Evolutionary events: Prokaryotes, Protists, Fungi, and Plants	Evolutionary events: Animals
Gymnosperms on the rise; adaptive radiation of orchids Modern distribution of angiosperms and gymnosperms	Neolithic to modern civilization *Homo erectus, Homo sapiens* (Neanderthal), *Homo sapiens sapiens,* modern mammals
	Large extinction of earlier mammals
Grasslands spread	First hominids (Ramapithecus), numerous grazing mammals
	All of today's mammal families
Angiosperms and gymnosperms dominate	Earliest cetaceans, all of today's mammal orders formed
Most present-day angiosperm families develop	Age of Mammals begins; modern invertebrates appear
Extinction of dominant phytoplankton (coccolithophorids)	Massive extinction: nearly all dinosaurs and 70% of all animal species according to some estimates
Rise of angiosperms	Dinosaurs reach peak and rapidly decline; birds persist
Earliest angiosperms (restricted to higher tropical elevations), conifers, ferns, ginkgos dominant	First birds, teleost fishes, modern crustaceans; Age of Reptiles begins
Gymnosperms (conifers, cycads, ginkgos), ferns dominant plant form	First dinosaurs, earliest mammallike reptiles
Extinction of many dominant life forms	Most Paleozoic invertebrates (including trilobites) extinct
First gymnosperms; coal age forests (tall trees: lycophytes, horsetails, ferns, seed ferns)	Earliest reptiles, first winged insects, Age of Amphibians
Large terrestrial plants, first seeds, first vascular plants (psilopsids, lycophytes, sphenophytes)	First amphibians, sharklike fishes, bony fishes, lung fishes, mandibulate arthropods
Algae give way to terrestrial plants; green, red, and brown algae	Age of Fishes: lobe-finned fishes, jawed fishes; first terrestrial invertebrates, coral reef building; first vertebrates: jawless fishes, chelicerate arthropods
Algal forms dominate	Marine invertebrates dominate; trilobites abound
Multicellular life (algae, fungi?)	Late Precambrian: first evidence of multicellular animals: soft-bodied coelenterates and other marine invertebrates, burrowing wormlike animals
First eukaryotes (probable): green algae, other protists	
Photosynthetic anaerobes (prokaryotes)	
Origin of life: first bacterialike prokaryotes (anaerobic heterotrophs)	
Organic synthesis	

BIOLOGY THE SCIENCE OF LIFE

SECOND EDITION

BIOLOGY THE SCIENCE OF LIFE

SECOND EDITION

ROBERT A. WALLACE
University of Florida

JACK L. KING
University of California, Santa Barbara

GERALD P. SANDERS

SCOTT, FORESMAN AND COMPANY
Glenview, Illinois   London, England

This book is dedicated to the memory of Jack L. King

Three people are listed as the authors of this edition, although Jack King died while the book was in preparation. He is shown as an author for two reasons. First, the remaining authors are perhaps not emotionally prepared to see the book appear any other way. Second, and more important, his name *should* appear because of the great influence he had on the book's development. That influence was particularly felt in the critical early stages when the course and tone of the book were being set. With his prodigious intellectual abilities and his keen interest in the project, he not only provided us with specific scientific material (often uncovered in this esoteric journal or that) but he often came up with ways of describing it so that it made sense and fit the "big picture." In essence, these pages are interlaced with Jack's thoughts, his ideas, and his work. He remains not only our good friend, in our minds, but our esteemed coauthor as well.

Cover: Computer-generated image of DNA molecule by Computer Graphics Laboratory, University of California, San Francisco.

Credit lines for other illustrations are placed in the "Acknowledgments" section at the end of the book, which is to be considered an extension of the copyright page.

456-RRW-90898887

Library of Congress Cataloging-in-Publication Data

Wallace, Robert Ardell
    Biology, the science of life.

    Bibliography: p.
    Includes index.
    1. Biology. I. King, Jack L.
II. Sanders, Gerald P. III. Title.
QH308.2.W34 1986   574   85-22113
ISBN 0-673-16657-0

PREFACE

One might think that writing a new edition of an existing book would be a relatively straightforward, if not simple, task. Isn't it just a matter of updating and rearranging a bit, based on reactions of readers of the earlier work? Not according to our experience. But perhaps the problem lies in our approach to new editions. We agreed early on not to settle for a cosmetic touch-up. Instead, we consulted many of those who used the first edition. We asked about what worked and how we could improve the book. We dismantled large parts of the material and carefully reconstructed it, with an eye to producing a clearer and more effective way of learning about biology.

And even as we were altering and reconstructing those pages, biology itself was changing. Scientists were telling us new things about life on the planet. Some of the things they said were admittedly troubling. We learned of new human population problems, of possible unexpected results from nuclear warfare, of new diseases, and of changes in weather patterns and water supplies. But we also were told of happier and more hopeful findings, such as new ways to extend human life, new kinds of crops, the wonders of genetic engineering, better understanding of the molecules of life, and even strange new aquatic communities we hadn't seen before. So as the revision progressed, there were changes in both organization and substance.

We should add that we are grateful for the enthusiastic response of those who actually used the first edition, the students themselves. Many have written to tell us the book is not only informative (as any text should be) but *interesting*. It is unfortunate that it should be considered unusual for a biology text to be interesting. After all, biology is the study of *life*—that intricate web of which we are a part. The consideration of life should rivet our attention, and we hope that we have let the fascination of it all shine through. To this end we have gone beyond a simple compilation of facts and have tried to "tell the story" of biology in a form weighing under 12 pounds.

We wish we could have written a shorter book and told the same story but we couldn't. The tale we tell here is obviously one of breadth *and* depth. We have tried to set the stage and to reveal some of the drama of modern science. In some cases we show how the scientists themselves arrived at their conclusions and we even describe experiments that didn't work. All this is done to attempt to give readers the "feel" of science. We think the feel is important because this book is written primarily for people who are likely to go on in the sciences. Such an introduction, whether covered in one or two academic terms, is particularly important to those who will continue training in the *life* sciences.

In this edition we have maintained the evolutionary theme of the first edition. The first eight chapters have been extensively reorganized in order to more effectively introduce the basic chemistry of life. We have also increased the depth of existing material, not just in chemistry and energetics but in evolution and botany as well. We have greatly increased our coverage of the nervous system and have expanded areas of ecology and population dynamics. We trust that the result is a more useful tool in the effort to explain the fascinating and often complex principles of life.

In this edition, we have divided the material into 42 manageable chapters and have paid special attention to pedagogical aids that help get the material across. Each chapter begins with a brief introduction that orients the student toward new topics. Within each chapter, key terms appear in bold type, and we have made an effort to summarize detailed discussions within the text, where appropriate. Each figure now carries a brief title (in addition to a complete caption) so that readers can quickly identify what is being illustrated.

At the end of each chapter, we first offer an effective summary called "Key Words and Ideas," a thorough study outline that integrates important terminology with restatements of major concepts. The "Applications of Ideas" are thought-provoking cases or questions that ask students to synthesize information from the chapter. Page-referenced review questions also appear after each chapter. Suggested readings are given at the end of each part of the book.

We have developed over 40 special essays that delve into particular facets of biology in some detail. And we are proud of our new summary illustrations, which are intended as a reminder and an overview of complex ideas, a way to put the details into perspective.

At the end of the book, you will find an appendix on the classification scheme used in the text, a particularly useful biological lexicon, an extensive glossary, and a thorough index. A geologic timetable appears in the front of the text, and metric conversions are listed inside the back cover.

There are a number of supplements that accompany the text itself. They have been prepared at some expense and with great care to assist in the teaching and learning of the principles of biology through these pages. They include an Instructor's Manual and Test Item File, a Student Study Guide that contains a special section on working genetics problems, and a Laboratory Manual with 31 exercises appropriate for biology majors. Also available to adopters are 150 acetate transparencies, or 250 35mm slides, of relevant art from the text and *Biosphere: The Realm of Life*. Our computerized testing system includes capabilities for grade-book management, an electronic calendar, and a self-testing study program that allows students to take quizzes on monitors.

We are always surprised at how tough writing can be and we must admit this edition, as well, proved to be much more difficult and time-consuming than anyone had imagined. However, we believe the result speaks for itself. We are also aware that biology continues to change and that we will soon begin to work out new ways to present the material next time. So with some trepidation, we say, Here's to next time!

R.A.W.
G.P.S.

ACKNOWLEDGMENTS

There are a number of people who helped bring the book to fruition and we would like to thank them here.

First, we are grateful to Jim Levy, the general manager of Scott, Foresman's College Division. Without his support the project would never have gotten off the ground. His managerial skills and energy level are indeed impressive, and he seemed to have a way of appearing on the scene just when he was needed. We are also thankful for the quiet but persuasive support of Dick Welna, editor-in-chief. He knows the pressure points in this business and dexterously maneuvered the project through what could have been sticky situations.

Again we had the opportunity to work with Jack Pritchard, the editorial vice-president for science. We have been through a lot together over the years and we have come to deeply trust his judgment and guiding hand. Many of the key decisions fell on Jack. Working closely with both Jack and us has been someone we consider a bright light in this business. Becky Strehlow came aboard as editor on one of our earlier projects, and we quickly came to admire not only her knowledge about getting a book out, but her intense interest in and concern for her projects. We feel lucky to be one of them.

Bill Poole has a peculiar New England accent but we were able to understand him when he worked so closely with us during development and the early stages of production. He cared about the project and his intelligence and diligence ironed out many a difficult problem. Our project editor, Carrie Dierks, was the one most likely to suffer a heart attack. However, she somehow maintained her balance and good humor even after some unsettling telephone conversations with one of us. Her ability to quickly assess a situation and to find ways around problems saved us more than once.

We are very impressed with the design and layout of this edition, and for that we thank our designer, Lucy Lesiak. Meeting last-minute challenges seems to be her forte. The overall appearance of the book also reflects the efforts of our photo editor, Mary Goljenboom. We were often truly amazed at what she was able to find for us. The illustration program was monumental and we were indeed fortunate to have the services of some of the finest artists in the business, particularly Jeff Mellander of Precision Graphics, Sandra McMahon, and Jean Helmer. And one of the most trying jobs, that of preparing the index, was well done by our longtime friend Ethel King. Again, our thanks to all.

REVIEWERS

Biology is far too wide-ranging these days for anyone to have a firm grip on all its facets, so we strongly relied on reviews from our colleagues as the book progressed. Some of them are specialists who know the fine details of certain areas of biology, and they kept us accurate and current regarding content.

Others are generalists who focus on how the details fit into the big picture. They helped us make some sense of it all and assisted in developing ways to effectively present the material. We are grateful to these people for their help and the chance to forge a few new friendships. Our reviewers include:

Robert A. Andersen
DePaul University

Marilyn D. Bachmann
Iowa State University

Thomas G. Balgooyen
San Jose State University

William E. Barstow
University of Georgia

Joseph J. Beatty
Oregon State University

Robert A. Bender
University of Michigan

David B. Benner
East Tennessee State University

Richard D. Bliss
Yuba College

Donald J. Defler
Portland Community College

William A. deGraw
University of Nebraska at Omaha

Fred Delcomyn
University of Illinois at Urbana-Champaign

Dennis E. Duggan
University of Florida

Lawrence A. Dyck
Clemson University

Susan A. Foster
Mt. Hood Community College

Michael S. Gaines
University of Kansas

James T. Giesel
University of Florida

Elizabeth A. Godrick
Boston University

Walter M. Goldberg
Florida International University

Judith Goodenough
University of Massachusetts at Amherst

Steven R. Heidemann
Michigan State University

John J. Heise
Georgia Institute of Technology

Carl Helms
Clemson University

Doris R. Helms
Clemson University

Robert J. Huskey
University of Virginia

Judith A. Jernstedt
University of Georgia

Thomas Kantz
California State University, Sacramento

Samuel Kirkwood
University of Minnesota

Paul M. Knopf
Brown University

Robert A. Koch
California State University, Fullerton

John C. Makemson
Florida International University

John C. Mallett
University of Lowell

David V. McCalley
University of Northern Iowa

Roger Milkman
University of Iowa

John E. Minnich
University of Wisconsin, Milwaukee

Joseph R. Murphy
Brigham Young University

Frank G. Nordlie
University of Florida

David T. Rogers, Jr.
University of Alabama, University

Daniel C. Schierer
Northeastern University (Boston)

John A. Schmitt
The Ohio State University

Robert D. Simon
State University of New York, Geneseo

Roger D. Sloboda
Dartmouth College

Kingsley R. Stern
California State University, Chico

Daryl C. Sweeney
University of Illinois at Urbana-Champaign

Samuel S. Sweet
University of California, Santa Barbara

Wallace A. Tarpley
East Tennessee State University

Jay Shiro Tashiro
Kenyon College

Roger E. Thibault
Bowling Green State University

Ray E. Umber
University of Wyoming

Joseph W. Vanable, Jr.
Purdue University

Stephen H. Vessey
Bowling Green State University

Richard E. Widdows
East Tennessee State University

Mary Wise
Northern Virginia Community College

Carl R. Woese
University of Illinois at Urbana-Champaign

Edward Yeargers
Georgia Institute of Technology

John L. Zimmerman
Kansas State University

Uko Zylstra
Calvin College

We are especially grateful for the level and quality
of assistance we received from Bill Barstow, Judy
Goodenough, Dori Helms, and Roger Thibault.

OVERVIEW

CONTENTS

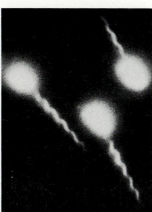

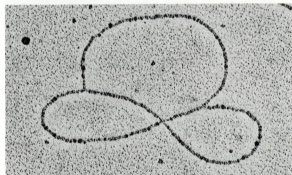

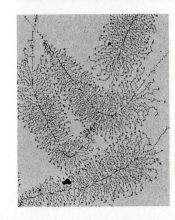

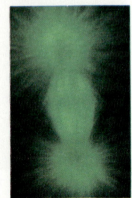

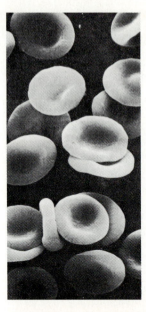
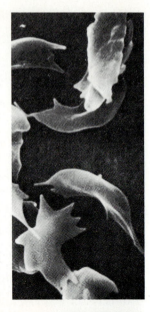

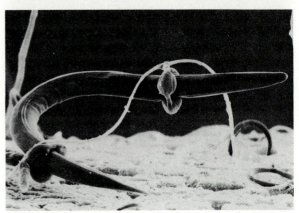

PART 5 THE PLANTS

PART 6 THE ANIMALS

PART 5 THE PLANTS

PART 6 THE ANIMALS

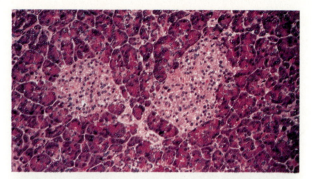

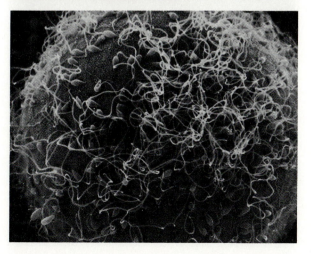

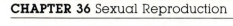

PART 8 ECOLOGY

David T. Rogers, Jr.
University of Alabama, University

Daniel C. Schierer
Northeastern University (Boston)

John A. Schmitt
The Ohio State University

Robert D. Simon
State University of New York, Geneseo

Roger D. Sloboda
Dartmouth College

Kingsley R. Stern
California State University, Chico

Daryl C. Sweeney
University of Illinois at Urbana-Champaign

Samuel S. Sweet
University of California, Santa Barbara

Wallace A. Tarpley
East Tennessee State University

Jay Shiro Tashiro
Kenyon College

Roger E. Thibault
Bowling Green State University

Ray E. Umber
University of Wyoming

Joseph W. Vanable, Jr.
Purdue University

Stephen H. Vessey
Bowling Green State University

Richard E. Widdows
East Tennessee State University

Mary Wise
Northern Virginia Community College

Carl R. Woese
University of Illinois at Urbana-Champaign

Edward Yeargers
Georgia Institute of Technology

John L. Zimmerman
Kansas State University

Uko Zylstra
Calvin College

We are especially grateful for the level and quality of assistance we received from Bill Barstow, Judy Goodenough, Dori Helms, and Roger Thibault.

Mr. Darwin and the Meaning of Life

Something was wrong. He couldn't put his finger on it, but he knew it just the same. The feeling had nagged at him before, but he had dismissed it. After all, it was hard to be troubled while standing on those grassy hills with the fresh winds of an exotic land brushing across his face.

The good ship *Beagle* lay at anchor off the coast of Argentina, and the young naturalist was 200 miles inland, glad to be ashore, crossing the Argentine pampas on horseback. Life had not always been a joyous adventure for Charles Darwin. His father and his grandfather had been among the wealthiest and most famous physicians in England. Charles had been expected to follow in their footsteps and, at the age of 16, was sent off to medical school. But he found that he became ill at the sight of blood, and he nearly fainted upon witnessing his first operation. He saved himself that disgrace only by rushing from the room. At one point Charles was mortified to be told by his father that "You care for nothing but shooting, dogs, and ratcatching, and you will be a disgrace to yourself and all your family."

So Charles tried again—this time law. But he had no aptitude for it and was soon shuttled into training for the clergy. He was duly enrolled in divinity school at Cambridge, where he promptly showed almost as little aptitude for divinity as for medicine. His curriculum included classics, which he loathed, and mathematics, which he couldn't understand. He once wrote a friend about his trouble with mathematics and said, "I stick fast in the mud at the bottom and there I shall remain." Still, his college experience had its pleasant aspects; he could keep up with his insect and rock collecting. At Cambridge he found a friend in one of his teachers, the Reverend John Henslow, who was a botanist as well as a clergyman. Often the two would take long walks in the countryside around Cambridge and discuss the natural history of the area. Darwin was even known by some as "the one who walks with Henslow."

When at last Charles surprised himself and his family by passing his final examinations at Christ's College, he came home shouting "I'm through! I'm through!" No more school. He was ecstatic. Of course, he was now expected to enter the clergy, but he spent his first postgraduate summer happily "geologizing" around the English countryside, away from difficult decisions and away from his family (Figure 1.1).

When Darwin returned home in the fall he found a letter from Henslow waiting. Henslow had been offered an appointment as naturalist on a British naval survey ship that was to sail around the world. Mrs. Henslow, however, had become so disconsolate at the idea of her husband being gone so long that he had reluctantly refused the offer. He

1.1 CHARLES DARWIN

Charles Darwin in 1860.

recommended that young Darwin go in his place. Darwin thought it was a great idea.

Unfortunately, Charles' father would have none of it. Darwin was disheartened, but continued to press for permission. Finally his father told him, "If you can find any man of common sense who would advise you to go, I will give my consent." Young Darwin didn't think anyone would give such advice and was prepared to decline the offer once and for all when his uncle Josiah Wedgwood (of pottery fame) said that he thought it would be a splendid thing for his favorite nephew. The two of them together persuaded Dr. Darwin to hold to his word.

On September 5, 1831, Charles was summoned to London to be interviewed by the captain of the *Beagle*, Robert Fitzroy. Darwin was only a year younger than the 23-year-old captain, who nevertheless had already distinguished himself as a seaman of remarkable abilities. There was an initial awkwardness between them (Fitzroy thought that the shape of Darwin's nose indicated a weak character). But soon Fitzroy's doubts about Darwin's nose evaporated and Darwin was accepted as the *Beagle's* naturalist (Figure 1.2). In fact, Charles shared quarters with the captain himself. There was no salary, and Darwin had to pay for his own room and board throughout the voyage.

Captain Fitzroy's major mission was to chart the waters off South America, but his private mission—his personal passion—was to find evidence that would establish once and for all the literal truth of the biblical account of the creation of the world. For that he needed a naturalist, and this amiable young divinity student who was hungry for adventure seemed just right for the job.

Now, as Darwin gazed across the lush grassland of South America, he felt that there was something unusual about the place, something not quite right. What was it? Why was he troubled? There was certainly no apparent reason to feel uneasy. The warm breeze gently smoothed the unkept grass, the sky was clear and blue, and his confidence and energy were high. But something was nagging him. What was it? Then it came.

There were no rabbits.

No rabbits. The phrase could be engraved in the consciousness of Western civilization along with $E = mc^2$ and *E Pluribus Unum*. No rabbits; such an innocuous phrase, but in a sense, it signalled the beginnings of a revolution.

Darwin was not thinking about revolution, though. He was thinking about rabbits. Where were they? Darwin knew rabbit country when he saw it, and this place was obviously a rabbit heaven. There was grass for rabbits to eat and bushes for rabbits to hide in and dirt for rabbits to dig in—still, there were no rabbits.

Where rabbits should have been there were other, strange little animals eating grass, digging holes, and hiding in bushes. They had long legs like rabbits, and large ears, and did many things that rabbits did, yet they were clearly not rabbits. They looked more like guinea pigs (Figure 1.3). But where were the rabbits?

On one level Darwin knew perfectly well why there were no rabbits. There were no rabbits on the Argentine pampas because he was in South America, and rabbits don't live there. That kind of answer usually satisfies almost everyone; however, it is really no answer at all. If the question *Why aren't there any rabbits in South America?* had been pressed, another naturalist might have answered that South America was really not the proper place for rabbits, that the land couldn't support them. But Darwin thought the land *could* support them. He continued to mull over the question and at last a partial answer formed in his mind.

Perhaps there are no rabbits in South America, he thought, *because rabbits can't swim across the Atlantic Ocean.*

That question and its apparently simple answer were to change our perception of the world forever.

AN EXPLANATION ABOUT RABBITS AND OCEANS

If Darwin's tentative answer doesn't seem earth-shaking, or even particularly interesting, perhaps it is because more than a century and a half have passed since Darwin looked across the lonely pampas. Our view of the world has changed dramatically since that time, largely because of the questions posed during that voyage and the answers Darwin suggested. Of course, rabbits cannot swim across the ocean. But that answer leads to new questions, those to still others, and eventually the whole line of questioning was to produce the logic of evolution. The absence of rabbits on the pampas meant that perhaps animals and plants are where they are, not because they were put there, but because their ancestors either made the journey or themselves originated there.

1.2 THE H.M.S. *BEAGLE*

The *Beagle* was one of many British vessels whose primary mission was to chart the oceans and collect oceanographic and biological information. The *Beagle* was a small vessel, just under 100 feet in length, but her fearless captain James Fitzroy, an expert navigator, guided her unerringly through a five-year voyage. Among her 74 voyagers was a young, very seasick divinity student, the ship's naturalist. There was little in his manner to suggest that his seagoing experiences would forever change the course of science. The *Beagle*'s travels took it around the globe, but most of its oceanographic and natural history studies were carried out along the coasts of South America and in the neighboring islands.

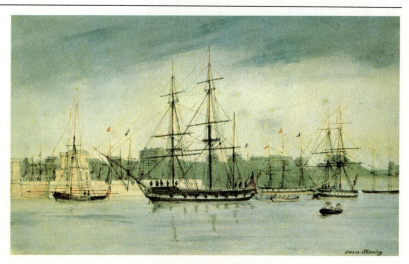

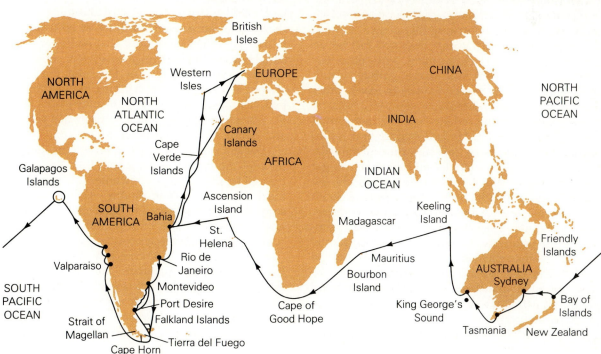

1.3 WHERE ARE THE RABBITS?

While rabbits are not native to South America, their ecological parallel is the Patagonian hare *Dolichotis patagonum,* or mara, as it is known. In spite of the apparent similarity, however, maras are not related to rabbits. Like squirrels and mice, they are rodents. The similarity is a special case of evolution (called convergent evolution), in which similar but isolated environments tend to favor the evolution of similar physical features.

Darwin also noticed that the rabbitlike rodents of the pampas were very similar to other South American rodents, such as guinea pigs. Why? The conventional wisdom was that South American rodents were similar because that general form of life was well adapted to (or designed for) life in South America. In Darwin's mind, a newer, different answer began to take shape: perhaps South American rodents were similar to each other because they were related.

It wasn't just the rodents and the rabbits that were stirring Darwin. He had dug up and reconstructed the bones of several extinct mammals, including a gigantic armadillo and some even larger giant ground sloths that were very much like the hippopotamus. Sloths are still found only in South and Central America, but the present-day sloths are small creatures. The extinct giants were clearly similar to—in fact, Darwin had to say, were apparently *related* to—the small burrowing armadillos and the tree-dwelling sloths that were still around. So he had bones of animals that no longer existed and living animals that were so similar that they seemed to be related.

Darwin found evidence that a number of South American mammals had become extinct. He speculated that they were driven to extinction by competition from invading North American species. But why hadn't the South American extinctions occurred earlier? Why had the North American animals taken so long to go south? Darwin thought that perhaps the narrow Isthmus of Panama, which connects the two great continents, had once been under water and that the animals had only advanced southward when a land bridge was formed. This was an excellent guess, and it suggested that conditions on the earth and the relationship of its inhabitants could change.

Darwin had taken with him on the *Beagle* a copy of the newly published first volume of his friend Charles Lyell's revolutionary book, *Principles of Geology.* By the time he had crossed the Atlantic, Darwin was a convert to the new geology. While in South America he received the second volume. Lyell had some rather startling things to say about the physical evolution of the earth. He said that the world was much older than anyone had imagined; that over long periods of time, continents and mountains rose slowly out of the sea; and that they just as slowly subsided again or were washed away. Most importantly, Lyell claimed that the very forces that had so changed the earth in the past were still at work and that the world was still changing.

Darwin's own observations of South American geology seemed to confirm Lyell's position at every hand. In his adventurous climbing of the Andes, he had found fossil clam shells at 10,000 feet. Below them, near an ancient seashore at 8000 feet, he found a petrified pine forest that had clearly once lain beneath the sea because it, too, was interspersed with seashells. In fact, the *Beagle* had arrived in Peru just after a strong local earthquake had destroyed several cities, in some places *raising the ground level by two feet.* The earth had changed and clearly was still changing.

Darwin was excited by his developing idea, but he kept the most revolutionary of his thoughts to himself because he was sometimes uneasy with his ideas and often full of doubts. After all, he had studied for the ministry and had believed in the literal truth of the Bible, but the evidence seemed to contradict the creation account in every detail.

Captain Fitzroy had no such doubts. To him the bones of extinct mammals merely proved the account of the Flood, if one simply allowed that perhaps Noah hadn't been able to round up all the animals. If there were no rabbits in South America, it was because rabbits did not belong in South America. There was a very good reason for everything. Fitzroy believed in laws and rules, and furthermore he knew what the laws and rules were. Darwin, on the other hand, was blessed (or cursed) with an ever-inquiring mind and was always ready to consider an alternative hypothesis. The hypothesis that was forming in his mind now was a dangerous one, and Darwin knew it could lead to trouble.

Darwin's thoughts began to come together during one momentous part of his five-year trip. The *Beagle,* on a dead run from the coast of South America, had reached a peculiar little group of islands and its anchor clattered into the quiet lee waters of an apparently insignificant island the English called Chatham. Chatham was one of the Galapagos Islands, a recently formed group of volcanic islands that lie astride the equator some 600 miles off the coast of Ecuador. Physically, the islands were quite unlike anything Darwin had seen on the mainland: black, bleak, dry, and hot. And isolated—there were relatively few species of plants and animals. In fact, there were no mammals other than those brought by European ships, and only a few species of birds other than sea birds (see Essay 1.1).

The Galapagos archipelago was, as Darwin wrote later, a little world unto itself. Yet Darwin noticed that there was something familiar about the plants and animals. Though most of the species were unique to the islands, Darwin had seen species like them only recently. They were similar to South American species and, as he was to learn, unlike those of Europe, Asia, North America, or Australia. Why should this be?

The usual explanation just wouldn't do. There was nothing in the physical environment that suggested that the islands were somehow appropriate for South American creatures. Indeed, the environment was almost identical to that of the Cape Verde Islands, volcanic islands where the *Beagle* had tied up for nearly a month early in its voyage. The Cape Verde Islands, however, lay off the coast of Africa, and its species were typically African.

Because rabbits can't swim across the Atlantic . . .

Darwin reasoned that it would be difficult, if not impossible, for Asian, Australian, and North American species to cross the Pacific Ocean to settle on these volcanic islands. But perhaps from time to time, a few drifting seeds, a few reptiles on floating logs, and a pair of birds blown off course might have traveled across the 600 miles from the South American mainland. Darwin wondered if, finding a hospitable island free from competition, they could have survived and increased in number.

THE WORKINGS OF SCIENCE

By now Darwin and Fitzroy were engaging in lively, if not heated, discussions about the nature and origins of life. Fitzroy would probably have been chagrined to think he was *helping* Darwin form his "heretical" ideas by providing a sounding board. Like most of his contemporaries, Fitzroy was convinced of the "immutability or fixity of species," which, in the English vernacular of the 19th Century, means their inability to change. In his mind species were as they had always been, ever since the Creator placed them here. Darwin probably was hesitant to reveal his true thoughts, for he was beginning to think that life does, in fact, change—that living species are modifications of earlier and different species. Other people, including Darwin's own grandfather, had said the same thing. But where their musings had been unsupported, Darwin was slowly gathering evidence. **Species,** we should point out, refers to any of the millions of unique kinds of organisms, such as humans, red maples, and rainbow trout. Members of any species are similar in enough ways to interbreed successfully (see Chapter 17).

The idea that Darwin was formulating was to be called **evolution,** or **descent with modification.** Evolution, in this context, has come to mean changes in populations through time. (We will soon see just how such changes come about.) Darwin began to try to apply his ideas to what he saw around him, and he now saw evidence for evolution everywhere. Could his observations prove that evolution is a fact of life on the planet? How are such facts established? For the answer we turn to the inner workings of science.

The Methods of Science

Science, in its most general sense, is the search for truth. That search, of course, can be conducted in a number of ways. If we witness a volcano rising from a seabed, we come closer to the truth about volcanoes. But simple observation is not enough. Today, most of what we call science is arrived at by the **scientific method.**

The scientific method has been explained many ways, yet it always eludes precise definition. Essentially, it is the process of establishing new facts and understanding mechanisms. Though there is no set algorithm—that is, no prescribed set of directions for accomplishing these things, we generally know

The Galapagos Islands

The Galapagos archipelago includes habitats ranging from dry, lowland deserts to wet, species-rich highlands. The islands are home to a bizarre collection of life, from grazing lizards and giant tortoises to a variety of bird life and shore dwellers. Darwin, who despised the place, was to make it famous because he saw it as an experiment of nature in progress.

(right) Opuntia cactus. The plant is a favorite food of the Galapagos tortoise.

(above left) Galapagos tortoise. The species vary according to the island on which they are found.

(above) The beautiful Sally Lightfoot crab, named for a dancer because of its quick and graceful movements.

(far left) Land iguana, a relatively rare land-bound cousin of the marine iguana.

(left) Marine iguana. They graze beneath the sea.

(left) Bartolome Island. Sharks are common in these shallow waters.

(below) Crown of Thorns, volcanic rocks that were once part of a larger island.

(above left) Courting blue-footed boobies.

(above) Courting frigate birds. The throat of the male is expanded as a display.

(far left) Waved albatrosses engaged in *billing,* a courtship behavior.

(below left) An oystercatcher. These birds feed on shellfish in intertidal areas.

(left) Red-footed booby.

the scientific process when we see it. The process usually begins and ends with observations about the real world. Between the first observation and the last occurs a fair amount of human mental activity. In a sense, then, science is the interaction of the human mind with the facts of nature. But where does it begin?

There are the initial observations, which are presumably not understood. Darwin, for example, observed that the land birds of the Galapagos are unique species with strong physical similarities to those of mainland South America. But what was he to do with such information?

What happens next is the least known aspect of science, and perhaps the most important. It begins with mulling over the observations and wondering, especially wondering *why*. Almost no one writes down very much at this stage, which may be why it is so little understood. Perhaps most scientists are wondering and mulling over facts most of the time, and most of the time nothing comes of it. Sometimes, though, wondering leads to speculation. Speculations are surmisings about untested ideas. We could no more do without speculation than we could do without facts, but even facts coupled with speculation do not add up to science.

At some stage the investigator must formulate a **hypothesis** to explain the facts. The hypothesis represents but one level of scientific activity. At a somewhat loftier level we find the **theory,** which is usually considered to be a larger generalization about some phenomenon—one that is supported by a considerable amount of evidence. All too often, the terms hypothesis and theory are used interchangeably, so let's try to make a clear distinction between the two.

A hypothesis is a tentative explanation of some observed phenomenon that can be used to make predictions that can then be tested. The hypotheses of others may be used because science is often a community effort. They can arrive fully formed, like Venus on a seashell, or in rough outline form so that modifications and refinements can be made to fit new observations. Keep in mind, though, that if a scientific hypothesis is to be useful it must lead to *predictions* and it must be *testable*. This is the crucial stage and it involves another observation—the observation that tests the prediction.

As an example, an investigator may believe that "hormone X" is responsible for some observed aspect of growth in plant embryos. This is his or her hypothesis. From the hypothesis, a prediction is made: "If I add hormone X to the growth medium

of the plant, I should be able to observe the following in the embryo. . . ." Such statements automatically suggest the next logical step—testing the prediction. The investigator does this by carrying out an **experiment,** deliberately setting up carefully specified conditions under which certain observations or results can be expected according to the prediction being tested. Further, the experiment must be set up in such a way that there can only be one explanation for the observations to come; frequently, this involves using **controls,** in which the crucial factor, sometimes called a **variable,** being tested is altered or left out. In other words, a control is a standard to which the experimental results can be compared. As such, the control must be identical to the experimental situation in every other way. Any unaccountable differences that creep in are called uncontrolled variables. In testing the hypothesis above, the experimenter might add a watery solution containing hormone X to the growth medium of some plants, but might add only a similar quantity of water to the control group. The water acts as a **placebo,** a substitute that cannot influence the experiment. (Figure 1.4 shows an example of an experiment using controls (called a **controlled experiment**).

The results of the experiment may or may not support the investigator's predictions. The next step will be to simply accept or reject the hypothesis accordingly. This step, in more formal language, is the investigator's **conclusion.** The conclusion can be tentative or firm, depending on the investigator's confidence in the strength of the evidence. Throughout the entire procedure the investigator must constantly review what others have said and written; someone else may have expressed the same ideas and made the same observations, but tested them and explained them in a different way.

We have said that a hypothesis is usually tested in an experiment, but it is important to note that not all experiments in biology are performed in laboratories. There have been many excellent field experiments—that is, those done under natural conditions. An example is a mark-and-recapture experiment, in which animals are captured, marked, released, and later recaptured, thus providing information about such things as their movement, growth, or longevity. Hypotheses can also be tested without experiments at all. An ecologist might predict that a certain relationship will be found in nature and then test the prediction by going out into the field to look for it. No matter which method one selects, however, it is still necessary to first make the prediction and then test it.

Finally, publication is an important part of the scientific method. Discovering something and keeping it to oneself is not in keeping with the goals of modern science. As you can see in Essay 1.2, there are rather formal guidelines about how one spreads the word.

In reality, the investigator's behavior is rarely as precise and straightforward as it would appear from reading a published article. For example, we know little about the phase that involves mulling, dreaming, or inspiration. We also know very little about the scientist's own motivation, his or her need to know and tell. The very best scientists are terribly curious, of course, but they also tend to be driving,

egocentric, proud enthusiasts who may covet fame, or titles and prizes, or the adulation of students. Perhaps the strongest and most common driving force of scientific investigation is the individual's very human need for the acceptance, approval, and admiration of other scientists.

Darwin was no exception. Near the conclusion of its voyage, when the *Beagle* arrived at Ascension Island in the South Atlantic Ocean, Darwin found a letter from his sisters waiting for him. In it they mentioned that Adam Sedgwick, an English scientist, had told Robert Darwin that on the strength of the letters and specimens that Henslow had already received from the *Beagle*, his son Charles would take a place among the leading scientists. Darwin wrote, "After reading this letter I clambered over the

1.4 A CONTROLLED EXPERIMENT

This experiment is intended to determine the effect of a herbicide (plant killer) on bird development prior to its use on plants. The first procedure **(a)** includes piercing the fertile egg (under sterile conditions) and injecting a specific concentration of the herbicide (dissolved in alcohol) into the air space below the shell. Following a desired period of incubation, the embryo is removed and checked for abnormalities.

As you may have surmised, the experiment contains several sources of potential error. If developmental abnormalities do occur, they might have been caused by the procedure, or perhaps the addition of the alcohol solvent had an effect. In any case, it is difficult to draw a meaningful conclusion from one embryo. To correct these errors a controlled experiment must be devised.

In the controlled experiment, four large groups of fertile eggs are prepared. The first is treated in the manner described above. The second is treated identically, except that the injection is a *placebo,* an alcohol solution identical to the first except that it contains no herbicide. A third group has been pierced and sealed, but not injected, and a fourth group has no special treatment, but is incubated under identical conditions to the others. Now the results of the four can be compared, and differences between the herbicide treated embryos and the others can be attributed to the presence of herbicide.

(a) Uncontrolled experiment
Herbicides injected into a number of fertilized eggs. These are incubated and studied at certain stages for undesirable effects.

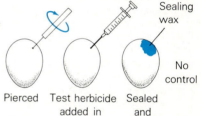

Pierced   Test herbicide   Sealed
     added in   and
     alcohol   incubated
     solution

Potential sources of error:
1. Does piercing affect development?
2. Does adding alcohol affect development?
3. Could the result be due to chance?
4. Could there be something wrong with the incubator?

(b) Controlled experiment

Group 1 Experiment: *Group 2 Control:*
herbicide in alcohol   alcohol solution
solution added to   added to
a number of eggs   a number of eggs

Group 3 Control: *Group 4 Control:*
a number of eggs   a number of eggs un-
pierced but   treated but handled
nothing added   in identical ways

All four groups are treated alike in all other ways

How Scientists Spread the News

The conventions of communication between scientists are as formal and intricate as an English country dance. It's not enough to just unravel the universe's little secrets. If new discoveries are to be taken seriously by other scientists, certain rigid criteria must be met. This means, among other things, that the experiments must be done right. Often, as it happens, the discovery comes first, not infrequently by (educated) accident. Still, before the new finding can be published and accepted, the experiment must be repeated and verified according to accepted procedures.

New findings are almost always communicated through a formal scientific paper or journal article. The article may appear in any one of several thousand scientific journals, usually one devoted to the narrow specialty of the investigator. The most important new findings may appear in a journal of general interest to all scientists, such as *Science* or *Nature*.

In either case, the article will not be published until it has undergone the scrutiny of the journal's editor and of two or three anonymous volunteer referees. This is one of the extensive safeguards of formal science. The referee system, however, is not without drawbacks. Many of the most important landmark papers in any scientific field have had to withstand an initial rejection by suspicious referees. Of course, these referees have also prevented the publication of innumerable allegedly grand ideas and supposed new paradigms that were, in fact, pure hokum.

A standard scientific article consists of six parts: the *summary, introduction, methods and materials, results, discussion,* and *literature cited.* The *summary* includes the principal finding, or conclusion, of the experiment being reported.

A short *introduction* reviews any previous relevant work and explains the reasons for proposing the hypothesis that is to be tested.

The *methods and materials* section tells exactly how the experiment was conducted. It is written with enough detail and clarity that anyone who is sufficiently interested can repeat the experiment. Repeatability is the only guarantee that the findings are legitimate.

The *results* section is the key part of most papers. It includes the observations made and experimental data compiled, along with any statistical analysis required to clarify the data. The investigator must accumulate enough data until it becomes extremely unlikely that the results could be due to chance.

In the *discussion* section, the author may be a little less formal, and can even indulge in speculation, comparisons, and suggestions for future research. Here, ambiguities in the data can be accounted for, potential objections explored, and persuasion and argument allowed. After all, it doesn't do much good to make a new discovery if you can't convince the rest of the world.

Every scientific paper is sprinkled with numerous parenthetical notes or footnotelike numbers referring the reader to the *literature cited,* a list of journal articles and other sources that concludes the presentation. These references alert the reader to other important work in the field, in case he or she wishes to pursue the subject. Just as important, they give credit to workers who have made previous contributions.

mountains of Ascension with a bounding step and made the volcanic rocks resound under my geological hammer!''

Going Home to Applause and Illness

After the Galapagos, the rest of the voyage was somewhat anticlimactic. The *Beagle* made for home as quickly as possible, which in this case meant continuing westward around the world. It called at Tahiti, New Zealand, Tasmania, Australia, Mauritius, South Africa, St. Helena, Ascension, and Brazil. As always when the *Beagle* was under sail, Darwin was wracked with seasickness.

By the time the *Beagle* returned to English waters, it was late in 1836 and Darwin was 27 years old. When the boat tied up, a grateful Darwin leaped ashore and immediately took a carriage home. Since he arrived at a late hour, he decided not to rouse his family and took a room at an inn nearby. The next morning he received a joyous greeting from his family, but he was especially delighted when his dog greeted him and immediately set off down the trail on which they had last enjoyed their morning walk five years before.

Darwin did indeed take his place among the leading scientists. His observations on geology and zoology were published and he was revealed as a keen observer of nature, undoubtedly one of the greatest natural historians of all time. He was a good storyteller as well. His journal, *The Voyage of the Beagle*, became successful popular reading in England and remains a classic today.

EVOLUTION: THE THEORY DEVELOPS

We will not be referring to "Darwin's hypothesis of evolution," except in its historical context, since today it is a full-blown theory. It achieved the status of theory by the weight of its increasingly supportive evidence. Whereas theories may never be proven, or provable, they do have solid evidence to support them and they do help explain the observations. The theory of evolution is supported by seemingly endless observations and not a few experiments, although experiments on evolution are admittedly difficult. Let's see how the theory developed.

For Darwin, the key event of the voyage on board the *Beagle* was the visit to the Galapagos. He was, from then on, convinced about the reality of evolution. He was especially impressed by the observation that each of the many small islands had different assemblages of species, and often their own unique species, even though the islands were all rather similar. This suggested that the animals and plants on each island were there because of past accidents of introduction and, more importantly, that the different island species had evolved from common ancestors. What Darwin needed now was to find the *mechanism of evolution.* Like any devoted observer of nature, he marvelled at how different organisms were so exquisitely adapted to their needs. Of one thing he was sure: *Any explanation of evolution that did not explain the adaptedness of species was no explanation at all.* It was the variety of little finches fluttering around the islands of the Galapagos that gave him one of his first clues.

These little birds were dark and drab and notably unspectacular. Darwin collected a number of them as specimens that would be examined later by specialists in England. Darwin noticed two things about them: they were all rather similar to species on the South American mainland, but they differed from each other in critical ways, such as bill size and foraging behavior (Figure 1.5). Darwin wondered if

1.5 DARWIN'S FINCHES

While they are now considered separate species, evidence from many studies supports the theory that the Galapagos finches evolved recently from a common finch ancestor. *(right)* The hypothetical relationship of Darwin's finches as they evolved from a common stock.

Ground finches, *Geospiza*

Tree Finches, *Camarhynchus*

Warbler finch, *Certhidea*

Cocos finch, *Pinaroloxias*

Original arriving pair of South American finches

the island species could have been somehow modified from an ancestral mainland stock. This idea did not fit with the prevailing idea of how species arose, but it was a cornerstone for Darwin's developing theory. We will come back to the Galapagos later (see Chapter 17).

The Theory of Natural Selection

Beginning in 1837, Darwin started to keep a journal titled *Transmutation of Species* and was soon making entries referring to "my theory." Reading that journal today, and the journal of the voyage of the *Beagle*, is like reading a detective story after you already know whodunnit—the suspense is in watching the detective sift through clues for the right answer. In his journal entries Darwin began to toy with the idea of **natural selection.** It is fascinating to watch the idea develop in his writings.

Keep in mind that Darwin was a country boy. He knew a lot about agricultural practices, and he knew about livestock breeding. Any good farmer knows that a breed can be improved by *selection*, or, in biological terms, **artificial selection**—selecting the best individuals of each generation for breeding (Figure 1.6). By longstanding folk tradition, the best of the breed (whether cattle, fowl, dog, or cucumber) were honored annually at country fairs and chosen for propagation. The reasoning was simple: like begets like, offspring tend to resemble their parents, and the "best" parents produce the "best" offspring.

But Darwin wondered, could selection operate *without* human intervention? And, if so, how? Agricultural selection involved the conscious choice of the breeder. If selection occurred in nature, who was the selector? This line of thought seemed at first to lead right back to a supernatural factor. Without a conscious selector, the inferior individuals were as free to breed as were the most superior, in which case no improvement or change or adaptation would occur.

Did selection have to be conscious, then? Perhaps not. Darwin was impressed with an essay on population that had been written by Reverend Thomas Malthus three decades earlier. Malthus stressed that all species had enormous reproductive capabilities and that their numbers tended to expand rapidly geometrically (2, 4, 8, 16, 32, and so on) unless held in check by starvation or disease. Natural populations, he argued, reached a balance in which all but a few of the young of each generation were forced to perish. *All but a few.* So, mused

Darwin, perhaps the inferior individuals were not free to reproduce after all, precisely because they were inferior! The environment, then, could select the individuals that were allowed to breed, and these would be the hardiest. And it all happened through natural means (Figure 1.7). Darwin's journal shows that by 1838 he had solved the major riddle of evolution. The mechanism of evolution, he said was *natural selection.* Briefly, natural selection includes (1) overproduction of offspring, (2) natural variation within a population, (3) limited resources, and the struggle for survival, and (4) selection by

1.6 ARTIFICIAL SELECTION IN HORSES

The most recent ancestor of the modern horse was probably similar to Przewalski's horse (top), a hardy little animal of the Asian steppes. Through artificial selection—selective breeding—variants as diverse as the thoroughbred racehorse (center) and the draft horse (bottom), have been developed (note the differences in leg structure). Such breeding programs must be maintained rigorously, however, because populations of horses left to their own devices will quickly revert to the original unselected type.

© Zoological Society of San Diego.

1.7 NATURAL SELECTION OF GIRAFFES

The giraffe's neck has long been a source of evolutionary speculation. The neck is apparently a feeding specialization for browsing on lofty tree foliage that is unavailable to most other mammals. How did this unusual specialization come about? Applying the principle of natural selection, we can speculate that the giraffe's ancestors had relatively short necks and probably had to compete for food and avoid predators along with many similar primitive browsing, hooved herbivores. Variants with slightly longer necks probably arose from time to time, but until long necks became important, the variants' impact on the giraffe population was minimal. Certainly, chance favored such variants. The longer-necked giraffes found untapped food sources in the higher branches. Because of their competitive edge, the longer-necked oddities thrived and thus were able to pass their novel genes on to more descendants than their shorter-necked contemporaries could.

the environment for those with traits that enable the individual to survive and reproduce.

A penciled manuscript in 1842 laid out the entire theory of the origin of species through natural selection. Darwin did not publish it, however. He knew that the idea would arouse fierce resistance and that he would have to back up every part of his idea with facts and experimental evidence. He set about preparing to defend his argument.

However, Darwin's robust health mysteriously began to fail. He began to vomit frequently and he complained of headache, nausea, fatigue, and heart palpitations. In time, because he tired quickly, he stopped seeing anyone, even his valued scientific colleagues. He resigned as secretary of the Geological Society. His father bought him a country house, and his wife became his nursemaid. Strangely enough, he continued to look completely healthy. His mind was as vigorous as ever, but his strength was gone.

The probable cause of Darwin's illness was not discovered until long after his death. It may have been the result of his great curiosity as a young man. In August 1835, he "experienced an attack of the Benchuga, the great black bug of the Pampas. . . ." It turns out that the bloodsucking Benchuga (now known as the barbeiros), *Panstrongylus megistus*, is the carrier of the protozoan *Trypanosoma cruzi*, which causes Chagas' disease. After a latency period of some years, the parasites usually invade the heart and the intestines, causing weakness, intestinal distress, and heart palpitations.

Gathering Evidence. When he was able to work, Darwin set about compiling information to back up his theory. He even suggested further tests. Darwin was also an experimental biologist, and he devised a number of ingenious laboratory experiments. He found through his experimentation that many kinds of seeds, as well as land snails, could have made the journey in sea water or perhaps be carried by birds or on driftwood from mainlands to oceanic islands. Such evidence was important because much of the foundation of his developing theory depended on the ability of species to disperse.

Working only a few hours a day, Darwin also published lengthy works on the systematics, or natural classification, of barnacles (still a major reference work), on his theory of the origin of coral islands, and on geology. But he continued to develop his theory of natural selection. He showed his early manuscript on natural selection to only one close friend, the botanist Joseph Hooker, who remained doubtful.

Twenty years passed, and finally in 1856 Darwin began writing his major work, to be called *Natural Selection.* It was planned as an enormous, six-volume monograph. But in 1858 his work on the manuscript was suddenly interrupted. Unexpectedly, another manuscript, a very brief one, arrived in the mail from A. R. Wallace, a young naturalist working in Malaya. Wallace asked politely whether Darwin would care to make any comments on the manuscript. The article was a short but well-written statement about evolution and natural selection—Darwin's ideas had been quite independently deduced by someone else! Darwin was mortified. In a letter to his friend Hooker he lamented, "So all my originality, whatever it may amount to, will be smashed. . . . Do you not think that this sketch ties my hands? . . . I would far rather burn my whole book, than that he or any other man should think that I had behaved in a paltry spirit." Darwin was about to be scooped; it is not an uncommon experience in science. Would his own work now be thought to have been based on the ideas of another?

Darwin's friends persuaded him to allow his previously unpublished 1842 summary and the text of a long letter describing his ideas to be presented together with Wallace's paper before the Linnean Society, a scientific association in London. Because of Darwin's much more substantial evidence his paper was given first. The two were presented in July and were published in August 1858. Now Darwin went furiously to work and, putting aside the idea of a huge, definitive monograph (which never was written), he quickly finished an "abstract" of it, the famous *Origin of Species,* published in 1859. The first edition sold out on the first day.

The Origin of Species

In the first chapter of *Origin of Species,* Darwin sought to establish that animals and plants under domestication are extremely variable, that the variation is heritable, and that the many domestic varieties have risen, under artificial selection, from wild ancestors. In the second chapter, he drew together what evidence he could find to show that plants and animals in nature were variable too, and that species differences were only one aspect of variation. In the third chapter, he discussed Malthus' idea of the *struggle for existence,* the idea that the natural reproductive capacities of living things greatly outreach the ability of the environment to support them, so that most organisms must perish by starvation or disease without even reproducing (Figure 1.8). In the fourth chapter, he introduced natural selection:

1.8 OPOSSUMS

Typically, organisms produce far more offspring than can possibly survive. According to Darwinian theory, only the "fittest" survive, with the environment acting as the selective agent. The theory has held up rather well, for while numbers in any population may fluctuate for many reasons, over the long run most organisms succeed only in replacing themselves.

How will the struggle for existence . . . act in regard to variation? Can the principle of selection, which we have seen is so potent in the hands of man, apply in nature? I think we shall see that it can act most effectively. . . . If such [variations] do occur, can we doubt (remembering that many more individuals are born than can possibly survive) that individuals having any advantage, however slight, over others, would have the best chance of surviving and of procreating their own kind? On the other hand, we may feel sure that any variation in the least degree injurious would be rigidly destroyed. This preservation of favourable variations and the rejection of injurious variations, I call Natural Selection. . . ."

Why Are There Mosquitoes? The theory of evolution through natural selection was immediately attractive to biologists because of its enormous explanatory power. Familiar and ancient observations, as well as more recent observations in natural history and paleontology (the study of fossils) were seen for the first time to make sense, and even philosophical questions were brought under a new light. The explanatory power of the theory was its strength but also, paradoxically, its weakness. It has the potential to become meaningless, since it tends to explain everything and thus is nearly impossible to falsify.

In a philosophical sense, the stage was set for life on the planet to be viewed in a new light. It seemed to some that the continued existence of life had come down to nothing more than survival and procreation. Those organisms successful in these matters would have continued to populate the earth, and the unsuccessful ones would have died out.

Of course, the very idea of reducing the grand pageant of life, including human life, to such terms seemed morally repugnant to many of Darwin's Victorian contemporaries and is equally repugnant to many people today. But nonetheless, the explanatory powers of the general theory could not be denied.

Testing Evolutionary Hypotheses

Even the brilliant debater Thomas Huxley, Darwin's most pugnacious defender in the 19th century, felt that the idea would remain untested and unproven until someone directly observed the experimental creation of a species. However, Darwin did not believe such a test was possible. He thought that such events simply take too much time. One of the key ingredients in his theory is time—a great deal of time—and a hypothesis cannot legitimately be tested with observations that were made before the hypothesis was formed; thus the theory's power to explain ancient observations did not provide a valid proof. It is not enough to have explanatory power; a hypothesis must have predictive power if it is to be tested and thereby to gain the status of a theory.

Can hypotheses about the past be tested? Yes. Predictive power means only the power to predict the outcome of observations before they are made. For instance, hypotheses and predictions in paleontology can be made and then tested by digging up fossil bones or by analyzing fossil bones that have previously been dug up.

Such experimental tests can be made of past evolutionary relationships, but it is not easy. The fossil record is spotty and ambiguous, especially when it comes to tracing transitional forms—those that form connections between existing groups. Perhaps the best large-scale test of the theory of evolution has come with the advent of **protein-sequencing** techniques in the last 15 to 20 years. It was predicted, on the basis of evolutionary theory, that species that appeared to be closely related on the basis of structure would have closely similar proteins, and that the sequences of these proteins would diverge in the same way the species were presumed to have diverged (Figure 1.9). Not too surprisingly, this prediction turned out to be true. Furthermore, the prediction was based on the theory of evolution. So in that sense the theory itself was tested.

Keep in mind that evolution and natural selection are not the same thing. The theory of evolutionary origins of species by repeated branching is one thing; the theory of natural selection is another, and the latter is exceedingly difficult to test experimentally. Of course, one can simulate natural selection in the laboratory. Darwin himself observed, "To keep up a mixed stock of even such extremely close varieties as the variously coloured sweet-peas, they must be each year harvested separately, and the seed then mixed in due proportions, otherwise the weaker kinds will steadily decrease in numbers and disappear" (Origin of Species). An experimental test of natural selection in moths, under more natural conditions, is described in Chapter 16.

The difficulty with the concept of natural selection is not that it predicts too little, but that it explains too much. As soon as it became apparent that natural selection explained the adaptations of organisms to their environment, natural selection began to be used in a lazy way to explain all kinds of biological phenomena. To explain any phenome-

1.9 TESTING EVOLUTIONARY HYPOTHESES

A computerized, treelike scheme shows evolutionary relationships based on differences in a protein called cytochrome *c*. Essentially, the organisms are arranged according to differences in the protein. The numbers on the chart are a measure of those differences.

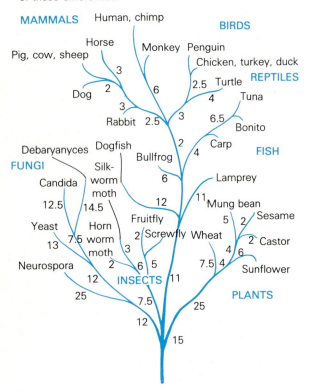

non, it began to seem that the speculating biologist need not prove anything new, but needed merely to dream up some halfway plausible way in which the phenomenon might benefit the organism. And if the imagination failed in this, it seemed adequate to state that even if the benefit of the phenomenon to the organism was not obvious, surely there must be one. For instance, if the Indian rhinoceros has one horn and the African rhinoceros has two (Figure 1.10), it might be argued that there must be something about the two environments that makes this arrangement the best one for all concerned.

This extravagant faith in the power of natural selection leads to the benign view that everything is always for the best and there is a reason for everything. Such a viewpoint elevates natural selection to the status of a new, all-powerful deity, and such faith is contradictory to science. It is certainly a contrast to Darwin's own conviction of the role of chance and **historicity** in evolution. Historicity is

the notion that things are as they are because of events that occurred in the past. Natural selection is a powerful phenomenon, but it is limited. The adaptations of organisms are marvelous, but they are never perfect, just as evolutionary change is always opportunistic and never predictable.

The Impact of Darwin

Of course, Darwin's ideas were immediately and bitterly controversial. And, unlike the once-controversial ideas of Newton and Einstein, his writings continue to resist resolution. Almost all biologists believe in the central notions of natural selection and evolution, but they still differ greatly among themselves on such questions as what constitutes a species, how species really change, why different species can't mate, and whether most evolutionary change comes in small, continuous increments or in larger and less regular leaps. Darwin, through his works, remains an active participant in the debate.

Darwin went on to publish other major theoretical works and continued his simple but first-rate experimentation. For example, he discovered plant hormones, as we'll see in Chapter 24. *The Expression of the Emotions in Man and Animals* (1872) was the foundation of the modern sciences of ethology and comparative animal behavior (see Chapter 38). *The Formation of Vegetable Mould Through the Action of Worms* (1881) established the importance of earthworms in soil ecology. Because of his work on orchids, climbing plants, and insectivorous plants, modern botanists claim Darwin as a fellow botanist.

But it is on the *Origin of Species* and a related work, *The Descent of Man*, that Darwin's reputation is based, and it is the idea of natural selection that has become the central concept of the science of biology.

There has been a longstanding argument over just how much of an intellectual achievement the theory of natural selection really was. Some biologists and historians of science have maintained, in all seriousness, that the idea is so simple as to be trivial. However, among Darwin's many gifts was an ability to see simple and obvious things that had previously escaped notice and to give them simple and obvious explanations; at least they seem simple and obvious to us, living in a post-Darwinian age. But again, much of science involves clever people pointing out something that, once explained, seems as if it must have been known all along. Our reactions must be somewhat irritating to the original scientists. Thomas Huxley, Darwin's contemporary and perhaps his best-known supporter, reacted in a blunt and disarming fashion when first presented with the idea. When the *Times* of London sent him a copy of the *Origin of Species* to review, he is said to have exclaimed of himself, "How extremely stupid not to have thought of that!"

Reducing and Synthesizing

While Darwin was able to focus on such detailed processes as earthworm diggings and snail longevity, he was a master at seeing the grand scale, the overall picture. His theory of evolution, in fact, could not have been crafted without this ability to generalize and deduce encompassing principles. Today, however, those who seek to test the grand old theory do not take the same approach; rather, they resort to testing fine detail. And so we see that there are different ways of approaching science.

Most of the scientific progress in biology in this century has not been achieved through such grand conceptual breakthroughs, but through what can be called **reductionism.** In reductionist science, the questions are small ones that can be stated and answered in specific, precisely defined terms. The reductionist tradition is based on the controlled experiment, an experiment in which only one variable at a time is allowed to change and all others are kept rigorously constant. The data from such experiments tend to be in the form of numbers, often plotted in graphical form. Cause and effect are determined, whenever possible, by eliminating all competing explanations until one is left. The reductionist seeks to find mechanisms, not reasons, for observed phenomena. In a sense, it is the tradition of physics and chemistry applied to biology.

Those dealing with overview and the big picture are called **synthesists.** They seek underlying order in other ways. In general, they seek to show that seemingly unrelated observations can be related after all. Synthesists are the emotional descendants of Darwin, clearly interested in forming grand rules and sweeping generalities. Of course, for science to work well, the reductionist and synthesist approaches should be interrelated. The synthesist, after all, is able to generalize because he or she has available so much detailed data produced by the reductionist. Furthermore, the synthesist's vision can be validated only through precise experimentation by the reductionist. A recent example of this interaction, in the area of biochemistry and cellular physiology, is the chemiosmotic theory of Peter Mitchell (see Chapter 6). Mitchell used the known pieces to construct a larger and previously unimagined system that explained many previously puzzling questions about the synthesis of high-energy compounds in cells.

Characteristics of Life: The Ultimate Reduction

As the reductionists take us spiralling inward to ask the most fundamental questions about life, they are finally confronted with the most basic question of all: What is life? How can we know the living from

---

1.10 IS NATURAL SELECTION THE ANSWER?

Why does the African rhinoceros have two horns and the Indian rhinoceros have only one? Remember that any explanation would have to be testable. What explanations come to mind? Can they be tested? Is it possible that there is no selective advantage of one horn over two? Or perhaps there is no longer a selective advantage. You won't find many biologists willing to say much about questions like this.

Signs of Life

Life is notoriously hard to define, but we know that it has certain properties. For example, life takes in and uses energy in order to retain its own highly organized state. Most energy enters the living realm through photosynthesis in the commonplace green plants.

While all organisms respond to their surroundings, some living things are often extraordinarily responsive to external stimuli. The chameleon's (*Chameleo senegalensis*) lightning "tongue-flick" response to the presence of a moth, a tasty morsel, clearly supports this assertion.

The aphid seen here (*below left*) is giving "live birth" to an offspring, an attribute of "higher life forms," but the real surprise is that the offspring developed *parthenogenetically*—from an unfertilized egg, just one more twist in diversity. Living things reproduce in an endless variety of ways. (*below right*) The embryos of some terrestrial vertebrates develop in a protective egg as did this emerging gavial (or gharial, *Gavialis gangeticus*), an endangered species from northern India.

It is important that living things adapt to their environment, and organisms do so in quite surprising ways. (*left*) For instance, chimpanzees are known to make and use simple tools. The chimp carefully strips a slender stem of its leaves and uses it to gently probe into a termite's nest. The stick is carefully withdrawn and any clinging termites are eaten with relish. (*below*) When not soaring through majestic mountain passes, these bald eagles pass the time freeloading in a garbage dump located on Adak Island in the Aleutians. Such visits occur on a regular basis when the usual food supplies become scarce.

With the physical environment in constant change, living things must also change, or evolve, if they are to remain in harmony with their surroundings. Much of what we understand about evolution comes from the study of fossils, but the fossil record is notoriously incomplete, especially in providing clues to transitional organisms—extinct forms representing steps in the evolution of today's lines. (*left*) The fossil insect seen here (*Sphecomyrma freyi*), is a rare and exciting find, a transition species representing a stage in the evolution of ants from waspy ancestors. Experts E. O. Wilson and Frank Carpenter state that its head, eyes, abdomen, and sting are typical of wasps, but its thorax (middle body) and waist are decidedly antlike and its antennas contain characteristics of both insects.

Organization is the catch-word of life. Even seemingly simple organisms such as the radiolaria (*below left*) are highly organized. The radiolarian skeleton is essentially of glass and its sculptured appearance is testimony to how beautifully complex life forms can become.

Life must exist within rather narrow limits and many forms have devised remarkable ways of regulating their internal environments. (*below right*) These bumblebees are vigorously fanning the nest, a common practice among bees when temperatures rise.

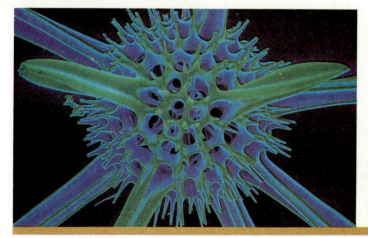

the nonliving? At many different levels, this is a simple question. It is easy to tell living things from nonliving—until the question falls into the hands of the reductionist. Then it can become most difficult, indeed.

In fact, there are disagreements over the characteristics that mark life, but there are some rather constant features that have been agreed upon, including these five (Essay 1.3):

1. Living things are orderly. The world is an unkempt and unruly place that tends to move toward further disorganization. For life to go against this trend it must be orderly and organized. The complex phenomena associated with life demands such coordinated organization.
2. Living things take chemicals and energy from the environment. With these chemicals and energy, they grow and maintain themselves.
3. Living things reproduce. By reproducing, living things pass on a chemical code of their organization to later generations.
4. Living things respond to their environment. Certain characteristics of the environment will be registered and, to some degree, reacted to by living organisms.
5. Living things adapt to the environment. Adaptation involves adjusting in a beneficial way. Individuals can adapt, as when a wolf grows thicker fur in winter. Populations can also adapt, as those individuals in the group with the traits that best fit the environment survive longer and reproduce more. In time, such adaptation by generations of populations results in *evolution*.

We should not rely too strongly on any such list alone since there are always exceptions. Further, many nonliving things possess at least one of these traits. Exceptions include such beings as viruses, whose responsiveness is pretty much limited to reproducing when they can and dying when they must. However, under some severe conditions, instead of dying they form crystalline structures. Some people have difficulty believing that anything that could crystallize could be alive. Regarding nonliving (we can't say "dead" because of the implicit assumption that it had once been living) entities mimicking life processes, we see that mineral crystals and oil droplets can grow and seem to reproduce by spontaneously breaking up, rusting iron takes energy from the environment, and many metals are highly ordered. Thus, we must admit that our definitions of life are inadequate and forge ahead. But keep in mind that the inadequacy is not just with our definition. The problem is fundamental. We really don't know much about the basic nature of life (although you may think that seems a bit improbable, considering the size of this book).

APPLICATION OF IDEAS

1. In explaining scientific and other intellectual achievement, someone once said, "Chance favors the prepared mind." What does this mean to you, and how do you define "prepared"? As you answer, think of the background and experiences of Charles Darwin, and consider how both chance and preparation fit in with what he accomplished.
2. Develop an organizational diagram that illustrates how science might work. Include both the intellectual and technological aspects of science and clearly distinguish these.
3. One important aspect of scientific progress is the rise of new technology. With new inventions and techniques, it is possible to test hypotheses that were heretofore untestable. Yet no technological innovation has replaced the human intellect. Consider the relationship between technology and intellect. How important might an advanced scientific technology have been to Darwin?
4. An antievolutionist derides Darwinian evolution, saying that the notion of new species arising is "unscientific" since such an event has never been observed. How would you counter such an assertion? Are there other, perfectly acceptable, scientific mechanisms that have never been observed?
5. Are humans in modern society subject to natural selection or have we managed to thwart the process as it occurs in other organisms? How might we interfere with the process? Could there be other selective forces at work?

KEY WORDS AND IDEAS

AN EXPLANATION ABOUT RABBITS AND OCEANS

1. Many of Darwin's ideas about evolution arose from observations made during his voyage in the *Beagle*. Many of these were made in South America and neighboring islands.

2. Darwin's observations of life in South America started him wondering about the absence of some species (rabbits) where one would expect them (grasslands). He pondered the meaning of fossils that were different from, yet vaguely similar to living species. He was also challenged by the similarity of different species within a common area.

3. Important to Darwin's growing notions about evolution was the revolutionary work of geologist Charles Lyell. Lyell's theories provided both a substantial time frame and far-reaching geological changes essential to an evolutionary scenario.

4. The prevailing opinion of the 19th century was that species were immutable (unchanging), but Darwin's observations suggested evolutionary descent with change.

5. The isolated Galapagos Islands offered Darwin a veritable laboratory of animals that had originated elsewhere and undergone change as they adapted to conditions there.

THE WORKINGS OF SCIENCE

1. Once Darwin had formulated the idea of **evolution,** or **descent with modification,** he began to see evidence of it everywhere.

The Methods of Science

1. While the **scientific method** can be elusive and difficult to describe, certain elements are standard. Most often, it begins with observations, followed by wondering, mulling over, and speculating about what an observation means.

2. A **hypothesis** is a provisional or tentative explanation of some observed phenomenon that must be tested. A **theory** is a larger generalization based on considerable evidence.

3. To be useful, a hypothesis must lead to *predictions* and be *testable*. The predictions suggest the test, which is often an **experiment,** or may simply call for more observations.

4. **Controlled experiments** include both an experimental, or variable, and a controlled element. The **control** becomes the standard of comparison and is used to prove that any difference between the two is due to the variable being tested and not to chance. The control must be identical to the experimental group to avoid introducing **uncontrolled variables.** In some kinds of experiments, a **placebo** is administered to subjects in the control group.

5. The **conclusion** is the investigator's decision about the hypothesis. Based on the experimental or observed results alone, the hypothesis will be accepted or rejected.

Going Home to Applause and Illness

1. Upon the completion of his voyage, Darwin was an immediate success as a scientist. He set to work, mulling over his notes and collections.

2. Darwin was convinced of the idea of evolutionary change, but lacked an explanatory mechanism.

EVOLUTION: THE THEORY DEVELOPS

The Theory of Natural Selection

1. A knowledge of livestock breeding and **artificial selection** suggested an evolutionary mechanism to Darwin, but it was the essays of Thomas Malthus that suggested the agent itself. Malthus observed that although more offspring were produced than could survive, natural populations reached a balance.

2. Malthus' ideas suggested that the environment was the agent of selection, and only the fittest individuals lived to reproduce. Darwin called selection by the evironment **natural selection.**

3. In addition to his observations and collections, Darwin relied heavily on experiments to provide evidence for his theory. He held off publicizing his theory for over 20 years until 1859, when *Origin of Species* was published.

The Origin of Species

1. In *Origin of Species*, Darwin presented his theory by first discussing artificial selection. He then described the great variability within species and followed this with Malthus' ideas on the **struggle for existence.** Finally, he introduced the idea of natural selection.

2. Natural selection includes (a) overproduction of offspring, (b) natural variation within a population, (c) limited resources and the struggle for survival, and (d) selection by the environment for those with traits that enable the individual to survive and reproduce.

3. As a hypothesis, natural selection can create intellectual laziness, since it has the potential to explain everything.

Testing Evolutionary Hypotheses

1. Evolutionary hypotheses can be tested by first making predictions and then, from the experiments or observations they suggest, accepting or rejecting the hypothesis.

2. Through new **protein sequencing** techniques, theoretical evolutionary relationships can be predicted and tested.

3. While evolution itself can be tested through observation, natural selection is far more difficult to test.

The Impact of Darwin

1. Darwin's major achievement was his theory of evolution through natural selection, but he also published in other areas of biology.

Reducing and Synthesizing

1. Science includes two aspects, **reductionism** and **synthesism.** Reductionism involves breaking problems down and investigating the smaller elements. Synthesism consists of putting many smaller ideas together into new grand schemes.

Characteristics of Life: The Ultimate Reduction

1. Living things have the ability to
   a. maintain a complex organization;
   b. take in matter and energy for growth and maintenance;
   c. respond to stimuli from the environment;
   d. adapt to a changing environment;
   e. reproduce their kind.

REVIEW QUESTIONS

1. In general, how did people in the 19th century perceive the species on the earth? What was their feeling about fossils that represented extinct life? (5)

2. In what way is the mara like the common rabbit? How do biologists explain the similarities? (2)

3. List three specific observations that might have been instrumental in Darwin's early thoughts about evolutionary descent. (4)

4. In what ways were Lyell's revolutionary ideas on geology important to the emerging theory of evolution? (4)

5. How does a hypothesis differ from a theory? (8)

6. What are the two most important characteristics of a hypothesis? (8)

7. Comment on the scientific validity of the statement, ''The experimenter then set out to prove the hypothesis.'' (8)

8. Suggest how a controlled experiment might be organized to test the usefulness of a new medicine. (8)

9. What is the purpose of an experimental control? (8)

10. Upon what, specifically, must a scientist base his or her conclusions? How are such conclusions usually reached? (8)

11. Was Darwin a careful scientist? Support your answer in detail. (9, 13–14)

12. List four important elements of the mechanism of natural selection. (12)

13. Why is the explanatory power of natural selection a problem? (15–16)

14. How can evolution, a gradual process, ever be tested? Be specific. (15)

15. In what way has the technique of protein sequencing been useful to evolutionary investigation? (15)

16. Distinguish between reductionism and synthesism. How is one dependent on the other? (17)

17. List four characteristics of life. Do any of these apply to nonliving entities? Explain. (20)

Small Molecules

In my hunt for the secret of life, I started research in histology. Unsatisfied by the information that cellular morphology could give me about life, I turned to physiology. Finding physiology too complex I took up pharmacology. Still finding the situation too complicated I turned to bacteriology. But bacteria were even too complex, so I descended to the molecular level, studying chemistry and physical chemistry. After twenty years' work, I was led to conclude that to understand life we have to descend to the electronic level, and to the world of wave mechanics. But electrons are just electrons, and have no life at all. Evidently on the way I lost life; it had run out between my fingers.

    Albert Szent-Györgi
    Personal Reminiscences

Szent-Györgi's wry comment on his search for the secret of life describes what has been called a reductionist's nightmare. This is because the "meaning of life" cannot be reduced, it seems, to basic understandable processes. He suggests that the nature of life will never be grasped as a pure, crystalline gem of truth. If the secret is ever unveiled it will probably be found somewhere in the very complexity that Szent-Györgi tried to discard. Nonetheless, the reductionist approach is valid; the processes of life depend ultimately on the behavior of lifeless molecules, atoms, and electrons moving mindlessly in space. The point is, in order to make sense of what we do know about life, we must know about its components. It is for this reason that we find ourselves immersed, from time to time, in the precise and measured world of biochemistry.

Of course, biology can be "done" without biochemistry. Darwin, for example, had very little knowledge of the subject, and even now, the woods are full of biologists with a love of the outdoors but a hatred of beakers, bases, and balanced equations. However, perhaps their love might even increase if they had a greater appreciation for the complexities of life at the molecular level. It cannot detract from the beauty of a delicately veined leaf to know that it can make food from carbon dioxide and water.

It may be true that some biochemists must periodically be convinced of the existence of the platypus, but only a biochemist can tell us how a fat little hummingbird is able to fly nonstop across the Gulf of Mexico. In this chapter, then, we will briefly enter the biochemist's world. We will begin with some basic information about atoms, inorganic molecules, and small organic molecules. At some point we'll cross the line between the lifeless world of chemicals and the vibrant world of life—but we're not sure where that line is.

ELEMENTS, ATOMS, AND MOLECULES

An **element** *is a substance that cannot be separated into simpler substances by purely chemical means.* There are 92 naturally occurring elements in the universe. Of these, only six—sulfur, phosphorus, oxygen, nitrogen, carbon, and hydrogen—make up approximately 99% of living matter. SPONCH (from the initial letters of their names) is an acronym that might help you to remember these six elements. Of course, these are not the only elements important to life. For instance, plants could not manufacture food without magnesium; and while a pine tree can live with only trace amounts of sodium, you cannot. Sodium is important in the functioning of nerves and muscles. Table 2.1 lists elements with significant roles in life. Most of the remaining elements are rare in nature and of less interest to biologists. Before we go on with our definitions, check below for a quick review of the meaning of the basic chemical symbols.*

An **atom** *is the smallest indivisible unit of an element* (Figure 2.1). An atom can be divided into smaller parts, as we'll see shortly, but if it is, it loses the special properties associated with the element it represents. Each element owes its special characteristics to the specific structure of its intact atoms. In other words, once separated, the subatomic parts no longer represent an element.

A **molecule** *is a unit formed by two or more atoms joined together.* The atoms of a molecule can be of the same kind or of different kinds; that is, they can be formed of the same or different elements. For example, a molecule of oxygen consists of two oxygen atoms bound together. The symbol for oxygen is simply O, but the molecule is symbolized by the chemical formula O_2. (The oxygen of the air is molecular oxygen.) Similarly, hydrogen gas consists of molecular hydrogen, H_2. And you are undoubtedly aware that two atoms of hydrogen and one atom of oxygen combine to form one molecule of water, H_2O.

---

*The number appearing *before* a chemical symbol shows how many units there are of whatever follows that number. The small subscript *after* a chemical symbol indicates the number of atoms of the element directly preceding. Thus in H_2O, we find two atoms of hydrogen and one of oxygen, combined to form one molecule of water. The symbol $12H_2O$, then, refers to 12 molecules of water. If a subscript number follows a molecular symbol that is in brackets, it indicates the number of molecules that precedes it and assumes they are part of an even larger molecule. Thus $(CH_2O)_3$ is such a molecule, composed of three CH_2O subunits.

2.1 A BIRD'S-EYE VIEW OF AN ATOM

Atoms as seen near the limits of magnification through the electron microscope.

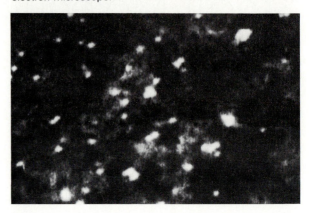

Water, by the way, is a **compound.** And what is a compound? *A compound is any pure molecular substance in which each molecule contains atoms of two or more different elements in specific proportions.*

Atomic Structure

Atoms are made up of subatomic particles. About a hundred distinct kinds of subatomic particles have been described, but most of them are short-lived and play no known role in biology.

The three stable subatomic particles that make up atoms are **neutrons, protons,** and **electrons,** and it is these that will concern us here. Protons and neutrons are about equal to each other in mass, and are much heavier than electrons. Thus, protons and neutrons make up most of the mass of the universe. Protons have a positive (+) electrostatic charge and electrons have an equally strong negative (−) electrostatic charge; neutrons, on the other hand, have no electrostatic charge (they are electrically neutral). Unlike charges attract and like charges repel; thus protons and electrons are attracted to one another through electrostatic force.

An atom is made up of an extremely small, incredibly dense cluster of neutrons and protons surrounded by orbiting electrons. The small, dense cluster of neutrons and protons is called the **atomic nucleus.** A force called the **strong nuclear force** holds clusters of neutrons and protons together, but it is poorly understood. Nonetheless, the strong nuclear force explains why the positively charged protons do not repel each other and tear the nucleus apart.

The number of neutrons plus protons determines an

atom's **atomic mass** (or **atomic weight,** an archaic but still common term). Atomic mass is measured in *daltons.* An atom of hydrogen, the lightest element, has a mass of about 1 dalton, while an atom of one of the heaviest elements, uranium 238, weighs just about 238 times as much. Thus the atomic mass of uranium 238 is 238 daltons (Figure 2.2). And, by the same token, *the* **molecular mass** *(or molecular weight) of a molecule is the sum of the atomic masses of the atoms that make up the molecule.*

Each element has its own **atomic number,** *which*

equals the number of protons in the atom. For instance, the atomic numbers of the six SPONCH elements are 16, 15, 8, 7, 6, and 1, respectively. (Notice that the acronym SPONCH lists the six elements in order of decreasing atomic number.) The number of protons in atoms also equals the number of electrons.

Neutrons, Isotopes, and Biology. From what we have just seen, determining the atomic mass of an element seems straightforward—we just add

TABLE 2.1

ELEMENTS ESSENTIAL TO THE PROCESSES OF LIFE

Element	% of SPONCH atoms in humans	Symbol	Atomic number	Atomic mass	Example of role in life
Calcium		Ca	20	40.1	Bone; muscle contraction
Carbon	10.50%	C	6	12.0	Constituent (backbone) of organic molecules
Chlorine		Cl	17	35.5	HCl in digestion and photosynthesis
Cobalt		Co	27	58.9	Part of vitamin B_{12}
Copper		Cu	29	63.5	Part of oxygen-carrying pigment of mollusk blood
Fluorine		F	9	19.0	Necessary for normal tooth enamel development
Hydrogen	60.90%	H	1	1.0	Part of water and of all organic molecules
Iodine		I	53	126.9	Part of thyroxin (a hormone)
Iron		Fe	26	55.8	Hemoglobin, (oxygen-carrying pigment of many animals); cytochromes (electron carriers)
Magnesium		Mg	12	24.3	Part of chlorophyll, the photosynthetic pigment; essential to some enzyme action
Manganese		Mn	25	54.9	Essential to some enzyme action
Molybdenum		Mo	42	95.9	Essential to some enzyme action
Nitrogen	2.47%	N	7	14.0	Constituent of all proteins and nucleic acids
Oxygen	25.60%	O	8	16.0	Molecular oxygen in respiration; constituent of water and nearly all organic molecules
Phosphorus	0.16%	P	15	31.0	Energy-rich bond of ATP
Potassium		K	19	39.1	Generation of nerve impulses
Selenium		Se	34	79.0	Essential to the workings of many enzymes
Silicon		Si	14	28.1	Diatom shells; glass sponge exoskeleton; arteries
Sodium		Na	11	23.0	Salt balance; nerve conduction
Sulfur	0.06%	S	16	32.1	Constituent of most proteins
Vanadium		V	23	50.9	Oxygen transport in tunicates
Zinc		Zn	30	65.4	Essential to the workings of the alcohol oxidizing enzyme

SPONCH shown in color

2.2 THE LIGHTEST AND HEAVIEST ATOMIC NUCLEI

Uranium 238 has a massive nucleus compared to that of hydrogen. The nucleus of uranium has 92 protons (+) and 146 neutrons (0). Hydrogen has just one proton and no neutrons (in its most common form). This tells us that the atomic mass of uranium is approximately 238 times that of hydrogen.

Nucleus of hydrogen	Nucleus of uranium-238
1 proton	92 protons
0 neutrons	146 neutrons
1 dalton	238 daltons

protons and neutrons. Hydrogen has one proton and no neutrons, so its mass is 1. Carbon has six of each, so its mass is 12, and oxygen's, with eight protons and eight neutrons, is 16. But, as it turns out, things aren't quite as neat as we might hope. Atomic mass varies among the atoms within most elements simply because the number of neutrons varies. The atoms of the elements are grouped according to their atomic masses and the groups are known as **isotopes** of that element.

The nuclei of some isotopes are unstable, or **radioactive.** "Unstable" refers to the decomposition or *decay* of the atomic nucleus, which releases radiation in some form—either an energetic sub-atomic particle (alpha or beta particles—helium nuclei or electrons, respectively), a highly energetic photon (gamma ray—similar to an X ray), or some combination of these. In the process of decay, some of the radioactive isotopes change from one element to another as they lose mass. The new element may or may not be radioactive.

The time required for half of the atoms of any radioactive material to decay is called the isotope's **half-life.** Half-lives can vary considerably. Some of the laboratory-produced radioisotopes have a fleeting half-life of seconds or minutes. On the other hand, most naturally occurring radioisotopes are extremely durable; some have half-lives of billions of years. Uranium 238, for instance, has a half-life of about one billion years, after which half

of its atoms would have formed an isotope called lead 206.

Isotopes are important to scientists in a number of ways. The longer-lived ones are often used in dating fossil-bearing samples from the earth's crust, or indeed, the earth itself. Such dating yields important clues to ancient geological and evolutionary events. In medicine, powerful radioisotopes are used in radiation treatment of cancer-ridden tissues. And, of course, scientists are vitally interested in the destructive effects of radiation on all life.

Research biologists also use shorter-lived, relatively benign, low-energy radioisotopes as **tracers,** to determine where certain chemicals go and how they behave in living cells. For example, the use of radioactive phosphorous, nitrogen, and hydrogen have been vital in determining the structure and function of DNA, the gigantic molecules that bear the hereditary information of a species.

Electron Orbitals and Electron Shells.   Electrons move, or occur, in definite regions outside the atomic nucleus. Their distance from the nucleus depends on each electron's energy. It requires energy for such lightweight negatively charged particles to resist being drawn into the positively charged nucleus. Electrons with minimum energy will be found closest to the nucleus, while their more energetic brothers will remain more distant. Further, electrons tend to distribute themselves as far from each other as possible.

The regions where electrons reside are called **orbitals** because of the way they were once depicted in drawings, namely as flat circles moving around the nucleus—much as the planets move in their orbits around the sun. There are many different ways of drawing electron orbitals, but probably none are truly representative of actual atoms. One common way of drawing a picture of an atom is to show the electron orbitals as "probability clouds," depicted as shaded areas in which the density of the shading is supposed to be proportional to the probability that an electron will be exactly at that place at any one point in time.

Figure 2.3 shows the orbitals and probability clouds of helium (which has two protons and two electrons) and of neon (which has ten protons and ten electrons). The two electrons of helium do not travel in flat circles, but can be found anywhere in a small, spherical orbital, fairly close to the nucleus. Two of the ten electrons of neon are also found in the same kind of small, spherical orbital, close to the nucleus, but the other eight electrons are found in one spherical and three odd, dumbbell-shaped orbitals that stick out at right angles to one another. Note that the electrons travel in pairs, so eight electrons are found in the four orbitals.

In another useful way of visualizing atoms, the **electron shell model,** the atom is drawn as a flat, two-dimensional structure, although, as we have stressed, real atoms exist in three dimensions. The nucleus is shown as a small circle labeled with the number of protons and neutrons it holds. Surrounding are a number of concentric rings representing energy levels or **electron shells.** Each, as we will see, contains specified numbers of electrons arranged according to their general energy levels (Figure 2.4). Although the electron shells represent energy levels, they are not to be confused with

2.3 ELECTRON ORBITALS

One way of depicting electron orbitals is with electron density clouds. An electron pair may be anywhere within the cloud at a given point in time. Here the cloudlike figures also represent the orbitals of helium and neon. **(a)** The two electrons of helium occupy a single, spherical orbital. With the exception of hydrogen, all elements have just two electrons in this first or innermost orbital. For example, neon has ten electrons, the first two of which move in a spherical orbital similar to that shown for helium. Of its other eight electrons, one pair moves in a somewhat larger spherical orbital and the other three pairs move in a pathway somewhat resembling a dumbbell **(b).** Neon's three dumbbell-shaped orbitals **(c)** are arranged as far from each other as possible.

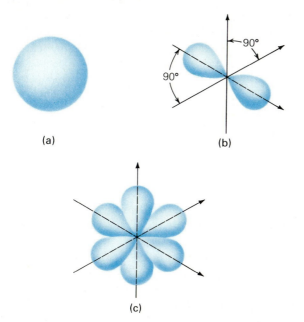

(a)

(b)

(c)

2.4 THE ELECTRON SHELL MODEL OF HELIUM AND NEON

Representing atoms as flat concentric circles of electrons surrounding the atomic nucleus is not consistent with what we know about atomic structure, but it is sometimes useful to convey other types of information. Note that the electrons in each shell have been depicted as pairs and that the pairs are separated from each other as far as possible. The concept of electron pairs is important in explaining how molecules are formed, which is one of the reasons for depicting the atoms in this manner.

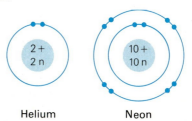

Helium   Neon

orbital paths, which we know do not take the form of concentric rings (compare Figure 2.4 with Figure 2.3). In the flat electron shell diagrams, the electrons are shown as paired black dots placed at the top, bottom, left, and right sides of the circle.

Actually, electrons do usually travel their orbitals in pairs, and, in fact, it requires a considerable amount of energy to pry apart two electrons that are sharing an orbital. Correspondingly, two unpaired electrons will release a considerable amount of energy if they are allowed to form a pair. The tendency of electrons to form pairs is one of the important forces in atomic and molecular structure and in chemical interactions.

Filled and Unfilled Shells. Now that we have some idea of atomic structure, let's see how such structure is important in the interactions or chemical behavior of the elements. We can begin by noting that there are three important forces, or **energetic tendencies,** that determine most of what happens in chemistry. These forces do not always operate in the same direction, however. In fact, conflicts between them are responsible for most of the properties of chemical interactions. We have already mentioned one force, *the tendency of orbiting electrons to travel in pairs.* We have also already noted that, because of positive and negative magnetic charges, *atoms show tendencies toward electrostatic attraction and repulsion, resulting in a balancing or nullifying of the opposite charges.* Now we'll consider a third energetic tendency that is probably the most important of all these competing forces when it

2.5 FILLING OUTER ELECTRON SHELLS

The electron shell model of sulfur, shown filling its shells according to the rules. Note the positioning of the electron pairs and also take note of the number of electrons in each shell. Atoms fill their shells from the inside out, and the maximum number of electrons in each shell (in the lighter elements) is 2, 8, 8. Only the outer shell remains unfilled, making it a good place for leftovers. This unfilled outer shell is what makes the elements interesting. You might say that chemistry is really the study of outer-shell electrons.

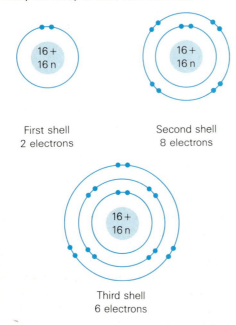

First shell
2 electrons

Second shell
8 electrons

Third shell
6 electrons

Sulfur: 16 electrons must fill shells according to the rules

comes to chemical reactions and molecular structure. This is the *tendency of atoms to establish completely full outer electron shells.*

What do we mean by "completely full outer electron shells"? Here, we resort to a rather dogmatic statement, sometimes called the "octet rule." The rule is most easily explained through the use of the electron shell atomic model. The innermost electron shell (designated "K") has room for just two electrons—one electron pair. The next shell (the "L" shell) is filled when it has eight electrons—four electron pairs. If there is a third (or "M") shell, it, like the second, has room for four pairs of electrons. So counting outward in our electron shell model, the first three shells are filled when they have two, eight, and eight electrons, respectively. Here we will be concerned only with those first three electron shells since most biochemical reactions involve elements with three or fewer shells.

Lighter atoms, like those of the SPONCH elements, always fill their electron shells progressively, completing the first before beginning the second, and then filling the second before beginning the third (Figure 2.5). Thus, if a SPONCH atom does not have enough electrons to fill all its electron shells, it will be the *outer shell* that ends up only partially filled. But, in fact, the atom usually finds a way to fill that shell—*and the strong tendency to fill the outer shell is what brings about chemical reactions and allows* **chemical bonds** *to form.*

As we mentioned, the different energetic tendencies, or needs, of the atom are sometimes in conflict. As an example, one tendency is to balance proton (+) and electron (−) charges. But when the protons and electrons are balanced, as they often are in the individual atoms of an element, the outer shell is frequently left unfilled. For example, each individual atom of oxygen has eight protons; thus is electrostatically balanced when it has eight electrons in its shells. But after the first shell is filled, there are only six electrons left to fill the second, or outer, shell, where eight is the stable number. Therefore, the outer shell of an electrostatically balanced oxygen atom is short two electrons. If the oxygen atom were to fill its outer shell with electrons, it would have ten electrons and only eight protons, and would thus have lost the balance between protons and electrons. How are these competing energetic tendencies accommodated? Before we answer this question, it will be helpful to understand the competing energetic tendencies better by looking closely at a class of elements, the **noble elements,** in which competition among the energetic tendencies just doesn't happen.

The so-called noble elements, including helium, neon, and argon, have no problems maintaining

2.6 ARGON, AN INERT ELEMENT

The electron shell diagram of argon shows filled first, second, and third (outer) shells. This element is very stable because it is electrically balanced, has a full outer shell, and has no unpaired electrons.

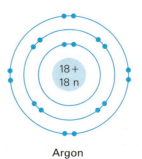

Argon

2.7 ELECTRONEGATIVITY IN THE SPONCH ELEMENTS

Here, some of the elements are arranged according to their electronegativity. The SPONCH elements and others important to life are indicated in color. At the extreme left, familiar elements including Li (lithium), Na (sodium), K (potassium), Mg (magnesium), and Ca (calcium) are only weakly electronegative and quite often form ionic bonds as they give up electrons. Those at the far right, including F (fluorine), Cl (chlorine), Br (bromine) and I (iodine), are strongly electronegative and also participate in ionic bonding, but they generally capture electrons from the first group. Note the value of C (carbon), one of the core elements of life. What behavior would you expect of its electrons? Why?

H 2.1																	
Li 1.0	Be 1.5											B 2.0	C 2.5	N 3.0	O 3.5	F 4.0	
Na 0.9	Mg 1.2											Al 1.5	Si 1.8	P 2.1	S 2.5	Cl 3.0	
K 0.8	Ca 1.0	Sc 1.3	Ti 1.5	V 1.6	Cr 1.6	Mn 1.5	Fe 1.8	Co 1.8	Ni 1.8	Cu 1.9	Zn 1.6	Ga 1.6	Ge 1.8	As 1.9	Se 2.4	Br 2.8	
																I 2.5	

Source: Pauling, Linus. *The Nature of the Chemical Bond,* 3rd ed. (Ithaca, N.Y.: Cornell University Press, 1960).

both balanced charges and full orbitals. A look at Figure 2.6 should quickly explain why. For example, helium has two protons and two electrons, so it is electrostatically balanced. Its two electrons are paired and its first electron shell, which is also its outer shell, is filled, so that requirement is satisfied too. Similarly, neon has ten electrons balancing the ten protons of its nucleus, and the first and second (outer) electron shells are nicely filled. Argon (atomic number 18) has its eighteen protons balanced by eighteen electrons, and its outer shell, the third shell in this case, is also fully satisfied. Why, then, are these elements "noble"? Atoms of helium, neon, and argon remain aloof and unreactive because as individual atoms they are both electrostatically balanced and have filled shells. And as we have mentioned, it is through reactions between and among atoms that the energetic tendencies of the elements are finally satisfied.

Chemical Reactions: Filling the Outer Electron Shell

The tendency for atoms to fill their outer shell leads us into some vital ideas in chemistry. An atom can fill its outer shell in one of only three ways: it can gain one or more electrons from another atom, filling up the hole or holes in its outer shell; it can lose one or more electrons to another atom, stripping its original outer shell bare and leaving a new, full outer shell at a lower energy level; or it can share one or

more electron pairs with another atom. Thus outer shells are filled through the loss, gain, or sharing of electrons.

In each instance, two or more atoms become involved, and through this involvement the atoms engage in what are called **chemical reactions.** Since the loss, gain, or sharing of electrons between or among atoms changes the properties of the elements involved, chemists often define chemical reactions in a more general way. Chemical reactions are chemical changes in which the starting substances—the **reactants**—are chemically changed into new substances—the **products.** Thus if we choose as reactants sodium, a silvery metal that reacts violently with water, and chlorine, a heavy, pungent, greenish-yellow gas, these elements will enter a chemical reaction in which the product is sodium chloride, simple table salt. During this chemical reaction, the reactants, you will agree, have undergone a dramatic change in their physical properties.

Whether interacting atoms gain, lose, or share electrons is predictable, and an atom's ability to hold its electrons and to attract more is known as its **electronegativity.** As we see in Figure 2.7, atoms of the different elements differ considerably in this characteristic. The range in electronegativity goes from calcium with a low of 0.7 to fluorine with a high of 4.0. (It may not surprise you to learn that when fluorine and calcium react, fluorine will attract electrons from calcium.)

The point here is that it is the difference in electronegativity between interacting elements that determines what happens to electrons. As a rule of thumb, if the difference is greater than 2.0, then the atoms of the element with greater electronegativity will completely capture electrons from atoms of an element of lesser electronegativity. If the difference is less than 2.0, then there will be a tendency for interacting atoms to share electrons. (We will see how this sharing is done shortly.) Actually, it's not an either-or situation. It is rather a continuum, with the complete loss or gain of electrons at one end and the totally harmonious sharing of electrons at the other. In between there are molecules in which the electron is not evenly shared but one atom holds the electron an inordinate amount of time.

Where electrons are gained or lost, the formerly electrically balanced atoms take on a net charge. That is, those gaining electrons become negative while those losing electrons become positive (their protons exceed their electrons). Such charged atoms are called **ions.** In situations where sharing occurs but in an unequal manner, the molecule formed has no real net charge on it, but takes on a *polar* nature. That is, the end of the molecule lacking electrons most of the time becomes positive, while the end holding the electrons most of the time becomes negative. (One of the best examples of a polar molecule is water, which, as we will see, owes most of its life-supporting qualities to the very fact that it is highly polar.) Finally, where electrons are equally shared, the molecules are nonpolar. In nonpolar molecules the charges are equally distributed throughout.

The interaction of electrons between and among atoms enables the atoms to form chemical bonds—those forces that hold molecules together. Of particular interest to us here are four kinds of chemical bonds: **ionic bonds, covalent bonds, hydrogen bonds,** and **van der Waals forces.** The first two involve electron transfer and sharing, respectively, while the latter two do not involve the transfer or sharing of electrons but are more temporary attractions between and among molecules with charge differences.

Let's consider the least known of these, the van der Waals forces, first and then return to discuss them in more detail in the next chapter. These forces are electrostatic attractions between molecules brought about by very temporary, slight shifting of charges. The asymmetry of the charges is created by the constant movement and shifting of the negatively charged electrons. Such movement creates momentary positive and negative areas around

the molecule, and thus the molecules are drawn together at places with opposite charges. The strength of such individually weak forces depends to a great degree on how much surface one molecule exposes to its neighbors. The van der Waals forces explain why some of the large, chainlike, nonpolar carbon compounds such as the heavy petroleum oils and sticky petroleum jellies and tars cling together tenaciously. In the next chapter, we will see how van der Waals forces help determine the final shape of the protein molecule. For now, let's take a more detailed look at the other three types of chemical bonds.

Ions and the Ionic Bond. The ionic bond is best demonstrated in the formation of common

2.8 IONIC BONDING IN SODIUM CHLORIDE

Sodium, at the left, has one unpaired outer-shell electron and an electronegativity of 0.9. Chlorine, at the right, has seven outer-shell electrons, one of which is also unpaired. It is strongly electronegative with a value of 3.0. Accordingly, with a difference of 2.1 we can predict that chlorine will capture sodium's lone outer shell electron. This will fill chlorine's outer shell and sodium's full second shell will now be its outer shell. When this occurs, both will become ions, sodium having a net positive charge and chlorine having a negative charge. The sodium and chlorine ions are then drawn together since negative and positive charges attract each other, and this attraction forms an ionic bond.

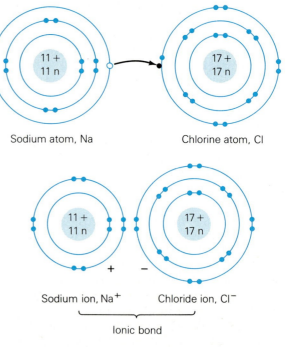

Sodium atom, Na   Chlorine atom, Cl

Sodium ion, Na$^+$ Chloride ion, Cl$^-$

Ionic bond

(Note the loss of the third shell in sodium)

table salt. Sodium (Na) and chlorine (Cl) differ considerably in electronegativity, with chlorine obviously being the candidate to capture electrons. But, it turns out that energetically, the two elements have complementary needs. As you see in Figure 2.8, sodium has 11 protons. In its elemental form it also has 11 electrons: two in the first shell, eight in the second shell, and only one in the third shell—seven electrons short of a completed outer shell, or one electron too many, depending on how you look at it. And that one extra electron is an energetic, unpaired electron as well. Sodium is only weakly electronegative, so there is no way for it to fill its outer shell—to gain seven electrons—at the expense of chlorine, but if it can get rid of one electron, the already full second shell will become the outer shell. On the other hand, the atom of chlorine, a strongly electronegative element, has 17 protons and 17 electrons in its free atomic state. Its third orbital has seven electrons, which is one short of a full shell—and one of the seven is an unpaired electron. Chlorine, therefore, can fill two of its three energetic needs by accepting one more electron.

Because of their special properties, atomic sodium and atomic chlorine react together very swiftly as an electron passes from sodium, the **electron donor,** to chlorine, the **electron acceptor.** As a result, sodium has only ten electrons to balance its eleven protons, which gives it a net positive charge of +1. But chlorine has 18 electrons as compared to its 17 protons, so it takes on a net negative charge of –1. The opposite charges attract, and sodium joins chlorine to form sodium chloride, common table salt (see Figure 2.8). Note that the name of the negatively charged atom has changed to *chloride* (although sodium is sodium, charged or not). As we mentioned, any charged atom or molecule, such as sodium (Na^+) or chloride (Cl^-), is called an ion.

The electrostatic attraction between the positively charged ion and the negatively charged ion forms the **ionic bond.** The ionic bond is defined as *a bond formed by electrostatic attraction after the complete transfer of an electron from the donor to the acceptor.*

The ionic bond is responsible for some interesting and complex structures. For example, sodium chloride forms crystals that can range from tiny grains to massive accumulations held together by ionic bonds between individual sodium and chloride ions (Figure 2.9). Since such linkages are quite fragile, if a sodium chloride crystal is placed in water, it will immediately dissociate (separate into its component ions); thus, we say that the salt *dissolves.* The net charge of the resulting *solution*

2.9 THE SODIUM CHLORIDE CRYSTAL

Sodium and chlorine form ionic bonds, but true discrete molecules do not form. Because of their electrostatic charges, the sodium and chloride ions attract each other, accumulating into crystalline formations that consist of alternating sodium and chloride ions. Crystals are of indefinite size and can vary from invisible to enormous. Regardless of the size, the geometry of a salt crystal is definite, as shown here.

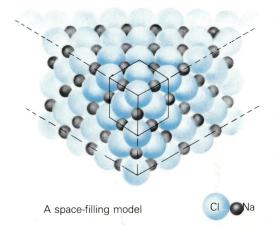

A space-filling model   Cl Na

formed remains approximately balanced—you can't get a beaker full of dissolved, positively charged sodium ions without approximately as many negatively charged chloride ions. Incidentally, salt water is an excellent conductor of electricity because of its mobile charged ions.

Before going on the covalent bond, let's take a moment to consider energetic tendencies and the formation of sodium chloride. What has happened to satisfy the competing demands? The demands are satisfied because

1. in sodium chloride there are no longer unpaired electrons;
2. charges are balanced in the sodium chloride crystal;
3. the outer shells of both sodium and chlorine become filled to capacity.

The Covalent Bond: Sharing Electron Pairs. As we know, the second way an atom can fill its outer electron shell is by sharing electron pairs with other atoms. This occurs when the electronegativity of either element is not great enough to allow one element to completely capture electrons from the other. Thus, two such atoms in close proximity can satisfy their shell requirements by simply sharing a pair of electrons. Atoms give up very little by sharing electrons, since such sharing allows them to meet their three energetic tendencies simultaneously. What they do give up, in energetic terms, is the freedom to go their separate ways. The sharing holds the two atoms together by what we called a **covalent bond** and thereby forms a molecule.

Consider the simplest covalent molecule, molecular hydrogen (H_2). It consists of two hydrogen atoms, each comprised of a nucleus with one proton and a single unpaired electron. In forming a molecule, the energetic tendencies of the element are satisfied as follows:

1. Charges between protons and electrons in the hydrogen molecule are balanced.
2. The two hydrogen atoms can pool their electrons to make a pair.
3. By pooling their electrons and then sharing the pair, they simultaneously satisfy the requirement of shell-filling for both atoms (two in the first, or innermost, shell).

Thus we have a molecule of hydrogen, as seen in Figure 2.10. Unlike the ionic bond in sodium chloride, the covalent bond produced in such a way is a rather powerful bond, not easily disrupted. Hydrogen molecules do not dissociate in water.

Since the two hydrogen atoms in the hydrogen molecule share their electrons in an equitable manner, the molecule contains a *nonpolar* covalent bond. Most molecules made up of one element only contain this type of bond. But, since the sharing of electrons often involves different elements with differing electronegativity, the sharing does not always occur on an equal basis. When the sharing is unequal, as in the water molecule, a *polar* covalent bond forms. The main reason water is such an excellent solvent is that it has polarity—a negative and positive end that readily interact with other substances to form solutions.

2.10 FORMATION OF COVALENT BONDS

In covalent bonding two or more atoms form a molecule as outer-shell electrons are shared. The simplest example is hydrogen gas (H_2). Covalent bonds are considerably stronger than ionic bonds because, as the checklist indicates, covalent bonding satisfies all the individual requirements of both partners in the molecule. Further, since electronegativity in the two hydrogens is identical, the resulting molecule is nonpolar.

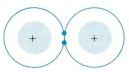

Checklist:
Electrons paired? √
Charges balanced? √
Outer shells filled? √

Molecular hydrogen, H_2

CHEMICAL BONDS AND THE SPONCH ELEMENTS

Now let's reconsider some of the SPONCH elements and see how they form covalent and ionic bonds. Then we can discuss some of the amazing characteristics of water before considering the peculiar hydrogen bond.

Carbon and the Covalent Bond

Molecules containing carbon, along with hydrogen and perhaps one or more other elements, are said to be **organic molecules.** There is an enormous range of molecules in this group and some are incredibly complex. However, we can learn a great deal about the chemistry of carbon compounds through a brief look at *methane,* one of the simplest.

Methane. The atomic number of carbon is 6, so free atomic carbon has two electrons in its inner shell and four in its outer shell. How can carbon fill its outer shell and yet retain a balance of charges? Carbon is only weakly electronegative, so it can't capture four electrons. And since it only has six protons in its nucleus to balance the charges, it can't give up its four electrons because a serious imbalance in charges in the opposite direction would result. So, carbon must share electrons to reach stability. Specifically, it forms four covalent bonds. When these four bonds are formed with hydrogen, we have a molecule of methane. **Methane** (CH_4) is a principal component of marsh gas, the waste product of certain primitive bacteria that rot organic material in the depths of oxygen-deficient bogs. There, it contributes to the delicate swampy aroma. More important, methane makes up most of the natural gas that helps fuel the world.

The usual way of writing the formula for methane is CH_4, but this is simply the **chemical formula.** It is often useful to use **structural formulas** to illustrate carbon compounds. Such formulas help indicate the geometric arrangement of atoms and reveal how electrons are shared. The structural formula of methane is written as:

The single lines indicate single covalent bonds, but remember that each covalent bond involves two

2.11 COVALENT BONDING IN METHANE

The electron shell diagram of methane (CH_4) illustrates how the four outer-shell electrons of carbon interact with those of four hydrogen atoms. The sharing of these electrons forms the four covalent bonds of CH_4. The difference in electronegativity between carbon and hydrogen is small (0.4), so the methane molecule, like many hydrocarbons, will be nonpolar.

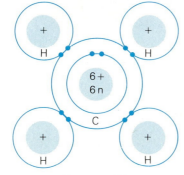

Methane, CH_4

$$H-\underset{\underset{H}{|}}{\overset{\overset{H}{|}}{C}}-H$$

Checklist:
Electrons paired? ✓
Charges balanced? ✓
Outer shells filled? ✓

shared electrons. The electron shell diagram is shown in Figure 2.11.

Actually, the electron orbitals of the second shell don't lie in a plane, but in three dimensions. If imaginary lines are drawn from the carbon to each of the four hydrogens, the angle between any two of them—the "bond angle"—is slightly more than 109 degrees, and the figure drawn is a tetrahedron.

Tetrahedrons are not easy to illustrate, but Figure 2.12 is an attempt. Note that there are several different "models" or ways of representing the methane molecule in three dimensions: the **electron cloud model,** the **ball-and-stick model,** and the **space-filling model.**

Actually, for our purposes, methane itself really isn't all that important, except as an example of how carbon compounds are organized. As we are about to learn, carbon is extremely varied in the ways in which it can join with itself and with other elements to form molecules of all sizes. As we will see, the chemistry of life is virtually the chemistry of carbon.

More on the Magic of Carbon. The biochemical magic of carbon, so essential to life, lies in the four electrons of its outer electron shell. As we saw with methane, a carbon atom can form four single

covalent bonds. It can also form double bonds, as it does in carbon dioxide. Carbon dioxide contains two sets of **double bonds** (each with two pairs of shared electrons) with the carbon in the middle, double-bonded to an oxygen atom on each side: O=C=O. (See Essay 2.1 for more on CO_2.) Carbon can even form *triple* bonds as in acetylene gas: H—C≡C—H; and in hydrogen cyanide, H—C≡N. Most importantly, carbon can form bonds with other carbon atoms, and these can take the form of long chains, rings, and other complex structures. There seems to be no limit to how large a carbon based molecule can be. A single giant DNA molecule, as we will see later on, can contain up to 50 billion atoms. These giants are aptly called **macromolecules** (*macro*, large).

2.12 FOUR VIEWS OF METHANE (CH_4).

The ball and stick model of methane (a) greatly exaggerates bond length but the bond angles are easily seen. The space filling model (b) is a much more accurate representation of the actual molecule. (c) In the geometry of methane, the one carbon nucleus lies at the very center, and the four hydrogen nuclei form the corners of an imaginary tetrahedron. (d) If its electron orbitals are emphasized, methane has four pear-shaped orbitals, each containing one electron from the carbon's outer shell and one electron from a hydrogen. The orbitals radiate from the carbon nucleus and envelop the hydrogen nuclei.

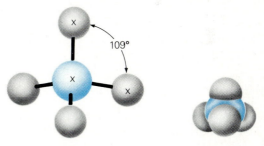

(a) Ball-and-stick model   (b) Space-filling model

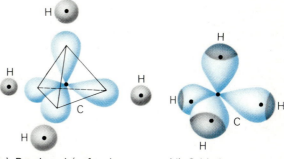

(c) Bond angles of carbon   (d) Orbitals

Carbon Dioxide and the Bicarbonate Ion

Carbon dioxide (CO_2) is an interesting and familiar substance. It makes soft drinks, beer, and champagne tingle, but where does it come from? In champagne and beer, the carbon dioxide is produced as a waste product of yeast metabolism. (Another waste product of yeast is alcohol, of course.) Your own body is loaded with CO_2, which is a waste product of your own metabolism. The carbon dioxide in soft drinks comes from steel cylinders.

There is a tiny amount of CO_2 in the air—not much, about ⅓ of one percent of air by weight, but this small amount is the only source of carbon for plants. Carbon dioxide is a symmetrical, linear molecule, which can be written as $O{=}C{=}O$. (Note the paired bonds when two pairs of electrons are shared. This is called a *double bond*.)

Carbon dioxide dissolves easily in water, and most of the CO_2 in beer and soft drinks occurs as CO_2 in simple solution. But CO_2 also reacts to some extent with the water to form *carbonic acid,* which is a weak acid that further dissociates into a *bicarbonate ion* and a *hydrogen ion* (a). The dissociation (separation) of hydrogen ions is what makes CO_2 solutions acidic (tart). Notice that the arrow is bidirectional. This indicates that the reaction can occur in either direction.

Sodium bicarbonate is a sodium salt of the bicarbonate ion. When you dissolve a spoonful of "bicarb" in a glass of water, some of the bicarbonate ions combine with H^+ ions from the water to reform carbonic acid, with the release of hydroxide ions (b).

In your stomach, sodium bicarbonate neutralizes some stomach acid (hydrochloric acid) and releases CO_2, which takes the form of an unseemly belch. Note that the sodium and chloride ions don't actually enter into the reaction (c).

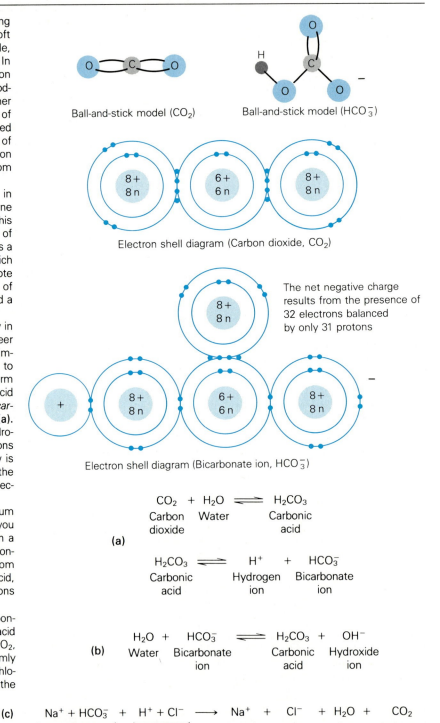

Ball-and-stick model (CO_2)

Ball-and-stick model (HCO_3^-)

Electron shell diagram (Carbon dioxide, CO_2)

The net negative charge results from the presence of 32 electrons balanced by only 31 protons

Electron shell diagram (Bicarbonate ion, HCO_3^-)

(a)
$$CO_2 + H_2O \rightleftharpoons H_2CO_3$$
Carbon dioxide   Water   Carbonic acid

$$H_2CO_3 \rightleftharpoons H^+ + HCO_3^-$$
Carbonic acid   Hydrogen ion   Bicarbonate ion

(b)
$$H_2O + HCO_3^- \rightleftharpoons H_2CO_3 + OH^-$$
Water   Bicarbonate ion   Carbonic acid   Hydroxide ion

(c)
$$Na^+ + HCO_3^- + H^+ + Cl^- \longrightarrow Na^+ + Cl^- + H_2O + CO_2$$
Sodium bicarbonate   Hydrochloric acid   Sodium ion   Chloride ion   Water   Carbon dioxide gas

2.13 THE GEOMETRY OF BUTANE

The ball-and-stick model of butane reveals its geometry. Note that butane's four carbon atoms do not lie in a straight line, but form angles. In addition, the shape of the chain can be changed by rotating the position of the carbon atoms on their bonds.

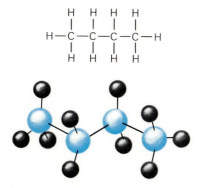

Carbon Backbones. Carbon atoms in long chains are often referred to as **carbon backbones** because they support various side groups that extend from the chains like ribs from a backbone. Such side groups include atoms of hydrogen, oxygen, nitrogen, phosphorous, sulfur, and others, forming mixed backbones of chains or rings. Some of the simpler backbones include the **hydrocarbons.**

A hydrocarbon is a compound that consists solely of carbon and hydrogen. We have already looked at the simplest hydrocarbon—methane. Propane and butane are three- and four-carbon chains, respectively. Both are familiar fuels, distilled from petroleum. Butane is seen in Figure 2.13.

Note that the ball-and-stick model of butane shows that the carbon backbone is not really straight. The single covalent bonds operate like little swivels, in which the carbon atoms can rotate freely about the axis of the chain.

The carbon-carbon linkages just mentioned are single covalent bonds, but carbon-carbon linkages can also be double or triple bonds (if two adjacent carbons share four or six electrons, respectively). Double and triple bonds don't swivel, and thus these molecules are more rigid than molecules with only single bonds.

There may be great biological differences between molecules in which carbons are joined by single bonds or multiple bonds. The two conditions therefore have their specific designations. Compare cyclohexane, C_6H_{12} (Figure 2.14), with benzene, C_6H_6 (Figure 2.15). In cyclohexane, all bonds are single bonds and each carbon has as many hydro-

gens as it can handle—this compound is therefore said to be **saturated.** In benzene, the carbons form double bonds between each other instead of with more hydrogens—this compound is said to be **unsaturated** with regard to hydrogen. We will come back to the chains and rings of carbon in the next chapter.

2.14 THE RING HYDROCARBON CYCLOHEXANE

Some hydrocarbons occur in symmetrical rings, as seen in the hydrocarbon cyclohexane, C_6H_{12}. This compound is a useful solvent but makes a poor quality, smoky fuel. Cyclohexane is also useful in the production of other ring hydrocarbons.

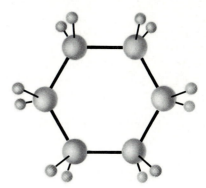

2.15 THE RING HYDROCARBON BENZENE

Benzene (C_6H_6) is a carbon ring similar to cyclohexane, but it is unsaturated. In drawings and in molecular models, every other carbon-to-carbon bond is depicted as a double bond. Reality is more complex: the molecular orbitals encompass the entire ring, and all of the carbon-to-carbon bonds are equivalent, "resonating" bonds. Benzene is a highly flammable compound often used as a solvent in industrial products such as pesticides, detergents, and special engine fuels. Other organic solvents such as toluene and xylene are derived from benzene.

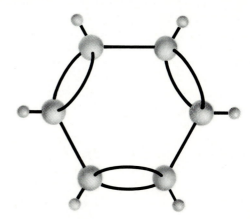

Nitrogen

Nitrogen is an essential component not only of proteins, but also of molecules such as the nucleic acids (DNA and RNA), some carbohydrates, some lipids (fats), and a number of other biological molecules. We will return to these classes of compounds in Chapter 3, but at this point, let's learn a few basic points about nitrogen. This is a peculiar element because it is so common (78% of air is nitrogen), yet so difficult for organisms to obtain in usable forms. This difficulty can present problems. Not only does a lack of suitable nitrogen compounds cause stunted growth in your garden plants, but lack of suitable organic nitrogen molecules results in severe malnutrition in people. Yet, curiously, while all animals require an intake of essential nitrogen—in the form of protein—once these compounds have undergone the usual chemical reactions in the body, some of the products are quite poisonous and must be isolated and flushed from the body.

Since air, a mixture of gases, is mostly molecular nitrogen (N_2), it would appear that this element is a readily available resource. However, most plants are perpetually starved for nitrogen, and plant growth is often limited by the availability of suitable forms of nitrogen. The reason for this peculiar shortage in the midst of plenty is that N_2 is very, very stable and tends not to react with other atoms or molecules under most conditions. Let's see why this is.

The atomic number of nitrogen is 7, so, allowing for two electrons in its inner shell, its outer shell has five proton-balancing electrons. Three more electrons are needed to fill the outer shell. Nitrogen usually forms three covalent bonds, as in molecular nitrogen (N_2) and in ammonia (NH_3). In some instances, however, nitrogen forms four covalent bonds, as it does in the ammonium ion (NH_4^+). Four-bonded *(quaternary)* nitrogen always carries a positive charge.

The nitrogen-nitrogen linkage is a triple covalent bond ($N \equiv N$), involving the mutual sharing of six electrons. It takes a lot of energy, such as lightning, or some very special chemical capability in cells to break this bond. Therefore, most organisms cannot use molecular nitrogen at all and must depend on other forms of the element such as the soluble ammonium (NH_4^+) or nitrate (NO_3^-) ions (Figure 2.16). However, a few species of bacteria have the chemical capability to *fix* nitrogen from atmospheric N_2, that is, to convert atmospheric

nitrogen into forms that can be used in living systems.

A few plants—notably the legumes such as peas, beans, peanuts, and alfalfa—meet their own nitrogen needs by harboring mutualistic (mutually beneficial) nitrogen-fixing bacteria within specialized root cells (Figure 2.17) (see Chapters 18 and 41).

2.16 IONS OF NITROGEN

Most plants cannot use atmospheric nitrogen for synthesizing protein but must depend on certain nitrogen ions. Principal among these are the soluble ammonium (NH_4^+) **(a)** and nitrate (NO_3^-) ions **(b)**. These ions are often incorporated into the ionic compound ammonium nitrate (NH_4NO_3), which is a crystalline substance and a valuable fertilizer.

(a) Ammonium ion, NH_4^+
The net positive charge results from the presence of 11 protons and 10 electrons

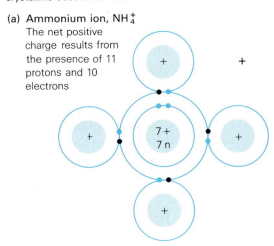

(b) Nitrate ion, NO_3^-
The net negative charge results from the presence of 31 protons and 32 electrons

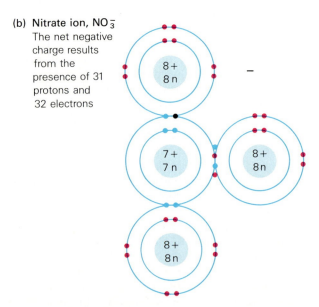

2.17 ROOT NODULES AND NITROGEN FIXERS

Root nodules are commonly found in leguminous plants. They contain populations of nitrogen-fixing bacteria, which live comfortably in a mutual relationship with the plant. Using atmospheric nitrogen (N_2), the bacteria produce nitrogen compounds that the plant uses to produce proteins and nucleic acids. The nitrogen-rich plants can be used directly or, as is often the case, they can be plowed under so that they will decay in the soil, releasing the nitrogen compounds for use by another crop.

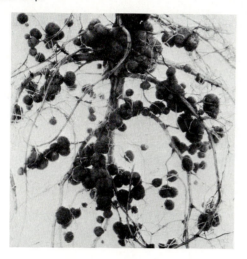

Phosphorus and Phosphates

Phosphorus is another of the essential SPONCH elements. In biological systems it is always combined with oxygen as a **phosphate.** A phosphate can be a free inorganic ion or it can be combined with a larger organic molecule to form a **phosphate group** (Figure 2.18). In the cellular and extracellular fluids of organisms, phosphate ions exist as HPO_4^{-2}, or as $H_2PO_4^-$ (with one or two hydrogen atoms and two or one negative charges, correspondingly).

In phosphate, the four oxygens are tightly bound to the phosphorus atoms, forming the four corners of a tetrahedron with the phosphorus inside. Since the phosphate often occurs as a free ion, it is given a symbol of its own, P_i *(inorganic phosphate).* Phosphate also forms covalent bonds, through its oxygen, with other molecules to form a phosphate group. Especially important is the **dehydration linkage** between two phosphates, in which a molecule of water is released when two phosphates are linked together. It takes a large amount of energy to remove the water; and later, breaking the bond by adding back a water molecule, a reaction known as **hydrolysis,** releases the same amount of energy.

2.18 PHOSPHORUS

(a) The electron shell model of phosphoric acid. (b) Phosphoric acid (nonionized state) and the phosphate ion.

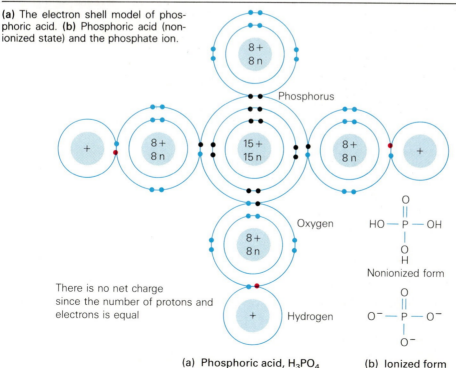

There is no net charge since the number of protons and electrons is equal

(a) Phosphoric acid, H_3PO_4

Nonionized form

(b) Ionized form

All living organisms use these *energy-rich* phosphate-to-phosphate bonds as a means of storing energy and shuffling it around to where it is needed. The energy rich bonds are found chiefly in molecules known as **adenosine triphosphate,** or more familiarly, **ATP,** and its less energetic product, **adenosine diphosphate,** or **ADP.** Phosphates are also found in certain fatty substances called **phospholipids,** which make up much of the structure of the many membranes found around and within cells. Finally, phosphate is also a part of the compound calcium phosphate, which is used to form the hard matter of bones and teeth.

Sulfur

Sulfur, our last SPONCH element, has several roles in the chemistry of life. It appears in some of the **amino acids,** small molecules that when joined together in chains form the familiar proteins of life.

The sulfur occurs in the amino acid as a **sulfhydryl group,** a sulfur atom covalently bonded to a hydrogen atom (–SH). When the amino acids are assembled into protein, pairs of sulfhydryl groups serve as a kind of interlocking "snap" or "hook" that can bond the long strands of protein together, aiding them in forming their special shapes. Two sulfhydryl groups give up their hydrogen and covalently bond together to form a reversible **disulfide bridge:**

amino acid—S—H + H—S—amino acid⟷

amino acid—S—S—amino acid + H_2

Key Functional Side Groups.   The –SH group is only one of several special functional groups or functional side groups, found in organic molecules.

A **functional group** is a specific chemical group that appears rather frequently in the many kinds of

---

TABLE 2.2

SELECTED FUNCTIONAL GROUPS AND THEIR CHARACTERISTICS

X—OH	Hydroxyl group X—O—H	Common in alcohols, slightly polar, tends to form H-bonds	X—COH	Aldehyde group	Slightly polar, soluble in water, common in sugars
X—NH₂	Amino group	Common in amino acids, weak base in water, accepts a proton:	X—SH	Sulfhydryl group X—S—H	Common in protein where it forms vital covalent linkages
			X—CO—X	Ketone group	Slightly polar, water soluble, common in sugars
X—COOH	Carboxyl group	Common in amino acids and other organic molecules, weak acid in water, releases a proton:	X—H₂PO₄	Phosphate group	Polar, weak acid, releases protons in water, common energy carriers of cells, un-ionized and ionized form:
X—CH₃	Methyl group	Common in organic molecules, nonpolar—rejected by polar molecules, insoluble in water			

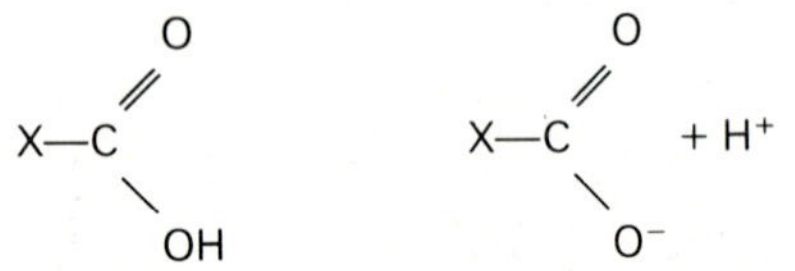

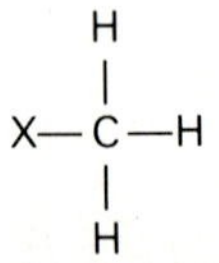

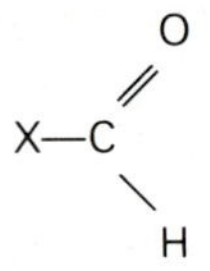

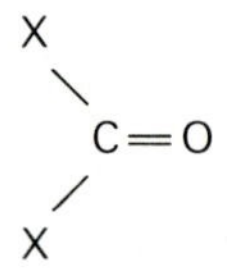

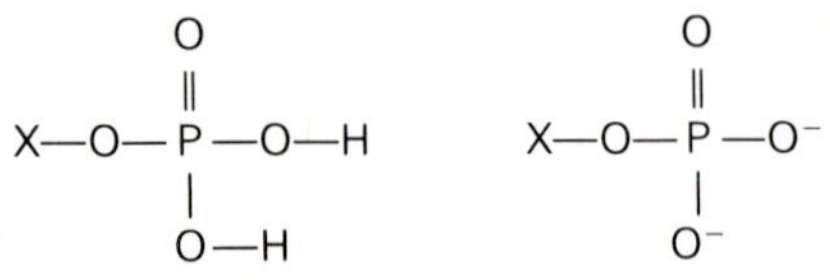

The letter *X* is just a convenience used to represent an undesigned or unnamed molecule to which the functional group is attached. Some of the functional groups listed above will ionize in water producing a charged condition, as we saw in the sodium and chloride ions.

2.19 WATER AND LIFE

Nowhere is the significance of water more dramatically evident than in a desert. Long periods of drought and dry, searing winds leave only the best-adapted plants to dot the landscape. Everything else lies dormant awaiting the seasonal rains. When water is available, the desert bursts into a riot of color as the annuals grow, flower, and drop their seeds, often in a matter of a few short weeks. Soon the bleak desolated condition will return as the cycle repeats.

organic molecules. They are important because their chemical behavior is pretty much the same, regardless of the kind of molecule to which they are attached. Some important functional groups are the carboxyl and amino groups, found in amino acids; phosphate and sulfhydryl groups (already discussed), methyl groups, hydroxyl groups, aldehydes, and ketones. Table 2.2 discusses the structure and characteristics of the most familiar functional groups.

THE WATER MOLECULE AND HYDROGEN BONDING

If you were to walk across some of our more arid and desolate deserts in the dry season, you might see no obvious signs of life for miles and miles. Then, in the distance, you may see green. Coming

closer, you find a few cottonwood trees and scattered tall bushes. You know why the plants have appeared. You might not be able to see it, but you know that it's there, beneath the surface at least: water. Where there's life, there's water. And just about any place where there is water, there is life (Figure 2.19).

On a grander scale, most biologists accept the proposition that life began in water, and certainly, the association between water and life endures. Hope of discovering life on Mars faded when we found that water was virtually absent on our celestial neighbor. After all, most of life's chemical reactions occur in aqueous solution; in fact most living organisms are between 50 and 90% water (Figure 2.20). *You* are about two-thirds water.

Thus, water is essential to life as we know it, and on our unique planet water is plentiful. Three fourths of the earth's surface is water, and enormous volumes of this essential liquid cycles into the atmosphere and back. Figure 2.21 shows the unending cycle of water between atmosphere and

2.20 WATER AND EMBRYOS

For many species, such as the salamander, the embryonic or larval forms are particularly high in water content. The early stages are often the most vulnerable for living things, thus emphasizing the critical role of water in the processes of life.

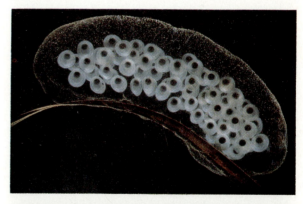

2.21 THE WATER CYCLE

The water cycle is a constant sequence involving evaporation and precipitation. Living organisms are a part of the cycle as plants absorb water through their roots and release it from their leaves in what is knows as transpiration. Some of this water is used to build the molecules of life. All organisms release excess water produced as they extract energy from foods. In the death and decay process, microorganisms complete the breakdown of molecules, returning water to the cycle.

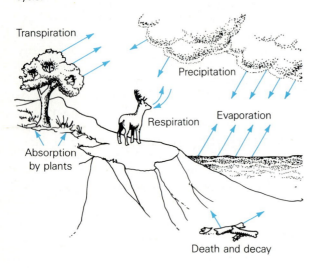

Transpiration
Precipitation
Respiration
Evaporation
Absorption by plants
Death and decay

oceans, known as the **hydrologic cycle.** The cycle is not simply a physical entity but a process that involves the organisms of the earth as well. Plants, animals, and the many other forms of life use water, incorporating it for a time in their cells as they go about their chemical activities. Eventually, all this water is returned to the cycle of the atmosphere and oceans as the organisms respire or die. In the later chapters, we will see the importance of water to life in many different ways, but for now, let's consider some of the molecular peculiarities of this vital, life-sustaining substance, and why its characteristics encourage chemical reactions.

The Characteristics of Water

The Polar Nature of Water. In each water molecule, two hydrogen atoms are covalently bonded to one oxygen atom. That is, the two hydrogens share electron pairs in two of the four electron-pair orbitals of the oxygen's outer shell. Figure 2.22 shows an electron shell diagram of water. Notice that the water molecule is quite polar—that is, it has positive and negatively charged sides. Further, the two hydrogen atoms

are close together, leaving the molecule a bit lopsided. (In certain views, the diagram of the water molecule reminds some people of Mickey Mouse. Now try looking at it without seeing him.) The polarity and lopsidedness of water are both important to its fascinating properties, as we shall see.

The two hydrogen atoms (the two Mickey Mouse "ears") of water occur at two corners of the same imaginary tetrahedron that we saw in the methane molecule. No matter which two of the four corners are occupied, the molecule is equally lopsided. More specifically, a pair of lines from the center of the oxygen to the centers of the two hydrogens would form an angle of approximately 109 degrees.

Water and the Hydrogen Bond

Now we will explore the last kind of bonding—the hydrogen bond—particularly with reference to its role in the behavior of water molecules. First we should note that because water molecules are polar, with positively charged and negatively charged ends, they readily interact with one another, each negative end attracted to the positive end of the next, and so on. Hydrogen bonds are formed from weak, electrical attractions between the positively charged, hydrogen-bearing part of one molecule and the negatively charged, oxygen-bearing part of another. This slight attraction between positive and negative parts of polar molecules tends to hold

2.22 POLARITY OF WATER

In water molecules, the electrical charge is unevenly distributed. The hydrogen end is positive, while the oxygen end is negative. This occurs because of oxygen's greater electronegativity—3.5 versus only 2.1 for hydrogen. The difference is not enough for the formation of oxygen and hydrogen ions to readily occur, but it does produce an uneven sharing of electrons which favor the oxygen. Thus, water is a polar molecule and, as such, it readily dissolves other polar substances.

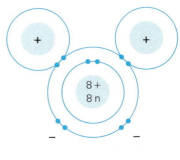

$$2H + O \longrightarrow H_2O$$

them together, even if just for an instant. Thus, the hydrogen bond is quite fleeting in nature.

Hydrogen bonds are quite common in the chemistry of organisms and perform vital roles in helping macromolecules hold their shape and remain intact. One of the best examples of this role is found in the famous DNA double helix, whose two long intertwined strands are held together by nearly countless, very weak, hydrogen bonds. In fact, the very weakness of the hydrogen bond is the key to its usefulness in biological reactions.

Ions, Hydration Shells, and Water as a Solvent. Water owes much of its excellent solvent qualities to its polar characteristics. We've already mentioned that salts such as sodium chloride dissociate into ions when dissolved in water. This dissociation is aided by the tendency of water to form **hydration shells** around charged molecules and atoms (don't confuse hydration shells with electron shells). For example, water molecules orient their positive (hydrogen) ends toward a negative ion, such as chloride, surrounding it with a hydration shell as shown in Figure 2.23a. This means that the water molecules of the innermost hydration shell have their negative (oxygen) ends pointing outward. This hydration shell, in turn, attracts the positive ends of other water molecules, and so on, forming progressively weaker concentric shells of oriented water molecules. The same kind of thing happens around positively charged ions, except that the orientation of the water molecules is reversed (positive end out). One of our more imaginative colleagues has compared the water molecules that surround an ion to groupies clustering around a highly charged rock star.

Water will form hydration shells around polar molecules as well as charged ions. For instance, sugars contain slightly polar, protruding hydroxyl functional groups (—OH; see Figure 2.23b and Table 2.2) with which water can build loose hydrogen bonds and form hydration shells. This keeps the somewhat polar sugar molecules from clumping together; in other words, sugar dissolves because hydration shells are formed.

Water and Nonpolar Molecules. If you mix a teaspoonful of water in a jar of salad oil, you might notice that the water will quickly form droplets that will eventually coalesce and isolate themselves from the oil. The reason for this is the strong mutual attraction of water molecules. Water has very little attraction for salad oil, which is a nonpolar compound that can't form hydrogen bonds.

If you try to mix a teaspoonful of salad oil in a

2.23 WATER FORMS HYDRATION SHELLS

Because of its polar structure, water interacts with sodium and chloride ions (a), and with slightly polarized oxygen and hydrogen on the glucose molecule (b), to form hydration shells. Note the specific manner in which the positive and negative ends of water orient to the negative and positive ions of chlorine and sodium and the positive and negative regions of glucose.

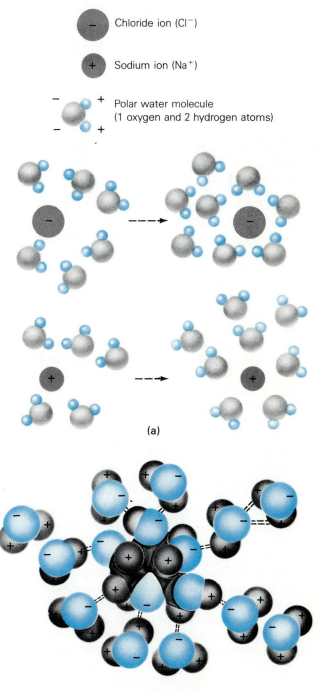

(a)

(b) Hydration shell about glucose
(space filling model)

jar of water, the results will be about the same. This time it is the oil that forms droplets that eventually coalesce. It would appear that the oil molecules too have a strong mutual attraction, but this is not the case. In fact, nonpolar molecules have very little attraction for each other. It is the strong mutual attraction among water molecules that excludes the oil: thus it is effectively isolated. Because the attraction among water molecules is strong, the forces tending to push nonpolar molecules together can be very strong in such a mixed system. This force is vital to the structure of proteins and of cell membranes.

Hydrophobia and Mayonnaise. Hydrophobia means "fear of water." Nonpolar molecules (or parts of molecules) are called **hydrophobic** because they mass together and do not mix with water (as in the salad oil and water example). The opposite of hydrophobia is hydrophilia—the "love of water"—a condition that would apply to ions or polar molecules. Some molecules, however, are polar, or **hydrophilic,** on one end and nonpolar (hydrophobic) on the other, and therefore can serve as links between water and fat. For instance, soaps and detergents are nonpolar on one end and charged on the other and can disperse nonpolar molecules in dishwater. The detergent molecules surround and infiltrate very small droplets of oil or grease, thus keeping them in solution and preventing them from rejoining. In another example, lecithin, a substance found in large quantities in egg yolk, is a natural detergent that is very useful in forming a bridge between dilute vinegar (which is polar) and salad oil (which is nonpolar). Beat well and the gel that results from your mixture of egg yolk, vinegar, and water is called mayonnaise. In the next chapter, we will come across molecules called phospholipids, which perform similar linking actions within the membranes of cells.

Water Is Wet. Water tends to get things wet. But what does this really mean? It means that it forms hydrogen bonds with the surface molecules of solid objects, except, of course, with objects made of oily or waxy substances that are composed entirely of nonpolar molecules. This wetting ability is the result of liquid **adhesion**—an attraction between two dissimilar substances. **Cohesion** is the attraction between similar substances; the hydrogen bonds between water molecules give water a considerable cohesion.

The tendency of water to adhere to and spread over solid surfaces is one of its special properties. If a thin glass tube is lowered into a beaker of water,

the water wets the inside of the tube and, as it does so, a column of water rises in the tube until it is higher than the water level in the beaker. If glass tubes of different diameters are put into the same beaker, the water will rise higher in the tube with the smallest bore. This is called **capillary action,** and it is due in part to the adhesion of water to glass and in part to the cohesion of water to itself. The two forces also explain the peculiar concave bend (meniscus) seen at the top of the water in a graduated cylinder (Figure 2.24).

Similar to capillary action, but on a finer scale, is **imbibition,** the movement of water into porous substances, such as wood or gelatin through **adsorption,** the adhesion to surfaces. The substances swell as the water moves in, and in fact, the swelling can generate a startlingly powerful force. Seeds can split their tough coats by the force of imbibition. And it has even been suggested that the great stones used in the construction of the Egyptian pyramids were quarried by driving wooden pegs into holes in the rock face and then soaking the pegs with water.

Water Has High Surface Tension. An old party trick involves carefully "floating" a needle in a glass of water. How is this possible—surely a needle cannot float? It works because of **surface tension.** Surface tension is one aspect of the cohesion of water. Where air and water meet, the water molecules at the interface have a much greater attraction to the water next to and beneath them

2.24 ADHESION AND COHESION

If glass tubes are placed into standing water, the water will rise in the tubes. The height it will rise is inversely proportional to the diameter of the tubing. Thus, if a piece of tubing is heated and drawn out into hairlike thinness, water will easily rise several feet above the container. Adhesion and cohesion are the forces responsible, and in the enlarged view their combined effect explains the formation of a curved meniscus at the top of the column. Such a meniscus is commonly seen in water being measured in graduated cylinders.

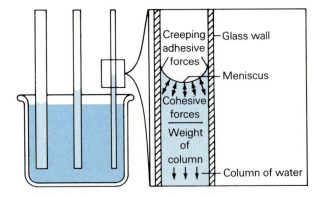

than to the air above them. While hydrogen bonding readily occurs among the surface water molecules, it cannot occur between water and air. They thus form a tough, elastic film of hydrogen-bonded water molecules. The familiar water strider (or water skater) shown in Figure 2.25 is able to walk on this film without breaking it. Surface tension is also the force that produces rain drops. Thus spring showers are delicate and pleasurable, rather than terrifying, as they would be if the same amount of water fell in disorganized and ponderous masses.

Water, Heat, and Life. One of the most important qualities of water as far as life is concerned is its temperature stability. It takes a large amount of heat to raise the temperature of water measurably, but once heat has been absorbed by water, it is just as stubbornly retained. Chemists are quite interested in the effects of heat on matter and refer to the amount of heat needed to raise the temperature of a given substance as its **specific heat.** Specific heat is generally stated in calories; in fact the calorie itself is defined as the heat needed to raise the temperature of one gram of water 1°C. Compared to most other substances, water has a high specific heat. For example, to raise the temperature of lead 1°C requires only 0.03 calories, while a one-degree rise in the temperature of table sugar requires 0.30 calories, and ethyl alcohol, 0.60 calories. But liquid ammonia (NH$_3$) has a greater specific heat than water, requiring 1.23 calories of heat to raise its temperature 1°C (for this reason it has been used as a refrigerant in refrigerators). Interestingly, molecules of ammonia, like water, tend to join by forming hydrogen bonds.

The high specific heat of both water and ammonia is closely related to the presence of hydrogen bonds. Chemists explain that temperature is a product of the rapid movement of molecules in a substance—more specifically, the average kinetic energy (energy of motion) of molecules. It is important to realize that under ordinary conditions molecules are always in very rapid random motion. But such motion in water and other substances that are subject to hydrogen bonding is greatly retarded. And, when hydrogen bonds are present, much of the heat input is required simply to rupture or counteract these countless attractive forces. Once this requirement is met, the heat can bring about increased molecular movement, and a thermometer will record this activity as a temperature rise. So, in a real sense, water simply captures and holds much of the heat to which it is exposed.

2.25 SURFACE TENSION

Some insects, such as this water strider, are able to walk about on the surface of water. This is because at the water-air interface, water molecules have a stronger attraction for each other than for the air molecules. Thus, a surprisingly tough water film forms at the interface. A close look at the appendages of such insects usually reveals that they are quite long and tipped with featherlike hairs, often coated with a nonwetting substance. Their weight is distributed on a large surface area because of their long legs, so their feet merely form indentations in the water surface.

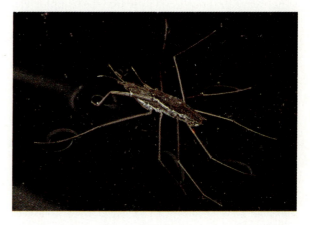

Incidentally, as you may expect, water resists evaporation better than do most liquids. Chemists (never at a loss for terminology—like biologists) say that water has a "high heat of vaporization." The heat energy required to convert one gram of water to water vapor is 540 calories. This is more than twice that required for ethyl alcohol, and nearly twice that required to vaporize liquid ammonia. The vaporization (or evaporation) of water is also retarded by hydrogen bonds, for the same reasons as stated above. Water must absorb a considerable amount of heat before its molecules move fast enough to escape as vapor.

Water's high resistance to vaporization is also significant to animals, such as ourselves, that rely on the evaporation of water from our body surface as a cooling device. It works well simply because evaporating water molecules remove a considerable amount of heat as they escape. On a much grander scale, in the cycling of water into the atmosphere and back, an enormous amount of heat is also carried aloft, and this is a key factor making the earth a habitable place. Further, the waters of the earth—particularly the larger bodies—are quite hospitable to life since their temperatures vary only slightly compared to temperatures on land.

The cooling of water reveals even more of its

peculiarities. Water freezes slowly because of the great amount of heat that must be withdrawn. As it cools, its molecules at first move closer together, increasing in their density and reaching maximum density at 4°C. The hydrogen bonds become more rigid at this time and the molecular latticework closes up. Thus, with the arrival of autumn in the temperate regions, layers of cooling lake water sink into the lake depths, helping to create a seasonal revolution or overturn that carries oxygen downward and nutrients upward (see Chapter 41). But as waters approach 0°C, the water lattice opens up and the molecules become widely separated, reaching their lowest density—becoming lightest—as ice crystals form. The separation is apparently consistent with the formation of four hydro-

gen bonds around each water molecule. (Figure 2.26 portrays the three states of water.) The fact that water at 0°C reaches its least dense state explains the buoyancy of the ice cubes in your drink and why ice skating is more popular than it might otherwise be.

Water: Ionization, pH, and Acids and Bases

Although water usually exists as a covalently bonded molecule, a very tiny fraction of the molecules in a drop of water will briefly and reversibly dissociate into a **hydrogen ion** (H^+) and a **hydroxide ion** (OH^-). That is, water molecules are continually breaking apart into ions, and the ions are continu-

---

2.26 THE THREE STATES OF WATER: GAS, LIQUID, AND SOLID

Because of their polar structure, water molecules attract each other, forming weak hydrogen bonds. The degree of attraction determines the physical state of water. **(a)** In its gaseous form (above 100° C or 212° F), thermal energy exceeds the attractive forces and molecular movement is quite rapid and random. Molecular distances are too great for the hydrogen bonds to hold. **(b)** In its liquid state (between 0 and 100° C, or 32 and 212° F), in the presence of less thermal energy molecular movement lessens. The association is still loose, however, resulting in sliding rows, or lattices, of molecules with some suggestion of geometric form. Because of this regular pattern, water has been called a "liquid crystal." **(c)** As thermal energy decreases and water reaches its freezing point (0° C or 32° F), the molecules become more rigidly held by hydrogen bonds. Eventually, crystals with definite geometric shapes will form. Now, the only molecular movement is vibrational. In all of this, note the relationships of the positive and negative ends of the molecule.

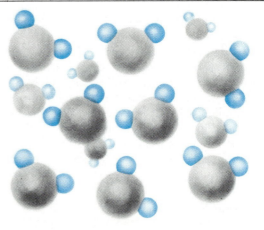

(a) Random arrangement as seen in water vapor or gas: limited hydrogen bonding, thermal energy exceeds hydrogen bond strength

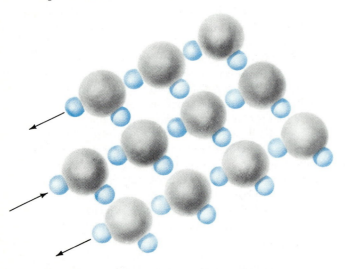

(b) Sliding lattice with geometric shape in liquid phase: greater hydrogen bonding, hydrogen bonds break and reform constantly

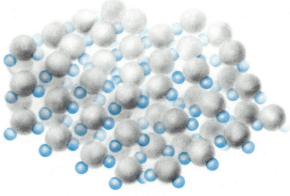

(c) Rigid crystal in solid (ice) phase with hydrogen bonding controlling the position of H_2O molecules

ally rejoining to make the neutral molecule again. In pure water, at any one instant, something like 1 molecule in 550 million will be dissociated into a pair of positively and negatively charged ions. An instant later, they will be back together again, but another approximately 1 in 550 million water molecules will dissociate in the meantime.

The *molar concentration* of hydrogen ions in pure water is 0.0000001 *mole per liter*. A **mole** of any substance is the weight *in grams* that equals the molecular mass in daltons of one molecule. Thus, a mole of hydrogen ions weighs 1 gram, and the concentration of hydrogen ions by weight in pure water is 0.0000001 *gram per liter*. The number 0.0000001, or the digit "1" seven places to the right of the decimal point, can also be written as 10^{-7}, which, in **scientific notation** is the equivalent of 1 divided by 10 million. The molar concentration of hydrogen ions in pure water, then, is 10^{-7} mole per liter, and the same concentration of hydroxide ions is also 10^{-7} gram per liter.

Actually, the hydrogen ion would be a naked proton, and while such things exist, they are rarely found dissolved in water. In reality, the hydrogen nucleus from a dissociating water molecule becomes bound up with another water molecule to make a **hydronium ion** (H_3O^+). Keep in mind that whenever we refer to a hydrogen ion, or write it as H^+, we mean a hydronium ion, H_3O^+. The dissociation and reassociation reaction may be symbolized as follows:

$$2H_2O \rightleftharpoons OH^- + H_3O^+$$

where the longer arrow pointing left indicates that most molecules are in the H_2O form at any one time. In the standard convenient fiction, of course, the same reaction would be written this way:

$$H_2O \rightleftharpoons OH^- + H^+$$

Some substances, when in water, release hydrogen ions in measurable quantities. We call solutions of such substances **acids.** Stomach acid, for instance, is dissolved hydrochloric acid, HCl. Pure HCl is a gas, but when it is dissolved in water, HCl ionizes to become paired H^+ and Cl^- (hydrogen ions and chloride ions).

Just how strong, or *acidic*, an acid solution is depends on the concentration of the hydrogen ions. The molecular weight of hydrochloric acid is 36 (1 for the hydrogen, 35 for the chloride). Thus, 1 mole of pure HCl gas weighs 36 grams, and 36 grams of HCl dissolved in 1 liter of water would create a concentration of 1 mole of HCl per liter, a decidedly strong acid.

TABLE 2.3

Molar concentration of H^+ ions	pH	Example	Molar concentration of OH^- ions
$1.0 = 10^0$	0	1 molar nitric acid	10^{-14}
$0.1 = 10^{-1}$	1	Gastric juices	10^{-13}
$0.01 = 10^{-2}$	2	Coca-Cola	10^{-12}
$0.001 = 10^{-3}$	3	Vinegar	10^{-11}
$0.0001 = 10^{-4}$	4	Tomato juice	10^{-10}
$0.00001 = 10^{-5}$	5	Urine (varies)	10^{-9}
$0.000001 = 10^{-6}$	6	Saliva	10^{-8}
$0.0000001 = 10^{-7}$	7	Pure water	10^{-7}
$0.00000001 = 10^{-8}$	8	Seawater	10^{-6}
$0.000000001 = 10^{-9}$	9	Baking soda	10^{-5}
$0.0000000001 = 10^{-10}$	10	Soap solution	10^{-4}
$0.00000000001 = 10^{-11}$	11	Ammonia solution	10^{-3}
$0.000000000001 = 10^{-12}$	12	Washing soda	10^{-2}
$0.0000000000001 = 10^{-13}$	13	Oven cleaner	10^{-1}
$0.00000000000001 = 10^{-14}$	14	1 molar sodium hydroxide	10^{-0}

Actually, since the HCl ionizes completely, the concentration of *chloride ions* would be 1 mole per liter and the concentration of *hydrogen ions* would also be 1 mole per liter—a concentration of hydrogen ions that is 10 million times greater than that of pure water. That's pretty acidic. A shorthand notation for the strength of acid solutions, or for acidity in general, is the **pH scale.** This scale uses scientific notation to express the H^+ concentration in a solution. Thus, if the acidity of some rather tart orange juice is 0.01 mole of H^+ ions per liter, which is the same thing as 10^{-2} moles of H^+ per liter, that orange juice has a pH of 2. We just leave out the "ten to the minus" part of the number. On the pH scale, then, pure water has a pH of 7. Table 2.3 gives the pH values of various substances. Note that the smaller the pH value, the more acid the solution.

A **base,** or **alkali,** is a substance that accepts protons and releases hydroxide ions (OH^-) when dissolved in water. Lye, or sodium hydroxide (NaOH), is a familiar example of a strong base. Ammonia is another base, although it contains no hydroxide ion itself, it accepts a proton from water to form an ammonium ion and a hydroxide ion:

$$\underset{\text{Ammonia}}{NH_3} + \underset{\text{Water}}{H_2O} \rightleftharpoons \underset{\text{Ammonium ion}}{NH_4^+} + \underset{\text{Hydroxide ion}}{OH^-}$$

A solution can be acidic, neutral, or basic (alkaline), but it cannot be acidic and basic at the same time. This is because hydroxide ions and hydrogen ions join spontaneously to form water. In pure water, which is neutral, the concentrations of hydrogen ions and hydroxide ions are both 10^{-7} moles per liter. But when the concentration of hydrogen ions rises to 10^{-1} mole per liter, as in gastric juice, the molar concentration of hydroxide ions falls to 10^{-13}, which is a very small number. In general, the exponents of the H^+ and OH^- ion concentrations in a solution always add up to -14. Thus, seawater, which is slightly basic, has a hydroxide ion concentration of 10^{-6} and a hydrogen ion concentration of 10^{-8}: $(-6) + (-8) = -14$. The seawater, then, has a pH of 8.0.

The chemical environment in which most life processes go on has a pH value of between 6 and 8. Human blood, for instance, has a pH of approximately 7.4, and the cell contents of most organisms have similar nearly neutral pH values. Exceptions include the stomach's digestive juices and fluids contained in citrus fruits, both of which are quite acidic.

APPLICATION OF IDEAS

1. Reread Szent-Györgi's lament that opens this chapter. Actually, he is an extremely successful scientist, a Nobel Prize winner, and the discoverer of many of life's secrets (for example, the structure of vitamin C and much of the biochemistry of respiration). In what sense, then, did "life" run out between his fingers?

2. Assume you are part of a space probe assigned the task of looking for signs of life as we know it on distant planets. Make a list of the compounds and elements you would search for and explain why each of these would be vital to any conclusion you might make.

3. Elements like hydrogen, sodium, and chlorine are rarely found in their elemental form. Explain why this is true. From a periodic table of the elements, list other elements with the same chemical characteristics and try to determine whether these are ever found in their elemental form. (The organization of the periodic table will tell you what the others are.)

4. It is interesting to compare silicon with carbon, since they both have four electrons in their outer shells. Silicon can form long chains called silicones, which are like the long chains of carbons called hydrocarbons. In addition, silicon combines with oxygen (SiO_2) to form crystals of silica or quartz and readily combines with fluorine to form highly soluble SiF_4. This similarity has not escaped the attention of science fiction writers who have described life based on silicon rather than carbon. What might such life forms be like? What similarities and differences would you expect between silicon- and carbon-based life forms? What might such alien scientists have to say about the "special properties" of silicon?

KEY WORDS AND IDEAS

ELEMENTS, ATOMS, AND MOLECULES

1. An **element** is a substance that cannot be separated into simpler substances by purely chemical means. Examples common to life include sulfur, phosphorus, oxygen, nitrogen, carbon, and hydrogen.

2. An **atom** is the smallest indivisible unit of an element.

3. A **molecule** is two or more atoms chemically joined.

4. A **compound** is a molecule in which two or more different elements occur in specific proportions.

Atomic Structure

1. Atoms consist of positively charged **protons,** uncharged **neutrons,** and negatively charged **electrons.**

2. Protons and neutrons make up the **atomic nucleus,** while electrons—minute particles—are in motion about the nucleus.

3. The combined mass of protons and neutrons account for the **atomic mass (atomic weight)** of an element. **Molecular mass** (mol. weight) is the combined atomic masses of the atoms in the elements making up that molecule.

4. The **atomic number** of an element is the number of protons (or electrons) in its atoms.

5. The number of neutrons in atoms of an element may vary, resulting in a number of **isotopes** of that element.

6. Radioactive isotopes are unstable isotopes that give off energy and/or matter in the form of radiation. The rate at which radioisotopes disintegrate is measured in **half-lives.**

7. Radioisotopes called **tracers** are used to determine the role of various elements and compounds in living organisms.

8. In electrostatically balanced atoms, the number of electrons and protons are equal.

9. **Orbitals** are regions where electrons are to be found. The distance of any electron from its nucleus is directly proportional to its energy.

10. In neon, two electrons occur in pairs with two in an innermost spherical orbital, while the other eight are found in one spherical and three dumbell-shaped orbitals.

11. In the **electron shell model** of atoms, electrons are depicted as being arranged in concentric **energy shells.**

12. Electrons have a strong tendency to occur in pairs, an important aspect of chemical activity.

13. Atoms have three chemically important **energetic tendencies:**
    a. Orbiting electrons tend to form pairs.
    b. Positive and negative charges tend to balance.
    c. Atoms tend to form full outer shells.

14. Atoms of the lighter elements fill their shells according to the octet rule, with the maximum numbers of 2, 8, and 8 in the first three shells.

15. The tendency to fill the outer shell brings about chemical reactions and allows chemical bonds to form.

16. The so-called **noble elements** (such as helium, neon, and argon) have filled outer shells in their elemental condition and thus are unreactive.

Chemical Reactions:
Filling the Outer Electron Shell

1. **Chemical reactions** are events in which **reactants** undergo changes into **products.** Chemical reactions involve losing, gaining, or sharing outer shell electrons.

2. Elements that are strongly electronegative tend to gain electrons when they react, while those less strong lose electrons. When **electronegativity** in the reactants is about equal, sharing of

electrons occurs. Where sharing is equal, *nonpolar* molecules form, but with unequal sharing, **polar molecules** (slightly positive or negative) result.

3. The forces that hold atoms together include **ionic bonds** (loss or gain of electrons), **covalent bonds** (sharing electrons), **hydrogen bonds** (attraction of negative and positive regions), and **van der Waals forces** (slight electrostatic attractions).

4. In their chemical reaction, sodium, an **electron donor,** loses an electron to chlorine, an **electron acceptor.** The resulting imbalance in proton and electron charges results in the formation of the positive sodium *ion* and the negative chloride ion.

5. The electrical attraction between negative and positive ions is the **ionic bond.** Ionically bonded substances tend to form crystals of indeterminate size. Ionic molecules separate readily into their ions in water.

6. When outer shells are filled through electron-sharing, a **covalent bond** results. Each atom has its outer shell filled part of the time. When two pairs of electrons are shared, a **double bond** is formed.

7. Covalently bonded molecules do not dissociate into ions in water.

8. Sharing is equal in H_2, so it is nonpolar, but in H_2O, the electrons are more attracted to oxygen, so water is polar, with positive and negative sides.

CHEMICAL BONDS AND THE SPONCH ELEMENTS
Carbon and the Covalent Bond

1. **Organic molecules** are produced naturally by organisms and contain carbon, hydrogen and other selected elements.

2. In **methane,** a simple hydrocarbon, one carbon atom shares its four outer shell electrons with four hydrogens, forming a molecule of **tetrahedral** shape.

3. **Chemical formulas, structural formulas, electron cloud, ball-and-stick models,** and **space-filling models** are all useful ways of portraying carbon (as well as other) compounds.

4. Carbon can form single, double, and triple bonds, and can form very long, straight or branched chains called **carbon backbones.** The backbones may also take the form of rings. Molecules composed mostly of carbon and hydrogen are called **hydrocarbons.**

Nitrogen

1. Nitrogen occurs naturally as stable, triple bonded N_2, making up 78% of the atmospheric gas.

2. Nitrogen gas is unavailable to most forms of life but can be converted to useful ammonia and nitrate by certain bacteria.

3. Nitrogen occurs in proteins, nucleic acids, and in some carbohydrates and fats.

Phosphorus and Phosphates

1. In life, phosphorus is utilized as phosphoric acid or the **phosphate** ion—designated as HPO_4^{2-} or as $H_2PO_4^-$ and symbolized as P_i. It is commonly attached to other molecules as a **phosphate group.**

2. Phosphate is used in **ATP (adenosine triphosphate)** and **ADP (adenosine diphosphate),** molecules used as energy sources for the chemical reactions of life. The energetic phosphate bonds are **dehydration linkages,** formed through dehydration, and broken through **hydrolysis,** the addition of water.

Sulfur

1. Sulfur occurs as **sulfhydryl (S—H)** groups of certain amino acids. When such amino acids are incorporated into protein, the S—H groups form important **disulfide linkages** within the molecule.

2. **Functional groups** are specific groups of atoms found in many molecules with each having its own chemical characteristics.

3. Amino, carboxyl, and phosphate functional groups ionize in water, while others such as hydroxyls, aldehydes, and ketones are slightly polar, and methyl groups are nonpolar.

THE WATER MOLECULE AND HYDROGEN BONDING

1. Water is so essential that without it life as we know it would not exist.

2. Water in the environment cycles between its liquid and gaseous phases in what is known as the **hydrologic cycle.**

The Characteristics of Water

1. Water is polar. The unequal sharing of electrons renders one part of the molecule negative, the other positive.

2. Water molecules interact, with weak, temporary **hydrogen bonds** forming between their positive and negative regions.

Water and the Hydrogen Bond

1. Water molecules tend to surround positive and negative ions forming **hydration shells.** This is the characteristic that makes water a good **solvent.**

2. Nonpolar molecules are repelled by water, so they cluster together in dense associations fostered by van der Waals forces.

3. Detergents are **hydrophilic** (water loving) at one end and **hydrophobic** (water fearing) at the other, so they form bridges between polar water and nonpolar oils.

4. **Capillary action** is the tendency for water to "creep" through minute spaces. **Adhesion** and **cohesion** are involved.

5. Water enters minute spaces through **imbibition,** which occurs through **adsorption** (the adhesion of molecules to surfaces).

6. **Surface tension** occurs where air and water interfaces occur.

7. Water has high **specific heat**—the amount of energy needed to raise its temperature. This is attributed to the resistance to molecular movement brought about by the presence of numerous hydrogen bonds. Water's ability to retain heat helps in climate moderation and as a medium for life, provides its inhabitants with relatively constant conditions.

8. Water reaches its greatest density at $-4°$ C, but below this, its density decreases rapidly to a minimum density with $0°$ C (freezing).

Water: Ionization, pH, and Acids and Bases

1. Pure water itself ionizes, but in minute amounts, spontaneously forming **hydrogen ions** (H^+) (protons or, more accurately, **hydronium ions**) and **hydroxide ions** (OH^-).

2. When H^+ ions outnumber OH^- ions, a solution is **acidic.** When OH^- ions outnumber H^+, a solution is **basic** or **alkaline.** Equal numbers of the ions produce a neutral condition.

3. The strength of acids and bases is represented by a **pH scale** from 0 to 14. Acids range from a pH of 7 down to 0 for the strongest acids, while bases range from 7 up to 14 for the strongest bases.

4. The chemical reactions of life generally occur at a pH near neutral (pH7).

REVIEW QUESTIONS

1. Write the names, symbols, and atomic numbers of the chemical elements abbreviated in the acronym SPONCH. (24–25)

2. Define the terms *atom*, *element*, *molecule*, and *compound*. (24)

3. Explain in terms of atomic structure how one element differs from another. (25)

4. Explain how the atomic number and the atomic mass of an element are determined. (25)

5. What are isotopes? Cite an example of a radioactive isotope with an extremely long half-life. (25–26)

6. List three energetic tendencies of atoms and explain what these have to do with chemical reactions. (27–28)

7. State the shell-filling (octet) rule and using potassium (atomic number 19), illustrate its operation. (28)

8. Using argon as an example, explain why the noble elements are unlikely to react with other elements. (28)

9. Define electronegativity and briefly explain what it has to do with the behavior of outer-shell electrons. (29)

10. Using potassium and chlorine (atomic numbers 19 and 17) as examples, show how they would form ions if they reacted together. Explain how an *ionic bond* would form between the two elements. (30–31)

11. Using a drawing, show in detail what would happen if the ionic compound potassium chloride were to be placed in water. (31)

12. Using two atoms of nitrogen (atomic number 7) as examples, illustrate how covalent bonding occurs. What is unusual about the bonding of the N_2 molecule? (32)

13. List three important uses to which plants put the element nitrogen. Why is using molecular nitrogen such a problem for plants? (36)

14. In what form is the element phosphorous usually found in living things and what are two important uses? (37–38)

15. In which of the molecules of life would one look for sulfur? What molecular purposes does it serve there? (38)

16. Draw structural formulas for the following molecules: ethane (C_2H_6), propane (C_3H_8), pentane (C_5H_{12}), and octane (C_8H_{18}). (32–33, 35)

17. Identify the following functional groups: X—CH_3, X—SH, X—NH_2, X—OH, X—$COOH$, X—O. (39)

18. Using water as an example, explain what a polar molecule is and discuss the nature of hydrogen bonding. (40)

19. What do hydration shells have to do with water's excellent solvent qualities? (41)

20. Define the terms *hydrophilic* and *hydrophobic*. What is the peculiar nature of soap molecules that enables them to interact with grease and water? (42)

21. Explain the meaning of specific heat, and discuss two ways in which the specific heat of water is important to life. (43–44)

22. What do hydrogen bonds have to do with water's resistance to temperature change? (43)

23. Which of the two ions of water predominates in acidic solutions? Basic solutions? Briefly describe the pH scale and cite examples of strong and weak acids and bases. (45–46)

Large Molecules

Life is not simple. How many times we've heard that, usually from weary or worried people trying to live out theirs and hoping their problems are only part of a larger, unpleasant phenomenon. But even those of us who don't focus on life's problems would have to agree that life is complex and often mysterious. In fact, it is complex in a number of ways, including its origins, mechanisms, and structures. Here, then, let's look at the molecular structure of living things, focusing on the complex building blocks of a complex phenomenon. Specifically, we will concentrate on the nature of the large molecules that comprise living things.

In a sense, we can apply what we learned about small molecules to the study of large ones. Although large molecules can appear dazzlingly complex, they are usually just **polymers**—that is, composed of many identical or similar small molecules, or **monomers.** The monomers, then, are assembled into polymers. The large polymers are also known as **macromolecules** ("large molecules").

In this chapter we will look at the structures and functions of the four major classes of molecules: the carbohydrates, lipids, proteins, and nucleic acids. Actually, such categorizing is not always accurate because the four classes aren't always completely distinct from each other. For example, along the way we will run into molecules that are part protein and part carbohydrate (glycoproteins and mucoproteins), some that are part protein and part lipid (lipoproteins), some that are combinations of protein and nucleic acid (nucleoproteins), and some that are part lipid and part carbohydrate (glycolipids).

THE CARBOHYDRATES

Carbohydrates are familiar to us as the sugars and starches in our diets. But, as we will see, they are important in other ways as well. Most carbohydrates have the empirical formula $(CH_2O)_n$. (The numbers in empirical formulas are reduced to their simplest terms. Here, the n means that there can be any multiple of CH_2O.) Actually, the simplest common carbohydrates are $(CH_2O)_3$, or 3-carbon compounds, but we will primarily be interested here in 6-carbon carbohydrates and the way they are linked together to form the large polymers. Let's begin by looking at the organization of the "single" and "double" sugars.

Monosaccharides and Disaccharides

All carbohydrates contain **simple sugars,** or **monosaccharides,** as they are called (*mono*, one; *saccharide*, sugar). There are many simple sugars, but the most familiar is the 6-carbon sugar, **glucose.** The next most complex type of carbohydrate is the **disaccharide,** which as its name suggests, contains two monosaccharides. The most familiar disaccharide is **sucrose,** or table sugar. Then, more complex molecules are formed from longer chains of simple sugars until we reach the **polysaccharides** (*poly,* many), which may contain hundreds or thousands of monosaccharide subunits, covalently linked into chains.

A monosaccharide consists of a short chain of carbon atoms (usually five or six), with nearly every carbon having a hydroxyl (—OH) functional group as well as a hydrogen (—H) side group (Figure 3.1). Note that the empirical formula $(CH_2O)_n$ for carbohydrates suggests one oxygen for every carbon. Often, one carbon at the *end* of each simple sugar forms a double bond with its oxygen, producing an **aldehyde** group:

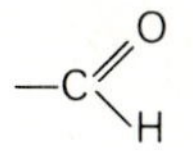

In other instances a carbon *within* the chain forms a double bond with an oxygen group, yielding what is called a **ketone** group:

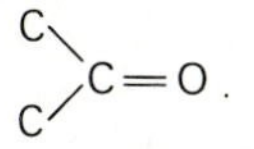

Simple sugars, we know, can exist as rings or open chains, but in polysaccharides they are all in rings. (See Essay 3.1 for a look at some of the ways glucose can be represented.)

As we said, the most common and most important monosaccharide is glucose (an aldehyde) the 6-carbon sugar that is not only fundamentally involved in energy metabolism and photosynthesis,

---

3.1 MONOSACCHARIDES

(a) Glucose is a simple sugar. Its chemical formula, $C_6H_{12}O_6$, can also be written as $(CH_2O)_6$, following the general formula for carbohydrates. Note the presence of hydroxyl and hydrogen groups and the aldehyde group, which is typical of many sugars. The numbers 1–6 identify the different carbons in the molecule. (b) Fructose has the same chemical formula as glucose but is quite different structurally. Further, although fructose contains many hydrogen and hydroxyl functional side groups, it is characterized by the presence of a ketone side group.

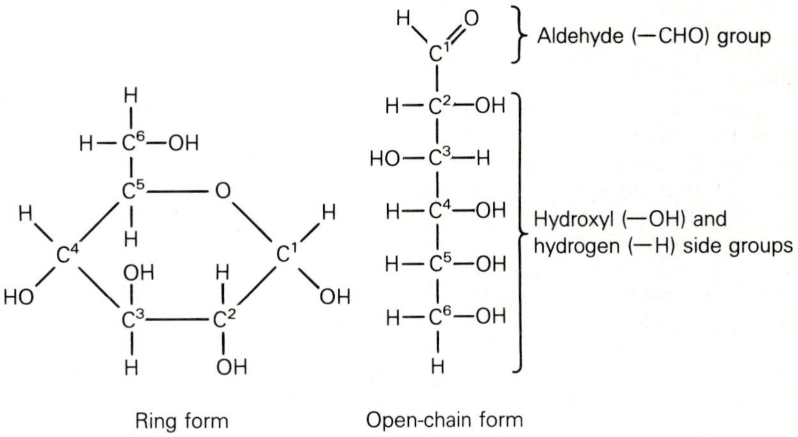

(a) Glucose, $C_6H_{12}O_6$

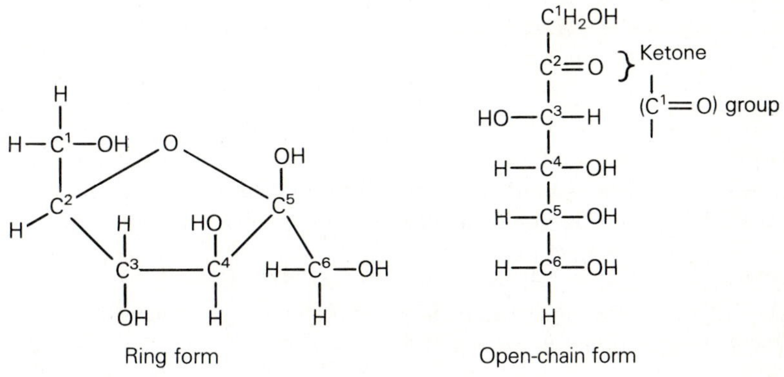

(b) Fructose, $C_6H_{12}O_6$

Reading Structural Formulas

A troublesome problem for many people when they are first introduced to the structural formulas of organic chemistry is the variable and seemingly inconsistent ways in which they are written. Why, they wonder, can't chemists pick some standard system of structural formulas and stick to it? The reason is, each of the systems used has its own strengths and weaknesses, and none will do for all purposes. In some cases, very fine detail is needed; for instance, if it is important to show where the electrons are, the electron shell diagram may be appropriate. But most of the time one doesn't need to know where the electrons are.

As larger and larger molecules are considered, the detail has to be suppressed more and more so that the important features can be seen. Side groups can be simplified, or perhaps left out altogether.

Examples may be helpful at this point, so consider some of the ways that can be used to represent glucose. There is the space-filling model (a). Then there are two types of complete structural formulas with tapered bonds to indicate the three-dimensional structure—a ring form and a "chair" form (b).

A skeleton structural formula, in which the carbons that form the ring are left out and hydrogen side groups are indicated by short lines, may also be used to depict the same glucose ring (c). And then there is the bare bones structural formula, showing

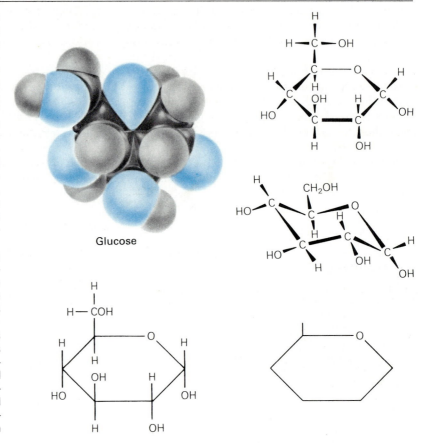

Glucose

only the ring, the oxygen in the ring, and the bond reaching to where the number 6 carbon should be (d).

The last configuration (e) is often used when glucose is incorporated into a polysaccharide, such as starch or cellulose. You can mentally fill in the carbons and the hydrogen and hydroxyl side groups because you know that the hexagon with the tail is supposed to represent glucose.

but is the building block from which all complex carbohydrates are built. It is also one of the two subunits of sucrose. Glucose is called blood sugar, corn sugar, or grape sugar, depending on its source. But whatever its source, glucose—or dextrose, as it is also called—is the simple sugar intravenously introduced into patients in emergencies and after surgery.

Sucrose, a disaccharide, is made up of one glucose and one fructose subunit (Figure 3.2). Technically, it is a 12-carbon sugar consisting of two 6-carbon sugars (glucose and fructose) linked together in a **dehydration linkage.**

Figure 3.2 illustrates what is meant by the term *dehydration linkage.* Two —OH side groups combine (R—OH + HO—R) to form an oxygen bridge (R—O—R) with the liberation of the two hydrogens and one oxygen as H_2O. Such a water-removing chemical reaction, called **dehydration synthesis,** requires the presence of biologically active proteins called **enzymes.** These macromolecules abound in the cell and are the catalysts—the activating agents—of most of the chemical reactions of life. (Enzymes are discussed in detail in Chapter 6, but let's just say here that enzymes are molecules that encourage certain chemical reactions within the body.)

In your gut (the biologist's indelicate term for gastrointestinal tract), the sucrose is broken back down into glucose and fructose by **hydrolytic cleavage** (*hydro,* water; *lysis,* rupture), which is just the opposite of dehydration synthesis. Here, with the assistance of an enzyme, a water molecule is added to the linkage, breaking the sucrose into its component parts. This is necessary because your gut can't absorb sucrose very efficiently and the cells of your body can't metabolize it. (If you gorge yourself on table sugar, some undigested sucrose may get into your bloodstream, but it will be excreted in your urine.)

The carbons of each sugar are numbered as we saw earlier (see Figure 3.1), so we can precisely describe where the linkages occur. For example, the dehydration linkage between glucose and fructose is a 1-2 linkage, since the number 1 carbon of glucose is linked by an oxygen bridge to the number 2 carbon of fructose. Biochemists also refer to the bond between adjacent glucose monomers as a **glycosidic linkage.**

Another disaccharide of interest is **lactose,** one of the so-called milk sugars. Lactose, in fact, is found *only* in milk (Figure 3.3). Lactose is not as sweet as sucrose, and whereas babies thrive on it, many adults (including nearly all non-Caucasian adults) who consume milk products suffer from **lactose intolerance.** They lack the enzyme that breaks lactose into its two subunits—glucose and galactose—and therefore can't digest lactose. Instead, the lactose, with other milk sugars, passes into the colon (bowel) where it is attacked by gas-forming bacteria, often causing an accumulation of painful gas.

Polysaccharides

Turning to the larger carbohydrates, the polysaccharides, let's note that they are formed from monosaccharide subunits covalently linked into lengthy chains. The polysaccharides have many forms and

3.2 FORMATION OF SUCROSE

Monosaccharides can be linked to form disaccharides (double sugars). The diagram illustrates the manner in which chemical bonds join simple sugars to form the double sugar sucrose. In the presence of specific synthesizing enzymes, one —OH group combines with the hydrogen from an adjacent —OH and a molecule of water is produced. The electrons of the remaining oxygen then pair up with those of the carbon to form an oxygen bridge. The process of enzymatic water removal, known as *dehydration synthesis,* is very common in the chemical activity of cells.

3.3 FORMATION OF LACTOSE

Lactose, or milk sugar, consists of one molecule of glucose and one molecule of galactose. Note the slight difference between the two. These sugars are covalently bonded by dehydration (losing H_2O), yielding lactose and water. The chemical formula and composition of lactose is identical to that of sucrose, but the molecular structures of the two compounds and their chemical behaviors are quite different.

Galactose   Glucose

Lactose, $C_{12} H_{22} O_{11}$

serve a large number of functions. Many familiar foods (potatoes, wheat and corn flour, whole grains, seed vegetables, fruits) contain large amounts of polysaccharides in the form of **plant starches. Glycogen** is another common polysaccharide, formed by animals as a means of storing glucose and often called "animal starch." Then there are **cellulose** and **chitin,** two polysaccharides used, not as storage carbohydrates, but as structural material. Cellulose is a structural polysaccharide in plant cell walls, and chitin is used by insects to form exoskeletons (skeletons surrounding the body), as well as by fungi to form the walls surrounding their cells.

Plant and Animal Starches. Starches form a large part of our diet, perhaps too large a part. Most snack foods ("junk foods"?) are high-calorie foods largely because of their high starch and sucrose content. Nevertheless, the starches are vitally important storage carbohydrates, which are readily metabolized for energy. The two most common starches are **amylose** and **amylopectin**—important food reserves for higher plants. All starches are made entirely of glucose subunits.

Amylose. *Amylose* is the simplest starch (Figure 3.4). Its molecules consist of unbranched chains of hundreds of glucose subunits, with 1-4 dehydration linkages between successive monomers (single units). Potato starch is about 20% amylose. In the gut or in a sprouting potato, breakdown (digestion) occurs in three steps: one enzyme breaks amylose into fragments of varying size, attacking the starch at random points; a second enzyme works on the ends of the fragments, cleaving off two glucose units at a time as disaccharides (glucose-1, 4-glucose, or *maltose*); and a third enzyme cleaves the disaccharides into glucose monomers.

Amylopectin. The other 80% of potato starch is *amylopectin*, which is a large molecule consisting of short 1-4 glucose chains *cross-linked* with occasional 1-6 linkages between chains (Figure 3.5). The individual straight chains are 20 to 30 glucose units long. The amylopectin in rice starch is made up of about 80 to 100 such chains cross-linked into a huge mesh, with one or two cross-links per chain. The same enzymes that cleave amylose also break up amylopectin, but yet another enzyme is required to cleave the 1-6 linkages.

3.4 AMYLOSE STARCH

(a) Polymers of glucose are exceedingly important to life. The polysaccharide amylose is a storage starch found in plants. It consists of hundreds of glucose molecules linked together at their number 1 and number 4 carbons. Amylose is insoluble in water and will form a helical structure as shown in part **(b)** as a result of its repulsion of water. Cooking amylose (boiling potatoes) fragments the chain, which becomes more soluble and easier to digest. The grains shown here have been photographed through polarizing filters, which reveal the regular structure of the starch molecules.

$HOCH_2$ $HOCH_2$ $HOCH_2$ $HOCH_2$ $HOCH_2$

(a) Unbranched chain (amylose)

(b) Helical structure (amylose)

3.5 AMYLOPECTIN STARCH

Amylopectin is a cross-linked polysaccharide. The short, helical chains consist of glucose subunits, joined at carbons 1 and 4 by enzymatic dehydration. The branching side chains are linked together by 1-6 linkages, which complicate digestion since additional enzymes are required to break these linkages. Both amylose and amylopectin are excellent sources of glucose for animals.

1-6 linkage

1-4 linkage

Glycogen.   **Glycogen,** the storage polysaccharide of animals, is usually a much larger molecule than are amylose or amylopectin, and is highly branched. The predominant linkage is 1-4, but there are many 1-6 linkages as well. Glycogen is a temporary, short-term storage unit in animal cells, particularly prevalent in the liver and muscles of vertebrate animals (Figure 3.6).

When glycogen is metabolized within cells, the dehydration linkages are enzymatically cleaved by

3.6 ANIMAL STARCH

Animals, using glucose from plants, synthesize their own starch, the polysaccharide glycogen. Glycogen is similar to amylopectin with 1-4 linkages in the straight chains and 1-6 linkages where branching occurs. It is, however, more highly branched, heavier, and more compact than its plant counterpart. Glycogen is stored in skeletal muscle and in the liver cells, where it forms large granules. Such storage is very temporary, however, since glycogen is rapidly converted to glucose-1-phosphate and used as an energy source for the cell. In the enzymatic conversion, phosphorolytic enzymes, rather than hydrolytic enzymes, are employed. The phosphorolytic enzyme utilizes inorganic phosphate in cleaving the 1-4 linkages between glucose subunits. The product is glucose-1-phosphate, which is valuabe to cells as an energy source and, unlike glucose, cannot diffuse out of the cell.

Glycogen (branched chain) of 1-4 and 1-6 linkages)

Phosphorolytic enzyme

Inorganic phosphate

Glucose-1-phosphate released from glycogen

phosphorylase. The cleavage is called a **phosphorolytic cleavage,** or **phosphorolysis.** The enzyme adds inorganic phosphate to the molecule (instead of water as in hydrolysis), and the product is not glucose but glucose-1-phosphate. (A phosphate functional group is attached to the number 1 carbon of glucose.) There is a great deal of energy in the bond between the glucose and phosphate, and by leaving this unit intact, the energy is conserved and can be recovered later by the cell.

Primary and Secondary Structure of Starches. Starches can be described according to their structures. The straight- and branched-chain polymers we've just described are called the *primary structures* of the starches. The primary structure of a polymer results from covalent bonding alone, and usually can be shown as a two-dimensional figure. When we referred to amylose as an unbranched chain, we had the primary structure in mind. In reality, amylose is a three-dimensional structure that loops and folds and forms helices (like winding staircases), as shown in Figure 3.4b. Amylose forms helices because of hydrogen bonding between subunits along its length. These helices make up the *secondary structure* of the molecule. Whereas we are fairly confident about the secondary structure of amylose, the secondary structure of amylopectin and glycogen are less well understood.

Structural Polysaccharides: Cellulose. **Cellulose** is a linear polymer of glucose subunits put together with 1-4 dehydration linkages. If that sounds familiar, it's because amylose is also a linear polymer of glucose subunits put together with 1-4 dehydration linkages. Nonetheless, amylose and cellulose are very different. Starch is fairly soluble, while cellulose is not; cellulose has great tensile (stretch) strength, while starch does not; starch is readily broken down by digestive enzymes, but cellulose is completely indigestible to all but a very few organisms. So, whereas the structural differences in the subunits of starch and cellulose are very slight (Figure 3.7), those slight differences have far-reaching effects.

Earlier, we mentioned that free glucose occurs in two forms, a linear (open) form and an alpha ring (closed) form. That was only partially true. Actually, there are three forms: linear, alpha ring, and beta ring. When the double-bonded oxygen of the number 1 carbon in the open, linear form becomes part of a hydroxyl group, there are two possible positions that the hydroxyl can take: pointing down (alpha-glucose) or pointing up (beta-glucose), as

3.7 BETA-GLUCOSE AND CELLULOSE

The primary structure of cellulose is similar to that of amylose, with straight chains of 1-4 linked glucose subunits. The significant difference is that the glucose utilized is beta-glucose rather than alpha-glucose. Because of the position of the number 1 —OH groups in each beta glucose, every other subunit is inverted as the cellulose chains are assembled.

Amylose

Alpha glucose

Cellulose

Beta glucose

shown in Figure 3.7. When pure glucose is in solution, about 34% of the molecules are in the alpha ring form, 65% in the beta ring form, and a fraction of a percent in the open, linear form.

Cellulose is difficult to digest precisely because of these linkages. In cellulose, glucose units are linked by what is called a **beta-glycosidic linkage.** Relatively few organisms produce the enzyme **cellulase** necessary to break suck a linkage. Garden snails can digest cellulose (such as in the leaves of whatever you are planting). Fungi and some bacteria also have cellulase. But what about animals such as grazers and termites? It turns out that they don't digest cellulose themselves, but they harbor helpful intestinal microorganisms that can break it down. The products of cellulose digestion are absorbed (and even the microorganisms themselves are ultimately digested and absorbed). What about the fact that people eat plants? People can eat all the cellulose they want and it will just go right through. In fact, it may go through pretty fast because it irritates the gut lining causing lubricating fluids to be released. Whole-bran cereals are very rich in cellulose.

Cellulose can form very strong materials because of the ways it can be arranged. As Figure 3.8 indicates, cellulose chains are organized into larger and larger units, like the strands of a cable, finally forming tough **microfibrils.** Although cellulose is insoluble and strong, it it also flexible and can even be soft. After all, facial tissues are almost pure cellulose but using them isn't like wiping your nose on a block of wood. The hardness of wood comes from another class of large molecules, the **lignins,** which infiltrate and harden some cellulose microfibrils of wood (see Figure 3.8). Lignins are not polysaccharides but are in a class of their own.

Other Structural Polysaccharides. In the tough walls that surround plant cells, cellulose

3.8 CELLULOSE, A STRUCTURAL POLYSACCHARIDE

Cellulose, a structural polysaccharide, forms a tough wall around nearly every cell in the plant body. **(top)** The microfibrils can be easily resolved by the electron microscope. Rows of microfibrils are laid down in laminated form, which explains the strength of cell walls. **(bottom)** Unlike other polysaccharides, cellulose forms lengthy *microfibrils.* These consist of cellulose chains embedded in cementlike substances including pectin, extensin (a glycoprotein), hemicellulose, and lignin. This combination forms the "wood" of woody plants.

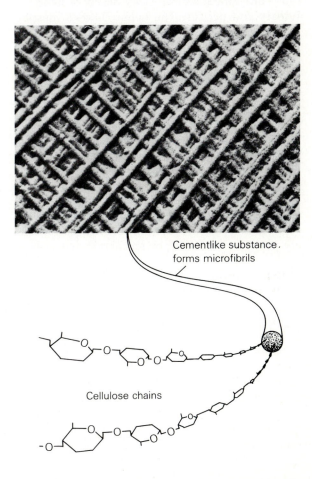

Cementlike substance, forms microfibrils

Cellulose chains

3.9 CHITIN, AN ANIMAL POLYSACCHARIDE

The lobster exoskeleton is composed of the complex carbohydrate chitin, a polymer of *N*-acetyl glucosamine. Insects commonly use this polymer as skeletal material, but in the lobster and other marine forms, additional firmness is provided by calcium carbonate.

N-acetyl glucosamine

Chitin

fibers are imbedded in a gluey matrix of **hemicelluloses** and **pectins.** The hemicelluloses, despite the name, are not structurally related to cellulose. Instead, they include polymers of some of the less common 5-carbon sugars. Pectin is the substance that gives jelly its consistency. The strong cell walls of the more common "higher" plants can be compared to iron-reinforced concrete, or resin-impregnated fiberglass, or any other materials that owe their strength to fibers imbedded in an amorphous (shapeless) matrix.

Other structural polysaccharides (such as algin, agar, and carrageenan) are found in various seaweeds from which they are extracted and sold to the food industry for such esoteric uses as thickeners in milk shakes. Also, many bacteria secrete slimy protective coats of polysaccharide material.

Chitin.   **Chitin** is a principal constituent of the **exoskeletons** (external skeletons) of insects and other arthropods, including lobsters and crabs (Figure 3.9). Chitin itself is rather flexible and leathery, as is found in the exoskeletons of grasshoppers and cockroaches, but can become very hard when impregnated with calcium carbonate, as in lobsters and crabs. It is a structural polysaccharide similar in many ways to cellulose, except that instead of glucose, the basic unit is ***N*-acetyl glucosamine.** Notice that this carbohydrate contains nitrogen as well as carbon, oxygen, and

hydrogen. [The *N*-acetyl part of the name of this compound means that an acetyl group ($-COCH_3$) is attached to the main structure through a nitrogen atom.] Chitin is indigestible to most animals.

Mucopolysaccharides.   Finally, another important group of nitrogen-containing polysaccharides of animals includes the **mucopolysaccharides,** which, as you might expect from the name, are a bit slimy. **Hyaluronic acid** is a mucopolysaccharide with an acid (carboxyl, COOH) side group. It shows up in quite a variety of places, such as the jellylike substance filling the eye, the lubricating fluid of joints, the umbilical cord, and the cementing substance holding the skin to the body. A closely related mucopolysaccharide is **chondroitin,** the principal component of cartilage and a significant part of adult bone as well as the cornea of the eye. Hyaluronic acid is made of repeating units of glucuronic acid and *N*-acetyl glucosamine; chondroitin is made of glucuronic acid and *N*-acetyl galactosamine.

THE LIPIDS

The **lipids** are a diverse group of molecules, defined by their solubility rather than by their structure. Lipids may be small molecules, large molecules, monomers, polymers, energy storage molecules, structural molecules, hormones, lubricants, or parts of proteins.

Lipids are organic molecules that tend to be fat-soluble rather than water-soluble. They are generally nonpolar and tend to dissolve in organic solvents, such as gasoline, chloroform, paint thinner, and salad oil. These are all substances that lack charges in their functional groups. Whereas carbohydrates and proteins are hydrophilic ("water-loving"), lipids are hydrophobic ("water-fearing"; see Chapter 2). It may seem odd that an important class of organic compounds is defined solely on the basis of a solubility characteristic, but it does appear that fat-solubility gives all lipids, no matter how different they are structurally, some common characteristics. For example, lipids tend to dissolve into one another and may often be found together in the same part of the cell.

Important groups within the category are fats, oils, sterols, waxes, and phospholipids. Most familiar to us are the fats (such as beef tallow) and the oils (such as corn oil, safflower oil, and cod liver oil). All are **triglycerides**—that is, compounds consisting of three **fatty acid** chains attached to a molecule of **glycerol** (see below). The difference between a fat and an oil is that a fat has a higher melting point; fat is solid at room temperature and oil, with its lower melting point, is liquid at room temperature. But oils, and even fats, tend to liquefy at the normal temperature of the living organism in which they occur. Because of this ambiguity, and to avoid confusion with petroleum oil and with other meanings of the word *fat*, biologists prefer to use the term *triglycerides* to refer to oils and fats.

Triglycerides

Glycerol, formerly known as *glycerine*, is a small molecule. It has three carbons and three hydroxyl side groups. Linked to the glycerol in a triglyceride are three fatty acid molecules. A fatty acid molecule consists of a hydrocarbon chain (that is, a chain consisting of a backbone of carbon atoms with only hydrogens as side groups) with a carboxyl group at one end (Figure 3.10). The synthesis of a triglyceride from glycerol and three fatty acids is a dehydra-

---

3.10 TRIGLYCERIDE SUBUNITS

Triglycerides consist of one molecule of glycerol and three of fatty acids. There are many different fatty acids, which explains the large number of triglycerides possible.

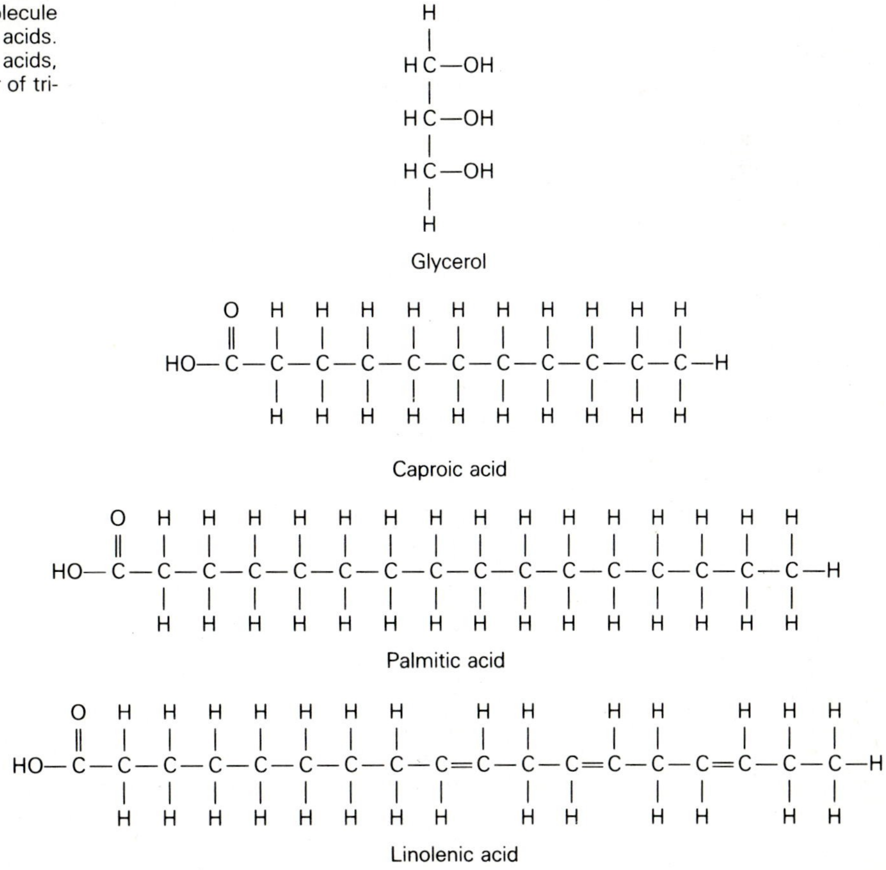

Glycerol

Caproic acid

Palmitic acid

Linolenic acid

tion process whereby three molecules of water are formed for each triglyceride produced. Note that enzymes—the organic catalysts of cellular reactions—are required for the reaction to go on (Figure 3.11).

Fatty acids can be of many different lengths and usually contain an even number of carbon atoms, most commonly in chains of 14, 16, 18, or 20. Free fatty acids are not very soluble in water, but in alkaline solutions the carboxyl group becomes ionized, making it strongly polar and thus soluble. Sodium and potassium salts of fatty acids are known as "soap" and are soluble in water. If you can get along without the added perfumes and deodorants, you can make your own soap from waste grease from the kitchen and sodium hydroxide (our forefathers got their sodium hydroxide or "lye" from wood ashes). Just obtain a recipe and be careful with the sodium hydroxide. You can put this soap in the guest bathroom.

Calcium salts of fatty acids, on the other hand, are not soluble in water, which is why soap forms disconcerting curds in hard (calcium-rich) water. Calcium salts also explain the "ring around the tub" and the dullness of hair washed in hard water.

Triglycerides are completely nonpolar because the carboxyl groups of fatty acids are bound up in linkages with the hydroxyl groups of glycerol. A dehydration linkage between a carboxyl group and an **alcohol** side group is called an **ester linkage.** An

alcohol is technically any substance with the formula R—C—OH. For example, the ethyl alcohol or ethanol of liquor is CH_3CH_2—OH. Thus, glycerol is an alcohol.

Saturated, Unsaturated, and Polyunsaturated Fats

Why all the diet talk about **saturated fats?** The obvious question here is, saturated with what? You may recall from an earlier discussion that the answer is hydrogen.

The carbon backbone of the hydrocarbon tails of fatty acids is composed primarily of carbon-to-carbon single bonds (one pair of shared electrons between successive carbons: C—C—C—C—C, and so on). But some fatty acids have double bonds between one or more pairs of successive carbons in place of side bonds to hydrogen (C=C—C—C—C=C, and so on). Therefore, they have fewer hydrogens (Figure 3.12). Such fatty acids, and the fats containing them, are said to be **unsaturated** because it is possible to add hydrogen to them if the double bonds are broken (with the aid of a platinum catalyst and a little pressure). By adding hydrogens to an unsaturated triglyceride, one can produce, for example, hydrogenated vegetable oil. When all the double bonds are gone, the molecule won't accept any more hydrogen, and it is then a saturated fat.

3.11 TRIGLYCERIDE SYNTHESIS

Through the action of specific dehydrating enzymes, three fatty acids are covalently linked to one molecule of glycerol. The emerging triglyceride is accompanied by three molecules of water.

Glycerol   +   Enzymes   Fatty acids   A triglyceride

3.12 SATURATED AND UNSATURATED FATTY ACIDS

Stearic acid is the main constituent of animal fat (see below left) and, like other saturated fats, has the maximum number of hydrogen atoms. Linoleic acid, from linseed and cotton-seed oils (below right), has two double bonds in the carbon chain, and, as the gaps in the rows of hydrogen atoms reveal, it is unsaturated.

Stearic acid glyceride

Linoleic acid glyceride

A fat with carbon-carbon double bonds is unsaturated, and one with more than one such double bond is called **polyunsaturated.** Common fatty acids with carbon-carbon double bonds are oleic acid (vegetable oil) and linolenic acid (linseed oil). Highly saturated fats tend to have rather high melting points (tend to remain solid, like lard, at room temperature), while polyunsaturated fats tend to have low melting points (tend to remain liquid or "oily"). Simple vegetable oils, such as peanut oil, are largely unsaturated, which explains why the oils in old-fashioned peanut butter separate, forming a layer at the surface (which usually spills when you open the jar). The problem is solved by the food industry through *hydrogenation*—partially saturating the oil with hydrogen, or adding a hydrogenated oil, thus forming a harder fat that does not separate at room temperature.

The membranes of our cells are rich in unsaturated fatty acids. But because the chemical processes of human metabolism can't form double bonds, it is necessary to include a small amount of unsaturated fats in the diet. In Chapter 29 we will discuss the dietary controversy that has grown over the increased use of unsaturated fats in many foods.

Triglycerides are concentrated energy storage compounds. In our bodies, they yield about twice as much available energy per gram dry weight as carbohydrates or protein. The difference is even more marked for calories per gram wet weight, since carbohydrates and protein absorb water and fat doesn't.

Because of their molecular properties, fats make ideal energy storage molecules. Plants store triglycerides in their seeds, and many kinds of animals store body fat as an adaptation for lean seasons or migration. Humans also store fat under the skin and around internal organs, although agriculture and food storage techniques have largely exempted us from the rigors of seasonal food depletion.

Stored fats also serve other purposes, such as insulation and flotation. The blubber of sea mammals serves both purposes. Storage fat is used as padding in your fingers and on your bottom, and makes a significant contribution to the general shapes of people.

Phospholipids

While the **phospholipids** are structurally closely related to the triglycerides, there are vital differences between them. In the phospholipids, two fatty acids are linked to a backbone of glycerol by ester linkages. In place of the third fatty acid found in triglycerides, however, phospholipids have a phosphate group, which is commonly linked to still other organic molecules (variable groups) as shown in Figure 3.13. Of the two fatty acids, one is commonly unsaturated and, in that one, the presence of the carbon-to-carbon double bond produces a peculiar but very significant bend in the tail as seen in the illustration. The bent tail has a lot to do with membrane fluidity, as we will see in Chapter 5. The phosphate or phosphate group is hydrophilic, so the charged phosphate heads mingle freely with water and other polar molecules. The tails themselves are hydrophobic, so they reject water but mingle freely with nonpolar substances. Actually, this gives phospholipids a detergent property in the sense that they can form bridges between water and oils (see Chapter 2). In addition they reduce the surface tension of water, thus acting as wetting agents.

Phospholipids can occasionally be found as food storage compounds (for example, lecithin in egg yolk), but they are more important in their roles as major components of the delicate membranes that surround and occur within living cells (Chapters 4 and 5). The **plasma membrane,** as the surrounding membrane is known, consists primarily of two layers of phospholipids, along with glycolipids, glycoproteins, and other dispersed proteins. It is believed that the hydrophilic phosphate ends associate with the protein, while the hydrophobic fatty acid tails dissolve into one another, giving an essentially liquid, oily core to the membrane.

In vertebrates, some nerves are sheathed in layers of plasma membranes, and therefore nervous tissue—especially that of the brain—has a high phospholipid content. (So perhaps "fathead" is an unintentional compliment.) In addition to the simple glycerol phospholipids already mentioned, brain and other nervous tissue have a wide array of marvelously named and complex phospholipids and glycolipids: sphingomyelin, dihydrosphingosine, cerebrosides, gangliosides, and many more just as poetic.

Other Lipids

Waxes, like neutral fats, are esters of fatty acids. However, the alcohol involved is not glycerol, but a long-chain alcohol with a single hydroxyl group. Beeswax consists largely of an ester of palmitic acid (16 carbons) with a straight-chain alcohol called myricyl alcohol (30 carbons). Because of its hydrophobic (water-repellent) properties, wax is useful to a variety of living things, especially those that need to save water. Insects generally have waxy cuticles. So do plants; the waxes of leaves, fruit skins, and flower petals contain very long fatty acids (up to 36 carbons), both as free fatty acids and as esters of long-chain alcohols. Many marine invertebrates manufacture various waxes, and so do your ears. Waxes are generally much harder and even more hydrophobic than fats—two reasons why commercial wax is used as a protectant. Heads would turn if you announced that you were going to fat your car.

Steroids are not structurally similar to fatty acid lipids, but since they are either partly or wholly hydrophobic, they qualify as lipids. All steroids contain a peculiar core consisting of four interlocking rings. The differences between various steroids depend on the variation in the side chains. Some steroids are highly hydrophobic, some less so. (By the way, steroids with —OH groups are also called **sterols**.) Let's consider two specific examples of steroids.

Lanolin (or **lanosterol**) is a greasy, almost waxy substance that is commercially refined from sheep's wool. Small amounts of lanolin are found in your own skin and hair, helping to keep them flexible.

Cholesterol is a substance we've all heard of because various food manufacturers have assaulted us over the airways with the message that their

3.13 PHOSPHOLIPIDS

Phospholipids are an essential component of plasma membranes. They differ from triglycerides in several important ways. Phospholipids have only two fatty acid tails, one saturated and the other unsaturated. Because of its double bond, the unsaturated fatty acid takes on a bent configuration as seen. Instead of the usual third fatty acid of triglycerides, phospholipids have a phosphate group, and this positively charged head is often attached to one of a number of charged organic molecules, designated here as a variable group. Since the phosphate and organic heads are charged, or polar, they readily interact with water, which is also polar. The fatty acid tails are uncharged so they have no affinity for water. Therefore the phospholipid heads are described as hydrophilic and the tails, hydrophobic.

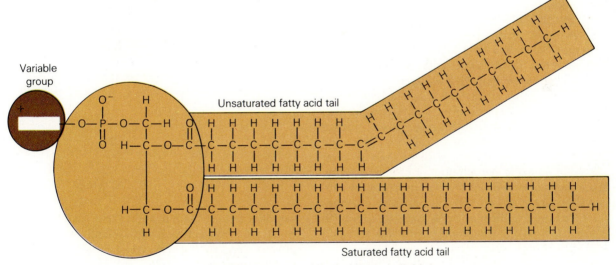

Variable group

Unsaturated fatty acid tail

Saturated fatty acid tail

Chemical Structure of Typical Phospholipid

Model of phospholipid

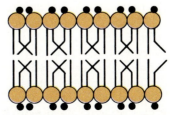

Arrangement in plasma membrane

products don't contain any. Why should we care? We might be interested because cholesterol has been accused of coating the lining of our blood vessels and thereby reducing their internal diameter and increasing blood pressure the way one might increase water pressure by placing the thumb over the end of a garden hose (Figure 3.14). But cholesterol is not all bad. In fact, it has a number of vital functions. For example, like phospholipids, cholesterol is a major constituent of plasma membranes. Also, bile acids, which are necessary for fat digestion, are modified cholesterol. When irradiated with ultraviolet light, cholesterol becomes vitamin D, which is necessary for normal bone growth and maintenance. You may have a greater appreciation for cholesterol when you realize that it can also be modified to make the various steroid hormones, including the sex hormones.

THE PROTEINS

For many reasons, proteins are incredibly important molecules. We've seen that carbohydrates function both as food reserves and as structural molecules. Proteins may function in these ways as well, but they also play other, equally vital, roles. For example, many proteins take the form of enzymes. We have mentioned the basic role of enzymes, but here let's stress that enzymes are biologically active proteins that work in cells as catalysts—agents that greatly speed up chemical activity and in some instances, bring about reactions that might never occur in their absence. As we proceed into the chemical reactions of cells you will soon develop an abiding respect, perhaps even an awe, for these amazing perpetrators of cellular activity, since they seem to be involved in nearly every conceivable aspect of life. We will return to enzymes later, but putting them aside for now, let's move to a more general consideration of proteins and perhaps pose some basic questions.

First, what *is* a protein? A **protein** is a macromolecule composed of one or more **polypeptides.** (It may also contain groups of sugars, lipids, or other small molecules.) Obviously, the next question is, what is a polypeptide? A polypeptide is a linear polymer made up of a chain of **amino acids** joined together by dehydration linkages known as **peptide bonds.** Actually, a polypeptide is quite large, most probably containing over 40 to 50 amino acids. On a considerably smaller scale is the **dipeptide,** two amino acids joined through peptide bonds, and the

3.14 STEROIDS

Steroids are complex four-ring molecules that have some lipid characteristics. One important steroid is cholesterol, a precursor or building block for many other important steroids. Cholesterol has also been implicated in the vascular disease atherosclerosis. Abnormal regions—thickenings and rough spots called plaques—form in the arterial linings and these may break away and clog vessels. In addition, the roughened lining may cause blood clotting within the circulatory system. Both situations are decidedly dangerous. Both clear (above) and partially clogged arteries are shown here.

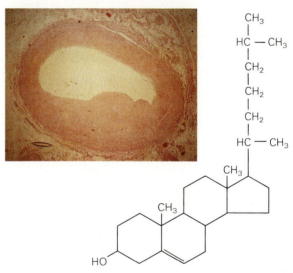

Cholesterol

peptide, a short chain of amino acids so joined. The distinction between polypeptide and protein is not always clear, by the way. Most proteins consist of two or more polypeptides, are biologically active, and take on complex shapes; however, polypeptides may also be active and complex in shape. Thus, some polypeptides are functional proteins, but for now, let's consider polypeptides as simple chains of amino acids. We will take up protein structure shortly. The range in molecular weight of the various proteins is enormous—from about 5700 in insulin to 7 to 8 million in some of the largest enzymes. But before going further into this, let's look more closely at the amino acids.

Amino Acids

Amino acids are the molecular subunits—the monomers—that become the "building blocks" of proteins. There are just 20 different amino acids (plus a few variants and modified versions) in proteins, and each has different and important properties. Twenty seems a small number when we consider how complex proteins are, but we might remember that *The Rise and Fall of the Third Reich* was written with only 26 letters. Let's have a closer look at these remarkable molecules.

The Structure of Amino Acids. First of all, each of the 20 amino acids has an amino group, which is basic, a carboxyl group, which is acidic, and a simple hydrogen group. It is in the R portion of the molecule that the 20 amino acids differ from one another. In Figure 3.15, the amino group and the carboxyl group are shown in the ionized state, which is the actual condition of amino acids in solution under normal physiological conditions. An amino acid is sometimes said to be a *zwitterion* (German for "half-breed ion"), since it is both neg-

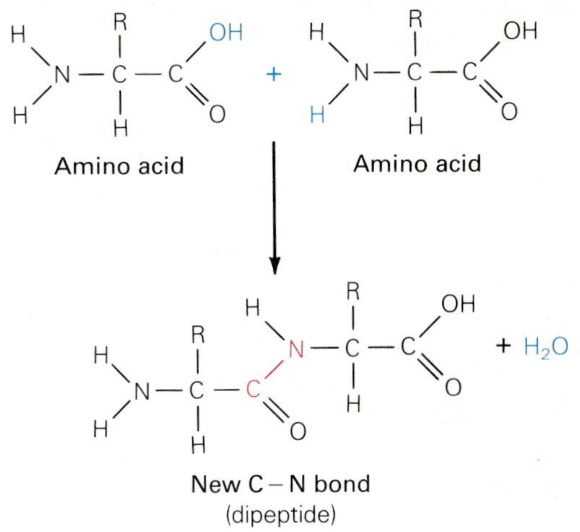

3.15 AMINO ACID STRUCTURE

The basic structure of each amino acid includes a short carbon chain with at least one amino and one carboxyl (acid) group attached. The amino acids differ from each other in the kinds of side groups (R) attached. The simplest amino acid, glycine, has one hydrogen atom at this location. The illustration shows a general amino acid in two ways: nonionized and ionized. In the ionized state (whenever water is present, which is almost always), the carboxyl group releases a proton, which tells us it is an acid group. The amino end collects a proton, thus acting as a base.

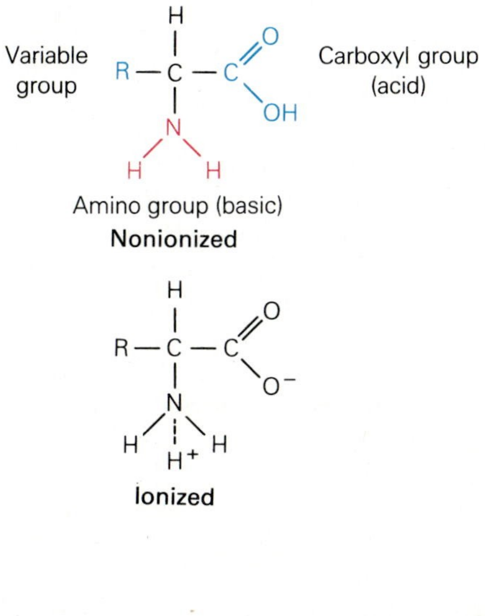

3.16 PEPTIDE BONDS

Amino acids are joined through peptide bonds in the simplest or primary level of protein organization. Peptide bonds always occur between the carboxyl and amino groups of adjacent amino acids. A polypeptide is formed when many amino acids have joined in this manner.

atively and positively charged. We will return to the amino acids for a closer look at their structure but let's first see how amino acids are incorporated into polypeptides.

Polypeptides and Proteins

In polypeptides, the carboxyl group of one amino acid is linked, through the familiar dehydration linkage, to the amino group of the next amino acid; the carboxyl group of the second amino acid is linked to the amino group of the third; and so on down the line. Each carboxyl-amino linkage is a peptide bond. The complete polypeptide has an *N*-**terminal,** or amino end (the first amino acid in the chain), and a *C*-**terminal,** or carboxyl end (the last amino acid). The structure of polypeptides is traditionally written from left to right with the terminal amino end on the left and the terminal carboxyl end on the right (Figure 3.16).

Proteins of any specific group are therefore far more uniform than are polysaccharides, which tend to branch randomly and be of random length. Further, proteins have precise, three-dimensional conformations determined by the amino acid sequence. (In solution, though, a protein molecule may be somewhat flexible and the enzymes may have "moving parts.") Proteins tend to be either long and stringy (fibrous proteins) or folded into nearly spherical balls, or globs (globulins).

Levels of Protein Organization. Proteins may have as many as four levels of organization and most have at least three. The first or **primary level** (Figure 3.17a) is determined by the number, kind, and order of the amino acids joined together by peptide bonds to form a simple polypeptide strand. The amino acid sequence, as we will see, is in turn, genetically determined.

The **secondary level** (Figure 3.17b) of organization forms spontaneously, as soon as the polypeptide has been synthesized. In many polypeptides, the chain of amino acids coils, forming what is called an **alpha helix** (right-handed coil). Coiling is brought about by numerous, weak hydrogen bonds that form between oxygen and hydrogen in the carboxyl and amino groups of every fourth amino acid in the chain. The individual hydrogen bonds are weak, but there are so many of them that the alpha helix is fairly stable. Another common secondary structure is a flattened zigzag folding called a **beta pleated sheet,** the final level of organization for many structural proteins.

The third or **tertiary level** (Figure 3.17c) of organization occurs when the alpha coil is coiled once more, and then folded back on itself in specific ways. It is particularly important to the structure of globular proteins such as enzymes and albumin (such as egg white protein). Tertiary coiling and folding is brought about by several kinds of interactions among amino acid R groups. Included are disulfide linkages, covalent bonds that form bridges between molecules of the sulfur-containing amino acid, **cysteine.** For this reason cysteine is of great importance in the diet of most organisms.

Some of the globular proteins reach a fourth or **quaternary level** (Figure 3.17d) of organization. In this level, two or more polypeptides join to form the finished protein. Needless to say, quaternary proteins are giants. Among them is the oxygen-carrying blood protein **hemoglobin.** It contains two pairs of interacting polypeptides (known as alpha and beta chains) attracted to each other by forces similar to those of the tertiary level. Also present are four very special iron-containing **heme groups** (see Figure 3.17), the actual sites of oxygen transport.

While proteins assume very specific three-dimensional shapes, and their shapes are highly important to function, they can readily lose both shape and function through **denaturation.** One common agent of denaturation is heat. Heat breaks some bonds and causes the random formation of others, and such events are generally irreversible. This is why it is hard to "unfry" an egg. Other denaturing agents are strong acids and bases and certain chemicals. Cold may also have a denaturing effect on protein, as will *urea,* a nitrogen waste produced by many animals. However, in both these instances, the denaturing effects are temporary and reversible, and the protein resumes its shape and function once normal conditions are restored.

The Amino Acids, Their R Groups, and How They Influence Protein Structure. Table 3.1 lists the 20 amino acids and shows their R group structures. The R groups of the amino acids determine the physical and chemical properties of the proteins in several ways. First, such general properties of the protein as solubility and electrical charge depend on the net effect of all the functional groups. Second, a few of the functional and side groups of an enzyme participate directly in some enzymatic reactions. Third, and most relevant to our discussion, the interactions between the different amino acid R groups of a given protein—especially such interactions as nonpolar interactions, ionic bonding, hydrogen bonding, and even covalent bonding—determine the three-dimensional shape of the protein. We see in Figure 3.17 how the R groups may interact in proteins. Now let's briefly review some of the distinguishing traits of the 20 amino acids. We should first note that there are three amino acids (arginine, histidine, and lysine) that have extra amino groups and are thus always positively charged at the usual range of acidity for living things (called physiological pH, the range being pH 6–8). Two other amino acids, aspartic acid and glutamic acid, have two carboxyl groups and thus are negatively charged at physiological pH.

Among the remaining 15 amino acids, a number are polar because their R groups contain hydroxyls or other functional groups containing slight charges. For this reason they are hydrophilic—that is, they interact readily with water and they readily form hydrogen bonds with other polar molecules. Such amino acids include asparagine, glutamine, serine, threonine, and tyrosine. The five ionic amino acids are also hydrophilic. Each of these has other special characteristics, as well.

Of the remaining nine amino acids, eight are hydrophobic. Just as oil and water separate, hydrophobic and hydrophilic side groups of a protein will not join. The hydrophobic side groups are forced together. Most globular proteins have hydrophobic centers (where the hydrophobic amino acid R groups get together) and hydrophilic exteriors (which form hydration shells in an aqueous medium). The eight hydrophobic amino acids are alanine, isoleucine, leucine, methionine, phenylalanine, proline, tryptophan, and valine.

That leaves just one amino acid, cysteine,

3.17 PROTEIN ORGANIZATION

(a) The primary level of organization involves the number and kinds of amino acid forming the polypeptides. **(b)** In the secondary level, an alpha helix, or right-handed coil, commonly occurs as polar groups within the amino acids form hydrogen bonds. **(c)** The tertiary or third level of organization occurs as the alpha helix coils and folds back upon itself and disulfide bonds connect polypeptide strands. The folding is due primarily to interactions between R groups. **(d)** Where the quaternary level of organization is found, two or more complete polypeptides join to form gigantic molecules such as the hemoglobin seen here. Note the special oxygen-attracting *heme* groups with their central atom of iron.

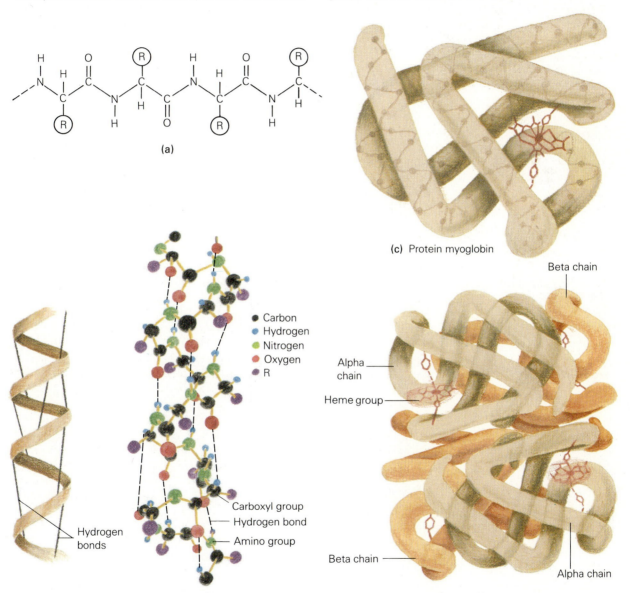

(a)

Carbon
Hydrogen
Nitrogen
Oxygen
R

Carboxyl group
Hydrogen bond
Amino group

Hydrogen bonds

(c) Protein myoglobin

Beta chain
Alpha chain
Heme group
Beta chain
Alpha chain

(b) Alpha helix with hydrogen bonds between carboxyl and amino group of amino acids

(d) Protein hemoglobin containing four polypeptides, two alpha chains, and two beta chains, along wih four iron-containing groups.

TABLE 3.1

THE 20 AMINO ACIDS* COMMONLY FOUND IN PROTEINS

Amino acids with nonpolar R groups	Amino acids with uncharged polar R groups

Alanine (Ala)

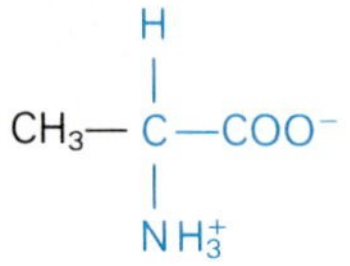

$$CH_3\text{—}\overset{\displaystyle H}{\underset{\displaystyle NH_3^+}{C}}\text{—}COO^-$$

Isoleucine (Ile)

$$CH_3\text{—}CH_2\text{—}\underset{\displaystyle CH_3}{C}H\text{—}\overset{\displaystyle H}{\underset{\displaystyle NH_3^+}{C}}\text{—}COO^-$$

Leucine (Leu)

$$\underset{\displaystyle CH_3}{\overset{\displaystyle CH_3}{CH}}\text{—}CH_2\text{—}\overset{\displaystyle H}{\underset{\displaystyle NH_3^+}{C}}\text{—}COO^-$$

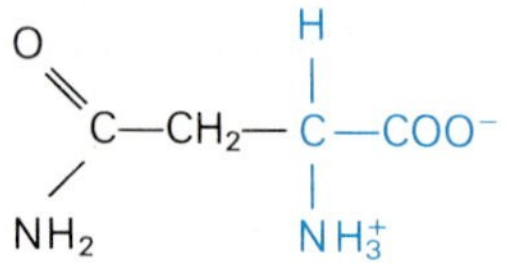

Methionine (Met)

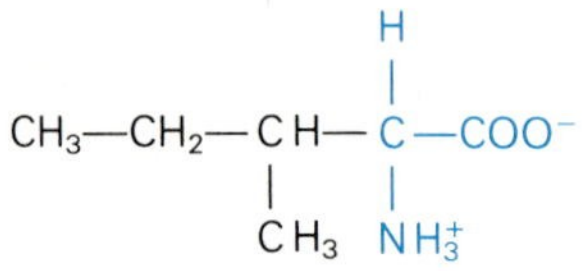

$$CH_3\text{—}S\text{—}CH_2\text{—}CH_2\text{—}\overset{\displaystyle H}{\underset{\displaystyle NH_3^+}{C}}\text{—}COO^-$$

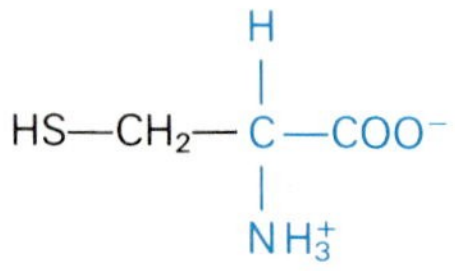

Phenylalanine (Phe)

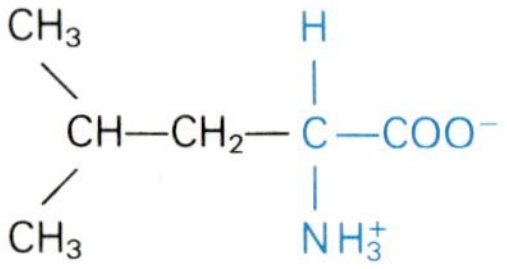

$$\text{⬡}\text{—}CH_2\text{—}\overset{\displaystyle H}{\underset{\displaystyle NH_3^+}{C}}\text{—}COO^-$$

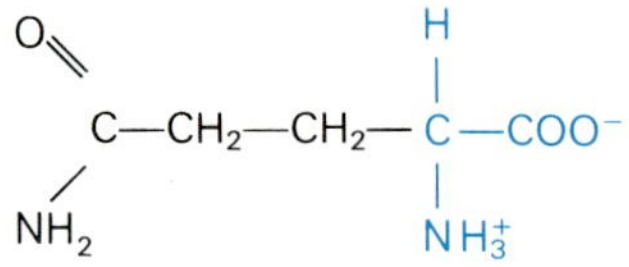

Proline (Pro)

$$\begin{array}{c} H_2C \diagup CH_2 \diagdown \\ H_2C \diagdown_{} \diagup C\text{—}COO^- \\ NH_2^+ \quad H \end{array}$$

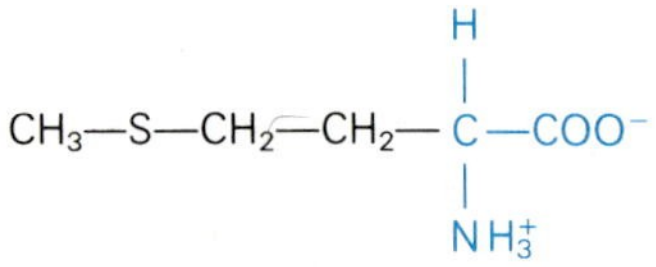

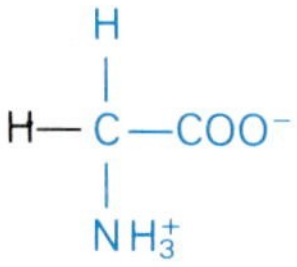

Tryptophan (Trp)

$$\begin{array}{c} \text{⬡}\overset{}{\underset{}{\diagup}}C\text{—}CH_2\text{—}\overset{\displaystyle H}{\underset{\displaystyle NH_3^+}{C}}\text{—}COO^- \\ CH \\ N \\ H \end{array}$$

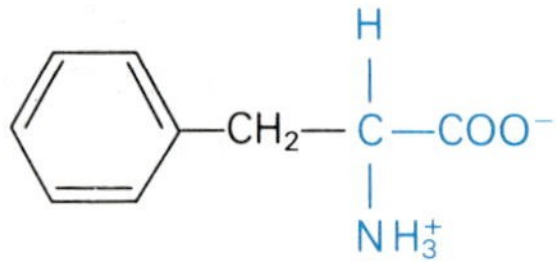

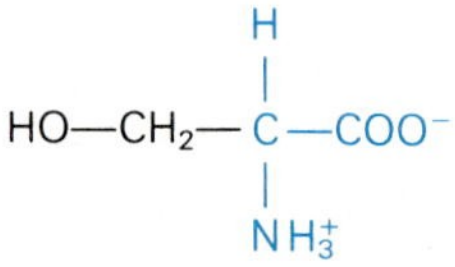

Valine (Val)

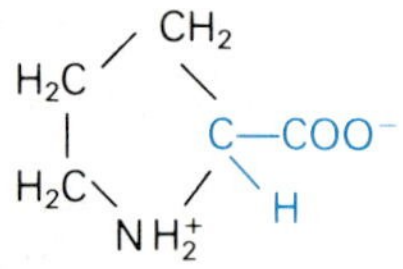

$$\underset{\displaystyle CH_3}{\overset{\displaystyle CH_3}{CH}}\text{—}\overset{\displaystyle H}{\underset{\displaystyle NH_3^+}{C}}\text{—}COO^-$$

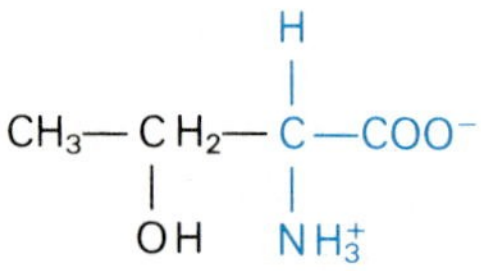

Asparagine (Asn)

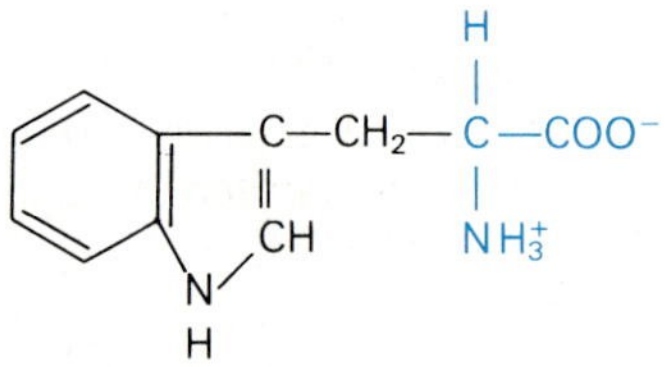

$$\overset{\displaystyle O}{\underset{\displaystyle NH_2}{C}}\text{—}CH_2\text{—}\overset{\displaystyle H}{\underset{\displaystyle NH_3^+}{C}}\text{—}COO^-$$

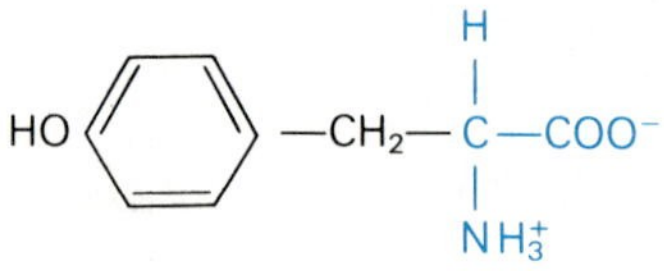

Cysteine (Cys)

$$HS\text{—}CH_2\text{—}\overset{\displaystyle H}{\underset{\displaystyle NH_3^+}{C}}\text{—}COO^-$$

Glutamine (Gln)

$$\overset{\displaystyle O}{\underset{\displaystyle NH_2}{C}}\text{—}CH_2\text{—}CH_2\text{—}\overset{\displaystyle H}{\underset{\displaystyle NH_3^+}{C}}\text{—}COO^-$$

Glycine (Gly)

$$H\text{—}\overset{\displaystyle H}{\underset{\displaystyle NH_3^+}{C}}\text{—}COO^-$$

Serine (Ser)

$$HO\text{—}CH_2\text{—}\overset{\displaystyle H}{\underset{\displaystyle NH_3^+}{C}}\text{—}COO^-$$

Threonine (Thr)

$$CH_3\text{—}\underset{\displaystyle OH}{C}H_2\text{—}\overset{\displaystyle H}{\underset{\displaystyle NH_3^+}{C}}\text{—}COO^-$$

Tyrosine (Tyr)

$$HO\text{—}\text{⬡}\text{—}CH_2\text{—}\overset{\displaystyle H}{\underset{\displaystyle NH_3^+}{C}}\text{—}COO^-$$

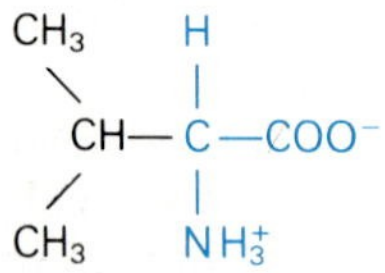

*The portion of the amino acid that is common to all is colored. Note that some of the amino acids contain more than one amino or acid group, giving them greater basic or acid qualities than the others. Cysteine contains a sulfur-hydrogen group at its R-terminal. This has special importance in determining the shapes of proteins. The abbreviations given in parentheses are used for convenience in writing protein formulas.

TABLE 3.1 (continued)

Amino acids with acid R groups (negatively charged at pH 6.0)	Amino acids with basic R groups (positively charged at pH 6.0)

Aspartic acid (Asp)

$$^-O{\diagdown}C-CH_2-\overset{\overset{\displaystyle H}{|}}{\underset{\underset{\displaystyle NH_3^+}{|}}{C}}-COO^-$$
(with $=O$ on the carbon)

Arginine (Arg)

$$^+H_3N-\overset{\overset{\displaystyle }{\|}}{\underset{\underset{\displaystyle NH}{}}{C}}-NH-CH_2-CH_2-CH_2-\overset{\overset{\displaystyle H}{|}}{\underset{\underset{\displaystyle NH_3^+}{|}}{C}}-COO^-$$

Glutamic acid (Glu)

$$^-O{\diagdown}C-CH_2-CH_2-\overset{\overset{\displaystyle H}{|}}{\underset{\underset{\displaystyle NH_3^+}{|}}{C}}-COO^-$$
(with $=O$ on the carbon)

Histidine (His)

$$HC{=\!=}C-CH_2-\overset{\overset{\displaystyle H}{|}}{\underset{\underset{\displaystyle NH_3^+}{|}}{C}}-COO^-$$
$$^+HN\quad NH$$
$$\diagdown C \diagup$$
$$|$$
$$H$$

Lysine (Lys)

$$^+H_3N-CH_2-CH_2-CH_2-CH_2-\overset{\overset{\displaystyle H}{|}}{\underset{\underset{\displaystyle NH_3^+}{|}}{C}}-COO^-$$

whose special role we have already mentioned. Cysteine has a **sulfhydryl** side group (–SH). Under mildly oxidizing conditions (which we'll consider later), two sulfhydryl groups will react to make a disulfide bridge:

$$R-SH + HS-R + \tfrac{1}{2}O_2 \longrightarrow R-S-S-R + H_2O$$

As we have seen, the disulfide linkages between cysteines help form the loops and winding shape of the molecule. **Keratins,** the structural proteins of skin and hair, are very rich in disulfide cross-linkages. Disulfide bonds can be broken by some of the concoctions used in permanents. They act by reversibly denaturing hair protein, allowing them to assume new, and hopefully more aesthetic, shapes.

Thus, it is basically the chemical characteristics of the amino acids and their strategic location in the polypeptide chains that determine the final shape a protein may take. To summarize, the forces determining the shape of a protein include (1) the disulfide covalent bonds between cysteine groups, (2) hydrogen bonding, (3) electrostatic or ionic interaction, and (4) hydrophobic interaction.

Each of these forces depends on the presence of specific amino acids in positions where the side groups are close together. In other words, the primary structure of a protein or polypeptide is instrumental in determining the kinds of folding a protein will undergo to assume its final shape. We should add a fifth factor that hasn't been brought up: the environment. The chemical and physical nature of the protein's surroundings can also strongly affect its structure.

Conjugated Proteins and Prosthetic Groups

A protein that consists only of amino acids, in one or more polypeptides, is called a **simple protein. Conjugated proteins** have something extra, either covalently bonded to the amino acid side groups or otherwise bound to the polypeptides. The something extra is called the **prosthetic group** (it might help to remember that a wooden leg, a glass eye, or a dental bridge is called a *prosthesis*, or added part). Among the conjugated proteins are glycoproteins (the prosthetic group is a carbohydrate), chromoproteins (*chromo* refers to color, so the prosthetic group is a pigment), and lipoproteins (the prosthetic group is fat-soluble). The hemoglobin molecule is a conjugated protein because each subunit contains a heme group which, in turn, contains an iron atom (Figure 3.18). Since heme is deeply colored, hemoglobin is a chromoprotein, although color has nothing whatever to do with its function. Blood just *happens* to be red.

The prosthetic group of a protein often determines function. The catalytic activities of many

3.18 PROSTHETIC GROUPS OF PROTEINS

Each of the polypeptides of hemoglobin contains a prosthetic group known as a *heme*. The heme groups contain iron and have the ability to bind with oxygen, forming *oxyhemoglobin*. The association of oxygen and hemoglobin is reversible, with oxygen released as blood circulates in oxygen-deficient tissues.

Heme group

enzymes, for instance, are dependent on **cofactors,** some of which are small prosthetic groups attached to the enzyme. Neither the factor nor the remainder of the enzyme alone has any catalytic properties. While some prosthetic groups are simple metals, others comprise the familiar dietary vitamins.

Binding Proteins

Proteins also perform a diversity of nonenzymatic, nonstructural functions. Many of these are related to protein's peculiar ability to form specific structures and shapes that enable them to *bind* to other substances. Hemoglobin, for instance, is a carrier protein; it binds oxygen in the lungs and releases it where it is needed in the body, and it binds carbon dioxide in the tissues and releases it in the lungs. There are a number of other circulating binding proteins, such as *transferrin*, which binds iron and *haptoglobin*, which binds hemoglobin. Egg white has a protein, *avidin*, that binds tightly to *biotin*, an important vitamin; fungi that attempt to invade eggs die of biotin vitamin deficiency.

One of the most critical functions of proteins is their ability to bind to other molecules as part of the antibody response. Some proteins are synthesized by white blood cells for the express purpose of binding with and ultimately destroying or inactivating

foreign molecules or particles. (See the discussion of immune responses in Chapter 31.)

We should add that there is good evidence that cells recognize each other during embryonic development through interactions of binding proteins on their surfaces. In addition, binding proteins on cell surfaces lend peculiarities to cells that enable us to smell and taste certain molecules.

Structural Proteins

Structural proteins are important in maintaining the physical form of organisms. They may be *intracellular* (inside cells) or *extracellular* (outside cells). In animals with backbones (vertebrates), common extracellular proteins include **collagen, elastin,** and **keratin.** These are long, fibrous molecules that clump together to make larger fibers. Collagen, the principal component of connective tissue such as tendons, ligaments, and muscle coverings, makes up about 25% of the protein in humans.

Elastin gets its name from its remarkable ability to stretch. Elastin fibers give elasticity to connective tissue, including that of your ears and skin. Ears are very rich in elastin, so if you reach over and tug on your neighbor's ear, it will snap right back to its original form (one would hope). You can also use pinching as an age test. If you pinch the skin on the back of your hand and it snaps back, you're young. If a ridge stands there, you're not. This is because with aging, elastin loses its elastic properties. The same phenomenon is responsible for the bagginess under the eyes and about the face and neck that in late life awaits nearly all of us.

Keratin is the protein of hair and of the outer layer of skin, as well as that of feathers, claws, nails, horns, antlers, and scales. In contrast to elastin or collagen, it is basically an intracellular protein. It is deposited as cells fill themselves with keratin, then dry up and die, leaving the keratin behind.

THE NUCLEIC ACIDS

Probably by now almost everyone has heard of the nucleic acids, particularly DNA, simply because some of the recent discoveries and technical breakthroughs in DNA research have been relentlessly and breathlessly touted by the popular press. Understanding the two nucleic acids, DNA and RNA, is so fundamental to biology that we will focus on it in three chapters (Chapters 9, 10, and 11). Here, we will be brief, introducing the basic structure and making a few comments on how the nucleic acids do their work.

The initials **DNA** and **RNA** refer to **deoxyribonucleic acid** and **ribonucleic acid,** respectively. Both molecules are present, with a very few exceptions, in all cells. DNA is often referred to as the molecular core of life since it contains the units of heredity we call **genes.** Essentially, genes are chemical codes that specify the kinds of protein an organism can produce, and these proteins include the enzymes that regulate most of the cell's activities. In addition, DNA is **self-replicating** which means that it can make copies of itself, as it does when the cell or organism reproduces.

RNA also has a critical role. It is essential to carrying out the protein specifying function of DNA. Later we will see that, in a sense, while DNA designates the specific protein to be produced, RNA is responsible for gathering the amino acids together and assembling them correctly into polypeptides.

Nucleic Acid Structure

Each DNA molecule consists of two extremely long polymers held together by hydrogen bonding and coiled into what is known as a **double helix** (Figure 3.19). Each of the two polymers consists of a great number of subunits called **nucleotides,** covalently bonded to form the long strands. Each nucleotide, in turn, consists of three parts: a phosphorus-containing group known as *phosphate,* a simple *5-carbon sugar,* and a *nitrogen base.* We will put off the molecular details for later, except to note that there are four different nitrogen bases used in DNA, so it follows that there are four different nucleotides. Their specific linear arrangement in the polymer constitutes the chemical coding (genetic coding) we mentioned earlier. In other words, the genes we all like to talk about at family gatherings are basically an arrangement of nucleotides in our DNA that simply determine what our proteins will be like.

RNA is a single polymer of nucleotides and does not take on the helical configuration. It is produced when a portion of the DNA helix unwinds and the chemical coding for that portion is followed as RNA is synthesized. Hence, RNA is much shorter than DNA, containing only enough nucleotides to code for the assembly of amino acids into a protein (or perhaps just a polypeptide). In addition, RNA's nucleotides utilize a slightly different sugar and a minor, but important, substitution of one of the four usual nitrogen bases. There are actually several types of RNA involved in protein synthesis, but we will get to these in later chapters.

3.19 DEOXYRIBONUCLEIC ACID

DNA consists of two very long polymers of nucleotides, coiled into a double helix configuration. Each nucleotide contains a phosphate or phosphoric acid group, a 5-carbon sugar (deoxyribose), and one of four different nitrogen bases (adenine, guanine, cytosine, or thymine). The one seen here is adenine. When assembled into a polymer, each nucleotide is covalently linked to the ones above and below via the phosphates and sugars, which form the backbone of the molecule. The nitrogen bases face inward, where weak but numerous hydrogen bonds attract the nucleotides of one polymer to those of the other. The combined forces produce the familiar helical configuration.

Hydrogen bonds

Nucleotides covalently linked into polymers

Linkage phosphate groups

A nucleotide

Phosphate

One of the four nitrogen bases (adenine)

5-carbon sugar

Double helix configuration

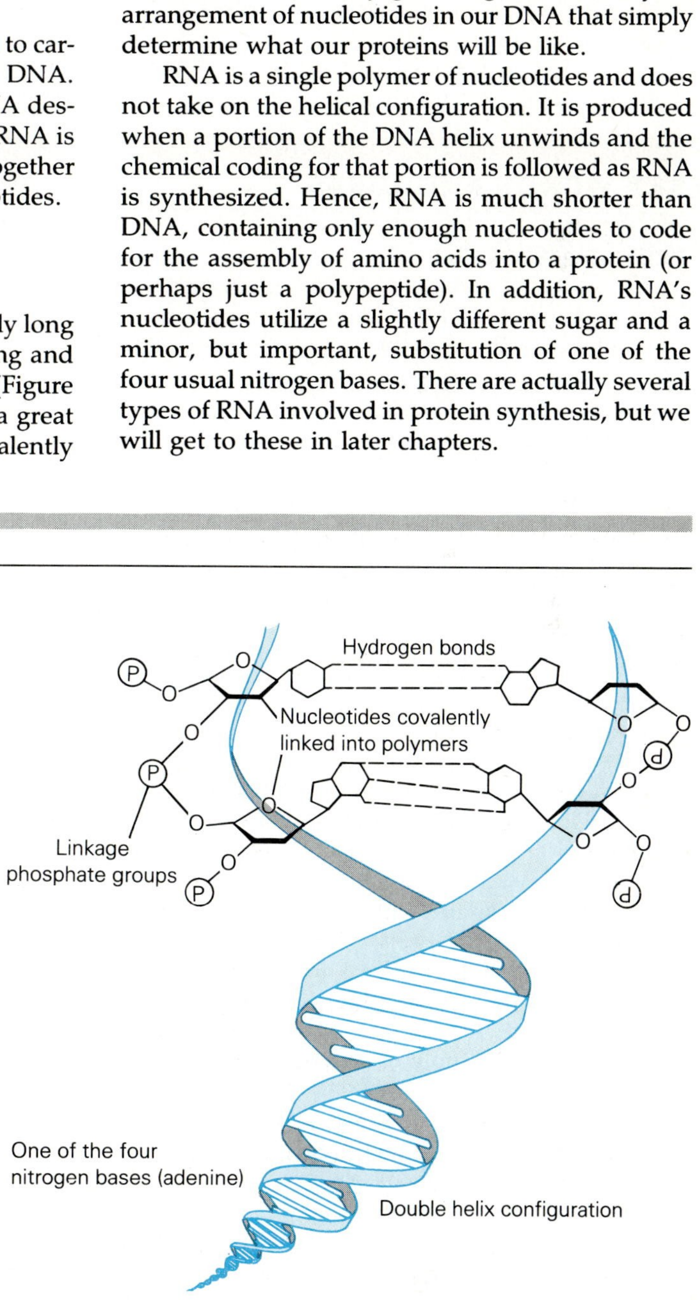

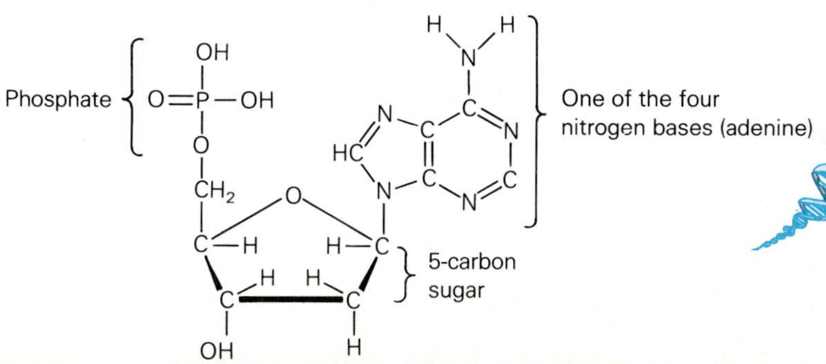

APPLICATION OF IDEAS

1. Nucleic acids and proteins are considered to be "informational molecules," but carbohydrates are not, even though they may be extremely large and complex. Why do carbohydrates fail to qualify? Could lipids be "informational molecules"?

2. Animals readily use enzymes to transform glucose and oxygen into water, carbon dioxide, and energy. Since all enzymatic reactions are reversible, can they transform water, carbon dioxide, and energy into oxygen and glucose? Explain your answer and plan to come back to this question after having read Chapters 6, 7, and 8.

3. Discuss the nutritional consequences of various fad diets, such as the all-banana or all-egg diet, a fat-free diet, or a low-carbohydrate, low-fat, all-meat diet. How might such diets affect one's health?

KEY WORDS AND IDEAS

THE CARBOHYDRATES

1. The empirical formula for **carbohydrates** is $(CH_2O)n$; thus all types of carbohydrates are multiples of this structure.

Monosaccharides and Disaccharides

1. The **monosaccharide glucose** is the subunit for many complex carbohydrates. **Sucrose,** a disaccharide, includes glucose and fructose as its subunits.

2. Functional groups in the simple sugars include **aldehydes** (glucose) and **ketones.**

Polysaccharides

1. Monosaccharides form disaccharides and polysaccharides through **dehydration synthesis,** an enzymatic dehydration process in which **dehydration linkages** (covalent bonds) form between adjacent monosaccharides. Water is also yielded. In the opposing process, **hydrolytic cleavage,** water is restored and the dehydration linkage is broken. In sugars, dehydration linkages are also called **glycosidic linkages.**

2. Plant starches include **amylose** and **amylopectin,** polysaccharides with hundreds of glucose subunits. The first is joined by 1–4 linkages, and the second by 1–3, 1–4, and 1–6 linkages.

3. **Glycogen,** a highly branched animal starch, contains both 1–4 and 1–6 linkages. It is cleaved through **phosphorolytic cleavage,** where phosphate is added.

4. While convalent bonding produces **primary structure** in starches, hydrogen bonding, ionic bonding, and van der Waals forces produce a coiling **secondary structure.**

5. **Cellulose,** composed of beta-glucose subunits, is insoluble and is digestible only to a few organisms that have the enzyme **cellulase.**

6. Cellulose forms strong structural elements (plant cell walls and fibers) because of its fibrous organization. Cellulose is strengthened through the addition of **hemicelluloses, pectins,** and **lignins.**

7. **Chitin,** a structural polysaccharide of insect **exoskeletons** and fungal cell walls, contains *N*-acetyl glucosamine.

LIPIDS

1. **Lipids** are fat-soluble, mainly hydrophobic molecules used for energy reserves, membrane components, and hormones.

Triglycerides

1. **Triglycerides** are nonpolar molecules consisting of one **glycerol** and three **fatty acids.** Triglycerides differ from each other in the kinds of fatty acids present.

2. Fatty acids themselves are polar and hydrophilic at their acid end and nonpolar and hydrophobic at their carbon chain.

Saturated, Unsaturated, and Polyunsaturated Fats

1. **Saturated** fats (chiefly the more solid animal fats) have no carbon-to-carbon double bonds and hold the maximum number of hydrogen atoms.

2. **Unsaturated** fats (chiefly liquid oils in plants) contain at least one carbon-to-carbon double bond, with fewer than the maximum number of

hydrogens. **Polyunsaturated** fats have two or more carbon-to-carbon double bonds.

3. Gram for gram, fats have twice the caloric value of carbohydrates and proteins.

Phospholipids
1. **Phospholipids,** components of cellular membranes, contain glycerol plus two fatty acids (one saturated, one unsaturated), and a polar phosphate group along with some variable group. Like fatty acids, the heads are polar and the tails nonpolar. Their nonpolar tails form an oily, water-resistant core in cell membranes.
2. **Waxes** lack glycerol and are more hydrophobic, serving as a waterproofing material in both plants and animals.
3. **Steroids** are lipid-soluble, multiple-ring molecules and include **lanolin** and **cholesterol.**
4. Cholesterol is a constituent of cell membranes, liver bile, and the starting molecule for many steroid hormones and for vitamin D.

THE PROTEINS
1. **Proteins** are important in cell structure, are chemically active as enzymes, are good sources of energy, and serve as hormones.
2. Proteins are composed of **amino acids,** linked via covalent **peptide bonds** arranged into **polypeptides.** In their final, active, form one or more polypeptides constitute a protein. The size and weight range of proteins is enormous.

Amino Acids
1. There are 20 amino acids in protein, each of which contains an amino and a carboxylic functional group, along with a variable R group. In the ionized state, the carboxyl group is negative, the amino group, positive.

Polypeptides and Proteins
1. Polypeptides are strands of covalently linked amino acids.
2. Proteins are either fibrous or globular and have four possible levels of organization.
   a. *Primary level:* determined by kinds and number of amino acids covalently linked into a polypeptide chain.
   b. *Secondary level:* polypeptide chains form an **alpha helix** through numerous weak hydrogen bonds.
   c. *Tertiary level:* folds and coils formed through covalent disulfide linkages and interactions between R groups.
   d. *Quaternary level:* the union of two or more polypeptides, through covalent bonding, hydrogen bonding, or van der Waals forces.
3. **Hemoglobin,** the oxygen-carrying pigment of many animals, is a quaternary protein. It contains four polypeptides and iron-containing **heme groups.**
4. The individual R groups of amino acids are significant to protein structure since they interact in nonpolar attractions, ionic bonding, hydrogen bonding and covalent bonding.
5. Some amino acids have extra amino groups and are positive, while others have extra carboxyl groups and are negative; both types are hydrophilic. Some have nonpolar clusters of hydrogen and are hydrophobic. One amino acid, cysteine, contains a **sulfhydryl group** (—SH) and forms the disulfide links within and between polypeptides.

Conjugated Proteins and Prosthetic Groups
1. Unlike **simple proteins, conjugated proteins** have non-amino acid groups such as carbohydrates, pigments, and lipids. The heme group of hemoglobin is a pigment. Some **prosthetic groups** act as **cofactors.**

Binding Proteins
1. **Binding proteins** are those that recognize and join with certain molecules.

Structural Proteins
1. **Structural proteins** in vertebrates include **collagen** (binding materials), **elastin,** (in flexible tissue), and **keratin** (hooves, nails, and hair).

NUCLEIC ACIDS
1. Nucleic acids include **DNA (deoxyribonucleic acid)** and **RNA (ribonucleic acid).**
2. DNA comprises the **genes,** forming a chemical or genetic coding. It is **self-replicating,** so copies can be sent from cell to cell and from organism to offspring. RNA is involved in translating the genes into protein products.

Nucleic Acid Structure
1. DNA is double stranded, consisting of two very long chains of covalently linked **nucleotides** intertwining into a **double helix.**
2. There are four different nucleotides and their linear order in DNA constitutes the genes and specifies the structure of the many kinds of proteins. While the specifications reside in DNA, they are copied into RNA which carries out the actual synthesis of protein.

REVIEW QUESTIONS

1. State the empirical formula for carbohydrate. Explain what such a formula means. (50)

2. To what do the terms *monosaccharide, disaccharide*, and *polysaccharide* refer? What do all three have in common? (50–51)

3. Write the straight chain and ring structural formulas for alpha glucose and number the carbons. (51–52)

4. Explain what is meant by a dehydration linkage or dehydration synthesis. What is the opposite process called? How are both important to most biological molecules? (52–53)

5. List the important differences between amylose, amylopectin, and glycogen. Where would each be found? (54–56)

6. Describe the secondary structure of amylose, and list the forces responsible. (56)

7. Compare the primary structure of amylose with that of cellulose. How does the difference affect the use of each carbohydrate as a food? (56–57)

8. What is it that makes cellulose strong? How is its strength further increased in plant fibers? (57)

9. What is the role of chitin and how is it different from other carbohydrates? (58)

10. List the essential parts of a triglyceride. Through what enzymatic process are triglycerides joined? What makes one triglyceride differ from the next? (59–60)

11. Distinguish between an oil and a fat and between a saturated fat, an unsaturated fat, and a polyunsaturated fat. (60–62)

12. Describe the general structure of phospholipids, and explain their arrangement into membranes. (62)

13. List the special uses of steroids and waxes. How do steroids differ structurally from other lipids? (63–64)

14. What are four significant uses for protein? (64)

15. Carefully define the term *protein*. How is a protein different from a polypeptide? (64)

16. What do all amino acids have in common? How do they differ? (65)

17. Using the amino acid as an example, describe the action of a zwitterion. (65)

18. Describe the levels of organization possible in a protein, and explain what forces are involved at each level. (66)

19. Discuss the following about the hemoglobin molecule: shape (globular or fibrous), level of organization represented, existence of prosthetic groups if any, size in comparison to other proteins, special function. (66)

20. List three different characteristics of R groups in the 20 amino acids. Which is useful in linking polypeptides together? (66)

21. Describe the general characteristics of the DNA molecule. What are its units of structure? (69)

22. What are two major roles of DNA and one of RNA? (71)

Cell Structure

Robert Hooke had just been appointed Curator of Experiments for the prestigious Royal Society of London, and he knew he had to come up with something good for the next weekly meeting. He was only too aware that the elite of British science would be there, and he wanted to present a demonstration that would enlighten, entertain, and impress them. It would not be an easy task.

Hooke had an idea. Perhaps he would use an exciting new technology of the 17th century—the casting and grinding of glass lenses. The world was buzzing with talk of lenses. With a pair of lenses in a frame the nearly blind were able to see—a miracle come true. Old men who had not been able to read for years had their books and letters restored to them. Earlier in the century, Galileo had pointed a lens to the sky and had started an intellectual revolution. The human eye has a voracious appetite and Hooke knew it. But Hooke himself was interested in a new use of the lens—to look at very small things. In fact, he had built his own microscope, one of the first in the world (Figure 4.1).

So for a scientific demonstration, Hooke thought of using the microscope to try to see why cork floats. Cork was a mystery. It appears to be solid, yet it floats. Perhaps it is not so solid after all, Hooke thought. He decided it was worth a look.

Hooke trained his microscope on a cork, and what do you think he saw? Nothing. It turns out that microscopes do not work very well on reflected light. So out of curiosity he took out his pen knife

and cut a very thin sliver of cork and shined a bright light *through* it. Then, when he peered into the microscope, he did see something. And it puzzled him. Hooke wrote at the time that the cork seemed to be composed of "little boxes." The little boxes, he surmised, were full of air, and that's why cork floats. He called the little boxes *cells* because they reminded him of the rows of monks' cells in a monastery, and a new scientific field was born.

CELL THEORY

The group that week was pleased by Hooke's demonstration, but a full century would pass before the scientific world would grasp the meaning and importance of Hooke's cells. One of the first people to move on the idea was the German naturalist Lorenz Oken, who wrote in 1805 what was to become known as the **cell theory:** "All organic beings originate from and consist of vesicles or cells." But in spite of this rather clear and encompassing statement, the formulation of the cell theory is usually credited to two other Germans, the botanist Matthias Jakob Schleiden and the zoologist Theodor Schwann. In 1839, they published a conclusion that they had reached more or less simultaneously; they said that all living things, from oak trees to violets, and from worms to tigers, were composed of cells. Either they were better at public relations than Oken, or perhaps the world was simply more ready to listen in 1839, but Schleiden and Schwann got the credit. Then, about 20 years after Schleiden and Schwann revealed their ideas, a fourth German, Rudolf Virchow, elaborated upon the proposal to add *"omnia cellula e cellula"*: all cells from cells. Under the conditions on earth today, cells only arise from preexisting cells. This addition was to become known as the biogenetic law, which, like the cell theory, remains a thoroughly respectable idea today. The provision, "under today's conditions," leaves room for a prominent theory proposing that life (cells) arose spontaneously in the early earth, but under much different environmental conditions (see Chapter 18).

We know now that not only must cells come from cells, but virtually every living thing is composed of cells. They are, indeed, the basic units of life. Once Hooke had described his little boxes, the art of microscopy blossomed, and nothing—literally nothing—proved sacred. People were peering through their handmade microscopes at everything. Of course, they went over the human body with avid interest, as they did other animal life and plant life as well—looking at and into everything. What they found were not just the dead cell walls that Hooke had described, but variable, changing cells of all descriptions. They could see some of these easily, but others needed to be stained before they were clearly visible. Living cells, they found, were full of a very active, changing, shifting fluid—a puzzling finding, indeed. In time, cell researchers would also find that the cells of every creature are unique, differing not only from one species to the next but from place to place within the same individual. We will delve into some aspects of this cell diversity, and you will find that in spite of the apparent diversity, cells are remarkably similar in many, quite fundamental ways.

WHAT IS A CELL?

We have said that the cell is the fundamental unit of life, but like most definitions, this one raises more questions than it answers. The phrase "fundamental unit" requires some explanation. What it really means is that the cell contains all of the structures and molecular constituents needed for life. With these entities in place, the cell can *(1) take in raw materials and from these, (2) extract useful energy, (3) synthesize its own molecules, (4) grow in an organized manner, (5) respond to stimuli from its surroundings, and, very significantly, (6) reproduce itself.* We will discuss the specific structures associated with these functions shortly, but Figure 4.2 shows how they are generally situated within the cell.

4.1 THE EARLY MICROSCOPE

Hooke's primitive microscope consisted of two convex lenses at either end of a body tube some 6 inches in length. Focusing was done by twisting the body tube along its spiral threads. The subjects were simply stuck on a pin attached to the base of the microscope. The light source was a flame, and its light was focused by a lens. The cork cells shown here are from a drawing by Hooke.

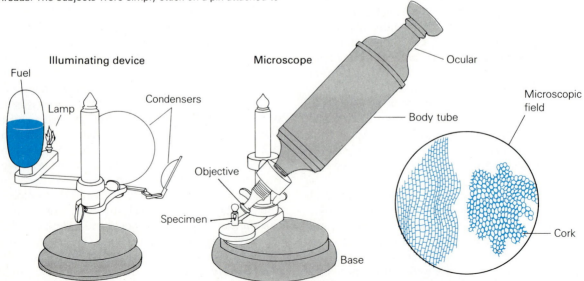

4.2 THE GENERALIZED CELL

Actually no one cell of either plant **(a)** or animal **(b)** has all the characteristics shown in these composite drawings. These are both *eukaryote* cells, which means each has a *nucleus*—the more or less spherical body with the double membrane or *nuclear envelope,* inside of which the genetic material and the dense *nucleolus* are separated from the rest of the cell. Inside the cell's *plasma membrane* is a semifluid mass called *cytoplasm* in which there are numerous organelles and internal membranes. Both plant and animal cells have *mitochondria* in which food molecules are oxidized, *Golgi apparatus* in which manufactured cell materials are collected, *ribosomes* on which proteins are synthesized, and an *endoplasmic reticulum* of variable internal membranes that communicate with both the nucleus and the plasma membrane.

But plant cells and animal cells are not alike. One of the most prominent features of the plant cell is its huge *vacuoles* filled with *cell sap,* a clear, often pigmented fluid. (Vacuoles occur in animal cells less frequently and are usually smaller.) Plant cells are encased in a thick, semirigid *cell wall.* The animal cell, in contrast, owes its more variable shape to a shifting, dynamic fibrous *cytoskeleton* of *microfilaments* and *microtubules.* (Plant cells may have microfilaments and microtubules at some stages of growth.) Centrioles are typically seen in animal cells, while plant cells often contain *plastids,* including the photosynthesizing *chloroplasts* and the *leucoplasts,* in which starch grains are stored, and *peroxisomes,* which neutralize hydrogen peroxide. The cell surface of the animal cell may be thrown up into absorbing *microvilli* or grasping *pseudopods;* some animal cells have motile, complex *flagella* or *cilia* with their associated *basal bodies.* Higher plants lack cilia or flagella, although ferns and lower plants have motile sperm cells with flagella. Organelles found exclusively in animal cells are *lysosomes,* small membrane-bound sacs in which powerful enzymes or other materials are stored. By definition, plant cells do not have lysosomes, although some plant cell vacuoles may perform the same functions.

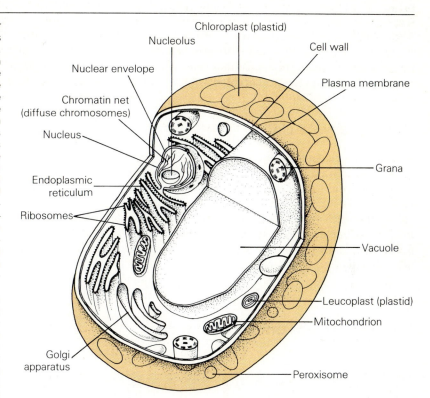

(a) Plant

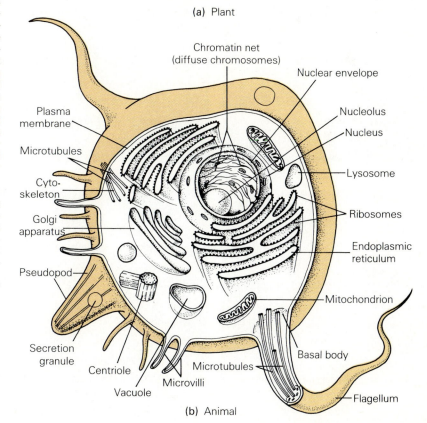

(b) Animal

We should add here that we will be focusing primarily on the cells of plants and animals. We see some of the variety among cells in Figure 4.3. Plants and animals, along with the various protists and the fungi (see Chapters 19 and 20), are known as **eukaryotes.** Their cells have certain traits that distinguish them from the **prokaryotes,** which include the bacteria (see Chapter 18).

The Eukaryote Cell

Basically, a cell consists of an outer boundary, the **plasma membrane,** a region within known as the **cytoplasm,** and a distinct **nucleus,** surrounded by a *nuclear membrane.* Since it is actually a double membrane, the nuclear membrane is known technically as the **nuclear envelope.** In addition, many cells, particularly those of plants, are surrounded by a rigid, protective, nonliving **cell wall.**

The plasma membrane marks the boundary of the cell and isolates it to some degree from other cells. It controls the passage of materials in and out, thus maintaining the internal integrity. The cytoplasm in most types of cells contains numerous minute structures called **organelles,** usually surrounded by complex membranes and each with a specific function. Also present are supporting elements that make up what is called the **cytoskeleton.** They include simple, hairlike, protein **microfilaments,** which help stiffen the otherwise semiliquid ground substance of the cytoplasm and support the organelles, and larger tubular protein elements called **microtubules.** Finally, there is the nucleus, a central body surrounded by a double membrane, or nuclear envelope. It contains the genetic material (DNA) whose role is to direct the activities of the cell and, prior to cell division, replicate itself (make copies) for distribution to new generations of cells.

4.3 DIVERSITY IN PLANT AND ANIMAL CELLS

These cells are from multicellular organisms, and each is well adapted to a specific role.

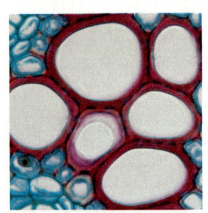
Thick-walled xylem

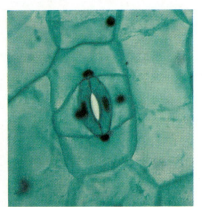
Guard cells (leaf)

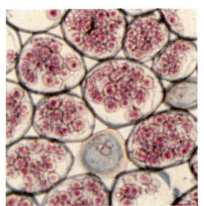
Storage cells of root

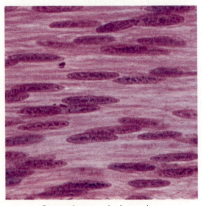
Smooth muscle intestine

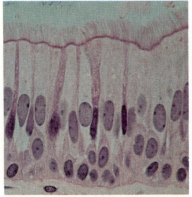
Ciliated epithelium trachea

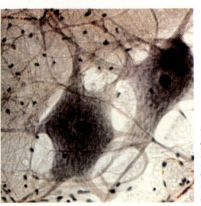
Spinal cord nerve cell

Small Units of Linear Measurement in the Metric System

The United States alone continues to use the peculiar English system of weights and measures in commerce and everyday life, but even in America the times are changing, as witnessed by a sure index: Ford and General Motors are shifting their production lines to the metric system. Track meets now regularly feature the 100-meter dash rather than the 100-yard dash, and the 1500-meter "metric mile" is displacing the traditional mile. (The mile will probably hang on as a special event because of its sentimental value.)

The sciences, including biology, have always used the metric system. Even if you did a good job of memorizing the metric system in grade school, you may well come across some unfamiliar units of measurement in your study of cell structure. You will not be alone. Trained biologists have been having the same problem in recent years because the international metric system itself has been changing. It is becoming even more regular, by decisions of international councils that have authority over such matters. The unit names in the metric system usually consist of a root word and a standard prefix. The root word gives the basic unit for a type of measurement: meter for linear measurement, liter for volume, gram for weight, and so on. The standard prefixes are added to these basic word roots to form larger or smaller

UNITS OF LINEAR MEASURE

Unit name	Symbol	Portion of a meter	Equivalent
kilometer	km	10^3 m	1000 m
meter	m	1 m	0.001 km
centimeter	cm	10^{-2} m	0.01 m
millimeter	mm	10^{-3} m	0.001 m
micrometer	μm	10^{-6} m	0.001 mm
nanometer	nm	10^{-9} m	0.001 μm

In addition to the above units, which are now official, three other metric units of linear measurement are in the process of being phased out. These units are still common in scientific books and articles, so we'll add them to our table:

micron	μ	10^{-6} m	1 μm
millimicron	mμ	10^{-9} m	1 nm
angstrom	Å	10^{-10} m	0.1 nm

The change from the name micron to micrometer means the symbol has changed from μ to μm. The change from millimicron (mμ) to nanometer (nm) also means a new name and a new symbol for 10^{-9} meter. The angstrom (Å) is being dropped altogether, to be replaced by 0.1 nanometer.

secondary units. Each prefix indicates an increase or decrease by a specific positive or negative power of ten. For instance, *micro* means one one-millionth; thus, a microliter is one one-millionth of a liter (10^{-6} liter), a microgram is one one-millionth of a gram (10^{-6} gram), a micrometer is one one-millionth of a meter (10^{-6} meter), and a micromole is one one-millionth of a mole (10^{-6} mole). Another regularity being imposed on the metric system is that each named unit of measurement is one one-thousandth (10^{-3}) of the next larger named unit. The table shows how this works out in practice for units of linear measurement. Note that the familiar centimeter, one one-hundredth of a meter, violates this regularity.

The basic unit of linear measurement is the meter. Other lengths relate directly to it, such as the kilometer (about 0.6 mile) and the centimeter (about 0.39 inch).

Plant and animal cells differ in a number of important ways, but the most obvious is the presence of a cell wall surrounding the plant cell. While the cell wall restricts movement, it provides the cell with considerable strength and protection. Since they lack cell walls, animal cells are much less confined, although their shape and movement are controlled by the cytoskeleton. The plasma membranes are anchored to the cytoskeleton, but so far no one has found out exactly how. By controlled chemical reactions of the cytoskeleton, animal cells can move their membranes extensively and thus change in shape.

The Size of Cells

How big are cells? Not very. Most cells are far too small to be seen with the naked eye. But, of course, there are exceptions; a chicken egg yolk is technically a cell, and a frog egg is somewhat more convincingly so. Neurons (nerve cells) may be over a meter long (such as those that run down a giraffe's leg), but they are still too thin to be seen. However, apart from such specialized cells, eukaryote cells fall within a surprisingly small size range. Very few are smaller than 10 μm (micrometers) in diameter or larger than 100 μm. (See Essay 4.1 for a discussion of metric units of measurement.) Plant cells tend to be somewhat larger than animal cells, but perhaps only because they contain large, water-filled internal cavities (or vacuoles); the average amount of cytoplasm is about the same in plant and animal cells. Bacteria, as we shall see, are much smaller than either, seldom exceeding a few micrometers in diameter or length.

Why Are Cells Small?   This all brings up an interesting question. Why are most cells small, and why hasn't cell size increased as organisms became larger? For that matter, why aren't there gigantic single-celled creatures? We have yet to hear of an 800-pound ameba lurking in the old swimming hole, and no one has reported sighting single-celled whales and elephants. The evolutionary trend has been to keep cell size small and simply increase the number to accommodate increases in size. Why?

The Surface-Volume Hypothesis.   The generally accepted hypothesis explaining the minute cell size has to do with some clear-cut physical problems. The size to which a cell can grow is dictated by its metabolic requirements, and these in turn are largely dependent on the ability of the membrane to serve the cytoplasmic volume within the cell. In other words, perhaps cells remain small because they are restrained by problems of area and volume. This is called the **surface-volume hypothesis,** and its operation would put close restraints on how large cells can be. Let's see why this should be.

The surface-volume hypothesis is explained mathematically by the observation that as the radius of a body, such as a sphere, is doubled, the *surface* area is *squared* while the *volume* is *cubed*. This relationship holds not just for spheres, but for any solid body with a well-defined shape (Figure 4.4). Therefore, we see that to keep the surface-volume ratio at a functional level, cells must be small.

Cells, especially growing cells, require nutrients and oxygen, and they must rid themselves of waste products. The nutrients and oxygen must move into the cell across the plasma membrane and must then pass through the dense cytoplasm to sites of chemical activity. The movement of materials within the cell quite often involves a process called **diffusion,** whereby molecules move randomly through heat energy they absorb (see Chapter 5). For example, in order for cells to use fuels such as glucose for energy, oxygen and glucose must diffuse through the cytoplasm and into organelles to interact with certain enzymes. The waste products

4.4 SURFACE AREA TO VOLUME RATIOS

The plasma membrane is vital to the transport of materials in and out of the cell. The total surface area determines how well the membrane can support the transport requirements. The small cell at the left has substantially more surface area per unit of volume than does the larger cell at the right. Small cells can thus operate more efficiently in membrane transport than can larger cells. At some point in growth, biologists believe, the increased workload on the membrane of a metabolically active cell becomes intolerable. Through natural selection, the answer for most cells has become cell division.

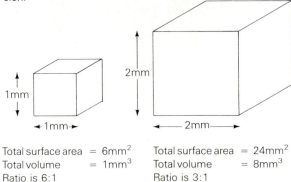

Total surface area = 6mm²	Total surface area = 24mm²
Total volume = 1mm³	Total volume = 8mm³
Ratio is 6:1	Ratio is 3:1

of respiration, carbon dioxide and water, likewise must pass by diffusion through the cytoplasm and across the membrane to leave the cell.

The problem of volume pertains to the nucleus as well. The genes reside in the nucleus, but one product of their action, protein synthesis, occurs in the cytoplasm, where the protein-assembling machinery and the numerous amino acids must be brought together to complete a protein. Thus, the sites of chemical activity in the cell must be fairly close to the source of reactants, whether from the outside or from the nucleus. Cells must remain small enough to ensure a high probability that these reactants will be able to get together.

The problems that restrict cell growth can therefore be restated: as a cell increases in size, its volume grows faster than does its surface area (the membrane). Thus, in large cells there may not be enough membrane area to accommodate the cytoplasmic mass. At some critical point, then, growth stops. This point is apparently reached very quickly by bacteria. It has been argued that eukaryote cells can become much larger than bacteria because they possess internal membranes and compartments. However, there is a limit even for eukaryotes. Plant cells, particularly those with large internal vacuoles, can stretch the limit a bit further by augmenting internal diffusion with a peculiar and puzzling process called **cyclosis** (cytoplasmic streaming), in which the cytoplasm whirls around the cell in more or less definite channels. So, cells must have a way of coping with the problem. For example, a nerve cell can grow to be very large but it grows mainly in one dimension, and all parts of the cytoplasm remain very close to the surface. Some cells in the kidney have highly convoluted borders, as do cells of the small intestine (Figure 4.5). This development greatly increases the surface area of these cells whose primary function involves moving material across their membranes. Large, spherical egg cells, such as the yolk cell of a chicken egg or an ostrich egg, have almost no metabolic activity except at the very surface—the rest of the volume is taken up by stored food reserves.

CELL STRUCTURE IN EUKARYOTES: THE ORGANELLES

Now that we have some idea about what cells are and how they vary, let's look closely at their structure. In keeping with the characteristics of life discussed earlier, the structures will be grouped according to basic cellular functions. A good place to begin is with those structures that support the cell and form the necessary boundaries that maintain the cell's integrity and provide for transport. From there we will progress to structures or organelles devoted to control and reproduction, synthesis and storage, energy transfer, and finally move-

4.5 SURFACE-TO-VOLUME STRATEGIES

The epithelial cells lining the small intestine absorb small molecules through a surface that is greatly expanded by the presence of innumerable *microvilli*, small projections of membrane supported by bundles of microfilaments (X50,000).

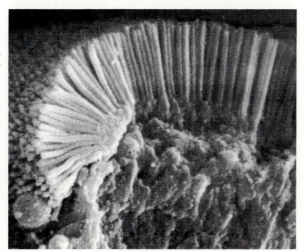

4.6 CELL WALLS

The plant cell wall is a dense, semirigid, but porous wall just outside the plasma membrane. The mature cell wall consists essentially of several layers of cellulose microfibrils, usually impregnated with strengthening and sometimes waterproofing materials. The outermost layer, or primary wall, is laid down first, shortly after cell division. Its fibers are reoriented as the cell lengthens, as are the fibers in each of the newer, inner layers of cellulose. The innermost and youngest still has its fibers laid down at right angles to the cell's long axis.

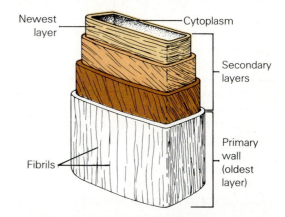

ment. Keep in mind, though, that there is often some unavoidable overlap among these categories. Cells simply don't fill our neat little categories as well as we might want.

Organelles of Support and Transport

The Cell Wall. As we've already noted, plant cells are surrounded by fairly rigid, nonliving walls composed primarily of cellulose and other polysaccharides (Figure 4.6). Microfibrils of cellulose chains are arranged in layers, with each layer lying at an angle with respect to the one below it. The result is a laminated, and very strong but porous covering over the cell (not unlike plywood). Once impregnated with lignin and pectin, hardening substances, the cell wall maintains the shape of the cell and has great strength and resiliency.

Actually, newly formed cell walls can't be too rigid, because each time a plant cell divides to form two cells, the new cells must be able to grow to the size of the original cell. As the cell divides, a new cell wall appears, effectively separating the two new cells. Walls of adjacent cells are cemented together with pectin to form a layer known as the **middle lamella.**

As the plant cells mature and differentiate into specialized tissues, other substances are deposited into the cellulose and pectin matrix. In the stems of woody plants, for instance, **lignin,** a complex chemical that is technically an alcohol, is continually secreted into the cellulose, forming a hard, thick, decay-resistant wall. **Suberin,** a waxy substance, is secreted into the outer layer of stem cells, forming a protective, waterproof, corky layer. Waterproofing of the upper surface of leaves, on the other hand, is accomplished by the secretion of **cutin,** a waxy substance added to the epidermal cell wall.

The Plasma Membrane. We have already mentioned this vital cell boundary, and vital is the right word because if anything is important to the integrity of the cell, the membrane is. Basically, it keeps the inside of the cell in and the outside out, and it lets through the materials that the cell must exchange with its environment. The membrane's complex structure permits it to accept effortlessly the passage of some substances, utterly reject others, and at times, expend energy to actively assist the transport of still others.

These important functions, and others, are made possible by the plasma membrane's intricate structure. It is essentially a bilipid film, consisting of two layers of phospholipids. Their fatty acid tails intertwine while their polar heads project outward (Figure 4.7). Interspersed in the lipid film are a number of different kinds of proteins, all essential to the movement of materials and other functions. We will look into plasma membrane function in detail in the next chapter, but we should point out that the basic bilipid-protein structure of the plasma membrane is also found in the membrane-bounded organelles within the cell.

The Cytoskeleton, a Microtrabecular Lattice. It is interesting to note how heavily our growing knowledge of cell structure relies on technological innovation. It is an understatement to say that cytologists rely heavily on the transmission electron microscope, but even though these powerful devices have been available for many years, biologists were still not able to learn much about the viscous cytoplasmic fluid in which the organelles seemed to float. Terms such as *cytoplasmic matrix* and *ground substance* were coined— terms that seemed to convey information but didn't. If there was a substructure in the cytoplasm, it was quite invisible through the electron microscope. But the newest technological innovation, the **high voltage electron microscope,** has begun to change our ideas on the cytoplasmic organization.

The high voltage electron microscope, as its name suggests, has greatly increased power— power, that is, to propel electrons through cells. While the preparation of tissue for the standard electron microscope requires that tissues be thinly sliced (see Essay 4.2) to permit electrons to pass through, this is not the case with the high voltage

---

4.7 THE PLASMA MEMBRANE

The plasma membrane is composed of a bilipid film—two layers of phospholipids with their hydrophobic tails intertwining and their hydrophilic heads facing outward. Distributed throughout the bilipid layer are many kinds of functional proteins that carry on transport and recognition, major functions of the plasma membrane.

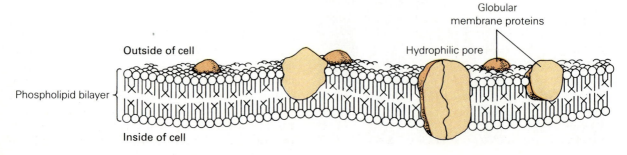

4.8 THE MICROTRABECULAR LATTICE

In this three-dimensional, high voltage electron micrograph, the cytoplasm is revealed as containing a weblike supporting framework of microtubules and microfilaments. Many of the cellular organelles lie suspended in the lattice, as seen in the artist's reconstruction.

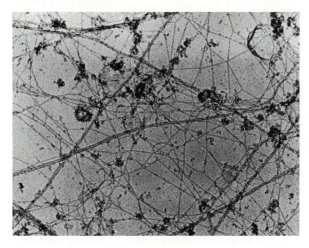

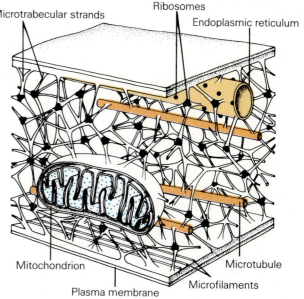

Microtrabecular strands

Ribosomes

Endoplasmic reticulum

Mitochondrion

Plasma membrane

Microtubule

Microfilaments

electron microscope. With its million-volt electron flow, thick slices of cells and, in some instances, entire cells can be viewed. The result is a three-dimensional view of the cell and the discovery of a previously unknown structural framework in the cytoplasm. Because of its resemblance to the spongy (or *trabecular*) network of bone, the newly discovered cytoplasmic substructure has been named the **microtrabecular lattice** (Figure 4.8).

The microtrabecular lattice, or simply the "lattice," appears as a mazelike network of hollow fibers extending throughout the cytoplasm, connecting with and suspending the organelles like a three-dimensional spider web. So pervasive is this network that its discoverers have suggested changing the name cytoplasm to *cytoplast*, denoting a structural body rather than an amorphous fluid. Researchers are already hard at work unlocking the secrets of this grand network and are now proposing both structural and metabolic roles.

The lattice appears to be an important structural component, supporting organelles and giving shape to the cell. Other hypotheses suggest that the lattice may help in organizing complex biochemical activities, those requiring teams of enzymes that carry out sequential reactions. These enzymes may be integrated into, or adhered to, the tubular framework; thus the product of one reaction would immediately be confronted by the next enzyme in line. Before the lattice was discovered, it was presumed that many enzymes and their reactants sim-

ply met through chance collisions in the fluids of the cytoplasm. The newer hypothesis is attractive because it suggests a more efficient system. The idea is fortified by the observation of *ribosomes*, protein synthesizing bodies, closely associated with the lattice. We have known for years that some ribosomes become bound to cytoplasmic membranes, and this has helped in understanding how they organize the incredibly complex task of protein assembly. Other ribosomes, however, appear to be free-floating in the cytoplasm, and again we are stuck with the idea of chance collisions between the free-floating ribosomes and the many components of protein synthesis. If it can be shown that the lattice helps in their organization, some difficult questions will have been answered. This is pretty speculative at the present, but speculation is sometimes a strong motivation to further research.

Control and Cell Reproduction: The Nucleus

The significant functions of control and cell reproduction go to the nucleus, the most prominent organelle of the cell. Because the nucleus is easily stained, and therefore easily viewed under the microscope, it was described early in the history of cell study. In fact, the word "nucleus," which means "kernel," was first used in 1831 by botanist Robert Brown about the time of the emergence of the cell theory. Nuclei have since been found in

Looking at Cells

The Light Microscope
The compound lens **light microscope** had achieved near technical perfection by the end of the 19th century. Of course, there have been some embellishments since, but the essentials were all there. Not only had the machine itself been developed, along with the theoretical optics to make it work, but the equally important techniques of tissue preparation and staining had been worked out. As a result of these early technical breakthroughs, almost all cell structures that can be seen with a light microscope had been named and catalogued before your grandparents were born.

A compound microscope consists basically of a tube with lenses at both ends. The **objective lens,** itself made up of a number of thinner lenses glued together, is brought very close to the object being viewed. The **ocular lens,** also made up of several thinner lenses, is brought close to the eye.

A third set of lenses is found in most quality microscopes. This is the **condenser,** and it lies beneath the specimen—that is, between the specimen and the source of light. This lens focuses light on the specimen. Finally, the condenser contains an **iris diaphragm** which cuts out stray light that would cloud the image and adjusts the light to accommodate various magnifications.

Fixing, Preparation, and Staining of Specimens for the Light Microscope
Small things tend to be transparent. Light simply passes through them without being absorbed. Special stains must therefore be used to distinguish cellular structures. Stains may be of a general nature, simply darkening the cytoplasm or nuclei of cells; or they may be highly selective for specific structures or molecules. Selective staining has been invaluable to the fields of **cytology** (the study of cells) and **histology** (the study of tissues), and many of the old techniques are still in use.

Generally, microscope specimens have to be quite small or very thin in order for light to pass through. Small, loose cells such as bacteria can be seen whole. They are killed with heat, **fixed** (treated with chemicals that coagulate their cytoplasm into a solid gel), affixed to a glass slide, stained, and viewed. Most biological material, however, must be cut into very thin slices, or **sectioned** (as Hooke found out with his cork) before staining and viewing.

Resolving Power: The Practical Limit
The limits to what can be seen with any type of microscope are described as the instrument's **resolving power.** Resolving power is a measure of how close two points can be and still be distinguished as being separate. The resolving power of the normal, unaided human eye at close range is about 0.1 mm, or just about the distance separating the dots that make up the image in a newspaper photograph. With even a very simple microscope or a hand lens of good quality, you have no trouble making out the dots in even the finest newsprint photo.

There is a theoretical limit to the resolving power of even the finest light microscope, and many important cellular structures are well below their resolving limits. Optical physicists have proven that no system can resolve points that are closer together than half the wavelength of the light

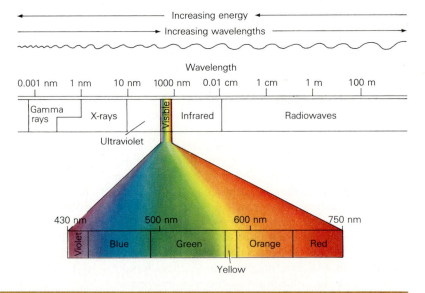

The electromagnetic spectrum. Our eyes are barraged by electromagnetic radiation of a variety of wavelengths, but we see only that portion with wavelengths between 430 and 750nm. Within that range, light of different wavelengths is discerned as color. The wavelengths in the visible spectrum also restrict microscope resolving power, since objects closer together than 250 nm will interfere with light passing between them and the separation will not be visible. The 250 nm is just about half the shortest wavelength in the visible spectrum.

Transmission electron microscope (upside down)

Film or screen

- Vacuum chamber
- Electron beam
- Viewing port
- Projector lens
- Objective lens
- Specimen on movable stage
- Condenser lens
- Electron gun

Light microscope

Film or eye

- Light beam
- Ocular lens
- Iris diaphragm
- Light source

The light microscope and the transmission electron microscope work on very similar principles. In both cases a focused beam, of photons or electrons, is transmitted through a very thin slice of the specimen and is absorbed differentially by different structures of stains. The emerging beam is then refocused to produce an image, either on a fluorescent screen that can then be seen by the viewer (electron microscope), directly into the eye (light microscope), or onto film (either system). The light microscope lenses are glass, and the electron microscope lenses are magnetic fields. Here, the electron microscope diagram is shown upside down to emphasize the correspondence between the two systems.

used to view them. Amazingly, the fine microscopes of the turn of the century approached this theoretical limit very closely; that is why there haven't been significant improvements since. The limitation is that of the nature of light itself.

The visible spectrum includes light with wavelengths between 750nm (red) and 430nm (violet). The theoretical and actual resolving power of the best light microscopes is therefore approximately 250nm (0.00025 mm)—about 400 times (400×) the resolving power of the human eye. It is possible to magnify images more than 400 diameters, but doing so will not appreciably improve the resolving power. Even photographic enlargement cannot overcome the limits of resolving power. (Have you ever had a snapshot taken with a cheap camera enlarged to portrait size?)

The Transmission Electron Microscope
The electron microscope has enormous powers of resolution, since the object is flooded with electrons rather than light waves and the wavelength of electrons is much smaller than the shortest light wave.

The **transmission electron mi**-croscope (TEM) provides one answer to the problem of overcoming the light microscope's limited resolving power. "Transmission" means that the electrons are directed through the specimen, rather than being reflected off its surface. The principle upon which the electron microscope works is relatively simple, although the actual technology required is not. In place of light, the electron microscope uses electrons shot from an electron gun similar to the electron gun in the picture tube of a television set. A vacuum is maintained in the chamber to prevent electrons from being scattered by molecules of gases. Electrons move in waves, just as do the photons of light, but the wavelength of moving electrons is much shorter than that of visible light, thus allowing much smaller structures to be resolved. In fact, while the useful magnification of the light microscope is in the neighborhood of 400×, that of the transmission electron microscope can reach 250,000×.

Wavelengths of visible light range from 430 to 750 nm. By comparison, the electrons in the electron microscope have an average wavelength of only 0.05 nm. Theoretically, this should produce a resolving power of about 0.025nm, but at present the best that can be accomplished is a resolving power of 0.2 to 0.3 nm. Compared with the 250nm resolving power of the best light microscopes, the electron microscope is 1000 times more powerful; however, a drawback of the electron microscope is that its picture is strictly black and white.

Focusing the electron microscope is done through electromagnetic condensers, which are only distantly related to the condenser of a light microscope. The negative charges of moving electrons make it possible for magnets to direct their paths in a way somewhat similar to the bending of light rays by glass lenses. Different magnifications are achieved by altering the strength of the magnetic lenses. As the electrons pass through the specimen, some are absorbed by the nuclei of heavier atoms and others pass through unhindered. Those that get through pass between additional electromagnetic lenses that spread them out. The electrons are finally focused either onto a phosphorescent screen or, if a permanent record is desired, onto photographic film. The pattern creat-*(continued)*

Heavy metal shadow casting. The objects in this photo (a) are viruses. The moonscape appearance is attributed to heavy metal shadow casting. Molecules of heavy metals are sprayed at an angle and thus deposited on one side of each virus. The result is a distinct shadowing that helps emphasize the shapes. This is a positive print of the original electron micrograph, so the heavy metal deposits are made to look light, strikingly like reflected sunlight. (b) Shadow casting is done by heating and vaporizing a small dab of gold or other heavy metal in a vacuum chamber. The gold "spatters" over the specimen.

(a)

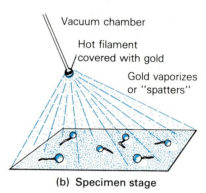

(b) Specimen stage

ed shows the image of the electron-absorbing parts of the specimen.

Fixing, Sectioning, and Staining for the Electron Microscope

The preparation and staining methods used for the transmission electron microscope are even more brutal than those for the light microscope. In fact, what is finally observed is often the image of a "casting" of the once-living object. The specimen is fixed, usually with an electron-dense metallic salt. The heavy metal not only solidifies the protein of the cytoplasm, but combines with it in specific ways. Thus, the metal selectively stains the specimen, since its atoms will later serve to absorb electrons.

The specimen is then embedded in hard plastic, and with a diamond blade (or the more traditional broken glass blade) is cut into exceedingly thin slices. The specimen has to be very thin, because electrons do not pass through solid material very well.

Small objects such as viruses, chromosomes, or bacterial flagella, are prepared somewhat differently. They are first spread onto a thin film of protein that is supported by a wire screen. Then the mounted specimen is placed in a vacuum chamber, and atoms of platinum or gold are spattered onto the specimen at an angle—a procedure called "shadowing." Since the specimen is not a flat section but consists of small three-dimensional objects, the metal atoms will fall on one side of the objects the

way evening sun falls on the tops and sides of a group of hills. Following this, the cellular material itself is dissolved away, leaving only the metal replica, which is then reinforced with carbon. (Has the word, "brutal," begun to sound appropriate yet?) The gold absorbs electrons, and the resulting image is reversed photographically. The result is a surprisingly lifelike photograph of three-dimensional bodies that look as if someone were shining a flashlight on them.

One of the most specialized methods of specimen preparation is of particular interest because it has been used so successfully to study the structure of cell membranes. This is a process called **freeze-fracturing,** or sometimes **freeze-etching.** The tissue is frozen to –100° C and, at that temperature in a vacuum, is broken apart. At –100° C, the aqueous portions of the cell are frozen much harder than the thin lipid sheet that is sandwiched in the plasma membranes, and these form natural weaknesses in the tissue. As a result, cell membranes may be split apart and opened up, exposing their inner structure from two aspects—the half of the membrane that faces out is broken apart from the half that faces in. Wherever water-binding materials, such as membrane proteins, are included within the layers of the ruptured membrane, indentations show on one inner side of the membrane and bulges show on the other inner side.

The Scanning Electron Microscope

The **scanning electron microscope** (SEM) was developed in the 1940s, but it was not perfected until the 1960s. While the resolving power of this instrument is far less than the standard electron microscope, it has the distinct advantage of producing the illusion of three-dimensional images, with unusually great depth of field. This capability makes the scanning electron microscope a unique tool for revealing surface features of a specimen.

In its operation, a thin scanning beam of electrons is passed back and forth across an object, causing it to scatter and reflect the electrons. The reflected electrons are captured on a positively charged plate, giving rise to a small electric current. This current is amplified and fed into a cathode ray tube whose own scanning beam is synchronized with the scanning beam in the specimen chamber. The image appears on a television screen. The specimen is mounted on a movable mechanical axis, so its position can be changed for varying views.

The High Voltage Electron Microscope

The most recent addition to the cytologist's growing arsenal of tools is the high voltage electron microscope, a gigantic, three-story version of the older electron microscopes. Because of its enormous energy output (up to 3 million electron volts, or 30 times

Vacuum chamber

Hot filament covered with gold

Gold vaporizes or "spatters"

The freeze-fracture technique. The TEM photograph shown here was made of a specimen prepared by freeze-fracture. This technique produces fracture lines within membranes that reveal their inner structure. Here, the nuclear membrane has been cleaved, revealing its pores. This answers some of the long-standing questions about transport in and out of the nucleus.

greater than other electron microscopes), tissues can be viewed without slicing, and the image produced is three-dimensional. Recently, the high voltage electron microscope's awesome penetrating power has been focused on the featureless cytoplasmic matrix, and an entirely new order of structure has become visible.

Ultracentrifugation

One of the most useful innovations applied to cell study, and an aid to electron microscopy, has been the **ultracentrifuge.** The term *ultra* indicates that these centrifuges are capable of much greater rotating speeds than ordinary clinical centrifuges. In fact, modern ultracentrifuges can attain rotating speeds of up to 70,000 revolutions per minute (rpm). The force exerted at this rotating speed may be up to 300,000 to 400,000 times normal gravity (g). The *heads,* or *rotors,* spin in a partial vacuum, thus avoiding air friction and turbulence. Recent models have optical systems for observing the cell suspensions at high speeds. Some models have even reached a technological state where substances can be added through the rotor while it is in motion.

Use of ultracentrifugation permits researchers to separate various components of tissues into layers in the centrifuge tubes. Cells are first disrupted so that their contents are free in a suspension. The preparation is placed in another fluid, such as a sucrose solution of a known density.

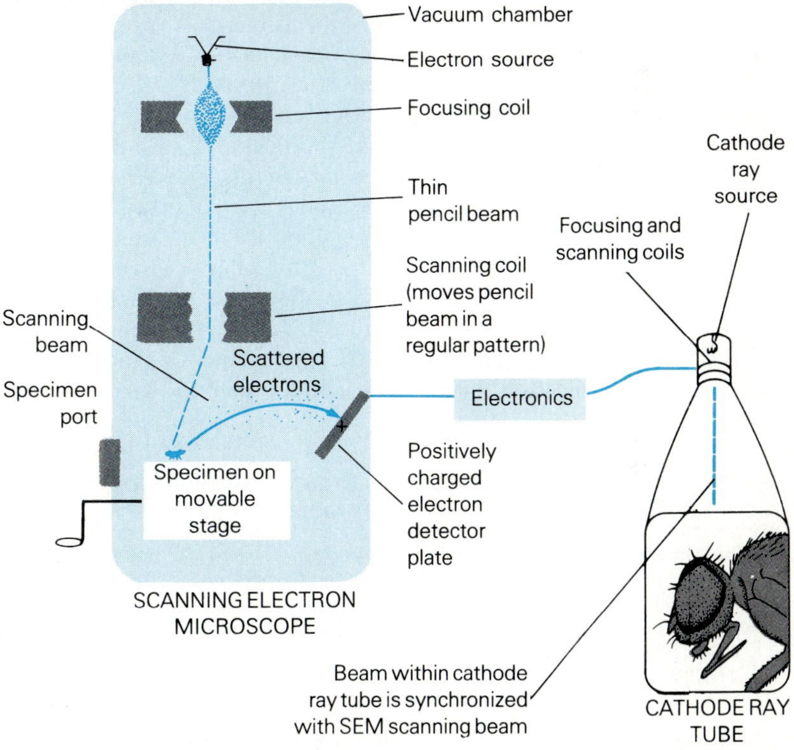

Diagram labels:

- Vacuum chamber
- Electron source
- Focusing coil
- Thin pencil beam
- Scanning coil (moves pencil beam in a regular pattern)
- Scanning beam
- Specimen port
- Scattered electrons
- Specimen on movable stage
- Electronics
- Positively charged electron detector plate
- Cathode ray source
- Focusing and scanning coils
- Beam within cathode ray tube is synchronized with SEM scanning beam

SCANNING ELECTRON MICROSCOPE

CATHODE RAY TUBE

The scanning electron microscope (SEM) is one of the newer innovations of electron microscopy. It involves more electronic wizardry than its predecessor, the TEM.

When the cell-sucrose suspension is spun at ultrahigh speeds, structures and molecules layer out according to their sizes and densities, with smaller and lighter molecules near the top, and larger and heavier cell fragments at the bottom, with zones of particles of other weights along the length of the centrifuge tube. This permits the investigator to remove zones and subject them to further biochemical tests or to electron microscope observation. Needless to say, having a relatively pure quantity of some cell organelle greatly simplifies the work of the cell biologist.

virtually every type of eukaryotic cell. In this chapter we'll consider the nucleus only briefly because it figures so prominently in chapters to come.

The nucleus contains the **chromosomes.** And chromosomes contain the organism's **genes,** which determine its hereditary traits. When chromosomes (DNA and associated protein) and the dramatic process of **mitosis** (chromosome movement and separation) were discovered, new attention was focused on the nucleus. It was soon reasoned that the nucleus and its chromosomes played an important role in cell reproduction. Mitosis is the process whereby the orderly and equal division of chromosomes occurs just prior to cell division. It is during mitosis that the chromosomes become condensed into tiny rodlike bodies which, with the application of dyes (staining), become readily visible (See Chapter 11).

Unless a cell is undergoing division, its chromosomes are not visible, but, nonetheless, basic stains will reveal a dark nucleus laden with the diffuse (loosely organized and spread out) DNA of chromosomes (Figure 4.9). In many cells, darker bodies can be seen inside the nuclei. These are the *nucleoli* (plural of *nucleolus,* "little nucleus"). The nucleoli are rich in a different nucleic acid, RNA (ribonucleic acid). Each nucleolus is a great lump of material from which new ribosomes are made at about the time of the next cell division (see Figure 4.2).

Between cell divisions, the nucleus is almost as featureless in an electron micrograph as it is under the light microscope. However, its membrane is more interesting. As we have mentioned, the nucleus is enclosed in a double membrane, dubbed the nuclear envelope by cytologists. Each of the membranes of the envelope is a regular lipid bilayer, so altogether there are four layers of lipid molecules and the usual associated proteins. The inner and outer membranes pinch together in scattered places over the nuclear surface to form **nuclear pores** (see Figures 4.9a and b). We might have expected such pores, since we know that very large molecules of RNA and the ribosomes must travel from the nucleus to the cytoplasm. We also know that all protein molecules that are found inside the nucleus—and there are many—are synthesized in the cytoplasm and must find their way in, presumably through the pores. It would make things a lot easier to understand if these pores were just holes, but they aren't. In fact, they are apparently filled with large globular proteins.

As we indicated earlier, the nucleus has two important functions: cell reproduction and control. First, it maintains the hereditary information of the cell. All the genetic instructions for development, metabolism, and behavior of the species are found in the DNA of chromosomes within the nucleus. Each cell nucleus of a multicellular organism has a complete copy of all the information needed to produce the organism. In its role as guardian and carrier of genetic information, the nucleus has the

4.9 ELECTRON MICROSCOPE STUDY OF THE NUCLEUS

(a) The dark area is the *nucleolus,* where RNA and protein are assembled into ribosomes. The slightly less dense regions are probably heavy concentrations of DNA in a tightly wound, inactive state. The lighter regions may be diffuse (unwound) and active DNA. Note the double nuclear envelope and several nuclear pores. **(b)** This freeze-fracture electron microscope study shows pores on the inner membrane. These are not simple holes, but may be filled with hydrated proteins. **(c)** This remarkable electron micrograph reveals the connection between the endoplasmic reticulum and the outer membrane. Some observers interpret this to mean that they are continuous channels. Others believe that the endoplasmic reticulum actually forms from protrusions in the nuclear envelope.

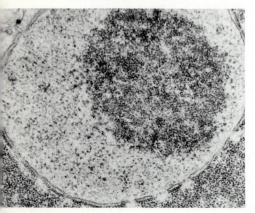
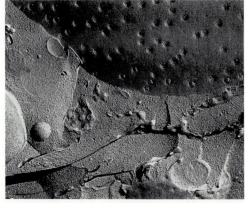
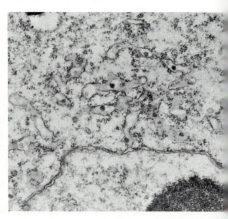

(a)   (b)   (c)

capability of preparing exact duplicates of this information for transmission to new generations of cells when cell reproduction occurs.

Second, the nucleus has the capability of controlling various cellular activities through a rather indirect mechanism. Its DNA directs the formation of enzymes, and these, in turn, direct the biochemical activities of the cell.

Organelles of Synthesis, Storage, and Cytoplasmic Transport

Here, we will look into the activities of several membrane-bounded organelles within the cytoplasm. Their roles will be chiefly synthesizing, producing, storing, and transporting the myriad chemicals that the cell must have to maintain itself and fulfill its special functions. As we will see, these organelles are closely related in both structure and function.

The Endoplasmic Reticulum and Ribosomes. The endoplasmic reticulum (ER) was unknown in the days when we were restricted to using light microscopes. In fact, in those days the contents of cells were thought of as a formless, soupy "protoplasm" (primary living material). But electron micrographs reveal a complex membrane system that takes up a large part of the cytoplasm of eukaryote cells, especially those cells that are engaged in significant protein synthesis. The membranes of the ER have many of the same components seen in the basic plasma membrane. Although the ER system was first seen in the mid-1940s, it wasn't until 1953 that Keith Porter of the Rockefeller Institute first suggested the name. A year later, Porter and George Palade suggested that the reticulum was a very dynamic, ever-changing structure.

Painstaking electron microscope studies of the ER leads researchers to believe that it is one continuous double sheet folded back and forth within the cytoplasm and forming a closed sac. Its internal space, the ER **lumen,** may account for up to 10% of the entire cytoplasmic volume, and the ER membrane itself, more than half the membrane of the entire cell. In effect, the ER divides the cytoplasmic space into two compartments, separated by a single membrane's thickness. It has been clearly demonstrated that the ER is continuous with the outer of the two membranes of the nuclear envelope, so actually the nuclear material is separated from the ER lumen by a single membrane as well. The ER plays a central role in the cell's synthetic activity. Newly synthesized materials such as protein, lip-

4.10 THE ROUGH ER

The rough endoplasmic reticulum (ER) consists of parallel rows of membranes forming flattened channels. The three-dimensional view suggests the immense amount of folding in the ER. The small round bodies along the ER membranes are ribosomes. Cells with large amounts of rough ER are known to be heavily involved in protein synthesis.

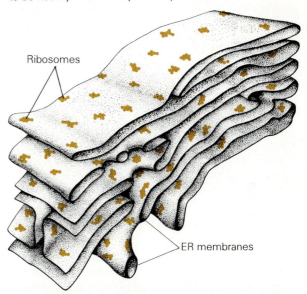

Ribosomes

ER membranes

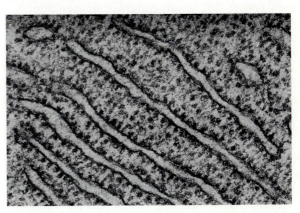

ids, and carbohydrates are transported within the ER lumen prior to storage or prior to their transport to the plasma membrane for release outside the cell.

The endoplasmic reticulum appears in two main forms: rough and smooth. The **rough endoplasmic reticulum** receives its name from the appearance of dense granules that tightly adhere to the outer sides of the membrane, making it look a little like coarse sandpaper (Figure 4.10). Channels are formed by two such membranes lying side-by-side with their rough surfaces out. The granules are ribosomes and as we have seen, they

4.11 THE SMOOTH ER

The smooth endoplasmic reticulum lacks the intimate association with ribosomes seen in the rough endoplasmic reticulum. It functions primarily in the assembly and storage of lipids and carbohydrates.

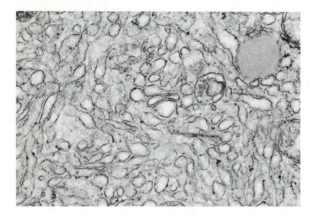

are essential to the synthesis of protein. Ribosomes are minute, two-part bodies, composed of protein and ribonucleic acid that act as organizing centers where the nucleic acids assemble amino acids into polypeptides according to genetic specifications (see Chapter 10). Rough endoplasmic reticulum is seen most commonly in cells that manufacture proteins to be secreted outside the cell, such as digestive enzymes and certain hormones.

Smooth endoplasmic reticulum (Figure 4.11) is primarily found in cells that synthesize, secrete, and store carbohydrates, steroid hormones, lipids, or other nonprotein products. The structure of smooth endoplasmic reticulum is more tubelike and saclike. The tubelike elements form intricate weblike associations. We find a lot of smooth endoplasmic reticulum in cells of the testis, oil glands of the skin, some hormone-producing gland cells, and absorptive cells lining the small intestine. There is strong evidence that the smooth ER in small intestine cells play an important role in the way animals take up the products of lipid digestion. During digestion, many triglycerides are hydrolyzed into fatty acids and glycerol (see Chapter 3) prior to being absorbed by the small intestine. Experimenters found that shortly after rats were fed a diet rich in triglycerides, the digestive products, the fatty acids and glycerol, ended up in the smooth ER of the intestinal cells. But by then, curiously, they had been once again assembled into triglycerides. The enzymes involved in lipid and steroid synthesis are now thought to be integrated into the smooth ER membranes.

Smooth endoplasmic reticulum in liver cells is

known to have another, quite different function: it contains oxidizing enzymes that detoxify certain drugs. Animals injected with large doses of the sedative phenobarbitol, for example, revealed a substantial increase in their liver cell smooth endoplasmic reticulum and in the associated enzymes. Also in the liver cells, we find glycogen (animal storage starch) closely associated with smooth endoplasmic reticulum. Studies reveal that the enzymes responsible for one of the steps in the breakdown of glycogen to glucose is present in the endoplasmic reticulum.

The Golgi Body or Apparatus.   In 1898, Camillo Golgi, an Italian cytologist, discovered that when he treated cells with silver salts, certain peculiar bodies showed up in the cytoplasm. The "reticular apparatuses," as he described them, had never been noticed before, and they didn't show up with other stains. But when other workers used Golgi's silver treatment, they found the bodies in a variety of secretory cells (cells that release their products outside). Still, many biologists continued

4.12 GOLGI APPARATUS

In this electron microscope view, the Golgi apparatus appears as flattened stacks of membranes. Note that the flattened membranous sacs that form cisternae, fill at either end, and "bud off" to form true closed membranous sacs called *vesicles*. The vesicles may contain products originally produced in the ER, but modified and packaged in the Golgi apparatus.

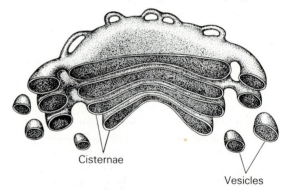

Cisternae

Vesicles

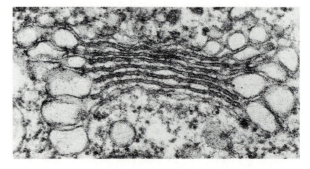

to deny their existence, and the problem wasn't resolved until the electron microscope was developed.

Modern studies, utilizing the great magnifying and resolving power of the electron microscope, reveal that these bodies have a characteristic and identifiable structure no matter what kind of cell they are found in. In every case, a Golgi body or apparatus (known as **dictyosomes** in plants) appears as a stack of flattened, baglike, membranous sacs or **cisternae** lying close to the nucleus (Figure 4.12). Typically, each stack contains about six cisternae, although fewer than six and several times that many are sometimes found. The number of stacks also varies, with the greatest numbers appearing in cells heavily involved in secretion. Each cisterna appears flattened toward its midregion and dilated at its ends. Periodically, the dilated ends are thought to round up and break away, forming what are called **Golgi vesicles.**

Thus the Golgi apparatus is a dynamic, ever-changing organelle, constantly forming from regions of the endoplasmic reticulum and constantly budding off spherical vesicles. Two types of vesicles form—smaller **coated vesicles** and much larger **secretory vesicles.** Coated vesicles appear to be involved in the intracellular sorting and transport of specific proteins. The large secretory vesicles, as you would expect, are quite numerous in cells that specialize in secretion, where their role is to store materials and carry them to the cell membrane for discharge. Let's look more specifically at what goes on within this busy structure.

It is generally accepted that the Golgi apparatus serves as a sort of molecular traffic-directing and packaging center for the cell. Many of the products of cellular synthesis, particularly those formed in the endoplasmic reticulum, end up in the Golgi apparatus, where they undergo modification and concentration. Included are many proteins, glycoproteins, membrane glycoproteins, and glycolipids. In plants, Golgi secretion vesicles release complex carbohydrates into the developing cell membrane, from where they are deposited into the cell wall.

The employment of radioactive tracers has permitted cytologists to trace the history of some substances from their initial synthesis to their release outside the cell (Figure 4.13). For example, studies of cells in the pancreas, a highly active secretory organ that produces both enzymes and hormones, clearly reveal the route taken. The copious production and secretion of protein begins in the rough endoplasmic reticulum. Then, as the membranes of the Golgi apparatus form, the protein finds its way into the cisternae. Following this, newly modified

4.13 CELL SYNTHESIS AND SECRETION

Polypeptides to be assembled into enzymes are synthesized by ribosomes bound to the rough endoplasmic reticulum. Upon entering the rough ER, the polypeptides undergo certain modifications and then enter transport vesicles budded from the rough ER membrane. The transport vesicles empty into Golgi cisternae and after further modification, the finished enzymes enter secretion vesicles formed on the Golgi's opposite face. The secretion vesicles eventually fuse with the plasma membrane, releasing the enzymes outside the cell.

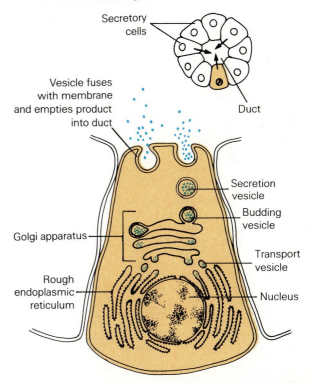

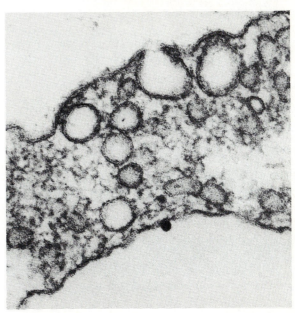

proteins are carried into secretion vesicles that bud from the cisternae. Secretion vesicles are then transferred to the plasma membrane, through which they are discharged either into ducts or into the nearby capillaries, depending on the product being carried. The interaction of secretion vesicles and the cell membrane can be quite intricate, as will be discussed in the next chapter.

Lysosomes. **Lysosomes** have been found in nearly all types of animal cells, although there is some disagreement about whether they exist in plant cells. They are membrane-bounded sacs that are roughly spherical in shape (Figure 4.14). But the simple appearance of the lysosome belies the rather startling role they play in the life of a cell. Lysosomes are bags of powerful hydrolytic enzymes, synthesized in the rough ER and packaged primarily by the Golgi apparatus. In all, about 40 different enzymes have been detected. Interestingly, the enzymes require an acidic condition to work efficiently, which helps prevent damage to

the cell cytoplasm should the lysosome leak. The lysosome creates its own acidic condition by actively pumping protons (H^+) from the cytoplasm to its interior.

Cytologist Christian de Duve predicted the presence of lysosomes from biochemical evidence and went looking for them with his electron microscope. When he found them, de Duve described lysosomes as little "suicide bags." His poetic fancy was not entirely unwarranted, since lysosomes are known to engage in **autophagy** (self-eating), a tidying up process whereby damaged or aged organelles within the cell are taken into a digestive vacuole and hydrolyzed by lysosomal enzymes. Actually, autophagy is not as destructive as it may seem. Sometimes, in fact, the destruction of cells is a normal part of metabolism. In other instances, lysosomes might destroy a superfluous cell that is not functioning well, or one in a part of the body that is undergoing reduction as part of a developmental process, such as the tissue between the fingers in a developing hand. Following the action of lyso-

4.14 THE ROLE OF THE LYSOSOMES

(a) Lysosomes, seen here as numerous large, dark spheres in the cytoplasm, are storage bodies for powerful hydrolytic enzymes. They may also serve as storage places for unwanted and potentially dangerous substances that cannot be excreted. (b) In this scenario, enzyme-laden lysosomes formed from Golgi vesicles have three fates. At the left, a lysosome fuses with a phagocytic vacuole that contains a captured protist. The protist will be digested and the products will

pass into the cytoplasm. At center, a vacuole containing part of an aging mitochondrion is joined by a lysosome that will digest the useless organelle, permitting its constituents to be recycled. At the right, several lysosomes become encased in a dense membrane originating from the plasma membrane. In this way, large lysosomes can store considerable amounts of hydrolytic enzyme for future use.

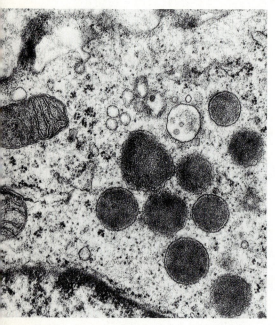

(a)

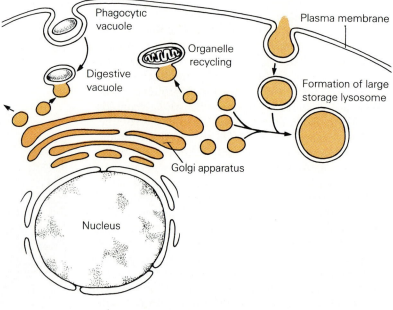

(b)

4.15 PEROXISOMES

Some peroxisomes have an unusual inner structure of crystallized enzymes. They are presently believed to function in enzymatic oxidizing reactions; for example, in some peroxisomes, hydrogen peroxide is oxidized to water and oxygen, rendering the peroxide harmless to the cell. The enzyme involved is catalase, which is found abundantly in liver tissue.

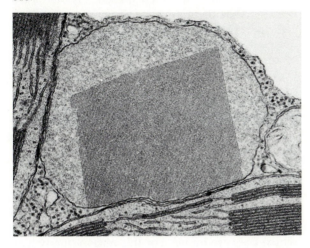

somes, phagocytic white cells—cells that engage in **heterophagy** (other-eating)—clean up what's left.

Lysosomes also have specific functions in certain healthy, active cells. Such cells take in solid substances and form *digestion vacuoles.* Afterwards, a lysosome will empty its contents into the digestive vacuole, thus aiding in the digestion of whatever has been captured.

There are a number of genetic disorders in which some essential lysosomal enzyme is altered or absent. This means that the substances in their unreacted, raw form accumulate in the cell, causing so-called **storage diseases.** Most such diseases are fatal in the first five years of human life. Tay-Sachs disease, a genetic disease that is relatively common among infants of European Jewish descent, is one such storage disease. In the disease, the absence of the critical lysosomal enzyme results in the buildup of lipids in brain cells, producing retardation, blindness, and, eventually, death.

Peroxisomes. The **peroxisomes** are membrane-bound bodies that often lie near mitochondria or chloroplasts (both of which are discussed below). Peroxisomes are found in a great variety of organisms, including plants, animals, and protists. In animals, they are most common in liver and kidney cells, where they are believed to originate from outpocketings of the smooth endoplasmic reticulum. Some peroxisomes appear as very dense bod-

ies with a unique crystalline array within, which causes the organelle to stand out unmistakably when it shows up in electron micrographs (Figure 4.15).

The crystals within peroxisomes are enzymes, and a principal enzyme of the liver and kidney peroxisomes is **catalase,** important in the break-down of hydrogen peroxide into water and oxygen:

$$2H_2O_2 \rightarrow 2H_2O + O_2$$

The role of the peroxisomal enzymes here is probably protective. Hydrogen peroxide is a rather dangerous, highly reactive oxidizing chemical, produced in considerable quantity as a product of the chemical activity in certain cells.

Vacuoles. The term **vacuole** is rather general. It refers merely to any membrane-bound body with little or no inner structure. Vacuoles generally hold something, but their contents vary widely, depending on the cell and the organism.

Plant cells generally have larger vacuoles than do animal cells. In many types of plant cells, the vacuole dominates the central part of the cell, crowding all other elements against the cell wall (Figure 4.16). Such vacuoles are filled with a fluid, the **cell sap,** which is primarily stored water with various substances in solution or suspension. The solutes may include atmospheric gases, inorganic salts, organic acids, sugars, pigments, or other materials. Sometimes the vacuoles are filled with

4.16 VACUOLES

Plant cells often have large central vacuoles filled with cell sap—water and dissolved metabolites. In leaves and soft stems, the water-filled vacuoles in most cells exert pressure against the cell walls, producing a firmness that helps maintain the shape of softer parts and keep young stems in an erect position.

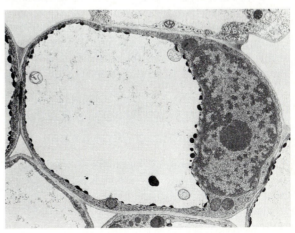

water-soluble blue, red, or purple pigments (the *anthocyanins*), which are responsible for many flower colors.

The large, watery vacuoles of plants serve another curious purpose. While all animals have elaborate excretory systems for getting rid of cellular and metabolic wastes, plants lack such systems. Instead, metabolic wastes and other poisons are sequestered in the central vacuoles, frequently forming crystals. Some plants store large amounts of waste poisons in their vacuoles, and it has been suggested that this serves as a protective device against herbivores. If a herbivore should eat the plant and crush the cell, it will get not only food, but stored poisons.

Of equal significance, the large, water-filled vacuoles of plants, particularly in cells of the leaves and in the stems and leaves of soft-bodied plants, hold the plant erect and firm. As we will learn in Chapter 5, the presence of solutes in the watery vacuoles is important to the continued uptake of water.

Other Storage Bodies: The Plastids. Plants also produce storage bodies known as **plastids.** They differ from vacuoles in that they are surrounded by a double membrane or envelope. Included in the plastids are the **chloroplasts,** one of the energy-generating organelles discussed in the next section. Storage plastids include the **leucoplasts** and **chromoplasts.** Leucoplasts (white plastids) are present in nearly all plant cells, but they can be most readily seen in leaf epidermal cells, onion or apple storage cells, and a variety of other white plant tissues specialized for storage. The principal role of leucoplasts is the storage of starch after it is formed from glucose. Chromoplasts (colored forms) are named for the pigments they contain; they impart many of the bright colors (other than green) that we see in plants, such as in their flowers. The colored pigments of chromoplasts include orange carotenes, yellow xanthophylls, and various red pigments. Chromoplasts can be important in attracting animals to the plant. While some animals transfer pollen from flower to flower, others help in seed dispersal by devouring the colorful fruit and depositing the seeds elsewhere.

Energy-Generating Organelles

The energy-generating organelles—the chloroplasts and **mitochondria**—are critical and interesting bodies. Their basic function is to convert energy to a form useful to the cell, and each does this in its own unique way. Their unique functions make the chloroplast and mitochondrion interesting on their own, but when one considers how these energetic little organelles may have gotten into cells in the first place, they can only be described as fascinating.

The Chloroplast. Chloroplasts are found in plants and algae, and their presence is one of the features that distinguishes plants from animals. They are large as organelles go and quite complex, with a double membrane, an intricate inner membranous structure, and, surprisingly, their own DNA and ribosomes. With DNA and ribosomes present, chloroplasts are able to reproduce themselves and to carry out the synthesis of some proteins, although many others are taken in from the cell cytoplasm where they were produced through the action of the nucleus. We will find that these characteristics are true of mitochondria as well. Studies of the chemical structure of chloroplast and mitochondrial nucleic acids and ribosomes reveal that they are much more closely related to certain bacteria than they are to the eukaryotic cells in which they are found. For this reason, many biologists conclude that both of these structures originated as bacterial invaders that soon established a mutual relationship with their eukaryotic hosts. As we will see in Chapter 18, this thinking has led to a novel hypothesis on the origin of eukaryotic cells.

Chloroplasts function in that critical and intricate process called **photosynthesis,** in which the energy of light is used to convert carbon dioxide and water to glucose and other important molecules. Among the vital participants in this process are active molecules known as the **chlorophylls.** *Chloroplast* means "green form," and the green comes from the photosynthetic pigment chlorophyll. We should note that many bacteria also contain chlorophyll pigments, but the chlorophyll in these prokaryotes has a somewhat different chemical structure and is not organized into chloroplasts.

Inside their double membranes, the chloroplasts consist primarily of layers of flattened, membranous disks known as **thylakoids.** Each stack of disks is known as a **granum** (plural, *grana*), and many grana are distributed in orderly rows throughout the chloroplast. Located along the thylakoid membranes are numerous granules, **CF1 particles,** the sites of ATP synthesis. Each granum is connected to its neighbors by **lamellae** (singular, *lamella*), membranous extensions of the thylakoids. The clear, watery area outside the thylakoids is called the **stroma** (Figure 4.17).

4.17 CHLOROPLASTS

(a) The chloroplasts in an intact plant cell appear as numerous minute, dark spheres. **(b)** At low electron microscope magnification the inner structure becomes visible. The dark, neat stacks are made up of thylakoids, each stack forming a granum. The thylakoid membranes extend from one granum to another. The clear substance suspending the grana is called the stroma. **(c)** This three-dimensional drawing shows a cross-sectional view of a chloroplast, with its stacks of disk-like thylakoids. **(d)** The thylakoid consists of membranes enclosing a fluid-filled lumen (cavity). Embedded within the membranes are large groups of proteins and pigments essential to photosynthesis. CF1 particles represent regions of ATP synthesis.

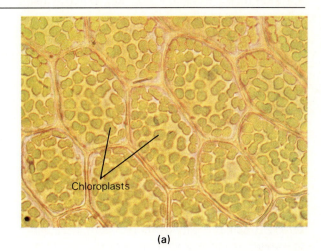

Chloroplasts

(a)

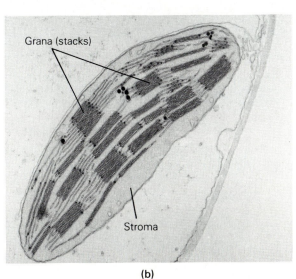

Grana (stacks)

Stroma

(b)

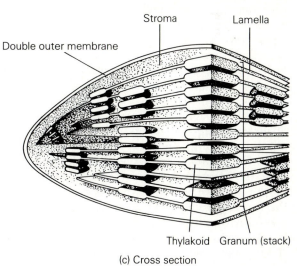

Stroma

Lamella

Double outer membrane

Thylakoid   Granum (stack)

(c) Cross section

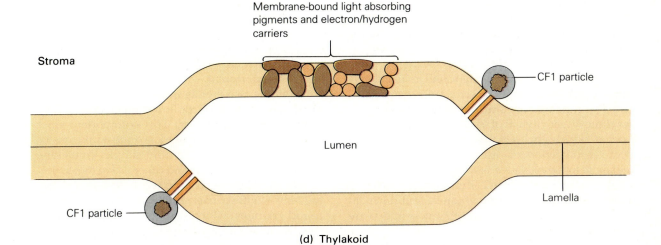

Membrane-bound light absorbing pigments and electron/hydrogen carriers

Stroma

CF1 particle

Lumen

CF1 particle

Lamella

(d) Thylakoid

In Chapter 7, we'll consider the biochemistry of photosynthesis in detail, but we would like to make one important point about it here. Each chloroplast is relatively autonomous, capable of carrying out photosynthesis quite on its own, as long as the raw materials, carbon dioxide and water, are made available by the plant. To retain this capability, the chloroplast must be intact. Although we can identify many of the enzymes and particular biochemical steps involved in the transformation of light energy to chemical energy, we also know that this mysterious process depends on the very precise structure and relationship of the chloroplast's components. If chloroplasts are disrupted and their components dispersed, photosynthesis abruptly ceases.

Mitochondria. Mitochondria (singular, *mitochondrion*) are complex energy-transferring organelles found in every eukaryote cell. Like chloroplasts, mitochondria (1) are enclosed in double membranes, (2) have their own circular DNA, (3) have their own ribosomes and other machinery of protein synthesis, and (4) are widely believed to be the descendants of a once-independent prokaryote ancestor.

To some extent, chloroplasts and mitochondria do exactly opposite things. Put simply, chloroplasts use energy and raw materials to produce carbon compounds and oxygen, while mitochondria use carbon compounds and oxygen to produce usable energy. We will deal with both these important biochemical processes in Chapters 7 and 8.

Mitochondria are considerably smaller than chloroplasts, especially in cross-section. In part, this is because chloroplasts tend to be more or less spherical, whereas mitochondria are usually long and slender. In electron micrographs, mitochondria usually appear as oval structures, with the inner of the two mitochondrial membranes show up as curious folds that extend partway across the inner cavity (Figure 4.18).

Although mitochondria are found in all eukaryotic cells, there are more in some cells than in others. The number varies in proportion to the cell's respiratory activity—that is, in proportion to the rate at which the cell uses up oxygen. This is because mitochondria are specialists in oxygen metabolism, or oxidative respiration. For example, a single hardworking liver cell may contain as many as 1000 mitochondria, but few can be found in a fat storage cell. As you might expect, mitochondria are abundant in muscle cells.

The inner membrane of the mitochondrion is highly folded. These folds are known as **cristae.** The

4.18 THE MITOCHONDRION

(a) Mitochondria are barely visible through the light microscope even at its highest magnification, but they show up clearly with low magnification in the electron microscope. In thin sections such as this, the long, tubular mitochondria appear as irregular ellipses and circles. Each is surrounded by two membranes. The inner one is folded repeatedly to form inner shelves known as cristae. (b) The artist's reconstruction of a mitochondrion clearly reveals its intricate membranous structure and the presence of F1 particles.

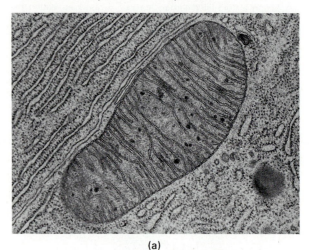

(a)

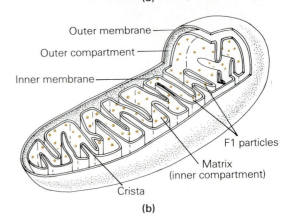

(b)

liquid-filled maze within the cristae is called the **inner compartment,** or **matrix,** and the space between the two membranes is the **outer compartment.** The folding of the inner membrane greatly increases its surface area. This is important since most of the biochemical work is done on the cristae themselves. They are far from simple, containing many versatile proteins and other active molecules. Special electron microscope techniques reveal that the tiny shelflike cristae are covered with small, round granules, which have been named **F1 particles.** These are similar in appearance to the CF1 particles. The F1 particles are also sites of ATP synthesis.

Organelles of Cellular Movement

A number of organelles are involved in cell movement. Some create movement within the cell, and others are involved in **locomotion,** moving the cell through its medium (or in stationary cells, moving the medium past the cell).

Centrioles. Under the electron microscope, each **centriole** (they are usually found in pairs) is seen to consist of two identical cylinders lying at right angles to each other (Figure 4.19). Each cylinder (actually a half-centriole) is made up of nine sets of triplet tubules, arranged like a pinwheel and surrounding a structureless center. Centrioles are self-replicating—a feat they accomplish after being separated during cell division. The term "replicate" is fitting, since each half-centriole apparently uses its existing structure to organize another just like itself. The new half then assumes the usual right angle position. (We haven't a clue as to how they do it.)

Centrioles are found in the cells of animals, most protists, and the more primitive plants. Apparently the more recently evolved plants and the fungi have lost their centrioles, and, importantly, eukaryotes without centrioles also lack basal bodies, cilia, and flagella—organelles of cell movement. This suggests a close relationship between centrioles and such organelles, and although evidence for such a relationship is mounting, the function of centrioles is still far from clear. They may organize basal bodies, and these in turn are known to have a role in the development of cilia and flagella. On the other hand, centrioles are present and active during the separation of chromosomes in cell division. They appear to have some function in organizing the cell for mitosis, but some researchers have pointed out that cells without centrioles do very nicely in organizing themselves for mitosis. We will return to the role of centrioles later (see Chapter 11).

Cilia, Flagella, and Basal Bodies. Cilia and **flagella** are fine, hairlike, movable organelles found on the surfaces of some cells. Although cilia and flagella appear to be outside the cell, they really are not; the cell membrane protrudes to cover each cilium or flagellum. Thus, we can consider them to be outpocketings of the cell proper.

Both cilia and flagella may either move the cell through some surrounding fluid or move some surrounding fluid past the surface of a stationary cell. For example, a sperm cell swims by undulations (wavelike movements) of its flagellum, a *Paramecium* (single-celled protist) moves by the coordinated beating of rows of cilia, and the cilia of the cells that line the passages of your lungs (unless you have permanently paralyzed them with tobacco smoke) sweep a cleansing film of mucus upward to your pharynx.

Cilia and flagella are structurally almost identical. They differ only in length, the number per cell, and their pattern of motion (Figures 4.20a and b). Cilia are short, numerous, and have a highly coordinated, unified rowing motion (like oars, except

---

4.19 CENTRIOLES

(a) Paired centrioles in the cell. **(b)** A thin section through the central axis of one of the paired bodies of a centriole. Note the nine sets of triplet microtubules. **(c)** In the three-dimensional reconstruction, each triplet consists of three fused microtubules. Each microtubule, in turn, consists of protein spheres arranged in a spiral form to produce the tubular unit.

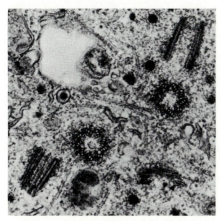
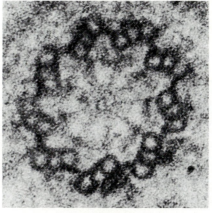
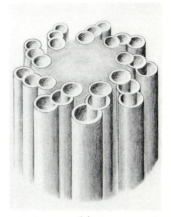

(a)   (b)   (c)

these "oars" must bend on the return stroke since they stay in the water). Flagella are long, few in number, and move by undulation. More specifically, most cilia are between 10 and 20 μm long, whereas flagella may, in exceptional cases, be several thousand micrometers long—that is, several millimeters long. The flagellum on a *Drosophila* (fruitfly) sperm, for instance, may be longer than the fly itself!

Prokaryotes have flagella, but these are built on an altogether different plan and are not related to eukaryote flagella (Figure 4.20c). Prokaryote flagella are solid crystals of protein that stick out through holes in the plasma membrane; they neither beat nor undulate, but spin like propellers.

Fine Structure and Movement in Cilia and Flagella. Under the electron microscope, the microtubular arrangement in eukaryote cilia and flagella is similar in cross section to that of centrioles.

While the half-centriole is made up of a circle of nine triplets of microtubules, both cilia and flagella have a "9 + 2" structure: a circle of nine pairs of microtubules with two single microtubules in the middle (Figure 4.21). Each of the nine doublets includes a complete microtubule and one that is only partially complete and fused to the other. Emerging from each complete microtubule are a pair of **dynein arms,** armlike structures composed of the protein dynein and essential participants in the mechanism of movement (dynein has the ability to cause an energy release in ATP, the energy storage molecule of cells). In addition, there are three important accessory proteins. The protein **nexin** forms connections between the microtubular doublets, and extending inward from each of the doublets are the **radial spokes,** which apparently contact an **inner sheath** of protein surrounding the two central microtubules. The entire active core of cilia and flagella—the microtubules, dynein arms,

4.20 ORGANELLES OF MOVEMENT

Cilia and flagella at work in the eukaryote and prokaryote. **(a)** Cilia, which occur only in eukaryotes, coordinate their movement to produce a highly orchestrated rowing action. Each cilium is extended during its power stroke but is bent on the return. **(b)** The eukaryotic flagellum may be used in a number of ways to direct movement. Usually the flagellum forms undulating, helical waves that can either push or pull the flagellated cell. **(c)** Bacterial flagella are solid, helically shaped filaments that cannot undulate. Surprisingly, they actually rotate in place like a boat propeller.

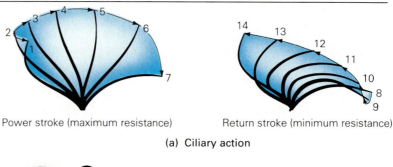

Power stroke (maximum resistance) Return stroke (minimum resistance)

(a) Ciliary action

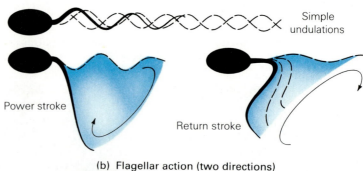

Simple undulations

Power stroke Return stroke

(b) Flagellar action (two directions)

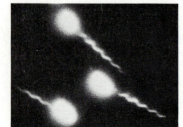

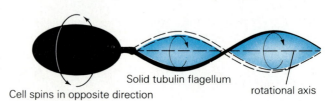

Cell spins in opposite direction Solid tubulin flagellum rotational axis

(c)

4.21 STRUCTURE OF CILIA

Electron micrographs of flagella—long a favorite subject among electron microscopists—aid in the three-dimensional reconstruction of this intricate organelle. The transverse section (a) is through the basal body of the flagellum, into which the nine doublet microtubules extend but the central pair do not. Basal bodies are quite similar to centrioles. The cross section (b) reveals the complex arrangement of microtubules, dynein arms, and accessory proteins. These are further clarified in the artist's reconstruction (c), where the axoneme is seen as a functional unit. The dynein arms are responsible for movement between adjacent doublet microtubules, while the accessory proteins and central microtubules bind the structure in such a way as to produce the characteristic bending and undulation seen in ciliary and flagellar movement.

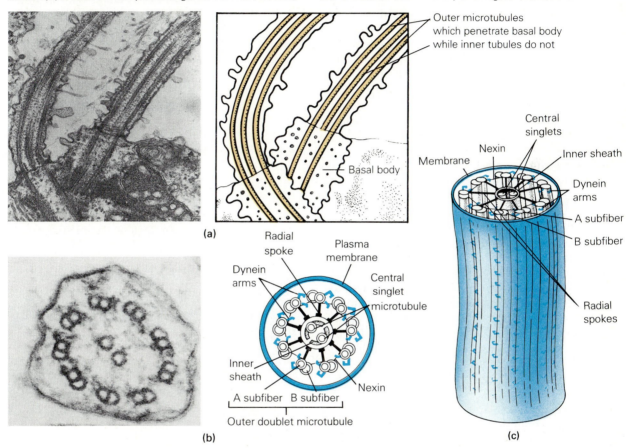

(a)

(b)

(c)

and accessory proteins—are known as the **axoneme** of the organelle. Let's see how these parts interact to produce the special kind of movement we find in cilia and flagella.

Using ATP as an energy source, the dynein arms of one microtubule apparently contact the adjacent microtubule and pull it downward in a ratcheting action. A very similar mechanism has been clearly identified in the contraction of muscle fibers (see Chapter 28). The sliding action in the intact axoneme is restrained by the accessory proteins and converted into the more unifying and directing bending response which is characteristic of these undulating organelles. Thus, while it is the dynein arms that actively move, it is the presence of the nexin, the radial spokes, and the inner sheath

that makes such movement useful. The sliding action is most easily seen in isolated axonemes—those that have had the accessory proteins digested away and have been activated by ATP and ions of calcium or magnesium.

What about the two central microtubules? Cytologists aren't certain, but it seems likely that they form a less movable base for the bending action, although some researchers report that they have an orienting function—determining the plane in which bending occurs. One thing is certain: they are essential to ciliary and flagellar action. In mutant organisms in which the central pair are absent, there is no bending action.

Each cilium and flagellum terminates beneath the surface of the cell at a basal body, which is structurally almost identical to a centriole. The nine pairs of microtubules of the cilium or flagellum, as they join the basal body, become nine triplet microtubules; the two single microtubules just end blindly. Incidentally, groups of basal bodies are usually joined by interconnecting fibers that presumably enable the cilia to coordinate their movements.

Cilia and flagella also apparently have some sort of primitive cellular sensory function. Many of our own sensory receptors are believed to have evolved from cilia. Amazingly, the 9 + 2 arrangement of microtubules can still be seen in (1) the rods and cones of the retina, (2) the olfactory fibers of the nasal epithelium, and (3) the sensory hairs of the cochlea and semicircular canals of the internal ear. It seems, then, that we see, smell, hear, and balance ourselves with highly modified cilia-bearing cells.

Microtubules and Microfilaments. You may have noticed that in our study of the parts of a cell we have been unable to keep the categories from overlapping. Thus, we mentioned ribosomes on the microtrabecular lattice before we told you what a ribosome was, and we have had several occasions to bring in microtubules before we got to this section. That's because the functions of different organelles are so closely interrelated. We have seen that microtubules are integral in the cytoskeleton of the cell and that they also make up the internal structure of centrioles, cilia, flagella, and basal bodies. We have mentioned asters and the mitotic spindle in passing, and you may recall that those structures are also made largely of microtubules.

It is generally accepted that all microtubules are made of a common protein, **tubulin.** Each tubulin molecule consists of two spheres of slightly different polypeptides linked together into a figure-8 shape. Under the electron microscope, the hollow tubular structure they form is evident. When microtubules are cut longitudinally, they appear as close, straight parallel lines. In cross section, they are tiny circles. But three-dimensional reconstructions reveal that each microtubule consists of a spiraling arrangement of the individual figure-8 tubulin molecules (Figure 4.22).

Microfilaments or actin filaments, as they are more specifically known, are also important in the cytoskeleton and in many structures of movement, such as animal muscle. The microfilament consists of a double strand of the globular protein, **actin,** with each strand wound about the other in the helical conformation so common in protein. We will come to actin again where vertebrate muscle is taken up in Chapter 28.

THE PROKARYOTE CELL: A DIFFERENT MATTER

The prokaryotes are bacteria—minute, single-celled forms that trace their roots to the most ancient forms of life known. Included in the prokaryotes are the familiar disease bacteria, along with common soil and water types. This group also includes the **cyanobacteria,** also known as *cyanophytes* ("blue plants"). Although these are cellular organisms, they lack many of the features that are seen in the more familiar eukaryotes (Figure 4.23).

In addition to their greater simplicity, prokary-

4.22 MICROTUBULES

Microtubules occur in many types of cells and apparently serve as a kind of cellular skeleton. Under the electron microscope they appear as long rods (vertical lines in the photograph), which, in cross section, are found to be hollow. The tubes consist of spherical units of polypeptides that comprise the globular protein tubulin. Each tubulin molecule actually contains two polypeptides—alpha and beta subunits, arranged in the tubule as shown.

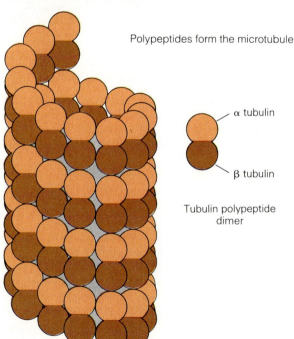

Polypeptides form the microtubule

α tubulin

β tubulin

Tubulin polypeptide dimer

4.23 TWO TYPES OF PROKARYOTES

The structural simplicity of prokaryotes is deceptive. They are often biochemically complex, carrying out many of the chemical functions of eukaryotes. This scanning electron micrograph shows a typical rod-shaped bacterium.

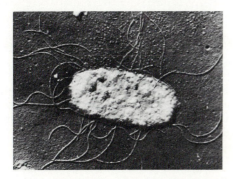

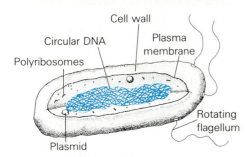

otes are also generally much smaller than the eukaryotes (although cyanobacteria tend to be larger than other bacteria). All prokaryotes have a plasma membrane; in fact, it has recently been established that in some groups, the cells are enclosed within two separate membranes. But prokaryotes lack an organized nucleus; that is, the genetic material or DNA is not enclosed by a membrane or nuclear envelope. Further, the DNA is not wound about protein as it is in eukaryotes, and the naked bacterial DNA usually takes the form of a circle. Because there is no nuclear membrane, there is no physical separation between the sites of gene activity in the cytoplasm.

In fact, prokaryotes lack nearly all signs of internal membranes and the membrane-bound organelles common to eukaryotes. However, cyanobacteria are again an exception. They have extensive internal membranes, the thylakoids, which are involved in photosynthesis. Further, during cell fission (division), some bacteria are seen to form a membranous structure called the **mesosome.** Finally, bacteria, like plants, are enclosed in cell walls, though the walls are chemically very different. In fact, the cell wall chemistry within different groups of bacteria is quite different. The differences between prokaryotes and eukaryotes are summarized in Table 4.1.

We have learned a great deal about those "little

TABLE 4.1

KEY DIFFERENCES AMONG THE CELLS OF PROKARYOTES, PLANTS, AND ANIMALS

Feature	Prokaryotic Cell	Higher Plant Cell	Animal Cell
Cell membranes	External only (two, separated by periplasmic space)	External and internal	External and internal
Supporting structure	None seen	Protein cytoskeleton	Protein cytoskeleton
Nuclear envelope	Absent	Present	Present
Chromosomes	Single, circular, DNA only	Multiple, linear, complexed with protein	Multiple, linear, complexed with protein
Membrane-bounded organelles	Absent except for mesosome and thylakoid	Many, including mitochondria, large vacuoles, and chloroplasts	Many, including mitochondria, lysosomes
Endoplasmic reticulum	Absent	Present	Present
Ribosomes	Smaller, free	Larger, some membrane-bound	Larger, somewhat membrane-bound
Cell wall	Peptidoglycan in some	Cellulose	None
Flagella or cilia (when present)	Solid, rotating	Never present*	Microtubular (9+2 pattern)
Ability to engulf solid matter	Absent	Absent*	Present, extensive movable membranes
Centrioles	Absent	Absent*	Present

*Although absent in higher plants, these features are found in more primitive plants. Apparently they have been lost in the course of evolutionary change.

boxes'' that Hooke brought to our attention some three centuries ago. They have yielded their small secrets, one at a time, over the years until their workings have come to be revealed as one of the most impressive symphonies in the concert of life.

Yet there remain those lapses into silence, and those dissonant chords of incomplete understanding, that send researchers on their continuing quest for the secrets of life's harmony.

APPLICATION OF IDEAS

1. Many exceptions to the cell theory have been noted. In view of this, how can the theory continue to be important and instructive? If you think it is not, offer reasons for this conclusion.

2. A biological principle called complementarity of structure and function holds that the two are very closely related. Cite five or six examples of how complementarity works at the cellular level.

KEY WORDS AND IDEAS

CELL THEORY

1. The **cell theory,** proposed in 1805 by Schleiden and Schwann, maintains that living organisms are composed of cells. Rudolf Virchow added that all cells come from preexisting cells.

2. Cells are extremely diverse, their differences reflecting specialized functions.

WHAT IS A CELL?

1. Cells are the basic units of life. They are capable of taking in materials, extracting energy, synthesizing molecules, growing, responding to environmental stimuli, and reproducing.

The Eukaryotic Cell

1. Eukaryotic cells are surrounded by a **plasma membrane,** and in plant cells, a **cell wall.** Its **nucleus** lies within a **nuclear envelope.** In the **cytoplasm** lie membrane-surrounded structures called **organelles,** along with supporting elements of the **cytoskeleton**—the **microtubules** and **microfilaments.**

The Size of Cells

1. Most cells range in size from 10 to 100 micrometers.

2. As cells increase in size, the volume increases proportionally faster than the surface area.

3. The **surface-volume hypothesis** maintains that cells must remain small in volume in order for the membrane, or cell surface, to provide a sufficient transport of materials in and out of the cell. The same critical problem is seen with the nuclear membrane and the cytoplasmic volume.

4. Cells can overcome the surface-volume limitations through growth in one dimension only (length or width) or by producing a highly folded plasma membrane.

CELL STRUCTURES IN EUKARYOTES: THE ORGANELLES

Organelles of Support and Transport

1. Dense cell walls surround the cells of plants, fungi, and many protists, maintaining shape and providing strength and flexibility.

2. In plants, the cell wall consists of cellulose fibers, laid down in several directions. A **middle lamella** of pectin holds adjacent walls together. **Lignin, suberin,** and **cutin** are secreted into some plant cell walls.

3. The plasma membrane controls the passage of materials in and out of the cell. It is composed of a bilayer of phospholipids and many surface and transmembranal proteins.

4. The **high voltage electron microscope** has resolved a detailed picture of the cytoskeleton, showing that the *cytoplasmic matrix* consists of a **microtrabecular lattice**—an interwoven tubular network. In addition to its support function, the lattice may help organize enzyme activity.

Control and Cell Reproduction: The Nucleus

1. The nucleus is a prominent, spherical body, containing the **chromosomes** which are made up of DNA (the **genes**) and protein.

2. Deep staining **nucleoli** contain the RNA of **ribosomes.**

3. The nuclear envelope is a complex double membrane containing protein-filled pores. Its lumen is continuous with the endoplasmic reticulum.

4. Genes within the nucleus are copied and transmitted through **mitosis** from one generation to the next. They contain the chemical instructions for assembling amino acids into protein including enzymes which direct the cell's chemical activity.

Organelles of Synthesis, Storage, and Cytoplasmic Transport

1. The **endoplasmic reticulum** includes extensive, dynamic, membrane-lined channels, the **lumen,** or *cisternae*, through the cytoplasm. It also occurs in a *vesicular* form when portions pinch away. The **rough endoplasmic reticulum** is named for the presence of numerous ribosomes (protein synthesizing bodies). The **smooth endoplasmic reticulum** occurs in cells where carbohydrates, lipids, and other nonprotein products are formed.

2. The **Golgi body** or **apparatus** (**dictysome** in plants) includes a number of membranous channels and sacs usually located near the nucleus. The Golgi complex is involved in chemical modification and concentration. Two types of **Golgi vesicles** pinch off from the main body, **coated vesicles** (involved in protein chemistry), and **secretory vesicles** (which carry materials to the plasma membrane for discharge).

3. **Lysosomes** are membrane-bounded sacs that contain hydrolytic enzymes for the hydrolysis of damaged or aging cell components **(autophagy),** or for the digestion of materials taken into phagocytic vesicles **(heterophagy).** In certain genetic **storage diseases,** critical enzymes are missing and an abnormal buildup of chemicals occurs in the lysosomes.

4. **Peroxisomes** are membrane-bounded bodies that contain **catalase,** a hydrogen peroxide-metabolizing enzyme.

5. **Vacuoles** are general saclike bodies that contain **cell sap,** water, certain chemicals, and pigments. When water-filled, they give firmness to softer plant structures.

6. **Plastids** include metabolically active **chloroplasts,** starch-storing **leucoplasts,** and colorful, pigment-filled **chromoplasts.**

Energy-Generating Organelles

1. Chloroplasts are complex, double-membraned plastids that carry on **photosynthesis:** they convert sunlight energy into chemical bond energy and, using water and carbon dioxide, use this energy to produce carbohydrates. **Chlorophylls,** the light-trapping photosynthetic pigments, are arranged in membranous **thylakoids,** which are interconnected by membranous **lamellae.** Enzymatic activity occurs in the clearer **stroma** between **grana,** the stacks of thylakoids.

2. **Mitochondria** are complex bodies that carry out cell respiration, the conversion of chemical bond energy in cellular fuels to high-energy bonds in ATP. Their outer membrane is simple and surrounds an **outer compartment.** The inner membrane is highly folded, forming the shelflike **cristae,** which surrounds an enzyme-rich **inner compartment,** or **matrix.** The cristae contain the **F1 particles,** sites in which ATP is synthesized.

Organelles of Cellular Movement

1. Cellular movement includes movement within the cell and **locomotion,** movement of the cell.

2. **Centrioles** are paired, rodlike, self-replicating bodies near the nucleus. Their function is not clear, but they are active during mitosis and cell division, and may organize **basal bodies.**

3. **Cilia** and **flagella** are involved in the movement of cells. They differ only in tail length, number, and movement pattern. They consist of **microtubules** in a 9 + 2 arrangement. **Dynein arms,** with their **radial spokes** and interconnecting **nexin,** and an **inner sheath** surrounding the central doublet, form the **axoneme.** The microtubules produce bending action by their energetic movement past each other. Cilia are also believed to be the forerunners of many sensory structures.

4. Microtubules are tubelike structures composed of spherical subunits of **tubulin** that are integral parts of cilia and flagella and are the spindle fibers of the mitotic spindle. **Microfilaments** are simpler fibers of the protein **actin** that are involved in cytoplasmic movement.

THE PROKARYOTE CELL: A DIFFERENT MATTER

1. Prokaryotes are the simplest forms of cellular life. While they contain a cell wall and plasma membrane, there is little cytoplasmic organization and organelles are generally absent. Since they lack a nuclear membrane, the DNA is free in the cytoplasm.

2. Photosynthetic, membranous thylakoids are found in **cyanobacteria,** and many bacteria produce a membranous **mesosome** at the time of fission.

REVIEW QUESTIONS

1. Summarize Robert Hooke's contributions to cell biology. (75)

2. Summarize the contributions to cell biology of Schleiden, Schwann, and Virchow. (75–76)

3. List five fundamental activities of cells. (76)

4. What is the usual size range of cells? Cite two exceptions in which cells are much larger than this. (79–80)

5. What happens to the relationship between surface area and volume when a cell doubles its size? (80)

6. Summarize the surface-volume hypothesis. How do large cells overcome such limitations? (80–81)

7. List and explain two important functions of the plant cell wall. (82)

8. What is the basic structure of the plasma membrane? What are its functions? (82)

9. Describe the microtrabecular lattice. What are two of its supposed functions? (82–83)

10. Summarize two important roles of the nucleus. Be specific. (83, 88)

11. Describe the nucleus and its contents, including the envelope, chromosomes, and nucleoli. (88)

12. Compare the structure and function of the rough endoplasmic reticulum with the smooth endoplasmic reticulum. (89–90)

13. What is the relationship between the endoplasmic reticulum and the Golgi apparatus? List two types of Golgi vesicles and their functions. (91)

14. Explain how the lysosomes function in autophagy and heterophagy. (92–93)

15. What is the relationship between lysosomes and storage diseases? (93)

16. Describe a common chemical reaction in peroxisomes and explain why it is important. (93)

17. Discuss two important roles of water vacuoles in plants. (93–94)

18. Using a simple drawing, illustrate the structure of a chloroplast including the following: envelope, grana, stroma, thylakoids, lamellae, CF1 particles. (94–95)

19. Generally discuss the function of chloroplasts, and explain briefly what happens in the thylakoids and stroma. (94)

20. How do evolutionists explain the presence of complex cellular organelles such as chloroplasts and mitochondria? (96)

21. Using a simple drawing, illustrate the mitochondrion. Include the outer membrane, inner membrane, outer compartment, matrix, cristae, and F1 particles. What is the function of the mitochondrion? (96)

22. Describe a centriole and summarize its function in mitosis. (97)

23. Using a drawing, explain the structure of a cilium. Be sure to include the 9 + 2 microtubular arrangement and the components of the axoneme. (98)

24. What are two differences between cilia and flagella? Carefully compare their styles of movement. (97–98)

25. Explain the bending action of a cilium or flagellum. (97–98)

26. Describe the structure of a microtubule. List three of its uses. (100)

27. Generally contrast the structure of a prokaryote and eukaryote cell. (100–101)

Cell Transport

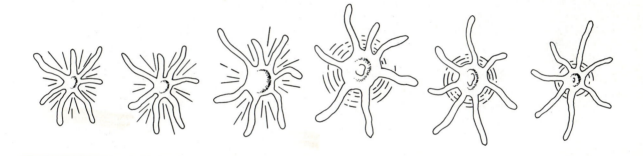

Most cells are so tiny they can't be seen, yet within them are systems so complex that they are not yet understood. One problem that has presented an enduring challenge is the way cells shift critical materials around so that the delicate processes of life can continue. After all, for cells to remain active they require a constant uptake of vital substances and a constant output of metabolic wastes—and, in many instances, the special substances they produce. Obviously, materials moving into or out of cells must pass through that vital envelope called the plasma membrane, so it is here that we can begin our search for the mechanisms of such movement.

THE PLASMA MEMBRANE

The plasma membrane is a critical structure, and rather taken for granted in its role as a living envelope around the cell. But for a long time some people refused to believe it existed. This is understandable; the plasma membrane is so thin that its existence, as well as its structure, was postulated entirely on circumstantial evidence until the advent of the transmission electron microscope (TEM). The cir-

cumstantial evidence was gathered from observations that something was regulating the passage of materials into and out of living cells. The circumstantial evidence even suggested the composition of the membrane. For example, it had long been postulated that the membrane has a lipid core simply because lipid-soluble materials and lipid solvents readily pass through. Since water also passes through the membrane, though at a slower rate, scientists believed that the membrane also had water-admitting pores, perhaps surrounded by proteins. As we will see, these ideas were generally on target.

With the transmission electron microscope we could see for the first time visual evidence that verified much of what the membrane theorists had proposed. Membranes, as viewed through the TEM, look like two dark lines separated by a clear line (Figure 5.1). The clear line, it turns out, is about 5.0 nanometers (nm) wide, which happens to be twice the average length of the hydrocarbon tails of membrane phospholipids. This suggested a two-layer structure with the tails fitting neatly back-to-back. The dark lines apparently represent the phosphorus-rich polar heads of phospholipids, along with proteins associated with the membrane.

The Fluid Mosaic Model

Figure 5.2 shows a diagram of what we now believe the plasma membrane structure to be. This is called the **fluid mosaic** model. The name "fluid mosaic" is purposefully vague. It refers to the fluidlike qualities of the phospholipid core and the dynamic behavior of proteins that seem to drift on the "lipid sea," some afloat, others partially or fully submerged. The phospholipids are shown as small, polar spheres of glycerol, phosphorus, and choline. (You may recall from Chapter 3 that phospholipids contain, in addition to phosphate, a variable group, which in the plasma membrane is often nitrogen-containing choline. This phospholipid is called **phosphatidylcholine.**) The polar (charged) heads have two hydrocarbon tails each, and these point inward. This arrangement forms a water-resistant barrier that admits only molecules that are lipid-soluble and rejects most water-soluble molecules. Recall from the last chapter that the polar heads of the phospholipids are hydrophilic but their nonpolar (uncharged) tails are hydrophobic. Thus, it is the very core of the membrane, where the hydrophobic tails intermingle, that rejects water and water-soluble molecules. How, then, do these molecules get past the barrier?

Large globular proteins are known to be embedded right in the membrane, and these proteins are believed to have channels through which water-soluble materials can pass. The anchoring of proteins to the phospholipid core is complex and involves individual amino acids in the protein structure. Some interact with the charged regions of the phospholipids, others with the uncharged fatty acid core.

The large globular proteins are believed to occur in three spatial arrangements. Some, the **transmembranal proteins,** pass through the membrane. The midregion of such proteins tend to be hydrophobic, like the lipid core with which they interact, but the ends protruding from both the outer and inner cell surface are not. The exposed regions are hydrophilic, like the phospholipid heads surrounding them. Other, smaller **peripheral proteins** have only one hydrophobic end and lie embedded only part way through the membrane. Then there are **surface proteins** that are entirely hydrophilic and only lie on the membrane surface, not penetrating the hydrophobic region at all.

Freeze-fracture preparations (Figure 5.3) have helped to substantiate the existence of such proteins. When frozen cells are split apart right between the tails of phospholipids, the membrane proteins remain as large globs on one of the two surfaces. The other surface is pitted with indentations where the proteins had been.

The membranal proteins have several important functions. Apparently, some of the proteins associated only with the outer membrane layer are the sites where hormones can interact with cells, and some of them may function in helping cells recognize others of their own type, as is discussed below. Still others are quite mobile, free to pass back and forth through the membrane, acting as carriers in the transport process, while others have enzyme characteristics, carrying out their reactions right in the membrane. Proteins associated only with the

5.1 EARLY MODEL OF THE PLASMA MEMBRANE

The TEM was invaluable in finally confirming the structure of plasma membranes. The pair of dark lines in the photo are two membranes close together. Each consists of a pair of parallel lines with lighter regions between. The structure was first suggested from studies of the membrane-forming properties of phospholipids. The drawing is adapted from one of the early models, which has now been substantially changed. It consists of two layers of protein with a bilayer of phospholipids between. Note the orientation of hydrophobic and hydrophilic ends of each phospholipid molecule.

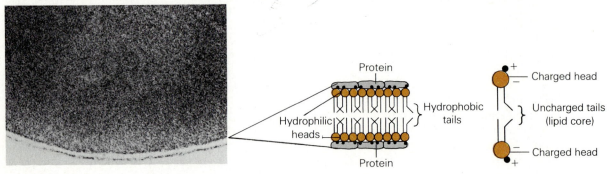

5.2 NEWER MODEL OF THE PLASMA MEMBRANE

The fluid mosaic model of the plasma membrane reveals its best-known components. The basic structure includes the well-substantiated phospholipid bilayer with the hydrophilic heads on the outside and the double hydrophobic tails pointing inward. Numerous globular proteins are embedded throughout the membrane. Hydrophilic pores occur in some. Glycoproteins project their branched heads above the surface and attach to subsurface proteins below. Other subsurface structures, including the microfilaments and microtubules, attach below, forming the cellular cytoskeleton.

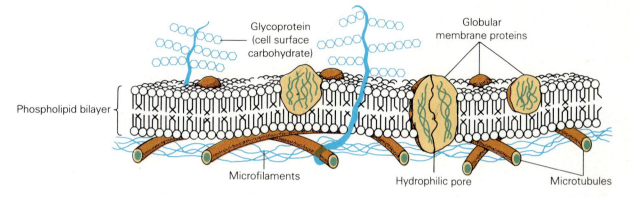

Glycoprotein (cell surface carbohydrate)

Globular membrane proteins

Phospholipid bilayer

Microfilaments

Hydrophilic pore

Microtubules

5.3 PROTEINS IN THE PLASMA MEMBRANE

Freeze-fracture studies help substantiate the globular protein component of the plasma membrane. These are clearly seen in the electron micrograph as tiny mounds of protein. Both globular proteins and the cavities they once filled are seen in the illustration where the upper lipid layer is torn away.

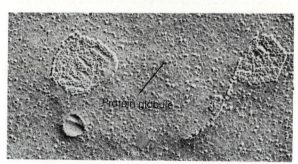

Protein globule

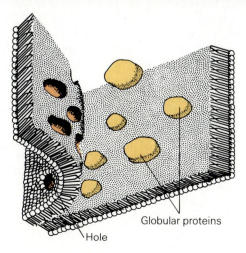

Globular proteins

Hole

inner surface of the membrane apparently interact with cytoplasmic proteins, and perhaps some of them even form bridges between the plasma membrane and the lengthy elements (microtubules and actin fibers) that make up the cytoskeleton. This is a bit conjectural at present, since the model is new and scientists are still looking into the protein functions.

The Glycocalyx.   The plasma membrane has an additional surface component—complex surface carbohydrates with either protein or lipid tails that pass through the lipid core. These are known as **glycoproteins** and **glycolipids,** the first of which is by far the most common (Figure 5.4). The glycoproteins contain up to 15 monosaccharides, often occurring in several branched chains. The surfaces of eukaryotic cells are rich in such carbohydrates, and the layer produced is known as a **glycocalyx.**

The cellular role of the glycocalyx has yet to be firmly established, but many cell biologists are convinced that it is important in cellular recognition— the ability of cells to identify and bind to each other when needed. This role is more clearly established in plants than animals. Cell recognition is important in many processes, including embryological development and immune responses (Figure 5.5).

To show the roles of proteins in cell recognition, cells from different tissues have been mixed together on a nutrient medium and then observed as they actually begin to move about. Surprisingly, each type appears to seek out others of its kind in such a way as to form individual masses of specific tissue

5.4 PLASMA MEMBRANE SURFACE FEATURE

(a) Dense membranal coats of glycolipid and glycoprotein are seen as the *glycocalyx* in electron micrographs of the plasma membrane. **(b)** Glycolipids, by virtue of their lipid tails, become integrated directly in the membrane. Their carbohy-. drate heads tend to be simple and unbranched. Glycoproteins are far more complex. Hydrophobic regions of their amino acid chain lie embedded in the lipid core, with the rest of the chain protruding above and below the membrane.

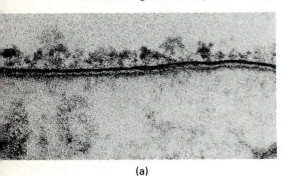

(a)

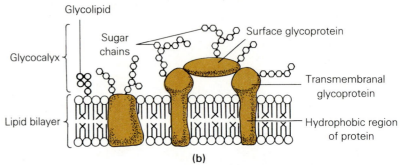

(b)

types. Presumably the recognition occurs through specific proteins on the cells' surfaces.

Cell-surface proteins have a number of roles in the maintenance of the human body. For example, certain proteins enable protective white blood cells to recognize most cancerous cells, which they then promptly destroy. And you undoubtedly have heard about the difficulties of rejection in organ and tissue transplants. Foreign tissues are rejected because their surface proteins and carbohydrates are not recognized by protective white blood cells, and are destroyed just as any invader of the body would be.

We have seen that the incredibly thin plasma membrane is a delicate but exquisitely complex structure. Its lipid, protein, and carbohydrate components have precise roles in the membrane's essential barrier and transport and recognition functions. Now that we have a clearer picture of the nature of the plasma membrane, let's consider how materials may move across it.

MECHANISMS OF TRANSPORT

We should begin by noting that the plasma membrane is highly selective in its transport role. Some substances pass readily through with little help or interference, while others can neither enter nor leave. The degree to which substances can pass through a membrane defines the membrane's **permeability.** Because the plasma membrane is more permeable to some substances than others, it is said to be **selectively permeable.** The admission or rejection of substances by the plasma membrane depends on a number of factors, but the most significant are size, polarity, and electrical charge. As a rule, small, nonpolar molecules (such as oxygen and nitrogen) and small, polar, but uncharged molecules (such as water, carbon dioxide, glycerol, and urea) pass rapidly across membranes. Somewhat larger molecules such as glucose and sucrose have great difficulty passing through the membrane on their own, and charged particles, such as sodium, potassium, and calcium ions are quite often rejected. Such rejected substances do cross plasma membranes, but they must receive an active boost from the cell, as explained below.

Cell transport can be divided into two categories, which we will call **passive transport** and **active transport.** The primary difference between the two is the source of energy for moving substances. In passive transport, as its name implies, the cell's own energy store, that is, its ATP energy, is not directly involved. However, we find that in some instances, the cell may actively transport one substance thereby encouraging the passive transport of another.

Examples of passive transport include processes called **diffusion, bulk flow, osmosis,** and **facilitated diffusion.** They all make use of external sources of free energy (energy that is available for work), particularly heat. Active transport, a process that is not very well understood, often involves ATP-powered **membranal pumps.** These are specialized proteins involved in the transport of ions and molecules in and out of the cell. In other instances, large portions of the membrane are involved as masses of solids and fluids are transported in and out through the assembly and disassembly of vacuoles. We will first discuss the passive processes, each of which involves certain physical processes or "laws" that govern the movement of matter—both inanimate (nonliving) and animate (living). Such phenomena govern the behavior of all matter, and in the course of evolution living cells have simply taken advantage of their existence.

Passive Transport

All forms of passive transport depend ultimately on energy in the system in which they occur. As mentioned, for the most part this energy is thermal or heat energy. Let's begin by finding out how the energy level of a system influences diffusion.

Diffusion.   The molecules or ions of any liquid or gas are in constant and random movement, forever bumping each other, rebounding, and taking new paths. The frequency with which these interactions occur is greatly influenced by the amount of heat in the surroundings. Since the movement of such particles is entirely random in gases and liquids, there is always a trend toward randomness in their distribution. The resulting movement is called **diffusion** and it is defined as the *net* movement of molecules or ions from regions of their greater concentration to regions of their lesser concentration until a state of **equilibrium**—a random distribution—is reached. The term "net" is important here since the movement of *individual* molecules or ions is fully random and continues irrespective of equilibrium. How can random movement result in net movement? Let's consider an example.

Movement Down Diffusion Gradients.   Everyone who has studied chemistry knows what happens when you uncork a container of hydrogen sulfide, permitting the "rotten egg" odor to penetrate the classroom (Figure 5.6). The next person entering the room may cast a suspicious eye about until he or she realizes the probable source of the odor. The

5.5 CELL-TO-CELL INTERACTION

Intercellular recognition is not a newly observed phenomenon, but has interested biologists for some time. Biologist H. V. Wilson experimented with the phenomenon in sponges in 1907, noting that sponge cells strained through fine cloth would regroup according to their tissue types. The phenomenon is readily seen in cells from the chick embryo (a) and (b). They had previously been separated with the aid of a digestive enzyme. One model of cell recognition proposes that two different macromolecules are involved at each recognition site. The clustering of cells within a tissue occurs as the paired surface macromolecules in the many cells interact (c).

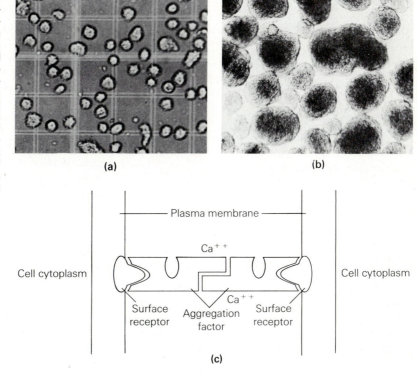

(a)

(b)

Plasma membrane

Ca^{++}

Cell cytoplasm

Cell cytoplasm

Surface receptor

Aggregation factor

Ca^{++}

Surface receptor

(c)

person may then locate the container by tracing the odor along an *increasing gradient* to its source. How does the odor spread through the room? And why is the container harder to find after some time has passed?

We should keep in mind that the original system, the uncorked bottle of hydrogen sulfide, is a highly ordered, energetic system. The molecules within the bottle, whether in a liquid or gaseous

5.6 DIFFUSION OF A GAS

A container of hydrogen sulfide (H_2S) has been uncorked at one corner of a room. The container contains a higher concentration of molecules than the room. Thus, the gas has great free energy. Molecules move in all directions, but the greatest movement in any direction is toward areas where there are fewer molecules of H_2S. As movement continues, a concentration gradient is established, with the greatest concentration still in the container. Finally, a state of equilibrium is reached when the distribution of molecules is random throughout the bottle and the room. The free energy level of the system is at a new low. Molecular movement continues, but no net directional movement occurs.

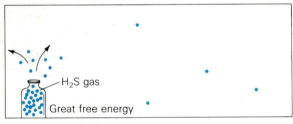

H₂S gas

Great free energy

Greatest free energy

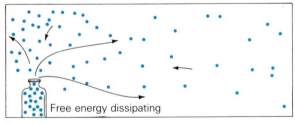

Free energy dissipating

Gradient established

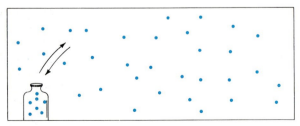

Equilibrium:
no free energy, no net movement in any direction

state, move constantly, colliding continually with each other and with the container walls, but there is no net movement in any direction.

When the cork is removed, some molecules chance to move out and through the room, blindly bumping into walls, nasal receptors, and each other. A few may even cross the room twice and then return to the bottle, but that isn't very likely. During this spreading process, the molecules take random paths. Thus, when the first hydrogen sulfide molecules have crossed the room, most of the rest are still jostling around somewhere near the bottle. The result is an increasing **concentration gradient,** between the source and the walls of the room. So, even though all of the molecular movement we have witnessed here is random, the *net* movement establishes a concentration gradient. Such gradients are always characteristic of diffusion. There is simply a greater degree of freedom in spaces where the molecules are scarce. What we have seen then is a random process by which molecules move under their own thermal energy, away from their place of higher concentration. The net movement will cease when the distribution of molecules is fully random, but as we have emphasized, the individual molecules continue to move.

Living cells frequently take advantage of diffusion, particularly in their ongoing exchange of gases. As an aerobic, or oxygen-requiring organism, you undoubtedly know that your cells require a continuous supply of oxygen, which is rapidly used up during respiration. Also during respiration, your cells continually produce carbon dioxide, a waste product of this process. Thus, the greatest concentrations of carbon dioxide are found within cells, and the greatest concentrations of oxygen occur outside. In both instances, the net movement of the two gases is in opposite directions, or, as biologists like to say, the two gases move *down their concentration gradients.* We are emphasizing gases here because their net movement in cells is usually straightforward. Clearly, however, with fluids such as water, the situation is somewhat more complicated.

Water Potential, Bulk Flow, and Osmosis. Because water is crucial to life, there has always been a keen interest in its passage into and out of cells. The general movement of water through cells is directed to some extent by diffusion, but it may also involve two other forces: **bulk flow** and **osmosis.** Whereas the movement of water through diffusion is generally a very slow process, bulk flow and osmosis can shift it more rapidly.

The bulk flow of water (or any other fluid) is that movement produced by pressure or gravity or both. The simplest example of bulk flow is the response of water to gravity (Figure 5.7). Because of gravity, water located high above the ground—in a tank on a hill or in a water tower, for example—contains a great deal of potential energy (the energy it took to pump it there in the first place). This energy is often referred to as **water potential,** although water potential can be created in other ways as well. The term *potential* is used here in its common context—that is, referring to the capacity to do work. As the water is permitted to run downhill through water mains and pipes, its potential energy is dissipated. In both instances, forcing water upward against gravity and allowing water to fall in response to gravity, water moved from a region of greater potential to one of lesser.

The concept of water potential is an important one to biologists seeking to predict the direction in which water will move. Keep in mind the point about water's moving from higher to lower potential. This may sound like diffusion in a general sense, but in bulk flow the water is moved by an energy source other than simple molecular motion.

We stated that the movement of water involved three forces—bulk flow, diffusion, and osmosis. Let's now turn to osmosis, which will bring us back to cells, where these concepts come together. Cells, as we have mentioned, are about 70 to 80% water, and within this water are many substances in solution, called **solutes.** Then there is the surrounding plasma membrane, which, as we have previously mentioned, is selectively permeable, so it plays a vital role in determining which substances move through its complex structure. Osmosis (Greek; *osmos;* impulse or thrust) refers to the diffusion of water through a semipermeable membrane, usually from a region of greater water potential to one of lesser water potential. We say ''usually'' because both gravity and pressure, the forces causing bulk flow, can also influence the direction of water movement in an osmotic system. The tendency for water to move in such a system is called, logically enough, **osmotic potential.** Osmotic potential depends upon the concentration of solutes on the other side of a membrane. Let's look at some examples of osmosis at work, beginning with a simple demonstration involving two containers, a thistle tube and a beaker. They are divid-

5.7 WATER POTENTIAL

Water in a reservoir has potential (free) energy that normally is depleted only when the water reaches sea level. As water passes through the dam and falls to a lower level, its potential energy becomes kinetic energy—energy in motion.

ed by a membrane that is *permeable* to water but *impermeable* to salt (Figure 5.8). Thus, water can cross the membrane but salt (actually sodium and chloride ions) cannot. Suppose the water in the thistle tube is a 3.0% salt solution, while that in the beaker is pure water. Which has the higher solute concentration? Obviously, it's the thistle tube. But which has the greater concentration of water? Not quite so obviously, it's the beaker, which is 100% water, compared to the tube which is 97.0% water and 3.0% (by weight) salt. Consequently, the beaker has greater water potential than the thistle tube. What can we predict about this model osmotic system?

We can predict that the net movement of water will be down its water potential and concentration gradient, moving through the membrane into the thistle tube. As a result, the volume of water in the tube will rise. How long will this net movement continue? Since the membrane is impermeable to salt, we might speculate that all of the water will move into the tube. But in this instance our guess may well prove wrong. Of course, if the sheer weight of water breaks the membrane, our demonstration will be finished, but what other forces are there at work? Remember that both gravity and pressure influence water potential. In this case, the weight of the rising water volume may well represent a substantial increase in water potential in the weakening salt solution. The point is that when the

5.8 DEMONSTRATION OF OSMOSIS

A thistle tube containing a salt solution is immersed in a container with distilled (pure) water. Separating the two fluids is a semipermeable membrane that will allow water to pass freely but restricts the passage of salt. The fluid in the thistle tube will rise to a considerable height, but the two solutions can never reach equilibrium because one will always be a salt solution.

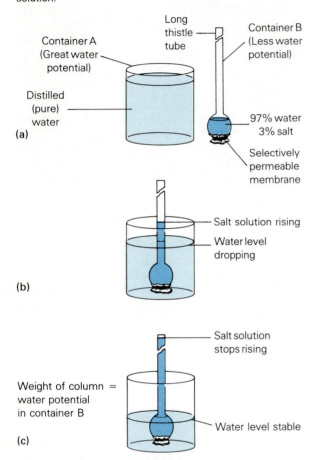

water potential in the salt solution equals the water potential of the pure water, the *net* movement of water molecules in the osmotic system will cease. You may have already noticed that the water and salt molecules can never reach equilibrium because of the membrane's impermeability to salt. Unlike simple diffusion, in which the net movement of substances continues as long as a gradient exists, osmosis involves other factors that determine whether net movement will continue until equilibrium is reached.

Osmotic pressure refers to the force that must be applied to stop the osmotic movement of water across a membrane. Osmotic pressure is generated when a cell receiving the osmotic flow of water can

no longer expand, as would happen in a plant cell whose tough cellulose walls would prevent unlimited expansion. If we could place a miniature version of a tonometer (the instrument used in measuring fluid pressure in the eyeball) against the cell wall and record the pressure within an expanding cell, we would have the osmotic pressure in that cell. Osmotic pressure varies, as you might suspect, with the solute concentration in the cell. A cell with 3% solute concentration would generate greater osmotic pressure than one with a 1% concentration. Another way of looking at it is to note that an increase in osmotic pressure (or any kind of pressure) also increases the water potential on that side of the membrane. This will oppose the inward flow of water brought about by osmosis.

In summary, then, osmotic systems greatly influence the movement of water by virtue of the semipermeable nature of membranes and the solutes commonly found in cells. The ability of water in osmotic systems to move from regions of high water potential to lower water potential is influenced by solute concentrations and limited by osmotic pressure. These factors will be more readily understood as we see them at work in cells, so let's turn to osmosis in plant and animal cells, where the impact of these physical forces is of interest to us.

Osmosis and the Cell.   Plant and animal cells are always subject to varying osmotic conditions, and through various adaptations most can exist within some range of these conditions. The two problems confronting both plant and animal cells are the possibilities of water loss and water gain. Plant cells can cope well with water gain by virtue of their tough, resilient cell walls, and, as we will find, they make use of steep water gradients to maintain shape and rigidity in their leaves and young shoots (Figure 5.9). Animal cells however, lack cell walls and are easily disrupted by excess water intake.

Because of their susceptibility to water intake, animal cells under study in the laboratory must be maintained in what is called an **isotonic** environment. An environment is isotonic if the surrounding solution contains the same concentration of water and solutes as does the cell. Red blood cells, for instance, will remain stable and maintain their shape for a time if they are placed in an isotonic 1.0% salt solution. Should these same blood cells be placed in a **hypotonic** environment—in a solution containing relatively *more* water (less solute) than is found within their cytoplasm—water gain through osmosis will be excessive and the cell will literally

5.9 THE ROLE OF TURGOR IN PLANTS

The softer parts of plants, the leaves and young shoots, are held erect by hydrostatic pressure when the plant is in a turgid condition. During this time, the water vacuoles are quite full and water is exerting pressure against the vacuole walls. Should a serious water loss occur, the plant may rapidly lose its turgid condition and the leaves and soft stems will wilt.

burst. The same life-threatening situation exists for the animal-like protists, tiny single-celled organisms living in fresh water. However, they are provided with mechanisms for getting rid of excess water, as we will see when active transport is discussed. At the other extreme, a **hypertonic** environment—one in which the water concentration is lower and the solute concentration higher than that of the cell, water is rapidly lost through osmosis. Such a cell will shrink in what is known as **plasmolysis.** Protist and plant cells are also susceptible to plasmolysis in a hypertonic environment (Figure 5.10). Later we will see how living organisms can exist in the marine environment, roughly a 3.0% solute concentration (Chapter 32).

Plants, as we have mentioned, encourage the uptake of water from what is usually a hypotonic environment. They do this by accumulating high solute concentrations in their large cell vacuoles. This, of course, decreases the water potential there and increases the osmotic potential. The subsequent intake of water creates great **hydrostatic pressure** (water pressure) against their cell walls. This

5.10 OSMOSIS AND THE CELL

Varying the osmotic environment of a plant (**a,b,c**) and an animal (**d,e,f**) cell. In each instance, the effects of an isotonic, hypertonic, and hypotonic environment are seen. Because they lack the restraining and supporting cell wall, plant cells are more affected than animal cells when osmotic conditions in the environment alter. In this figure, the blue dots represent water and the open dots represent solute.

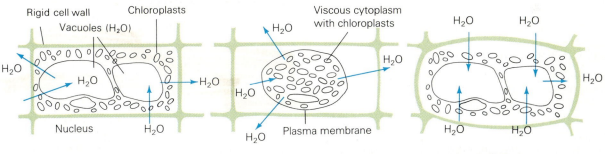

(a) Leaf in normal isotonic environment

(b) Leaf in 3% NaCl solution, which is hypertonic to the cell (loss of turgor)

(c) Leaf in distilled water (hypotonic maximum turgor)

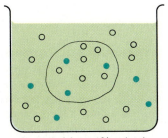

Isotonic (about 1% solute)

(d) Animal cell remains intact

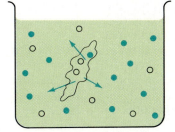

Hypertonic (3% solute)

(e) Animal cell shrinks as water leaves

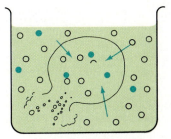

Hypotonic (100% H_2O)

(f) Animal cell swells up and bursts

5.11 MODEL OF FACILITATED DIFFUSION

In facilitated diffusion no cell energy source, such as ATP, is used. It is likely that carrier proteins undergo reversible conformational (shape) changes rather than total movement across the membrane. In this highly schematic, so-called "ping-pong" model, the "pong" conformational state exposes molecule binding sites to the outside. When the sites are filled, the carrier protein changes shape, assuming the "ping" conformation, and exposing the molecules to the cytoplasm, into which they diffuse.

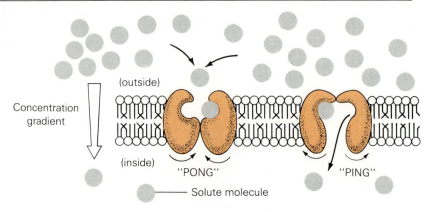

Concentration gradient

(outside)

(inside)

"PONG"   "PING"

——— Solute molecule

pressure partly explains why even spindly tomato plants can stand upright. Botanists refer to the water-swelled condition as a state of **turgor** and often refer to hydrostatic pressure as **turgor pressure.** Turgor is maintained only when the soil water potential exceeds the water potential in the plant cells.

But, if the tomato plants are not watered you may soon see the effects of a *shifting* water potential. As the soil water is lost, the solute concentration in soil water increases, reducing water potential there. (Adding too much dry fertilizer, which forms solutes in the soil water, can create the same effect.) When the water potential in the plant cells, particularly those of the root, is greater than that of the soil, the movement of water through osmosis is reversed, and water exits down the new gradient. In a word, the plant wilts. Fortunately for the plant, the massive plasmolytic effects of wilting are reversible, up to a point, and with timely watering they will again become turgid.

Facilitated Diffusion and the Role of Permeases. Facilitated diffusion, our last example of passive transport, refers to the movement of molecules across a membrane with the assistance of special **carrier molecules** embedded within the membrane. It is similar to simple diffusion in that the energy involved is thermal energy and the net movement of molecules is always from regions of higher concentration to regions of lower concentration. It differs in that in facilitated diffusion only certain kinds of molecules can cross plasma membranes, and these cross much more readily than do similarly sized, similarly charged competing molecules. How does this happen? Facilitated diffusion

is not well understood, but certain carrier molecules, proteins called **permeases** embedded in the membrane, are believed to greatly increase the membrane's permeability to specific substances. Some permeases have even been isolated, but no one knows just how they work. The hypothetical scheme seen in Figure 5.11 shows a permease carrier at work, but the scheme is highly tentative. It relies on the ability of some proteins to change their molecular conformation (tertiary shape).

Active Transport Across Plasma Membranes

So far, the various kinds of movements we have considered are all passive. Passive transport does not involve the expenditure of the chemical energy stored in the molecules (ATP) of cells. Now we will focus on active transport, which requires chemical bond energy. There are many instances in which substances are actively transported across the plasma membrane. The process is characterized by (1) the movement of materials against the existing diffusion gradient and (2) the associated expenditure of ATP energy. Moving molecules against a gradient is a bit like rolling boulders uphill: work has to be done, and energy has to be expended.

Active Transport and Membranal Carriers. The most familiar mechanism of active transport is the **sodium/potassium ion exchange pump** common to many cells, including nerve cells and red blood cells. The sodium/potassium ion exchange pumps continually move sodium ions out of cells and force potassium ions in, opposing what would otherwise be the diffusion or concentration gradient.

As seen in Figure 5.12, the sodium/potassium pump consists of a transmembranal protein capable of conformational changes. The pumping cycle begins as three sodium ions are attracted to a site on the protein, which then interacts with ATP. This interaction triggers the first change in the protein's shape, exposing the sodium ion to the outside, where it then is released. In the second phase two potassium ions from the outside fill their sites, and upon the loss of phosphate by the carrier protein, a second conformational change occurs, dumping the potassium ion inside the cell.

Endocytosis and Exocytosis. Endocytosis and **exocytosis,** as the terms suggest, refer to the incorporation and the expulsion of substances, respectively. They were first observed in the feeding and digestive activity of amebas and other single-celled organisms. They were later seen in certain white blood cells of multicellular animals, as well as other cells. Both endocytosis and exocytosis are testimony to the versatility of the plasma membrane, which plays a highly active role in both processes.

Endocytosis. Endocytosis begins as a depression in the plasma membrane, eventually forming an

5.12 ACTIVE TRANSPORT: THE SODIUM/POTASSIUM EXCHANGE PUMP

Through an energy-requiring process, sodium (Na^+), and potassium (K^+) ions cross the plasma membrane against the diffusion gradient. No one is quite certain how it occurs, but one theoretical mechanism, the sodium/potassium ion exchange pump, is believed to be common to many cells. It works through highly specific, energy-driven, conformational changes in certain active membranal proteins. Four such pumps are seen here, each carrying out part of a four-step process. Each pump has three Na^+ carrying sites and two K^+ sites. In each complete action, three sodium ions are pumped out of the cell and two potassium ions are taken in. Note the minute changes in the carrier sites and the major conformational change when ATP acts. **(a)** The sequence begins with three sodium ions joining the pump. **(b)** Next, energy is provided by ATP (a common source), which brings about a major shape change **(c)** resulting in the three sodium ions being released outside. At this time, smaller changes permit two potassium ions to join the pump, and a final change **(d)** causes the release of the potassium to the interior of the cell. Following this, the sequence will be repeated.

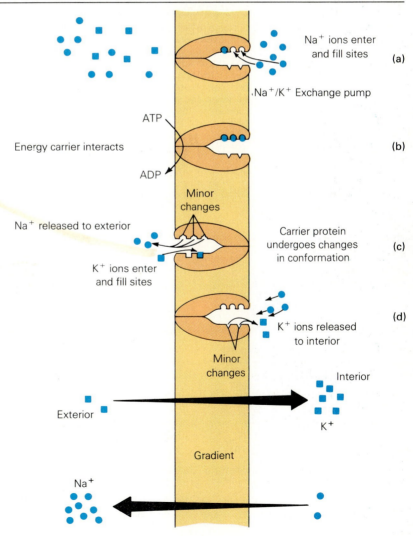

Na^+ ions enter and fill sites **(a)**

Na^+/K^+ Exchange pump

ATP

Energy carrier interacts **(b)**

ADP

Minor changes

Na^+ released to exterior

Carrier protein undergoes changes in conformation **(c)**

K^+ ions enter and fill sites

K^+ ions released to interior **(d)**

Minor changes

Interior

K^+

Exterior

Gradient

Na^+

"impouching" and finally a spherical, saclike vacuole as the pouch deepens and is finally pinched off from the surface. Vacuoles formed this way are actually portions of the plasma membrane turned inside out. Their inner surface is derived from the outer surface of the original plasma membrane. In this peculiar manner, the captured material that was originally *outside* the cell ends up *inside* the cell, within a vacuole, of course. The cell will eventually digest the contents within the vacuole, and the products of digestion will pass, by active transport, through the vacuole membrane into the cell's cytoplasm. We saw how this happens in the last chapter; lysosomes fuse with food vacuoles and release their hydrolytic enzymes, which then digest the contents. If the forming vacuole engulfs visible solid material, the process is called **phagocytosis** (cell eating); if it engulfs only dissolved materials, such as proteins, the process is called **pinocytosis** (cell drinking). Both processes are illustrated in Figure 5.13.

Receptor-Mediated Endocytosis.   More recently, cell biologists have identified another form of endocytosis, one that is **receptor-mediated.** That is, specific receptor molecules in the plasma mem-

---

5.13 ACTIVE TRANSPORT IN AN AMEBA

A very busy ameba demonstrates both endocytosis and exocytosis. At the upper region, (also see photo) it is engulfing a small food particle by phagocytosis (cell eating). At the right, a channel has surrounded a solution of large molecules in a process called pinocytosis (cell drinking). Eventually, both endocytic processes will create a vacuole from the plasma membrane. At the lower side of the ameba, the undigested residue from a food vacuole is being expelled by exocytosis. Actually, the food vacuoles merge or fuse their membranes with that of the cell, so there is never a time when the cytoplasm is exposed to the vacuole's toxic contents. The two insets (above and below) reveal some of the details of endocytosis and exocytosis. In endocytosis a portion of the plasma membrane first impouches, and then closes up, capturing substances from the external medium. The membrane-surrounded sphere and its contents pinch off and move into the cytoplasm as a vacuole. In exocytosis, the vacuole or vesicle reaches the plasma membrane, where its outer lipid layer fuses with the membrane's inner lipid layer. The fused region opens up, and the entire vacuole or vesicle everts—turns inside out, as it were—dumping its contents out of the cell.

Endocytosis

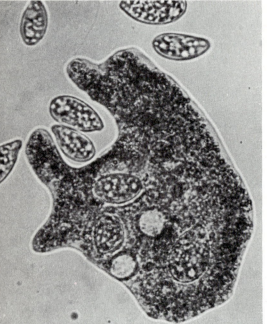

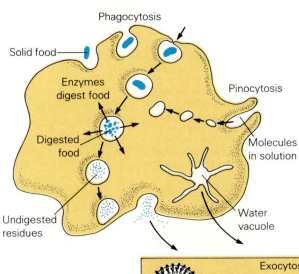

Phagocytosis

Solid food

Enzymes digest food

Pinocytosis

Digested food

Molecules in solution

Undigested residues

Water vacuole

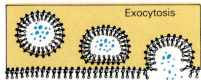

Exocytosis

brane bind with matching substances from the external environment for transport into the cell. As more and more sites are filled, a surface indentation forms, followed by a complete inpouching, and finally separation of the spherical, membrane-bound droplet, which takes on the characteristics of a storage vesicle. The cell then acts upon the material taken in, generally through the interaction of the newly formed vesicle with a lysosome. But things can get somewhat more complicated, as the scenario in Figure 5.14 suggests. There we find that the vesicle produces a new region of membrane, which pinches off from the old receptor-bearing portion. The latter then moves back to the plasma membrane, rejoining it to act again in receptor-mediated endocytosis. We should point out here that some authorities consider receptor-mediated endocytosis to be responsible for many instances of pinocytosis.

Receptor-mediated endocytosis was first reported in the early 1960s when Thomas Roth and Keith Porter determined that this process helped bring protein into birds' eggs and into the immature eggs (oocytes) of the mosquito for storage. In the latter, they identified 300,000 pitted sites on the oocyte surface, each of which was capable of forming a protein-laden vesicle. Other substances entering the cell through this process include cholesterol, iron, hormones, and mammalian antibodies being transported from the mother to the fetus late in fetal life. Workers at the Salk Institute, Stanford University, and the University of Colorado have succeeded in identifying a specific glycoprotein receptor responsible for moving **transferrin,** an iron-containing protein (see Figure 5.14). The receptor is a paired strand of amino acids with a total molecular weight of about 180,000, with 800 amino acids in each strand. The receptor bridges the membrane and contains two transferrin-capturing active sites that face the outer side.

Actually, most eukaryotic cells are constantly involved in endocytosis. For the most part this includes pinocytosis as cells take in larger molecular entities, but phagocytosis is also common. If you think carefully about the effects of endocytosis, you may recognize a problem: cells constantly lose bits of membrane that end up forming vacuoles. How much do they lose? Some of the larger phagocytic white blood cells are so heavily involved in endocytosis that they can literally use up their plasma membrane in about one hour. However, such cells retain a stable volume and surface in spite of this loss. Obviously, plasma membrane is replaced as fast as it is used. Typically we find that much of the

5.14   RECEPTOR-MEDIATED ENDOCYTOSIS

In receptor-mediated endocytosis, receptor sites on the plasma membrane fill with substances to be transported inward. As filling occurs, vesicles form, transporting the substance into the cell where they are often acted upon by enzymes from lysosomes. During their transit through the cytoplasm, the receptor bearing portion of the vesicular membrane separates and recycles, joining the plasma membrane to act once again in their unique transport function. The material laden vesicle then fuses with a lysosome containing hydrolyzing enzymes. Following this, the materials will be digested and released into the cytoplasm.

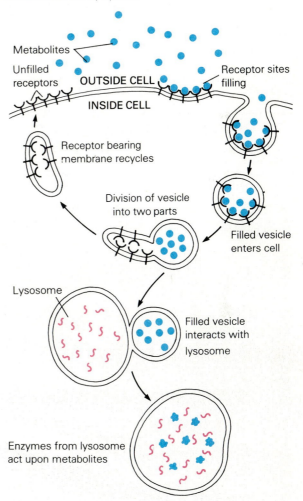

Metabolites

Unfilled receptors

OUTSIDE CELL

Receptor sites filling

INSIDE CELL

Receptor bearing membrane recycles

Division of vesicle into two parts

Filled vesicle enters cell

Lysosome

Filled vesicle interacts with lysosome

Enzymes from lysosome act upon metabolites

plasma membrane lost through endocytosis is constantly restored through the opposite process of exocytosis. And as we have just seen, the plasma membrane is rapidly recycled during receptor-mediated endocytosis.

Exocytosis.   In exocytosis, materials are expelled from the cell. While this is the first time we have encountered the specific term, we discussed several instances of exocytosis in the last chapter. For example, it is through exocytosis that the Golgi vesicles expel their contents outside secretory cells. For exocytosis to occur, a vacuole containing such material must fuse with the plasma membrane. The fusion is a complex and little-understood process, but some studies reveal that contractile microfilaments are involved. They apparently become attached to the vacuole and the membrane and draw them together. Then, membranal elements of the two fuse, the site of contact opens up, and as the vacuole is turned inside out, its contents are deposited outside the cell. As discussed in Chapter 4, this very process is used in emptying the contents of the large Golgi secretion vesicles into the surrounding ducts.

Water Transport by Contractile Vacuoles

As a final example of active transport, we will consider the **contractile vacuole,** a water-regulatory device found in protists. We mentioned earlier that single-celled organisms in the hypotonic freshwater environment have an ongoing problem with water uptake. The problem must be solved in special ways because there are no known instances of membranal carriers transporting water out of the cell. The freshwater protozoan *Paramecium* (Figure 5.15) for example, has more or less permanent water vacuoles specializing in the expulsion of water. While the process is not very well understood, it is apparent that bundles of microfilaments

5.15 ACTIVE TRANSPORT AND THE CONTRACTILE VACUOLE

(a) Contractile vacuoles are readily seen in protists such as *Paramecium,* a complex ciliate (cilia bearer). **(b)** A contractile vacuole fills and its membrane fuses with the plasma membrane, which then ruptures, releasing its contents. Microfila-

ments, contractile filaments composed of the protein actin, assist in the filling and emptying process. As you might suspect, this requires the expenditure of energy.

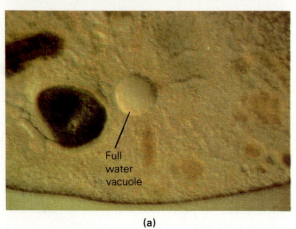
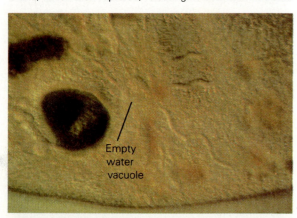

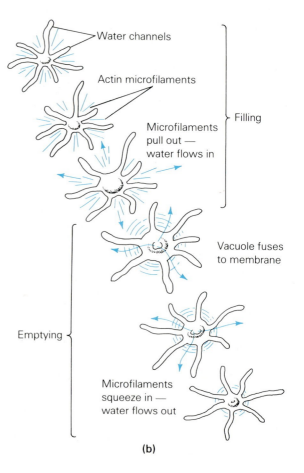

(a)

(b)

first pull on the water vacuoles, enlarging them and reducing the pressure within them. Water, following the pressure gradient, enters and fills the vacuole. Then other microfilaments contract, squeezing the vacuole and suddenly reversing the pressure. This forces the water out through a pore to the exterior. Both filling and emptying require the expenditure of ATP.

Auxiliary Mechanisms and Structures of Transport

A number of processes may aid the transport of materials into and out of cells. We have portrayed the cytoplasm as a relatively passive entity in transport, but in reality the cytoplasm may move about quite actively and help distribute materials. Further, we find that certain groups of cells in plants and in many fungi communicate freely through special junctions in their walls where the cytoplasm of one cell is continuous with the next. Let's see how these conditions can facilitate active transport.

Cytoplasmic Streaming.   The movement of cytoplasm, known as **cytoplasmic streaming,** or **cyclosis,** is a well-documented but largely unexplained phenomenon. It is easily seen in some plant cells, where the cytoplasm circulates around the cell, commonly around a large central vacuole, carrying along with it chloroplasts, mitochondria, and other organelles. Cytoplasmic streaming has also been observed in the cytoplasm of certain protists, where food vacuoles are seen to move in apparently predetermined paths. It is also common in the amebas, where, it turns out, the ability of the cytoplasm to flow enables them to move about and capture prey. In our own lengthy nerve cells, and those of other animals, materials produced by the cells are rapidly transported within thin, tubelike extensions of nerve cells that reach distant parts of the body.

Transport and Cell Junctions.   If we have left the impression that cells in multicellular organisms tend to be isolated from each other, let's correct that now. Communicating channels of several kinds have been found between the closely packed cells of multicellular tissues. Included are the intricate **gap junctions** of animal cells and the slender **plasmodesmata** of plant cells, both of which are very common.

Gap junctions are formed by rosettes of protein that firmly hold adjacent membranes together but at the same time form simple, minute pipelines or tubes that make the cytoplasm of adjacent cells effectively continuous (Figure 5.16). The passages are apparently about 200 nanometers in diameter and will freely admit substances up to a molecular weight of 1000 daltons. This would include most ions, water, amino acids, sugars, vitamins, hormones, and even nucleotides. Gap junctions are believed to be important to heart muscle, where they facilitate the spread of electrical waves, and in the embryo, where they help with the passage of materials through layers of cells.

The protein-lined channels of gap junctions are not simply passive openings but are known to disjoin from the neighboring cells if an injury occurs. This effectively isolates the damaged cell, preventing the loss of essential materials from the continuous cytoplasm. The researchers who first observed this phenomenon also determined that it was probably an influx of calcium ions from the outside that caused the uncoupling. Calcium ions are generally in short supply in the cytoplasm of cells, but when these ions are injected into the cell, the junctions are rapidly uncoupled. Eventually, a cell treated in this manner will recover, isolating the calcium ions in its mitochondria, whereupon the gap junctions will be reestablished.

Junctions known as plasmodesmata (singular, *plasmodesma*) have been recognized in multicellular plants for some time. In fact, there are very few plant cells that do not communicate through these fine channels. Plasmodesmata are fingerlike cytoplasmic extensions that pass from cell to cell through plasma membrane-lined channels in adjacent cell walls (Figure 5.17). The diameter of plasmodesmata ranges between 20 and 40 nm. Also extending through the porelike openings in adjacent walls are elements of the endoplasmic reticulum (ER) of the two cells. They form slender channels, the **desmotubules,** through the pores, which means that the lumen of the ER in the two cells is also continuous. Researchers have discovered thickenings at either end of the plasmodesmata and have suggested some valvelike role, permitting some control over the passage of cytoplasmic materials between adjacent cells.

If interpretations of the plasmodesmal structure are correct, then it is certain that these cytoplasmic connections are important in intercellular transport. They may be especially significant to the passive processes of diffusion and bulk flow, facilitating the passage of water and other materials through dense tissues and over great distances. One theory of plant transport proposes that the process of cytoplasmic streaming and the presence of plasmodesmata may explain how sugar solutions are rapidly transported through living cells in plants. The abundance of plasmodesmata in secretory cells provides further indirect evidence of their transport function.

Creative plant physiologists have also gathered evidence of a more direct nature. For example, when certain dyes that cannot normally cross the plasma membrane are introduced into plant tissues containing plasmodesmata, the dyes spread rapidly from cell to cell. Further, tissues whose cells abound with plasmodesmata will conduct electrical currents more readily and with less loss than those without plasmodesmata. Actually, plasmodesmata, like their counterparts in animals, the gap junctions, are not entirely passive but apparently have considerable control over what passes through. For example, in spite of their generally larger diameter, the plasmodesmata will not admit molecules as large as those that pass through the gap junction, apparently being limited to those lighter than 800 daltons. Nevertheless, the combined effect of countless plasmodesmata in the plant body is to create a virtually continuous cytoplasm, thereby greatly facilitating coordination of the plant's activities.

5.16 INTERCELLULAR COMMUNICATION

(a) Model of a *gap junction,* an organelle that allows the *direct* exchange of nutrients and intracellular hormones through channels that pass between cells. Each channel is created by a pair of "pipes," each pipe consisting of six dumbbell-shaped protein subunits. (b) Gap junctions make the cytoplasm of many multicellular tissues effectively continuous. (c) If some cells are injured, however, the gap junctions quickly seal off the wound, possibly in response to materials released by rupturing cytoplasmic lysosomes. (d) The injured cells are hydrolyzed (digested) by the free lysosomal enzymes. (e) Contact between the surviving cells is reestablished within 30 minutes.

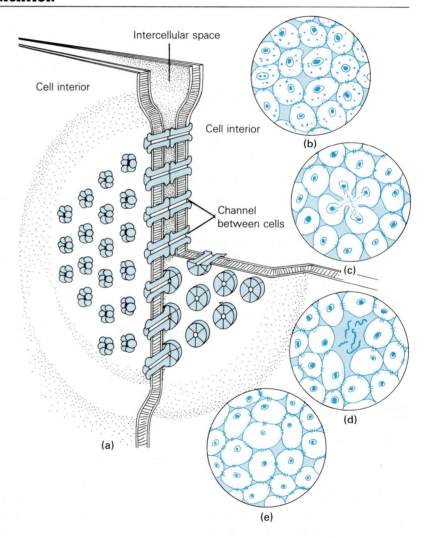

Intercellular space

Cell interior

Cell interior

Channel between cells

(a)

(b)

(c)

(d)

(e)

5.17 PLASMODESMATA

(a) The presence of plasmodesmata in so many types of plant tissue makes much of the plant's cytoplasm effectively continuous, facilitating the transport of water and nutrients and aiding in other necessary intercellular communication. (b) In a closer view plasmodesmata appear as pores in the cell wall.

Passing through each pore are the plasma membranes of the adjacent cells, the cytoplasm, and desmotubules (extensions of the endoplasmic reticulum). The electron micrographs show plasmodesmata in cross section (c) and longitudinal section (d).

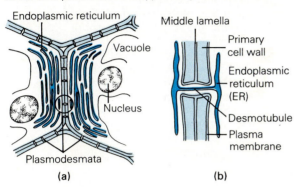

Endoplasmic reticulum
Vacuole
Nucleus
Plasmodesmata

(a)

Middle lamella
Primary cell wall
Endoplasmic reticulum (ER)
Desmotubule
Plasma membrane

(b)

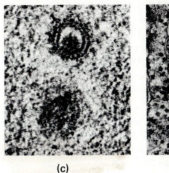

(c)

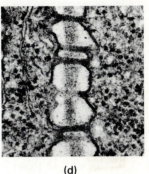

(d)

APPLICATION OF IDEAS

1. It was once thought that frogs could not survive in distilled water—that is, until someone actually kept some in a pan of distilled water for months without ill effect. Theoretically, what should have happened to the frogs? Why do you think it did not?

2. Early methods of preserving food included drying, pickling, smoking, salting, and sugar-curing. Offer a physiological explanation for the failure of spoilage bacteria and molds to grow readily on food treated in these manners.

3. In an experiment, the leaves of the water plant *Anacharis* are treated in several ways and observed under the light microscope. The cells are normally rectangular in shape, with a large central vacuole and a thin layer of chloroplast-containing cytoplasm between the vacuole and plasma membrane. Explain each microscopic observation and compare it with the others.
   a. Leaves in pond water show no change.
   b. Leaves in distilled water show no change.
   c. Leaves in 1% and 2% glucose solutions show no change, but in 3% glucose, the membrane-surrounded chloroplasts form a tight sphere near the cell center.
   d. Leaves in 1% NaCl solution show no change, but those in a 1.5% NaCl solution resemble those in the 3% glucose solution.

KEY WORDS AND IDEAS

THE PLASMA MEMBRANE

1. The presence of a plasma membrane and its specific structure at first were only theorized on the basis of permeability studies.

2. The earliest models of the plasma membrane included a double layer of phospholipids and a covering of protein on either side.

3. The sandwichlike structure of the plasma membrane was confirmed by early electron microscope studies.

The Fluid Mosaic Model

1. In the **fluid mosaic model** of the plasma membrane the bilayer of phospholipids of earlier models is retained, but the arrangement of

membrane proteins is much different. The two layers of phospholipids are arranged with their polar heads outward and their nonpolar tails intermingling. Some proteins are **transmembranal proteins,** passing through the entire membrane, some are **peripheral proteins,** partly submerged in the lipid core, and others, the **surface proteins,** appear on the surface only.

2. The function of the oily core is to restrict the movement of water and other polar molecules. The proteins serve several functions, including actively transporting substances in and out, recognition and binding with other cells, recognition of messenger molecules, and activation of the cell.

3. Other surface components include **glycoproteins** and **glycolipids,** which make up the **glycocalyx.** Included among these are cell-surface carbohydrates such as those that determine blood type, along with others involved in cell recognition.

MECHANISMS OF TRANSPORT

1. Membranes that admit some substances and not others are **selectively permeable.**

2. Factors influencing the membrane's **permeability** include particle size and electrical charges, if any.

3. There are two known mechanisms of transport. **Passive transport** does not require ATP energy output by the cell while **active transport** does.

Passive Transport

1. **Diffusion** is the net movement of molecules or ions from areas of their greater concentration to areas of their lesser concentration.

2. Molecule or ion movement never ceases, but diffusion ends when a net movement in one direction stops. At this time, the particles are said to be in **equilibrium.**

3. The energy of diffusion is in the molecules or ions themselves.

4. Substances in diffusion are said to move *down the gradient* or **concentration gradient.** Such gradients exist wherever the distribution of molecules or ions is unequal.

5. The movement of water down its gradient is referred to as **bulk flow,** or in special circumstances, as **osmosis.**

6. Bulk flow occurs because of differences in potential energy as would occur if gravity or pressure were involved. Such energy is often referred to as **water potential.**

7. Osmosis is the movement of water across a membrane from an area of greater concentration to one of lesser. **Osmotic potential** is the tendency or capability for such movement.

8. **Osmotic pressure** is a measure of the force that must be applied to stop the osmotic movement of water across a membrane.

9. The net movement of water in and out of cells through osmosis is determined by the comparative amounts of water and solutes on either side of the plasma membrane
   a. when the solutes are equal, the system is **isotonic** and no net movement of water occurs;
   b. when the concentration of water outside is greater (and solutes less) than inside, the surroundings are **hypotonic** and water will move in, bursting the cell;
   c. when the concentration of water outside is less (and solutes greater) than inside, the surroundings are **hypertonic** and water will move out, causing **plasmolysis.**

10. When plant cells fill with water, **hydrostatic pressure** and a state of firmness called **turgor** are produced. A loss of hydrostatic pressure results in wilting.

Facilitated Diffusion and the Role of Permeases

1. In facilitated diffusion, no ATP energy is used, but special membrane **carrier molecules,** or **permeases,** speed the passage of specific molecules or ions through the membrane.

Active Transport Across Plasma Membranes

1. In active transport, materials move against an established gradient and ATP is consumed.

2. In the **sodium/potassium ion exchange pump,** a transmembranal protein undergoes energy-driven conformational (shape) changes, carrying sodium ions into the cell and potassium ions out. ATP is converted to ADP and P_i.

3. In **endocytosis,** bulk substances are taken in by an inpouching portion of the plasma membrane which forms a vacuole. Solid material is taken in by **phagocytosis,** and fluids are taken in by **pinocytosis.**

4. In **receptor-mediated endocytosis,** receptor molecules in the plasma membrane are activated by a substance, and when enough molecules react, a vacuole forms and the material enters the cell. One of the suspected receptors is **transferrin,** an iron-containing protein.

5. In **exocytosis,** a cellular vacuole or vesicle attaches to the cell membrane, whereupon the two membranes fuse and the materials are expelled from the cell.

Water Transport by Contractile Vacuoles

1. Many protists form **contractile vacuoles,** which fill periodically with cytoplasmic water, and then, with the aid of microfilaments, contract, forcing their contents outside.

Auxiliary Mechanisms and Structures of Transport

1. **Cytoplasmic streaming,** or **cyclosis,** aids in the distribution of materials within cells.

2. **Gap junctions,** protein-lined pores between adjacent animal cells, permit the movement of materials between cells.

3. **Plasmodesmata,** slender strands of cytoplasm that extend through pores in adjacent plant cell walls, help in the intracellular transport of materials. The pores are lined by the plasma membrane which is continuous from one cell to the next; portions of the ER—the **desmotubules**—also communicate between cells.

4. Plasmodesmata may be involved in the rapid transport of sugars and water through the plant body.

REVIEW QUESTIONS

1. Prepare a simple drawing of the fluid mosaic model of the plasma membrane. Label the phospholipids and the surface, transmembranal, and peripheral proteins. (106–107)

2. Carefully explain the significance of the phospholipid arrangement and that of the transmembranal proteins. (106)

3. List four functions of membranal proteins. (106–107)

4. What molecules occur in the glycocalyx and what are their functions? (107–108)

5. Why is it that molecules such as glycerol, oxygen, carbon dioxide, and steroid hormones pass readily through the plasma membrane, while glucose and many essential ions have difficulty? (108)

6. Specify two significant differences between passive and active transport. (109, 114)

7. Write a precise definition of the term *diffusion* and, using an example, explain how it works in cells. (109–110)

8. Under what conditions will diffusion begin? When is it over? Why is the term "net" included in discussions of molecular movement in diffusion? (110)

9. What is the energy source behind the bulk flow of water? Cite two examples of bulk flow. (111)

10. The term *osmosis* was coined to cover a special case of diffusion. Explain what this case is. (111)

11. Explain what would happen if cells averaging about 1% solute concentration were placed in a 3% salt solution. What term describes the osmotic conditions in the cell? Outside the cell? (112)

12. Distinguish between the terms *permeable, impermeable,* and *selectively permeable.* (108)

13. Explain why microorganisms might fail to grow on food that has been preserved by salting. (112–113)

14. In what way do plants rely on high hydrostatic or turgor pressure in their cells? (113–114)

15. In what way does facilitated diffusion differ from ordinary diffusion? (114)

16. Explain how a sodium/potassium ion exchange pump works. (115)

17. What is the role of the plasma membrane in endocytosis? How does this differ from its role in exocytosis? (115–116)

18. How does phagocytosis differ from pinocytosis? (116)

19. Explain what *receptor-mediated endocytosis* is. (116–117)

20. What problem does the *Paramecium* solve through the operation of its water vacuoles? How do the vacuoles function? (118–119)

21. What are gap junctions? In what way do they serve the animal cell? (119–120)

22. Plasmodesmata are common in plant cells. Fully describe their structure. What important questions about transport might they help answer? (119–121)

Cell Energetics

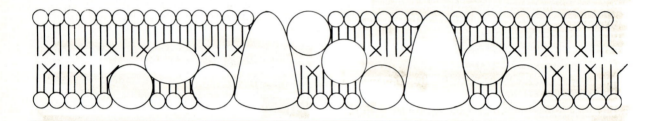

Flowering plants growing peacefully on a flower-strewn hillside are often engaged in a battle for life or death. They quietly fight for survival as they compete with other plants for the sun's energy and the water and minerals in the soil below. But success itself can spell new problems. Once a plant has successfully stored energy in the food it makes, it may suddenly have to yield its hard-won gains to some voracious plant eater who rearranges the plant's molecules and energy to suit its own needs. The plant eaters, in turn, often fall to sharp-toothed predators, bent on fulfilling their own requirements (Figure 6.1). This can, of course, go on, as the links in the chain of life are set into place. But the chain is not endless. As far as life is concerned, the flow of energy will cease when the microbes of decay have their way. In this, the final episode, simple life forms extract the last remnants of useful energy as they rearrange the molecules of plant and animal corpses.

The name of life's game, then, is energy. Energy begins and ends in the cells of organisms and this is where we will now direct our attention. But as we go along, we should not lose sight of the "big picture." After all, these various shifts of molecules and energy are but a part of the grand scheme of life, providing the basis for the grander drama. The subject of energy is indeed immense, and a good place to begin is with the word itself.

ENERGY

What is energy? What are its characteristics and how does it behave? Like life, energy is an elusive concept; in fact, we can only perceive its existence through its effect on matter. Nonetheless, let's begin with a functional, if somewhat creaky, definition. Put in its simplest terms **energy** *is the ability to do work.* Defining "work" can get complicated, but if you can find a physicist in a good mood, you might hear something about "the transfer of energy from one system to another," or "the transfer of energy by applying force"—just before the inevitable equations begin to bog you down. Since the question can easily fall into the "sorry I asked" category, let's move from defining to describing, an approach with which we can all feel more comfortable.

Energy—the ability to do work—occurs in two states, **potential energy** and **kinetic energy.** The first really defines itself, implying that energy is "stored" and not doing anything. For example, the five-ton boulder on the hill above your house represents a considerable store of potential energy. It was raised to that level in some forgotten time when geological forces carried out a great deal of work to get it there. Should something happen to the soil holding the rock, you may quickly come to grasp the difference between potential and kinetic energy.

Another term for potential energy, one we will use frequently, is **free energy.** The term is usually applied to the potential energy in the chemical bonds of molecules, which, like the rock, are at rest and not reacting. Kinetic energy, as you probably have determined by now, is energy of motion. We know this because of its effect on matter. Not only does the boulder leave a trail through the brush as its potential energy becomes kinetic, but it might have a substantial effect on the matter that is your house.

Forms of Energy

Energy can take various forms. Thus, we speak of chemical bond energy, electrical energy, magnetic energy, mechanical energy, and radiant energy. The latter includes the familiar electromagnetic radiations, such as heat, light, ultraviolet rays, and so on. We will soon see that all of these forms of energy are interchangeable; that is, under certain conditions, one form of energy can become another.

The behavior of energy has been greatly simplified into time-honored principles known as the **laws of thermodynamics.** Let's see how these laws work and how they apply to living things.

The Laws of Thermodynamics

There are certain concepts in science that appear to hold up no matter how often they are tested. We call these ideas *laws,* and we like to think that these natural laws can't be broken under any circum-

6.1 ENERGY TRANSFER

In the intricate webs of life, the passage of energy from organism to organism can at times be violent.

stances. Biology, because of the almost infinite variety of life, is actually rather short of laws. But living systems are made of matter that interacts chemically and physically, and the laws of physics and chemistry are as inviolable in living systems as anywhere else. The laws of thermodynamics certainly fit into this category. We will confine our discussion to the first and second laws, those most relevant to biology.

The First Law. The first law of thermodynamics states that *energy can neither be created nor destroyed.* It is also called "the Law of Conservation of Energy." What this means is that the total amount of energy present in a system remains constant. Let's qualify that by making it an imaginary, **closed system**—that is, one in which matter and energy can neither enter nor leave.*

Energy can, as we have mentioned, change form, and it does so in an often bewildering variety of ways. Energy in the chemical bonds of gasoline, when combined with oxygen and a spark of electricity inside a lawnmower engine, can be transformed into heat and motion. Not only has the potential energy in the chemical bonds of gasoline molecules become kinetic energy as the bonds break, but the sudden release of heat expands gases in the cylinder, moving the piston. The energy conversions go from chemical bond energy to heat energy, and then to mechanical energy as the piston moves. If the lawnmower were in a closed system, all of the original chemical bond energy released could finally be detected as heat. But then, what happens to heat? The answer falls into the province of the second law.

The Second Law. The second law of thermodynamics (sometimes called *the law of entropy*) states that *the free energy of a system is always decreasing.* In other words, systems tend to move toward randomness. Thus, highly organized systems, those with a complex, organized molecular nature and great free energy, move toward a simple, random molecular organization and a decrease in free energy. The technical term describing the disorganized state is **entropy.** Actually, entropy is a measure of the disorder or randomness in any system.

---

*Closed systems—those in which matter cannot enter or leave—do not really exist (except perhaps for the universe), but are contrived as models by scientists who wish to test their ideas under hypothetical conditions that can be limited and controlled.

A state of entropy can be found in the fuel we expended in the lawnmower. Gasoline molecules are long, combustible, hydrocarbon chains, so a tank full of gasoline represents both a high degree of molecular organization and great free energy. But in the end, the hydrocarbons, having reacted with oxygen, are broken down into simple carbon dioxide and water. The great free energy store has been dissipated as heat. The energy didn't disappear, it just became evenly distributed in the much simpler system. One can say, then, that both matter and energy reached a state of maximum entropy. By the way, it should be apparent that there is no way to organize the dissipated heat into potential energy again. Entropy, in this situation, is irreversible.

If doting on entropy seems pessimistic, we can note that the second law does have its more cheerful aspects. For example, in order for the lawnmower to run, the ordered gasoline molecules *must* move to a lower energy state, releasing their free energy as they do. (At least you didn't have to use a push-mower, thereby increasing your own state of entropy.) Work can be done only when entropy

increases and available free energy diminishes. This brings up a vital point about chemical reactions in general. *No chemical transfer of energy is totally efficient;* some energy is always lost as entropy increases. But since it is the tendency for molecules to change from higher to lower energy states that makes the reaction happen in the first place, it is essential that some energy be lost.

Entropy and Life

We will relate the laws of thermodynamics to life by posing two questions. First, if the trend toward maximum entropy is real, how can one explain the great molecular complexity of life? Second, is life on earth headed toward disorganization and energy dissipation? The answer to the second question is no, and the answer to the first explains why. The laws of thermodynamics apply to closed systems. And while the universe may be a closed system, the earth is not. The earth receives a constant input of free energy, in the form of sunlight. And as long as energy reaches the earth, living things can be expected to use that energy to decrease entropy. As molecules reach simple and disorganized states, energy will be expended to continuously reorganize them into the complex and ordered systems of life (Figure 6.2). So, it turns out that the second law of thermodynamics will have to wait another 10 billion years or so, when the sun is scheduled to burn out, before the earth's teeming life is reduced to simple molecules and heat.

6.2 LIFE AND FREE ENERGY

An organism can only resist the trend toward entropy through the constant intake and harnessing of energy. Once this ceases, simplification and disorganization begins until the molecules and energy that once defined its life become randomly distributed.

CHEMICAL REACTIONS AND ENERGY STATES

We often find very strange and dramatic examples of the laws of thermodynamics at work. Consider the case of the airship *Hindenburg.*

The *Hindenburg,* the most famous of the giant zeppelins, was, unfortunately, filled with molecular hydrogen gas. Back in the 1930s, Germany's Third Reich intended to impress America with its hydrogen-bloated airship. It indeed impressed America. It blew up. Somehow it caught fire while landing in New Jersey, and the hydrogen gas combined violently with the molecular oxygen of the atmosphere to produce water. The rest is history (Figure 6.3).

But why was the water formed so explosively? It was a matter of energy states. Water (H_2O) is in a lower energy state than an equivalent amount of H_2 and O_2. We know this because a mixture of hydro-

6.3 THE *HINDENBURG* DISASTER

On May 6, 1937, the hydrogen-filled zeppelin *Hindenburg* exploded during its approach to the mooring site at Lakehurst, New Jersey. We will never know the actual cause of the disaster, but the vast quantities of pure hydrogen gas required to keep the airship aloft represented an immense source of free energy—just waiting for a spark. The spark may well have come from an atmospheric discharge, igniting a minor hydrogen leak. The heat and light of the explosion represent excess energy released, as hydrogen and oxygen atoms took part in an immense chain reaction to form water. Whatever the chemistry of the *Hindenburg* disaster, the lives of 36 passengers and crew were suddenly snuffed out. Surprisingly, 61 survived to tell about their experience. The loss of the *Hindenburg,* the world's largest lighter-than-air ship, and other similar disasters spelled the end of an era.

gen gas and oxygen gas confined in a space and ignited by a spark will release heat. More specifically, if one mole (see Chapter 2) of H_2 and a half mole of O_2 (the quantities needed to produce water with no hydrogen or oxygen left over) are placed in a device called a calorie bomb or calorimeter, and a spark is added, one mole of water (actually steam) will be produced and a heat rise of just about 58 kilocalories (kcal) will occur. This reminds us that the mole of water thus produced contains less free energy than the mix of hydrogen and oxygen gases. On a gargantuan scale, the exploding *Hindenburg* illustrated this same principle. When the *Hindenburg* burned, the chemical mix went quickly to a lower free energy state, producing low-energy water and releasing the excess energy as an enormous fireball of heat and light.

The molecular reaction of the tragedy may be written as:

$$2H_2 + O_2 \rightarrow 2H_2O + Energy$$

If you try to impress your friends by mixing a little oxygen with hydrogen before their very eyes, however, they will be totally unimpressed. Nothing will happen; the molecular hydrogen and molecular oxygen will just sit there with their outer electron shells nicely filled. What now? Perhaps a spark of intuition.

Heat of Activation

It turns out that with a tiny spark, the mixture will blow their hats off. But why? Why didn't the mixture blow up without the spark? The reason molecular oxygen and molecular hydrogen don't react with each other at room temperature is that before they can, the atoms of each molecule must first be forced apart. The H_2 will then become 2H, and the O_2 will become 2O. This frees the electrons that were tied up in the covalent bonds, permitting individual hydrogen and oxygen atoms to join, forming water. The spark provides the intense amount of energy required to bring such events about. This energy is referred to as the **heat of activation,** or **energy of activation.**

The heat of activation must exceed the bond energy that originally held the molecules together. Actually, the spark, or heat of activation, only gets things started. Once the first molecules are disrupted, the energy liberated sets off a chain reaction, an avalanche of reactions; we might say the system catches fire. As mentioned, $2H_2O$ is in a lower energy state than a mix of $2H_2 + O_2$, so as that system goes to its lowest energy state, the energy difference will be revealed as heat and light. The heat and light released in the *Hindenburg* disaster was considerable.

The explosive formation of water from H_2 and O_2 is a good example of an **exergonic reaction,** one that releases energy. The sudden release of energy in the *Hindenburg* disaster dramatizes what can happen as such a reaction proceeds. But reactions can also be **endergonic,** that is, they can absorb energy. For example, when two reactive hydrogen atoms (2H) form the hydrogen molecule (H_2), energy is actually absorbed into the H—H covalent bond. Thus, an endergonic chemical reaction is one in which the product ends up with more free energy than its constituents had to begin with. This might seem to be a violation of the second law, but it really isn't. Endergonic reactions proceed at the expense of some other source of energy greater than that in the new bond. Many of the synthetic (molecule-building) reactions in cells fall into this category.

We have seen that chemical reactions can be quite violent when the energy states of molecules are disturbed abruptly. Although supplying the heat of activation is a common way to stimulate chemical reactions, there are quieter ways as well.

6.4 ACTIVATION ENERGY

The graph compares the activation energy requirement for a reaction with and without an enzyme. The reactant (A) is stable at ordinary temperatures. To form the product (B), the activation energy barrier must be overcome. In the absence of enzymes, heat must be applied, but in their presence, the reaction can occur at considerably lower temperatures.

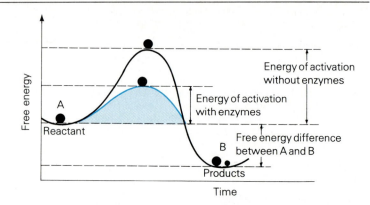

Quieter Changes in Energy States. Hydrogen and oxygen will combine to form water at room temperature if a **catalyst** is present. Catalysts are substances that provide shortcuts between the higher and lower energy states of a molecule, greatly lowering the heat of activation required to get things started. For instance, hydrogen and oxygen will combine readily, even at room temperature, if powdered platinum is present. The hydrogen first combines with the platinum and then with the oxygen, leaving the platinum in its original state; by definition, catalysts themselves are not changed by the reactions they facilitate. Almost all new cars in the United States have platinum catalyst devices in their exhaust systems so that smog-producing unburned gasoline, other hydrocarbons, and carbon monoxide can be combined with oxygen to form harmless carbon dioxide and water vapor. In biological systems, catalysts are a special class of proteins called **enzymes.**

Enzymes are so vital to chemical reactions in the cell that life without them is difficult to imagine. We have mentioned them several times before, but now is the time for a more serious look.

ENZYMES: BIOLOGICAL CATALYSTS

If you were to leave hydrogen and oxygen gases together, some of them would randomly unite to form water, but not to any significant degree. Nevertheless, the process would go on. A piece of wood left on the forest floor will gradually react with oxygen from the air, but not as fast as it would in a forest fire. In both instances, a spark is necessary to speed up processes that would have occurred anyway. However, in a structure as delicate as a cell, one cannot expect to find sufficient heat to initiate biochemical processes, and the cell can't wait for them to occur slowly over time. Thus the cell's activities are greatly dependent on the presence of enzymes.

Enzymes, the catalysts of living cells, are a special class of proteins that are biologically active. This means that they actively participate in cellular activities, specifically by acting as catalysts that accelerate chemical reactions. There are a great number of enzymes in any cell's biochemical arsenal, each of which functions in only one kind of biochemical reaction; that is, each enzyme reacts only with its specific **substrate** (any substance altered by an enzyme).

Thus, from the beginning we should be aware of three important characteristics of enzymes: (1) they are proteins, (2) they are very specific in their reactions, and (3) they do not require substantial heat to bring about chemical reactions. Figure 6.4 compares the energy of activation required in an enzyme-catalyzed reaction with that needed in a typical heat-activated reaction.

A Matter of Shape

Proteins make efficient enzyme catalysts because their shapes are very precise. A crucial part of any enzyme is its active site, which is usually a groove or depression on the surface of the protein molecule. The shape of the active site and the configuration of bonding forces present there cause the enzyme to interact specifically with a certain substrate and no other. The complex of enzyme and substrate is called, appropriately enough, the **enzyme-substrate complex,** or simply **ES.**

In 1890, the famed chemist Emil Fischer described the structural relationship between enzyme and substrate as a "lock-and-key" arrangement. According to his analogy, the unique way a lock and its matching key fit together are similar to the specificity between an enzyme and its substrate. This concept has served well over the years, but like almost all such early simplifications, it has failed in the face of newer knowledge. For example, we now know that enzymes are not the rigid structures they were once thought to be. Further, unlike a lock and its key, the active site of an enzyme does not exactly fit its substrate molecule. Biochemists now propose the **induced fit hypothesis,** which states that as the substrate binds to the enzyme's active site, the site changes to fit the substrate. Thus, the recognition process between enzyme and its matching subject is quite dynamic. The induced fit hypothesis includes

one more critical factor. When the substrate assumes its somewhat strained position to fit the active site of the enzyme, it becomes stressed. This stress helps in the chemical disruption of specific bonds. At least this seems to be true in hydrolytic reactions, those water-utilizing reactions that cleave the substrate into parts. With all this in mind, let's look more closely at the chemical events.

When the enzyme and substrate are positioned correctly, one or more side groups of the substrate may react directly with one or more amino acid side groups in the enzyme (Figure 6.5). Or, the enzyme may only bring two reactive substrate groups close together so that they can react on their own. The intricacies of enzyme action are quite complex, but they occur with dazzling speed. Enzyme reactions can be summarized as:

$$S \;+\; E \;\rightarrow\; E\text{-}S \;\rightarrow\; P \;+\; E$$

Substrate Enzyme Enzyme- Product Enzyme
                    substrate

6.5 ENZYME ACTION: A HYDROLYTIC CLEAVAGE

(a) Enzymes commonly admit substrate molecules into an active site—a flexible cleft in their molecular conformation. The shape of the cleft is generally very specific for the substrate, but, importantly, the fit between enzyme and substrate is inexact. In this instance, the bond to be broken by the enzyme occurs between R_1 and R_2. Note that one of the amino acid side groups of the enzyme projects a hydrogen into the active site. This hydrogen will participate in the enzymatic reaction. **(b)** The amino acid side group of the enzyme moves to form an energetically favorable but temporary bond with the substrate. It "attacks" and breaks the bond between R_1

and R_2. The hydrogen (H) from the enzyme forms a covalent bond with R_1, while the enzyme covalently bonds with R_2. Next the enzyme "moves," actually forming a new shape that physically divides the substrate into two products. The first product of the reaction is R_1—H, which is no longer bound and drifts away. At this point water enters into the hydrolytic reaction. **(c)** The temporary bond between R_2 and the enzyme is broken and an H group from the water joins and restores the enzyme. An —OH group from water joins the R_2 product, which is then repulsed by the enzyme and drifts away. The enzyme is restored to its original state, ready to act again.

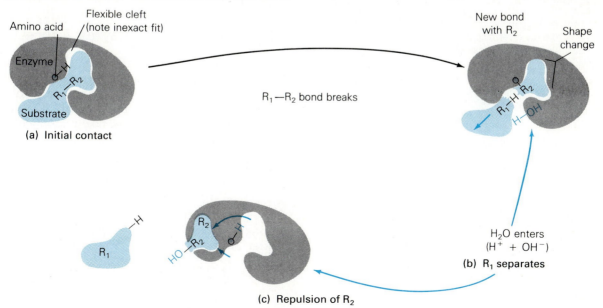

Flexible cleft (note inexact fit)
Amino acid
Enzyme
R_1—R_2
Substrate

(a) Initial contact

R_1—R_2 bond breaks

New bond with R_2
Shape change
R_1—H R_2
H—OH

H_2O enters
(H^+ + OH^-)

(b) R_1 separates

—H
R_1
R_2
HO—R_2

(c) Repulsion of R_2

Since the enzyme comes out unscathed, the net reaction is simply:

$$S \rightarrow P$$

When the enzyme and substrate finish their business, they drift apart. The substrate has been chemically altered, but the enzyme emerges unchanged, ready to repeat the reaction. The ability of enzymes to react with substrate molecules again and again is another of their important characteristics and provides a good example of the cell's frugal use of valuable resources. Actually, enzymes are not as immortal as they may seem. Like all biologi-

cal molecules and macromolecules, they are "biodegradable"—they break down in time and lose their potency.

We have concentrated so far on the chemically disruptive, or **catabolic** aspects of enzyme action. But enzymes also participate in constructive, or **anabolic** activity—the building of molecules and macromolecules. Let's consider the synthesis of starch from its glucose subunits, since this brings up an important point. You may recall from our earlier discussion of starch that as the glucose units are linked together, hydrogen and hydroxyl groups are removed to form water. We called this process dehydration synthesis. Such a reaction results in a product of greater free energy than is found in individual glucose molecules. This is by definition, then, an endergonic reaction, one that, as we said earlier, may seem to violate the second law of thermodynamics. The question then arises, how can free energy in a system be increased?

The answer is, free energy is not increased. Starch cannot be produced from glucose without an external supply of free energy. This supply comes from an important source in cellular chemistry, the *ATP* (adenosine triphosphate) mentioned briefly in earlier chapters. We'll have much more to say about ATP a little later, but the important point here is that a combination of glucose, dehydrating enzymes, and energy-rich ATP molecules represents a considerably greater amount of free energy than we find in the products of the reaction. Thus, starch can only be synthesized at the expense of cellular energy. In the overall picture, then, seemingly *endergonic* reactions in the cell are in larger terms *exergonic* reactions (Figure 6.6).

6.6 HIDDEN COSTS IN BIOCHEMICAL ACTIVITY

(a) At first glance it seems obvious that the conversion of simple glucose into complex starch is an endergonic reaction. In other words, through biochemical "magic," glucose at a lower free energy level is converted to starch at a higher free energy level. **(b)** However, look at the overall process of starch synthesis. We see that in the overall reaction, ATP is used. The ATP is at a higher free energy level than starch and is degraded to ADP. Therefore, in terms of the entire event, starch synthesis is exergonic. The long view of this tells us that life goes on at the expense of some energy source with a free energy level that is higher than the sum total of free energy for all living organisms. What is that source?

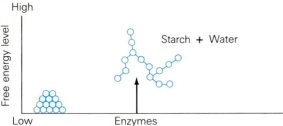

(a) An endergonic reaction? Glucose at a lower free energy level is converted to starch at a higher free energy level. *Biochemical magic?*

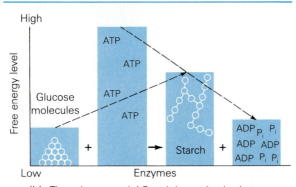

(b) *There is no magic!* Starch is synthesized at a considerable expense of ATP energy. The overall reaction is exergonic.

Characteristics of Enzyme Action

It may seem, so far, as if enzymes are in control of the cell. However, enzymes really don't control anything; they are merely active but subservient molecules whose activity is itself controlled by surrounding conditions. Let's now take a closer look at some principles of enzyme action.

Rate of Reaction.   First, the speed at which products are formed from substrate in an enzymatic reaction often depends on the quantity of substrate present. There is no mystery here, since the joining of an enzyme and its substrate is purely a matter of chance collision. Therefore, when the number of substrate molecules increases, the chances of collision are increased. This suggests

6.7 RATE OF ENZYME ACTIVITY

(a) The graph represents the amount of product formed when a fixed amount of enzyme and an excess of substrate are placed in a single reaction tube. The tube is checked at regular time intervals for the amount of product formed (perhaps in a colorimeter, which measures color change). As the graph indicates, the amount of product formed per unit of time increases at first, but the reaction rate soon levels off. We can conclude that the enzyme molecules become totally saturated—their active sites are all occupied—and that the rate of new product formation depends on the time each enzyme requires to complete its action.

  (b) In this experiment 11 reaction tubes are prepared and each is checked after the same time interval. All tubes contain the same amount of enzyme, but the concentration of substrate in each varies, with the lowest concentration in tube 1 and the highest in tube 11. The points on the graph represent the initial velocity at the different substrate concentrations. As the graph indicates, increasing the substrate concentration raises the initial velocity, but only up to a point. The data indicate that increasing the quantity of substrate improves the chances of enzyme-substrate complexes forming. This is logical, since their formation relies on random movement and chance collision. At the right on the graph, however, the initial velocity stabilizes. This tells us that the enzyme molecules have reached a "turnover" point. In other words, *every* enzyme molecule is busy reacting with a molecule of substrate as fast as conditions permit. What might happen in tubes 8 through 11 if heat were added?

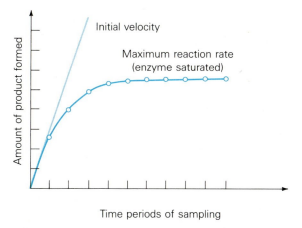

(a) Fixed amount of enzyme and excess of substrate

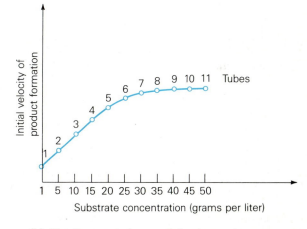

(b) Fixed amount of enzyme, but increasing amounts of substrate carried out in eleven reaction tubes

that if conditions such as pH and temperature remain constant, a maximum rate of product formation will be reached (Figure 6.7). At this rate, the time required for the enzyme to act (an incredibly brief instant) will then determine the rate of product formation. When the substrate concentration is low, not only will product formation be slowed, but in some instances the reaction itself will be reversed, with the product now forming the substrate. The effect of substrate or product concentration on enzyme action leads us to **Laws of Mass Action.**

The Laws of Mass Action: A Two-Way Affair.
One of the notable characteristics of many enzyme-catalyzed reactions is their reversibility. In other words, the products can react with the enzyme, recreating the substrate. The direction of the reaction has nothing to do with what we think the cell needs; it simply follows the Laws of Mass Action. These laws state that the direction taken by

many enzyme-mediated reactions depends on the relative amounts of substrate and product. If a great deal of substrate is present but very little product, then the reaction proceeds toward more product formation. When the opposite is true, that is, when there is much more product present than substrate, the reaction reverses itself. When substrate and product are in balance, there is no net direction of reaction. Things simply reach an equilibrium.

We should hasten to point out that such reversibility is not always possible, for two important reasons. First, if the product is immediately **metabolized** (changed chemically) by the cell, which is often the case, the reaction cannot reverse and mass action favors the product. Second, the process can't be reversed if the reaction is highly exergonic. If the product has very little free energy, then reversibility is energetically difficult if not impossible unless the cell can invest new energy reserves to make it happen.

The Effect of Heat.   We have said that enzymes apparently overcome the need for high activation temperatures, yet heat has important implications to enzyme-mediated reactions. As a rule of thumb, an increase of 10°C doubles the rate of most chemical reactions, including enzyme activity (Figure 6.8). This is because heat in a system is expressed as molecular motion; thus an increase in the movement of the substrate and enzyme molecules increases the chances of their collision.

Of course, there is a limit to how much heat can usefully be added to any system. After all, enzymes are made of protein, and heat in excess can denature protein—that is, break the fragile bonding forces that maintain the tertiary shape of enzymes. And to an enzyme, shape is everything. Above the normal operating temperature, then, many enzymes become misshapen and their catalytic activity ceases. In some cases, however, the damage is reversible.

The Effect of Acids and Bases.   A third physical factor influencing enzyme activity is pH—the acidity of the surroundings. While a few enzymes, such as **pepsin** (a stomach enzyme that digests protein), perform optimally at a strongly acidic, very low pH, most require a more neutral condition such as that found in most cells (Figure 6.9). Apparently, the presence of an excessive number of hydrogen or hydroxide ions can interfere with the tertiary shape of enzymes. In some instances,

6.8 HEAT AND THE ENZYME

The addition of heat accelerates the action of enzymes—up to a point. As the graph indicates, once a critical temperature has been reached, enzymatic activity halts abruptly. What has happened is that the enzyme has been denatured, its shape permanently altered by the action of excessive heat.

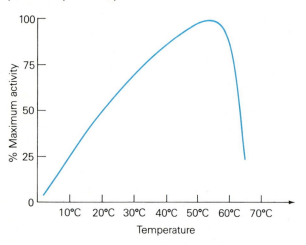

6.9 pH AND THE ENZYME

Most enzymes operate optimally at a pH near neutral, as we see in trypsin, an intestinal enzyme. A few, such as the stomach enzyme pepsin, require a strongly acidic environment to function optimally.

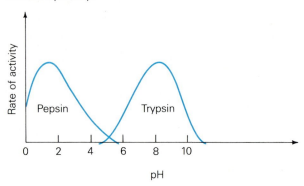

an enzyme's actions may be reversed by changes in cellular pH. For example, the same enzyme that catalyzes the synthesis of glycogen from glucose at a high pH will reverse the process at a low pH, breaking glycogen down to glucose.

Teams of Enzymes: Metabolic Pathways. As you might expect, enzyme activity in the cell can become quite complex. While single-step reactions frequently occur, it is just as likely that a single enzymatic reaction may be just one link in a long sequence of such reactions. Such a chain of reactions is called a **metabolic pathway,** and there are many of these at work in the cell. A metabolic pathway includes teams of different enzymes, each involved in its own chemical change. In effect, the product of one enzyme becomes the substrate of the next, and so on along the pathway. Metabolic pathways are quite like industrial assembly lines, in which each worker makes some change in the product passing by.

Allosteric Sites and Enzyme Control

In addition to mass action, enzymes are controlled in more subtle ways—through inhibition, for example. In addition to its active site, an enzyme may have a second binding site, called an **allosteric site.** The allosteric site will accept a specific molecule that fits. When the site is filled, the shape of the enzyme's primary active site is changed, and the enzyme will no longer bind to its usual substrate. The allosteric site is sometimes keyed to one of the enzyme's products, whereupon the amount of product accumulating determines how many enzymes will be inhibited. More likely, enzymes with allosteric sites will be part of a team of enzymes that act sequentially on the same substrate. In such an instance, the final product of the team would be the one to fill the allosteric site (Figure 6.10). It is important to know that the binding is not permanent, and as soon as product becomes scarce—is used up or enzymatically altered to a new product—the inhibitory effect is lost and the enzyme can again act.

The operation of the allosteric site is an example of an important biological regulatory phenomenon: **negative feedback,** or **feedback inhibition.** We will come across it again and again as we proceed, but a brief definition will suffice here. In negative feedback, the product of an action (not necessarily the action of an enzyme, as we will see) directly or indirectly inhibits the action that produced the product in the first place. In our example, as the quantity of product became excessive, it began to inhibit the enzyme that produced it. Such regulatory mechanisms are elegant in that they are both simple and automatic. As we will see, negative feedback mechanisms work at all levels of life.

The following summarizes some of the characteristics of enzymes and their functions:

1. Enzymes are proteins with highly specific shapes.
2. Each kind of enzyme has a specific role, reacting with its own specific substrate.
3. Enzymes may have slightly flexible active sites, which fit the substrate's molecular configuration.

6.10 ALLOSTERIC CONTROL OF ENZYMES

Some enzymes have two different binding sites. The primary site binds with the substrate molecule as usual, but the second (allosteric) site can bind with a small product molecule. When the allosteric site is free, the enzyme is active, but when it becomes occupied, the shape of the enzyme changes enough to render it incapable of forming an enzyme-substrate complex. Thus, the allosteric site becomes a control mechanism that inhibits action as long as product is plentiful. **(a)** The uninhibited enzyme produces P_1 and P_2. **(b)** Product P_2 rejoins the enzyme at an allosteric site; note the effect of the change in shape. When products P_1 and P_2 are used up in other reactions, the enzyme will once again be free to carry out its role in the cell.

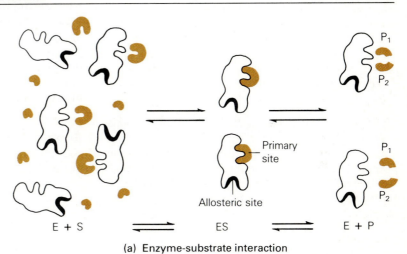

(a) Enzyme-substrate interaction

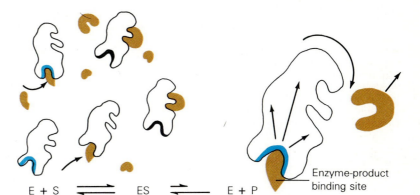

(b) **Inhibition begins:** product binds the allosteric site and configurational changes alter the active site — the substrate no longer fits the enzyme

4. Enzymes, like other catalysts, emerge unaltered from their reactions and, allowing for normal molecular degradation, can be used over and over.
5. Enzymes can act in both catabolic and anabolic reactions.
6. Enzymes lower the heat of activation required for a reaction to occur.
7. The rates at which enzymes function is determined by substrate and product concentrations, heat, and pH.
8. Enzyme action may be reversible, often depending on the relative amounts of substrate and product.

Now let's consider some specific characteristics of energy transfers within cells. We can begin by introducing a molecule that seems to be involved in nearly all the work the cell does. This is the molecule called *ATP, adenosine triphosphate,* a name we will frequently encounter as we continue to explore the chemical complexities of the cell.

ATP: THE ENERGY CURRENCY OF THE CELL

ATP can accurately be called the "coin" of the cell's energy transactions. Like the penny, our smallest monetary unit, an ATP molecule represents the smallest energy unit, or smallest useful amount of energy in the cell. When larger amounts of energy are needed, then more than one ATP must be invested.

All living cells, by the way, use ATP. It is a universal molecule of energy transfer. If energy is made available by some cellular process, such as photosynthesis or respiration, it is stored in ATP. If that energy is then needed, it is released by ATP. If you recall from Chapter 3, carbohydrates, lipids, and even proteins contain energy reserves the cell can use. But energy cannot be taken directly from these molecules. First it must be transferred to the special bonds of ATP.

The Molecular Structure of ATP

To understand how ATP works, it is helpful to know about its molecular structure. As Figure 6.11 reveals, each ATP molecule consists of three parts: an **adenine base,** a 5-carbon ring sugar named **ribose,** and three phosphates. The most obvious part is adenine, a double ring of carbon and nitrogen atoms. Ribose forms a connecting link between adenine and the three phosphates that make up the "tail" of the molecule. Each phosphate is linked to the next via an oxygen atom in what is called a **pyrophosphate bond.** As we will see in later chapters, this arrangement of a base, sugar, and phosphate tail is also common in the nucleic acids, the giant macromolecules known as DNA and RNA.

Energy-Rich Bonds. Now look closely at the pyrophosphate bonds. You can see that two bonds are shown as curved lines. These curved lines indicate energy-rich bonds, a designation that means that an unusually large amount of energy must be used to form the bond. It follows, therefore, that a

---

6.11 ATP STRUCTURE

Adenosine triphosphate (ATP) is a triphosphate nucleoside, identical to that found in RNA. It consists of adenine covalently bonded to the number 1 carbon of ribose. Ribose, in turn, is covalently bonded at its number 5 carbon to three phosphates. The last two contain high-energy bonds that, upon cleaving, yield about 8000 calories per mole.

large amount of energy is released when such bonds are broken.

We can gain some insight into the nature of energy-rich pyrophosphate bonds if we compare their energy value to that of the bond between the first phosphate and ribose. It has been found that breaking the ribose-phosphate bond will liberate about 2200 calories per mole of ATP. However, breaking either of the two pyrophosphate bonds will yield 8000 calories of energy per mole. Incidentally, if you are a calorie counter, you are familiar with the larger caloric unit, the kilocalorie (kcal). If we apply this unit here, the yield in the two types of bonds would be 2.2 kcal and 8 kcal, respectively. These are easier numbers to handle so we will use the kilocalorie in future discussions.

Why do some chemical bonds require and release more energy than others? We can offer only a rough explanation. Apparently, it is because the oxygens of the phosphate group draw electrons away from the phosphorus atom, rendering it positive. The charged condition produces a repulsion between phosphorus atoms, and so significant bond energy is required to bring the reluctant phosphorus atoms together. This energy is released when these unusual bonds are broken.

ATP and Chemical Reactions in the Cell

One of the cell's most essential reactions involves the recycling of ATP when its energy is spent. Anytime ATP's terminal pyrophosphate bond is broken, it is changed to **ADP (adenosine diphosphate),** or if its second pyrophosphate bond is broken, **AMP (adenosine monophosphate.)** The names tell us that ADP has two phosphates remaining and AMP has only one. AMP, as you may have guessed, is the fully spent form since it has lost both of the energy-rich pyrophosphate linkages. The most common energy transferring reaction, in which ATP becomes ADP, can be summarized in its simplest terms as:

$$ATP + H_2O \rightleftharpoons ADP + P_i + energy$$

or:

$$\text{A-P} \sim \text{P} \sim \text{P} + H_2O \rightleftharpoons \text{A-P} \sim \text{P} + P_i + energy$$

Here ATP reacts with water to form ADP, P_i, and energy. (The phosphate ion is symbolized as P_i, or inorganic phosphate.) What happens to the water? Its components (H^+ and OH^-) join the ADP and P_i when the pyrophosphate bonds are broken. This may have a familiar ring to it since this is a hydrolytic reaction; the role of water is precisely the same

6.12 ADENOSINE TRIPHOSPHATE (ATP) CYCLE

The ATP molecule stores energy for the cell's activities. Energy is released when the bond holding the third phosphate is broken, releasing the phosphate (P_i) and changing ATP to ADP. The ADP is reconverted to ATP during photosynthesis and respiration.

as it is in the disruption of carbohydrate, lipid, and amino acid linkages (see Chapters 2 and 3).

The bidirectional arrows in these reactions are also important. They tell us that the reactions are reversible, but be aware that proceeding from ADP + P_i to ATP is energetically "uphill," and some outside energy source must be available. The regeneration of ATP, an ongoing process in all cells, is often portrayed as we see in Figure 6.12, the ATP cycle. The two sources of energy seen there are photosynthesis and cellular respiration, two complex processes we will describe in Chapters 7 and 8.

The restoration of ATP's third phosphate (or the second *and* third, as is required when ATP is broken down to AMP) is an example of a **phosphorylation reaction**—the adding of phosphate. Phosphorylation reactions are important in other ways as well. Quite often when ATP is broken down, the terminal phosphate, along with its energy-rich bond, is transferred to some substrate. If we let R stand for the substrate, this phosphorylation can be represented as:

$$ATP + R \rightleftharpoons ADP + R\,P$$

or:

$$\text{A-P} \sim \text{P} \sim \text{P} + R \rightleftharpoons \text{A-P} \sim \text{P} + R\,P$$

Transferring the energy-rich phosphate group to the substrate is usually an intermediate step leading to other reactions. During synthesis (molecule building), for example, phosphorylation of a substrate may be the first step in linking two substrate molecules together. With its free energy increased, the phosphorylated substrate can more readily enter into reactions that are useful to the cell.

To summarize the roles of ATP in the life of a cell, we can first say that much of its energy is used to do work, including chemical synthesis, the building of the cell's molecules and macromolecules, some of which are used to form structure and others that are used for storing energy.

ATP is also used for pumping molecules and ions across membranes, contracting muscles, and conducting neural impulses (messages). These are all forms of work, energy expenditures that organisms must make to resist and delay the inexorable trend toward entropy. Organisms can maintain an ordered existence only by continuing to bring in new energy, much of which is then used for cycling ADP back to ATP. Now we will consider the ways cells handle this critical molecule. We will first set the stage by introducing some other kinds of molecules.

COENZYMES: NAD, NADP, AND FAD

Coenzymes, as their name implies, work together with enzymes. But unlike enzymes, coenzymes are not proteins. They are much smaller than proteins, and as a matter of fact, the components of some are similar to those of ATP. Coenzymes are involved in many kinds of cellular reactions, including molecular synthesis, but one of their major roles is to participate in reactions that lead eventually to the transfer of chemical bond energy to the energy-rich bonds of ATP.

The two most common coenzymes are **NAD** and **NADP (nicotinamide adenine dinucleotide** and **nicotinamide adenine dinucleotide phosphate).** We will also be concerned to a lesser degree with a third one, **FAD (flavin adenine dinucleotide).** These are usually pronounced as "nad," "nad-phosphate," and "fad." NADP operates in photosynthesis, and NAD specializes in cell respiration. FAD functions in both processes.

Chemically, NAD and NADP are quite similar, but NADP contains one more phosphate. Like ATP, both contain the nitrogen base adenine, along with ribose sugars and phosphate. But in addition, they contain a compound known as **nicotinic acid** (Figure 6.13), a component of vitamin B_6. Nicotinic acid is the active component in both coenzymes. FAD is

similar to the other two coenzymes, but its active group is **riboflavin,** a vitamin B_2 derivative.

In general, the three coenzymes carry out similar functions. They team up with enzymes that are involved in the **oxidation** and **reduction** of substrates. Both of these chemical processes are essential to the generation of ATP in photosynthesis and cell respiration. We can understand the roles of these coenzymes better if we take a brief look at the processes of oxidation and reduction.

Oxidation and reduction are common chemical reactions, but their importance here is related to the fact that the energy of any cellular fuel lies in the chemical bonds that hold its atoms together. For example, the energy of glucose is contained chiefly in the carbon-hydrogen bonds. One might wonder how anything so elusive and volatile as energy could be transferred from the simple bonds of glucose to the energy-rich bonds of ATP. But the bonds are simply a function of the behavior of paired electrons, and those electrons can themselves be transferred, bond energy and all. This is exactly what is done by oxidizing enzymes and their coenzymes. But what is oxidation?

Oxidation is the removal of electrons from a substance and reduction is the opposite, the addition of electrons to a substance. Since electrons cannot exist alone, it follows that when something is oxidized (loses electrons), something else is reduced (gains electrons). We must point out that, in some cases, when an electron is removed, a proton tags along. In these instances, oxidation may involve the removal of hydrogen atoms (a proton and an electron together). Similarly, reduction may involve the addition of hydrogen atoms. Let's return now to the coenzymes.

Coenzymes in Action

Unlike enzymes, coenzymes do not emerge from reactions unchanged. In the oxidation of a substrate, it is the coenzyme's function to accept the electrons or hydrogens, thereby becoming reduced. The reduction of a coenzyme is a temporary change, representing half of its role. A reduced coenzyme will, in turn, pass its electrons or hydrogens on to a different substrate. In reducing that substrate, it will itself become oxidized, and ready to act again. Because of their role, coenzymes are often referred to as **electron carriers** or **hydrogen carriers.** They take electrons or hydrogen from one molecule and deposit them in another.

At the pH levels normally found in the cell, oxidized NAD and NADP bear positive charges, and are generally designated as NAD⁺ and NADP⁺ Usually, both NAD⁺ and NADP⁺ can take on two electrons, one of which attracts a proton (H⁺) from the surrounding medium, while the other neutralizes the positive charge. Thus, we usually show the reduced forms as NADH and NADPH. Since the reduction of the two coenzymes often involves *two* protons, as well as two electrons, some people prefer to write the reduced forms as NADH + H⁺ and NADPH + H⁺, to indicate that the second proton will be found in the surrounding medium (Figure 6.14). We can write the reduction of NAD (or NADP) as:

$$NAD^+ + 2e^- + H^+ \rightarrow NADH$$

and the oxidation of NAD (or NADP) as:

$$NADH \rightarrow NAD^+ + 2e^- + H^+$$

If we wanted to show the reduction of NAD⁺ (or NADP⁺) as indicating the acceptance of two hydrogen atoms (which often happens), we would have to show it as:

$$NAD^+ + 2e^- + 2H^+ \rightarrow NADH + H^+$$

and the oxidation as:

$$NADH + H^+ \rightarrow NAD^+ + 2e^- + 2H^+$$

6.13 THE ELECTRON AND HYDROGEN CARRIERS NAD, NADP, AND FAD

NAD and NADP are nearly identical, except that NADP includes a phosphate group attached to the number 2 carbon of ribose. Note that each contains adenine nucleotide—one of four basic units of a nucleic acid (in this instance RNA, since the sugar is ribose). The active group of both NAD and NADP is nicotinic acid (seen here in its oxidized state). FAD also contains adenine nucleotide, but its active group is known as an isoalloxazine ring (the triple carbon, nitrogen rings seen above the nucleotide portion—also seen in the oxidized state), which is part of riboflavin.

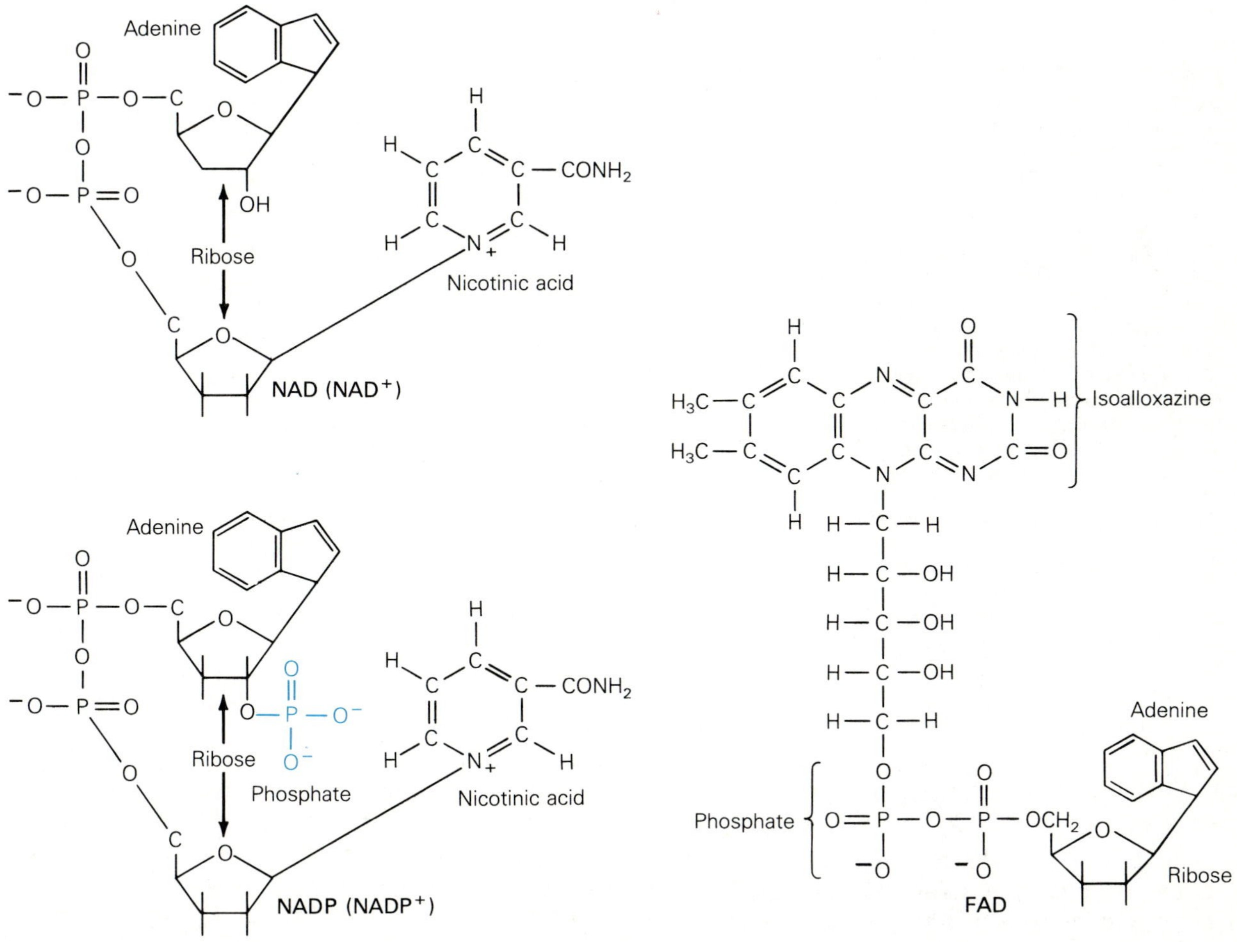

What about FAD, you're probably anxiously muttering—what happens to FAD? You will be glad to hear that FAD doesn't give us the same problems since it readily takes on two electrons and two protons (the equivalent of two hydrogen atoms), so in its reduced state FAD is written $FADH_2$ (see Figures 6.13 and 6.14).

To return to the combined action of an enzyme, a coenzyme, and a substrate, we can symbolize a reaction in this manner:

$$\text{Substrate} + NAD^+ \xrightarrow[\text{enzyme}]{\text{Oxidizing}} \text{Oxidized substrate} + NADH$$

The enzyme isn't changed in the oxidation reaction, so by tradition it is listed under the arrow.

It is important to note that the reduction of a coenzyme increases its free energy state. As a result, it becomes more reactive. Its usual reaction is to pass its electrons and hydrogens off to some molecule that is in a lower free energy state. The ability to reduce other substances is called **reducing power**. NADH and NADPH may or may not require an enzyme to reduce a substrate, depending on how receptive the recipient is. For example, a substrate that has been previously phosphorylated (had phosphate added) by ATP might readily release its phosphate to accept electrons or hydrogens. A simplified representation of such a reduction might be:

$$NADH + \text{Substrate-P} \rightarrow$$
$$\text{Substrate-H (reduced substrate)} + NAD^+ + P_i$$

The entire purpose of coenzymes, then, is to receive electrons or hydrogens and use the newly acquired reducing power to pass them along to some substrate. This is why the term "electron" or "hydrogen" carrier is so fitting. But as we will now discover, the three coenzymes are not the only molecules with this ability. While NAD, NADP, and FAD are small mobile carrier molecules, we find that the transport of electrons and hydrogens through oxidation and reduction also occurs in nonmobile carriers. These fixed carriers are also important to the energy transfers that lead to ATP production.

Stationary Electron Carriers of Mitochondria and Chloroplasts. Stationary electron and hydrogen carriers are found in strategic locations in the membranes within mitochondria and chloroplasts. Actually, we find them in the *inner membrane* of the mitochondrion and in the *thylakoid membranes* of the chloroplast, structures that were discussed in Chapter 4. We will get to the significance of this shortly, but first, let's describe these membrane-embedded carriers and see what they do. Unlike the tiny NAD, NADP, and FAD molecules, these carriers are often large. Many fall into a

6.14 REDUCTION OF COENZYMES

Both the oxidized forms and the reduced forms of **(a)** nicotinic acid (NAD and NADP) and **(b)** the isoalloxazine ring (FAD) are seen. The nicotinic acid could have come from either NAD^+ or $NADP^+$. In the reduction of either coenzyme, two hydrogen atoms are involved, with one proton and two electrons joining the nicotinic acid ring. The second proton is released into the aqueous medium. However FAD is different; it is reduced by two protons and two electrons—two hydrogen atoms—so in its reduced form it is called $FADH_2$.

6.15 ELECTRON/PROTON TRANSPORT SYSTEMS

(a) In this hypothetical situation, fixed electron and proton carriers are arranged in a manner similar to those of the mitochondrial and thylakoid membranes. Their general role is to receive energetic electrons, which then pass along from carrier to carrier, sequentially reducing the next carrier in line. As an energetic electron passes through the electron transport system, its free energy level diminishes until, at the end, it is greatly reduced. In the meantime, the energy is used to accomplish work. **(b)** Electron energy is used at critical points (transmembranal proteins) to power the pumping or ferrying of protons (H^+) across the membrane in one direction. Note that the proton-transporting carriers collect electrons and protons together, thus actually becoming hydrogen (H) carriers for a brief time. The eventual accumulation of protons in one compartment generates a system of considerable potential energy (like water behind a dam). This free energy is later put to work synthesizing ATP.

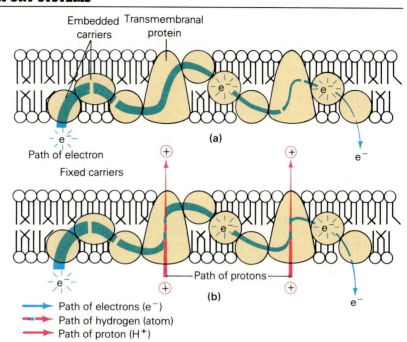

Path of electrons (e^-)
Path of hydrogen (atom)
Path of proton (H^+)

family of iron-containing proteins known as **cytochromes,** although larger, less understood proteins are also included. We will name some of these embedded carriers in the next chapter.

Imagine a series of such carriers arranged in such a manner that they run in zigzag fashion across a membrane and back. Beginning with the first, each carrier in the system, when reduced by an electron, has a slightly greater reducing power than the next in line. Thus carrier 1 can reduce carrier number 2, and number 2 can reduce carrier number 3, and so on through the system. If the first carrier is reduced by a mobile carrier with great reducing power, such as NADH or NADPH, the members can begin to pass electrons, like buckets of water in a bucket brigade, down the system. And ''down'' is the right word since the sequential action occurs in an energetically downhill direction (Figure 6.15a). In any such system, the free energy of electrons being transported is depleted as they move through the carriers.

The sequential carriers grouped in the inner membrane of mitochondria and in the thylakoid membranes of chloroplasts are called *electron transport systems,* although their real work is to transport protons (hydrogen ions, H^+) at selected points (see Figure 6.15b). In the next chapter, we'll see that the

movement of electrons and hydrogens through such transport systems can prepare the way for the conversion of ADP to ATP. We have been leading up to this for some time, but it is important that each aspect be in place. Now we can tie together the action of oxidizing and reducing enzymes, the mobile coenzymes, and the membrane-bound carriers of electrons and protons. We will bring these entities together in what is called the **theory of chemiosmotic phosphorylation.**

ATP PRODUCTION BY CHEMIOSMOTIC PHOSPHORYLATION

The theory of chemiosmotic phosphorylation was proposed by Peter Mitchell in 1961 (Essay 6.1). Mitchell, it seems, had his own peculiar way of looking at things. While other biochemists were trying to understand ATP synthesis by analyzing each specific component in its manufacture, Mitchell took a broader view and looked at ATP production in terms of the intact chloroplasts and mitochondria. His first clue was a basic and well-substanti-

Peter Mitchell and the Overthrow of an Enshrined Scientific Dogma

A new scientific truth does not triumph by convincing its opponents and making them see the light, but rather because its opponents eventually die, and a new generation grows up that is familiar with it.

Max Planck
Scientific Autobiography

A whole generation was reared on a number of accepted "basic facts" about the mitochondrial electron transport system. The electron carriers were known to operate in a specific sequence, visualized as a kind of bucket brigade, and the energy released in such transactions were believed to be used directly to form ATP and ADP. Unfortunately, the researchers were working with reacting molecules in free solution. And they ran into a few problems.

For example, although carriers floating free in solution (no longer bound to the mitochondrial membrane) would readily perform the individual transfers, no ATP would be made. For ATP production to take place, the mitochondrion must remain intact. Most scientists shrugged their shoulders at this inconvenient and mysterious fact, and continued to work out the details of the carrier-to-carrier reactions. Furthermore, some

ingenious experiments seemed to indicate just where the ATP was generated in the intact electron transport chain. The experiments involved poisoning specific reactions in intact mitochondria with enzyme inhibitors and noting how ATP production was affected.

That the individual reactions were involved in "coupled phosphorylations" directly involving both ATP and specific electron carriers was never thought to be a hypothesis; it was

Peter Mitchell

considered an established fact of science. It had a strong appeal to the scientific (or human) love of precision. The energy relationships were just right, and with three sites of coupled phosphorylation along the electron transport pathway, it was readily calculated that every molecule of reduced NADH that was oxidized yielded exactly three ATPs from ADP. The total energy yield of the oxidation of one molecule of glucose—including glycolysis, the citric acid cycle (see Chapter 8), and the electron transport pathway—was precisely 36 ATPs. "Energy yield budgets" based on this analysis filled even basic biology textbooks with rather precise numbers for students to memorize. The electron pathway of photosynthesis was similarly dissected.

But then in 1961, Peter Mitchell (at the Glynn Research Laboratories in England) presented a radically different idea, which he called the *chemiosmosis hypothesis*. He suggested that ATP synthesis in the mitochondrion depends on a hydrogen ion gradient across the mitochondrial membrane and that the electron transport system operates according to the physical and spatial relationships of its elements in the membrane. Mitchell challenged the notion that hydrogens and electrons were

ated observation: neither chloroplasts nor mitochondria can make ATP unless they are physically intact. Mitchell devised a scheme in which ATP synthesis was dependent on intact, closed membrane systems and membrane-separated organelle compartments.

In Mitchell's system, hydrogen ions (H^+), or protons, as we will refer to them, are pumped through membranes, building up on the other side within the compartments of organelles. Such a buildup produces what Mitchell called a **chemiosmotic differential,** which is also known as an **elec-**

trochemical proton gradient (Figure 6.16). Such gradients contain great free energy, and this energy can be used to produce ATP in chloroplasts, mitochondria, and bacterial cells (see Essay 6.2). Let's look more closely at the energetics of the chemiosmotic differential.

The term "electrochemical proton gradient" can be translated as "an electrical and a chemical gradient produced through the buildup of sizable pools of protons within certain membrane-bound compartments in the cell or its organelles." The gradient results from there being more protons inside the

freely interconvertible. Indeed, he saw the difference in free electrons and hydrogen-bound electrons as crucial to his new idea. Mitchell hypothesized that the key steps of "coupled ATP synthesis" were in fact the steps at which hydrogen ions were pumped across the mitochondrial membrane. These electron transport reactions were involved with ATP formation, all right, but he saw the coupling as being indirect and imprecise.

This holistic way of looking at the problem was quite new and seemed bizarre to biochemists steeped in the established tradition of analyzing isolated chemical reactions of enzymes in artificial solution. Actually, Mitchell himself had originally only been interested in what had seemed to be a different problem, namely the active transport of hydrogen ions (H^+), or "protons," across membranes using ATP as an energy source. But then it occurred to him that the hydrogen ion pump reaction might be reversible and that ATP could be generated by a flow of hydrogen ions *back* across the membrane.

At first there was not much of a controversy; Mitchell was simply not taken very seriously. But he was more than just a theoretician. He also knew his way around the laboratory, and he had formed a very fortunate relationship with a particularly gifted colleague, Jennifer Moyle. Together they set out to do the experiments that would test the hypothesis. Over a period of 15 years they built up a body of experimental evidence that made the chemiosmotic interpretation increasingly plausible and the traditional isolated molecular "coupled phosphorylation" idea increasingly unlikely. Along the way, it became apparent that the chemiosmotic hypothesis could explain photosynthetic generation of ATP in chloroplasts as well as oxidative generation of ATP in mitochondria.

Unlike most scientists, Mitchell and Moyle now work at home; specifically, they do their work in a modern research laboratory that Mitchell built in his own home, which happens to be a medieval castle in Cornwall. The aristocratic Mitchell supports his laboratory through the unlikely combination of government research grants and proceeds from his dairy herd.

Other investigators became interested in Mitchell's idea and contributed to its scientific validation. A recent experiment strongly supported the chemiosmosis hypothesis. It consisted of isolating small, closed vessels made up of inner mitochondrial membrane. In the process of disrupting the mitochondria and isolating the vessels, the membranes were turned inside out, with the F1 particles on the outside, just as the CF1 particles are on the outside in living chloroplast membranes (thylakoids). The small vessels were soaked in an acid solution until their contents became acidic. Then they were quickly separated and plunged into a basic medium. As the hydrogen ions passed back out through the tiny, isolated membranes, ATP synthesis could be measured. The experiment was repeated after the membrane vessels had been treated in a way that stripped off the F1 particles (as could be verified in electron micrographs). This time no ATP was generated.

Scientists may often be hostile to heretical new ideas. Perhaps it's just as well; otherwise, science would constantly be inundated by half-baked notions. But scientists are generally receptive to well-done research, and the chemiosmosis hypothesis has won increasing acceptance. There are now very few biochemists who would not agree that it is the most plausible explanation of mitochondrial and chloroplast function so far. Peter Mitchell was awarded the Nobel Prize in chemistry in 1978.

compartment than outside it. The primary locations of chemiosmotic systems in eukaryotes are the chloroplast and the mitochondrion, while in prokaryotes the gradient occurs between the cell and its surroundings. In the chloroplast, electrons are activated by light energy, while in the mitochondrion, the source is cellular fuel—that is, the carbohydrates, fats, and proteins committed to that purpose.

As its name suggests, the electrochemical proton gradient is actually two gradients in one, depending on what is being measured (see Figure 6.16). The electrical part of the chemiosmotic gradient is a product of the preponderance of positive charges (protons) on one side of the membrane resulting in a preponderance of negative charges on the other. One can easily appreciate the potential energy of such a polarized system. In purely chemical terms the "chemical" gradient is actually a pH gradient—the product of a high concentration of protons (H^+) on one side of the membrane and hydroxide ions (OH^-) on the other (see the discussion of pH in Chapter 2). It is important to realize that the pH gradient has great potential energy, and we can expect that energy to be used somewhere

6.16 HIGH FREE ENERGY SYSTEMS

System (a) shows the hydrogen and hydroxide ions at equilibrium on either side of a membrane. Free energy is low, as it would be in any random system. In (b), the molecular pumps have been at work, actively transporting ions. Now there are two systems of high energy separated by the membrane. This potential energy can be tapped if the ions are permitted to cross in a controlled manner, as they are in the chloroplast and mitochondrion.

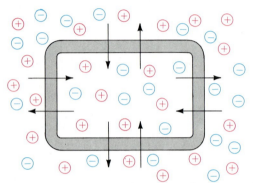

(a) Equilibrium (randomness)
Low free energy

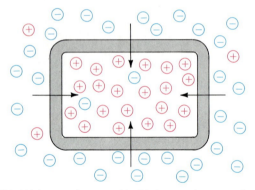

(b) H⁺ (protons) pumped inside by membrane carriers
High free energy

down the line. In fact, the energy can be released if the protons are permitted to move back across the membrane to a lower energy state.

This is exactly what happens in chemiosmotic phosphorylation. While the electron/proton transport systems constantly build the chemiosmotic gradient, protons are permitted to pass back across the membrane to lower energy states. But this return passage is closely controlled. The protons return through special channels that connect with enzyme-laden particles (CF1 particles in the chloro-

plast and F1 particles in the mitochondrion; see Chapter 4). Protons passing through these particles are confronted by ADP, P_i, and phosphorylating enzymes, and it is within these particles that some of the proton energy is transferred to the energy-rich bonds of ATP (Figure 6.17).

Little is actually known about specific activity in the F1 and CF1 particles except that the phosphorylating enzymes are present and ADP and P_i must be available. We can gain one clue about the amount of energy released if we consider that it is within this complex that protons and hydroxide ions come together to form water. Essentially, this involves adding an acid to a base ($H^+ + OH^-$), and such combinations give off a considerable amount of heat. In fact, bringing together large concentrations of acids and bases can produce a dangerously explosive condition.

We will look more closely into these systems in the next two chapters, where photosynthesis and respiration are discussed in greater detail.

6.17 CHEMIOSMOSIS

The chemiosmotic system in the mitochondrion reveals a buildup of protons (H^+) between the inner and outer membrane (in the outer compartment). These ions were pumped across the inner membrane using the energy of electrons from fuels in the mitochondrial matrix. The great free energy of the chemiosmotic gradient (electrochemical proton gradient) is then used to synthesize ATP in the F1 particles.

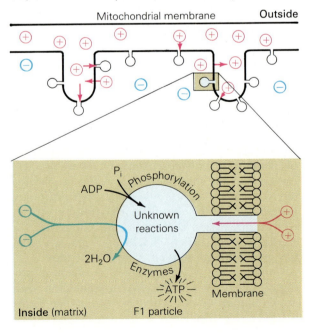

Bacterial Chemiosmosis and Liposome Models of Chemiosmotic Phosphorylation

The clearest models of chemiosmotic phosphorylation are seen in some of the photosynthetic bacteria that use a light-powered proton transport mechanism to establish their chemiosmotic differential. Bacteria do not contain much in the way of membrane-bound organelles, so they establish their chemiosmotic gradients between their cytoplasm and their immediate external environment.

The well-known salt-loving bacteria of genus *Halobacterium* carry out the clearest and simplest known natural form of chemiosmotic phosphorylation in this manner. When grown in the absence of oxygen, *Halobacterium* develops patches of purple pigment on its plasma membrane, and these patches consist of *bacteriorhodopsin,* a light-activated pigment that contains *retinal,* a visual pigment identical to that found in the rod cells (light-sensitive cells) of the vertebrate retina.

When light is absorbed by bacteriorhodopsin, a transmembranal protein, each photon powers the transport of two protons out of the cell. The continued transport of protons to the immediate surroundings sets up an electrochemical proton gradient similar to those in the chloroplasts and mitochondria of the eukaryote. And, as we find in these organelles, protons travel down their osmotic and electrical gradient to return into the tiny bacterial cell. Their return, however, is directed through phosphorylating enzyme complexes, where the proton's free energy is used to power the formation of ATP.

Cell biologists have taken advantage of the simplicity of *Halobacterium* to produce an even simpler system—a synthetic model that has helped fully establish much of the chemiosmotic theory. In this synthetic model, ATP is synthesized in the absence of a cytochrome-containing electron transport system, once widely believed to be directly involved in the phosphorylation of ADP.

In 1974, researchers borrowed bacteriorhodopsin proteins from *Halobacterium,* and F1 particles (phosphorylating enzyme complexes) from mitochondria, and using a strange synthetically produced model cell called a *liposome,* they produced the simplest complete chemiosmotic model. Liposomes are synthetically produced hollow lipid spheres—membranes consisting of a bilayer of phospholipids, similar to plasma membranes and ranging in size from 1 micrometer down to 25 nanometers. They have been used as model membranes to test many ideas about membranes and transport. In the chemiosmotic model, bacteriorhodopsin and F1 particles were incorporated right into the phospholipid bilayer membrane of liposomes. A source of hydrogen ions, ADP, and P_i were added to the suspension of liposomes, and the system was exposed to light. The bacteriorhodopsin began transporting protons into the liposome, and the protons moved down their gradient and out through the F1 particles. Incredibly, the outflow of protons through the particles powered the synthesis of ATP! Thus, a synthetic system of chemiosmotic phosphorylation proved capable of ATP production without the intervention of other membrane components. This clearly indicates that ATP synthesis can be structurally independent and separate from other photosynthetic and respiratory reactions.

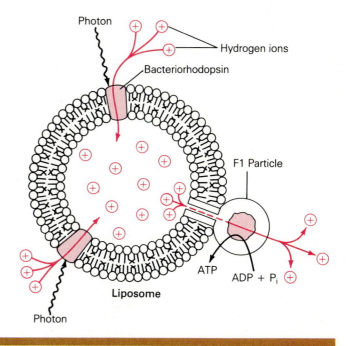

Liposome

APPLICATION OF IDEAS

1. Many inventors have attempted to design "perpetual motion machines." All failed. The reasons for such consistent failure were not understood until the laws of thermodynamics were formulated. Explain how they apply to the problem.

2. When asked in an essay exam to define life, a thoughtful student once wrote an answer totaling six words. "Life is an interruption in entropy." Is this true? Using the laws of thermodynamics for background, comment on this statement.

3. Reviewing what has been presented so far on chemiosmosis in mitochondria and chloroplasts and the electrical differential generated across their membranes, develop an analogy using a rechargeable battery. How are they similar, and where does the analogy fail?

KEY WORDS AND IDEAS

ENERGY

1. **Energy** is often defined as the ability to do **work,** and work can be defined as the transfer of energy to a body through the application of force.

2. **Potential energy,** also called **free energy,** is energy stored, or energy available for work; **kinetic energy** is energy in motion, or energy being transferred.

Forms of Energy

1. Energy can take various forms, including chemical bond, mechanical, magnetic, and electromagnetic energy—all of which are interchangeable.

The Laws of Thermodynamics

1. The behavior of all matter in the universe is described by the **laws of thermodynamics.**

2. The *First Law* (the law of conservation of energy) states that energy can neither be created nor destroyed. The sum of energy in a **closed system** remains the same but energy can be transformed to another state.

3. The *Second Law* (law of entropy) states that the free energy of a system is always decreasing. Complex, organized systems tend to reach simplicity and randomness, and a measure of the disorganization and randomness is called **entropy.** Of great biological importance, transfers of energy are always accompanied by a loss of free energy.

Entropy and Life

1. The molecular complexity of life in the earth is possible only because of readily available light energy. Without it, the overall state of entropy would soon increase and life would cease.

CHEMICAL REACTIONS AND ENERGY STATES

1. The *Hindenburg* disaster illustrates the nature of an **exergonic reaction**—one in which energy is released. Hydrogen gas, in the presence of a spark, rapidly united with oxygen to form water, and both gases reached a lower energy state. The energy difference was the light and heat of the explosion.

Heat of Activation

1. At usual temperatures, molecules remain stable unless activated by energy. Such energy is called **heat of activation** or **energy of activation.** Such energy sets off a chain reaction whereby the energy of one reaction provides the heat of activation for another, etc.

2. In **endergonic reactions,** the products are at a higher free energy level than the reactants.

3. **Catalysts** are agents that lower the required heat of activation; thus chemical reactions can occur at lower temperatures than is otherwise possible.

ENZYMES: BIOLOGICAL CATALYSTS

1. **Enzymes** are biological catalysts that (a) are proteins, (b) are highly specific for their **substrate,** and (c) bring about reactions without the usual heat of activation requirement.

A Matter of Shape

1. Enzymes are usually large, globular, quaternary proteins with **active sites** that closely fit the shape of their substrate. When joined they form an **enzyme-substrate complex.**

2. The older "lock and key" analogy of enzyme and substrate shape has been replaced with the **induced fit hypothesis.** Rather than a perfect

fit, the substrate fits imperfectly, thus assuming a stressed condition when in the active site.

3. Side groups of enzyme and substrate, along with water, interact chemically. The substrate emerges chemically altered, but the enzyme emerges fully restored.

4. Enzymes facilitate both **catabolic** (cleaving) reactions and **anabolic** (synthesizing) reactions.

5. While the synthesis of starch from glucose increases its free energy level and is considered endergonic, that energy increase is possible only because of the availability of even greater free energy from ATP.

Characteristics of Enzyme Action

1. The rate and direction in which enzymatic reactions occur follows the **Law of Mass Action.** Greater amounts of reactant force the reaction to go toward product formation at an increased rate. When the product exceeds the reactants, the reaction may reverse itself unless one of the products is being removed and made unavailable—that is, **metabolized.**

2. Heat increases the chances of collision between substrate and active sites, thus speeding up enzymatic reaction. At some point heat denatures the enzyme and the reaction ceases.

3. With few exceptions, most enzymatic reactions occur at a near neutral pH, and strong acids or bases slow reactions or denature the enzyme.

4. **Metabolic pathways** include teams of enzymes that perform sequential reactions. The product of the first enzyme becomes the substrate of the second, and so on through the pathway.

Allosteric Sites and Enzyme Control

1. Enzymes with second binding sites, called **allosteric sites,** are subject to **negative feedback** or **feedback inhibition.** When the product is in excess, it enters the allosteric sites of enzyme molecules, slowing the activity. As the product is further metabolized, the sites are freed and activity resumes.

ATP: THE ENERGY CURRENCY OF THE CELL

1. The available energy in each ATP molecule represents the smallest amount of energy useful to the cell.

2. All energy-requiring chemical and physical activity in the cell employs ATP or some other nucleoside triphosphate as its energy source.

The Molecular Structure of ATP

1. Each ATP contains an **adenine base**, a **ribose** sugar, and three phosphates with two **pyrophosphate bonds.**

2. Upon hydrolysis, the pyrophosphate bonds in each mole of ATP yield 8000 calories (8 kcal).

ATP and Chemical Reactions in the Cell

1. As its energy-rich bonds are spent, ATP is degraded to ADP (the diphosphate) and P_i (inorganic phosphate)—or sometimes to AMP, the monophosphate form.

2. ADP and AMP are converted to ATP in an ongoing cycle utilizing photosynthesis and respiration.

3. In some reactions, high-energy phosphates from ATP are transferred to the substrate in a **phosphorylation reaction.** This prepares the substrate for further transfers.

COENZYMES: AGENTS IN ENERGY TRANSFER

1. **Coenzymes** are small, nonprotein molecules that work in cooperation with enzymes.

2. The most familiar coenzymes are **NAD, NADP,** and **FAD,** which operate in oxidation-reduction reactions, accepting or transferring electrons and protons (hydrogen ions). They are derived from vitamins—**nicotinic acid** from vitamin B_6 and riboflavin from vitamin B_2.

3. **Oxidation** is the chemical removal of electrons or hydrogens while **reduction** is their addition. Such reactions are necessarily coupled so that the oxidation of one substance results in the reduction of another.

Coenzymes in Action

1. When reduced, coenzymes may in turn pass their electrons or hydrogens to other coenzymes, thus called **electron** or **hydrogen carriers.**

2. The coenzymes in their oxidized and reduced forms, respectively, are:
   a. $NAD^+ \longrightarrow NADH + H^+$
   b. $NADP^+ \longrightarrow NADPH + H^+$
   c. $FAD \longrightarrow FADH_2$

3. A reduced coenzyme is at a greater free energy level than an oxidized coenzyme and is said to have **reducing power.**

4. Although the coenzymes above are often mobile, many electron-hydrogen carriers are stationary, or fixed in place, usually within membranes such as those of the thylakoid and the cristae of the mitochondrion.

5. Stationary carriers include **cytochromes**—protein complexes and coenzymes.

6. In their passage through the carriers of **electron transport systems,** electrons gradually lose free energy, and this energy is harnessed in useful ways.

ATP PRODUCTION BY CHEMIOSMOTIC PHOSPHORYLATION

1. The **theory of chemiosmotic phosphorylation** proposes that carriers in the membranes of mitochondria and thylakoids use electron energy

to actively pump protons into isolated compartments, concentrating them so as to produce a **chemiosmotic differential** of great free energy. Then, through the controlled transit of such protons back through the membrane, the free energy of the system is used to form energy-rich ATP bonds.

2. Such gradients are also called **electrochemical proton gradients,** indicating both an electrical and chemical nature. In either case, there is considerable potential energy established. It is electrical because positive charges (H^+) accumulate on one side of the membrane and negative

charges (OH^-) accumulate on the other. It is chemical (a pH gradient) because the charges are present on hydrogen ions (acid) and hydroxide ions (base).

3. Protons recrossing the membranes of mitochondria and thylakoids pass through F1 and CF1 particles, respectively.

4. Both particles have phosphorylating enzymes, ADP, and P_i present, lacking only an energy source to restore the energy-rich pyrophosphate bond. That energy is provided by protons passing down the electrochemical gradient to reach a lower free energy level outside.

REVIEW QUESTIONS

1. Using the terms "work" and "force," define energy. (124)

2. Distinguish between the terms potential, kinetic, and free energy. (124–125)

3. What are some of the forms in which energy is expressed? Cite real-life examples of energy going through several conversions. (125)

4. Using the First and Second Laws, explain why life is improbable in a closed system. (125)

5. Explain the statement, "Entropy is the inverse of free energy." (125–126)

6. Using the *Hindenburg* disaster as an example, and considering the free energy of the reactants and products that were involved, explain how an exergonic reaction occurs. (126–127)

7. What does *heat of activation* have to do with chemical reactions? How does a catalyst affect this requirement? (127–128)

8. Explain an endergonic reaction and provide an example. (127)

9. List three important characteristics of all enzymes. (128)

10. Briefly summarize the events within the enzyme-substrate complex when a hydrolytic reaction occurs. (129)

11. Compare the "lock and key" analogy with the "induced fit concept" of enzyme action. (129)

12. Do enzymes actually enter into chemical reactions? Explain. (129–130)

13. Explain how the law of mass action applies to enzymes. (131)

14. Summarize the effect of heat on enzymatic reactions. What rule applies and what are the limits? (132)

15. Using the allosteric site as an example, explain how negative feedback operates in biological systems. Why is such an automated process valuable and necessary? (132–133)

16. Carefully describe the structure of ATP and point out specifically where the energy-rich bonds are located. (134)

17. Compare the energy yield in pyrophosphate and other bonds. How is the difference explained? (135)

18. Describe a phosphorylation reaction. Why are these used by the cell? (135–136)

19. Carefully define oxidation and reduction and explain the role of a coenzyme in these processes. (136)

20. List the three common mobile coenzymes of cells and write their formulas in the oxidized and reduced states. (137–138)

21. What are "stationary carriers"? What do they carry and what purpose do they serve? (138–139)

22. What is an electron transport system? What function do such systems serve in the mitochondrion and chloroplast? (138–139)

23. What is a chemiosmotic differential and why is it considered both electrical and chemical? (140–141)

Photosynthesis

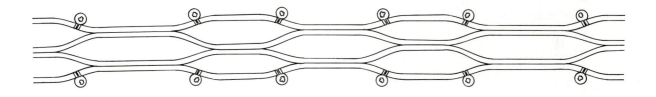

Our frequent bantering about the merits of solar energy has an almost touching naivete about it. We seem to treat solar energy as a recent idea, a new concept. But, of course, the sun is an ageless source of energy. The fossil fuels—oil, gas, and coal—are simply releasing solar energy stored away in the bodies of long-dead plants and algae. So while we continue to wrestle with the maze of engineering problems associated with harnessing the sun's energy to produce electricity, perhaps we should turn to the real experts for some ideas. And the real experts are likely to be green (Figure 7.1).

It is no secret that plants get their energy from sunlight. But it's less well known that plants long ago evolved the ability to convert **photons** of radiant energy into *electrical* energy. (Light energy is often described in terms of photons—discrete packets of energy traveling in a wavelike manner.) This happens in those tiny, highly organized bodies called chloroplasts. We mentioned earlier that chloroplasts convert sunlight energy into chemical bond energy by first passing this energy to electrons, which, in a new high-energy state, pass along pathways known as electron transport systems. However, we did not point out that this flow of electrons has many of the characteristics of a current of electricity moving along a conductor. We will return to this point shortly, but first let's step back for a broader look at the process of photosynthesis.

PHOTOSYNTHESIS: AN OVERVIEW

Photosynthesis is the process by which the energy of sunlight is used to bond certain molecules together to produce carbohydrates, usually the monosaccharide glucose. The raw materials of photosynthesis are carbon dioxide and water. The products are glucose, water, and oxygen. The process can be expressed in the general equation:

$$6CO_2 + 12H_2O \rightarrow C_6H_{12}O_6 + 6O_2 + 6H_2O$$

The reaction, when translated, simply tells us that water and carbon dioxide, in the presence of light and chlorophyll, are used to produce glucose, water, and oxygen. While this now-ancient formula is worth remembering, it obviously hides a vast amount of detail.

For example, while light is essential, photons of certain wavelengths of light are preferred over others. Water itself plays a highly active role in photosynthesis, yielding hydrogen for glucose-building, but only in the presence of powerful oxidizing agents. Thus, the water at the right side of the equation is not the same water as that at the left, and the oxygen at the right comes only from the water at the left, not from the carbon dioxide.

Carbon dioxide's role is more straightforward. It

7.1 SOLAR ENERGY

Plants and other photosynthesizers are masters at solar energy capture.

is incorporated, with slight changes, into the glucose molecule. Finally, the equation hides the fact that before any glucose can be synthesized, light must be captured and shunted into high-energy compounds such as ATP and NADPH, which are not even mentioned in the venerable old formula. Much of what we have referred to here represents hard-won information from discoveries spanning the past 200 years. The highlights are presented in Essay 7.1.

THE TWO PARTS OF PHOTOSYNTHESIS

The many events of photosynthesis are more readily understood if they are divided into two parts, those that require light and those that do not. The two are called the **light reactions** and the **light-independent reactions**. In a sense, we can consider the first to involve "charging up" the system and the second, to letting the charged-up system "run back down" (Figure 7.2). As we have emphasized, a system of free energy, when running down, can accomplish a great deal of useful work. And a great deal of work is necessary, since the prospect of low-energy carbon dioxide molecules joining to form

high-energy glucose is highly improbable and unfavorable as far as energetics are concerned.

In the light reactions, sunlight energy is captured, and this energy is used in two ways: to convert ADP and P_i (inorganic phosphate) to ATP and to reduce $NADP^+$ to NADPH. Both substances will be needed in the light-independent reactions, ATP for its energy-rich bonds and NADPH for its hydrogen. By introducing ATP and NADP into the general formula and considering only the light reactions, we have:

$$18ADP + 18P_i + 12NADP^+ + 12H_2O \rightarrow$$
$$18ATP + 12NADPH + 6O_2 + 18H_2O$$

In this formula we are getting a little closer to what actually happens in the light reactions. The numbers preceding the reactants (left side of arrow) should not concern you just yet, but they do represent the numbers of ATPs, NADPs, and H_2Os needed for a yield of one glucose molecule at the end of photosynthesis. Further, we should mention that the extra water appearing at the right of the arrow is produced when ADP is joined by P_i, a process that yields water through dehydration synthesis.

The light-independent reactions can also be more accurately summarized by varying the general formula to include ATP and NADPH:

$$18ATP + 12NADPH + 12H_2O + 6CO_2 \rightarrow$$
$$C_6H_{12}O_6 + 12NADP^+ + 18ADP + 18P_i + 6H^+$$
(glucose)

7.2 PHOTOSYNTHETIC ENERGY CURVE

In the light reactions the photosynthetic mechanisms charge the system, using sunlight energy to produce molecules with great free energy in their chemical bonds. In the light-independent reactions, the energetic products of the light reactions go to lower energy states, but in the process, energy is conserved in the chemical bonds of carbohydrates and other products.

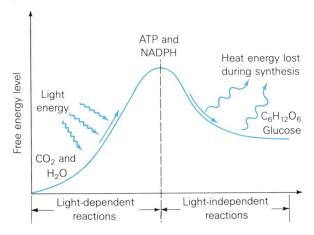

From Willows, Mice, and Candles to Electron Microscopes

Tree: 5 lbs.

Tree: 169 lbs., 3 oz.

Soil: 200 lbs.

Soil: nearly 200 lbs.

Prior to the time of Jan Baptiste van Helmont, a Belgian physician of the 17th century, it was commonly accepted that plants derived their matter from materials in the soil. (Probably, many people who haven't studied photosynthesis would go along with this today.) We aren't sure why, but van Helmont decided to test the idea. He carefully stripped a young willow sapling of all surrounding soil, weighed it, and planted it in a tub of soil that had also been carefully weighed. After five years of diligent watering (with distilled water), van Helmont removed the greatly enlarged willow and again stripped away the soil and weighed it. The young tree had gained 164 pounds. Upon weighing the soil, van Helmont was amazed to learn that it had lost only 2 ounces. Van Helmont's conclusion was inescapable: "It's the water!" he cried, but he was only half right. (Carbon dioxide hadn't been discovered yet.)

The next investigative breakthrough occurred late in the 18th century, around 1771, when British clergyman Joseph Priestley, made an astute observation. In a simple but classic series of experiments, Priestly learned that plants growing in a sealed container would support the burning of a candle and breathing of a mouse longer than either would go on without the plant present. What Priestley had found, in addition to the reciprocity of gases in plants and animals, was that plants produce oxygen.

Priestley's discovery was followed almost immediately by the ambitious effort of a Dutch engineer named Jan Ingenhousz. Ingenhousz carried out some 500 experiments (some say, all within one year), which led to proof that light was essential for photosynthesis. Ingenhousz determined that plants gave off oxygen only in the light and that under darkened conditions, they even produced some carbon dioxide. By the mid-1800s, the basic equation for photo-

synthesis was well established. Toward the end of the 1800s, Theodore Engleman determined that the chloroplasts were the sites of photosynthesis. He observed the peculiar manner in which bacteria clustered about the ribbon-shaped, oxygen-evolving chloroplasts of certain filamentous algae.

Many of the mysteries in the biochemistry of photosynthesis were resolved in the 1930s, particularly through the efforts of C. B. van Niel. In van Niel's time, it was held that during photosynthesis, light energy broke down carbon dioxide, with oxygen escaping as a gas and carbon joining water to form CH_2O—the empirical carbohydrate. Van Niel knew from his studies of sulfur bacteria that photosynthesis could occur without the production of oxygen. In sulfur bacteria, the source of hydrogen is hydrogen sulfide, and when oxidized, the byproduct is elemental sulfur rather than oxygen. This fact suggested to van Niel that the source of oxygen released during plant photosynthesis was water rather than carbon dioxide. He proposed a general equation for all photosynthesis:

$$CO_2 + 2H_2A \longrightarrow (CH_2O) + H_2O + 2A$$

In this equation the H_2A represents the source of electrons, which could be either hydrogen sulfide, water, or

some other similar substance—but not carbon dioxide.

While van Niel's arguments were certainly persuasive and logical, persuasion and logic are not proofs. There was, at the time, no way of knowing whether bacterial photosynthesis was similar enough to photosynthesis in algae and plants for van Niel to be certain of his hypothesis. In fact, the final proof had to await the development of radically new research techniques, which in this instance turned out to be products of the newly emerging field of nuclear physics. In 1941, certain radioactive tracers became available, and at the University of California, Samuel Ruben and Martin Kamen used oxygen 18 (^{18}O) to carry out the critical experiment. If van Niel was correct, then if plants or algae were given water containing the radioactive isotope of oxygen, ^{18}O ($H_2^{18}O$), rather than the usual H_2O, the oxygen released during photosynthesis would be radioactive. Further, there would be no radioactivity present in the carbohydrates produced by the plants at this time. On the other hand, still testing van Niel's hypothesis, if the carbon dioxide given plants or algae contained radioactive oxygen ($C^{18}O$), then the carbohydrate would contain radioactive oxygen, and the oxygen gas given off

(Continued on next page)

would be free of radioactivity. Using the green alga *Chlorella,* Ruben and Kamen were able to show that van Niel's hypothesis was quite correct. In the first instance, the oxygen gas emerging from the photosynthesizing alga was radioactive, while the carbohydrate was not. In the second, as expected, the oxygen gas given off was nonradioactive, while the carbohydrate was shown to contain ^{18}O, the radioactive isotope.

In the late 1930s, British biochemist Robert Hill provided strong experimental evidence that supported van Niel's theorizing. Hill was among the first to successfully isolate and experiment with chloroplasts. From his experiments it was determined that oxygen production in chloroplasts could occur independently of carbon dioxide fixation. In fact, it occurred in the absence of carbon dioxide. All that was required was the presence of an electron acceptor. From these observations, it became clear that photosynthesis occurs in two separate series of reactions: the light reactions involving water and the light-independent reactions involving

carbon dioxide fixation. From Hill's discoveries, the experimental focus turned to the role of light in creating a flow of electrons and the pathways taken by the electrons. By 1951, the role of $NADP^+$ (first called TPN^+) as an electron acceptor was known. In 1954, Daniel Arnon was able to demonstrate the entire photosynthetic process in isolated chloroplasts. His accomplishment was a milestone in the saga of research in photosynthesis.

Arnon and his associates at the University of California at Berkeley, taking their clues from Hill's brilliant work, used intact chloroplasts as their "experimental organism." Much of the essential material in this chapter has come down from Arnon's initial discoveries. For example, in an early experiment, he determined that if carbon dioxide was withheld, chloroplasts could carry out all of the known photosynthetic reactions except carbohydrate synthesis. In other words, the process yielded only ATP, $NADPH + H^+$, and oxygen—but no glucose. From this observation, Arnon was able to propose that photo-

synthesis occurred in two distinct phases, the light and dark reactions.

Arnon then determined that if he withheld both carbon dioxide and NADP, providing his chloroplasts with ADP, P_i, water, and light only, then the only photosynthetic product would be ATP. There was, as expected, no glucose, no $NADPH + H^+$, and, perhaps surprisingly, *no oxygen*. The lack of oxygen was significant in that it indicated that a simpler, cyclic process may be going on in the absence of NADP that could yield ATP. We, of course, now call this independent process the cyclic light reaction.

In a final experiment, elegant in its conceptual simplicity (although technically very difficult), Arnon placed his chloroplast suspension in the dark for a time, washed away any late-forming products, and waited for synthesis to stop. He then supplied the chloroplasts with CO_2 along with the products normally produced in the light-dependent reactions: ATP and $NADPH + H^+$. Once again, the amazing little organelles became active and went on to produce carbohy-

Through the use of freeze fracturing, a thylakoid membrane is sheared and peeled back between its bilipid layers, and various protein complexes are revealed. Cytologists, reconstructing the thylakoid membrane from such views, think that the particles represent photosystems I and II within stacked regions and photosystem I in the unstacked regions. The largest clustered particles seen here (center) represent proteins of photosystem II.

drate. As you can see, we owe much of what we understand about photosynthesis to the efforts of Robert Hill and Daniel Arnon and his associates.

Research in photosynthesis has been consistently intense over the years, but recently the interest has moved from the specific chemical reactions to the structures in which they occur. Motivating the switch has been the chemiosmotic theory of Peter Mitchell, the British biochemist (see Essay 6.1). The chemiosmotic theory has become well entrenched in both photosynthetic and respiratory biochemistry today.

Recent studies of the thylakoid membrane, using freeze-fracture techniques, reveal details that help reinforce the chemiosmotic hypothesis. As Mitchell's hypothesis predicted, the CF1 particles are structurally independent of all other membrane proteins. Mitchell had predicted that phosphorylation was a separate process from the photoactivation of electrons and their subsequent transport through electron transport systems. Today's efforts have also revealed an elaborate protein array in the thylakoid membranes, and cytologists are almost certain that they have identified specific proteins and other elements of photosystem I and photosystem II.

Freeze-fracture studies of the thylakoid membrane (see the electron micrograph) reveal that the molecular constituents found in stacked membranes (of the grana) differ considerably from constituents found in unstacked regions (the lamellae between grana and the uppermost and lowermost membranes of the grana). This freeze-fracture study reveals details at four depths in a pair of membranes from the stacked region, but such details must be reconstructed before they can be interpreted. Such reconstructions by cytologists have led to the artist's representation seen here.

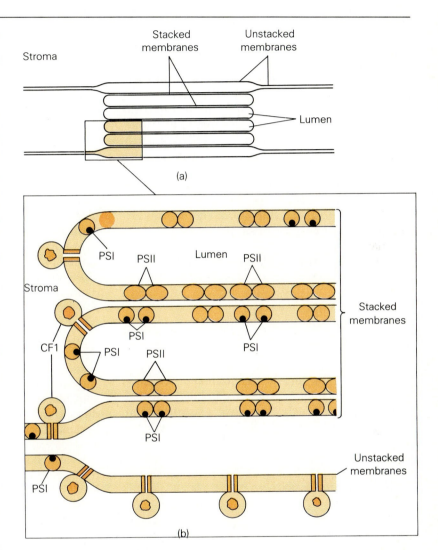

(a)

(b)

In the series, the first (a) helps distinguish the stacked and unstacked regions of the thylakoids. In the second (b) it becomes obvious that two types of molecular complexes occur in the unstacked regions: the CF1 particles and the Photosystem I components, including their light-harvesting antennas and the components of the electron/proton transport system (see legend). Their isolation from other components supports the belief that ATP phosphorylation is independent of light capture and electron/hydrogen transport, and that the cyclic events may occur quite independently of other photosynthetic events. Finally, in the stacked region, where adjacent thylakoids come close together, we find the molecular constitutents of both photosystems, I and II. Note that each photosystem II (P680) system is opposed by a photosystem I (P700) system—just as one might predict in considering the noncyclic events.

As we can see, energetic ATP and reduced NADPH make their contributions to glucose-building and are returned to their original condition (actually recycled). The carbon dioxide and the hydrogen from NADPH end up in glucose. The extra water entering is used in the hydrolysis of 18ATP and eventually forms the —H and —OH groups in ADP and P_i. Finally, as we will see, there are many more steps in the reactions than the formula indicates, and each is directed by specific enzymes.

Now that we are aware of some of the basic chemistry of photosynthesis, let's turn to the structures in which the process occurs. Modern research in photosynthesis is focused on the structure of the cell, and we are beginning to make clear and incredibly fine associations between biochemistry and structure. The structure we are most concerned with is, of course, the chloroplast and its intricate membranous inner structures, the thylakoids. Understanding the structure of the thylakoid will take us a long way toward understanding the light reactions.

CHLOROPLASTS AND THE LIGHT REACTIONS

We described the general organization of the chloroplasts in Chapter 4, where we pointed out that they contain relatively clear areas of **stroma,** and many, many membranous thylakoids, stacked into **grana.** Each granum, you may recall, is attached to its neighbor via membranous extensions called **lamellae.** Figure 7.3 reviews the structure of the chloroplast and introduces the thylakoid membrane.

The Thylakoid

The disklike shape of a single thylakoid, viewed in cross-section, is revealing (Figure 7.3b and c). The double membrane bulges out to enclose a central space called the lumen. Each lumen formed is thus isolated by its membranes from all others and from the stroma outside the thylakoid. But our real interest here is in the special structures imbedded into the membranes. It is here that light is absorbed, and here that its energy is used to produce ATP and NADPH.

The membranous structures include numerous chlorophyll-rich **light-harvesting antennas** and their all-important **reaction centers, electron transport systems,** and **hydrogen carriers**—all grouped into what are known as **photosystems.** There are two kinds of photosystems, designated **photosystem I** and **photosystem II,** (or PSI and PSII) each with its own light-harvesting antenna and electron transport system. The reaction center in each photosystem absorbs light of a somewhat different wavelength, but the two photosystems pair up to work cooperatively in the light reactions. Separate from the photosystems are the CF1 particles, where, as you may recall from the last chapter, ATP is actually synthesized (see Figure 7.3c). We will look into each of these in greater detail, beginning with the chlorophyll pigments and their role in absorbing light.

Chlorophyll and the Light-Harvesting Antennas. As their name implies, the role of the light-harvesting antennas is to intercept light, to gather it in, as it were, and concentrate the energy. The antennas consist essentially of three kinds of pigment molecules: **chlorophyll** *a,* **chlorophyll** *b,* and **carotenes.** Hundreds of these molecules are found in each antenna, bound into a pattern by a framework of protein. Their arrangement results in the absorbed light energy being shunted to a reaction center which usually contains one molecule of chlorophyll *a* (see Figure 7.3d).

Chlorophyll *a* and *b* are quite similar, but they differ completely from the carotenes (Figure 7.4). Because of their molecular differences, each absorbs wavelengths of light in its own way. Thus, chlorophyll *a* absorbs primarily in the violet and red region of the visible spectrum, while chlorophyll *b* absorbs more in the blue and red region. The carotenes absorb light primarily in the blue and green wavelengths. For more on this subject, see Essay 7.2.

The reaction centers form the heart of the photosystems, with the reaction center of photosystem I absorbing light, as we mentioned, at somewhat different wavelengths from the reaction center of photosystem II. In the first, light is best absorbed at wavelengths of 700 nm (nanometers), while the second best absorbs light at 680 nm. For this reason, the reaction center of PSI is often referred to as P700 and that of PSII, as P680. We will find the photosystems I and II have slightly different roles in the light reactions.

When light is shunted to chlorophyll *a* of the reaction center, the energy is absorbed by its elec-

7.3 FINDING THE PHOTOSYSTEMS

Moving through several levels of organization, we find that the plant leaf contains countless light-capturing structural units. **(a)** Each leaf contains layer upon layer of photosynthetic cells, each with many chloroplasts present. **(b)** Within each chloroplast lie the thylakoids, stacked into grana. Each thylakoid balloons out to form a hollow lumen. **(c)** The thylakoid membrane is complex, with two kinds of photosystems and the ATP-producing CF1 particles embedded in its bilipid structure. **(d)** The heart of each photosystem is a light-harvesting antenna with its central reaction center.

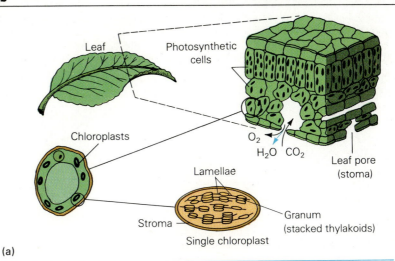

(a)

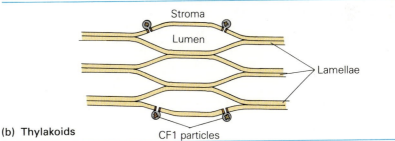

(b) Thylakoids

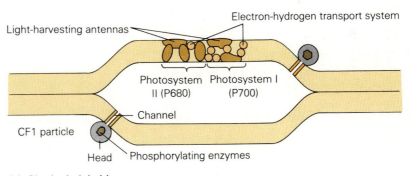

(c) Single thylakoid

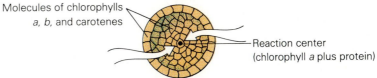

(d) Light-harvesting antenna (top view)

7.4 PHOTOSYNTHETIC PIGMENTS

The chlorophylls consists of a complex head with a lengthy hydrocarbon tail. Chlorophyll *a* differs only slightly from chlorophyll *b*. Two common carotenoids, beta-carotene and lutein, are shown here. The carotenoids are accessory pigments to the chlorophylls. They absorb light only in the blue-green range. Their function is believed to be that of a light-absorbing antenna, shunting their energy into more active regions of chlorophyll.

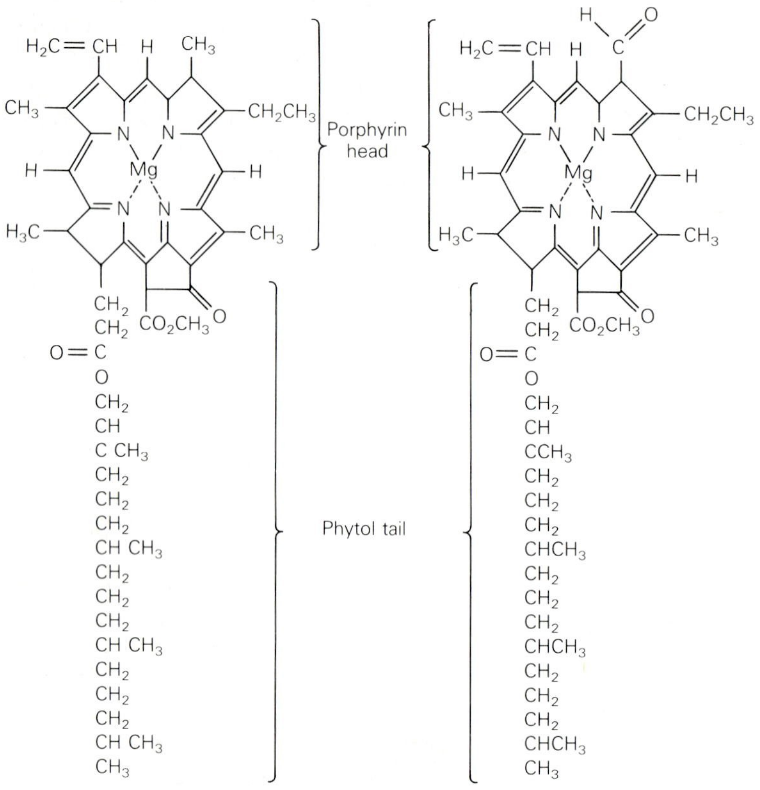

Chlorophyll *a*

Chlorophyll *b*

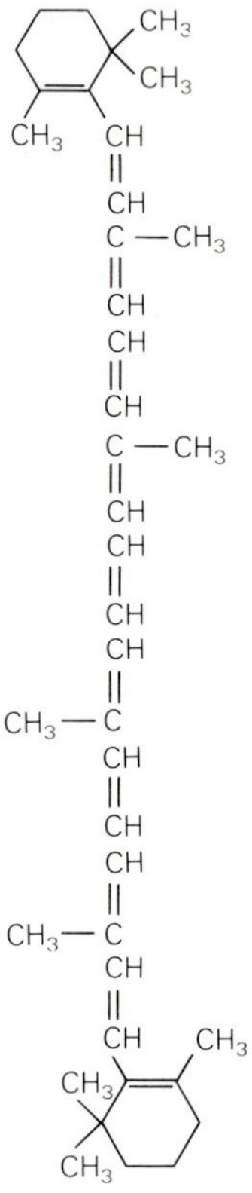

β-Carotene

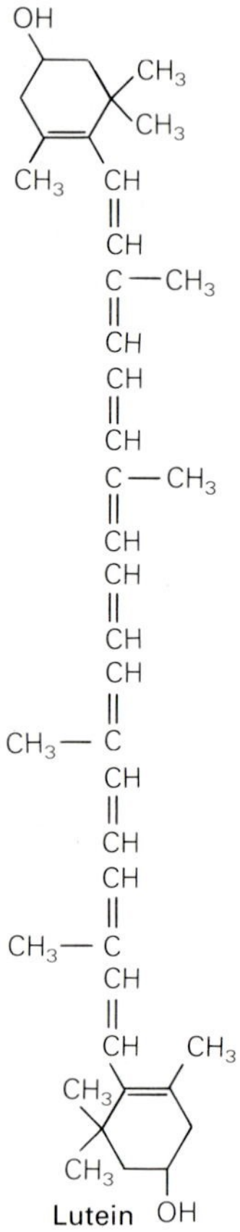

Lutein

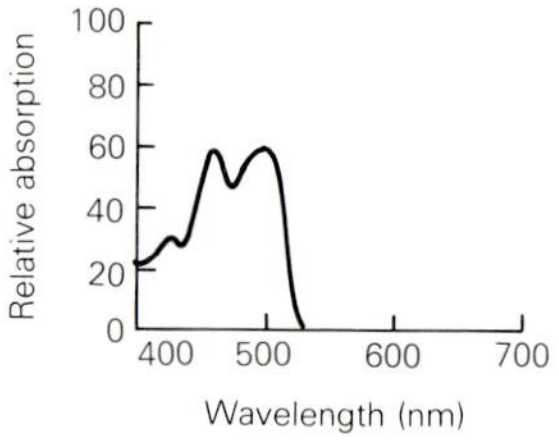

Carotenoids absorb in the blues and greens

trons. What happens next is the very essence of photosynthesis, providing the link between light energy and the energy soon to be found in the chemical bonds of glucose. An electron from chlorophyll *a* absorbs the radiant energy and then, with its energy level substantially raised, it leaves its orbit. In the world of molecules, escaping electrons are not unusual and they generally just fall back into their orbitals, releasing their temporary energy as light and heat. But in the photosystem, this does not happen (Figure 7.5). The energized electron is captured at once by the electron transport system, which will drain it of its energy and thereby power the light reactions of photosynthesis. Of course, one energized electron cannot accomplish much photosynthetic work. But considering the many photosystems at work in each thylakoid, and multiplying this by the great number of thylakoids in

each chloroplast, and then by the number of chloroplasts per leaf cell, and so on, and you can build a new appreciation for the unapplauded green leaves that grace our lives.

Electron Transport Systems and Proton Pumps. Also embedded in the thylakoid membrane and closely associated with the two photosystems are the electron transport systems and the proton, or hydrogen ion, pumps. We must emphasize that while much is known about the members of electron transport systems in the mitochondrion, little is known about these systems in the thylakoids. This account therefore represents some of what is known along with some conjecture on our part.

The electron transport systems of the thylakoid have two important roles. As we pointed out in the

7.5 TRAPPING ELECTRON ENERGY

In nonphotosynthetic situations (a), the absorption of light energy by an atom produces a temporary shift in an electron's energy. It may leave its orbit, give off heat and light, and return to its former energy level. When light is absorbed by the photosystems (b), energetic electrons are similarly ejected, but they do not give off all their excess energy. Rather, the excited electron joins an electron acceptor, which then starts it through the electron transport system.

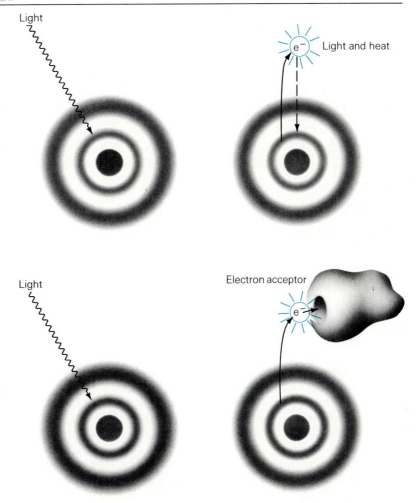

The Visible Light Spectrum and Photosynthesis

The earth is constantly bathed in radiation emanating not only from the sun, but from a host of other celestial bodies. Part of the radiation that reaches us is **visible light.** Visible light, however, is only part of an **electromagnetic spectrum** that includes (from lower to higher energy) radio waves, microwaves, infrared radiation (heat), visible light, ultraviolet (UV) radiation, X rays, and gamma rays.

At the low-energy end of the spectrum are very long waves, while at the other extreme are very short, highly energetic waves. Near the middle of this continuum is visible light. Visible light is visible because it interacts with special pigments (light-absorbing molecules) in our eyes. It also interacts with pigments such as chlorophyll, the molecule that absorbs the energy of light and provides power for photosynthesis.

Light is very hard to describe in nontechnical terms. One reason is that it can be considered in two ways: as particles called **photons,** or as waves. Arguments have raged over these two concepts for years. Here,

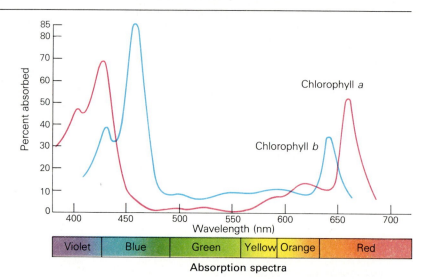

Absorption spectra

we will assert that light is composed of photons that move like waves. This side step will enable us to describe the energy of a photon in terms of its wavelength.

The specific light-absorbing qualities of the chlorophylls and carotenes can be determined by using a device

known as a spectrophotometer. First, the pigments are extracted and dissolved in a solution. Next, light of a known wavelength is passed through the solution and whatever light is not absorbed is detected on the other side. The wavelength of the entering light can be varied to see which wave-

last chapter, these systems use the energy of photoactivated electrons to pump protons across the membranes of mitochondria and thylakoids. In the thylakoid, the protons gather in the lumen, where their presence produces the electrochemical proton gradient of the chemiosmotic system. The second role of the electron transport systems is to reduce $NADP^+$ to NADPH, which, as we have noted, is needed for glucose production in the light-independent reactions.

The CF1 Particles.   Recall that ATP is produced in the CF1 particles. These are embedded in the thylakoid membrane but are separated from other membranal elements (see Figures 7.3b and c). The protruding, spherical bodies contain the all-important phosphorylating (phosphate-adding) enzymes for forming the energy-rich ATP bonds. The specific orientation of the particles and their chan-

nels is significant to chemiosmotic phosphorylation.

Now let us see how the various elements of the photosystems do their work. The antennas are ready, the electron transport systems and proton pumps are idling, the CF1 particles have gathered in ADP and P_i, and the $NADP^+$ is waiting. We also have a supply of water on hand. So what do we need? We need light.

THE LIGHT REACTIONS OF PHOTOSYNTHESIS

Light falling on a green leaf passes into its photosynthetic cells and enters thousands upon thousands of green chloroplasts. Within the many thylakoids, millions of light-harvesting antennas re-

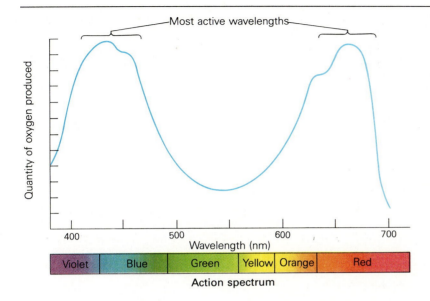

Action spectrum

length is most absorbed by the solution. Finally, the data are usually plotted on a graph to form what is called an **absorption spectrum**. Note the absorption spectrum for chlorophylls *a* and *b* on the graph shown here. The peaks represent light that is absorbed by the pigment, while the valleys represent light that passes through the solution. As you can see, the green and yellow hues are least absorbed and are transmitted or reflected, although the yellows are often masked by the darker green color. Thus, these are the colors we see when we look at a chlorophyll solution—or when we look around us at the greenery we treasure.

Certain wavelengths, such as violet-blue and orange-red, are strongly absorbed by chlorophyll. This indicates that these wavelengths are used in photosynthesis, but the evidence is circumstantial. It is possible, however, to obtain more direct kinds of evidence. One way is to discover the rate at which some product of photosynthesis is produced when groups of plants are subjected to different wavelengths of visible light. Plants produce oxygen gas at a rate proportional to the rate of photosynthesis: for every glucose produced, six oxygen molecules are released. Measuring the volume of oxygen gas produced under varied wavelengths can determine which wavelengths are most effective in photosynthesis.

With these data we can plot what is known as an **action spectrum** (as shown here). Action spectra for the chlorophylls turn out to be very similar to absorption spectra, indicating that the wavelengths of light absorbed by chlorophyll are the wavelengths that drive photosynthesis.

spond. As we take a close look at only one pair of photosystems, the events of the light reactions begin to unfold. There are two ways in which photosystem I and II can act: (1) the two can work cooperatively to build the chemiosmotic differential and reduce NADP⁺ to NADPH, an effort known as **noncyclic photophosphorylation;**[1] or (2) photosystem I can act alone, simply building the chemiosmotic differential for ATP synthesis in what is called **cyclic photophosphorylation.** We will begin with noncyclic photophosphorylation, since it appears to be the most common event and since it is more significant to the light-independent reactions.

Photosystems II and I and the Noncyclic Events

The noncyclic reactions (Figure 7.6) begin as light is absorbed by the photosystem I and II light harvesting antennas. While the events following occur simultaneously in both photosystems, we will begin with photosystem II and its P680 reaction center. As light energy is absorbed by the antenna and shunted to chlorophyll *a* of the reaction center, an electron is energized and, resisting the powerful pull of its own nucleus, it quickly passes to a nearby electron acceptor molecule, a large protein desig-

[1]We should note that the terms *noncyclic* and *cyclic photophosphorylation* were developed before the chemiosmotic hypothesis appeared, when ATP synthesis was believed to occur during electron transport. It will become obvious that the terms are not totally accurate in describing the light reactions since phosphorylation is no longer believed to occur during the noncyclic and cyclic events, but is now thought to be a separate process. For this reason, we will refer to "chemiosmotic phosphorylation" in our descriptions of ATP synthesis.

7.6 PHOTOSYSTEMS AND NONCYCLIC EVENTS

(a) Photosystems II and I contain light-absorbing antennas and electron/proton transport systems. Each carrier is designated by letters representing its longer name (see legend).

(b) In the noncyclic events involving photosystem II, electrons from water end up in NADPH, but their route is complex and circuitous. ① Light is absorbed by the reaction center (P680) of photosystem II. ② An excited electron leaves chlorophyll *a*, passing through the electron transport system. ③ During the transit, a proton is attracted into the system and ferried across to the lumen of the thylakoid. ④ The electron continues through the electron transport system to ⑤ photo-

system I. ⑥ Following the ejection of the P680 electron, its chlorophyll *a* is reduced (its electron restored) by an electron from water.

(c) Events in photosystem I are more direct. ⑦ Light is absorbed by the P700 reaction center and ⑧ an excited electron is ejected (making room for the P680 electron). ⑨ The electron passing through the photosystem I electron transport system reaches FAD, and as soon as FAD has collected two electrons, it captures two protons from the stroma, being reduced to $FADH_2$. ⑩ $FADH_2$ in turn reduces NADP, and the light reactions end.

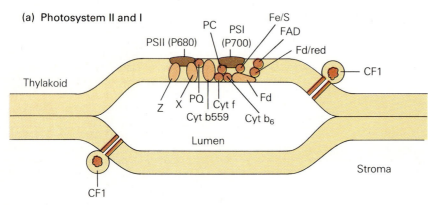

(a) Photosystem II and I

Z - water oxidizing system
X - unknown receptor
PQ - plastoquinone
Cyt b559 - cytochrome b559
Cyt f - cytochrome f (or C_{552})
PC - plastocyanin
Fe/S - iron/sulfur protein
Fd - ferredoxin
Fd/red - ferredoxin reductase
Cyt b_6 - cytochrome b_6 or (b_{563})
FAD - flavin adenine dinucleotide

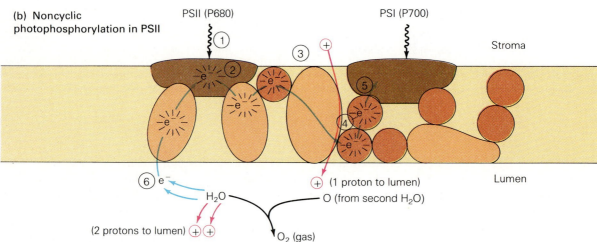

(b) Noncyclic photophosphorylation in PSII

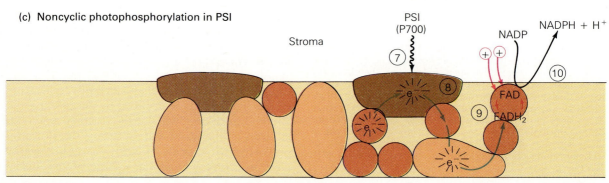

(c) Noncyclic photophosphorylation in PSI

nated X in Figure 7.6a. The symbol X simply means that this carrier has not been completely identified. Upon accepting the electron, protein X's reducing power is greatly increased, and it will soon pass the electron to the next carrier in the electron transport system. Before tracing its path through the system, let's look at another important event associated with chlorophyll *a* which involves water.

Water and the Reduction of P680. The loss of an electron leaves chlorophyll *a* in a highly oxidized, "electron hungry" state. The electron needs of oxidized chlorophyll *a* will be satisfied by electrons from nearby *protein Z*, which, in turn, will obtain its electrons from water. Protein Z is quite essential, but biochemists know little about its structure.

Water is quite reluctant to give up its electrons, and only a very powerful oxidizing agent or agents will do the job. The actual chemistry is quite complex (Figure 7.7) and incompletely understood, but it is set in motion by the light-induced oxidation of chlorophyll *a*. Electrons are literally pulled from the hydrogen atoms of water molecules, to be passed to protein Z and on to chlorophyll *a*, reducing it and permitting it to act again. For every two electrons entering Z, two hydrogen protons and an oxygen atom are left over.

The oxygen atoms join oxygen from other similarly disrupted water molecules, forming O_2, which simply escapes from the plant as the oxygen gas given off during photosynthesis. The fate of the two protons is more important to photosynthesis. As you can see in Figure 7.6, these are released directly into the thylakoid lumen, and thereby add directly to the chemiosmotic differential.

Proton Pumping and the Reduction of P700. Returning to our energy-rich P680 electron, we follow it through the photosystem II electron transport system. As was mentioned, the function of an electron transport system is to power the pumping of protons across the membrane, in this case from the stroma outside to the thylakoid lumen within. Following the route of an energized electron in Figure 7.6, we see that it is handed off from the reaction center to a small, mobile carrier (protein Q) and then to a large protein complex, which captures a proton from the stroma and then pumps it to the lumen. The electron, now having lost much of its free energy, enters photosystem I. The most important event in the photosystem II electron transport chain has been the pumping of a proton into the lumen, thus increasing the chemiosmotic differential.

7.7 THE OXIDATION OF WATER

Basing the chemistry of this event on the formation of one molecule of oxygen (O_2) water is oxidized two molecules at a time. The Z protein, fully reduced when it has received four electrons from water, becomes a strong reducing agent. It then reduces (passes its electrons to) chlorophyll *a* of the P680 reaction center, as the reaction center is itself oxidized by light energy, passing its electrons to "protein X." Note that for each complete oxidation sequence (two water molecules), four protons are added to the chemiosmotic gradient and one molecule of oxygen is released as a gas.

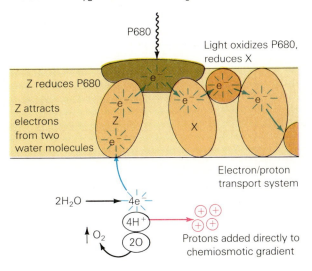

Events in Photosystem I. P680 electrons can only enter P700 when it is ready to receive them, which means that the chlorophyll in that reaction center must be oxidized. This occurs as light is absorbed and an excited P700 electron enters the first acceptor in the photosystem I electron transport system. Thus, while oxidized chlorophyll *a* in the P680 reaction center is reduced by electrons from water, it is the P680 electron that reduces the reaction center of P700.

The transit of electrons through the carriers of P700 is direct (see Figure 7.6), with no pumping of protons. The only purpose of photosystem I in the noncyclic events is to use the energy of excited P700 electrons to reduce $NADP^+$ to NADPH. Actually, the final membranal carrier is coenzyme FAD, which, after receiving two electrons from the P700 carriers, attracts two protons from the stroma, becoming $FADH_2$. It is reduced FAD that finally reduces $NADP^+$ to NADPH. This is the final step in the noncyclic events.

In summary, we have seen that the chemiosmotic differential is increased twice—once during the oxidation of water and once in the photosystem II electron transport system. We have also

7.8 REDUCING POWER AND THE Z-SCHEME

In the Z-scheme diagram, the vertical scale represents the oxidation reduction potential, in which the negative numbers indicate high reducing power. Events begin with the oxidation of a water molecule, in which two hydrogen ions are released into the interior space of the thylakoid, two electrons enter the electron transport chain, and one oxygen atom escapes to join another, forming O_2. The path of the electrons is followed from left to right through a series of carriers. While electrons flow toward the most oxidized (least reduced) carriers, at two points—P680 and P700—photon (light) energy is used to boost the electrons to a more excited state with increased reducing power. Note that during their transit through the two photosystems, the energy of the excited electrons is used to pump hydrogen ions into the thylakoid lumen, and at the end to reduce $NADP^+$ to NADPH.

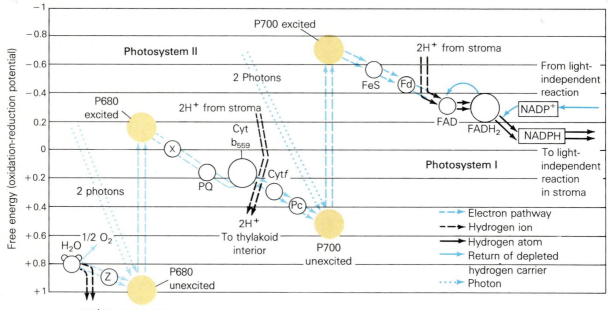

seen how electrons from photosystem I are used to reduce $NADP^+$ to NADPH. Further, the noncyclic events can be viewed as a continuous process, with electrons flowing from water to $NADP^+$, like the electrical current alluded to earlier.

Traditionally, the shifting energy levels of electrons and carriers and the currentlike flow through the photosystems has been shown somewhat differently, in what is called the **Z-scheme** (Figure 7.8). It may help to look this over and gain a somewhat different perspective on the process. Note that in the Z-scheme, electrons move and protons are shifted as pairs.

Finally, if we return to our general formula for the light reactions (above), we see that the process must occur many times to provide enough NADPH and ATP for the synthesis of one glucose in the light-independent reactions.

Summing Up the Noncyclic Events. Now let's briefly recap what goes on in the light depen-

dent reactions. In the order of occurrence, the events are

1. Absorption of photons by P680;
2. Movement of excited P680 electrons into the electron transport system;
3. Oxidation of water, restoring of electrons to P680, adding of protons to the lumen, and releasing of oxygen gas;
4. Use of the free energy of P680 electrons to power proton transport from the stroma to the lumen;
5. Absorption of photons by P700;
6. Replacement of P700 electrons by those of P680;
7. Movement of excited P700 electrons into the electron transport system;
8. Use of P700 electron energy to reduce $NADP^+$ to NADPH.

The most important point to keep in mind is that the light-dependent reactions provide ATP and NADPH for the light-independent reactions.

Photosystem I and the Cyclic Events

As we indicated earlier, the photosystems can act in a cooperative, noncyclic manner, or photosystem I can act on its own, in a cyclic manner. As we see in Figure 7.9, the term *cyclic* refers to the repeating pathway of P700 electrons. When activated by light, electrons leave P700, pass through the carriers in their related electron transport system, and return to P700, where they reduce its chlorophyll *a*. At the moment, we can only make informed guesses about the exact path of cycling electrons, so the scheme in Figure 7.9 is hypothetical. Unlike the noncyclic events, cyclic activity does not produce NADPH. Instead, the decreasing energy of P700 electrons is used strictly in pumping protons from the stroma to the lumen. Thus, the cyclic events help only in producing ATP.

No one is quite sure how the cyclic events came to be, but there are some prevailing ideas. One of these is that the cyclic events occur when $NADP^+$ is in short supply. In other words, photosystem I goes into "idle," switching to proton pumping, until the light-independent reactions can cycle oxidized NADP ($NADP^+$) back to the light reactions. This effort is certainly not wasted since any increase in the chemiosmotic differential means a potential gain in ATP. We might also point out that in very recent electron microscope studies of the thylakoid, it was discovered that the elements of photosystem I frequently occur independently of and at some distance from those of photosystem II. The cyclic events may thus prove to be a routine part of the light reactions (see Essay 7.1).

Chemiosmotic Phosphorylation

Now that we have seen how the chemiosmotic differential is built up during the noncyclic and cyclic events, let's turn to chemiosmotic phosphorylation and the production of ATP. We will begin by reviewing some of what is known about the chemiosmotic system and mentioning a few specific things about the system in chloroplasts. There are important differences between the chemiosmotic differential in chloroplasts and mitochondria.

As positively charged protons accumulate in the lumen, they tend to attract negatively charged chloride ions (Cl^-) from the stroma. Chloride ions move rather freely across the thylakoid membranes, and as a result, the lumen becomes a pool of concentrated **hydrochloric acid** (HCl). Further, as protons are pumped out of the stroma, this outer region takes on an alkaline or basic characteristic, simply because of the loss of protons or hydrogen ions (H^+). Thus, the chemiosmotic system in the thylakoid becomes a concentrated acid solution (about pH 4 to pH 5), separated by a membrane from a concentrated alkaline solution (pH 8 to pH 8.5). We characterized this as a pH gradient in Chapter 6. Such a system has great free energy, and this energy can be used to do work if the system is permitted to run down or reach equilibrium.

7.9 CYCLIC LIGHT REACTIONS

In the cyclic events of the light reactions, photoactivated electrons from P700 ① begin their journey through the photosystem I electron transport system as usual ②, but when they reach a compound called ferredoxin (FD), they follow an alternative route through a pigment known as cytochrome b_6 ③. In so doing they do not reach FAD, so no NADPH is produced. The alternate, cyclic route takes the electron through cytochrome *f*, back to plastocyanin (PC) ④, and along this route a proton is attracted from the stroma and ferried across to the lumen ⑤ (at the present time the proton's actual route is unknown). The electron then returns to P700, completing its cycle. Thus, the final product of the cyclic events is simply the enrichment of the chemiosmotic differential by one proton per electron from P700.

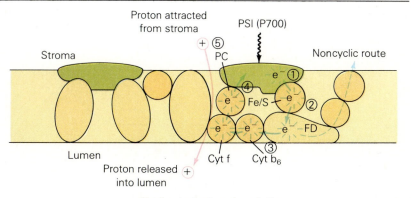

Cyclic photophosphorylation

In the laboratory, permitting a mixture of acids and bases to come together would produce a sudden heat rise—perhaps even an explosion. But in the thylakoid, things are much more gradual and certainly better controlled. Thus, as the acidic protons (H^+) and basic hydroxyl groups (OH^-) come together, the heat rise will be minimal and much of the free energy released as the system runs down will be used to generate energy-rich ATP bonds (Figure 7.10). This happens as protons pass down through the channels of the CF1 particles to the heads, where ADP, P_i, and phosphorylating enzymes are waiting. No one yet knows the actual events in this step, but somehow, the free energy is trapped in new energy-rich ATP bonds.

We have portrayed the light reactions in a rather mechanical, and slightly unrealistic, manner. For one thing, the process seems to be a rather plodding event when, in reality, it occurs with dazzling speed and in countless repetitions in innumerable chloroplasts. Perhaps the light reactions are more accurately visualized as a constant flow of electrons and energy through the photosystems. P680 and P700 work in both a "pushing" and "pulling" manner, with P680 providing the push and P700 providing the pull. The overall flow of electrons can best be portrayed as going from water to NADPH with the products made available for light-independent reactions. Thus, water must be constantly available,

entering along with sunlight at one end, with oxygen, ATP, and NADPH leaving at the other. The entire process is summarized in Figure 7.11.

THE LIGHT-INDEPENDENT REACTIONS

Through the light reactions, the chloroplast builds up a considerable store of potential energy and reducing power in ATP and NADPH, respectively. Thus, with an input of carbon dioxide, and more water, glucose production can actually begin. The glucose, as we have mentioned, is synthesized in the unstructured region of the chloroplast, the stroma. While there is no membranous organization in the stroma, it does contain a battery of synthesizing enzymes. The essentials of the light-independent reactions can be summarized as:

$$6CO_2 + 12NADPH + 18ATP + 12H_2O \rightarrow$$
$$C_6H_{12}O_6 + 12NADP^+ + 18ADP + 18P_i$$
(glucose)

Basically, the second half of photosynthesis involves incorporating, or **fixing,** inorganic carbon dioxide into glucose or other organic molecules that are useful to the plant. This process shouldn't seem too difficult since essentially all that must be done is

7.10 CHEMIOSMOTIC PHOSPHORYLATION

ATP is synthesized as protons reach a lower energy level by passing through the CF1 particles.

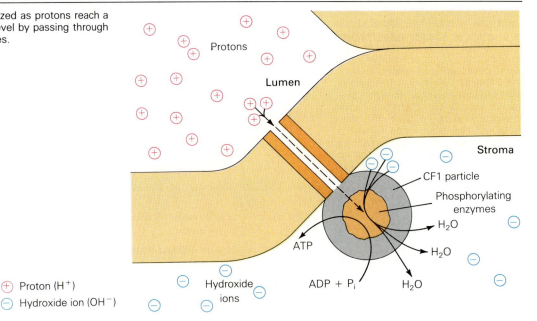

Protons

Lumen

Stroma

CF1 particle

Phosphorylating enzymes

H_2O

H_2O

ATP

ADP + P_i

H_2O

⊕ Proton (H^+)

⊖ Hydroxide ion (OH^-)

Hydroxide ions

7.11 LIGHT REACTIONS SCENARIO

The total picture of the light reactions of photosynthesis is seen as a continuous flow of electrons from water to NADPH. The currentlike flow is boosted by light absorbed in the reaction centers of the two photosystems. Note that the reaction seen here is actually only a "half reaction." A complete light reaction brings about the release of one molecule of oxygen (O_2), so it requires two molecules of water. Here we see two photons bringing about the release of two pairs of electrons from P680, and these are replaced by electrons from one molecule of water. The electrons will bring about the enrichment of the chemiosmotic gradient by four protons, two from water and two ferried across in photosystem II. In addition a reduced NADP (NADPH + H$^+$) will be formed at the end of photosystem I. The energy of the steep chemiosmotic gradient being formed is then tapped for producing ATP (lower left).

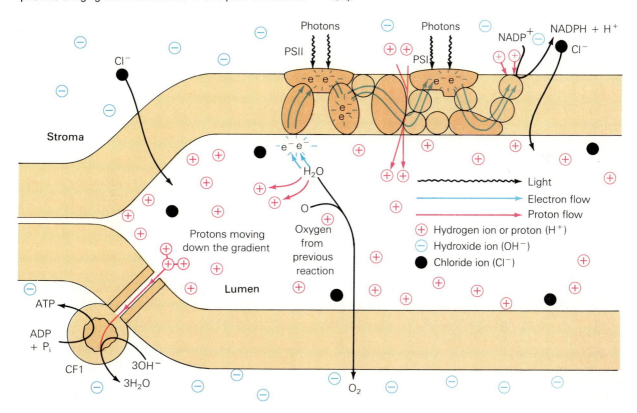

to put six CO_2 molecules together and rearrange some hydrogen and oxygen groups. But biochemistry is never that simple; the energetics of such straightforward events would simply be too costly. As stated earlier in the chapter, carbon dioxide is at a far lower free energy level than is glucose, and therefore would be quite reluctant to form glucose spontaneously. Only the great free energy of ATP and NADPH from the light reactions, and a battery of synthesis-directing enzymes, makes such an event possible. And these must be used in a highly ordered manner to achieve the right results.

With the aid of a key enzyme, which we will come to shortly, CO_2 molecules are ushered in and each is then chemically joined to a 5-carbon molecule already in existence. This all happens in a repeating pathway called the **Calvin cycle.**

Discovery of the Calvin Cycle

The Calvin cycle is named in honor of Melvin Calvin, who in 1945, along with his colleagues, began a lengthy series of investigations into the light-independent or carbohydrate-synthesizing reactions of photosynthesis. Specifically, they wanted to understand the chemical pathway by which carbon dioxide is incorporated into carbohydrate. Calvin chose a green alga called *Chlorella*, a single-celled, photosynthetic protist, as his subject. He also used newly available radioactive tracers, specifically carbon dioxide, that contained radioactive carbon 14 (^{14}C).

Calvin's procedure consisted essentially of injecting carbon 14 dicxide ($^{14}CO_2$) into active suspensions of the algae and, at certain intervals, killing the algae suddenly and extracting the carbohydrates. The time periods of each trial were critical, since the object was to stop the chemical pathway at various places and then locate, isolate, and identify any compounds that contained the radioactive carbon. After ten years of exacting analysis, Calvin was finally able to identify the chemical constituents of the carbohydrate-synthesizing pathway. This remarkable achievement won him international recognition and the coveted Nobel Prize. Let's see what Calvin found.

The Calvin Cycle

Notice in Figure 7.12 that the Calvin cycle essentially consists of a sequence of reactions in which teams of enzymes, working in a biochemical pathway, take their turns at altering the substrate moving through. With each change, a substrate molecule takes on a new form and identity. Notice that at a key point in the cycle, some of the products are drawn away to enter pathways where glucose and other final products are made. Other products stay in the Calvin cycle, returning to the starting place and keeping the cycle going.

Each modification of the substrate requires the presence of a specific enzyme as atoms or side groups are added and removed and bonds are rearranged. Note also that carbon dioxide enters the cycle at one step only, and that ATP and NADPH are brought in at very specific steps. We will go through the more significant steps, one at a time, beginning with step ① at the left of Figure 7.12.

Phosphorylation of Ribulose Phosphate. In the first step, one of the ATPs generated in the light reactions is converted to ADP when its terminal phosphate is transferred to **ribulose-5 phosphate (RuP),** a 5-carbon sugar already present in the cycle. The product, now containing two phosphates, is called **ribulose-1,5-bisphosphate (RuBP).** What has happened here is that a phosphorylating enzyme has added phosphate to the substrate, priming the molecule for changes to come. Phosphorylation simply sets the stage, making a future substitution reaction energetically favorable, so we can expect some other side group eventually to replace the phosphate.

Enter Carbon Dioxide. It is at this point that carbon dioxide first enters the cycle. In reaction ②, carbon dioxide is added to RuBP through the action of a **carboxylating** (carbon-adding) **enzyme.** We won't be naming many enzymes in this discus-

sion, but this one is an exception. The carboxylating enzyme, **ribulose-1,5-bisphosphate carboxylase,** is one of the most significant enzymes on earth, since it is responsible for the initial fixing of carbon into compounds upon which nearly all life depends. Further, some biochemists acknowledge that the enzyme is probably the most abundant protein on the earth. This brash-sounding statement is based on the fact that plants and other photosynthesizers make up the vast majority of living material on the planet.

We now have a 6-carbon sugar, but it's the wrong one, and a highly unstable one besides, breaking down spontaneously, without the help of an enzyme (reaction ③). In addition to a release of heat—this step is highly exergonic—the breakdown produces two molecules of the 3-carbon **phosphoglyceric acid,** known in its ionized form as **3-phosphoglycerate (3-PG).**

On to PGAL. Following their formation, the two 3-PGs react with two ATPs, emerging as doubly phosphorylated, **diphosphoglyceric acid,** or **1.3 diphosphoglycerate (DPG)** (reaction ④). The two emerging ADPs will recycle back to the light reactions. In view of what was said above about phosphorylation, the formation of two energy-rich phosphate bonds in each DPG should warn us that something significant is about to happen.

What happens is that NADPH enters the Calvin cycle. Two NADPHs react with the two DPGs, replacing a phosphate on each with a hydrogen; that is, the two DPGs are *de*phosphorylated and reduced (reaction ⑤). The newly oxidized NADP (NADP$^+$) cycles back to the light reactions along with the two P$_i$s that were replaced by hydrogen. The products of this key step are two molecules of **phosphoglyceraldehyde,** or PGAL. This is a key step since PGAL is actually a 3-carbon sugar—a carbohydrate. It is also a vital intermediate that can be used to produce many of the molecules of life. Before we see where the production of glucose comes in, let's first summarize the cycle.

Summarizing the Calvin Cycle. The major events of the Calvin cycle are

① Ribulose-phosphate (RuP) is phosphorylated by ATP, and ribulose-bisphosphate (RuBP) is formed;

② Carbon dioxide is joined to ribulose-bisphosphate, forming an unstable 6-carbon intermediate;

③ The 6-carbon intermediate is split into two molecules of 3-phosphoglycerate (3-PG);

7.12 THE CALVIN CYCLE

The light-independent reactions that comprise the Calvin cycle use the ATP and NADPH generated in the light-dependent reactions, plus carbon dioxide from the atmosphere. There are five key steps involved in the abbreviated version shown here: ① RuP is phosphorylated by ATP. ② Carbon is added to RuBP, ③ forming an unstable intermediate that is cleaved into two molecules of 3-PG. ④ Another phosphorylation converts each 3-PG to DPG. ⑤ Finally, NADPH enters, reducing the two DPG molecules of PGAL, a 3-carbon carbohydrate. The ADP remaining from the phosphorylation and the oxidized NADP will be recycled through the light-dependent reactions. ⑥ Quite often, PGAL is used in producing glucose. Two molecules of PGAL are joined enzymatically to form fructose diphosphate. ⑦ Later, other enzymes remove a phosphate and change the fructose to glucose-1-phosphate.

Upon dephosphorylation, simple glucose emerges, or it can immediately be converted to starch. PGAL must also recycle most of the time to keep the Calvin cycle going.

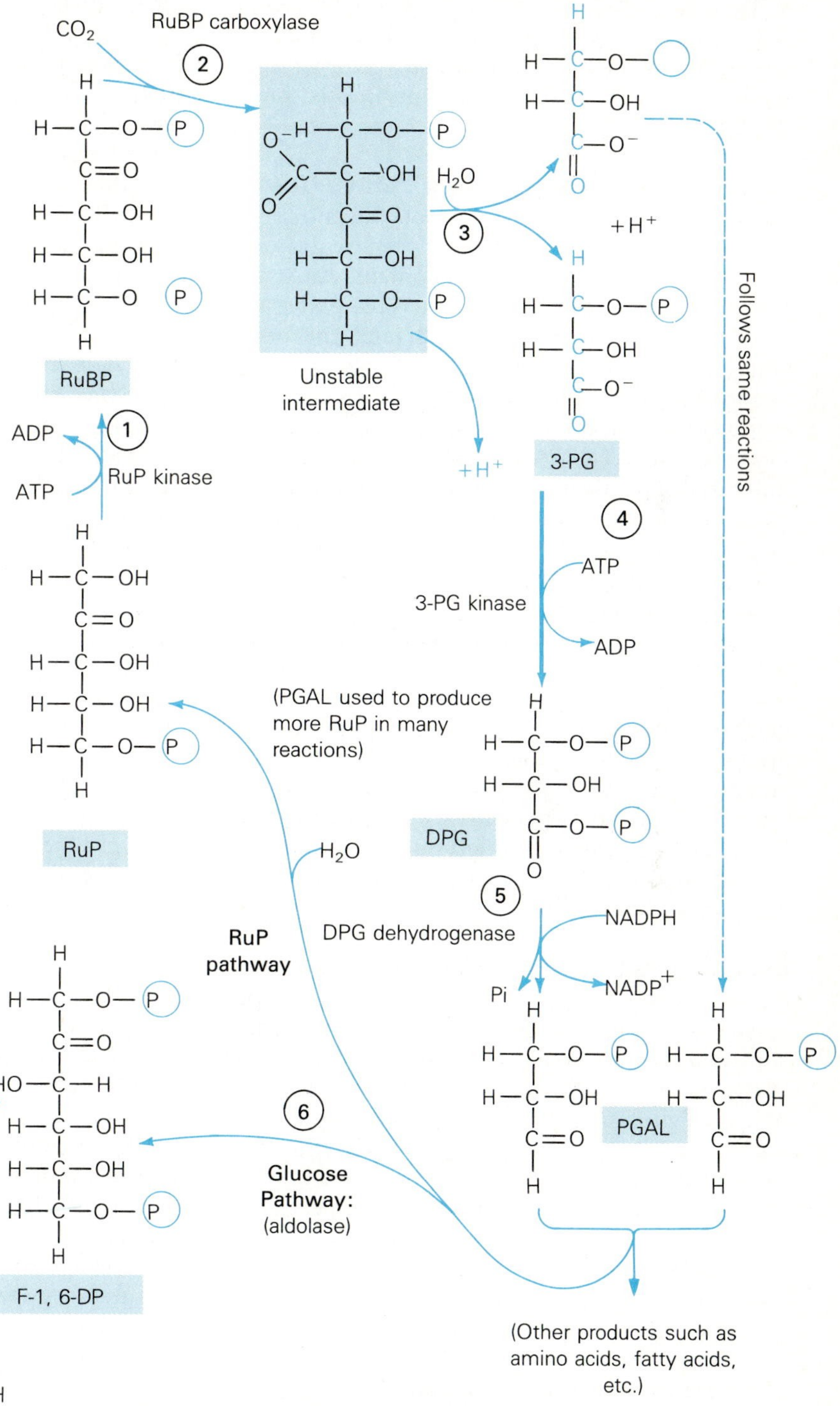

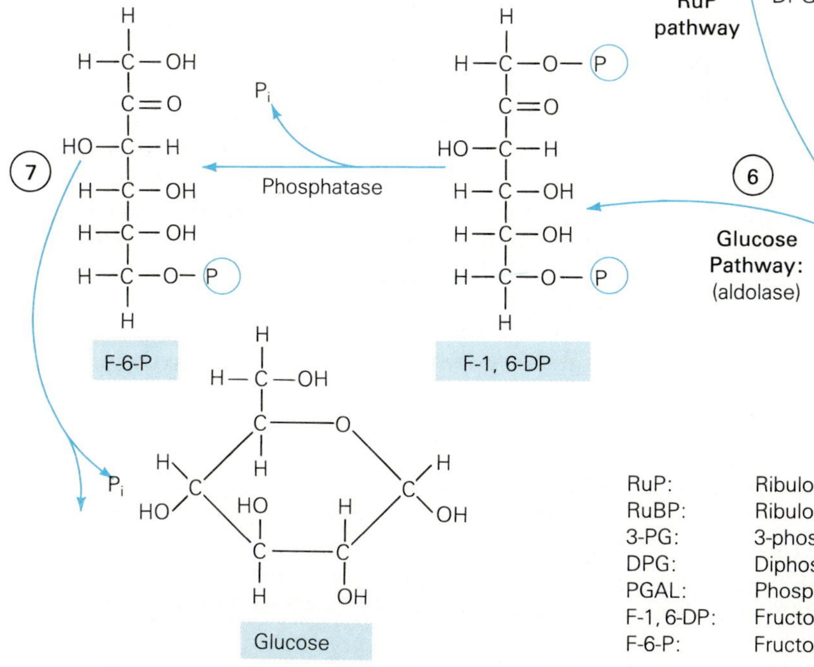

CO_2

RuBP carboxylase

② ③ ④ ⑤ ① ⑥ ⑦

RuBP

Unstable intermediate

$+H^+$

3-PG

Follows same reactions

ADP

RuP kinase

ATP

$+H^+$

④

ATP

3-PG kinase

ADP

(PGAL used to produce more RuP in many reactions)

RuP

H_2O

DPG

⑤

NADPH

$NADP^+$

P_i

RuP pathway

DPG dehydrogenase

PGAL

F-6-P

P_i

Phosphatase

F-1, 6-DP

⑥

Glucose Pathway: (aldolase)

P_i

Glucose

(Other products such as amino acids, fatty acids, etc.)

RuP: Ribulose-phosphate
RuBP: Ribulose-bisphosphate
3-PG: 3-phosphoglycerate
DPG: Diphosphoglycerate
PGAL: Phosphoglyceraldehyde
F-1, 6-DP: Fructose-1, 6-diphosphate
F-6-P: Fructose-6-phosphate

④ Two ATPs phosphorylate the two 3-PGs, forming two diphosphoglycerates (DPG);

⑤ Two NADPH + H⁺ reduce the two DPGs to two phosphoglyceraldehydes (PGAL), and with the appearance of these two 3-carbon carbohydrates, the Calvin cycle synthesis is complete.

PGAL, Glucose, and Keeping Things Going.
We have seen that PGAL represents the end of the Calvin cycle, but we have yet to see any glucose. Let's focus on the PGAL for a moment. After it is produced, PGAL can take two possible paths: the main pathway continues cycling, and the other finally leads to the assembly of glucose from the two PGALs (see Figure 7.12). The PGALs each lose a phosphate and are assembled into the 6-carbon sugar **fructose-1,6-diphosphate,** which is then converted into **glucose-1-phosphate.** From there, free glucose may be formed, or many glucose-1-phosphates can be combined to form starch through dephosphorylation and dehydration synthesis (see Chapter 3).

PGAL, it turns out, can be used for more than the production of glucose and other carbohydrates. For example, it can be modified and assembled into fatty acids and glycerol, thus permitting lipid synthesis. It can also be used in the synthesis of amino acids for assembly into protein. Alternatively, PGAL can be modified and sent into the mitochondria, where, through cell respiration and according to the plant's needs, its energy can be used to produce ATP.

Recycling PGAL.   Following the main cycle from PGAL, we see that much of the PGAL is simply recycled back to ribulose phosphate, the starting substrate of the Calvin cycle. If you think about it, this makes sense. We have described only one turn of the Calvin cycle, which means that only *one* carbon dioxide was brought in. Should the two PGALs produced be removed, the cycle will obviously stop for lack of intermediates. The Calvin cycle cannot yield a 6-carbon glucose molecule until six molecules of carbon dioxide have been ushered in, or until six complete turns of the Calvin cycle have occurred. A quick look at the general formula presented earlier will also confirm this. There we see that $6CO_2$ are required (along with 12 NADPH and 18 ATP, most of which have yet to be used).

Looking at the yield of the Calvin cycle in terms of PGAL, we can note that for every 12 PGALs produced, two can be removed to form glucose. The other 10 must be recycled to form more ribulose phosphate. This is enough for the six RuPs needed for six turns of the Calvin cycle.

Summing Up the Light-Independent Reactions.   We see, then, that the light-independent reactions are a sequence of biochemical events that bring the light reaction products together with carbon dioxide to produce glucose and other products. Glucose production is not a simple, straightforward process, but is complex, involving many intermediate steps in which very small molecular changes are made.

We have also seen that ATP is used to phosphorylate the substrate, preparing it for the chemical changes to follow. NADPH is used at key steps to reduce the substrate as new carbohydrates are formed. Most of the reactions in the Calvin cycle are directed by specific enzymes, as we have come to expect. Finally, the Calvin cycle must repeat six times for each glucose molecule formed.

THE PROBLEM OF PHOTORESPIRATION AND C3 AND C4 PLANTS

When the Calvin Cycle Goes Awry

The picture of the light-independent reaction in plants, as we have presented it, is one of marvelous efficiency. After all, everything balances (perhaps with a few ATPs left over). It turns out that the thermal efficiency of the process is 38%, which means that about 38% of the energy in the original photons is eventually captured in the chemical bonds of glucose and the rest is lost as heat. Unfortunately, real life interjects itself to ruffle our formulas once again. A formal efficiency of 38% is quite admirable by any standards, but *in actuality*, the photosynthetic efficiency of most plants can at times be abysmal. Often, less than 1% of the light energy absorbed by a plant is converted into carbohydrates.

The reason, quite simply, is that the reactions do not always go the way we have described them. We have described an ideal system, but we do not live in an ideal world. The carbon-fixing ability of the carboxylating enzyme, 1,5-ribulose bisphosphate carboxylase, depends to a large extent on the amount of carbon dioxide present. In other words, the "mass action law" (see Chapter 6) rears its head again. When carbon dioxide levels are low, or when carbon dioxide is used faster than it can be provided by diffusion, the efficiency of the carboxylating enzyme—the one responsible for joining carbon dioxide to ribulose phosphate—falls off drastically, and strange things begin to happen. Keep in mind that under the best of circumstances, carbon dioxide is not very abundant; it occurs at about 3 parts per 10,000 in air, and since it must diffuse into leaves and cells, the concentration of carbon dioxide available to the Calvin cycle is usually very low.

When carbon dioxide levels fall below critical levels, many plants react by undergoing what is called **photorespiration.** In photorespiration, oxygen begins to compete successfully with carbon dioxide for RuBP, especially when the O_2 is high and the CO_2 concentration is low. When this happens, oxygen is added to RuBP, which then breaks down into 3-PG and **phosphoglycolate,** which leaks out of the chloroplasts as **glycolate.** This is further broken down to carbon dioxide by cytoplasmic peroxisomes, the same ones that catalyze the breakdown of hydrogen peroxide (see Chapter 4). The reactions are summarized in Figure 7.13. From all we can determine about photorespiration, it is a

waste of the plant's precious resources. It appears to accomplish absolutely *nothing* except to use up the valuable NADPH and ATP so diligently produced in the light reactions.

Dedicated plant physiologists have tried to determine the function of photorespiration. It is known that it occurs under hot, dry conditions when plants must resist water loss by closing or restricting the size of leaf pores (stomata). When this happens, the CO_2 intake is greatly restricted. So far, however, no one has been able to fully explain this paradoxical process.

Photosynthesis in C4 Plants

Photorespiration is not a universal problem in plants. In fact, many tropical plants have evolved pathways that seem to overcome the carbon dioxide concentration problem altogether. It seems that in these plants, carbon dioxide entering the leaf is not sent directly into the Calvin cycle but is first run through an entirely different pathway.

In the mid-1960s, M. D. Hatch and C. R. Slack,

7.13 PHOTORESPIRATION

When shortages of CO_2 occur in many plants, the usual Calvin cycle ceases, and photorespiration ensues. In reaction 2, the carboxylating enzyme that usually adds carbon dioxide to ribulose-bisphosphate, reacts with oxygen instead (note the concentration of oxygen rather than carbon dioxide). The two products are phosphoglycerate (PGA), a normal Calvin cycle constituent, and phosphoglycolate (or phosphoglycolic acid), an abnormal 2-carbon product (above). Phosphoglycolate is dephosphorylated to glycolate, whereupon it leaves the chloroplast to enter a peroxisome. There it is further altered by the enzymes to form useless waste products.

7.14 C4 PLANTS

At least 10 families and over 100 genera include some C4 plants. The three examples here are **(a)** sorghum, **(b)** sugar cane, and **(c)** corn. Although these are all grasses, many broadleafed C4 plants show the same biochemical and structural adaptations, which apparently have evolved independently several times.

(a)

(b)

(c)

working in Australia, announced their discovery of an alternative pathway of CO_2 fixation. It seemed that in maize, sugar cane, and a number of other tropical and subtropical plants (Figure 7.14), the initial fixation involves a 3-carbon compound called **phosphoenolpyruvate,** or simply, PEP. With the addition of CO_2, PEP is transformed into into a 4-carbon acid called **oxaloacetate (oxaloacetic acid).** Oxaloacetate is then reduced by NADPH to another 4-carbon acid known as **malate:**

$$CO_2 + PEP \longrightarrow oxaloacetate$$

$$oxaloacetate + NADPH \longrightarrow malate + NADP^+$$

Plants with this capability were soon referred to as **C4 plants,** in deference to the 4-carbon acids formed. All other plants, those in which CO_2 fixation leads to the formation of 3-carbon acids (3-phosphoglycerate) in the Calvin cycle, were of course renamed the **C3 plants.** The special carbon-fixing pathway of C4 plants is known either as the **PEP cycle** or the **Hatch-Slack pathway** (Figure 7.15).

This strange manner of CO_2 fixation was surprising, and the next finding was just as unexpected. The 4-carbon malate diffuses out of the chloroplast and even out of the **leaf mesophyll cell** (loosely arranged photosynthetic cells) in which it was formed, passing into a **bundle-sheath cell** (cells surrounding leaf veins; see Figure 7.16), where it undergoes additional reactions. (Note how the C4 and C3 leaf organizations differ in Figure 7.16.)

Upon entering the chloroplasts of the bundle-sheath cells, the malate is chemically changed to a 3-carbon acid known as pyruvate (pyruvic acid), and in the reaction, $NADP^+$ is reduced to NADPH, and carbon dioxide is released:

$$Malate + NADP^+ \longrightarrow Pyruvate + CO_2 + NADPH$$

The pyruvate then diffuses back into the mesophyll cells, where it is phosphorylated by ATP and converted back to phosphoenolpyruvate (PEP). The PEP can then react with more CO_2 to form oxaloacetate once again, and the cycle continues. But the important point is that carbon dioxide, and useful NADPH, have been released into the bundle-sheath cell, where they then enter the Calvin cycle.

The overall formula for the C4 light-independent reaction is the same as that for the C3 light reaction, except that 12 more ATPs are used for each glucose molecule:

$$6CO_2 + 12NADPH + 30ATP \rightarrow$$
$$C_6H_{12}O_6 + 12NADP^+ + 6H_2O + 30ADP + 30P_i$$
(glucose)

This means that the *theoretical efficiency* of C4 photosynthesis is less than that of C3 photosynthesis. But in reality, C4 photosynthesis is usually much more efficient. The reason is that the extra ATPs aren't wasted; instead, their energy is used to move CO_2 and electrons from one cell to another, and this, it turns out, is critical.

A serious problem with C3 photosynthesis, as we have seen, is that the concentration of CO_2 is often too low for the system to operate with full efficiency. The carboxylating enzyme in the C4 pathway is far more efficient at taking in CO_2 at low concentrations than is the carboxylase of the C3 pathway. Thus, in C4 plants, the initial fixation in the mesophyll cells goes quickly and efficiently. Then in the bundle-sheath cells, where the C3 (Calvin) cycle takes over, the refixation of CO_2 becomes efficient simply because of a steady input of 4-carbon malate.

Over 100 species of C4 plants have been identified. And the actual, net efficiency of some of them is phenomenal, compared with that of C3 plants. For example, fully 8% of the sunlight energy falling on an acre of sugar cane (a C4 plant) can later be harvested as carbon compounds.

7.15 THE C4 PATHWAY

The C4 pathway concentrates carbon dioxide in the vicinity of the carboxylating enzyme of the Calvin cycle. This makes it possible for plants to incorporate carbon dioxide rapidly, improving their photosynthetic efficiency. The major step is the addition of CO_2 to phosphoenolpyruvate in the mesophyll cells, forming oxaloacetate. This is then converted to malate through the reduction of NADP to NADPH. The malate moves to the bundle-sheath cells, where it is oxidized and decarboxylated, forming pyruvate. The decarboxylation makes CO_2 available to the Calvin cycle going on in the bundle-sheath cells. Pyruvate is then changed to phosphoenolpyruvate-utilizing ATP. The cycle can then repeat.

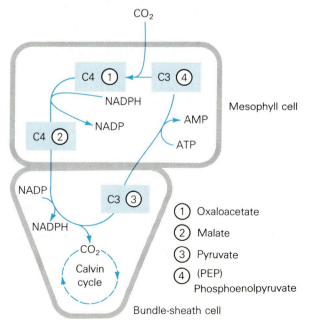

7.16 LEAF MESOPHYLL AND BUNDLE-SHEATH CELLS

(a) A comparison of the leaf anatomy of C3 and C4 plants is quite revealing. The fixation of carbon in the Calvin cycle of C3 plants takes place in most of the cells, including the leaf mesophyll or spongy layer. In C4 plants, the leaf mesophyll cells act mainly in transporting carbon dioxide within malate, to the bundle sheath cells. It is in the bundle sheath cells that the Calvin cycle occurs. (b) Some of the special features of C4 plants are easily seen in *Sorghum sudanense*. The bundle-sheath cells (left) are prominent with their large, elongated chloroplasts. Compare these to the mesophyll cell chloroplasts. The central group of cells comprise the vein with xylem, phloem, and supporting cells. The arrangement of mesophyll and bundle-sheath cells in C4 plants is an integral part of their special biochemistry.

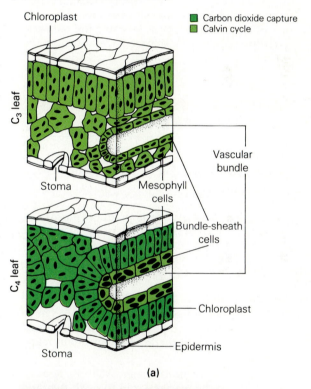

(a)

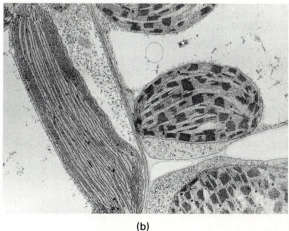

(b)

In general, plant species from temperate climates tend to be C3 plants, and tropical and desert plants tend to be C4 plants. Considering this relationship, we would surmise that tropical and desert plants have plenty of photons and must turn their efforts to the fully efficient capture and use of that rare resource, carbon dioxide. In plants from temperate climates, the input of light energy may be the limiting factor, and the more photon-efficient C3 system will do. However, this is merely speculation. It may simply be that the C4 system is always more efficient but has evolved relatively recently in only a limited number of lines.

APPLICATION OF IDEAS

1. Earlier in this century, plant physiologists determined the precise role of water and carbon dioxide in photosynthesis by using the radioactive isotopes carbon 14, oxygen 18, and tritium (hydrogen 3). Using your new knowledge of photosynthesis, suggest how these isotopes might be used in such determinations, and what you might expect in the results.

KEY WORDS AND IDEAS

PHOTOSYNTHESIS: AN OVERVIEW

1. The general equation for **photosynthesis** is:

$$6CO_2 + 12H_2O \rightarrow C_6H_{12}O_6 + 6O_2 + 6H_2O$$

Although photosynthesis produces glucose as shown, much of the process is not seen in the general formula (including the roles of light, chlorophyll, enzymes, electron carriers, coenzyme NADP, and ATP).

THE TWO PARTS OF PHOTOSYNTHESIS

1. Free energy in the thylakoid is increased in the **light reactions** and decreased as the synthesis of glucose occurs in the **light-independent reactions.**

2. In the light reactions, water, ADP, P_i, and NADP are the reactants and through the capture of light energy the products oxygen, ATP, NADPH, and H^+ are produced.

3. In the light-independent reactions, ATP, NADPH, H^+, water, and carbon dioxide are the reactants while glucose, ADP, P_i, NADP, and water are the products.

CHLOROPLASTS AND THE LIGHT REACTIONS

1. Chloroplasts include a clear area of *stroma,* membranous *thylakoids* stacked into *grana* that are interconnected by *lamellae.*

The Thylakoid

1. Thylakoids have a large number of **photosystems** embedded in their membranes. Each includes **light-harvesting antennas,** a **reaction center,** and an **electron (and hydrogen) transport system** (ETS). Photosystems designated I and II (PSI and PSII) work in pairs, although PSI also works alone.

2. The antennas contain **chlorophylls** *a* and *b* and **carotenes.** In each, one molecule of chlorophyll *a* along with an associated protein is designated as a reaction center.

3. Chlorophylls *a* and *b*, and the carotenes each absorb light of slightly different wavelengths.

4. The reaction centers of PSI and PSII are designated P700 and P680, respectively, according to their light absorption.

5. Photosystems shunt energy to their reaction centers where it is absorbed by electrons of chlorophyll *a*. Instead of energized electrons dropping back to their old energy level, they escape the reaction center to begin reducing carriers of the electron transport system (ETS).

6. The passage of electrons through the ETS yields free energy that is used in pumping protons from the stroma to the thylakoid *lumen*.

7. The phosphorylation of ADP occurs in the CF1 particles, which abound in the thylakoid membrane but are not associated directly with the photosystems.

THE LIGHT REACTIONS OF PHOTOSYNTHESIS

1. The light reactions include both **noncyclic** and **cyclic photophosphorylation.**

Photosystems II and I and the Noncyclic Events

1. Light absorbed in P680 of PSII produces an electron flow that first reaches protein Q and is then passed into the PSII ETS.

2. Oxidized chlorophyll *a* in P680 is reduced by electrons from protein X, which is itself reduced by electrons from water molecules within the lumen.

3. As water molecules are disrupted, oxygen gas is liberated, electrons reduce protein Z, and protons are released into the lumen, adding directly to the chemiosmotic differential.

4. Each electron passing from P680 to P700 powers the transport of a proton from the stroma to the lumen.

5. Light-excited electrons in P700 leave PSI and pass to NADP, reducing it to NADPH + H$^+$. These electrons are replaced by those from P680.

6. The noncyclic light reactions enrich the chemiosmotic gradient and reduce NADP, thus providing for ATP formation and a supply of hydrogen for glucose synthesis.

Photosystem I and the Cyclic Events

1. Photosystem I also acts independently, capturing light energy and sending its electrons through its ETS from which they return to chlorophyll *a* (thus water is not involved). Protons are pumped into the lumen enriching the chemiosmotic differential.

Chemiosmotic Phosphorylation

1. The chemiosmotic diffential of the thylakoid is a pH differential produced by a buildup of protons and chloride ions in the lumen and an increase in hydroxide ions in the stroma. Its electrical characteristics are lost because of the influx of chloride ions, which form **hydrochloric acid.**

2. As protons escape the lumen through channels in the CF1 particles, their free energy is used in the phosphorylation of ADP to ATP. Upon entering the stroma, the protons join hydroxide ions to form water.

THE LIGHT-INDEPENDENT REACTIONS

1. In the second half of photosynthesis, ATP and NADPH are used in the reduction, or **fixing** of carbon dioxide into carbohydrate. This occurs in the unstructured stroma where synthesizing enzymes are located.

Discovery of the Calvin Cycle

1. Using the alga *Chlorella,* Calvin discovered the chemical steps and enzymes of the synthetic pathway that became known as the **Calvin cycle.**

The Calvin Cycle

1. The Calvin cycle is a pathway where several enzymes make subtle changes in resident molecules, bringing in ATP, NADPH, and carbon dioxide to produce carbohydrate.

2. The Calvin cycle includes the following chemical steps:
   a. *Phosphorylation:* **RuP** + ATP → **RuBP** + ADP
   b. **Carboxylation:** RuBP + NADPH + H$^+$ → 6-C sugar + NADP$^+$
   c. *Cleaving:* 6-C sugar → two **3-PG**
   d. *Phosphorylation:* (2) 3-PG + two ATP → (2) **1,3-DPG** + 2 ADP
   e. *Reduction:* (2) 1,3-DPG + 2NADPH + H$^+$ → **2PGAL** + 2NADP$^+$
   f. *Recycling:* NADP$^+$, ADP, and P$_i$ are recycled to the light reactions

3. The product PGAL can be used to produce glucose or other essential molecules, but some must be recycled to keep the Calvin cycle going. Glucose pathway: 2PGAL → **fructose-1,6-diphosphate** + 2P$_i$
   fructose-1,6-diphosphate → glucose-1-phosphate → glucose + P$_i$

4. Since only one carbon dioxide is brought in per turn of the cycle, six turns must occur before two PGALs can be withdrawn to form one glucose. The other 10 PGALs are recycled to form ribulose phosphate, the starting molecule. This recycling keeps the cycle going.

THE PROBLEM OF PHOTORESPIRATION AND C3 AND C4 PLANTS

When the Calvin Cycle Goes Awry

1. Photosynthesis in C3 plants becomes inefficient when carbon dioxide gas is in low concentration, a time when the enzyme *1,5-ribulose bisphosphate carboxylase* cannot readily incorporate carbon dioxide.

2. When carbon dioxide is unavailable, **photorespiration** ensues. During this process, **phosphoglycolate,** which forms from RuBP, leaks from the chloroplast, as **glycolate** to be broken down to carbon dioxide in peroxisomes. This is a waste of light reaction products.

Photosynthesis in C4 Plants

1. The problem of photorespiration is avoided in the **C4 plants** which have an alternative pathway for incorporating carbon dioxide that involves certain 4-carbon acids. This alternative is called the **PEP cycle** or **Hatch-Slack pathway.**

2. Carbon dioxide enters **leaf mesophyll cells,** where incorporation occurs:
   a. CO_2 + **3-C phosphoenolpyruvate** → **4-C oxaloacetate**
   b. 4-C oxaloacetate + NADPH → **4-C malate** + NADP$^+$

3. Malate enters the **bundle sheath cells,** where
   4-C malate + NADP → **3-C pyruvate** + CO_2 +
   NADPH

4. The carbon dioxide then enters the Calvin cycle
   and the pyruvate is converted to phosphoenolpyruvate, completing the cycle.

5. **C3 plants** appear to be adapted to temperate climates while most C4 plants are desert dwellers.

6. Comparative studies of C3 and C4 plants reveal
   that in intense sunlight, C3 plants undergo photorespiration while C4 plants continue photosynthesis.

REVIEW QUESTIONS

1. Write the general formula for photosynthesis
   and list several factors missing from this simple
   representation. (147)

2. Briefly explain why disruption of the chloroplast
   and thylakoids stops photosynthesis. (148)

3. Explain specifically what is meant by this statement: In the light reactions the system is
   charged up, while in the light-independent reactions it runs back down. (148)

4. Write a more detailed balanced equation for the
   light reactions including water, NADP, ATP,
   and oxygen. (148)

5. Write a detailed balanced equation for the light-
   independent reactions including glucose, water,
   ATP, and NADP. (148)

6. Prepare a large drawing of a granum, showing
   the relationship of the thylakoids, their membranes and lumen, the lamellae, and the surrounding stroma. (152)

7. Prepare a simplified drawing of paired photosystems I and II, carefully labelling the following: lumen, stroma, light-harvesting antennas,
   reaction centers, and ETS. (152)

8. Compare the absorption of light among chlorophylls *a* and *b* and the carotenes. In what way is
   the broad combined range of these pigments
   adaptive? (152)

9. Summarize the events of the noncyclic light reactions, beginning with the absorption of light
   by P680.
   a. path of electrons
   b. role of water
   c. protons gained in (a) and (b)
   d. final electron acceptor (159-160)

10. In general, what is gained by the noncyclic
    events? (160)

11. Summarize the cyclic events:
    a. path of electrons;
    b. role of water, if any;
    c. protons gained;
    d. role of NADP, if any. (161)

12. Explain how and when elements of the electron
    transport system become hydrogen transporters.
    Be specific. (159, 161)

13. Describe the CF1 particles—their location, orien-
    tation, structure, and spatial relationship to the
    photosystems. (162)

14. Using a simple diagram, illustrate the phosphor-
    ylation of ADP in a CF1 particle. (162)

15. Summarize the results of the light reactions.
    What are the products and what, in general, are
    they used for in the light-independent reac-
    tions? (162)

16. Write a short paragraph summarizing the bio-
    chemical pathway known as the Calvin cycle as
    though you were explaining it to a novice. (162)

17. The Calvin cycle contains only five major steps
    (excluding the restoration of ribulose phosphate.
    List these from memory and add one word for
    each step that defines the kind of reaction oc-
    curring (such as dehydration, hydrolysis, and so
    on). (164)

18. What, specifically, are the roles of carbon diox-
    ide, NADPH, ATP, and enzymes in the Calvin
    cycle? (164)

19. Briefly explain why it requires six turns of the
    Calvin cycle to generate one molecule of glu-
    cose. (166)

20. Summarize the steps used in converting PGAL
    to glucose. What alternative products are possi-
    ble? (166)

21. From a plant's point of view, what is wrong
    with photorespiration? Under what conditions
    does it occur? (166-167)

22. Describe the four steps of the C4 cycle, listing
    the reactants and products of each. Include the
    specific cell types in which each reaction occurs.
    (168)

Glycolysis and Cell Respiration

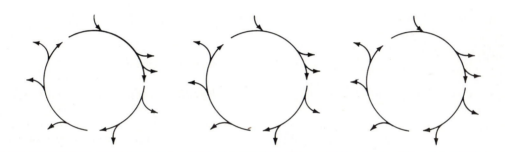

We have seen that life can only exist through intense and precise efforts to keep its molecules organized. As soon as the tendency to remain organized ceases, so does life. A corpse is a once-organized entity, gradually becoming disorganized as it decays until finally, its molecules have no more to do with each other than they do any other molecules. At this point they are behaving randomly. They are no longer organized, and the corpse is gone.

In the last chapter, we saw how living things manage to capture the energy of sunlight and to use that energy to reorganize matter into molecules that are essential to life. Now let's take a look at how living things go about using that energy to stay organized.

Energy is used for such vital tasks as movement, active transport, and synthesis. Cells store much of their energy in the molecule called adenosine triphosphate, or ATP. Because of the many roles of energy in the processes of life, cells need a constant supply of ATP. Most of the earth's organisms produce their ATP by using chemical bond energy in organic molecules—carbohydrates, fats, and proteins—which we can refer to as cellular fuels. Here, we will focus primarily on the carbohydrate called glucose, the most familiar and common of the cellular fuels.

In most organisms, the chemical energy of glucose is converted to the high-energy bonds of ATP through two processes: **glycolysis** and **cell respiration.** Glycolysis precedes cell respiration and is **anaerobic**—that is, it does not require oxygen. Cell respiration, on the other hand, is **aerobic**, or oxygen-requiring. In addition, a significant number of bacteria and fungi obtain their ATP through **fermentation,** an anaerobic process quite similar to glycolysis. We will sort out the three processes, but first let's compare the basic energy-yielding processes of photosynthesis and respiration.

COMPARING PHOTOSYNTHESIS AND RESPIRATION

In the last chapter we saw how glucose is produced in photosynthesis, and we will shortly see how, through respiration, it is broken down as its energy is released. In many ways the two processes are simply the reverse images of each other. Photosynthesis begins with water, carbon dioxide, and energy, and results in the production of glucose. On the other hand, glycolysis and cell respiration begin with glucose and end up with carbon dioxide, water, and energy.

8.1 THE ENERGETIC RELATIONSHIP BETWEEN PHOTOSYNTHESIS AND RESPIRATION

Photosynthesis is portrayed as an uphill (endergonic) process. Through the use of solar energy, carbon dioxide and water, with little free energy and simple molecular organization, are reorganized into glucose, which has greater free energy and complexity. Respiration begins with molecules of greater free energy and proceeds downhill (exergonic), with the transfer of some energy into ATP bonds, but with much of it released as heat. The end products are low-energy carbon dioxide and water, again with little free energy and very simple organization. The small bump in the curve at the right of glucose represents the energy of activation needed to get stable glucose to react.

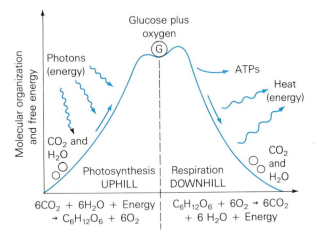

Compare the overall net photosynthetic reaction for the *synthesis* of glucose:

$$6CO_2 + 6H_2O + Energy \rightarrow C_6H_{12}O_6 + 6O_2$$
Glucose

to the overall net reaction for the *metabolism* of glucose:

$$C_6H_{12}O_6 + 6O_2 \rightarrow 6CO_2 + 6H_2O + Energy$$
Glucose

The reactions appear to be the same, with only the direction reversed. There are other differences, though. For example, energy enters the photosynthesis equation in the form of photons and is converted to chemical bond energy. Energy enters glycolysis and respiration in the form of the chemical bonds of fuels and is converted to heat and the high-energy bonds of ATP.

Energetically, the fact that the two reactions are somewhat reversed is highly significant. Photosynthesis is essentially an *uphill* process, with the final chemical products (oxygen and glucose) possessing much more free energy than the starting chemical materials (water and carbon dioxide). This is accomplished by incorporating the energy of photons.

Conversely, glycolysis and cell respiration are *downhill* reactions. Here, glucose and oxygen combine and fall to the substantially lower free-energy state of carbon dioxide and water, with some of the energy captured and conserved in the high-energy bonds of ATP and the rest lost as heat. Figure 8.1 illustrates the relationship between the two processes in terms of energy and molecular organization.

THE THREE PARTS OF RESPIRATION

Our discussion of glucose metabolism will be divided into three parts, as shown in Figure 8.2:

Part I Glycolysis (literally, the breaking of sugar), including fermentation;

Part II Cell respiration: the citric acid cycle (or Krebs cycle);

Part III Cell respiration: electron transport and chemiosmotic phosphorylation.

(Note that Part I, glycolysis, occurs in the cytoplasm. Parts II and III take place in the mitochondria.)

A Word About ATP

ATP is produced in two ways during the metabolism of glucose. One is by chemiosmotic phosphorylation, which we have already encountered with photosynthesis, and the other has been called **substrate level phosphorylation.** Our discussions of chemiosmotic phosphorylation in the last two chapters will help here, but there are some differences between the process in chloroplasts and in mitochondria, as we will see.

In glycolysis, ATP is formed only through substrate level phosphorylation. The term reflects the fact that ADP receives its terminal high-energy phosphate *directly from the substrate* (fuel) molecule. Substrate level phosphorylation is simple and direct, and is common to all forms of life. Even in aerobic organisms, as we mentioned, the metabolism of glucose begins with anaerobic glycolysis and includes direct phosphorylation of the substrate. In aerobic organisms, after glycolysis yields the first few ATP molecules through substrate level phosphorylation, the products of this reaction proceed into the aerobic process of cell respiration, where much more ATP is produced through chemiosmotic phosphorylation.

GLYCOLYSIS: METABOLIZING GLUCOSE WITHOUT OXYGEN

Since anaerobic metabolism of glucose is common to most of the earth's organisms, it has been suggested that both glycolysis and fermentation were early evolutionary developments. This notion is supported by evidence that there was little oxygen available in the early atmosphere (see Chapter 18). Cells of most life forms today have retained the glycolytic pathway, while many simpler organisms such as bacteria, protists, and fungi, and a few primitive worms, use the similar process of fermentation.

Figure 8.3 describes the essentials of glycolysis and substrate-level phosphorylation. The glycolytic reactions, beginning with glucose, are a bit complicated, but as we saw with the Calvin cycle (see Chapter 7), cells often do things in a roundabout way. In glucose metabolism, the primary biochem-

8.2 GLYCOLYSIS AND CELL RESPIRATION

The metabolism of glucose may be divided into three sequential parts. The first is glycolysis, which occurs in the liquids of the cytoplasm. Glucose is broken down to pyruvate, but along the way ATP is produced and NAD^+ is reduced to $NADPH + H^+$. Glycolysis is followed by cell respiration, which includes parts II and III, both of which occur in the mitochondrion. Part II, the citric acid cycle (or Krebs cycle), occurs in the matrix (inner compartment), yielding a quantity of $NADH + H^+$ and

$FADH_2$ along with a small amount of ATP and carbon dioxide. Part III includes electron transport and chemiosmotic phosphorylation, accounting for most of the ATP yield of respiration. Electrons are sent through the electron transport system, whereupon their energy is used to pump protons into the outer compartment. The free energy produced by the system is used to phosphorylate ADP, yielding ATP.

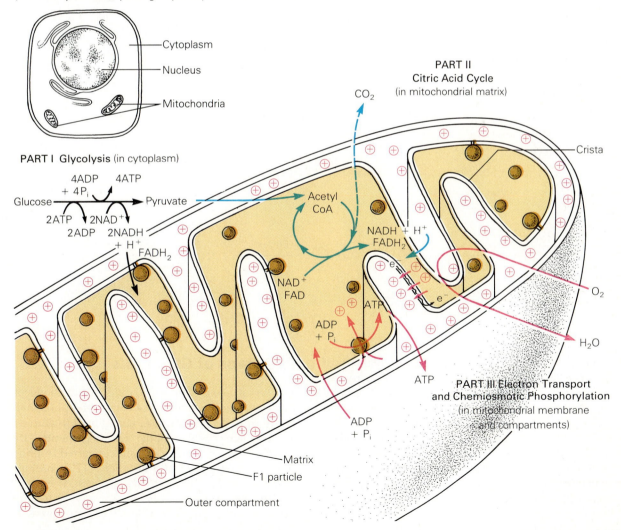

8.3 THE GLYCOLYTIC PATHWAY

In glycolysis, the phosphorylation of ADP occurs at the substrate level. This means that the bond energy in the substrate or fuel molecule (such as glucose) is used directly to form the terminal high-energy phosphate bond of ATP. There is no electron transport system or chemiosmotic gradient involved. There are a total of nine major events, each with its own specific enzyme. Two molecules of ATP must be invested to complete reactions ① and ③. These are returned further along in the phosphorylation of ADP during reaction ⑥. ATP is again produced further along in reaction ⑨. In all, glycolysis represents a net gain of 2 ATPs and 2 NADH + H$^+$. The final reaction yields two molecules of pyruvate, each containing a considerable amount of potential energy.

ical trick is for the cell to break the chemical bonds of glucose in such a way as to avoid the sudden release of energy, since this would result in the wasteful production of heat. After all, the rapid combining of glucose and oxygen with the production of massive amounts of heat would produce the crackle of burning cytoplasm—obviously an unsatisfactory situation for cells. In cells, the transfer of bond energy from glucose to ATP is efficient, but the process proceeds through very indirect, gradual, and roundabout means.

The Steps of Glycolysis

As you can see in Figure 8.3, the first step in glycolysis involves the transfer of a terminal phosphate group from ATP to glucose, changing it to glucose-6-phosphate. With subtle rearrangements, the phosphorylated glucose is then converted to phosphorylated fructose, and another ATP is brought into the picture. The fructose-6-phosphate receives another phosphate and is thus converted to **fructose-1,6-diphosphate** (in other words, fructose with phosphate groups on the number 1 and number 6 carbons). This change involves three steps, each requiring its own specific enzyme.

Enzyme
① Glucose + ATP ⟶ Glucose-6-phosphate + ADP

Enzyme
② Glucose-6-phosphate ⟶ Fructose-6-phosphate

Enzyme
③ Fructose-6-phosphate + ATP ⟶
Fructose-1,6-diphosphate + ADP

[*Note*: Numbers in circles also refer to Figure 8.3.]

At this point, the cell has spent two ATP molecules, degrading them to ADP. This may seem paradoxical since, if the object of respiration is to build ATP from ADP and phosphate, things seem to have started off exactly backwards. Why, then, is ATP being expended when the idea is to gain ATP? In a sense, the first two ATPs can be considered to be an investment—a priming of the pump, as it were.

There are several explanations for these pump-priming steps, and they are not mutually exclusive. One idea is that the charged phosphate groups serve as "handles" and that these serve to bind the intermediates to their appropriate enzymes. Another hypothesis notes that simple sugars, such as glucose, are free to diffuse out of the cell, but that phosphorylated sugars are unable to cross the plasma membrane. This suggests that the attached phosphates serve to keep glycolysis localized within the cell. A third hypothesis proposes that the addition of phosphates makes the stable glucose molecule suddenly unstable and reactive. Once the stability of its bonds has been weakened, the glucose can be broken apart, and at the right moment the phosphates can be removed along with some of the chemical bond energy that was invested in them.

To resume, let's look at reaction ④, in which fructose-1,6-diphosphate is cleaved into *two products*, phosphoglyceraldehyde (PGAL) and dihydroxyacetone phosphate. The latter is then transformed into PGAL in a separate reaction, thus yielding two molecules of PGAL in all. These enzyme-mediated reactions do not require the input of additional energy. You may recall that PGAL also functions in the Calvin cycle, the light-independent reactions of photosynthesis.

Enzyme
④ Fructose-1,6-diphosphate ⟶ 2PGAL

With the two molecules of PGAL, our fuel is now thoroughly primed for one of the most impor-

tant steps: a coupled oxidation-phosphorylation reaction. In a kind of hydrogen and phosphate switch, an enzyme coupled with NAD removes two hydrogens and substitutes an inorganic phosphate group (P_i taken from the cytoplasmic fluids):

Enzyme
⑤ 2PGAL + 2NAD + 2P_i ⟶
(2) 1,3-diphosphoglycerate + 2NADH

The removal of hydrogens by such oxidations could release a great deal of energy. However, the oxidation is coupled with a nearly simultaneous phosphorylation reaction (addition of phosphate). The *coupled* reaction here is important. And *uncoupled* phosphorylation reaction would require a considerable input of energy, but since the energy is available from the oxidations, the net reaction goes on without an input or loss of energy. Since the coupled reaction is energetically self-sustaining, it is also readily reversible.

While this reaction (step ⑤) does not result in ATP production, the new phosphate bond has what is called "high transfer potential." This means that it can readily be passed over to ADP when the opportunity arises. Three things have been accomplished in this coupled reaction:

1. In aerobic organisms, highly energetic reduced NAD (NADH + H$^+$) is made available for use in the chemiosmotic synthesis of ATP—a step we will come to later.
2. The oxidations may diminish the bond stability of the fuel molecule, setting it up for coming reactions.
3. The phosphate bond that has been added serves as an *energy store*. Some of the energy of the fuel molecule is absorbed in the phosphate bond.

The next reaction brings ADP into the picture, and with the aid of the appropriate enzyme, the phosphate, along with an appropriate amount of energy, is transferred to ADP, forming ATP with its energy-rich terminal phosphate bond. Keep in mind that everything doubled after the splitting of fructose-1,6-diphosphate, so actually, two ADPs and two 1,3-diphosphoglycerates are involved. This means that two ATPs are now produced. Remember also that two ATPs were invested to get glycolysis started in the first place, so this step is really just a "payback" and the net gain in ATP so far is zero.

Enzyme
⑥ (2) 1,3-diphosphoglycerates + 2 ADP ⟶
(2) 3-phosphoglycerates + 2ATP

Now our two 3-carbon fuel molecules have been changed to two molecules of 3-phosphoglycerate, a stable molecule that must again be prepared for substrate-level ATP phosphorylation. Two preparatory reactions follow. In the first reaction, an internal rearrangement, the phosphate is moved from the number 3 carbon to the number 2 (end to center), where it becomes a potential energy store. The second event, a **dehydration reaction** rather than an oxidation, removes a water molecule (two hydrogens and an oxygen) from each 3-phosphoglycerate. Thus, a second disruption and another redistribution of energy within the molecule occurs, changing the phosphate bond, which had previously been a relatively low-energy bond, to a high-energy bond (one with high transfer potential).

⑦ (2) 3-phosphoglycerate $\xrightarrow{\text{Enzyme}}$

(2) 2-phosphoglycerate

⑧ (2) 2-phosphoglycerate $\xrightarrow{\text{Enzyme}}$

2 Phosphoenolpyruvate + 2H$_2$O

Finally, in both phosphoenolpyruvates, the phosphate with its high-energy bond is transferred to two ADPs, forming two ATPs. The products of this final step are two **pyruvates** (two *pyruvic acids*, but we will continue to use the ionic ("-ate") form, as in earlier chapters). Depending on the organism, the two pyruvates can either enter aerobic cell respiration or they can enter one of several fermentation pathways.

⑨ 2 Phosphoenolpyruvate + 2ADP $\xrightarrow{\text{Enzyme}}$

2 Pyruvate + 2ATP

With the final reaction of glycolysis, we see that substrate level phosphorylation has returned the original ATP investment and produced a gain of two ATPs. Let's see what the cell has gained in energy so far (also in Table 8.1).

Energy Yield of Glycolysis. The high-energy bonds of ATP have a caloric value of 8 kcal/mole. Ignoring the NADH + H$^+$ and pyruvate for the moment, the net yield through the glycolytic metabolism of one mole of glucose to the pyruvate stage is two moles of ATP, worth 16 kcal. Now, one mole of glucose weighs 180 grams (or about 6.3 ounces), and if it is burned in the presence of oxygen, it can release 680 kcal. On the molecular level, then, the energy stored in the two ATP high-energy bonds represents about 2.4% of the oxidizable energy of a molecule of glucose (that is, gly-colysis is 2.4% efficient). This is a low yield compared to cell respiration. But cell respiration is an aerobic process, and if the cell cannot use oxygen, it must rely on an anaerobic process.

The Fates of Pyruvate

We have seen that glycolysis is a common process in many organisms and that its end product, pyruvate, can enter aerobic cell respiration in the mitochondrion. There, much more energy is removed and the final products are carbon dioxide and water. However, as we see in Figure 8.4, this fate is but one of several possibilities.

Other possibilities may involve fermentation pathways. Fermentation is commonly defined as the formation of ethyl alcohol from glucose, although other final products are also possible.

TABLE 8.1

ENERGY YIELD IN GLYCOLYSIS AND FERMENTATION

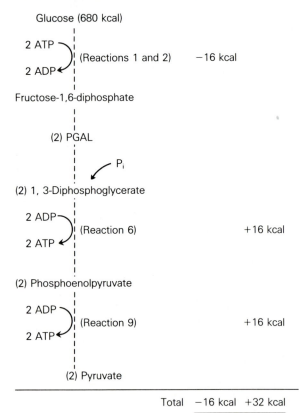

Total	−16 kcal	+32 kcal
Net gain	2 ATP	16 kcal
Percentage efficiency	$\frac{16}{680} =$	.024 or 2.4%

8.4 THE METABOLISM OF PYRUVATE

Three common fates of pyruvate include two anaerobic fermentation processes, as well as the aerobic pathway. **(a)** During muscular activity the pyruvate produced during glycolysis enters the lactate fermentation pathway, where the pyruvate is reduced by NADH to lactate. Eventually much of the lactate will be converted back to pyruvate and sent through aerobic pathways or converted back to glucose. **(b)** In another anaerobic process, some organisms carry on alcohol fermentation. Pyruvate is reduced by $NADH + H^+$ (which was formed during glycolysis), and carbon dioxide is removed, yielding ethyl alcohol. **(c)** Under aerobic conditions, pyruvate enters the mitochondrion, where it is converted to acetyl CoA and sent through a complex metabolic pathway called the citric acid cycle. In its conversion to acetyl CoA, pyruvate is oxidized and decarboxylated—that is, some of its hydrogen reduces NAD^+ to $NADH + H^+$ and CO_2 is released. Acetyl CoA is a 2-carbon acetyl group joined with coenzyme A.

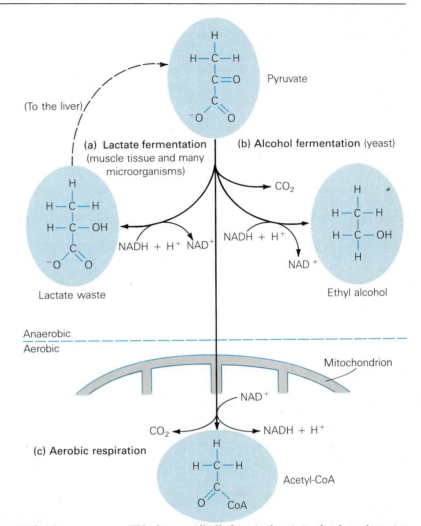

(a) **Lactate fermentation** (muscle tissue and many microorganisms)

(b) **Alcohol fermentation** (yeast)

Pyruvate

(To the liver)

CO_2

$NADH + H^+$ NAD^+

$NADH + H^+$

NAD^+

Lactate waste

Ethyl alcohol

Anaerobic

Aerobic

Mitochondrion

NAD^+

CO_2 $NADH + H^+$

(c) **Aerobic respiration**

Acetyl-CoA

Except for the final products, however, the fermentation pathway is identical to that of glycolysis.

Here we will focus on two fermentation pathways (see Figure 8.4). The first is **ethyl alcohol fermentation.** It occurs in yeasts and some bacteria, and the waste products are carbon dioxide and ethyl alcohol. The second is **lactate fermentation,** which occurs in many bacteria and in animal muscle during exertion. Its final product is **lactate (lactic acid).** Neither of these anaerobic pathways yields any additional energy, but, as we will see, they serve to recycle desperately needed oxidized NAD (NAD^+). Why is the presence of NAD^+ so important? Its continued availability is always a potential problem for cells since, in the absence of NAD^+, the key oxidation step in glycolysis (or fermentation)

will halt, as will all chemical activity further along in the pathway.

Alcohol Fermentation.   As mentioned, in alcohol-fermenting organisms, including yeasts and certain bacteria, pyruvate is degraded to ethyl alcohol and carbon dioxide (Figure 8.5). Ethyl alcohol, or **ethanol,** as it is also known, is the alcohol of beer, wine, and liquor. Fermentation by yeast is also important to the baking industry, but bakers are primarily interested in the carbon dioxide, which causes bread to "rise." (The ethyl alcohol produced simply evaporates in the oven.)

The transformation of pyruvate into ethyl alcohol occurs in two steps, each mediated by a specific enzyme. The pyruvate is first acted upon by a

8.5 THE ENERGY HILL AND FERMENTATION

As yeasts metabolize glucose anaerobically, only a small amount of energy is transferred to ATP. The end products are carbon dioxide, which has little free energy, and ethyl alcohol, which has a great deal; thus it comes to rest only part way down the "energy hill." More of its energy can be released when oxygen is available.

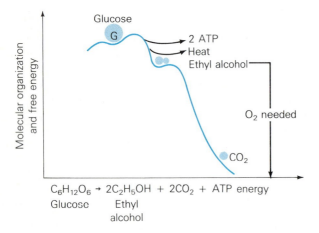

$$C_6H_{12}O_6 \rightarrow 2C_2H_5OH + 2CO_2 + \text{ATP energy}$$
Glucose   Ethyl
                   alcohol

decarboxylase enzyme, which removes the carboxylic acid group (COOH or COO−), converting it to carbon dioxide. The nearly depleted fuel molecule that remains is known as **acetaldehyde.** Next, with the aid of the enzyme **alcohol dehydrogenase,** the NADH formed earlier in glycolysis adds hydrogens to each molecule of acetaldehyde, reducing it to ethyl alcohol and regenerating oxidized NAD (NAD^+) (see Figure 8.4).

As we have said, the last step permits NAD^+ to recycle back into glycolysis. Actually, in the absence of oxygen, the hydrogen of NADH is only a waste product anyway. The final step in alcohol fermentation, then, is the combining of two waste products, hydrogen and acetaldehyde, to form one waste product, alcohol (Figure 8.6).

Yeasts in general are **facultative anaerobes;** that is, depending on conditions, they can switch the metabolism of pyruvate between anaerobic alcohol fermentation and fully aerobic respiration, where carbon dioxide and water are the end products. The aerobic cell respiration of yeast, by the way, is biochemically identical to human aerobic respiration.

Lactate Fermentation. We have noted that in some cases, the end product of fermentation is lactate, a 3-carbon derivative of pyruvate. Among the lactate fermenters are certain bacteria of genus *Lactobacillus* and *Streptococcus*, whose lactate secretions are used to create the unique flavors of rye

breads, yogurt, and some cheeses. And as we mentioned, lactate is also produced in the active muscles of many kinds of organisms, including the vertebrates. As in the fermentation of ethyl alcohol, the conversion of pyruvate to lactate is essential to the continuation of glycolysis, since it frees NAD^+ from its reduced state, whereupon it can then be recycled and continue its role in the oxidation of PGAL. The enzyme responsible for reacting with pyruvate and reduced NAD is **lactate dehydrogenase,** (see Figure 8.4; note the transfer of two hydrogens from NADH + H⁺ to lactate). The story of lactate fermentation in our own voluntary or skeletal muscle is an interesting one, since it explains a lot about exercise, fatigue, and physical conditioning. Let's look at the details.

Glycolysis and Lactate Fermentation in Muscle Tissue. During heavy exertion, a great deal of ATP can be used up quickly. Skeletal muscle can operate aerobically, carrying out oxidative cell respiration, but in larger animals at least, there is no way for the circulatory system to bring in oxygen fast enough for oxidative phosphorylation to replace the ATP being used during such activity. However, muscle cells have two backup systems.

8.6 WINE-MAKING, AN ANAEROBIC PROCESS

The strictly sanitary and controlled conditions of the modern winery are a sharp contrast to the anaerobic swamp, where nature has its way. Anaerobic conditions are maintained by the careful exclusion of oxygen, and pure cultures of yeast are used to convert grape sugars to ethyl alcohol.

The first backup system is a store of high-energy bonds in a molecule that is abundant in muscles, **creatine phosphate** (or **phosphocreatine,** as it is also known).

High-energy bond

Creatine phosphate
(Phosphocreatine)

Creatine phosphate is a much more compact molecule than ATP, and the cell can store a great deal of readily available energy in this form. Creatine phosphate doesn't provide muscle contraction energy directly, but it can transfer its own energy-rich phosphate to ADP to regenerate ATP:

Creatine phosphate + ADP ⟶ ATP + Creatine

As the creatine phosphate is gradually used up, the muscle tissue falls back to yet another quick energy source—glycolysis, the process described earlier and in Figure 8.3. So, in addition to pyruvate, the end products include a modest supply of ATP and NADH. Of course, pyruvate is still energy-rich and can be sent into the mitochondrion to undergo aerobic cell respiration. This would provide a good source of ATP and a continuous supply of recycled NAD (as we will see in the next section). But when the oxygen required for aerobic respiration in muscle cannot be supplied fast enough, the continued recycling of NAD becomes threatened, and as we have seen, without it glycolysis cannot proceed. The answer, at least temporarily, is to recycle NAD another way. Using the hydrogen of NADH + H$^+$, pyruvate is reduced to lactate. The conversion to lactate is, in a sense, a dead end, since this product cannot be used directly by muscle. (Compare the production of lactate with that of ethyl alcohol in Figure 8.4.)

As a result, lactate accumulates rapidly during intense muscular activity, and sooner or later something must be done with it. Much of the lactate is carried out of the muscles by the blood stream and directed to the liver (and to a much lesser extent to the kidneys). Some of it may remain in the muscles until the amount of oxygen being brought in by the circulation exceeds the muscle's current needs, in which case the lactate is converted back to pyruvate, which then proceeds into the oxidative respiration pathway. There it is broken down to carbon dioxide and water, and its chemical bond energy used to produce ATP.

The lactate sent to the liver has a different fate. It is also converted back to pyruvate, but the pyruvate is then sent through a special biochemical pathway known as **gluconeogenesis.** In several steps, a few of which are clearly a reversal of glycolysis, the lactate is converted to glucose. Some of this same glucose finds its way back to the muscles again. The conversion is not without cost, however, because glucose has greater free energy than lactate. Thus, much of the initial pyruvate produced must be oxidized to provide energy to run the gluconeogenic pathway. (Figure 8.7 summarizes each aspect of muscle activity.)

Oxygen Debt.   After a period of heavy exertion, the muscle tissues in humans and other vertebrates will be depleted of creatine phosphate, and both the liver and the muscle will be loaded with lactate. Perhaps you have experienced this condition; fatigue hurts and most people try to avoid it, although a few eccentrics, like marathon runners, claim to find it exhilarating. The final stage is sheer exhaustion. But marathon running or not, following any strenuous muscle activity, it takes a period of time and a large amount of oxygen and ATP for the lactate to be metabolized and for the creatine to be regenerated as creatine phosphate. During this time you can expect to find yourself continuing to breathe hard, taking in as much oxygen as the lungs can handle. The state of oxygen and creatine phosphate depletion creates what is known as the **oxygen debt,** which can be expressed as the amount of extra oxygen needed to restore the system to its preexertion equilibrium. How long it takes to repay the oxygen debt obviously depends on a person's physical condition. Physical conditioning, in turn, involves increasing respiratory and circulatory capacity, as well as increasing the storage capacity and possibly the mass of the muscles themselves. Conditioned runners have enlarged lung capacities, hearts that pump more blood with each stroke, and an increased ability to use oxygen. Some may even have a high tolerance for pain—at least they act as though they do (Essay 8.1).

The Metabolism of Starch and Glycogen. Starch and glycogen are polymers of glucose in which the glucose units are held together by 1-4

8.7 THE ENERGY RESERVES OF MUSCLE

① In sustained muscular activity, the ready ATP reserves are depleted in the first few minutes. ② To refurbish this supply, a much larger reservoir of creatine phosphate is tapped. Its high-energy phosphate is transferred to ADP, producing more ATP. As activity continues, even the creatine phosphate becomes depleted. ③ To provide a continued supply of ATP for the restoration of creatine phosphate and the replenishment of ATP reserves, the muscle relies mainly on the fast but limited anaerobic process of glycolysis, in which each glucose is converted to two pyruvates, two ATPs and two $NADH + H^+$. Some of the pyruvate enters the mitochondrion ④ where it is used in oxidative respiration, but oxidative respiration cannot keep up with the demands of vigorous muscular activity. As glycolysis proceeds, a shortage of NAD^+ sets in. To compensate, NADH unloads its hydrogen onto pyruvate, reducing it to lactate ⑤ and freeing NAD^+ for recy-

cling back to glycolysis. Lactate starts to accumulate, but most of it is quickly removed by the circulatory system and carried to the liver ⑥, where it is converted back to glucose and conserved.

What we have been describing is often summarized in the phrase "oxygen debt," which refers to the extra quantity of oxygen used to bring about the complete recovery of the resting state in the muscle. This requires time since it involves the much slower aerobic respiratory process ⑦ in the mitochondrion. As usual, glucose is converted to pyruvate, which is then sent through the aerobic process, where a rich supply of ATP ⑨ is made available, eventually restoring the muscle to its previous resting condition. During this recovery period, runners breathe heavily as their respiratory process gradually overcomes the oxygen debt. When all three reservoirs are replenished, recovery is complete.

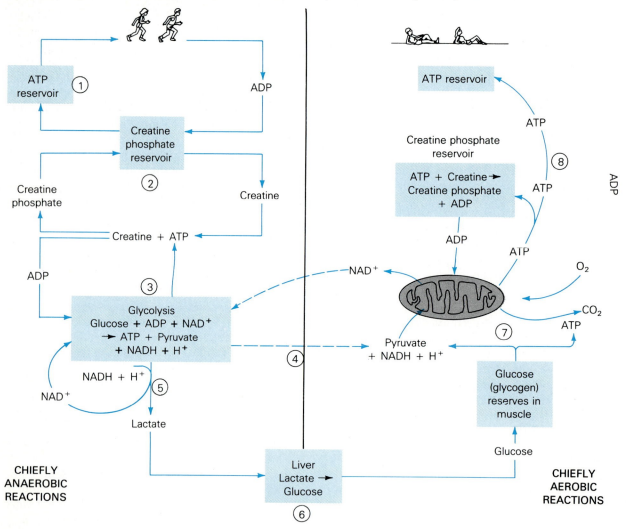

A Marathon Runner Meets the "Wall"

There is a point in the marathon when many runners hit the dreaded "wall." The wall usually looms before them at about mile 18 or 20, and it is here that many flounder, falling prostrate, staggering, or simply walking. It is here that other, better-trained runners begin to pass those whose dust they had been eating, on their way to the conclusion of the 26-mile, 385-yard race.

There has been a lot of discussion over the years about what the wall really is, but everyone who has encountered it can attest to its reality. Physiologically, it seems to correspond to the period when muscle glycogen, the most readily available reserve of glucose, is depleted. Glucose is normally depleted after two to three hours of slow running, at about an eight minute per mile pace. The runner at this point must switch to metabolizing fat molecules. The blood pH lowers and the transition is, for some reason, exceedingly stressful. It has also been suggested that women switch to fat metabolism more easily than men, a suggestion supported by their increased relative numbers in 50- and 100-mile "ultramarathons."

Interestingly, when one encounters the wall, sheer willpower has little effect. You simply can't go on. The muscles refuse. Any effort is extremely painful and fatiguing. In addition, blood sugar levels drop and you may suffer the psychological depression associated with hypoglycemia.

Marathoners have been able to beat the wall by training so hard that it no longer exists for them. This training includes long runs of 20 miles or more, during which muscle glycogen is repeatedly depleted and then restored in even greater abundance during the following week. Some runners believe they are able to move the wall back by a depletion run one week before the race followed by three days of low-carbohydrate dieting and three days of carbohydrate "loading," when it is theorized that the deprived muscles overcompensate and store glycogen in large quantities.

---

linkages (see Chapter 3). The energy in this type of bond is relatively small, but the presence of 1-4 linkages means that these molecules contain more energy than does free glucose. In fact, few cells contain very much free glucose at any one time, and their carbohydrate energy stores are usually in the form of starch or glycogen. The polymers are broken down into single glucose units by *phosphorolysis*, in which the 1-4 bond is split by the enzyme phosphorylase, which substitutes a phosphate ion on the number 1 carbon. The resulting breakdown product of both starch and glycogen is glucose-1-phosphate, which can be changed to glucose-6-phosphate by another enzyme. Interestingly, ATP is not required in this instance, and when the glucose-6-phosphate is run through glycolysis, only one ATP, rather than the usual two (see Figure 8.3), is required to "prime" the molecule. As a result, the net gain in glycolysis using starch or glycogen is three ATPs per glucose subunit rather than just two (Figure 8.8).

Hardworking muscle tissue is not the only animal tissue in which glycolysis takes place. Other cell types can also use the lactate fermentation pathway. Red blood cells, for instance, lack nuclei and mitochondria and respire anaerobically. Many tumors and cancers metabolize glucose anaerobically, throwing off large amounts of lactate into the blood stream. The reason for this shift in respiratory pathways is not known, but researchers have suggested that it is of basic importance in understanding cancer. One prominent hypothesis is that the shift to anaerobic metabolism comes first and that this change causes the tissue to be cancerous.

8.8 IMPROVING THE GLYCOLYTIC YIELD

When the starting fuel for glycolysis is glycogen rather than simple glucose, a higher ATP yield is made possible. In the conversion of glycogen to glucose by hydrolysis (adding water), the 1-4 linkages are broken by phosphorolysis (adding phosphate) rather than hydrolysis (adding water), and the product is glucose-1-phosphate. As seen here, the enzyme phosphorylase cleaves the glycogen into glucose units and at the same time adds inorganic phosphate to each number 1 carbon. Note that ATP is not required. Because of this simple phosphorolysis step, no ATP is invested in glycolysis until step 3 (see Figure 8.4) and the net yield at the end of glycolysis is three ATPs instead of the usual two.

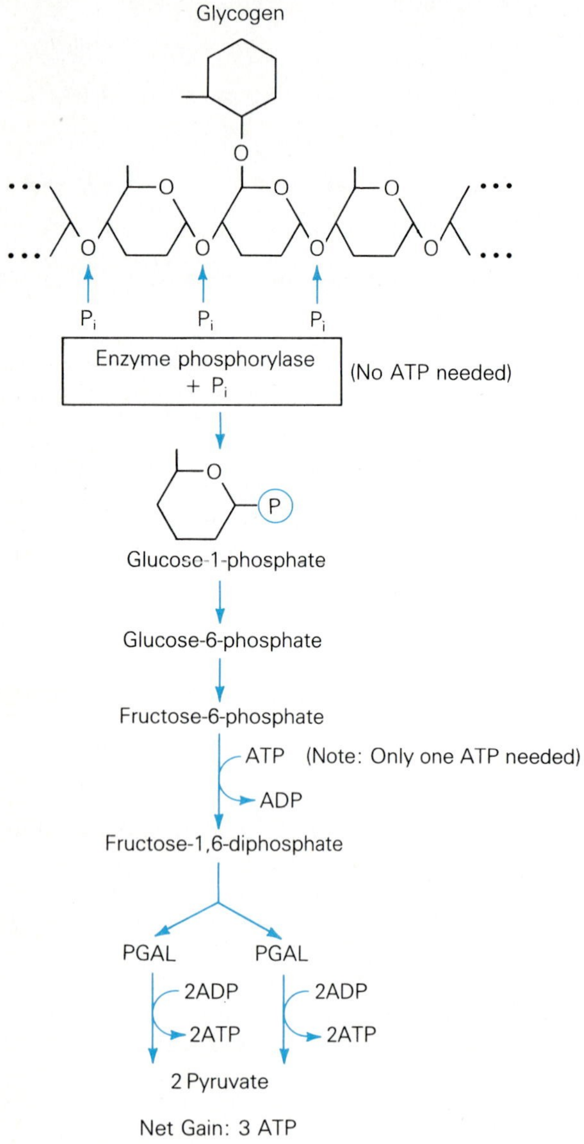

Glycogen

Glucose-1-phosphate

Glucose-6-phosphate

Fructose-6-phosphate

Fructose-1,6-diphosphate

PGAL   PGAL

2 Pyruvate

Net Gain: 3 ATP

CELL RESPIRATION

The other fate of the pyruvate from glycolysis is the most important one in all eukaryote cells—the one that involves aerobic or oxidative cell respiration. We can begin our account of cell respiration where we left off with glycolysis—with pyruvate and NADH, products that are formed in the cytoplasm. The site of activity then changes to the mitochondria, where cell respiration takes place.

Pyruvate to Acetyl CoA

NADH cannot enter the mitochondria, but as we will discuss later, there is a mechanism, a shuttle carrier, as it were, to carry its energetic electrons inside. The pyruvate, however, can enter the mitochondria, where it is altered so that it can take part in the citric acid cycle. This alteration turns out to be a fairly complex process. The overall reaction is:

$$2 \text{ Pyruvate} + 2\text{NAD}^+ + 2 \text{ Coenzyme A} \longrightarrow 2 \text{ Acetyl CoA} + 2\text{NADH} + 2\text{H}^+ + 2\text{CO}_2$$

Coenzyme A (CoA) is covalently linked by a high-energy bond to what's left of our fuel molecule, which has now been diminished from a 3-carbon pyruvate to a simple 2-carbon acetyl group (CH_3CO) to form **acetyl CoA** (Figure 8.9).

Acetyl CoA is a vital intermediate in the utilization of cellular fuels other than glucose. In other metabolic pathways, amino acids (from proteins) and fatty acids (from fats) are also broken down into acetyl CoA (Essay 8.2). Let's now see what happens to acetyl CoA in the citric acid cycle.

The Citric Acid Cycle

The **citric acid cycle** (or **Krebs cycle**) was first postulated in 1937 by H. A. Krebs. Ironically, although his paper has become a classic in biochemistry, it was rejected by the disbelieving editors of *Nature*, the prestigious British journal to which it was first submitted. However, when science was finally ready for Professor Krebs, he was awarded the Nobel Prize for his efforts.

Before we look into some of the details of the citric acid cycle, let's take note of the fact that nearly its entire purpose is to provide energetic electrons to the electron transport system, and that their energy will be used to increase the chemiosmotic differential of the mitochondrion and permit ATP to be synthesized.

The citric acid cycle occurs in the matrix, or inner compartment, of the mitochondrion. As soon as acetyl CoA enters this inner chamber, it reacts with a 4-carbon molecule known as **oxaloacetate** and an enzyme called **citric synthetase.** The high-energy bond between coenzyme A and the acetyl group causes the reaction to go quickly, with the release of a considerable amount of free energy. The coenzyme A is simply released to recycle, forming more acetyl-CoA, but the acetyl group combines with the oxaloacetate to form the first new molecule of the cycle, the 6-carbon compound citric acid—the same citric acid found in lemons. One molecule of water is used up in the citric acid synthesis reaction.

An Overview of the Citric Acid Cycle. The cycle begins as the acetyl group of acetyl CoA joins with oxaloacetate to form citrate and the CoA is

8.9 THE ACETYL CoA STEP

Upon entering the mitochondrion, pyruvate must first be converted to acetyl CoA before it can join the citric acid cycle. The pyruvate joins an enzyme complex, is decarboxylated, and oxidized (with NAD^+ reduced to $NADH + H^+$), and the acetyl group remaining is joined by a high-energy bond to coenzyme A. For each molecule of glucose entering respiration, two molecules of acetyl CoA are eventually formed.

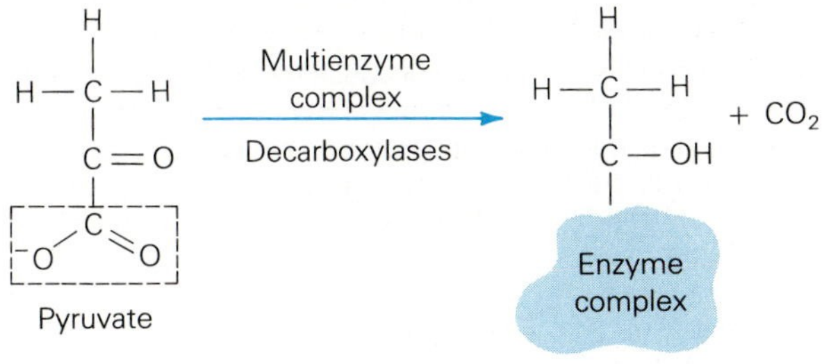

(a) Removal of CO_2

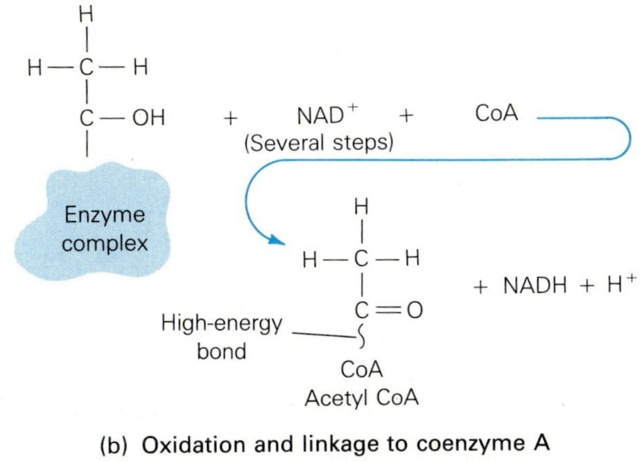

(b) Oxidation and linkage to coenzyme A

released. Figure 8.10 illustrates the entire cycle, showing the series of events as our fuel molecule fragment is acted upon sequentially by a battery of enzymes. The reaction numbers in the following discussion refer to the numbers in the diagram.

As you can see, in the course of the cycle, a 6-carbon citrate is internally altered twice and broken down to a 5-carbon compound, and the first of two carbon dioxides formed in the cycle is released (④). The 5-carbon molecule is short-lived and in the next reaction (⑤), the second carbon dioxide is released (this is the carbon dioxide we exhale from our lungs with each breath). The 4-carbon molecule emerging is altered in four reactions and the final product is oxaloacetate (⑥ through ⑨), which is ready to begin the cycle anew. Thus, with each complete cycle, two carbons enter as the acetyl group, and two carbons leave as carbon dioxide.

The energy of the acetyl group is obtained from the cycle in repeated removals of hydrogens and electrons. These are transferred to NAD^+ and another hydrogen carrier, flavin adenine dinucleotide (FAD), reducing them to NADH and $FADH_2$ (steps ④, ⑤, ⑦, and ⑨). Every hydrogen and electron removal is another oxidation of the fuel molecule, so the glucose is very gradually "burned." At one point along the way, an additional bit of energy is picked off in a high-energy substrate-level phosphorylation (⑥). Notice also that three water molecules enter the cycle at various points, including the water that enters into the synthesis of citrate; one water molecule leaves the reaction (①, ②, ③, and ⑧). The net reaction of one turn of the cycle, leaving out the various cofactors (such as NAD^+ and FAD), is:

$$C_2H_4O_2 + 3H_2O \rightarrow H_2O + 2CO_2 + 8H$$
Acetic
acid

In case you are wondering how eight hydrogens can be obtained from one acetyl group—they can't. Some of the hydrogens come from the water that enters the reaction.

If we include the coenzymes, one turn of the cycle can be written as:

Acetyl-CoA + $3NAD^+$ + FAD + $3H_2O$ + GDP + $P_i \rightarrow$
CoA + 3NADH + $3H^+$ + $FADH_2$ + GTP + H_2O + $2CO_2$

(GDP and GTP are the triphosphate nucleotides guanine diphosphate and guanine triphosphate, respectively, both chemically similar to ATP.) Of course, if we want to continue following both of the acetyl CoA molecules that result from the break-

Alternative Fuels in Cell Respiration

Physiologists talk about glucose so much that it might appear that this is the only cellular fuel. While carbohydrates are important and common cellular fuels, they are not the only ones, and quite often they are not even the most important ones. For instance, we all rely heavily on stored fats as a primary energy supply. We maintain a sizeable supply of glucose in the form of glycogen, but only enough to maintain us for about a day of normal activity. Following brief periods of fasting our body fats become mobilized, ready for use by cells. By morning, for instance, our bodies have switched to fats so that most of the ATP energy produced in the mitochondria has come from the oxidation of fatty acids. It should be good news to dieters that body fats (particularly those stored in places we would rather not draw attention to) can also be sent through the respiratory mill. In fact, fats are excellent energy sources, yielding about twice the amount of energy as carbohydrates on a gram per gram basis. Proteins can also be used as energy sources, but their use is complicated by their much more varied composition. Interestingly, fatty acids are the preferred chemical fuel of heart muscle, but they are utterly rejected by the brain, whose chief energy source is glucose.

As all of us are aware, the three basic foods, carbohydrate, lipid (fat), and protein, must be broken down into their chemical subunits to be of use to our bodies as fuels and for other purposes. But for use in respiration, even the subunits must undergo considerable modification (as seen in the illustration) before they can enter the aerobic respiratory process of the mitochondria.

Fatty acids must be activated by coenzyme A (CoA) in order to be transported across the mitochondrial membranes to the matrix, where they are prepared for the citric acid cycle. Activation takes place on the outer membrane and consists of joining the fatty acid to coenzyme A. The activated fatty acid is then ferried across the membrane by special carriers, ending up in the matrix. Once there a 2-carbon acetyl-CoA is split away to enter the citric acid cycle. Then another coenzyme A joins the remaining fatty acid fragment, and the sequence is repeated, freeing a second 2-carbon acetyl-CoA, and so on until the fatty acid chain has been completely fragmented, two carbons at a time, into acetyl CoAs.

For example, **palmitic acid,** a 16-carbon fatty acid, goes through seven reactions yielding eight acetyl-CoAs. But there is a special bonus in all of this: each reaction includes an oxidation and a reduction, so along with the acetyl-CoAs, seven NADH + H$^+$ and seven FADH$_2$s also form. The reduction products, of course, greatly enrich the chemiosmotic differential of the mitochondrion, leading to the production of ATP. This explains why fats or lipids are such high-energy food.

Proteins are first hydrolyzed by digestive enzymes into the usual 20 amino acids. They then enter the citric acid cycle in several different ways. All must first have their amino groups (NH$_3^+$) removed, a process called **deamination,** and these groups must be rendered harmless. Unattended, the amino groups would accumulate as dangerous ammonium ions. Five of the 20 amino acids are simply converted to pyruvate, which is then converted to acetyl CoA. Six more of the amino acids skip the pyruvate step and enter as acetyl CoA itself. Fourteen amino acids are actually converted into specific acids of the citric acid cycle, joining it directly in four places, as seen in the illustration. If you add this up, you will find a disturbing total of 25 different amino acids entering the citric acid cycle, but five amino acids can enter at either of two places, still leaving 20 in all.

The amount of energy available from each type of amino acid depends, as you might suspect, on where its conversion product joins the citric acid cycle. This makes sense because the point of entry also determines how much reduced NAD is generated. The five amino acids converted to pyruvate have the most to offer, since NADH + H$^+$ is yielded each time pyruvate is formed. Those entering as acetyl CoA are next in yield, while those entering as acids of the cycle are, of course, the least productive.

Fatty acid conversion to acetyl-CoA yields a bonus—a number of NADHs and FADH$_2$s. Note the cleavage sites in palmityl-CoA, an intermediate in fatty acid oxidation.

$$\text{Activated palmityl-CoA} + 7\text{FAD} + 7\text{NAD} + 7\text{H}_2\text{O} + 7\text{CoA} \longrightarrow 8 \text{ Acetyl-CoA's} + 7\text{NADH} + 7\text{H}^+ + 7\text{FADH}_2$$

The amino acid alanine, one of the simplest, is deaminated in a reaction with NAD^+ and water, which yields pyruvate, the same end product formed when glucose is metabolized in glycolysis. Pyruvate is then acted upon in the mitochondrion to form acetyl-CoA and enters the citric acid cycle. In our own bodies, the amino group removed during deamination will follow another pathway, where, in a very different cycle, it will be converted from the reactive ammonium ion (NH_4) to a relatively harmless molecule called urea. Urea can be safely transported in the blood to the kidneys where it is excreted from the body in the urine.

Alanine $+ NAD^+ + H_2O \xrightarrow{\text{deamination}}$ **Pyruvate** $+ NADH + H^+$

$NH_3 \dashrightarrow$ Different Pathway
Ammonia

Pyruvate $+ NAD^+ + CoA \longrightarrow$ **Acetyl-CoA** $+ CO_2 + NADH + H^+$

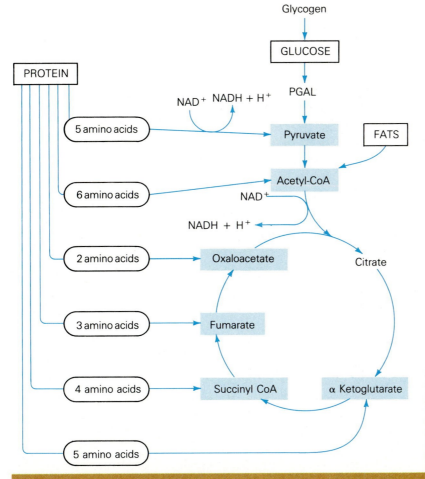

Glycogen
GLUCOSE
PGAL

PROTEIN

5 amino acids
NAD^+ $NADH + H^+$
Pyruvate   FATS

6 amino acids
Acetyl-CoA
NAD^+
$NADH + H^+$
Citrate

2 amino acids
Oxaloacetate

3 amino acids
Fumarate

4 amino acids
Succinyl CoA   α Ketoglutarate

5 amino acids

Amino acids enter the citric acid cycle at various points depending on kind of amino acid concerned. The point of their entry determines the amount of reduced NAD or FAD to be produced and whether or not ATP can be generated directly in the cycle (the step following succinyl CoA).

8.10 THE CITRIC ACID, OR KREBS CYCLE

The citric acid cycle has nine major enzymatic steps. Each turn of the cycle begins and ends with oxaloacetate. In the first reaction, oxaloacetate joins with acetyl CoA and water, and Coenzyme A is released. The product formed is citric acid. As each enzyme does its job in the succeeding steps, molecules change, CO_2 is released, NAD and FAD are reduced, H_2O enters and leaves, and one ATP is formed. However, the main function of the citric acid cycle is to make energetic electrons available to the electron transport systems in the inner membrane. As the electrons pass through these systems, their energy will be used to pump protons across the membrane, thus increasing the free energy of the chemiosmotic differential, and making chemiosmotic phosphorylation possible.

down of one glucose molecule, the terms in the formula have to be doubled.

In Figure 8.10, we have included the names of all the enzymes and intermediates and have given the structural formulas as well, to fill you in on the details of what's happening. In our discussion, however, the names of these substances are secondary. We are more interested in following the events.

Let's proceed through the cycle clockwise from step ①, which we have already discussed, on through steps ② and ③, in which water is removed and then added, and the citric acid molecule is subtly rearranged, forming first cis-aconitate and then isocitrate. Step ④ is far from subtle. Here, isocitrate is oxidized to oxalosuccinic acid, as NAD^+ is reduced to NADH. Oxalosuccinate, while still joined to the isocitrate dehydrogenase enzyme, loses a carbon dioxide. The remaining 5-carbon compound, called alpha-ketoglutaric acid, is then released. Let's review the first four steps:

① Acetyl-CoA + Oxaloacetate + H_2O $\xrightarrow{\text{Enzyme}}$ Citrate + CoA

② Citrate $\xrightarrow{\text{Enzyme}}$ Cis-aconitate + H_2O

③ Cis-aconitate + H_2O $\xrightarrow{\text{Enzyme}}$ Isocitrate

④ Isocitrate + NAD^+ $\xrightarrow{\text{Enzyme}}$ Oxalosuccinate + NADH + H^+

and,
Oxalosuccinate $\xrightarrow{\text{Enzyme}}$ Alpha-ketoglutarate + CO_2

The enzyme alpha-ketoglutarate dehydrogenase then enters the picture and performs a dazzling display of chemical virtuosity. Essentially, what happens is that NAD^+ is reduced, carbon dioxide is released, and the remaining 4-carbon skeleton is covalently linked to coenzyme A as succinyl CoA. Note in Figure 8.10 that the succinyl CoA bond is a high-energy or energy-rich bond, that is, it has high transfer potential—enough to phosphorylate ADP.

⑤ Alpha-ketoglutarate + CoA + NAD^+ $\xrightarrow{\text{Enzyme}}$

Succinyl-CoA + CO_2 + NADH + H^+

In the next reaction (step ⑥ in Figure 8.11), the high energy represented by the succinyl-CoA bond is picked up by guanosine diphosphate (GDP) to form guanosine triphosphate (GTP), as coenzyme A is released. We don't know why the cell uses GDP here instead of ADP, but in any case, the high-energy bond of GTP is readily transferred to ADP:

⑥ Succinyl CoA + P_i + GDP $\xrightarrow{\text{Enzyme}}$

Succinate + CoA + GTP

GTP + ADP $\rightleftharpoons$ GDP + ATP

Proceeding to step 7, we find another oxidation (or dehydrogenation) reaction. But instead of the hydrogen being passed to NAD^+, as we have come to expect, coenzyme FAD does the job.

⑦ Succinate + FAD $\xrightarrow{\text{Enzyme}}$ Fumarate + $FADH_2$

In step ⑧, the 4-carbon skeleton adds another water to become malate. Malate dehydrogenase then oxidizes the malate (step ⑨) and produces our old friend oxaloacetate, to start the cycle again. As you see, NAD^+ is again reduced to NADH in the process.

⑧ Fumarate + H_2O $\xrightarrow{\text{Enzyme}}$ Malate

⑨ Malate + NAD^+ $\xrightarrow{\text{Enzyme}}$ Oxaloacetate + NADH + H^+

Let's stand back now look at the entire process, going all the way back to glucose in glycolysis. That same glucose molecule has now been oxidized completely to carbon dioxide. And what is there to show for it? So far, there has been a net gain of only four ATPs, two from glycolysis and two from the citric acid cycle (via GTP), both occurring through substrate level phosphorylation. This may not seem quite right, since the citric acid cycle was touted as the major energy producer. So far, it is no more efficient than glycolysis. But keep in mind that we have accumulated a total of 24 hydrogens in the form of $FADH_2$ and NADH + H^+. That's a pretty good trick if you come to think about it, since glucose has only 12 hydrogens to start with. The other 12 hydrogens come from the six molecules of water that enter the cycle as the two acetyl CoA molecules (from glucose) go through the citric acid cycle (Table 8.2). These hydrogens are going to have a great deal to do with making more ATP. (Figure 8.11 presents an abbreviated view of the citric acid cycle.)

ELECTRON TRANSPORT SYSTEM

We have already been through an electron transport system in the thylakoid membrane of the chloroplast (see Chapter 7). And as we first saw in Chapter 6, the electron transport chain of the mitochondrion has many similar features, which should not be too surprising. Both systems are believed to have evolved from a common ancestor long ago.

TABLE 8.2

ADDING UP THE HYDROGEN

The total hydrogens gained per glucose from all of respiration is 24:

Glycolysis	$2NADH + 2H^+ \rightarrow$		$2FADH_2$*
Pyruvate to acetyl-CoA	$2NADH + 2H^+$		
Citric acid cycle			
Step ④	$2NADH + 2H^+$		
Step ⑤	$2NADH + 2H^+$		
Step ⑦			$2FADH_2$
Step ⑨	$2NADH + 2H^+$		
Total		24 Hydrogens	

*Note: Cytoplasmic NADH is converted to $FADH_2$ at the outer mitochondrial membrane.

You may wish to compare Figure 7.11 with Figure 8.12.

We saw in Chapter 4 that the mitochondrion is a cell organelle with two membranes. The outer membrane is simple in structure, is highly permeable, and has the form of a closed sac. The inner membrane, however, is highly convoluted and forms extensive folds or shelflike cristae that reach into the interior of the organelle. The presence of the two membranes produces an outer compartment and an inner compartment, or matrix, as the latter is known. As in the case of the chloroplast, all mitochondrial electron and proton carriers are integrated into a membrane—in this instance the inner membrane.

The inner membrane is, essentially, a bilipid membrane in which large proteins are bound together by phospholipids, more or less the way bricks in a wall are bound together by mortar. The enzymes and electron carriers of the electron transport system are tightly bound within the inner membrane. Stalked F1 particles, which appear to be identical to CF1 particles, are also located on the inner membrane, projecting into the matrix or inner compartment. Keep in mind that it is within the bulbous particles that ADP and P_i are combined to form ATP. We will be making additional comparisons between mitochondria and chloroplasts further along.

As you would expect, the arrangement of the mitochondrial membranes, the electron transport systems, and the F1 particles has special significance. What we see here is an active transport system, two compartments separated by a membrane, and a system of ATP-synthesizing enzymes—all

familiar ingredients for an energy-packed chemiosmotic differential to form. As with the chemiosmotic differential in the thylakoid, it will be the energy of electrons passing through electron carriers that will be used, at critical points, to pump protons through the membrane. In this case, the protons will concentrate in the outer compartment. Their isolation will produce an electrochemical gradient whose energy can be harnessed in the synthesis of ATP. Let's go now to some of the details.

Building the Chemiosmotic Differential. Earlier we referred to the membranal carriers as both electron *and* proton (H^+) carriers. The electrons of the mitochondrial chain, traveling in pairs, start on the inside and end on the inside, making at least two, probably three, round trips across the inner mitochondrial membrane. In each round trip, the electrons travel out as hydrogen atoms in one kind of carrier and return as electrons in a different kind of carrier. Each time a hydrogen carrier reaches the outside surface, it liberates protons into the external medium, but the electrons are passed on to an electron carrier in the membrane. Then each time an electron carrier reaches the inner surface, the electrons are passed to a hydrogen carrier, which combines them with protons taken from the internal medium. For each pair of electrons completing its round trips, the system pumps six protons across the membrane.

Of course, the electrons and hydrogens follow a thermodynamically predictable pathway, and

8.11 SHORT VERSION OF THE CITRIC ACID CYCLE

The cycle begins with the addition of a 2-carbon molecule (acetyl-CoA) to an existing 4-carbon molecule (oxaloacetate). The products seen are reduced NAD and FAD, ATP, and carbon dioxide.

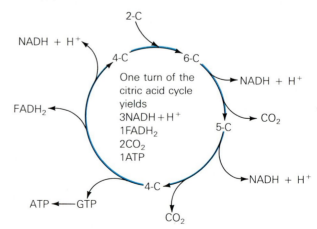

One turn of the citric acid cycle yields
$3NADH + H^+$
$1FADH_2$
$2CO_2$
$1ATP$

8.12 ELECTRON TRANSPORT IN MITOCHONDRIA

The enzymes of the citric acid cycle are in the inner compartment (matrix) of the mitochondrion; the enzymes and carriers of the electron and proton transport system are embedded in the inner membrane.

(a) Building the gradient: Activity in the electron transport system begins with the transfer of two hydrogen atoms ① from NADH to the FMN-enzyme complex. The hydrogens are moved across the membrane, where they are dissociated into electrons and protons. The protons are released into the outer compartment of the mitochondrion ②, and the two electrons reduce, first FeS_1 and the FeS_2, the next electron carriers. FeS_2, next reduces enzyme Q (of CoQ complex III) ③, which attracts two protons from the matrix becoming QH_2. The reduced coenzyme carries its two hydrogen atoms across the membrane, releasing two protons into the outer compartment ④ and the two electrons to the next electron carrier, cytochrome b (cyt_b). The electrons then reduce a second cytochrome b, and in passing to cytochrome c_1, provide the energy to somehow shuttle a third and final pair of protons ⑤ into the outer compartment. The electrons then pass through the final two cytochromes—cyt c and cyt a, a_3 ⑥. Finally, the electrons reach oxygen, which, when combined with protons from the matrix, produces water ⑦. Altogether, the electrons of NADH power the transport of six hydrogen ions across the inner mitochondrial membrane. The $FADH_2$ from the citric acid cycle (⑧, see inset) enters the system further along at the CoQ complex III, so its electrons power the transport of only four protons across the membrane.

(b) Using the gradient to produce ATP: the resulting proton gradient (a pH and electrical gradient) represents a store of free energy which is tapped in two ways. In the F1 particle, two hydrogen ions ⑨ are admitted into the mitochondrial matrix and the energy is used to phosphorylate ADP to ATP (k). A proton-powered active transport system ⑩ uses a third proton to power a shuttle that pumps out an ATP molecule and simultaneously brings in an ADP molecule.

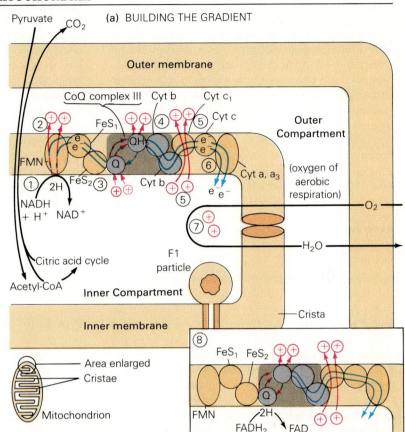

(a) BUILDING THE GRADIENT

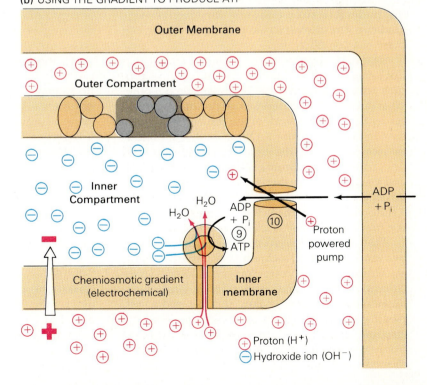

(b) USING THE GRADIENT TO PRODUCE ATP

the free energy of each carrier state is less than that of the one before. When the electrons complete their zigzag pathway, they are almost depleted of energy; most of it has been lost as heat or transferred into the free energy of the chemisomotic differential (buildup of protons on one side of a membrane). What becomes of the energy-depleted electrons? We will soon see, but for now keep this clue in mind: we have not yet mentioned the role of oxygen in cell respiration.

Electron and Proton Transport. Electron and proton transport begins inside the inner membrane with NADH (which was reduced in the citric acid cycle). When NADH reaches the inner membrane, it reduces a large transmembranal complex, **FMN (flavin nucleotide,** complexed with a NAD^+ oxidizing enzyme). The FMN, now reduced to $FMNH_2$, carries the hydrogen across the membrane, where it releases two protons into the outer compartment of the mitochondrion. The two electrons pass through a pair of iron-sulfur proteins and the FMN recycles. Back at the interior side of the membrane, the iron-sulfur proteins pass the electrons to a complex known as **CoQ complex III.** Within this complex, a mobile hydrogen carrier, coenzyme Q (CoQ) receives the electrons and at the same time picks up two protons from the inner medium, forming hydrogen atoms once again. Q is thus reduced to QH_2, whereupon it moves to the outer surface of the membrane, where it releases the protons, further enriching the chemiosmotic gradient. The electrons pass to the next carrier, but there is a complication.

The next step isn't clearly known except that CoQ complex III is capable of pumping four protons across the membrane. A reasonable hypothesis states that the second pair of protons is ferried across through the work of the three cytochromes next in line. As Figure 8.12 indicates, QH_2 hands its electrons off to the first of two cytochromes, designated as **cytochrome *b*.** It reduces the second, and the electrons are then passed to **cytochrome c_1.** It is in the association of the three cytochromes that a third pair of protons is ferried across, but again the precise pumping mechanism isn't known.

From this site, the electrons pass down through two more carriers, **cytochrome *c*** and **cytochrome *a*, a_3,** but no further proton transport occurs. Cytochrome *a*, a_3, the final carrier and a large transmembranal complex, makes the last reduction, passing the two electrons to an oxygen atom. The electrons and oxygen are joined by two protons from the inner medium and a water molecule is formed (see Figure 8.12).

The Role of Oxygen in Cell Respiration.   We have now answered the basic question about the role of oxygen in cell respiration. It enters the scene at the very end of the process, collecting the energy-poor electrons after their reducing power has been removed. One question that has bothered biologists for decades, and that remains unanswered, is just how the oxygen molecule finally enters the picture. One theory is that the cytochromes work in pairs, or possibly in quartets, and that the final reaction is between one O_2 molecule, four electrons, and four hydrogen ions to make two molecules of water. Another possibility, for which there is some evidence, is that every second reaction involves two hydrogen ions and two electrons, which break up an oxygen molecule into two highly reactive particles called radicals; the other reactions then supply two more hydrogen ions and two more electrons to turn the free radicals into harmless water. For various reasons, no one is really satisfied with either hypothesis.

The Role of $FADH_2$.   There are several important points to make about the mitochondrial electron transport system. First, most of the electrons from the citric acid cycle are carried as NADH so they start at the beginning of the chain as described, and each electron will serve to pump three hydrogen ions out against the gradient. One of the reactions in the citric acid cycle, however, passed a pair of hydrogens directly to a lower-energy FAD coenzyme. $FADH_2$ doesn't have the reducing power to start at the beginning of the cycle, but must interact with the system further down the line. As you see in Figure 8.12 (inset), $FADH_2$ passes its hydrogens to the CoQ complex III, reducing Q to QH_2. From there on, the events are the same as we described earlier. Thus, the lower-energy electrons of $FADH_2$ miss out on the first proton transporting reaction and will contribute only two hydrogen ions each to the chemiosmotic gradient.

There is another source of FAD reducing power (free energy) for the electron transport system—the two NADH + H^+ left over from glycolysis. But this reduced NAD^+, as was mentioned, is in the cytoplasm and cannot enter the mitochondrion. In another minor miracle of nature, however, a special "shuttle" system oxidizes NADH to NAD^+ and carries the electrons and protons across to the inner mitochondrial membrane, where they reduce FAD. The $FADH_2$-bound hydrogens can then enter the transport chain at the CoQ complex as seen, and can account for another contribution of four protons to the chemiosmotic gradient. The shuttle, we should mention, also provides a means for recycling NAD^+ back to the glycolytic pathway, where it can act again.

The Electrochemical Gradient and ATP

The importance of all of this electron movement and proton pumping, once again, is in its contribution to the chemiosmotic differential, or electrochemical gradient, as it is also called (see Chapter 6). A high concentration of protons on one side of a membrane represents both an electrical gradient (a charge difference on either side) and a chemical pH gradient (a lower pH on one side than the other). But whichever way we characterize it, what is important is the free energy of the differential. As we have seen in photosyntheisis, this energy can be harnessed to phosphorylate ADP.

Phosphorylation occurs as protons rush down their gradient, through the channels leading into the F1 particles. The F1 heads contain the phosphorylating enzymes and a supply of ADP and P_i (inorganic phosphate). Again, the specifics of energy transfers accompanying phosphorylation are incompletely known, but as the protons lose energy, ADP is phosphorylated, gaining in energy as ATP (see Figure 8.12b).

The ATP Balance Sheet in the Metabolism of Glucose

One of the favorite activities of biochemists is to try to balance the books in a complicated series of reactions such as we have seen. Today, however, we are less sure of these numbers. Remember, the development of the chemisomotic view of chloroplast and mitochondrion function has made any exact accounting difficult—at least at our present level of understanding—and has put something of a damper on any tit-for-tat view of the process. Even if it were possible to determine exactly how many protons were transported across the membrane for each reaction and how many protons were passed back across the membrane to produce each ATP, we'd still have problems calculating precise numbers. For one thing, laboratory measurements indicate that the membranes leak. For another, some of the free energy of the hydrogen ion gradient is siphoned off for uses other than ATP formation. For instance, some of the gradient's energy is used to power the active transport of essential molecules across the mitochondrial membrane.

The estimates today vary, ranging from two to four protons reentering the matrix for each ATP

phosphorylation. The highest number, four protons, reflects a different theory of phosphorylation in the F1 particle, but let's stay with Peter Mitchell's proposals (see Essay 6.1). Estimates from the earlier days of the Mitchell hypothesis maintained that chemiosmotic phosphorylation requires the passage of only two hydrogen ions through the F1 particle for each ATP generated. This fits well with still older calculations, which maintained that ATP was generated within the mitochondrial electron transport system itself. The chemiosmosis theory, of course, maintains that no such direct coupling occurs. In addition, we noted in our consideration of photosynthesis that each pair of hydrogen ions pumped across the thylakoid membrane accounts for the synthesis of only two thirds of an ATP molecule in that system.

According to thermodynamic measurements, there isn't quite enough energy in transferring a pair of hydrogen ions to account for one ATP synthesis, if all reactants are present in equal molar concentrations. Apparently, the reaction occurs only because there is a high concentration of ADP within the inner cavity of the mitochondrion. This ADP concentration is there only because of a proton-powered active transport exchange shuttle that brings in an ADP and a hydrogen ion while simultaneously removing an ATP. So it appears that each ATP synthesis, in the long run, may require the movement of three hydrogen ions across the membrane after all—one to bring in ADP and two to phosphorylate it.

If we assume that from two to three hydrogen ions are involved, the total number of ATPs produced is 36 to 25 per glucose (Table 8.3). Actually, it is more likely that the efficiency of respiration is not constant but varies according to a number of factors. We can assume only that the average yield is somewhere between the two numbers.

About 680 kcal are released in the complete oxidation of one mole of glucose. The caloric value of the energy-rich bond in ATP is about 8 kcal/mole. The 25 to 36 ATPs produced by the total metabolism of glucose represent a range in efficiency of 29 to 43%. Even 29% is quite respectable by any standards, and you can be certain that your mitochondria are more efficient in energy conversion than is your car.

Comparing the Mitochondrion with the Thylakoid

A comparison of the orientation of F1 particles in the mitochondrion and CF1 particles in the thylakoid has shown they are opposites. The knobby ends face the inner compartment in the mitochondrion and the outer compartment in the thylakoid. There are also differences in the electron transport system of chloroplasts and mitochondria. In particular, the photosynthetic electron transport chain *ended* with the electrons being passed from the last membrane-bound carrier to $NADP^+$; the mitochondrial electron transport system, on the other hand, *begins* with electrons being passed from NADH to the first membrane-bound carrier in the chain. In addition, the electron in the Z-scheme of photosynthesis *began* with the splitting of water into hydrogen ions, electrons, and molecular oxygen; in the mitochondrial system, the flow *ends* with the combining of hydrogen ions, electrons, and molecular oxygen into water. Thus, mitochondrial and chloroplast electron transport systems are not only inside out but also chemically backwards relative to one another.

In summary, we have seen that the molecules of food undergo extremely delicate, complex, and rigorous changes in living systems. The energy that has been stored there by plants is gradually released to do the work required in order to keep the molecules of life organized.

TABLE 8.3

ATP BALANCE SHEET (ATPs PER GLUCOSE)

Glycolysis	$2NADH + H^+ \rightarrow$	2FADH$_2$	2ATP direct (substrate level)
Acetyl-CoA	$2NADH + H^+$		
Citric acid cycle	$6NADH + H^+$		2ATP direct (substrate level)
		2FADH$_2$	
	$8NADH + H^+$ + $4FADH_2$ + 4ATP		

Each NADH provides the energy to enrich the differential by 6 protons, while each FADH$_2$ provides the energy to enrich it by 4 protons. Therefore:

$$8NADH = 48 \text{ protons}$$
$$\underline{4FADH_2 = 16 \text{ protons}}$$
$$\text{Total} \quad\quad 64 \text{ protons}$$

Estimate 1: If ATP synthesis requires 2 protons: 64 protons = 32 ATP (+4 ATP direct) = *36 ATP per glucose.*

Estimate 2: If ATP synthesis requires 3 protons: 64 protons = 21 ATP (+ 4 ATP direct) = *25 ATP per glucose.*

Estimate 3: If ATP synthesis requires 4 protons: 64 protons = 16 ATP (+ 4 ATP direct) = *20 ATP per glucose.*

APPLICATION OF IDEAS

1. What does it mean for a runner to "go anaerobic"? Who depends more on anaerobic respiration, a sprinter or a marathon runner? Explain your answers.

2. Is anaerobic respiration really inefficient? Consider the question in terms of the free energy actually available to dwellers of the anaerobic environment. Also consider the answer in terms of whether or not anaerobic organisms are successful, well-adapted organisms.

KEY WORDS AND IDEAS

COMPARING PHOTOSYNTHESIS AND RESPIRATION

1. Cells produce ATP through **anaerobic glycolysis** (and **fermentation**), and through **aerobic cell respiration.**

2. The general formula for photosynthesis appears to be precisely opposite to the general formula for respiration, and the energy sources are quite different.

3. Light energy drives the photosynthetic reactions while chemical bond energy drives those of respiration.

4. Photosynthesis is energetically uphill while glycolysis and cell respiration are downhill.

THE THREE PARTS OF RESPIRATION

1. Glucose metabolism includes glycolysis and cell respiration. The latter includes the **citric acid cycle, electron transport,** and **chemiosmotic phosphorylation.**

A Word About ATP

1. ATP is produced through two processes: **substrate level phosphorylation** and chemiosmotic phosphorylation.

2. In substrate level phosphorylation, the high energy phosphate bonds of ATP are aquired directly from the fuel.

3. In chemiosmotic phosphorylation, energy for new ATP bonds comes from the free energy of the chemiosmotic differential.

GLYCOLYSIS:
METABOLIZING GLUCOSE WITHOUT OXYGEN

1. Glycolysis is an **anaerobic** process occurring in the free cytoplasm of the cell. The starting molecule is glucose and the phosphorylation of ADP occurs only at the substrate level.

The Steps of Glycolysis

1. Glycolysis includes nine major enzymatic steps, beginning with glucose and ending with **pyruvate (pyruvic acid)** and NADH. Along the way two ATPs are invested and four ATPs are returned. The principal reactions are:
   a. *Phosphorylation* (priming): Two ATPs are used to phosphorylate glucose, which is converted to **fructose-1,6-diphosphate.**
   b. *Cleavage:* Fructose-1,6-diphosphate is broken down into two 3-carbon PGALs (phosphoglyceraldehyde).
   c. *Phosphorylation* and *Oxidation:* The PGALs are phosphorylated with P_i, and oxidized to form two 3-carbon 1,3 diphosphoglycerates. The hydrogens from oxidation reduce NAD^+ to $NADH + H^+$.
   d. *Substrate level phosphorylation of ADP:* Two ADPs react with the 1,3 diphosphoglycerates, yielding two ATP's and two 3-phosphoglycerates, which are changed to two 2-phosphoglycerates (the phosphate moved to the number 2 carbon).
   e. *Dehydration:* Water is removed from the 2-phosphoglycerates producing two 3-carbon phosphoenolpyruvates.
   f. *Substrate level phosphorylation of ADP:* ADP reacts with the phosphoenolpyruvates, yielding two ATPs and two pyruvates.

2. During glycolysis two ATPs are consumed and four ATPs are recovered with a net gain of two ATPs. Considering the free energy available in glucose, this is about 2.4% efficient.

3. Pyruvate is taken into the mitochondrion in aerobic cells, and there, considerably more energy is removed through oxidation and chemiosmotic phosphorylation.

The Fates of Pyruvate

1. In anaerobic forms, glycolysis is replaced by **fermentation,** which is identical except for the fate of pyruvate. Fermenters produce energy-rich alcohols and organic acids as waste products, but in so doing they are able to recycle the NAD^+ needed to keep fermentation going.

2. Yeasts are well-known **ethyl alcohol fermentation** organisms. In their metabolism, they **decarboxylate** pyruvate (remove CO_2), forming **acetaldehyde,** and then, using the enzyme **alcohol dehydrogenase** and NADH, they reduce it to **ethanol.**

3. Yeasts are **facultative anaerobes,** capable of switching to aerobic respiration when conditions are right. (Low oxygen, excess glucose → alcohol fermentation; abundant oxygen, limited glucose → aerobic oxidation in mitochondrion.)

4. **Lactate fermenters** include a number of bacteria and the muscle tissue of animals. In both, pyruvate is reduced by NADH to 3-carbon **lactate (lactic acid),** permitting NAD^+ to recycle. In bacteria, the lactate is a waste product but in muscle tissue, lactate is removed and more energy extracted.

5. During muscle exertion ATP is provided as follows:
   a. A limited reserve of ATP gets the muscle going.
   b. ADP is at first phosphorylated by energy rich phosphate bonds from **creatine phosphate (phosphocreatine).**
   c. Creatine phosphate is replenished by ATP produced during glycolysis (lactate fermentation).
   d. During glycolysis, lactate is built up in the muscle tissue, but much of it is carried away by the blood to be metabolized in the liver.
   e. When the muscle rests, and oxygen supplies increase, its aerobic respiration begins to catch up, thus it replenishes its ATP and CP reserves.
   f. Lactate in the liver is first converted back to pyruvate. It then undergoes **gluconeogenesis,** where it is reconstituted into glucose.

6. The amount of extra oxygen needed to restore ATP of the resting state is known as the **oxygen debt.**

7. When glycolysis begins with glycogen or starch, the polymers are phosphorylated by P_i as the linkages are broken, producing glucose-1-phosphate. Thus the first phosphorylation is "free" and the net gain of glycolysis is three ATPs.

Pyruvate to Acetyl CoA

1. In aerobic organisms, pyruvate enters the mitochondrion where it is converted to **acetyl CoA.** Additional products are carbon dioxide and $NADH + H^+$.

2. Acetyl CoA then enters the **citric acid cycle** or **Kreb's cycle,** where a number of oxidations begin that yield reduced NAD^+ and FAD. The purpose for most citric acid cycle activity is to yield electrons and protons for the chemiosmotic process to come.

3. The citric acid cycle includes 10 major steps, beginning with the union of 2-carbon acetyl CoA joining 4-carbon **oxaloacetate** to form 6-carbon **citrate.** The major events are:
   a. *Oxidation:* Once formed, citrate is converted to isocitrate which is oxidized to form oxalosuccinate. NAD^+ is reduced to $NADH + H^+$.
   b. *Decarboxylation:* Oxalosuccinate loses CO_2 to become alpha-ketoglutarate.
   c. *Oxidation* and *Decarboxylation*: Alpha-ketoglutarate is oxidized to succinyl-CoA, and loses a CO_2. NAD^+ is reduced to $NADH + H^+$.
   d. *Substrate level phosphorylation*: Succinyl-CoA is phosphorylated and dephosphorylated and converted to succinate. An ATP is yielded.
   e. *Oxidation:* Succinate is oxidized to fumarate. FAD is reduced to $FADH_2$.
   f. *Oxidation*: Succinate is converted to malate and malate is oxidized to oxalosuccinate (end of cycle). NAD^+ is reduced to $NADH + H^+$.
4. For each glucose (two acetyl CoAs) entering the citric acid cycle, eight oxidations have occurred. Six $NADH + H^+$ and two $FADH_2$ are formed. In addition, two substrate level phosphorylations yield two ATPs.

The Electron Transport System in Mitochondria
1. In the mitochondrion, numerous electron and hydrogen transport systems are embedded in the inner membrane (cristae). The channels leading to the F1 particles originate in the outer compartment and lead into the inner compartment or matrix.
2. Electrons enter the mitochondrial ETS from the matrix and return to the matrix. Protons enter from the matrix and exit into the outer compartment.
3. The purpose of the ETS is to use electron energy to pump protons into the outer compartment to build the chemiosmotic differential.
4. NADH from the citric acid cycle delivers its electrons and protons (reduces the first carrier) at the start of the ETS, but $FADH_2$ delivers its electrons and protons further along. NAD^+ and FAD then recycle back to the citric acid cycle.
5. Spent electrons finally return to the matrix to react with oxygen and protons there to form water. This is the reason all aerobic forms of life require oxygen.

The Electrochemical Gradient and ATP
1. ATP is generated from the free energy of protons from the outer compartment. They pass through the F1 particles, where their energy is used to phosphorylate ADP, forming ATP.

The ATP Balance Sheet in the Metabolism of Glucose
1. From two to three protons must be pumped into the outer compartment, and leak back through an F1 particle to generate one ATP.
2. At a ratio of three protons per ATP, 25 to 36 ATPs can be produced per glucose molecule entering metabolism.
3. On a calorie basis, in the total metabolism of glucose, the efficiency is 29 to 43%.

REVIEW QUESTIONS

1. Contrast glycolysis with cell respiration, citing such factors as locale, oxygen use, energy yields, and type of phosphorylation used. (173)
2. Discuss three significant ways in which the chemistry of photosynthesis differs from the chemistry of glycolysis and respiration (energy source, direction, starting and ending products). (173–174)
3. Briefly describe the two means by which ADP is phosphorylated. (174)
4. How does glycolysis differ from fermentation? How is it similar? (179)
5. List three different kinds of chemical reactions that go on during glycolysis and mention what the system gains by each. (177–178)
6. What is the purpose of NAD in glycolysis? Why must it be constantly recycled? (178)
7. Suggest reasons for the phosphorylation of glucose at the start of glycolysis. (177)
8. How many energy-consuming steps occur in glycolysis? Energy-yielding steps? What is the net gain in ATP? (177–179)
9. What does a fermenter gain by the reduction of pyruvate at the end of fermentation? (179)
10. Briefly describe the chemical steps that occur in alcohol fermentation from pyruvate. (180)
11. Describe what is meant by the term *facultative anaerobe* and cite an example. (180)
12. Explain the chemical steps leading from pyruvate to lactate. In which of the organisms mentioned is lactate a waste product? (180)
13. List three steps involved in providing ATP for vigorous physical activity in our muscles. (181)

14. Specifically, what is the fate of lactate in our own bodies? (181)

15. Explain how an "oxygen debt" arises and how it is paid off. (181)

16. How is it possible for cells to gain extra energy by starting glycolysis or fermentation with glycogen rather than glucose? (183)

17. Briefly summarize the preparations needed for pyruvate to enter the citric acid cycle. (184, 188)

18. What is the overall purpose of the citric acid cycle? Name the two molecules that play a central role in carrying out this purpose. (184–185)

19. What is the total yield in hydrogens in one turn of the citric acid cycle? List these as NADHs and $FADH_2$s. What other valuable products does the citric acid cycle yield? (185)

20. With what molecule does the citric acid cycle start and end? Explain what happens to the carbon as the cycle goes on. (185)

21. Briefly describe the organization of the inner membrane components of the mitochondrion. Include the electron and hydrogen transport system and the F1 particles. How does the orientation of F1 differ from that of CF1? (190)

22. Describe the path of electrons and protons from NADH as they cross the membrane. Where do they start and end? How many protons cross the membrane in each episode? (190, 192)

23. Compare the proton pumping capabilities of FAD and NAD^+. Why is there a difference? (192)

24. What is the ultimate fate of all electrons passing through the ETS of the mitochondrion? (192)

25. Briefly describe the chemiosmotic differential in the mitochondrion. Where do protons accumulate? What is the nature of the differential? (193)

26. Describe the events surrounding the phosphorylation of ADP. How many protons participate? (193)

SUGGESTED READING

Alberts, B. et al. 1983. *Molecular Biology of the Cell.* New York: Garland Publishing Co.

Calvin, M., ed. 1973. *Organic Chemistry of Life: Readings from Scientific American.* W.H. Freeman, San Francisco. A collection of *Scientific American* reprints, featuring their superb technical illustrations and written for the intelligent lay person.

Darwin, C. 1859. *On the Origin of Species through Natural Selection.* A facsimile of the first edition. Harvard University Press, Cambridge, Mass. We strongly urge all serious biology students to take the time to read it as it was written; Darwin makes much better reading than any of the innumerable books that have since been written about him. Although the writing style is early Victorian, the ideas are fresh and modern and the spirit is infectious.

de Beer, G. 1965. *Charles Darwin: A Scientific Biography.* Doubleday, New York. A sober but intelligent account of Darwin's life and work, with a strong emphasis on the evolution of his scientific thought.

Dickerson, R. E., and I. Geis. 1969. *The Structure and Action of Proteins.* Harper and Row, New York. After more than a decade, this remains the best source book on protein structure. A supplement is available that gives three-dimensional stereograms of important proteins.

Folsome, C. E., ed. 1979. *Life: Origin and Evolution.* A *Scientific American* book. W. H. Freeman, San Francisco.

Gould, S. J. 1977. *Ever Since Darwin: Reflections in Natural History.* W. W. Norton, N.Y. Essays on evolutionary theory and natural history presented in a lucid, and often amusing manner by one of today's most interesting evolutionary theorists.

Hall, D. L., P. D. Tessner, and A. M. Diamond. 1978. "Planck's Principle." *Science* 202:717. Do younger scientists accept new scientific ideas—in this case, the theory of natural selection—more readily than older scientists?

Ifftt, J. B., and J. E. Hearst, eds. 1974. *General Chemistry.* A *Scientific American* book. W. H. Freeman, San Francisco.

Jensen, W. A. 1970. *The Plant Cell.* 2d ed. Wadsworth, Belmont, Calif.

Kennedy, D., ed. 1974. *Cellular and Organismal Biology.* A *Scientific American* book. W. H. Freeman, San Francisco.

King-Hele, D. 1974. "Erasmus Darwin, Master of Many Crafts." *Nature* 247:87. Charles' wonderful grandfather, poet, inventor, and essayist, made some amazingly clear and sensible statements about evolution and sexual selection. Unlike Charles, he never bothered to collect the facts that would back them up.

Kozlowski, T. T. 1964. *Water Metabolism in Plants.* Harper and Row, New York.

Laüger, P. 1972. "Carrier-mediated Ion Transport." *Science* 178:24. Experimental work on the transport of potassium ions by valinomycin, and a general consideration of how ion carriers function across plasma membranes.

Lehninger, A. L. 1975. *Biochemistry.* 2d ed. Worth, New York. The current standard of excellence in biochemistry texts.

Margulis, L., L. To, and D. Chase, 1978. "Microtubules in Prokaryotes." *Science* 200:1118. Margulis found in 1978 just what she had predicted in 1968: some spirochaete bacteria have tubulin-based microtubules.

Moorehead, Alan. 1969. *Darwin and the Beagle.* Harper and Row, New York. We think this book is the most thoroughly enjoyable account of Darwin's seminal years aboard the H.M.S. *Beagle.* Lavishly illustrated.

Porter, E. 1971. *Galápagos.* Ballantine, New York. All evolutionary biologists dream of making the trip to the Galápagos—and some of them make it. A different world.

Evolution. A *Scientific American* book. 1978. W. H. Freeman, San Francisco. The September 1978 special issue on evolution, reissued as a book. Superior articles include R. E. Dickerson, "Chemical Evolution and the Origin of Life"; J. Maynard Smith, "The Evolution of Behavior"; J. W. Valentine, "Multicellular Plants and Animals"; R. M. May, "Ecological Systems"; and J. W. Schopf, "The Earliest Cells."

Singer, S. J., and G. Nicolson. 1972. "The Fluid Mosaic Model of the Structure of Cell Membranes." *Science* 175:720. The original description of the now-accepted fluid mosaic model.

Stephenson, W. K. 1978. *Concepts in Cell Biology.* John Wiley, New York. A very good study aid with a succinct, self-teaching approach.

Van Valen, L. 1971. "The History and Stability of Atmospheric Oxygen." *Science* 171:439. Although oxygen-producing photosynthesis evolved prior to the change from a reducing atmosphere to an oxygen-rich atmosphere, Van Valen argues that the oxygen atmosphere was almost certainly not created by biological forces, but by long-term spontaneous dissociation of water molecules in the ionosphere and the subsequent loss of hydrogen ions to space.

Weissmann, G., and R. Clairborne, eds. 1975. *Cell Membranes: Biochemistry, Cell Biology and Pathology.* HP Publishing Company, New York. Superbly illustrated articles on all aspects of plasma membranes, incorporating the latest research findings. Notable are E. Racker on the inner mitochondrial membrane and the chemiosmosis hypothesis; W. R. Lowenstein and G. D. Pappas on cell-to-cell communication via membrane junctions; and J. F. Danielli on the Danielli lipid bilayer concept as it has evolved over the years.

White, E. H. 1964. *Chemical Background for the Biological Sciences.* Prentice-Hall, Englewood Cliffs, N.J.

Whittingham, C. P. 1977. *Photosynthesis.* Carolina Biological Supply Company, Burlington, N.C.

Zelitch, I. 1971. *Photosynthesis, Photorespiration, and Plant Productivity.* Academic Press, New York. What is accomplished by the apparently wasteful process of photorespiration?

Scientific American articles. San Francisco: W. H. Freeman Co.

Bassham, J. A. 1962. "The Path of Carbon in Photosynthesis." *Scientific American,* June. The Calvin cycle illustrated.

de Duve, C. 1983, "Microbodies in the Living Cell." *Scientific American,* May.

Dustin, P. 1980. "Microtubules." *Scientific American,* January.

Govindjee and R. Govindjee. 1974. "Primary Events of Photosynthesis." *Scientific American,* December.

Hayflick, L. 1980. "The Cell Biology of Aging." *Scientific American,* January.

Hinkle, P. C., and R. E. McCarty. 1978. "How Cells Make ATP." *Scientific American,* March. A thoroughly convincing presentation of the chemiosmosis hypothesis and the evidence for it in both photosynthetic and oxidative phosphorylation. We have drawn heavily on this article in preparing Chapters 6 and 7.

Margaria, R. 1972. "The Sources of Muscular Energy." *Scientific American,* March.

Margulis, L. 1971. "Symbiosis and Evolution." *Scientific American,* August. Still the best presentation of her once startling theory of the origin of eukaryotes.

Porter, K. R., and J. B. Tucker. 1981. "The Ground Substance of the Living Cell." *Scientific American,* March.

Satir, P. 1974. "How Cilia Move." *Scientific American,* October. An intriguing but still speculative model of how sliding microtubules might be responsible for the movement of cilia and flagella.

Schopf, J. W. 1978. "The Evolution of the Earliest Cells." *Scientific American,* September. Biochemical pathways clearly demonstrate that basic life processes were evolved in an anaerobic world, and that steps involving free molecular oxygen have been "tacked on" to previously evolved systems.

Shulman, R. G. 1983. "NMR (nuclear-magnetic-resonance) Spectroscopy of Living Cells." *Scientific American,* January.

DNA and the Central Dogma

Not very long ago, as recently as the early 1950s, the chemical basis of heredity was unknown. In fact, they were widely considered to be so mysterious and complex as to be quite literally beyond human comprehension. Scientists had generally agreed that the gene had a chemical composition, but what genes were, how they were put together, and how they worked seemed hopelessly beyond grasp. As early as the 1940s there was some good evidence that the genetic material was **deoxyribonucleic acid (DNA),** but at mid-century, most biologists, if they allowed themselves an opinion at all, would have said that genes were probably made of protein. After all, they knew that enzymes were proteins, and they believed that genes must be much more complex even than enzymes. And before that time, enzymes themselves seemed to be so enormous and complex that their own structure could never be known. (It was not until 1953 that Frederick Sanger determined the amino acid sequence of insulin, one of simplest proteins.) How could we ever hope to understand genes? It had been pointed out that genes not only had to direct all the life processes of the cell and the development of the organism, but also they somehow had to be able to make exact copies of themselves every cell generation.

The more pessimistic souls must have been shocked when two vigorous young researchers, James Watson and Francis Crick, showed that the structure of genes and the mechanism of gene function and replication were essentially not so difficult to understand after all. The secrets lay in the simple structure of DNA, a molecule whose presence in the cell had been known for about 80 years. The gene itself turned out to be a surprisingly simple chemical coding of protein-synthesizing instructions. Furthermore, the code contained only four elements. These four elements determined the order in which the 20 amino acids are arranged in any protein. And, after all, it is the order of amino acids that makes one protein different from the next.

One might ask, what does protein synthesis have to do with genes and heredity? The answer is, everything: remember that proteins form much of the structure of living things. They also comprise the many pigments that provide color. Many of the hormones that quietly nudge their target cells into action are proteins, and most significantly, so are the many enzymes that guide all cellular activity. Logically, then, much of heredity can be expressed through protein.

As you will see, the revelations of Watson and Crick turned out to be one of those classical and retold stories of biology. In this chapter, we will review the essential points Watson and Crick found out about DNA in those early days. It makes a compelling story, and one that is pleasingly logical and understandable when it is told one step at a time. So let's take it that way, beginning with the structure of DNA. Try to keep in mind that many scientists of the 1950s found it difficult to accept what we are about to say. To many, the mechanical simplicity of the ideas threatened to take away the central mystery of life.

THE CENTRAL DOGMA?

Since we are dealing with the processes of life here, you may be surprised to see the word *dogma*. Dogma usually refers to something that can't be questioned and is generally reserved for matters of religion and politics. In a sense, the only proper dogma in science is that there is no dogma, and anything can be questioned. The term **central dogma** was originally meant as a joke of sorts, initiated by Francis Crick, but the name stuck.

The central dogma completed an intellectual process begun by Gregor Mendel, often called the "father of genetics," some 100 years earlier, and gave our concept of life (including human life) a rigorous chemical and physical basis. Of course, there are still important details to be worked out. Many smaller mysteries remain to be solved, and some questions, perhaps, may never be answered. But *the essence of this life process is understood.* What, then, is the central dogma (Figure 9.1)? Simply this:

1. Genetic information—the message of the gene—is encoded and maintained in the DNA molecule and is interpreted via RNA into protein structure.
2. DNA is copied from other DNA (a process called **replication**).
3. RNA is copied from DNA (**transcription**).
4. Protein synthesis is directed by one class of RNA (messenger RNA, or mRNA) in such a way that three sequential nucleotide bases specify one amino acid in a protein chain. The genetic code followed is the same for nearly all organisms **(translation)**.

As it happens, scientists quickly found exceptions to the rules above, but as far as we know, the exceptions occur only in certain viruses:

9.1 THE CENTRAL DOGMA

A very diagrammatic scenario of DNA activity reveals the basic core of the central dogma. Included are its three primary mechanisms: *replication, transcription,* and *translation.* ① DNA, which remains in the nucleus, contains the genetic instructions for each species, and the chemical coding of these instructions is copied faithfully through generations, assuring survival through the process of *replication.* ② Within the active cells of an organism, the genetic instructions are copied into RNA in a highly organized and specific manner. This is *transcription.* ③ Several types of RNA leave the nucleus and then, in the cytoplasm and in concert with the ribosomes, assemble proteins according to the coding system transcribed from DNA. Converting the genetic code into proteins is known as *translation.*

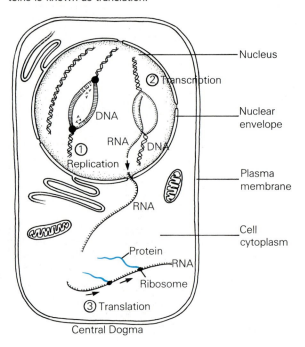

Central Dogma

5. For some viruses, viral RNA is copied from viral RNA (**RNA replication**).
6. For other viruses, viral DNA is copied from viral RNA (**RNA-dependent DNA synthesis** or **reverse transcription**).

The Principle of Colinearity

The statements so far, it would seem, are hardly the stuff for an earthshattering intellectual revolution, but the effects were far-reaching. Let us not dwell on those effects just yet. For now, be aware that the essence of the dogma is this: *All genetic information* (here, "information" refers to how certain molecules direct other molecules) *is (1) contained in linear sequences of* **nucleic acid bases** (the bases are arranged in a line, one above the other), (2) *retained when these sequences are copied into other nucleic acid sequences, and (3) expressed in the synthesis of proteins with specific linear amino acid sequences.* This statement is known as the **principle of colinearity** (Figure 9.2). All of this may be befuddling at present, but our task here is to unfuddle it.

DNA STRUCTURE

We have already indicated that there are two kinds of nucleic acids, DNA (deoxyribonucleic acid) and RNA (ribonucleic acid). As the "deoxy" implies, there is one less oxygen atom in each unit of DNA than in RNA. In this chapter we will concentrate on DNA, leaving the details of RNA structure and activity for Chapter 10.

The DNA Nucleotides

We saw in Chapter 3 that the basic unit of DNA is called a **nucleotide** (or sometimes, a **deoxynucleotide**). The nucleotides can exist as free-floating molecules in the cell fluids, but when they are strung together into long polymers, they form **DNA strands**, or **nucleic acids**.

The free-floating nucleotides are usually found as triphosphates or, technically, **nucleoside triphosphates,** meaning that they have an appendage of three phosphates strung together. In Chapter 6, we described ATP as the universal energy currency in the cell because of its two energy-rich phosphate-phosphate bonds. ATP is a ribose nucleotide (what we would expect to find in RNA), not a deoxynucleotide. Nevertheless, the free-floating deoxynucleotides also have high-energy bonds, and the DNA equivalent of ATP is logically designated dATP. In addition to the phosphate subunits, the deoxynucleotide consists of a sugar, **deoxyribose,** and any one of four different **nitrogenous bases.**

Deoxyribose is a 5-carbon sugar, exactly like the more common sugar ribose except that one of the oxygens has been removed. In other words, the —OH side group on one of the five carbons is replaced by a —H. Deoxyribose attaches to one of the nucleotide bases at its number 1' carbon. The five carbons of deoxyribose are traditionally numbered as 1' through 5' (read "one prime through five prime").

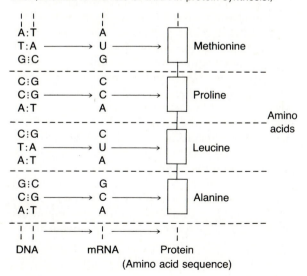

Deoxyribose

The nucleotide bases are also called **nucleic acid** bases, or nitrogenous bases (the latter because they consist of nitrogen-containing rings). In DNA, two of the bases, **thymine** and **cytosine,** are **pyrimidines,** meaning that they consist of a single six-cornered ring. The other two, **adenine** and **guanine,** contain two rings, a five- and a six-cornered ring connected together. Adenine and guanine are **purines.** (Actually, pyrimidine and purine are specific ring compounds from which the nitrogen bases are derived. It may help to keep the purines and

---

9.2 DNA, RNA, AND PROTEIN

The colinearity principle describes the relationship between DNA, messenger RNA, and a protein. Within DNA, each three nucleotides code for three nucleotides in RNA, which in turn code for the placement of a specific amino acid in a polypeptide. (Omitted is the role of tRNA in protein synthesis.)

DNA	mRNA	Protein
A:T	A	
T:A →	U →	Methionine
G:C	G	
C:G	C	
C:G →	C →	Proline
A:T	A	
C:G	C	
T:A →	U →	Leucine
A:T	A	
G:C	G	
C:G →	C →	Alanine
A:T	A	

Amino acids

DNA   mRNA   Protein
                    (Amino acid sequence)

pyrimidines straight if you recall that the smaller molecules have the longer name.) The colored N—H groups shown below are the sites of reaction with the deoxyribose.

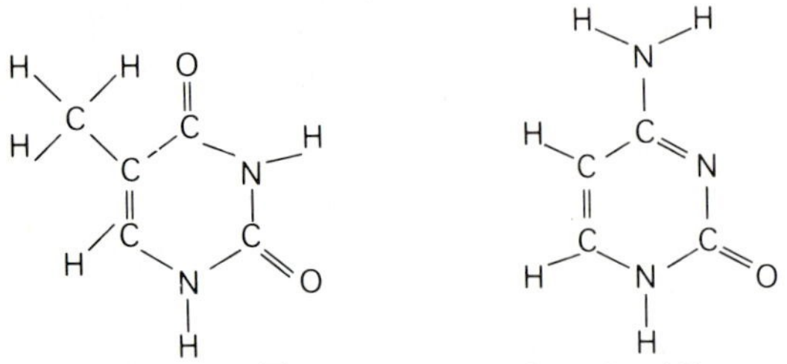

Thymine (T) Cytosine (C)

Pyrimidines
(single ring)

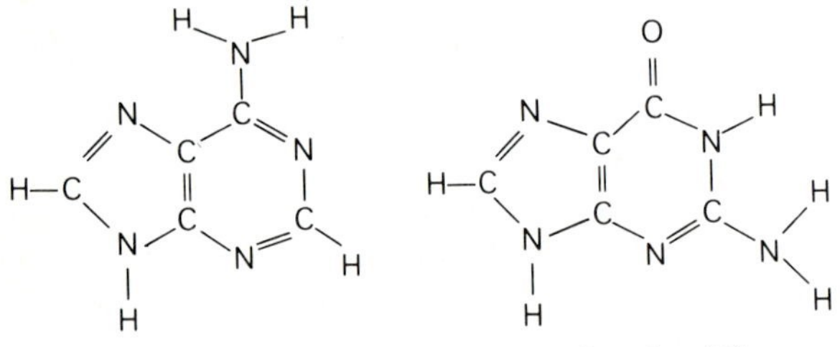

Adenine (A) Guanine (G)

Purines
(double ring)

Adenine, guanine, thymine, and cytosine are also known simply by the letters A, G, T, and C, respectively. The corresponding nucleotide triphosphates are dATP, dGTP, dTTP, and dCTP; again, the *d* designates "deoxyribose." (*Adeno-* means gland; adenine and thymine were first isolated from the thymus gland. *Cyto-* means cell, and, as you may have guessed, cytosine is present in all cells. The least romantic is guanine, first isolated from guano, or feces, usually from bats or seabirds). In a free nucleotide triphosphate, the string of three phosphates is covalently linked through the oxygen group of the 5' carbon, and the nucleotide base is covalently linked by one of its nitrogens directly to the 1' carbon.

The triphosphate nucleotide (dATP) shown here consists of phosphates, sugar, and a nucleotide base linked together. The 1' carbon of deoxyribose is also linked, through an oxygen, to the 4' carbon, making the sugar a pentagonal ring. The 2' carbon is the one without any —OH side groups. Note that in the free triphosphate nucleotide below, only the 3' carbon has an —OH side group. Let's see why this is important.

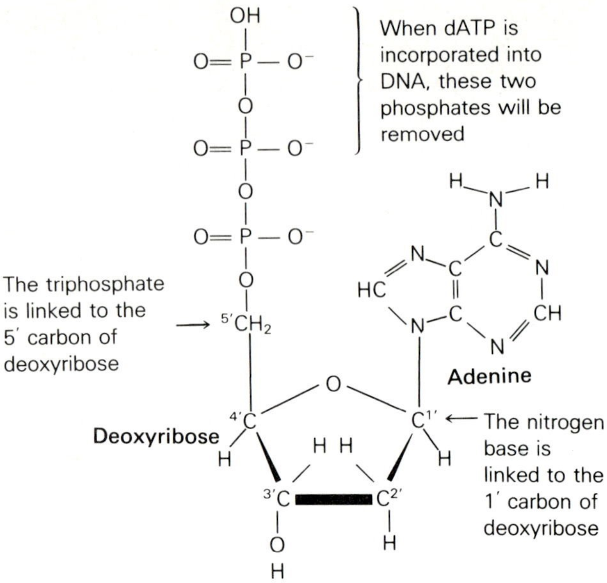

When dATP is incorporated into DNA, these two phosphates will be removed

The triphosphate is linked to the 5' carbon of deoxyribose

Deoxyribose

Adenine

The nitrogen base is linked to the 1' carbon of deoxyribose

Deoxyadenosine triphosphate (dATP)

DNA Bonding. In the synthesis of DNA chains, the high-energy bond between two of the phosphates is used to form a strong covalent bond between nucleotides. In the process, the two outermost phosphate groups are liberated as *pyrophosphate*, leaving the innermost phosphate attached. This remaining phosphate forms a link between two deoxyribose subunits as follows: The free 3' —OH group of deoxyribose in the first nucleotide reacts with the first 5' phosphate of the second nucleotide, forming a sugar-phosphate-sugar linkage. The 3' —OH group of the second nucleotide is free to interact with the 5' phosphate group of the third nucleotide, and so on in (Figure 9.3).

Eventually, a long polymer of nucleotides is produced. The backbone of the chain consists of alternating sugars and phosphates linked by shared oxygen atoms. The nucleotide bases are side groups hanging off the sugars and are not part of the backbone. The end of the polymer that has the free 5' phosphate is known as the **5' end,** and the end with the free 3' —OH group is called the **3' end.** Synthesis of a nucleic acid chain (DNA or RNA) always proceeds from the 5' end to the 3' end.

DNA is almost always found as a double strand. The nucleotides of each strand are linked by strong covalent bonds, but the two strands are only weakly attracted to each other. In the inactive state the two DNA strands are wrapped around each other, held by numerous individually weak hydrogen

9.3 THE COVALENT LINKAGES OF DNA NUCLEOTIDES

(a) Nucleotides are added to a growing DNA chain by the enzyme DNA polymerase. Although not shown here, there is always a preexisting strand that serves as a template or guide for new strands. As each nucleotide is added, two phosphates are cleaved from the nucleoside triphosphate, and the hydrogen from the—OH group of the 3′ sugar carbon is transferred to the liberated diphosphate molecule. Newly added nucleotides are subsequently attached at the 3′ carbon of the

deoxyribose, forming 3-5′ phosphate linkages as the strand grows. **(b)** The single strand of DNA grows in length to become an incredibly long and thin polymer. All four nucleotide bases must be present in great abundance to accommodate nucleic acid synthesis. During this synthetic period of the cell cycle, nearly all the cell's resources are devoted to this immense task.

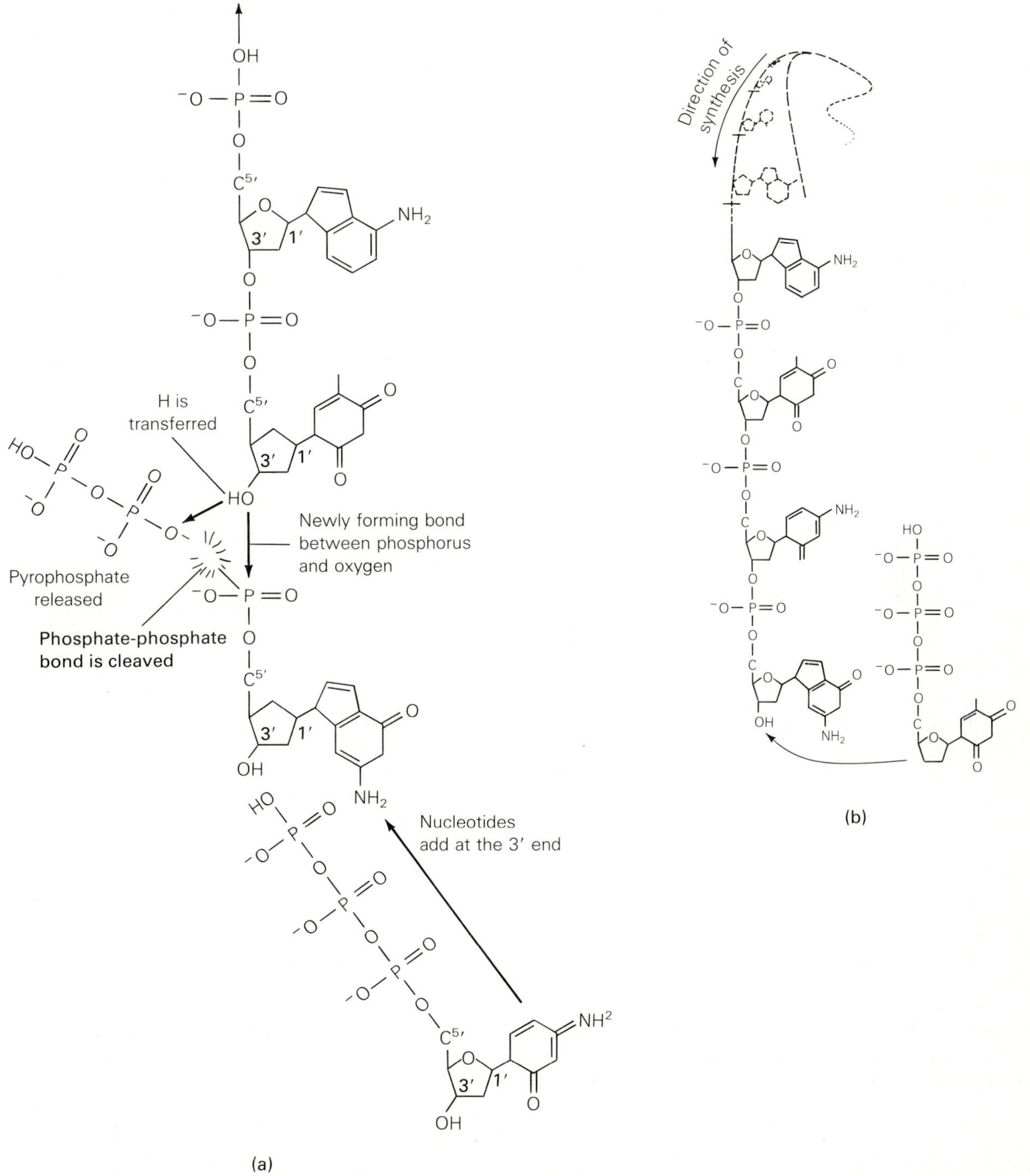

(a)

(b)

bonds. The configuration of the two strands of DNA wound around each other is that of the famous **double helix.** (A double helix is best visualized by imagining two wires twisted around each other.)

Base Pairing. As the sugar-phosphate-sugar chains of DNA curl around each other, the nucleotide bases point inward, much as if someone had twisted a ladder, and the pairs of bases are the rungs. Watson and Crick determined that the bases pair up in a curious fashion but one that turns out to have critical functional implications, which will be discussed later. The specific manner in which the nucleotide bases join, forming the rungs of the ladder, is known as **base pairing.**

The structures of the bases are such that ade-

nine will form *two* hydrogen bonds with thymine, and guanine will form *three* hydrogen bonds with cytosine (Figure 9.4). Thus, in DNA, each adenine in one strand is paired with a thymine in the other strand (and vice versa); each guanine in one strand is paired with a cytosine in the other strand (and vice versa). This results in the peculiar coiling of the molecule. The geometry of the inside of the DNA molecule is very precise. Furthermore, the A:T, T:A, G:C, and C:G (the dots represent bonds) pairs lie perfectly flat, one pair on top of the next like a stack of pennies.

Let's look more closely at the specific base pairing of the nucleotides. The two purines, adenine and guanine, are so large that they can't pair because they would overlap; the pyrimidines, thymine and cytosine, are so small that they can't reach each other; and side groups of pairs of adenine and cytosine or guanine and thymine don't match. Thus, the four base pairs A:T, T:A, C:G, and G:C are the only pairs that *can* match up.

The Double Helix Within the double helix, the two chains of DNA run in opposite directions. That is, if you pictured a DNA molecule vertically on a page, one of the chains would run from top to bottom in its 5'-to-3' direction and the other would run from bottom to top in its 5'-to-3' direction (Figure 9.5). It is traditional to show the 5' to 3' polymer on the left and the 3' to 5' polymer on the right.

DNA molecules can be quite long. In fact, the average DNA molecule in a human cell nucleus consists of about 160 million nucleotide pairs (two strands of 160 million nucleotides each)! However, because molecules are so small, if such a DNA double helix were stretched out, it would be only 5 cm long. (The 46 DNA molecules in each human cell nucleus, placed end to end, would measure almost exactly 2 m. One might argue that that's pretty big to be contained in a microscopic cell. It's all a matter of perspective—and some pretty amazing packaging as well.)

9.4 THE CONCEPT OF BASE PAIRING

Simply put, pairing in DNA always occurs between A and T and between G and C. As you see, hydrogen bonding (dashed lines) between adjacent bases is very specific (two bonds between T and A, and three between C and G); thus it will occur only when the base pairing is correct. Any time a mistake occurs, such as thymine pairing with guanine, there is a misfit, and bonding will not readily occur. (In reality, it does sometimes occur, but produces problems. Mispaired bases are believed to be spontaneously removed by repair systems in the nucleus.)

Thymine

Adenine

Cytosine

Guanine

DNA REPLICATION

Synthesis means making something and *replication* means making an exact copy of something. With very few exceptions, DNA synthesis and DNA replication are exactly the same thing, since DNA molecules are made only by copying other DNA molecules. Historically, it was realized very early that

9.5 A COMPLETE DNA MOLECULE

In a fully assembled double polymer of DNA, one strand appears upside down relative to the other. Each polymer is held together by covalent bonds within the sugar-phosphate backbone, while the two strands are held in place by numerous hydrogen bonds between adjacent base pairs.

one of the properties that the unknown genetic material must have is the ability to replicate when a cell divides. A complete set of genes (within the chromosomes) has to be passed down to each new cell, and any reproducing cell or organism must pass along a full set of genes to descendents. Watson and Crick were aware that the specific pairing of DNA nucleotides suggested how DNA itself might be replicated. "It has not escaped our notice that the specific pairing we have postulated immediately suggests a possible copying mechanism for the genetic material," they said. And the suggestion was soon shown to be correct.

The very special way in which nucleotide bases pair in DNA suggests the analogy of a positive and a negative photographic film. The same information is present in both; you can reproduce a positive from a negative and a negative from a positive. Similarly, DNA replication enzymes can produce a "Watson" strand from a "Crick" strand, and a "Crick" strand from a "Watson" strand. In so doing, a new Watson strand and a new Crick strand are produced; thus there are two DNA molecules where there was one before. ("Watson strand" and "Crick strand" do not represent official terminology, but this is a little inside joke that caught on as a handy way of referring to the two DNA polymers.)

We must confess that all the details of DNA replication aren't known, such as what initiates the process, why it goes out of control in cancer, and why it ceases forever in cells that have reached a terminal state of specialization. But we do know most of the essentials (Figure 9.6).

Replication is carried out by **replication complexes,** which consist of several types of enzymes. We are aware that, first, an **unwinding enzyme** called **helicase** unwinds the two strands of DNA so that there are regions, each a few hundred nucleotides long, in which the two strands are unpaired, with their nucleotides sticking out in an exposed and reactive manner. Then, using supplies from the nuclear fluid, the enzyme **DNA polymerase** goes to work, matching the four kinds of nucleotide triphosphates—dATP, dTTP, dGTP, and dCTP—to the bases exposed by the opened-up, single-stranded DNA.

Working in the 5'-to-3' direction of the new chain being synthesized and utilizing dehydration synthesis, the enzymes combine the reactive 5' phosphates of free nucleotide triphosphates with the free 3' —OH groups of the growing chain. DNA replication requires a great deal of energy, and we find that while the energy for unwinding the helix comes from ATP, the energy needed for adding phosphates comes from the deoxynucleotide triphosphates themselves.

As the nucleotides of the unwound segments find new partners, the replication complexes move along the strand, unwinding more DNA and creating two moving **replication forks** (Figure 9.6). Since the original strands run in opposite directions, the two newly replicated strands are synthesized in opposite (but still 5'-to-3') directions as well. Replication must proceed in the 5'-to-3' direction on each strand because DNA polymerase, the enzyme that catalyzes the process, must have the free 3' —OH group of deoxyribose on which to add nucleotides. This presents a problem for one end on each growing strand, but cells apparently have the answer.

Because of the special 5'-to-3' replication requirement, and because replication proceeds in both directions in each strand, each end of a grow-

9.6 DNA REPLICATION

(a) Prior to the onset of replication, DNA is wound tightly in its helical configuration. (b) Replication begins when the double helix is opened by two replication complexes moving in opposite directions and forming two moving replication forks. Each replication complex contains the unwinding enzyme, *helicase* and the nucleotide-adding enzyme, DNA polymerase. (c) During replication, each new strand is assembled according to the bases present in the preexisting strand, which acts as a template. However, things become complicated. Nucleotides are added to both ends of each of the newly forming strands, yet the assembly of nucleotides can only occur in the 5' to 3' direction. As a result, the smooth sequential "one-after-the-other" addition can only occur at what amounts to the *leading ends*. At the *lagging ends,* nucleotides are first assembled by DNA polymerase into short *Okazaki fragments,* and then, with the help of the enzyme ligase, such fragments are spliced into the growing strand. (d) Following replication, each new double strand assumes the usual double helix configuration. The two newly produced DNA molecules will be identical, but, as we see, each replicated molecule contains either an original Watson strand or an original Crick strand, both fully intact. This is known as *semiconservative replication,* since elements of the old coding are maintained.

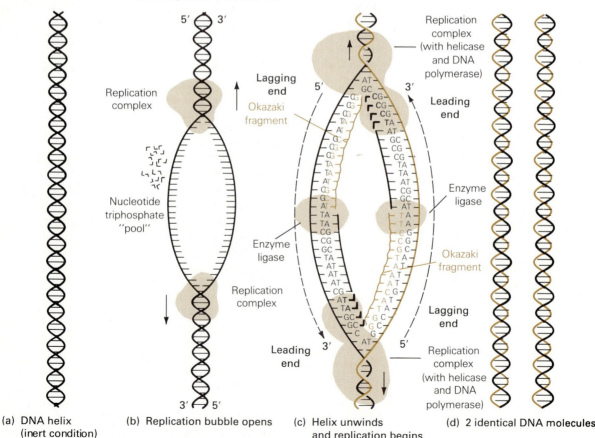

(a) DNA helix (inert condition)

(b) Replication bubble opens

(c) Helix unwinds and replication begins

(d) 2 identical DNA molecules

ing polymer must be synthesized in a different manner. At the exposed 3' ends, or **leading ends**, nucleotides are added one at a time in a smooth, uninterrupted manner, just as you might expect. But at the opposite 5' or **lagging ends** of each growing polymer, the nucelotides cannot be added this way since replication would then follow a 3' to 5' direction—an impossibility. Instead, they are first joined together away from the old strand in the required 5'-to-3' direction in what are called **precursor fragments,** or **Okazaki fragments** (named in honor of one of their discoverers, Reiji Okazaki, who found them in the colon bacterium *Escherichia coli*). These fragments consist of a couple of hundred nucleotides (thousands in bacteria), assembled according to the coding on the preexisting lagging end. Finished Okazaki fragments are added one after the other, but following the 5'-to-3' direction. Once in place, the enzyme *ligase* joins together the ends of the fragments (see Figure 9.6).

In spite of such complications, we see that in the replication or synthesis of DNA, each individual strand remains intact and each receives a new fully matching strand. Such synthesis is termed **semiconservative,** which, as the name implies, means that part is retained and part is new.

There was a time, not long ago, when biologists were doggedly trying to determine the nature of replication; that is, whether it is semiconservative as opposed to either conservative or dispersive (the latter two suggest that either the DNA molecule remains completely intact while being copied, or is completely dismantled and copies somehow reassembled). Essay 9.1 reviews one of the truly elegant experiments of this period. Later we will find that the processes of DNA repair and DNA synthesis have much in common. Both synthesis and repair of damaged or mutated DNA involve the matching of free nucleotides against an intact DNA single strand.

Comparing DNA Replication in Eukaryotes and Prokaryotes

We will be comparing eukaryote and prokaryote (bacterial) replication next, but first let's stand back and take a longer view of the process. Replication begins with the two sides of the twisted ladder of DNA pulling apart at some point along the length of the strand and unwinding, forming a "bubble." Each end of the bubble, where the two single strands of DNA emerge from the doublestranded

9.7 SIMULTANEOUS REPLICATION

In eukaryotes, many replication events occur simultaneously along each DNA molecule. Individually, each event occurs in a bubblelike affair, with replication forks moving in opposite directions. Rewinding of each new replica into its final helical configuration occurs as replication proceeds.

DNA, contains a replication fork. Newly synthesized DNA strands are formed enzymatically on both of the unwound single strands, following the Watson-Crick pairing rules. As replication proceeds, further unwinding moves the replication forks in opposite directions, causing the bubble to become larger.

In eukaryotes, each of the numerous chromosomes contains one very long, linear DNA molecule prior to DNA replication. At the initiation of DNA replication, hundreds of bubbles are formed along the length of each DNA molecule (Figure 9.7). As the bubbles enlarge, the replication forks of different bubbles run into each other, and the bubbles merge. One replication fork runs off each end of the molecule, and eventually two long, linear DNA molecules exist where there was one before.

Prokaryotes are different in that they require only two replication forks. It's a matter of speed. Multiple replication sites are essential to the eukaryote since its replication complexes work much slow-

Meselson and Stahl

Although the structure of DNA was clarified by Watson and Crick in 1953, the details of replication remained something of a mystery. Watson and Crick's information strongly suggested a *semiconservative* mechanism, in which the two strands of the double helix separate during replication but each remains intact, acting as a template for the assembly of a new partner. It was not inconceivable, however, that DNA replication might be *conservative,* or perhaps even *dispersive.* According to the conservative replication hypothesis, the entire molecule would remain intact and another double strand of new material would be replicated alongside. The dispersive replication hypothesis stated that the strand would be dismantled piece by piece and replication would occur along the pieces.

In 1957, Matthew Meselson and Franklin Stahl set out to determine how replication was accomplished. Their procedure utilized two modern tools of biology, isotope-labeled biological chemicals and the ultracentrifuge. Using special techniques for isolating DNA from bacteria, they were able to centrifuge the DNA in a cesium chloride (CsCl) solution, which bands each molecule according to its specific gravity (density). Since bacteria are normally exposed only to nutrients containing common nitrogen 14, a ^{14}N reference line was available. In other words, DNA containing ^{14}N settled in the centrifuge tubes at a certain level, forming what was to be the first reference line. Next, a second reference line was established, this time for bacterial DNA extracted from cells that had been grown for many generations in culture media containing ^{15}N-labeled nucleotides. Since ^{15}N is heavier than ^{14}N, the ^{15}N DNA settled at a somewhat lower position in the centrifuge tube. With these reference points established, the experiment could begin.

Predictions of competing hypotheses of DNA replication

		Before replication	After 1 round of replication	After 2 rounds of replication
^{14}N chain ^{15}N chain				
1 Hypothesis: conservative replication				
Prediction: original double helix remains intact, while all new DNA lacks any of the original molecule (color indicates the two strands of the original molecule).				
2 Hypothesis: dispersive replication				
Prediction: original molecule becomes increasingly diluted with new material.				
3 Hypothesis: semiconservative replication				
Prediction: the two strands of the double helix come apart, but each strand remains intact.				

Bacteria were grown in a culture medium containing ^{15}N nucleotides for several generations so that DNA containing ^{15}N would be present in the vast majority of bacterial chromosomes. Bacteria from this culture were then removed, washed, and resuspended in a culture medium containing only nucleotides with ^{14}N. *One round of replication was permitted.* Cells from this stage were then removed, and the DNA was extracted and centrifuged in the CsCl gradient.
Prediction:
1. If DNA replication is *conservative,* then two regions of DNA will be seen in the centrifuge tube. One will contain only ^{14}N DNA, while the other will contain only ^{15}N DNA.

Time	Observation		Interpretation	
	Observed density gradient centrifuge bands	Quantitative results	~~~ ¹⁵N DNA strand ~~~ ¹⁴N DNA strand	
	^{14}N (light)　　hybrid (intermediate)　　^{15}N (heavy) 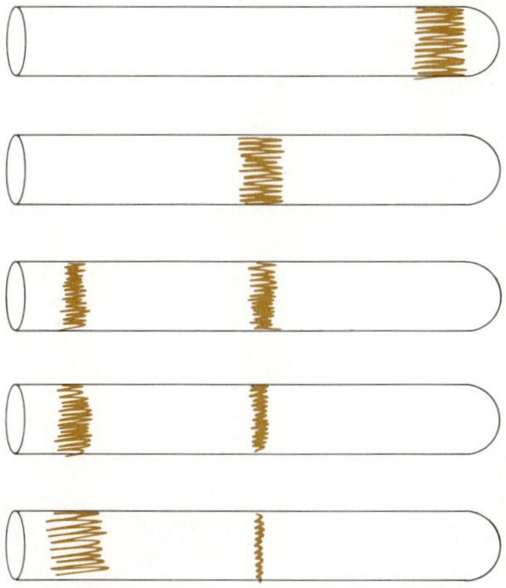			
Before replication		100% ^{15}N DNA		
After 1 round		100% hybrid DNA		
After 2 rounds		1/2 hybrid DNA 1/2 ^{14}N DNA		
After 3 rounds		1/4 hybrid DNA 3/4 ^{14}N DNA		
After 4 rounds		⅛ hybrid DNA ⅞ ^{14}N DNA		

2. If replication is *dispersive,* all DNA after one round of replication will be halfway between the reference densities of ^{14}N and ^{15}N DNA.
3. If DNA replication is *semiconservative,* then only one region of DNA sediment will be found. This region will contain *hybrid* DNA, each moelcule containing a ^{14}N strand and a ^{15}N strand. The sedimentation will occur halfway between the two reference lines established earlier.

In other words, one round of replication could not distinguish between the dispersive and semiconservative hypotheses. But other bacteria were allowed to continue into *a second round of replication*. The cells were then removed, and the DNA was isolated and centrifuged in the CsCl gradient.

Prediction:
1. If DNA replication is *conservative,* then the same two sedimentation bands will appear as did in prediction 1, but the ^{14}N band will contain three times as much DNA as the ^{15}N band.
2. If DNA replication is *dispersive,* then all DNA after two rounds of replication will be uniform; namely, 25% will be ^{15}N and 75% will be ^{14}N. These should appear as a single band appropriately spaced between the two reference points.

3. If DNA replication is *semiconservative,* then two equal sedimentation bands will appear. One will contain *hybrid* DNA and will form a band at the same location as did hybrid DNA in prediction 2. The other will contain ^{14}N DNA only and will settle out at the ^{14}N reference line established earlier.

Results: The results, as diagrammed here, were clear. What Meselson and Stahl saw were the bands shown on the left. Their interpretation is shown on the right—only the semiconservative hypothesis could be supported. Note that the experiment was followed for four rounds of replication, just to be sure.

er than those of prokaryotes. While bacterial replication can proceed at a rate of 1 million base pairs added per minute, eukaryotic rates only range from about 500 to 5000 base pairs per minute. At that rate it would require about one month for a eukaryotic cell to complete replication, but because of multiple replication forks, each moving bidirectionally, replication in eukaryotes averages just a few hours. *Drosophila*, the fruit fly, may be a record holder, since the newly fertilized egg forms some 50,000 replication sites and its four pairs of chromosomes can be replicated in about three minutes.

In prokaryotes, most or all of the DNA is in a single circular molecule that is not nearly as long as a eukaryote DNA molecule (you have about 1400 times more DNA per cell than an average bacterium). There is no end and no beginning of a circle, of course, but replication in the bacterium *Escherichia coli* always begins at a single, specific initiation point (Figure 9.8). Again, unwinding proteins form a bubble of single-stranded DNA, and the unpaired strands are replicated with new "Watson and Crick" strands in the usual way. The two replication forks—and in prokaryotes there are only two—travel in opposite directions around the circular DNA molecule, eventually meeting each other at a specific termination point halfway around the circle. In Figure 9.8, a smaller *E. coli* chromosome, known as a **plasmid,** is caught in the act of replicating.

9.8 REPLICATION IN PROKARYOTES

In prokaryotes, where circular chromosomes are seen, replication begins at a specified point of origin and two replication forks (and only two) travel in opposite directions, eventually producing the two replicas.

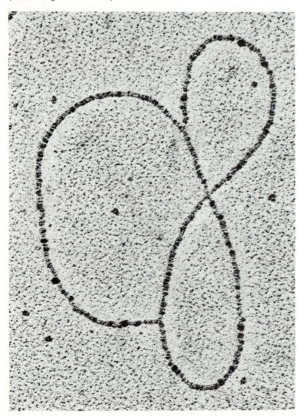

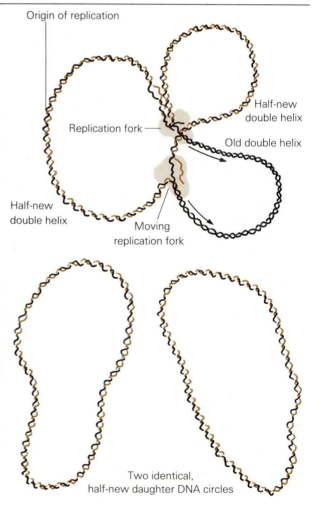

Origin of replication

Half-new double helix

Replication fork

Old double helix

Half-new double helix

Moving replication fork

Two identical, half-new daughter DNA circles

A BRIEF HISTORY OF THE DISCOVERY OF DNA

One of the great scientific arguments of this century centered over what cellular structures were the "units of inheritance," how hereditary factors were physically transmitted. Was it by DNA? Protein? Or was it some incredibly complex substance as yet undiscovered? In one sense, the genetic material was discovered in the last century, about the time Darwin was formulating his ideas on natural selection. Precise and brilliant genetic experimentation by Gregor Mendel, the Austrian monk, strongly indicated that genes were physical, particulate entities he called **factors** (Chapter 12). These factors replicated in cell division, segregated as sex cells (sperm and egg) formed, were transmitted from generation to generation, and then, in some way, directed the development of the organism.

But what, precisely, was this material? Although there were strong indications of ties between peculiar bodies called **chromosomes** and inheritance, what were these chromosomes exactly? The word *chromosome* means "colored body." Chromosomes are not naturally colored, but the phosphate groups of their DNA bind to certain basic dyes very strongly, so chromosomes were first seen as brightly stained objects—actually, brightly stained DNA.

The Early Efforts

Surprisingly, DNA itself was discovered while both Darwin and Mendel were working. In 1869, Friedrich Miescher, a Swiss chemist, used the enzyme pepsin to digest the proteins of the nucleus of white corpuscles taken from pus and showed that a strange, phosphorus-containing material remained. In a private letter, discovered much later, Miescher actually speculated that this material might be the stuff of heredity. Later, in 1914, a German chemist named Robert Feulgen invented a still widely used staining procedure, **Feulgen staining,** that is specific for DNA. The Feulgen staining procedure has the advantage of staining DNA more or less strongly according to how much DNA is present. Thus, the amount of DNA present can be calculated by measuring the strength of the color. A few key experiments revealed that virtually every cell nucleus in a given plant or animal has the same amount of DNA, except for gametes (eggs and sperm), which have *half* that amount.

Once the existence of DNA was known, most biologists still couldn't bring themselves to take it seriously, for a few very convincing reasons. In the first place, the structure of DNA is very simple: just four different nucleotides are present. How could anything so simple be the physical basis of anything so wonderful as the gene? How could only four nucleotides produce the complex variations of life? In the second place, DNA didn't seem to *do* anything. It just sat there, some scientists said, probably holding the chromosome together, or making it acidic, or doing some even more trivial thing.

But chromosomes also contain proteins. Proteins! Now *there* was a likely source of variation. Proteins are wonderfully complex and do all kinds of marvelous things. So all bets were on proteins, providing one was willing to believe that genes were chemicals at all. Not every biologist had given up ideas about such things as "vital forces" and other mystical, unexplainable entities. Genes were thought of simply as developmental information, and the idea of *informational molecules* hadn't yet been worked out. If this all seems absurd to us now, remember that even today no one knows the biochemical basis of such things as thoughts and memories, and indeed, any biochemical basis of such things remains to be proven.

Genes, Enzymes, and Inborn Errors of Metabolism.   The world was not ready for the first major contribution to the biochemistry of the gene. In 1908, A. E. Garrod, a physician influenced by Mendel's work, published a book called *Inborn Errors of Metabolism*. His subject was *Homo sapiens,* which, at the time, was an unusual experimental organism for genetics research. Garrod was interested in metabolic defects, breakdowns in the complicated biochemical processes of life. He searched for the abnormal products of such defects in the urine, where many metabolic products end up. Of special interest to Garrod was **alkaptonuria,** a disease in which the urine contains **metabolites** (products of metabolism) called **alkaptones**—substances that happen to turn black upon oxidation, and so are easily revealed. Infants with alkaptonuria are usually detected as soon as their diapers start turning black. As the child grows older, black pigments begin to settle in cartilage and other tissues, blackening the ears and even the whites of the eyes. Another more serious effect is a form of arthritis, caused by the accumulation of the metabolite in the cartilage of the joints.

Garrod observed that the disease tended to be found in several brothers or sisters in a single family. By studying family histories, he correctly inferred that alkaptonuria and certain other inborn errors of metabolism were genetic in origin. The problem, he deduced, is caused by the absence of specific enzymes that are necessary for the long chains of biochemical reactions to occur. If an enzyme for a particular reaction is absent, no reactions can take place past the point where it normally enters the chain, and the substance that the enzyme acts on builds up.

Other inborn errors of metabolism produce albinism (a complete or partial lack of melanin pigment in the hair, skin, iris, and sometimes the retina) and phenylketonuria (which also affects hair and skin pigmentation but has a much more severe effect on mental development because of the accumulation of toxic metabolites in the nervous system). As it turns out, albinism, phenylketonuria, and alkaptonuria are all caused by defects in enzymes that act on the metabolism of the amino acids phenylalanine and tyrosine (Figure 9.9).

Significantly, Garrod discovered that heredity

9.9 METABOLIC PATHWAYS OF TYROSINE AND PHENYLALANINE

Seven genetic disorders, including two forms of albinism, phenylketonuria, and alkaptonuria, result from enzyme deficiencies in the pathways. An enzyme deficiency means that the DNA segment that codes for the enzyme has undergone a chemical alteration resulting in the absence of a key enzyme, the accumulation of some unmetabolized substrate, and the loss of important products further down the pathway. **(a)** Phenylketonuria, the accumulation of phenylalanine and loss of

the subsequent product tyrosine, occurs when the enzyme *phenylalanine hydroxylase* is absent. **(b)** Albinism, the absence of the pigment melanin, occurs when the enzyme *tyrosinase* is absent. **(c)** Alkaptonuria, near the end of the pathway, occurs when the enzyme *homogentisase* is absent, leading to an accumulation of homogentisate (alkapton) in the body.

played a role in enzyme activity and that there was a definite connection between heredity and the presence of normal or abnormal enzymes—quite an accomplishment for his day. But he was ignored. Biochemistry was still an infant science, and geneticists of the time were more interested in how the genes influenced *morphology*—visible structure.

H. J. Muller: A Man Ahead of His Time.

In the 1920s, H. J. Muller worked with **mutations,** rare, inherited changes in the gene, providing strong evidence that genes are indeed chemical structures of some kind. He devised a way of measuring the **mutation rate** of lethal genes in the *Drosophila melanogaster*, the common fruit fly (also known as the garbage fly or vinegar fly). Not only did he show that he could create mutant genes with X rays, but he also measured the rate of occurrence of spontaneous mutations. In addition, he showed that the rate of spontaneous mutations changed with temperature at about the same rate that known chemical reactions changed with temperature. For example, in most instances an increase in temperature of $10°$ C doubles the rate of chemical reactions, and it also doubled the mutation rate in Muller's experiments.

Transformation.

In 1928, bacteriologist Fred Griffith conducted what seemed to be an oddball experiment but one that proved to be a classic. He was studying the **virulence** (disease-producing capability) of two strains of *Pneumonococcus*, a bacterium that causes pneumonia. One strain was dangerous and one harmless. The virulent (disease-producing) strain synthesized a smooth, gummy polysaccharide coat that seemed to protect it from the host's defenses; the harmless strain (a laboratory curiosity) did not. When grown in the laboratory, the virulent strain produced "smooth-looking" colonies; the harmless strain lacked the proper enzymes to coat themselves and produced colonies that appeared "rough."

When Griffith injected smooth-strain bacteria into mice, the mice died. When he injected rough-strain bacteria into mice, the mice did not die. He then killed some smooth-strain bacteria by heating them and injected their bacterial corpses into more mice. The mice did not die. So far, all he had shown was that the smooth polysaccharide itself didn't kill the mice when the bacteria were dead. But then Griffith killed some smooth-strain bacteria and mixed them with live rough-strain bacteria,

both of which had proven to be harmless, and injected the mixture into still more mice. These mice came down with severe pneumonia and died. (At this point, why not pause and make your own best guess about what was happening?) Did the chemical remains of the smooth-strain bacteria help the rough strain do its dirty work? To further confuse things, autopsies showed that the dead mice were full of virulent, living, smooth bacteria! Where did they come from?

Griffith thought that perhaps he had erred in his experimental technique, so he repeated the experiment with great care, again and again. The results were clearly not due to faulty techniques, nor were they due to accidental contamination; the dead smooth-strain *Pneumonococci* were indeed dead. As Sherlock Holmes said, when you have eliminated the impossible, whatever is left must be the truth, no matter how improbable. It seemed unlikely, but apparently the living rough-strain *Pneumonococcus* had somehow been *transformed*. That is, they had incorporated hereditary material from an outside source and, in so doing, expressed a new (smooth) trait (Figure 9.10).

Others improved on the experiment, trying to discover what was behind these results. They found that **transformation** could occur in test tubes, as well as in living mouse hosts. Various materials from the dead smooth bacteria were isolated and purified and then injected to see if they were the mysterious transforming substance. Not until 1944 was it demonstrated that pure DNA extracted from smooth-strain *Pneumonococcus* could transform rough-strain bacteria, giving them the ability to synthesize the necessary enzymes for making the smooth polysaccharide coat. After years of research, we now know that the harmless rough cells actually take in pieces of smooth-cell DNA. With a low but measurable frequency, the DNA-repairing enzymes of the rough cell insert the deadly smooth-strain DNA fragments into the cell's own chromosome, replacing the rough strain's defunct gene. The rough bacteria are then able to synthesize new enzymes and become smooth.

By 1944 it was becoming clear that DNA was the genetic material—clear, that is, to only relatively few of the more "visionary" biologists. Most researchers were still far from convinced, even after some 30 different instances of bacterial transformation by purified DNA had been reported. But perhaps a bit of caution and a critical attitude aren't all that bad in science.

9.10 TRANSFORMATION (1928)

Twenty-five years before the central dogma was finally resolved, Griffith's experiments clearly laid the groundwork for the idea that DNA was the genetic material. Griffith worked with two strains of *Pneumonococcus*, the agent of bacterial pneumonia. His smooth strain was virulent, while his rough strain was rendered harmless by mouse body defenses. Griffith found that some substance or factor from heat-killed smooth *Pneumonococci* could be transferred to the rough strain, transforming the rough form to the smooth form and rendering the bacteria virulent. He had no idea why this happened, but today we know that bits of chromosome (DNA) were being taken in by the rough bacteria and copied during replication. Descendants of the rough bacteria had incorporated smooth DNA genes and become smooth and deadly.

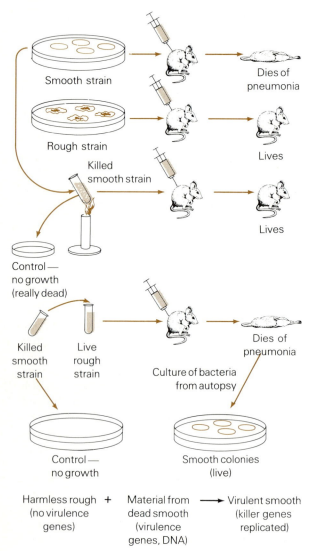

Smooth strain

Dies of pneumonia

Rough strain

Lives

Killed smooth strain

Lives

Control — no growth (really dead)

Killed smooth strain

Live rough strain

Culture of bacteria from autopsy

Dies of pneumonia

Control — no growth

Smooth colonies (live)

Harmless rough (no virulence genes) + Material from dead smooth (virulence genes, DNA) → Virulent smooth (killer genes replicated)

More Recent Events

One Gene, One Enzyme. The slogan "one gene, one enzyme" was an electrifying bit of public relations in its day, some 30 years after Garrod's work. It referred to the Nobel Prize-winning studies of George Beadle and Edward Tatum, who, like Garrod, looked to heredity for an explanation of metabolic pathways. But instead of drawing inferences from human abnormalities or crossing mutant flies, Beadle and Tatum imaginatively chose what was until then a very unusual experimental organism: the fungus *Neurospora*, the same pink mold that may have ruined your bread on occasion. *Neurospora* was to be the first in a series of important microorganisms that would be used in the upstart field of molecular biology.

Neurospora crassa (green mold), *Escherichia coli* (the colon bacterium), *Saccharomyces cerevisiae* (brewer's yeast), and bacteriophages (viruses that infect bacteria) were soon to become the standard tools of latter-day geneticists and molecular biologists. Such microorganisms have two huge advantages for genetic study over mice, flies, peas, humans, and other higher organisms: (1) they can be grown cheaply in enormous numbers in a very short time and (2) unlike higher organisms whose genes come in pairs, through most of their life cycle these organisms have only one copy of each gene, so many genetic complications can be avoided altogether. Further, the effects of mutation (sudden genetic changes) can be seen immediately, since they aren't hidden by a corresponding normal copy.

Beadle and Tatum produced random mutations by using X rays to irradiate *Neurospora* **spores** (dormant, resistant cells important to survival and reproduction; see Chapter 20). They then grew the irradiated spores and screened them for biochemical mutations (Figure 9.11); that is, they looked for strains that could not grow unless certain simple biochemical compounds were added to the medium (the food on which the fungus was grown). These simple compounds were the metabolites (intermediate products), such as vitamins and amino acids, that are routinely present in biochemical pathways under normal conditions. Their idea was that if a mutant gene was not producing a certain enzyme, then the enzyme's usual product would not be produced and the biochemical pathway would be brought to a lethal halt. The biochemical

9.11 EXPERIMENTAL PROCEDURES OF BEADLE AND TATUM

(a) The procedure for producing and isolating nutritional mutants in *Neurospora* begins with the irradiation of tiny fungal spores (see chains of spores in photo). Irradiation produces a number of mutations, some of which will be nutritional mutants. **(b)** The spores are then placed in a simple, unsupplemented food medium (minimal medium), where only the normal, nonmutated spores (wild type) grow, forming the usual long fungal filaments. **(c)** The medium is then poured through cheesecloth, which traps the filaments while letting the mutated spores pass through. **(d)** The spores are then placed on agar that is supplemented in various ways with vitamins, amino acids, or other metabolites. Spores that germinate are classified according to which supplement they require.

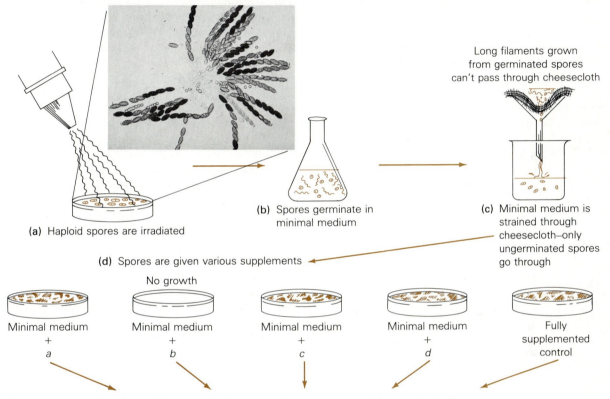

Long filaments grown from germinated spores can't pass through cheesecloth

(a) Haploid spores are irradiated

(b) Spores germinate in minimal medium

(c) Minimal medium is strained through cheesecloth—only ungerminated spores go through

(d) Spores are given various supplements

No growth

| Minimal medium + *a* | Minimal medium + *b* | Minimal medium + *c* | Minimal medium + *d* | Fully supplemented control |

Spores that grow are collected and become mutant stock for experimental crosses

pathway could be said to be *blocked* at a critical step. But adding the missing product of the blocked step would unblock the pathway and allow it to proceed to completion, allowing the fungus to thrive.

Once the nutritional mutants had been identified, the strains could be maintained and used to determine the hereditary basis for enzyme deficiencies. All of the enzyme deficiencies turned out to be the result of simple single gene changes; hence the slogan "one gene, one enzyme," or, in translation, "It requires the action of one gene to produce one enzyme." Garrod's idea had been rediscovered and confirmed: biologists were now more confident that specific genes were responsible for the presence or absence of specific proteins. Actually, the "one gene, one enzyme" slogan has had to be revised since that day. As we will learn in the next chapter, genes actually code for polypeptides. You will recall that many proteins, and the complex enzymes in particular, consist of two or more polypeptides, and these are considerably modified before being incorporated into protein.

Hershey and Chase.   Alfred Hershey and Margaret Chase performed another classic experiment that, in retrospect at least, firmly established that DNA was the genetic material. They used a new tool in genetic analysis, the **bacteriophage** (called *phage* for short; also see Chapter 18). If you have ever wondered whether a germ can get sick, you will be glad to learn that it can. It can become infected with phage (a kind of virus) and may even die.

The phage Hershey and Chase worked with destroys *Escherichia coli*, the common, rod-shaped bacterium that lives harmlessly in your intestine. This phage consists of a DNA chromosome contained in a body made of protein. The whole thing looks like a moon lander and operates like a hypodermic needle. When the phage touches down, tail first, on the surface of its bacterial host, it makes a small hole in the bacterial wall and shoots its DNA inside. There the phage DNA takes over the synthetic machinery of the host cell, causing it to make a hundred or so new viruses and then to rupture, releasing a myriad of newly constructed viruses (Figure 9.12).

The important point, with respect to the experiment that Hershey and Chase were going to perform, is that only the DNA of the phage enters the host cell; the protein portion of the phage stays outside. Hershey and Chase didn't know this at the time, but discovered it themselves through experimentation. They did this by growing bacteriophage on a medium containing radioactive sulfur (^{35}S) and radioactive phosphorus (^{32}P). Since proteins contain sulfur but no phosphorus, and nucleic acids

9.12 THE BACTERIOPHAGE

This low magnification electron micrograph shows "moonlander" phage particles attached to the bacterium *E. coli*. New phage particles are being assembled within the bacterium, and when this insidious activity has been completed, the cell will rupture, releasing numerous new particles to invade other cells.

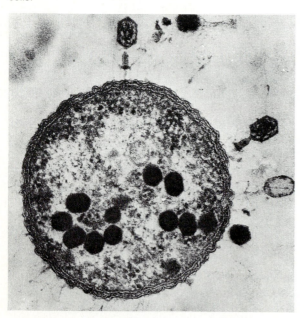

contain phosphorus but no sulfur, the protein and nucleic acids conveniently labeled themselves by incorporating the radioactive sulfur and phosphorus, respectively.

Hershey and Chase infected "cold" (normal, nonradioactive) bacteria with their "hot" (radioactive) bacteriophages and allowed enough time for the phages to attach themselves and inject their DNA—but not enough time for the production of new phage. Then they put the phage and bacteria into a kitchen blender. The empty "ghosts" of the bacteriophage were dislodged from the bacterial surface and could be separated by centrifugation. It turned out that all the radioactive sulfur was found in the empty bacteriophage ghosts; all the radioactive phosphorus was found inside the infected bacteria. Thus, Hershey and Chase proved that only the DNA entered the cell. More important, they showed that naked DNA alone has all the information necessary to enable the host to make new bacteriophage DNA, and new bacteriophage protein as well (Figure 9.13).

Thus, it appeared that DNA was the stuff of which genes were made—at least the genes for smooth coats in *Pneumonococcus* and the genes of certain bacterial viruses. Actually, Hershey and Chase were lucky because some viruses use RNA as their genetic material, and in certain viruses, some or all of the protein does in fact enter the host cell. But at the very least, they demonstrated that DNA was *capable* of being the genetic material.

Still, no one had very much of an idea of how DNA worked. Genes were known to produce enzymes, which are complex proteins. The big question was, how could a molecule with only four different subunits determine the specificity of proteins, which are composed of 20 different amino acids? The next task was to look at the DNA molecule as closely as possible, and from every possible angle.

Chargaff's Rule. For a long time it had been thought that the nitrogen bases, adenine, thymine, cytosine, and guanine (A, T, C, and G), occurred in equal frequencies. In fact, it was once believed that the basic unit of DNA was a simple, repeating tetranucleotide consisting of one each of the four different nucleotides.

Then in 1950 Erwin Chargaff showed that DNA from different sources had *different base ratios*, that is, different frequencies of the four subunits. *Escherichia coli*, for instance, has about 25% A, 25% T, 25% C, and 25% G, fitting the original theory—but the DNA of humans and other mammals is

9.13 PROTEIN IS NOT THE GENETIC MATERIAL (1950)

In the early 1950s, Hershey and Chase followed a hunch that phage viruses injected only DNA into their hosts and that DNA had the genetic instructions for producing more viruses. At this time, people weren't sure whether it was the DNA or the protein that carried the instructions. The researchers grew phage viruses on radioactive sulfur (^{35}S) and phosphorus (^{32}P). After allowing the labeled phage particles to infect their hosts, but permitting no time for replication, Hershey and Chase succeeded in dislodging the empty shells of the phage. The rest is history. Only the radioactive ^{32}P entered the host cells, showing that DNA was almost certainly the genetic material. Another link in the central dogma had been forged.

Incubated to allow phage injection, but not long enough for phage replication

Bacteriophage culture   Normal *E. coli* culture

Phages inject their nucleic acids, which contain ^{32}P, while protein coat with ^{35}S remains outside

Blender separates phage coats from bacterial cells

Centrifuging settles cells to bottom, phage coats remain in liquid

about 21% C, 21% G, 29% A, and 29% T. As Chargaff looked at the DNA of more and more organisms, he found increasingly different ratios, but he also discovered one general rule: The amounts of A and T in his samples were always equal, and the amounts of G and C were always equal (Table 9.1). Or, in shorthand, A = T and G = C. In addition, one could also state that A + G = T + C = 50%. That is, regardless of the source of DNA, exactly half of the nucleotide bases are purines (adenine and guanine) and exactly half are pyrimidines (thymine and cytosine).

We now know that the reason for these equalities is the specific pairing of nitrogen bases, but to Chargaff they were only intriguing, mysterious observations. Nevertheless, they were key observa-

TABLE 9.1

CHARGAFF'S RULE (1949–1953).[a]

	Base composition (mole percent)			
	A	T	G	C
Animals				
Human	30.9	29.4	19.9	19.8
Sheep	29.3	28.3	21.4	21.0
Hen	28.8	29.2	20.5	21.5
Turtle	29.7	27.9	22.0	21.3
Salmon	29.7	29.1	20.8	20.4
Sea urchin	32.8	32.1	17.7	17.3
Locust	29.3	29.3	20.5	20.7
Plants				
Wheat germ	27.3	27.1	22.7	22.8
Yeast	31.3	32.9	18.7	17.1
Aspergillus niger (mold)	25.0	24.9	25.1	25.0
Bacteria				
Escherichia coli	24.7	23.6	26.0	25.7
Staphylococcus aureus	30.8	29.2	21.0	19.0
Clostridium perfringens	36.9	36.3	14.0	12.8
Brucella abortus	21.0	21.1	29.0	28.9
Sarcina lutea	13.4	12.4	37.1	37.1
Bacteriophages				
T7	26.0	26.0	24.0	24.0
λ	21.3	22.9	28.6	27.2
φX174, single strand DNA[b]	24.6	32.7	24.1	18.5
φX174, replicative form	26.3	26.4	22.3	22.3

[a]By determining the composition of nitrogen bases in the DNA of a variety of organisms, Chargaff and his contemporaries were able to provide vital information as the central dogma emerged. Pay close attention to the relative quantities of A and T, and G and C here. (But note that the values are not exactly equal due to experimental error.)

[b]Note that this virus has single-stranded DNA, which does *not* follow Chargaff's rule. Why not?

Adapted from A. L. Lehninger, *Biochemistry,* 2d ed. (New York: Worth, 1975).

tions that enabled Watson and Crick to work out the structure and function of life's key molecule.

X-Ray Diffraction and More Puzzles. X-ray **crystallography** helps physical scientists explore the fine structure of crystals. Essentially, the technique involves aiming X rays at a crystal and noting how the rays are bent (diffracted) by the regular, repeating molecular structures within the crystal. The closer together the regularly repeated structures, the greater the angle through which diffracted X rays are bent. The pattern produced on a photographic plate consists of whorls and dots, with those farthest from the center of the plate representing the most closely spaced repeating structure (Figure 9.14). In this way, the relatively simple structures of inorganic crystals have been deduced.

Interestingly, organic chemicals can also be crystallized. In recent years, the three-dimensional structures of many proteins and of some RNA molecules have been worked out. In Watson and Crick's time, however, only preliminary work on a few proteins had been done. People were just starting to look at DNA, among them Maurice Wilkins (who received a Nobel Prize with Watson and Crick) and Rosalind Franklin (who died before her work was fully acknowledged). Their studies revealed a few repeated intramolecular distances, namely, 2.0 nm, 0.34 nm, and 3.4 nm, numbers that showed up again and again in the X-ray image but didn't make complete sense at the time. Wilkins and Franklin also recognized in the patterns a helical molecule with the phosphates on the outside. Franklin even argued that there were probably two strands, not one or three, as had been suggested by others. Although Wilkins and Franklin had a general idea of what the molecule was like, they still didn't know the specifics. It was at this point that Watson and Crick appeared on the scene.

The Watson and Crick Model of DNA

Watson and Crick knew several things when in 1953 they tackled the DNA problem. They knew that DNA was a polymer consisting of four different nucleotides, they understood the chemical structure of the nucleotides, and they realized that the nitrogen bases dangled off the sides of the ribose-phosphate backbone. From Chargaff's data they knew that somehow the number of adenines had to equal the number of thymines, and that the number of guanines had to equal the number of cytosines. They also knew of Wilkins and Franklin's ideas

9.14 X-RAY DIFFRACTION STUDIES OF DNA

(a) Even to many biologists, the X-ray diffraction pattern of DNA might as well be a Rorschach ink blot intended for detecting personality disorders. But to Watson, Crick, and others schooled in X-ray crystallography, this picture was a thing of beauty, actually suggesting the helical structure of DNA. The dark upper and lower patterns represent the dense packing of atoms in the nitrogen bases at the core of DNA, while the X-shaped central pattern is interpreted as a helical structure. From patterns like this, scientists have deduced the structure of DNA, RNA, and many complex proteins.

(b) In X-ray diffraction, an X-ray beam is deflected by repeating structures in the regular crystalline array of closely packed molecules. The closer together two repeating structures are, the farther from the center the beam will be deflected. There are now computers that can be fed the raw data and will immediately draw three-dimensional pictures of the molecule, but keep in mind that Watson and Crick didn't even have a pocket calculator.

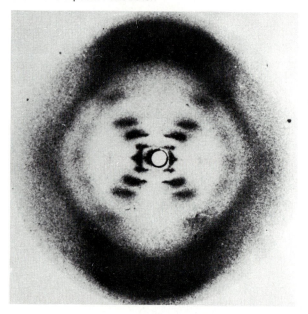

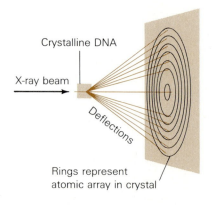

Crystalline DNA

X-ray beam

Deflections

Rings represent atomic array in crystal

about DNA and the results of their X-ray diffraction studies—the intramolecular measurements 2.0 nm, 0.34 nm, and 3.4 nm.

Watson and Crick's experimentation was based on real models made of wire, sheet metal, and nuts and bolts (Figure 9.15). Their models provided a graphic, "hands-on" representation of the emerging molecule (sometimes the fingers can grasp what the mind cannot). Biological intuition, along with fitting the pieces this way and that, seemed to indicate that there might be two strands wrapped around one another, with the phosphate-sugar backbone on the outside and the nucleotide bases inside, facing one another. But how were the bases arranged inside?

Wilkins and Franklin's numbers began to make sense. The 2.0 nm measurement represented the total width of the double helix. The 0.34 nm represented the thickness of the nucleotide bases if they were laid perfectly flat. That is, if the bases were stacked one on top the other, like pennies in a roll, each layer would be 0.34 nm thick. The 3.4 nm measurement was a tough one, but it turned out that if each base was set slightly off from the one above, like steps in a circular staircase, the double backbone would make one complete twist every 3.4 nm along the axis of the molecule. Further, with a little

9.15 WATSON AND CRICK'S DNA MODEL

Using wire, bits of metal, nuts and bolts, and a great deal of intuition, Watson and Crick put what information they had to work deducing the structure of DNA.

9.16 A SPACE-FILLING MODEL OF DNA

In the DNA double helix, the sugar-phosphate groups stand out clearly as the backbone of the molecule. Within lie the bases, represented here as horizontal rows of spheres, stacked one atop the other within the helix. Each base pair is 0.34 nm in thickness, and ten such pairs produce one full turn of the helix, a distance of 3.4 nm. The overall width, including the base pairs and the backbone, is 2.0 nm. Each measurement was determined through X-ray diffraction studies.

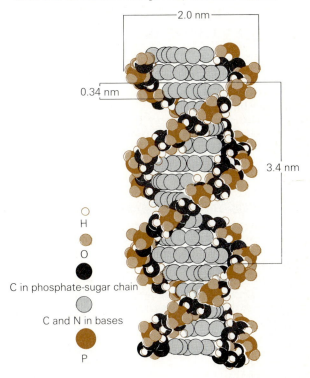

arithmetic (3.4 nm ÷ 0.34 nm = 10), we find that there would be exactly ten nucleotide pairs in each helical turn (Figure 9.16).

It was Watson who insightfully grasped the true meaning of Chargaff's strange data about the ratios of A, T, G, and C. In the sheet metal and wire model, two purines would not fit opposite one another within the 2.0 nm confines of the double helix, and two pyrimidines would leave a gap. But one purine and one pyrimidine *could* fit opposite one another. Watson saw that adenine and thymine would form two hydrogen bonds, while guanine and cytosine would form three hydrogen bonds (see Figure 9.4). This was the only way the DNA molecule could fit together. Thus, the specifics of base pairing represented one of Watson and Crick's major findings. The linear order of nucleotides in any one strand might be perfectly arbitrary, but whatever nucleotides were in one strand rigidly fixed the sequence of nucleotides in the other strand.

The successful analysis of DNA structure by Watson and Crick opened "the golden era of molecular biology." Things happened fast after the publication of their short paper in April of 1953. Before the decade was over, Crick, whose awesome intellect had come to dominate molecular biology, was able to formulate what he called "the central dogma." We have seen one aspect of the central dogma at work in the process of replication. In the coming chapter, we will look at other aspects as we describe the genetic code, the role of DNA in transcription, and that of RNA in translation.

APPLICATION OF IDEAS

1. Before the discovery of RNA-dependent DNA synthesis (reverse transcription), the central dogma had included the provision that DNA arises only from other DNA. The new knowledge was then incorporated into the dogma. Discuss how such a flat contradiction affects a theory. Is the theory invalidated?

2. In the prehistory of life, nucleic acids might have come before proteins, or proteins before nucleic acids (this is still being argued). How could a purely protein organism store information or reproduce such information for the next generation? Is there some other way around the problem?

KEY WORDS AND IDEAS

THE CENTRAL DOGMA?
1. The **central dogma** is an explanation of the chemical and physical basis of heredity and gene expression through the operation of DNA and RNA. Its tenets are:
   a. Genetic information is stored in DNA.
   b. DNA is copied into new DNA during **replication.**
   c. RNA is copied from DNA during **transcription.**
   d. RNA uses its transcribed coding of nucleotides to carry out protein synthesis during **translation.**
   e. Exceptions occur in viruses where viral DNA is sometimes copied from viral RNA (**RNA-dependent DNA synthesis,** or **reverse transcription**).

The Principle of Colinearity
1. The *colinearity principle* correlates the linear arrangement of *nucleic acid bases* in DNA and their corresponding linear arrangement in messenger RNA with the linear arrangement of amino acids in the primary structure of proteins.

DNA STRUCTURE

The DNA Nucleotides
1. Four different **deoxynucleotides,** or **nucleotides,** the structural units of DNA, are assembled into long polymers of **DNA strands,** or **nucleic acids.** Prior to assembly, they are in the form of *nucleoside triphosphates* similar to ATP.

2. Each nucleotide contains the three parts: *phosphate,* **deoxyribose,** and a **nitrogenous base,** in that order.

3. The four bases of DNA, their designations and their triphosphate form are **adenine** (dATP), **guanine** (dGTP), **thymine** (dTTP), and **cytosine** (dCTP).

4. When incorporated into DNA, a pyrophosphate is released from each nucleoside triphosphate. Nucleotides are joined by their phosphates and sugars, which form the backbone of the polymer with the nitrogen bases projecting off to the side.

5. Synthesis of DNA polymers proceeds from the **5' end** to the **3' end.** In its finished form, DNA is a double strand of nucleotides wound into a **double helix.** Adjacent bases are attracted by numerous weak hydrogen bonds.

6. In **base pairing,** adenine is only opposed by thymine (A-T or T-A), and guanine with cytosine (G-C or C-G).

7. In the double helix, the two polymers run in opposite directions (5'-to-3' and 3'-to-5'). Many millions of nucleotides may be present.

DNA REPLICATION

1. Replication is the preparation of DNA copies prior to reproduction of the cell or organism.

2. Because of specific base pairing, upon separation of the DNA double strand, each strand can reproduce the other ("Crick" strands can form "Watson" strands and "Watson" strands "Crick" strands).

3. Replication is carried out by **replication complexes,** which include the **unwinding enzyme helicase** and the nucleotide adding enzyme **DNA polymerase.**

4. Working from the 5'-to-3' direction, the helix is unwound, the separated strands forming **replication forks,** and new nucleotide triphosphates are added according to base pairing.

5. The addition of bases to the 3' or **leading end** of a polymer occurs smoothly, one base at a time, but at the 5' or **lagging end, Okazaki** or **precursor fragments** must first be assembled in the 5'-to-3' direction, and then, utilizing the enzyme **ligase,** they are added in.

6. Because each new polymer is base-paired to an old, DNA replication is called **semiconservative** (half is conserved).

Comparing DNA Replication in Eukaryotes and Prokaryotes

1. Replication sites "bubble out" as they form, and bubbles lengthen as replication proceeds in both directions.

2. In eukaryotic replication, multiple replication forks work simultaneously, forming the many bubbles. In prokaryotes, only two replication forks (one bubble) form along the circular chromosome, but replication in prokaryotes is much faster.

A BRIEF HISTORY OF THE DISCOVERY OF DNA

1. Mendel helped establish that heredity was controlled by "particulate factors" and **chromosomes** were soon suspected of carrying the factors (genes).

2. Miescher identified DNA in 1869 and in 1914, Feulgen perfected a specific DNA stain **(Feulgen stain);** however the connection between DNA and heredity was not made for many years.

Early Efforts

1. Garrod identified metabolic disorders such as **alkaptonuria** by the presence of abnormal **metabolites** such as **alkaptones** in the urine. He determined that such conditions were inherited and surmised that they involved abnormalities in the enzymes of metabolic pathways. He correctly associated the abnormal metabolites with abnormal enzymes and such enzymes with abnormal genes.

2. In 1920 Muller determined that X rays could create chemical changes in the genes, and found ways of measuring natural **mutation rates.**

3. In 1928 Griffith, experimenting with **virulence** in *Pneumonococcus,* determined that nonvirulent strains could be **transformed** (genetically changed) to virulent strains if the remains of dead virulent bacteria were made available. In 1944, the transforming material was finally determined to be DNA.

More Recent Efforts

1. Using X rays to create nutritional mutants in the fungus *Neuospora,* Beadle and Tatum determined that the presence of enzymes in metabolic pathways was controlled by genes. They identified X-ray-damaged genes by locating blocked steps (enzyme failures) in the metabolic pathways. This was done by determining which nutrient had to be supplied to get the mutant strain to grow.

2. Using a **bacteriophage** and the bacterium *Escherichia coli,* Hershey and Chase were able to show that only viral DNA entered the host; thus it was DNA that directed the production of new viral particles. This strongly suggested that DNA was the genetic material.

3. In 1950, Chargaff developed the principle of base-pairing. He determined the relative amounts of A, T, G, and C in a variety of cells, proving that A = T and G = C and that there is exactly as much purine in the nucleus as there is pyrimidine.

4. Through the use of **X-ray crystallography,** Wilkins and Franklin determined that DNA was double stranded, probably formed a helix, and had intramolecular measurements of 2.0 nm, 0.34 nm, and 3.4 nm.

The Watson-Crick Model of DNA

1. In 1953, having used critical information from the work of others and by constructing models of their own, Watson and Crick determined the structure of DNA, including its phosphate-sugar backbone, specific (A-T, G-C) base-pairing of purines and pyrimidines, and the meaning of the intramolecular distances. From their models, mechanisms for replication, transcription, and translation were proposed (the central dogma).

REVIEW QUESTIONS

1. List the four main ideas of the central dogma. (200)

2. Summarize the concept of colinearity. (200-201)

3. Using a simplified drawing, identify each component of a nucleotide triphosphate and explain where they are bonded together. (201-202)

4. List the four kinds of nucleotides in DNA, and identify whether they are purines or pyrimidines. (202)

5. Explain how a nucleotide triphosphate is added to a growing polymer of DNA. Specifically, where do the new nucleotides attach and in what direction is the chain synthesized? (202)

6. Describe the final structure of DNA. Include its geometric form, what holds it together, and some idea of its length. (204)

7. Using the terms *hydrogen bonding, pyrimidine, purine*, and the letters *G, A, C,* and *T,* explain the manner in which nitrogen bases fit together in the completed DNA molecule. (204)

8. Explain why base-pairing is so essential to the process of replication. (204)

9. List the three components found in a replication complex, and briefly explain what each does. (205)

10. In what direction *must* replication proceed? Since replication on each DNA strand proceeds in both directions, what problem does this introduce? How is the problem solved? (206-207)

11. Explain the term *semiconservative replication.* What are the two alternatives? (207-209)

12. Compare prokaryote and eukaryote replication in terms of replication forks and bubbles. Suggest reasons why the *rate* of replication in prokaryotes may be so much faster than it is in eukaryotes. (207)

13. Explain the link Garrod found between abnormal substances in the urine and the role of genes in metabolic pathways. How did Garrod arrive at the notion that such metabolic disorders are heriditary? (211)

14. In what way did Muller's work indicate that genes have a discrete chemical structure? (213)

15. Briefly summarize Griffith's observations. What was his conclusion, and what would we add to this conclusion today? (213)

16. Explain the statement, "One gene, one enzyme." Is this really accurate? Explain. (214-215)

17. Summarize Beadle and Tatum's procedure for producing, recovering, and identifying nutritional mutants of *Neurospora.* What was their final conclusion? (214)

18. Describe how the use of radioactive tracers by Hershey and Chase led to identifying DNA as the hereditary agent of the phage virus. (216)

19. If Chargaff had worked with the bacterium *E. coli* only, how would this have effected the discovery of base-pairing? Describe what he actually found. (217)

20. What did the measurements 2.0 nm, 0.34 nm, 3.4 nm mean to Wilkins and Franklin? What do they actually represent? (218)

21. List three aspects of DNA structure, previously discovered by others, that led Watson and Crick to construct an accurate model of DNA. Which of these led to Crick's prediction of how replication would work? (218-219)

DNA in Action

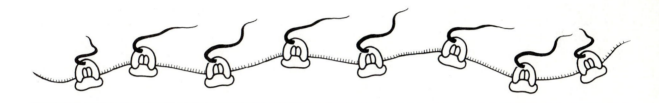

The story of the search for the "stuff of heredity" and the unveiling of its structure indeed make a good tale. But there is another story to tell—the one about how DNA works. Here, we will consider its role in some very specific cellular activities. We will see how DNA dictates the formation of proteins, especially enzymes. Later we will consider in more detail just how the enzymes assume their important roles in the intricate processes of life. Two major points will be established: (1) DNA contains all of the chemical instructions needed to produce new cells and even entire individuals, and these instructions are faithfully replicated prior to cell division, and (2) DNA dictates the business of the cell by transcribing its information into RNA, which in turn translates the genetic code into the synthesis of polypeptides. The polypeptides will be used to produce enzymes and other proteins.

The discovery that genes are composed of DNA and that their information lies in the linear arrangement of DNA nucleotides was indeed an enormous breakthrough, but it also leaves many questions unanswered. For example, what is the precise organization of the genetic instructions? And how are these instructions used to determine the order of amino acids in polypeptides? It didn't take researchers long to determine that in eukaryotes, DNA is not involved in the actual process of protein synthesis. They found that in eukaryotes, DNA remains in the nucleus, while protein assembly occurs in the

cytoplasm. In fact, even the proteins found in the nucleus are produced in the cytoplasm and transported through the nuclear membrane. How does the DNA direct the synthesis of proteins some distance away? The answer lies in the roles of RNA.

RNA STRUCTURE AND TRANSCRIPTION

Comparing RNA to DNA

Let us begin our discussion of RNA with a brief review of its structure. Although there are some similarities, RNA differs from DNA in a number of important ways (Figure 10.1):

1. The 5-carbon sugar in RNA is **ribose** instead of deoxyribose.
2. While both RNA and DNA contain adenine (A), guanine (G), and cytosine (C), the fourth nucleotide base differs in the two molecules. DNA contains thymine (T) and RNA contains **uracil** (U), a closely related but slightly different nitrogen base.
3. DNA almost always occurs as a double-stranded helix. RNA almost always occurs as a single-stranded molecule, which often has complex, twisted and folded *secondary* and *tertiary structures.*

4. DNA molecules are almost always much longer than RNA molecules—typically, 1000 to 1 million times longer.
5. DNA is generally more stable than RNA; it is more resistant to spontaneous and enzymatic breakdown and damage can be repaired. RNA is more reactive partly because of the additional reactive —OH side group of ribose, and repairs are not possible.
6. Although there is only one type of DNA, there are several different kinds and sizes of RNA, each with its own function.

Transcription: RNA Synthesis

In **transcription,** the chemical instructions encoded in DNA are copied into RNA. Only one strand of the double-stranded DNA is transcribed, and it is called, logically enough, the **transcribed strand.** The other, then, is the **nontranscribed strand** (or

noncoding strand). The segment of the DNA molecule on which a single RNA molecule is transcribed is, in a sense, equivalent to a **gene.** A gene can also be defined as a length of a DNA molecule that specifies a polypeptide. As we will learn, only portions of any DNA molecule are transcribed.

We have seen that DNA replication is *semiconservative*. RNA transcription, on the other hand, is **conservative**—that is, the molecule being copied is conserved, not changed by the process of transcribing RNA.

Transcription requires the presence of **RNA polymerase,** which happens to be among the largest known enzymes. Eukaryotes contain three different RNA polymerases, each of which helps form a specific type of RNA.

RNA polymerase can begin assembling new bases only after it identifies and binds to a specific DNA sequence known as a **promoter;** a different kind of promoter is involved for each of the three polymerases. Following this binding, the giant

10.1 RNA NUCLEOTIDES

The RNA nucleotides differ from those of DNA in two ways. First, the RNA nucleotides contain the sugar ribose rather than deoxyribose. Ribose contains an —OH group on the 2' carbon rather than the —H group found in deoxyribose. Second, in RNA, the nitrogen base uracil replaces the thymine of DNA. Note that uracil and thymine are quite similar, except for the presence of a methyl group (—CH₃) in thymine where a hydrogen appears in uracil.

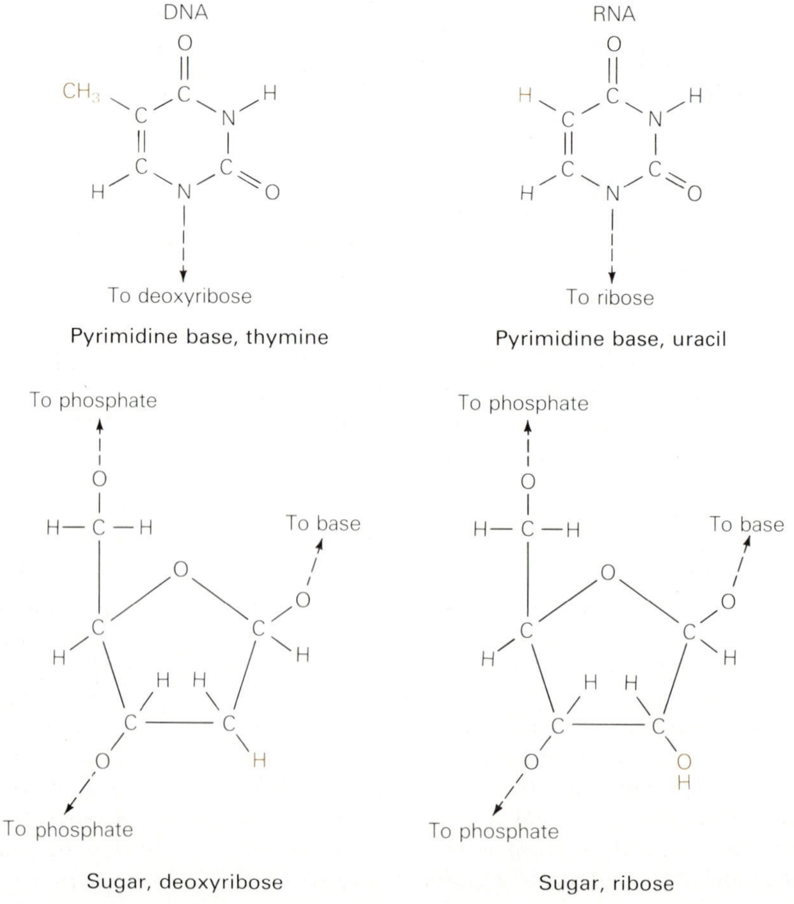

DNA

Pyrimidine base, thymine

RNA

Pyrimidine base, uracil

Sugar, deoxyribose

Sugar, ribose

10.2 RNA TRANSCRIPTION

(a) In the synthesis of RNA during transcription, the giant enzyme RNA polymerase moves down the DNA polymer, unwinding small portions of the helix as it goes. The unwinding opens about one full turn of DNA at a time, and once the RNA strand has been assembled along the exposed bases, the helix immediately rewinds. **(b)** and **(c)** Base pairing in transcription is very similar to that found in replication. Cytosine pairs with guanine and uracil pairs with adenine (remember, uracil replaces thymine in RNA). Once the phosphate-sugar bonds have formed, the RNA polymer dislodges from DNA. It will perform its function as a single strand.

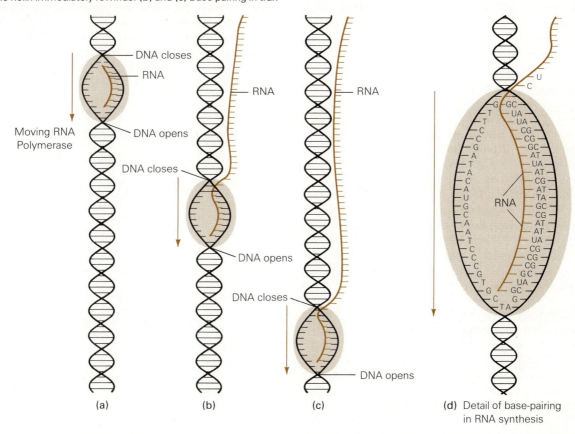

(a)   (b)   (c)   (d) Detail of base-pairing in RNA synthesis

enzyme begins to unwind the DNA. It unwinds about one full turn of the DNA helix, exposing a segment of freed DNA bases. It is here that the base-pairing will occur. As a segment is transcribed, the opened helix rewinds while a new segment ahead unwinds (Figure 10.2). Interestingly, as the new RNA bases are brought in and base-paired with the transcribing DNA strand, they temporarily form a double polymer, as though replication were occurring. However, the attractive forces between the two strands are weak, and soon the original DNA helix reforms, with the RNA replaced by the nontranscribing DNA strand.

Except for the substitution of uracil for thymine, the base-pairing rules of RNA synthesis are very similar to those of DNA replication. The RNA strand is synthesized in the usual 5'-to-3' direction, using ribose triphosphate nucleotides, or ribonucle-

otides (rCTP, rGTP, rATP, and rUTP; the "r" stands for "ribose"). By way of summary, then, for every C in the transcribed DNA strand, RNA polymerase puts in a G ribonucleotide; for every G, a C; for every T, an A; and for every A, a U. When the process is completed, the new RNA has the same order of bases as the appropriate stretch of the nontranscribed strand of DNA, with U substituting for T (see Figure 10.2).

Transcription of a DNA strand continues until the moving polymerase encounters what is called a **termination signal.** Like the promoter, this consists of a special sequence of DNA bases, but it acts to dislodge the growing RNA strand and to release the bound RNA polymerase.

Many RNA molecules can be transcribed from different parts of the same gene simultaneously, for as soon as the first few bases of a sequence have been copied, they are free to begin a new RNA strand (Figure 10.3). The number of simultaneous transcriptions possible depends ultimately on the availability of RNA polymerase. In the bacterium *E. coli*, each cell contains between 3000 and 6000 RNA polymerase molecules (compared to only 10 to 20 DNA polymerases), and some of the more active eukaryotic cells contain as many as 70,000 RNA polymerases. With so much polymerase around, simultaneous transcription is the rule rather than the exception.

Varieties of RNA: Physical and Functional Classes

There are three kinds of RNA in most cells: **ribosomal RNA (rRNA), messenger RNA (mRNA), and transfer RNA (tRNA).** Each is transcribed from DNA as described above and each plays a role in protein synthesis. Ribosomal RNA contributes significantly to ribosomal structure, while messenger RNA, as its name implies, contains the coded message that will determine the polypeptide to be produced. Transfer RNA plays a key intermediary role, identifying amino acids and delivering them to the ribosome, the site of polypeptide synthesis.

Ribosomal RNA and the Ribosome. Ribosomal RNA, as we have just seen, is found in ribosomes. While this isn't too surprising, its transcription is quite special. In the eukaryotes, ribosomal RNA, unlike other RNA, is transcribed within the *nucleolus*. In addition, the DNA involved in the transcription is a special form. Loops of DNA within the nucleolus contain what are called **nucleolar organizer regions,** each consisting of multiple copies of the gene responsible for rRNA transcription. Why are so many identical genes needed? It seems that at certain times the demand for rRNA is much greater than at other times. Thus, when large amounts of rRNA are needed, thousands of RNA polymerases can travel along copies of the transcribing genes, spinning off strand after strand of rRNA. Nowhere is the vast synthetic activity of the nucleolus more dramatically shown than in the feathery **lampbrush chromosomes** found in the oocytes (immature eggs) of amphibians. As you see in the electron micrograph in Figure 10.4, the term "feathery" is quite fitting even if you have never seen a lampbrush. Each slender branch of the feather's rib represents a loop of DNA along which rRNA is being transcribed.

Following their production, the long rRNA transcripts are immediately processed to yield the specific shorter strands of ribosomal RNA needed for

ribosome assembly. These strands are of three types, called 18S, 5.8S, and 28S rRNAs. A fourth, 5S rRNA, is prepared outside the nucleolus. ("S" is a sedimentation or density unit used in describing the results of ultracentrifugation.) While the processed rRNA will form the skeleton of the ribosome, the remainder will consist of special ribosomal proteins assembled in the cytoplasm. Such proteins enter the nucleus and find their way to the nucleolus, where they join the rRNA. But while ribosomal assembly begins in the nucleolus, it must be completed in the cytoplasm. Cytologists suspect that this keeps the ribosomes from becoming active

10.3 SIMULTANEOUS TRANSCRIPTION

Typically, transcription of a gene will occur simultaneously along the DNA strand. Each of the four "bubbles" represents a moving RNA polymerase and an independent transcription event. The electron micrograph shows a central strand of DNA with feathery branches of RNA emerging. The dark dots along DNA are RNA polymerases.

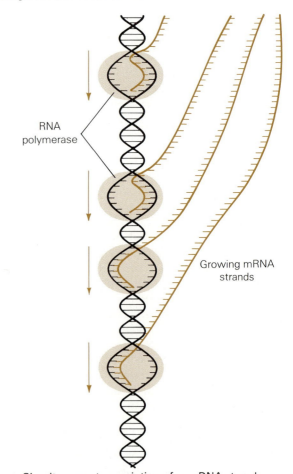

RNA polymerase

Growing mRNA strands

Simultaneous transcription of one DNA strand

10.4 LAMPBRUSH CHROMOSOMES

(a) Ribosomal RNA transcription is seen occuring along a feathery *lampbrush chromosome,* an active region of DNA seen here in an amphibian egg cell nucleolus. **(b)** Each lampbrush loop is an exposed segment of DNA, along which rRNA transcription is occurring. The loops emerge from the main chromosome axis, which is actually a denser region of DNA and protein.

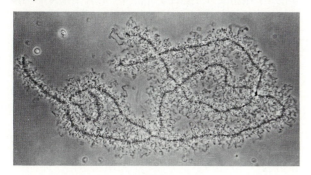

(a)

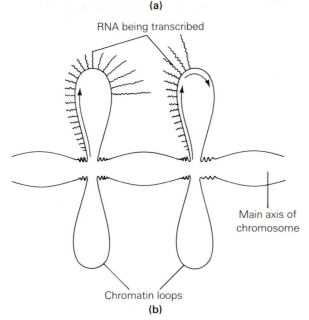

(b)

and starting to function in the nucleus. The transcription of rRNA and assembly of ribosomes is reviewed in Figure 10.5.

Completed ribosomes are made up of two different sized subunits (Figure 10.6). The smaller subunit fits into the larger in an elaborate interlocking manner, and when assembled will clamp itself over a messenger RNA molecule. The larger—60S—subunit contains the 28S, 5.8S, and 5S rRNA, while the smaller—40S—subunit contains the 18S portion. (You may have noticed that the numbers of the S

units don't add up. This is for technical reasons we won't discuss here.) Proteins make up about half the ribosomal mass, with the smaller subunit containing some 30 different proteins and the larger 45 to 50, all tightly bound to the RNAs. Prokaryote ribosomes, by the way, are somewhat smaller than those of eukaryotes.

It should be stressed that ribosomes are the only places where proteins are synthesized. A ribosome can be compared to a kind of workbench or to the reading head of a tape recorder. The ribosome is where mRNA, tRNA, and the growing polypeptide chain of a protein all work together to create their

10.5 RNA SYNTHESIS IN THE NUCLEOLUS

Synthesis of both ribosomal RNA and ribosome synthesis occurs in the nucleolus. There, multiple copies of the rRNA transcribing genes located on nucleolar organizer DNA transcribe the rRNA in vast amounts. The rRNA is then processed, and along with 5S rRNA from the nucleus, is combined with ribosomal protein from the cytoplasm to form the ribosomal subunits. Ribosome synthesis is completed in the cytoplasm.

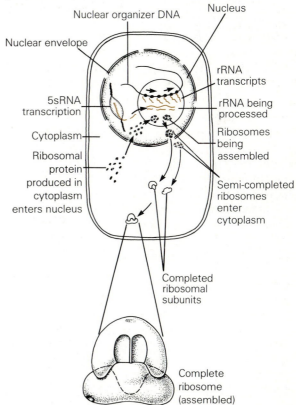

10.6 RIBOSOME DETAIL

Ribosomes are composed of two subunits that join during protein synthesis. Here the eukaryotic ribosomes, an 80S body when intact, consists of a larger 60S and a smaller 40S subunit. (The S is the *Svedberg unit*, a measure of the speed of sedimentation. Note that one can't add the S values of the parts to get the S value of the whole.) Each ribosome has three attachment sites, identified here as two tRNA binding sites and a site to which the leader of an mRNA will attach when translation begins. In addition, each ribosome has a site where the enzyme *peptidyl transferase* operates. This enzyme is instrumental in joining new amino acids to the growing strand.

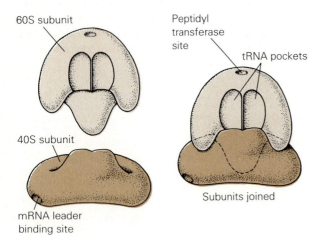

magic: making proteins. The ribosome itself is not passive in this process. Its surface has three **attachment sites** (specifically shaped cavities)—one site for a short, three-nucleotide stretch of mRNA that is located on the leading end of the molecule and one site each for a new and a used tRNA, as we will see. There is also a site where the enzyme **peptidyl transferase** works to form peptide bonds between adjacent amino acids (see Figure 10.6).

Exons and Introns: The Tailoring of Messenger RNA. Messenger RNA (mRNA) emerging from the DNA template in eukaryotic cells contains the instructions for polypeptide synthesis. In its *primary* or "raw" form, however, it is not quite ready to perform its task. On the average, **primary mRNA** transcripts may contain up to 6000 nucleotides, although transcripts of 50,000 nucleotides are known to exist. However, before they can leave the nucleus and begin polypeptide synthesis, the RNA transcripts will undergo what is called **posttranscriptional modification**, a process whereby the raw RNA transcripts are reduced to 500 to 3000

nucleotides. (This does not occur in prokaryotes, in which mRNA is produced in a form ready to go to work—an important basic difference between the two forms of life.)

Eukaryotic mRNA contains long noncoding nucleotide sequences called **intervening sequences,** or **introns.** These must be identified and enzymatically removed before the mRNA can become active. The remaining portions of RNA are called the **expressed sequences,** or **exons.** Following the removal of the introns, the exons are spliced together to form the finished RNA (Figure 10.7a). Individual introns may contain anywhere from 100 to 10,000 nucleotides. Their removal is a precise and complex process, with little room for error, since the loss or gain of even one nucleotide would alter the genetic message, resulting in a faulty protein product when translation occurs.

But on with the story. After posttranscriptional modification, each mature eukaryotic mRNA consists of three regions, a **5′ leading region,** a **cistron,** and a **3′ trailing region** or **poly-A tail.** The leading region receives special treatment known as **capping,** whereby the 5′ end becomes linked to a special **methylated** version of a triphosphate guanine nucleotide in which one of the nitrogens in the base has a methyl group added, as does the 2′ carbon of first ribose sugar. The trailing region contains about 200 adenine nucleotides. Capping at the leading region is essential to mRNA's interaction with the ribosome, but the trailing region has no known function. The cistron—roughly the equivalent of a gene—carries the coding that will determine the sequence of amino acids in one polypeptide.

Prokaryotic mRNA is quite different. It lacks the methylated "cap" and the poly-A tail of the eukaryote. In addition, it is common for prokaryotic mRNA to be **polycistronic**—to contain the coding for producing more than one polypeptide and thus actually represent several transcribed genes. Proteins coded by a polycistronic gene often work in closely related functions, so producing them simultaneously is useful. In essence, polycistronic mRNA is the transcription product of two or more genes. Figure 10.7b compares eukaryotic and prokaryotic mRNA.

Messenger RNA and the Genetic Code. Having explored some of the processes of transcription, let's look more closely at its product, messenger RNA (mRNA). As its name suggests, messenger RNA carries the message—in this case, information regarding the sequence of amino acids of the polypeptide to be produced. The linear sequence of every protein a cell produces is encoded in the DNA of a specific gene on one of the chromosomes. But DNA, as we have noted, does not make proteins directly—it can only direct the synthesis of RNA or of copies of itself. In eukaryotes, in fact, protein synthesis occurs in the cytoplasm, while DNA remains in the nucleus (recall the exception of mitochondrial and chloroplast DNA discussed in Chapter 4). Messenger RNA is the physical link between the gene and the protein. The mRNA molecule is synthesized on DNA and faithfully incorporates the information necessary to specify a protein. The information contained in mRNA is written in the **genetic code.** Each code word, or **codon,** is made up of three adjacent nucleotides. The three nucleotides specify one of the 20 common amino acids (Table 10.1). The sequence GAG in mRNA, for instance, specifies the amino acid glutamic acid. (Therefore, its nucleotide equivalent in DNA would be CTC.)

Since there are four different RNA nucleotides that can occur in any of the three positions of a codon, there are $4 \times 4 \times 4 = 64$ different codons. Three of these, UAA, UAG, and UGA, are **stop codons** that specify the end of a protein, like the period at the end of a sentence. The remaining 61 codons specify the 20 amino acids. Obviously, there are more types of amino acid-specifying codons than there are types of amino acids, so most amino acids are coded by more than one codon. For instance, codons GGU, GGC, GGA, and GGG all code for one amino acid, glycine. These are called **synonymous codons.**

One codon, AUG, is seemingly ambiguous in that it has more than one role. It can either specify the amino acid methionine in the middle of a protein or serve as an **initiator** or *start* signal that directs the enzymes of the cell to begin polypeptide synthesis. Apparently, a ribosome's reaction to an AUG codon during translation depends on the initiator codon's proximity to the 5' end of the mRNA, and possibly to the methylated cap. Given this positioning factor, there is really no ambiguity anywhere in messenger RNA.

Transfer RNA.   As we have seen, the mRNA carries the coded message (a long sequence of nucleotides) to the ribosome, where it is decoded. The decoding determines the sequence of amino acids in the developing polypeptide. But, except for initiator and terminator codons, the ribosome

10.7 PROKARYOTE AND EUKARYOTE mRNA

(a) Messenger RNA in prokaryotes differs from eukaryotic mRNA in two important ways. Prokaryotes produce polycistronic RNA (copies of two or more genes), which does not require posttranscriptional modification before becoming active. In fact, the leading end of mRNA being transcribed by the bacterial chromosome may start translating while the trailing end is still being transcribed. Eukaryotic mRNA always contains one cistron, and in addition, in its primary form **(b),** contains a large number of noncoding introns, which must be excised prior to its leaving the nucleus and for translation to begin.

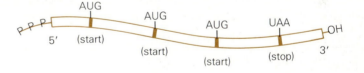

(a) Prokaryote (polycistronic) mRNA

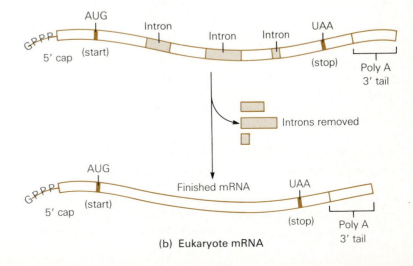

(b) Eukaryote mRNA

itself cannot tell one codon from the next. Deciphering the codons, one at a time, is the job of our last type of RNA: transfer RNA, or tRNA.

Amino acids and tRNAs are linked by special **charging enzymes.** Such enzymes are quite specific and there is a different one for each amino acid. Further, as would be expected, there are also different tRNAs for each of the 20 amino acids found in proteins. It follows that we are dealing with 20 different enzymes and at least 20 different tRNAs (there may be more than 20 tRNAs because of synonymous codons). The charging enzymes recognize each particular type of tRNA and link it to its own special amino acid with a high-energy, covalent bond. Thus, one such enzyme, called a **phenylalanine tRNA charging enzyme,** is joined by a

phenylalanine tRNA in one of its receptive sites, a phenylalanine amino acid in another site, and, with the aid of ATP, joins the specific amino acid to its specific tRNA (Figure 10.8).

Each tRNA molecule is a relatively short length of RNA consisting of about 90 nucleotides. In its primary or raw form, when it is first transcribed from DNA, the tRNA is somewhat longer, but before it becomes active it undergoes some post-transcriptional modification (Figure 10.9a). Many raw tRNAs contain introns, which must also be excised. In addition, the molecule is "tailored" as special enzymes remove segments from each end. Other enzymes make chemical modifications of some of the bases in special places on the different tRNAs so that the completed molecule contains

TABLE 10.1

THE GENETIC CODE

The genetic code can be described in terms of the codons in mRNA. The table is read in the following manner: There are 64 possible codons. The left-hand column contains the first letters of the codons. Across the top are the second letters, and at the right are the third letters. If you wanted to know which amino acid was coded by CAU, you would find C at the left, A at the top, and U at the right. Where the three letters intersect in the table, you will find CAU and the abbreviation His, which stands for the amino acid histidine.[a] Note that the code is *synonymous*, meaning that there is more than one codon for each amino acid (with two exceptions). UAA, UAG, and UGA do not code for amino acids; they signal *stop*, and are known as *terminators*. And one more irregularity needs to be mentioned. AUG has two purposes—it codes for methionine and it also means *start*. Every mRNA begins with AUG, so it is an *initiator*.

SECOND LETTER

FIRST LETTER	U	C	A	G	THIRD LETTER
U	UUU UUC } Phe / UUA UUG } Leu	UCU UCC UCA UCG } Ser	UAU UAC } Tyr / UAA stop / UAG stop	UGU UGC } Cys / UGA stop / UGG Trp	U C A G
C	CUU CUC CUA CUG } Leu	CCU CCC CCA CCG } Pro	CAU CAC } His / CAA CAG } Gln	CGU CGC CGA CGG } Arg	U C A G
A	AUU AUC AUA } Ile / AUG Met start	ACU ACC ACA ACG } Thr	AAU AAC } Asn / AAA AAG } Lys	AGU AGC } Ser / AGA AGG } Arg	U C A G
G	GUU GUC GUA GUG } Val	GCU GCC GCA GCG } Ala	GAU GAC } Asp / GAA GAG } Glu	GGU GGC GGA GGG } Gly	U C A G

[a]Amino acid abbreviations: alanine, Ala; arginine, Arg; asparagine, Asn; aspartic acid, Asp; cysteine, Cys; glutamic acid, Glu; glutamine, Gln; glycine, Gly; histidine, His; isoleucine, Ile; leucine, Leu; lysine, Lys; methionine, Met; phenylalanine, Phe; proline, Pro; serine, Ser; threonine, Thr; tryptophan, Trp; tyrosine, Tyr; valine, Val.

"exotic" RNA bases in addition to the usual four. These exotic RNA bases may serve in part to help preserve the molecule by retarding enzymatic degradation. Yet another enzyme adds three more nucleotides to the 3' end of every tRNA so that all completed tRNAs end with the sequence —CCA. (Here, then, *protein* enzymes determine a *nucleic acid* sequence. Does this seem to violate the central dogma?)

The tRNA molecule is precisely coiled and loops back on itself in a characteristic conformation. The folded tRNA has three loops and a stem (Figure 10.9b), and, finally, the whole molecule is held in a twisted, L-shaped configuration by hydrogen bonds between its nucleotide bases and between the bases and the free —OH side groups of the ribose units (Figure 10.9c). Such contortions would be quite impossible for DNA because deoxyribose lacks the free —OH group.

Transfer RNA can do three things:
1. As we have mentioned, tRNA can bond to its specific amino acid. One end, the —**CCA stem,** is covalently bonded to an amino acid by a charging enzyme.
2. It can connect to a ribosome. A portion of at least one side loop can make a specific

10.8 CHARGING ENZYMES

In the charging of alanine-tRNA, four components are required: alanine, alanine tRNA, ATP, and a specific charging enzyme. **(a)** With the components in place, AMP is transferred to the amino acid and pyrophosphate (P ~ P) is released from the site. **(b)** The phosphate bond energy is used in forming a covalent bond between the amino acid and its specific tRNA, with AMP released. **(c)** Following the bonding event, charged alanine-tRNA is released from the enzyme ready to act.

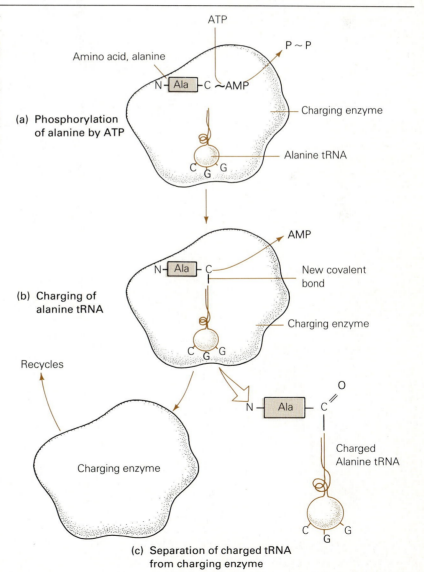

(a) Phosphorylation of alanine by ATP

(b) Charging of alanine tRNA

(c) Separation of charged tRNA from charging enzyme

10.9 LEVELS OF tRNA STRUCTURE

Transfer RNA molecules contain about 90 nucleotides and are transcribed similarly to mRNA. **(a)** Upon leaving the nucleus, the crude tRNA is "tailored"—modified by enzymes. These modifications include removing ends and clipping out introns, leaving about 90 nucleotides, some of which are chemically modified from the U, C, A, and G of the original transcript. **(b)** In its modified form, tRNA contains three larger loops and one smaller (variable) loop. These are called the *D loop, T loop, anticodon loop,* and *variable loop.* The 3' end always contains the codon CCA and is the attachment site for amino acids.

The *anticodon loop* contains the nitrogen bases that base-pair with codons on mRNA. **(c)** tRNA achieves its final form by folding from its cloverleaf shape into a twisted form resembling an inverted L. **(d)** The exact sequence of an alanine tRNA. The unfamiliar symbols stand for various modified bases. The base sequence is called *primary* structure; the helical base pairing is called *secondary* structure; and the final folding into a complex three-dimensional shape is called *tertiary* structure.

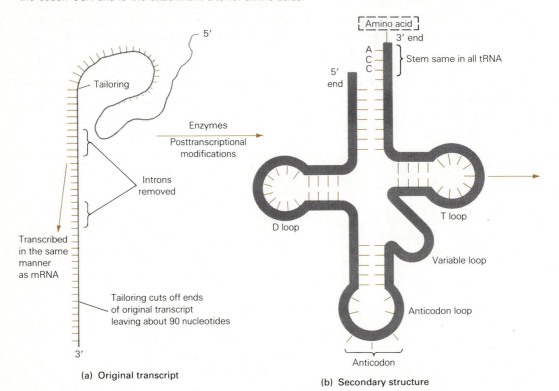

(a) Original transcript

(b) Secondary structure

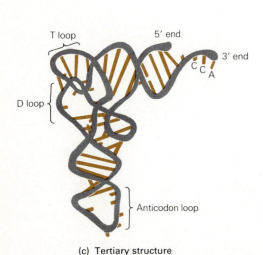

(c) Tertiary structure

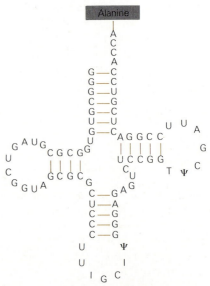

(d) Alanine tRNA showing primary and secondary structure

10.10 CODON AND ANTICODON

One of the highly specific regions of tRNA is the anticodon loop. Here, we see the anticodon 5'-AGC-3' base pairing with the mRNA codon 5'-GCU-3', a perfect match, which will code for alanine (see Table 7.2). The arrows indicate the 5'-to-3' direction. In order to obtain a proper match, the codon and anticodon must be antiparallel—that is, oriented in opposite directions. Since the diagram shows the codon reading from left to right, like English, the anticodon must read from right to left, like Hebrew.

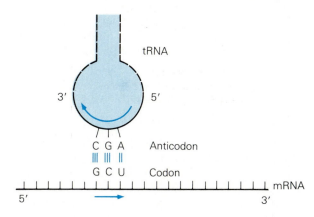

Watson-Crick pairing (that is, C:G, A:U, and so on) with one of the RNA constituents of the ribosome. This portion of the loop is the same for all tRNAs because they all need to connect to ribosomes in the same way.

3. It can base pair with mRNA. One end, the **anticodon loop,** can recognize specific codons of mRNA. This is a major point so let's look at the details (Figure 10.10).

The "recognition" between codon and anticodon is yet another example of the Watson-Crick type of pairing in which U in the codon pairs with A in the anticodon, C pairs with G, and so on. Thus, the tRNA anticodon 5'-AGC-3' pairs with the mRNA codon 5'-CGU-3' (which codes for the amino acid *alanine*, Figure 10.10). Remember that strands run in opposite directions in Watson-Crick pairing.

TRANSLATION: HOW POLYPEPTIDES ARE ASSEMBLED

Translation, converting the coded genetic message to work in synthesizing polypeptides, includes three parts: **initiation, elongation,** and **termination.** All we have said about the RNAs and the genetic code now come together and we begin with the ribosome and initiation.

Initiation

Initiation begins when the leading end of the mRNA winds around the smaller 40S ribosomal subunit and fits into a special attachment site (Figure 10.11). This step is quite precise and if all goes well, the initiator codon AUG will have aligned itself in the left hand receptor site, or pocket, as we will call it. In eukaryotes, such as ourselves, the first amino acid to be incorporated into a polypeptide is always methionine. (In prokaryotes, it is *N*-formyl methionine.) A special **initiator tRNA** with a 5'-CAU-3' anticodon recognizes and pairs with the initiation codon 5'-AUG-3'. It also binds with the smaller of the two major ribosome subunits, as mentioned, in an energy-utilizing reaction involving at least three specific initiation proteins. Only after this **initiation complex** (the three initiation proteins, the smaller ribosomal subunit, a charged methionine tRNA, and the mRNA initiator codon) is formed can the two subunits of the ribosome combine to form a functional, intact ribosome (Figure 10.11).

The methionine will form the *N*-**terminal end** of the growing polypeptide. This simply means that the amino group (NH_2) of the first amino acid in a polypeptide will be exposed. Further, the amino groups of any newly arriving amino acids will all attach to the carboxyl groups of existing amino acids. Thus, the final amino acid of any polypeptide will have its carboxyl group ($—COO^-$) exposed, making up the *C*-**terminal end** of the polypeptide. By tradition, the *N*-terminus of the polypeptide is written at the left and the *C*-terminus at the right.

Elongation

That is how a protein starts, but how does it grow? How are additional amino acids added to the chain? Let's take a look at the process about halfway along. Polypeptides come in all lengths, but a length of 250 amino acids is typical, so let's say that the first 125 amino acids have already been incorporated, and the 126th is about to be added on.

Figure 10.12 shows the situation. A length of mRNA lies in the groove between the two halves of the ribosome—nine bases reading —AAA—GGC—UUA—, which are three codons specifying lysine, glycine, and leucine, respectively. Let's say that these will be the 125th, 126th, and 127th amino acids in the protein. Lysine, the 125th, is already part of the chain.

As we have seen, the ribosome has two large, adjacent pockets, which are the tRNA attachment sites. The messenger RNA runs along the bottom of the two pockets, with codon AAA in one pocket and the next codon, GGC, lying in the bottom of the other pocket. At this particular moment (captured in Figure 10.12a), a lysine tRNA lies in the pocket on the left.

The other tRNA pocket is empty, except for the next mRNA codon that lies along the bottom of it. This condition lasts for a few microseconds, which is a short time for us but a fairly long time for a chemical reaction. During this time, all sorts of small molecules randomly bump in and out of the nearly empty pocket in the ribosome, including any tRNA molecules that might happen to be in the neighborhood. Sooner or later, a charged glycine tRNA will wander in, and it will fit so well that it sticks. The good fit results from the combination of the shape of the pocket, which fits all charged tRNAs, and the matching of the glycine tRNA anticodon —CCG— with the GGC codon of the mRNA.

Figures 10.12b and 10.12c show the two tRNA pockets in the ribosome filled with the lysine and glycine tRNAs. Part (c) shows that the entire 125

amino acids, ending with lysine, have been transferred from the stem of the lysine tRNA to the amino side group of the glycine, as a peptide bond is formed between the carboxyl group of the lysine and the amino group of the glycine. The energy is provided by the lysine tRNA high-energy bond, and the union is guided by the enzyme peptidyltransferase. The lysine tRNA is now uncharged, and the growing polypeptide is 126 amino acids long. The process is not completed until the system is ready to begin another cycle.

The uncharged lysine tRNA has lost its affinity for the ribosome, which only binds to charged tRNAs. It will soon drift out of its pocket and will continue to bump around in the cell fluid until it runs into its own special charging enzyme, which will charge it with another lysine so that it can enter the process again.

Translocation: A Key Step. In Figure 10.12c, the glycine tRNA has moved from one pocket to the other, bringing along with it both the growing polypeptide and the mRNA, which is still bound to its anticodon. This crucial step is called **transloca-**

10.11 PART 1 OF PROTEIN SYNTHESIS: INITIATION

(a) The required elements are mRNA, the two ribosomal subunits, and charged methionine tRNA. (b) The initial event is base pairing between the mRNA initiator, AUG, and the anticodon UAC, which is only found on methionine tRNA. As the base pairing occurs, the smaller ribosomal subunit joins the RNAs. (c) Only then can the larger subunit move in and complete the polypeptide assembling complex.

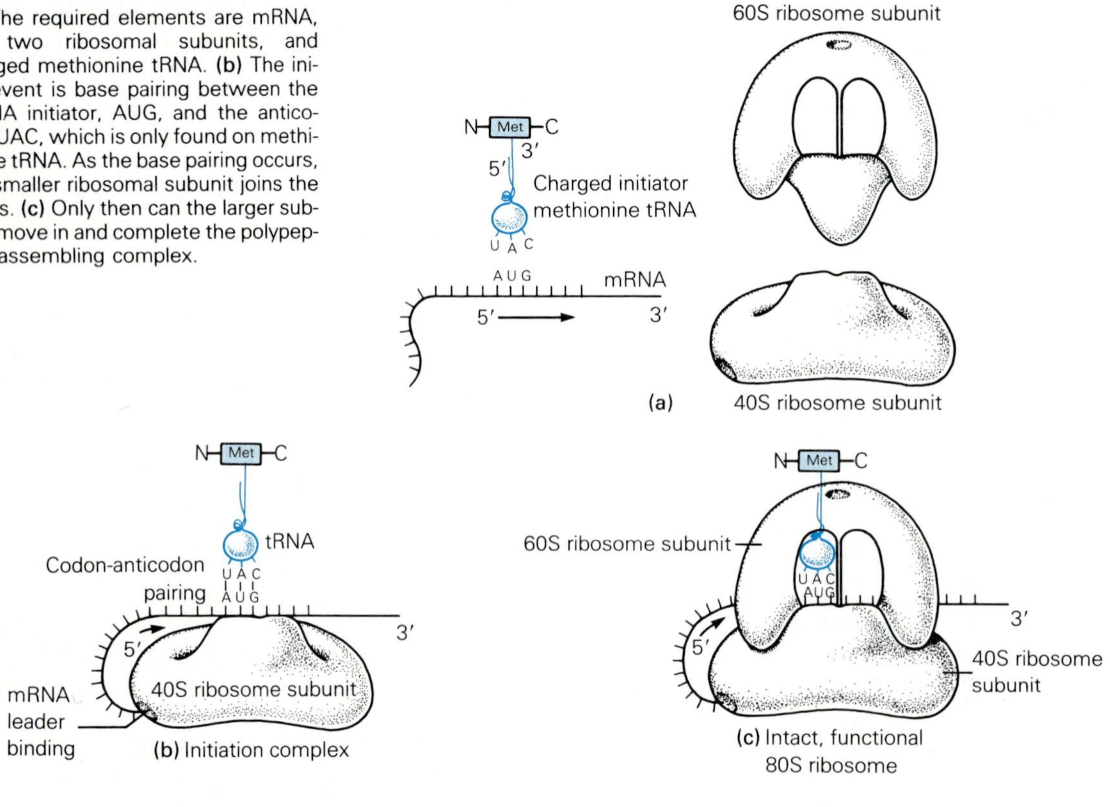

10.12 PART 2 OF PROTEIN SYNTHESIS: ELONGATION

(a) The activity represents the translation process midway through the synthesis of a polypeptide. Three code groups of mRNA are shown, with the left one occupied by lysine tRNA. Its amino acid, lysine, has already covalently bonded to the growing polypeptide chain. (The previous occupant is shown tilting away for recycling.) Glysine tRNA has landed in the right-hand pocket. (b) Next, with the aid of the enzyme peptidyl transferase, a covalent bond forms between the two amino acids. Such bonding sets the stage for the orderly movement down the mRNA strand—an event called translocation. (c) During translocation, the ribosome moves along one codon to the right. Following this, lysine tRNA is released to recycle, glycine tRNA occupies the left pocket, and the right pocket is again empty. Leucine tRNA, a newcomer, is about to drop into the empty pocket where its anticodon is a match for the mRNA codon UUA. Then bonding and translocation will occur once more as the elongation process continues.

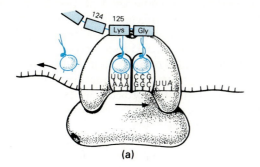

(a)

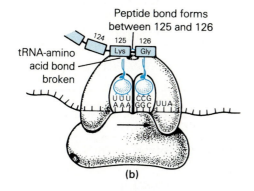

Peptide bond forms between 125 and 126

tRNA-amino acid bond broken

(b)

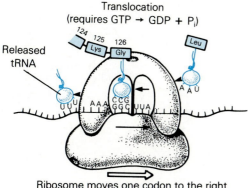

Translocation (requires GTP → GDP + Pᵢ)

Released tRNA

Ribosome moves one codon to the right along the mRNA and is now ready for the next charged tRNA

(c)

tion. Translocation, you will notice, moves the ribosome three nucleotides to the right along the mRNA. (The high-energy bond of guanosine triphosphate (GTP) provides the energy for this process.) As a consequence, the UUA codon now lies along the floor of the newly empty tRNA pocket. Soon a charged leucine tRNA will randomly bump into place, and the elongation process will proceed. This, then, is why we compared the ribosome with the head of a tape recorder: the ribosome not only "reads" the mRNA, but moves it along. (It is arbitrary whether we visualize the mRNA moving past the ribosome, or the ribosome traveling down the length of the mRNA.)

The question most often asked when translation is explained is, how does everything know where to go? The scheme just described is indeed straightforward and elegant, but it prompts questions. For example, the movement of the molecules appears to be totally random and there is every reason to believe that it is. Does this mean, then, that the base pairing of codon and anticodon is just a matter of chance contact? Of course, no combination but the correct one will work and all others will be rejected. One way of improving the odds is for a great deal of charged tRNA to be around and in motion. Actually, it seems that most chemical events in cells depend on random motion and accidental but predictable collision. Does this explain why some protein synthesis is such a slow process? A ribosome of *E. coli* requires about six seconds to produce a polypeptide, but our own comparatively sluggish cells take two to three minutes.

Polypeptide Chain Termination

Well before the ribosome reaches the end of the mRNA molecule, it runs into a chain **terminator,** or *stop* codon. There are three of these: UAA, UAG, and UGA. Sometimes there are double stops (for example, UAA—UAG), apparently just to be sure that the ribosome gets the message.

None of the tRNAs have anticodons that are complementary to any of the three stop codons. Instead, there are specific proteins that apparently occupy the tRNA site once a stop codon has been reached and clog the works, grinding it to a halt. Then, yet another protein factor frees the *C*-terminal carboxyl group from the last tRNA. Following this, the completed polypeptide is released, and the ribosome falls apart into its two components (Figure 10.13). The entire translation process is reviewed in Figure 10.14.

Polyribosomes

High-speed centrifugation of crushed cells can separate cell contents into various fractions, according to the size and specific gravity of the solid particles (see Essay 4.2). Ribosomes appear in two such fractions. Under the electron microscope it can be seen that single ribosomes are found in one group, **polyribosomes** are found in the other. The heavier polyribosomes (also called **polysomes**) consist of several ribosomes, usually 5 to 10 (up to 40 in some cells), bound together by an mRNA molecule. They look like little strings of beads (Figure 10.15). Why are they bound together in groups? It seems that different ribosomes are reading the same mRNA molecule and are spaced along it at appropriate intervals—a minimum of 25 nucleotides apart. Each ribosome will travel the whole length of the cistron of the messenger, from the initiator codon to the terminator codon; then each will fall apart and drop off. Meanwhile, other ribosomes will assemble themselves at the initiator codon and begin moving along the mRNA.

The several ribosomes reading the same mRNA are like a group of ancient Talmudic scholars all reading different parts of the same long scroll. While the last scholar to arrive begins with reading Genesis, another may be just finishing Deuteronomy, while others are working on Exodus, Leviticus, and Numbers. When the first scholar is finished, he can take a break, begin a new scroll, or go back to "In the beginning. . ." Thus, the different ribosomes in a polysome can be producing different copies of the same polypeptide simultaneously, each working on a different portion of the sequence.

Free and Bound Ribosomes

In Chapter 4, we noted that ribosomes occur in two general places, floating free in the cytoplasm or bound to membranes. In higher organisms, **bound ribosomes** are attached to one side of the membranes of the rough endoplasmic reticulum; in fact, their pebbly appearance gives the rough endoplasmic reticulum its name (see Figure 4.10). You may recall that the endoplasmic reticulum, or ER, specializes in transporting (and modifying) newly synthesized substances and also forms the Golgi apparatus, which helps transport materials to the plasma membrane for exocytosis. Bound ribosomes are also seen in bacteria, but prokaryotes lack an ER, and the ribosomes are bound to the inner surface of the plasma membrane itself.

The polypeptides that are produced along the ER experience a different fate from those produced by the free ribosomes. Whereas polypeptides produced by free ribosomes are simply released into

10.13   PART 3 OF PROTEIN SYNTHESIS: TERMINATION

(a) In the first event, the ribosome has reached a terminator codon (UAA). There is no opposing anticodon, since no AUU-bearing tRNA exists. The amino acid just added will be the last. (b) The second event is a "derailing" of the ribosome. The details of this process aren't yet clear, but there appear to be special proteins involved that have been named "releasing factors" (this term is deliberately vague in anticipation of new information). Somehow, the presence of UAA triggers a change in the ribosome, and its subunits separate as the final tRNA is released. Our polypeptide moves away for final shaping into secondary, tertiary, and perhaps quaternary form, and the ribosomal subunits go back to "start" for another round of translation.

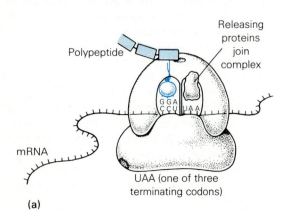

(a)

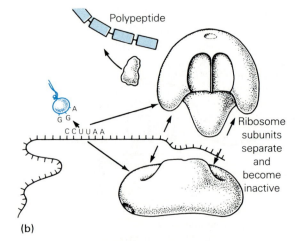

(b)

10.14 SCENARIO OF POLYPEPTIDE SYNTHESIS

The entire translation process is summarized here, with initiation at the left, elongation toward the center, and termination at the right. While the events seen here appear to be highly ordered, everything happening is thought to be completely random, and the proper interaction of all molecules is really just a chancy affair. The odds are improved by large numbers of charged tRNAs and the amazingly high speed at which events occur.

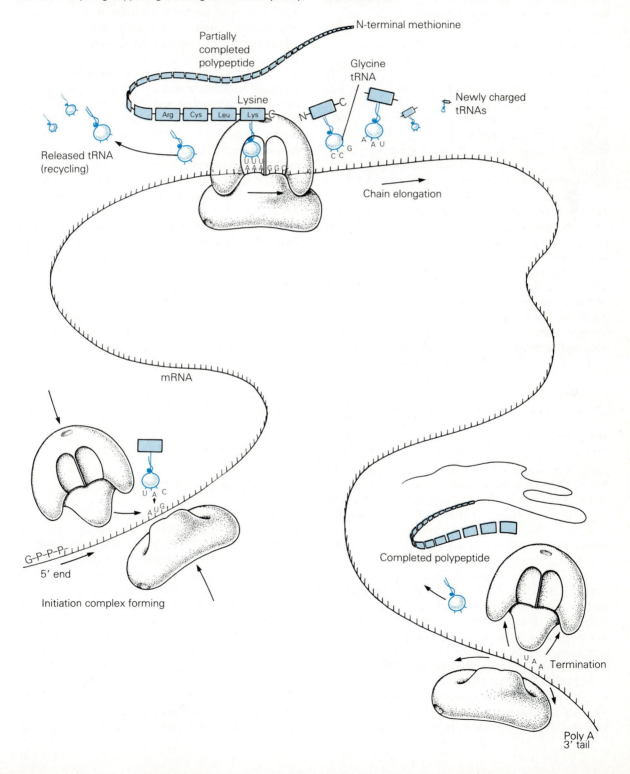

N-terminal methionine

Partially completed polypeptide

Glycine tRNA

Newly charged tRNAs

Lysine

Arg Cys Leu Lys

Released tRNA (recycling)

Chain elongation

mRNA

G-P-P-P

5' end

Initiation complex forming

Completed polypeptide

Termination

Poly A 3' tail

10.15 POLYSOMES IN A PROKARYOTE

Polyribosomes (or polysomes) are shown in this electron micrograph. The amount of protein synthesis is greatly increased when several ribosomes read the mRNA code at the same time. In *Escherichia coli*, polyribosomal translation is so rapid that ribosomes often move along mRNA that is still being transcribed along the chromosome.

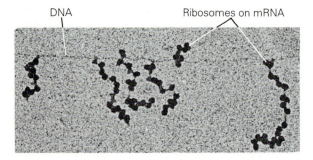

DNA   Ribosomes on mRNA

the cytoplasm, those produced on the bound ribosomes are moved across the membrane to reappear within the lumen (Figure 10.16).

The most recent thinking by cytologists is that bound ribosomes are no different from free ribosomes, and that their presence on the ER is a function of the polypeptide they happen to be producing rather than any specialization of their own. In other words, if they are producing polypeptides destined to enter the ER, then they become bound to the ER. Cytologists have long wondered how the polypeptide gets from the bound ribosome into the lumen or cisternal space of the ER, and the new observations seem to provide an answer.

Signal Peptides. Polypeptides destined for the ER contain at their *N*-terminus a special short segment of amino acids, or **leader sequence,** like the leader strip on a home movie reel. The particular amino acids involved are hydrophobic (water-fearing; see Chapter 3), so they will have a natural affinity for the lipid-rich membrane of the ER. The leader sequence acts as a **signal peptide,** identifying with a specific **receptor site** on the ER. Once contact is made, the polypeptide is actively transported, bit by bit, through the membrane into the lumen. There the signal peptide, which is of no further use, is clipped off by an enzyme called **signal peptidase.** The growing polypeptide strand continues to move into the ER as fast as it is synthesized, keeping both the ribosome and the associated mRNA bound tightly to the ER surface.

The question arises, how does the ribosome, with its mRNA and special polypeptide, find its way to the endoplasmic reticulum? Cytologists

believe that whenever a ribosome-mRNA complex begins the synthesis of an ER-destined polypeptide, the signal peptide attracts and temporarily binds with a cytoplasmic protein known as a **signal recognition protein.** This binding apparently jams the ribosome and mRNA, halting further synthesis until the signal peptide, probably through random movement, enters a receptor site on the ER, whereupon the signal recognition protein is released and synthesis resumes.

The evidence for the role of signal peptides and receptor sites is compelling. For example, proteins or polypeptides that are customarily synthesized and used in the free cytoplasm never contain the leader sequences. Further, when polypeptides known to enter the ER for packaging are synthetically produced outside the cell, they always contain the special leader region. These same proteins normally synthesized in the cell and found in the ER lack the leader or signal peptide, which, as we mentioned, has been enzymatically removed. Geneticists, always eager to apply their own tools toward the solution of such problems, have isolated mutant bacteria and provided some answers. Certain mutant strains of *E. coli* produce polypeptides with faulty signal peptides, and these polypeptides remain in the cell cytoplasm instead of becoming integrated into the plasma membrane as they normally would. In addition, geneticists have succeeded in hybridizing membranal proteins containing the signal peptide with cytoplasmic proteins, and these hybrids find themselves inserted into the membrane.

MUTATION

The Stability of DNA

The DNA molecule is well adapted to its function as a repository of genetic information, primarily because it is a relatively stable molecule. Only the sugar-phosphate backbone is exposed to outside influences, and the deoxyribose sugar has all its potentially reactive side groups already bonded. Most of the potentially reactive side groups of the nucleotide bases are immobilized by hydrogen bonds. In addition, the nucleotide bases are safely tucked inside, and rather immobilized by the geometric tightness of the molecule. Consequently, DNA does not readily react with other chemicals, although an exception may be the highly reactive free radicals (aberrant chemical groups with unpaired electrons) that often form in the cell. Substances such as radicals that chemically alter DNA are called **mutagens.** In addition to chemicals, mutagens occur in the form of heat, ultraviolet light, and ionizing radiation such as gamma radiation and manmade X rays.

In spite of its protection, DNA is subject to some spontaneous denaturation—the random alteration of chemical bonds. Such a change can inactivate an enzyme, but spontaneous changes in DNA are potentially far more dangerous than spontaneous changes in other kinds of molecules. They alter the genes themselves and therefore fall under the category of **mutations.** Mutations are heritable, irrepa-

rable changes in DNA. Mutations in the tissue that produces sperm or eggs can result in seriously ill (or dead) descendants; mutations in the other cells of the body can cause cancer or cell death. Obviously, then, it is best for so critical a molecule as DNA to be relatively resistant to spontaneous changes.

It turns out that DNA in many eukaryotes has additional protection from chemical damage. The molecule is tightly wound around **nucleosomes,** spheres of proteins called **histones** (Figure 10.17). The tightly binding histones have strong positive charges and form ionic bonds with DNA by attaching to the negatively charged phosphates. Histones

10.16 BOUND RIBOSOMES

Recent discoveries presented in this scenario strongly suggest that the ribosomes studding the rough endoplasmic reticulum are at least partially held there by the specific polypeptide they are synthesizing. **(a)** A free-floating ribosome has begun the synthesis of a polypeptide. (Note that in this instance it is the mRNA, not the ribosome, that does the moving.) At its leading (*N*-terminal) end, the polypeptide contains a signal peptide that will bind with a specific surface receptor protein on the membrane of the rough ER. Once bound, the signal peptide is drawn into the rough ER. **(b)** A bound ribo-

some continues translating its polypeptide, which is continually drawn into the lumen of the rough ER. Within the lumen, an enzyme called *signal peptidase* cleaves the signal peptide from the growing polypeptide, and the remainder can then take on its usual function. Polypeptides accumulate in the rough ER, which will later form vesicles of the Golgi body where final protein-forming modifications will occur. **(c)** Once a polypeptide nears completion and mRNA reaches its *stop* codons, termination occurs and the ribosomal subunits separate and are freed from the rough ER.

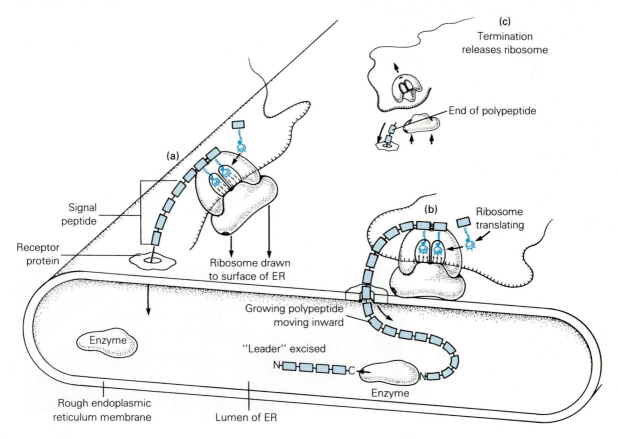

also bind DNA by specific foldings and thereby protect it from thermal and chemical damage.

The most important protection that DNA has against random mutation, however, depends on the specific A:T and G:C pairing that we have discussed. The biological information that must be preserved lies in the **specific order** of the base pairs. That information is redundant (present more than once) in the DNA molecule in the sense that the specific order of bases in one chain determines the specific order of bases in the other chain. This redundancy, then, helps maintain the integrity of the information. For example, if one of the two chains is accidentally altered (say, by ultraviolet radiation), the cell discards the damaged strand and makes a new, perfectly good one by using the remaining intact partner as a template, following the A:T and G:C format.

DNA Repair. Several kinds of **DNA repair systems** are present in the cells of bacteria and probably most other organisms. The first, and probably most general of these is a "recheck" function of DNA polymerase. Here, the polymerase simply reverses its direction and picks up and repairs mismatches (say, a guanine paired with a thymine). The DNA polymerase is able to repair the potential damage because parent DNA strands have methyl groups added to their bases but new strands do not, and the polymerase is able to detect the parent template strand repairing the mismatch correctly instead of perpetuating it.

The next kind of repair, called **excision repair,** consists of excising or cutting out the bases immediately on either side of a damaged region so that replication or transcription would not be blocked. DNA polymerase then comes along and recopies the template strand. Then a new enzyme, **ligase,** sutures the new piece into the strand being repaired (Figure 10.18).

Because of DNA's ability to be repaired, we should note that the term *mutation* does not refer to initial changes in the DNA molecule; these are **primary lesions.** Mutation refers to the very small proportion of those primary lesions that cannot be repaired, or that are repaired incorrectly. (Even DNA repair complexes make mistakes sometimes.) Still, experiments with microorganisms and fruit flies indicate that something like 99.9% of all primary lesions in DNA are completely repaired by the repair enzymes.

Mutations at the Molecular Level

Chromosomal mutations involve entire chromosomes, in which errors in cell division produce daughter cells with extra or missing chromosomes. Or, when chromosome breakage occurs, small bits of chromosome can be lost or chromosomes can be rejoined improperly, leading to all kinds of genetic

10.17 NUCLEOSOMES

(a) Because of its critical role, DNA must be protected. The double helix with its bases turned inward is one level of protection. In addition, DNA in eukaryotes is tightly wound about large, globular protein spheres of histone. The association forms *nucleosomes,* giving DNA an inert quality that protects it from chemical and physical damage. **(b)** Nucleosomes resemble nothing more than beads on a string.

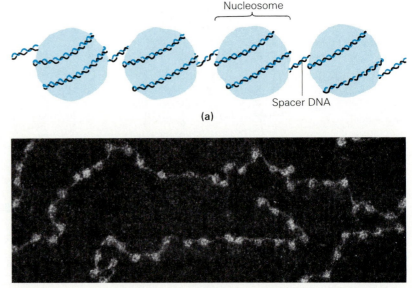

(a)

(b)

10.18 DNA REPAIR

In their role as safeguards against spontaneous damage, teams of roving repair enzymes travel along the DNA strands, somewhat like railroad inspection crews. Any damaged portions are "clipped" out upon discovery, part of the repair process. Following this, bases from the nucleotide pool are properly paired into a short strand. Then, with the help of the enzyme ligase, the short strands are fit into place and covalently bonded. Thus, a potentially harmful genetic change is eliminated.

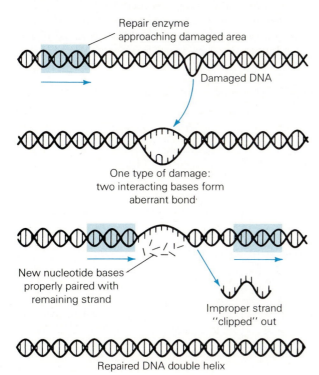

Repair enzyme approaching damaged area

Damaged DNA

One type of damage: two interacting bases form aberrant bond

New nucleotide bases properly paired with remaining strand

Improper strand "clipped" out

Repaired DNA double helix

and developmental problems. For the most part such gross abnormalities spell the death of the cell or organism, so most such changes in DNA are not passed on to future generations. We will look more closely at chromosomal mutations in Chapter 13.

Mutations involving minor changes in DNA sequences—those that geneticists call **point mutations**—are far more subtle, less destructive, and stand a better chance of being passed on from generation to generation. Actually, the effects of these changes may be neutral or even beneficial. Neutral mutations produce slight, harmless changes in the amino acid sequence of a protein and for that reason are known as **silent mutations.**

We should point out that most DNA (of higher organisms at least) does not code for polypeptide sequences. It has been estimated that only about 1% of human DNA actually codes in this way. An even smaller proportion codes for RNAs other than mes-

senger, such as ribosomal RNA and transfer RNA. Of the remaining 98 or 99% of human DNA, some may form so-called "spacer DNA" (untranscribed segments between genes); and a great deal, as we have seen, produces the introns of mRNA. Some, it seems, has no function at all ("junk" DNA), but generally it is assumed that most DNA has some kind of "controlling" function, such as regulating the transcription of the structural genes. We don't know how this control information is coded, or whether "coded" is the right word to describe it. We do know that at least some gene control involves complex sequence-specific binding interactions with controlling proteins. We can assume that permanent changes (mutations) in the controlling regions of DNA are important, but we do not know exactly how they might work or how much influence, if any, single base changes might have in the binding of controlling proteins. We understand much better the effects of changes in the DNA that codes for polypeptides.

Point Mutations. The three most common types of point mutations are

1. **Base substitutions**—the number of nucleotide base pairs in a length of DNA is unchanged, but one base pair is changed from one of the four types to another. This is the most common type of mutation.
2. **Insertion**—one or more bases are added.
3. **Deletion**—one or more bases are lost and the gap closed.

Let's consider each of these in more detail, focusing especially on their effects on polypeptide synthesis.

Base Substitutions. The most obvious effect of single base substitutions is that they alter a codon from one of the 64 possible types to another. DNA base substitutions that cause polypeptide amino acid substitutions are often called **missense mutations.** Of course, not all such codon changes would alter the amino acid sequence in a polypeptide. For example, a change from GAA to GAG would have no effect, since both of these codons specify glutamic acid. In fact, synonymous changes in the third base are very common in evolution; since they would not be weeded out by natural selection, they accumulate randomly over time. As you can see, the presence of synonymous codons offer a certain amount of protection against mutations.

But other single base changes may alter the amino acid sequence and change the nature of the polypeptide. For example, a change of GAA to GUA would change the amino acid somewhere in the polypeptide from glutamic acid to valine (Figure 10.19). Such amino acid substitution may have no measurable effect at all or may prove lethal. An example of a severe change is the mutation that changes glutamic acid to valine in the sixth position in the beta chain of human hemoglobin. This change causes the hemoglobin molecule to form long, sickle-shaped crystals that distort the red blood cells and cause painful, eventually fatal, blood clots in small blood vessels. This genetic disease is **sickle-cell anemia** (see Essay 12.1).

Some of the most interesting base substitutions involve termination codons. If the new codon produced by a mutation is one of the three termination codons, the base substitution mutation is called, not surprisingly, a **chain-termination mutation.** Unless the terminating codon is very close to the end of the structural gene, chain-terminations result in totally nonfunctional polypeptides. The majority of *lethal* mutations, at least in microorganisms, are of the chain-termination type.

Sometimes, the normal termination codon mutates to some other form, say UAA to GAA. Unless a "double stop" message is present, the ribosome

10.19 BASE SUBSTITUTION MUTATIONS

A base substitution in DNA can have serious implications. **(a)** The proper coding is shown for a short segment of one of the polypeptides of hemoglobin. A simple substitution of A (adenine) for T (thymine) in the DNA **(b)** changes the corresponding codon in mRNA from GAA to GUA. Instead of glutamic acid appearing in the corresponding position in the polypeptide, we now see valine. The chemical structure in the R-groups of valine and glutamic acid is quite different, and the R-groups of amino acids often help determine the shape of the protein. This particular change sets up a chain of events, drastically changing the nature of hemoglobin and leading to the crippling genetic disease, sickle-cell anemia.

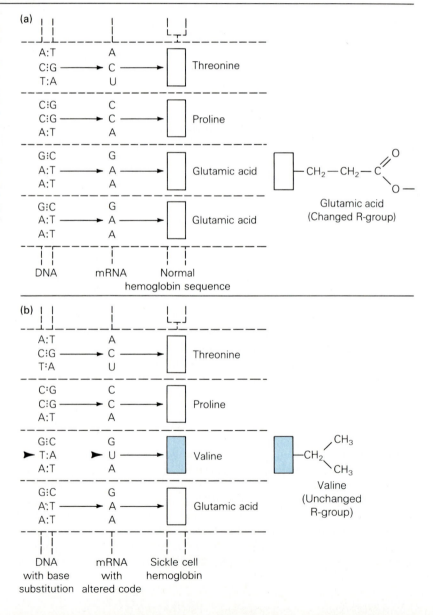

10.20 FRAME SHIFT MUTATION

In the frame shift mutation, nitrogen bases are either added or deleted, and unless this happens in multiples of three, an entirely new coding is produced "downstream" from the incident.

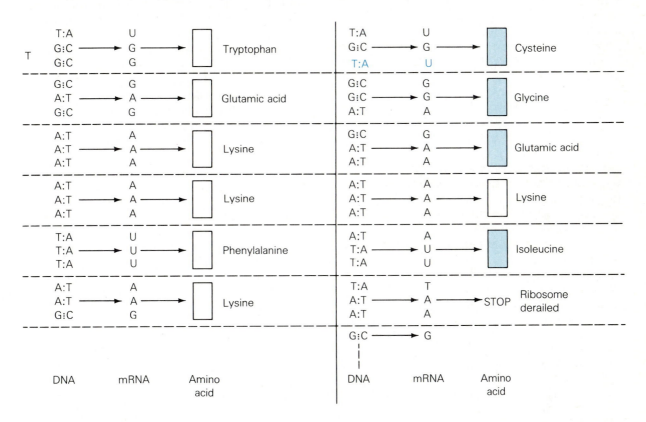

DNA   mRNA   Amino acid

DNA   mRNA   Amino acid

simply adds in the newly coded amino acid and continues to read out amino acids bringing up the trailing end of the mRNA. Following the normal stop codon are (probably) a random assemblage of mRNA, followed by the poly-A tail mentioned earlier. Since 3/64 of all random codons are chain terminators, the ribosome will eventually hit the combination that allows it to let go and fall apart. Obviously the presence of double stop codons or double terminators significantly reduces the incidence of chain termination mutations.

Chain termination mutations are known to exist in DNA coding for both the alpha chain and the beta chain of human hemoglobin protein. Those inherited defects can result in 12 extra amino acids being tacked onto the C-terminal end of the abnormal alpha chain, and 31 extra amino acids onto the end of the abnormal beta chain. But remember, genes come in pairs. Persons who carry the defect on only one member of their hemoglobin-specifying gene pair are not very healthy, but they may be able to survive.

Insertions and Deletions. What happens when extra nucleotides are inserted or deleted? Clearly, if three nucleotides are inserted or deleted, a single change in the amino acid chain will occur.

However, insertions and deletions usually don't come in threes. Single additions or deletions are more likely and, interestingly enough, these can have far greater effects than changes involving triplets. Inserting or deleting a single nucleotide (or any number of nucleotides not divisible by three) will cause what is called a **reading-frame shift** (as we see in Figure 10.20, which shows the result of a uracil nucleotide insertion into a series of codons). The ribosomes, not noted for their intelligence, simply plod along, reading the mRNA nucleotides three at a time. However, the insertion or deletion throws off the nucleotide sequence, which leaves the mRNA reading the wrong triplets. Consequently, the amino acids assembled from the point of the frame shift onward will constitute an incorrect sequence. It is not surprising that frame shift mutations are usually lethal.

MECHANISMS OF GENE CONTROL

With the refinement of the central dogma, biologists finally knew the chemical structure of the gene and the basic mechanism by which it expresses its information. But a major question remained. What controls transcription? This question has proven difficult indeed, partly because there are in fact numerous mechanisms among different organisms.

Human chromosomes carry an estimated 40,000 to 50,000 pairs of protein-coding genes (for the record, *E. coli* carries about a tenth that number of genes or about 1/700 as much DNA as in the human genes). These genes cannot all be active at the same time; otherwise, all the cells in the body would look the same and be doing the same things. In order for specialization to occur, some genes must be active while others are shut down.

How do the genes of eukaryotes become turned off or on in the course of cell specialization and in response to environmental stimuli? We don't know. Answers are beginning to take shape, however. For instance, it is known that some genes are activated by hormones, and a fair amount is known about hormone action. But the overall picture of gene control in eukaryotes is hazy at best. The mechanisms of gene control in prokaryotes, such as bacteria, are much better understood, so we must turn once again to our small contemporaries.

Gene Organization and Control in Prokaryotes: The Operon

Shortly after Crick put together the idea of the central dogma, and while molecular biologists were still trying to work out the genetic code and the mechanism of protein synthesis, two French microbiologists were concluding a long experimental program of their own. In 1961, Francois Jacob and Jacques Monod unveiled their model of bacterial gene organization and control. They called their system, which included a set of genes and the things that influenced them, the **operon.** Molecular biology suddenly took a huge leap forward.

Inducible Enzymes in *E. coli*. Jacob and Monod knew that some of *Escherichia coli's* many enzymes were produced constantly. But they also knew that the synthesis of other enzymes was under some kind of control, and that the tiny bacterium could make these enzymes when it needed them and could stop production when it didn't.

For instance, if *E. coli* is grown on a medium that does not contain lactose (milk sugar), it will not bother to produce the specific enzymes that are needed for lactose metabolism. That makes sense from the standpoint of evolution and energetics; protein synthesis is expensive, and producing unneeded enzymes would be wasteful.

On the other hand, if these same bacteria are placed in a medium that does contain lactose, they will almost immediately begin to produce enzymes that break lactose down into its constituents, glucose and galactose. These lactose-metabolizing enzymes are said to be *inducible,* because their production can be induced or stimulated by an appropriate substrate.

It turns out that there are three enzymes whose synthesis is induced by the presence of lactose—**beta galactosidase, galactose permease,** and **thiogalactoside transacetylase,** but Jacob and Monod found it easier to refer to them simply as enzymes *z, y,* and *a.* The same letters were used to designate the structural genes responsible for the production of the three enzymes. (A structural gene is any gene that codes for proteins, including enzymes.) Jacob and Monod's question was an ambitious one: just how did the presence of lactose in the medium activate or induce activity in genes *z, y,* and *a*?

The Lactose Operon. The inducible lactose operon (or lac operon) was found to consist of the three structural genes along with a fourth gene and two very special segments of DNA that are not genes, but have important control functions in transcription (Figure 10.21). The three structural genes, *z, y,* and *a,* are located in a row on the *E. coli* chromosome, while the fourth, a key controlling element, is located some distance away. The structural genes are transcribed into messenger RNA together; that is, the mRNA transcript for genes *z, y,* and *a* is polycistronic. (Recall that polycistronic messenger RNA is a feature found only in bacteria. We eukaryotes never have more than a single cistron on one mRNA molecule.)

The two special DNA segments include one that is dubbed the **operator,** or **o** for short, and another called the **promoter,** or **p**. The operator lies between the promoter and *z,* the first structural gene. The term "promoter" applies to a region of DNA to which the transcription enzyme, RNA polymerase, binds in preparation for the synthesis of RNA. Once it binds there, it moves along to an initiation site, opens the DNA helix, and begins to assemble RNA. The operator, as we will see, is a very special region that can, under certain condi-

10.21 THE JACOB-MONOD OPERON MODEL

The production of the three inducible enzymes responsible for the metabolism of lactose is under control of a system known as the lac operon. The operon consists of three principal parts, as shown in the diagram: the regulator gene *(i)*, the promoter/operator region, and the structural genes, *z, y,* and *a.*

  (a) The production of enzymes is shut down. The repression of the structural genes is accomplished by an interaction between regulator gene *i* and the operator. ① The regulator gene transcribes messenger RNA that ② is translated into a repressor protein, which ③ coats the operator, blocking the

action of RNA polymerase along the promoter. Thus the transcription of mRNA in the structural genes ④ is inhibited.

  (b) When lactose enters the cell ⑤ it acts as an *inducer,* ⑥ tying up the repressor protein. ⑦ Once the operator is unblocked, RNA polymerase can then begin transcription. ⑧ The polycystronic mRNA produced translates into the three lactose-metabolizing enzymes. When all of the lactose has been metabolized ⑨, newly produced repressor protein can once again bind the operator and the lac operon shuts down.

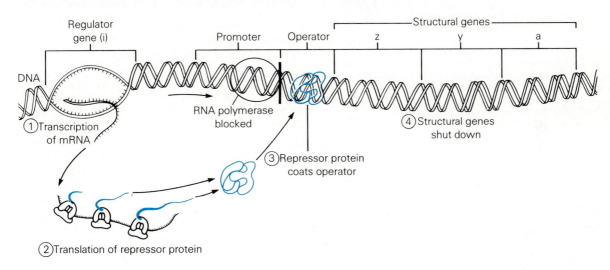

(a)

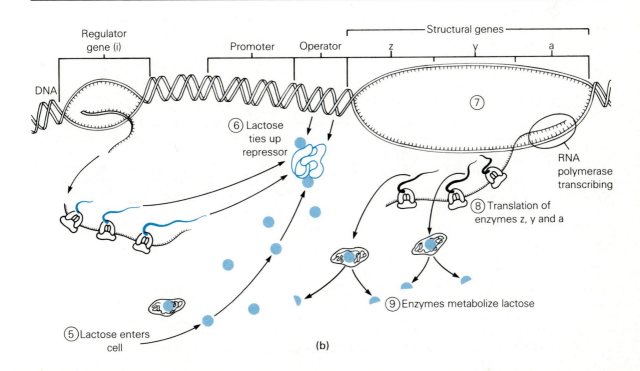

(b)

tions, prevent RNA polymerase from completing its task. Without these conditions, the operator gene does not interfere with the promoter and transcription goes on. The final element, the fourth gene, is the key to the operon's functioning.

The fourth gene of the lactose operon is the **regulator gene,** designated *i* for its role as an inhibitor.

The *i* gene is the key to the control process. In all cells, induced or not, it constantly but very slowly transcribes mRNA. The mRNA codes for a **repressor protein,** which is produced in very small quantities (about 20 to 40 molecules per cell). It is called a repressor because it binds to the operator, blocking the action of RNA polymerase along the promoter

10.22 THE TRYPTOPHAN OPERON

(a) The repressible tryptophan operon works in the opposite way of the inducible lactose operon. In this system, cells grown in glucose continually produce the enzyme tryptophan synthase ①, which is essential in producing the amino acid tryptophan. Although the regulator gene produces the repressor protein ②, it cannot bind the operator gene in its present form. Therefore, the system remains turned on.

(b) If the cells are fed the amino acid tryptophan ③, the system immediately shuts down. Tryptophan joins the repressor protein ④ to form a tryptophan-repressor complex. This complex, in turn, coats the operator, blocking the action of RNA polymerase along the promoter. ⑤ This shuts down the structural genes ⑥ that produce tryptophan synthase. As you can see, this is a conserving process. Why produce the enzyme when its product is abundant?

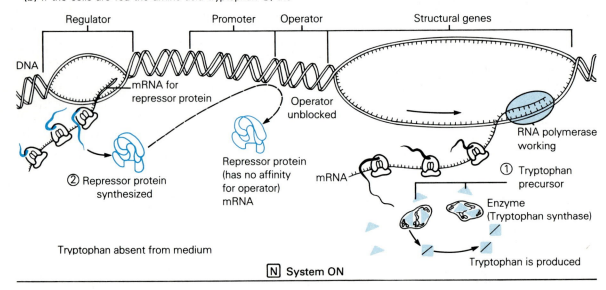

Tryptophan absent from medium

N System ON

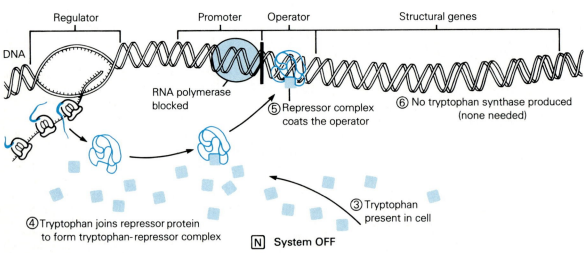

N System OFF

and inhibiting transcription. In effect, the structural genes are turned off while the repressor protein is present and active in the cell.

However, the repressor protein has an even greater affinity for the sugar lactose. When lactose is present in the cell, some of it will bind to the small number of repressor protein molecules, effectively removing them from the operator. This frees the promoter region and RNA polymerase becomes functional and transcription of the structural genes begins. The messenger RNA transcribed becomes translated on the bacterial ribosomes into the three enzymes. This, then, is the way lactose acts as an *inducer* of gene action (see Figure 10.21). (Actually, today we know that the inducer is a variant of lactose called *allolactose*. But since this is formed from lactose, the explanation is still valid.)

Perhaps it will help to consider a fanciful analogy. Suppose RNA polymerase is a train that must run along a DNA track. The repressor protein is an elephant that has a penchant for sitting on the track. However, when peanuts (lactose) are available, the elephant lumbers off to eat the peanuts, and the train may at last proceed. Thus, the availability of peanuts controls whether the train will run or not.

The next question is, how is the lactose-operon system shut down? When the lactose-metabolizing enzymes hydrolyze the lactose molecules into glucose and galactose, lactose is no longer available to tie up the repressor still slowly being produced. Thus, the repressor protein is once again free to bind the operator gene, and so, transcription of the three structural genes is halted. As you see, the lactose operon exercises negative control; the inducer molecule, lactose in this case, makes something happen by repressing a repressor.

The Tryptophan Operon. The second operon to which Jacob and Monod turned their attention was the **tryptophan operon,** which controls a series of enzymes that are needed for the synthesis of the amino acid tryptophan. They were interested because the cell needs a constant supply of tryptophan for protein synthesis. They found that the tryptophan operon functions continually *except* when there is ample free tryptophan already in the medium (Figure 10.22).

Jacob and Monod learned that the tryptophan operon is also under negative control; however, in this case the repressor protein will *not* bind to its operator *unless* it first binds to a tryptophan molecule. So the tryptophan operon is a *repressible* operon and tryptophan is its *repressor*.

Many other operons have been discovered since 1961. Alas, each seems to operate on its own principles. Some operator sequences, for example, have multiple binding sites for repressor molecules, allowing for graduated responses; that is, the gene doesn't necessarily have to be either "turned on" or "turned off," but is in effect controlled by a "dimmer switch."

Gene Control in Eukaryotes

For the most part, gene control systems in higher organisms tend to be quite different. In recent years, however, experimenters have found some control systems in higher organisms that are similar to gene control in bacteria. For example, steroid hormones can control gene activity rather directly.

The oviducts (egg-conducting tubes) of baby chicks have been found to be responsive to the steroid estrogen, the female sex hormone. In the presence of estrogen the chick oviduct will begin to produce albumin and other egg white proteins. Radioactively labeled estrogen molecules have been allowed to enter the cell, where they are bound by a specific cytoplasmic receptor protein. The protein-steroid complex then passes into the nucleus, where it binds tightly to a nonhistone chromosomal protein. This complex then initiates mRNA transcription through positive control.

The system is similar to that of a bacterial operon. But even this, the simplest of all known eukaryote gene control systems, is much more complex than the lactose operon. To illustrate, estrogen doesn't induce the synthesis of egg white proteins in any other kind of chick cell; apparently only oviduct cells have the specific chromosomal receptors. Also, estrogen may have very different effects on different genes, suppressing the synthesis of some proteins as it induces the synthesis of others.

We have come a long way since Sir Francis Crick first gave us our "central dogma." The euphoria that swept the scientific community in those days, when a simple system of genetic coding was unveiled, has been dampened by the realization that the system is not so simple after all. But, the very complexity of life that discourages some people stimulates others. They see the complexity, the variety of life, as providing not just challenges, but opportunities. Some of the greatest opportunities lie before us now. We now have the technical capability to manipulate genes—to turn them on, to turn them off, to move them about, and to make them work for us in very critical and specific ways, as we will see in chapters to come.

APPLICATION OF IDEAS

1. It has recently been shown that UGA codes for tryptophan and CUA codes for threonine in the mitochondrial translation systems of yeasts and hamsters. What do these codons usually specify? What does that do to the universal code? Comment on the significance, if any, of these minor departures from the code.

2. With the increased use of nuclear power, the subject of harmful ionizing radiation has once more become a public issue. Such radiation is known to alter DNA in random ways. Its effects, which are accumulative, have two aspects, medical and genetic. Distinguish between the two in terms of potential dangers. Is human life the only concern? Comment on the long-term meaning of small accumulative effects.

3. Proponents of nuclear energy are comparing the risk of a nuclear power plant accident with the risk of airplane and highway travel and industrial fires. What, if any, is the flaw in this kind of argument? Consider accumulative effects of radiation and the fact that we are only one of millions of species.

KEY WORDS AND IDEAS

RNA STRUCTURE AND TRANSCRIPTION

Comparing RNA to DNA

1. RNA differs from DNA in that it contains the sugar **ribose,** substitutes **uracil** for thymine, is single-stranded, readily breaks down, and exists in several forms.

Transcription: RNA Synthesis

1. Copying the code from DNA **(transcription),** occurs on the **transcribed strand,** leaving the opposing **nontranscribed,** or **noncoding strand,** idle. Transcription is **conservative.**

2. In transcription, **RNA polymerase,** the principal enzyme, binds to a DNA **promoter** region, the helix unwinds, and, proceeding in the 5'-to-3' direction, RNA triphosphate nucleotides are paired to the exposed DNA bases. U is always substituted for T.

3. Transcription ends when RNA polymerase encounters a **termination signal.** Multiple transcription along a DNA strand is common.

Varieties of RNA:
Physical and Functional Classes

1. The three kinds of RNA are **ribosomal (rRNA), messenger (mRNA),** and **transfer (tRNA).** All must undergo modification from the **precursors** to the final mature form.

2. Ribosomal RNA makes up most of the ribosome, a two-part, interlocking organelle that is the site of protein synthesis. Each ribosome has attachment sites for mRNA and tRNA.

3. In eukaryotes, the long **primary mRNA** transcript undergoes **posttranscriptional modification,** where noncoding segments called **intervening sequences,** or **introns,** are removed, leaving only coding segments called **expressed sequences,** or **exons.** Introns are absent in prokaryotes.

4. Eukaryote mRNA contains a **capped** or **methylated 5' leading region** and a **3' trailing region** or **poly-A tail** (many adenine nucleotides). Prokaryotic mRNA lacks the cap and tail and is **polycistronic** (contains several **cistrons,** polypeptide coding regions).

5. The linear arrangement of bases in mRNA constitutes the **genetic code.** The amino acids are specified by **codons**—nucleotides in groups of three. There are 64 possible codons, so most amino acids have more than one **(synonymous)** specifying codon. One codon, the **initiator,** specifies both *start* and an amino acid, while three codons specify *stop.*

6. **Charging enzymes** bond tRNAs to their specific amino acids. Each tRNA has a specific amino acid binding site, a ribosomal binding site, and an **anticodon loop.** The anticodon matches a codon on mRNA, assuring that the amino acid carried by a tRNA will be inserted correctly in the polypeptide.

7. The primary tRNA transcripts are first *tailored,* then folded into a cloverleaf secondary shape, and finally folded again into the tertiary "L" shape.

TRANSLATION: HOW POLYPEPTIDES ARE ASSEMBLED

1. Translation includes **initiation, elongation,** and **termination.**

Initiation

1. Initiation requires the smaller ribosomal subunit, an **initiator tRNA** (with a 5'—CAU—3' anticodon), and the mRNA initiator codon (5'—AUG—3'), all of which form the ribosomal **initiation complex.**

2. When each component is in place, the larger ribosomal subunit joins the complex and a second amino acid can be inserted.

3. Polypeptides in eukaryotes all begin with methionine, and each has an *N*-terminal (NH$_2$) **end** and a *C*-terminal carboxyl (COO$^-$) **end.**

Elongation

1. The elongation of polypeptides occurs through **translocation**—the formation of the peptide bond and the movement of a tRNA from the right to the left ribosomal pocket.

2. As a polypeptide grows, a charged tRNA whose anticodon matches the mRNA codon in the right pocket becomes attached. A peptide bond forms between its amino acid and the last one in the polypeptide above, and translocation occurs again.

3. Following translocation, the tRNA in the left pocket is released and drifts away to recycle.

Polypeptide Chain Termination

1. When the ribosome reaches a chain **terminator** (stop) codon, proteins block the pockets and the final tRNA is released along with the completed polypeptide.

Polyribosomes

1. **Polyribosomes,** or **polysomes,** occur in clusters on one mRNA molecule, where they carry on simultaneous translation.

Free and Bound Ribosomes

1. **Bound ribosomes** are located along the rough endoplasmic reticulum (and on the plasma membrane of prokaryotes). Their polypeptides enter the ER.

2. Polypeptides destined to enter the ER are tipped by a **leader sequence** or **signal peptide.** It binds to a **signal recognition protein** and together they contact a **receptor site** on the ER surface. The recognition protein is released and the polypeptide enters the lumen, where its leader is removed. As the polypeptide grows it binds the ribosome to the ER.

MUTATION

The Stability of DNA

1. Because it lacks reactive side groups and its bases are tucked into the double helix, DNA is not very reactive. Irreversible changes that do occur in DNA are called **lesions,** and unrepaired lesions become **mutations.** Agents that cause mutations are called **mutagens.**

2. The presence of **histones** in eukaryote chromosomes and the redundancy of base-pairing reduce the incidence of mutation.

3. Most lesions are repaired by **DNA repair systems.** Two ways in which repairs are made are through a DNA polymerase "recheck" mechanism and through **excision repair** with **ligase.**

Mutations at the Molecular Level

1. **Chromosomal mutations** involve entire chromosomes, while **point mutations** are minor changes in DNA sequences. Many are **silent mutations**—harmless changes in amino acid sequences.

2. Many mutations are undetected or not well understood, since most DNA is made up of noncoding, so-called spacer DNA.

3. Point mutations include **base substitutions, insertions,** and **deletions.**

4. Base substitutions change a codon, and unless the change produces a synonymous codon, an amino acid substitution will occur in the polypeptide (a **missence mutation**). The result may be negligible or it may produce a serious condition such as **sickle-cell anemia.** In this disorder a single base substitution brings about the collapse of red blood cells.

5. Base substitutions producing an unwanted *stop* codon, or **chain-termination mutations,** stop polypeptide synthesis too early. If it is the *stop* codon that is altered, the mRNA poly-A tail will add amino acids to the polypeptide and it may "lock up" the ribosome as termination fails.

6. Base insertions and deletions can produce highly abnormal proteins with lethal effects on the cell. They cause **reading-frame shifts,** in which all codons from the insertion or deletion onward will be altered.

Retroviruses and Transposable Elements

1. **Retroviruses** and **transposable elements** represent cases of random insertions and deletions of base sequences. In an exception to the central dogma, they have the effect of transcribing RNA into DNA, which can then be reinserted into the chromosome. The DNA transcribing enzyme is **reverse transcriptase.**

2. Both retroviruses and transposable elements can be passed through generations and some include cancer-causing **oncogenes.**

3. They are considered to be mutations since their presence causes the gene to fail.

MECHANISMS OF GENE CONTROL

1. Of the 40,000 to 50,000 protein coding genes in humans, and 4000 to 5000 such genes in bacteria, many are active only part of the time, suggesting a controlling mechanism.

2. While prokaryote gene controlling mechanisms are well understood, not much is known about them in eukaryotes.

Gene Organization and Control in Prokaryotes: The Operon

1. The **operon** model of prokaryotic gene control was reported by Jacob and Monod in 1961.

2. The **lactose operon** in *Escherichia coli* is inducible, its three lactose-metabolizing enzymes (symbolized, *z*, *y*, and *a*) are only produced when a substrate is present in the cell.

3. Elements of the operon include the structural genes for producing enzymes *z*, *y*, and *a*, the operator (o), the **promoter (p)** and the **regulator gene (i).**

4. The lactose operon functions as follows:
   a. Enzyme synthesis repressed. The regulator produces a **repressor protein** that binds with the operator, blocking RNA polymerase so that it cannot interact with the promotor and begin transcription.
   b. Enzyme synthesis induced. Lactose enters the cell and binds the repressor protein, freeing the operator. RNA polymerase interacts with the promotor, transcription begins, and mRNA is translated into the three enzymes.
   c. Enzyme synthesis ends. The lactose-metabolizing enzymes hydrolyze lactose, breaking it down, thus releasing the repressor protein, which once more binds the operator and stops synthesis.

5. The **tryptophan operon** controls the production of enzymes that synthesize that amino acid. In this system the enzymes are synthesized continuously, even though a repressor protein is produced. Here, the repressor protein cannot bind the operator unless it first joins a tryptophan molecule.

Gene Control in Eukaryotes

1. Little is known about eukaryote gene control, but it is known that the steroid hormone estrogen enters oviduct cells in baby chicks, where it joins a cytoplamic receptor protein. Together they enter the nucleus, where they activate transcription of an albumen-specifying gene.

REVIEW QUESTIONS

1. List four ways in which RNA differs from DNA. (223-224)

2. If the sequence of nucleotides in a transcribed strand of DNA was A-A-G-C-C-T-T-A-G-G-C-A, what would be the sequence of nucleotides in the RNA transcript? (225)

3. Briefly describe the process of transcription. Include mention of the primary enzyme and the role of the promotor and termination signal. What, actually, are the last two elements? (224, 225)

4. What determines how many simultaneous transcriptions occur along a gene? Does this factor seem limiting in *E. coli?* Explain. (226)

5. Describe the organization of the ribosome as we now see it. (227)

6. Briefly discuss what happens to eukaryote mRNA during its posttranscriptional modification. What are introns and exons? (228)

7. Describe the three regions of the mature eukaryote mRNA transcript. In what three ways does prokaryote mRNA differ? (228)

8. Following the key points mentioned, describe the organization of the genetic code. (a) What is a codon? (b) How many codons are there? (c) Since only 20 codons are required, what happens to the extras? In what molecule do we find the code written? (229)

9. List the steps and the molecules involved in the charging of a tRNA. (230)

10. Briefly describe the form taken by tRNA in its three levels of organization and identify three key regions. (230, 232)

11. List the different elements of the initiation complex. Explain how they get together to initiate translation. (233)

12. Using a simple drawing, show your understanding of chain elongation by illustrating the three steps involved. (233-234)

13. Briefly explain how chain termination occurs. (235)

14. What are polysomes? How does their presence affect the efficiency of protein synthesis? (236)

15. Are bound ribosomes really part of the ER as they seem to be? Explain fully. (236)

16. Briefly describe the relationship between signal peptides, signal recognition proteins, and ER receptor sites. (238)

17. In what ways does the double helix itself help prevent lesions or mutations from occurring? How do histones help? (240)

18. Describe the activity of DNA repair systems. What characteristic of DNA makes such a system feasible? (240)

19. How does a point mutation differ from a chromosomal mutation? Which tends to be more serious? (241)

20. Describe an instance in which a base substitution would have no effect on a polypeptide. What, if any, adaptive significance does this suggest? (241)

21. Using an example, explain how a single base substitution could be disastrous. (242)

22. Why would a base insertion or deletion tend to be more serious than a base substitution? (243)

23. What effect on a polypeptide would a mutation in the *start* or initiator codon have? Could this be detected? Explain. (242)

24. How does a repressible operon differ from one that is inducible? Be specific. (244)

25. Carefully explain the role of the following in the lactose operon of *E. coli:* (a) promoter, (b) regulator, (c) operator, (d) structural genes, (e) repressor protein. (244, 246-247)

26. Describe the gene-arousing action of the steroid hormone estrogen in baby chicks. (247)

The metabolically active composite cell is involved in seemingly endless activity, beginning with ciliary movement and transport. Materials cross the plasma membrane through streaming plasmodesmata via the sodium/potassium exchange pump and through endocytosis. Energy for the support of these and other chemical activities is supplied by chloroplasts and mitochondria, where ATP is produced through chemiosmosis. Nuclear activity includes both replication and transcription, the latter

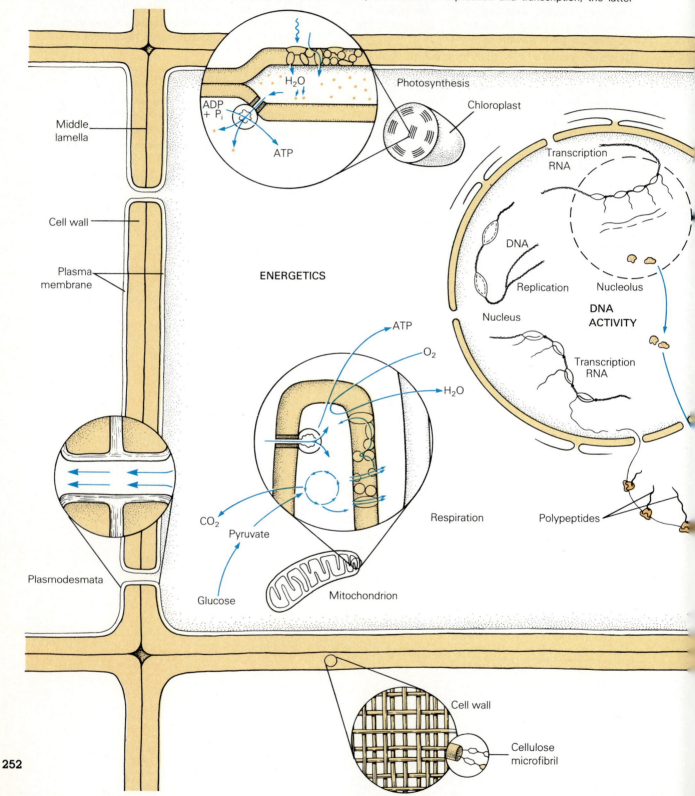

Middle lamella

Cell wall

Plasma membrane

Plasmodesmata

H_2O

$ADP + P_i$

ATP

Photosynthesis

Chloroplast

ENERGETICS

ATP

O_2

H_2O

CO_2

Pyruvate

Glucose

Respiration

Mitochondrion

Transcription RNA

DNA

Replication

Nucleolus

Nucleus

DNA ACTIVITY

Transcription RNA

Polypeptides

Cell wall

Cellulose microfibril

occuring in both the nuclear fluids and within the nucleolus. Ribosomes and RNA produced there are put to work in the cytoplasm, where bound ribosomes are seen spinning raw polypeptides into the rough ER channels. The products then enter transport vesicles, which join the forming faces in the Golgi membranes. Following final chemical modification, the finished products enter secretion vesicles pinched off the opposite face and travel to the plasma membrane for exocytosis.

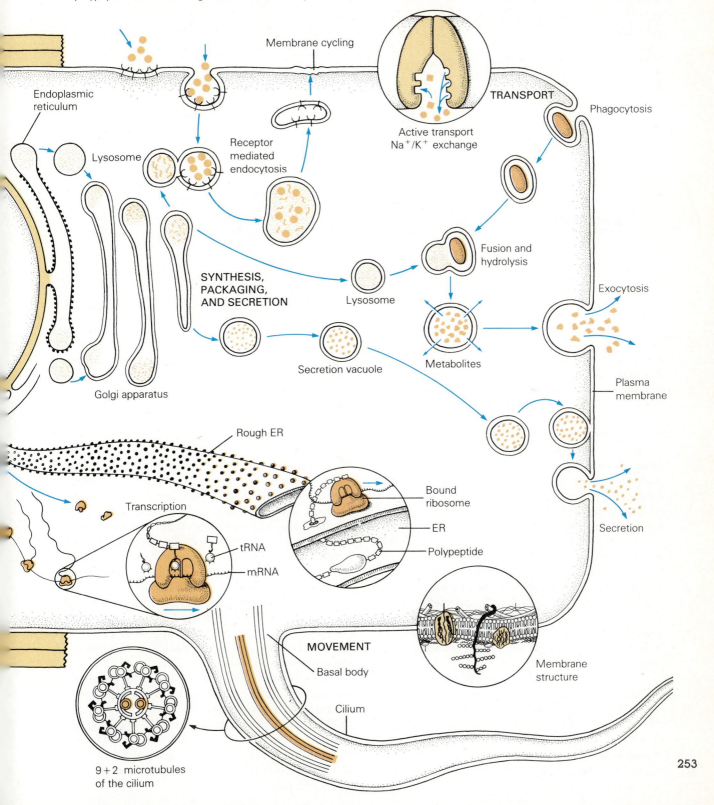

Membrane cycling

TRANSPORT

Active transport
Na+/K+ exchange

Phagocytosis

Endoplasmic
reticulum

Lysosome

Receptor
mediated
endocytosis

Fusion and
hydrolysis

SYNTHESIS,
PACKAGING,
AND SECRETION

Lysosome

Exocytosis

Secretion vacuole

Metabolites

Plasma
membrane

Golgi apparatus

Rough ER

Bound
ribosome

Transcription

ER

tRNA

Polypeptide

mRNA

Secretion

MOVEMENT

Basal body

Membrane
structure

Cilium

9 + 2 microtubules
of the cilium

Mitosis, Meiosis, and the Cycle of Life

You are not the same person you were a few years ago. In fact, you aren't the same person you were a few seconds ago. This is not to say that the first sentence is so profound that you will never be the same for having read it. It's just that even as you sit here reading, your cells are growing, dying, multiplying, and dividing. Your body is coping with a rigorous world and it must change or perish. Some cells, of course, are more active in such changes than are others. For example, the palms of your hands, and any other parts of your body that receive a great deal of friction, must constantly replace worn cells. Other parts of your body, however, have very low cell turnover. And some, such as the skeletal muscles and nervous system, are never replaced in adults. Nonetheless, at this very moment, your cells are quietly obeying their genetic overlords, many of them slowly and methodically reorganizing themselves and dividing again and again, duplicating themselves ever so precisely in their unending pageant until that day when it all stops.

At this point, we will change our focus and concentrate on just how these things occur. We have had a look at the functioning of the DNA molecule—its structure, replication, and molecular role in cellular activity. We have also seen how its unique structure assures faithful replication in preparation for cell division. Now we will look more closely at the process of cell division, and for that we need to know a bit more about the nucleus. We will also leave the molecular details of the chromosome and begin to pay more attention to the chromosome as a whole.

NUCLEAR STRUCTURE

We considered the gross nuclear structure in Chapter 4, so we will only briefly review some major points here. Recall that the fluid interior of the nucleus, the **nucleoplasm,** is surrounded by the double layered membrane called the **nuclear envelope.** The outer membrane of the nuclear envelope is continuous with the endoplasmic reticulum, and the space between the outer and inner membrane, the **perinuclear space** (*peri,* surrounding), is continuous with the ER lumen. The close relationship of the nuclear envelope and the ER may explain the envelope's peculiar ability to disappear quickly during cell division, and to reappear just as rapidly at the end of the process.

Because of the great selectivity of the envelope, materials within the nucleus are usually quite different from cytoplasmic materials. However, when the nuclear envelope disappears from view during cell division, materials in the nucleoplasm mix with those of the cytoplasm. Apparently, at this time in the life of a cell, it is not so important that the pristine nucleoplasm be separated from the rough-and-tumble world of the cytoplasm. The chromosomes are somewhat protected because now they are tightly coiled into the rather inert form they assume during cell division.

DNA, Chromatin, and the Eukaryote Chromosomes

In discussing the life of a cell, terminology is important, so some definitions are in order. DNA was defined earlier, but now we see its relationship to **chromatin** and **chromosomes.** Chromatin is DNA that is complexed (joined) with proteins. These proteins are of two types—the **histones** and others known, logically enough, as **nonhistone chromosomal proteins.**

Chromosomes are chromatin that is condensed (tightly coiled) into large, highly visible bodies. This is the form they take during cell division. When the chromosomes "relax" (or, more properly, become diffuse), they form very long, thin threads (Figure 11.1). In fact, if a DNA molecule from the largest human chromosome were straightened out, it would be about 12 centimeters long. Most of the time, DNA is in the diffuse form.

11.1 THE EUKARYOTE CHROMOSOME

In eukaryotes, the DNA strand is wrapped around numerous globules of histones, creating a beadlike appearance. Each bead, or nucleosome, contains about 140 nucleotide pairs on its surface, while the strand between beads consists of 50 nucleotide pairs. The one long DNA molecule, its histone complex, and its associated nonhistone proteins constitute a single eukaryotic chromosome. The chemical substance of eukaryotic chromosomes in general is termed *chromatin.*

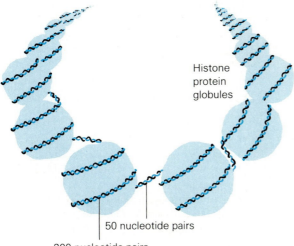

Histone protein globules

50 nucleotide pairs

200 nucleotide pairs

11.2 THE CHROMATIN NET

The only structural features of the interphase nucleus are the dark-staining nucleoli and the netlike distribution of chromatin, representing the diffuse chromosomes.

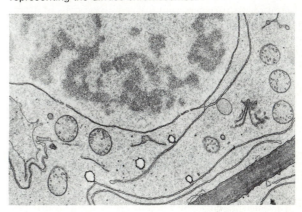

When the DNA is associated with histones, together they form tiny, beadlike globules called **nucleosomes,** each consisting of about 140 DNA base pairs wound around each cluster of histone. Between the nucleosomes is a string of DNA about 50 nucleotide pairs in length. Each chromosome, then, consists of one long DNA molecule wrapped around countless beads of histone. (Later, however, we will see that there is a time following replication when each chromosome will consist of two DNA molecules.)

The mass of chromatin in nondividing cells, which resembles a bundle of yarn subjected to the attentions of a demented kitten, is sometimes referred to as the **chromatin net** because of its netlike appearance under the electron microscope (Figure 11.2). Parts of the chromatin net are thin and diffuse, while other parts are thicker and more tangled. Only the DNA in the diffuse portions of the net functions in gene activity—that is, transcribing RNA. The condensed portions are temporarily inactive.

The Cell Cycle: Growth, Replication, and Division

A cycle is endless, and the **cell cycle** is an endless repetition of **growth, chromosomal replication, mitosis, cytokinesis,** and then, more growth, chromosomal replication, and mitosis. Not all cells follow a cycle continuously. Some cells eventually break out of their cycle, but when they do, they pay dearly. As an example, fingernail cells become filled with keratin (a tough protein) and then die, only to be pushed along by dividing cells further back. The nuclei of red blood cells in birds, reptiles, and fish also are permanently inactivated—turned off, so to speak—so that once formed, these cells will die without undergoing another division. It should be even more apparent that the red blood cells of mammals, which have no nuclei, are also living on borrowed time. Thus, there are exceptions (biology is full of exceptions), but in this chapter we will focus on cells that cycle.

Stages of the Cell Cycle. The cell cycle is traditionally divided into four phases: the **M phase, G$_1$ phase, S phase,** and **G$_2$ phase** (Figure 11.3). The M stands for **mitosis** (and includes cell division), S for **synthesis** of DNA, and G$_1$ and G$_2$ for **gap one** and **gap two,** respectively.

Briefly, in the M phase (mitosis), the chromosomes condense, the nuclear envelope and nucleoli disappear, the weblike **mitotic spindle** forms, the two identical sister strands of each chromosome separate and go to opposite poles, two new nuclear

envelopes and two new sets of nucleoli form, and the cell divides into two daughter cells.

The other three phases G$_1$, S, and G$_2$, are collectively called **interphase.** The chromosomes decondense as each new daughter cell enters G$_1$, the first stage of interphase. This is generally a very active period, the time when the cell synthesizes the enzymes and structural proteins necessary for cell growth. In this stage, each chromosome consists of only a single unreplicated DNA strand with its associated histones and other chromosomal proteins.

The S phase is the period in which the DNA and chromosomal proteins are replicated. This phase typically lasts a few hours and can be distinguished in radioactive labeling experiments as the stage in which the cell nucleus will incorporate radioactive nucleotides such as thymine. Recall that thymine is found in DNA and not in RNA, so its uptake by the nucleus tells the experimenter when replication is occurring.

The gap 2, or G$_2$ phase is an important preparatory period preceding mitosis. The proteins of the

11.3 THE CELL CYCLE

This scheme represents two generations of cells, which follow a cyclical series of events as they divide, synthesize and grow, replicate, and prepare to divide again. Each cycle is divided into four major parts: mitosis (M), protein synthesis (G$_1$), DNA replication, or synthesis (S), and microtubular synthesis (G$_2$). Some cells are arrested in G$_1$, differentiating into specialized types that will no longer cycle.

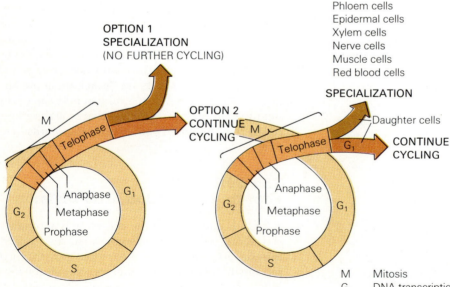

Phloem cells
Epidermal cells
Xylem cells
Nerve cells
Muscle cells
Red blood cells

SPECIALIZATION

OPTION 1
SPECIALIZATION
(NO FURTHER CYCLING)

OPTION 2
CONTINUE
CYCLING

Daughter cells

CONTINUE
CYCLING

M
Telophase
Anaphase
Metaphase
Prophase
G$_1$
G$_2$
S

M	Mitosis
G$_1$	DNA transcription–translation, growth
S	DNA replication–new chromosomes
G$_2$	DNA transcription–translation, spindle proteins

mitotic spindle are synthesized at this time, in preparation for the coming nuclear division. The mitotic spindle is an elaborate structure that is involved in chromosome movement during mitosis. It is constructed anew for each cell cycle, then dismantled after it has been used once (the name "spindle" comes from a fancied resemblance to the spindle of a hand loom). You may recall from Chapter 4 that the mitotic spindle is composed of tubulin, a protein consisting of two dissimilar spheres of polypeptides in a figure-8 form (see Figure 4.23). The tubulin spheres are used in the construction of the tubelike microtubules of the spindle.

In G_2, each chromosome consists of two DNA molecules and their related proteins joined at the **centromere** (to be discussed shortly). Each individual strand in this connected state is referred to as a **chromatid.** However, the two chromatids of a G_2 chromosome are so tightly bound that they cannot be distinguished visually until they appear during mitosis.

The relative amount of time the cell spends in interphase varies greatly according to the cell's type and stage of development. The shortest periods generally are found in the rapidly growing embryo. The large, newly fertilized frog egg, for example, will complete a cell cycle every 30 minutes, thereby producing about 8000 cells in 12 hours. In the cell proliferation of many early embryos, there is no G_1 phase, so these large cells become progressively smaller, up to a point, with each division.

MITOSIS AND CELL DIVISION

It is important to realize that cell division has two parts: **mitosis,** the condensation and division of the chromosomes, and **cytokinesis** (cytoplasmic division), the actual division of one entire cell into two **daughter cells.**

In the interphase nucleus—that is, during G_1, S phase, and G_2—there is a great deal of biochemical activity, but not much of interest is seen with the light microscope. Presumably, this is why earlier microscopists called interphase the "resting phase." However, observations through the phase contrast microscope (see Essay 4.2) reveal a very busy cytoplasm, churning and seething, with bodies appearing and disappearing in a cauldron of activity. There appears to be almost total anarchy in the cytoplasm at this time, but organization is the key to life and we can be sure that all this activity is highly regulated.

Mitosis (M phase) is itself divided into four subphases: **prophase, metaphase, anaphase,** and **telophase,** briefly described as follows:

1. In *prophase*, the chromosomes condense, the nucleoli and nuclear envelope fade from view, and the **mitotic spindle apparatus** forms.
2. In *metaphase,* the chromosomes are brought to a well-defined plane, the **metaphase plate,** in the middle of the mitotic spindle. Their movement is believed to be produced by the microtubule spindle fibers, which attach to each chromosome at specific sites known as **centromeres.**
3. In *anaphase*, the centromere of each chromosome separates and the two daughter chromosomes of each pair travel to opposite poles (ends) of the spindle.
4. In *telophase*, new nuclear envelopes form around each group of daughter chromosomes, the nucleoli appear, the chromosomes decondense, and finally the cytoplasm of the cell divides to form two daughter cells (Figure 11.4 and Table 11.1). (Incidentally, in some fungi and protozoa, the nuclear envelope does not break down during mitosis, but pinches into two daughter nuclear envelopes near the conclusion of cell division.)

Now for the details.

Prophase

Interphase ends and mitotic prophase begins with the onset of two events: chromosome condensation and the first steps in the organization of the mitotic spindle apparatus.

Chromosome Condensation.   From what one can tell with a microscope, chromosome condensation essentially involves coiling and supercoiling in each DNA strand (Figure 11.5). Supercoiling can be illustrated by the tungsten filaments of those large, clear light bulbs. The filament is basically a helix whose strand is composed of a much smaller, tighter helix. In fact, in mitotic chromosomes, microscopists have described a helix within a helix within a helix.

Why do chromosomes condense so tightly in mitosis? ("Why" questions in biology usually mean, what good is it?) There is a logical answer, though it would be hard to prove. It is probably because chromosomes that are condensed into tight

11.4 MITOSIS IN ANIMALS AND PLANTS

(a) Mitosis in animals. Mitosis is a highly organized process in which the products of replication, the chromatids, are separated at their centromeres and exactly and equally divided between the emerging daughter cells. The process is divided into five phases, each with its specific events, as shown here with light microscope photographs and drawings. The cells shown here are from an early embryo stage in the whitefish.

① In interphase, the nucleus shows no hint of impending events, but a vast amount of biochemical activity precedes mitosis. Each DNA molecule replicates and gradually condenses into visible chromosomes.

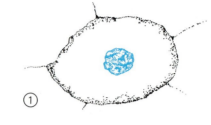
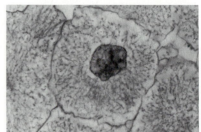

② Chromosomes begin to condense in early prophase.

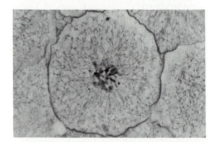

③ In midprophase the nucleolus has disappeared and chromosomes are clearly visible.

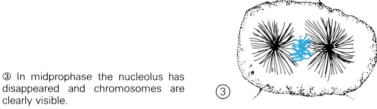
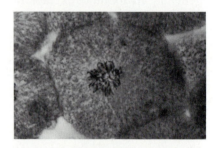

④ Centromeres in metaphase are aligned on the metaphase plate.

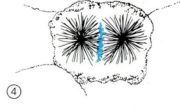
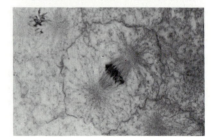

⑤ Anaphase chromosomes move rapidly apart as the fibers connected to their centromeres shorten. Can you distinguish the chromatids?

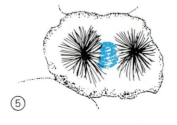
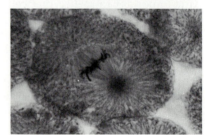

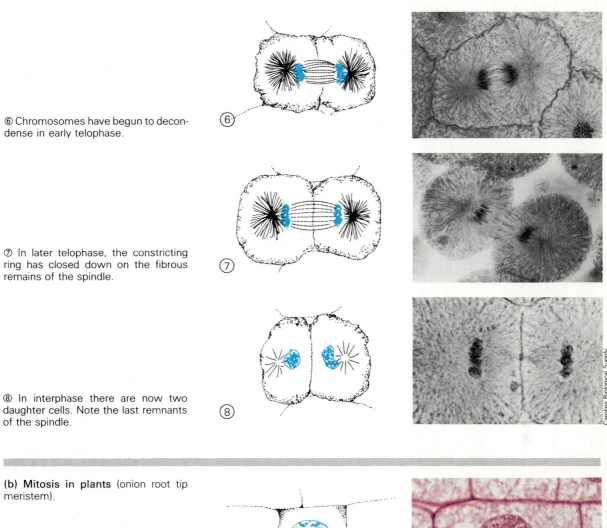

⑥ Chromosomes have begun to decondense in early telophase.

⑦ In later telophase, the constricting ring has closed down on the fibrous remains of the spindle.

⑧ In interphase there are now two daughter cells. Note the last remnants of the spindle.

Carolina Biological Supply.

(b) Mitosis in plants (onion root tip meristem).

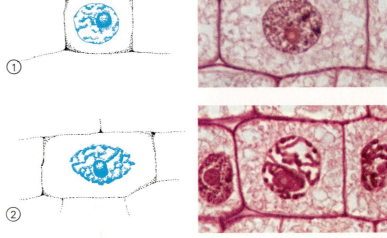

① In interphase the chromatin is diffuse and little detail can be seen. The dark-staining nucleoli stand out sharply within the nucleus.

② In early prophase the nuclear envelope is still in place, but the chromatin has condensed into distinct chromosome bodies. In later prophase the nuclear envelope and nucleoli will no longer be seen and chromosome condensation has continued.

11.4 (continued)

③ The chromosomes are very distinct in metaphase as they line up on the metaphase plate. This is a brief interlude and some of the centromeres have already divided.

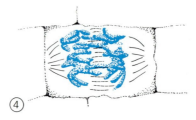

③

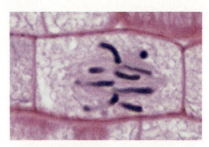

④ Centromere division is complete in early anaphase.

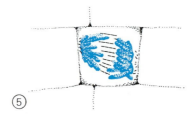

④

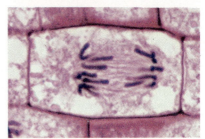

⑤ In later anaphase the daughter centromeres have been drawn to opposite sides of the cell.

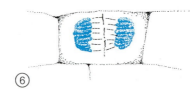

⑤

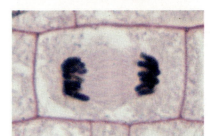

⑥ In early telophase the individual chromosomes begin losing their identity as decondensation begins. The newly forming cell plate can be seen between the two nuclei.

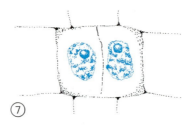

⑥

⑦ The chromosomes continue their decondensation and nucleoli are just visible. Rigid cell walls have now formed.

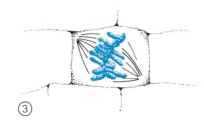

⑦

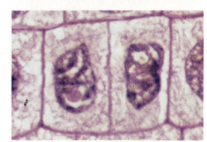

sausagelike bundles (Figure 11.6) are simply easier to move around the cell than the long, wispy strands of uncondensed chromosomes. You may well ask, then, why the chromosomes don't simply remain in their shortened, coiled form all the time, even through interphase. The answer to that seems to be that uncoiling is a prerequisite to action. Like many of us, chromosomes can't do their job when they're all wound up. Their job, of course, is to direct protein synthesis.

Chromatids, Chromatin, and Centromeres.
By the time a cell enters prophase, its single strands of chromatin comprising each G_1 chromosome have doubled (in the S phase) so that each chromosome now consists of two chromatids (see Figures 11.5a and 11.6). The doubled structure is still considered to be one chromosome (consisting of two chromatids) until anaphase, when it divides into two chromosomes, each with one strand of chromatin. There is no such thing as a chromosome with only one chromatid, since the term *chromatid* is defined as one subunit of a doubled chromosome. Keeping this in mind will avoid some of the common pitfalls in the language of mitosis. Problems arise when one refers to the number of chromosomes in the cells of this species or that. The species chromosome number, like many of its characteristics, is highly specific, as you can see in Table 11.2. Chromatids are never individually counted in determining that number. (One trick to counting chromosomes is to count centromeres.)

The two chromatids of the chromosome are not visibly distinguishable in early prophase and, in the cells of many tissues, they do not become distinct until metaphase or even early anaphase. In some cells, however, the two chromatids become quite distinct by late prophase. Each is supercoiled individually, and, by metaphase, they are held together only at the centromere.

TABLE 11.1

MITOSIS AND CELL DIVISION

Interphase	Prophase	Metaphase	Anaphase
Chromosomes are decondensed	Chromosomes condense	Spindle is fully formed	Centromeres divide
S phase: DNA and chromosomal proteins synthesized	Separate chromatids may become visible	Chromosomes are aligned with their centromeres on the *metaphase plate*	Sister chromatids separate to become chromosomes
Spindle proteins and other mitotic proteins formed in G_2 phase	Asters and mitotic spindle begin to form	Centromeres begin to divide	Daughter chromosomes go to opposite poles
Cell increases in volume	*Nuclear membrane* breaks down		Centromeric spindle fibers shorten; the spindle as a whole elongates
	Nucleolus disperses		
	Centromeres become attached to the *centromeric spindle fibers*		
	Chromosomes migrate toward the *metaphase plate*		
Telophase	Cytokinesis (cytoplasmic cell division)[a] in animals	Cytokinesis (cytoplasmic cell division)[a] in plants	Interphase
Chromosomes begin to decondense	Microfilaments associated with the cell membrane form a circular band around the cell	Small membrane vesicles fuse in the plane of the previous metaphase plate to form the *cell plate*	Daughter cells are in the G_1 stage of interphase (one DNA molecule per chromosome) after mitosis
Nuclear membranes reform	Microfilaments contract, pinching apart daughter cells	Cell plate grows to separate daughter cells	Chromosomes become diffuse
Nucleoli reappear			
Spindle disappears		*Middle lamella* and *cell walls* are laid down	
Cytokinesis (cytoplasmic cell division) usually occurs			

[a]Not a feature of mitosis proper.

11.5 SUPERCOILING AND THE LEVELS OF CHROMOSOMAL ORGANIZATION

The condensed chromosome (a), easily visible through the light microscope, is a supercoil with several levels of organization, each a coil of its own. With the greater magnification in part (b), the chromosomal arms are seen to consist of a thickened helix. (c) Examining a small segment of this supercoil reveals that it consists of an incredibly long thread of DNA coiled back and forth as it is packed into the supercoil. (d) Yet more coiling of the chromosome strand. Here a short length coils back on itself. (e) The nucleosomes: DNA coiled around histone molecules. Structure at this level of organization can be resolved by the electron microscope, but its final helical organization cannot. The familiar DNA helix must be determined using techniques such as X-ray diffraction. (f) Thus, the last level of coiling is the double helix of DNA.

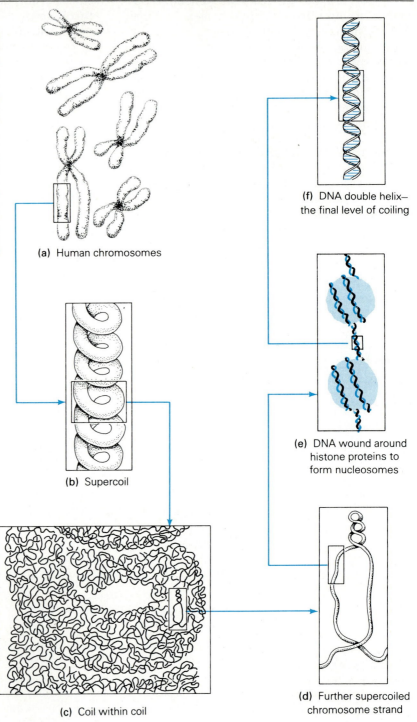

(a) Human chromosomes

(b) Supercoil

(c) Coil within coil

(d) Further supercoiled chromosome strand

(e) DNA wound around histone proteins to form nucleosomes

(f) DNA double helix— the final level of coiling

11.6 METAPHASE CHROMOSOME

The protrusions sticking out of the sides of this metaphase chromosome are loops coiled back on themselves, and not broken pieces. This chromosome consists of two chromatids joined at the centromere, permanent regions on chromosomes where spindle fibers attach.

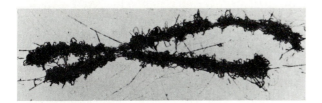

The centromere, as seen under the electron microscope, is a constriction between the thickened chromatids, seemingly squeezing them together, rather like a napkin ring holding two napkins. Some cytologists suggest that the centromere may simply be a region where DNA replication has not been completed, so only a single strand exists. should this idea be substantiated, it would explain the constricted appearance of the centromere. In stained preparations, the centromere appears under the light microscope as a prominent construction separating the chromosome into "arms." The spindle fibers attach to the centromere, and their actual sites of insertion appear to be two thin, discrete plates of protein on either side of the centromere, known specifically as **kinetochores** (Figure 11.7).

The chemical structure of the centromere is unknown, but it is considered to be a permanent part of the chromosome. It always appears at exactly the same place in any given chromosome, presumably under the direction of DNA. The constancy of the placement of the single, highly visible centromere is one of the most reliable characteristics used to identify specific chromosomes.

The Mitotic Spindle.   The mitotic spindle is architecturally at the very core of mitotic movement. It consists primarily of the protein tubulin, common in cilia and flagella. In fact, in the late G_2 phase, tubulin accounts for as much as 10% of the total protein of the cell. These proteins assemble themselves into **microtubules** (see Figure 4.23) at some unknown signal (Figure 11.8). The spindle, then, is a delicate structure composed of microtubules.

During interphase, the *centrioles* (see Figure 4.21) look like a pair of bright dots under the light

TABLE 11.2

CHROMOSOME NUMBERS[a]

Alligator	32	*Hydra*	32
Ameba	50	Lettuce	18
Brown bat	44	Lily	24
Bullfrog	26	Magnolia	38, 76, 114
Carrot	18	Marijuana	20
Cat	32	Onion	16, 32
Cattle	60	Opossum	22
Chicken	78	*Penicillium*	2
Chimpanzee	48	Pheasant	82, 81
Corn	20	Pigeon	80, 79
Dog	78	Planaria	16
Dogfish	62	Redwood	22
Earthworm	36	Rhesus monkey	42
Eel	36	Rose	14, 21, 28
English holly	40	Sand dollar	52
Fruit fly	8	Sea urchin	40
Garden pea	14	Starfish	36
Goldfish	94	Tobacco	48
Grasshopper	24	Turkey	82, 81
Guinea pig	64	White ash	46, 92, 138
Horse	64	Whitefish	80
House fly	12	Yucca	60
Human	46		

[a]There is no apparent significance to chromosome number as far as biologists can determine. If you feel good about your 46 (see human being), check the turkey, ameba, cattle, tobacco, and yucca. Note the variation in some plant and animal species. Plants undergo spontaneous doubling and tripling of chromosome number.

11.7 SPINDLE FIBER ATTACHMENT

In this high-magnification electron micrograph, spindle fibers, hollow microtubules, are seen attached to the curved, plate-like kinetochores, part of the centromere vaguely outlined here.

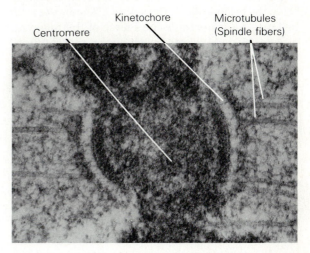

Centromere   Kinetochore   Microtubules (Spindle fibers)

11.8 THE MITOTIC SPINDLE

The mitotic apparatus in the young embryo of a whitefish (pictured at left) includes two prominent asters—numerous microtubules radiating outward from the spindle poles, along with the mitotic spindle. At the center of the spindle are the dark-staining chromosomes, clustered at the metaphase plate. The centromeres, clearly seen in the drawing, are not very visible at this magnification.

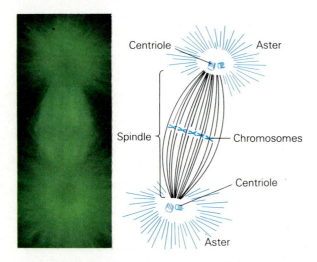

microscope. But under the electron microscope each is revealed as a pair of complex short cylinders also made up of microtubules.

At the onset of mitosis the two centrioles, which lie close to each other near the nucleus, separate and migrate to opposite sides of the nucleus. They come to rest in what are known as the **spindle poles,** areas from which the microtubules of the spindle radiate (see Figure 11.8). The position of the two centrioles was traditionally believed to determine the orientation of the spindle. They were also believed for years to be the source of the microtubules of the mitotic spindle. However, most authorities agree today that there is no direct relationship between the centrioles and the spindle. In fact, the higher plants that lack centrioles (primarily the conifers and flowering plants—those that also lack any form of motile cells), still have well-organized spindles and spindle poles. This ambiguity has remained a source of irritation to generations of biologists. But recent investigations clearly indicate that centrioles are not essential to spindle formation. For example, even if the centrioles are burned away with lasers, the mitotic spindle will form in a normal manner. For this reason, cytologists have turned to the clear, structureless region surrounding the centrioles as the site of microtubular assembly, although little is known about actual mechanisms. Nonetheless, in anticipation of new find-

ings, this region has been named the **microtubular organizing center (MOC).**

If centrioles are not responsible for the mitotic spindle, what are they doing in the middle of things and what do they do for a living? There is evidence that centrioles give rise to the **basal bodies** from which cilia and flagella emerge. (Basal bodies are very similar to centrioles in that they both are composed of a ring of nine groups of tubules; see Chapter 4.) But then, why do the centrioles migrate to the future poles of the spindle? The best answer we can come up with is that the process ensures that each daughter cell gets one.

Spindle Fibers. There are three different arrangements of microtubules in the spindle. One, the **centromeric microtubules** (or **centromeric spindle fibers**) originate in the microtubular organizing centers (near, but not attached to, a centriole). The centromeric spindle fibers extend out to attach to the chromosomes at the kinetochore, the specialized outer part of the centromere (see Figure 11.7). Note that a number of spindle fibers are attached to each kinetochore.

The second group of spindle fibers are known as **polar microtubules** (or **polar spindle fibers**). These fibers originate at opposite poles and cross the cell, whereupon opposing pairs of the microtubules overlap each other for a short distance before ending short of the other side. They may be closely associated with each other in the overlapping region, a point that has functional significance, as we will see. The polar spindle fibers were previously known as *continuous spindle fibers,* a misnomer from a time when they were thought to be single, continuous fibers. This was cleared up through electron microscope studies in the mid-1970s.

The third arrangement of microtubules is into two **asters,** whose **astral rays,** looking like starbursts, project outward from the area of the centriole. What do they do? Their chemical structure is the same as that of the spindle but we can only guess at their role. Perhaps they do nothing. Maybe they are just leftover microtubules that didn't get into the spindle or perhaps they anchor the spindle to the plasma membrane.

Asters, like centrioles, are absent in higher plants, so the search for their function becomes immediately more complicated. But perhaps the differences in plant cells and other cells suggest the role of the spindle. Flowering plants have relatively rigid cell walls, so perhaps the asters serve to make the cytoplasm of animal cells temporarily rigid. It is known that the spindle, asters, centrioles, and chromosomes do form a rather rigid mitotic apparatus. Two researchers, Daniel Mazia and S. K. Dan, astonished their colleagues in the late 1950s by isolating the mitotic apparatus intact. This was surprising because the spindle and asters had always been thought to be far too fragile. However, when the researchers carefully washed away the plasma membrane and the other loose cytoplasm with weak detergents, they were surprised to find the mitotic apparatus remaining as a feeble, but intact, framework.

Summary of Mitotic Prophase.   As we see, prophase is marked by the condensation of chromosomes and the formation of the mitotic spindle. In addition, we have noted that following the migration of the centrioles, the breakdown of the nuclear envelope, and the attachment of the microtubules to the centromeres, the mitotic apparatus is set to do its work. As a final part of prophase, the chromosomes are pulled by the spindle fibers from their more-or-less random prophase distribution to line up on a plane called the metaphase plate. The metaphase plate lies in the middle of the spindle and perpendicular to its axis. Actually, only the centromeres are in the plane of the metaphase plate; the arms of the chromosomes tend to hang off in all directions.

Metaphase and Anaphase

Metaphase is a brief pause in the continuous sequence of events that characterizes mitosis. The chromosomes are aligned on the metaphase plate, and to each are attached centromeric spindle fibers from each pole. Metaphase ends and anaphase begins when each chromosomal centromere divides. At one time it was believed that the centromeric separation was caused by tugging microtubules, but studies show that separation occurs even if the spindle fibers are cut. Upon separation, the newly divided chromatids—now referred to as chromosomes—are pulled rapidly to opposite poles by the spindle fibers. Since the point of attachment is the centromere, this region leads the way, dragging the chromosomal arms behind.

Some of the hypotheses about chromosome movement during anaphase are discussed in Essay 11.1. Whatever the process, the end result is that one copy of each chromosome ends up at each pole of the mitotic spindle. Each of the two nuclei that will form there will contain one of these copies. Thus, the daughter cells will contain identical genetic instructions. This, after all, is what mitosis is all about.

Telophase

Telophase is somewhat the reverse of prophase, and at its conclusion there will be two nuclei where there was one. The chromosomes decondense, unwinding from their superhelices and becoming diffuse chromatin once more. The mitotic apparatus is disassembled, its microtubules broken back down into actin and tubulin. Finally, the nuclear envelope is reassembled and the nucleoli reappear.

The sudden reappearance of the complex nuclear envelope is one of the ongoing mysteries of mitosis. It appears that during prophase, the nuclear envelope is actually disassembled into small, closed fragments closely resembling bits of ER. Then, during telophase, the fragments associate with the newly separated chromosomes, wrapping partially around each one and forming a tight mass. Then the fragments join together, the chromosomes tightly held within, and as the membranes expand, new nuclear pores are formed. Finally, the new nuclear envelope is complete. It has been suggested that the close association of envelope fragments and chromosomes may help in excluding cytoplasmic elements from the nucleoplasm as the new nucleus is forming. In any case, with the reappearance of the nucleus, telophase and mitosis come to an end and cytokinesis begins.

Cytokinesis: Cytoplasmic Division

Cytoplasmic Division in Animals.   In animals, cytokinesis—division of the cytoplasm—generally begins with an equatorial furrow. This is an indentation in the surface of the cell, usually at about the plane of the metaphase plate. It looks as though someone were squeezing the cell in half with a tight but invisible band, cleaving it in two. For this reason, animal cytokinesis is often referred to as **cleavage.** The furrow is created by the contraction of a ring of actin microfilaments in the cytoplasm just beneath the plasma membrane. In eukaryotes (other than flowering plants, most algae, and the fungi), plasma membranes generally have the power of movement thanks to these actin microfilaments. The furrow continues to form, to become a deep groove still being drawn in from below. The remains of the spindle may be caught by the tightening ring of contracting microfilaments, but no matter; the constriction continues until eventually the cell is pinched in two (Figure 11.9).

Let's consider some exceptions to this general pattern. Mitosis in plants is a case that quickly comes to mind and we will discuss it next. But there

How the Spindle Works

We know that the spindle serves to move the daughter chromosomes to opposite poles during anaphase. How it accomplishes this has been a popular subject of debate among biologists for half a century or more. We now have a pretty good idea of how it's done.

The basic observations are simple. During anaphase (the period of chromosome movement), the spindle drastically changes its shape in two ways: (1) the centromeric spindle fibers (the ones attached to the chromosomes) become shorter and (2) the spindle itself becomes longer, with the poles moving farther apart.

These two classic observations have given rise to a long-standing controversy over the forces at work during chromosome movement. The "pull" hypothesis holds that chromosome movement is caused by the shortening of the centromeric spindle fibers, which reel in the chromosomes like trout on a line. The "push" hypothesis holds that the chromosomes are forced apart by the seeming elongation of the spindle, which in turn is caused by the elongation of

the polar spindle fibers; the centromeric spindle fibers simply bind each set of chromosomes to its proper pole. The argument, each side with its distinguished proponents, has continued for decades. But we do have some new information from electron micrographs.

It now seems that both sides are declaring victory. Yes, indeed, one side argues; the centromeric spindle fibers do get shorter, apparently exerting some pull on the chromosomes. Yes, indeed, says the other; the spindle itself does get longer, pushing the poles apart. So who's correct? Maybe both. Evidence for two different, seemingly redundant forces is becoming strong. Each of the separate processes, it turns out, is sensitive to different chemical treatment. For example, when mitotic cells are exposed to a dilute solution of the anesthetic chloral hydrate, the polar spindle fibers fail to lengthen, but the centromeric fibers shorten and the newly separated anaphase chromosomes move apart as usual. It also appears that spindle elongation is considerably more important in some

cells than in others—so important in some that the poles increase their distance from each other by as much 15 times! Let's look more closely at each of these ideas.

In considering the "pull" idea, the **subunit disassembly hypothesis,** as it is known today, it is important to keep in mind that the spindle fibers are actually microtubules. They consist of spherical subunits of the protein tubulin stacked into the tubular structure. It seems that the molecular subunits leave the microtubule—are disassembled, that is—without affecting the integrity of the part that remains so that the microtubule grows shorter while somehow remaining attached at both ends.

There are many variants of this hypothesis. According to some theorists, either the centromere, or the microtubular organizing centers around the centrioles, or both have the ability to actively disassemble molecular subunits from the ends of the microtubules while still keeping a grip on the fibers. This reminds us of one of the favorite games of the old frontier parties in which each of two

are other examples as well. Many algal and fungal cells undergo a number of mitoses (nuclear divisions) without cytoplasmic division, thereby becoming multinucleated. Other kinds of cells may normally undergo rounds of chromosome replication with neither nuclear nor cytoplasmic division. One such example is found in the giant **polytene chromosomes** in the salivary glands of flies, including *Drosophila*. Their size results from repeated chromosomal replication without subsequent separation.

Cytoplasmic Division in Plants. Let's see why we have had to exclude plants from our generalizations about cell division. How are they so different? To begin, plant cell walls are not very flexible and they are not given to bending and folding, so cleaving is simply out of the question. Thus

plant cells, with their rather rigid cell walls and relatively immobile plasma membranes, must divide in a different manner. At about the time of telophase, the assembly of a new cell wall begins in the plane of the metaphase plate. Small membranous vesicles formed from the Golgi apparatus move along the polar microtubules and accumulate along a plane in the center of the cell. There they join and form a virtually continuous double membrane, but one that contains dense materials that will later be used in wall construction. This dense, double-membraned structure is called the **cell plate.**

The cell plate begins forming in the middle of the cytoplasm and then, as more Golgi vesicles join, it grows outward to fuse with the periphery of the cell, unlike the animal cell division furrow, which begins at the periphery and extends toward the middle.

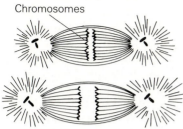

Anaphase:
Spindle as a whole gets longer
Centromeric fibers get shorter

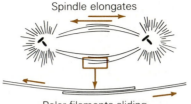

Sliding filament hypothesis
(continuous fibers)

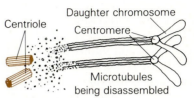

Subunit disassembly hypothesis
(centromeric fibers)

players held one end of a long piece of molasses taffy in his or her mouth; then they chewed their way to each other's lips. If you can imagine that the two players are the MOC and the centromere, and the taffy is a spindle fiber, you are in a position to envision a thoroughly absurd analogy.

According to the push explanation, now called the **sliding microtubule hypothesis,** the polar microtubules do not themselves grow longer; instead, the two overlapping sets slide apart from one another, leaving a smaller and smaller region of overlap. The vexing question raised by this observation is, "What provides the motive force for such movement?" The concept of filaments forcefully sliding past each other is certainly not without precedent. A well-known system of sliding fibers or filaments is clearly responsible for muscle contraction. But the system in spindle elongation must be different, since the experimental application of muscle inhibitors has no effect on spindle elongation. For a different approach, cytologists have now turned to the protein dynein, an active constituent of the cilium and flagellum (see Chapter 4). In these organelles, it generates its motive force through the hydrolysis of ATP. And it turns out that the same chemical inhibitors that stop ciliary and flagellar action by interfering with the ATP hydrolyzing enzyme stop the sliding of polar spindle fibers. In addition, cytologists have been able to see crossbridges in the spindle that are clearly similar to the dynein arms of cilia and flagella. Thus, a dynein bridge-based system is now the favored explanation of how the polar fibers slide and the spindle elongates.

The cell plate becomes impregnated with the tough material called pectin and ultimately forms the **middle lamella,** a gummy layer that lies between the cell walls of adjacent cells. The new plasma membranes then grow in over the middle lamella, secreting new cell walls as they grow (Figure 11.10). The details of plant cell wall production are described in Chapter 22.

The Adaptive Advantages of Mitosis. Mitosis, we see, is a critical step in the process that allows cells to divide, to produce two daughter cells where there was one. It is a fundamental property of many kinds of life, but its mechanisms are basically the same wherever is occurs. But what are the advantages of mitosis?

In many single-celled organisms, cell division by mitosis is a way of reproducing **asexually** (without combining genetic material from two sources). Since mating is not involved, the organisms can make copies of themselves at a prodigious rate, swamping the competitors and spreading out into its available spaces.

In multicellular organisms, mitosis permits growth, the formation of new cells. These can then replace older, worn cells, or they can provide new material for cell specialization. In this case, the cells take different developmental pathways toward specialization for different tasks. In humans, for instance, billions of new red blood cells must be produced daily to keep up with the rapid attrition. In addition, the growing organism can achieve relatively large sizes while maintaining a favorable surface-to-volume ratio (see Chapter 4). Because the cytoplasm is never far from the plasma membrane, it can communicate with outside resources.

In some species, mitosis permits very precise wound healing, and in others, regeneration of lost body parts is possible through complex processes involving mitosis. Finally, mitosis is involved in the long developmental progression toward the formation of eggs and sperm. Obviously, mitosis is a critical and tightly controlled process in the lives of cells. Now, however, we will consider another process, seemingly similar and also tightly controlled, but with an entirely different function.

MEIOSIS

In both plants and animals, individuals are **diploid.** This means that every cell in the organism has a double set of genes and chromosomes, one set from each parent. In sexual reproduction, the two part-

ners produce **gametes** (sex cells, usually eggs and sperm), which join through fertilization, to produce an offspring. Thus sexual reproduction ensures that the offspring will receive genes and chromosomes from each of the two parents.

Two related questions come to mind: first, if fertilization joins the genetic material of two parents, why is it that the amount of genetic material is not doubled in every generation? Second, if each parent has two sets of genes and chromosomes, how is it that the offspring receives only one set from each parent?

Logically enough, the answer to the second question also answers the first. Gametes are **haploid**—that is, they have only one set of chromosomes. Thus, each gamete (sperm or egg) enters into fertilization with only one set of chromosomes instead of two; thus, there will be no doubling in each generation. The problem then becomes,

11.9 CYTOKINESIS IN ANIMAL CELLS

(a) Following the division of chromosomes and their migration to opposite sides of the cell, cytokinesis (or cytoplasmic division) begins. **(b)** Typically, shallow indentations form and then deepen around the cell, producing a definite furrow. **(c–f)** The furrow deepens and eventually the cytoplasm is pinched in two. The structures responsible for the formation and deepening of the cleavage furrow are the actin microfibrils, which, in a manner not entirely understood, pull the plasma membrane in until separation is complete.

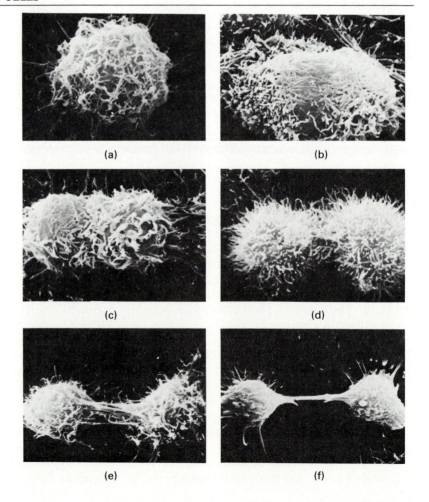

(a)   (b)

(c)   (d)

(e)   (f)

11.10 CYTOKINESIS IN PLANTS

Rigid cell walls prevent any form of cleavage from occurring. Instead, a new cell wall is constructed between daughter cells. In the earliest events **(a)**, vesicles originating from the Golgi apparatus migrate along the overlapping microtubules to coalesce at the cell equator. **(b)** As the vesicles extend across the cell, they form the early *cell plate,* and the carbohydrate precursors within are converted to wall materials such as pectin and hemicellulose. Such materials form the gummy *middle lamella,* which acts to cement the new cells together when the *primary cell wall* has formed. **(c)** As events continue, the cell plate is extended outward, reaching toward the existing plasma membrane. This growth is accomplished by microtubules of the spindle, which reorient in such a manner as to encourage the migration of vesicles to the edges of the cell plate. There they both physically join the plate and add their contents. Once the cell plate is established **(d)**, the daughter cells on each side begin laying down cellulose microfibrils, which will form the primary cell wall.

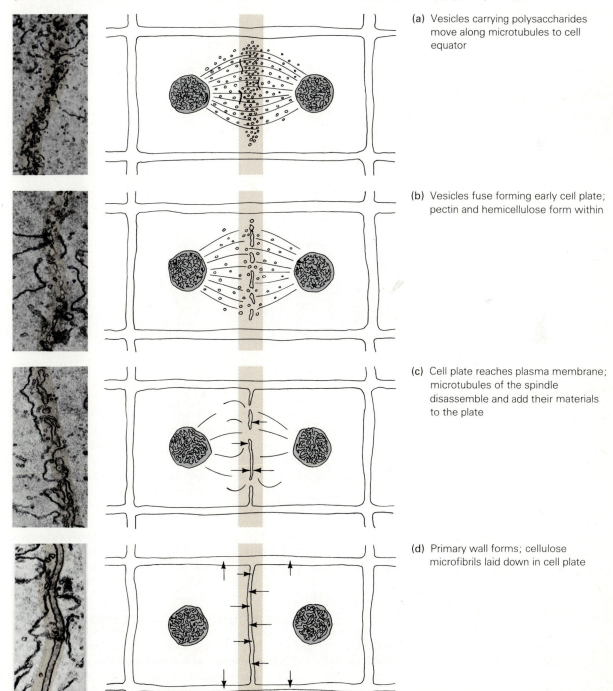

(a) Vesicles carrying polysaccharides move along microtubules to cell equator

(b) Vesicles fuse forming early cell plate; pectin and hemicellulose form within

(c) Cell plate reaches plasma membrane; microtubules of the spindle disassemble and add their materials to the plate

(d) Primary wall forms; cellulose microfibrils laid down in cell plate

11.11 SIMPLIFIED SCHEME OF MEIOSIS

(a) Prior to meiosis, the chromosomes replicate, and at the start of meiosis **(b)**, eight chromatids are present. **(c)** The first of two successive meiotic divisions then occurs. Note that the centromeres do not divide as they did in mitosis and that the members of each pair of chromosomes have been separated. Thus, each daughter cell receives two half-pairs (but with a total of four chromatids). **(d)** With the second meiotic division, the centromeres divide and the daughter cells receive two chromosomes per cell. This means they are now haploid (compare these to the diploid cell at the top). In animals, the haploid cells develop directly into gametes, which remain haploid until fertilization, whereupon they resume the diploid condition.

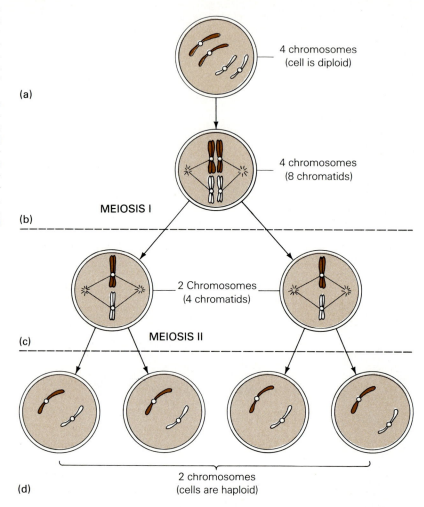

4 chromosomes (cell is diploid)

4 chromosomes (8 chromatids)

MEIOSIS I

2 Chromosomes (4 chromatids)

MEIOSIS II

2 chromosomes (cells are haploid)

how is the number of chromosomes halved during gamete formation from two complete sets per cell to one complete set per cell? The answer involves a complex and orderly set of steps called **meiosis.**

We can define meiosis as the process whereby a diploid (double or paired) set of chromosomes is reduced to a haploid (single or unpaired) set of chromosomes in a cell, as part of the process that produces a gamete. In diploid organisms, meiosis guarantees that the chromosome number will remain stable from generation to generation and that each sexually reproduced offspring will get two complete sets of genetic instructions, each set being a random mix taken from the parents' own two sets of genetic instructions (Figure 11.11). Humans, for instance, have 46 chromosomes in each cell of the body—23 originating from the father's sperm and 23 from the mother's egg. In meiosis, the 46 chro-

mosomes per cell will be reduced to 23 chromosomes per gamete. The two sets of 23 chromosomes are sometimes referred to as **paternal** (father's) and **maternal** (mother's), but we will soon see that the meiotic process also virtually guarantees that each gamete will be different from every other gamete.

It is important to note that most of what we will say about meiosis applies to animals, most plants, and a few protists—specifically to those organisms that spend most of their life as diploids. The fungi and many of the protists differ in many aspects of sexual reproduction, as we will see later.

Homologous Chromosomes

Each chromosome within a cell of an individual is distinctive and can be identified through a process called **karyotyping** (Essay 11.2). In the human karyotype, each of the 23 different chromosomes is

Karyotyping

A *karyotype* is a graphic representation of the chromosomes of any organism, in which the chromosomes are systematically arranged according to size and shape. Each species has its own particular karyotype; thus we know the number and kinds of chromosomes found in cabbages, fruit flies, and people. Karyotyping is done according to an established method.

Consider, for example, human karyotyping. (1) A blood sample is withdrawn and (2) a few drops are put into distilled water. The relatively fragile red blood cells burst under the resulting osmotic stress, but the white blood cells merely swell and can be separated by centrifugation. They are transferred to (3) an isotonic cell culture medium, containing cell nutrients and two plant-derived chemicals, **plant lectins** and **colchicine.** The plant lectins, for some unknown reason, induce the white

blood cells to enter *S* phase and proceed on to mitosis. The colchicine, which acts on the chemical structure of the mitotic spindle, arrests all cells in mitotic metaphase, when the already doubled chromosomes are maximally condensed. (4) The cells are again suspended in distilled water, which again causes them to swell. This swelling increases the distances between the chromosomes so that individual chromosomes can more readily be seen. (5) A drop of stain is added, and the swollen cells are squeezed flat between a microscope slide and a cover slip, further spreading the chromosomes. (6) A favorable cell—one showing all the chromosomes spread out and separated from each other—is photographed through a light microscope. (7) The photograph is enlarged to an 8 X 10 glossy print. (8) Here comes the most technically sophisticated trick of

all: the investigator uses scissors to cut out each chromosome from the photograph. (9) The cut-out images of the chromosomes are sorted by size and shape. The homologous pairs are matched up, and the chromosomes are laid out in one or several lines, from the largest to the smallest, and fastened down with rubber cement. This is the karyotype. (10) Once the karyotype has been constructed, the investigator can count the chromosomes and can identify individual ones. By comparing the karyotype with a karyotype of a normal individual, the investigator can detect even minor chromosomal abnormalities. (The X shape of the chromosomes results from the fact that the two condensed chromatids of each chromosome have almost completely separated and are held together only by the centromere.)

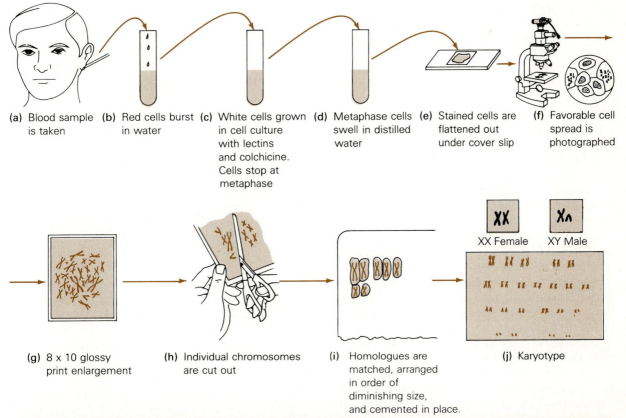

(a) Blood sample is taken

(b) Red cells burst in water

(c) White cells grown in cell culture with lectins and colchicine. Cells stop at metaphase

(d) Metaphase cells swell in distilled water

(e) Stained cells are flattened out under cover slip

(f) Favorable cell spread is photographed

(g) 8 x 10 glossy print enlargement

(h) Individual chromosomes are cut out

(i) Homologues are matched, arranged in order of diminishing size, and cemented in place.

XX Female   XY Male

(j) Karyotype

identified by its size, the position of its centromere, and its unique pattern of dark-staining bands.

We see, then, that the 46 chromosomes in each human diploid cell come in 23 pairs; one member of each pair comes from the father's sperm and one comes from the mother's egg. The two members of each pair are called **homologues,** or **homologous chromosomes.** The two homologues are functionally equivalent in that they have genes that code for the same feature and they are arranged in the same order. Thus, if a gene on a chromosome from the father's sperm influences eye color, the homologous chromosome from the mother's egg will have a gene that also influences eye color, and it will be located in the same position on the chromosome. The two homologues differ from each other only in ways in which two individuals within a species might differ from one another. For instance, the chromosome from the father might carry a form of the eye color gene that tends to make eyes blue, and the homologue from the mother might carry a form of the gene that tends to make eyes green.

An Overview of Meiosis. During meiosis, the homologous chromosomes are involved in three critical events. First, the two homologues pair, or undergo **synapsis,** and second, they become, in a sense, scrambled. These two processes are called **pairing** and **crossing over.** Briefly, what happens is that in pairing, the two homologues come to lie alongside each other. In crossing over, the homologues break apart at specific places and exchange chromosomal parts. In this way, they come to have unique combinations of genes not present in the parents. Such an exchange is known as **genetic recombination.** We will say more about this intriguing process shortly.

The third critical event follows the crossing over. During this phase, the homologues separate and go to different daughter cells; thus, each of the developing daughter cells will receive only one copy of a gene from each gene pair. In our hypothetical blue eye/green eye example, some of the daughter cells would become gametes with the blue-eye form of the gene and some would carry the green-eye form of the gene, but none of the gametes would normally carry both types. The homologous genes, then, would exist unpaired until fertilization.

With these points in mind, let's review the meiotic process in more detail. At the start of meiosis, the DNA of cells entering meiosis will have doubled, and each chromosome will consist of two chromatids. This means that in the human meiotic

cell, there are actually 92 DNA molecules or chromatids at this stage. The cell will then go through two divisions, called **meiosis I** and **meiosis II,** in order to produce gametes with 23 chromosomes apiece. (In the arithmetic of meiosis, one cell with 92 DNA molecules produces two with 46 DNA molecules each, which, in turn, produce four cells with 23 DNA molecules each.)

Meiosis I: The First Meiotic Division

Meiotic Prophase and Its Substages. Prophase of meiosis I is longer and more complex than mitotic prophase. It has been divided into five substages for simplicity's sake, each with its own peculiar events. As we see in Figure 11.12, the substages are

1. **Leptotene,** the earliest visible condensation of the chromosomes.
2. **Zygotene,** when homologous chromosomes pair up, or undergo synapsis.
3. **Pachytene,** when the paired chromosomes condense further and crossing over occurs.
4. **Diplotene,** when the results of crossing over are first visible as X-shaped **chiasmata,** (singular, *chiasma*), and in which the four chromatids of each homologous pair are sometimes visible. The chiasmata represent regions where crossovers have occurred.
5. **Diakinesis,** in which the chromosomes are highly condensed, and in which **repulsion** occurs with homologues held together only at their ends.

As in mitotic prophase, the chromosomes of leptotene slowly condense (see Figure 11.12a). Perhaps one reason the process takes so long is that in the early stages, the chromosomes are shifting about in that murky soup of the nucleus. It is now that each chromosome somehow joins its chromosomal counterpart, its homologue in synapsis.

How do homologues find each other in zygotene? Chromosomes can't move under their own power (as far as we know), and this early movement and pairing occurs before the mitotic spindle has even formed. The only clue to such movement is that the nuclei of most organisms actually rotate during the early stages of meiosis, so that with time-lapse photography they look rather like dryer windows in a laundromat. Pairing or synapsis can be seen to occur when homologous chromosomes touch each other in the right places, but first there seems to be a lot of random groping (see Figure 11.12b).

11.12 SUBSTAGES OF MEIOTIC PROPHASE

Meiotic chromosomes from the lungless salamander appear remarkably clear. **(a)** In leptotene, the chromosomes begin to condense and are visible as long and spindly single strands. Actually, replication has already occurred, but individual chromatids will not be visible for some time. The granular appearance of the chromosomes reflects the presence of numerous beadlike nucleosomes along their length. **(b)** In late zygotene, the chromosomes have completely synapsed, with all homologues paired up. While the actual pairing isn't completely clear, fewer individual strands occur than were visible in leptotene. The dense spherical bodies seen here and there are centromeres. The fuzziness about the chromosomes represents free loops of DNA that are actively transcribing RNA. **(c)** With pachytene, the paired chromosomes now consist of four aligned chromatids, or tetrads, comprising two homologous pairs of chromosomes. While this isn't entirely visible, we do see the products of greater condensation. Pachytene is also the time of crossing over, although this cannot be determined through visual means. **(d)** In diplotene the homologous chromosomes have begun to separate. Individual chromatids are now visible (see the highly magnified view), as are chiasmata, clinging, X-shaped regions representing places where homologous chromatids have previously undergone crossover. In all other regions, a general repulsion of homologues is seen. **(e)** With diakinesis, prophase will end. The homologues are now widely separated except for remaining points where crossover occurred during pachytene. It is in this form that they will become arranged on the metaphase plate.

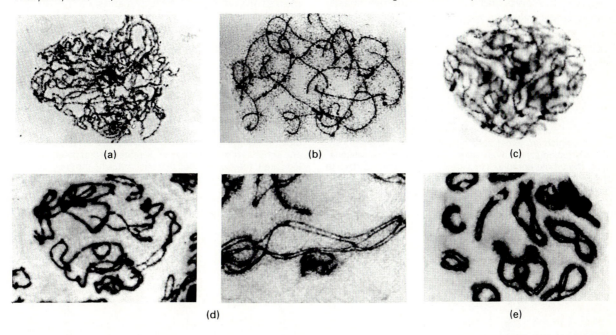

(a)   (b)   (c)

(d)   (e)

Ironically, since fertilization, the two homologues of the chromosome pairs have ignored each other, each going through mitosis after mitosis completely independent of the other, and relating, if at all, only through the biochemical interactions of their gene products. But in early meiotic prophase all that changes, and suddenly chromosome number 13 begins to have an affinity for the other chromosome number 13, and number 7 for number 7, and so on. With this new affinity, when two homologues happen to touch in the same region, they hold. The result, over a period of time, is that the two homologues seem to come together like the two halves of a zipper. It turns out that in synapsis, there is a specialized structure, the **synaptonemal complex** (made up of protein and RNA), that bridges the two homologues in meiotic prophase, and it even *looks* like a zipper (Figure 11.13).

With the pairing up of chromosomes in zygotene, each grouping (joined homologues) actually consists of four chromatids, each of which is called a **tetrad.** The meiotic cell of a human has 23 such tetrads. Each tetrad, however, has two centromeres, so when the dust clears we are still looking at just 23 pairs of chromosomes.

A Closer Look at Pachytene and Crossing Over. In pachytene, the four chromatids become scrambled by undergoing crossing over (see Figure 11.12c). It turns out that at any point along the DNA backbone of a chromatid, special enzyme complexes, forming gigantic **recombination nodules,** can cause the chromatids to unwind so as to reveal two single strands of DNA. A complex and poorly understood series of events then occurs.

It was long held that at each crossover event,

several thousand base pairs from a chromatid of one chromosome are exchanged for their counterparts from a chromatid of the homologous chromosome (thus not between sister chromatids, which are exact replicas of each other). More recently, however, other ideas have been proposed. In one popular model, the double helix itself opens up in homologous chromatids and the four individual strands are brought close together. We'll call the strands *a*, *b*, *c*, and *d*. Then an enzyme from the family of **endonucleases** (nucleic acid-hydrolyzing enzymes) produces a break, or "nick," as it is called in, let's say, the *b* and *c* strands. One of *c*'s resulting free ends and a length of unpaired bases then crosses over to a correlating region of nearby

strand *b*, and displaces it. The displaced portion of strand *b* then moves over to replace the missing section of *c*. Actually, a number of the newly paired bases will not match, since homologous chromatids are expected to differ as a natural part of genetic diversity. But this isn't a problem. Any mismatched bases or any unpaired segments would be corrected by enzymes similar to those used in the DNA repair process (see Chapter 10). The newly exchanged *b* and *c* strands would act as the template and the correct pairing of bases would be carried out. Thus, when the homologues separate during anaphase, each of the chromatids involved will bear portions of the other's DNA.

11.13 CROSSING OVER AND CHIASMA FORMATION

(a) In pachytene, following the zipperlike pairing of homologous chromosomes in the previous zygotene stage, the maternal and paternal chromosomes are intimately associated and bound together by the synaptonemal complex. Each of the two chromosomes consists, in turn, of two sister chromatids, which are even more intimately associated and cannot be distinguished even under the electron microscope. Thus the whole complex consists of four strands. Breakage and reunion of the DNA strands is known to occur in pachytene, although it cannot be seen under the microscope. **(b)** In diplotene, the homologous maternal and paternal chromosomes separate but the sister-strand regions are still tightly

held together as indicated. Wherever crossing over (physical rejoining of homologous chromatids) has occurred, the chromosomes are still held together. The visible evidence of this exchange is a cross-shaped conformation called a *chiasma*. Here, two chiasmata are shown as they would appear in the diplotene stage. **(c)** If the four chromatids could be unwound and separated, and the regions of maternal and paternal origin indicated, they would look like this. Actually, the four chromatids will separate in later stages of meiosis. Any one exchange involves only two of the four chromatids, but ordinarily there are many exchanges between each pair of homologues and all four chromatids become scrambled.

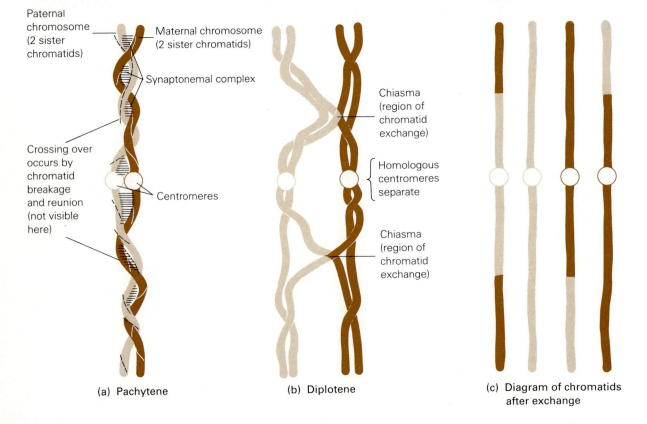

(a) Pachytene

(b) Diplotene

(c) Diagram of chromatids after exchange

It is quite possible that the synaptonemal complex plays an important, though as yet unknown, role in crossing over. It always seems to be present where crossover is occurring, and absent where it is not. For example, crossovers are never detected in the males of *Drosophila melanogaster*. Close examination of the synapsing chromosomes reveals that the synaptonemal complex is absent in the male's chromosomes. The same can be said for females of the silk moth *Bombyx*, in which neither the synaptonemal complex nor crossing over has been seen.

Whatever the recombination mechanism, the final result is that at randomly located but equivalent places along the chromatids, the DNA of one of the original four chromatids is covalently bonded with, and continuous with, the DNA of one of the other three chromatids. The process is reciprocal, so that at the point where one chromatid leaves off being paternal and begins being maternal, the opposite chromatid leaves off being maternal and begins being paternal (see Figure 11.13).

In human cells, there is an average of about ten crossovers for every meiotic tetrad. In other species, there may be fewer. But one thing is clear: after meiotic prophase, one cannot refer to individual chromatids as maternal or paternal. The crossovers produce new kinds of chromatids, and it should be mentioned that this is an important source of genetic variation in a population. This constantly renewed variation is crucial to the evolutionary process.

Evidence of crossing over having occurred is seen first in the diplotene stage (see Figure 11.12d). The sister chromatids of each of the two homologues are still held tightly together, but the homologous chromosomes begin to separate. Like a true zipper, the synaptonemal complex "unzips," permitting the homologues to begin parting. However, where a chromatid exchange or crossover has taken place, the homologues seem to cling together, forming the cross-shape configurations we earlier called chiasmata. So for a brief period, the chiasmata tell the observer where crossing over has occurred.

In diakinesis (see Figure 11.12h), the individual chromatids finally become visible. Recall that this occurred very early in prophase of mitosis. The chromatids of individual chromosomes are joined at their centromeres, just as you would expect. However, while the homologous chromosomes actually reject or repel each other along most of their length, they still manage to cling together at the crossover points, just as they did in diplotene. This clinging will continue into metaphase I.

11.14 METAPHASE OF MEIOSIS I

When the chromosomes move to the metaphase plate in meiosis I, the repulsion of homologous chromosomes is very pronounced, as though the pairs were trying to get away from each other. They eventually touch only at the tips of the chromosome arms. In this preparation of metaphase I salamander chromosomes, the spindle and the centromeres cannot be seen, but homologous centromeres will be attached by centromeric spindle fibers from opposite poles.

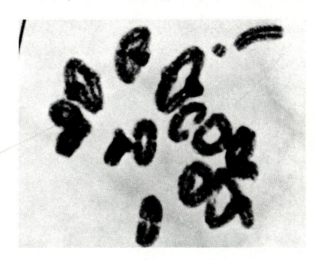

Metaphase I. As we have noted, meiosis actually consists of two cell divisions, complete with two metaphases, two anaphases, and two telophases. In the first metaphase (Figure 11.14), the tetrads are brought to the metaphase plate, but as you might expect, the appearance and behavior of the chromosomes and the ensuing events are quite unlike any mitotic metaphase. The centromeres of homologous chromosomes appear to be separated from one another as far as possible, with only the ends of the chromosome arms touching. And, most significantly, the attachment of the centromeric spindle fibers in metaphase I is quite different from that in metaphase of mitosis. In mitosis, you'll recall, centromeric spindle fibers from each pole attached to centromere. In metaphase I of meiosis, however, centromeric spindle fibers from just one pole attach to each centromere (see Figure 11.11b). This peculiar orientation is quite important, but its implications only become clear when anaphase I begins.

Anaphase I. Anaphase I follows quickly (Figure 11.15), but because of the unique way in which the centromeric spindle fibers are attached, the results are quite different from those of mitotic anaphase. In mitosis, the pull on each chromosome came from two directions, and as the centromeres divided, each chromatid—then a full fledged chromosome—moved to opposite poles. In metaphase I of meiosis, the pull is from one direction only and the centromeres do not divide. When the migration begins, homologue separates from homologue all down the line, and each chromosome, still composed of sister chromatids, is drawn to a pole. In humans, this means that 23 chromosomes go to each pole and that the homologues—the chromosome pairs—are forever separated. So, following the events of anaphase I, we can see that the number of chromosomes arriving in each daughter cell has been reduced by half. For this reason, the first meiotic division is sometimes known as the **reduction division.**

11.16 TELOPHASE OF MEIOSIS I

(a) Telophase of meiosis I in the grasshopper testis: The cell retreats from meiosis as chromosomes cluster tightly together, becoming individually indistinct. Accompanying this will be the reconstruction of the nuclear envelope, nucleoli, and duplication of centrioles.

11.15 ANAPHASE OF MEIOSIS I

(a) Unlike mitotic anaphase, the first meiotic anaphase proceeds without the division of centromeres and the separation of chromosomes containing two chromatids is clearly seen. **(b)** The homologous chromosomes have fully separated and form two clusters. The centromeres in each chromosome are quite visible, as are the four arms of each chromatid. Along with the anaphase separation go all of the crossover products formed during prophase.

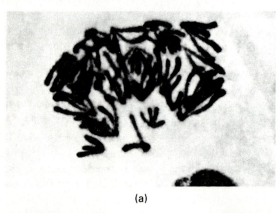

(a)

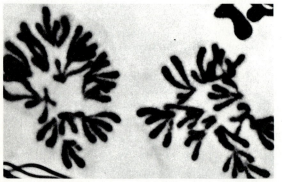

(b)

Telophase I and Meiotic Interphase.   Anaphase I is followed by telophase I (Figure 11.16), complete with the reorganization of nuclei, decondensation of chromosomes, duplication of centrioles, and cytoplasmic division. The cell then enters a **meiotic interphase.**

Meiotic interphase is unlike other interphases in one very important detail: there is no S phase. In other words, there is no DNA replication as there would be between mitotic divisions. Meiotic interphase may be extremely brief and the chromosomes may or may not completely decondense, or it may be exceedingly long, depending on the species.

Meiosis II: The Second Meiotic Division

Prophase II.   Meiosis II is a bit easier to follow since it is reminiscent of the simpler process of mitosis. Prophase II, for example, is rather like a mitotic prophase. The chromosomes condense and each contains two genetically recombined chromatids, still attached by their centromeres. This time the centromeric microtubules become attached in standard mitotic fashion. The kinetochores of each chromosome attach to fibers from opposite poles, just as in mitosis. With this spindle orientation, the chromosomes are moved to the metaphase plate.

Metaphase II and Anaphase II.   In metaphase II, the chromosomes line up on the metaphase II plate in preparation for separation. As anaphase II begins, the centromeres finally divide and as the chromatids separate, they become full-fledged chromosomes. These chromosomes are then pulled apart as they are in mitotic anaphase, with the same forces and attachments. It is important to remember that the chromatids that have separated in anaphase II are not sisters and are not genetically identical. So, actually, both anaphase I and anaphase II are distinctly different from mitotic anaphase.

Telophase II. As anaphase II proceeds into telophase II, nuclear envelopes once again form around the four groups of decondensing chromosomes. Unlike the centrioles in mitotic telophase, the newly isolated centrioles do not replicate. (Why do you suppose this is?) The four resulting cells then enter interphase, where they are destined to remain in the G_1 phase. DNA replication is over until fertilization triggers new cell cycles (Figure 11.17).

Summing Up Meiosis. To sum up, then, each of the four newly scrambled, unique chromatids of the prophase I tetrad ends up in one of the four cells that are the products of meiosis. As in mitosis, the amount of genetic material was doubled before the process started (during the S phase), but in two divisions, the products were distributed to four cells. Each cell now has one fourth of the normal G_2 amount of DNA. Each of these is now haploid—that is, each cell has only one of each of the different types of chromosomes of the species (for example, 23 chromosomes in human gametes). In animals meiosis is also called **gametogenesis,** and the four products of meiosis are called **gametocytes,** cells that must undergo cellular differentiation to become gametes. (Figure 11.18 and Table 11.3 compare meiosis with mitosis in animals.)

11.17 MEIOSIS II

Events in meiosis II are strikingly similar to those of mitosis. (a–c are salamander, d and e are grasshopper) **(a)** Prophase II: The simple prophase of mitosis is recalled here as chromosomes condense and the nuclear envelope fades from view. **(b)** Metaphase II: In meiotic metaphase, the chromosomes, consisting of sister chromatids attached at their centromeres, are drawn from their prophase positions and align along the metaphase plate. Centromeric fibers extend from both sides of each centromere. **(c)** Anaphase II: Now the centromeres have separated. Each chromatid, now a chromosome, has separated from its formerly attached member, and the two migrate in opposite directions. (Note the seemingly "reluctant" chromosomes, whose longer arms have just parted at the midline.) **(d)** Telophase II: The complete product of metaphase II is four cells, which in telophase II are represented by four haploid clusters of chromosomes. **(e)** With the reconstruction of the nuclear membrane, four haploid cells will form. These will undergo maturation into gametes.

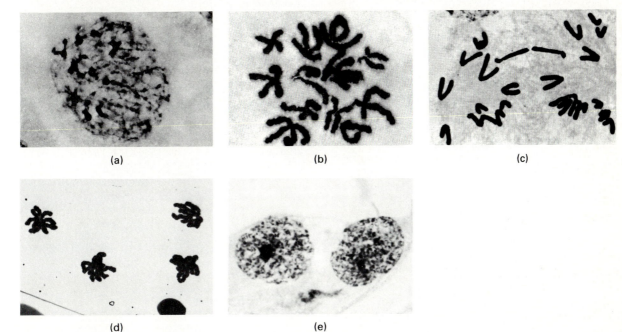

(a) (b) (c)

(d) (e)

11.18 A COMPARISON OF MITOSIS AND MEIOSIS

The differences between mitosis and meiosis become apparent when they are seen together. We have already described the longer and more elaborate prophase of meiosis I, so here we will simply point out that in mitosis, homologous chromosomes have no particular interest in each other and are arranged randomly in the nucleus, while meiotic homologues have paired up.

At metaphase, the alignments are also different. Mitotic chromosomes align randomly with the spindle fibers attached on each side. In meiotic metaphase I, attachments form on one side only.

At anaphase, the mitotic centromeres divide and chromatids separate, while in anaphase I of meiosis they do not. Sister chromatids remain attached and homologous chromosomes, still doubled, move apart.

With telophase, mitosis is completed. The cell enters interphase and its DNA will be replicated in the S phase. In addition, the manner of division at anaphase assures that the chromosomes in each daughter cell will be identical in both number and genetic makeup to those in the cell that produced them. The meiotic daughters, however, do not enter an S phase and no DNA replication occurs. Further, because of crossing over and genetic recombination, the chromosomes are now quite different from what they were at the start of meiosis.

At this point, the meiotic event is only half finished. A second division will occur with centromeres now dividing and sister chromatids (now chromosomes) moving to opposite poles, just as happened in mitosis. Unlike mitosis, however, the four daughter cells will be haploid, with exactly half the chromosome number of the mitotic daughter cells, and they will contain a mix of maternal and paternal genes.

MITOSIS

1. **Interphase**
Chromosome not visible; DNA replication.

2. **Prophase**
Centrioles migrated to opposite sides; spindle forms; chromosomes become visible as they shorten; nuclear membrane, nucleolus fade in final stages of prophase.

3. **Metaphase**
Chromosomes aligned on cell equator. Note attachment of spindle fibers from centromere to centrioles.

4. **Anaphase**
Centromeres divide; single-stranded chromosomes move toward centriole regions.

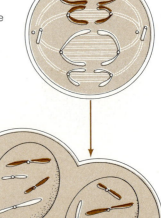

6. **Daughter cells**
Two cells of identical genetic (DNA) quality; continuity of genetic information preserved by mitotic process.

These cells may divide again after growth and DNA replication has occurred.

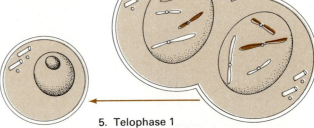

5. **Telophase 1**
Cytoplasm divides; chromosomes fade; nuclear membrane, nucleolus reappear; centrioles replicate (reverse of prophase).

11.18 (continued)

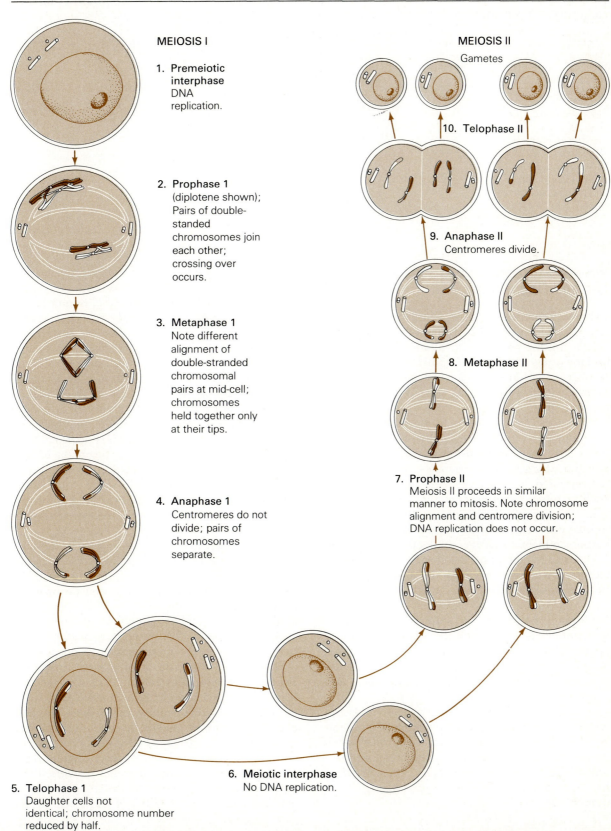

MEIOSIS I

1. **Premeiotic interphase**
   DNA replication.

2. **Prophase 1**
   (diplotene shown); Pairs of double-standed chromosomes join each other; crossing over occurs.

3. **Metaphase 1**
   Note different alignment of double-stranded chromosomal pairs at mid-cell; chromosomes held together only at their tips.

4. **Anaphase 1**
   Centromeres do not divide; pairs of chromosomes separate.

5. **Telophase 1**
   Daughter cells not identical; chromosome number reduced by half.

6. **Meiotic interphase**
   No DNA replication.

MEIOSIS II
Gametes

10. **Telophase II**

9. **Anaphase II**
   Centromeres divide.

8. **Metaphase II**

7. **Prophase II**
   Meiosis II proceeds in similar manner to mitosis. Note chromosome alignment and centromere division; DNA replication does not occur.

TABLE 11.3

DNA DUPLICATIONS AND CELL DIVISION (FOR HUMANS N = 23)

Stage	Number of chromosomal DNA molecules per cell	Number of centromeres (= number of chromosomes) per cell	Number of DNA molecules attached to each centromere
Haploid gamete	N	N	1
New zygote	$2N$	$2N$	1
Diploid cell in G_1	$2N$	$2N$	1
Diploid cell in G_2	$4N$	$2N$	2
Meiotic prophase I	$4N$	$2N$	2
Meoitic interphase and prophase II, per cell	$2N$	N	2
Haploid gametocyte	N	N	1

Where Meiosis Takes Place in Higher Organisms

In higher organisms, meiosis is restricted to **germinal tissues** in specialized organs. In animals, these organs are the **gonads**—the **ovaries** of the female and the **testes** of the male. In flowering plants, meiosis occurs within the flower, specifically, in **ovules** located within the plant ovaries and in cells within the **anthers.**

The term *germinal* refers to reproduction, from the Latin word for seed. Germinal tissue is sometimes referred to as **germ-line** tissue, and the cells that can later give rise to gametes are said to be in the germ line. In plants, unspecialized active tissue, such as the *meristem* described in Chapter 22, can give rise to reproductive organs (flowers) and is therefore at least potentially germ-line tissue. In higher animals, however, a group of cells is set aside early in embryonic development, and germinal tissue will arise only from these cells. All other cells are **somatic cells** (*soma*, body).

Meiosis in Animals.   In animals, the germ line, once established, goes through several phases. It begins with the early group of germinal cells that is set aside in the embryo. These will become germinal tissue of the gonads, then meiotic cells, and finally the gametes.

The germinal tissue of the gonads (ovaries and testes) is called the **germinal epithelium** (Chapter 36). (*Epithelium* is a general term for surface tissue, either internal or external.) The germinal epithelium of the testes and the ovaries, like any other tissue, undergoes repeated mitosis during development. Meiosis will occur later.

During fetal development in human males, the germinal epithelium of the testes forms the lining of long, convoluted tubes—the **seminiferous tubules** (Chapter 36), but after they form, both mitosis and meiosis in this tissue are suppressed until puberty. Then, the germinal epithelium undergoes a type of asymmetrical mitotic division that is common in epithelial tissues of all types. With each mitotic division, one of the daughter cells remains a germinal epithelium cell, capable of further mitoses, but the other daughter cell becomes differentiated as a **primary spermatocyte,** a specialized diploid cell ready to begin that once in a life-cycle event, meiosis. It will proceed through two divisions and eventually form four sperm cells. Billions of sperm will be produced in the germinal epithelium daily for the rest of the man's life.

During fetal development in human females, germinal epithelium forms in the ovaries. Following this, the germinal cells begin to undergo meiosis. They complete most of the first prophase of meiosis, more or less simultaneously during the final months of fetal development, but then meiosis stops suddenly. Thus, when a baby girl is born, her germ-line cells, now referred to as **oocytes,** are well along in the process of gamete formation. They will become suspended in this state for 10 to 12 years without completing meiosis. Then, when the menstrual cycle begins, they will be released, one or sometimes two at a time, during each monthly cycle for the next 35 years or so. Meiosis will not be completed unless the oocyte is fertilized. Although a woman may have several thousand oocytes, only about 400 to 500 will become mature, and of course most of these will not be fertilized. Unlike germinal cells in males, then, those in females begin to form gametes before the individual is born.

Since all oocytes are suspended in the diplotene stage of meiotic prophase just before metaphase I, the implication in human females is that meiotic prophase can last 45 or perhaps 50 years. Actually, the long prophase I in females is typical of many animals, including all vertebrates. In spite of being suspended at one stage, however, the paired meiotic chromosomes are metabolically very active, producing large amounts of ribosomes and mRNA for the developing oocyte. During this period of intense activity, the female prophase I chromosomes become greatly puffed up and form large numbers of side loops. These are the *lampbrush chromosomes* shown in Figure 10.4.

By the time the oocyte is ready to be released, it is a relatively large cell, visible to the naked eye and full of yolk, mRNA, ribosomes, and other essentials for starting a new life.

Meiosis in Females, a Special Case. The chromosomes behave the same way in meiosis in males and females, so most of our discussion of meiosis applies equally to both sexes. But major differences are seen in the orientation of the two meiotic metaphase plates and in the cytoplasmic events to follow. That is, meiotic cytokinesis is quite different in the sexes.

In females, the valuable cell constituents are not divided evenly among the products of meiosis (Figure 11.19). Instead, most end up in one cell. Only that one cell gets to be the egg, and it is relatively fat and opulent and swollen with nutrients. About the time of ovulation, the oocyte's meiotic spindle forms *off to one side of the oocyte,* just under the surface. A normal first meiotic chromosome separation and cell division occurs, but one of the two daughter cells is very small and is called a **polar body.** The other cell is known as the **secondary oocyte.** A second meiotic division occurs, pinching off another tiny cell as a second polar body, which emerges just under the first. Polar bodies are clearly visible with light microscopes, as seen in Figure 11.19.

Our natural love of numerical symmetry would be satisfied if all the divisions of meiosis were completed, and if the first polar body joined in with a metaphase II of its own to give four products of meiosis, however unequal they might be. But reality strikes again, and we learn that the first polar

11.19 MEIOSIS IN HUMAN FEMALES

Meiosis in human females begins during embryonic life, but the egg cells are frozen in prophase I until puberty. When an oocyte resumes meiosis, cytokinesis is quite unequal, resulting in the production of one large cell and one tiny functionless polar body in each division. In humans, the second meiotic division begins at ovulation but will not be completed unless fertilization has occurred. Male and female pronuclei will fuse soon afterwards, and a second polar body will arise below the first. The photo captures a human oocyte at the moment of fertilization. One polar body is seen at the top and several sperm are seen near the egg surface.

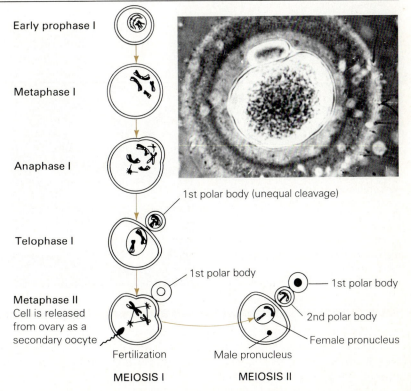

When Meiosis Goes Wrong

Meiosis is a much more complicated process than mitosis. When you consider the lengthy prophase with all the subphases of chromosome pairing, crossing over, and repulsion, in addition to the presence of two cell divisions, you shouldn't be surprised to learn that frequently something goes wrong. In humans, for instance, about a third of all pregnancies spontaneously abort in the first two or three months. When the expelled embryos can be examined, it turns out that most of them have the wrong number of chromosomes. Failure of the chromosomes to separate correctly at meiosis is termed **nondisjunction.**

But not all failures of meiosis result in early miscarriage. There are late miscarriages and stillbirths of severely malformed fetuses. Even worse, about one live-born human baby in 200 has the wrong number of chromosomes, accompanied by severe physical and/or mental abnormalities.

Apparently, nondisjunction can occur with any chromosome. Having only one **autosome** of a pair instead of the normal two is invariably fatal. An autosome is any chromosome other than the X and Y chromosomes, which determine sex. Having three instead of two is almost always fatal, resulting in spontaneous abortion or death in infancy. There are a few exceptions. About one baby in 600 has three copies of tiny chromosome 21. Such persons may grow into adulthood, but they have all kinds of abnormalities. The syndrome is known both as **trisomy-21** and **Down's syndrome,** after the 19th century physician who first described it. Characteristics of the syndrome are general pudginess, rounded features, and a rounded mouth in particular, an enlarged tongue which often

protrudes, and various internal disorders. Often a peculiar fold in the eyelids is seen and in the past this was erroneously equated with the characteristic eye fold of Asians (thus the earlier name "mongoloid"). Trisomy-21 individuals also have a characteristic barklike voice and unusually happy, friendly dispositions. The "happiness" is a true effect of the extra chromosome and not a result of their usually extremely low IQs; those with serious mental impairment are usually, by most indications, miserable.

Trisomy-21 occurs most frequently among babies born to women over 35 years old. At that age the incidence is about 2 per 1000 births. By age 40, this climbs to about 6, and by age 45, 16 Down's children are born for each 1000 births. The age of the father apparently has a much smaller, if any, effect. We can guess that the much-prolonged prophase I of the human oocyte might have something to do with this.

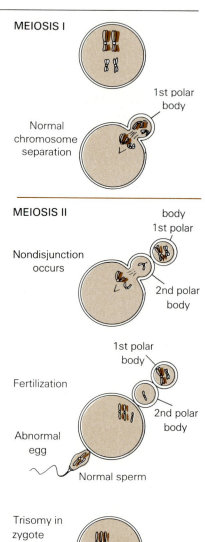

MEIOSIS I

Normal chromosome separation

1st polar body

MEIOSIS II

Nondisjunction occurs

body
1st polar

2nd polar body

Fertilization

1st polar body

Abnormal egg

Normal sperm

2nd polar body

Trisomy in zygote

body usually gives up and dies without bothering to divide first. What would be the use, anyway? In fact, what is the function of polar bodies? The polar bodies can be regarded as no more than convenient little garbage cans into which an unwanted three quarters of the genetic material of the primary oocyte can be dumped, leaving one large oocyte with its haploid complement of newly scrambled chromosomes and the lion's share of its predecessor's cytoplasm and nutrients.

Why Meiosis?

Meiosis, we see, is a difficult and complicated process. Furthermore, its complexity means that all too frequently something goes wrong (see Essay 11.3). So why do organisms bother? Maybe because of sex. (Perhaps you haven't noticed that we have been dealing with sex all along.) Meiosis, after all, is part of sex. So the next question is, what is sex good for in the long run? (You probably have an answer about the short run.)

In the long run, of course, we are all dead. But our genes live on. When we leave offspring ("we" includes all us sexual beings), we pass our reproductively "successful" genes on to our descendants. The problem is, our descendants may not find the world to be the same as the one *we* lived in; the weather changes, new diseases show up, and unexpected competitors eat our food. Logically, no one particular combination of genes, including ours, will be adequate to meet all new situations. So, according to evolutionary theory, sexual reproduction comes to the rescue. Sexual unions keep reshuffling the genes of all the successful individuals in the population so that virtually infinite possible combinations of genes are produced. Most of the new combinations will be less successful than the old ones because, whereas the old ones are tried and true, the new ones are genetic experiments. That is unfortunate, but in the long run it's the cost of success. The winning evolutionary strategy is to cover as many bets as possible by having highly variable offspring. Then, when the weather changes or the Creeping Purple Flu shows up, perhaps not all of our descendants will be wiped out. Some of them are likely to have the right combinations of genes to be able to live under these conditions.

We can see this principle at work in certain organisms that can reproduce either sexually or asexually. For instance, *Daphnia*, the little water flea, reproduces asexually for generation after generation, females producing only females. In fact, sometimes one can see an asexual female embryo within an asexual female embryo within an asexual adult! The little animals keep up this kind of reproduction as long as conditions are stable. But when their pond begins to dry up, or things otherwise get dangerous and unpredictable, they change strategies. They begin to produce both males and females. These go on to mate and reproduce sexually in the usual way. Thus, in hard times, a few of their highly variable offspring may have that particular combination of genes that will allow them to cope with their new environment.

There are two major take-home lessons from the story of meiosis. First, meiosis halves the chromosome number of eggs and sperm, making fertilization feasible. Second, meiosis provides a means of shuffling and reorganizing chromosomes, thus increasing genetic variation in offspring. This shuffling and reorganization takes place in at least five ways: (1) in the prophase I scrambling of chromosomes through crossing over; (2) in the random lining up of homologous chromosomes in metaphase I, so that maternal and paternal centromeres are randomly distributed to the two poles; (3) in females, the random distribution of chromosomes into polar bodies; (4) in the random ascendancy of one oocyte in the ovary to begin development while the rest remain frozen in prophase I; and (5) in the chancy competition of billions of individually unique sperm for a single unique egg. This is not to mention the somewhat random way two sexual individuals often get together in the first place.

APPLICATION OF IDEAS

1. Single-celled organisms that reproduce asexually through mitosis die in enormous numbers. But potentially, at least in some sense, such organisms are immortal. Multicellular organisms that can reproduce by fission (splitting) or budding (asexually producing miniatures of themselves) are also potentially immortal in the same sense. What do these statements mean? What kind of a tradeoff might there have been in the evolution of sex? Is death more of a reality to sexual beings than to asexual beings?

KEY WORDS AND IDEAS

NUCLEAR STRUCTURE

1. The nuclear structures include the general **nucleoplasm,** a double membrane called the **nuclear envelope,** the **perinuclear space** between membranes, and the nuclear pores that selectively admit the passage of materials.

2. The nuclear envelope disappears from sight during much of cell division.

DNA, Chromatin, and the Eukaryotic Chromosomes

1. **Chromosomes** consist of **chromatin,** which includes DNA and nuclear proteins known as **histones** and **nonhistone chromosomal proteins.**

2. The DNA of chromosomes is wound about numerous beadlike **nucleosomes,** which consist of histone.

3. In cell division, chromosomes are highly condensed and DNA is inactive. When active, they are uncoiled and spread out, forming the **chromatin net.**

The Cell Cycle:
Growth, Replication and Division

1. The **cell cycle** includes **growth, replication, mitosis,** and **cytokinesis.**

2. Some cell specialization requires that cells leave cycling. Many such cells no longer reproduce so their life span is usually limited. Others survive the lifetime of the organism.

3. The **M phase** (mitosis) of the cell cycle includes formation of the **mitotic spindle,** separation of chromosome replicas, and division of the cell into two. The remaining phases constitute **interphase.**

4. **G_1 phase** is a period of protein synthesis, while the **S phase** involves DNA replication, and the **G_2 phase** includes synthesis of the mitotic spindle protein.

5. At the end of G_2, each chromosome consists of two **chromatids** joined at their **centromere.**

MITOSIS AND CELL DIVISION

1. Cell division includes **mitosis** (chromosome separation) and **cytokinesis** (cytoplasmic division).

2. Mitosis includes four subphases: **prophase, metaphase, anaphase,** and **telophase.**

Prophase

1. Chromosome condensation is a matter of coiling and supercoiling in at least three orders.

2. Condensation facilitates the movement of chromosomes during mitosis.

3. In its proper use the term *chromatid* only applies to DNA replicas held together at the centromere in a chromosome. Upon centromeric division, chromatids become chromosomes.

4. Spindle fibers attach to centromeres at a region called the **kinetochore.**

5. The mitotic spindle consists of **microtubules** made of the protein **tubulin.**

6. Centrioles are short cylinders of microtubules arranged in nine groups of three each.

7. At the onset of mitosis, the centrioles move to opposite **spindle poles.** Their precise function is unknown, but they are not essential to spindle formation. Studies suggest that centrioles give rise to **basal bodies.**

8. The featureless area surrounding centrioles is called the **microtubular organizing center (MOC).**

9. Three types of spindle fibers include **centromeric microtubules** (pole to centromere—attached to kinetochore), **continuous microtubules** (pole to midcell—overlapping), and **astral rays** (part of the two **asters** that radiate outward like a starburst).

Metaphase and Anaphase

1. At metaphase, the chromosomes pause after aligning on the **metaphase plate.** Centromeric microtubules are seen on each side of each centromere.

2. At anaphase, the centromeres divide and the newly separated chromosomes migrate to opposite poles.

3. The mechanism of chromosome movement is not known, but the *sliding filament* and *subunit disassembly* hypotheses attempt to explain its action.

Telophase

1. Telophase is the reverse of prophase, including: chromosome uncoiling, reestablishment of nuclear envelope, and the formation of nucleoli. Newer data suggest that bits of the old nuclear envelope are conserved to be reassembled at this time.

Cytokinesis: Cytoplasmic Division

1. In animals, cytokinesis, or **cleavage,** involves the contraction of microfilaments below the cell membrane forming a cleavage furrow that tightens, dividing the cytoplasm.

2. In some fungi and algae, mitosis occurs without cytokinesis, forming multinucleate cells. Repeat-

ed replications without mitosis occur in some insects, producing giant **polytene chromosomes.**

3. Because of the cell wall, cleavage does not occur in plants, but cytoplasmic division requires that a new wall form between daughter nuclei.

4. Following telophase in plants, vesicles from the Golgi apparatus coalesce to form a **cell plate** between daughter nuclei. It fills with pectin, forming a gummy **middle lamella,** around which the new plasma membranes secrete new cell walls.

5. Mitosis and cell division are a primary means of **asexual** reproduction in single-celled organisms. In multicellular organisms, mitosis accommodates growth, cell replacement, repair, and cell specialization.

MEIOSIS

1. **Diploid** organisms must undergo **meiosis** to form **haploid** sex cells. Meiosis reduces the number of chromosomes in half (by the separation of pairs) and permits new gene combinations to arise.

Homologous Chromosomes

1. Chromosomes occur in pairs (in the original fertilized cell, one from each parent) and each pair can be identified through **karyotyping.**

2. Members of a chromosome pair are called **homologues,** or **homologous chromosomes.** Each member of a pair has the same array of genes although there may be a difference in the way each expresses itself.

3. During meiosis, homologues undergo **pairing (synapsis)** and **crossing over** (exchange of genes). Following this, the homologues are separated, ending up in different daughter cells.

4. Meiosis requires two divisions, **meiosis I** and **meiosis II,** which in humans reduce the 23 pairs of chromosomes (actually 46 chromatids because of replication) in a diploid cell to 23 chromosomes in each haploid gamete.

Meiosis I: The First Meiotic Division

1. Meiotic prophase includes five substages:
   a. **Leptotene:** earliest condensation
   b. **Zygotene:** synapsis, or pairing up
   c. **Pachytene:** crossing over
   d. **Diplotene:** X-shaped **chiasmata** form (regions still touching represent points of crossover).
   e. **Diakinesis:** homologue **repulsion**

2. During synapsis, homologues are bridged by the **synaptonemal complex** which is made up of protein and RNA. Its function is unknown, but it is absent in homologues that do not cross over.

3. Synapsis is also called the **tetrad** (foursome) state.

4. In crossing over, enzyme complexes called **recombination nodules** overlie the homologues. Regions along chromatids unwind, the double helix opens and single homologous strands break and cross over. Strands still remaining, and now unmatched, are broken down and resynthesized by base pairing.

5. Crossover produces **genetic recombination,** the creation of new gene associations in the chromosomes.

6. Chiasmata reveal where crossovers have occurred.

7. In metaphase I the highly condensed zygotene chromosomes are aligned on the metaphase plate. Centromeric spindle microtubules from one pole attach to *only one side of each centromere.*

8. In anaphase I, no centromeric division occurs, and because of the unique spindle attachment, each set of homologues separates, going to opposite poles. The subsequent reduction of chromosomes is called **reduction division.**

9. During telophase, the reorganization of the nucleus occurs, followed by cytoplasmic division.

10. There is no DNA replication during meiotic interphase.

Meiosis II: The Second Meiotic Division

1. Meiosis II closely resembles mitosis, and in anaphase II, centromeres divide and single chromosomes (former chromatids) are drawn to opposite poles.

Where Meiosis Takes Place in Higher Organisms

1. Meiosis occurs in **germinal tissues: gonads** in animals, and **ovules** and **anthers** in plants. Such tissues completely differentiate in the embryo of animals, but in plants can arise from simple undifferentiated tissues whenever plants flower.

2. **Germ-line** tissue gives rise to gametes. All other cells are called **somatic cells.**

3. In the human male embryo, **germinal epithelium** forms in the **seminiferous tubules** of the **testes,** but actual meiosis does not begin until puberty.

4. In the human female embryo, germinal epithelium develops in the **ovaries** and by the time of birth meiosis has begun in all germ-line cells or **oocytes.** Meiosis ceases during prophase I, and remains suspended in each oocyte until it actually undergoes ovulation.

5. Though meiosis is suspended in oocytes, some DNA uncoils into **lampbrush chromosomes,** transcribes RNA, actively produces protein.

6. While each meiosis in males produces four functional sperm, meiosis in females results in only

one oocyte forming. The other cells are lost as **polar bodies** following unequal cleavages in telophase I and II.

Why Meiosis?

1. Meiosis provides variation through (a) random genetic exchange during crossing over, (b) random chromosome lineup at the metaphase plate, (c) random discarding of chromosomes into polar bodies, (d) random selection of oocytes for development and ovulation, and (e) random success of genetically different sperm.

REVIEW QUESTIONS

1. Using the terms *DNA, chromatin, histone,* and *nucleosome,* describe the organization of a chromosome. (255)

2. Why is isolation of the nucleoplasm of less importance during mitosis than during interphase? (255)

3. What is a chromatin net? Of what significance are the dense and diffuse regions? (255)

4. Mention two kinds of cells that cease cycling and die and explain how their death makes their function possible. (255)

5. List the four stages of a cell cycle, and briefly describe the events of each. (256)

6. What is the relationship between the chromosome and the chromatid? In what case is the latter term appropriate? (257)

7. Clearly distinguish between mitosis and cytokinesis. (257)

8. In a few sentences, clearly explain the function of mitosis. (267–268)

9. List the four phases of mitosis, and summarize the events of each.

10. Using a simple illustration, suggest how four levels of coiling occur in the chromosome. What is the function of such coiling or condensing? (257, 261)

11. Describe the centromere. What observation suggests that it is a permanent part of the chromosome? Upon which part of a centromere do the spindle microtubules attach? (261, 263)

12. How do the centrioles behave during mitosis? What observation suggests that they are not essential to mitosis? (264)

13. Name and describe the orientation of the three different spindle microtubules. (265)

14. Describe the arrangement of chromosomes and the centromeric spindle fibers at mitotic metaphase, and also explain the significance of the arrangement. (265)

15. Summarize the events of mitotic anaphase. What do these events assure about the genome of daughter cells? (265)

16. Briefly describe cytokinesis in animal cells. What part of the cytoskeleton is involved? (265)

17. Explain how a new cell wall forms between daughter cells in the plant. Why is cleavage impossible? (266)

18. Discuss the importance of mitosis to single-celled organisms such as algae and protozoans. (267)

19. Humans have 46 pairs of chromosomes. How many chromosomes would one find in a cell in G_1, a cell in mitotic metaphase, a sperm cell, a cell in meiotic prophase I, a cell in meiotic telophase I, and a cell in meiotic telophase II? (280)

20. Using humans with their 46 chromosomes as an example, explain the number problem solved through meiosis. (268, 270)

21. In a few words, summarize the events in the following: leptotene, zygotene, pachytene, diplotene, diakinesis. Be sure to use the terms *repulsion, synapsis, crossing over, condensation* and *chiasma.* (272)

22. Summarize the hypothetical events in crossing over. What does crossover accomplish? (274–275)

23. Compare the attachment of centromeric microtubules in metaphase I of meiosis with that of mitosis. What specific difference will this make in anaphase? (275–276)

24. Summarize the events of meiosis II, comparing each phase to those of mitosis. (276–277)

25. Briefly summarize the development of germinal epithelium in the human female. How does this development differ in the male? (280)

26. Using a diagram, suggest how meiotic cytokinesis in the female vertebrate differs from cytokinesis in the male. Offer a logical reason for the difference. (280, 283)

27. What is the functional significance of lampbrush chromosomes? (281)

28. List four ways in which genetic variability is increased through sexual reproduction. (283)

Mendelian Genetics

Occasionally you may hear someone just back from a trip to Spain boasting of having "fought a bull." In fact, that very noun may come to mind when you learn that the "bullfight" was actually with a heifer (a young cow). At certain times of the year, guests are invited to the ranches where fighting bulls are bred. There they can watch the ranchers test the young cows to determine which stock will be used for breeding. As part of the festivities, the guests may be invited to try their hand at caping a heifer. The ranchers seem to think that nothing is quite so festive as watching gringos trying to stand their ground with an angry cow.

Professionally, the ranchers are interested in how the cows fight because it is traditionally accepted in such circles that a fighting bull gets its strength from its father, but its courage from its mother. The origin of such beliefs is lost in tradition, but it is evident that people applied the principles of genetics long before they knew anything about genes.

Charles Darwin, as the son of a gentleman farmer, was well aware of the basic principle of inheritance, namely that offspring tend to resemble their parents. The offspring of heavier animals, he knew, tended toward heaviness; among cows, the daughters of high-yielding milk producers tended to produce more milk. Certain traits tend to be transmitted from generation to generation.

WHEN DARWIN MET MENDEL (ALMOST)

Darwin was probably the 19th century's greatest biologist, and one of its greatest geneticists. Even so, most of his ideas about heredity were wrong. Darwin's chief mistake was to accept the only intellectually respectable theory of his day, which was **blending inheritance.** It was believed that the "blood," or hereditary traits, of both parents blended in the offspring, just as two colors of ink blend when they are mixed. The blending hypothesis appeared to work reasonably well for some traits, such as height or weight, which have a continuous gradation of possible values, but it couldn't account for other observations.

It is always surprising, in hindsight, to recognize the degree to which a strongly held belief will blind its proponents to obvious contradictions. The blending theory would predict that the offspring of a white horse and a black horse should always be gray, and that the original white or black should never reappear if gray horses continued to be bred. But it turns out that the offspring are not always gray, and black or white descendants do appear. Similarly, if a honey-colored cocker spaniel were crossed with a black one, the offspring would usu-

287

ally be black or honey, not some kind of golden black. Something obviously was wrong with the theory, but no one seemed to notice, or if they did, they tried to ignore it.

In fact, the blending theory doesn't work even for continuous, graduated traits like height and weight. If it did, the offspring in every generation would always be less extreme than their parents for every trait. Every time a fat horse mated with a thin horse, the offspring should always have been an "average" horse. In that case the world would by now be full of average, gray horses and average everything else, but in fact, in a group of horses that eat about the same amounts, some will be fat and others thin. Horses are also still found in many different colors, and even expert breeders usually can't predict the color of a foal.

When Darwin finally got around to publishing his theory of natural selection, some of his sharper critics used the weakness of his genetic arguments against him. They pointed out, quite correctly, that natural selection will not work with blending inheritance because any variation will be blended away. Try as he might, Darwin could not come up with an answer for them, and the tired old man, so sickened in his later years, began to backtrack. In later editions of the *Origin of Species*, he began to give more and more weight to the possibility of the inheritance of acquired traits, but his desperate explanations were so bad that we won't bother you with them. The idea of "inheritance of acquired traits," which stated that traits acquired during the lifetime of an individual (muscular development, criminal tendencies, knowledge, and so on) could be passed on to offspring, is actually ancient but remained quite popular and respectable in the 19th century. A leading proponent was Jean Baptiste Lamarck, the famous 19th century French naturalist. The basic tenets of Lamarck's hereditary notions have largely been discarded.

His shortcomings notwithstanding, Darwin was a keen enough observer to realize that some parental traits could, in fact, disappear in the offspring and reappear in the following generation. Actually, while he was still working on *Origin*, he did some very good experiments to try to solve this nagging puzzle. He performed almost exactly the same experiment that the Austrian monk, Gregor Mendel, was performing at just about the same time, and the two great scientists, unknown to each other, obtained almost exactly the same results. But only Mendel was able to understand what had happened. This almost incredible historical incident points up an aspect of Darwin's science that is

sometimes overlooked: he was an experimenter of dazzling abilities.

In his set of experiments, Darwin crossed two **true-breeding** strains of snapdragons (true-breeding means that when individuals of a strain are crossed, the offspring are the same, generation after generation). The two strains differed in only one respect: one strain had abnormal, radially symmetrical flowers (what botanists called **peloric** flowers), and the other had the normal, bilaterally symmetrical snapdragon flowers (Figure 12.1). Of the progeny of this first cross, Darwin wrote, "I thus raised two great beds of seedlings, and not one was peloric." Darwin called this tendency of one trait in a cross to suppress another **prepotence**. We now call it **dominance**. We will leave Darwin's genetics experiment now, but will return to it later.

And, as we know, while Darwin was working on his snapdragon experiment, scientific history was being made across the English Channel, deep in Europe. The self-effacing but incredibly bright and dedicated Gregor Mendel was crossing strains of garden peas.

The Abbot Gregor Johann Mendel (1822–1884), a member of an Augustinian order in Brunn, Moravia (now part of Czechoslovakia), is often depicted as a kindly old man of the cloth who, while puttering around his monastery garden, somehow stumbled onto important genetic laws. Again, our tendency to embellish the memory of already worthy individuals has led us astray. The real Mendel was remarkable enough in his own right. In many

12.1 DARWIN, THE GENETICIST

One of Charles Darwin's many excellent experiments included breeding snapdragons. In one observation he crossed true-breeding *peloric*-flowered snapdragons (a) with true-breeding bilaterally symmetrical flowers (b). He found that the peloric flower trait was lost in the progeny. In describing what we would call the dominance of one trait over another, he coined the term *prepotence*.

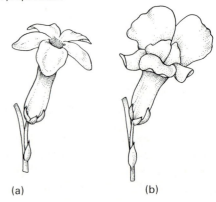

(a)　　　(b)

12.2 GREGOR MENDEL (1822–1884)

Bringing together an innate curiosity, keen observational powers, and mathematical training, Mendel developed the basic laws of heredity. He was undoubtedly the first mathematical biologist. His work, published in 1866, went unnoticed for many years until it was rediscovered about the turn of the century. Once his ideas were understood, they opened the door to 20th century genetics.

ways he seems more like a 20th century biologist somehow displaced into the wrong century (Figure 12.2). His first published paper is a landmark in its clarity and a model modern experimental report. Many scientists believe that it is perhaps the best scientific paper ever written.

Early in his life Mendel began training himself, and he became a rather competent naturalist. To support himself during those years, he worked as a substitute high school science teacher. The professors at the school, noting his unusual abilities, suggested that he take the rigorous qualifying examination and become a regular member of the high school faculty. Mendel took the test and did reasonably well, but he failed to qualify. Mendel joined a monastic order instead.

His superiors, confident of his abilities, sent him in 1851 to the University of Vienna for two years of concentrated study in science and mathematics. There, he learned about the infant science of statistics, which was to serve him well.

When Mendel returned to the monastery he began his plant-breeding studies in earnest. He impressed his fellow monks with his intelligence and his vigor, and he applied these qualities to the study of plant breeding. He developed new varieties of fruits and vegetables, kept abreast of the latest developments in his field, joined the local science club, and became active in community affairs.

Mendel's Crosses

Mendel began to experiment by observing the effects of crossing different strains of the common garden pea. To begin with, he based his research on a very carefully planned series of experiments and, more important, on a statistical analysis of the results. The use of mathematics to describe biological phenomena was a new concept. Clearly, Mendel's two years at the University of Vienna had not been wasted.

The careful planning that went into Mendel's work is reflected in his selection of the common garden pea as his experimental subject. There were several advantages in this choice. Since others had successfully used the garden pea in experimental crosses, he wasn't proceeding entirely on guesswork. Furthermore, pea plants were readily available and fairly easy to grow, and Mendel was able to purchase 34 true-breeding strains. These strains differed from each other in very pronounced ways so that there could be no problem in identifying the results of a given experiment. Mendel chose seven different pairs of traits to work with:

1. round or wrinkled seed form;
2. yellow or green seed contents;
3. white or gray seed coat;
4. green or yellow unripe seed pods;
5. inflated or constricted ripe seed pods;
6. short (9-18 inches) or long (6-7 feet) stem;
7. axial (along the stem) or terminal (at the end of the stem) flowers.

The old description, "alike as two peas in a pod," then, is not always genetically valid.

Knowing that they yield more solid data, Mendel chose to work with large numbers. His choice of the pea was fortunate since it produces large numbers of offspring. Keep in mind that each pea in a pod is a seed—essentially a new plant, an individual with its own **genotype** and **phenotype.** An organism's genotype is its total combination of **genes,** and the phenotype is the combination of its *observable traits.* Thus, the phenotype of a plant's offspring—traits such as seed form and coat color—can easily be determined by simply examining the peas.

Mendel's approach was to cross two true-breeding strains that differed in only one characteristic, such as seed color. Peas ordinarily self-fertilize, so to cross two strains, the pollen had to be transferred by hand. We now call the original parent generation P_1 and designate the first generation offspring F_1 (first filial) generation. When the F_1 plants are crossed with each other or are allowed to self-pollinate, the resulting offspring are called the F_2 generation, and so on.

The Principle of Dominance. When Mendel crossed his original P_1 plants, he found that the characteristics of the two plants didn't blend, as prevailing theory said they should. When plants grown from round seeds were crossed with plants grown from wrinkled seeds, their F_1 offspring were not intermediate seeds. Instead, all of them were round seeds. Mendel termed the trait that appeared in the F_1 generation the *dominant* trait, and the one that had failed to appear the **recessive** trait. But he was now left with a vexing question. What had happened to the recessive trait? It had been passed along through countless generations, so it couldn't have just disappeared.

Mendel then allowed his F_1 pea plants to pollinate themselves. And, about that time in England, Darwin was allowing his two beds of snapdragon seedlings to pollinate themselves.

In his book, *Variation in Plants and Animals under Domestication*, Darwin recorded his results. Of the offspring of the F_1 snapdragons that bore flowers, he had found that 88 had normal flowers, 37 had abnormal (peloric) flowers, and 2 had a few flowers he couldn't classify. Thus he could classify 125 of the F_2 plants into two discrete groups of unequal size. But where had the peloric flowers come from? Weren't they bred out in the F_1 generation? What about the blending theory? The trait that had disappeared in the first generation had reappeared in the second. Now, here was something that had to be explained! Darwin was greatly puzzled but finally came up with a suitable explanation. He said the phenomenon was due to **latency.** He said that the latent character had "gained strength by the intermission of a generation." Apparently satisfied, he directed his attention to other things. Darwin went to his grave unaware of how close he had come to arriving at yet another great biological principle.

Mendel, however, stayed with the problem. In the offspring of the F_1 generation, which we call the F_2 generation, Mendel found that roughly one fourth of the peas were wrinkled and that about three fourths were round. He repeated the experiment with other pea strains and obtained similar results. He crossed a yellow-pea strain with a green-pea strain. All of the F_1 peas were yellow, but in the F_2 generation about one fourth of the peas were green again. The ratios did not escape the tenacious Mendel, determined to badger the problem until he could make some sense of it.

Darwin, when he recorded his own numbers, probably did not notice that 37/125 is fairly close to one fourth. Even if he had noticed, it would probably have been of no more significance to him than one third or one fifth. Darwin apparently had a very different kind of mind than Mendel. Mendel was not only one of the first mathematical biologists familiar with statistics, but he also knew how to isolate small parts of great problems. Darwin's genius, on the other hand, was in the enormous breadth and scope of his ideas and the ability to fit seemingly unrelated details into a grand scheme.

Mendel expanded his experiments. He continued to use the trusty garden pea in spite of the fact that his artificial crosses relied on his manipulation. Left to themselves, most pea plants will simply self-fertilize and produce plants just like themselves. To get a cross between two plants or two strains, it was necessary to open the normally closed pea flower, remove the pollen-producing anthers (to prevent self-pollination), and apply the foreign pollen with a small paintbrush. This way, Mendel could control his crosses, prevent accidental contamination, and allow self-pollination only when it suited his needs.

One of Mendel's early observations was critical to the development of his genetic model. Mendel realized that there were two kinds of round peas: the true-breeding kind, like the original parent stock, which would grow into plants that would bear only round peas; and another type, which when grown and self-pollinated would produce pods containing both round and wrinkled peas. At this point we can almost hear Mendel musing, " . . . two kinds of round peas—one pure-breeding, one not." Were there also two kinds of wrinkled peas? He found that there were not. When wrinkled peas were cultivated and allowed to self-pollinate, they always bore only wrinkled peas. We now call the true-breeding traits **homozygous,** and the other kind **heterozygous.** If an organism is homozygous for a trait, then both of its genes for that trait are identical. If it is heterozygous for that trait, then the two genes are different.

Mendel realized that it was impossible to tell the difference between the two kinds of round peas unless they were allowed to mature and reproduce. He decided he needed an F_3 generation, which he obtained through a procedure called **progeny testing.** This test would allow two seemingly identical round seeds to show just how different they really were. The kind we call homo-zygous, if allowed to self-pollinate, would produce one type of pea, while the heterozygous plant would produce two types of peas. He planted the round peas of the F_2 generation, waited another year for them to grow and bear pods, and then opened the pods to determine whether the plants were true-breeding or not. He found that one third of the F_2 round peas had been true-breeding (were homozygous) and that two thirds had not (and thus were heterozygous). More numbers! Darwin would have fallen asleep by this time.

But Mendel was intrigued. He determined that three fourths of the F_2 generation had been round, and one third of the round F_2 had been true-breed-ing. One third of three fourths is one fourth. Fur-ther, when he considered all the F_2 peas together, he found the following (Figure 12.3):

¼ of the total F_2 were round and true-breeding (homozygous)

½ of the total F_2 were round and not true-breeding (heterozygous)

¾ round

¼ of the total F_2 were wrinkled and true-breeding (homozygous)

¼ wrinkled

The same thing worked for yellow and green peas. There were two kinds of yellow and one kind of green. Mendel looked at all seven traits (Table 12.1), and the same principle worked in each case; roughly the same ratios were generated. In every case, one form, bearing what Mendel called the *dominant* trait, was present in 100% of the F_1 peas or pea plants. In every case, three fourths of the F_2 peas or pea plants showed the dominant form of the trait (we would say the dominant phenotype),

TABLE 12.1

MENDEL'S F_2 GENERATIONS

The dominant and recessive traits analyzed by Mendel are shown along with the results of F_1 and F_2 crosses. Note the large numbers he worked with. How does a large sample size help with the conclusions? How well do his numbers in the last two columns agree with what probability tells us to expect in the crosses? The proportion of the F_2 generation showing recessive form is in the last column.

Dominant form	Number in F_2 generation	Recessive form	Number in F_2 generation	Total examined	Ratio	Proportion of F_2 generation
Round seeds	5474	Wrinkled seeds	1850	7324	2.96:1	0.253
 Yellow seeds	6022	 Green seeds	2001	8023	3.01:1	0.249
Gray seed coats	705	White seed coats	224	929	3.15:1	0.241
 Green pods	428	 Yellow pods	152	580	2.82:1	0.262
 Inflated pods	882	 Constricted pods	299	1181	2.95:1	0.253
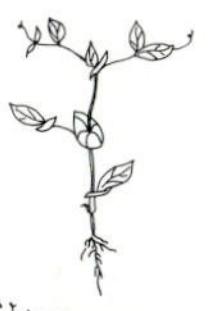 Long stems	787	 Short stems	277	1064	2.84:1	0.260
 Axial flowers (and fruit)	651	 Terminal flowers (and fruit)	207	858	3.14:1	0.241

12.3 ROUND X WRINKLED PEAS

(a) In one of his early experiments, Mendel cross-pollinated pea plants that were true-breeding for round peas (seeds) with those that were true-breeding for wrinkled peas (seeds). These represented the P_1 or parental generation. (b) This cross produced only round seeds, which occur in the pods as the F_1 generation. (c) When the round F_1 seeds were grown into mature F_1 plants and allowed to self-pollinate, the F_2 progeny that appeared in the pods included both round and wrinkled. (d) These occurred in a near perfect 3:1 ratio: ¾ of the pea seeds were round and ¼ were wrinkled. (d) Mendel realized he was confronted with two kinds of round peas. There were those that always produced offspring exactly like themselves and those that produced both offspring like themselves and offspring with an alternative trait. (e) In an effort to better understand his F_2 results with its 3:1 ratio, Mendel planted the F_2 seeds and let the adult plants self-pollinate once again. While all wrinkled F_2 seeds were true-breeding, round F_2 seeds were of two kinds: some were like the original true-breeding parental strain and bore pods with only round seeds; others were like the F_1 plants in having both round and wrinkled seeds in their pods. The true F_2 ratio, then, was 1:2:1—namely, ¼ true-breeding round, ½ not true-breeding round, and ¼ true-breeding wrinkled.

Round seed   Wrinkled seed

Only this cross requires hand-pollination

F_1 seeds planted, allowed to grow, and self-pollinate

F_2 progeny counted, planted, and allowed to self-pollinate

Mendel's progeny testing

(a) P_1 or parental generation: cross between two true-breeding strains

(b) F_1 generation progeny: all round seed in pods

(c) F_2 generation progeny: both round and wrinkled seeds in pods

(d) Ratio: ¾ round, ¼ wrinkled

(e) F_3 generation progeny test

⅓ of round-seed F_2 plants (¼ of all F_2 plants) are true-breeding: only round seeds in pods

⅔ of round seed F_2 plants (½ of all F_2 plants) are *not* true-breeding: both round and wrinkled seed in pods

All wrinkled-seed F_2 plants (¼ of all F_2 plants) are true-breeding: only wrinkled seeds in pods

and one fourth of the F_2 peas or pea plants showed the contrasting, recessive trait.

In every case, a breeding test revealed that in actuality, one fourth of the F_2 were the dominant form and true-breeding, one half of the F_2 were the dominant form but not true-breeding, and one fourth of the F_2 were the recessive form and true-breeding.

Clearly, something determining the recessive form was passed down from the true-breeding recessive parental strain, through the hybrid F_1, to the true-breeding recessive F_2; but whatever it was, it was skipping the F_1 generation. Mendel thought he could use algebraic symbols to express his dilemma (Figure 12.4).

He let **A** represent the factor that determines the dominant form and he let **a** represent the factor that determines the recessive form. The F_1 hybrid, and in fact all the non-true-breeding plants, must have both factors present, as represented by the symbol **Aa**. Since there are two parents, Mendel figured that in the hybrid, **A** comes from one parent and **a** from the other. Mendel determined experimentally that it didn't matter which parental strain bore the peas and which provided the pollen. He let the capital letters signify dominance and the lowercase letters signify recessivity, which merely meant that the **Aa** individuals would look exactly the same as the **A**-bearing parental stock.

If the heterozygous plants get an **A** from one parent and an **a** from the other parent, and are symbolized **Aa,** it makes sense that the true-breeding dominant forms get two **A** factors—one from each parent—and can be symbolized **AA**. In the same way, the true-breeding recessive forms get **a** factors from both parents and can be symbolized **aa**. We can use **AA** to symbolize the dominant **homozygote** (when the factor from each parent is identical), **aa** to represent the recessive homozygote, **Aa** to represent the **heterozygote** (when the factor from each parent is different), and **A___** to symbolize those plants with the dominant phenotype whose complete genotype is not known.

Briefly, then, the basic F_2 ratio of Mendelian genetics is:

Cross: **Aa** × **Aa**
Offspring: ¼ **AA**
½ **Aa**   } ¾ dominant phenotype, **A___**
¼ **aa**   } ¼ recessive phenotype, **aa**

Now Mendel was ready for some conclusions. First, he said, each of the seven characteristics was controlled by transmissible factors that came in two

12.4 MENDEL'S SYMBOLS

From his many trials Mendel knew that his heterozygous F_1 peas carried two factors responsible for producing the pea shape round or wrinkled. He represented these two factors as **A** and **a**. The capital **A** represented dominance, while the lowercase **a** represented recessiveness. Each individual in his crosses can be represented with a pair of letters. As you can see, the progeny of an F_2, represented this way, applies itself to algebraic factoring to obtain the genotypes of the F_1 that produced it. Likewise, you can begin with the F_1 and simply multiply the algebraic factors **Aa** × **Aa** together to predict the F_2 results.

AA × aa P_1 true-breeding strains
Aa F_1 progeny
Aa × Aa F_1 cross
AA : 2Aa : aa F_2 progeny

¼ round (**AA**)   = homozygous round
½ round (**Aa**)   = heterozygous round
¼ wrinkled (**aa**) = homozygous wrinkled

forms, dominant and recessive. Second, every hybrid individual had two different copies of the factor controlling each character, one from each parent. Third, if the factors were the same (if the individual were homozygous), the individual would be true-breeding. Fourth, if the factors were not the same (if the individual were heterozygous), the dominant factor would determine the appearance, or phenotype, of the plant. Today, these factors—the alternate forms of a gene on two homologous chromosomes—are known as **alleles**. Mendel considered these important conclusions to be preliminary. Let's now look at some of the implications of Mendel's work in more detail.

Mendel's First Law: The Segregation of Alternate Alleles

Mendel's first law specifies that *any trait is produced by at least a pair of factors, or alleles, which segregate (separate into different gametes) during pollen or ovule (or sperm and egg) production.* Each pair of alleles received may be heterozygous or homozygous. Where two alleles differ, the dominant one will be expressed over the recessive. Let's look further into these ideas.

Mendel was the first to realize that *when a heterozygote reproduces, its gametes* (in plants the sperm-containing pollen and egg-containing ovule) *will be of two types in equal proportions.*

Now, how did he come up with that? He did it by playing with the numbers his experiments were generating. He figured that when the heterozygous plant was allowed to pollinate itself, it would produce half **A** ovules and half **a** ovules. It would also produce half **A** pollen and half **a** pollen. When it self-pollinated then, what proportion of all the fertilized ovules would be **aa?**

The probability of any given ovule being **a** is ½. The probability of this ovule being fertilized by an **a** pollen is also ½. One of the basic **laws of probability,** the **multiplicative law,** states that *the probability of two independent events both occurring is equal to the product of their individual probabilities* (Figure 12.5). Since an ovule being **a** and the pollen being **a,** have

12.5 THE MULTIPLICATIVE LAW

The probability of two independent events both occurring is equal to the product of their individual probabilities. The law can be tested with coins. In this scheme, two coins are used to represent two independent events. Toss two pennies 100 times. Record the results on a graph. Explain why the probability of getting one head and one tail is $2 \times \frac{1}{2} \times \frac{1}{2} = \frac{1}{2}$. What does this have to do with Mendel?

When each penny is tossed: Probability of H = ½
Probability of T = ½

Heads (H)   Tails (T)   Heads (H)   Tails (T)

When both pennies are tossed together:

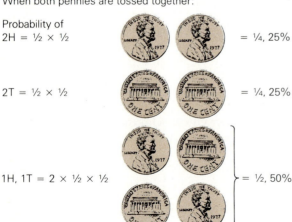

Probability of
$2H = \frac{1}{2} \times \frac{1}{2}$ = ¼, 25%

$2T = \frac{1}{2} \times \frac{1}{2}$ = ¼, 25%

$1H, 1T = 2 \times \frac{1}{2} \times \frac{1}{2}$ = ½, 50%

no influence on each other, the probability that both of these things will happen for any given pea is ½ × ½ = ¼. (If the probability of your flunking your next exam is ⅙ and the probability of rain on that day is ⅕, then the probability that both these things will happen is ⅟₃₀, unless your test performance influences the weather, or vice versa.)

The same reasoning applies to the quarter of the F_2 progeny that are **AA:** a ½ chance of **A** pollen times ½ chance of **A** ovule gives a ¼ chance of a pea (a newly fertilized ovule) being **AA.**

But *half* of Mendel's F_2 peas were **Aa.** How do we account for that? We find that there are two different ways that **Aa** zygotes are formed in an **Aa** × **Aa** cross. Either an **a** pollen fertilizes an **A** ovule, or an **A** pollen fertilizes an **a** ovule. The probability of the first combination of events is ½ × ½ = ¼, and the probability of the second combination of events is ½ × ½ = ¼.

There is no other way of getting an **Aa** zygote from such a cross, and the two possibilities are *mutually exclusive* (they can't both happen to the same zygote). Furthermore, Mendel had shown that the **Aa** heterozygotes were identical regardless of which parent contributed which factor.

This leads to another basic law of probability, the **additive law:** *The probability of either one or another of two mutually exclusive events occurring is equal to the sum of their individual probabilities.* If there is a ½ probability of a coin coming up heads, and a ½ probability of it coming up tails, then you can be 100% sure (½ + ½ = 1) that it will come up either heads or tails.

But back to our question: Why were half the F_2 peas **Aa?** We know that the two mutually exclusive possibilities are **(1) a** pollen and **A** ovule, and **(2) A** pollen and **a** ovule. If either one or the other of these events occur, the zygote will be an **Aa** heterozygote. The probability of a zygote from such a cross being an **Aa** heterozygote is ¼ + ¼ = ½.

We have already seen another example of the additive law in probability: the events "zygote is **AA**" and "zygote is **Aa**" are mutually exclusive. The probability of the first in an **Aa** × **Aa** cross is ¼, and the probability of the second is ½. Thus, the probability that an individual will have the dominant phenotype **A**___ is ¼ + ½ = ¾, which, of course, is what Mendel observed.

Early in the 20th century, these relationships were put into a graphic form by a fan of Mendel named Reginald Crandall Punnett. Figure 12.6 shows a Punnett square used to keep track of a cross between two F_1 round individuals. The letters **R** and **r,** have been used to represent the dominant

12.6 PUNNETT SQUARES

Punnett squares are often used as a means of keeping track of the factors being multiplied in a cross. In the simple cross illustrated here, the factors **R** (round) and **r** (wrinkled) are separated (as they are in meiosis) from each parent and placed on the squares. We derive the genotypes of the offspring where the factors intersect in the squares. In this example, each square represents one fourth of the offspring.

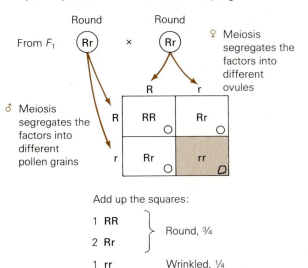

Add up the squares:

1 RR
2 Rr } Round, ¾

1 rr   Wrinkled, ¼

round and recessive wrinkled alleles, respectively (another convention in genetics). Each little square represents the simultaneous occurrence of the event directly above it and to its left. Or we might simply say that gametes from one parent are written at the left and from the other parent at the top (as shown). The symbols within the square represent possible fertilizations. Note that there are two **Rr** squares, which must be added to account for all the **Rr** individuals in the F_2 progeny.

We have called Mendel a mathematical biologist not just because he was trained both in mathematics and biology, or because he was among the first biologists to use statistical analysis in his work, but also because of the way he arrived at his conclusions. So, we might now ask, what is a mathematical biologist and how does he or she work? Usually he or she starts with some set of observations. In Mendel's case it was the dominance of one trait in the first generation and the reappearance of the recessive trait in some following generation. By some mental process involving both intuition and logic, the mathematical biologist then constructs a **model.** The model is an imaginary biological system, based on the smallest possible number of assumptions, and it is expected to yield numerical data consistent with past observations. New exper-

iments are then performed to test further predictions of this model. If the new data don't fit the predictions, the model is discarded or adjusted to fit the new observations so that further experiments can be done. A model, then, is a biological hypothesis with mathematical predictions (although the term is often applied in a nonmathematical manner).

It is essential that one understand that in science, models and hypotheses cannot be *proven* with experimental data. We can only say that the data are consistent with the model. Mendel did not *prove* his model of the segregation of alternate traits, but the simplicity of the model and the excellence of the fit enabled him to make new predictions and test them. His success came very close to a proof, at least as far as he was concerned. However, others were unconvinced until after the discovery of chromosomes and meiosis. Have you noticed how well Mendel's findings fit with what you already know about meiosis? Imagine how elated Mendel would have been if meiosis had been discovered in his own lifetime.

Mendel's Second Law: Independent Assortment

In his second law, the law of independent assortment, Mendel observed that *the inheritance of any one pair of factors (alleles) in pollen or ovule formation occurred independently of the simultaneous inheritance of any other pair.* Let's see how he arrived at this.

We have noted that Mendel succeeded in breaking his problem down to its smallest parts, partly by studying only one genetic characteristic at a time. The next step was to study two characters at a time, so he decided to cross a true-breeding strain that bore round, yellow peas with another true-breeding strain that bore wrinkled, green peas.

The F_1 offspring were all round and yellow. We can symbolize this as follows:

$$RRYY \times rryy \longrightarrow RrYy$$

where **R** are **r** are symbols for the two *alleles* of the round-wrinkled gene and **Y** and **y** are symbols for the two *alleles* of the yellow-green gene. Although we are using Mendel's basic symbols and concepts, we can now introduce another modern term. We defined *allele* earlier as "a particular form of a gene"; now let's note that each gene resides at a specific **locus.** The term *locus* (plural, *loci*) derives from our knowledge that each gene occupies a specific place,

12.7 INDEPENDENT ASSORTMENT OF TWO PAIRS OF ALLELES

Mendel developed his second law by crossing pea plants for two traits simultaneously. The scheme here illustrates the crosses from P_1 through F_2, using the Punnett square to show the results of the F_1 self-pollinated cross. The traits being considered are shape (round or wrinkled) and color (yellow or green), both characteristics of seeds. (a) When true-breeding round-yellows (RRYY) are crossed with true-breeding wrinkled-greens (rryy) (b), the F_1 is heterozygous round-yellow (RrYy), since these are the dominant traits. (c) Inbreeding the F_1 individual produces four different types of gametes. The F_2 generation is diagrammed in the Punnett square. It is composed of four distinct phenotypes, which include all the possible color and shape combinations. (d) These occur in the 9:3:3:1 ratio. Mendel determined this ratio by counting and classifying 556 pea seed offspring.

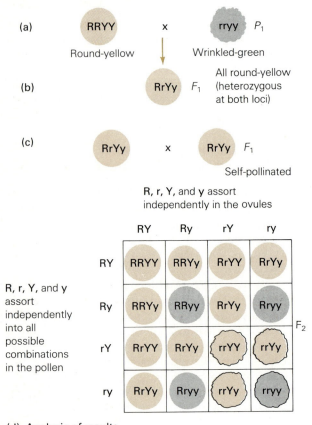

(d) Analysis of results

Round and yellow (like the F_1)	9	(9/16)
Round and green	3	(3/16)
Wrinkled and yellow	3	(3/16)
Wrinkled and green	1	(1/16)
Total	16	

or locus, on the chromosome. For Mendel's purposes, it was sufficient to suppose that different features were affected by alternate factors.

Now let's see what happened in the F_2 generation when Mendel conducted a dihybrid cross—that is, crossed plants that were different in two ways (Figure 12.7). First, as we have said, the F_1 peas were uniformly round and yellow. In the F_2—the offspring of $F_1 \times F_1$ (RrYy × RrYy)—Mendel classified 556 peas into four groups (the R__ and Y__ indicate that although the individual is dominant, the second allele isn't known).

315	round and yellow	R__ Y__
101	wrinkled and yellow	rrY__
108	round and green	R__ yy
32	wrinkled and green	rryy

Note that 133 peas altogether were wrinkled and 140 peas altogether were green. Both numbers are close to 139, which is one fourth of 556.

The results supported Mendel's first law: about one fourth of the F_2 peas were wrinkled and one fourth were green, while three fourths were round and three fourths were yellow. But the data indicated more than that. Mendel now suspected that the two pairs of contrasting characters were inherited independently. We have already mentioned the multiplicative law of probability, which gives the probability of two independent events both occurring. If the probability of being round is 3/4 and the probability of being yellow is also 3/4, the probability of being both round and yellow is 3/4 × 3/4 = 9/16. And 9/16 of 556 peas is 312.75 peas. When Mendel counted he found 315 round, yellow peas—remarkably close! Researchers rarely reach such good agreement with their hypotheses (Table 12.2).

Mendel thought the fit between expected and observed numbers was impressive. He went on to determine which of these peas were true-breeding and which were not. Here again, the fit was good.

TABLE 12.2

MENDEL'S PREDICTIONS AND RESULTS FOR F_2 PHENOTYPE

Phenotype of F_2	Fraction predicted	Number predicted out of 556	Number actually observed
Round and yellow	9/16	312.75	315
Wrinkled and yellow	3/16	104.25	101
Round and green	3/16	104.25	108
Wrinkled and green	1/16	34.75	32

He expected one fourth to be **RR,** half to be **Rr,** one fourth to be **rr:** one fourth to be **YY,** half to be **Yy,** and one fourth to be **yy,** so that:

$$\frac{1}{4} \times \frac{1}{4} = \frac{1}{16} \text{ should be } \mathbf{RRYY}$$
$$\frac{1}{2} \times \frac{1}{4} = \frac{1}{8} \text{ should be } \mathbf{RrYY}$$
$$\frac{1}{2} \times \frac{1}{2} = \frac{1}{4} \text{ should be } \mathbf{RrYy}$$

and so on. Mendel was able to get 529 of his 556 F_2 peas to bear F_3 progeny. The breakdown of their genotypes are listed in Table 12.3.

Again, Mendel felt that the numbers fit his model quite well. He tried combinations of other traits; he even tried three traits together. In each case, the different pairs of alternative traits behaved independently. We know why, of course: maternal and paternal chromosome pairs line up and separate independently during meiosis (Figure 12.8). But Mendel had never heard of meiosis (except that in a sense, he discovered it). In Mendel's words, "The behavior of each pair of differing traits in a hybrid association is independent of all other differences between the two paternal plants . . . ." This is known as Mendel's second law, or the law of independent assortment. It can be restated: *If an organism is heterozygous at two unlinked loci, each locus will assort independently of the other.*

The more recent statement is based on Mendel's work as modified by subsequent findings. But how did the word *unlinked* get in there? It was added many years later. What Mendel didn't know was that the law of independent assortment works only for genes that are not on the same chromosome. If **R** and **Y** are on the same chromosome, they are said to be linked, and Mendel's law of independent assortment simply doesn't hold for such loci. We will come back to all this in the next chapter. As it happened, none of the three gene loci that Mendel used in his multifactor crosses were linked. Perhaps Mendel was just lucky. In fact, Mendel had been very lucky.* With a little different luck, he might have become completely confused and given up, or he might have discovered gene linkage!

---

*In 1936, Mendel's data were reanalyzed by R. A. Fisher, a noted statistician and geneticist. The data fit Mendel's model, all right. But it fit *better* than it should have by random chance; his data were literally too good to be true. Either Mendel had fudged his data, consciously or unconsciously, or he had presented only his best results and had left out other, less favorable experiments. Or, we could say that the man was absurdly lucky to an unlikely degree. You can take your choice. Perhaps it is worth noting that Mendel recorded several thousand pea and pea plant phenotypes, and, unlike many other researchers, including Darwin with his snapdragons, never once found a pea he felt he couldn't classify one way or another.

TABLE 12.3

MENDEL'S PREDICTIONS AND RESULTS FOR F_2 GENOTYPE

Genotype of F_2	Fraction expected	Number expected according to hypothesis	Number actually observed
RRYY	$\frac{1}{16}$	33	38
RRyy	$\frac{1}{16}$	33	35
rrYY	$\frac{1}{16}$	33	28
rryy	$\frac{1}{16}$	33	30
RRYy	$\frac{1}{8}$	66	65
rrYy	$\frac{1}{8}$	66	68
RrYY	$\frac{1}{8}$	66	60
Rryy	$\frac{1}{8}$	66	67
RrYy	$\frac{1}{4}$	132	138

Mendel's Testcrosses

It is not sufficient for a mathematical biologist to build a model that is consistent with prior observations. He or she must make a *testable* prediction from the hypothesis—and then test it. And Mendel did this. In his words:

> It seems logical to conclude that in the ovaries of hybrids as many kinds of germinal cells—and in the anthers as many kinds of pollen cells—are formed as there are possibilities. . . . Indeed, it can be shown theoretically that this assumption would be entirely adequate . . . if one could assume at the same time that the different kinds of germinal and pollen cells of a hybrid are produced, on the average, in equal numbers. In order to test this hypothesis, the following experiments were chosen. . . .

In his test of the hypothesis, Mendel used what is now known as a **testcross.** In this instance such a test is called a **backcross.** This means the heterozygous F_1 individuals are crossed back to parental stock—that is back to individuals of the two pure-breeding parental strains. Figure 12.9 illustrates such a cross with coat color in sheep. Mendel chose to use his two-factor cross, in which one strain was **RRYY** (round and yellow) and the other strain was **rryy** (wrinkled and green). As we have seen, all the F_1 progeny were round and yellow, **RrYy.** There were two kinds of backcrosses to make, and Mendel made both of them: **RrYy × rryy** and **RrYy × RRYY.** Let's consider them one at a time.

Backcross to the Recessive Parent. Mendel's first backcross was with the F_1 progeny and the recessive parent, **RrYy** × **rryy** (see table at right). His prediction held!

"A favorable result could hardly be doubted any longer," wrote Mendel, "but the next generation would have to provide the final decision." He was referring to the other backcross—with the double dominant parent.

	Gametes from **RrYy** parent			
	RY 25%	Ry 25%	rY 25%	ry 25%
Gametes from rryy parent				
ry 100%	RrYy	Rryy	rrYy	rryy
Phenotypes	Round, yellow	Round, green	Wrinkled, yellow	Wrinkled, green
Expected (208)	52	52	52	52
Observed (208)	55	51	49	53

12.8 MEIOSIS AND MENDEL'S FACTORS

The behavior of two pairs of chromosomes during meiosis provides the chromosomal basis for Mendel's law of independent assortment. The cell entering meiosis with two pairs of genes on two different chromosome pairs has two possible ways of aligning its chromosomes at metaphase of meiosis I ("Possibility one" or "two" in the diagram), and the chances of either are equal. Of course, once a meiotic cell goes to either alternative, that's it for that cell, so two or more cells must go through meiosis for both alternatives to be possible. (Actually, it is necessary to count large numbers of individuals or gametes—as did Mendel—to test probability.) Following the alternatives as each continues through meiosis reveals that there are four equally likely allele combinations possible in the gametes. All four must be taken into account when predicting the outcome of a cross involving two allele pairs on two pairs of chromosomes.

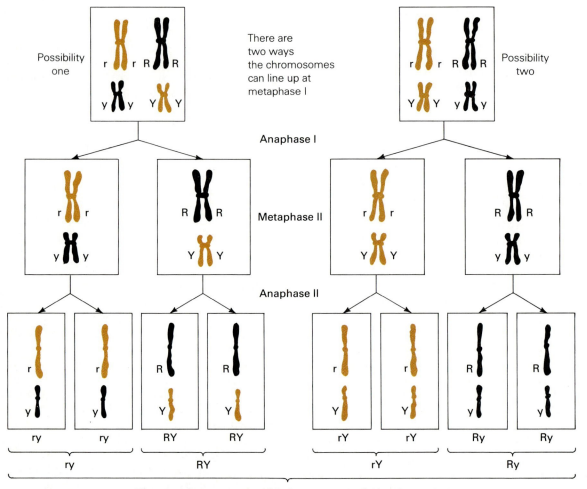

These are the gametes possible from genotype **RrYy** when meiosis occurs in two or more cells with crossing over

12.9 THE TESTCROSS

Sir Beauregard Thickfuzz is a prize ram by all ram-judging standards. He will be useful for breeding, but only as long as we are sure he is homozygous for white wool. White is dominant over black in sheep. The genotypes for color are **WW** = white, **Ww** = white, and **ww** = black. To test our prize white ram, we backcross it to some homozygous black ewes. **(a)** If Sir Thickfuzz is homozygous, all his sperm will carry the dominant **W** allele. All the eggs of the black ewes will carry the reces-sive **w** allele, but the offspring will be white lambs. If he passes the test, our prize ram will probably father hundreds of lambs. **(b)** However, if he is a heterozygote, half of his sperm will carry the allele **w** for black wool and about half of the test progeny will be black. Just one black sheep in the family will be enough to condemn Sir Beauregard Thickfuzz to a life of celibacy!

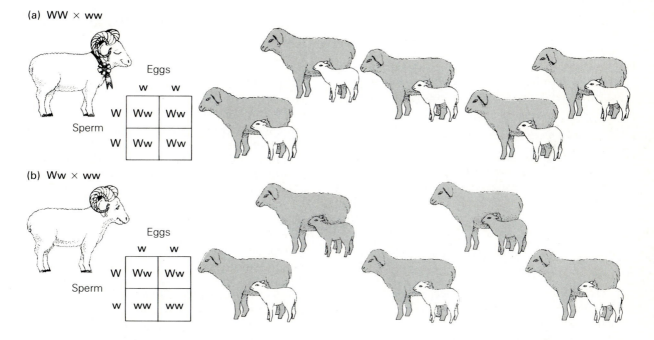

(a) WW × ww

(b) Ww × ww

Backcross to the Double Dominant Parent.
Here the cross was **RrYy** × **RRYY**. We can analyze it the same way:

	Gametes expected from **RrYy** parent			
	RY 25%	Ry 25%	rY 25%	ry 25%
Gametes from **RRYY** parent				
RY 100%	**RRYY**	**RRYy**	**RrYY**	**RrYy**
Phenotypes	Round, yellow	Round, yellow	Round, yellow	Round, yellow
Expected	100% round, yellow			
Observed	100% round, yellow			

The only phenotype expected among the progeny of such a cross was **R__Y__**, and that was the only one found. However, there were four different *genotypes* that Mendel expected to occur in equal numbers (as shown in the table above), and these could be determined by progeny testing. Mendel planted his round, yellow peas, allowed them to

mature and self-pollinate, and then looked into their pods. Even without counting he could tell whether the plant was **RRYY** (it would have only round, yellow peas in its pods) or **RrYy** (it would have some green and some wrinkled peas in its pods as well), or any of the other genotypes. His results are given in Table 12.4.

Mendel had been able to predict the *genotypes* of his all-round, all-yellow backcross progeny. It was quite an achievement. He was able to use his theory to predict something new, and then to test his prediction. It held.

A Final Word About the Testcross.   Before going on, let's review what a testcross can tell us. We've seen, for instance, that if we are not sure whether something that shows the dominant phenotype is homozygous or heterozygous, we can perform a testcross by mating it with a homozygous recessive individual. This applies to organisms other than peas, as we saw in Figure 12.9. If the unknown is homozygous, all its offspring will

TABLE 12.4

PROGENY TEST OF BACKCROSS TO RRYY

Found in pods	Expected	Observed
Round, yellow peas only (RRYY)	44	45
Round, yellow and green (RRYy)	44	42
Round and wrinkled, yellow (RrYY)	44	47
Round and wrinkled, yellow and green (RrYy)	44	43

have the dominant phenotype; if it is heterozygous half will have the recessive phenotype:

$$AA \times aa \rightarrow 50\% \ AA, \ 50\% \ Aa \ (100\% \ A\_)$$
$$Aa \times aa \rightarrow 50\% \ Aa, \ 50\% \ aa$$

The testcross can also be used to determine whether two genes are linked. How? In this case, a known double heterozygote is crossed with a test stock that is homozygous recessive at both gene loci. If there is independent assortment, the four kinds of offspring occur in equal proportions:

$$AaBb \times aabb \rightarrow 25\% \ AaBb + 25\% \ Aabb$$
$$+25\% \ aaBb + 25\% \ aabb$$

However, if the two gene loci are on the same chromosome, this 1:1:1:1 ratio of offspring phenotypes is *not* likely to occur, and some other distribution of progeny will be seen. The multiplicative law cannot be used because the two loci will not assort independently. In other words, Mendel's second law does not apply when genes are linked (are on the same chromosome).

Somehow Mendel figured this out, and he performed the appropriate test to see whether his differing traits assorted independently or not. As we have seen, they did.

The Decline and Rise of Mendelian Genetics

In 1865, after seven years of experimentation (at the very time Darwin was beating his brains out over the enigma of heredity), Mendel presented his results to a meeting of the Brunn Natural Science Society. His audience of local science buffs just sat there politely, probably not understanding a word of what they were hearing. In fact, they wanted to discuss the intriguing new idea of natural selection. His single paper was published in the society's proceedings the following year and was actually distributed rather widely. However, the learned scientists of the day were just as baffled and just as uninterested as Mendel's original audience. Eventually a German botanist included an abstract of Mendel's work in an enormous encyclopedia of plant breeding. Again, the world responded with silence. Apparently, no one had the foggiest notion of the importance of the Austrian monk's experiments and analysis; 1865 minds were just not ready for 20th century mathematical biology.

Historians have come up with a small, sad, but remarkable piece of information. In Darwin's huge library, which is still intact, a one-page account of Mendel's pea work appears in that German encyclopedia of plant breeding. Some relatively obscure work is described on the facing page, and it is covered with extensive notes in Darwin's handwriting. The page describing Mendel's work is clean. Darwin must have seen the paper that would have explained his own snapdragon work and, more important, could have clarified his theory of natural selection, saving him years of agony and uncertainty. But even Darwin was not ready for mathematical biology, and he too failed to grasp Mendel's simple but profound ideas.

Mendel's work continued to be ignored—and the ignorance lasted until 1900. In that year three biologists in three different countries, all trying to work out the laws of inheritance, searched through the old literature and all came up with Mendel's paper. All three immediately recognized its importance. Science had changed in 35 years. The obscure monk became one of the most famous scientists of all time. But he had been dead for 16 years.

Later in the 20th century, Mendelian genetics was applied to the theory of natural selection, and Darwin's reputation, which had faded considerably, ascended to new heights. Darwin had been right all along, if only he had gotten his genetics straight. It is often too easy for those of us who have just had something explained to us to say, ". . . but of course." However, breaking new conceptual ground, even a little, can be stultifyingly difficult. Mendel's work and his reasoning, like Darwin's, are so obvious to us now that we often forget how formidable the monk's task was as he placed those first pea seeds in the carefully cultivated soil of that monastery garden. You may well have found that even when Mendel's careful reasoning is all laid out, it can still be difficult to follow. Biologists can usually be divided into two groups: those who dote on abstract reasoning, who eagerly devour Mendel's work and find that solving genetics problems is fun, and those who don't.

MENDELIAN GENETICS IN THE 20TH CENTURY

A Closer Look at Dominance and Recessivity

The various ways in which two alleles at one gene locus can affect the phenotype are called **dominance relationships.** As we saw in Chapters 9 and 10, genes are part of informational molecules, lengthy polymers of DNA, that direct the production of proteins and determine the course of development. In many organisms, including humans, they come in pairs, one member of each pair coming from the father and one coming from the mother. This double dose of information provides, among other things, a backup system in case one of the genes isn't working properly.

But what happens when the gene from the father and the corresponding gene from the mother give conflicting information? Suppose an allele for brown eyes comes from one parent and an allele for blue eyes comes from the other. There are many possible results, but most commonly, the information from one allele will appear to be ignored. The allele that is expressed no matter what its partner does, we have called the **dominant allele,** and the allele that is suppressed in the heterozygotes, the **recessive allele.**

But so far we have given only observations and definitions, not explanations. What really happens when two different alleles occur together? How is one suppressed? And how does the organism "choose" between its two sets of information? Actually, there are several different ways that dominance can occur.

The Inoperative Allele. Dominance usually occurs simply because the recessive allele isn't doing anything. An allele that isn't making an enzyme will result in a phenotype based on the absence of that enzyme's function. This is clearest in rare medical disorders, where the absence of an enzyme can have a severe and often lethal effect. In an example of a relatively benign disorder, **albinos** lack an enzyme that is necessary to make melanin pigments. Both their fathers and their mothers provided what can be called an **inoperative allele,** so their cells cannot make the pigment. Such individuals are homozygous for the recessive, inoperative alleles. Heterozygotes for albinism or other enzyme deficiencies, on the other hand, have one working allele and one that is inop-

erative. They therefore produce only half the normal amount of enzyme, but in most cases this is still enough to metabolize all the enzyme's substrate, and the phenotype of the heterozygote will be perfectly normal.

We should add here that one problem with studying such relationships is that sometimes dominance is in the eye of the beholder. For instance, many mutant genes have been found that create bizarre changes in the eye color of *Drosophila* (see Figure 13.2). Almost all of them are recessive to the normal alleles, which are involved in the synthesis of one or both of two strong eye color pigments. For some of the gene loci involved, dominance is once again just a matter of half the enzyme being enough to complete some necessary step. For other genes, however, half as much gene activity results in exactly half as much eye pigment being produced. This can be determined easily enough if the pigments are extracted and quantified. But in the living fly, the observer cannot tell the difference between a fly with its full, normal amount of eye pigment and a heterozygote with only half as much pigment, so the normal allele is recorded as being dominant. It's as though the three genotypes were represented by a jar full of black India ink, a jar full of water, and a jar with a 50/50, but black, mixture of ink and water. All anyone can see is that there are two jars with black fluid and one with clear fluid. There's an important lesson here: *dominance depends on how the phenotypes are classified.*

In some cases, the recessive allele is involved in some normal function and the dominant allele makes some kind of gene product that prevents that function. For example, true-breeding white leghorn chickens are homozygous for a dominant allele, **I,** that inhibits melanin (color) formation. That's why they're white. Other chicken breeds are homozygous for the alternative allele, **i,** and can be all sorts of colors. In a cross between a homozygous white leghorn chicken and a chicken of another color, all the offspring are F_1 heterozygotes:

$$II \times ii \longrightarrow Ii$$

And, because the inhibitor allele is dominant, all the offspring of such a cross will be white. A similar example of dominant color inhibition exists in sheep, where the black recessive allele shows up occasionally in spite of attempts to breed the gene out (see Figure 12.9).

Other Dominance Relationships. Whatever its basis, full dominance is common, but it is not universal. There are other ways in which alleles with conflicting instructions can interact. Let's review a few particularly interesting dominance relationships.

Partial Dominance. Whenever the phenotype of the heterozygote is somewhere between those of the two homozygotes, and not exactly like either one of them, we can call the relationship **partial dominance.** A classic example of partial dominance is found in snapdragons, in crosses between strains with red and white flowers. Here, the two alleles can be symbolized by **r** and **w**. (We have deliberately avoided upper case letters since there is no true dominance.) When the homozygous **rr** red-flowered snapdragons are crossed with homozygous **ww** white-flowered snapdragons, the plants in the F_1 generation, the **rw** heterozygotes, all have pink flowers. If the pink-flowered F_1 plants are self-pollinated to produce an F_2 generation, we have an interesting Mendelian ratio. Instead of a 3:1 ratio, the F_2 snapdragons are 25% red, 50% pink, and 25% white, yielding a 1:2:1 ratio. Since all three genotypes are easily distinguished, the F_2 phenotypic ratio is exactly the same as the F_2 genotypic ratio (Figure 12.10).

Sometimes the phenotype of the heterozygote is not simply a compromise between the two homozygous phenotypes, but has some unique character of its own. Consider Roy Rogers' horse Trigger, for instance. Trigger was a beautiful palomino horse, with a golden coat and blond mane and tail. All palomino horses are heterozygotes (Figure 12.11). Crosses between palomino horses yield brown, palomino, and white foals in an approximate 1:2:1 ratio.

Rare Dominants (Lethal-Recessive Dominants). Here we return to the realm of abnormal, mutant alleles. **Rare dominants**—since they are dominants—do not have to become homozygous to work their terrible ways. A single dose is enough to alter the phenotype substantially. But we also call these mutants **lethal-recessive dominant,** an awkward phrase that describes a situation that may be difficult to grasp. Lethal-recessive dominants are in one sense considered recessive because two alleles are required to produce death. Yet the abnormal allele is dominant since it is always expressed when opposed by its normal counterpart.

Probably the most familiar lethal-recessive dominant trait is **achondroplastic dwarfism,** or **achondroplasia.** The heterozygous carriers **(Aa)** of the defective allele appear normal at birth, but they don't grow normally. The arm and leg bones, in

particular, stay very short, and there are usually facial bone abnormalities as well. Affected persons have nearly normal bodies and normal-sized heads, and are generally in good health and often extremely athletic, although they are conspicuously stunted because of their short limbs (Figure 12.12).

Achondroplastic dwarfs are usually fertile, although studies have shown that they tend not to marry. If they do marry, and their mates happen to be normal homozygotes **(aa),** about half their chil-

12.10 PARTIAL DOMINANCE

Snapdragons represent a good example of *partial dominance*. When white-flowered snapdragons are crossed with red, the offspring are pink, as shown in the first cross. In the second, pink heterozygotes have been crossed. The result is the appearance of red, pink, and white in the progeny. The ratio of the three colors is 1:2:1. Since we can identify the heterozygous progeny by their color, we have derived both the phenotypic and genotypic ratios. Is it possible to produce a true-breeding strain of pink snapdragons?

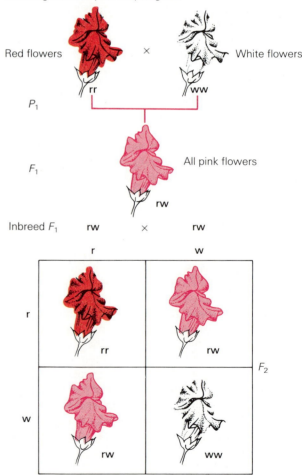

1. What colors appear in the F_2 generation?
2. What is the phenotypic ratio?

12.11 A PALOMINO MARE WITH HER FOAL

The beautiful golden color is the result of heterozygosity of alleles for brown and white coat color. Palomino horses are not true-breeding.

dren will be dwarfs like themselves (which is in keeping with Mendel's first law). Sometimes, two achondroplastic dwarfs will marry each other (**Aa** × **Aa**). Such matings might be expected to produce children who are homozygous for the dwarf allele; apparently, however, the double dose of the allele is usually not compatible with survival of the embryo and it dies.

Dominant traits, unlike recessive traits, do not skip generations (except for the special case of alleles with *incomplete penetrance*, which we will deal with in the next chapter). An individual cannot carry the allele without showing its effects. Thus if you are normal but your mother and two brothers are dwarfs, you don't have to worry about passing the dominant gene on to your own children: you don't have it.

There are many less common or less noticeable dominant traits. In hereditary night blindness, the affected carriers have no functional **rods** in their retinas; they have only **cones**. (Rods are retinal cells that perceive black and white and provide night vision, while cones perceive color and are inactive at night.) In one family pedigree, the trait has been traced through 300 years, with each affected person having either an affected mother or an affected father.

Codominance. Sometimes one homozygote will show one phenotypic trait, the other homozygote will have a different phenotypic trait, and the heterozygote will show both traits. This condition is called **codominance.** A simple example occurs in spotted housecats. One gene locus produces spotting as a simple recessive trait. However, another gene locus determines the color of the spots. Homozygous **BB** cats have black spots. Homozy-

gous **RR** cats have orange spots. Heterozygous **BR** cats have black spots *and* orange spots.

Codominance is actually rather rare among traits that are easily seen, but it is common among genetic traits that can only be measured by biochemical tests (called *in vitro* [in glass] phenotypes). Codominance is often encountered, for instance, in the genetics of blood groups. Let's consider one of these as another example.

In the **MN** blood group system, two codominant alleles, **M** and **N** account for three genotypes, **MM, MN,** and **NN.** The *in vitro* phenotype is revealed when blood samples are tested with two kinds of antisera, anti-M and anti-N. The two antisera are produced from blood serum removed from animals that have been previously sensitized against M or N blood. When blood samples react with an antiserum, the red cells agglutinate or

12.12 A DOMINANT MUTATION

Achondroplasia is a dominant genetic trait in which an individual's arms and legs remain very short. Two pedigrees are shown here: **(a)** an achondroplastic dwarf (**Aa**) marries a normal person (**aa**); on the average, the couple can expect about half their children to be affected **Aa** heterozygotes. **(b)** Occasionally two dwarfs marry (**Aa** × **Aa**). One fourth of the zygotes they produce will be **AA**, which apparently is lethal in embryonic development. Of their surviving children they can expect, on the average, that about two thirds will be affected **Aa** heterozygotes. About one third will be normal and will carry no genetic tendency for the condition.

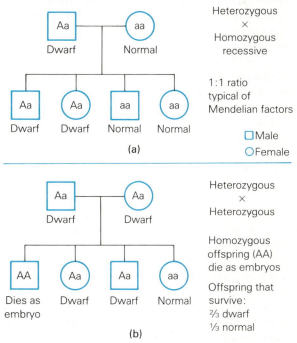

clump together forming visible clusters. Here is how the three genotypes are determined:

Genotype	Reaction with anti-M	Reaction with anti-N
MM	+	−
MN	+	+
NN	−	+

Here, + means that the red blood cells agglutinate, or clump, and − means that they do not. Look at the reaction with anti-M. Considering these three reactions alone, it is clear that allele **M** is dominant. Now look at the reaction with anti-N. For this test, **N** is clearly dominant. When both antisera are used, alleles **M** and **N** are codominant.

Overdominance. If the phenotype of the heterozygote is beyond the range of the phenotypes of the two corresponding homozygotes, **overdominance** is said to occur. For instance, we see overdominance if one homozygote is short, the other homozygote is tall, but the heterozygote is even taller than the tall homozygote. Overdominance with regard to Darwinian fitness (the ability to survive and reproduce) is of great interest to biologists. When overdominance increases fitness, both of the alleles can become common in the population.

Pigeons have two common alleles for slightly different forms of a protein called **egg-white transferrin.** The alleles are simply codominant by the *in vitro* criterion of gel electrophoresis, in which the two protein variants can be characterized by their rate of movement in an electric field. But they are overdominant by another criterion, namely egg hatchability. One allele appears to give the egg a greater resistance to fungal infections, while the other appears to give greater resistance to bacterial infections. The two together give the egg a greater chance of surviving than either allele alone can provide, a clear case of overdominance.

Multiple Alleles

Fortunately for Mendel, he had to deal with only two alleles at any gene locus; otherwise he may have become hopelessly entangled in complex genetic systems. Most of our examples so far have involved only two alleles. However, there is a great deal of genetic variation among the normal alleles. Often, many forms of a gene or **multiple alleles** can occur at a given gene locus, although any one individual can have only two—one from each parent. A Harvard research group recently did an intensive study of a randomly chosen enzyme locus in a North American fruit fly population. They uncovered 37 different alleles at one locus, out of a sample of only 146 flies. As far as could be determined, all 37 variants functioned normally, although they produced proteins with different amino acid sequences.

In a much larger sample, even more alleles of human **beta hemoglobin** (the beta polypeptide strand in hemoglobin; see Chapter 3) have been found. One variant at this locus is the **S** allele, which causes sickle-cell anemia in homozygotes, although it gives heterozygotes some protection against malaria (Essay 12.1). Most people are homozygous for the normal **A** allele of the beta hemoglobin locus, even in Africa, where the **S** sickle-cell allele is fairly common. In other parts of the world where malaria is a problem, other alleles of the hemoglobin locus may be common. For example, in parts of Asia, **C** and **E** alleles occur. Like the sickle-cell allele, each of these variants differs from the normal hemoglobin by a single amino acid, and each appears to offer heterozygotes some protection from malaria. Homozygous **CC** and **EE** individuals are somewhat anemic, but the effects are not as severe as in sickle-cell anemia.

Another set of alleles of the hemoglobin locus causes the various forms of **thalassemia,** found in moderately high frequencies in parts of Italy. Some of them fail to produce beta chain protein at all; others produce only 20 to 30% of the normal amount of gene product. In homozygous persons, many of these alleles give rise to a severe genetic disease called **thalassemia major.** Children with thalassemia major usually die in infancy. Heterozygotes have half the normal amount of hemoglobin, which makes them somewhat anemic—their condition is known as **thalassemia minor**—but apparently they also have some kind of protection against malaria.

Blood Groups

We are aware, of course, that there are different types of blood, but we may not be familiar with the genetics behind them. Let's take a look. The most important and best-known blood group system is the ABO system, and the corresponding four **blood types** to which people belong are A, B, O, and AB. The blood types are determined by identifying cell-surface **antigens** carried by persons of different genotypes.

The term *antigen* refers to any substance that produces a response in the body's immune system. This includes the production of active proteins called **antibodies.** Antibodies bind to antigens, clumping them together for easier destruction by phagocytic white blood cells. The antigens of the ABO system happen to be polysaccharides. The red blood cells of a type A person, for instance, have A antigens on their cell surface, and the blood cells of a type B person have B cell-surface antigens. Type AB people have both A and B antigens, so the genes for A and B are codominant. The two alleles are usually written I^A and I^B, to distinguish the allele that makes the antigen from the antigen itself. Individuals with the fourth blood type, type O, have neither the A antigen nor the B antigen on the surfaces of their red blood cells. A third allele, then, is I^O; this allele is recessive to both I^A and I^B. The possible genotypes and their corresponding phenotypes are shown in Figure 12.13.

The three alleles can combine into six different genotypes, but since $I^A I^A$ and $I^A I^O$ are both type A, and $I^B I^B$ and $I^B I^O$ are both type B, there are only four ABO blood types or phenotypes.

The ABO genotypes have another, more indirect effect on the phenotype. The immune systems of all individuals, for reasons that are not entirely understood, always produce antibodies against the A and B antigens, whichever is lacking on the individual's own cell surfaces. Thus, type A persons produce anti-B antibodies and type B persons produce anti-A antibodies; type O persons produce both kinds of antibodies. Since type AB persons lack neither—that is, they have both A and B antigens present—they produce neither the A nor the B antibody.

The antibodies are specific, two-headed binding proteins. By *two-headed* we mean the protein has two active sites. Each antibody is able to bind to two of the specific molecules that it is directed against. The anti-B antibody, for instance, can bind tightly to the B molecule (antigen) on the surface of one red blood cell and at the same time bind tightly to another B molecule on the surface of another red blood cell. Since each red blood cell is covered with these antigens, the antibodies will cause susceptible cells to agglutinate, or clump. Tests of blood types are carried out with drops of blood and antisera on glass microscope slides (Figure 12.13).

As a result of the A and B antibodies produced by persons of various genotypes, the blood types of donors and recipients have to be very carefully matched. Type A blood transfused into a type B person will sensitize that person to the A antigen, and should this be repeated, fatal clumping may occur. In some cases, type O blood can be transfused into persons of other blood types, but even this is unwise, because the introduced type O blood brings with it some anti-A and anti-B antibodies.

12.13 ABO BLOOD TYPING

The presence of A and B antigens in one's blood can readily be determined in a simple test using anti-A and anti-B test reagents. (These reagents are usually extracted from plants and are usually color-coded for convenience.) As the scheme shows, two drops of blood and two drops of reagent (one of anti-A and one of anti-B) are used. The reaction (or the lack of reaction) is then used to determine the blood type. A positive reaction, called agglutination, is characterized by a distinct clumping of cells in the mixture. The red cells eventually form a patchwork pattern in the serum. Where there has been a negative reaction (no agglutination), the blood remains homogeneous, with an even texture.

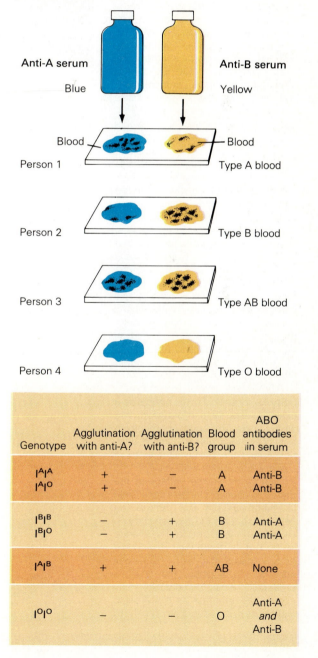

Genotype	Agglutination with anti-A?	Agglutination with anti-B?	Blood group	ABO antibodies in serum
$I^A I^A$	+	–	A	Anti-B
$I^A I^O$	+	–	A	Anti-B
$I^B I^B$	–	+	B	Anti-A
$I^B I^O$	–	+	B	Anti-A
$I^A I^B$	+	+	AB	None
$I^O I^O$	–	–	O	Anti-A *and* Anti-B

Sickle-Cell Hemoglobin, an Example of Almost Everything

Dominance relationships are in the eye of the beholder. For instance, normal eye color in *Drosophila* can be a dominant if you just look at the eye, but a partial dominant if you measure the amount of pigment. The Manx (tailless) allele in cats is a dominant if you consider only the presence or absence of a tail, but a recessive if you consider the presence or absence of the whole cat. The **M** allele is a dominant by one agglutination test and a recessive by another, or a codominant if both tests are considered together. Pigeon transferrins are codominant by electrophoretic criteria but overdominant for egg hatchability. And so on. Just to drive this important principle home, we'd like to consider one all-inclusive example—the normal and sickle-cell alleles of the beta hemoglobin gene.

Human adult hemoglobin (the red pigment of blood) is a large protein consisting of four subunits: two **alpha chains** and two **beta chains.** Different gene loci code for alpha and beta hemoglobin chains. The beta hemoglobin gene locus actually has hundreds of different alleles, most of them rare. We will only consider the normal allele, usually written **Hb^A,** and the most common mutant allele, the sickle-cell allele, which is written **Hb^S.** We'll shorten the symbols down to **A** and **S** and consider the three genotypes **AA, AS,** and **SS.**

The **S** allele makes a hemoglobin beta chain identical to the normal beta chain, except for one out of its 146 amino acids. The normal hemoglobin beta chain has a glutamic acid at position 6, and the sickle-cell beta chain has a valine at that position.

The valine on the surface of the molecule causes it to form large, sickle-shaped crystals that deform the cell and reduce the oxygen-carrying capacity of the molecule. The change in the red cells and their loss to the body explains the term "sickle-cell anemia." In addition, the **S** allele produces less hemoglobin—only about 40% as much as the **A** allele (although this varies from family to

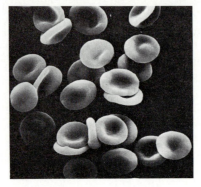

Normal red blood cells

Sickle-shaped cells

Threonine   Proline   Glutamic acid   Glutamic acid

Normal hemoglobin (A) positions 4–7

Threonine   Proline   Valine   Glutamic acid

Sickle-cell hemoglobin (S) positions 4–7

family). The homozygous **SS** person, then, has a severe anemic disease. His or her blood will also tend to form small clots that clog the smaller arteries of the muscles, heart, and brain, causing pain, heart trouble, and brain damage. The condition is usually lethal for **SS** homozygotes, but **AS** heterozygotes enjoy essentially normal health. So **A** appears, clinically at least, to be dominant over **S**. However, the **AS** carrier may have only about 70% as much hemoglobin as a normal **AA** person, and is somewhat more subject to anemia when placed under stress. Thus, the **A** allele is only partially dominant over the **S** allele.

The sickle-shaped cells are seen in the blood of heterozygous carriers when a sample of that blood is deprived of oxygen and viewed under the microscope. Carriers can also be detected by gel electrophoresis (see illustration), since the two proteins have different electrophoretic mobilities. **AA** individuals show a fast band on the electrophoretic apparatus; **SS** individuals show a slow band; and **AS** heterozygotes show both bands, indicating codominance.

The **S** allele is common in parts of Africa where malaria is endemic, and there it shows overdominance with the **A** allele with respect to Darwinian fitness. Where malaria is endemic, *everyone* is bitten by malaria-carrying mosquitoes and *everyone* contracts malaria. Heterozygotes (**AS**) are much less seriously affected by malaria because blood cells influenced in any way by sickling do not provide a suitable home for the malarial parasite. Thus, in these areas, heterozygotes survive better, live longer, and have more children than homozygous (**AA**) persons; in other words, there is overdominance for Darwinian fitness. This means that, in these areas, there is selection *for* sickling allele so that in some places nearly 40% of the population are carriers for the sickle-cell gene.

Is the sickle-cell allele a dominant allele, a recessive allele, a partial dominant allele, an overdominant allele, a codominant allele, or a lethal-recessive dominant allele? It is all of these, depending on how the phenotype is classified.

1. *Dominant:* The **S** allele is dominant over **A** for the trait *sickle-shaped cells under the microscope.*

2. *Recessive:* The **S** allele is recessive to the **A** for the severe anemic disease syndrome.

3. *Partial dominant:* The **A** and **S** alleles are partial dominants for the amount of hemoglobin, the **AS** genotype being just halfway between **AA** and **SS.**

4. *Overdominant:* Under conditions of endemic malaria, the **A** and **S** alleles are overdominant with regard to Darwinian fitness. The fitness of the **AS** heterozygote exceeds that of either homozygote.

5. *Codominant:* The **S** and **A** alleles are codominant for their electrophoretic patterns in vitro.

6. *Lethal-recessive dominant:* The **S** allele is dominant for the above trait, and effectively a recessive-lethal under primitive conditions.

Our use of the **S** and **A** alleles as examples of dominance, codominance, partial dominance, and overdominance may have seemed confusing or even perverse, but it was deliberate, and for a good reason. The moral of the sickle-cell anemia tale is that dominance is, in fact, in the eye of the beholder.

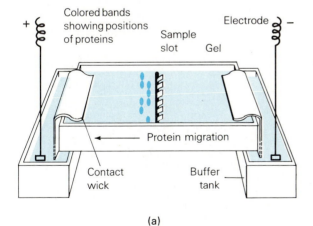

(a)

Phenotype	Genotype	Hemoglobin electrophoretic pattern Origin ⟶ +	Hemoglobin types present
Normal	AA		A
Sickle-cell trait	AS		S and A
Sickle-cell anemia	SS		S

(b)

Using gel electrophoresis (**a**), it is possible to compare the hemoglobin in the three genotypes in question, **AA, AS,** and **SS (b).** Normal hemoglobin (**AA**) forms a single fast band and is seen to migrate the farthest. **SS** sickle-cell hemoglobin forms a single slow band, which lags behind the normal hemoglobin. The heterozygote (**AS**) hemoglobin produces a smaller **A** band and an **S** band, revealing codominance.

The Rh Blood Group.   Many other blood group loci also have multiple alleles. For example, there is the Rh (from rhesus, the monkey in which it was found) blood group system, which has eight fairly common alleles and many rarer ones. As in the case of the ABO system, the different alleles can be identified according to how the blood types react to known antibodies in an antiserum. You are probably aware that you're referring to the Rh system when you add the words "positive" or "negative" to your blood type. One Rh antigen is particularly reactive with other types, and the Rh blood groups are classified as *Rh positive* and *Rh negative* on the basis of the presence or absence of this one—the most potent—Rh antigen.

People don't naturally have anti-Rh antibodies, but transfusing antigen-containing Rh positive blood into an Rh negative person would sensitize that person against the Rh positive factor. The recipient would produce anti-Rh antibodies, but in first exposures the reaction is generally minor. However, should such a transfusion occur again, the next reaction would be rapid and massive and the effects quite dangerous. (Once the immune system has "learned" to produce an antibody for a specific antigen, any subsequent invasion is met almost at once by a veritable army of rapidly synthesized antibodies; see Chapter 31.) Mismatching blood types in transfusions isn't too common but there is another, more familiar Rh problem—one that confronts mothers.

Let's see why Rh negative women were once not kindly disposed toward mates who were Rh positive (Figure 12.14). Following a pregnancy in which the baby turns out to be Rh positive, Rh negative women *sometimes* build up antibodies against the Rh positive antigen. The word "sometimes" is emphasized because such a reaction requires a leakage of the baby's Rh positive blood across the placenta into the mother's blood, and this seldom

12.14 RH INCOMPATIBILITY

In the mating of an Rh positive man (**RR** or **Rr**) and an Rh negative woman (**rr**), any offspring stand at least a 50% chance of being Rh positive. A serious Rh incompatibility may arise in a newborn Rh positive infant (**Rr**) from this mating, resulting in a severe and sometimes fatal anemia. The incompatibility problem does not affect the first Rh positive infant, but can arise during second and subsequent pregnancies. **(a)** The problem begins if, during the time of delivery of the first baby, blood cells from the Rh⁺ fetus should leak across the placenta into the Rh⁻ mother's circulatory system. **(b)** The baby will be normal, but in the meantime, the mother's immune system responds by producing Rh antibodies against the Rh⁺ cells. Such cells are destroyed, but more critically, the mother has been *sensitized* against the RH⁺ antigen. **(c)** In second or subsequent pregnancies, a similar leakage would then trigger a massive production and release of Rh antibodies in the sensitized mother (just as the immune system responds to the second onset of a disease). The new Rh antibodies then enter the baby's circulatory system and begin to attack its red blood cells, destroying them. (The disease is called *erythroblastosis fetalis.*)

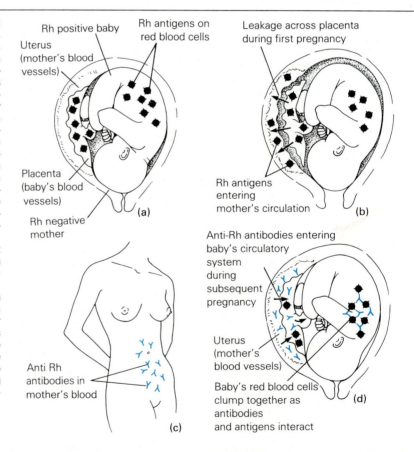

Rh positive baby
Uterus (mother's blood vessels)
Rh antigens on red blood cells
Placenta (baby's blood vessels)
Rh negative mother
(a)

Leakage across placenta during first pregnancy
Rh antigens entering mother's circulation
(b)

Anti Rh antibodies in mother's blood
(c)

Anti-Rh antibodies entering baby's circulatory system during subsequent pregnancy
Uterus (mother's blood vessels)
Baby's red blood cells clump together as antibodies and antigens interact
(d)

happens. But, when it does, Rh negative women can become sensitized against their Rh positive fetuses. This can cause trouble in subsequent pregnancies because the mother's new antibodies can enter the blood of the next fetus shortly before birth and destroy its red blood cells, a condition called **erythroblastosis** (Figure 12.15).

However, Rh negative women no longer have to search out Rh negative men. Those who have just given birth to an Rh positive baby are now routinely given injections of anti-Rh serum, which destroys any fetal red blood cells that may have leaked into the mother's body before her immune system starts building up antibodies against them. This prevents Rh incompatibility problems with any future pregnancies.

Organ Transplants. Careful genotype matching must also be done in the case of heart and other organ transplants. Cells other than red blood cells have **histocompatibility antigens** on their surfaces; these are controlled principally by four closely linked **histocompatibility loci.** In spite of there being only four loci, matching a donor and a recipient is extremely difficult because each histocompatibility gene locus has hundreds of alleles, and none is common. Virtually every person is unique for his or her cell-surface antigens, identical twins being the exception, and immediate family members being more likely to show similarities. But if you should ever need a heart transplant, you may wish you were not so special. One technique that facilitates transplantation is to chemically inhibit the body's immune system. The side effects, however, are traumatic, and treated individuals are vulnerable to a host of invading disease organisms.

Gene Interactions and Modified Mendelian Ratios

You will recall that Mendel found that the F_2 of the round-yellow and wrinkled-green cross yielded a 9:3:3:1 ratio. After Mendel's work was rediscovered, his enthusiastic followers gleefully produced this ratio again and again. Their work is a textbook standard and clearly indicates some of the ways in which genes at different loci can interact to produce different phenotypes.

This all worked out very nicely for Mendel and his followers because the round, wrinkled characters do not influence the inheritance of the yellow, green characters, but this is not true of all pairs of gene loci. For instance, it doesn't work for loci that regulate coat color in mice. In this case, we find that

Erythroblastosis, the destruction of red blood cells in the fetus, is now uncommon in medically sophisticated societies. The child from which this blood sample was taken inherited its father's Rh positive phenotype. Its mother, an Rh negative person, was probably sensitized by a former pregnancy. Her anti-Rh antibodies found their way into the baby's bloodstream and began the systematic destruction of red cells shown here. Mass transfusions were necessary to save the baby.

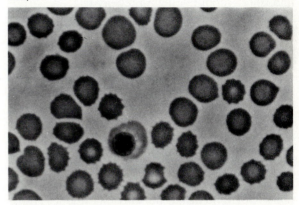

at one gene locus, **B** is dominant to **b**, such that **BB** and **Bb** mice are black and **bb** mice are brown. A cross between a homozygous black **(BB)** mouse and a homozygous brown **(bb)** mouse will produce nothing but black heterozygotes **(Bb).** A cross between two heterozygous black mice produces an F_2 generation with three fourths black mice and one fourth brown mice. So far so good.

But at another gene locus, **C** is dominant to **c,** such that **CC** and **Cc** mice can make pigment (black or brown) and **cc** mice cannot, and are thus white albinos. The **C** allele, in effect, allows for coat color. Hence, the two gene loci at **B** and **C** control two different steps in the biochemical pathway that produces the pigment normally present in mouse fur. Thus, the **c** locus is said to be **epistatic** to the **B** locus. Epistasis refers to one gene's interfering with the expression of another.

Now consider a mating between a true-breeding white mouse and a true-breeding brown mouse. What would you expect? You might *not* expect the litter to be entirely black. Mendel, however, wouldn't have been surprised. After all, here is one possibility in a mouse cross:

CCbb (brown) × **ccBB** (white) → **CcBb** (black)

So the black F_1 are heterozygous at two gene loci. The real surprise comes at the next cross, however. If you mate two such double heterozygotes, you will find that in the F_2 generation one fourth are **cc** (white), regardless of what's happening at the **B** locus. Of the remaining colored mice, three fourths are black and one fourth are brown. The phenotypic classes of the F_2 are $9/16$ black; $3/16$ brown; and $4/16$ *white* (Figure 12.16). This is just the old 9:3:3:1 ratio with the last two terms combined (9:3:4), because once a mouse is white you can't tell whether it is brown or black. Hmmm. Perhaps we should say this another way. One cannot distinguish between **BBcc, Bbcc,** and **bbcc** without doing a progeny test.

Similarly, one can combine two *different gene loci* that have *identical effects.* For instance, **pp** mice are also white, while **PP** and **Pp** mice are normally pigmented. These genes control yet another step in the biochemical pathway that produces pigment as its end product. The F_1 offspring of **Ppcc** and **ppCC** (both white phenotypes) are **PpCc,** a genotype that gives normally pigmented mice (let's say black). In the F_2 generation, about half are white and about half are black. If we have large enough numbers, we may be able to show that the ratio is not really half black or half white, but $9/16$ black to $7/16$ white. This is the 9:3:3:1 ratio again, but this time the last three groups are lumped together:

$9/16$ **P__C__**	Black	$9/16$ black
$3/16$ **P__cc**	White	
$3/16$ **ppC__**	White	$7/16$ white
$1/16$ **ppcc**	White	

Or, put another way, once a mouse is white, it can't be any whiter, but as long as it has both a **P** allele and a **C** allele, it will produce the normal black coat color.

12.16 EPISTATIC INTERACTION

In this scheme we follow the path of two pairs of genes that influence the coat color of mice. The **B** (black) and **b** (brown) alleles occur at one locus, and the **C** (color) and **c** (albino) alleles occur at another locus. The two pairs assort independently in typical Mendelian fashion but produce strange ratios when heterozygous F_1 are inbred (**CcBb** × **CcBb**). The resulting ratio is 9:3:4 because of an epistatic interaction.

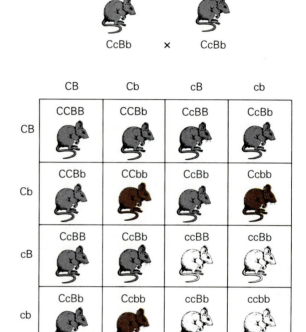

APPLICATION OF IDEAS

1. Promiscuous Alice had five lovers in one week, became pregnant, and doesn't know who the father is. Alice received marriage proposals from her five former lovers, but she has decided that only the true father of her child can have her hand in marriage. Scientific Alice determines that her blood type is B negative (that is, has B antigen, no A antigen, no Rh positive antigen). Her five suitors have volunteered blood samples of their own:

Henry	O negative
Arthur	A negative
William	B negative
Schuyler	A positive
Buddy	AB positive

The child is born before indecisive Alice can make up her mind, and proves to be O positive. Which of the five men could have been the child's father? Prove your answer. Should the lucky candidate renege, would vengeful Alice's evidence be conclusive in court? Explain.

2. In one breed of chickens, *pea* comb (**PP, Pp**) is dominant to *single* comb (**pp**). In another breed of chickens, *rose* comb (**RR, Rr**) is dominant to *single* comb (**rr**). When a homozygous *pea* comb chicken (**PPrr**) is mated with a homozygous *rose* comb chicken (**ppRR**), the F_1 progeny all have yet a fourth phenotype, *walnut* comb (**P__R__**). What are the expected phenotypic ratios among the F_2 progeny of two F_1 *walnut* comb (**PpRr**)

chickens? In the progeny of an F_1 *walnut* comb chicken (**PpRr**) crossed to a *single* comb chicken (**pprr**)?

3. In another breed of chickens, **BB** chickens are *black*, **Bb** chickens are *blue*, and **bb** chickens are *white*. A rooster from a true-breeding strain of *black, single* comb chickens (**BBpp**) is mated with several hens from a true-breeding strain of *white, pea* comb chickens (**bbPP**). What is the expected phenotype of the F_1 offspring, and what kinds of chickens can be expected in the F_2 generation in what expected frequencies? Construct a Punnett square of the $F_1 \times F_1$ (**BbPp** $\times$ **BbPp**) cross.

4. The Japanese geneticist, Hagiwara, crossed two true-breeding strains of Japanese morning glories, both of which had blue flowers. The plants of the F_1 generation all had purple flowers. When Hagiwara crossed the purple-flowered F_1 back to one of the parental strains, the progeny of the backcross were approximately 50% blue-flowered and 50% purple-flowered. When the purple-flowered F_1 was crossed to the other parental strain, the results were the same. But when the F_1 purple-flowered progeny were crossed to each other, the F_2 offspring included purple-flowered, blue-flowered, and scarlet-flowered plants. Hagiwara concluded that two gene loci were involved, an **Aa** locus and a **Bb** locus. The scarlet-flowered plants, then, must have been **aabb**.

   a. What were the genotypes of the parental strains, the F_1 hybrid, and the progeny of the two backcrosses?

   b. List all phenotypes and all genotypes that appear in the F_2.

c. Below is a list of individual crosses between pairs of F_2 plants, together with the ratios of phenotypes found in the progeny. Determine the genotypes of the parents in each cross.

Male parent	Female parent	Progeny		
		Blue	Purple	Scarlet
purple	scarlet	—	100%	—
purple	scarlet	50%	50%	—
purple	scarlet	25%	50%	25%
blue	blue	25%	50%	25%
blue	blue	75%	—	25%
blue	blue	100%	—	—
blue	blue	—	100%	—
blue	blue	50%	50%	—

5. Baur crossed two true-breeding strains of *Antirrhinum majus*, the same snapdragon that Darwin had experimented with. One strain had regular (bilaterally symmetrical) white flowers, and the other had peloric (radially symmetrical) red flowers. The F_1 progeny, which had regular pink flowers, were crossed to each other to produce an F_2 generation. Here is Baur's data on the F_2, in numbers of plants of each flower phenotype:

regular pink	94
regular red	39
regular white	45
peloric pink	28
peloric red	15
peloric white	13

   a. Explain the results.
   b. For the observed total of 234 F_2 progeny, compute the *expected* number of each phenotype, and compare with the observed number.

KEY WORDS AND IDEAS

WHEN DARWIN MET MENDEL (ALMOST)

1. Darwin subscribed to the idea of **blending inheritance,** which held that traits from both parents are ''blended'' in the offspring.

2. Critics of natural selection pointed out how blending would destroy variations—a key part of Darwinian evolution.

3. Darwin crossed **true-breeding** strains of snapdragons, in which he observed **dominant** traits, which he then called **prepotent.**

Mendel's Crosses

1. Mendel used carefully planned experiments and applied statistical analysis to his data.

2. Mendel reported on experiments with seven different characteristics of garden peas. These included seed form, color of contents, color of coats, color of pods, shape of ripe pods, length of stem, and position of flowers.

3. His approach was to cross **true-breeding** strains, manipulating pollen by hand to avoid self-pollination.

4. In peas, each pea in a pod has its own **genotype** and **phenotype.** (*Genotype* is the total combination of an organism's genes; *phenotype* is the combination of observed or measured traits, generally what is readily visible.)

5. The symbols P_1, F_1, and F_2 are used to designate first and subsequent generations in crosses.

6. Both Mendel and Darwin perceived the same results in their crosses. There was no blending, but the **recessive** trait disappeared in the F_1 generation and reappeared in the F_2. Darwin called this **latency** and pursued the problem no further, but Mendel noted that the reappearance of a *recessive* trait occurred with a definite frequency in one fourth of the F_2.

7. He determined that for each dominant trait there were two kinds (genotypes), those that produced two kinds of offspring **(heterozygous)**, and those that produced one **(homozygous)**.

8. To determine whether an individual with a dominant trait is heterozygous or homozygous requires **progeny testing.** Mendel did this by breeding an F_3 generation from F_2 round peas. He determined that one third of the round F_2 peas were true-breeding, two thirds were not. From this he determined that one fourth of the total F_2 were round and true-breeding; one half of the F_2 were round and not true-breeding; and one fourth of the F_2 were wrinkled and true-breeding.

9. From his work so far, Mendel concluded that:
a. Characteristics were controlled by factors in two forms, dominant and recessive.
b. **Hybrids** had two different factors, one from each parent.
c. If the factors were the same, the individual was true-breeding (homozygous).
d. If the factors were not the same (heterozygous), the dominant factor would prevail (produce the phenotype).
e. In modern terms, factor is replaced by **allele,** an alternate form of a gene.

Mendel's First Law:
The Segregation of Alternate Alleles

1. Heterozygous **(Aa)** individuals produce two kinds of gametes (sex cells) in equal proportions:

2. Following this, the multiplicative law from the laws of probability can be applied to crosses: "The probability of two independent events both occurring is equal to the *product* of their individual probabilities."

3. The F_1 cross **Aa × Aa** can be stated (1/2A + 1/2a) × (1/2A + 1/2a). Multiplying produces 1/4AA + 1/2Aa + 1/4aa. The 1/2Aa is determined algebraically but can be explained genetically. The combination **A + a** can occur two ways, **A + a** or **a + A.**

4. The *additive law* is applied: "The probability of either one or another of two mutually exclusive events occurring is equal to the *sum* of their individual probabilities."

5. The probability of an **A** pollen and an **a** ovule combining is one fourth, as is the probability of **a + A.** Since they are mutually exclusive events, the probability of a heterozygous F_2 individual is 1/4 + 1/4 = 1/2.

6. Punnett illustrated Mendel's principles using squares:

	A	a
A	AA	Aa
a	Aa	aa

When summed up the results are:
Genotype: 1/4**AA** + 1/2**Aa** + 1/4**aa**
Phenotype: 3/4 dominant +
                    1/4 recessive

7. Mendel can be considered a mathematical biologist because he constructed a **model** that yielded numerical predictions consistent with observations.

Mendel's Second Law: Independent Assortment

1. Mendel crossed two alleles of the round **locus, R** and **r,** and two alleles of the yellow locus, **Y,** and **y.** *Locus* refers to a specific gene location on a chromosome.
     RRYY × rryy (P_1 cross)
     all **RrYy** (F_1 offspring)
     RrYy × RrYy (F_1 **dihybrid cross**)

2. To predict the results, consider the following:
a. You know that **Rr × Rr** produces 1/4**RR,** 1/2**Rr,** and 1/4**rr.** Likewise, **Yy × Yy** produces 1/4**YY,** 1/2**Yy,** and 1/4**yy.**
b. To predict the results when both are considered simultaneously, follow the multiplicative law for all possible **R** and **Y** combinations. These are as follows:

Genotype	Separate probabilities		Combined probabilities	Grouped into phenotypes
RRYY	¼ × ¼	=	¹⁄₁₆	Round and yellow ⁹⁄₁₆
RrYY	½ × ¼	=	⅛	
RRYy	¼ × ½	=	⅛	
RyRy	½ × ½	=	¼	
rrYY	¼ × ¼	=	¹⁄₁₆	Wrinkled and yellow ³⁄₁₆
rrYy	¼ × ½	=	⅛	
RRyy	¼ × ¼	=	¹⁄₁₆	Round and green ³⁄₁₆
Rryy	½ × ¼	=	⅛	
rryy	¼ × ¼	=	¹⁄₁₆	Wrinkled and green ¹⁄₁₆
			—	¹⁶⁄₁₆
		¹⁶⁄₁₆		

3. In addition to being consistent with his expectations, Mendel's findings indicate that two traits, pea shape and color, are inherited independently. If they were not, the multiplicative law wouldn't have worked.

4. Today we know that all of the characters Mendel studied this way were located on different chromosomes. Because of the random way pairs of chromosomes align at metaphase and separate at anaphase, pairs of factors separate independently of each other.

5. Mendel's second law can be restated: "If an organism is heterozygous at two unlinked loci, each locus will assort independently of the other."

Mendel's Testcrosses

1. Mendel used **backcrosses** to determine whether a dominant type was homozygous or heterozygous, and whether assortment was independent. Suspected heterozygotes are crossed with homozygous recessive and homozygous dominant individuals. These are called backcrosses because the P_1 is the source of the types needed.

2. Mendel's backcrosses were as follows:
RrYy × **rryy** and **RrYy** × **RRYY**

3. If the test individual were heterozygous, the expected result from the first testcross would be all four categories—round, wrinkled, yellow, and green—appearing in equal numbers. If the test individual were partially or fully *homozygous,* or if assortment was not independent, other results would be found.

4. In the second cross, back to a homozygous dominant, all the offspring would be dominant round and yellow. To complete the test, each offspring would have to be **progeny tested.**

5. Test-crossing back to recessive types can help determine two things:
a. Whether the test subject is homozygous or heterozygous;
b. Whether genes are linked. When test crossed, double heterozygotes will produce a 1:1:1:1 ratio in the offspring if there is independent assortment. If the gene pairs are linked, some other ratio will show up.

The Decline and Rise of Mendelian Genetics

1. Mendel's findings, which were not understood in his time, were rediscovered about the turn of the century.

MENDELIAN GENETICS IN THE 20TH CENTURY
A Closer Look at Dominance and Recessivity

1. The double dose of genetic information provides organisms with a backup system in case of failure.

2. When the information is in conflict, the **recessive allele** is usually ignored, and the **dominant allele** is expressed.

3. Recessivity may mean that a gene is simply not functioning. An example of this is **albinism,** in which a recessive cannot fulfill its role in producing the pigment melanin. Two alleles failing to produce pigment results in an albino individual.

4. Recessive alleles may be functioning, but to a lesser degree than dominant genes.

5. In some instances, dominance is apparent visually, but on closer examination or measurement there is a difference between having one or two genes functioning normally. In fruit flies, for example, there may be less pigment produced by one gene than by two.

6. Some dominant alleles are also known as *inhibitor alleles,* since they prevent recessive alleles from expressing themselves. The dominant white of white leghorn chickens is an example.

7. Four specific **dominance relationships** are **partial dominance, lethal-recessive dominance, codominance,** and **overdominance.**

8. Partial dominance in snapdragons is seen when red and white are crossed. The offspring are pink. Phenotypic and genotypic ratios are the same (1/4:1/2:1/4).

9. Lethal-recessive dominance, or **rare dominance,** is seen in **achondroplasia** or **achondroplastic dwarfism.** Homozygous offspring die as embryos, while heterozygotes survive. For lethality to be expressed, two abnormal alleles must be present; thus the name lethal-recessive dominance.

10. Codominance occurs when one homozygote expresses a trait differently from the other homozygote, but the heterozygote shows both traits. In cats, black spotting *or* orange spotting represent the two homozygotes. Black *and* orange spotting represents the heterozygote.

11. The blood groups **M** and **N** are another example of codominance. When tested with **antisera,** *agglutinations* show that each genotype can be biochemically identified. Both **M** and **N** are clearly expressed.

12. In overdominance, the heterozygote expresses the dominant trait to an even greater degree than does the homozygous dominant. In pigeons, one allele helps fight off fungi while the other helps fight off bacteria. Together they increase hatchability or *Darwinian fitness.*

Multiple Alleles

1. The term *multiple alleles* means that even though each individual in a population gets two alleles for a trait, more than two different alleles are present in the population's genes.

2. The gene that translates into **beta hemoglobin** (one of the polypeptides of the hemoglobin tetramer) can have the **A** or **S** allele, but there are others. (Examples are the **C** and **E** alleles from Asia and the **th** allele from the Mediterranean region.)

Blood Groups

1. The blood types of the ABO system are determined by cell-surface **antigens**. Antigens are large molecules that are capable of reacting with specific **antibodies**. In terms of dominance, A and B are codominant and both are dominant over O. A, B, and O are multiple alleles.

2. From the alleles A, B, and O there are six possible genotypes. From these six genotypes there are only four blood groups or phentotypes.

3. The immune systems of individuals produce either anti-A or anti-B antibodies. The type A person produces anti-B, type B produces anti-A, type AB produces neither, and type O produces both.

4. When antibodies meet opposing antigens on red cell surfaces, they bind to these and form bridges to other red cells nearby. The massive binding of these cells is called agglutination, or clumping.

5. Because of antigen-antibody reactions, blood transfusions have to be preceded by careful blood-matching tests.

6. Rh blood groups can be divided in to **Rh positive** and **Rh negative.** The Rh positive antigen is very potent. In blood transfusions, this factor must be considered along with the ABO factors.

7. Rh antibodies are only present when a person receives Rh$^+$ antigens. The presence of antigen induces an immune reaction and the antibodies are produced.

8. The immune reaction of Rh$^-$ people to Rh$^+$ antigens extends into reproduction. Rh$^+$ fathers pass the antigen-producing gene to their offspring. When Rh$^-$ mothers bear Rh$^+$ children, there can be mixing of maternal and fetal blood, causing the mother to produce antibodies. Subsequent pregnancies are potentially dangerous, since the antibodies can enter the fetal blood and cause agglutination. The immune response can be clinically suppressed if the incompatability is known.

9. Organ and tissue transplants are difficult because of four **histocompatability loci** that produce **histocompatability antigens.** Because of the large number of multiple alleles, matching between people is extremely difficult and the immune system of the recipient must be suppressed to avoid tissue rejection.

Gene Interactions and Modified Mendelian Ratios

1. Mendel's two-factor crosses worked well because the pairs of genes involved did not interact with each other.

2. In many instances the 9:3:3:1 ratio Mendel observed is not produced when other loci or organisms are studied in similar crosses.

3. Sometimes pairs of alleles control different steps in producing the same trait. These are known as **epistatic** genes or alleles.

4. An example of epistatic genes at work is seen in mouse hair color. One pair of alleles produces brown or black, but another pair permits or prevents the presence of color. When the color preventer (a recessive) is homozygous, the mice are white.

5. In the epistatic cross **BbCc** × **BbCc,** the phenotypic ratio in the offspring turns out to be 9:3:4 instead of the classic 9:3:3:1 because every offspring with a **cc** is white.

6. In the epistatic cross **PpCc** × **PpCc,** two color preventors are at work and the phenotypic ratio in the offspring becomes 9:7 because either **pp** or **cc** in the offspring produces white (or prevents color).

REVIEW QUESTIONS

1. What is blending inheritance? How did Darwin's use of this concept open him to severe and justifiable criticism of his natural selection hypothesis? (287)

2. Describe Darwin's experiments in the heredity of snapdragons. What did he mean by prepotence? (288)

3. What kind of preparation did Mendel have for mathematical biology? What other characteristics led to his success? (289)

4. Describe Mendel's general procedure from P_1 to F_2. What kinds of crosses did he make? (289)

5. What important question confronted Mendel as he observed his F_1 generation in each cross? (290)

6. Distinguish between the terms *genotype* and *phenotype*. (289)

7. Carry out Mendel's round versus wrinkled crosses from F_1 to F_2 (round and wrinkled) and verify his ratios in the F_2. State the phenotypic ratio of the F_2. (290)

8. Distinguish between the terms *heterozygous* and *homozygous*. (290)

9. Explain how Mendel used progeny testing to determine the *genotypes* of his F_2 peas. What did he learn about the F_2 round peas? (290)

10. State Mendel's first law. (293)

11. Using the cross **Aa** × **Aa** and applying the multiplicative law, answer the following: (294)
    a. What is the probability of an **A** sperm fertilizing an **A** egg? Why?
    b. What is the probability of an offspring carrying the **Aa** combination? Explain carefully.
    c. What is the probability of an **a** sperm fertilizing an **A** egg? Why?
    d. How does the *additive law* apply to predicting the **Aa** offspring?

12. Carry out Mendel's two-character cross from P_1 through F_2. Write the *phenotypic* ratio of the F_2. (295–296)

13. State Mendel's second law and explain what it has to do with the results of the above cross. (297)

14. Using diagrams of chromosomes to represent the cross **AaBb** × **AaBb,** show how independent assortment really works. (Review meiosis in Chapter 11.) *Hint:* There are two ways the homologous pairs of chromosomes can align on the metaphase plate. (298)

15. Show the two types of testcrosses or backcrosses Mendel made for the double heterozygote **AaBb.** What is the purpose of a *testcross?* (297–298)

16. How would a testcross tell you whether a white ram carried a recessive gene for black? Prove your answer. (299)

17. Why do you need to do progeny testing if your testcross was to a homozygous dominant individual? (299)

18. Explain in terms of pigmentation how dominance works in albinism. (301)

19. What is meant by the statement: Dominance depends on how the phenotypes are classified? Give an example. (301)

20. Define and give an example of each of the following: (302–304)
    a. Partial dominance;
    b. Lethal-recessive dominance (rare dominance);
    c. Codominance;
    d. Overdominance.

21. List the responses of the **MM, MN,** and **NN** blood groups to anti-M and anti-N antisera. Use the term *agglutination.* (303–304)

22. How can the alleles controlling the protein transferrin be both codominant and overdominant? (304)

23. Review the following aspects of sickle-cell anemia (Essay 12.1): (306–307)
    a. Clinical symptoms;
    b. Visual effect on red cells at low oxygen levels;
    c. Transmission in families.

24. How is sickle-cell anemia an example of (Essay 12.1):
    a. Simple dominance;
    b. Partial dominance;
    c. Codominance;
    d. Overdominance.

25. Carefully explain what is meant by multiple alleles. (304)

26. List both the blood phenotypes (types) and the genotypes possible in the ABO blood system. (305)

27. Why is blood type O considered to be a universal donor? (305)

28. Review the problem of *Rh* incompatibility between *Rh* positive males and *Rh* negative females. What are the clinical solutions? (308–309)

29. What are histocompatibility antigens and what do they have to do with organ transplants? Why is it that identical twins can trade organs without problems? (309)

30. Carry out the epistatic cross between two **BbCc** mice (as described in the text). From your results, derive the 9:3:4 ratio and explain why the familiar 9:3:3:1 phenotypic ratio did not show up. (309–310)

Genes, Chromosomes, and Sex

The 19th century was a very comfortable one for the Victorian spirit before Darwin so reluctantly and abruptly jarred its sensibilities. The 20th century, however, descended on our small planet like a crazy quilt tossed on a globe. The patterns of the quilt suggested what was indeed to be a patchy century. As technology began to blossom, small, isolated groups of researchers worked feverishly to bring their own hard-won areas of expertise to their culmination. Among the many advances in those wee years of the century was the incredibly rapid development of good light microscopes. We haven't improved upon them much to this day.

Not only were microscopes vastly improved, but new stains were developed and people were seeing things that had never been seen before. Interest in the cell and its constituents was growing and the questions were becoming more sophisticated. Perhaps, though, in this proud age, one of the greatest discoveries was . . . the 19th century monk, Gregor Mendel.

Once his writings were brought to light, a chorus swelled, "Of course!" One could almost hear the heels of hands slapping against foreheads. Suddenly an army of experimenters emerged to duplicate old Mendel's work, and 9:3:3:1 became as familiar a number as 1776.

SUTTON AND MORGAN: MEIOSIS AND SEX

The Chromosomal Basis for Mendel's Laws

Other advances were not long in coming. One of them originated in the thinking of Theodore Boveri, a German who, working with sea urchin eggs and sperm, determined that the chromosomes were an essential part of fertilization and development. He is also credited with the discovery of the centriole.

Another important advance came from a bright young graduate student at Columbia University. His name was Walter Sutton and in 1902 he published a paper in which he attested that he could see a relationship between Mendelian inheritance and meiosis. Keep in mind that Mendel didn't know about meiosis and that those studying it weren't sure how it fit into the big picture. The field was obviously patchy, and one of the most difficult chores of scientists is to merge patches.

Finding a relationship between meiosis and Mendel's laws required that investigators observe the actual *random* lineup of homologous chromosomes on the metaphase plate and their subsequent *random* separation during anaphase. However, this would only be possible if pairs of homologous chromosomes were distinguishable from each other. In other words, the researchers needed to "see" chromosomes engaged in segregation and in independent assortment.

This is exactly what they did. Investigators found **heteromorphic** chromosome pairs in the meiotic cells of male grasshoppers. In heteromorphic chromosome pairs, the two homologous chromosomes have some visible difference, such as an extra knob of chromatin on one end, or a dark or faint band. The sex chromosomes, X and Y, which are clearly distinguishable in most species, are an example of a heteromorphic chromosome pair. In the germinal cells of the grasshopper testis, there is just one X chromosome, and it could be seen to segregate to only half the gametes. This was a visual confirmation of Mendel's first law: that paired factors segregate into different daughter cells when gametes are formed.

Further, when two different heteromorphic chromosome pairs were present in the same organism, they were seen to segregate independently in different meiotic cells, giving visible proof to Mendel's second law (independent assortment: the segregation of one pair of factors has no effect on the segregation of any other pair). Thus, while Mendel drew his inferences from statistical analysis of numbers, the cytologists could actually *view* alternate segregation and independent assortment in the meiotic process through their light microscopes.

Sutton went on to predict gene linkage. He reasoned that since there are only a small number of chromosomes in any one cell, and there are many hereditary factors, each chromosome must carry many genes. This prediction was rapidly tested and verified in the early 20th century.

Chromosomes and Sex

In 1910, an innocuous little insect entered the world of genetics. Thomas Hunt Morgan, also at Columbia University, began a program of breeding experiments with the fruit fly, *Drosophila melanogaster*. This is the tiny brown fly you find quietly hovering around your bananas.

Drosophila turned out to have many advantages for genetic studies. They are easy to maintain in the laboratory, they mate readily, each female lays hundreds of eggs, the young develop and mature in 10 days (Figure 13.1), they can be anesthetized for easy inspection, they have only four pairs of chromo-

13.1 *DROSOPHILA* LIFE CYCLE

The life cycle of *Drosophila melanogaster* is short. Metamorphosis from egg to emerging adult is completed in about 10 days. Each female is capable of producing hundreds of offspring.

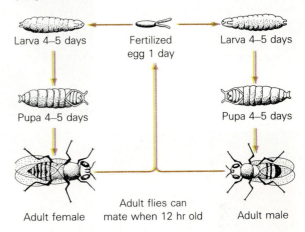

somes, and in the large salivary glands of the larvae these become giant (polytene) chromosomes on which specific loci are readily distinguishable under a low-powered microscope. In addition, mutants are readily recognized (Figure 13.2). People have now identified about 1000 gene loci, many with a number of known mutant alleles. *Drosophila melanogaster* has little economic importance, and except for occasionally floating belly-up in your cider or flying up your nose, they don't seem to bother people much. In fact, it has been said that God must have invented *Drosophila melanogaster* just for Thomas Hunt Morgan.

Now, in order to study inheritance one must have variation. At first, all Morgan's flies looked alike, except that males were visibly different from females. However, as Morgan carefully scrutinized each new generation, he eventually turned up one variant—among a group of flies with normal brick-red eyes, Morgan found a single male with white eyes. He correctly surmised that this was caused by a *mutation*, or spontaneous change in a gene.

Morgan decided to apply Mendel's newly appreciated techniques. He carefully nurtured his little white-eyed specimen and crossed it with several of its red-eyed virgin sisters. (In the laboratory, a fruit fly will mate with any other fruit fly of the opposite sex, if given no other choice. But female fruit flies, if given a choice, prefer to mate with strangers rather than their own brothers, although they will accept their brothers if no other males are around. Male fruit flies don't seem to care one way

13.2 SOME MUTANTS OF *DROSOPHILA*

The wing and eye mutations shown here were created in the laboratory using X-radiation. In doing this, geneticists are able to create variants from the normal gene and subsequently study the genetic control of the trait. In addition, they have learned a good deal about the nature of mutation and radiation damage.

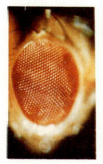

Carolina Biological Supply. (6)

or another and will even attempt to mate with each other.) From these matings, all the F_1 were red-eyed, to the surprise of none of the "new Mendelians." Obviously, *white eyes* was a recessive trait.

The experiments continued, and when the F_1 flies were mated with one another to produce an F_2 generation, sure enough, about one fourth of the F_2 flies were white-eyed and about three fourths were red-eyed. The actual numbers were not as close to this expected ratio as Morgan had hoped because, as it turned out, the white-eyed flies have a somewhat lower rate of survival than the normal flies. But there was something more disturbing about the F_2 flies. Every single white-eyed fly was a male! In fact, the F_2 ratio approximated:

¼ red-eyed males: ½ red-eyed females: ¼ white-eyed males

At this point, you may have decided that only males can be white-eyed. You would be wrong. Morgan discovered this for himself when he first did a testcross, mating his original, now-geriatric, white-eyed male to its own red-eyed F_1 daughters. A simple testcross should have provided a 1:1 ratio of dominant to recessive, and this one did. In fact, the testcross offspring consisted of approximately equal numbers of red-eyed males, red-eyed females, white-eyed males, and white-eyed females. So females could have white eyes. But even more surprises were in store. When Morgan mated

white-eyed females to red-eyed F_1 males, in what is called a **reciprocal testcross,** again, half of the offspring were red-eyed and half were white-eyed. But now every single one of the males was white-eyed, and every female had red eyes.

If you have a penchant for puzzles, can spare a few minutes, and think you're smart, try to figure out how this could have happened. Keep in mind what you learned about sex chromosomes in the chapter on meiosis. The rest of us can refer to Figure 13.3.

Morgan was aware of Sutton's suggestion that a single chromosome may carry a number of hereditary factors, although up to this time no one had *proven* that chromosomes carry anything at all. Morgan surmised that the sex-determining factor and the eye-color factor are somehow linked together, since these traits did not follow the law of independent assortment in the F_2 or in the testcrosses. Hence, he reasoned, as the X chromosome segregates at anaphase, so do the genes on it, including the red-eye or white-eye alleles.

Sex Chromosomes in *Drosophila*. By this time, Morgan knew something about the chromosomes that determine gender. He knew that male and female *Drosophila* were different with regard to one of their four chromosome pairs, and he guessed that this chromosome difference was the

13.3 MORGAN'S CROSSES

In his P_1 cross, Morgan mated his newly discovered white-eyed male with a normal red-eyed female, starting a revealing set of experimental crosses. All the F_1 offspring were red-eyed. Inbreeding the F_1 produced an F_2 that was three fourths red-eyed and one fourth white. But all the white-eyed flies were males.

The first testcross was between an F_2 red-eyed female and the P_1 white-eyed male. The offspring clearly showed that the F_2 female was heterozygous and that females could be white-eyed. Note the testcross results.

A reciprocal testcross was done using an F_1 red-eyed male and one of the white-eyed testcross females. This time, white eyes appeared in all the male progeny, and only in males. From these crosses arose the concept of sex-linkage.

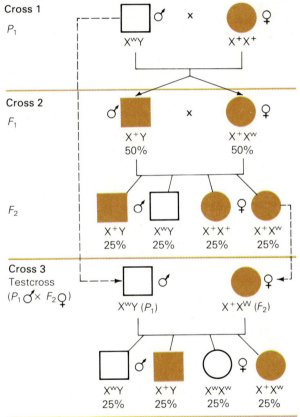

Cross 1
P_1

Cross 2
F_1

F_2

Cross 3
Testcross
($P_1 \male \times F_2 \female$)

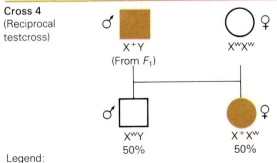

Cross 4
(Reciprocal testcross)

Legend:
X^wY = White-eyed male
X^+Y = Red-eyed male
X^+X^+ = Homozygous red-eyed female
X^+X^w = Heterozygous red-eyed female
X^wX^w = Homozygous white-eyed female

cause and *not the result* of sex differences. The X chromosome had been named and described earlier by H. Henking, who couldn't figure out what his finding meant (hence the name "X," for unknown). Female flies have two X chromosomes, while the males have one X and one Y (the same as with humans). At metaphase in *Drosophila melanogaster*, the X appears as a long, rod-shaped chromosome and the Y, as a shorter, J-shaped chromosome (Figure 13.4). (The human X and Y chromosomes are shown in Essay 11.2.)

We have said repeatedly that you have two of every gene—one from your father and one from your mother. You may realize by now that we have been lying to about half of our readers. While females are truly diploid, males are only partly diploid. Actually, at about 10% of their gene loci, men have one gene from their mother—period. Although the X and Y chromosomes behave like homologues in meiosis, lining up together, they do not carry the same genes. While the X chromosome has thousands of perfectly functional genes—those determining growth patterns, enzymes, and so on—the stunted little Y chromosome bears almost nothing other than a few genes relating to male sexual development. In *Drosophila*, the Y chromosome carries about six genes, all having to do with male fertility, while the X chromosome has approximately 1000 genes.

A gene on an X chromosome has no homologous gene on the Y chromosome to interact with or yield to. Thus, the Y chromosome behaves as if it has a recessive allele for virtually all the X chromosome loci. However, for a recessive allele to be expressed in a female, it must be present on both of her two X chromosomes. It follows that since a male has only one copy of any X-linked gene, it will express itself, whether it is recessive or dominant in

13.4 *DROSOPHILA* CHROMOSOMES

Drosophila has four pairs of chromosomes. Each of the four pairs is homologous in the female, but in the male only three pairs are homologues. The fourth pair consists of an X chromosome, which is identical in shape and content to that of the female, and a J-shaped Y, which is not. The Y is almost devoid of loci, and those that are present deal with male fertility only. In contrast, the X chromosome is known to have about 1000 loci.

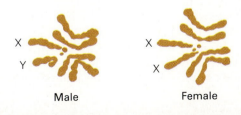

Male   Female

Sex Chromosome Abnormalities

As with trisomy-21, or Down's syndrome, abnormal numbers of sex chromosomes are brought about by nondisjunction—the failure of chromosomes to assort properly during meiosis. A leading factor causing this problem is believed to be aging in the oocytes. While most nondisjunctions in autosomal chromosomes are fatal to the embryo, wrong numbers of sex chromosomes in humans result in live babies and abnormal adults. There are many varieties of sex chromosome conditions. The shorthand designations of the normal individuals and the most common abnormalities are:

XX   Normal female (two X chromosomes)
XY   Normal male (one X, one Y)
XO   Turner's syndrome female (one X, no homologue)
XXY Klinefelter's syndrome male (two Xs, one Y)
XYY Extra Y, or XYY syndrome male
XXX Trisomy-X, or XXX female

In addition to the above, there are many more extreme situations, such as XXYY, XXXY, XXXYY, XXXX, XXXXX, XXXXYY, and so on, each syndrome having its own distinguishing characteristics. However, we can make four generalizations: first, one must have at least one X chromosome to live. Second, the presence of a Y causes the individual to develop as a male, and the absence of a Y causes the individual to develop as a female. Third (and this is probably why these syndromes are not fatal), all but one of the X chromosomes will condense into heterochromatin and be visible as a Barr body when stained, so that XO females lack a Barr body, XXY males have one, XXX females have two, and XXXXXYY males have four, and so on. Fourth, the more sex chromosomes a person has, the taller he or she will be, so that XO females are tiny, while XXX, XXY, and XYY individuals are usually much taller than chromosomally normal men and women, and XXYY men are huge.

XO (Turner's) individuals are phenotypically female but do not develop ovaries. They remain sexually immature as adults unless given hormones. XXY (Klinefelter's) males are tall and have small, imperfect testes and low levels of male hormones. They may have femalelike breast development and somewhat feminine body contours. XXX females are tall and frequently sterile but otherwise appear normal. XYY males appear normal except for their extreme height and for a tendency toward severe acne. They are also generally sterile. On the average, they have somewhat reduced IQs and, in common with other low-IQ groups, they average significantly increased criminal arrest records. At one time there was speculation that XYY males had "genetic criminal tendencies," but other analysis suggested that an XYY male is no more likely to be arrested than an XY or XXY male of the same IQ.

---

females. For X-linked genes in males, the term **hemizygous** (*hemi* means half) is used instead of homozygous or heterozygous.

Thus, Morgan had discovered X-linked inheritance, a discovery with vast implications. But more important, at the same time, he helped establish that genes are on chromosomes.

Sex Linkage in Humans

Human Sex Chromosomes. Although normal females are XX and normal males are XY in both *Drosophila* and humans, the physiological mechanism determining sex is somewhat different in the two species. In *Drosophila*, sex is determined by the number of Xs: two Xs, female; one X, male. Abnormal XXY flies are fully functional females, and XO flies (one X, no Y) are sterile males that look normal. In humans and other mammals, the *active* presence of a Y determines the development of testes, which in turn determines male development.

Thus, XO humans are sterile, abnormal females, and XXY humans are sterile, abnormal males. You will recall from Essay 11.3 that such abnormalities are the result of nondisjunction during meiosis, although there we were considering **autosomal** (non-sex chromosome abnormalities.) For more on X and Y chromosome abnormalities see Essay 13.1.

X Chromosomes and the Lyon Effect. The "saliva test" is sometimes used in women's athletic competitions as a test of whether a competitor is in fact a woman. The test has nothing to do with saliva and a great deal to do with X chromosomes. A swab of the inner surface of person's cheek will carry away a few cells from the mucous membrane lining. When these cells are stained, the cell nucleus in females shows a dark-staining body, the **Barr body,** absent in male cells. The Barr body is actually a condensed X chromosome. The condensed X chromosome can also be viewed microscopically in certain white blood cells, where it forms a characteristic projection from the cell nucleus called a **drumstick.** Whereas the Barr body was named after Murray L. Barr, its discoverer, the drumstick got its name because it reminded someone of a chicken leg (Figure 13.5).

Since every female has twice as many X-linked genes as are present in male cells, we can presume that there would be a physiological imbalance if all these genes produced twice as much gene product in one sex as in the other. In any case, evolution has solved the problem by permanently inactivating one of the two X chromosomes in each XX cell during embryonic development; each cell, that is, except for the germ-line cells. Which of the two X chromosomes is inactivated—the one from the father or the one from the mother—seems to be a matter of chance. Every tissue in the adult female is a mosaic of cell lines in which one or the other X has been inactivated. The highly condensed X is replicated normally and is passed from cell to daughter cell, but it is inactive with regard to genetic func-

13.5 THE CONDENSED X

One of the X chromosomes in each human female somatic cell is represented by permanently condensed heterochromatin. **(a)** In the nerve cell of the rhesus monkey, it is visible as a Barr body. **(b)** In some white cells, the heterochromatin takes the form of a ''drumstick.''

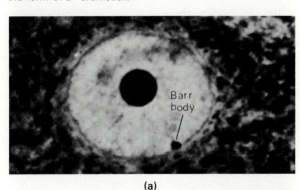

(a)

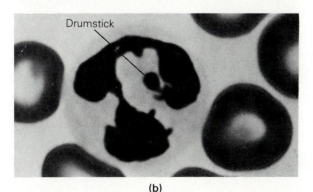

(b)

13.6 A COLOR BLINDNESS TEST

Color vision is tested using colored plates such as the one seen here. Actually, several plates are required for the complete test. If you have trouble seeing the number 9 (in red and orange) you may want to take the complete test.

tion. This peculiar behavior of the X chromosome was first established by an English geneticist, Mary Lyon, and is therefore known as the **Lyon effect.**

Incidentally, the Lyon effect probably accounts for the fact that extra numbers of sex chromosomes do not have the serious effects that other chromosome imbalances have. XXX females, for instance, have two Barr bodies and two drumsticks in the cheek and blood cells. As you will see in Essay 13.1, humans with sex chromosome number abnormalities survive rather well, although they are usually sterile.

Color Blindness. **Color blindness** in humans is usually caused by a recessive allele at either of two closely linked gene loci on the X chromosome. Human color vision depends on the differential sensitivity of three groups of receptors in the retina, called **cones.** One group of cones is maximally sensitive to blue light, one to red light, and one to green light. Perception of other colors and of subtle hues depends on the relative stimulation of these three types of cones. Persons homozygous (or hemizygous) for a recessive allele at one of the two X-linked loci lack the cones that are most sensitive to green, and homozygotes for recessive alleles at the other X-linked locus lack the cones that are maximally sensitive to red. By the way, the locus controlling the blue-sensitive group is autosomal, and blue-insensitive color blindness is very rare. Both X-linked defects are called **red-green color blindness** (Figure 13.6) because neither type of color blind people can distinguish between red and green. The defect was first described in a little boy who couldn't learn how to pick ripe cherries. He always brought home a mix of red and green fruit.

Somebody in the British Navy was aware of this situation well over a century ago and set about developing "running lights" on boats that even color blind men could tell apart. The green (starboard) side had a touch of blue and the red (port) side had a touch of orange. When traffic lights were introduced on railroads and later on city streets, these readily recognizable hues were the logical choice and we now see them daily.

About 8 percent of American men have one form or another of X-linked color blindness—about 6% are hemizygous for a recessive allele at one of the two loci and about 2% are hemizygous for a recessive allele at the other locus. Women are affected far less often—only about 0.4% of American women are red-green color blind. The reason for the sex difference is that, to be affected, a man need only receive one recessive allele from his mother. But an affected woman must receive recessive alleles from both her mother *and* her father. The chance that both these things will happen is much smaller than the chance that just one of them will. Recall our probability discussion in Chapter 12: "The likelihood of two independent events occurring is equal to the product of their individual occurrences."

A woman who is heterozygous for color blindness shows no symptoms of the condition. Among her children, however, she can expect half her sons to be color blind and half her daughters to be carri-

ers like herself—assuming that she marries a man with normal vision. What could she expect if she marries a color blind man? Figure 13.7 shows the appearance of color blindness in one family.

Other Sex-Linked Conditions. Many sex-linked genetic conditions have had great impact on people's lives. Two of the most common, but also dramatic, of all genetic disorders are **hemophilia**, the bleeder's disease, in which the blood doesn't clot normally, and **muscular dystrophy,** in which muscle tissue breaks down in late childhood. The usual forms of both of these are sex-linked. There are actually two common forms of X-linked hemophilia, governed by different X-linked loci. Most hemophilic males formerly bled to death in their youth. But in recent years, modern medicine has allowed affected hemophiliacs to survive and reproduce, thanks to blood transfusions and to infusions of a blood-derived substance known as **antihemophilic factor,** which supplies the critical substance missing in hemophiliacs. So now we find adult hemophilic males, a new situation in human history. In fact, even hemophilic females occur occasionally (in spite of the fact that homozygous females must receive a recessive allele from each parent). Hemophilia has had interesting implications in European history, as we find in Essay 13.2.

There is, as yet, no effective treatment for muscular dystrophy, so boys with this genetic disease still die before reaching adulthood. Muscular dystrophy has a delayed age of onset so that hemizygous boys appear perfectly normal as infants and toddlers, only to begin wasting away some time

13.7 COLOR BLINDNESS IN A FAMILY

In this hypothetical family tree, color blindness can be traced back to the great-grandfather and great-grandmother, although the problem was intensified by the marriage of a

color blind man and a carrier woman as shown at the right. We did not complete the great-grandmother's circle because we want you to decide whether she was a carrier.

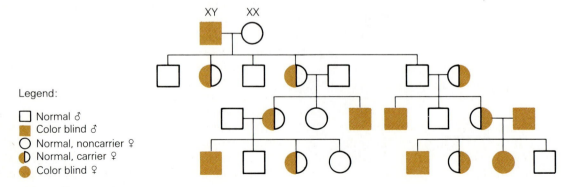

Legend:

☐ Normal ♂
■ Color blind ♂
○ Normal, noncarrier ♀
◐ Normal, carrier ♀
● Color blind ♀

The Disease of Royalty

Because it was the practice of ruling monarchs to consolidate their empires through marriage alliances, a highly restricted "royal mating population" was created, and hemophilia was transmitted throughout the royal families of Europe. Hemophilia is a sex-linked recessive condition in which the blood does not clot properly, so that any small injury can result in severe bleeding and, if the bleeding cannot be stopped, in death. Hence, it has sometimes been called the *bleeder's disease.*

The hemophilia of European royalty has been traced back as far as Queen Victoria, who was born in 1819. One of her sons, Leopold, Duke of Albany, died of the disease at the age of 31. Apparently, at least two of Victoria's daughters were carriers, since several of their descendants were hemophilic.

Hemophilia also played an impor- tant historical role in Russia during the reign of Nikolas II, the last czar. Alex- is, the only son of Nikolas II, was hemophilic, and his mother, the czari- na, was convinced that the only one who could save her son's life was the monk Rasputin, known as the "mad monk." Through this hold over the reigning family, Rasputin became the real power behind the disintegrating throne.

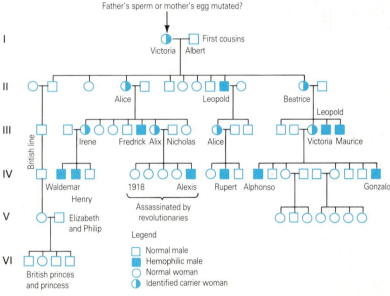

during their elementary school years. Their hetero- zygous mothers appear to be unaffected. Interest- ingly, however, microscopic tissue samples of females heterozygous for muscular dystrophy show that clusters of muscle cells accounting for about half the total number of muscle cells have atrophied by adulthood. We assume that the affect- ed muscle cells are simply demonstrating the Lyon effect, and the normal X happened to become con- densed. The women do not weaken because the remaining muscle cells expand in response to the increased load they must carry—another example of how dominance can work.

Sex-Linked Dominants. A few sex-linked mutant alleles are dominant, or are rare dominants. **Dom- inant brown spotting** of the teeth, for instance, is passed from affected men to *all their daughters* and *none of their sons;* affected women pass the condi- tion on to half their daughters and half their sons (work that one out). Dominant brown spotting affects about twice as many women as men, since they have two chances of receiving the X chromo- some with the brown-spotting gene. Another curi- ous X-linked genetic disease is known as the **oral- facial-digital syndrome,** which involves irregulari- ties of the mouth, face, fingers, and toes. This con-

323

dition affects only women, who get it only from their mothers. Such women pass the condition on to half their daughters but to none of their sons. It turns out that as a group, affected women have twice as many daughters as sons. We can therefore surmise that the allele in the hemizygous state is a lethal one.

More Variations on Mendelian Traits

We have already discussed several ways genes interact: dominance relationships, multiple alleles, epistasis (see Chapter 12), and sex linkage. We know how genes can interact to upset our neat $3:1$ and $9:3:3:1$ ratios in the F_2 generation. But you should be aware that there are other factors that can influence the way genes are expressed, such as the environment in which the gene appears.

Environmental Interactions. Environmental interactions may be very obvious or very subtle. Let's consider a straightforward example: the Siamese cat. One of the enzymes in its pigmentation pathway is temperature-sensitive; it won't function when it is warm. As a result, pigmentation of the fur occurs primarily in the colder extremities of the cat—the ears, the tail, the feet, and the nose. While these parts are black or dark brown, the rest of the cat is tan or almost white, which is why Siamese cats look like Siamese cats (Figure 13.8a). If you keep your Siamese cat in the warm house, it may grow to be almost white, but if you put it out at night, it will get to be quite dark.

Another strikingly dramatic example of the effect of environmental temperature on phenotype is seen in the *Drosophila* wing mutant called *curly.* Fruit flies raised under temperatures exceeding 16°C lose their ability to fly effectively as the wings curl upwards over the back (Figure 13.8b).

Incomplete Penetrance. If there is sufficient variation in a mutant phenotype, it may overlap the normal phenotype. In this case an individual may have an abnormal genotype without showing it, a condition known as **incomplete penetrance.** For instance, a rare dominant trait in human genetics is **polydactyly,** the tendency to have extra fingers or toes. Persons carrying this dominant gene show variable expressivity in that all four extremities may be affected, or only the hands, or only the feet, or perhaps only one hand or one foot (Figure 13.9). Both hands and both feet of a given carrier have the same genes and the same environment,

but may have either normal or abnormal numbers of digits, indicating that some form of developmental chance is at work. Of course, this means that a person carrying the gene may just be lucky enough to have only five toes on each foot and only five fingers on each hand. In such a case, fortune has smiled four times and the person would be unaware of carrying this gene if it weren't for the fact that unless fortune continued smiling, about half of his or her children will have extra fingers and toes.

13.8 GENE-ENVIRONMENT INTERACTION

(a) Siamese cats produce pigment at the cooler extremities of the body only, those regions where circulation is limited. (b) The wings of fruit flies are normally straight under any environmental temperature in which the insect lives. One type of wing mutation, known as *curly wing,* causes the wings to curl up when temperatures in their surroundings exceed 16°C.

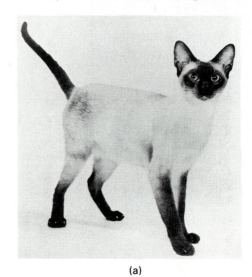

(a)

Normal fly at either temperature

Mutant at 25°C

Mutant at 16°C

(b)

13.9 POLYDACTYLY AND INCOMPLETE PENETRANCE

(a) The individuals shown here have inherited a dominant gene that expresses itself in the development of extra fingers or toes, or both. Although the gene is dominant, its expression is somewhat incomplete. Thus, while some carriers might not express the trait, their children may not be so lucky. **(b)** In the family protrayed in the chart, the numbers accompanying the individuals represent the number of digits on the left and right hands (at the top) and left and right feet (at the bottom). The dominant gene expressed itself in either the hands or feet of every carrier except for the two males in the third generation. Although both males were normal, three children of the male on the left were afflicted, thus, the father was a carrier, although the gene did not penetrate. The male on the right *may* have been genotypically normal (why "may?").

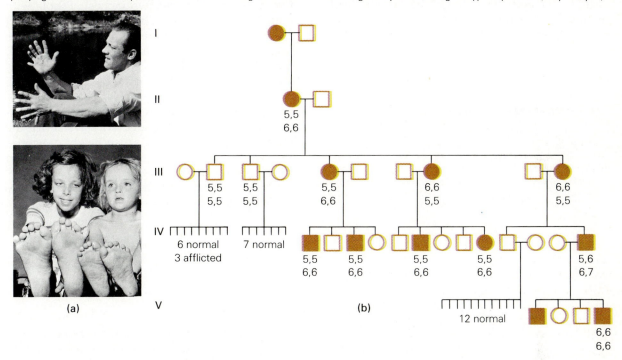

(a) (b)

Sex-Limited and Sex-Influenced Effects. A dominant gene is known to be responsible for a rare type of cancer of the uterus. Since, needless to say, the gene affects women only, it controls a **sex-limited** trait. **Sex-influenced** traits are another matter. The most common kind of middle-aged baldness, for example, is caused by a dominant gene, which produces only thinning of the hair in women. It doesn't affect eunuchs, either—unless they have been given injections of male sex hormones. (At last, a cure for baldness!) Sex-influenced genetic conditions then, affect one sex more often, or more severely, than the other. Another example is **pyloric stenosis,** a common and serious tumorous malformation of the digestive tract; it runs in families but affects five times as many boy babies as girl babies. As a final case, we might mention that men are more frequently affected by stomach ulcers than are women.

Variable Age of Onset. Baldness and muscular dystrophy both have delayed, **variable ages of onset.** Muscular dystrophy can begin at very different ages even in affected brothers, who would have received the same abnormal X-linked allele from their mother. Variable age of onset is also seen in **Huntington's chorea,** or **Huntington's disease,** as it is now known, a severe neuromotor disease caused by a dominant gene. The disease begins as a personality disorder and progresses through muscular shakiness, with symptoms similar to intoxication, on to complete paralysis and death in about 15 years. Huntington's disease does not begin to show its ultimately lethal effects until some time in adulthood (Figure 13.10).

Continuous Variation and Polygenic Inheritance

Many of the phenotypic traits that are most important to biologists, and especially to plant and animal breeders, do not fit into "either-or" categories. Instead, these traits occur in a gradient with more than two phenotypes, a situation called **continuous**

variation. Alleles still occur in pairs, and they still segregate and assort according to Mendelian law, but alleles at more than one locus are involved. When alleles at more than one locus contribute to the same trait, this is called **polygenic inheritance** and the trait, a **polygenic trait.** (How does this differ from multiple alleles and epistasis? See Chapter 12.) Examples in humans include skin color, foot size, nose length, birth weight, height, and intelligence. Let's look at one example.

A great many gene loci determine human height, but to simplify things we'll assume that height is determined by only three loci (three gene pairs at three locations). Also, in reality multiple alleles may be possible at each gene locus, but in our example we'll assume only two alternatives are available: "short" alleles and "tall" alleles. Further, we will assume that the presence of a "tall" allele, rather than a "short" allele increases adult height by five centimeters (about two inches). People with only the "short" alleles (six in all) grow to be about 160 cm (5'3"), while those with only "tall" alleles (again, six) grow to 190 cm (6'3"). In the middle, with three "short" alleles and three "tall" alleles, are the average individuals about 175 cm tall (5'9").

Table 13.1 summarizes the seven height categories possible with this model. We have added the relative frequencies of the seven height categories that would be predicted in the offspring from a large number of heterozygous couples (**AaBbCc × AaBbCc**). Note that the distribution approximates a "bell-shaped curve" or normal distribution.

13.10 VARIABLE AGE OF ONSET

Huntington's disease is a progressive nervous disorder that is genetic in origin. Its onset (appearance of symptoms) can occur at any time between ages 15 and 60, but most of its sufferers experience its onset between the ages of 30 and 50, as this graph reveals. Unfortunately, people may be well into their reproductive years before Huntington's disease is diagnosed, thus leaving the unfortunate legacy to another generation.

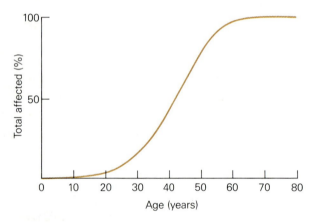

LOCATING THE GENES: LINKAGE, CROSSING OVER, AND CHROMOSOME MAPPING

In science, it is often said, the answer to one question gives rise, inevitably, to new questions. And this was certainly the case with early 20th century genetics. Sutton, Morgan, and others soon decided that chromosomes bore the hereditary factors and that many such factors, or genes, were contained in

TABLE 13.1

POLYGENES CONTROLLING HEIGHT: A MODEL OF THREE LOCI

Genotypes	Number of "Tall" Alleles	Number of "Short" Alleles	Height in Centimeters	Distribution
AABBCC	6	0	190	1/64 (1.6%)
AaBBCC, AABbCC, AABBCc	5	1	185	6/64 (9.4%)
aaBBCC, AAbbCC, AABBcc, AaBbCC, AaBBCc, AABbCc	4	2	180	15/64 (23.4%)
aaBbCC, aaBBCc, AabbCC, AaBBcc, AaBbCc, AAbbCc, AABbcc	3	3	175	20/64 (31.3%)
aabbCC, aaBBcc, AAbbcc, AaBbcc, AabbCc, aaBbCc	2	4	170	15/64 (23.4%)
aabbCc, aaBbcc, Aabbcc	1	5	165	6/64 (9.4%)
aabbcc	0	6	160	1/64 (1.6%)

Variation in adult human male height could be explained by supposing three equivalent gene loci, each with two alleles. Each "tall" allele (capital letters) adds five centimeters to height. Many different genotypes may have the same phenotypic effect. The distribution is calculated on the assumption that the "tall" and "short" alleles are equally common in the population. The same distribution would be expected in the F_2 of a hypothetical controlled cross between **AABBCC** and **aabbcc** individuals of, say, corn plants. Although models like this involve many unrealistic assumptions, they nevertheless have proven to have good predictive power.

each chromosome. The problem they soon encountered, however, was that Mendel's principle of independent assortment didn't work out with all genes. This, they decided, was because genes on the same chromosome stayed together during segregation, moving with the chromosomes as part of a **linkage group.** A linkage group came to be defined as any group of genes on the same chromosome.

Such a linkage would, of course, confound the law of independent assortment. And soon enough, some exceptions to the law began to show up. William Bateson and Reginald Punnett (of Punnett square fame), while trying to confirm Mendel's findings, got some puzzling results. They started with two true-breeding strains of sweet peas: one with blue flowers **(BB)** and long pollen grains **(LL),** the other with red flowers **(bb)** and round pollen grains **(ll).** The F_1 offspring of this cross had blue flowers and long pollen grains, hence "blue" and "long" were known to be dominants. Their P_1 cross was:

$$P_1 \qquad\qquad F_1$$
$$\text{BBLL} \times \text{bbll} \longrightarrow \text{BbLl}$$

So far there were no surprises. Then, following Mendel's now-established procedures, they sought to reconfirm the law of independent assortment by crossing the doubly heterozygous F_1 back to the doubly recessive P_1 stock, a standard testcross:

$$\text{BbLl} \times \text{bbll}$$

Mendel's second law predicted that they should get equal numbers of all four possible phenotypes—a 1:1:1:1 ratio of blue-long, blue-round, red-long, and red-round (you may want to confirm this for yourself), but this is not what Bateson and Punnett observed. Their ratio was approximately 7:1:1:7 for the four phenotypes. Their expectations and results were as follows:

Phenotype	Genotypes	Expected	Observed
Blue, Long	**BbLl**	25.0%	43.7%
Blue, round	**Bbll**	25.0%	6.3%
red, Long	**bbLl**	25.0%	6.3%
red, round	**bbll**	25.0%	43.7%

As you can see, there is significant disagreement with the outcome predicted by Mendel's second law. Perhaps, then, Sutton's new concept of gene linkage would apply. But according to the linkage concept, where the alleles for blue and long were on one chromosome, and the alleles for red and short on its homologue, the results should have been 50% blue-long, 50% red-round (Figure 13.11).

13.11 THE BEHAVIOR OF LINKED GENES

If the allele pairs for flower color and pollen shape are on the same chromosome, they cannot assort independently. At least, that's what Bateson and Punnett thought when confronted with gene linkage. With this being the case, the phenotypic ratio of offspring in their cross should have been 1:1, or 50% blue-long and 50% red-round. In actuality, not many linked genes behave quite this way.

Assumption:

Two pairs of alleles linked on the same chromosome pair will not segregate in a Mendelian fashion thus

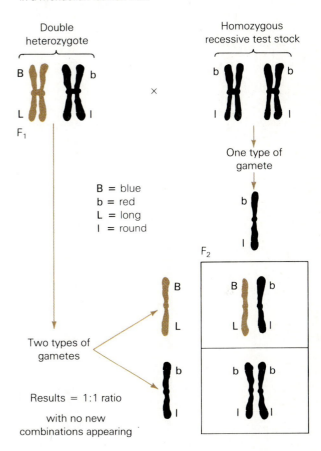

B = blue
b = red
L = long
l = round

Blue flowers would always appear with long pollen grains and red flowers with round pollen grains. Bateson and Punnett's results were inconsistent with either model. They were never able to figure out what was going on. What *was* going on? Bateson and Punnett were dealing with linked genes all right, but in a small proportion of meiotic events, the genes had become "unlinked" and represented genetic recombination. Crossing over had occurred (Figure 13.12). It's small wonder that Bateson and Punnett gave up in disgust. However, a new generation of geneticists were soon to unravel the mys-

13.12 NEW GENE COMBINATIONS

Continuing from Figure 13.11, we look at the same testcross, but with crossover occurring. Note that with crossover, the types of gametes double. The result is the appearance of two new classes of offspring, but not necessarily or even generally in a frequency equal to the expected phenotypes.

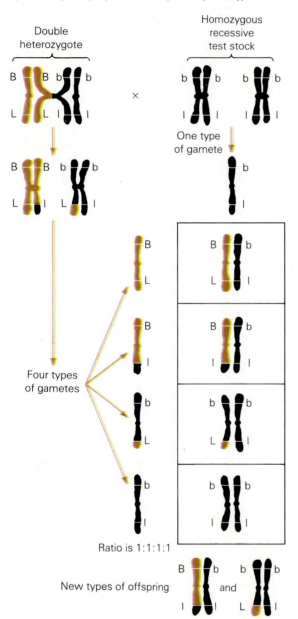

Double heterozygote

Homozygous recessive test stock

One type of gamete

Four types of gametes

Ratio is 1:1:1:1

New types of offspring   and

tery. We will leave the history for a time and review some things we already know about crossing over from Chapter 11.

Crossovers and Genetic Recombination

The recombination of two different genes that are physically parts of homologous chromosomes—in modern terms two homologous regions of DNA—requires the actual breaking of chemical bonds in the two molecules, at the same loci, and the exchange of these regions (see Chapter 11). The early geneticists had a hard time of it, but they managed to work out the concept of crossing over using only numbers from their crosses, testcrosses, and progeny counts. Part of the confusion at first was that some pairs of genes showed deviations from the law of independent assortment, but each of these pairs seemed to have its own degree of deviation. Two genes that tended to assort together in a testcross (such as **B** and **L**, and **b** and **l**) were called "linked," but some gene pairs showed stronger tendencies to "link" than others. In the Bateson/Punnett example, 12.6% of the testcross progeny were recombinants, but in other crosses the recombinant progeny might represent only a fraction of a percent. Gene pairs that had very low percentages of recombination came to be known as "tightly linked genes"; those with higher percentages, "loosely linked." These testcross numbers had none of the appeal of Mendel's wonderfully precise ratios.

Crossing over, then, allows linked gene loci to segregate independently. In fact, crossing over is so frequent that genes located far apart on the same long chromosome once again obey Mendel's second law. Although they are physically part of the same molecule at the beginning of meiosis, there is such a strong likelihood of one, two, three, or even more exchanges occurring between the two homologous chromosomes that the probability of the genes ending up in the same gamete is just about the same as if they had been on separate chromosomes to begin with—50%.

The situation is different with genes that lie close together on a chromosome. They tend to be shunted around as a unit during meiosis, and thus tend to be inherited as a unit in testcrosses. For example, let's say that two alleles, **A** and **B,** are so close together on a chromosome that there is only a 10% chance that they will be separated in crossing over and a 90% chance that they won't. That is, there is a 10% chance of recombination in the region between the two loci.

As an exercise, start with a double heterozygote, with one chromosome carrying **AB** and with its homologue carrying **ab.** You can see that if there were no crossing over at all, half the gametes would carry the **AB** chromosome and half would carry the **ab** chromosome; none would carry **aB** or **Ab.** But only 90% of the chromosomes are **nonrecombinants.** Of the other **recombinant** chromosomes, half will be **aB** and half will be **Ab.** Thus, we can expect the following kinds of gametes (Figure 13.13):

Nonrecombinants, 90% 45% **AB** 45% **ab**

Recombinants,   10% 5% **Ab** 5% **aB**

In a testcross, all of the gametes from the double homozygote test stock parent would be **ab,** so the distribution of the testcross progeny should be the same as the distribution of the gametes:

45% AaBb
45% aabb   Nonrecombinants = 90%

5% Aabb
5% aaBb   Recombinants = 10%

Therefore, the total frequency of recombination between the **A** locus and the **B** locus is 10%. In general, let R represent the frequency of recombination between two gene loci. Then, from the progeny of the testcross **AaBb** × **aabb,** R is the number of recombinant offspring (**Aabb** and **aaBb**) divided by the total number of offspring. The value of R is *related* to the probablility that a crossing-over event will occur between two loci on a given chromatid. However, it is *not exactly equal* to this probability. If two crossing-over events occurred between **A** and **B** on the same chromatid, then the first would separate **A** and **B** and the second would put them back together again, so the gamete getting that chromatid would show no evidence of any recombination at all (Figure 13.14). Thus, R, the probability of net recombination between the **A** locus and the **B** locus, is equal to the probability of an *odd number* of recombination events occurring between these two markers for any given chromatid.

Mapping Genes.   Soon after geneticists stumbled upon the phenomenon of crossing over and inferred that the **recombination frequency** depended at least in part on the physical distance between two gene loci, geneticist A. H. Sturtevant

13.13 10% CROSSOVER

In this example of crossover, the frequency of occurrence is 10%, which means the noncrossover frequency is 90%. We end up with four types of gametes, but each gamete has a percentage that must be considered. When tabulated, the results are a 9:9:1:1 ratio. (Divide each value in the offspring by the smallest, 5, to get this ratio.) This is a far cry from the 1:1:1:1 phenotypic ratio one might expect when independent assortment occurs, or the 1:1 ratio if the alleles were unable to assort independently. Perhaps you can see what geneticists were up against when first confronted with such ratios.

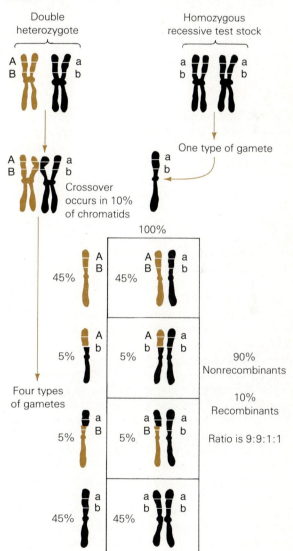

13.14 DOUBLE CROSSOVER

If a crossover occurs between **A** and **B**, what happens if another crossover should occur between them? The answer is above. There will be no new recombinants in the gametes if both crossovers occur in the same two chromatids. Thus, the observer would not even be aware of a crossover having occurred.

A double crossover or *any even number* of crossovers between two loci of one chromatid

No net recombinants between A and B loci

realized that recombination frequency data could be used to construct **genetic maps.** A genetic map shows where specific genes lie along a chromosome, based on the recombination frequencies observed between loci.

A map yields information, regarding not only the order in which gene loci occur on the chromosome, but the distances between the loci as well. For instance, if testcrosses of *Drosophila* indicate 13% recombination (meaning they are transmitted together 87% of the time) between *bar eye* (the dominant trait for eye shape) and *garnet* (recessive bright red eye), 7% recombination between *garnet* and *scalloped wings,* and 6% recombination between *scalloped wings* and *bar eye,* we can infer that *scalloped wings* is located between the other two gene loci (Figure 13.15). Mapping enables the discoverer of a new mutant allele to determine whether the discovery is a variant of a known gene locus. And it makes it possible to determine which genes belong to which chromosomes. Even gene loci that are too far apart on one chromosome to show any significant linkage in a two-locus recombination test can be related to other gene loci between them on the chromosome, and be shown to belong to the same linkage group.

Following Sturtevant, other ways were developed for mapping genes. We have mentioned that the chromosomes of some *Drosophila* larval organs, in particular those of the larval salivary glands, are very large and are banded in very regular ways. Actually, each such polytene or "giant chromosome" is a tight bundle of about 1000 sister strands lying side-by-side, the result of about 10 rounds of

DNA doubling without any nuclear or cytoplasmic division. The giant salivary chromosomes are interesting in their own right—they seem to be necessary for the production of the huge amounts of saliva needed by the little maggots—but they are even more interesting to geneticists because they provide a physical picture of each chromosome that can be related to its genetic content. One can draw physical chromosome maps (Figure 13.16) showing a total of about 5000 cross-bands of different widths that occur in recognizable patterns in the four chromosome pairs of *Drosophila.*

Individual gene loci can be correlated with individual giant chromosome bands through a process known as **deletion mapping.** Deletion mapping uses chromosomes that have been broken apart by X-radiation. Any segments missing as a result can be identified under the microscope by noting which salivary chromosome bands aren't there. A chromosome with a small missing segment, or deletion, behaves in genetic crosses as if it had recessive missing bands. Flies with missing segments can then be crossed with flies carrying known mutant alleles at all the gene loci that normally occur in the alleles to find out which genes belong with which bands. Such maps are called **deletion maps,** or **cytological maps,** to distinguish them from the recombination genetic maps we just mentioned.

Comparisons between physical chromosome maps (based on direct observations of altered chro-

13.15 GENE MAPPING

By studying the crossovers between groups of genes, their relative locations on the chromosome can be determined and their distance apart in map units decided. Thus genetic maps of chromosomes can be prepared. In this example, the locus of *scalloped wings* was determined to be between *bar eye* and *garnet eye,* all mutant traits of *Drosophila.* Can you follow the logic used?

Test crosses indicate

13% recombination between bar eye and garnet eye
7%   recombination between garnet eye and scalloped wings
6%   recombination between scalloped wings and bar eye

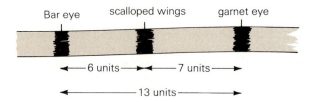

13.16 TWO CHROMOSOME MAPS

This illustration contains two kinds of maps of *Drosophila*. Parts labeled **(a)** are genetic maps. These were constructed by exhaustive studies of crossovers among the mutant loci. The numbers refer to the percentage of recombinations that occur between the loci, which are represented by the letter abbreviations. Parts labeled **(b)** are cytological maps that have been prepared from microscopic observations and micrographs of the giant salivary gland chromosomes obtained from *Drosophila* larvae. The giant chromosomes represent numerous replications of DNA rather than a single strand. The bandings are very distinct, and geneticists have identified bandings and associated them with the actual loci for specific mutants. Shown here are the tiny fourth chromosome and the left and right arm of chromosome 3.

mosomes) and recombination maps (inferred from counts of recombinant offspring) have turned up some interesting things. First, the different gene loci appear in exactly the same order in both kinds of maps, which is a relief since it proves the validity of years of incredibly painstaking recombination mapping.

Second, the relative distances between genes are not the same on the two maps, which means that the frequency of crossing over is not the same all along the DNA molecule, but varies from place to place on the chromosome. For instance, there are inactive chromatin regions on either side of the centromere that are actually quite large, but since almost no recombination takes place there, these regions appear to be short on the recombination map.

Third, and perhaps most interesting, is that the great majority of gene loci that have been mapped do not affect things like the shape or color of eyes and bristles, but are **vital genes** that can have *lethal recessive* alleles that kill the fly before it can grow up and be counted. These are called vital genes because the wild-type (normal) alleles of the recessive lethal loci are necessary for fly survival. Recessive lethals can be mapped like any other recessive gene, but there are a lot more of them. Comparing the physical chromosome map of vital gene loci with the recombination map of vital gene loci produces a striking result: with rather few exceptions, there is one vital gene in each of the dark bands that mark the salivary chromosomes.

In an intensive study of a limited region of one *Drosophila* chromosome, Burke Judd used a deleted X chromosome that was missing a segment just 14 bands long. Female flies that were heterozygous for

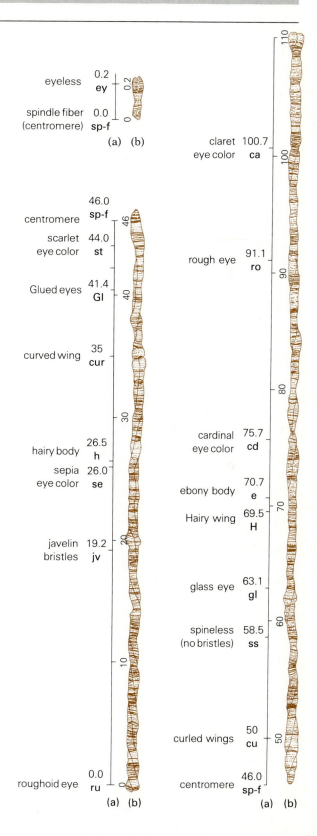

this test stock X chromosome and a normal X chromosome would survive because the normal X chromosome covered for the deleted region. However, any lethal recessive mutation that occurred in the corresponding region of the undeleted X chromosome would not be covered and could be detected. After collecting hundreds of strains of flies heterozygous for recessive lethal mutations in this region, Judd made crosses that showed that there were precisely 14 different vital gene loci involved: every one of their hundreds of lethal recessive mutations were alleles of one or another of these vital gene loci. Deletion mapping further showed that each vital gene locus was in a different chromosome band.

This seemed to indicate that there is one vital gene in each chromsome band, and that when we look at the bands on a giant salivary gland chromosome, we are in effect looking at row after row of genes. It has since been shown that some bands may contain two different vital genes, so the relationship is not absolute. However, using biochemical techniques, geneticists have estimated that there are about 5000 different kinds of mRNA in *Drosophila*. Assuming that most functioning genes produce a single kind of mRNA, we can conclude that there are about 5000 genes in the organism. This estimate is identical to estimates based on mutation rate studies and is also identical to the number of microscopically visible salivary chromosome bands. Thus, the evidence for the existence of about 5000 *Drosophila* genes is pretty good. Furthermore, the good match between the number of bands and the estimated number of vital genes helps establish the believability both of the gene number estimate and of the one-band-one-gene hypothesis.

Human cells in tissue culture have been studied in the same way, and it has been calculated that such cells have about 35,000 different kinds of mRNA, most of them represented by only a few molecules. This estimate is in agreement with work done by J. L. King, who had estimated an upper limit of about 40,000 vital genes in humans, based on mutation rate studies. You may be relieved to know that you have seven times as many genes as a fruit fly.

Bands and Puffs. The giant salivary chromosomes of the *Drosophila* larva, bizarre as they are, are good functioning chromosomes, and some of their genes are busy transcribing mRNA, as one would expect. But in these cases we have an exceptional opportunity to leave our theorizing and simply watch the system working. Because we can identify specific bands, we can actually see genes doing their job. At specific stages in *Drosophila* larvae, highly localized and predictable bands of giant chromosomes loosen up and send great loops of chromosomal material into the nuclear sap. In such places, chromosomes are no longer dense and compact, but diffuse, with tiny individual strands of DNA looking like the tangled line on an overly ambitious flycaster's reel. These areas are called **puffs** (Figure 13.17). Experimentation has shown that radioactive uracil becomes concentrated in these puffs, indicating that mRNA transcription is taking place. The largest puffs show the most RNA transcription activity. As the larvae mature, they need different kinds of enzymes, and hence they need new RNA to organize them. Thus, one area puffs and then recedes as the strands of another unwind to expose the four critical bases that are the foundations of life as we know it. Once the puffs recede, transcription ceases, the chromosomes rewind and settle down to lie quietly side-by-side, and the ordinary banding pattern once again appears in that place. Then the puffing begins in a new and different set of bands. These early observations, by the way, were among the first confirmations of the central dogma that chromosomal genes function by transcribing RNA. (Also see the lampbrush chromosome in Figure 10.4.)

CHROMOSOMAL MUTATIONS

We have seen in past chapters that although DNA is a relatively stable molecule, it is nonetheless subject to wear and tear and random change. As we noted, any permanent change in the DNA molecule is called a *mutation*. While mutations can occur within DNA, with deletions, insertions, and base substitutions (see Chapter 10), they also occur at a much greater level, involving whole chromosomes.

Actually, whole chromosomes can mutate, sometimes in comparatively spectacular and bizarre ways. Chromosomal mutations are better known as chromosomal **rearrangements**. They all involve the breaking of the double backbones of one or more DNA molecules and the rejoining or fusion repair of these molecules in unusual and abnormal ways. Let's begin with a close look at fusion repair.

Breakage and Fusion Repair

Breakage is just that, it occurs when both of the phosphate-sugar backbone strands of a double-stranded DNA molecule break and the chromosome falls into two pieces. You'd think that this would be fatal to the cell because only one of the pieces would have a centromere and the other would surely be lost in the next mitotic division. But in fact, chromosome breakage seems to happen all the time, usually with no permanent effect. The two broken ends simply come back together and are healed by the same repair enzyme systems that heal other kinds of DNA damage. Under the microscope it looks as if the two broken ends were "sticky." In careful studies of such breaks in *Drosophila* sperm chromosomes, it has been shown that more than 99.9% are healed by normal fusion repair.

Problems arise when two chromosome breaks occur in the same nucleus at the same time. Two breaks means *four* "sticky," unhealed broken ends. The fusion repair enzymes aren't choosy and will rejoin any two ends, sometimes rightly and some-

times wrongly. When the products of such events end up in gametes, and such gametes succeed in fertilization, the effect on an embryo can be disastrous.

Consider what happens when both breaks occur in the same chromosome. This means that there will be three fragments, one of them with two sticky ends. The two end fragments may rejoin and leave out the middle fragment, resulting in a **deletion** of genetic material. A small deletion affecting only one or a few genes may be passed on to the next generation, but most larger deletions are immediately lethal. The middle fragment with two broken ends will eventually form into a ring, and if this fragment is large and contains a centromere, it may become an abnormal **ring chromosome** (Figure 13.18a). There are reasonably healthy people walking

13.17 PUFFS AND BANDS

Chromosome puffs, a general loosening and looping of DNA from a region on the giant chromosomes, occur with regularity in fruit fly larvae. These are known to be regions of DNA transcription of RNA. **(a)** Microphotograph of one puff after experimental exposure to the molting hormone, ecdysone. **(b–d)** Puff is diagrammed in increasing magnifications. The granules are thought to be messenger RNA being transcribed.

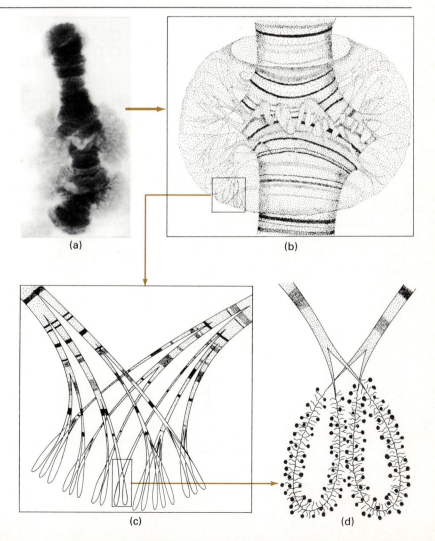

(a)

(b)

(c)

(d)

around today with ring chromosomes in every one of their cells. Another possible outcome is that the middle fragment may flip over before being rejoined to the two end fragments, resulting in a chromosome with an *inversion,* or inverted segment (Figure 13.18b).

If the two breakpoints occur in separate but homologous chromosomes or chromatids, and in different positions on the two homologues, abnormal fusion repair can result in one chromosome being too short and the other being too long. The short chromosome will have a deletion and the long homologue will have a *duplication* of genetic material; it will have an extra copy of all of the genes that occur in the duplicated region. Believe it or not, such duplications are sometimes beneficial and have played an important role in evolution. Gene duplication has resulted, for instance, in the highly productive, multiple hemoglobin genes that each of us carries on certain of our chromosomes.

13.18 INVERSION

A. INVERSION

(1) Inversion occurs

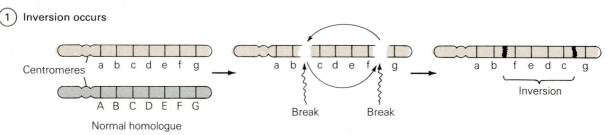

Centromeres
a b c d e f g

A B C D E F G

Normal homologue

Break   Break

a b f e d c g

Inversion

(2) Meiosis with inversion chromosome

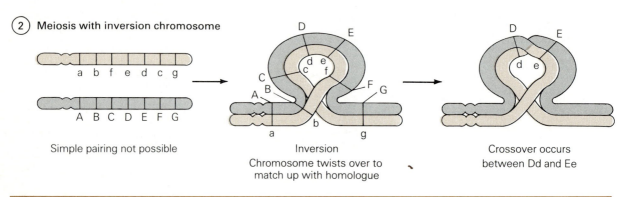

a b f e d c g

A B C D E F G

Simple pairing not possible

Inversion
Chromosome twists over to match up with homologue

Crossover occurs between Dd and Ee

(3) Products of crossover

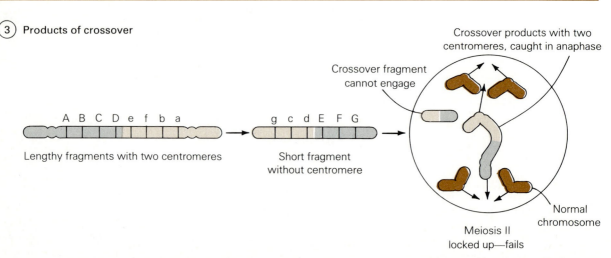

A B C D e f b a

Lengthy fragments with two centromeres

g c d E F G

Short fragment without centromere

Crossover products with two centromeres, caught in anaphase

Crossover fragment cannot engage

Normal chromosome

Meiosis II locked up—fails

Translocations. As a final example, consider what can happen when two nonhomologous chromosomes break in the same nucleus at the same time. An incorrect fusion repair results in a **translocation** of genetic material from one chromosome to the other. If both chromosomes break more or less in the middle, such misrepair can result in two reciprocal translocation chromosomes. When the breaks happen to be close to the ends in both chromosomes, the tiny end fragments may become lost

and the only surviving chromosome will be a **fusion chromosome** containing most of the genetic material of both chromosomes. The best-known example occurs with chromosomes 14 and 21 and produces what is called **translocation Down's syndrome.** We described nondisjunction Down's syndrome—trisomy-21—in Chapter 11. Although the outcome in the Down's child is the same translocation Down's differs in that it is inherited. The bizarre pattern of inheritance is reviewed in Figure 13.19.

13.18 (continued)

B. RING

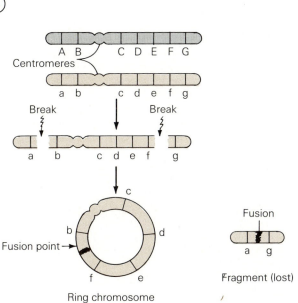

Ring chromosome

Fusion

Fragment (lost)

(a) An inversion is just what it sounds like; ① a chromosome undergoes breakage at two points along its length and the broken ends fuse back together but in an end-to-end position. Inversions are usually harmless in the somatic cells, since no genetic material has actually been lost. But trouble begins when inversions undergo meiosis. ② During synapsis (pairing up) in prophase I, the homologous chromosomes tend to take on a looping configuration that enables homologous loci to pair. When crossovers occur during these contortions, chromosomes can end up with double doses of genes or deletions. Further, some chromosomes end up with two centromeres and others, none. As you might expect, ③ such events produce highly abnormal daughter cells (gametes), and upon fertilization, gross developmental abnormalities. (b) Rings. An alternative to the inversion is the ring chromosome, ④ where the sticky ends of the broken chromosomal segment containing a centromere join to form a ring, and the other segments join to form a linear chromosome. The linear fragment without a centromere is lost during cell division, resulting in the loss of its genes. Should a cell bearing a ring chromosome attempt to complete meiosis ⑤ a crossover could produce results similiar to those seen in the inversion.

---

⑤ Meiosis and pairing up with normal homologue

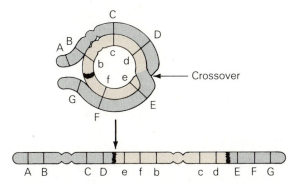

Crossover

One chromosome: Two centromeres and two copies of genes—
same meiosis II problem as inversion

Translocations and other chromosomal rearrangements, like individual gene mutations, aren't always harmful. Most are, but rare harmless or beneficial ones are the stuff of evolution. We've already noted that gene duplications have played an important role in evolution. Even closely related species *as a rule* show substantial chromosomal differences that must have originated as mutations. For instance, while we (normal) humans do indeed have 46 chromosomes per cell, our closest relatives, the chimpanzees, gorillas, and orangutans, have 48. A close study of the chromosomes of the four species shows that one of our treasured human chromosomes arose as a fusion translocation mutation in some possibly apelike ancestor.

This chapter has covered a lot of ground and, it must be admitted, genetics is not the easiest of topics to deal with, but it is becoming increasingly important to understand its basic principles. Some people may be interested in genetics for very specific reasons, such as how to make the sandy soil of Georgia yield peanuts or how to breed guard dogs that can't be stolen. The population in general, however, is rapidly approaching the need to understand genetics for broader reasons. We want to know why those miracle crops haven't yet begun to feed this hungry planet, as once promised. We also want to know about the inheritance of our dispositions. Are we naturally aggressive or docile? We want to know the answers to a host of questions, and many of them may well be answered as we learn more about how the gene works.

13.19 TRANSLOCATION 14/21

Translocation Down's syndrome is brought about through an inherited chromosomal mutation. The translocation and fusion of chromosomes 14 and 21 are found in a carrier mother **(a)**, either having arisen in her gametes during meiosis or been passed on to her by her mother. The translocation has no effect on the carrier since the correct number of homologues is present. **(b)** The problems arise during meiosis, where there are two possible ways for the chromosomes to align themselves in metaphase I. **(c)** From these alignments, four kinds of eggs are possible. After fertilization **(d)** four kinds of zygotes can result **(e)**. One will have a normal chromosome complement. A second will be a 14/21 carrier, like the mother. A third will contain a deletion of chromosome number 21 and will die as an early embryo. The fourth contains the trisomy-21 condition that will produce Down's syndrome in the child.

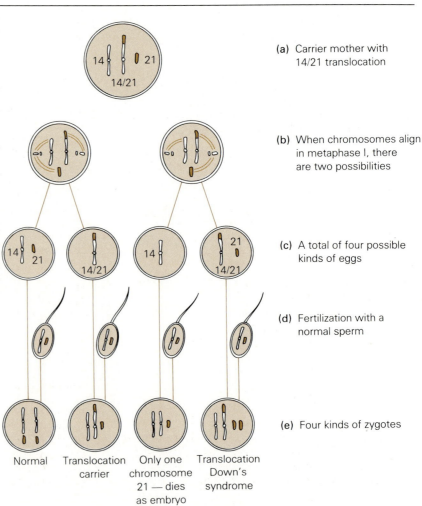

(a) Carrier mother with 14/21 translocation

(b) When chromosomes align in metaphase I, there are two possibilities

(c) A total of four possible kinds of eggs

(d) Fertilization with a normal sperm

(e) Four kinds of zygotes

Normal   Translocation carrier   Only one chromosome 21 — dies as embryo   Translocation Down's syndrome

APPLICATION OF IDEAS

1. Eleanor Perkins is phenotypically normal, but her family has its share of sex-linked abnormalities. Her husband, Garvey, has the X-linked dominant allele for brown spotted teeth but is otherwise normal. Her brother, Arthur, and her son, Little Chester both suffer from hemophilia A, the most common type of sex-linked bleeders' disease. Her father, Grandpa, is not hemophilic, but he is color blind.
   a. Draw and label a pedigree diagram of this family.
   b. From this information alone, Eleanor knows that she herself is a carrier. Which X-linked mutant gene or genes does Eleanor carry?
   c. Once Grandpa came to breakfast wearing one red sock and one green sock and couldn't understand why Eleanor, Garvey, and Arthur were all laughing at him. But when Little Chester didn't get the joke, Eleanor knew that her son was both hemophilic and color blind, and she suddenly realized what must have happened in one of her own oocytes shortly before her birth. Explain.
   d. Eleanor and Garvey are expecting another child. If it is a daughter, what is the probability that it will be hemophilic? Color blind? Have brown teeth?
   e. If their child is a son, what is the probability that he will be hemophilic? Color blind? Have brown teeth?
   f. What is the probability that such a son will be normal—that is, that he will have none of the traits mentioned? (Note: This last question requires more information. Haldane and Smith calculated that the recombination frequency between the red-green color blindness and hemophilia A locus is about 10%. List the expected frequencies of all possibilities for Eleanor's and Garvey's sons, and be sure they add up to one.)

2. Thomas Hunt Morgan, discoverer of X-linkage in fruit flies, later studied inheritance in an insect, *Euchistus*. The bug also proved to have XX females and XY males. In one species, *Euchistus variolarius*, all males have a black spot on the abdomen. In a closely related species, *Euchistus servis*, the males lack the spot, just as do the females of both species. A mating between *E. variolarius* females and *E. servis* males produced numerous F_1 hybrid progeny in which all the males had spots and the females did not. When the F_1 were allowed to inbreed, Morgan found 97 spotless females, 32 spotless males, and 84 spotted males in the F_2 generation.
   a. How many gene loci are involved in the spotting phenomenon?
   b. Is spotting a dominant or recessive trait?
   c. Is the spotting gene on an X chromosome, a Y chromosome, or on an autosome? Explain.
   d. If the original cross had been between a spotted male *E. variolarius* and an unspotted female *E. servis*, what would have been seen in the F_1 and F_2 generations?

3. In many plant species, male and female floral parts are borne on separate plants. Bauer and Shull took pollen from a narrow-leaved male *Lynchnis alba* and dusted it on the flowers of a broad-leaved female plant. In the F_1, both males and females were all broad-leaved. The F_1 male and female plants were crossed to each other and in the F_2, the females were broad-leaved but the males were both broad- and narrow-leaved.
   a. Explain these results.
   b. Is breadth of leaf sex-linked or sex-limited?
   c. What do you expect would happen in the F_1 and F_2 of a cross between a male of a true-breeding broad-leaved strain and a female of a true-breeding narrow-leaved strain?

4. T. H. Morgan crossed a white-eyed, yellow-bodied *Drosophila* female with a red-eyed, brown-bodied male. In the F_1 progeny, the females were red-eyed and brown-bodied, while the males were white-eyed and yellow-bodied. Allowing these to cross to produce an F_2 generation, Morgan counted the following:

Eyes	Body	Sex	Numbers
white	yellow	male	474
white	yellow	female	543
red	brown	male	512
red	brown	female	647
white	brown	male	11
white	brown	female	6
red	yellow	male	5
red	yellow	female	7

   a. What happened?
   b. Of the total of 2205 F_2 flies, what proportion were white-eyed? Yellow-bodied? Male? Recombinant?
   c. Do these values depart from the expected 50% of Mendel's first and second laws? Why?
   d. What is the map distance between the white and yellow loci?

KEY WORDS AND IDEAS

SUTTON AND MORGAN: MEIOSIS AND SEX

The Chromosomal Basis for Mendel's Laws

1. Sutton proposed that the hereditary factors were contained on the chromosomes and that each chromosome carried many factors.

2. Establishing the relationship between Mendel's first and second laws and meiosis required following two **heteromorphic** pairs of chromosomes through meiosis.

3. In 1910, Morgan began experimenting with *Drosophila melanogaster*, the fruit fly, which became an important research organism. It is easily maintained, has a short generation time, and produces hundreds of offspring. It has four pairs of readily identifiable chomosomes, which occur in polytene form in the larva.

4. Morgan's discovery of a mutant white-eyed male led to the discovery of sex linkage. In his experimental crosses he observed the following:
   a. P_1: White-eyed males × red-eyed females → F_1: all red-eyed.
   b. F_1: Red-eyed males × red-eyed females → F_2: ¾ red, ¼ white, but *all white-eyed flies were males*.
   c. Testcross: P_1 white-eyed male × F_1 red-eyed female → ½ red, ½ white-eyed offspring evenly distributed between males and females.
   d. **Reciprocal testcross:** Red-eyed male × white-eyed female → ½ white, ½ red, but *all females were red-eyed and all males were white-eyed*.
   e. The solution to the above was the linkage between the white eye allele and the alleles that determined sex: both were on the X-chromosome.

5. In *Drosophila* the sex chromosomes are heteromorphic: the X is rod-shaped and the Y is "J-shaped. Females have two X chromosomes, while males have an X and a Y.

6. The *Drosophila* X chromosome carries about 1000 genes, while the Y has only six, and these control fertility. Thus, all X-linked genes are **hemizygous** in males, acting as dominants.

Sex Linkage in Humans

1. In *Drosophila*, sex is determined by the number of Xs, thus the abnormal XXYs are functional females in the fly, but sterile males in humans (*Klinefelter snydrome*). So, the presence of a Y chromosome in humans produces maleness. Further, the absence of a second sex chromosome (XO) in fruit flies produces a sterile male, while this combination in humans results in a sterile female (*Turner's syndrome*).

2. In each cell of the human female, one X chromosome remains condensed. It is detected as a **Barr body** in cheek cells and a **drumstick** figure in the nucleus of certain white cells. The selection of such Xs is random and a mosaic of maternal and paternal Xs occurs in the tissues, producing the **Lyon effect.**

3. **Red-green color blindness** in humans, an abnormality in the **cones** of the retina, is caused by a recessive allele at either of two closely linked loci on the X chromosome. **Color blindness** occurs in 8% of American males and 0.4% of American females.

4. Women heterozygous for a sex-linked trait do not express the trait, although half their sons do.

5. Other sex-linked traits include **hemophilia** and **muscular dystrophy.** Because of the availability of **antihemophilic factor,** bleeders survive and reproduce. Because of the mosaic Lyon effect, half the muscle cells of women heterozygous for muscular dystrophy show signs of deterioration.

6. Brown spotting of the teeth is a sex-linked dominant that occurs more frequently in daughters (all the daughters, but none of the sons of an afflicted male). There are no homozygous females since two such alleles are lethal to an embryo.

7. In the **oral-facial-digital syndrome,** mothers pass the condition on to daughters only, and they have twice as many daughters as sons, since the hemizygous allele is lethal.

More Variations on Mendelian Traits

1. Many alleles are not necessarily expressed. They may be subject to environmental influence, sexual influence, or those as yet not understood.
   a. Dark hair pigment in Siamese cats increases in response to cold.
   b. The dominant gene for **polydactyly** (extra digits), as an example of **incomplete penetrance,** may not be expressed at all, or the degree of expression (number of affected digits) may vary widely.
   c. **Sex-limited** traits (for example, a tendency toward uterine or prostate cancer) can only affect one sex, while **sex-influenced** traits (such as baldness) tend to be expressed in one sex and not in the other.
   d. **Huntington's disease,** a neuromuscular disorder, has a **variable age of onset,** appearing most often near middle age.

Continuous Variation and Polygenic Inheritance

1. In many instances a trait is produced by the accumulative effect of genes from different loci. This is called **polygenic inheritance,** and its effect is to produce **continuous variation** in a trait. Examples include body height and skin color.

2. When polygenic traits are measured and plotted, the data commonly produce normal distributions which take the form of bell-shaped curves.

LOCATING THE GENES: LINKAGE, CROSSING OVER, AND CHROMOSOME MAPPING

1. Linkage, the occurrence of genes in linkage groups, and crossing were discovered when expected Mendelian ratios failed to appear. In a testcross with garden peas, Bateson and Punnett found odd ratios and seemingly impossible combinations of traits in the progeny.

Crossovers and Genetic Recombination

1. In the testcross **AaBb** X **aabb,** the predictable Mendelian ratio in the progeny would be 1:1:1:1, or 25% **AaBb,** 25% **Aabb,** 25% **aaBb,** and 25% **aabb.** If the alleles are on different chromosomes, no other combinations are possible.

2. If, however, the genes in the above testcross were linked on the same chromosome, then the ratio in the progeny would be 1:1, or 50% **AaBb** and 50% **aabb.** If they are fully linked, no other combinations are possible.

3. Frequently, even when linkage is established, testcrosses can reveal "impossible" new combinations and strange ratios. For example, the progeny might be 9:9:1:1, or 45% **AaBb,** 45% **aabb,** 5% **Aabb,** and 5% **aaBb.** Apparently, an exchange of alleles between homologous chromosomes has occurred.

4. In the situation above, crossover in meiosis occurred often enough between the **A** and **B** loci for 10% of the gametes to bear the results of recombination (and did not occur the other 90 percent of the time). The gametes can be represented as: 90% **nonrecombinants:** 45% **AB** and 45% **ab,** and 10% **recombinants:** 5% **Ab** and 5% **aB.**

5. Actually, more crossovers occur than recombinant genotypes reveal, since double crossovers

(or any even number) put the alleles back on their respective homologues.

Mapping Genes

1. **Recombination frequencies** are used to construct **genetic maps.** Map distances are measured in recombination percentages between loci.

2. **Recombination genetic maps** identify the chromosome in which a locus is found, and relative locations of genes and distances between them.

3. **Cytological maps** are based on actual chromosome topography. The polytene chromosomes from salivary glands in the larvae of *Drosophila* contain distinct banding patterns that are useful in the identification of loci. In **deletion mapping,** X-ray damage can be seen in the bandings and correlated with observable changes in traits.

4. Cytological and recombination maps show close correlation in the relative location of loci, but distances differ because of large *inactive chromatin* regions about the centromere, where crossing over does not occur.

5. The giant, banded chromosomes of *Drosophila* are known to loosen up and produce synthetically active loops or **puffs,** which have been shown to be active in the transcription of RNA.

CHROMOSOMAL MUTATIONS

1. Mutation at the chromosomal level involves **rearrangements** of chromosomal segments after breakage.

Breakage and Fusion Repair

1. Chromosome breakage and fusion repair is very common, and most repairs occur without problems.

2. When the wrong sticky ends are rejoined, **deletions, ring chromosomes, inversions,** and **translocations** can result.

3. In translocations, the greater part of a broken chromosome is fused to a normal chromosome. Generally, minor portions are lost.

4. **Translocation Down's syndrome** is produced by the fusion of part of chromosome number 21 with normal number 14. When a translocation egg is fertilized by a normal sperm, the resulting extra chromosome 21 produces a syndrome identical to trisomy-21. Unlike trisomy-21, a product of nondisjunction, translocation Down's syndrome is inherited.

REVIEW QUESTIONS

1. Review the contributions of Sutton and explain why they were valuable to understanding heredity. (316–317)

2. Why was the presence of two pairs of heteromorphic chromosomes needed in order to understand the cytological basis for Mendel's laws? (317)

3. List several ways in which *Drosophila* is well suited for the study of heredity. (317)

4. To test your understanding of sex linkage, carry out the calculations that explain the behavior of the white-eye trait in fruit flies. Cross the white-eyed male with a homozygous red-eyed female. Inbreed the F_1 and explain the results in the F_2. (318–319)

5. Can females express sex-linked traits? Using fruit fly eye color, prove your answer. (318)

6. Compare the shapes, gene complement, and sex-determining behavior of the X and Y chromosome in humans and *Drosophila*. (319–20)

7. Using a Punnett square and the symbols XX (female) and XY (male), determine the probability of male or female offspring. (320)

8. Explain why males always inherit X-linked traits from their mothers and never from their fathers. (320)

9. Explain the statement: Males are hemizygous for X-linked traits. (320)

10. What is the probability of color blindness in the children of a color blind carrier woman and her normal-visioned mate? (321–322)

11. Suggest two reasons why hemophilic females did not appear in the geneology of European royalty. (323)

12. What observations of women who carry muscular dystrophy suggest that the allele for the condition is really codominant to the normal allele? (323)

13. Suggest a reason why women with oral-facial-digital syndrome have twice as many daughters as sons. (323)

14. Briefly describe two examples of how physical conditions in the surroundings determine whether or not a trait will be expressed. (324)

15. Using polydactyly as an example, explain how incomplete penetrance seems to work. (324)

16. How do sex-limited traits differ from sex-influenced traits? (325)

17. How does the expression of the dominant allele for Huntington's disease differ from the usual? (325)

18. Using skin color as an example, explain what is meant by continuous variation. If two pairs of alleles were responsible, how many different skin colors would there be? What does your answer tell you about the genetics of this trait? (326)

19. How does polygenic inheritance differ from that of multiple alleles? Epistasis? (326)

20. What two aspects of their garden pea testcross data thoroughly confused Bateson and Punnett? (327)

21. Using the testcross **GgOo** × **ggoo,** what genotypes in what ratios might one expect if (a) the genes were unlinked, (b) the genes were linked but could not cross over, (c) the genes were linked but crossed over 20% of the time? (328–329)

22. Using a diagram, carefully explain why only odd numbers of crossovers are revealed in recombination frequencies. (329)

23. What is the basis for map distances in genetic or recombination mapping? Explain the logic of this. (330)

24. What is deletion mapping? Why is *Drosophila* well suited for this methodology? (330)

25. Construct a simple recombination map from the following alleles and recombination frequencies: Genes **P** and **L**, 30 map units; genes **L** and **X**, 5 units; and genes **X** and **P**, 35 map units. (330)

26. In general how do the relative position and relative distance of loci compare in recombination and deletion maps? (331)

27. How do chromosomal and point mutations differ? Which tends to be the more disastrous? (332)

28. Explain how a chromosome inversion might form. What happens when an inversion and its normal homologue pair up in meiosis? (334)

29. Why do chromosome fragments commonly fail to segregate properly during meiosis? What would the effect on the gamete and next generation be if they did segregate properly? (333)

30. Beginning with a premeiotic germinal cell carrying a normal 14 chromosome, a normal 21 chromosome, and a 14-21 translocation, show the four different possibilities in the gametes following meiosis. (335)

New Frontiers of Genetics

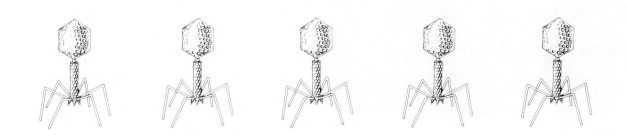

Some scientists these days are disturbed and others elated by our newly emerging abilities to manipulate genes. The grounds for the disagreement are apparent when one considers the possibilities that may lie before us. No one cares much that we can insert rat growth-hormone into developing mice to produce giant mice. And probably few would object to our ability to make human hormones in bacterial cells. But dare we try to change *people?* Not just to bring abnormal people to normality, but to make normal people "better"? Taller? Stronger? Smarter? Let's take a look now at the promise and the threat of our expanding ability to manipulate genes.

MICROBES ENTER THE GENETICS REALM

Over the years, different organisms have tended to dominate the study of genetics. First we had Mendel's true-breeding pea plants, then the hardy, prolific, and amazingly versatile fruit fly. Later, the cutting edge of genetic research focused on the corn plant, *Zea mays;* then Beadle and Tatum brought the pink mold *Neurospora crassa* into the spotlight. Then, in the late 1940s, a surprising organism became vogue: an exceedingly commonplace, but invisible microbe. Geneticists began to focus on *Escherichia coli*, the common colon bacterium, and the viruses with which it is infected.

Perhaps the emergence of *E. coli* should not have been so surprising. After all, the thrust of molecular biology has been to reduce problems to their simplest terms, and bacteria, when all is said and done, are much simpler than fruit flies, corn, or even mold. The viruses that infect bacteria are simpler still. It was thought that the wisest move would be to try first to understand these simpler organisms and then to progress to more complex ones.

E. coli did, indeed, offer a number of advantages as an experimental subject. Not only were they genetically simpler (they are essentially haploid), but bacterial cells have little internal structure and are easily broken open so that their cellular machinery can be isolated and analyzed biochemically. More important, bacteria and viruses can be grown in enormous numbers in very short periods of time—*E. coli* can double in number in 20 minutes—so experiments can be done quickly (Figure 14.1). Billions of such organisms can be grown in a test tube of fluid, so statistical sampling is never a problem, and even very rare events (such as the occurrence of specific mutations) will occur at a dependable frequency. Of course, finding one out of 1 million bacteria does offer a few logistical problems.

14.1 BACTERIAL GROWTH CHAMBER

Through the use of modern fermenters, fully automated devices that maintain ideal growth conditions, bacteria can be readily grown by the pound. This permits sizeable quantities of bacterial products to be harvested.

Locating Mutant Bacteria

Genetic experiments with bacteria obviously require the identification and collection of subjects. Among the most readily located are nutritional mutants—those unable to grow on simple foods readily used by wild-type, or unmutated bacteria. Such growth media, like those used by Beadle and Tatum when they collected nutritional mutants of *Neurospora* (see Chapter 9) are known as minimal media. Techniques for isolating bacterial mutants are different, and one involves adding **penicillin** to the minimal medium.

Penicillin is a powerful antibiotic that kills cells by interfering with cell wall synthesis. In this procedure, the experimenter puts a large, known quantity of wild-type bacteria into a minimal medium containing penicillin (Figure 14.2). Among the bacteria are a small but unknown quantity of nutritional mutants. The normal cells metabolize the medium, grow, try to divide, and die. But the mutants survive. They remain inactive since they are unable to metabolize and cannot grow and enter cell division, and in this state, the penicillin doesn't kill them. Thus through the use of penicillin an experimenter can substantially increase the frequency of nutritional mutants in the subject population, and these mutants can then be used. The recovered cells are then transferred to a gel-like agar medium in Petri dishes, where this time the medium is supple-

mented with a richer supply of nutrients. Now the mutant cells can grow—and they find themselves surrounded by others of their kind.

Another ingenious way of recovering mutations is by **replica plating,** developed by Esther and Joshua Lederberg. As you can see in Figure 14.3, replica plating is a sort of "now you see it, now you don't" process. Here, the bacterial colonies that *don't* grow after being transferred are the nutritional mutants. Among other things, the replica-plate experiment proved for the first time that nutritional mutants occurred spontaneously on supplemented medium.

Recovering Recombinants. Different strains of bacteria are able to recombine in various ways that we will discuss shortly. However, they have never mastered the process of meiosis, so the recombination doesn't involve sex as we know it. What passes for sex in bacteria is varied, bizarre, and infrequent—so infrequent that the odds of a bacterium having sex is about one in a million. In order to study such infrequent behavior, it's obviously necessary to take some shortcuts. No one wants to check millions of bachelor bacteria before finding one in the act of mating.

Suppose you have two nutritionally mutant strains of *Escherichia coli,* each of which requires a different amino acid in its nutrient medium. One strain needs tryptophan and another needs arginine (the first is symbolized trp^-, arg^+; the second, trp^+, arg^-); neither can grow on minimal medium. To find that one-in-a-million recombinant, you would mix 50 million or so bacteria of each type, let them mate (or undergo whatever other kind of genetic recombination event they can manage), and transfer them to minimal medium. The trp^-, arg^+ bacteria and the trp^+, arg^- bacteria will both just sit there. Those of the two strains that have managed to recombine to produce trp^+, arg^+, however, will form colonies (Figure 14.4).

The Life and Times of the Bacteriophage

The bacteriophage or phage virus, a parasite of bacteria, has already been mentioned with respect to Hershey and Chase's classic experiment, which showed that only the viral DNA entered the bacterial host cell while the empty coat stayed outside (see Figure 9.14). Let's look more closely at the bacteriophage and then see what happens when it finds its host. We will consider a typical **T-even bacteriophage,** one that commonly invades *E. coli.*

14.2 RECOVERING MUTANTS WITH PENICILLIN

Bacterial mutants can be recovered through a technique involving antibiotics. When a population consisting primarily of nutritionally normal wild-type bacteria and a few nutritionally mutated individuals are grown together in a medium containing penicillin, the normal individuals attempt to grow but soon die. The nutritional mutants are unable to grow in an unsupplemented medium so they remain dormant but living. Then when they are placed in the correctly supplemented medium, they grow and divide, producing a mutant clone ready for further study.

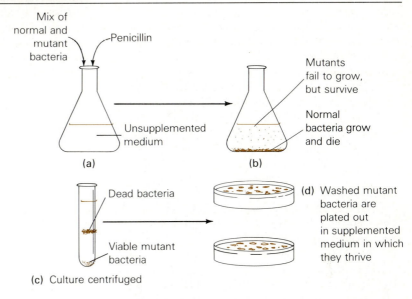

14.3 IDENTIFYING MUTANTS BY THEIR ABSENCE

Replica plating is a method of locating nutritionally mutant bacteria. Plate I, containing a completely supplemented medium that can support any growth, is coated with a mix of nutritionally normal bacteria and unidentified nutritional mutants. The resulting bacterial colonies form a random pattern of dots on the agar. After such growth has occurred, transfers are made to a minimal medium in plate II, as shown, and a comparison of plate I and II soon reveals the nutritional mutants. Such colonies can be recovered from plate I and grown for further study.

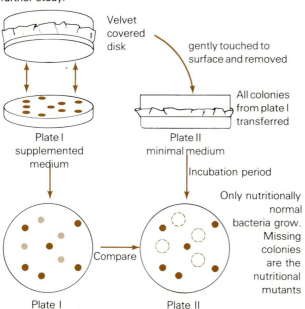

14.4 BACTERIAL SEX DISCOVERED

Biochemical and genetic evidence for sexual reproduction in bacteria was first gathered with the techniques shown in this scheme. Two strains, trp$^+$, arg$^-$ and trp$^-$, arg$^+$, which were selected through replica plating, were grown together in a supplemented medium. Then samples were plated out on minimal medium that lacked tryptophan and arginine. If colonies grew, they represented genetic recombinants with the trp$^+$, arg$^+$ genotype.

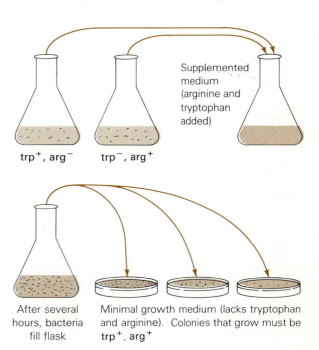

As viruses go, T-even phages are complex, consisting of a DNA core surrounded by a polyhedral protein head, which ends in a cylindrical, enzyme-laden tail. Attached to the tail are a peculiar set of leglike, protein tail fibers, completing a form that is strikingly like one of NASA's moonlanders. Upon "touching down" on its host, the enzymes go to work breaking down the bacterial cell wall, whereupon the tail firmly attaches itself and the viral DNA is injected into the cell. Once the viral DNA gets inside, it may immediately run into trouble with one of the host's protective mechanisms. Many bacteria have cytoplasmic **restriction enzymes** that chop up foreign DNA sequences. Each restriction enzyme recognizes a specific, short DNA sequence (eight to 12 nucleotides long) and chops it in the middle. (The bacterium has other enzymes that protect its own DNA.) Not all restriction enzymes kill all phage. As in all host-parasite relationships, the bacterial host and its specific bacteriophage parasites evolve together. In terms of restriction enzymes, natural selection favors any variation that does not contain the fatal DNA sequence recognized by such enzymes; such variants are the survivors. On the other hand, the gene coding for restriction enzymes also changes, and sooner or later a variant there will produce the right enzyme to cope with the new phage DNA sequences. Thus, those phage variants whose genotype puts them one jump ahead of their host are successful. The others do not survive host defenses.

The product of such ongoing changes is a see-saw battle between parasite and host, but more interestingly the changes promote host-parasite specificity. In other words, every strain of bacteria is subject to infection by only a few bacteriophage strains, and every strain of bacteriophage is able to infect just a few host strains. Later we will find out about some new and fascinating uses molecular biologists have found for restriction enzymes.

If the viral DNA gets past the restriction enzymes, it may encounter a friendlier host protein: *RNA polymerase*. RNA polymerase, unlike the restriction enzymes, isn't able to distinguish between its own DNA and foreign DNA. It reacts to the viral DNA as if it were its own and transcribes mRNA from it, producing *viral* mRNA. The viral mRNA is then blindly translated by the host's ribosomal machinery into viral proteins.

Then the virus turns tables on the host. Some of its own newly made enzymes readily distinguish between host DNA and viral DNA. These enzymes chop the host DNA into fragments, presumably so that the nucleotides can be recycled into viral DNA.

Other viral enzymes make copies of the viral DNA; some even make more viral mRNA to speed up the process. The virus genes also make **coat proteins,** the leglike "landing gear," and viral penetration enzymes. The final act of the viral proteins is to **lyse** (rupture) the envelope of the host cell, now hardly more than a bag of virus particles. Lysis of an *E. coli*

14.5 CELL LYSIS CYCLE OF A BACTERIOPHAGE

The T-even phages inject their DNA into a host cell, whereupon they take over all synthesizing activity and use the host's enzymes and protein-producing machinery to produce many more phage particles. Eventually the cell will lyse, releasing new phages that infect other bacterial cells.

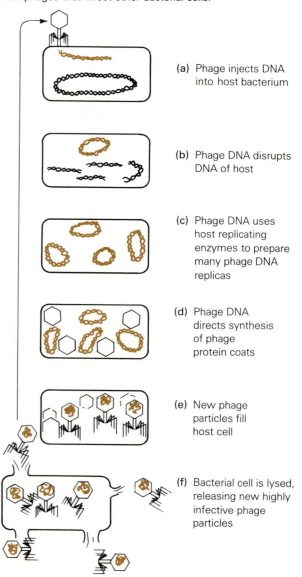

(a) Phage injects DNA into host bacterium

(b) Phage DNA disrupts DNA of host

(c) Phage DNA uses host replicating enzymes to prepare many phage DNA replicas

(d) Phage DNA directs synthesis of phage protein coats

(e) New phage particles fill host cell

(f) Bacterial cell is lysed, releasing new highly infective phage particles

14.6 BACTERIOPHAGE LYSOGENY

(a) In this series of events, a phage injects its DNA into the host cell. (b) The viral DNA joins the much larger circular bacterial DNA in an event that requires one crossover. (c) Eventually, the host will replicate both its own DNA and that of the viral segment. (d) The host cell then divides and the host/viral genome ends up in generation after generation of cells. (e) In some instances, the phage will break out of the lysogenic cycle and enter a lytic phase, in which many viral particles are produced and the host cell is lysed (see Figure 14.5).

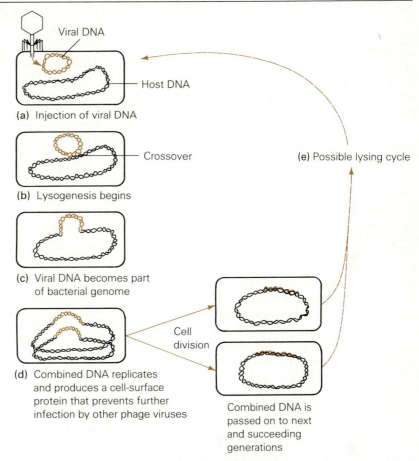

Viral DNA

Host DNA

(a) Injection of viral DNA

Crossover

(b) Lysogenesis begins

(c) Viral DNA becomes part of bacterial genome

(d) Combined DNA replicates and produces a cell-surface protein that prevents further infection by other phage viruses

Cell division

Combined DNA is passed on to next and succeeding generations

(e) Possible lysing cycle

cell may release about 100 new, infectious particles (Figure 14.5).

Lysogeny, an Alternative Cycle.   That is the basic bacteriophage life cycle, at least for the group known as the T-even phages. There are also other important phage life cycles. For example, sometimes the phage doesn't immediately kill its host. Instead of destroying the host chromosome, the phage DNA may cut into and join the host DNA. The phage is said to **lysogenize** the cell, and the whole phenomenon is called **lysogeny** (Figure 14.6). The viral DNA undergoes replication with the rest of the bacterial chromosome and is passed on to all the bacterium's descendants. While in the chromosome, it does very little except to put out a repressor protein that protects the cell from further attack by the virus' close relatives. Then, in some future generation, the viral DNA may *excise*—that is, break away from the host DNA—and the cycle

will resume where it left off. Again, viral proteins will be produced, viral replication will occur, and the cell will be lysed, releasing a hundred or so active virus particles.

Counting Viruses, or Holes in the Lawn. Viruses are much smaller than bacteria and can be found in very great concentrations. They are completely parasitic and can only grow in living cells; thus they do not grow on laboratory medium. However, they can be grown in the laboratory if they are diluted and then spread on agar plates that have been especially prepared with a "lawn"—a continuous surface layer—of susceptible bacteria. Each active virus particle infects just one bacterium, but after lysis its progeny spread to the neighboring bacteria until there is a clear hole in the lawn for each virus. These holes, called **plaques,** are then counted to determine how many viruses were originally present (Figure 14.7).

BACTERIAL RECOMBINATION

Sex in bacteria is so different from sex in even the lowliest eukaryotes that you may not think they really have sex at all. But it's just a matter of semantics. If the broadest possible definition of sex is *any mechanism that combines genetic material from two cells into one cell,* then bacteria have sex. In fact, they have a highly varied sexual repertoire.

Transformation

The first kind of recombination to be observed in bacteria was **transformation** (see Griffith's experiment, Chapter 9). It seems that bacteria can, with extremely low efficiency, take up fragments of DNA from other bacteria into their cells and incorporate it into their own chromosome. Interestingly, the fragment is incorporated into the bacterial chromosome by what amounts to double crossing over (Figure 14.8). The fragment aligns with the corresponding section of the host's chromosome, and the two sections are exchanged. The leftover host segment is then simply dismantled by host enzymes. The new addition is passed along to the cell's descendants when the cell divides.

14.7 GROWING THE BACTERIOPHAGE

Unlike bacteria, viruses will not become active or reproduce in an organic food medium. To do this they must find their way into a living host. In one procedure, dormant bacteriophage are spread out over a previously prepared "lawn," a layer of bacterial cells on the gel-like surface of nutrient agar in a Petri dish. After a period of incubation, the dishes are examined for the presence of plaques (clear circles) in the lawn. The circles are areas where bacteria are now absent, an indication that they have been lysed by the phage virus which continues to invade the cells in an ever widening circle.

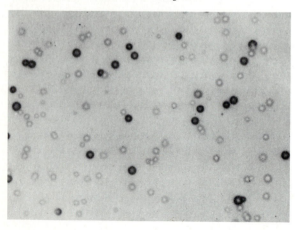

14.8 BACTERIAL TRANSFORMATION

In the transformation process, bacteria take in a fragment of DNA from other bacteria. As the illustration shows, once inside, the fragment aligns itself along the opposing segment of the circular DNA and becomes incorporated through double crossing over. All the bacterium's descendants will then carry the new recombinant chromosome.

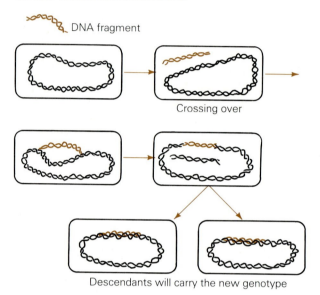

DNA fragment

Crossing over

Descendants will carry the new genotype

Transduction

Transduction involves the transfer of genetic material from one bacterium to another using a bacteriophage as the carrier. What happens is this: sometimes, when the phage coat protein envelope has already formed and is beginning to fill with DNA, there may still be fragments of the host's DNA floating about. The developing viruses may take up this bacterial DNA in place of viral DNA. The developing virus coat doesn't distinguish between them. After lysis, the viral particle with the bacterial DNA attacks a new host, injecting the piece of foreign bacterial DNA. When it encounters the DNA of the host, recombination can take place. Essentially, then, the protein body of the virus has brought bacterial DNA from one strain into another (Figure 14.9). It has not escaped the attention of biologists that this means of genetic transfer by such omnipresent agents as viruses could have important evolutionary significance.

Plasmids and the Plasmid Transfer

Many bacteria may contain, in addition to their principal chromosome, one or more very small circular chromosomes called **plasmids.** Certain kinds of plasmids were previously known as **episomes** (*epi*, upon; *soma*, body). The plasmids contain genes and generally replicate in synchrony with the main chromosome and are passed on to progeny of the cell at the time of cell division.

Some plasmids have genes that code for the proteins that make up special structures called **pili** (singular: *pilus*). While most pili are simple proteinaceous spines, some are larger and tubelike. The simpler pili aid the cell in maintaining contact with surface layers of water, cell surfaces (in parasites), and other surfaces. The larger pili are believed to be instrumental in sexual reproduction, drawing cells together and forming a bridge or **conjugation tube,** a hollow tube through which the transfer of genes can occur. For this reason, the larger pili are often called **sex pili** (Figure 14.10a).

14.9 BACTERIAL TRANSDUCTION

Transduction is a sort of DNA hitchhiking transfer. During phage lysing, a segment of the host's fragmented DNA is incorporated into the viral coat. When that virus infects a new host, the segment of bacterial DNA that it carries is then added to the new host's DNA through recombination.

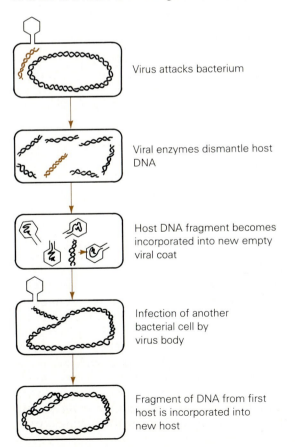

Virus attacks bacterium

Viral enzymes dismantle host DNA

Host DNA fragment becomes incorporated into new empty viral coat

Infection of another bacterial cell by virus body

Fragment of DNA from first host is incorporated into new host

If a plasmid-containing, or **plasmid-plus** bacterium and a plasmid-free, or **plasmid-minus** bacterium come close to one another, a sex pilus is formed, and the plasmid-plus bacterium draws up against the plasmid-minus bacterium. The plasmid DNA undergoes an extra round of replication, and as it does so, one of the new copies opens up and is transferred (as a linear DNA molecule) through the sex pilus into the plasmid-minus bacterium. Once inside its new host, the DNA becomes circular again. Other plasmid genes then produce a cell-surface substance that prevents further plasmic invasions. Both cells are now plasmid-plus. Plasmid transfer, then, represents a fourth mechanism for genetic recombination in bacteria (Figure 14.10b).

Some plasmids transfer themselves quite easily from one species of bacteria to another. This is unfortunate for us because some plasmids, as we will learn, carry genes that make bacteria simultaneously resistant to many different antibiotics.

F Plasmid. Our understanding of the fourth kind of genetic recombination is based directly on the work of Joshua Lederberg and Edward Tatum. They worked out the behavior of a certain kind of plasmid that was originally known as the *F* epi-some. The *F* episome got its name from a *fertility plus strain* of bacteria. As you might expect, there was also an opposing strain, the *fertility minus strain.* These bacteria were particularly interesting because genetic material from the positive strain always moved across the sex pilus into the negative strain. Furthermore, after such a union, the negative strain changed, itself becoming positive. But we are getting ahead of ourselves. Let's go back a few years and review a bit of history that shows why this plasmid is so important to modern geneticists.

In 1946, Lederberg and Tatum performed a rather simple but imaginative experiment. They mixed two nutritionally deficient strains of colon bacteria together to see if they could get any wild-type progeny. The experiment was unusual because, at that time, everyone was convinced that bacteria were confirmed celibates.

The two strains were:

Strain A: met⁻, bio⁻, thr⁺, leu⁺, thi⁺
Strain B: met⁺, bio⁺, thr⁻, leu⁻, thi⁻

The minus means that the nutrient is required and the plus means that it is not, and the nutrient symbols refer to methionine, biotin, threonine, leucine, and thiamine, respectively. Neither strain could grow on minimal medium.

14.10 THE PLASMID AND BACTERIAL RECOMBINATION

(a) The slender tube between the two bacteria of different strains seen in this electron micrograph is known as a *sex pilus*. Its proteins are produced in a *plasmid-plus* bacterium by genes on the plasmid chromosome. These slender pili are believed to draw bacteria together for conjugation to occur. (b) Once the sex pilus forms, the plasmid undergoes a round of replication. The new plasmid chromosome passes through the pilus into the plasmid-minus bacterium. The arrival of a plasmid changes the recipient into a plasmid-plus bacterium, which is capable of repeating the process.

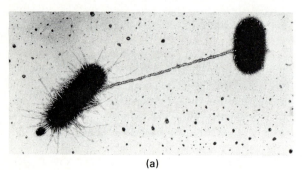

(a)

Plasmid-plus bacterium

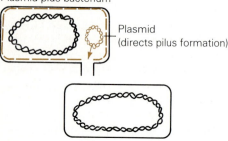

Plasmid (directs pilus formation)

Plasmid-minus bacterium

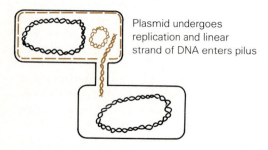

Plasmid undergoes replication and linear strand of DNA enters pilus

Both bacteria are now plasmid-plus

Plasmid produces a membrane protein that prevents further plasmids from entering the bacterium

(b)

It turned out that about one cell in 10 million grew perfectly well on minimal medium and therefore had the following *recombinant* genotype:

Recombinants: met$^+$, bio$^+$, thr$^+$, leu$^+$, thi$^+$

Lederberg and Tatum next tried to work out linkage relationships, as geneticists are prone to do when mapping chromosomes (see Chapter 13). They had a terrible time. Genes that appeared to be closely linked in one experiment were evidently distantly linked in another. Years later, after any number of ingenious (but erroneous) hypotheses were discarded, biologists concluded that the bacterial chromosome was circular. Imagine the problems anyone would have in trying to work out linkages with circular chromosomes!

Other investigators began to employ Lederberg and Tatum's techniques, and as new data came in, a strange and confusing picture began to emerge. For example, it was discovered that bacterial strains could be divided into two groups, which were promptly (perhaps too promptly) dubbed "male" and "female." The names seemed appropriate because males mated only with females, and mixtures of two male strains or two female strains never produced recombinant progeny.

What distinguished male from female was that gene transfer was always in one direction, from the male type to the female type. But, surprisingly, the male usually transferred only a few genes at a time. This is obviously a very different situation from that of the truly sexual eukaryotes, whose zygotes receive an equal input of genes from both parents.

A Closer Look at Bacterial Sex. How was it discovered that only a few genes from the male generally are transferred to the female? Let's look at a typical cross between two strains, as it was done in the early days. The fact that a combination of strains with different nutritional requirements can give rise to completely wild-type progeny doesn't tell us much other than that recombination does take place. To get to the root of the matter, it was necessary to look at the transfer of *unselected* loci—that is, the inheritance of nutritional mutants under conditions in which either the plus or the minus allele could survive. Such unselected genes were called **markers** because they were simply used to mark places (loci) on the chromosomes and did not affect whether or not the bacterium would survive. To be sure that recombination is taking place, a medium supplemented with only some of the nutrients is used. Then the recombinant progeny are looked at more closely, with replica plating, on other kinds of partially supplemented agar plates.

For example, suppose the original strains,

Strain A: met⁻, bio⁻, thr⁺, leu⁺, thi⁺
Strain B: met⁺, bio⁺, thr⁻, leu⁻, thi⁻

are mixed and plated on a medium supplemented with biotin, leucine, and thiamine but lacking methionine and threonine. All the surviving progeny would have to be met⁺, thr⁺, indicating that a mating has taken place. But what are their genotypes with regard to bio, leu, and thi, which are acting as the three genetic markers? There are actually eight different possible genotypes among the recombinant colonies (Table 14.1). To diagnose the genotypes and to count the number of colonies of each type, three replicate plates are made. Each replicate plate is supplemented with two out of three nutrients. Linkage relationships can be worked out from the relative numbers of colonies of each of the eight types. For instance, if leu⁺, thi⁺ and leu⁻, thi⁻ combinations were common, while leu⁺, thi⁻ and leu⁻, thi⁺ were both rare, one could conclude that the leucine locus and the thiamine locus were close together on the chromosome.

The results of experiments such as these also showed that the males contributed relatively little to the recombinant progeny. For instance, if strain A were male and strain B were female, most of the recombinant progeny in the above experiment would be met⁺, bio⁺, thr⁺, leu⁻, thi⁻—just like their strain B mothers for all unselected marker loci.

What recombinant genotype would be the most common if strain A were female and strain B were male? In line with the above reasoning, the most common of the met⁺, thr⁺ recombinants would be bio⁻, leu⁺, and thi⁺—just like their strain A mothers. Let's move on now to some very clever related work and find out how some of the riddles seen so far were solved.

William Hayes and the *F* Episome, or Making *E. coli* Sexier. In the early 1950s, William Hayes devised some unnerving experiments. In separate experiments, he treated either the male or the female strain of *E. coli* with the antibiotic streptomycin before mixing them together. Streptomycin prevents protein synthesis but not conjugation. The streptomycin-treated females produced no viable progeny because they couldn't divide, but the streptomycin-treated males could and did leave progeny of their own by transferring some of their genes into the receptive, untreated females. (In fact, this bizarre experiment was originally used to distinguish male from female strains.)

Then Hayes showed that the "maleness" of the bacteria was *catching*. Male and female bacterial strains (with easily distinguishable genotypes) were mixed together and plated. The resulting colonies were then tested for maleness or femaleness. It turned out that the male bacteria were all still male, but now about a third of the bacteria of the previously all-female strain were now male. Hayes thought that was a curious kind of sex. Of course, we know from our earlier discussion that maleness is due to a plasmid, the *F* plasmid, and that this plasmid is readily transferred from a male (plasmid-plus) to a female (plasmid-minus) bacterium. Once the new host receives the sex plasmid, it becomes a male (plasmid-plus).

Electron microscopes were not well enough developed in the 1950s for anyone to know about bacterial conjugation, but because both sexual recombination and contagious masculinity required direct physical contact, Hayes had a pretty good idea of what was going on. Something, originally called the **fertility factor,** was apparently transferred from individual to individual. This is why the male strains were renamed *F*⁺, short for **fertility-positive,** and the female strains were renamed *F*⁻, short for **fertility-negative,** although in fact both types were necessary for successful fertility (an obvious case of male chauvinism). Soon the fertility factor itself was identified as a circular piece of DNA and was then dubbed the *F* episome. It wasn't until many years later that it was realized that the *F* episome was only one of a large class of plasmids.

TABLE 14.1

EIGHT POSSIBLE GENOTYPES AMONG THE RECOMBINANT COLONIES

Genotype (all met⁺, thr⁺)	Growth on replicate plates[a]		
	Lacking only biotin	Lacking only leucine	Lacking only thiamine
bio⁺, leu⁺, thi⁺	+	+	+
bio⁺, leu⁺, thi⁻	+	+	−
bio⁺, leu⁻, thi⁺	+	−	+
bio⁺, leu⁻, thi⁻	+	−	−
bio⁻, leu⁺, thi⁺	−	+	+
bio⁻, leu⁺, thi⁻	−	+	−
bio⁻, leu⁻, thi⁺	−	−	+
bio⁻, leu⁻, thi⁻	−	−	−

[a] + grows on medium; − does not grow on medium

The *F* plasmid is one of many plasmids that ensure their own survival by inducing the formation of sex pili and transferring copies of themselves to bacteria. Like other plasmids, it also causes its host to produce a substance that makes it immune to infection from other *F* plasmids. The "contagiousness" of the F^+ condition, then, is similar to phage lysogeny.

As we mentioned earlier, the *F* plasmid is normally a closed circle of DNA, but following replication one daughter chromosome temporarily becomes a linear structure and is transferred to an F^- cell. Thus, the new host becomes infected with masculinity and is able to initiate conjugation.

The *F* Plasmid as an Agent of Gene Transfer. What about the transfer of bacterial genes? After all, that's what makes the *F* plasmid so interesting to scientists in the first place. It turns out that while simple transfer is very common, in rare cases the *F* plasmid may become inserted into the chromosome of its host. The plasmid has no preferred site of insertion and can show up anywhere in the host's circular chromosome. After insertion, the *F* DNA is replicated right along with the host DNA and can be passed on to all the bacterial offspring. So far, this sounds like phage lysogeny. But later, when the plasmid forms a conjugation tube and attempts the transfer to an F^- bacteria, strange things happen. First, the *F* plasmid opens up, as usual. But since it is now part of a larger circle, *the entire structure—host chromosome and all—is changed into a linear DNA molecule.* Then, when one end of the *F* plasmid goes through the conjugation tube, it starts to pull the entire host chromosome in after it. Because the chromosome is very long, it takes about 89 minutes for the whole thing to get through. In fact, the tube usually breaks apart before transfer is completed. In this case, since the recipient cell usually gets only a piece of the F^+ sequence, it isn't transformed into an F^+ bacterium, but the bacterial DNA that is brought in can recombine, by crossing over, with the DNA of the recipient ("female") bacterium (Figure 14.11).

It wasn't long before Hayes and others were able to isolate clones (identical individuals) in which the *F* plasmid had been integrated into the host chromosome. Then they didn't have to depend on rare, random insertions. Every bacterium in such a clone had the plasmid integrated in exactly the same place. These strains showed a potential for recombination that was 1000 times greater than that of the ordinary F^+ strains because in these strains

14.11 TRANSFERRING THE INTEGRATED PLASMID

In the rare cases where the *F* plasmid joins the entire bacterial chromosome, the sex act changes. When the newly converted F^+ bacterium mates with an F^- type, the entire chromosome replicates, episome and all. A small portion of the episome replica starts through the sex pilus, dragging the lengthy bacterial chromosome behind it. Usually, the host chromosome replica breaks before the entire transition is completed. Any portions that do make the transfer may be incorporated into the F^- host through crossing over and recombination—*a truly sexual act.* Since most of the replicated episome rarely gets through, the second cell must retain its F^- status.

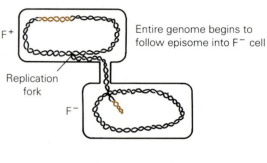

F+ Entire genome begins to follow episome into F⁻ cell

Replication fork

F⁻

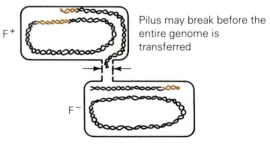

F+ Pilus may break before the entire genome is transferred

F⁻

F⁻ is still F⁻ Crossing over and recombination occur between fragment and F⁻ chromosome

each *F* plasmid was integrated, a rare occurrence in other F^+ strains. The new clones were called *Hfr,* for **high-frequency of recombination.**

Mapping the *E. coli* Chromosome: The Great Kitchen Blender Experiment. Progress in molecular biology often follows the development of new technology. In this case, the technology was provided by Fred Waring, a popular band leader of the 1950s. When he was not leading his group, The Pennsylvanians, in song, he was busy inventing the Waring blender. The blender, of course, beats food into a mush, and in so doing, it disrupts cells. Molecular biologists needed a good way to separate mating cells, so they swiped the blenders from their kitchens.

In one of the first experiments, Francois Jacob (of the Jacob and Monod team) and Elie Wollman broke apart mating bacteria with such a blender. They let *Hfr* and *F⁻* bacteria mate for precisely measured lengths of time before pushing the button, rudely interrupting the transfer in a primitive form of coitus interruptus.

As we have mentioned, it takes about 89 minutes for the *Hfr* chromosome to transfer to the *F⁻* bacterium. Thus, if the transfer is interrupted after just a few minutes, only a few of the known gene loci would make it across while most of the others would remain behind. As the periods of mating were extended, more and more loci would be transferred and could be recovered in the progeny, although at lower efficiencies because of spontaneous breakups of the bacterial couples. The last genetic loci come through, but with extremely low efficiency, toward the end of the 89 minutes.

Jacob and Wollman discovered that they could rapidly map *Hfr* bacterial chromosomes by using their blender. They found that on a standard *Hfr* chromosome the first locus passed through in about two minutes, and each subsequent marker was given a position appropriate to the time it entered the recipient cell. Figure 14.12 shows the genetic map of *Escherichia coli*.

One question that might have occurred to you is whether the donor *Hfr* cell survives the transfer of its chromosome. Recall that plasmid transfer is associated with replication. Apparently, replication occurs simultaneously with transfer, with one of the replication fork strands going through the tube and the other staying behind. The donor cell always keeps a copy of the entire *Hfr* chromosome. The answer then is yes, the *Hfr* cell survives, and sex is not a suicidal act for a male bacterium.

The early geneticists forced all sorts of sexual interpretations upon their originally baffling data. By now it seems clear that "bacterial sex" is, in fact, a rare aberration of plasmid transfer and has little to do with sex as we eukaryotes know it. But it was, and remains, a very useful tool in molecular biology. It may also prove to play a significant role in long-term bacterial evolution.

Sexduction

Sexduction is admittedly an awful term—a combination of sex and transduction. In sexduction, a small bit of the large host chromosome somehow becomes incorporated into the free *F* plasmid. It can then be transferred to any *F⁻* cell. This means that the recipient cell is functionally *diploid* for those bacterial genes that are present in the episome (that is, it has two copies of each such gene, one on its own chromosome and one on the *F* plasmid). In such partially diploid bacteria, gene loci can be heterozygous, and one can even investigate dominance and recessivity relationships in organisms where such things don't normally occur. It was through the use of sexduction that Jacob and Monod were able to work out the DNA control mechanisms, known as operons, in bacteria (see Chapter 10). Also, sexduction clearly suggests a means by which the molecular biologists could manipulate genes. If they could somehow get an episome—a plasmid, that is—to take up a strand of selected DNA, it might be possible to get bacteria to do some work for them. Sound far-fetched? Keep sexduction in mind as we go on.

Plasmids and Plasmid Genes

The *F* plasmid is only one of many kinds of bacterial plasmids that have been discovered. None of the rest can transfer the host's chromosomal genes to other bacteria, but they do play roles in bacterial genetics. They have genes of their own, often including genes for making pili and for coating the cell surface to protect the host (and the plasmid) from infection by other plasmids. In addition, some have genes that protect their hosts (and thus themselves) in other ways. For example, plasmids are responsible for the deadly toxins produced by disease-causing bacteria.

Some plasmids even carry genes that make their hosts resistant to antibiotics. One plasmid, known as **R6,** endows its host with resistance to six important antibiotics (Figure 14.13). R6 can also be transferred between distantly related species of bacteria. It was previously believed that plasmids were a kind of parasite, and that plasmid infection resembled a disease infection in some ways; we have to modify that now since some plasmids protect their hosts (and thus themselves) against the ravages of antibiotics.

Now that we know about "infectious antibiotic resistance," some mysteries have been cleared up. Time and again physicians have successfully treated bacterial infection with antibiotics, only to find that the same disease turns up again and is no longer susceptible to the same antibiotic. What has happened is that natural selection has been at work, and our very success at battling microbes has paved

14.12 GENETIC MAPPING OF *E. COLI*

This genetic map of the circular chromosome of *Escherichia coli,* produced by recombinant studies of interrupted *Hfr* chromosomes, is divided into 89 parts, corresponding to the 89 minutes required for a complete chromosome transfer. Each gene locus is identified by code letters that represent the ability or inability to synthesize various substances. Note that many of the codes represent amino acids using the standard abbreviations. The technique used in developing the map includes the sequential interruption of gene transfer and, following this, identifying the genes that made it through.

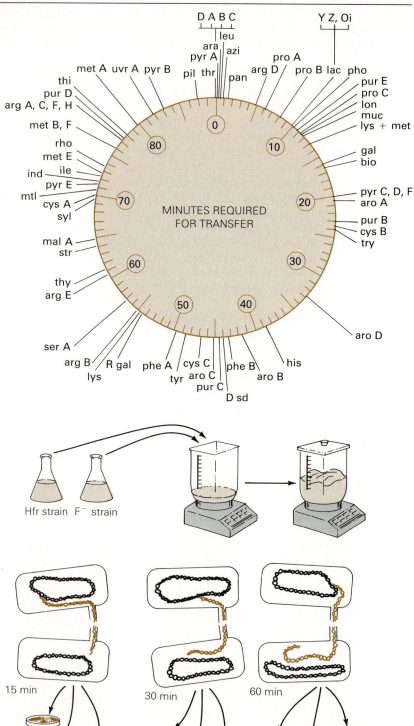

Bacteria grown together but conjugation disrupted after specified time intervals.

Hfr strain   F⁻ strain

Length of time determines number of markers transferred

15 min   30 min   60 min

Test for recombination determines which markers have made it through.

the way for the rise of new, more potent strains. Fortunately, the drug industry has more or less kept up with the challenge by continually developing new or altered antibiotics.

Because of the resistance phenomenon, the indiscriminate use of antibiotics is today a controversial issue in medicine. Antibiotics are routinely used in livestock feed to accelerate growth rates and to protect against disease, and those antibiotics can be transferred to meat-eating consumers. (Some livestock breeders have recently agreed to forego the use of antibiotics in this way.) Fortunately, the use of antibiotics to retard spoilage in foods such as milk is illegal and is discouraged by the constant surveillance of public health agencies. It is also clear that too many physicians routinely prescribe antibiotics for all manner of minor ailments, in part to guard against secondary infections and malpractice suits, but mainly to satisfy their patients' demands that something dramatic be done to end the sniffles. It is now believed that this indiscriminate overuse of antibiotics has had a positive selective effect on monstrous superresistant bacterial strains.

On the other hand, there is evidence to back up a different view, which is that the widespread use of antibiotics by American and European physicians may have had a net beneficial effect. It has been maintained that such widespread usage has managed virtually to eliminate rheumatic heart disease and salmonellosis in those parts of the world where antibiotics are readily available. (Rheumatic heart disease, which is caused by a *Streptococcus* bacterial infection, was until recently, a leading cause of death among young and middle-aged adults. Salmonellosis, until recently, was a major debilitating disease. Both diseases are still raging in less developed parts of the world, but both are susceptible to standard antibiotics.) Antibiotics have also greatly reduced the prevalence of syphilis, once the leading cause of insanity. Unfortunately, resistant strains of the syphilis organism are on the loose, and some strains of *Salmonella typhimurium*, the agent for salmonellosis, has picked up the R6 plasmid, with some strains showing resistance to antibiotics.

Plasmids have recently taken on a new significance in molecular biology. Many of the most recent advances in **genetic engineering** and **gene cloning** have involved plasmids. Their small genomes are readily isolated and manipulated, and they are widely used as vectors (carriers) of spliced genes in DNA experiments. Let's take a closer look at this new and fast-growing area of biological progress.

GENETIC ENGINEERING

Molecular biology has suddenly become interesting to big business. The reason is simple: there are now opportunities to apply some of this technology to practical ends. Or, in the dim view of some critics, science is about to be perverted because there are now opportunities to make big bucks.

The Gene-Splicing Controversy

The opportunities arise because a new generation of molecular biologists has learned to manipulate genes with amazing ease through **recombinant DNA technology,** which includes **gene splicing, gene cloning,** and **gene sequencing.** In gene splicing, a specific segment of DNA can routinely be removed from cells (including human cells) and be inserted into a tamed plasmid. This plasmid can then be cloned in large quantities, removed again, analyzed by gene sequencing techniques for its nucleotide sequence, perhaps altered to some new specification, and inserted into bacterial or yeast cells, where it can produce unlimited quantities of its protein product. Already produced this way are human insulin, human growth hormone, and several human types of a promising substance called **interferon,** an antiviral substance naturally produced by virus-infected cells. Such capabilities stagger the imagination. We may soon have the ability to easily and cheaply manufacture any of the body's chemicals.

14.13 THE R6 PLASMID

The circular object is the R6 plasmid, a circle of DNA which contains genes that endow the bacterial cell with resistance to certain antibiotics.

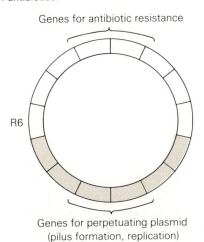

Genes for antibiotic resistance

R6

Genes for perpetuating plasmid
(pilus formation, replication)

When the idea of using plasmids for these gene-splicing techniques was introduced by Paul Berg and others in the late 1970s, the possibilities were so dramatic and so bizarre that the new technology immediately spawned a raging controversy. Trouble started when the very people who invented the techniques called for a research moratorium to give them time to consider any possible dangers associated with their work. Did they say *possible dangers?* The scare was on.

If there were possible risks, why was the work continuing? What new benefits did the technique offer? Supporters of the new gene-splicing technology claimed that it held great promise for humanity and that some of our more pressing problems might soon be solved. They spoke glowingly of cancer cures and of a possible end to all genetic disease. Some even visualized a possible end to hunger with the synthesis of new plants that combine, say, the productivity of maize with the nitrogen-fixing ability and protein yield of peanuts.

The principal benefit, in the minds of such proponents, is that scientists like themselves will be able to do more experiments and to learn things faster—and that's nothing to scoff at. There are also the more practical benefits, some of which have been realized and some of which are on the horizon. For example, there are thousands of growth hormone-deficient people who can be made to grow to normal height with the help of human growth hormone synthesized in bacterial cells.

Before the new techniques were developed, growth hormone had to be extracted from the pituitary glands of human cadavers, and was prohibitively expensive. Even though the glands of 50 cadavers are needed for one dose, a few thousand seriously undersized adolescents have been treated to some extent, as well as one man who grew to normal height after having been only four feet tall until the age of 35. Very soon, no one will have to be any shorter than he or she wants to be. Does that seem to you like a blessing, or like cavalier tampering with nature?

One enormously valuable feat has already been accomplished: engineered *E. coli* now grows the antigen for hoof-and-mouth disease, making possible a relatively inexpensive vaccine against that costly disease for the first time. That bit of genetic engineering alone might help shore up the economy of Mexico.

With such possible benefits, the risks must have been seen as great indeed to suggest halting the work. In fact, alarmed critics of the new techniques saw nothing but disaster ahead. Their fears ranged from the possible release of newly created disease organisms—genetic monsters capable of creating uncontrollable plagues—to new kinds of cancer, to constructing new kinds of humans. Nonsensical scare fiction began to appear (such as the film *The Boys from Brazil*). Some critics accused the new biologists of "playing God," of messing around with primal forces, or of seeking demonic new power. Politicians tried to capitalize on the scare; the city officials of Cambridge, Massachusetts, passed an ordinance prohibiting recombinant DNA research within the city limits (which include Harvard University and MIT). The National Institute of Health set up rigid "guidelines" as to what kinds of research were to be allowed and what precautions were to be taken. Finally, molecular biologists themselves took steps to assure that their bacterial subjects could not survive outside the laboratory.

Then, just as suddenly, the controversy died. Interestingly, this happened at about the time that gene splicing became big business, with new developments being offered for sale to the highest bidder. One would hope we never find out whether there was ever any real basis for the scare. But in any case, Cambridge relented, the N.I.H. guidelines were relaxed, and the cry of the critics diminished. No new epidemics have been unleashed and no new forms of cancer have been created—so far. And the promises have begun to be realized.

How Gene Splicing Is Done

Gene splicing depends on the availability of restriction enzymes, which, as we know, are a normal part of the bacterial cell's defenses against the foreign DNA of viruses and plasmids. Restriction enzymes have the ability to cut DNA at specific places along its length. They are easy to produce and commercial companies now purify dozens of kinds of restriction enzymes and sell them to laboratories by the bottle. Each variety recognizes a certain short nucleotide sequence in DNA and cleaves the molecule wherever the sequence is found.

For example, a restriction enzyme called **Eco-R1** recognizes the following sequence in double-stranded DNA, and cuts (*) it at a particular place:

$$\overset{*}{}$$
-X-X-X-C-T-T-A-A-G-X-X-X-
-X-X-X-G-A-A-T-T-C-X-X-X-
$$\underset{*}{}$$

(X stands for any nucleotide.) Note that the two strands are not cut directly opposite each other, but that the cuts are offset (asterisks). This leaves new free ends of complementary, single-stranded DNA:

The Gene Machine

If you aren't convinced of the almost bizarre possibilities of scientific technology, consider one of the latest additions. The *gene machine,* a device whose name has definite Madison Avenue overtones, is now available for about $30,000, and it does just what it says. An operator can create any DNA sequence he or she wants simply by imagining it and typing it out. The computerized machine does the rest.

For instance, one might have a protein amino acid sequence but be unable to isolate the corresponding mRNA. From a knowledge of the genetic code, the investigator can predict what a short segment of the gene might be, type it into the gene machine, and use a *probe* to locate the rest of the gene from a *DNA library* (cloned chromosome segments of the species in question).

If all one has is a tiny amount of protein, a highly sensitive protein-sequencing "accessory" is available. Samples of protein placed in the *automatic sequenator* are dismantled chemically, one amino acid at a time, from the *N*-terminal end; these are identified automatically and the amino acid sequence is printed out.

Once a probable nucleotide sequence is decided on, it is typed into the gene machine. The gene machine goes through all the steps of DNA synthesis. In about a day's time a small quantity of the artificial gene is available. In order to get a sufficient quantity of absolute purity, it is cloned into a bacterial plasmid, which is then taken up by a receptive bacterium, and standard biological gene cloning methods take over. Potentially, any enzyme the experimenter can imagine or any other kind of protein or gene can be made to order.

(left) The gene machine. *(right)* DNA sequencing. After treating DNA with base-specific restriction enzymes, the operator compares electrophoretic patterns, recording A, T, G, or C as the nucleotides are identified.

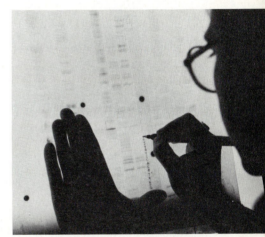

-X-X-X-C-T-T-A-A-   -G-X-X-X-
-X-X-X-G-   A-A-T-T-C-X-X-X-

The free, single-stranded ends will recognize and base-pair with one another, given the opportunity and the right conditions. For this reason they are called "sticky" ends. After such base pairing, the DNA can be *healed* again with another enzyme called *ligase.*

Specific sequences, such as the one shown above, are found in many life forms. The same restriction enzyme will break any of these, and if the DNA of one life form, broken in this way, should encounter the DNA from another life form, the two forms of DNA can heal together, mixing the genes of two species. Do you want an organism that combines bacterial genes with genes from, say, a chicken? It's been done.

The gene splicers' favorite trick is to use a restriction enzyme to cut open the circular DNA of a plasmid and then to splice in a fragment of DNA from some other organism (Figure 14.14). After healing, the plasmids with foreign DNA inserts are allowed to infect susceptible bacteria. As the newly infected bacteria rapidly multiply, copies of the altered plasmid and its new insert are passed on to all descendants. If the altered bacteria are infused with genes that make human insulin, the bacteria will make human insulin. And bacteria multiply so rapidly that it isn't long before significant amounts of this hormone can be recovered.

But why did some people find this so alarming? Perhaps because it's possible to put the gene for botulism toxin into *E. coli,* a bacterium that is already adapted for thriving in your gut. It is also possible to clone the entire genome of a cancer-

355

14.14 GENE SPLICING

Gene splicing permits researchers to insert segments of foreign DNA into bacteria, where it can transcribe mRNA and produce proteins in huge quantities. The techniques and skills required are not technically sophisticated by today's laboratory standards, and much of the procedure relies on chance events. With bacteria, the number of cells raised is unlimited, so the problems of chance are not serious, just tedious. (a) The procedure makes use of bacterial plasmids for the splicing of genes. These are obtained by removing the bacterial walls and differentially centrifuging the plasmids from other cell elements. (b) The relatively pure plasmid fractions are then cleaved with restriction enzyme (EcoR1). (c) Eukaryote DNA is then cleaved with the same restriction enzymes, add-

ed to the cleaved plasmids, and the combination treated with ligase, the suturing enzyme. (d) The plasmids with successful splices are then separated from unspliced plasmids and linear fragments of DNA. What remains are many circular spliced plasmids, part bacterial and part eukaryote. (e) The bacterial-eukaryote plasmids are then added to bacterial protoplasts (cells without walls), taken in, and grown in large numbers on regular nutrient agar plates. Transfers can then be made to broth culture tubes where even larger numbers are grown. The gene products can then be harvested or the eukaryote DNA can then be extracted, purified, and used as desired (for example, sequenced, or inserted into a bacterial operon by additional trickery).

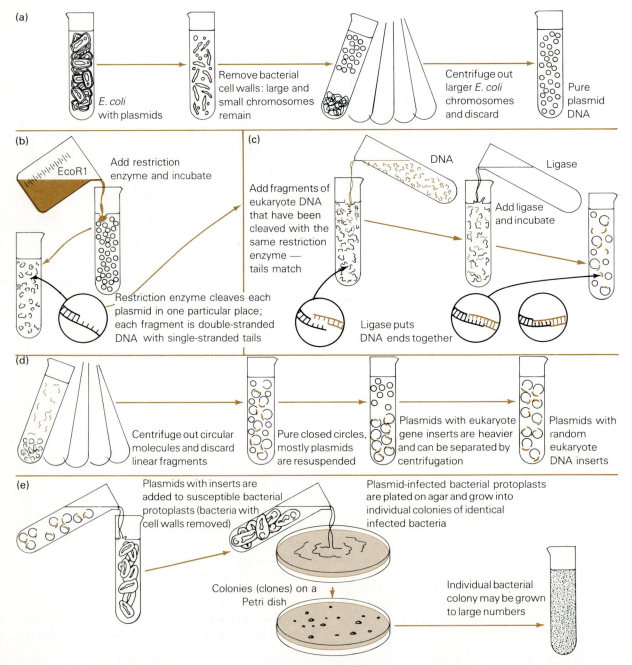

(a) *E. coli* with plasmids → Remove bacterial cell walls: large and small chromosomes remain → Centrifuge out larger *E. coli* chromosomes and discard → Pure plasmid DNA

(b) EcoR1 Add restriction enzyme and incubate — Restriction enzyme cleaves each plasmid in one particular place; each fragment is double-stranded DNA with single-stranded tails

(c) Add fragments of eukaryote DNA that have been cleaved with the same restriction enzyme — tails match — DNA — Add ligase and incubate — Ligase — Ligase puts DNA ends together

(d) Centrifuge out circular molecules and discard linear fragments → Pure closed circles, mostly plasmids are resuspended → Plasmids with eukaryote gene inserts are heavier and can be separated by centrifugation → Plasmids with random eukaryote DNA inserts

(e) Plasmids with inserts are added to susceptible bacterial protoplasts (bacteria with cell walls removed) → Plasmid-infected bacterial protoplasts are plated on agar and grow into individual colonies of identical infected bacteria → Colonies (clones) on a Petri dish → Individual bacterial colony may be grown to large numbers

causing virus, or the specific cancer-causing genes. Unlikely, you think? This, too, has been done. That possibility was one of the scare items, but now that it has actually happened, the isolation of cancer-causing genes, or **oncogenes,** has been hailed as the research breakthough that may lead science to a final understanding of the cancer process, and possibly from that to a dependable cure. Not that anyone can afford to be careless: we mustn't forget that the last two minor epidemics of smallpox in Europe—including the last death caused by this virus—were started by escaped laboratory strains. (The deadly smallpox virus, which only two decades ago infected millions of people, is now extinct except for laboratory cultures.)

Gene Sequencing

Other new techniques, based on plasmid transfer, have permitted molecular biologists to determine the exact nucleotide sequence of any piece of DNA or RNA they can isolate—and they can isolate just about any piece they want to (Essay 14.1). Through gene sequencing techniques, the entire base sequence of DNA in bacteriophage, plasmid, polio virus, and human mitochondria have now been

determined. They fill pages and pages with the nucleotides, A, T, C, and G; for example, there are exactly 5315 in the first bacteriophage sequence to be discovered. Your own DNA sequence of some 6.4 billion nucleotide pairs, if it were ever determined, would fill a thousand volumes the size of this one.

Incredible as it sounds, human gene sequencing is already well under way. Tom Maniatis, a Harvard molecular biologist, has chopped human DNA into fragments a few tens of thousands of nucleotides long, inserted them into the DNA of a bacteriophage, and is now maintaining 3 million separate clones of cultured human DNA. Such a collection is called a DNA library.

Genetics has come a long way from the days when Gregor Mendel began counting his peas. He might well be delighted today at the strides modern genetics has made since his rediscovery. On the other hand, like any good scientist, he might well be concerned about the pace and the direction genetic research has taken in recent years. Indeed, neither the pace nor the direction could have been anticipated. Yet our abilities grow. We can only assume that with care and thoughtfulness, the great promise of modern genetics will be realized.

APPLICATION OF IDEAS

1. Biologists are sometimes (often?) thought of as being quite whimsical in their interests. What kind of people intentionally grow spiders, flies, or skunk cabbage; chase cockroaches or butterflies; or count fish scales? Needless to say, they have trouble convincing people that their experimental subject may have been chosen for its usefulness in uncovering some new concept or principle, and that they themselves aren't all that excited about keeping company with cockroaches. Considering the historical development of modern genetics, offer your own convincing argument about how and why organisms are chosen for research. Use a liberal sprinkling of examples.

KEY WORDS AND IDEAS

MICROBES ENTER THE GENETIC REALM
1. The experimental subject *Escherichia coli* is useful because of its biological simplicity, rapid generation time, and simple requirements.

Locating Mutant Bacteria
1. Mutations that affect nutritional requirements of bacteria are readily identified.
2. Nutritional mutants in a population are isolated by first eliminating the wild types through the use of **penicillin,** a cell wall inhibitor, in a minimal growth medium. The nutritional mutants are survivors that failed to grow at all in the minimal medium. They can then be removed and grown in supplemented medium.
3. **Replica plating** includes transferring colonies from a supplemented to a minimal medium and looking for colonies missing from the pattern.
4. Recombination through sexual means is normally a rare event in bacteria. Originally, nutritional mutants were used to detect such events. If after growing different nutritional mutants togeth-

er, some descendants reveal no nutritional deficiency, then recombination between the two has occurred.

The Life and Times of the Bacteriophage

1. Bacteriophage are viruses that inject their DNA into a bacterial host. The bacterium's defense includes **restriction enzymes,** each of which recognizes and hydrolyzes specific DNA sequences.

2. Evolution between virus and host goes on simultaneously. Variations in viral DNA sequences that help the virus avoid restriction enzymes are selected for, as are changes in the bacterial host enzymes that can act against the new viral DNA sequence.

3. If the phage succeeds in avoiding the restriction enzymes, its next act will be to use the host's RNA polymerase to transcribe the viral DNA and synthesize viral enzymes.

4. Viral enzymes disrupt host DNA and recycle its nucleotides for use in replicating many copies of viral DNA. In addition viral DNA is transcribed into RNA to be used in producing viral **coat proteins.** When viral reproduction is complete, the host cell will be **lysed,** and many infectious viruses released.

5. An alternative to lysis is **lysogeny.** The viral DNA is incorporated into the host chromosome, and the viral genes are replicated along with the host's as reproduction occurs. Then, in some future generation, the virus may again enter its lytic cycle.

6. Because of their manner of reproduction, the culturing of viruses requires living cells. Phage activity in a bacterial plate can be recognized as **plaques**—clear areas in an agar plate that otherwise has heavy bacterial growth.

BACTERIAL RECOMBINATION

1. Although bacterial sex is unusual and infrequent, true sexual reproduction—the combination of DNA from two individuals—can occur through several mechanisms.

Transformation

1. **Transformation** is the uptake and incorporation of fragments of stray DNA into the bacterial chromosome.

Transduction

1. In **transduction,** stray DNA fragments carried into the bacterial cell by a phage virus are incorporated into the bacterial chromosome.

Plasmids and Plasmid Transfer

1. **Plasmids** are small, supplemental DNA molecules in bacteria; one group, described early, was labeled **episomes.**

2. One kind of plasmid codes for the synthesis of the **sex pilus,** which unlike other solid spikelike **pili,** are hollow and act as **conjugation tubes.**

3. Bacteria with the plasmid are labeled **plasmid-plus,** others, **plasmid-minus.** Such plasmids can be transferred to plasmid-minus bacteria, changing them to the plus form.

4. The *F* **plasmid,** one of several kinds of plasmids, represents the first thoroughly investigated sexual activity in bacteria.

5. In 1946, Lederberg and Tatum demonstrated sex in bacteria by crossing two nutritionally mutant strains and obtaining wild-type progeny. Their failure to work out gene linkage relationships led biologists to later conclude that the bacterial chromosome was circular.

6. It took many years for biologists to determine the mechanism of bacterial sexual exchange. Linkage studies from crosses of nutritional mutants revealed that during sexual exchange, only a few genes are transferred at a time, and that transfer occurs in one direction only.

William Hayes and the *F* Episome, or Making *E. coli* Sexier

1. Through his experiments, William Hayes added the notion that bacterial sex was catching. Maleness, the ability to transfer genes, could be passed from one individual to another. (It was later determined that it was a sex plasmid that was transferred.)

2. "Maleness" was at first called the **fertility factor;** "males" were termed F^+ **(fertility positive),** and "females" F^- **(fertility negative).** It was after this that the term "*F* episome" was coined.

3. The *F* episome, or plasmid, codes for the sex pili, which permits copies of itself to be transferred. After replication, it opens to a linear form to pass through the sex pilus into the F^- recipient.

4. In rare instances, the *F* DNA will insert into the host's main chromosome (and, as in phage lysogeny, can be replicated and passed along to progeny).

5. Once integrated into the host chromosome, the *F* plasmid can then open to the linear form and the *host's main chromosome* passes through the sex pilus (usually just a portion makes it). Bacteria that did this more commonly were isolated and grown in pure cultures and termed *Hfr* **(high-frequency of recombination).** Thus, a large amount of recombination is potentially possible. This also explains how bacterial recombination often works.

6. The peculiar chromosome transferring ability of the *Hfr* strain made genetic mapping of the *E. coli* chromosome possible. Jacob and Wollman grew cultures of *Hfr* and *F⁻* individuals containing genetic markers. Then they interrupted the transfer at different time periods and determined the recombinant types that had developed. In this manner, the linear order of genes was established.

Sexduction

1. In **sexduction,** a small portion of main chromosome is incorporated into an *F* plasmid. When transferred, it confers a unique partial *diploid state* on the recipient. Jacob and Monod used partial diploids in working out their operon model (see Chapter 10).

Plasmids and Plasmid Genes

1. Recently, the **R6** plasmid, one that confers its bacterial host with resistance to antibiotics, has been discovered. There is concern with the R6 plasmid since its appearance in a dangerous pathogen could make that bacterium impossible to control. The R6 plasmid may already be responsible for the increases in many antibiotic-resistant strains around today.

2. The R6 risk factor is intensified by the heavy use of antibiotics in the cattle industry and their overprescription by many physicians.

3. The intense use of antibiotics has, however, all but eliminated certain important diseases from developed nations.

4. The most recent significance of plasmids has been as an agent for **genetic engineering—gene splicing, gene sequencing,** and **gene cloning.**

GENETIC ENGINEERING

1. Gene splicing involves producing chosen DNA sequences (genes), inserting them into plasmids, reintroducing the plasmids to bacterial cells, cloning the cells, and harvesting large amounts of the gene products (such as insulin, growth hormone, and **interferon**).

2. Genetic engineering remains controversial, although industry now employs some of the technical procedures. There is an element of medical and genetic risk, but much of the controversy centers around the moral and philosophical issues.

3. A major gene splicing technique involves
   a. obtaining plasmids from bacteria through ultracentrifugation;
   b. using restriction enzymes to cleave plasmid DNA, introducing foreign DNA strands to be cloned, and using the enzyme *ligase* to reattach the sticky ends.
   c. preparing recipient protoplasts by dissolving the cell walls, and permitting the new plasmids to be taken in.
   d. producing large clones and harvesting the gene product.

4. Biologists can now determine the nucleotide sequence for any segment of DNA, and such sequences can be produced by cloning and stored for future reference.

5. The technology is available for determining the amino acid sequence of any protein, then synthetically producing a DNA strand that will code for that sequence. The DNA can then be spliced into a plasmid and the protein produced in quantity.

REVIEW QUESTIONS

1. List five important experimental organisms used in genetics research. (341)

2. Suggest several advantages for using the bacterium *E. coli* in studies of genetics. (381)

3. Briefly outline the general procedure for isolating nutritional mutants of bacteria. (342)

4. Suggest a procedure that could be followed to determine whether genetic recombination had occurred in bacteria. (342)

5. Suggest several ways in which sexual reproduction in bacteria is different from sexual reproduction in the eukaryote. (342)

6. In what specific way are the natural restriction enzymes important to bacteria? (344)

7. How do bacteria respond to the rise of genetic variants in their viral parasites? (344)

8. Explain how a bacteriophage reproduces itself. (344)

9. Distinguish between lytic and lysogenic cycles in bacteriophage. (344–345)

10. Outline a procedure for visibly detecting the presence of bacteriophage. (345)

11. In what way is transformation similar to transduction? How are they different? (346)

12. Specifically, what is a plasmid? How might a plasmid get from one bacterium to another? (347)

13. Using the terms *F⁺* and *F⁻*, explain how conjugation occurs in *E. coli*. (348)

14. What two goals were Lederberg and Tatum trying to reach? Which turned out to be the most difficult? Why? (348)

15. List two peculiar things that happen to a "female" or F^- (fertility minus) bacterium that has undergone recombination with an F^+ (fertility plus) bacterium? (349)

16. Explain briefly how nutritional mutants can be used to determine the relative distances from one locus to another on the bacterial chromosome. (349)

17. What, exactly, is a fertility factor? What are F^+ and F^- bacteria? (349)

18. Briefly explain how the F^+ episome was able to transfer genes from the larger bacterial chromosome. How common was this phenomenon? (350)

19. How does your answer to review question 18 supplement your answer to number 17? (350)

20. What special capabilities were there in Hayes' *Hfr* strain? (350)

21. Explain the kitchen blender experiment, including the purpose, the timing, the reasoning behind it, and the choice of experimental organism. (351)

22. In what way was the earlier observation of sexduction suggestive of what was coming to molecular biology in the 1980s? (351)

23. What is an R6 plasmid? In what way can such plasmids be threatening to us? (351)

24. How does the presence of the R6 plasmid relate to the problems brought on by the intensive use of antibiotics in clinical practice and in the raising of livestock? (353)

25. List three gene products now available through genetic engineering techniques. (353)

26. Briefly discuss what some people consider to be potential risks in the new recombinant DNA technology. (354)

27. What is the difference between gene splicing, gene sequencing, and gene cloning? (353)

28. List the general steps that are involved in obtaining a gene product such as interferon or growth hormone. (354–355)

29. What is a gene library? What is the potential value of such a collection? (357)

SUGGESTED READING

Avery, O. T., C. M. MacLeod, and M. McCarty. 1944. "Studies on the Chemical Nature of the Substance Inducing Transformation of Pneumococcal Types." *Journal of Experimental Medicine* 79:137. The classic experiment that proved to the world—at least in hindsight—that the genetic material is DNA.

Chedd, G. 1981. "Genetic Gibberish in the Code of Life." *Science 81*, November.

Crick, F. H. C. 1979. "Split Genes and RNA Splicing." *Science* 204:264. The codiscoverer of the structure of DNA was active in the race to decipher the genetic code and has some cogent things to say about those mysterious newly discovered entities, gene inserts or introns.

Darnell, J. E. 1978. "Implications of RNA—RNA Splicing in the Evolution of Eukaryote Cells." *Science* 202:1257. Another thoughtful attempt, by one of the leaders in the field of eukaryote gene function, to make some sense out of those puzzling introns.

DuPraw, E. J. 1970. *DNA and Chromosomes*. Holt, Rinehart and Winston, New York (paperback).

Goodenough, U. 1978. *Genetics*, 2d ed. Holt, Rinehart and Winston, New York. Of the current general genetics texts, Ursula Goodenough's has the best coverage of recent developments in molecular genetics and is particularly strong in the molecular genetics of higher eukaryotes.

Hamilton, W. D. 1967. "Extraordinary Sex Ratios." *Science* 156:477. Why have some species departed from the usual one-to-one ratio of males to females?

Jacob, F., and J. Monod. 1961. "Genetic Regulatory Mechanisms in the Synthesis of Proteins." *Journal of Molecular Biology* 33:318. A modern classic and an example of fine scientific writing and reason, this is an account of the revolutionary experiments that demonstrated the existence of operons, the operator, cytoplasmic regulating molecules, and messenger RNA.

Lewin, R. 1983. "A Naturalist of the Genome." *Science* 222:402

Maynard Smith, J. 1971. "What Use Is Sex?" *Journal of Theoretical Biology* 30:319. One major difference between prokaryotes and eukaryotes is the invention, in the latter, of regular biparental reproduction. Being widespread and of ancient origin, it presumably serves some function, but what is it?

Mendel, G. 1965. "Experiments in Plant Hybridization (1865)." Translated by Eva Sherwood. In

The Origin of Genetics, ed. C. Stern and E. Sherwood. W. H. Freeman, San Francisco. In addition to the full text of Mendel's classic paper, this volume includes some of Mendel's letters and minor works, the three "rediscovery" papers of 1900, and a fascinating exchange between R. A. Fisher and Sewall Wright on the question of whether or not Mendel drylabbed the whole thing.

Menosky, J. A. 1981. "The Gene Machine." *Science 81,* July/August.

Meselson, M., and F. W. Stahl. 1958. "The Replication of DNA in *E. coli.*" *Proceedings of the National Academy of Sciences* (U.S.) 44:671. The brilliant and influential experiment that proved that DNA unwinds and replicates semiconservatively, as foreseen by Crick. All biology students must study this work in detail at some time or another.

Mourant, A. E., et al. 1978. *The Genetics of the Jews.* Clarendon (Oxford University Press), New York and Oxford. Ashkenazi and Sephardic Jews form a genetically distinct racial group after all, according to blood group and enzyme polymorphisms. Basically Palestinian, they show surprisingly little evidence of past mixing with European groups but rather more (5–10%) negroid admixture, presumably from the time spent in slavery in Egypt.

Okazaki, R. T., et al. 1968. "Mechanism of DNA Chain Growth: Possible Discontinuity and Unusual Secondary Structure of Newly Synthesized Chains." *Proceedings of the National Academy of Sciences* (U.S.). On the Okazaki fragments.

Rensberger, B. "Tinkering with Life." *Science 81,* November.

Sanger, F., et al. 1977. "Nucleotide Sequence of Bacteriophage φX174 DNA." *Nature* 265:687. The first publication of the entire genome of any organism and a *tour de force* of molecular biology.

Shine, I., and S. Wrobel. 1976. *Thomas Hunt Morgan: Pioneer of Genetics.* University of Kentucky Press, Lexington, Ky. Interesting narrative and lively anecdotes of the early days of genetics in America.

Strickberger, M. W. 1976. *Genetics,* 2d ed. Macmillan, New York. Of the current general genetics texts Strickberger's has the clearest treatment of classical Mendelian genetics. In addition, we feel that the 120 pages that are included on population genetics and quantitative genetics happen to constitute the best textbook in print on these difficult subjects.

Van Valen, L., and G. W. Mellin. 1967. "Selection in Natural Populations. 7. New York Babies." *Annals of Human Genetics* (London) 31:109. Among newborns, it's better to be average, because small and large babies are both at risk.

Watson, J. D. 1968. *The Double Helix.* Atheneum, New York. Deftly hidden in the narrative of this witty, often hilarious and picaresque account of the personal triumph of a young scientist and an old graduate student is a surprising amount of solid scientific information. Certainly the most enjoyable account of how "the scientific method" actually works in practice.

Watson, J. D., and F. H. C. Crick. 1953. "Molecular Structure of Nucleic Acids. A structure of deoxyribose nucleic acid." *Nature* 171:737. This is the one that started it all: the most influential single page in scientific history.

Wilson, E. O., and W. H. Bossert. 1971. *A Primer of Population Biology.* Sinauer, Sunderland, Mass. Students tell us that this self-teaching approach to the elementary mathematics of population genetics and population ecology is more helpful than most formal courses on the subjects.

Scientific American articles. San Francisco: W. H. Freeman Co.

Aharonowitz, Y., and Cohen, G. 1981. "The Microbiological Production of Pharmaceuticals." *Scientific American,* September.

Allison, A. C. 1956. "Sickle Cells and Evolution." *Scientific American,* August. The first demonstration of overdominance: the sickle-cell allele is actually beneficial to carriers in certain environments.

Anderson, W. F., and Diacumakos, E. G. 1981. "Genetic Engineering in Mammalian Cells." *Scientific American,* July.

Bishop, J. M. 1982. "Oncogenes." *Scientific American,* March.

Brill, W. J. 1981. "Agricultural Microbiology." *Scientific American,* September.

Campbell, A. M. 1976. "How Viruses Insert Their DNA into the DNA of the Host Cell." *Scientific American,* December.

Chambon, P. 1981. "Split Genes." *Scientific American,* May.

Chilton, M. 1983. "A Vector for Introducing New Genes into Plants." *Scientific American,* June.

Cohen, S. N., and Shapiro, J. A. 1980. "Transposable Genetic Elements." *Scientific American,* February.

Crick, F. H. C. 1962. "The Genetic Code." *Scientific American,* October.

―――. 1966. "The Genetic Code III." *Scientific American,* October.

Dickerson, R. E. 1972. "The Structure and History of an Ancient Protein." *Scientific American,* April. How cytochrome *c* has evolved its present shape and amino acid sequence over the last 2 billion years.

Friedman, T. 1971. "Prenatal Diagnosis of Genetic Diseases." *Scientific American*, November. A primer of transabdominal amniocentesis and a valuable discussion of the moral implications involved in the interruption of pregnancy.

Grivell, L. A. 1983. "Mitochondrial DNA." *Scientific American*, March.

Hopwood, A. 1981. "The Genetic Programming of Industrial Microorganisms." *Scientific American*, September.

Howard-Flanders, P. 1981. "Inducible Repair of DNA." *Scientific American*, November.

Hunter, T. 1984. "The Proteins of Oncogenes." *Scientific American*, August.

Kornberg, R. D., and Klug, A. 1981. "The Nucleosome." *Scientific American*, February.

Kretchmer, N. 1972. "Lactose and Lactase." *Scientific American*, 227:70. Milk produces flatulence in adults of Oriental or African ancestry because they lack the enzyme needed to break down milk sugar.

Lake, J. A. 1981. "The Ribosome." *Scientific American*, August.

McKusick, V. A. 1965. "The Royal Hemophilia." *Scientific American*, February. Queen Victoria was heterozygous for the X-linked recessive allele and through political marriages of her daughters managed to pass it on to all the leading royal families of Europe.

Miller, O. L. 1973. "The Visualization of Genes in Action." *Scientific American*, March. Some remarkable electron micrographs of transcription and translation, looking almost exactly like diagrams that had originally been made on biochemical evidence alone.

Nomura, M. 1984. "The Control of Ribosome Synthesis." *Scientific American*, January.

Novick, R. P. 1980. "Plasmids." *Scientific American*, December.

Pestka, S. 1983. "The Purification and Manufacture of Human Interferons." *Scientific American*, August.

Evolution and Changing Alleles

The theory of evolution is one of the most pervasive and explanatory themes in modern biology. As an intellectual fulcrum it has been used to pry loose countless gems from the complex matrix of life. In fact, it is so useful that one wonders how biology could ever have been done without a clear understanding of its principles. Of course, as we know, much biology *was* done without it, but, as we also know, much of it was wrong. So let's take a close look at the venerable old idea. We will see that some parts of it have weathered, aged, hardened, and cured, while other parts have been changed and new parts—parts that Darwin could never have imagined—have been added, as the concept of evolution has itself evolved.

We might set the stage here by saying that most evolution proceeds by natural selection acting on variation. Variation arises simply with the appearance of individuals that differ from other members of the species. Some of these differences will inevitably confer reproductive advantages to some individuals over others. The descendants of those with such advantages can be expected to increase in the population. In a sense, then, nature selects the most successful reproducers, just as a farmer selects the best layers to produce the next generation of chickens. By such means, a population changes through time and this, we will see, is the essential idea of evolution.

ALLELES AND ALLELE FREQUENCIES

Now let's begin to refine the idea. It is important to remember that *individuals do not evolve*—at least not in the sense that we will consider evolution here. *Populations evolve.* Such evolution is evidenced by changes in the **gene pool,** which includes all the genes of any population at any given time. More precisely, evolution involves changes in **allele frequencies.** As the relative number of one form of a gene, or one allele, increases, the relative number of an alternate form of the same gene decreases. Such a change can spread, perhaps slowly, perhaps rapidly, through a local population or an entire species. The result is evolution. Evolution, by the way, is not always the result of natural selection. It can also proceed somewhat randomly as this or that gene accumulates by mere chance. For now, though, we will focus on the effects of natural selection—how "good" genes are preserved. Where do "good" (beneficial) or "bad" (harmful) genes come from? In large part from existing genes that have changed through mutation, a permanent, random chemical change in the DNA molecule itself. A gene may also become more, or less, beneficial if, while it remains stable itself, the environment changes, thereby altering the level of its benefit.

Darwin knew about evolutionary changes in physical appearance (what we now call phenotype), and he knew that these changes must reflect changes in the hereditary makeup. But Darwin did not know about alleles, Mendel's "alternate factors." We must keep in mind that understanding the mechanism of evolution requires a meshing of Darwinian and Mendelian thinking, for Darwinian evolution ultimately depends on Mendelian genes. As we proceed to explore the relationship between the two, we will see first what happens to gene frequencies in populations that, theoretically, are not affected by natural selection and then what happens to populations that are.

THE CASTLE-HARDY-WEINBERG LAW

Godfrey Hardy, an eminent mathematician, had few professional interests in common with Reginald Punnett, the young Mendelian geneticist, but they frequently met for lunch and tea at the faculty club of Cambridge University. One day in 1908, Punnett was telling his colleague about a small problem in genetics. Rumor had it that Gudny Yule, a strong critic of the Mendelians, had said that if the allele for short fingers were dominant and the allele for normal fingers were recessive, then short fingers ought to become more and more common each generation. Within a few generations, Yule thought that no one in Britain should have normal fingers at all. Punnett didn't think this argument was correct, but he couldn't explain why.

Hardy said he thought the problem was simple enough and wrote a few equations on his napkin. He showed that, given any frequency of alleles for normal fingers and alleles for short fingers in a population, the relative numbers of people with normal fingers and people with short fingers ought to stay the same for generation after generation as long as there was no natural selection involved. Today, we say that the population is in **genetic equilibrium** for the gene.

Punnett was excited and wanted to have the idea published (on something besides a napkin) as soon as possible, but Hardy was reluctant. The idea was so simple and obvious, he felt, that he didn't want to have his name associated with it and risk his reputation as one of the great mathematical minds of the day. But Punnett prevailed, the equation was published, and the relationship between

genotypes and phenotypes in populations quickly became known as *Hardy's law.* Hardy, who was indeed one of the great mathematical minds of the day, is now known almost solely for this modest contribution. Yule, incidentally, denied ever having said that dominant traits should increase from generation to generation, so this little incident in the history of science was based on a misunderstanding from the outset.

In Germany, within weeks of the publication of Hardy's short paper, the same law was described by the German physician Wilhelm Weinberg. Eventually, the formula became known as the **Hardy-Weinberg law,** and then later as the **Castle-Hardy-Weinberg law,** in recognition of the belated discovery that an American, William Castle, had published a neglected exposition of the same observation in 1903. In any case, the Castle-Hardy-Weinberg law is the starting point for studying the population genetics of diploid sexual species. The law, restated, is that both gene and genotype frequencies will remain unchanged—in equilibrium—unless outside forces change those frequencies.

The Implications of the Castle-Hardy-Weinberg Law

To approach such questions as Britain's "finger length problem," we must begin by considering how genes behave in populations. We have been using the term **population,** and you may have taken it to mean a group of individuals. You were not wrong, but at this point we can give it a more precise meaning. In biology, a population designates a group of interbreeding or potentially interbreeding individuals. With this in mind, let's consider the ratios (or frequencies) of different alleles for a specific characteristic in a population.

The notion of **frequency** is fundamental to the genetics of evolution, and here the term has a special meaning. It has nothing to do with how frequently something happens or how often something occurs in time, such as when we refer, say, to the frequency of tornados in the spring. In genetics, frequency is a *proportion*, namely a proportion of items of a particular kind in a more general class. For example, if there are 10,000 registered voters in town, and 4300 are Republicans, the *frequency* of Republicans in this town is $4300/10,000 = 0.43$. Frequencies are always numbers between 0 and 1 because they are always a fraction consisting of a part divided by the whole. Both the numerator and the denominator are counts of individual items, and any individual that appears in the numerator must also appear in the denominator. For instance, the 10,000 registered voters *includes* the 4300 registered Republicans. Now let's see why allele frequencies for some trait such as finger length might tend to remain stable.

First, imagine a population of only two individuals in which the male is homozygous for dominant trait **A** and the female is homozygous for recessive trait **a**. We know that all their F_1 offspring will be heterozygous **(Aa)** for that characteristic. Next, assume that the F_1 individuals mate and produce an F_2 generation. The Punnett square shows us that in the F_2 generation, three out of four individuals will show the dominant trait and only one will show the recessive trait. It might appear, then, that we are on the way to eliminating the recessive allele from the population, since the frequency of the trait has decreased from ½ to ¼. However, if we plot the succeeding generations (F_3, F_4, F_5, and so on), we

TABLE 15.1

POSSIBLE MATINGS AMONG THE F_2 GENERATION

Father	Mother	Frequency of mating	Frequency combined
AA (¼)	AA (¼)	1/16	1/16
AA (¼)	Aa (½)	1/8	
Aa (½)	AA (¼)	1/8	¼
AA (¼)	aa (¼)	1/16	
aa (¼)	AA (¼)	1/16	1/8
Aa (½)	Aa (½)	¼	¼
Aa (½)	aa (¼)	1/8	
aa (¼)	Aa (½)	1/8	¼
aa (¼)	aa (¼)	1/16	1/16

will find that the proportion of dominant and recessive alleles in the population has not changed at all (Figure 15.1). In fact, if we think again of the F_1 and F_2 generations, we see that there is a $1:1$ ratio of the two alleles even at these stages:

F_1 All **Aa** (Ratio of **A** to **a** is $1:1$)
F_2 ¼ **AA**
    ½ **Aa** } (Ratio of **A** to **a** = $1:1$)
    ¼ **aa**

A population of two individuals is unrealistic, but it serves to point out how the phenomenon occurs in larger populations. We can show by means of Punnett squares that the frequency of alleles for any characteristic will remain unchanged in a population through any number of generations—unless this frequency is altered by some outside influence such as natural selection.

Thus, according to the Castle-Hardy-Weinberg law, if there is no natural selection, mutation, or any other force that changes gene frequencies in populations, and if random mating is permitted, the frequencies of each genotype will remain constant through the following generations.

Note the frequencies of the combinations **AA, Aa,** and **aa** in the F_2 generation in Table 15.1, which is our first opportunity to see all the possible combinations. We see in the F_2 that one fourth of the population is **AA,** one half is **Aa,** and one fourth is **aa.** In order to see what will happen in the next generation, we can list all the different kinds of matings and how often these should be expected to occur. With three different genotypes, there are three kinds of males and three kinds of females, or nine types of matings altogether. If we combine

15.1 AN F_3 GENERATION

In the F_2 generation of a Mendelian cross, both males and females occur in three genotypes **AA, Aa,** and **aa,** and the genotype frequencies are ¼, ½, and ¼, respectively. When all nine possible matings are considered, the offspring are represented in the rectangles bounded by heavy black lines (each of which is actually a Punnett square for an individual cross). Within each type of mating, Mendelian genotypic ratios again occur: $1:0$, $1:1$, and $1:2:1$, depending on the mating. But if all offspring are considered together, the F_3 generation again has the same distribution of genotypes as the F_2: ¼ **AA,** ½ **Aa,** and ¼ **aa.**

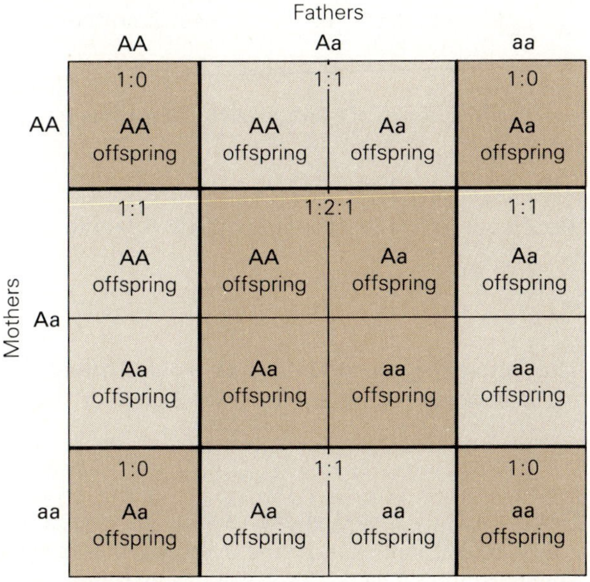

Fathers

	AA	Aa		aa
AA	1:0 AA offspring	1:1 AA offspring	Aa offspring	1:0 Aa offspring
Aa	1:1 AA offspring	1:2:1 AA offspring	Aa offspring	1:1 Aa offspring
	Aa offspring	Aa offspring	aa offspring	aa offspring
aa	1:0 Aa offspring	1:1 Aa offspring	aa offspring	1:0 aa offspring

Mothers

TABLE 15.2

EXPECTED F_3 GENERATION

Mating	Frequency	Offspring expected		
		AA	Aa	aa
AA × AA	1/16	1/16	—	—
AA × Aa	1/4	1/8	1/8	—
AA × aa	1/8	—	1/8	—
Aa × Aa	1/4	1/16	1/8	1/16
Aa × aa	1/4	—	1/8	1/8
aa × aa	1/16	—	—	1/16
Total	1	1/4	1/2	1/4

reciprocal matings, such as **AA × aa** and **aa × AA,** there are still six different types. Each type of mating can be expected to produce certain kinds of offspring in the usual Mendelian ratios.

These six different types of matings, together with the resulting proportions of each of the three offspring types, are listed in Table 15.2. For instance, one fourth of the matings are **AA × Aa** (or **Aa × AA**). Since this should produce a 50:50 Mendelian ratio of **AA** and **Aa** children, among the F_3 a total of one eighth will be **AA** children and one eighth will be **Aa.**

From Table 15.2 and Figure 15.1, we can see that under the required Castle-Hardy-Weinberg conditions, random mating of the F_2 produces an F_3 generation that is ¼ **AA,** ½ **Aa,** and ¼ **aa,** just as in the F_2. The F_4 and F_5 will also have the same genotypic ratios. And that is why recessive traits continue to exist in a population.

An Algebraic Equivalent

In the example above, the two alleles start out at the same frequency—half **A** and half **a.** However, the Castle-Hardy-Weinberg distribution also maintains stable genotype frequencies in populations in which the allele frequencies are not the same. For those burning with a fierce love of mathematics, let's put it all into algebra. First, let p be the allele frequency of allele **A,** while q is the allele frequency of allele **a.** If there are only two different alleles, $p + q = 1$ (1= all).

The Castle-Hardy-Weinberg distribution says that the expected genotype frequency of **AA** is p^2, the expected genotype frequency of **Aa** is $2pq$, and the expected genotype frequency of **aa** is q^2. We can show that these frequencies will also be stable. As

before, we list the six (combined) types of matings, the frequencies in which they should occur, and the distribution of offspring of each type of family (Table 15.3). The genotype frequencies of the offspring will be p^2, $2pq$, and q^2, just as in the parents' generation. Thus we have the well-known Castle-Hardy-Weinberg formula or distribution. Where p *represents the dominant allele and* q, the recessive, we have:

$$p^2 + 2pq + q^2 = 1$$

The Castle-Hardy-Weinberg distribution is easier to understand if we forget about random mating of diploid individuals and just consider the *random association of gametes* (Figure 15.2). It can be proven that this amounts to the same thing. After all, when an egg and sperm meet—either in the open sea, as with sea urchins, or in the dark confines of a human oviduct—the parents' diploid genotypes no longer really matter. All that matters is the haploid genotypes of the two gametes, and if there is random mating, the probabilities for each gamete will be p and q of being **A** or **a,** respectively.

The Castle-Hardy-Weinberg law has very specific and important implications. For example, if we know the prevalence of a recessive trait such as albinism (the absence of normal melanin pigment) in the population, we can predict, within limits, the probability that any couple in that population will have an albino baby. Here's how it works. Normal skin and eye pigment in humans, **A,** is dominant over the albino condition **a.** The genotype **aa** occurs in about one of every 20,000 people. According to the Castle-Hardy-Weinberg equation, the frequency of genotype **aa** is represented by q^2 so that

$$q^2 = 1/20{,}000$$

and the frequency of a single allele **(a)** for this trait is thus:

$$q = \text{square root of } 1/20{,}000 = 1/141$$

The frequency of the dominant allele **A** would then be

$$p = 1 - q$$

or

$$p = 1 - 1/141 = 140/141$$

The heterozygous condition **Aa** would, therefore, occur in the population with a frequency of

$$2pq = 2 \times 140/141 \times 1/141$$
$$= 1/70, \text{ or about } 1.4\%$$

Since 1.4% of 20,000 is 280, about 280 people in every 20,000 will be carrying a recessive allele for albinism, while, as we have seen, only one will be

TABLE 15.3

EXPECTED ALLELE FREQUENCIES

Mating ♀ ♂	Mating frequency	Offspring frequencies		
		AA	Aa	aa
AA × AA	p^4	p^4	—	—
AA × Aa ⎫ Aa × AA ⎭	$4p^3q$	$2p^3q$	$2p^3q$	—
AA × aa ⎫ aa × AA ⎭	$2p^2q^2$	—	$2p^2q^2$	—
Aa × Aa	$4p^2q^2$	p^2q^2	$2p^2q^2$	p^2q^2
Aa × aa ⎫ aa × Aa ⎭	$4pq^3$	—	$2pq^3$	$2pq^3$
aa × aa	q^4	—	—	q^4
Total	$(p^2 + 2pq + q^2)^2 = 1$	$p^2(p^2 + 2pq + q^2) = p^2$	$2pq(p^2 + 2pq + q^2) = 2pq$	$q^2(p^2 + 2pq + q^2) = q^2$

affected. Hence, in the absence of a family history of this characteristic in either parent, the chance that any couple will have an albino child is very slim.

A Closer Look at the Castle-Hardy-Weinberg Model

The Castle-Hardy-Weinberg model indeed helps answer a wide range of problems about the real world. But any such model is liable to have some basic restrictions and exceptions. It turns out that this model is most useful when several very specific conditions are met:

1. Single gene loci must be involved and the alleles must segregate according to Mendel's first law, eliminating, for example, sex-linkage, polygenic inheritance, incomplete penetrance, and so on;
2. Loci must have only two alternate alleles (or classes of alleles); more complex mathematical forms of the law must be applied where multiple alleles are present;
3. Mating must be completely random in the population; there can be no preferences;
4. There can be no migration into or out of the population;
5. There can be no gene change through mutation;
6. There can be no differences in viability and reproductive capacity among genotypes in the population; all survive and produce the same numbers of offspring;
7. The population (and samples) must be of infinite size.

15.2 THE CASTLE-HARDY-WEINBERG EQUILIBRIUM

If there are two alleles, **A** and **a**, occurring in relative frequencies p and q, respectively, and mating is random, a proportion of p of the sperm will carry the **A** allele, and p of the eggs will also carry the **A** allele. For each zygote formed, the probability that the egg and sperm will both carry **A** alleles is $p \times p$ or p^2. Similarly, the probability that both of the uniting gametes will carry **a** alleles is q^2. There are two ways that an **Aa** zygote can be formed: **A** sperm uniting with **a** egg and **a** sperm uniting with **A** egg. The total probability of one of these two events occurring is $pq + pq$, or $2pq$. (The arrow connects the two boxes that represent the same **Aa** genotype.) The relative frequencies of the three kinds of genotypes in the population will be equal to the individual probabilities of each kind of event: p^2, $2pq$ and q^2 will be the frequencies of genotypes **AA, Aa,** and **aa** respectively.

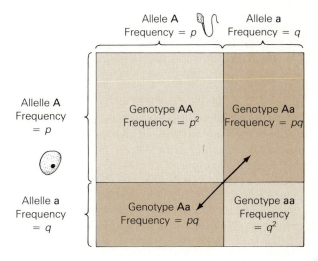

Allele **A** Frequency = p Allele **a** Frequency = q

Allele A Frequency = p

Allele a Frequency = q

Genotype **AA** Frequency = p^2 Genotype **Aa** Frequency = pq

Genotype **Aa** Frequency = pq Genotype **aa** Frequency = q^2

Of these stipulations, the last is the most blatantly unrealistic, since all populations and all samples are finite. In practice, then, one seldom gets, or really expects to get, the exact Castle-Hardy-Weinberg expectations. Let's look at a specific example of how problems can come up.

An Application of the Castle-Hardy-Weinberg Model

Albinism has been a favorite textbook example for decades, in part because everyone is familiar with the striking phenotypes of affected homozygotes and has some idea of what is being discussed. We have just used it ourselves to show how the Castle-Hardy-Weinberg law can be applied. Now we'll show just how easily it can be misapplied.

One day, not too many years ago, an albino man and his albino wife produced a child who was normally pigmented (in fact, black). Their doctor, who remembered his introductory biology course, cheerfully informed the parents that their child was illegitimate, since two recessive homozygotes could not possibly have a child with the dominant phenotype. This point of view did not favorably impress the albino father, who fumed that any fool could tell that he and his wife were different kinds of albinos—for instance, she could not stay out in the sun for even five minutes without burning badly, while he could last a couple of hours out of doors.

A geneticist investigated the situation and found that the father's intuition was perfectly correct. Mutant alleles can block pigment formation at several different points in the pathway of melanin development. By different points we refer to the different gene loci that code for the various enzymes involved in melanin production. The husband and wife were recessive homozygotes at different loci— let us say that the father was **aaBB** and the mother was **AAbb,** in which case their child was **AaBb** and thus (phenotypically) perfectly normal. The geneticist then developed a simple biochemical test that differentiated the two types of albino. Tyrosine is an amino acid from which melanin is synthesized, and in some albinos a defective gene prevents its synthesis. If a few hairs are pulled out by the roots and incubated in a tyrosine-rich medium, one type of albino hair root produces dark pigment and the other type of albino hair root does not.

The hair-root test was performed on a sample of albinos from different families. About half responded to the test with pigment formation and about half did not. Thus, there are *at least two* different gene loci that can produce the albino phenotype. Actually, there are probably quite a few, some of which have several alleles that can produce varying degrees of albinism.

The Castle-Hardy-Weinberg law, like all mathematical models, has certain restrictions. In this case, the model gave incorrect estimates of the allele frequencies of albinos because it does not allow for the possibility of more than one gene locus controlling a trait.

What Good Is the Castle-Hardy-Weinberg Model? Physicists often quite casually begin some study by making absurd assumptions: frictionless surfaces, falling bodies without air resistance, mass concentrated at a single point, and so on. Just because these things don't really happen does not mean that the formulas derived from them aren't any good. Similarly, the Castle-Hardy-Weinberg expectations are based on a set of idealized circumstances that may seldom occur in nature, but the expectations serve to help us make predictions all the same. In fact, it turns out that there are three ways in which the Castle-Hardy-Weinberg model is useful:

1. *Developing a population genetics theory:* Much of the theory of the genetics of populations is built on *hypothetical models* of what we think is going on. In many of these models, the Castle-Hardy-Weinberg population model acts as the ground floor on which everything else is built. (Algebraic models in biology are useful, of course, only if they make predictions that can be tested.)

2. *Estimating recessive allele frequencies when there is complete dominance:* This special use of the Castle-Hardy-Weinberg distribution depends on all the assumptions being at least nearly true. As we saw in the case of recessive albinism, it is very easy to reach incorrect conclusions when you don't know all the facts. An extension of this is using the Castle-Hardy-Weinberg distribution to test whether some specific condition *is* or *is not* a simple Mendelian trait. If the frequencies of supposed carriers and alleged homozygotes don't match expectations, or if the *ratios* in the offspring of certain types of matings don't match Mendelian laws, the investigator is forced to try another hypothesis.

3. *Explaining deviations in natural populations:* For many purposes, the Castle-Hardy-Weinberg predictions are the most useful when they don't come true. We have listed several condi-

tions in natural populations that will give rise to departures from Castle-Hardy-Weinberg expectations. These departures themselves are interesting and can sometimes tell us such things as how much mutation, natural selection, or plain random change may be going on. These, of course, are the primary forces of evolution. The only way evolution can proceed is through such events; thus a population *out of equilibrium* is immediately interesting.

MUTATION AND GENETIC VARIATION: THE RAW MATERIALS OF EVOLUTION

Natural selection is the force that determines the direction of evolution, but it must have something to work with. The raw material is genetic variation. Of course, the genetic recombination that occurs in meiosis continually reshuffles genes to insure some new variation, but the source of new variations in both sexual and asexual species is *mutation,* the chance alteration of DNA. In other chapters we have seen how mutations arise through rare, unrepaired changes in base sequences (see Chapter 10) and through chromosome breakage and rejoining (see Chapter 13). But now we are interested in the fate of mutations as they enter the gene pool of populations.

Mutation Rates: Constant Input of New Information

In any population, a given gene will mutate at one time or another (usually to a nonfunctional or even harmful form). In fact, each gene undergoes mutation in a statistically regular and predictable manner. Thus, there is a constant and measurable input of new genetic information into the gene pool. Typically there is about one new mutation per gene locus per 100,000 gametes. Some of the most severe mutations go unnoticed, however, because no offspring are produced. It has been estimated that *a third* to *a half* of all human zygotes fail to develop because of dominant chromosomal mutations. The potential parents are usually quite unaware that anything untoward has happened, even as the lethal gene carries the embryo into oblivion.

It is likely that most of the mutations that are transmitted from generation to generation are

recessive, since recessive alleles either fail to function at all or function at a reduced level. Many of these recessive mutations are also lethal, but only in the homozygous state—and this is not likely to happen until many generations after they first enter the gene pool.

The easiest mutations to study, of course, are dominant mutations that cause visible changes. An *invisible* change, for example, might be an alteration in some enzyme that had little or no effect on the organism. A *visible* change, on the other hand, might be dwarfism.

A peculiar type of human dwarfism, identified earlier as achondroplasia (see Chapter 12), occurs in one out of about 12,000 births to normal parents (Figure 15.3). In achondroplasia the head and trunk are of normal dimensions, but the arms and legs are stunted or fail to grow at all. Since the condition is readily visible and also dominant, each dwarf born

15.3 A DOMINANT MUTATION

Each occurrence of achondroplasia (hereditary dwarfism) in the children of normal people can be traced directly to a new mutation in the father or mother of the afflicted individual. The allele or alleles responsible are dominant, so they cannot be hidden by the heterozygous condition as is the situation for most mutant alleles. We know the trait is dominant since half the children of unions between achondroplastic dwarfs and normal individuals are achondroplastic. What might we expect if the allele were recessive?

to normal parents represents a newly mutated gene.

With an average mutation rate of about 1 mutation per gene locus per 100,000 gametes, and with an estimated 80,000 genes per human zygote, everyone is likely to be carrying a new gene mutation. Since many of these will accumulate and be passed on to future generations, the total number of people with less than ideal genotypes is very large indeed—in fact it includes all of us if we include the potentially devastating recessive alleles that we all carry. The more minor the genetic defect, in general, the more readily it is passed on, and the more people it will affect. Some of the more common of these less-then-ideal genetic conditions are the familiar nuisances of missing teeth, malocclusion (such as buck teeth), nearsightedness, deviated nasal septum, and so on.

Of course, some genetic conditions are more severe, and to conditions we have already discussed, such as achondroplasia, hemophilia, albinism, sickle-cell anemia, and phenylketonuria, we can add schizophrenia, manic-depressive syndrome, early-onset diabetes, hereditary deafness, cystic fibrosis, Tay-Sachs disease, and literally thousands of other known genetic disorders. (Keep in mind, however, that not all mutation is bad. After all, mutation is one of the reasons you are not an ameba, or even a random collection of methane, ammonia, and mud.)

The Balance Between Mutation and Selection

Natural selection can cause a beneficial mutation to increase in a population, but we must admit that this is an extremely rare event. Few mutations will make a gene work better than it did before; the great majority will make the gene work less well or not at all. It's as if you raised the hood on your Porsche and let your neighbor's kid randomly bang the internal workings with a hammer. He may make precisely the adjustment needed to make the car run better, but what are the odds of that? (As a completely irrelevant aside, when the Apollo astronauts were having trouble with a delicate, sophisticated mechanism on their moon rover, Mission Control deliberated 12 hours before coming up with a solution: try banging it on the side with a hammer. It worked.) The point of all of this is simply that most mutations are harmful. Still, randomly occurring changes provide the variability on which natural selection will ultimately act.

It is important to realize that, genetically, muta-

tions that prevent an individual from reproducing ultimately have the same effect, whether they kill an embryo or an adult, or simply render the individual sterile. In either case, that person's genes are not passed along; they die with him or her. Thus, while lethal alleles are constantly fed into the gene pool by mutation, they remain rare because they are not passed along. However, if the mutant allele is only *partially* limiting in its effect, afflicted individuals may reproduce but at a reduced rate, and the mutant form may become much more common. It will still be held in check, though, by natural selection. Population geneticists tell us that eventually an equilibrium will be attained, so that the number of affected individuals in the population will be *directly proportional* to the mutation rate, but *inversely proportional* to the individual's loss of reproductive fitness (Figure 15.4). That is why minor genetic abnormalities are so common: a mutant allele for, say, buck teeth is passed from generation to generation, producing whole families with malocculusion but with otherwise good health, most of the affected individuals still managing to marry and to raise numerous, toothy children. (Essay 15.1, a tongue-in-cheek story about Manx cats, may help you gain some perspective about how aberrant genes reach equilibrium.)

15.4 THE BALANCE BETWEEN MUTATION AND SELECTION: A SCHEMATIC ANALOGY

Water in the beaker represents mutant genes in the gene pool. Water enters the beaker at a constant rate (spontaneous mutation). Water flows out of the beaker (natural selection) at a rate that depends on the current level in the beaker. **(a)** At *mutation equilibrium,* the flow of mutant alleles into the gene pool through spontaneous mutation exactly equals the loss of mutant alleles through natural selection. **(b)** If the mutation rate is increased, perhaps because of radiation or other mutagens in the environment, the level of mutant alleles in the gene pool will rise until the outflow again equals the inflow. **(c)** Similarly, if natural selection against the mutant alleles is reduced, for instance, by improved medical treatment of affected individuals that allows them to reproduce, the level of mutant alleles in the gene pool will rise until a new equilibrium is reached between the inflow by mutation and the outflow through natural selection.

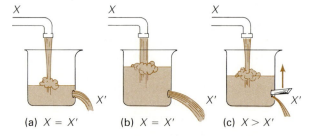

(a) $X = X'$ (b) $X = X'$ (c) $X > X'$

The Tail of Two Genotypes

To see how natural selection can act in rather unexpected ways, consider what might happen in the relatively simple case of *lethal* alleles. If an individual harbors a harmful allele, death removes the individual and the allele from the gene pool. To illustrate, consider Manx cats (see photo).

Manx cats are peculiar genetic anomalies. They have rather large hind legs and no tails (or very short tails). No one has ever been able to develop a strain of true-breeding Manx cats for the simple reason that the tailless animals are all heterozygotes. Normal cats, with tails, are TT homozygotes; Manx cats, without tails, are Tt heterozygotes. The homozygous tt genotype is an embryonic lethal; that is, it kills the embryo. So already we see a strange thing. The t allele is dominant for one trait, and recessive for another. It is dominant for the absence of a tail, and it is recessive for the absence of a kitten. Thus, two Manx **(Tt)** cats, mated, produce ¼ normal cats, ½ Manx cats, and ¼ dead (lost or "absorbed" as early embryos). Actually, then, the only litter you would see from such a cross would be ⅔ Tt Manx and ⅓ TT alleycat.

Now suppose that someone should populate a remote island with a whole shipload of Manx cats, which would then run wild, yowling and scratching and mating randomly, as cats are wont to do. What would happen? The frequency of the Manx allele, **t,** starts out at $q = ½$. In one generation it is reduced to $q = ⅓$. What happens then? With random mating, the third generation of *zygotes* will be $p^2 = 4/9$ TT, $2pq = 4/9$ Tt (Manx), and $q^2 = 1/9$ tt homozygous lethal. (See the accompanying text for a detailed explanation of the algebra.)

But when the recessive homozygotes are removed by natural selection, the remaining cats are now half Manx and half alleycat. The frequency of the recessive lethal t allele has gove from ½ to ⅓ to ¼ in three generations, and in succeeding generations it will fall further to ⅕, ⅙, and so on. Meanwhile, the proportion of homozygous lethal **tt** zygotes will decrease accordingly (applying the Castle-Hardy-Weinberg formulas): ¼, 1/9, 1/16, 1/25, 1/36 and so on. Eventually Manx cats will be fairly rare on our hypothetical island. But, significantly, the severe selection against the t allele will diminish. The accompanying graph plots the course of the genotypes over 70 generations. Note that the allele frequency of t has fallen to 1.37% (1/70) by the end of the 70 generations, and that about 2.8% of the cats will then be Manx. It would take another 70 generations to bring the recessive allele frequency down to 0.7%.

Selection doesn't have to be so severe, of course. As a general rule, the speed of gene change is proportional to the amount of selection against the unfit genotype. For instance, in the above example, suppose that the recessive genotype tt wasn't lethal, but merely reduced the individual cat's reproductive ability by one tenth. Then it would take 700 generations, rather than 70, to go from $q = 50\%$ to $q = 1.37\%$.

You might wonder why there are any Manx cats at all. The truth is that people are impressed by anything bizarre in cats, and tend to keep the Manx kittens while disposing of the alleykittens. So in the final analysis, it is human intervention—what can be called *artificial selection*—that keeps the Manx gene going.

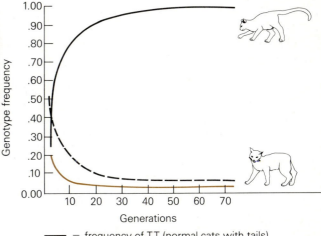

— = frequency of TT (normal cats with tails)

- - - = frequency of Tt (Manx cats)

— = frequency of tt (homozygous lethal zygotes)

The Genetic Future of *Homo sapiens*

This brings us to a philosophical and moral question that inevitably arises about genetic variation in future human populations. If better medical and social care saves the lives of persons with adverse genetic conditions, thus allowing them to reproduce, what is to become of us?

It's not hard to find examples that illustrate the problem. The genetic condition **pyloric stenosis,** an abnormal overgrowth of a stomach valve muscle, was once invariably fatal in infancy. Since the 1920s a simple surgical procedure has saved the lives of nearly all affected infants (in developed countries). But about half the offspring of the saved people are also affected, and also need surgery. And since there are just as many new mutations as ever, the genes for the condition have actually increased in frequency. Where will our species be in 10,000 years?

What has happened in effect is that a *severe* genetic condition for "certain death in infancy by intestinal obstruction" has been transformed to a relatively *mild* genetic condition for "simple abdominal surgery needed in infancy." Suppose that the remaining risk of the condition, including possible delayed diagnosis and surgical mishap, is such that now about one affected infant in 20 fails to survive. The mutant genes for the condition will continue to increase until, many millenia from now, a new equilibrium between mutation and selection will be reached. At that time there will be exactly 20 times

as many affected individuals as there were before the operation was developed. If after all that time the condition still takes the life of one in 20, the total number of persons dying from the effects of the gene will have once again reached its 1920 level. In the meantime, many lives will have been saved and many people with slightly aberrant genotypes will have joined the human population. Is that good or bad?

In fact, this sort of thing has already happened repeatedly; every species adapts to changes in its environment, even such changes as the availability of skilled surgeons. In other cultures, such as those of primitive hunting and gathering tribes, **myopia** (nearsightedness) must be a condition that is often lethal. Presumably, nearsighted aborigines can't find roots and berries efficiently, let alone a zebra. But long before the invention of corrective lenses, the stable social conditions that came with villages and agriculture allowed myopics to survive, possibly even giving them a measure of protection as male myopics stayed closer to the home base and dealt with matters that did not require distance vision. In any case myopia is much more common among people with a long history of agriculture and urban civilization than among groups that have more recently given up nomadism or the hunting-gathering life. American Indians and American blacks are blessed with much better visual acuity, on the average, than are Americans of European or Oriental ancestry, people long involved in agriculture.

APPLICATION OF IDEAS

1. In a herd of wild mustangs, Greg Meddlesome counted 10 palominos, 29 dark (brown or black) and one white. (a) If the genotypes are **Aa, AA,** and **aa,** respectively, what are the respective genotype frequencies? (b) What is the allele frequency of **a?** (c) Is the group in approximate Castle-Hardy-Weinberg equilibrium? Calculate the CHW expectations. (d) Greg further noticed there was only one stallion, which happened by pure chance to be the white **(aa)** horse. The rest constituted a harem of mares—not an unusual situation with groups of wild horses. What will be the approximate allele frequency of **a** in the next generation, assuming that the group remains isolated? (e) Would you expect the next generation to be in CHW equilibrium? Explain.

2. Among Americans of European descent, about 70% find weak solutions of the chemical phenylthiourea (also called phenylthiocarbamide, or PTC) bitter and distasteful. The remaining 30% are unable to taste the chemical unless it is extremely concentrated. The difference in ability to taste the chemical is genetically determined by a single gene locus with two alleles. Tasting the chemical is dominant over not tasting, so tasters are designated as **TT** and **Tt,** while nontasters are **tt.** What is the frequency of the each allele (**T** and **t**) in the above population? What are the frequencies of the three genotypes (**TT, Tt, tt**)? What is the probability that any taster in the population is homozygous?

KEY WORDS AND IDEAS

ALLELES AND ALLELE FREQUENCIES

1. Evolution occurs in populations, not in individuals. It begins with changes in the **gene pool** (all of the population's genes).

2. Changes in genes occur through mutation, while changes in **allele frequencies** occur through **natural selection** and random events.

The Castle-Hardy-Weinberg Law

1. The **Castle-Hardy-Weinberg law** predicts the behavior of alleles and allele frequencies in a model population. It explains why in the absence of selective forces, allele frequencies cannot change.

The Implications of the Castle-Hardy-Weinberg Law

1. A **population** is defined as a group of interbreeding individuals, and **frequency** refers to a fraction or proportion of one type of item to all of the items in a general class.

2. In the absence of outside influences, allele frequencies remain constant. This can be proven by using an individual cross of **Aa** × **Aa,** and carrying the cross through any number of generations. The frequencies of **A** and **a** remain ½.

An Algebraic Equivalent

1. The algebraic equivalent of a population in terms of the alleles for any gene can be stated as: $p^2 + 2pq + q^2 = 1$. In the formula, p^2 equals the frequency of the homozygous dominant genotype, $2pq$, the frequency of the heterozygote genotype, and q^2, that of the recessive genotype.

2. From the CHW law, it is possible to predict, within its limits, the probability and outcome of matings in a population.
   a. This requires, first, that the allele frequencies of the gene in question be determined.
   b. Such frequencies can be determined through application of the formula, $p + q = 1$. Here p is the frequency of the dominant allele, and q is the frequency of the recessive allele.
   c. Together their frequencies make up all of those alleles in the population, or, 1.
   d. Allele q is usually detectable simply because it is recessive. First, all of the recessives (qq) in a population are identified and counted. Since the number equals qq or q^2, q is the square root of that number.
   e. When q is known, p is readily determined, since $p = 1 - q$.
   f. When both p and q are known, the genotype frequencies can be determined by substituting the numbers for factors in the general CHW formula. $p^2 + 2pq + q^2 = 1$.

3. The CHW formulas can be used for predicting the probability of any couple in a population producing certain genotypes in their offspring. Albinism is an example:
   a. recessive genotype frequency: $q^2 = \frac{1}{20,000}$
   b. recessive allele frequency: $q = \frac{1}{141}$
   c. dominant allele frequency: $p = 1 - \frac{1}{141} = \frac{140}{141}$
   Since knowing p and q permits the genotype frequencies to be determined, applying the multiplicative law enables predictions of certain matings to occur.

A Closer Look at the Castle-Hardy-Weinberg Model

1. The CHW model fits only an ideal population, so there are restrictions: single loci only, two contrasting alleles, random mating, no loss through migration, no mutation, equal reproductive fitness, infinite size population.

2. Consequently, allowances must be made and variables considered in applying the CHW law.

3. Uses for the CHW formula include population modeling, spotting and explaining deviations, and estimating allele frequencies.

MUTATION AND GENETIC VARIATION: THE RAW MATERIAL OF EVOLUTION

1. Natural selection operates on variation originating from mutation (the raw material).

Mutation Rates: Constant Input of New Information

1. The input by mutation is constant: 1 per gene locus per 100,000 gametes. Most humans carry a new mutation.

2. A large proportion of human embryos fail because of mutated genes, most of which are recessive.

3. The effects of mutations vary from simple conditions like deviated septum to cystic fibrosis.

The Balance Between Mutation and Selection

1. The vast majority of mutations are harmful. If they prevent an individual from reproducing, they have no lasting impact on the gene pool.

2. Abnormalities are strongly selected against at first, but the degree of selection tapers off as the alleles (and thus the homozygous recessive individuals) become less common.

The Genetic Future of *Homo sapiens*

1. The usual attrition of harmful mutations is thwarted where genetically derived abnormalities are medically corrected. Such action permits once doomed or reproductively limited humans to survive and reproduce, and the mutant gene frequency to increase.

REVIEW QUESTIONS

1. Explain the statement, "Individuals do not evolve, populations evolve." (363)

2. What, exactly, constitutes a *gene pool?* (363)

3. Explain what the term *frequency* refers to in population genetics. (364)

4. Carefully define the term *population.* (364)

5. Using the logic of Hardy, explain why, in the absence of selection, the frequency of a recessive gene remains constant. (365)

6. Using Punnett squares and the alleles **B** and **b,** prove that the recessive **b** remains constant through three generations. (365–366)

7. Define each of the terms of the CHW expression: $p^2 + 2pq + q^2 = 1$. (366)

8. A certain recessive genotype appears in 16% of the individuals in a population. Determine the frequency of its two alleles, **T** and **t.** (366–367)

9. What is the probability of two heterozygotes mating in the population in number 8? What is the probability of two heterozygotes mating and producing an offspring that is also heterozygous? (366)

10. The CHW distribution applies to model populations. List the seven assumptions that must be made to make the application valid. (367)

11. Considering your answer to the last question, how can the CHW law be of any use to geneticists? (368)

12. List three specific ways the CHW law can be applied. (368–369)

13. Explain the relationship between evolution, variation, mutation, and natural selection. (369)

14. What is the usual mutation rate? What are the chances that you carry a new mutation? (369–370)

15. How important is mutation to the survival of human embryos? Why aren't we fully aware of the size of the problem? (369)

16. List five human maladies that can be traced to mutations. (370)

17. Why is it that nearly all mutations are harmful? (370)

18. Considering the Manx cat story, characterize the way, over time, that natural selection works against a harmful allele. List two reasons why all such alleles aren't completely eliminated from the gene pool. (370–371)

19. Suggest three or four instances in which humans in modern society seem to overcome selection against harmful alleles. In a larger sense, do we really thwart this important evolutionary force—that is, in these instances will natural selection simply work in some more subtle manner? (372)

Evolution and Natural Selection in Populations

One of the more fascinating observations about life on earth is how greatly individuals vary from each other. Sometimes the variation is obvious, as in human faces. Sometimes it is less apparent, as in colonies of gulls whose members look identical to us, even though they sort themselves out quite nicely. In other cases, the variation may be virtually invisible without technical assistance, as we find in individuals with slight differences in their body chemistry.

Life is indeed highly variable, and we have just seen some of the ways this variation can arise. In this discussion, we will try to see more precisely how such variation is important to the processes by which populations change through time.

MICROEVOLUTION AND MACROEVOLUTION

Evolution can be viewed in two ways—at two distances, in a sense. Up close, and in fine detail, we can consider the processes of **microevolution**—that is, the evolutionary changes within populations. These changes are the result of shifts at the molecular level, and they lead to the sorts of variation that are expressed within species.

One can also take the broader view. **Macroevolution** involves the changes that occur on a grander scale, above the species level. These changes must be viewed over geologic time as we consider the evolutionary development of groupings such as species, genera, orders, classes, and even phyla (also called divisions).

As usual, we will first discuss the smaller changes and use this information to build the larger case. We will begin by considering how evolutionary change might proceed in those genetically simple groups called clones.

NATURAL SELECTION IN "CLONES"

A **clone** is composed of a population descended from one ancestor, a group in which all individuals are genetically identical. Theoretically, because there is no variation in a clone, no evolution can take place. But even in clones some variation can appear through the process of random mutation. If mutation results in the formation of, say, two groups that vary only slightly, each will be present at a certain frequency. Evolutionary change occurs when the frequency of one type increases and the frequency of another type decreases. Such changes in frequency can be due to chance, but it is vastly more interesting to most biologists when the changes are the result of differential survival or reproduction—in other words, of natural selection.

Antibiotic Resistance

A familiar example of natural selection in a clone is the evolution of antibiotic resistance in a population of bacteria. *Antibiotics* are natural or synthetic substances that, in minute concentrations, kill or inhibit the growth of microorganisms. For instance, penicillin kills susceptible bacteria by interfering with the synthesis of bacterial cell walls. But some populations are naturally resistant to penicillin, their resistance genes having arisen by mutation. They may appear in people who too frequently treat their illnesses with penicillin (see Chapter 14).

Penicillin-resistant organisms can also be identified in laboratories. We could begin with a clone of susceptible bacteria and cover them with a dilute solution of penicillin. If it is a true clone, and if the dose of penicillin is large enough, all the bacteria will die.

But bacteria occur in large numbers, and we can predict that in a culture of any size there will be one or more spontaneous genetic mutants to the penicillin-resistant state. The *frequency of such mutants*, in the absence of penicillin, may be something like one resistant individual among 100 million. We can symbolize the frequency by p, thus:

$$p = 1/100,000,000 = 0.00000001 = 10^{-8}$$

If we then kill all the susceptible bacteria with penicillin, perhaps only one or two resistant cells will remain, but the *frequency of the resistant type among the survivors* will have increased to $p = 1$. (That's one way of producing *fast* evolution!) Because of their rapid reproduction, the descendants of the resistant bacteria will then fill an entire culture vessel in a short time. For a time, this strain will be a clone.

If the penicillin is diluted so that it only kills *some* of the susceptible bacteria, the change in frequency is not so great, and evolution is not so fast. Suppose that the penicillin kills half the susceptible bacteria. (This, of course, means that at least three strains were present: susceptible, resistant, and partly susceptible.) Then the frequency of the resistant type goes from one in 100 million to approximately two in 100 million—that is, from 10^{-8} to 2×10^{-8} (0.00000002). That's a large *relative* increase—a doubling—though the *absolute* increase is very small.

However, if the difference in survival went on for 10 generations, the frequency of the resistant type would increase about a thousandfold, from 10^{-8} to 10^{-5} (one resistant type per 100,000 susceptible types).

The same relationship holds whether the entire bacterial population is actually growing, staying the same, or shrinking, as long as the *relative fitness* of one type is greater than the other. In this case, the *fitness* of the resistant type is 2 relative to the fitness of the susceptible type; conversely, the fitness of the susceptible type is 0.5 relative to that of the resistant type. The difference can be in death rate or in rate of reproduction, as long as the average number of descendants of each of the resistant bacteria is twice that of the susceptible types after some specified period of time. Soon, of course, the descendants of the resistant bacteria would outnumber and eventually supplant those of all the susceptible ones.

We might do yet another experiment with a still more dilute solution of penicillin. Suppose the effect were such that the resistant bacteria grew at a rate that was just 10% greater per generation. The rate of change in the frequency of the resistant type would be slower, but again the favored type would become relatively more common and the unfavored type would become relatively less common. If we made a graph of the frequency of the penicillin-resistant type over time, it would show a **sigmoid** or S-shaped curve (Figure 16.1).

Notice that the absolute change in frequency is initially very slow. It then increases, reaching a maximum *rate of change* at $p = 0.5$. After this point, it slows down again to a *fixation* level, where it very slowly nears the value $p = 1.0$ (when every individ-

---

16.1 NEW MUTANTS IN AN ASEXUAL POPULATION

If a beneficial mutation occurs in a strain of asexual organisms, over a number of generations it will increase in frequency at the expense of the less favored type. A graph of the frequency as it changes through time takes the form of a smooth, S-shaped curve (sigmoid curve).

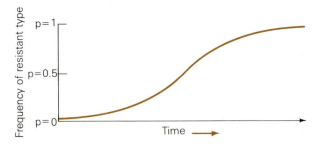

16.2 TWO MUTANTS IN A POPULATION OF ASEXUAL ORGANISMS

When two unrelated mutant strains occur in the same population of asexual organisms and both are superior to the original type, they will be in competition with each other. Eventually, only the one strain with the higher reproductive potential will survive (type 2), and the other potentially beneficial mutation will be lost (type 1).

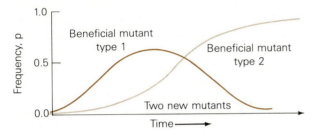

ual in the population is penicillin-resistant), without ever quite reaching it, producing an asymptotic curve. In a real population, of course, infinitely small frequencies don't occur, and eventually the last surviving nonresistant type would bite the dust.

Selecting for Two Characteristics at Once

So far, we have considered changes in a single trait controlled by one gene locus. Now let's see what sort of interactions are possible when natural selection operates on two phenotypes, and two gene loci, at once.

In populations of asexual organisms, there may be simultaneous selection for two or more different factors. For instance, one new type might have a selective advantage because, let us say, it has a new mutant enzyme that enables it to metabolize arginine more efficiently. Another strain of the bacterium may have a selective advantage because a different gene gives it better resistance to damage from ultraviolet radiation. What happens now?

With asexual organisms, it's always a matter of "may the best strain win." If the strain with better ultraviolet resistance can outreproduce either of the other two strains (with and without the mutant enzyme), it will drive both of them to extinction, and the new and improved arginine enzyme would be lost. No matter, perhaps, because bacteria are so numerous and reproduce so rapidly that a very similar mutation will probably show up again in our hypothetical ultraviolet-resistant strain. The point is, in asexual organisms, only one new evolutionary

change can be fixed at a time, so progressive evolutionary changes are *sequential* (Figure 16.2).

Natural Selection in Sexual Populations.

In sexual organisms, two different favorable mutations can arise in separate individuals and go to fixation simultaneously. For instance, an ultraviolet-resistant mutation in one organism and a mutation with more efficient metabolism in another organism would not be competing against each other, as they were in the asexual population. Genetic recombination can free genes from competing with each other, as each gene locus behaves as if it were independent. Recombination will create individuals bearing both favorable mutations (Figure 16.3).

We should remind ourselves of a small but important point here. It is only in sexually reproducing populations that unlinked genes behave as independent units in evolution. They can be separated by various "shuffling" processes discussed earlier and be brought together independently of each other. Gene A need not always appear with gene B. Because of this, natural selection can mold many different aspects of an evolving organism at the same time, and every change that arises can be combined into the evolving phenotype. According to people who have given the question much thought, this is the chief advantage of recombination and perhaps the basis upon which meiosis and sex evolved in the first place.

One claim for the evolutionary advantage of sex is that sexual organisms can evolve more rapidly, since many genes can be selected for simultaneous-

16.3 NEW MUTANT GENES IN A POPULATION OF SEXUALLY REPRODUCING ORGANISMS

With sexual recombination, alleles at different loci are not in competition with one another. If two or more beneficial mutant alleles are increasing in the same population, recombination will allow the emergence of individuals that carry the best allele at each gene locus, and different beneficial mutations can go to fixation independently and simultaneously. This is one of the long-term advantages of sexual reproduction.

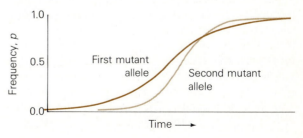

ly. Opponents of this view object that asexual organisms can evolve just as rapidly or even more rapidly, at least when only one trait is involved. In fact, asexual organisms can evolve *too* rapidly, leading a line of asexual organisms down a blind alley. For instance, a succession of hot, dry years would favor the expansion of asexual organisms (clones) adapted to such conditions, and quickly drive to extinction those lines that are better adapted to cooler, moister climates. But what would happen if the dry spell ended and cooler weather set in? The hot-weather organisms would follow the cool-weather organisms into oblivion. Sexual combination, on the other hand, ensures some genetic variation in a population—perhaps as unexpressed recessive genes in heterozygotes, but there nonetheless. This hidden variation allows sexually reproducing species to change more slowly, giving them a certain resilience against this sort of devastation.

The last advantage of sex that we will mention is related to the constant reshuffling of genes. Just as any harmful ones sometimes ride along harmlessly, their ill effects being swamped by good genes, as chance would have it, someone occasionally winds up with a load of bad genes and dies. Thus, a large number of bad genes are removed from the population by the death of only one individual.

An interesting study illustrates a case of rapid evolution in nature resulting from sexual recombination. Actually, since two species are adapting to each other it is a case of rapid **coevolution.** The study involves a species of pine and the scale insect that infects it. Pines, like other plants, have a wide variety of chemical defenses against insect pests, but in pines, as in other plants, one or a few insect parasites are usually able to overcome these barriers. In this case, the pines are much more long-lived than the insects. Thus, the short-lived insects undergo many generations on a single host. As the tree ages, so does the population of scale insects, with the best-adapted parasites surviving. In time, the pest population becomes increasingly adapted to its venerable host.

In fact, the insect becomes so adapted to its specific host that it can't survive anywhere else. In experiments, scale insects can successfully be transplanted from one part of a tree to another part, but they nearly all die if they are moved onto a different tree, even one of the same species. In fact, for the most part, the scale insects cannot even infect the young seedlings that grow from the parent tree to which they are adapted. The reason may well be sexual recombination. The young seedlings have different genotypes from those of their parents, and

thus they have new and different combinations of defenses against insect attack, defenses for which the insects aren't adapted. Most of the young pine seedlings are therefore relatively free of scale insects, but even on the hardiest young hosts, a few invading insects manage to survive. Then, as the tree ages, descendants of the first colonists become increasingly resistant to their host's defenses through natural selection that favors new combinations of the insect's own counterdefenses. With such recombination, adaptation is fairly rapid. In this continuing relationship, sexual recombination is important to the scale insects. Thus, sex is important to both the tree and the insect.

The problem of how sex ever arose to begin with has plagued biologists for years. It has been pointed out, for example, that if the name of the "fitness" game is simply to leave one's kinds of genes, the best way would be to make replicas of oneself rather than to produce offspring that carried only half of one's genes. The advantages of reshuffling genes and distributing them among variable offspring would have to be enormous to offset the advantage of simply replicating oneself. Nevertheless, questions regarding fundamental advantages of sex have never been answered satisfactorily. We will return to the question in later chapters.

The Peppered Moth: Natural Selection in a Natural Population

Biston betularia is a British moth, commonly called the **peppered moth.** It occurs in two **morphs;** that is, it may have either of two distinct forms or appearances, as shown in Figure 16.4. One morph is light and mottled (or peppered) and the other morph is black. The British have a long tradition of butterfly and moth collecting, and records on the peppered moth go back two centuries. The black morph, whose color is controlled by a single dominant allele, originally showed up in 18th century collections as a rare, highly prized variant. In the early stages of the industrial revolution (in the 1840s), the black form began to show up in greater frequencies in collections, especially near cities. The black morph continued to become more and more common in industrialized areas, until it greatly outnumbered the light peppered morph. In Manchester, England's industrial center, the dominant black morph achieved a frequency of 98%. Meanwhile, the light peppered morph remained the predominant form in rural areas.

The environment had changed and the species, through differential mortality and a change in allele

16.4 THE PEPPERED MOTH

In this "bird's eye view," the black and peppered morphs of *Biston betularia* are seen on the natural lichen-covered, and unnatural soot-covered backgrounds. Their frequencies are drastically affected by predatory birds.

frequencies, adapted to it. The environmental factor was soot from burning coal. Industrial England, as the 19th century proceeded, quietly submitted to its dark cloak of carbon. Meanwhile, the *Biston betularia* adapted. Bird predation is probably the species' principal cause of death. Over the long course of evolutionary time, the moth had achieved a camouflaging coloration that blended well with the light, peppered appearance of lichen-covered tree trunks. But pollution killed the lichens and blackened the trees, making the light peppered morph highly visible and extremely vulnerable to predation. In industrialized areas, the black morph achieved a significant selective advantage because it was less easily spotted by birds.

From a graph of the frequency increase of the black form in the historical data, J. B. S. Haldane (one of the founders of population genetics) calculated its relative fitness in an industrial environment to be twice that of the more conspicuous peppered form. But a British naturalist, H. B. D. Kettlewell, performed the crucial experiment. He released known numbers of marked black and light peppered moths in unpolluted woodlands and in polluted, soot-blackened woodlands. In each habitat, after a period of time had elapsed, he recaptured a portion of the released moths. Kettlewell's mark-and-recapture data are seen in Table 16.1.

For the first set, released in the unpolluted woodland, almost exactly twice as many light forms survived as black forms. That's equivalent to a 100% advantage of the light type in a brief exposure to predation. The frequency of the black morph in this sample fell from $p = 0.488$ to $p' = 0.326$ (that is, from 473 out of 969 to 30 out of 92).

TABLE 16.1

KETTLEWELL'S MARK-AND-RECAPTURE EXPERIMENT WITH MOTHS

Dorset, England unpolluted woodland	Peppered morph	Black morph
Marked and released	496	473
Recaptured after predation	62	30
Percentage recaptured	12.5%	6.3%
Relative survival	1.00	0.507

Birmingham, England soot-blackened woodland	Peppered morph	Black morph
Marked and released	137	447
Recaptured later	18	123
Percentage recaptured	13.1%	27.5%
Relative survival	0.477	1.00

In the second data set, selection was against the light peppered morph. Almost exactly twice as great a percentage of the favored black type survived. The frequency of the black morph rose from $p = 0.765$ to $p' = 0.872$.

Incidentally, England has been doing pretty well of late in its battle against air pollution. The woodlands near the cities are once again becoming covered with lichens, and the soot is disappearing. As one might predict, the black morphs of *Biston betularia* are now declining in frequency.

Genetic Variability in Higher Organisms

The main difference between natural selection in diploid, sexually reproducing organisms, and in haploid, asexual organisms like bacteria and unicellular algae is the much greater amount of genetic variability in populations of diploid organisms. Whereas a population of bacteria may often be essentially a clone of identical individuals, or at most a mix of a very few clones, in diploid populations every individual is unique. This genetic variability, as we now know, allows sexual diploid populations to respond very rapidly to natural selection.

We humans are quite aware of this variability in our own species. We are incredibly sensitive to individual differences, especially when it comes to facial features (Figure 16.5). And facial differences in humans are almost entirely genetic, as we are reminded when confronted with identical twins. It turns out that although individuals of other species might look very much alike on casual observation, they may be quite variable (Figure 16.6). As we will see when we consider animal communication in Chapter 39, other species are fully aware of these individual differences among their own kind, and, for all we know, they might think that all humans look alike.

Our uniqueness and variability are not limited to our faces; we are utterly unique, down to our very chemistry. This can be demonstrated by simple chemical classification of blood proteins and other biochemical differences. Also, you are probably aware of the difficulties of organ transplants resulting from tissue incompatibility. It is no accident that Darwin began his book on evolution by discussing variation and differences between individuals; this is the raw stuff of natural selection.

Maintaining Genetic Diversity. What maintains all this genetic variability in diploid sexual species? We don't know. People have come up with many ideas, but as yet we don't know which ones are closest to the truth. A generation of population geneticists and evolutionary theorists are devoting their professional lives to just this question. Part of the basis for variability lies in dominance relationships, specifically the tendency of recessive traits to "go into hiding" whenever they

16.5 VARIATION IN A POPULATION

Human beings, like all large populations of sexually reproducing organisms, are extremely variable. No two individuals are genetically alike, with the exception of single-egg twins.

16.6 VARIATION ISN'T ALWAYS OBVIOUS

Penguins, like humans, form large populations of sexually reproducing individuals. No two penguins are exactly alike—at least to another penguin. Every king penguin here knows its mate, its offspring, and all its nesting-ground neighbors.

occur in low frequencies. Another part of the answer is overdominance, in which the heterozygous genotype has a greater reproductive fitness than either homozygote. We saw an example of overdominance in the sickle-cell and normal alleles of beta hemoglobin in populations where malaria is endemic (see Essay 12.1). Since each of the two homozygous types is inferior to the heterozygote, natural selection keeps both alleles in the population in intermediate frequencies.

APPLYING THE RULES OF ARTIFICIAL SELECTION

In developing his idea about natural selection, Darwin drew on what he knew about **artificial selection,** in which humans determine which plant or animal variant is to live and reproduce. As we will see, artificial selection mimics natural selection, at least in the short term. In fact, the results of artificial selection experiments can be startling because it is so rapid and so effective. Considering how we have, in a very brief time, successfully created the vast array of domestic dogs that range from 200-pound behemoths to hairless, mouselike creatures, through artificial selection. (A visitor from outer space might well conclude that *Homo sapiens* is, above all, a whimsical creature.)

That visitor would have other evidence as

well—such as tall chickens. Figure 16.7 shows the result of a famous experiment done in Berkeley, California, in which a population of chickens was selected for long legs. Note, by the way, that some of the long leg genes are associated with more general effects and also cause increases in neck and tail length. It seems that only another century of artificial selection would suffice to produce chickens with legs like storks and necks like giraffes, requiring major revisions in henhouse architecture.

In another set of experiments, the British population geneticist D. S. Falconer selected for body weight in mice for 23 generations. A large group of mice was divided into 18 separate groups. In six such groups, only the heaviest mice in each generation were allowed to breed. In six other groups only the lightest mice were allowed to breed. In the remaining six populations, which served as a control for environmental variation, mice to be bred were chosen by lot. As shown in Figure 16.8, selection for greater or smaller body weight in mice was successful. Individual lines appear jagged because of sampling error or slight, subtle variations in the environment from generation to generation, but on the average all lines responded in a predictable and repeatable way. These results, in fact, are quite typical of those from hundreds of similar experiments with mice, chickens, corn, and fruit flies. The response to artificial selection is immediate and substantial, and the rate of change is constant, or lin-

16.7 ARTIFICIAL SELECTION OF CHICKENS

University of California geneticists have selected a line of chickens for longer and longer legs. The female on the left is a representative bird from the unselected control line; the female on the right is from the line selected for long legs. Artificial selection for this character has obviously been successful; in addition, there has been a related response in neck, tail, and body length.

16.8 ARTIFICIAL SELECTION OF MICE

Eighteen populations of mice were involved in an artificial selection program in which body weight at six weeks was measured. The experiment was carried out for 23 generations. Six replicate populations were selected for increased body weight, six replicate populations served as unselected controls, and six replicate populations were selected for smaller body weight. Although there were random variations between replicates, the general trends are clear. Note that there is much less progress in the selected lines in the last 12 generations, compared with the rather substantial progress made in the first 12.

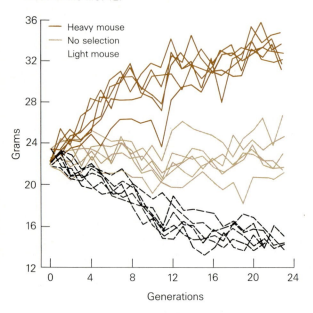

ear. There are limits to its power, however. After a dozen or two generations, the rate of phenotypic change gets much smaller, until finally there is no more response to selection.

Not reflected in the graph of mouse body weight but present in Falconer's experiment (and in virtually *all* artificial selection experiments) is the fact that each of the experimental mouse populations got progressively weaker, had smaller and smaller litter sizes, and showed increasing infertility and higher infant mortality. In other words, their own natural fitness decreased under artificial selection, no matter whether the artificial selection was for larger or for smaller mice. This pattern is seen again and again, no matter what phenotypic character is being selected for. In agriculture this presents a great risk to people who grow new miracle grains and other highly selected and inbred strains.

But just what is going on? First, the mouse studies and most other selection experiments deal with a **polygenic** character—that is, a trait in which selec-

tion operates on many gene loci. Also, the strong and linear initial response indicates that the experimenter is not dealing with rare alleles. You will recall that initial responses are very small when alleles are rare. In the case of the mice, the alleles for variable body weight must already be present in the original population, with both "lightweight" and "heavyweight" alleles in intermediate frequencies. Thus, by the end of the experiment, the low-line, intermediate, and high-line populations still have the same alleles, but in different frequencies. The low line has accumulated a large number of lightweight alleles, while the high line has accumulated many heavyweight alternatives of these alleles. The intermediate line remains more variable than either the high or the low line, as shown in Figure 16.8.

At this point you may be jumping up and down and screaming, "What causes the rate of selection response to slow down after 10 or 20 generations?" Actually, a number of factors conspire to give this result. One might be that some of the alleles become fixed (reach equilibrium) in the population, or they may attain very high frequencies, so that we are seeing the high end of the sigmoid curve. In other words, the experimental lines run out of genetic variability, and without genetic variability there is no evolution. Another factor is the decreasing fitness of the selected lines. As litters get smaller and fewer, it is more and more difficult for the experimenter to pick parents that are notably heavier or lighter than their littermates. And frequently, the most extreme animals are the least fertile. Let's see why this happens.

Selected lines lose vigor and reproductive fitness for several reasons. One is simple: the delicate physiology of the animal is adjusted to a certain body weight range, and outside that range there may be physiological problems of various sorts. Another reason is that some of the combinations of alleles that are selected for by the experimenter are simply harmful to the mouse. Most genes have effects on many systems, and a lightweight gene might well have an effect on, say, hormone production or thermoregulation.

Finally, by repeatedly choosing just the most extreme animals for mating, the experimenter inevitably begins to mate related animals, and the selected lines become increasingly inbred. Inbreeding of normally outbred organisms always causes drastic reductions in fitness, primarily because it allows harmful recessive alleles to become homozygous (also see Essay 16.1).

Natural Selection Compared with Artificial Selection. To some extent, artificial selection experiments have given us misleading ideas about how natural selection works in the long-term evolution of species. In these experiments, all the genetic variation is present in the beginning population. We thus see only the (admittedly crucial) stage in microevolution in which allele frequencies are changing within a population, which of course alters the average phenotype of members of the population. But one vital ingredient in long-term evolution is missing—the unpredictable appearance of new genes by random mutation. Mutation is a slow process; beneficial mutations, in particular, occur so rarely that they can't be expected to show up in the course of laboratory experiments on higher organisms. Artificial selection, then simulates only one part of the whole complex process of adaptive evolution.

POLYGENIC INHERITANCE AND NATURAL SELECTION

The story of peppered moths in polluted woodlands is instructive, but it can also be misleading. To be sure, evolution does indeed sometimes proceed by the rapid sweep of a dramatically advantageous allele through a population. However, most of the genetic differences between individuals in a population, such as height, weight, blood pressure, length of limbs, skin color, and swiftness of foot and mind, are not due to a few genes with dramatic effects but to numerous genes with individually small effects.

The cumulative effect of such **polygenic inheritance** on the phenotypic variation in a population is enormous. And it is on this available variation that natural selection works. The individual genes obediently follow Mendel's rules, but there are so many effects of gene interaction and other complications that the response of a population to natural selection is not simply one of allele increase and replacement. So let's now consider how natural selection works with continuous characters. (You may want to review our previous discussion of polygenic inheritance in Chapter 13.)

Three Patterns of Natural Selection

How does natural selection work with traits such as height, which vary continuously and gradually between individuals? To begin with we should be aware that most individuals will be about average for most traits. Thus, for any given measurement, there will be relatively few individuals with extremely high or low values. If we group all of the individuals in a population according to one trait, they will almost always form the bell-shaped curve—the statistician's **normal distribution.** In a normal distribution, most of the measurements cluster around the mean, or average, dropping off to two long tails on either side (Figure 16.9). The question then becomes one of how the individuals in the middle of the distribution thrive compared with those on either extreme. The answer is, it depends. In general, however, we find three

16.9 NORMAL DISTRIBUTIONS

(a) When plotted, continuous traits form the familiar "bell-shaped curve." Such distributions are common where polygenic traits are recorded. **(b)** The accompanying photograph shows a good example. The men are arranged in rows according to height.

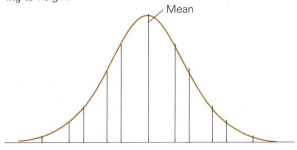

Selection for Going Up or Down

The accompanying graph shows response to selection, in two directions, for *geotaxis* in *Drosophila*. Positive geotaxis is the tendency to move up. Jerry Hirsch invented an ingenious maze that enabled experimenters to select flies for geotaxis (see accompanying illustration). The maze consists of a series of interconnected chambers. As the fly enters each chamber, it finds only two exits: one straight up and one straight down. To get from one end of the maze to the other, the fly must make an up-down choice 14 times. At the end of the maze are 15 collection vials with fly food in them. The flies that end up in the top vial have to have made the up choice 14 times in a row; the flies that end up in the bottom vial have to have made the down choice 14 times in a row. The flies that go up seven times and down seven wind up in the middle vial (vial 8). The experimenter can simply put a population consisting of hundreds of flies in one end of the maze and collect the high scorers and low scorers from the top and bottom vials at the opposite end of the maze. In a selection experiment, only the high-scoring flies are kept in the up line, and only the low-scoring flies are kept in the down line.

In the original experimental runs, the flies showed no particular preference—as a population, that is. But within the original population, some flies showed a very slight preference for going up and some showed an equally slight preference for going down. This behavior is a polygenic trait; there are many gene loci that are variable for alternate up and down alleles in the *behavioral phenotype*. A selection response in both directions continued for 12 to 16 generations. By that time, almost all the flies in the down line always ended up in the extreme down vials (numbers 14 and 15). The average of the up line was around vial number 4, which corresponds to three down choices and 11 up choices.

As usual for selection experiments, the fitness of the extreme up and extreme down lines declined markedly. While it is normal for a fly population to have a mix of both kinds of alleles, it is not normal to accumulate too many of one type and too few of another. We have no idea of what other things these alleles are doing, but they evidently are harmful in the wrong balance and combinations. After 20 generations, the researchers who ran this particular experiment stopped selecting the most extreme flies, and just let their populations mate and reproduce as they pleased. Every few generations the "relaxed" populations were tested. Natural selection had taken over, and the balance of up and down alleles returned toward normal in both populations.

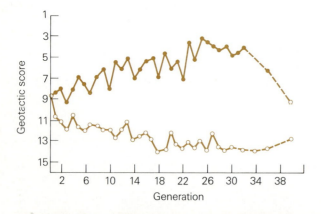

(a) The results of selection for geotaxis in *Drosophila*. Open circles indicate the progress of populations selected for negative geotaxis (down); solid circles indicate the progress of populations selected for positive geotaxis (up). The dashed lines indicate the progress of the same populations after selection had been "relaxed." Note that the up line, in particular, rapidly lost its selected behavior.

trends: **directional selection, stabilizing selection,** and **disruptive selection.**

Directional Selection. Directional selection favors one extreme of the phenotypic range—that is, one end of the curve. This is the kind of artificial selection practiced by dairy breeders who want only the offspring of the cows that give the most milk. In nature, directional selection may be a response to a change in the environment. The response may be to a new predator or parasite, for example. Or a small population may find itself in an unfamiliar territory, with new challenges. Or the species may suddenly be able to expand its **ecological niche** (the way it interacts with its environment—obtains food, interacts with other organisms, and affects its surroundings). In such cases, the formerly aberrant individuals at one tail of a

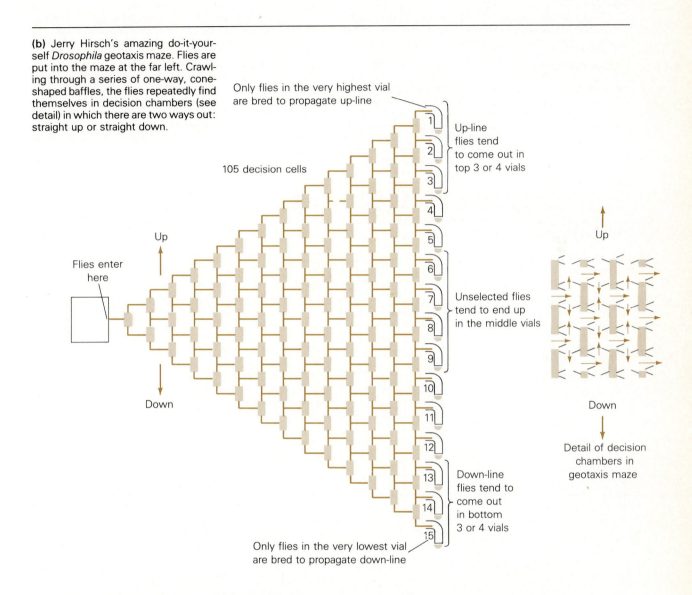

(b) Jerry Hirsch's amazing do-it-yourself *Drosophila* geotaxis maze. Flies are put into the maze at the far left. Crawling through a series of one-way, cone-shaped baffles, the flies repeatedly find themselves in decision chambers (see detail) in which there are two ways out: straight up or straight down.

Only flies in the very highest vial are bred to propagate up-line

105 decision cells

Flies enter here

Up

Down

Up-line flies tend to come out in top 3 or 4 vials

Unselected flies tend to end up in the middle vials

Down-line flies tend to come out in bottom 3 or 4 vials

Only flies in the very lowest vial are bred to propagate down-line

Up

Down

Detail of decision chambers in geotaxis maze

curve become the most fit. Their previously unfavored traits may become the new optimum. The evolution of the long neck of giraffes is often cited as a classical case of directional selection (Figure 16.10).

Stabilizing Selection. Stabilizing selection is usually associated with a population that has become well adapted to its particular surroundings. As these surroundings stabilize, the population's ecological niche becomes well established and any genetic change is unlikely to be helpful. While genetic variability still exists, selection favors the mean, or average, individual. Since most populations are rather well adapted to their environments most of the time, stabilizing selection is the most common kind of natural selection.

Consider those giraffes, for instance. As far as

16.10 DIRECTIONAL SELECTION FAVORS THE EXTREME

In this example we consider the *past* evolution of the giraffe, an animal that browses on tree leaves. Along with each drawing is a population frequency distribution, which characterizes the mean and spread of the population with respect to an important giraffe character, height. **(a)** Among the antelope-like ancestors of the giraffe, height is variable, as is any other character (bell-shaped distribution). **(b)** The tallest individuals with the longest necks are best able to reach the foliage of trees on the African veldt; these individuals survive longer and reproduce more offspring than shorter animals. The average height of the surviving/reproducing individuals is greater than that of the population as a whole, as can be seen in the frequency distribution. **(c)** The offspring of the survivors tend to resemble their successful parents, although there is some regression to the mean. The average height increases over the course of one generation (exaggerated here). Over many generations, giraffes become taller and taller. (And so, incidentally, do the trees, as only the tallest trees escape defoliation by giraffes.)

16.11 STABILIZING SELECTION

Phenotypes are usually already well adapted to the needs of the organism and are not under directional selection. **(a)** In this scheme, we assume that the giraffe population is already at its optimal tallness—on the average. But there is still some variation, with some giraffes being too short to browse well and some being too tall for their own good. (The tallest giraffes may have trouble drinking efficiently, or perhaps may be too tall for the trees, or may be subject to high blood pressure or enormous sore throats.) **(b)** The most successful giraffes are no longer the tallest individuals but the most average individuals, and these leave the most offspring. **(c)** The population mean is not expected to show any further change under these conditions. Because of genetic recombination, the distribution (spread) of phenotypes also remains the same from generation to generation (curves **a** and **c**).

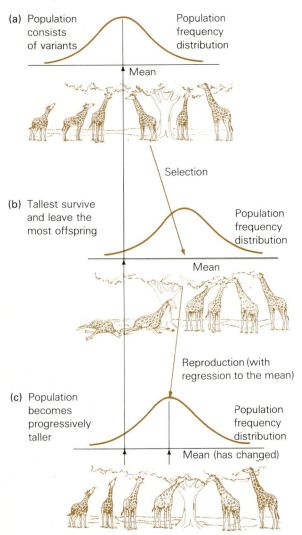

(a) Population consists of variants — Population frequency distribution — Mean

(b) Tallest survive and leave the most offspring — Selection — Population frequency distribution — Mean

(c) Population becomes progressively taller — Reproduction (with regression to the mean) — Population frequency distribution — Mean (has changed)

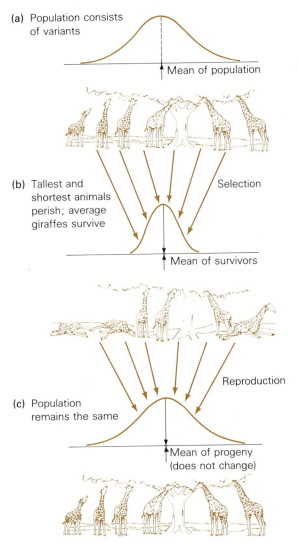

(a) Population consists of variants — Mean of population

(b) Tallest and shortest animals perish; average giraffes survive — Selection — Mean of survivors

(c) Population remains the same — Reproduction — Mean of progeny (does not change)

necks and legs go, giraffes are well adapted to their niches and are no longer subject to directional selection. And they haven't changed much in 20 million years (Figure 16.11).

Perhaps the best-studied example of selection for an intermediate condition is related to birth weight in human babies. The studies are based on data provided from the obstetric wards of major hospitals. If we plot survival rate against birth weight, we find that abnormally small babies have relatively low rates of survival, which is not too surprising. But we find that abnormally large babies also have lower survival rates (Figure 16.12). The highest survival rate is for babies around 3.4 kg (7.5 lb). In this case, the optimal birth weight is almost exactly the average birth weight. Selection is against genes for both large and small birth weight. Similar situations are found for almost any continuous polygenic trait. In essence, the average is the best, and "survival of the fittest" becomes "survival of the most common."

Although stabilizing selection reduces both extremes from the population, it is also true that mutation, genetic recombination, and other effects are sufficient to reestablish the range of phenotypes in each new generation. Thus, both the mean and the range of phenotypes remain about the same over time.

Disruptive Selection. Disruptive selection is produced when the extremes of a population are favored and those with an intermediate, or "aver-

16.13 THE BIMODAL DISTRIBUTION

Plotting the range of body weights for male and female elephant seals produces a decidedly bimodal curve, although the sexual differentiation is rather obvious without such treatment.

age," condition are selected against. Such selection can produce bimodal (or two-humped) curves. For example, it may be advantageous for males to be large and females to be small, as is the case in elephant seals (Figure 16.13). In such a case individuals of intermediate sizes would be weeded out of the population.

Frequency-Dependent Selection

Frequency-dependent selection occurs when the fitness of a genotype depends on its frequency in the population. If the success of a genotype is dependent on how frequently it appears in a population, the results may be a stable polymorphism (*poly*, many; *morph*, shape)—that is, a variety of distinctly different phenotypes. This happens if a genotype has a net advantage when it is rare and a net disadvantage when it is more common. The problem is, net advantage when rare will result in an increase in allele frequency; when this occurs, the genotype becomes less rare and its advantage decreases.

For example, some tropical freshwater fish populations are polymorphic for a common gray morph and a relatively rare red morph. The red fish, by their color alone, intimidate the other fish and almost always win out in fish-to-fish competitions. On the other hand, the red morph is easier to see and is more subject to predation by birds. Thus, when the red fish are rare, they have a net benefit because of their powers of intimidation. However, if the red form becomes common, their gray competitors encounter them frequently and become

16.12 STABILIZING SELECTION FOR HUMAN BIRTH WEIGHT

The highest rate of survival or optimal birth weight range appears to be between 2.7 and 3.6 kg (6 and 8 lb).

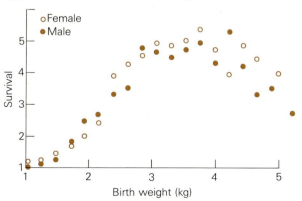

aware that those red fish aren't so tough after all. The gray fish that are harder to intimidate produce more offspring than the more timid gray fish, so a bolder gray morph evolves. Thus, each morph has its advantages, but their relative numbers are important. The red forms are kept at an *equilibrium frequency* at which the benefits of being red just balance the disadvantages (Figure 16.14).

GRADUALISM VERSUS PUNCTUATED EQUILIBRIUM

One of the assumptions of natural selection in Darwinian terms has been that evolution proceeds gradually. Such **gradualism** encompasses the idea that populations change slowly over time because of the accumulation of small shifts in gene frequencies. This notion, you can see, is very compatible with the mechanisms that would cause, for example, directional selection.

However, there is now some very good evidence that some evolutionary changes do not proceed so slowly. Niles Eldredge at the American Museum of Natural History has pointed out that fossil records reveal evolution proceeding in a different manner entirely. He notes that organisms seem to remain about the same for thousands or even millions of years and then are suddenly replaced by a clearly different form. Eldredge argues that evolution proceeds, not gradually, but in fits and starts. Thus, the very descriptive term **punctuated equilibrium** was coined. No one knows what causes these rapid and drastic changes, but the idea of punctuated equilibrium has generated a great deal of argument in scientific circles and most biologists now seem willing to accept it.

In the fossil record, whole groups appear suddenly, evidence of a period of rapid specialization and change. Then there may be extended periods in which no change is seen. Most species (according to what we can learn from fossils) survive for millions

16.14 FREQUENCY-DEPENDENT SELECTION

(a) Among certain species of tropical fish, rare red individuals may have a selective advantage over the more common gray form because their bright coloration tends to intimidate other fish of the same species. The red forms get the best breeding territories and produce more offspring. (b) If the red form becomes too common, though, it no longer has its advantages. Not only do birds learn to search for the easily spotted red fish, but the common gray fish learn not to be intimidated. In the end, a balance is struck between advantages and disadvantages, and the red forms stabilize in numbers and continue to persist at a low frequency.

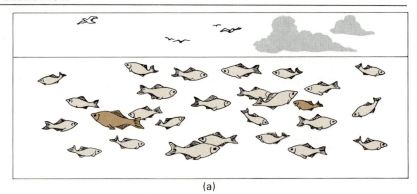

(a)

(b)

16.15 THE BOTTLENECK EFFECT

(a) In any large, diploid population, most individuals will be heterozygous carriers for rare recessive alleles at several different gene loci. **(b)** At the time of a population bottleneck, only a relatively few individuals survive. The survivors will carry a random sample of the rare alleles that were present in the formerly large population. **(c)** After the bottleneck, the few survivors will produce large numbers of descendants. Many of these progeny will carry some of the same recessive alleles, which will no longer be rare. Numerous individuals will become homozygous for these alleles (shading). On the other hand, many rare recessive alleles from the original population will not occur at all in the new population.

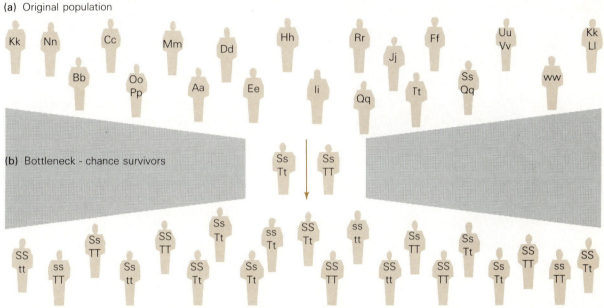

(a) Original population

(b) Bottleneck - chance survivors

(c) New population with more individuals homozygous for recessive alleles

of years without any visible change, only to be replaced suddenly with new species or new groups of species. Such appears to be the nature of evolution.

GENETIC DRIFT, POPULATION BOTTLENECKS, AND THE FOUNDER EFFECT

Genetic drift refers to *random* changes in gene frequency—that is, changes due simply to chance. The phenomenon is most easily observed in small populations. For example, should some catastrophe, such as a flood or plague, suddenly wipe out most of a population, the survivors would then begin to reproduce. However, the frequency of alleles in this small population might be different from that of the former large group. Thus, the allele frequency in the renewed population would depend on just which alleles happened to have been carried by the few survivors. In the jargon of genetics, the population would have gone through a **bottleneck** (Figure 16.15). As a result, the once rare harmful alleles may now be not so rare.

Endangered species whose numbers have been restored through comprehensive protection programs provide good examples of genetic bottlenecks. The northern elephant seal discussed earlier is probably the best-known example since its numbers were reportedly reduced to about 100 at the turn of the century. Today there are many thousands of descendants, but so far there have been no reported increases in genetic defects. But perhaps the tule deer population of California, which also experienced a genetic bottleneck, has not been so fortunate (Figure 16.16).

Bottlenecks also occur when a few individuals stray out of their normal species range and establish a successful colony in a new habitat, any rare genes they happen to carry would become common among their decendants. This special kind of bottleneck is referred to as the **founder effect.**

There are many examples of genetic bottlenecks

16.16 POPULATION BOTTLENECK

During the 1950s, mammologist Lloyd Ingles noted that the range of the dwarf, or tule elk, an unobtrusive member of the deer family, had been reduced from hundreds of miles through California's central valleys and mountains to a fenced 1100-acre reserve in Kern County. Today National Park Service zoologists suspect that one outcome of the apparent genetic bottleneck is an increased incidence of a shortened lower jaw, a condition that creates grazing difficulties.

and founder effects in human history. The numerous Afrikaaners of South Africa all descend from some 30 17th-century European families. As you know, all humans carry their share of recessive mutations. Most of these are harmful, but they are very rare and are seldom expressed in homozygotes. However, the Afrikaaners' genes have gone through a bottleneck of only 30 families, so present-day Afrikaaners suffer from a unique set of recurrent recessive genetic diseases that are seldom seen in other populations. For example, a normally rare condition called **porphyria variegata** is common among the South African settlers of Dutch ancestry. It is a metabolic disorder that is characterized by excess iron porphyrins in the blood (the heme group of blood hemoglobin; see Figure 3.16), red urine, acute sensitivity to light, and eventual liver damage. On the other hand, Afrikaaners are almost completely free of other recessive genetic diseases; the 30 families obviously did not carry those genes to Africa. Similarly, Jews of Eastern European ancestry harbor a different but equally distinctive array of recessive genetic diseases, one of which is Tay-Sachs syndrome, a storage disease discussed earlier (see Chapter 4). This can be taken as circumstantial evidence for one or more population bottlenecks in the history of this group.

Neutralists Versus Selectionists

In recent years there has been some controversy over how much natural variation is due to genetic drift and to **neutral mutations,** harmless changes in the amino acid sequences of proteins (see Chapter 10). The **neutralists** say that much variation on the molecular level is simply incidental and unadaptive. For example, there are frequently several slightly different allelic forms of any enzyme in a population, differing from each other by only one or two amino acids. Neutralists hold that most of this structural variation has no effect on the function of the enzyme. They think that functionally equivalent alleles just happen by chance mutation and that the allele frequencies drift around meaninglessly. **Selectionists,** on the other hand, prefer to believe that virtually all variation results from natural selection and has some adaptive basis, even if we don't happen to know what it is. Take note that the disagreement centers over normally *invisible traits*, not over obvious differences in phenotypes. Neither side maintains that any *visible* phenotypic variation is likely to be meaningless. The neutralist and selectionist hypotheses are restricted to the question of whether normally *invisible* details of molecular structure are always subject to differential natural selection.

In *Origin of Species,* Darwin himself summed up the neutralist argument:

> Variations neither useful nor injurious would not be affected by natural selection, and would be left either a fluctuating element, as perhaps we see in certain polymorphic species, or would ultimately become fixed. . . .

It can be shown mathematically that a completely neutral mutation would usually disappear by chance, but that it does have a finite probability of spreading through the population and thus creating a molecular polymorphism (variations in specific proteins). There is also a somewhat smaller chance that such a neutral mutation would drift to fixation in the species, effortlessly ousting its fully equivalent predecessor. Let's look at some examples of genetic drift at the molecular level.

Molecular Evolution. Horse beta hemoglobin and human beta hemoglobin are exactly the same at 129 out of 146 amino acid positions, but are different at the 17 remaining positions. Since the time of the last common ancestor of the two species, which was a little insectivore that lived about 80 million years ago, the beta hemoglobin gene has undergone 17 evolutionary changes—17 occasions on which a new mutant has replaced an older allele, in one line of descent or the other. Has all this change been adaptive? Are these changes examples of the replacement of an inferior allele by a superior new one through natural selection? Or have some or most of the changes been the result of meaningless neutral mutations and chance alone? We don't know.

The remarkable fact that has emerged from molecular evolutionary studies is that the rate of change in proteins (an indication of genetic change) is nearly constant in different lines of descent. For instance, since the time of the last common ancestor of a human and a carp—an ancestor that was surely a bony fish—the rate of change in hemoglobin molecules has been the same in the carp line of descent and in the human line, about one amino acid change per protein every 7 million years. There are, of course, thousands of proteins, and other proteins change at different rates, but each protein type has its characteristic evolutionary pace.

Now, this is all a bit unexpected. One amino acid change every 7 million years? Yet species change astoundingly rapidly. Doesn't a great morphological change reflect a great genetic change? Perhaps not. First of all (as we have seen in Chapter 10), some genes control great blocks of other genes, and a single change in a controlling gene could therefore alter the expression of the rest. Also, many genes continue through generation after generation, maintaining an intermediate optimum and playing a minor role in genetic variation. When conditions change and the genes are suddenly needed, they increase in frequency and the phenotype quickly changes.

APPLICATION OF IDEAS

1. It is always fascinating to speculate on the future course in the evolution of life, particularly of human life. Using what you have learned so far, prepare a scenario depicting humans (or what humans will have become) a few million years into the future. Try to base you assumptions on what you know about the directions of selection today. Which of the human attributes (strength, muscular coordination, intelligence, craftiness, and so on) would one expect to persist and perhaps be further emphasized? Which will be lost? Will the pace of evolution have increased over what it was in the past few million years? Why?

KEY WORDS AND IDEAS

NATURAL SELECTION AT WORK IN THE CLONE
1. A **clone** is a genetically identical group of organisms descended from one individual (commonly seen in bacteria, fungi, and protists).
2. Evolution in a bacterial clone can be demonstrated by exposing it to an antibiotic. In a high concentration, all but extremely rare, mutated, antibiotic-resistant individuals die. Using a lesser concentration may reveal the presence of individuals with some tolerance to the antibiotic.

Antibiotic Resistance
1. In low concentrations of antibiotic, the resistant bacteria would increase rapidly at first, but the increase would taper off (a graph would reveal a **sigmoid,** or S-shaped curve). While the increase would continue, it would slow and many generations might pass before the last susceptible individual died.

Selecting for Two Characters at Once
1. In asexually reproducing organisms, only one advantageous mutation can be fixed in the population at a time. If more than one occurs, their bearers become competitive.

NATURAL SELECTION IN SEXUAL POPULATIONS
1. Sexual organisms are more likely to survive environmental changes because of their greater variability and their ability to fix more than one genetic change at a time.
2. Sexual recombination permits the scale insect to overcome the pine defenses—an example of **coevolution**—but it also becomes too specialized to infest other trees.

The Peppered Moth: Selection in a Natural Population

1. A classic case of natural selection is seen in *Biston betularia,* the **peppered moth.** A black variant, or **morph,** is rare in rural areas, but predominant in industrial regions where it blends in with soot-covered trees. The peppered morph is much more common in the clean forests, where it blends in with lichen-covered trees.

2. Release and recapture studies of both moths in both environments clearly revealed that background coloration and predation were the factors affecting survival. In more recent times, air pollution control has meant a decrease in the black moth in industrial regions.

Genetic Variability in Higher Organisms

1. A major difference between simpler haploid asexual organisms and more complex diploid sexual organisms is in the amount of genetic variability. Individuals in the latter group are often genetically unique from their physical appearance through their biochemistry.

2. Darwin recognized the uniqueness of sexual individuals and made variation a key part of his evolutionary theory.

3. Variability in diploid, sexual organisms caused by many factors, including the tendency for recessives to be hidden by the dominant phenotype and overdominance, or greater fitness of heterozygotes.

APPLYING THE RULES OF ARTIFICIAL SELECTION

1. **Artificial selection** mimics natural selection in the short term only.

2. Where certain characteristics are artifically selected in a population, changes are dramatic and rapid for a time, but soon level off as variability decreases.

3. In populations undergoing artificial selection, the extremes of the population develop a large range of problems that affect survival.

4. The survival problem is attributed to a loss in genetic variability and to a subsequent increase in undesirable homozygous recessive traits.

5. In the long run, artificial selection is not very comparable to natural selection. Further, unlike the situation in natural populations, all variations are preexistant in the experimental population.

POLYGENIC INHERITANCE AND NATURAL SELECTION

1. Most genetic variability in populations is the result of **polygenic inheritance,** the combined effect of many alleles at a locus, which is the rule rather than the exception.

Three Patterns of Natural Selection

1. When their frequency is graphed, polygenic traits form what is called a **normal distribution** (a bell-shaped curve).

2. Selection occurs in three kinds of trends: **directional selection, stabilizing selection,** and **disruptive selection.**

3. In directional selection, one extreme of the phenotypic range is favored, and that phenotype increases.

4. In stabilizing selection, selection favors the mean or average phenotype. In humans, for example, natural selection favors the average birth weight.

5. Disruptive selection occurs when the environment favors individuals at either extreme of the population. Selection against intermediate types can produce a bimodal population curve.

Frequency-Dependent Selection

1. In **frequency-dependent selection,** selection favoring a phenotype is intense when it is rare, but decreases when the phenotype becomes common.

GRADUALISM VERSUS PUNCTUATED EQUILIBRIUM

1. **Gradualism** is the belief that evolution occurs very gradually or in such small increments as to be nearly unobservable.

2. **Punctuated equilibrium** describes evolutionary change occuring in relatively sudden spurts, following long periods of minor change or no change.

GENETIC DRIFT, POPULATION BOTTLENECKS, AND THE FOUNDER EFFECT

1. **Genetic drift** refers to the absence of natural selection. Random changes in gene frequencies occur, including the random loss of alleles.

2. Random changes in the gene pool can be produced by catastrophic events where there are few survivors. Following such events, allele frequencies in the population may become quite different through chance alone. Such events are sometimes called population **bottlenecks.**

3. A bottleneck effect can be created where a few people colonize a new territory. This is known as a **founder effect.** In some instances, the incidence of recessive abnormalities such as the disease **porphyria variegata,** is greater than in the parent population.

NEUTRALISTS VERSUS SELECTIONISTS

1. **Neutralists** propose that most unseen (biochemical) variation is the product of random change **(genetic drift** and **neutral mutation)** and has nothing to do with natural selection.

2. **Selectionists** maintain that any variability seen is affected by natural selection.

GENE CHANGES IN EVOLUTION

1. Molecular evolution as measured by amino acid changes is extremely slow. Humans and carp are comparable, both experiencing only one amino acid change per protein every 7 million years.

2. Great morphological changes may be caused by mutations in genes that control large blocks of other genes.

REVIEW QUESTIONS

1. Suggest how the frequency of a mutant in a clone could go from 1/100,000,000 to *one* in a single generation. (376)

2. What is it that determines just how fast new mutants can be fixed in a population? (376)

3. Explain why mutants can only be fixed one at a time in asexual populations. (377)

4. Why in sexual species is it possible for one or more mutations to become fixed simultaneously? (377)

5. How might change that is too rapid send a population of organisms into oblivion? How might sexual diploids avoid this? (378)

6. Using the example of the coevolution of the pine and the scale insect, suggest how sexuality continually aids an organism to adapt. (378)

7. Why do biologists believe that sexuality has enormous adaptive value? (378)

8. In the case of the peppered moth, what evidence suggests that predation was a significant factor in survival? (378–379)

9. Which of the peppered moth variants seems to be on the increase today? Suggest why. (380)

10. In what way does Kettlewell's experiment fail to prove that the moths were evolving?

11. Suggest a number of ways in which each human individual is genetically unique. Cite evidence from organ transplants. (380)

12. How does overdominance encourage genetic variability? In what way is sickle-cell anemia an example? (381)

13. Briefly characterize, from start to finish, the observations made in artificial selection studies of mice and chickens. (381–382)

14. What results seem to be characteristic of highly selected populations? How is this explained? (382)

15. What are polygenic traits? What are their effects on genetic variability and how significant are these effects in evolution? (382–383)

16. Draw three graphs, showing the frequency of a certain allele undergoing directional, stabilizing, and disruptive selection. (384–385)

17. Suggest a situation that might tend to encourage disruptive selection. (387)

18. What kinds of environmental changes might encourage directional selection? What other influences might favor this? (384–385)

19. Using the red and gray fish as an example, explain how selection can be frequency dependent. (387)

20. Explain what is meant by punctuated equilibrium. How might the geological record be supportive of this theory? (388)

21. Specifically, how does genetic drift differ from selection? (389)

22. Mention two ways that populations can go through a "bottleneck." How might such occasions affect the allele frequencies of the parent population? (389–390)

23. What do the neutralists maintain about variation? The selectionists? At what level do their ideas apply? (390)

Evolution and Speciation

One of the simplest questions in biology is also one of the most confounding: What is a species? You would think that the question would have been decided and dismissed ages ago, but it lingers on and continues to vex. It is like a great, swollen pillow, able to yield and to absorb attacks from any direction, and then fluff back up to its original shape. One reason that the question hasn't been answered to anyone's satisfaction is that although *everyone* has an answer, almost no one agrees.

Here, then, we will consider a few of the arguments about what a species is. We will first come up with an acceptable definition for our purposes, and then we will look at some of the mechanisms that give rise to species (whatever they are). We will also look briefly into an argument that didn't hold up and thus had disastrous consequences; see Essay 17.1. Finally, we will take a look at evolutionary relationships among living things from several angles, and consider some of the implications. Here, of course, we are dealing with the principles of *macroevolution*, the sorts of changes involved with evolution above the population level. Before we can discuss how species change, though, we must first approach the old question regarding the definition of a species.

WHAT IS A SPECIES?

As we have said, we don't have a definite answer to the question, but we should at least see what some of the authorities are saying. Often, a **species** is defined as a group of organisms that interbreed or that *could* interbreed and produce viable and fertile offspring. If organisms are to interbreed, they must have certain things in common, such as their morphology, behavior, ecology, and general physiology. Right away then, we see the assumption that members of a species must be very similar in critical aspects.

In general, a simple definition like that works fine because it tells us, for example, that African golden jackals comprise a species. They breed among each other, but they show no sexual interest in wildebeests or lions (perhaps a good thing for them). Golden jackals, then, are one of the distinct *species* you might see on a safari.

The jackals are certainly a different species from the wildebeests and lions. But are they a different species from the timber wolf? That seems like an easy enough question. In the first place, jackals don't look at all like wolves. They are adapted to hot, dry, open country, and timber wolves are adapted to cold, damp forests. Jackals live in Africa, and timber wolves live in northern Europe, Siberia, and North America. They are apparently two different species.

But maybe we should take another look. When we do, we learn that jackals can mate with domestic dogs, and the offspring are fertile. Furthermore, domestic dogs can and do interbreed with wolves, and for that matter—given the opportunity—with coyotes and even Australian dingos.

Thus, we see that according to the criterion of "potential interbreeding," timber wolves, jackals, coyotes, and dingos would all be one big, highly variable species, because all of them can interbreed successfully with domestic dogs (Figure 17.1).

But it's even more complicated than that. It turns out that there are three quite distinct species of jackals: the golden jackal *(Canis aureus)*, the side-striped jackal *(Canus adjustus),* and the black-backed jackal *(Canus mesomelas).* And while they will all breed with dogs, they apparently don't produce

hybrids with each other. In the Serengeti of Eastern Africa, the ranges of all three jackals overlap, and there they simply ignore each other. A highly territorial jackal of one species won't even bother one of a different species that wanders through its personal domain. On the other hand, the range of the coyote overlaps slightly with that of the now-rare red wolf *(Canis rufus)* in Texas, and there, wolf-coyote hybrids have been found (Figure 17.2). Still, wolves remain wolves and coyotes remain coyotes, and the two are physically quite distinct animals. Aside from this and the promiscuity of "man's best friend," then, the interbreeding situation among the eight species of the genus *Canis* appears to be more one of unwillingness rather than inability. The eight species are similar enough *physiologically* to be able to produce fertile hybrid offspring. But they are true species, reproductively isolated in nature, partly because of geographical separation and partly because their behavioral tendencies do not encourage interbreeding.

Should we just say, then, that those animals

17.1 **THE GENUS** *CANIS*

Certain problems with the species concept are shown here. Each of the *Canis* species here is considered separate, but each is known to mate readily with the domestic dog. However, they generally do not mate with the others in the group, even if they are given the opportunity. If these matings were to occur, the offspring would presumably be fertile hybrids. So are they all members of one highly variable species? The arrows between recognized groups indicate well-documented interbreeding.

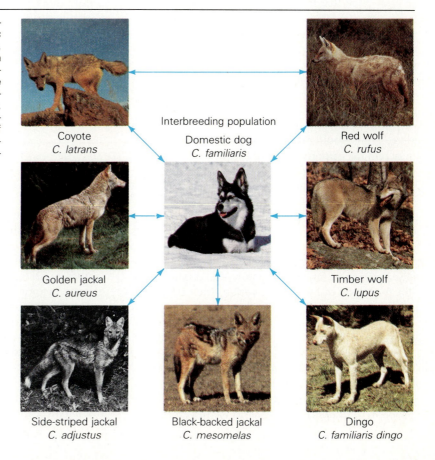

Coyote
C. latrans

Interbreeding population
Domestic dog
C. familiaris

Red wolf
C. rufus

Golden jackal
C. aureus

Timber wolf
C. lupus

Side-striped jackal
C. adjustus

Black-backed jackal
C. mesomelas

Dingo
C. familiaris dingo

17.2 HYBRIDS IN THE DOG GROUP

(a) Hybrids sometimes occur between species in nature. In Texas, where the range of the red wolf *(Canis rufus)* overlaps with that of the coyote *(Canis latrans),* natural hybrids are common, such as this 16-month-old male. (b) Interbreeding between wolves and domestic dogs is fairly common, even though they are recognized as separate species. The parents of these superb animals were a dog *(Canis familiaris)* and a timber wolf *(Canis lupus).*

(a)   (b)

that *do* interbreed successfully are of the same species, and let it go at that? It would certainly be the safest bet, but it's a bit arbitrary. Furthermore, it is so restrictive that it would be misleading. The timber wolves of Siberia and Alaska are identical, and no one has suggested that they are different species, but actual interbreeding is prevented by their geographic isolation. Among birds, the large, glistening, and raucous grackles of Texas are, in almost every important way, very similar to the smaller grackles of the West Indies. Except for size, their physical, vocal, and behavioral traits are very much alike. They would probably manage to interbreed if they had the opportunity, but the physical barrier of water keeps them apart. Though it seems inappropriate to classify them as different species, there is a total absence of evidence to the contrary.

For a natural experiment, we can turn to a group of salamanders in California. Their range extends over a circular route, and at the southern end, two variants overlap (Figure 17.3). They breed continually along their range, but where the variants from the two ends of the circle overlap, they ordinarily do not interbreed. In fact, they don't even look alike, and they treat each other as different species. Because of the continuity of their breeding range, they are considered one species containing a number of distinct geographical units called *subspecies.* So again, our definition gives us problems.

Obviously there isn't any easy answer to the problem of defining species, and we haven't begun to mention the problem of speciation in plants. We will save that tangle for further along. Actually, we could go with Darwin on the species problem and simply deny their existence, but that is not a valid scientific position today. Most biologists would not go along with it anyway because the species concept is very useful in dealing with evolution. For now perhaps the best working definition, at least for animal species, is one by the respected zoologist, Ernst Mayr: *A species is a group of actually or potentially interbreeding natural populations that is reproductively isolated from other such groups.*

REPRODUCTIVE ISOLATION

When a population of animals cannot mate with individuals of another population, the two are said to be in a state of **reproductive isolation** from each other. Such isolation may result from a number of factors such as incompatible mating structures, genetic incompatibility, differences in behavior, or separation by physical or geographic barriers. The reproductive isolation of many species of insects, for example, is due to their very specific genitalia. The penis is often bent, hooked, and barbed, and it simply won't fit into the female genitalia of any species other than its own.

Of course, if populations are separated by physical barriers, they won't be able to mate. A river can very effectively separate species of mice or other terrestrial animals that are poor swimmers. In one classic case, the Kaibab and Abert squirrels of the Grand Canyon have long been isolated by the Colorado River.

Individuals can also be reproductively isolated by genetic means. In most instances the sperm and egg are simply incompatible. Even if offspring are successfully produced by a mating between members of different species, the offspring are commonly sterile. The problems arise later when the offspring reach maturity and gamete production begins. Since their meiotic cells contain the chromosomes of two different species, they usually cannot pair up or synapse properly during prophase I of meiosis (when homologues usually find and join homologues). So, either the resulting gametes or

17.3 *ENSATINA* AND THE SPECIES ARGUMENT

Populations of the salamander, *Ensatina eschscholtzi* are found in California along the coast and in the inland mountains. Skin color and size varies more or less continuously along the range, which forms a circular shape, as shown on the map. In the southernmost end of the range, two variants overlap, sharing the same territory. While individuals of each type may live just a few yards from each other (see *E.e. eschscholtzi* and *E.e. klauberi*), there is no evidence of interbreeding between types. At the same time, studies made along the range of the salamander indicate that other neighboring variants can and do interbreed. If clinal variants are of the same species but individuals in the extremes of the range do not interbreed, are they two of the same species? Many scientists disagree on this issue; others believe the question to be meaningless, arguing that the species concept itself is invalid.

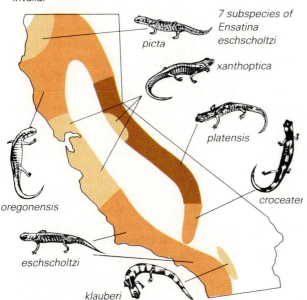

7 subspecies of *Ensatina eschscholtzi*

picta

xanthoptica

platensis

oregonensis

croceater

eschscholtzi

klauberi

the embryo, should one form, will most likely fail. Thus, horses and donkeys—very similar species—are able to produce strong, healthy offspring in the form of mules, but for the reason just mentioned, mules are sterile. (Nevertheless, a case of mule reproduction was reported in 1984, and gained a great deal of attention.) Later we will see that plants sometimes overcome the genetic limitations imposed on animals.

Behavior can reproductively isolate different species should they come in contact. Mating may involve very precise signals, and the signalling of one species simply may not trigger a mating response in another species.

How Species Are Named

Humans come in two types: those who like asparagus and those who don't. Or those who like biology and those who don't. The point is, it's only human to divide the world up, pigeonholing everything.

Pigeonholing living things has proven difficult, however, because people disagree on how large any hole is, if it exists at all, and what should go into it. We have already mentioned part of the problem: living things vary. How much can an animal vary from the others before it goes into a different hole (that is, before it gets a different name)? Another problem is that we don't know enough about relationships between organisms. How strongly related are organisms with very similar appearances? We need to know about these things, because in each hole are yet smaller holes and these have holes that are smaller yet. It's not too difficult to place most things in the largest holes. For example, if it has leaves and roots, we can safely place it in the great pigeonhole labeled "plants." If it has ears and it bites, it's definitely an "animal." But not all plants have leaves and not all animals have ears, so the story already grows complicated.

What scientists have done is divide living things into categories, and more categories within these categories. As we proceed from larger to smaller categories, the plants and animals in each grouping have more and more traits in common until they are so much alike that they can interbreed. These, then, are the species. We will say more about such categories shortly, but first we should say something about the **binomial system.**

Binomial means "two names." This simply refers to the way scientists label living things. Thus, humans are *Homo sapiens.* Our genus (or our *generic* name) is *Homo.* Our species (or specific) name is

Homo sapiens. (Note that when species are called by their scientific names—generic plus specific names—both are italicized. Sometimes, however, the generic name can be abbreviated. For example, *H. sapiens* is acceptable when the reader knows what the *H.* stands for.)

The two-name system of classification was developed by Karl von Linne (1707–1778), who latinized his own name to Carolus Linnaeus. Linnaeus is responsible both for establishing categories within categories and for naming things with two Latin names. Latin was chosen because it is a dead language, not commonly spoken anywhere outside a few academic halls or in religious ceremonies. Thus, it isn't likely to change much. Also, because it is the root of a number of present-day languages, latinized names can transcend many language barriers.

In a monumental undertaking, Linnaeus single-handedly classified and named many of the earth's creatures; today, however, the rules for assigning names are rigidly enforced by international commissions. In fact, Charles Darwin sat on the very first such commission. There are now international nomenclature commissions for zoology, botany, bacteriology, and virology (not to mention enzymology, organic chemistry, and just about any other branch of science in which names are important).

TAXONOMIC ORGANIZATION: KINGDOM TO SPECIES

The scientific discipline responsible for the assignment of organisms into species, and species to higher groupings such as classes and kingdoms is called **taxonomy,** while the determination of the evolutionary relationships is the role of **systematics.** Most taxonomists today accept the organization of species into five great categories known as **kingdoms.** (We will learn in the next chapter that a sixth kingdom is now being seriously considered.) We have loosely used the term "kingdom" from time to time, so let's now be more precise.

Kingdoms are the largest taxonomic category. If we were describing political organization, kingdoms would be the equivalent of nations. They are generally subdivided into what taxonomists call **phyla** (singular *phylum*), roughly equivalent to our own states if we continue the political analogy. Each phylum is itself subdivided into **classes** (counties?), and so the subdivisions continue. It is easier at this point to look at the total organization as we make

our way down to the species level (Figure 17.4). The increasingly finer divisions, along with a mnemonic to help remember them, are considered below.

Taxonomic Categories	Memory Aid
Kingdom	"King
Phylum*	Philip
Class	Came
Order	Over
Family	From
Genus	Greece
Species	Singing
Subspecies	Songs"

*In botanical terms—plant taxonomy—the phylum is replaced by the *division* (whereupon the mnemonic becomes, "King *David* came over . . .).

(The nonsense about King Philip, or some ribald version of it, has been memorized by generations of biology students, some who find, after many years, that it is one of the few things about biology they still remember!)

In essence, these are just names applied to groups of organisms assumed to be related to one another. Each group, from subspecies to kingdom, is called a **taxon** (plural *taxa* from *taxis*, "to put in order"). Naming, of course, is necessary if we are to talk or write about living things. But we must keep in mind that names are human contrivances, created for human purposes, and have no other importance and no separate reality in themselves. At the same time, if a name is going to be useful, it should not be totally arbitrary, but should reflect some kind of reality.

The Five Kingdoms

We should note that the organization of life into kingdoms is constantly being revised. The prevailing notion is that there are five kingdoms as proposed by R. H. Whittaker, and the five kingdoms are related to each other as seen in Figure 17.5. Other schemes have shown anywhere from two to six kingdoms. For now though, let's take a closer look at the five-kingdom scheme.

1. **Prokaryota** or **Monera:** The prokaryotes or bacteria. As we will see in Chapter 18, biologists are now in the process of reorganizing the bacteria into two kingdoms, Archaebacteria and Eubacteria.
2. **Protista** or **Protoctista:** Various simple, mostly single-celled eukaryotes, including both *protozoa* (animallike protists) and algae (photosynthesizing algal protists).

17.4 CLASSIFICATION OF TWO ANIMALS AND A PLANT

The two animals whose classification from kingdom to species are listed here represent the largest and smallest members of a diverse class, the mammalia. The great blue whale *(Balenoptera musculus)* grows to about 100 feet in length, can weigh over 150 tons, and devours 8 tons of food per day. Kitti's hog-nosed bat *(Craseonycteris phonglongyai)* weighs about 2 grams (0.07 oz), and has a wing span not exceeding 17mm—altogether about the size of a bumblebee. Both these animals have hair, are warm-blooded, suckle their young, and nourish their embryos through a placenta. As it happens, neither species is divided into subspecies. The red maple *(Acer rubrum)* is a flowering plant of considerable size and is rather closely related to those other species that are called maples.

Kingdom:	Animalia	Animalia	Kingdom:	Plantae
Phylum:	Chordata	Chordata	Division:	Anthophyta
Subphylum:	Vertebrata	Vertebrata	Subdivision:	
Class:	Mammalia	Mammalia	Class:	Dicotyledones
Order:	Cetacea	Insectivora	Order:	Sapindales
Family:	Mysticeti	Craseonycteridae	Family:	Aceraceae
Genus:	*Balenoptera*	*Craseonycteris*	Genus:	*Acer*
Species:	*B. musculus*	*C. phonglongyai*	Species:	*A. rubrum*

3. **Fungi:** Mostly multicellular, parasitic and scavanging organisms (the molds and mildews, mushrooms and toadstools, yeasts, and elements of the lichens).
4. **Plantae:** Multicellular, photosynthetic organisms from mosses to oaks.
5. **Animalia:** Multicellular animals, including sponges (parazoa) and other animals (metazoa).

There are, however, some problems with this organization. You may have noted that there is no mention of the viruses. They should probably have their own kingdom since they are unlike anything else (Chapter 18), but kingdoms are supposed to reveal evolutionary relationships, and since their evolutionary origins are unknown, we can't yet justify this step.

Within the five-kingdom scheme we find further problems, particularly but not exclusively in the protists. Kingdoms are supposed to be monophyletic units—that is, all of their members should be traceable to a common evolutionary ancestor. Yet there is no evidence that the protozoa and algal protists are descended from a single line. In fact, they are definitely polyphyletic, having evolved from different ancestors. However, they are lumped together provisionally until their evolutionary origins can be sorted out.

THE MECHANISMS OF SPECIATION

Let's assume that species exist and review some of the prevailing ideas about how species are formed. There was a time, according to prevailing theories, when the simplest forms of life arose spontaneously (see Chapter 18), but the conditions favoring these momentous events were short-lived, and today biologists are convinced that life only originates from preexisting life. Thus, species today must stem from existing ones. However, the process takes place in rather precise ways and under special conditions. We'll begin with a more detailed look at some of the influences of geography.

Ranges and Clines

The geographic area over which a species or population may be found is called its **range.** A range may be limited and compact, as we see in a population of field mice living in an isolated field. Or it may be quite extensive, covering entire continents. We see this in some bird populations. The larger ranges, of

17.5 FIVE-KINGDOM SCHEME

The five kingdoms are arranged in a manner that suggests their phylogenetic relationship to each other. The underlying kingdom, known as Prokaryota and Monera, is now being reorganized into two kingdoms, Archaebacteria and Eubacteria, containing two quite diverse groups of prokaryotes. Above this is the large assemblage provisionally called Protista or Protoctista. Some of its members are descended from the same lines that gave rise to plants, fungi, and animals.

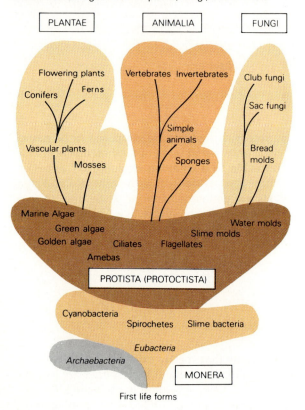

course, may encompass many different kinds of habitats, and in such cases it is possible to see the effects of natural selection as the population changes from one kind of place to the next—for example from cooler northern to warmer southern climates.

Natural selection will have predictable effects where such climatic differences exist across a range. For example, we know that among the warm-blooded animals, the birds and mammals, those living in cooler areas will be larger, which reduces the relative amount of heat-radiating surface compared to mass. (Surface-volume relationship was discussed in Chapter 4.) They will also have shorter extremities, since these, too, dissipate body heat (the Arctic fox has these characteristics while its relative in the desert, the kit fox, has huge ears, long legs, and a long tail.

The results of such differential selection *within* a species will depend ultimately on the amount of **gene flow** throughout the range. Gene flow refers to the free exchange of genes throughout a range. It can occur when individuals move from the place where they were conceived to the place where they mate, or when pollen is transported from one flower to another, or when a seed is dispersed.

If there is a great deal of gene flow throughout the species range, the species is likely to have a uniform phenotype—its genotypes will be well mixed. But when species do not tend to mate freely across a range, subpopulations may become closely adapted to their own habitat. The entire group may thus establish a **cline,** in which members of a species vary gradually and continuously across their range.

In some cases, however, the range does not vary gradually. Instead it may be marked with sharp disjunctions. Here, also, phenotypes are more likely to be discontinuous. Further, should the flow of genes be cut off between populations at one extreme or the other of the range, such as, for example, by the rise of a mountain range or the formation of a river, gene flow can be almost entirely stopped. Such isolated groups, subjected to different selective factors, are commonly the raw material of speciation. Eventually, geographic isolation may lead to physiological or behavioral reproductive changes as the isolates become more and more different. Eventually, then, should they be rejoined and the opportunity to interbreed arise, interbreeding might be impossible. Here then, we have the formation of two species where there was one.

There are numerous cases of geographic factors

A Politician Helps Explain Speciation: The Saga of T. D. Lysenko

The story of Soviet biology from 1937 to 1964 is among the most unfortunate in all of science. During this period, one man (not a good scientist, and apparently not even very bright) imposed his own crackpot, neo-Lamarckian ideas on the nation. He destroyed Soviet biological science and set back Soviet agriculture by several decades. At the height of his reign, official Soviet publications regularly bleated denunciations of Gregor Mendel, T. H. Morgan, and anyone else so corrupt as to believe in genes.

The Soviet Union had been quite advanced in genetics in the 1920s and 1930s, but this man, T. D. Lysenko, was able to send the most prestigious geneticists to slave labor camps and to silence the rest. Lysenko is now regarded as an illiterate and fanatical charlatan who was allowed absolute dictatorship over Soviet biology and agriculture.

Lysenko's scientific credentials initially rested on his 1926 report of finding that under mild winter conditions, planting peas before a cotton crop buffered the cotton against the ravages of winter. This was a modest but useful finding which Lysenko, who did have a fine instinct for public relations, tried to sensationalize. He was ignored by his colleagues then, as well as when he later claimed credit for the "discovery" of *vernalization,* a process whereby the germination of winter wheat is speeded by soaking the seeds in ice water. Peasants had been using the process for at least a century, but Lysenko claimed that *one such treatment would permanently change a winter wheat strain to a spring wheat.* The paper in which he made his announcement drew criticism because of its false claims to originality, its dependence on Lamarckian reasoning, and its aura of inflated importance. But in the Soviet Union at that time, only one opinion counted—that of Josef Stalin—and Stalin was impressed.

Pravda carried the banner headline: "It is possible to transform winter into spring cereal: an achievement of Soviet Science." It seems that the idea that the environment could change the genetic condition of organisms was in line with Marxist dogma, which stresses the idea that people are the creation of their environment. At the time, racism and claims of genetic superiority were rampant in Hilter's Germany; and Stalin, who often expressed his contempt for the traditional Russian character and "our slavish past," was trying to create what he termed "the new Soviet man." Lysenko's rejection of genetics and his claims for the influence of the environment on heredity fit well into Stalin's philosophical framework and political needs. Lysenko became Stalin's man, and Lysenkoism became the official science. In one sense, the dictator was being idealistic in his rejection of genetics, a fact we can't ignore if we want to understand the true horror of what was going on. The imposition, by brutal force, of "correct" ideology and wishful thinking on science in fact destroyed the very essence of science, and brought only farce and disaster.

Lysenko, as director of the Soviet Institute of Genetics, promoted more and more bizarre "findings." He and his growing band of opportunistic followers published "results" that showed, for instance, that wheat stalks sometimes produced oat grains, and that animal hybrids could be formed by injections of blood or plant hybrids by intraspecific grafts.

Now, one of science's most cherished but vulnerable rules is that other scientists accept reported observations at face value; we must trust each other. But this was too much. It was clear to every scientist outside the Soviet Union that Lysenko was a fraud. In official communist publications, however, angry ranting at G. Mendel and T. H. Morgan continued unabated. For a while it looked bad for Lysenko when Stalin died in 1953, but Lysenko's grip on science, biology, and argiculture increased under the rule of Stalin's eventual successor, Nikita Khrushchev.

I. A. Rapoport was among the most prestigious of the older Russian geneticists to have escaped the purges. He survived by dropping his genetics studies and moving into other areas of biology. Then one day in 1964, nearly 30 years after Lysenko's rise to power, Rapoport received an urgent call from the Kremlin. Could he write an encyclopedia article favorably describing Mendelian genetics? The startled Rapoport said that he certainly could. Could the manuscript be ready in 24 hours? Well, yes. Actually it took Rapoport 30 hours, but the next day, the first page of *Pravda* carried his detailed scientific article on the accomplishments of Gregor Mendel, complete with a laudatory banner headline. Ordinary readers may have been perplexed, but the more astute knew that this could mean only one thing: Nikita Khrushchev had been toppled from power.

Lysenko and his by now numerous followers were also stripped of power, but they retained their academic ranks and the privileges that go with them. T. D. Lysenko died in 1976 at the age of 78, probably still dreaming of a return to authority.

influencing gene flow in nature. For example, consider two Australian birds, *Rhipidura* and *Seisura* (Figure 17.6). In both species, the birds are larger in the cooler southern part of their range. (Remember, in the land "down under," the south is colder.) *Rhipidura* has a continuous distribution, with considerable gene flow throughout, and it shows a gradual cline in size (and, presumably, in the frequency of the many genes involved in determining size). *Seisura*, which is divided into three discrete populations without much gene flow between them, simply has three different size phenotypes.

Allopatry and Speciation

The formation of new species through the geographic isolation of once continuous populations is known as **allopatric speciation**. There are two primary ways that populations of a species might become isolated. One possibility is that a small group of individuals, or a seed, or even one inseminated female, might find itself in a new but hospitable place—perhaps by some accident or a navigational error in migration. Ocean islands, for instance, are occasionally populated by the descendants of unwilling, drenched, and thoroughly disgusted passengers on driftwood logs. Birds and flying insects may also be blown to some island by particularly violent storms.

The other possibility is that a geological change might divide a previously continuous range. The most fundamental earth changes occurred some 230 million years ago when the worldwide process of continental drift began, and again about 65 million years ago when Africa and South America parted company. The uniqueness of species communities in the world's present-day continents, particularly those of Australia and South America, has been the result of millions of years of allopatric speciation originating from continental drift (Essay 17.2).

The Galapagos Islands, Home of Darwin's Finches. The finches of the Galapagos Islands (see Chapter 1 and Figure 17.7) provide a well-known example of allopatric speciation. Compared to the giant tortoises, strange flightless cormorants, and impish sea iguanas living there, Darwin's 13 species of finches are not particularly

---

17.6 SIZE CLINES AND GENE FLOW IN TWO GROUPS OF AUSTRALIAN BIRDS

In both cases, the southern form, which lives in the colder climate, is larger than the corresponding northern form. In addition, the effect of gene flow on phenotypic clines can be seen by comparing variation in the discrete (separate) populations of *Seisura inquieta* with the continuous population of *Rhipidura leucophrys*. In the species with separated populations, the southern form is 22% larger than the northern form, while in the species with a single continuous population the most extreme southern form is only 11% larger than the most extreme northern form.

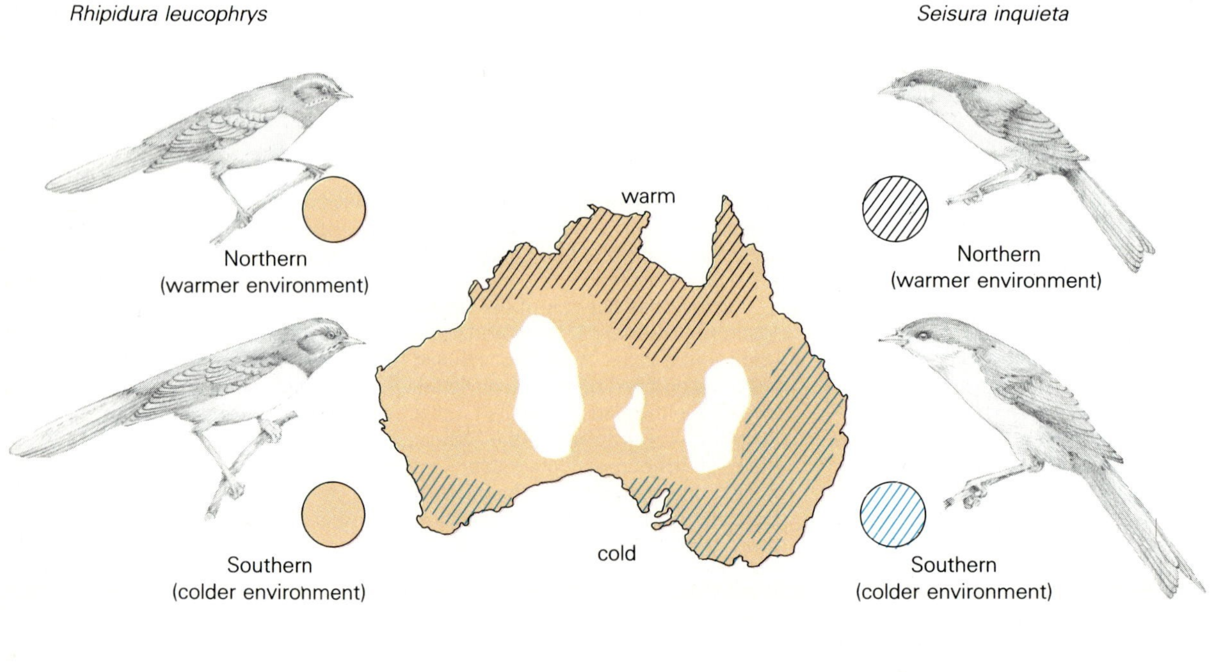

Rhipidura leucophrys

Northern
(warmer environment)

Southern
(colder environment)

warm

cold

Seisura inquieta

Northern
(warmer environment)

Southern
(colder environment)

Continents Adrift

Lower Triassic period
225 million years ago

Lower Cretaceous period
90 million years ago

Paleocene epoch 65 million years ago
END OF MESOZOIC ERA

Have you ever noticed that the bulge of Eastern South America would fit rather nicely against the western coast of Africa? So has almost every fourth-grader who ever played with a globe. But after the initial observation no one usually pursues the matter further.

No one, that is, except people like Alfred Wegener. Wegener was captivated by the shape of the continents surrounding the Atlantic Ocean, and he too noted the particularly good fit between South America and Africa. In work first published in 1912, he was able to coordinate this jigsaw-puzzle analysis with other geological and climatological data to propose the idea of continental drift. He proposed that about 200 million years ago, all of the continents were joined together into one enormous land mass which he called *Pangaea*. In the ensuing millenia, according to Wegener's hypothesis, Pangaea broke apart and the fragments began to drift northward (by today's compass orientation), to their present location.

Wegener was not treated well in his lifetime. His geologist contemporaries attacked his naivete as well as his supporting data, and the theory was pretty much discarded until about 1960. About that time, a new generation of geologists revived the idea and found new data to support it. The most useful data has been the determination of magnetism in ancient lava flows. When a lava flow cools, it permanently fossilizes a small sample of the earth's magnetic field, recording for future geologists both its north-south orientation and its latitude. Detailed maps of the positions of the continents through the ages can then be made. It now seems that Wegener's (and all those fourth-graders') insight was absolutely right. Not only did continental drift occur as Wegener hypothesized, but it continues to occur today.

Geologists have long maintained that the earth's surface is a restless crust, constantly changing—sinking and rising through incredible unrelenting forces below. These constant changes are now known to involve large distinct segments of the crust known as *plates*. At the edges of these immense masses, new ridges are constantly built up and, in response, some edges are ground down. Where continents or pieces of continents are slammed together, mountain chains have formed. When ridges are built up in the ocean floor, the oceans expand. For example, astoundingly precise satellite studies reveal that the Atlantic Ocean is growing 5 cm wider each year.

In addition to its fascinating geological implications, the theory of continental drift is vital to our understanding of the distribution of life on the planet today. It helps explain the presence of fossil tropical species in Antarctica, for example, and the uniqueness of animal life in the Australian continent and South America.

As the composite maps indicate, the disruption of Pangaea began some 230 million years ago in the Paleozoic era. By the Mesozoic era, the Eurasian land mass, now named *Laurasia,* had moved away to form the northernmost continent. Gondwanaland, the mass that included India and the southern continents, had just begun to divide. Finally, during the late Mesozoic, South America and Africa completed their separation. For a time Australia and South America remained connected through Antarctica, which was more temperate then. During this time the marsupials spread over all three continents, but were driven to near-extinction by placental mammals in all but Australia. Both the North and South Atlantic Oceans would continue to widen considerably up into the Cenozoic, a trend that is continuing today. So we see that whereas the bumper sticker "Reunite Gondwanaland" has a trendy ring to it, it's an unlikely proposition.

17.7 DARWIN'S FINCHES

The darker birds on the ground and standing on the low cactus are ground finches (9–14); those in the trees are tree finches (1–8). What suggestions can you make about the diets of the different finch species by looking at the size and shape of their bills? The tree finches are **(1)** *Camarhynchus pallidus* (the woodpecker finch), **(2)** *C. heliobates*, **(3)** *C. psittacula*, **(4)** *C. pauper*, **(5)** *C. parvulus*, **(6)** *C. crassirostris*, **(7)** *Certhidea olivacea* (the warblerlike finch), and **(8)** *Pinaroloxias inornata* (the Cocos Island finch). The ground finches are **(9)** *Geospiza magnirostris*, **(10)** *G. fortis*, **(11)** *G. fuliginosa*, **(12)** *G. difficilis*, **(13)** *G. conirostris*, and **(14)** *G. scandens*.

interesting—that is, not until the saga of their evolution is revealed.

The finches are all 10 to 20 cm long, and both sexes are drab-colored browns and grays. Six are ground species, each feeding on different, appropriately sized seeds or cactus, while six are tree finches. In each species the beak has become modified for its specific diet. One of the strangest tree-dwellers is the woodpecker finch. Lacking the long, piercing tongue of the woodpecker, it uses long, barbed cactus spines to pry insect grubs out of cracks and crevices in the trees. The 13th species at first looks more like a warbler than a finch. This species lives isolated on distant Cocos Island, a miserable, mosquito-ridden, outlying lump of an island in the Galapagos group.

On the *Beagle's* historic visit to the Galapagos Islands, Darwin collected everything he could find or catch, including the ordinary little brown finches. He took no special interest in them, but a London bird taxonomist later examined the specimens and noted that the inhabitants of different islands, though very similar to each other, were clearly different species. Ordinarily, birds living in one locality tend to be very different from one another. Whereas these birds were all clearly finches, Darwin had seen many of them doing things that finches don't ordinarily do.

Why have these little birds become so important to our understanding of speciation? Darwin concluded (and his conclusions are backed by years of careful research by others since) that the different species of birds were all descended from the same stock (Figure 17.8). Long ago, about 10,000 years ago by current estimates, the volcanic islands were colonized by South American finches that probably were blown out to sea by a storm. Apparently conditions on the islands were favorable and the "castaways" flourished. Their descendants eventually populated all the islands by occasional island hopping. However, the island hopping was rare enough to ensure the virtual isolation of each population. What followed then is referred to as **adaptive radiation**—the branching of an evolutionary line through the invasion of new environments, or through the partitioning of an existing environment.

According to our scenario, the little birds had the islands to themselves, as far as they were concerned. There was a great variety of food. They were already well adapted for foraging for small seeds on the ground, but there were other plentiful untapped food resources—food not ordinarily eaten by finches.

Soon enough the expanding populations were depleting the available supply of small seeds. Thus, natural selection began to favor birds that could also cope with larger seeds and with other food resources. In time, the birds' bill sizes began to change as each population began to adjust itself more closely to the different kinds of food found on each island. Natural selection was working its way. Eventually the isolated birds of the different islands differed genetically to the extent that any island hoppers would find themselves to be reproductively incompatible with the residents of other islands. Clearly, speciation, the formation of new species, had occurred.

As differences leading to speciation appeared, there would have been reduced competition. If the genetic differences led to differences in resource utilization, competition might become so low that two emerging species might be able to coexist on the same island. Their coexistence would also set the stage for further change. For example, with two species of finches trying to survive on one island, natural selection would favor the individuals in *each* population that were as different as possible from those in the other population, thereby further reducing competition between them. The tendency for differences in competing species to become exaggerated as each specializes is called **character displacement** (Figure 17.9).

After thousands of years of finches occupying

17.8 DIVERGENCE OF DARWIN'S FINCHES

The numbers refer to the birds in Figure 17.7. The scheme shows the division of tree finches, ground finches, warbler finch, and isolated Cocos finch into separate branches.

Ground finches, *Geospiza*

Tree Finches, *Camarhynchus*

Warbler finch, *Certhidea*

Cocos finch, *Pinaroloxias*

Original arriving pair of South American finches

17.9 CHARACTER DISPLACEMENT

Character displacement is revealed in beak size changes in three species of Galapagos finches, *Geospiza fuliginosa, G. fortis,* and *G. magnirostris.* Where all three species are found together (graph *a*), the ranges in beak sizes are clearly separated with no overlap. Where *G. magnirostris* is absent (graph *b*), its beak size range is overlapped by *G. fortis,* but notice that it and *G. fuliginosa* remain distinct. Where *G. fortis* and *G. fuliginosa* exist separately and alone (graphs *c* and *d*), their beak size ranges completely overlap.

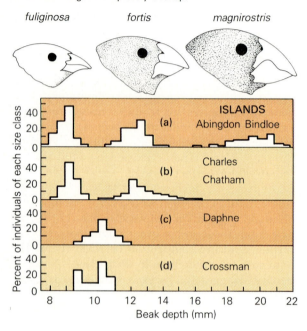

the dismal Galapagos Islands and separating, changing, specializing, and rejoining, the different populations today are totally unable to interbreed. This means that several of the species can exist side-by-side on every island. Each species uses the resources of the island in its own unique manner, in some cases filling niches that are occupied by other kinds of birds on the mainland. (A similar case of adaptive radiation is seen in the birds of another archipelago—the honeycreepers of Hawaii.)

Darwin's finches, then, nicely illustrate both allopatric speciation and the important evolutionary process of adaptive radiation as they spread out into new ranges, each group adapting to its own habitat. Allopatric populations, as we have seen, are clearly separated and it is under such conditions that most speciation takes place. The separation, of course, would clearly impede or prevent interbreeding and, likewise, gene flow between the populations. The separation would also encourage increased differences as each adapted to its own conditions and as different ran-

dom allele-frequency changes occurred in each group. However, although geographic separation certainly facilitates speciation, we will now see that under special circumstances, speciation can occur in the absence of isolation.

Sympatry and Speciation

Sympatry refers to populations occupying the same geographical area; allopatry, to those occupying different geographical areas. While allopatric speciation—the formation of new species in geographically isolated groups—is by far the most common route of speciation, **sympatric speciation**—speciation within a single habitat—can indeed occur. The best examples are found among plants.

Sympatric Speciation in Plants. Sympatric speciation is, in fact, quite common in some plants. Among the flowering plants, in particular, there are many examples of new species arising by the hybridization of existing species. We have mentioned that, in animals, hybrid offspring, such as mules, are usually sterile. The mule's sterility can be traced to gametogenesis, where during prophase I of meiosis, the horse and donkey chromosomes are different enough so that they fail to synapse (see Chapter 11). Pairing up is essential for both crossing over and for the proper alignment and separation of homologous chromosomes in the first division of meiosis. Such a failure results in abnormal, random chromosome separations in anaphase, and eventually in abnormal gametes.

Why don't hybridizing plants have the same problem? First of all, plant species that appear to be very different physically may be genetically similar enough that hybrids between them can undergo normal or nearly normal meiosis. Where the ranges of such plant species overlap, there may be extensive mixing and partial hybridization, thus forming highly variable **hybrid swarms.** Fertile hybrids, of course, allow the exchange of genetic material between species and stretch the traditional working definition of species. However, such hybridization between "good" (established and recognized) plant species is common, and, surprisingly enough, doesn't seem to result in the breakdown of either species or their merging into a single species. This could be because hybrid swarms occasionally find their own niche to fill and become new, distinct species.

Polyploidy: Autotetraploids and Allotetraploids. Plant hybrids can succeed dramatically through **polyploidy,** in which whole sets of chromosomes become doubled. This happens spontaneously from time to time in the mitotic divisions of the growing plant. In the process, the chromosomes double in preparation for cell division, but for some reason the cell and the cell nucleus fail to divide. Thus, the abnormal cell and all of its progeny will have four complete sets of chromosomes. The abnormal cell is now called a **tetraploid** cell, and its progeny will sometimes form the tissue of an entire tetraploid branch, complete with tetraploid flowers.

When this happens in an ordinary diploid plant, the resulting tissues (or whole plants) are called **autotetraploids** ("self-tetraploid" or "self-four-genomes"). Meiosis in autotetraploid flowers is, once again, abnormal. The chromosomes will still try to pair two-by-two, but each chromosome will now have *three* other homologues. So clumps of homologous chromosomes form, and meiotic segregation becomes jumbled and imprecise. Perhaps a few balanced seeds will be produced, but autotetraploids are comparatively infertile. They probably have an extremely limited role in evolution.

Tetraploidization in hybrid plants is different. As we just saw, the reason most hybrids fail in meiosis is that the chromosome sets from the parental species fail to recognize one another and to synapse properly. Because the hybrid plant cell nucleus has two different haploid chromosome sets, some chromosomes simply can't find a homologue. But when spontaneous tetraploidization occurs, there will suddenly be two different but complete *diploid* chromosome sets. Such a tetraploid is called an **allotetraploid** ("other-tetraploid"). When flowers form in the allotetraploid hybrid, there is no longer a compatibility problem in meiosis, since every chromosome now has a homologue (a matching pair member) and meiosis can then proceed normally (Figure 17.10).

The pollen and ovules of an allotetraploid will be diploid, but the plant can pollinate itself to reconstitute new, entirely allotetraploid seeds and new, entirely allotetraploid and entirely fertile plants. But the allotetraploid plants are fertile only with themselves and with each other; if they are crossed to either parental species, with its haploid pollen and ovules, they produce only *triploid* seeds, which are infertile and often inviable. This reproductive isolation means that the new self-fertile allotetraploid plant constitutes, in fact, *an instant species*. The new species will soon fail, of course, unless it happens to fill a new niche or unless it can outcompete another plant species—perhaps one of its parents—in its own territory.

Polyploidy can occur repeatedly. Thus, we find **hexaploids** (six genomes, or three doubled copies of three different parental genomes), **octaploids,** and so on (the term *polyploid* covers any level of chromosome duplication higher than diploid). All of this may seem freakish, but in fact it is not even a rare process in flowering plants. Many plant species, both

17.10 HYBRID STERILITY

While a plant hybrid may readily be produced in nature, it is often sterile. Sterility arises when, during meiosis, unmatched chromosomes fail to find their homologues. Here, species A is successfully pollinated by species B, producing a hybrid. Since there is no match in the chromosome complement, the hybrid cannot carry out meiosis and is therefore sterile. However, should a spontaneous doubling of chromosomes occur, as it sometimes does, normal pairing-up and meiosis occurs and normal gametes are produced. Following self-fertilization, a new viable species is produced.

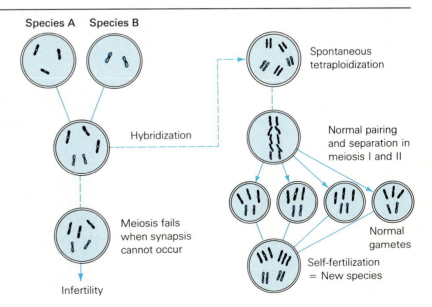

Species A   Species B

Spontaneous tetraploidization

Hybridization

Normal pairing and separation in meiosis I and II

Meiosis fails when synapsis cannot occur

Normal gametes

Self-fertilization = New species

Infertility

wild and domestic, are polyploids. Wheat, for example, is actually an allohexaploid, a genetic combination of the chromosome complements of three entirely different Middle Eastern species of wild grasses. *Triticale* (Figure 17.11) is a human-engineered application of the tendency of wheat and rye grasses to form allotetraploids. In the final analysis, it turns out that polyploidy is an excellent mechanism for increasing plant diversity through rapid speciation.

For reasons not fully understood, polyploid species may have much greater tolerance for harsh climatic conditions. In one study, for instance, only 26% of the plants sampled in lush tropical regions were polyploids, while 86% of all the flowering plant species in the raw environs of northern Greenland were polyploid.

Thus, hybridization followed by polyploidization creates instant, sympatric species of flowering plants, ready to be tested by the forces of natural selection. This versatility helps explain, hypothetically at least, how flowering plants arose rather abruptly in evolution and very quickly radiated out over the landscape to create the incredible diversity of plant species that dominate our world.

Sympatric Speciation in Animals? While there are polyploid species of animals, hybridiza-tion is not likely in species that have sex chromosomes because, in polyploid meiotic cells, the Xs tend to pair with Xs and the Ys with Ys, instead of the normal XY pairing. This means the gametes produced might be XX or YY, so they are unbalanced for sex chromosomes. There are many polyploid fish, reptile, and amphibian species, however, since gender, in many of these species, is determined by factors such as temperature and not by chromosome ratios.

Evolutionary theorists argue over whether sympatric speciation could occur in animals in ways other than polyploidy, but nowhere among the animals is the situation as clear-cut as in the polyploid plants and hybrid swarms. In fact, there is no solid evidence that speciation in animals ever follows a path other than that of allopatry.

MAJOR EVOLUTIONARY TRENDS

How can one put the concept of speciation into the larger context of evolutionary history? Let's begin by taking a look at some of the major trends in evolution as revealed by the fossil record and by comparing living animals.

17.11 MIRACLE GRAIN

Triticale is a hybrid grain produced by crossing wheat and rye. (The name is a combination of *Triticum* and *Secale,* the genus names of the two.) Actually, the hybrid formed is sterile because of the usual chromosome incompatibility. But agricultural scientists solve the problem by treating the hybrid with *colchicine,* a mitosis inhibitor that causes a doubling of chromosomes—synthetic tetraploidy. Following this treat-ment, the treated plant will carry out a normal meiosis and produce viable gametes and a viable F_2, F_3, and so on. Triticale combines the vigor of rye with the high grain yield of wheat. Further, the genes of the rye add the amino acid lysine to the usual high quality of wheat protein, resulting in a more nutritious product.

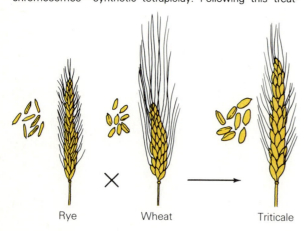

Rye   Wheat   Triticale

17.12 DARWIN'S TREE OF LIFE

The growing tips of twigs on Darwin's phylogenetic tree represent living species. These branch from recent ancestral types, which in turn diverge from more primitive ancestors. The trunk itself represents the original ancestral form of life. Dead twigs are extinct species, and some of the lower branches are entire extinct groups. The progressive division into trunks, limbs, and branches may correspond to higher taxonomic groupings. The concept is useful for illustrating relationships among various groups or species and is frequently used for this purpose.

Divergent Evolution

We used the term *adaptive radiation* to describe the manner in which speciation occurred in the Galapagos finches. The finches spread into new ecological niches, and in the process of adapting they became less similar to one another. Thus, one way in which adaptive radiation expresses itself is through **divergence** or **divergent evolution,** which simply means that newly emerging species tend to become increasingly different from each other over time. The idea of divergence has strong ecological overtones since it often occurs when a species finds some new and different way of using the environmental resources, thus establishing a new ecological niche or, in Darwin's phrase, a new "place in the polity of nature." The opening of new niches is greatly encouraged by natural selection, since it relieves (for a time) the unrelenting competition for resources. Such divergence has suggested a scheme called the **phylogenetic tree,** as we see in Figure 17.12.

The long-range outcome of divergence in evolutionary history can be visualized through the analogy of the highly branched phylogenetic tree. The outermost, green twigs represent species that are alive today, while the branches from which the twigs originate are their last common ancestors—the ones that founded the different genera. Moving inward, we find older branches, which themselves have originated from previous divergences. All branches, with the exception of the green twigs, represent ancestors that no longer exist—that have become extinct—both the ones that have speciated into more advanced forms and the dead-end species that produced no living descendants. These tree analogies are very commonly used to represent specific evolutionary hypotheses about the origin of species and of higher taxonomic groups.

Monophyletic, Polyphyletic, and Paraphyletic Groups. In **cladistics,** which is one version of the ideal evolutionary taxonomy, every taxon (any taxonomic group, such as kingdom, or species) is represented by a part of a branching tree, and any branch can be, metaphorically, broken off at a single point and waved around as a single, discrete unit. In such a system, each taxon would have a single ancestral species that *would also* (if it were still alive) *fit the definition of the group,* and if any branch were broken off, there would be no members of this group left on the tree. *Every* living member of the taxon would be more closely related to every other member than *any* would be to the member of another taxon. Such an ideal group is said to be **monophyletic.**

For instance, the vertebrate class *Aves* (birds) is a monophyletic group (Figure 17.13). As far as we can tell, all birds have descended from some single species, and that ancestral species was itself a true bird. Thus, any two birds are more closely related to each other than either is to any other animal. Similarly, the mammals appear to constitute a monophyletic group. However, some dissenters theorize that the monotremes (egg-laying mammals, such as the platypus and the echidna) arose from reptilian ancestors independently of the other mammals. If, indeed, the last common ancestor of the monotremes and the other mammals should prove to have been a reptile, then we would have to say that the mammals constitute a **polyphyletic** group—a group with more than one point of origin. That is, they now have similar characteristics that cause us to label them as mammals, but they arose from different reptilian lines and independently evolved similar features (see Chapter 27).

17.13 THE PHYLOGENETIC TREE OF TERRESTRIAL VERTEBRATES

The branches bearing the mammals and birds illustrate what is meant by *monophyletic* groups. Each group has a common ancestor of its own that gave rise to no other groups (*a for the mammals, *b for the birds). The reptiles are an entirely different matter, as you can see by the multiple branching. The reptiles are a *paraphyletic* group: all are descended from a common ancestor that was also a reptile (*c); but this common ancestor also gave rise to birds and mammals, which are not reptiles. Today's amphibians represent a mystery, but may be a *polyphyletic* group with several fishlike ancestors.

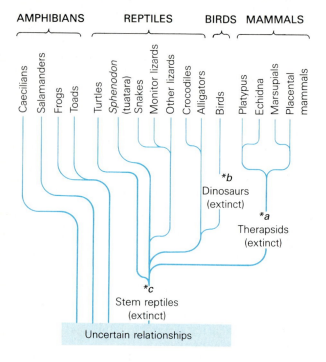

A clearer example of a polyphyletic group is the protists. Some members are often included among the plants, others with the animals. Still others have been and still are carrying on the roles of the fungi. The incorporation of such diverse organisms into kingdom Protista is an uneasy truce at best until they are further sorted. Protista is almost certainly a polyphyletic taxon.

Rigid adherence to the ideal of monophyletic groupings clearly doesn't always work. Even if convenient polyphyletic groups could be broken up successfully, we would still need another kind of group, the **paraphyletic** taxon. In a paraphyletic group, every member is descended from a common ancestor that also belongs to the group, just as in any monophyletic taxon. However, a paraphyletic group is one that has itself given rise to other groups that have diverged to the extent that they no longer belong to the parent group.

A familiar paraphyletic group is the vertebrate class *Reptilia*, the reptiles. Reptiles have a number of features in common, and they are all derived from one ancestor that was also a reptile. But some reptilian ancestors gave rise to nonreptiles, namely, the birds in one case and the mammals in another. In Figure 17.13, note that one line of descent leads to the crocodilians, dinosaurs, and birds, while other, earlier departures lead to turtles, mammals, and snakes and lizards.

If we quantify "closeness" of evolutionary relatedness in terms of recentness of common ancestry, then birds are more closely related to alligators than they are to lizards. More to the point, alligators are more closely related to birds than they are to lizards, even though alligators *look* a lot more like lizards. Appearances can be deceiving.

While we're on the subject of lizards, they too form a smaller paraphyletic group. The monitor lizards are believed to be more closely related to snakes than they are to other lizards. On the other hand, the "glass snake" is not really a snake at all, but a legless lizard.

Paraphyletic taxa are common in the plant kingdom. The two major groups among the flowering plants are the **monocots** (grasses, palms, lilies, orchids, and so on) and the **dicots** (beans, apples, magnolia, clover, geraniums, oaks, avocado, and many more). The two groups are named for the number of **cotyledons** ("seed leaves") in the plant embryo. The monocots are a monophyletic taxon. Although the origin of the monocots is obscure, it is believed that the monocots branched off from one group of the dicots, making the dicots a paraphyletic taxon. Thus, some dicots are more closely related, in time at least, to the monocots than they are to certain other dicots. Among the nonflowering plants, the living gymnosperms (ginkgo, cycads, and conifers) are also a paraphyletic group.

Tracing Evolutionary Origins Through Homology and Analogy. One of the important ways in which evolutionary origins can be traced is by comparing the physical structure of different species, an area of study known as **comparative anatomy.** The goal is to see how closely related the species are by noting similarities in their physical traits. One may begin, for instance, by studying the segmented musculature of a tiny fish and then follow these segments through increasingly complex specializations, finally to arrive at mammalian musculature with a clearer idea of how such a system could have evolved. The task is not always easy, nor are the conclusions immediately evident.

Often only the fossil record retains the necessary clues. *Homologous* bones, muscles, and other organs have undergone extensive modifications in different lines. Homology is suggested if there are intermediate forms between two existing species, or if the fossil record indicates that particular structures on two living forms have descended from a common ancestral structure. Homology is also indicated if structures being considered in two species have a common embryonic origin. The bones of the forearm, for example, show a fairly clear homology (Figure 17.14). Even obscure homologies can be traced by these methods. The tiny bones of the middle ear are homologous with reptilian jaw bones, which in turn evolved from the bony supports of the gills of ancient fish.

Thus, the forelimbs of a whale, on close examination, reveal this great beast to be not so distantly related to a bat or a human. Since the bones of the forelimbs of these disparate creatures form embryologically in similar ways, they are said to be homologous.

Can we consider the pectoral fin of a shark or a perch as homologous with the flipper of a whale? Figure 17.15 shows that the superficial resemblances of the structure and the similarity in how they are used are deceiving. They have embryologically different origins and thus are not evolutionarily related. Such structures are called **analogous** and analogy is not very useful in constructing phylogenetic trees. Insect and bird wings, though serving similar functions, are clearly analogous.

17.14 HOMOLOGOUS STRUCTURES

The forelimbs of several representative vertebrates are compared here with those of the suggested primitive ancestral type. In each case, individual bones have undergone modification, although they can still be traced to the ancestor. Since they all have the same embryological origin, they are said to be *homologous*. The most dramatic changes can be seen in the horse, bird, and whale. Many individual bones have undergone reduction and, in some instances, fusion. Another interesting modification is seen in the greatly extended finger bones of the bat.

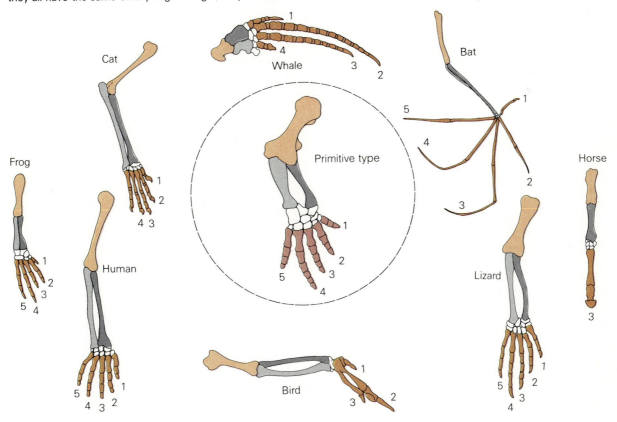

Modern Construction of Phylogenetic Trees.
The animal and plant kingdoms are so vast and
varied that we are often at a loss to describe the
pedigree of a living species. The fossil record is
simply incomplete, or absent. Therefore, constructing a phylogenetic tree is quite a task. If we
simply lump living species, we end up with incorrect trees. For example, lizards and crocodiles
would be positioned on the same limb, which, as
we mentioned earlier, wouldn't be correct. A
somewhat better procedure is to construct all possible trees and choose the one that requires the
smallest number of evolutionary changes. We
could call this the **minimum mutation tree.** The
problem is, nature does not necessarily minimize
the number of necessary evolutionary changes.
Further, for groups of any size, the number of possible trees is astronomical; for only 15 living species, there are exactly 7,905,853,580,625 theoretically possible trees! For 16 species there are 27
times that many. Even computers can't evaluate
that many trees. Thus, taxonomists must take
some shortcuts, such as starting with smaller trees

and adding a branch at a time. Even then, it takes a
high-speed computer to consider just the more
plausible relationships. Figure 17.16 is based on
such a computer analysis.

Convergent Evolution

While one may be struck with the trend toward
divergence in long-term evolution, it is not the only
trend. In some cases quite different species may
grow more alike. This should not be unexpected in
species that are adapting to similar environments.
In fact, the selective action may be so similar that
quite different species can come to resemble each
other in many ways. The process is called **convergent evolution.** Convergence occurs on all levels,
from the biochemical to the morphological. Thus,
similar enzyme systems, or perhaps photosensitive
pigments may arise; or, alternatively, similar limb
modifications, coloration, or body shape may be
seen.

Darwin was struck by the evidence of convergent evolution in his lengthy travels—evidence that

17.15 ANALOGOUS APPENDAGES

The pectoral fin of the shark and the flipper of the whale serve
similar functions and are generally similar in appearance.
However, they are not homologous limbs. In the whale, the
paddlelike flippers contain highly modified hand bones of the
tetrapod, (four-footed), land-dwelling ancestor—with five finger bones, two of which are greatly extended. In the shark's

pectoral fins, the supporting elements are of cartilage, and
bear no resemblance to the hand bones of the whale. Sharks
evolved fin structures long before the first tetrapods appeared
on the land. Structures of similar function in animals, but of
different origin are described as analogous.

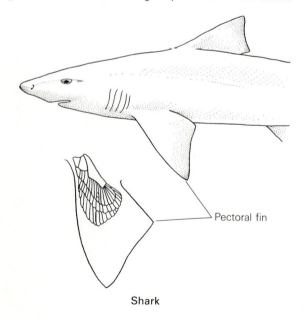

Pectoral fin

Shark

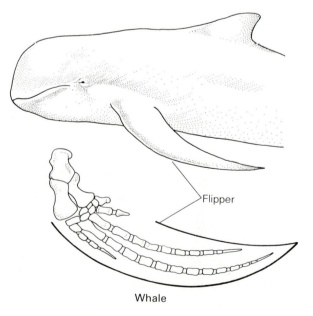

Flipper

Whale

17.16 CYTOCHROME c PHYLOGENY

This computer-generated phylogenetic tree represents amino acid differences in the polypeptide sequence of respiratory pigment cytochrome c, which is found in all aerobic organisms. Each amino acid difference represents at least one DNA mutation that has been "accepted" in evolution. The estimated number of DNA changes (numbers on the tree) determine the evolutionary distance from one organism to another. In general, the tree derived from one short protein is consistent with other trees based on morphology or the fossil record, although the computer seems to think that the turtle is a bird.

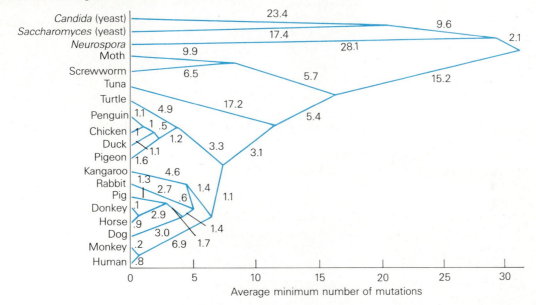

Average minimum number of mutations

suggested the whole idea of evolution in the first place. He recorded in his journal that the South American mara or Patagonian hare, *Dolichotis patagonum*, was quite similar in both appearance and ecological niche to the European rabbit (see Figure 1.3). Close examination of the mara revealed that it was not a rabbit at all, but a rodent, and a particularly South American rodent at that. Thus, it is actually more closely related to guinea pigs.

Convergence is dramatically illustrated in comparisons between placental mammals and the distantly related, geographically isolated marsupial (pouched) mammals of Australia. Marsupials have established niches very similar to those of many placental mammals of other continents. Through a long period of isolation, selection, and adaptation, pairs of unrelated species have often taken on a striking resemblance to one another. Thus in Australia we have the rabbit bandicoot, the marsupial mouse, the marsupial mole, the flying phalanger (which very much resembles a flying squirrel), the Tasmanian wolf, and the banded anteater (Figure 17.17). All have a placental counterpart on other continents.

Coevolution

Quite often the direction of evolution in one species is strongly influenced by what is happening in the evolution of another, particularly if either species is dependent on what the other has to offer. The most obvious examples are evolutionary directions seen in predator-prey relationships. As natural selection improves the predator's skill, the prey also adapts. Selection on the prey favors behavior that will improve ability to escape or to escape detection. As the prey develops new ways to avoid the predator, the predator population also changes in ways that increase its probability of hunting success. Such reciprocal influence by two species, in which they influence each other's adaptive directions, is called **coevolution.**

We see similar adaptations between parasite and host species. An animal may harbor a number of parasite worms without obvious ill effect. Such parasites have succeeded in infiltrating the host's defenses and the host no longer exhibits the drastic immune system responses seen when the two species first came together. But should a parasite find

17.17 CONVERGENT EVOLUTION

The concept of convergent evolution is well supported by observations of Australian mammals. Although phylogenetically unrelated to their counterparts in other continents, they bear a striking resemblance to many. All of these animals are marsupials—pouched mammals whose young complete their development in the pouch. Notably absent in Australia is a marsupial bat—there's no such animal. The niche is filled by numerous placental types, clearly related to bats in other continents. Unlike other terrestrial mammals, bats are not easily isolated by water barriers, and apparently they made an island-hopping migration from Asia sometime in the past.

A placental flying squirrel *(above left)* and a marsupial sugar glider *(above)*.

Placental two-toed sloth *(far left)* and marsupial cuscus *(left)*.

The placental wolverine *(right)* and the Tasmanian devil *(far right)*.

Queensland tube-nosed bat *(below)*.

Placental hare *(below center)* and marsupial rabbit bandicoot *(below right)*.

itself in the wrong host, as is seen when the dog tapeworm finds itself in a human, the immune reaction is immediate and widespread, and the person becomes quite ill. Both predator and prey, and parasite and host relationships reveal unending adaptive adjustments.

Some coevolution leads to more peaceful ends. Evolution of the great array of flowering plants was influenced by the coevolution of insect pollinators, so that today some flowering plant species are selectively pollinated by only one insect species. The yucca is entirely dependent on the yucca moth for pollination, and the yucca moth in turn is entirely dependent on the yucca flower, where it lays its eggs and where its larvae grow on a diet of yucca seeds. The yucca moth goes to a considerable amount of trouble to assure pollination, since only properly pollinated flowers will produce the seeds the moth larvae need.

An entertaining example of herbivore-plant coevolution is that between passion-flower vines and a kind of butterfly (*Heliconius*) whose larvae specialize in eating its leaves (Figure 17.18). This plant has succeeded in manufacturing poisons that prevent most other insects from devouring its leaves and young shoots, but the butterflies have evolved the necessary biochemistry to detoxify the poisons. The female lays bright yellow eggs on the young leaves and shoots. This bright color is a warning to other female butterflies that it would be wiser for them to find their own fresh leaves or shoots, and the other butterflies have evolved the response to avoid laying eggs where there are eggs already, thereby cutting down the competition their own larvae will face.

Infestation by toxin-resistant butterfly larvae is destructive to the passion vine, and in response, it has undergone the evolution of a number of its own defenses. The plant takes advantage of the butterflies' selective egg-laying behavior by producing perfectly good *mimic* eggs on the leaf surfaces—little round lumps of bright yellow tissue scattered randomly on the leaves and shoots. The mimic eggs tend to persuade the female butterfly to hunt elsewhere. The passion vine has gone one step further in its coevolution with the destructive butterflies. The same vines produce variable leaves, leaves with very different shapes, apparently confusing the butterflies. Further, some of the leaf shapes mimic the leaf-shriveling that usually accompanies larval infestations—and female *Heliconius* avoid shriveled leaves.

The concept of evolution, we see, encompasses a range of complex and interactive ideas. Since its inception, it has also provided us with a host of questions, enduring puzzles that continue to resist our intellectual attacks. Someday we may know the answers to all our vexing questions, about how and why life changes and how the various forms are related. And yet, perhaps not. It is possible that some of the events of our biological past will never be unraveled, and that these same questions will remain forever as unweathered monuments to human limitation.

17.18 COEVOLUTION

The passion-flower vine (*Passiflora*), continues to evolve new strategies to resist the parasitic activity of the *Heliconius* larva. The yellow spots, randomly arranged on some leaves (*below left*), are actually displaced nectar glands that resemble the butterfly's eggs (*bottom*), and discourage females from laying eggs on leaves so adorned. In addition, the passion-flower produces a variety of leaf shapes, imitating those of other nearby forest plants. Such mimicry may fool the butterfly into continuing her search for the right plant. These defenses could limit the *Heliconius'* success and conceivably lead to its extinction since it is highly specific for the type of plant food eaten by its larvae. However, predators are also subjected to selection, and it wouldn't be surprising to learn of behavioral countermeasures evolving in the butterfly populations.

APPLICATION OF IDEAS

1. The fossil record for nearly all forms of multicellular life on earth begins rather abruptly shortly before the Cambrian period of the Paleozoic era. This raises numerous questions and in the minds of some, adds fuel to the creationist's arguments. Forgetting "special creation" for the moment, offer two different hypothetical explanations for the seemingly abrupt appearance of such life.

2. Until recently, phylogenetic trees were hypotheses that could only be tested in terms of the characteristics contrived by the tree's originator. This is no longer true. Describe two important discoveries of this century that offer other ways of testing these hypotheses and explain how such testing is done.

3. In assigning organisms to various taxa (or in determining which taxa they have previously been assigned), a dichotomous key is commonly used. Find an example of a dichotomous key in the literature and explain how one is used.

KEY WORDS AND IDEAS

WHAT IS A SPECIES?

1. The usual criterion for grouping organisms into a **species** is the ability to interbreed. For this they must share many characteristics such as those related to physiology, anatomy, behavior, and ecology.

2. Members of the genus *Canis*—the "dog" species—cloud the species issue for although they are considered distinct species, a certain amount of interbreeding is possible.

3. Members of a species from the extreme ends of long ranges may differ enough to fail to interbreed, even when given the opportunity.

4. Mayr's definition of species represents a truce among biologists: "A species is a group of actually or potentially interbreeding natural populations that is reproductively isolated from other such groups."

REPRODUCTIVE ISOLATION

1. Populations that cannot interbreed are said to be **reproductively isolated.** Many factors contribute to this condition.

2. Hybrid offspring such as mules may be produced by the mating of two species, but the offspring are commonly sterile. During the meiosis I of gametogenesis, the parental chromosomes fail to synapse and separate properly and any gametes formed are abnormal.

How Species Are Named

1. Organisms are named according to the **binomial system,** which includes two names: **genus** and **species.** The first is capitalized and both are latinized and italicized.

2. The binomial system was originated by Carolus Linnaeus in the 18th century. Today, the naming of species is supervised by commissions.

TAXONOMIC ORGANIZATION: SPECIES TO KINGDOMS

1. **Taxonomy** is the science of classification, and **systematics** is the science of evolutionary relationships.

2. The largest **taxon** is the **kingdom.** In our scheme, there are five kingdoms and these are subdivided into descending levels of taxa: **phylum, class, order, family, genus, species,** and **subspecies** (sometimes **varieties** and **races**).

3. The kingdoms include (a) **Prokaryota** or *Monera* (bacteria; now *Eubacteria* and *Archaebacteria*), (b) **Protista** (protists), (c) **Fungi** (molds, mildews, mushrooms), (d) **Plantae** (plants), and (e) **Animalia** (animals).

MECHANISMS OF SPECIATION

Ranges and Clines

1. A **range** is the geographic area a species inhabits. In large ranges where climatic differences occur, populations of the species tend to vary from one region to the next.

2. In colder regions, species tend to be larger, but with shorter extremities.

3. Differences within species depend on extent of **gene flow.** Where the gene flow is great, uniformity results.

4. In ranges that are geographically disrupted, and gene flow limited, variation within a species increases, and a **cline** may form. Isolated members may undergo speciation.

Allopatry and Speciation

1. **Allopatric speciation** is the formation of new species through the geographic isolation of groups from the parent population (as occurs through colonization or geological disruption).

2. Historically, the drifting apart of continents produced major instances of allopatric speciation.

3. Darwin's Galapagos finches represent a classic case of allopatric speciation, since all 13 species are believed descended from a single population on mainland South America.

   a. Each finch species originated from groups that became established and isolated on various islands where their adaptations differed somewhat. The formation of new species, arising from a common ancestor, resulting from their adaptation to different environments is known as **adaptive radiation.**

   b. Today's finch species are reproductively isolated and are said to have undergone **speciation.** Where similar finches coexist, minor genetic differences tend to become exaggerated in what is called **character displacement.**

Sympatry and Speciation

1. **Sympatric speciation** occurs within a population and without geographical isolation. It is rare in animals but not in plants.

2. An important mechanism of sympatric speciation is the formation of hybrids and subsequent arising of **polyploids.** Large numbers of partial hybrids are called **hybrid swarms.**

3. In an ordinary diploid plant, **tetraploidy,** or **autotetraploidy,** will cause meiosis to fail since the four homologues in each group will synapse abnormally.

4. When a spontaneous chromosome increase occurs—chiefly a doubling through failure of the chromosomes to separate in mitosis—otherwise infertile hybrids can become fertile. Doubling produces the tetraploid (or **allotetraploid**) condition, which means that the chromosomes then have homologues. Thus, meiotic synapsis and crossing over occur and normal gametes can be produced.

5. Allotetraploids are reproductively isolated. They are incompatible with other members of their species and must self-fertilize. They are essentially new species.

6. Polyploid species tend to tolerate harsh climatic conditions and in such regions form the majority of species.

MAJOR EVOLUTIONARY TRENDS

Divergent Evolution

1. In **divergence** or **divergent evolution,** newly formed species become increasingly different from each other.

2. Divergent evolution is readily illustrated in a **phylogenetic tree.** In such trees, all but the outermost branches represent extinct types. Dead branches represent extinct lines, but those with living tips represent extinct types that underwent speciation, eventually forming the living species.

3. According to ideal **cladistic** theory, each natural taxonomic group (organisms grouped by similarities) is **monophyletic**; that is, all members can trace their ancestors to one species that gave rise to all of the group. Birds are a good example.

4. Natural taxa such as class Mammalia may be **polyphyletic;** there may be more than one evolutionary ancestor.

5. **Paraphyletic** groups are those that have given rise to other groups that have changed radically from the parent group (reptiles gave rise to birds and mammals, but the reptile is still a clearly identifiable group). In the plants, paraphyletic taxa include the **monocots** and **dicots,** groups distinguished in part by their numbers of **cotyledons.**

6. Phylogenetic trees are in part based on **comparative anatomy** and embryology. The presence of **homologous** structures—those of similar embryological origin, whatever their present function may be—demonstrates relatedness (such as the forelimb of horse, human, whale, cat, and so on).

7. **Analogous** structures in different species, have similar functions and appearances, but are of very different embryological origin (such as the wings of birds and insects).

8. Computers, evaluating identified mutations (for example, amino acid substitutions in a common protein), now help establish phylogenetic trees. **Minimum mutation trees** are those that require the smallest number of evolutionary changes.

Convergent Evolution

1. Species subjected to the same adaptive pressures and selective action, can undergo **convergent evolution,** whereby they come to resemble each other (such as marsupials and their mammalian counterparts who occupy similar environments).

Coevolution

1. **Coevolution** occurs when the evolutionary directions taken by one species strongly influence those followed by a different but dependent species.

2. Coevolution is common between predator and prey and herbivore and plant.

REVIEW QUESTIONS

1. Using the interbreeding principle, define the term *species*. Briefly criticize such a definition. (394)

2. Using examples, explain why the species concept runs into trouble in the genus *Canis*. (395)

3. In what way can the size of a species' range affect its ability to interbreed? (396)

4. List four ways in which members of a population can become reproductively isolated from each other. (396)

5. Explain how chromosome differences prevent hybrids from successfully reproducing. (397)

6. Using an example, explain how the binomial system is used in naming species. Why has Latin been retained when it is a dead language? (398)

7. Beginning with the largest and most encompassing, name the taxonomic categories. Darwin used the analogy of "boxes within boxes," to describe these. Explain what he meant. (398)

8. List the five kingdoms, and comment on why some people are now considering extending this to six. (398–399)

9. Ideally, kingdoms should be monophyletic. Explain this and comment on which kingdoms, if any, qualify. (399)

10. Explain, using examples, how the physical characteristics of a species might be affected by a range with sharply differing climates. (400)

11. Suggest a logical explanation for the short ears, legs, and tail of the Arctic fox, and the long ears, legs, and tail of desert foxes. (400, 402)

12. Distinguish allopatric from sympatric speciation. Which is believed to be more common? (402, 406)

13. Using the Galapagos finches as an example, explain how adaptive radiation is believed to occur. (402, 405)

14. What is character displacement? Using two species of Galapagos finches as an example, characterize this phenomenon. (405)

15. Briefly explain how a tetraploid cell might arise. (407)

16. Explain why an autotetraploid is infertile yet an allotetraploid could be quite fertile. (407)

17. Following allotetraploidy, what future is likely to be in store for the plant? Upon what does its success depend? (407)

18. Where do we find most polyploid plant species today? Suggest reasons for their being better able to succeed under such conditions and why they fail to displace diploid species under less extreme conditions. (408)

19. Explain the organization of a phylogenetic tree. How are the extant species shown? The extinct species? The dead-ended extinct species? What form of evolution does the tree best represent? (409)

20. Define and cite examples of monophyletic, polyphyletic, and paraphyletic groups of animals. (409–410)

21. Using placental and marsupial mammals for examples, describe four instances of convergent evolution. What are the forces behind convergent evolution? (412–413)

22. Distinguish between homologous and analogous organs. Which is of the greatest use in establishing phylogenetic relationships? (410–411)

23. What is coevolution? Cite three examples of situations that might encourage such evolutionary directions. (413, 415)

SUGGESTED READING

Banks, H. P. 1975. "Early Vascular Plants: Proof and Conjecture." *Bioscience* 25:730. On the origin of one of the five kingdoms.

Darwin, C. 1859, 1966. *On the Origin of Species.* A facsimile of the first edition. Harvard University Press, Cambridge, Mass.

Dawkins, R. 1976. *The Selfish Gene.* Oxford University Press, New York and Oxford. Dawkins is an enthusiastic and often persuasive writer who, in this popular paperback, puts modern evolutionary theory into clear and vivid language. His approach is logical and nonmathematical, with some emphasis on theories of the evolution of animal behavior.

de Beer, G. 1974. "Evolution." In *The New Encyclopedia Brittanica*, 15th ed. 7:7. A concentrated synopsis of evolutionary thought, with an extensive bibliography.

Dobzhansky, T. 1963. "Evolutionary and Population Genetics." *Science* 142:3596.

_____. 1970. *Genetics of the Evolutionary Process.* Columbia University Press, New York. A great compendium of observations and interpretations of the genetics of natural populations by an influential evolutionary geneticist.

Dodson, E. O. 1974. "Phylogeny." In *The New Encyclopedia Brittanica,* 15th ed. 14:376. Dodson includes a full historical account of the various one-, two-, three-, four- and five-kingdom schemes as well as an overview of phylogenetic relationships within the plant and animal kingdom.

Gilbert, L. E., and P. H. Raven. 1975. *Coevolution of Plants and Animals.* University of Texas Press, Austin, Tex. Such fascinating esoterica as flowers that mimic insects and insects that mimic flowers, as well as such important topics as the coevolution of plants and their herbivores.

Gould, S. J., and Eldredge, N. 1977. "Punctuated Equilibria: The Tempo and Mode of Evolution Reconsidered." *Paleobiology,* 3:115. Do the mechanisms of evolution proceed in a slow and orderly manner, or are they rapid and sporadic?

Grant, V. 1963. *The Origin of Adaptations.* Columbia University Press, New York. A balanced view of evolutionary adaptations in plants and animals.

Haldane, J. B. S. 1932. *The Causes of Evolution.* Cornell University Press, Ithaca, N.Y. (paperback). All the difficult math is relegated to the appendix in this rather old—but very wise—discussion of evolutionary genetics. A must for advanced students of population biology.

Joravsky, D. 1970. *The Lysenko Affair.* Harvard University Press, Cambridge, Mass. Fascinating reading about the man whom biologists—especially geneticists—love to hate.

Kettlewell, H. B. D. 1956. "Further Selection Experiments on Industrial Melanisms in the Lepidoptera." *Heredity* 10:287. Here you can read for yourself one of the most often cited experiments in modern evolutionary biology.

King, J. L., and T. H. Jukes. 1969. "Non-Darwinian Evolution." *Science* 164:788. The authors suggest, among other things, that most evolutionary changes on the molecular level may be meaningless noise, the result of mutation and random drift; that most DNA in higher organisms is not genetic material, and that no more than about 1% codes directly for proteins.

Lack, D. 1947. *Darwin's Finches.* Cambridge University Press, New York. Lack investigates the coevolution of competing species on different islands in the Galápagos archipelago.

Leedale, G. F. 1974. "How Many Are the Kingdoms of Organism?" *Taxon* 23:261.

Levin, D. A. 1979. "The Nature of Plant Species." *Science* 204:381. The "biological species concept" quickly breaks down when plant species are considered.

Margulis, L. 1968. "Evolutionary Criteria in the Thallophytes: A Radical Alternative." *Science* 161:1020. A cogent synthesis of the problem of the origin of eukaryotes and the kingdoms of organisms.

Maynard Smith, J. "Group Selection and Kin Selection." *Nature* 201:1145.

Patterson, C. 1978. *Evolution.* Cornell University Press, Ithaca, N.Y. (paperback). This short textbook makes a useful supplement to a general biology course or stands on its own. It is a good secondary reference for Darwin's finches and industrial melanism.

A *Scientific American* book. 1978. *Evolution.* W. H. Freeman, San Francisco. Reprinted from the September 1978 special issue on evolution.

Silver, W. S., and J. R. Postgate. 1973. "Evolution of Asymbiotic Nitrogen Fixation." *Journal of Theoretical Biology* 40:1.

Stebbins, G. L. 1971. *Processes of Organic Evolution.* Prentice-Hall, Englewood Cliffs, N.J. Somewhat dated, but still a superior textbook on evolution.

Volpe, P. 1981. *Understanding Evolution.* 4th ed. Wm. C. Brown, Dubuque, Iowa.

Whittaker, R. H. 1959. "On the Broad Classification of Organisms." *Quarterly Review of Biology* 34:210. Whittaker's influential revolt against the classic plant–animal dichotomy.

Scientific American articles. San Francisco: W. H. Freeman Co.

Bishop, J. A., and Laurence M. Cook. 1975. "Moths, Melanism and Clean Air." *Scientific American,* January. How the peppered moth is readapting to the improving atmosphere of postindustrial England.

Eigen, M. et al. 1981. "The Origin of Genetic Information." *Scientific American,* April.

Gilbert, L. E. 1982. "The Coevolution of a Butterfly and a Vine." *Scientific American,* August.

Margulis, L. 1971. "Symbiosis and Evolution." *Scientific American,* August.

Patterson, C. 1978. *Evolution.* Cornell University Press, Ithaca, N.Y. (paperback). This short textbook makes a useful supplement to a general biology course or stands on its own. It is a good secondary reference for Darwin's finches and industrial melanism.

Stanley, S. M. 1984. "Mass Extinction in the Ocean." *Scientific American,* June.

Ia. Microevolution: Genetic Variation Evolution requires the constant rise of variation, and the initial source of variation is mutation. Variation is greatly increased through a number of means in both prokaryotes and eukaryotes. **Ib. Microevolution: Changes in gene frequencies** Another form of microevolution involves increased change in alleles through natural selection and genetic drift. New alleles, and new combinations of alleles, are continuously tested by the forces of natural selection. While the effects of natural selection are adaptive, genetic drift is a completely random mechanism whereby allele frequencies change simply through chance. **II. Macroevolution: The Development of New Species** Speciation occurs chiefly through allopatry and sympatry. In the first, new species arise in recently divided populations where, through the continuous action of different selective forces acting on old and new variants, the two isolates evolve along different paths. Should they find themselves sharing a habitat, reproductive isolation will prevent interbreeding. In sympatry, chiefly a flowering plant phenomenon, "instant species" most commonly arise within a population through spontaneous polyploidy. **III. The Fossil Record: An Example of Macroevolution** The products of 50 million years of variation, selection, and speciation are clearly seen in the fossil history of *Equus,* the modern horse. The history is paralleled by the increase in grasslands to which the horselike ancestors adapted.

Ia. MICROEVOLUTION: Genetic Variation

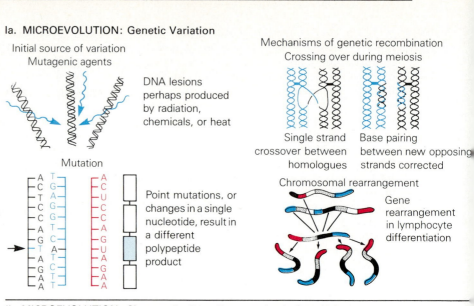

Initial source of variation
Mutagenic agents

DNA lesions perhaps produced by radiation, chemicals, or heat

Mutation

Point mutations, or changes in a single nucleotide, result in a different polypeptide product

Mechanisms of genetic recombination
Crossing over during meiosis

Single strand crossover between homologues

Base pairing between new opposing strands corrected

Chromosomal rearrangement

Gene rearrangement in lymphocyte differentiation

Ib. MICROEVOLUTION: Changes in Gene Frequencies in Population

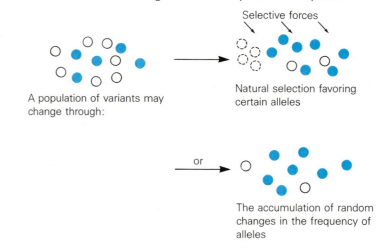

A population of variants may change through:

Selective forces

Natural selection favoring certain alleles

or

The accumulation of random changes in the frequency of alleles

III. THE FOSSIL RECORD: An Example of Macroevolution

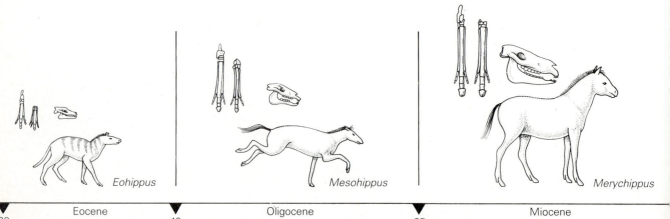

Eohippus

Mesohippus

Merychippus

| Eocene | Oligocene | Miocene |

60 Millions of years   40   25

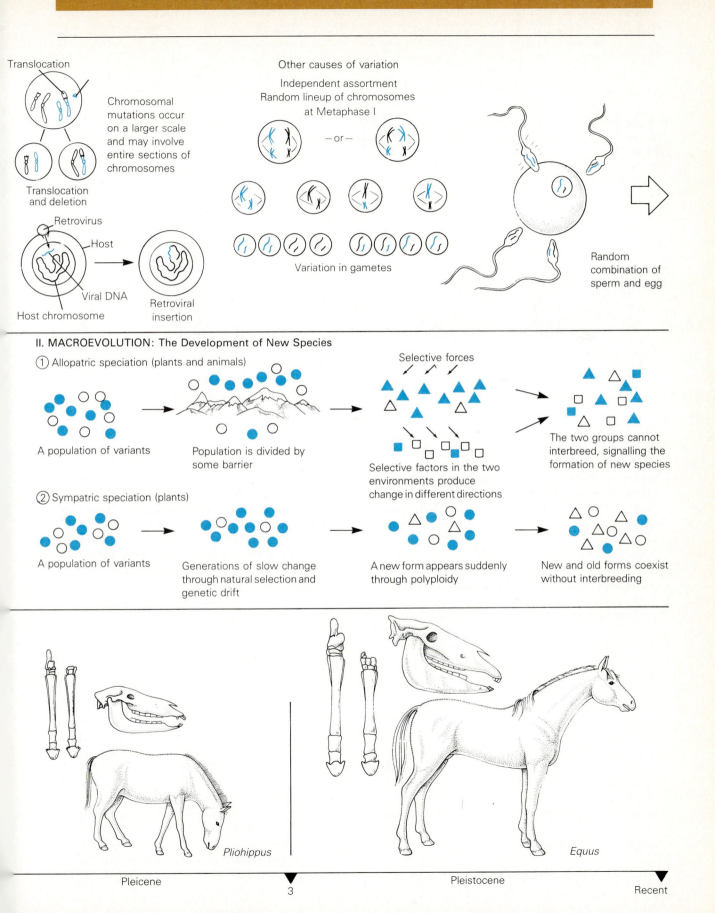

Translocation

Chromosomal mutations occur on a larger scale and may involve entire sections of chromosomes

Translocation and deletion

Retrovirus

Host

Viral DNA

Host chromosome

Retroviral insertion

Other causes of variation

Independent assortment
Random lineup of chromosomes at Metaphase I

— or —

Variation in gametes

Random combination of sperm and egg

II. MACROEVOLUTION: The Development of New Species

① Allopatric speciation (plants and animals)

A population of variants

Population is divided by some barrier

Selective forces

Selective factors in the two environments produce change in different directions

The two groups cannot interbreed, signalling the formation of new species

② Sympatric speciation (plants)

A population of variants

Generations of slow change through natural selection and genetic drift

A new form appears suddenly through polyploidy

New and old forms coexist without interbreeding

Pliohippus

Equus

Pleicene

Pleistocene

Recent

Prokaryotes, Viruses, and Origin of Life

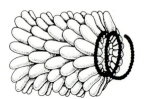   

We have now had a look at some of the principles governing evolution, and we have explored its classical triad of forces: mutation, variation, and natural selection. There is indeed a substantial body of information on such processes, yet many of the most fundamental questions remain. The ultimate question, of course, is how did life begin? The question is as old as humanity itself. The ancients throughout the world were absorbed by this mystery, and their conclusions have lingered in our consciousness, often forming the basic premises of many philosophies and religions.

Undoubtedly, many of the ancient cultures had great faith in their ideas. But faith in an idea, while it may serve as some sort of motivating force, is only a starting point. The faith must stand ready to be tested, and the burden of proof falls on the faithful.

But how does one investigate an improbable event that may have happened only once, several billion years ago? It is clear that we can never *prove* how life came to be on this planet. We can, however, examine any number of seemingly plausible notions of how life *might* have arisen. All scenarios or speculations about the origin of life must involve suppositions. These suppositions can lead to predictions, and it is the predictions that can be tested. Many such speculations have been trotted out, and

Note: Geological time is charted inside the front cover.

many of the predictions based on their suppositions have been tested. Most of the proposed schemes of how life might have originated couldn't possibly have worked; they have depended on assumptions about nature that have proven to be untrue. But through the process of elimination, we have narrowed the range of possibilities and have highlighted the most promising avenues for further examination. At this time, scientists generally agree that the preponderance of evidence indicates that *life successfully arose only once* in the earth's history and that the conditions at that time were unique, probably never to occur again.

A HYPOTHESIS ON THE ORIGIN OF LIFE

Although Darwin speculated that life might have arisen in a warm, phosphate-rich pond, the first serious proposals concerning the spontaneous origin of life (those that were based on sound biochemical and geological information) began to appear some 50 years ago. Such schemes were presented by J. B. S. Haldane, a Scottish biochemist, and by A. P. Oparin, his Russian counterpart. They proposed that soon after the earth's formation, when conditions were quite different from those existing today, a period of spontaneous chemical synthesis began in the warm ancient seas. During this era, amino acids, sugars, and nucleotide bases—the structural subunits of some of life's macromolecules—formed spontaneously from the hydrogen-rich molecules of ammonia, methane, and water. Such spontaneous synthesis, explained Haldane, was only possible because there was little oxygen in the atmosphere, an important condition since oxygen will react spontaneously with many organic substances. The energy for this synthesis was, at this time, readily available in the form of lightning, ultraviolet light, heat, and higher energy radiations. Since there were no organisms to degrade the spontaneously formed organic molecules, they accumulated until

the sea became a "hot, thin soup" (of, perhaps, about the consistency and nutritive quality of chicken boullion (Figure 18.1).

Both Haldane and Oparin suggested that polymerization—the joining of the structural subunits into macromolecules—was greatly encouraged by the ever-increasing concentrations of precursor molecules. Amino acids joined to form the first polypeptides and eventually the first complete proteins. Active versions of these proteins became the earliest enzymes. The rate of activity must have been very slight at first, but there was plenty of time. The active proteins catalyzed more building activity, producing more proteins like themselves.

18.1 SCENARIO OF EARLIEST EARTH CONDITIONS

Torrential rains formed the young seas while volcanoes released gases into the developing atmosphere. Heat and electrical discharge provided energy for chemical synthesis.

Collections of these new catalysts then became enclosed by simple membranelike, water-resistant protein or lipid shells, all the while perpetuating themselves by making use of the energy-rich nutrients of the ancient sea.

To continue with Oparin and Haldane's scenario, at first these conglomerates divided simply because they became too large and fell apart. But eventually the crude membrane-surrounded droplets, or **coacervates** as Oparin called them, took in new types of molecules, *autocatalytic* molecules that in some way induced the formation of molecules somewhat like themselves. These were the precursors of genes, and with such systems the coacervate can be called a **protocell** (prototype cell). Keep in mind that in Haldane and Oparin's time, not much was known of DNA or of its crucial genetic role. These first protocells, with their crude genetic mechanism, persisted and increased in numbers. The presence of this genetic mechanism enhanced the continuance of the chemical characteristics of the successful protocells. They also set the stage for the inexorable forces of natural selection. The survival of the fittest had begun. According to Haldane and Oparin, from these early and persistent successes, the forces of natural selection molded and winnowed the tiny, membranous droplets and eventually produced the first simple cellular life.

The **Haldane-Oparin hypothesis** relies heavily on a number of assumptions about the physical conditions of the early earth, as well as the occurrence of largely random events. Nonetheless, although some of the assumptions are now known to be incorrect and the original hypothesis has had to be modified in many ways, the overall hypothesis is yet to be disproven.

Haldane's and Oparin's hypothetical scheme remained a neglected intellectual curiosity for more than a quarter of a century. But in 1952, Nobel laureate Harold Urey (the discoverer of deuterium, a heavy isotope of hydrogen) and Stanley Miller, his graduate student, began testing some of the assumptions in earnest.

The Miller-Urey Experiment

The crucial proposition of the Haldane-Oparin hypothesis was that the precursors of the molecules of life would form spontaneously in the primitive atmosphere. Miller and Urey created a laboratory apparatus at the University of Chicago that attempted to simulate what were then believed to be some of the primitive conditions of the earth (Figure 18.2). They introduced a small amount of water and a mixture of gases including methane, ammonia, water vapor, and hydrogen—but no free oxygen—into the apparatus. As an energy source, they produced repeated electrical discharges (lightning?) through the atmosphere of the upper flask. After a week, they analyzed the sediments that collected in the lower flask. Among the various molecules they found aldehydes, carboxylic acids, and, most interestingly, amino acids. All of these are commonly found in living cells.

Although these small molecules were a far cry from anything alive, their production under the simulated primitive conditions provoked a lively revival of interest in the Haldane-Oparin hypothesis. But before we discuss the work that followed Miller and Urey's breakthrough, let's review what scientists today believe the early earth was like.

The Early Earth

The best estimates suggest that the earth took form about 4.6 billion years ago, along with the sun and the other planets of our solar system. Prior to this, the precursor of the solar system was a vast, flattened cloud of gases, dust, and other debris (Figure 18.3). Recent theories maintain that the cloud was cold, but that as the sun and planets coalesced, a great deal of heat was released from gravitational energy, supplemented with heat from radioactivity. When it took form, the earth's crust was a molten semiliquid. Some 600 to 800 million years were to pass before it solidified.

When the crust finally cooled to below the boiling point of water, torrential rains began to fall, initiating the formation of the oceans. The atmosphere, of volcanic origin, consisted largely of water vapor, methane, carbon monoxide, carbon dioxide, ammonia, hydrogen, and hydrogen sulfide. However, computer simulations now show that the reducing chemicals, methane, ammonia, and hydrogen sulfide, would be rapidly broken down by ultraviolet radiation, and that most of the liberated hydrogen would be lost to outer space. According to current theory, then, the principal constituents of the early atmosphere were water vapor, carbon dioxide, carbon monoxide, molecular nitrogen, and possibly some free hydrogen. This development has actually supported the hypothesis, since recent experiments, similar to those of Miller and Urey but using the revised version of the probable primitive atmosphere, produced even greater yields of small organic molecules.

Energy sources abounded in the primitive atmosphere. Although the sun was not as bright as it is now, ultraviolet light was plentiful, since the ozone layer, which screens out much of this energy, had yet to form. Other plentiful supplies of energy included lightning, heat, and what geologists call shock energy (from earthquakes and tremors). So the major requirements of chemical evolution, reactive gases and sources of high energy, were in abundance. At least that is the consensus from scientists today.

From recent NASA studies of the moon and Mars, we know that the earth and all of the planets of the solar system continually received massive bombardments of meteors and asteroid fragments for a long while. This meteor rain ceased abruptly about 3.8 billion years ago and many hold that life must have begun about this time, although it could have begun earlier (Figure 18.4). Actually, the oldest fossils of reliable identification date back some 3.5 billion years. According to their discoverer, Stanley Awramik, they resemble today's complex, photosynthetic cyanobacteria.

18.2 THE CLASSIC MILLER-UREY EXPERIMENT

Heated gases of the theoretical primitive atmosphere were subjected to electrical discharges in a sealed, sterile environment. Residues were collected in the lower chamber and periodically analyzed. Results indicated that some of the simple monomers of life could be produced spontaneously under the test conditions.

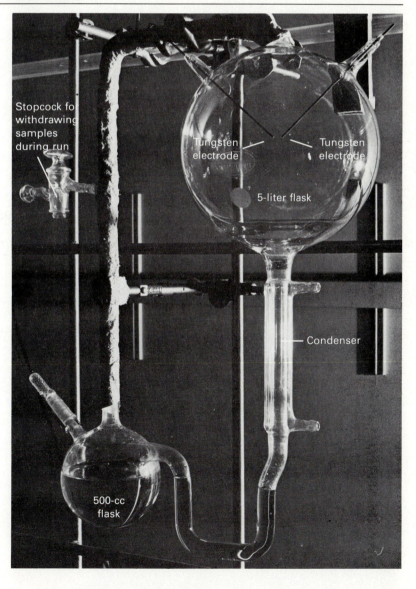

18.3 THE YOUNG SOLAR SYSTEM

The hypothetical events in the formation of the solar system from a flattened cloud of cold gases: *(top)* the flattened dust cloud, *(center)* sun and planets forming, and *(bottom)* the completed solar system.

The Hypothesis Today: How Strong Is It?

Scientists today continue to test the hypothesis of the spontaneous generation of life. In particular, five aspects of the hypothesis must receive attention if the whole is to be taken seriously:

1. Verification and clarification of the physical conditions of the primitive earth must continue. Scientists must continue with their efforts to understand the conditions of the primitive earth. Without such understanding, any hypothesis becomes a house of cards, ready to tumble if its foundations are disturbed (status: very good and getting better).
2. It must be shown that essential monomers, such as amino acids, nitrogen bases, and simple sugars, can be produced under primitive conditions, in the absence of enzymes or other biological activity (status: well established).
3. It must be shown that the familiar polymers of life—proteins, nucleic acids, and so forth— can form spontaneously from monomers (status: not well established).
4. The spontaneous formation of active, well-defined, cell-like bodies with isolating membranes or borders must be verified (status: fairly good).
5. It must be shown to be at least possible that all of this will result in the production of simple, self-replicating systems, with repositories of genetic information and the ability to maintain metabolic processes from generation to generation (status: not yet established).

The Monomers.   The early work of Miller and Urey gave rise to a host of similar experiments. The atmospheric constituents and the energy sources have been varied, especially as our knowledge of primitive conditions has been refined. The results have been rewarding. The list of laboratory-synthesized monomers now includes all of the nucleotide bases of DNA and RNA, along with the essential sugars, all of the amino acids, and most

18.4 AN EARTH CALENDAR

Here, the earth's history is shrunk into a period of 12 hours—from midnight to noon. The events are in chronological order and the spacing indicates lapsed time.

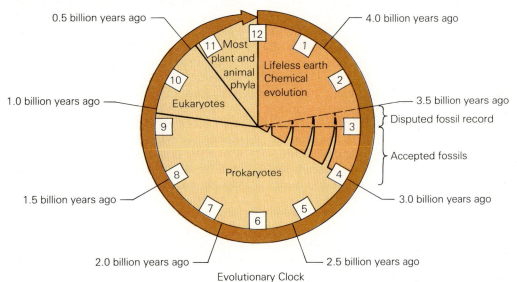

Evolutionary Clock

12 Midnight —	Earth forms	11:30 AM —	Age of dinosaurs
3:00 AM —	First undisputed life	11:50 AM —	Age of mammals
3:00 AM-9:15 AM —	Prokaryotes	11:59:00 —	1st hominids
9:15 AM —	1st eukaryotes	11:59:40 —	1st humans
10:45 AM —	Primitive animal phyla evolving	11:59:59 —	All of human history
10:54 AM —	First terrestrial plants	12:00 Noon —	Present
11:00 AM —	1st vertebrates		

essential vitamins. (As an aside, it turns out that adenine, the most widely occurring nucleotide base in today's life forms, has been the easiest to synthesize.) Thus, research supports the idea that many of the monomers associated with life were produced through spontaneous generation on the early earth.

The Polymers. It is one thing to produce such monomers and quite another to induce them to join, forming polymers, without the assistance of enzymes. It is here that the original hot, thin soup hypothesis is weakest. All biological polymerizations involve *dehydration linkages* between the monomers—that is, removal of water to produce the linkage (see Chapter 3). Researchers agree that such reactions would *not* have been energetically favorable in the primitive sea, or indeed in any aqueous medium, because of the mass action law. In water, biological polymers slowly dissociate back into monomers, and heat just accelerates the process. Without the catalyzing effects of enzymes, spontaneous polymerization is possible only when the concentration of monomers in water is very high.

With such problems, then, how could polymerization have been achieved? There have been many suggestions. For example, biologist Carl R. Woese, now with the University of Illinois, has proposed that life began, not in the sea, but in the hot, extremely dense atmosphere of the *very* early earth. In addition, new ideas have sprung from the experiments of Sidney Fox of the University of Miami. Fox has demonstrated that polymerization of amino acids occurs readily under hot, drying conditions such as might be found along the edges of volcanoes or even on the hot beaches of ancient seas. Pools of organic precursors, rich in amino acids, could have been concentrated by evaporation and heated to allow the spontaneous formation of polypeptides. Fox has succeeded in producing polymers of 200 or more amino acids under hot, drying conditions. Fox calls aggregations of these spontaneously generated polymers **proteinoids** (Figure 18.5). (Interestingly, Sidney Fox's spontaneously polymerized proteinoids also form coacervatelike encapsulated spheres in water.)

One of the greatest problems in reconstructing the development of life lies in explaining how those essential molecules called nucleotides might have formed. Investigators have even stacked the deck in trying to form such molecules. They have boiled and dried concentrated energy-primed nucleotide triphosphates in the presence of single-stranded

18.5 ORGANIZATION WITHOUT CELLS

Proteinoids are polypeptides that polymerize spontaneously from evaporating concentrations of amino acids. Such bodies could have formed in small, hot pools of spontaneously formed monomers located in regions of intense volcanic activity.

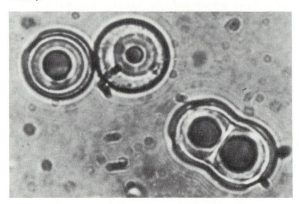

templates of DNA, producing an energetically favorable direction of reaction. However, without the appropriate enzymes, no second strand of DNA is produced. The units do not link to form polymers. Linkages can be forced, but they occur in the wrong places. To date, there is no evidence that nucleic acid polymers can be produced spontaneously, hence some scientists are now theorizing that the first synthesis of nucleic acids might have come long after the actual origin of life.

Coacervates: The Active Droplets. After presenting his hypothesis in the 1920s, Oparin spent most of the next 50 years experimenting with versions of the bodies he had named *coacervate droplets*. As we have seen, coacervates arise when proteins or other polymers form dense clusters in water. We noted earlier that, like cells, coacervates tend to divide when their mass reaches a critical size. In addition, their thickened surface behaves in some ways like a plasma membrane (Figure 18.6).

Oparin and his coworkers produced coacervate droplets by mixing various molecules, including proteins and nucleic acids. For instance, coacervate droplets containing simple nucleic acids and nucleic acid replicating enzymes were immersed in a medium containing triphosphate nucleotides (see Chapter 9). The coacervates "grew," "divided," and "replicated" themselves. But of course we must remind ourselves that coacervate droplets are not living entities, and their more significant feats depend on their being supplied with enzymes that had previously been extracted from living organisms. Nevertheless, they help us envision a critical stage in the origin of life.

Self-Replicating Systems.   One of the questions that bothers some investigators is which macromolecules formed first: nucleic acids or proteins? The problem is that contemporary organisms use nucleic acids—DNA, mRNA, tRNA, and rRNA—to synthesize the polypeptides incorporated into enzymes, yet at the same time, enzymes appear to be necessary to synthesize nucleic acids. It's the ultimate chicken-and-egg problem. Those who favor proteins as primitive macromolecules point out the relative ease with which peptide bonds are produced and the great generality of protein's enzymatic activity. Those who favor nucleic acids dwell on the information-carrying feature of the genetic material and the fact that some enzymelike characteristics are associated with certain RNA molecules. They suggest that rRNA and tRNA are the remnants of a once-large class of nucleic acid "enzymes" or enzymelike molecules. It's possible that the two systems originated together.

The Earliest Cells

By taking the giant step from metabolically active colloidal aggregates to self-reproducing **protocells** (leaving huge gaps for future theorists to deal with), we can apply some informed speculation to many

questions about early cellular life: what were the earliest cells like? What were their energy sources? How do we get from the earliest stages to Awramik's 3.5 billion-year-old cyanobacteria, a complex photosynthetic prokaryote?

There is considerable disagreement as to the metabolic characteristics of the earliest cells. Some hold that the first life forms were **phototrophic** or **chemotrophic.** Both are **autotrophs,** deriving their energy from light or inorganic chemicals, respectively, and living independently of other organisms. Others propose that the earliest cells were primitive versions of today's *Clostridium*, a soil bacterium that is rapidly killed by oxygen. These life forms probably relied heavily on the comparatively simple anaerobic process of fermentation and substrate-level phosphorylation (see Chapter 8). (Remember that there was no available oxygen or oxidative respiration.)

The original energy supply for the early heterotrophic cells may well have come from the spontaneously produced monomers that were still available in the ocean. But we can surmise that expanding populations of the new, living cells soon began to use up the available resources, and the resulting competition led natural selection to favor those cells able to exploit new energy sources or to exploit old ones more efficiently. Cells could prey on one another, but this simply redistributed the limited and dwindling supply of organic nutrients. The next major advance was to break out of this limited food chain altogether. The cells that could begin to do this, even somewhat inefficiently, would begin a line that would replace others. Thus we have the rise of autotrophy.

The Early Autotrophic Cells.   The primitive phototrophs today obtain their hydrogen from dissolved hydrogen sulfide and their energy from sunlight. It's a good bet that the earliest successful autotrophs also used light energy in the simplest of photosystems to pry the hydrogen away. But the number of places in the world that provide both hydrogen sulfide and abundant sunlight is severely limited. At some point, some ancestral cyanobacterium began to obtain its photosynthetic hydrogen supply from an energetically less favorable but far more abundant source: water. This step required a complex photosystem, as we discussed in detail in Chapter 7. But the accomplishment was a success, and it was to change the earth forever.

Where water is the source of hydrogen in photosynthesis, the waste product is molecular

---

18.6 THE COACERVATE

When proteins or other polymers are introduced into water, they tend to cluster together into distinct droplets called coacervates. The coacervate surrounds itself with a boundary layer that is selective in admitting kinds of molecules. When coacervates reach a critical size and mass, they divide spontaneously, a process characteristic of cells.

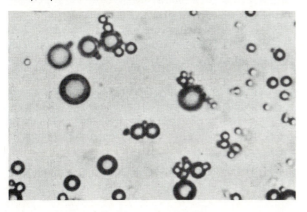

18.7 THE HISTORY OF THE EARTH'S OXYGEN

The accumulation of oxygen in the atmosphere of the early earth was at first an extremely slow process, requiring about 2 billion years to approach the concentration that now exists. There are several reasons for this, including the union of atmospheric oxygen with elemental metals in the earth's crust. Only after these oxygen sinks were saturated could oxygen begin to increase in the atmosphere.

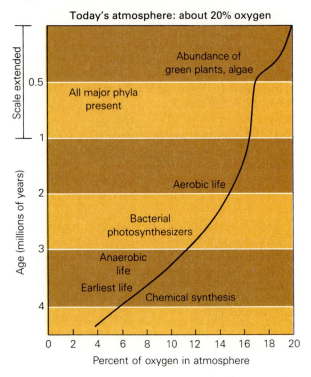

Today's atmosphere: about 20% oxygen

It took about 2 billion years, but eventually photosynthesis changed the strongly reducing atmosphere of the early earth to the oxygen-rich, strongly oxidizing atmosphere of today. The early earth contained many **oxygen sinks** that absorbed oxygen as fast as it could be produced. For example, the elemental iron, elemental sulfur, and abundant iron sulfide of the early earth's crust took in enormous amounts of oxygen as they were transformed into iron oxide and various sulfates. Only when these oxygen sinks were finally saturated could free oxygen exist in the air. By the time oxygen became a significant atmospheric gas, organisms of a new kind—the eukaryotes—were already making their presence felt.

New modes of nutrition arose, too. The burgeoning cyanobacteria themselves represented an abundant new source of food for any heterotroph able to engulf and assimilate their stored energy. Thus the world saw the emergence of the first herbivores. Those anaerobes unable to cope with oxygen and eat other organisms were soon relegated to the backwaters of life's great progression. They could exist only if hidden away from poisonous oxygen in pockets of the earth's crust, in nutrient-rich muds and in deep recesses of stagnant waters. And it is in such places that they remain to this day.

In our imaginative scenario, we have seen two metabolic forms of life emerging: the photosynthetic, oxygen-producing phototroph and the aerobically respiring, oxygen-utilizing heterotroph (Figure 18.8). Now that we have arrived at the time of the first fossil evidence of life on earth, we can leave this hypothetical world and begin to consider the known world. We leave many questions unanswered, and we must put away those explanations based on conjecture and on an imperfect knowledge of the primitive earth. But this is the way of science; those questions will not lie untouched on some intellectual shelf. They will be repeatedly brought out, dusted off, tested again—and perhaps altered by the weathering effect of new data.

We will move on now to learn more about the prokaryotes, the organisms believed to be the most direct descendants of the earliest forms of life on earth. We have visited this remarkable group before, specifically, in our comparisons of cells and our discussion of prokaryotes as experimental subjects. Here, though, we will consider the prokaryotes as our first subjects in our consideration of the diversity of life.

oxygen. As the early cyanobacteria flourished, exploited new niches, and multiplied, the amount of oxygen they released became significant, "poisoning" the water for their anaerobic competitors (Figure 18.7). At first, the regions of oxygen poisoning would have been local, just a thin layer of oxygenated water in the sea or in a shallow pond, but this gradually changed.

Although many oxygen-sensitive organisms undoubtedly became extinct, being literally driven into the mud, new forms less sensitive to the poisonous gas were to emerge through mutation and natural selection. At first, they probably only developed ways of detoxifying oxygen. Later, however, the corrosive power of oxygen was actually utilized, put to work in extracting the energy from organic foodstuffs. Thus, oxidative aerobic respiration came into being.

THE PROKARYOTES

Origins

There is no longer much doubt that the prokaryotes preceded all other forms of life. Their ancient history is clearly written in stone—in fact, in deposits 3 to 3.5 billion years old. The most widespread evidence of their antiquity is seen in the strange columnlike deposits known as **stromatolites** (Figure 18.9). These are highly laminated deposits of sedimentary rock, with each layer produced by dense populations of bacteria (usually cyanobacteria) that deposited mineral particles around their cells.

Some paleontologists doubted that these strange geological formations were true evidence of

early life forms until living stromatolite-forming mats of bacteria were discovered. The living versions were found in Yellowstone National Park in the United States, and along the shores of Shark Bay, Australia.

Organization of the Prokaryotes

As recently as 1975, biologists were quite comfortable with the organization of prokaryotes into kingdom Monera, which contained two major groups, the bacteria and the cyanobacteria. But as you are well aware, opinions about how things should be grouped can change rapidly in biology. Intensive research pioneered by Carl R. Woese, who used new molecular techniques for establishing phyloge-

18.8 THE EARLIEST CELLS

A history of the evolution of cells from the simple anaerobic heterotroph, sopping up spontaneously formed nutrients, to the more sophisticated aerobic, photosynthetic cells that preceded known life.

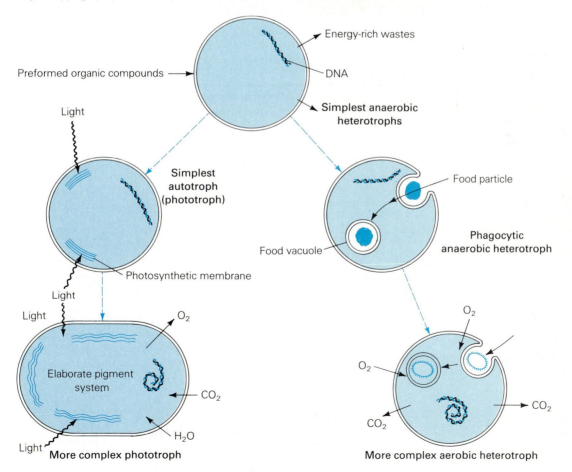

18.9 THE OLDEST FOSSILS

Fossilized stromatolites are thought to represent the most ancient evidence of life on the earth. They arose on rocky ledges in the tidepools of shallow Precambrian seas. Their peculiar, layered construction was the product of constantly growing populations of bacterial cells—probably cyanobacteria—that were continually infiltrated by tide-carried sand particles. As new populations arose at the surface, their predecessors underwent fossilization below, leaving a permanent record of past struggles.

netic relationships, has changed the way we classify microorganisms. Some aspects of this work are described in Figure 18.10.

It was Woese who introduced the idea that the prokaryotes actually include two distinct, unrelated groups—*Archaebacteria* and *Eubacteria* (the "first or ancient bacteria" and "true bacteria," respectively). They are now appearing as separate kingdoms in a number of schemes. While both groups are definitely prokaryotic, they are different in enough ways to indicate separate origins from the earliest forms of life (Table 18.1). The two kinds of bacteria look much alike when viewed through the light microscope, but the electron microscope and chemical analysis reveal basic differences in the structure of their cell walls. While Archaebacteria incorporate a variety of proteinaceous substances into their walls, nearly all Eubacteria use *amino sugars* organized into a wall of **peptidoglycan**. We will look at the peptidoglycan in more detail shortly and characterize both groups further, but first, it might be useful to describe some of the common characteristics of prokaryotes.

18.10 THE BACTERIAL KINGDOMS

New evidence from molecular biology has led biologists to reconsider older concepts of prokaryote taxonomy. The outcome is a division of the older kingdom, Monera, into two kingdoms, the Archaebacteria and the Eubacteria. The first includes thermoacidophiles, extreme halophiles, and methanogens, a group with strange biochemical adaptations. All of the other bacteria, including the cyanobacteria and the familiar decomposers, parasites, and some photosynthesizers, fill the second kingdom. According to one prominent theory, some of the Eubacteria became symbionts in primitive eukaryotic cells (dashed lines), establishing themselves permanently and providing the early eukaryotes with chloroplasts, mitochondria, cilia, flagella, and centrioles.

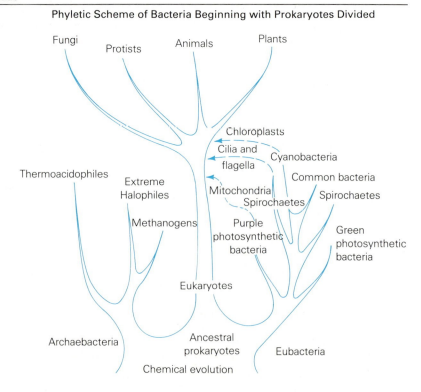

Phyletic Scheme of Bacteria Beginning with Prokaryotes Divided

TABLE 18.1

DIFFERENCES BETWEEN ARCHAEBACTERIA AND EUBACTERIA

	Archaebacteria	Eubacteria
Cell wall	Variety of substances, often proteinaceous	Peptidoglycans
Plasma membrane lipids	Modified branched fatty acids	Straight-chain fatty acids
Nuclear membrane	Absent	Absent
DNA	Naked, circular	Naked, circular
Membrane-bounded organelles	Absent	Absent (except for mesosome and thylakoid in some)
Ribosomes	30S, 50S subunits; structural similarity to eukaryotic	30S, 50S subunits, unlike archaebacteria and eukaryotic
Flagella	Unknown	Solid, rotating (protein flagellin)
Photosynthetic pigments	Bacteriorhodopsin	Bacteriochlorophyll
Cell division	Fission	Fission

Prokaryote Characteristics

One of the most obvious differences between prokaryotic and eukaryotic cells is that the latter are generally much larger. In fact, eukaryotic cells may be 1000 times larger than those of prokaryotes. Prokaryotes also lack a nuclear membrane, as well as the other elaborate membrane-bound cytoplasmic organelles that are typical of eukaryotes (see Figure 18.11). A number of other traits also mark the prokaryotes. For example, they have circular chromosomes of nearly naked DNA, relatively free of the extensive framework of nuclear protein present in the eukaryote chromosomes. Prokaryotic cells do not undergo mitosis or meiosis. Commonly, DNA replication is followed by a kind of cell division called **binary fission** (some undergo budding, others fragmentation). In binary fission (Figure 18.12), the plasma membrane and cell wall simply grow inward, constricting the cell and dividing it in two. Each half will contain similar cytoplasmic constituents and a replica of the chromosome. The **mesosome,** an odd, quite mysterious, membranous extension of the plasma membrane, may be involved in the separation of the circular chromosomes during fission. However, the role of the mesosome has not been completely clarified, and some microbiologists believe it is instead involved in cell transport.

Occasionally, as we saw earlier (Chapter 14), prokaryotes undergo a primitive kind of sexual exchange in the form of conjugation, during which a generally incomplete genetic recombination oc-

18.11 THE PROKARYOTE CELL

Compared to the typical eukaryote cell, the prokaryote cell is quite small. Further, it lacks the membrane-bound organelles of the eukaryote, including the organized nucleus. Prokaryotic DNA (the nucleoid) is naked and circular, lacking the protein complex of eukaryote chromosomes. A dense cell wall, quite different chemically from eukaryote cell walls, surrounds the membrane and is often itself surrounded by a slimy sheath. Free ribosomes and polyribosomes are common, as are tube-like, cytoplasmic projections known as pili. Membranous mesosomes are often seen in cells undergoing fission, but their function has not been established. Where flagella appear, they are solid, rotating entities, anchored in the cytoplasm and cell wall, and quite unlike those of eukaryotes.

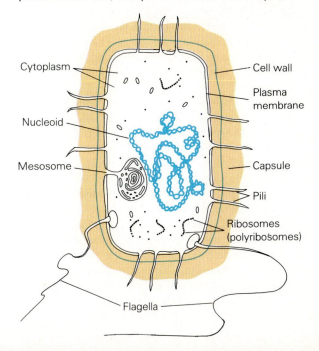

18.12 ASEXUAL REPRODUCTION IN BACTERIA

Cell division in bacteria occurs through a process called fission. Little is known about the mechanisms at work. Note the inward growth of the plasma membrane, followed by the synthesis of a new wall. The events are clear enough, but we are at a loss to explain how DNA replicas are properly divided between daughter cells.

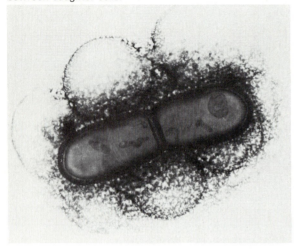

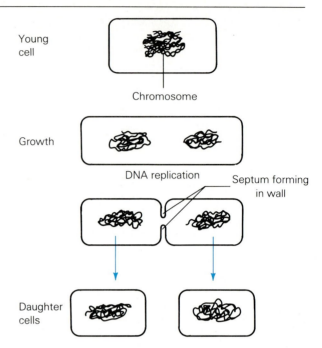

Young cell

Chromosome

Growth

DNA replication

Septum forming in wall

Daughter cells

curs through the transfer of DNA replicas. The transfer is believed to occur through tubular sex pili (see Figure 14.10).

In addition to the unique cell wall described earlier, many prokaryotes are surrounded by a slimy layer or capsule consisting of complex carbohydrates. In many human disease-causing bacteria, the capsule enables the bacterial cell to resist being engulfed by the phagocytic white blood cells. Certain of the carbohydrates invoke antibody reactions, and are thus considered antigens. The polysaccharide coat also aids bacteria in adhering to surfaces, an important ability for both parasitic and free-living bacteria.

Finally, while many prokaryotes move via a flagellum, it is structurally and functionally different from the eukaryotic flagellum. (In fact, the motor apparatus powering the flagellum appears to be different in Eubacteria and Archaebacteria.) The flexible eukaryote flagellum is composed of microtubules of the protein tubulin, is surrounded by the plasma membrane, and undulates to propel the cell (see Figure 4.21). In contrast, the bacterial flagellum is a stiff, solid, rodlike structure, permanently bent into an S-shape and composed of a different protein, called **flagellin.** The flagellum protrudes through the bacterial cell wall, makes a sharp turn (the "hook"), and, when it moves, rotates on its axis like a ship's propeller (Figure 18.13).

Archaebacterial Life

Archaebacteria are not very easy to grow in the laboratory and therefore are not as familiar to bacteriologists as the Eubacteria. When it first became apparent that prokaryotes tend to fall into two distinct groups, it was assumed that the Archaebacteria were rare, "primitive," and possibly relics of the earliest form of bacterial life. They were also considered to be quite bizarre in that the most familiar Archaebacteria lived in strange and improbable habitats, such as near-boiling hot springs and very acidic or salty ponds (harsh conditions not unlike those of the primitive earth). But the largest group of Archaebacteria, the **methanogens** (methane generators), are found in habitats where carbon dioxide and hydrogen are readily available, but where there is little or no oxygen. We find them in anaerobic marshes, in sewage treatment plants, in the mucky, anaerobic sea and lake bottoms (such as the Black Sea), and, as we have seen, in the oxygen-deficient bowels of animals, including humans. The familiar *Escherichia coli*, an aerobic Eubacterium, is better known than our archaebacterial gut inhabitants only because it is so much easier to grow in laboratory cultures. There are no hard and fast rules about oxygen utilization by prokaryotes, but some Eubacteria as well as most Archaebacteria are **obligate anaerobes**—that is, they cannot survive in the presence of oxygen. Members of both groups, however, have the ability to use oxygen in their metabolism.

Methanogens use hydrogen gas to reduce carbon dioxide, producing methane gas (CH_4, or marsh gas, as it was first known) and water. The reaction (which requires a battery of enzymes) is:

$$4H_2 + CO_2 \longrightarrow CH_4 + 2H_2O$$

Incidentally, sewage treatment plants can supply energy by using the methane gas they produce as a fuel to power electrical generators.

Other archaebacterial types include the aptly named **extreme halophiles** ("salt lovers"), **extreme thermophiles** ("heat lovers"), and the **thermoacidophiles** ("heat and acid lovers"), names suggesting rather drastic living conditions. The halophiles thrive in Great Salt Lake, the Dead Sea, and in salt-

18.13 THE BACTERIAL FLAGELLUM

(a) The prokaryotic flagellum is a solid structure permanently bent into a helical configuration and composed of the protein flagellin. While it appears to undulate, this is illusory, for unlike any other known cellular organelle, the bacterial flagellum *rotates*. When groups of bacterial flagella rotate counterclockwise, they join to produce a smooth synchronous movement that propels the cell along in a definite direction. But when they spin clockwise, the individual flagella fly apart, and their spinning sends the cell into a tumbling movement that is more random in direction. (b) High magnification by the electron microscope reveals that the flagellum is anchored to a hooklike shaft, which penetrates the cell wall and is anchored in two ring-shaped bases. The innermost ring, composed of 16 spherical proteins, rotates, while the outer, similarly constructed ring is fixed in place. Movement is believed to be powered by an influx of protons (H^+) generated in a chemiosmotic gradient between the cell and its immediate surroundings.

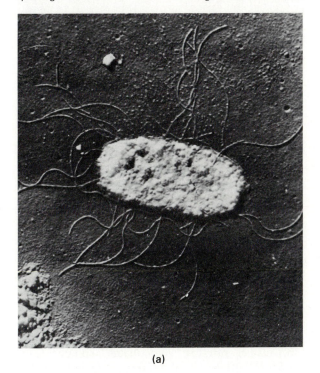

(a)

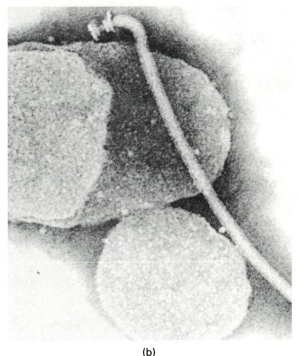

(b)

Smooth
directional
swimming

Counterclockwise
rotation

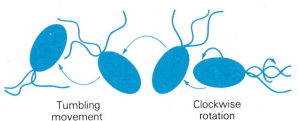

Tumbling
movement

Clockwise
rotation

18.14 EUBACTERIAL FORM AND ARRANGEMENT

The three common eubacterial forms are the bacillus (rod-shaped), coccus (spherical), and spirillum (spiral). Bacilli may occur singly (**a**) or in chains (**b**), as seen here, and they may be enclosed in sheaths. The coccus form also occurs singly (**c**) and in chains (**d**) (streptococcus), but in addition are seen in pairs (**e**) (diplococcus), in eights (**f**) (sarcina), and in clusters (**g**) (staphylococcus). The spirillum form (**h**) occurs singly with great variation in size. Each growth form is useful in identifying specific bacterial groups and sometimes species.

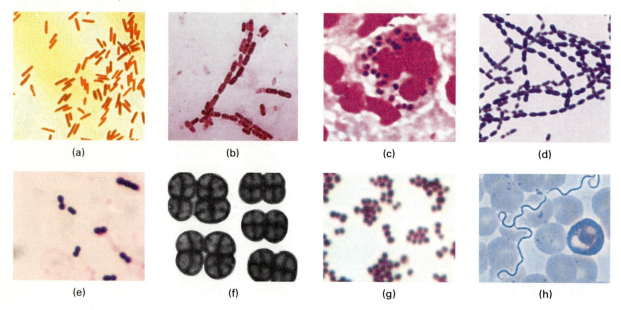

(a)   (b)   (c)   (d)

(e)   (f)   (g)   (h)

evaporation facilities, where they carry on a unique version of photosynthesis. Instead of the usual bacteriochlorophylls found in the Eubacteria, the halophiles use a pigment called **bacteriorhodopsin** to harness the sun's energy. As we learned earlier (see Essay 6.2), bacteriorhodopsin uses light energy to pump protons out of the bacterium, creating a unique prokaryotic version of the ATP-generating chemiosmotic differential found in chloroplasts.

The thermophiles live under incredibly harsh conditions, some of them thriving at 90° C (194° F, near boiling). In the hot springs of the deep-sea Galapagos rift, bacteria survive at even higher temperatures. (Recent studies of *Pyrodictium*, a rift bacterium, reveal that it thrives at 105° C.) The thermoacidophiles live under even harsher conditions, some thriving in strong acids heated to near boiling. Surprisingly, the internal pH of such Archaebacteria remains close to neutral.

The Eubacteria

The Eubacteria are more familiar and include species that tend to live under less drastic conditions than those tolerated by Archaebacteria. While many are heterotrophic—requiring organic nutrients—

three groups are phototrophs and a number fall into a category we have seen in the Archaebacteria, the chemotrophs. The phototrophic Eubacteria all utilize chlorophylls, although the specific type may vary from the chlorophyll *a* and *b* used by plants and algae. The chemotrophs are bacteria that utilize inorganic substances such as iron and sulfur compounds from the earth's crust. Of the heterotrophs, some are pathogens—disease-causing parasites—of animals (including humans) and plants. Others become pathogenic when accidentally introduced into other organisms. But the vast majority are independent, free-living soil and water bacteria; these are the **reducers** or **decomposers,** vital recyclers of essential ions and molecules. Through their fermenting and decomposing processes, carbon dioxide, water, nitrates, phosphates, sulfates, and numerous other critical substances are released to be used again by the autotrophic forms of life.

Eubacterial Characteristics. Traditionally, Eubacteria have been classified according to their shape and their reactions to certain stains. Today, it is known that shape is not a very useful phylogenetic character. As we mentioned, the phylogeny now emerging is based more on biochemical

characteristics than simple morphology. Still, shape is useful in identifying certain kinds of bacteria. The three common shapes of bacteria are rodlike, spherical, and spiral, known respectively as **bacillus, coccus,** and **spirillum.**

The rodlike bacilli occur as single cells and in chains (Figures 18.14a and b). Some chain formers are enclosed in sheaths. Many bacilli are also known for their ability to form highly resistant, thick-walled **endospores** (Figure 18.15) in response to unfavorable conditions. These dehydrated bodies contain the cellular components held in a state of dormancy, ready to absorb water and resume their metabolic activities when conditions improve. The spherical or coccoid forms occur singly, in pairs (diplococcus), beadlike chains (streptococcus), grapelike clusters (staphylococcus), and in groups of eight (sarcina) (see Figures 18.14c through g). The spiral-shaped cells, which form a group called the **spirochetes,** commonly have long flagella (see Figure 18.14h).

As was mentioned, Eubacteria are characterized by peptidoglycan cell walls made up of amino sugars. In many, the wall is quite dense, consisting of several layers of the amino sugars interconnected by peptide side chains. In addition, the cell walls of many Eubacteria are coated by a polysaccharide

18.15 THE ENDOSPORE

Many bacteria, particularly those of the soil and water, survive unfavorable conditions by forming tough-shelled, resistant endospores. Within each endospore is the nucleoid—the naked chromosome—and a bit of dehydrated cytoplasm. Under more ideal conditions, the endospore will take in water and the cell will resume activity. Bacterial endospores are found everywhere and some are so heat resistant that they survive boiling temperatures. Where sterile conditions are vital, endospores can be killed by successive boilings, by heat and pressure, or, where practical, by the use of chemical agents called bacteriocides.

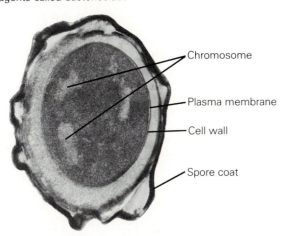

Chromosome

Plasma membrane

Cell wall

Spore coat

called **teichoic acid,** which appears to play a role in the transport of ions into the cell.

In other Eubacteria, the peptidoglycan wall is much thinner and far more fragile, and the coating of polysaccharides is absent. In its place, surrounding the wall, is a complex, membranelike structure containing lipoprotein, phospholipid, and lipopolysaccharide. This "outer membrane," as some call it, acts as a barrier to many substances, including penicillin and certain dyes. We mention that because the reaction to dyes is important in the classification of bacteria, and it is medically important to know which bacteria are not susceptible to penicillin. (The two types of cell walls are compared in Figure 18.16.)

The staining characteristics of Eubacteria with thick peptidoglycan walls are quite distinct from those without. In particular, the two groups react differently to **Gram's stain,** a deep purple dye (named for its discoverer, Hans Christian Gram). Bacteria with dense peptidoglycan walls readily absorb the bright purple stain and are referred to as **Gram positive** bacteria. Because of the outer membrane, bacteria of the second group reject the stain and are dubbed **Gram negative** (Figure 18.17).

In the presence of penicillin, Gram positive bacteria—those with the dense peptidoglycan layer and teichoic acids—cannot synthesize the peptidoglycan wall properly, and newly divided cells rapidly die. The cytoplasm and plasma membrane of the bacterial cell are under considerable osmotic tension, and only the presence of a tough, flexible wall keeps the cell from overfilling with water and bursting. Being without a wall spells trouble for most organisms in the rough and tumble bacterial realm. Thus, penicillin is often the antibiotic of choice for controlling Gram positive bacteria. However, penicillin is useless against infections by Gram negative bacteria. Those cells with the complex outer membrane go right on dividing and synthesizing cell walls; for that reason, other antibiotics such as tetracycline and streptomycin are used since they both stop growth by inhibiting protein synthesis.

Bacterial Villains.   The worst of the pathogenic bacilli include the agents of such dread diseases as leprosy, typhus, black plague, diptheria, and tuberculosis. These diseases are so dangerous that they have been the focus of a great deal of research attention. As a result, researchers have developed such effective prevention and curative treatment throughout most of the world that most physicians will never be called upon to treat patients with these diseases.

18.16 THE EUBACTERIAL CELL WALL

(a) While eubacterial cell walls contain peptidoglycan, there are two basic differences in their organization. In one group the peptidoglycan is very dense and is coated by an outer layer of teichoic acid. In the second group, the peptidoglycan is quite thin and is covered by a complex outer membrane. These differences are reflected in how diagnostic stains are taken in and how the cell reacts to antibiotics. (b) The structural units of peptidoglycan are two amino sugars, N-acetylglucosamine and N-acetylmuramic acid. To form the cell wall, the subunits are interconnected by peptide bridges, forming dense rows of the tough conglomerate.

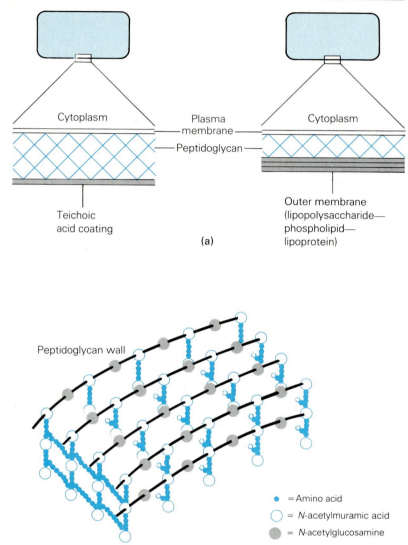

Cytoplasm

Plasma membrane

Peptidoglycan

Cytoplasm

Teichoic acid coating

Outer membrane (lipopolysaccharide— phospholipid— lipoprotein)

(a)

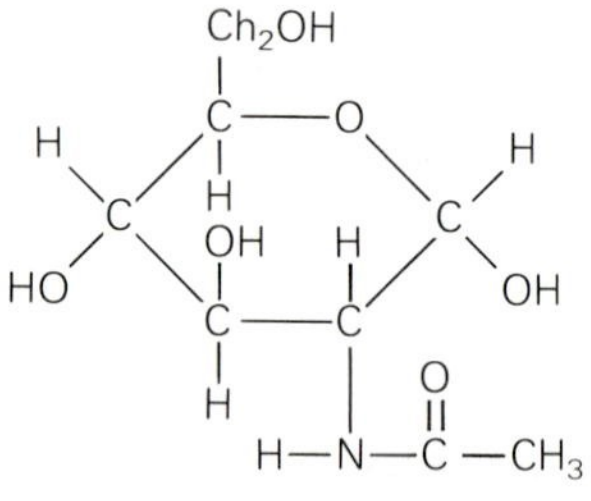

N-acetylglucosamine

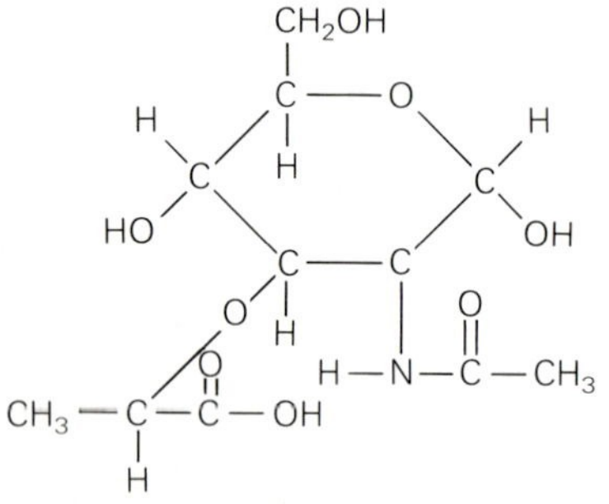

N-acetylmuramic acid

Peptidoglycan wall

- = Amino acid
- = N-acetylmuramic acid
- = N-acetylglucosamine

(b) Amino sugars

Among the bacteria still capable of inflicting misery on us are relatively common anaerobic soil bacilli from the genus *Clostridium: Clostridium tetani, Clostridium botulinum,* and *Clostridium perfringens. C. tetani* thrives in oxygen-free pockets of the soil, where it decomposes dead organic material. Unfortunately it can also thrive in deep oxygen-poor wounds, from which its toxins spread to the nervous system producing **tetanus,** an excrutiatingly painful condition with spasmotic muscle contractions that leave the body rigid and arched for prolonged periods. *C. botulinum,* which can grow in improperly canned food, causes a food poison-

ing called **botulism,** which brings on muscle paralysis and respiratory arrest. *C. perfringens* is the agent of **gas gangrene,** literally a rotting of the flesh surrounding an infected wound. Untreated, gas gangrene is invariably fatal.

Coccoid pathogens, such as certain staphylococci, are commonly involved in minor skin infections, boils, and pimples, but under some conditions can create enormous, dangerous infections. One notorious group crops up occasionally in hospitals and creates troublesome, dangerous, and resistant infections, especially among the newborn. **Strep throat** can be brought on by a strepto-

18.17 GRAM REACTIONS

The bacteria on the left are Gram positive (purple), while those on the right are Gram negative (red). The reaction to Gram staining is an important diagnostic tool, particularly in clinical use, since Gram negative bacteria are generally not inhibited by penicillin. In the Gram reaction, a purple stain is taken into the cytoplasm and cannot be washed out with ethyl alcohol. Conversely, the purple stain cannot enter the Gram negative bacterium and is readily washed from the wall with alcohol. These bacteria have an outer coat of mucopolysaccharides that resist the purple stain. The red coloring within these cells is a "counterstain" that does penetrate, making them easier to spot.

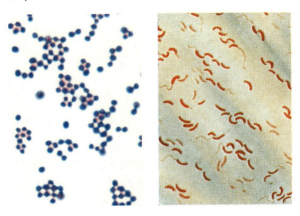

coccus. Because of the availability of antibiotics, strep throat isn't nearly as serious a health threat it once was in the United States. Actually, the body's own reaction to the streptococcus is the real threat; hyperallergic reactions in the form of scarlet fever, rheumatic heart disease, and rheumatic nephritis are still significant causes of death, especially in undeveloped regions.

Neisseria gonorrhoeae is a socially unacceptable diplococcus that causes the venereal (sexually transmitted) disease **gonorrhea** (Figure 18.18a). Signs of gonorrhea are readily apparent in men (painful urination along with a pus discharge), but often go undetected in women since the infection is more internalized. Should the infection spread into the pelvic region it can cause sterility. Since gonorrhea can be transmitted during birth and there is a risk of its causing blindness in the baby, it is routine to treat the eyes of newborns with drops of silver nitrate. For a time, gonorrhea was well controlled by penicillin therapy, but today, the disease is on the rise. It seems that evolution permits bacteria to adapt to the challenges of their changing environment, and now we are forced to deal with new, antibiotic-resistant strains.

18.18 VENEREAL DISEASE

Two common agents of sexually transmitted disease are *Neisseria gonorrhoeae* (left) and *Treponema pallidum* (right), the eubacteria of gonorrhea and syphilis, respectively. Gonorrheal bacteria are diplococci, occurring in pairs within a capsule. The agent of syphilis is a spirochete. Both gonorrhea and syphilis are almost always transmitted through sexual contact, although both occur in infants born to infected mothers. Infant gonorrhea is contracted, particularly in the eyes, during

the birth process when the baby passes through the infected cervix and vagina. For this reason, the eyes of newborns are routinely treated with an antibacterial agent whether gonorrhea is suspected or not. The spirochetes of syphilis are another matter. They can cross the placenta after the 18th week of development, infecting the baby and causing traumatic, often fatal developmental defects.

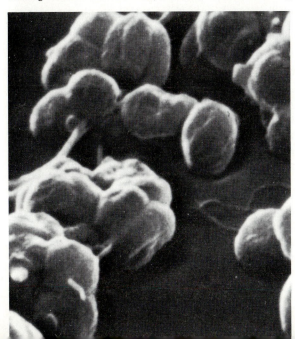

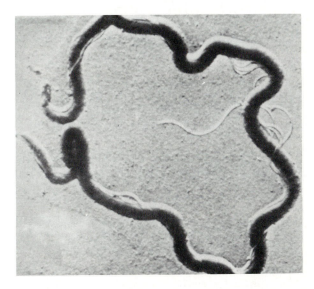

The most notorious of the spirillum form of bacteria is *Treponema pallidum,* the corkscrew-shaped spirochete of **syphilis,** another common sexually transmitted disease (Figure 18.18b). An early sign of syphilis is usually the appearance of a "chancre," a small, ulcerated sore on the genitals or mouth. However, it soon disappears and an unwary person might believe that all is well. It isn't. In its more progressive stages, syphilis has been called the "great pretender" since its widespread effects mimic many other diseases. If it is allowed to go untreated, the spirochetes will eventually enter the nervous system, permanently damaging brain tissue and causing blindness, insanity, and death. The spirochete is also known to cross the placenta, infecting the embryo, and producing serious birth defects in the baby. Because of their disastrous effects, many of the bacterial villains are quite well known, but let's now leave them and shift our attention to a few less familiar groups of bacteria.

Other Eubacteria

Myxobacteria. The **Myxobacteria** include the **slime** or **gliding bacteria** (Figure 18.19). They secrete a slime, or mucus, that forms a slippery layer on solid surfaces, allowing the bacteria to "creep" or "glide" over it. Some investigators

18.19 REPRODUCTIVELY ADVANCED BACTERIA

The Myxobacteria are more commonly known as the *slime bacteria* because of the layer of slippery material they secrete. These bacteria are unusual in that some produce fruiting bodies in which spores form. Because of this apparent specialization, they are considered to be multicellular.

18.20 SHEATHED BACTERIA

Chlamydobacteria form sheaths around themselves. The sheath generally attaches to some object in its watery surroundings or to the water surface itself. This particular organism is *Spaerotilus,* a bacterium capable of oxidizing iron compounds.

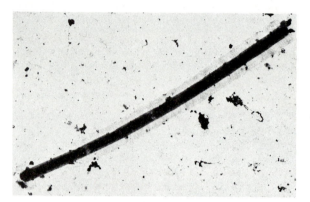

believe that the mucus is secreted in a concentrated form and that it expands as it takes up water, thereby pushing the organism along. (The eukaryotic diatoms apparently move in a similar fashion.)

The reproductive behavior of these bacteria is interesting because it is so advanced, much like that of the eukaryotic protists. Individual cells converge into a slippery mass, from which a multicellular **fruiting body** grows, producing a **cyst** in which many spores form. When the cysts rupture, the spores are released. Myxobacteria thus can reach a degree of cellular specialization very similar to that of simple multicellular organisms.

Chlamydobacteria (Mycelial Bacteria). The **Chlamydobacteria** or **mycelial bacteria** appear to be intermediate between Eubacteria and fungi. They are sheathed, filamentous bacteria that occur in colonies, often in moist soil. The sheath consists of extracellular, secreted material in which the bacterial cells are lined up end-to-end (Figure 18.20). One species is common in polluted water; its filaments actually attach to the underside of the water surface. In its dispersal phase, swarms of motile, flagellated cells are released.

One of the best-known groups, the **Actinomycetes,** have highly branched filaments without the cross-walls that occur in some other species. These filaments may give rise to thin, clublike, asexual reproductive structures that stand vertically from the mass. The swollen tips break off and release sporelike reproductive cells. These may either be flagellated or nonmotile and windborne, and each may potentially start a new colony. Actinomycetes may also reproduce by fragmentation in which the filaments break into individual bacteria, as well as by binary fission.

Chlamydobacteria are medically important as the source of several antibiotics, notably streptomycin, aureomycin, and actinomycin, produced by *Streptomyces*, *Aureomyces*, and *Actinomyces*, respectively.

Mycoplasmas.   The **mycoplasmas** are among the smallest living things, with the cells of some species being less than 0.16 μm in diameter, far smaller than most bacteria. In fact, they are smaller than some viruses. Oddly, they lack a rigid cell wall and therefore have no definite shape. Almost all are parasites of animals including humans, but some mycoplasmas, or something very much like them, have recently been found in plant cells. They are completely resistant to penicillin, presumably because penicillin normally kills by interfering with the growth of bacterial cell walls. They sometimes form filamentous colonies resembling certain fungi. Reproduction is by binary fission. One form is responsible for a relatively mild form of human pneumonia (Figure 18.21).

18.21 THE SMALLEST CELLS

The structure of mycoplasmas indicates that they are true cells, although they lack the cell wall found in other bacteria. It is extremely doubtful whether organisms smaller than the mycoplasmas exist, other than viruses. Anything smaller than the size of the mycoplasmas would not have the volume to carry out the synthetic processes necessary for life.

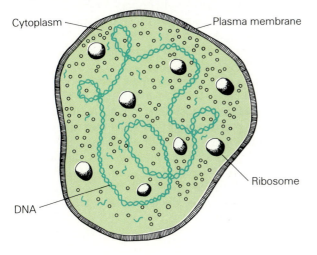

18.22 THREE COMMON CYANOBACTERIA

A few of the diverse cyanobacteria include the filamentous, beadlike *Nostoc* (a), the slowly undulating, filamentous *Oscillatoria* (b), and the spherical *Gloeocapsa* (c), enclosed in its gelatinous wall. All are inhabitants of fresh water. The larger, thick-walled cells of *Nostoc* are known as *heterocysts,* cells that specialize in nitrogen fixation.

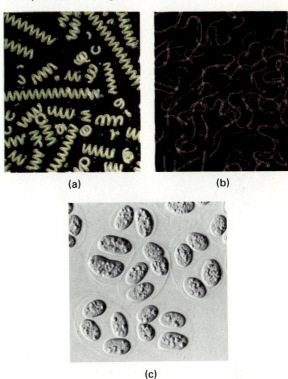

(a)   (b)

(c)

Photosynthetic Eubacteria: The Cyanobacteria.   The **cyanobacteria** are widely distributed. There are some 2500 species, and they live in aquatic environments including oceans, ponds, lakes, tidal flats, moist soil, swimming pools, and even around leaky faucets. Some flourish in hot springs where temperatures reach 75° C (167° F). They can often be recognized by their blue-green color, but they also may be black, purple, brown, or red. Cyanobacteria abound in the marine environment where, through their photosynthetic activity they release significant amounts of oxygen and contribute to the organic matter on which marine life depends.

Cyanobacteria occur as single cells, as colonies of cells, and even in a simple multicellular state (Figure 18.22). Many cyanobacteria occur in long chains of cells organized into filaments, which may or may not be branched. In other instances, cyano-

18.23 CYANOBACTERIAL STRUCTURES

Electron microscope studies of cyanobacteria reveal that their cellular organization can be quite complex. Unlike other eubacteria, they do have some membranous structure in the cytoplasm. Numerous chlorophyll-containing thylakoids form a prominent part of the cytoplasm. This is, of course, where the light reactions of photosynthesis occur. Although the thylakoids are not organized into grana, as we find in the chloroplasts of eukaryotes, their complexity is startling for a prokaryote.

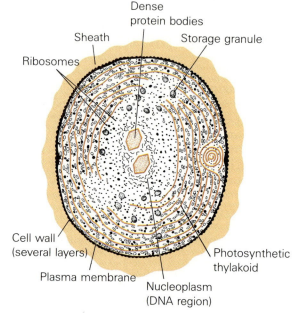

bacteria form gelatinous masses of no definite shape. The cyanobacteria can move: filamentous forms such as the common *Oscillatoria* rotate in a screwlike manner, while the gelatinous forms glide along in a mucuslike slime they produce, like the slime bacteria mentioned earlier. Some species of cyanobacteria have a degree of cell specialization, their filaments containing specialized vegetative cells that manufacture food which is transferred to reproductive cells through thin connecting strands. Similar connections carry food to other cells specialized for a vital process called **nitrogen fixation** (see below). Such specializations may seem surprising, since prokaryotes are generally thought of as single-celled and primitive. (It once more emphasizes the problem of building hard-and-fast rules in biology.) Finally, as far as it is known, reproduction in the cyanobacteria is entirely asexual, most commonly by fission.

The cells of cyanobacteria reveal an amazing degree of complexity (Figure 18.23). Their chlorophyll is located on numerous internal membranes—the photosynthetic thylakoids. Actually, the entire photosynthetic cell is comparable to a eukaryotic chloroplast. Photosynthesis in the cyanobacteria is nearly identical, biochemically, to that of algae and plants. Like the algae and plants, their photosynthetic pigments include chlorophyll *a* and the accessory pigment beta carotene, although they lack chlorophyll *b*. In addition, cyanobacteria produce blue pigments called **phycobilins,** which are important in the capture of light energy and contribute to the characteristic color of these bacteria. The glucose produced through their photosynthesis is stored in their own form of starch, which is similar to animal glycogen.

A number of cyanobacteria produce specialized, nitrogen-fixing cells called **heterocysts.** Their role is to incorporate atmospheric nitrogen into a form useful for producing amino acids and other nitrogen-containing molecules. Interestingly, the formation of nitrogen-fixing heterocysts is inhibited in many species when alternate sources of nitrogen—ammonia or nitrates—are added to their medium. We will look into the details of nitrogen fixation in Chapter 41.

THE EUKARYOTES: A DIFFERENT MATTER

The origin of eukaryotes is shrouded in mystery. While it seems apparent that the earliest eukaryotes must have been protists—relatively simple, unicellular forms—there is no fossil record to support the idea. The earliest undisputed fossil records of eukaryote life consist of worm tracks from the ancient sea floor, estimated at about 800 million years old. There are also some arguably eukaryotic fossils, thought to be red and green algae, in the 1.3 billion-year-old Bitter Springs limestone deposits of central Australia.

In any case, the prokaryotes appeared first and it seems that the eukaryotes evolved from them. There are many ideas about how the required dramatic changes came about, but one, the **endosymbiosis hypothesis,** is particularly convincing. Part of its plausibility stems from evidence from organisms that are still with us today.

The Endosymbiosis Hypothesis

The endosymbiosis hypothesis is an ingenious and widely accepted explanation of the origin of the eukaryote cell. It maintains that the eukaryote line arose when cells with various specific capabilities were incorporated as **endosymbionts** into other cells. (*Symbiosis* simply refers to a close relationship between two species.) The invaders lived somewhat independently at first, but soon an interdependence or **mutualism** was established. It was through these incorporations, goes the hypothesis, that the newly emerging eukaryote came into possession of mitochondria, chloroplasts, flagella or cilia, and centrioles.

The endosymbiosis hypothesis is not new. In fact, some aspects of it were suggested early in this century. But more recently, the hypothesis has been revived and new evidence brought to light. A large part of the revival has resulted from the interest and efforts of Lynn Margulis of Boston University. She has used information from a wide variety of fields to enrich the hypothesis. In her version of endosymbiosis, she maintains that the primitive eukaryote cell was derived by at least three separate events that involved the union of four separate prokaryotic lines (Figure 18.24).

Line A, which she calls the **protoeukaryote,** had evolved the ability to move its plasma membrane, and thus was able to engulf particles and form digestive vacuoles and other internal membranous structures. It, then, was the first predator. It was capable only of anaerobic respiration (glycolysis), but it may have had multiple chromosomes and a nuclear membrane. There is no such organism alive today.

Line B was an *aerobic* bacterium, not too dissimilar to *E. coli,* that was engulfed by the line A cells. Eventually, a **mutualistic symbiosis** (mutually beneficial coexistence) developed so that engulfed line B cells were not digested, but instead began to help break down other foodstuffs ingested by the larger cell. In time, the two became completely dependent on one another. Finally, line B cells lost the ability to live outside their hosts and, Margulis hypothesizes, their descendants exist today as mitochondria. Over time, much of the mitochondrial genetic material (from the former line B bacteria) was shifted to the host chromosomes, but even today, mitochondria retain a complete functioning set of tRNAs, bacterialike ribosomes, and a bacterialike circular chromosome of nearly naked DNA. That, we are told, was the first symbiotic event.

The second symbiotic event, she suggests, perhaps involves a bit more faith. Here, a new organism enters the picture: *line C.* According to Margulis, line C resembled a modern prokaryote, the **spirochete,** in that it was long, thin, and highly motile. Margulis assumed that it contained microtubules in a 9 + 2 arrangement. The line C organism first attached to the outer surface of the line AB complex to become the first flagellum. It introduced the protein *tubulin,* which was to give rise to cilia, flagella, basal bodies, centrioles, and spindle fibers. Acquisition of cilia and flagella gave further mobility to the evolving eukaryote cell. With the latter, the stage was set for mitosis and meiosis.

The third symbiotic event was the acquisition of cells of *line D.* Margulis proposes that line D cells were simply ancient cyanobacteria. Note that in none of these steps did it pay the host to be too voracious. The eukaryote cell able to engulf photosynthetic bacteria without digesting them acquired a reliable source of energy—and gave its guests mobility, nutrients, and protection. Part of the evidence for this last step is that chloroplasts retain their own prokaryote type of ribosomes and circular DNA, and even have their own tRNAs. Margulis argues that there are hundreds of known cases in which modern host organisms have taken photosynthetic symbionts into their cytoplasm.

The endosymbiosis hypothesis as stated here is not completely accepted by all biologists. There admittedly are numerous holes to be filled. Evolutionist Peter Raven suggested that the real ancestral symbiont of green plants was not a cyanobacterium but a different photosynthetic prokaryote, one that used chlorophylls *a* and *b* and carotenoids, but not phycobilins. Having supposed that this hypothetical ancestral group was now extinct, he must have been pleasantly surprised when a living prokaryote that fit his description was found, in an unlikely place—living as a symbiont within the saclike body of an animal called a tunicate (see Chapter 27) in Baja California (Figure 18.25). Thus, a new prokaryote group, called the **chloroxybacteria,** was promoted to the position of ancestor of the group from which the chloroplasts of green plants evolved.

Margulis' case for endosymbiosis seems strongest for the origin of chloroplasts. Their internal structure and biochemistry are very similar to that of cyanobacteria. Like cyanobacteria and chloroxybacteria, chloroplasts store starch, and they arise

18.24 THE ENDOSYMBIOSIS HYPOTHESIS

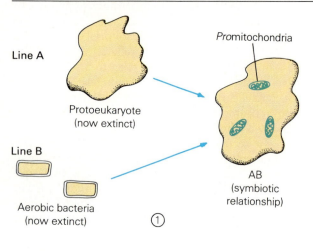

Line A

Protoeukaryote
(now extinct)

Line B

Aerobic bacteria
(now extinct)

Promitochondria

AB
(symbiotic
relationship)

①

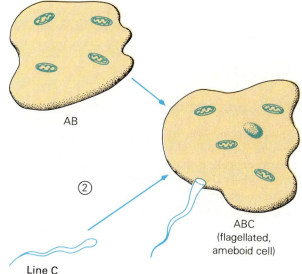

AB

Line C

ABC
(flagellated,
ameboid cell)

②

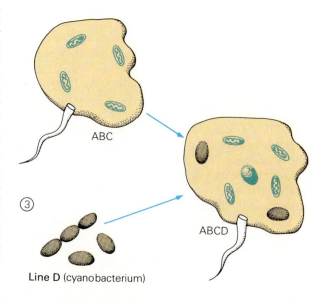

ABC

③

Line D (cyanobacterium)

ABCD

① The union of prokaryote lines A and B. Two hypothetical prokaryotes, now extinct, included a protoeukaryote (line A), which was a simple anaerobe with the ability to phagocytize its food. One organism it fed upon was an aerobic bacterium from line B. Eventually, line B cells became incorporated, and a symbiotic relationship developed. They became interdependent, with the mitochondrialike bacteria carrying on aerobic respiration within the line A cells. The line B cells lost their ability to live outside their hosts. As the host reproduced, the line B cells did likewise, just as mitochondria do in eukaryotic cells today.

② The union of prokaryote lines AB and C. The new AB symbiont now took one step closer to a eukaryotic status by incorporating into its cytoplasm a cell from line C. These new cells brought with them microtubule proteins and the flagellate structure. In addition, it is hypothesized, the tubular centriolar structure arose from this new incorporation. The new development improved the ability of the symbiont to move about, becoming a more efficient heterotroph.

③ The union of prokaryote lines ABC and D. In the final step, the highly improved prokaryote incorporated a cyanobacterial cell with efficient photosynthesizing structures. By this step, the prokaryote had several of the organelles of today's eukaryotes, including mitochondria, contractile proteins, centrioles, cilia, and now chloroplasts. All that remained was for the membranous endoplasmic reticulum and the nuclear membrane to evolve and a modern eukaryote would emerge.

18.25 TUNICATE AND CHLOROXYBACTERIA SYMBIONT

It was recently discovered that this colonial tunicate, *Diplosoma virens,* harbors a photosynthetic bacterial symbiont that contains typical eukaryotic chlorophylls (*a* and *b*) and carotenoids. The photosynthetic symbiont is now indicated as a possible ancestor of today's chloroplasts. Each colony of filter-feeding *D. virens* is about as big as your fingernail and consists of hundreds of individuals.

only by division of other chloroplasts. Both chloroplasts and mitochondrial DNA are susceptible to certain antibiotics that do not affect the eukaryote nucleus—further evidence of the similarity between the two organelles and prokaryotes. When the chloroplasts of the protist *Euglena* have been destroyed by such antibiotics, the parent cell line can be kept growing indefinitely in a nutrient broth, but the chloroplasts never reappear.

The case for the bacterial origin of mitochondria is almost as good. True, mitochondria have internal membranes (cristae) unlike anything seen in bacteria, and they seem to have very few functional genes. But at least one mitochondrial protein, cytochrome *c*, is recognizably similar in shape and amino acid sequence to the cytochromes of certain photosynthetic bacteria. According to Carl Woese, it now seems likely that the original mitochondrion was a photosynthetic symbiont, and that it later became restricted to respiratory tasks.

The case is weakest for the symbiotic origin of

flagellar structures. No flagellar DNA or protein-synthesizing machinery have been reported, and no free-living prokaryotes with the characteristic 9 + 2 microtubule organization are known. However, Margulis has argued for the origin of flagella by symbiosis with an impressive analogy that is truly a parade of symbioses within symbioses. She notes that termites eat wood but have no enzymes capable of digesting cellulose. Instead, all termites harbor in their guts various symbiotic protozoans, that ingest and digest the wood particles. One such symbiotic protozoan, *Myxotricha paradoxa*, lives in the gut of certain Australian termites. In addition, the protozoan harbors as endosymbionts no fewer than *three* bacterial species! One bacterial species lives in the protozoan cytoplasm and aids its host by digesting wood. The other two bacteria live on the surface of *Myxotricha* and provide the protozoan with a unique form of locomotion. What were once thought to be *Myxotricha* flagella turned out to be spirochaete bacteria, wriggling away furiously with their basal ends firmly embedded in the host's membrane (Figure 18.26).

We will leave the bacteria now and consider a group of even smaller creatures, so small and unusual that some researchers have pronounced them as nonliving, or "bits of heredity looking for a chromosome." No matter what one thinks of their lifestyle, however, the viruses can certainly not be considered insignificant.

18.26 ENDOSYMBIOSIS TODAY

The presence in termites of the protozoan symbiont *Myxotricha paradoxa* illustrates that the type of complex symbiotic relationships suggested by Margulis can occur. In this example, a symbiont harbors its own symbionts. Part of the *M. paradoxa's* mobility is provided by numerous spirochetes, which affix themselves to its surface. In addition, it harbors a number of cellulose-digesting bacteria within its cell and additional bacteria living symbiotically on its surface.

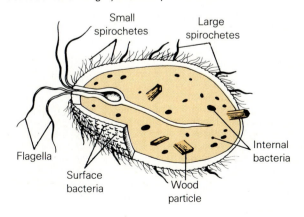

THE VIRUSES

Viruses are minute, biologically active particles made up of a nucleic acid core, a covering of protein, and sometimes an enzyme or two. For this reason we must take a bit of a conceptual leap here because viruses don't fit in any organizational scheme of living organisms. In fact, since their discovery and description, scientists have questioned whether they are actually alive at all since they can crystallize and remain inactive for a seemingly indefinite period.

Adding to the biologist's dilemma, there are many apparently unrelated viruses that share only a few features, such as their extremely small size and highly limited mode of life. Viruses range in size from 20 to 300 nm (a range that extends from the size of large molecules to that of the smallest bacterial cells; see Figure 18.27). But, while they certainly cannot be considered cells, they do have at least one of the fundamental capabilities of living organisms: they can, under the proper conditions, reproduce.

Of course, our interest in viruses is not entirely academic. After all, we humans are susceptible to several hundred viral diseases, not to mention those that infect our crops and domestic animals. The list of human maladies includes smallpox, polio, German measles, chickenpox, mumps, and the many forms of influenza that periodically sweep through our population. More recently, viruses known as oncogenic viruses, long known to be related to cancers in laboratory animals, have been implicated in certain human cancers.

Most of what we know about viruses was fairly recently discovered—a product, more or less, of 20th century research efforts. But the story begins long before that.

The Discovery of Viruses

The discovery of viruses came at a time when the newly emerging field of bacteriology was making its first great gains. By the late 1800s, Louis Pasteur, Robert Koch, and others had convinced the world that bacteria were the agents of some diseases. But other diseases apparently were caused by something else. The methods that usually proved so successful in finding the bacterial villains simply didn't work for these diseases. Oddly enough, the stymied bacteriologists, although unable to find the culprits, were able to produce vaccines that were effective against some of them. For example, vaccines for the prevention of smallpox and rabies in humans and hoof-and-mouth disease in cattle were developed long before the viral agents of these devastating diseases were found. A number of causative factors were proposed, some rather imaginative. As an example, the term *virus* itself means "poison," and for ages, people attributed viral diseases such as yellow fever and smallpox to poisons carried by the "deadly night air."

By 1892, studies of a contagious disease in tobacco, soon dubbed the **tobacco mosaic virus,** revealed several important clues to the nature of viruses. Dimitri Iwanowski, the Russian biologist, extracted juices from infected plants, strained the liquid through an extremely fine filter (the type used to remove bacteria from growth media too delicate to sterilize by heating), and using the filtrate, succeeded in spreading the disease to healthy plants. From his work, Iwanowski was able to establish two facts about the disease agent: it was smaller than a bacterium and it could be transferred from plant to plant—that is, it was contagious. By the turn of the century, the filtration technique had been broadly applied, and the list of these **filterable viruses** grew substantially. But such progress was

18.27 VIRAL SIZES

By comparing several representative viruses with the familiar bacterium *E. coli*, we get some idea of their minute size and the range of sizes found. *E. coli* measures about 3000 nm in length, (about half the diameter of one of our red blood cells). The smallest virus, one of the bacteriophages, is quite minute (about 20 nm), while one of the largest in diameter, the tobacco mosaic virus, is just visible in the light microscope.

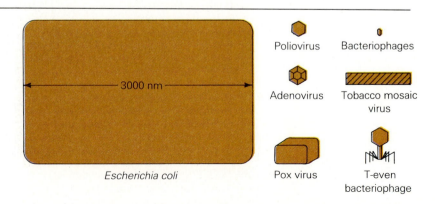

3000 nm

Escherichia coli

Poliovirus Bacteriophages
Adenovirus Tobacco mosaic virus
Pox virus T-even bacteriophage

not without its problems. Despite persistent efforts, the early bacteriologists were unable to grow the filterable agents in the usual bacterial culture media. This meant that the techniques so useful in identifying bacterial disease agents could not be fully applied to the tiny viral agents. Further, they still could not tell whether the filterable agent was a living entity or just a chemical of some kind. The puzzle persisted.

Then, in 1935, the American microbiologist Wendell Stanley trained his light microscope on a drop of filtrate from infected tobacco plants and became the first person to see a virus. Actually, what he saw was the crystalline form—the long, slender crystals of the dormant tobacco mosaic virus (Figure 18.28). (In that same year, the large, rod-shaped virus particle itself was first observed through the newly invented electron microscope.) Stanley also showed that while the tobacco mosaic virus appeared in many ways to be nonliving and could not be grown by any usual bacterial culture methods, it could somehow reproduce if it was present in healthy tobacco plants. Stanley's procedure was straightforward. He introduced a small drop of dilute viral filtrate into a healthy plant. The plant would then develop the disease and from its tissues he could recover a much greater quantity of the viral substance. The search was on to learn more about these peculiar little entities. Let's now focus on some of the things researchers have been able to tell us in the past few decades.

The Biology of Viruses

All viruses are parasites. The reason they cannot be grown on a bacterial growth medium is that they lack the enzymes and metabolic pathways commonly found in living cells; therefore they must invade cells and use the metabolic machinery of their hosts to complete their life cycles.

Each virus particle, or **viron** as the individual particles are known, consists of a core of nucleic acid (either DNA or RNA) and a protein coat called a **capsid.** Whereas an inactive viron contains an array of genes, it cannot carry out even the simplest requirements for living without a living host. So viruses must be considered the most specialized of parasites, lacking all structure except that which facilitates the invasion of a host and the production of more viruses.

Viral Geometry.   Most viruses are either helical or polyhedral (many-sided) in shape, but a few are cubic, and some are brick-shaped. The shape of a viron depends ultimately on the arrangement of

18.28 THE TOBACCO MOSAIC VIRUS

(a) The leaves of an infected tobacco plant are mottled in color and wrinkled in texture. (b) Under the electron microscope, virus particles on an infected plant are seen to form a multi-layered herringbone pattern. (c) In this TEM view, the crystallized virus takes on a cylindrical or rod-shaped appearance.

(a)

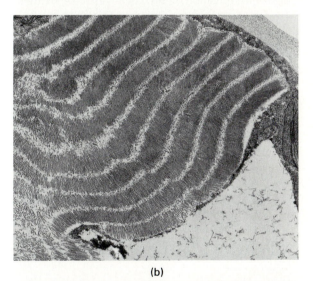

(b)

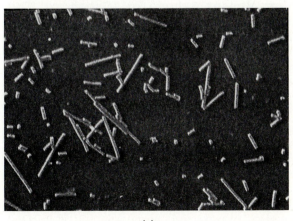

(c)

protein subunits, or **capsomeres,** in the capsid. The helical viruses, for example, consist of a spiral of nucleic acid surrounded by a capsid whose capsomeres follow the nucleic acid spiral. Included among them is the tobacco mosaic virus, the first virus to be described. Influenza viruses also have the spiral form. The polyhedral viruses may sometimes appear spherical in shape, but a close look reveals that the spheres are actually 20-sided polyhedrons, known as *icosahedrons*. Included in the polyhedral viruses are the familiar "even numbered" or T-even bacteriophages (discussed in Chapters 9 and 14), with a polyhedral head and a cylindrical tail. Helical and polyhedral viruses may be enclosed in a complex covering or envelope (Figure 18.29). Among the polyhedral viruses that are surrounded by an envelope are members of the disreputable *Herpes simplex* group. Among the cubic or brick-shaped viruses is the agent of smallpox, which, thanks to the efforts of the World Health Organization, is now believed to be extinct. (The last reported case of smallpox occurred in 1977.)

Biochemical Differences. In addition to these variations in form and covering, viruses can also differ in the kind of nucleic acid core they contain. For example, in some viruses the genetic material is standard double-stranded DNA, just as it is in all metabolizing organisms. But in other viruses, we may find (1) single-stranded DNA, (2) double-stranded RNA, (3) a single large molecule of single-stranded RNA, or (4) several small strands of RNA. The nucleic acid can be circular or linear, and the single-stranded nucleic acid can be either the transcribed strand or the nontranscribed strand. (Many of these differences are listed in Table 18.2.)

How Viruses Behave or Misbehave: The Bacteriophage Revisited. Viruses begin their attack on cells by attaching to the cell surface by use of specialized enzymes. They then penetrate the cell and take over the host's metabolic machinery, as we saw in Chapters 9 and 14. The site of penetration is often quite specific, requiring the biochemical recognition of certain host cell surface proteins by the virus prior to attachment. In some viruses, notably the T-even bacteriophages, the bacteriophage tail releases *phage lysozyme,* an enzyme that digests its way through the bacterial cell wall, preparing the way for the next event (see Figure 9.12). The capsid in the T-even phages remains outside and the nucleic acid is injected into host cytoplasm.

Once a T-even phage enters host cytoplasm, it is called a **prophage.** The prophage DNA can take either of two routes; it can enter either a short, deadly **lytic cycle** or an extended **lysogenic cycle** (see Figures 14.5 and 14.6). In the first, you may recall, the invading prophage disassembles the host's DNA and uses it for its own replication. It then uses the host transcription and translation machinery for producing many new viral coats, tail and tail fibers (landing gears), and viral enzymes. The amino acids to be assembled into these viral proteins are, of course, also "borrowed" from the host (like your freeloading Uncle Charlie, who, having eaten all of your groceries, borrows your overcoat as you show him out the door). Finally the host bacterium is lysed, literally bursting, and the new infectious virons are released.

In the lysogenic cycle, the viral DNA inserts into the host chromosome and in this inert form is passed through many generations of bacteria. Unfortunately, the apparently harmless lysogenic phase can change to the disruptive lytic cycle. The transition from a lysogenic to a lytic prophage can be brought about by a number of agents such as ultraviolet light or certain chemicals. Such appears to be the case in one human virus, which causes fever blisters to appear after a day's exposure to bright sunlight at the beach. The lysogenic and dormant **Type I herpes simplex** virus, hiding away in our nerve cells, rapidly enters a lytic cycle, bringing on the familiar blisters on the lips. (We will look into the activities of the more disconcerting **Type II herpes simplex** shortly.)

Infection by Animal Viruses. The manner in which animal viruses penetrate their hosts is generally similar to the action of T-even phages. But, as you might expect, the differences between prokaryotic and eukaryotic cells will have some influence on the way viral infection occurs. The capsids of animal viruses, like those of the T-even phages, contain specific sites that have complementary target sites on the host plasma membrane. In the adenoviruses (the viruses of respiratory infections), active sites are at the corners of the polyhedron, while in the enveloped myxoviruses (measles, mumps, and some flu viruses) they are on "spikes" on the envelope.

There are no injecting mechanisms in animal viruses, so they must enter the cell in a different manner (Figure 18.30). In those viruses lacking an envelope, the entire viron passes through the host's plasma membrane. In the enveloped viruses, the envelope itself sometimes fuses with the membrane at a complementary site, and the

18.29 **VIRAL DIVERSITY**

(a) In the naked, spiral or helical-shaped tobacco mosaic virus, the capsomeres follow the spiral form of the nucleic acid core. **(b)** The spiral-shaped influenza virus is surrounded by a dense envelope, complete with a number of spiked projections, important to its ability to penetrate cells. **(c)** The adenovirus appears roughly spherical in this electron micrograph, but it actually takes the form of the icosahedron drawn here. The tiny spheres are capsomeres, the proteinaceous units of the capsid. **(d)** The herpes simplex virons seen here are polyhedral viruses that are surrounded by an envelope. **(e)** One of the most complex of the viruses is the ''moon-lander'' T-even bacteriophage, with its polyhedral head and complicated tail. The head contains the nucleic acid core, while the tail is important in attaching to a host bacterial cell.

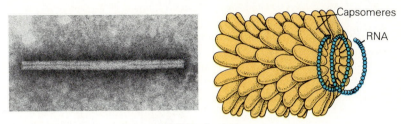

(a) Naked spiral
(tobacco mosaic virus)

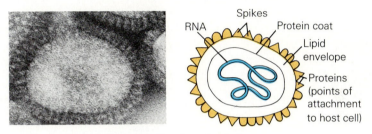

(b) Enveloped spiral

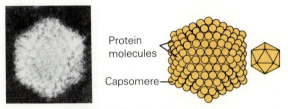

(c) Naked polyhedron

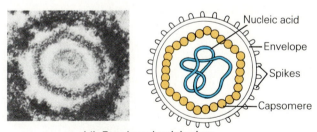

(d) Enveloped polyhedron

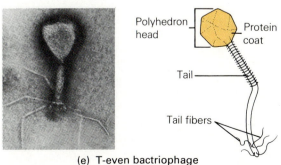

(e) T-even bactriophage

TABLE 18.2

VIRUSES AND HUMAN DISEASES

DNA Viruses	Diseases
Poxviruses (Brick-shaped)	
Vaccinia Type	Smallpox
Paravaccinia	Nodules on hands
Herpesviruses (Icosahedral, enveloped)	
Group A	
Herpes simplex 1	Fever blisters, respiratory infections, encephalitis
Herpes simplex 2	Genital infections, cervical cancer (?)
B virus	Encephalitis
Group B	
Varicella-zoster	Chicken pox
Cytomegalovirus	Jaundice, liver/spleen damage, brain damage
Ungrouped	
Burkitt lymphoma	Burkitt lymphoma, Hodgkin's disease, infectious mononucleosis
Adenoviruses (Icosahedral, naked)	
Humanadenoviruses	Respiratory infections
Papoviruses (Icosahedral, naked)	
Human papilloma	Warts

RNA Viruses	Diseases
Picornaviruses (Icosahedral, naked)	
Enteroviruses	
Poliovirus	Poliomyelitis
Coxsackie A	Muscle and nerve damage, common cold, meningitis
Coxsackie B	Meningitis, paralysis
ECHO	Paralysis, diarrhea, meningitis
Rhinoviruses	Common cold
Arboviruses (Helical, enveloped)	
Dengue	Fever, rash
California encephalitis	Encephalitis
Myxoviruses (Helical, enveloped)	
Influenza viruses	Influenza
Type A_O, A_1, A_2	
Type B_O, B_1, B_2	
Type C	
Paramyxoviruses	
Sendai virus	Common cold
Mumps virus	Mumps
Pseudomyxoviruses	
Measles	Measles, pneumonia, common cold
RNA Tumor Viruses	Breast cancer(?)

Miscellaneous Viruses	Diseases
Rubella virus	German measles
Hepatitis Viruses	
Type A	Short incubation hepatitis (infectious)
Type B	Long incubation hepatitis (serum)
Type nonA-nonB or C	Hepatitis

viral nucleic acid enters there. In other instances the host cell obliges the virus by engulfing it through phagocytosis, enclosing the virus in a food vacuole. The host's lysosomal enzymes are then released into the vacuole and attack the viron just as it would a food particle. However, the enzymes are only able to hydrolyze the protein capsid. This releases the active, infectious core of the virus and the invasion is complete.

Once the viral DNA is released, reproduction can begin. Just how replication and capsid synthe-

sis occurs depends on the organization of nucleic acids in the virus. In the double-stranded DNA viruses, the first viral genes to be transcribed code for the replication enzyme DNA polymerase and viral DNA replication follows. Next, transcription and translation occur and the protein capsids are produced. In the single-stranded RNA viruses, replication often begins with the synthesis of a special, double-stranded RNA, which becomes the template for the replication of more viral single-stranded RNA. In the polio virus, the single-

18.30 ANIMAL VIRUSES ATTACKING HOST CELLS

(a) To penetrate its host cell, the herpes simplex virus first adheres to special sites on the cell. Then its envelope fuses with the plasma membrane and the nucleic acid core and resident enzymes are dumped into the cell. (b) In other

instances, animal viruses are taken into the host cell by active phagocytosis. Once inside, the protein envelope and coat may be digested away by lysosomal enzymes and the nucleic acid core released.

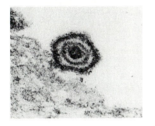

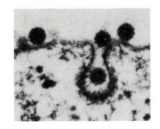
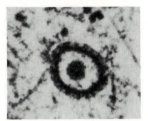

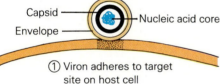

Capsid — Nucleic acid core
Envelope —

① Viron adheres to target
site on host cell

Capsid — Nucleic acid core
Envelope —

① Viron adheres to target
site on membrane

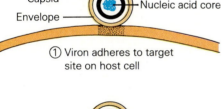

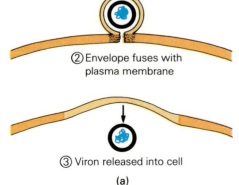

② Envelope fuses with
plasma membrane

③ Viron released into cell

(a)

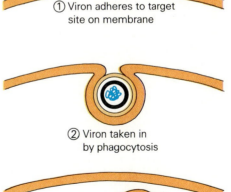

② Viron taken in
by phagocytosis

③ Lysosome fuses,
releases enzymes

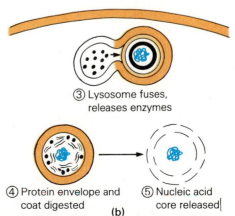

④ Protein envelope and
coat digested

⑤ Nucleic acid
core released

(b)

18.31 PUTTING THE HOST CELL TO USE

(a) The virons of the myxoviruses are assembled in the host cell, but the envelope is formed in an odd manner. As the electron micrograph indicates, the virons reach the cell surface where they join the membrane and are "budded" from the cell. Thus, a bit of the plasma membrane becomes the viral envelope. (b) A single-stranded RNA virus, the Semiliki Forest virus, or SFV (named for the Uganda rain forest in which it was first found), emerges in this manner, providing itself with a complex spiked envelope. After replication and capsid synthesis have occurred in the host cell, the virus enters a final phase of preparation for its emergence, whereupon the spike proteins are synthesized. The spike proteins pass into host endoplasmic reticulum for modification, and are transported to the Golgi apparatus, where the spikes are assembled and transported to the host plasma membrane. Upon arrival both the Golgi membrane and the finished spikes are incorporated into the plasma membrane's protein-phospholipid layers. When the newly produced SFV particles reach the spike-studded plasma membrane, budding can proceed.

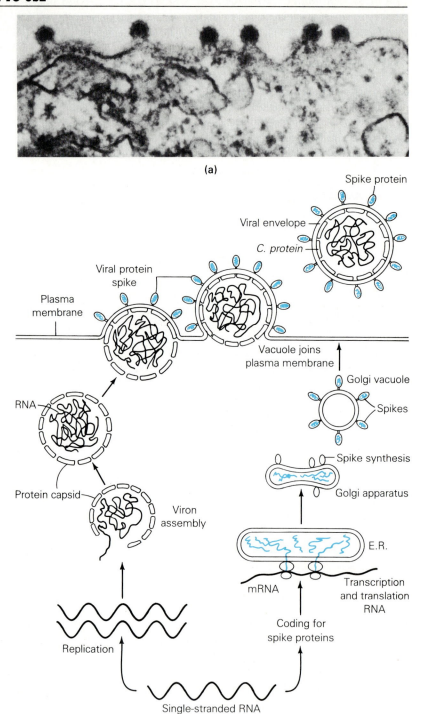

(a)

(b)

stranded RNA doubles as messenger RNA, attaching to host ribosomes and producing both replicating and synthesizing enzymes. The translation product, it turns out, is an enormous polypeptide that is later cleaved into viral enzymes and capsid proteins.

The double-stranded RNA viruses include the poorly understood **reoviruses,** common agents of respiratory infection. In these the genome, the protein-synthesizing genes, is divided into several double-stranded RNA fragments. Each is replicated much like DNA, using enzymes carried in by the virus itself. In the case of the myxoviruses (influenza, measles, and mumps), the new viruses are assembled inside the cell but their envelopes come from an unusual source. The newly replicated RNA segments, surrounded by capsids but lacking their envelopes, merge with and use bits of the host plasma membrane for their outer coverings. The new virons, sporting a "borrowed" envelope, are then budded off the surface of the cell (Figure 18.31).

A group of single-stranded RNA viruses, the **retroviruses,** infect humans in a remarkable way. Using the enzyme *reverse transcriptase*, the invading viral RNA strand can transcribe from its RNA sequence a double-stranded DNA/RNA "hybrid." Next the RNA is enzymatically removed and replaced by a matching DNA strand (Figure 18.32). The final product, a regular, double-stranded DNA molecule, is then inserted into one of the host chromosomes. As the infected cell divides, the DNA replica of the RNA virus is proliferated right along with it. It may even be passed down through the generations in the eggs and sperm. Some retroviruses even ensure the rapid proliferation of their host cells by *transforming* them into fast-growing cancer cells. Further, in recent times, a retrovirus has become strongly suspect as the agent responsible for AIDS (acquired immune deficiency syndrome; see Chapter 31).

At some later time, the incorporated viral DNA cuts loose from the host chromosome and reverts to the other strategy—taking over the cell. It simply transcribes many infectious, single-stranded RNA molecules, produces the necessary capsids, and kills the cell. Such RNA viruses will switch to this behavior whenever something goes wrong with the host cell's replicating activity (as if the virus recognizes that the host is in trouble and it is time to jump ship).

These rather bizarre but fascinating viral capabilities have not escaped the attention of that new generation of biologists, the genetic engineers. Using reverse transcriptase, they have streamlined

18.32 REVERSE TRANSCRIPTION

Acting on a single-stranded RNA molecule, the enzyme reverse transcriptase first produces a hybrid DNA/RNA molecule. The RNA is then broken down and a new DNA polymer is assembled in its place. The double-stranded DNA then contains the viral genetic coding. It can be transcribed into mRNA for viral protein synthesis or into single-stranded RNA for more virons.

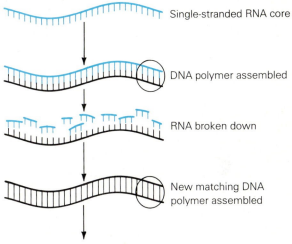

Single-stranded RNA core

DNA polymer assembled

RNA broken down

New matching DNA polymer assembled

Inserted into host chromosome

the task of isolating and identifying specific DNA sequences for cloning in bacterial plasmids (see Chapter 14). In the usual procedure, great numbers of randomly broken DNA segments are inserted and cloned with the hope of later isolating those that code for the desired protein product. While this "shotgun" technique works, it is quite cumbersome. Recall from our previous discussions that much of DNA consists of introns, or intervening sequences, that do not code for protein. When messenger RNA is transcribed, these introns are excised as posttranscriptional modification goes on, and the finished mRNA consists mainly of the polypeptide coding sequence. In using reverse transcriptase, the procedure is greatly simplified by first isolating and collecting finished messenger RNA that codes for the desired polypeptide, and then, using the viral enzyme, synthesizing correlating segments of DNA. The clearly identified, intron-free DNA segments are then used in the cloning procedure.

A Word About Plant Viruses.   While the list of animal viruses is long and impressive, plants are certainly not exempt from these insidious invaders. In most instances plant viruses are spread by insects feeding from one plant to another, or by insect pollinators carrying infected pollen from plant to plant. Viral infections are particulary dangerous to crops and are common in potatoes, sugar beets, wheat, soybeans, and tobacco. Infected plants are easily recognized by the presence of mottled and wrinkled leaves, loss of color, tumors, dying tissue, and stunted growth. The tobacco mosaic virus does so well in the domestic tobacco plant that the minute viral particles can account for one tenth of an infected plant's dry weight!

Herpes Simplex II: Scourge of the '80s

We cannot end our discussion of viruses without saying a bit more about the **herpes simplex II** virus. It is one of "the" diseases of the '80s, along with the much more severe AIDS (the '60s had ulcers, and in the '70s it was high blood pressure). Strain I Herpes simply causes cold sores, but strain II causes the famed genital herpes. So what is herpes all about? The herpes viruses (both I and II) reside in nerve cells, which are apparently unharmed by the invading dormant form. But when the host's defenses are down, perhaps because of stress, the virus buds off from its sanctuary deep in the spinal nerves and infects cells of mucous membranes (especially the moist membranes lining the mouth and genitals). Here they liberate numerous infectious new virus particles. It is at this time that the herpes victim is highly contagious and can readily pass the genital herpes virus on to a sex partner.

The problem is, when the person's antibodies build up, the viruses simply retreat into the nerve cells where they are safe from immunological detection. To make things worse, genital herpes has been associated with higher rates of cervical cancer, although the link isn't completely understood. Herpes infections are currently permanent and incurable, although intensive research has recently led to treatment that reduces the length and severity of symptoms.

APPLICATION OF IDEAS

1. There are strong arguments against the liberal use of antibiotics. Numerous kinds of bacteria live on and inside the body, often in an innocuous manner. Many are heterotrophs while others are marginal parasites, and still others are a constant threat as virulent parasites. What, if anything, does the frequent use of antibiotics have to do with these normal populations? Why not kill all the bacteria possible?

2. Despite the great care taken in the commercial canning of foods, occasional cans of spoiled food appear in supermarkets. A common form of spoilage can be identified by a general bulging of the can. What causes the bulging? Suggest ways in which bacteria might survive the canning process. What kind of bacteria might live in a sealed environment?

3. By gathering together pertinent information about bacteria, develop a table or chart that could be used in identifying individual groups. Use such characteristics as form, arrangement, nutritional requirements, and other requirements of a biochemical nature.

KEY WORDS AND IDEAS

A HYPOTHESIS ON THE ORIGIN OF LIFE

1. Compelling evidence suggests that life arose spontaneously on the earth, at a time when unique conditions were present.

2. Haldane and Oparin were the first to speculate on the spontaneous origin of life. The **Haldane-Oparin hypothesis** proposed that the precursors of life's molecules formed from inorganic sources and these underwent polymerization to form macromolecules. Such macromolecules included primitive versions of enzymes. Droplets, or **coacervates,** formed and were enclosed by protein or lipid shells. A next hypothetical step required the formation of genelike **autocatalytic** molecules that could assure the faithful reproduction of such enzymes.

The Miller-Urey Experiment

1. Miller and Urey tested the Haldane-Oparin hypothesis with a device that simulated the primitive environment. Using a mix of gases and applying an electrical discharge, they succeeded in synthesizing amino acids, aldehydes, and carboxylic acids, all known to be monomers of cells.

The Early Earth

1. New knowledge of ultraviolet light has changed ideas on the primitive earth atmosphere. The atmosphere in which life arose was more likely to have consisted of water, carbon dioxide, carbon monoxide, molecular nitrogen, and some free hydrogen.

2. Energy sources in the primitive atmosphere may have included ultraviolet light, lightning, heat, and geological shock energy.

3. Estimates place the origin of life some 3.8 billion years ago; fossils of cyanobacteria are believed to be 3.5 billion years old.

The Hypothesis Today: How Strong Is It?

1. For the hypothesis to remain viable, scientists must be confident about the following: proposals about early conditions, the spontaneous formation of monomers of life and their polymerization into the polymers of life, the spontaneous formation of cell-like bodies, and the formation of simple, self-replicating chemical systems.

2. Scientists are still debating over which came first, the nucleic acids or the proteins.

3. Recent experiments on the revised atmospheric conditions have been successful in producing the usual monomers and a few that were not formed in the Miller-Urey experiment.

4. The mass action law suggests that polymerization is not likely to have occurred in the sea, but more likely in heated and highly concentrated pools of monomers. Treating amino acids in this manner, Sydney Fox produced polymers that aggregate into what he called **proteinoids.**

5. Efforts to polymerize functional nucleic acids from nucleotide triphosphates, even when using single stranded DNA for templates, fail in the absence of polymerizing enzymes.

The Active Droplets: Coacervates and Proteinoids

1. Coacervates simulate life processes such as growth and reproduction. By taking in selected materials, they grow to a critical mass and then divide.

2. Experimenting with coacervates, Oparin constructed simple biochemical systems such as those containing replicating enzymes and nucleic acid elements. He managed to get the coacervate to duplicate some elementary biochemical mechanisms.

The Earliest Cells

1. There is disagreement on the metabolic characteristics of the early **protocells.** Some suggest they were chemotrophic or phototrophic, while others claim they were simple anaerobic heterotrophs. The continued autocatalytic synthesis of monomers may have provided carbon compounds for early heterotrophs.

2. Keen competition for energy-rich monomers may have encouraged variants that could extract hydrogen from inorganic sources, using light energy in simple photosystems. The utilization of water as a hydrogen source probably came later since it requires complex photosystems.

3. When the use of water in photosynthesis occurred, oxygen joined the gases of the atmosphere for the first time.

4. The buildup of significant amounts of oxygen in the atmosphere required an enormous time period because of the presence of **oxygen sinks—** elemental substances such as sulfur and iron that readily combined with the gas.

5. As cell populations grew, new modes of heterotrophic nutrition arose. Food chains emerged as cells began devouring cells through phagocytosis.

THE PROKARYOTES

Origins

1. Among the earliest evidences of life are 3 to 3.5 billion-year-old columnlike **stromatolites,** laminated deposits produced by dense bacterial populations.

Organization of the Prokaryote

1. Recent revisions of prokaryote taxonomy have divided the older kingdom **Monera,** into two groups (kingdoms), **Archaebacteria** and **Eubacteria.** The two differ significantly in their cell wall chemistry (proteinaceous substances in Archaebacteria, **peptidoglycan** in Eubacteria) and in other aspects of biochemistry.

Prokaryote Characteristics

1. Prokaryote cells are smaller than those of eukaryotes, they lack membrane-bound organelles, and their chromosomes are circular DNA molecules, lacking in protein. They divide by **binary fission** and occasionally undergo genetic recombination through conjugation. They often have slimy capsules, and the flagellum, where present, consists of a stiff, S-shaped, rotating rod, composed of the protein **flagellin.**

Archaebacterial Life

1. Archaebacteria commonly live in hot, acidic, and salty environments. Most are **obligate anaerobes.**

2. Examples of archaebacteria include **methanogens**—anaerobic methane generators, **extreme halophiles, extreme thermophiles,** and **thermoacidophiles,** bacteria that live in dense salt concentrations, very hot water, and hot, acidic conditions, respectively.

3. Where found, photosynthetic archaebacteria use the pigment **bacteriorhodopsin** for the capture of light energy used in building the chemiosmotic gradient and ATP synthesis.

The Eubacteria

1. Eubacteria include phototrophs (which utilize bacteriochlorophyll), chemotrophs (which obtain energy from inorganic compounds), and heterotrophs (including **reducers** or **decomposers** and parasites).

2. Eubacteria occur as the rod-shaped **bacillus,** the spherical **coccus,** and the corkscrew-shaped **spirillum.**

3. Bacilli occur singly or in chains, and commonly form dormant, resistant **endospores.** The coccus forms are arranged in pairs (diplococcus), chains (streptococcus), and clusters (staphylococcus). Spirillum forms—**spirochetes**—occur singly.

4. Eubacterial cell walls contain peptidoglycan, which is very thick in some and thin in others. Some are coated with a polysaccharide called **teichoic acid,** while others have an outer lipid membrane.

5. Vital stains are used to color cells for microscopic viewing and for classification. **Gram's stain** is used for the latter purpose; cells that retain the dark purple stain are **Gram positive** and those that don't are **Gram negative**. Gram positive bacteria are susceptible to penicillin, which inhibits wall synthesis.

6. Many serious bacterial diseases have been controlled in developed nations. Three species of soil bacteria, *Clostridium tetani, botulinum,* and *perfringens,* agents of **tetanus, botulism** and **gas gangrene,** respectively, are still a threat. Other common pathogens include strains of staphylococcus and streptococcus, which cause resistant skin infections and **strep throat.**

7. Venereal (sexually transmitted) diseases include **gonorrhea** and **syphilis.** Some strains of the former are very resistant and do not respond to antibiotics.

Other Bacteria

1. **Myxobacteria (slime** or **gliding bacteria)** utilize slime secretions in movement. They produce multicellular **fruiting** bodies and aerial **cysts** that contain spores.

2. **Chlamydobacteria (mycelial bacteria)** are sheathed and filamentous, some producing swarms of flagellated cells. **Actinomycetes** produce branched filaments with clublike spore-forming bodies. Chlamydobacteria are the source of some antibiotics.

3. **Mycoplasmas** are the smallest of cells, lacking both cell walls and definite shape. All may be animal parasites, and one causes a form of pneumonia in humans.

4. The **cyanobacteria** are found in all aquatic environments, including the oceans, where they are important to marine food chains. They occur as single cells or in colonies and in simple multicellular states. Some produce specialized reproductive cells and **nitrogen fixing** cells known as **heterocysts.**

5. Photosynthesis in cyanobacteria is carried on in highly organized, membranous thylakoids containing chlorophyll *a*, carotenes, and light-capturing **phycobilin** pigments. Their starch is like animal glycogen.

THE EUKARYOTES: A DIFFERENT MATTER

The Endosymbiosis Hypothesis

1. The **endosymbiosis hypothesis** proposes that eukaryote cells arose from a number of prokaryotic cell lines that became **endosymbionts** through invasions and incorporations. Such events explain the presence of such organelles as mitochondria, chloroplasts, basal bodies, flagella, cilia, and centrioles.

2. According to the endosymbiosis hypothesis,
   a. *Line A* **protoeukaryotes** were phagocytic predators capable of anaerobic respiration only.
   b. *Line B,* an aerobic bacterium, was taken into line A, and was the forerunner of the mitochondrion.
   c. *Line C* cells, when incorporated, brought in the 9 + 2 microtubular flagellum containing the protein *tubulin,* giving rise to centrioles, basal bodies, and cilia and flagella.
   d. The final incorporation involved photosynthetic cells of *line D,* possibly a cyanobacterium or a **chloroxybacterium.** This incorporation established the chloroplast.

3. Evidence supporting the endosymbiosis hypothesis comes from several sources. The eukaryote mitochondrion and chloroplast have critical similarities to bacteria. Included are ribosomal RNAs, cytochrome *c*, and amino acid sequences. The link to a flagellated bacterium is weak, although flagellate symbionts are known to exist in certain protists.

THE VIRUSES

1. Taxonomic relationships within the **viruses** and between viruses and other life is unknown. They range in size from that of large molecules to the smallest bacteria. Viruses are responsible for many human, animal, and plant diseases.

The Discovery of Viruses

1. Pasteur and Koch had developed vaccines against viral diseases long before anyone knew what viruses were. This is reflected by the name *virus* since it means "poison."

2. Iwanowski experimented with the **tobacco mosaic virus** in the late 1800s, transferring a virulent filtrate from plant to plant. By 1900, viruses were regularly isolated by filtration and they became known as **filterable viruses.**

3. W. M. Stanley identified tobacco mosaic virus crystals in 1935, the same year they were first seen in the newly developed electron microscope. The fact that the virulent material increased while in the plant supported the idea that it was alive.

The Biology of Viruses

1. Viruses are **obligate parasites**—they cannot carry out life processes outside a host cell.

2. Each virus particle or **viron** contains a nucleic acid core and a protein coat or **capsid.**

3. Viruses are helical (tobacco mosaic virus), polyhedral **(herpes simplex),** or cubic and brick-shaped (smallpox). Shape depends on the arrangement of coat proteins or **capsomeres.**

4. Some viruses have the standard double-stranded DNA, others double-stranded RNA or single-stranded RNA.

5. Viruses enter the host, disrupt its DNA, undergo their own DNA or RNA replication, and transcribe their genes into mRNA for producing viral proteins. All materials—synthesizing enzymes, ribosomes, and energy—are provided by the host.

6. The T-even bacteriophages use **phage lysozyme** to digest their way into a bacterium. **Prophages** (bacteriophages that have penetrated a cell) can enter a **lytic cycle** (reproductive and cell-destroying stage) or a **lysogenic cycle** (dormant stage). In the latter, the viral DNA is inserted into a host chromosome and is replicated and passed along through generations of cells.

7. Viruses in a lysogenic cycle can break out and become lytic when specifically stimulated. Exposure to bright sunlight can activate the herpes simplex I (fever blister) virus.

8. Some animal viruses either pass right through the animal plasma membrane or are engulfed through phagocytosis. In the latter, the virus ends up in a food vacuole which receives lysosomal hydrolytic enzymes. Digestion frees the indigestible nucleic acid core and the virus then takes over the cell and multiplies.

9. Single-stranded RNA viruses first synthesize a double-stranded RNA, which then replicates more viral RNA. In some it can act as a messenger RNA for producing protein.

10. Double-stranded RNA viruses replicate more double-stranded RNA in a manner similar to DNA replication.

11. Some viruses use the host plasma membrane directly in producing a new coat.

12. Using their single strand of RNA as a template, **retroviruses** employ the enzyme **reverse transcriptase** to assemble an opposing strand of DNA nucleotides. The RNA of the RNA/DNA hybrid is then disassembled, and the DNA is next converted to a double-stranded DNA, which is inserted into a chromosome (just like a lysogenizing prophage would do). Later it may transcribe many single-stranded RNAs, produce coat proteins, assemble new viruses, and lyse the cell.

13. The enzyme reverse transcriptase is now used in genetic engineering as a shortcut method for synthesizing DNA from messenger RNA. This avoids the usual problem of "reading" mRNA with the usual cumbersome intervening sequences present.

Herpes Simplex II: Scourge of the '80s

1. Herpes viruses are dormant and protected in the spinal nerves but emerge periodically to infect the genitals and the mucous membranes of the mouth. As the immune system responds, the viruses retreat to the spinal nerves once again. Genital herpes may be associated with cancer.

REVIEW QUESTIONS

1. Briefly describe Haldane and Oparin's hypothesis on the origin of life. (422–424)

2. According to most biologists, the spontaneous origin of life occurred only once in the earth's history. What is the basis for this restricted statement? Why not several times? (423)

3. What steps are involved in going from inorganic matter to the coacervate level? What, according to Haldane and Oparin, were the energy sources? (423–424)

4. What is the status of the original Haldane-Oparin hypothesis today? (427)

5. Describe the experimental apparatus and materials used by Miller and Urey in testing the spontaneous origin hypothesis. What did Miller and Urey actually establish? (424)

6. In what specific ways has our understanding of the primitive atmosphere changed since the days of the Miller-Urey experiment? What effect on the original hypothesis has subsequent experimentation produced? (424–425)

7. List five requirements that must be met if the spontaneous origin of life hypothesis is to continue to hold up. With which of the five requirements are scientists making satisfactory progress? (427–428)

8. What problems, if any, do researchers find with attempts at spontaneously synthesizing nucleic acids? Did nucleic acids necessarily precede enzymes? Explain. (428)

9. Briefly describe the polymerization experiments of Sidney Fox. Why didn't he use a water solution such as might have been found in the early oceans? (428)

10. Describe Oparin's experimental work with coacervates. Did he actually produce protocells? Has he proven the spontaneous origin hypothesis? Explain. (428)

11. Why might the earliest phototrophs have used hydrogen sulfide as a hydrogen source rather than water? In what way did the later shift to water "change the earth forever"? (429–430)

12. According to the latest thinking, it took 1 billion years for significant oxygen to accumulate in the atmosphere. Why was such a long period of time required? (430)

13. Describe stromatolites and their formation. What evidence best proves that they were produced by living cells? (431)

14. In what ways do the cell walls of Archaebacteria and Eubacteria differ? (432–433)

15. Using the following categories, compare prokaryote cells with eukaryote cells: size, chromosomes, organelles, cell division, sexual reproduction, cell walls, flagella. (433–434)

16. Briefly describe the environments of methanogens, extreme halophiles, and thermoacidophiles. (434–435)

17. Briefly discuss the most ecologically significant activity of bacteria. Why is this activity so vital? (436)

18. Using simple illustrations, describe the major shapes and arrangements seen in bacteria. (437)

19. What is an endospore? In what way is its formation adaptive to the bacterium? (437)

20. How does penicillin kill bacteria? In which group of bacteria is it ineffective? How are such bacteria clinically controlled? (437)

21. Name and describe the activities of the three important species of *Clostridium*. (438)

22. Describe the complications that can occur with streptococcal infections. (438–439)

23. Compare the manner in which syphilis and gonorrhea infections occur in babies. Which disease presents the more difficult problem? Why? (439–440)

24. In what ways are the myxobacteria and the chlamydobacteria similar to fungi? Where might this tend to place such groups in taxonomic schemes? (440)

25. List three unique features of the mycoplasmas. (441)

26. List three or four ways in which the cyanobacteria are biochemically similar to algae and plants. (442)

27. Briefly explain what problem the endosymbiosis hypothesis addresses, and list the eukaryotic organelles concerned. (443)

28. Beginning with the protoeukaryote cells in line A, list the hypothetical events leading from the prokaryotic to the eukaryotic cell. (443)

29. For which eukaryotic organelles is the endosymbiosis hypothesis most satisfactorily supported? Briefly summarize the supporting evidence. (445)

30. What, precisely, are viruses? How does this explain why viruses require a living host? (446–447)

31. What was Iwanowski's contribution to our understanding of viruses? (446)

32. How did Stanley prove that viruses could reproduce? (447)

33. List the three general forms taken by viruses. What determines the form? (448)

34. List four variations in the nucleic acids occurring in viruses. (448)

35. Briefly summarize the events in the lytic and lysogenic cycles of the phage virus. (448)

36. Review two ways in which animal viruses enter the host cell. (448, 451)

37. Compare replication in viruses with single-stranded and double-stranded RNA. (453)

38. Explain how replication occurs in the retroviruses. (453)

Diversity in Protists

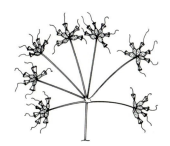 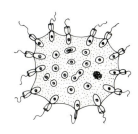 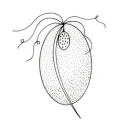 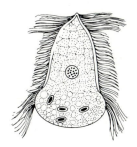

Here, we will describe the kingdom Protista. The problem is, we're not sure that there *is* a kingdom Protista. That is, we're not sure the organisms that fall into this kingdom are any more closely related to each other than they are to some organisms in other kingdoms. They are lumped together by default and for negative reasons, since they are all eukaryotes yet quite unlike other eukaryotes. Recently, some researchers have suggested that the name *Protoctista* replace the term Protista. They suggest that Protoctista be a larger, more encompassing group. Whereas Protista includes the unicellular organisms only, Protoctista would make room for many exceptions and would help define more clearly the three higher kingdoms as monophyletic. In any case, as we will see, the kingdom is best defined by "exclusion." That is, its members share the commonality of being neither animal, plant, fungus, nor prokaryote.

The protists, or protoctists, if you will, are clearly eukaryotes. They are commonly single-celled or unicellular, but some have colonies of cells and some species within the algal protists are clearly multicellular. There is great diversity in feeding, although among the animallike protists endocytosis and simple absorption, the diffusion of digested food across the body wall, are common. Many of the protists are photosynthetic, so both autotrophs and heterotrophs are found. Both asexual and sexual reproduction are common, and most protists carry on mitosis and meiosis. Movement, where it occurs, is chiefly through microtubular cilia and flagella, although some groups form **pseudopods** (false feet), which produce a sort of flowing movement.

A look back at Figure 17.20 shows that the relationships of one kingdom to another are indicated by position. The protists form a broad polyphyletic group underlying the three other eukaryote kingdoms. The protists are located directly under the roots of the other kingdoms because those kingdoms are believed to have protist ancestors. Thus, below the Plants are the plantlike protists—the filamentous green algae from which the plants arose. Below the Fungi are the protists that are most like fungi, and below the Animal kingdom are the ciliate protists that are believed to share ancestry with the animals.

Should the protists then be divided into several separate kingdoms? Maybe. Or should they perhaps be parceled out among the three other existing eukaryote kingdoms? As you can see, there is always room for debate in discussions of phylogenetic relationships, but we've made our choices; now let's look at the protists themselves in more detail.

We can begin by dividing the protists into three groups. By tradition, the protists that resemble animals are called **protozoans** and those that resemble plants are called **algae.** There are also the **slime molds** and other funguslike creatures that were, until recently, classified as fungi. Subdividing the protists in this manner is quite useful, but it doesn't begin to cover all the cases. For example, there are the puzzling little *Euglena* (see below). Like the algae, they contain chlorophyll and can be photosynthetic. But if they are grown in the dark so that they can't photosynthesize, their green area fades and they become heterotrophic, like an animal. *Euglena* also has a very close relative called *Astasia,* to which it is morphologically almost identical, but which is nonphotosynthetic and **saprobic** (living on dead matter; see Chapter 20). Biochemical studies show that *Euglena* is also related to *Crithidia,* a parasitic protozoan. So, what is *Euglena*? Because of their flagella, some taxonomists place both *Euglena*

and *Astasia* in the protozoan phylum Mastigophora—the flagellates. *Euglena* and *Astasia* are markedly different from the other flagellates, however; for example, they don't reproduce sexually.

The task of sorting all this out must fall on the evolutionists of the future. For now, let's keep in mind that our categorizing is tenuous and subject to change should new information become available.

THE ANIMALLIKE PROTISTS

The "animallike" protists, the protozoans, were formerly in the phylum called *Protozoa* and were described as single-celled animals. The term *single-celled* leaves much to be desired, since it implies that these organisms are like single animal cells. But many of the protozoan protists are far more complex than the most elaborate cell of any animal or plant. It is probably better to refer to them as small and **acellular** (without cells).

There are six phyla of protozoan protists, but here we will concentrate on the four most significant ones: Mastigophora, Sarcodina, Sporozoa, and Ciliophora. The distinctions are based mainly on their manner of locomotion (or lack of locomotion).

Mastigophora: The Flagellates

The Mastigophora are the flagellated protozoans (Figure 19.1). Most members of this phylum propel themselves by whiplike flagella, which occur singly, in pairs, or in greater numbers. The flagella have the 9 + 2 microtubule structure typical of eukaryotes. Movement through undulation of the flagella permits the flagellate protozoan to move efficiently in any direction (see Figure 4.21).

Flagellated protozoa feed in a variety of ways. They may hunt and capture prey or simply absorb nutrients through their body covering. Reproduction is primarily by asexual cell division. Protozoans are eukaryotes, so we expect to see nuclear division with the typical movement of chromosomes and action of the mitotic apparatus. One might also expect meiosis and sexual fusion, but in fact, our knowledge of sexual reproduction in the flagellates is indeed scant, as it is with many other protists.

Each of the four classes of protozoa has its parasitic forms. One of the most notorious flagellates, *Trypanosoma gambiense,* is the causative agent of

---

19.1 THE FLAGELLATES

A small sample of the Mastigophora includes the colonial *Protospongia*, which may share ancestors with the sponge, and the branching colonial, *Codosiga botrytis*. Also included are *Trichomonas vaginalis,* whose species designation is quite descriptive. When it occupies the human vagina it can cause considerable discomfort as the body's immune response begins. The multiflagellated *Trichonympha campanula* lives as a symbiont in the termite gut, where it digests cellulose for its host.

Codosiga botrytis

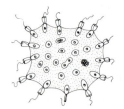
Protospongia haeckeli

Trichomonas vaginalis

Trichonympha campanula

19.2 A FLAGELLATE PARASITE

The flagellum of *T. gambiense,* attached along the body by a thin, membranous flap, produces an undulating motion as the parasite makes its way through the host's blood cells. When humans are infected, the trypanosome reproduces in the blood and lymph glands and eventually enters the cerebrospinal fluid where it brings on the symptoms of African sleeping sickness *(trypanosomiasis)*: irregular fever, swollen lymph nodes, painful edema. As the parasites invade the central nervous system, tremors, headache, apathy, and convulsions commence, leading to coma and death. When a tsetse fly draws blood from an infected animal, the parasite enters the fly's body and reproduces in the midgut. The immature offspring then make their way into the salivary glands, where they become mature and capable of infecting a new host.

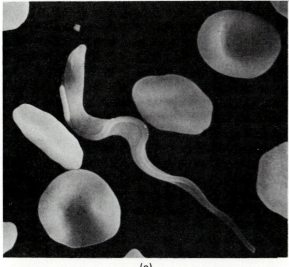

(a)

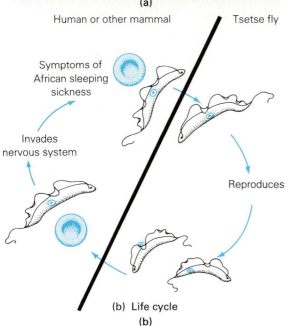

(b) Life cycle
(b)

African sleeping sickness. The trypanosome is carried by the equally infamous *tsetse fly,* whose bite injects the parasite into the mammalian victim. Control of the disease is very difficult, since nearly all large mammals in tropical Africa harbor the parasite. Figure 19.2 depicts the life cycle of the sleeping sickness trypanosome.

Sarcodina: Ameboid Protists

Imagine a toughened, determined, well-trained army on the move into some primitive land. Then imagine that army brought to a dead halt, as soldiers drop their weapons, clutch at their bellies, and dart for the nearest cover. It has happened. Armies have literally been stopped by a tiny protozoan called *Entamoeba histolytica,* the carrier of amebic dysentery (Figure 19.3). The ameba is transmitted by infested human feces entering the food or water system. That, of course, sounds highly unlikely in civilized countries such as our own, but keep in mind that hepatitis, a common enough problem in the United States, is often transmitted the same way. The problem with amebic dysentery, though, is that it can be transmitted by people who carry the disease but show no symptoms.

These impressive amebas are just one organism in the phylum *Sarcodina.* The sarcodines are the ameboid protozoans. They are best known for their physical plasticity, for they have no definite form and can send out a pseudopod—a temporary extension of the cell—in any direction as they move.

19.3 *ENTAMEBA HISTOLYTICA*

The agent of ameboid dysentery is shown here in both the active state and in its protective cyst. The spread of ameboid dysentery *(amebiasis)* occurs through contaminated water supplies, food, utensils, and so on. The ameboid cyst ruptures in the intestine of the person who has eaten it and there the ameba reproduces. The progeny then penetrate the wall of the colon, where they create ulcerous erosions.

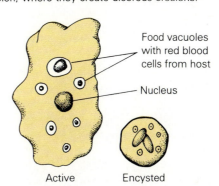

Food vacuoles with red blood cells from host

Nucleus

Active   Encysted

19.4 AMEBOID MOVEMENT

(a) Movement in amebas is accomplished through rapid changes in the consistency of parts of the cytoplasm. As a pseudopod or "false foot" begins its formation, the more liquid inner cytoplasm, known as *endoplasm*, flows towards the leading edge. As it flows to the sides, it is converted to a stiffer, gellike *ectoplasm*, which forms a tubelike border, lending shape to the enlarging pseudopod. Note that at the trailing edge, the stiffer ectoplasm is converted back to the free flowing endoplasm, so that the protist can sort of "keep up with itself" as it moves forward. Pseudopods are used in both movement and in feeding **(b)**, which occurs through phagocytic action.

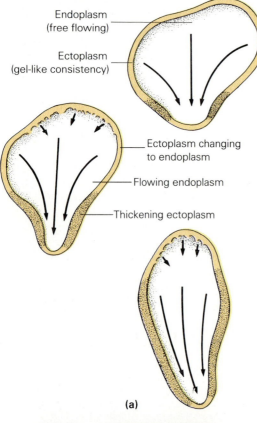

Endoplasm (free flowing)

Ectoplasm (gel-like consistency)

Ectoplasm changing to endoplasm

Flowing endoplasm

Thickening ectoplasm

(a)

(b)

Basically, they do this simply by shooting out a foot and then flowing into it. Ameboid movement is accomplished by dynamic reorganization of the cytoplasm between a firm "gel" state and a flowing "sol" state (Figure 19.4). Certain white blood cells are creeping through the tissues of your body in just such a way this very moment.

Sarcodines typically feed through phagocytosis. Food particles or prey are surrounded by converging pseudopods, or they may stick to the surface of the sarcodine and become engulfed.

One group of sarcodines, the **heliozoans,** produces a hardened capsule of silicon dioxide (glass) from which extend long, slender **axopodia**. The axopodia are actually needlelike pseudopods upon which the cytoplasm creeps back and forth, gathering food through phagocytosis. The axis of each axopodium includes a large number of microtubules in a spiral arrangement. The microtubules permit the axopod to be extended or withdrawn from the main body (Figure 19.5). Another group, the **radiolarians,** also produce coverings of silicon dioxide. They have been on earth a long time, and their siliceous corpses on the ageless ocean floor form deposits known as the **radiolarian ooze.** The **foraminiferans** also form skeletons, elaborate shells of calcium carbonate. This "foram" shell is pitted with numerous pores through which the pseudopods move in and out as the organism feeds. Foraminiferans also contribute to the dense ooze of the ocean floor. In fact, some land areas that were once under the sea are still composed of dense deposits of foraminiferan shells. The white cliffs of Dover, England, are a good example. Oil geologists use the presence of certain ancient foram bodies to predict where oil might be found.

Some species of the Sarcodina reproduce both asexually and sexually, some only asexually through mitosis and cell division. So far, sexual reproduction in the most familiar sarcodines, the amebas, has been seen only rarely, and then whole individuals fuse together. In the foraminiferans that reproduce sexually, the basic event involves diploid cells that undergo meiosis and then form haploid sex cells, which are often flagellated. These can join in fertilization to produce a diploid zygote that can then reproduce asexually for many generations. In the heliozoans, individuals withdraw into cystlike states, undergo meiosis, form polar bodies which are discarded, and then undergo fusion, restoring the diploid state. When growth conditions are poor, some sarcodines produce protective cysts—tough, dehydrated bodies that allow them to remain dormant until conditions become more hospitable. A few types of Sarcodina are shown in Figure 19.6.

19.5 A HELIOZOAN

Actinosphaerium has a spherical body covered with long, fine axopodia, made up of numerous microtubules, surrounded by streaming cytoplasm. Waves of cytoplasmic movement carry trapped food particles into the main body, where they may enter the cell by pinocytosis or phagocytosis. Food vacuoles in the cytoplasm are the sites of digestion. Contractile vacuoles aid in maintaining a suitable water balance.

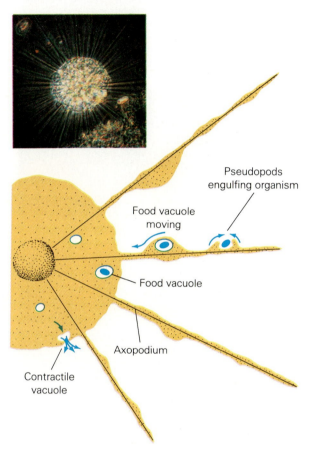

Pseudopods engulfing organism

Food vacuole moving

Food vacuole

Axopodium

Contractile vacuole

Sporozoa: Spore Formers

From one point of view, the **sporozoans** may be among the most important of all protozoa to humans. That's because one group can cause **malaria,** a disease that has been called humanity's greatest curse. Sporozoans have three major characteristics: (1) they form spores, (2) they live as parasites, and (3) in their mature stages, they have no special means of locomotion. Sporozoans exhibit complex life cycles, with both sexual and asexual phases. They often complete different portions of their parasitic life cycle in different hosts.

Malaria is spread from person to person by the female *Anopheles* mosquito, a tiny insect harboring the tinier parasite, *Plasmodium vivax* (the most common of several species that infect humans). This

decidedly dangerous mosquito—recognized by the habit of standing on its head when it feeds—is alive and well throughout the world (Figure 19.7). One of the most effective mosquito control agents has been the pesticide DDT, but its use is highly controversial and today we are confronted by DDT-resistant mosquitoes. Because of the dangers of DDT, our best tactics against malaria may be draining ponds, putting mosquito-eating fish in streams, or filling in ditches to eliminate the mosquito's breeding grounds. This is all quite feasible in developed nations, but quite unrealistic in many parts of Asia and Africa where malaria can be epidemic.

During its life cycle (see Figure 19.7), *P. vivax* reproduces in both the mosquito and the vertebrate host. The cycle begins with the injection of mosquito saliva containing **sporozoites** (the ameboid stage). The sporozoites make their way through the bloodstream and enter the liver cells, where they

19.6 SARCODINE DIVERSITY

(a) The radiolarians are typically spherical, with highly sculptured skeletons. **(b)** The foraminiferans tend to form spiral shapes reminiscent of tiny snails. The skeletons of marine sarcodines contribute to the thick ooze common on the ocean bottom.

(a)

(b)

19.7 PLASMODIUM LIFE CYCLE AND MALARIA

The female *Anopheles* mosquito is a known carrier of the malarial parasite, *Plasmodium vivax*. The life cycle of *P. vivax* takes place partly in the mosquito and partly in the reservoir animal, in this case a human. **(a)** The cycle begins when the mosquito injects saliva containing sporozoites into the bloodstream. **(b)** From there, the sporozoites enter the liver and produce large numbers of merozoites, which **(c)** reenter the bloodstream and **(d)** invade red blood cells to reproduce once again. **(e)** The sudden, coordinated release of numerous parasites brings on a period of fever and chills. As the fever sub-

sides, a new infection of red cells occurs, ultimately bringing about a new release of the parasites. Then fever and chills occur again. **(f)** Eventually, gamonts (sexual forms) are produced. When these are taken into the intestine of a mosquito, they begin a sexual cycle, forming sperm and eggs. After fertilization, the zygote develops in the intestinal wall of the mosquito. **(g)** Eventually, the maturing sporozoites venture into the salivary gland, from where they move into the saliva that may then be injected into another victim.

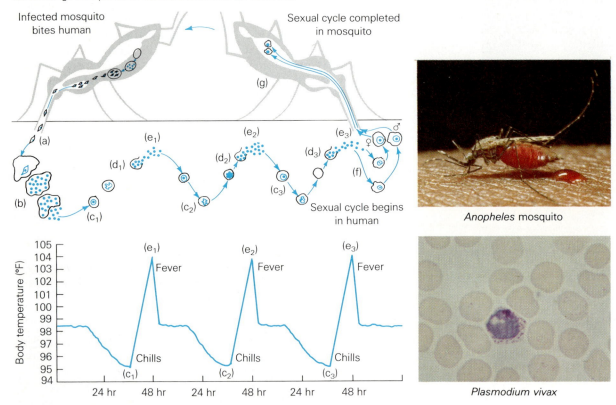

Anopheles mosquito

Plasmodium vivax

undergo a quiet asexual phase. At this point, the host still shows no reaction. However, the parasite is reproducing rapidly and changing to the **merozoite** phase. Eventually, large numbers of merozoites enter the bloodstream. Their task is to invade the red blood cells, where they reproduce simultaneously, rupturing the blood cells and releasing poisons throughout the host's body. The poisons bring on the familiar fever and chills of malaria. The cycle repeats itself every 48 hours, 72 hours, or longer periods, depending on the species of *Plasmodium* involved.

Eventually, some merozoites develop into potential gamete producers known as **gamonts**, and

Plasmodium enters a sexual phase. The gamonts, which are either male or female, are quite impotent within the vertebrate host. To produce gametes, they must first find their way into a mosquito. When an infected person is bitten and the mosquito sucks a few gamonts into its gut, the gamonts change into sperm or eggs and fertilization occurs. The zygotes then develop in the walls of the intestine. The fertilized cell, here known as an **oocyst,** divides asexually, producing large numbers of sporozoites that migrate to the salivary glands of the mosquito. When the mosquito bites, she (only the females bite) injects a little of her infected saliva and the infection cycle begins anew.

Ciliophora: Ciliates

The **ciliates** are a highly diverse group of protozoans, and some species undoubtedly represent the most complex single cells on earth. In size alone they range from about 10 micrometers to 3000 micrometers (approximately the same relative difference between shrews and blue whales, see Figure 19.8).

Ciliates are identified by the cilia that appear at some stage in their life cycle, and that are used in movement, or feeding, or both. The cilia occur in rows, either longitudinal or spiral, or they can occur in fused tufts. Each cilium can perform a precise rowing motion (see Chapter 4), but for efficiency in both locomotion and feeding, a large number of cilia must perform in a coordinated manner, moving in sequence much like rows of wheat bending before gusts of wind. This coordination is made possible by an elaborate network of nervelike fibers that connect one basal body with the next. In *Paramecium*, a group of well-known ciliates, some of the cilia are fused into a kind of membrane that lines the **cytostome**, a mouthlike structure. In other genera, such as *Euplotes*, the cilia may be arranged in tufts, which serve as many "legs," enabling this species to scramble over the bottom sediment or paddle along through the water above.

The body covering of the ciliate is often a tough but elastic **pellicle**, a thin, translucent envelope of secreted material outside the plasma membrane. Because it is elastic, the ciliates can bend and wriggle and contort and manage to get past or through all sorts of obstructions.

Alternating with the ciliary basal bodies are vaselike structures called **trichocysts.** Trichocysts can be forcefully discharged from the body surface en masse when the *Paramecium* is disturbed. Each discharged body is a long, threadlike cylinder with a barblike head. Some ciliates have toxic trichocysts that are used in capturing prey or in defense, while others use the trichocyst to moor themselves in place while feeding.

Ciliates also have **contractile vacuoles** with which they actively pump out excess water (see Chapter 5). This is an important means of maintaining proper osmotic conditions, since the cells often live in a hypotonic (watery) environment. Both the filling and emptying cycles appear to operate by ATP-powered contractions of cytoplasmic microfilaments.

Most ciliates feed in an animallike manner, taking bulk food such as bacteria or other protists, rath-

19.8 CILIATE DIVERSITY

The ciliates are enormously varied in both size and structure. One of the largest is *Spirostomum*, which is 3000 μm (or 3 mm) in length, and easily visible to the unaided eye. *Diplodinium*, one of the smallest ciliates, is 1/300 the size of *Spirostomum*. *Vorticella*, a funnel-shaped cell, is attached to a long contractile stalk. Its open end has a ring of cilia that creates a vortex, drawing in its food particles. A peculiar variation is *Euplotes*, with fused cilia forming tufts, which it uses to scurry along or to row with. *Balantidium*, one of the few parasitic ciliates, invades the colon of humans, where it causes ulceration of the lining. *Paramecium* is one of the best-known ciliates—a laboratory favorite.

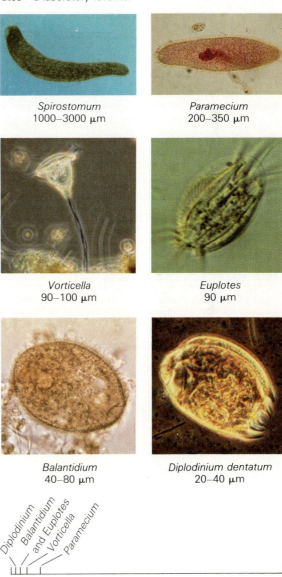

Spirostomum
1000–3000 μm

Paramecium
200–350 μm

Vorticella
90–100 μm

Euplotes
90 μm

Balantidium
40–80 μm

Diplodinium dentatum
20–40 μm

Diplodinium
Balantidium
and Euplotes
Vorticella
Paramecium

Spirostomum

er than by simple absorption. Some are seemingly "ravenous" predators; one species, *Didinium*, although relatively small, can "swallow" ciliates considerably larger than itself. *Paramecium* feeds less conspicuously, principally on bacteria, which it sweeps along its **oral groove** leading to its cytostome and on into **food vacuoles** that continually form. The site of food vacuole formation, as one might expect, is a thin region of the body wall, containing only the plasma membrane. Once filled, the food vacuole breaks away and moves through the cytoplasm in a well-defined path. As digestion occurs, the nutrients diffuse out of the vacuole into the cytoplasm along the way. The undigested residues are then expelled from the body through exocytosis. If the cytostome is the protozoan's mouth, then the **cytopyge** is its anus. It appears periodically—always in the same place—when solid wastes are ready to be expelled (Figure 19.9).

Reproduction in ciliates is extremely complex and rather fascinating. Outwardly, the process seems simple enough; ciliates usually reproduce asexually by mitosis and cell division (Figure 19.10). But that's the easy part. It is when they become involved with sex that their lives become complicated (but we will draw no analogies.) Let's look at

Paramecium caudatum as a protist with a complex sex life. In *P. caudatum,* sexual reproduction occurs through a process called **conjugation.** There are eight types of *Paramecium,* and any two individuals must be of different mating types in order to conjugate. These types are not really sexes; by definition, there are only two sexes, in the strict sense, among all the living things. However, no one knows why there are only two sexes or why *Paramecium* has eight mating types.

Paramecium is unusual in that its genetic material is separated into a germ line (involved in reproduction) and a somatic line (involved in the maintenance of the individual). The germ-line DNA is confined to the diploid **micronucleus.** The somatic DNA, from which all RNA is transcribed, occurs in the polyploid **macronucleus** (having multiple copies of each chromosome). In any case, during conjugation, meiosis in both partners produces several haploid micronuclei (Figure 19.11). A pair of these are then exchanged through a cytoplasmic bridge that appears between the partners. In each partner, the incoming micronucleus fuses with a resident micronucleus. The result is a new diploid combination of genetic material in each of the conjugates when they draw apart. One puzzling aspect of all

19.9 VACUOLES IN A CILIATE

Ciliates such as *Paramecium* use vacuoles for the digestion of food and for maintaining water balance. **(a)** Food particles taken into the *cytostome* (gullet) enter food vacuoles for digestion. As the food vacuoles move through the cytoplasm, digestion occurs and the products diffuse into the cytoplasm.

The vacuole empties its waste contents at a special region in the pellicle known as the cytopyge. **(b)** Water regulation is managed by the contractile vacuoles, which occur at specific sites. They balloon out when filling and contract suddenly as they empty their contents to the outside.

Food is swept into oral groove and cytostome (gullet)

Digestion proceeds (activated by enzymes and acid secreted into the vacuole)

As digestion is completed, amino acids and sugar diffuse from the vacuole into the cytoplasm

Enzymes enter the vacuole

Indigestible material remaining in the vacuole is carried toward the cytopyge

Cytostome

The vacuole detaches and moves away

At the cytopyge, waste is ejected

Vacuole filling

Vacuole emptying

(a) Food vacuole

(b) Contractile (water) vacuole

19.10 ASEXUAL REPRODUCTION BY MITOSIS AND CELL DIVISION IN *PARAMECIUM*

During its interphase, the *Paramecium* replicates its chromosomes in preparation for mitosis. Mitosis occurs with the usual events, including the condensation of chromosomes, formation of a mitotic spindle, separation of chromatids, and nuclear division. The nucleus is separated into two bodies—the *macronucleus*, which simply pinches in half, and the *micronucleus*, which goes through replication and mitosis. During cell division, the oral groove of *Paramecium* closes and cytoplasmic division occurs across that region. As division proceeds, new grooves begin to form. After the two resulting cells separate, all the cilia and other cytoplasmic organelles are reproduced.

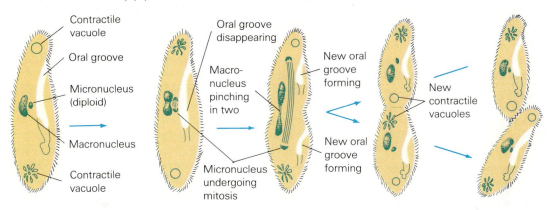

this is that some ciliates are capable of a form of self-fertilization wherein two haploid micronuclei fuse. The evolutionary advantage of such behavior is unclear. After conjugation, the macronucleus disintegrates while the micronucleus undergoes mitosis, forming two daughter micronuclei. One of these enters the germ line; the other undergoes multiple DNA replications and becomes the new polyploid macronucleus. Ciliates, by the way, are the only organisms that have managed such a complete segregation of germ-line and somatic DNA.

THE FUNGUSLIKE PROTISTS

Gymnomycetes: Slime Molds

The poetic name *slime molds* (Phylum Gymnomycota) is perhaps unfortunate, since by most schemes, these admittedly slimy creatures are not molds. Some biologists include these with the fungi, but for reasons that will soon become obvious, we will go along with the classification here. Actually, there are two distinct groups, the acellular slime molds and the cellular slime molds, and they may not be closely related to each other.

One of the best-known acellular slime molds, *Physarum polycephalum*, has all of the characteristics of a huge ameba except that it is multinucleate, its many nuclei the product of numerous mitotic events without cell division. *P. polycephalum* may commonly be found creeping over the moist underside of rotting tree trunks. As it creeps along, it feeds by phagocytizing bits of organic matter. The mass shows some sensitivity and avoids obstacles and dry areas (Figure 19.12).

At some point in the creeping plasmodium stage of its life cycle, *P. polycephalum* will seek out a drier habitat (or perhaps its moist habitat will dry out). It is only then that it shows the characteristics of a fungus. Like many fungi, the drying mass produces slender vertical props—**sporangiophores**—topped by spore-forming **sporangia**. Each sporangium undergoes a number of meiotic divisions, producing numerous haploid spores. The spores emerge to be carried aloft by air currents. If a spore lands in a suitable place, it begins to divide and produce either an amoeboid **myxameba** or a flagellated **swarm cell**. If conditions aren't quite suitable, either the myxameba or swarm cell can suspend activity, forming a dormant cyst from which it will emerge later. Otherwise the myxameba may divide a number of times, producing many ameboid descendants. Finally, either the myxamebas or the swarm cells can pair off and fuse, producing the diploid state once more. From there, either can produce the plasmodium state and again repeat the life cycle.

The Oomycetes: Water Molds

Members of one group of protists helped change the course of history, at least for the Irish. If you are of Irish ancestry, you just might owe your American citizenship to the activity of a highly destructive funguslike species of **water mold.** A heavy period of Irish immigration to the United States occurred between 1843 and 1847 as a direct result of the famous *potato famine.* The culprit responsible for this great exodus was *Phytophthora infestans,* better known as *late blight* of potatoes. Let's look at the **oomycetes,** or water molds, as a group and see what else they have been up to.

The **oomycetes** (*oo,* egg: *mycetes,* threadlike growth) receive their name from their sexual cycle, in which large egg cells are produced inside a special structure called an **oogonium.** Because of the threadlike growth in many, the oomycetes were classified as fungi, but they were always considered primitive because of two characteristics. First, some species consist of very few cells; second, they commonly form flagellated spores, a decidedly protistan characteristic. Actually, such organisms are

19.11 SEXUAL REPRODUCTION IN *PARAMECIUM*

The *Paramecium* reproduces sexually through a process of conjugation involving two individuals of opposite mating types. **(a)** These partners fuse together at their oral grooves and the process begins. **(b)** The diploid micronuclei undergo meiosis, producing four haploid micronuclei in each partner. **(c)** Three of these disintegrate and the remaining one undergoes mitosis, forming two haploid micronuclei. These are the equivalent of gametes in other sexually reproducing organisms. **(d)** Following this, one of each pair of haploid micronuclei is exchanged across the cytoplasmic bridge. **(e)** In each partner, the exchanged micronucleus fuses with the one remaining to produce a new diploid micronucleus. At this time, the macronucleus is disassembled. **(f)** In each partner the new micronucleus undergoes mitosis without cell division. This is followed by several rounds of DNA replication in one of the daughter nuclei, to form the new large, polyploid macronucleus. **(g)** Following conjugation, each partner goes through two more rounds of mitosis, producing four offspring each.

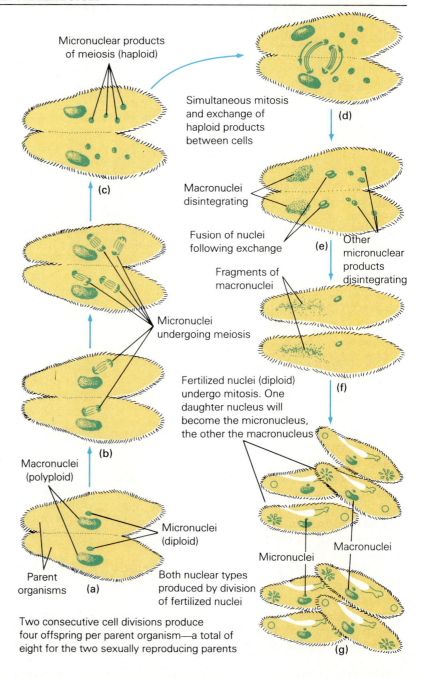
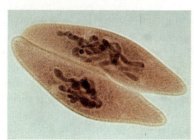

19.12 LIFE CYCLE OF AN ACELLULAR SLIME MOLD

Acellular slime mold life cycles have essentially two phases. These are a creeping *plasmodium* phase, seen in the photo at left, and a fruiting stage (center, right) in which sporangia rise up on short stalks and release spores into the environment. As you can see in the drawing, the creeping stage begins with the zygote, which forms a large *plasmodium* as the cytoplasm increases and nuclei reproduce mitotically, without cell divi-

sion. This is the feeding stage. The plasmodium stage ends with the production of sporangia in which spores form by meiosis. The second half of the life cycle begins as spores are released and germinate into small, single-nucleate ameboid cells or flagellated cells, some of which form the gametes of the acellular slime mold. Gametes then fuse, forming zygotes, and the cycle repeats.

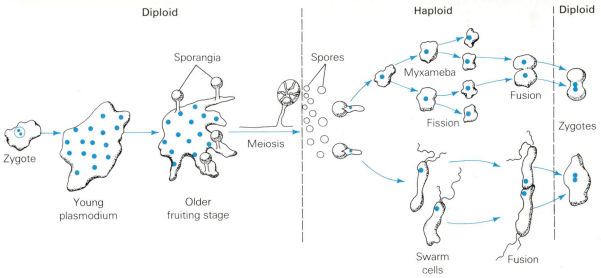

quite interesting to biologists since they tend to span or bridge taxonomic groups, indicating how one may have given rise to another in the dim past.

Some water molds actually live in water, where many species invade injured or diseased plants or animals. Other members, the terrestrial species of the oomycetes, are dangerously parasitic. This group includes **downy mildews** and **blights,** both of which can be seen on the undersides of infected leaves. You can expect them on your house or garden plants if you tend to overwater or undercultivate. Many crops, such as potatoes, beans, melons, sugar beets, and grains, are especially susceptible to infection. *Phytophthora,* the villain of the Irish potato famine mentioned above, grows on moist leaves by

penetrating the numerous stomata (leaf pores) with its **mycelium.** These natural openings on leaves permit the fungus to penetrate the photosynthesizing cells, absorbing the nutrients as they are produced. Eventually, an asexual reproductive stage begins and the filaments grow back out through the stomata, producing spore-forming structures at their tips (Figure 19.13).

In their asexual cycle, the water molds produce sporangia. From these, flagellated spores emerge and begin to swim about. If they encounter a food source, growth occurs and a new colony develops (Figure 19.14a).

Sexual reproduction in water molds (Figure 19.14b) begins with the emergence of unusually thick filaments that produce egg cells in spherical oogonia. The sperm reach the egg cells in an unusual way. Filaments growing near an oogonium send fingerlike branches over the spherical body. Within these branches are the **antheridia,** the sperm-producing structures. The branches form fertilization tubes that penetrate the oogonia, permitting the sperm to reach the eggs. Some of the events are clearly reminiscent of fertilization in higher plants; however, the similarity may be a simple coincidence—a case of convergent evolution.

19.13 INFESTATION OF THE POTATO LEAF BY *PHYTOPHTHORA INFESTANS*

When any downy mildew or late blight infests a plant, it can be readily seen on the underside of the leaves. *Phytophthora* sends its mycelial mass throughout the spongy tissue of the leaf, where it can absorb the plant's photosynthetic products. This eventually weakens and kills the plant, but not before *Phytophthora* completes its asexual cycle. Hyphae of *Phytophthora* emerge through the leaf pores, sending sporangiophores out into the air. At their tips, numerous sporangia develop to produce immense numbers of haploid spores, which can then spread to other plants to produce a new mycelium (as shown in the photograph).

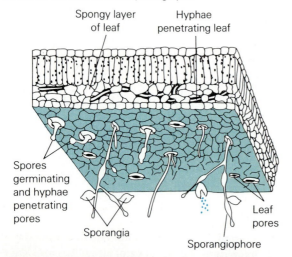

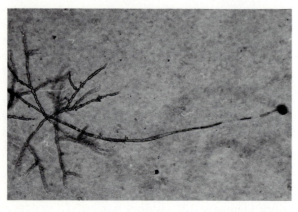

19.14 ASEXUAL AND SEXUAL REPRODUCTION IN THE WATER MOLDS

(a) During asexual reproduction, sporangia form and produce numerous flagellated spores. Some eventually reach a new food source, grow filaments, and repeat the asexual cycle. (b) Sexual reproductive activity in water molds begins when rounded thickenings form in certain filaments. Other filaments send fingerlike projections called antheridia around the enlarging spheres. Eventually, the swellings become an egg-producing oogonia, forming clusters of haploid eggs within. The antheridia produce sperm cells that reach the eggs through fertilization tubes. After fertilization, each diploid zygote will form a nonmotile, resistant zygospore, and under proper conditions, the zygospores will germinate and the asexual and sexual cycles will repeat.

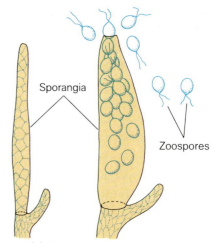

(a) Asexual reproduction

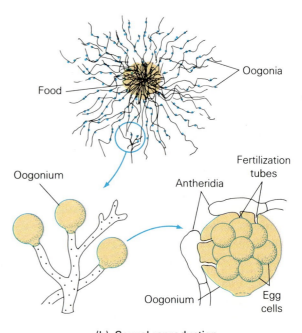

(b) Sexual reproduction

THE ALGAL PROTISTS

As was mentioned, the algae are photosynthetic, plantlike protists. They are extremely widespread, living in all aquatic habitats and a few terrestrial ones. Significantly, algal protists make up much of the floating **phytoplankton,** which provide a source of energy for many marine food chains. Along with the **seaweeds** and **kelps** of the marine environment, they produce more new living material each year than all other photosynthetic forms together. In addition, many algal protists live as intracellular photosynthetic symbionts in the cells of corals and other animals.

Life Cycles: Alternating Haploid and Diploid Phases

As we look into the life history of some of the algae, you will come to what at first might appear to be a peculiar twist in how life cycles go on. We will clarify this point, but let's back up for a moment. You are probably most familiar with the life cycles of animals, including ourselves, which are spent almost entirely in the diploid chromosome state (when chromosomes within cells occur in pairs). That is, nearly all of our cells are diploid and the haploid state is restricted to gametes which, as you may recall, arise from meiosis during gametogenesis. This haploid state ends abruptly when gametes meet in fertilization. But there are many exceptions to this familiar condition. For example, the fungi are essentially opposite to animals in that they are haploid except for brief diploid interludes following sexual fusion. And these usually lead directly into meiosis and a return to the haploid state. The life cycles of some of the protozoans and some of the algae are similar to that of the fungi—that is, they spend most of their lives in the haploid state. But what about other algae, and the plants?

Some algae and all plants exhibit what is known as an **alternation of generations** (Figure 19.15). The alternation of generations seems to be a compromise between the diploid and haploid extremes of animals and fungi, respectively. It involves *two phases* within the life history of *one individual,* one diploid and the other haploid. So far this doesn't sound particularly earth shattering, but we still must mention that in some species, the two phases occur separately. That is, the life of one individual occurs in two independent and separated "subindividuals," for want of a better word. For this reason, the term "generations" is perhaps misleading,

since the two phases still occur within one individual's life. For example, we will see later that the graceful garden fern, while capable of copious spore production, is really a celibate diploid. It carries on meiosis but never forms gametes. Both gametogenesis and fertilization occur in a tiny, but very sexual, groundhugging haploid individual that emerged from one of the spores.

The two alternating phases are known as the **sporophyte phase** (or generation) and **gametophyte phase** (or generation), respectively. The diploid sporophyte begins life as a zygote and produces haploid spores through meiosis, ushering in the gametophyte phase in which gametes are produced by mitosis. (This may seem strange, but we will see many instances of gametes being produced by mitosis before long.)

As we examine the algae in which there is a pronounced alternation between the sporophyte and gametophyte phases, we find that this phenomenon can be accented in two ways. In the first, known as **isomorphic alternation,** the sporophyte and gametophyte generations are identical in appearance, differing only in chromosome number. In the second, **heteromorphic alternation,** one of the two phases is quite dominant in the life cycle. It is usually larger, longer lived, and nutritionally independent. The other phase is greatly reduced, often to just a few cells, and is often nutritionally dependent on the first. We now will review the life histories of three algal protists that do *not* show a clear alternation of generations.

Pyrrophyta: Dinoflagellates

Visitors to tropical waters may be surprised and delighted as they are rowed back to their anchored ship in a dugout canoe after a rousing night ashore. Each time the paddle slides into the water, the water seems to explode with tiny iridescent lights. Even the wake of the canoe is aglow. Objects tossed overboard leave a shimmering trail as they disappear into the briny depths. It seems like magic, but it is the magic of **pyrrophytes,** or **dinoflagellates,** the "fire plants."

Some of the pyrrophytes are also responsible for another dramatic and far more dangerous phenomenon: the dreaded **red tide.** When conditions are right, such as with sudden increases in mineral nutrients, certain pyrrophytes multiply to incredible densities called "blooms." The species with reddish pigments may also contain a potent nerve poison, and the results of their periodic blooms spell death for enormous numbers of fish. The water turns a rusty or blood color, and fish by the thousands appear belly up (Figure 19.16). Clams, oysters, and mussels are unaffected by the dinoflagellates, but they can accumulate the poisons, making their flesh toxic to humans.

Let's look more closely at these fascinating little creatures. Figure 19.17 shows a variety of dinoflagellates. They all have contractile vacuoles and two flagella. Most have chloroplasts and cellulose cell walls. Some eat other organisms, while others are

parasites, and yet others are rather formless photosynthetic symbionts within larger organisms. But most dinoflagellates are free-living and photosynthetic. Their large chromosomes are permanently condensed and remain attached to their nuclear membranes. The nuclear membranes themselves are unusual, in that they are single membranes, as opposed to the double nuclear membranes found in other eukaryotes. The nuclear membrane does not break down during dinoflagellate mitosis (reminis-

19.15 ALTERNATION OF GENERATIONS

The term "alternation of generations" refers to the presence of a diploid sporophyte state and a haploid gametophyte state in the life history of a species. Definite generations are clearly seen in some of the algae and in all plants. Either generation may be dominating, in both size and tenure, or they may be equal in both. The gametophyte generation arises when cells in the sporophyte undergo meiosis, forming haploid spores. The spores then give rise to cells that form gametes through mitosis. When gametes fuse in fertilization, the diploid sporophyte stage emerges once again. The variation in life histories is seen in **(a)** an animal and some of the animallike protists, **(b)** a fungus, or in some cases an alga, **(c)** a plant or alga with equal sporophyte and gametophyte stages, **(d)** a plant or alga with a dominating sporophyte state, and **(e)** a plant or alga with a dominating gametophyte state.

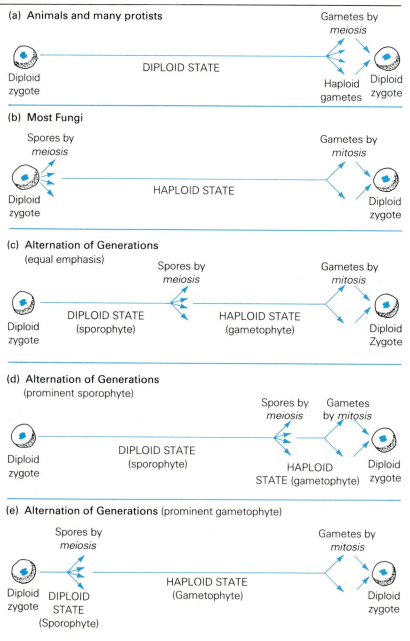

19.16 RED TIDE

The dead fish seen floating here were killed by the highly toxic metabolic products of dinoflagellates from the "red tide"—a sudden population bloom. The dinoflagellates are always present, but they tend to increase seasonally due to enrichment of the waters by upwellings of nutrients from deep sediments.

cent in some ways of the fission process in prokaryotes). Meiosis does occur, but it too is unusual. Apparently it involves only one division, not two as in other eukaryotes. In general, dinoflagellates have many unique cellular features and seem to have diverged from most other eukaryotes very early in the evolution of life. Some specialists consider them to be intermediate to prokaryotes and eukaryotes, and prefer to call them **mesokaryotes.**

Euglenophyta: Euglenoids

Among the Euglenophyta is the *Euglena*. We have mentioned *Euglena* before as an example of why classification can be so difficult. But it's also interesting for other reasons, as we will see.

The name Euglenophyta comes from *eu-*, "true"; *glena-*, "eye"; and *-phyton*, "plant"—the plant suffix referring to their former association with the plant kingdom. The visible "eye-spot" of euglenoids (Figure 19.18) is really not an eye at all, but is a shield of red pigment lying next to a light-sensitive area called a **photoreceptor.** Because of the shield's arrangement, euglenoids can detect light and move toward it. When the sensitive area is not shielded, the organism simply begins to move. The result is that it continually approaches the light—an adaptation for photosynthesis. Actually, only about a third of euglenoid species are photosynthetic.

Euglenoids are unicellular, flagellated, and asexual. *Euglena* is the most widespread genus and

can serve as an example. Like all euglenoids, *Euglena* lacks rigid cell walls and has a flexible body, but tends to maintain a flasklike shape. A single long flagellum arises from a mouthlike depression. There is actually no evidence that *Euglena* ingests solid food particles, but other genera of euglenoids are known to do so. A second very short, rudimentary flagellum is present in the gullet, but its function is not known. *Euglena* contains chlorophyll *a* and *b* and carotenoids, incorporated into a chloroplast—all quite similar to what we find in plant cells. It stores its starches in the form of *paramylon*, which is quite different from plant and algal starches. So far, no one has observed sexual activity in *Euglena*, but asexual reproduction, as you might expect, occurs by mitosis and cell division. Euglenoids can reproduce so rapidly that they may impart a green color to a pond, except for one ruddy species that colors it red.

Chrysophyta: Yellow-Green and Golden-Brown Algae

Both the yellow-green algae and the golden-brown algae live in both marine and fresh waters, where they form a vital part of the food chains. The names are derived from the color of their light-absorbing pigments. All **Chrysophyta** ("golden plants") con-

19.17 SEVERAL COMMON DINOFLAGELLATES

There is considerable variation in size and shape. *Gymnodinium* (bottom left) and *Gonyaulax* (bottom right) are red tide species.

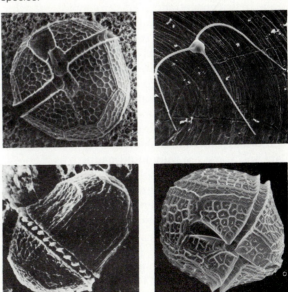

19.18 *EUGLENA GRACILIS*, A WELL-KNOWN EUGLENOID

Euglena's chloroplasts are quite like those of plants—highly organized and surrounded by a double membrane. Each chloroplast is associated with a *paramylon body* where starch is stored. The large, membrane-bound, circular structure at the center is the nucleus, which, with its nucleolus, is typically eukaryotic. At the base of the cell is the flagellum, and enclosed in the flagellar membrane is the photoreceptor. The photoreceptor, along with its *pigment shield,* permits the protist to orient itself toward light, which it requires for photosynthesis.

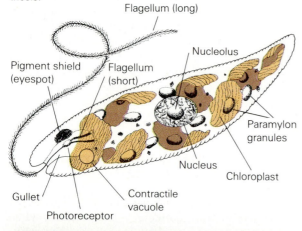

Flagellum (long)

Pigment shield (eyespot)

Flagellum (short)

Nucleolus

Nucleus

Chloroplast

Paramylon granules

Gullet

Photoreceptor

Contractile vacuole

tain high concentrations of carotenoids, of which the most abundant is called **fucoxanthin.** Fucoxanthin absorbs much of the light, transferring it to molecules of chlorophyll *a.* The chrysophytes store their carbohydrates in the form of **leucosin.** Chrysophytes are a diverse group. Some surround themselves with cell walls that are essentially glass (silicon dioxide) boxes. Some chrysophytes are flagellated, others not; most are solitary but a few genera produce filamentous colonies; some are multinucleate and some are truly multicellular.

Although they are photosynthetic, some ameboid species of golden-brown algae can ingest solid food. Many yellow-green algae can be blown about by the wind and are often found growing on tree trunks, rocks, or soil.

The Diatoms. Perhaps you are wondering about the glass boxes. These are best seen in the **diatoms,** the most common photosynthetic marine organism. The boxes are indeed very fragile and

peculiar structures. The diatom shell consists of an inner box and an outer lid, both made of the colorless silicon dioxide mentioned earlier. The boxes strongly resemble Petri dishes, of all things.

When they reproduce asexually, diatoms undergo mitosis and divide within their shell, whereupon the two old halves of the shell each become the outer lid of a daughter cell as new inner half-boxes are secreted (Figure 19.19). Because glass

19.19 ASEXUAL REPRODUCTION IN DIATOMS

The two halves of the skeleton separate with the daughter cells. Following mitosis (a) and cell division, one new cell gets the outer lid of the parent and the other, the inner half (b,c). Diatoms only synthesize inner halves, as shown here, meaning that the one receiving an inner half (c) produces another inner half, thereby becoming smaller than its parent cell. When this smaller diatom divides, one of its offspring will be even smaller (d). This keeps happening until the diatoms reach a critical minimum size; then they must enter a sexual process. The photos show a diatom at the beginning and end of cell division.

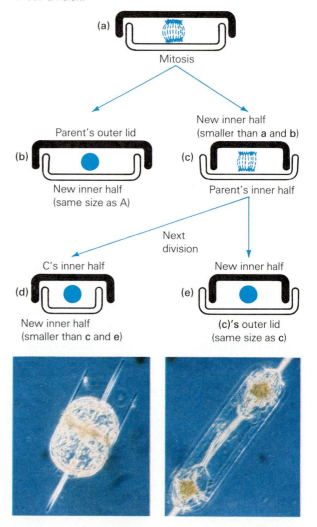

(a) Mitosis

Parent's outer lid

New inner half (smaller than a and b)

(b) New inner half (same size as A)

(c) Parent's inner half

Next division

C's inner half

(d) New inner half (smaller than c and e)

New inner half

(e) (c)'s outer lid (same size as c)

19.20 SEXUAL REPRODUCTION IN DIATOMS

In this species of diatom, one ameboid gamete fuses with an identical gamete from another diatom, becoming a diploid auxospore. When the auxospore grows and matures it produces a new shell, large enough to go through many asexual divisions before the sexual process must be repeated.

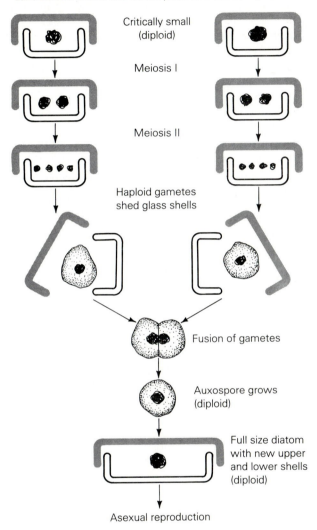

Critically small (diploid)

Meiosis I

Meiosis II

Haploid gametes shed glass shells

Fusion of gametes

Auxospore grows (diploid)

Full size diatom with new upper and lower shells (diploid)

Asexual reproduction

boxes don't stretch very well, one of the daughter cells, the one that inherited the inside half of the parent cell, must be smaller than the other. This presents the diatom with a problem. After all, it wouldn't do to grow smaller with every generation. So, what's the solution?

Eventually the smallest diatom undergoes meiosis in its box, producing haploid gametes. In many species (chiefly those in fresh waters), the gametes are ameboid and are not differentiated. In others (mainly marine), a small flagellated sperm and a larger egg are formed. In either case, the haploid gamete sheds its box and becomes the free and naked sexual form. Gametes meet and fuse to produce the diploid zygote, called an **auxospore** (*auxo*, "increasing") because it increases in size for a while before reaching the adult size. The auxospore then matures, secretes a new glass house, and the cycle continues (Figure 19.20).

The walls of marine diatoms are highly ornamented and beautiful to the human eye. The intricate designs are produced by the arrangements of tiny holes through which gas and water are exchanged. Diatoms can be radially or bilaterally symmetrical (Figure 19.21).

Ocean floor sediments consist largely of diatom shells. Deposits of ancient diatom shells form **diatomaceous earth,** which is used in toothpaste, swimming pool filters, and insulating material. The minute glassy particles have an abrasive quality which is useful in polishing. The thickest deposits of diatomaceous earth (about 1 km in depth) occur off the coast of Lompoc, California.

19.21 DIATOM DIVERSITY

(a) Diatoms are a highly diverse group in terms of shape and design. The fine sculpturing of their glass skeletons varies from one species to another. The etched surfaces are so fine that they make excellent objects for testing microscope lens systems. (b) The scanning electron micrograph is a closeup of the two halves of the skeleton.

(a)

(b)

19.22 RED ALGAE

Most red algae are small compared to the brown algae, measuring from a few centimeters to perhaps a meter in length. Most species are marine and are found in warm waters, usually attached to rocks by their *holdfasts*. Some of the branched bodies form widened, flat blades, but most are frilly and delicate, as seen here.

Rhodophyta: The Red Algae

The **Rhodophyta** (*rhodo-*, "red"; *-phyta*, "plants") include about 300 species, all called **red algae.** All contain the pigments **phycocyanin** (meaning "algal blue-green") and **phycoerythrin** (meaning "algal red"), as well as chlorophyll *a,* and many have chlorophyll *d* (an oxidation product of chlorophyll *a*). The pigments of the red algae are strikingly similar to those of the prokaryotic cyanobacteria (see Chapter 18).

Flagella and cilia, typically found in the rest of the eukaryotes, are not found in the red algae. Even the sperm (called *spermatia*) lack tails and must float passively to receptive female cells.

All but a few red algae are aquatic, and nearly all of them are marine (Figure 19.22). You have undoubtedly seen red algae if you've ever looked into a tide pool. In fact, red algae comprise a large portion of what we commonly call *seaweed.* You might not recognize them as red algae, since they are often green or black, and some are blue or violet, but those growing well below low tide are often red. Actually, color can't be used to identify species. Individuals of the same species often are found to be of different colors at different depths, in what is believed to be an adaptive response to the changing intensity of light available for photosynthesis.

Marine red algae grow primarily on rocky coasts, where they attach firmly to the seabed by specialized structures known as **holdfasts.** Holdfasts are simply anchors, not roots. After all, algae grow in water and therefore have no need of a root system for the extraction of water and minerals.

Red algae store their foods in the form of a starch called **floridean starch** and produce a number of other polysaccharides. One of these, **agar-agar** (or **agar** for short), is used by Indonesians to thicken soup and by biologists as a jellylike medium on which to grow bacteria. The bacteria cannot metabolize the agar but use only the nutrients that have been added to it. Although red algal polysaccharides cannot be metabolized by most microorganisms or by anything else, we may eat them anyway. A species of red alga known as *Irish moss* is harvested in enormous quantities for its polysaccharide **carrageenan,** which gives a fake richness to such things as chocolate-flavored dairy drink and milkshakes.

The common red algae *Polysiphonia* presents us with an unusually clear example of alternating sporophyte and gametophyte phases, as do many of the seaweeds and kelps (Figure 19.23). In this genus, the sexual haploid and the asexual diploid are both multicellular forms with branching growth forms that closely resemble one another—what we described earlier as isomorphic alternation of generations. The phases can be distinguished only by inspecting the reproductive structures under the microscope (or by chromosome counts). Neither generation is dominant—that is, neither is decidedly larger or more common than the other.

The sporophyte begins with the union of gametes into a zygote called a **carpospore,** a number of which form in a pouchlike **carposporangium.** It is in the sporophyte that meiosis takes place, but again, unlike meiosis in animals, in which gametes are produced, meiosis results in the production of spores. The role of the meiotic spore in these instances is to disperse—to be carried away—and to germinate and grow by mitosis into a multicellular haploid organism, a gametophyte. The gametophyte then produces gametes through the simple process of mitosis (its cells are already haploid), and these will join in fertilization to produce the next sporophyte.

Some species of red algae are heteromorphic, with either the sporophyte or the gametophyte generation clearly dominant. In some, the gametophyte generation is reduced to a single-cell stage, and eggs and sperm develop directly by meiosis, just as in animals. There aren't too many hard and fast rules about alternating phases, even in a group of only 300 species.

The fossil history of the red algae is not very complete, but it seems likely that they shared the earth with the green algae (as well as with the far older cyanobacteria) as far back as the Cambrian period at the beginning of the Paleozoic era, some half billion years ago.

Phaeophyta: The Brown Algae

The **Phaeophyta** includes about 1000 named species. The **brown algae** are distinguished by their characteristic brown pigment, **fucoxanthin.** Brown algae at least have the decency to be brown—all of them. All of them also store carbohydrates in the form of **laminarin** and **mannitol,** and they have characteristic structural polysaccharides as well (notably **algin,** an important constituent of commercial ice cream and frozen custards). Unlike the red algae, the brown algae have flagellated sperm, and strangely enough, some female reproductive cells are flagellated as well.

The brown algae live only in the ocean, particularly in cold coastal waters. An exception of sorts is the genus *Sargassum,* which is found in great masses in the warm water that flows through the middle of the Atlantic Ocean, an area called the Sargasso Sea (Figure 19.24). Even *Sargassum,* however, is really a coastal seaweed. The individuals in that floating mass have been uprooted by storms and have drifted out to sea. Though they blissfully continue to photosynthesize, grow, and release gametes, the zygotes that are produced fall to the bottom of the ocean and fail to develop, succeeding only in adding to the marine food chain.

The brown algae come in all shapes and sizes and have a wide range of life histories. The strongly heteromorphic life cycle of one of these, *Fucus,* is shown in Figure 19.25. With its greatly reduced gametophyte phase, occurring within specialized **conceptacles** in the blade, *Fucus* is quite similar reproductively to the more advanced terrestrial plants, demonstrating that evolution sometimes takes similar directions in unrelated groups. Some brown algae are microscopic filaments, while others

19.23 THE LIFE HISTORY OF A RED ALGA

Polysiphonia's life cycle consists of separate but isomorphic sporophyte and gametophyte generations. The *sporophyte is diploid,* produced by growth of a *carpospore.* The sporophyte generation ends with meiosis and the production of *haploid spores,* marking the beginning of the *gametophyte* generation. Female gametophytes produce egg cells, each with long slender filaments that protrude outward. The male gameto- phyte produces *nonmotile* sperm cells, which when released adhere to the female filament. Their nuclei penetrate this structure and migrate down to fertilize the egg cells. Fertilization ends the haploid gametophyte generation, starting a new, diploid sporophyte generation. The zygote divides mitotically within the pouchlike *carposporangium,* which ultimately releases many diploid *carpospores.*

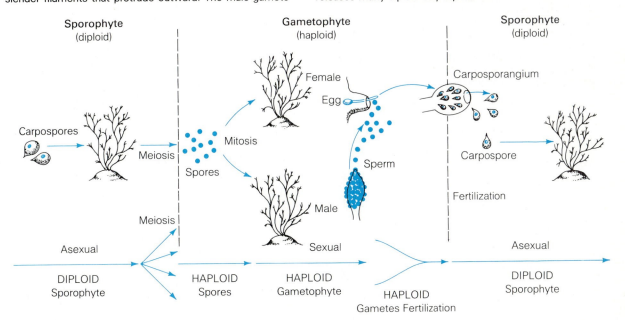

19.24 THE BROWN ALGA *SARGASSUM*

Sargassum is a floating seaweed commonly found in the warm southern Atlantic waters. *Sargassum* forms branching stipes and flat, leaflike *blades*. The spherical objects attached to the stipes, seen in the bottom photo, are bladders (floats) that help the seaweed remain near the surface where light is plentiful.

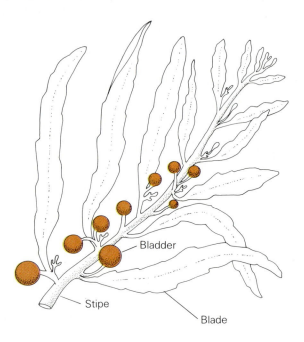

Bladder

Stipe

Blade

19.25 THE BROWN ALGA *FUCUS*

This alga is surprisingly like an advanced seed plant in its reproductive life. Its highly dominating diploid sporophyte forms specialized blade areas called *conceptacles*—hollow chambers within which oogonia and antheridia form. Meiosis leads to a very brief haploid state which remains within the sporophyte tissue, and through mitosis produces egg and sperm cells. These are eventually released into the water, where fertilization and the resumption of the sporophyte generation occur.

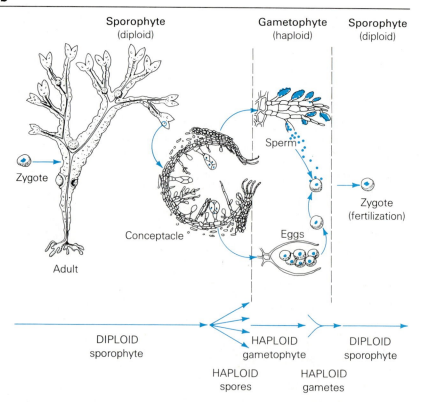

Sporophyte (diploid)

Gametophyte (haploid)

Sporophyte (diploid)

Zygote

Sperm

Zygote (fertilization)

Conceptacle

Eggs

Adult

DIPLOID sporophyte

HAPLOID gametophyte

DIPLOID sporophyte

HAPLOID spores

HAPLOID gametes

such as the giant kelps, are enormous. One of the largest kelps, *Nereocystis*, can grow to 30 m (100 ft) in length rivaling the height of many terrestrial trees.

The kelps appear to be the most advanced brown algae. For example, *Macrocystis*, another giant kelp (Figure 19.26), has highly specialized tissues and organs (see Figure 19.25b). These include the convoluted *holdfast*; **stipes**, which are stemlike structures supporting the **blades;** blades themselves, which resemble flattened leaves; and spherical, hollow **bladders** (or floats), which keep the photosynthetic cells of the kelp near the surface. The brown alga *Postelsia* even has specialized conducting tissue that closely resembles that of some plants. But in general, algae do not possess elaborate conducting systems.

Chlorophyta: The Green Algae

There are about 7000 named species of **green algae,** and whereas most of them are freshwater forms, there are a respectable number of marine species as well. A few terrestrial/airborne species appear on the surface of melting snow, on the moist sides of trees, or free in the soil. The chlorophytes include both unicellular and multicellular species. Many single-celled green algae have become photosynthetic symbionts in lichens, ciliates, and invertebrates. Biochemically, the green algae closely resemble plants. They contain chlorophyll *a* and *b* and carotenes. A chief storage carbohydrate is starch and their cell walls contain cellulose, hemicellulose, and pectin.

The green algae occur as single-celled flagellated or unflagellated forms, as chains or filaments, as inflated "fingers" (*Codium*, also called "dead man's fingers"), and as delicate flattened blades (*Ulva*, the delicate *sea lettuce* of tide pools). The group is so diverse that it is difficult to find a representative species, so we'll arbitrarily choose a few examples.

Single-Celled Green Algae. The most primitive of the green algae are the *single-celled* and *colony-forming* types. As far as anyone can figure, this group, although related to the multicellular green algae, is not descended from them and thus simply has never gotten past the cellular level of organization.

19.26 THE GIANT KELPS

The giant kelp *Macrocystis* thrives in cold oceans. **(a)** Seen here are its highly branching stipe and smaller bladders at the base of each blade. Kelps like these occupy large submerged beds along the continental shelf and provide protective cover for many other marine species. **(b)** This view from the Monterey Bay Museum near San Francisco gives an idea of the great size of *Macrocystis*. The giant aquarium simulates the marine environment with a motor-driven cylinder that circulates seawater from which the kelp obtains mineral nutrients.

Chlamydomonas (Figure 19.27) is a favorite organism of many biologists. It is easily grown, and its genetics and physiology have been studied in detail. The cells have two flagella, a single light-sensitive **red eyespot,** and one cup-shaped chloroplast. Like *Euglena, Chlamydomonas* is positively phototactic (it moves toward light), and if kept in a transparent container of water, individuals will congregate on the side closest to the light. *Chlamydomonas* is usually seen in a haploid chromosome state and its alternation of generations is not noticeable. In its haploid stage it reproduces asexually by mitosis and cell division and builds huge clone populations. It will not enter into a sexual cycle unless conditions are right.

The right conditions include the presence of opposite mating types (the plus and minus strains) as well as some form of environmental stress, such as absence of nitrogen compounds. When both mating types are present, the haploid *Chlamydomonas* produce **isogametes** mitotically. (Isogametes are gametes in which the two types are identical in appearance.) The two isogametes fuse by butting, head on, and entering a wildly spinning "nuptial dance" as their nuclei slowly fuse. Following the union, the zygote, now representing the diploid state, forms a tough, resistant **zygospore.** The zygospore may remain dormant for a considerable time or, if conditions are right, it may immediately enter into meiosis, producing four haploid **meiospores.** This restores the haploid state and the individuals mature into the familiar haploid population. As we see, then, alternation of generations is not heavily emphasized in this green alga, and its life cycle is very similar to that of the typical fungus.

Colonial Green Algae. Another remarkable green alga is *Volvox* (Figure 19.28a). Although it is not related to any truly multicellular form, its history tantalizingly suggests what the earliest beginnings of multicellularity *might* have been like. *Volvox* is composed of a spherical colony of virtually identical cells, each one, in fact, very similar to an individual *Chlamydomonas.* So, in a sense, *Volvox* is a group of individuals in a sphere behaving as an organism. Or is the *sphere* really the individual? The seeming simplicity of this sort of question is

19.27 LIFE HISTORY OF *CHLAMYDOMONAS*

The diploid zygospore is a brief interlude in an otherwise lengthy haploid history. Zygospores enter meiosis and produce flagellated meiospores. These can enter either a sexual phase, producing many isogametes, or an asexual phase, producing enormous cloned populations through repeated mitotic divisions. Sexual reproduction occurs when different (+ and −) mating strains meet. Flagellated isogametes fuse head-to-head, producing diploid zygospores once again.

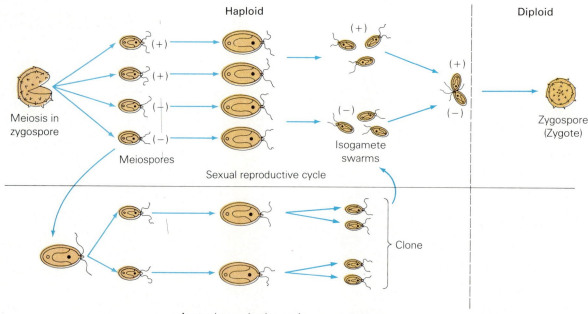

Meiosis in zygospore

Meiospores

Haploid

(+)

(+)

(+)

(−)

(+)

(−)

(+)

(−)

Isogamete swarms

Diploid

Zygospore (Zygote)

Sexual reproductive cycle

Clone

Asexual reproduction cycle
(may continue indefinitely or may enter sexual phase above)

19.28 *VOLVOX*, A COLONIAL ALGA

Volvox, an alga, is a spherical colony of tiny, interconnected, flagellated cells. The dense spherical bodies seen within the colony are concentrations of cells that were produced asexually. **(a)** During asexual reproduction, individuals of the sphere increase in size, divide, grow, and divide again, but eventually they are all reduced down to the original size. The result at first is a miniature of the original colony, except that the flagella are all pointed inward. Next the new colony undergoes inversion—it actually turns inside out—and a more typical

flagellated colony emerges, soon to escape and live on its own. **(b)** In sexual reproduction, flagellated cells of the sphere differentiate into sperm- and egg-producing structures. Sperm are released and may then enter and fertilize the egg cells of another colony. With fertilization, a zygospore is produced, complete with roughened, resistant casing. *Volvox* colonies are haploid, so zygospore germination is immediately followed by meiosis and the emergence of haploid, flagellated spores that grow into new colonies.

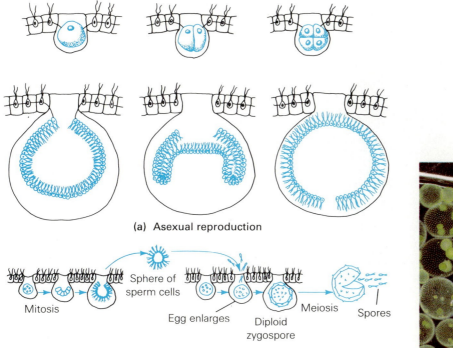

(a) Asexual reproduction

Mitosis · Sphere of sperm cells · Egg enlarges · Diploid zygospore · Meiosis · Spores

(b) Sexual reproduction

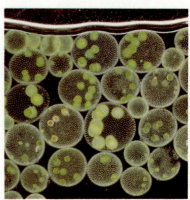

deceptive. Biologists have been trying to answer it for years.

Volvox is interesting not only because it suggests a way multicellularity could have arisen, but also because it shows a kind of rudimentary differentiation within the colony. The sphere swims in an organized manner, with the flagellated cells in front pulling and those in the rear pushing. In the sexual phase some cells become specialized for reproduction, producing a few large eggs or many smaller sperm. In asexual reproduction, pockets of the sphere depress inward, and new spheres of cells pinch off and come to lie inside the parental sphere. The new, young spheres are inverted, however, with their flagella directed inward. The daughter *Volvox* colony has to turn itself inside-out within the parent structure. Eventually it is released through a hole in the parent colony (Figure 19.28b).

Siphonous Green Algae. Green algae of another group, the **siphonous algae,** have taken off in a different evolutionary direction. Old botany texts once called it "an evolutionary dead end," but this is probably unfair because, after all, they're still around—and doing quite well in fact. Algae of this group aren't exactly multicellular, but they're not exactly unicellular either. In fact, they present a direct challenge to the famous cell doctrine of Schleiden and Schwann. A few organisms are clearly not cellular, though some individuals grow to a considerable size and may produce branched **thalli** (bodies). Although the nuclei undergo mitosis, the cytoplasm doesn't divide and cell walls are not laid down between the newly separated nuclei. The result is a multinucleate mass of cytoplasm (or *coenocyte*) within an enveloping plasma membrane and cell wall. The group includes *Acetabularia*, a weird little alga that looks

19.29 THE SIPHONOUS ALGAE, *ACETABULARIA* AND *CODIUM*

(a) *Acetabularia*, romantically called the "mermaid's wine-glass," is a most unusual organism. Although *Acetabularia* can grow as long as 9 cm, it is single-celled with only one (compound) nucleus, located at the base of the stalk. These delicate organisms are found in warm, tropical and subtropical waters along the Gulf of Mexico and the Florida coast. (b) *Codium* grows in the manner of some fungi in that cell walls are not formed after mitosis. In effect, its cytoplasm is continuous and multinucleated. *Codium* produces a branching fingerlike growth, from which it gets its common name, "dead man's fingers." It can often be found growing on the shells of mollusks.

(a)   (b)

like a toadstool or parasol. Its single cell stands about 5 to 9 cm tall. Some *Acetabularia* are multinucleate coenocytes, but in other species, the multiple nuclei are not dispersed through the cytoplasm; instead, they clump into a single **compound nucleus** in the foot, so that the structure of the whole organism is very like that of a single cell, but an extraordinarily large one (Figure 19.29).

Filamentous Green Algae. Among the structurally more complex **filamentous algae** are *Ulva* and *Ulothrix.* The thallus of *Ulva* is leaflike, broad, and thin (only two cells thick), the product of growth in two dimensions rather than in only one as in *Ulothrix* and many other filamentous algae. *Ulva* has a distinctive alternation of generations and the gametophyte and sporophyte stages are isomorphic—that is, about equally prominent, as in the red alga *Polysiphonia*. In *Ulothrix,* we see the nearly total domination of the haploid state over the very brief diploid state. Following meiosis in specialized cells called **gametangia,** isogametes emerge and fuse in pairs to form diploid zygospores. The zygote immediately undergoes meiosis to restore the haploid state as we saw in *Chlamydomonas* (Figure 19.30).

The themes of alternation of generations and of the tendency of one phase to become dominant continue in the evolution of the land plants. While

19.30 *ULOTHRIX*, A FILAMENTOUS ALGA

Like *Chlamydomonas, Ulothrix* has an extremely brief diploid state. The diploid zygospore enters meiosis, producing numerous zoospores. In an asexual phase, the zoospores produce filaments, which, in turn, develop sporangia and more motile, tetraflagellate spores, which settle to begin the growth of new filaments. In the sexual phase, some filaments develop gametangia, which produce biflagellate isogametes. Different mating types fuse to produce diploid zygospores once again.

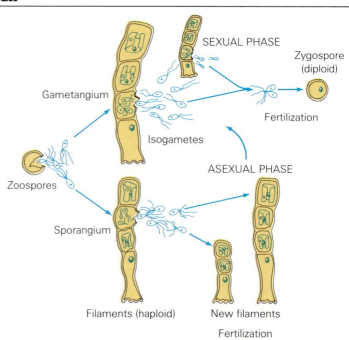

SEXUAL PHASE

Zygospore (diploid)

Gametangium

Fertilization

Isogametes

ASEXUAL PHASE

Zoospores

Sporangium

Filaments (haploid)   New filaments

Fertilization

there is no fossil evidence suggesting that the ancestors of the land plants were closely related to present-day multicellular green algae, most authorities agree that this is the case. We will soon see that some of the filamentous green algae probably share ancestors with the plants.

The green algae were probably the dominant form of aquatic life in the Cambrian period and throughout the Ordovician and Silurian, too (405 to 570 million years ago). Most authorities believe that the first terrestrial plants descended from the filamentous green algae, possibly during the Lower Devonian period (some 400 million years ago). The brown and red algae remained in the sea specializing in other directions.

APPLICATION OF IDEAS

1. The use of DDT has been greatly curtailed for important ecological reasons, yet it has proven in the past to be the greatest malaria deterrent known in parts of the world where other measures have been ineffective. Many organizations would like to see this form of mosquito control resumed. Discuss the issue of human health versus preservation of the environment. (Read *The Silent Spring*, by Rachel Carson, for some insight into the DDT controversy.)

2. Euglenoids, and in particular *Euglena gracilis*, have been classified as plants by botanists, animals by zoologists, and protists by others. Review the characteristics of *Euglena* and discuss the basis for the taxonomic disagreement. In what ways does the protist designation help? What does such a problem reveal about the categories contrived by taxonomists?

3. One of the problems of controlling African sleeping sickness has been the movement of nomadic tribes and their cattle in and out of endemic areas. Your job as an official attempting to control the spread of the disease is to explain to the tribes why they can no longer move about freely and why some of their cattle have to be disposed of for health reasons. How would you explain the life cycle of the trypanosome and the spread of sleeping sickness to these primitive people?

4. Some protists, particularly the very large ciliates, seem to represent exceptions to the cell theory, and the term *unicellular* or *single-celled* seems inappropriate. Using one of the larger ciliates as an example, make a case for using the term *acellular*. What might substitute for the term "organelles" in an acellular organism?

KEY WORDS AND IDEAS

1. Because Kingdom Protista is polyphyletic and extremely diverse, some consider the very old name, *Protista*, to be inadequate and suggest it be replaced with *Protoctista*.

2. Protists are often single-celled, but some are colonial and a few are definitely multicellular. Autotrophs are common and all common feeding modes occur among the heterotrophs. Movement is through cilia, flagella, and **pseudopods.**

3. Protista is divided into the animallike **protozoans,** plantlike **algae,** and a few funguslike forms. Each is believed to have given rise to the higher groups they resemble.

THE ANIMALLIKE PROTISTS

1. The animallike protists or protozoans were formerly Phylum Protozoa of the animal kingdom. Since some are far larger and more complex than most cells, they are sometimes called **acellular** (without cells).

Mastigophora: Flagellates
1. Most flagellates have the 9 + 2 flagellum. They may capture food or simply absorb it through the body wall.

2. *Trypanosoma gambiense*, the agent of **African sleeping sickness,** is one of several important flagellate parasites.

Sarcodina: Ameboid Protists
1. *Entameba histolytica*, an ameboid protist is the water borne agent of ameboid dysentery.

2. Most ameboid protists move about by the formation of pseudopods (false feet), temporary extensions of the cell into which the body flows. They feed mostly through phagocytosis.

3. Sarcodines have retractible microtubular **axo-**

podia, spinelike extensions covered by a moving ameboid cytoplasm used in feeding.

4. **Radiolarians** and **foraminiferans** have hardened skeletons, of silicon dioxide and calcium carbonate, respectively. Both contribute to ocean floor deposits, including the **radiolarian ooze.**

5. Asexual reproduction in sarcodines is through mitosis and cell division. Sexual reproduction is not often seen but usually involves meiosis and cell fusion. Foraminiferans produce flagellated haploid cells that fuse. In heliozoans, three fourths of the gametes are discarded as polar bodies during meiosis, and fertilization is through simple fusion.

Sporozoa: Spore Formers

1. **Sporozoans** are simple, spore-forming, nonmotile parasites.

2. Medically, the most important sporozoan is *Plasmodium vivax,* an agent of **malaria.** It is carried by the female *Anopheles* mosquito. The most effective control agent is the highly controversial insecticide, DDT.

3. The life cycle of *P. vivax* includes sexual stages in the mosquito and asexual stages in the host. The mosquito injects **sporozoites** into the host and these enter the host's liver, where many **merozoites** form. When freed, they invade the red blood cells and release toxins that bring on the alternating fever and chills of malaria. Some enter a sexual stage, forming **gamonts,** which when taken back into the mosquito form sperm and eggs. The zygote, called an **oocyst,** divides asexually to form many sporozoites which are then injected into the next host.

Ciliophora: Ciliates

1. **Ciliates** have an enormous size range. Cilia are commonly used in movement and in feeding. The ciliate covering, or **pellicle** is made up of a tough, secreted material. Some ciliates release threadlike **trichocysts** for capturing prey and for mooring. Water balance is maintained by **contractile vacuoles** that collect and force water out of the cell. Food taken in through the **oral groove** and the **cytostome** is commonly digested within **food vacuoles,** and wastes expelled through the **cytopyge.**

2. Germ-line DNA (for replication and mitosis) in *Paramecium* is maintained in one or more **micronuclei,** while somatic-line DNA (for transcription) is maintained in the **macronucleus.**

3. Asexual reproduction occurs through mitosis and cell division. Sexual reproduction **(conjugation)** involves meiosis, cell fusion, and the exchange of haploid micronuclei. There are no gametes or sexes but different mating types occur.

THE FUNGUSLIKE PROTISTS

The Gymnomycetes: Slime Molds

1. The acellular **slime mold,** *Physarum polycephalum,* has a giant multinucleate, creeping and feeding ameboid stage that ends with a spore-forming stage. Haploid spores are formed through meiosis in **sporangia,** which develop atop erect **sporangiophores.** Such spores germinate into an ameboid **myxameba** or flagellated **swarm cell.** Either can form a dormant cyst and either can fuse with others, becoming diploid and forming another feeding stage.

Oomycetes: Water Molds

1. A parasitic **oomycete,** *Phytophthora infestans,* also called **late blight,** created the Irish potato famine by infecting the potato plant.

2. Oomycetes receive their name from the large egg cell produced in an **oogonium.** The **water molds** are funguslike in appearance, but they are **coenocytic** (multinucleate cytoplasm) and they produce flagellated cells.

3. The aquatic water molds are nonparasitic or only weakly parasitic. However, terrestrial species include the parasitic crop-destroying **downy mildews** and **blights.** *Phytophthora* invades the leaf through the stomata and returns through these pores to produce spores.

4. In asexual reproduction, aquatic water molds produce flagellated spores. In sexual reproduction, spherical oogonia produce egg cells. Fingerlike sperm-producing **antheridia** invade the oogonia, permitting sperm and egg to meet.

THE ALGAL PROTISTS

1. Algae—photosynthetic protists—range from single-celled to complex multicellular forms. Most are aquatic, living in all of the earth's waters, and in a few instances on land. Included is the vast, floating, microscopic marine **phytoplankton** and the marine **seaweeds** and **kelps.**

Life Cycles: Alternating Haploid and Diploid Phases

1. While animals are diploid except for their gametes, algae, fungi, and primitive plants can have extensive haploid stages in their life cycles and very brief diploid stages.

2. Some algae and all plants undergo an **alternation of generations,** whereby diploid phases and haploid phases occur in separate individuals. These phases are known as the **sporophyte** (spore-producing) and **gametophyte** (gamete-producing) phases, respectively. The diploid sporophyte phase begins with fertilization and ends with meiosis. The gametophyte phase begins with meiosis and ends with fertilization.

3. The two phases may be **isomorphic**—identical in appearance, or **heteromorphic**—quite different. They may be independent of each other or one may be nutritionally dependent on the other.

Pyrrophyta: Dinoflagellates

1. **Pyrrophytes** (the "fire-plants," **dinoflagellates**) produce the famous **red tides,** which bring about the death of fish and other marine life.

2. Dinoflagellates are mainly flagellated, single-celled photosynthesizers. Some are free-living and others live in the cells of other organisms as symbionts. They have a primitive form of cell division in which the single nuclear membrane does not break down and only one division occurs during meiosis. Some regard them as intermediates between prokaryotes and eukaryotes, calling them **mesokaryotes.**

Euglenophyta: Euglenoids

1. *Euglena*, a single-celled representative euglenophyte, is a flagellated, very flexible swimmer that orients to light through a light-sensitive **photoreceptor.** Its chloroplast contains chlorophyll *a* and *b*, and carotenoids in which its starches are stored as **paramylon.** It reproduces asexually by mitosis and transverse cell division—no sexual reproduction has been observed.

Chrysophyta: Yellow-Green and Golden-Brown Algae

1. **Chrysophytes inhabit fresh and salt water, and owe their color to concentrated light absorbing carotenoids (mainly fucoxanthin).** Carbohydrate storage is in the form of **leucosin.**

2. Some chrysophytes have glassy coverings and some are flagellated, while others produce colonies, a few are multicellular, and some are even terrestrial.

3. Because of their glassy, boxlike coverings, **diatoms** must solve the problem of getting progressively smaller as they undergo mitosis and cell division. Their solution is sexual reproduction. Many produce ameboid gametes that abandon the covering and fuse, forming **auxospores.** Following this, a full-size covering is produced and the cycle repeats. The boxes fall to the seabed to form **diatomaceous earth.**

Rhodophyta: The Red Algae

1. The pigments (phycocyanin and phycoerythrin) and chloroplasts of **red algae** are similar to those of cyanobacteria.

2. Most red algae are marine seaweeds that grow on rocky coasts, using **holdfasts** to fasten themselves in place. They produce **floridean starch** and other polysaccharides, including **agar-agar** and **carageenan.**

3. The alternation of generations of **Polysiphonia** is isomorphic but in others it is heteromorphic. Meiosis in the sporophyte results in the production of haploid spores and the start of the gametophyte phase. The gametophyte produces haploid gametes that fuse in fertilization, restoring the sporophyte phase. There are no motile cells in the red algae.

4. Red algae fossils date back to the Cambrian period.

Phaeophyta: The Brown Algae

1. **Brown algae** contain the pigment fucoxanthin and store their carbohydrates as **laminarin** and **mannitol.** They have flagellated sperm, and sometimes, flagellated eggs.

2. The Sargasso Sea contains great masses of the floating brown algae *Sargassum*.

3. *Fucus* is a reproductively advanced alga. Its alternation of generations is heteromorphic with a very prominent sporophyte and reduced gametophyte.

4. Brown algae include the giant kelps, *Nereocystis* and *Macrocystis*. Their specialized bodies include anchoring *holdfasts,* stemlike **stipes,** and large, leaflike **blades.**

Chlorophyta: The Green Algae

1. **Green algae** occur as single cells, coenocytic strands, filaments of cells, and flattened multicellular blades. They contain chlorophyll *a* and *b*, and produce cell walls of cellulose and hemicellulose in a pectin matrix—all plant characteristics.

2. *Chlamydomonas*, a representative single-celled green alga, is motile with two flagella and has a **red eyespot.** It exists primarily in the haploid state, where it reproduces through mitosis. When opposite mating types meet, **isogametes** form, fertilization occurs, and a brief diploid **zygospore** forms. Zygospores undergo meiosis, producing **meiospores.**

3. *Volvox* forms a spherical colony of small flagellated cells. Some specialize in movement, some as eggs and sperm, while others form small asexual spheres.

4. **Siphonous green algae** are coenocytic (multinucleate). They undergo repeated mitosis and become quite large, but little cell division occurs.

5. **Filamentous green algae** are the most complex algal forms. *Ulva*, a multicellular marine form, has a distinct isomorphic alternation of generations. *Ulothrix* has a heteromorphic alternation with a prominent gametophyte phase like that of *Chlamydomonas*.

6. Filamentous green algae are important since they are believed to represent the ancestral group from which plants arose.

REVIEW QUESTIONS

1. Review the taxonomic problems of Kingdom Protista. Is the kingdom monophyletic? Polyphyletic? In what way is the kingdom conditional and what are some of the main problems to be worked out? (459)

2. The protists are said to be grouped together for negative reasons. What does this mean? (459)

3. In what ways does *Euglena* seem to characterize the taxonomic problems we have with the protists? (460)

4. Discuss how the protists form an underlying or ancestral group for the other kingdoms. (459)

5. Briefly review the characteristics of Mastigophora. (460)

6. Briefly review the life history of the agent of African sleeping sickness. Why has it been so hard to control its spread? (461)

7. Describe the best-known sarcodine parasite, its effect on the human host, and the symptoms it produces. (461)

8. Explain the formation and functioning of a pseudopod. What are its two functions? (462)

9. Explain what axopodia are, where they are found, and how they are used. (462)

10. In what ways are the skeletal parts of radiolarians different from those of foraminiferans? What evidence suggests that they were incredibly numerous in former times? (462)

11. Describe an example of sexual reproduction in the sarcodines. (462)

12. List three important characteristics of the sporozoans. (463)

13. Briefly summarize the life cycle of *Plasmodium vivax*, and where appropriate, relate events to the symptoms of malaria. (464)

14. Describe the organization of the cilia in *paramecium* and in *Euplotes*. What are its two principal uses? (465)

15. Describe the trichocysts and name their functions. (465)

16. Summarize the steps involved in feeding in *Paramecium*. (466)

17. Discuss the peculiar manner in which DNA is divided in *Paramecium*. (466–467)

18. Summarize the events involved in sexual reproduction in *Paramecium*. (466–467)

19. In what way is *Physarum* a typical protist? How does it get from a spore state to a multinucleate state again? (467)

20. Describe the manner in which late blight infects potato plants. (469)

21. Explain how sexual reproduction takes place in the aquatic water molds. (470)

22. Using a simplifed diagram, explain how alternation of generations between sporophyte and gametophyte might appear in an isomorphic species. What event always starts a gametophyte phase? A sporophyte phase? (471)

23. Explain in detail how mitosis and meiosis differ between a dinoflagellate and a *Paramecium*. (472)

24. List several animallike and several plantlike characteristics of *Euglena*. (473)

25. Describe the problem diatoms have with continued asexual reproduction, and explain how it is solved. (475)

26. List several examples of how the red algae are similar to the cyanobacteria. (476)

27. Briefly summarize the events in the alternation of gametophytes and sporophytes in *Polysiphonia*. (476)

28. List three polysaccharides of the red algae and name some uses for these products. (476)

29. List the characteristic pigments and polysaccharides of the brown algae. (477)

30. Briefly characterize the life history of the brown alga *Fucus*. Why is it considered an advanced form? (477–478)

31. List several specialized structures of *Macrocystis* and describe their functions. (479)

32. List several important characteristics of the chlorophytes. (479)

33. Summarize the life history of *Chlamydomonas*. Is this protist considered primitive or advanced? Why? (480)

34. Using *Volvox* as an example, explain the colonial form of life. (481)

35. Describe the organization of the siphonous algae. Why aren't they seriously considered as ancestral to plants? (481–482)

36. Briefly describe the life history of *Ulva*, the sea lettuce, characterizing its alternating phases. (482)

37. Of the several groups of chlorophytes, which seems to be most closely related to the plants? Why? (483)

Diversity in Fungi

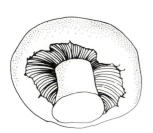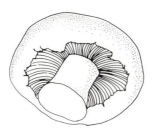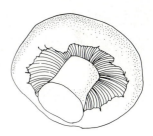

You may have been walking in a damp woodland one day and suddenly come upon a large mushroom on the forest floor. It can be an intriguing experience, especially if you don't know much about mushrooms, because their size, shape, and even coloration may seem rather outlandish (Figure 20.1). They have a distinct sense of mystery about them, possibly because they are so often associated with dark, wet forests—pensive places. If the area is remote, the day is overcast, and you are alone, it is not hard to secretly believe in the "little people," if only for a moment.

If you have experienced a moment like this, you may not be pleased to hear that the mushroom is a fungus. A fungus! The word *fungus* is simply *not* associated with beauty and mystery. A fungus, to many people, is bad; it grows on foods we like; it lurks on shower floors. You won't be any happier to learn that, as a group, fungi tend either to feed as *parasites*, from living material, or as *saprobes*, which sounds even worse—feeding off the dead. Nonetheless, mushrooms, along with baker's or brewer's yeast and a number of other friendly organisms, are fungi. Fungi can be quite useful and even delicious, and with a little objectivity, we might even perceive some as beautiful.

WHAT ARE THE FUNGI?

In the past, the fungi have been classified as plants, and even today, in some schemes they are considered protists. However, in the system we have chosen to follow, they are placed in their own kingdom (see Figure 17.5).

Many of the fungi are multicellular. Although most lack tissue organization in their feeding or vegetative structures, they often have highly elaborate reproductive structures. These are often as complex as those of any plants and animals, but they are very different from them. Although many fungi are organized into cells just as we find in most plants and animals, some are quite different in that their "cells" are long tubular filaments containing a flowing cytoplasm. Within the moving cytoplasmic stream are numerous nuclei. This type of multinucleate organization is termed a *syncytium* or *coenocyte*, and such fungi are described as *coenocytic*. You may recall this organization in the siphonous green algae *Codium* and *Acetabularia*. In addition to such noncellular organization, most fungi have cell walls containing **chitin**, a nitrogenous carbohydrate that also occurs in the exoskeletons of arthropods (such as insects and crustaceans; see Chapter 3). While

20.1 FUNGAL ENCHANTMENT

The circle of mushrooms probably originated when spores from some distant mushroom were carried by air currents to this ideal location, where dead vegetation, moisture, and warmth made growth possible. Beneath the surface is a vast fungal growth that spread from a central point in the circle.

chitinous cell walls are most common, a number of fungi have cell walls of cellulose.

Nutritionally, fungi are heterotrophs. They require an organic food source much the same as other heterotrophs, although some are able to use simple organic substances while others require food sources as complex as our own. Like bacteria, the fungi feed by **absorption**—they secrete digestive enzymes into their surroundings, which happen to be their foods, and the products of digestion are allowed to diffuse back into the cell. Both the parasites and the saprobes feed by absorption.

Fungi carry on a primitive form of mitosis and meiosis seen earlier in some protists (Chapter 19). The nuclear membrane is not dismantled during nuclear division but remains intact, constricting between the developing daughter nuclei. Further, although mitotic or meiotic spindles form, centrioles are lacking, as they are in the higher plants (see Chapter 11). This is not all that surprising, however, since fungi are nonmotile, lacking flagella and cilia, and of course, basal bodies. The presence of cilia and flagella, and especially the basal bodies of these active organelles, is a prerequisite to the presence of centrioles. Finally, it was recently discovered that the fungal chromosomes have extremely small quantities of histones and some lack the typical eukaryotic chromosome/protein association altogether. All of these features should reinforce the

assignment of fungi to their own kingdom, but it turns out that most fungi characters are shared with one or another group of protists. This means that there is still much work to be done if we are to resolve the phylogeny of the protists and fungi.

The fungi were once believed to have evolved from algal stock, but most biologists now believe they sprang from the colorless, heterotrophic, flagellated protists. This distinctive ancestry further separates the fungi from the plant line. Whether the fungi are a monophyletic or a polyphyletic group is still unsettled, and we will later review the arguments for both cases. While the phyletic relationships of the fungi present a challenging problem to biologists, other aspects of this group are equally fascinating. As we mentioned, a considerable number of fungi are economically important to us in a number of ways, not always pleasant.

The list of parasitic activities among the fungi is long and infamous and ranges from rose mildew to jock itch, with a number of serious troublemakers in between. And if you occasionally have "itchy feet" don't start packing. You might examine the basis of that itch since a common fungus, *Trichophyton mentagrophytes* is the agent of "athlete's foot."

There is, however, another side to the fungal world. They can be quite beneficial. For example, they share with bacteria the role of reducer or decomposer. Using dead organisms as a source of organic nutrient, they aid in the cycling of important mineral nutrients (for example, ions of nitrogen, sulfur, iron, phosphorous, and calcium) that would otherwise be tied up in corpses. In addition, many plants depend on a mutualistic (mutually beneficial) fungal association in their roots known as **mycorrhizae** ("fungus-roots"). Here, the fungus absorbs and supplies to the plant certain minerals while deriving carbon compounds that the plant produces (see Chapter 23). We also make use of fungi in such commercial activities as the manufacture of cheeses, and antibiotics, linen, bread, wine, and beer.

You will notice, as we consider the fungal phyla, that their names end in *-mycota* or *-mycetes*. These suffixes are derived from the Greek word for fungus, *myketos*. The root word is also our source for the term **mycelium**, which is the body of the fungus. The mycelia of many of the fungi are made up of threadlike masses of individual filaments known as **hyphae** (Figure 20.2). The hyphae may be **septate**—composed of individual cells—or **nonseptate**, or coenocytic.

Typically, fungi develop spore-producing organs called **sporangia**, which through mitosis produce tiny spores by the millions. Each spore consists of a haploid nucleus, a cytoplasm greatly reduced by dehydration, and a protective covering or spore case. The spore enables fungi to get from one locale to another, since the mycelium is stationary. Spores are most commonly borne by air currents, but may also be disseminated by moving water and by animals. Spores, then, afford fungi an asexual reproductive stage, a way of surviving harsh conditions, and a means of dispersal.

Fungi reproduce not only asexually by spore formation, but also sexually. In fact, the fungi remain haploid through most of their life cycle, reaching a diploid state only when fertilization occurs. Typically, their diploidy is brief and followed immediately by meiosis and the return to the haploid state. Such an extensive haploid state is not unique to fungi, but is common in the algae, as we have seen. But you may recall this is quite different from what we find in animals.

Biology is a science of exceptions, and a peculiar exception to the typically brief diploid interlude is called the **dikaryotic** ("two nuclei") state. Following sexual union, the emerging cell and its descendants retain two *separated* nuclei for a considerable time before the actual fusion of nuclei (fertilization) occurs. When this finally does occur, it is immediately followed by meiosis, in true fungal fashion.

20.2 FUNGAL GROWTH

Most fungi grow by producing filamentous hyphae, fine tubes of cytoplasm that penetrate the substrate to obtain food. The entire mass of hyphae is known as the mycelium. When conditions favor such growth, the sporangiophores rise up and produce sporangia at their tips, eventually releasing numerous haploid spores into the environment.

Before describing the specific kinds of fungi, let's review some of the characteristics we have elaborated upon. Fungi are chiefly multicellular organisms that are sometimes coenocytic, lacking typical cellular organization. Their cell walls are usually composed of chitin. As a group, they lack the common microtubular flagellum or cilium and although they carry on mitosis and meiosis, they do so without the presence of centrioles. Reproduction, both asexual and sexual, is through spore formation, and nearly all of the life cycle is spent in the haploid chromosome state.

KINGDOM FUNGI

Kingdom Fungi consists of four phyla: **Zygomycota** (conjugating molds, including bread molds), **Ascomycota** (sa fungi, including *Neurospora* and some yeasts), **Basidiomycota** (club fungi, including mushrooms), and **Deuteromycota** (also called Fungi Imperfecti). We will refer to them by their more familiar names, **Zygomycetes, Ascomycetes, Basidiomycetes,** and **Fungi Imperfecti.** Some schemes also include the water molds and slime molds, but we have placed them in the protists (see Chapter 19).

The Zygomycetes: Conjugating Molds

The Zygomycetes are sometimes called the "lower fungi" because they lack regularly spaced cell crosswalls and are coenocytic. Most members of the phylum are saprobic. They are named after the **zygospore**, a fungal version of a zygote and the diploid product of sexual reproduction in this division. *Pilobolus* (or the "spitting mold"), one of the zygomycetes, has a bizarre way of disseminating its spores, as we see in Essay 20.1.

***Rhizopus*, A Bread Mold.**   The common bread mold, *Rhizopus stolonifer*, turns up in everyone's refrigerator at one time or another (Figure 20.3). But don't necessarily look for it on bread because *Rhizopus* grows vigorously on many foods—and besides, most of our bread is so full of preservatives that fungal growth is inhibited. Spores that land on any suitable medium will first absorb water and then germinate. Hyphae sprout in all directions, branching time and again until the mycelial mass has fully penetrated the food. The favored conditions for the growth of *Rhizopus* and

A Fungus that Spits

There isn't much question that life has many meanings, and nowhere can we see its meaning to be more bizarre than in *Pilobolus*. To succeed in this world, *Pilobolus*, a zygomycete fungus, depends on the ability to spit in an accurate and timely manner. You see, spitting is its primary way of assuring that its offspring—spores contained in water droplets—will find their way into the same fortunate circumstance that previously led to the success of the spitter. Success, it turns out—and the true meaning of life to *Pilobolus*—is managing to get from one manure pile to another. Let's see what is going on here.

After gorging itself on delicacies in the dung of a herbivore, *Pilobolus* shifts from feeding to asexual reproduction. It produces a typical upright sporangiophore, tipped by a sporangium filled with hardy, resistant spores. But below the sporangium lies a swelling, bulbous structure, the *subsporangial swelling*, not seen in other zygomycetes (see photo). Further, this fungus doesn't wait for wind or water to carry its spores away to shift for themselves. Under the right conditions, its subsporangial swelling, water-filled to the bursting point, will literally explode, ejecting the sporangium for a distance of two meters or more. If fortune smiles on the sporangium, it will land in and adhere to a blade of grass. If fortune smiles twice, the grass will be eaten by a herbivore and the spores will survive the rigors of the digestive tract to emerge with the animal's feces, guaranteed of a food supply and a good prospect for future success. The adaptive value of this aiming system seems obvious. It is far more advantageous to send the spores upward and outward than it would be to shoot them down into the exhausted food reserves.

It turns out that sporangium ejection is not a random process but is closely oriented to the direction from which it receives light. In aligning itself, the stalklike sporangiophore bends first in one direction and then in another, but eventually aims the sporangium into the light. At this point, an explosion in the subsporangial swelling is triggered, and off goes the sporangium into the light.

Such a description leaves us with many questions, particularly about the light-seeking growth of the sporangiophore. And precisely what triggers the rupturing of the subsporangial swelling? Researchers have suggested a lens effect whereby light is focused directly on a photoreceptor that controls the direction of growth in the sporangiophore. When the sporangium is aimed directly at the light, the photoreceptor receives maximum stimulation. However, this is all speculation, and researchers have yet to put their finger on the precise mechanisms involved.

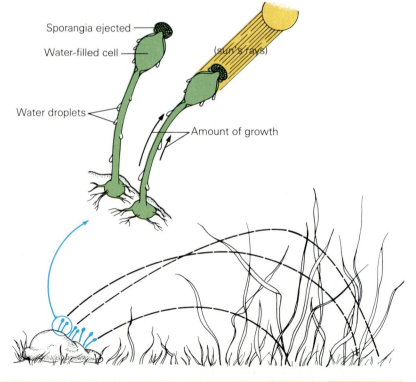

Sporangia ejected

Water-filled cell

(sun's rays)

Water droplets

Amount of growth

20.3 *RHIZOPUS*, THE COMMON BLACK BREAD MOLD

The mycelium of *Rhizopus* grows in a tangled mat throughout the medium and is not easily visible, but its sporangia, which are black when mature, form the dots on the large mass seen in the photograph. Like most saprophytic fungi, *Rhizopus* secretes enzymes from the hyphae into the food, where digestion occurs extracellularly. The simplified products of digestion are then absorbed into the cells.

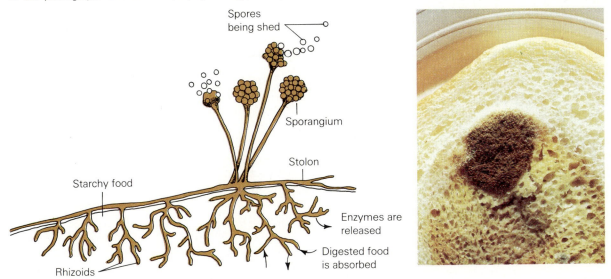

many other fungi—in addition to a supply of proper nutrients—are moisture and warmth, although they can do rather well in one's refrigerator. When the bread mold has grown for a time, it develops long horizontal hyphae known as **stolons**, which venture along the surface of the food, sending rootlike growths called **rhizoids** back down into the food. It is the tangled growth of stolons that often gives moldy food a fuzzy appearance.

You may have noticed the tiny black dots on moldy bread. These are the sporangia. They appear at the tips of upright, aerial hyphae called **sporangiophores,** and are a sure sign that asexual reproduction and spore formation is occurring. Numerous haploid spores arise through mitosis on each sporangia. Each spore contains several haploid nuclei, a dehydrated and metabolically inactive or dormant cytoplasm, and a surrounding protective coat. When released, the spores will be lofted by the air and the rest is up to chance. If the proper conditions are encountered, a spore will germinate; that is, it will become metabolically active and begin taking in water, breaking through the spore case, and starting a new mycelium. Since many of the spores fall right back near the sporangia that bore them, the food may become covered with *Rhizopus* overnight.

Sexual Reproduction.   *Rhizopus* also has a fairly simple sex life (Figure 20.4). It begins with **conjugation,** the fusing of two cells and the subsequent union of nuclei. There are no true males and females, so we refer to compatible mating types as *plus* (+) and *minus* (–). When different mating types are grown in the same medium, specialized club-shaped hyphae are produced. The hyphae of plus and minus mating types can fuse to form closed chambers known as **gametangia,** which contain many haploid nuclei. The gametangia then fuse and their adjoining walls break down, permitting haploid nuclei from each strain to meet and fuse. The common chamber, now containing a number of diploid nuclei, then forms a tough resistant wall, becoming a *zygospore*. The diploid period may be short or the tough, resistant zygospore may remain dormant for a considerable period. Under favorable conditions meiosis soon occurs within the zygospore, and many haploid nuclei—four times the number of diploid nuclei—are formed. Next, the zygospore germinates and a hypha emerges, immediately producing a sporangium. The haploid nuclei produced earlier are then incorporated into spores to be released. As the spores germinate, each new organism will have its genes rearranged because of genetic recombina-

20.4 SEXUAL REPRODUCTION IN *RHIZOPUS*

Rhizopus will enter into a sexual phase if opposing mating types are present in its medium. These strains are simply called *plus* (+) and *minus* (−). Neighboring plus and minus hyphae form clublike gametangia, each of which contains many haploid nuclei. The gametangia join and the plus and minus nuclei fuse, producing many diploid nuclei. Next, a tough, thick-walled zygospore forms around the diploid nuclei. The nuclei then undergo meiosis, producing many haploid nuclei once again. Upon germination, hyphae will emerge forming sporangia. The haploid nuclei will then be distributed into spores which, when released, will repeat the cycle.

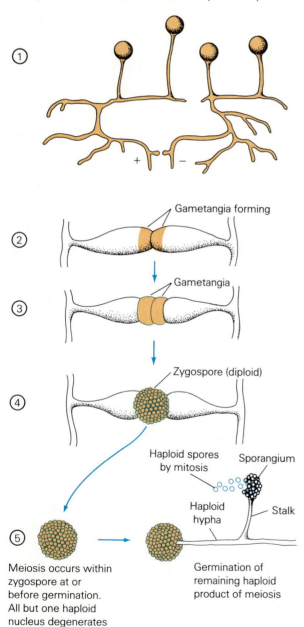

①

② Gametangia forming

③ Gametangia

④ Zygospore (diploid)

Haploid spores by mitosis   Sporangium

Haploid hypha   Stalk

⑤ Meiosis occurs within zygospore at or before germination. All but one haploid nucleus degenerates

Germination of remaining haploid product of meiosis

tion during meiosis. Keep in mind this general pattern: sexual reproduction always requires the presence of plus and minus (+ and −) mating types and is often associated with times of environmental stress, long periods of dormancy, and eventually, the exploitation of newly favorable environments.

Ascomycetes: Sac Fungi

Phylum Ascomycota and phylum Basidomycota are often referred to as *higher fungi*. This elevation in status is primarily due to their septate condition—that is, except in yeasts, cell divisions are routinely accompanied by the formation of crosswalls, or **septa**. Generally, in haploid hyphae, each resulting cell compartment has one nucleus, but the septa have perforations through which cytoplasm and nuclei can pass. Thus, in spite of the crosswalls, some hyphal cells may become multinucleate.

The Ascomycetes, or **sac fungi**, include some familiar members such as the edible morels and truffles, the powdery mildews (not to be confused with "downy mildew" mentioned in the last chapter), and the blue and green molds of citrus fruit. Some yeasts are also placed in this group. The sac fungi include the infamous chestnut blight, *Endothia parasitica*, an Asian mildew that brought about the extinction of vast forests of the graceful American chestnut. A sac fungus with rather ghastly implications for our own species is one known as *Claviceps purpurea*, a parasite of rye and other grasses.

Claviceps produces a plant disease called **ergot**, which in rye is characterized by the formation of a compact, black mycelium. Bread that has been made from "ergoted" flour contains certain alkaloids with the peculiar ability to constrict blood vessels in the body extremities. People who continue to eat bread made from such flour suffer from a condition known medically as **ergotism**, but historically as *Saint Anthony's fire*. The old name reflects the common symptoms of ergotism: burning sensations in the hands and feet accompanied by hallucinations. The restricted blood flow to these parts can bring on gangrene, often requiring amputation of the affected part. Ergot poisoning can even prove fatal, since respiratory and heart failure may occur. Such horrors were common in past centuries but are now rare.

Today, derivatives of ergot are used clinically as tranquilizers, for migraines, and to control difficult cases of internal bleeding after childbirth. Perhaps the most famous ergot-derived pharmaceutical is lysergic acid diethylamide (LSD).

Reproduction in Sac Fungi. During asexual reproduction, the Ascomycetes characteristically produce spores known as **conidia**. Conidia are produced in long chains at the ends of numerous specialized hyphae known as **conidiophores** (Figure 20.5).

Sexual reproduction among the Ascomycetes is quite diverse. However, there are some underlying similarities among the various types. The name *sac fungi* originates from the production of **ascospores** in a saclike container called the **ascus** (plural, *asci*). A group of asci are usually found in a fruiting structure known as the **ascocarp**, which occurs in a variety of shapes. What goes on inside the asocarp is most unusual (Figure 20.6). The diploid ascospores are produced through the unusual process we referred to earlier as delayed fertilization and the dikaryotic state. Let's review the events of sexual reproduction in more detail.

When the hyphae of opposing mating types of the sac fungus *Peziza* come into contact, each produces large multinucleate swellings. One, the **ascogonium**—the fungal version of a female sexual structure—will contain many haploid nuclei that remain in place. The other, the **antheridium**—an equivalent male structure—will also contain numerous haploid nuclei, but these will later leave the antheridium. Soon a bridgelike conjugation tube, the **trichogyne**, will form between the two bodies, and haploid nuclei from the antheridium will cross to enter the ascogonium. So far, this sounds pretty typical of sexual reproduction, but at this point the typical aspects end. This is because *the two newly associated haploid nuclei fail to fuse.* As new fungal cells form, each ends up with two nuclei, plus and minus, that do not join in fertilization. Such cells are referred to as **dikaryons** ("having two nuclei"). From these odd unions, or "non-unions," as it were, dikaryotic hyphae emerge to produce the asci, asexual structures in which the plus and minus nuclei will finally fuse. Let's come back to this in a moment.

Following the formation of dikaryotic hyphae, the parent fungi—the plus and minus individuals that got together in the first place—cooperatively produce hyphae to form the ascocarps that characterize sexual reproduction in the sac fungi (see Figure 20.6). Since these growths are not produced sexually, they are simply referred to as **sterile hyphae**. Their role, then, will be to produce the structure that will house the asci, the emerging dikaryotic sexual growths.

Finally, after a long delay, plus and minus nuclei fuse in the asci. But, as usual, the diploid state is quite temporary. Nuclear fusion is *immediately* followed by meiosis. The four meiotic products undergo one round of mitosis each to produce eight haploid ascospores. When the ascospores are fully mature, the ascus breaks open and the spores disperse. Unlike the asexual spores, however, each will contain a unique genome, the product of the genetic recombination that accompanied meiosis. Each will be capable of producing new fungal growths. These growths remain haploid until another fertilization produces the brief diploid interlude.

20.5 ASEXUAL REPRODUCTION IN THE SAC FUNGI

In each conidiophore, the terminal cell rounds out and matures into a spore called a conidium, followed by the one below and so on, until a long chain of such haploid spores forms. They will then break away and become airborne, some landing on new food sources.

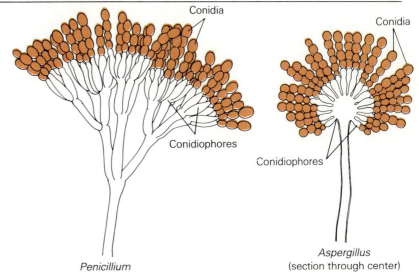

Conidia

Conidia

Conidiophores

Conidiophores

Penicillium

Aspergillus
(section through center)

The Lichens. Another well-known group of sac fungi lives with certain algae or cyanobacteria. Together, they form **lichens**. (It should be mentioned that some of the other fungi are also capable of this relationship.) It is generally believed that the algae provides sugars to the fungus in exchange for moisture, housing, and anchoring. This symbiotic association is usually considered to be mutualistic—beneficial to both parties. But in reality, one benefits more than the other because the algal or cyanobacterial partner can get along without the fungus. In fact, studies show that it does much better on its own. The fungus, however, thrives only if supplied with the complex nutrients it normally gets from its partner. For this reason, some authorities consider this often-touted example of pure mutualism to actually border on parasitism. (If they are correct, we have just lost another long-cherished notion.)

About 17,000 combinations of lichens are known. Ecologically, the lichens are important as soil builders, eroding rock surfaces and harboring bits of organic matter in their crusty bodies. The efforts of these hardy pioneers pave the way for a succession of organisms in many desolate, rocky environments. You probably have seen these organisms on rocks, although they may also be found on trees and on the soil. One lichen, the reindeer moss, represents an important primary food source in the ecology of some arctic regions (Figure 20.7).

Both the fungi and the algae that comprise lichens reproduce asexually through spore production. They are also capable of growing from frag-

20.6 SEXUAL REPRODUCTION IN *PEZIZA*, A SAC FUNGUS

Plus and minus strains approach each other and ascogonia and antheridia develop. Haploid nuclei from each antheridium migrate across the trichogyne—a conjugation tube—to an ascogonium. The plus and minus nuclei do not fuse but lie side by side, entering what is called a dikaryotic state. As growth proceeds, the dikaryotic hyphae produce asci, where nuclear fusion finally takes place. But much of the cuplike ascocarp is produced cooperatively by the two parent mycelia, whose haploid hyphae are referred to as "sterile." The plus and minus nuclei in the dikaryotic asci finally fuse in fertilization, but this event is immediately followed by meiosis, with four haploid nuclei appearing. These nuclei will double through mitosis and will become the ascospores after tough, resistant spore coats have formed. The ascospores are then released, whereupon each can begin a new vegetative growth.

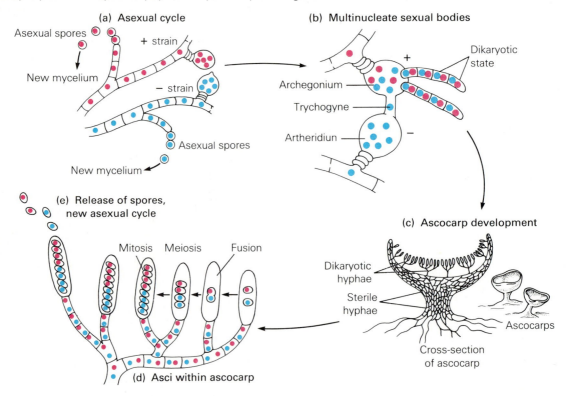

(a) Asexual cycle
Asexual spores
+ strain
New mycelium
− strain
Asexual spores
New mycelium

(b) Multinucleate sexual bodies
Dikaryotic state
Archegonium
Trychogyne
Artheridiun
+
−

(c) Ascocarp development
Dikaryotic hyphae
Sterile hyphae
Ascocarps
Cross-section of ascocarp

(e) Release of spores, new asexual cycle

Mitosis Meiosis Fusion

(d) Asci within ascocarp

20.7 LICHENS, A MUTUALISTIC ASSOCIATION

Some lichens form a crusty mat on the surfaces of a rocky outcropping, where they play an important part in erosion and soil building, while others are much more luxuriant in their growth. The diversity of lichens is suggested by the types shown here.

ments that may wash away and settle in new habitats. Often, the algal spores that are released are associated with short lengths of fungal hyphae, and so, if conditions are right when they settle, they are set to go into business.

Yeasts. Sac fungi also include many **yeasts**. A yeast is a single-celled fungus; the cells may be round or oval and may occasionally be found in short, branched, hyphalike chains. Yeasts grow by **budding**, a mitotic process with unequal cytoplasmic division. The buds (the parts receiving the lesser amount of cytoplasm) enlarge and finally separate from the parent cell. A complete organism grows from these tiny, dispersing bodies.

As you may now be aware, seemingly simple fungi may not lead very simple lives. To drive the point home, consider the sex lives of the yeasts. Common bakers' yeast can be diploid or haploid, and either form can undergo asexual reproduction by budding (Figure 20.8). The diploid form can also undergo meiosis. The four haploid products of meiosis become separate ascospores inside the original cell wall of the parent, which thus becomes the saclike ascus. (That, by the way, is why yeasts are classed as ascomycetes or "sac fungi.")

Two of the four ascospores will be the *alpha* mating type and two will be the *a* mating type. They can undergo immediate fusion to regain the diploid state, or sexual fusion between alpha and *a* cells can occur much later, as long as the two types are present. Generally, diploid strains do not undergo

meiosis when conditions are good; meiosis usually is induced by food shortages and drying.

Like other eukaryotes, yeasts have mitochondria, and when oxygen is available, they can oxidize sugar completely to carbon dioxide and water. But mitochondria cannot function under anaerobic conditions, and in the absence of oxygen, yeasts derive their energy from anaerobic glycolysis. When food and oxygen are both abundant, the yeast may still utilize the anaerobic process.

As you may recall, under anaerobic conditions, glycolysis in yeast is followed by the fermentive process in which ethyl alcohol and carbon dioxide are the final waste products (see Chapter 8). Both products are immensely important to humans and have been throughout history. The carbon dioxide is what makes bread rise and gives some beverages their bubbles. The ethyl alcohol is what brewers and vintners are after when they make beer and wine. As far as the yeasts are concerned, they continue to ferment until their own alcoholic waste poisons them. This occurs when the concentration of alcohol reaches about 13%, the percentage of alcohol in unfortified wines. It is possible to get a higher concentration of alcohol from fermentation products, of course, but you need a still to get rid of the water.

Not all yeasts are so compatible with humans. Some yeasts can produce rather nasty vaginal infections, for instance. Yeast infections are most common in women who have had a prolonged exposure to antibiotics or certain steroid hormones, which kill off the bacteria that normally inhabit the

20.8 YEAST LIFE CYCLE

Yeasts have both a haploid state and a diploid state, which resemble each other rather closely and are without cellular specialization. ① In the diploid state, a cell reproduces asexually by mitosis and budding (photo), or may undergo meiosis, ② producing four haploid ascospores within the ascus. Two will be alpha and two will be *a*. ③ These ascospores emerge from the cell and carry out the usual yeast activities, including budding for a time, so that large haploid populations are produced. ④ If both alpha and *a* mating strains are present, a fusion will occur, returning the yeast to the diploid state of the cycle.

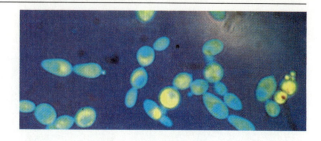

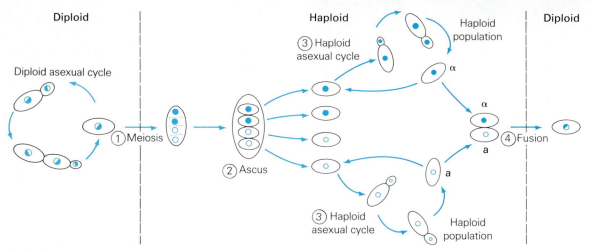

vagina. While the infectous yeasts are also a part of the normal flora of the body, under normal conditions competition with the harmless bacteria holds their numbers in check.

Basidiomycetes: Club Fungi

Basidiomycetes, or **club fungi**, are significant to us in a number of ways. Although the group includes such familiar varieties as mushrooms, toadstools, and shelf fungi, it also includes some devastating parasites. Among these are wheat rust and corn smut. Considering the worldwide dependence on wheat and corn, it is easy to appreciate the concern over these parasites. In fact, the development of rust-resistant strains of wheat is one of the great success stories of applied genetics. Saprobic members of the Basidiomycetes are also partly responsible for the decomposition of dead trees and litter on forest floors (Figure 20.9).

Like the sac fungi, the club fungi are septate. They also reproduce sexually and produce complex sexual structures. The group can be divided roughly into two groups that differ both nutritionally and

structurally. We will use the saprobic mushroom to represent one group and the parasitic wheat rust to represent the other.

Sexual Reproduction in the Mushroom.

The mycelial mass of the common mushroom is not seen above ground; it lies deeply embedded in organic matter, breaking it down and using its stored nutrients. What we see growing on the forest floor is the sexual, spore-producing **basidiocarp** (fruiting body), which produces **basidiospores**, the direct products of meiosis. (You may have eaten "cream of basidiocarp soup" or "basidiocarps sautéed in butter.")

Below the basidiocarp lies a branching and extensive mycelium, which secretes enzymes, breaks down organic matter, and absorbs nutrients. The mycelium may also form a symbiotic relationship with the roots of higher plants, essentially providing minerals in exchange for organic nutrients (see Chapter 23).

The basidiocarp emerges above ground after the mycelia of plus and minus strains have joined (Figure 20.10). As we found in the sac fungi, the actual union of plus and minus nuclei is delayed, leaving each cell in the basidiocarp a dikaryote. Fertilization finally occurs in cells of the many **basidia** ("little clubs"), minute clublike structures that line the thin **gills** found on the underside of the mushroom's umbrella-shaped **cap**. Fertilization is immediately followed by meiosis. Each haploid cell becomes a spore, and these are released in enormous numbers to be scattered to the winds.

We can't leave the subject of mushrooms without a word about picking your own wild specimens for the table. People who do this are often referred to as "critically ill" and sometimes as "departed." As you probably know, some species of mushrooms (called toadstools) are extremely poisonous. But which ones? Even the experts can have difficulty distinguishing certain poisonous species from the edible ones. Your best bet is to obtain your mushrooms from the same source as do experienced mycologists—the corner supermarket.

Wheat Rusts. Now let's consider wheat rust, *Puccinia graminis*, a species with a truly complex life cycle, most of which is spent in the dikaryotic state, and which involves two different hosts and several kinds of spores. One host, of course, is wheat but the parasite also must infect the common barberry plant, *Berberis vulgaris*.

While in the wheat, the rust fungus produces dikaryotic spores of two types. One—a red spore, or **uredospore**—simply infects other wheat plants, while the other—a black spore, or **teliospore**—is the rust's investment in future generations. Teliospores survive the winter season and germinate in moist soil in the spring. Until then the black spores remain in a dormant state. As such they are not active and cannot infect the wheat or the barberry. Then, just prior to germination, each diploid nucleus undergoes meiosis in the usual fashion, producing what become the four basidiospores. Upon their release, the basidiospores infect the wild barberry.

It is only in the barberry leaf that the sexual phase occurs. As usual in fungi, plus and minus strains must be present. When they meet a union occurs, with plus and minus nuclei remaining separated, thus restoring the dikaryotic state. Finally, another round of spore formation occurs, this time

20.9 DIVERSITY IN CLUB FUNGI

Basidiomycetes can be large and colorful, as is the edible mushroom *Lepiota* **(a)**, and the large, wrinkled bracket fungus **(b)**. Although many are simple saprobes, living on dead organisms, others such as the corn smut **(c)**, are devastating parasites. Another mushroom, *Amanita phalloides* **(d)**, looks harmless enough, but its deadly toxins have led people to dub *A. phalloides,* the "death cap." Poisonous mushrooms are commonly called toadstools.

(a)

(b)

(c)

(d)

20.10 LIFE CYCLE OF THE MUSHROOM

(a) Below the soil surface, plus and minus hyphae meet and fuse, producing a line of dikaryotic cells. **(b)** After sufficient growth has occurred, the basidiocarp emerges above ground and the cap expands and opens, exposing numerous gills on its underside. **(c)** Each gill contains numerous tiny, clublike basidia, where fertilization or fusion of plus and minus nuclei finally occurs. Each fused nucleus then undergoes meiosis, producing four haploid basidiospores to be released into the environment.

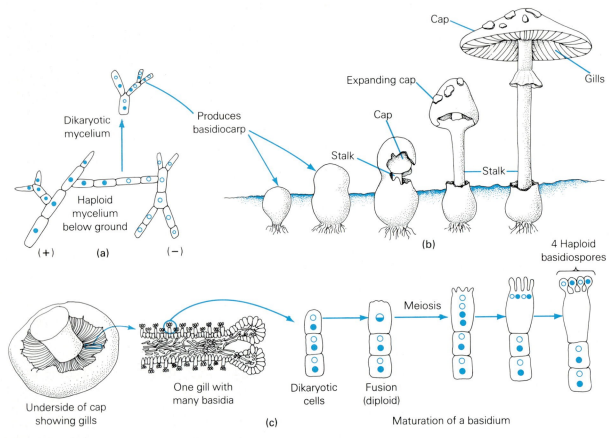

with dikaryotic **aeciospores** emerging. It is the aeciospore that can infect the next wheat crop.

Wheat infected by *Puccinia graminis* is invariably weakened or killed. After the wheat rust's life cycle was discovered, it became common practice to break the infection cycle by burning contaminated fields and by systematically destroying barberry bushes.

THE FUNGI IMPERFECTI

A fungal group with the name **Fungi Imperfecti** will need explaining. *Imperfect* is a botanical term referring to a lack of sexual reproduction, and the approximately 24,000 named species of Fungi Imperfecti have no known sexual phase. Actually, the Fungi Imperfecti still require a lot of taxonomic work. Undoubtedly, many will eventually be placed with the sac and club fungi. But a vital part of fungal classification is based on sexual stages, and if these phases have not yet been observed, the fungi can't be classified.

This group harbors a number of human parasites, including the common parasite of the skin, *Trichophyton mentagrophytes*, which causes athlete's foot. But on a more positive note, we also find some harmless and even useful species. Consider *Penicillium roquefortii* and *Penicillium camembertii* as examples, whose names give them away as agents used in the flavoring of famous cheeses. Then there is *Aspergillus oryzae*, which, along with lactic acid bacteria, is responsible for the special flavor of soy sauce and, in part, for the fermenting of saki. *A. oryzae* is also important in the fermentation and nutritional enrichment of livestock feeds.

Some species are predators that actually catch certain tiny roundworms that live in the soil (Figure 20.11). They develop ringlike mycelial structures that constrict like an inflated noose when a hapless roundworm inadvertently passes through. Then fungal mycelia enter the victim and digestion begins.

ANOTHER LOOK AT FUNGAL ANCESTORS

Tracing the origin of fungi from ancestral stock is at best a conjectural exercise. There are many possibilities, but none, so far, is based on the fossil record or an elaborate biochemical analysis of amino acids, or other compounds. The principal guesses include the following:

1. Fungi are monophyletic, derived from green algal stock. The derivation was accompanied by a loss of the photosynthetic pigments of algal forerunners, and the retention of cell walls;
2. Fungi arose monophyletically from protozoan stock. The connection is represented by the slime molds, which, evolving from sarcodines, developed aerial spore formation as an adaptation to land. Higher fungi then developed from slime molds;
3. Fungi are a polyphyletic group, some having been derived from algal ancestry and others from protozoans. Aerial spore formation is an example of convergent evolution, since it represents a very successful adaptation to life on land.

Fossil fungi are dated tentatively back to the Precambrian era, some 900 million years ago, but these aren't very reliable finds. Fossils from the Ordovician period, between 450 and 500 million years ago, have been clearly identified as fungi and mycorrhizial associations have been preserved in Silurian deposits dating back over 400 million years.

As far as taxonomic association go, we can presently do little more than compare structures and life cycles, look for possibly favorable comparisons, and arrange the phylogenetic tree accordingly. Recent cytochrome *c* sequence studies indicate that cellular slime molds are closer to the protist *Euglena* than to the true fungus *Neurospora*, so in spite of some of

20.11 A PREDATORY FUNGUS

Arthrobotrys, a microscopic fungus, bears many snares (open loops). These swell rapidly upon touch, trapping anything inside. Gotcha! The nematode worm becomes ensnared by a rapidly closing noose. Following the capture, hyphae from the fungus will penetrate the worm's body and digestion and absorption will begin.

the decidedly fungal characteristics of the slime molds, they have been reassigned to the protists. There are many, many questions about fungal origins, but until more fossil or biochemical evidence is available, the questions will remain unanswered.

MULTICELLULARITY

We have seen that the some of the "simple" protists and fungi are not so simple after all. Their life cycles are complex and elusive; they appear in all sorts of forms and colors, and they are rather complex in structure. Some are not only multicellular, but contain a variety of cell types. Each cell type, of course, plays its own role in the life of the organism, and such specialization enables the protist and fungus to exist under quite precise and demanding conditions.

Let's consider for a moment the biological implications of multicellularity and cell specialization. Why are so many successful species composed of many cells of different types? The answers may provide an important insight to the way living things have adapted to such a complex planet.

The Origin of Multicellularity

We can begin by defining multicellularity. *A multicellular organism is one that is composed of a number of cells that cooperatively carry out the functions of life.* This definition is meant to distinguish the multicellular organisms from aggregations and colonies of single-celled organisms, although the distinction is not always clear. For our purposes, multicellular organisms are made up of cells that cannot survive independently under natural conditions.

We assume that multicellularity began when aggregations of cells became interdependent. This would have happened when different cells became adapted for different roles. For example, some cells might have become specialized for producing or obtaining food, others for transport, and still others for moving the aggregate to a more favorable place. We might imagine that, finally, some cells came to specialize in reproduction.

As you know, the organisms of the earth are divided into taxonomic groups according to shared characteristics. Two of these groups, the bacteria and the protists, are commonly referred to as unicellular. This description should exclude these taxa from membership in the multicellular club. However, the taxonomy is not that clear-cut. Some bacteria and protists are clearly multicellular. Further, some of the fungi are unicellular.

Fungi help bridge the gap between unicellular-colonial and multicellular organisms. While yeasts are unicellular, Fungi Imperfecti form hyphae that can be considered colonies of identical cells; they also form specialized dispersal cells (spores). The reproductive structures of sac and club fungi may be quite complex, some species of the latter containing specialized cells for stalk, cap, gills, and covering. The noose of the predatory fungus *Dactylaria* is one exceptional example of fungal cell specialization in nonreproductive tissue.

But keep in mind that for the most part there is little specialization in the fungal mycelium. Cell specialization is almost entirely reserved for the reproductive structures. It is the plants and animals that have made the transition to somatic cell specialization—to the tissue, organ, and organ system levels of organization. We can't let this transition slip by unnoticed.

It seems obvious that when the necessary labors of life are doled out to specialized cells in a colony, then each will become better at performing its particular task and the result will be greater efficiency. It follows, then, that multicellular organisms must somehow be better off than the single-celled organisms. But this is not necessarily so. Single-celled organisms have survived very well through evolutionary history without being displaced by multicellular life. We have emphasized that some of the protists, such as the ciliates, are highly complex and nicely adapted creatures. Their lineage has certainly continued without the drastic adaptive changes that have marked ours. And they have become successful without following the multicellular trend. If single cells are so successful, then, why did multicellular life develop?

The answer has already been stated: in a word, it's *specialization*. This term has two connotations, however. First, cells become specialized and interdependent so that the success of one type depends on the success of others. In addition, the relative numbers of each type, the absolute numbers of the cells (the size of the multicellular body), and the direction of their specialization have permitted various kinds of living things to specialize by adapting to particular kinds of environments. For example, some multicellular bodies may have disproportionate numbers of contractile (muscle) cells so that they can survive under environmental conditions that require strength. However, they may not have many cells that are specialized to detect soundwaves, so in environments where hearing is important, they must yield before other organisms that perhaps have fewer muscles but better hearing. Thus, cellular specialization means that organisms may come to occupy different niches, and may branch out, specialize, and diversify, creating the staggering array of living things we see around us.

APPLICATION OF IDEAS

1. The manner in which wheat rust is controlled—removing the barberry plants where it spends part of its ife cycle—is an important example of intelligent ecological management. What are some other examples of this more "natural" approach to agricultural problems? Compare the desirability of these to the use of fungicides or some other chemical method.

2. The fungi are still considered plants in some taxonomic schemes. In what ways are they plantlike? If they are plants, do they represent primitive, advanced, or perhaps degenerate forms? How might one find out? Be sure you define these evolutionary terms as you prepare your answer.

3. In discussing evolution and evolutionary milestones, such as the evolution of the first land life, we should avoid using terms that seem to credit the organism with conscious forethought, purpose, or intent. Present your own hypothetical explanation of the colonization of land, but use terms such as variation, adaptation, competition, and natural selection.

KEY WORDS AND IDEAS

Fungi get a lot of bad press because they are either *parasites,* invaders of living organisms, or *saprobes,* feeders of dead materials such as our foods and organic goods.

WHAT ARE THE FUNGI?

1. Fungi are multicellular, nonmotile heterotrophs, lacking tissue organization except in their reproductive structures. Some are coenocytic, others cellular; most have chitinous cell walls, while a few have walls of cellulose.

2. Fungi feed through **absorption**, secreting enzymes outside and absorbing the digested food.

3. Fungi lack centrioles (also basal bodies, cilia and flagella), and the nuclear membrane remains intact during mitosis. Their chromosomes contain very little protein, and in some it is absent.

4. Fungi may have evolved from heterotrophic, flagellated protist ancestors.

5. While fungi are commonly parasites, many are reducers, joining soil bacteria in breaking down nutrients and recycling essential mineral ions. One group forms a mutual symbiosis with plant roots called **mycorrhizae,** in which they assist the roots in mineral and water absorption. We use fungi as food and in food production.

6. The suffixes -*mycota* and -*mycetes* refer to the fungal body or **mycelium.** The mycelium is often made up of threadlike tubular cells called **hyphae.** Hyphae may be cellular (**septate**) or coenocytic (**nonseptate**). The mycelium is essentially a feeding structure that gives rise to reproductive structures.

7. Asexual reproduction often occurs through mitotic spore production in **sporangia.** Spores are easily dispersed and highly resistant.

8. Each fungus is haploid except for a brief diploid state following fertilization. An exception is the **dikaryotic** state, in which following sexual reproduction, the two haploid nuclei in each cell remain separated during the growth of elaborate sexual structures. Eventually they fuse, but the diploid cell then enters meiosis, producing haploid spores.

KINGDOM FUNGI

Kingdom Fungi contains phyla **Zygomycota, Ascomycota, Basidiomycota,** and **Deuteromycota.**

Zygomycetes: Conjugating Molds

1. Zygomycetes are coenocytic and most are saprobic.

2. The ideal growth conditions for *Rhizopus stolonifer* (bread mold), a common contaminant of foods, are sufficient nutrients, moisture, and warmth. Its **stolons** (horizontal hyphae) send **rhizoids** (penetrating hyphae) into food. **Sporangiophores** (upright hyphae), are tipped by sporangia in which mitotic spore formation occurs.

3. Sexual reproduction occurs through a form of **conjugation** involving plus and minus strains. Haploid nuclei gather in **gametangia,** which then join, permitting the nuclei to fuse. Several diploid nuclei form a thick-walled **zygospore,** which may become dormant. A sporangium containing the diploid nuclei emerges and with meiosis, the diploid nuclei undergo genetic recombination and reduction division. The haploid products then form spores.

Ascomycetes: Sac Fungi

1. Ascomycetes, a group of higher fungi, is septate (cellular), and its members include powdery mildews, citrus molds, and certain species of yeasts.

2. *Claviceps purpurea* produces **ergot**—a compact, black mycelium—in the rye plant. When ergoted rye flour is eaten, the symptoms of **ergotism**—restricted blood flow to extremities, burning sensations, hallucinations, and gangrene—may develop. Derivatives of ergot are used in tranquilizers, to control migraines, and to slow bleeding after childbirth.

3. In asexual reproduction, **sac fungi** form **conidiophores** on which spore-forming **conidia** emerge through mitosis.

4. Sexual reproduction in *Peziza* requires plus and minus strains. One produces an **ascogonium,** the other an **antheridium,** and these are joined by a tubular **trichogyne,** through which nuclei migrate. The plus and minus nuclei join in cells, but enter a dikaryotic state. The nonsexual or **sterile hyphae** of the two strains cooperatively produce an **ascocarp,** in which the dikaryons develop saclike **asci.** Fusion of plus and minus nuclei occurs in the asci, followed immediately by meiosis and the formation of haploid **asco-**

spores. They then germinate to form a new mycelium.

5. The **lichens** include some 17,000 specific associations, each of which is an intimate mutualistic symbiosis, made up of an alga or cyanobacterium and a fungus. The phototroph supplies organic nutrients, while the heterotroph reportedly provides moisture and anchorage. Recent studies suggest that only the fungus benefits. Lichens are important soil builders, and one type, reindeer moss, forms the energy base of certain arctic food chains. In asexual reproduction, spores are produced by each member.

6. **Yeasts** are single-celled sac fungi that increase through **budding** (mitosis and highly unequal division). A colony can be diploid or haploid and the diploids may undergo meiosis producing ascospores within the old cell. The meiotic spores can immediately fuse, restoring the diploid state, or this may happen after many mitotic divisions and budding events. Some yeasts are important to us in producing bread, wine, beer and liquor, while others are parasites, causing serious infections.

Basidiomycetes: Club Fungi

1. **Club fungi** include saprobic mushrooms and shelf fungi, along with important parasites such as wheat rust and corn smut. Most of the mushroom mycelium is below ground, while above is the **basidiocarp,** a fruiting body that produces the **basidiospores.**

2. Following the subterranean union of plus and minus mushroom hyphae, the dikaryotic products form the **basidia** that line the many **gills** underlying the mushroom **cap.** The basidiocarp is produced through the cooperative efforts of sterile hyphae of the two strains. Fusion of plus and minus nuclei occurs in the basidia, followed by meiosis and basidiospore formation.

3. The wheat rust, *Puccinia graminis,* requires two

hosts, the wheat and the barberry plant. While in wheat, it produces dikaryotic red **uredospores.** which infect other wheat, and dikaryotic black **teliospores,** which winter over. In the spring, the winter dormancy ends and the diploid nucleus of each teleospore undergoes meiosis. The emerging plus and minus haploid basidiospores then infect the barberry. Sexual reproduction between the plus and minus hyphae follows. More dikaryotic cells emerge and enter another round of spore formation. The dikaryotic **aeciospores** produced then infect the new wheat crop.

THE FUNGI IMPERFECTI

1. **Fungi Imperfecti** are unclassified fungi with no known sexual phase. Many are probably sac and club fungi. Members include the athlete's foot fungus and several *Penicillium* species used in flavoring foods. Another species, *Dactylaria,* captures and feeds upon soil roundworms.

ANOTHER LOOK AT FUNGAL ANCESTORS

1. Fungal ancestors are in question, but guesses include a single green algae, a sarcodine protozoan or, if the fungi are polyphyletic, both. Their fossil record, which is very tentative, can be traced back 900 million years (Precambrian).

MULTICELLULARITY

The Origin of Multicellularity

1. Multicellular organisms are composed of many cells cooperatively carrying out the life functions. They may have originated from casual aggregations that became interdependent. The gap between unicellularity and multicellularity is bridged by the fungi (and some of the algae).

2. While in multicellular organisms, the ability of cells and tissues to specialization is vital, this does not mean that unicellular organisms are inefficient.

REVIEW QUESTIONS

1. List four characteristics shared by most fungi. (487–488)

2. Describe the absorptive feeding mode. How does this differ from the manner in which digestion occurs in animals? (488)

3. What is unusual about the cell walls of most fungi? How might this be used as an argument against including fungi in the plant kingdom? (487)

4. Distinguish between the terms *coenocytic, multi-*

nucleate, and *nonseptate.* To which group or groups of fungi do these apply? (488)

5. Among the protists, which is the best guess for a fungal ancestor? What protist groups may have sprung from this ancestor as well? (488)

6. Explain how a mycorrhizal association serves as an example of a mutualistic symbiosis. (488)

7. Distinguish between the terms *mycelium, hypha, stolon,* and *rhizoid.* (488)

8. What is the function of a sporangium? How is this function adaptive to the fungus? (488–489)

9. In what way does timing in the union of fungal haploid nuclei (fertilization) seem unusual when compared to plants and animals? (489)

10. From what reproductive characteristic is the name *Zygomycete* derived? (489)

11. Describe the asexual growth and reproduction in the common bread mold, *Rhizopus stolonifer.* (491)

12. List the steps taken during conjugation in *Rhizopus.* (491)

13. What is the ''sac'' of the sac fungi? (493)

14. What is ergoted wheat? Ergotism? (492)

15. List three ways in which the activities of *Claviceps purpurea* have benefitted humans. (492)

16. How does conidiospore development differ from the development of the asexual spores in bread molds. (493)

17. Trace the events in sexual reproduction in the sac fungus, using such terms as *ascogonium, dikaryon, trichogyne, sterile hyphae, plus and minus strains, antheridium, asci,* and *ascospore.* (493)

18. What are lichens? What is the most recent thinking about the symbiotic relationship in lichen associations? (494)

19. List two ecologically significant activities of lichens. (494)

20. Describe the specific activity in yeasts that places them in the sac fungi. What are the two fates of haploid meiospores? (495)

21. List two economically important uses for yeast, and tell which products of yeast metabolism are important to each. (495)

22. Trace the events leading to the production of the mushroom basidiocarp. What reproductive event happens in the basidiocarp? (496–497)

23. In what way(s) is sexual activity in the mushroom similar to that of the sac fungi? (496–497)

24. In which host does the sexual phase of the wheat rust life cycle occur? Trace the events of this phase. (497)

25. At what point in the life cycle of *Puccinia graminis* can the terms *diploid* and *haploid* be applied? Explain the events surrounding this. What nuclear term applies the rest of the time? (497)

26. Why would one combine burning the wheat fields with removal of barberry plants as a rust preventing measure? Why not just remove the barberry plants? Would burning have to be repeated? (498)

27. What is the basis for assigning fungi to the Fungi Imperfecti? List three familiar examples. (498–499)

28. List three possible explanations of fungi origins. (499)

SUGGESTED READING

Alexopoulos, C. J. 1979. *Introduction to Mycology.* John Wiley, New York. For enthusiasts of things fungal.

Dagley, S. 1975. ''Microbial Degradation of Organic Compounds in the Biosphere.'' *American Scientist* 63:681. Microbes have ecology, too.

Fox, G. E., et al. 1980. ''The Phylogeny of Prokayotes.'' *Science,* 209:457.

Miller, S. L. 1935. ''Production of Some Organic Compounds under Possible Primitive Earth Conditions.'' *Journal of the American Chemical Society* 77:2351. The first *experimental* approach to the question of the origin of life on earth.

Raven, P. H. 1970. ''A Multiple Origin for Plastids and Mitochondria.'' *Science* 169:641.

Salle, A. J. 1979. *Fundamental Principles of Bacteriology.* Not for leisure reading, but if there is anything you want to know about bacteria you'll find it in here.

Schopf, J. W., and D. Z. Oehler. 1971. ''How Old Are the Eukaryotes?'' *Science* 193:47. Perhaps they are not as old as had been thought, but originated only a few hundred million years before the first oldest known fossil animals.

Scientific American articles. San Francisco: W. H. Freeman Co.

Barghoorn, E. S. 1971. ''The Oldest Fossils.'' *Scientific American*, May.

Dickerson, R. E. 1978. ''Chemical Evolution and the Origin of Life.'' *Scientific American*, September. We think that this is the best no-nonsense review of this primarily speculative field.

Ptashne, M., et al. 1982. ''A Genetic Switch in a Bacterial Virus.'' *Scientific American*, November.

Simons, K., et al. 1982. ''How an Animal Virus Gets into and out of Its Host Cell.'' *Scientific American*, February.

Vidal, G. 1984. ''The Oldest Eukaryotic Cells.'' *Scientific American*, February.

Woese, C. R. 1981. ''Archebacteria.'' *Scientific American*, June.

Evolution and Diversity in Plants

 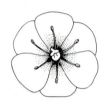  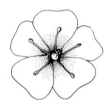

With our arrival in the plant kingdom we come to those organisms that are familiar to everyone. These are the life forms that grace our lives by lending soft hues to our surroundings, by providing our shelter, and by warming our hearths. But perhaps most important, plants are our ultimate source of oxygen and food. What, then, are plants? The question seems simple, but it is not. Formally, the plants are multicellular, photosynthetic organisms that have evolved specialized tissues, organs, and systems. That said, we can safely assume that an elm is a plant. Reputable scientists in the past have argued over whether slime molds, fungi, and even bacteria belong in this group as well, but the five-kingdom scheme we have been following (see Figure 17.5) will help us handle all this diversity more easily.

Let's begin with some other characteristics that help distinguish the plants from all other organisms—in particular the algal protists with which they share a number of features. We will briefly mention these here and come back to them as we proceed. One very dependable difference is that plants develop from enclosed and protected embryos, which undergo a great deal of tissue differentiation during development. But although plant tissues are quite diverse and specialized, this is a mat-

---

Note: Geological time is charted inside the front cover.

ter of degree rather than an absolute difference between plants and algal protists. For example, we have seen that some of the marine algae have specialized tissues. Plants contain the photosynthetic pigments chlorophyll *a* and *b*, the latter of which is absent in many other phototrophs. Finally, plants usually produce starches as their storage carbohydrate and cellulose as their cell wall material. As usual, our list is not truly exhaustive, but when considered in its entirety, it, in fact, defines the plants. With this in mind, let's consider a major aspect of the plant life cycle. This will then lead to some interesting ideas about the evolution of plants.

In Chapter 19 we saw the alternating gametophyte and sporophyte phases in the life histories of some algae. In this chapter we will see that, in most terrestrial plants, there is a definite trend toward a more prominent sporophyte. In fact, in the most recently evolved group of plants, the flowering plants, the sporophyte encompasses roots, stems, leaves, and flowers—everything except for a few cells produced by meiosis and mitosis within the anthers and ovaries of the flower itself. This is the case in nearly all plants, but there is an important exception, a group in which the gametophyte remains dominant. We will begin our consideration of plant evolution and diversity with these "holdouts." These are the bryophytes.

DIVISION BRYOPHYTA: NONVASCULAR PLANTS

The **bryophytes—mosses, liverworts,** and **hornworts**—are all multicellular plants with simple tissues, but the tissues are differentiated enough to divide the tasks of life efficiently. Although the bryophytes are terrestrial, they retain a reproductive feature common in the aquatic algae, a swimming sperm that must travel through water to the egg.

Bryophytes lack well-developed **vascular tissue** (water- and food-conducting tissue), a trait that readily distinguishes them from the other terrestrial plants. The others—those with well-developed conducting tissues—are collectively called the **vascular plants.** As we will see, vascular tissue contains tough fibers that lend great strength to roots, stems, and branches. Because bryophytes lack such tissues, there are no tall liverworts or mosses; instead they cling close to the earth or hang from the supporting trunks and branches of vascular plants.

Nevertheless, bryophytes have been very successful—surprisingly so—since their origin some 350 million years ago. They fill an ecological niche that was seemingly ignored by the vascular plants. While vascular plants underwent the evolutionary changes associated with largeness and adapted to drier and drier climes, the bryophytes became specialists in inhabiting the regions unsuited to the other plants. They weren't entirely restricted to moist habitats, however, since bryophytes have successfully penetrated the hot, dry deserts, the windswept, rocky outcroppings of high mountains, and the frigid polar regions. Further, along with lichens, these simple plants contribute to the tundra ground cover of the arctic region. It seems that the only conditions bryophytes cannot tolerate are those produced by humans. For example, they are notably absent in areas of severe air pollution. There are as many as 23,000 named species of bryophytes, most of them mosses. The other bryophytes, mainly liverworts, are much less common. And you probably can't recall ever having seen a hornwort.

How do the bryophytes fit into the evolution of plants? We can report with confidence that no one today seriously believes that the bryophytes were ancestors of the vascular plants. But here the concensus ends. Some botanists maintain that the mosses, liverworts, and hornworts evolved independently of the vascular plants from their own green algal ancestor. They cite a number of similarities of bryophytes to certain green algae, including early growth stages in mosses that resemble filamentous green algae, almost identical pigments including chlorophyll *a* and *b* and xanthophylls, as well as the *pyrenoids,* algal starch-storage bodies curiously present in the photosynthetic cells of hornworts.

There are other theories, however. Some botanists maintain that the bryophytes and vascular plants share a common green algal ancestor. They propose that the bryophytes diverged from the early vascular plant line about 50 million years after it had begun. The bryophyte's lack of vascular tissue, they suggest, is a product of evolutionary simplification—a loss of whatever vascular development had occurred in their immediate ancestors. Further, the unique characteristics of bryophytes are the product of evolutionary specialization, so the bryophytes may not be as primitive as some people think. In fact, they may be relative newcomers. Recognizable fossils of mosses can only be traced back to the Devonian period, about 350 million years ago, while those of vascular plants can be traced back to the Silurian, some 400 million years.

One of the best guesses is that the common ancestor of bryophytes and vascular plants resembled today's green alga *Coleochaete* (Figure 21.1). One line of evidence is the manner in which mitosis and meiosis occur. As in the plants, the nuclear envelope in *Coleochaete* is dismantled during cell division and following this, formation of the cell plate between daughter cells occurs through the coalescing of Golgi vesicles (see Chapter 11).

The bryophyte contains various tissues that have become somewhat specialized for different tasks such as water absorption, anchorage, foliage support, and photosynthesis. Although bryophytes do not develop true roots, stems, and leaves, which contain vascular tissue, and are exclusive to vascular plants, they do produce analogous structures that carry out similar functions. In the bryophytes, anchorage in the soil is provided by the **rhizoids,** elongated threads of cells that emerge from the base of the leafy gametophyte (Figure 21.2). Unlike a true root, however, the rhizoid does little to bring in water and minerals. The "stem" is a complex cylinder of many cell layers surrounded by a photosynthetic **epidermis** (covering layer). The epidermis surrounds a region of **cortical cells,** which in turn contains a **central cylinder** of thick-walled, toughened cells. Many mosses contain cells that resemble the conducting tissues of vascular plants, although there is disagreement as to their importance to transport.

21.1 AN ALGAL ANCESTOR

Green algae similar to *Coleochaete* are candidates for the ancestor of both bryophytes and the vascular plants. Although it is still around and doing well in the shallows of lakes and ponds, *Coleochaete* is thought to be quite primitive, exhibiting a number of characteristics that were present in its ancestors. Its growth form today includes a disklike single layer of cells, containing hair cells for which it is named.

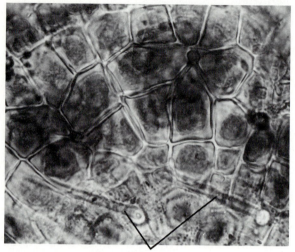

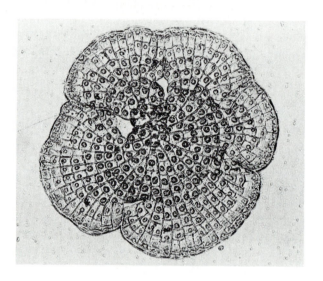

Hair cells

21.2 BRYOPHYTE ADAPTATIONS TO LAND LIFE

Bryophytes are well adapted to land life and the problem of desiccation (dehydration), although in some ways they are still tied to the watery environment. In some species specialized structures, the *rhizoids,* penetrate the soil, anchoring the plant. In many species, a layer of waterproofing cutin protects the plant from excessive water loss. In the liverworts, water loss is avoided somewhat by the presence of numerous pores that admit air but retard the excessive escape of water. Further, the embryo is protected in the *archegonium,* surrounded by cells of the gametophyte. Although specialized food- and water-conducting cells have not evolved in the bryophytes, a central cylinder of tissue may function in water conduction. Spores—dormant cells released into the environment—contain hardened, protective coats. In at least one important way, however, bryophytes have not fully adapted to land. The sperm must depend on rainwater or dew to reach and fertilize the egg.

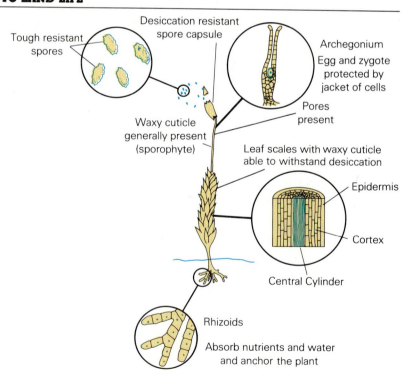

Tough resistant spores

Desiccation resistant spore capsule

Archegonium
Egg and zygote protected by jacket of cells

Waxy cuticle generally present (sporophyte)

Pores present

Leaf scales with waxy cuticle able to withstand desiccation

Epidermis

Cortex

Central Cylinder

Rhizoids
Absorb nutrients and water and anchor the plant

In many bryophytes, the "leaves" make up most of the simple, rather undifferentiated gametophyte body or **thallus,** as it is called. Although they lack the complexity of true, vascular leaves, they serve the same function. Each "leaf" is composed of a flattened, ribbonlike array of photosynthetic cells. In some species, the "leaf" cells secrete a covering or **cuticle** containing waxy **cutin** (see Chapter 3), a moisture barrier.

In the bryophyte life cycle, the gametophyte is quite dominant, and the reduced sporophyte is often dependent. As we have mentioned, the direction of evolution in vascular plants has been the reverse, with the progressive reduction of the gametophyte and its final incorporation into the sporophyte. Let's move along now for a brief look at the three bryophyte classes.

Class Hepaticae: Liverworts

The name liverwort dates from the ninth century, when the plant was popularly used for the treatment of liver ailments. (The liverwort's lobelike growth was thought to resemble the liver.)

Liverworts produce a rather simple gametophyte, which can be either "leafy" or **thallose,** the latter tending to be flat and resembling (for lack of a better analogy) branching, green cornflakes. Just about everyone's favorite liverwort is *Marchantia,* a thallose liverwort. Its dark green gametophyte thallus is a branching, ribbonlike growth that remains flattened against the earth. Each branch is notched at its tip, and within the notches are a number of **apical** (growing) cells. Simple single-celled rhizoids and multicellular **scales** anchor the plant and absorb water from the soil. On the surface of the thallus are cuplike structures within which are the **gemmae,** small, vegetative, budlike growths that function in *Marchantia's* special asexual reproduction. When the gemmae are dislodged, they are capable of growing into an entire gametophyte thallus. Independent individuals also form when portions of the older thallus die off, leaving newer growths to exist on their own.

Marchantia has a sexual cycle as well, with a life history typical of bryophytes (Figure 21.3). Sexual reproduction begins when the gametophyte thallus produces stalked structures, the **archegoniophores** and **antheridiophores,** that bear **archegonia** and **antheridia** (egg- and sperm-producing structures, respectively). These gamete-producing structures are borne on stalks within structures that resemble miniature umbrellas. Liverworts can either be **monoecious** ("one house") in which both sperm and eggs are produced by one individual, or **dioecious** ("two houses"), in which sperm and eggs are produced in separate individuals. Depending on the species, female and male structures are produced either on different branches of the same thallus or on different thalli. When the gametes are mature, swimming sperm will be literally splashed by raindrops from the antheridia to the archegonia. The sperm then enter the archegonia and fertilize the eggs. With this union, a new diploid sporophyte generation begins.

The liverwort sporophyte is dependent on the gametophyte and remains attached to the archegonium. Eventually, each new sporophyte will develop a sporangium, whose diploid cells go through meiosis and produce numerous haploid spores. This ends the brief diploid sporophyte phase of the life cycle. Spore dispersal is assisted by the presence of odd, spiral cells called **elators.** They respond to minute changes in humidity by coiling and uncoiling, and this twisting movement dislodges and frees the spores. Each spore, under suitable conditions, can produce a new gametophyte.

Class Anthocerotae: Hornworts

The *hornworts* are a relatively minor group, with only 100 or so named species. Although the gametophyte thallus somewhat resembles that of the liverworts, its lobed growth forms a rosette rather than a ribbon. Further, in *Anthoceros,* our representative genus, each thallus cell contains one large chloroplast along with a starch-storing pyrenoid. These are very primitive traits, characteristic of green algae. Antheridia and archegonia are produced directly in the thallus (Figure 21.4). Fertilized egg cells develop into unusual diploid sporophytes—long, slender, hornlike growths (from which the name *hornwort* was derived) that form lengthy sporangia. When the sporangia split, spores are released.

Two other aspects of the sporophyte's growth are unusual. First, it can grow continuously, producing spores over a long period of time. This is made possible by an active **meristematic** region, an area of simple, undifferentiated, actively dividing cells at its base. Second, under exceptional conditions it can survive on its own should the supporting gametophyte die. Like the young moss sporophyte, it is photosynthetic. Following meiosis in the sporophyte, the haploid spores are carried away to begin a new gametophyte generation. Some hornworts form a symbiotic relationship with nitrogen-fixing cyanobacteria, hosting them in their body cavities. This association provides the hornwort with a rich source of nitrogen compounds.

Class Musci: Mosses

The *mosses* are the most numerous and common of the bryophytes, with about 9500 named species (second only to the flowering plants). They grow best in moist conditions but are found in a variety of habitats. A few mosses even thrive in deserts, having made a simple adaptation to the drying conditions. Those that lack a cutin layer do not resist dehydration as do most other land plants. When their habitat becomes dry, these mosses simply dry out too, and, as you would expect, all metabolic activities cease. The tinder-dry moss is not dead, however. With the next rainfall or heavy dew, water diffuses into the dry tissues and the moss literally "comes to life," seemingly resuming where it left off.

Interestingly, a certain terrestrial animal can do

21.3 ALTERNATION OF GENERATIONS IN *MARCHANTIA*, A LIVERWORT

The diploid sporophyte in *Marchantia* is briefly surrounded by the archegonium, from which it develops. The sporophyte consists of a foot, seta, and sporangium. The gametophyte generation begins when meiosis and spore formation occur in the sporangium. When the haploid spores mature, they are propelled from the sporangium by springlike, humidity-sensitive *elators,* which constantly stir the sporangium. Successful spores germinate to produce the gametophyte thallus. *Marchantia* frequently undergoes an asexual cycle, producing new thalli from reproductive buds known as *gemmae* (see photo, left). Each gemma, when freed from its cup, can produce a regular thallus. Sexual reproduction involves the development of stalked *archegoniophores* (divided, umbrella-like heads with archegonia on the underside) and *antheridiophores* (disklike heads with antheridia on the top side), on separate thalli (see photo, right). Sperm from mature antheridia are splashed by rain to a neighboring archegonium and fertilize the egg cell within, beginning the diploid sporophyte generation once again.

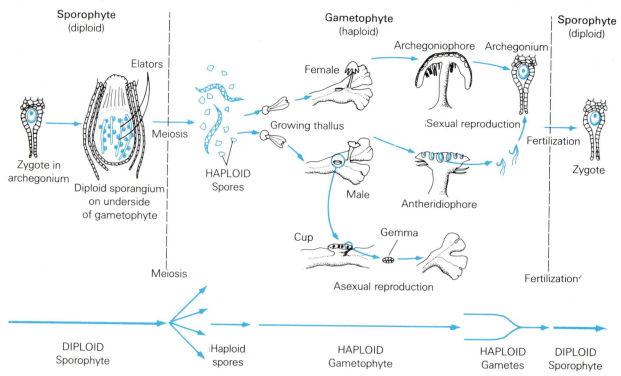

the same thing. One group of *Tardigrades* (obscure, microscopic animals that somewhat resemble arthropods), are also capable of being desiccated (sometimes for years) and of then rehydrating successfully. It is no coincidence that the tardigrade species are found in an appropriate locale—living and browsing among the mosses.

Mosses are of considerable ecological importance since they provide food for a number of animals, particularly in Arctic regions where, along with lichens, they provide much of the ground cover. As with lichens, they are also well known as pioneer plants, species responsible for colonizing previously uninhabited regions and producing the first soils. Thus, we often find mosses growing on bare rocky outcroppings. Gardeners commonly use mosses for decorative purposes, and *Sphagnum* (peat moss) is used as a mulching or bedding material for lawns and gardens. In Ireland and a few other places, *Sphagnum* from peat bogs is dried and used as heating and cooking fuel (and is also burned to provide the smoky flavor of Scotch whisky).

The Moss Life Cycle. Let's look at some of the details of the moss life history, beginning with the haploid spore and the gametophyte phase (Figure 21.5).

Once a haploid spore has established itself in a suitable place, it will absorb water and sprout into a young threadlike moss gametophyte stage called a **protonema** (plural, *protonemata*). The protonema is very similar to certain filamentous green algae. If conditions are favorable, the protonema will bud in several places, finally producing the mature leafy gametophyte. At maturity, the moss gametophyte enters into sexual reproduction. Most moss species are monoecious, and at their gametophyte tips they bear clusters of sperm-producing antheridia or egg-producing archegonia. Both are surrounded by a protecting layer of nonreproductive cells called **sterile jacket cells.** These cells are an important terrestrial adaptation, providing protection for the embryo while in the archegonium.

If water is available, sperm emerge from the mature antheridium, enter a neighboring archegonium, and fertilize the egg. As with the liverworts, the water often takes the form of raindrops, which literally splash the sperm from the antheridia to the archegonia. Some researchers maintain that the sperm are directed into the archegonium by chemical attractants.

Fertilization ushers in the diploid sporophyte generation. The typical moss sporophyte consists of

21.4 THE HORNWORT

In *Anthoceros,* a hornwort bryophyte, the gametophyte is dominant but the sporophytes are quite conspicuous and "hornlike." Each of the several sporophytes commonly borne by the gametophyte consists of a foot, a very conspicuous (hornlike) seta, seen emerging from the gametophyte, and a stalklike sporangium. (There is no spore capsule as seen in mosses and liverworts.) Spores are continuously formed within the lengthy, stalklike sporangium, and mature spores are released as the upper region of sporangium dries and splits open. This lengthy spore-forming process is made possible by continuous upward growth from the base.

a broad **foot** firmly anchored in the archegonium and a slender, elongating, stalklike **seta,** which serves to elevate an enlarged capsule containing the multicellular **sporangium.** Most sporophytes are quite small, but in some species the stalk may become as long as 20 cm (about 8 in). At first, the moss sporophyte is green and photosynthetic, so, unlike the liverwort sporophyte, is not dependent on the gametophyte. Later it will lose its chlorophyll and take on a brown hue. Interestingly, the moss sporophyte contains numbers of **stomata,** pores surrounded by **guard cells**—structures generally associated with the leaves of vascular plants.

Within the sporangium, cells undergo meiosis, producing numerous haploid spores. When the capsule is mature and filled with spores, its cap, the **operculum,** falls away, and the tiny spores are scattered to the winds. The likelihood of landing in a suitable place is not very good, but the low odds are countered by the great number of spores. With spore formation, the moss gametophyte generation begins once again.

21.5 ALTERNATION OF GENERATIONS IN MOSSES

Mosses have dominating gametophyte generations. The sporophyte consists of a stalked growth emerging from the gametophyte. It is topped by a capsule containing the sporangium where meiosis and spore formation occur. With meiosis and spore formation, the haploid gametophyte generation begins. Spores germinate, producing threadlike *protonemata,* which mature into the familiar, green gametophytes. Male gametophytes produce antheridia and female gametophytes, archegonia, The sperm from the antheridia swim, in a film of rainwater or dew, to the archegonia, fertilizing the eggs. Upon fertilization, the diploid chromosome number is restored and the sporophyte generation begins again. Although the sporophyte grows from the gametophyte, it is actually independent. Seen in the photos are stalked sporophytes of *Polytrichum* moss growing from the low-lying gametophytes (left) and spores being released from the spore case (right).

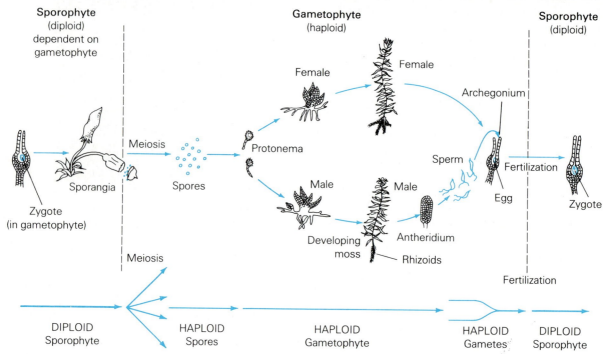

THE VASCULAR PLANTS

Bryophytes are interesting, but if someone were to ask you to think of a plant, you would probably think of a vascular plant, such as an oak, rose, or onion, and not a hornwort. Most people simply are far more familiar with vascular plants than with the others.

The vascular plants are often referred to by the more technical term **tracheophyte (tracheo-,** tubes; **-phyta,** plants), which refers to the tubelike vascular elements, **xylem** and **phloem,** which transport water and foods, respectively. Our comments will be general here, since these tissues are discussed in

detail in Chapter 22. The most important point we can make is that vascularity was a smashing evolutionary success, providing an answer to some of the evolving plants' terrestrial problems. First, vascular tissue, including the woody, hardened xylem and tough accompanying supporting tissues, allowed plants to grow to immense sizes. Apparently, large size provided a novel way of adapting to the terrestrial environment—one not available to the bryophytes. Second, and closely related, once the problem of desiccation was overcome, the successful transition to terrestrial life also depended on getting water up from the soil to the photosynthetic tissues. Thus, in the presence of increasing competition among rapidly emerging species, natural selection often favored root systems that could penetrate deeply and spread widely in the soil, stems that could hold the foliage higher, and in some instances, leaves that grew longer and broader.

The origin of vascular plants is lost in antiquity, but the earliest fossils are to be found in the 400 million-year-old rocks of the Silurian period of the Paleozoic era. Among the earliest vascular plants is an extinct group called the **rhyniophytes.** They made their debut along with the first vertebrates— those clumsy, unimpressive armored fishes with jawless, sucking mouths. (But even these creatures were probably not very impressed by the rhyniophytes.) Consider *Rhynia*, whose fossils are well known. It grew in foul marshes, probing the air with simple leafless stems arising from **rhizomes,** a form of stem that grows horizontally (Figure 21.6). The peculiar growth of the fingerlike stems was important for two reasons. First, their branching and rebranching provided a vascular framework for the later evolution of the flattened leaf blade. But equally significant, the primitive stems contained a system of tiny tubelike, nonliving, water-filled elements we call **tracheids,** a part of the water-conducting xylem common to today's vascular plants. With the tracheids came the promise of new horizons for, as we have seen, xylem not only conducts water, but its tough, fibrous elements provide vital support in many plants. Xylem forms the wood of woody stems, providing the strength that permitted vascular plants to grow tall and alter the appearance of the earth.

Today, vascular plants inhabit nearly every part of the planet that protrudes above water, and a few that do not. As they radiated out over the landscape, they adapted to their environments in different ways and thus produced a wide variety of forms. Most developed true (vascular) leaves, stems, and roots (and some evolved **seeds**). A finishing touch was the development of mechanisms for fertilization that did not require the flagellated sperm or a watery medium for sperm travel. You may recall that the more recently evolved plants, and the fungi, lack centrioles and the 9 + 2 microtubular flagella that characterized the swimming sperm of their ancestors.

Vascular plants have been classified in a number of ways, including one in which the tracheophytes are grouped under division[1] (phylum) *Tracheophyta.*

---

[1]The taxonomic term *division* is used by botanists instead of phylum, a term that has been used elsewhere.

21.6 THE EARLIEST VASCULAR PLANT

Rhynia, reconstructed from fossil studies, represents one of the oldest known vascular plants. The rhyniophytes flourished nearly 400 million years ago. The aerial stems, emerging from rhizomes, branched dichotomously, ending in terminal sporangia.

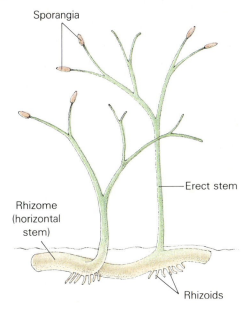

Sporangia

Erect stem

Rhizome (horizontal stem)

Rhizoids

For simplicity's sake we will use the system that organizes the vascular plants into nine divisions, six of which probably will be unfamiliar to everyone except perhaps a few knowledgeable gardeners. Be aware that the names will differ from one source to another. Also note that the vascular plants are further divided into those that do not produce seeds and those that do. The nine divisions are:

Vascular plants without seeds

Psilophyta ("naked plants"): *Psilotum*
Lycophyta ("spiderlike plants"): ground pine or club mosses
Sphenophyta ("wedge plants"): horsetails
Pterophyta ("winged plants"): ferns

Vascular plants with seeds

Coniferophyta ("cone bearers"): the pines and other conifers
Cycadophyta: the cycads
Ginkgophyta: the ginkgo
Gnetophyta: the gnetophytes
Anthophyta ("blossom or flower plant"): the flowering plants (also called **angiosperms**)

TABLE 21.1

SUMMARY OF THE PLANT KINGDOM

Kingdom Plantae: Phyla (Divisions)	Examples	Characteristics	Alternation of Generations
The Nonvascular Plants		Little or no organized tissue for conducting water and food (xylem and phloem)	
Division Bryophyta (7000 species)	Mosses, liverworts, hornworts, *Polytrichum, Riccia, Anthoceros, Marchantia*	Terrestrial and fresh water aquatic, multicellular with tissue specialization, chlorophyll *a* and *b*, many adaptations to land	Separate generations, gametophyte usually dominant, swimming sperm, egg and zygote protected
The Seedless Vascular Plants		Organized vascular tissue, limited root development, many primitive traits, chlorophyll *a* and *b*	
Division Psilophyta (4 species)	Whisk ferns, *Psilotum*	Terrestrial, tropical, primitive vascular tissue, no true leaves	Separate generations, dominant sporophyte, minute gametophyte, swimming sperm
Division Lycophyta (1000 species)	Ground pine or club moss, *Lycopodium, Selaginella*	Terrestrial, widespread, well-developed vascular tissue in some, true leaves	Separate generations, dominant sporophyte, minute gametophyte, swimming sperm
Division Sphenophyta (12 species)	Horsetails, *Equisetum*	Terrestrial, widespread but preferentially marsh dwellers, underground stems (rhizomes), simple, hollow aerial stems, well-developed vascular tissue	Separate generations, dominant sporophyte, minute gametophyte, swimming sperm
Division Pterophyta (11,000 species)	Ferns: *Sphaeropteris, Platycerium, Osmunda*	Terrestrial, moist soils to dry, underground stems, highly developed variable leaf, well-developed vascular tissue	Separate generations, dominant sporophyte, minute gametophyte, swimming sperm
The Seed-Producing Vascular Plants		Seeds, organized vascular tissues, extensive root, stem, leaf development	All with dominant sporophyte, which retains the gametophyte as a specialized tissue. All pollen producers

(Also see Table 21.1.) The first three divisions were once an important part of the earth's greenery, but their heyday is over and they are now relatively minor groups. However, we will consider them anyway because they are interesting and because some of their characteristics may represent significant stages in the evolution of the seed plants. Of the estimated 261,000 named species of vascular plants, only about 1000 species comprise the first three groups, and of these, all but 28 are found in division Lycophyta. Thus, in considering these divisions, we will be looking at a very few survivors of another time, before the ferns and then the seed plants became the dominant forms of plant life on earth. Of the second group—the seed plants—the first four divisions are commonly grouped into the **gymnosperms,** while the fifth group or flowering plants (Anthophyta) are usually called the **angiosperms.**

TABLE 21.1 (continued)

SUMMARY OF THE PLANT KINGDOM

Kingdom Plantae: Phyla (Divisions)	Examples	Characteristics	Alternation of Generations
Gymnosperms		Seeds without surrounding fruit, tracheids only for water conduction, terrestrial	
Division Ginkgophyta (1 species)	Maidenhair tree, *Gingko biloba*	Native of China, widespread by introduction, large tree, fan-shaped leaf	Swimming sperm (utilizes plant fluids)
Division Cycadophyta (100 species)	Cycads: *Zamia, Cycas*	Widespread but mainly tropical, thick, partially subterranean stem, palmlike foliage, very large pollen and seed cones	Swimming sperm (utilizes plant fluids)
Division Gnetophyta (71 species)	*Gnetum, Welwitschia, Ephedra*	Warm-temperate regions, deserts, angiosperm features: bladelike leaf, two types of water-conducting tissue (tracheids and vessels), flowerlike pollen cones	Nonmotile sperm
Division Coniferophyta (550 species)	Pines, redwoods, firs, junipers, larch, cypress, hemlock	Widespread in temperate and subarctic areas; needle leaves common, evergreen, multiple cotyledons	Nonmotile sperm wind-pollinated
Angiosperms		Flowers, seeds with fruit surrounding	
Division Anthophyta Class Dicotyledonae (about 200,000 species)	Dicots: magnolia, cabbage, tobacco, cotton, willow, apple, oak, maple, bean	Worldwide, widely cultivated, net-veined leaf, floral parts in 4s and 5s or their multiples, two cotyledons	Nonmotile sperm, mainly insect pollinated
Class Monocotyledonae (about 50,000 species)	Monocots: iris, orchids, grains (grasses)	Worldwide, widely cultivated, parallel-veined leaf, floral parts in 3s or its multiples, one cotyledon	Nonmotile sperm, both wind and insect pollinated

Division Psilophyta: Whisk Ferns

Division **Psilophyta,** a remarkably persistent group of plants, is represented today by two genera and only four species, all found mainly in the tropics. Sporophytes in *Psilotum nudum,* the "whisk fern," (Figure 21.7) produce hairy rhizoids similar to those of the bryophytes, but no true roots. The stems are fairly well developed, with simple, primitive vascular tissue. The stems are photosynthetic and bear

simple, primitive sporangia above tiny scalelike appendages. The sporophyte phase in the life cycle of the psilophytes is quite dominant. The gametophyte, a tiny, separate plant, produces both antheridia and archegonia. Following fertilization, the sporophyte emerges from the gametophyte in a manner reminiscent of the mosses. However, it soon becomes independent and the tiny gametophyte withers and dies.

The psilophyte ancestors are well represented in

21.7 A LIVING FOSSIL

Psilotum, a vascular plant with many primitive traits, produces an erect, highly branched stem with very simple vascular tissue. The sporophyte generation (shown here) is quite dominant. Spores are produced meiotically in the scattered sporangia (seen as spherical bodies).

the fossil record of 375 million years ago. They may represent an early stage in the evolution of vascular plants, but we can't be sure because there are fossils of other, more advanced plants that date even further back.

Division Lycophyta: Club Mosses

The most familiar **lycophytes** are those in genus *Lycopodium* (Figure 21.8). You may have seen *Lycopodium* sporophytes without knowing it, since they are often mistaken for pine seedlings. In some parts of the United States they are known as **ground pine** or **club moss.** They are green year-round and quite conspicuous in the winter when other ground cover has died back. It was once considered quaint to gather them to make a Christmas wreath, but the practice has actually endangered this interesting little plant and is now prohibited by law in many places.

Lycopodium sporophytes are dominant, and they have vascular roots, stems, and leaves. In the adult plant, the roots are of the **adventitious** type; that is, they emerge from the stem rather than from a primary root. The vascular system of *Lycopodium* is well developed, extending from the adventitious roots through stem and leaves. The leaves, however, are small and arranged in a whorl (a spiral) around the stem. They are called **microphylls** (small leaves), and are believed to have evolved as simple outgrowths from the stem epidermis (outer-

most covering tissue). They contain a single vein of vascular tissue. In contrast, the leaves of higher vascular plants are called **megaphylls** (large leaf). They contain many veins and owe their evolutionary origin to the fusion of branching stems (such as those seen in *Rhynia*). Some lycophyte species produce horizontal runner stems, which eventually form a mat over the forest floor, with occasional roots and vertical stems springing from the runners.

In lycophytes, spores are formed in **sporophylls,** which are located either in the base of the leaflike microphylls or atop the plant in enlarged conelike clusters called **strobili.** In either case, when the haploid spores are released they germinate and grow into monoecious gametophytes, which in some species live underground. Fertilization occurs when there is sufficient water over the gametophyte for the sperm to swim to the archegonium. The diploid sporophyte then emerges from the gametophyte to repeat the life cycle.

***Selaginella*: A Reproductively Advanced Lycophyte.**   *Selaginella* is a lycophyte genus of considerable evolutionary importance. While its sporophytes are not unlike those of *Lycopodium*, a very basic and important difference exists in the kind of spores they form.

21.8 CLUB MOSSES

Lycophytes, the club mosses, are not mosses at all but vascular plants with true roots, stems, and leaves. Their roots are adventitious, emerging from numerous rhizomes (horizontal stems) that lie on the soil surface. The leaves are primitive microphylls, emerging from the stem epidermis. In some species, spores are produced in the axils of sporophylls distributed along the stems, but in others the sporophylls are clustered into strobili at the tips of leafy upright stems.

Plants have taken two different directions with regard to spore formation. For example, *Lycopodium* is **homosporous,** which means that only one kind of spore is produced. Homosporous species are found in the psilophytes, sphenophytes, lycophytes, and in nearly all pterophytes (ferns). *Homospory,* by the way, is the only type of plant spore formation we have discussed so far. However, *Selaginella* is an example of a **heterosporous** species. It produces two types of spores—**microspores** (tiny) and **megaspores** (large). We will use these terms again in discussing the seed plants since heterospory is an advanced condition invariably seen in the gymnosperms and angiosperms.

In spite of their literal meaning, the terms *microspore* and *megaspore* are not always related to size (especially in seed plants), but refer to function. Microspores give rise to male gamete-producing individuals—the **microgametophytes**—while megaspores give rise to those that produce female gametes—the **megagametophytes.** In heterosporous species such as *Selaginella,* the microspores are produced in sporophylls suitably named **microsporophylls,** while megaspores are produced in **megasporophylls.**

As the microsporophylls of *Selaginella* develop, certain diploid cells within undergo meiosis, each forming four haploid microspores. Each microspore then undergoes mitosis and develops into a two-celled microgametophyte (part of the gametophyte generation). One of these cells develops into an antheridium containing both sterile cells and sperm cells. Similar events occur in the megasporophylls. Following meiosis, each megaspore develops into a megagametophyte (also part of the gametophyte generation), within which the archegonia develop that will contain sterile cells and female gametes. The gametes are brought together in a seemingly odd, but apparently effective manner: microsporangia break loose and sift down along the sporophyte to the megasporophylls of the plant, reaching the locale of the megasporangia. Then, when water is present, the microsporangia break open and the sperm swim to the egg cells.

The lycophytes were not always the humble, ground-hugging plants we see today. One order, the Lepidodendrales, produced great treelike plants over 50 meters tall and 2 meters in diameter. In fact, 300 to 400 million years ago, the lycophytes were the dominant plants of the Devonian and Carboniferous forests (Figure 21.9). But change is inexorable, and with the drying climate of the Permian period, the primitive giants died, possibly replaced by the newly evolving gymnosperms, but leaving only a few remnants of a once prominent group. Their ghosts now haunt us—the corpses of these great plants became partly decomposed to form vast coal deposits, fossil fuels that influence modern global politics as we squabble over their remains. Let's move along for a closer look at another survivor.

Division Sphenophyta: Horsetails

The **sphenophytes,** like the lycophytes, had their day in the late Paleozoic, when they contributed significantly to the lush Carboniferous forests. Today there is only one genus, *Equisetum* (equi-, "horse"; -setum, "bristle"), with 20 or so species. Most species grow to less than a meter in height. One tropical giant, *E. giganteum,* is reported to be 6 m tall, but its scrawny stems are less than 3 cm (1.25 in) in diameter. It is commonly supported by neighboring trees.

Equisetum is known as "horsetail," which it tends to resemble, or "scouring rush," a name reflecting its pre-Brillo use as a pot cleaner. Its abrasiveness comes from heavy deposits of glassy silica in the stem epidermis.

Two kinds of shoots arise from the sporophyte rhizome (Figure 21.10). Vegetative shoots bear scalelike microphylls and whorls of short, lateral branches, while unbranched reproductive shoots produce haploid spores meiotically in a prominent strobilus. As we saw earlier, the sphenophytes are homosporous. The spores develop into separate, monoecious gametophytes, often about the size of a pinhead. The sperm must then have sufficient water in order to swim to the egg.

Division Pterophyta: Ferns

Pterophytes—ferns—are undoubtedly among the most enchanting of plants. The spring forest, dripping with cool rain, is accented by delicate ferns rising with tiny bowed heads from the damp floor. Later, the plants will lend an exotic touch to the woods as they stand full grown, their leaves splayed, as if placed there as decoration.

Ferns have survived in great numbers since Paleozoic times, apparently having adapted more readily to changing environments than have the other primitive vascular plants. Today some 11,000 named species of ferns are widely distributed over the earth, including both tropical and temperate regions; some even live in arid climates (Figure 21.11).

21.9 HISTORY OF VASCULAR PLANTS

(a) Paleozoic times. The earliest forests became established in the Devonian period of the Paleozoic era. By the Carboniferous period, huge forests of lycophytes, psilophytes, sphenophytes, and primitive seed ferns flourished. In this scene, some of the giants of these now insignificant plant groups are seen. The luxurious growth seen here is now represented on earth by fossil fuel deposits.

(b) Early Mesozoic times. The Permian period of the Mesozoic era saw changes in both the land and the plants and animals that inhabited it. While there were still moist areas, much of the land at this time was dry and windswept, conditions unfavorable to primitive vascular plants, but favorable to the new seed plants, principally gymnosperms. The ferns persisted, mostly in the remaining moist lowlands. The great age of reptiles also began. The ginkgos, conifers, and cycads, which had barely begun their evolution in the Permian, increased in importance during the Triassic and Jurassic periods.

(c) Late Mesozoic times. By the Cretaceous period, the flowering plants had established themselves and were becoming dominant. Along with the flowering plants came the insects needed to pollinate them. Reptiles gave way to mammals, and birds underwent rapid evolutionary radiation. In another 60 million years the first hominids would join the descendants of these mammals.

① Eusthenopteron	⑯ Cynognathus
② Eogyrinus	⑰ Podokesaurus
③ Diplovertebron	⑱ Camptosaurus
④ Meganeuron	⑲ Compsognathus
⑤ Eryops	⑳ Allosaurus
⑥ Seymouria	㉑ Archaeopteryx
⑦ Limnoscelis	㉒ Stegosaurus
⑧ Varanosaurus	㉓ Ramphorhynchus
⑨ Ophiacodon	㉔ Brontosaurus
⑩ Sphenacodon	㉕ Anatosaurus
⑪ Araeoscelis	㉖ Ankylosaurus
⑫ Dimetrodon	㉗ Tyrannosaurus
⑬ Edaphosaurus	㉘ Pteranodon
⑭ Saltoposuchus	㉙ Triceratops
⑮ Plateosaurus	㉚ Struthiomimus

Ⓐ Eospermatopteris	Ⓜ Matonidium
Ⓑ Calamites	Ⓝ Schizoneura
Ⓒ Lepidodendron	Ⓞ Neocalamites
Ⓓ Sigillaria	Ⓟ Palmetto
Ⓔ Cordaites	Ⓠ Pandanus
Ⓕ Araucarioxylon	Ⓡ Sassafras
Ⓖ Bjuvia	Ⓢ Ginkgo
Ⓗ Macrotaeniopteris	Ⓣ Cornus
Ⓘ Wielandiella	Ⓤ Quercus
Ⓙ Williamsonia	Ⓥ Magnolia
Ⓚ Cycadeoidea	Ⓦ Sabalites
Ⓛ Araucarites	Ⓧ Salix

Devonian Period | **Carboniferous Period**

(ferns, seed ferns, lycophytes, and sphenophytes)

Jurassic Period

(gymosperms and ferns)

Permian Period

iest gymnosperms)

Triassic Period

(conifers, cycads, cycadeoides, and ginkgos)

Jurassic Period

Cretaceous Period

(flowering plants and gymnosperms)

21.10 THE HORSETAIL

Equisetum, the horsetail, produces two types of shoots: one that is vegetative with whorls of photosynthetic branches and another—a fertile branch—that in this species is nonphotosynthetic, unbranched, and bears a spore-producing strobilus at its tip. Both shoots have tiny, scalelike, nonphotosynthetic leaves at regular nodes along their stems and both belong to the dominant sporophyte generation.

The fern sporophyte (Figure 21.12a) typically consists of a thick rhizome from which arise many fine roots. Vascular tissue is well developed. The large divided leaves are true megaphylls, having evolved from branch systems. In many species, the leaves emerge above ground from the rhizomes as **fiddleheads,** each of which uncoils into a large and often frilly compound leaf (that is, a leaf subdivided into leaflets). The leaves come in all sorts of sizes and shapes. A few, such as the tree ferns, have tall stems supporting a leafy rosette.

The fern's sporophyte is decidedly dominant and produces copious numbers of spores. Spores are usually produced on the underside of the leaves in sporangia, which are commonly clustered into structures called **sori** (singular, *sorus*). In their younger stages, the sori are sometimes hidden under a scalelike cover called an **indusium** (Figure 21.12c). You have undoubtedly seen the sori on the underside of ferns. They often occur as rows of brown dots and may be mistaken for an insect or fungal invasion. Their sudden appearance has sent many an alarmed gardener scurrying to the nearest plant nursery for pesticides.

21.11 DIVERSITY IN FERNS

A small sample of the diversity in ferns includes **(a)** the tree fern *(Sphaeropteris),* **(b)** the staghorn fern *(Platycerium),* **(c)** the maidenhair fern *(Adiantum),* and **(d)** the floating fern *(Salvinia).* Ferns are a diverse group, perhaps because they were not seriously diminished by geological changes as were the other primitive vascular plants.

(a)

(b)

(c)

(d)

21.12 FERN SPOROPHYTE

(a) Anatomy of a fern sporophyte. Below the soil, an extensive rhizome (underground stem) sends its leaves (megaphylls) above ground to form the cluster seen here. Leaves emerge from the rhizome as highly curled *fiddleheads,* each of which unwinds into its final complex, divided form. The rhizome also produces hairy true roots that absorb water and minerals. **(b)** In many species, spore-forming structures called sporangia are borne in sori many species, spore-forming structures called sporangia are borne in sori on the underside of the leaf. **(c–e)** Spores are produced meiotically in the sporangia and are catapulted out when lip cells opposite the annulus split.

Labels in (a): Leaf; Sori (dots); Fiddlehead; Rhizome; Roots

Labels in (c): Lip cells; Leaf; Spore; Annulus; Sporangium; Indusium (cover)

(a)   (b)

(c)   (d)   (e)

The sporangia may be quite complex, containing elaborate mechanisms for spore release. In some species, spore release—actually, forceful ejection—occurs when rapid changes in the thick-walled cells of the **annulus** (a row of cells along one wall of the sporangium) split open the thin-walled **lip cells.**

Figure 21.13 illustrates the life history of a typical fern. Haploid spores formed by meiosis in the sporangia are released into the breeze. When one settles into a moist, well-protected surface, it germinates into a delicate, heart-shaped gametophyte known as the **prothallus.** This photosynthetic haploid plant produces rhizoids on its lower surface, and for a time it lives as a completely independent, if seemingly insignificant, individual. Eventually, however, each prothallus produces antheridia and archegonia in which haploid gametes are produced through mitosis. If at least a film of water is available, sperm cells from the antheridia will become active, swim to the archegonia, and fertilize mature egg cells. The diploid embryo, housed and pro-

tected for a brief time by the prothallus, develops into the large sporophyte plant.

One might wonder whether ferns can self-fertilize by using such a system. After all, it usually isn't very far from an antheridium to the nearest archegonium on the wet surface of a single fern gametophyte. Actually, self-fertilization is not uncommon, but cross-fertilization is more adaptive from a genetic point of view. Accordingly, many ferns have evolved mechanisms for inhibiting self-fertilization. The first prothallus to mature produces a hormone known as **antheridogen** (similar to the hormone gibberellic acid, discussed in Chapter 24). The hormone *suppresses* male (antheridial) development in the gametophyte in which it is produced but *stimulates* male development in neighboring gametophytes. Thus, the hormone's effect is to produce separate male and female prothallia, assuring cross-fertilization. In other species, sperm and eggs from one gametophyte are physiologically incompatible and only fertilizations involving different genotypes are successful.

THE SEED PLANTS

The primitive vascular plants and their less conspicuous contemporaries, the earliest seed plants, persisted through the Devonian, Carboniferous, and Permian periods. Toward the close of the Permian period, at the end of the Paleozoic era (about 225 million years ago), the plant life of the earth experienced a strange and dramatic change. It may have been the same sort of drastic change that brought on the extinction of the dinosaurs some 100 million years later. Powerful geological movements produced a completely different landscape and climate. The earth had been a rather smooth globe, and its surface had permitted countless warm and shallow lowland seas to cover the planet. Rather quickly (in geological terms), things changed. Soggy lowlands were uplifted to form vast mountain ranges. The monotonous warm climate gave way to a general cooling, even as the new uplands dried. Thus, the primitive, marsh-loving giants of the late Paleozoic era perished, leaving only a few scattered remnants of the once-prominent groups.

With the demise of the ancient forests, the competitive edge in the plant world passed to the inconspicuous seed plants, which were better prepared to survive on this new kind of earth. They continued to evolve, becoming larger and better able to invade the land, and soon appeared everywhere over the new Mesozoic landscape.

Among the primitive *gymnosperms* were the ancestors of today's conifers. The secret of their newfound success was their ability to draw water from deep in the dry surface of the earth, transfer it in efficient vascular systems, and conserve it in water-resistant stems and leaves. They also developed a different mode of reproduction, with the female gametophyte fully enclosed within the sporophytic tissues. In primitive vascular plants, as we have seen, the gametophyte is separate, often small and ground-hugging, and altogether seeming quite vulnerable. It may well have been the new drier and harsher conditions that brought about the loss of too many gametophytes and thus the great reduction of these early vascular plants. But there was still the problem of transporting the sperm to the egg. Let's see how this was solved in the newly emerging seed plants.

21.13 ALTERNATION OF GENERATIONS IN A FERN

The sporophyte is highly dominant and the gametophyte is greatly reduced. The two are clearly separated and independent, except for the brief period when the sporophyte begins its growth. The mature sporophyte produces haploid spores meiotically, and upon germination the spores produce a heart-shaped *prothallus,* which bears antheridia and archegonia. Following fertilization, the young diploid sporophyte begins its growth from the archegonium, soon producing its own roots and photosynthetic tissue as the prothallus withers and dies.

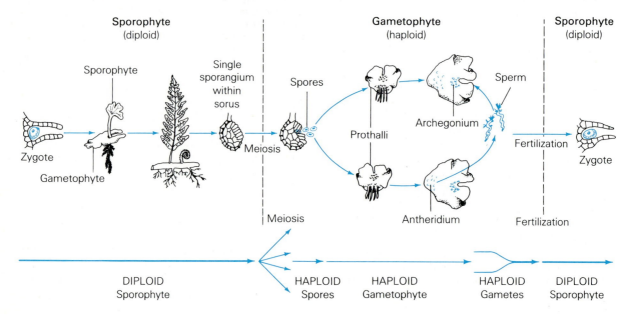

21.14 POLLEN AND SEEDS

Two reproductive developments helped the gymnosperms adapt to the new environment. First came the pollen grain, with a tough, resistant wall—the young male gametophyte. Following pollination, the gametophyte produced the male gametes. The pollen shown here in the scanning electron micrograph is from the northern white cedar. Second, the embryo came to be protected in the seed, another tough con-tainer. The seed, such as those of the pine seen here, contains the embryo of the plant, some stored food, and a surrounding seed coat, which is hard and water-resistant. The protective seed coat makes it possible for the embryo to remain intact for considerable periods of time, until conditions are right for growth.

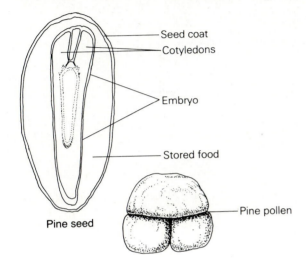

Pine seed — Seed coat, Cotyledons, Embryo, Stored food, Pine pollen

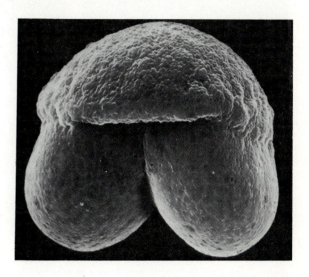

Reproduction in the New Seed Plant

In the emerging seed plant, the male gametophyte was greatly reduced in size and was housed and transported in the form of tiny, but tough and resistant, **pollen** grains. These could easily be transported from plant to plant via winds, insects, other animals, and even water. At first, the newly evolving seed plants retained the swimming sperm to be released near the egg. Eventually, though, the motile, flagellated sperm cell was replaced in the modern gymnosperm by nonmotile sperm, released in the female gametophyte after pollination (transfer of pollen) had successfully occurred. Although a minute quantity of fluid is required for pollen germination, the sperm no longer have to swim from one plant to the next. Thus, the last link to the ancient, watery plant environment was finally broken. The venture was successful and gymnosperms came to dominate their new environment (see Figure 21.9).

Pollen is very effective, indeed, at protecting and transporting the male gametophyte. In some species, in fact, it is so resistant that it can remain viable for years. Some pollen is carried worldwide in air currents (which isn't exactly news to hay fever victims). It is even found far out at sea, as well as drifting in the uppermost layers of the atmosphere.

Pollen grains fossilize well, and their fossil records help reconstruct the nature of the flora of bygone periods. In fact our limited knowledge of flowering plant origins is based heavily on the analysis of fossil pollen from the soil and rocks. (As an aside, those who study fossil pollen are called *palynologists*.)

The second reproductive innovation in this changing world—the seed—consists of an **embryo** and stored food enclosed within hardened **seed coats** (Figure 21.14). Most gymnosperms produce seeds in cones; in contrast, the *Angiosperms*, the flowering seed plants, produce a fruit around the seed. Seeds are often quite resistant and can lie dehydrated and dormant in the soil for long periods of time until conditions are right for germination.

The Gymnosperms

The name *gymnosperm* literally means "naked seed," and it refers to plants that have seeds without fruit. The term is not an official taxonomic grouping, although in past use it has designated both a division and class. Today, however, the

21.15 *GINKGO:* ONE OF A KIND

Although its leaves resemble those of angiosperms, *Ginkgo biloba* is a gymnosperm—albeit a strange one. The leaves of the *Ginkgo* are fan-shaped, divided slightly into two lobes, and its seeds have odorous, fleshy coverings. Trees are either male or female. *G. biloba* has been maintained for years in Japanese and Chinese temple gardens, although it is extremely rare in the wild. It is remarkably fungus-, smog-, and insect-resistant, and some living ginkgos are 1000 years old.

gymnosperms are considered to encompass four divisions, the Ginkgophyta, Cycadophyta, Gnetophyta and Coniferophyta. The first three are not very widespread today, but the fourth division includes the familar conifers. We will return to the conifers after a brief look at the other gymnosperm divisions.

The Ginkgo.   The one **ginkgophyte** species that still exists is often referred to as a "living fossil." *Ginkgo biloba*, the **maidenhair tree** (Figure 21.15) was once known to Europeans only from fossils and was thought to be extinct. But the living *Ginkgo* was subsequently found on the grounds of Oriental temples, and finally, in 1946, growing wild in China. They are now commonly cultivated as decorative plants throughout the world. Today it seems strange that great parts of the earth were covered by Ginkgos in the early Mesozoic.

From its appearance you might think that *G. biloba* was an angiosperm—it even sheds its leaves in autumn, after they turn from green to lovely hues of yellow. However, its fruitless seeds reveal its membership in the gymnosperm club. Reproductively, the *Ginkgo* retains a primitive reproductive trait; the swimming sperm. Male *Ginkgo* trees (the species is dioecious) produce air-borne pollen that germinate near the ovules in female trees. The microgametophyte then grows toward the egg cell. On reaching the vicinity of the egg cell, it releases two motile, multiflagellated, swimming sperm cells. Fluids for the very short distance the sperm must travel are produced by the receptive sporophyte itself. Following fertilization, the female *Ginkgo* produces seeds in the manner of a typical gymnosperm.

The Cycads.   Plants resembling modern cycads had their day during the late Triassic period of Mesozoic, some 200 million years ago. We know from fossils that they were among the most common plants in those ancient forests. They may even have been an important part of the diet of giant reptiles of that era. Today, 100 or so remaining species are found mainly in tropical regions (Figure 21.16). You may have seen cycads in museums and parks, and you may have mistaken them for palms or ferns.

In the cycads, pollen is produced in conelike strobili. Sexes are separate, and pollen is carried by the wind to the female cones. The windborne pollen produces sperm cells that, like those of the *Ginkgo*, are flagellated, representing our last look at this primitive mode of sperm transport in plants.

The Gnetophytes.   The ancestors of the 70 or so named species of gnetophytes alive today may have been related to the ancestors of today's flowering plants. We can't be sure because information on the evolutionary origin of flowering plants is

21.16 A CYCAD

Except for the prominent, brightly colored cones such as the pollen cone seen here, many cycads look like a cross between a palm and a pineapple. These interesting gymnosperms grow wild in the tropical and subtropical regions of most continents and are garden favorites wherever they can be grown.

woefully scarce. One gnetophyte, *Gnetum*, reproduces as a true gymnosperm, but has blade-like leaves resembling those of the cherry tree. Another, *Welwitschia*, produces flowerlike pollen cones. Within the stems of gnetophytes are water-conducting **xylem vessels.** While such vessels are common in the angiosperms and gnetophytes, other gymnosperms produce only the more primitive, water-conducting tracheids.

Welwitschia, by the way, may be the world's most bizarre plant (Figure 21.17). It produces only two leaves, but they grow continuously, splitting and resplitting, spilling their twisted, tentaclelike growth over the ground, as if the object of some relentless torture. *Welwitschia* lives in the very dry deserts of coastal southwestern Africa, drawing its water almost entirely from fog that spreads inland nightly.

The Conifers.   The **conifers** (class Coniferinae) include nine families containing only about 550 named species, not a large number in spite of their great populations in the coniferous forests of the world. The most common and best known are in the pine family Pinaceae. In addition to the familiar pines, this family includes firs, spruces, hemlocks, Douglas firs, junipers, and larches.

The conifers are the evergreens, the cone bear-

21.17 DESERT GNETOPHYTE

The splitting, twisted, and tortured foliage seen here belongs to *Welwitschia mirabilis,* a desert-dwelling gnetophyte. The leaves would extend for many meters if they had not split and become gnarled.

21.18 CONIFEROUS FORESTS

Coniferous forests are found chiefly in the cold regions of the earth. They form a continuous belt across North America, northern Europe, and the northern Soviet Union. Some types of conifers also thrive in more temperate regions.

ers whose corpses decorate your house at Christmas. Most species are found in the cold northerly climates of the earth (Figure 21.18). The conifers form a vast, worldwide belt of forests found in North America, Europe, and Asia called the **taiga** (see Chapter 40). Cold weather species are also found in the higher altitudes of high mountains further south. But conifers are not restricted to the colder climates. For example, there are great pine forests over much of the southeastern United States, and one of the tallest plants known, the redwood, *Sequoia sempervirens*, thrives in the mild, foggy climate of coastal California and Oregon.

Typically, conifers produce their needlelike or scalelike leaves seasonally, but unlike other seasonal plants, the shedding of old dead leaves is gradual and continuous so that the trees remain green ("evergreen") all year round. The unique leaf form is an adaptation to arid conditions brought on by both limited precipitation and extreme cold (where water is tied up as ice and snow). Because of the needle- or scalelike form, conifer leaves have little surface area from which water can be lost. In addition, each leaf is surrounded by a thick waterproof cuticle, with the stomata (leaf pores) safely recessed.

Reproduction in a Conifer. Conifers typically bear separate male (pollen-bearing) and female (ovule-bearing) cones. So while the tree is monoecious, a cone is either male or female. Not surprisingly, each species has distinctive cones. Let's look at some of the complex reproductive events in the pine, a typical conifer.

As you can see in Figure 21.19, male gametophytes develop from haploid microspores within the microsporophylls of a pollen cone. Within the hardening walls of the microspore, the haploid cell undergoes mitosis to form the male gametophyte. At first, the gametophyte includes two cells—a **generative cell** and a **tube cell.** Thus the mature pollen grain is not a simple spore, even though pollen is produced and released in a manner rather similar to the spores of ferns and mosses. The pollen grain is the young male gametophyte. Before fertilization is accomplished, the generative cell in the tiny male gametophyte will have undergone mitosis and two sperm cells will be produced. One of these will disintegrate and the other will fertilize the egg nucleus. The tube cell, by the way, controls the growth of a **pollen tube,** whose role is to form a channel through which the sperm will be carried to the female gametophyte.

The female gametophyte in the female pine cone develops entirely within a sporophyte structure known as the **ovule.** An ovule is a megasporangium surrounded by the **integuments** or **seed coats.** As you would expect, four megaspores are produced in each meiotic event, but only one succeeds. It grows into a mass of haploid megagametophyte tissue. Two or more haploid archegonia, each bearing a single haploid egg, differentiate in one end of this tissue. The surrounding sporophyte tissue of the ovule differentiates into the hard seed coat. A small pore in one end of the seed, the **micropyle,** serves as the opening through which the pollen tube enters.

Upon fertilization of the egg nucleus, the young sporophyte—the pine embryo—begins development. It will be nourished by the surrounding starchy gametophyte tissue, which will later nourish the young germinating seedling. Interestingly, pines can take two years to complete reproduction: gamete development, pollination, and fertilization can require a year's time, and the development of the embryo and seed another year. Pine cones in each state of development can be found on a single tree.

THE END OF AN ERA

The lives of the gymnosperms and the great dinosaurs were once strangely intertwined. For eons, the great beasts foraged through immense Mesozoic forests of gymnosperms, but then, for some reason, they both began to die off. The dinosaurs had marched to extinction by the end of the Mesozoic, but some gymnosperms, we know, managed to survive. With the arrival of the Cenozoic era about 65 million years ago, a new kind of plant gained dominance, the upstart anthophytes or flowering plants—the angiosperms as we will call them. With their greater adaptability to fluctuating conditions, the angiosperms fanned out over much of the planet in those distant days. Although the colder north temperate and mountainous regions remained the domain of the hardy gymnosperms, the angiosperms abounded throughout the tropical and temperate regions.

The most recent chapter in the saga of the interaction between gymnosperms and angiosperms is an interesting one. The gymnosperms seem to be on the comeback trail! In the last few million years, extensive northern spruce and pine forests have displaced tracts of hardwood angiosperms. We

appear to be in a period marked by rapid speciation and adaptive radiation of conifers. No one knows why this is happening, though.

The Angiosperms: The Rise of the Flowering Plants

The astounding diversity of flowering plants stands in stark relief against the sameness of the conifers. This diversity may help explain the dramatic rise of the angiosperms at the close of the Mesozoic. There seemed to be a flowering plant ready to exploit any developing habitat. Nonetheless, there are many unanswered questions regarding their sudden ascent. When did flowering plants first appear? Were they inconspicuously sprinkled among the primitive gymnosperms that dominated Mesozoic? Or were they absent altogether in those days? And

21.19 ALTERNATION OF GENERATIONS IN A CONIFER

With the exception of cells hidden in their cones and an occasional yellow cloud of wind-borne pollen, all of the conifer we see is the sporophyte. The conifers, like angiosperms and other gymnosperms, incorporate their gametophyte within the sporophyte tissues. Pines are typically monoecious, producing both male and female cones. Cells in the male cones—the microspore mother cells—enter meiosis and form four-celled microspores, each of which forms a winged pollen grain. Only two of the cells in each pollen grain has a future: one gives rise to a cell that controls pollen tube growth while a descendant of the other will much later produce two sperm cells. One sperm fertilizes the egg; the other may or may not engage in fertilization, depending on how many archegonia

are present. Cells in the female cone—megaspore mother cells—enter meiosis and produce megaspores. One of these produces a female gametophyte, which after many mitotic divisions has formed a mass of gametophyte cells surrounded by sporophytic cells. These form archegonia, each with an enclosed egg. When pollen, the male gametophyte, lands on the female cone, a pollen tube grows into the tissue surrounding the egg cell within an archegonium. The two sperm cells penetrate the egg cytoplasm, with one fertilizing the cell and the other disintegrating. The embryo and surrounding seed of the pine may take as long as two years to mature and be ready for germination and growth.

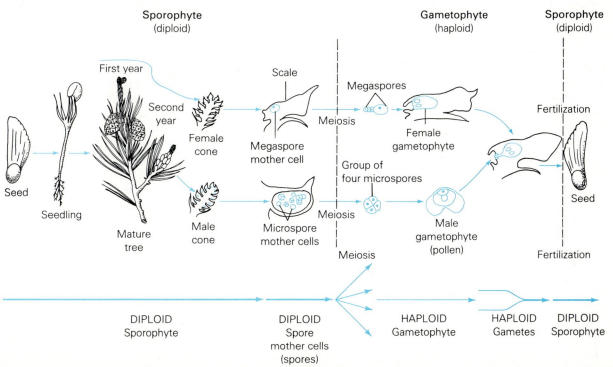

especially, what conditions would have promoted the sudden explosion of flowering plants?

Perhaps the important question is, if there *were* flowering plants during the Mesozoic, *where* were they? The fragmentary evidence seems to point to the drier upland regions of the earth (see Figure 21.9). The lowland gymnosperms were in an ideal location for formation of fossils. However, the upland regions were too dry, which may be why we find few fossils of flowering plants from the Mesozoic. The most ancient fossil remains of the angiosperms are fossil pollens estimated to be about 127 million years old.

The sudden emergence of flowering plants has been linked, hypothetically, to geological changes at the end of the Mesozoic and drastic climate changes during the early Cenozoic (a startling drop in the average temperature of the earth of perhaps 20° C). The changes in climate were brought about by two geological events. The first, as mentioned, was a period of mountain building. The second was a series of events that geologists are only now beginning to understand, called **plate tectonics** or **continental drift.** According to evidence from studies of the ocean floor, the major land masses of the earth began shifting at about the start of the Mesozoic era, some 200 million years ago (see Essay 17.1). Most of the world's land mass broke apart and the pieces drifted northward, so that by the Cenozoic the continents had roughly assumed their present positions. Apparently the angiosperms were versatile enough to have survived both these changes. Let's look at one more possibility.

The end of the Mesozoic era appears to have been marked by a major catastrophe—the collision of a gigantic asteroid with the earth. The dust it raised, according to what has become known as the **Alvarez theory** (after paleontologist Walter Alvarez and his physicist father, Louis Alvarez), blocked the sun, causing massive weather changes and large-scale extinctions. Such drastic changes on the earth's surface would have marked the demise of many living things and provided countless opportunities for other organisms to expand and diversify. As fantastic as the asteriod theory may seem, it has received a great deal of attention and is now provisionally accepted by a number of prominent scientists. (Atmospheric scientists are quite concerned with a similar scenario, called *nuclear winter,* which many believe would follow nuclear war.) We will take a closer look at the Alvarez hypothesis in Chapter 27.

Were climatic and geological events the only factors responsible for the rise of angiosperms? Perhaps not. Perhaps the great diversity of flowering plants itself suggests another factor. Perhaps flowering plants have a greater variability and can undergo more rapid speciation than can many other living things. There is some compelling evidence that this is true. For example, the ability of angiosperms to form fertile hybrids is well known. Hybrid flowering plants can even form "instant species" through *polyploidy,* an increase in chromosome number through irregularities in the meiotic process following hybridization (see Chapter 17).

Perhaps the most important factor in the rise of the flowering plants was a developing partnership with the insects. Insects became pollen carriers as they were drawn to flowers by the lure of sweet nectar and nutritious pollen. The mutual adaptation presents many classic cases of coevolution. Wind pollination is quite effective in grasslands and in coniferous forests where there are great numbers of the same species. But wind is certainly an inefficient vehicle in a mixed forest where the next individual of one's species may be quite distant. Not only is insect pollination more efficient than wind pollination under these conditions, but it is much more selective; once a bee finds nectar in one flower, it will return to that kind of flower. Such specificity on the part of insects makes it easier for new species to reproduce and for many different species to coexist. We will return to this point shortly.

The Angiosperms Today

So the angiosperms have inherited the earth—for the time being. Today they indeed constitute the vast majority of plant species. We really don't know how many there are, although estimates range from 200,000 to 250,000 named species.

In spite of such diversity, the angiosperms can be divided taxonomically into two major classes—the Monocotyledonae (monocots) and the Dicotyledonae (dicots). The first is in the minority, with about 50,000 named species. The two represent different evolutionary lines, and they vary from each other in several important aspects, as shown in Figure 21.20.

We won't dwell on the structure of angiosperms here since this group is the subject of the next three chapters. Let's look instead at some of their evolutionary aspects. Figure 21.21 presents a typical angiosperm life history.

21.20 MONOCOTS AND DICOTS

Monocots and dicots differ in four obvious ways, in their leaf venation, seed cotyledons, flower parts, and vascular systems.

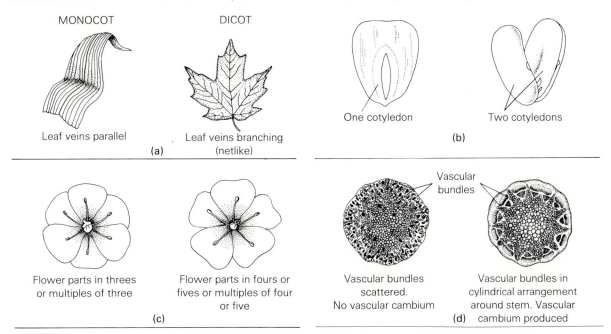

MONOCOT | DICOT

Leaf veins parallel | Leaf veins branching (netlike)
(a)

One cotyledon | Two cotyledons
(b)

Flower parts in threes or multiples of three | Flower parts in fours or fives or multiples of four or five
(c)

Vascular bundles

Vascular bundles scattered. No vascular cambium | Vascular bundles in cylindrical arrangement around stem. Vascular cambium produced
(d)

Origin and Phylogenetic Relationships. There are a number of taxonomic schemes that attempt to show evolutionary relationships among the angiosperms. There is some agreement that the family Magnoliaceae (the magnolias) in the order Ranales, is a living representative of the ancestral group. The fossil record isn't much help in resolving the issue since lower Mesozoic rock is devoid of angiosperm fossils. Such fossils begin to appear in the strata of the Cretaceous period and in greater numbers in the early Cenozoic. Most of our presumed plant relationships, therefore, are based on the study of today's plants, particularly the comparison of their flowers. As an example, the magnolia family is suggested as the ancestral type because of its primitive flower structure, as we will see.

When botanists use the floral structure to determine taxonomic relationships, they must first decide what is primitive (original equipment) and what is advanced (new)—and that presents problems.

The concepts of *primitive* and *advanced* are always troublesome. We must keep in mind that every living organism has exactly as long an evolutionary history as every other living organism—some 3 to 3.5 billion years—even though some groups may have changed rapidly while others appear to have been marking time. So perhaps it is best not to label any living organism, or even any current taxon, as either primitive or advanced. Instead, we may consider individual *features* within a group to be primitive (present in the original founders of the group) or advanced (evolved subsequently by only some members of the group and different from the earlier condition). If only specific features are considered as primitive or advanced, many semantic and philosophical problems disappear. We might, then, keep in mind that every living organism has a mix of primitive and advanced features.

In keeping with this reminder, botanists have worked out what they consider to be "primitive" floral features. Any departure from these types represents an advancement or divergence from the ancestral type. Let's consider the magnolia *(Magnolia grandiflora),* to illustrate the primitive condition (Figure 21.22):

21.21 LIFE HISTORY OF AN ANGIOSPERM

The sporophyte is clearly dominant in the angiosperm life cycle. The cycle begins with fertilization of the egg and development of the embryo and seed. The seed germinates and grows into a mature plant, which produces the flower. The gametophyte will develop in the ovules and anthers of the flower. Following meiosis in the female reproductive structure, a megaspore goes through three rounds of mitosis and the haploid female gametophyte (megagametophyte) generation commonly emerges as a seven-celled (eight-nucleate) embryo sac. Following meiosis in the anthers (the male counterpart to the ovary), each resulting haploid microspore enters a round of mitosis, producing the young male gametophyte (microgametophyte). It contains a tube cell and a generative cell. The male gametophyte or pollen grain is enclosed in the tough, resistant wall. Following pollination, the pollen germinates and a pollen tube grows toward the ovules. The generative nucleus divides through mitosis, producing two sperm. Upon reaching the embryo sac, one sperm fertilizes the egg cell and the other the nearby binucleate cell. The first becomes the zygote while the second becomes the triploid endosperm. Next, the zygote develops into an embryo, enclosed in a seed. The ovary itself continues development, emerging as the fruit.

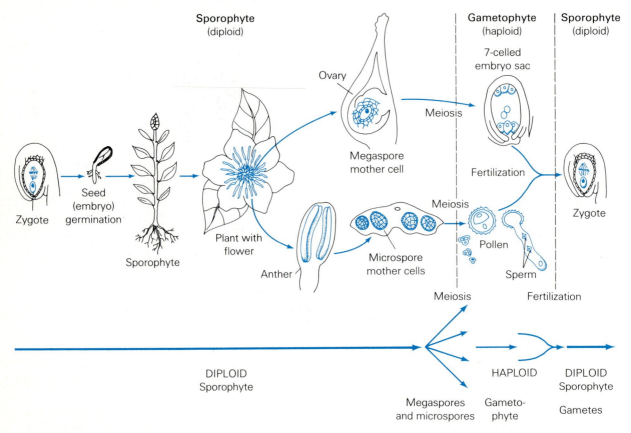

1. floral parts are arranged in spirals;
2. **carpels** (ovule-containing parts) are always superior to (that is, above) the **receptacle** (the enlarged end of floral stem) and other floral parts;
3. carpels and **stamens** (pollen-producing parts) are numerous and not fused;
4. **petals** are separate or completely divided, never fused;
5. the symmetry of the flower is radial (*regular*), not bilateral (*irregular*);
6. the flower is *complete* and *perfect* (has all parts, including both accessory and reproductive).

A Last Look at Angiosperm Diversity.   Now that we have seen a hypothetical evolutionary pathway for the flowering plants, we can return to our old question: Why is there so much diversity in the flowering plants? One answer involves the co-evolution of plants and insects mentioned earlier. We saw that insects may "specialize," that is, be attracted to one kind of flower. A pollinator that has been rewarded with nectar even once may become "keyed" to that particular type of flower, ignoring all others, and will transfer pollen only between flowers of the same species, or even of the same strain. Natural selection may then tend to favor plants with the most distinctive flowers. A new strain with a slightly modified flower can therefore become genetically isolated by the behavior of the pollinating insect, and will not have to become geographically isolated in order to take a new evolutionary pathway. Actually, there are many examples of insect specialization that have led to angiosperm diversity. You may recall from Chapter 17 our discussion of coevolution between the passion vine and a butterfly that lays its eggs in

21.22 A PRIMITIVE FLOWER

The flower of the magnolia, *Magnolia grandiflora,* is considered to be primitive. Like the ancestral angiosperm, the magnolia flower is complete and perfect (has all of the floral parts, including microgametophyte and megagametophyte), and also contains numerous divided parts, including many sepals, petals, stamens, and carpels. In addition, it is radially symmetrical (a line drawn through the center of the flower in any plane will divide it into two equal halves; see inset) and insect-pollinated.

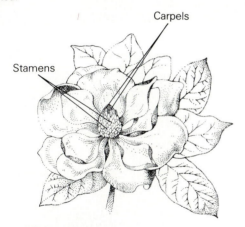

(a) Many divided parts

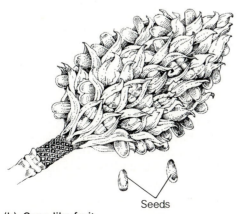

(b) Cone-like fruit

21.23 THE YUCCA AND YUCCA MOTH

When in bloom, the *Yucca,* a desert plant, sends up a long shoot that bears large clusters of white flowers at its end. The *Yucca* depends exclusively on the yucca moth for pollination. As the moth pays its nightly visits to the yucca flowers, it gathers pollen and rolls it into small balls. At successive stops, it pierces the ovary using its sharp ovipositor (egg-burying structure) and injects a number of eggs among the young ovules. It then places a ball of sticky pollen into the opening (perhaps a form of "sealing behavior" since many insects entomb their eggs). Upon hatching, the larvae eat their way through the carpels and reproductive tissue, but they emerge before they destroy too much. The remaining undamaged ovules form seeds and drop to the ground about the same time the moth larvae become pupae. Both moth and plant have evolved to a state where they are entirely dependent on each other. The moth cannot survive without the plant, and the plant is not naturally pollinated by any other means.

the vine's leaves (see Figure 17.18). In another instance, the *Yucca* requires a specific insect, the *Yucca moth,* to accomplish pollination, but the interaction goes far beyond that, as explained in Figure 21.23. Another example of specialization is seen in the familiar "skunk cabbage" (*Symplocarpus*). This marsh plant produces a flesh-colored structure, which bears flowers within. Its pungent, unpleasant odor, like that of rotting flesh, will have you searching through your garden looking for an unfortunate cat. Honeybees avoid this flower, but flies swarm over it. Flies, in fact, lay their eggs in the flower and in so doing pollinate it. Some large, drably colored, nectar-rich, night-opening flowers are pollinated by bats. They have a strong fruitlike or fermenting odor, which attracts their bat pollinators. The coevolving relationship has gone even further. In one species, the pollen has a much higher than usual protein con-

21.24 INSECT POLLINATION DEVICES

Some plants have evolved elaborate devices for assuring pollination. Many flowers, such as *Wedelia,* have markings known as *nectar guides.* These are pigment lines that are easily visible to insects and guide them into the nectaries. In some cases, the guides are not visible to us **(a)**, but show up under ultraviolet light, which is visible to bees **(b).** Perhaps the most intricate and elaborate mechanism is found in orchids. **(c)** One orchid, *Oncidium,* which has specialized anthers with very adhesive surfaces, imitates male bees, and transfer of the pollinium (pollen sac) occurs when an aggressive, territorial male bee is deceived into picking a fight with a flower. **(d)**

Note the pollinium on the bee forehead. When a bee enters the orchid, its head or body brushes the adhesive projections, which become firmly attached. **(e)** As the bee backs out, these break away. The pollinia are then deposited in the next orchid visited by the bee, and new pollen sacs may be picked up. These devices are all excellent examples of how interacting species coevolve, usually, but not always, to their mutual benefit. Sometimes only one of the species is benefited. For example, the insect can be deceived. **(f)** Flowers of the orchid *Ophrys speculum* mimic female wasps. Male wasps try to copulate with them but manage only to pollinate the orchid.

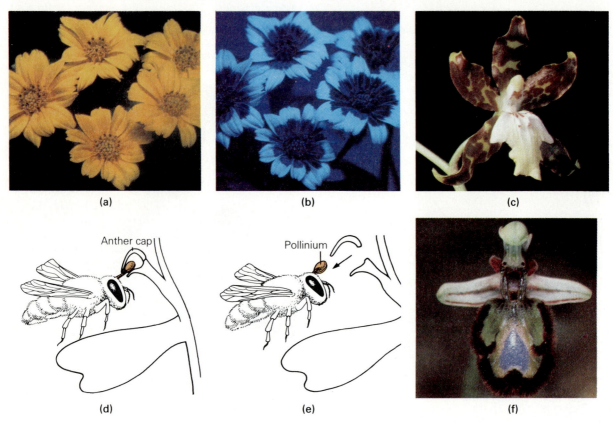

(a) (b) (c)

Anther cap

Pollinium

(d) (e) (f)

tent, which, as it turns out, is an important dietary supplement to the bat pollinator of that species.

This is not to say that all flowering plants are such specialists. Some take a different adaptive route and depend on a variety of animal species. They thus develop traits that are attractive to a number of pollinators. For example, many insect pollinators are attracted to flowers with distinctly separated petals as well as aromatic secretions. The adaptations for attracting a variety of pollinators are, of course, tempered by other factors such as environmental influences (Figure 21.24). It

wouldn't do to have big, floppy flowers on dry, windy slopes.

Among the greatest evolutionary trends has been the remarkable development of plants. They have silently but relentlessly winnowed their way into virtually every available nook and cranny, constantly changing and adapting with such a pervasive influence that they have also markedly affected a wide range of other forms of life. Among the most successful of the plants has been the fascinating group that produces flowers, as we will see in the next chapters.

APPLICATION OF IDEAS

1. Present an argument for dividing the plant kingdom into two kingdoms. Why would one consider this in the first place? What would the kingdoms contain and from what ancestral types would they have emerged?

2. Describe the conditions (climatic, topographic, and so on) that would have to prevail for the bryophytes to return to the prominence they once held. Which animal groups might persist under these conditions?

3. The fossil record clearly demonstrates the presence and even prominence of fernlike but seed-producing plants in the Devonian forests. The significance of these plants to the evolution of today's seed producers is unknown. Present two competing hypotheses that might account for seed evolution having occurred twice. Support the more plausible of the two.

4. The involvement of animals in plant reproduction may have been a key factor in the rise and spread of angiosperms. Discuss this proposition and explore two aspects of angiosperm reproduction that involve animals today. How might these have influenced animal evolution?

KEY WORDS AND IDEAS

Plants are characterized as being multicellular and photosynthetic, most with tissue, organ, and system organization (that is, roots, stems, leaves, flowers). They develop from protected embryos, contain chlorophylls *a* and *b*, produce starches, and have walls of cellulose. There is a pronounced tendency toward a dominant sporophyte phase of life cycles.

DIVISION BRYOPHYTA: NONVASCULAR PLANTS

1. **Bryophytes** include **liverworts, hornworts,** and **mosses,** all of which require water for reproduction. Each has limited tissue differentiation, and all lack significant **vascular tissue** (water-conducting and supporting tissue) which distinguishes them from **vascular plants.** The gametophyte phase of the life cycle is dominant.

2. Bryophytes originated some 350 million years ago. While many are restricted to a moist habitat, some have adapted to desert and rocky conditions and some live in the tundra.

3. There are several theories of bryophyte origin. One suggests a different ancestor from the plants. Another more recent one is that bryophytes branched off the vascular line and through evolutionary simplification lost their vascular tissue and took on other bryophyte characteristics. The fossil record supports this view. The common ancestor to both groups might have resembled the green alga, *Coleochaete*.

4. Bryophytes have large, stationary eggs, produced and protected in the **archegonium.** The swimming sperm are often transported to the archegonium by splashing rain.

5. Bryophytes lack true (vascular) roots, stems, and leaves, but there are organized tissues. The **rhizoids** anchor the plant, and a photosynthetic "stem" **epidermis** surrounds a cylinder of **cortical cells,** which contain thick-walled cells of a supporting **central cylinder** (which conduct water to some extent). The "leaves" are a flattened array of photosynthesizing cells, which in some species secrete a **cuticle** containing waterproofing **cutin.**

Class Hepaticae: Liverworts

1. Haploid liverwort gametophytes originate from spores. They can be leafy or **thallose** (ribbonlike). *Marchantia* has a thallose, branched gametophyte and each branch contains an **apical** (growing) region. Anchorage is provided by simple rhizoids and multicellular **scales.**

2. In asexual or vegetative reproduction, fragments of the body or cuplike **gemmae** break away and grow.

3. Marchantia's sexual structures include separate egg-producing **archegonia** and sperm-producing **antheridia.** They may appear on the same plant **(monoecious)** or different plants **(dioecious)** borne on **archegoniophores** and **antheridiophores.** Sperm are rain-splashed from antheridia to archegonia, where fertilization occurs. A nutritionally dependent diploid sporophyte develops. It produces a sporangium, where spores are produced through meiosis, beginning a new gametophyte phase. Springlike **elators** aid in spore dispersal.

Class Anthocerotae: Hornworts

1. The gametophyte of the hornwort *Anthoceros* is a low rosette form. Each cell contains one large chloroplast and a pyrenoid (similar to green algae). The sporophyte is hornlike and undergoes continuous growth and spore formation, its growth emerging from a **meristematic** base. It may survive independently.

Class Musci: Mosses

1. Mosses make up the great majority of bryophytes. They grow in many habitats including deserts, where, through dormancy, they withstand long droughts. Mosses are important in food chains, and as pioneer plants they are active soil builders. Peat moss (*Sphagnum*) is used for mulching and forms peat, an important fuel.

2. The moss life cycle includes a haploid spore, a threadlike **protonema** stage, and the leafy gametophyte. The latter produces antheridia and archegonia, which are surrounded by protective **sterile jacket cells.** Rain splashes sperm to the archegonia, where fertilization occurs. The nutritionally independent, diploid sporophyte—consisting of a **foot, seta,** and **sporangium,** extends from the gametophyte. The sporophyte is unusual in that it contains numerous **stomata,** surrounded by **guard cells.** Spores are released as the caplike **operculum** breaks away. A new cycle starts as haploid spores are produced through meiosis in the sporangium.

THE VASCULAR PLANTS

1. Vascular plants, often called **tracheophytes,** have tubelike vascular tissues, the **xylem** and **phloem,** a highly dominant sporophyte, and a greatly reduced gametophyte.

2. The presence of efficient, water-conducting vascular tissue made it possible for vascular plants to grow very large, to draw water from deep in the soil, and to transport it up to the leaves.

3. The **rhyniophytes** of the Silurian, among the earliest vascular plants, had leafless, upright shoots arising from **rhizomes** (horizontal stems). These simple stems contained tough, supporting xylem tissue with water-conducting **tracheids.**

4. Evolutionary developments in the rapidly spreading vascular plants included true roots, stems, and leaves. In some, evolution produced nonmotile sperm, new mechanisms for fertilization, and an embryo protected within a **seed.**

5. Vascular plant divisions include the seedless plants—**Psilophyta, Lycophyta, Sphenophyta,** and **Pterophyta** (ferns), and the seed plants—**Coniferophyta** (conifers), **Cycadophyta, Ginkgophyta, Gnetophyta (gymnosperms),** and **Anthophyta, (Angiosperms** or flowering plants).

Division Psilophyta

1. The dominant sporophyte of psilophytes, or whisk ferns, has anchoring rhizoids but no true roots, photosynthetic true stems, and primitive sporangia. The tiny separate gametophyte produces sperm and eggs in antheridia and archegonia, and upon fertilization the young sporophyte becomes independent. Psilophyte fossils date back 375 million years, but their phylogeny is not well known.

Division Lycophyta: Club Mosses

1. The lycopod sporophyte **(ground pines** or **club mosses)** has true and **adventitious** roots (arising from the stem), true stems, and true leaves. The leaves are **microphylls** (derived from the epidermis, as opposed to larger **megaphylls,** derived from branching stems). Spores form in many **sporophylls** in the leaf axils or in a single, terminal **strobilus.** Separate gametophytes produce the gametes, and after fertilization the young sporophyte becomes independent.

2. Spore formation is either primitive and **homosporous** (spore of one type) or advanced and **heterosporous** (spores of different types). The latter includes **microspores** (small) and **megaspores** (large). The first produces **microgametophytes,** which give rise to male gametes, the second, **megagametophytes,** which give rise to female gametes. Microspores and megaspores are produced in **microsporophylls** and **megasporophylls,** respectively. Heterospory is characteristic of advanced plants.

3. In *Salaginella*, events in the microsporophylls are as follows:
   a. meiosis forms four microspores;
   b. microspores divide mitotically, forming two-celled microgametophytes;
   c. microgametophytes form antheridia, which produce sperm.

Events in the megasporophylls include the following:

   a. meiosis forms four megaspores
   b. each megaspore forms a megagametophyte;
   c. megagametophytes form archegonia, where egg cells develop.

4. During the Devonian and Carboniferous periods, lycopods formed forests of large trees, the remains of which form fossil fuels.

Division Sphenophyta: Horsetails

1. One surviving genus of Sphenophyta, *Equisetum,* forms slender photosynthetic, branched or unbranched stems with whorls of microphylls. Spore formation occurs in a strobilus. The gametophyte is tiny and separate. Sphenophytes contributed to the lush Carboniferous forests.

Division Pterophyta: Ferns

1. Ferns are still widespread and numerous. The sporophyte consists of thick rhizomes (underground stems), which give rise to **fiddleheads** that uncoil to form large, highly divided leaves (megaphylls).

2. Spores are produced in sporangia, often clustered into **sori** and sometimes covered by an **indusium.** Elaborate spore-ejecting mechanisms involve the **annulus** and **lip cells** of sporangia.

3. When a haploid spore germinates, it forms a photosynthetic gametophyte—the **prothallus,** which forms archegonia and antheridia. In some, the hormone **antheridogen** helps prevent self-fertilization. Following fertilization, the young independent sporophyte emerges.

4. Ferns were among the dominant species of the Paleozoic forests.

THE SEED PLANTS

1. The demise of the ancient forests and the rise of the seed plants may have been brought about by vast geological changes—upheavals that raised the land and subjected it to drying and cooling trends. The gymnosperms were efficient at gathering and conserving water; they incorporated the gametophyte into their protective tissues.

2. The male gametophyte took the form of tough, resistant **pollen** grains, transported to the female gametophyte via winds, water, and animals. The swimming sperm was eventually replaced by a nonmotile form that was carried to the egg in fluids produced by the plant.

3. Seeds consist of an **embryo, stored foods,** and hardened **seed coats.** They are resistant and can remain dormant for long periods, or until specific, favorable conditions are present.

4. The new evolutionary trends in plants included heterospory, a highly dominating sporophyte, and an incorporated gametophyte.

The Gymnosperms

1. **Gymnosperm,** now an informal term, includes ginkgos, cycads, gnetophytes, and conifers.

2. Ginkgophytes are composed of a single, broad-leaved species, *Ginkgo biloba,* the **maidenhair tree.** It is wind-pollinated, produces seeds in cones, and retains the swimming sperm for which it provides a fluid environment.

3. The **cycads** are palmlike plants with separate sexes. Pollen is produced in large conelike strobili. Flagellated sperm persist. Cycads were prominant in the Triassic, some 200 million years ago.

4. Gnetophytes reproduce as do gymnosperms, though some have angiosperm characteristics. One species has broad leaves, one produces flowerlike cones, and another has advanced water-conducting elements called **xylem vessels.**

5. **Conifers** form great northern and mountain forests in the northern hemisphere. Pines are the most common tree. The leaves, typically needle- or scalelike, are replaced gradually. The leaf form is an adaptation to dryness. Conifers bear separate male (pollen-producing) and female (ovule-producing) cones.

6. After meiosis and microspore production in the male pine cone, mitosis in each produces a two-celled gametophyte with a **generative cell** and a **tube cell.** The gametophyte becomes enclosed in a tough covering, forming a pollen grain. Before fertilization, the generative cell divides to form two nonmotile sperm. The tube cell guides the development of a **pollen tube,** which penetrates the female gametophyte, and one sperm fertilizes the egg.

7. In the female pine cone, the gametophyte develops in the **ovule**—a megasporangium surrounded by **integuments** or **seed coats.** After meiosis, a single surviving megaspore produces the megagametophyte, which develops archegonia, each with one egg. Sperm enter through a **micropyle.**

THE END OF AN ERA

1. The gymnosperms and the dinosaurs fell from prominence together at the close of Mesozoic. The period is marked by the rise and divergence of the angiosperms, but in recent times the conifers appear to be making a comeback.

The Angiosperms:
The Rise of the Flowering Plants

1. Flowering plants are far more diverse than conifers. Flowering plants probably originated in the Mesozoic, but until the Cenozoic they were apparently restricted to drier uplands where fossils rarely formed.

2. The sudden rise of flowering plants is attributed to a rapidly changing climate accompanied by mountain-building and **plate tectonics,** or **continental drift.** The latter began some 200 million years ago and represented a northerly shift of the continents.

3. According to the **Alvarez theory,** a third great geological event, the collision of a gigantic asteroid with the earth, darkening the atmosphere with dust and creating massive weather changes, brought about many extinctions.

4. The replacement of gymnosperms by angiosperms may have also been due to the angiosperms' greater variability, adaptiveness, and high rate of speciation. An additional factor was efficient insect pollination and the coevolution of pollinators and plants.

The Angiosperms Today

1. The quarter million angiosperm species form two major classes, the Monocotyledonae (monocots) and Dicotyledonae (dicots). They differ in seed structure (number of cotyledons), flower anatomy, and vascular tissue distribution in stem and leaf.

2. The ancestral angiosperm is represented by family Magnoliaceae and specificially by the magnolia. The fossil record is scant and of little help, so plant phylogeny is based mainly on comparative features of living angiosperms.

3. The primitive floral condition includes spiral flower parts, **carpels** above the **receptacle** and other parts, numerous and free carpels and **stamens,** separate **petals** (not fused), **radial** symmetry *(regular), perfect* and *complete* flowers (has all parts, including male and female).

4. The great diversity of flowering plants may be attributable in part to the continuing adaptation (coevolution) of the plant and its animal pollinator. In the ongoing competition for pollinators and the increasing efficiency of pollination, natural selection may have favored variants.

REVIEW QUESTIONS

1. List five significant characteristics of plants. Which if any is (are) unique to plants only? (504)

2. In what way are the bryophytes "tied" to the ancestral watery environment? (505)

3. Has the absense of vascular tissue meant failure to the bryophytes? Explain. (505)

4. List four different environments inhabited by bryophytes. (505)

5. The current theory of bryophyte origins suggests that they shared common ancestors with the vascular plants. How, according to this, does one account for the simplicity of bryophytes and their lack of vascular tissue? (505)

6. What evidence recommends the green alga *Choleochaete* as a common ancestor to plants? (505)

7. What are "true" roots, stems, and leaves? What characteristic of bryophytes permits them to get along fine without these? (505)

8. Characterize the life cycle of the bryophyte. How does this compare with the general trend in other plants? (507)

9. Describe the body form of *Marchantia*. What purpose do the gemmae serve? (507)

10. In what structures do the liverworts produce sperm and egg? What kind of help do sperm get in reaching the eggs? (507)

11. How does the hornwort sporophyte differ nutritionally from the liverwort sporophyte? How does its growth differ from that of other bryophyte sporophytes? (507)

12. Describe a specific desert adaptation of mosses. (508–509)

13. List two ecologically important activities of mosses. (509)

14. Briefly summarize the life cycle of a typical moss, beginning with the germination of haploid spore. (509)

15. Explain how the presence of vascular tissue has influenced the distribution and the size of vascular plants. (510–511)

16. Describe the body form of the rhyniophytes. Why are they biologically important? (511)

17. List four divisions of vascular plants that do not produce seeds. In what two ways is reproduction in these groups primitive? (511)

18. Briefly describe the body form of the whisk fern, *Psilotum nudum*. Which of its features would one consider primitive? (513)

19. Describe the sporophyte of *Lycopodium*. How do its leaves, called *microphylls*, differ from *megaphylls?* (514)

20. In what way is spore formation in *Selaginella* advanced over spore formation in *Lycopodium?* How does the location of the gametophyte phase differ? Which of the two species appears to be the most similar to seed plants? (515)

21. List the steps involved in the formation of sperm in *Selaginella*. Include the terms *microsporophyll, microspore,* and *microgametophyte.* (515)

22. List the steps involved in the formation of eggs in *Selaginella*. Use terms equivalent to those in question 25. (515)

23. Describe the body form of a typical horsetail. Is its reproductive cycle advanced or primitive? Explain. (515)

24. Characterize the distribution of ferns. What does this indicate about their evolutionary success? (515, 518)

25. Describe the body form of a typical fern, using the terms *rhizome, fiddlehead,* and *divided megaphylls.* (518)

26. Using the terms *sorus, indusium, sporangia, annulus,* and *lip cells,* describe the structures responsible for spore formation and dissemination in a typical fern. (518–519)

27. Briefly summarize the life cycle of a fern, starting with the haploid spore. (519)

28. Summarize the geological changes that may have prompted the end of the ancient Paleozoic forests. How might the small independent gametophyte have added to the survival problems of the early vascular plants? What plants replaced the lycopods, ferns, and psilophytes? (520)

29. Explain how a nonmotile sperm gets from the pollen grain to the egg cell. (521)

30. What constitutes a seed? How might the evolution of seeds given the gymnosperm an advantage over earlier vascular plants? (521)

31. Which of the gymnosperm divisions is most prominent today? In what regions are most of these to be found? (524)

32. In what ways are the ginkgo and cycad reproductively primitive? (522)

33. List three characteristics of certain gnetophytes that make this group seem related to the angiosperms. (522–523)

34. What specific adaptation do the needle- and scalelike leaves of the conifer represent? In what other ways are conifer leaves adapted to this situation? (524)

35. Briefly summarize the life history of a pine. (524)

36. When did the angiosperms rise to prominence? Where were their forebearers living during the time gymnosperms were prominent? (525)

37. Briefly describe three factors that might have produced vast changes in life at the close of Mesozoic. (526)

38. Name the two groups of angiosperms and list several characteristic differences between them. (526)

39. In what way do the evolutionary terms "primitive" and "advanced" present usage problems? Are individual species entirely primitive or entirely advanced? How might one avoid this semantics problem? (527)

40. List four flower characteristics that would be considered primitive. Suggest a specific flower to which they apply. (529)

41. Explain how the ongoing coevolution of insect pollinator and plant might increase diversity. Describe an extreme case of specialization between flower and insect. (529)

Reproduction and Development in Plants

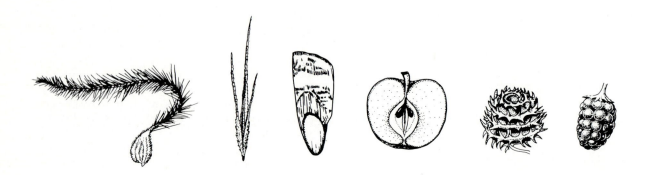

As plants began to expand over the earth's often parching surface, they were met by a host of unexplored opportunities. At the same time, they encountered perhaps as many real threats to their very existence. Among those threats was the problem of reproducing in the dry air; as we will see here, that problem was largely solved by flowers.

In the flowering plants—the angiosperms—the role of sexual reproduction falls on the flower itself. The flower is, to many of us, the most notable part of an angiosperm, and we extol its beauty and fragrance. We must keep in mind, however, that they are not simply decorations; they have a job to do. Furthermore, we should keep in mind that after fertilization, the flower's petals drop off, and part of the flower enlarges to become the seed-bearing *fruit*. The seed, we know, is the flowering plant's investment in the future. So let's begin our consideration of sexual reproduction in the angiosperms where it begins—with the flower.

THE FLOWER

A typical flower is composed of as many as four organs, usually occurring in **whorls** around the **receptacle** or **base** (Figure 22.1). The term **whorl** means simply that the floral parts are repeated in a circle. Actually, each member of a whorl is a modi-fied leaf, but some are much more modified than others. The outermost whorl consists of the **sepals,** which are usually green and leaflike. They surround and protect the flower bud before it opens and are commonly photosynthetic. The whorl of sepals is collectively known as the **calyx** (meaning "cup").

Just within the calyx lie the **petals;** these are a second whorl of floral parts. They are often large and colorful, but most are still somewhat leaflike. A whorl of petals is known as a **corolla** (meaning "garland").

Within the corolla is found a third region, the **androecium** ("house of man"), a whorl of **stamens** called the **microsporophylls,** which, in this case, are roughly the equivalent of a male reproductive structure (see Chapter 21). That is, they have the capacity to produce cells that later form the male gametophyte that produces the sperm. Stamens are also leaf derivatives, but they are so modified that their ancestry is not apparent. Each stamen includes a slender stalk known as a **filament,** and cross sections through the filament reveal the presence of a leaflike vascular system. Atop the filament is the **anther,** where microspores (future pollen grains) are produced.

The fourth region of the flower, the **gynoecium** ("house of women"), contains one or more **megasporophylls** or **carpels,** as they are known in flowering plants. Because of its shape, the gynoecium is traditionally known as a **pistil** (a corruption of the term "pestle" of the chemist's "mortar and pestle," although today we might think of it more as bowling pin-shaped). The pistil is commonly a whorl of parts, but it is often too modified to be recognized as a leaf-derived structure unless one considers its evolutionary and embryonic development. In the grand diversity of flowers, pistils may be single or multiple, fused or separate. A pistil comprises one or more carpels. Further, each of the carpels consists of three parts: the **ovary** (called the *ovulary* by some, since it functions quite differently from the animal ovary), the **style,** and the **stigma.** The ovary produces the ovules. Extending from the ovary is the style, a stalk that supports and elevates the stigma, an enlarged structure at its tip. The stigma is often hairy or sticky and specializes in receiving pollen.

In summary, the flower consists of a calyx (whorl of sepals), corolla (whorl of petals), androecium (whorl of stamens), and gynoecium (whorl of pistils containing carpels). The last two are the most intimately involved in reproduction. Sepals and petals are called **accessory parts,** since they play no direct role in sexual reproduction. However, the petals can be quite important to the process, particularly in insect-pollinated species (also see Chapter 21).

22.1 A GENERALIZED FLOWER

Flowers generally consist of four major parts supported by a **receptacle:** the *calyx* contains the leaflike *sepals;* the *corolla,* larger colorful petals; the *androecium,* a whorl of *stamens,* each with a *filament* and *anther;* and the *gynoecium* or *pistil,* one or more *carpels,* each with a *stigma, style,* and *ovary.* **Nectaries** are located at the petal bases.

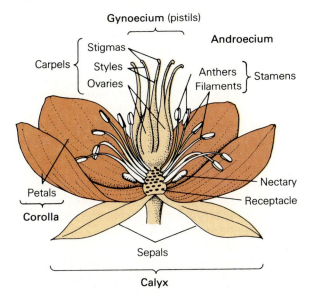

Gynoecium (pistils)

Androecium

Carpels { Stigmas / Styles / Ovaries

Anthers / Filaments } Stamens

Petals

Corolla

Nectary

Receptacle

Sepals

Calyx

22.2 A FLY-POLLINATED FLOWER

Rafflesia flowers are among the world's largest, up to a meter in diameter. They have the odor of rotting meat and attract flies and other carrion insects, which act as pollinators.

Variations in Flower Structure

Flowers can be beautiful, elegant, and fragrant. Some can also produce sweetness of another sort—**nectar.** Color, fragrance, and nectar all have a role in flowers that depend on animals such as insects or birds for transferring pollen from one flower to another. The color and fragrance attracts pollinators to the flowers with the promise of sweet, energy-rich nectar (usually secreted from **nectaries** near the base of the flower). We, too, are attracted to flowers for the same reason. It is interesting that we are attracted to the same structures as are the insects, despite the vast differences in our perceptive abilities. We do not see what they see. For example, insects can perceive ultraviolet light, so they may find purple patterns in flowers that to us appear white. And since most insects can't see red at all, a beautiful red rose appears to them as black. Flower structure, appearance, and scent are often modified to attract and assist specific kinds of pollinators, whether beetles, flies, moths, butterflies, bats, or other animals (Figure 22.2 and Table 22.1).

Natural selection has molded, warped, changed, hidden, and amplified flower parts in endless ways. For example, there tend to be clear differences even between monocots and dicots. As mentioned in the last chapter (see Figure 21.20), most dicots have floral parts arranged in fours and fives or multiples of these numbers, while monocot flowers have parts arranged in threes and multiples of threes. The most primitive floral types have well-defined parts; the sepals, petals, carpels, and stamens occur in a **radially symmetrical** pattern—that is, circular and disklike so that any line drawn through the center of the disk produces equal halves. Highly evolved flowers, in contrast, are often **bilaterally symmetrical** or **irregular;** they have right and left halves that are essentially mirror images (as is true of humans and most other animals). A few highly advanced flowers seem to have lost their symmetry altogether.

Other variations include flowers with missing parts. A flower with one or more basic parts absent is called **incomplete,** as opposed to the **complete** flower, which has all of the usual parts. If either (or both) of the sexual parts, the pistils or stamens, is absent, the flower is referred to as **imperfect.** If they are both present, the flower is **perfect.** Obviously, flowers that are imperfect must also be incomplete, and flowers that are complete must be perfect. Cornflowers are imperfect and are either **pistillate flowers** (female) or **staminate flowers** (male), but both occur on the same plant. In date palms, pistillate and staminate flowers grow on separate trees, and to provide for pollination a date-grower usually plants one staminate tree per 10 or so pistillate trees.

In some instances flowers are grouped together into clusters, or **inflorescences,** that take various forms such as the heads of daisies, marigolds, and sunflowers. Thus a single sunflower is really a bouquet of flowers. **Composite flowers** are inflorescences composed of tiny, individual flowers, each of which produces a single seed-bearing fruit. In the daisy, marigold, and sunflower head, the outer rows of flowers are in fact imperfect, sexless flowers, serving mainly to attract pollinators.

There is also a very large group of plants that are wind-pollinated. These include the grasses, such as wheat, rye, and corn. It may seem strange that grasses produce flowers, but keep in mind that grasses are angiosperms. Further, grass flowers are similar to those of insect-pollinated flowers. They have stamens and pistils, but they are plain, small, and inconspicuous (Figure 22.3). There would be little adaptive value in such energy-costly amenities as colorful petals and sweet nectaries in the wind-pollinated flowers. Instead, these flowers produce copious amounts of pollen and tend to grow in dense groupings, traits that would enhance the effect of the wind.

Sexual Activity in Flowers

Earlier we noted that a few of the primitive vascular plants (for example, *Selaginella;* see Chapter 21), and all seed plants, are heterosporous. That is, they produce two kinds of spores, megaspores and microspores. We also saw that as plants evolved, the gametophyte phase of the life cycle became more and more reduced and dependent on the sporophyte. In the flowering plants, the female gametophyte and at least the younger phase of the male gametophyte are tucked away in the ovaries and anthers of the flower. With these points in mind, let's look into the process of sporogenesis, the formation of spores, and their development into the gametophyte in flowering plants.

The Ovary, Megaspores, and the Megagametophyte.   Within the soft tissues of the flower's ovary are found the young ovules. The ovules originate from a surrounding ovarian tissue known as the **placenta** and for a time remain attached to it by a stalklike **funiculus.** Each ovule consists mainly of a megasporangium, or **nucellus,** as it is known in flowering plants, and one or two **integuments**—skinlike protective coverings that will much later form the seed coat. At the exposed end of the integument is a small opening called the **micropyle.** Each ovule contains one large cell called the **megaspore mother cell,** and it is this cell that will concern us here.

The megaspore mother cell, like its male counterpart in the stamen, the **microspore mother cell,** will undergo meiosis, producing haploid spores.

TABLE 22.1

FLOWERS AND ANIMAL POLLINATORS

Animal	Visual Cues	Chemical Cues
Beetles	Not significant, flowers dull colored or white	Strong odors—fruity, spicy, or foul
Bees	Bright colors; yellow or blue (ultraviolet perception); highly divided floral parts	Strong fragrance
Flies	Large flowers; dull, flesh-colored	Musky to rotting odors
Moths and butterflies	Bright colors—reds, oranges, yellows, blues	Strong fragrance
Birds	Bright colors—reds and yellows	Copious, sugary nectar; little odor
Bats	Color not significant (night flyers)	Copious nectar; fruity, fermenting odors

22.3 WIND-POLLINATED FLOWERS

An obvious difference between insect-pollinated flowers and the wind-pollinated flowers is the absence of color and odor in the latter. Also absent are sepals and petals. Upon opening, the flower reveals feathery, plumelike, pollen-catching stigmas, stamens usually in threes, and pollen that does not tend to stick together as it does in insect-pollinated species. Wind pollination is only efficient where many plants of the same species are clustered.

Wild oat

Bog cotton

Typical of plants, these meiotic products are not gametes, but instead they will enter mitosis to produce the gametophyte, which, in turn, will produce the gametes. Let's follow these events more closely (Figure 22.4).

Meiosis in the megaspore mother cell involves the usual two divisions, so at the end of the process there are four haploid spores—the megaspores. Usually, three of the megaspores simply disintegrate. The surviving megaspore enlarges considerably before its haploid nucleus undergoes three successive mitotic divisions, giving rise to the gametophyte or, more accurately, the **megagametophyte** (the female gametophyte generation of the flowering plant). If our arithmetic is correct, the megagametophyte will contain eight identical haploid nuclei, all in one greatly enlarged cell (one divides into two, two divide into four, and four divide into eight). With the mitotic events completed, the megagametophyte becomes known as an **embryo sac,** which may be a relief to those having trouble with pronunciation.

The next events may vary according to the species, but we will consider what commonly happens. In most cases the eight nuclei are equally distribut-

ed to the two ends of the embryo sac, after which two of the nuclei, one from each end, migrate to the center of the cell. Finally, new cell walls are laid down and the cytoplasm is divided into seven separate, unequal cells. The result of all this is that one cell, often the largest one, ends up in the center with two nuclei. This binucleate cell will later give rise to the **endosperm** (a tissue containing stored food) in the seed, but for now it is known as the **central cell** (and sometimes as the **endosperm mother cell**), and its nuclei are called **polar nuclei.** Of the remaining cells, one will become the **egg cell,** the single female gamete of the megagametophyte. There will be no more nuclear divisions in the embryo sac until after fertilization. At that time, both the egg cell and the central cell will participate in a unique and impressive event known as **double fertilization.**

Events in the Anthers: Microspores and the Microgametophyte.

While the embryo sac has been undergoing its development, similar but less complex events have been occurring in the anthers, the sites of **microsporogenesis** (Figure 22.5). Typically, each anther contains four microsporangia—chambers that contain numerous diploid **microspore mother cells,** each of which will enter meiosis. By the end of meiosis, each microspore mother cell will have produced four microspores. The haploid nucleus within each microspore will then undergo mitosis, producing two cells, a **generative cell** and a **tube cell,** both of which remain within the original microspore. The two cells represent the young male gametophyte, or **microgametophyte.** The cells of a microgametophyte are still haploid since mitosis doesn't change the final chromosome number. Meanwhile, each microgametophyte develops a tough, resistant coating and becomes a grain of pollen. In many species, there will be no further cellular activity until the pollen finds its way to the stigma of a flower.

Pollination and Double Fertilization.

Technically, **pollination** occurs when pollen is deposited on a receptive stigma. The source of the pollen may be the male parts of the same flower, other flowers on the same plant, or another individual. We are aware of the dangers of inbreeding, and since self-fertilization is obviously inbreeding of the severest sort, it is not surprising to learn that some plants have ways of avoiding self-pollination, such as the position of their anthers, the timing of microspore and megaspore production, or

22.4 MEGASPOROGENESIS

(a) The female gametophyte develops within the ovules. The ovule, enlarged in the inset, contains the megaspore mother cell and its accompanying tissues, the nucellus and enfolding integuments, all arising from the placenta. **(b–e)** The megaspore mother cell undergoes meiosis, producing four haploid megaspores. **(f)** Three megaspores disintegrate while the fourth enlarges. **(g–i)** The surviving haploid nucleus then undergoes mitosis three times, forming an embryo sac with eight nuclei. With cell wall formation, the embryo sac becomes a seven-celled structure. The large, binucleate cell with the polar nuclei is the central cell, which will become the endosperm mother cell. One of the smaller cells becomes the egg cell.

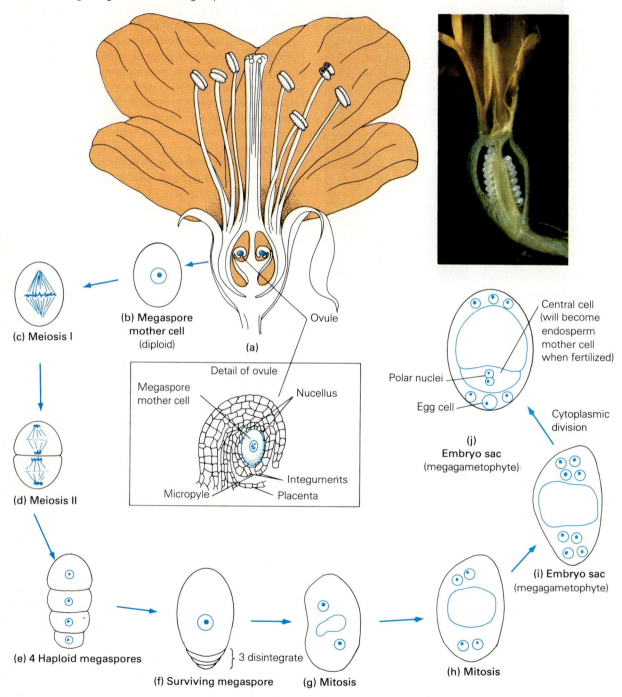

(c) Meiosis I

(b) Megaspore mother cell (diploid)

Ovule

(a)

Detail of ovule

Megaspore mother cell

Nucellus

Integuments

Micropyle

Placenta

(d) Meiosis II

(e) 4 Haploid megaspores

(f) Surviving megaspore

} 3 disintegrate

(g) Mitosis

(h) Mitosis

(i) Embryo sac (megagametophyte)

Cytoplasmic division

Central cell (will become endosperm mother cell when fertilized)

Polar nuclei

Egg cell

(j) Embryo sac (megagametophyte)

even physiological incompatibilities. Nonetheless, strangely enough, self-pollination is a regular event in some species (some hypothetical reasons for this are explored in Essay 22.1.)

When pollen germinates, the tube nucleus produces what is called a **pollen tube,** a long slender tube of cytoplasm surrounded by a growing wall. The pollen tube grows through the stigma and into the long style (Figure 22.6). The tube cell produces enzymes that actually digest the tissue ahead of the tube. It is generally during this advance (earlier in some species), that a vital act occurs: the generative cell undergoes mitosis, producing two **sperm cells.** With this event, the male gametophyte is fully mature and there is no further mitosis.

Finally, the tube penetrates the ovule at the tiny, porelike micropyle. The two sperm cells then move from the tube and enter the embryo sac, where that unique *double fertilization* mentioned earlier occurs. One sperm fertilizes the egg cell and a diploid zygote is formed. This event ushers in the next sporophyte generation in the life cycle of flowering plants. The fertilized egg, or zygote, will develop into the plant embryo. The second sperm penetrates the central cell, where its chromosomes fuse with those of the two polar nuclei already present, producing a *triploid* **primary endosperm nucleus.** (While a triploid primary endosperm nucleus is common in many species, it may in others be diploid, or even tetraploid, depending on the number of polar nuclei.) The endosperm provides food for early development after seed germination occurs. Double fertilization, as seen here, represents one of the basic differences between angiosperms and gymnosperms. You may recall from the last chapter that only one sperm is functional in the gymnosperms.

After fertilization, the flower begins to change, and its beauty may fade. Commonly, those parts that do not sustain the seeds begin to wither and fall off. Then the ovary itself will swell, sometimes to enormous proportions. This begins the development of **fruit,** another structure characteristic of angiosperms. In everyday usage, "fruit" usually refers to a sweet and juicy structure such as the

22.5 MICROSPOROGENESIS

(a) Development of microspores occurs within the microsporangia, each of which contains many microspore mother cells. (b) Each microspore mother cell undergoes meiosis, producing four haploid microspores. (c) Individual microspore nuclei divide once through mitosis, producing a microgametophyte made up of a tube cell and a generative cell. These become enclosed in tough coverings, forming a pollen grain. Later, with the division of the generative cell, the male microgametophyte will be complete.

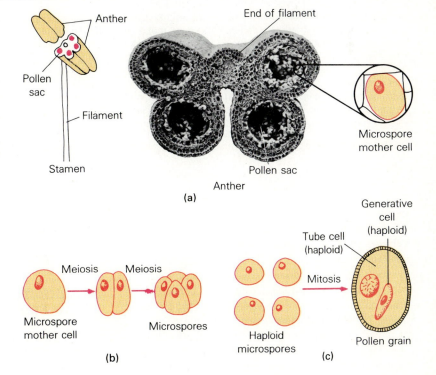

Apomixis: Seed Production Without Sex

Sexual reproduction in higher plants is always through seeds, but interestingly, not all seeds are produced sexually. In some instances they are produced asexually—that is, without the union of sperm and egg nuclei. This form of asexual reproduction in plants is known as *apomixis,* and the phenomenon can occur in several ways. Usually, however, the megaspore mother cell simply skips meiosis, thus retaining its diploid number, and the seed embryo begins development without the intervention of the male gamete. Of course, the new embryo has exactly the same chromosomes and genetic characteristics as the single parent sporophyte on which it is borne. Thus, all of the descendants of an apomict line are essentially clones.

It is interesting that while completely asexual plant species are very common, none of these species appears to have been around very long in evolutionary terms. Unless the environment remains exceptionally stable, none of them can be expected to hold out in the evolutionary long run. Most apomicts, however, can be expected to continue because they can also reproduce sexually when necessary. It's as though these species are hedging their bets, going the asexual route when it is advantageous but always maintaining its main evolutionary sexual line. Let's see how the theorists explain this versatile but odd behavior.

Natural selection is indeed opportunistic and seems to proceed on the profound and unbiased principle that whatever works, works. And sometimes asexual reproduction works extremely well for a plant. For example, a particularly successful genotype can be saved from the vagaries of genetic recombination that occur during meiosis and when unrelated haploid cells are brought together in fertilization. An apomict can pass 100 percent of its genes along to all its offspring, and this favored genotype can quickly spread through a habitat. But, one might ask, under what conditions could a lack of variability be adaptive? Isn't variability in a population the key to success?

Plants living under drastic conditions, such as those in arctic regions, cling to existence through the most precarious and fragile adaptations. Any disruption of its precise genotypes could spell disaster for a population. Further, in such rigorous environments, plants may be so far apart as to make pollination difficult, so they have adapted to the problem by simply skipping sex for a time.

Apomixis may also speed up evolution by increasing the chances of successful hybridization—the union of two species into one. Plants readily hybridize, but all too often the hybrid is sterile. Its efforts to reproduce fail during meiosis, when, early in prophase I, the chromosomes are unable to find compatible partners or homologues with which to pair (see Chapters 11 and 17). One answer to the problem is straightforward. After the successful union of two species, the plant simply becomes an apomict, producing seeds without troublesome meiosis. This has happened time and again with the highly variable Kentucky bluegrass.

The problem with apomixis should be apparent. When the environment changes, as it inevitably will, the apomictic population may not contain the variability needed to assure even minimal survival. Those plants in sexual phases, however, will be shuffling genes, producing new combinations, and leaving variable offspring. Thus, these will have a good chance of success.

Since sexual organisms have more variability built into their genetic systems, they have the short-term advantage and will once again take over the habitat. Then, in response to changing environmental stimuli, the newly succeeding variants may then spin off a new apomict variety as conditions warrant. But from the fact that strictly apomictic groups generally don't have long evolutionary histories, we can cheerfully conclude that sex is the safer bet in the long run.

---

familiar apple, grape, or banana. Botanically, however, the fruit is the ripened or mature ovary. Whereas it does commonly consist of a sweet fleshy structure, it may also include any other mature ovary such as the familiar string bean, cereal grain, pumpkin, or tomato. It even includes certain hardened parts of the coconut, walnut, and the so-called "seed" of the sunflower. For a more complete look at fruits and their development, see Essay 22.2.

The United States Supreme Court at one time was required to decide whether the tomato was a fruit or a vegetable. It seems that Congress had passed a 10% tariff on vegetables, but there was no tariff on fruit. The nine learned judges, wise men all, decided that the tomato was not sweet enough to be a fruit, and was cooked for dinner often enough to be a vegetable.

Development of the Embryo and Seed

After fertilization, the new diploid zygote, the primary endosperm nucleus, and all the associated maternal tissues begin mitosis and the differentiation that will produce the plant **embryo** and its life support systems. The finished product—the

22.6 POLLINATION AND FERTILIZATION

(a) Pollination occurs when pollen lands on the stigma. It is followed first by pollen germination, then production of a pollen tube. The pollen tube is a prerequisite to fertilization. (b) As the pollen tube grows down through the style, the generative cell undergoes mitosis, producing two sperm cells. The tube cell then penetrates the ovule and the two sperm enter the embryo sac. (c) Typically, one sperm fertilizes the egg cell, forming a zygote, while the other unites with the diploid nucleus of the central cell, forming the triploid endosperm. (d) Nuclear fusion completes fertilization.

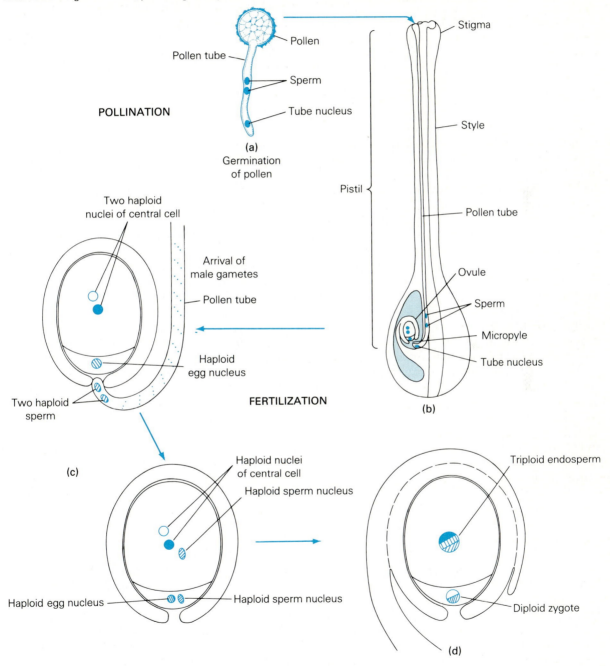

POLLINATION

Pollen
Pollen tube
Sperm
Tube nucleus

(a)
Germination
of pollen

Stigma
Style
Pistil
Pollen tube
Ovule
Sperm
Micropyle
Tube nucleus

(b)

Two haploid
nuclei of central cell
Arrival of
male gametes
Pollen tube
Haploid
egg nucleus
Two haploid
sperm

FERTILIZATION

(c)

Haploid nuclei
of central cell
Haploid sperm nucleus
Haploid egg nucleus
Haploid sperm nucleus

Triploid endosperm
Diploid zygote

(d)

seed—will consist of (1) an embryo, (2) some kind of food supply, and (3) a protective **seed coat.**

Seed development varies widely among flowering plants, so we will follow the process in only a few cases, beginning by contrasting the events in dicots and monocots.

Seed Development in the Dicot.   The most familiar dicot seeds are probably foods such as peas, beans, and peanuts—seeds often eaten without much alteration. The embryo, a miniature sporophyte sporting two large food-laden cotyledons (a storage organ), takes up most of the dicot seed. You can see this for yourself in the peanut. Remove the shell (fruit) and examine the seed within. Most of the structure consists of the two cotyledons. Separate the cotyledons very carefully and you can see the rest of the embryo, a miniature plant, still attached to one side. Let's now look into the events that lead up to seed formation.

After fertilization, the ovule contains an embryo sac with a diploid zygote, a primary endosperm nucleus, and several layers of surrounding and supporting diploid cells of the maternal tissue (Figure 22.7a). For convenience, what follows can be divided into three events: developments in the endosperm, growth of the cotyledons, and growth of the rest of the embryo, or **embryo axis,** as it is called.

Within the primary endosperm, the nuclei divide continuously until a multinucleate mass surrounds the zygote (see Figures 22.7b and c). (This stage is carried to an extreme in the coconut, a monocot, since "coconut milk" is nothing but a mass of white, fluid endosperm cytoplasm with free nuclei floating in it.) Ultimately, in most dicot seeds, cell walls are laid down around the endosperm nuclei.

The zygote itself does not usually begin to develop into an embryo until the endosperm has grown to a considerable size. Then the zygote begins a series of cell divisions that will eventually produce the embryo. The first few divisions produce a vertical row of cells. Then, at the inner end of the row, cell divisions speed up and, more important, begin to occur in many planes, producing a three-dimensional ball of cells (see Figure 22.7d). Unlike the cells in animal embryos, which lack walls and grow between, around, and over each other as the embryo takes form, plant embryo cells must do this chiefly through changes in cleavage planes. At any rate, it is this ball of cells that becomes the embryo. A **suspensor** is formed by cells at the other end of the row that divide much more slowly. One becomes a large **basal cell,** swol-

len by the uptake of water. The suspensor anchors the embryo in place and aids in the transfer of food to the rapidly growing embryo.

Up to this point, development has been about the same in monocots and dicots, but now we see a fundamental difference between the two groups. In the rapidly growing dicots, the cells of the embryo form a two-pronged heart-shaped structure (see Figures 22.7e and f). In the monocots, equally rapid cell division produces a simple, elongated structure. The two lobes of the heart will produce the two cotyledons of dicot plants; the single growth will form the cotyledon of monocots.

In beans and many other dicots, the cotyledons continue to grow, absorbing all or nearly all of the endosperm and filling most of the space of the growing embryo sac. As they grow, the planes of division change once more, and the result is that the cotyledons form a U-turn, growing back toward the base of the seed (see Figure 22.7g).

Meanwhile, just below the cotyledons, the remainder of the embryo axis begins to take shape. Where the cotyledons divide, a naked dome of tissue, the **shoot apical meristem,** forms. All **meristematic tissue** in plants is made up of cells that for now remain simple and uncommitted, a source of cells from which new tissues can be derived as growth continues. In some embryos, the shoot apical meristem is surrounded by a **plumule,** tiny embryonic leaves that emerge from below. Since the plumule develops above the base of the cotyledons it is also referred to as an **epicotyl.** At this time the embryo's outer surface consists of a layer of cells called the **protoderm,** a name that suggests its future role in producing the epidermis (outer covering). Toward the central axis of the embryo lies a line of cells known as the **procambium.** Later it will form the all-important vascular tissues, the primary phloem and xylem. Much of the tissue between the procambium and protoderm is nondistinctive and is simply referred to as **ground meristem.**

In the bean, the **hypocotyl** ("below the cotyledons") is very prominent. Upon seed germination, it will erupt in a great surge of rapid growth. In some species it will lift the embryo out of the soil and expose its first delicate leaves to the sunlight. At the embryo's base is a second area of uncommitted tissue, the **root apical meristem,** just above the young root tip. Its role will be to provide cells for root growth throughout the active life of the plant. In some embryos, the root apical meristem will have already produced some embryonic root growth. This embryonic root is called the **radicle.** Upon germination, the radicle will grow rapidly, reaching toward precious moisture.

Flowers to Fruits

A fruit is a ripened ovary that sometimes exists in association with certain floral parts. There are three basic types of fruits: simple, aggregate, and multiple, depending on the number of ovaries in the flower or the number of flowers in the fruiting structure.

Simple fruits may be derived from a single ovary or, more commonly, from the compound ovary of a single flower. They can be divided into two groups according to their consistency at maturity: simple fleshy fruits and simple dry fruits.

Simple fleshy fruits include the **berry, pome,** and **drupe.** The berry has one or several united fleshy carpels, each with many seeds. Thus the tomato **(a)** is a berry, and each of the seed-filled cavities is derived from a carpel. Watermelons, cucumbers, and grapefruits are also berries (but, oddly enough, blackberries, raspberries, and strawberries technically are not berries). Pome **(b)** means "apple," and the group includes apples, pears, and quinces. In the pome, only the inner chambers (roughly, the "core") are derived from the ovary, and most of the flesh comes from the calyx and corolla. A drupe—what a wonderful word—is also derived from a compound ovary, but only a single seed develops to maturity. The ripened ovary consists of an outer fleshy part and a hard, inner stone, containing the single seed. Peaches and cherries are drupes.

There are many kinds of simple

(a) Young, simple fleshy fruit (berry) of the tomato with only sepals remaining. The flower is seen at right.

Mature fleshy berry of the tomato. A cut at right angles to its axis reveals five fused carpels, each containing the seed-bearing, fan-shaped parts of the ovary.

(b) Apple flower

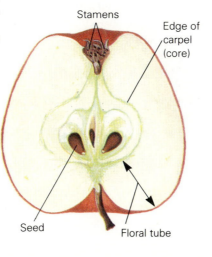

Stamens

Edge of carpel (core)

Seed

Floral tube

(left) The organization of pome becomes apparent in the young fruit, as the bases of corolla and calyx form the *floral tube* surrounding the ovary.

(right) In the mature fruit, most of the floral parts have withered away, but the floral tube has greatly enlarged, producing the sweet, edible portion of the fruit that contains the ovary (core). In the cross section we see the remnants of the flower, including the united carpels.

dry fruits, but they are neatly categorized as follows: (1) those with many seeds, which split open and release their seeds, and (2) those with few seeds, which do not split open or release seeds. The first group is called **dehiscent (c),** from the verb *dehisce,* to split or to open, and includes poppies, peas, beans, milkweed, snapdragons, and mustard. The second group is called **indehiscent (d)** (nonsplitting). Its members include sunflowers, dandelions, maples, ash, corn, and wheat.

Aggregate fruits (e) are derived from numerous separate carpels of a single flower. Blackberries, raspberries, and strawberries are aggregate fruits, Aggregate fruits consist of many simple fruits clumped together on a common base.

Multiple fruits (f) are formed from the single ovaries of many flowers joined together, as seen in the mulberry, fig, and pineapple. The pineapple starts out as a cluster of separate flowers on a single stalk, but as the ovaries enlarge, they coalesce to form the giant multiple fruit. (The commercial variety, the kind we most commonly see, is a seedless hybrid.)

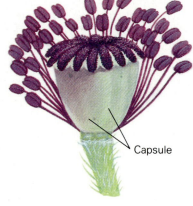

Capsule

(right) **(c)** Flower of poppy

(far right) The dry dehiscent fruit takes the form of a capsule surrounding the maturing seeds.

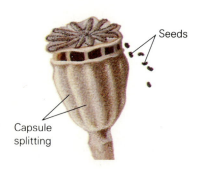

Seeds

Capsule splitting

Fused carpels from capsule

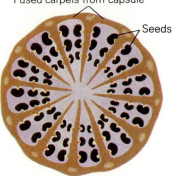

Seeds

Mature fruit. The capsule, free of other floral parts, becomes a seed-dropping machine as it splits. In cross section, the united carpels are visible in this simple fruit. Each carpel contains numerous seeds in rows along its length.

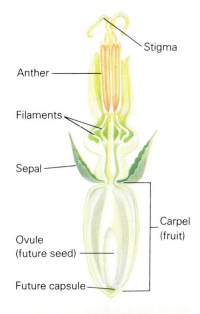

Stigma

Anther

Filaments

Sepal

Carpel (fruit)

Ovule (future seed)

Future capsule

(d) The sunflower, a composite form, consists of many individual disk flowers, making up the *head*. Each flower is simple, consisting of one carpel that will hold a single seed. As the dry indehiscent fruit matures, it will become surrounded by the familiar hardened "shell."

Individual carpels

Stamens

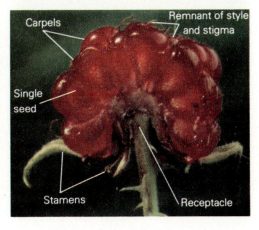

Carpels

Remnant of style and stigma

Single seed

Stamens

Receptacle

(e) *(far left)* Flower of the blackberry

(left) Maturing aggregate fruit of the blackberry. Each of the small spheres, the carpels of this aggregate, is actually a simple fleshy fruit (drupes, in this case) containing a hard seed.

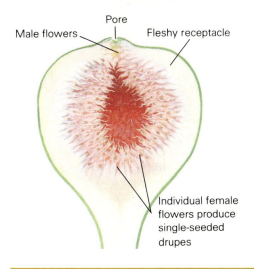

Pore

Male flowers

Fleshy receptacle

Individual female flowers produce single-seeded drupes

(f) In the common fig, functional male and female flowers both develop within vase-like receptacles, but on different trees. The male receptacle remains tiny (and inedible), producing staminate flowers within its porelike opening. However, it also contains nonfunctional female flowers that serve only to attract the female fig wasp who lays her eggs within. When the young hatch, the males fertilize the females and then usually die, but the females crawl out, becoming dusted with pollen as they leave. Later, as they prepare to lay their eggs, some will inadvertently enter female flowers. Since the pistillate flowers are not suitable the wasps soon leave, but not before pollinating all of the flowers within. Each flower then matures into a simple drupe, but forms a swollen, fleshy mass characteristic of the edible fig.

22.7 SEED AND EMBRYO DEVELOPMENT IN THE DICOT

(a) The development of both the seed structures and the embryo in dicots begins soon after fertilization. (b) Early activity in the endosperm includes numerous mitoses, but at first, no actual cell division. (c) The zygote divides several times, forming a growing chain of cells. (d) Varied planes of cell division at one end of the embryo result in three-dimensional growth, and the embryo takes the form of a ball of cells. (e) Below the embryo, a single file of cells with a bulbous basal cell form the suspensor. As events progress, the spherical embryo takes on a heartlike shape, marking the two cotyle-dons. (f) Changing planes of cell division again do their work and the enlarging cotyledons form a U-shape that is directed back toward the base of the embryo. (g) Early signs of differentiation in the embryo include a domelike region of apical meristem (between the cotyledon bases), a lengthy hypocotyl, and developing root tip with its own root apical meristem. Tissue types include a central procambium, surrounding ground meristem, and outer protoderm, meristematic tissues that, respectively, will give rise to the vascular system, cortex, and epidermis.

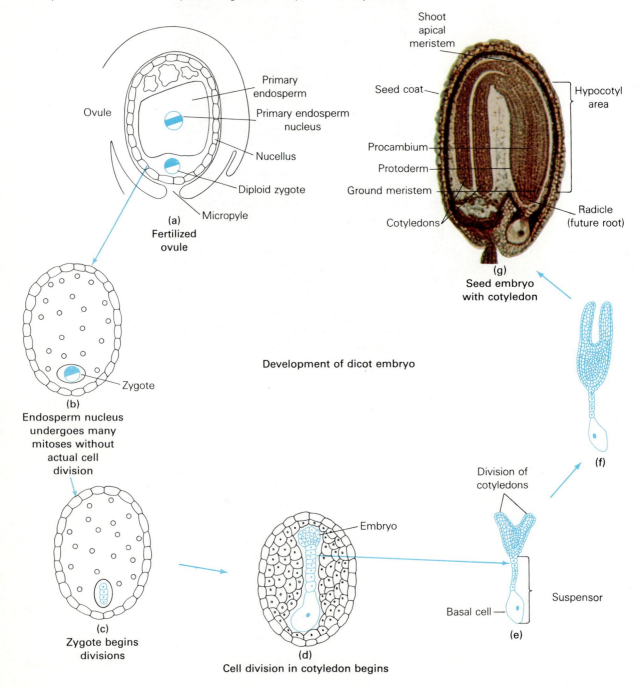

Development of dicot embryo

Shoot apical meristem

Seed coat

Procambium

Protoderm

Ground meristem

Cotyledons

Hypocotyl area

Radicle (future root)

(g)
Seed embryo
with cotyledon

Ovule

Primary endosperm

Primary endosperm nucleus

Nucellus

Diploid zygote

Micropyle

(a)
Fertilized ovule

Zygote

(b)
Endosperm nucleus undergoes many mitoses without actual cell division

(c)
Zygote begins divisions

Embryo

(d)
Cell division in cotyledon begins

Division of cotyledons

Basal cell

Suspensor

(e)

(f)

Fruit and Seed Development in the Monocot. The fruits of monocots such as corn and other grasses are rather different from those of dicots and are often confused with seeds (Figure 22.8). A corn kernel or wheat grain, for instance, may look like a seed, but it is actually a hard fruit encasing a single large seed. In other words, each kernel of corn is a single fruit—the equivalent of a peach or cherry, which are simple dicot fruits.

Food storage in the monocot seed differs from that of dicots. In the monocots food is stored primarily in undifferentiated endosperm tissue. In wheat or corn, the white or yellow starchy bulk of the kernel is endosperm tissue, and the wheat germ and corn germ are the embryos and the associated cotyledon. The thin coverings of wheat and corn grains are far more complex than they seem. They include the outer **pericarp,** which arises from diploid parental tissue in the ovary wall, and the inner triploid **aleurone,** which arises from the endosperm.

The monocot embryo of corn and other grains (Figure 22.9) consists of a plumule (embryonic shoot) and a radicle (embryonic root), separated by a short length of undifferentiated stem tissue. The radicle is surrounded by a protective sheath called the **coleorhiza.** The plumule contains partially developed embryonic leaves and is itself surrounded by a closed, tubular protective sheath known as the **coleoptile.** Both the coleorhiza and coleoptile are distinctive features of monocots. The coleoptile will first be visible as a tiny, simple, shaft-like growth as the green shoot emerges from the soil. In a short time, it will split open as the first leaf emerges. The monocot cotyledon, or **scutellum,** functions in digestion. Upon germination, the starchy food reserves of the endosperm will be digested by enzymes from the aleurone, and the resulting sugars will be absorbed by the scutellum and passed into the growing embryo.

Seed Dispersal

In order for seeds to germinate successfully, they must begin their growth under suitable conditions—generally away from the competition of parents and other seeds. Seed dispersal is a particularly critical factor in the life cycles of all plants. Plants have therefore developed intricate means of dispersal. Annuals, plants that live for one growing

---

22.8 MONOCOT AND DICOT SEEDS

(a) In the monocot corn kernel (a fruit), food reserves occur in the form of an unorganized starchy endosperm, taking up most of the kernel's volume. During germination and early growth, the lengthy scutellum—the single cotyledon—will act as a food-absorbing structure, taking in digested nutrients from the endosperm. (b) In the dicot bean seed, food is stored in cells of the paired cotyledons, which occupy much of the seed. Upon germination the bean cotyledons, which emerge from the soil, develop chlorophyll and for a time carry on photosynthesis.

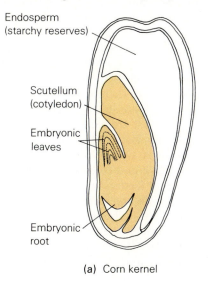

Endosperm
(starchy reserves)

Scutellum
(cotyledon)

Embryonic
leaves

Embryonic
root

(a) Corn kernel

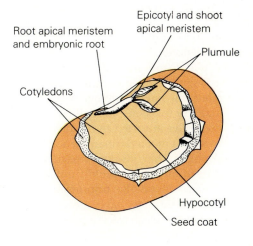

Root apical meristem
and embryonic root

Epicotyl and shoot
apical meristem

Plumule

Cotyledons

Hypocotyl

Seed coat

(b) Bean seed

22.9 MONOCOT EMBRYO AND FRUIT

Each corn kernel is actually a fruit—the mature ovary of corn. The corn kernel consists of tough outer coatings, a large starchy endosperm, and the embryo, which includes a single cotyledon—the scutellum. Other parts of the embryo include the plumule, coleoptile, and radicle with its overlying coleorhiza.

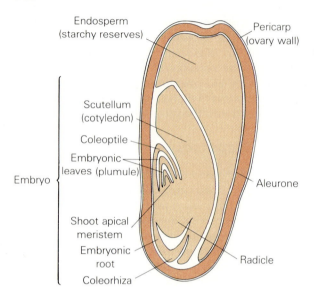

Endosperm (starchy reserves)

Pericarp (ovary wall)

Scutellum (cotyledon)

Coleoptile

Embryonic leaves (plumule)

Embryo

Aleurone

Shoot apical meristem

Embryonic root

Radicle

Coleorhiza

Corn kernel

season, have a brief time to produce and disperse their seeds. Many live in disturbed habitats and their seeds must be well scattered to assure a few of success in the transitory conditions to which they have adapted. Perennials, plants that live on for many seasons, particularly the large trees, have a different kind of problem. Undispersed seeds that germinate beneath their parent's boughs must compete with it for water, mineral resources, and, especially, sunlight. If a seed germinates close to the parent plant, the new seedlings, may be subject to heavy attack by the same insect predators and fungal root parasites attracted to the parent.

Many seeds enter a dormant period during their dispersal phase. They lose most of their water and enter a state of suspended activity until germination occurs. They will not lie unprotected, however, because their seed coats have already formed. The walls of the outer cells harden and thicken, forming resistant, protective casings over the entire seed. Further, in their dehydrated state, seeds readily resist the growth of bacteria and fungi.

The dormant period varies greatly, and apparently some seeds can remain viable for many, many

years. For instance, some lotus seeds found in peat deposits near Tokyo germinated after 2000 years of dormancy. The record, though, is held by a delicate flower of the Yukon, *Lupinus arcticus*. One group of seeds grew into fine flowers after having lain dormant in frozen soil for over 10,000 years.

How do seeds, bearing the young embryos, get around? The primary natural seed carriers are water, wind, and animals (Figure 22.10). Perhaps the best-known example of a water-borne seed is the coconut. Coconut palms have become established on even the most remote and tiny islands of the Pacific and are found throughout the tropical and subtropical regions of the earth. Wind dissemination is common in the maples and elms, which produce winged seed-bearing fruits. As they fall from trees they gain speed and at a certain rate of descent they begin to fly sideways, spiraling to new frontiers away from their parents. Dandelion fruits are carried in the wind on fine plumes derived from the calyx. An odd adaptation to wind dispersal is the tumbleweed: the whole plant forms a large and very light ball that dries up, breaks off at its base and rolls across the prairie, scattering seeds and giving rise to songs as it goes tumbling along.

Seeds can be carried by animals in two ways—on the outside and on the inside. Those that are carried on the outside often have spiny, barbed fruit (the geranium and burr clover are familiar examples) or barbed or sticky seed coats. What dog owner hasn't had to remove cockleburs or foxtails from the animal's ears or paws? But how can seeds be transported inside?

Seeds are transported on the inside largely because of one quite successful adaptation involving the evolution of tempting and digestible fruit surrounding indigestible seeds. When eaten, the seeds are carried for a while in the animal's digestive tract to be deposited, intact and ready to germinate, in a mound of excrement. This is what bright-colored, fleshy, sweet-tasting fruits and berries are all about—at least from a plant's point of view. Many such fruits also contain powerful laxatives to help the process along. In some cases, the seed itself is attractive and nutritious to the dispersing animal. (Birds at feeders are attracted to the seeds.) Since these seeds are small, a plant can produce many of them and, although they can be digested, a number will escape the animal's digestive processes to germinate later in a nitrate- and phosphate-rich dropping. Some animals also store seeds for later use, and many an oak has grown from an acorn that some absentminded squirrel has buried and forgotten.

22.10 DIVERSITY IN SEED DISPERSAL

Seeds and seed-bearing fruits are dispersed by wind, water, or animals. Plumes and wings help in lofting the seed-bearing structures of *Clematis,* dandelion, and pine into breezes that will carry the offspring well away from their parents. The cranesbill, foxtail, and burr clover have adaptations for clinging to the fur of animals. The seeds within blackberry and strawberry fruits are carried in an animal's digestive tract for a time, but are eventually deposited in the feces. The coconut seed, well protected by its tough fruit (husk) is uniquely adapted for drifting in the sea.

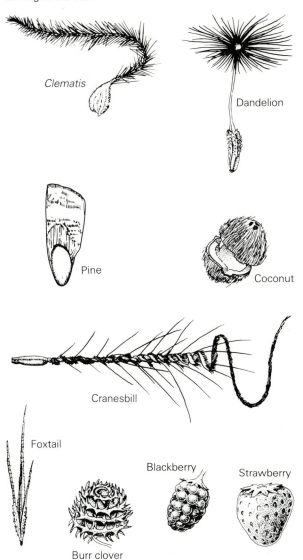

Clematis

Dandelion

Pine

Coconut

Cranesbill

Foxtail

Blackberry

Strawberry

Burr clover

GROWTH AND DEVELOPMENT IN FLOWERING PLANTS

Germination and Growth of the Seedling

Germination, the beginning of growth from a seed, is essentially a resumption of metabolic activity—a sort of "reawakening" of the once active but now dormant embryo. Most plants have the same requirements for germination: an adequte supply of water, favorable temperatures, and the availability of oxygen. Metabolism resumes when water is absorbed into the dry interior of the seed and the many essential enzymes are mobilized.

While all plants have these requirements, many also require special conditions to get germination going. The seed coats of some species are especially tough and resistant. Some seeds require a period of freezing temperatures to split their coats, while some require the heat of fire. A few require the abrasive action of sand particles carried by flowing water, and still others require the hydrolyzing action of animal digestive enzymes. On the island of Mauritius there are 11 huge Tambalacoque trees, a very rare species. All of the trees are about 300 years old; there are no younger trees. Every year they put out a crop of huge seeds with very thick seed coats. But the seeds never germinate. It seems that normal germination requires that the seed be passed through the crop of another native of the island, a bird called *Raphus solitarius,* better known as the dodo. But, thanks to us, the dodo has been extinct for about 300 years, so the seeds just lie around waiting for the bird that will never come. (You may be pleased to learn, however, that recently, the naturalist who worked out this curious relationship managed to get several of the seeds to germinate by passing them through a turkey.) We can speculate that each variation in germination requirements has some special adaptive value to the species and is related to ideal conditions for survival.

Following the uptake of water in monocots such as the oat or other grains, germination begins with the mobilization of the large starchy endosperm. This is a roundabout process involving hormones known as **gibberellins** (see Chapter 24). The target of the gibberellins is the aleurone cell layer of the endosperm, just below the seed coats. The aleurone cells respond to the hormone by producing alpha-amylase, a starch-digesting enzyme that hydrolyzes the starch to mobile, soluble sugars that the embryo can put to use.

As germination continues in the monocot, the embryonic root or radicle emerges first, breaking through the coleorhiza and pushing down into the soil (Figure 22.11a). Next, the shoot, surrounded by the protective coleoptile, elongates rapidly, pushing upward through the soil. Soon after, the first leaf breaks through the coleoptile and emerges into the sunlight. The rapidly growing young root below is soon accompanied by more roots. Oddly, in corn and some other monocots, these newcomers will emerge not from the primary root, but from the stem region above. As these new roots grow and mature, the primary root will shrivel up and die and newer root growth will take over.

Dicots vary considerably in their early growth, but the early visible signs of germination include the swelling of the seed and the splitting of the restrictive seed coats. As germination in the bean proceeds, the young root emerges and quickly penetrates the soil, forming secondary roots and anchoring itself securely. The hypocotyl then elongates rapidly (Figure 22.11b). (The intake of water into cells is an important factor in elongation, since it is turgor pressure that stretches young cell walls

22.11 CORN AND BEAN GERMINATION

(a) As corn germinates the rapidly elongating radicle emerges first, breaking through the coleorhiza to form the primary root. The kernel itself remains in the soil. Following primary root growth, the coleoptile, with the plumule enclosed, pushes through the soil. As the first leaves unfurl, they tear through the coleoptile and emerge into the light. Meanwhile, the primary root has been joined by numerous adventitious roots, which emerge directly from the shoot. (b) Events in the germinating bean seedling are quite different. Following the emergence of the primary root, the rapidly elongating hypocotyl forms a curved bumper, grows upward, and as it proceeds, draws the cotyledons behind it. Upon emerging from the soil, the hypocotyl straightens and the cotyledons and young plumule are fully exposed. Just below the plumule a newly elongating region, the epicotyl, lifts the young unfurling leaves even higher. As their food reserves diminish, the cotyledons wither and eventually drop from the shoot. By then, the young primary root has produced many secondary roots as the root system matures.

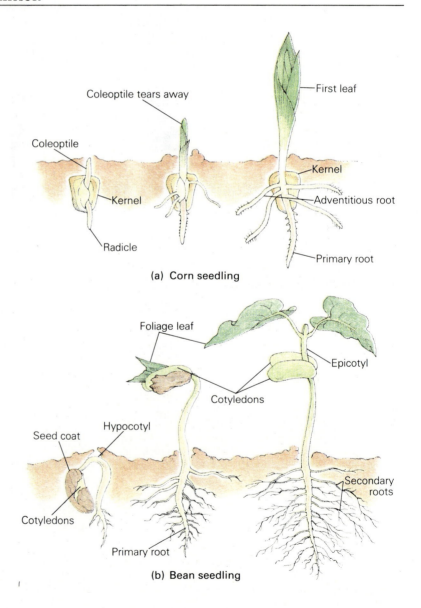

(a) Corn seedling

(b) Bean seedling

and lengthens the cell.) As the hypocotyl grows it forms a sharp curve—the **hypocotyl hook**—which acts as a protective bumper to the delicate parts below. In its upward growth, the hypocotyl draws the cotyledons along behind it. Upon breaking through the surface the hypocotyl straightens, the cotyledons open up, and the young plumule is exposed to the light. At this point the epicotyl begins its elongation, lifting the expanding leaves up to the light.

When the seedling breaks the surface, its cells begin producing their first chloroplasts, turning the plant a deep green. The embryonic leaves are exposed to energy-laden light as their growth continues. At this time, the food reserves in the cotyledons diminish and they wither and fall away, leaving telltale scars on the young shoot. The young dicot seedling is then fully on its own.

Primary and Secondary Growth

Seed germination and the early growth of the shoot are examples of **primary growth.** Primary growth goes on throughout the life of a plant. It is responsible also for continued growth in the length of shoots and roots, and for the production of leaves, flowers, and new branches. In contrast, **secondary growth** accounts for growth in girth or thickness. While all of the conifers are capable of secondary growth, within flowering plants, secondary growth is seen mainly in dicots. However, not all dicots undergo extensive secondary growth. In annuals, for instance, it may be absent or quite restricted. They germinate, grow, develop, and reproduce in a single season. As you might expect, then, extensive secondary growth in flowering plants occurs chiefly in the perennials, such as long-lived shrubs and trees.

Perennial plants are also unique among living things in that they exhibit what is called **open growth,** or **indeterminate life span.** This means that barring fire, injury, or disease, theoretically they can live forever. There are the usual arguments over which plant holds the record, and for years the documented champion was a 4900-year-old conifer, a bristlecone pine from the White Mountains of California (Figure 22.12). Today, we're not so sure. Some cottonwoods may have reached an age of 8000 years, and the age of certain desert creosote bushes was recently estimated at 11,700 years. In any case, we can be sure that somewhere on earth grows a plant that is the world's oldest living thing.

22.12 ONE OF THE OLDEST PLANTS

At the timberline of the White Mountains, a minor range in eastern California, is found one of the oldest living organisms on earth. One specimen of the bristlecone pine *(Pinus longaeva)* is known to have reached 4900 years of age.

In a sense, plants with indeterminate or open growth never completely mature. They always have a reserve of uncommitted, undifferentiated, meristematic tissue. In addition, many cells that have differentiated and begun to function in a specific role are capable, in a manner of speaking, of having their "clocks set back" by **dedifferentiation.** They return to an immature state, from which they have the potential to mature or differentiate all over again, perhaps along new routes. We will return to this point, but let's prepare for a closer look at primary and secondary growth by learning something about plant tissues.

Tissue Organization in the Plant

The plant body is certainly a complex entity, composed of many types of cells organized into seemingly countless layers within the roots, stems, and leaves. However, all these are composed of only a few types of tissues. There is the perpetually young apical meristem, which produces three types of **primary meristems:** protoderm, ground meristem, and procambium. We noted earlier that these are the first specific tissues formed in the embryo. Each of these, in turn, gives rise to more differentiated tissues that carry out the many specialized functions that characterize plant life. This organization is illustrated in Table 22.2.

Protoderm and the Dermal System.   Mature protoderm contributes to the **dermal system**—the outermost layers of cells, or epidermis. In the leaf and young primary stem, epidermal cells are commonly flattened and irregular in shape. They are often covered by a waxy **cuticle** or **cutin,** which provides waterproofing and prevents water loss from the more delicate tissues within. In addition to the simple epidermal cells, pairs of more intricate **guard cells** surround pores, together forming **stomata** (singular, *stoma*), which permit the exchange of gases between photosynthetic cells and the atmosphere. Epidermal cells just above the root tip have the vital job of absorbing water. They produce numerous, lengthy **root hairs** that greatly increase the absorbing surface area. Where secondary growth occurs, the epidermis is replaced by a multilayered tissue called **periderm.** The periderm is made waterproof by the addition of **suberin,** another waxy substance.

Ground Meristem.   Ground meristem differentiates into three important tissues: **parenchyma, collenchyma,** and **sclerenchyma.** Parenchyma is widely distributed throughout the plant. It is an often loosely arranged tissue consisting of large, thin-walled, irregularly shaped cells. Parenchyma takes in a considerable amount of water, producing the turgid condition important to leaves and the shoots of young plants. It is also a food storage tissue, often burgeoning with starches. Leaf parenchyma (also called *chlorenchyma*) contains most of the chloroplasts and is the most important site of photosynthesis. Parenchyma tissue is capable of cell division and differentiation into other tissue and is responsible for the growth of new tissues in wound healing.

Collenchyma is a supporting tissue. Its thickened cell walls form tough strands of tissue below the epidermis of young stems and alongside vascular tissues in the supporting ribs of some leaves. Sclerenchyma tissue takes the strengthening and toughening process one step further. It provides the plant with **fibers** and **sclereids.** The tough, tapered fibers, which provide tensile strength in the plant, are the functional substance of jute, used in making sacking and cord. Tough strands of sclerenchyma tissue are removed from celery stalks by considerate hosts before serving them to their guests. (Incidentally, one goal of agricultural geneticists has been to selectively produce strains of celery that are easier to chew, but still strong enough to stand erect in the soil.) The thick-walled form of sclereids called **stone cells** make up the hard parts of cherry stones, peach pits, coconuts, and nut shells, along with the gritty material in the flesh of pears.

Procambium and the Vascular Tissues. Procambium, the third product of meristem, has only one role: to produce **primary xylem** and **phloem,** the vascular tissue of the young plant. Since we now have some idea of the evolution and basic functions of these conducting tissues, we can now focus more tightly on their development and organization.

Xylem.   As we know, water and minerals move upward from the root to the stem and leaves through the xylem. Xylem in angiosperms includes *tracheids* and *vessels,* along with nonconducting fibers and parenchyma. (You may recall that with the exception of angiosperms and gnetophytes, vessels are absent in vascular plants; thus vessels are considered to be an advanced evolutionary structure.) Both tracheids and vessels are of little use while alive, but after they die, their cytoplasm disappears and the elongated cell walls form the extensive, lengthy water-conducting structures.

Tracheids are long and slender with tapered ends (Figure 22.13a). They do not lie precisely end to end; their ends form staggered, overlapping connections. One might expect the end walls to be absent where the elements fuse together. The end walls do persist, but they contain very thin, porous regions called **pits** or **pit pairs,** which consist of primary cell wall material only. The pits of adjacent tracheids are aligned so that water readily passes through from one tracheid to the next. The side walls of tracheids are commonly pitted as well; thus fluid can move laterally. Note in Figure 22.13 that pit pairs can be either simple or bordered (recessed).

Vessel elements (derivatives of individual cells) tend to be shorter and much wider than the cells of tracheids (Figure 22.13b). Tapering, if it occurs at all, is much less noticeable in the end walls. The end walls themselves may be heavily perforated or even entirely absent. As a result, vessel elements lying end to end form uninterrupted pipelines, the **vessels,** through which water readily flows. Like tracheids, the side walls of vessel elements are heavily pitted and readily admit water. The xylem fibers,

TABLE 22.2

PRIMARY MERISTEMS AND THEIR TISSUE DERIVATIVES

In primary growth, apical meristematic tissue gives rise to:

Protoderm

Protoderm differentiates into covering tissues, including the epidermis of roots, stems, and leaves. More specialized examples include guard cells, leaf hairs, and roots hairs.

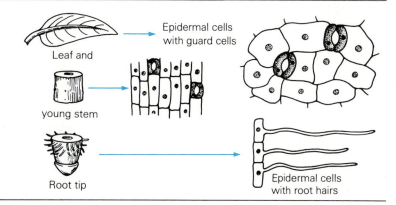

Leaf and

young stem

Root tip

Epidermal cells with guard cells

Epidermal cells with root hairs

Ground meristem

Ground meristem differentiates into three basic tissue types:

Parenchyma is widely distributed in the stem and root and makes up the photosynthetic tissues of the leaf. The cells are large and thin-walled, often involved in storage.

Collenchyma is primarily involved in support. Its thick-walled cells form tough strands of tissue below the epidermis, within vascular tissue, and in the supporting portions of the leaf.

Sclerenchyma, in its *fiber* form, strengthens young shoots. In its *sclereid* form, it provides hardness for seed coverings and shells.

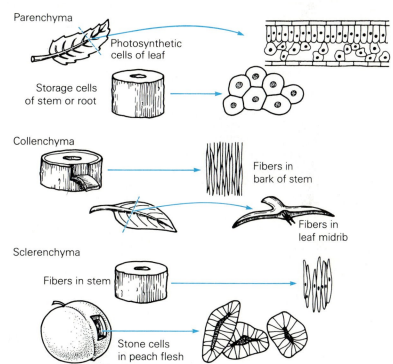

Parenchyma

Photosynthetic cells of leaf

Storage cells of stem or root

Collenchyma

Fibers in bark of stem

Fibers in leaf midrib

Sclerenchyma

Fibers in stem

Stone cells in peach flesh

Procambium

Procambium differentiates into xylem and phloem, the conducting (vascular) tissue of roots, shoots, and leaves. Xylem specializes in water and mineral transport; phloem, in food transport.

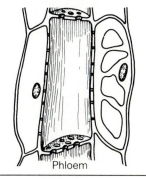

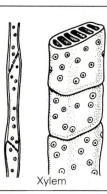

Phloem

Xylem

22.13 XYLEM TISSUE

Xylem includes conducting elements known as tracheids and vessels. **(a)** Tracheids are long, slender elements with pitted side and end walls. **(b)** Pits, or pit pairs as they are more properly known, consist of thin-walled regions where only the two adjacent primary cell walls and the middle lamella are intact. This thin region is called a pit membrane. Pit pairs may be simple, as described above, or may be bordered, which means that the secondary wall overlaps the pit membrane, forming a chamberlike recess around the pit. **(c)** Although vessel elements have side wall pits, their end walls are completely perforated. Also, they are usually much larger and less tapered than tracheids.

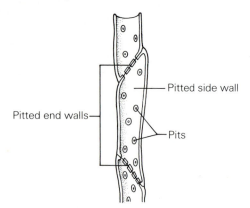

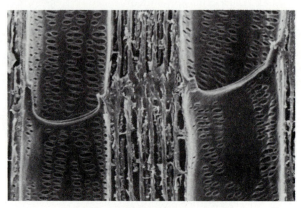

(a) **Tracheids** (cutaway view)

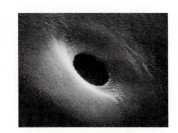

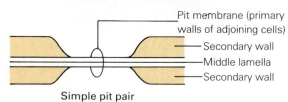

Simple pit pair

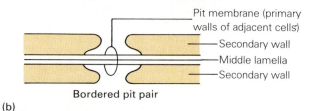

Bordered pit pair

(b)

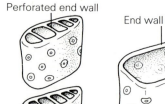

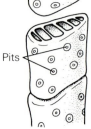

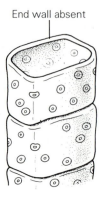

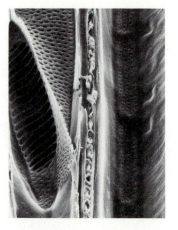

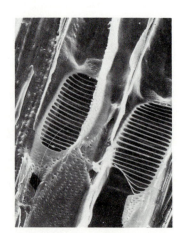

(c) **Vessels**

which can be long and tough, have a role in support, although some, called **xylem parenchyma,** serve as storage cells.

Phloem. While xylem conducts water, phloem is responsible for the movement of foods, primarily sugars. Unlike xylem, phloem must remain alive to carry out its function. This is a complex tissue, composed of **phloem parenchyma, companion cells,** and **sieve tube elements** (Figure 22.14). (A somewhat different, less specialized conducting element, the *sieve cell,* is found in gymnosperms.) The actual conducting cells are the sieve tube elements. (Note that the term "cell," as with the xylem, is again replaced by the term "element.")

Sieve tube elements lie end to end in the phloem, forming lengthy **sieve tubes.** They are named for the prominent pores that pock their walls. The pores occur both in the side and end walls, but the largest are often in the end walls where the **sieve plates** are found. Plasmodesmata, thin streams of cytoplasm extending through the enlarged pores of sieve plates, can conduct substances from one sieve tube element to the next (see Chapter 5).

Companion cells are aptly named since they lie adjacent to the sieve tube elements and also function in conduction. The exact relationship between the two isn't entirely clear, but we know that while the companion cell contains a nucleus, the sieve tube elements do not. Presumably then, the companion cell takes care of any needs related to protein synthesis for both types of cells. There is also considerable evidence that the companion cells function importantly in lateral transport. Specifically, they actively transport sugars in and out of the sieve tube elements—a vital part of the overall food transporting process, and one we will look into in the next chapter. The third element of the phloem triad, the simpler phloem parenchyma cell, is involved chiefly in storage but also actively transports foods in and out of the sieve tube elements. Supporting fibers are also incorporated in phloem.

Primary Growth in the Root: The Root Tip

Primary growth—growth in length—in the plant root takes place at the **root tip.** The root tip is indeed the business end of the root, for its tasks include growth, penetration of the soil, and the absorption of water and minerals.

A look at Figure 22.15 will reveal the organization of a typical root tip. The apical meristem is located at the very end, just above a region of larger, loosely arranged cells called the **root cap.** The root tip is commonly divided into three functional zones, the first of which includes the meristem and forms the region of cell division (note the many nuclei involved in mitosis). Above this lies the region of elongation and further back, the region of maturation. We can now consider each in greater detail.

In the apical meristem, the region of cell division provides a constant supply of cells for the root cap below and for the zone of elongation above. In addition, some cells are held in reserve, not specializing in either direction, to assure that these provisions continue. The root cap acts as a protective bumper, shielding the delicate meristem behind. As the root tip penetrates the abrasive soil, the cap cells are sloughed away and must continually be replaced. Further, the rupture of the cap cells releases a slimy substance, a polysaccharide that acts as a lubricant to the lengthening root tip. Root cap cells, as we will see in Chapter 24, may also be important in causing the downward growth of roots.

While new cells added below the apical meri-

22.14 PHLOEM TISSUE

Phloem includes sieve tube elements bordered by companion cells and phloem parenchyma. Sieve tubes conduct the phloem stream, while the other two provide physiological support and active lateral transport. The end walls of sieve tubes contain sieve plates with plasmodesmata, where the cytoplasm of one element is continuous with the next.

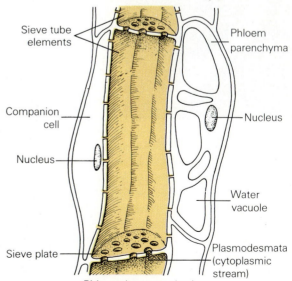

Phloem (cutaway view)

22.15 THE ROOT TIP

Primary growth is clearly seen in the
root tip. Its major regions include the
root cap and regions of cell division,
elongation, and maturation. The root
cap and region of elongation receive a
continuous supply of cells from the apical meristem between. As cells complete elongation they mature into the
three primary meristematic tissue
types: protoderm, ground meristem,
and procambium. Within the region of
maturation, cells of the young epidermis, derived from protoderm, produce
extensive growths of root hairs (see
inset of corn seedling). Toward the center, procambial cells mature into xylem
and phloem, while in the area between
these vascular tissues and the epidermis, ground meristem matures into
parenchyma tissue that makes up the
cortex. .

(a) Radish seedling

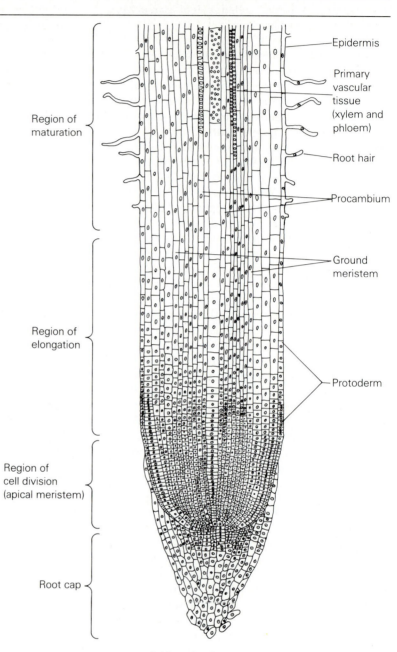

(b) Longitudinal section

stem replace those lost by the root cap, new cells added above will differentiate into the three basic tissue types: protoderm, ground meristem, and procambium. But first they will play an important role in the zone of elongation. It is within this zone that the cells elongate, providing the "downward push" we associate with root growth.

Elongation requires changes in the cell wall. In the young cells above the root meristem, the primary wall is still thin and pliable. In water-filled cells, the resulting turgor pressure causes the cell to elongate. As a result, the first layers of cellulose fibrils circling the cell change their orientation, first becoming diagonal and then longitudinal. Any new fibrils laid down during this process will be at right angles to the original fibers (Figure 22.16). Once the cell wall is completed and hardened, no further elongation is possible. As we will see (Chapter 24), plant hormones called **auxins** are involved in cell elongation in the stem, but their role in root cell elongation is not so clear.

Once their elongation has been completed, the older cells, which now form the region of maturation, are left behind the growing root tip to mature and begin differentiation. The outer tissues, the protoderm, will form the epidermis, and as we saw in Figure 22.15, many of these young covering cells will produce lengthy root hairs. Ground meristem just within the protoderm forms the **cortex**, a region of thin-walled parenchyma. Toward the central region of the root, some of the ground meristem will mature into a cylinder of cells known as the **endodermis** (inner skin). The endodermis performs an active role in the vital uptake of water by the plant (see Chapter 24). Just within the endodermis lies the **stele**, which houses the developing vascular system. This region deserves special attention.

The Stele. The region containing the vascular tissue is known as the stele (Figure 22.17). The stele is surrounded by a cylinder of cells called the **pericycle**, just inside the endodermis. The pericycle conducts water and minerals inward to the vascular tissue, but in dicots it has another function as

22.16 MECHANICS OF CELL ELONGATION

Cell elongation is a highly precise process. **(a)** The rapid uptake of water causes lengthening in young cells that have yet to produce their secondary walls. As lengthening occurs, the soft cellulose fibrils of the primary cell wall are reoriented **(b)** to follow the long axis of the cell. As elongation proceeds **(c)**, the secondary wall is laid down within the primary wall. As new cellulose layers are assembled, their fibrils also become reoriented but take on a more diagonal direction. **(d)** A final layer of fibrils is incorporated into the secondary wall as elongation ceases, and these remain oriented in a transverse direction.

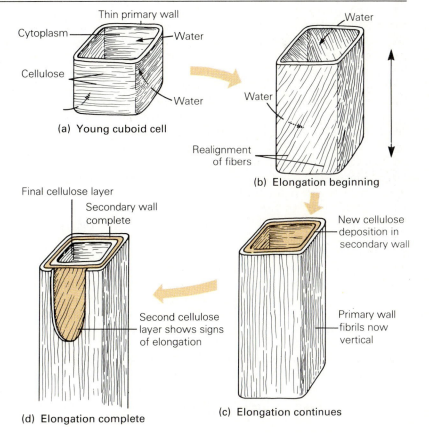

(a) Young cuboid cell

(b) Elongation beginning

(c) Elongation continues

(d) Elongation complete

22.17 THE MATURE PRIMARY ROOT

In the region of maturation, the primary meristematic tissues mature into the epidermis, the cortex (including the endodermis), and the stele. The latter begins with a ring of pericycle and includes the xylem (large, thick-walled cells) and phloem (smaller, thin-walled cells) within. In many dicots a region of primary cambium forms between primary xylem and phloem. This cambium will provide for secondary growth.

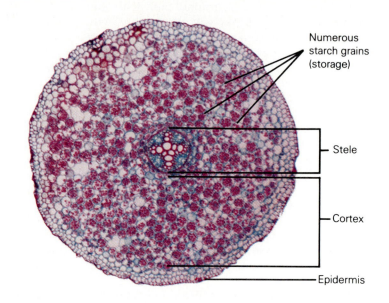

Numerous starch grains (storage)

Stele

Cortex

Epidermis

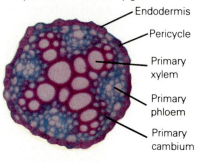

Endodermis
Pericycle
Primary xylem
Primary phloem
Primary cambium

well. It has the capacity to produce secondary roots called **branch roots.** These young secondary roots emerging from the pericycle closely resemble primary root tips in their organization. Their growth has a brutal aspect since they must push their way through the endodermis and cortex, crushing many cells and finally bursting through the epidermis of the primary root (Figure 22.18). Eventually, each branch root will develop its own vascular tissue that will join that of the primary root.

In some dicots, the organization of the vascular tissue within the stele looks (in cross-section) like a four-armed star (see Figure 22.17). The broad arms contain the primary xylem, which is made up of large, thick-walled, water-conducting cells. The primary phloem, with its conspicuous sieve tubes (food-conducting structures), is found between the arms of the star. In plants capable of secondary growth, undifferentiated procambium produces a strip of **vascular cambium** between the xylem arms and phloem regions. Vascular cambium will later produce the **secondary xylem** and **secondary phloem.**

Root Systems. Root systems differ between dicots and monocots. While both begin by producing a fast-growing main or primary root, the primary root in monocots does not usually persist as it does in dicots. For instance, in corn, the primary root is replaced early in life by **adventitious roots,** which emerge from the stem. These in turn produce branches that originate from the root peri-

cycle in the usual manner of secondary roots. You may have seen exposed adventitious roots in corn arch downward to the soil. These are often called "prop roots" and are important to supporting the heavy mature corn plant. Adventitious roots are also seen in the dicots, and some well-known examples include those of English ivy, the tropical red mangrove, and the banyan tree.

Roots can be organized into either **tap root systems,** or **diffuse root systems,** although combinations of the two exist. Tap root systems tend to have a main root and many lesser secondary roots. Diffuse root systems, which are common in monocots and best known in the grasses, are characterized by a great many roots of similar size. Figure 22.19 compares the two systems.

Primary Growth in the Stem

The organization of the apical meristem of the shoot is somewhat more complex than that of the root (Figure 22.20). As we mentioned earlier, a small mound of tiny, actively dividing apical meristem is found at the very tip of the shoot. As cell division occurs, daughter cells at the lower side of the meristem undergo extensive elongation, carrying the meristem upward or outward. Below this region of elongation, cells that have completed the process undergo differentiation, but the pattern is quite different from that of the root tip. In the center, thin-walled parenchyma cells will form the **pith.** Just outside the pith, strands of procambium differenti-

22.18 GROWTH OF SECONDARY ROOTS

Secondary or lateral roots are produced by the pericycle of mature primary roots. Cells in the pericycle undergo repeated division, producing what at first appears to be a formless mass. The mass elongates, growing through the cortex and bursting through the epidermis. During this period, an apical meristem develops at the tip, producing a root cap ahead and typical primary tissues in its wake.

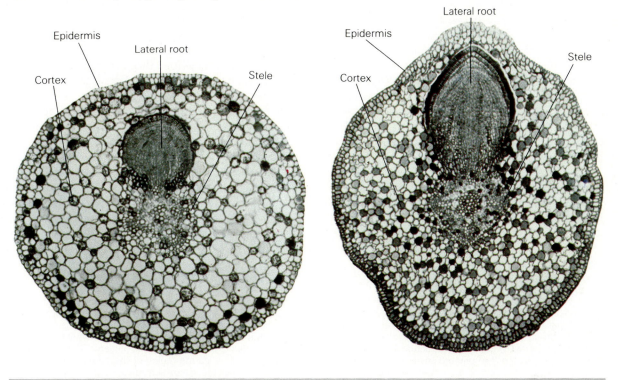

22.19 TWO KINDS OF ROOT SYSTEMS

(a) Diffuse root systems are common in monocots such as the bluegrass seen here. This vast, spreading system consists of adventitious and secondary roots. (b) In tap root systems, the primary root maintains dominance, leading the way and producing a profusion of secondary or branch roots.

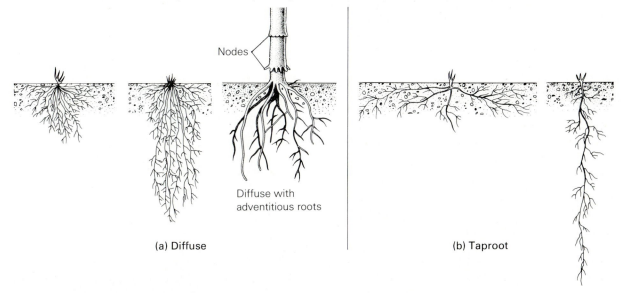

Nodes

Diffuse with adventitious roots

(a) Diffuse

(b) Taproot

ate into the primary xylem and phloem. At the close of primary growth in dicots, the bundles of vascular tissue will have become arranged in a ring around the perimeter of the pith. The vascular bundles are surrounded by cortex, which extends to the young differentiating epidermis. In many young stems, the cortical cells just below the epidermis contain chloroplasts and carry on photosynthesis.

The distribution of vascular tissue in the dicot and monocot shoot, as you may recall from the last chapter, are fundamentally different. While the vascular bundles commonly form a rather neat ring within the dicot stem, in monocots they typically are scattered throughout the stem. Figure 22.21 compares the arrangement in corn and sunflower plants.

As the shoot grows, it leaves behind it differentiating tissue as we described above, but it also leaves behind patches of meristem that give rise to **leaf primordia** and **branch primordia.** The primordia are patches of embryonic tissue that give rise to

new tissues. General regions of potential leaf production are known as **nodes.** The stem region between successive nodes, is logically enough, called an **internode.**

Nodes and internodes are most easily seen in the growing tips of deciduous woody plants—seasonally growing plants that shed their leaves in the autumn, such as oak or hickory trees (Figure 22.22). When a leaf falls from such a plant, it leaves a scar that remains visible for a time on the young branch. Above each **leaf scar** is a tiny **lateral branch bud,** or **axillary bud,** which in the younger region may produce leafy stems in the spring. The age of a young woody stem can be determined by counting the number of terminal **bud scale scars** below the **terminal bud** at the stem tip.

Lateral branch buds may be inhibited from sprouting into lateral branches by the action of the hormone auxin, which originates in the apical meristem. Under such inhibition, a tree may take on a triangular shape or at least will have somewhat

22.20 THE SHOOT TIP

The shoot tip contains the shoot apical meristem, a dome-shaped mass of tiny, undifferentiated cells. The small projections rising from the meristem are the newest leaf primordia. Older leaf primordia (large, upright structures) rise up to cover the meristem. In the leaf primordium axil, two patches of dark tissue mark the location of lateral bud primordia. These buds have the potential for producing branches. Further down the shoot are two more lateral bud primordia. Note the general elongation of cells below the apical meristem and on down through the stem. Most of this tissue is ground meristem, but further down are islands of procambium that will later give rise to primary xylem and phloem. The cross section (right) through a mature region of the primary shoot reveals the circle of vascular tissue surrounding a region of pith parenchyma.

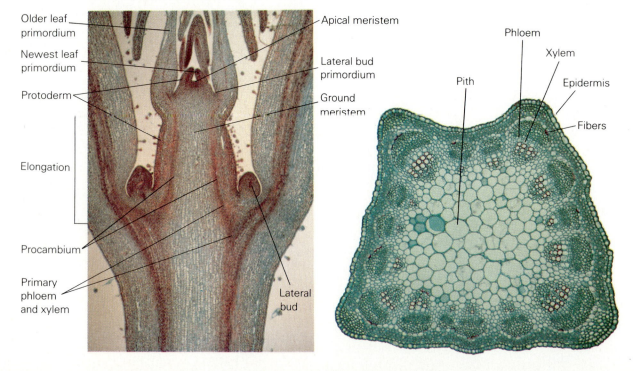

Older leaf primordium
Newest leaf primordium
Protoderm
Elongation
Procambium
Primary phloem and xylem
Apical meristem
Lateral bud primordium
Ground meristem
Lateral bud
Phloem
Xylem
Pith
Epidermis
Fibers

22.21 MONOCOT AND DICOT STEMS

(a) A cross section through a corn stem reveals a scattered distribution of vascular bundles that is typical of monocots. Each bundle (see detail) contains xylem and phloem surrounded by tough, fibrous sclerenchyma tissue, important in providing support to the stem. **(b)** The vascular bundles of the sunflower, a dicot, occur in an orderly row around the perimeter of the stem, which is similar to the way they are arranged in young woody dicots. Note (in the detail) how each vascular bundle is capped by a collection of tough sclerenchyma cells. The primary phloem and primary xylem are separated by a region of procambium. Most of the stem's interior is soft pith, so for strength the sunflower relies on a layer of collenchyma below the epidermis and on the fibers of the vascular bundles.

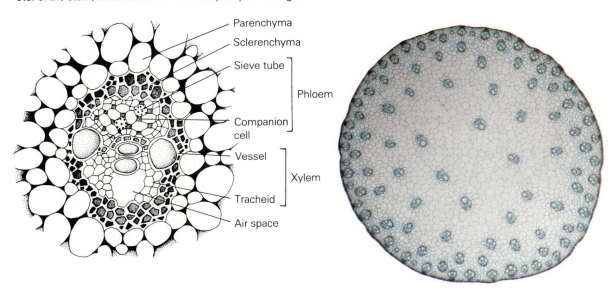

(a) Monocot (corn) vascular bundle

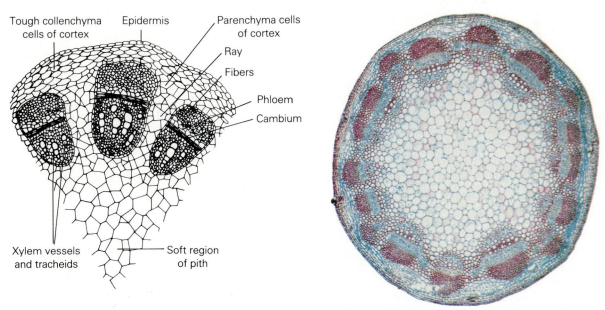

(b) Dicot (sunflower) vascular bundles

22.22 SHOOT ANATOMY

This dormant deciduous woody stem tip reveals two years of growth history. The dormant terminal bud will emerge as the next season's growing tip. The last season's growth can be determined by measuring the distance between the terminal bud and the first terminal bud scale scar. The distance between the two terminal bud scale scars marks the growth of the season before.

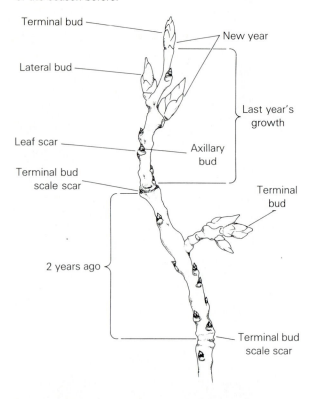

staggered branches. But under certain conditions, depending on the position of the bud relative to the apex and other factors, the lateral branch bud may be released from its dormancy and sprout into a lateral branch. Or, if a principal branch is lost (or removed through pruning), a nearby dormant apical meristem, suddenly released from its auxin-inhibited state, will begin a growth surge, thus replacing the branch.

It should be noted that grasses do not follow the path of growth that we have described here, perhaps, it is thought, because grasses evolved under strong pressure from animal grazers. If grasses grew from their tips, it would be difficult for them to recover from grazing. But new growth in grasses continually emerges from a meristematic reserve near the roots called the **basal meristem.** Thus, the loss through grazing does not interfere with growth.

Growth of Leaves

In its development, each leaf requires contributions from the basic tissue types formed by primary meristem. Protoderm contributes to the highly specialized epidermis, which, you will recall, has the conflicting tasks of slowing water loss yet admitting air. The light-trapping photosynthetic cells within the leaf are essentially parenchyma, arising from ground meristem. Also originating from ground meristem are the many collenchyma cells that form important supporting elements. And finally, because of the activity of the procambium, the leaf contains an extensive vascular system—xylem and phloem elements—which brings in water and minerals and takes away newly produced foods.

Leaf Anatomy.   The typical dicot leaf (Figure 22.23a) is attached to the stem by a twiglike **petiole,** which extends to the flattened **leaf blade.** The vascular system of the stem passes into the leaf through the petiole and into the blade along a large central vein called a **midrib.** Typically in dicots, the central midrib has a number of major branches or veins that penetrate the blade on either side, branching and rebranching so that no cell in the blade is very far from an extension of the vascular system. The smaller veins are surrounded by specialized cells that form what is called the **bundle sheath,** and all materials passing in or out of the veins must pass through the bundle sheath cells. (You may recall the special role of the bundle sheath cells in C4 plants discussed in Chapter 7.)

Looking at the remaining tissue organization of the dicot leaf (Figure 22.24), we see that the **upper epidermis** (that side most exposed to the sun) consists of large, simple, flattened cells whose outer walls are typically shiny—coated with a waxy layer of cutin. These cells lack chloroplasts and are very transparent. Within the leaf are several layers of photosynthetic parenchyma, or **leaf mesophyll,** the tightly packed, vertically arranged, **palisade parenchyma,** and below this, the very loosely arranged **spongy parenchyma.** The latter is aptly named since the loose arrangement contains numerous spaces filled with water or very moist air. The spaces are vital avenues of carbon dioxide diffusion.

Carbon dioxide enters and water vapor exits the leaf through numerous porelike stomata. Each stoma includes a tiny pore surrounded by two guard cells, which, through changes in shape, can open and close the pores (see Chapter 24). The stomata are found primarily in the **lower epidermis**—the underside of the leaf. Lower epidermal cells also often contain lengthy **leaf hairs,** which give the undersurface a fuzzy appearance. Leaf hairs tend to impede the movement of air over the pore-marked surface, cutting down on evaporation in windy weather. In addition, some hairs are sharp and hooked and can repel voracious insect larvae.

That's the dicot leaf. The organization of the monocot leaf (see Figure 22.23b) is quite different. Monocots, such as grasses, have no petioles. The leaf consists of two parts; the leaf sheath, which surrounds the stem, and the blade, which extends outward. The sheath is extremely tough, as you can

22.23 DICOT AND MONOCOT LEAVES

(a) Dicot leaves are typically net-veined, with a major, central vein giving rise to smaller veins at either side. The smaller veins subdivide to form finer and finer veins (see inset). **(b)** Monocot leaves are parallel-veined, with a number of similar-sized veins running alongside each other. These are connected by a fine network of smaller veins (see inset). Details of vein structure in each reveal that no cell in the leaf is far from the vascular system.

(a) Dicot—net venation

(b) Monocot—parallel venation

22.24 LEAF ANATOMY

Cross sections through a dicot leaf clearly reveal the tissue organization within. The tightly packed layer of long cells is the palisade parenchyma, the first photosynthetic cells to receive incoming light. Below this lies the spongy parenchyma, a loosely arranged layer that contains many air- and water-filled spaces. These spaces communicate directly with the stoma in the lower epidermis. Note the presence of smaller veins in the thinner leaf region. The midrib contains the major leaf vein along with regions of thick-walled, supporting collenchyma, which gives strength to the midrib and petiole.

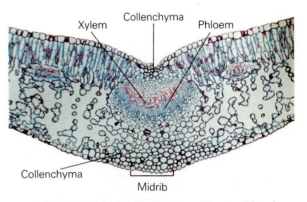

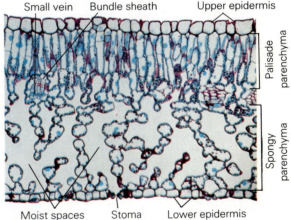

attest if you have ever tried to tear off a corn or bamboo leaf. In addition, the veins do not branch from a central midrib but emerge in parallel rows, interconnected by many small veins.

Secondary Growth in the Stem

We have been considering the products of primary growth in plants; now we will look at what happens in those plants capable of secondary growth and development. While primary growth provides the growing plant with its basic tissues and accounts for growth in length, such growth is highly limited.

Secondary growth is necessary if the plant is to continue to thrive season after season. Secondary growth begins where primary growth ends, in the maturing root and shoot. It has some fascinating, and perhaps even bizarre, aspects.

Our discussion here can be simplified by noting that secondary growth in roots and stems is similar in many ways, so we will confine our discussion to the stem. Secondary growth arises from regions of undifferentiated cambium that become active. As we have mentioned, secondary growth among angiosperms is characteristic of dicots and is most notable in woody shrubs and trees.

The Transition to Secondary Growth.   As we have seen, primary growth in the dicot stem ends with the vascular bundles becoming arranged into a ring or cylinder. A tiny region of procambium remains between patches of xylem and phloem elements. Remaining between the bundles are areas of cortex consisting of versatile parenchyma cells. We will simply mention the highlights of events to follow since the details are more easily described with illustrations, as in Figure 22.25.

As the procambium between the xylem and phloem becomes active, rapid cell division ensues, and the expanding tissue links up with the cortical

22.25 FROM PRIMARY TO SECONDARY GROWTH

Because woody dicots grow continuously, it is possible to examine all stages of growth in the same plant. Primary growth occurs near the stem tip **(a)**, where the newly forming vascular bundles are seen. **(b)** In a slightly older region further down, primary growth is complete and the vascular bundles have formed a distinct ring. Each bundle contains a region of phloem and a region of xylem separated by remaining procambium. **(c)** In the next oldest region, evidence of secondary growth is seen in the continuous ring of vascular or secondary cambium that has formed. **(d)** The events of secondary growth will continue until the separate bundles become one

complete ring of secondary phloem and xylem. The newly emerging phloem will crowd older tissues outward as it progresses toward the stem's perimeter. The secondary xylem will crush most of the inner tissues, leaving a small amount of pith and some primary xylem. **(e)** Further down the stem, in a two-year-old region, one complete annual ring is seen and the formation of another one is in progress. In the latter, the vascular cambium has formed second-year xylem within and second-year phloem outside. Note the mature periderm with its cork cambium and cork cells.

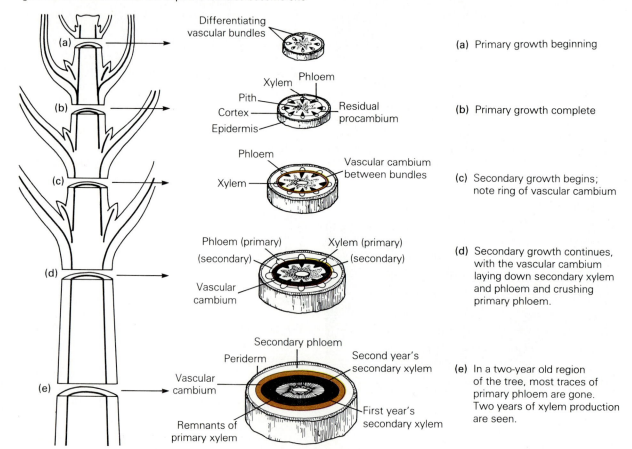

(a) Differentiating vascular bundles

(a) Primary growth beginning

(b) Xylem   Phloem   Pith   Cortex   Epidermis   Residual procambium

(b) Primary growth complete

(c) Phloem   Xylem   Vascular cambium between bundles

(c) Secondary growth begins; note ring of vascular cambium

(d) Phloem (primary) (secondary)   Xylem (primary) (secondary)   Vascular cambium

(d) Secondary growth continues, with the vascular cambium laying down secondary xylem and phloem and crushing primary phloem.

(e) Secondary phloem   Periderm   Vascular cambium   Second year's secondary xylem   First year's secondary xylem   Remnants of primary xylem

(e) In a two-year old region of the tree, most traces of primary phloem are gone. Two years of xylem production are seen.

22.26 LENTICELS

Lenticels are eruptions in the periderm that admit oxygen into the metabolically active stem and permit carbon dioxide to escape.

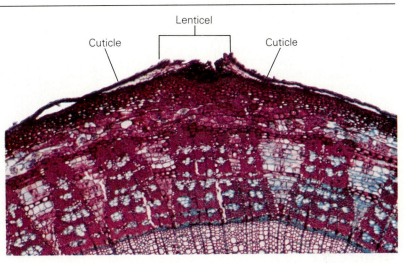

cells between the vascular bundles. This forms a cylinder of active tissue, now called the **vascular cambium,** or **secondary cambium.** As the cells in the ring of newly formed vascular cambium divide, some will differentiate into new xylem and others into new phloem, just as the procambium did in primary growth, although it is more accurate to now refer to these tissues as **secondary xylem** and **secondary phloem.** Vascular cambium also contributes cells to the rays—spokelike lines of cells discussed below.

As growth continues, the vascular cambium forms a ring of rapidly expanding tissue, with secondary xylem forming inside the ring and secondary phloem forming outside. As you might expect, pressure resulting from the rapid production and growth of new cells in internal regions of the stem can cause mechanical problems. As new phloem and xylem emerge, their outwardly expanding front reaches delicate older phloem and parenchyma tissues, crushing them and eventually obliterating this primary organization. Soon, only the tougher primary xylem remains in the pithy center to mark the earlier stages. The newly emerging xylem enlarges and matures, and as it grows it pushes the vascular cambium outward. Likewise, the new phloem pushes outward, crushing the fragile parenchyma tissue of the cortex. Eventually the ring of vascular cambium and phloem comes to rest near the perimeter of the stem. Thus, we see that secondary growth also has its brutal aspects.

Secondary growth in the vascular tissues within the stem is accompanied by changes in the shoot epidermis, a primary tissue. Cells in the neighboring cortex take on a new role, producing what becomes the **cork cambium** (see Figure 22.25). The cork cambium, like the vascular cambium, undergoes continuous cell division, contributing layer after layer of cells that expand outward, rupturing and replacing the old epidermis. This new tissue constitutes the shoot's periderm, whose function is to provide a tough, water-resistant covering over the stem. The outer cells of the periderm will be impregnated with waterproofing suberin and become the plant's **cork.** Cork is sometimes called **bark,** but technically, *bark* refers to all stem tissue outside the secondary xylem, including the vascular cambium, phloem, and periderm.

For the record, the cork in wine bottles is true cork. But it doesn't occur naturally, not even on the cork oak from which it is obtained. Cork growers must remove the natural periderm of the oak, causing the tree to respond to the injury by "obligingly" forming a new, smoother regrowth. This is then stripped away periodically, cut into cylinders and placed into wine bottles in such a way that they almost invariably break or crumble when you try to get them out.

The cork layer is highly water-resistant, but it is not entirely sealed. The living tissues below require a constant supply of oxygen, and they must release carbon dioxide, just as we would expect from metabolically active cells. It turns out that this vital gas exchange occurs through minute openings in the periderm known as **lenticels.** Lenticels (Figure 22.26) are most numerous over metabolically active stem regions, and they are also found in the outer skin of pears, apples, and some other fruits.

22.27 OLDER WOODY STEM

Cross sections through woody stems, such as the two-year-old stem seen here, reveal several different tissues. The outermost region is the periderm, which contains a layer of suberized cork followed by cells produced by the cork cambium just within. A narrow band of cortex follows, and just inside this is the phloem region. Notice the triangular patches of phloem separated by widening rays of phloem parenchyma. Just inside the phloem is the all-important vascular cambium, which seasonally produces new growths of phloem and xylem. Finally, beginning just within the ring of vascular cambium is the xylem or wood of the tree. Each of the distinct annual rings within represent a season's xylem production. At the very center is a small region of residual pith.

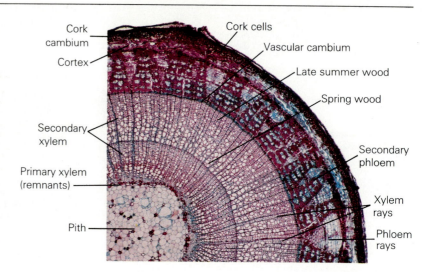

The Older Woody Stem.   Older regions of the woody stem, those that have gone through several seasons of secondary growth, are mostly composed of nonliving xylem tissue—the "wood." The vascular cambium, phloem, and periderm—all lying outside the xylem—form a rather thin ring of living material around the perimeter of the stem. In fact, while many large trees can survive a burned out interior or woody region, removing a strip of bark all the way around the trunk (called girdling) will kill even the most robust forest tree. Why is this? Consider what the loss of the phloem might mean to the roots of a tree.

Cross sections of older woody stems reveal a definite pattern in trees that undergo seasonal growth. The patterns are easily seen as the familiar **annual rings** of temperate zone trees (Figure 22.27). Annual rings form because growth conditions change during each season. In the north temperate zone, trees grow rapidly in spring and summer, tapering off in autumn and stopping growth in winter. The earliest layers of xylem cells are very large, while those produced later in the season tend to be smaller. There is normally less water available and less time for cell enlargement as summer passes and autumn draws near. The tree's annual rings can be counted to determine the age of the tree.

Also in the woody stem, another growth pattern is seen in the form of radiating spokes that extend through the wood. These are called **vascular rays,** or simply **rays.** They consist of strands of parenchyma, produced by the vascular cambium. Vascular rays provide a means of lateral transport and help accommodate the stress created by rapidly expanding cylinders of xylem as the tree grows in girth. Without the rays, such stress could readily split the trunk of a tree.

APPLICATION OF IDEAS

1. Characteristically, the great diversity in angiosperm floral parts can partly be explained through the dependence of plants on animals for pollen transfer. However, one large group has become adapted for pollination by wind. Which group is this? Would you expect much diversity in the flowers of this group? In what other ways might wind-pollinated flowers differ from those pollinated by insects? Compare the specific structures in a typical wind-pollinated angiosperm flower with that of an insect-pollinated flower.

2. An important characteristic of many plants is indeterminate growth. What does this mean and how does the organization of plants provide for such growth? Do any animals share this characteristic? Is there anything comparable in the other kingdoms? (Consider a clone of protists.)

3. A young boy carved his initials at about eye-level in a small tree. Returning as a man, he noted that both he and the tree had grown and that the initials were still there. In what position were the man's eyes when he read the initials—looking up, down, or straight ahead? Explain your answer with a brief but technical discussion of growth in the woody stem.

4. Dendrochronology involves the study of growth rings in wood stems—both intact and fossilized. Of what do growth rings consist and what kinds of information might a dendrochronologist gain by their study in plant fossils? How might this information be useful?

KEY WORDS AND IDEAS

SEXUAL REPRODUCTION IN THE FLOWERING PLANT

The Flower

1. The flower consists of modified leaves arranged into **whorls,** which include:
   a. **Accessory parts:** the **receptacle,** the **calyx (sepals),** and **corolla (petals).**
   b. Sexual parts: the **androecium (stamens** or **microsporophylls,** consisting of **filaments** topped by **anthers),** and the **gynoecium** or **pistil** (one or more **carpels** or **megasporophylls,** each of which contains an **ovary,** a necklike **style,** and a **stigma).**
2. Floral structure varies with the pollinating agent. Attractants include, odor, **nectar** (within **nectaries),** pollen, color, and structural pattern.
3. Floral parts in monocots occur in threes and its multiples, while dicot floral parts occur in fours or fives or their multiples. More primitive flowers are **radially symmetrical,** while advanced types are often **bilaterally symmetrical** or **irregular.**
4. **Complete flowers** have all four basic parts, but **incomplete flowers** may have one or more missing. **Perfect flowers** have both pistils and stamens but **imperfect flowers** are **pistillate** (female—lacking stamens) or **staminate** (male—lacking pistils).
5. Some flowers are clusters, or **inflorescences,** and if they contain individual flowers, each with a single fruit, they are **composite flowers.**
6. Wind-pollinated flowers such as grasses are inconspicuous, lacking colors and odors.

Sexual Activity in Flowers

1. Flowering plants are heterosporus, producing megaspores and microspores, with the gametophyte restricted to cells in the ovules and anthers.
2. Ovules originate in **placental** tissue to which they are connected by a **funiculus.** An ovule is a megasporangium or **nucellus,** containing a **megaspore mother cell** and accompanied by skinlike **integuments.** An opening, the **micropyle,** occurs at one end.

3. In **megasporogenesis,** megaspore mother cells undergo meiosis, producing four haploid products. Three disintegrate but the fourth divides mitotically three times, producing an eight-nucleate **megagametophyte,** later called an **embryo sac.** Upon cell division, six small, single-nucleated cells emerge, one of which is the **egg cell.** A seventh, large **central cell** (or **endosperm mother cell**), containing two **polar nuclei,** will later give rise to the **endosperm.**
4. Microsporogenesis begins in the **anthers,** where each **microspore mother cell** undergoes meiosis, producing four haploid **microspores.** The nucleus of each divides mitotically, producing a **generative cell** and a **tube cell,** thus giving rise to the male gametophyte or **microgametophyte,** commonly referred to as a **pollen** grain.
5. **Pollination** is the transfer of pollen onto a stigma. Many plants have structural or physiological adaptations for avoiding self-pollination.
6. Germinating pollen produces an enzyme-containing **pollen tube,** which digests its way into the ovary. At the time of pollen tube growth (or earlier) the generative cell undergoes mitosis, producing two sperm, one of which fertilizes the egg and the other, the central cell. In a **double fertilization,** the first fertilization gives rise to the diploid zygote, and the second, the triploid **primary endosperm nucleus.**
7. Following fertilization, the ovary matures into the seed-bearing **fruit.**

Embryo and Seed Development

1. Mitosis and differentiation in the primary endosperm and zygote will produce the **embryo,** which along with a food supply and **seed coat,** make up the seed.
2. Developmental activity often begins with the endosperm, which undergoes repeated mitosis, becoming multinucleated. Then the embryo begins its mitotic divisions. The early embryo is a ball of cells anchored by the large **basal cell** of the **suspensor.**
3. In the dicot, the embryo takes on a heart shape as two cotyledons take form. In monocots, the

cotyledon is single. As the bean cotyledons grow they make a turn and enlarge greatly, enclosing the **embryo axis** and comprising much of the seed.

4. Continued growth in the embryo produces a **shoot apical meristem** at its tip (made up of simple, uncommitted **meristem tissue**), which in some species is surrounded by a **plumule.** Any growth above the origin of the cotyledons is called **epicotyl.** Primary tissues include the outer **protoderm,** central **procambium,** and **ground meristem.** Most of the embryo axis in the bean is **hypocotyl.** At its base is the **root apical meristem,** which produces the **radicle.**

5. In monocots such as corn, each kernel develops from an ovary so it is technically a fruit. The fruit includes a starchy endosperm, an embryo, and a digestive structure, the **scutellum** (cotyledon). The endosperm forms an **aleurone,** which is surrounded by the **pericarp.** The corn radicle and plumule are protected by a **coleorhiza** and **coleoptile,** respectively.

Seed Dispersal
1. Successful dispersal of seeds reduces competition and improves the chances of survival.

2. Many seeds enter a dormant period in which they harden through dehydration. Some dormant seeds have remained viable for thousands of years.

3. Seeds are dispersed by water, wind, and animals. Wind-dispersed seeds may have winglike structures or plumes. Some seeds are eaten by animals and are distributed in the animals' wastes.

GROWTH AND DEVELOPMENT IN FLOWERING PLANTS
Germination and Growth of the Seedling
1. For many seeds, **germination** (the emergence from dormancy) simply requires the uptake of water, suitable temperatures, and the availability of oxygen, but others require special conditions such as exposure to freezing temperatures, fire, abrasion, or exposure to animal digestive enzymes.

2. Following the uptake of water by monocot grains, hormones called **gibberellins** stimulate the aleurone to release enzymes that hydrolyze the endosperm starches into sugars. Next the radicle emerges from the coleorhiza and grows downward, absorbing water and anchoring the seedling. The coleoptile-enclosed shoot grows upward, breaks free, and unfurls the first leaf.

3. Seedling growth in the bean, a dicot, begins with the emergence of the hypocotyl, which forms a bumperlike **hypocotyl hook.** Its growth raises the seed out of the soil, whereupon the epicotyl rapidly elongates and spreads the plumule out to face the sunlight.

Primary and Secondary Growth
1. **Primary growth** includes growth in length in the shoot and root along with the production of leaves, flowers, branches, and branch roots. **Secondary growth,** seen primarily in dicots (and in conifers), provides for growth in thickness.

2. Perennial plants have **open growth,** or **indeterminate life span** (unlimited growth); thus they have no maximum life span. Continuous growth is made possible by a reserve of undifferentiated meristematic tissue, but mature plant cells can also undergo **dedifferentiation** and then differentiate into some other cell type.

Tissue Organization in the Plant
1. Apical meristem produces three kinds of **primary meristems:** protoderm, ground meristem and procambium. Each gives rise to three specialized tissue systems.

2. Protoderm produces the **dermal system,** including leaf epidermis and **guard cells,** which surround the pores, forming the **stomata,** root epidermis with **root hairs,** and the complex **periderm.** Waterproofing materials include **suberin** (in the periderm) and **cutin** of the leaf or stem **cuticle.**

3. Ground meristem differentiates into the very common thin-walled **parenchyma** of leaves, stems, and roots; the tough supporting **collenchyma fibers** of stems; and the **sclerenchyma,** including **sclereids,** which in the pits of fruits occur as **stone cells.**

4. Procambium produces **primary xylem** and **phloem.**
    a. Xylem includes water-conducting **tracheids** and **vessels** and supporting xylem fibers and **xylem parenchyma.** The tracheids are long and slender and have pitted end and side walls (**pits** are thin regions containing primary wall only). **Vessel elements**—the cellular units of vessels—are wider and shorter with open end walls and pitted side walls.
    b. Phloem consists of food-conducting **sieve tube elements, phloem parenchyma,** and **companion cells.** Sieve tube elements end in **sieve plates,** which contain plasmodesmata. When assembled they form **sieve tubes** with a continuous cytoplasm. Since sieve tube elements lack nuclei, nuclear functions are carried out by companion cells, which are also involved in the active transport of nutrients into sieve tubes. Phloem parenchyma specializes in storage and active transport.

Primary Growth: The Root Tip

1. The **root tip** contains a mitotically active root apical meristem, which gives rise to replacement cells for the **root cap** region ahead and primary tissues behind.

2. Behind the meristem, cells stimulated by the hormone **auxin** elongate, pushing the root tip through the soil. In the primary tissues, protoderm cells produce the young epidermis, which forms numerous root hairs, while those of the ground meristem begin formation of the root **cortex** and **endodermis.** Within the endodermis, procambium forms the xylem and phloem of the **stele,** a cylinder of conducting and supportive tissue.

3. The stele includes an outer **pericycle** and regions of primary xylem and primary phloem. In plants capable of secondary growth, procambium forms regions of **vascular cambium,** which later gives rise to **secondary xylem** and **secondary phloem.** The pericycle gives rise to **branch roots.**

4. Monocot root systems often include **adventitious roots** that emerge from the stem. **Tap root systems** have a prominent main root and many branching or secondary roots, while all of the roots in **diffuse root systems** are similar in size.

Primary Growth in the Stem

1. A dome-shaped shoot apical meristem gives rise to cells that elongate, producing shoot growth. When the young primary tissues differentiate, they form a thin epidermis along the outside and islands of vascular tissue which surround an inner region of thin-walled parenchyma called the **pith.**

2. Meristem also gives rise to leaves, **leaf primordia,** and **branch primordia. Nodes** give rise to leaves and mark the ends of **internodes. Lateral branch buds** or **axillary buds** give rise to branches unless suppressed by hormones. **Terminal buds** form **terminal bud scale scars** at the end of each growth season.

3. Grasses emerge continuously from a **basal meristem.**

Growth of Leaves

1. While protoderm contributes to the complex leaf epidermis, ground meristem differentiates into photosythetic parenchyma **(leaf mesophyll)** and procambium gives rise to the leaf vascular tissues.

2. The structures of leaves include a supporting **petiole,** the flattened **leaf blade,** and a central vein called a **midrib.** Smaller veins are surrounded by **bundle sheath cells,** through which materials must pass.

3. A cross-section through the leaf reveals a simple **upper epidermis,** then **palisade parenchyma,** followed by **spongy parenchyma** with air spaces, and then the more complex **lower epidermis.** The lower epidermis includes cells with extensive **leaf hairs** and guard cells that surround the porelike stomata.

4. While veins in dicot leaves tend to be branching (net-veined), those in monocots generally run parallel (parallel-veined).

Secondary Growth in the Stem

1. During secondary growth, secondary xylem and phloem arises from vascular cambium **(secondary cambium).** The growth of newly produced **secondary xylem** and **secondary phloem** expands outward, forming a continuous ring and crushing older growth in its path.

2. In secondary growth, the epidermis is replaced by a periderm or **cork,** originating from **cork cambium** that forms in the cortex. The outer cells are suberized, but gases can be exchanged through the **lenticels.**

3. Older woody stems are composed mainly of xylem or "wood," with the living tissues restricted to a thin outer layer called **bark. Annual rings** in the wood represent seasonal growth patterns. Spokelike patterns—the **vascular rays,** or **rays**— aid in transport and help the stem expand.

REVIEW QUESTIONS

1. List the four major regions of the flower. Which are accessory and which are the "male" and the "female" parts? (536–537)

2. What are equivalent floral terms for microsporophyll and megasporophyll? What is the function of each? (536)

3. What colors and odors would you expect in flowers that are pollinated by the following: bats, carrion flies, bees? (537)

4. What is an imperfect flower? Can an imperfect flower be complete? Explain. (538)

5. From the viewpoint of natural selection, explain why grasses lack colorful, fragrant flowers. (538)

6. Describe events in a megaspore mother cell that lead to the production of the eight nucleate megagametophytes. (539)

7. List the parts of an embryo sac. Which participate in fertilization? (539)

8. Trace the events that carry a microspore mother cell through microsporogenesis, ending with a microgametophyte. What is another more common name for the latter? (539)

9. Beginning with pollination, describe fertilization in the flowering plant. (539, 541)

10. Give a technical definition of the term *fruit*. Describe the fruit of corn and string beans. (542)

11. What are the three parts of a seed? (544)

12. Trace the development of the dicot embryo. Include a description of the suspensor, the heart-shaped embryo, and the cotyledons. What are the three primary tissues? (544)

13. Using a simple drawing of an embryo, locate the epicotyl, a hypocotyl, radicle, and plumule. From what groups of cells does each arise? (544)

14. What are two basic differences between the monocot and dicot embryo and seed? (549)

15. List three agents of seed dispersal. Why is dispersal an important problem to plants? (549–550)

16. Of what strategic value to the plant is a sweet, fleshy, and nutritious fruit? (550)

17. Explain how the growing grain embryo is provided with a supply of simple sugars. (551)

18. Compare the early growth of the corn seedling with that of the string bean. (552–553)

19. Distinguish between primary and secondary growth. List two groups of plants in which secondary growth is possible. (553)

20. How do plants make provision for indeterminate growth? (533)

21. In general, what does the protoderm provide? List three of its tissue derivatives. (544)

22. List three tissues derived from ground meristem, and mention a function of each. (554)

23. Compare the structure of tracheids and vessel elements. Which would seem better able to transport water? Describe the structure that makes provision for lateral transport among tracheids and vessel elements. (554)

24. List the three kinds of cells in phloem tissue and state a function for each. (557)

25. What is the function of the sieve tube? List two provisions that make transport possible from one sieve tube element to the next. (557)

26. How is the sieve tube element able to function without a nucleus? (557)

27. Prepare a diagrammatic drawing of a root tip, labeling the following: root cap, rood hairs, epidermal cells, apical meristem, stele, cortex, regions of elongation and maturation, procambium, ground meristem, and protoderm. (557, 559)

28. Draw a cross section of a mature region in a root that has completed primary growth, labeling the stele, epidermis, endodermis, pericycle, cortex, primary xylem, vascular cambium, and primary phloem. (559–560)

29. Contrast the tap root system with the diffuse root system. In what plants would one most likely find the latter? (560)

30. Compare the arrangement of the vascular system in the young shoot with that of the root. (562)

31. What might determine whether a lateral branch bud will sprout? (562, 564)

32. Draw a simple representation of a dicot leaf and monocot leaf, labeling midrib, blade, petiole, leaf sheath, net venation, and parallel venation, as appropriate. (564–565)

33. Visualizing a cross section of a dicot leaf, name the tissues beginning with the upper layer. List two special features of the lowermost tissue layer. (564–565)

34. Explain how secondary growth changes the arrangement of vascular tissue in the stem from bundles to rings. (566–567)

35. What is the role of cork cambium? From what tissue source does it arise? (567)

36. Construct a "pie-slice" drawing of a three-year-old stem, and label the important tissue areas. Include the cork and cork cambium, cortex, secondary phloem, phloem rays, vascular cambium, secondary xylem, and annual rings. (568)

Mechanisms of Transport in Plants

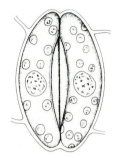 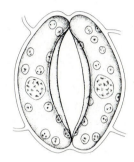 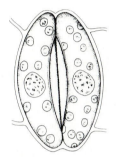 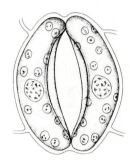

As living things go, we can count ourselves among the giants. The average size of organisms on the planet is closer to that of a mouse. However, there are many living things far larger than we are and the greatest of these are plants—towering, invincible plants, great trees that have resisted every threat except our need for lawn furniture. Why have so many of the plants become so large? One reason is that there are decided advantages to being tall, especially if one is a forest plant, since here there is intense competition for sunlight. The lofty giants are able to expose their leaves to unobstructed light, thereby outcompeting other species. Also, the vast root systems typically produced by large plants allows them good access to water and minerals and provide firm anchorage. Furthermore, tall trees can disperse their seeds farther from home than can their shorter competitors, thus lessening competition with their own offspring. There are, then, a number of advantages to being a forest giant.

Largeness, however, has its price. The scaly leaves of a giant redwood (Figure 23.1) constantly lose water to the atmosphere. That water must be replaced by a root system in the soil far below. On the other hand, those roots must be fed by carbohydrates manufactured in the distant leaves. The plant must somehow bring water and minerals from the soil up to its leafy canopy and food from the leaves down to the roots. Thus, large plants have had to pay for their large size by developing extensive vascular systems for the transport of water and nutrients.

But the cost of large size does not end there. The high forest canopy must be supported, sometimes through strong winds and heavy rains or even snow. Therefore, much of the plant tissue must be devoted to physical support. The forest giants, and other higher plants as well, solve the problems of both support and water transport with the same tissue—the woody **xylem.** Xylem is well adapted for its functions. Wood is marvelously strong supporting material, and enables plants to raise water many meters above the earth. This ability has long mystified human engineers since the water is lifted with *little apparent direct expenditure of plant energy*. It is important to point out that there are no known active transport mechanisms that move individual water molecules.

The other part of the vascular system is the **phloem,** which specializes in carrying food molecules. Phloem tissue is living. Even its **sieve tube elements** are living, although they lack nuclei. The distribution of food molecules does require the plant to expend energy, but there are mysterious aspects to this sort of movement as well. Food molecules do not haphazardly pass from the leaves downward, but are somehow routed according to the specific requirements of the different tissues below. Furthermore, the distribution can shift as conditions change. Food can even be moved upward from storage regions in the root; in spring, trees and bushes that have gone through the winter with bare, leafless branches routinely move food molecules up from their roots to the newly forming leaves and twigs. There are a number of riddles regarding the movement of nutrients through the phloem. We will come back to these later, but first let's look closely at the role of xylem.

23.1 A KING-SIZED TRANSPORT PROBLEM

This towering California coast redwood, *Sequoia sempervirens* is the world's tallest tree (107m or 350 ft). If you know of the problems of water transport, your appreciation for these giants can only increase as you visit one of the remaining stands. The tree is also awe-inspiring to foresters who stand in reverence at visions of board-feet and profits after taxes.

THE MOVEMENT OF WATER AND MINERALS

Water from the soil enters the plant through countless thin-walled root hairs, crossing the root cortex mainly along the porous cell walls between cells but some passing through the cytoplasm as well. The water then passes through the endodermis and moves into the **stele** (the root vascular cylinder that contains both xylem and phloem; see Chapter 22). Upon entering the stele, water passes into the xylem, the vessels and tracheids of the root vascular system. It then continues through the xylem of the stem and on to the foliage. There, it passes out of the xylem, moving through numerous cells to eventually enter the air spaces in the leaf's spongy mesophyll. It is in the moist air spaces that evaporation occurs, and water, in its gaseous state, diffuses out of the plant through the leaf stomata (Figure 23.2), except for a relatively meager but vital amount that is used in photosynthesis. Water loss from the leaf through evaporation and diffusion is called **transpiration.**

A great deal of water is shifted from the soil to the atmosphere through transpiration. A single potato plant, for instance, can lose about 95 liters of water in one growing season, and in the same amount of time 206 liters of water can evaporate from the leaves of a single corn plant (see Table 23.1). In Chapter 41 we will find that the transpiration from a forest has immense ecological and meteorological significance, affecting both organisms and climate. (We are learning this too late in the Amazon rain basin where vast changes in rainfall patterns are in store as the forests are ravaged.)

How is so much water moved through a plant? And more to the point, *why* is so much water moved? Answering the second question is relatively simple. Plants require water to carry on photosynthesis and to provide the turgor pressure that holds leaves and young stems erect. The first question is a little tougher. The problem of water moving up through, say, a tall tree, boils down to two basic explanations. Either something is *pushing* the water up or something is *pulling* it up. The first explanation that often comes to mind is that somehow the root provides the force needed to push water up into the leaf. After all, water must enter the root first. Let's see if this is true.

23.2 PLANT WATER POTENTIAL GRADIENT

The movement of water and minerals through a plant follows an established water potential gradient from soil water to air surrounding the leaf. The transport of water is passive, while that of minerals is often active. In addition, the active transport of ions into the plant and from cell to cell often sets up water potential gradients.

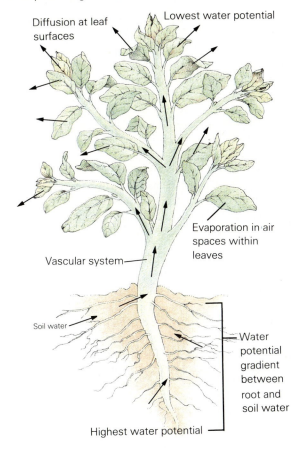

Diffusion at leaf surfaces

Lowest water potential

Vascular system

Evaporation in air spaces within leaves

Soil water

Water potential gradient between root and soil water

Highest water potential

TABLE 23.1

TRANSPIRATION RATES IN LITERS PER DAY FOR SELECTED PLANTS

	Liters per day
Cactus	1/50
Tomato	1
Sunflower	5
Ragweed	6
Apple	19
Coconut palm	75
Date palm	450

Root Pressure: Is the Root a Kind of Pump?

If you've ever celebrated into the wee hours of the morning, you may have gotten your knees wet as you crawled home across your lawn. At the time, you may have thought the moisture was dew. Perhaps it was, but often the presence of water drops on the grass is the result of its having been pushed up by the roots through the plant's tiny vascular system. The blades of grass have special openings near their tips, where excess water can escape. Such loss of water in its *liquid* phase is known as **guttation,** not to be confused with transpiration, which involves evaporation and water vapor. Guttation occurs mainly at night, when evaporation is reduced and water excesses occur. The water has been forced up to the tips of the grass blades by **root pressure** from below.

Root pressure, a water-moving force developed in the root through osmosis, occurs when ions are actively secreted into the stele from nearby living cells. The presence of ions in the xylem fluids sets up a water potential gradient (see Chapter 5), and water moves by osmosis into the stele. The endodermis, we know, has the ability to actively transport ions in one direction, resisting their diffusion back into the cortex. If the root is a pump, then the pressure the pump generates can be traced back to an energy-requiring mechanism, the active transport of ions, which creates an osmotic condition that encourages the passive inward movement of water.

Root pressure can be demonstrated by removing the stem from a small plant and attaching a mercury manometer in its place (Figure 23.3). A rise in the manometer's column of mercury is used as a measure of atmospheric pressure. With this apparatus it has been possible to demonstrate that roots can generate pressures of about 3 to 5 atmospheres (3 to 5 times greater than that of the atmosphere). So the immediate question is: Is this sufficient to push water to the top of a tall tree? *The answer is no. The weight of a 100-meter column of water is at least two times too great to be supported by a root pressure of even 5 atmospheres.* The weight of such a column ought to push water right out of the vascular cylinder, back into the earth. Yet trees continue to stand there with water constantly moving upward through their enormously tall stems and out the lofty foliage.

One more observation should dispel the notion of root pressure as a total explanation of upward

23.3   DETECTING ROOT PRESSURE

Root pressure can be measured by substituting a manometer for the stem and foliage of a potted plant. Mercury in the S-shaped tube rises as water is exuded from the vascular system of the stem. The force originates in the root where solutes in the cells produce an inward water potential gradient. But although the existence of root pressure can be demonstrated, measurements reveal that it cannot account for the rise of water to the leaves in taller plants.

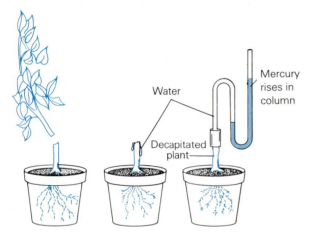

water transport. During periods of peak water loss, the pressure inside the trees is less than the pressure outside. If the roots were generating great pressure, cutting into the xylem would produce a *spurt of water*. Actually, however, cutting produces a sucking action as air is drawn *in*. In summary, then, root pressure apparently plays a role in water transport, but it cannot be considered to be a major force, at least not in tall trees.

Most plant physiologists now believe that root pressure, when it occurs, is only an indirect effect of the active transport of inorganic nutrients, such as nitrates, potassium, and phosphates, into the roots. So we find that roots can provide only a modest push, certainly not enough to explain water transport through tall plants. Therefore, we must consider the alternate explanation, the possibility of a pulling force. Is it possible to pull water? This is an unusual prospect, and one for which we will have to lay out some background.

We might first review the characteristics of water and its movement. In a truly literal sense, the vascular plant has evolved "around" such characteristics. The special peculiarities of root hairs, the root endodermis, the xylem vessels and tracheids, the leaf mesophyll, and the stomata are all adapted to take advantage of the special properties of water.

Water Potential and the Vascular Plant

All the forces that cause water to move depend on water potential. However, water moves from regions of higher water potential to regions of lower water potential, regardless of which forces are at work. Water potential within cells is dependent on how dilute the cell's fluids are. The *more* solutes there are dissolved in the water, the *lower* its water potential will be. Conversely, if the solute concentration is *low*, the water potential will be *high*. Since water moves from the soil into the root and from the root through the stem to the leaves, and finally out of the leaves to the surrounding air, we can infer that the water potential is high in the soil, somewhat lower in the plant and lowest in the surrounding atmosphere.

When, for any reason, this gradient is disrupted, the plant is in trouble. This can happen, for example, when solutes in the soil water (such as nitrates and phosphates in fertilizers) are too concentrated around the plant. Either situation would increase the osmotic potential and decrease the water potential in the soil water outside the root. Thus, water would tend to move out of the root. In response, the leaves would rapidly lose water and with the subsequent loss of leaf cell turgor, they would collapse or wilt. Fortunately, the loss of turgor also occurs in guard cells, which would thereby change shape, closing the stomata and slowing transpiration. If the normal water potential gradient is not disrupted too long, the leaf changes would be reversed.

So far we have only described the movement of water along a water potential gradient. We have not seen how the gradient is established and how it is maintained in the plant. But most important, we have yet to see how such a gradient can be translated into the substantial *pull* needed to move water to the foliage of tall trees. These are some of the problems we will approach now, beginning with the leaf.

The Leaf, Transpiration, and the Pulling of Water

To conduct photosynthesis, plants must permit carbon dioxide to diffuse in from the surrounding air, but because of the normal prevailing water potential gradient, they lose water when carbon dioxide is admitted. The problem arises because of the tissue organization of the leaf (see Chapter 22), which includes a loosely arranged, spongy parenchyma. The open spaces between cells are avenues of diffusion, both for incoming carbon dioxide and for outgoing water vapor.

Evaporation of water is indeed a problem for any terrestrial organism and it is often minimized through the process of natural selection. However, plants may have opportunistically developed a means of using the forces of evaporation to help move water. Let's see how this is possible.

We can begin by noting that water evaporating from the leaf spaces is replaced by water from the cells surrounding the spaces (Figure 23.4). This leaves these cells with a reduced water potential, and a considerably increased osmotic potential. The result is that water will move into these cells from others even closer to the source. The source, of course, is the water-filled xylem of the leaf veins. Thus the continued evaporation of water establishes a steep water potential gradient between the vascular system and the spongy region of the leaf, and water tends to move rapidly down its gradient. In addition to its movement along porous cell walls, water movement through the leaf cells is facilitated by the presence of numerous plasmodesmata—

minute cytoplasmic streams passing through pores in adjacent cell walls and connecting the cytoplasm of one cell to that of another (see Chapter 5).

Transpiration at the leaf surface translates into a powerful water potential gradient within the leaf, which produces a major pulling force. Measurements of the leaf forces reveal that the osmotic pressure of water-depleted leaf cells can reach 12 atmospheres—enough to lift a xylem-sized column of water 130 m high! This force is therefore quite sufficient to raise water from the soil to the tops of the tallest trees.

We have seen that the movement of water from cell to cell in the leaf is always along a gradient from higher water potential to lower water potential, and that it is a *passive* process, not a direct result of biochemical work done by the plant. The process is powered by the water potential gradient set up by transpiration, the evaporation of water from the air spaces of the leaf. The **free energy of evaporation,** as it is called, is an indirect form of solar energy since the sun affects the wind, humidity, and temperature, all of which influence the rate of transpiration. Let's now focus on the effect of these forces, and others, on the xylem.

23.4 WATER POTENTIAL IN THE LEAF

Evaporation from the leaf sets up a water potential gradient, with water moving from the xylem to the moist spaces in the spongy parenchyma. Since water potential is highest in the xylem and lowest in the air outside the leaf, its loss is inevitable as long as the leaf pores (stomata) are open. The loss of water through evaporation constitutes an energetic force, with the energy supplied by the sun. Water movement out of the leaf creates a pull on continuous columns of water in the xylem, so water moves upward from the root.

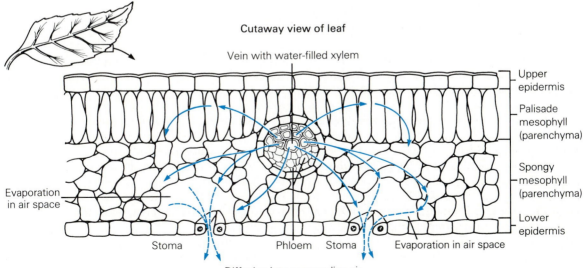

Cutaway view of leaf

Vein with water-filled xylem

Upper epidermis

Palisade mesophyll (parenchyma)

Spongy mesophyll (parenchyma)

Lower epidermis

Evaporation in air space

Evaporation in air space

Stoma

Phloem

Stoma

Diffusion into surrounding air

Water Movement in the Xylem Through the TACT Mechanism

Four important forces influence the behavior of water in the xylem elements: **transpiration, adhesion, cohesion,** and **tension,** which we will call the **TACT** forces. We have already discussed transpiration as a major force in creating the water potential gradient, so let's briefly look at the roles of the others. The second force, adhesion, refers to the attraction of water molecules to materials such as glass or the cellulose in plant cell walls. Cohesion is the attraction of certain molecules to each other. (We considered both in detail in our discussion of water in Chapter 2. You may recall that it is the polarity of water that attracts the molecules to each other.) Tension, the fourth force, refers to the stress produced by exerting a pull on a narrow column of water—namely, the pull on water in the vessels and tracheids of xylem during transpiration.

The adhesion of water to cellulose cell walls may be a major water-moving force throughout the plant since it greatly facilitates water's creeping movement from cell to cell and through the cellulose-lined vessels and tracheids of the xylem. Evidence suggests that it is adhesion that causes water to leave the leaf mesophyll cells and enter the surrounding air spaces where evaporation occurs. Accordingly, some authorities claim that it is these adhesive forces that set in motion the "pull" of transpiration-pull. In support, they point to the immensity of the combined surface area of cells surrounding the air spaces of a leaf. Is the force of adhesion enough to lift long columns of water to the leaves of the tallest trees? It may be, but there is much more to the process.

Cohesion—the attractive force between the molecules of a substance—is particularly evident in water. The positive (hydrogen) ends of one molecule are attracted to the negative (oxygen) ends of the next, forming countless, shifting hydrogen bonds. It is cohesion that gives water its physical viscosity ("liquid-stickiness"). When water is present in a cylinder of narrow diameter such as that of the vessel or tracheid, its cohesive forces are greatly magnified. Such a column has significant tensile strength—the ability to be pulled without breaking. For example, steel wire has more tensile strength than cotton string, which has more tensile strength than cooked spaghetti. Water would seem to have even less tensile strength than cooked spaghetti, but the tensile strength of a column of a fluid depends on all sorts of complex factors and prominent among these is its diameter. By some calcula-

tions, the tensile strength of a column of water in a tiny xylem vessel approaches that of a steel wire of the same diameter!

Tension, as we mentioned, is the stress placed on an object by a pulling force. As you might expect, when a plant is heavily transpiring, the tension on the water columns within the many xylem vessels and tracheids can be enormous. In fact, the pull has a measurable squeezing effect on the trunks of trees; their trunk diameters actually decrease during intense transpiration! (Recall our mention of the faint sucking observed when the xylem is punctured.) Essay 23.1 explains how such delicate measurements are made. The important point is that as tension increases, the cohesive strength of water in the xylem elements prevents breakage, and the pull on the column is transferred downward, creating a decrease in water potential in the root below. In response, water moves into the stele, permitting the upward movement to continue.

Thus, water movement in the plant is caused by the combined effect of four cooperating "TACT" forces. In review, we've seen that transpiration produces a water potential deficit that through a chain reaction produces a water potential gradient, which along with adhesive forces creates a powerful pull on minute water columns in the leaf xylem. Because of incredibly strong cohesive forces characteristic of water in minute columns, the columns remain intact and water is literally lifted through the entire vascular system. This all works very well, but as you might expect, it can only continue if water intake by the root can keep up with the movement of water through the stem and out of the leaf.

Water Transport in the Root

We have already seen that osmotic conditions in the root can provide a certain amount of push in the transport of water. But let's emphasize the point that during intense transpiration, root pressure is totally overwhelmed by transpiration pull and is not considered a major force in the movement of large quantities of water at this time. Root pressure as a force in water transport is probably most important when transpiration is slow or when water potential in the soil water is low. At those times, a proper osmotic gradient in the root would be essential to the continued movement of water into the root and stele. It may also be important much of the time to ground-hugging plants.

Within the root, the ongoing pull occurring in the xylem creates a water potential gradient in the

Testing the TACT Theory

Parts of the transpiration, adhesion, cohesion, and tension (TACT) theory can be readily tested. To demonstrate cohesion and tension, simple techniques similar to those used to demonstrate root pressure can be applied. The stem of a plant is cut (under water so as not to break the water columns) and connected to a water-filled glass tube. (If dye is added to the water, it will soon appear in the leaves and flowers; florists sometimes use this trick to dye flowers odd and unnatural colors.) If the lower end of the water-filled glass tube is put into a dish of mercury, the column of mercury tracks the water, rising in the tube substantially, and indicating a strong pull. However, the analogy fails in some ways, since the glass tubing is far greater in diameter than are vessels and tracheids.

But how can it be demonstrated that it is evaporation and not some other force that is involved in transpi-

ration? A clever mechanical model seems to support the evaporation principle. The setup is the same as before, but this time the top of the glass tube is attached to a porous clay *potometer,* instead of the crown of a plant. There is no air in the potometer, only water. The wetted microscopic pores of the clay potometer permit water to literally creep out to the surface of the clay cylinder through the forces of adhesion and cohesion (much like what is believed to go on in the spongy tissue and air spaces of the leaf). Moisture evaporates from the damp outer surface and the column of water rises, creating a partial vacuum in the tubing that is filled by the rising column of mercury. The rise is much faster if a fan is used to blow air around the potometer, just as transpiration would be more rapid on a dry, windy day.

A very clever experiment lends strong support to the TACT theory.

D. T. MacDougal, a plant physiologist, wondered about the enormous tension which must be placed on water columns in the immense xylem of a tree if the transpiration pull idea were correct. He reasoned that such a tension might even affect the stem diameter, perhaps causing an actual decrease during peak activity. MacDougal tested this idea by devising an instrument that could measure minute changes in the stem. The instrument, a *dendrometer* (later called a *dendrograph* when a recording drum was added), did in fact detect the expected changes. Not only did the stem increase in diameter, but it did so in regular day-night cycles. It is common knowledge that plants transpire much less at night when evaporation rates are low, and MacDougal's measurements coincided with this very nicely.

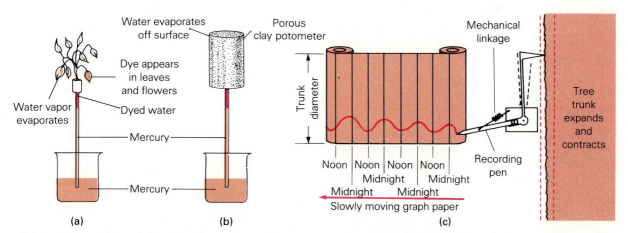

(a) In a demonstration of the pulling force of transpiration, a fresh leafy branch is attached to a vertical glass tube that is filled with dyed water. The other end of the tube rests in a beaker of mercury. As the living branch transpires, dye appears in the leaves and water is drawn into the tube with a force that lifts the mercury up the column. Transpiration pull exists. Furthermore, energy is expended. But how?
(b) In an analogous but completely nonliving system, a porous clay potometer is substituted for the leafy branch. As water evaporates from the surface of the potometer, more water is pulled in from the tube and the column of mercury rises. Tran-

spiration does not require the expenditure of cellular energy. (c) Changes in the diameter of a tree over several days can be registered on a drum recording. D. T. MacDougal, testing the hypothesis that transpiration pull places enough tension on columns of water within the stem to compress the trunk, found that his intriguing data support the hypothesis. The 12-hour variations in trunk diameter correspond nicely with measured transpirational activity. Plants transpire most at midday and least at night. Thus the transpiration pull hypothesis is supported.

23.5 THE MOVEMENT OF WATER IN THE ROOT

While some water passes through the cytoplasmic route in root cells, most moves along the porous walls of cells. This continues until water reaches the endodermis, whereupon it must pass through the endodermal cytoplasm. From there it resumes its transit along cell walls to enter the xylem. During periods of intense transpiration, the pulling force exerted on the xylem is believed to overwhelm all other forces, drawing water through the root to the vascular system.

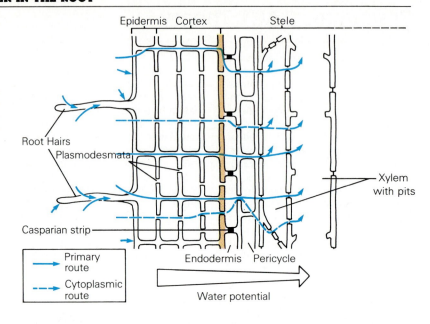

23.6 THE CASPARIAN STRIP

(a) Looking down at a cross section through tissues of the young root (see inset), we see how the partial suberization of endodermal walls forms the Casparian strip. While water readily moves through and along cell walls in the other root cells, it cannot enter the stele in this manner. (b) The Casparian strip forms a watertight barrier between adjacent endodermal cells, so water must pass through the endodermal cell cytoplasm on its way to the stele.

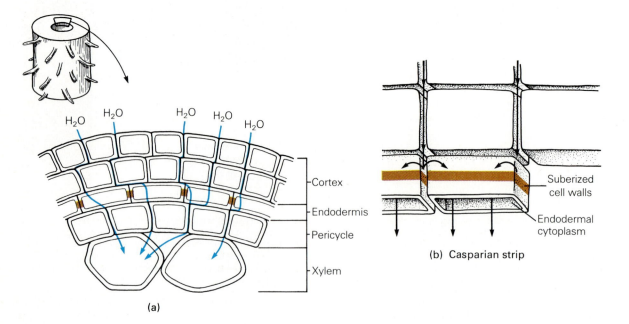

other cells of the stele that continues on across the cortex to the extensive root hairs of the epidermal cells (Figure 23.5). So, as long as soil water is plentiful, water will passively enter the epidermal cells and pass across the root to the xylem. In its transit through the cortex, water passes mainly along the highly porous parenchyma cell walls. But once it encounters the endodermis, its path changes.

Because there is no recourse, water must pass through the cytoplasm of the root endodermal cells. The endodermal walls contain a waxy region—the **Casparian strip**—arranged in such a way as to direct the water flow across the plasma membrane and through the cell cytoplasm rather than along its cell walls (Figure 23.6). The cytoplasm, of course, is a living and highly regulated substance, so the passage of water may not be a simple process there. Because of the Casparian strip, it is possible that the endodermis may exercise both a physical and osmotic directional control over the movement of water, although the importance of this during intense transpiration isn't clear. It is possible that one function of the endodermis is to help maintain an inward water potential gradient by actively transporting mineral solutes into the stele. This is important in preventing water loss when transpiration slows or when water potential in the soil drops. Of course, the endodermis also transports ions into the stele, where they can be carried to the leaves for use in metabolic activity there.

Guard Cells and Water Transport

When aquatic plants made the evolutionary transition to land, an immediate problem was the obvious risk of drying. The evolution of a watertight cuticle helped, but with the terrestrial plants' dependence on atmospheric carbon dioxide, it wouldn't do for them to become completely sealed. In time, the problem was largely solved by the development of the versatile *stoma* (see Chapter 22), formed by an opening between two specialized *guard cells*. The guard cells can open and close the pore, thereby admitting gases but controlling water loss from the plant interior. (Note that the term stoma designates a complete unit: two guard cells and the pore they form.)

Guard cells are located mainly on the underside of leaves (Figure 23.7) but are also found on the green stems of many herbaceous plants and leafless plants such as cacti. They vary greatly in density, from just a few to many thousands per square centimeter (12,000 cm^2 in tobacco).

The size of the stomatal opening is regulated by

23.7 GUARD CELLS

Guard cells are most often found on the leaf undersides, although they also appear on green, photosynthetic stems. **(a)** They occur in pairs, with their borders forming the stomata. Unlike other epidermal cells, guard cells contain chloroplasts and carry on photosynthesis. **(b)** The scanning EM view of a guard cell is as it would appear from inside the leaf.

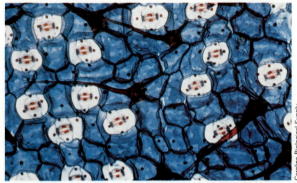

(a)

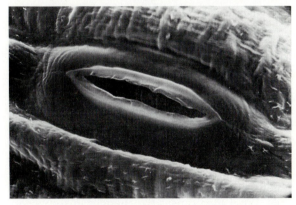

(b)

the turgor of the guard cells (Figure 23.8). Put simply, a pair of guard cells pressed together reduce the stoma, and when pulled apart they open the stoma, allowing gases to diffuse more freely in and out of the plant.

At first glance, it might appear that an increase in turgor should close the stomatal opening. But because the cells' inner walls, those that border the stoma, are much thicker than the outer, the guard cells bend when turgor increases. A loss of turgor permits the thick, resilient walls to straighten, decreasing the stoma.

In general, the turgor increases during daylight and decreases at night. In this manner, gases are allowed to pass through the stomata when the plant is engaged in photosynthetic activity, and water loss is restricted when it is not.

23.8 OPENING AND CLOSING MECHANISM

Guard cells look somewhat like twin sausages, their curved shapes forming the stoma. The control of the stoma depends on the turgidity of the guard cells. As their water content increases, the guard cells assume a greater curvature, increasing the stomatal opening. Note the thickened inner cell walls in both the relaxed and stressed condition. The thick walls are elastic, and when turgor pressure within the cell falls, they straighten, constricting the stoma.

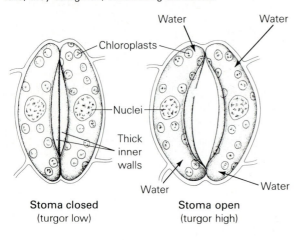

Stoma closed
(turgor low)

Stoma open
(turgor high)

The Guard Cell Mechanisms. How does turgor change in guard cells? As you might expect, water enters and leaves guard cells in response to changes in solute concentrations, which affects water potential. When solutes increase in guard cells, water potential lessens and water enters from surrounding cells. When solutes decrease, water potential increases and water leaves the guard cells. What then causes the changes in solute concentration? The answer is not clear, but several independent factors may be involved. For a time the presence of chloroplasts in the guard cells, which are usually lacking in other epidermal cells, seemed to provide a major clue. The reasoning was that the chloroplasts carry on photosynthesis early in the day and the resulting carbohydrate molecules increase the solute concentration in the guard cell, thus reducing the water potential there. Water passing down the newly established gradient enters the guard cell, increasing its turgor and opening the stoma. Then, when the sun goes down, the processes slow down, sugar is metabolized, and the gradient is reversed. Water leaves the guard cell, turgor decreases and the stomata close. It's a rather neat explanation and it fits our expectations for day/night behavior of guard cells. But it's wrong.

As usual, some restless soul came along and carried out one experiment too many. The sugar-osmosis hypothesis didn't hold water, as it were. It turns out that changes in turgor occur much too rapidly to be explained by the synthesis of sugar solutes in photosynthesis.

In the late 1960s, physiologists found that plants in direct light show an increase in the active transport of potassium ions into the guard cells. (Actually, they made use of an observation first reported in Japan in 1943, but we weren't holding very many conventions with Japanese scientists that year.) It turns out that only the blue part of the light spectrum is effective because the process is triggered by a blue-sensitive pigment. The entrance of potassium increases the solute concentration of the cell, speeding the uptake of water and bringing about stomatal opening (Figure 23.9).

Unfortunately, the **potassium transport mechanism,** which is now well-substantiated, has not answered all the questions. The link between blue light and the guard cell opening may be direct, with the pigment itself powering potassium transport, or it may be indirect, with the light reactions of photosynthesis providing a coded signal for potassium transport. The latter explanation is preferred at the moment, although the relationship between light reaction activity and potassium pumping is not known.

Furthermore, the general water loss by the plant apparently overrides the effects of the potassium transport mechanism. If water loss is severe, plants undergo what is called "water stress," and in some species a hormone known as abscisic acid (or ABA) is released by nearby cells. In the presence of ABA, guard cells rapidly lose turgor and the stomata close. Plant physiologists have also studied simpler variables affecting the activity of guard cells, such as carbon dioxide concentrations, temperature, and light. Experiments with corn reveal that stomata close rapidly when carbon dioxide levels are increased and that they open, even in the dark, when CO_2 levels are experimentally lowered. Stomata also close when plants are exposed to abnormally high temperatures, certainly an appropriate response since high temperatures hasten water loss. But strangely, if carbon dioxide is depleted, the response fails, suggesting (but not proving) that the two factors—heat and carbon dioxide—could be related. We can speculate that under normal conditions, increasing temperatures accelerates respiratory activity in the cells, thereby increasing the carbon dioxide output.

Then there are the **CAM plants,** a group that seems to break all the rules.

CAM Photosynthesis, or How To Hold Your Breath All Day.

In some plants the opening and closing cycle of the stomata is reversed. Included are many cacti and other succulents (fleshy-bodied plants common to the desert), and members of the family Crassulaceae—the "stonecrops." In such plants, the stomata are *closed in the daytime and opened at night.* This strange adaptation helps prevent intolerable water loss in the excessively arid desert environment in which the plants live. It also means that access to atmospheric carbon dioxide is cut off during photosynthesis, when it is needed. Yet the plants carry out the light-independent or sugar-synthesizing reactions of photosynthesis in a manner similar to other plants. How do they handle their carbon dioxide requirements?

The plants use what is called **crassulacean acid metabolism,** or **CAM.** Carbon dioxide is admitted at night and temporarily incorporated into certain organic acids. Then, in the daylight, the chemical reactions are reversed and the carbon dioxide is released within the cell. From there it is used in the production of glucose according to the usual biochemcial pathway—the Calvin cycle of the light-independent reactions (see Chapter 7). The energy and hydrogen sources, ATP and NADPH, are provided, as usual, from the light reactions. Such chemical activity means that they must use precious ATP reserves, but that's one price they pay for living in a less competitive environment.

FOOD TRANSPORT IN PLANTS

In 1671, just six years after he described the cells in cork, Robert Hooke and a colleague, Robert Brotherton, "girdled" a tree. That is, they removed a ring of bark around the trunk of the tree, including the soft, moist layer beneath the dead outer cork. The tree eventually died, but they observed that it continued to grow for some time, putting out new leaves and branches. They also noted something odd. The trunk of the tree increased in diameter above the ring *but not below it.* The roots did not grow, and in fact began to shrivel. Hooke and Brotherton concluded that the material a plant receives from the air is transported downward in the bark, while the material it receives from the roots is transported upward in the wood. Hooke and Brotherton's conclusions have remained unaltered for 300 years.

Actually, sugar and other nutrients move through the phloem in either direction, down toward the roots when the leaves are photosynthesizing and up toward the branches at other times, especially in spring when new leaf buds are growing. It is now clear that some form of active transport is involved in the movement of sugar and other nutrients through the phloem. We will look at the role of active transport after making a few general observations about food transport in the phloem.

Mechanisms of Phloem Transport

We learned earlier that in the movement of water from roots to foliage through the xylem, plants make use of the energy of evaporation, water potential gradients, and those molecular peculiarities called adhesion and cohesion. The TACT forces, we found, provide what is essentially a pulling force that creates great tension within the tracheids and vessels. Only the thickness and strength of the cellulose xylem walls prevent these conduct-

23.9 POTASSIUM TRANSPORT HYPOTHESIS

Studies of potassium transport in guard cells indicate that a light-mediated process brings about changes in turgor. When protoplasts (cells with walls removed) are subjected to light, particularly certain blue wavelengths, they rapidly take in potassium ions (K^+). The uptake of K^+ is followed immediately by water, with cell size increasing by as much as 50%.

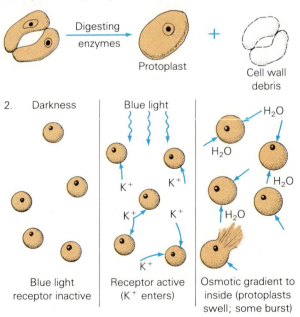

1. Preparation of protoplasts

Digesting enzymes → Protoplast + Cell wall debris

2. Darkness | Blue light

H_2O

K^+ K^+

K^+ K^+

K^+

H_2O

H_2O

H_2O

| Blue light receptor inactive | Receptor active (K^+ enters) | Osmotic gradient to inside (protoplasts swell; some burst) |

ing elements from collapsing under this great force. The situation is quite different in the phloem, though. The content of the sieve tubes, the phloem **sap,** is instead, *pushed* along. The pushing force, as we will see, is provided by *hydrostatic pressure,* which can be considerable in the sieve tubes.

A Little Help from the Aphids. Aphids take advantage of the high hydrostatic pressure of phloem sap by using their long, hollow mouthparts to drill tiny holes in individual sieve tubes, essentially creating little artesian wells. The tiny insects then relax and let the nutrient-laden sap flow into their bodies. In fact, the sap often flows right through their bodies and accumulates in drops at the other end, where it is euphemistically called "honeydew"—a delicacy among ants. Plant physiologists sometimes take advantage of the aphids' drilling technique—a feat they have technical difficulty achieving—in studying phloem sap. The trick is to let the aphids drill, then anesthetize them, cut the head away from the imbedded mouth parts, and use the resulting "artesian wells" as a source of pure sap for analysis (Figure 23.10).

It turns out that sap is actually a rather dense fluid, especially when the plant is photosynthetically active. Nonetheless, it moves rather rapidly through the phloem stream, up to one meter per hour. Sap is mainly water and dissolved solutes with sucrose making up about 90% of the latter. Other sugars, along with nutrients, hormones, and amino acids, are also transported in the phloem. The movement of such substances in the plant is referred to as **translocation.**

Flow from Source to Sink. As we have mentioned, phloem sap flows in either direction along the vascular system, according to the plant's needs. It first flows into developing leaf primordia, and then, when the leaf matures and begins photosynthesizing, it moves out of the leaf and down to the roots, the maturing fruit and growing seeds, the apical meristem, or very often into storage regions of the stem and root. We call these places where food is destined for use or storage **sinks** (a term borrowed from engineering, as in "heat sink"). The place where moving food originates, whether the site of photosynthesis or storage, is called the **source.** The flow of sap, therefore, proceeds from *source to sink.* Active transport is essential in both regions.

The active transport of nutrients through the phloem is often against the concentration gradient. Sugars in the leaf cells, for instance, are often at a

23.10 THE APHID AND PHLOEM CONTENT ANALYSIS

An ingenious method of obtaining samples of pure sap from the phloem stream involves the use of aphids. These insects use their fine, tubelike mouthparts to pierce the phloem and suck out sugars, an accomplishment researchers find technically difficult to duplicate by other means. Taking advantage of a method that works, physiologists simply anesthetize the aphid and cut the mouthparts away from the head. When successful, fine capillary tubes can be used to catch the phloem sap as it oozes out the mouth parts.

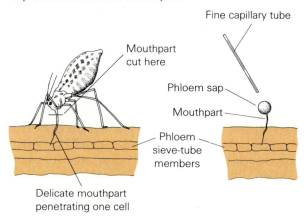

Fine capillary tube

Mouthpart cut here

Phloem sap

Mouthpart

Phloem sieve-tube members

Delicate mouthpart penetrating one cell

much lower concentration than they are in the nearby phloem elements into which they are being loaded. In turn, such sugars may be unloaded from the phloem into food storage cells with even higher concentrations of sugars. Moving materials against their concentration gradient takes energy, provided, as usual, by ATP. Ordinarily glucose, the principal sugar, is converted to starch for storage and hydrolyzed into glucose once again for transport in the phloem. Because of their size, the large starch molecules are more apt to stay put than are the very small, more mobile, glucose molecules.

The Pressure Flow Hypothesis. We have seen that active transport is used in loading nutrients into the sieve tubes at the source and unloading them at the sink, but what accounts for the flow of such materials in between? How do they pass through the sieve tubes? Many hypotheses have been proposed to account for phloem sap flow. Interestingly, the most favored one—the **pressure flow hypothesis**—is far from the newest. Proposed back in 1927 by Ernst Munch, a German plant physiologist, it is based on differences in water potential between the phloem and xylem—differences that are created by the active transport of nutrients.

The idea is straightforward: the active transport of sugars into the phloem stream greatly decreases the water potential there in comparison to the high water potential in the nearby water-filled xylem elements. Water responds by moving down the gradient into the phloem. The inward movement of water raises the hydrostatic pressure within the phloem sieve tubes. This pressure forces along the phloem sap with its load of nutrients and water. Then, at the target tissues—the nutrient sink—foods are moved out of the stream by active transport. The loss of solutes increases the water potential within the phloem and the water escapes from the phloem stream, usually finding its way back into the xylem (Figure 23.11).

Thus, water apparently circulates in plants, moving from xylem to phloem at the source, and moving from phloem back to xylem at the sink. The sieve tube elements themselves may play a passive role in transport, with the companion cells and phloem parenchyma providing the energy for active transport and somehow determining whether nutrients are to be loaded or unloaded. A cellular scheme of nutrient transport is seen in Figure 23.12.

A half-century-old idea that explains such a vital

23.11 MODEL OF PRESSURE FLOW

In this model, (a) a length of dialysis tubing (permeable to water) containing a sugar solution is connected by glass tubing to (b), dialysis tubing of similar length but containing water alone. The three represent the source, phloem sieve tubes, and sink, respectively. The apparatus is immersed in distilled water, which enters (a), considerably increasing its hydrostatic pressure. The sugar solution is pushed out of (a) through the glass tubing to (b). In the model, the onward movement of the sugar solution will continue until sugar molecules are equally distributed, whereupon the system will reach equilibrium. In the plant, the active transport of sugar into the phloem at (a) and its active transport out of the phloem at (b) would prevent the system from reaching equilibrium and the pressure flow would continue.

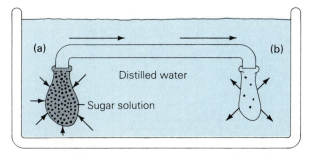

Sugar molecules

Water movement

23.12 THE PRESSURE FLOW HYPOTHESIS

The movement of sugars in the phloem begins at the source (a), which in this example is photosynthetic tissue in the leaf. The sugars are actively transported to the sieve tube through the leaf parenchyma, bundle sheath cell, and companion cell (or phloem parenchyma). Loading of the phloem sets up an osmotic gradient that facilitates the movement of water into the dense phloem sap from the neighboring xylem (b). As hydrostatic pressure in the phloem sieve tubes increases, pressure flow begins (c), and the sap moves through the phloem. Meanwhile, at the sink (d), incoming sugars are actively transported out of the phloem elements and into storage cells. The loss of solute produces a high water potential in the phloem, and water passes out (e), returning, eventually, to the xylem.

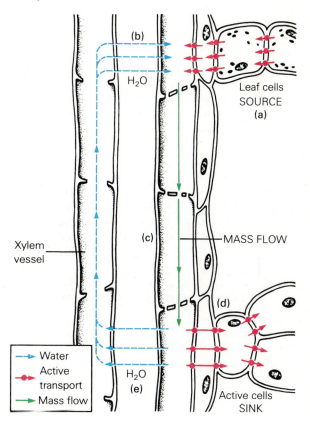

aspect of plant biology as food transport reminds us that our own generation has no monopoly on creativity. But, venerable though it is, the pressure flow hypothesis is not without its problems and critics. Scientists would still like to know, for instance, how the phloem sap is isolated from the phloem cytoplasm. How does it move through plasmodesmata without disturbing the cytoplasmic extensions there? And what determines whether nutrients are to be loaded or unloaded? In other words, what controls translocation? Obviously, science still needs bright and inquiring minds.

GAS TRANSPORT IN PLANTS

Plants continually exchange gases with the environment while carrying out both photosynthesis and respiration. But in general plants don't have specific systems to deal with gases. Few plants have developed ways to move gases around in their own tissues. In monocots, the vascular bundles commonly contain an air channel along with the xylem and phloem and supporting fibers. So at least some plants have made provision for the passage of air. Otherwise, gas is usually transported by diffusion where thin-walled cells, such as the those of the root tip epidermis, those just within the lenticels (pores along the stem), and the spongy parenchyma bordering the air spaces within the leaves, contact the air or water. If oxygen and carbon dioxide are to be transported in the xylem, however, they must be dissolved since air bubbles destroy cohesion and tension in these columns. Should this happen, water will move into adjacent columns through pits in the walls of the tracheid or vessel (see Figure 22.13).

Root hairs are good gas exchangers when soil conditions permit. Since they arise near areas of intense metabolic activity—growth, maturation, and active transport—their oxygen requirement is great. The exchange of gases takes place across the

23.13 GAS EXCHANGE IN THE BALD CYPRESS

The bald cypress *(Taxodium distichum)*, with its roots submerged in stagnant water, has adapted to its oxygen-poor environment by producing "knees." These porous root extensions rise well above the surface of the water and absorb air through their spongy tissues. The air can then diffuse into the metabolically active root tissues.

23.14 PASSAGE OF IONS THROUGH THE ROOT

Unlike the movement of water, the passage of ions through the root is primarily through the cell cytoplasm. Such movement is facilitated by the presence of numerous plasmodesmata, and active transport is believed to play a major role.

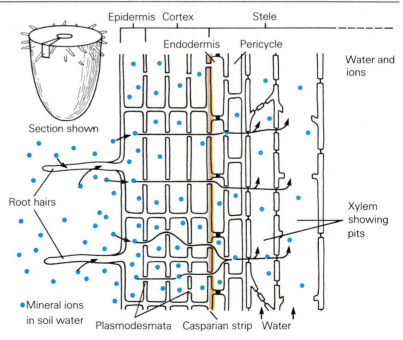

cells of the root tip and the root hair surfaces. The exchange is easy here because the thin-walled cells have no resistant cuticle, and the tremendous surface area of root hairs provides a large interface for gas exchange.

Where roots are submerged in water, gas exchange and oxygen availability are adversely affected. For this reason, the roots of plants that are grown in tanks of water must be aerated just as one would do for aquarium fishes. This is also why you can kill your potted plants by overwatering them: without gaseous oxygen in the soil, they simply drown. Some plants, however, have adapted to flooded, oxygen-poor soils. These include marsh and swamp plants and domesticated plants such as rice. It turns out that many of these plants have large, hollow air channels in their stems (not to be confused with the much smaller channels associated with the vascular bundles of monocots). The "knees" of cypress trees extend above the surface of the stagnant swamp waters, permitting air to diffuse into their loosely arranged tissue, and then down into the roots (Figure 23.13).

Lenticels, common in woody stems, provide an avenue of gas exchange for the very active tissues in the bark. Recall that both the vascular cambium and the cork cambium are intensely active during growth. Further, as we have just seen, the phloem elements and their associated cells, also part of the bark, are metabolically active, using ATP for transporting food materials into and out of the phloem stream. Thus an adequate supply of oxygen is essential.

MINERAL NUTRIENTS AND THEIR TRANSPORT

Dissolved minerals entering the root hairs and epidermal cells follow a route (Figure 23.14) that takes them through the cytoplasm of parenchymal cells in the cortex. Apparently most of their movement involves active transport and requires a considerable investment of ATP. Active transport in the endodermal cells passes the ions to the pericycle, which then secretes them into the water columns for transport upward.

The uptake of mineral ions is not always straightforward. In many instances the plant gets a little help from a friend—a friendly fungus, that is. An association of plant root and fungal mycelium known as a *mycorrhiza* (Figure 23.15) is one of those

23.15 A FUNGUS–ROOT ASSOCIATION

Mycorrhizae, associations of certain fungi with plant roots, can be quite essential to the health and vitality of the plant. In this experiment, all of the seedlings were first grown in a nutrient solution with all essential materials provided. The group on the left were then transplanted directly into prairie soil. The group on the right were planted first in forest soil where mycorrhizal associations are common, and then, after such associations had formed, transplanted to prairie soil.

rare instances of true mutualism: both members benefit. However, the relationship may not be clear at first. After all, plants have a lot to offer a fungus, but what does a fungus have that a plant could possibly need?

It turns out that the mycorrhiza efficiently absorbs and concentrates certain ions, notably phosphate. In some instances the fungal mycelium with its mineral load actually penetrates the root cortex and deposits ions there. In other cases, the mycelium simply surrounds the root epidermis and brings the plant into close association with the ions. In either case, the fungus extends its many mycelial fingers out into the soil, acting like a second root system for the plant. In transporting phosphates directly into the plant or concentrating them near the epidermis, the fungus makes this valuable ion more available to the plant. The fungus, for its part, absorbs complex carbon compounds such as sugars and amino acids—plant products that are initially produced through photosynthesis. Logically enough, mycorrhizal associations appear to be most common to those plants that live where certain minerals are scarce, such as in the tropics, where mineral recycling is so fast that little residual nutrient remains in the soil.

23.16 IRON DEFICIENCY

Plant mineral requirements are determined through deprivation studies carried out under rigidly controlled conditions. Depriving plants of one or more mineral nutrients produces obvious growth deficiencies. The iron-deprived tomato plants clearly reveal problems of iron deficiency through stunted growth and foliage lacking the deep-green color of the control plants.

Plant scientists have used many approaches to determine the mineral requirements of plants. These include deprivation experiments in which plants are grown in aerated distilled water of rigidly controlled purity. The only mineral ions present are those added by the experimenter. Once the minimal nutrients needed for growth are known, researchers can determine the particular effect of any mineral simply by witholding it from the solution (Figure 23.16).

From such experiments botanists have accumulated a list of important mineral nutrients essential to plant growth (Table 23.2). Notice that the table is divided into **macronutrients** and **micronutrients** (trace elements). These terms simply reflect the relative quantities required. For example, ions such as chloride (Cl^-) are required in such minute quantities that they defy measurement, yet photosynthesis cannot proceed without them. Trace elements often cause problems for plant scientists in trying to control experiments—if you can't find it, you can't always keep it out.

Many of the mineral nutrients of plants are made available through cycles that are often quite intricate. These are usually referred to as **biogeochemical cycles,** some of which are described in Chapter 41. Key elements such as nitrogen, sulfur, phosphorus, and calcium, as well as carbon, oxygen, and hydrogen, are constantly recycled between the living and nonliving realms of the earth. When these cycles fail, the result can be infertile soil and limited plant growth.

Perhaps the strangest way a plant obtains nutrients is seen in the **insectivorous plants,** which Darwin studied extensively. These plants supplement their nitrogen needs by capturing and digesting small insects. Examples include the sundew (*Drosera*), the Venus flytrap (*Dionaea*), and the pitcher plant (*Sarracenia*) (Figure 23.17). The most unusual is probably the Venus flytrap, since it actively captures its prey. Insect-catching by plants is an adaptation to life in nitrogen-poor swamps and bogs, where many nutrients are leached away by water or consumed by anaerobic bacteria. Thus, trapped insects can provide the missing nutritional requirements. (Even though such predation is unnecessary when nitrogen supplies are added, these plants go right on killing flies.)

TABLE 23.2

ELEMENTS AND NUTRIENTS ESSENTIAL TO PLANT GROWTH

Element	How Taken In	Examples of Use
Macronutrients		
Calcium	Calcium ion (Ca^{2+})	Cell wall, plasma membrane, coenzyme activity
Carbon	Carbon dioxide (CO_2)	Proteins, lipids, carbohydrates
Hydrogen	Soil water (H_2O)	Proteins, lipids, carbohydrates
Magnesium	Magnesium ion (Mg^{2+})	Chlorophyll molecule
Nitrogen	Nitrate ion (NO_3^-), Ammonium ion (NH_4^+)	Amino acids, purines, pyrimidines, protein
Oxygen	Atmospheric oxygen (O_2)	Cell respiration
Phosphorous	Phosphate ion ($H_2PO_4^{2-}$)	Nucleic acids, phospholipids, ATP
Potassium	Potassium ion (K^+)	Plasma membrane, enzyme activity, guard cell mechanism
Silicon	Silicate ion ($HSiO_3^-$)	Cell walls
Sulfur	Sulfate ion (SO_4^{2-})	Proteins, coenzyme A
Micronutrients		
Boron	Borate ion (BO_3^-), Tetraborate ion ($B_4O_7^{2-}$)	Cell elongation, carbohydrate translocation
Chlorine	Chloride ion (Cl^-)	Accumulates as HCl in chemiosmotic photosynthesis
Copper	Cupric ion (Cu^{2+})	Enzyme activity
Iron	Ferrous ion (Fe^{2+}), Ferric ion (Fe^{3+})	Enzyme activity, chlorophyll synthesis
Manganese	Manganese ion (Mn^{2+})	Enzyme activity
Molybdenum*	Molybdenum ion (Mo^{3+})	Enzyme activity, including N-fixation
Zinc	Zinc ion (Zn^{2+})	Hormone activity

*Indirectly important in the nitrogen cycle

23.17 INSECTIVOROUS PLANTS

A few plants are able to supplement their nitrogen supplies by capturing and digesting insects. **(a)** The bladderwort *(Utricularia),* a water plant, consists of a saclike chamber with a door. When the trap is tripped, the door opens inward, the prey is sucked in by inrushing water, and the door shuts behind it. **(b)** The pitcher plant *(Sarracenia)* has evolved a one-way passage into its trap. Insects venturing into the vaselike structure find their return blocked by downward-pointing hairs. **(c)** The sundew *(Drosera)* makes use of a thick, sticky fluid that traps the insect. Insectivorous plants grow in swamps and bogs, habitats that are notoriously poor in nitrogen sources.

We see, then, that as life became more complex, that complexity often involved an increase in size. With this growth came a need for transporting the molecular necessities of life from one part of the organism to another. This need has been met in plants in a variety of ways that apparently employ simple, basic laws of physics. But in some cases these laws don't seem so simple after all, and the search for the precise mechanisms of transport in plants goes on.

APPLICATION OF IDEAS

1. The statement is made in this chapter that vascular plants have literally "evolved around the characteristics of water." Discuss the full meaning of this statement. Begin by explaining exactly how vascular plants have taken advantage of water's peculiarities. Then rephrase the statement in such a way as to introduce the terms *natural selection* and *adaptation*. Complete your discussion by considering how this evolutionary direction made great increases in size possible, and how this result has been advantageous to evolving plants.

KEY WORDS AND IDEAS

1. A major problem in plant physiology has been to explain how water passes from the roots to the foliage of tall trees with little direct expenditure of plant energy.
2. The vascular system of plants has the dual function of transporting water through its **xylem** and foods through its **phloem.** In the latter, the plant uses active transport and expends considerable energy.

THE MOVEMENT OF WATER AND MINERALS
1. Water travels from the soil to the air around the leaf via root hairs, root cortex, stele, root xylem, stem, leaf, leaf mesophyll, air spaces, stomata, and outside.
2. The evaporation and subsequent diffusion of water from the leaf is called **transpiration.** In many plants, the quantity transpired daily can be measured in liters.

Root Pressure: Is the Root a Kind of Pump?
1. **Root pressure,** a force orginating from osmosis conditions in the root, is responsible for **guttation,** the forcing of water in its liquid phase from the tips of grass blades.
2. Root pressure is produced when ions are actively transported into the stele, setting up a water potential gradient.
3. Root pressure is not a major force in the movement of water to the tops of tall plants. When the xylem of a tree is tapped during active transpiration, a sucking action rather than a spurt occurs, indicating a negative rather than a positive root pressure.

Water Potential and the Vascular Plant
1. Water always moves from regions of greater water potential to regions of lesser water potential. Water potential is the inverse of the solute concentration in cells.
2. The normal water potential gradient of a plant is from high potential in soil water to low potential in air surrounding the leaf.
3. A substantial loss of water potential in the soil leads to a loss of water from the root and wilting in the foliage.

The Leaf, Transpiration, and the Pulling of Water
1. Water loss is unavoidable if CO_2 is to be available for photosynthesis. Plants make use of the continued water loss to initiate the mechanism that raises water from the roots.
2. Transpiration creates a steep water potential gradient through the cells of the leaf, and water enters from the leaf xylem. The cell-to-cell movement is facilitated by numerous plasmodesmata.
3. The water potential gradient in the leaf produces a pulling force of up to 12 atmospheres, enough to raise water through the tallest trees.
4. The movement of water through the stem and leaves is provided by the **free energy of evaporation.** Indirectly, this energy comes from the sun.

Water Movement in the Xylem Through the TACT Mechanism

1. **Transpiration** and **adhesion,** the first two **TACT** forces, set up the pulling force that raises water through the xylem.

2. Adhesion, the attraction of one type of molecule to another (such as water to those of surrounding surfaces), may be a major force, since the adhesion of water to the cellulose walls of leaf cells may initiate the pulling action. It is also a factor in cell-to-cell movement.

3. **Cohesion,** the mutual attraction of similar molecules, is critical in the tiny water columns of the xylem. It provides the tensile strength needed to hold them together when under tension.

4. **Tension** is the stress placed on the water columns by the pulling forces above. During intense transpiration, tension can be sufficient to cause a narrowing of tree trunks.

5. These combined forces provide and maintain the pull needed to raise water through the plant.

Water Transport in the Root

1. During intense transpiration, root pressure is overwhelmed and the pull from above is simply transmitted across the root. Root pressure may be significant when transpiration slows or when water potential outside the root is low.

2. Because water is directed around the **Casparian strip** and through the endodermal cytoplasm, the endodermis may somehow influence water movement. It does transport ions into the stele.

Guard Cells and Water Transport

1. Paired guard cells form porelike stomata on the under surface of leaves. In the turgid condition the thicker inner walls of the cells are bent into a curve that opens the stoma, but when turgor is lost, the thick walls straighten, closing the stoma.

2. The stomata are generally open in daylight and closed at night.

3. Turgor changes in guard cells are brought about by changing solute concentrations. An early, erroneous hypothesis proposed that sugars produced in guard cell chloroplasts were the solutes.

4. A favored hypothesis today is the **potassium transport mechanism,** which proposes that the solute responsible is the potassium ion. In daytime, light falling on a photoreceptor sets in motion the active transport of K^+ ions into the guard cells. The increased solute concentration speeds the uptake of water and the stoma opens.

5. The stomatal mechanism is complex. During periods of water stress, a hormone, abscisic acid, overrides the potassium transport mechanism and the stomata close. Corn plant stomata close when carbon dioxide levels increase. Desert dwelling **CAM** (Crassulacean acid metabolism) **plants** manage to close their stomata in daytime and open them at night when they take in CO_2. They store CO_2 in organic acids, releasing it for glucose synthesis in the daytime.

FOOD TRANSPORT IN PLANTS

1. In 1671, Hooke and Brotherton determined that substances produced in leaves were transported downward while those collected by the roots were transported up.

Mechanisms of Phloem Transport

1. While water is moved by a pulling force, a push in the form of hydrostatic pressure is responsible for food transport.

2. Aphid mouthparts are used in sampling phloem sap, a dense solution containing sucrose (about 90%), other nutrients and hormones. The movement of foods in the plant is called **translocation.**

3. Foods are commonly moved from **source** to **sink.** They are actively transported into and out of the phloem stream.

4. The favored explanation of phloem transport is the **pressure flow hypothesis.** Sugars are actively transported into the phloem stream at the source. This increase in solute decreases the water potential in the phloem elements and water enters, increasing hydrostatic pressure. The increased pressure forces the dense phloem sap along to the sink, where the solutes are actively transported out. The loss of solute raises the water potential of the phloem and water reenters the xylem.

GAS TRANSPORT IN PLANTS

1. Plants must carry on gas exchange in their metabolically active tissues. Commonly, in woody plants, oxygen enters and carbon dioxide leaves the stem via the porous lenticels. In some monocots the passage of gases occurs through open air channels.

2. In the metabolically active root, gases are readily exchanged across the extensive root hairs in the root tip epidermis.

3. Marsh plants may have a problem with providing oxygen to the roots. Some have hollow, air-conducting channels in their stems. Cypresses have very porous "knees" that extend above the water surface, permitting gases to diffuse into the roots.

MINERAL NUTRIENTS AND THEIR TRANSPORT

1. The transport of mineral ions through root cells is primarily active.

2. Many plants are assisted in mineral uptake by a fungus-root association called a **mycorrhiza.** The fungus absorbs and concentrates ions near the root hairs or within the root, and in turn absorbs useful compounds from the plant.

Mineral Requirements of Plants

1. Mineral nutrient requirements are commonly determined through deprivation experiments. A mineral is withdrawn and the plant carefully studied for deficiency symptoms.

2. Mineral nutrients are commonly provided through **biogeochemical cycles,** where the ions cycle between organisms and the physical environment. They are divided into **macronutrients** and **micronutrients.**

3. **Insectivorous plants** obtain their nitrogen from the digestion of insects.

REVIEW QUESTIONS

1. Name the two different tissues of the vascular system and summarize their roles in transport. (573–574)

2. In what major way does the mechanism of food transport differ from the mechanism of water transport? (573–574)

3. Starting with the epidermis, trace the movement of water through the plant, naming the important tissues through which it passes. (574)

4. What is transpiration? How could transpiration possibly affect the weather? (575)

5. Explain how root pressure is created by the plant. In what way is root pressure visible in grasses? (575)

6. Explain why root pressure cannot be responsible for the rise of water to the tops of tall trees. What is more likely to be the purpose of root pressure? (575)

7. What is the general rule about water potential and the direction in which water will flow? What effect do solutes have on water potential gradients? (577)

8. Explain transpiration again, this time by describing activity within the air spaces of a leaf. What effect does evaporation have on cells bordering the leaf spaces? On cells further in? (577)

9. Describe the water potential gradient of the leaf, beginning with the air outside a stoma and ending with the water-filled xylem. (577)

10. It has been said that the energy for transpiration is unlimited. Explain what this means. (577)

11. Name and briefly describe the TACT forces. (578)

12. Assuming that transpiration and adhesion create the necessary pulling force, what is the role of cohesion in water transport? In what way is the diameter of xylem vessels and tracheids related to cohesion? (578)

13. What two kinds of evidence do physiologists find that suggest the presence of tension in the transpiring tree? (578)

14. Under what two conditions might root pressure be significant to the transport of water? (578, 581)

15. Why is it reasonable to suspect that the endodermis has a role in the control of water uptake by plants? (581)

16. What is the usual cycle of stomatal opening and closing in the plant? (581–582)

17. Describe the shape of a pair of guard cells in the turgid condition. What factor apparently influences turgor? (581)

18. Explain the potassium transport mechanism of guard cell function. (582)

19. Describe two situations in guard cells that indicate that their operation is not at all straightforward. (582)

20. Describe the peculiar stomatal cycling of the CAM plants. How can they do this and still produce carbohydrates in the daytime? (583)

21. Compare the pressures in the phloem stream with that of the xylem stream. (584)

22. Explain how physiologists make use of the aphid in studying phloem sap. Summarize their findings on the general makeup of phloem sap. (584)

23. Briefly describe how events at the source produce a flow in the phloem sap. Is this a push or a pull? (584)

24. Explain what happens to the nutrients and to the hydrostatic pressure when the phloem stream reaches the sink. (584)

25. What function do lenticels, cypress knees, root hairs, and rice stem channels have in common? (587)

26. Explain how the formation of a mycorrhizal association is beneficial to the plant. How is it beneficial to the fungus? (587)

27. Describe a deprivation experiment and explain what information can be gained by this procedure. List two examples. (588)

Regulation and Control in Plants

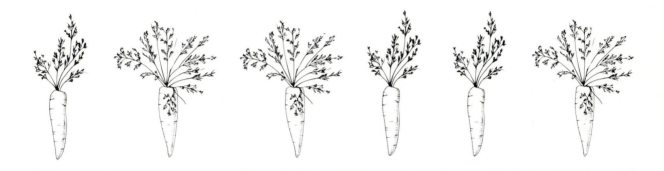

Charles Darwin's investigations into evolution led him into a wide array of scientific fields. One such field was the area of behavior. In fact, his publication *The Expression of the Emotions in Man and the Animals* (1872) was to become the foundation of animal psychology. But Darwin realized that the area was so complex that many of the most basic questions remained unanswered. He sought to reduce behavior to its simplest terms. And what, he thought, could be simpler than studying the behavior of plants? Thus Darwin became the first plant psychologist and, some would say, the last.

Of course, one is tempted to ask: Do plants actually *behave?* The answer is, they do—if behavior is considered to be a response in the form of a movement.

Darwin began his studies with the most interesting of all plant behavior, that of the insect eaters such as the Venus flytrap. In his publication *Insectivorous Plants* (July 1875) he described in exquisite detail the triggering of the plant and its response, but he was at a loss to explain just *how* such rapid movement was accomplished. So he changed the focus of his research; he looked into the slower and seemingly simpler behavior of climbing plants.

The tips of climbing plants, he found, move about in circles while they grow, "as if seeking something to twine around." If no object is encountered, the tip eventually grows straight upward. But if a twig or a beanpole is encountered, the slow-

ly growing tip begins to spiral around it and to tighten up. In this way, vines and climbing plants can reach great heights without producing thick woody supporting structures such as those found in trees (Figure 24.1).

Darwin also found that on a hot day a new shoot of a hop plant can revolve once in 2 hours, 8 minutes. After 27 such revolutions without encountering anything, the tip of Darwin's plant was describing a circle of 19 inches in diameter and, by looking closely, Darwin found the movement was discernible to the human eye.

Darwin described many variations of this basic pattern and (not surprisingly) interpreted them in terms of natural selection. He showed that the underlying mechanism was the elongation of first one and then another part of the stem below the growing shoot, causing it to bend first one way and then another (*Climbing Plants*, September 1875). But what brought about this elongation?

THE DISCOVERY OF AUXIN

Darwin, now 71 years old, worked on the problem with his son, Francis (Figure 24.2). Their experimental organism was canary grass. Like other grasses, in its early stages of embryonic development it produces a tubular sheath, the **coleoptile.**

As it emerges from the soil, it surrounds the primary (first) leaf. Darwin found that when the coleoptile is illuminated from one side, it bends toward the light. The bending, they also noted, does not occur at the tip, but in the elongating part of the leaf, well below the tip. Was light acting directly on the bending part? Apparently not. Darwin illuminated only the growing tip of the coleoptile, and it bent as before. With a small piece of foil, he covered the tip and exposed the rest of the plant to light. Nothing happened; it didn't curve. Darwin then knew that *something was happening in the growing tip that caused certain areas of the stem beneath to elongate.* In *The Power of Movement in Plants* (1880), Darwin ascribed this fundamental discovery to "some matter in the upper part which is acted upon by light, and which transmits its effect to the lower part . . . . When seedlings are freely exposed to a lateral light some influence is transmitted from the upper to the lower part, causing the latter to bend." Thus, in a way, Charles Darwin and his son were the first to propose the existence of a **plant hormone.**

The Experiments of Boysen-Jensen and Paal

Early in this century, a Dane, Peter Boysen-Jensen, and a Hungarian, A. Paal, extended Darwin's observations. Boysen-Jensen decapitated a coleoptile of an oat seedling a few millimeters from the tip, put a tiny block of gelatin on the stump, and replaced the tip, which he then illuminated from one side. The coleoptile bent as before, showing that Darwin's "influence" was something that

24.1 CLIMBING PLANTS

Climbing plants are adapted to rapid upward growth at little metabolic expense, using other plants or objects for support.

24.2 DARWIN'S PLANT TROPISM EXPERIMENT

Charles Darwin and his son Francis were the first to report studies of the response of plants to light. They subjected canary grass seedlings to light from one side, noting the bending growth **(a)**. To determine where the photosensitive region was, they used thin metal caps to cover various parts **(b)**. Their results indicated that something from the tip was influencing bending in the shoot region below.

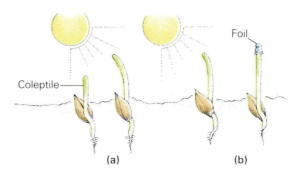

(a)   (b)

could move through gelatin. To verify this, Boysen-Jensen placed tiny slivers of impermeable mica between the coleoptiles and the stumps. There was no bending. When he inserted mica into slits cut partway through coleoptiles and then illuminated them from different sides, he found that the active substance did not move down the illuminated side, but only down the side away from the light (Figure 24.3). Thus, the stimulus that caused the normal cell elongation and bending was present on the darkened side of the coleoptile only.

In his efforts, Paal decapitated oat coleoptiles and then restored the tips on the growing stubs, but off to one side or the other. The shoot always bent away from that side, even in the dark. This clearly suggested that there was a material substance emanating from the tip, and that this substance stimulated the elongation of cells just below. The next logical step was to isolate this material and see precisely what it could do.

Fritz Went and the Isolation of Auxin

By 1926, Fritz Went, a Dutch scientist, had succeeded in doing just that. Went's technique consisted of decapitating oat seedlings and placing the tips on agar blocks (Figure 24.4). He then took tiny squares of the agar and placed them on the side of the decapitated seedlings. Went kept his subjects in a darkened environment throughout the experiment, so that any bending response could be attributed to substances in the agar block alone. In other words, he permitted only one variable at a time in his experiment. Typically, bending occurred within an hour after the blocks were applied, proving that an active, collectable agent was the cause of bending. Although he did not chemically characterize the active substance, Went named it **auxin** (from the Greek word *auxein*, "to increase"). Later, when more of its chemistry became known, it would be called **indoleacetic acid,** or **IAA.**

Any biochemical isolation requires an **assay system,** a way of measuring whether a substance is present, and in what concentrations. An assay system for the coleoptile growth substance was soon worked out with oat seedlings, using Went's basic methods. Hormone concentrations are calculated by the degree of bending seen in the oat seedlings. The greater the angle, the more hormone present.

24.3 BOYSEN-JENSEN EXPERIMENT

Boysen-Jensen elaborated upon Darwin's experiments. In one series (a), he decapitated seedling tips, placed a small gelatin block on the cut, and replaced the tip. The plant bent toward the light in the usual manner, indicating that a stimulating substance was passing through the gelatin. In a second series (b), the gelatin was replaced by a thin sheet of mica, a light-admitting mineral through which the substance could not pass. There was no curving. Third, (c), slivers of mica were inserted halfway through the tip, blocking the travel of any stimulating substances. The lack of curvature in the subject on the left and the definite curvature of the subject on the right indicate that the substance responsible for curvature passed down the unlighted side of the stem only.

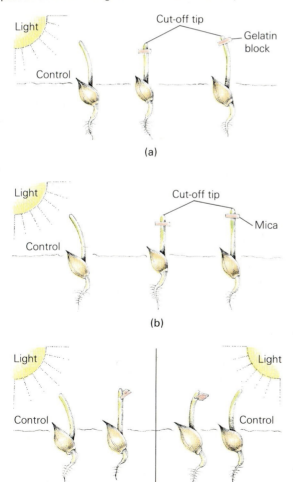

(a)

(b)

(c)

AUXIN

Unknown to Boysen-Jensen, Paal, or Went, the active substance of auxin had been isolated from fermentations back in 1885 by a pair of biochemists, E. and H. Salkowski. However, the Salkowskis hadn't a clue about its biological significance. The substance was not successfully isolated again until 1934, when it was finally purified from, of all things, human urine. We now know that the mysterious coleoptile growth substance is actually a widespread and fundamental plant growth hormone. The structure of auxin is very similar to the amino acid tryptophan; in fact, tryptophan is its primary building block:

Indole   Acetic acid group

Auxin, or indoleacetic acid (IAA)

Tryptophan

Auxin doesn't cause cells to proliferate by mitosis, but rather promotes cell enlargement through elongation (Figure 24.5). Cell elongation, a primary plant growth mechanism, occurs during the growth of stems after the tiny, tightly packed daughter cells form through mitosis in the apical meristem (see Chapter 22).

Although the exact mechanisms through which auxin works are still being investigated, it is currently believed that elongation requires a loosening of the closely bound microfibrils of cellulose throughout the cell walls. This permits turgor pressure to expand the cells, generally producing an increase in length, but with the direction of expansion depending on the orientation of the cellulose in the wall (see Figure 22.16). In most cases, the actual

24.4 WENT'S AUXIN ASSAY

(a) Fritz Went was able to collect the growth-stimulating substance in agar and use tiny agar squares to study its effects on growth. His experiments were carried out in the dark. If you have ever seen oat seedlings you might have some appreciation for the technical skill required to carry out his procedures. Following the removal of the coleoptile, a tiny agar block with hormone added is placed on the coleoptile's cut edge alongside the young shoot. **(b)** Went's assay procedure is used to determine the presence of a specific hormone and to determine the quantity if it is unknown. The quantity is determined by measuring the amount of bending in the shoot and comparing it to a standard.

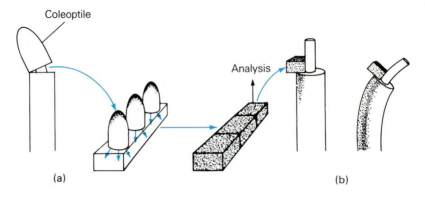

Coleoptile

Analysis

(a)

(b)

24.5 THE ACTION OF AUXIN

Auxin is believed to promote cell elongation in the growing tips of stems. The region called the apical meristem continually divides, producing cells that then elongate. **(a)** When light is multidirectional, elongation proceeds equally around the stem. **(b)** But when light is available from one side only, only the cells on the unlighted side elongate. Studies show that auxin concentrates on the unlighted side.

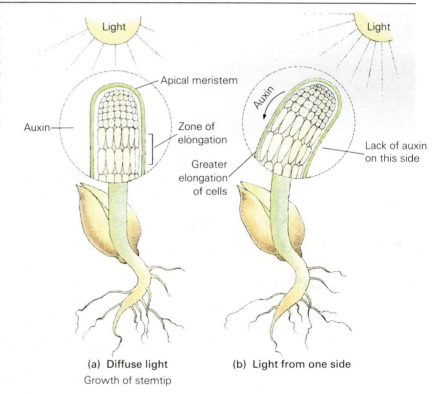

Light

Light

Apical meristem

Auxin

Auxin

Zone of elongation

Greater elongation of cells

Lack of auxin on this side

(a) Diffuse light

(b) Light from one side

Growth of stemtip

cytoplasm of the cell doesn't increase appreciably; instead, the central vacuole enlarges. Interestingly, some evidence suggests that auxin may not act directly on cell walls but rather on the genetic mechanism of RNA transcription. Under experimental circumstances, auxin has been shown to induce RNA synthesis in cells. Other studies indicate that in the presence of auxin, the pH of the cell wall decreases, activating latent enzymes responsible for fiber-loosening reactions. As you can see, we are still some distance from understanding this fundamental aspect of auxin's action in the plant.

Auxin is also closely involved in leaf fall or **abscission** (a role formerly ascribed to abscisic acid). Where each leaf attaches to the stem a special layer of cells called the **abscission zone** forms at the base of the petiole (Figure 24.6). Changes in this zone bring about the normal separation of leaves. The fall of leaves in deciduous perennials is a common seasonal event, but it can cause problems for the plant, nevertheless. The primary problem is the risk of water loss and parasite invasion at the site where the leaf was attached. You may recall that the vascular system of plants extends directly into the midrib of the leaf; therefore, such exposure could be hazardous and the stem must seal off the wound as rapidly as possible. In the natural fall of leaves, this begins when the reduced auxin levels in the leaf brings on the events of abscission. As auxin diminishes, many valuable materials within the leaf are mobilized and withdrawn into the plant. Next, cellulases—cellulose-splitting enzymes—break down the cell walls within the abscission zone. In some instances, new cells with a highly suberized content are formed, further protecting the plant. Eventually, only a few strands of vascular tissue actually hold the leaf in place, but finally, an enlargement of thin-walled parenchyma weakens this last link, triggering the separation. Then comes the raking.

THE TROPISMS

It now appears that auxin and other plant hormones are responsible for much of the specialized growth responses and movement that so intrigued Darwin. These are collectively known as **tropisms.** Growth that is influenced by light is called **phototropism.** Growth *toward* light is **positive phototropism;** growth away from light is **negative phototropism.** Growth that is influenced by gravity, such as that seen in root tips, is called **geotropism.** Perhaps

24.6 THE LEAF ABSCISSION LAYER

Before a leaf falls, the cells of the abscission zone (arrows) die and harden. Eventually, the leaf petiole will break away and fall. The dead abscission zone forms a sealing scar, averting the loss of water and danger from invading microorganisms. (The dark object in the leaf axil is a bud.) Abscission is closely related to falling levels of auxin in the leaf.

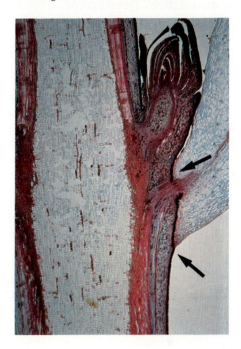

the strangest of all plant responses is **thigmotropism**—the response of plants to touch. The rapid closing of the Venus flytrap is a most notorious example, but far more familiar is the growth of plant tendrils, the wiry, coiled growths we see in garden squash and grapes.

Phototropism

Let's take a closer look at phototropism as an example of tropism in general. Most researchers believe that light stimulates something in the growing tip and causes the auxin to move laterally in the apical meristem. The process appears to be controlled by a photoreceptor, probably a yellow pigment. Although this initial photoreceptor has not yet been identified, it is known to be most responsive to blue light. The entire growing tip produces auxin in quantity, but as it diffuses down the shoot it crosses over to the unlighted side, where cells respond by elongating. Such unequal growth produces the typical bending of the shoot (see Figure 24.5).

In common with many hormones, auxin is

short-lived. Instead of accumulating in the stem as it migrates downward, it is inactivated by specific enzymes in the lower parts of the plant. The inactivation process tightly regulates the level of auxin within the plant. Interestingly, excess auxin can actually inhibit growth.

Auxin and Apical Dominance. The familiar triangular shape of conifers (Figure 24.7) is the result of a phenomenon known as **apical dominance.** Apical dominance operates through inhibition. It turns out that auxin, produced in the dominant growing tip, inhibits growth in lateral branches below and even stops the development of any new branches. Thus, the plant's energies are devoted to upward growth. If a plant is "topped"—its dominant tip removed—the inhibitory effects of apical dominance are removed. The highest remaining lateral branches then begin to grow upward, seeming to vie with each other for dominance. But what is happening is that auxin suppression has not yet been established in the lateral branch tips. Ultimately, all but one of the lateral branches will be suppressed, and that one—the new suppressor—will be transformed into the main trunk of the tree. Their many centers of auxin production and their ability to replace dominant growing tips enable plants to survive extensive damage from browsing animals, windstorms, and inept gardeners.

24.7 APICAL DOMINANCE IN CONIFERS

The growing tip releases auxin that retards growth in the lateral branches below. The older branches, however, have had longer to grow, even at a reduced rate, so the result is the familiar triangular appearance in plants of this group.

24.8 APICAL DOMINANCE IN BROAD-LEAVED PLANTS

Broad-leaved plants often lack a single dominant tip **(a)** and produce a spreading growth. Each branch contains its own apically dominant tip. **(b)** Very dense decorative hedges are produced through constant pruning. With repeated loss of growing tips, each stem continues to produce new lateral branches.

(a)

(b)

Apical dominance is very dramatic in conifers but less pronounced in broadleaved flowering trees. Flowering trees may have a single trunk up to a certain height and then branch out into several large limbs, with no one limb truly dominant. Some form of apical dominance may occur within each of these major limbs (Figure 24.8a).

Continued removal of growing tips, as is done in pruning, increases the number of lateral shoots. This results in a decorative bushy plant growth (Figure 24.8b). Pruning can also stimulate a response in fruit trees whereby the plant's resources are put into production of fruit rather than wood.

Roots and Geotropism

When a seedling is placed on its side, the shoot tip bends upward and the root tip bends downward. We have seen how light affects the activity of auxin in promoting cell elongation, and we can therefore see how the shoot bends upward towards the light. But what causes the root tip to grow downward, a phenomenon known as the positive geotropic response?

The question of what causes the downward growth of roots is still awash in a sea of contradictory hypotheses. Root cells are known to be extremely sensitive to artificially applied auxin, which inhibits their elongation. One hypothesis is that auxin moves downward away from light as it does in the shoot, and inhibits the lengthening of the lowermost root cells. The normal elongation of the cells above causes root curvature. Another, less probable, hypothesis is that auxin moves *upward* in the root and that small amounts of auxin stimulate cell elongation. Both hypotheses are weak in that there are no auxin gradients in the root and auxin has never been found in the very tip of the root. Researchers have, however, found another hormone in the root cap, leading to a very different third hypothesis.

The third hypothesis maintains that auxin is not involved in root tip curvature at all, but that another hormone, **abscisic acid,** moves upward from the root cap and inhibits elongation in the underside of root cells (Figure 24.9). It is known that abscisic acid has a wide spectrum of activities and can, under some conditions, inhibit growth. Even so, it isn't

24.9 THE ABSCISIC ACID HYPOTHESIS

Recent hypotheses maintain that abscisic acid (ABA), produced in the root cap, travels up the root tip to the region of elongation. If the seedling root is oriented horizontally, abscisic acid concentrates on the lower side, inhibiting elongation there. As a result, normal elongation on the other side of the root tip sends it curving toward the soil.

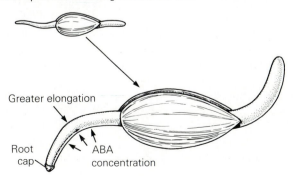

Greater elongation

Root cap

ABA concentration

24.10 PLASTIDS AND ROOT CURVATURE

One hypothesis for explaining geotropic responses in the root is that hormones may be distributed according to gravity. Starch-filled plastids containing the growth hormone always settle on the underside of cells. Should the root tip assume a horizontal position, the plastids—with elongation inhibiting hormone adhered—sink to the lower side in each cell. Their concentration there, the hypothesis suggests, inhibits elongation on that side and curvature follows.

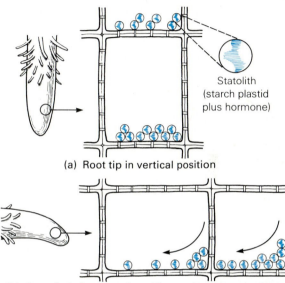

Statolith (starch plastid plus hormone)

(a) Root tip in vertical position

(b) Root tip in horizontal position undergoes curvature

immediately obvious what controls the distribution of the hormone.

There have been several attempts to describe gravity-sensing mechanisms in plants. For instance, many plant cells contain fairly large, heavy, starchy plastids that lie on the lower end of large central vacuoles. When the plant is laid on its side, the plastids settle within a few minutes to the downward side of the cells (Figure 24.10). This movement is soon followed by root curvature, and physiologists are looking for a cause and effect relationship between the two. It may be that the plastids contain high enough levels of root-inhibiting hormone to inhibit elongation if the plastid contacts the cell wall. But there are mechanical difficulties with this idea, and the questions surrounding geotropism remain largely unanswered.

Thigmotropism

The ability of some plants to respond quickly to touch has fascinated laypeople and frustrated plant physiologists. The physiologists are frustrated because they can't figure out how the plants do it. The basic mechanism of rapid movement is thought to be through very sudden changes in turgor, but how is this achieved and how does it result in movement? Two plants that can illustrate the response for us are the mimosa and the Venus fly-trap.

Touching or pressing any of its many fine leaflets sends the highly sensitive *Mimosa pudica* (Figure 24.11a) into a spasmlike reaction. The leaflets fold up and the leaf petioles suddenly droop. Some theorists suggest that this response may discourage browsing animals, while others maintain that it simply helps the plant avoid excessive water loss when hot, dry winds blow. The second hypothesis is supported by the finding that *M. pudica* is heat sensitive, responding to heat as it does to touch.

The trap of the Venus flytrap, *Dionaea muscipula*, (Figure 24.11b), actually a modified leaf, lies open when at rest. Each half of the trap has three tiny, hairlike triggers that spring the trap when brushed by a hapless insect. When triggered the toothed leaf, now tightly closed, presses the insect against digestive glands on the inner surface. The rest is history as the plant fills its nitrogen requirements.

Responses in both plants probably involve rapid changes in turgor. *M. pudica* has specialized thickenings—**pulvini**—at the base of leaf petioles and leaflet joints. Apparently, sudden changes in turgor within these structures permit the leaf and its leaflets to fold up and the petioles to droop. The movement is preceded by the transport of potassium ions out of cells that lose turgor and into cells that gain in turgor. The uptake of solute decreases the water potential and water enters by osmosis, bringing on rapid turgor changes needed for movement. Similar changes occur when the Venus flytrap is stimulated, but the trap isn't easily fooled. The triggers have a built-in code; two hairs must be touched in succession, or one hair must be touched twice before the trap goes into action. Generally, this means that a falling twig or inert object brushing the trap will have no effect. A wandering insect, on the other hand, is more apt to stumble across the code.

OTHER PLANT GROWTH HORMONES

A number of plant growth hormones have now been identified, and some of their stories are quite fascinating. We will look into the action of four of these.

Gibberellins

The **gibberellins,** a family of some 57 known molecules, received their name from the source of their discovery, *Gibberella fujikuroi,* a fungus. The fungus, which once threatened rice harvests in Japan, causes the stem to elongate strangely and the failure of the rice plant to produce normal flowers.

24.11   TOUCH RESPONSES IN PLANTS

Thigmotropisms are known to occur through swift changes in turgor, brought about by the active transport of ions such as potassium. **(a)** *Mimosa,* the touch-sensitive plant, visibly "cringes" when touched along its leaflets. **(b)** *Dioneae,* the Venus flytrap, springs its trap on a hapless insect if its triggers are touched in a certain way.

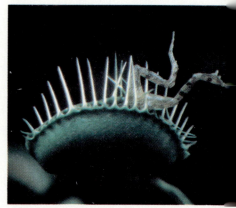

(a)

(b)

Between 1926 and 1935, Japanese botanists had already isolated and purified the active substances produced by the fungus, but it wasn't until the 1950s that other nations took an interest in these strange molecules. Today, we are continuing to study the natural production of gibberellins to learn more about how they operate in conjunction with auxins to control cell elongation. Some of the results are finding their way into agricultural use.

A gibberellin

Gibberellins are formed in young leaves around the growing tip, and possibly in the roots of some plants. (However, we don't know what role they play in root activity.) The power of gibberellins in stem elongation has been most dramatically illustrated in experiments with genetic dwarfs. Dwarf corn, for instance, can be induced to grow to normal height after the application of gibberellins. This indicates that a hormonal failure causes the plants' shortness and, except for this abnormality, dwarf corn has the potential to grow tall. Incidentally, the degree of growth in a dwarf depends on the quantity of gibberellins applied. Thus the botanist has an excellent method of gibberellin hormone assay (Figure 24.12).

Gibberellins also have another role. They act as chemical messengers to stimulate the synthesis of an enzyme called **alpha-amylase** in grains such as barley and corn. As these grains germinate, the embryo secretes gibberellins, which move to the cell layer that surrounds the starchy endosperm, the aleurone. The cells of the aleurone, apparently stimulated by the hormonal messenger, begin to produce alpha-amylase, which then breaks down starch and makes glucose available to the growing plant (Figure 24.13). There may be many steps or only a few between the arrival of the hormone messenger and the transcription activity required for producing the enzyme. As you may recall from past discussions (see Chapter 10), the specific hormonal gene-activating mechanisms are still under investigation.

Cytokinins

Most of what we know about the family of hormones called **cytokinins** springs from work begun in the 1950s, when it was found that plant growth

24.12 GIBBERELLINS AND STEM GROWTH

Gibberellins have a dramatic effect on stem growth, as seen in these five cabbage plants. The control plants at the left and the experimental plants at the right were treated identically except that gibberellins were applied to the group at the right. Notice the flower development at the stem tips, a phenomenon that normally occurs only in cabbage's second year of growth.

could be influenced by something from corn kernels. In 1964, the first of the molecules, **zeatin**, was described; since then, three others have been identified.

Zeatin

Prior to the discovery of zeatin, research scientists had found that in order to grow cultures of plant tissues, they had to add coconut milk to the medium. What was so special about coconut milk and corn kernels? They began to isolate chemicals found in both sources and the mysterious substance turned out to be a group of hormones they called cytokinins.

The first thing biologists found out about cytokinins was that they stimulated cell division in plants, although they had to be coupled with other plant hormones to do this. Using tissue taken from tobacco plants, researchers mixed **kinetin** (one of the cytokinins) and auxins in different ratios and then grew tobacco pith cells in the various media. The interaction of the two hormones is quite complex. The growing pith at first produced only an undifferentiated **callus,** the same tissue that forms around wounds. Then, when hormones were administered, it turned out that auxin encouraged root growth while kinetin encouraged shoot growth—but only when the opposing hormone was in low concentration. Further, when very low concentrations of both hormones were present, little callus growth and no differentiation occurred. At

24.13 GIBBERELLINS AND SEED GERMINATION

Gibberellins bring about the conversion of starch to glucose, as seen here. **(a)** Water is absorbed by the seed. **(b)** The embryo secretes gibberellins. **(c)** The hormone reaches the aleurone cell layer, stimulating these cells to release alpha-amylase. **(d)** Alpha-amylase, released into the starchy endosperm, hydrolyzes the starch to glucose, and the seedling has an energy-rich food supply for growth and development.

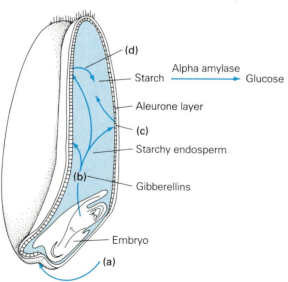

24.14 CYTOKININS AND DIFFERENTIATION

In this experiment, the pith is treated with varying combinations of auxin and cytokinin, each of which produces some variation in growth and differentiation.

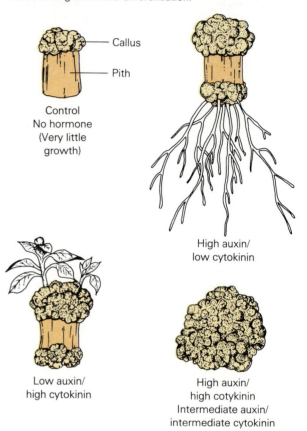

Callus

Pith

Control
No hormone
(Very little
growth)

High auxin/
low cytokinin

Low auxin/
high cytokinin

High auxin/
high cotykinin
Intermediate auxin/
intermediate cytokinin

high concentrations of both, the callus grew but no differentiation occurred (Figure 24.14).

The studies indicated that the undifferentiated and ordinary cells of the pith, the parenchyma cells, contain all the genetic information necessary to develop into a number of other kinds of plant cells, producing a variety of tissues and organs. Even differentiated tissue, grown on the proper medium, would sprout into complete, normal plants. The potential value of such studies staggers the imagination—whole individuals grown from a piece of tissue! This capability in cells is referred to as **totipotency**—a phenomenon that first involves **dedifferentiation,** the return to a simple undifferentiated state, and then differentiation, usually along a different path from that of the original tissue. The concept of totipotency implies that cells, no matter how specialized, never lose the genetic capability of the fertilized egg cell. More on this subject and studies of plant totipotency are found in Essay 24.1.

The third role of the cytokinins is associated with **senescence,** or aging, in plants. For example, senescence occurs in leaves prior to their seasonal fall from the branches of deciduous trees. Leaves also age if they are picked while in the prime of life. Then, of course, they die. When picked leaves are treated with cytokinins, however, aging is retarded. The chlorophyll does not disintegrate (so the leaves stay green), protein synthesis continues, and carbohydrates do not break down. Synthetic cytokinins have been applied to harvested vegetable crops such as celery, broccoli, and other leafy foods in order to extend their storage life.

We have a few ideas about how cytokinins do their work. It is known, for example, that some of the tRNA molecules contain cytokinins as a functional part of their structure, so it is possible that some cytokinins may facilitate protein synthesis. Frankly, however, plant physiologists are groping at this point, the precise answer to how cytokinins work constantly eluding them.

Ethylene

Ethylene is a relatively simple compound when compared to other plant hormones:

$$H \diagup{C}=C \diagdown H$$

Ethylene

Ethylene is a gas—one you can smell around ripening fruit. In fact, it controls the ripening process. It is synthesized by altering the amino acid methionine, commonly found in cells. Since it is a gas, it readily diffuses out of plants; thus the concentration of ethylene in plant tissues depends on its rate of production and escape.

In addition to its role in initiating the ripening process, ethylene is believed to play an important part in the emergence of seedlings from the soil. As some seeds sprout, the upper portion forms a sharp curve, sheltering the young fragile leaves underneath (Figure 24.15), as the tough, thickened curve of the shoot plows its way up through the soil. Somehow, the presence of ethylene prevents the plant from straightening and keeps the fragile leaves from unfolding to the sky until the shoot is free of the ground. Once this happens, the light ethylene gas readily escapes, lowering its concentration, permitting the stem to straighten and the leaves to expand.

The concentration of ethylene in plants is important, as is the case with all plant hormones. When the concentration rises to an abnormal level, the result can be disaster. For example, ethylene in the form of air pollution can cause defoliation, and eventually death, in plants. Plant physiologists suggest that ethylene promotes the production of the cellulases that normally increase just before regular abscission occurs.

Ripening fruits produce ethylene naturally, but ethylene gas is also a petroleum product—the same one that is polymerized to make polyethylene plastic. The synthetic gas has long been used to initiate the ripening of fruits in transit to the world's markets. The banana industry in particular owes much of its success to this simple chemical. It is now possible to ship unripened fruit to world markets in prime condition without worrying about untimely ripening. The release of ethylene in warehouses or in ship cargo holds full of green bananas will produce ripe, yellow bananas at just the time they reach the consumer.

Ethylene may also help determine sex in certain flowers. In plants that bear separate male and

24.15 ETHYLENE'S ROLE IN SHOOT CURVATURE

(a) Many dicot seedlings form a curved, protective "bumper"—the *epicotyl hook*—as they push their way through the soil. The curve is maintained by a surrounding concentration of ethylene as long as the shoot is in the soil. When the seedling breaks through the soil, the curve straightens. (b) The effects of ethylene on curvature in pea seedlings are demonstrated by this experiment, in which seedlings on the right have increasing ethylene concentrations.

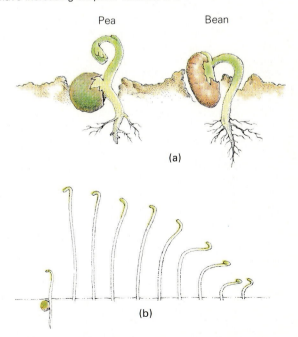

Pea   Bean

(a)

(b)

The Problem of Differentiation

Scientists have indeed uncovered detail after detail about plant development until it seems that we surely must have the big problems solved. But one haunting question prowls the attics of our minds: "How do plant cells differentiate?" This, of course, is a most fundamental question, so we might wonder, what do we really know? The answer is essentially the same for plants and animals: we haven't yet developed entirely adequate explanations. Perhaps none exist, but biologists are making progress on a number of fronts.

Botanists in particular have provided valuable basic information. This may well be because in some ways plants make ideal subjects for studying cell growth and proliferation under controlled conditions. For one thing, some of them can regenerate from bits and pieces. Many plants can regenerate roots from stem cuttings, stems from bits of root, and even entire plants from leaves. This knowledge has been invaluable in agriculture through the ages, but for biologists it suggests clues to the puzzle of differentiation. It supports the idea, for example, that plant tissues are *totipotent*—that is, their cells retain the capability needed to produce the entire organism from which they come. If this is true, then, can individual plant *cells,* acting alone, duplicate the regenerative feat we see in cuttings? This question has been partly answered through studies using tissue culture techniques.

Plant researchers turned to tissue culture methods in the 1930s and 1940s. A pioneer in these efforts, Johannes van Overbeek, discovered that a suitable medium was coconut milk. Besides important nutrients, coconut milk contains some critical hormones needed for plant growth. Van Overbeek succeeded in growing individual cells that he had separated out from young carrot embryos (a). He cultured them in his coconut milk and later planted them in soil. Lo! They produced normal adult carrot plants. The results indicated that cells in the embryo, at least, were totipotent.

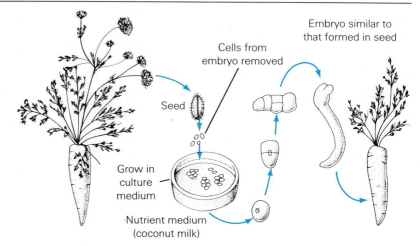

(a) Van Overbeek's experiment

Embryo similar to that formed in seed

Cells from embryo removed

Seed

Grow in culture medium

Nutrient medium (coconut milk)

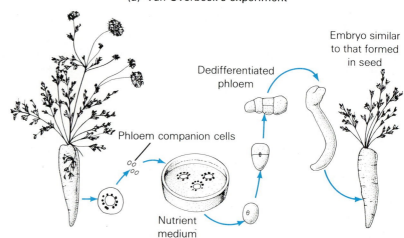

(b) F.C. Steward's experiment

Embryo similar to that formed in seed

Dedifferentiated phloem

Phloem companion cells

Nutrient medium

It wasn't until the late 1950s, however, that mature tissue was first cultured in a similar manner by F. C. Steward at Cornell University. Mature carrot cells were successfully removed and grown in a nutrient medium. The mature cells grew into root-like structures that, when planted, produced entire carrot plants (b). More recently, there has been excellent progress in culturing redwood trees and orchids. Plant physiologists have also succeeded in producing clones of potato varieties from cells in the leaf. In so doing, they have provided agriculture with a potentially economical way of growing potatoes, and biology with yet another "experimental organism." The potato clone promises to lend itself to investigations into the unyielding mysteries of development.

Extending this work, researchers at Kansas State University obtained experimental material from leaf cells that were first converted to protoplasts (c). Enzymes were used to remove cell walls and intercellular materials, leaving the naked cells behind. The new protoplasts then massed together, entered cell division, and produced new cell walls.

The resulting tissue was then transferred sequentially through several types of culture media. Each medium contained certain mixes of synthetic hormones, each of which had its own effect on differentiation. The first mix produced a large undifferentiated callus. The second encouraged shoot growth and differentiation, and in response to a third mix, the tissue produced roots. Newly differentiated plants were then transferred to regular soil beds, where they were grown to maturity. This procedure proves once again that mature plant cells lose none of their original genetic potency.

Though the work is in its early stages, the potential benefits are clear. Through cloning, agriculturists could produce highly selected, unvarying crop plants. (Who needs variable potatoes?) In keeping with the new era of genetics and molecular biology, potato protoplasts can be used in gene splicing and recombinant DNA studies. The Kansas group has succeeded in fusing nuclei in potato and tomato protoplasts, producing a hybrid that would not readily occur in nature or through sexual, genetic crosses. Such readily manipulated experimental organisms hold great potential for the ongoing study of development.

A more immediate goal in such efforts is to confer the disease resistance of one species on another less resistant one. For example, the tomato plant used in the fusion described above carries with it genes that will enable the potato to resist the fungus of *late blight* (see Chapter 19). Not only could this eliminate a constant and serious threat to an important food supply, but it also could eliminate the widespread use of certain dangerous chemicals—the fungicides now used to control late blight.

Let's digress here and consider the seemingly facetious question, "Who needs variable potatoes?" The answer is, we do. In fact, we would do well to begin developing "gene banks" to preserve the variation in

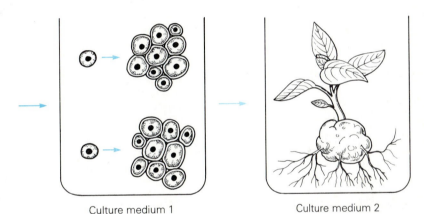

① Potato plant

② Cell walls digested, treatment causes cytoplasm to withdraw and ball up

Youngest leaves

Cellulose digesting enzyme

Free protoplasts

Culture medium 1

Culture medium 2

③ Protoplasts divide, produce new cells that grow into callus

④ Callus treated with hormones, differentiates into shoot and root: ready for planting

(c) Kansas State University experiment

the original phenotypes. Once we drive these older varieties into extinction through selective breeding for our favorite traits, there will be no going back. Biologists are becoming alarmed at what might happen if our highly selected food strains should fall to some new fungal or bacterial mutant or perhaps some environmental variable that we can't control. We will need those original strains just to start again.

We see, then, that any cell retains the capability necessary to produce the entire plant. Second, when removed from their cell associations, cells can regress to an undifferentiated state from which they can then proceed through normal embryonic development. How does this answer the question about differentiation? It tells us both that genetic information is not lost as development proceeds, and that it is not irreversibly repressed.

Such findings are important from the standpoint of pure research, but they may gain in importance as we seek to provide for an increasingly hungry world.

female flowers, treatment with gibberellins encourages male flowers to develop while treatment with ethylene encourages female flowers. The artificial induction of female flowers in the fields can be quite profitable if one is marketing the fruit.

Abscisic Acid

Our last example of a plant hormone, one we have already mentioned, is abscisic acid. Its molecular structure consists of a carbon ring with a short carbon chain containing a carboxyl (acid) group:

Abscisic acid

It is generally believed that abscisic acid induces winter dormancy by suppressing mRNA production, which would inhibit the growth-promoting hormones, auxin and gibberellin. The suppression is especially important to deciduous plants of the temperate regions. Dormancy, we might say, is the deciduous plant's answer to water shortages that occur when freezing temperatures lock water up as ice.

Applications of Plant Hormones

Investigation of plant hormones has led to the development of increasingly precise analytical techniques such as the hormone assay. The improved techniques have enabled us to regulate and understand (sometimes in that order) a wide range of hormones. Such knowledge has been extremely useful to people in a variety of ways. Not only can we produce the specific plants we want, but we can also destroy them—including some weeds and at one time, the trees of Vietnam's forests (which hid the activities of the Viet Cong and North Vietnamese soldiers).

Since the discovery of auxins, chemists have developed an entire family of analogous compounds, some of which have even greater growth-promoting effects than the original auxin. One such compound is 2,4-dichlorophenoxyacetic acid, known mercifully as 2,4-D. While 2,4-D promotes growth at very *low* concentrations, it kills plants at higher concentrations. This compound is a very common and potent agent of weed control. Dicots,

including the broad-leaved plants we often think of as "weeds" are much more sensitive to the herbicide than monocots (wheats, oats, and so on). This means that you can kill the dandelions in your lawn with 2,4-D while sparing the rye and bermuda grass. Of major ecological importance, 2,4-D offers the advantage of rapid **biodegradability** (it rapidly ceases biological activity when exposed to the elements and organisms of the soil).

Weed control agents are not uncontroversial, however. For example, there have been serious questions about the effects of a chemical called 2,4,5-T (2,4,5-trichlorophenoxyacetic acid), closely related to 2,4-D but used in controlling the growth of trees. It reportedly has caused birth defects in mammals—and we are mammals. For this reason, its broad use for valid purposes such as fire prevention has to be measured carefully against other risks.

The problem is, we create a weed problem by clearing the land, building roads, and otherwise disrupting the natural habitat. Then we solve the problem by applying unnatural controls, which in turn may create additional, often unexpected problems. Actually, the real risk, as we now see it, does not arise so much from the control agents themselves as from a lack of quality control in the manufacturing process. In the past, 2,4,5-T is known to have been contaminated by **dioxin,** the infamous and frightening environmental pollutant common to illegal (and legal) chemical dumps. We now know that the 2,4,5-T used in the defoliant **agent orange** of Vietnam fame was also heavily contaminated with dioxin. The potential consequence of the frequent exposure of American military personnel (and who knows how many unfortunate Vietnamese citizens) to agent orange during the Vietnam conflict is still a volatile issue. Veterans' groups have claimed that such exposure has caused them a number of drastic health problems and increased birth defects in their children.

LIGHT AND FLOWERING IN PLANTS

The response of organisms to changing lengths of the day and night is known as **photoperiodism.** It is a familiar phenomenon in many forms of life. Birds, for instance, respond to lengthening days through the appearance of their spring plumage and an amazing increase in gonadal size as the breeding season approaches. The mating activities of other vertebrates, and many invertebrates as well, are

24.16 PHOTOPERIODICITY

Each of the plants here has its own specific photoperiod requirement for flowering. **(a)** The chrysanthemum is a short-day plant that produces flowers in autumn. Flowering can be stalled if the grower applies light for an hour during the night. **(b)** Poinsettias are also short-day plants and will bloom when the night is longer than 13 hours. Their blossoms and the bright red color in the surrounding leaves can also be easily controlled by night illumination; thus poinsettias can be brought into bloom just in time for Christmas. **(c)** For its flowering, the henbane requires a long day (short night), usually in excess of 12 hours.

(a)   (b)   (c)

also governed by day and night length. But certainly, a familiar and delightful seasonal event is the response of plants to the changing days—the flowering response.

Every flower has its time. Throughout much of the Northern Hemisphere, the early-flowering crocus is the famous herald of spring. As the days lengthen, species after species bursts into bloom in a rather predictable sequence. As summer gives way to fall and days again begin to grow shorter, still more species will unfurl their colors. Hardly any other good news in this life is so dependable.

But one wonders, how does the crocus "know" that spring has arrived? And how does the ragweed "know" that fall is here? Or, to regain some scientific objectivity, what physiological processes stimulate the tissues of the plant to shift to flower production? The questions are not simple ones. Biologists would like to have a better idea of how to both stimulate and inhibit the reproductive process. If the process could be controlled, florists could have their product ready just in time for Mother's Day, and even more important, farmers could regulate the production of fruit and develop a more steady market. As a matter of fact, botanists are learning to control these very things, and they are getting better at it with the passing years.

Still, the answers are not all in by any means. For example, flowering seems to be under hormonal control, but no specific flowering hormone has ever been discovered. Auxin is at least indirectly involved, but in a negative way. The concentration of auxin drops in plants that are about to go into flower; furthermore, the application of auxin can sometimes be used to prevent or delay flowering.

Photoperiodicity

As to how a plant knows what the season is, we do have some answers. The critical factor for many species is the *length of the night*, rather than the length of day, as was once believed. Of course, the length of night varies with location and with the season, but plants somehow are able to "count the hours." When the period of darkness reaches a certain length, the plant responds in a predictable manner, as though it "knows" April, June, or January has arrived.

The responses of plants to changes in the length of night can be roughly organized into three categories. Plants that begin the flowering process before the summer solstice, June 21, are called **long-day plants** because they flower only when the shortening nights reach a critical brevity. Plants that do not flower until after the summer solstice are called **short-day plants,** because these plants respond to the lengthening nights. The third category of flowering plants are the **indeterminate** or **day-neutral** plants, which appear to be indifferent to the light cycles. They may flower continuously or in response to other stimuli (Figure 24.16).

How was it determined that it was the duration

of *darkness* that triggers flowering, and not the duration of *daylight*? Actually, the demonstration is rather simple and can be done in laboratories, where the light and dark periods can be regulated. Under such conditions it was found that the period of light can be changed without causing any flowering changes, but if the length of darkness is tampered with, the plant responds. Quite dramatically, a single, relatively brief exposure to intense light in the middle of the night can "trick" a long-day (short-night) plant into responding as if the short nights of spring had arrived. Such a plant can then be made to bloom in any season. The same treatment, if done every night, can prevent a short-day (long-night) plant from flowering at its normal late summer or autumn time, since it will react as if the nights were still too short for such activity.

This bit of academic tinkering was immediately put to practical use by chrysanthemum growers. For many years they had extended the chrysanthemum season into early winter, artificially lengthening the days by keeping bright lights turned on for several hours after sunset. But now they found they needed only to turn on the lights for a few minutes each night in order to get the same results. Similarly, but much more important economically, sugar cane growers can stall flowering in their fields by turning floodlights on at night. Sugar-laden sap that might otherwise be used to produce commercially useless flowers can then continue to accumulate in the stems.

The cocklebur has proved to be an ideal organism for studying the initiation of flowering. It is so hardy that researchers have been able to alter it drastically and still count on its survival. The cocklebur is a short-day (long-night) plant, and will put out homely little green flowers if exposed just once to a period of darkness longer than 8 ½ hours.

Armed with knowledge, plant physiologists began to toy with the plant. One of the first things they found was that the stem and even the insipid flowers were insensitive to the midnight flash effect; only the *leaves* were sensitive.

The scientists also found that varying the wavelengths of the light had curious results (Figure 24.17). The most effective light for establishing the photoperiod was found to be in the red or orange-red region. Far-red light, with its longer wavelength, had the opposite effect; that is a flash of far-red light actually reversed the effect of either white or red light if it immediately followed either of them. However, if the far-red flash *preceded* the red flash, or was *delayed* for more than 35 minutes after the red flash, the red flash had its full effect. Curious indeed.

The midnight flash experiments gave rise to the hypothesis that there was some receptive pigment involved. The receptive pigment could not be chlorophyll (which is abundant in the photosensitive leaves) because it responds to different wavelengths from those produced by the floodlights. The scientists suggested that perhaps there is another pigment that occurs in two forms: P_r, the red-absorbing form and P_{fr}, the far-red-absorbing form. Absorption of light of the appropriate wavelength would change this hypothetical pigment from one form to another. During the day, the predominance of red light over far-red would convert all the pigment from the P_r form into the P_{fr} form, but at night there would be a spontaneous reversion of the pigment back to the P_r form. A midnight flash of red light would immediately convert the pigment once again into the P_{fr} form, but in this state it could again be reversed to P_r by absorbing far-red light.

Light-induced change:

$$\text{Far-red light} \quad P_{fr} \rightarrow P_r$$

$$\text{Red Light} \quad P_r \rightarrow P_{fr}$$

Spontaneous dark reaction: $\quad P_{fr} \rightarrow P_r$

There was joy in the streets when the hypothetical pigment was found. It was named **phytochrome,** and it turned out to be a large membrane-bound protein complex. It now seems that phytochrome is involved in a number of other light-induced phenomena, such as the turning of leaves toward light and the rapid orientation of chloroplasts as they move broadside to the light source.

Phytochrome indeed occurs in the two forms hypothesized, but apparently it also goes through many other conformational changes, which are not at all simple or well understood. At one time it was thought that the conversion of one form of phytochrome to another was a slow process that might "time" the duration of the night but in fact these conversions are relatively rapid. There is clearly some kind of "dark clock" involved, but the phytochrome itself is apparently only the detector of light and the trigger that sets the dark clock. According to this idea, a flash of red light in the middle of the night resets the dark clock back to zero, just as it is set to zero at the end of a normal day.

But just what *is* the dark clock, and how does it work? Again, we don't know. It is presumably some kind of chemical reaction, but unlike most chemical reactions it is almost completely insensitive to temperature differences within normal ranges. It is a most peculiar phenomenon indeed.

Transmitting the Stimulus. The dark clock and its photoreceptor, phytochrome, are located in the leaves. But the signal that actually induces flower development has to be delivered from the leaves to other parts of the plant. In the resilient cocklebur, an isolated, amputated leaf can be subjected to an appropriate flower-inducing photoperiod regimen and subsequently grafted back onto a plant, causing the plant to flower. The signal is evi-

dently hormonal, but, as we said, no flower-inducing hormone has been found. In some experiments using plants other than the cocklebur, it appears that there are *repressing* hormones produced by the leaves at all times *except* when the dark clock indicates that the appropriate time for flowering has arrived.

The hormonelike transmission of the photoperiod effect has been demonstrated most convinc-

24.17 PHOTOPERIODS IN THE COCKLEBUR

(a,b) The cocklebur, a short-day plant, will flower when the nights become long. Its requirement of a specific dark period to initiate flowering is demonstrated at **(c)**, where the period is interrupted by a flash of red light. **(d)** A similar flash of far-red light has no effect. **(e)** Preceding the red flash by a flash of far-red light apparently has no effect on the inhibiting effect of red. **(f)** But when the flashes are reversed, with far-red following red, the effect of red is reversed and the flowering occurs. **(g)** The timing of these flashes is critical, since a lapse of 35 minutes or more between red and far-red flashes is apparently enough time for the effect of red light to reset the "dark clock," inhibiting flowering.

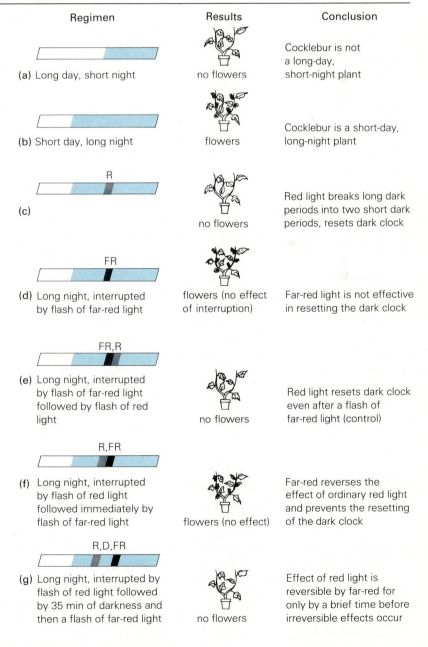

Regimen	Results	Conclusion
(a) Long day, short night	no flowers	Cocklebur is not a long-day, short-night plant
(b) Short day, long night	flowers	Cocklebur is a short-day, long-night plant
(c) R	no flowers	Red light breaks long dark periods into two short dark periods, resets dark clock
(d) FR — Long night, interrupted by flash of far-red light	flowers (no effect of interruption)	Far-red light is not effective in resetting the dark clock
(e) FR,R — Long night, interrupted by flash of far-red light followed by flash of red light	no flowers	Red light resets dark clock even after a flash of far-red light (control)
(f) R,FR — Long night, interrupted by flash of red light followed immediately by flash of far-red light	flowers (no effect)	Far-red reverses the effect of ordinary red light and prevents the resetting of the dark clock
(g) R,D,FR — Long night, interrupted by flash of red light followed by 35 min of darkness and then a flash of far-red light	no flowers	Effect of red light is reversible by far-red for only by a brief time before irreversible effects occur

24.18 THE LEAF: A PHOTORECEPTOR

Of the many experiments that reveal the photoperiod effect, this is among the most dramatic. It establishes conclusively that the photoreceptor is in the leaf. Only one leaf in this series of six grafted cocklebur plants is exposed to the proper photoperiod, yet all six plants produce flowers. It is almost certain that some kind of chemical messenger is being transferred from the exposed leaf to the other plants, but thus far, no specific hormone has been found.

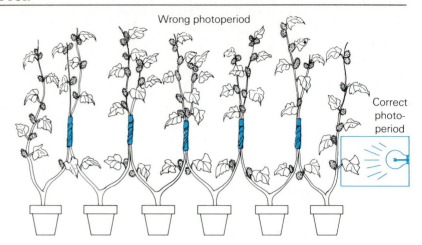

Wrong photoperiod

Correct photo-period

All six plants produce flowers

ingly by an experiment in which six cocklebur plants were grafted together in a row (Figure 24.18). One leaf of the plant at one end of the line was enclosed in a box and given the appropriate photoperiod treatment, after which *all six plants* flowered, one after the other, right down the line. In other grafting experiments, the cambia of the two plants were separated by a sheet of paper, but the message got through anyway, just as we would expect if the signal were hormonal.

The hypothetical flower-inducing hormone has somewhat presumptuously been named **florigen.** However, because it has proven to be so elusive, some plant physiologists believe that there is no special flower hormone at all but that the message is conveyed by particular levels and combinations of auxin, gibberellins, and other known plant growth hormones.

We are aware that, paradoxically, auxin has an inhibitory effect on flowering. Since growing tips of shoots produce auxin in quantity, flowering can sometimes be induced or increased by simply cut-

ting off the shoot tips. Rose growers have discovered this, and the clicking of pruning shears can be heard over the countryside at the beginning of the floral season.

Pineapples and Auxin. Pineapples are different: in pineapples, auxin helps induce flowering. Pineapples are also tropical plants, living where there is little seasonal change in day length. Not surprisingly, they are indifferent to photoperiods and normally bloom and bear fruit on irregular schedules of their own, all year long. In the past, pineapple pickers had to roam the fields daily, looking for ripe pineapples—an expensive and time-consuming process. Now, growers synchronize the flowering of whole fields by the application of artificial auxins. The plants dutifully flower and ripen all at once, and can be efficiently harvested by machines. That's why pineapples are cheaper than they used to be. (But do you suppose we're paying hidden costs by spraying artificial auxins into the environment?)

APPLICATION OF IDEAS

1. Formerly, plant physiologists attributed the success of plant roots in reaching water to *hydrotropism,* a water-initiated growth response. While the idea has now been discarded, root growth is known to be most dense in moist areas of soil. Suggest an alternative hypothesis to explain this observation. How do roots happen to find water? How might your hypothesis be tested?

2. Many angiosperms respond to photoperiods in their flowering activities. What might be some advantages to complex flowering systems that relate to day or night length? If photoperiodism is adaptive to plants, how can the success of the day-neutral plants be explained?

KEY WORDS AND IDEAS

Charles Darwin was among the first to experiment with plant growth mechanisms, and his observations involved insectivorous plants and climbing plants.

THE DISCOVERY OF AUXIN

1. Working with canary grass, Darwin found that the young **coleoptile** covered shoot would respond to light by bending toward it. By experimenting, he found that the agent responsible was formed in the tip and passed downward.

The Experiments of Boysen-Jensen and Paal

1. Boysen-Jensen decapitated oat seedlings, mounted tiny blocks of gelatin, a very permeable substance, on the stump, and then placed the tips on the gelatin. On others, he used a tiny chip of impermeable mica to separate the tip from the stump. The response toward light occurred only in the group where gelatin was used, indicating something from the tip had to pass downward for the bending response to occur.

2. Paal decapitated oat seedlings and used the tip alone for experiments. He placed the tip to one side of the stump and the bending response always occurred on that side. He concluded that some influence from the tip caused cell elongation.

Fritz Went and the Isolation of Auxin

1. Fritz Went placed oat seedling tips on agar for a time and then used tiny agar blocks in his experiments. Arranging the agar blocks on the seedling stumps in various ways, and keeping his seedlings in darkness, he observed the usual bending, indicating that the active substance had diffused into the agar.

2. Went named the active substance **auxin** (later identified as **indoleacetic acid** or **IAA**). He also worked out an **assay system** relating bending to auxin concentration.

AUXIN

1. Auxin is derived from the amino acid tryptophan. Its primary action is to stimulate cell elongation, the mechanism of primary growth. It may function by loosening cellulose microfibrils, thereby permitting turgor pressure to lengthen the cell. The mechanism may involve lowering the cell wall pH and thus activating microfibril loosening enzymes.

2. Auxin is involved in **abscission** (leaf fall). Prior to leaf fall, changes in cells at the base of the petiole produce an **abscission zone.** Then as auxin diminishes, materials in the leaf are reclaimed by the plant, cell walls in the abscission zone break down, and cells nearer the stem are sealed off with suberin.

TROPISMS

1. Auxin is involved in *tropisms*, or growth responses. They include **positive phototropism** (toward light), **negative phototropism** (away from light), **geotropism** (gravity), and **thigmotropism** (mechanical or touch).

Phototropism

1. In phototropism, a blue-sensitive photoreceptor detects light direction and, through processes still unknown, the migration of auxin is directed to the unlighted side of the shoot where elongation then occurs.

2. Auxin plays an inhibitory role in **apical dominance,** whereby one growing shoot elongates and the rest are inhibited to some degree. Loss of the dominant tip releases inhibition and another branch becomes dominant. The triangular shape of conifers is a good example of apical dominance.

Roots and Geotropisms

1. A seedling laid on its sides will reveal phototropism in the shoot and geotropism in the stem. One explanation of the different response of roots is that auxin has an inhibiting effect on root cells; thus its usual movement away from the light inhibits elongation only in the lowermost cells. A second hypothesis suggests that **abscisic acid,** produced in the root cap, moves upward and subsequently inhibits the lowermost layer of cells. It is further suggested that plants have gravity-sensing mechanisms—large plastids, perhaps, with a root-inhibiting hormone that settles to the lowermost cells when the root is laid on its side.

Thigmotropism

1. Touch responses, seen in the *Mimosa pudica* and *Dionaea muscipula*, are attributed to sudden changes in turgor brought on by the rapid transport of potassium ions from one tissue to another, changing water potential gradients.

OTHER PLANT GROWTH HORMONES

Gibberellins

1. **Gibberellins** are known to cause dramatic stem elongation. They restore normal growth to genetic dwarf corn. When released by the embryo of grains, they stimulate the aleurone to produce starch-digesting **alpha-amylase.**

Cytokinins

1. **Cytokinins** such as **zeatin** and **kinetin** stimulate cell division. Together, auxin and kinetin influ-

ence differentiation in the **callus** formed from tobacco pith cells. The relative proportions and concentrations of the two hormones determine whether the growth will form shoots, roots, both, or neither.

2. The capability of plant cells to **dedifferentiate** from a specialized state and then redifferentiate into a new tissue type is called **totipotency.**

3. Cytokinins retard **senescence,** or aging, in plants and can be used to extend the storage life of leafy vegetables. These hormones may interact with tRNA, facilitating protein synthesis.

Ethylene
1. **Ethylene** is a gas derived from the amino acid methionine. It hastens fruit ripening and helps retain the hypocotyl hook in some emerging seedlings.

Abscisic Acid
1. **Abscisic acid** inhibits auxin and gibberellins, bringing on plant dormancy in preparation for winter.

Applications of Plant Hormones
1. Many plant hormones or their analogs are now produced synthetically. 2,4-D is a **biodegradable** dicot weed control agent. 2,4,5-T, a defoliant, is used in controlling tree growth but may be hazardous to mammals. Its contamination with **dioxin** when used in **agent orange** during the Vietnam conflict has raised legal and moral issues.

PHOTOPERIODISM AND FLOWERING IN PLANTS
1. **Photoperiodism** is any response to changing lengths of night or day. Flowering in plants is often photoperiodic, but so far no specific hormonal mechanism has been found.

Photoperiodicity
1. In their response to photoperiods, plants fall into three categories: **long-day, short-day,** and **indeterminate** or **day-neutral.**

2. It is now known that plants respond to the length of night rather than the length of day. A brief exposure to light during the night will induce flowering in long-day plants and stall flowering in short-day plants.

3. The most effective light for interrupting flowering is in the red region of the spectrum, while far-red reverses the red effect if it follows within 35 minutes. Two interconvertible variations of a light receptor, P_r and P_{fr} have been proposed. The receptor has been identified as **phytochrome,** a large, membrane-bound protein, but its operation as a "dark clock" isn't known.

4. Experiments with the cocklebur strongly suggest a hormonal mechanism that may work through the repression of flowering at all times except during the proper photoperiod. The hormone has been presumptuously called **florigen.**

REVIEW QUESTIONS

1. Describe Charles Darwin's experiments with phototropism. What conclusion did he reach? (594)

2. What information did Boysen-Jensen and Paal add to what Darwin had already determined? (594)

3. Describe the experimental technique used by Went, and explain how his assay procedure worked. (594–595)

4. Describe the specific action of auxin on plant cells and the related enzymatic mechanism. (595, 597)

5. List the main steps involved in leaf abscission. What role does auxin play? (597)

6. Using the idea of a photoreceptor and the pattern of auxin diffusion, present an explanation of phototropism. (597)

7. Briefly explain how pruning affects apical dominance in stems. (598)

8. Describe two different hypotheses explaining geotropism. (599)

9. What is thigmotropism? Describe the mechanism responsible. (599–600)

10. Briefly describe how gibberellins were discovered. What effect do they have on dwarf corn? (600–601)

11. Explain the role of gibberellins in the germination and growth of grain embryos. (601)

12. Explain the concept of totipotency. How do agriculturalists make use of this capability? (602)

13. What effect do cytokinins have on senescence? How might cytokinins interact with auxins in this role? (603)

14. How might abscisic acid help temperate zone plants survive? (606)

15. Name two auxin analogs and describe how they are used. (606)

16. List and explain the three photoperiodic categories of plants. (607)

17. Briefly explain how plant physiologists determined that the period of darkness was the significant factor in the photoperiodic flowering responses. (608)

18. Describe the experimental observations that led to the conclusion that the flowering photoreceptor had two forms, P_r and P_{fr}. Write a simple equation showing the effects of red and far-red light. (608)

19. Explain how a repression system might work to instigate flowering. (609)

20. What is the status of the flowering hormone hypothesis today? (610)

SUGGESTED READING

Baker, H. G. 1963. "Evolutionary Mechanisms in Pollination Biology." *Science* 139:877. The always fascinating sex lives of plants.

Epstein, E. 1972. *Mineral Nutrition of Plants: Principles and Perspectives.* John Wiley, New York.

Esau, K. 1977. *Anatomy of Seed Plants,* 2d ed. Wiley, New York. The authoritative sourcebook by one of the world's most distinguished botanists.

Galston, W. W., and P. J. Davies. 1970. *Control Mechanisms in Plant Development.* Prentice-Hall, Englewood Cliffs, N.J.

Hardy, R. W. F., and V. D. Havelka. 1975. "Nitrogen Fixation Research: A Key to World Food?" *Science* 188:633. Basic research in plant biochemistry has impressive prospects for practical application in the world food crisis.

Hutchinson, J. 1969. *Evolution and Phylogeny of Flowering Plants.* Academic Press, New York and London.

Jensen, W. A. 1973. "Fertilization in Flowering Plants." *Bioscience* 23:21.

Lehner, R., and J. Lehner. 1962. *Folklore and Odysseys of Food and Medicinal Plants.* Tudor, New York. Did you know that crabgrass was deliberately introduced into the United States by Polish immigrants? In 19th century Poland people made bread from crabgrass seed.

Raven, P. H., et al. 1981. *Biology of Plants,* 3rd ed. Worth, N.Y.

Rutter, A. J. 1972. *Transpiration.* Oxford University Press, Oxford.

Sporne, K. R. 1971. *The Mysterious Origin of Flowering Plants.* Carolina Biological Supply Company, Burlington N.C.

Temple, S. 1977. "The Dodo and the Tambalacoque Tree." *Science* 197:885. On Mauritius there are some geriatric trees whose seeds normally germinate only when passed through the gizzard of a dodo. But the last dodo died in the 17th century.

Wade, N. 1974. "Green Revolution (1): A Just Technology, Often Unjust in Use." *Science* 186:1093. All fundamental changes, even of the most promising kind, end up hurting someone. Peasants of the third world end up getting it in the neck, it seems, no matter what happens.

Wardlaw, I. F. 1974. "Phloem Transport: Physical, Chemical or Impossible?" *Annual Review of Plant Physiology* 25:515. What *does* make sap flow? Although the answer is still not known, at least some hypotheses can be ruled out. Unfortunately, it sometimes seems that all hypotheses can be ruled out.

Wareing, P. F., and I. D. J. Phillips. 1970. *The Control of Growth and Differentiation in Plants.* Pergamon, New York.

Went, F. W. 1963. *The Plants.* Life Nature Library. Time, Inc., New York. Although it is now getting along in years, this beautifully illustrated volume remains a superbly readable, informative, and intelligent nontextbook approach to botany.

Scientific American articles. San Francisco: W. H. Freeman Co.

Albersheim, P. 1975. "The Walls of Growing Plant Cells." *Scientific American,* April. How cellulose fibers are laid down in careful patterns.

Brill, W. J. 1977. "Biological Nitrogen Fixation." *Scientific American* 236:68.

Epstein, E. 1973. "Roots." *Scientific American,* August.

Groves, D. J., et al. 1981. "An Early Habitat of Life." *Scientific American,* October. (Stromatolites).

Kaplan, D. R. 1983. "The Development of Palm Leaves." *Scientific American,* July.

Shepart, J. F. 1982. "The Regeneration of Potato Plants from Leaf-Cell Protoplasts." *Scientific American,* May.

Sigurbjörnsson, Björn. 1971. "Induced Mutations in Plants." *Scientific American,* January. The rapid progress in plant breeding that has led to the so-called "green revolution" has depended in part on deliberately caused mutations.

SUMMARY FIGURE REVIEW OF PLANT BIOLOGY

The outwardly serene and passive appearance of a plant is deceptive, for within its organs and tissues a veritable symphony of physiological acts is ongoing. Proceeding in a counterclockwise direction from the top we begin with ① the foliage, the site of photosynthesis and the origin of the vital TACT forces, responsible for lifting water many feet from root to foliage. During transpiration, water diffusing out of the leaves through the versatile stomata sets up the water potential gradient that is reflected in the pull on countless xylem water columns. ② At the growing tips, primary growth and differentiation provide a constant source of new tissues and account for lengthening of the stems. ③ Below the growing tips, the transition to secondary growth goes on with simple islands of vascular tissue coalescing to form the familiar ringed appearance of the perennial woody stem. ④ The source of continued secondary growth, a provision for increased girth and strength, is the vascular cambium, which gives rise to new xylem and phloem. Further out in the stem or trunk, cork cambium continually replaces the water- and disease-resistant cork. ⑤ Within the root system, primary growth occurs in countless root tips, where mitotic cell division provides an unending source of new tissues. In the maturing primary root, a well-defined stele arises just within the water-controlling endodermis. Within, newly emerging primary xylem will provide water conduction, and primary phloem will provide a food source for the actively growing root. ⑥ The vital uptake of water and minerals occurs in the young root, where lengthy root hairs emerge from epidermal cells, vastly increasing the absorbing surfaces. ⑦ Water transport through the stem is the task of xylem, with its slender tracheids and squat vessels. In addition, xylem contains countless fibers that lend resiliency and strength to the woody parts. ⑧ The transport of sugars occurs in the phloem, where sugar concentrations, created through active transport at the source, encourage water to enter and provide the pressure needed to transport the sugars to various sinks. ⑨ Finally, there are micro- and megagametogenesis, the meiotic events that lead to gamete production in the flower. These are closely followed by fertilization and seed development.

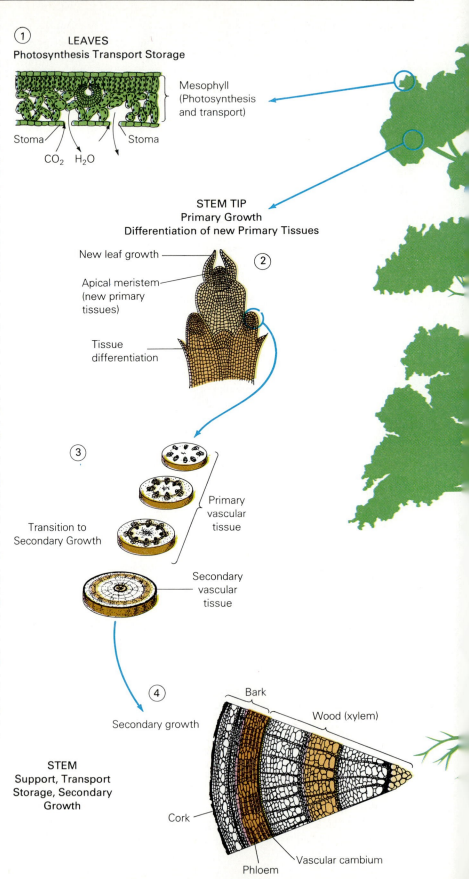

① LEAVES
Photosynthesis Transport Storage

Mesophyll (Photosynthesis and transport)

Stoma   Stoma

CO_2 H_2O

STEM TIP
Primary Growth
Differentiation of new Primary Tissues

New leaf growth

Apical meristem (new primary tissues)

②

Tissue differentiation

③

Primary vascular tissue

Transition to Secondary Growth

Secondary vascular tissue

④

Secondary growth

STEM
Support, Transport
Storage, Secondary
Growth

Bark

Wood (xylem)

Cork

Phloem

Vascular cambium

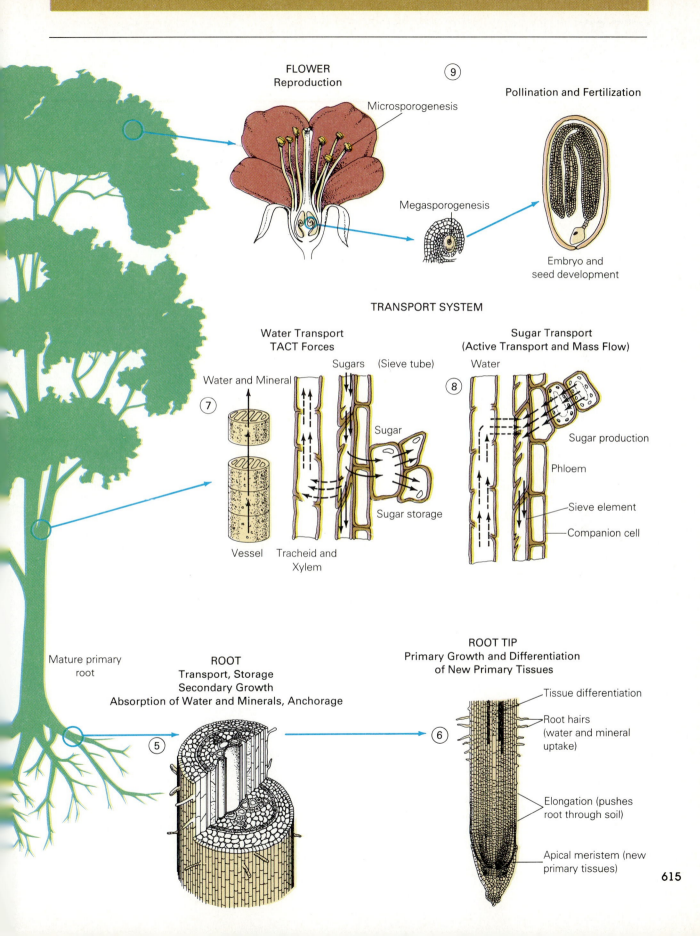

FLOWER
Reproduction

⑨

Microsporogenesis

Pollination and Fertilization

Megasporogenesis

Embryo and
seed development

TRANSPORT SYSTEM

Water Transport
TACT Forces

Sugar Transport
(Active Transport and Mass Flow)

Water and Mineral

Sugars   (Sieve tube)

Water

⑦

Sugar

Sugar production

⑧

Phloem

Sugar storage

Sieve element

Companion cell

Vessel   Tracheid and
Xylem

Mature primary
root

ROOT
Transport, Storage
Secondary Growth
Absorption of Water and Minerals, Anchorage

ROOT TIP
Primary Growth and Differentiation
of New Primary Tissues

Tissue differentiation

Root hairs
(water and mineral
uptake)

⑤

⑥

Elongation (pushes
root through soil)

Apical meristem (new
primary tissues)

Evolution and Diversity: Lower Invertebrates

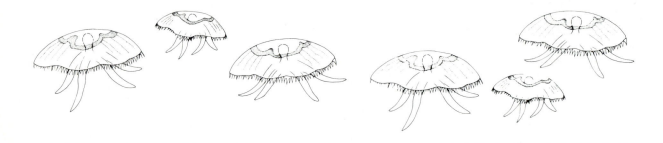

If you were driving across the countryside of Kenya and passed a troop of baboons, would it be safe to stop, get out, and walk among them, or are they dangerous? Would you be afraid of wolves if you walked across Isle Royale in Michigan? Why don't people use watchcats instead of watchdogs? Can a horse be trusted to guard the pasture? Should you pick up a centipede? Obviously, our fellow animals on the planet are quite diverse. Yet through all of them there runs the common thread of origin. We will now consider the diversity of animal life that shares this place. We will focus not only on the bewildering, fascinating, and even perplexing array of forms, but also on those great underlying themes that bind us all together.

WHAT IS AN ANIMAL?

If someone asked you what an animal is, you could probably come up with a fair definition. But you would probably be wrong just the same. The reason is that animals are so varied that exceptions can be found to almost every rule. So what kinds of rules can we make? Some would simply say that animals move around and have mouths and eat things. That's not a bad start. If we add "and are multicellular," we've pretty much covered the animal king-

dom. But some animals would virtually shout, "Not me!"

Because of the difficulties of finding commonality among such diversity, any description of the group must be both broad and precise. Such descriptions vary among scientists, but our list here will at least be generally acceptable and fill our needs for this discussion:

1. *Animals are multicellular, eukaryotic heterotrophs.* They are composed of numerous eukaryotic cells and their nutritional requirements include complex substances such as carbohydrates, lipids, proteins, and vitamins.
2. *Animals have cell and tissue differentiation and well-organized systems that carry out different life functions.* In other words, systems are composed of organs, organs of tissues, and tissues of cells. All except the simplest phyla (and a few types that have lost theirs) have a nervous system, digestive system, and reproductive system. Nearly all have skeletal frameworks and muscles and most have some form of excretory system.
3. *Between the cells of animals is an extracellular matrix of the complex protein* **collagen,** which acts as the "cementing substance" holding them together. In vertebrates, collagen is by far the most common body protein.

4. *Most animals reproduce sexually, with large, non-motile eggs and small, flagellated sperm.* Animals are diploid organisms and all cells except the gametes are diploid. The majority of species in every animal phylum is capable of sexual reproduction. However, each phylum also contains species that reproduce asexually.

There are some very conspicuous exceptions to these general rules, as we will see. Some of the exceptions, in fact, can be instructive because of their novelty, interest, or fascination.

Most of us are probably familiar with at least a few representatives of the major phyla. For example, we ourselves are chordates (phylum Chordata.) But so are some invertebrates such as sea squirts and salps. Everyone has seen insects (phylum Arthropoda), earthworms (Annelida), and snails (Mollusca). You probably know that your dog has roundworms (Nematoda), and if you live near a rocky coastline, you're likely to know about jellyfish (Coelenterata), starfish (Echinodermata), sponges (Porifera), and flatworms (Platyhelminthes), all of which can be found in the isolated tidepools that remain when the tide goes out.

With these examples, we have just named the nine major animal phyla. Each phylum has certain characteristics that set it apart from others, and each has a range of diversity within it (Table 25.1). There are at least as many rare animal phyla as the common ones listed above—all quite fascinating and, at times, perhaps even weird. Furthermore, all the animal phyla can be found as fossils in rocks that are at least half a billion years old.

TABLE 25.1

CHARACTERISTICS OF THE MAJOR ANIMAL PHYLA

Phylum	Level of Organization	Body Symmetry	Cleavage Pattern	Larval Type	Digestive Tract	Body Cavity	Segmentation
Subkingdom Parazoa							
Porifera	Cellular	Radial	NA	Flagellated	NA	Primitive spongocoel	NA
Subkingdom Metazoa							
Coelenterata	Tissue–Organ	Radial	NA	Ciliated planula	Gastrovascular cavity	NA	NA
Platyhelminthes	Organ-system	Bilateral	NA	Similar to trochophore	Gastrovascular cavity	None	No
Nematoda	Organ-system	Bilateral	Spiral cleavage	None	Complete gut (protostome)	Pseudocoelom	Yes, but greatly reduced in many mollusks and arthropods
Annelida	Organ-system	Bilateral	Spiral cleavage	Trochophore	Complete gut (protostome)	Coelom (schizocoel)	Yes, but greatly reduced in many mollusks and arthropods
Mollusca	Organ-system	Bilateral	Spiral cleavage	Trochophore	Complete gut (protostome)	Greatly reduced in mollusks	Yes, but greatly reduced in many mollusks and arthropods
Arthropods	Organ-system	Bilateral	Spiral cleavage	Nauplius in some crustaceans	Complete gut (protostome)	Greatly reduced in mollusks	Yes, but greatly reduced in many mollusks and arthropods
Echinodermata	Organ-system	Penta-radial	Radial cleavage	Dipleurula or similar	Complete gut (deuterostome)	Coelom (enterocoel)	No
Hemichordata	Organ-system	Bilateral	Radial cleavage	Dipleurula or similar	Complete gut (deuterostome)	Coelom (enterocoel)	No
Chordata	Organ-system	Bilateral	Radial cleavage	Unique where occurring	Complete gut (deuterostome)	Coelom (enterocoel)	Yes

25.1 A POSSIBLE METAZOAN ANCESTOR

According to some theorists, the metazoans arose from multinucleate ciliates. One line of evidence is the similarity between certain living multinucleate ciliates (protozoans) **(a)** and certain very simple, ciliated flatworms (metazoans) **(b).** Much of the flatworm's body is noncellular, and further, it resembles the ciliate in feeding, food sources, behavior, physiology, and ecology. This doesn't prove the ciliate hypothesis, but it does offer compelling though indirect support.

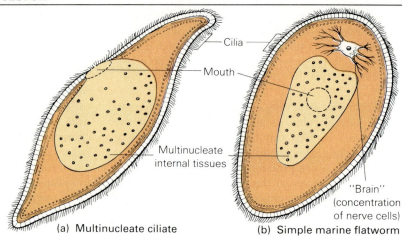

(a) Multinucleate ciliate

(b) Simple marine flatworm

ANIMAL ORIGINS

The questions "Who are we?" and "Where did we come from?" have long interested scientists, as well as philosophers, mystics, theologians, primitives, and drunks. It seems certain that today's plants, animals, and fungi sprang from ancient protists. The consensus among biologists is that the animal kingdom has its roots in the ancestors of two different protist lines. One of these, a group of flagellate protozoans (see Chapter 19), produced the subkingdom **Parazoa,** consisting of one phylum, **Porifera,** the sponges. The other gave rise to the **Metazoa.**

The Metazoa—the rest of the earth's animals—are all probably derived from a single metazoan line, which in turn was derived from a protozoan of some sort. No one knows exactly what sort, but there are two leading hypotheses. One is that the unicellular ancestor of the metazoans was a flagellated protozoan, although one different from the parazoan ancestor. Flagella in today's animals are found in sperm. The other leading hypothesis is that animals are derived from a ciliate protozoan. Today, most animals have tissues with cilia.

Proponents of the ciliate hypothesis stress the fact that the cilia of animals and those of ciliate protozoans are perfectly identical in size, in structure, and in their complex, asymmetrical pattern of beating. They also note the striking similarity between some advanced multinucleate ciliate protozoans and certain primitive, ciliated, gutless flatworms that live in the ocean today (Figure 25.1). However, the issue remains unresolved since there is biochemical evidence indicating that at least some metazoan proteins are demonstrably more similar to those of flagellates than to those of modern ciliates.

Unfortunately, there is no fossil evidence that bears on the issue at all. In fact, the oldest eukaryote fossils are similar to some of today's metazoa. These animals, which lived some 680 million years ago, were not very different from modern jellyfish. This leads us to the question, "How did the first multicellular animal come into being?" Let's see if we can marshal our informed ignorance to develop a scenario that might put together some of these ideas.

How does one go from a single-celled state to a multicellular one? Perhaps by single-celled forms aggregating—joining to form a coordinated group that became somewhat interdependent. There are, in fact, a number of living protists that show various stages of aggregation that suggest, by analogy, how things might have gone for our own ancestors.

As aggregating cells became more dependent on each other, the stage would have been set for specialization. That is, different cells could come to have different functions, some, for example, specializing in movement, others in feeding, and still others in reproduction. We know from observing living colonial forms that this specialization can be quite efficient and successful.

Most biologists agree that the earliest developments toward multicellularity in animals must have occurred in the ancient seas, among marine protozoans. Perhaps the simplest crawling or stationary forms were followed by floating or swimming versions, or maybe animals anchored by long stalks—any adaptation that could enable the organism to begin to exploit new food resources. Admittedly, much of this scenario remains hypothetical. But

TABLE 25.2

GEOLOGIC TIMETABLE

Eras (Years since Start)	Periods	Extent in Millions of Years
Cenozoic	Quaternary	
	Holocene (present)	last 10,000 years
	Pleistocene	.01–2
	Tertiary	
	Pliocene	2–6
	Miocene	6–23
	Oligocene	23–35
	Eocene	35–54
65,000,000	Paleocene	54–65
Mesozoic	Cretaceous-Paleocene discontinuity	65
	Cretaceous	65–135
	Jurassic	135–197
225,000,000	Triassic	197–225
Paleozoic	Permian	225–280
	Carboniferous	280–345
	Devonian	345–405
	Silurian	405–425
	Ordovician	425–500
570,000,000	Cambrian	500–570
Precambrian	All invertebrate phyla present	570–4500

Origin of earth, 4.5 billion years

intellectual rigor demands that we consider the problem of origin, and that we consider as many kinds of evidence as possible. With this in mind, let's see what the rocks can tell us.

The Animal Fossil Record

The earliest reliable fossil record of animals is found in Precambrian deposits dating back 580 to 680 million years. (See Table 25.2 for a review of geological time.) Still older rocks bear only what seem to be fossil tracks and burrows that some authorities suggest were made by marine worms. Whatever these holes and tracings are, they are about 700 million years old. But the oldest clearly identifiable animal fossils, the **Ediacara fauna** from southern Australia, are rich in jellyfish, soft-bodied corals and worms (Figure 25.2).

If the Ediacara fauna are at all representative of animals in their time period, then Precambrian evolution must have proceeded with incredible speed. An extremely rich fossil deposit, dated at about 570 million years, occurs in the **Burgess Shale Formation** of western Canada. Its remarkable fossils, many of which are of soft-bodied animals, include representatives from *all* of today's major animal phyla, including animals that were much more complex than those of the Ediacara fauna just 10 to 100 million years older (Figure 25.3). However, some evolutionists consider the Ediacara fauna to include aberrant, "backwater" forms of life, not rep-

25.2 PRECAMBRIAN ANIMALS

The Ediacara fossils from the late Precambrian seas are dominated by the fossilized remains of thin-bodied coelenterates. Among these are **(a)** the jellyfish—not unlike those seen today—and **(b)** stalked, featherlike corals, nearly identical to today's "sea pens." The bottom dwellers include several species of annelids **(c)**, segmented worms that are identifiable by the lines crossing the body. Arthropods **(d)** are also represented, as are a few shelled mollusks **(e)** and some fossils from phyla that are now extinct **(f)**. The egglike mass **(g)** is believed to be algae.

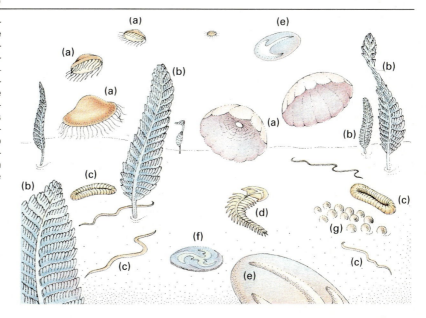

25.3 ANIMALS OF THE BURGESS SHALE

Burgess Shale fossils include some hauntingly familiar animals, whose complexity suggests a long evolutionary history. The "pineapple slices" (a) are coelenterates, an animal group not nearly as common here as in the earlier Ediacara fauna. The large stalked animals (b) are primarily filter-feeding sponges, while the many-legged creatures scurrying across the ocean floor (c) are arthropods, possibly ancestors to today's crabs and lobsters. Tube-dwelling annelids (d), the predecessors of today's marine worms and terrestrial earthworms, are also seen. The large wormlike creatures in the U-shaped burrows (e) are priapulids, very similar to some bizarre species that burrow in the ocean floor today. The marine onychophoran (f) closely resembles the modern terrestrial onychophorans, puzzling animals that share traits with the annelids and arthropods. The Burgess Shale fossils include primitive chordates (members of our own phylum), closely resembling today's *Branchiostoma* (g). The first vertebrates were not to make their appearance for many millions of years. Some species are unknown today (h), but the strange, spiny creature (i) is a mollusk and the stalked animal with the feathery tentacles (j) is an echinoderm.

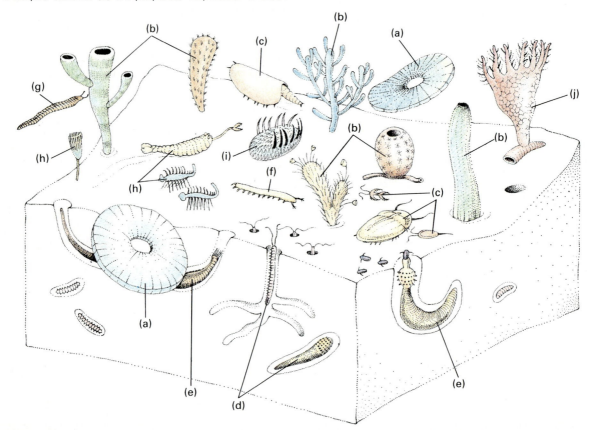

resentative of most of today's animal phyla. If this is true then we can make no statement about the rate at which evolution occurred in those times.

As you can see, there are still great voids in our knowledge of animal evolution. Nonetheless, we find that while the animals of the Ediacara deposits were not particularly complex, they had come a long way from any hypothetical ciliated ancestor. The Burgess shale animals are more complex yet. In fact, the Burgess shale itself was a rare and lucky find since the soft tissues of animals rarely fossilize, and it has filled a great hole in the existing data. Almost certainly other equally intriguing and informative early fossil beds await discovery.

Phylogeny of the Animal Kingdom

The evolutionary history and relationships among animal phyla can be represented by a phylogenetic tree (see Chapter 17). The tree shown in Figure 25.4 is only one of several possibilities, but it does indicate what are believed to be some of the major milestones of animal evolution. As with all such trees, it uses several sources, including the fossil record, comparative anatomy, physiology, and embryology.

Notice in particular the tree's major branches. They represent important evolutionary events that led to the diversification we see today. Also notice the names enclosed in boxes. They reveal two major divisions, the **protostomes** and **deuterostomes**. The terms mean "mouth first" and "mouth second" and refer to the embryological origin of the mouth. We will say more about all this later. Keep in mind that the branches of a tree such as this imply nothing about evolutionary *time*. However, it may be that nearly all of the branching occurred very rapidly, between the time of the Ediacara and Burgess Shale Fauna. With these relationships in mind, let's now go to the individual phyla.

PHYLUM PORIFERA: SPONGES AND THE LEVELS OF ORGANIZATION

We can begin by showing why sponges (phylum Porifera) are different. Simply put, they have not reached the **organ-system level of organization.** But what does this mean? We mentioned earlier that most animals have well-developed organs and organ systems. From an anatomical viewpoint, the

25.4 PHYLOGENETIC TREES

This scheme is one of several reasonable possibilities. The branches of the tree represent the major animal phyla, along with important evolutionary designations (boxed items). No specific time frame is suggested, but the main trunk represents the ancestral animal stock, and the lower branch-points are the earliest evolutionary divergences. Each phylum, although shown far out on its limb, may have its evolutionary roots quite far down in its main branch. The scheme reflects the finding that the Parazoa and Metazoa arose from separate protist ancestors. The earliest metazoan animals apparently were radial (circular or spherical animals, represented today by the coelenterates and ctenophores). The bilateral (left- and right-sided) metazoans presumably arose from the radials, diverging into the two major branches seen in the trunk. These two branches are designated as deuterostomes and protostomes, referring to important embryological differences in the higher animal phyla. The left limb, the deuterostomes, gave rise to several phyla, including the chordates, which in turn produced a subphylum containing the vertebrates—our own group. The right limb gave rise to most of the familiar invertebrate animal phyla. The designations *acoelomate, pseudocoelomate,* and *eucoelomate* refer to the presence or absence of certain body cavities, which also represent important evolutionary milestones. Theorists are not in agreement as to whether the coelom arose just once (**a**) or whether it arose independently a second time (**b**).

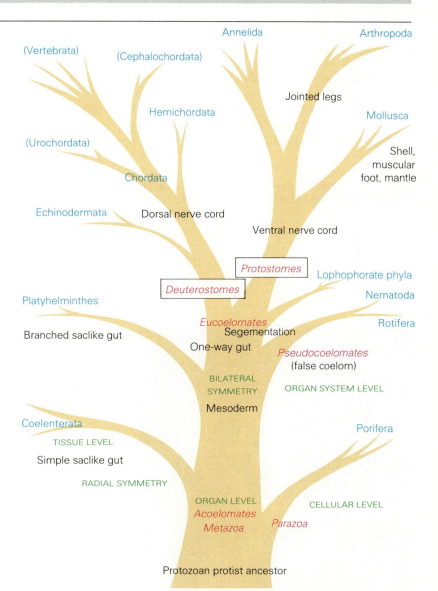

fundamental level of an organism is the **cell** (putting aside sublevels such as organelles and macromolecules). The second level is the **tissue,** which consists of a group of specialized cells with some common function. The lining of the intestine, for instance, is a tissue that specializes in food absorption. Tissues, in turn, are organized into **organs,** such as the intestine itself, which perform all or part of a major function. The intestine contains absorbing tissue along with muscle, nerve, secretory, and other tissues. Organs generally interact with other organs to form the *organ-system*. The digestive system, including the mouth, esophagus, stomach, intestine, and so on, is an organ system, as are the respiratory and reproductive systems.

From an evolutionary viewpoint, prokaryotes and protists have achieved only the cellular level of organization. Although the parazoans (sponges) are animals, they too can really only claim a cellular level of organization. Sponges consist of only a few specialized but loosely organized types of cells with little coordination among them. In fact, sponges can be completely disrupted into individual cells and

they will reaggregate (crawl back together) and reestablish themselves into a new organism. Let's see how such simple cellular organization works for the sponges.

The Biology of Sponges

Since sponges evolved separately from other animals, we can expect them to be different. However, there is no doubt that sponges are animals since they are multicellular heterotrophs with all of the requirements and reproductive characteristics of animals. Most sponges are marine, living in shallow ocean waters throughout the world, although a few live in fresh water. The adults are sessile, living quietly attached to the bottom in clear water without apparent movement.

In the simpler sponges, the body is vaselike, with a large central cavity or **spongocoel,** with an opening, the **osculum,** at the top (Figure 25.5). The outer wall of the thin body is riddled with porelike cells called **porocytes,** which admit water into the central cavity. A jellylike, unstructured region, the **mesohyl,** separates the outer cell layer from the

25.5 SPONGE ANATOMY

In the simple, vaselike sponges, the body wall consists of two organized cell layers. Between lies the gelatinous mesohyl, which contains amebocytes and calcareous, supporting spicules. Flagellated choanocytes (collar cells) lining the body cavity create water currents, drawing in water from which microorganisms and bits of organic debris collect in mucus on the collars. The mucus passes to the cell below, where phagocytosis and digestion occur.

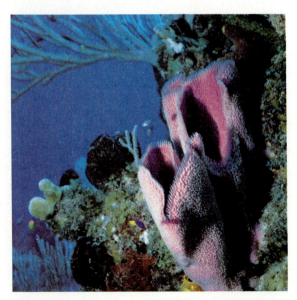

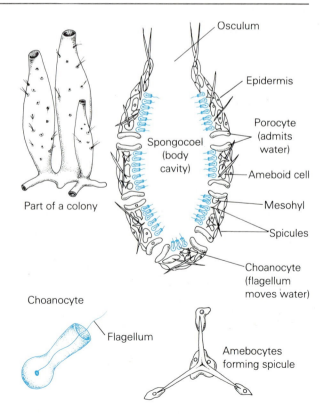

inner and houses wandering **amebocytes.** The inner wall contains numerous flagellated collar cells, the **choanocytes,** which line the central cavity and there create inward water currents. The incoming water brings with it fresh supplies of oxygen and food. As filter-feeders, sponges feed on microscopic organisms and bits of organic matter that become trapped in mucous secretions of the choanocytes. These trapped particles pass down the collar of the choanocytes to the cell below, where they are phagocytized and digested in food vacuoles. The water currents that brought in the particles then pass out through the osculum, laden with carbon dioxide and other cellular wastes.

In more complex sponges, the body wall may be relatively dense and the body cavity highly branched. The pores open into canal systems that frequently widen into chambers lined with collar cells before leading into a reduced central cavity.

Sponges maintain their body shape by **spicules** or fibers that are scattered through the body wall and that act as skeletal elements. The spicules, produced by amebocytes, differ in each of the three classes of sponges. In **calcareous sponges** the spicules contain calcium carbonate, while those of the beautiful **glassy sponges** consist of silicon dioxide (glass). In a third group, the **proteinaceous sponges** or ''bath sponges,'' the skeletal material is **spongin,** a fibrous protein. (Most cleaning sponges sold today are synthetic and the once thriving sponge fishing industry is no more.)

Sponges are usually **hermaphroditic,** with each individual producing both sperm and eggs. Sperm release (often in clouds) is aided by the currents from the collar cells. The egg cells are often located just beneath the collar cells, where the sperm can easily penetrate. A fertilized egg escapes as a highly flagellated, swimming larva, which after a brief period settles to the bottom and begins to sit out its adult life. Some sponges can also reproduce asexually, releasing **gemmules,** clusters of cells surrounded by a resistant wall. The gemmules of fresh water sponges may remain dormant through hard times, such as cold or dry spells, before becoming active and producing a new sponge body.

As adults, the sponges' ability to respond to the environment is extremely limited and is similar, in fact, to responses by colonial flagellates. There is no accumulation of nerve cells that could be considered a central nervous system. Restriction to the cellular level of organization means that functions are carried out by individual cells or aggregates of cells, with little of the coordination of activity seen in the tissue level.

THE RADIAL PLAN AND A STEP UP IN ORGANIZATION

The body plans of metazoan animals reveal two basic kinds of symmetry, **bilateral** and **radial** (Figure 25.6). Both are conceptually simple but have far-reaching evolutionary implications. The bilateral body plan has only *one* plane of symmetry, which basically divides the body into right and left sides. We are a good example of a bilateral animal. The radially symmetrical form is quite different. The body is essentially disklike or cylindrical (sometimes spherical), perhaps with radiating arms. Theoretically, there are unlimited planes of symmetry. In other words, any plane that passes through the center of the body from top to bottom will divide the body into roughly equal halves.

Radial symmetry in today's animals exists in only a few phyla, including the coelenterates and the ctenophores. (Later, we will find that adult echinoderms such as sea stars, sea urchins, and sand dollars have an odd, **pentaradial** or five-part radial symmetry.) Let's now look at the radially symmetrical coelenterates and ctenophores.

Phylum Coelenterata

The **coelenterates,** or **Cnidaria,** as they are also known, include the **hydroids, jellyfish, corals,** and **anemones.** For the most part, coelenterates have arrived at a tissue level of organization but include a few examples of organs as well. An example would be the tentacle, which contains specialized covering cells, an abundance of nerve processes, and contractile fibers. Most coelenterates have tentacles armed with stinging cells called **cnidocytes** that release, in harpoon fashion, stringlike stinging structures called **nematocysts.** The coelenterates are very delicate and extremely thin-walled animals, composed essentially of an outer and inner layer of cells. Sandwiched between these layers is acellular, jellylike matter—the **mesoglea.** Wandering ameboid cells, nerve processes, and contractile fibers are found in the mesoglea. The outer cells (or epidermis) are mainly protective, while the inner cells (or gastrodermis) form a saclike gut called the **gastrovascular cavity** because it serves both digestive and circulatory functions. A simple opening serves as both mouth and anus—or, to put it another way, coelenterates have a mouth but no anus and must spit out the undigested remnants of whatever they swallow.

As with other radial animals, there is little cen-

25.6 ANIMAL SYMMETRY

In radial symmetry the body is essentially spherical, disk-shaped, or cylindrical. Any plane drawn through the center produces roughly equal left and right halves. Bilateral symmetry, on the other hand, means that the body can be equally divided by one plane only. This division produces mirror-image left and right halves that contain similar structures. Perfect examples of radial or bilateral symmetry are not always found. Can you think of structures in your own body that are asymmetric—that is, neither paired nor composed of mirror-image right and left halves?

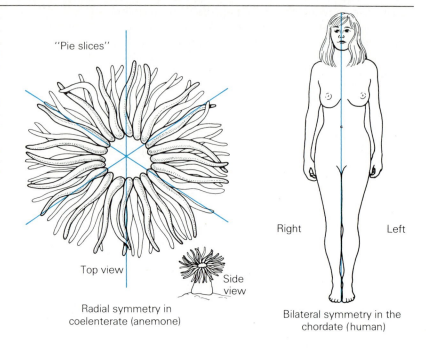

"Pie slices"

Top view

Side view

Radial symmetry in coelenterate (anemone)

Right   Left

Bilateral symmetry in the chordate (human)

25.7 LIFE CYCLE OF A HYDROZOAN

In its life cycle, *Obelia,* like many other hydroids, develops a number of polyps, which in turn specialize in either feeding or reproduction. The reproductive polyps form swimming medusae, which are released to swim away. When mature, the medusae produce eggs and sperm, which are released into the surrounding water where fertilization occurs. The zygote develops into a ciliated larva called a planula. Eventually the planula settles to the ocean bottom and begins its transition into the adult hydroid state, whereupon the alternating cycle will repeat.

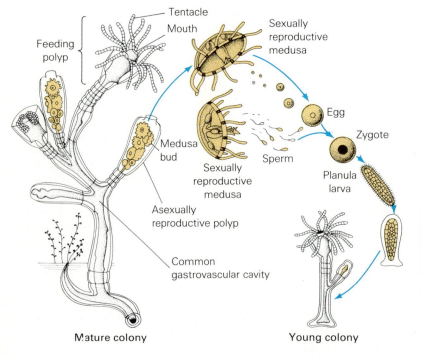

Feeding polyp

Tentacle

Mouth

Sexually reproductive medusa

Medusa bud

Sexually reproductive medusa

Sperm

Egg

Zygote

Planula larva

Asexually reproductive polyp

Common gastrovascular cavity

Mature colony

Young colony

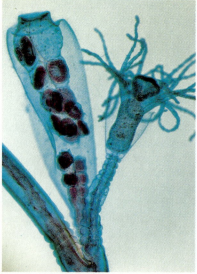

tralization of the nervous sytem. In fact, it is quite diffuse and takes the form of a **nerve net,** covering the entire animal. In addition, the neurons or nerve cells of coelenterates are quite unusual in that nerve impulses can travel in either direction from one cell to the next. In bilateral animals, neural conduction is generally one way.

Coelenterates may take either of two forms: the **polyp** or the **medusa.** Polyps are usually sedentary, attached to rocks, wharf pilings, and other objects, whereas medusas (jellyfish) can swim. In some coelenterate species, both polyp and medusa stages exist in an unusual alternation of generations. Unlike alternation of generations in plants, however, both polyp and medusa forms are diploid, true to their animal mode of life.

Class Hydrozoa. Most **hydrozoans** are found as colonies of polyps with specialization for feeding and reproduction. One well-known hydrozoan, *Obelia*, is seen in Figure 25.7. The colonies grow asexually by simply budding more polyps. Sexual reproduction begins when minute medusae are produced. Sexes are commonly separate in the free-swimming medusae, and sperm released into the water enter the female medusa to fertilize the eggs. The hydrozoan zygote first develops into an intermediate or larval state known as a **planula larva.** The planula larva is a ciliated, swimming form that soon settles to the bottom, attaches, and there begins its strange metamorphosis into a colony of polyps. The transition from egg to larva, prior to the emergence of an adult, is commonly seen in the invertebrate phyla.

One familiar, but aberrant, hydrozoan is the tiny, free-living *Hydra*. *Hydra* and a closely related genus *Chlorhydra* are unusual in several ways, one being that they are freshwater animals in an almost exclusively marine phylum. And unlike many marine coelenterates, *Hydra* is a solitary individual. Further, there is no medusa stage and no planula larva. The *Hydra* polyp reproduces asexually by budding new polyps. As in so many other minute freshwater forms, sexual reproduction occurs only under harsh conditions, and the zygote is protected by a toughened, resistant case. When conditions improve, the encased zygote germinates into a tiny polyp. Most hydra are brown in color but *Chlorohydra* is bright green due to the presence of an intracellular photosynthetic symbiant, a green alga, upon which it depends for nutrition.

Other hydrozoans are quite unlike these. For example, the large Portuguese man-of-war *(Physalia)* looks like a jellyfish, but it is actually a floating hydrozoan colony. Each colony includes a gas-filled float and feeding and reproductive polyps with long trailing and stinging tentacles. A fish, perhaps seeking food or refuge in the dense "foliage" of the hydroid's tentacles, is at once paralyzed by the venom of thousands of ejected nematocysts as it brushes past the graceful, often brightly colored lures. The fish becomes totally ensnared and is then subjected to digestive enzymes that prepare its body for absorption as the feeding polyps take over.

Class Scyphozoa. In **scyphozoans,** the alternation of generations is just opposite to that of most hydrozoans. Here, the swimming medusae are the dominant generation while the polyps are greatly reduced or, in some species, entirely absent. Most medusae have a bell-shaped jellyfish body with the mouth centrally located on the underside, surrounded by tentacles. For the most part, they are drifters, but they can swim a bit by contracting muscle fibers around the bell. While most are modest in size, some, such as *Cyanea*, attain dramatic proportions—up to 2 meters in diameter with tentacles 70 meters long. These enormous jellyfish are indeed dramatic sights when seen drifting in clear, blue waters, far out at sea.

Sexes are separate in the jellyfish. Sperm, produced in the testes of males, swim out through the mouth and into the gastrovascular cavity of the female, penetrating the ovary, where fertilization occurs. A swimming planula larva develops and settles to the ocean floor, where an unusual transition occurs. Usually hanging upside down from the underside of a rock ledge, the attached polyp, or **scyphistoma,** begins to bud off miniature jellyfish, or **ephyra,** in assembly line fashion (Figure 25.8). These mature into adults.

Class Anthozoa. Anthozoans—the corals and anemones—exist only as polyps (Figure 25.9). It was mentioned earlier that polyps can reproduce asexually by simply budding new polyps. In their sexual reproduction, female and male polyps produce eggs or sperm separately and fertilization occurs in the water. The zygotes develop into swimming planula larvae that form more anthozoans.

The most familiar anthozoans in temperate waters are the heavy-bodied sea anemones. Anemones abound in rocky tidepools and in shallower coastal waters. Corals are colonial anthozoans whose polyps secrete surrounding walls of limestone (calcium carbonate). The lovely coral formations sold in curio shops are actually their dead skeletons. The cumulative effects of coral secretions

25.8 LIFE CYCLE OF A SCYPHOZOAN

Jellyfish are free-swimming coelenterates with a highly reduced polyp stage. Most are relatively small, a few inches in diameter, but many are much larger. One North Atlantic species, *Cyanea capillata*, reaches 2 m in diameter, with tentacles trailing some 70 m below, and can weigh up to a (metric) ton.

(a) In its life cycle, the adult scyphozoan jellyfish produces egg or sperm cells. The sperm are shed into the water and fertilization is internal. The zygote is released and develops into a ciliated planula larva that attaches as a polyp to the underside of an overhanging rock. There it feeds and develops into a budding stage, producing many new jellyfish through an asexual process known as strobilation. Each individual thus produced (an *ephyra*) will mature into an adult jellyfish. In the scyphozoan life cycle, the medusa stage **(b)** is dominant, and the polyp stage is small and brief.

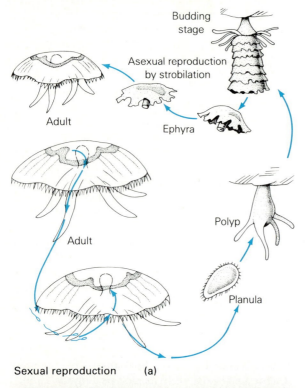

Budding stage

Asexual reproduction by strobilation

Adult

Ephyra

Adult

Polyp

Planula

Sexual reproduction   (a)

(b)

(along with significant contributions by coralline algae) over millions of years have produced coral atolls (islands), as well as the barrier reefs that protect many islands in the Pacific. Although all corals can feed with tentacles, most tropical corals play host to photosynthetic, symbiotic algae that provide most of their nutrients.

Phylum Ctenophora

The **Ctenophora** (Figure 25.10), commonly called "comb carriers" or "comb jellies," appear to be closely related to the coelenterates. Like the coelenterates, comb jellies are radially symmetrical; they have a gastrovascular cavity formed primarily by two cell layers separated by a jellylike mesoglea. These marine animals are characterized by fused cilia that form eight rows or **combs.** The undulation of these combs pushes the animals along. The Ctenophora have tentacles, but rather than nematocysts, they are armed with "glue cells" that snare and capture prey. Their sticky secretions capture plankton (small drifting invertebrates) and, to the dismay of oyster hunters, oyster larvae. The Ctenophora are well known for their intriguing luminescence. One species, *Cestus veneris*, or "Venus' girdle," has a glowing, ribbonlike body nearly a meter long.

THE BILATERAL TREND AND THE VERSATILE MESODERM

In the animal kingdom, the radial species are the minority. Most animals are bilaterally symmetrical. With the evolution of bilateral symmetry, a significant trend in body organization known as **cephalization** arose. Animals developed a head or leading end and a tail or trailing end. At first the head may have been simply a concentration of muscles, an adaptation for burrowing in the soft sea beds of ancient oceans, which is where the oldest fossils of bilateral animals are found. But leading ends soon became equipped with sensory structures for the detection of food, light, sound, and other stimuli. With senses concentrated in the leading end, an animal could more quickly perceive what sort of environment it was moving into. Such sensory structures require neural support and integration; thus, clusters of nerve cells were concentrated at the head end and the evolution of the brain was underway.

25.9 REPRESENTATIVE ANTHOZOA

Sea anemones and corals are members of class Anthozoa. **(a)** Corals are colonial, living in extensive limestone coverings that they secrete. Their secreted skeletons often produce extensive reefs and atolls. Corals are polyp forms, similar to the anemones but generally much smaller. Here, the polyps are seen both closed (left) and open (right). **(b)** Anemones are more solitary, although they are often seen in clusters, attached by their pedal disks to the substrate. They can move about but most of the time remain attached to one place. In their feeding posture they are extended, but when disturbed they quickly retract into a short, rounded body.

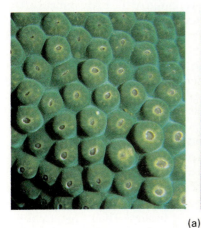

(a)

(b)

25.10 COMB JELLIES

These delicate comb jellies are propelled through the water by rows of cilia found along their combs. The combs are the dense vertical ridges around the body wall. The central pouch-like object is the stomach or gastrovascular cavity. Comb jellies feed by snaring small prey in gluey secretions from their tentacles (not seen here). The tentacles can trail along behind the animal or may be retracted into sheaths.

The Three-Layered Embryo

In the coelenterate and ctenophoran embryos, only two types of embryonic tissue ever appear, and these two tissues, or **germ layers,** as they are known, give rise to all of the structures to come. During development the first layer, the **endoderm** (inner skin), forms many internal linings, including those of the gut and the many glands and blood vessels. The **ectoderm** (outer skin) contributes chiefly to outer linings and forms much of the nervous system. All other metazoan embryos have, in addition to these two layers, a third germ layer, the **mesoderm** (middle skin). It is from the mesoderm that muscle, blood, and the internal skeleton, or **endoskeleton,** are derived. The first group we will consider whose embryos form from three germ layers is the flatworms.

Phylum Platyhelminthes: The Flatworms

Phylum **Platyhelminthes,** our first bilateral phylum, includes one class of free-living species and two classes of parasites, the flukes and the tapeworms. Some species in this group are highly complex, with fully developed organs and organ systems; others are more simple. As bilateral animals, flatworms have definite anterior (head) and posterior (tail) orientation. Their nervous systems are not well developed, but their nerves are concentrated at the anterior end as with other bilateral animals.

Flatworms, like coelenterates, lack special respiratory and circulatory structures, and they depend on diffusion for their oxygen needs. Because of their flattened structure, all their cells are close to the surface. The organ systems are best developed in the free-living flatworms, some being entirely absent in the parasites.

Class Turbellaria. Although many of the **turbellarians**—the free-living flatworms—are marine, the group is well represented by the very common freshwater species, *Dugesia*, better known as **planaria** (Figure 25.11). These animals have been the subject of anatomical, developmental, and behavioral studies for many years. We will describe these findings in more detail in later chapters.

Planarians have clearly arrived at an organ-system level of development, yet they retain some primitive characteristics. For example, like the coelenterates they carry out digestion in a blind sac, the gastrovascular cavity, but the cavity in flatworms can be highly branched and complex. As we mentioned earlier, the simple nervous system is concentrated at the anterior (head) end, where large **ganglia** (clustered nerve cells) and light-receptors are located. Two ventral (frontal) nerves run down the length of the body sending out lateral branches from a number of smaller ganglia in what has often been described as a ladderlike nervous system (see Figure 25.11).

In freshwater flatworms, **osmoregulation**—the maintenance of water and ion balances in the body—is accomplished by the action of the **protonephridia,** also known as the **flame cell system.** A protonephridium consists of two highly branched systems of tubules with a number of blind sacs containing the **flame bulbs.** Each bulb is made up of a group of ciliated **flame cells** (the waving, or "flickering" of the cilia, as seen under the microscope, inspired their name). Their primary function is getting rid of excess water that continually enters the body because of a natural osmotic gradient.

Planarians swim in a curious sort of crawl. They secrete a layer of mucus underneath them and then beat their way through it with the cilia that abound on their ventral body surface. They are predators and scavengers, feeding through a protrusible, muscular **pharynx,** while pinning the unlucky victim in the slimy mucous layer below. Food enters the highly branched gastrovascular cavity, where digestion begins (see Figure 25.11), but bits of food are also phagocytized by cells lining the cavity and digestion is completed in food vacuoles within the

cells. Thus digestion is both extracellular and intracellular.

Some planaria can reproduce asexually by a fission process—they simply pinch in half just behind the pharynx, each half regenerating the missing structures. They also have tremendous regenerative power and can recover from drastic experimental surgery as long as enough of the body remains. The reproductive system is quite complex and advanced (see Figure 25.11). Each planarian has well-defined ovaries and testes. They are hermaphroditic, each individual having both a penis and a vagina, but they do not generally self-fertilize. During sexual reproduction, copulation is reciprocal, with simultaneous penetration and exchange of sperm.

Class Trematoda. The **Trematoda,** or **flukes,** are parasites. They are not ciliated and the gastrovascular cavity is divided into two parts. Flukes are equipped with large suckers, one in front and one on the bottom side, both of which they use to attach to their host's tissues. Their epidermal tissue layer is resistant to digestive fluids.

Flukes invade many kinds of animals and make their home in various parts of the host, including the gills, lungs, liver, and intestines. One species has even found its niche in the urinary bladder of frogs. One human liver fluke (*Chlonorchis sinensis*), common to Asia, has an enormously complex life cycle involving three very different hosts—human, snail, and fish. Note in Figure 25.12 the ongoing change to **miracidia, sporocysts, redia,** and **cercaria,** each a specialized state in one bizarre life cycle.

An important fluke in the Nile Valley of Egypt today is the human blood fluke, *Schistosoma mansoni*, which has recently become widespread in the area. The extensive irrigation canals fed by the Aswan Dam have allowed aquatic snails to flourish. The snails serve as an intermediate host to the fluke, which causes the debilitating disease **schistosomiasis.** Generally, it weakens the host to the point that other diseases cause early death. The parasite enters the body directly through the skin when people wade in the water. It is estimated that well over 60% of the population in the Nile delta is now infested with the blood fluke, but the disease is also common in other tropical regions.

Class Cestoda. Class **Cestoda,** the **tapeworms,** are quite different from other flatworms. These parasites have no digestive cavity and must absorb digested food of the host directly across the body wall. To this end, the outermost external membrane is highly folded, greatly increasing the sur-

25.11 ANATOMY OF A FLATWORM

The apparent simplicity of *Dugesia,* a free-living planarian, is highly deceptive. Within its tiny, flattened body lie several complex organ systems that permit the animal to feed, reproduce, and respond to external stimuli. **(a)** Digestion occurs in a gastrovascular cavity, which, unlike the gastrovascular cavity of the coelenterates, is highly branched and includes a complex, muscular pharynx. **(b)** The reproductive system includes well-defined testes and ovaries, along with related ducts and yolk glands. It further includes distinct copulatory organs (penis and vagina) that serve in a manner typical of more complex animals. *Dugesia* is hermaphroditic, so both male and female structures are present. **(c)** Light-sensitive eyespots help the flatworm avoid light, and a ladderlike nerve network coordinates movement of its muscles.

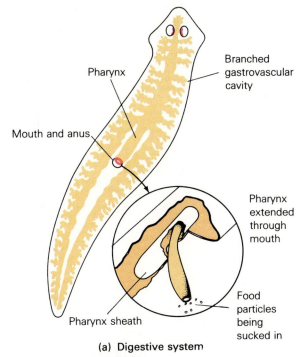

Pharynx

Branched gastrovascular cavity

Mouth and anus

Pharynx extended through mouth

Pharynx sheath

Food particles being sucked in

(a) Digestive system

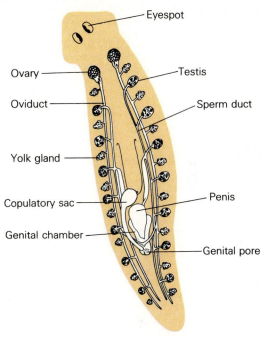

Eyespot

Ovary

Testis

Oviduct

Sperm duct

Yolk gland

Copulatory sac

Penis

Genital chamber

Genital pore

(b) Reproductive system

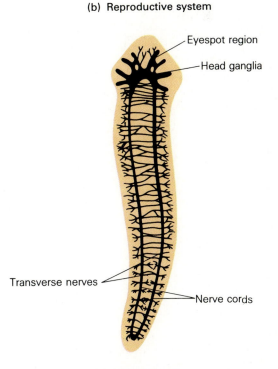

Eyespot region

Head ganglia

Transverse nerves

Nerve cords

(c) Nervous system

25.12 LIFE CYCLE OF A PARASITE

The human liver fluke *(Chlonorchis sinensis),* common in many rural regions of Asia, lives part of its life cycle in the bile passages of the liver, where heavy infestations can bring on cirrhosis and death. Its life cycle is one of the most complex known, requiring three separate hosts: human, snail, and fish. Its primary or sexual stage occurs in humans **(a)** where the eggs are fertilized and the first of several intermediate stages, the *miracidium,* forms. The egg cases then pass into the intestine and out of the host with the feces. When the miracidia are eaten by a freshwater snail **(b)** (in rice paddies or other polluted waters), they hatch from the egg case and bore into the snail's tissues, forming sporocysts. The sporocysts then enter an asexual phase, producing numerous *redia,* each

of which in turn produces many swimming *cercaria.* The cercaria escape from the snail and seek out a fish, the next host. **(c)** They bore into the fish's muscles and secrete protective cysts around themselves. The encysted cercaria (or *metacercaria)* remain there until some hapless human eats the fish raw or partially cooked. The digestive enzymes weaken the metacercarian capsules and the young flukes emerge and make their way up into the bile duct to the bile passages of the liver. Then the cycle repeats. You can probably see several ways in which the chain of infestation could be broken, the most obvious of which would be to thoroughly cook the fish. But customs of hundreds of generations of rural Asians are not so easily changed.

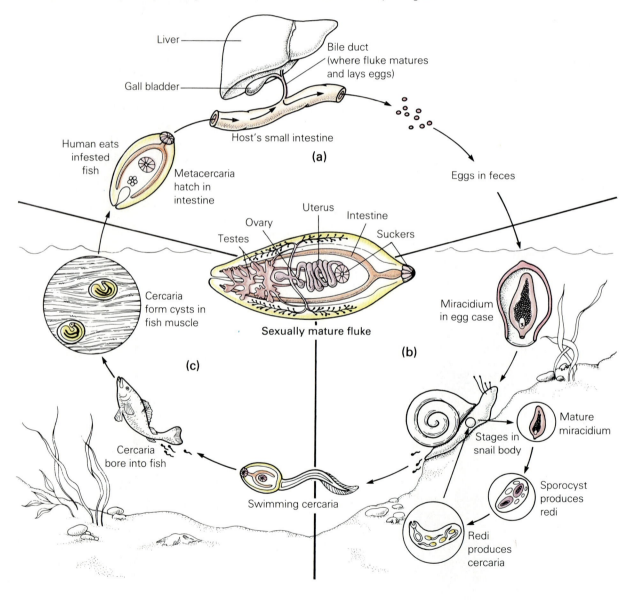

face area. Virtually all vertebrates harbor tape-worms, generally with little harm to the host animal. Humans are host to seven different species.

Typically, the sexually mature tapeworm consists of an extremely small, rounded **scolex** located anteriorly and equipped with suckers and hooks for attaching to the host's intestinal wall. Extending from this is a long row of segments known as **proglottids.** (This type of segmentation is a form of asexual reproduction by budding, and is not to be confused with the true segmentation we will see later.) The proglottids are produced continuously from a region next to the scolex. As they mature, they begin to swell with eggs as they are moved back by the growing chain. Each proglottid contains both ovaries and testes (Figure 25.13), and self-fertilization within and between proglottids is common, as is cross-fertilization between two worms. Afterward, greatly enlarged, mature proglottids containing thousands of fertilized eggs break away from the organism and pass out with the host's feces. The fertilized eggs must then lie about until they are ingested by the next host.

In addition to the **primary host,** where sexual reproduction occurs, most tapeworms have one or two **intermediate hosts,** a different species in which they may lie dormant for a while or undergo asexual reproduction. When an egg is introduced into an

25.13 TAPEWORM ANATOMY

The business end of a tapeworm is the scolex (a), which is equivalent to a head. Most sensory structures are absent, but the scolex is equipped with a ring of hooks and several suckers for holding on to the host's intestinal wall. The remainder of the tapeworm's body consists of reproductive units known as proglottids (b) and (c). These form continuously at the scolex, enlarging as they mature. Each contains well-defined ovaries and testes, along with related ducts and glands. Following fertilization, the developing eggs crowd other structures aside, filling and enlarging the proglottid.

(a)

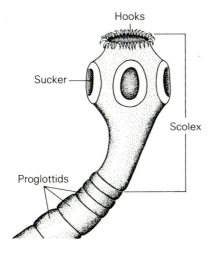

(b)

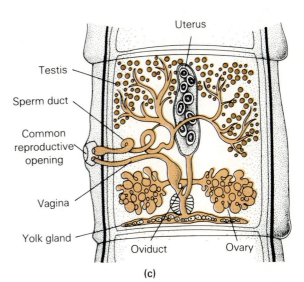

(c)

intermediate host it will be transformed into a tunneling creature and begin boring its way through the host's tissues (often muscle). Then it will form an encapsulated, protective cyst around itself as it develops into a stage called a **bladderworm,** which will lie dormant for a time. Pigs, cattle, and fish serve as intermediate hosts to some human tapeworms. Uncooked or undercooked pork, beef, or fish, then, may carry the bladderworms to a human. (The uncooked fish in sushi has been implicated in some parts of the world.)

When humans serve as an intermediate host to a tapeworm, the results can be quite serious. In such a case, the parasite does not remain in the intestine but ventures into other body parts, where it may secrete dangerous toxins and produce cysts that resemble large tumors.

BODY CAVITIES, A ONE-WAY GUT, AND A NEW BODY PLAN

The bilateral body plan in animals brought other changes. Two of these were to go hand-in-hand—the formation of internal body cavities and the formation of a tubelike, one-way digestive tract, including a mouth, gut, and anus (keep in mind that "gut" is not an indelicate term in biology). With a gut inside a body wall, the body plan resembled a sort of "tube within a tube" (Figure 25.14). The body cavity between the tubes was to become known as a **coelom** (pronounced sea'-loam). Animals with a coelom are known as **eucoelomates** ("having a true coelom") or simply **coelomates,** while those lacking such cavities are **acoelomates.** If there is a "true coelom" can there be a "false coelom"? We've begged the question, so let's see.

The body cavity first appears in that varied group, the roundworms, where it is called a **pseudocoelom** or "false coelom." As you see in Figure 25.14c, the true coelom is completely lined by tissues that were derived from embryonic mesoderm. For instance, in the earthworm (and other coelomates, such as ourselves), both the inner body wall and the gut are covered by a mesodermally derived lining known as the **peritoneum.** In the roundworm, the pseudocoelom is lined by tissues derived from ectoderm and mesoderm (see Figure 25.14b). Furthermore, the gut is completely endodermal, lacking a muscular wall and peritoneum. It is usually only one cell layer in thickness and simply absorbs digested food from the host. Let's look more closely at animals with these kinds of bodies,

beginning with a brief visit with some strange animals that can claim a complete gut although they have never evolved a body cavity.

Phylum Rhynchocoela (Nemertina): The Proboscis Worms

You quite likely have never seen a **Rhynchocoela,** or **nemertine** worm. Although they aren't really rare, they usually don't make headlines either. But at least they don't burrow into your soft tissues; they are free-living marine predators. Nemertines don't seem to be closely related to any other group. They have some characteristics in common with flatworms, but they share a few characteristics with most other animal phyla, and even possess a few that are theirs alone. Like flatworms, nemertines are acoelomate, lacking a coelom or pseudocoelom. But unlike flatworms, the nemertines have a one-way gut—food goes in the mouth and feces leave through the anus. Their thin, ribbonlike, ciliated bodies are solid, and they have well-developed protonephridia.

Nemertines have a unique feeding adaptation: a long, retractable **proboscis** that lies in a cavity just above the mouth. The hollow proboscis can be rapidly everted (turned inside out) by fluid pressure (like blowing into a rubber glove to pop out an inverted finger). A sharp stylet on the extended proboscis pierces a prey organism, and strong retractor muscles reel it in.

Phyla Nematoda, Rotifera, and Priapulida

Now we come to the **nematodes, rotifers,** and **priapulids.** The first are the "roundworms" while the second includes a number of tiny, almost microscopic animals that look as if they have wheels. The third group resembles rotifers but only in their larval state.

The Nematodes. The nematodes, or roundworms, are best known for their parasitic members, some of which are truly horrendous. Simply reciting the life cycle of some of them can cause nightmares. But let's stay calm and objective. Roundworms are slender, cylindrical, and usually tapered at both ends. Sexes are separate. They are bilaterally symmetrical, although from outside appearances, they lack a distinctive head. Most are surrounded by a strong and flexible cuticle. Since there is no skeleton, the turgid, fluid-filled body cavity acts as a **hydrostatic skeleton,** helping the

25.14 BODY CAVITIES

The coelom is a mesoderm-derived and mesoderm-lined body cavity between the gut and the body wall. Since flatworms lack a coelom, (a), they are referred to as acoelomates. An extensive body cavity is present in the roundworms (b), but it is designated as a pseudocoelom, since it does not contain the completely mesodermal linings found in the true coelom. Thus the roundworms are designated as pseudocoelomates.

In eucoelomates such as the earthworm (c), the body cavity is fully lined by mesodermally derived tissue, including the gut, which usually contains mesodermally derived layers of muscle. A glistening membrane, the peritoneum, commonly lines the coelomic cavity. Since the coelom is seen in both protostomates and deuterostomates, it may have evolved independently twice in evolutionary history.

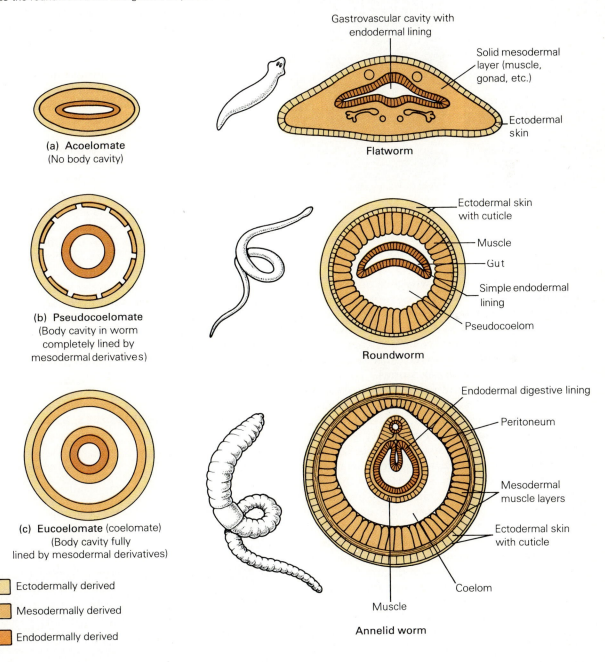

Gastrovascular cavity with endodermal lining

Solid mesodermal layer (muscle, gonad, etc.)

Ectodermal skin

(a) Acoelomate
(No body cavity)

Flatworm

Ectodermal skin with cuticle

Muscle

Gut

Simple endodermal lining

Pseudocoelom

(b) Pseudocoelomate
(Body cavity in worm completely lined by mesodermal derivatives)

Roundworm

Endodermal digestive lining

Peritoneum

Mesodermal muscle layers

Ectodermal skin with cuticle

Coelom

Muscle

(c) Eucoelomate (coelomate)
(Body cavity fully lined by mesodermal derivatives)

Ectodermally derived

Mesodermally derived

Endodermally derived

Annelid worm

25.15 THE NEMATODES

Nematodes are common in most soils but abound in moist, fertile places. Many are scavengers or predators of soil organisms and are considered free-living, but a great number infest plants and are significant agricultural pests. Shown at the right is a nematode cyst, containing hundreds of eggs, attached to a potato root.

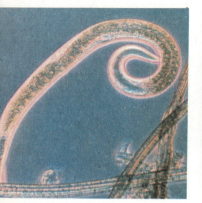 

worm maintain its shape and acting as a resistant base for muscle action. Nematodes move by flexing longitudinal muscles first on one side and then on the other, producing a wriggling or undulating motion. This wouldn't get one very far in open water, but many nematodes live in the soil and the action helps them thread their way between the soil particles. Nematodes come in a range of sizes but they are usually easy to recognize since they all look like smooth, glistening worms.

Nematodes thrive in nearly every conceivable moist and aquatic habitat, from the soil of flower gardens to the world's oceans. Most of the estimated half million species of nematodes are free-living. Winnowing their way among moist soil particles, they feed on protozoans, rotifers, small earthworms, and each other. Some capture their prey with a paralyzing saliva. Others make use of a tiny bladelike device that impales the hapless prey while the sucking lips and pumping pharynx drain its juices. Because of their astounding number, free-living nematodes are an essential link in the earth's ecological organization (Figure 25.15).

Many nematodes are very important parasites of plants. Plant nematodes may have a devastating effect on agriculture and require constant control at great expense. In fact, sometimes the farmer must completely shift crops in order to make infested land usable again.

Parasitic roundworms probably infest all of the vertebrates, including humans. In fact, at least 10

species are dangerous to humans, and perhaps 50 more live in humans without doing great harm. There are two types that are particularly dangerous. *Ascaris lumbricoides,* a giant among members of its phylum, inhabits the bodies of both hogs and humans (Figure 25.16). Females often exceed 20 to 35 cm (8 to 14 inches) in length. The males, shorter and narrower, are easy to identify by their hooked posterior extremity. At the tip of the hook is located a pair of bristlelike spicules, which males use to hold the female's genital pore (reproductive opening) open during copulation.

Ascaris reproduces in the host's intestine and its resistant egg cases pass out with feces. The tough egg cases containing developing embryos may survive in the soil for many years. The odds against any embryo finding the proper host are very great, so the *Ascaris* counters this by producing incredible numbers of offspring. A female may contain about 30,000,000 eggs, which can be released at a rate of 200,000 a day! There is no intermediate host so the eggs are passed from one person to the next.

The larvae hatch in the small intestine and begin a strange odyssey through the wall of the gut, to the heart and lungs, up and out of the breathing passages, to the pharynx, where they are swallowed, finally coming to rest again in the intestine. By the time they reach the intestine they may be 2 or 3 millimeters long. But once there, they grow to immense proportions. *Ascaris* survives in its host by neutralizing trypsin, a key enzyme in protein digestion. Without the enzyme, heavily parasitized chil-

25.16 *ASCARIS LUMBRICOIDES,* MALE AND FEMALE

Female ascarid worms grow to 35 cm (14 in) in length, while the males are shorter and more slender. Males can immediately be distinguished from females by the presence of a curled tail with fine bristles emerging from the tip. The adults are commonly found in the small intestine of infected people and hogs. The *Ascaris lumbricoides* of humans and hogs are identical in their physical features, but they are apparently physiologically distinct species. They do not develop well in the wrong host, so we can't blame the hogs for this one.

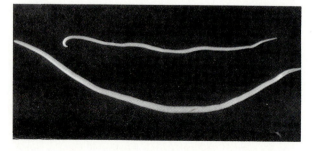

25.17 AN INVITATION TO TRICHINOSIS

The light microscope photograph shows the *trichina* worm encysted in pork. If such meat is eaten without proper cooking, these worms will leave their cysts and venture through the body of the flesh-eater. Their favorite regions are the diaphragm, ribs, tongue, eye muscles, and larynx. The infestation is known as *trichinosis,* and its severity depends on how many worms have been ingested. The first rule of prevention is thorough cooking of all pork products. Apparently, deep-freezing at –15°C for 20 days also kills the worms. Most cases of trichinosis can be traced to locally produced pork products, particularly from farms where uncooked garbage is (illegally) fed to hogs.

dren may suffer dietary protein deficiency resulting in impaired growth and mental capacity.

The roundworm *Trichinella spiralis* is a common inhabitant of swine where, following sexual reproduction, females burrow into the intestinal wall and remain there, where they each produce about 1500 live young. The young then begin to migrate through the lymph channels, eventually finding their way into the voluntary muscles, tissue rich in blood and oxygen. There the worms mature and surround themselves with a protective cyst (Figure 25.17). The cycle ends there unless the hog is eaten by another hog or some other carnivore, such as ourselves. If we eat infested pork or pork products without first killing the worms by cooking or deep freezing (–15°C for 20 days), the worms may travel through our own lymph channels, to encyst in our own muscles. It has been estimated that an ounce of heavily infected sausage can contain 100,000 or more encysted larvae, which could produce at least 100 million venturesome offspring in their new host. The symptoms of this infestation, called **trichinosis,** include muscle pain, fever, blood disorders, edema, and gastrointestinal disturbances. There is no way of killing the parasites, but most people survive the invasion unless other medical complications set in.

The Rotifers.   The rotifers, we can report, are not parasitic. In fact, their elaborate feeding system is unlike that of all other pseudocoelomates. They feed by sweeping microscopic organisms into their mouths through the action of double rings of cilia around their head area. Food is swept into a grinding gullet, or **gizzard,** which is unlike anything else in the animal world. The digestive system contains other specialized organs, including a stomach, two digestive glands, intestine, and anus (Figure 25.18). Rotifers are equipped with protone-

25.18 THE "WHEEL ANIMALS"

Rotifers, extremely common freshwater animals, are about the same size as many protists but are much more complex. Note the well-defined "brain" and the complex digestive and reproductive systems. Cement glands at the base of the foot permit the animal to temporarily stop its whirling motion and anchor itself to the bottom.

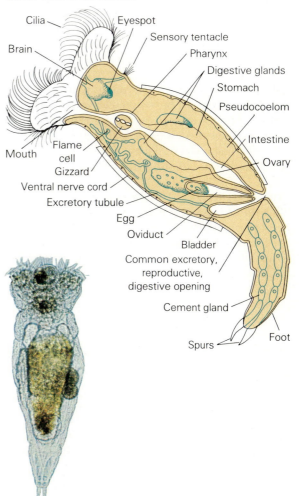

phridia (flame cell systems) like those of the flatworms. However, in the rotifer, the protonephridia form a true excretory system, solving the problems of both nitrogen waste disposal and osmoregulation. In addition, waving cilia direct fluids along a system of ducts to a **urinary bladder,** which subsequently empties through an opening to the outside. As you see, the digestive and excretory systems are quite advanced in such otherwise simple animals.

Female rotifers are much larger than the males. Males are scarce, however, and, in some species, they have never been found. One reason may be that the male digestive system is so rudimentary and inadequate that, apparently, they die soon after mating. The short interlude of the male's existence would seem to threaten the survival of these animals were it not for a surprising ability of the females to compensate. The females, it turns out, are of two types: those that mate with males and those that don't. Before you begin to extrapolate to humans, we hasten to explain.

Rotifers reproduce sexually or asexually. In the latter, chaste females produce offspring **parthenogenetically** (the eggs develop without fertilization). These offspring are all females. Apparently, environmental conditions determine whether these daughters will be of the sexual or asexual type. So, in the world of rotifers, males are reduced to the least common denominator. All they contribute is an occasional input of genes, adding some genetic variability to an otherwise all-female population.

The Priapulids. The **priapulids** were named after the Roman god Priapus, noted for his exaggerated phallus which the worm is said to resemble (Figure 25.19). Although there are only eight living species of priapulids, this group is common in Cambrian fossil strata. Priapulids are more complex than other pseudocoelomates, and researchers wonder whether, since the presence of a peri-

25.19 PRIAPULIDS

The adult priapulid is a wormlike marine predator that may or may not be related to aschelminthes. Larval priapulids are almost indistinguishable from adult rotifers.

toneum has been reported, the priapulids are actually eucoelomates. One line of evidence regarding their relationship is that larval priapulids share some characteristics with adult rotifers.

We will now leave the peculiar and fascinating realm of perhaps some of the more unfamiliar animals on the planet and begin to consider those that are, in many ways, more like ourselves.

APPLICATION OF IDEAS

1. Interestingly, one group of animals and one group of plants represent an aside from the main line of animal and plant evolution. Both emerged from protist ancestors, although very different ones, and each is doing well today but in restricted environments. Describe these two groups and how they relate to the main lines. In what way does each differ from main line organisms?

2. Explain the importance of animal embryology to the development of evolutionary theories and phylogenetic trees. Give examples. List some specific taxonomic problems that might arise if embryos were ignored.

3. Radially symmetrical animals are all aquatic. Suggest reasons why this type of symmetry is not found in the terrestrial environment. If it were, how might such an animal function?

KEY WORDS AND IDEAS

WHAT IS AN ANIMAL?

1. Animals are multicellular, eukaryotic heterotrophs, usually organized at the organ-system level with specialized organs, tissues, and cells. Animals produce a cellular matrix of **collagen.** They reproduce sexually, producing large, stationary eggs and small, flagellated sperm.

ANIMAL ORIGINS

1. While one of the two animal subkingdoms, **Parazoa,** probably arose from flagellate protist ancestors, the other's origin is in question. **Metazoa** may have ciliate or flagellate roots in the protists. Animals today often have both flagella and cilia.

2. The transition from single-celled to multicellular life probably began with casual aggregations in which cells became interdependent, then specialized as they reached the colonial level of organization.

The Animal Fossil Record

1. Reliable animal fossils date back to Precambrian some 580 to 680 million years ago, although fossil worm tubes and tracks may be as old as 700 million years.

2. The **Ediacara fauna** fossils included jellyfish and worms; the slightly newer **Burgess Shale formation** includes fossils from most of today's phyla.

Phylogeny of the Animal Kingdom

1. The phylogenetic tree representing animal evolution reveals the separate origins of metazoans and parazoans. An early major split produced the **protostomes** and **deuterostomes,** which include the higher invertebrates and the vertebrates, respectively.

PHYLUM PORIFERA:
SPONGES AND THE LEVELS OF ORGANIZATION

1. Within organisms, the **cell** represents the fundamental level of organization, followed by the **tissue** level, the **organ,** and the **organ system.** The sponges of phylum **Porifera** represent the cellular level.

The Biology of Sponges

1. Sponges are all aquatic, and mostly marine. Their body consists of a few kinds of cells, including **porocytes, choanocytes,** and **amebocytes,** the latter occurring in a gelatinous **mesohyl.**

2. The amebocytes secrete skeletal elements—**spicules**—which in **calcareous sponges** are of calcium carbonate, in the **glassy sponges,** silicon dioxide, and in the **proteinaceous sponges,** the protein **spongin.**

3. Sponges are filter-feeders that draw water in through the body wall and phagocytize tiny food particles in the body cavity or **spongocoel,** before sending it out through the **osculum.**

4. As **hermaphrodites,** sponges produce sperm and egg, and upon fertilization, a ciliated larva forms. In asexual reproduction, **gemmules,** clusters of cells, break away to form a new body.

5. Sponges lack organized nerve cells, so all responses are carried out by individual cells.

THE RADIAL PLAN AND A STEP UP IN ORGANIZATION

1. Most animal phyla are **bilateral**—that is, the body can only be divided symmetrically by forming left and right halves. A few are **radial,** meaning that *any plane* passing through will divide the body into right and left halves. Echinoderms have a five-part **pentaradial** symmetry.

Phylum Coelenterata

1. Phylum **Coelenterata (Cnidaria)** is organized primarily on the tissue level with some organ development. They are thin-walled, saclike animals with tentacles and stinging cells—**cnidocytes**—that release **nematocysts.** The body wall contains an inner **mesoglea** that contains nerve cell processes, contractile fibers, and ameboid cells. The central **gastrovascular cavity** carries on digestion and respiration. Responses are facilitated by a widespread **nerve net** of nerve cells and processes.

2. Coelenterates include two basic forms, the stationary, attached **polyp** and the swimming **medusa** (jellyfish).

3. Most **Hydrozoa** are marine, forming a stationary feeding and asexually reproducing colonial polyp. A swimming medusa stage carries out sexual reproduction, and after fertilization a swimming **planula larva** forms a new polyp colony. *Hydra,* a freshwater colenterate, is solitary and reproduces asexually by budding and sexually by sperm and egg. It has no medusa or swimming larval stage. *Physalia,* the Portuguese man-of-war, is a floating hydroid colony.

4. **Scyphozoans** (jellyfish) drift in the ocean, capturing prey with stinging tentacles. Scyphozoans produce sperm and eggs, and following fertilization a swimming planula larva forms a fixed **scyphistoma** (polyp), producing many **ephyra** (miniature jellyfish) asexually.

5. The **Anthozoans** (corals and anemones) are strictly stationary polyps. Following fertilization, the usual planula larva forms, but it develops

directly into a new polyp. Coral polyps secrete limestone coverings, forming coral beds and reefs.

Phylum Ctenophora
1. **Ctenophora,** the comb jellies, are organizationally similar to coelenterates but for the fused cilia or **combs** and tentacles bearing glue cells.

THE BILATERAL TREND AND THE VERSATILE MESODERM
1. Part of the bilateral trend included **cephalization,** an emphasis on a head or leading end, bearing feeding and sensory structures.

The Three-Layered Embryo
1. Metazoans have two embryonic **germ layers, endoderm** and **ectoderm,** from which some body tissues are derived. A third layer, the **mesoderm,** first seen in the flatworms, makes many new tissues possible, including organized muscle, blood, and internal skeleton **(endoskeleton).**

Phylum Platyhelminthes: The Flatworms
1. Phylum **Platyhelminthes,** the **flatworms,** includes free-living and parasitic members, many with the organ and organ-system level of organization.
2. The free-living **turbellarians** include *Dugesia* **(planaria),** a complex species with the following characteristics:
   a. an extensive, branched gastrovascular cavity;
   b. a ladderlike nervous system with large anterior **ganglia** and light-sensitive structures;
   c. **osmoregulation** (water management) through **protonephridia**—the **flame cell system**—in which excess water is forced through collecting tubules to the outside by ciliated **flame cells** within **flame bulbs;**
   d. movement by cilia, which push against mucus released by the body;
   e. feeding through a complex protrusible **pharynx,** with digestion beginning in the gastrovascular cavity and being completed within the lining cells;
   f. asexual reproduction through fission and sexual reproduction involving copulation and complex sexual organs. Planarians are hermaphroditic.
3. Class **Trematoda,** the **flukes,** are all parasitic, using a sucker mouth to feed on the host. Important human parasites include the human liver fluke and human blood fluke, the latter of which causes **schistosomiasis.** The life cycle of some may occur in several hosts.
4. Class **Cestoda,** the **tapeworms,** are found in most vertebrates. Their body is greatly reduced to a small hooked and suckered **scolex.** While in the **primary host,** the tapeworm produces numerous **proglottids** in which the sex organs and eggs and sperm develop. Ripened proglottids filled with fertilized eggs pass out with the host's feces. When eggs are taken into an **intermediate host,** they rupture and a burrowing form enters muscle tissue, where it encysts as a **bladderworm,** remaining dormant until the muscle is eaten by the primary host.

BODY CAVITIES, A ONE-WAY GUT, AND A NEW BODY PLAN
1. **Eucoelomate** animals have a body cavity—the **coelom**—and a tubelike, one-way gut, complete with mouth and anus. In roundworms, the cavity is a **pseudocoel** or false coelom. The true coelom is mesodermally lined, and the gut is usually surrounded by mesodermally derived muscle tissue. Animals lacking either are **acoelomates.**

Phylum Rhynchocoela (Nemertina): The Proboscis Worms
1. The **nemertean,** or **proboscis worms** are a minor phylum but have evolutionary importance in that they have characteristics of both flatworms and roundworms.

Phyla Nematoda, Rotifera, and Priapulida
1. The **Nematoda,** or roundworms, are slender, cylindrical pseudocoelomates with little cephalization. They are soft-bodied but are made firm by fluid pressure forming a **hydrostatic skeleton.** Most are free-living predators, forming an important part of soil fauna, but others are important parasites of plants and animals.
2. The largest nematode, *Ascaris lumbricoides,* infests humans and hogs, where males and females sexually reproduce. The fertilized eggs pass out with the host's feces. When swallowed, the eggs hatch in the intestine, make a circuit through the lungs and back down to the intestine, where the life cycle repeats.
3. *Trichinella spiralis,* or the trichina worm, reproduces sexually in the hog intestine and fertilized females bore into the muscle and reproduce asexually, forming great numbers. When the muscle is eaten by a human or another hog, the cycle repeats, producing a disease called **trichinosis.**
4. Phylum **Rotifera** is made up of tiny but complex, free-living aquatic pseudocoelomates. Their digestive system includes several specialized organs, such as a grinding **gizzard.** A simple flame cell system carries out osmoregulation and excretion, and a **urinary bladder** stores liquid excretory wastes. Reproduction is commonly through **parthenogenesis**—development of eggs without fertilization.
5. **Priapulids** resemble other pseudocoelomates, but the presence of a peritoneum suggests that they may be eucoelomates.

REVIEW QUESTIONS

1. List four important characteristics of animals. Of the four, which is (are) exclusive to animals? (616–617)

2. List four systems found in most animals and suggest the primary function of each. (616)

3. List several differences between the Parazoa and the Metazoa. (618)

4. What evidence suggests that the metazoan animals evolved from ciliate ancestors? (618)

5. Approximately when does the undisputed animal fossil record begin? What animals were present then? Are these the first multicellular animals? Explain. (618–619)

6. What do the Burgess Shale fossils tell us about the state of animal evolution at the end of the Precambrian? (619–620)

7. List the four levels of organization found in organisms and define each. (621–622)

8. Prepare a drawing of a simple sponge and label the osculum, spongocoel, and mesohyl. (622–623)

9. List four specific kinds of sponge cells and explain their functions. (622–623)

10. Discuss two reasons why the sponge is relegated to the cellular level of organization. (623)

11. Describe the manner in which sexual reproduction occurs in sponges. (623)

12. Define and give examples of bilateral and radial symmetry in animals. (623)

13. Describe the typical coelenterate body plan and include the terms *mesoglea, gastrovascular cavity, tentacles, cnidocytes, nematocysts,* and *nerve net.* (623, 625)

14. What is the difference between a polyp and a medusa? In which colenterate class is each of these prominent? (625)

15. Describe the peculiar way in which hydroids carry out sexual reproduction. How is this similar to alternation of generations in plants? What is the fundamental difference? (625)

16. Describe the life cycle of a typical scyphozoan coelenterate. (625)

17. In what two significant ways does the anthozoan life cycle differ from that of most hydrozoans and scyphozoans? (625)

18. List two differences between ctenophorans and coelenterates. (626)

19. With the advent of bilateral symmetry, what new ways of exploiting the environment may have been made available? What evidence is there that bilateral symmetry may be as old as 700 million years? (626)

20. List the three embryonic germ layers. What structures does the flatworm owe to the presence of a third layer? (627)

21. Describe the osmoregulatory organ system in the turbellarian and explain how it functions. (628)

22. Compare the digestive systems of planaria and hydra, listing primitive and advanced characteristics where applicable. (625, 628)

23. Briefly review the life cycle of the human liver fluke, *Chlonorchis sinensis.* How might such a complex cycle both help and hinder our control of the parasite? (628)

24. Describe the body form of a typical tapeworm. In what ways is it less complex than other flatworms? Is the lack of structure an advanced or primitive condition? (628, 631)

25. Briefly review the life cycle of a two-host tapeworm. (631)

26. Using simple drawings, carefully distinguish between the terms *eucolomate, acoelomate,* and *pseudocoelomate.* (632)

27. Describe a taxonomic peculiarity of the proboscis worm and comment on its evolutionary significance, if any. (632)

28. Describe the hydrostatic skeleton of the roundworm and explain how it helps in movement. (632–634)

29. In what habitat are most nematode species located? Why are they ecologically important? (634)

30. Describe the life cycle of *Ascaris.* (634)

31. At what stage would *Trichinella spiralis* produce the symptoms of trichinosis? Describe these and explain briefly how such a problem can be avoided. (635)

32. For their minute size, rotifers are amazingly complex. Using the digestive system as an example, support this comment. (635–636)

33. Describe the usual method of reproduction in a rotifer population. Under what conditions do males become important? (636)

Evolution and Diversity: Higher Invertebrates

We now move to those animals with true and beautiful body cavities. This is not to imply that pseudocoeloms are false and ugly, but the prefix *eu* does indeed mean true and beautiful, and it is here that we begin to consider the **eucoelomates.** Since many of these species will have a number of advanced traits, as a group they are referred to as "higher" invertebrates.

In specific terms, the gut of eucoelomates is often complex and includes a specialized, endodermally derived inner lining that functions in digestion and absorption; a muscular, mesodermally derived layer for churning, mixing, and moving food along; and a smooth outer epithelium, also mesodermally derived, that allows the gut to slide around freely within the coelom or body cavity.

The eucoelomate's advanced, one-way gut is usually divided into specialized regions for swallowing, grinding, digesting, and absorbing food, and for concentrating wastes. This means that food can be processed continuously, undergoing different stages of digestion in different areas of the gut.

PROTOSTOMES: THE INVERTEBRATE LINE

Two Ways of Producing a Gut

We learned earlier that in the evolution of animal life, probably in the late Precambrian, a great split occurred and two great animal groups were formed—the **protostomes** and the **deuterostomes.** The protostome line includes the annelids, arthropods, mollusks, and lophophorates, which together comprise the vast majority of today's animal species. The deuterostomes include the echinoderms— sea stars and their relatives—our phylum, chordata, and a few minor phyla such as the urochordates.

Both the protostomes and the deuterostomes are essentially bilateral, and both have the tube-within-a-tube body plan, with a one-way gut and a true coelom. But the gut forms differently in the embryos of these two groups, and there is reason to believe that they evolved quite independently as well. That is, the guts in the two groups are analogous, but not homologous.

In embryos of both groups, the gut originates as an infolding of the early embryonic tissues, at the time that the embryo, a **blastula,** is essentially a hollow ball of cells. When during **gastrulation** the wall of the hollow ball of cells sinks inward to form the **primitive gut,** the depression that results is called the **blastopore.** At this stage, again in both groups, the primitive gut is blind and saclike. To achieve the tube-within-a-tube structure, the gut must develop a second opening to the outside. As we know, the protostomes and deuterostomes both manage this but in different ways (as we see greatly simplifed in Figure 26.1).

In the protostome embryo, the blastopore remains and ultimately becomes the mouth opening. (Thus the term, which means "first, the mouth.") The second opening of the gut to the outside, namely the anus, develops later. We deuterostomes develop the opposite way. In the deuterostomes, the blastopore becomes the location of the future anus; the mouth forms much later at the opposite end of the embryo (thus the name "second, mouth"). Other differences are described in Essay 26.1. With this reminder let's go on to the protostomes.

The Lophophorate Phyla

You may have never heard of the **lophophorates,** but they are important in the scheme of things because they represent our first look at eucoelomate animals. We might mention that there are some authorities that favor placing the lophophorates with the deuterostomes since they have characteristics in common with this group as well.

Lophophorates include three minor phyla: **Ectoprocta, Brachiopoda,** and **Phoronida** (Figure 26.2). The ectoprocts, also known as bryozoans, are easily seen along beaches at low tide. They form branched, reddish fronds that look something like seaweed but are actually colonies formed by asexual budding. Each individual bears tentacles and feeds in the fashion of a coral, but there the similarity ends. The tentacles are attached to a curved ridge called the **lophophore,** and each tentacle is covered with a ciliated lining. Some feathery ectoprocts are dried and sprayed with green paint, and sold in grocery stores as "living air ferns—requiring no care."

Brachiopods are shelled lophophores that superficially resemble a clam or scallop. However, the

26.1 TWO WAYS OF FORMING A GUT

During the development of an embryo, an invagination of cells occurs during a process called gastrulation, producing the first outlines of the gut. In the protostomes, the initial site of gastrulation becomes the mouth as development proceeds. In deuterostomes, however, this invagination marks the site of a future anus. These embryological distinctions, along with other considerations, form the basis of the two phylogenetic divisions of animals.

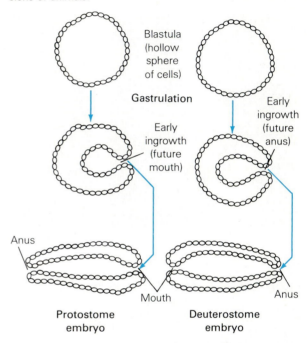

two valves or shells hinge so as to form a top and bottom, rather than two sides (dorsoventral rather than lateral). The shell encloses a coiled ridge of ciliated tentacles that comprise the lophophore.

The phoronids are wormlike lophophorates that live in tubes penetrating the oxygen-starved bottoms of estuaries and bays. They also bear ciliated lophophores. Phoronids, brachiopods, and ectoprocts all produce larvae very similar to the larvae of primitive annelids and mollusks.

While today's lophophorates are only minor phyla in terms of their species numbers, they were once more important. In fact, they are well represented in the oldest fossils. Both brachiopods and ectoprocts are found in Cambrian deposits over 500 million years old. Some rock strata are composed almost entirely of their fossilized shells. The brachiopods have dwindled from about 3000 named species in the Ordovician fossil beds to about 200 named species alive today.

26.2 LOPHOPHORATES

The lophophorates include three phyla that all have a structure known as the lophophore. It consists of a series of ciliated tentacles that are used in feeding and providing water currents for respiration. **(a)** One of the phyla, Ectoprocta, forms either crusty colonies (found on rocks) or seaweedlike fronds. Close examination reveals tentacles thrusting in and out of the body. **(b)** Members of the second phylum, Brachiopoda, are often mistaken for clams, but the shell halves are arranged dorsoventrally rather than laterally. The dominating feature of brachiopod anatomy is the numerous tentacles of the lophophore. Some of the brachiopods live in tubelike burrows, attaching themselves to the ends of the burrows by a long contractile stalk. When disturbed, they draw themselves into the burrow. **(c)** The third phylum is Phoronida. These wormlike creatures feed by extending their lophophores from their burrows like a spirally coiled fan.

(a)

(b)

(c)

The Segmented Body Plan

Along with the coelom and a newly specialized digestive system, a third evolutionary milestone appeared in most of the eucoelomates: the **segmented body plan** or **metamerism.** Here, both the body and the coelom are divided by transverse septa (cross walls) into a sequence of units, or **metameres.** In the simplest forms of segmentaton the units are more or less repetitious. We see such structural repetition in most annelids and in arthropods such as millipedes and centipedes. A more complex version of the theme is also found in other eucoelomate animals. In vertebrates, for example, we find a form of segmentation in the repeated vertebrae, ribs, spinal nerves, and trunk muscles. However, vertebrate segmentation is believed to have evolved independently of that of invertebrates. Nonetheless, the theme prevails in eucoelomate animals, thereby attesting to its evolutionary importance. As we will see, many adaptive variations have sprung from this basic body plan.

Phylum Annelida

The extreme segmentation in the annelids is believed to be an adaptation for burrowing. The trait developed hand in hand with the fluid-filled coelom, which produces the hydrostatic pressure found in both earthworms and roundworms. The hydrostatic skeleton was an early alternative to the hardened skeleton that arose later.

The annelids include three classes: the terrestrial **Oligochaeta** (earthworms), the **Hirudinea** (leeches), and the marine **Polychaeta** (segmented marine worms).

Class Oligochaeta: The Earthworms. **Earthworms** are obviously segmented as viewed from the outside, but the repetition is found in many of their internal structures as well. Internally, each segment contains elements of the circulatory, digestive, excretory, and nervous systems. Their circulatory system, like our own, is "closed." A **closed circulatory system** is one in which the blood is virtually always enclosed in blood vessels, not allowed to percolate freely through tissue spaces as is so common in invertebrates. The earthworm's circulatory system includes five pairs of **aortic arches** (hearts), which are essentially pulsating vessels, along with arteries, veins, and capillaries. The blood of many annelids contains hemoglobin, an oxygen-carrying protein, but it is not bound within blood cells, as is our own hemoglobin. Instead, it is dissolved in the circulating fluid.

The excretory system is also highly developed

in the terrestrial worms. Nearly all segments have paired **nephridia.** A nephridium is considered to be quite advanced over a protonephridium (see Chapter 25) because it not only takes up fluids, ions, and nitrogen wastes, but it recovers most of the water and valuable ions, sending them back into the blood. (As a terrestrial animal, the earthworm must be a water conserver.) Thus, the nephridia perform functions similar to the vertebrate kidney, maintaining water and ion balances and removing nitrogen-containing metabolic wastes from the coelomic fluids (Figure 26.3).

The earthworm is hermaphroditic, with complex male and female reproductive organs present in each individual, but is not self-fertilizing. When earthworms copulate, they exchange sperm reciprocally, and each stores the sperm in its **sperm receptacles** until the proper time for fertilization. The **clitellum,** the smooth, whitish cylinder of external tissue on earthworms, secretes a mucous cocoon that will slide down the body, receiving eggs and sperm from special reproductive pores in the body. Fertilization occurs within the cocoon, which is then shrugged off. It will house the embryos while they develop. There is no larval stage and the young hatchlings resemble the adults.

The digestive system is complete, following the tube-within-a-tube plan. The tube is subdivided in the first 20 or so segments into swallowing, storing, and grinding regions (see Figure 26.3). The intestine, a digestive and absorptive structure, continues to the anus. Earthworms feed on organic matter in the soil.

The earthworm's nervous system, like that of other protostomes, is characterized by a **ventral nerve cord.** This major nerve extends from the brain to the last posterior segment, giving rise to paired ganglia (clusters of nerve cells) in each segment and

26.3 EARTHWORM BODY PLAN

Most of the organ systems in the earthworm are quite complex. Note the prominent segmental body plan, interrupted only by the smooth, glandular clitellum. The body wall contains circular and longitudinal muscle groups, used in extending and in contracting the body. The transport of food, oxygen, and carbon dioxide is carried out by a closed circulatory system that includes five paired "hearts" and an extensive system of blood vessels. The muscular gut, which is suspended in the coelom, contains several specialized regions (pharynx, esophagus, crop, gizzard, and intestine) for the batch-processing of food. Earthworms are hermaphrodites, so both male and female gonads are seen. The copulatory structures are simple, consisting of pores in the body wall. Following fertilization, the eggs are enclosed in a cocoon of slime, which is produced by the clitellum. The nervous system consists of a pair of enlarged ganglia (clusters of neurons) above the esophagus and a lengthy ventral nerve that gives rise to ganglia in each segment (not seen here). Nearly all segments contain paired nephridia, complex tubular structures that take in coelomic fluids, selectively eliminate nitrogen wastes, and reabsorb fluids, thus regulating body water.

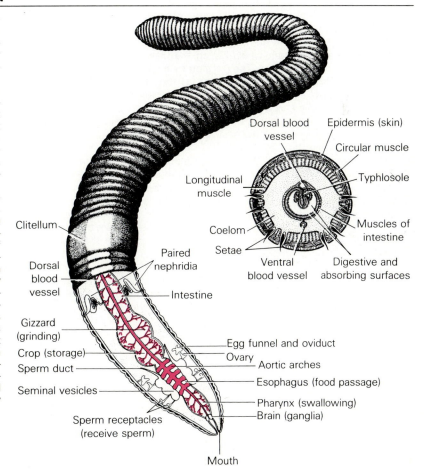

Embryos and Evolution:
Protostome and Deuterostome Characteristics

Embryos often provide interesting clues to evolutionary history, sometimes filling in the many gaps of the fossil record. We can, for instance, usually distinguish between the two great invertebrate groups—the protostomes and deuterostomes—on the basis of several distinctive embryological differences. The two groups express these differences almost from the start of life. (Protostomes include the bryozoans, phoronideans, brachiopodians, molluscans, annelids, onychophorans, arthropods, tartigradians, and others. Deuterostomes include the chaetognathans, echinoderms, hemichordates, and chordates.)

(a) Spiral cleavage
(annelids, mollusks, arthropods)

(b) Radial cleavage
(echinoderms, chordates)

Cleavage Patterns

Shortly after fertilization the zygote begins the first of many cell divisions or cleavages that will provide the immense bank of cells from which the embryo's form will emerge. The zygote first divides into two cells, then into four, then eight, sixteen, and so forth. If the animal is a protostome, by about the third round of divisions the new cells come to lie on the dividing line of the cell layer below, rather than directly over them **(a)**. Thus we have *spiral cleavage*—that is, the cleavage planes are diagonal to the polar axis of the egg. With the deuterostomes, the cell divisions occur in radial fashion, called *radial cleavage*. Radial cleavage places each new cell neatly atop another **(b)** so that the cleavage lines are nicely aligned.

Mouth Development

A second important distinction explains the terms *protostome* and *deuterostome* ("first the mouth" and "second the mouth," respectively). During the normal course of events in an embryo's development, the cleav-

ages produce a hollow ball of cells called a blastula. Then, certain cells begin to sink in, or invaginate, producing the blastopore **(c)**. In the protostomes the blastopore marks the origin of the embryonic mouth, but in the deuterostomes, this region marks the site of the anus. The mouth develops from a secondary opening called the *stomadeum*.

Schizocoels Versus Enterocoels

A third distinction is seen in the manner in which the coelom develops following the ingrowth of cells mentioned earlier. Although protostomes and deuterostomes are both eucoelomates, their coeloms are believed to have evolved independently; thus, as you might expect, their mesodermal lined body cavities form quite differently. In the protostome, the *schizocoelous* process is seen. The term "schizocoelous," not surprisingly, refers to a "splitting"—in this instance the coelom forming as a split in the mesoderm, which itself has formed in patches near the blastopore **(d)**. The split produces two hollow

patches of mesoderm that expand by cell division, filling the old cavity and forming the mesoderm of the coelom. In the deuterostomes, the *enterocoelous* process occurs. In this instance **(e)**, the mesoderm forms two pouches along the hollow center of the embryo, and cell division in the pouches spreads the mesodermal lining into the cavity, forming a new cavity or coelom within.

Swimming Larvae

A final distinction can be seen in the appearance of swimming larvae in those protostomes and deuterostomes that produce this embryological stage. Nemerteans, bryozoans, mollusks, and annelids all produce a pear-shaped larva generally referred to as the *trochophore* **(f)**. While the larvae in these phyla are certainly not identical, they bear important similarities, especially the mollusks and annelids. The deuterostomes, when represented by the echinoderms, produce several types of larvae which, according to evolutionary theory, are based on an ancestral type called the *dipleurula larva*. In the sea star, the specific dipleurula is the *bipinnaria larva* (g), while in the brittle stars, sea urchins, and sand dollars, the larval derivative is called a *pluteus*. In each echinoderm class, the larvae vary from a basic hypothetical ancestral form that no longer exists. The hypothetical dipleurula and the early bipinnaria are also quite similar to the larval form of the acorn worm, another recognized deuterostome. Note the similarity in body plan and arrangement of the gut and ciliated bands between the bipinnaria and the *tornaria larva* (h) of the acorn worm.

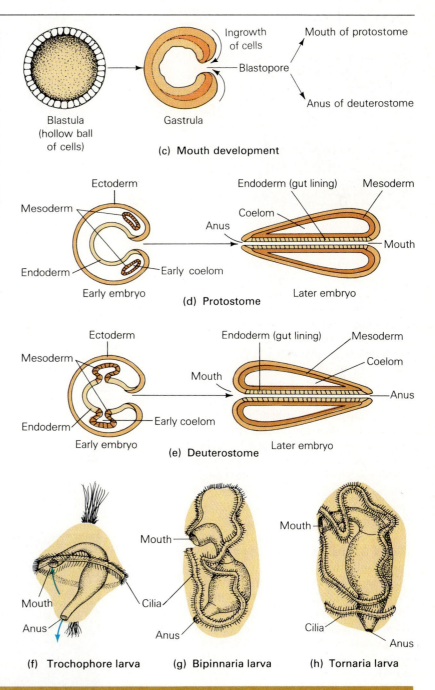

Blastula (hollow ball of cells)

Gastrula

Ingrowth of cells

Blastopore

Mouth of protostome

Anus of deuterostome

(c) Mouth development

Ectoderm

Mesoderm

Endoderm

Early coelom

Early embryo

Endoderm (gut lining)

Mesoderm

Coelom

Anus

Mouth

Later embryo

(d) Protostome

Ectoderm

Mesoderm

Endoderm

Early coelom

Early embryo

Endoderm (gut lining)

Mesoderm

Coelom

Mouth

Anus

Later embryo

(e) Deuterostome

Mouth

Anus

(f) Trochophore larva

Mouth

Cilia

Anus

(g) Bipinnaria larva

Mouth

Cilia

Anus

(h) Tornaria larva

26.4 THE LEECH

Leeches are armed with sharp, piercing mouthparts and suckers for holding onto their hosts and drawing blood. As a leech fastens onto the skin of its host, its secretes into the small wound an anticoagulant known as *hirudin*. Leeches feed voraciously once attached: they are able to ingest three times their body weight at one time. After feeding, they drop off and fast for long periods.

to segmental branches that serve the organs therein. The brain consists of greatly enlarged, paired ganglia above the esophagus and smaller paired ganglia below. The presence of large nerve cell masses in the head region, although modest as brains go, is another example of increasing cephalization in bilateral animals.

Burrowing and other movements are accomplished by two layers of body wall muscles—*circular* and *longitudinal*—which extend and shorten the body, respectively. Acting as a hydrostatic skeleton, the turgid, fluid-filled body provides a flexible but resistant base for muscle action while maintaining the earthworm's shape. During burrowing, anchorage is provided by the **setae** (singular, *seta*), chitinous bristles found on most segments, which can be inserted into the burrow to hold some segments fast while other parts of the body are extended or contracted.

Class Hirudinea: The Leeches.   Leeches are commonly thought of as parasitic, but there are many predatory and scavenging species. They are primarily found in fresh water, although some species are found in marine and moist terrestrial habitats. The parasitic species are **ectoparasites**—that is, they attach themselves to the skin of the warm-blooded hosts, including humans, and draw blood.

The segmented body of the parasitic leech is usually flattened, with suckers at the anterior and posterior ends (Figure 26.4). Suction is applied by a muscular pharynx, and some species have horny teeth in the anterior sucker that cut through the skin of the host organism. When they have made an incision, they secrete an anticoagulant called **hirudin** into the wound. You may know that leeches were once sold in pharmacies as a popular remedy for the swelling and "blackness" of black eyes, and in earlier times to "bleed" patients as a common treatment for illness.

The body of the leech is formed of modified segments. The segmentation is apparent in the nervous, reproductive, and excretory systems, but the coelom itself is not divided. Like the earthworm, leeches are hermaphroditic but not self-fertilizing. They are also similar to earthworms in their copulation, egg laying, and development.

Class Polychaeta: The Segmented Marine Worms.   The **polychaetes** include a number of filter-feeders as well as the predatory clamworm, *Nereis*, which is a marine carnivore, although it feeds on algae and decaying matter as well (Figure 26.5). Clamworms make their homes in the sandy and muddy tidal flats. Although they tend to burrow reclusively in the bottom, they can be roused to swim when startled or when mating.

Externally, the most striking features of the clamworm are its powerful, retractable jaws, the numerous pairs of fleshy **parapods** along its segments, and the sensory appendages at its anterior end. The parapods, used both in movement and in gas exchange, are muscular extensions of the body wall that usually bear setae.

26.5 *NEREIS*, A MARINE POLYCHAETE WORM

The clamworm is an aquatic relative of the earthworm but is highly specialized for its marine existence. Its anterior has eyes, sensory projections, and a set of retractable grasping jaws, which reveal it as a voracious predator.

Unlike earthworms and leeches, sexes are usually separate in marine worms. Gametes are produced seasonally by the coelomic lining. Some species simply release their gametes into the surrounding water, but in others the females swell with eggs and then literally burst. In one particularly exotic species, the Samoan palolo worm *(Eunice viridis)*, a large number of posterior segments become filled with mature gametes. Then during the night, on some unknown cue, the ripened, specialized regions of countless worms break away simultaneously and swim to the surface (each segment has light receptors). At about dawn, with the surface literally crawling with the odd creatures, each bursts, and sperm and egg find each other in an orgy of fertilization. The fertilized eggs then sink to the bottom. As with other marine worms, the zygotes develop into swimming **trochophore larvae.** The trochophore is typical of many other marine invertebrates, including flatworms, nemerteans, brachiopods, bryozoans, and mollusks. It gets its name from the spinning motion (Greek; *trochos*, "wheel") of its ciliated, pear-shaped body.

Among the polychaetes are a number of tube-dwelling species. Some construct tubes of sand particles, cemented with mucus or calcium secretions. They are easy to see in their burrows because their feathery plumes (which are actually feeding and respiratory devices) constantly pop in and out. These colorful ciliated tentacles have inspired common names such as "fan worm," "feather worm," and "peacock worm."

The tube-building worms have left important clues to the annelid history. Fossilized tubes have been found along with what seem to be tracks made by the ancient worms in Precambrian strata. The earliest tracks are random and dispersed, but those appearing later are aggregated, showing a possible change in the social behavior of the worm. The fossilized worm tracks are among the earliest indications of the eukaryote life.

Phylum Mollusca

The earth is burgeoning with **mollusks.** In fact, we already know of about 100,000 species, which, in sheer numbers, puts them behind only the nematodes and insects. Some mollusks are minute, living in tiny inconspicuous shells, while the giant North Atlantic squid that swims confidently through the cold ocean waters reaches over 18 meters (59 feet) in length. Of the seven classes, we will consider only the four larger classes as representatives of the phylum.

The term *mollusk*, derived from Latin, means soft-bodied, though for an invertebrate, this isn't very descriptive. Aside from being soft-bodied, all mollusks are distinguished by a muscular **foot** that contains sensory and motor systems and may be used for creeping, digging, holding on, or capturing prey. Some have external **shells** produced by secretions of a fleshy covering known as the **mantle,** which is also present in nonshelled members of the phylum. The **mantle cavity** (the space enclosed by the mantle) houses feathery respiratory **gills** in aquatic mollusks. In terrestrial mollusks, the lining of the mantle cavity is highly vascular and serves as a respiratory membrane across which oxygen and carbon dioxide can pass (Figure 26.6).

Mollusks are coelomates, but whereas the coelom is conspicuous in embryos, it is very reduced in adults, often present only as an open region surrounding the heart with some remnants in the nephridia and gonads. Except for the cephalopods (octopuses and squid), an **open circulatory system** is present, with blood leaving vessels to enter open, spongelike sinuses before returning to the heart from where it is again pumped to the vessels. The digestive system is, of course, tubelike, with both a mouth and anus. Most classes of mollusks have a rasping tonguelike structure called the **radula** with which they file their food into small particles. Mollusks, like annelids, have nephridia that carry out osmoregulation and excretion.

Whether mollusks are actually related to other protostomes is a subject of controversy, but two factors support just such a relationship. First, like marine annelids, many mollusks produce a trochophore larva. Second, fossils tell us that some mollusks were segmented and apparently had a well-developed coelom, similar to that of annelids. For many years this theoretical relationship was based primarily on fossils of *Pilina*, whose shell revealed evidence of segmental muscle attachments. But in 1952 a Danish research vessel dredged up 10 "living fossils" from a depth of 3 kilometers. They were the remnants of a class known as Monoplacophora, long believed to be extinct. The specimens, dubbed *Neopilina*, are clearly segmented, and the coelom of the adult is well developed and extensive.

Class Amphineura: The Chitons. Chitons, probably the least specialized mollusks, are thought to retain many of the features of the ancestral mollusks. Their flattened bodies and long foot are protected by a shell composed of eight simple plates (see Figure 26.6a). The gut, beginning with a radula-equipped mouth, extends the length of the

26.6 MOLLUSK BODY PLAN

The basic body plan of the mollusks is best seen in the primitive chitons, which have a simple foot, mantle, and shell. The other classes shown here have each evolved from the basic plan in a different way. The greatest change appears to have occurred in the cephalopods (the octopuses and squid), which have lost all or nearly all of the shell and modified the food into grasping tentacles. Note the divergent evolution of the shell, mantle, gill, foot, mantle cavity, and gut in these four classes.

Chiton

Scallop

Tree snail

Squid

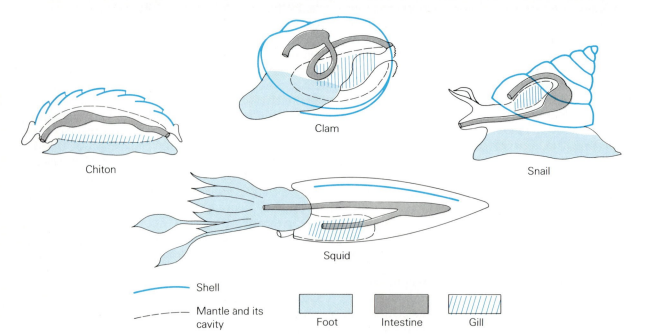

Chiton

Clam

Snail

Squid

——— Shell

— — — Mantle and its cavity

Foot   Intestine   Gill

body. The mantle cavity forms a groove entirely around the foot and houses numerous feathery gills along each side.

Chitons are slow-moving, rather uninspiring, dull creatures. They creep along over the rocky bottoms of tide pools, grazing as they go, by using their sharp radula to scrape the algae from the surfaces of the rocks. When disturbed, their defense is to grip the rocks with powerful foot muscles and flatten so that it is very difficult to remove them. These primitive beasts have been around a long time; chitons very similar to those of today show up in the fossil record in Ordovician deposits (500 million years old).

Class Gastropoda: The Snails. The **gastropods,** or "stomach-foot" animals, slide along on a large, extendable, muscular foot, which lies just below most of their digestive organs. Some have rather successfully invaded the land, and a host of beautifully shelled creatures glide over the seabed. In **pulmonates,** terrestrial and freshwater snails and slugs, the gills have been replaced by a simple, vascularized mantle cavity that serves as a lung. The freshwater snails, and a few marine species, were apparently terrestrial at one time and have returned to an aquatic habitat. However, their descendants still retain lungs instead of gills and come to the surface to breathe.

The basic organization of gastropods is rather primitive. The body plan is superficially similar to that of the chitons, except that coiled gastropods have undergone **torsion** (twisting or rotation) as they developed. The result is a twisted hump matching the shape of the shell. This twisting causes all sorts of internal changes and no one is quite sure what the advantages might be. As you can see in Figure 26.7, during larval development the body undergoes a 180° twist, bringing the mantle cavity forward and the anus around to a position just above the head (certainly not the most esthetic arrangement possible). The humped arrangement provides considerable space for retraction of the entire head-foot into the shell. Once the tough, horny, doorlike cover, the **operculum** is in place across the opening into the shell, the snail is safe from many predators.

There are a lot of gastropods around, some 50,000 named species in all. They vary from those with elaborate twisted shells to shell-less creatures such as the slug, nudibranch, and sea hare (Figure 26.8).

Class Bivalvia (Pelycypoda): The Bivalves. The **pelycypods**—bivalve (two-shelled) mollusks—differ from gastropods primarily in that they have two shells, and the head, with its sensory appendages, is greatly reduced. The foot is still fleshy and highly extensible, and some species (clams) use it for burrowing. The hinged shell is drawn closed by two powerful muscle groups.

Bivalves lack the radula and are chiefly filter-

26.7 TORSION AND ITS EFFECTS ON THE GASTROPOD

In larval snails, the internal organs undergo a twisting as the animals develop. As you can see in the young ciliated larva, the digestive tract is simply U-shaped. As time passes, it completes a 180° twist that brings the anus up over the mouth. This twist is related to other developmental events as well.

For example, organs on one side of the body are compressed and fail to develop. The nervous system contorts into a figure 8. Spiral development of the shell and body mass, which occurs in many snails, is a separate, later developmental effect.

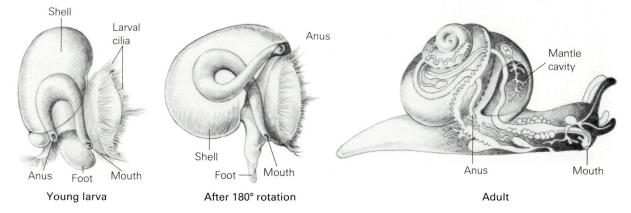

Young larva   After 180° rotation   Adult

26.8 REPRESENTATIVE MOLLUSKS

There are 50,000 named species of gastropod mollusks. These include (a) the beautiful nudibranch, (b) the cone snail, seen here eating a paralyzed fish, and (c) the keyhole limpet.

(a)

(b)

(c)

feeders, screening small particles of food from the water. Currents of water, bearing algae and other minute organisms, are drawn into the mantle cavity through an **incurrent siphon** by the action of cilia that line the large gills. The food particles are then trapped in heavy mucus secretions and moved by liplike, ciliated **labial palps,** which separate out grains of sand and permit the bits of food to move into the mouth. The gill surface also serves in the exchange of carbon dioxide for oxygen with water passing on its way to the **excurrent siphon.** The circulatory system of bivalves is typical of mollusks. The heart is located in its coelomic cavity. As you can see in Figure 26.9, the gut also passes through the cavity; in fact, the heart is actually wrapped around the gut, so the gut appears to go through the heart.

Class Cephalopoda: The Squid and Octopus. The **cephalopods,** "head-foot" mollusks, include the squid, octopuses, cuttlefish, and nautiluses. All but the nautiluses lack the external shell.

26.9 ANATOMY OF A CLAM

The clam, a bivalve mollusk, has highly specialized feeding and digestive systems. In this diagram, one shell, one pair of gills, and part of the body wall have been cut away. The head is reduced, with none of the elaborate sensory features found in many other mollusks. The foot makes up most of the body. Clams are filter-feeders; they bring in a steady flow of water through the incurrent siphon, pass it through pores in the gills, and then send it back out through the excurrent siphon. As water passes through the gills, as shown by the arrows, rows of cilia sweep any particles forward to the mouth. Sensitive labial palps near the mouth detect whether an object is edible. The mouth opens into a short gullet that leads into a stomach pouch. Food then moves into a lengthy intestine, which is twisted back and forth through the foot. The intestine actually passes through the heart and back to the anus. The heart is a simple pumping chamber that forces blood out into an open circulatory system.

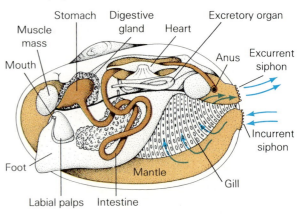

26.10 CEPHALOPOD VISION

The eyes of the cephalopods (octopus and squid) are remarkably similar to those of vertebrates. The squid is believed to have binocular vision, like humans and a few other mammals, giving it the ability to judge distances with great accuracy.

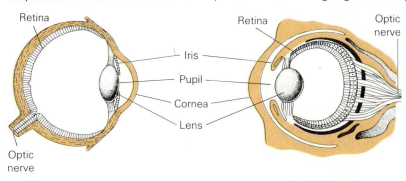

Vertebrate eye

Cephalopod eye

In many ways, they are highly specialized creatures. They are active and voracious predators, feeding on fast-moving invertebrates and vertebrates. Their circulatory system is closed, an advanced condition with the efficiency needed for large fast-moving animals. There are two pairs of hearts; one boosts blood through the gills, the other through the body. Water is moved over the gills by the pumping of the muscular mantle, in contrast to the ciliary currents found in other marine mollusks. In cephalopods, the head is greatly enlarged and is covered by the mantle itself. The thick, muscular molluscan foot is highly modified into tentacles, some complete with suckers. The mouth is armed with a horny beak used in ripping and tearing prey.

Because of their active behavior the cephalopods tend to attract the attention of other carnivores. Their defense is their speed and agility. Octopuses and squid move by forcefully ejecting water through their siphons in a jetlike action. The siphon can be turned in any direction. They may also quickly change colors or release a cloudy "ink" to foil predators. Octopuses often hide in burrows and can squeeze their large bodies through and into tiny cracks and crevices. Like other cephalopods they have highly developed sensory receptors, including eyes that are very similar to those of vertebrates (Figure 26.10). The octopuses also have a large, well-developed brain and are surprisingly intelligent invertebrates, as we will see in Chapter 38. Unlike many invertebrates, their behavior is not entirely genetically programmed, and they are capable of a significant amount of learning.

Phylum Arthropoda

There are a lot of **arthropods.** You can get the idea by imagining Noah trying to load his ark with animals. If, by chasing, swatting, and kicking, he could have gotten one pair aboard per minute, he would have spent 18 months, night and day, simply loading his quota of crickets, crabs, lice, flies, centipedes, aphids, wasps, weevils, dragonflies, lobsters, ticks, and the like into the vessel.

To date, some 900,000 species of arthropods have been described, and some authorities estimate that about 1 million more await identification. We must conclude that as a phylum, **Arthropoda** is the most successful on earth.

So what are these pervasive little creatures? The name *arthropod* tells us that they have "jointed feet," actually jointed legs. They are also segmented, but these segments are not simply repeating units as in earthworms; they may instead be highly specialized for different tasks. Arthropods have an **exoskeleton** (*exo*, outside) made of chitinous material, often hardened with calcium salts. All arthropods have open circulatory systems. Their nervous system consists of major ganglia, clustered nerve cells in the head, and a ventral nerve cord giving rise to smaller paired ganglia in most segments. Except for a few marine forms, arthropods are small because of restrictions placed on them by the physics of this plan. (At least we don't have to deal with 80-pound mosquitoes.)

Arthropods have successfully invaded most of the earth, often developing amazingly narrow specializations. (Consider the adult mayfly, a delicate creature that emerges without mouthparts with

26.11 PHYLOGENY OF ARTHROPODS

The phylum Arthropoda can be organized into two subphyla on the basis of mouthparts and appendages. The Chelicerata, shown in the left branch, lack jaws, but produce clawlike *chelicerae* (which in the spiders are venomous fangs). In addition, they have four pairs of walking legs and other appendages that are generally related to feeding. The Mandibulata, branching off to the right, have jaws and three or more pairs of walking legs. Most have antennas. The subphylum Mandibulata includes the enormous class Insecta.

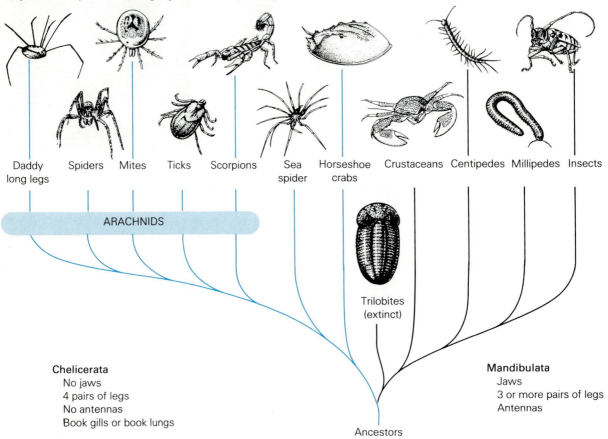

Daddy long legs   Spiders   Mites   Ticks   Scorpions   Sea spider   Horseshoe crabs   Crustaceans   Centipedes   Millipedes   Insects

ARACHNIDS

Trilobites (extinct)

Chelicerata
No jaws
4 pairs of legs
No antennas
Book gills or book lungs

Mandibulata
Jaws
3 or more pairs of legs
Antennas

Ancestors

which to feed, and which must mate and leave offspring in the few precious hours of life allowed it.)

Arthropods have varying diets and include omnivores, herbivores, carnivores, scavengers, ectoparasites, and endoparasites. They live on, under, and above the surface of land and water, from deep ocean trenches to the highest mountain peaks. In many cases, because of their highly specialized niches, numerous species can exist side-by-side without seriously competing with each other.

Arthropod success is due to other important factors as well. The reproductive capacity of many arthropods is phenomenal. In addition, many produce larvae that have an entirely different diet from the adults, thus expanding the niche and avoiding competition between parent and offspring. Because of the large brood size the opportunities for genetic recombination, particularly in insects, are great and there is a great deal of genetic variation in most populations. This helps the insect to overcome all sorts of environmental deterrents. (For example, pesticides; DDT-resistant mosquitoes have evolved in only a few generations.)

Subphylum Chelicerata

Arthropods have been divided into two subphyla, **Chelicerata** and **Mandibulata** (Figure 26.11). The more ancient group, the Chelicerata, includes horseshoe crabs, sea spiders, and class Arachnida: true spiders, daddy longlegs, ticks, mites, and scorpions. They lack jaws and have six pairs of appendages, four of which are legs. The first two pairs are sensory palps and **chelicerae** (fangs). None of the Chelicerata have antennas.

Class Arachnida. The **arachnids** don't win much affection with their habit of sucking the juices of other organisms. Spiders, equipped with hollow fangs and venom, are all carnivores. Typically, when a spider bites its prey, it injects venom, which weakens the unfortunate animal. Next it injects digestive enzymes into its host's body that liquify the tissues there. Then it sucks the fluid out.

Spiders also differ from other arthropods in that they respire through a structure known as the book lung. Book lungs are internal sacs that open to the outside through slits. The sacs are lined with leafy folds, similar to the pages of a book. Blood passing through the thin "pages" exchanges gases with air in the sac.

Most spiders are equipped with silk-producing glands connected by ducts to external devices called **spinnerets** (Figure 26.12). Both the **web** and

the **cocoon** of the spider are produced by this system. The silk of spiders (and insects) is a protein known as **fibroin,** composed of the amino acids glycine, alanine, and tyrosine. Spiders as a group are known to produce seven different kinds of silk, each with its own function. Webs have both sticky parts with which to ensnare prey and safe, dry parts along which the spider can run.

Subphylum Mandibulata

Members of the subphylum Mandibulata have **mandibles** (jaws), as well as antennae and various numbers of paired appendages, including three or more pairs of walking legs. There are four major classes and two rather minor ones. The major classes are the Crustacea (including marine and freshwater crabs, lobsters, copepods, barnacles, and sowbugs), Chilopoda (centipedes), Diplopoda

26.12 SILK GLANDS OF THE SPIDER

The orb spider, a master web-maker, can produce several kinds of silk. The silk glands, as shown here, are located in the abdomen. They open into the spinnerets (below right), which emit threads of different diameters. The web in the photograph belongs to an orb spider. The spider can move about quickly on the web, but it must know where the sticky parts are or it will become ensnared in its own trap.

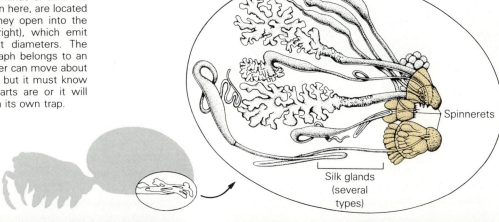

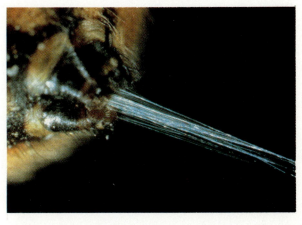

26.13 REPRESENTATIVE MANDIBULATES

The Mandibulata are such a diverse group that it's difficult to find a representative example. These species can give you some idea about the diversity of their size, appearance, and habitat. There are crustaceans such as **(a)** the long-arm crab, seen here in a defensive posture, and **(b)** the red cleaner shrimp. **(c)** The oak treehopper is an attractive insect found in North American deciduous forests. **(d)** The exotic Javanese leaf insect, a native of Indonesia and New Guinea, contrasts the Hercules beetle, **(e)** whose long jaws serve to attract females. **(f)** *Cathedra serrata* is an unusual inhabitant of the tropics.

(a) Long-arm crab   (b) Red cleaner shrimp   (c) Oak treehopper

(d) Javanese leaf insect   (e) Hercules beetle   (f) *Cathedra serrata*

(millipedes), and Insecta (flies, grasshoppers, bees, and so forth). Figure 26.13 illustrates the wide variety within this subphylum. The crustaceans are not closely related to the other Mandibulata, and the tendency in recent years has been to elevate them to a subphylum of their own.

Class Crustacea. **Crustaceans** live in both marine and fresh water. Their size ranges from microscopic ostracods and copepods to giant crabs from the western North Pacific. Species in this diverse group, such as the familiar pillbugs (or sowbugs) clearly demonstrate their relationship to the primitive annelids, while others (like the crabs) have become much more advanced, departing dramatically from the ancient plan. The segments of the most primitive crustaceans bear paired appendages, each somewhat like the next. In fact, they are reminiscent of the polychaete worms in this respect.

Crustaceans are gill breathers, with gill cavities partly covered by folds of the body wall. The active movement of certain appendages creates water currents that pass over the feathery gills (see Chapter 30). Because they are encased in a semirigid, secreted exoskeleton, to allow for growth crustaceans frequently molt, shedding their exoskeleton in a hormonally regulated manner (see Chapter 33).

Among their highly developed sensory structures is a **compound eye,** often borne on a moveable or retractable stalk. The eye, similar to that of insects, is composed of a large number of visual units called **ommatidia.** Each has a **corneal lens** and pigmented, light-sensitive **retinal cells** (Figure 26.14). The compound construction and stalked position enable the animal to obtain a nearly 360° view of its surroundings.

Many of the marine forms of the diverse crustaceans are of great ecological importance. For example much of the ocean's minute floating life, the **plankton,** consists of microscopic crustaceans. They are so numerous that one order, Euphausiacea, commonly known as "krill," is the main diet of the largest whales. But they are also an important part of the ocean's longer food chain.

Classes Chilopoda and Diplopoda. Centipedes (Chilopoda) are sometimes confused with **millipedes (Diplopoda).** However, millipedes are harmless herbivores, while centipedes are more fearsome, and perhaps even dangerous, predators and scavengers. The centipede's first pair of legs are modified into perforated claws with which they inject poison into prey. The sting of common temperate zone centipedes is painful but probably no more life-threatening to humans than a wasp sting. However, the sting of larger tropical species can put an adult in bed with a fever for several days.

Class Insecta. The majority of arthropods and, indeed, the majority of animals, are **insects.** Except in the sea, where the crustaceans hold sway,

insects rule the earth in terms of numbers and kinds. We are continuously engaged in conflict with them and, even today, we often lose. Insects are of such importance ecologically and economically that we literally could not have reached our own place in nature without understanding something about them.

The segmented body of insects has undergone considerable modification from the ancestral form, primarily through the fusion and alteration of segments. The modification has been so sweeping, in fact, that only in the abdominal region and in the embryo is it possible to see a clearly segmented body. The exoskeleton, like that of many other arthropods, is basically of chitin, but it retains a flexibility not seen in the crustaceans.

The bodies of insects have three major regions: the **head,** the **thorax,** and the **abdomen** (Figure 26.15). The thorax gives rise to three pairs of legs and, in some species, wings. The legs are modified in different species for running, jumping, catching, holding, or simply resting. The appendages on the abdomen have been modified to form the genitalia and anal structures. In many species, females have

---

26.14 THE COMPOUND EYE

(a) Crustaceans have compound eyes, as do the insects. The compound eye consists of many individual light-receiving units called *ommatidia.* **(b)** Each unit has a compound lens (a thickened cuticular lens and a crystalline cone lens), a mirror-lined channel and a cluster of long light-sensitive cells. The mirrored channel is an efficient light collector and is most abundant in crustaceans that live in dim light. The unit is surrounded by pigment cells that screen it and prevent light from scattering from one unit to another. The eye is apparently well adapted for detecting movement, but may not present as clear an image as the vertebrate eye.

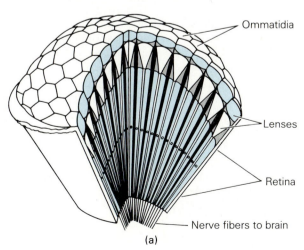

Ommatidia

Lenses

Retina

Nerve fibers to brain

(a)

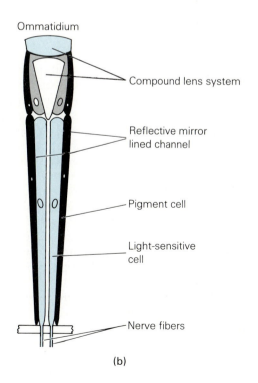

Ommatidium

Compound lens system

Reflective mirror lined channel

Pigment cell

Light-sensitive cell

Nerve fibers

(b)

26.15 INSECT BODY PLAN

The insect body plan (represented by the grasshopper) consists of three main parts: *head, thorax,* and *abdomen.* Typically in insects, segmentation is most apparent in the abdomen and is obscured in the thorax and head because the segments there are fused. The grasshopper's head bears a pair of compound eyes and two or three simple eyes, in addition to a pair of sensory antennas. Its mouthparts include the jaws and sensory *labrum* and *labium* (lips). The largest accumulation of nerves is located in the head, and comprises the insect brain. The thorax commonly gives rise to three pairs of walking legs and two pairs of wings, while the abdomen contains 12 hinged segments. Paired *spiracles* open into each segment, admitting air to the branching system of tracheae that extend throughout the body. The last segments bear the reproductive structures and the anal opening. In the female, some of these segments are modified into *ovipositors,* structures for aiding the process of egg laying.

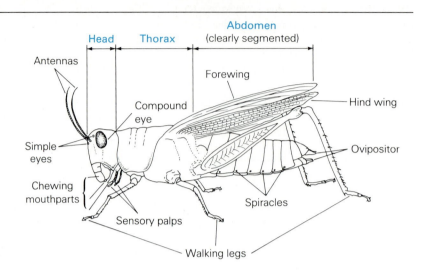

an **ovipositor,** a hollow appendage that is used to dig into the soil or bore holes in plants, in which they lay their eggs.

Insects breathe by admitting air into a complex **tracheal system** through external openings known as **spiracles.** The air passes into highly branched **tracheae,** which subdivide into smaller **tracheoles,** finally reaching individual cells where gas exchange occurs. Some tracheae end in balloonlike air sacs.

The major sense organs and mouthparts are in the head, where you would expect them to be. Consider, for example, the head of the grasshopper. It has one pair of sensory antennas, a pair of compound eyes, and three simple eyes, or **ocelli.** Its mouthparts consist of a single **labrum** (upper lip) and a **labium** (lower lip) bearing **sensory palps,** and there are two laterally movable **mandibles** (jaws) that do most of the chewing. These are assisted by a pair of **maxillae** also bearing sensory palps. The entire apparatus is magnificently adapted for biting and chewing leaves. The sensory palps detect texture and flavor, the jaws rip off and chew bits of vegetation, and the lips hold the food in place during chewing.

Mouthparts in other insects are often highly specialized to fit their specific feeding niche. For example, the maxillae of butterflies are modified

into a long, flexible sucking device (neatly coiled when not in use). On the other hand, the cicada has a piercing and sucking mouth, while the housefly has a large, complex, tonguelike apparatus used for licking (Figure 26.16).

Returning to our grasshopper, let's examine its locomotive structures. First, although some grasshoppers are great jumpers, they are also flying insects. They have two pairs of wings; one pair actually propels the insect, and the other serves as a protective wing cover. The desert locust, a close relative of the grasshopper, can fly over 200 miles in a day. We will look into the details of insect flight in Chapter 28.

The legs of insects are also highly specialized for specific modes of life. The grasshopper, for example, has two pairs of legs of similar size for walking and grasping, while its third pair is greatly enlarged. The combination of extremely powerful muscles in the third pair and the light exoskeleton produces a great mechanical advantage, enabling the grasshopper to jump great distances from a standing start. The praying mantis has very strong, spiked forelegs that snatch and hold prey, which is then ripped apart by powerful jaws (Figure 26.17). Honeybees have walking legs modified for pollen collection. The first pair is equipped with a pollen

26.16 FEEDING VARIATIONS IN INSECTS

Variation in insect mouthparts is seemingly endless, but a few examples will be impressive. **(a)** The grasshopper's chewing mouth is nicely adapted for eating foliage. Two sets of palps contain taste receptors while biting is done with the paired mandibles and maxillae. The labium and labrum help by holding the food. **(b)** Butterflies use their lengthy mouthparts in sucking up plant juices. The long, extendable paired *maxillae* are held together like the sides of Ziploc bags, forming a long drinking tube. When not in use, the tube is kept tightly coiled by flexible rods. **(c)** The housefly has its own feeding strategy. It salivates on its food, stirring it with the large spongelike labium. Its labrum, part of a tubelike device, sucks up the partially digested mess.

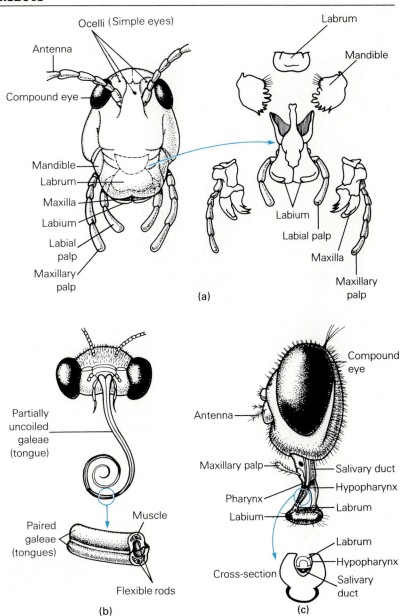

26.17 PRAYING MANTIS

The praying mantis is a voracious predator. This mantis has torn its insect prey open with its powerful barbed forelegs and is feeding with its strong jaws.

brush for collecting pollen from the body and a circular indentation in one segment for cleaning the antenna. (Pollen gathering is a messy business.) The posterior legs contain hollow grooves, or pollen baskets, for compacting pollen (Figure 26.18). And then consider the sensitive legs of roaches, which detect delicate air currents set up by anything that might be after them, such as an irate apartment dweller.

Of course, these are only a few examples of insect diversity. We will discuss additional aspects of their biology elsewhere, but Table 26.1 should underline the immense variation in this group.

Arthropods have been around a very long time. Fossils of marine forms are found in Cambrian strata and possibly in the Precambrian. By the Ordovician period, some members had left the sea and had begun to explore the terrestrial environment. This escape from their marine enemies was only temporary, though; by the Devonian, the first amphibians had struggled ashore and have been eating insects ever since. But the arthropods did not fall easily before the hungry vertebrates. They ran, flew, and crawled away, invading every part of the globe and establishing all sorts of new niches.

26.18 SPECIALIZED LEGS OF THE HONEYBEE

The first and last pairs of legs on the honeybee are specialized for collecting pollen, which the bee uses for food. The first pair is used to collect pollen and to clean the eyes and antennas. Note the eye brush and antenna cleaner. The second pair of legs is relatively unspecialized, but does have spurs that the bee uses to remove wax "plates" from its abdomen (for use in hive building). The hind legs contain a pollen comb and a

pollen basket. These utensils are used to remove pollen from the other legs, crush the pollen into compact masses, and carry it home. As the bee goes about its business, plants get pollinated because some pollen is lost, but most goes to form "bee bread," a food made of a combination of nectar and pollen.

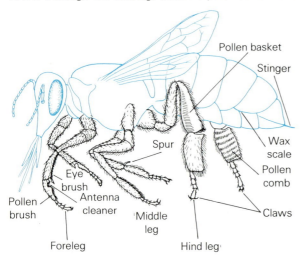

26.19 A CONNECTING LINK

The phylum Onychorphora has many characteristics of both the annelids and the arthropods. Its thin-walled, segmented body resembles the earthworm's, and the serial duplication of internal structures such as the paired nephridia is also remi- niscent of the annelids. On the other hand, onychophorans have jointed legs, tipped with claws, and an open circulatory system, both of which are arthropod characteristics.

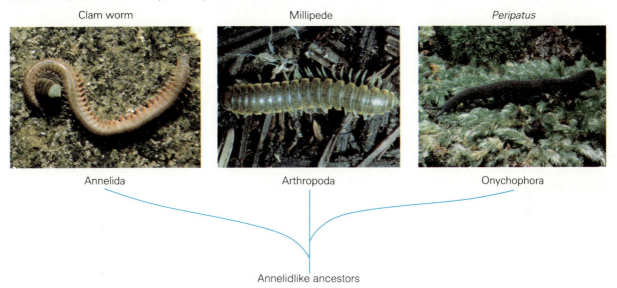

Clam worm

Millipede

Peripatus

Annelida

Arthropoda

Onychophora

Annelidlike ancestors

Phylum Onychophora

Phylum Onychophora, which includes only about 65 to 70 named species, is so minor that it could well be ignored if it weren't for the fact that some researchers consider it an evolutionary link be- tween the Annelida and Arthropoda. The ony- chophorans are believed to represent an ancient animal that arose from the primitive arthropod line very shortly after the arthropods diverged from the early annelids. After leaving the arthropod line, they made the difficult transition to land where they assumed the rather restricted existence we see today (Figure 26.19).

The best-known member of the onychoporans is the peculiar *Peripatus,* which lives among the damp leaves of the tropics in Africa, Southeast Asia, New Zealand, Australia, and the West Indies. *Peripatus* has a thin, soft cuticle covering a muscular body wall. It has two muscle layers that run in circular and longitudinal directions. The animal is highly segmented, revealing its annelid heritage, but the segmentation is largely confined to its appendages and internal structure; its body from the outside is fairly smooth. However, it also has several features characteristic of arthropods. Tiny claws or small pincers arm its soft appendages, and its head even has antennas. It breathes through a series of spira- cles along its body, which open into tiny pits con- nected to a number of tracheae. This arrangement permits air to reach the internal organs easily, but there is no way of controlling the size of the open- ing, so *Peripatus* avoids water loss by living in moist habitats.

DEUTEROSTOMES: THE ECHINODERM-CHORDATE LINE

At this point let's again consider the importance of the evolutionary divergence of the protostome and deuterostome lines. With relatively few exceptions, the protostome line was to emphasize great diversi- ty, small bodies, extremely rapid reproductive cycles, and great population size, all characteristics that more than compensate for limited learning abil- ity.

The early deuterostomes weren't much differ- ent—if anything, they were less intellectually gifted and less adventurous. As we will see, they were also small, simple creatures with extremely limited

TABLE 26.1

MAJOR INSECT ORDERS[a]

Order (number of named species within order)	Examples		Characteristics
Coleoptera (300,000)	Japanese beetle Stag beetle		Two pairs of wings, horny and membranous Heavy, armored exoskeleton Biting and chewing mouthparts Complete metamorphosis Herbivores, predatory carnivores, scavengers Serious agricultural pests
Lepidoptera (140,000)	Moths Butterflies		Two pairs of wings Hairy bodies Long coiled tongue for sucking Complete metamorphosis Serious agricultural pests
Hymenoptera (90,000)	Bees Wasps Ants		Two pairs of membranous wings Head mobile Well-developed eyes Chewing and sucking mouthparts Stinging Complete metamorphosis Many are social Important as pollinators and predators of other insects
Diptera (80,000)	Flies Mosquitoes		One pair of wings and halteres Sucking, piercing, lapping mouthparts Complete metamorphosis Medically dangerous as carriers of diseases (malaria, yellow fever, sleeping sickness, encephalitis, filiariasis)
Hemiptera (55,000)	True bugs: Bed bug Assassin bug Homoptera: Aphids Scale insect Cicada Plant lice		Two pairs of wings, horny and membranous Piercing mouthparts Many agricultural pests
Orthoptera (30,000)	Roaches Grasshoppers Mantis Crickets		Two pairs of wings, horny and membranous Adults, biting and chewing mouth Incomplete metamorphosis Herbivores (except mantis)
Odonata (5000)	Dragonfly		Two pairs of wings Biting mouth food basket (legs and spines) Incomplete metamorphosis Predator

[a]The number of named species in each of the thirteen orders listed ranges from a very modest 1000 in the Dermaptera to 300,000 in the Coleoptera. In fact, by conservative estimate, it is accurate to say that one in four animals roaming the earth is a beetle. Most of the remainder of the animals are butterflies, moths, flies, mosquitoes, bees, ants, and wasps (and their near relatives).

TABLE 26.1 (continued)

Order (number of named species within order)	Examples	Characteristics
Neuroptera (4000)	Ant lion Dobson fly Lacewing	Two pairs of membranous wings Biting mouthparts Complete metamorphosis Silk cocoon
Anoplura (2400)	Lice	Wingless Sucking or biting mouthparts Small with flattened body, reduced eyes Legs with clawlike tarsi (for clinging to skin) Highly host-specific (in humans, head, body, and pubic lice are of different species) Medically important (carry typhus and cause relapsing fever) Incomplete metamorphosis
Isoptera (2000)	Termites	Two pairs of wings, but some stages are wingless Chewing mouthparts Social, division of labor for reproduction, work, defense Incomplete metamorphosis
Siphonaptera (1200)	Fleas	Small, wingless, laterally compressed Piercing and sucking mouthparts Jumping legs Complete metamorphosis Medically important (carry bubonic plague, typhus)
Ephemeroptera (1000)	Mayfly	One pair of wings Vestigial mouthparts in adults (do not feed) Few days of adult life (reproduce and die) Incomplete metamorphosis
Dermaptera (1000)	Earwig	Two pairs of wings, leathery and membranous Biting mouthparts Large pincerlike cerci in males Incomplete metamorphosis

nervous systems. Yet somehow there was amazing potential in this kind of animal. The deuterostomes were to produce not only the echinoderms, such as the witless starfish, but the chordates which would in turn produce the largest and brainiest creatures of all. However, hundreds of millions of years separated the early deuterostomes from the vertebrates.

We've already seen that the first embryonic opening—the *blastopore*—forms the anus in deuterostome embryos and the mouth in protostome embryos. The two groups also differ in the origin of the skeleton. In the protostomes, the exoskeletons of the arthropods, the shells of the mollusks, and the bristles of the annelids consist of nonliving, noncellular materials secreted by the ectoderm-derived epithelium. In contrast, most deutero-

stomes produce **endoskeletons**—stiff, internal skeletons that are formed by mesodermal cells. In most vertebrates the endoskeleton is bone, a hardened matrix containing numerous living bone cells.

The important question for evolutionists is, how could both protostomes and deuterostomes have arisen on the earth? One would think that the first animal species to have developed such a major advantage as the tube-within-a-tube body plan would have taken over the world. Yet these two coexisted. Both these evolutionary lines underwent massive adaptive radiations in the Ediacara period of the late Precambrian, and the phyla that descended from each are today in competition with one another. However, ecological and evolutionary theory suggests that the earliest protostomes and deuterostomes must somehow have failed to fall

26.20 REPRESENTATIVE ECHINODERMS

One of the most striking features of the echinoderms is their five-fold radial symmetry, most easily seen in the sea stars and the brittle stars. In addition, these marine animals have a spiny skin, with the longest spines found in the sea urchins. The echinoderms are unique in that they possess a water vascular system, used in movement and feeding. Shown here are **(a)** sea star, **(b)** brittle star, **(c)** sea urchin, **(d)** crinoid, and **(e)** sea cucumber.

(a)

(b)

(c)

(d)

(e)

into competition with one another at the time they first became established. There are at least two ways this might have happened.

The first hypothesis is that the early protostomes and early deuterostomes were protected from direct competition because they were *ecologically* isolated. Whereas the earliest protostomes were burrowing worms, the primitive deuterostomes were upright filter-feeders. So while the annelid-arthropod-mollusk ancestors were crawling about in the muck, the echinoderm-chordate ancestors—our own forebears—were sitting on stalked rear ends sucking seawater. Who could have predicted which of the two lines would eventually produce eagles and astronauts?

The second hypothesis is that the earliest protostomes and deuterostomes were protected from direct competition because they were *geographically* isolated. You will recall from our discussions of continental drift (see Chapters 17 and 21) that the present-day land masses were produced by the breakup and drift of the supercontinent, Pangaea, which began about 230 million years ago. Studies of magnetic lines and forces in ancient lava beds indicate that continental drift also occurred at an even earlier time. Current thinking is that Pangaea itself

was formed through the union of two great land masses or supercontinents. The northern supercontinent, **Proto-Laurasia,** consisted of what is now North America, Siberia, and China; the southern supercontinent, **Proto-Gondwana,** included everything else. But what has this to do with protostomes and deuterostomes? Simply that some early fossil evidence suggests that the protostomes began in the shallow shelf waters of the southern supercontinent and that the deuterostomes radiated out from the shallow waters of the northern supercontinent. Thus, they were firmly established, diverse, and occupying different niches before the shifting land masses brought them the pleasure of each other's company. With these ideas in mind, let's look closely at our first deuterostome, one of the splendid phyla whose anus forms first.

Phylum Echinodermata: The Spiny-Skinned Animals

Your first look at **echinoderms** may led you to believe that you are on the wrong road to the vertebrates. Sea stars, brittle stars, urchins, sea cucumbers, and sand dollars (Figure 26.20) appear unlikely relatives to amphibians, fishes, mammals, and birds, not to mention humans. The comparative anatomy of echinoderms and vertebrates won't instill much confidence in you either. To see the relationship between the echinoderms and the vertebrates, we must look at their earliest embryonic development, when the first cleavages occur, and later at the developing mouth. Here the echinoderms are unveiled as deuterostomes (see Essay 26.1). Their embryos indicate that they are related, not directly to the vertebrates, but to their soft-bodied cousins, the hemichordates and urochordates, which in turn show clear relationships with the chordate phylum. We will review this relatedness in the next chapter.

Echinoderms are unusual in many respects. Their spiny, crustéd covering is actually an endoskeleton, even though it seems to be on the outside. Echinoderms are covered by an epidermis, beneath which lie the calcareous plates that are secreted by the dermis, which is of mesodermal origin. (Recall that the shells of mollusks and the exoskeletons of arthropods are both secreted by epidermal cells that stem from ectoderm.)

The body of adult echinoderms is essentially a pentaradial (five-part radial) construction, a scheme found in no other phylum. The larvae are bilateral, so the pentaradial condition turns out to be only a peculiar secondary state. Echinoderms show no obvious segmentation, indicating that they are not closely related to the annelid line.

One of the most unusual features of the echinoderm is the **water vascular system,** which is composed of a series of water-filled canals, which by hydrostatic pressure help extend their numerous muscular **tube feet.** In many echinoderms, each tube foot is equipped with a terminal sucker that is used in grasping. We will come back to this unique characteristic shortly.

Echinoderms are exclusively marine animals. Many species are found in the shallow waters of the continental shelf, particularly just below the reaches of the lowest tides. However, one group, the brittle stars, have been found in the deepest trenches of the ocean.

The phylum consists of five classes: the **Crinoidea** (sea lilies), **Holothuroidea** (sea cucumbers), **Echinoidea** (sea urchins and sand dollars), **Asteroidea** (sea stars), and **Ophiuroidea** (serpent stars and brittle stars). Each class differs considerably from the others, with perhaps the greatest departure being the soft-bodied, wormlike sea cucumbers, which have only scattered elements of an endoskeleton.

Class Asteroidea: The Sea Stars.   As a representative of the phylum, we will concentrate on one class of echinoderms—the sea stars (see Figure 26.20). Sea stars (or starfish, if you insist) are well known for their voracious appetite when it comes to gourmet foods such as oysters and clams. Obviously, they are the sworn enemy of oystermen, although these same oystermen may have inadvertently helped the spread of the sea stars. At one time, when they caught a starfish, they chopped it apart and vengefully kicked the pieces overboard. But they were unfamiliar with the regenerative powers of the starfish. The central disk merely grows new arms, and a single arm can form a new animal (Figure 26.21).

Since sea stars are slow-moving predators, their prey must be even slower-moving or immobile. Their ability to open an oyster shell is a testimony to their persistence. When an oyster or clam is discovered by a sea star, it clamps its shell together tightly, a tactic that discourages most would-be predators. But not the sea star—it bends its body over the oyster, attaches its tube feet to the shell, and then begins to pull. Finally, the oyster can no longer hold itself shut, and it slowly opens—only a tiny bit, but this is enough. The star then protrudes its thin-walled stomach out through its mouth. The soft stomach slips into the slightly opened shell, surrounds the oyster, and digests it in its own shell. Let's see what powers the tireless tube feet.

26.21 REGENERATION IN THE SEA STAR

Sea stars are endowed with considerable regenerative ability. Most species can regenerate an entirely new body from any portion of the central disk. (One species, *Linckia*, can even regenerate a new body, complete with disk and arms, from very small pieces of any arm.) While their regenerative powers are well known, the cellular details of their regeneration are not. It is generally believed that regeneration occurs through a dedifferentiation of cells and a return to the earlier embryological sequence.

26.22 WATER VASCULAR SYSTEM

The water vascular system of echinoderms consists of a supply system composed of a *sieve plate,* a number of *canals,* and numerous *tube feet.* Water is used to fill the *ampullae,* which in turn are used to extend the tube feet. This is accomplished by contraction of the ampullae, which fill the tube feet with water, greatly lengthening them. Each tube foot ends in a sucker tip. When several of the tube feet contact a surface, they can contract by muscular action, forcing water out and sucking at the surface (water is permitted to reenter the ampulla and canal system by a series of valves). In this way, the tube feet, working in series, pull the sea star along the sea bottom. The tube feet are also used in opening oysters.

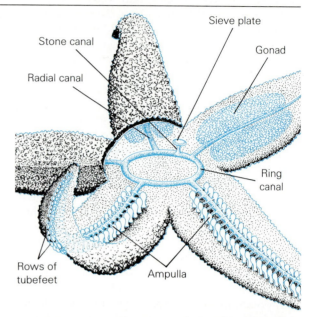

Tube feet are located in double rows in the grooved underside (the oral side, since the starfish mouth is on the bottom) of each arm (Figure 26.22). Each tube foot is attached to a short **radial canal,** which in turn connects with a pipelike **central canal** that extends along the arm. The central canals from each arm all connect to the hollow **ring canal** in the central disk. Water enters and leaves the water vascular system through a vertical tube, the **stone canal,** which has a sieved opening called a **madreporite** (mother pore) on the aboral (upper) side of the star.

Each tube foot contains longitudinal muscles that can contract, shortening the delicate structure. Above each foot is a rounded sac, an **ampulla,** which is similar to a squeeze-bulb. When muscles contract a water-filled ampulla, water is forced into

the foot, extending it hydrostatically. The sucker end of the foot attaches to a surface and the longitudinal muscles then contract, shortening the tube and causing suction on the undersurface. In this way, the animal gets its "footing." Thus, the tube feet, contracting in a highly coordinated fashion, can open an oyster or draw the animal over the rocky bottom of its habitat.

The sea star, like all echinoderms, lacks a centralized nervous system. The greatest concentration of nerves is in a ring around the mouth, with branches extending into the arms. Sensory receptors are limited in the sea star to one sensory tentacle and a light-sensitive region at the tip of each arm.

Echinoderms are either male or female, but it is hard to tell which is which. Sperm or eggs, produced in the well-defined testes or ovaries found in each arm, are shed into the water through tiny openings in the arms. Fertilization takes place in the water. As described earlier, the sea star, a deuterostome, undergoes radial cleavage in its early embryological development (as opposed to the spiral cleavage of protostomes). The embryo differentiates into a free-swimming **bipinnaria larva** that is distinctly different from the trochophore larvae of other invertebrates but similar to that of other deuterostomes (see Essay 26.1). Thus, whereas adult sea stars don't resemble vertebrates at all, echinoderm embryology suggests that our ancestors were probably rather closely related.

The echinoderms, represented by fossil crinoids, are found in Cambrian deposits along with fossil protostomes. These primitive echinoderms, called **eocrinoids,** were stalked animals somewhat different from their living descendants. Other classes of echinoderms appear first in the Ordovician, the next oldest period.

With the echinoderms, then, we have set the stage for our entrance into a more familiar world, that of backboned animals. However, the first few animals we will next encounter will not have backbones and will probably not be familiar.

APPLICATION OF IDEAS

1. A popular post-Darwinian concept of evolution placed humans at the pinnacle of a long, gradual progression of forms, all evolving toward the human line. (One such scheme placed 19th century English gentlemen at the very tip. Guess where this idea originated.) Other animals were represented as failures or cases of arrested progress. Present a strong argument against this point of view. Does evolution really lead anywhere? Have any of today's animals "given rise" to any other of today's animals? Are humans and other species still evolving?

2. Using the mollusks as examples, explain the evolutionary terms *primitive, specialized, generalized,* and *advanced.* Why are such terms commonly inappropriate when applied to the entire organism?

KEY WORDS AND IDEAS

PROTOSTOMES: THE INVERTEBRATE LINE

Two Ways of Producing a Gut
1. A major division in animal evolution produced the **protostomes** and **deuterostomes.** Each has bilateral symmetry, a one-way gut, and a true coelom, although the latter two features are probably analogous and not homologous. During **primitive gut** formation in protostome embryos, the **blastopore** forms the mouth, while in deuterostomes this is the anal area.

The Lophophorate Phyla
1. **Lophophorates** have characteristics of both protostomes and deuterostomes. Three phyla include **Ectoprocta** (bryozoans), **Brachiopoda,** and **Phoronida.** The **lophophore** is a group of ciliated tentacles attached to a curved ridge. Brachiopod and ectoproct fossils occur in Cambrian strata.

Segmented Body Plan
1. In the **segmented body plan,** or **metamerism,** the body is composed of repeated segments or **metameres.** Segmentation is also seen in vertebrates.

Phylum Annelida
1. In annelids, extreme segmentation and a hydrostatic skeleton may have evolved as an adaptation to burrowing.
2. In the **earthworm,** an **oligochaete,** each segment contains elements of most organ systems. Earthworms have a **closed circulatory system** with five pairs of pumping, **aortic arches** and blood that contains hemoglobin. Excretory and osmoregulatory structures called **nephridia** are greatly advanced over the protonephridium. Earthworms are hermaphroditic, each with **sperm receptacles.** Fertilized eggs are released into a

slime cocoon, secreted by the **clitellum.** The gut includes specialized areas for swallowing, storing, grinding, digesting, and absorbing. The nervous system includes a brain (enlarged ganglia) and a **ventral nerve cord** with branches in each segment. The body wall contains circularly and longitudinally arranged muscles.

3. **Hirudinea,** the **leeches** are **ectoparasites.** They attach themselves via toothed suckers to warm-blooded hosts. A secretion, **hirudin,** prevents clotting. Although segmentation is modified, its systems are similar to those of the earthworm.

4. **Polychaetes** are marine annelids. *Nereis,* the clamworm, has prominent jaws and fleshy, vascular, **parapods** that function in swimming and gas exchange. In marine worms, sexes are separate, and following fertilization, a swimming **trochophore larva** forms, similar to that of many other marine invertebrates. Many polychaetes are tube-dwellers, and the fossils of this group are the oldest traces of eukaryote life.

Phylum Mollusca

1. **Mollusks** are among the most numerous invertebrates, ranging in size from microscopic **bivalves** to the giant squid. All have a variation of a **foot, mantle, mantle cavity, shell,** and in most aquatic forms, a **gill.** The coelom is highly reduced, an **open circulatory system** is present in most, and many feed with a rasping **radula.** Nephridia are present.

2. The evolutionary relatedness of mollusks to other protostomes is in question, but they do produce a trochophore larva, and some fossil mollusks reveal segmetation and a well-developed coelom.

3. **Chitons** have a primitive body plan, with traces of segmentation in the eight shell plates and rows of simple gills.

4. Most **gastropods** are snails or snaillike, but some lack shells. While fully aquatic snails have gills, **pulmonate** snails use a vascularized mantle-cavity for lungs. Many snails undergo **torsion** (a 180° twist) as they grow. A horny **operculum** covers the retracted body.

5. The **bivalves,** or **pelycypods,** clamlike mollusks with paired, hinged shells, often have a powerful, retractable digging foot. Clams are filter-feeders, using cilia to draw water into the body through an **incurrent siphon,** whereupon food is removed and taken in by **labial palps,** gases are exchanged in the gills, and water leaves by an **excurrent siphon.**

6. **Cephalopods** differ considerably from other mollusks. They have a well-developed brain, image-forming eyes, a closed circulatory system, a muscular mantle, and a foot highly modified

into grasping tentacles. The octopus and squid move rapidly by jetting water from the mantle.

Phylum Arthropoda

1. Nearly 1 million **arthropod** species are known. Common characteristics include segmented body, jointed legs, ectodermally secreted **exoskeleton,** open circulatory system, and ventral nerve cord. They live in all major habitats and produce great numbers of genetically variable young, which often live on a different diet from adults.

Subphylum Chelicerata

1. **Chelicerata** are arthropods that lack antennas and jaws, and have six pairs of appendages. They have venom-injecting fangs known as **chelicerae.** The largest group are **arachnids,** the spiders, all of which are venomous carnivores. They exchange gases in book lungs, produce **fibroin** silk for **webs** and **cocoons** with **silk glands** and **spinnerets.**

Subphylum Mandibulata

1. **Mandibulates,** arthropods with jaws, have antennas and three or more pairs of walking legs.

2. **Crustaceans** are mainly aquatic, both marine and freshwater. Their body plans range from the highly segmented primitive condition to advanced states with little visible segmentation. They are mostly gill breathers, with a chitinous exoskeleton made rigid with calcium salts. They have **compound eyes,** consisting of multiple **ommatidia,** each with a **corneal lens** and **retinal cell.** Ecologically important crustaceans include the floating marine and freshwater **plankton** that are important to the aquatic food chains.

3. **Chilopods** and **diplopods** are represented respectively by the venomous, predatory, **centipede** and the herbivorous **millipede.**

4. The majority of animal species are **insects.** Their characteristics include a flexible chitinous exoskeleton, segmentation chiefly in the abdomen, three pairs of legs, one pair of antennas, and sometimes wings.

5. Other features include a three-part body with **head, thorax,** and **abdomen,** greatly modified legs, and complex genitalia with a specialized **ovipositor.** Gases are exchanged through a **tracheal system** composed of branched **tracheae** and **tracheoles** that sometimes end in **air sacs. Spiracles** control the external openings.

6. Sensory structures in grasshoppers include compound eyes and simple **ocelli.** Mouthparts include the **labrum** and **labium** (lips), chewing **mandibles, maxillae,** and **sensory palps.** Insect mouthparts vary greatly.

7. Some leg specializations include modifications for jumping, grasping, pollen collecting, and de-

tecting intruders by movement of air.

8. Arthropod fossils extend back to Precambrian times.

Phylum Onycophora

1. The **onycophorans,** typified by *Peripitus*, have the bidirectional body wall muscles and pronounced internal segmentation of the annelid, yet, like arthropods, they also have clawed appendages and antennas, and exchange gases through a tracheal system complete with spiracles. They may have diverged from ancient arthropods shortly after that group diverged from annelids.

DEUTEROSTOMES:
THE ECHINODERM-CHORDATE LINE

1. Protostomes tend to occur with great diversity and have small bodies, rapid reproduction, and limited learning. Although some deuterostomes are similar to protostomes, an opposite trend began with the chordates. Further, in deuterostomes, the exoskeleton is replaced by a mesodermally derived **endoskeleton.**

2. The divergence of protostomes and deuterostomes occurred in the late Precambrian. Two hypotheses seek to explain how such divergence occurred:
   a. In one, *ecological* isolation occurred as one group specialized in burrowing for food and the other formed stalks and filtered food from the sea.
   b. A second hypothesis proposes that isolation was *geographical*. Prior to the formation of Pangaea, there existed **Proto-Laurasia** and **Proto-Gondwana,** two supercontinents in which, respectively, the deuterostomes and protostomes arose.

Phylum Echinodermata:
Spiny-Skinned Animals

1. The phylogenetic link between **echinoderms** and vertebrates is seen in the embryo, where similarities in cleavage planes and mouth development are seen.

2. The spiny skin of the echinoderm is secreted by mesodermally derived tissue and is covered by an epidermis. Although adult echinoderms are pentaradial, their larval forms are bilateral. The **water vascular system** with its **tube feet** is an exclusive feature.

3. Five echinoderm classes include **Crinoidea** (sea-lilies), **Holothuroidea** (sea cucumbers), **Echinoidea** (sea urchins), **Asteroidea** (sea stars), and **Ophiuroidea** (serpent stars).

4. Asteroidea (sea stars), predators of bivalves, use their tube feet and water vascular system in moving about and in feeding. When they succeed in prying a bivalve open, their thin-walled stomach is everted between the shells to digest the prey. Tube feet occur in double rows within grooves in each arm. Each connects to a **radial canal,** then to a **central canal,** which in turn connects to the **ring canal.** The ring canal gives rise to a **stone canal** that opens to the outside through a sieved **madreporite.** Tube feet are extended by water pressure from an **ampulla,** whereupon the sucker end fastens to an object and minute muscle fibers shorten the foot.

5. The simple nervous system consists of a ring and branches into each arm.

6. Separate individuals produce eggs and sperm and fertilization is external. The swimming **bipinnaria larva** is unlike the protostome trochophore.

7. The earliest echinoderms were stalked Cambrian **eocrinoids.**

REVIEW QUESTIONS

1. List the main features of the eucolomate body and gut. (640)

2. List the major animal phyla represented by the protostomes and deuterostomes. (640)

3. Describe the difference in mouth formation in the embryos of protostomes and deuterostomes. (641)

4. What is the leading characteristic of the lopophorate? List the three phyla and name a member of each. (641)

5. Briefly explain metamerism and give an example of a highly metameric animal. (642)

6. Explain how a hydrostatic skeleton is used in movement. (642)

7. Describe the earthworm's circulatory system and blood. Would this be considered primitive or advanced? Explain. (642–643)

8. Use a simple drawing to explain the functioning of the earthworm's nephridia. (643)

9. What role does the earthworm's clitellum play in reproduction? (643)

10. List examples of specialization in the earthworm's gut. (643)

11. What evidence of cephalization is found in the earthworm? (646)

12. Explain how the earthworm makes use of its muscles and setae in burrowing. (646)

13. Describe the feeding specializations that occur in the leech. (646)

14. List three anatomical differences between the clamworm and the earthworm. (646–647)

15. In what phyla is the trochophore larva found? What is the phylogenetic significance of this? (647)

16. List the four molluskan structures that vary greatly from class to class. (647)

17. What two lines of evidence link the mollusks with the annelids and arthropods? How did the discovery of *Neopilina* help support this relationship? (647)

18. Why is the chiton body plan considered to be primitive? (647, 649)

19. Explain what torsion is and how it affects the final body plan of the gastropod mollusk. (649)

20. Describe functional specializations in the foot, mantle, and gills of the clam. (649)

21. Describe the anatomical and functional specializations seen in the foot and mantle of the cephalopod mollusk. (651)

22. List several ways in which octopuses and squid are well adapted to their roles as predators. (651)

23. Briefly describe the characteristics of arthropod appendages, skeleton, nervous system, and size. (651)

24. Suggest how the following may be significant to the success of arthropods: rate of reproduction, potential variability, life cycle with distinctly different stages, small size. (652)

25. List three characteristics of the chelicerates that are not present in the mandibulates. (652)

26. Briefly explain how spiders produce and design their webs. (653)

27. List the four major classes of mandibulates and cite an example from each. (653–654)

28. How does the crustacean exoskeleton differ from that of the insect or arachnid? (654)

29. Describe the structure of the compound eye and explain what it actually enables arthropods to see. (654)

30. List two vital ecological roles played by the marine crustaceans. (655)

31. What are some ways in which chilopods differ from diplopods? (655)

32. Name the three main body parts of insects. In which of these is the evidence of segmentation most obvious? (655)

33. List three major characteristics of insects. (655)

34. Compare the structures of gas exchange in spiders and terrestrial insects. (653, 655)

35. Describe the mouthparts of the grasshopper. For what are they specialized? (656)

36. Describe three specializations in the legs of insects. (656, 658)

37. List both the arthropod and annelid characteristics of the Oncophyora. Where do these strange animals seem to fit in invertebrate phylogeny? (659)

38. Briefly summarize an ecological hypothesis and a geographical hypothesis that explain how the protostome and deuterostome evolutionary lines became established without conflict on the early earth. (662)

39. List two embryological characteristics of echinoderms that suggest their evolutionary relationship to chordates. (663, 665)

40. List the elements of the water vascular system and briefly explain how the tube feet extend, fasten on, and shorten. What are two roles of this unique system? (663–664)

41. How does the spiny covering of the echinoderm qualify as an endoskeleton? (663)

Evolution and Diversity: Vertebrates

The hairy-nosed wombat cannot run very fast. Neither can its cousin, the naked-nose wombat. But coyotes can, and cheetahs can outrun coyotes. None of these animals jumps from the tops of trees, but flying squirrels and birds do. And a bird may go up instead of down. Both birds and chimpanzees build nests in trees, but only one gives milk and neither can hold its breath as long as a turtle. And most fish don't have to come up for air at all. Fish, however, do have certain things in common with turtles. In fact, all these animals have enough traits in common that they are placed in the same phylum: the **Chordata.** In addition they are all in the same subphylum: the **Vertebrata;** that is, they all have backbones.

The chordates are a fascinating and highly diverse group of animals that, while seemingly quite different, have certain fundamental traits in common. We will get to them directly but first let's have a look at an unusual, distantly related phylum that yields some interesting clues about chordate ancestors and the evolution of deuterostomes.

PHYLUM HEMICHORDATA

The **hemichordates** ("half-chordates") are represented by the acorn worm, a burrowing marine animal with a conical (acornlike) **proboscis** or nose, and just behind it a **collar** (Figure 27.1a). Note the rows of **gill slits** in the wall of the pharynx, behind the collar. These slits reveal the worm's affinity to the chordates, since chordates have gill slits at some time during their development—even if only as embryos. In the acorn worm, the apparatus acts as a kind of gill and helps in feeding as well.

While the gill slits link the hemichordates with the chordates, another seemingly insignificant trait links the acorn worms to the echinoderms—the close resemblance of their larvae (Figure 27.1b). In addition, the embryos of echinoderms, hemichordates, and chordates undergo radial cleavage (see Chapter 26). It seems that in the distant past, the ancestors of echinoderms, chordates, and hemichordates were rather closely related. This theoretical relationship is explained in Figure 27.2.

PHYLUM CHORDATA

The chordates are customarily divided into three subphyla: **Urochordata, Cephalochordata,** and **Vertebrata.** In addition to the gill slits just mentioned, chordates have **dorsal, hollow** (tubular) **nerve cords,** a peculiar structure called the **notochord,** and a **postanal tail.** In primitive chordates, the gill slits become sieves for feeding, and in the fishes, the structures between the slits called **arches** will later support the gills. In other vertebrates gill slits appear only in the embryo.

Our own spinal cord is a good example of a dorsal, hollow nerve cord. (Recall that the higher bilateral invertebrates in the last chapter had *ventral solid* nerve cords.) We will learn in Chapter 37 how the dorsal nerve cord is formed during embryological development.

Chordates also have, at least at some point in their development, a notochord. The notochord is a flexible, turgid rod running along the back that serves as kind of skeletal support. It consists of large cells, apparently under considerable hydrostatic pressure, encased in a tight covering of connective tissue. In nearly all living chordates the notochord exists only in the embryo or larva. In vertebrates it begins to dissipate as the backbone appears, early in embryonic development.

The final defining characteristic of chordates, the postanal tail, is present at least in some stage of development in all chordates. Whereas adults of one chordate group, the **tunicates,** generally do not have such a tail, it is found in their larva, just as it is in human embryos.

Subphylum Urochordata

Tunicates, also known as **ascidians** or "sea squirts," are the most common urochordates. The group also includes the bizarre, transparent, free-swimming **salps** and a minor but highly significant group, the **Larvacea.** The larvaceans are planktonic forms that live in tiny, jellylike shells. The name *tunicate* refers to the transparent, tough covering, or **tunic,** on the outside of the soft, saclike body.

Tunicates are revealed as chordates by their tadpolelike larvae, which bear a notochord, a dorsal, hollow nerve cord (which enlarges anteriorly), gill slits, and a postanal tail (Figure 27.3). In its transition to the adult form, the tunicate loses its obvious chordate characteristics. Its tail, notochord, and nerve cord are absorbed into the body, leaving only the enlarging **gill sac** (or gill basket) as a clue to its chordate relationship. Tunicates are filter-feeders, using their gill clefts as strainers. The adult circulatory system is open, consisting of little more than a bizarre heart that pumps blood first one way, then the other. To further confuse the biologist, their tunic contains cellulose. Cellulose, a common cell wall material in plants, is rarely found in animals.

Though the tunicate larva changes to a saclike adult form, the larvacean retains the juvenile tadpole body plan, with its gill slits, tail, and notochord. Thus, except for sexual maturity, it never reaches what is the adult stage seen in other urochordates. The retention of the juvenile form in reproductively mature animals is called **neoteny;** to evolutionary theorists, this odd phenomenon suggests how the main chordate line might have originated (see Figure 27.2).

27.1 A HEMICHORDATE

(a) The acorn worm has a large number of pharyngeal *gill slits,* a chordate trait. It burrows in mud flats, using its proboscis somewhat in the manner of an earthworm. As it tunnels along, water and organic debris enter the mouth, and the nutrients in the debris are sorted and digested by the gut, while the water passes out through the gill slits. The hemichordate affiliation with the echinoderms may be difficult to imagine, but it becomes very clear when the larvae are compared.

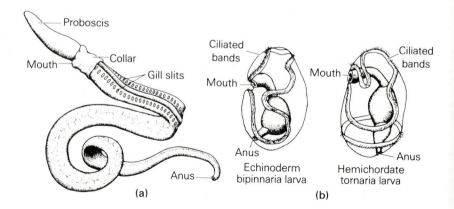

27.2 DEUTEROSTOME RELATIONSHIPS

According to one theory, echinoderms, hemichordates, primitive chordates, and vertebrates arose from a sessile deuterostomate. Like so many of today's marine invertebrates that are fixed in place, these simple animals produced a swimming larva. From these early creatures arose the echinoderms and hemichordates, both of which still retain the free-swimming larva. Following these divergences, the deuterostome line divided once more, but in a most unusual manner. One branch produced the urochordate line, today's tunicates and salps, whose members persist today. The other arose from certain larvae that had undergone neoteny: they had somehow retained their juvenile body form yet reached sexual maturity. From this evolutionary experiment arose the cephalochordates and the vertebrates.

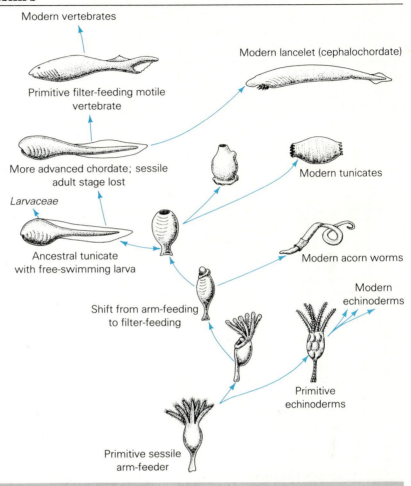

Modern vertebrates

Primitive filter-feeding motile vertebrate

More advanced chordate; sessile adult stage lost

Larvaceae

Ancestral tunicate with free-swimming larva

Modern lancelet (cephalochordate)

Modern tunicates

Modern acorn worms

Shift from arm-feeding to filter-feeding

Modern echinoderms

Primitive echinoderms

Primitive sessile arm-feeder

27.3 THE TUNICATE AND ITS LARVA

The adult tunicate is hardly what we expect in a chordate. The adults commonly live in a colonial fashion on rocks and wharf pilings and are often found encrusted on boat hulls. The larval form of the tunicate places the animal securely in the chordate phylum and reminds us of the need for studying the entire life cycle of an organism when establishing phylogenetic relationships. As a larva, the tunicate has the pharyngeal gill slits, a lengthy notochord, and a dorsal, hollow nerve cord—all hallmarks of the chordates. As you can see, the body form of the adult is rather simple and has no specialized sensory devices, although it maintains the pharyngeal gill slits throughout adult life. Currents of water are produced by cilia in the gill structure drawing food particles into the digestive tract. Respiration occurs as water exits through the body wall.

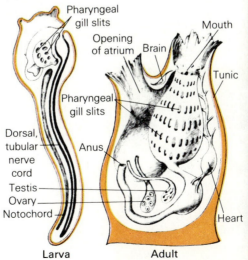

Pharyngeal gill slits
Opening of atrium
Brain
Mouth
Tunic
Pharyngeal gill slits
Dorsal, tubular nerve cord
Anus
Testis
Ovary
Notochord
Heart

Larva   Adult

According to the theory, all cephalochordates and vertebrates can trace their roots to the fortunate evolutionary accident that led to the arrival of these perpetual juveniles. From this primitive line arose species that at first were awkward, slow, jawless creatures, sucking up nutrients from the mucky sea bottom sediments. But in 500 million years, evolution has time to exert its powerful effects, and from these humble beginnings arose a species that would someday change the very face of the earth. We will come to this species shortly, but first let's consider a subphylum whose members more closely resemble their Precambrian ancestors.

27.4 THE LANCELET

This lancelet is *Branchiostoma*, an inhabitant of sandy regions of the shore in most tropical and subtropical parts of the earth. The adult body is clearly similar to the vertebrate body plan, and its appearance is fishlike. The lancelet is a burrower without eyes. The notochord runs the length of the body and functions as an anchor for the paired rows of *myotomes* (muscle segments). In addition, there is a dorsal, hollow nerve cord and pharyngeal gill slits. Its filter-feeding apparatus is similar to that of the tunicate and is enclosed in an atrium. The gill basket strains food particles from water that is drawn into the mouth, passes through the gill slits, and finally exits through the posterior atrial opening. The anus opens separately, ahead of the tail (another chordate characteristic). In some ways, the lancelet resembles a tunicate larva; in other ways, a fish.

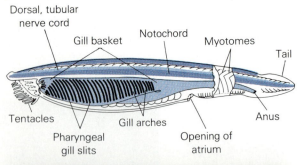

Subphylum Cephalochordata

You may recall that cephalochordate fossils, similar to today's species, have been found in the Burgess Shale of Precambrian origin, so they are a truly ancient group. The cephalochordates, or **lancelets,** as they are commonly known, are nonetheless members of our own phylum. Whereas they look like fish (Figure 27.4), the notochord is prominent in adults. In fact, the notochord protrudes beyond the brain. (This is how the creature got its name.) The relationship between cephalochordates and vertebrates is best seen by comparing the lancelet to the larval lamprey, a primitive vertebrate (Figure 27.5). The adult cephalochordates retain the pharyngeal gill slits and develop a dorsal, hollow nerve cord. Another prominent feature is the body musculature, composed of repeating units called **myotomes** (an example of segmentation, a trait found in most fish and all vertebrate embryos), which aid in both swimming and burrowing. Lancelets are filter-feeders. Beating cilia draw water into a complex mouth whose tentacles separate out food particles to be taken in for digestion. Water taken into the mouth passes through the gills where oxygen and carbon dioxide are exchanged, enters an outer chamber, the **atrium,** and exits through an opening near the tail.

SUBPHYLUM VERTEBRATA: ANIMALS WITH BACKBONES

Most **vertebrates** have several traits in common in addition to the standard chordate characters. They include a vertebral column (backbone), cranial brain development, a ventrally placed heart and dorsal aorta, gills or lungs for gas exchange, a maximum of two pairs of limbs, one pair of eyes, paired kidneys, and separate sexes.

There are seven classes of living vertebrates. The first three are fishes, while the rest includes the familiar amphibians, reptiles, birds, and mammals. We will also consider one extinct class of fishes.

Class Agnatha: Jawless Fishes

The oldest well-defined vertebrate fossils on earth are a group of **agnathans** called the **ostracoderms.** These were heavy, armored, jawless fishes (Figure 27.6). Their fossils date back half a billion years, occurring in the sediments of the Ordovician, Silu-

27.5 AN AMMOCOETE LARVA

The ammocoete larva of the lamprey bears many similarities to the lancelet (and to the tunicate larva).

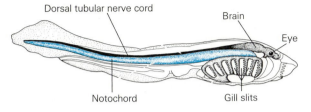

27.6 THE EARLIEST FISHES

The ostracoderms were jawless fishes of the Ordovician period. *Hemicyclapis* was large; its head contained the gill region and was enclosed in a hard bony shield with a single olfactory opening (nostril) on top. The mouth was a fleshy, sucking device. The heavy head may have been used to plow through the muddy bottom, where soft-bodied worms and mollusks could be found and sucked into the mouth. The *Anglapsis* was more typically fishlike and was probably more adapted to swimming than bottom dwelling.

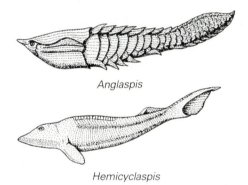

Anglaspis

Hemicyclaspis

rian, and Devonian periods (see Table 25.2 p. 619). Their fin structure, which does not include the two pairs of fins seen in later fishes, tells us that they were sluggish swimmers. They probably depended on their armor and bottom-dwelling habits for protection against the large invertebrate predators of the day. Without jaws, the ostracoderms were probably not predacious but more likely strained their food from the water.

Today's agnathans, boneless remnants of the armored fishes (which died out at the end of the Devonian period), remain jawless. The lampreys and hagfish are members of the subclass **Cyclostomata** ("round mouths"). Their bodies are long and cylindrical, with simple median fins adapted for wriggling along the ocean bottom. They lack the ventral fins of other fishes, their skeleton is cartilaginous, and, oddly, their notochord persists throughout life. Parasitic lampreys feed with the aid of a rounded sucker mouth armed with horny spikes and a rasping tongue (Figure 27.7). The hagfishes, or slime hags, as some species are called, lack the sucker mouth and feed by a rasping device, boring into the bodies of dead or dying animals. By some standards they aren't a particularly admirable group.

The lamprey life cycle includes a filter-feeding larva that resembles the adult lancelet (see Figure 27.5). During the very long larval period, the animal inhabits fresh water, but the adult form of most species normally lives at sea until it is time to spawn. The sea lamprey has shown its physiological adaptability by surviving in the fresh waters of the Great Lakes, where it now completes it life cycle and is a very successful parasite.

27.7 THE CYCLOSTOMES

Today's jawless fishes consist of two groups, the lampreys and the hagfish. Both lack a bony skeleton, and a notochord persists in the adult along with a rudimentary cartilaginous skeleton. **(a)** The parasitic lamprey uses rasping and sucking mouthparts in feeding, fastening to a fish and drawing in its body fluids. This is only possible because the lamprey has a separate water inlet for ventilating the gills. The hagfish **(b)** is a bottom scavenger known to attack weakened or injured fish.

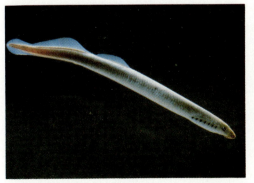

(a)

(b)

Class Placodermi: Extinct Jawed Fishes

The **placoderms** are believed to have descended from one line of ostracoderms, but the evidence is poor. They originated back in the Silurian period, their numbers swelling during the Devonian and dying some 150 million years later. Clearly, they must be considered a successful group in terms of their tenure on earth.

Placoderms were armored, with hinged jaws. There is little doubt that the evolution of moveable, biting jaws was one of the most significant events in vertebrate history, quickly permitting the establishment of many new predatory feeding niches and the broadening of others. (It also gave vertebrates a way to make life miserable for each other.)

The placoderm jaw is known to have derived from the gill arches (the bony structure that supports the gills). This, of course, required a considerable modification in both the position and strength of the gill arches, as well as the pharynx. Today, the upper and lower jaws of most fish form a hingelike structure that is only loosely attached to the cranium (Figure 27.8). The hinge improved jaw mobility and enabled the primitive vertebrates to become predators, a role formerly dominated by invertebrates. One of the more fascinating species was the gigantic *Dunkleosteus* (Figure 27.9), about the size of a modern gray whale and possibly one of the most fearsome predators that ever lived.

From the placoderm line arose the two large classes of jawed fishes we see today—the cartilaginous fishes, (class Chondrichthyes) and the bony fishes (class Osteichthyes).

Class Chondrichthyes: Sharks and Rays

Class **Chondrichthyes** includes the cartilaginous fishes, a fascinating group of predators and scavengers made up mainly of sharks and rays. Many of these modern predators have replaced the protective armor and the heavy skeleton of their ancestors with a tough skin, light frame, and great speed. Their origin remains a puzzle, but their fossils first appear in the early Devonian period. By the Mesozoic era, such fossils dwindled as the numbers of bony fishes burgeoned. But beginning in the Jurassic period, they again increased gradually, and since then they have held their own.

The cartilaginous skeleton of sharks and rays was once taken as evidence of their primitive nature. Evolutionists more recently have established that the cartilaginous skeleton of sharks is a *derived* condition—that is, the cartilaginous skeletons were developed from bony skeletons. Some biologists believe the evolution of a lightweight cartilaginous skeleton is an adaptation to deep-water life. In support, they note that sharks and rays lack the hollow, buoyant swim bladder found in bony fishes, although some buoyancy is provided by a large store of lightweight lipids. More recently, scientists have found exciting new evidence of residual bony tissue in the spinal column of sharks.

In addition to their cartilaginous skeleton, sharks and rays are unique in several other respects. For example, their body and tail shape is unlike that of most bony fishes (Figure 27.10). The shark's tail is asymmetrical, more like that of the extinct placoderms than those of modern bony fish-

---

27.8 GILL ARCHES TO JAWS

The jaws of today's vertebrates are believed to have evolved from the primitive gill arches. In this hypothetical series, the events unfold in a sequence from the ancient jawless fishes to a modern shark. **(a)** In ostracoderms, the gill arches are all similar and unspecialized. **(b)** In the placoderms, the first gill arches have been modified into the primitive upper and lower jaws, accompanied by related muscle development. **(c)** In a modern shark, the gill arches are greatly modified into strong, hinged upper and lower jaws. One gill slit has become the spiracle, a small, somewhat dorsal opening. In terrestrial vertebrates, this gill slit becomes the middle-ear passage, and the bones of the second and third arch become the bones of the inner ear (not shown).

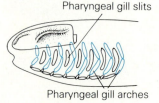

Pharyngeal gill slits

Pharyngeal gill arches

(a) Jawless ostracoderm
(unspecialized gill arches)

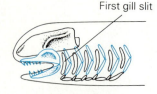

First gill slit

(b) Primitive jaw gill arches
(modified into weak jaws)

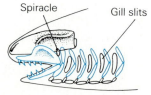

Spiracle

Gill slits

(c) Jaw of shark (gill arches modified into strong jaws and their supporting elements)

27.9 THE JAWED FISHES

The placoderms were fierce predators. Note the teeth and hinged jaws of *Dunkleosteus*. Important advances in fin development undoubtedly helped with the predatory mode of life. The placoderms were armored, some with massive plates surrounding the head and other portions of the body. *Dunkleosteus* was 10 m long.

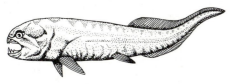

Dunkleosteus

es. The body, particularly at the anterior, is dorso-ventrally flattened, rather than laterally as in typical bony fishes.

The skin of sharks and rays is very rough, embedded with miniature "teeth" known as **placoid scales** (see Fig. 27.10). Each is anchored into the dermis by a basal plate. Embryologically, these scales form the same way as teeth, and in fact, the shark's teeth are simply larger versions of the same structure. The teeth themselves develop continuously throughout life, with newly formed teeth migrating forward from rows inside the mouth to replace older or broken teeth. The rays and several kinds of sharks use heavy, flattened "paved" teeth for feeding on shellfish. Surprising to some, the largest sharks, the whale shark and the basking shark, are harmless filter-feeders that eat small zooplankton they strain from the sea.

A curious feature in the shark's short digestive system is the **spiral valve** (also seen in lampreys and some bony fishes), which consists of a spiralling flap of absorptive tissue that greatly increases the surface area of the intestine (see Figure 29.8). The spiral valve slows the movement of large chunks of food, permitting more time for the powerful digestive enzymes to work. The digestive system ends in a **cloaca,** a posterior chamber that also receives ducts from the reproductive and excretory systems and ends with the **vent.** Thus the cloaca not only receives feces and urine, but is also a sperm receptacle and a birth canal. The cloaca is found in all vertebrate classes, but exists only in the embryo in most mammals (see Chapter 37).

Sharks have keen senses and locate prey by smell, by sight, and by certain patterns of vibrations (distress movements) that are picked up by the **lateral line organ.** The lateral line organ, characteristic of all fish, is formed of canals running the length of the body, containing groups of sensory cells that

respond to water movements. Other sensory canals on the head are sensitive to bioelectrical disturbances or weak electrical fields created by nearby animals, although little is known of the receptor mechanism.

Contrary to the tales of some scuba divers, the toothsome, gaping grin on the mouth of an approaching shark is not necessarily anticipatory. It is generally accepted that by constantly swimming with its mouth open, the shark is simply avoiding suffocation. This assures a continuous flow of oxygen-laden water into their mouths, over their gills, and out through the gill slits.

Fertilization in sharks and rays is internal but embryo development varies, from simple, primitive, egg-laying in some species to a more advanced regime of protecting the independent egg and embryo in the uterus until hatching in others. A few shark species are quite reproductively advanced, retaining the egg and embryo in the uterus and pro-

27.10 SPECIAL FEATURES OF SHARKS

Sharks are often described as veritable "eating machines" because of their ravenous habits when hungry. There is no doubt as to their success as ocean predators, and their bodies are well adapted to this niche. **(a)** The body shape is an excellent lesson in streamlining. Sensory structures include an extensive lateral line organ that detects nearby movement, along with keen olfactory and visual senses. **(b)** The skin is tough and flexible, consisting of placoid scales, each a miniature toothlike structure. Within the intestine, a flaplike spiral valve provides additional surface area and slows the movement of great chunks of food the shark swallows whole.

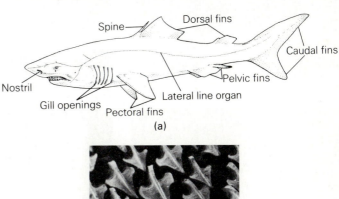

(a)

(b)

viding nourishment, removing wastes, and exchanging gases, just as mammals do. In some shark species, although embryos are produced in a steady assembly line, only the oldest survive. They feed by devouring their younger brothers and sisters in the uterus as fast as they appear.

Class Osteichthyes: The Bony Fishes

Like many modern animal groups, class **Osteichthyes,** the bony fishes, have developed just about every conceivable aquatic niche. They are a very diverse group (Figure 27.11), with well over 20,000 known species. It should be made clear that bony fishes are not so closely related to sharks as a first glance might indicate. There are a number of important differences in these animals. For example, the fins of many bony fish are much more refined. Sharks are great swimmers, but they can't stop short or make quick darting turns. The body surface of the bony fish is well adapted for swimming. The specialized, scaly skin has numerous mucous glands that produce the familiar slimy covering that can reduce drag, or water friction, by nearly 70%.

As we mentioned, bony fish have a swim bladder, a hydrostatic organ that improves balance and permits the animal to remain stationary at varying depths. In its most advanced form, the swim bladder has two specialized regions, a **gas gland,** which secretes gas into the bladder, and a **reabsorptive area,** which removes gas. In a more primitive state, the swim bladder wall is not so specialized in the release and uptake of gases, but is connected by a duct to the pharynx, as in the air-gulping lungfishes. In what is believed to be its most ancient state—in the primitive gar pike, for instance—the bladder has the spongy nature of a lung, suggesting that today's swim bladder may once have been a breathing device.

27.11 CLASS OSTEICHTHYES

Bony fish come in a great diversity of shapes. Most of the structural differences are in fin and body shape, but many of the rules are seemingly broken by such odd species as the lionfish, and even more dramatically by the creatures of the abysses such as the angler fish.

Bay blenny

Butterfly fish

Blue tang

Psychedelic fish

Angler fish

Lionfish

27.12 A FAMILY TREE OF FISHES

Included are three living classes (Agnatha, Chondrichthyes, and Osteichthyes) and one extinct class (Placodermi). The Osteichthyes, or bony fish, are divided into three groups: ray-finned fishes, lungfishes, and lobe-finned fishes. Most living fish belong to one of the three orders of ray-finned fish. Three living genera of lungfish remain, but there is only one species of lobe-finned fish—the coelacanth *Latimeria*. The lobe-finned fishes gave rise to the terrestrial vertebrates.

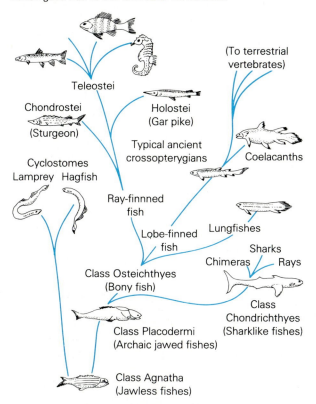

Many bony fish actively draw water into the mouth and pump it over the gills. The gill structures include five rows of gills on each side, located in a common **gill chamber** and protected by the bony **operculum,** a moveable external flap. Each gill consists of a supporting **gill arch** with toothed **gill rakers** (that keep food out of the gills) and rows of **gill filaments.** Gas exchange occurs in the gill filaments, which are extremely thin-walled with a rich supply of fine blood vessels just beneath the surface. Respiration is closely supported by the circulatory system, which includes a two-chambered heart. In its circuit through the body, blood is pumped from the heart by the single, muscular **ventricle** (larger pumping chamber) directly to the gills, where it exchanges carbon dioxide for oxygen before being circulated to the body and returned to the heart (see Chapters 30 and 31).

Reproduction is as varied in bony fishes as it is in the cartilaginous fishes. Most are egg-layers that employ external fertilization. Some species release enormous numbers of eggs freely into the environment, while others release far fewer eggs into sheltered places or nests they have built. Others do not lay eggs but maintain the embryo within the female body and bear relatively developed young.

Bony fish are believed to have evolved from some kind of air-breathing, freshwater ancestor that had already developed lungs. From this distant stock, the bony fishes split into three groups: the **Actinopterygii** (ray-finned fish), from which most of today's bony fishes—those with thin, flattened fins with radiating supportive spines—evolved, **Dipneustei** (lungfishes), and **Crossopterygii** (lobe-finned fishes), those with flattened fins attached to fleshy lobes (Figure 27.12). While ray-finned fishes make up the vast majority of species, only three genera today represent the lungfish. They live in swamps in Australia, Africa, and South America (Figure 27.13). These strange fishes have highly vascularized lungs that connect to the pharynx, which unlike that of any other living fish, is con-

27.13 LUNGFISH DISTRIBUTION

(a) The presence of lungfish on three continents long troubled evolutionists. What was once believed to be a triple case of convergent evolution can now be explained in a simpler manner. The ancestral lungfishes evolved and were widespread before the continents drifted apart. The lungfishes are presently found in tropical regions where seasonal droughts are common. **(b)** All have the ability to withstand long periods of drought by breathing air or by closing themselves up in burrows, where they become dormant through the dry season.

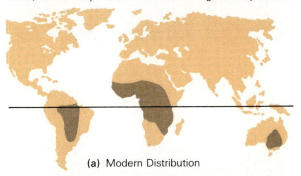

(a) Modern Distribution

(b) *Neoceratodus*

27.14 A LIVING FOSSIL FISH

The coelacanth, *Latimeria*, is not a fossil reconstruction, but a recently discovered specimen. The group to which it belongs, long believed extinct, is now known to exist off the Comoro Islands, which lie between the northern tip of Madagascar and the East African coast. Since their discovery in 1939, a number of *Latimeria* have been caught and subjected to intense study. Note the fleshy, lobed-fin bases. Interestingly, the adults retain the notochord throughout life and have vestigial, fat-filled lungs.

27.15 THE EARLIEST AMPHIBIANS

The amphibian line is believed to have arisen from certain lobe-finned fishes. These are then somehow connected to the labyrinthodonts. The labyrinthodont shown here *(Diplovertebron)* may have been the primitive amphibian from which both reptiles and amphibians emerged, although the lineage remains obscure.

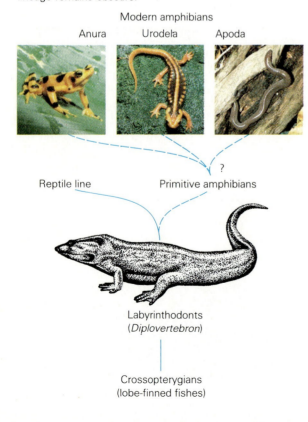

Modern amphibians

Anura   Urodela   Apoda

Reptile line   Primitive amphibians

?

Labyrinthodonts
(*Diplovertebron*)

Crossopterygians
(lobe-finned fishes)

nected by openings to the nostrils. Lungfish survive in highly stagnated water by filling their lungs with oxygen gulped from the surface. Some species can even survive periods of pond drying by lying dormant in the mud.

Lobe-finned species were always few in number, but they have had enormous evolutionary influence. They gave rise to the amphibians—the first animal land invaders—and from these arose the reptiles, and from these, the birds and mammals. Today, The Crossopterygii include just one species—the rare, deep-sea **coelacanth** (*Latimeria chalumnae),* the sole survivor of a once rather large and diverse group that was long believed to be extinct (Figure 27.14). Latimerians themselves are not ancestral to anything alive today, and do not even belong to the specific crossopterygian group from which land vertebrates evolved. *Latimeria* is a distant relative indeed, but it is nonetheless more closely related to us than is any other living fish.

Class Amphibia: The Amphibians

Amphibians today are represented by three orders: **Urodela** (the salamanders), **Anura** (frogs and toads), and **Apoda** (the wormlike caecilians). Nearly all amphibians reproduce and develop in aquatic habitats, though some are well adapted to drier environments and a few even live in deserts. Both modern amphibians and modern reptiles are believed to have evolved from members of one ancient terrestrial group, the **labyrinthodonts** (Fig-

27.16 FISH TO LAND VERTEBRATES

The legs of amphibians evolved from the bony, muscular fins of their crossopterygian ancestors. The lobe-finned fish probably began adapting to land by crawling from pond to pond or possibly by climbing out of the water to avoid aquatic predators.

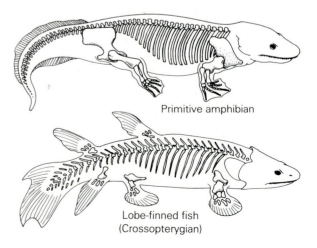

Primitive amphibian

Lobe-finned fish
(Crossopterygian)

27.17 VERTEBRATE LIMBS AND POSTURE

The adaptation to terrestrial life is seen as a transition in the positioning of limbs in the tetrapods. **(a)** Amphibians like the salamander (shown here) and newt have their thin, lightly muscled legs splayed out to the side, where the body weight is borne on flexed joints. **(b)** Reptiles show a trend toward placing the legs alongside the body, but the limbs remain flexed as they carry the body weight. **(c)** Mammals, the fastest moving land creatures, have their bodies raised above the ground with the limbs essentially below them. The positioning of the limbs is clearly seen in the comparison of heavyweights such as crocodiles and rhinos.

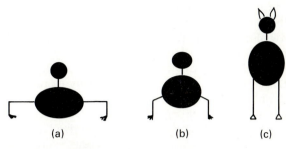

(a)   (b)   (c)

(a)   (b)   (c)

ure 27.15), whose fossils are found in Carboniferous rocks.

Note the differences in the skeletons of a lobe-finned fish and a fossil amphibian in Figure 27.16. Among the skeletal modifications necessary for terrestrial life is increased development and size in the vertebral columns and the shoulder and hip girdles. In the evolution of other terrestrial vertebrates, particularly those to whom speed was important, the legs gradually came to be located alongside or under the body and projected downward rather than sideways. Locomotion came to depend on the muscles of the limbs and limb girdles, rather than on lateral undulations of the entire body (Figure 27.17).

Amphibians have three-chambered hearts and go the fish one better by having a blood circuit through the lungs that is separated from the rest of the body. This arrangement permits oxygenated blood to be returned to the heart for a second pump before entering the body circuit. The separation is imperfect, however, and some mixing of oxygenated and deoxygenated blood occurs in the single ventricle (see Chapter 31).

Amphibians have moist, highly vascularized skin, which, along with the lungs, is an important organ of respiratory exchange. (Amphibian lungs have little surface area since they are essentially vascularized sacs and not spongy like those of other

vertebrates.) A moist skin is essential for gas exchange, but if gases can cross the skin, so can water. Thus, the amphibian is always at risk of dehydration, and accordingly, many live a semi-aquatic life. There are important exceptions, though. The desert spadefoot toad, for example, survives its desert habitat by spending long dry spells in a fully dormant state, buried in the soil.

Like the desert spadefoot, most other toads and frogs fertilize the eggs externally, in water, but fertilization in caecilians and most salamanders is internal. Male salamanders release their sperm in small gelatinous packets called **spermatophores.** The females crawl over the packets and squat, picking them up in their cloacas, where fertilization occurs before the eggs are laid. Other caecilians actually copulate, the male introducing sperm directly into the female's cloaca.

Some amphibians have managed to escape a water-bound existence. *Pipa pipa*, the Surinam toad of South America, carries the fertilized eggs in moist pouches on its back, assuring the young a relatively safe developmental period. In *Rhinoderma*, a small Chilean frog, the male scoops up the eggs in his mouth, where the embryos then complete their development. But in spite of such evolutionary adaptations to drier environments, amphibians remain a semiaquatic group. The first vertebrates to live a truly terrestrial life were to be the reptiles.

Class Reptilia: The Reptiles

Eventually, backboned animals conceived, lived, and died entirely on the dry earth—and these were the scaly creatures called **reptiles.** The new land dwellers continually adapted to the dry environment and, in so doing, created a wide variety of specializations. Changes in their sexual reproduction and development were especially significant. Reptiles retained the cloaca, but males developed a penis for efficient copulation and internal fertilization, a prerequisite for full-time land dwelling.

A highly significant change was the evolution of the **amniote egg,** in which the embryo is surrounded by a fluid-filled, membranous sac known as an **amnion** (characteristic of reptiles, birds, and mammals). The amniote egg of the reptile is a porous, leathery case, containing a food supply, vascularized membranes that transport food and oxygen into the embryo and remove wastes, and sufficient fluid to cushion it from injury and protect it from drying. These delicate, extensive **extraembryonic membranes,** as they are known, are produced by the embryo and grow along with it.

Adult reptiles avoid water loss in a number of ways. Like their embryos and like the birds and terrestrial insects, they produce a crystalline nitrogen waste that doesn't require much water for elimination. Water loss is minimized by a dry skin covered with protective scales and few mucus-secreting glands. The dry, scaly skin lacks elasticity and must be shed periodically as the animal grows.

Reptile lungs, unlike the simple saclike lungs of amphibians, are composed of many smaller compartments and thus have a spongy texture and a greatly increased surface area. Like that of the amphibian, the heart in most reptiles is three-chambered, but a partial septum divides the ventricles. In crocodiles and alligators, the ventricular septum is complete, forming a four-chambered heart. Reptiles are **ectothermic** (commonly known as "cold-blooded")—that is, their body temperature is not regulated and thus changes with surrounding temperatures. Those in temperate climates, however, have evolved a number of simple behavioral strategies for escaping both heat and cold (see Chapter 32).

Most of the reptiles are carnivorous, except for a few herbivorous lizards, tortoises, and turtles. Because their metabolism is slower, reptiles require considerably less food than do birds and mammals of the same size. Reptiles locate whatever food they do eat by a variety of means, including vision, heat detection, olfaction (smell), and hearing, and some use venom to subdue prey.

Reptile History.   The earth doesn't harbor as many reptiles now as it once did. In fact, the Mesozoic era, which lasted over 160 million years, is known as the "Age of the Reptiles." However, the reptiles actually originated much earlier, branching off from the amphibian line in the Carboniferous period of the Paleozoic era. They peaked during the Jurassic and dwindled down to the comparatively few species of today (see Table 25.2).

From the earliest reptiles, called the **cotylosaurs** or "stem-reptiles," there emerged a fascinating array of species. The land, for a time, was theirs. There were both giants and dwarfs. They climbed, hunted, hid, lurked, and threatened. As their numbers increased, competition and predation probably led some to return to the water from which they came. Others protected themselves with armor or sheer speed, striding around on their long hind legs. At least two lines of flying reptiles developed, one of which led to modern birds. The other, a much earlier and immediately successful group, were the **pterosaurs,** the first flying vertebrates (Figure 27.18). They were persistent, surviving from the Jurassic period to the end of the Cretaceous, but like so many reptile lines, they are not ancestral to anything alive today. Their bodies commonly reached about 1 meter in length. Some recently discovered fossils contained a surprise: *pterosaurs, at least the smaller ones, were furry*, indicating that reptiles, like mammals and birds, had made inroads into **endothermy**—maintenance of a constant body temperature.

By the start of the Cretaceous period, some 135

27.18 FLYING REPTILES

The pterosaurs represented an experiment in flight. Their fossils have left us with a host of questions. Some had a wingspan of 7 to 10 m, yet the wing was merely membranous skin, attached to an enormously elongated digit (finger). There doesn't appear to be much evidence of large, strong wing muscles. Most paleontologists believe that they did more soaring and gliding then active flying. The problem is in determining how they became airborne. The legs seem too fragile for them to have made a running start. One idea is that they climbed trees or cliffs and then jumped. Most pterosaur fossils are in marine strata, indicating that the animals may have hunted for fish from cliffs along the seashore.

27.19 RULING REPTILES

The "ruling reptiles" of the Mesozoic included some of the largest animals ever to rove the land. Contemporaries during the Cretaceous, the closing period, included the legendary *Tyrannosaurus rex* and the even larger though gentler herbivore, *Diplodocus. Triceratops,* another herbivore, was smaller than these giants but heavily armored. Great size was even a factor in the flying reptiles such as the pterosaurs, where wingspans in some species reached 10 m.

million years ago, large was "in." This was the age of the "ruling reptiles," the great dramatic beasts known as **dinosaurs** (Figure 27.19). Even the pterosaurs produced a giant—*Pteranodon,* the size of a small airplane. The capstone of these great beasts was *Tyrannosaurus* ("tyrant lizard"), probably the largest terrestrial carnivore ever. But even larger dinosaurs stalked the earth. Great lumbering herds of plant-eating beasts devastated the taller trees, stripping them of their foliage. One called *Diplodocus,* holds the record of 30 meters long and weighing in at 30 metric tons.

The reptiles were hugely successful, as witnessed by their prominence on earth for so many millions of years. Why did they die off? Their demise came with the closing events of the Cretaceous period, and no one is sure what brought it about. We do know, however, that the delicate webs of life are highly vulnerable to change, and the loss of one member affects all. Perhaps a change in the climate started the process. We know, for example, that their habitat was cooling at about this time. Some scientists blame the emergence of the early mammals, which might have preyed on the eggs of the giant reptiles. An exciting and well-substantiated theory has to do with the vast effects brought about by a giant asteroid that collided with the earth at the close of Mesozoic. This idea, the Alvarez hypothesis, is discussed in Chapter 22 and in Essay

27.1. Another recent theory suggests that the sun has a companion star, named *Nemesis,* that, in cycles of about 26 million years, causes a rain of comets that destroys most life on earth. The geologic record roughly suggests such a cycle but the star has not been found.

Modern reptiles probably descended from four separate branches of the stem-reptiles and today form four orders (Figure 27.20). The turtles and sidenecked turtles form one line of descent, the alligators and crocodiles another, the lone tuatara constitutes a third order, and a fourth includes the lizards and snakes. Snakes are believed to have evolved away from one branch of the lizard line about the start of the Jurassic, midway through the reptile era. The four lines survived the rigors of the Cenozoic era and are represented today by 6000 to 7000 species.

Class Aves: The Birds

The fossil ancestors of modern birds can be traced back to the early Jurassic period, about 180 million years ago. *Archaeopteryx,* the oldest known fossil bird, had feathers and presumably could fly (Figure 27.21). The skeleton of this early bird is, interestingly, almost indistinguishable from that of the thecodonts, the reptilian stock that produced crocodiles.

27.20 SURVIVORS OF THE REPTILE AGE

Today's reptiles are believed to have evolved through four separate lines. The snakes and lizards descended from a line of smaller reptiles that somehow survived despite the ruling clique of reptilian giants. The lizardlike tuatara of New Zealand is the lone remnant of another group. Right out of the central line of dinosaurs emerged the crocodilians, forming their own line in the second half of the Mesozoic era. Note that the side-necked turtles (shelled reptiles with long, crooked necks that fold sideways) have a long, separate history of their own.

Both kinds of turtles, like the snakes and lizards, sidestepped the giants and emerged from ancient stock early in the Mesozoic. From this proposed lineage, it is apparent that although lizards and crocodiles are both reptiles, they are really not closely related. Many extinct groups are also shown here. The widths of the colored lines roughly indicate the relative numbers of species at any point in time; dashed lines indicate our best guesses in the absence of fossil data.

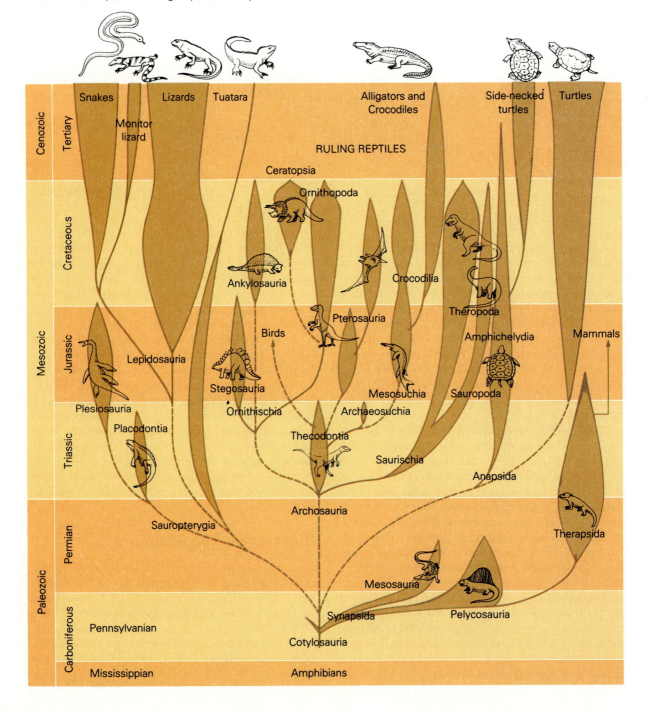

Exit the Great Reptiles

The demise of the great reptiles has long been a puzzle. Hundreds of hypotheses have been considered. Some people once believed that dinosaurs were simply outcompeted by brainier mammals, or that the little rascals ate the dinosaur eggs. Others say that drastic climate changes marked the close of the Mesozoic or that a companion star to the sun causes comets to rain down every 26 million years. The problem is, none of these ideas could account for the truly dramatic suddenness of the mass extinction.

Perhaps the answers can be found in the rocks. There are places on the earth where the 65-million-year-old rock stratum that marks the boundary between the Mesozoic and the Cenozoic can be found. In the waters off Gubbio, Italy, the marine deposits are particularly clear. What can they tell us? Below the boundary, layer after layer of Mesozoic carbonate rocks show a clear assemblage of Mesozoic plankton skeletons, dominated by very large foraminifera (protists; see Chapter 19). But above the Mesozoic rocks, in the strata that were deposited in the Cenozoic, the fossil life forms are dramatically different. In layer after layer of Cenozoic rocks, we find fewer, much smaller species. Even more interesting, at the boundary between the two layers of

plankton deposits is a single layer of clay about a centimeter thick.

In 1980 the paleontologist Walter Alvarez and his Nobel Prize-winning physicist father, Luis Alvarez, reported that this boundary layer has about 50 times more of the elements iridium and platinum than would be expected. Iridium is extremely rare in the earth's crust, but is a common metal in meteors and asteroids. The Alvarez hypothesis, as it has come to be known, is that an asteroid collided with the earth 65 million years ago, and that's what ended the Mesozoic. The idea is not so far-fetched as it seems; astronomers calculate that enough asteroids pass through the earth's orbit that we can expect one to hit our planet every 100 million years or so. From the amount of iridium they found in that thin layer, they calculated that the asteroid was about 10 kilometers in diameter. A rock six miles across is enormous; it could knock you down. But this one may have done more than that. Computer simulations tell us that the dust thrown up by the impact brought the world into total darkness—much darker than the darkest night—for about three months. The effect of such darkness would be enormous. Since the sun's rays could not reach the earth, photosynthesis would stop and plants begin to wither and die.

Many of the earth's animals would have been blinded and even the weather would have changed. Tropical species would have felt subfreezing temperatures for the first time. Further research has corroborated the Alvarez hypothesis, and the iridium-rich Mesozoic-Cenozoic boundary layer has been found in many places.

Few organisms were equipped to survive such a catastrophe. Seed plants and marine plankton that formed resistant cysts were seemingly unaffected. But animal species changed significantly; for example, no animal species weighing more than 26 kg (57 lb) survived into the Cenozoic. Most of the earth's great dinosaurs, then, would have been killed by the earth's great darkness.

The demise of the reptiles was obviously not quite complete, since we have a considerable number with us today. As you watch turtles going about their business, keep in mind that they are descendants of survivors of the asteroid impact. But other species are descended from those Mesozoic reptiles too. One had great prospects in store. That species would build highways, develop fascinating ways to annihilate each other, vote twice and devise elaborate justifications for all its behavior. But that was still a long way off.

27.21 THE FIRST BIRD

Archaeopteryx, the famous feathered, winged reptile of the Mesozoic, is not a direct ancestor to today's birds, but rather represents a divergent line from common thecodont origins. Note the presence of numerous teeth and the persistence of clawed fingers in the reconstruction. Without the clear, fossilized impressions left by the feathers, *Archaeopteryx* would probably have been identified as simply another small dinosaur.

Modifications for flight are found throughout the modern bird's body (Figure 27.22). The skeleton is light and strong. Many bones are hollow, containing extensive air cavities crisscrossed with bracings for strength. The reptilian jaw has been drastically lightened, and teeth have been replaced by a light horny beak. The neck is long and flexible, and the bones of the trunk (pelvis, backbone, and rib cage) have become fused into a semirigid unit. The breastbone (sternum) is greatly enlarged and pos-

sesses a large keel to which the large flight (breast) muscles attach in flying species. The tail is greatly reduced and, with the exception of a few individual bones, has become fused. The feet are often specialized, for example, for perching or grasping.

Feathers, another important modification for flight, are believed to have evolved from reptilian scales. Feathers are unusually strong for their weight. Part of their strength results from their interlocking barbs (branches). Softer, noninterlock-

27.22 DESIGN FOR FLIGHT

Modifications for flight are seen in nearly every aspect of the bird's anatomy and physiology, from the streamlined form to the elevated metabolic rate. In spite of the demands placed on it, the skeleton is extremely light. The frigate bird, for instance, has a wingspan of just over 2 m (7 ft), yet its skeleton weighs an average of just 113 g (4 oz). In general, the slender, hollow bones of birds have a deceiving delicate appearance; in fact, they are strong and flexible, containing

numerous triangular bracings within (see X-ray image). Part of the skeletal strength is due to fusion, as is seen in the hip girdle, tail vertebrae, and, most spectacularly, in the long finger bones. Flight feathers, which may weigh more than the skeleton, owe their extreme strength and flexibility to numerous vanes. These have an interlocking arrangement of hooklike barbules.

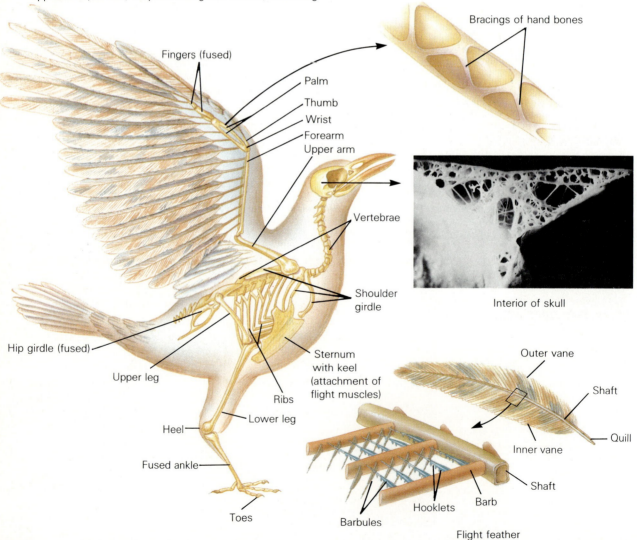

Flight feather

27.23 THE LAND EGG

The amniotic egg of birds and reptiles represents the complete transition from amphibian reproduction to the terrestrial form of life. The egg provides conditions similar in some ways to the aquatic environment, in which most amphibians still develop. The *amniotic cavity* formed by the *amnion* is fluid-filled, protecting the embryo. The *yolk sac* grows out over the food mass of the *yolk,* traversed by blood vessels that bring the vital food supply into the embryo. The *allantois,* an extension of the urinary bladder, collects the embryo's nitrogen wastes. The vascularized *chorion* assists the embryo in its exchange of respiratory gases by providing the necessary surface area. Finally, the egg case itself, often leathery in the reptile and calcified in the bird, protects the contents while permitting gas to be exchanged with the surroundings. Seen here is the bird embryo, which is similar to that of a reptile.

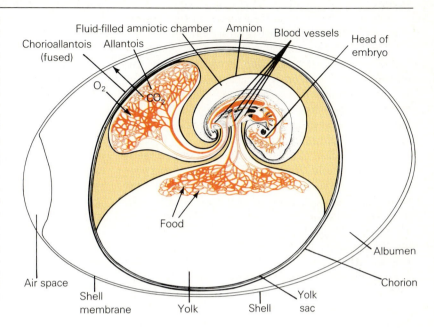

ing down feathers provide insulation in all young and many adult birds.

Birds, like mammals, are endothermic, which permits adaptation to a range of climates. This innovation is quite costly in colder climates, however, since considerable food reserves must be converted to heat energy. Not only does endothermy require a greater intake of food, but it also requires a plentiful supply of oxygen for cell respiration. To meet the greater oxygen demand, both birds and mammals have highly efficient four-chambered hearts, which provide two fully separated circuits: the pulmonary circuit for oxygenating the blood and the systemic circuit for gas exchange in the rest of the body. In the bird lung, we find an extremely extensive, spongy respiratory surface, with numerous air sacs extending from the complex lung. The sacs serve primarily as reservoirs and as bellows for inflating and deflating the passive lungs, as we will see in Chapter 30.

Water conservation is important to most birds since the diet is often dry and drinking opportunities are limited. Most birds lack a urinary bladder. The excretory waste, uric acid, is simply added to the digestive wastes in the cloaca and most of the accompanying urinary water is reabsorbed. The water-conserving regime and lack of a bladder help keep the bird's weight down. And many female birds have only one ovary, perhaps further trimming for flight.

Birds have continued the reptilian tradition of producing large, resistant, amniotic eggs, with basically the same supporting extraembryonic membranes and food supply (Figure 27.23). As a rule, birds produce fewer eggs than do reptiles, a reflection of the parental care so characteristic of nearly all bird species. (With such care, a greater proportion of eggs are likely to hatch.)

There are about 8600 named species of birds, divided into about 27 orders. In spite of such extensive speciation, however, all birds have rather similar general structures. Apparently, the demands of flight have superimposed a specific set of design requirements. Birds differ most strikingly in the development of their beaks and feet, both of which are often related to their feeding behavior (Figure 27.24).

Class Mammalia: The Mammals

Mammals are hairy animals with glands that give milk. In fact, it is the **mammary gland,** or **mamma,** for which the class—and one of our parents—are named. The hair or fur of the mammals is produced in follicles and is lubricated by numerous oil glands. Hair provides an insulating covering that helps the mammal maintain a relatively constant body temperature. The young of most mammals develop in the uterus, where they are nourished and sustained by a **placenta** (Chapter 37). All have a muscular diaphragm that helps move air in and out of the lungs.

While the limbs of amphibians and reptiles are commonly directed outward and then down, the mammal's (and the bird's) limbs are directed generally downward, raising the body clear of the ground. Variations on this basic theme have produced mammals that swim, run, climb, burrow, leap, and even fly. In some species, such a limb arrangement also permits greater efficiency in catching, holding, and killing.

The evolution of the mammalian jaws and teeth has been equally dramatic. Except for the crocodiles, only the mammals have socketed teeth. And only the mammals have a fully mobile, chewing jaw. The teeth, of course, vary with the diet. Herbivores generally have massive, grinding molars, while carnivores have large canine teeth and high-ridged, bone-crushing molars. The chisellike incisors of rodents are specialized for gnawing, while insectivorous mammals have sharp, pointed teeth. The massive tusks of walruses (modified incisors) are used both defensively and for raking up clams. Elephants also use their tusks (also modified incisors) in defense. The great peglike teeth of the killer whale are specialized for grasping and crushing large prey such as other whales and seals. Human teeth are among the least specialized, so we are capable of eating about anything we come across, from lettuce to spider crabs. Incredibly, the vagaries of evolution have transformed some of the bones of the reptilian lower jaw, step by step, into the tiny bones—the **ear ossicles**—of the inner ear (see Chapter 34).

The Mammalian Brain. Above all, the hallmark of a modern mammal is its relatively large and versatile brain. Today's mammals are smarter than other vertebrates; that is, they (1) rely less on genetically programmed instinct and (2) adjust more readily to their environment, basing more of their behavior on individual experience and learning. Their remarkable characteristics are reflected in the structure of their brains. Not only has the cerebrum (the part associated with learning and conscious thought) become larger, but new parts have been added, such as the **corpus callosum,** which integrates the mental activities of the left and right halves of the brain (see Figures 35.4 and 35.14).

27.24 BIRD DIVERSITY

The general organization of birds is similar throughout the class, probably because of the demands of flight. However, there is marked diversity in the beaks and, as seen here, in the feet. These variations permit birds to explore many food sources and habitats. For example, the flicker's hooked hind toes easily cling to the bark of trees, while the hawk uses its sharply curved talons to capture prey. In contrast, the lobed feet of the coot and webbed feet of the duck are well suited to their aquatic habitats. The greatest divergence from general organization is seen in the flightless birds, such as the emu, which relies on its speed afoot to escape enemies.

Emu

Hawk

Flicker

American coot

Duck

Keep in mind, however, that increased learning capacity is simply one evolutionary alternative to the challenge of survival. For example, insects have a comparatively limited ability to learn, yet they are our chief competitors for the earth's resources. Their alternative path of evolution obviously works well for them.

Mammal Origins. The mammals had their reptilian origin somewhere in the Permian period, branching off very early from a large group that preceded the ruling reptiles, known as the **therapsids,** sometimes known as the "mammallike reptiles." The transition into the full mammalian condition was gradual, with no sharp dividing line seen in the fossil record. Some early mammals are said to have been doglike in appearance, but this is not very accurate and you could probably clear out a bar if you walked in with one on a leash (Figure 27.25).

The therapsids were flesh eaters that developed the forelegs and musculature necessary for running. The angle of their elbows was quite different from that of the first terrestrial reptiles, whose elbows angled out to the side. Other indications of their mammalian association include tooth structure as well as skull and jaw features.

Survival in the developing mammals may have been enhanced by the ability to carry embryos within the body and to bear live young, rather than deserting the eggs to predators or having to defend a stationary nest. Another advancement was the production of milk, a very convenient food that provided the young with high-protein, high-calorie nutrition during the critical stages of growth. In turn, the young would have tended to remain with at least one parent in order to be fed, and this continued association would have permitted the offspring to learn from the parent. This opportunity would have enhanced the tendency toward braininess.

The advancing Cenozoic era brought the "age of mammals." By the end of its Cretaceous period, three types of mammals had diverged from the original line. Today's descendants are grouped into three subclasses: **Prototheria, Metatheria,** and **Eutheria.** Respectively, these are (1) **monotremes,** order Monotremata (the egg-laying mammals), (2) **marsupials,** order Marsupalia (the pouched mammals), and (3) placental mammals (Figure 27.26).

The monotremes never became important and today are restricted to a few minor species, such as the duckbilled platypus and the echidna or spiny anteater. The marsupials reached prominence in geographically isolated Australia and, earlier, in the

27.25 THE EARLIEST MAMMALS

The mammals arose from the therapsids, an early reptile line. The earliest therapsids retained many reptilian traits, including scales, but by the Triassic period (197 to 225 million years ago), features such as the limbs and stance had already changed. Raising the entire body from the ground was an adaptation for running and leaping, but it introduced new problems in balance and coordination—problems that required simultaneous adaptive changes in the brain.

Americas. Marsupials differ from placental mammals in that they lack a true placenta, although a "pseudoplacenta" develops from the yolk sac. The young emerge from the uterus very early and complete their development in the pouch.

With the exception of the opossum, marsupials were replaced by placental mammals in all continents except Australia. This replacement began in North America, where marsupials evolved, but soon spread to South America, and by 50 million years ago, the marsupials were restricted to subtropical Antarctica and to Australia, then joined by a land bridge. It seems possible that replacement might have been completed were it not for a final episode in continental drift—the separation of Antarctica from Australia (see Essay 17.2).

The three subclasses radiated from an explosion of species in the early Cenozoic era. By the Eocene period, which began 54 million years ago, the main lines of mammalian evolution had already become established. The Cenozoic era was a disruptive time indeed, marked by drastically fluctuating climatic patterns and intense selection. Many Cenozoic mammals didn't survive the severe stress; however, those that did underwent rapid speciation and divergence, filling many new and highly specialized niches. Today's 4500 species are organized into 18 orders (or 19, depending on whether you consider the pinnipeds—walruses, seals, and sea lions—to be in the carnivores or in their own order) each representing a major line that appeared in the early Cenozoic (Figure 27.27 and Appendix).

27.26 THREE TYPES OF MAMMALS

Three lines of mammals probably evolved from the ancestral therapsids. These were the primitive Prototheria (monotremes), the Metatheria (marsupials), and the Eutheria (placentals). The reproductive differences are summarized in these three modern representatives. (a) Monotremes are hairy egg layers. Their young feed by simply lapping milk off the mother's fur. (b) The marsupials give birth to immature young, which climb to the pouch where they fasten to a nipple and remain there as they complete their development. (c) The placental mammal completes its embryonic development in the uterus, supported by membranes that form the placenta. It also nurses for a time after birth and, like the others, requires a period of time for growth and learning before it ventures out on its own. This gives the complex brain time to develop, and intelligence has been important to the success of mammals.

(a)

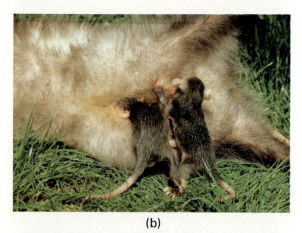

(b)

(c)

27.27 THE EMERGENCE OF MAMMALS

Today's mammals occupy 18 or 19 orders and are dispersed into innumerable niches over the earth. They live below the earth's surface, on its surface, in trees, in oceans, streams, and lakes, and have even taken to the air as gliders or as true flyers. They feed on everything, from minute plankton to great

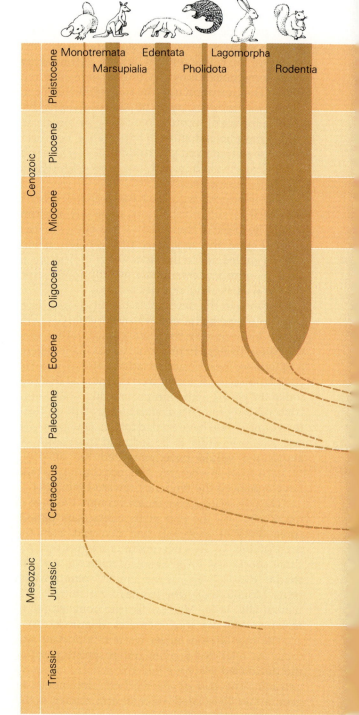

trees and, of course, each other. Mammalian roots reach back to the early Mesozoic era but do not appear to branch out significantly until the Cenozoic, where they became the dominant vertebrates. The Cenozoic, an era of great climatic changes, saw the end of the ruling reptiles. A multitude of mammals arose and died off, but many persisted to give rise to today's forms. Dashed lines represent hypothetical relationships; note that nearly all Mesozoic mammalian relationships are hypothetical.

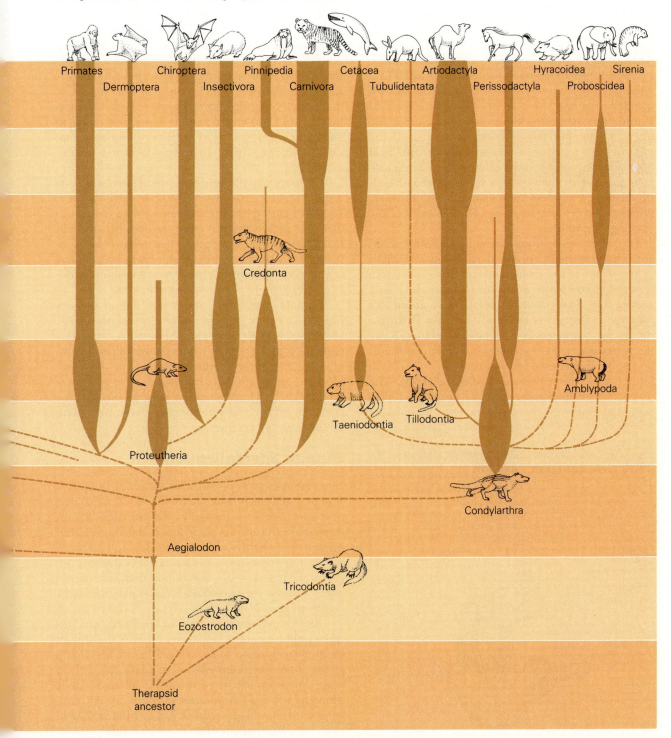

The Primates

Modern primates, order **Primates,** are divided into two major groupings, the first consisting of the more primitive tree shrews, tarsiers, lemurs, and the loris (Prosimii), and the second, the monkeys, apes, and humans (Anthropoidea). The latter is further subdivided into the New World monkeys (Ceboidea), Old World monkeys (Cercopithecoidea), and apes plus humans (Hominoidea). New World, western hemisphere monkeys are those with the prehensile (grasping) tails and nostrils set apart by a wide nasal septum, typified by the familiar "organ-grinder monkey" or spider monkey. The Old World, Eurasian and African monkeys, are more likely to have short tails and doglike snouts with the nostrils close together, as in the baboons. The great apes (gibbons, orangutans, chimpanzees, and gorillas), along with humans, are referred to as **hominoids,** but members of our own family, **Hominidae,** whether extinct or extant, are called (careful of the spelling) **hominids.**

Primates can be traced through the fossil record as far back as the Paleocene epoch, some 65 million years. The earliest primates were unlike modern types in many ways. For example, they had long snouts and claws, and actually looked more like rodents. In fact, it is believed that they occupied nearly the same niches as modern rodents, some living in trees, some scurrying about on the ground, and some in burrows feeding on grubs. Our best-known example is *Plesiadapis* (Figure 27.28). The divergence of the various primate groups is represented in Figure 27.29. Note how recently humans

have diverged from common ancestry with the gorillas and chimpanzees.

How do primates differ from other mammals? This would seem to be an easy question, and it is unless one attempts to become too specific. Primates are often described as generalized mammals, meaning that they have many unspecialized features, some even quite primitive.

The limbs and body of *most* primates are well adapted for arboreal (tree-dwelling) life. For example, all have prehensile (grasping) hands divided into clawless (nailed) fingers. Most have long arms and some have prehensile tails. Primates have also developed binocular, stereoscopic vision with eyes located in a forward position. If you are going to swing on limbs, you need good eye-hand coordination—and the ability to judge distances (Figure 27.30).

Of course, all of these anatomical specializations require a substantial degree of brain development, with strong emphasis on centers dealing with coordination and vision. But more significantly, the primate brain is highly adapted to learning. This ability is especially developed in chimpanzees and gorillas (and the capuchin), but humans are without parallel in intelligence and relative brain size.

Primates tend to be omnivores, eating all sorts of food, so their teeth are relatively unspecialized. However, with the exception of humans, primates, especially males, have rather large canine teeth, used primarily in fighting, threat, and defense. Humans have reduced canines and have substituted tools such as rocks, baseball bats, and cruise missiles. An officer and a gentlemen would never bite the enemy; his canines are too short.

Human Specialization

In some cases the specializations that are unique to humans are really only refinements of traits that also exist in other animals. Several primates have an opposable thumb that can touch the fingers, but ours has a slightly different musculature and is much more dexterous. With this manual ability we can make precise tools, and although other animals make and use very primitive tools, none can build a watch (Figure 27.31).

And then there is our upright posture, which is quite unlike any other modern primate's. Savannah chimpanzees stand to see over tall grass and many chimpanzees walk bipedally (on two legs) a short way, but normally they move on all fours. It is amusing to see a chimpanzee or other primate running on its hind legs because it is so humanlike but ungainly. Humans, however, are beautifully adapt-

---

27.28 THE EARLIEST PRIMATE

Fossils of the extinct *Plesiadapis*, a rodentlike tree mammal from the Paleocene epoch, have been found in North America and in Europe. Its type is believed to be ancestral to the primate line. *Plesiadapis* was rather small, about the size of a housecat—not very impressive but it had a bright future. Its general body structure and long tail indicate that it was arboreal.

27.29 PRIMATE HISTORY

This tentative phylogeny of the primates places *Plesiadapis* at the base. The earliest branches produced the tree shrew and prosimian lines, followed next by the New World monkeys. These two groups represent the greatest divergence in the living primates. Old World monkeys diverged some 10 million years later, subsequently followed by the great apes. The most recent divergence occurred between humans and chimpanzees.

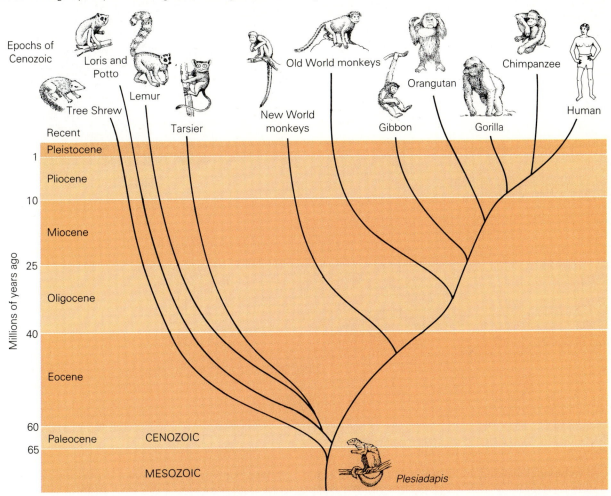

27.30 A COUPLE OF SWINGERS

Good brachiation (swinging from limb to limb) requires prehensile hands, long arms, and keen hand-eye coordination. Stereoscopic vision is also a must. Prehensile feet and tails don't hurt either. **(a)** The gibbon is one of the most talented brachiators and a big attraction at zoos. Note the lengthy arms, the four-fingered grasp, and the position of the thumb, lower than our own and out of the way. **(b)** Both the hands (outside) and feet (center) of the orangutan are well suited for grasping and swinging.

27.31 THE HUMAN EDGE

(a) While apes can assume a bipedal posture, neither hip nor leg structure supports this very well. In their attempts at bipedal walking, the weight is borne on the outer edge of the archless foot. **(b)** The hands of humans and chimpanzees appear generally similar, but there are important differences. Chief among these are the length and musculature of the thumb. In the chimpanzee, the thumb doesn't quite reach the base of the forefinger, while in humans the thumb extends nearly to the middle joint. This "out of the way" location in the chimp is important to a brachiator but makes handling tools and other human objects far less precise. Compare the precision grip of the two, noting that in using its thumb, the chimpanzee must flex its fingers. (Try buttoning your shirt that way!)

(a)

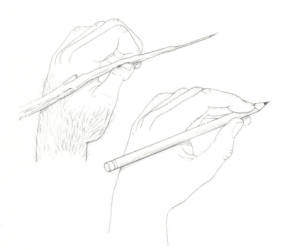

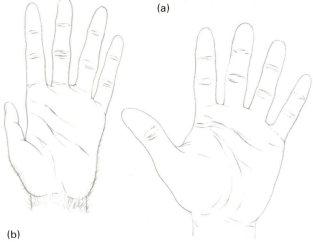

(b)

ed to bipedal walking and running, surpassing any other vertebrate in this accomplishment. We owe this ability partly to our highly specialized foot, with its arched construction, and partly to our enlarged *gluteus maximus* (buttock) muscles.

We pride ourselves on our intelligence, but its evolution has not been without cost. The large skull that houses our brain produces all sorts of birth difficulties. And, even more than most other mammals, we are born in a helpless state. A baby hare will lunge and hiss at an intruder, and some baby birds will crouch and "freeze" at the mother's warning call. But we are more defenseless, and our newborns seem totally witless.

In a sense, humans have sacrificed a "prewired" brain, one that is programmed to react in certain ways under given situations, for one with greater potential. That is, we rely less on genetically based instincts and more on raw intelligence than do other species. Certainly this makes it possible for us to cope with a more variable and unpredictable environment. But we must undergo long periods of learning and therefore have evolved extended parental care. The fact that we have adapted to our world through intelligence, learning, and nurturing (rather than, say, having thousands of children and leaving their survival to chance, as some organisms do) has undoubtedly altered our social patterns to an enormous degree.

Aside from our magnificently overgrown brain, the principal and unique specialization of the human species is our capacity for language. Language is not easily separated from intelligence because much of our brain is directly related to this function. Despite highly publicized and controversial studies, there is no evidence whatsoever that dolphins or apes can tell each other about their experiences in any but the most basic terms. The capacity for creating abstract symbols to communicate concrete ideas, which is what language is, appears to be absent in even the most highly trained animals.

With language, evolution entered a new dimension. Humans are capable of *cultural* evolution in addition to genetic evolution. We can tell our children the truths (and lies) that we learned from our parents. Factual information can be transmitted via word of mouth over tens of generations; the historically accurate account of the siege of Troy, handed down by illiterate singing bards for hundreds of years before being committed to paper, is but one example. Language also allows for politics: even unlettered tribes manage social organization far above the level that could ever be achieved by gibbering apes. Written language, of course, has extended the scope and power of cultural evolution by yet another order of magnitude.

HUMAN EVOLUTIONARY HISTORY

As is usual when humans try to assess themselves, their genes, or their heritage, the air becomes full of indignation and accusations. The issue has been a volatile one since Darwin's day, when the very idea that humans had an evolutionary ancestry at all was offensive to many. Even Alfred Wallace, the codiscoverer of the principle of natural selection, couldn't bring himself to believe that human beings had originated in the same way other species had. Darwin himself firmly believed that humans and apes had evolved from a common ancestor, but there was no fossil evidence for this contention. Opponents of the idea of human evolution made much of the absence of a "missing link" and continued to make much of it long after many hominid (humanlike) fossils had been discovered.

In the early 1960s, biochemists Allan Wilson and Vincent Sarich began comparing primate proteins and DNA sequences. Their data clearly indicated that humans were more closely related to the chimpanzees and gorillas of Africa than to the orangutans of Asia. They were aware that molecules in these substances change at a steady and dependable rate and thus provide us with a kind of "molecular clock." This clock suggested that humans and African apes diverged from a common ancestor not more than 6 million years ago—a startling notion at the time. Anthropologists had routinely assumed a much more ancient divergence and were mostly hostile to the new ideas. Another furious,

often boring scientific feud was underway. However, new data, both biochemical and paleontological, back up the biochemists rather than the anthropologists, causing us to adjust our notions of human lineage. Let's review the current theory.

Our Ancestral Scenario

About 2 to 3 million years ago, a number of humanlike forms lived in relatively dry open grasslands of Africa. Their fossils have been given a great number of contradictory names, but when the dust settles four groups can be recognized: *Australopithecus robustus, Australopithecus africanus, Australopithecus afarensis*, and *Homo habilis*. It is not conclusively established that any of these forms was directly ancestral to *Homo sapiens*, but it is clear that the first two *Australopithecus* species were widespread, successful species that persisted almost unchanged until at least 1 million years ago, long after the genus *Homo* was well established. Still, our own ancestors could have evolved from early offshoots of the *Australopithecus* line.

The Australopithecines. The **australopithecines** (members of the genus *Australopithecus*) were rather small-boned, light-bodied creatures, about four feet tall. They had humanlike teeth, jaws with small incisors and canines, and an upright posture (Figure 27.32). They were hunters, apparently preying on baboons, gazelles, hares, birds, and even giraffes.

Their weapons were clubs made from the long bones of their prey. Near some *Australopithecus* bones, anthropologist Raymond Dart found a fossil baboon skull with a fossil antelope leg bone jammed into it. Their cranial capacity, somewhat larger than the gorilla's, ranged from 450 to 650 cubic centimeters (compared to 1200 to 1600 for modern adult humans).

Australopithecus robustus, as the name suggests, was larger-boned, though not taller than the *A. africanus*. Most strikingly, *A. robustus* had huge jaws and greatly expanded cheek bones to accommodate massive jaw muscles. Their teeth were very large, though more human than apelike in most respects, and the tooth wear suggests that the australopithecines were vegetarians. We should point out that some anthropologists now question whether *A. robustus* and *A. africanus* were actually two different species, since there seems to be an array of intermediate forms.

The first specimen of *Australopithecus afarensis,* dubbed "Lucy" (after the Beatles' song of the 1960s), was found in the northern Ethiopian desert

27.32 EARLY HOMINIDS

The australopithecines were small creatures by modern human standards, from 1 to 1.5 m (3.5 to 5 feet) in height and up to 68 kg (150 lbs). They were heavy-boned, suggesting strong muscularity. Though their bodies appeared very humanlike, their heads were more apelike than human in appearance, with a low cranial profile and little or no chin. However, the jaws were large and forward-thrusting. There is no evidence that they produced tools, although they may have made use of materials at hand, as chimpanzees are known to do.

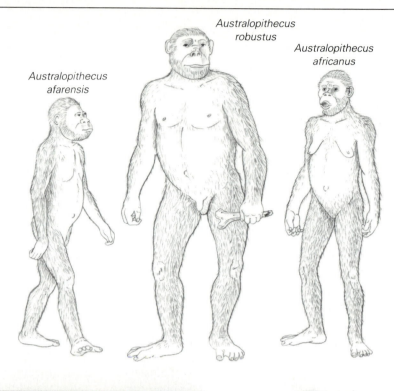

Australopithecus robustus

Australopithecus afarensis

Australopithecus africanus

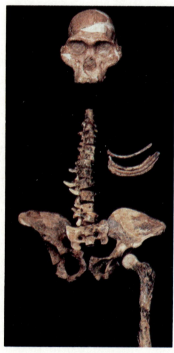

A. africanus

A. robustus skull

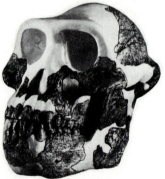

A. afarensis
(reconstructed from
parts of several individuals)

region in 1974 by Donald Johanson. It consisted of more than half a fossilized skeleton of an upright-walking female, calculated to be about 3 million years old. The following year Johanson unearthed the fossilized remains of 13 additional "Lucy" types, clustered together in one area. Johanson concludes that his australopithecines were more primitive than *A. robustus* and *A. africanus*, and has tentatively placed them as a separate species in the main line of hominid evolution, suggesting that they were ancestors to both the australopithecine and human line (Figure 27.33).

Homo habilis. *Homo habilis* is the name given by famed anthropologist Louis Leakey to certain fossils from the Olduvai Gorge of East Africa. As the name indicates, these fossils seem to be close to the modern human form, with an average cranial capacity of about 656 cc (cubic centimeters). Critics have argued that the new find is just a variant of *Australopithecus* and that Leakey was unjustified in trying to place his fossils within the genus *Homo*.

More recently, Leakey's son Richard and his widow Mary found a hominid skull of unusual interest at Koobi Fora. At 1.6 million years old, it is older than some *Australopithecus* fossils, but it appears to be much closer to the human line than is *Australopithecus*. For one thing its cranial capacity—over 800cc—is greater than those of either *Australopithecus* or the earlier disputed *Homo habilis* finds. The Leakeys, apparently fed up with the sterile arguments that have raged over the naming of hominid fossils, simply identified their unique find by its arbitrary field identification: fossil skull 1470.

Homo erectus. The fossil beds of the eastern shore of Lake Turkana have more recently yielded fossils that are closer to the modern human. The new species has been called *Homo erectus* and is considered to be an extinct member of our own genus. The earliest fossil African *Homo erectus* skulls are known to be more than 1.5 million years old. They are most interesting in that they appear to be of the same species as some of the first hominid fossils ever found—those once called "Java Ape Man" and "Peking Man" and designated *Pithecanthropus*. But the Lake Turkana fossils were only about half a million years old, and some *Homo erec-*

tus fossils may be even less than 200,000 years old. In other words, *Homo erectus* flourished, relatively unchanged, for well over a million years. During its first 300,000 years it coexisted with other, more primitive hominid species, including the australopithecines and possibly *Homo habilis*. In at least the last 100,000 to 200,000 years of its existence, *Homo erectus* overlapped with yet another upstart species, *Homo sapiens*. Fossils of *Homo erectus* have been found in China, Europe, and southern Africa, and of course, East Africa (Figure 27.34). The evidence left by *Homo erectus* provides fascinating grist for the mills of our imagination.

Homo erectus had a small brain, heavy brows, and strong jaws and teeth. But from the neck down, they apparently were very much like us. In addition to their fossilized bones, *Homo erectus* left behind crude stone tools. The tools, consisting of hand axes and scrapers, are designated **Lower Paleolithic** or **Lower Old Stone Age.** There is evidence that these Lower Old Stone Age people built shelters, hunted small game, and gathered plant foods. The era of *Homo erectus* arrarently ended just after the first glaciers of the Pleistocene receded.

27.33 HOMINID HISTORY

The known hominid history spans nearly 4 million years. According to David Johanson, the most recent ancestor to the human line was *Australopithecus afarensis,* the type represented by "Lucy." He suggests that *A. afarensis* produced two lines of descent, the Australopithecines and *Homo,* with *Homo habilis* provisionally representing the earliest human, although not in the main line. Next we find *Homo erectus,* who had a fairly long history, coexisting toward the end with the first *Homo sapiens,* an offshoot of the *H. erectus* line. The earliest *H. sapiens,* as far as we know, were the familiar Neanderthals, a diverse but successful group, finally replaced entirely by modern humans a mere 40,000 years ago.

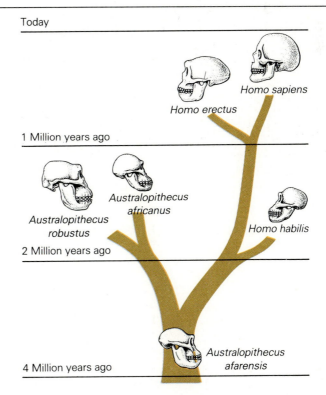

Today

Homo erectus

Homo sapiens

1 Million years ago

Australopithecus africanus

Australopithecus robustus

Homo habilis

2 Million years ago

4 Million years ago

Australopithecus afarensis

Recent findings at a well-known site at Zhoukoudian, China, reveal that one large cave was continuously inhabited by *H. erectus* for over 200,000 years, and finally abandoned 230,000 years ago. During this incredibly long period of continuous residence, both the anatomy and the technology of *H. erectus* changed significantly. Skulls upearthed in the oldest debris of the cave revealed a cranial capacity of about 915 cc, while those in the most recent layers had reached 1140 cc. The stone tools had progressed from large, crude, hastily fashioned choppers and scrapers in the oldest, deepest layers, to smaller, much more refined tools in the newer surface layers. The cave inhabitants used fire from the start, hunted both large and small game, and ate a variety of nuts, fruits, seeds, and other plant matter.

Neanderthals and Us. Even while Darwin was studying and writing at his country estate, workers in a steep gorge in the Valley of Neander (in German, *Neanderthal*) were pounding at something that turned out to be a skeleton, inadvertently smashing it to bits but leaving enough for researchers to see evidence of a new and different kind of human (a similar skull had been unearthed at Gibraltar a few years earlier but had not created much of a stir).

Scientists are rarely at a loss for words, and an explanation was immediately forthcoming. A professor Mayer of Bonn examined the heavy-browed skull and proclaimed that the skull and bone fragments belonged to a Mongolian cossack chasing Napoleon's retreating troops through Prussia in 1814; an advanced case of rickets had caused him great pain and his furrowed brow had produced the great ridges; he was so distraught he had crawled into a cave to rest, but alas, had died there. This scenario has been rejected in favor of the idea that the bones were those of a member of an early form of human that became extinct—the **Neanderthal Man.**

Members of *Homo sapiens neandertalensis* thrived, or survived, until around 40,000 years ago. They left a rich record of their fossil remains and some indication of their tools and culture. Their geographic range was large, extending at least from Portugal in the west to Soviet Central Asia in the east (see Figure 27.34).

Neanderthals had large brow ridges and sloping foreheads but were not as radically different from modern humans as was once believed. Their chins did not protrude as far as those of *Homo sapiens*, but their necks and bodies—when they didn't have rickets—were like ours and indeed like those of *Homo erectus* (Figure 27.35). There is no way to

27.34 DISTRIBUTION OF EARLY HOMINIDS

The richest Australopithecine finds are located along Africa's Rift Valley in Ethiopia, Tanzania, and Kenya. *Homo erectus* was far-ranging, with fossils recovered in eastern and southern Africa, Europe, China, and Indonesia. Neanderthal fossils have been unearthed throughout much of Europe, the Mid-East, North Africa, and China.

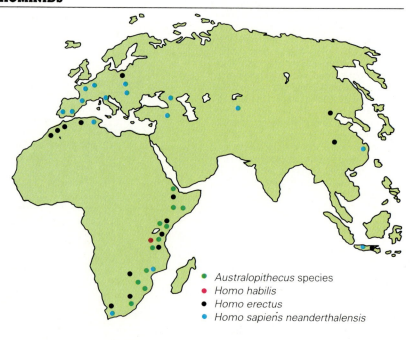

● *Australopithecus* species
● *Homo habilis*
● *Homo erectus*
● *Homo sapiens neanderthalensis*

27.35 A "FACE-TO-FACE" CONFRONTATION

A modern human and Neanderthal reveal many general similarities and some striking differences. **(a)** Notable in Neanderthal is the slope of the cranial vault (forehead), the large brow ridges and receding lower jaw. The jawbone itself has less of an angle where it curves upward to form its articulation with the skull. **(b)** The facial reconstruction produces an image quite different from the traditional view of Neanderthal, long considered an apelike creature as it is still seen in some museum paintings and restorations. With the exception of the obvious differences in skull structure, Neanderthals had fairly modern skeletons.

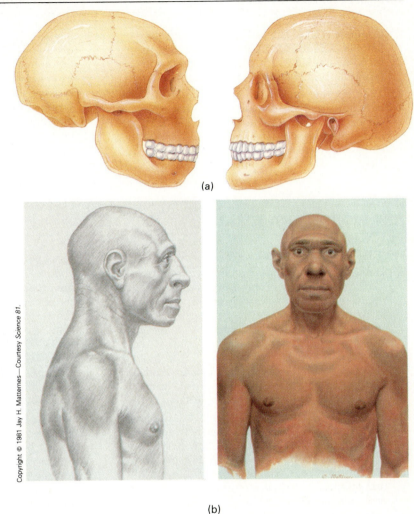

Copyright © 1981 Jay H. Matternes—Courtesy *Science 81.*

(a)

(b)

know whether they had a spoken language. However, we do know, from fossil pollens found in their remains, that Neanderthals sometimes covered their dead with flowers before burial. Whether or not that's sophisticated, it certainly is human—touchingly so.

Neanderthal-type fossils are not found in rocks younger than about 40,000 years. Why did the Neanderthals disappear? One school has proposed that the classic Neanderthal line and the line of modern humans diverged as long ago as 250,000 years, perhaps when early *Homo sapiens* was separated geographically by the great glaciers of the Pleistocene. About 40,000 years ago, the scarce *Homo sapiens sapiens*—the modern humans—suddenly burgeoned and Neanderthal became extinct, possibly by being wiped out by *Homo sapiens sapiens,*

a subspecies famous for its eagerness to go to war.

A second theory minimizes the differences and states that *Homo sapiens* in the Neanderthal age was one highly variable species that included both the classic Neanderthal types and individuals much like ourselves. This theory places the Neanderthals squarely in the mainstream of human evolution. Modern humans, the theory maintains, simply evolved from somewhat archaic Neanderthal ancestors. Therefore the Neanderthals didn't die out, but merely changed. Those who believe this interpretation suggest that earlier anthropologists simply put too much emphasis on some extremes of normal variation. Much of the remainder of human evolution is called *history* and is within the domain of historians and archaeologists.

APPLICATION OF IDEAS

1. Compare a bony fish (a vertebrate) with the lancelet (a cephalochordate). In the comparison try to determine which characteristics in each represent primitive or advanced states.

2. The Cenozoic era is often referred to as the Age of Mammals. Did the mammals evolve in the Cenozoic? If not, why the label? When did mammals first appear and what did some of the earliest representatives look like?

3. Suggest three anatomical differences that clearly distinguish humans from other primates. Are the human characteristics abrupt departures from general primate characteristics or are they differences in degree? Cite several examples. Comment on how some of these differences may have been responsible for the uniqueness of human accomplishments.

KEY WORDS AND IDEAS

PHYLUM HEMICHORDATA

1. **Hemichordates** include the burrowing acorn worm, named for its conical **proboscis** protruding from a **collar.** While its **gill slits** suggest its relatedness to chordates, its larvae are quite similar to those of echinoderms.

PHYLUM CHORDATA

1. Chordates—the **Urochordata, Cephalochordata,** and **Vertebrata**—have gill slits, **dorsal, hollow nerve cords,** a **notochord,** and a **postanal tail.**

2. In primitive chordates, the gill slits are used as sieves, while in fishes, the **arches** between them support the gills. The dorsal, hollow nerve cord in chordates contrasts to the ventral, solid nerve cord of many invertebrates. The notochord, a firm supporting rod, is seen mainly in the chordate embryo. The postanal (after the anus) tail is present in most adult chordates and in the embryos of the rest.

Subphylum Urochordata

1. **Tunicates** (**ascidians** or sea squirts) are the best-known urochordates, but the subphylum includes **salps** and **larvaceans.**

2. The name *tunicate* comes from the saclike body covering, or **tunic.** While the swimming, tadpolelike, larval tunicate has all of the chordate traits, the only one remaining in the stationary adult is the **gill sac.**

3. Larvaceans retain the larval form even when sexually mature, a condition called **neoteny.**

Subphylum Cephalochordata

1. **Cephalochordates (lancelets)** are fishlike in shape and exhibit all of the chordate characteristics. Segmentation is revealed in the **myotomes,** repeated muscle units in the body wall.

2. Lancelets use cilia for filter-feeding and have a complex mouth surrounded by tentacles. Gases are exchanged when water, leaving the mouth, passes over the gills and out of an opening in the surrounding **atrium.**

SUBPHYLUM VERTEBRATA: ANIMALS WITH BACKBONES

1. **Vertebrates** have a vertebral column, cranial brain, gills or lungs, no more than two pairs of limbs, one pair of eyes, kidneys, and separate sexes.

Class Agnatha: Jawless Fishes

1. The oldest vertebrate fossils are **agnathans,** half-billion-year-old armored jawless fishes called **ostracoderms.** Living remnants of the class are the eellike jawless and boneless **cyclostomes**—the lampreys and hagfish.

2. Parasitic lampreys feed via a rasping sucker mouth. Normally, the adults live at sea but reproduce in rivers and streams, where a long larval stage develops.

Class Placodermi: Extinct Jawed Fishes

1. The extinct **placoderms,** armored fishes with hinged jaws, arose during the Silurian, giving rise to the cartilaginous and bony fishes.

Class Chondrichthyes: Sharks, Rays, Skates, and Chimaerma

1. **Chondrichthyans** have cartilaginous skeletons, an asymmetrical tail, a dorsoventrally flattened body, and unique, **placoid scales**—miniature but true teeth.

2. The digestive, reproductive, and excretory systems of sharks and rays open into a common **cloaca,** which leads to the external **vent.** Digestive surface in the gut is increased by a **spiral valve.**

3. Sharks have a highly developed **lateral line organ,** which senses water movement. They swim continually, passing water over the gills where gases are exchanged.

4. Sharks utilize internal fertilization, and while some shark species are egg-layers, others retain the embryos throughout development. A few species provide the embryo with nutrients and remove metabolic wastes.

Class Osteichthyes: The Bony Fishes

1. The diverse bony fishes of class **Osteichthyes** are often highly maneuverable, with a mucus-covered body surface that helps eliminate drag. Most have an adjustable **swim bladder** for maintaining buoyancy at various depths. Its gases are generated in a **gas gland** and removed by a **reabsorptive area.**

2. Gas exchange occurs in thin-walled **gill filaments,** supported by **gill arches** with toothed **gill rakers,** located in two **gill chambers,** each protected by an **operculum.** In most bony fishes water is pumped over the gills. Blood flowing through fine vessels in the gills receives oxygen, which is then distributed to the body. Respiration is supported by circulation in a two-chambered heart.

3. Most bony fishes are egg-layers and external fertilization is common. A few retain the eggs during development.

4. The bony fishes arose from an air-breathing ancestor to form the **Actinopterygii** (ray-fins), **Dipneustei** (lungfishes), and **Crossopterygii** (lobe-fins). The vast majority are ray-finned fishes, while only three species of lungfish persist. They breath by gulping air and forcing it into the swim bladder. Only one species of lobe-finned fish, the **coelacanth** (*Latimeria chalumnae*) is known. The lobe-finned line is ancestral to the terrestrial vertebrates.

Class Amphibia: The Amphibians

1. Amphibians today occur in three orders: **Urodela** (salamanders), **Anura** (frogs and toads), and **Apoda** (ceacilians). Amphibians and reptiles arose in the Carboniferous period.

2. Most amphibians have lungs, but to some extent also use their moist, vascularized skin for gas exchange. Gas exchange is also facilitated by the presence of a new circuit through the lungs, made possible by a third heart chamber.

3. Reproduction in amphibians ranges from simple external fertilization and development in water to internal fertilization, by actual mating or by the transfer of sperm packets **(spermatophores).**

Class Reptilia: The Reptiles

1. Specific adaptations of **reptiles** for dry terrestrial life include internal fertilization and development in the independent, self-supporting, **amniote egg** named for the **amnion,** a membrane surrounding the embryo. It contains a food supply and supporting **extraembryonic membranes** developed by the embryo. There are a dry, scaly skin; complex, spongy, protected lungs; and, although reptiles are **ectothermic,** a behavioral repertoire for avoiding the effects of extreme temperatures.

2. Reptiles arose from **cotylosaurs** ("stem-reptiles") to become the prominent vertebrates during the Mesozoic (the "Age of Reptiles"). The highly diverse group included **pterosaurs,** large flying reptiles, some of which may have been **endothermic** (warm-blooded).

Class Aves: The Birds

1. The earliest known bird, *Archaeopteryx,* appeared in the early Jurassic. Reptilian traits include scales on the legs and the general reptilian skeletal framework.

2. Flight modifications in birds include forelimbs modified into wing bones, a light, strong skeleton with fused units and reduced tail, a light beak (no teeth), enlarged breastbone with a keel, perching and grasping feet, and scales modified into feathers used for flight surfaces and insulation.

3. Birds are **endothermic,** a condition supported by efficient circulatory and respiratory systems. The heart is four-chambered and the pulmonary (lung) and systemic (body) circuits are completely separated. The lungs are complex and numerous air sacs are present.

Class Mammalia: The Mammals

1. Mammals, named for the **mammary glands** (milk glands), produce hair and are endothermic. Nearly all young develop in the uterus supported by the **placenta.** Breathing is assisted by a muscular diaphragm. The jaws are quite mobile and contain socketed teeth that vary widely with feeding and dietary habits and are thus important to classification.

2. The mammalian brain, with its enlarged cerebrum and newly evolved **corpus callosum,** is large and versatile and provides for more learning than that of other vertebrates.

3. Mammals originated from the **therapsids,** flesh-eaters whose limbs were modified for running on all fours.

4. By the Cenozoic, the Age of Mammals, three groups had diverged: **Prototheria** (egg-layers),

Metatheria (marsupials), and **Eutheria** (placentals). The first is minor today with only a few species, while the second, the marsupials, are concentrated in Australia. Placental mammals make up the majority.

The Primates
1. Order **Primates** includes the more primitive **Prosimii** and the **Anthropoidea.** The latter is subdivided into New World monkeys, Old World monkeys, and apes and humans. Apes and humans make up the superfamily **Hominoidea,** or the **hominoids.** The human family, **Hominidae,** including extinct forms, are the **hominids.**
2. Primates are generalized mammals whose characteristics include long forelimbs (arms) and **prehensile** hands with fingernails rather than claws. Some have prehensile tails. Vision is binocular and stereoscopic, and the brain is highly adapted to learning. Most are omnivores.

Human Specializations
1. Human specializations include an opposable thumb that is more versatile than that of other primates, a naturally upright posture, arched foot, and bipedal gait. Most striking is human intelligence, the capacity for abstract thought, learning, and a facility for language and mathematics. These are coupled with a very long and dependent developmental period. Humans experience cultural as well as physical evolution.

HUMAN EVOLUTIONARY HISTORY
1. Theories of human origins have a history of controversy. Newer data from the study of hominoid protein amino acid sequences suggest that humans and apes diverged no more than 6 million years ago.

Our Ancestral Scenario
1. The known fossil hominid line begins with the *Australopithecus robustus, A. africanus,* and *A. afarensis.* The first two lived up to about 1 million years ago.
2. The **australopithecenes** were about four feet tall, had humanlike teeth, walked upright, and hunted with crude bone weapons. Their cranial capacity was somewhat larger than the gorilla's. *A. robustus* had a larger frame and larger jaws and teeth than the others.
3. Anthropologist Donald Johanson maintains that *A. afarensis* is the most primitive of the genus, suggesting that it is ancestral to both the australopithecenes and humans.
4. The Leakey discovery, *Homo habilis,* is tentatively placed near the start of the human line. Somewhat closer to the human line is skull 1470, another Leakey find.
5. Fossils of *Homo erectus* are found throughout much of the Old World, persisting for as long as 1.3 million years and overlapping *H. sapiens.* The body was like a modern human's, but the brain was smaller (about 900 cc) and the brows and jaws larger. Their tools are classified as **Lower Paleolithic (Lower Old Stone Age).**
6. The earliest *H. sapiens* were the **Neanderthals** (*Homo sapiens neanderthalensis*), who ranged across the Old World from 40,000 to 70,000 years ago. Their bodies were like those of modern humans and their heads were similar except for a receding lower jaw and a sloping crown. All fossils since the Neanderthal's demise are *Homo sapiens sapiens* (modern humans). The fate of Neanderthal is unknown, but they may have simply become *H. sapiens.*

REVIEW QUESTIONS

1. What characteristics of the acorn worm suggest that it is related to both the echinoderms and the chordates? (669)
2. List and briefly describe the four leading chordate characteristics. Which appears in the embryos of humans but not in the adults? (670)
3. Considering what tunicates look like, what is the strongest evidence that they are really chordates? (670)
4. Using an example from the urochordates, explain the concept of neoteny. How does this concept help theorists bridge the gap between the earliest chordates and the vertebrates? (670)
5. Compare lancelets, juvenile lampreys, and larvaceae. What does the comparison suggest? (672)
6. List five important characteristics of vertebrates. (672)
7. List several characteristics of the ostracoderms. When were they prominent? (672–673)
8. Name the two survivors of the Agnatha. Describe their mode of feeding. (673)
9. Why are the placoderms so significant in the history of vertebrates? Briefly explain how their major evolutionary innovation came about. (674)
10. According to recent thinking, what is the origin

of the cartilaginous skeleton? How might it be adaptive to the shark or ray? (674)

11. Describe the digestive system of the shark, beginning with the teeth and ending with the vent. Include the spiral valve. (675)

12. Characterize the wide range of reproductive modes in the shark. Which is most advanced? (675–676)

13. What is the function of the advanced swim bladder? How does the fish compensate for various depths? (676)

14. Compare the gill structure of bony fishes and sharks and explain how gills are ventilated in each. (675, 677)

15. Name and briefly describe the three groups of bony fishes and rate them in terms of their numbers and importance today. (677–678)

16. In what way were the lobe-finned fishes significant to vertebrate evolution? (678)

17. Name the three groups of living amphibians and list common examples of each. (678)

18. Beginning with the salamanders, trace changes in body stance and limb position in vertebrates. (679)

19. In what way is the three-chambered heart important to improved respiration in the amphibian? Would you expect pulmonary branches of the circulatory system to enter the skin? Why? (679)

20. List two examples of adaptations that permit amphibians to reproduce out of water. (679)

21. List four specific ways in which reptiles have adapted to the terrestrial environment. One should involve the embryo. (680)

22. Trace the history of reptiles. When did they originate and in what geological period were they most prominent? Briefly, how is the demise of the ruling reptiles explained? (680–681)

23. List five specific ways in which the bird's body is modified for flight. (684–685)

24. Describe the four-chambered heart and complex respiratory system of birds and explain why both were important in the evolution of birds. (685)

25. List three characteristics that are most unique to the mammals. (685)

26. Describe five significant variations in the jaws and teeth of mammals, and explain how they relate to food-getting or defense. (686)

27. Suggest ways in which the mammalian brain and behavior differs from that of other vertebrates. (686)

28. How might the production of milk for the young by mammals have influenced learning and how might this have continued to influence the evolution of the brain? (687)

29. List the three subclasses of mammals and describe reproductive differences among them. (687)

30. List five ways in which the primates are different from other mammals. Why is it difficult to be specific? (690)

31. Trace the evolutionary history of the primates, mentioning the original type, and listing the main branches as they occurred. (690)

32. List four significant ways in which our anatomy is different from all other primates. (690)

33. Name the australopithecene species, briefly describe them, and suggest how they fit into the hominid evolutionary line. Where does *Homo habilis* fit in? Is it properly named? (693)

34. Describe *H. erectus*, its range, its technology, and degree of success. What happened to *H. erectus*? (695–696)

35. Compare the physical features of Neanderthal to modern humans. (697)

36. Suggest two possible fates for Neanderthal. (696)

Support and Movement

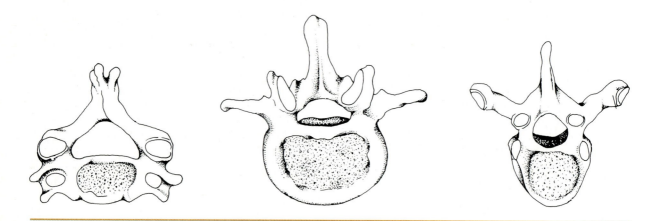

Compared to other celestial bodies, the earth is a rather unspectacular place. Here, for example, we lack those fascinating black holes through which they tell us time and space can be stretched. Temperatures are moderate compared to those of other planets. Even the surface here is relatively quiet compared to the great storms on Jupiter. Perhaps it is because the earth is so moderate and unspectacular that life has formed here. However, this is not an entirely benign place. The earth does present its guests with problems. Two of these problems relate to locomotion and support.

One challenge arises because the earth is not only large, it differs greatly from one place to another. It also changes. This means that some areas are more suitable to life than others, and areas that are quite hospitable at one time may be quite unsuitable (even deadly) at another time. Thus, many animals have had to develop the ability to move from one place to another. Some travel easily, such as the plover, whose migratory distance between its wintering and breeding grounds would take it around the world each year. Other animals, such as the lowly chitons, creep slowly over wave-washed rocks, scraping off morsels of slimy food as they go (Figure 28.1). Nonetheless, movement is as important to one as it is to the other.

The problem of support arises because, for some reason, celestial bodies with great mass tend to pull objects toward their surfaces. Not only do apples fall, but animals find it hard to jump. Gravity exerts a continuous tug on our bodies, seeking to flatten us against our great globe. But we resist with stiff skeletal systems that are able to withstand the persistent pull and, often, with muscles that contract and defy that pull. These structures help us not only to move from one place to another, but also to behave in an adaptive manner wherever we are. So let's have a closer look at the structures of support and movement among the animals that live on this small planet we call home.

INVERTEBRATE SUPPORT AND MOVEMENT

We are aware that many animals have skeletons of one type or another, and it is equally apparent that there are many soft-bodied animals that have no visible means of support. These animals gain some support from the cytoskeleton within each cell (see Chapter 4), but some are also supported by a **hydrostatic skeleton** that maintains form and provides a firm base for muscle movement.

The Hydrostatic Skeleton

A hydrostatic skeleton, as the name implies, involves fluids. These fluids exert pressure by pushing against a restraining outer wall. This stretches muscle fibers that can then contract against the fluids. Although hydrostatic pressure is important in many ways to most animals, the best examples of hydrostatic mechanisms of support and movement are found in the saclike coelenterates and in wormlike annelids and nematodes.

Coelenterates, you will recall, are comparatively simple animals with hollow, water-filled, saclike bodies. In the sea anemone's body, for example (see Chapter 25), the flow of seawater in and out is controlled by circular, valvelike, contractile filaments at the opening to the sac. With its body cavity filled with water and the "mouth" closed, the coelenterate body will remain firm in any position. The contraction of circular muscle fibers and the relaxation of longitudinal fibers extend the body into a slender, vaselike shape. But contraction of longitudinal fibers and relaxation of the circular fibers pulls the body down into a wider, flatter squat, as happens when the animal is startled. Muscles exert force by contraction only—relaxation is a passive process—so animal muscles usually work in opposing groups such as these.

The earthworm is a segmented animal, so the fluid mechanism is a bit more complicated than that of the anemone. The dense muscle fibers in the earthworm are arranged in an outer circular layer and an inner layer that runs along its length (Figure

28.1 ANIMAL MOVEMENT

Both the chiton and the plover use their skeletal and muscular structures in movement—but with quite different results. The chiton's travels, as it creeps along in its rocky marine world, are measured in centimeters. The plover, in contrast, is a world traveler capable of navigational feats that stagger the imagination.

28.2 BODY MOVEMENT IN THE EARTHWORM

The earthworm, an annelid, is capable of a wide range of movements. Like the anemone, the earthworm uses hydrostatic forces to maintain its body shape and as a resistant base for muscle action. The muscles are arranged in longitudinal bundles and in a circular sheet. Without hydrostatic pressure, the earthworm's wide range of movement would be seriously curtailed.

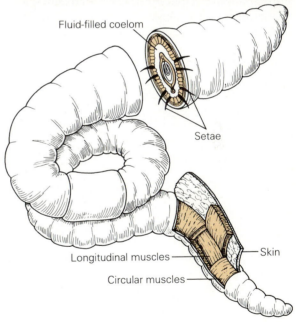

Fluid-filled coelom

Setae

Longitudinal muscles

Circular muscles

Skin

28.2). This arrangement permits a wide range of movements, but nevertheless, movement occurs according to the same simple principle described for the anemone. During burrowing, contraction of the circular muscles extends the body, whereupon the bristlelike setae dig in and take hold. Then the contraction of longitudinal muscles shortens and thickens the body, drawing the posterior end up and preparing the anterior end for another extension.

The fluid located in the body cavity, or coelom, makes the inching motion possible.

The roundworm (phylum Nematoda) has longitudinal muscles only; however, it uses a similar principal in its movement. The body fluid is held in one long, unsegmented body cavity, but the body wall is criss-crossed with numerous collagen fibers, enabling it to resist a high hydrostatic pressure within. Because it has only longitudinal muscles, the roundworm can't inch its way around. It can only whip back and forth, pushing against fine soil particles as it moves through the moist earth.

Exoskeletons

Now let's leave the concept of watery skeletons and talk about those that are more familiar: hard skeletons. These can lie outside the soft flesh or inside, but wherever they are they serve in protection, support, and locomotion. **Exoskeletons** are on the outside of the organism and are formed from secretions of the epidermal cells (cells derived from embryonic ectoderm). **Endoskeletons** are inside, surrounded by living tissue (and produced by mesodermally derived tissue).

Hardened outer exoskeletons are a characteristic of the phylum Arthropoda, the gigantic phylum that includes insects, crustaceans, spiders, and other joint-legged types. In additional to providing a firm muscle base and protection for the animal, the arthropod exoskeleton, or **cuticle** as it is more specifically known, helps retard water loss in terrestrial species. You may recall that its principal constituent is chitin, a complex carbohydrate. In crustaceans (lobsters, shrimp, crayfish, crabs, and so on), the chitin is hardened with calcium carbonate. The cuticle at the joints remains thin and flexible, but tough like leather.

Muscle action in arthropods is more or less the opposite of that in vertebrates. Note in Figure 28.3 that a muscle that *extends* a limb in an arthropod is located about where one would be that would flex the limb in a vertebrate. An arthropod leg is a hollow, jointed tube, and its activating muscles and tendons extend down its length to move the extremity from within. The muscles of arthropods are commonly attached to skeletal projections known as **apodemes,** ingrowths of the exoskeleton that serve as points of attachment.

It seems that exoskeletons should be rather cumbersome, but you have most likely observed that insects are quick, exceedingly agile little beasts.

28.3 ARTHROPOD MUSCLE ATTACHMENTS

The attachment of muscles in the arthropod limb (top) is the opposite of that in a vertebrate limb. In both, the muscles originate in the upper region and attach in the lower. The (a) muscles in both extend the limb, and the (b) muscles flex the limb. Note their opposite attachment with respect to the skeleton. Note also the apodeme shown in the insect exoskeleton.

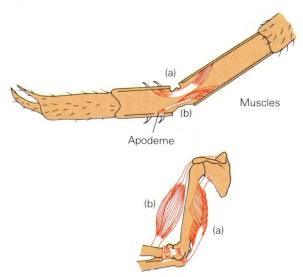

Also, we have all been impressed with the strength of insects, perhaps having watched an ant drag a dead insect many times its size. The strength and agility of insects, it turns out, is not due to a superior structure, but an inferior size. They can move so strongly and swiftly with an exoskeleton precisely because they are small. It has been calculated that an arthropod as large as a human would hardly be able to move at all. Thus, we don't have to be afraid of encountering science fiction's giant ants. We would be too fast and strong for them.

Insect Wings and Their Muscles.   The movement of thin, chitinous, sheetlike insect wings is somewhat unusual. There are no flight muscles in the wing itself, but there are two pairs of powerful muscles in the insect thorax. In some species these muscles move the wings *directly*, through a complex lever arrangement. In others, the movement is *indirect*, and depends on distorting the cuticle of the thorax, which, in turn, moves the wing. Both kinds of movement are explained in Figure 28.4.

Small insects can beat their wings at incredible rates. Midges ("no-see-ums") have been recorded at over 1000 wing beats per second—far faster than

28.4 INSECT FLIGHT MUSCLES

In some instances, the wings of insects are moved by pairs of *direct flight muscles* that attach to the wings themselves via a complex "seesaw" pivot **(a).** One set of muscles raises the wing, while the other lowers it. In other insects, *indirect flight muscles* move the wings by acting on the thorax **(b).** Vertically arranged muscle pairs contract, pulling the roof of the thorax down. As you can see by their attachment, this raises the wings. Downward movement is provided by longitudinal muscles, whose contraction pops the thoracic roof back up. The actual "lift" in insect wing action is provided by the constant changing of wing angles, which describes a figure 8 in each complete cycle. The wings tilt downward on the downbeat, scooping air over the leading edge. During the upbeat, the wings tilt upward, spilling the air over the trailing edge. So far, no one has been able to explain how hovering insects do their stunts.

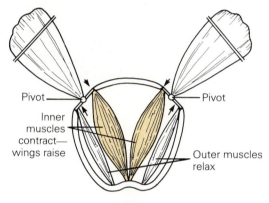
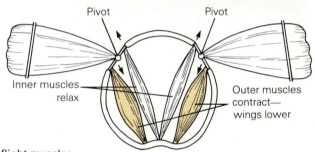

(a) Direct flight muscles

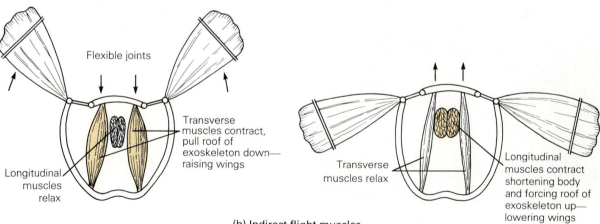

(b) Indirect flight muscles

nerve impulses can be generated. How are the wings activated? Experiments on blowflies reveal that although their wings beat at 140 cycles per second, the muscles receive only about 12 impulses per second, or one impulse for every 11 or 12 wing beats. Apparently, the muscles themselves are capable of maintaining the cycle between neural impulses through some kind of resonating mechanism.

Mollusk Shells: A Type of Exoskeleton.
Mollusks, such as snails, clams, and nautiluses, generally have both hydrostatic skeletons and hard exoskeletons. The exoskeletons are what we generally refer to as their "shells" and are secreted by the fleshy mantle. The shells of clams and many other bivalves are used chiefly for protection. These animals move by using muscles to extend the foot and hydrostatic pressure (blood) to enlarge its tip, forming an anchor. The muscles then draw the body forward. On the other hand, snails use their shell as a muscle base and protective house into which they withdraw when threatened.

28.5 CUTTLEFISH AND CUTTLEBONE

The cuttlebone seen here is actually an internalized exoskeleton of the cuttlefish. This highly porous flotation device is located just behind the head and along the inside of the dorsal surface. Squid, close relatives of the cuttlefish, also have a cartilaginous structure, but it has only a skeletal function.

28.6 EXOSKELETON GROWTH IN MOLLUSKS

Unlike the arthropods, the mollusks grow continuously without molting. This is accomplished through the addition of shell material, secreted by the mantle. Limpets and bivalves form a cone as they grow. The *Nautilus*, on the other hand, adds new shell material unevenly, forming a continuously enlarging, logarithmically spiraled cone. The *Nautilus* secretes cross walls as it grows, and it lives in its most recent compartment.

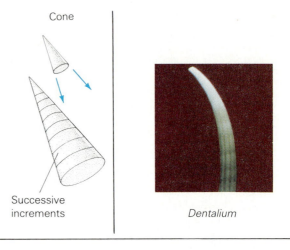

Cone

Successive increments

Dentalium

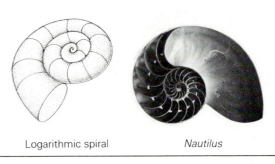

Logarithmic spiral

Nautilus

One cephalopod mollusk, the cuttlefish, has developed a new use for its internalized skeleton (Figure 28.5). Its skeleton is cuttlebone—the beak sharpener of parakeets and canaries. The cuttlebone in the live cuttlefish is used as a flotation device. It has many hollow chambers supported by strong calcium carbonate walls. Water is removed from the cuttlebone chambers, leaving only a vacuum behind—a rather peculiar situation.

Although the shells of mollusks, like arthropod exoskeletons, are secreted by epidermal tissue, unlike arthropods, mollusks don't shed their exoskeletons as they grow. Instead, their shells grow by increments at the edges. This puts a particular kind of constraint on the sorts of shells mollusks can develop. It turns out that the problem is one of geometry. There are only a few rigid surfaces that can grow by increments and still retain the same shape (think about it). One is a special kind of spiral called the **logarithmic spiral;** another is a **simple cone.** Both of these shapes, it turns out, are common in mollusks (Figure 28.6).

28.7 THE ENDOSKELETON OF ECHINODERMS

(a) In the starfish, the skeleton consists of numerous plates composed of calcium carbonate. In fact, the familiar spines are actually overgrown plates. The crusty skeleton appears to be outside the animal, but it is actually an endoskeleton since it is covered with a thin epidermis and is secreted by mesodermally derived cells. **(b)** The sea urchin "test" reveals the plates very clearly.

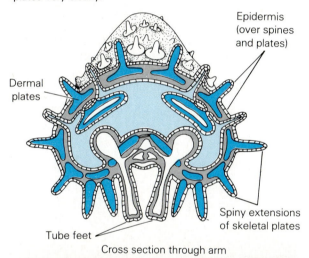

Epidermis (over spines and plates)

Dermal plates

Spiny extensions of skeletal plates

Tube feet

Cross section through arm

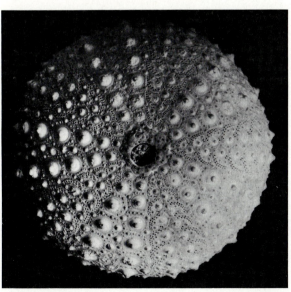

Endoskeletons

The spiny-skinned echinoderms, such as starfish, also have endoskeletons, just as do the vertebrates. This may not be too surprising if you remember that both groups are deuterostomes, quite distant from the protostomes with their exoskeletons. Under the echinoderm epidermis, the thick dermis (of mesodermal origin) secretes a series of plates that fit closely against each other. The plates themselves are composed of many tightly interlocked little spined bodies, formed chiefly of calcium carbonate, a common material in marine animals. As the skeletal plates form, they erupt over the entire upper surface of the starfish, creating short, blunt, immovable spines. The spines of sea urchins, on the other hand, are very long, often very sharp, and movable (Figure 28.7).

VERTEBRATE SUPPORTING STRUCTURES

Before we get into vertebrate supporting structures, we will look briefly at how the vertebrate body is organized. The vertebrate body is often quite complex, and there is a great amount of variation among and even within the various classes. Yet, interestingly, vertebrates have just four basic tissue types that contribute to the many organs and organ systems. Let's look more closely at these tissues.

Vertebrate Tissues

Tissue, you will recall, is an aggregation of similarly specialized cells, together with their noncellular *matrix* (supporting substance). Vertebrate tissues and those of all other metazoan animals fall into four major categories: **epithelium, connective tissue, muscle,** and **nerve.** Muscle tissue is explored in detail later in this chapter, and nerve tissue is the subject of chapters to come, so here we will concentrate on the first two.

Epithelial Tissue.   **Epithelium,** or epithelial tissue, forms most surface linings or coverings of the organism, both interior and exterior. Thus the *epidermis*—the outermost layer of the skin—is epithelial. Epithelial tissue also lines the mouth, the nasal cavities, the respiratory system, the coelom, the tubes of the reproductive system, the gut, and the body cavities, and forms the interior, or **endothelium,** of blood vessels. The lining tissues of glands contain a thickened epithelium called **glandular epithelium.** The cells of the glandular epithelium both line the ducts (openings) and secrete whatever that gland secretes. As you might expect, epithelial tissues differ greatly in different parts of the body (Figure 28.8).

Connective Tissues. Connective tissue is found throughout the vertebrate body. This is probably a good thing since its role is essentially to hold the body together. Connective tissue consists of cells that produce a noncellular **connective tissue matrix** (Figure 28.9). Examples of familiar connective tissues include bone, cartilage, ligaments, tendons, and even blood. (The matrix of blood is called plasma).

The principal and most abundant substance in most connective tissue matrices is **collagen,** the fibrous protein that accounts for at least a third of the total body protein in the larger vertebrates. Collagen is the "glue" of the animal body and forms extremely tough structures such as the cornea of the eye, the tendons and ligaments, and the all-impor-

tant disks that cushion the spine. Collagen is even the principal nonmineral component of bone.

Collagen is produced and secreted by cells known as **fibroblasts.** As the protein is secreted, it takes the form of tough, lengthy fibers that form the connective tissue matrix (Figure 28.10). These fibroblasts are vitally involved with wound healing and the mending of broken bones. In wound healing, a number of different kinds of connective tissue cells cooperate, reforming normal structures and, when necessary, forming tough collagenous **scar tissue.**

In the mending of bones, fibroblasts form a thin sheet of connective tissue around bones and temporarily fill the fracture site with tough fibers of collagen. Eventually, the new bone forms and hardens, permanently fusing the break.

Vertebrates: Endoskeletons of Bone

As vertebrates, when we think of supporting structures, we normally don't think of cuticles, shells, or spiny plates. We think of bone. Bones are the light, hardened structures that hold us erect (keeping us

28.8 EPITHELIAL TISSUES

Epithelial cells vary depending on their location and tasks. Simple linings are formed by flattened squamous cells **(a)**, which may be one or more layers. Many layers are more common in regions of rapid wear and replacement. The respiratory linings—nasal passages, trachea, and bronchi—contain columnar cells **(b)**, commonly including glandular, or secretory, and ciliated cells. Glands may also be formed by epithelial, secretory cells **(c)** that empty their secretions into a common duct.

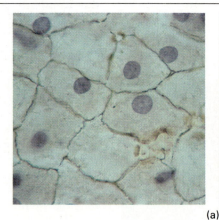
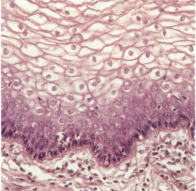

(a)

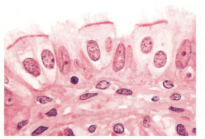
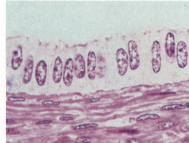

(b)

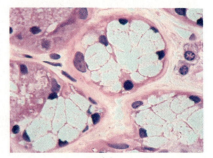

(c)

28.9 CONNECTIVE TISSUES

In loose connective tissues (a), the cellular elements lie scattered in a loosely arranged matrix of dense collagen and fine elastic fibers. Loose connective tissue occurs beneath the skin and epithelial linings, and surrounds blood vessels and nerves. Dense connective tissue (b), such as that forming tendons and ligaments, contains much more of the fibrous matrix arranged in parallel rows, with the fibroblasts crowded among individual collagen fibers.

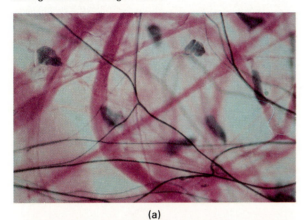

(a)

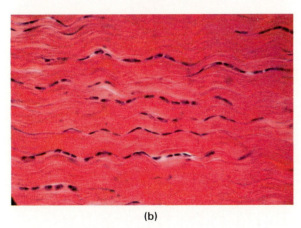

(b)

from lying about, dotting the landscape in great heaps of protoplasm) and shield our more delicate parts from damaging environmental objects. What is bone, and what gives it its toughness?

Except for the cartilaginous fishes (see Chapter 27), vertebrates have skeletons composed primarily of bone (although some cartilage usually persists, as we shall see). The bony skeleton is important in at least five ways: (1) it supports the body and (2) serves as an attachment for muscles. (3) Certain parts of the skeleton are protective. In particular, the skull and rib cage protect the delicate structures within; the skull is nature's version of the motorcycle helmet. Further, some bones (4) produce red blood cells in their interior spaces, and (5) all bones act as a natural reservoir for the body's calcium.

Let's look at the structure of bone and see why it works so well.

The Structure of Bone. Typically, the long bones of the limbs are surrounded by a thin membrane of connective tissue called the **periosteum.** Within, we find two kinds of bone, **spongy** and **compact** (Figure 28.11a). Spongy bone is well named since it is porous and spongelike, made up of a weblike structure of hard bone (laid down along lines of stress so that this seemingly fragile bone is very strong). The spaces in the web are filled with soft tissue. Spongy bone is found at the enlargements at each end of long bones. Some bones, such as the ribs, sternum, vertebrae, and hip bones, contain **red marrow,** the site of red blood cell production.

Compact bone makes up the shaft—the cylindrical, lengthy portion of the long bone—and is

28.10 COLLAGEN

(a) Collagenous fibers form a tough binding material that literally holds the vertebrate body together. Logically enough, collagen is the primary material of tendons—strong, cordlike structures attached to bones. (b) Individual collagen fibers in tendons form a wavy pattern, with distinct cross-banding of dense protein.

(a)

(b)

From *Tissues and Organs—A Text-Atlas of Scanning Electron Microscopy,* by Richard G. Kessel and Randy H. Kardon. W. H. Freeman and Company, copyright © 1979.

28.11 THREE VIEWS OF BONE

(a) The sectioned long bone consists of a hard shaft and a central cavity of yellow marrow. Spongy bone is seen near the joints. **(b)** A cross section of bone reveals its inner organization. **(c)** Numerous Haversian canals, responsible for carrying blood vessels and nerves, are surrounded by hardened con-centric rings, or lamellae, containing bone cells (osteocytes). Such cells are entombed in lacunae, minute cavities that communicate with each other via minute crevices called canaliculi. Such structures are grouped into Haversian systems.

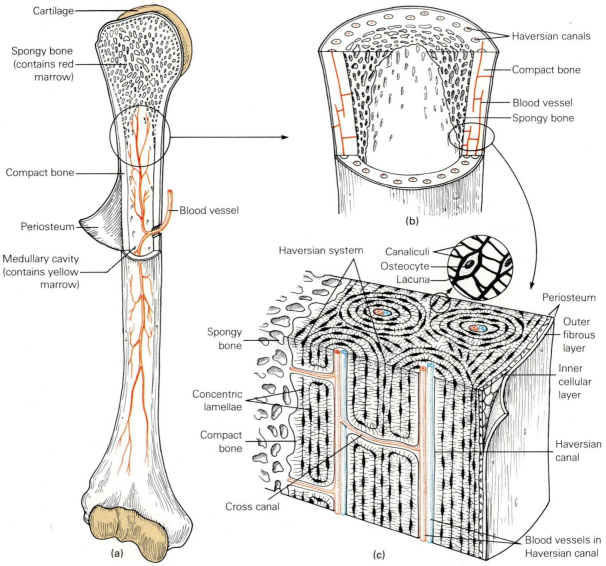

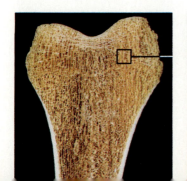

Light microscope view of bone tissue

SEM of Haversian system

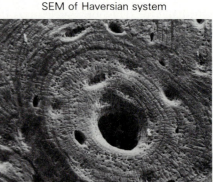

From Tissues and Organs—A Text-Atlas of Scanning Electron Microscopy, by Richard G. Kessel and Randy H. Kardon. W. H. Freeman and Company, Copyright © 1979.

thick and dense. Its central cavity contains **yellow marrow,** a dense, fatty material. Despite its stone-like appearance, compact bone is definitely a living tissue. It contains numerous metabolically active bone cells, as well as cells of nerves and blood vessels. Most of the bony mass, however, consists of hardened calcium phosphate in a collagen matrix.

The Microscopic Structure. Under the microscope, thin sections of compact bone (Figure 28.11b) reveal intricate, repeated units of structure called **Haversian systems,** or **osteons.** Each Haversian system consists of a **central canal** that contains blood vessels and nerves, and a surrounding region of concentric rings or **lamellae,** consisting of calcified bone. The rings are, in effect, tubes within tubes—a laminated tubular construction that imparts great strength and resiliency to the bone.

It is within the lamellae that we find the **osteocytes**—the bone cells. The main body of each osteocyte resides in a tiny cavity, or **lacuna** (plural, *lacunae).* Each lacuna is interconnected with others via minute canals, aptly named **canaliculi,** which pass throughout the hardened bone. The osteocytes touch each other via long projections through the canaliculi. Some of these extensions reach the central canal, permitting the cells to carry on an exchange of materials with the circulatory system. The osteocytes are important in calcium deposition and withdrawal, processes that are controlled by hormones (see Chapter 33).

The Skeleton

Vertebrate skeletons are usually divided, for convenience, into **axial** and **appendicular** components. The axial skeleton includes the **cranium** (skull), the **vertebral column,** and, when present, bones of the **thorax** (chest). The appendicular consists of the limbs (always two or fewer pairs) and the two girdles, **pelvic** (hip) and **pectoral** (shoulder). Much of the appendicular skeleton, by the way, is absent in most snakes, which is why they are not broad-shouldered. Before getting into the details of all this, let's pause to consider a few joints.

The Joints. The various bones of the body meet at **joints.** Each joint allows only a certain kind and degree of movement and some, such as the **sutures** between the bones of the skull, allow no movement at all (try flexing your skull). The more movable joints, though—for example, the knee, elbow, or hip—are both complex and elegant. The articulating (contacting) surfaces are covered with a thin layer of exceedingly smooth cartilage. The bones

28.12 JOINTS IN THE HUMAN SKELETON

The range of body movements in humans and other vertebrates is made possible by joints in the skeleton. Most movement occurs in the following types of joints: **(a)** gliding—a slight back-and-forth movement, **(b)** pivotal—turning back and forth in one plane, **(c)** ball-and-socket—free rotation, and **(d)** hinge joints—extending and retracting in one plane.

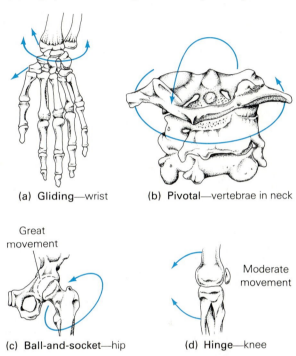

(a) Gliding—wrist

(b) Pivotal—vertebrae in neck

Great movement

(c) Ball-and-socket—hip

Moderate movement

(d) Hinge—knee

are held together by short bands of white fibers called **ligaments,** which can be torn or "sprained." Where the ligaments form an enclosing capsule they produce a **synovial joint.** The entire joint is constantly lubricated by **synovial fluid,** the secretion of a **synovial membrane** that lines the capsule. The presence of smooth cartilage and lubricating synovial fluid is highly advantageous. It helps keep the joint from wearing out by reducing friction and allows for quieter movement. As you can imagine, a hungry lion creaking and squeaking around in the grass is likely to come up short.

The movable joints are generally named according to the type and degree of their movement. Figure 28.12 depicts the **gliding, pivotal, ball-and-socket,** and **hinge** joints.

The Axial Skeleton. The axial skeleton includes the skull and jaw, structures that have traditionally yielded a great deal of information about the evolution of vertebrates. In addition, the axial skeleton includes the vertebral column and the rib cage. Let's have a closer look, now, at the skull.

The Skull. Skulls from the vertebrate classes make a rather nice comparative series. In the fishes the cranium consists of a large number of loosely connected, individual plates. The jaws are simple and form a single-directional hinge, which means they can bite but they can't chew by moving the jaws sideways. In fish, both the top and bottom jaws move, a feature still found in some birds but one that has been lost in the mammals; our upper jaw, the *maxilla*, is fused to the skull.

As we compare the skulls of other vertebrates

28.13 COMPARISON OF THE SKULL STRUCTURE IN VERTEBRATES

In the evolution of the vertebrate skull, the trend has been toward the fusion of bones. We see less fusion in the skulls and jaws in fishes, amphibians, and reptiles, and more rigid skulls in birds and mammals due to fusion. Note that the mammalian jaw, while comparatively simple, is very movable. Its great mobility permits chewing—a seemingly common act, but actually unusual among vertebrates.

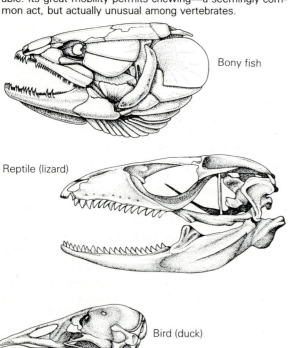

Bony fish

Reptile (lizard)

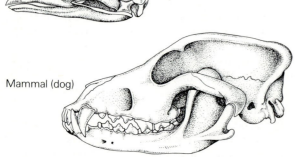

Bird (duck)

Mammal (dog)

28.14 THE HUMAN SKULL

The human skull consists of 28 individual bones, many of which fuse together in joints called sutures. At the base of the skull is a large opening known as the *foramen magnum*, through which the spinal cord passes into the brain. (The two smooth surfaces, the occipital condyles, articulate with the first cervical vertebra, permitting "yes" movements.) Other external openings include the auditory canal, optic foramen, nasal cavities, and various minor openings for blood vessels and nerves.

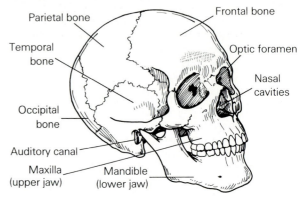

Parietal bone
Frontal bone
Temporal bone
Optic foramen
Occipital bone
Nasal cavities
Auditory canal
Maxilla (upper jaw)
Mandible (lower jaw)

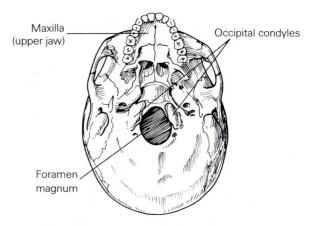

Maxilla (upper jaw)
Occipital condyles
Foramen magnum

in the order in which we believe they progressed through their evolution, the trend is toward fusion of parts, resulting in fewer bones (Figure 28.13). (For an account of an interesting evolutionary problem with the vertebrate skull, see Essay 28.1). Amphibians and reptiles have more mobility in the jaws than do fishes, but like birds, they swallow food whole. In fact, the jaws of snakes unhinge to swallow prey that is often larger in diameter than the snakes themselves. They are able to do this because the lower jaw is hinged to the quadrate bone, which is loosely attached to the skull and can be pulled away from it.

In vertebrates the skull not only protects the

Evolution of the Mammalian Skull

The history of the mammalian skull provides an unusual illustration of evolution at work. We could call this example "How to repair a mistake, or you can't hardly get here from there." Originally, most of the bones of the skull weren't part of the axial skeleton at all. The axial skeleton consisted of the vertebrae, the ribs, the gill arches (later, jaws), and a tiny *braincase,* suitable for a tiny brain. Most of the bones of the head were *dermal bone,* protective plates of bone developed in the lower layers of the skin, as we see in fish today.

When fish became terrestrial, the dermal bone plates fused and hardened over time to become the skull of the ancient amphibians and early reptiles. There were holes for the eyes and a simple, membranous, partly open braincase deep inside. There is one problem with this design: the jaw muscles are inside the skull, between the skull and the open, membranous braincase (a). One result is that when the jaw muscles contract and therefore bulge, there is no place for the bulge to go except to squeeze the brain. It would be much better to have the jaw muscles on the outside of the skull.

Evolution, as Darwin argued tirelessly, tends to proceed by small increments. Thus, it would seem that there could be no way to get the heavy jaw muscles from the inside of the skull to the outside without leaving a string of maladapted monsters halfway between.

But we now know that, in the evolutionary descent of reptiles, mem-

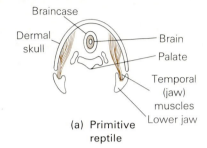

(a) Primitive reptile

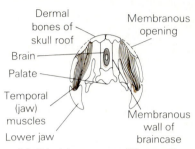

(b) Primitive mammallike reptile

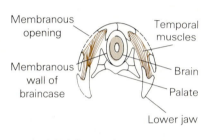

(c) Advanced mammallike reptile (therapsid)

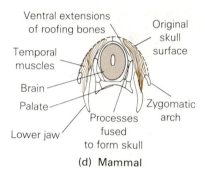

(d) Mammal

branous openings gave the bulging jaw muscles someplace to go, relieving pressure inside the skull. Meanwhile, flangelike projections from the dermal bones of the roof of the skull developed between the jaw muscles and the brain, further protecting it. At the same time, similar processes grew up from the floor of the braincase (b).

Eventually, the protective flanges, or side processes, that developed over the brain fused, creating a new and more solid encasement for this delicate—and enlarging—organ. And the openings in the dermal skull grew

larger and larger (c,d), until eventually all that was left of the original skull outside the jaw muscle was the thin, fragile, *zygomatic arch* (cheekbone). In mammals, the zygomatic arch served as the origin of a new muscle, the *masseter,* which allows side-to-side chewing. It took many tens of millions of years to accomplish this by evolution and natural selection, with each step along the way a slight improvement over the one before. But in the end, the jaw muscles were outside the skull where, logic seems to tell us, they belonged in the first place.

brain but also houses a complex group of receptors. As you know, many of the senses are "up front," including the organs of sight, hearing, balance, smell, and taste. The sensory structures are actually isolated from the brain itself, but the major nerve pathways enter the braincase via openings known as **foramina** (singular, *foramen*). The largest of these, the **foramen magnum,** is at the base of the skull and

serves as an entrance for the spinal cord. These and other features of the vertebrate skull are shown in Figure 28.14.

The Vertebral Column. Throughout the evolution of vertebrates, the backbone has carried out its supporting role in a number of ways. For example, in the vast majority of terrestrial animals that walk on all fours, the backbone has often had to support the great weight of the belly. On the other hand, aquatic species are buoyed by water, and their reduced weight places far less stress on the vertebral column. Thus, the backbone or vertebral column can be quite varied and specialized.

The backbone consists of a series of bones that form a flexible axis for the body. Each bone is individually known as a **vertebra,** and their number varies greatly from one class of animal to another. The reptiles have the record, with as many as 400 (most of them very similar to each other). The birds, on the other hand, have comparatively few, partly because the posterior ones are fused. All mammals, from giraffes to shrews to whales, have exactly seven vertebrae in the neck, although the number in the remainder of the column varies.

Humans have 32 to 34 vertebrae, including three to five that fuse to make up the tiny, finger-like **coccyx** (tailbone). All vertebrae, other than those that have fused, are separated by **intervertebral disks** of **fibrocartilage.** Fibrocartilage is cartilage containing numerous reinforcing strands of collagen. Each disk has a compressible "soft cen-

28.15 THE HUMAN VERTEBRAL COLUMN

The vertebral column in humans, allowing for some variation in the coccyx, consists of about 33 vertebrae. Note that there are seven cervical, twelve thoracic, five lumbar, five (fused) sacral vertebrae, and three or four in the coccyx. The lateral view reveals that the vertebral column is curved. Also notice surfaces (in color) that articulate with adjacent vertebrae or with the ribs. At the right, the vertebrae from three regions are shown in detail.

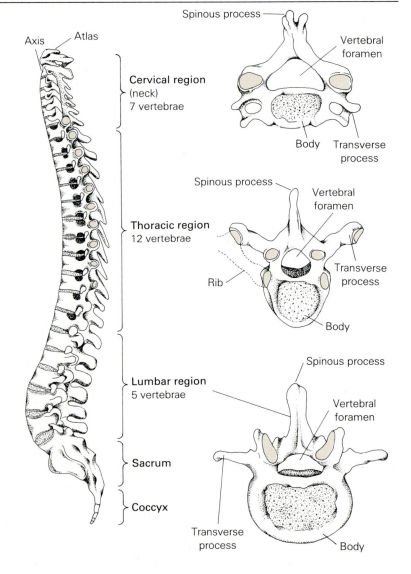

ter'' (sometimes unblinkingly described as a ''chewy center''), which provides cushioning and permits lateral movement. Unfortunately, intervertebral disks can and do rupture, causing the softer center to protrude outward, which can result in debilitation and great pain.

In addition to its supportive role, the vertebral column protects the **spinal cord.** The spinal cord passes through the **neural canal,** which is formed by a series of vertebral foramena. Certain bones of the vertebral column also provide attachments for most of the ribs.

The human vertebral column contains five regions (Figure 28.15). The bones of the neck comprise the **cervical** region, which contains the seven vertebrae common to all mammals. The first and second cervical vertebrae, called the **atlas** and **axis,** respectively, support the skull and assist the yes/no motions of the head. Next are the 12 **thoracic** (chest) vertebrae. Each of these has well-developed lateral processes that articulate with the ribs. A third spinal region contains the five **lumbar** vertebrae, which lie between the thoracic region and the sacrum. These are the largest, with a more massive central body and heavier spines. Their large size is to be expected since the vertebrae carry an increasing load as the column descends. The lumbar region is followed by the **sacrum,** a triangular group of fused vertebrae wedged between the right and left hip bones, thereby completing the pelvic girdle. Finally, below the sacrum is the **coccyx,** a vestigial tail.

Besides the skull and the vertebral column, the axial skeleton includes the ribs and the **sternum** (breastbone). The sternum is composed of three parts: the upper **manubrium,** middle **body,** and lower **xiphoid process.** The rib cages of many species have become highly modified, producing some interesting variations. For example, frogs and toads have very little rib development. They also lack a diaphragm and breathe by moving the muscles in the mouth and throat. Another unusual variation of rib structure is found in the turtle, whose ribs and vertebrae are joined to the platelike bones of the **carapace** (upper shell). In birds, the sternum gives rise to the **keel,** which has a large, flat vertical projection to which the powerful flight muscles are attached. Humans have 12 pairs of ribs. All articulate with the thoracic vertebrae, but not all connect to the sternum (Figure 28.16).

The Appendicular Skeleton. The appendicular skeleton includes the two girdles (rings) of bone and bones of the limbs (arm and leg bones).

28.16 THE HUMAN RIB CAGE

The human rib cage consists of the sternum (three bones), twelve pairs of ribs, and their related cartilages. Note that only the first seven pairs of ribs articulate directly with the sternum. The eighth, ninth, and tenth pairs (''false ribs'') articulate with the cartilage of the seventh pair. The eleventh and twelfth pairs do not attach at all in the front, but are ''floating ribs.'' Each rib articulates with one of the twelve thoracic vertebrae to complete the rib cage in the back.

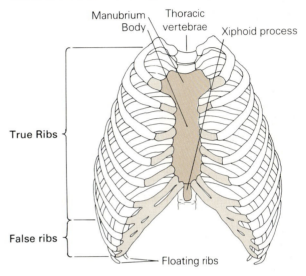

The Pectoral and Pelvic Girdles. Humans (and birds) are unusual among vertebrates in that they are bipedal (walk on two feet). Therefore, our pectoral girdles, greatly relieved of their weight-supporting burden, are considerably reduced. The human pelvic girdles are also different even from the other primates, which are not well adapted to an upright gait. As you can see in Figure 28.17, our upright posture is associated with a distribution of weight that is unusual among primates. The upper torso rests entirely on the **sacroiliac joint** (the joint between the *sacrum* and the *ilium*), with some help from certain back muscles that distribute the weight to the ilium.

Each half of the pelvic girdle is formed from three fused bones: the **ilium** (hip), the **ischium** (the one you sit on), and the **pubis** (the pubic bone—in front, where you would imagine). The two halves join ventrally but remain slightly separable, forming the **pubic symphysis.** Dorsally, the halves join the sacrum, forming the sacroiliac joint. The joint transfers considerable weight from the vertebral column and upper torso to the pelvic girdle, as we have seen. Where the three pelvic bones join, they form the **acetabulum,** or hip socket, a deep depres-

28.17 THE APPENDICULAR SKELETON OF THE HUMAN AND THE GORILLA

Of all the primates, only humans are truly bipedal. Compare the pectoral and pelvic girdles, position of skull, relative length of arms and legs, and the shape of the foot in the human and the gorilla. Note the horizontal position of the pelvis in the gorilla compared to its vertical position in the human. The ischium is quite long in the gorilla and flares in a forward direction. The pectoral girdles are more generally similar, with greater mass in the scapula of the gorilla, as might be expected in a heavy, brachiating primate. Note the relative lengths of arms and legs in each.

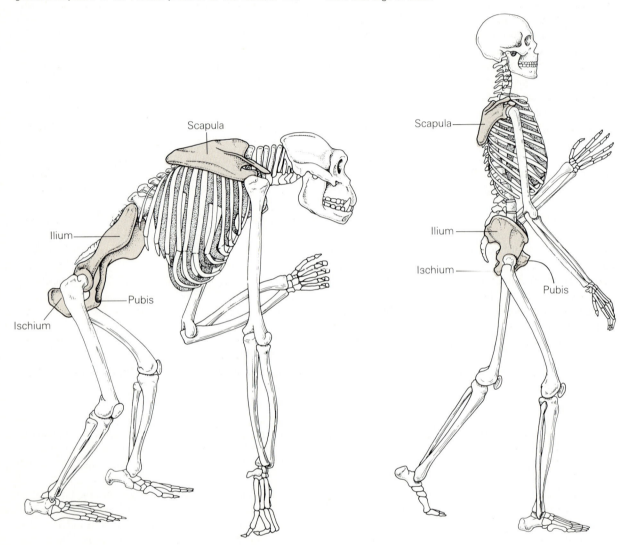

sion that receives the spherical head of the thigh bone, forming the ball-and-socket joint between the upper leg and hip. Note in Figure 28.18 that the pelvic bones of women are shaped somewhat differently than those of men.

The pectoral girdle consists of two **scapulae** (shoulder blades) and two **clavicles** (collar bones). The shoulder blade is a triangular, flattened bone whose outer margin is thicker and forms part of the shoulder joint, a ball-and-socket arrangement. In movements such as the "bear hug," the scapula glides outward over the rib cage. The clavicles are attached to the ventral (front) side of the shoulder joint (Figure 28.19a). You can feel the clavicles move as you shrug your shoulders.

Vertebrate Limbs.   As we have mentioned, vertebrates have no more than two pairs of legs. Those with fewer than two are rather rare and include snakes, whales, dugongs, and kiwis. We see in the flying vertebrates limbs that have been greatly modified for locomotion. In the two living groups, birds and bats, the forelimbs and especially the **digits** (fingers) have been greatly modified for flight (see Figure 17.14).

With the exception of the thumb and the arch of the foot, human appendages are rather generalized (see Figure 28.19). In Chapter 27 we described in detail our famous opposable thumbs, our arched foot, and well-developed gluteus maximus (buttock muscles), each of which distinguishes us from other primates.

The mammalian forelimbs, such as our arms, are made up of three long bones: the large **humerus** of the upper arm and the **radius** and **ulna** of the lower arm. The elbow is a complex, pivoting hinge joint while the wrist forms a flexible hinge (see Figure 28.19a).

The **phalanges** (the bones of the fingers and toes) are attached to the **metacarpals,** or hand bones. The finger joints are essentially hinges, but they are capable of some lateral movement, as we see when the index finger is wagged back and forth as a warning. The metacarpals join the wrist, which is composed of eight small bones, the **carpals.** These are all somewhat blocky or *cubic* in shape and form gliding joints (see Figures 28.19a and b).

The human leg, like the arm, consists of a single upper bone, or **femur** and two lower bones, the tibia and **fibula,** as shown in Figure 28.19c. Most vertebrates have this general plan. The upper and lower bones meet at the knee joint, but only the tibia articulates with the femur. The fibula articulates with the tibia just below the knee. A fourth bone, the chestnut-shaped **patella,** or kneecap, which is embedded in a tendon, covers and protects the juncture of the tibia and femur to complete the knee joint. The two rounded heads of the femur rest in a seemingly precarious manner on the flattened head of the tibia. But the presence of several strong ligaments (the ligamentous capsule) and the attached muscle tendons make the joint exceptionally strong. As strong as it is, though, the knee joint cannot withstand very much twisting.

At the foot, the tibia and fibula articulate with a large ankle bone called the **talus,** forming a versatile hinge joint that permits the foot to be flexed, extended, and rotated. The **tarsals** form the ankle. There are seven tarsals in all, the largest being the **calceneous,** a long bone that forms the protruding heel and receives the large **Achilles tendon** from the calf muscle. This joint works on the principle of a lever and fulcrum, but has rather poor mechanical advantage since it requires considerable muscle power to raise the body up on the toes.

The **metatarsals** comprise the rest of the foot, and the phalanges make up the toes. We might add here that the arch of the foot, unique to humans, is not formed by the bone structure. Instead, it is formed by the bindings of its ligaments.

28.18 THE HUMAN PELVIS, MALE AND FEMALE

Some of the differences between male and female pelvises can be detected at a glance. The flared pubic arch and broader pelvic inlet in females are accommodations to child-bearing.

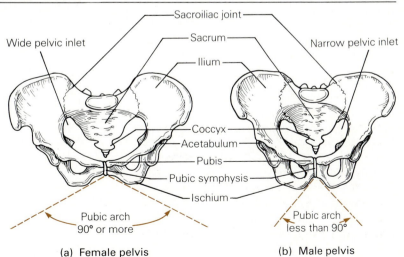

(a) Female pelvis   (b) Male pelvis

28.19 THE HUMAN APPENDAGES

(a, b) The arrangement of the arm bones is such that the elbow is both a hinge and a slightly rotating joint. The wrist is also a hinge, but does not in itself rotate. Grasping the wrist firmly and moving the hand will reveal a "back and forth" wigwag motion in the hand. (c) The hip bones form a versatile ball and socket joint, while the knee is a simple hinge with very limited rotation. The ankle is quite mobile as a hinge and rotating joint. (d) The bones of the foot form a definite arch, but its curve is maintained by ligaments. (e) Note the complex arrangement of ligaments and tendons about the otherwise flimsy-appearing knee joint. Its ligaments make it one of the body's strongest joints.

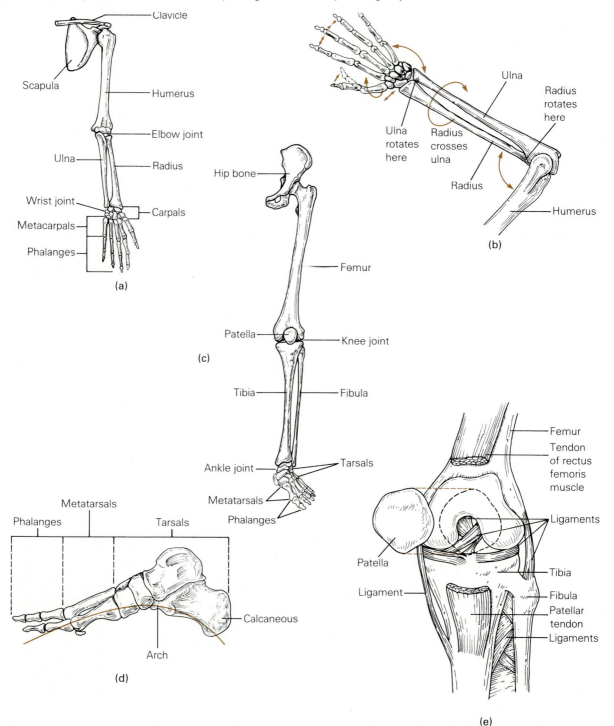

MUSCLE: ITS ORGANIZATION AND MOVEMENT

In a sense, the ability of a cell to alter its shape has altered the shape of our world. Muscular movement, after all, involves no more than altering the cell shape.

Types of Muscle Tissue

Vertebrate muscles can be divided into three types according to their microscopic structure, the nerves that activate them, and where they occur. The three types are **smooth muscle, cardiac muscle,** and **skeletal muscle.**

Smooth Muscle.   Smooth muscle is also called **involuntary muscle** and **visceral muscle,** terms that tell us that we have little voluntary control over its action and that it can be found in the gut. In addition to the gut, smooth muscle occurs in various internal organs: in the walls of blood vessels, at the base of every hair, in the iris of the eye, and in the uterus and other reproductive organs. You may have noticed that your hair rises involuntarily under eerie conditions, or that you can't stop your stomach from growling when meeting your date's parents, or that you can't call back a blush once it has begun. Smooth muscle even influences sex since it is the muscular control of blood entering and leaving the genitals that accounts for the responses in erectile tissue. Smooth muscle is under the control of the **autonomic nervous system,** which controls many activities that occur below the conscious level.

Smooth muscle tissue is, logically enough, smooth in appearance, lacking the highly patterned striations or cross stripes seen in the other types of muscle (Figure 28.20a). It tends to occur in sheets rather than in bundles. The individual cells are tapered (or spindle-shaped) and have a single nucleus. In spite of its differences from cardiac and skeletal muscle, smooth muscle contains the same contractile proteins.

Cardiac Muscle.   Cardiac muscle is heart muscle. While it is unique, cardiac muscle has some things in common with both smooth and skeletal muscle. For example, like smooth muscle, the heart is largely under control of the autonomic nervous system. Like skeletal muscle, the cardiac muscle cells are cylindrical and striated (striped). But cardiac muscle fibers are branched; thus, heart muscles contract in a twisting motion that "wrings" the chambers with each beat. Also, unlike skeletal or smooth muscles, the individual cells are separated by **intercalated disks**—prominent interlocking cell borders where the membranes are folded and tightly compressed, permitting an efficient transmission of the neural impulses that control contraction (see Figure 28.20b). The most impressive feature of cardiac muscle is its tireless, rhythmic beating.

Skeletal Muscle.   Skeletal muscle is also known as **voluntary muscle** and as **striated mus-**

28.20 HUMAN MUSCLE TYPES

Smooth muscle (a) is involuntary, consisting of long, spindly cells, each containing a single nucleus. Smooth muscle often occurs in sheets. Cardiac muscle (b) is also involuntary, but unlike other types, the tissue branches and rebranches and is interrupted by intercalated disks. Like skeletal muscle it is heavily striated (cross-banded). Skeletal muscle (c) is, of course, largely voluntary—moved at will. It is heavily striated and multinucleate.

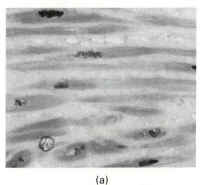

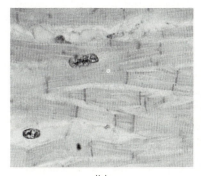

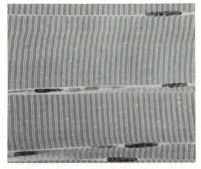

(a)   (b)   (c)

cle. It is controlled by motor neurons that are part of the **somatic** (voluntary) **nervous system.** As for its three names, it is *skeletal* because these are the muscles that move the skeleton; it is *striated* because of its molecular organization; and *voluntary* because it is *possible* to move the muscles at will. Skeletal muscle can contract much more rapidly and forcefully than the other types, but it cannot sustain such contraction without tiring.

Skeletal muscle cells are striated, unbranching, cylindrical, and multinucleate, with the nuclei lying just beneath the cell surface (see Figure 28.20c). Because of their multinucleate condition and their great length (several centimeters in some instances), they are usually referred to as fibers rather than cells. Each fiber is packed with precisely arranged contractile proteins. We will return to the details of this arrangement shortly, when we look at muscle ultrastructure and the contractile mechanism.

Gross Anatomy of Skeletal Muscle

Skeletal muscle fibers are bound into bundles known as **fascicles,** which are surrounded by a tough sheath of connective tissue. The fascicles in turn are bound together to form the muscle, and it is the fascicles that give meat its stringy appearance. Blood vessels run throughout the fascicular bundle, supplying oxygen and nutrients and carrying off wastes. Nerves also penetrate the bundle, their branches dividing ever more finely until they reach each fiber. These nerves will carry the messages that cause the fibers to contract.

Skeletal muscles are enclosed in a tough casing of connective tissue, the **fascia,** which is continuous with the inner connective tissue sheaths. (*Fascia* is Latin for a bundle of sticks fastened together, emphasizing the strength that comes from binding. It may help to remember that *fascism* is a rather binding form of government.) At the ends of the muscle, the fascia coalesces into denser collagenous tissue, forming the cordlike **tendons,** which may be broad and short or thin and long. Tendons, in turn, integrate with the periosteum covering the bones, thus insuring the continuity of force from muscle to bone. Each muscle has what is called an **origin** and **insertion.** When the muscle contracts, one end usually moves a great deal (the insertion) and one moves much less (the origin). As a general rule, the origin is more stationary and it is closer to the midline of the body, while the insertion is on the part—commonly a bone—being moved. Consider the biceps muscle in Figure 28.21. Where is the origin? The insertion?

Not all skeletal muscles move bones. For instance, there are the facial muscles, which in humans are quite significant since we rely so heavily on facial expressions in communicating. Tongue muscles also have complex origins and insertions. Thus, the tongue has considerable dexterity, which is essential for eating, cleaning the teeth in expensive restaurants, and speaking. Some muscles form **sphincters,** rings that surround passages within the gut and at the anus and mouth. Other muscles may form into flattened sheets that join with broad, thin tendons called **aponeuroses.** A familiar example is the sheet of abdominal muscles that hold your paunch in.

Muscle Antagonism.   Unlike the hydrostatic skeleton, in which the incompressibility of fluids can return the body to its former position, bones are returned to their original positions by opposing muscle action (or by gravity). Thus, skeletal muscles usually have, on the other side of a bone, opposing muscles. A pair of opposing groups are known as **antagonists,** an ill-fitting but descriptive term. In fact, unless something goes wrong, there is a great deal of cooperation and coordination between the antagonists.

The Ultrastructure of Skeletal Muscle

Now let's take a look at the details of muscle ultrastructure and contraction. We will look into the finer structural levels of muscle and find out how it contracts and how ATP actually works at the contractile site.

Muscle structure is best understood by mentally dissecting it down through its levels of organization, beginning with the gross tissue itself (Figure 28.22a). We have already discussed the organization of muscle down to the fiber level. We saw that the muscle itself is subdivided into bundles, each containing numerous fibers. Describing the individual muscle fibers in detail (Figure 28.22b) brings us into the ultrastructural level of organization, where most of our knowledge comes from electron microscope studies. Let's review a few aspects about muscle fiber first and then look at its contractile structure.

Skeletal muscle fibers range from a few millimeters in length to several centimeters—enormously long as cells go. Each muscle fiber (or cell) is surrounded by its equivalent of a plasma membrane, the **sarcolemma.** The sarcolemma receives the endings of **motor neurons** at **neuromuscular junctions.** Motor neurons relay impulses from the central ner-

28.21 THE HUMAN MUSCULATORY SYSTEM

The externally visible muscles of the human illustrate the various types and arrangements and give us some idea about origins and insertions. Only the major muscles have been named. By finding tendons of origin and insertion, and the general orientation of a muscle, you can determine just what it does. Keep in mind that although muscles can contract forcefully, they cannot extend with force. For this reason, they work in opposing units known as antagonists.

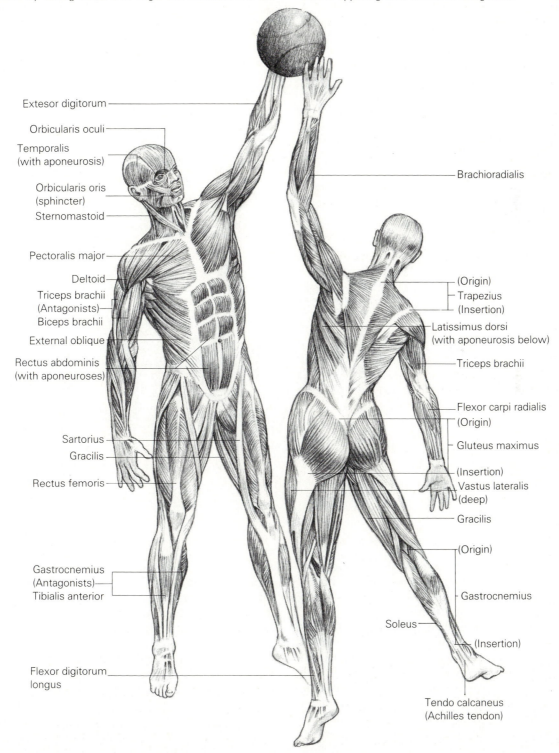

Extesor digitorum

Orbicularis oculi

Temporalis
(with aponeurosis)

Orbicularis oris
(sphincter)

Sternomastoid

Pectoralis major

Deltoid

Triceps brachii
(Antagonists)

Biceps brachii

External oblique

Rectus abdominis
(with aponeuroses)

Sartorius

Gracilis

Rectus femoris

Gastrocnemius
(Antagonists)

Tibialis anterior

Flexor digitorum
longus

Brachioradialis

(Origin)
Trapezius
(Insertion)

Latissimus dorsi
(with aponeurosis below)

Triceps brachii

Flexor carpi radialis

(Origin)

Gluteus maximus

(Insertion)
Vastus lateralis
(deep)

Gracilis

(Origin)

Gastrocnemius

Soleus

(Insertion)

Tendo calcaneus
(Achilles tendon)

28.22 SKELETAL MUSCLE ORGANIZATION

(a) Skeletal muscle had at least five levels of order or organization. The first three, the entire muscle, the fascicular bundles, and the fiber (or cell), are shown. Whole muscles are surrounded by a fascia, which is continuous with the tendons of origin (the anchor) and insertion (the part to be moved). The thicker midregion is commonly called the belly. Each bundle consists of many fibers bound together in a connective tissue matrix. **(b)** Continuing down the levels of organization, each fiber is surrounded by a plasma membrane called a sarcolemma, which contains numerous T-shaped extensions called T-tubules. Within the sarcolemma, numerous nuclei are visible along with a membranous sarcoplasmic reticulum and numerous mitochondria. Each fiber contains many myofibrils that form the contractile units. And each myofibril has an even more minute level of organization. The myofibrils reveal the banding that gives skeletal muscle its other name, striated muscle. These same striations are visible in cardiac muscle as well.

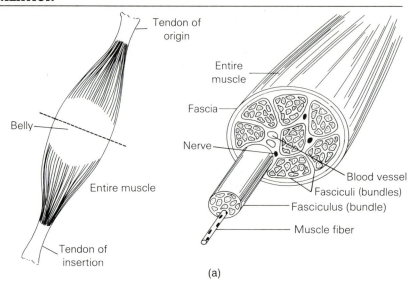

(a)

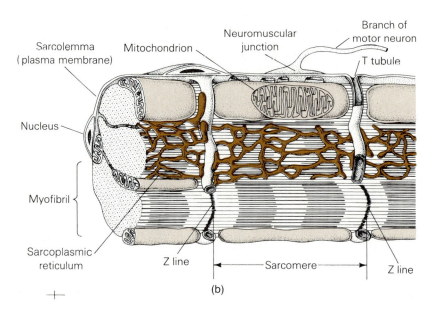

(b)

vous system (the brain and spinal cord) to the muscles. The neuromuscular junctions are the nerve-muscle interfaces through which impulses pass, instigating muscle contraction.

Within the muscle fiber, just below the sarcolemma, are found a number of nuclei and mitochondria and many glycogen granules—just what one would expect for such an active tissue. Lying just beneath the sarcolemma is the **sarcoplasmic reticulum,** which, like the endoplasmic reticulum of other cells, is a membranous, hollow structure. Extending from the sarcolemma or plasma membrane are numbers of T-shaped tubules, also hollow

and membranous but somewhat larger than the sarcoplasmic reticulum. These are called **transverse tubules,** or **T-tubules.** The T-tubules tend to mark the boundaries of each contractile unit. So far, we have named only supporting cellular structures. The active contractile elements of the muscle fiber are the rod-shaped **myofibrils,** or simply **fibrils.** The myofibrils form the visible pattern of striations that gives skeletal muscle one of its names.

Isolating one myofibril permits us to obtain a clear view of the highly visible striations of skeletal muscle. To understand the pattern, we must first realize that each myofibril contains numerous rod-

28.23 CLOSEUP OF THE MYOFIBRIL

Each sarcomere, or contractile unit, is bordered by two Z lines. The sarcomere's darkest band, containing a lighter zone at its center, is known as the A band. It consists of thick myosin filaments alternating with thin actin filaments. The actin filaments don't quite reach the center when the muscle is at rest, so the light H zone in the center of the A band is myosin

only. Note that the actin filaments reach all the way to the Z lines, and where they leave the A band they form the lighter I band. The cut end of the myofibril reveals that it is composed of myofilaments of actin and myosin. The cut has been made through the region of overlap and shows the arrangement of actin around myosin.

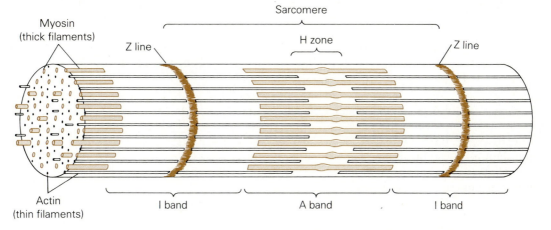

shaped filaments, or **myofilaments,** of the protein **myosin,** and an even greater number of thinner filaments of the protein **actin.** When viewed in cross section, it is common to observe each thick filament surrounded by six thinner filaments.

The two kinds of filaments are also arranged in a highly specific longitudinal manner, as we see in Figure 28.23.

Let's look at the longitudinal view from the surface, beginning with the very prominent **Z lines.** The region between the Z lines is the contractile unit, also known as the **sarcomere.** Moving inward from either Z line, we come to a very light zone called the **I band.** It consists of actin only, with no overlapping of myosin. Further inward is a broad, dark **A band,** with a smaller and lighter central strip, the **H zone.** The two darker parts of the A band consist of overlapping actin and myosin myofilaments. The lighter central H zone represents myosin alone unless the muscle is contracted. When muscle contracts, the H zone tends to disappear, as we shall see.

Contraction, an Ultrastructural Event

When the muscle fiber is stimulated, the Z lines move closer together (the sarcomere is shortened). This changes the banding, as you can see in Figure 28.24, producing a dark line where the H zone was. What actually happens is that actin myofilaments slide inward through the myosin. When the muscle

28.24 MUSCLE CONTRACTION

Two electron micrographs of muscle show what happens at contraction. **(a)** The muscle in a relaxed state: note the distance between Z lines, the width of the I bands, and the density of the A band. (b) Now observe the muscle in a contracted state: the dashed lines have been included as guides to reveal changes in Z line distance and I band width. Note these changes and the increase in density in the A band. Has the A band width changed? What has happened in the H zone? How would these changes affect an entire muscle?

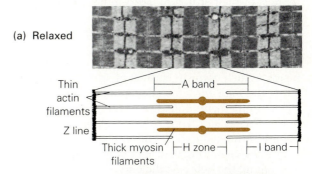

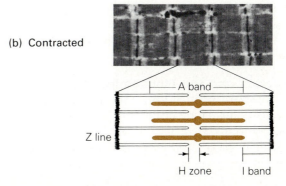

is fully contracted, the light I bands are greatly reduced because the actin and myosin are then almost completely overlapping. Notice that the width of the A band remains constant—it is exactly equal to the length of a myosin filament. It's important to keep in mind that the filaments themselves don't change in length, although sometimes the actin appears to pile up in the center of the sarcomere.

All this leaves us with a host of questions. What causes actin filaments to slide? What is actin, anyway? What is the relationship between its movement and ATP? Do all muscles work on the sliding filament principle? In all organisms? Is this the mechanism of movement in cilia, flagella, and the mitotic spindle? Only since the perfection of electron microscopes have biologists been able to answer some of these questions. For example, at very high magnification, researchers have found tiny bridges between the actin and myosin filaments (Figure 28.25). These bridges apparently extend from the myosin filaments and fasten to the actin. During contraction, they act as tiny rachets that attach at an angle and then pull themselves perpendicular to the actin fibers; then they reattach at a point further along the actin filament and pull themselves up again. The bridges thus rapidly pull the actin in toward the center.

Muscle cytologists have concentrated on studying isolated actin and myosin. Using magnifications of nearly half a million diameters, they have observed the filaments in enough detail to construct models of their structure. Myosin has been shown to be constructed of long, rodlike proteins spirally wound into each myofilament. Where individual proteins end they turn outward, forming the club-headed myosin heads (Figure 28.26). These myosin heads, which occur in clusters at specific distances along the filament, are believed to be the bridges.

Actin filaments consist of three protein components. The first of these, actin, is globular, with its spheres arranged end to end in a helical strand. A second protein, long and slender **tropomyosin,** follows along the helix. At the end of each tropomyosin the third protein, **troponin,** forms a globular tip. With these molecular considerations in mind we will move on to some biochemical aspects of contraction.

Calcium and the Biochemical Mechanism of Muscle Contraction.   The study of muscle chemistry has a long history that reaches back into the 19th century. In 1883, it was determined that calcium ions (Ca^{2+}) are required for the contraction

of frog muscles. A considerable body of biochemical and physiological information has emerged during this century, but, curiously, nearly 80 years passed before calcium's exact role was determined. Only in the past 20 or so years have we been able to relate the role of Ca^{2+} to what we have recently learned about muscle structure. Let's see what this synthesis has produced.

28.25 ARRANGEMENT OF MYOFILAMENTS AND CROSS-BRIDGES

(a) The electron micrograph of a cross section through the myofibril shows the arrangement of myofilaments actin and myosin with respect to each other. The drawing shows the bridges between myosin and the surrounding actin. (b) Notice the relationships of the myofilaments and their bridges in the three-dimensional drawing. (c) During contraction, these bridges actively pull the actin fibers inward, thus shortening the contractile unit.

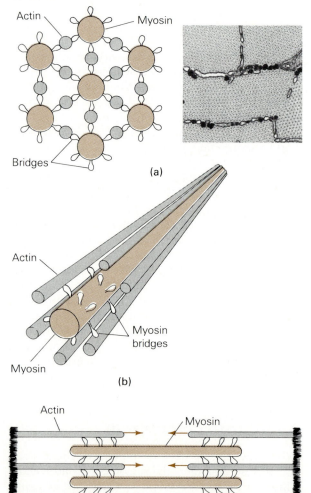

(a)

(b)

(c) Myosin bridges slide actin inward

Recall that in a resting muscle filament, the Z lines are widely separated and that there is only a small amount of overlapping myosin and actin. In this position, a few myosin bridges are near—but *not touching*—the slightly overlapping actin. Each bridge contains ATP, as we might expect. In the relaxed state, the troponin-tropomyosin-actin complex inhibits bridge formation with the myosin heads. For contraction to occur the bridges must attach, but as we will now see, this requires the calcium ion.

Muscle contraction begins when the sarcolemma receives an impulse from a motor neuron. The impulse is transferred across the neuromuscular junction to the sarcolemma, where a change in polarity begins, spreads across the membrane, and travels down the T-tubules. From the T-tubules it reaches the sarcoplasmic reticulum. Actually, the shift in polarity sweeps through all of the muscle fibers so quickly that they react more or less simultaneously. The sarcoplasmic reticulum, which is loaded with calcium ions that have been actively transported in, suddenly releases these ions and they flood the muscle contractile units (Figure 28.27a).

Once in the contractile units, the calcium ions join with the troponin-tropomyosin-actin complex. This breaks the inhibition, thus permitting bridge formation. Second, the myosin bridge attachment triggers the enzymatic hydrolysis of ATP, forming ADP and P_i. Third, energy released from the breakdown of ATP causes the bridge to move, drawing the actin filament along (see Figure 28.27b).

As the actin moves inward, more bridges attach to the actin and new ATP reserves join the spent bridges, which are still attached. Adding ATP restores the bridges, permitting them to act again. The repeated tugging by the bridges brings the muscle into its fully shortened state. As long as calcium ions are present and ATP is available, the contracted state will hold.

Calcium ions, then, apparently are the controlling factor in muscle contraction. However, their presence depends on the permeability of the sarcoplasmic reticulum to calcium. This permeability is, in turn, under the influence of the depolarizing nerve impulse.

In the relaxation process, the actions reverse themselves (Figure 28.27c). This occurs when the motor neuron that began the process ceases to stimulate the muscle. The membrane than becomes less permeable to the Ca^{2+} ions, so they can no longer leave the sarcoplasmic reticulum. Further, in the absence of neural stimulation, the *active transport* quickly sends the calcium ions back into the sarco-

28.26 THE MOLECULAR STRUCTURE OF ACTIN AND MYOSIN

(a) The thicker myosin myofilament consists of long, slender proteins spirally wound into the filament. At the end of each rod there emerges a thick, club-shaped tip—the myosin bridge. (b) The thinner actin myofilaments contain three kinds of protein. The major protein is globular actin. The individual actin spheres are arranged in a long, slender, helical filament. Tropomyosin, a short filamentous protein, follows the curve of the helix for a distance, ending with a molecule of troponin. The tropomyosin-troponin-actin complex of the actin filament acts as a controlling mechanism for the myosin bridges.

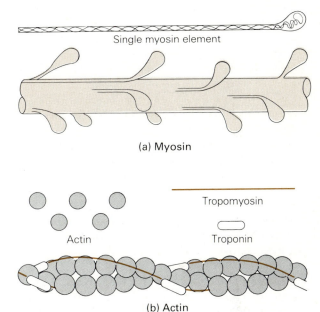

Single myosin element

(a) Myosin

Actin

Tropomyosin

Troponin

(b) Actin

plasmic reticulum, where they will be stored. Without calcium in the contractile units, the troponin-tropomyosin-actin complex again inhibits the myosin bridges from attaching to the actin. The muscle is now at rest.

We have seen how individual skeletal muscle fibers react to an impulse from a motor neuron, but we should point out that individual muscle fibers rarely act alone. In fact the minimum number of fibers that might react simultaneously is about 10 or 12. One motor neuron usually branches into at least that many muscle fibers and 10 or 12 are probably the minimum needed to accommodate the simplest task. When an impulse arrives, all of the fibers served by that motor neuron will react simultaneously; thus the reacting group is aptly called a **motor unit.** Motor units of just a few muscle fibers are usually found where precision is important, as in the muscles of the hands or eyes. On the other hand, motor units in the calf muscle, which tends to react as a whole, may include hundreds of muscle fibers. Generally, however, large muscles contract

28.27 CALCIUM AND MUSCLE CONTRACTION

Muscles will contract when a motor neuron transmits an impulse to the muscle surface. In the resting state **(a)**, calcium ions are pumped into and stored in the sarcoplasmic reticulum and the muscle is polarized (positive charges outside, negative inside). In this state, myosin-actin bridge formation is inhibited. **(b)** When a motor neuron transmits an impulse and the neural impulse reaches the muscle surface, a depolarizing wave is created, passing down the T-tubules and into the sarcoplasmic reticulum. There it stops the calcium pumps momentarily and calcium leaks out flooding the contractile units. The Ca^{2+} removes the inhibition and the bridges form.

Movement begins at ATP present in the bridges is enzymatically cleaved to ADP and P_i in the usual energy-releasing action. Each bridge shortens, doing its part in moving the actin filament inward. Spent bridges take in a second ATP and are restored, reattaching further down in the actin filament to tug again. In this manner, entire contractile units shorten, leading to the shortening of entire muscles. **(c)** When neural stimulation ceases, the actions reverse themselves. Calcium ions are again pumped into the sarcoplasmic reticulum. As these ions are again inhibited, contraction ceases and the muscle returns to its resting state.

●Ca^{2+} (calcium ions)

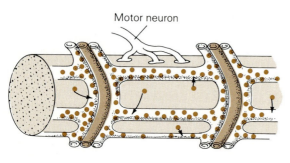

Motor neuron

(a) Resting state Ca^{2+} pumped into sarcoplasmic reticulum

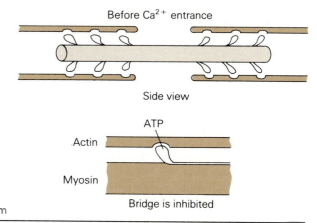

Before Ca^{2+} entrance

Side view

Actin

ATP

Myosin

Bridge is inhibited

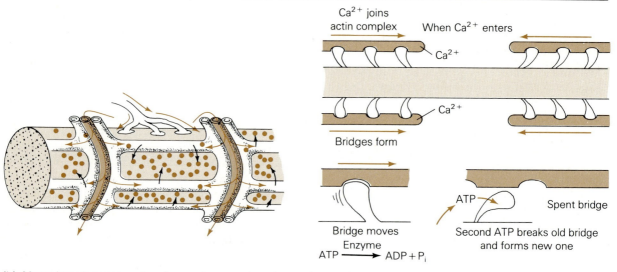

Ca^{2+} joins actin complex

When Ca^{2+} enters

Ca^{2+}

Ca^{2+}

Bridges form

Bridge moves

Enzyme

$ATP \longrightarrow ADP + P_i$

ATP

Spent bridge

Second ATP breaks old bridge and forms new one

(b) Motor impulse occurs   Impulse reaches muscle and spreads to surface into T-tubules, then to sarcoplasmic reticulum. Ca^{2+} leaves sarcoplasmic reticulum and enters contractile units.

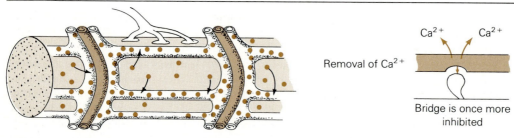

Ca^{2+} Ca^{2+}

Removal of Ca^{2+}

Bridge is once more inhibited

(c) Muscle recovers   When impulse ceases, the calcium-ion pumps again transport Ca^{2+} into the sarcoplasmic reticulum.

in a graded manner, with only the minimum number of motor units required undergoing contraction. For instance, shooing a fly from your arm through a slight movement of the triceps involves far fewer motor units than might be used in doing pushups. The arrangement of muscle fibers into motor units permits the frugal expenditure of precious energy reserves, and if nothing else, the body is frugal.

In this chapter, then, we have considered skeletal systems and the muscles that move them. As usual, many questions remain, particularly pertaining to muscle contraction. But let's continue now with our look at animals, keeping in mind that whereas the answers may be fascinating, so, in many cases, are the questions themselves.

APPLICATION OF IDEAS

1. The evolution of skeletons has occurred in two distinctly different directions in the protostomes and deuterostomes. How do skeletons in the two groups differ? What causes the drastic terrestrial size restriction in one of these trends? Is there a restriction in the other? Are such restrictions present in aquatic species? Explain.

2. The bird skeleton probably has more specialized features than are seen in any other vertebrate. Describe several specific examples. Also cite examples of the use of triangular forms. What is the value of this? How do humans make use of such a form in architecture?

3. Describe how muscle studies have involved the merging of different scientific disciplines. For example, the visual aspects of contraction have now been interpreted on the biochemical level.

Explain where one leaves off and the other begins in the contractile process.

4. Levers are classified according to the position of the fulcrum, the force applied, and the weight moved. *First*, *second*, and *third class levers* are described respectively as follows: weight and downward force at opposite ends, fulcrum near weight (pry bar lifting a stone); fulcrum at very end, weight near fulcrum, upward force at other end (wheelbarrow); fulcrum and weight at very ends, lifting force near the center (hand shovel). Illustrate the levers with simple drawings, and then find examples of each in the human skeleton and muscles. Determine which, if any, offers the best mechanical advantage (the least amount of force needed to move a part the greatest distance).

KEY WORDS AND IDEAS

INVERTEBRATE SUPPORT AND MOVEMENT

Hydrostatic Skeleton
1. The **hydrostatic skeleton** firms the animal body through fluid pressure against a resistant wall. It also provides a base for muscle action.

2. The coelenterates move and maintain their body shape and position by contractions of muscle fibers in the water-filled gastrovascular cavity.

3. Earthworms have complex muscle tissue arranged in two directions around their fluid-filled coelom. The body is extended by the squeezing action of circular muscles and shortened by contraction of longitudinal muscles.

4. Roundworms are fluid-filled with muscles directed longitudinally, permitting lashing movements only.

Exoskeletons
1. **Exoskeletons** are outside, produced by ectodermally derived epidermis, while **endoskeletons** are inside and mesodermally derived.

2. The arthropod chitinous exoskeleton or **cuticle** provides a muscle base, protects the soft parts, and prevents desiccation in terrestrial species. In crustaceans, the cuticle is hardened by calcium salts, but it is leathery and flexible at joints.

3. Muscle attachments and action in arthropods and vertebrates are reversed and opposite with respect to each other. Arthropod muscles are often attached to projecting **apodemes.**

4. The great strength and agility of arthropods is only possible because of their minute size.

5. Wing movement in flying insect species is either *direct* or *indirect*. In the first, opposing *direct*

flight muscles simply move the wings up and down, while the second, *indirect flight muscles* are vertically and longitudinally arranged and move the thorax up and down and the wings respond through a complex hinge. A tilted angle provides lift. Wingbeat rates can exceed neural impulses because of a resonating mechanism in the muscles.

6. Mollusk shells protect the body, but parts such as the fleshy foot and eyestalks are often moved through hydrostatic mechanisms. Air is withdrawn from spaces in the spongy cuttlebone, a shell remnant in cuttlefish, making it an effective flotation device.

7. In mollusks, external shells are continuously secreted by the mantle and take one of three forms: a **logarithmic spiral, simple cone,** or **spiralled cone.**

Endoskeletons
1. The dermis of echinoderms secretes calcareous plates and spines that originate below the epidermis.

VERTEBRATE SUPPORTING STRUCTURES

Vertebrate Tissues
1. The vertebrate body consists of four tissue types: **epithelial, connective, nerve,** and **muscle.**
2. **Epithelium** or epithelial tissue covers the body as **epidermis** and lines internal organs as well. Blood vessels have an **endothelium,** while glands have a **glandular epithelium.**
3. Connective tissue contains cells in a noncellular **connective tissue matrix** as seen in bone, cartilage, ligaments, tendons, and blood. **Collagen** is a common binding or tough matrix material. Cells called **fibroblasts** secrete collagen and are active in bone mending.

Vertebrates: Endoskeletons of Bone
1. The bony endoskeleton provides support, a muscle base, protection, a source of blood cells, and acts as a calcium reservoir.
2. Bone is surrounded by the sheathlike **periosteum.** Within are regions of weblike **spongy bone,** containing **red marrow** where red cells form, and regions of **compact bone,** containing fatty **yellow marrow.** Bone consists of hard calcium phosphate in a collagen matrix containing numerous bone cells, blood vessels, and nerves.
3. Compact bone is organized into **Haversian systems** or **osteons,** each with a **central canal** and hard, concentric, laminated **lamellae. Osteocytes** (bone cells) reside in hollow **lacunae,** interconnected by weblike **canaliculi.**

The Skeleton
1. Vertebrate skeletons consist of two divisions. The **axial** contains the **cranium** (skull), **vertebral column** (backbones), and **thorax** (rib cage), and the **appendicular** contains the two **girdles—pelvic** (hip bones) and **pectoral** (shoulder bones and limbs).
2. Immovable joints include the fused **sutures** in the skull. Movable joints have articulating surfaces of cartilage, and the bones are held in place by **ligaments** that often form the **capsules** of **synovial joints.** Lubrication is provided by **synovial fluid** secreted by a **synovial membrane** lining the capsule.
3. Movable joints include **gliding** (wrist bones; slight back and forth), **pivotal** (head and neck; rotation in one plane), **ball-and-socket** (shoulder; free rotation), and **hinge** (knee; hingelike extension and flexion in one plane).

The Axial Skeleton
1. The skull or cranium in modern fishes includes many separate bones with little fusion. Fusion increases in the amphibians, reptiles, and birds, and reaches its peak in mammals. Jaw mobility is also greatest in the mammals, although the **maxilla** (upper jaw) is fused to the skull. Sensory receptors located on the cranium communicate with the brain through **foramina.** The **foramen magnum** at the base of the skull admits the spinal cord.
2. The number of **vertebrae** in the vertebral column varies among vertebrate classes. Reptiles have the most, while in birds many have fused. Mammals all have seven neck vertebrae.
3. The 32 to 34 human vertebrae include three to five in the **coccyx.** The free vertebrae are separated by compressible, soft-centered **intervertebral disks** of **fibrocartilage.** The vertebral column houses and protects the **spinal cord,** which passes through the **neural canal.**
4. The human vertebral column is divided into five regions: **cervical** (neck), **thoracic** (chest), **lumbar** (trunk), sacrum (hip), and coccyx (tail). The first two vertebrae, the **atlas** and **axis,** support the skull and permit pivotal movements. Thoracic vertebrae attach to the ribs, while lumbar vertebrae are massive and support the upper body. **Sacral** vertebrae join the vertebral column to the pelvic girdle, while the coccyx is a vestigial tail.
5. The thoracic bones include the rib cage and the three-part **sternum (manubrium, body,** and **xiphoid process).** Amphibians lack rib development, while in the turtle the ribs are fused to the **carapace.** In birds a large **keel** is fused to the sternum.

The Appendicular Skeleton

1. Each half of the human pelvic girdle contains an **ilium, ischium,** and **pubis.** On each side, the three bones form the **acetabulum,** or hip socket. The two pubi form the **pubis symphysis** in front, and the two ilia and the sacrum form the **sacroiliac joint** in the back. The pectoral girdle includes the two **scapulae** and **clavicles,** a loose assemblage that forms the shoulder joint.

2. Human forelimbs are very generalized, except for subtle thumb modifications. The **humerus** (upper arm) joins the ball-and-socket shoulder joint, while the **radius** and **ulna** (lower arm) form a complex pivoting hinge joint at both ends. The hands include the **phalanges** (fingers), long **metacarpals** (hand bones), and cubic **carpals** (wrist bones).

3. The human leg contains the large **femur** (upper leg) and the **tibia** and **fibula** (lower leg). The hinged knee joint, formed by the femur, tibia, and **patella** (knee cap), depends on a capsule of many tough ligaments for its strength. The ankle forms a rotating hinge joint where the **talus** meets the leg bones. The largest of the seven **tarsals** in the ankle, the **calcaneous,** forms the heel and receives the **Achilles tendon** from the calf muscle. The slender **metatarsals** and the phalanges (toes) complete the foot. The arch is produced by ligaments binding the bones.

MUSCLE: ITS ORGANIZATION AND MOVEMENT

Types of Muscle Tissue

1. *Smooth muscle* (also **involuntary** or **visceral**), is slow moving and rhythmic, occurring in the gut, blood vessels, hair roots, iris, uterus, and other places under the unconscious control of the **autonomic nervous system.** Individual cells occur in sheets, are long and tapered, and have a single nucleus.

2. **Cardiac muscle** (heart) is involuntary. Its fibers are cylindrical and branching, with interlocking borders called **intercalated disks,** important to contraction. It has a rhythmic and tireless contractile characteristic in which isolated cells contract spontaneously.

3. **Skeletal muscle** (also **voluntary** or **striated**), is controlled at the conscious level through the **somatic nervous system.** Cells, called **fibers,** are striated and multinucleated.

Gross Anatomy of Skeletal Muscle

1. Muscle fibers form bundles called **fascicles,** and a number of fascicles bound by connective tissue form muscle bundles. Muscles contain a rich supply of blood vessels and nerves. A sheathlike **fascia** surrounds the muscle, forming the **tendons** at its ends. Tendons connect mus-

cle to bone, forming **insertions** on the part to be moved and **origins** on the stationary part.

2. Muscles also move the face and the tongue, and others form ringlike **sphincters** in the gut. Sheetlike muscles in the abdomen originate and insert via flattened tendons called **aponeuroses.**

3. Muscle **antagonists** (opposing sets) account for opposite movements of the skeleton.

The Ultrastructure of Skeletal Muscle

1. Muscle fibers are surrounded by the **sarcolemma,** which receives **motor neurons** at **neuromuscular junctions.** Each fiber contains several nuclei, many mitochondria, and an extensive **sarcoplasmic reticulum.** Hollow **transverse tubules,** or T-tubules, form boundaries for each contractile unit. The pattern of contractile **myofibrils,** or **fibrils,** produces the striations.

2. Myofibrils contain **myofilaments,** or **filaments,** made up of two kinds of protein: **myosin** and **actin.** Each myosin filament is surrounded by six actins. The striations are seen in the longitudinal view. Prominent **Z lines** mark the **sarcomere,** or contractile unit. Within the Z lines lies a light **I band** (actin alone), and further in lies a darker **A band** (actin and myosin), ending in a lighter central **H zone** (myosin alone).

Contraction in the Ultrastructure

1. Upon contraction, the Z lines move closer together, the I bands are reduced, and the H zone darkens. However, the A band remains constant, indicating that it is the actin that moves, while the myosin is stationary. The patterns change because minute myosin bridges contact each actin filament, pulling it inward.

2. The myosin filaments are composed of spirally wound protein rods with regularly clustered, club-shaped heads. Actin contains the assembled spheres of the protein actin and slender fibers of the protein **tropomyosin,** ending with the globular protein **troponin.**

3. In resting muscle, the troponin-tropomyosin-actin complex inhibits bridge formation and the myosin knobs do not touch the actin.

4. In contraction, a neural impulse reaches the muscle fibers and travels along the T-tubules to the sarcoplasmic reticulum, which responds by suddenly releasing calcium ions into the contractile unit. The ions alter the inhibition state, and the bridges connect and straighten sequentially along the actin filaments, pulling them inward as contraction continues.

5. Restoration of the resting state begins when neural stimulation stops and the calcium ions are actively transported back into the sarcoplasmic reticulum. The action occurs simultaneously throughout the **motor unit**—the fibers served by a motor neuron.

REVIEW QUESTIONS

1. Knowing that coelenterates have contractile fibers running around and along their soft, sac-like bodies, explain why water is necessary for them to make useful movements. (702–703)

2. Describe the muscle arrangement of the earthworm, and explain how it makes use of this in burrowing. Of what specific use is hydrostatic pressure in the coelom? (703–704)

3. Discuss the size restrictions of terrestrial arthropods brought about by the exoskeleton. Are there similar restrictions in vertebrates? (704)

4. Explain how the molluskan shell grows. To what geometric forms is such growth restricted? (706)

5. Describe the endoskeleton of the echinoderm. Considering that much of it protrudes through the epidermis, what qualifies it as an endoskeleton? (707)

6. What is the general role of epithelial tissue? List four different kinds and mention where they are found. (707–708)

7. List five human structures that consist primarily of connective tissues. (708)

8. What are two roles of collagen? How is it produced? (708)

9. List five important functions of the vertebrate skeleton. (709)

10. Compare spongy and compact bone. In which are the red and yellow marrows found? (709–711)

11. With the aid of a simple diagram, illustrate an osteon, labeling lamellae, canaliculi, lacunae, osteocytes, and the central canal. Which is (are) nonliving? (711)

12. List four types of joints, and using examples from the human body, describe the movement possible in each. (711)

13. Describe a typical synovial joint, including the role of ligaments, cartilages and the synovial membrane. (711)

14. Describe the organization of the human vertebral column, naming the main regions and listing characteristics of vertebrae in each. Also describe the structure of the disks separating the vertebrae. (713–715)

15. List the bones that make up the pelvic girdle and name the three joints they form. (715–716)

16. List the bones of the pectoral girdle. In what way is the human pectoral girdle different from those of most mammals? (716)

17. Name three groups of bones in the hands and in the feet. Briefly discuss a unique human characteristic in each appendage. (717)

18. List one unique characteristic each for skeletal, cardiac, and smooth muscle tissue. (719–720)

19. List two ways in which cardiac and smooth muscle are similar, and one way in which cardiac and skeletal muscle are similar. (719–720)

20. How do the roles of ligaments and tendons differ? (720)

21. Distinguish between tendons of origin and tendons of insertion. Do tendons always insert or originate on bones? Explain. (720)

22. Starting with the complete muscle, write the following terms in a logical descending order: Complete muscle, myofilament, troponin, fiber, actin, myofibril, actinomyosin, sarcomere, myosin, fascicle. (720, 723)

23. Using a simple drawing, explain the organization of a contractile unit. Begin with two Z lines, and fill in between. Explain the composition of each part of the pattern. (723)

24. Describe the arrangement of actin and myosin in a contractile unit, and in so doing explain how the actin moves during contraction. (723–724)

25. How does the absence of Ca^{2+} affect the contractile unit? Where are these ions when the muscle is at rest? (725–726)

26. Beginning with an impulse moving along a motor neuron, describe the sequence of events leading to contraction. (725)

Digestion and Nutrition

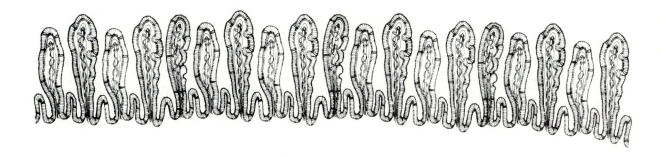

It's always a little risky to generalize about evolutionary trends since evolutionary change runs in many directions and at very different rates. Some traits change rapidly, some change slowly. If a species happens to harbor many rapidly changing traits, then the species itself changes rapidly. By the same token, a species with a great proportion of slowly changing traits will likely be very similar to its distant ancestors. Even within the same kind of animal, different organs may evolve at different rates and show different levels of specialization. And in some cases, "advanced" species can harbor some very primitive traits. For example, we humans consider ourselves advanced, but we have primitive mouths. The specialized bill of the seemingly primitive platypus, on the other hand, is considered one of the most advanced mouth structures among vertebrates. But the gut! We humans can be proud of our gut (though we rarely boast about it). We can be proud because we have a very highly evolved gut—not as highly evolved as that of the cow with its four-chambered stomach, but still very efficient and rather complicated. It is only fitting, then, to learn more about the specializations of such a nice gut, but first we'll look at those of other animals and see how they have done.

DIGESTIVE SYSTEMS

Digestion refers to the chemical breakdown of complex nutrients (food molecules) into their component parts so that they can be absorbed and used as energy sources and as building blocks for other complex molecules. **Nutrition** is an encompassing term that refers to any process involved in nourishment or being nourished. Here we will consider not only nutrients, but their sources, digestion, chemical characteristics, and roles in metabolism.

Some organisms do not actually ingest food but use **extracellular digestion.** Fungi and bacteria simply secrete digestive enzymes into the surrounding nutrients and break the food materials down into simpler subunits that then move into the cell. On the other hand, many eukaryotes are capable of taking food particles into their cells, where they are digested **intracellularly** in **digestive vacuoles.** You can readily see digestive vacuoles in large protozoa such as *Paramecium* and *Amoeba* (see Figure 19.10).

We will now briefly survey digestion and digestive systems in a few representative animals, beginning with the invertebrates. Then we'll move on to the vertebrates, concentrating on humans.

Saclike Systems in Invertebrates

Food-getting in the sponges—animals that lack organ systems—is the responsibility of the flagellated **collar cells,** or **choanocytes** (Figure 29.1). These active cells line the body canals, and their beating causes seawater to swirl and eddy throughout the body of the sponge. The water carries in tiny food particles, which are trapped by a layer of mucus on each collar, passing down to enter the cell below. From here the food is then moved to other cells, where it becomes enclosed in digestive vacuoles. Molecules of digested food are then transported throughout all the cells of the sponge. It is believed that peculiar, wandering cells called **amebocytes** aid in the distribution processes by carrying food from one place to another. So the sponge, a simple multicellular animal, carries on digestion (intracellular) without a well-defined digestive system.

The coelenterates (Figure 29.2) and flatworms (Figure 29.3) have somewhat more complex digestive structures, including a prominent saclike gut or **gastrovascular cavity** with a complex lining of enzyme-secreting, phagocytic (food-engulfing), and ciliated cells. Digestion begins extracellularly when enzymes are released into the gastrovascular cavity, but after the food is partially broken down, digestion is intracellular. Particles are engulfed by phagocytic cells lining the gut, and digestion is completed in food vacuoles. In both coelenterates and flatworms, the presence of a blind gastrovascular cavity means that indigestible material must eventually be ejected through the organism's mouth. It is not surprising that such an arrangement is considered primitive.

Feeding and Digestive Structures in Higher Invertebrates.

Beyond the flatworms and coelenterates, nearly all other animals have the tube-within-a-tube body plan (Figure 29.4) and use extracellular digestion. (Recall from Chapters 25 and 26 that this plan evolved with the pseudocoelomates and eucoelomates.) Yet some intracellular digestion persists. The so-called "liver" of many invertebrates, such as mollusks and crustaceans, actually consists of intestinal side passages lined with phagocytic cells, and some food is digested within these cells after having been broken down into small particles.

As an example of digestive systems in invertebrates, we can consider two quite different groups: earthworms and insects. As you can see in Figure 29.5, the gut in earthworms is well defined and associated with several specialized organs. As the earthworm burrows through the soil, it feeds on decaying organic matter. Its mouth is a simple opening, followed by a very muscular **pharynx** that swallows the organic matter and forces it along. A short **esophagus** directs the mass into the thin-walled **crop,** which is a temporary storage organ. The ingested material is next moved into the thick-walled, muscular **gizzard.** The gizzard is filled with bits of gravel and, like the gizzard in seed-eating birds, is used to grind the food.

29.1   INTRACELLULAR DIGESTION IN SPONGES

Sponges obtain their food by filtering seawater and engulfing bits of organic matter. Food particles pass down the collar cells and enter the cell below through phagocytosis, ending up in food vacuoles. As digestion continues, some food vacuoles enter nearby amebocytes that move about and distribute the food to other parts of the body.

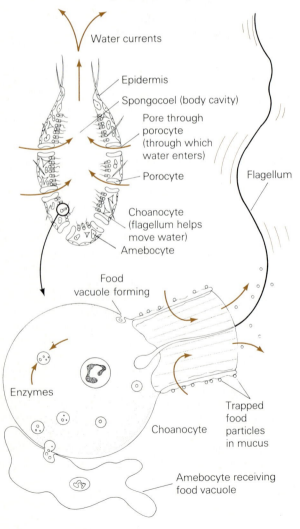

Water currents

Epidermis

Spongocoel (body cavity)

Pore through porocyte (through which water enters)

Porocyte

Flagellum

Choanocyte (flagellum helps move water)

Amebocyte

Food vacuole forming

Enzymes

Choanocyte

Trapped food particles in mucus

Amebocyte receiving food vacuole

29.2 FEEDING IN *HYDRA*

Hydra, a freshwater coelenterate, demonstrates the manner in which many coelenterates feed. **(a)** A minute crustacean is captured by paralyzing it with toxins from stinging cells in the tentacles. **(b)** Then it is brought into the mouth and gastrovascular cavity, where extracellular digestion begins. **(c)** Partially digested particles are then incorporated into food vacuoles where digestion becomes intracellular.

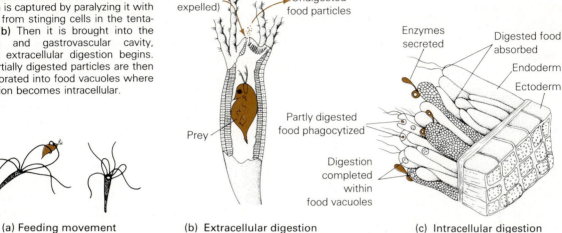

(a) Feeding movement

(b) Extracellular digestion

(c) Intracellular digestion

29.3 FOOD-GETTING AND DIGESTION IN PLANARIANS

Planarians' highly branched gastrovascular cavity is more of a distribution system than a digestive system. They obtain food by sucking with a protrusible pharynx. Bits of food are distributed throughout the gastrovascular cavity, where they are phagocytized and digested intracellularly.

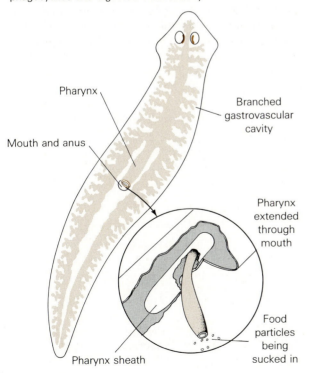

29.4 THE TUBE-WITHIN-A-TUBE PLAN

The gut shown here is highly diagrammatic. The outer tube is the body wall, while the inner represents the gut. Between the tubes is the body cavity. Regions of the gut are specialized in carrying out the various tasks of digestion and absorption.

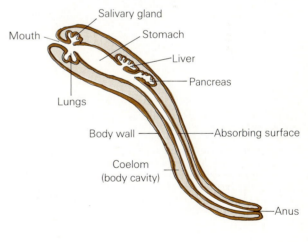

29.5 SPECIALIZATION IN THE TUBE

The complex earthworm digestive system is "complete," which means that food moves onward through a tube from mouth to anus. The gut has several specialized regions, including the pharynx, crop, gizzard, and lengthy intestine. The absorbing surface is greatly increased by an intestinal fold known as the typhlosole.

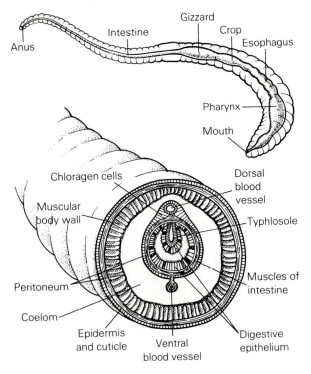

From the gizzard the food moves into the **intestine**, where it is digested by enzymes secreted by cells in the intestinal lining. The digestion and uptake of food occur along the entire length of the intestine. The **typhlosole**, a large, foldlike extension of the gut wall, speeds up absorption by greatly increasing the surface area (see Figure 29.5). From the absorptive surface, the food molecules pass into the circulatory system for distribution throughout the body. Although the earthworm has nothing to compare with the vertebrate liver, groups of skin cells known as **chloragen** cells perform some of the same functions, including the conversion of glucose to glycogen, the **deamination** (removal of nitrogen) of amino acids, and the formation of urea. Worms get their color partly from the accumulation of some of these products. In addition (strangely enough), roaming phagocytes pick up whole bits of soil, which pass through the earthworm's intestine and then move to the skin, with the resulting earthy coloration presenting a major visual problem to the early bird.

It is difficult to generalize about insect feeding structures. In fact, the group serves as a lesson in digestive variation, one that illustrates the tremendous range of adaptations to the varied energy sources of the earth. It is difficult to think of anything that at least some insects don't eat. This versatility has been particularly troublesome to humans since the dawn of agriculture; the availability of food in large patches has made it possible for some insect species to increase in number dramatically.

29.6 THE INSECT GUT

The grasshopper's digestive system is similar to that of many other insects. While there is enormous variation in mouthparts, the digestive tract typically consists of three main divisions, the fore-, mid-, and hindgut.

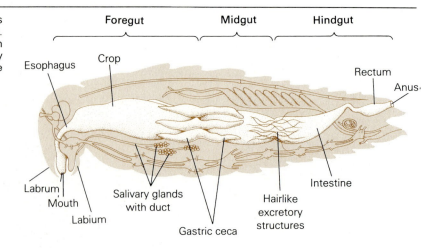

Biting and chewing mouthparts are very common in insects, particularly in predators and foliage eaters (see Figure 26.16). They can be clearly seen in the grasshopper, which has, essentially, a five-part mouth. The grasshopper tastes its food with its **palps,** pairs of sensory structures attached to the mouthparts. Biting and shearing are done with the large saw-toothed **mandibles,** while the smaller **maxillae** do the finer grinding. The upper and lower lips, the **labrum** and **labium,** assist by holding food so that the shearing and grinding can occur efficiently. Saliva from well-formed salivary glands in grasshoppers helps with the grinding and begins digestion.

The grasshopper's digestive system, like that of many insects, consists of three parts, the **foregut, midgut,** and **hindgut** (Figure 29.6). The foregut is divided into a pharynx, esophagus, and crop. The crop functions primarily in food storage. In some insects this region has a roughened surface that is used to further grind the food. The midgut is a digestive structure where both digestion and absorption occur.

Surrounding the junction of the fore- and midgut are 12 **gastric cecae** (singular, *cecum*)—pouch-like accessory structures that secrete digestive enzymes into the midgut. The hindgut is mainly a water-absorbing structure that receives both digested food and cellular nitrogenous wastes. As digestive residues pass through, the water is removed and conserved, and the feces leave in a semidry condition, an important adaptation to the dry terrestrial environment.

The digestive system of the housefly is slightly different: it carries on some digestion outside the gut. You should be aware that the housefly walking across your chocolate cake is doing two things. First the fly is tasting the icing with its feet—a place where many insects have taste receptors. Second, when it bends its head down, it is disgorging its enzymes onto the surface and stirring it all around with its hairy labium. But don't worry—it will suck up any mess it makes with its hollow, tubelike mouthparts. The rest is yours.

Plant eaters, by the way, frequently have piercing and sucking mouthparts. Recall the aphids that suck up the phloem sap (see Chapter 23). There are also fruit-eating moths that tear the fruit first, then suck up the exposed fluids. Of course, piercing and sucking mouthparts are found among the blood-sucking and predatory species as well. When the ant lion sinks its huge curved mandibles and maxillae into the ant's vulnerable abdomen, it doesn't chew. It uses channels formed between the paired mouthparts to suck out the ant's body fluids.

Feeding and Digestive Structures in Vertebrates

The jaws, teeth, and beaks of vertebrates provide us with clues to what they eat. For example, if you find a large skull with conical teeth, you've probably found a fish eater. If the skull is very large, though, the conical teeth may have once graced a killer whale who used them to crush the bodies of large prey such as seals. On the other hand, if the skull is very small and the teeth more pointed, the animal probably ate insects. You can rest assured, however, that the creature with the pointed teeth was not a grazer. Let's find out about feeding and digestive specializations in one of the sharper-toothed types from the briny depths.

Sharks and Bony Fishes. Sharks are certainly carnivores, despite the fact that they swallow practically anything. One of your authors, while on an expedition, helped dissect a 12-foot tiger shark and found a Polaroid negative that had been thrown overboard six days earlier and 200 miles away! Finding a picture of oneself in a shark's stomach can lessen the pleasures of seafaring life.

Two aspects of the shark's food-getting and digestive apparatus are especially noteworthy. First, its teeth occur in rows, most of which are around the periphery of the mouth (Figure 29.7). As the teeth grow, the rows migrate outward so that the outermost ones fall out (and sink to the ocean floor, thereby providing tourists with necklaces). The jaw itself is a hinge joint, moving in one plane only. Instead of chewing, the shark uses the sideways thrashing and tearing action typical of shark attacks.

29.7 AN "EATING MACHINE"

The fearsome mouth of the shark is lined with rows of teeth. The outer rows are the oldest and are often broken or lost as the shark feeds. However, new teeth soon move forward and outward to replace them.

The second unusual aspect of the shark's digestive system is the **spiral valve** in its relatively short intestine. The valve (Figure 29.8) resembles a circular stairway. Apparently, its function is to slow the passage of food through the shark's short intestine to give digestive enzymes time to act. It also provides additional absorptive surface.

Bony fish are extremely diverse, with over 20,000 species described, several of which will be neglected here. Like the sharks, the jaws of the bony fishes can only move in one plane and are used for biting or sucking only. The teeth are numerous, and in many species they grow from the roof of the mouth as well as from the upper and lower jaws.

The herbivorous fishes have interesting mouthparts. For one thing, they lack tearing teeth. However, you need not fear being gummed by an old carp, perch, or minnow, since they also lack that ridge. Instead, their teeth are on the floor of the mouth, where they grind bits of plant material against leathery patches on the roof. Other herbivorous fishes filter minute particles from the water with highly branched gill rakers through which water passes as it enters the gill chambers. Herbivorous fishes usually have very long, highly coiled intestines, typical in animals that extract nourishment from cellulose-enclosed plant cells.

Amphibians, Reptiles, and Birds. The amphibians have a digestive system similar to that of bony fishes but with some important differences. For example, whereas fish have rather immobile tongues, the tongue in amphibians is often highly movable. In fact, frogs use their tongue to capture insects. The tongue is attached to the front of the mouth, permitting the sticky organ to be flicked far out with considerable speed and accuracy. The prey is then crushed against a peculiar patch of teeth on the roof of the mouth and swallowed whole—amphibians can't chew either.

The reptiles are a rather diverse group and boast a wide variety of techniques in detecting and catching prey. For example, some reptiles have many teeth, while in others, the teeth are fewer in number but more specialized. Crocodiles and alligators are unusual in that their teeth are seated in sockets in the bone, much like our own. Evolution has provided some venomous snakes, including the North American pit vipers, with retractible, grooved, or hollow **fangs** with intricate injecting structures. Other venomous snakes—cobras, kraits, and mambas, for instance—have permanently erect fangs and must "chew" the venom into their prey. All snakes, by the way, are carnivorous and have undergone a unique modification of the jaw for swallowing whole prey (Figure 29.9; also see Chapter 27).

Snakes and lizards find food with the aid of well-developed eyes and a specialized olfactory (odor sensing) organ at the roof of the mouth. This structure, called **Jacobson's organ,** detects molecules (odors) from the air, which are brought to it by the moist, flicking tongue. The pit vipers orient to warm-blooded prey through the use of acutely sensitive heat-detecting receptors in the facial pits (see Chapter 34).

The dentition of birds is as scarce as hen's teeth since all of today's birds have bills rather than teeth. The bill or beak is a bony structure with a tough, continuously growing covering of keratin, a horny protein also found in fingernails and hair. While the basic structure of the bill is similar in all birds, there is considerable variation in its shape and size. Each type strongly reflects the feeding habits of the bird (Figure 29.10).

29.8 THE SPIRAL VALVE

While the shark gut is short and abrupt, its absorbing surface is greatly expanded by the presence of the spiral valve. Spiral valves are also found in some of the more primitive bony fishes.

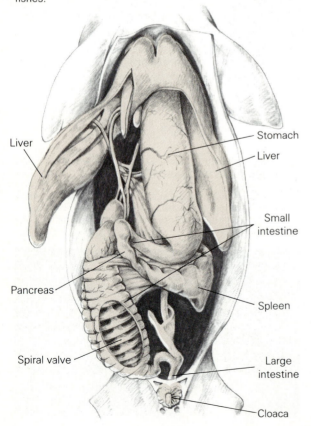

Liver
Stomach
Liver
Small intestine
Pancreas
Spleen
Spiral valve
Large intestine
Cloaca

29.9 A DOUBLE-HINGED JAW

Snakes cannot tear or chew their food. Captured prey are swallowed whole, a surprising feat since the prey is often larger than the snake's mouth. The mouth is extremely flexible because of the arrangement of bones in the head and jaw. The lower jaw is loosely attached to the quadrate bone, and even the bones of the palate are movable, all helping draw the prey into the gaping mouth. The esophagus and stomach can stretch considerably as can the body wall. There is no sternum, so the ribs can move freely as the prey passes through the gut.

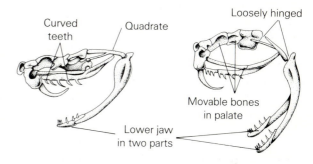

29.10 BIRD FEEDING SPECIALIZATIONS

The bills of birds provide clues to their feeding habits.

Keel-billed toucan

Oystercatcher

Blue-winged teal

Northern cardinal

White ibis

Bald eagle

In some instances, the feeding habits are reflected in the feet as well. Typically, raptorial birds, or birds-of-prey, have long, curved talons for grasping prey. Ground foraging species such as grouse and pheasants have heavy, strong feet for scratching the soil, while the feet of the ostrich and emu can be deadly weapons.

The digestive system of seed-eating birds includes a large crop for storing and moistening food, followed by a two-part **stomach.** The first portion—the very glandular **proventriculus**—secretes gastric juices into the coarse food. It then passes into a very muscular **ventriculus** (gizzard), where it is pulverized with the aid of a hardened lining and abrasive sand grains the bird routinely swallows (Figure 29.11).

Foraging and Digestive Structures in Mammals.   Mammalian teeth grow in sockets in the jaws, a trait that distinguishes them from almost all other vertebrates. Generally, the mammal begins life with temporary teeth, or milk teeth, which are later replaced by permanent teeth. The basic structure of each permanent tooth is the same in all mammals (Figure 29.12).

While the teeth of most fishes and reptiles show little specialization, mammals have four types of teeth, each usually with its own specialized function. The front teeth, or **incisors,** highly developed

in the rodents and rabbits, are chisel-shaped for gnawing and cutting. The second group, the **canines,** are used for capturing and killing prey, tearing food, and defense. They are very prominent in carnivores and completely absent in some of the herbivores. (Imagine feeding a carrot to a horse with fangs.) The last two groups, the **premolars** and **molars,** are specialized for grinding. They are large and flat in the herbivores and are used to break down the cellulose walls and fibers of plants. Some carnivores, such as dogs, have large, sharp-edged premolars specialized for shearing and crushing bones.

Dentition in humans is rather generalized, lacking the prominent, specialized, exaggerated features often found in other mammals. Because of the lack of specialization, some zoologists regard human dentition as primitive (see Figure 29.12).

Digestion in Grazing Mammals: The Ruminants. Plant eaters have the formidable task of digesting cellulose. Cattle, horses, and other herbivores begin the process with heavy, flat teeth that are specialized for grinding. The digestive tract of cattle and other grazers, or **ruminants** (including deer, giraffes, antelope, and buffalo), have four-chambered "stomachs" including the **rumen** from which the name *ruminant* is derived, and the **reticulum, omasum,** and **abomasum** (Figure 29.13). Cattle don't produce enzymes that can digest cellulose but instead, like termites, harbor certain protozoans and bacteria in their rumen. These organisms *can* break down cellulose, forming nutritious products that, along with the microorganisms themselves, can begin digestion in the abomasum, the actual stomach, in the usual manner. The unusual aspects of digestion in the four-part stomach, including "regurgitation" and "cud chewing," are described in Figure 29.13.

THE DIGESTIVE SYSTEM OF HUMANS

The human digestive system (Figure 29.14) is rather representative of the mammalian system in many respects. We will now follow the enchanting path of food through the human tract, discussing the specialized structures along the way. We will then take a close look at the chemistry of digestion and some aspects of nutrition.

29.11 DIGESTION IN BIRDS

Herbivorous birds have organs specialized for handling plant foods, such as crops and gizzards. In addition, they have a special stomach region called the proventriculus, which saturates food with digestive enzymes before it enters the ventriculus, or gizzard.

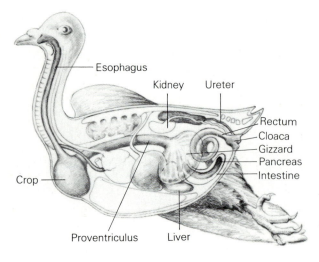

Esophagus

Kidney   Ureter

Rectum

Cloaca

Gizzard

Pancreas

Intestine

Crop

Proventriculus   Liver

29.12 MAMMALIAN TEETH

Nearly all mammals share the same basic tooth structure **(a)**: a hardened layer of enamel surrounding a softer dentin region. Within the dentin is a pulp cavity, penetrated by blood vessels and sensory nerve endings. The arrangement and individual shapes vary considerably, however, according to the diet to which the species has adapted. **(b)** Note the simple pointed teeth of the insect-capturing insectivore. Compare the gnaw- ing and grinding teeth of a rodent **(c)** with the tearing and crushing teeth of a canine **(d)**. Also note the nipping lower incisors of the ruminant **(e)** and the absence of upper inci- sors—a horny covering meshes with the lower incisors for nipping foliage. Human dentition **(f)** is quite uniform and rela- tively unspecialized.

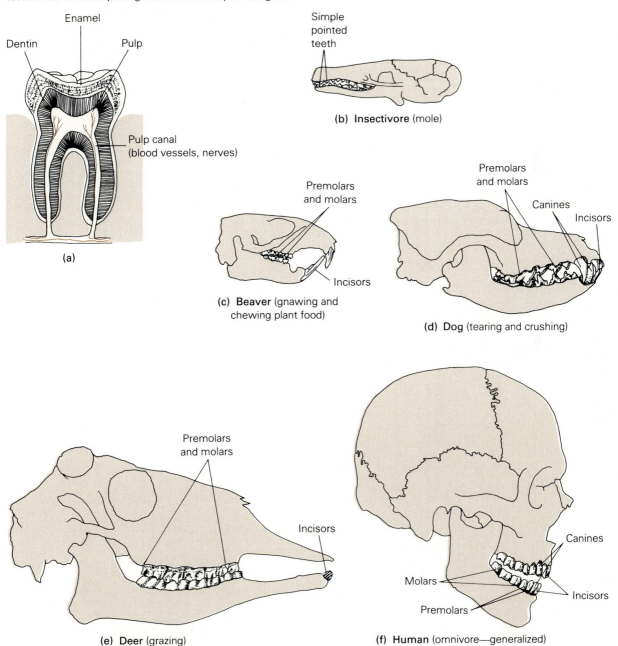

Enamel

Dentin   Pulp

Pulp canal
(blood vessels, nerves)

(a)

Simple
pointed
teeth

(b) Insectivore (mole)

Premolars
and molars

Incisors

(c) Beaver (gnawing and
chewing plant food)

Premolars
and molars

Canines

Incisors

(d) Dog (tearing and crushing)

Premolars
and molars

Incisors

(e) Deer (grazing)

Canines

Molars

Incisors

Premolars

(f) Human (omnivore—generalized)

29.13 A FOUR-PART STOMACH

Ruminants such as the cow have special digestive structures for the digestion of cellulose. Digestion begins in the large rumen, where vast numbers of protozoans and bacteria begin the breakdown of cellulose under anaerobic conditions. The products are regurgitated as the *cud,* rechewed and swallowed, this time entering the **reticulum** (follow the dashed arrow). The partially digested mass then enters the **omasum** and finally the abomasum (the true stomach) where digestion of the microorganisms themselves begins. Digestion is completed in the small intestine.

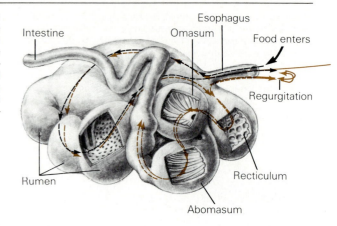

29.14 THE HUMAN DIGESTIVE SYSTEM

The three salivary glands (inset) open into the oral or mouth cavity. The pharynx and esophagus are primarily simple, muscular passages secreting only lubricating mucus. The stomach is quite complex, secreting hydrochloric acid and protein digesting enzymes. The small intestine is both secretory and absorptive, its villi and microvilli greatly increasing its absorbing surface. It also receives the secretions of the liver and pancreas. The colon, or large intestine, is chiefly involved in water absorption and the compacting of digestive wastes. It also provides a suitable environment for enormous populations of bacteria, some of whose waste products are useful to the host.

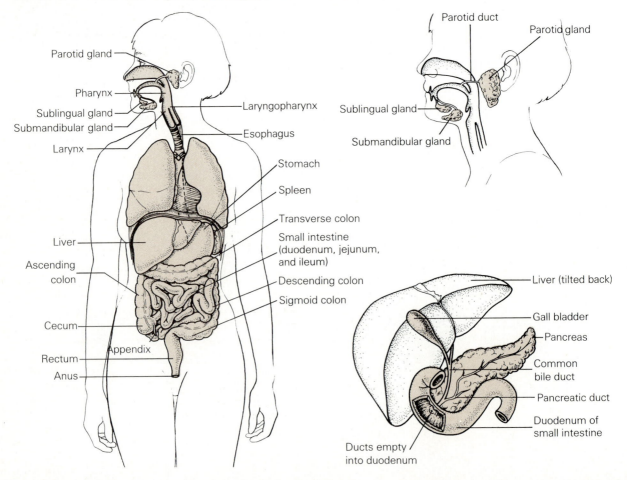

The Oral Cavity and Esophagus

We have already discussed human dentition, so now let's concentrate on some of our other feeding structures, such as the lips and tongue. In all mammals, the lips have a rather essential role in eating. If you don't know what that role is, try eating a meal without closing your lips, but do it somewhere else. Fortunately, even those crass souls who chew with their mouths open must close them to swallow.

The tongue is also important to eating. In addition to its role in moving food into position for chewing, swallowing, and sorting out fish bones, it constantly monitors the texture and chemistry of foods. This is a valuable chore, since it warns us about peculiarities in time to prevent our swallowing something we shouldn't. Further, the tongue *tastes,* that is, it can distinguish certain chemicals by specialized chemoreceptors clustered in sense organs called **taste buds.** These receptors distinguish four basic tastes: salty, sour, sweet, and bitter (Figure 29.15). In addition to informing us of flavor, the stimulation of taste buds enhances the flow of **saliva.**

There are three pairs of salivary glands, the **parotids, submaxillaries,** and **sublinguals** (see Figure 29.14). Each salivary gland empties its secretions into the mouth. These secretions, collectively, form the saliva, which consists mainly of water, ions, lubricating mucus, and the starch-splitting enzyme **amylase.** Salivary amylase, however, is probably less important in digestion than it is in oral hygiene. It helps break down starchy food particles caught between the teeth. Saliva is mainly important for moistening and lubricating food.

The pharynx, in the rear of the oral cavity, forms a common passageway with the nasal cavity. Just below the base of the tongue the pharynx divides, forming the anterior **larynx** and the posterior **laryngopharynx.** During swallowing, food is pressed by the tongue against the **soft palate,** closing the nasal passageway; the larynx is then raised, bending the **epiglottis** over the **glottis** (laryngeal opening), thus closing the air passageway to the lungs and directing food into the esophagus (Figure 29.16). Should this action fail, food will enter the larynx, producing violent spasms of coughing.

The esophagus, like the pharynx, has no digestive function but simply moistens food and moves it to the stomach. Its upper portion contains skeletal (voluntary) muscle, while the rest has smooth (in-

29.15 THE TONGUE

The tongue is an important organ for sensing the taste and texture of food. In addition, it directs food to the teeth for chewing and to the pharynx for swallowing, actively helping in that complex process. Taste buds on the tongue surface tend to specialize in detecting one of the four flavors. The scanning electron microscope study reveals flattened, columnlike projections that contain the taste buds.

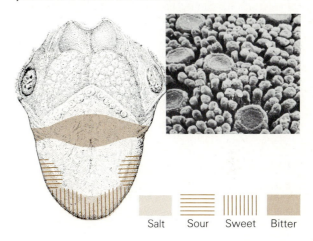

Salt   Sour   Sweet   Bitter

voluntary) muscle. Thus, swallowing begins as a conscious act but soon becomes automatic.

The lower esophagus, by the way, is very similar in its tissue organization to the rest of the digestive tract. As you can see in Figure 29.17, there are three principal layers of tissue and a fibrous outer coat, or **serosa.** The innermost layer of tissue, the **mucosa,** contains a mucus-secreting epithelium (covering), followed within by a region of connective tissue. The second tissue layer, the **submucosa,** contains more connective tissue with blood and lymphatic vessels, nerves, and glands. The third layer, the **muscularis,** is a region of smooth muscle with an inner circular layer and an outer longitudinal layer.

The muscles of the esophagus move food along by wavelike **peristalsis.** Food being swallowed is pressed into a **bolus** (clump) by the tongue and pharynx. As the bolus approaches the involuntary muscles of the esophagus, the circular layer just ahead of it relaxes and the muscles behind the bolus contract, pushing food along. The muscles are coordinated by nerve endings that are activated by the presence of food. Because peristalsis is under the control of the autonomic nervous system, it is an involuntary action, and unless you have an embarrassingly noisy intestine you are probably not aware of the process.

29.16 SWALLOWING

The important structures of swallowing are the tongue, pharynx, soft palate, epiglottis, larynx, and laryngopharynx. Swallowing is initially voluntary and then becomes involuntary or reflexive. It begins as a food mass is pressed upwards by the tongue against the palate and back toward the pharynx. It is here that the involuntary stages begin. The soft palate elevates, closing off the nasal cavity. The larynx elevates, bending the epiglottis over the glottis (larynx opening) and effectively sealing off the breathing passage. Food then enters the laryngopharynx, where continued reflexive muscular action forces it into the esophagus.

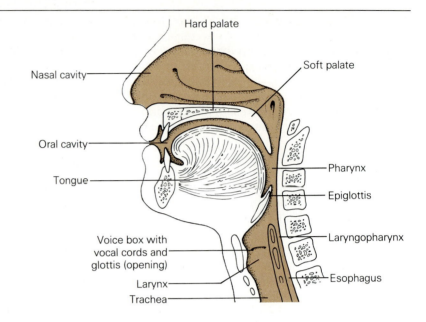

29.17 ORGANIZATION OF THE DIGESTIVE TUBE

The basic structure of the tube includes three tissue layers. The innermost mucosa surrounds the lumen and functions in lubrication, secretion, and, except for the esophagus, absorption. The middle submucosa contains connective tissue and is the avenue for blood vessels, nerves, and lymphatic vessels. The outer muscularis contains circular and longitudinal smooth muscles and is covered outside by a tough fibrous sheath called the serosa (the stomach has a third muscle layer that runs roughly diagonally).

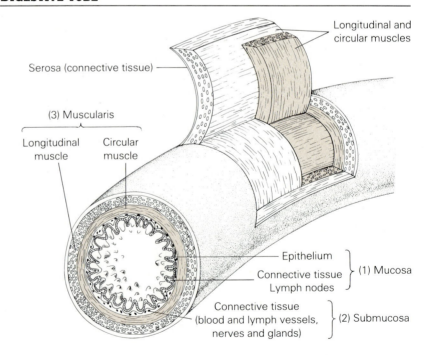

The Stomach

The human stomach is structurally and functionally similar to those of most other vertebrates. It temporarily stores food and begins its digestion. In addition, its acids and enzymes kill any microorganisms we might swallow. The stomach can be closed off at either end by two muscle sphincters (Figure 29.18a). When contracted, these circular groups of muscles permit the stomach to churn and liquefy food without forcing it back into the esophagus or into the intestine before it is ready. You may have noticed that the upper ring of muscle, the **cardiac sphincter,** sometimes fails when the stomach is overfull or filled with gas, and allows acidic fluids to enter the esophagus, creating "heartburn." The overproduction of these acids is also related to emotional stress.

The muscle layers of the stomach are more complex than those of the esophagus. Here, we encounter a third, diagonal layer of muscle just inside the circular layers. This one produces a twisting action that accompanies the usual wringing (by circular muscles) and shortening (by the longitudinal muscles). Also, peristalsis in the stomach is nondirec-tional, and moves food back and forth in a sequence of contortions until it reaches the well-churned state required by the intestine.

The stomach lining (mucosa) contains many long tubular glands that secrete the gastric juices. The glands contain **chief cells,** which secrete the protein **pepsinogen,** and **parietal cells,** which secrete **hydrochloric acid** (HCl) (see Figure 29.18b). The acid activates the pepsinogen, forming the digestive enzyme **pepsin.** Other glands secrete water, mucus, and small quantities of **gastric lipase,** a fat-splitting enzyme. The presence of HCl produces a very low pH of 1.6 to 2.4 in the stomach fluids. The acidity of the stomach and the potency of its enzymes could (and sometimes does) endanger the lining itself. One reason we don't digest our stomachs is that a layer of **mucin,** an insoluble mucoprotein, forms a coating over the stomach lining.

The Small Intestine

Once the food reaches a liquefied state, now referred to as **chyme,** the **pyloric sphincter,** another ring of muscle, relaxes a bit, allowing a small

29.18 THE HUMAN STOMACH

(a) The stomach is a J-shaped pouch with a sphincter at either end. Its muscular layers run in three directions and can produce powerful writhing and wringing motions. The complex lining bears cells and glands that secrete water, mucus, acids, and enzymes. **(b)** A close look at the lining reveals that mucus-secreting cells abound at the tips of long tissue columns. In the deep crevices between the columns, pepsinogen-secreting chief cells and acid-secreting parietal cells release their products.

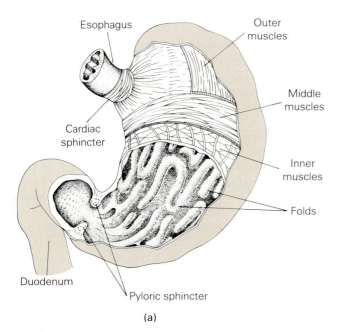

(a)

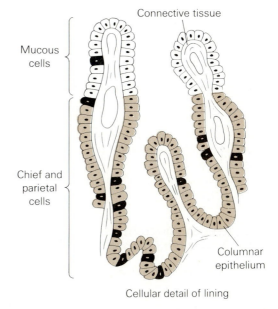

Cellular detail of lining

(b)

amount of liquefied food to move into the **small intestine,** where digestion continues and absorption begins. The small intestine is about 6 m (20 ft) long in humans, consisting of three regions: the **duodenum, jejunum,** and **ileum** (see Figure 29.14).

As we have emphasized, a principal role of the small intestine is absorption, which as we will see includes a great deal of active transport and some passive transport as well. The small intestine is uniquely adapted for absorption (Figure 29.19). To begin with, it has an enormous surface area (estimated at about 700 m² or the floor area of four or five three-bedroom houses). Its inner surface has the appearance of a badly wrinkled bath towel. Closer examination of the surface reveals that there are many folds, each covered with tiny projections called **villi.** Each villus, as you see in Figure 29.19c, is in turn covered with columnar epithelial cells bristling with **microvilli,** fingerlike projections of the plasma membrane.

Within each villus is a **capillary bed** and a **lacteal** (a lymphatic vessel) (Figure 29.19b). The villi are capable of rather vigorous movement, like millions of wiggling fingers. They help in mixing food and enzymes in the gut and assist in absorption by creating pressures that tend to concentrate digested food in the many recesses of the lining.

Until recently, it was believed that digestive enzymes were free in the intestinal lumen. But careful studies of the epithelial cells have led physiologists to believe that most of the enzymes are actually bound to the plasma membrane, forming a frilly surface they have named the **glycocalyx** (see Figures 29.19c and d and Chapter 5). This precise positioning of the enzymes presumably helps prevent the lining of the small intestine from being digested.

The cellular lining of the small intestine is thought to be very temporary and always in a state of replacement. This isn't surprising if you consider the amount of frictional wear that it suffers. Abrasion scuffs away an estimated 17 million cells each day. However, new cells are continuously produced at the base of each villus, and these migrate upward in an orderly fashion, replacing those worn away at the tip.

Accessory Organs: The Liver and Pancreas

The duodenum of the small intestine receives the secretions of two accessory organs, the **liver** and the **pancreas.** The liver, an organ of many functions,

aids in digestion by secreting a slightly alkaline fat emulsifier known as **bile** that breaks fats up into minute globules. Bile is a complex substance containing cholesterol, bile salts and pigments, water, and modified amino acids. The bile salts are actually steroids and are important in dissolving fats. The bile pigments are products of hemoglobin destruction, since the liver (along with the spleen) is the red blood cell graveyard. These breakdown products of hemoglobin become part of the digestive wastes.

After being produced in the liver, the bile is stored in the **gall bladder.** The release of bile is brought about by a hormone whose name defies pronunciation, **cholecystekinin-pancreozymin,** which causes the gall bladder to contract (and also stimulates the release of pancreatic enzymes). Bile reaching the duodenum via the **bile duct** is joined by highly alkaline fluids from the merging **pancreatic duct,** just before it enters the intestine (see Figure 29.14).

In humans, the pancreas is a long glandular organ lying nestled in the first turn of the small intestine (see Figure 29.14). One of its products, **sodium bicarbonate,** neutralizes the acid accompanying partially digested food from the stomach, protecting the small intestine and raising the pH, thereby enhancing the activities of the next series of enzymes. The pancreas secretes an entire battery of digestive enzymes that are involved in the breakdown of fats, carbohydrates, protein, and nucleic acids. The pancreatic enzymes, together with those of the small intestinal lining, carry out most of the digestive process.

The Large Intestine

The small intestine joins the **large intestine,** also called the **colon** or **bowel,** on the right side of the abdominal cavity. The large intestine consists of the **cecum, ascending colon, transverse colon, descending colon,** and **sigmoid** (S-shaped) **colon** (see Figure 29.14). The last portions of the digestive tract are the **rectum,** the **anal canal,** and the **anus**— where the story ends.

Below the point where the two parts of the intestine join, the large intestine forms a blind pouch, the cecum. Protruding from the cecum is the **appendix**—a hollow, fingerlike extension. At the union of the small and large intestines is the **ileocecal valve,** a one-way valve that prevents a backflow of the food residues into the small intestine.

The primary function of the colon is to absorb water and minerals into the blood and to prepare the feces to leave the digestive tract. The mucosa of

29.19 THE INTESTINAL LINING

The inner surface or mucosa of the small intestine contains a number of folds (a), each of which is covered by the fingerlike intestinal villi. (b) Each villus consists of a column of cells containing a capillary network and a lacteal (a blind ending of a lymph duct), into which digested foods move. (c, d) The surface cells of the villus have their membranes folded into numerous microvilli, as seen in the electron micrograph. The folding provides an enormous surface area and the oscillating movement of the villi further assure their exposure to foods. At the surface, membrane-bound enzymes form the hazy surface glycocalyx.

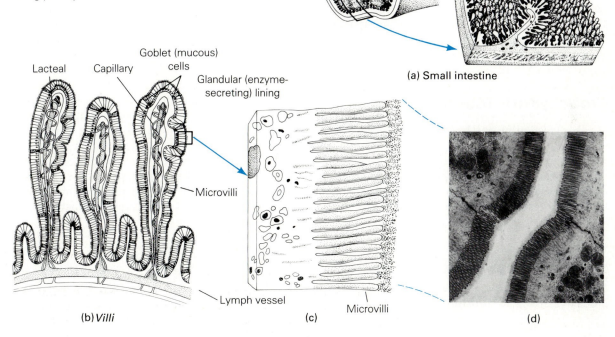

(a) Small intestine

Lacteal   Capillary   Goblet (mucous) cells   Glandular (enzyme-secreting) lining

Microvilli

Lymph vessel

Microvilli

(b) *Villi*   (c)   (d)

the colon is glandular and secretes mucus, which is used in the production of feces and lubricates the lower portions of the colon. Very little, if any, digestive activity takes place in the colon itself, although a rapid digestion of food residues is carried on by teeming populations of microorganisms. *Escherichia coli*, along with a number of methanogens (see Chapter 18), are prominent inhabitants. Strange as it may seem, a sizable amount of fecal material, perhaps a third by dry weight, is composed of bacteria. *E. coli* in the intestine play an important and rather unexpected role in nutrition. A flourishing bacterial colony can help prevent deficiencies in such vitamins as vitamin K, biotin, and folic acid.

The final region of the digestive system is the rectum. Its structure is basically the same as that of the rest of the colon, except for the presence of folds in its walls. These are the **rectal valves,** which help support the weight of feces so that gravity doesn't

become the enemy of propriety. The anus is the external opening. The movement of wastes through the anal sphincter is usually, thankfully, under voluntary control.

THE CHEMISTRY OF DIGESTION

You may recall from our previous discussions that proteins, carbohydrates, and fats are often very large, complex molecules composed of repeating molecular subunits (see Chapter 3). Except for some of the neutral fats, these macromolecules can't be absorbed through the lining of the gut. So in the digestive process, enzymes attack the linkages of the molecular subunits, breaking them down into their components.

The bonds are broken through a process called **hydrolysis.** Essentially, water is added enzymatically to the linkages, disrupting them and yielding the molecular subunits. Following digestion, the simplified foods cross the gut lining, but physiologists are now convinced that while water and some ions move by osmosis and diffusion, respectively, most foods must be actively transported through membranal carrier mechanisms across the intestinal villi into the blood or lymphatic fluids within.

In review, during digestion, carbohydrates are hydrolyzed into simple sugars, fats into fatty acids and glycerol, proteins into their various amino acids, and nucleic acids into free nucleotides. The enzymes and their functions are listed in Table 29.1.

Carbohydrate Digestion

Starch digestion begins in the mouth, where salivary amylase breaks some linkages, producing maltose and some larger molecular fragments. Starch digestion is then temporarily stalled by the acidity of the stomach. It resumes, in full swing, in the small intestine, where **pancreatic amylase** (from the pancreas) converts all the starch into maltose. But maltose is still too large to pass into the bloodstream. Finally, it is broken down by **maltase** into the monosaccharide glucose, which can be absorbed (Figure 29.20). Sucrose (table sugar) and lactose (milk sugar), both disaccharides, are also split into monosaccharides by enzymes from the mucosa of the small intestine.

The enzyme **lactase,** which breaks down milk sugar, is present in the intestines of all normal human babies and nearly all Caucasian adults. Black and Asian adults and their older children usually lack intestinal lactase, and milk can cause excessive gas buildup and diarrhea, symptoms of what is called **lactose intolerance.** Cheese is more digestible because its lactose has already been broken down by microorganisms used in its manufacture. It has been suggested that the continued production of lactase in Caucasian adults is a product of natural selection; milk has long been a major part of the Caucasian diet.

Glucose and fructose are transported into the capillaries of the villi in the small intestine, and from there directly to the liver where they are recombined into long chains of glycogen. The active trans-

TABLE 29.1

DIGESTIVE ENZYMES AND THEIR FUNCTIONS

Sources and Enzymes	Substrate	Product
Salivary glands		
Salivary amylase	Starch	Maltase and starch fragments
Stomach lining		
Pepsin	Protein	Peptides
Rennin	Casein	Insoluble curd
Gastric lipase	Triglyceride	Fatty acids + glycerol
Pancreas		
Trypsin	Peptide linkage	Shorter polypeptides
Chymotrypsin	Peptide linkage	Shorter polypeptides
Ribonuclease	RNA	Nucleotides
Deoxyribonuclease	DNA	Deoxynucleotides
Pancreatic amylase	Starch	Maltose
Pancreatic lipase	Triglyceride	Fatty acids, glycerol
Carboxypeptidase	C-terminal bond	Shorter peptide and one free amino acid
Intestinal lining		
Aminopeptidase	N-terminal bond	Shorter peptide and one free amino acid
Tripeptidase	Tripeptide	Dipeptide + amino acid
Dipeptidase	Dipeptide	Two amino acids
Nuclease	Nucleotide	5-carbon sugar + nitrogen base
Maltase	Maltose	Two glucose units
Sucrase	Sucrose	Glucose + fructose
Lactase	Lactose	Glucose + galactose

29.20 STARCH DIGESTION

Amylose, a straight chain starch, is readily digested by amylase, a starch splitter that breaks 1-4 linkages thereby forming maltose, a double sugar. Amylopectin digestion is more complex because it is a highly branched chain. Amylase cannot attack the 1-6 linkages; thus its products include both 1-4 linked maltose and short, branched chains containing both 1-4 and 1-6 linkages. All of the products are hydrolyzed to glucose by enzymes of the small intestine, but additional enzymes are required for the hydrolysis of the 1-6 linkages.

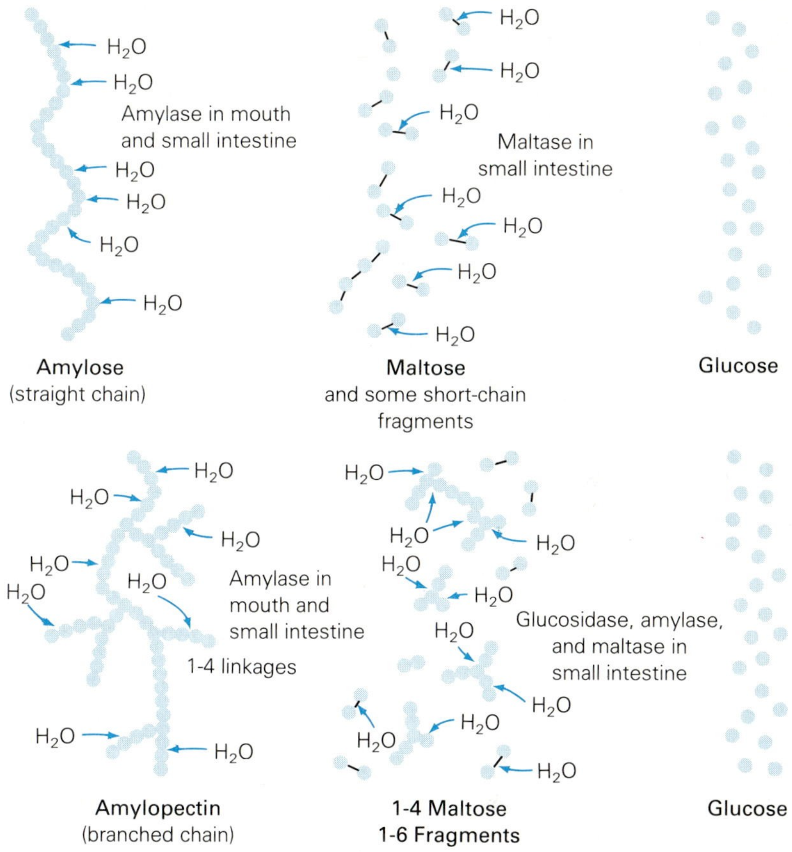

Amylose
(straight chain)

Amylase in mouth
and small intestine

Maltase in
small intestine

Maltose
and some short-chain
fragments

Glucose

Amylopectin
(branched chain)

Amylase in
mouth and
small intestine

1-4 linkages

Glucosidase, amylase,
and maltase in
small intestine

1-4 Maltose
1-6 Fragments

Glucose

port of glucose is carried out by the same ATP-powered carrier that transports sodium ions into the intestinal cells. From these, glucose diffuses into nearby capillaries.

Fat Digestion

Fats, for the most part, reach the small intestine with little chemical change. This is a bit surprising since we know that the stomach secretes lipase. Once in the small intestine, bile salts emulsify fats, or separate them into tiny droplets. These droplets are further broken down by lipase into fatty acids, glycerol, monoglycerides, and diglycerides (Figure 29.21), which simply diffuse across the lipid-soluble core of the plasma membrane of lining cells. Once inside a lining cell, these products are reassembled into triglycerides that, along with absorbed cholesterol, then gather into **chylomicrons,** minute bodies enclosed by a thin protein envelope. In this form, they pass across the cell and enter the lacteals of the villi, from which they are transported via lymph vessels to the circulatory system.

Protein Digestion

Proteins are the most complex of the food molecules. It is not surprising, therefore, that their digestion is also complex. As we have seen, a first step occurs in the stomach, where pepsinogen is secreted by chief cells and activated by HCl to form the enzyme pepsin. Pepsin has a broad-range action, lacking the specificity of most other enzymes. As you can see in Figure 29.22, it attacks here and there *within* the strand of protein; therefore, it is called an **endopeptidase.** Its common targets, however, are the peptide bonds of the amino acids methionine, leucine, tyrosine, phenylalanine, and tryptophan. As a result, pepsin breaks proteins down into shorter peptides, but seldom into single amino acids.

The action of the pancreatic enzyme **trypsin** is more specific. It hydrolyzes only the bonds on the carboxyl side of arginine and lysine. The second pancreatic enzyme, **chymotrypsin,** is also specific, attacking only the peptide bonds on the carboxyl side of the amino acids phenylalanine, tyrosine, and tryptophan. The combined action of pancreatic

29.21 FAT DIGESTION

Fats are digested by the enzyme lipase which attacks the bonds between each of the fatty acids and the glycerol. Three water molecules are consumed in the process. Some fats bypass the digestive process in the intestine and are taken into the lacteal through pinocytosis.

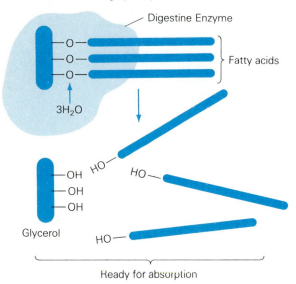

Ready for absorption

Nucleic Acid Digestion

Almost everything we eat contains at least some nucleic acids. The enzymes that hydrolyze nucleic acids are called **nucleases,** generally classed as **DNAases** and **RNAases.** They are divided into two classes on the basis of how they attack the large molecules. **Exonucleases** cleave off the end nucleotides, while endonucleases attack bonds within the molecule. Exonucleases, also called **phosphodiesterases,** specialize in working either from the 3′ end or the 5′ end of the nucleotide, hydrolizing the bonds between phosphates and ribose or deoxyribose sugars and thus splitting off terminal nucleotides as they go. Endonucleases also specialize in hydrolyzing the bonds between the phosphates and sugars but do not require free ends upon which to work. In their combined action, the nucleases break down nucleic acids into small units of either single nucleotides or small chains of three or four.

Integration and Control of the Digestive Process

The release of digestive enzymes and related substances must be very precisely timed. It turns out that such timing is under at least three types of control—mechanical, neural, and hormonal—with some interaction among the three types. For example, saliva flow can be stimulated mechanically by chewing a tasteless substance like paraffin. Yet, even the thought of eating chocolate cake can evoke salivation. Further along the digestive tract, gastric secretions can also be stimulated by our simply sensing the presence of food. A neural message is sent along a branch of the **vagus nerve** from the brain to the stomach lining.

The presence of food in the stomach mechanically evokes gastric secretions by stimulating sensory neurons in the stomach wall itself. Impulses reach the brain and again the brain sends down the word via the vagus nerve to resume the flow of gastric secretions.

A third mechanism is hormonal and independent of the nervous system. For example, when food, particularly concentrated protein, is present in the stomach, the hormone **gastrin** is released into the bloodstream by cells of the stomach wall itself. Once in the blood, the hormone circulates freely but is ignored by all cells except highly specific target cells, which have plasma membrane or cytoplasmic receptors for that specific hormone (see Chapter 33). Once the hormone has been received and recognized, the reaction is very precise. The target cells, in this case the nearby gastric glands, respond

enzymes hydrolyzes the peptide fragments into some individual amino acids as well as small peptides consisting of perhaps two to ten amino acids.

Finally, enzymes known as **exopeptidases** cleave the remaining peptide bonds, chopping off single amino acids or pairs of amino acids from the ends of the peptide fragments. This leaves the individual amino acids ready for absorption and transport to the liver. The exopeptidases can be divided into **carboxypeptidases** (pancreatic in origin), which take single amino acids off the carboxyl terminal end of the peptide, and **aminopeptidases** (intestinal), which take them off the amino terminal end, and **dipeptidases,** (also intestinal), which specialize in breaking up amino acid pairs.

Amino acids are actively transported by several kinds of membranal carriers into the intestinal lining cells and from there enter the blood, probably through simple diffusion. They are removed from the blood by liver cells. Once in the liver, the amino acids can be used in a number of ways. Most are deaminated and used for energy. Others are reconstituted into serum proteins (the proteins of blood plasma) and released into the blood so that individual cells can fill their amino acid needs. Except for the blood vessels between the digestive system and the liver, free amino acids do not occur in high concentrations in the blood.

29.22 PROTEIN DIGESTION

The digestion of protein begins in the stomach with the action of pepsin, an endopeptidase that breaks the peptide bonds of five specific amino acids. **(b)** The peptide fragments are then attacked in the small intestine by the pancreatic endopeptidases, chymotrypsin and trypsin, which also act at specific sites. While some amino acids are released at this time, peptide fragments of varying lengths must again be hydrolyzed, this time by exopeptidases and dipeptidases from the small intestine itself **(c).** The first releases terminal amino acids while the second splits dipeptides. Finally, a residue of single amino acids emerges and protein digestion is complete.

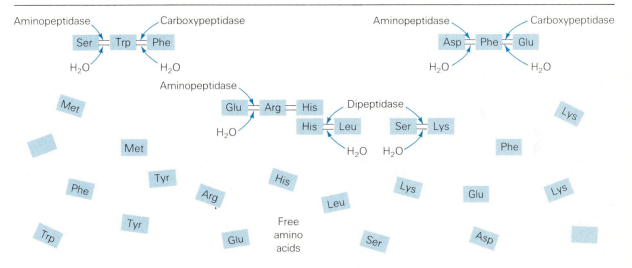

(a) **Stomach**—pepsin (an endopeptidase) breaks down proteins to polypeptides

(b) **Small intestine**—pancreatic enzymes (endopeptidases) break down polypeptides to smaller peptides

(c) **Small intestine**—intestinal enzymes (exo- and dipeptidases) break down peptides to amino acids

to gastrin by releasing gastric juices. The gastrin mechanism, like many that are hormonally regulated, works through a negative feedback principle. The release of gastric juices, which include HCl, lowers the stomach pH. When it reaches pH2, gastric secretion slows greatly. Thus, the product of the hormone's action—the HCl—has a negative effect, shutting down the hormone's release. We will examine negative feedback in more detail in Chapter 33. The five major hormones involved in the integration and control of digestion are shown in Table 29.2.

SOME ESSENTIALS OF NUTRITION

The subject of nutrition is of increasing interest to the general public. Unfortunately, interest in food doesn't necessarily imply knowledge about nutrition. Many food faddists are almost ignorant of the subject. Some do not seem to realize that "organically grown" food is no better for you than food grown with chemical fertilizers; that rose hip vitamin C is no different from the commercial vitamin C produced by bacteria; that large doses of vitamins are not healthful and may be harmful; that dietary protein in excess of body needs is simply deaminated and converted to glucose or fatty acids in the liver—and fat on the hips. Actually, most Americans eat far more protein than they need, and tests with long-distance runners show that athletes have more stamina on a high-carbohydrate diet than on a high-protein diet. On the other hand, some health faddists prescribe diets so low in protein as to produce virtually the only remaining cases of protein-deficiency disease in America. All this is interesting but confusing, so let's have a closer look at diet and see if the scientific findings have any relevance for us.

Carbohydrates

Carbohydrates are common sources of energy. As we saw in Chapter 8, they are used in generating ATP, but they are used in other ways as well. For instance, once glucose makes its way from the gut into the circulatory system, it is immediately removed and stored as glycogen by the liver. From there it is meted out according to need. Thus, the blood glucose level remains rather constant, with a temporary 10 to 20% increase immediately after a high-carbohydrate meal. Glycogen is also stored in the muscles, where it is broken down into glucose and used as an energy source in anaerobic glycolysis. Long-distance runners build additional glycogen reserves in muscles by continually depleting their reserves in long training runs and letting them build back up at ever higher levels. For those who have no intention of becoming involved in heavy exercise, it is important to know that excess glucose is readily converted into body fat.

Fats

Fats in the diet are not all bad. The polyunsaturated fats remain an essential nutrient to humans as well as to other animals, although the quantity needed is open to debate. Fats are excellent concentrated energy sources with, gram for gram, twice the energy value of carbohydrates. In addition, they contain the essential fat-soluble vitamins A, D, E, and K. Unsaturated fats are necessary constituents of all plasma membranes, yet vertebrates and most

TABLE 29.2

DIGESTIVE HORMONES

Hormone	Stimulation	Function
Gastrin	Protein in stomach Vagus nerve	Stimulates release of gastric juice
Enterogastrone	Acid state in intestine Fats in intestine	Inhibits gastric juice release and slows stomach contractions
Secretin	Acid state in intestine Peptides in intestine	Stimulates release of sodium bicarbonate
Cholecystokinin–Pancreozymin	Food entering small intestine	Stimulates secretion of pancreatic enzymes and contraction of gall bladder
Enterocrinin	Stomach materials in intestine	Stimulates intestinal secretions

Cholesterol and Controversy

Cholesterol is a vital constituent of plasma membranes. Our own liver can make it readily. Nevertheless, the amount of cholesterol circulating in the blood is greatly influenced by dietary intake. Not only is cholesterol in food taken up directly, but the amount of cholesterol synthesized and its concentration in the blood can be dramatically decreased by reducing either the amount of cholesterol or the amount of saturated triglycerides in the diet, and especially by increasing the consumption of unsaturated fats.

How did cholesterol get its unsavory reputation? The *plaques* that plug arteries in chronic arteriosclerosis are heavily infiltrated with cholesterol. The fact that persons with abnormally high levels of circulating cholesterol have abnormally high rates of arterial disease and heart disease suggests a cause-and-effect relationship. Thus, it seems perfectly reasonable to conclude that anyone will benefit from reducing cholesterol intake and replacing saturated (animal and hydrogenated vegetable) fats with unsaturated (nonhydrogenated vegetable) fats. Some doctors still give such advice, but new information is suggesting other routes to good health.

In a recent study, a large group of middle-aged men was divided into an experimental population and a control population. The experimental population was a given a low-cholesterol, high-unsaturate diet. Sure enough, their blood cholesterol levels went down, and over the years, their rates of arterial disease and heart disease were lower than those of the control population. Case proven? Not quite. While the death rate due to heart attacks was lower in the experimental group, the overall death rate of the experimental group was significantly *higher* than that of the control group. It turns out that the experimental group had a higher cancer death rate.

What happened? One possibility is that the higher levels of cholesterol in the control group gave them healthier plasma membranes that protected them from cancer and other disease, or perhaps gave them a better-balanced steroid metabolism.

Another, even more likely, possibility is that the higher levels of unsaturated fats in the experimental group were actively toxic. Unsaturated fats spontaneously form *free radicals,* highly reactive molecules that contain unpaired electrons. Free radicals tend to react with and degrade other molecules in a random manner. Interestingly, ionizing radiation damages cells in a similar manner. It creates free radicals that attack DNA and other cellular constituents. In any case, the study indicates what can happen to those who radically change their diets in response to even a seemingly well-founded scientific opinion. A generation or two of Americans have avoided cholesterol like the plague and, overall, may have gained nothing from their efforts.

---

other animals cannot synthesize all the types needed. Some, known as essential fatty acids, must be provided in food. Enlightened dieters, then, include at least some fats in the diet. However, fats can be synthesized from carbohydrate and protein excesses, so cutting down on these nutrients is also required. For this reason, high-carbohydrate or high-protein weight-reducing diets may defeat their own purpose if the intake is not carefully measured.

You are probably aware of the persistent controversy over saturated fats and cholesterol, and may faithfully avoid these in favor of the unsaturated fats heavily touted in advertisements for margarine and cooking oils. Essay 29.1 reveals a startling twist to some prevalent ideas on the subject.

Protein

Since our bodies don't store protein well, we should include it in our daily diet. The amino acids of protein are used in several ways: they are the building blocks of our own protein, some kinds are used to form the nitrogen bases of the nucleic acids DNA and RNA, they can be oxidized for energy, and, as mentioned, they can be converted to fats and carbohydrates.

It is interesting that humans (along with all other animals, and protozoa as well) can convert some amino acids to others. In fact, we can synthesize 12 of the 20 amino acids we require for maintenance and growth. The remaining eight must be supplied, intact, in the diet. These are referred to as the **essential amino acids,** a phrase that delights advertising agencies, although it is something of a misnomer, since all 20 are biologically essential at some level.

Dietary proteins vary in "quality." High-quality proteins contain more kinds of amino acids than do low-quality proteins. Protein synthesis in your own cells cannot proceed unless all of the constituent amino acids are present, so the usefulness of a protein is limited to the relative concentration of its scarcest essential amino acid. In general, animal proteins are of higher quality than plant proteins. Although plants do store proteins in their seeds for the development of new seedlings, these proteins are usually of comparatively low quality for human consumption. Plant storage proteins are typically deficient in tryptophan, methionine, or lysine and lysine deficiency in particular can become a problem for vegetarians.

Vitamins

In addition to essential fatty acids and essential amino acids, a variety of substances, known as vitamins, are also essential in the diets of animals. Most of the vitamins have been clearly identified by biochemists so that we know both their molecular structure and their precise function (Table 29.3). A few are only vaguely understood, however. All we can say about them is that if you don't have them,

you will develop something-or-other, and it will probably be bad. Shortages of niacin (vitamin B$_6$) and riboflavin, for example, can lead to serious illness because the coenzymes NAD and FAD, necessary for glycolysis and cell respiration, are derived from these vitamins. Vitamins C and E have a somewhat more general function: both are **antioxidants** that remove spontaneous free radicals that would otherwise cause damage through random oxidation reactions.

Vitamins can be categorized as fat-soluble or water-soluble. The two behave quite differently in the body. For example, water-soluble vitamins function as coenzymes, fat-soluble vitamins do not. Whereas water-soluble vitamins function in most animals, the fat-soluble function only in verte-

TABLE 29.3

VITAMINS

Vitamin	Source	Function	Daily requirement	Result of deficiency
A, retinol	Fruits, vegetables, liver, dairy products	Synthesis of visual pigments	1500–5000 IU[a] 3 mg	Night blindness, crustiness about eyes
B$_1$, thiamine	Liver, peanuts, grains, yeast	Respiratory coenzyme	1–1.5 mg	Loss of appetite, beriberi, inflammation of nerves
B$_2$, riboflavin	Dairy products, liver, eggs, spinach	Oxidative chains in cell respiration	1.3–1.7 mg	Lesions in corners of mouth, skin disorders
Niacin, nicotinic acid	Meat, fowl, yeast, liver	Part of NAD and FAD cell respiration	12–20 mg	Skin problems, diarrhea, gum disease, mental disorders
Folic acid	Vegetables, eggs, liver, grains	Synthesis of blood cells	0.4 mg	Anemia, low white blood count, slow growth
B$_6$	Liver, grains, diary products	Active transport	1.4–2.0 mg	Slow growth, skin problems, anemia
Pantothenic acid	Liver, eggs, yeast	Part of coenzyme A of cell respiration	Unknown	Reproductive problems, adrenal insufficiency
B$_{12}$	Liver, meat, dairy products, eggs	Red blood cell production	5–6 µg	Pernicious anemia
Biotin	Liver, yeast, intestinal bacteria	In coenzymes	Unknown	Skin problems, loss of hair and coordination
Choline	Most foods	Fat, carbohydrate, protein metabolism	Unknown	Fatty liver, kidney failure, metabolic disorders
C, ascorbic acid	Citrus fruits, tomatoes, potatoes	Connective tissues and matrix antioxidant	40–60 mg	Scurvy, poor bone growth, slows healing (colds?)
D	Fortified milk, seafoods, fish oils, sunshine	Absorption of calcium	400 IU[a]	Rickets
E	Meat, dairy products, whole wheat	Uncertain	Unknown	Infertility, kidney problems
K	Intestinal bacteria	Blood-clotting factors	Unknown	Blood-clotting problems

[a]IU = International units.   mg = milligram (1/1000)   µg = microgram (1/1,000,000)

TABLE 29.4

MAJOR MINERAL NUTRIENTS

Mineral	Food source	Function	Daily requirement	Result of deficiency
Calcium	Dairy foods, eggs	Growth of bones and teeth, blood clotting, muscle contraction, nerve action	1 g	Tetany, rickets, loss of bone minerals and muscle coordination
Cobalt	Common in foods, water	Vitamin B_{12}	1 mg	Anemia
Copper	Common in foods	Production of hemoglobin, enzyme action	2 mg	Anemia
Fluorine	Most water supplies	Prevents bacterial tooth decay	Unknown	Tooth decay, bone weakness
Iodine	Seafood, iodized salt	Thyroid hormone	0.25 mg	Hypothyroidism
Iron	Meat, eggs, spinach	Hemoglobin (oxygen transport)	10 mg (men) 18 mg (women)	Anemia, skin problems
Magnesium	Green vegetables	Enzyme function	350 mg	Dilated blood vessels, irregular heartbeat, loss of muscle coordination
Manganese	Liver, kidneys	Enzyme function	Unknown	Loss of fertility, menstrual irregularities
Phosphorus	Dairy foods, eggs, meat	Growth of bones and teeth, ATP nucleotides	1.3 g	Loss of bone minerals, metabolic disorders
Potassium	Most foods	Nerve and muscle activity	2–4 g	Muscle and nerve disorders
Sodium	Most foods, salt	pH balance, nerve and muscle activity, body fluid balance	0.5 g	Weakness, muscle cramps, diarrhea, dehydration
Zinc	Common in foods	Enzyme action	Unknown	Slow sexual development, loss of appetite, retarded growth

brates. Finally, excess water-soluble vitamins are excreted, but the fat-soluble ones are stored in the body fat. This is why it is dangerous to overdose on fat-soluble vitamins.

Mineral Requirements of Humans

In addition to the organic nutrients mentioned, animals require a variety of inorganic ions that are generally referred to as **minerals** (Table 29.4). Some of those needed in substantial amounts are calcium (a major constituent of bone and of many cellular processes), magnesium (necessary for many enzyme activities), and iron (a constituent of hemoglobin and of the cytochromes of cell respiration). Sodium, potassium, and chloride are also needed in fairly substantial quantities; they are involved in ion balance, and sodium and potassium are also involved in nerve cell conduction.

Trace Elements. As we mentioned with plants, elements that are necessary for life but in only very small amounts are called *trace elements*. Iodine is perhaps the best-known trace element, since an iodine deficiency produces **goiter,** a highly visible overgrowth of the thyroid gland (see Figure 33.11). Iodine is essential in producing thyroxin, the thyroid hormone, and the overgrowth is the thyroid gland's odd way of compensating for its failure. Many other minerals are constituents of coenzymes (for example, cobalt in vitamin B_{12}), are otherwise necessary for enzyme function, or are involved in synthetic processes (for example, zinc is needed for insulin synthesis). The functions of many trace elements are unknown—for instance, it is not known why small amounts of fluoride retard cavities in teeth.

Among the more bizarre recently discovered mineral requirements are arsenic, silicon, and selenium. Arsenic is an extremely deadly poison, yet laboratory animals have died of arsenic deficiency! Silicon is one of the most abundant elements (sand and rocks). Yet, silicon deficiencies may also be widespread. The mineral is needed as a cross-linking agent in the elastic walls of major arteries. Finally, selenium is known to be necessary for the functioning of at least one enzyme. Human selenium deficiency disease is common in some parts of China. (In other parts of China the population suffers from selenium poisoning.)

So, what's a body to do? We've seen that low-cholesterol and high-cholesterol diets are both harmful, that the balance between saturated and unsaturated fats mustn't be tipped too far in either direction, that too much protein is harmful but too little is worse. Vegetarianism can be dangerous, but so can meat. A tiny amount of arsenic is deadly, but an even tinier amount may help keep you alive; you can get sick from too much vitamin A or D, or from not enough. Most people consume too much sodium, but everyone needs some. Too little selenium is bad, but so is too much. How can one make intelligent decisions in the face of conflicting evidence?

There is no simple answer. We must resort to generalities and cliches, such as "moderation in all things" or "variety is the spice of life" or "don't eat so much." No one ever suffered from eating sensible amounts of fresh fruits and vegetables. Get enough roughage. Avoid faddism. Avoid fats, but don't be a fanatic about it (unless you are overweight, in which case a little antifat fanaticism may be in order). Part of the problem is, we live in a food-laden environment. Most of us can choose from a range of foodstuffs, some traditional, others recently and chemically contrived. With the opportunity for choice, however, comes the responsibility of being informed. And since much of what we know has just been learned recently, the responsibility remains a continuing one.

APPLICATION OF IDEAS

1. Select five distinct variations in mammalian dentition (including its absence) and relate these variations specifically to the food consumed by each animal. Suggest how such variation might arise.

2. The feeding habits of animals are highly varied. Omnivores commonly feed on a wide variety of foods, while more specialized species, whether herbivore or carnivore, may feed mainly on one specific item with little deviation. Discuss examples of each and suggest short- and long-term evolutionary advantages and disadvantages in either direction.

3. Careful studies of herbivorous animals reveal that they rarely produce the enzyme cellulase, yet a considerable amount of their diet consists of cellulose. Using examples, explain how mammals extract nutrients from cellulose. Considering that a large part of the carbohydrate produced on earth is in the form of cellulose, suggest reasons why most animals haven't evolved the ability to synthesize the enzyme. What advantage would such a capability offer humans?

KEY WORDS AND IDEAS

DIGESTIVE SYSTEMS

1. **Digestion** is the chemical breakdown of nutrients into absorbable parts, while **nutrition** refers to nourishment and the characteristics of essential nutrients.

2. **Extracellular digestion** refers to digestion outside of cells as seen in bacteria and fungi. **Intracellular digestion** occurs in cells, generally in **digestive vacuoles.**

Saclike Systems in Invertebrates

1. Sponges sort food particles out of seawater brought in by **choanocytes (collar cells).** Particles caught in the mucus-lined collar cells are phagocytized by the cells below for intracellular digestion and transported about by **amebocytes.**

2. Coelenterates and flatworms begin digestion extracellularly in the **gastrovascular cavity,** with

food phagocytized and digestion completed intracellularly.

Feeding and Digestive Structures in Higher Invertebrates

1. Higher invertebrates have the tube-within-a-tube body plan and digestion is mainly extracellular. Intracellular digestion occurs in the invertebrate liver, which is lined by phagocytic cells.

2. The earthworm gut contains specialized regions. Food is swallowed by the **pharynx** and passes through the **esophagus** to the **crop** for storage. Grinding occurs in the **gizzard,** and digestion and absorption occur in the long **intestine.** The **typhlosole,** a deep fold in the intestinal wall, increases surface area. **Chloragen** cells convert glucose to glycogen and **deaminate** amino acids to be used as fuels.

3. The grasshopper's chewing mouthparts include sensory **palps,** shearing **mandibles** and **maxillae,** the liplike **labrum** and **labium,** and **salivary glands** that secrete saliva. The digestive system includes a **foregut** consisting of a pharynx, esophagus, and crop for swallowing and grinding, a **midgut** for digestion and absorption, **gastric cecae** for enzyme secretion, and a water-absorbing **hindgut.** Insect mouthparts are also adapted for piercing and sucking.

Feeding and Digestive Structures in Vertebrates

1. Vertebrate jaw and tooth structure varies with feeding specializations. Carnivorous sharks have continuously growing rows of teeth. In the gut, a winding flap, the **spiral valve,** increases the surface area for digestion and absorption. Teeth in bony fishes vary from numerous sharp teeth to patches on the roof of the mouth. Herbivorous fishes generally have lengthy coiled intestines suitable for the time-consuming cellulose digestion process.

2. Some amphibians have a protruding tongue for capturing insects. The jaw, like that of the reptile is not suitable for chewing, so prey is swallowed whole.

3. Crocodiles and alligators have socketed teeth, while some venomous snakes have hollow or grooved retractable **fangs** for injecting venom. Snakes and lizards capture air molecules with the tongue and analyze them with the olfactory **Jacobson's organ.** Pit vipers use their heat sensitive pits to orient themselves to their prey.

4. The beak (and feet) of birds are often specialized for their source of food. Digestive specializations in seed-eaters include a storage crop and a two-part **stomach** composed of the glandular, enzyme-secreting **proventriculus** and the gravel-filled and muscular **ventriculus,** or gizzard.

5. Mammalian teeth are socketed and include temporary, or milk teeth, and permanent teeth. Four types of teeth include **incisors, canines, premolars,** and **molars.** The shape and size of each group commonly relates to diet.

6. **Ruminants** are grazing mammals whose four-part stomach includes a **rumen,** where microorganisms digest cellulose. The products and microorganisms are themselves digested by the **reticulum, omasum,** and **abomasum** (true stomach).

THE DIGESTIVE SYSTEM OF HUMANS

The Oral Cavity and Esophagus

1. The lips and tongue assist in eating and swallowing. **Taste buds** located on the tongue detect salt, sour, sweet, and bitter flavors. **Saliva,** a watery solution of ions, mucus, and **amylase,** a starch-digesting enzyme, are produced in the **parotid, submaxillary,** and **sublingual** salivary glands.

2. The pharynx divides to form the larynx and **laryngopharynx.** Some of the structures involved in swallowing are the tongue, esophagus, **soft palate, epiglottis, glottis,** and **larynx.** The larynx is raised during swallowing, preventing food from entering the air passage.

3. The esophagus, a food-conducting tube, is made up of an innermost, secretory **mucosa,** a vascular and glandular **submucosa,** and a circular and longitudinal layer of smooth muscle, the **muscularis,** all of which are surrounded by a fibrous **serosa.** Wavelike contractions called **peristalsis,** coordinated by the autonomic nervous system, move the food **bolus** to the stomach.

The Stomach

1. The stomach stores and mixes food, and its acidity helps destroy microorganisms. The **cardiac** and **pyloric sphincters** close off the stomach during its churning peristalsis. A third, oblique muscle layer aids the churning by producing a twisting action.

2. The glandular lining contains **pepsinogen**-secreting **chief cells** and **parietal cells,** whose **hydrochloric acid** secretions activate pepsinogen to the active form, **pepsin.** Other enzymes include **gastric lipase** and milk-digesting **rennin.**

The Small Intestine

1. The **small intestine** consists of the **duodenum, jejunum,** and **ileum.** Its role is digestion and absorption. In addition to what is provided by length, the surface area of the intestine is increased by folding, by projections called **villi,** and by cellular projections called **microvilli.** Each villus contains a **capillary bed** and a **lacteal,** and is capable of movement.

2. The small intestine surface contains bound enzymes forming the **glycocalyx.** New cells continually replace those lost by abrasion.

Accessory Organs: The Liver and Pancreas

1. The **liver** secretes **bile,** an alkaline secretion that emulsifies fats. Its contents include hemoglobin breakdown pigments (bilirubin and biliverdin).

2. Bile is stored in the **gall bladder,** and secreted through the **bile duct,** which joins the **pancreatic duct** before entering the intestine. Bile secretion is a response to the hormone **cholecystekinin-pancreozymin,** which is released when fats are present in the gut.

3. The glandular **pancreas** secretes **sodium bicarbonate,** which neutralizes stomach acids in the intestine. It also secretes a variety of enzymes that act on carbohydrates, fats, proteins, and nucleic acids.

The Large Intestine

1. The **large intestine,** or **colon,** begins with the **ileocecal valve,** which opens into a pouchlike cecum (to which the **appendix** is attached). Included are the **ascending colon, transverse colon, descending colon,** and **sigmoid colon.** Further along are the **rectum, anal canal,** and **anus.**

2. The colon absorbs water, concentrates the feces, and provides a suitable environment for bacteria that secrete useful vitamin K, biotin, and folic acid. Within the rectum, **rectal valves** help support the feces, and the anus controls defecation.

THE CHEMISTRY OF DIGESTION

1. The bonds connecting subunits of complex foods are broken through **hydrolysis,** releasing the simple subunits that can be absorbed or actively transported into cells lining the intestine.

Carbohydrate Digestion

1. Starches are hydrolyzed into maltose by salivary amylase in the mouth and by **pancreatic amylase** in the small intestine. **Maltase** hydrolyzes maltose into glucose, which is then actively transported by a sodium carrier into the blood.

2. **Lactase** hydrolyzes milk sugar, but adults in some races fail to produce the enzyme. Upon taking in milk they experience the symptoms of **lactose intolerance.**

3. Glucose is transported to the liver, where it is converted to glycogen and gradually released again as glucose.

Fat Digestion

1. Fats are hydrolyzed by lipase into fatty acids, glycerol, monoglycerides, and diglycerides. These diffuse into the lining cells, reform as triglicerides, and join cholesterol to form **chylomicrons,** which pass into the lacteals.

Protein Digestion

1. Protein digestion begins in the stomach, where pepsin, an *endopeptidase*, hydrolyzes inner peptide bonds forming shorter peptides.

2. In the small intestine, pancreatic **trypsin** and **chymotrypsin,** also endopeptidases, hydrolyze specific amino acid linkages, and finally, intestinal **exopeptidases** break the remaining peptide bonds. The latter includes **carboxypeptidases** from the pancreas and **aminopeptidases** and **dipeptidases** from the intestine.

3. Amino acids enter the blood for transport to the liver, where some are deaminated and used for energy, while others form serum proteins for transport to needy cells.

Nucleic Acid Digestion

1. **Nucleases,** nucleic acid hydrolyzing enzymes, include **exonucleases** that work on the ends and **endonucleases** that work within the chain. **Phosphodiesterase** breaks sugar-phosphate linkages.

Integration and Control of Digestive Enzymes

1. Digestive enzymes are released through mechanical (physical presence of food), neural (thoughts and detection of food), and hormonal mechanisms. Neural messages travel to the stomach via the **vagus nerve** when the gut lining senses the chemical presence of food. The cells react by releasing a hormone such as **gastrin** into the blood. When hormones reach their specific target cells or tissues, they stimulate them to release enzymes into the gut.

2. Hormonal systems generally work through negative feedback, in which a stimulus evokes the release of a hormone, which removes the stimulus and thereby stops its own release.

SOME ESSENTIALS OF NUTRITION

Carbohydrates

1. Carbohydrate is a vital energy source, but in excess it is converted to body fat.

Fats

1. Fats are excellent energy-providing foods and sources of fat-soluble vitamins. Unsaturated fats are needed for plasma membranes. Although unsaturated fats must come from the diet, most foods can be converted to saturated fats.

Protein

1. Protein must be provided daily because of its constant turnover. Excess amino acids are deaminated and used for energy or converted to fats and carbohydrate.

2. Eight of the amino acids—**essential amino acids**—must be provided by the diet, while the remaining 12 are interconvertible by the body. High-quality dietary protein contains sufficient

amounts and kinds of amino acids to provide all that is needed.

Vitamins

1. Fat-soluble vitamins are generally essential in forming coenzymes (NAD and FAD), and water-soluble vitamins as **antioxidants** (removing free radicals) and other uses. The former are harmful in overdoses.

Mineral Requirements of Humans

1. Major essential **minerals,** inorganic ions, include calcium, magnesium, iron, sodium, potassium, and chloride, used in structure, enzyme action, respiration, and osmotic regulation.
2. Trace elements include iodine, cobalt, and silicon. A shortage of iodine can cause **goiter,** an overgrowth of the thyroid.

REVIEW QUESTIONS

1. Distinguish between extracellular and intracellular digestion and provide an example of each from the invertebrates. (731)
2. Explain how the sponge takes in nutrients, carries on digestion, and distributes the products. (732)
3. Describe digestion in the gastrovascular cavity of a coelenterate. List three digestive functions served by the cavity. (732)
4. List the specialized regions of the earthworm digestive tract and give a function of each. Include absorbing structures. (732, 734)
5. List the three parts of the insect gut and explain what each does. (735)
6. Describe two anatomical feeding specializations in the shark and explain how they operate. (735–736)
7. List four anatomical feeding specializations in the amphibians and reptiles. (736)
8. Explain the functions of the bird crop, proventriculus, and ventriculus. (738)
9. List the four types of mammalian teeth and describe specializations in each. (738)
10. List the parts of the four-chambered stomach of cattle and trace the path followed by food. (738)
11. Since cows cannot digest cellulose, how do they obtain needed nutrients from grasses? (738)
12. Draw a simple human taste map and label the specialized areas. (741)
13. List the three salivary glands and explain how saliva functions in digestion. (741)
14. Explain how peristalsis works and how it is controlled. (741)
15. Describe the muscularis of the stomach and its action in moving food. (743)
16. Describe the stomach lining and state the specific functions of the parietal and chief cells. Name a secretion that protects the stomach lining. (743)
17. Describe the four levels of structure that provide the small intestine with its enormous surface area. (744–745)
18. Draw a simple villus. Label its parts, including any specialized cell surface features. (744)
19. What is the glycocalyx? How is this an improvement over the simple release of enzymes into the lumen? (744)
20. What is the digestive function of the liver? How does the release of its secretion take care of two problems at once? (744)
21. List two general functions of the pancreatic secretions. (744)
22. Briefly discuss three functions of the colon. (744–745)
23. Name and describe the common chemical mechanism of digestion. What is the opposite process called? (746)
24. List the steps in the digestion of starch. What is the end product? (746)
25. Explain the peculiar manner in which fatty acids are absorbed. (747)
26. What is an endopeptidase? Name three of these and state where they are produced. (747)
27. Where are the exopeptidases produced? What steps in protein digestion do they accomplish? (748)
28. List three different mechanisms controlling enzyme release. (748)
29. Explain how gastrin is released and what stops its action. (748, 750)
30. List several essential reasons for including saturated and unsaturated fats in the diet. (750–751)
31. What determines the quality of dietary protein? What vegetable foods contain high-quality proteins? (751–752)
32. List two specific ways in which vitamins are used by the body. (752)
33. List four major and two trace mineral requirements and state their uses. (753–754)

Respiration

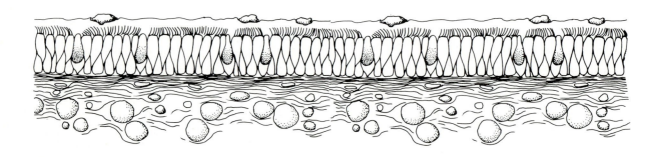

This may not be the best of all possible worlds, but it's the only one we've got, and it does have a way of presenting us with an array of offerings. In this world, in a very real sense, the processes of life are involved with overcoming various problems imposed by our environment and with taking advantage of the opportunities that we find before us.

For example, one such opportunity has been afforded by oxygen, but it did not always fall under the category of an opportunity; it was once a dire threat to life. In fact, much of the oxygen present today may have arisen as a form of atmospheric pollution—a toxic product of early photosynthetic life (see Chapter 18). It took many millions of years for living things to adapt to the increasing levels of atmospheric oxygen. At first it must have been a dangerous game indeed, since oxygen can play havoc with the usual chemical reactions of life unless its effect is somehow neutralized. At the time oxygen first appeared, there was little in the atmosphere to absorb ultraviolet radiation, so atmospheric oxygen was quickly converted to ozone (O_3), a highly poisonous oxidizing agent. (Ozone still arises spontaneously in the upper atmosphere.) The advent of oxygen, then, transformed the surface of an already unpredictable earth into an even more dangerous place.

Today, however, ozone helps protect us from dangerous ultraviolet radiation from the sun. Further, most organisms can actually use molecular oxygen (O_2). Interestingly, perhaps because of common evolutionary descent, these species tend to use oxygen in the same general way—as an acceptor of spent electrons during cell respiration. (Recall that energy-depleted electrons leave the respiratory electron transport systems of the mitochondrion to join oxygen and eventually form water, a metabolic waste product; see Chapter 8.)

Many species have even developed very elaborate systems to distribute oxygen to the cells. In many animals, including humans, the **respiratory system** exchanges carbon dioxide for oxygen, while the **circulatory system** carries these gases, as well as nutrients, wastes, and hormones, throughout the body.

In this chapter, we'll look at the respiratory systems of a variety of animals, focusing on both their unity and diversity. We will find that solutions to problems of gas exchange cut across all taxonomic lines, and that similarities occur not only because of evolutionary relatedness but because of adaptations to similar environments. In other words, we will be looking at examples of both convergent and divergent evolution of the respiratory system.

30.1 METHODS OF GAS EXCHANGE

(a) In skin breathers a simple exchange of O_2 and CO_2 occurs across the moist body wall. (b) External gills provide a greatly increased surface area for gas exchange. (c) The gill becomes more elaborate in fish, with oxygen-carrying water entering through the mouth, moving across the gills, where oxygen is exchanged for carbon dioxide, and out through the gill openings. (d) In insects, highly branched, thin-walled tracheae penetrate the fluid-filled body cavity, exchanging gases through-out. (e) The vertebrate lung produces a great surface area in a relatively small space by the presence of numerous blind, thin-walled sacs where gases are exchanged. (f) The bird respiratory system is quite unusual, sporting a one-way flow of air through the lung tissue itself and a number of air sacs that act as bellows and reservoirs for inflating and deflating the lungs.

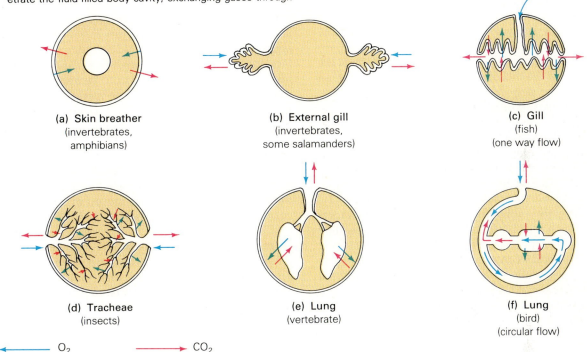

(a) **Skin breather**
(invertebrates,
amphibians)

(b) **External gill**
(invertebrates,
some salamanders)

(c) **Gill**
(fish)
(one way flow)

(d) **Tracheae**
(insects)

(e) **Lung**
(vertebrate)

(f) **Lung**
(bird)
(circular flow)

O_2 CO_2

GAS EXCHANGE SURFACES

Let's begin by emphasizing that the exchange of gases requires structures with two basic characteristics.[1] First, they must have a permeable surface area, with a thin-walled membrane of sufficient size to meet the animal's requirements. Second, this surface area must be moist, since gases can't normally cross dry membranes (such as our skin, which contains an outer layer of dry, dead cells). These requirements have been met through evolution in an interesting variety of ways. Most familiar are the spongy lungs of air-breathing vertebrates and the frilly gills of fishes and numerous aquatic invertebrates. But there are others (Figure 30.1).

The problem of maintaining moistness in the terrestrial environment, in many species, is solved by the secretion of copious amounts of mucus by cells of the exchange membrane. We will refer to the exchange membrane as the **respiratory interface** as we proceed since it is the respiratory structure that is exposed to the environment. The respiratory interface can vary from lungs and gills to body surfaces. Our survey will begin with some of the animals that use their body surfaces for respiration. Then we will consider specific structures that are more complex.

---

[1]The requirements have been formally stated in what has been called Fick's Law: $F = K A (P_1 - P_2)$, where F is the diffusion rate, K is a constant called the diffusion coefficient, A is the surface area, and $(P_1 - P_2)$ is the partial pressure gradient between blood and medium. The value of K is influenced by the presence of moisture on the respiratory surface and by barrier thickness. The value of A depends on the size of the respiratory surface and by barrier thickness. This is just Fick's way of saying that increased membrane area and moistness favor more rapid gas diffusion.

The Simple Body Interface

Use of the body surface or skin as a respiratory interface occurs in a number of unrelated groups. It is an apparently simple solution to the problem of gas exchange but one that has severe restrictions as well. You may recall the general structure of some of the simpler sponges discussed earlier. Their thin-walled, vaselike bodies permit a simple exchange of gases with both the surrounding sea water and a current of sea water carried through the hollow body by flagellated collar cells. However, we should remind ourselves that not all sponges are small and the more complex sponges may reach an impressive size indeed. How do gases penetrate their mass? While the body surface is still the exchange interface, mass is no problem since the body, rather than being vaselike, is riddled with small canals. These canals are lined with collar cells that create strong currents of oxygen-laden water that reaches all parts of the sponge body before being expelled through the osculum (Figure 30.2).

Similarly, although some species of coelenterates (such as jellyfish) attain considerable size, they require no specialized respiratory structures. Their saclike bodies consist essentially of two cell layers that are virtually always in contact with the external environment. Their outer body wall and extensive gastrovascular cavity provide an adequate exchange interface (see Figures 25.7 to 25.9).

30.2 GAS EXCHANGE IN THE COMPLEX SPONGES

In *Euspongia*, the respiratory surface is increased substantially by an extensive array of interconnecting canals. The canals lead into chambers lined with flagellated collar cells that create strong water currents. Rather than a single body cavity and one ostium, the complex sponges have numerous excurrent canals leading to many ostia through which water leaves the sponge body.

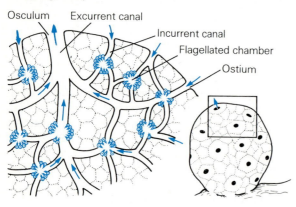

Osculum   Excurrent canal
Incurrent canal
Flagellated chamber
Ostium

Though flatworms are considerably more dense than coelenterates, they are, nevertheless, flat, and this shape yields a large surface area for its mass. Flatworms also have an extensive highly branched gastrovascular cavity (see Figure 25.11). Because of these characteristics, much of its body mass is exposed to its watery environment, and no cells are far from it. It can, therefore, use cell-by-cell diffusion as a means of exchanging gases with the environment. Of course, as long as it relies on diffusion alone to transport gases, it can never grow very large—its volume would quickly outgrow its surface area. So if you ever see an 8-foot planarian gliding toward you in your favorite swimming hole, ignore it. It doesn't exist. Some marine flatworms grow to as much as 10 cm wide and 60 cm long but they are only a few millimeters thick.

In the spiny echinoderms, an expansion of the respiratory interface has been accomplished through the **dermal brancheae,** outpocketings of the coelomic wall that protrude through pores in the endoskeleton. Fluids are moved in and out by ciliated action in the coelom (Figure 30.3).

Complex Skin Breathers. A few terrestrial animals, such as the earthworm and small lungless salamanders, manage to use their moist skin as a respiratory surface (the salamander also uses its highly vascular pharynx). But these complex animals are very special cases. For example, the earthworm's habitat is really only semiterrestrial since it restricts itself to moist soil. The lungless salamanders are also limited to damp places. They spend their days underground (or under logs), venturing out mainly at night when risk of desiccation is lessened. The skin of earthworms and lungless salamanders, by the way, is kept moist by the secretion of a slimy layer of mucus.

Although such skin breathers have no special structures with which to exchange gas with their environments, their great mass prevents them from relying on the cell-by-cell diffusion of gases. The skin breathing is enhanced by efficient circulatory systems that readily transport gas once it passes into their bodies. Both the earthworm and the salamander have closed circulatory systems containing hemoglobin, the oxygen-carrying protein pigment. The salamander's hemoglobin is contained within special blood cells, as is our own, but the earthworm's hemoglobin is dissolved in the blood. Even so, the worm's blood is a far more efficient oxygen carrier than is simple plasma. Thus skin breathing works for some complex terrestrial

30.3 GAS EXCHANGE IN AN ECHINODERM

The dermal brancheae of the starfish—numerous extensions of the skin that protrude through the skeletal plates—provide much of the respiratory exchange surface. The dermal brancheae are continuous with the coelom, and the ciliated lining of that cavity helps in gas exchange by keeping the coelomic fluids moving. Note the many pincerlike devices that help keep the sea star's surface clear of would-be encrusting organisms so common in the marine habitat. A significant amount of gas exchange also takes place across the numerous tube feet.

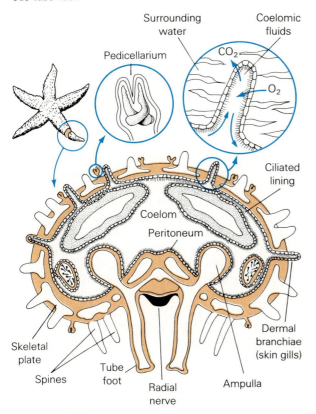

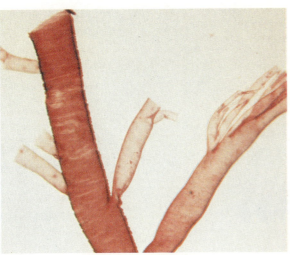

organisms—if they restrict themselves to a moist habitat and if they have an efficient circulatory system. The earthworm and the lungless salamander, then, share an interesting combination of behavior, anatomy, and physiology that permits their simple skin respiratory systems to work.

Expanding the Interface: Tracheae

Most terrestrial animals have developed more specialized breathing surfaces. These usually involve specialized infoldings of the gut or body surface. By internalizing their respiratory surfaces in this way, land dwellers reduce the problem of excessive water loss. Furthermore, these internal pouches are usually highly folded and convoluted, providing an increased area for gas exchange. The lungs are the most obvious example of such an arrangement.

Among the variations of this internal arrangement are those of the arthropods, a group that does most things differently. Insects, for example, have a **tracheal system.** The insect body is riddled with a series of tiny, highly branched tubules called **tracheae** and **tracheoles** (Figure 30.4). These tubes open to the outside via tiny valvelike openings known as **spiracles,** commonly situated along the insect's

30.4 THE TRACHEAL SYSTEM OF INSECTS

Insects have an elaborate respiratory system that includes spiracles along the body that open into extensive, thin-walled tracheal tubes. The tracheae branch throughout the body and some end in blind pouches that increase the exchange surface area. The extensive, branching tracheal tree permits gas exchange to occur with all body cells.

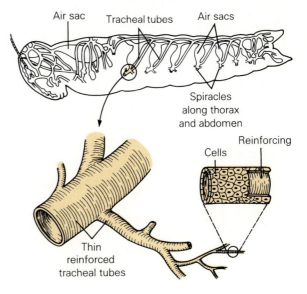

sides. In grasshoppers and some other larger insects, air is pumped in and out of the spiracles by a bellows action of the abdomen, the spiracles opening and closing in synchrony with the pumping action. In grasshoppers, bees, and others, the branching tracheae terminate in elastic **air sacs** that expand and contract and aid in the exchange processes. In other species of insects, air simply diffuses through the tracheal system. Oxygen and carbon dioxide are exchanged through the walls of very fine tracheoles, which permeate the body and carry oxygen to the immediate vicinity of every active tissue. The finer branches of the tracheae are filled with fluid through which the respiratory gases can diffuse.

In aquatic insects, the tracheae may terminate in **tracheal gills**—thin-walled extensions of the exoskeletons that contain finely-branched tracheal networks. Although insect circulatory systems are efficient, they appear to serve primarily to distribute food molecules and metabolic products; the respiratory functions are thus left to the tracheal system.

Complex Interfaces: Gills

Aquatic mollusks, arthropods, and chordates are often highly active animals, and their increased respiratory demands are met with a more complex system. Here we find the **gill,** which, like the lung, is a combination respiratory-circulatory exchange surface. We might point out here that the presence of similarly constructed gills in widely divergent animal groups does not indicate common ancestry, but rather the rigorous restrictions for gas exchange imposed by the environment.

Gills are thin-walled, finely divided, and feathery structures with extensive capillary beds through which blood flows, carrying carbon dioxide from the body to be exchanged for oxygen from the surrounding water. Capillaries will be described later, but here we can say that they are the smallest blood vessels, with walls only one cell thick, across which gases can freely pass.

Gills in Mollusks and Arthropods. Most mollusks rely on gills. In bivalves (such as the clam) the gill serves in both food filtering and respiration, as was mentioned in Chapter 26. The clam gill (Figure 30.5a) consists of sheetlike folds that protrude into the mantle cavity. Water is drawn by ciliary action into the incurrent siphon. From there, it passes through microscopic pores in the gills, entering channels that direct it out through the excurrent siphon. As water passes through the channels, oxygen and carbon dioxide are exchanged through tiny blood vessels that permeate the gill partitions. Oxygenated blood is then returned directly to the heart for distribution to the

30.5 INVERTEBRATE GILLS

Both the clam and the lobster make use of gills for gas exchange, although the manner in which water is moved across the gill surface is quite different in the two. **(a)** In the clam, a stream of water created by ciliary action enters through an incurrent siphon, passes over the gills, and leaves by an excurrent siphon. Each gill contains numerous pores that direct water through passages leading to the excurrent opening. The incoming flow brings both food and oxygen to the filter-feeding clam. **(b)** The lobster makes use of a pair of "bailers," specialized, paddlelike appendages, to create a current of water, drawing it forward under the protective carapace (part of the exoskeleton) where the gills are located. The lobster circulatory system is vital to the transport and exchange of gases with the body tissues.

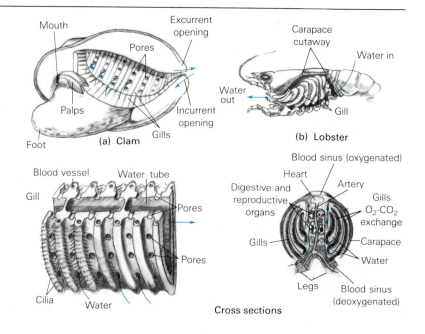

(a) Clam

(b) Lobster

Cross sections

body. Mollusks lack both red blood cells and hemoglobin but they do have respiratory pigments. Strangely, they make use of a copper-containing protein called **hemocyanin** for transporting oxygen. Hemocyanin is also found in crustaceans, spiders, and scorpions. In hemocyanin oxygen is bound to copper rather than iron as in hemoglobin, and on oxygenation the pigment turns from colorless to blue.

The arthropod gills are varied, but those of the lobster, a crustacean, will serve as a good example. The lobster gills are feathery extensions of the body wall, attached partly to the bases of the legs (Figure 30.5b). They are covered by a **carapace,** which is open at both ends. Water is drawn under the carapace, across the gills toward the head by the action of a paddlelike appendage near the jaws known as the **bailer.** Blood from a hemocoel (an open blood passage) in the thorax floor passes through channels in the gills, exchanging gases as it goes. Upon leaving this vital circuit, it returns to the heart to be pumped through the body again. As we saw earlier, such a circulatory system—with blood leaving the vessels and percolating through open spaces—is called an **open circulatory system.** Further, the blood of crustaceans contains hemocyanin, and since this pigment readily combines with oxygen, it is ideal for creatures such as crayfish that live in stagnant, oxygen-poor water.

Gills in Fishes. Gills in fishes arise from a number of supporting **gill arches** that contain rows of feathery **gill filaments.** Each gill filament contains a series of platelike **lamellae,** each of which houses a network of capillaries (Figure 30.6). Carbon dioxide-laden blood from the body enters the gill filaments in thin-walled capillaries emerging from **afferent** (incoming) **vessels** in the gill arch. The capillaries branch and rebranch in each lamella, forming the network in which the exchange of carbon dioxide for oxygen occurs. Oxygen-rich blood leaving the gill filament enters an **efferent** (outgoing) **vessel** in the gill arch and, upon leaving the gill, joins with the dorsal aorta to be distributed throughout the fish's body.

Two factors facilitate the exchange of gases within the minute capillary networks of the lamellae. First, since the walls of both the capillaries and the filaments are extremely thin, there is little distance between the blood and water passing over the gills. Second, the flow of blood opposes the flow of water, setting up a highly efficient **countercurrent exchange.** In the gill, blood passing over the exchange network is constantly confronted by a fresh supply of water; thus, as long as blood is

30.6 THE FISH GILL

(a) The supporting structures of the fish gills are the cartilaginous gill arches. Inside the curve of the arch are the rakers, which screen foreign particles out of the gills. At the outer curve, two rows of gill filaments protrude from each arch. **(b)** The surface area of each filament is greatly increased by the presence of lamellae, as shown. Each has a rich supply of capillaries that branch from afferent vessels, carrying deoxygenated blood. Crossing a lamella, the blood loses its carbon dioxide and picks up oxygen before entering the efferent vessel leaving the filament. The opposing movement of blood and water in each lamella sets up a countercurrent exchange, greatly enhancing the exchange of gases.

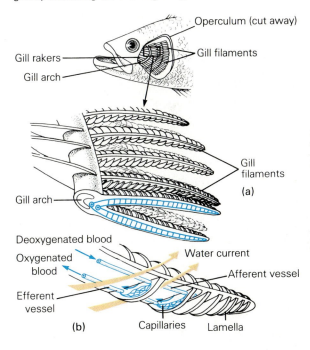

passing across the lamellae, there is an inward diffusion gradient for oxygen and an outward diffusion gradient for carbon dioxide. Such efficiency is especially important to active animals in aquatic environments where getting enough oxygen is difficult.

Even if one has gills, underwater breathing presents problems. Because of oxygen's low solubility, its concentration in water is rather low. The solubility of oxygen in water is inversely proportional to temperature, so in general, cold water contains more oxygen than warm water. But even the coldest saturated water contains less than 1% oxygen by volume. This is paltry compared to the oxygen content of air—21% by volume. Further, the diffusion rate of oxygen through water is vastly slower than in air.

Water's low oxygen content and high viscosity mean that active aquatic animals must devote considerable time and energy just to moving large volumes of water over their respiratory surfaces.

The faster-swimming bony fishes solve their respiratory problem much as do the sharks—they keep moving. Others rely on a pumping action. They keep a continuous flow of water moving over the gills by the subtle action of pharyngeal muscles. To "breathe," the fish closes its opercula (gill covers), opens its mouth, and by expanding its mouth and gill chambers, draws water in. The fish then closes its mouth, contracts the oral cavity, and opens its opercula to let the water flow out across the gills. That's why motionless fish look as though they are gulping.

Complex Interfaces: Lungs

Evolution of the Vertebrate Lung.   The vertebrate lung has had a curious evolutionary history. Early in vertebrate history, perhaps as early as the Precambrian, there lived the ancestor of modern bony fishes, a freshwater inhabitant that obtained oxygen by gulping air. A pocket of highly vascularized tissue became specialized in the pharynx of this ancestral fish as a place to hold its oxygen-laden air between swallows. Eventually, the air pocket grew and branched, and evolved into the vertebrate lungs.

Some of the descendants of that early ancestral fish species retained the lungs as accessory organs, although they developed gills as well. One such group is the lungfishes (see Figure 27.13), a peculiar assemblage of freshwater fishes that are uniquely adapted for survival in ponds that tend to dry up. A different group, the lobe-finned fishes, was later to leave its freshwater habitat and become the ancestor of the terrestrial vertebrates. For this group, the possession of lungs was a vital **preadaptation**—an adaptation to one way of life that, by coincidence, enabled the organism to survive in another.

Those ancient lobe-finned fish then achieved a second evolutionary breakthrough—literally. The nasal cavities in most fish are blind pouches used only for smelling the water. But the ancestor of terrestrial vertebrates evolved a pair of **internal nares**—openings between the nasal cavity and the mouth. It's not clear that they actually did use their internal nares for breathing. (Lungfish today are unusual among bony fishes, in that they have internal nares but also gulp air through their mouths.) Still, this feature proved to be yet another preadaptation that helped greatly when certain descendants invaded the land.

To this day, the internal nares of amphibians and most reptiles open directly into the mouth cavity, but these animals can breathe with their mouths closed. In the evolution of mammals, a bony ledge, the **palate,** finally separated the mouth cavity from the nasal cavity, displacing the internal nares back to the region of the pharynx. Thus we can chew and breathe at the same time, as can crocodiles.

Most bony fish have kept the primitive lung structure, but its function has changed. In these species, the lung became a flotation device—the swim bladder—and the gills regained their status as the organism's sole gas-exchange organ. (In an entirely different role, the swim bladder also aids in sound reception.)

The Vertebrate Lung.   The lung in vertebrates is an inpocketing, branching tube (basically an extension of the gut) that in some species ends in a multitude of tiny air sacs (alveoli), where blood and air are separated by a thin, moist membrane. Air is usually pumped in and out through a bellows mechanism. By muscular control of its breathing apparatus and the valves of its mouth and pharynx, the vertebrate can control the amount of air that passes through its lungs. Of course, moisture loss is unavoidable; the exhaled air is laden with water vapor.

The evolutionary pathway from an aquatic to a terrestrial respiratory system is perhaps suggested in today's amphibians. Nearly all amphibians have lungs (the lungless salamanders we mentioned apparently once had lungs, but lost them along the way), and nearly all exchange gases in two ways: through rather simple, baglike lungs and through their moist, highly vascularized skin. The ability to breathe through the skin enables many amphibians to "hibernate" in mud through cold or dry seasons, a time when a restricted oxygen supply requires them to remain inactive. (It's probably hard to be very active while encased in mud anyway.) Not only are the lungs simple, but even breathing movements in amphibians are quite primitive. Since there is no diaphragm or well-developed rib cage, air is forced in and out of the lungs by the raising and lowering of the floor of the mouth. This coarse pumping action is aided by the presence of "check valves" in the nostrils and the **glottis** (the opening of the trachea) (Figure 30.7).

Since the dry reptilian skin is impervious to air, reptiles, including the many aquatic species, are strictly lung breathers. Perhaps *strictly* is not the right word. The inevitable exception this time is

30.7 BREATHING MECHANISM OF THE FROG

The frog breathes through lungs, but gases also pass across its moist, highly vascularized skin. Air is forced in and out of the lungs by the bellows action of its throat and body wall (there is no diaphragm). Note the four-stroke system and the role of check valves in the nostrils and trachea. Like all vertebrates, the frog has a closed circulatory system with red blood cells that contain hemoglobin. As you can see, frogs breathe with their mouths closed.

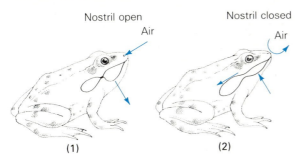

Nostril open — Air (1)   Nostril closed — Air (2)

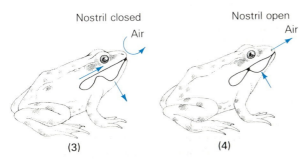

Nostril closed — Air (3)   Nostril open — Air (4)

found in some water-breathing turtles that use a large, heavily vascularized cloaca as a supplemental respiratory surface—a kind of water lung. (You may have trouble picturing this if you remember that a cloaca is the common posterior chamber through which digestive and urinary wastes and eggs pass from the body; see Chapter 27.) The lungs of reptiles are essentially saclike, but some are more complex than those of amphibians. For example, some species, such as the monitor lizards, have more subdivided air passages. Breathing movements involve the muscles around the entire body cavity since there is no diaphragm and no separate abdominal and thoracic (chest) cavities.

The Unique Bird Lung. You will recall that one of the major advances in the evolution of the digestive system was the development of a one-way gut that allowed continuous processing to replace the batch processing of coelenterates and flatworms. A similar breakthrough occurred in the evolution of the lung in birds. While the lung itself is quite spongelike with many passageways, the passages

do not end in blind sacs as they do in the lungs of reptiles and mammals, but *pass right through the lung.* There is no new opening to the outside; birds still breathe through their mouths and nostrils and the air passes in and out through the main bronchus (essentially a trachea, albeit a lengthy one). But from the main bronchus on, things become quite different, and these differences produce a highly efficient exchange system. The blood and air move in different directions in the bird lung in what is called **crosscurrent flow.** Here, the air passes through the lung, anterior to posterior, and the blood moves around it at right angles, the vessels being similar to ribs around a cylinder through which blood flows.

In addition to the lengthy main bronchus, the bird's respiratory system includes two groups of hollow air sacs, three pairs anterior and two pairs posterior to the lung (Figure 30.8a). The air sacs are quite extensive, some even passing into the major long bones. But extensive as the air sacs are, they have little to do with gas exchange, acting instead as a combination bellows and reservoir for air entering and leaving the lung between. Figure 30.8b traces the breathing movements and the flow of air through the bird respiratory system.

The important point is that air flows in a one-way path through the lung—from the posterior air sacs, through the lung, to the anterior air sacs. This one-way flow provides for a most efficient exchange since it crosses the flow of blood through the lung capillaries. Such an arrangement required dramatic evolutionary changes from the simpler reptilian lung. How were such vast changes adaptive?

The primary advantage of the bird lung is that it is well adapted for flight at high altitudes where oxygen levels are low. Maintaining rigorous activity at higher altitudes requires the most efficient gas exchange possible, and this, apparently, is where the unique bird respiratory system is most significant. High altitude flying is particularly important to birds that migrate and while many birds migrate at an altitude of only 1200 to 1500 m, many fly higher. Pilots have reported birds flying at 6000 m (19,685 ft), and radar tracking has indicated that they can fly at 7000 m (about 23,000 ft). Mammals, including the flying bats, subjected to activity at this altitude would have great difficulty functioning at all and would quickly fall into a metabolic stupor. As you might expect, bats are low altitude migrators.

We now turn to a more detailed look at a respiratory system in the mammals, concentrating on humans.

30.8 ONE-WAY AIR FLOW IN THE BIRD LUNG

The extensive bird respiratory system includes posterior and anterior air sacs (a) that act as reservoirs and bellows for filling and emptying the lungs. Air flows *through* the lungs (b) rather than *in and out,* as in other air-breathing vertebrates. The one-way passage establishes a crosscurrent exchange of gases between the air and blood since they move in perpendicular directions. Oddly enough, the bird lung contracts on inhalation and expands on expiration (exaggerated here), but this is in response to expansion and contraction in the air sacs, rather than to air entering or leaving the body. (c) In the scanning EM of the lung tissue, cylinder-shaped parabronchi (branches of the bronchi) are surrounded by highly branched walls containing the capillaries.

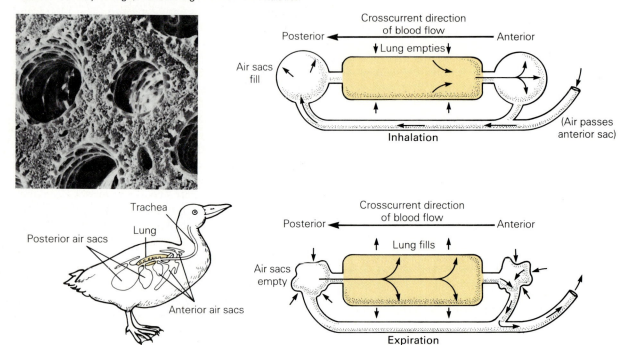

(a)

(a)

(b)

(b)

THE HUMAN RESPIRATORY SYSTEM

The respiratory system in humans (Figure 30.9) is fairly typical of those of other mammals. The major structures include the mouth and nasal passageways, the **pharynx, larynx, trachea, bronchi,** and **lungs.** Air first passes through the nasal passages (unless one is a chronic mouth breather). In addition to their role in directing the air flow, these passages are important in filtering, warming, and moistening the air prior to its entering the lungs. The nostril hairs act as an initial filter, trapping dust particles. (A nose full of hair may not be esthetically pleasing, but your lungs undoubtedly appreciate it.)

In all mammals, the nasal cavity is separated from the mouth by the **palate,** a bony shelf in the roof of the mouth ending in a softer muscular region, the **soft palate.** The nasal cavity is lined with mucous membrane, which includes mucus glands, mucus-secreting **goblet cells,** and cells bearing cilia (Figure 30.10). These glands and cells produce a film of mucus, which is constantly swept toward the throat by the ciliary action. In addition, the lining of the nose contains many dense capillary beds that warm the air before it enters the lungs. This countercurrent flow of air and capillary blood is very effective; even on very cold days, the air entering the lungs is warmed almost to body temperature.

The inhaled air moves from the nasal passages

30.9 THE HUMAN RESPIRATORY SYSTEM

In humans and other mammals, the air passages include the nostrils, nasal cavity (and mouth), pharynx, larynx, trachea, bronchi, and bronchioles. The bronchioles terminate in numerous blind alveoli (see inset). The spongy lungs are subdivided into lobes, with three in the right lung and two in the left lung. The muscular, shelflike diaphragm seals the thoracic cavity off from the abdominal cavity below.

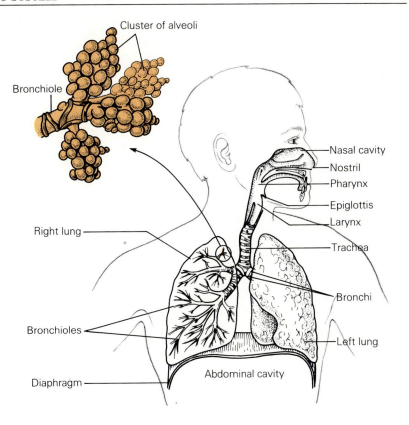

Cluster of alveoli

Bronchiole

Nasal cavity
Nostril
Pharynx
Epiglottis
Larynx
Trachea

Right lung

Bronchi

Bronchioles

Left lung

Diaphragm

Abdominal cavity

30.10 MICROSCOPIC STRUCTURE OF THE RESPIRATORY PASSAGES

(a) Much of the surface of the respiratory passages contains large numbers of goblet and ciliated cells that, respectively, secrete dust-trapping mucus and sweep it toward the throat. What happens then depends on your upbringing. (b) The scanning electron microscope view shows such a ciliated lining.

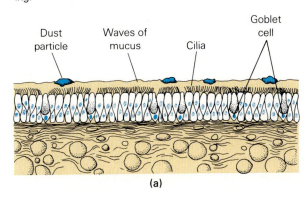

Dust particle   Waves of mucus   Cilia   Goblet cell

(a)

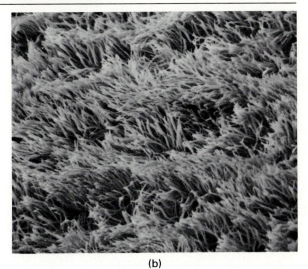

(b)

into the pharynx and from there through the larynx into the trachea—unless you are swallowing, whereupon the laryngeal opening is closed (Figure 30.11). You can't breathe while you swallow (of course, immediately everyone tries). The larynx contains the **voice box** and vocal mechanism—the **vocal cords.** Below the larynx begins the trachea, a tube that contains many C-shaped rings of stiff car-tilage that hold the airway open. As the trachea enters the chest, it branches into right and left **primary bronchi,** which branch again and again into the **bronchioles** that form the **respiratory tree.**

The trachea and some of the other bronchial structures have a secreting and sweeping lining similar to that in the nasal passages, so the air is cleaned of dust and debris once again. (The air

30.11 THE LARYNX

The larynx consists of cartilage and muscle. Toward the front is a protruding cartilage commonly called the "Adam's apple." **(a)** A principal function of the larynx is to guard the entrance of the trachea, preventing food from entering as it is swallowed. **(b)** In swallowing, the larynx is elevated (you can check this with your fingers) and the epiglottis is folded over the respi-ratory openings, directing food into the esophagus. In addi-tion, the larynx **(c)** houses the vocal cords. As we breathe, the vocal cords form a triangle; but as we speak, the cords tighten to form a slitted opening that is varied in breadth to produce different pitches **(d)**.

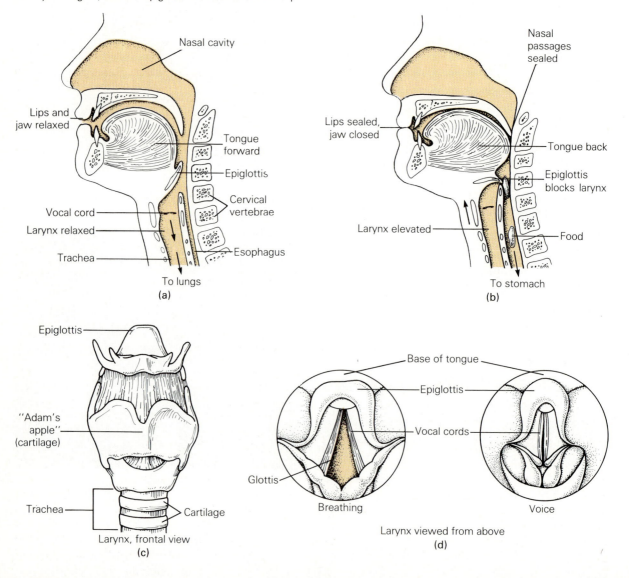

30.12 A HISTORY OF LUNG CANCER

The normal ciliated epithelium **(a)** of the respiratory passages includes columnlike, ciliated, and mucus-secreting goblet cells. In the smoker's respiratory lining **(b)**, the cilia become partially paralyzed, and the mucus accumulates on the irritated lining. Where an early cancerous state exists, the lowermost basal cells divide more rapidly and begin to displace normal columnar cells. As the cancer progresses **(c)**, most of the normal columnar cells are replaced by simple cancer cells that form a spreading tumor. In advanced cases, clusters of cancer cells may be carried away in the lymphatic system, spreading to other parts of the body. The contrast between normal and cancerous cells is seen at right in the SEM of human lung tissue.

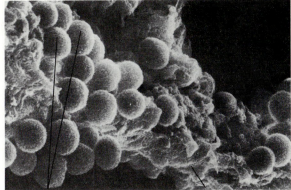

Cancer cells   Normal lung tissue

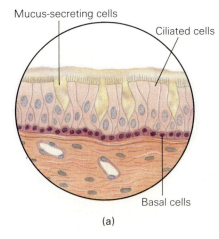

Mucus-secreting cells

Ciliated cells

Basal cells

(a)

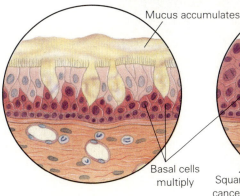

Mucus accumulates

Basal cells multiply

(b)

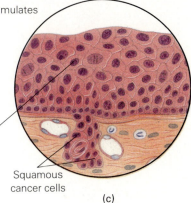

Squamous cancer cells

(c)

going out is probably cleaner than the air coming in, so in a sense, humans help clean the earth's air.) Tiny particles of inhaled dust become trapped in the mucus, which is continually being moved up into the throat and swallowed. This capability is lost to heavy smokers, whose respiratory linings are subject to drastic change as the years and cigarette packs go by (Figure 30.12).

The bronchioles, the tiniest branches of the respiratory tree, end in grapelike clusters of air spaces known as **alveoli** (Figure 30.13). Each cluster is enclosed in a dense capillary bed, and here the atmosphere and blood are only a membrane apart. The clusters, of course, provide an enormous surface area. In fact, the total surface area of the approximately 300 million alveoli of the human lungs has been estimated at nearly 70 m² (750 ft²).

When viewed intact, the lungs appear lobed and roughly triangular in shape, with a broad base. Two airtight, moist baglike membranes, the **pleurae,** enclose the lungs. The inner pleura is attached firmly to the spongy surface of the lung. The outer pleura forms the tough lining of the **pleural cavity,** which is the space between the lungs and chest wall. (*Pleurisy* is an inflammation of the pleura.)

The pleural cavity is bounded at its lower portion by that dome-shaped muscular shelf, the **diaphragm.** The diaphragm, a mammalian characteristic, separates the body cavity into its thoracic and abdominal portions.

The Breathing Movements

Breathing, or **ventilation,** in humans involves both the diaphragm and muscles of the rib cage, the **intercostal muscles** (Figure 30.14). In the relaxed condition, the diaphragm rises into a dome shape and protrudes into the pleural cavity. Inhalation is accomplished by contracting (lowering and flattening) the diaphragm and contracting the external intercostals, which raises the rib cage. This in turn increases the volume of the airtight pleural cavity, creating a partial vacuum. It is the partial vacuum that causes the lungs to fill with air. The weight of the atmosphere itself forces air down the trachea to inflate the lungs. Were it not for the lungs, the air

30.13 CIRCULATION IN THE ALVEOLI

The bronchioles terminate in grapelike clusters of alveoli—blind, thin-walled sacs whose surfaces contain extensive capillary beds. Deoxygenated blood, laden with carbon dioxide, passes through the capillary beds where the waste gas diffuses into the air-filled alveolar spaces. From there it is carried away by exhalation. Simultaneously, oxygen in the alveolar spaces enters the bloodstream, to be carried out of the lungs to the body.

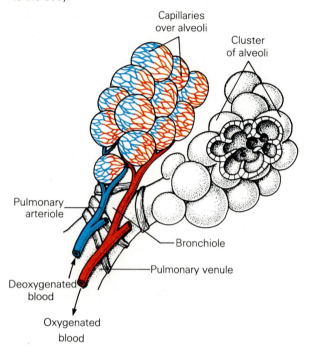

Capillaries over alveoli

Cluster of alveoli

Pulmonary arteriole

Bronchiole

Pulmonary venule

Deoxygenated blood

Oxygenated blood

would simply replace the vacuum in the thoracic cavity—the lungs just happen to be in the way.

Exhalation, when passive (not forced), is produced by the relaxation of the rib cage and diaphragm, actions that decrease the pleural cavity volume. The elastic lungs resume their former state, and air is forced out through the respiratory passages. Forced expiration involves the contraction of internal intercostal and abdominal muscles, which further exhaust the lungs.

While we are at rest, about half a liter of air is moved in and out with each breath we take. But when we fill our lungs to their greatest capacity, about four liters are moved in and out. This maximal measurement, the **vital capacity,** depends mainly on body size, and varies considerably among individuals and between the sexes. In a well-trained male athlete, the vital capacity can exceed six liters. But no matter how hard one tries to expel all of the air in the lungs, about 1.5 liters always remains. This **residual air** is highly signifi-

cant since, as we will see, it contains carbon dioxide that is essential in maintaining an adequate respiratory rate. Even in the superefficient bird lung, the main bronchus (trachea), by virtue of its great length, helps retain a critical amount of carbon dioxide in the respiratory system.

The Exchange of Gases

Partial Pressure. To understand some of the basic aspects of the transport of oxygen and carbon dioxide, it is helpful to know some things about the behavior and characteristics of gases in general. As you know, our atmosphere is composed of a mixture of gases—primarily nitrogen and oxygen, with a much smaller amount of carbon dioxide. Together, the gases of the atmosphere exert a pressure—the total atmospheric pressure. The pressure exerted by any one gas in the mixture is logically called its **partial pressure,** designated as P_g (the letter g represents any specific gas). At sea level, and under what chemists call **standard conditions,** the total atmospheric pressure is known to be 760 mm Hg (mercury). For instance, air is about 21% O_2; therefore, the partial pressure of O_2 at sea level would be 21% of 760, or 160 mm Hg ($P_{O_2} = 160$). Carbon dioxide gas accounts for only about 0.04% of the atmosphere; thus it exerts a partial pressure of 0.3 mm Hg ($P_{CO_2} = 0.3$) (.04% of 760). Most of the remaining total atmospheric pressure is due to nitrogen gas. Since atmospheric pressure decreases with altitude, so do the partial pressures of the atmospheric gases. On a mountaintop, the air is still 21% O_2 and 0.04% CO_2, but since the total atmospheric pressure is much less, so, therefore, are the partial pressures of the two gases. (At 14,000 feet, the total atmospheric pressure may be only 450 mm Hg, while the P_{O_2}, would be 95 and the P_{CO_2}, 0.18.) For this reason, we must breathe faster and deeper up there.

The concept of partial pressure is important to our understanding of gas exchange because diffusion of dissolved gases always occurs from regions of higher partial pressure to regions of lower partial pressure. For instance, suppose we placed an open container of a fluid that was rich in CO_2 and poor in O_2 (as measured by the partial pressures of these two gases) inside a closed container with a mixture of gases that was rich in O_2 and poor in CO_2. In due time the fluid and the gas would equilibrate, so that the partial pressure of both gases would be the same in both the fluid and the air. The actual amount, in contrast to the immediate availability, of O_2 and CO_2 in each part of the system would depend on the solubility of each gas in the fluid, which in turn would depend on such factors as temperature and pH. Partial pressure differences are important in setting up diffusion gradients, but as we will see, evolution has added a few twists to the simple exchange of gases.

Exchange in the Alveoli. We breathe in air rich in O_2 and breathe out air rich in CO_2. The exchange takes place on the moist inner surfaces of the alveoli, primarily through simple diffusion (Figure 30.15). The blood that enters the lung from the heart has previously been routed through the body tissues, where mitochondrial respiration has depleted the oxygen supply, creating a very low partial pressure of oxygen (P_{O_2} = about 40). At the same time, metabolic activity has increased the partial pressure of CO_2 in the body tissues. The blood has tended to equilibrate with the body tissues. Hemoglobin (the carrier of oxygen in the blood) is deoxygenated in the body tissues and picks up molecules of carbon dioxide. A typical partial pressure for carbon dioxide in blood entering the alveoli would be about 45.

So, compared to partial pressures in the atmosphere, blood entering the lungs has a low partial pressure of O_2 and a high partial pressure of CO_2. Conversely, in the alveoli, the partial pressure of oxygen is high—about 100 mm Hg—and the partial pressure of carbon dioxide is about 40 mm Hg. (In case the greater than expected partial pressure of carbon dioxide bothers you, recall that alveolar air is not completely replaced during inhalation, but is a mixture of old and new air.) In the brief time that the blood and air are in near contact on opposite sides of the thin alveolar membrane, nearly complete equilibrium takes place. That is, the partial pressures of O_2 and CO_2, in the blood and air become almost equal as molecules of O_2 diffuse in and molecules of CO_2 diffuse out. Then the equilibrated air is exhaled and fresh air is inhaled while the blood flows continuously through the lung. Blood leaving the lung is relatively high in oxygen, with a partial pressure of about 95 mm Hg. The partial pressure of carbon dioxide is about 40 mm Hg.

The partial pressure of oxygen in metabolically active tissue shifts dramatically as oxygen is consumed. There its partial pressure may fall to 25. The

30.14 BREATHING MOVEMENTS IN HUMANS

The combined action of muscles in the rib cage and the diaphragm forces air into and out of the human lung. Inspiration **(a)** occurs when intercostal rib muscles contract, elevating the rib cage and expanding the chest cavity. Simultaneously, the diaphragm contracts, pulling downward and further increasing the chest volume. Both actions decrease air pressure in the chest cavity to less than the atmospheric pressure outside. Consequently, air rushes into the respiratory passages and inflates the lungs, which fill the chest cavity. In expiration **(b)**, the rib muscles and diaphragm relax, and the rib cage falls, decreasing the volume of the chest cavity. This produces internal pressure that exceeds the atmospheric pressure, forcing air back out of the shrinking lungs. The lungs have no muscles of their own, but their elasticity assists in expiration.

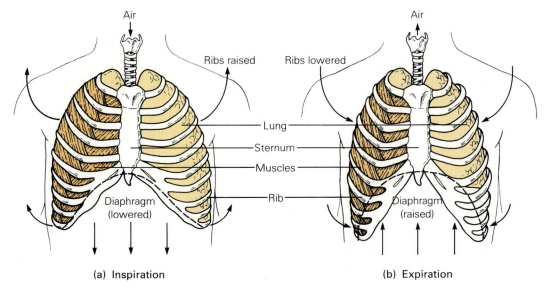

Air

Ribs raised

Ribs lowered

Air

Lung

Sternum

Muscles

Rib

Diaphragm (lowered)

Diaphragm (raised)

(a) Inspiration

(b) Expiration

partial pressure of carbon dioxide, as you would expect, increases, perhaps reaching 46.

Thus, the respiratory gases diffuse down their pressure or concentration gradient and the organism takes advantage of this "free" means of exchange to keep its oxygen and carbon dioxide at optimal levels. For some organisms, this is about all there is to respiration, but for many, including the vertebrates, there is much more. Passive diffusion is quite important, but it is only a part of the total gas exchange process. Other aspects of the process are directed by complex biochemical mechanisms. Let's consider the process in the human respiratory system.

Oxygen Transport

Because of hemoglobin's great affinity for oxygen, it can carry 60 times as much as can an equal weight of water alone. Thus, if we didn't have a specific oxygen carrier, our circulation would have to be 60 times as fast, or our activity would have to be greatly curtailed.

So what precisely is this magic molecule called hemoglobin and how does it work? First, it consists of four polypeptide chains and four heme groups (Figure 30.16), one heme group for each polypeptide chain. Each heme group contains an iron atom that can bind with one molecule of oxygen (O_2). (Thus, each hemoglobin molecule can hold eight oxygen atoms when saturated.) The association and dissociation of O_2 and hemoglobin (Hb) is usually simplified as:

$$Hb + O_2 \rightleftharpoons Hb \cdot O_2$$

The left side of the formula shows **deoxyhemoglobin** and the right side, **oxyhemoglobin**.

Once key to the efficiency of our respiratory system is the behavior of hemoglobin itself. Hemoglobin is highly specialized to associate and dissociate with oxygen under certain conditions. Whereas the affinity of an ordinary passive fluid (such as water) for O_2 is simply dictated by partial pressure, hemoglobin has the remarkable ability to "change its mind" about how much oxygen it will accept. To put it a bit more scientifically, hemoglobin can change its affinity for oxygen depending on such factors as pH, the partial pressure of CO_2 and tem-

30.15 PARTIAL PRESSURE IN THE ALVEOLI

Exchange of gases across the alveolus follows a partial pressure gradient. The net movement of molecules of CO_2 and O_2 is shown in the capillary network (a) and in a small segment of a capillary (b). In the capillary network, blood entering the alveolar region has an increased CO_2 partial pressure and greatly lowered partial pressure of O_2. But at the other side of the capillary bed, the situation is reversed. Carbon dioxide has diffused out of the blood into the air sac while O_2 has diffused in. In the enlarged view, note the thin membrane of the capillary wall and alveolar membrane separating the blood from the inhaled air. Red blood cells (drawn to scale) are instrumental in the exchange as they easily give up CO_2 and attach O_2 to their hemoglobin to form oxyhemoglobin.

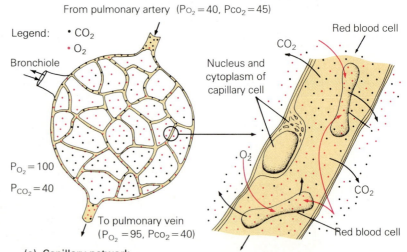

(a) Capillary network—
flow of blood across an alveolus

(b) Segment of a capillary

Partial Pressures in Lung

Gas	Atmosphere	Alveoli	Blood entering	Blood leaving
Oxygen	158	100	40	95
Carbon dioxide	0.3	40	45	40

perature. A shifting affinity for oxygen allows the hemoglobin to give up much of its oxygen in its journey through the body and to become quickly saturated again in the lungs.

One more point about oxygenation of the blood: at sea level, and under normal conditions, blood passing over the alveoli is nearly saturated with O_2. That is, most of the hemoglobin molecules pick up their full allotment of four O_2 molecules. Increasing the O_2 concentration in the lungs has very little effect on the amount that crosses into the blood. There is a considerable safety margin in this relationship because it also works the other way: reducing the oxygen concentration doesn't easily change the amount of oxygen reaching the blood either. At half the partial pressure of O_2 at sea level, the blood will still be 80% saturated. This is why humans can live comfortably at varying altitudes, including high on mountains where O_2 partial pressure is low.

Furthermore, at high altitudes, additional red blood cells are produced—a long-term adaptation. The rate of red cell production is controlled through a feedback system originating in the kidney. Certain cells there detect the oxygen level in passing blood and when it is not up to a certain standard, the hormone **erythropoietin** is released. Its targets are blood-forming elements in red bone marrow, which respond by producing more red blood cells. Added red blood cells mean an increase in blood oxygen and the kidney cells, no longer stimulated, slow the erythropoietin release.

Carbon Dioxide Transport

Now let's look at a related phenomenon: carbon dioxide transport. It turns out that where partial pressure of CO_2 is high, hemoglobin has a lower affinity for O_2. When the CO_2 level is low, hemoglobin holds its O_2 more tightly. This peculiar influence by carbon dioxide, known as the **Bohr effect** (Figure 30.17), clearly has adaptive significance. The presence of carbon dioxide indicates that oxidative metabolism is occurring, increasing the oxygen demand. The most metabolically active regions produce the greatest quantities of CO_2, which in turn hastens the release of O_2 from the passing bloodstream. Conversely, oxygen-rich blood passing less metabolically active tissue will release less of its oxygen.

The transport of carbon dioxide in the blood is far more complex than that of oxygen. About 8 percent simply goes into solution in the water of blood plasma. The remainder enters the red cells, where it

30.16 HEMOGLOBIN

Hemoglobin is a complex, heavyweight protein containing four polypeptides linked into one globular unit. Each hemoglobin macromolecule contains two alpha chains and two beta chains, and each of the four polypeptides bears a heme group. Each heme group is capable of accepting one oxygen molecule. In vertebrates, vast numbers of hemoglobin molecules are bound to each red blood cell, and the oxyhemoglobin produced in oxygen saturation gives blood its bright red color.

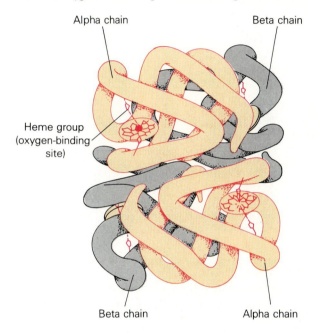

Alpha chain

Beta chain

Heme group (oxygen-binding site)

Beta chain

Alpha chain

follows one of two paths. Some combines in a loose association directly with the hemoglobin. Instead of combining with the heme groups as O_2 does, CO_2 reacts with amino side groups in other parts of the protein. The combination of carbon dioxide and hemoglobin is called **carbaminohemoglobin:**

$$Hb + CO_2 \rightleftharpoons Carbaminohemoglobin$$

The remaining carbon dioxide reacts with water in the red cells forming carbonic acid (H_2CO_3). Carbonic acid in turn dissociates into hydrogen ions (H^+) and bicarbonate ions (HCO_3^-). This reaction can occur in the plasma, but only very slowly. Red cells, however, contain the enzyme **carbonic anhydrase,** which not only speeds up the formation of carbonic acid but can also rapidly convert carbonic acid back to carbon dioxide and water. This is quite important since carbon dioxide must be reformed quickly if it is to leave the body during the transit of the blood through the alveoli of the lungs.

When carbonic acid dissociates in the red cells, the hydrogen ions are buffered (pH changes resist-

30.17   THE BOHR EFFECT

In the Bohr effect, hemoglobin surrenders its oxygen load more readily in the presence of increasing amounts of carbon dioxide. Actually, it is the increasing acidity brought about by the formation of carbonic acid as carbon dioxide goes into solution that produces the Bohr effect. In the organism (a), this chemical behavior means that metabolically inactive cells will receive less oxygen than metabolically active ones, regardless of the oxygen gradient. From an adaptive point of view, this peculiar behavior of hemoglobin makes sense, since the oxygen need is much greater in the active cells. (b) The Bohr effect is readily seen in graphs where the percent saturation of hemoglobin with oxygen is plotted against oxygen partial pressure at varying pH levels. As blood CO_2 increases, pH decreases (becomes more acidic), and oxygen dissociates more readily from hemoglobin.

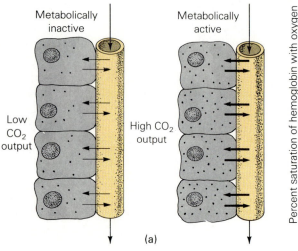

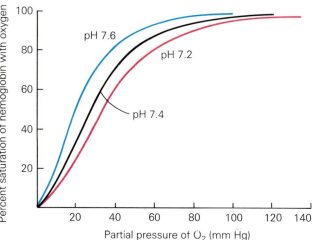

(a)

ed) by the protein hemoglobin itself. The bicarbonate ions diffuse out into the plasma, where they are balanced by sodium ions, forming sodium bicarbonate (NaHCO$_3$, or more precisely, the ionic forms, $Na^+ + HCO_3^-$). In addition to providing a means of transporting carbon dioxide, sodium bicarbonate forms an important part of the body's **acid-base buffering system**—that is, it helps neutralize any acids or bases that might form in the blood, keeping the pH constant—near neutral. (The sodium and bicarbonate in salt form, by the way, is identical to commercial baking soda and to the main ingredient in many stomach acid neutralizers.) The reactions so far can be summarized as

$$1) \quad Hb + CO_2 \rightleftharpoons \text{Carbaminohemoglobin}$$
$$2) \quad CO_2 + H_2O \underset{\substack{\text{carbonic} \\ \text{anhydrase}}}{\rightleftharpoons} H^+ + HCO_3^-$$
$$3) \quad Na^+ + HCO_3^- \rightleftharpoons NaHCO_3$$

As we see, the reactions are all reversible, and in each, it is the quantity of carbon dioxide present that dictates the direction of the reaction (an example of the mass action law at work; see Chapter 6).

So, in active tissues, where carbon dioxide levels are high, the direction of the reactions is toward carbaminohemoglobin and toward the formation of hydrogen and carbonate ions and sodium bicarbonate. But in the alveolar capillaries, any free carbon dioxide diffuses out of the blood, so its concentration is low. This prompts a speedy cascade of reversing chemical events: (1) The carbaminohemoglobin releases its carbon dioxide, (2) the bicarbonate of sodium bicarbonate in the plasma reenters the red cells, (3) it joins hydrogen ions to form carbonic acid, and (4) with a boost from carbonic anhydrase there is the rapid conversion back to carbon dioxide and water. The carbon dioxide then diffuses out of the capillary and into the alveolus for expiration out of the body.

The Control of Respiration

The perpetual, incessant changes that are the hallmark of living things are usually only minor adjustments—fine tuning—in response to shifting conditions within and without. For example, the need for O_2 and the production of CO_2 are always changing in every part of the body—now a little more here, now a little less there. Considering all such adjustments, however, the body's task is enormous and its ability to respond is admirable, even dazzling. How can it handle such a complex chore as regulating oxygen and carbon dioxide levels in literally millions of places at once?

First, we should be aware that both the respiratory and circulatory systems must respond in a coordinated way to changing oxygen levels. After all, increasing the breathing rate would be of little value unless one also increased the rate of blood flow. Second, it is important to know that respiratory control is exceedingly complex, involving many types of sensors and a variety of complex interactions. Let's now look at the role of the nervous system in respiratory control.

Neural Centers.   There is little question that the respiratory system is under involuntary control. We can, of course, vary the rate and depth of breathing at will, but only up to a point. If your little brother holds his breath to get his way, don't worry. He may begin to lose his rosy complexion for a time, but the ruse won't work. As his CO_2 level rises, his autonomic nervous system will win out and he will be forced to breathe although if he is stubborn enough, he may faint first.

The partial pressure of carbon dioxide and the closely related pH (hydrogen ion or H^+ concentration) in the blood turn out to be the most important factors in controlling respiration. This may be surprising since it is oxygen that we require for cell respiration. However, we now know from experiments that increases in the carbon dioxide or acidity in our blood bring about a rapid acceleration in the breathing rate. On the other hand, similar increases in oxygen uptake have little effect on the breathing rate.

Breathing movements are coordinated by a **respiratory control center** deep in the medulla (brainstem) that responds to changes in CO_2 and H^+ levels and, to a much lesser extent, oxygen levels in the blood. Such changes are detected by the **aortic** and **carotid bodies,** chemical sensors located in major arteries. Other sensors in the **fourth ventricle,** a fluid-filled space in the medulla (brainstem) itself, detect pH changes in the cerebrospinal fluid therein (Figure 30.18).

Any increase in the partial pressure of CO_2 or decreased pH in the blood, or any decrease in the pH of the cerebrospinal fluid, produces a barrage of neural impulses to the respiratory control center. It

30.18 RESPIRATORY CONTROL

The primary respiratory control center is located in the medulla of the brain. Its mission is to continually receive neural input from outside sensors, and to respond by altering the rate and depth of breathing. Because of the respiratory center's action, the body's oxygen demands can be fulfilled under a variety of conditions, from rest to strenuous activity. The monitoring sensors are located in the fourth ventricle of the brain, the carotid arteries, and the aorta.

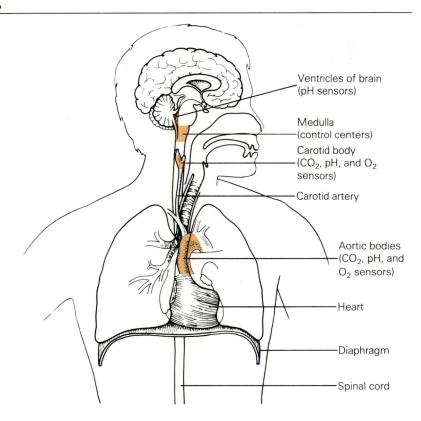

Ventricles of brain (pH sensors)

Medulla (control centers)

Carotid body (CO_2, pH, and O_2 sensors)

Carotid artery

Aortic bodies (CO_2, pH, and O_2 sensors)

Heart

Diaphragm

Spinal cord

responds, in turn, by directing motor impulses to the respiratory muscles which increase their activity. As the breathing rate and depth increases, more CO_2 is exhaled and the blood pH increases. Thus the blood is quickly restored to its optimal condition. On the other hand, decreases in these stimulating elements produce the opposite effect on the control center, and breathing rate and depth slow. These mechanisms are quite delicate and constantly adjust breathing to maintain carbon dioxide levels in the blood.

You may have noticed that our discussions of gas exchange have often led to discussions of the circulatory and nervous systems. It is quite difficult, if not impossible, to discuss one system to the exclusion of all others. In large part this is because animal systems function in an integrated and interdependent manner, and no system functions alone. As we now focus more specifically on the circulatory system, we will again weave in our findings from other discussions as we continue to try to understand the basic fabric of this thing called life.

APPLICATION OF IDEAS

1. All aquatic mammals, birds, and reptiles are air breathers. No matter how well they have adapted to the water in other ways, they must come to the surface to breathe. What does this clearly indicate about their ancestors? Given enough time, is it probable that these animals would evolve a gill-like device? Explain your conclusion. What might some disadvantages be if gills were present in an endotherm?

KEY WORDS AND IDEAS

Since oxygen can disrupt many life processes and reactions, organisms have had to adapt to the rise of oxygen as an atmospheric gas. Its presence as ozone in the upper atmosphere protects organisms from ultraviolet radiation. Oxygen is essential in aerobic life as an electron acceptor in the cell respiratory process. Oxygen and carbon dioxide are exchanged in the **respiratory system** and transported in the **circulatory system.**

GAS EXCHANGE SURFACES

1. Gas exchange in animals requires extensive, thin, moist membranes, such as those seen in the gill and lung. Such membranes are called the **respiratory interface.**

The Simple Body Interface

1. In complex sponges, gas exchange across the body interface works well because of the extensive canal system. It also works in coelenterates because of the extremely thin (two-cell layered) body wall. Denser flatworms increase the respiratory interface through the highly branched gastrovascular cavity, but this severely restricts their size. The surface of the echinoderm is increased by the presence of **dermal brancheae.**

2. Earthworms remain in moist soil most of the time, and the salamander is nocturnal, so they can safely exchange gases through the skin interface. Further, both have an efficient closed circulatory system and blood containing hemoglobin.

Expanding the Interface: Tracheae

1. Most terrestrial animals have internalized respiratory interfaces that help resist desiccation. Insects have a **tracheal system,** which includes body openings called **spiracles,** highly branched tubes called **tracheae** and **tracheoles,** and, in some, thin-walled **air sacs.** Gas exchange occurs in the fine branches, which are often fluid filled. In aquatic insects, the tracheae terminate in external **tracheal gills.**

Complex Interfaces: Gills

1. The **gill** is similar in invertebrates and vertebrates, with the respiratory interface intimately associated with the circulatory system. Feathery structures contain extensive capillary beds where blood and surrounding water are separated by a single membrane.

2. Through the use of cilia, clams draw water in through an incurrent siphon. Water crosses the

vascular gill exchange surface and exits through an excurrent siphon. Oxygen transport is aided by the copper-containing protein **hemocyanin.**

3. The lobster gills lie in chambers below a protective **carapace.** Water is drawn across the gills by an appendage called a **bailer.** The circulatory system is open in most of the body, but blood vessels occur in the gills.

4. The fish gill includes supporting **gill arches** with rows of **gill filaments,** each containing many capillary beds in platelike **lamellae.** Blood enters the filament from **afferent vessels,** crosses the lamellar capillaries and exits the filament to **efferent vessels.** During its transit, CO_2 is exchanged for O_2 across the thin walls. Water flow across the lamellae opposes blood flow, setting up an efficient **countercurrent exchange.**

5. Water holds less oxygen than air and diffusion is slower; thus aquatic animals must expend energy to move a large volume of water across the gills (or to move themselves through water).

Complex Interfaces: Lungs

1. The earliest fishes gulped air at the surface, using a vascularized pharynx for exchange. This system, an example of **preadaptation,** underwent modification into the lung in the first terrestrial vertebrates. In addition, **internal nares,** openings between the nasal cavity and mouth, evolved in the early terrestrial animals. In mammals, the **palate** came to separate the pharynx into mouth and nasal cavities. In fishes, the primitive lung evolved into the swim bladder, a flotation device.

2. The vertebrate lung is a highly branched inpocketing that, in many species, contain numerous air sacs, intimately associated with capillaries, and inflated and deflated by a bellows action.

3. The amphibian lung is a simple paired saclike affair, inflated and deflated by a pumping action in muscular mouth floor. Check valves in the nostrils and **glottis** help coordinate the filling and emptying cycles. Amphibians also use the skin as an exchange surface.

4. Reptile lungs are spongy and complex and, aside from the vascular cloaca of aquatic turtles, represent the only respiratory interface. Breathing occurs through a bellowslike action of the entire body wall.

5. In birds, air passes *through* the lung rather than in and out. Air moving through the trachea bypasses the lung to enter posterior air sacs and from there enters the lung. The passage of air through the lung is crosscurrent to the flow of blood through lung capillaries. The bird respiratory system represents a specific adaptation to flight at higher altitudes where oxygen is limited.

THE HUMAN RESPIRATORY SYSTEM

1. Major structures of the human respiratory system include the **mouth, nasal passages, pharynx, larynx, trachea, bronchi, bronchioles, alveoli** and **lungs.** The **palate** separates the mouth from the nasal passages which moisten, warm, and filter air entering the lungs. **Goblet cells** secrete dust-trapping mucus in the nasal passages and cilia move the mucus film toward the pharynx for swallowing. Warming of incoming air is aided by a countercurrent blood flow.

2. Incoming air moves through the larynx, into the trachea and **primary bronchi,** through branching **bronchioles,** and finally into the **alveoli.** The larynx contains the **voice box** and **vocal cords.**

3. Much of the **respiratory tree** contains a mucus-secreting and ciliated epithelium that traps dust particles and sweeps them upward, out of the system.

4. The lungs lie in the **pleural cavity,** enclosed by double membranes, the **pleurae.** The dome-shaped **diaphragm** forms the lower part of the cavity.

The Breathing Movements

1. In inspiration, **intercostal muscles** contract, elevating the rib cage, and the diaphragm flattens, creating a partial vacuum in the pleural cavity. The lungs passively expand in response to the inward movement of air. In exhalation the rib cage drops, the diaphragm resumes its relaxed dome shape, and the elastic lungs resume their former shape, forcing air out.

2. While the minimal (resting) exchange of air is about 0.5 liters, the maximum, or **vital capacity,** is about 4.0 to 6.0 liters. Some **residual air** always remains in the lungs.

The Exchange of Gases

1. While all gases of the atmosphere contribute to total atmospheric pressure, each gas exerts a **partial pressure**—P_g. Under **standard conditions** (total pressure 760 mm Hg) $P_{O_2} = 160$ mm Hg, and $P_{CO_2} = 0.3$ mm Hg. P_g decreases with altitude.

2. The diffusion of a gas follows its partial pressure gradient.
   a. In blood arriving at the alveolar capillaries, $P_{O_2} = 40$ mm Hg, while $P_{CO_2} = 45$ mm Hg.
   b. In alveoli air, the $P_{O_2} = 100$ mm Hg and $P_{CO_2} = 40$ mm Hg. Thus CO_2 leaves the blood and O_2 enters the blood.
   c. In blood leaving the alveoli, $P_{O_2} = 95$ mm Hg and $P_{CO_2} = 40$ mm Hg.
   d. In very metabolically active tissues, these values can change to 25 and 46 mm Hg respectively.

Oxygen Transport

1. Hemoglobin can carry 60 times as much oxygen as water. Its four polypeptide chains contain four heme groups and each **deoxyhemoglobin** can reversibly associate with one O_2, forming **oxyhemoglobin.**

2. The association and subsequent dissociation of hemoglobin and oxygen is complex, depending on pH, P_{CO_2}, and temperature.

3. Hemoglobin has a built-in safety factor. At sea level the blood leaving the lungs is nearly saturated with oxygen, and increasing the P_{O_2} has little effect. But at half the sea level P_{O_2} (high altitude), the blood will be 80% saturated. In addition, when a consistently low P_{O_2} occurs in the blood, the hormone **erythropoietin** is released, stimulating new red cell formation, a slower, long-term adjustment.

Carbon Dioxide Transport

1. Because of the **Bohr effect,** the affinity of hemoglobin for oxygen is inversely proportional to the partial pressure of carbon dioxide. Thus, in blood passing metabolically active tissues with a high CO_2 concentration, substantially more O_2 will dissociate.

2. In the transport of CO_2,
   a. about 8 percent of the CO_2 goes into solution in plasma;
   b. the remainder enters the red cells. Some associates with amino acids in hemoglobin, forming **carbaminohemoglobin,** and the rest forms carbonic acid with water in the red cell;
   c. the speed of dissociation reactions depends on the action of the enzyme **carbonic anhydrase;**
   d. within the red cell, carbonic acid dissociates into bicarbonate and hydrogen ions (acid).
   e. some hydrogen ions are buffered by hemoglobin while the carbonate ions diffuse out to the plasma, where they are buffered by sodium ions, forming sodium bicarbonate—a part of the body's **acid-base buffering system;**
   f. all reactions are reversible, according to the mass action law.

3. The reactions of carbon dioxide reverse in the lungs where the P_{CO_2} in the alveolus is low. The restoration of ions into CO_2 is greatly speeded by the action of carbonic anhydrase.

The Control of Respiration

1. Respiration is under the control of the autonomic nervous system and is ultimately involuntary.

2. Respiration is strongly influenced by the P_{CO_2} and pH of the blood. A **respiratory control center** in the medulla reacts to changes in these factors. Other sensors are located in the **aortic** and **carotid bodies,** and **fourth ventricle** of the brain.

REVIEW QUESTIONS

1. Explain why the circulatory and respiratory systems in complex animals are closely integrated. (758)

2. What are two basic characteristics required of a gas exchange structure? What additional restriction is imposed by the terrestrial environment? (759)

3. List three invertebrates that use the simple body interface for gas exchange, and explain how each satisfies the basic requirements. (760)

4. Explain in specific terms how the flatworm's respiratory requirements restrict its size. (760)

5. The surface area of the earthworm and lungless salamander alone are not sufficient to provide an adequate gas exchange for their dense bodies. What else is required? Explain in detail. (760–761)

6. Describe the insect tracheal system. What is its relationship to the circulatory system? (761–762)

7. Describe the manner in which the clam gill functions. Name its respiratory pigment. (762–763)

8. Describe the structure of the fish gill using the following terms: *gill arch, gill filament, lamellae, afferent* and *efferent blood vessels,* and *capillary beds.* (763)

9. The passage of blood and water in the fish gill is said to be a countercurrent flow. What does this mean and how does it affect gas exchange efficiency? (763)

10. Discuss the problem of availability of oxygen in the aquatic environment and describe solutions in the bony fishes. (763)

11. Summarize the main events in the evolution of the vertebrate lung. (764)

12. Compare the amphibian and reptilian lung. How does the amphibian supplement its lung surface area, and why haven't the reptiles taken advantage of this? (764–765)

13. Describe the four-stroke breathing mechanism in the frog. (765)

14. List several ways in which the bird lung differs from that of the reptile or mammal. (765)

15. Describe the movement of air through the bird respiratory system, beginning with air entering the trachea. (765)

16. What special requirements of birds does its extremely efficient respiratory system satisfy? (765)

17. List the structures through which air passes in the human respiratory system, beginning with the nostrils and ending with the alveoli. (766)

18. Describe the structure of the nasal passages and explain their role in respiration. (768)

19. Describe the structure of the alveoli, including their structural relationship to the circulatory system. (769)

20. Summarize the movements of the rib cage, diaphragm, and lungs in inspiration and expiration. What actually causes the lungs to fill? To empty? (770)

21. Carefully define partial pressure and state the partial pressures of oxygen, carbon dioxide, and nitrogen at sea level. How does this change at higher altitudes? (771)

22. What factors affect the movement of gases down a partial pressure gradient? (771)

23. State the partial pressures of oxygen and carbon dioxide in blood *entering* the alveoli, and the partial pressures of the two gases in the alveolar air. In what direction do the two gases move as a result? (772)

24. Write a formula representing the reactions between hemoglobin and oxygen. What four factors influence the rate and direction of the reactions? (772–773)

25. Describe the response of the body to a prolonged reduction in P_{O_2} in the arteries. (773)

26. Describe the Bohr effect and explain how it is adaptive. (773)

27. What happens to carbon dioxide when it enters the blood? What is the role of carbonic anhydrase? (774)

28. What is the function of sodium bicarbonate formation in the blood plasma? (774)

29. Explain how mass action affects the blood carbon dioxide chemistry in the lung. (774)

30. What two factors in the blood seem to have the most influence on respiratory activity? (775)

31. List three major regions where neural sensors measure conditions in the blood. (775)

32. Explain how the respiratory control center responds to increased pH, increased P_{CO_2}, and decreased pH. (775)

Circulation and Immunity

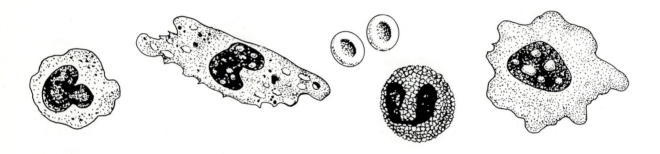

Watching the early development of a vertebrate embryo is a fascinating and often reaffirming experience, even to crusty old biologists who have spent their adult lives considering the various manifestations of life. It might seem that the feeling of wonder would eventually diminish, but somehow it doesn't. The sight of the quick pulses of a tiny, unformed heart signalling a new life is perpetually intriguing. The heart is, in a sense, the very symbol of life.

In its earliest days, the vertebrate heart gives little indication of how complex it will become since at that time it is only a simple, twitching tube. Curiously, the beating begins hours before there is any blood to pump. In fact, the circulatory system is among the first systems to take form. Why should this be?

Actually, if we consider that the embryo needs an inflow of nutrients and oxygen and the prompt removal of metabolic wastes, we should expect the circulatory system to develop early. But the circulatory system—that is, the heart and the vessels it serves—is important throughout the life of the animal. In this chapter, we'll see why. We can begin by reminding ourselves that the primary mission of the circulatory system is transport: transport of oxygen, carbon dioxide, nutrients, water, ions, hormones, antibodies, and metabolic wastes. The circulatory system, along with the lymphatic system, which we

will discuss shortly, constantly shifts water, ions, and proteins about, thus helping to regulate osmotic conditions in the body. In addition, the two closely related systems play a vital role in supporting a third—the **immune system.** The immune system is that vast, loosely organized army of highly specialized white blood cells and their products, responsible for combating bacteria, viruses, and other invaders. We will return to the lymphatic and immune systems later in the chapter, but first let's look into circulation and transport.

INVERTEBRATE TRANSPORT SYSTEMS

The circulatory systems of vertebrates and invertebrates can vary in a number of basic ways. We will begin with the invertebrates, where we will encounter systems in which the blood leaves the vessels.

Open Circulatory Systems

In closed circulatory systems, the blood remains within vessels of some sort through virtually its entire circulation; in open systems blood does not remain in vessels. Typically, in the latter case, the blood is pumped from the heart into a major vessel that eventually empties into channels leading to sinuses or cavities called **hemocoels.** The blood percolates through such hemocoels and then returns to other vessels that lead to a cavity surrounding the heart. From the cavity, the blood is drawn into the heart and pumped out again into the vessels. If you know about boats, this kind of heart is reminiscent of a bilge pump, the submergible water pump that draws water from its surroundings and directs it into hoses.

It is difficult to make generalizations about the many species of arthropods, but we can say that probably all of them have open circulatory systems. If you compare the circulatory systems of the crayfish and the grasshopper in Figure 31.1, you will see that the system is more complex in the gill-breathing crayfish. Its blood is involved in respiration as well as transport, whereas in the insect, the two systems are independent. The hearts of the two arthropods are also quite different. Note that the grasshopper heart is no more than a long dorsal blood vessel with muscular thickenings along its posterior region. In the crayfish, the heart is larger, more central, and more complex.

Closed Circulatory Systems

Other invertebrates have closed circulatory systems. As you may recall, the annelids have a remarkably well-developed closed system with distinct vessels, tubular hearts, and hemoglobin-rich blood. In addition, they have a fluid-filled coelom that assists in the distribution of materials. Note in Figure 31.2 the major vessels and the five pairs of hearts, or **aortic arches.** The valves in the aortic arches assure a one-way flow of blood.

TRANSPORT IN THE VERTEBRATES

All vertebrates have a closed circulatory system (if we ignore the fact that blood percolates through the liver) and a centrally located heart. The most complex of these is the four-chambered heart of crocodiles, mammals, and birds. But let's first consider the simpler two- and three-chambered arrangement in fish, amphibians, and reptiles.

31.1 OPEN CIRCULATORY SYSTEMS

Both the crayfish and the grasshopper have open circulatory systems. In these systems, blood vessels empty out into coelomic cavities known as sinuses or hemocoels. **(a)** The system of vessels is more extensive in the crayfish, with the blood serving a respiratory function as it passes through the gills. **(b)** The grasshopper has one dorsal vessel that contains several pumping regions or "hearts." Blood is collected by the hearts and pumped forward, first to sinuses in the head and then through the body sinuses, eventually returning to the dorsal vessel. Transport of the sluggish fluid is assisted by body movements and diffusion.

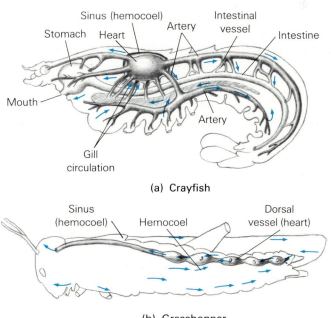

(a) Crayfish

(b) Grasshopper

Fishes and the Two-Chambered Heart

Fish are the most simple-hearted of all vertebrates. Their heart essentially consists of two pumping chambers, a single **atrium** (also called an **auricle)** and a single **ventricle.** The atria are generally thin-walled structures that receive blood and transfer it to ventricles. Ventricles are larger, thick-walled chambers that send blood into arteries with considerable force. In fish, incoming blood is first received by an enlarged vein, the **sinus venosus,** prior to entering the atrium, and blood leaving the ventricle first enters an enlarged artery, the **conus arteriosus.** These enlarged chamberlike vessels are also found in the three-chambered hearts of amphibians and reptiles but are lost in the four-chambered hearts of birds and mammals.

Notice in Figure 31.3 that blood leaving the fish's heart passes through the **ventral aorta** going directly into the blood vessels and capillaries of the gills. From the gills, freshly oxygenated blood flows through the body and from there returns to the heart. Blood returning to the heart moves sluggishly, since much of it has passed through two major capillary beds, one in the gills and the other in the body tissues, where frictional resistance is great.

Some of the blood has passed through other capillary beds, further decreasing its pressure. In some cases, blood that has already passed through the gills enters capillary beds in the intestine, where it picks up the products of digestion. The capillaries then rejoin, forming the **hepatic portal vein,** which goes to the liver. Here it divides once more into capillaries. It then enters a large vessel, the **common cardinal vein,** and returns to the heart.

In summary, the fish's blood flows from heart to gills via the ventral aorta, through the gill arches, from the gills to the tissues via the dorsal aorta (and some through the portal circuit in the liver), and from the tissues to the heart via the common cardinal vein. Again, during each cycle, all the blood flows through the gills and continues on through the other body tissues. Oxygen is picked up in the gills and released in the tissues prior to the blood's return to the heart.

Amphibians, Reptiles, and the Three-Chambered Heart

In the evolution of animals that, however temporarily, could survive on land, a new emphasis on lungs and a deemphasis on gills meant that the simple one-cycle circulatory system of the fish was no longer enough. As oxygen demands increased, greater blood pressure and a new way of oxygenating the blood were in order. New circuits in the system arose, with the single atrium becoming divided into two in the amphibians and reptiles. Also, some of the blood pumped from the heart was directed through a new **pulmonary circuit** to the lungs, which became separate from the **systemic system** to the other body tissues. Today, we find the three-chambered heart in all amphibians and nearly all reptiles.

As an example we can consider the frog heart. We see in Figure 31.4 the large sinus venosus (also present in fish) but here we see *two atria*. The third chamber makes possible the two-circuit system mentioned above. Let's follow the flow of blood to see how the system works.

Deoxygenated blood returning from the body enters the **right atrium** and from there is pumped into the single ventricle. Nearly simultaneously,

31.2 CLOSED CIRCULATORY SYSTEMS

The circulatory system in the earthworm and other annelids is closed: blood remains within vessels. Blood is pumped along by five pairs of aortic arches, complete with valves. The large dorsal vessel contracts to push blood forward into the pumping arches; these send it into the ventral vessel, which directs it to the body and head. The two large vessels are connected in each segment by extensive capillary beds that serve the digestive, excretory, reproductive, and respiratory regions.

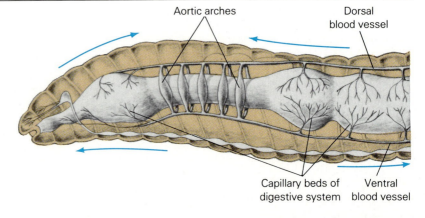

Aortic arches

Dorsal blood vessel

Capillary beds of digestive system

Ventral blood vessel

31.3 CIRCULATION IN THE FISH

Blood follows a simple circuit in the fish: from heart to gills to body and back to heart. The heart has two pumping chambers—one atrium and one ventricle—but in addition there is a receiving sinus venosus and a sending conus arteriosis. The hepatic portal system contains two capillary beds connected by a portal vein, so, counting its circuit through the gills, blood passing into the hepatic vein has been through three capillary beds.

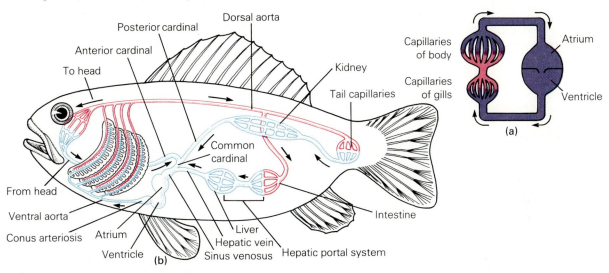

31.4 THE AMPHIBIAN THREE-CHAMBERED HEART

The three-chambered heart of the frog, shown here in both ventral (front) and dorsal (back) view, includes two atria and one ventricle. Deoxygenated blood arriving at the heart from the body enters the right atrium after being collected in the sinus venosus (dorsal view). The right atrium pumps its contents into the single ventricle. Simultaneously, the left atrium, which has received oxygenated blood from the lungs and skin, empties its contents into the single ventricle. An intricate system of flaps and valves helps reduce the mixing of oxygenated and deoxygenated blood. The ventricle then contracts, sending its contents into the conus arteriosus and then out into the divided aorta. Branches of the aorta carry blood into the lungs and skin through the pulmonary circuit and back to the heart again. Other branches supply blood to capillary networks in the head and body as seen in the schematic diagram. Blood returning through the veins reenters the heart through the sinus venosus once again.

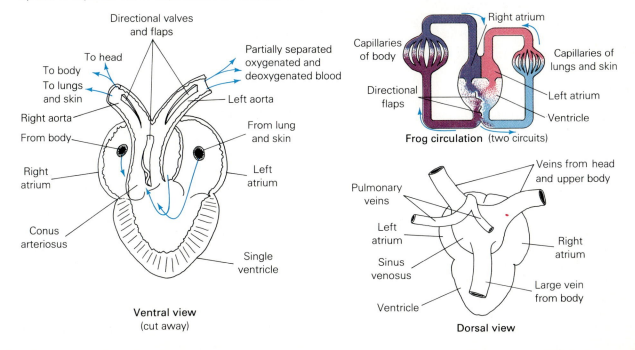

31.5 THE FOUR-CHAMBERED HEART

Mammals, birds, and crocodiles have four-chambered hearts. The four-chambered heart essentially allows two separate blood circuits, shown more diagrammatically in the inset. The right side receives deoxygenated blood from the body and pumps it through the lungs in one pulmonary circuit. The left side receives blood from the lungs and pumps it through the body in a separate circuit. The two circuits actually unite in the capillary beds, where arteries from the systemic circuit branch into capillaries, which rejoin to form veins, which return blood to the heart to be pumped through the pulmonary circuit.

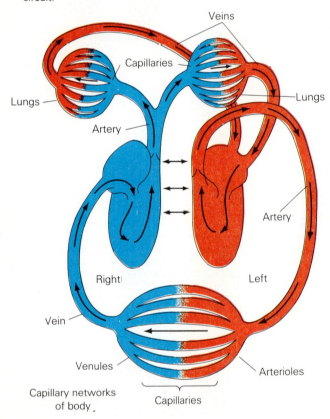

Veins

Capillaries

Lungs

Lungs

Artery

Artery

Right

Left

Vein

Venules

Arterioles

Capillary networks of body

Capillaries

oxygenated blood returning from the lungs and skin enters the **left atrium** and is also pumped into the ventricle. This system has produced a vexing question for biologists. Is the oxygenated blood from the lungs mixed with the deoxygenated blood from the body? The evidence indicates that it does mix, but only partially. In the amphibian heart, total mixing is prevented by flaps and partial valves that separate the oxygenated and deoxygenated blood somewhat and direct it into the proper arteries. Reptiles also have a partial **septum,** or partition, between the ventricles that helps keep the two kinds of blood separated. Even a hint of such mixing would be physiologically unacceptable for birds and mammals—endothermic creatures with much greater oxygen demands.

In summary, let's recall that a major structural change is seen in the circulation of amphibians and reptiles. We now see a two-part system because of the added pulmonary circuit. As a result, the extra boost provided by the single ventricle after oxygenated blood returns to the heart from the lungs keeps the blood pressure strong as the blood is sent to the tissues.

The Four-Chambered Heart

The evolution of the four-chambered heart was not a sudden thing. It occurred by a succession of tiny steps. We can see indications of some of these intermediate steps among living amphibians and reptiles. First there was the three-chambered heart with its directing valves. Then a partial septum evolved in the single ventricle, further dividing oxygenated and deoxygenated blood. Then, in the crocodiles, the ventricular division became complete, and birds and mammals were to share this evolutionary innovation.

The four-chambered heart of birds, crocodiles, and mammals includes a right atrium and ventricle and a left atrium and ventricle. The conus arteriosus and sinus venosus, so important to other vertebrates are finally lost in the mammals and birds. In the four-chambered heart, some of the blood is oxygenated in the lungs and some is deoxygenated in the body tissues in each complete circuit. The crucial difference between this complete circuit and that of the fish is that the pulmonary circuit is completely separate. Thus, a complete circuit in birds and mammals involves two trips through the heart, once through the left side and once through the right side. In fact, it is convenient to think of the heart as two separate pumps, as is illustrated in Figure 31.5.

THE HUMAN CIRCULATORY SYSTEM

Humans, like all mammals, have a four-chambered heart and a closed circulatory system. The system consists of the heart, arteries, capillaries, and veins. The arteries are muscular vessels that always carry blood away from the heart and, with the exception of the **pulmonary arteries,** always carry oxygenated blood; hence, arterial blood is usually bright red. Pulmonary arteries direct deoxygenated blood from the heart to the lungs. Capillaries are thin-walled vessels that branch from **arterioles** (smaller arteries). All exchanges between the blood and the cells of the body occur through the capillaries. **Venules** (smaller veins) receive the blood from capillaries and flow into ever-larger veins; these direct blood toward the heart. With the exception of the **pulmonary veins,** all veins carry deoxygenated blood.

Arteries tend to be round and thick-walled, and invested with heavy layers of smooth muscle and elastic connective tissue. Constriction or dilation of arteries and arterioles controls the blood flow in specific tissues. Veins, on the other hand, tend to be thin-walled and flattened, with much larger openings in cross section. They also generally lie nearer the surface, so most of the vessels you see under the skin are veins.

Circulation Through the Heart

The Right Side. Circulation through the four-chambered heart begins much like that in the three-chambered heart. Deoxygenated blood arrives at the right atrium and, simultaneously, in an oxygenated condition at the left atrium. Blood arriving at the right atrium is carried by two major veins, the **superior** and **inferior vena cavae** (Figures 31.6 and 31.7). (They are named "superior" and "inferior" because in humans one is higher than the other—a product of our upright posture.) From the right atrium the blood flows into a chamber below it, the muscular **right ventricle.** About

31.6 THE HUMAN HEART

The human heart consists of four chambers. The large, muscular right and left ventricles make up the bulk of the heart. The right and left atria form muscular caps at the top of each ventricle. Note the large arteries and veins entering and leaving the heart. The heart muscle has its own vessels, the coronary arteries and veins. The coronary arteries branch directly from the aorta, sending their branches deep into the heart muscle. The coronary veins emerge from heart muscle to form the coronary sinus, which empties into the right atrium. The coronary arteries are often involved in heart attacks, where blockages stop the critical blood supply to areas of the heart muscle.

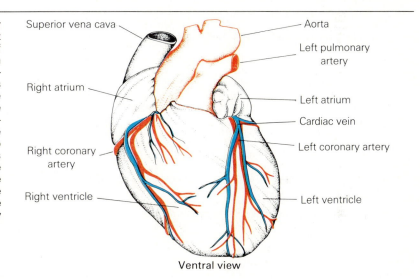

Ventral view

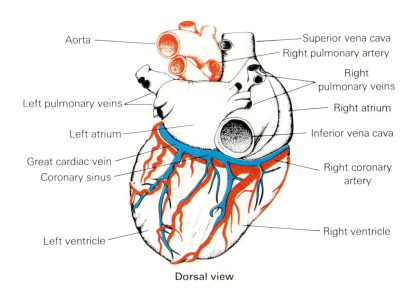

Dorsal view

80% of the blood entering the right ventricle gets there by flowing into the expanding chamber; the rest is pumped by contraction of the right atrium. When the right ventricle contracts, it forces blood into the pulmonary arteries and on to the lungs for gas exchange.

One-way valves between the chambers, and between the chambers and the arteries leaving the heart (Figure 31.8), ensure that the blood flows in the right direction. The valve between the right atrium and ventricle, the **tricuspid valve,** checks any backward flow created by the ventricular contraction. When the tricuspid is closed, it resembles a three-part parachute; the "shroud lines" are a series of **chordae tendineae,** tendonous cords that prevent the three flaps from collapsing backward like an

umbrella in a strong wind. As blood passes into the muscular **pulmonary artery,** its backflow is prevented by the **pulmonary semilunar valve,** a three-part curved flap that closes when the artery is full and expanded. The name *semilunar* refers to the half-moon shape of each of the three flaps.

The Left Side.   Blood from the right and left branches of the pulmonary arteries finally wends its way into the capillaries surrounding the alveoli of the lungs, where gases are exchanged. The capillaries then rejoin to form the **pulmonary veins,** which return oxygen-rich blood to the left atrium. When the left atrium contracts, blood passes through the double-flapped **bicuspid,** or **mitral valve** into the **left ventricle.** This valve has two

---

31.7 CIRCULATION THROUGH THE HEART

(a) Blood enters the right and left atria almost simultaneously. The superior and inferior venae cavae bring in deoxygenated blood from the body, emptying into the right atrium. Oxygenated blood entering the left atrium comes from the pulmonary circuit via the pulmonary veins. Blood flows from the atria to the ventricles. The contraction of the ventricles sends blood from the right ventricle into the pulmonary arteries (to the lungs) and from the left ventricle into the aorta (to the head and body). Backflow into right and left atria is prevented by the tricuspid and bicuspid valves. Blood entering the pulmonary artery and aorta is prevented from flowing back into the ventricles by the pulmonary semilunar and aortic semilunar valves, respectively. The pumping action is summarized in (b).

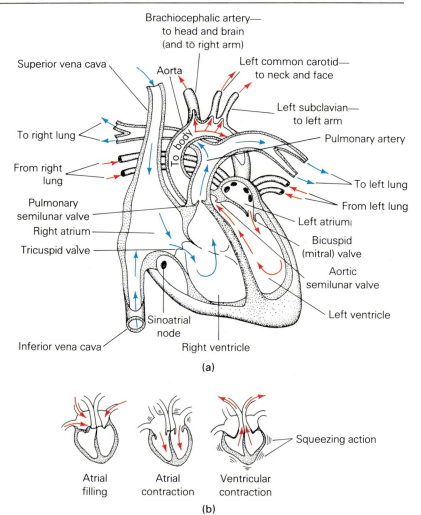

(a)

(b)

flaps, as opposed to the tricuspid of the right side.

The walls of the left ventricle are much thicker than those of the right, which probably gave rise to the myth that the heart is on the left side of the chest. (Actually, it is in the center, but tilted.) The left ventricle is large and muscular enough to force the oxygenated blood throughout all the tissues of the body and into the veins that eventually return it to the right side of the heart—an enormous task. Contraction of the left ventricle sends blood through the **aortic semilunar valve** and on into that great artery, the **aorta**. The aorta makes a U-turn to the left, giving rise to paired and single side branches that serve the head, arms, and digestive organs as it turns and passes down through the trunk. Below, it splits into the **iliac arteries** which branch into the legs.

Control of the Heart

Some might say that the subject of the human heart has been overworked. After all, not only have volumes been devoted to describing its relentless and tireless activity, but we have eulogized and venerated it as the center of love and emotion. In reality, though, it needs no romanticizing, since no discussion can be cold and clinical enough to drain it of its wonder. In Chapter 28 we mentioned that the heart muscle is unique because of its unusual contractile tissue. This was an extreme understatement. There are a number of physiological differences between heart muscle and other muscle types, including its branching fibers, interlocking membranes (intercalated disks), and tireless contractions. One of its most interesting features is that the contraction itself and its rhythmicity are inherent in the heart muscle cells. The heart, in other words, is capable of beating without outside regulation. This is dramatically illustrated in transplanted hearts, which sometimes go on beating for many years with no nerve connections whatever.

Extrinsic and Intrinsic Control.   In no sense, however, is the normal heart functionally isolated, beating merrily away, independent of the rest of the body. The body's constantly changing demands wouldn't allow that. The heart is, in fact, greatly affected by a number of conditions, even including the emotions. We will find that the heart rate is under both *extrinsic* (from without) and *intrinsic* (from within) control.

Extrinsically, the heart can be fine-tuned by the autonomic nervous system. The nerves that serve

31.8 THE VALVES OF THE HEART

The larger valves are between the atria and ventricles. The bicuspid (mitral) valve has two flaps of tissue, while the tricuspid has three. The aortic and pulmonary valves have heavier and smaller flaps and lack the support of chordae tendineae. The stringlike chordae can be seen in the photograph. These attach to papillary muscles projecting from the base of the ventricles. When the ventricles contract, these cords are tightened, holding the thin flaps in place as blood surges against them.

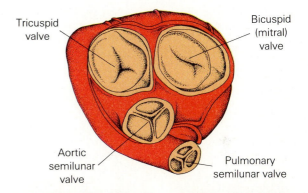

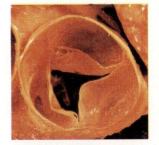
Heart valve open

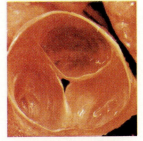
Heart valve closed

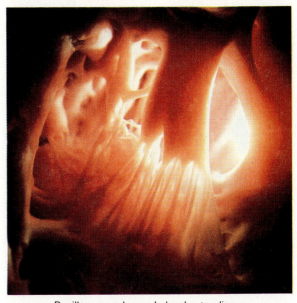
Papillary muscles and chordae tendineae

31.9 EXTRINSIC CONTROL OF THE HEART

(a) The contraction cycles of the heart originate in the sino-atrial (SA) node, more familiarly known as the pacemaker. **(b)** Impulses generated in the SA node spread across the atria, causing contraction of the muscle walls. **(c)** The same impulse reaches a second nodal area known as the atrioventricular (AV) node. **(d)** From this region, the impulse is regenerated and spread down the ventricular septum. **(e)** The impulse spreading over the ventricular surfaces initiates the contraction of the ventricles, first at the tip and then moving rhythmically upward. The arrangement of the muscle fibers causes a squeezing and wringing action in the ventricles. Interestingly, conduction through the AV node is slow. In fact, there is enough delay for the atria to complete their contraction before the ventricles begin.

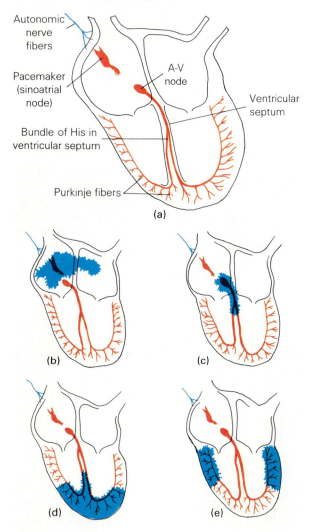

the heart originate in the spinal cord and in the medulla of the brain. One group, the **sympathetic nerves,** accelerates the heart rate. The second, the **parasympathetic nerves,** acts to slow the heart-beat. (We will have a closer look at these nerves in Chapter 35.) The two kinds of nerves permit tight regulation of this vital organ so that it operates in concert with the rest of the body. The heart may be further influenced by hormones from the endocrine system. **Epinephrine (adrenalin)** from the adrenal gland has the same effect on heart rate as does the accelerator nerve.

One may wonder, how does the heart direct itself if it is influenced by all these nerves? It directs itself through intrinsic control.

The origin of the heartbeat is in the modified muscle of the right atrium, at a region known as the **sinoatrial node (SA node)** (Figure 31.9). More commonly called the **pacemaker,** this is also the region influenced by the autonomic nerve fibers. The pacemaker transmits impulses across the atrial walls, causing the two atria to contract simultaneously, and on to a second node—the **atrioventricular node (AV node).** There is enough delay in the transmission of the impulse for the atria to complete their contraction before the ventricles begin theirs. The AV node initiates contraction of the two ventricles. Impulses from the AV node pass through a strand of specialized muscle in the ventricular septum known as the **bundle of His** (the name is the same in both sexes, and anyway, it's pronounced *hiss).* The bundle branches into right and left halves and travels to the pointed apex of the heart, where they branch into **Purkinje fibers** that initiate contraction there. The ventricular muscle fibers are arranged in a spiralling fashion so that their contraction produces a twisting, wringing motion that squeezes the blood out.

The Working Heart

There are two major heart sounds, usually described as *lub dup, lub dup.* The first is that of the sudden closing of the tricuspid and bicuspid valves as they shudder under the tremendous force of the contracting ventricles. The second sharp sound occurs as the aortic and pulmonary valves are snapped shut by arterial backflow and pressure when the ventricles relax. Then there is a brief pause. Then (one hopes), another *lub dup.* The period of ventricular contraction in the heart, heralded by the first sound, is known as **systole,** while the longer period beginning with the second sound,

31.10 ANATOMY OF AN ARTERY

Arteries are thick-walled, muscular vessels with considerable elasticity. In cross section they are shown to consist of three regions. The innermost is a smooth endothelium surrounded by a connective tissue sheath. The central region consists of a layer of smooth muscle that circles the artery. Outside the smooth muscle is an elastic membrane that gives the artery its resiliency. The outer region is a sheath of loose connective tissue.

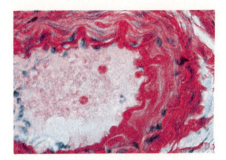

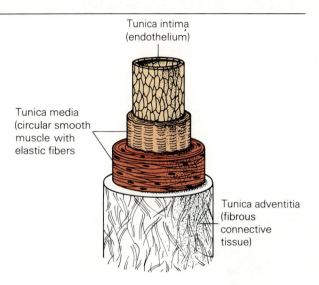

Tunica intima (endothelium)

Tunica media (circular smooth muscle with elastic fibers

Tunica adventitia (fibrous connective tissue)

during which the chambers of the heart fill up once again, is **diastole.**

It has been calculated that the resting heart rate is generally 72 contractions per minute, and that each contraction forces about 80 ml (2.7 oz) of blood into the aorta. The amount of blood passing through the heart with each heartbeat is the **stroke volume,** comparable to engine displacement. **Cardiac output,** the amount pumped each minute, is about 5 to 6 liters (11 to 13 pints). During activity, cardiac output can be increased up to 30 to 35 liters per minute—a remarkable reserve power! In case you have sensed an arithmetic problem here, we hasten to explain that your heart doesn't have to increase its beat rate by six to get a six-fold increase in cardiac output—if you are in good condition. During vigorous activity an athlete's heart rate climbs rapidly, but so does the stroke volume. A more sedentary person, on the other hand, must maintain cardiac output primarily through increased heartbeat, with less increase in stroke volume. This is one reason an unconditioned person tires quickly and recovers slowly. In trained athletes, the heart rate may go up to 180 per minute, but the difference is, theirs comes down faster (to about 120 within three minutes).

Blood Pressure. The aorta and major arteries receive the full impact of the powerful surges of blood from the left ventricle. A great many body functions depend on a rather constant pressure in the circulatory system, so the structure and function of arterial walls is vital (Figure 31.10). The sud-

den swell of blood during systole (heart contraction) expands the elastic walls of arteries, accommodating the increased volume. As the blood moves onward through the arterial tree, the expanded arterial walls contract through their elasticity, squeezing the rapidly decreasing blood volume. The speed of this recoil allows some lowering of blood pressure but prevents a drastic pressure drop. For example, there is a pressure drop in diastole of about 35%, or, as this would be usually expressed, about 40 to 60 mm of mercury (mm Hg). A typical systolic pressure is 120 mm Hg; a typical diastolic pressure is 80 mm Hg, and your doctor would call this set of blood pressure measurements "120 over 80." You've probably had your blood pressure taken with a **sphygmomanometer** (Figure 31.11).

A common problem in older persons is **arteriosclerosis,** or "hardening of the arteries." Arteriosclerosis has several causes, including calcium deposits in the arterial connective tissue, general loss of arterial elasticity, and especially **atherosclerosis**—arterial lesions or thickenings (plaques) characterized by localized growth of smooth muscle cells and deposits of lipids, especially cholesterol (see Figure 3.14). As vessels lose their elasticity, they fail to yield before the surging blood, creating serious burdening of the heart as resistance to flow increases. Normally elastic arteries can maintain pressure within their channels in spite of intermittent blood volume surges. Atherosclerotic plaques, however, impede blood flow and further stress the arteries.

31.11 MEASURING BLOOD PRESSURE

(a) Blood pressure is commonly measured in the brachial artery with a device known as a sphygmomanometer. It measures *systolic pressure,* the arterial pressure between contractions, and *diastolic pressure.* To make the measurements, a cuff is wrapped around the upper arm and inflated. At the same time, a stethoscope is inserted just under the cuff above the joint. As the cuff is inflated, the brachial artery is collapsed and no sound is heard. Pressure in the cuff is slowly reduced, and when a beating sound is first heard the pressure on the dial is recorded. Then, as the pressure continues to be released, all sound ceases (the diastolic pressure). **(b)** In the graph, the peaks represent systolic pressure and the valleys diastolic pressure. The shaded area is the pressure range at which the arterial sounds are heard. The diagonal line represents pressure decreasing in the sphygmomanometer.

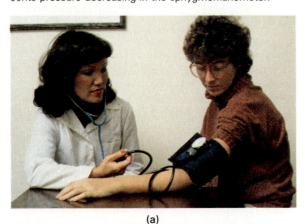

(a)

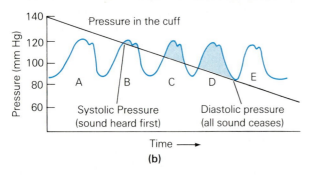

(b)

Vasoconstriction and Vasodilation. The smooth muscle lining the walls of arterioles plays a vital role in maintaining and directing the flow of blood. The lack of surge in these smaller vessels testifies to the precision of their regulation. Opening up and closing down of the arterioles, known as **vasodilation** and **vasoconstriction,** is partly under the control of the autonomic nervous system and partly under the control of hormones. In addition, oxygen, carbon dioxide, hydrogen ions, and a variety of drugs can bring about changes in the diameter of arterioles. There is some evidence that blood pressure can be consciously controlled, to a degree, as well.

Circuits in the Human Circulatory System

Though we are not about to discuss all the pathways of the human circulatory system here, we would like to consider a few interesting circuits. A circuit can be loosely defined as a distinct and major pathway of blood supply and return, generally where some function is performed. Figure 31.12 shows the four major circuits: pulmonary, hepatic portal, renal, and systemic (general body circuit which includes muscles, skin glands, and so on). In addition to these, there is the vital **cardiac circuit,** the circulatory system of the heart itself (see Figure 31.6). The pulmonary circuit, already described, is a good example of a relatively simple circuit.

Hepatic Portal Circuit. In the **hepatic portal circuit,** arteries branching from the abdominal aorta spread across the intestinal membranes and enter the digestive organs. Those entering the small intestine form capillaries within the intestinal villi. Digested foods are collected in the capillaries, which then merge together and, along with vessels from the rest of the intestine, form the **portal vein** leading to the liver. Upon entering the liver, the portal vein divides to form a second capillary bed. Actually, the liver contains not only capillaries, but many branched **sinusoids** (minute cavities) through which the blood passes. (As was mentioned, it could be argued that this makes the human system a partially open one.) From there, the blood, now free of its cargo of food and cleansed by the liver of toxic materials, returns directly through the inferior vena cava to the heart.

Renal Circuit. As it passes through the abdominal cavity, the aorta sends right and left branches, the **renal arteries,** into the kidneys. There, in one of the most complex filtering systems imaginable, the nitrogenous waste urea, excess water, miscellaneous metabolic byproducts, and some salts are removed from the blood. Of equal significance, water and salt levels are carefully adjusted, a vital process in the maintenance of the precise osmotic conditions required by the body. Blood returns from the kidneys via the **renal veins** to the inferior vena cava, and then to the heart.

Cardiac Circuit.   The cardiac circuit includes the arteries, capillaries, and veins of the heart muscle. The heart muscle receives first crack at oxygenated blood leaving the left ventricle since the first branches of the aorta are the right and left **coronary arteries.** They emerge just beyond the aortic valve, coursing separately to the right and left as they follow a furrow around the outer surface of the heart, finally joining toward the posterior side (see Figure 31.6).

As the arteries branch and rebranch, they continually rejoin each other through what anatomists call **anastomoses.** The resulting weblike network with its accompanying capillaries is quite significant, since it helps assure that no sizable part of the heart muscle will be without a blood supply should a blockage, such as a **coronary thrombosis** (a blood clot), occur in some vessel. Unfortunately, this arrangement is not very effective against coronary arteriosclerosis and atherosclerosis. Apparently nature's "fail-safe" backup systems cannot keep up with our sedentary but stressful lifestyles and diets rich in animal fats. (Or perhaps we are just living too long.)

Following its transit through the capillary networks of the heart, blood enters the **coronary veins,** whose routes parallel those of the arteries. The veins finally join each other to form the **coronary sinus,** a receptacle that opens into the right atrium.

Systemic Circuit.   The systemic circuit includes the rest of the body. Here we refer to capillary beds in muscles, glands, bones, the brain, and so on. In these places CO_2 and O_2 are exchanged, required substances for cell maintenance are delivered, wastes are removed, and so on. Capillaries are so profusely distributed that no cell lies far from the blood supply. The walls of the capillaries are usually composed of only one layer of flattened epithelial cells. They are fascinating little vessels and deserve a closer look.

31.12 THE MAJOR CIRCULATORY CIRCUITS

Circulation in the heart itself occurs in the cardiac circuit (see Figure 31.6). Closest to the heart is the pulmonary circuit, where CO_2 and O_2 are exchanged. Note the relationship of this circuit to the heart. The second circuit is the hepatic portal circuit and serves the digestive organs and the liver. Note that a major branch of the aorta goes directly to capillary beds in the digestive system, forms the portal vein, and then forms another capillary bed in the liver before returning to the heart. The third circuit is the renal circuit, which includes branches to and from the kidneys. After blood passes through the renal circuit, it returns immediately to the heart. Finally, there are the upper and lower systemic circuits, which carry blood to general areas of the body.

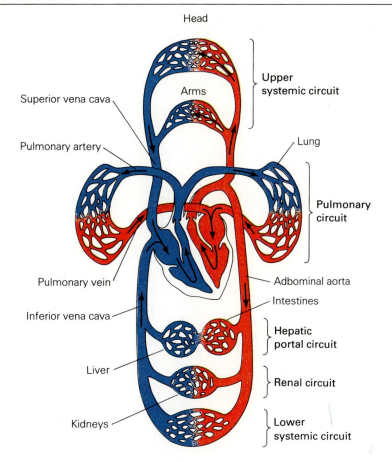

The Role of the Capillaries

As we learned earlier, the capillaries are, in a sense, the functional units of the circulatory system—all exchanges take place in these vessels. One might even say that the role of every other part of the system is simply to assist them. Their walls are made up of interlocking cells in single thickness, so the surroundings are never more than one cell away from the blood (Figure 31.13a).

One of the more interesting traits of the circulatory system is its ability to open and close selected capillary beds (Figure 31.13b). Such changes are made possible by the presence of smooth-muscle **precapillary sphincters** located in the tiny arterioles just about where they divide into the capillaries. Thus, for example, in an emergency situation, the vasoconstriction of arteriole sphincters in the diges-

tive area will partially close down circulation in that system so that more blood can be sent to the muscles and brain.

As blood passes through a capillary bed, the transport of water and various molecules to and from the capillary depends on several opposing forces. (Measurements of these forces are summarized in Figure 31.14.) Blood entering a capillary bed will lose nutrients, ions, water, and oxygen. Blood leaving a capillary bed will have picked up ions, water, carbon dioxide, and other metabolic wastes. As you can see, water and ions leave the capillaries at one end of the bed, only to return at the other. (Recall that the factors involved in the exchange of oxygen for carbon dioxide in capillaries are more complex; see Chapter 30.) Diffusion gradients can account for some of this peculiar behavior, but hydrostatic pressure is also important. Hydrostatic

31.13 ORGANIZATION OF CAPILLARIES

(a) Capillaries are composed principally of endothelial cells. Because their walls are only one cell thick, materials pass rapidly through them. **(b)** Capillary beds arise where arterioles branch. The capillaries, in their delicate profusion, serve tissues directly before they rejoin into their thicker-walled ven-

ules. The beds can be partially shut down in various regions by action of precapillary sphincters, muscular valves in the arteriolar walls. Thus, blood can temporarily be diverted to organs where the need is greatest. In this instance, the left side is open while the right side is shut down.

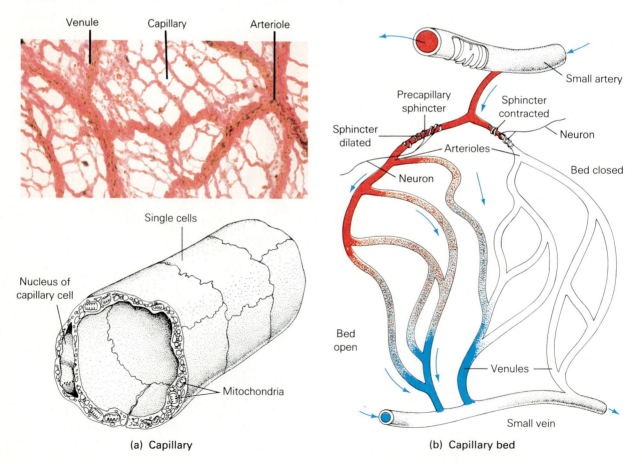

(a) Capillary

(b) Capillary bed

31.14 PRESSURE RELATIONSHIPS IN CAPILLARY BEDS

The movement of water and solutes in and out of the circulatory system occurs in the capillary beds. This movement, in turn, is controlled by pressure relationships in the bed. The two forces at work are hydrostatic pressure and osmotic pressure, and both operate inside and outside the capillary.

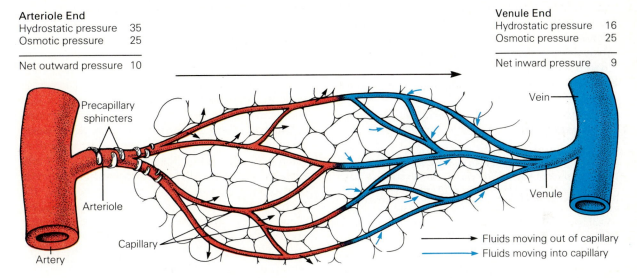

Arteriole End

| Hydrostatic pressure | 35 |
| Osmotic pressure | 25 |

| Net outward pressure | 10 |

Venule End

| Hydrostatic pressure | 16 |
| Osmotic pressure | 25 |

| Net inward pressure | 9 |

Precapillary sphincters

Vein

Arteriole

Venule

Capillary

Artery

→ Fluids moving out of capillary
→ Fluids moving into capillary

pressure is great at the arteriolar end of a capillary bed, and many substances are simply forced through the thin capillary walls in a filtration process. But things change at the venule end. Hydrostatic pressure is reduced, and since water has been forced out of the capillary, the protein-dense plasma left behind becomes a hypertonic medium and an osmotic gradient arises. Thus, most of the water simply returns to the blood through osmosis and the ions diffuse down their gradient as well. Any excess outside the capillary will eventually be returned to the blood by the lymphatic system.

But the capillaries are also actively involved in transport. Electron micrographs of capillaries show that the flat, thin cells that make up the capillary wall are engaged in extensive pinocytosis (Figure 31.15).

Work of the Veins

Capillaries join, forming venules, which in turn coalesce to form veins. Because of frictional resistance in capillary beds, blood pressure is lowest in the veins, where it can measure as low as 5 mm Hg. Thus, although blood volume leaving the arteries is equal to the volume entering the veins (except for water loss to the lymphatic system), the force of its movement is greatly depleted. Consider an analogy: if you block the water leaving a hose with your thumb, the pressure in the hose (arteries) is considerable, but as the spray (capillaries) strikes

31.15 PINOCYTIC VESICLES IN THE CAPILLARY WALL

As the electron micrograph of a capillary wall clearly indicates, capillaries are far more than minute tubes. The flattened cells comprising their walls accommodate simple diffusion, but their versatile plasma membranes carry on a considerable amount of active transport. Note the numerous pinocytic vesicles in the capillary cells.

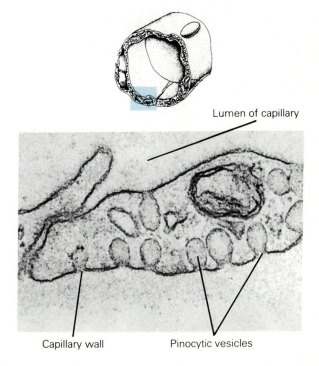

Lumen of capillary

Capillary wall   Pinocytic vesicles

the surface of the ground and the water trickles down the gutter (veins), the high energy level dissipates.

Because of the greatly reduced blood pressure in veins, the blood needs help in getting back to the heart. Many of the veins have one-way, flaplike valves, that allow blood to move in one direction—toward the heart. The walls of the veins, though much thinner than arterial walls, do contain some smooth muscles that can contract and help push the blood along (Figure 31.16). But venous flow

31.16 THE VEINS

(a) Veins are commonly thinner-walled than arteries. They also tend to be larger. In cross section, we see that a vein contains some of the regions we saw in the artery in Figure 31.10. But while the inner and outer layers of tissue are similar to those in the artery, the middle tissue layer typically contains much less muscle and is not as elastic. **(b)** The onward movement of blood in the veins is assisted by one-way valves.

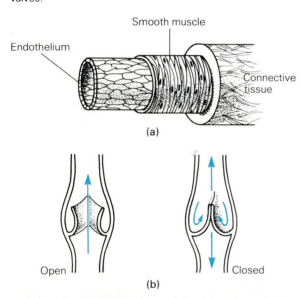

(a)

(b)

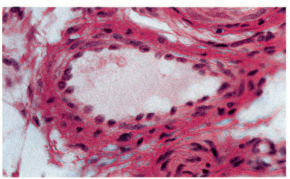

receives assistance in other ways as well. In the simple movement of the limbs, for instance, the muscles squeeze and massage the veins, moving blood along. In the chest area, breathing movements squeeze the walls of the vessels, sending the blood toward the heart.

The Blood

By definition, blood is a tissue. That is, it is composed of several types of cells in more or less resident (unchanging) status. More precisely, it is a connective tissue, with **plasma** as its matrix (ground substance). Gentle centrifugation of blood easily divides it into three parts: plasma on top, a thin, clear band of **leukocytes** (white blood cells) and **platelets** in the middle, and the heavy **erythrocytes** (red blood cells) on the bottom. By volume, the erythrocytes comprise about 45% of the whole blood in men. In women, the average percentage is about 43%. The straw-colored plasma consists of about 90% water. The remaining 10%, the plasma solids, includes a long list of substances; by weight, about 70% of the plasma solids are proteins, and the remaining 30% includes urea, amino acids, carbohydrates (mostly glucose), organic acids, fats, hormones, and various inorganic ions.

Three of the major plasma proteins are the **albumins, globulins,** and **fibrinogen.** The albumins are large proteins that bind miscellaneous impurities and toxins in the blood and aid in the transport of certain hormones, fatty acids, and metal ions. They are also involved in maintaining the osmotic potential of the blood, so important to capillary functions. The globulins include the **antibodies,** or **immunoglobins,** and certain proteins involved in the transport of lipids and fat-soluble vitamins. Fibrinogen is important in the blood clotting process.

The Red Blood Cells. Red blood cells are quite small (about 6 to 8 μm in diameter), are biconcave and disklike in shape, and, when mature, lack nuclei (Figure 31.17). Normally, there are about 5 million red cells per cubic millimeter of blood. Red and white blood cells are produced continuously in the **red bone marrow.** In adult humans, red cell formation occurs almost exclusively in the axial skeleton. Red cells have about a four-month life expectancy. When they age, some rupture, spilling their hemoglobin into the blood. Most aging or damaged red cells are phagocytized by certain white cells in the spleen and liver.

31.17 RED BLOOD CELLS

Erythrocytes make up about 45% of total blood volume in men and 43% in women. We each have about 25 billion red cells at any one time. In this SEM sequence of the formation of a red cell, a six-day process, the cell expels its nucleus on day three (top), then continues to twist into its final disk shape.

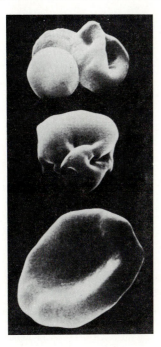

The White Blood Cells.   Unlike red cells, white cells (leukocytes) do not discard their nuclei as they mature. As was indicated, the white blood cells are active in the immune system, which will be discussed in detail shortly. For now it is important to know that the **neutrophils,** whose numbers make up the majority of leukocytes, are important phagocytic cells. They aggregate at infection sites, engulfing invading microorganisms. **Basophils** are involved in the inflammatory response, releasing histamines and serotonin at infection sites. **Eosinophils** are also involved in inflammatory responses, but in addition, they help destroy larger parasites. **Lymphocytes** are the backbone of the immune system. In lymphatic tissue they carry out many tasks that aid in fighting off infectious organisms, destroying foreign molecules and providing longer-lasting immunity from disease. The **monocytes,** once activated at infection sites, develop into **macrophages** which, like neutrophils, specialize in phagocytizing foreign cells and any cellular debris they encounter.

Platelets.   Blood platelets, or **thrombocytes** as they are known technically, are tiny, extremely numerous structures. Unlike erythrocytes, which are whole cells that have since lost their nuclei, platelets are only fragments of cells. The small, disk-shaped vesicles are formed directly from **platelet mother cells (megakaryocytes)** when these large cells mature and fragment without mitosis—a most peculiar process. Platelets play an important role in blood clotting.

Figure 31.18 illustrates the cellular origin of the blood cells and platelets. Curiously, each of the formed elements of the blood—the cells and platelets—has the same origin, the **hemocytoblast,** or simply **stem cell.**

Clotting.   Blood clotting is essential to any organism that relies on circulating fluids. The world is full of objects that can pierce our bodies and cause our fluids to leak out, and clotting minimizes such leakage.

Blood clotting is an extremely complex and only partially understood process. At least 15 substances are involved, some of which are part of an intricate arrangement that prevents accidental clotting. Two proteins are basic to the process: **prothrombin,** an inactive clotting protein present in the plasma, and fibrinogen, one of the major plasma proteins. Platelets and damaged cells also play a role. Generally, the clotting process proceeds as follows:

1. A vessel is damaged.
2. Platelets attach at the wound site, form lengthy extensions, and adhere to collagen fibers, forming what is essentially a "plug."
3. The platelets rupture, releasing (a) vasoconstrictors that cause nearby vessels to constrict, reducing blood loss, and (b) **thromboplastins** (enzymes).
4. In the presence of thromboplastins and calcium ions, prothrombin in the plasma becomes **thrombin,** a specific endopeptidase.
5. Thrombin breaks apart the large fibrinogen molecules of the plasma, the smaller pieces forming a fibrous, sticky protein called **fibrin.**
6. Fibrin fibers, along with damaged platelets, form a network that solidifies, becoming a clot that stops the bleeding.
7. The clot contracts, pulling the wound together, further preventing bleeding and encouraging healing.

31.18 DEVELOPMENT OF BLOOD CELLS

Stem cells in the red bone marrow give rise to leukocytes, erythrocytes, and platelets, the formed elements of the blood.

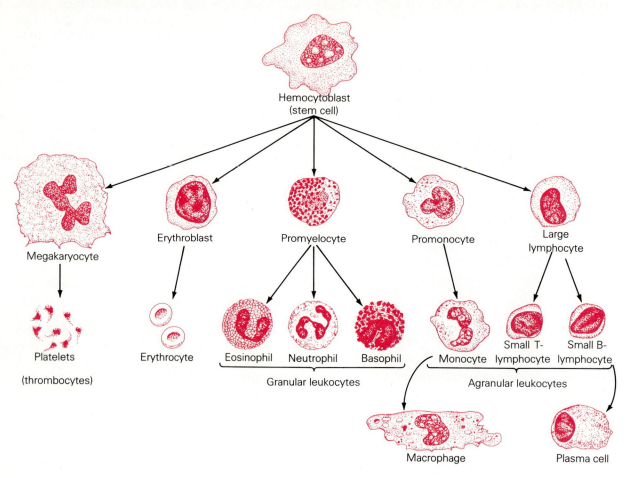

Hemocytoblast (stem cell)

Megakaryocyte

Erythroblast

Promyelocyte

Promonocyte

Large lymphocyte

Platelets

(thrombocytes)

Erythrocyte

Eosinophil   Neutrophil   Basophil

Granular leukocytes

Monocyte   Small T-lymphocyte   Small B-lymphocyte

Agranular leukocytes

Macrophage

Plasma cell

THE LYMPHATIC SYSTEM

The lymphatic system has three essential roles: it maintains fluid and electrolyte (ion) balances in the body, it transports certain fatty acids from the intestinal villi to the blood, and it assists with the work of the immune system. In its first role, the lymphatic system drains tissue spaces and cavities of fluid and ions that have not been recovered by the capillaries. Such fluids are returned to the bloodstream through ducts near the heart.

The lymphatic system (Figure 31.19) consists of **lymph vessels** and clusters of **lymph nodes** (often called "lymph glands"). Lymph vessels include countless tiny blind endings called **lymph capillaries** (which include the lacteals of the intestinal villi; see

Chapter 29). Lymph capillaries collect fluids, solutes, and foreign materials from tissue spaces, emptying into **lymphatic collecting ducts,** which join to form the **lymphatics,** major vessels that carry the lymph to the bloodstream. The watery lymph is pushed along by the squeezing action of muscles and changes in pressure within the thoracic cavity. Like the veins, lymph vessels have one-way check valves that help keep the lymph from backing up so that the flow continues toward the heart. Along the way, much of the lymph filters through lymph nodes.

The lymph nodes are located throughout the body, but their greatest concentrations are in the head, neck, armpits, abdomen, and groin (see Figure 31.19). Each node is a compartmentalized mass of tissue that harbors multitudes of lymphocytes—

primary cellular agents of the immune system. Foreign materials, bacteria, and viral particles carried into the vessels are swept into the nodes, where they are attacked by resident white cells. Activated lymphocytes cause the nodes to enlarge, so swollen lymph nodes are a telltale sign of infection.

Occasionally, the ducts and vessels in the lymphatic system are turned into deadly avenues for the spread of cancer. While cancer cells entering the lymph nodes are commonly attacked and killed by highly specialized lymphocytes, often some survive and continue their rapid cell division as they are carried by the lymphatic stream throughout the body.

THE IMMUNE SYSTEM

Simply stated, the immune system recognizes and eliminates any foreign substance or invading organism that finds its way into the body. Foreign molecules that initiate a reaction by the immune system are called **antigens.**

All animals have in common one very effective first line of defense: primitive ameboid cells that phagocytize foreign particles and invading microorganisms. Some of these cells, the macrophages, are fixed in place, for instance in the spleen and the lymph nodes, and simply engulf anything suspicious that passes their way. Other defensive cells

31.19 THE LYMPHATIC SYSTEM

(a) The lymphatic system consists of a number of lymph vessels and nodes. The lymph vessels eventually empty into large veins near the heart. (b) Fluids are directed by a number of one-way valves, similar to those seen in veins. (c) Lymph nodes are located generally throughout the body but cluster in several regions, including the groin, abdomen, armpits, neck, and head.

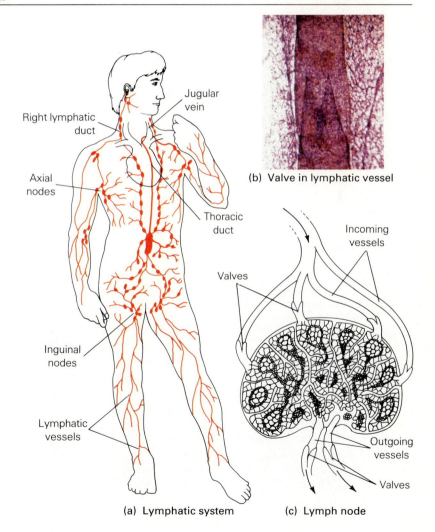

(b) Valve in lymphatic vessel

(a) Lymphatic system

(c) Lymph node

roam about freely as if on patrol, which, in a sense, they are. In addition to the huge macrophages, human phagocytes include the somewhat smaller neutrophils. The patrolling macrophages slip in and out of the bloodstream, but they can quickly congregate at the site of a wound or lesion, apparently attracted by substances released from damaged platelet cells. If there is a sizable infection, arriving phagocytes will literally eat themselves to death, engulfing and digesting the bacteria. Eventually the macrophages are killed by the accumulation of toxic products and their dead bodies form **pus.** Such phagocytes are considered primitive because they are found in all animals, including those that lack the more sophisticated immune responses provided by our next group of cells, the lymphocytes.

Lymphocytes are the second-most plentiful type of cell in the immune system. (One estimate is that the human body contains about 2 trillion of them.) They are involved in a complex second line of defense called, simply, the **immune response.** The immune response includes two different mechanisms, the **humoral immune response** (involving the production of free immunoglobins, or antibodies) and the **cell-mediated response** (direct cellular action). Each requires the action of a specific type of lymphocyte, the **B-cell lymphocyte** and the **T-cell lymphocyte,** respectively.

Both types of lymphocytes originate from stem cells in the red marrow, the cells that produce all blood cells. The B-cell lymphocytes also mature in other places, such as the bone marrow (the "B" in the name is derived from the *bursa of fabricus* in the chicken, where B-cells were first isolated). The T-cells are so named because they complete their development in the **thymus,** a lymphoid organ located above the heart in the thoracic cavity. In the inactive state, B- and T-cells are nearly impossible to distinguish visually, but once activated, the B-cells contain an unmistakably extensive rough endoplasmic reticulum while the T-cells contain large concentrations of free ribosomes (Figure 31.20). Biologists, using cell surface protein differences, can now differentiate between the inactive B- and T-cells. This capability has led to much of the knowledge we now have about their different roles.

Unlike most other systems, which consist of distinct, fairly centralized organs, the immune system is loosely organized and widespread, although many areas of **lymphoid tissue**—where lymphocytes aggregate—can be readily identified. We see in Figure 31.21 that these include the bone marrow and thymus—**central lymphoid tissue** where the cells originate and where development occurs, respectively. Once they have matured, the lymphocytes migrate to the many **peripheral lymphoid tissues,** where they become active. These widespread places include the lymph nodes, spleen, adenoids, tonsils, and *Peyer's patches* in the small intestine. In carrying out their function, the lymphocytes move freely in and out of the bloodstream, squeezing

31.20 LYMPHOCYTES

(a) Virgin B-cells and T-cells are identical in appearance. Following activation, however, the B-cell lymphocyte **(b)** produces an extensive rough endoplasmic reticulum, a sign of extensive protein synthesis, which presumably would include its specific immunoglobin. The T-cell lymphocyte **(c)** produces very little rough endoplasmic reticulum and its numerous ribosomes are free.

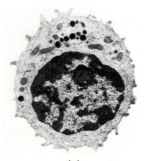

(a)

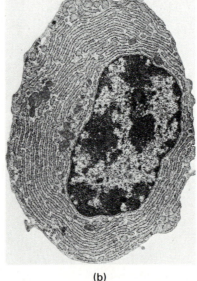

(b)

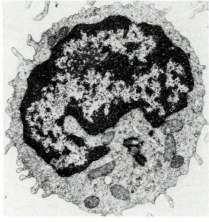

(c)

31.21 THE IMMUNE SYSTEM

The human immune system is dispersed throughout the body, but central lymphoid tissue is found in the bones and thymus. Also significant are the peripheral lymphoid tissues in the lymphatic system, adenoids, tonsils, spleen and Peyer's patches of the small intestine. Lymphocytes develop in the bones and thymus and interact with antigen in the other tissues.

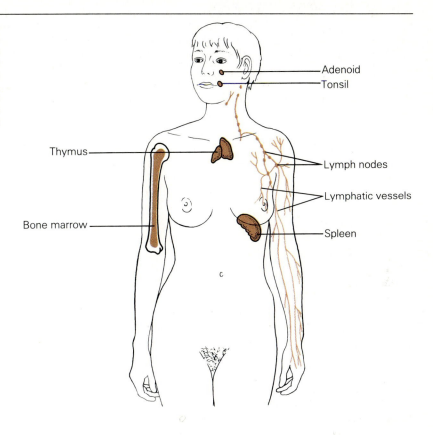

through cell junctions in capillaries and following the lymphatic stream back to the circulatory system again. Their constant movement isn't surprising when you consider that an important part of their task includes seeking out and identifying invaders.

B- and T-Cell Lymphocytes and Clonal Selection

The B-cell lymphocytes specialize in the production and release of antibodies, protein molecules that have the amazing ability to recognize and bind to foreign molecules—those we have called antigens. The T-cell lymphocytes also produce antibodies but in relatively minor amounts, and these are integrated into their membranes, where they help recognize matching antigens on the surfaces of invading cells. The amazing part is that there are antibodies against practically any foreign substance imaginable—even some synthetic molecules that are not found in nature. We will see how the lymphocytes come by this enormous capability later, but for now let's see what happens when a B- or T-cell encounters a specific antigen for the first time. The events

are encompassed in the **clonal selection theory** (Figure 31.22).

Virgin B- and T-cell lymphocytes, those that have yet to encounter antigens, carry specific antigen recognition proteins (similar to but not antibodies) integrated into the outer face of their plasma membranes. According to the clonal selection theory, each virgin cell is a specialist, bearing one and only one kind of recognition protein. Since we are talking about such incredibly large numbers, there is, hypothetically, a specialized recognition protein for every conceivable antigen. Certainly, the great majority of virgin lymphocytes will never find that matching antigen. But when a match occurs, things begin to happen.

The antigen forms a reversible association with the receptor site on the virgin lymphocyte's plasma membrane. The B-cells then enlarge and undergo mitosis and differentiation, giving rise to clones of two cell types, **plasma cells** and **memory cells** (see Figure 31.22).

The plasma clone becomes active, synthesizing an antibody that matches the captured antigen. Plasma B-cells are short-lived, lasting perhaps for a few days, but during that period they secrete copi-

31.22 LYMPHOCYTE ACTIVATION

According to the clonal selection theory, an antigen can only cause a response in a lymphocyte that has already been committed to it. This assume that there are millions of different predisposed lymphocytes in the immune system, each waiting for its confrontation with a specific antigen. Upon this confrontation, the lymphocyte is activated, dividing rapidly and producing clones of plasma and memory cells. Activated plas- ma cells produce and release copious amounts of the specific antibody, while activated T-cells wear protein recognition molecules on the cell surface and attack cells with matching antigen. Some B-cell and T-cell descendants, the "memory cells," are long-lived and remain inactive until they have a second encounter with the same antigen.

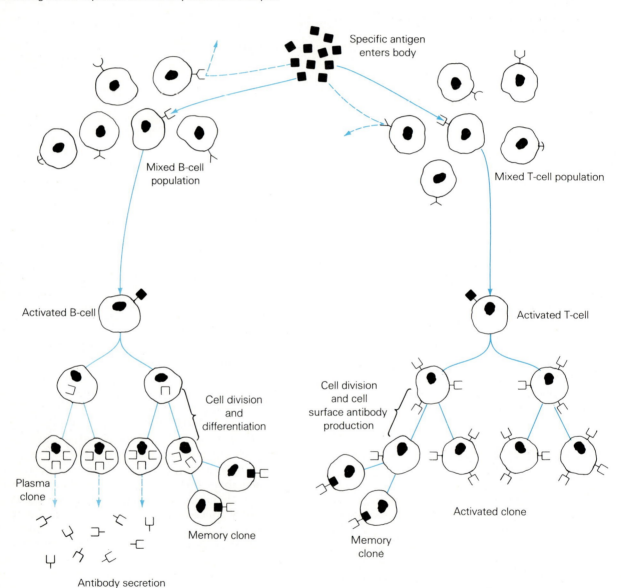

ous amounts of antibodies, reportedly up to 2000 molecules per minute. The antibodies then join with and immobilize the offending antigen molecules and the antigen-antibody complexes are cleaned up by phagocytic cells as mentioned earlier. In the meantime, we may become quite ill from the presence of the antigens or invading organisms and their toxic secretions. It takes time, but if all goes well, the offending antigens will be dealt with and we will recover. This early reaction of the immune system to an invader is called the **primary immune response.** (We will look into the specifics of anti-body-antigen reactions further along.)

What about memory cells? They don't seem to do anything exciting at this time, but as the name indicates, they remember—and this turns out to be vital to the immune process. Memory cells store the instructions necessary to construct the antibody that their sister plasma cells are so busy making and releasing. In addition, they give rise to clones. But unlike the plasma cells, memory cells have extreme-ly long lives—perhaps years. They are kept in reserve and should a second invasion by the same antigen or antigen-bearing invader occur, the clone will respond on a more massive scale, producing many more plasma cells than was possible in the primary immune response. This is the **secondary immune response,** a reaction that occurs much more rapidly than the primary response. In fact, we may be completely unaware of the reaction since the symptoms are either absent or quite mild.

Incidentally, there is a shortcut to establishing the cellular memory banks that prepare us for the secondary immune response—immunization through vaccination. Injections of weakened or killed disease agents are used to promote a primary immune response without the misery and risk of the actual disease. During the primary immune response, banks of memory cells are produced as usual, and should we later be confronted with the active disease agent, our immune systems will go right into the secondary immune response. Such a response is limited, however, by the longevity of memory cells, and for that reason, "booster shots" are often recommended.

Let's now look more closely at the antibody itself, and how so many different varieties can be produced.

The Structure of Antibodies

Antibodies, as suggested earlier, are found in the class of plasma proteins known as immunoglobins. There are a number of different general forms of antibodies (Table 31.1), but each has the same basic

TABLE 31.1

ANTIBODIES

Immunoglobin class		Characteristics
IgA	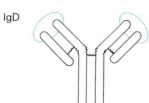	Occurs as a dimer (two immunoglobin molecules) but forms a monomer in blood. Transported to sur-faces. Common in moth-er's milk, respiratory mu-cus films, saliva, tears, and the intestinal lumen. First line of defense in secondary immune re-sponse.
IgD		Occurs as a monomer (single immunoglobin) on the surface of B-cell lym-phocytes. Functions as re-ceptor.
IgE	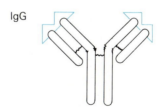	Occurs as a monomer (single immunoglobin). Acts in allergic reactions caus-ing the release of hista-mine, which is associated with tissue inflammation (hay fever symptoms).
IgG		Occurs as a monomer (single immunoglobin). Common in blood during late primary infections and throughout secondary in-fections.
IgM	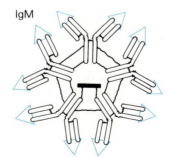	Occurs as a pentamer (five immunoglobins). Op-erates in blood as part of the initial response to in-fection.

31.23 STRUCTURE OF IMMUNOGLOBINS

Antibodies, or immunoglobins, have a common organization. **(a)** Included in the complex proteins are two heavy and two light amino acid chains. The chains are joined to each other by disulfide bridges, and the heavy chains have a hingelike region that permits flexibility **(b)**. Each immunoglobin has a nonspecific constant region and two highly specific variable regions. The variable regions of each antibody are unique, and millions are known to exist.

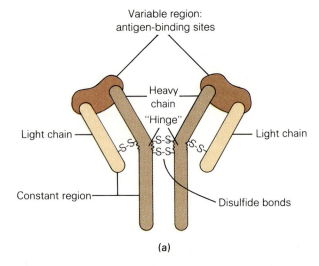

(a)

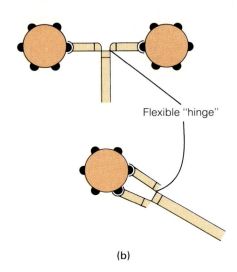

(b)

four-part structure. Each consists of two identical heavy (long) chains of amino acids and two identical light (shorter) chains. Interestingly, the heavy chains each contain a hingelike region that gives the molecule considerable flexibility. As we will see, this flexibility permits groups of antibodies to interact with antigens and cells. The light chains have about 220 amino acids each and the heavy chains contain about 450, so the whole protein has about 1340 amino acids (Figures 31.23 and 31.24). The four chains are bound together by disulfide linkages. In some classes of immunoglobins, small polymers of two to five of the individual immunoglobin units are formed.

Each of the chains is divided into two functional regions: a **constant region** and a **variable region.** The constant regions of both the light chains and the heavy chains are the same in thousands of different antibodies, but the variable regions are specific to different antigens. The variable regions of the light chain and the heavy chain together form a cleft, or **antigenic determinant,** that binds tightly to the surface of at least some part of the antigen. Since the molecule is flexible and has two identical variable regions, two antigens can be bound simultaneously by the same antibody, which then forms a bridge between them. When many antibodies attach to many antigens, the foreign molecules agglutinate into great clumps (Figure 31.25). The specificity of the antigenic determinants explains why each antibody attacks only a certain kind of antigen.

As we indicated earlier, antibodies bound to the surface of a foreign cell induce rapid phagocytosis. The antibody-antigen complex also binds, and activates, a group of circulating blood plasma proteins that collectively are called **complement.** When some such activated complement ruptures the plasma membrane of the invading cell, this releases additional foreign substances that act as antigens, heightening the immune reaction.

Monoclonal Antibodies. In 1976, immunologists learned to produce in the laboratory clones of any B-cell lymphocyte plasma cell desired, and as a result, to obtain pure samples of the specific antibody produced by that B-cell. Plasma cells don't survive well in tissue culture, so it is first necessary, using recombinant techniques, to fuse the desired line of B-lymphocytes with B-lymphocyte myeloma cells, a kind of tumor cell that has lost its ability to produce antibodies. Tumor cells do especially well in tissue culture—so well that their cell lines are sometimes referred to as "immortal." The fused cells, called **hybridomas,** then go on through their cell cycles, producing a clone.

And as is the case with plasma cells, each produces only one specific antibody, called a **monoclonal antibody.**

The medical potential for monoclonal antibodies is staggering. Theoretically, one could order a supply of antibodies against any protein, whether in a viral coat, bacterial cell wall, or cancer cell plasma membrane. In addition, monoclonal antibodies show great potential in getting specific cytotoxic (cell killing) drugs to their specific target cells. Getting drugs into the desired cells and tissues is often a problem for the clinician, and using specific carriers helps avoid the large dosages often needed to assure success.

The Problem of Antibody Diversity

How Specific Antibodies Are Produced: The Germ Line and Somatic Variation Theories. All mammals and birds, and perhaps all vertebrates, can produce antibodies against any large molecule. This ability suggests some interesting questions. If you inject a rabbit with crocodile hemoglobin, the rabbit will make hundreds of different specific antibodies against crocodile hemoglobin. How does the immune system do it? Can rabbits be *preprogrammed* to resist crocodile proteins? Are there enough antibody-coding genes to go around to genetically enable rabbits to resist

31.24 SPACE-FILLING MODEL OF AN IMMUNOGLOBIN

This view of an immunoglobin impresses us by its vast number of amino acids. The general Y shape is visible with the two antigen binding sites (the variable ends) at the tips of the Y.

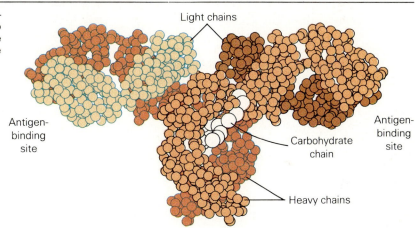

31.25 AN ANTIBODY–ANTIGEN REACTION

When antibodies encounter antigens, a massive coupling or agglutination occurs. Because of its hingelike arrangement and two binding sites, an antibody is capable of binding to two antigens at once. Following agglutination, the mass can be readily attacked by phagocytic white cells.

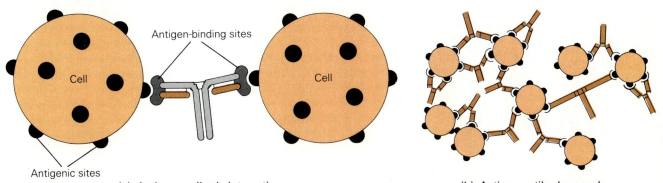

(a) Antigen-antibody interaction

(b) Antigen-antibody complex

crocodile molecules? More significantly, are there enough genes to code for antibodies against millions upon millions of potential antigens?

An older theory, developed some 20 years ago by Macfarlane Burnet, the brilliant Australian immunologist, suggests there are. This is the **germ line theory.** It proposes that the genetic information required to generate all variable regions is inherited. That is, it is encoded in the DNA of the zygote and passed from cell to cell, just like the genes for any protein. So when the immune system develops, it is quite ready for any antigen it might confront.

Today, we know a little more, and the **somatic variation theory** has developed from the older idea. The somatic theory relieves us of the problem of numbers of genes, since it proposes that genes coding for the variable regions in antibodies are themselves generated from about 300 DNA segments located throughout the chromosomes. These segments are rearranged during the immune system's development. That is, the genes coding for the immunoglobin variable regions are actually broken apart and recombined in a nearly infinite number of ways (estimated at about 18 billion). (You may recall that we've said that all the cells of the body have the same genetic information. Obviously, a correction is in order. You can now see that each clone of differentiated lympho-

cytes had rearranged its DNA so that it has a pair of tailormade genes that other cells don't have.)

In summary, then, it is now believed that during lymphocyte maturation, the DNA of the chromosome itself is permanently rearranged to create new variable region genes (Figure 31.26). Since there are many ways to break and rearrange the chromosome, any one of a vast array of possible antibody genes can be created in a given lymphocyte.

Recognizing Self. The somatic variation theory satisfactorily explains how great diversity in antibodies arises, but the potential to produce an antibody against every conceivable substance raises another question. What keeps the immune system from reacting against proteins and other molecules in the very body in which it resides? There are two answers. First, the immune system must learn *not* to react against "self," a process called **tolerance.** Second, sometimes the immune system does react against self, an abnormal condition that we will consider shortly. Let's first see how the immune system learns.

During embryonic development, newly formed virgin lymphocytes begin to move through the bloodstream, becoming familiar with their parent organism. When the wandering cells find a chemical match with their surface antigen recognition

31.26 THE MAKING OF A VARIABLE REGION GENE

During lymphocyte maturation, the DNA in uncommitted cells is recombined, and the cells become genetically unique. In producing new combinations of genes for variable regions, genes from several pools are excised and rejoined to produce unique combinations. With this enormous possible variation, the body can produce immunoglobins for practically any antigen. In this diagram, a variable region gene (V3) is selected, excised from the chromosome, and joined to one of several "joining genes" (J4), which has also been excised. Following the removal of any introns that may have been present, the new variable region gene along with its joining gene is affixed to a constant region gene (C), and one polypeptide coding segment for an immunoglobin can then be transcribed into mRNA.

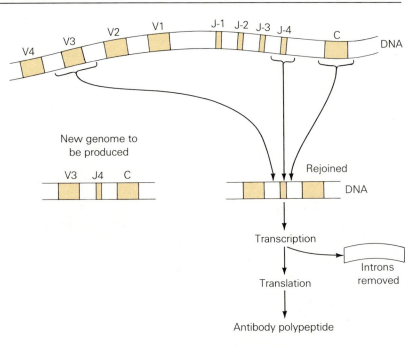

31.27 IDENTIFYING SELF

Experiments with larval tree frogs have verified that a developmental period of "self-identification" occurs that suppresses autoimmune responses. In this experiment, the pituitary gland of one tadpole has been transplanted to another early in development. Later, the pituitary is returned but is rejected as foreign tissue. Hypothetically, during the pituitary's absence, the young lymphocyte population completed its identification of self. In the control, only half the pituitary was removed, and when returned to its original owner, fails to provoke an immune response and is accepted as "self." This shows that the rejection was not based on changes that may have occurred while the pituitary was growing in the temporary host.

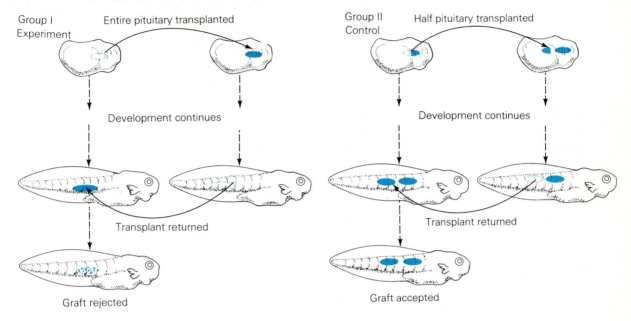

Group I Experiment — Entire pituitary transplanted — Development continues — Transplant returned — Graft rejected

Group II Control — Half pituitary transplanted — Development continues — Transplant returned — Graft accepted

proteins, rather than becoming activated, as we described earlier, they become permanently suppressed. The exact mechanisms of this self-recognition and suppression process are unknown, but that it occurs has been firmly established through experimentation.

In an elegant experiment with tree frog larvae (Figure 31.27), the pituitary gland was removed and transplanted to a host larva to keep it alive. Later, when the gland was implanted into its original owner, the graft was rejected. One conclusion would be that the recipient frog's immune system, which had since matured, had missed its chance to become familiar with the missing tissue. Thus it did not recognize its old pituitary as "self" and treated it as an antigen. But how do we know that something did not simply happen to the pituitary when it was in the recipient, rendering it "foreign" upon its return to the host? This question was answered in follow-up experiments. The procedure was altered so that just half of the pituitary was removed and maintained in a new recipient. Later, the transplanted half was returned to its original owner. The graft was not rejected.

But how does an organism recognize itself? One hypothesis seeking to explain the suppression of antibody production against self involves a **suppressor lymphocyte.** During the self-recognition process, certain T-cell lymphocytes take on a suppressing characteristic. Throughout the remaining life of the organism, the suppressor lymphocytes interact with other B-cell and T-cell lymphocytes in such a way as to prevent them from acting against the body's own molecules. Interestingly, this recognition of self can be transferred from one organism to another by a transplantation of suppressor lymphocytes. The continuing development of this idea will be interesting to follow, but let's look now at what happens when the suppression mechanisms fail.

Autoimmunity: Attack Against Self. Considering the enormous complexity of the immune system, it should be no surprise that it sometimes goes awry. A line of lymphocytes may begin reacting against self, that is, against one of the organism's own proteins or tissues as though it were a foreign invader. Such a reaction is called an **auto-**

immune response, or **autoimmune disease.** Among the many known or suspected autoimmune diseases are arthritis, nephritis, rheumatic fever, systemic lupus erythromatosis, various hormone disorders, certain forms of diabetes, and possibly schizophrenia.

Some disease organisms are particularly insidious in that their surface antigens are sufficiently like our own that the invaders are recognized as self, which lets them slip by our defenses. Eventually, however, they will be detected and the immune system will come up with antibodies that will knock out these impostors. Problem solved? Not quite. Unfortunately, the antibodies created may cross-react with our own tissues, causing a severe autoimmune reaction. The *Streptococcus* bacteria that cause strep throat are notorious for this. An infection in the throat creates antibodies that can attack tissue elsewhere, notably in the kidneys or heart valves, with serious and sometimes fatal results. But perhaps the most extreme case of the immune system gone awry is in the disease **AIDS (acquired immune deficiency syndrome),** a truly exotic and frightening disease we will look into shortly.

The Work of T-Cell Lymphocytes

Our knowledge of T-cell activities is far less precise than that of the B-cells, although what we are finding is intriguing. We know that T-cell lymphocytes are different from B-cells in a number of significant ways. They do not, as we have mentioned, release antibodies into the blood, but instead they wear them. The antibodies they produce are integrated into glycoprotein complexes in the cell surface. The T-cells have three principal functions: (1) In vertebrates, they destroy any foreign cells they encounter, including cellular parasites, virus-infected cells, and tumor cells, all of which have a cell surface chemistry that the T-cells recognize as antigenic. (2) They assist certain B-cell and other T-cell lymphocytes in their response to antigen and activate macrophages and certain other white cells. (3) They act, as we have seen, to suppress immune reactions in some T-cell and B-cell lymphocytes. These functions are carried out by different specialized populations of T-cells.

A number of different T-cell subpopulations of specialists have been identified, each fulfilling a certain part of the cell-destroying role. Included are two groups—**effector T-cells** and **regulatory T-cells.** The first includes the **cytotoxic T-cells** and **encapsulating T-cells.** The second group is made up of **helper T-cells** and **suppressor T-cells.**

The Role of Effector T-cells. The cytotoxic T-cells carry out the first task—killing foreign or virus-infected cells. Typically, when T-cells come across and identify such cells, they cluster about them, their antigen recognition surface molecules forming associations with antigenic sites on the invader. These associations apparently signal their presence to the immune system in general, and before long numerous macrophages arrive on the scene and phagocytize the cell masses.

The manner in which cytotoxic T-cells attack virus-infected cells is worthy of mention (Figure 31.28). Cytotoxic T-cells are able to identify a virus-infected body cell by changes that the virus induces in the cell surface. (Viruses customarily cause surface changes that prevent invasion by any other kind of virus; see Chapters 14 and 18.) The crucial task of the T-cell is to identify and kill the host cell before viral particles can be replicated and their new coats synthesized. While the killing mechanism isn't well known, cytotoxic T-cells are known to release toxins that destroy both themselves and the invader. Once the cell has been disrupted, phagocytizing cells quickly clean up the debris.

The encapsulating T-cell lymphocytes, with the assistance of accessory cells such as macrophages and fibroblasts, work quite differently. They encapsulate foreign cells or antigens, localizing the infection, and preventing its spread to other parts of the body.

The Regulatory T-cells. The helper T-cells, also called **inducer T-cells,** assist both other T-cells and B-cells, in responding to antigens, and they activate other types of leukocytes, such as macrophages. The helper T-cells are essential to the responses of B-cell lymphocytes to many, but not all, antigens. The helper T-cells may be stimulated by the presence of antigenic material attached to a macrophage. After engulfing and partially digesting foreign cells, the macrophage will incorporate antigenic materials into its plasma membrane, thus becoming **antigen-presenting cells.** Such cells rove the body, touching lymphocyte after lymphocyte until they contact a helper T-cell with a matching antigenic recognition complex in its membrane. Such contacts activate the helper T-cell, which then activates effector T-cells and B-cells. Just how this communication takes place is a difficult question and one that immunologists are trying to answer. The suppressor T-cells, as mentioned earlier, specialize in suppressing lymphocyte reactions that might lead to attacks against self.

31.28 A CYTOTOXIC T-CELL ON THE ATTACK

One response of the cytotoxic T-cell is to attack virus-infected cells. When a T-cell encounters such a cell, it recognizes changes in its surface chemistry, and destroys the infected cell before the virus can replicate and produce new viral coats. Reportedly, the cytotoxic T-cell interferes with the viral mRNA action.

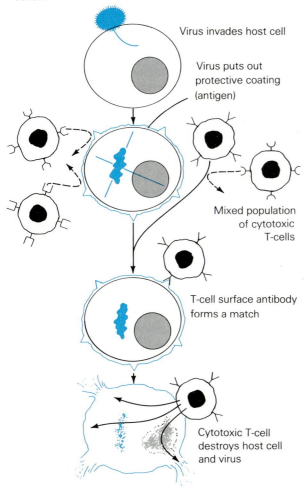

Virus invades host cell

Virus puts out protective coating (antigen)

Mixed population of cytotoxic T-cells

T-cell surface antibody forms a match

Cytotoxic T-cell destroys host cell and virus

T-Cells and Tissue Grafts. One source of information on the work of T-cells has been the recent surge in organ transplant surgery. Since grafts bring foreign cells into direct contact with the recipient's cells, graft rejection is primarily a problem of T-cell reaction. Following the graft of any foreign tissue, T-lymphocytes whose antigen recognition surface molecules match the proteins or antigens of the graft are aroused to a highly activated state. Much of the rejection reaction occurs as cytotoxic T-cells attack the foreign tissue. Unless they are somehow suppressed, such lymphocytes aggregate at the graft site, attach themselves to the exposed graft cells, and disrupt their plasma mem-

branes. In addition, the cytotoxic T-cells attract numerous macrophages to the graft site, where they readily engulf the injured graft cells. The macrophages are then joined by other leukocytes, and soon the badly damaged graft tissue fails and sloughs away.

The rejection of grafts, a problem of tissue incompatibility, has a clear genetic basis. Simply put, the tissues of unrelated individuals differ in a large group of cell-surface antigens. These antigens are grouped into what is known as the **major histocompatibility complex,** or simply **MHC.** In humans these are more specifically known as the **human leukocyte-associated antigens** or, mercifully, **HLA antigens.** The MHC antigens are actually glycoproteins that stud the surface of plasma membranes. Coding for their production are some 20 different gene loci, each with many different alleles. At one locus, for example, there are at least 50 known alleles. The point is, with this many alleles, there is nearly an infinite number of possible genotypes. It is for this reason that, except for identical twins, the chances of any two people being completely compatible is incredibly minute. Thus we reject each other's tissues and organs.

AIDS and the Defunct Immune System

Over the past few years, a new and terrible disease called aquired immune deficiency syndrome (AIDS) has gained worldwide attention. The problem is a suppressed immune system, the results of which tragically remind us of this system's vital importance. The symptoms may begin as a simple but persistent cold. But as the disease progresses, AIDS victims grow increasingly susceptible to all sorts of diseases, even very rare ones. In fact, the appearance of certain rare diseases such as **Kaposi's sarcoma** and **pneumocystic pneumonia,** assists in identifying the condition in its later stages. By this time, unfortunately, such information is often more useful to the pathologist or medical statistician than to the patient since both diseases are usually fatal.

The first cases of AIDS were recognized late in 1979. By late 1984, the disease had spread to over 40 states and 20 countries, and the number of victims was in the thousands and climbing. Certain high-risk groups, primarily highly promiscuous male homosexuals, hemophiliacs, and intravenous drug users, account for 96% of AIDS cases. It is not known why these particular groups are at risk. There is also increasing evidence that promiscuous heterosexuals are at risk; in fact, in some countries, the disease is acknowledged as a primarily hetero-sexual problem. There is also evidence that AIDS can be transmitted by blood transfusions, although public health officials tell us that the relatively small numbers of AIDS carriers and sufferers makes the risk from transfusion infinitesimally small. (Just the same, cautious surgical patients in some cities are bringing their own blood donors along to the hos-pital with them.)

Fewer than 14% of AIDS victims survive more than three years after onset of the condition. No treatment has been successful. Intensive research into AIDS is continuing, however, and some clues to its mysteries are now revealing themselves. Sci-entists now have evidence that the agent is a human retrovirus (of the HTLV group) that specifically attacks human helper T-cells. Retroviruses, you

may recall (see Chapter 18), integrate their own genetic code into the host chromosome, playing havoc with the host's genetic mechanisms. The virus may be a variant of one implicated in leuke-mia. However, in leukemia, the helper T-cells grow in an uncontrolled way, while in AIDS, they are killed. Scientists are looking for a mutant form of the leukemia retrovirus.

AIDS researchers have currently placed their hope in three possibilities: interferon, a new drug called interleukin-2, and the development of a vac-cine. Interferon doesn't do much for the defunct immune system, but it does seem to bring on remis-sion in the Kaposi's sarcoma tumors. Interleukin-2 seems to revive the virus-combating ability of the immune system cells; and the development of a vaccine is increasingly likely because of the success in isolating the HTLV retrovirus.

APPLICATION OF IDEAS

1. Tracing the embryological development of the four-chambered heart in humans reveals stages when the heart appears as a simple tube, as two-chambered, as essentially three-chambered, and finally, four-chambered. What does this suggest about the genetic framework upon which the human heart is organized? Could your hypothesis be tested? In what way does the fate of the pharyngeal arches (see Chapter 37) in humans help support your hypothesis?

2. The gravest danger in coronary embolism (blockage) immediately follows its onset, a time when tissue death is occurring. This is true even when only a small amount of heart tissue is destroyed. Why is such loss significant to the normal heart function? What characteristic of the cardiac arteries helps minimize such dam-age?

KEY WORDS AND IDEAS

INVERTEBRATE TRANSPORT SYSTEMS

Open Circulatory Systems
1. In **open circulatory systems,** blood is pumped into vessels but leaves them to percolate through spaces called **hemocoels.** Blood returns to the heart to be drawn up for another circuit. In **closed circulatory systems,** the blood ele-ments remain within vessels.
2. The gill-breathing crustacean uses its circulatory system to transport oxygen and carbon dioxide, and accordingly the system is more complex than in the insect where it does not have a res-piratory function.

Closed Circulatory Systems
1. Annelids have closed circulatory systems, dis-tinct vessels, and five pairs of tubular **aortic arches** (hearts) with one-way valves.

TRANSPORT IN VERTEBRATES

Fishes and the Two-Chambered Heart
1. The fish heart consists of one **atrium (auricle)** and one **ventricle.** Blood entering the atrium is first collected in a large vein, the **sinus venosus,** and blood leaving the ventricle enters an en-larged artery, the **conus arteriosus.**
2. Blood leaving the fish heart enters the **ventral aorta,** which directs it to the gills, where gas ex-

change occurs. The blood then passes through the head and body. All blood returning to the heart has traveled through at least two capillary beds. Most blood is returned to the heart via the **common cardinal veins.**

Amphibians, Reptiles, and the Three-Chambered Heart

1. Terrestrial vertebrates have a second atrium and a **pulmonary circuit,** which carries blood to the lungs and back. In both amphibians and reptiles, some mixing of oxygenated and deoxygenated blood occurs in the single ventricle.

2. In the frog's circulation, deoxygenated blood from the body enters the sinus venosis, then the **right atrium** and single ventricle. At the same time, oxygenated blood from the lungs and skin enters the **left atrium** and then the ventricle. Partial separation is provided by flaps and partial valves in the heart and the conus arteriosus, through which blood passes on its way back to the lungs, skin, and body.

3. Reptiles have only a partial **septum** within the single ventricle, so mixing of oxygenated and deoxygenated blood occurs in these animals.

The Four-Chambered Heart

1. The four chambered heart of crocodiles, birds, and mammals includes a right and left atrium and a right and left ventricle, but the conus arteriosus and sinus venosus are absent. The pulmonary circuit is completely separated.

THE HUMAN CIRCULATORY SYSTEM

1. The human circulatory system consists of the four-chambered heart, arteries, capillaries and veins. The muscular arteries carry blood away from the heart, and except for the **pulmonary arteries,** this blood is oxygenated. All exchanges occur in the thin-walled capillaries, following which blood returns to the heart through the veins. While arteries tend to be thick-walled and muscular, veins tend to be larger, to have thinner walls, and to contain less smooth muscle.

Circulation Through the Heart

1. Deoxygenated blood from the body enters the right atrium from the **superior vena cava** and **inferior vena cava.** From the right atrium, blood enters the muscular **right ventricle,** which pumps it to the lungs for gas exchange. Backflow into the right atrium is prevented by the **tricuspid valve,** a one-way valve whose thin flaps are held in place by **chordae tendineae.** Backflow from the pulmonary artery to the right ventricle is prevented by the **pulmonary semilunar valve.** Deoxygenated blood from the pulmonary artery enters the capillaries of the lung, where its gases are exchanged.

2. Oxygenated blood returns from the lungs via the **pulmonary veins,** enters the left atrium, and moves on to the thick-muscled **left ventricle,** which pumps it into the **aorta.** Backflow into the left atrium is prevented by the **bicuspid valve (mitral valve),** while backflow into the left ventricle is prevented by the **aortic semilunar valve.**

Control of the Heart

1. While heart muscle has an inherent contractile nature, control of its rate and effort is both *extrinsic* (external) and *intrinsic* (internal). Extrinsic control is through the autonomic nervous system, which accelerates heartbeat via **sympathetic nerves** and slows it via **parasympathetic nerves. Epinephrine** from the adrenal glands also accelerates the heart.

2. Intrinsic control originates in the **sinoatrial node (SA node),** or **pacemaker,** which sends contractile impulses across the atrial walls, causing contraction there. The impulse reaches the **atrioventricular node (AV node)** and is relayed to the **bundle of His** in the ventricular septum. The bundle's two branches pass to the base of the ventricles and up their outer walls, giving rise to many branched **purkinje fibers.**

The Working Heart

1. The *lub* of the *lub dup* heart sound is that of the tricuspid and bicuspid valves shutting, while the *dup* is that of the shutting of the aortic and pulmonary valves. The period of ventricular contraction is **systole** while the period between contractions is **diastole.**

2. The amount of blood per contraction is the **stroke volume,** while the **cardiac output** is the rate output per minute.

3. Elasticity in the major arteries maintains blood pressure during diastole. Clinically, blood pressure is measured with a **sphygmomanometer** and is recorded as systolic over diastolic pressure (for example, 120/80 mm Hg).

4. Problems in maintaining pressure arise in **arteriosclerosis** (loss of elasticity) and **atherosclerosis** (plaque formation), arterial wall diseases that increase resistance, burdening the heart.

5. Arterioles are capable of **vasodilation** (opening) and **vasoconstriction** (closing), shunting blood where need is greatest.

Circuits in the Human Circulatory System

1. Circuits are circulatory pathways where some special function is performed by the blood. They include the following:
   a. **pulmonary circuit** (gas exchange);
   b. **hepatic portal circuit** (food carried from gut to liver for storage and distribution via the **portal vein,** which forms **sinusoids** in the liver);

c. **renal circuit** (**renal arteries** and **veins** bring blood to and from the kidneys where wastes, excess water, and other substances are removed from the blood);

d. **cardiac circuit** (blood supply to heart muscle, includes **coronary arteries** and **coronary veins.** Arteries form netlike **anastomoses,** which help limit the effect of **coronary thrombosis.** The coronary veins empty into the right atrium via the **coronary sinus);**

e. **systemic circuit** (a catch-all for the rest of the body).

The Role of the Capillaries

1. All transport functions are carried out by the capillaries, which are composed of a single thickness of interlocking cells. Blood is directed into or away from capillary beds by smooth muscle **precapillary sphincters** in arterioles.

2. The functioning of capillaries depends on diffusion gradients, hydrostatic pressure, and active transport (commonly, pinocytosis).
   a. Hydrostatic pressure and to a lesser amount, diffusion gradients, account for losses of water, ions, and nutrients at the arteriolar end of a capillary bed.
   b. Steep osmotic and diffusion gradients bring water and ions back into the capillaries at the venule end.
   c. The exchange of oxygen and carbon dioxide follows the diffusion gradient, and water not reclaimed is recycled by the lymphatic system.

Work of the Veins

1. Blood pressure is lowest in the veins, but although force is reduced, volume in the veins must nearly equal that of the arteries. The onward movement of venous blood is assisted by one-way valves, muscular squeezing, and breathing movements.

The Blood

1. Blood (a connective tissue) consists of cells (mainly **erythrocytes,** but also **leukocytes** and **platelets**) and plasma (a watery matrix with proteins, nutrients, ions, hormones, and wastes).

2. Plasma proteins include **albumins,** which aid in transport, **globulins,** which include **antibodies** or **immunoglobins,** and **fibrinogen,** which functions in blood clotting.

3. Red blood cells are constantly replaced by production in the **red bone marrow.** Aging cells are phagocytized by **macrophages** in the spleen and liver. Hemoglobin is converted to **bilirubin,** which joins the bile. Its buildup in the blood produces **jaundice** (yellow coloring).

4. White blood cells include **neutrophils, basophils, eosinophils, lymphocytes, monocytes,** and **macrophages,** each of which plays a role in the **immune responses,** which include recognition, phagocytosis, inflammation, and antibody formation.

5. Platelets, or **thrombocytes,** are cellular fragments formed from **platelet mother cells (megakaryocytes).**

6. All blood cells originate from one cell type, the **hemocytoblast** or **stem cell.**

7. Clotting is quite complex with some 15 steps. Following damage to a vessel, platelets gather at the wound, form a collagen plug, and release vasoconstrictors and enzymes called **thromboplastins.** The latter converts **prothrombin** to active **thrombin.** Thrombin cleaves fibrinogen, forming **fibrin,** which becomes the clot.

THE LYMPHATIC SYSTEM

1. The lymphatic system maintains fluid and ion balances, transports lipids, and cooperates with the immune system.

2. Lymphatic structures include **lymph vessels,** and **lymph nodes.** The vessels begin as **lymph capillaries,** which lead to **collecting ducts** and then the **larger lymphatics.** Fluids move through the squeezing action of muscles and by breathing movements.

3. Lymph nodes are the sites where disease organisms and cancer cells are destroyed by lymphocytes and other cells of the immune system. Cancer commonly spreads via the lymphatic system.

THE IMMUNE SYSTEM

1. The immune system recognizes and eliminates **antigens,** foreign substances or organisms that enter the body.

2. Ameboid, phagocytic cells, the macrophages and the neutrophils engulf particles and invaders, massing at infection sites and forming **pus.**

3. **B-cell lymphocytes** and **T-cell lymphocytes,** the most common immune system cells, participate in the **immune response,** both **humoral** (involving antibodies or immunoglobins) and **cell-mediated** (cellular). When activated, T-cells contain free ribosomes, while the B-cells have bound ribosomes (rough endoplasmic reticulum).

4. The immune system is widespread. **Central lymphoid tissue** includes the bone marrow and thymus, while **peripheral lymphoid tissue** includes the lymph nodes, spleen, adenoids, tonsils, and **Peyer's patches.**

B- and T-Lymphocytes and Clonal Selection

1. The manner in which lymphocytes identify specific antigens is proposed by the **clonal selection theory.** Virgin cells each bear a specific antigen

recognition protein. When one finds an antigen match, the virgin B-cell undergoes mitosis, producing **plasma cells** and **memory cells,** while the T-cell produces memory cells and several kinds of T-cells.

2. In the **primary immune response,** plasma cells produce antibodies, immobilizing the antigen-bearing cells or molecules. Memory cells store antibody-synthesizing instructions, living quite long and producing clones. T-cells produce memory cells and several other kinds of T-cells. If the same antigen appears again, a **secondary immune response** occurs as the clone produces many plasma cells that rapidly release antibodies. The immune response can be brought on artificially by vaccination.

The Structure of Antibodies
1. Antibodies, or immunoglobins, all contain two identical heavy amino acid chains and two identical light chains. Each chain also has a **constant** and a **variable region.** The latter contains an **antigenic determinant,** which is specific to each antigen. Following their reaction, agglutinated antibody-antigen complexes are readily phagocytized.

2. Recombinant DNA techniques now make it possible to produce clones of specific antibody-producing B-cell lymphocytes. The B-cell line is fused with a tumor cell forming hardier **hybridomas,** and, when cloned, they produce copious amounts of specific **monoclonal antibodies** that can be harvested and used to treat disease.

The Problem of Antibody Diversity
1. The question of how the B-cells can produce antibodies against all conceivable antigens has produced a number of hypotheses, but two major ideas have emerged.
   a. The **germ line theory** maintains that there are enough genes to code for all antibody variable regions.
   b. The more recent **somatic mutation theory** proposes that only a few genes exist, and two or more of these code for the variable region of any antibody. These genes are rearranged in a great many ways as antibodies are produced, accounting for the enormous potential of B-cells.
2. Lymphocytes go through a self-recognition stage in the embryo, developing a **tolerance** for the body's own chemical makeup. At this time, lymphocytes bearing self-matching surface molecules are inactivated.

3. Newer studies suggest that certain T-cells, **suppressor lymphocytes,** interact with both T- and B-cells during the sensitive period, suppressing their ability to react.

4. The failure to recognize self is seen in **autoimmune responses** or **autoimmune diseases.** For instance, the antibodies against certain *Streptococci* may cross-react with body tissues.

The Work of T-Cell Lymphocytes
1. T-cell lymphocytes specialize in reacting against foreign cells, cancer cells, and virus-infected cells, all of which are destroyed. They also assist in antigen responses and the activation of macrophages, and they act to suppress reaction to self.

2. **Effector T-cells** include **cytotoxic T-cells** and **encapsulating T-cells,** the first of which identify virus-infected cells, killing them before viral reproduction can occur. Encapsulating T-cells seal in foreign cells or antigens.

3. **Regulatory T-cells** include **helper T-cells** and **suppressor T-cells.** Helper or **inducer T-cells** activate macrophages and assist some B-cell responses. Macrophage cells become **antigen-presenting cells** that activate helper T-cells.

4. Transplanted organs and tissues activate the immune system's T-cells unless suppressed by drugs. When cytotoxic T-cells identify graft antigens, they disrupt the graft plasma membranes, attracting macrophages that engulf the injured cells.

5. Graft rejection involves tissue incompatibility based on glycoproteins from a group called the **major histocompatibility complex,** or **MHC (human leukocyte-associated antigens,** or **HLA antigens),** coded by 20 gene loci with many alleles.

AIDS and the Defunct Immune System
1. **AIDS,** or **acquired immune deficiency syndrome,** involves a near total failure of the immune system, characterized by rare diseases such as **Kaposi's sarcoma** and **pneumocystic pneumonia.** Originally AIDS was a disease predominately of promiscuous homosexuals and a few others, but it is increasing in the general population.

2. No successful treatment for AIDS is known, but it is now believed that a retrovirus is the causative agent. Possible treatments include the use of interferon, interleukin-2, and vaccines that are now being developed.

REVIEW QUESTIONS

1. List five substances commonly carried in the blood. (780)

2. Explain how open and closed circulatory systems differ and list an example of each from the invertebrates. (780–781)

3. In what way is the circulatory system of a crustacean more complex than that of an insect? (781)

4. Trace the flow of blood through the earthworm's circulatory system. (781, 782)

5. Draw a simplified scheme of the fish circulatory system and explain the major circuits. What is a portal circuit? (782, 783)

6. Name a separate blood circuit in the amphibian and reptile not found in the fish. Is this separation perfect? Explain. (782)

7. Beginning with the sinus venosus, trace the flow of blood through the amphibian heart, naming each major structure through which it passes. (782, 784)

8. In what groups of animals has the heart progressed to four chambers? Prepare a simple diagrammatic drawing of the four-chambered heart, naming the chambers and (with arrows) illustrating the flow of blood. (784)

9. Compare the structure of arteries, veins, and capillaries. Which is directly involved in exchanges with the tissues? (785)

10. Trace the flow of blood from the vena cavae to the lungs, naming chambers, valves, and vessels along the way. (785–786)

11. Trace the flow of blood from the pulmonary veins to the aorta, naming chambers, valves, and vessels along the way. (786, 787)

12. Starting with the SA node, describe the pathway followed by a contractile impulse as it passes through the heart. (788)

13. Describe the dual control of heart rate produced by the autonomic nervous system. (788)

14. Distinguish between stroke volume and cardiac output, and compare changes in these factors in a well-conditioned person and a poorly conditioned person during vigorous exercise. (789)

15. Explain how blood pressure is maintained during diastole. (789)

16. Describe the renal circuit and hepatic portal circuit and explain the special function of each. (790)

17. Describe the vessels of the cardiac circuit. What special arrangement occurs in the smaller arteries and how is this adaptive? (791)

18. Discuss the cellular construction of the capillaries and relate this to their function. (792)

19. List the forces at work at either end of a capillary bed and explain how these affect the materials in the blood. (792–793)

20. What happens to blood pressure when blood reaches the veins? In view of pressure changes, explain how blood is moved back to the heart. (793–794)

21. List six components of blood plasma. Which is the most common? (794)

22. Discuss the red blood cell, including its size, shape, life expectancy, and what happens to it when it ages. (794)

23. List five types of white blood cells and briefly describe their functions. (795)

24. Summarize the clotting process, mentioning the role of platelets, thromboplastins, prothrombin, and fibrin. (795)

25. Describe the composition of the lymphatic system. What are its three main functions? (796)

26. Distinguish between humoral and cell-mediated immune responses. In which do B- and T-cell lymphocytes specialize? (798)

27. According to the clonal selection theory, how are virgin B-cell lymphocytes activated, and what is the role of each kind of clone produced? (799, 801)

28. List five regions of lymphoid tissue. (798–799)

29. Distinguish between a primary and secondary immune response. How are these artificially stimulated? (801)

30. Prepare a simple drawing of an immunoglobin and label the heavy chain, constant region, light chain, variable region, disulfide linkage, hinge, and antigenic determinant. (801–802)

31. Explain the somatic variation theory of antibody diversity. (804)

32. Explain how the immune system learns to avoid reacting with "self." What experimental evidence suggests that this happens in the embryo? (804–805)

33. List four specialized kinds of T-cell lymphocytes. Which attacks virus-infected cells and what effect does it have on them? (806)

34. Describe the effect of unsuppressed T-cells on an organ transplant. (807)

35. What is the genetic basis for tissue rejection? What are the chances of one's finding a suitable donor from unrelated persons? (807)

36. Briefly summarize the problem of AIDS. What is it, who most often gets it, and what is a likely causative agent? (807–808)

Homeostasis: Thermoregulation, Osmoregulation, and Excretion

About 200 years ago, Charles Blagden, then the secretary of the Royal Society of London, proved that he was one of the most persuasive people on earth. He managed to talk two friends into joining him in a small room in which the temperature had been raised to 126°C (260°F). Being aware that water boils at 100°C, they naturally were a bit reluctant. But Blagden prevailed, and the men, taking along a small dog and a steak, entered the room. They emerged 45 minutes later and were surprised to find that they were not only alive, but in good shape. The dog was fine, too. But the steak was cooked!

Thus, Blagden demonstrated the remarkable abilities of living organisms to adjust to severe environmental conditions by regulating their internal processes. These regulatory abilities should be expected, though, as the delicate processes of life demand extremely constant and specific conditions.

Let's consider another example. A young woman had thought she was in good shape living in the city, but now she is finding it pretty tough going while backpacking high in the mountains. Like the dog in the hot room, she too is panting, but for a different reason. The level parts of the terrain are easy enough, but when climbing, she has to stop every 100 m or so just to catch her breath. Her heart races, she feels light-headed and sick to her stomach, and has a sharp pain in her right side. After each rest, though, her heavy breathing subsides, she feels more comfortable, and is able to resume climbing. Her companion, who has been in the mountains all summer and is having no problem, tells her that she'll get used to the altitude in a week or two. Her companion is right. In fact, by the next morning she is already feeling a bit better, and in a couple of weeks she will hardly notice the altitude at all.

HOMEOSTASIS AND FEEDBACK MECHANISMS

Both Blagden and the climber have demonstrated what are called homeostatic mechanisms. The *Random House Dictionary* defines **homeostasis** as "the tendency of a system, especially the physiological systems of higher animals, to maintain internal stability, owing to the coordinated response of its parts to any situation or stimulus tending to disturb its normal condition or function." Homeostasis, then, is one of the pervasive themes in biology.

Homeostasis can operate at a number of levels. In Blagden's experiment, the dog probably sought to leave the room—a behavorial response. Then he probably gave up and lay there panting, producing physiological changes. Regarding the novice mountain climber, she paused by the trailside to "catch her breath" in response to falling levels of oxygen in her blood. Neither the dog nor the climber has much control over the panting; it's an almost completely automatic adjustment to the stimuli. The sharp pain in the climber's side is caused by the spontaneous muscular contraction of her spleen as it becomes emptied of its reserve blood supply, while the nausea is due to the shunting of blood from her gut to her oxygen-starved brain and muscle tissue. On a different time scale, continued warm days would have caused the dog to shed some of his heavy coat, and continued exposure to high altitudes would have stimulated the climber to produce more red blood cells. These are familiar responses, but what may not be familiar is the way such changes are coordinated. How does the body

know when it is cool enough, or that there are sufficient numbers of red cells? As you might expect, such compensating activities are rather automated; they are usually controlled by what is called a **negative feedback loop.**

Negative and Positive Feedback Loops

The negative feedback loop, also described in Chapter 29, simply refers to a situation in which a stimulus produces a reaction that ultimately reduces the stimulus. Thus the negative feedback loops are inherently stabilizing, maintaining the organism in a steady state. Let's look at a somewhat homey analogy. A baby cries when it is hungry; the parent feeds it; the baby is no longer hungry and stops crying. From the baby's point of view, the stimulus was the hunger, and its response was crying, which ultimately made the hunger go away. From the parent's point of view, the stimulus was the crying and the response was feeding the baby, which made the crying go away. A classic nonbiological example of negative feedback control is the thermostat on a

32.1 HOUSEHOLD HOMEOSTASIS

The typical home temperature-regulating thermostat provides an example of a negative feedback system. In this instance, two thermostats regulate the temperature of a home through alternate heating and cooling, keeping the temperature within a 15°F range. The thermostats sense the air temperature in the room and activate the furnace or air conditioner when temperatures fall below or above their settings. These same

thermostats, like so many self-regulating systems in organisms, are subject to the effects they produce. In other words, they shut down when air temperatures fall within the limits set. And, like living mechanisms, they have their limits, or thresholds, at which they begin to fail. In the system shown here, the control mechanisms fail when the outside temperatures fall below 25°F or above 110°F.

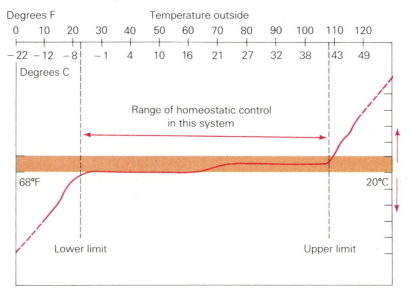

house's heating or cooling system, which is explained in Figure 32.1.

Sometimes, the loop fails in one way or another. The baby may cry not because it is hungry, but because it is in pain from diaper rash. If the parent tries to feed it, it cries harder, so the parent tries harder to feed it, so it cries harder, and so on. This constitutes a **positive feedback loop,** in which the stimulus evokes a response that further increases the stimulus instead of decreasing it. So unlike negative feedback systems, positive feedback loops are inherently unstable.

Positive feedback loops are uncommon in nature and are usually associated with illness. High blood pressure, for instance, can damage arteries and arterioles, and the damaged vessel walls can become infused with lipid materials, scar tissue, and cellular growth (see Figure 3.14). This in turn restricts the size of the vessel opening, further increasing blood pressure and arterial disease. And we all know of people who are depressed because they are overweight, and then eat because they are depressed.

In this chapter, we'll consider two important and highly complex examples that involve multiple, interacting homeostatic controls. One is **thermoregulation,** the animal's control over its internal temperature. The other is **osmoregulation,** which is closely related to **excretion** in most animals.

Osmoregulation and excretion involve processes through which animals control their osmotic conditions, internal salt balances, and pH, and rid themselves of toxic metabolic wastes. Actually, we have discussed other homeostatic mechanisms elsewhere; for example in the hormonal control of digestive enzyme release and in the body's ability to regulate heart rate and blood pressure. It is difficult to confine such a pervasive idea as homeostasis to any one discussion in biology.

THERMOREGULATION

Thermoregulation is the ability to maintain body temperatures within a certain range. The thermoregulatory capability of animals ranges from essentially no control to virtually complete control. In Chapter 27, we used the terms **ectothermic** (or **poikilothermic**) and **endothermic** (or **homeothermic**) in reference to regulatory abilities in vertebrates. Traditionally, the latter refers to temperature regulation in birds and mammals—the best thermoreg-

ulators—and the former to the rest of the animal kingdom—those that are "cold-blooded" and at the mercy of their surrounding temperatures. We would now like to modify that point of view.

Recent studies reveal that there is no neat dividing line between ectothermic and endothermic groups. At least there is no clear dividing line between taxonomic groups. We now know that thermoregulation is common in certain bony fishes, sharks, and some reptiles, and that some birds and mammals exhibit short-term ectothermy under specific environmental conditions. As more of the earth's creatures are placed under close scrutiny, the list of line-crossers will probably grow. The most recently discovered ones are, of all things, certain beetles. Their efficient metabolic thermoregulation may help explain the extraordinary evolutionary success of this group with its 300,000 or so species.

Why Thermoregulate?

Animals are faced with two alternatives when environmental temperatures fall below optimum. First, they can go into a metabolic stupor, wherein their biochemical processes slow down and they become sluggish or immobile. Second, they can take measures to conserve metabolic heat and retain it in those parts of the body that are more immediately critical. But since the primary source of heat is the process of cell respiration, the animal faces a paradoxical situation. As the body cools, greater metabolic heat is required, yet the loss of body heat slows the metabolic or chemical activity required to produce heat. The result, under frigid conditions, is an accelerated (positive feedback) cooling of the animal unless it can take measures to counteract the process. And that, in a nutshell, is what thermoregulation is all about.

Thermoregulation, of course, involves problems other than heat loss. In fact, heat gain can cause even greater difficulties. Physiological damage created by excessive heat is often irreversible, whereas damage by low temperatures, except in extreme cases, is more likely to be only temporary. (Enzymes are destroyed by heat but only inactivated by cold.) We should also keep in mind that whether the problem is heat loss or heat gain, combined strategies, involving behavior as well as physiology, are very common. In some cases in which physiological mechanisms are not well developed, behavioral strategies become essential. Let's look at some specific examples.

32.2 COUNTERCURRENT HEAT EXCHANGE

Heat exchangers are practical ways of hastening a cooling process. Two pipelines, arranged in a closely parallel fashion and carrying hot and cold liquid, respectively, are efficient countercurrent exchangers because the moving flow of hot liquid is continually opposed by a moving flow of cold. Thus, heat is lost to the cold water along the entire length of the pipelines—as long as the pipes are parallel and as long as there is a difference in temperature between the two.

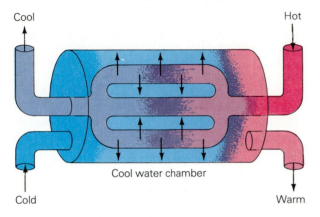

Cool   Hot

Cool water chamber

Cold   Warm

Countercurrent Heat Exchangers, or How the Tuna Keeps Moving in Cold Water

Many animals with closed circulatory systems have evolved a simple mechanism that prevents excessive heating of blood deep in the body and cooling of blood in the extremities. The mechanism, earlier described as a **countercurrent exchange,** borrows its name from engineering. We find examples at work in cooling systems that employ **countercurrent heat exchangers** (Figure 32.2). In living organisms, countercurrent exchanges are usually involved in retaining body heat. They assist in preventing heat loss throughout the body and retaining heat in specific and essential core areas.

As a specific case, let's see how a countercurrent exchange helps a fast-moving predator of the sea—one that was formerly considered to be an ectotherm. The bluefin tuna is among the fast-swimming fishes, in spite of its preference for *very* cold water. Let's see how an animal, long believed to be an ectotherm, can manage to stay active under such extreme conditions.

A vitally important adaptation in the bluefin tuna is its specialized circulatory arrangement. Compare in Figure 32.3 the circulatory systems of colder- and warmer-bodied species of fish found in similar habitats. Notice that in the colder-bodied fish, the major vessels are centrally located, with

branches moving out in all directions to serve the propulsion muscles and outer portions of the body. Such an arrangement can do little to retain body heat in a central place. On the other hand, the major arteries and veins of the bluefin tuna tend to run parallel to each other with blood flowing in opposite directions, thus acting as countercurrent heat exchangers.

Furthermore, the tuna and other warm-bodied fishes have another heat-conserving and directing arrangement of vessels. Figure 32.4 shows two types of swimming muscles, dark lateral muscle and surrounding light muscle. (You won't find the dark muscle in higher-priced cans of tuna; it's oily and has a strong flavor.) The darker muscle is a region of higher temperature—as much as 10°C higher than the skin temperature. One reason it is darker is that the muscles are surrounded by immense networks of blood vessels. This highly vascular region, consisting of parallel arteries and

32.3 COLD-BODIED AND WARM-BODIED FISHES

(a) In fishes that do not thermoregulate, the major vessels are centrally located along the body axis. The branching vessels supply muscle beds in the body and tail in such a way that heat is lost. (b) In the warm-bodied fishes, the major vessels branch out into the muscles, where countercurrent networks are arranged. As a result, these fishes are able to thermoregulate, keeping body heat concentrated in the swimming muscles.

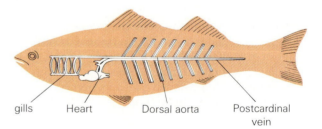

gills   Heart   Dorsal aorta   Postcardinal
                                                  vein

(a) Slow-moving fish (cold-bodied)

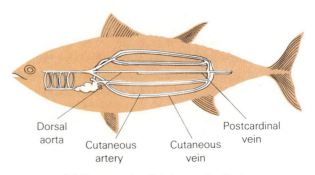

Dorsal   Postcardinal
aorta   vein
        Cutaneous   Cutaneous
        artery   vein

(b) Fast-moving fish (warm-bodied)

32.4 THE COUNTERCURRENT NET OF THE TUNA

Within the darker muscle regions is found an extensive countercurrent exchanger known as the *rete mirabile.* Veins and arteries arranged side-by-side set up a heat exchange that makes this one of the warmest regions in the tuna's body. Measuring temperatures all around the body at various distances below the skin demonstrates the effectiveness of this form of thermoregulation. The results are plotted here in a *thermocline,* similar to what we see in weather maps. The tuna was immersed in water cooled to 19°C before measurements were begun. As you can see, the body temperature deep in the darker muscle region, where countercurrent exchange is greatest, is over 12°C warmer than the water temperature.

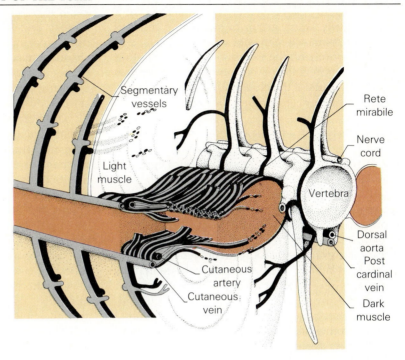

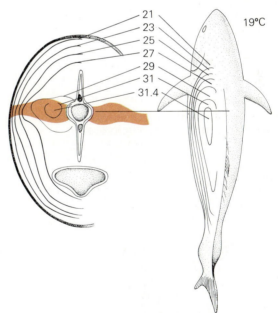

veins, is known as the **rete mirabile** ("wonderful net," and it truly is). The net, in fact, is a very dense countercurrent heat conserver. The tuna thus has a comparatively warm group of swimming muscles and can remain active in cold water.

At the moment, the extent to which most other fishes thermoregulate isn't known. But what has been learned about the bluefin tuna should warn us about taking our categorizing too seriously.

Regulating Behaviorally

Many reptiles also have interesting strategies for maintaining body heat, most of which are behavioral (although some lizards make use of their skin pigments as well). It turns out that the degree to which reptiles can regulate their body temperature depends on where they live. Reptiles from the tropics, for example, can't regulate their body temperature nearly as well as those from temperate regions, where weather changes more drastically. In fact, historically, the success of reptiles in invading temperate zones from the tropics has depended heavily on their ability to thermoregulate behaviorally. The strategies of temperate zone lizards are apparently very successful since these reptiles maintain remarkably constant body temperatures as they move about.

Both heating and cooling can be achieved through specific orientation to the sun's rays. Reptiles, for example, absorb a great deal of heat by basking in the sun, and the temperate desert lizards have developed the technique into an art. They avoid excessive heat gain by avoiding the heat of midday and by simply facing the sun's direction, thus exposing as little surface as possible. Heat is absorbed maximally by presenting more of the body to the sun. So, by constantly adjusting its amount of exposure to the sun, the lizard keeps a rather constant body temperature (Figure 32.5).

The horned lizard (or "horny toad," as some insist) is also a behavioral thermoregulator, but it supports these efforts through a physiological mechanism. It gains or loses the sun's heat by altering its color. At low temperatures, the pigmented cells in its skin expand and darken so that less solar energy is reflected and more is absorbed. This ability improves the efficiency of basking considerably (Figure 32.6). When body heat is high, the pigmented cells contract, producing a much lighter, heat reflecting color.

Questions about the extent of thermoregulation in reptiles also extend to species that are no longer around. Since endothermy is not taxonomically restricted there is an ongoing controversy over whether dinosaurs were endotherms or ectotherms, and the current scarcity of dinosaurs only adds to the problem. It has been suggested, in any case, that the strange, highly vascularized backbone plates of the stegosaurs were excellent heat radiators and absorbers (Figure 32.7). Other dinosaurs (for example, *Dimetrodon*) had high dorsal fin struc-

32.5 BEHAVIORAL THERMOREGULATION

Lizards and other desert reptiles must develop strategies of thermoregulation to cope with drastic temperature changes. A lizard buries itself in the sand at night to conserve body heat. To become active in the morning, it must crawl out and absorb the sun's rays. Early in the day, the lizard exposes its head to the sun, slowly warming up, without exposing its sluggish body to predators. During midday, when the lizard must escape from the heat, it finds shadows to linger in as it moves from one place to another. Later in the day, as cooling begins, the lizard basks in a position parallel to the sun's rays, absorbing the last remnants of the day's energy supply before digging in for the night. These are all behavioral mechanisms that assist the body in making physiological adjustments to changing temperatures.

32.6 THERMOREGULATION IN THE HORNED LIZARD

The horned lizard (like many reptiles and amphibians and some fishes) is able to change its color through pigment migration—an automatic response. **(a)** When exposed to cooler temperatures in experimental situations, the lizard's color darkens. **(b)** At higher temperatures, it lightens considerably.

(a)

(b)

tures that probably served both to capture solar heat and to radiate unwanted heat, depending on the animal's orientation to the sun and surrounding temperatures.

There are many other examples of heat regulation in the so called "cold-blooded" animals—those that were once believed to be nonregulators. The point here, in addition to introducing the subject and pouncing on some popular old ideas, is to illustrate the continuous diversity of form and function in living things.

Thermoregulation in the Endotherms

Mammals and birds are endotherms that have, indeed, made important inroads into maintaining a constant body temperature. Actually, we should be referring to *internal* body temperature, since the temperature of the extremities and skin may be quite different from the temperature of the deeper tissues. Your skin temperature, for example, may vary widely and is usually cooler than the temperature of your internal organs.

Endotherms are identified by their ability to regulate their internal temperature by varying the heat produced through oxidative metabolism, but they also control of heat loss and gain through radiation, conduction, and convection. These two general mechanisms work in close harmony to maintain optimal internal temperatures. Let's now look at the outward manifestations of temperature control and then consider the more complex regulatory process-

32.7 THERMOREGULATION IN *STEGOSAURUS*

Stegosaurus had a prominent row of large plates covering its dorsal surface. They are now thought to have been thermoregulating structures, containing capillary beds that helped in radiating body heat out of the animal and, alternatively, absorbing heat as part of a basking behavior. Exposing their large surface area to the early morning sun would have hastened the warm-up period many reptiles require to get themselves going each day.

es themselves. In particular, we'll focus on the processes common to humans.

Removal of Heat Energy.   The most important avenues of heat escape in humans are the skin and respiratory passages, areas from which heat is lost through radiation, conduction, and convection. Heat, we know, radiates into cooler surroundings. Radiation, therefore, is important in dissipating heat from the peripheral circulation, through the skin. In addition, heat escapes through conduction (in which heat passes by contact to a cooler solid, such as a tile wall or cool grass) and convection (in which heat energy is carried away by the cooler air molecules). Generally, the circulatory system itself can help heat escape. For example, when internal temperatures rise, muscle sphincters in arterioles near the skin relax and the vessels dilate, permitting blood to flow into skin capillaries, increasing heat loss and producing the familiar flushed appearance (Figure 32.8). Cooling the body is also accelerated by the return of blood from the extremities. Blood returning through the arms and legs is shunted to surface veins, where it is cooled.

The rate of heat escape by any means is greatly enhanced by the evaporation of moisture on the skin and in the respiratory passages. Water, as you may recall, can absorb substantial amounts of energy, and its evaporation from the skin is an effective cooling process. When air temperature exceeds body temperature, the skin can be cooled by sweating (see Figure 32.8). However, when high air temperature is combined with very high humidity, even this mechanism fails. At this time, humans may resort to behavioral strategies, slowing their activity, bathing their bodies in cool water, or perhaps turning on an air conditioner. The cessation of activity is important, and humidity can be a far

more important factor than you think. Some time back, one of your authors (Wallace) suffered heat stroke in a marathon. He reports:

The temperature was 75°F and the humidity was 80%. All seemed well until the 24th mile. My time was about 6:25 minutes per mile when, suddenly, I felt strange—remote, dizzy and spacy. I slowed down, thinking I would quickly recoup. But I didn't recoup. My goal was anything under 2:50 and with 2 miles to go my time was 2:36. I tried to mentally subtract 36 from 50 to see how fast I would have to run the final 2 miles, but any mental calculations were impossible. I began to realize I was in trouble and when I felt my arm I found it was dry and chilled. A wave of dismay swept over me because I knew continued stress could cause kidney or brain damage, but I wanted to finish, so I slowed down even more. Then I began almost to fall asleep again and again while running. I managed to stay awake and trotted in at 2:55. Recovery, with water and mineral laden drinks, was rapid. Strangely enough, a blanket was needed to fend off what felt like a chill wind. Physiologists later told me that I had possibly altered my thermoregulatory mechanisms so that I had lost too much heat and that the sleepiness could have been the same sort described by people who are freezing.

32.8 THERMOREGULATION BY THE SKIN

The human skin is an excellent thermoregulatory organ. Sweat glands secrete water through ducts that empty into skin pores. When evaporating from the surface, water has a cooling effect. Capillary beds below the surface can bring blood near the skin for cooling or, alternatively, shut down, directing blood to deeper regions in order to conserve heat. Temperature-sensing neural structures detect heat *(Krause's end bulbs)* and cold *(Ruffinian corpuscles),* passing this information along to the brain for the proper response. Insulation is provided by layers of fat below the skin.

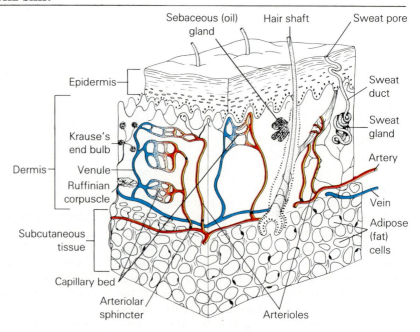

32.9 **HUMAN HEAT EXCHANGE**

Countercurrent heat exchangers are common in human extremities. In the arm, for instance, the larger arteries and veins travel close together deep in the muscle. When the arm is cold, blood returns through the deeper veins, where the countercurrent exchange of heat between the arteries and veins conserves internal body heat. An alternate route for returning blood is through the veins that lie near the surface, just below the skin. When the arm is warm, returning blood follows the surface route, increasing the radiation of heat from the skin.

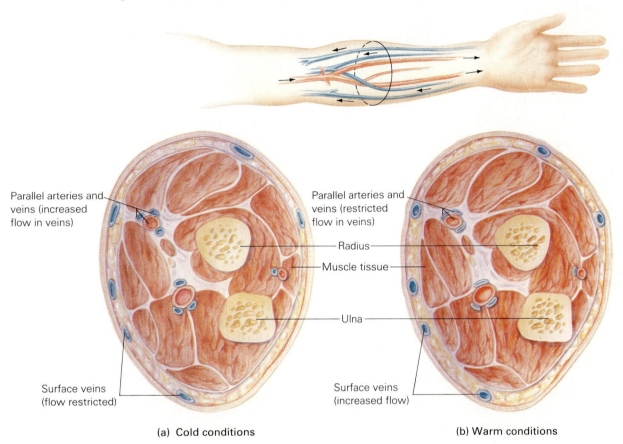

Parallel arteries and veins (increased flow in veins)

Parallel arteries and veins (restricted flow in veins)

Radius

Muscle tissue

Ulna

Surface veins (flow restricted)

Surface veins (increased flow)

(a) Cold conditions

(b) Warm conditions

The point is, the thermoregulatory mechanisms are in delicate balance, and if shifted off balance, the results may not be merely drastic, but unexpected.

Conserving Heat. What happens when *retaining* body heat is the problem? At very low external temperatures, constriction of the muscular sphincters of arterioles near the skin can minimize skin circulation and redirect the flow into deeper regions (see Figure 32.8). Blood returning from the arms and legs is shunted into deeper veins that run parallel to the arteries, thus setting up a countercurrent heat exchange (Figure 32.9).

Shunting blood deeper into the body to conserve heat is a costly process. Cooler muscles do not work as efficiently, and the extremities can become numb. But more significantly, people subjected to extreme cold, such as mountain climbers suffering from exposure, will experience freezing in the fingers, toes, and ears before their internal temperatures begin to fall. These parts may be sacrificed as the vital organs will receive first priority in the ensuing battle for survival.

Actually, birds and mammals conserve heat by a number of means. Perhaps you've seen chickadees perched at your bird feeder on a cold winter day, with their feathers all fluffed up and looking like tennis balls with beaks. The hair of mammals can also stand on end, and the process, in both cases, is called **piloerection.** Insulating air is trapped under the hair or feathers and buffers the animal from the

32.10 PILOERECTION

Piloerection may aid as an insulator in birds and mammals. The feathers or hair are raised above the skin by contraction of erectile muscles, producing air spaces that serve as insulation, which helps the animal resist both warming and cooling.

+30°C

+10°C

−20°C

environment (Figure 32.10). In humans, the action produces the familiar goosebumps.

Piloerection may not mean very much to humans in terms of heat conservation, but shivering does. Actually, since shivering involves rapid contraction of muscles, the work itself produces heat. Interestingly, neither shivering nor piloerection are reactions solely to temperature. Did you ever see anything that made your hair stand on end, or sent a shiver through you? (Try looking a politician right in the eye.)

The Internal Source of Thermoregulation

We see, then, that an organism can change in many ways to meet the challenge of variable environmental temperatures. The question then is: How are these responses regulated and integrated? In particular, how are they handled in humans? Humans, after all, maintain an average internal temperature of about 37°C (98.6°F) within an amazingly narrow range. Surely some form of sophisticated internal control is required since we have so many independent ways to regulate heat.

First, we need to know how the body measures

temperature. Essentially, it is done two ways. One is external, through the sensory neurons of the skin. We react voluntarily to this information by doing whatever is necessary to bring the skin temperature back to its desired level. We may move to a warmer place, take off some of our clothes, stomp our feet, or fan ourselves.

The second method of measuring heat is internal. The brain constantly monitors the temperature of the blood flowing through it. Furthermore, the responses initiated by the brain override all other controls; hence, the brain can be thought of as the real thermostat of the body.

The heat-monitoring part of the brain is the **hypothalamus.** It receives temperature information by intercepting impulses from thermoreceptors in the skin on their way to conscious centers of the brain, and, more directly, by sensing the temperature of blood passing through from the body core.

Actually, the hypothalamus behaves very much like an ordinary home thermostat, but with far more precision. Physiologists have been able to detect neural activity in the hypothalamus when blood temperature varies by as little as 0.01°C.

The hypothalamus, in its constant monitoring of blood temperature, can elicit a variety of responses (Figure 32.11). For example, through its influence on the autonomic (sympathetic and parasympathetic) nervous system, it can increase the rate of heat production by stimulating the **adrenal medulla** (a hormone secreting body), causing it to release the hormone **epinephrine** into the bloodstream. Epinephrine speeds up the conversion of glycogen to glucose in the liver. The release of glucose into general circulation and its subsequently increased availability to cells enables cells to increase their respiratory activity, which creates heat. In addition, the hypothalamus can prompt the autonomic nervous system to reduce blood flow in the skin.

The hypothalamus also stimulates the **pituitary gland** to release **thyroid-stimulating hormone,** which in turn causes the **thyroid gland** to increase the output of its hormone, **thyroxin** (see Chapter 33). The effect of thyroxin is to increase cellular respiratory activity and thereby produce body heat. Increases in body heat are subsequently detected by the hypothalamus, which then eases off on the pituitary, slowing heat output—another example of negative feedback control.

Thus, information arriving from peripheral and internal sensors results in a considerable variety of hypothalamic actions. As a homeostatic control structure, the hypothalamus keeps the organism in a finely tuned and responsive state, maintaining body temperature at an optimal level. One might wonder, though, just how does the hypothalamus measure blood temperature; how does the thermostat work?

No one is quite sure how the hypothalamus works in thermoregulation. In fact, after all is said and done, perhaps the hypothalamus doesn't actually measure heat. Maybe it measures heat-related chemical changes in the blood instead. A leading hypothesis suggests that it monitors calcium ions.

It is known that the thermostat can be artificially altered by suffusing the brain with calcium solutions. Whereas it is not known whether calcium directly affects the thermostat, the effect of the presence of calcium cannot be argued. In fact, calcium suffusion works so well that with it, body temperature can be chemically lowered enough to permit **cryogenic** (deep-cooled) **surgery,** duplicating the effect of submerging the body in ice water. The use of calcium removes the need to lower body temperature by removing body heat; the thermostat is merely set to a lower level.

OSMOREGULATION AND EXCRETION

Now let's look at another problem of regulation in animals—the problem of how animals regulate the water and ion concentrations in their cells. Most cells are about two-thirds water, but in some cases, "about" won't suffice. In certain cells, the amount of water is critical, as is the relative abundance of various ions in the cell fluids. The problem of maintaining the proper water and ion balance is called **osmoregulation.** And for many animals, this necessarily leads us into the methods by which excesses of water or ions are removed from the body by the processes of **excretion.** Excretion refers to the removal of metabolic wastes—the byproducts of cellular reactions—from the body. First let's see how these wastes are produced.

Producing Nitrogen Wastes

Amino acids, you recall, are the constituents of proteins. Typically, proteins must be broken down into individual amino acids during digestion so that

32.11 THERMOREGULATION BY THE HYPOTHALAMUS

The hypothalamus receives thermal information both from thermoreceptors in the skin and by direct sensing of the temperature of blood arriving from the core of the body. Its response to changing temperatures is carried out in two ways: through activation of the autonomic nervous system and of the endocrine system. Negative feedback occurs through the continued monitoring of blood temperature.

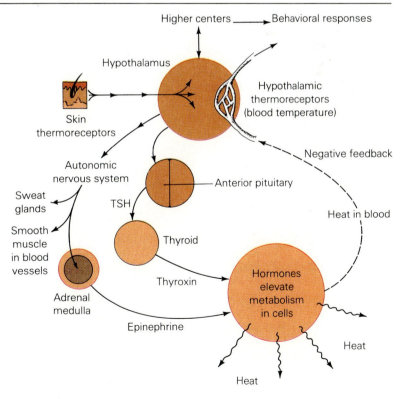

they can be absorbed across the gut wall and into the bloodstream. Those same amino acids can be used to build new proteins in the cells; they may also be converted to fatty acids or carbohydrates for storage, or used as fuel in respiration. Some of these changes produce leftover fragments of nitrogen, and accumulations of these nitrogen fragments can be extremely poisonous.

During **deamination** (Figure 32.12), the amine group,—NH_2, is removed from the amino acid as NH_3, or ammonia. In the presence of protons from water, it readily forms ammonium ions (NH_4^+). (Note in the figure the role of the cellular coenzyme NAD here—another example of the multiple use of molecules.) Ammonia is highly toxic, but many organisms can safely handle it if they live in water and are small enough to exchange materials with their environment easily. However, in many complex animals and especially those that live on land, the ammonia can't be removed fast enough to keep from poisoning the organism. In these cases, ammonia is changed to something more manageable.

Alteration of ammonia generally yields one of two compounds. Insects, reptiles, and birds produce a solid, insoluble waste known as **uric acid**

or related compounds such as the nitrogen base guanine. In many invertebrates, some fishes, amphibians and all mammals, the primary nitrogen waste is the highly soluble compound **urea**. Frogs are interesting in that their aquatic young, the tadpoles, excrete their nitrogen as ammonia. Only when they metamorphose into the partially terrestrial, water-conserving adults do they switch to urea production. A certain salamander is even more versatile; it produces urea while on land, but when it reenters the water to breed, it quickly switches to the less costly route of excreting ammonia. So, the precise way that nitrogen wastes are handled is also dictated by the need to conserve water, as we will see shortly.

The Osmotic Environment

Each kind of environment presents its own osmoregulatory problems for animals. Over the eons, as the species attempted to explore new environments, they faced a multitude of water-regulation problems. The results today are varied. Let's see how animals that live in salt water, fresh water, and on land have solved their water regulation problems.

32.12 DEAMINATION

In the metabolic breakdown of alanine, pyruvic acid is formed, along with the waste product, ammonia. Pyruvic acid is then free to enter one of several pathways (such as the citric acid cycle). Ammonia may be excreted directly, or it may be converted to uric acid or urea.

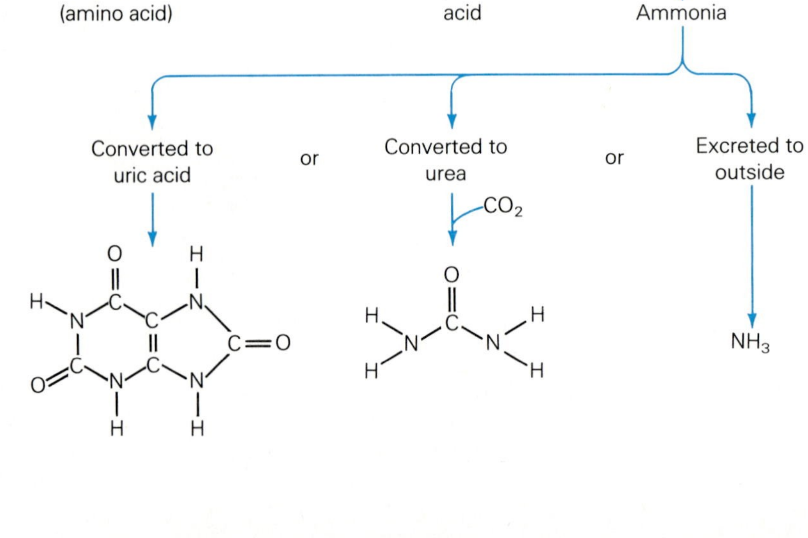

32.13 RESPONSES TO SALINITY

Organisms living in the marine environment solve the problem of changing salinity of the seawater in two ways: they are either passive osmoconformers or active osmoregulators.

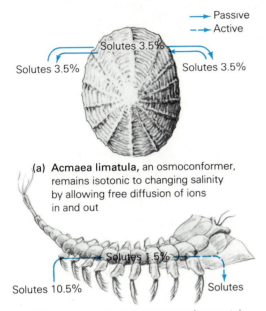

→ Passive
⇢ Active

Solutes 3.5%
Solutes 3.5%
Solutes 3.5%

(a) **Acmaea limatula**, an osmoconformer, remains isotonic to changing salinity by allowing free diffusion of ions in and out

Solutes 1.5%
Solutes 10.5%
Solutes

(b) *Artemia salina*, an osmoregulator, retains a stable solute balance by actively transporting Na$^+$ across gill membranes

The Marine Environment. Animals of the sea live immersed in a complex solution of ions, including sodium, potassium, calcium, magnesium, chloride, and others. We usually think about the 3.5 percent sodium chloride content of the solution, but we can't neglect the others. In their totality, the presence of such solutes creates osmoregulatory problems for the animal inhabitants. To survive in the marine environment, organisms must adapt to two conflicting osmoregulatory conditions. First, they are in a *hypertonic* medium—one in which the solute concentration of the surroundings is greater than the solute concentration within their cells and tissues and, conversely, one in which the water concentration is greater inside the cells and tissues than in the surroundings. As a result, there are two opposing concentration gradients. The tendency, therefore, is for marine animals to continually gain ions and to lose water as the two move down their respective gradients. If unchecked, either can be disastrous.

To survive this double threat, marine animals have evolved a variety of adaptive strategies, most of which place the animal in one of two broad categories: they are either **osmoconformers** or **osmoregulators.**

Osmoconformers. As the term *osmoconformer* implies, the solute concentration within the body simply conforms, changing passively, as it were, to that of its surroundings. Thus it simply matches the solute concentration of, or is *isotonic* to, the seawater. Many marine animals are osmoconformers, as this method requires little energy expenditure.

As an example, the limpet *Acmaea limatula* is a resident of the intertidal zone (Figure 32.13). As such, it commonly experiences changes in salinity as the tides change or as runoff from the land dilutes its surroundings. Rather than fighting to maintain any particular internal salinity, the limpet simply conforms to these salinity changes. Remarkably, its body fluids can remain isotonic in a salinity range of 1.5 to 5%. For some reason, these changes have little effect on the animal. A similar change in our body solutes would produce instant death.

While most vertebrates are osmoregulators, sharks and rays are conformers, or at least partly so. Their strategy for coexisting in the hypertonic marine environment is to retain sufficient urea in the blood and tissue fluids to create an isotonic condition with the seawater outside. The nitrogen waste, then, is put to use. If this seems peculiar, keep in mind that as far as osmosis is concerned, it doesn't make much difference what the solutes are as long as the concentrations of *water* inside and outside the organism are equal. When they are unequal, there will be a net movement of water down its gradient, in or out, until an equilibrium is reached. Incidentally, sharks and rays help solve the problem of excessive salt intake by actively transporting it back out. The site of transport is a special gland opening into the rectum (Figure 32.14).

Osmoregulators. Osmoregulators maintain a relatively constant internal solute concentration despite changes in salinity around them. Osmoregulation, though, involves work—active transport. Many of the marine arthropods are osmoregulators. One of the best-known regulators is the crustacean *Artemia salina* (see Figure 32.13), which lives in salt ponds. It can maintain constant solute concentrations in its body when environmental salinity varies from less than 1% up to 30%! Only a highly specialized and efficient osmoregulator could survive under such varying conditions. The osmoregulatory organ of *Artemia salina* is the gill, which has salt glands that actively transport salt out of the blood. Interestingly, the same structures are used by marine fishes.

Most marine vertebrates are osmoregulators.

32.14 OSMOREGULATION IN MARINE FISHES AND SHARKS

(a) The bony marine fish survives its salty environment by actively transporting salt and ammonia directly across the gills. It obtains its needed body water by drinking seawater. **(b)** The shark, however, uses a different strategy. It increases its solute concentration by maintaining a high urea concentration in the blood. In addition, it actively transports salt through special glands in the rectum. Does this make the shark an osmoconformer? An osmoregulator? Both?

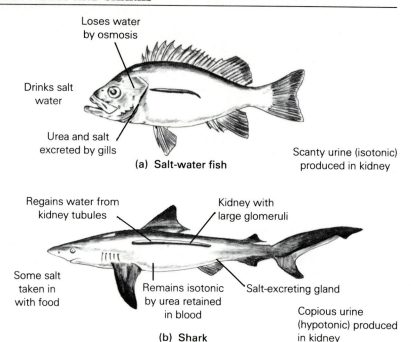

Loses water by osmosis

Drinks salt water

Urea and salt excreted by gills

Scanty urine (isotonic) produced in kidney

(a) Salt-water fish

Regains water from kidney tubules

Kidney with large glomeruli

Some salt taken in with food

Remains isotonic by urea retained in blood

Salt-excreting gland

Copious urine (hypotonic) produced in kidney

(b) Shark

The bony fishes have the same problem as the invertebrates: they lose water and gain salt. They make up for water losses by drinking seawater, but only enough to replace what water they lose. A terrestrial mammal (such as a human) that drank seawater would subsequently require huge quantities of fresh water to remove the salt from its body. But because of the special salt-secreting cells in its gills, the fish has no such problem. With salt actively transported from the body, seawater becomes perfectly suitable for drinking. The other ions are either excreted by the kidneys or passed through the gut unabsorbed.

Marine birds and reptiles also take in seawater with their food. The solute concentrations of their bodies, however, are about the same as those of terrestrial species. So what do they do with all those salts? Unlike their terrestrial cousins, they do not rely on the kidney; instead, they actively excrete salt through special glands located near the eye that drain through a duct into the nose. Their kidneys, in fact, are not very efficient salt removers. Studies of gulls reveal that the salt concentration is about 3.5% in the seawater and only about 0.3% in their urine. The salt glands, however, secrete a 5% sodium chloride solution.

Unlike marine fishes and birds, marine mammals have highly efficient kidneys and can actively transport ions into their urinary collecting ducts. For them, then, the kidney is the salt-excreting gland. However, most marine mammals avoid drinking seawater and rely on relatively low-solute fluid from the body fluids of the fish they eat.

The Freshwater Environment. Just as salt water presents osmoregulatory problems, fresh water, too, has its share of hazards. Fresh water is *hypotonic* to its inhabitants, which tend to take in the fresh water through osmosis. If the organism isn't able to keep excess water out, its cells may swell and rupture. But there is another problem. Fresh water doesn't provide its denizens with the ions they need. In particular, sodium, potassium, and magnesium are in short supply. Thus, these particles tend to obey the laws of diffusion and leak out of the organisms into the environment. The organisms must expend energy to pump these errant molecules back in by active transport. Such transport may occur across the membranes of the skin, gills, or kidney tubules.

A freshwater existence also has certain advantages. For example, the animals there generally don't have much of a problem with handling nitrogenous wastes. In fact, many of them simply produce ammonia and flush it out with water.

One of the simplest organized osmoregulatory

systems is the **flame cell system,** or **protonephridium,** of the planarians, freshwater flatworms. This system is chiefly an osmoregulatory device and possibly an excretory system as well (Figure 32.15). (Interestingly, protonephridia are absent in marine flatworms, which do not have the problem of excess water.) The flame cell system consists of numerous minute blind sacs leading into a complex network of tubules that empty to the outside via excretory pores. Water from intercellular fluids is transported into the flame cells by pinocytosis ("cell drinking"). Each flame cell contains numerous cilia facing the tubular lumen of its **flame bulb,** and the cilia set up currents that carry the fluids through the tubules to the pores. The cilia gave rise to the name, "flame cell," since their waving action resembles that of a flickering flame.

The kidneys of freshwater fishes, as you might expect, are not adapted for conserving water, and they produce copious amounts of highly dilute

32.15 THE FLAME CELL SYSTEM

Planarians have a simple but extensive osmoregulatory system composed of protonephridia. Numerous ciliated blind sacs—the flame cells—lead into tubules throughout the animal's dense body. Each flame cell transports water from nearby cells into its lumen. The waving cilia of individual flame cells create a current that carries the water out of the body through an excretory pore.

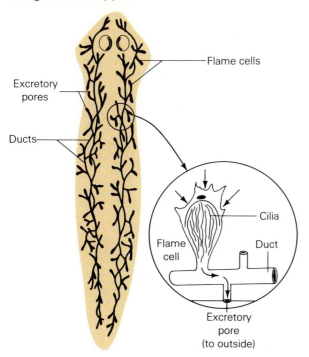

Flame cells

Excretory pores

Ducts

Cilia

Flame cell

Duct

Excretory pore
(to outside)

urine. Their kidneys are, however, extremely efficient at retaining salts, which would otherwise be quickly depleted. As we will see in the next section, in the filtering action of the kidneys, the first filtrate that leaves the kidneys contains many valuable materials that must be recovered before the urine is finally formed. But despite its efficiency, the kidney cannot prevent some loss of salts. To counter this loss, the freshwater fish must transport salt into its body against its gradient, using active transport. Interestingly, once again the active transport of salt takes place in the gill, but in the opposite direction of what we saw in marine fishes.

The excretory system of amphibians is also rather highly developed. In frogs, the tadpole produces and excretes its nitrogen waste in the form of ammonia, much the way a fish does. This is possible because of its watery habitat and because most kinds of tadpoles exist on vegetarian diets light on proteins and heavy on carbohydrates. As mentioned earlier, adult frogs shift to water-conserving urea as the principal nitrogen waste, but even so, most frogs drink and excrete huge amounts of water—up to 30% of their body weight daily. Like fishes, amphibians tend to lose body salts in the urine. Also like fishes, they actively transport salts from their watery surroundings back into the body, but the site of the active transport in amphibians is the skin.

The Terrestrial Environment.   In one important way, the terrestrial environment presents a problem similar to that of the oceans: how to avoid water loss. Terrestrial animals solve their water-retention problem through high water intake, highly efficient water-conserving excretory systems, watertight skin, and behavioral patterns that help deal with the problem of exposure. We will first consider a few land-dwelling invertebrates.

Invertebrates of the Terrestrial Environment. Though the earthworm is technically terrestrial, it lives in a perpetually moist environment. Nonetheless the earthworm is a water conserver. The fluids in each segment are constantly filtered and recycled by rather complex structures known as **nephridia.** (In some ways, the nephridium is similar to the nephron, the filtering structure of the vertebrate kidney, which we will consider shortly.) As you can see in Figure 32.16, each paired nephridium cleans the fluid in the segment just ahead of it. Any fluid moving through the ciliated tubule must pass a rich supply of capillaries in the tubule wall, where materials can be exchanged

32.16 THE NEPHRIDIUM

The excretory system in the earthworm includes paired nephridia in nearly every segment. Each has a ciliated funnel that extends into the next anterior compartment. Coelomic fluids are drawn into the funnel and pass through a meandering tubule with several U-shaped turns before reaching an excretory pore. The nephridium has a dense, overlying capillary bed that is responsible for reclaiming much of the water, salts and minerals, and other useful products before the fluid wastes are released to the outside. The chief nitrogen products of the earthworm are urea and ammonia.

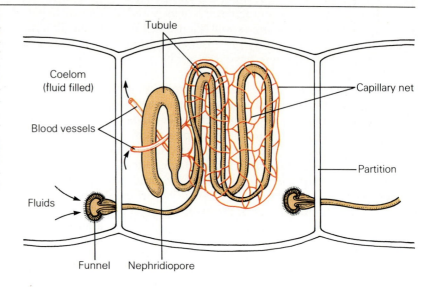

between the fluids and the blood. Water, minerals, and other essential materials are reabsorbed into the blood, while dissolved wastes pass to the outside through an opening known as the **nephridiopore.** Here, then, osmoregulation and excretion are accomplished by the same organ, just as in the vertebrates. But before leaving the invertebrates, let's consider some specializations in that enormous group called arthropods.

The arthropods of dry land have developed some very distinct ways of solving their osmoregulatory and excretory problems while reducing water loss. Their adaptations involve a water-resistant, waxy cuticle, a well-protected respiratory surface, and the elimination of nitrogen wastes in the form of semisolid uric acid.

The excretory system in insects consists of many blind, hollow, tubular structures known as **Malpighian tubules.** They emerge at about where the mid- and hindgut join (Figure 32.17). The tubules extend into the coelomic fluids and here absorb body wastes. Inside the tubules, nitrogen wastes from the coelomic fluids are converted to insoluble uric acid crystals which pass through the lumen of the Malpighian tubules directly into the gut. Once in the gut, the uric acid joins the digestive wastes, passing to the outside at defecation. This method of excretion conserves water in two ways. First, the nitrogen waste is a solid that doesn't require water for dilution. Second, fluids from the Malpighian

32.17 MALPIGHIAN TUBULES

In insects, the excretory system consists of a number of threadlike Malpighian tubules that lie in the coelomic fluids. They are believed to actively transport some ions out of the fluid and permit the passive transport of water and nitrogen wastes. The tubules empty into the gut at the junction of the midgut and hindgut. The cellular lining of the hindgut reclaims most of the water and much of the useful mineral content of the waste materials passing through. In this manner, the insect avoids undue water and mineral loss.

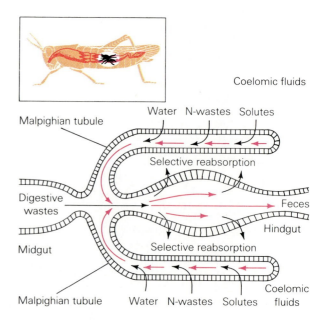

tubules can be reabsorbed by cells in the lining of the hindgut. Thus the insects produce a dense, fairly dry digestive and excretory waste.

Vertebrates of the Terrestrial Environment. Vertebrates that roam the land must constantly replace water that is lost during breathing and during the removal of nitrogen wastes; thus their movements are restricted by the availability of water. Aside from an essentially waterproof skin their most important means of conserving water is a specialized kidney. Two vertebrate groups, the reptiles and the birds, conserve water through the production of uric acid, which can be eliminated in a semisolid state. Uric acid production is also an adaptation to development within an egg since it can be safely isolated from the embryo. (Imagine the problem in the bird or reptile egg if ammonia collected.) The primary nitrogen waste in mammals, however, is urea (with some uric acid also produced), which requires a constant supply of water for its elimination. This is why a water-conserving kidney is so important in mammals. The human kidney serves as an example of this efficient organ.

THE HUMAN EXCRETORY SYSTEM

Whereas humans in developed nations are very wasteful of water, their kidneys are not. (Consider that in the home, Americans generally use about three gallons of highly purified water to flush away a few ounces of urine.) Like other mammals, humans have centralized, complex, and efficient kidneys, whose primary roles are the excretion of the nitrogen waste urea and the conservation and control of the body's water and mineral content. Let's first concentrate on the structure of the kidneys themselves and then discuss how they work and how their function is controlled.

Anatomy of the Human Excretory System

The human excretory system (Figure 32.18) includes the **kidneys, ureters, urinary bladder,** and **urethra** and the blood vessels of the **renal circuit** (see Chapter 31). The renal circuit includes the

32.18   THE HUMAN EXCRETORY SYSTEM

In humans, the excretory system consists of the paired kidneys, their blood vessels, the ureters, the bladder, and the urethra. The system has both an excretory and an osmoregulatory function. The blood vessels of the kidney, which form the renal circuit, consist of the paired renal arteries, a complex capillary network, and the paired renal veins.

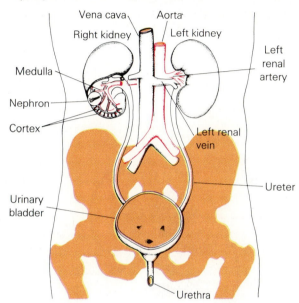

paired **renal arteries,** along with a great number of special branches, and the paired **renal veins.** Because it comes directly from the descending aorta, blood in the renal arteries is under considerable pressure. This pressure, as we will see, is vital to the kidney's function, and should it fall drastically, kidney function will immediately be threatened.

Figure 32.19a illustrates the major features of the human kidney. Its anatomical regions include the **cortex,** the **medulla,** and the **renal pelvis.** The dense cortex consists of the filtering units of the kidney—the **nephrons** and their related blood vessels. Each nephron consists of a spherical capsule and a long tubule. The tubules form long loops that venture down into the medulla and return to the cortex. These in turn are connected to numerous **collecting ducts** that join to form the dense, fan-shaped **pyramids.** The pyramids narrow down, leading into funnel-shaped **calyces** that empty into the renal pelvis.

The Urine-Forming Pathway. Briefly, the urine-forming pathway begins when the nephrons

32.19 ANATOMY AND MICROANATOMY OF THE KIDNEY

(a) The kidney contains an outer cortex, an inner medulla that contains the pyramids, several large funnellike calyces, and a final collecting region known as the renal pelvis. **(b)** Note the relationship between the cortex and medulla of the kidney and the loop of Henle. The medulla plays an important role in salt reclamation. **(c)** The functional units of the kidney are the nephrons. These are microscopic tubules that selectively remove urea, excess salt, and excess water from the circulatory system. Each nephron consists of four anatomical regions: Bowman's capsule, proximal convoluted tubule, loop of Henle, and distal convoluted tubule. Each nephron is joined

to a nearby collecting duct. Note the blood supply to the nephron. Blood enters the nephron at the Bowman's capsule, where an afferent arteriole has branched into a mass of smaller vessels that comprise the glomerulus. Emerging from the glomerulus is an efferent arteriole that immediately branches to form the extensive peritubular capillary network over the entire nephron. At the other side of the network, the capillaries join to form a venule that will eventually join other venules to form the renal vein. **(d)** The *juxtaglomerular region* of the nephron is essential to the control of sodium reclamation.

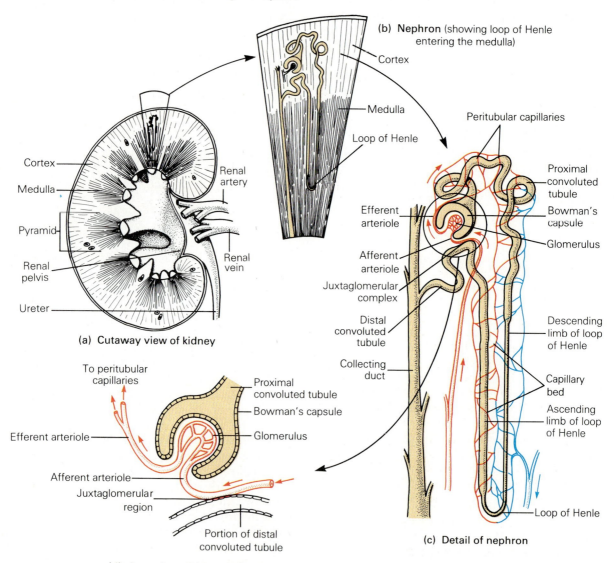

(b) Nephron (showing loop of Henle entering the medulla)

Cortex

Medulla

Loop of Henle

Peritubular capillaries

Proximal convoluted tubule

Bowman's capsule

Glomerulus

Efferent arteriole

Afferent arteriole

Juxtaglomerular complex

Distal convoluted tubule

Collecting duct

Descending limb of loop of Henle

Capillary bed

Ascending limb of loop of Henle

Loop of Henle

(c) Detail of nephron

Cortex

Medulla

Pyramid

Renal pelvis

Ureter

Renal artery

Renal vein

(a) Cutaway view of kidney

To peritubular capillaries

Proximal convoluted tubule

Bowman's capsule

Glomerulus

Efferent arteriole

Afferent arteriole

Juxtaglomerular region

Portion of distal convoluted tubule

(d) Juxtaglomerular complex

receive materials that have been filtered out of the blood, and from this filtrate the urine forms. Once formed, the urine passes into the collecting ducts to empty into the renal pelvis. This is the first place where pools of urine collect. Urine passes from there to the bladder via the ureters. From the bladder, urine is voided from the body through the urethra.

Microanatomy of the Nephron

The nephrons, which number about 1 million per kidney, are the functional or filtering units. Their structure reflects their function, as is so often the case in life (see Figure 32.19b). The nephron begins with **Bowman's capsule,** a hollow bulb or cup surrounding a ball or tuft of capillaries known as a **glomerulus.** Extending from the capsule is a lengthy tubule with a peculiar hairpin loop in its midsection. We will return to the loop, but first let's take a more detailed look at the glomerulus and the other blood vessels of the nephron.

The glomerular capillaries arise from the **afferent arterioles**—tiny branches of the renal artery. As the blood finally emerges from the glomerulus, it enters the **efferent arteriole.** The efferent arteriole branches into a second capillary network, the **peritubular capillaries,** which form a fine network over the entire nephron. The peritubular capillaries coalesce to form a venule, which is subsequently joined by venules from the other nephrons. These merge to form the renal vein. This peculiar routing of the blood vessels into two capillary beds is essential to glomerular filtration and tubular reabsorption, the two major processes in the nephron.

Bowman's capsule leads into a lengthy tubule that makes up most of the nephron. The first region, the **proximal convoluted tubule,** winds a meandering route and then forms the long hairpin bend called the **loop of Henle.** This loop is found in all water-conserving kidneys and is extremely prominent in the kidneys of desert-dwelling mammals for reasons we hope to make clear. After forming the loop of Henle, the tubule again begins to twist and contort into another convoluted section, this time called the **distal convoluted tubule.** It then joins a collecting duct, which also receives tubules from a number of other nephrons.

Note in Figure 32.19b that the afferent arteriole forms a passing connection with the distal convoluted tubule. This is the **juxtaglomerular complex,** an association that is vital to the kidney's role in sodium reabsorption, as we will see shortly.

In summary, the elements of the nephron proper, in the order of urine flow, are (1) Bowman's capsule, surrounding the glomerulus, (2) the proximal convoluted tubule, (3) the descending limb of the loop of Henle, (4) the ascending limb of the loop of Henle, and (5) the distal convoluted tubule, which leads to the collecting duct (not an anatomical part but shared by several nephrons).

The Work of the Nephron.   To see how all of this works, let's return to Bowman's capsule and the glomerulus, where blood is first filtered (Figure 32.20). Arterial blood is pushed into the glomerulus with considerable force and is thus under great hydrostatic pressure. About one fifth of its volume is forced through the capillary walls (the process being enhanced by the presence of small pores in the capillary walls) and into the cavity of Bowman's capsule. This filtering process is called **force filtration,** a truly descriptive term.

Force filtration is totally nonselective, and leaving the blood are water, ions, glucose, amino acids, urea, and other small molecules. Of the blood plasma, only the large protein consitituent (molecular weight above 30,000) escapes force filtration. Since many of the substances in this filtrate are extremely valuable, this may seem to be an extreme measure for getting rid of urea and a small amount of excess water and salt. It is hard to account for the evolution of such a gross process, except to remind ourselves that in nature, what works, works.

We might point out that while the crude filtrate entering the glomerulus is considerable, not much of this ends up as urine. For instance, while the adult kidneys on an average day (depending on fluid intake) filter about 180 liters of fluid from the blood, usually no more than 1.2 liters of urine are formed. Thus more than 99% of the water filtered out of the blood (and large quantities of salt) is returned to the blood before urine leaves the collecting ducts.

Reabsorption in the Proximal Tubule.   The filtrate, in its primary, highly dilute state, enters the nephron, passing into the proximal convoluted tubule, and some rather viscid, concentrated, hypertonic blood is carried off by the efferent arteriole that leads from the glomerulus. This hypertonic condition results in a steep water gradient that encourages the reentry of water from the dilute primary urine; thus, much of the water from the proximal tubule reenters the blood by simple osmosis. In addition, valuable substances in the filtrate, such as sodium, glucose, and amino acids, are actively transported out of the tubule to enter the surrounding cells and then cross the walls of the peritubular capillaries, returning to the blood. Chloride ions also leave the tubule, but passively follow sodium. As the filtrate leaves the proximal convoluted tubule it contains urea, miscellaneous toxic substances (small amounts of ammonia, creatine, and uric acid), some salt, and, still, much of the water.

Activity in the Loop of Henle.   It is in the loop of Henle and in the collecting ducts that most of the remaining water and salt will be returned to the blood. As far as we know, water cannot be actively transported by membranal pumps in biological systems, yet somehow most of the water returns to the blood. Let's see how.

Anatomically, the descending limb of the loop of Henle, the ascending limb of the loop, and the

32.20 MECHANISMS IN THE NEPHRON

In this diagrammatic view, the glomerulus and convoluted tubules are seen in the renal cortex, while the two limbs forming the loop of Henle and collecting ducts pass down into the renal medulla. Following nonselective force filtration in the Bowman's capsule, each of the remaining three nephric regions and the collecting duct perform highly specific roles in reabsorption, tubular secretion, and final urine formation. Valuable substances must be reclaimed, wastes removed, and a delicate water and ion balance retained in the blood. Note the steep osmotic gradient established in the medulla through the countercurrent activity of the loop of Henle and collecting duct.

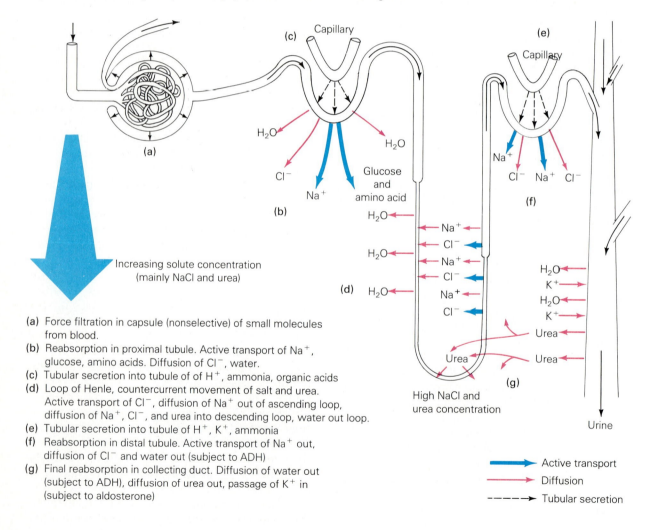

(a) Force filtration in capsule (nonselective) of small molecules from blood.
(b) Reabsorption in proximal tubule. Active transport of Na$^+$, glucose, amino acids. Diffusion of Cl$^-$, water.
(c) Tubular secretion into tubule of of H$^+$, ammonia, organic acids
(d) Loop of Henle, countercurrent movement of salt and urea. Active transport of Cl$^-$, diffusion of Na$^+$ out of ascending loop, diffusion of Na$^+$, Cl$^-$, and urea into descending loop, water out loop.
(e) Tubular secretion into tubule of H$^+$, K$^+$, ammonia
(f) Reabsorption in distal tubule. Active transport of Na$^+$ out, diffusion of Cl$^-$ and water out (subject to ADH)
(g) Final reabsorption in collecting duct. Diffusion of water out (subject to ADH), diffusion of urea out, passage of K$^+$ in (subject to aldosterone)

collecting tubules form a curious S-shape, with all three portions lying more or less side-by-side. The glomeruli and the convoluted tubules lie in the cortex (outer part) of the kidney, while the loop and the collecting tubules extend into the medulla (inner part). The tissue fluids of the medulla are relatively salty (hypertonic), so as the dilute urine flows down the descending limb of the loop, it loses water and gains salt, both by passive diffusion. At the bottom of the loop, the urine, still carrying with it the unwanted wastes, has lost much of its water, but it has picked up a lot of salt, and salt must be conserved.

In the ascending limb of the loop of Henle—the portion headed back toward the cortex—salt is actively transported from the nephron into the surrounding tissue. (Actually, only chloride (Cl^-) is pumped out, but sodium (Na^+) and potassium (K^+) ions follow the negative chloride ions.) This active transport both enables the blood to recapture the salt and maintains a high salt concentration (hypertonicity) in the kidney medulla, thus encouraging the movement of water out of the nephron. Incidentally, the transfer of salt in and out of oppositely moving columns of fluid in the two limbs of Henle's loop constitutes a countercurrent mechanism, one we have seen several times but under different circumstances. Its role here is to help concentrate salt in the medulla.

Activity in the Distal Tubule and Collecting Ducts. In the distal tubule, as in the proximal tubule, sodium ions are actively transported out, followed by chloride ions, adding still more solute to the surroundings. As before, water follows the increased gradient. Following the loss of salt and water, filtrate in the tubule tends to become isotonic with blood in the nearby peritubular capillaries, so there is no further tendency for water to move in or out. But, the newly formed dilute urine next flows through the collecting ducts and has to traverse the salty medulla one more time. As it again enters regions with an increasing outward osmotic gradient, the urine may lose yet more water. We say "may" since, as we will see, a special hormonally regulated mechanism is at work in the collecting ducts. In addition, according to recent research, urea itself diffuses out of the collecting duct and into the medulla, further increasing the hypertonic state therein. Urea also diffuses back into the loop at its bottom region, thus setting up its own countercurrent mechanism with this substance.

Tubular Secretion. While reabsorption, involving both active and passive transport, is responsible for reclaiming valuable materials and getting them back into the peritubular capillaries, another process, **tubular secretion,** also occurs. But unlike reabsorption, tubular secretion transports substances *out of* the peritubular capillaries and *into* the renal tubule. Not all of the blood's water (with its low molecular weight solutes) is forced into Bowman's capsule during filtration. Certain molecules and ions left in the blood after it leaves Bowman's capsule, such as ammonia, hydrogen ions, potassium ions, organic acids, and creatine, are actively transported from the capillaries to the nephron. Tubular secretion, then, is a last mechanism for getting rid of additional wastes. Table 32.1 reviews the structures of the nephron, the forces at work in each structure, the contents of the nephron at key points, and the contents of the peritubular capillaries.

Control of Nephron Function

The hypothalamus, in addition to its many other homeostatic functions, measures the osmotic pressure, or water content, of the blood passing through its capillaries. If the blood is hypotonic (has excess water), the kidney can begin to release more water as urine. If the blood registers as hypertonic (meaning the body is beginning to run short of water), the kidney is put on a water-rationing regime. In such cases, neurons of the hypothalamus secrete **antidiuretic hormone,** or ADH into the posterior lobe of the pituitary and the hormone is then released into the blood. Its targets are the epithelial cells of the collecting ducts of the nephron (Figure 32.21a). As a clue to ADH's function, consider that a **diuretic** is an agent that increases urine flow; thus an **antidiuretic** does the opposite.

As we mentioned earlier, ADH increases the permeability of these cells to water. Thus, more water leaves the tubule to reenter the blood. When the osmotic consistency of the blood is restored to a normal range, the stimulation of the hypothalamus slows and ADH secretion falls off, producing an increase in urine volume. The stimulation of the hypothalamus also creates the sensation of thirst. As the individual drinks, the blood becomes diluted, and again, the hypothalamus stops sending out antidiuretic hormone and the kidney increases its output of fluid. Eventually another sensation is produced, this time from the stretch receptors of the bladder. And a new behavior is initiated.

A hormonal mechanism also controls the retention of sodium chloride (Figure 32.21b). In this case, the agent of control is the hormone **aldosterone,** a steroid (see Chapter 33) released by the cortex of the adrenal gland when salt reabsorption drops. The link between sodium chloride and aldosterone release is indirect and not entirely understood. Apparently, certain cells in the nephron's distal convoluted tubule, part of the juxtaglomerular complex, monitor sodium chloride transport and communicate their information to cells in the adjacent afferent arteriole. In response, the arteriolar cells secrete **renin,** an intermediary hormone, into the blood. Renin's target is the adrenal cortex, which responds by releasing aldosterone into the blood. The target of aldosterone is the distal convoluted tubule, where it increases the reabsorption of salt from nephron filtrate. As the active transport of sodium ions (closely followed by chloride) increases, the juxtaglomerular complex slows its renin release, aldosterone secretion slows, and sodium transport is once again readjusted. In the absence of aldosterone, then, the urine becomes saltier. As with water, the quantity of salt excreted depends on the uptake.

Sodium chloride reabsorption in the distal tubule is tied to the secretion of potassium ions, also regulated by aldosterone. While aldosterone increases salt reabsorption it decreases that of potassium, which is secreted out of the blood and into the tubule. In fact, the sodium-potassium exchange pump (see Chapter 5) is at work in the distal tubule, moving the ions in opposite directions. Aldosterone's regulation of potassium ions extends to the collecting duct, where the ions also enter the urine path.

We should add that the excretory system is also involved in other homeostatic mechanisms—for

TABLE 32.1

THE NEPHRON AT WORK

Region	Process	Substances in Transit
Bowman's capsule	Force filtration from blood	Water, ions, glucose, urea, amino acids
Proximal convoluted tubule	Active transport out	Glucose, amino acids, K^+, Na^+, HCO_3^-, Ca^{2+}
	Diffusion out	Cl^-, water
	Tubular secretion capillary to tubule	H^+, ammonia, organic acids, K^+, creatine
Ascending loop of Henle	Active transport out	Cl^-
	Diffusion out	Na^+
	Diffusion in	Urea
Descending loop of Henle	Diffusion out	Water
	Diffusion in	Na^+ and Cl^-
	Countercurrent concentration of salt and urea in kidney medulla	
Distal convoluted tubule	Active transport out	Na^+
	Diffusion out	Cl^- and water
	Secretion into tubule	H^+, NH_3, K^+ (varies with aldosterone)
Collecting duct	Diffusion out of tubule	Water (varies with ADH), K^+, urea (joins salt in kidney medulla)
	Secretion into tubule	NH_3, H^+, K^+ (K^+ varies with aldosterone)

32.21 **HORMONAL CONTROL**

(a) water reabsorption is controlled in part by the hypothalamus, which secretes ADH via the posterior pituitary. Negative feedback is direct, since the osmoregulators in the hypothalamus respond to the increased water content of the blood. **(b)** The control of salt reabsorption is more complex. Although aldosterone influences the active uptake of sodium ions by the nephron, release of aldosterone by the adrenal cortex is initiated by renin secretion from the nephron itself. Negative feedback, therefore, is between the salt-sensing cells of the distal convoluted tubule and renin-secreting cells in the glomerular arteriole.

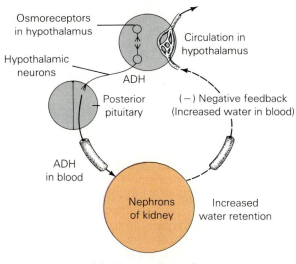

(a) Water reabsorption

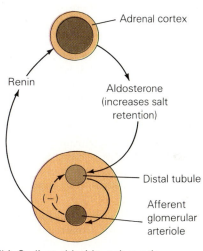

(b) Sodium chloride reabsorption

instance, in the control of blood acidity. The pH of our blood can change, depending on what we eat, but it is generally kept within fairly narrow limits. Since the acidity of urine can vary tremendously, from pH 4 to 9, excess hydrogen and hydroxide ions in the blood, like other toxic substances, are eliminated by secretion through the kidneys. Aldosterone, by the way, also speeds the movement of hydrogen ions into the urine.

In this chapter we have seen only a few of the many delicate, interacting, and highly coordinated mechanisms that keep the body's internal environment within the extremely precise limits critical to life. Remember, we live in what is essentially a disruptive environment. To remain organized in the face of potential disruption requires an ongoing, uphill battle against entropy-increasing forces. That battle is best fought under optimal physiological conditions, and the body's homeostatic mechanisms help insure those conditions.

APPLICATION OF IDEAS

1. Many of the animal body regulatory functions are automated, occurring with little, if any, conscious intervention. What are some adaptive advantages to this kind of control? What is the alternative, and how might it affect the general efficiency of an animal's metabolic activities?

2. Homeostatic mechanisms involve far more than thermoregulation and osmoregulation. Describe examples of homeostasis in other systems, such as digestive, respiratory, and circulatory, and in the action of skeletal muscle.

3. Once the problem of cross-matching tissue types is solved, transplanting a kidney is a comparatively simple and straightforward procedure. Knowing what you do about kidney function, explain this statement. Exactly what is required for a kidney to function, and why are its needs so simple?

4. Is thermoregulation in an endotherm more likely to fail under conditions of extreme heat or extreme cold? Explain fully.

KEY WORDS AND IDEAS

HOMEOSTASIS AND FEEDBACK MECHANISMS

1. **Homeostasis** is defined as the tendency for physiological systems to maintain internal stability through the coordinated response of its parts to anything tending to disturb such stability.

2. Homeostasis can operate through short term, immediate responses or through long term, cyclic responses.

Negative and Positive Feedback Loops

1. Many homeostatic mechanisms are regulated by **negative feedback loops,** whereby a stimulus creates a response that in turn alters or removes the stimulus, thus lessening or stopping the response.

2. Positive feedback loops also occur, but they are generally a sign of physiological trouble. In this case a stimulus creates a response that in turn intensifies the stimulus, thus intensifying the response, and so on.

3. Two homeostatic mechanisms in animals are **thermoregulation** and **osmoregulation.** The latter is often involved with another function, **excretion,** which occurs in the **excretory system.**

THERMOREGULATION

1. Thermoregulation is the act of maintaining body temperatures within a certain range. Animals that maintain a relatively constant internal temperature are **endothermic** (or **homeothermic**), while those animals whose body temperatures vary with surroundings are **ectothermic** (or **poikilothermic**).

Why Thermoregulate?

1. Under severe cold, the alternative to thermoregulation is metabolic inactivity. With excessive heat loss a positive feedback loop begins, as cold slows the metabolic activity required to produce body heat.

2. Surviving high temperatures is often more difficult than surviving low ones since the first may destroy essential enzymes while the second only temporarily inactivates them.

Countercurrent Heat Exchangers, or How the Tuna Keeps Moving in Cold Water

1. Excessive loss of body heat is prevented by **countercurrent exchange,** more specifically, **countercurrent heat exchangers** in the circulatory system, whereby warmed blood from deep in the body passes through vessels closely paralleling those carrying cooler blood from the extremities. The cooler blood takes up the heat as it returns, thus keeping heat out of the extremities where it would be lost.

2. Countercurrent exchangers abound in the major blood vessels of the bluefin tuna and in the **rete mirable,** a dense region of countercurrent vessels that keeps the muscles warm and active in spite of the cold surrounding water.

Regulating Behaviorally

1. Animals actively seek out areas where temperatures are optimal. Reptiles, generally considered ectotherms, use basking behaviors, in which they expose their bodies maximally to warm up and minimally to cool down. Color changes through pigment migration in some species help in either the absorption or reflection of sunlight. Physiological mechanisms of thermoregulation in reptiles are suspected.

Thermoregulation in the Endotherms

1. Endotherms thermoregulate by using the metabolic heat produced by the body's oxidative processes.

2. The circulatory system hastens cooling by shunting blood to the skin, where heat is lost through radiation (passage through air), conduction (passage to cooler solids) and convection (being carried away by perspiration).

3. The evaporation of sweat greatly accelerates heat loss because water holds heat. Evaporative cooling fails where the relative humidity is very high.

4. In retaining heat, most blood is shunted away from the surface areas, particularly into deeper veins that run parallel to the arteries. The countercurrent exchange internalizes much of the heat but decreases activity in the extremities. Upon exposure to extreme cold, the less vital extremities freeze before the general body temperature falls to critical levels.

5. Birds and mammals conserve heat through **piloerection,** improving insulation by fluffing the down feathers or body hair.

The Internal Source of Thermoregulation

1. The body also regulates metabolically—increasing or decreasing heat output by varying the rate of cell respiration.

2. Temperature changes are detected by sensory neurons, and reactions may include increased heat production, shivering, piloerection, or behavioral responses.

3. Heat measurements are also made by the **hypothalamus** of the brain. In response it may
   a. stimulate the **adrenal medulla** to release the hormone **epinephrine,** which prompts the liver to release glucose for increased cell respiration and heat output;

b. prompt the **pituitary** to release **thyroid-stimulating hormone,** which in turn causes the **thyroid gland** to release the metabolism-elevating hormone, **thyroxin.**

4. The hypothalamal thermostat may measure calcium levels rather than heat. Suffusing the brain with calcium ions for **cryogenic surgery** will cause a reduction in metabolic activity and a cooling of the body.

OSMOREGULATION AND EXCRETION

1. Osmoregulation is the ability of animals to regulate ions and water in the body. Osmoregulation is closely related to excretion, the removal of metabolic wastes.

Producing Nitrogen Wastes

1. During **deamination,** the amine groups are removed from amino acids as toxic ammonia, NH_3, which is excreted as is by some aquatic animals.

2. In insects, reptiles, and birds, the primary excretory waste is **uric acid,** while for many other invertebrates, and the fishes, amphibians, and mammals, the primary excretory waste is **urea.**

The Osmotic Environment

1. The marine environment is a **hypertonic** medium in which organisms tend to lose water and gain ions. The tissues of **osmoconformers** conform to the surroundings, becoming **isotonic. Osmoregulators** have mechanisms for actively removing ions.
   a. Osmoconformers include the limpet, *Acamea,* along with sharks and rays and the coelocanth. The latter three maintain their isoosmotic condition by retaining urea in the blood and body fluids. Sharks and rays also osmoregulate by actively pumping salts out of their bodies.
   b. Osmoregulators such as the crustacean *Artemia* and the marine bony fishes actively secrete salts out through the gill. Such secretion permits the bony fish to drink seawater. The nitrogen waste of bony marine fishes is ammonia, which is excreted across the gill. Marine birds and reptiles secrete salts from glands located near the eye, while marine mammals use the kidney for this purpose.

3. The freshwater environment is **hypotonic,** so water tends to enter the body through osmosis. Conversely, ions tend to diffuse out of the body. Since water conservation is not necessary, nitrogen wastes can be readily flushed out with water.
   a. Some freshwater invertebrates use the **flame cell system** or **protonephridium.** Excess fluids are transported into the **flame bulb** by pinocytosis, and the waving cilia push the fluids through the tubules and out through pores in the body wall.
   b. The kidney of the freshwater fish recovers some ions, but some must also be actively transported in through gill structures.
   c. The tadpole excretes ammonia, but the adult frog shifts to urea excretion. Amphibians actively transport ions in through the skin.

4. Osmoregulatory problems in the terrestrial environment primarily involve conserving water.
   a. The earthworm's many **nephridia** remove nitrogen wastes and excess salts, releasing them through the **nephridiopore,** but reclaim water and other essential materials. Arthropods have a watertight cuticle and often rely on metabolic water. Their **Malpighian tubules** collect nitrogen wastes from coelomic fluids and produce uric acid, which is excreted into the gut. The hindgut reabsorbs essential water, producing a semidry waste.
   b. All terrestrial vertebrates have specialized, water-reabsorbing kidneys. While reptiles and birds and their embryos produce uric acid, which can be excreted in a semidry state, mammals produce urea, which requires water for its excretion.

THE HUMAN EXCRETORY SYSTEM

Anatomy of the Human Excretory System

1. The human excretory system consists of the **kidneys, ureters, urinary bladder, urethra,** and the **renal circuit (renal arteries** and **renal veins).**

2. The kidney includes an outer **cortex,** middle **medulla,** and the **nephrons.** The nephrons include a capsule and a looping tubule that joins others to form the **collecting ducts,** making up the **pyramids.** The pyramids empty into the **calyces,** which lead into the **renal pelvis.**

3. The nephrons form urine, which passes from the collecting ducts to the renal pelvis. The renal pelvis empties into the ureters, which conduct urine to the urinary bladder, and the urethra voids the urine from the body.

Microanatomy of the Nephron

1. The nephron begins with **Bowman's capsule,** which surrounds the **glomerulus,** a ball of capillaries arising from an **afferent arteriole** of the renal artery. Leaving the glomerulus is an **efferent arteriole,** which forms the **peritubular capillaries,** where reabsorption takes place. These spread over the nephron to later form a venule that joins others to make up the renal vein.

2. Bowman's capsule leads to the **proximal convoluted tubule,** the **loop of Henle,** and the **distal convoluted tubule,** which joins a collecting duct. The afferent arteriole also connects with the distal convoluted tubule, forming the **juxtaglomerular complex.**

3. Each part of the nephron functions as follows:
   a. *Bowman's capsule.* **Force filtration** in Bowman's capsule causes much of the water and ions and smaller molecules to leave the blood and enter the proximal convoluted tubule.
   b. *The proximal tubule.* The peritubular capillaries contain blood in a hyperosmotic state, so much of the water filtrate reenters the blood by osmosis. Active transport also returns sodium (chloride follows passively), glucose, and amino acids to the blood.
   c. *The loop of Henle.* The ascending loop actively transports chloride ions (sodium ions follow passively) into the surrounding area, recycling salt and creating a hyperosmotic state in the kidney medulla. The hypertonic state is further increased by urea, which diffuses out of the collecting ducts.
   d. *The distal tubule.* The active secretion of sodium ions occurs with chloride ions and water passively following. Potassium ions enter the tubule.
   e. *Collecting ducts.* Water leaves the collecting ducts in response to **antidiuretic hormone (ADH),** which is secreted by the posterior pituitary in response to osmotic conditions in the blood (actually detected by the hypothalamus).

4. **Tubular secretion** forces ammonia, hydrogen ions, potassium ions, organic acids, and creatine into the tubule.

Control of Nephron Function

1. Nephron control is hormonal, with water reabsorption controlled by ADH from the posterior pituitary and sodium chloride reabsorption controlled by **aldosterone** from the adrenal medulla. Sodium chloride transport is monitored by the juxtaglomerular complex. The arteriolar cells secrete **renin,** which stimulates the adrenal cortex to secrete aldosterone. Aldosterone increases the reabsorption of sodium chloride and the excretion of potassium.

REVIEW QUESTIONS

1. Define the term *homeostasis* and list several homeostatic mechanisms (outside of thermoregulaton and osmoregulaton). (813)
2. Using the household thermostat example, explain how a negative feedback loop works. (814)
3. State an example of positive feedback. In biological systems, what do most instances of positive feedback indicate? (815)
4. Define the terms *endothermic* and *ectothermic.* Cite examples of animals with each type of system. (815)
5. Explain how a positive feedback loop operates when an individual freezes to death. (815)
6. Describe the operation of a countercurrent heat exchanger in the circulatory system. (816)
7. What is the "rete mirable" of the bluefin tuna? What does it enable the tuna to do that might otherwise be impossible? (818)
8. Describe a behavioral thermoregulatory response in a reptile. (818)
9. What are the two general physiological ways in which endotherms thermoregulate? (819)
10. Cite examples of radiation, conduction, and convection as agents cooling the body surface. (819–820)
11. Explain how the skin aids both in losing heat and retaining heat. (820)
12. Using blood vessels in the human arm as an example, explain how we make use of the countercurrent heat exchange principle in retaining body heat and in cooling the body. (821)
13. In what way does piloerection aid the animal in thermoregulation? How valuable is this to humans? (821–822)
14. Where in the human is body temperature controlled? What does this region actually measure? (822–823)
15. What is the role of the thyroid in maintaining a constant internal body temperature? (822)
16. Define osmoregulation. What substances are involved? (823)
17. List the three common nitrogen wastes. Through what chemical activity are such wastes produced? What kind of diet would tend to produce the most nitrogen waste? (824)
18. What are two osmoregulatory problems of animals living in the marine environment? (825)
19. Describe two ways in which marine animals respond to their hyperosmotic surroundings, and cite examples of each. (825)
20. Compare the means of osmoregulation in the following: marine birds, reptiles, bony fishes, and mammals. (826)

21. What osmoregulatory problems confront the freshwater animal? (826)

22. Describe the protonephridia of planarians. What aspect of osmoregulation is this system concerned with? (827)

23. How is it that freshwater fishes can safely form and excrete the nitrogen waste ammonia? (827)

24. Compare the method and form of nitrogen waste excretion in the earthworm and insect. Which, if either, is better adapted to dry terrestrial habitats? (827–829)

25. Why must most terrestrial mammals drink copious amounts of water on a regular basis? (829)

26. Describe the gross anatomy of the human excretory system, listing five significant parts. Trace the flow of urine through these parts. (829, 831)

27. What microstructures make up most of the cortex, the medulla, and the pyramids of the kidney? (829)

28. List the five parts of a nephron in their func-

tional order, and carefully explain the arrangement of related blood vessels. (831)

29. List the forces at work in each part of the nephron. (834)

30. Returning to the microanatomy, describe what happens to each of the following in each part of the nephron: urea, water, sodium ions, chloride ions, amino acids, glucose. (831–832, 833)

31. In general, what is the function of the loop of Henle? What is the significance of its hairpin shape? What does the collecting duct contribute? (832–833)

32. What are the targets of antidiuretic hormone? How do the target cells respond? How is antidiuretic hormone regulated? (833)

33. Describe the process of tubular secretion. What are the materials involved and in which direction are they secreted? (833)

34. Explain how the secretion of aldosterone is regulated. (834)

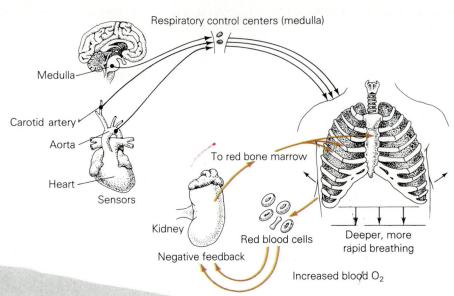

RESPIRATION

Increased oxygen demands of rapid oxidative respiration and lowered oxygen partial pressure because of increased altitude (neural and hormonal responses):

*Carbon dioxide, pH, and oxygen sensors activated: respiratory control centers respond by increased rate and depth of breathing, spleen releases red blood cell reserves.

*Kidney sensors activated by lowered oxygen load, stimulate homeopoietic mechanism, increasing red blood cell production.

METABOLIC RATE

Great energy demands of body (hormonal response):

*Hypothalamus prompts pituitary which activates thyroid and adrenals— the first releases thyroid hormone, elevating metabolic rate and the second releases steroids that increase glucose release by the liver.

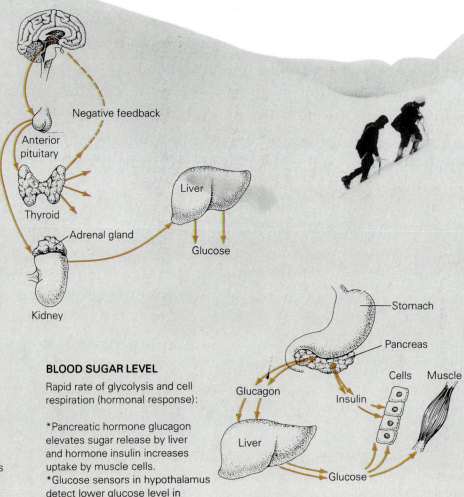

BLOOD SUGAR LEVEL

Rapid rate of glycolysis and cell respiration (hormonal response):

*Pancreatic hormone glucagon elevates sugar release by liver and hormone insulin increases uptake by muscle cells.

*Glucose sensors in hypothalamus detect lower glucose level in blood, stimulate hunger sensation.

Legend:
→ Circulating hormones
→ Neural impulse
→ Negative feedback

CIRCULATION

Increased oxygen demands, and need for carbon dioxide clearance (neural response):

*Sympathetic nerves increase heart rate, blood pressure.
*Precapillary sphincters shunt blood away from noncritical areas to muscles and brain.

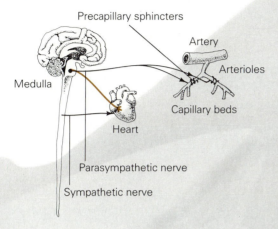

THERMOREGULATION

Decreasing temperature in surroundings, loss of body heat in exposed parts, gain of heat in protected areas (neural and hormonal responses):

*Skin sensors and blood temperature activate nervous system, which responds by stimulating hypothalamus, which activates pituitary. Pituitary activates thyroid and adrenal as in metabolism above, with adrenal also releasing epinephrine, which speeds sugar release into blood.
*Blood shunted to deeper vessels in cold surface areas and to skin and surface vessels in heated surface areas.

OSMOREGULATION AND EXCRETION

Loss of body fluids and salts through heavy activity and limited water intake (hormonal response):

*Osmoreceptors in hypothalamus activated by hypertonic blood: posterior pituitary secretes ADH, more water reabsorbed by kidney. Thirst sensation begins.
*Kidney senses sodium loss, prompts adrenal gland to secrete salt conserving aldosterone.

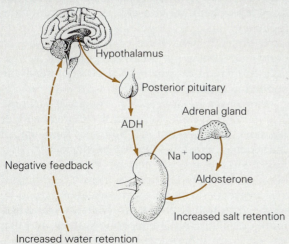

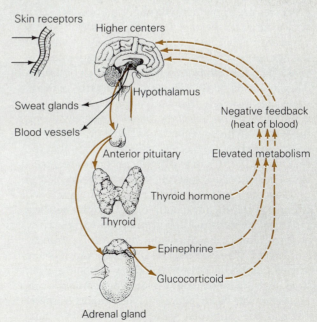

Hormonal Control

Life. A simple term. We intuitively know what it is. And we seem to have some sort of great respect for it (possibly because of its alternative). We see life around us and our interest in it is intense, but no one really knows what it is. Its very nature, and even its definition, escape us. One of the reasons is its complexity. Another reason is that some of its processes are apparent and others are not. We see a bird wading so we say, "Some life wades." That's apparent. What we don't see, though, are the strange little molecules silently leaking into the bird's bloodstream from scattered clumps of tissues deep inside its body. If we traced the paths of those molecules through that body, we would see that some of them were ignored by all the tissues along the way until they reached cells that were extremely sensitive to those molecules. We would see that the cells respond to them in a highly programmed manner. These unseen reactions may even be so vital as to keep the bird alive. The presence of some such chemicals might, for example, induce the wading bird to work in its feeding ground more intensively and store fat that will sustain it on a long journey southward. The bird may be utterly ignorant of the impending flight, but nevertheless, at just the right time, it will simply begin. Even then, much of its behavior will be dictated by these invisible molecules as they interplay through its body in a daz-

zling and dissolving concert. So let's look more closely at these molecules, these chemical messengers that help direct the symphony we call life.

THE NATURE OF HORMONES

Hormones are molecules that are formed in one part of the body and travel to other parts of the body, where they cause changes. As such, they are part of an array of active substances, collectively known as **chemical messengers.** They constitute regulatory and developmental agents, acting to make the body behave in a coordinated fashion. In vertebrates, many hormones are produced in well-defined glandular organs, collectively called the **endocrine system.**

Hormones are not alone in regulating the body. The nervous system, for example, shares many of the same functions but in a different, more abrupt time frame. In fact, it's difficult to consider the hormonal system and the nervous system separately because their efforts are often closely coordinated and interdependent. If a generalization has to be made, it is that the nervous system usually acts more quickly than the endocrine system, and that its effects are shorter-lived.

HORMONES AND THE INVERTEBRATES

While hormone-secreting endocrine systems exist in invertebrates, they are quite different from those of vertebrates. In view of the long evolutionary divergence of the two groups, this is certainly to be expected. For example, in many invertebrates, the sources of most hormones are nerve cells or clusters of nerve cells. (This is also true of vertebrates, but to a much lesser extent.) The hormones of two classes of arthropods—the crustaceans and the insects—have been rather intensively studied.

Molting Activity in Crustaceans

Crustaceans produce a hardened exoskeleton, a dense, calcium-impregnated cuticle secreted by cells of the underlying epidermis. While the exoskeleton of a crab or lobster is a fine suit of armor, its presence creates one physiological problem: the animal can grow but the exoskeleton can't. Thus the cuticle must periodically be discarded to give the living organism within it space to expand.

The process of molting, known technically as **ecdysis,** is preceded by **preecdysis,** in which the crusty exoskeleton becomes thinner and more fragile as calcium and organic substances are reabsorbed into the body. At the same time, cells in the underlying epidermis grow and divide rapidly, forming a new, soft and extendable cuticle. Then ecdysis begins; the old exoskeleton splits down the back, and the animal arches its back and pushes its way out, leaving a ghostly shell of its former self. As soon as the confining shell is discarded, the animal quickly swells by absorbing water, thereby stretching the new, flexible cuticle. Ecdysis is followed by **postecdysis,** in which the new cuticle hardens through both a "tanning" process and by impregnation with calcium salts. The time between moltings is **interecdysis,** when the animal can go about its business without worrying about changing its armor. During the animal's early growing stages, the whole process is repeated every few weeks. As the animal matures, interecdysis may lengthen to five or six months. Mature individuals of some species may even stop molting altogether.

Researchers began to suspect that hormones were involved in ecdysis when it was discovered that crayfish molted every two to three weeks if their eyestalks were removed, and that they returned to a normal molting cycle when the eyestalks were restored. Clearly, something in the eyestalk was regulating ecdysis. A search turned up the tissue that was responsible, tentatively called the **X-organ** (Figure 33.1). Further surgical experimentation led to the discovery of another mysterious patch of tissue located below the base of the eyestalk. Removal of this organ created an opposite effect: molting activity ceased altogether. The second organ was promptly named the **Y-organ** (both names remain in use today). The researchers believed that the two organs functioned together in controlling the process and did so through the release of hormones. However, proving this was another matter. Testing for hormone action is a rigorous business—one with very demanding rules. It is generally agreed that the role of any hormone can be described only after the following criteria are met:

33.1 MOLTING HORMONES

Ecdysis, the periodic shedding of the exoskeleton in crustaceans is under the control of two hormones. The molting hormone, crustecdysone, is secreted by the Y-organ, while molt-inhibiting hormone (MIH) is secreted by the nearby X-organ. Both of these endocrine structures are located at the base of the eyestalks. Light appears to be an important factor in controlling ecdysis.

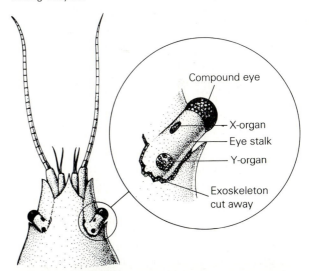

Compound eye
X-organ
Eye stalk
Y-organ
Exoskeleton cut away

33.2 METAMORPHOSIS IN A BUTTERFLY

The silkworm, or Hyalaphora moth, passes through a complete metamorphosis in its development. The fertilized egg develops into a larva, which is an active feeding and food-storing stage. The larval stage is followed by pupation. The pupa remains dormant through winter but in the spring undergoes important developmental changes within the cocoon. From this stage emerges the adult moth. Its primary activity is mating and reproduction. Each stage is directed by hormonal action.

Egg   Larva   Pupa   Adult

1. removal or damage to the suspected organ of production must cause a deficiency symptom;
2. the substance in question must be isolated, purified, and, if possible, chemically identified;
3. reintroduction of the substance must alleviate the symptom;
4. halting the restorative treatment must result in the reappearance of the symptom.

At the moment, physiologists believe there are two hormones (or groups of hormones) involved in ecdysis. The Y-organ is believed to secrete the hormone **crustecdysone,** and the X-organ is believed to secrete a **molt-inhibiting hormone** known simply as **MIH.** These two hormones appear to interact antagonistically, with molting activity determined by their relative concentrations. Increased crustecdysone levels are believed to increase molting activity, while higher MIH levels suppress the activity.

The factors controlling the levels of the two hormones aren't known, but several environmental conditions seem to have an effect. One is light. For example, the burrowing land crab *Geocarcinus* will not molt if kept in continual darkness. Other possible factors include the general health of the animal, its reproductive state, its diet, and the water temperature.

Hormonal Activity in Insects

Insect development has received a great deal of attention over the years, partly because some people would like to bring much of it to a screeching halt. It is difficult to overemphasize the importance of this research area, since our ability to control insects—probably our leading ecological competitors—depends on an exact knowledge of all aspects of their life. Much of what we know about insect development has come from a few in-depth studies such as those on the silkworm moth, *Hyalophora cercropia*. As it grows, this moth undergoes a complete metamorphosis; that is, it passes through all the developmental stages of insects: **egg, larva, pupa,** and **adult** (Figure 33.2). The structural changes from one stage to another are strikingly dramatic.

In its normal metamorphosis, the silkworm larva goes through several molts before entering pupation. It spends a winter in the pupal stage, emerging from the pupa case in the soft light of spring. It has carefully conserved its stored energy through the winter by existing in a very low metabolic state known as **diapause.**

Insect metamorphosis is known to be influenced by hormones from two paired brain extensions, the **corpus cardiaca** and the **corpus allata,** along with a pair of endocrine glands called the **prothoracic,** or **ecdysial glands.**

During the larval period the corpus allatum dominates. Its secretion, **juvenile hormone,** or **JH,** controls molting as the larva progresses through several growth periods (Figure 33.3). As long as JH levels are high, the young insect will remain in its larval stage, growing larger and larger. But eventually, the secretion of JH begins to dwindle. Meanwhile, secretory cells in the brain produce **brain hormone (BH),** also known as **prothoracicotrophic hormone,** or **PTTH,** which is actually secreted by the corpus cardiacum. Its target, the prothoracic gland, responds by secreting **ecdysone,** which favors the development of adult structures. With ecdysone levels high (relative to JH levels), the larval stage abruptly ends and the pupal stage begins.

The continued dominating effect of ecdysone eventually influences the adult emergence.

Some rather ingenious experiments with insect growth hormones have led to some fascinating findings. For example, when several larval corpora allata are transplanted into larvae, the insects begin molting as usual, but they don't pupate—they keep growing into larger and larger larvae! Since the corpora allata secrete juvenile hormone, the experiment illustrates the overriding effect of juvenile hormone over brain hormone and ecdysone, when it is present in sufficient levels.

In another series of experiments, threads were tied around the larvae just behind the thoracic gland (Figure 33.4). The part of the insect anterior to

33.3 HORMONAL CONTROL OF METAMORPHOSIS

Metamorphosis in the insect is under the control of four secretory endocrine structures. These include parts of the brain, the paired corpora cardiaca and corpora allata, and the prothoracic gland. During the larval stages, growth and molting are under the control of juvenile hormone (JH), secreted by the corpus allatum. As long as this hormone dominates, the larval stage will persist. When growth is sufficient, the prothoracic gland, under the influence of brain hormone (BH), produces prothoracic gland hormone (PGH, or ecdysone) while the JH diminishes. A pupa is then formed, which is followed by emergence of the adult.

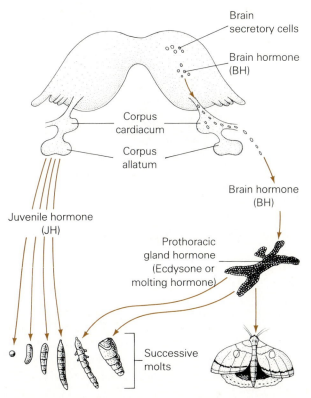

33.4 EXPERIMENTAL RESTRICTION OF PGH

If prothoracic gland hormone (PGH) is restricted from reaching the posterior parts of an insect larva, those parts will remain in the larval stage while the head and thorax undergo normal adult development. If the body is tied off just behind the thoracic gland, the head area will mature normally, but the rear area, not reached by hormones, will molt unchanged.

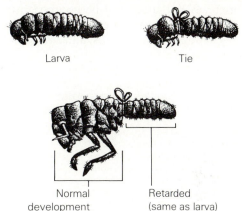

Larva   Tie

Normal development   Retarded (same as larva)

the ligature pupated, while the posterior region remained in its larval state. Apparently, the difference was due to the failure of growth hormones to reach the hindpart of the insect, leaving it in a juvenile condition when it molted.

VERTEBRATES: THE HUMAN ENDOCRINE SYSTEM

As the vertebrates evolved, both the nervous system and hormonal system became increasingly important as a means of maintaining homeostasis. The hormones work in a rather similar fashion in many of the vertebrates, so we will again draw on that representative mammal with the generalized teeth and the untalented toes.

Anatomically, the human endocrine system consists of a number of distinct glands as well as less organized tissues in various parts of the body (Figure 33.5 and Table 33.1).

The endocrine glands are often referred to as **ductless glands,** since their secretions are deposited directly from the gland into the blood. Others, such as the salivary glands, are **exocrine glands;** they release their secretions through ducts. Once released, the hormonal messengers circulate throughout the body, passing unnoticed through most tissue but causing dramatic changes when they encounter their specific target cells.

Hormones are rapidly degraded; in fact, there are enzymatic mechanisms to assure their breakdown. Thus, they do their job and disappear. Their instability is important—it wouldn't do to have a hormone hanging around in a delicately adjusted body, repeating the same monotonous message even after it was no longer needed.

Vertebrate hormones can be grouped into six major classes: **steroids, proteins, polypeptides, peptides, modified amino acids,** and a recently discovered group known as **prostaglandins.** Examples are seen in Figure 33.6, and we will look at the structure of some others as we proceed. Recall from Chapter 3 that the steroids contain four interlocking carbon rings with various side groups, while the proteins, polypeptides, and peptides consist of sequences of amino acids. One of the larger of these is insulin, which is actually a lightweight protein containing two polypeptides linked together by two disulfide bridges.

33.5 THE HUMAN ENDOCRINE SYSTEM

The hormone-secreting structures in humans consists of several distinct ductless glands and a number of less organized tissues in the stomach, small intestine, and placenta. Endocrine glands secrete their hormones directly into the circulatory system. The secretions are released in very minute quantities and are often controlled by delicate negative feedback loops.

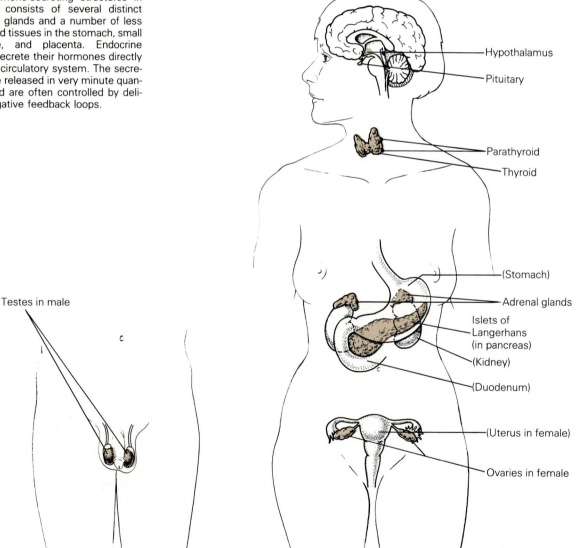

Testes in male

Hypothalamus

Pituitary

Parathyroid

Thyroid

(Stomach)

Adrenal glands

Islets of Langerhans (in pancreas)

(Kidney)

(Duodenum)

(Uterus in female)

Ovaries in female

33.6 HORMONE CHEMISTRY

The chemical classes of hormones seen here include proteins, amino acids, and lipids (prostaglandins and steroids).

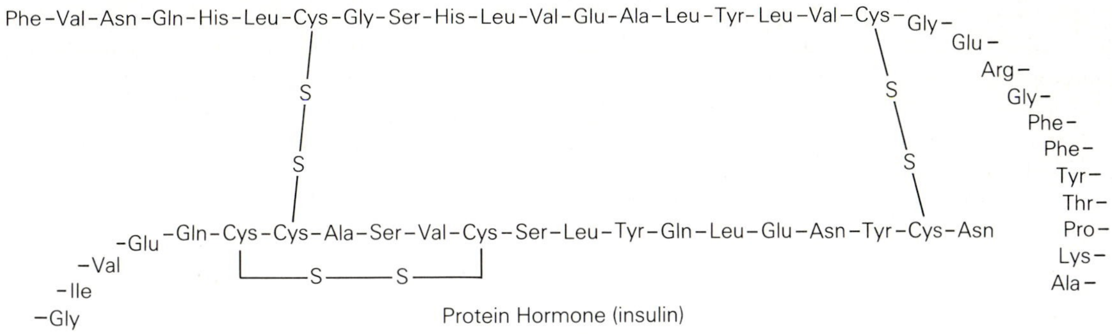

Phe–Val–Asn–Gln–His–Leu–Cys–Gly–Ser–His–Leu–Val–Glu–Ala–Leu–Tyr–Leu–Val–Cys–Gly–
                                                S   S Glu–
                                                S   S Arg–
         Gly–
        Phe–
        Phe–
        Tyr–
        Thr–
–Glu–Gln–Cys–Cys–Ala–Ser–Val–Cys–Ser–Leu–Tyr–Gln–Leu–Glu–Asn–Tyr–Cys–Asn   Pro–
–Val   S——S   Lys–
–Ile   Ala–
–Gly   Protein Hormone (insulin)

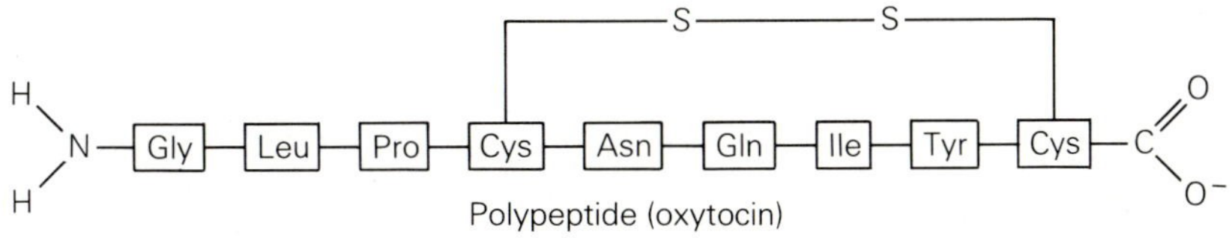

Polypeptide (oxytocin)

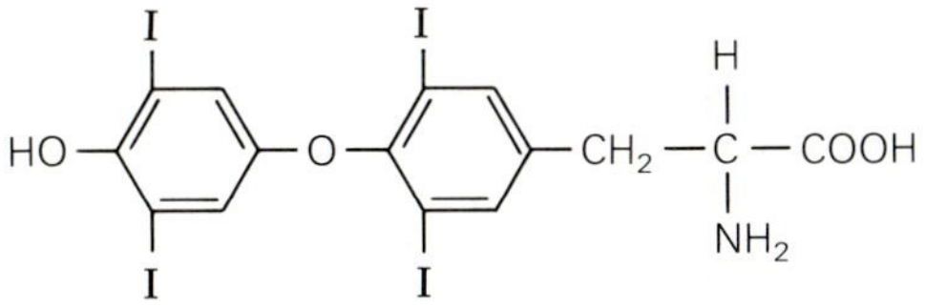

Modified amino acid tyrosine (thyroxin)

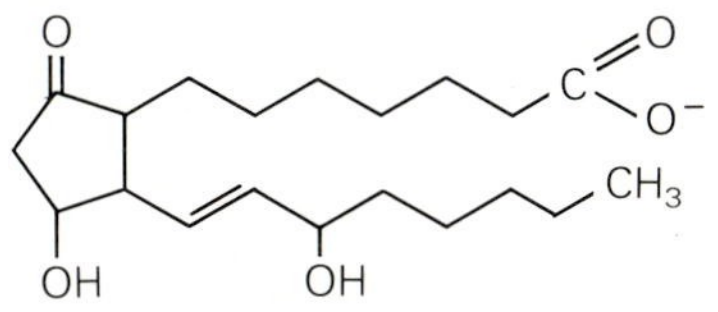

Fatty acid derivative (prostaglandin E$_1$)

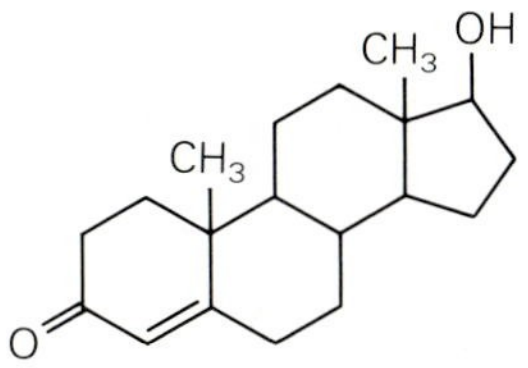

Steroid (testosterone)

TABLE 33.1

HUMAN ENDOCRINE SYSTEM

Endocrine structure and major hormones	Target	Action
Hypothalamus		
Releasing factors	Anterior pituitary	Stimulate or inhibit hormone release
Oxytocin*	See posterior pituitary	See posterior pituitary
Antidiuretic hormone (ADH)*	See posterior pituitary	See posterior pituitary
Anterior pituitary		
Adrenocorticotropic hormone (ACTH)	Adrenal cortex	Secretes glucocorticoid hormones
	Fat storage regions	Fatty acids released into blood
Growth hormone (GH)	General (no specific organs)	Stimulates growth; amino acid transport; spreads breakdown of glycogen
Thyroid-stimulating hormone (TSH)	Thyroid	Secretes hormones
Prolactin	Breasts	Promotes milk production
Follicle-stimulating hormone (FSH)	Ovarian follicles	Stimulates growth of follicle and estrogen production in females, spermatogenesis in males
Luteinizing hormone (LH)	Mature ovarian follicle	Stimulates ovulation, conversion of follicle to corpus luteum, and production of progesterone
	Interstitial cells of testis	Stimulates sperm and testosterone production
Posterior pituitary		
Oxytocin	Breasts	Stimulates release of milk
	Uterus	Contraction of smooth muscle in childbirth and orgasm
Antidiuretic hormone (ADH)	Kidney	Increases water uptake in kidney (decreasing urine volume)
Thyroid		
Thyroxin ⎫ Triiodothyronine ⎭	General (no specific organs)	Increases oxidation of carbohydrates; stimulates (with GH) growth and brain development
Calcitonin	Intestine, kidney, bone	Decreases blood calcium level; increases excretion of calcium by kidney, slows absorption by intestine, inhibits release from bones
Parathyroid		
Parathormone	Intestine, kidney, bone	Increases blood calcium level; decreases excretion of calcium by kidney and speeds absorption by intestine and release from bones

*The hypothalamic hormones, oxytocin and ADH, are released into the bloodstream via the posterior pituitary.

The Pituitary

As we consider the various endocrine glands of the human body, some will be familiar but some may not be. Certain endocrine glands may even help explain some of your less admirable behavior. We will begin our survey with the **pituitary** gland, or **hypophysis,** as it is also known.

The hormones of the pituitary activate other glands, which in turn ultimately send messages back to the pituitary, moderating its behavior. Actually, most of its moderation stems not directly from such feedback but through the interceding action of a distinct part of the brain, the **hypothalamus,** which both initiates and terminates pituitary actions. The relationship of the pituitary and the hypothalamus is part of a direct link between the nervous system and the endocrine system. We will return to all this shortly, but for now keep in mind that the neural and endocrine systems are tightly interrelated, both functionally and morphologically.

The pituitary (Figure 33.7) is a tiny, bilobed structure 12 to 13 mm in diameter, or about the size of a kidney bean. If you point your finger directly between your eyes and stick your other finger straight into your ear, you will not only gain the attention of other people on the bus, but the lines will intersect at about the location of the pituitary.

TABLE 33.1 (continued)

Endocrine structure and major hormones	Target	Action
Islets of Langerhans		
Alpha cells: glucagon	Liver	Stimulates liver to convert glycogen to glucose; elevates glucose level in blood
Beta cells: insulin	Plasma membranes	Facilitates transport of glucose into cells; lowers glucose level in blood
Delta cells: somatostatin, GH releasing hormone	Unknown	Unknown
Adrenal cortex		
Mineralocorticoids—aldosterone	Kidneys	Increased recovery of sodium and excretion of potassium and hydrogen ions; uptake of chloride ion and water
Glucocorticoids—cortisol	General (no specific organs)	Increases glucose synthesis through protein and fat metabolism; reduces inflammation
Sex hormones—androgens and estrogens	General (many regions and organs)	Promotes secondary sex characteristics
Adrenal medulla		
Epinephrine } Norepinephrine	General (many regions and organs)	Increases heart rate and blood pressure; directs blood to muscles and brain; "fight or flight mechanism"
Ovaries		
Estrogen	General (many regions and organs)	Development of secondary sex characteristics; bone growth; sex drive (with androgens); regulates cyclic development of endometrium in menstruation; maintenance of uterus during pregnancy
Progesterone (ovarian source replaced by placenta during pregnancy)	Uterus (lining)	
Testes		
Testosterone	General (many regions and organs)	Differentiation of male sex organs; development of secondary sex characteristics; bone growth; sex drive
Pineal gland		
Melatonin	Uncertain in humans—perhaps hypothalamus and pituitary	May have some influence over hypothalamus or pituitary in cyclic activity
Thymus		
Thymosin	Lymphatic system	May stimulate development of T-cell lymphocytes

The **anterior lobe** of the pituitary is primarily glandular tissue and, as a matter of fact, produces at least six important hormones. It develops in the embryo from an outpocketing of the primitive mouth. The anterior pituitary has an interesting blood supply, with another of those portal systems—those with dual capillary beds described earlier (see Chapter 31). The first capillary bed forms in the hypothalamus above and the second in the pituitary below (see Figure 33.7). The arrangement is important because it is into its first capillary bed that the hypothalamus secretes **releasing hormones** (RHs), or **releasing factors** (RFs), substances that stimulate the release of pituitary hormones. These controlling secretions make the hypothalamus, in a sense, the "master" of the "master gland." The second capillary bed delivers the releasing hormones to the anterior pituitary and receives hormones that are produced there.

The **posterior lobe** of the pituitary originates in the embryo from an outpocketing of the developing forebrain. Its downward growth meets the upward progress of the anterior pituitary until they join, finally forming the bilobed structure. Since the posterior lobe is really a part of the brain, it communicates with the hypothalamus via secretory neurons and not through the blood. These neurons deliver two hormones—**oxytocin** and **antidiuretic hormone (ADH)**—directly into the posterior pituitary. Thus, the posterior lobe doesn't really manufacture hormones. Instead, its cells simply act as storage reservoirs for hormones produced in the hypothalamus above. Upon stimulation, the hormones are released into capillaries passing through the posterior lobe.

Control by the Hypothalamus. The relationship between the hypothalamus and the pituitary is so critical that before we go into the role of the pituitary hormones we should clarify a few points about hypothalamic control.

As we have noted, the hypothalamus exercises control over the anterior pituitary through its releasing hormones or factors that enter the blood and are carried to the anterior pituitary through the connecting portal blood vessels. Each releasing hormone brings about the secretion of a specific anterior pituitary hormone, so they have such names as **growth hormone releasing hormone (GHRH), follicle-stimulating hormone releasing hormone (FSHRH),** and **thyrotrophic hormone releasing hormone (THRH).** In addition to the releasing hormones, some neurosecretions in the

hypothalamus act in an inhibiting manner, although only a few of these are known and their action is poorly understood. A list of releasing and inhibiting hormones is found in Table 33.2.

Hormones of the Anterior Pituitary. Now lets consider the hormones that are produced by the busy little bean that rests so securely in the middle of our skulls. A good hormone to start with is **adrenocorticotropic hormone**—ACTH—because of its clear regulation through our famed negative feedback principle. The target of the ACTH is the cortex (outer tissue layer) of the **adrenal gland,** an endocrine structure located on top of the kidney. When stimulated by ACTH, the adrenal cortex itself secretes a number of hormones called **glucocorticoids,** a name that tells us they are involved with glucose, are cortical in origin, and are steriods. As the levels of such steroids rise in the blood, they inhibit the hypothalamic stimulation of ACTH production in the pituitary (Figure 33.8). In addition to activating the adrenal cortex, in many animals ACTH also regulates the metabolism of fats. Under its influence, the body releases fatty acids into the bloodstream for redistribution through the organism.

Growth hormone, or GH (also called **somatotro-**

33.7 THE PITUITARY

The anterior pituitary is embryologically, structurally, and functionally distinct from the posterior. The anterior pituitary has its own capillary circuit and produces a number of hormones which it secretes directly into that capillary network. The posterior pituitary receives and stores hormones transported to it by neurons originating in the hypothalamus, releasing the hormones on demand.

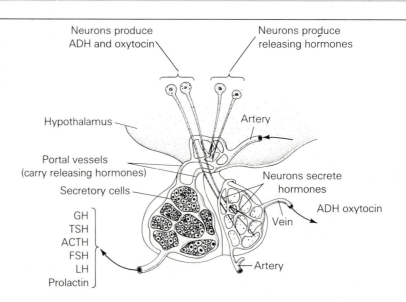

Anterior pituitary
(receives chemical signals;
produces and releases hormones)

Posterior pituitary
(stores and releases hormones
produced in the hypothalamus)

TABLE 33.2

HYPOTHALAMIC RELEASING AND INHIBITING HORMONES

Name of Hormone	Abbreviation	Role
Thyroid-stimulating hormone-releasing hormone	TRH	Release of TSH
Adrenocorticotropic hormone-releasing hormone	CRH	Release of ACTH
Follicle-stimulating hormone-releasing hormone	FSHRH	Release of FSH
Luteinizing hormone-releasing hormone	LHRH	Release of LH
Growth hormone-releasing hormone	GHRH	Release of GH
Growth hormone-inhibiting hormone (also called somatostatin)	GHIH	Inhibition of GH
Prolactin-releasing hormone	PRH	Release of prolactin
Prolactin-inhibiting hormone	PIH	Inhibition of prolactin

pin), is very broadly targeted, activating many tissues. It stimulates growth but more specifically, GH assists amino acids across plasma membranes, which results in an acceleration of protein synthesis within the cells. In addition, GH stimulates the release of stored fats and the breakdown of liver glycogen into glucose. In the latter, GH increases blood glucose levels and stimulates the release of another hormone, insulin.

Proper levels of GH in the body are essential to normal growth. This becomes distressingly obvious when something goes wrong. Excessive secretion (such as might occur with a pituitary tumor) during the early years can produce **giantism (pituitary giants).** Many people with this condition range from seven to nine feet tall. Conversely, in young people, an underactive pituitary with lower than normal GH levels produces **pituitary dwarfs,** often called **midgets** (Figure 33.9a). There are several thousand pituitary dwarfs in the United States today. Until recently, treatment for growth deficiencies was difficult and expensive. Stimulating growth in even one pituitary dwarf required a daily injection of all the GH that could be isolated from several human bodies. But since GH is a protein, new sources are now available through recombinant DNA technology (see Chapter 14).

Should GH levels rise suddenly in an adult human, the resulting condition is known as **acromegaly.** In this abnormality, growth that has stopped on time mysteriously resumes, but it is

33.8 A NEGATIVE FEEDBACK LOOP

In systems with negative feedback control, the product of an action has a suppressing effect on that action. Here the hypothalamus has stimulated the pituitary to release ACTH. In response, the target organ—the adrenal gland—secretes hormones from its cortex (outer region). The rise of glucocorticoids in the blood acts as a negative feedback loop. When sufficient quantities are present, they inhibit the hypothalamus and the system slows or shuts down.

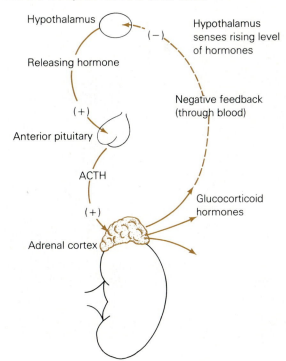

33.9 GROWTH HORMONE ABNORMALITIES

(a) When GH secretion is insufficient, the result can be growth retardation and the development of a pituitary dwarf. Oversecretion during adolescence can have the opposite effect, producing a pituitary giant. (b) Sudden increases in GH after maturity and when bone lengthening has ended produce growth in certain body parts such as the face and hands. Here we see the result, a condition known as acromegaly.

(a)

(b)

restricted to areas where cartilage persists, such as the hands, feet, jaw, nose, tongue, and some internal organs (Figure 33.9b).

Thyroid-stimulating hormone, or TSH (also known as **thyrotropin**) is responsible for stimulating the **thyroid,** another endocrine gland, to release the hormones **thyroxin** and **triiodothyronine.** These hormones regulate the metabolic rate of the body, and we will have a closer look at their functions when we consider the thyroid gland. Like GH, thyroxin and triiodothyronine are widely targeted, so many tissues respond.

Prolactin promotes milk production in mammals, including humans. Toward the end of pregnancy, the blood levels of prolactin increase to 28 times that in nonpregnant females. But milk is still not produced because it is inhibited by high estrogen and progesterone levels associated with pregnancy. These levels fall off dramatically at birth; then milk flow begins. The release of milk is in part under the influence of **oxytocin** from the posterior pituitary. Thus while milk *production* is influenced by prolactin, milk *release* is brought on by oxytocin.

Follicle-stimulating hormone (FSH) and **luteinizing hormone** (LH) are both involved in stimulating the production of gametes and sex hormones, and thus are called **gonadotropins.** But since FSH and LH activity is discussed in Chapter 36, we will move along to the last two pituitary hormones—those stored in the posterior lobe.

Hormones of the Posterior Pituitary. Two hormones, both short peptides, are stored in the posterior lobe of the pituitary. These are oxytocin and antidiuretic hormone, or ADH (mentioned earlier). The two are chemically similar, each consisting of nine amino acids:

Oxytocin: Cys-Tyr-**Ile**-Glu-Asp-Cys-Pro-**Leu**-Gly
ADH: Cys-Tyr-**Phe**-Glu-Asp-Cys-Pro-**Arg**-Gly

Both hormones are manufactured in the neural cell bodies of the hypothalamus and transported to the posterior pituitary through the axons of those cells. A variety of conditions—some we know about and some we don't—trigger their release. Antidiuretic hormone, for instance, is released when the osmotic potential of the blood falls below a certain level.

In females, oxytocin is released by, among other things, the stimulation of nipples and by sexual intercourse, and causes uterine contractions during orgasm. The uterine activity, known as "tenting," may help assure fertilization by actively drawing semen into the uterus (see Chapter 36). And we've seen that oxytocin functions with prolactin to stimulate milk release.

Oxytocin plays an important role in bringing about the contractions of the uterine muscles during labor and delivery (see Chapter 37). But once the baby is born, suckling stimulates oxytocin production, which helps bring the distended uterus back down to its normal size. Oxytocin is sometimes administered by obstetricians to stimulate uterine contractions if labor has been unusually protracted, or the contractions too feeble to be useful. Apparently, the hormone increases the permeability of uterine myofibrils to calcium ions, an essential part of the contraction process. Like prolactin, all the known activities of oxytocin have to do with female functions, although the hormone is found in males also. Perhaps it has other roles that remain to be discovered.

The antidiuretic hormone, or ADH, was discussed in the last chapter where we noted that its targets are the collecting ducts of the kidney, which respond by conserving water.

The Thyroid

In humans, the **thyroid gland** is shaped somewhat like a bow tie and is located in an appropriate place for one, slightly below the larynx (Figure 33.10). The thyroid weighs about 30 grams (1 oz). It produces two very similar hormones, thyroxin and triiodothyronine, both modified versions of the amino acid tyrosine. A third hormone, **calcitonin,** is a peptide composed of 32 amino acids. In the embryo, the thyroid develops as a ventral outpocketing of the throat that pinches off and embarks on its own developmental pathway.

Chemically, thyroxin (also called T_4) and triiodothyronine (T_3) differ by one iodine atom only (they carry four and three iodines, respectively). In their production, two molecules of the amino acid tyrosine are covalently bonded and joined by iodine (see Figure 33.6). Both hormones apparently have the same function, and thyroxin is believed to be converted to triiodothyronine when it reaches or enters its target cells.

Thyroid hormone is important in bringing about metamorphosis in amphibians. Young tadpoles have very little thyroxin, but at an appropriate stage, thyroxin is generated, causing dramatic changes. As the larva's tail and gills are reabsorbed, the legs and other adult structures emerge under the direction of thyroxin. The hormone is so

33.10 THE THYROID

The thyroid gland is a bilobed structure nearly surrounding the trachea at a point just below the larynx. Its hormones include thyroxin, triiodothyronine, and calcitonin. In general, these hormones increase the rate of body metabolism. Note the position of the parathyroid glands on the posterior side.

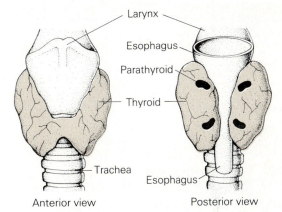

potent that even if it is administered to newly hatched tadpoles, they will promptly turn into tiny "adult" frogs the size of flies.

In humans, thyroxin influences the biochemical pathways that determine the rate at which carbohydrates are oxidized in the cells. Although we have known this much for decades, just how thyroxin works is not entirely resolved. We know that overactive and underactive thyroid glands produce **hyperthyroidism** and **hypothyroidism,** respectively. The symptoms of hyperthyroidism are nervousness, hyperactivity, insomnia, and weight loss. Skinny, active people are often accused of being hyperthyroid, but in fact most skinny, active people are perfectly normal— they're just skinny and active. In Grave's disease, a hyperactive thyroid produces, in addition to the usual symptoms, a characteristic bulging of the eyes brought about by fluid accumulation in the tissues behind them (Figure 33.11a).

A **goiter** is unsightly (see Figure 33.11b), but it is actually a normal response to low levels of iodine in the diet. The prominent, bulging, overgrowth of the gland increases the efficiency of iodine utilization and retention so that more nearly normal levels of thyroxin can be maintained. Before the introduction of iodized salt early in this century, such goiters were common where iodine-rich seafood was not a common part of the diet.

Hypothyroidism (underactive thyroid) causes a general slowing of the metabolic rate, bringing on a condition in adults called **myxedema.** Such individ-

33.11 THYROID-RELATED ABNORMALITIES

A malfunctioning thyroid can cause a number of important clinical problems. **(a)** Many of these Peruvian villagers suffer from iodine deficiency. As a result, their thyroids have become enlarged, forming goiters that help the body some-what in utilizing what iodine is available. **(b)** A hyperactive thyroid may produce rapid metabolic rate, general nervousness, and a failure to gain weight irrespective of diet. In Graves disease, the symptoms also include bulging eyes.

(a)

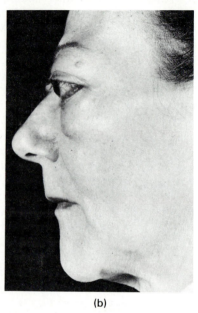

(b)

uals are usually overweight, physically and mentally sluggish, and have a puffy, bloated appearance due to fluid and protein accumulation beneath the skin.

In some cases, genetic abnormalities can cause a lack of thyroxin from birth. If left untreated, **cretinism,** a condition marked by extreme mental and physical retardation, may develop. At present, if the low thyroid condition is diagnosed early enough, cretinism can be treated effectively by the administration of a daily dose of thyroxin.

Let's run through thyroxin control again. Releasing hormone from the hypothalamus reaches the anterior pituitary, which responds by secreting TSH. Its target, the thyroid, reacts by secreting thyroxin, which spreads throughout the body, elevating metabolic activity. However, rising thyroxin levels in the blood form a negative feedback loop, inhibiting the hypothalamus and slowing the system down (Figure 33.12).

The other thyroid hormone, the peptide calcitonin, regulates the concentrations of calcium ions (Ca^{2+}) in the bloodstream. They do this in an interesting way involving the **parathyroid glands,** as we will now see.

33.12 NEGATIVE FEEDBACK CONTROL OF METABOLISM

When thyroid hormonal level in the blood is low, the hypothalamus responds by stimulating the pituitary via a specific releasing factor (TSH-RF). In response, the pituitary secretes TSH, which in turn stimulates the thyroid to secrete its hormones. The body responds with an elevated metabolic rate. While this is happening, the hypothalamus senses the rise in the level of thyroid hormone in the blood and at some precise threshold is itself inhibited, slowing its stimulating action on the pituitary.

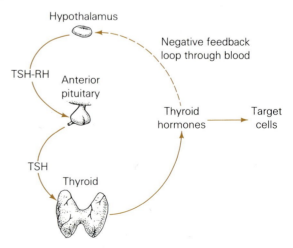

Hypothalamus

Negative feedback loop through blood

TSH-RH   Anterior pituitary

Thyroid hormones → Target cells

TSH

Thyroid

The Parathyroid Glands and Parathormone

The parathyroid glands are pea-sized bodies embedded in the tissue of the thyroid, two to three in each wing of the bow tie (see Figure 33.10). They produce **parathormone,** or **parathyroid hormone** (PTH). Parathormone is a polypeptide of 81 amino acids, large enough to be called a protein. It helps regulate the calcium ion levels of the blood, as does calcitonin. Actually, however, PTH and calcitonin produce opposing results; PTH increases the concentration, and calcitonin decreases it (Figure 33.13). The parathyroid and thyroid cells are sensitive to blood calcium, releasing PTH when its levels are lower than normal and calcitonin when calcium levels are higher than normal.

These two opposing hormones have three known target tissues: the intestine, the kidney, and the bone cells. In the intestine, PTH increases the absorption of calcium and calcitonin does the opposite. Thus, dietary calcium is absorbed when it is needed and passed through the gut when it is not.

The main buffer for the body's calcium, however, is the extensive reserve of calcium phosphate in the bones. Here, PTH increases the activity of **osteoclasts,** peculiar bone-eating ameboid cells that dissolve calcium phosphate and release calcium ions

33.13   CALCIUM REGULATION

Parathyroid hormone (PTH) and calcitonin cooperatively regulate body calcium. PTH secretion increases when blood calcium drops. It brings that level up by withdrawing calcium from the bones, increasing intestinal uptake of calcium, and slowing the excretion of calcium by the kidneys. Calcitonin has the opposite effect. When blood calcium is high, it decreases the rate at which calcium leaves the bone, reduces intestinal uptake, and increases excretion of calcium through the kidneys.

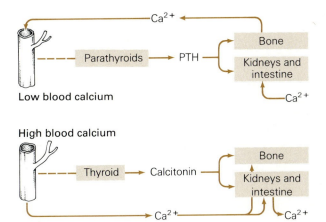

33.14   THE ISLETS OF LANGERHANS

The islets are endocrine tissues clustered in the pancreas (an exocrine gland). Note the differences in the islet cells and the surrounding pancreatic cells. Cells in the islets secrete insulin, glucagon, and somatostatin, the first two of which are instrumental in regulating glucose levels in the blood. The role of pancreatic stomatostatin is not clear.

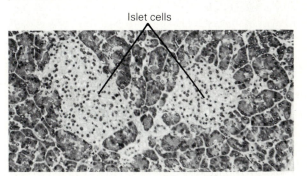

Islet cells

into the bloodstream. Calcitonin, on the other hand, increases the activity of the **osteoblasts,** the bone-forming cells, and increases the excretion of calcium through the kidneys, thus reversing the action of the osteoclasts (see Chapter 28).

Abnormally low levels of PTH mean low blood calcium, a situation that can cause muscle convulsions and eventual tetany (uncontrolled muscle spasms). Should the level fall low enough, the larynx will go into spasms, causing asphyxiation and death. Abnormally high levels of PTH result in **osteoporosis,** a severe decalcification of bone, with the subsequent formation of fibrous cysts in the skeleton. The high blood calcium level, in turn, causes irregularities in heart contraction and depresses skeletal muscle contraction.

The Pancreas and the Islets of Langerhans

Scattered through the pancreas, an exocrine gland, are groups of endocrine cells known as the **islets of Langerhans,** or simply the **islets** (Figure 33.14). Each islet contains three types of cells, **beta, alpha,** and **delta,** listed in the order of frequency of occurrence. The **insulin**-secreting beta cells are more numerous than the alpha cells, which produce the hormone **glucagon.** The delta cells are few and scattered, and they are believed to secrete the **growth hormone inhibiting hormone** (GHIH), also known as **somatostatin.** Hypothalamic somatostatin (mentioned earlier; see Table 33.2) inhibits the release of growth hormone (GH) produced by the pituitary. At this point, we might ask, why is GHIH produced by both the pituitary and the delta cells of the pancreas? Perhaps the pancreatic somatostatin has a role in the nearby digestive processes. For example, there is sketchy evidence that it inhibits the release of insulin and glucagon, and other evidence suggests that it suppresses the release of the digestive hormones such as gastrin. Directly or indirectly, somatostatin also reduces the uptake of triglycerides in the intestine. None of these functions is well understood. However, the other two hormones, glucagon and insulin, have well-defined roles in the regulation of carbohydrate metabolism.

Glucagon is a polypeptide consisting of a single chain of 29 amino acids. Its principal role is to stimulate target cells in the liver to break down glycogen into glucose. Glucagon is released when blood glucose levels fall below optimum levels.

Insulin has the opposite effect on blood glucose—that is, it decreases blood glucose levels. The two islet hormones complement each other in maintaining the delicate balance of glucose in the blood (Figure 33.15).

Insulin starts off as a protein that contains just 81 amino acids. It is inactive in this form and must be altered by enzymes before it can do its job. The enzymes remove a number of the amino acids from the middle of the sequence, producing active insulin composed of two short chains with disulfide cross-linkages (see Figure 33.6). Insulin's best-known role is to help move glucose across plasma membranes, but it is also thought to encourage glycogen synthesis in the liver. It is believed that target cells have very specific sites in their membranes—places at which insulin will bind so that the membrane becomes more permeable to glucose.

Sugar Diabetes.   Deficiencies in insulin activity produce the familiar clinical symptoms of **hyperglycemia** (high blood sugar), more commonly known as **sugar diabetes** or **diabetes mellitus.** In some diabetics, the ultimate cause of the disease is an actual deficiency of insulin, which can be caused by insufficient insulin production or by increased levels of **insulinase,** an enzyme that destroys insulin. However, most adult diabetics have normal levels of insulin but a deficiency of receptor sites on the membranes of the target cells. In some cases, the target cells have enough receptor sites but simply fail to respond properly (we will discuss target cells in more detail shortly).

The symptoms of diabetes include glucose in the urine, greatly increased urine volume, dehydration, constant thirst, excessive weight loss, exhaustion, fatty liver, ulcerated skin, local infections, blindness, and many other problems. Some symptoms are the result of glucose starvation of the cells—though glucose may be present, an insulin deficiency prevents it from crossing the plasma membranes. In addition, the untreated diabetic has a remarkable halitosis. Detectable in the exhaled breath are chemicals called ketones, especially acetone (one of the strong-smelling chemicals in some fingernail polish removers).

Mild cases of some forms of diabetes in adults can be controlled by a well-regimented diet in which carbohydrates and fats are carefully regulat-

33.15 INSULIN VERSUS GLUCAGON

The beta and alpha cells of the islet of Langerhans closely coordinate the glucose level of the blood. High blood sugar prompts the beta cells to release insulin, which helps cells take up glucose from the blood, thereby lowering the blood glucose level. Low blood sugar, however, sets the alpha cells in action, releasing glucagon into the blood. Its target, the liver, responds by converting glycogen to free glucose, which diffuses out into the blood, elevating its glucose level.

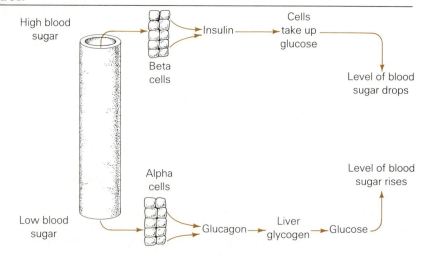

ed. More advanced cases can be controlled by insulin injections. (Diabetes in young people is essentially a different disease and is likely to be very serious indeed.)

Cells may also be glucose-starved because of **hypoglycemia,** or low blood glucose. A pancreatic tumor may be at the root of this problem. As the pancreatic tissues grow out of control, too much insulin enters the blood. Any blood glucose is cleared away rapidly and sequestered as intracellular glycogen. The brain cells, which depend on glucose as a fuel, are the first to be affected. Low blood sugar in the brain results in dizziness, tremors, violent temper, and sometimes blurred vision and fainting.

In the absence of pancreatic tumors, some people can have milder (but sometimes serious) types of chronic hypoglycemia. In such cases, one of our much-praised negative feedback mechanisms goes haywire. An increase of blood glucose brings about an increase in insulin secretion, as you might suspect. The insulin in turn facilitates the conversion of blood glucose into tissue glycogen, bringing the glucose level back down. Unfortunately, there is a time lag, and bodies may overproduce insulin, causing the blood glucose levels to fall sharply.

The Adrenal Glands

In most vertebrates, the **adrenal glands** are closely associated with the kidney. (In Latin, *ad* means upon and *renal* refers to kidney.) In humans the glands are perched atop each kidney, each adrenal being essentially two glands in terms of origin, structure, the chemical class of their hormones, and embryological origin. There is an outer layer called the **cortex** and an inner one called the **medulla** (Figure 33.16).

Hormones of the Adrenal Cortex. Some of the adrenal cortical hormones are known by the peculiar name of **mineralocorticoids,** which reflects their role in mineral (ion) regulation. Each of these hormones has the basic 4-ring steroid structure described earlier. Perhaps the best known is **aldosterone. Cortisol,** a glucocorticoid, is also an important adrenal cortical hormone, as we will see.

We saw in Chapter 32 that the target cells of aldosterone are the cells of the renal tubules of the kidney. The hormone increases the kidney's recovery of sodium and water by the circulatory system and decreases the uptake of potassium and hydrogen ions. Thus, aldosterone and antidiuretic hor-

Testosterone: Why Boys Are Different

Certain sex hormones are vital to the development of the sex organs in embryos—and their presence explains why baby boys aren't like baby girls. In XY (genetically male) individuals, the male hormone, testosterone, is produced in the rudimentary testes and first appears at about the eighth week of development, a time when the external genitalia, the sex organs, are identical in both sexes. Upon the appearance of testosterone, the penis and scrotum begin their differentiation. Where testosterone is absent or present in limited amounts, as normally occurs in XX (genetically female) individuals, the same structures that formed the penis and scrotum automatically form external female genitals—the vulva. so, for females at this time, it is the absence of testosterone that makes the difference in sexual development.

If the testes are removed from a developing XY rat or monkey embryo, the embryo will develop phenotypically as a female. It seems that there are always both female and male hormones present in normal individuals of both sexes, and normal female or male development depends on the relative balance between the two. In the absence of either testes or ovaries, female hormones, normally produced in low amounts by the adrenal glands, predominate (in normal females, the bulk of the female hormones are produced by the ovary).

Curiously, male hormones do not appear to interact directly with the genetic material. In the target cells that are sensitive to male hormones, specific enzymes alter the circulating testosterone into an active form—either dihydroxytestosterone or, sometimes, estradiol. This means that estradiol, which is a female hormone, acts as a male hormone once it is within the nucleus of certain male target cells.

While experimentation on human embryos just isn't done, the treatment of male rats and monkeys described above is more or less mimicked in certain abnormalities. An odd, rare human genetic condition called *androgen-insensitivity syndrome* dramatically illustrates the role of sex hormones in the development of sexual morphology. The syndrome is caused by the failure of a gene on the

---

The second well-known steroid of the adrenal cortex is cortisol, a glucocorticoid. In general, cortisol increases blood levels of glucose, amino acids, and fatty acids. Cortisol is also an anti-inflammatory agent. A pharmaceutical preparation, **hydrocortisone,** is available as an over-the-counter drug for this purpose. It suppresses lymphocytes that bring about the release of the histamines causing the inflammation. Such lymphocytes have very specific cortisol receptor sites on their membranes, and when cortisol binds to these, the cells die at once. For this reason, cortisol has been used in massive doses as an immunosuppressant following transplant surgery and for the treatment of autoimmune disease.

Cortical steroids also include the sex hormones. They play an important role in sexual development, particularly in the male fetus. In fact, the adrenal cortex makes the very same hormones as do the gonads (ovaries and testes). Hormones from the adrenals and gonads promote the appearance of secondary sex characteristics, such as beards and breasts (usually on different people), and help keep the reproductive system functioning. The principal adrenal sex hormones are **androgens** (male), with **estrogens** (female) making up only a minor part of the total output.

Hormones of the Adrenal Medulla. The adrenal medulla produces two fascinating modified amino acid hormones: **epinephrine** and **norepinephrine** (commonly called **adrenalin** and **noradrenalin,** old trade names originally coined by a pharmaceutical company). The two differ chemically only in that epinephrine has an added methyl ($—CH_3$) group. Their actions in the body are nearly identical, but epinephrine is widespread in the body while norepinephrine concentrates its efforts in regions served by the sympathetic nervous system (see Chapters 34 and 35).

Epinephrine causes a wide variety of dramatic changes in the body when an emergency arises. Their release, usually a reaction to danger, fright, or anger, brings on what is called the "fight or flight" response. The two hormones affect circulation by accelerating heart rate, thus increasing cardiac output, and by shunting blood into the skeletal muscles and away from other organs such as those of digestion and reproduction. This is accomplished by selected vasoconstriction and vasodilation, which closes and opens capillary beds, respectively. Blood flow in the capillaries near the body surfaces is usually restricted (causing the skin to pale and reducing the extent of bleeding). In addition, blood pressure increases as glucose is

human X chromosome. In normal males, this gene codes for the enzyme that converts testosterone to its active form. When an XY individual has inherited, from "her" mother, an X chromosome with a nonfunctional form of this gene, her tissues will be unable to respond to male hormones.

We say "her" because such individuals are indeed females, although, as you're aware, normal XY individuals are males. Persons with the androgen-insensitivity syndrome have a completely bald pubic area. Internally, they lack the ovaries, oviducts, and uterus and, curiously, have testes in the labia majora (outer lips of the female genitalia).

Androgen-insensitive individuals are biochemically unlike normal females since they have high levels of circulating testosterone (to which they are totally insensitive). Like normal XY males, and unlike normal XX females, they lack Barr bodies and do have a cell-surface antigen found only in males (the Y-linked H-Y cell-surface antigen). They are, as you might expect, sterile.

Several investigators of this strange but informative phenomenon have noted that the affected individuals appear to have exaggerated female secondary sex characteristics, being unusually voluptuous in comparison with their own XX sisters. Although their adrenals produce only the same, relatively small amounts of estrogen normally found in males, the balance between male and female determinants is, functionally, 100% on the side of femaleness. The reason they lack pubic hair is that the development of this hair in both males and females is a response to androgens, the male sex hormones.

The fact that sexual differentiation in females is a negative event, that is, it occurs only in embryos without testes or those otherwise deprived of testosterone, has led some researchers to conclude that the female form is more "primitive" than the male. (In this instance it's difficult to interpret what is meant by "primitive"—since the forerunner of the separate sexes was probably hermaphroditic, males have been around exactly as long as females!)

suddenly released into the blood stream from the liver. We have all heard stories of people exhibiting superhuman feats of strength at these times, and you may have had such an experience yourself.

The Gonads: Ovaries and Testes

The endocrine function of the ovary and testis is discussed in Chapter 36, so let's just briefly note some of the major points here. First, the two hormones of the ovary—**estrogen** and **progesterone**—and the principal hormone of the testis—**testosterone**—are all steroids. Estrogen, by the way, includes a whole family of molecules, among them **estradiol, estriol,** and **estrone.** Estradiol is the major estrogen secreted by the ovary. For a look at the significance of sex hormones in development, see Essay 33.1.

In addition to their reproductive function, sex hormones have an important role in skeletal development. Their sudden increase in the blood at the onset of puberty stimulates the lengthening of the long bones in humans, and causes a marked spurt of growth. (Interestingly, it also causes a simultaneous spurt in mental growth as measured by test performance). The onset of puberty is variable: some of us are precocious and some are definitely late bloomers. Toward the end of puberty, the sex hormones have the opposite effect. They promote the final fusion of growth regions in the long bones, after which no further increase in height is normally possible. Thus late-maturing adolescents, those in which bone fusion is delayed, may grow to be unusually tall. Low levels of estrogen secretion may delay puberty and produce tall girls. The same relationship, by the way, holds for testosterone and boys. Remember that tough, well-muscled kid who matured way ahead of everyone else—and then quit growing, so that by high school he was a "short guy" and quit picking on you after your own hormones finally got going?

The Thymus and Pineal Body

The **thymus,** located just above the lungs, is considered to be an endocrine gland because its secretion, **thymosin,** is believed to stimulate the development of lymphatic tissue and especially the T-cell lymphocytes (see Chapter 31).

In humans, the **pineal body** is located just above the brainstem, connected by a stalk to a structure known as the **posterior commissure** (Figure 33.17). Today, some biologists have serious reservations about whether the pineal has an endocrine func-

33.17 THE PINEAL BODY

While the pineal body is strongly suspected of having endocrine functions, its specific roles remain obscure. Its only identified product, melatonin, acts as a pigment-controlling and gonad-inhibiting hormone in other animals, but in humans its only role—an unproven one at best—is the regulation of daily cycles in the body. The pineal body is a minute stalked structure, connected to the posterior commissure in the brain.

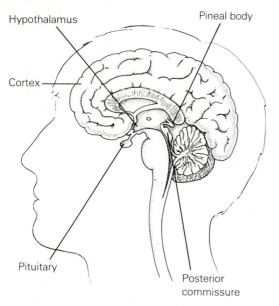

Hypothalamus

Pineal body

Cortex

Pituitary

Posterior commissure

tion, but many others are becoming convinced that it does. The first strong evidence of an endocrine function resulted from investigations in the 1950s, when an active substance was extracted from the pineal body of cattle. When injected into frogs, the substance, now known as **melatonin,** caused the frogs to change color! A cattle extract caused frogs to change color? Now *there* was an interesting bit of news. People began to toy with the phenomenon and found that the color changes resulted from the clumping together of pigment granules, as melatonin does in the pigment bodies (melanophores) of fish and amphibians. It is also known to inhibit gonadal development and to play a role in the control of **circadian rhythms** (daily behaviorial and physiological changes).

Melatonin may also be a gonadal inhibitor in mammals. In rats exposed to long periods of light, melatonin secretion is slowed and the rat enters reproductive readiness. In continuous light, the rat remains in constant readiness. The connection between light and the pineal is indirect, with nerve pathways relaying their impulses several times between the retina of the eye and the pineal gland. In birds, the light receptor is not the eye at all but some unknown region of the brain. This receptor has been located in frogs and lizards and is called the "third eye," or "median eye." Removal of the median eye results in an increase in sexual activity in these vertebrates.

People have continued to speculate about the pineal, particularly about its evolutionary history. The answer is a bit startling. It turns out that it may be the remains of some ancient and vestigial third eye!

Here is a case in which an understanding of evolutionary origins was a real help in working out the physiology and function of an organ. The fossil skulls of the earliest ostracoderms (jawless fishes) have holes for three eyes. The third eye is in the middle, right on top of the skull. Only a remnant of the eye remains in modern fish (and humans as well), but some amphibians and reptiles have retained a rudimentary but recognizable eye. In the long, geographically isolated New Zealand reptile *Sphenodon punctatus,* called the tuatara, the small, median or third eye even has an easily recognizable lens and retina. Some very common American lizards also have a tiny, degenerate third eye on the top of their heads, complete with light receptors and an eyehole in the skull covered by a special translucent scale. This third eye can't form an image, but it is sensitive to light. It is, in fact, the light receptor by which the lizard sets its biological clock, behaving one way when days are long and another way when short, winter days draw on.

Physiologists are notably vague in discussing the role of the pineal body in humans. It is known to secrete melatonin, and the quantity released follows a day-night cycle, rising in the night and falling in the daytime. Some physiologists suggest a gonad-suppressing effect in humans, but this has not been substantiated.

Other Chemical Messengers

We have deliberately omitted a number of other hormone-producing tissues, in some cases because they are discussed elsewhere. The digestive control hormones, for example, and those of the kidney were discussed earlier. There is also the placenta, a temporary endocrine structure that will be discussed later (see Table 33.1). Let's move along to chemical messengers that don't seem to clearly fall into any endocrine category.

Prostaglandins. The **prostaglandins** make up the most recently discovered class of vertebrate hormones. In fact, it was only in 1982 that the Nobel Prize for this discovery was awarded to Sune K. Bergstrom and Bengt Samuelsson of Sweden and John R. Vane of England. Yet some of the prostaglandins are among the most potent biological materials known. Perhaps the reason for their late discovery is that they are not produced by specialized organs, although the first prostaglandins discovered were isolated from sheep prostate glands (hence the name). Prostaglandins of some sort are produced by most kinds of tissues.

The release of prostaglandins can be brought about by other hormones or by almost any irritation of the tissues, including mechanical agitation. Not surprisingly, some prostaglandins are involved in inflammatory responses and in the sensation of pain. Aspirin may be effective against pain, inflammation, and fever because it inhibits the synthesis of prostaglandins.

Prostaglandins have been found to have other actions. For instance, one type causes uterine contractions. Since this type is also present in semen, it may serve as something of a pheromone (chemical signal between different individuals). In the 1970s, Samuelsson found that blood platelets, which are involved in blood clotting reactions, produce a prostaglandin called **thromboxane** that causes platelets to stick together and the walls of arteries to squeeze shut. But Vane also discovered another prostaglandin, **prostacyclin,** that has exactly the opposite effects—it prevents clots, keeping blood fluid, and prevents arteries from closing. Since prostacyclin is produced by the cells that line the blood vessels, it seems that prostaglandins may be important in maintaining normal circulation.

We mentioned that aspirin inhibits prostaglandin synthesis. It turns out that thromboxane synthesis is inhibited by very low levels of aspirin, while such levels have little effect on prostacyclin. So is aspirin good or bad for us? It can be both. First, there is evidence that some heart attacks, arterial disease, and stroke are caused by an imbalance in thromboxane, where spontaneous clotting (embolus formation), and arterial constriction bring on the familiar symptoms. Seemingly, one answer is lowering thromboxane levels, and here is where aspirin can help. On the basis of circumstantial evidence, it seems that aspirin in low doses (one tablet a day or every other day) is *beneficial*. Habitual aspirin users have *far lower* rates of arterial disease and heart attacks than do nonusers.

THE MOLECULAR BIOLOGY OF HORMONE ACTION

So far we have described what hormones do, what adaptive changes they bring about in the organism, and what happens in cases of their under- or overabundance. Now we will take a closer look at *how* they work—how they act on their target cells.

Cyclic AMP: The Second Messenger

Much of our information about the molecular biology of hormone action is somewhat recent, and there are still many unanswered questions. For example, it was only in 1971 that a Nobel Prize was awarded for the discovery of **adenosine 3′, 5′-cyclic monophosphate,** or more simply, **cyclic AMP** or **cAMP,** as a participant in hormone activity. Cyclic AMP (Figure 33.18) is so important that it is now referred to as a **second messenger.** In light of this concept, we could refer to many, but not all, hormones we've been talking about as "first messengers," since it now appears that many of them bring their message only as far as the target cell's plasma membrane. Within the cell, the actual response is often triggered by cAMP. This intracellular second messenger is, in fact, widespread in nature; it is the same molecule that causes the curious solitary cellular slime molds to aggregate.

The importance of cAMP came to light as scientists were trying to discover how epinephrine works by studying its activity in the liver. They found that, although many hormones operate by stimulating the transcription of specific kinds of mRNA in the target cell nucleus, epinephrine works entirely in the cytoplasm.

You may recall that one of the effects of epinephrine is a sudden increase in the blood glucose level, which is accomplished through the enzymatic breakdown of glycogen reserves in the liver. The chain of events leading to the elevated glucose level is rather well understood (Figure 33.19).

Epinephrine, released from the adrenal medulla, is carried by the blood to its target cells, such as those of the liver. Like all target cells, liver cells have very specific hormone receptor sites along their plasma membranes. Once fixed to its specific receptor in a liver cell plasma membrane, epinephrine activates an enzyme known as **adenylate cyclase.** This enzyme is part of the membrane system itself and remains inactive until the epinephrine is in place. But when adenylate cyclase is activated, it immediately converts cytoplasmic ATP to cAMP, starting a sequence that activates critical enzymes and eventually leads to the breakdown of glycogen into glucose, which then escapes into the blood.

Apparently, at least a dozen other hormones stimulate the production of cAMP and use it as a second messenger. The list includes glucagon, ACTH, LH, FSH, TSH, PTH, ADH, and calcitonin. But the question arises: How is the activity of so many hormones regulated if they all activate the same second messenger? The answer incorporates a number of important points. For instance, each cell has specific receptor sites and thus are target cells to specific hormones (although they may contain receptor sites for more than one hormone). But equally important, cAMP, perhaps in all cases, activates a specific starting enzyme called a **protein kinase.** Protein kinases are a class of enzymes that activate other enzymes through the transfer of phosphate from ATP to that enzyme. So, in a sense, the situation is a "set-up"; that is, the specificity of hormonal action depends on which protein kinases are present in the cell. This ultimately depends on which structural (enzyme-encoding) genes are active. Because of these factors, a number of hormones can utilize cAMP as a second messenger.

Other Cell-Activating Mechanisms

At a time when laboratory after laboratory was reporting cAMP involvement in hormone after hormone, it began to appear that all polypeptide hormones might function in this way. This turned out not to be the case, at least with insulin and a number of mitosis-prompting agents called **mitogens.** Such hormones do interact with membrane receptors, but then things get a bit more complex. Each is known to form a permanent covalent linkage with its membrane receptor. When a number of these receptors have been bound to the appropriate circulating hormone, all the receptors of that type cluster at one site on the surface of the cell. Then, in what seems to be one "great, slow gulp," the patch of receptors and their hormones sink beneath the surface of the cell, engulfed in a vacuole that pinches off from the plasma membrane. You may recall that this form of transport is called **receptor mediated endocytosis** (see Figure 5.14).

Apparently, these hormones are then digested, and in some manner still unknown, the nucleus is stimulated to undergo mitosis (in the case of the mitogens). Other second messengers may be involved in this process, but so far they have not been identified.

Steroid Hormones: Direct Action at the DNA Level. Steroid hormones apparently have no membrane receptors. Since they are lipid-soluble, they readily pass through plasma membranes and into cells. No matter if the cells are not target cells; such hormone molecules, once inside, find themselves unemployed and are denatured or simply diffuse right out again. In appropriate target cells, however, the steroid hormones are caught and tightly bound to a receptor called a **cytoplasmic binding protein.** Then the newly formed protein-hormone complex moves into the nucleus, where it joins with yet other proteins that are part of the chromatin complex. The chromosome-bound receptor might be thought of as a tiny organelle of cellular machine that has control of the genetic program. In any case, the complex of hormone, cytoplasmic receptor, and chromosomal receptor initiates gene activation, with the production of appropriate mRNA and protein (Figure 33.20).

33.18   CYCLIC AMP

Cyclic AMP (adenosine 3,5-cyclic monophosphate, or simply cAMP) is called a *second messenger* because it is activated by hormones, or "first messengers." Cyclic AMP has a variety of specific roles in the cell, which range from the activation of enzyme systems to the activation and perhaps deactivation of genes.

ATP

Cyclic AMP

Pyrophosphate

33.19 THE ACTION OF CYCLIC AMP

The action of epinephrine begins with epinephrine binding to its specific receptor site on the membrane of a liver cell. **(a)** The enzyme adenylate cyclase, which is also incorporated into the membrane, is activated by the receptor-hormone complex, and ATP is converted to cAMP within the cytoplasm. **(b)** Following this, the cAMP (the "second messenger") joins with an inactivated enzyme complex; with the aid of ATP, it activates the enzyme by removing a blocking poly-

peptide. **(c)** The activated enzyme phosphophosphorylase kinase converts another enzyme, phosphorylase *b*, to phosphorylase *a*, again with help from ATP. **(d)** Phosphorylase *a* then begins cleaving and phosphorylating the bonds between glucose molecules in glycogen. **(e)** Eventually, the phosphates are removed from glucose and the molecule crosses the liver cells' plasma membranes into the circulatory system.

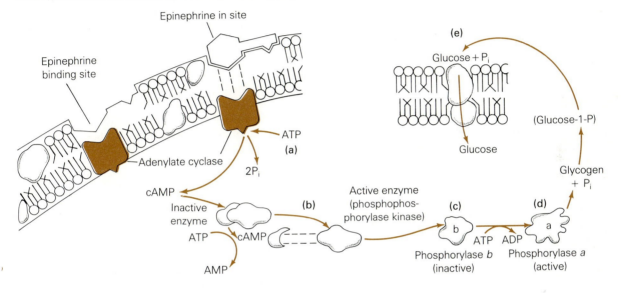

33.20 THE ACTION OF A STEROID HORMONE

Progesterone, a lipid-soluble steroid, readily passes through the plasma membrane and joins a receptor called a cytoplasmic binding protein. Together they penetrate the nucleus where the hormone-protein complex joins a chromosomal receptor which then activates a DNA segment, RNA is transcribed, and a protein is produced along ribosomes.

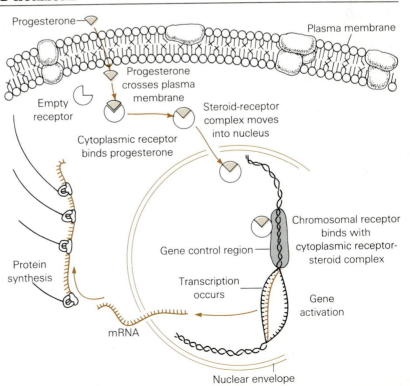

The system was first worked out by Bert W. O'Malley and his co-workers, then at Vanderbilt University. It was known that estrogen and progesterone treatments could cause the oviducts of female baby chicks to produce egg-white albumin and other proteins. These researchers labeled steroid hormones with radioactive tags and followed them through the sequence described above. At each step, the various intermediate compounds were isolated and identified. And at last something was known directly about how hormones turn on genes.

So, the continuous changes occurring within the living multicellular bodies are sometimes mediated by chemical messengers as the bodies constantly adjust and readjust to meet the demands of their specific worlds. One wonders why a complex system of messengers arose at all. Wouldn't it be simpler for cells to initiate change simply in response to environmental cues rather than to be directed by secretions from another part of the body—a part that was itself responding to the environment? Perhaps so, but nature doesn't always operate according to the rules of simplicity. It goes with what it has, and it may have, in fact, been simpler for existing tissues to respond to the environment by incidentally leaking fluids, fluids that caused other tissues to change. The kinds of changes that were adaptive were kept and refined. As time passed, the various systems would have interacted ever more precisely and become more regulated and coordinated, even as the target areas became increasingly sensitive. The result is an orchestrated array of chemical changes within living bodies as first this hormone, and then that, surges and ebbs, causing living things to change adaptively in response to their varied surroundings.

APPLICATION OF IDEAS

1. As two 19th century physiologists attempted to determine the function of the pancreas, they observed flies swarming around the urine of dogs from which the pancreas had been removed. This observation was an initial step in our understanding of the role of insulin in sugar metabolism. What might the next steps have been in isolating insulin and what rigorous rules should the investigators have followed in establishing its function?

2. Assuming that insect hormones could be synthesized in sufficient quantity, suggest a specific hormone that might be applied to insect control. Explain your choice, how it might work, and reasons why this method of insect control might be more ecologically suitable than the use of chemical insecticides.

3. While the anterior pituitary and thyroid glands are known to control growth, malfunctions in any of the endocrine glands can also affect growth. Considering each of the endocrine glands, explain why this is so.

4. The controlling functions of the endocrine system often overlap those of the nervous system. Cite examples of this overlap and comment on the adaptive significance, if any, of this.

KEY WORDS AND IDEAS

THE NATURE OF HORMONES

1. **Hormones,** and other **chemical messengers,** are chemically active substances that cause changes in cells and tissues. The **hormonal system** often shares regulatory functions with the nervous system.

HORMONES AND THE INVERTEBRATES

1. The source of hormones in many invertebrates is the nervous system.

Molting Activity in Crustaceans

1. **Molting,** or **ecdysis,** begins with **preecdysis** (absorption of exoskeletal materials), followed by ecdysis (loss of the old cuticle, stretching of the new), **postecdysis** (hardening of new cuticle), and **interecdysis** (variable amount of time between molts).

2. In crayfish, the **Y-organ** secretes **crustecdysone,** the molting hormone, while the nearby **X-organ**

secretes **molt-inhibiting hormone (MIH).** Light is an essential part of regulating ecydysis.

3. Establishing the validity of a hormonal substance includes (a) removing the organ and observing the appearance of deficiency symptoms, (b) isolating the substance in pure form, (c) removing deficiency symptoms by introducing the substance, and (d) withdrawing the substance and observing the reappearance of the deficiency symptoms.

Hormonal Activity in Insects

1. Many insects pass through a complete metamorphosis, including **egg, larva, pupa,** and **adult** states. Insect metamorphosis is influenced by the **corpus cardiacum, corpus allatum,** and the **prothoracic,** or **ecdysal gland.** The juvenile state is maintained by **juvenile hormone (JH).** Final molting and the end of the larval state occur when **brain hormone (prothoracicotrophic hormone—PTTH)** from the corpus cardiacum stimulates the prothoracic gland to secrete ecdysone.

2. Transplanted larval **corpora allata** cause an extension of the larval state. If the transport of hormones is stopped, parts of the body mature while other parts remain in a larval state.

VERTEBRATES:
THE HUMAN ENDOCRINE SYSTEM

1. The human **endocrine system** consists of about nine distinct glands and a number of less organized tissues. The endocrines are **ductless glands,** whose secretions enter the blood (as opposed to **exocrine glands,** whose secretions empty through ducts). Hormones act in minute amounts and are rapidly degraded.

2. Vertebrate hormones are classified as **steroids, proteins, polypeptides, peptides, modified amino acids,** or **prostaglandins.**

The Pituitary

1. The **pituitary gland** (or **hypophysis**) influences other endocrine glands and body tissues through negative feedback loops. The **hypothalamus** of the brain acts as an intermediary, monitoring feedback activity and stimulating the pituitary.

2. The pituitary includes two major lobes, the **anterior** and **posterior pituitary.** The anterior lobe receives **releasing hormones (releasing factors)** from the hypothalamus via the blood, while the posterior lobe receives hormones from the hypothalamus, transported within connecting neurons.

3. The hypothalamus influences the anterior pituitary through secretions of six releasing hormones, and two inhibiting hormones. The hormones are peptides that are secreted in trace amounts.

4. The anterior pituitary hormones include the following:
   a. **Adrenocorticotropic hormone** (ACTH) regulates fat metabolism and stimulates the adrenal cortex to secrete its **glucocorticoids,** which are involved in glucose metabolism. Negative feedback begins when rising levels of the adrenal hormone are sensed by the hypothalamus, which responds by lessening its pituitary stimulation.
   b. **Growth hormone** (also, GH or **somatotropin**) accelerates the cellular uptake of amino acids, decreases carbohydrate metabolism, and stimulates insulin release. Excesses of GH in childhood result in **giantism (pituitary giants),** while insufficiencies cause **dwarfism (pituitary dwarfs).** Excesses in the adult cause **acromegaly,** growth in cartilaginous areas of the body.
   c. **Thyroid-stimulating hormone** (TSH or **thyrotropin**) stimulates the thyroid to release its hormones.
   d. **Prolactin** promotes milk production, while **oxytocin** promotes milk release. Prolactin increases in males and females during sexual excitement.
   e. **Follicle-stimulating hormone** (FSH) and **luteinizing hormone** (LH) are **gonadotropins** that stimulate gamete and sex hormone production in the gonads.

5. The posterior pituitary releases the hormones oxytocin and **antidiuretic hormone (ADH or vasopressin).** In addition to milk release, oxytocin stimulates the uterine muscles to contract during sexual excitement and during labor. ADH, a **diuretic,** regulates water reabsorption in the collecting ducts of the kidney.

The Thyroid

1. Two of the three thyroid hormones, **thyroxin** (T_3) and **triiodothyronine** (T_4) differ only in their iodine content. In amphibians thyroid hormone stimulates metamorphosis, while in humans it influences carbohydrate metabolism.

2. **Hyperthyroidism** produces weight loss and nervousness, while a shortage of iodine produces a **goiter,** a visibly enlarged thyroid gland. **Hypothyroidism** causes **myxedema**—sluggish behavior, body puffiness, and fluid retention. In children, hypothyroidism leads to **cretinism,** severe mental and physical retardation.

3. **Calcitonin,** the third thyroid hormone, works with the **parathyroid glands** to regulate calcium.

The Parathyroid Glands and Parathormone

1. **Parathormone (parathyroid hormone** or **PTH),** a polypeptide, cooperates with calcitonin in regulating the distribution of body calcium. Parathormone increases calcium levels in the blood by increasing intestinal absorption and kidney reabsorption, and by stimulating the bone decal-

cifying activity of **osteoclasts.** Calcitonin performs roughly the opposite actions including the stimulation of bone calcifying activity by **osteoblasts.**

2. Low levels of PTH produce blood calcium deficiencies resulting in severe muscle spasms and loss of muscle control. High PTH levels cause severe bone calcium loss **(osteoporosis)** and irregularities in muscle contraction.

The Pancreas and Islets of Langerhans

1. The **islets of Langerhans** are clusters of insulin-secreting **beta cells,** glucagon-secreting **alpha cells,** and **delta cells** that are believed to secrete **somatostatin,** or **growth hormone inhibiting hormone** (GHIH).

2. Glucagon stimulates the liver to convert glycogen to glucose, which elevates the blood level. Insulin stimulates cells to take up glucose, thus lowering blood levels.

3. Deficient insulin levels cause **hyperglycemia** (high blood sugar), commonly called **diabetes mellitus.** Diabetes is most often caused by a shortage of or defects in insulin receptor sites, but it can be caused by a shortage of insulin or an overabundance of the insulin-deactivating enzyme **insulinase.**

4. The symptoms of diabetes are varied, but glucose in the urine is common. It is often successfully controlled by diet and insulin injection.

5. **Hypoglycemia** (low blood sugar) is often a result of insulin overproduction. Commonly, the blood regulating feedback mechanism goes awry, and a time lag occurs between increased blood sugar and insulin release.

The Adrenal Glands

1. The **adrenal glands** have two layers—the outer **cortex** and inner **medulla.** The adrenal cortex secretes **mineralocorticoids,** including **aldosterone** and **cortisol (hydrocortisone).**
   a. Aldosterone increases potassium excretion by the kidney and slows sodium excretion. Hyperactivity causes **alkalosis,** whose symptoms include **edema** (fluid retention). Aldosterone hypoactivity causes excessive salt and water loss and potassium retention, causing a potentially fatal **acidosis.**
   b. Cortisol increases blood levels of glucose, amino acids, and fatty acids and is an anti-inflammatory agent, suppressing and killing histamine-releasing lymphocytes.
   c. Adrenal steroid sex hormones include **androgens** (male) and **estrogens** (female). They play a role in fetal sex development.

2. Adrenal medullary hormones include **epinephrine** and **norepinephrine** (adrenalin and noradrenalin). Their general effect is to prepare the body for emergencies (for example, increasing

heart and breathing rate, elevating blood sugar, and shunting blood away from areas such as the digestive tract).

The Gonads: Ovaries and Testes

1. The ovaries produce two kinds of steroids, **estrogens (estradiol, estriol,** and **estrone)** and **progesterone,** while the testes produce the steroid hormone **testosterone.** Sex hormones influence reproduction, but also influence body development and bone growth.

The Thymus and Pineal Body

1. The thymus secretes **thymosin,** which stimulates T-cell lymphocyte development.

2. The only known secretion of the **pineal body** is **melatonin,** which causes color changes in frogs and inhibits gonadal development. It may inhibit sexual activity in mammals and control **circadian** (daily) rhythms.

Other Chemical Messengers

1. Other hormones include those that control digestive enzymes and those of the placenta.

2. Prostaglandins are produced almost universally in the body. Some are involved in inflammation, others cause uterine contractions; one **(thromboxane)** aids blood clotting, while another **(prostacyclin),** slows clotting. Aspirin inhibits thromboxane synthesis, and new studies on aspirin suggest that this may be related to reduced incidences of heart and arterial disorders.

THE MOLECULAR BIOLOGY OF HORMONE ACTION

Cyclic AMP: The Second Messenger

1. **Adenosine 3,5-cyclic monophosphate (cyclic AMP or cAMP)** is called a **second messenger** (hormones are the first messengers).

2. Epinephrine attaches to a specific receptor site on its target cell, activating the membranal enzyme **adenylate cyclase.** Adenylate cyclase converts ATP to cAMP, which begins a series of reactions that converts glycogen to free glucose.

3. Other hormones use the cAMP mechanism. Wide use is made possible by specific cell receptors for different hormones and preorganization of specific enzyme pathways involving the **protein kinases** (enzyme-activating enzymes).

Other Cell Activating Mechanism

1. *Receptor mediated endocytosis* may be responsible for insulin and **mitogens** entering the cell. When enough are bound to membrane receptor sites, they are taken in by phagocytosis. Once inside the cell, they are able to stimulate activity.

2. Steroid hormones are lipid-soluble and may pass right through the plasma membrane. Inside the cell, they join **cytoplasmic binding proteins.** Upon entering the nucleus, the complex joins chromatin proteins and activates genes.

REVIEW QUESTIONS

1. Characterize the hormone—its role, quantity required, and longevity. (842)

2. Describe the interplay between the X-organ and Y-organ of the crayfish. (843–844)

3. Name and briefly describe the four stages of molting. (843)

4. List the steps that an endocrinologist might follow when identifying a new hormone. (844)

5. List the secretions of the insect corpus cardiacum, corpus allatum, and prothoracic gland. Which promotes the final larval molt and emergence of the pupa and adult? (844–845)

6. What happens when PTTH is present in the insect head and thorax but is prevented from reaching the larval abdomen? (845)

7. In what basic way does the endocrine gland differ from the exocrine gland? (845)

8. List the six chemical classes of hormones. (846)

9. Compare the functional relationship of the anterior and posterior pituitary lobes with the hypothalamus. What does this tell you about the posterior lobe as a gland? (848–849)

10. List several releasing hormones, and generally explain what they do. (850)

11. Review the negative feedback loop established between the hypothalamus, pituitary, and adrenal cortex. (850–851)

12. How does the role of somatostatin differ from that of somatotropin? (851)

13. What are the targets of the anterior pituitary hormones ACTH, TSH, GH, and prolactin? (850–851, 852)

14. What are the effects of too much GH in the child? In the adult? (851–852)

15. Under what conditions do prolactin levels rise in the blood? (852)

16. What is the relationship between oxytocin and prolactin? (852)

17. What is the true source of oxytocin and antidiuretic hormone? What is the target and action of the latter? (852)

18. How does the chemical structure of triiodothyronine differ from that of thyronine? Compare their actions. (853)

19. Compare the symptoms of an overactive thyroid with those of an underactive thyroid. Under what conditions does cretinism occur? (853–854)

20. Compare the calcium-regulating action of calcitonin with that of parathormone. What are the symptoms of an overabundance of the latter? (855)

21. Name the three types of islet cells, compare their abundance, and name their secretions. (855)

22. List several symptoms and several specific causes of diabetes mellitus. (856)

23. Briefly discuss the interaction of glucagon and insulin in the precise regulation of blood sugar. (856)

24. What are some possible causes of hypoglycemia? What symptoms accompany the abnormality? (857)

25. Name the two regions of the adrenal gland and list their hormones. How does the chemical structure of the two groups of hormones differ? (857–858)

26. Briefly describe the manner in which sodium and potassium are regulated. (857)

27. Describe the action of the hormone cortisol. Which of its actions are also produced by the drug hydrocortisone? (858)

28. List five body reactions to epinephrine. Under what conditions do these reactions normally occur? (858–859)

29. Other than their reproductive functions, what roles do the sex hormones have? (859)

30. What are the two known actions of melatonin, the pineal secretion? (860)

31. Where are the prostaglandins produced? List two actions of prostaglandins in the circulatory system. What might a low level of prostacyclin cause in the blood? (861)

32. Outline the events that occur when epinephrine reaches one of its target cells. Include membranal and cytoplamic events and the final outcome. (861–862)

33. How is it possible for several hormones to use the same second messenger? (862)

34. Describe the manner in which receptor mediated endocytosis brings about the entry of chemical messengers into the cell. (862)

35. At what level do the hormones progesterone and estrogen seem to work? Briefly review how they enter and activate the cell. (862)

Neural Control: Neurons and Their Roles

Our planet is a lively place, and one of mixed blessings. It provides us not only with opportunities, but dangers. In fact, that is what much of life is about: taking advantage of the opportunities and avoiding the dangers. This means, of course, that living things must be able to accurately assess the nature of the environment and to respond to it appropriately. Deer, rabbits, and grasshoppers are able to sense the presence of young tender plants. As the plants quietly submit to the ravages of the plant eater, they may be avenged by some sharp-toothed predator peering from behind a rock. The predator is able to detect and evaluate signals emanating from the prey as surely as the plant eaters can find grass. Both kinds of animals share the same environment, but their sensory abilities are specialized to react to different aspects of it. Here we will see not only how signals are detected but also how they move through the body, activating a pattern of responses that help the animal succeed in a complex and changing world.

Essentially what we will see is a network of highly specialized cells that are devoted to the task of transmitting information. These are the nerve cells, or **neurons,** and they do three things. They *detect* events in their surroundings and send a signal to coordinating centers. These centers, which con-

sist of other neurons, *integrate* the information and formulate a response. Additional neurons carry this decisive message to some body part, where a *response* is initiated. As you can see, the three key terms are "detection," "integration," and "response." The environmental cue that started all of this is referred to as a *stimulus*. Figure 34.1 illustrates the three aspects at work in a real-life situation.

THE NEURON

Cellular Structure

Neurons come in many sizes and shapes, but they are characterized by having a **cell body** from which a number of processes extend (Figure 34.2). These processes can be extremely long, since some neurons must carry messages to a distant part of the body (such as those that reach from your foot to your spinal cord).

The cell body contains the nucleus and, generally, most of the cell's cytoplasm. The cytoplasm contains the typical organelles such as ribosomes, an endoplasmic reticulum, and numerous secretory bodies. The cell body can produce **neurotransmit-**

34.1 ANIMAL SENSITIVITY AND RESPONSE

The ability to respond rapidly to incoming environmental cues is an outstanding characteristic of animals. This ability is vital to finding food and recognizing and avoiding danger. Specialized receptors detect events in their surroundings, and this "raw information" is directed to the animal's brain for processing and integration. The animal's response, which depends on a variety of factors, may range from automatic and innate responses to complex actions based on intelligence.

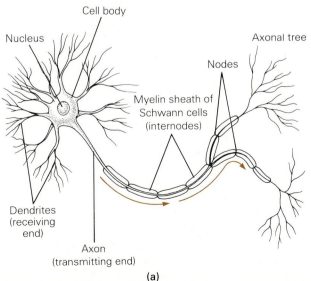

ters—chemical messengers that ordinarily must be present for a neural impulse to be relayed from one neuron to another. Neurotransmitters are transported within the slender neural processes along a highly organized cytoskeletal framework. Eventually they are released at specific sites, where one neuron associates either with another neuron or with other kinds of responding structures called **effectors.** Effectors can be any of the muscles, including those of skeleton, heart, blood vessels, and intestine. An effector can also be one of the many glands, such as a salivary gland, which upon stimulation pours its secretions into the mouth.

Among the unusual features of a neuron are the lengthy cellular extensions, or neural processes (see Figure 34.2). There are two types: **dendrites** and **axons.** A dendrite (which means "little tree") is the receiving end of the neuron. It receives stimuli from the environment or from other neurons, depending on its specialization, and converts this information into a neural impulse that is then transmitted *toward* the cell body. (At times the cell body itself receives impulses from the environment or from other neurons, and dendrites are not involved).

An axon transmits the neural impulse *away from* the cell body. It may communicate with other neurons or directly with an effector. A neuron often has

34.2 THE NEURON

(a) Each neuron has a receiving end consisting of slender, highly branched processes called dendrites; the cell body, an enlarged region with many typical cellular structures, including the nucleus; and an axon, the lengthy portion which branches at its terminus. Axons may be enclosed in a fatty, myelin sheath consisting of numerous Schwann cells or oligodendrocytes. These cells are interrupted by minute gaps called nodes. In most instances, neural impulses move one way only—from dendrite to axon. (b) A primary function of the cell body is the production of neurotransmitters, specialized molecules that are essential in the relaying of impulses from one neuron to the next. Supporting this secretory function are a prominent Golgi complex, numerous mitochondria, and an extensive endoplasmic reticulum. Vesicles containing neurotransmitters pass along an extensive cytoskeleton in the axon to be sequestered in the axonal tree.

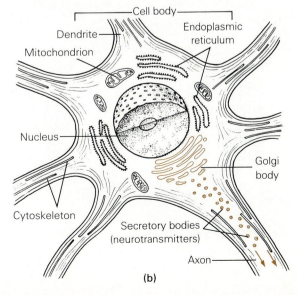

(a)

(b)

many dendrites, but it usually has only one axon. The single axon, however, may branch at any point along its length. An axon commonly divides and redivides at its tip, forming a terminal **axon tree** (also called an arborization), each branch of which forms a knoblike ending. It is from these endings that the axon releases a neurotransmitter that either causes an impulse to begin in the next neuron or activates an effector. Where the tips of the axons innervate a muscle fiber, they spread to form **neuromuscular junctions** (see Figure 28.27).

The axon seen in Figure 34.2 is surrounded (jelly-roll fashion) by a **myelin sheath,** a flattened sheath of fatty material typical of that found in many, though not all, types of vertebrate neurons. Like any lipid, myelin has great electrical resistance and acts as an insulator. But the sheath is interrupted at frequent intervals by the **nodes,** (also called nodes of Ranvier), where the axon is in direct contact with the surrounding intercellular fluid. The nodes are small spaces between the end of one wrapping cell, or **internode** as the wrapped regions are called, and the beginning of the next. The fatty myelin sheath is composed of layers of the flattened and rolled plasma membrane of a specialized axon-encasing cell. Outside the central nervous system—

the brain and spinal cord—myelin sheaths are formed from **Schwann cells** (Figure 34.3), while within the central nervous system, they are formed from somewhat similar cells called **oligodendrocytes.** The myelin sheath and the nodes have much to do with the speed of neural impulses, as we will see shortly.

Types of Neurons

Neurons fall into three functional categories: **sensory neurons, interneurons,** and **motor neurons** (Figure 34.4). Sensory neurons (or **afferent neurons**), as their name implies, conduct impulses from **sensory receptors,** structures in the body that in turn are responsive to stimuli such as air movement, gravity, touch, light, heat, chemicals, or other more subtle environmental stimuli. Sensory receptors, as we will see, are quite specialized; thus, the cells in the retina of the eye are sensitive only to light, while others deep in the skin respond only to pressure. Generally, sensory neurons carry information concerning not only the external environment, but the internal as well. Thus we also receive "situation reports" from our internal organs.

Interneurons are nerve cells that communicate

34.3 THE MYELIN SHEATH

In this electron micrograph, a cross section through a myelinated axon reveals several layers of wrappings surrounding the axon itself. The wrappings are encircling membranes of Schwann cells or oligodendrocytes, whose cytoplasm and nucleus are seen as an enlarged region. The membranal wrappings produce an effective insulation, which plays an important role in impulse propagation. During development **(b)**, a sheath-forming Schwann cell begins its enveloping action by simply surrounding part of an axon. Then one free end of the plasma membrane grows under the other, advancing along in a "burrowing" action until it has wrapped itself about the axon several times. This growth must be completed by numerous cells for a complete sheath to form.

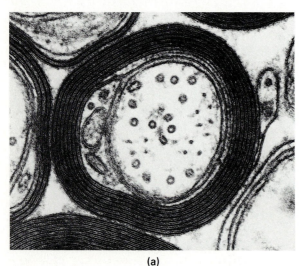

(a)

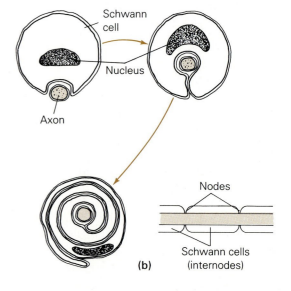

(b)

34.4 THREE KINDS OF NEURONS

Nerve cells generally fall into three categories—sensory neurons, interneurons, and motor neurons. Interneurons **(a)** and **(b)** are the most variable of the three types; they may be quite small with numerous, highly branched processes, as we see in the pyramid cells from the cortex of the brain. Purkinje cells are perhaps the most branched, each communicating with as many as 100,000 other neurons. Many interneurons make their connections within the spinal cord. Many motor neurons **(c)** originate in the cord, sending their axons out over great distances to reach effectors such as muscles of the skeleton, heart, blood vessels, and intestines. It is typical in vertebrates for sensory neurons **(d)**, particularly those of the skin, to have exceedingly long dendritic processes, since they extend all the way to just outside the spinal cord, where their cell bodies accumulate in ganglia. In these instances, the axons may be quite short since they enter the cord where they communicate with interneurons located there.

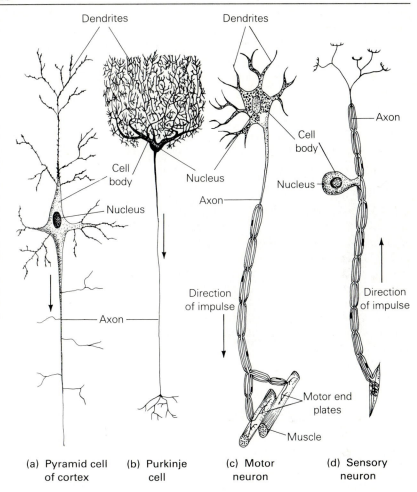

(a) Pyramid cell of cortex

(b) Purkinje cell

(c) Motor neuron

(d) Sensory neuron

only with other nerve cells. They receive input from other neurons such as the sensory neurons and are largely responsible for integrating the stimuli and responses in the nervous system. In our own nervous system, interneurons make up much of the spinal cord and brain.

The motor neurons (or **efferent neurons**), as their name implies, are responsible for the body's reaction to its environment. They receive information from the interneurons and transmit this information to effectors, structures that are able to react.

In addition to the neurons, the nervous system—specifically the central nervous system—contains vast numbers of supporting cells known as **neuroglia.** In fact, neuroglial cells, or **glial cells,** comprise up to 90% of the cellular matrix of the brain. The function of neuroglia is poorly understood. We do know, though, that the most common

type of neuroglia, the star-shaped cells called **astrocytes,** appear to function in transporting materials from the capillaries to the nerve cells. We will return to the neuroglia in the next chapter.

Nerves

The axons and the peripheral processes or branches of many neurons often travel together as they extend throughout the body. In such parallel arrangements, they form larger structures called **nerves** (Figure 34.5). A nerve can be likened to a telephone cable carrying many individual lines, each insulated from the others. The large glistening nerves are surrounded by their own coverings of tough connective tissues. As you are probably aware, damaging or severing a major nerve can produce, in the organs served, such devastating effects as paralysis or loss of sensitivity, or both.

34.5 NERVE STRUCTURE

A nerve, as seen in cross section through the scanning electron microscope, contains numerous neural processes, possibly both axons and dendrites. Each contains its own insulating sheath, and the entire ensemble is surrounded by dense connective tissue that binds it into its cablelike structure. Blood vessels penetrate the nerve, providing the exchanges needed to maintain the neurons.

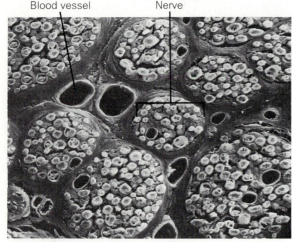

Blood vessel   Nerve

From *Tissues and Organs: A Text-Atlas of Scanning Electron Microscopy*, by Richard G. Kessel and Randy H. Kardon, W. H. Freeman and Company, copyright © 1979.

THE NEURAL IMPULSE

While neurons can be said to conduct impulses from one part of the body to another, it's important not to confuse the terms "conduct" with the usual electrical connotation. Neurons are far more than simple conductors, and a moving **neural impulse** is quite different from an electrical current passing through a wire conductor. Electrical current passing through a conductor travels at a much greater speed than the neural impulse. But most important, while, because of resistance, an electrical current in a copper wire diminishes over time and distance, neural impulses, once started, do not diminish.

Once the **threshold of stimulation** (minimum stimulation required) has been reached, an impulse traveling along a fiber will *regenerate* itself so that it is as strong at the end of its journey as it was at the beginning. To understand precisely how the neural impulse is generated, how it moves along, and how conditions for the succeeding impulses are restored, we must look more closely at the special nature of the neural membrane and at the roles of certain ions. We can divide the complex events of the neural impulse into three significant parts:

1. the role of ions in establishing a **resting state,** or **polarized state** (perhaps better called a "ready state");
2. creation of the neural impulse, also dramatically called an **action potential** and described as a **depolarizing wave;**
3. reestablishment of the polarized or resting state, or **repolarization.**

The Resting State: A Matter of Ion Distribution

The resting state in a neuron refers to the period when no impulses are being generated. But more than that, it is a time when ions inside and outside the neuron reach a precarious distribution—one that, if disturbed, will bring about a sudden, very significant rush toward equilibrium. Let's see what this is all about.

The ions important to neural activity are sodium ions (Na^+), potassium ions (K^+), and larger, less mobile, negatively charged proteins. During the resting state, the negatively charged proteins are immobile and remain within the neuron's interior, so although they are quite important, we can think of them as stationary and inactive. The membrane itself, while at rest, is impermeable to sodium, most of which remains outside. Potassium ions are freer to diffuse down their concentration gradient, so they are found on both sides.

There is one more essential aspect of this distribution. Like many cells, the neuron maintains steep sodium/potassium gradients through numerous ATP-driven **sodium/potassium ion exchange pumps.** You may recall this active transport mechanism from Chapter 5 (see Figure 5.12). For every molecule of ATP hydrolyzed, the ion exchange pump moves three sodium ions out of the axon and two potassium ions into the axon. This work helps maintain the two opposite concentration gradients of ions: sodium ions (Na^+) become heavily concentrated outside the neuron, while potassium ions (K^+) accumulate inside. But since potassium ions are able to diffuse across the membrane and down their gradient, they readily do so. In fact, the potassium ions would probably reach equilibrium were it not for the negatively charged proteins within the neuron. Negative charges attract positive charges, therefore, much of the potassium is held back, and an equilibrium of sorts is reached between potassium's outward diffusion gradient and the electrical attraction from within. Finally, with potassium reaching its tricky equilibrium, and with sodium accumulating outside the neuron, the net effect is that the charges between its interior and exterior are unbalanced—and this becomes a vital part of neural conduction (Figure 34.6).

The Resting Potential. With some positive ions excluded from the neuron and others leaving through diffusion, the result is that during the resting state, the interior of the cell becomes *negatively charged* relative to the outside. In other words, the neuron is *polarized*.

The slightly polarized state can be verified experimentally by inserting one tiny electrode into an axon and placing another on its outer surface. This will permit the measurement of the voltage potential developed across the membrane by the differently charged regions. If the electrical charges inside and outside the resting axon's membrane are compared, ordinarily we find that there is an

34.6 THE POLARIZED, RESTING NEURON

When not conducting neural impulses, a neuron **(a)** maintains a sodium ion (Na$^+$) concentration outside the axon, and a potassium ion (K$^+$) concentration inside. **(b)** Within the axon are many immobile negatively charged proteins. While the membrane is impermeable to sodium, potassium can freely diffuse back and forth. Some of the positively charged potassium ions diffuse out, leaving the interior slightly negative in relation to the exterior. This is characteristic of the resting neuron. Help in establishing the sodium/potassium gradients, and especially in moving sodium out of the neuron after an impulse, is provided by special membranal carriers known as ATP-dependent sodium/potassium exchange pumps. They transport sodium out of and potassium into the neuron. In each ATP-powered revolution, three sodium ions are pumped outside and two potassium ions are carried in.

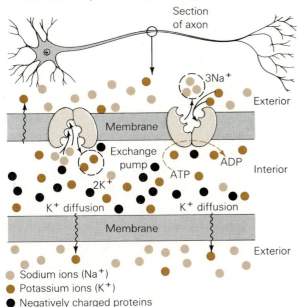

Sodium ions (Na$^+$)
Potassium ions (K$^+$)
Negatively charged proteins

electrical potential difference across the membrane, or **resting potential,** of –60 millivolts (mV: $\frac{1}{1000}$ of a volt; the –60 mV varies with different information sources). Thus the resting neuron is electrically polarized (Figure 34.7a).

The resting potential represents a significant amount of potential energy. To tap this energy, some of the opposing net charges on either side of the axonal membrane must be allowed to join. When this happens, a small current flow is produced and some energy is dissipated. With this in mind, let's return to the neural impulse.

The Action Potential

When a neuron is sufficiently stimulated, the point of stimulation becomes suddenly *depolarized,* and the depolarization spreads rapidly along the length of the neuron, followed within milliseconds by repolarization. The depolarization is created by rapid change in membrane permeability and a corresponding shift in the precarious balance of ions maintained during the resting state. This shift of ions and the accompanying shift in electrical charges is the neural impulse, or action potential (see Figure 34.7b, c, and d). What causes the action potential? The answer is quite complex, so let's begin with *what* happens and return for the "whys" shortly.

To begin, when a nerve impulse passes any point in an axon, the plasma membrane at that point suddenly becomes permeable to sodium ions, and the resting potential is rapidly lost as sodium ions rush down their steep gradient into the cytoplasm of the axon. The –60 mV differential is abruptly shifted to an electrical peak of +40 mV—hence the depolarization of the neuron. A shift of 100 mV may seem a lot, but in actuality such a change involves relatively few ions. Neurobiologists estimate that only one ion in 10 million actually moves across the membrane and most of the axon's interior is undisturbed. The depolarizing event and those following are illustrated in Figure 34.7e.

Action potentials last briefly. The region just inside of the neuron becomes positively charged only for milliseconds. But as one area along the neuron experiences this shift in ions, it triggers the next area to do the same. Once started, the shift in ions continues in a cascading manner—in a *wave of depolarization*—down the length of the axon. But, in its wake, an immediate recovery or repolarization begins and the resting potential of the neuron is quickly reestablished. Again, relatively few ions are involved in any single action potential, so a given

34.7 THE ACTION POTENTIAL

Neural impulses are depolarizing waves that sweep along the axon, immediately followed by repolarization. During its passage, the resting potential of –60 mV jumps to +40 mV and this peak is called an action potential. **(a)** Events begin in the resting neuron where, because of the distribution of ions, the resting potential is about –60 mV. **(b)** When an action potential begins, an inrush of sodium ions occurs as the membrane permeability changes. **(c)** The permeability to sodium ceases abruptly as the membrane potential reaches +40 mV, followed by a rapid recovery. **(d)** Recovery or repolarization occurs as potassium ions rush outward, once again restoring the resting potential of –60 mV. **(e)** Each of the events is plotted against the mV scale.

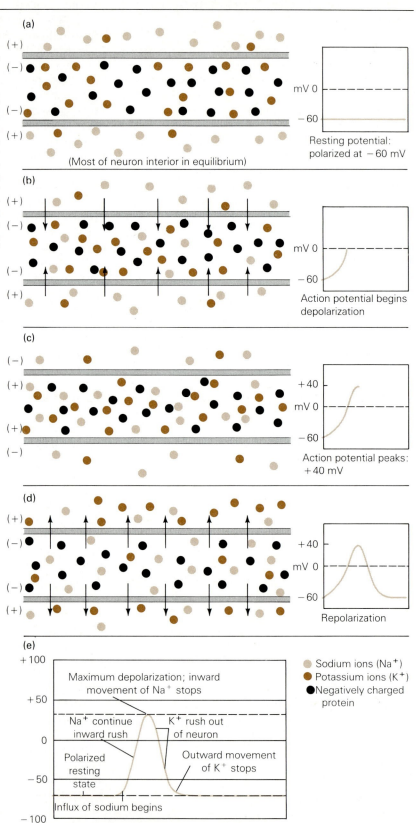

(a)

(+)
(−)

(−)

(+)

mV 0

− 60

(Most of neuron interior in equilibrium)

Resting potential:
polarized at − 60 mV

(b)

(+)

(−)

(−)

(+)

mV 0

− 60

Action potential begins
depolarization

(c)

(−)

(+)

(+)

(−)

+ 40

mV 0

− 60

Action potential peaks:
+ 40 mV

(d)

(+)
(−)

(−)

(+)

+ 40

mV 0

− 60

Repolarization

(e)

+ 100

+ 50

0

− 50

− 100

Maximum depolarization; inward
movement of Na⁺ stops

Na⁺ continue
inward rush

K⁺ rush out
of neuron

Polarized
resting
state

Outward movement
of K⁺ stops

Influx of sodium begins

- Sodium ions (Na⁺)
- Potassium ions (K⁺)
- Negatively charged
  protein

neuron could react thousands of times without a complete redistribution of the sodium and potassium ions.

Reestablishing the Resting Potential. As an action potential peaks, the membrane's permeability to sodium drops sharply and sodium is again rejected. Then potassium becomes the essential ion of recovery. With the newly arrived positive sodium ions balancing the negatively charged proteins inside the neuron, there is nothing to stop potassium from diffusing down its own gradient. Thus potassium ions move out of the neuron until their number balances the sodium ions that moved in. If our arithmetic is correct, this should restore the resting potential. Interestingly, as soon as the potential voltage again reaches –60 mV, that portion of the neuron may fire again—but not before. We will see why shortly.

It might have occurred to you that while the neuron is now polarized, the distribution of specific ions is different from before, and we left sodium trapped inside the neuron. Again, because of the meager number of ions involved in an action potential, this is irrelevant for a time. But for the sake of logic and symmetry, let's note that the sodium/potassium ion exchange pumps eventually restore the earlier distribution of ions. Sodium is pumped out and potassium is pumped in, and quite soon potassium resumes its old balancing act.

This is about all we knew until just a few years ago. But neurobiologists now understand a great deal more about the events underlying the behavior of ions during both the resting state and the action potential. For instance, you may have wondered why the membrane permeability changes so suddenly and why an action potential moves continually along an axon. And how is it that recovery happens so fast? Let's see if the newer findings provide some answers.

Ion Channels and Ion Gates

What is it that makes the axonal membrane suddenly permeable to sodium ions when the nerve impulse begins? The studies reveal that, in addition to the sodium/potassium exchange pumps, the membrane contains a large number of very special voltage-sensitive **ion channels.** These channels are selective. Some will admit sodium, some potassium, and others, as we will see, admit calcium, but the ions are admitted only under certain conditions. Controlling the ion channels are versatile mem-

brane proteins appropriately called **ion gates** or more specifically **sodium ion gates** and **potassium ion gates.** As their name implies, the gates control the passage of ions by opening and closing; actually, the proteins that comprise the gates undergo changes in shape. As we mentioned, the sodium and potassium gates in axonal membranes are voltage sensitive, or voltage dependent—thus they are activated by electrical disturbances.

Whereas potassium channels have but one voltage-activated gate, every sodium channel has two, an **activation gate** and an **inactivation gate,** both of which must be open before the rapid ion movement seen in an action potential can occur. As it happens, the two gates are both open only at critical times (Figure 34.8). In the resting axon, the inactivation gate is open, but the highly voltage sensitive activation gate is closed. When an action potential arrives, bringing with it a change in voltage, the activation gate also opens, and with both gates open the sodium ions rush into the cell. The inrush of sodium ions, as we have seen, depolarizes the neuron. Then the accompanying voltage shift reaches voltage-dependent sodium gates further along, and when these act, more sodium gates further along respond, and so on. Thus, we speak of a depolarizing wave moving along the neuron. Let's look at the ion gates and the recovery process.

Depolarization by an action potential is brought to its peak at +40 mV when the inactivation gates on the sodium channels spontaneously close and the sodium ion influx ceases. Then the potassium channels come into play. The potassium gates, which began opening earlier, are much slower acting than the sodium gates. But now the potassium ions rush out, moving down their gradient, and even with sodium ions still trapped inside, the voltage differential quickly reverses to the resting potential.

It is at this point that the roles of the two types of sodium gates becomes apparent. If only a voltage-sensitive activation gate were present, it would be "frozen" open as the action potential peaked. But the inactivation gate responds to different cues—we aren't sure what they are—however, it closes during the fully depolarized state. The sodium inactivation gates remain shut until the resting potential is reached, which is why at this time no new action potential can proceed in this part of a neuron—there is no way for another sodium inrush to occur. This very brief period when no stimulus can elicit a response is known as the **refractory state.**

With repolarization and the resting state established, the sodium activation gates finally close, and

34.8 ION CHANNELS AND GATES

The rapid change in the axonal membrane's permeability during a neural impulse can now be explained in terms of gated sodium and potassium channels, whose behavior changes to accommodate an action potential. Here the oscilloscope tracings reveal activity in the sodium and potassium gates. In the resting state (a), the sodium inactivation gate is open, but the pore is still blocked by a voltage-sensitive activation gate. The single potassium gate is closed, and only the usual resting state potassium leakage occurs. (b) With the arrival of an action potential, the voltage-sensitive activation gates open and a sudden inrush of sodium ions begins. At the same time, the slower acting voltage-sensitive potassium gates also start to open. (c) Then the sodium inactivation gates close, stopping the influx, and the action potential peaks. Simultaneously, potassium rushes freely down its gradient, leaving the axon and restoring the resting voltage. (d) Within milliseconds from its start, the action potential is completed (and has moved along the axon to the next series of gates). Like a well-charged battery, the neuron can support many, many action potentials before running down.

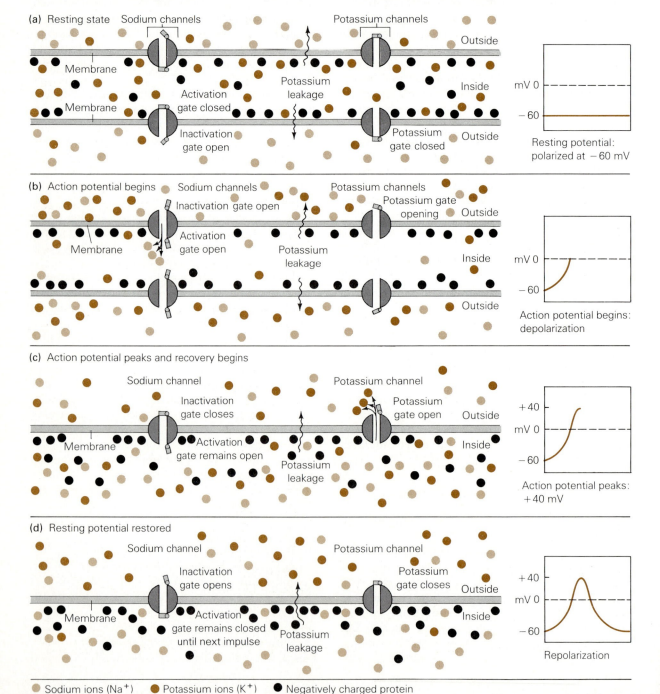

(a) Resting state

Resting potential: polarized at −60 mV

(b) Action potential begins

Action potential begins: depolarization

(c) Action potential peaks and recovery begins

Action potential peaks: +40 mV

(d) Resting potential restored

Repolarization

● Sodium ions (Na⁺)   ● Potassium ions (K⁺)   ● Negatively charged protein

the inactivation gates open. Simultaneously, the potassium gates close, and the neuron is ready to act again. Of course, the eventual redistribution of specific ions will require more work by the sodium/potassium ion exchange pumps. The exchange pumps do the housekeeping chores, restoring the sodium/potassium distribution over the long run. Some have used the analogy of bilge pumps, pumping water out of a leaky boat. Should the pumps stop, the boat would eventually fill with water and sink, but this would take a while. Similarly, the sodium/potassium pumps would have to be inactive for some time before the neuron became hopelessly depolarized. At this point, then, let's briefly summarize what we have seen.

1. The resting potential (–60 mV) and polarized state are produced by the presence of many negative ions inside the neuron, the specific manner in which potassium and sodium ions distribute themselves, and the work of the sodium/potassium ion exchange pumps.
2. An action potential or neural impulse occurs when voltage-activated sodium ion gates open, permitting an inrush of sodium ions and a depolarization (to +40 mV) to occur.
3. When the depolarized state reaches about +40mV, the sodium inactivation gates close, stopping the inrush of sodium ions. With the potassium gates open and potassium diffusing out, the resting potential (–60 mV) is restored.
4. The sodium inactivation gates open and the voltage activated gates close, permitting action potentials to occur again. The potassium gates close, reducing the potassium outflow to the usual leakage that maintains the resting state. Eventually the sodium is pumped out of the neuron, readjusting its distribution.

Myelin and Impulse Velocity

As we mentioned, axons in vertebrates are commonly surrounded by myelin sheaths—the plasma membranes of Schwann cells or oligodendrocytes—that act both as insulators and as a mechanism for speeding up impulses. In those human neurons that are myelinated, for example, impulses are conducted at a speed of up to 100 meters per second, while the speed of conduction in our nonmyelinated neurons and those of invertebrates have much slower rates. But how do myelin sheaths function? We can see how the fatty sheaths would insulate neurons, but how do they increase impulse velocity? And for that

matter, how can an action potential be generated in an axon that is insulated from the surrounding, sodium-rich extracellular fluid?

The answer to both questions lies in the arrangement of the myelin sheath. We must admit now that the smooth-flowing, wavelike nerve impulse that we described in such detail only pertains to nonmyelinated fibers. In the myelinated neurons, the action potentials occur only at the nodes, those gaps in the myelin found at regular intervals along the axon.

We find, then, that the increased speed of the impulse in myelinated fibers occurs because the neural impulse "jumps" from one node to the next all along the axon, in what is called **saltatory propagation** (Figure 34.9). The generation of an action potential in one node produces a minute current or electrical field that spreads *instantly* throughout the internode. When this field reaches the voltage-dependent sodium and potassium gates in the next internode, those gates become activated and another action potential begins further down the axon. Like a skillfully thrown rock skipping along a lake surface, saltatory propagation permits a greatly increased velocity of transmission. And, as you can imagine, this method of propagation is more energy efficient since there are so few ions to later be pumped in and out. In one comparison between myelinated and nonmyelinated axons conducting at the same speed, it was estimated that the nonmyelinated axon required 5000 times as much ATP energy to restore the distribution of sodium and potassium ions.

The speed of impulses is also increased in axons of larger diameter. For example, the fast-reacting giant axons of the squid that permit it to "jet-propel" itself away from danger are several millimeters in diameter. But even the squid's giant axon conducts impulses at only about 30 meters per second, still far below vertebrate capabilities.

Communication Among Neurons

The place where the axon of one neuron communicates with the dendrite or cell body, or with the membrane of an effector, is called a **synapse.** Neurons do not ordinarily make physical contact with one another, and action potentials do not ordinarily continue simply from the axon of one neuron to the dendrite or cell body of a second. Between the connecting processes of most neurons is a tiny space (some 20 nm wide) known as a **synaptic cleft** (Figure 34.10). The minute gulfs between neurons are a vital part of neural control and coordination. (We said "not ordinarily" because there are exceptions—some neurons make direct cytoplasmic connections with other neurons, and there the action potentials move right along over the junction.)

Since action potentials do not ordinarily cross the synaptic cleft, how does one neuron stimulate the next? Commonly, the neural activity moves *chemically* from one cell to the next by means of neurotransmitters, highly specific messengers that, when released into the synaptic cleft, stimulate the next neuron to fire. There are many different kinds of neurotransmitters, depending on the functions of the neurons involved. Outside the brain and spinal cord, the most common are **acetylcholine** and **norepinephrine** (others are discussed in the next chapter). We will look closely at activity in acetylcholine-activated neurons.

Action at the Synapse.   The axonal knobs, or **synaptic knobs,** as they are also known, are filled with tiny vesicles, or sacs, that contain the neurotransmitter substance, in this instance, acetylcholine. When the action potential reaches these synaptic knobs, voltage-sensitive calcium ion gates are activated, and upon opening, calcium ions (Ca^{2+}) diffuse from the extracellular fluids into the axonal knob. Their presence brings on contractions in the cytoskeleton, which in turn draws the vesicles against the plasma membrane which is here referred to as the **presynaptic membrane,** whereupon the vesicles empty their neurotransmitter molecules into the synaptic cleft. The molecules then diffuse across the narrow cleft and become attached to specialized receptors on the **postsynaptic membrane** (the plasma membrane of the next neuron or effector; see Figure 34.10).

34.9 SALTATORY PROPAGATION

In myelinated neurons, action potentials occur only at the nodes—minute spaces between Schwann cells. Thus the neural impulse leaps from node to node along the axon **(a)**. This type of transmission, known as saltatory propagation, is considerably faster and much less costly in terms of ATP than transmission in nonmyelinated neurons. The internodal "leaps" **(b)** are the result of an electrical current flow set up at each node as depolarization occurs. The current generated at one node depolarizes the next nodal region and so on down the axon. Saltatory propagation allows myelinated neurons to transmit impulses up to 20 times faster than the most efficient nonmyelinated neurons.

(a) Movement of ions

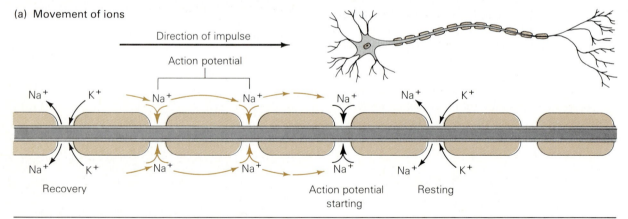

(b) Current flux

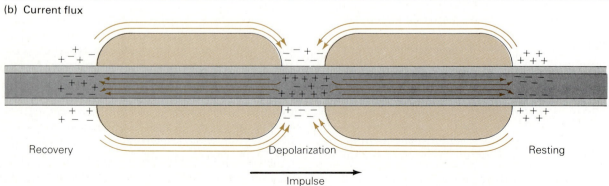

34.10 THE SYNAPSE

(a) Each axonal knob is filled with saclike synaptic vesicles, containing neurotransmitter molecules such as acetylcholine. When an impulse arrives at the axonal knob, an influx of calcium ions occurs, causing the vesicles to fuse with the presynaptic membrane, spilling their molecules into the synaptic cleft. Traveling across the cleft, the neurotransmitter molecules will fill receptor sites in the postsynaptic membrane. When a critical number have been filled, this initiates a new depolarizing event, producing a new action potential that sweeps along the receiving neuron. (b) The synaptic cleft, as viewed through the electron microscope, is a minute space between the axonal knobs of one neuron and the receptor membrane of another.

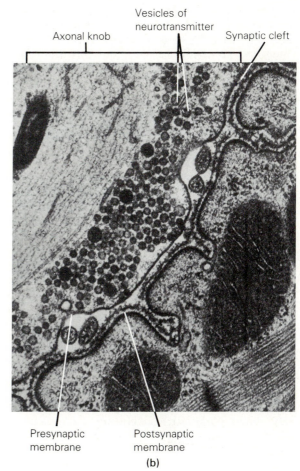

Axonal knob — Vesicles of neurotransmitter — Synaptic cleft

Presynaptic membrane   Postsynaptic membrane

(b)

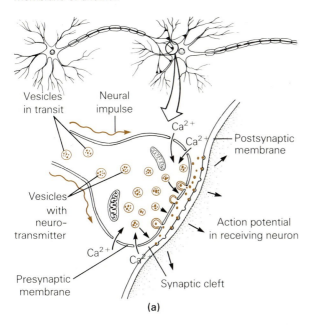

Vesicles in transit

Neural impulse

Ca^{2+}
Ca^{2+} — Postsynaptic membrane

Vesicles with neurotransmitter

Ca^{2+}
Ca^{2+}

Action potential in receiving neuron

Presynaptic membrane

Synaptic cleft

(a)

In acetylcholine-activated neurons, the postsynaptic membrane receptors are actually integrated into **chemically gated channels,** those activated by the presence of specific chemicals. When a critical number of receptor sites have been filled in the postsynaptic membrane, the gates open and there is an inrush of sodium and sometimes, potassium ions. (Note that the ions admitted depends on the specific type of synapse.) The entry of positive ions into the next neuron starts the cascading depolarization effect, creating an action potential that starts in the dendrite and passes down the nerve cell body and through the axon. If it were to end at an effector such as a skeletal muscle fiber, the depolarization would result in muscle contraction (see Chapter 28).

Transmission across the synaptic cleft not only explains how neural impulses travel from one neuron to another, but also explains why nerve impulses usually travel in one direction only (there are also exceptions here). Simply stated, the postsynaptic membrane (dendrite) has no neurotransmitters to release, and the presynaptic membrane (axon) lacks neurotransmitter receptor sites.

The neurotransmitter released into the synapse doesn't remain long. Most types are simply returned to the presynaptic membrane (the unlikely term "reuptake" is applied to this recycling), but in the case of acetylcholine, an enzyme, **acetylcholinesterase,** breaks the neurotransmitter down into its components, choline and acetyl groups. The breakdown or hydrolysis is so rapid that acetylcholine's presence is quite fleeting, lasting for less than a millisecond.

Significantly, some synapses do not activate the neuron that receives the neurotransmitter. Instead, they *inhibit* it by **hyperpolarization** of its neuronal membrane. When the receptor sites at inhibitory synapses receive a neurotransmitter, they may respond by opening **chloride ion gates** that allow the negatively charged ions to move into the neuron. The resulting increase in negative charges inside the neuron increases the polarization beyond that produced by the ordinary resting potential. In other instances, potassium ion gates open and potassium escapes, but with the same net effect and increase in negativity within. Such hyperpolarization raises the stimulus threshold of the neural membrane; therefore the cell requires a larger stimulus (a greater number of incoming impulses) before an action potential can occur.

A dendrite may form synapses with a number of axonal endings—some from excitory neurons, some from inhibitory neurons. Whether the affected neuron will fire depends on the net effect of the inhibitory and excitory synapses. Inhibitory neurons also generally decrease the likelihood of uncontrolled massive neural excitation such as that which occurs in epileptic seizures.

The Reflex Arc:
A Simple Model of Neural Activity

Now let's see how a system of neurons might operate to produce a simple behavioral response. Consider the **reflex arc.**

In a reflex arc, a receptor receives an impulse and transmits it to a coordinating center, which then sends an impulse to an effector, such as a muscle. Such an arc may directly involve as few as two or three individual neurons. A familiar example is that of a doctor tapping the tendon below your knee in order to watch your leg jump. (Doctors aren't just easily entertained; actually, they are trying to rule out about a dozen neurological disorders.) What happens is that the blow causes the tendon to stretch slightly, which also allows the **quadriceps muscle** to lengthen a bit. When this happens, certain **stretch receptors**—specialized neurons within muscles—flash a neural message to the spinal cord. Once the impulse reaches the spinal cord, it is relayed to a tiny interneuron, which in turn stimulates the appropriate motor neurons to act. The quadriceps contracts slightly, and your knee "jerks" (Figure 34.11). The reflexive response is sudden and involuntary, and the brain is only indirectly involved (as a passive and surprised observer).

Stretch receptors abound in the muscles. The ones in the muscles at the back of the neck are what cause your head to snap up when you start to doze during lectures. A delicate interplay of reflex responses is also essential in maintaining posture

34.11 REFLEXIVE ACTION

In the knee-jerk reflex, striking the tendon causes the muscle above to extend slightly. Stretch receptors, specialized sensory neurons in muscles and tendons, sense this change, transmit impulses to the spinal cord, where synapses with appropriate motor neurons are present. The response is direct, and the level of contraction in the muscle is restored without intervention by the brain. Such direct actions are known as reflex arcs.

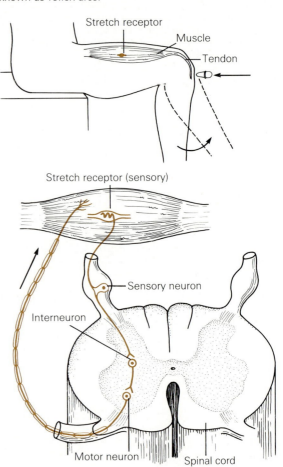

and limb position. Highly coordinated contractions in opposing muscles of the trunk and legs are constantly adjusted by reflexive neural centers in the spinal cord and brain without much conscious intervention. (When's the last time you thought to yourself, "I really must contract my left sternocleidomastoid; my head is leaning to the right"?)

We will now move to other aspects of neural function, looking particularly at sensory neurons. These are the cells with which organisms perceive their surroundings—perceptions that sometimes require rapid responses.

ANIMAL SENSORY RECEPTORS

Animals depend heavily on information received through their sensory receptors; they are the link to the outside world, providing information upon which adaptive short- and long-term responses are made. As we mentioned earlier, sensory receptors are highly varied and often quite specialized. Here, we will look briefly at how these receptors respond to stimuli, first in the invertebrates and then in the vertebrates, with special emphasis on humans.

Sensory receptors can be quite simple with little more than a specialized dentritic region, as with free neural endings in the skin, or incredibly complex, as are the light-sensitive retina of the vertebrate eye or the sound receptors of mammals.

Sensory receptors have several characteristics in common. They are **transducers**—that is, they convert the varied stimuli they receive into electrical signals—action potentials in the associated sensory neurons. Receptors themselves undergo depolarization, but through **generator potentials.** Unlike the "all or none" ("go or don't go") action potentials, generator potentials can increase in intensity. If you think about it, such "grading" of stimuli can be quite useful, since it is through such input that we make very fine distinctions between bright and dim light, soft and loud sounds, and faint odors and damaging toxic fumes. In addition, receptors are highly specialized for certain stimuli, so that a taste receptor cannot respond to light wavelengths and vice versa. But if the generator potentials created by stimuli in sensory structures precipitate action potentials in sensory neurons, and all action potentials are the same, how on earth do we know what we are sensing?

Neural Codes

It is likely that sensory neurons, while activated by a variety of receptors, don't really sort anything out. They simply generate action potentials. The sorting, identifying, interpreting, and integrating is done elsewhere—specifically, in the interneurons that comprise the spinal cord and the brain. One way the brain can decode incoming impulses is through their frequency. A mild stimulus, such as a light touch, might produce just a few impulses per second and might even be ignored by the brain. A much stronger stimulus, such as hitting one's finger with a hammer, produces a virtual barrage of impulses in a very short time. Further, the *number* of neurons involved is another means of conveying information. The hammer will undoubtedly stimulate many sensory neurons at once, making the stimulus difficult to ignore (you can test this for yourself).

Decoding, however, involves more than just distinguishing between degrees. It involves making distinctions in "kind" as well. Distinguishing specific kinds of incoming information—touch, light, taste, sound, temperature, pain—depends on how the sensory neurons and the integrating interneurons are connected, or "wired." Partly through learning and experience, regions of the brain specialize in interpreting sensory information, and each sensory neuron connects via specific tracts to its corresponding brain region. Some centers of the brain deal only with sound, for example, and when they are stimulated, "hearing" results. Interestingly, stimulating such centers directly with electrodes will likewise be interpreted as sound (or memories of sound). Finer levels of distinction and integration also exist; thus most of us can distinguish a reasonably accomplished tenor from an alley cat hitting the same note. Just how neurons connect to the proper brain center is a fascinating part of the developmental processes in the embryo, as we will see in Chapter 37.

In our discussion of sensory receptors, we will be interested in the receptors specialized for touch, heat and cold, chemicals, sound, position, balance, and light, although, as we will see, some of these are very closely related. Let's begin with the receptors that respond to touch, the **tactile** or **mechanoreceptors.**

Tactile Reception

Tactile receptors, or mechanoreceptors, are extremely sensitive, fast-firing neurons that respond to any force that deforms or alters the shape of their plasma membrane. In some cases, nonliving bristles or hairlike structures protruding from the surface of the plasma membrane touch objects in the environment before the organism reaches the objects.

Touch in Invertebrates. A number of arthropods, from caterpillars to spiders, are fuzzy. One adaptive advantage of their covering is related to an increased sensitivity to touch, since the bristles extending from the body are connected to tactile receptors.

Tactile sensitivity may be important for a variety of reasons. For example, web-building spiders

34.12 SENSORY HAIRS

Touch receptors in spiders are often associated with hairs. When a hair is bent or moved, it activates a sensory neuron that transmits its impulse to an associated ganglion for integration. Some sensory hairs are fine enough to be moved by light air currents.

have hairy legs that respond to vibrations set up by prey caught in the web (Figure 34.12). Cockroaches have tiny hairs protruding from their abdomen that, when stimulated by very light air currents, send them scurrying for safety. Many aquatic insects have bristles on their heads that are sensitive to water currents. R. S. Wilcox has accumulated a fascinating body of evidence that indicates how certain aquatic insects use water vibration in communicating with each other. Touch sensitivity, in general, is common among a wide range of invertebrates and is usually employed in finding food, mating, and avoiding predators.

Touch in Vertebrates.   Vertebrates have two kinds of tactile receptors. **Distance receptors,** such as the lateral line organs of fishes (Figure 34.13), are sensitive to water motion. And then there are **contact receptors.** In humans and other mammals, these are of two types: pressure and touch. The pressure receptors are located deep in the skin and seem to consist of encapsulated nerve endings called **Pacinian corpuscles.** Light touch, on the other hand, is believed to be registered in **Meissner's corpuscles,** which lie near the surface of the skin (Figure 34.14). Among primates, sensitivity to touch is greatest around the lips, nipples, eyes, and fingertips. In addition to touch receptors, the skin abounds with simple, unspecialized, free nerve endings that register pain, and thermorceptors that respond to heat. A special kind of mechanoreceptor found in the arteries, the **baroreceptor,** is activated by blood pressure within the vessel.

In humans, as in most mammals, touch sensitivity is particularly great around hairy areas. As you may have noted, this includes the hairline around the face, and the genitals. The "whiskers" of many animals are especially sensitive to touch. While the scant body hairs of humans are of limited value in keeping us warm, they are good sense organs: try moving a body hair without feeling anything.

Until 1982 it was generally believed that hairs, in their sensory function, were simply dead, mechanical levers that when touched would jostle the sensory nerves surrounding their roots. Then another theory was developed based on information that the hair protein keratin is so highly structured as to be essentially crystalline. In fact, it was found that each hair comprises a single **piezoelectric crystal.** A piezoelectric crystal discharges electricity when it is deformed. (Cheaper phonograph pickups use piezoelectric crystals to translate needle movements into electric currents.) Researchers found that like other transducers, a perfectly dead hair generates a small electrical discharge when it is bent, and that nerve endings respond to this electricity.

Incidentally, the outer layer of skin is also made primarily of keratin, and it too has a piezoelectric effect. Bending or depressing the epidermis generates detectable electric discharges that are picked up by the touch-sensitive free nerve endings.

Thermoreception

Thermoreception is important in a wide range of animals. Many of the processes of life are possible only within a specific and often narrow range of

34.13 LATERAL LINE ORGANS

The lateral line organs of bony fishes and sharks can detect distant movements in water. Such movement creates disturbances that are picked up by receptors in the lateral line canal. The receptors, known as *neuromasts,* are clusters of sensory hairs embedded in a gelatinous mass. Movement of water in the canal moves the hairs, which stimulates associated neurons to fire, sending impulses over a nerve to the brain.

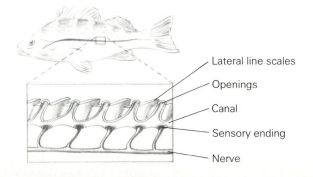

Lateral line scales

Openings

Canal

Sensory ending

Nerve

34.14 THE SENSITIVE SKIN

The skin contains a variety of sensory structures, each specialized for detecting certain stimuli. Interestingly, the hair shaft itself is a sensory structure. A slight bending of the hair causes it to discharge a very small electrical impulse, which is picked up by neurons that surround the hair root.

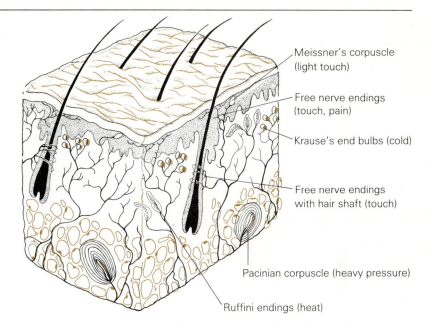

Meissner's corpuscle (light touch)

Free nerve endings (touch, pain)

Krause's end bulbs (cold)

Free nerve endings with hair shaft (touch)

Pacinian corpuscle (heavy pressure)

Ruffini endings (heat)

temperatures, and animals must be able to sense the temperature of their environment in order to place themselves under the most nearly optimal conditions.

Heat Receptors in Invertebrates.   While thermoreception may be important to many invertebrates, little is known about its mechanisms. Heat detection is particularly important to the ectoparasites of birds and mammals: leeches, fleas, mosquitoes, ticks, and lice—parasites that must find a warm-blooded host. Their thermoreceptors are generally located on the antennas, legs, or mouthparts.

Heat and Cold Receptors in Vertebrates. Many vertebrates have specific heat receptors. The pits of the poisonous pit viper are a pair of indentations between the eyes and nostrils. These are loaded with heat (infrared) receptors that tell the snake when it is facing a living thing that is generating metabolic heat (Figure 34.15).

There is some disagreement over whether humans and other mammals have specific receptors for detecting heat and cold, or whether free nerve endings register these stimuli. It has even been suggested that a single neuron can register both heat and cold, simply by firing in different patterns. Some physiologists believe that heat receptors are located deeper in the skin than cold receptors. Further, whether the brain registers "warm" or "cold"

34.15 THERMORECEPTION IN PIT VIPERS

Heat sensors, used for detecting prey by the pit viper (in this case, a rattlesnake), are located in depressions near the eyes. Each pit consists of an outer chamber that ends in a thin membrane covering an inner chamber below. Delicate, highly branched, heat-sensing neurons extend over the membrane, where they detect minute changes in temperature in nearby objects. By moving its head back and forth, the pit viper is able to use incoming thermal cues from the two pits to zero in on its warm-bodied prey.

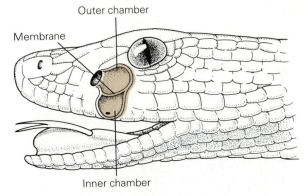

Outer chamber

Membrane

Inner chamber

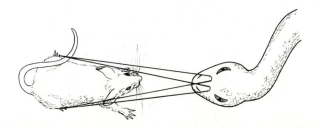

(within limits) depends on the immediate previous experience of the receptors. If the skin has been warm and touches something less warm, the object will seem cold, but if the skin has been cold and touches something cool, the second sensation will be exaggerated warmth. This can be demonstrated by switching your hands between cold water and ice water.

Other physiologists ascribe thermoreception to the free nerve endings and to specialized skin receptors called **Ruffini corpuscles** and **Krause corpuscles** (or **Krause end bulbs;** see Figure 34.2). The latter are far more numerous and are stimulated by a greater temperature range than are the former.

Chemoreception

Chemoreception is the ability to perceive specific molecules. These molecules are often important cues to the presence of specific entities in the environment. Nearly all animals and a great number of protists exhibit chemoreception. It is essential to many animals in finding food, locating a mate, and avoiding danger.

Chemical Receptors in Invertebrates. Planarians locate food by following chemical gradients in their aquatic surroundings. Their simple chemoreceptors are located in pits on their bodies over which they move water with their beating cilia. Certain insects have taste and smell receptors of astounding sensitivity. Among insects, chemoreceptors may be found in the body surface, mouthparts, antennas, forelegs, and, in some cases, the ovipositor (since many insects lay their eggs only on certain plants).

Among the incredible stories of insect olfactory (sense of smell) abilities we find the tale of the silk moth. The silk moth smells with its antennas and adult males have enormous antennas with thousands of sensory hairs (Figure 34.16). About 70% of the adult male receptors respond to only one complex molecule called **bombykol,** a sex attractant released by females. As the molecules drift downwind finally to touch the antennas of a wandering male, they enter the tiny pores of a "hair," or **sensillum** that houses his olfactory receptors. The molecules then dissolve in the fluid that moistens the receptor and interact with its membrane. Upon perception of the molecule, the male becomes excited, reorients his flight, and heads upwind, a behavior that sooner or later should lead him to the waiting siren.

Chemical Receptors in Vertebrates. Vertebrates detect chemicals in a number of ways,

---

34.16 CHEMICAL DETECTION IN MOTHS

The antennas of the male *Hyalaphora* moth are remarkably large, complex, and sensitive. They are able to detect a single molecule of sex attractant released by a receptive female. Each large bristle of an antenna has numerous hairlike sensillae extending outward. Within each sensillum is a fluid-filled cavity containing sensory neurons. Airborne molecules enter the cavities through pores and, when in solution, stimulate dendritic endings, producing impulses that are sent to the brain.

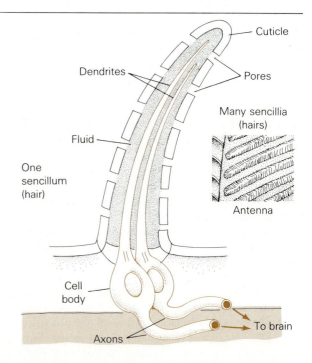

34.17 OLFACTION IN HUMANS

The olfactory receptors in the human nose connect to the olfactory bulb of the brain. These receptors are able to distinguish a wider variety of stimuli than are the taste buds. The olfactory neurons are part of the nasal epithelium dispersed with other cells. Each neuron has numerous olfactory "hairs," actually modified cilia, that protrude from the epithelium.

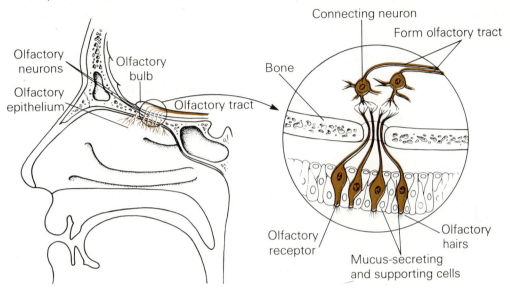

employing three major means: general receptors, and two types of specialized receptors, **gustatory** (taste) and **olfactory receptors.** For example, many aquatic vertebrates have generalized chemical receptors scattered over the body surface. Olfaction and gustation in vertebrates are usually accomplished by moving chemical-laden water or air into a canal or sac that contains the chemical receptors.

In land vertebrates, the keenest olfactory senses are found among the mammals, especially the carnivores and rodents. Good olfaction is essential to carnivores; wolves and lions, for example, often locate their prey by smell. It is less clear why rodents have evolved such strong olfactory ability, since their plant food sources are not usually widely dispersed and certainly do not run away. However, their sense of smell may be associated less with food-finding than with their complex social organization. Rats, for instance, may use their sense of smell to discriminate among families (or tribes) and the individuals within them.

Nearly all vertebrates, including most mammals (but not primates), have olfactory receptors not only in their noses but also in the roofs of their mouths, known as **Jacobson's organs.** In amphibians, Jacobson's organ simply gives the animal a better notion of what it's eating, but snakes and lizards actually smell with the structure. As they flick their forked tongues in an out, the moist tips capture scent molecules that are delivered into the twin openings of their Jacobson's organ inside the mouth.

Compared to many other mammals, olfaction in primates is rather poor. For example, chimpanzees smell about like you do (no offense). The olfactory neurons of primates, like those of other mammals, are located in nasal epithelium, specialized here as **olfactory epithelium,** where molecules in the air encounter the moist surface produced there (Figure 34.17). Scattered among the epithelial cells are specialized olfactory neurons that synapse with neurons of the **olfactory bulb** of the brain, an extension located just above the bony roof of the nasal cavity.

In all land-dwelling vertebrates, taste receptors, or **taste buds,** are confined to the mouth area. As we mentioned earlier in our discussions of digestion (see Chapter 29), humans experience four basic tastes: sweet, sour, salty, and bitter. Humans are omnivorous; that is, we'll eat almost anything, plant or animal. So it has undoubtedly been important to survival that through our long evolutionary history, we have become able to distinguish among a variety of tastes and smells. To illustrate, many alkaloid poisons taste bitter, and unless we deliberately train ourselves otherwise, we tend instinctively to avoid it. Sweet, on the other hand, is the taste of carbohydrates and many kinds of fruit taste sweet, especially when they are ripe and have their highest nutritional value. Sour is not one of our favorite tastes, being the taste of acid and of unripened fruit that will benefit us more if we wait until it ripens. Gustation, then, is not only a matter of taste, but of survival.

Proprioception

Proprioception is the ability determine the position of the body, or the position of one part of the body relative to another. It is made possible by sensory receptors in the muscles, joints, and tendons.

Proprioceptors in Invertebrates.   It is particularly important for animals with many body parts, such as arthropods, to be able to coordinate those parts—to know what each is doing. Sensors in arthropods may be located peripherally, for example, at the base of ''hairs,'' or deep within the muscles. Cockroaches, which are running animals, have proprioceptive hairs on the inside of their legs. These respond to flexion of the knees, so they are aware of whether a leg is extended or flexed. Other arthropods—the crabs and lobsters—have proprioceptors located within the muscles themselves.

Proprioceptors in Vertebrates.   Some proprioceptors in vertebrates, as well as in some invertebrates, take the form of stretch receptors (mentioned earlier in our discussion of the reflex arc). These respond to the stretching of skeletal muscle and tendons that is produced as limbs are extended or flexed. You may not be able to see your foot under the table, but you could probably point to it because your proprioceptors tell you where it is. Proprioception is well developed in mammals such as the primates, which must move with incredible agility through their forest canopy.

Auditory Reception

Audition, or hearing, is similar in some respects to lateral line reception in fishes since in both a distant stimulus is transmitted to the receptors via a medium (air or water). In fact, evolutionary theory holds that the structure of balance and hearing in the vertebrates evolved from increasingly complex lateral line organs.

Hearing in Invertebrates.   Most invertebrates don't have specialized sound receptors, but many are sensitive to vibrations in the air, water, or soil in which they live. One notable exception among the invertebrates is the insects, a group that boasts several kinds of sound receptors. For example,

most species have sensory hairs that respond to low-frequency vibrations of air, but others have a specialized organ, located in a leg, that is sensitive to movements of whatever the insect is standing on. Yet others, including grasshoppers and crickets, have **tympanal organs** that respond to high-frequency vibrations, much like the human eardrum. For a special case of insect hearing and its adaptiveness, see Essay 34.1.

Hearing in Vertebrates.   The sound receptors of most species of vertebrates are located in the **inner ear;** however, vibrations reach the inner ear in a variety of ways. In many fishes, sound vibrations in water are conducted directly to the inner ear by vibrating water, but in others (such as the minnows, catfishes, and suckers), a set of tiny bones—the **Weberian ossicles**—connect the swim bladder to the inner ear. Sound vibrates the air-filled bladder, which moves the bones and stimulates the receptors in the inner ear. The fishes that have such an apparatus can detect a much wider frequency of sound than those that lack it.

Land vertebrates have an **auditory canal** and one to three **middle ear** bones that transmit vibrations of the **tympanic membrane** (eardrum) to the inner ear. The tympanic vibrations usually stimulate auditory receptors, which then carry the impulses to the brain.

In humans and other mammals, the auditory apparatus consists of the **external, middle,** and **inner ear** (Figure 34.18). The external ear includes the **pinna** (''ear''), the **auditory canal,** and the **tympanum** (eardrum). Most mammals are able to move the pinna so as to maximize sound input and locate its source. (Humans have largely lost this ability and those who can move their ears are often in great demand for social events).

The three bones of the middle ear are the **malleus, incus,** and **stapes** (which translate to *mallet, anvil,* and *stirrup,* respectively). Acting as a jointed lever, they transfer vibrations of the eardrum to a thin membrane in the **cochlea** of the inner ear.

The inner ear of mammals consists of the cochlea and the **vestibular apparatus.** The cochlea is a lengthy fluid-filled tube, doubled back on itself and then coiled in the manner of a snail shell. It is hard to visualize how the structure works, but if we can imagine it to be straight—as it is, in fact, in birds—we see what is actually a U-shaped tube containing the sensory neurons involved in hearing (see Figure 34.18b and c). One end of the U-tube holds the **oval window,** to which the stapes is affixed, while the other end contains the highly flexible **round window.**

As sound waves strike the eardrum, the resulting vibrations are transferred by the three middle ear bones to the oval window, which, in response, vibrates rapidly. This creates a wave movement in the fluid of the outer tube and, at its far end, a compensating bulging of the round window (so the round window serves to dissipate the sound energy). As fluid pulsates within the tube, it activates what is called the **organ of Corti.** As you see in Figure 34.18c, the organ of Corti consists of a **basilar membrane,** which bears sensory neurons called **hair cells** (modified cilia). The hairlike tips of these cells are embedded in the gelatinous **tectorial membrane.** As the basilar membrane moves, the sensory hairs are bent, creating generator potentials, which in turn activate neurons leading to the cochlear nerve. This nerve leads to the midbrain, where the neurons synapse with those that carry impulses to be processed in the hearing centers of the cortex.

34.18 STRUCTURES OF HEARING

(a) When sound waves vibrate the human eardrum (tympanic membrane), they set in motion three tiny leverlike bones: the malleus, incus, and stapes. The stapes, attached to the oval window, sets fluids in motion within the snail-shaped cochlea. **(b)** The cochlea is actually a U-shaped tube, divided by the basilar membrane. **(c)** Sensory cells of the membrane are embedded in the gelatinous tectorial membrane. The two membranes and the sensory cells are called the organ of Corti. The sound impulses pass inward over one surface of the basilar membrane, turn a corner, and pass outward over the opposite surface of the membrane to be dissipated at the round window. Whenever incoming and outgoing sound waves are in phase (vibrating together) on the two sides of the basilar membrane, sympathetic vibrations in the membrane excite the sensory cells of the organ of Corti. Different regions of the basilar membrane are sensitive to different sound frequencies.

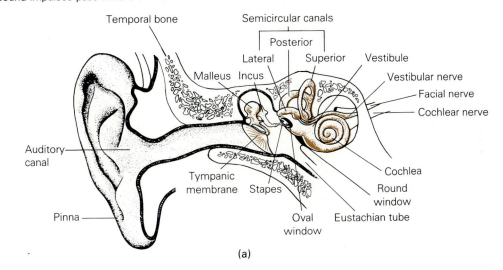

(a)

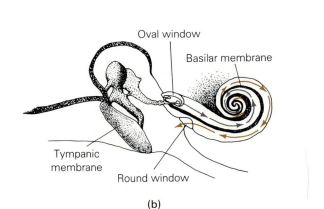

(b)

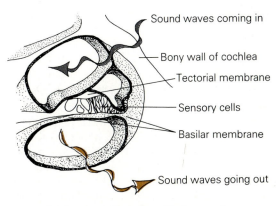

(c) Cross-section of vestibular canal

Predator Strategies and Prey Defenses

The study of insect hearing has provided us with some good stories about scientific detective work, and one of the best ones is about the coadaptation of moth and bat hearing. It seems that noctuid (night flying) moths are a favorite prey of certain bats. These bats fly swiftly and can turn on a dime to capture their prey on the wing. The bats locate the prey by emitting high-frequency sounds that bounce back from any structure in the environment—the principle of sonar. The echos provide the bat with information on the flying insect, and the bat simply intercepts the hapless moth. In fact, the term "echolocation," familiar in submarine jargon, was first coined to describe the use of sonar by bats.

But just as bats have evolved ways of catching moths, noctuid moths have developed ways of avoiding bats. When they hear the cry of a bat, they take evasive measures. As you might suspect when dealing with bats, this is easier said than done. It turns out, though, that the moth does it with a simple hearing apparatus.

Noctuid moths, like many insects, have paired *tympana*, one on either side of the thorax (the inner midsection). Kenneth Roeder found that each tympanum has only two receptors. One, called the *A1 cell,* is sensitive to low-intensity sounds. The other, the *A2 cell,* responds only to loud sounds. Surprisingly, neither kind of receptor is very good at distinguishing frequencies (high versus low notes)—a sound of 20,000 hertz (cycles per second) elicits the same neural action potential as one of 40,000 hertz, a much higher sound.

As sound becomes louder, however, the A1 cell fires more frequently and with a shorter lag time after the stimulus. The A1 cell also shows a greater firing frequency in response to pulses of sound than to continuous sounds. And it just so happens that bats emit pulses of sound.

In a sense, the moth has beaten the bat at its own game. Its very sensitive A1 neuron is able to detect bat sounds long before the bat is aware of the moth. The moth can not only detect the distance of the bat, but it can tell whether the bat is coming nearer, as the sound of an approaching bat grows louder.

In addition, the moth is able to detect the direction of the bat. The mechanism is simple. If the bat is on the left side **(a)**, the left thoracic receptors of the moth will be exposed to the sounds while the receptors on the right will be shielded (see illustration). Therefore, the left receptor fires sooner or more frequently than the right if the bat is on the left. If the bat is directly behind **(b)**, both neurons will fire simultaneously. Thus, the moth can determine the distance and direction of the bat. But what about its altitude?

If the bat is above the moth **(c)**, the bat sounds will be deflected by the upward beat of the moth's wings. If the bat is beneath the moth, however, the wing beats will have no effect on the pattern of neural firing.

The moth, then, decodes the incoming data, probably in its thoracic ganglion (from which the auditory neurons emerge) so that it pretty well has the bat pinpointed.

But what does it do with this information? If the bat is some distance away, the moth simply turns and flies in the opposite direction, thus decreasing the likelihood of ever being detected. The moth probably turns until the A1 cell firing from each ear is equalized. When the bat changes direction, so does the moth.

Bats fly faster than moths, though, and if a bat should draw to within 2.5 m (8 ft) of the moth, the moth's number is up—at least if it tries to outrun the bat. So it doesn't. If the bat and moth are on a collision course, that is, if the moth is about to be caught, the sounds of the onrushing bat will become very loud. At this point, the A2 fiber begins to fire—the signal of imminent danger. These messages are relayed to the moth's brain, which then apparently shuts off the thoracic ganglion that had been coordinating the antidetection behavior. Now the jig is up and the moth changes tactics. Its wings begin to beat in peculiar, irregular patterns or not at all. The insect itself probably has no way of knowing where it is going as it begins a series of unpredictable loops, rolls, and dives. But it is also very difficult for the bat to plot a course to intercept the moth. If all goes well, the erratic course will take the moth safely to the ground, where the echoes of the earth will mask its own echoes.

Sounds, we know, vary in intensity and pitch. The perception of intensity, or loudness, seems to depend on the *number* of auditory neurons that fire, as well as the *frequency* of their firing. Differences in pitch (highness and lowness) depend on which auditory neurons are stimulated. The basilar membrane is narrow at the broad base of the cochlea and wide at its apex end. It responds to high-pitched tones better at its thinner and more rigid beginning, while low-pitched tones stimulate the wider and more flexible apex of the snail-shaped chamber. As we have described, sound waves travel up one side of the U-shaped tube and down the other. Each particular sound frequency traveling up one side of the basilar membrane will be exactly in phase with the same frequency traveling down the other side of the membrane only in one region. The inphase resonance (or reinforced vibration) at that region vibrates the basilar membrane, producing the sensation of pitch.

The noctuid moth's evolutionary response to the hunting behavior of the bat serves as a beautiful example of the adaptive response of one organism to another. It also shows clearly that the sensory apparatus of any animal is not likely to respond to elements that are irrelevant to its well-being. It is not important for moths to be able to distinguish frequencies of sound, but it is important that they are sensitive to differences in sound volume. Anyone who tried to train a moth to respond to different sound frequencies could only conclude that moths are untrainable.

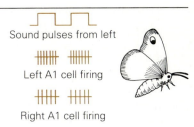

(a) Bat on left

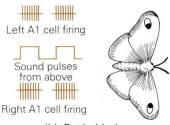

(b) Bat behind

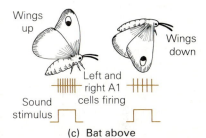

(c) Bat above

The relationship of sound pulses from a hunting bat and auditory neural firing in the hunted moth depends on whether the bat approaches the moth from (a) the side, (b) behind, or (c) above.

Sensing Gravity and Movement in Invertebrates and Vertebrates

Gravity sensors are important both in determining body position and maintaining balance or equilibrium. In most invertebrates, this information is conveyed by **statocysts.** A statocyst is an opening, a chamber, that contains sensory neurons and loose grains, usually of fine sand. In crayfish, as long as a sand grain is resting on the sensory hairs that line the bottom of the chamber, the animal will remain in an upright position. If the animal is turned over so that the grains fall away from the hairs, it will immediately right itself. If, on the other hand, the sand grains are replaced by a bit of iron, the animal will behave normally until a magnet is brought near the statocyst. In that case, as far as the crayfish is concerned, the magnetic field represents gravity.

In humans and other mammals, body movement, position, and balance is detected by the inner

34.19 SEMICIRCULAR CANALS

The inner ear is very sensitive to the body's position and movement. The semicircular canals lie at right angles to each other so that body movement in any direction shifts the fluid in at least one of them. The saccules and utricles contain tiny crystals of calcium carbonate, whose movement excites sensory hairs which, in turn, transmit impulses to the brain. In this way the brain is informed about changes in position or gravity.

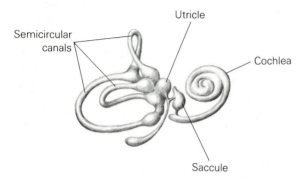

Utricle

Semicircular canals

Cochlea

Saccule

ear by the vestibular apparatus (Figure 34.19). It is composed of the **semicircular canals,** the **saccule,** and the **utricle.** These three structures are closely associated with those of hearing. Each semicircular canal lies in a different plane, at right angles to each of the other two. This arrangement permits the sensing of movement—acceleration or deceleration—in any direction. Each canal is filled with fluid and its movement jostles sensory hairs that extend into the canals. As the fluids move, they bend the hairs, creating generator potentials, which in turn activate sensory neurons that send impulses to the brain.

The saccules and utricles contain sensory hairs coated with fine granules of calcium carbonate. As in the crayfish statocyst, shifts in these granules pressing on the sensory hairs change the rate of neural impulses, providing information about the position of the head with respect to gravity. Some impulses travel to the spinal cord, where body position can be adjusted by reflex action; others are sent to the cerebellum, where other reflexive muscular coordination is orchestrated; and yet others move on to higher centers involved with the control of eye movement. Input from the eyes is important in maintaining balance. (Try to close your eyes and stand on one leg.)

Visual Reception

Light receptors are sensitive to a particular part of the spectrum of electromagnetic energy—the part we call visual light, although some animals can detect ultraviolet light. Visible light wavelengths range from about 430 to 750 nm. As far as is known, considerably shorter wavelengths, such as X rays, beta rays, and gamma rays, can't be detected by any animal. Neither can the very long ones, such as radio waves, although many animals can detect infrared (heat).

Vision in Invertebrates. The planarian flatworm (see Figure 25.11) has two eyespots that are shaded on opposite sides. It can therefore tell which direction light is coming from according to which eye is being stimulated. Such ability is, of course, important to these bottom-dwelling and dark-seeking animals.

Among invertebrates, the cephalopod mollusks—the octopus, squid, and others—are unique in that they have image-forming eyes, quite like those of vertebrates (see Figure 26.14). Arthropods, such as spiders, crayfish, and insects, have exceptionally good vision, although their eyes are adapted to detecting movement rather than producing sharp images. Spiders have eight eyes (two rows of four each), but in most species the eyes lack enough photoreceptors to form clear images. However, the jumping spiders (Salticidae) have a relatively large number of photoreceptors. These are undoubtedly beneficial; the spiders leap from a distance onto their prey and it wouldn't do for a weak-eyed spider to leap onto a hungry bird.

Crustaceans, such as crayfish, have two kinds of eyes: **simple eyes** and **compound eyes.** The simple eyes lack lenses and are found in the larval stages. In many species, the adult's sensitive compound eye rests on movable stalks. The term "compound" refers to the numerous visual units known as **ommatidia** (Figure 34.20). The ommatidia can't move to follow an image, so each stares blankly in its own direction until something moves into its visual field. Then it fires a signal to the central nervous system. Any movement across the animal's field of vision stimulates a series of ommatidia in turn, so that even very slight movements are detected. The convex surface of such eyes may give a visual field of 180°, but the world is probably perceived as some sort of shimmering mosaic.

Vision in Vertebrates. Let's now consider vision in humans as representative of vertebrates. Humans, after all, rely on vision to help them avoid danger, coordinate movement, maintain equilibrium, read the faces of other humans, avoid stepping in unpleasant things, and to experience a great deal of pleasure and creativity. While much is

34.20 THE ARTHROPOD EYE

Compound eyes are common among insects and other arthropods. Each eye consists of many units known as ommatidia.

known about the eyes and their operation, how the brain handles all this is an enormously complex process, and many questions remain.

The Eye. We will begin with a brief discussion of the anatomical aspects of vision, where things tend to be straightforward. The human eyeball (Figure 34.21) is roughly spherical, but protruding somewhat in front. It is surrounded by a tough, white outer layer, the collagenous **sclera,** except for the anterior bulge, which is covered by the transparent **cornea.** The sclera maintains the shape of the eyeball and serves as an attachment for the voluntary muscles that move the eyes. Just within the sclera is the highly vascularized **choroid.** At its anterior portion the choroid supports the **iris** of the eye, a thin diaphragm of smooth muscle. At the center of the iris lies the **pupil,** an opening whose size the iris regulates. The ringlike iris contains pigments we recognize as eye color. Behind the iris is the dense, transparent **lens,** held in place by another ring of muscle, and then a large space filled with transparent gel. Finally, there is an innermost layer of tissue at the back of the eye—the **retina,** a complex region of light-sensitive and supporting neu-

34.21 THE HUMAN EYE

The structure is multilayered, with two major fluid-filled chambers separated by the lens. The amount of light entering the eye through the pupil is regulated by the iris, which is able to change the diameter of the pupil. The light is focused on the retina by changes in the shape of the pliable lens, as various degrees of tension are applied by the ciliary muscles. The retina contains light-sensitive cells (the rods and cones) that lie beneath two layers of associated neurons. The highest concentration of receptors is in the fovea. Neural impulses travel to the brain via the optic nerve, which forms a "blind spot" where its fibers leave the eye.

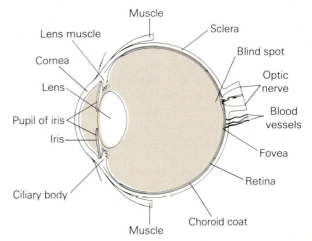

rons. The neural fibers form the **optic nerve,** a common pathway out of the eye. Let's now see how each of these structures work.

Light entering the eye passes through the transparent cornea, and then through the **anterior chamber,** filled with a fluid called the **aqueous humor,** and through the pupil of the iris. The iris adjusts the pupil size according to the brightness of entering light—like the light-controlling diaphragm of a microscope or camera but far more delicate and responsive. It does this through two very fine muscle layers, one circular and the other longitudinal. The first constricts the pupil while the second dilates it, and each is under the control of a different part of the autonomic nervous system. From the pupil, light traverses the lens and then passes through the clear gels and fluids—called the **vitreous humor**—of the **posterior chamber** to fall finally upon the retina.

The eye is often compared to a camera—the lens to the camera lens and the retina to camera film. However, the camera lens adjusts to close or distant objects by moving back and forth (interestingly, sharks focus the same way), but our lenses accommodate for distance by changing shape. This is done by a ring of precisely coordinated **ciliary muscles** and their delicate supporting ligaments, which attach directly to the lens. Relaxing the ciliary muscles flattens the lens, while contraction forces it to assume a more rounded form. As we grow older, the flexibility of the lens decreases, and many of us must wear reading glasses in order to see the details of nearby objects.

The Retina. The retina consists of four layers of cells (Figure 34.22). The deepest layer, attached to the inner surface of the choroid coat, is pigmented. It absorbs light that might otherwise reflect about inside the eye and create visual problems. In many vertebrates, including dogs and cats, the pigment-

34.22 THE RETINA

The retina is a field of two types of photoreceptor cells: the rods and cones. The rods especially react to shades of gray in dim light. Rods tend to be localized near the periphery of our fields of vision. There are at least three types of cones, each with its own type of pigment that responds maximally to either red, green, or blue. If a red-sensitive and a blue-sensitive cone are stimulated, the colors are summed at their syn-

apses and we see purple. Actually, a great deal of stimulus filtering, summing, blending, and inhibiting goes on in the retina, mediated by the bipolar and ganglionic cells. The brain therefore receives preprocessed stimuli, which helps in rapid integration of the varied information arriving from the environment.

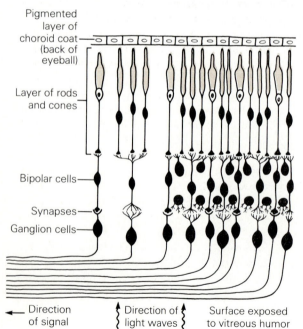

Pigmented layer of choroid coat (back of eyeball)

Layer of rods and cones

Bipolar cells

Synapses

Ganglion cells

← Direction of signal   Direction of light waves   Surface exposed to vitreous humor

ed layer is modified at night, revealing a reflective layer that increases their sensitivity to low levels of light (and helps them avoid being run over at night). Overlying the pigmented tissue are the **rods** and **cones,** the actual light receptors. You might expect them to be in the direct path of incoming light, but they are covered by two more layers of rather transparent neurons. When the rods and cones are stimulated by light, they activate the overlying **bipolar cells,** which in turn synapse with the layer of **ganglion cells** just above. Ganglion cells gather from all parts of the retina to form the optic nerve tract, which carries impulses to the brain.

The Rods and Cones.   The rods are slender cells, sensitive to all wavelengths (that is to say, all colors) of light between 430 and 750 nm. By no coincidence, we call this very same spread of wavelengths the **visible spectrum.** While rods have a wide range of responses, they cannot distinguish one color from the next. Rods form a clear majority of light receptors, numbering about 125 million, as compared to only 7 million cones in the human eye.

It may come as no surprise to learn that cones are cone-shaped (see Figure 34.22). Unlike the rods, each cone responds optimally to only its specific portion of the visible spectrum. Cones are also capable of much finer discrimination of detail in bright light than are rods, although they hardly respond at all to dim light (which is why we can see light but not much color at night). Vision is most acute in the **fovea centralis,** or simply **fovea,** a slight depression directly in the path of light focusing on the retina. The fovea in mammals contains a rich concentration of cone cells with few other retinal cells. Since cones are not very functional at night, it is easier to see a distant object, such as a faint star, if we look to the side of that object, avoiding the cone-rich fovea.

According to the most prevalent theory of color vision, there are three types of cones: red-sensitive, green-sensitive, and blue-sensitive. We can see so many variations of color because the sensitivity ranges of these receptors overlap. The color sensitivities of the cones depend on what sorts of visual pigments (light absorbers) they contain—or, more specifically, on variations in the proteins that join with visual pigments.

Rods operate on an entirely different principle.

Each contains **rhodopsin,** a light-sensitive pigment. When rhodopsin absorbs light, it bleaches out, losing its color and chemically breaking down into two subunits: **opsin,** a colorless protein, and a derivative of vitamin A called **retinal.** It is the chemical change itself that initiates a generator potential in the rod, which, as we have seen, activates the bipolar and ganglionic cells, and information is then relayed to the visual centers of the brain. The manner in which the rod cell interacts with the bipolar neurons has not been clarified, but one hypothesis proposes that when they are *not* photoactivated, rod cells are actually very active secreting neurotransmitters that inhibit the bipolar neurons above. Then when light is absorbed, the rod cell becomes hyperpolarized, slowing its neurotransmitter release, which allows the neurons to fire. Thus vision may begin through a negative mechanism.

Opsin and retinal eventually enter into a chemical pathway in which they are restored to rhodopsin. Apparently, dietary vitamin A must be supplied continually to keep the pathway going in the right direction. If the body is deprived of vitamin A, severe night blindness can result. Going from a very brightly lit room into near total darkness can produce a more temporary form of night blindness. Under intense light, most of the rhodopsin in the cones is in the opsin/retinal form, and a certain period of time is required for the conversion to catch up. In the meantime, the cones, which aren't very useful at night, help you peer dimly into the darkness to see whatever brought you to the door in the first place. While humans have reasonably good night vision, many day-flying birds lack rods altogether and are almost totally night blind, which explains why so many birds come to roost at twilight. Owls and bats, on the other hand, avoid daylight and are quite at home in the dark. It is not surprising, therefore, to learn that their eyes have only rods.

We see, then, that it is the chemical, mechanical, and electrical interplay of specialized neural cells that enables us to sense some small part of the world around us and to respond appropriately to our situation. We will now see just how these sorts of cells are organized to form the great and mysterious central nervous system that has been so crucial to the development of the dramatic role of humans on this earth.

APPLICATION OF IDEAS

1. Compare the functioning of the image-producing human eye with that of the compound eye of arthropods. We know what *we* see, but what might the arthropod actually see? What are some of the advantages of each kind of eye?

2. Compare the degrees of development in special senses in various vertebrates. Explain how an emphasis on certain of these can adapt the animal to its specific environment.

KEY WORDS AND IDEAS

THE NEURON

Cellular Structure

1. **Neurons** consist of a **cell body** with the usual cellular organelles, and receiving and sending processes called **dendrites** and **axons,** respectively. One neuron activates a second, or an **effector,** through chemical substances called **neurotransmitters** that are produced in the cell body and released at knobby ends of the **axon tree.** Neurotransmitters that activate muscles are released at **neuromuscular junctions.**

2. Many vertebrate neurons are wrapped in **myelin sheaths,** produced by **Schwann cells** outside the brain and spinal cord and **oligodendrocytes** within. The sheaths or **internode** regions, contain minute gaps called **nodes.**

Types of Neurons

1. Neurons include **sensory neurons** (also **afferent neurons), interneurons,** and **motor neurons** (also **efferent neurons**). The first receive stimuli such as light and heat and communicates with the second, which integrates the response and transmits it to the third, which produces the response. **Neuroglia** (or **glial cells**), especially **astrocytes,** are extremely common in the brain.

Nerves

1. **Nerves,** composed of numerous axons and dendrites, form cablelike structures surrounded by connective tissue.

THE NEURAL IMPULSE

1. Unlike an electrical impulse, the **neural impulse** travels comparatively slowly and is continually regenerated, so it is as strong at the end as it was at the beginning.

2. The parts of a neural conduction include the **resting state** or **polarized state,** the **action potential** or **depolarizing wave,** and reestablishment of the resting state, or **repolarization.**

The Resting State: A Matter of Ion Distribution

1. Ions important to the neuron are negatively charged proteins within the neuron and positive sodium and potassium ions. In the resting state, the neural membrane is impermeable to Na^+, which is pumped out by the **sodium/potassium ion exchange pumps.** The membrane is permeable to K^+, which is pumped in. K^+ diffuses out but is also attracted in by the negative charges, so an outward concentration gradient is maintained.

2. In the resting neuron, the positive charges predominate outside and the negative inside, producing a polarized state and **resting potential** of −60 mV.

The Action Potential

1. Stimulation of a neuron causes depolarization, a shift in ion concentrations and electrical charges. Membrane permeability changes, admitting Na^+, which shifts the potential to a peak of +40 mV, whereupon the membrane loses its permeability to Na^+.

2. Immediately, K^+ exits the neuron, repolarizing the neuron, thus restoring the resting potential of −60 mV. Na^+ is eventually pumped out by the ion exchange pumps, and K^+ restores its resting equilibrium. As long as the membrane is in a depolarized state, it remains impermeable to Na^+ and no new action potential can proceed.

Ion Channels and Ion Gates

1. Studies of axonal membranes reveal the presence of specific voltage-sensitive sodium and potassium **ion channels** and **sodium** and **potassium ion gates.** While potassium channels have one gate, sodium channels have an **inactivation gate** and an **activation gate.** The gates work as follows:

   a. Resting state: Sodium activation gates are closed, inactivation gates are open, and potassium gates are closed.

   b. Action potential: Sodium activation gate opens and Na^+ rushes into the axon, but as the electrical peak occurs, the inactivation gates close and remain closed until the resting potential is restored. This is the **refractory period** when new impulses cannot occur.

c. Repolarization: Potassium gates open (earlier), and K^+ moves outward, soon repolarizing the region and restoring the resting potential. At this point the potassium gates close.

d. Resting state: Upon repolarization, the sodium activation gates close, the inactivation gates open, and the region is ready for another action potential.

Myelin and Impulse Velocity

1. Myelinated neurons conduct impulses much faster than nonmyelinated neurons, since depolarization occurs only at the nodes.

2. Action potentials at the node create enough current flow to activate the sodium gates in the next node, so the impulse jumps from node to node in what is called **saltatory propagation.** Electrical current flow through internodes is almost instantaneous.

Communication Among Neurons

1. Commonly, axons communicate with dendrites and effectors via **synapses,** where minute spaces (about 20 nm) called **synaptic clefts** occur.

2. Neurotransmitters, such as **norepinephrine** and **acetylcholine,** are released from the **presynaptic membrane,** diffuse across the cleft to the **postsynaptic membrane,** and there activate the next neuron. Transmission across the synapse occurs as follows:

   a. When an action potential reaches the **synaptic knobs,** voltage-sensitive calcium gates open, admitting Ca^{2+} into the knob, where the ions cause vesicles to merge with the membrane, exocytotically releasing neurotransmitter into the cleft.

   b. The postsynaptic membrane has specific neurotransmitter receptors associated with **chemically gated channels.** When activated, the gates open, admitting positive ions that start an action potential in the receiving neuron.

   c. Activity ceases when the neurotransmitter is cycled back to the presynaptic membrane, where it is recovered or inactivated by enzymes. Acetylcholine is inactivated by the enzyme **acetylcholinesterase** and only choline is recycled.

3. In inhibitory synapses, specific neurotransmitters activate either chloride gates or potassium gates, letting Cl^- in or K^+ out, thus causing **hyperpolarization** and resulting in a much greater stimulus necessary to activate that neuron.

The Reflex Arc:
A Simple Model of Neural Activity

1. A **reflex arc** may involve as few as two or three neurons. In the "knee jerk reflex," a tap on a tendon just below the knee cap causes the **quadriceps muscle** above to relax. Neurons

called **stretch receptors** are activated, sending an impulse to an interneuron in the cord, which relays it though a motor neuron back to the muscle, which contracts slightly.

ANIMAL SENSORY RECEPTORS

1. Sensory receptors are highly specialized for the stimuli they receive, but all act as **transducers,** converting external stimuli first into **generator potentials** and then into action potentials. Generator potentials are graded, increasing from threshold level to an intense level, thus providing greater information about the stimulus.

2. While all action potentials are the same, the number and speed of impulses can act as a code in the central nervous system. Interpretation of sensory information depends on neural organization in the brain, along with experience, memory, and learning.

Tactile Receptors

1. **Tactile** (touch) or **mechanoreceptors** respond to deforming force. In invertebrates, they are often associated with sensory hairs. They detect moving solid objects and air and water currents.

2. In vertebrates mechanoreceptors include the **distance receptors** (such as the lateral line organ of fishes) and **contact receptors** that specialize in touch. In humans, **Pacinian corpuscles** are activated by pressure, while **Meissner's corpuscles** respond to light touch. **Baroreceptors** in the arteries respond to changes in blood pressure. Body hairs and the outer dead skin layer act as **piezoelectric crystals,** discharging current when deformed.

Thermoreception

1. **Thermoreception** is the detection of changes in temperature. Heat detection in invertebrates is especially important to ectoparasites of the warm-bodied mammals and birds.

2. In vertebrates, heat sensors include the pit of pit vipers, and **Ruffini corpuscles** and **Krause end bulbs** in mammals. Thermoreception in humans is poorly understood and, up to a point, the perception of heat and cold may depend on the last experience of the receptors involved.

Chemoreception

1. **Chemoreception** is widespread in animals. Insects have chemoreceptors on many body parts, but the **sensillum** of certain moths is the most acute known and can be activated by a single molecule of **bombykol,** the sex attractant.

2. Vertebrate chemoreceptors include general receptors and **gustatory** and **olfactory receptors** (taste and smell).

3. In rodents, a keen olfactory sense aids in social interactions. Reptiles sample chemicals with their forked tongue, which is inserted into the

paired **Jacobson's organ** in the mouth where receptors are located. In mammals, an **olfactory epithelium** contains sensory neurons that respond to chemicals, synapsing with neurons in the **olfactory bulb** above.

4. In fish, taste receptors may be scattered over the body surface, while in terrestrial vertebrates they are in the mouth. In humans, taste receptors are located in **taste buds** on the tongue, where there is some specialization for sweet, sour, salty, and bitter.

Proprioception

1. **Proprioception**—sensing the position of the body or its parts—in invertebrates occurs through sensors in the body hairs and muscles. In vertebrates, proprioceptors include the stretch receptors involved in the reflex arc.

Auditory Reception

1. **Audition** in insects includes the specialized **tympanal organ,** a drumlike device.

2. Some fishes have an **inner ear** containing tiny bones, **Weberian ossicles** that move in response to sound, stimulating action potentials in nearby senors. In land vertebrates, the hearing organs include an **auditory canal,** and one to three **middle ear** bones that transmit sound from the **tympanic membrane** to the inner ear.

3. The human hearing organ includes the **external ear (pinna,** auditory canal and **tympanum).** The middle ear bones include the **malleus, incus,** and **stapes,** which connect the tympanum with the **cochlea.** The inner ear includes the snail-shaped cochlea and the **vestibular apparatus.**

4. The cochlea is a fluid-filled, U-shaped tube with the **oval window** (stapes attachment) on one end and the **round window** at the other. Sound sets up motion in the fluids, which activates the **organ of Corti**—a **basilar membrane** containing **hair cells** embedded in the **tectorial membrane.** Bending of the hair cells creates generator potentials, which in turn start action potentials in neurons from the cochlear nerve. Differences in intensity and pitch are produced by activity at various parts of the basilar membrane.

Sensing Gravity and Movement

1. Some invertebrates have **statocysts,** chambers containing grains that move against sensory hairs when changes in body position occur.

2. In humans and other mammals, movement, position, and balance are detected by the vestibular apparatus. Three fluid-filled **semicircular canals,** lying in three planes, detect acceleration and deceleration as the fluids move against sensory hairs. Fine granules in the **saccule** and **utricle** shift with movement that activates sensory hairs, which inform the brain about position.

Visual Reception

1. Visual receptors in arthropods include **simple eyes** and **compound eyes,** the first of which lack lenses. Compound eyes contain immovable units called **ommatidia** that register the field of vision as a mosaic.

2. The vertebrate eyeball consists of tough, outer **sclera,** with a forward, transparent **cornea** and more internal **choroid.** The choroid supports the **iris,** which surrounds the **pupil.** The **lens** separates two fluid-filled chambers. The light-sensitive **retina** within the eyeball communicates with the **optic nerve.**

3. Light entering the eye passes through the **anterior chamber (aqueous humor),** lens, and **posterior chamber (vitreous humor),** to the retina. Focusing is done through changes in lens shape brought about by the **ciliary muscles.** The amount of light entering is altered by the iris, which determines the size of the pupil.

4. The retina consists of four cell layers, including an innermost pigmented layer, a layer of **rods** and **cones,** the sensory receptors, a layer of **bipolar cells,** and a final layer of **ganglion cells.** The bipolar cells and ganglion cells are neurons.

5. Rods respond to all wavelengths of the **visible spectrum** and function well at night, while each cone responds to either red, green, or blue light wavelengths and has little night function. Cones are most concentrated in the **fovea.**

6. Upon exposure to light, **rhodopsin** in rods bleaches to **opsin** and **retinal,** and the chemical change activates the synapses with the neurons above, where action potentials begin. The two products recycle, using vitamin A to produce rhodopsin. Since much of the rhodopsin is in the opsin/retinal form in bright light, temporary night blindness occurs when the surroundings are suddenly darkened.

REVIEW QUESTIONS

1. List the three things neurons do. (868)
2. Name the three functional parts of a neuron and explain what each does. (868, 869–870)
3. Identify the neural structure described by Schwann. (870)
4. List four examples of effectors. How are effectors activated? (869)

5. Describe the structure of a nerve. (871)

6. In what ways is the flow of an electrical current unlike a neural impulse? (872)

7. Describe the manner in which positive sodium and potassium and negative protein are arranged in the resting neuron. What complicates the distribution of potassium? Specifically, what is the effect of its distribution on the voltage potential across the membrane? (872–873)

8. Describe permeability changes in the membrane at the start of an action potential. How does this affect the distribution of the ions? The voltage potential across the membrane? (873)

9. Explain how repolarization occurs. What does this do to the distribution of sodium and potassium? How does the neuron eventually recover? (875)

10. Why is a neural impulse often referred to as a depolarizing wave? (873)

11. What is the general role of the sodium/potassium ion exchange pump in neural activity? (875)

12. Describe the position of the inactivation and activation sodium gates and the potassium gates when the neuron is at rest. (875)

13. Explain what happens to the ion gates that brings on an action potential. What specific characteristic of the gates makes this possible? (To what do they respond?) What causes sodium entry to cease? (875)

14. Describe the recovery process in terms of the ion gates. (875, 877)

15. Explain how a neural impulse proceeds through a myelinated neuron. Why is it faster and energetically less costly than in a nonmyelinated neuron? (877)

16. Using a simple drawing, describe the organization of a synapse, labeling the synaptic knob, postsynaptic membrane, synaptic cleft, neurotransmitter, vesicle, receptor sites, calcium gates, and presynaptic membrane. (877, 879)

17. Briefly describe the activity at the synapse, starting with the arrival of a neural impulse at a synaptic knob and ending with the opening of gates in the next neuron. How do these gates differ from the sodium and potassium gates farther along? (878–879)

18. What keeps the neurotransmitter from continually activating a neighboring neuron? What special situation exists for acetylcholine? (879)

19. Suggest two ways in which a neurotransmitter can inhibit the generation of an action potential in the next neuron. (879–880)

20. Explain the organization of a reflex arc. Of what adaptive value are such simple arrangements? (880)

21. List three important characteristics of sensory receptors. (881)

22. In what major way does generator potential differ from an action potential? How is this difference adaptive? (881)

23. List the two types of mechanoreceptors found in the human skin and state the specialization of each. (882)

24. What is a piezoelectric crystal? What tactile structures qualify as such structures? (882)

25. In what specific instance might sensitive thermoreceptors be highly adaptive to invertebrates? (883)

26. What are the specializations found in Ruffini and in Krause corpuscles? (884)

27. Describe the structure of the moth antennas, stressing the organization of an individual sensillum. Comment on their sensitivity. What is the adaptive value of this elaborate sensory arrangement? (884)

28. Explain how Jacobson's organ functions in the reptile. (885)

29. Describe the mammalian olfactory epithelium and its connections to the brain. (885)

30. Discuss ways in which the human and primate taste specializations may be adaptive. (885)

31. List the parts of the human external, middle, and internal ear, and include a description of the cochlear organization. (886, 887)

32. Describe the events of hearing, beginning with air waves buffeting the tympanum, and ending with an action potential in the auditory nerve. (886, 888)

33. Explain how a crayfish senses its body position. In what way is this similar to the operation of the saccule and utricle of humans? (889)

34. List the structures or regions of the eye through which light passes on its way to the retina. Where appropriate, explain how each structure functions. (891–892)

35. Describe the organization of the retina. In which region is color vision most acute? Why is this? (892–893)

36. In what way is the retina specialized for receiving red, blue, and green wavelengths? Other colors? (893)

37. Briefly summarize the chemical events occurring in the rod cells. When is rod vision most essential? Why? (893)

Neural Control: Nervous Systems

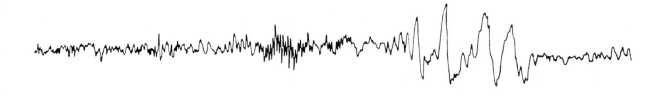

We humans pride ourselves, rightly or not, on our intelligence, and we are aware that the seat of that intelligence resides in that great gray structure we call the brain. We are also aware that the brain and the long stem that descends from it, the spinal cord, are very complex associations of the sorts of neural cells that we have just discussed. So here, then, we will focus on the structure and activities of both the brain and the spinal cord, which together compose what is called the **central nervous system.**

Before we launch into a discussion of brains and intellect, we might first note that many kinds of animals seem to do just fine without either. So perhaps we should begin by placing all this in some kind of evolutionary perspective. The brain, after all, is the result of an overdevelopment of one end of the nervous system. Note that in talking about ends, we must be talking about animals with longitudinal body plans. Starfish don't have ends. Neither do sea anemones. Being radially symmetrical, these animals are as likely to move off in one direction as another. Worms, on the other hand, have ends and they tend to move along in the direction of their body axis. In the course of evolutionary history, it seems that it would have become advantageous for

animals with elongated bodies, as they move about, to lead with the same end. Logic and evolutionary expediency thus seem to dictate that this leading end would be the site of a concentration of receptors that could tell the animal the nature of the environment into which it was moving before its whole body was exposed to that environment. It wouldn't do to back into an unfamiliar place. So **cephalization,** the development of an anterior end, set the stage for the formation of the brain by localizing neurons in the leading end of a longitudinal body. The brain was probably a subsequent specialization of a highly sensory and reactive area.

Let's now consider a few of the specializations in a selected sequence of animals, moving from those that don't have brains to those that do. We will arrange these species from simple to complex with the tacit assumption that we may be tracing, albeit roughly, the evolutionary development of the central nervous system. The assumption is a common one, but keep in mind that each representative animal has had its own long evolutionary history, and does not necessarily represent some sort of evolutionary stepping stone. We are only looking for hints and suggestions that might help lead us through an unchronicled history.

INVERTEBRATE NERVOUS SYSTEMS

Simple Nerve Nets and Ladders

The simplest metazoan neural organization appears in the coelenterates, such as the hydra. This is a diffuse system in which there is no clustering of nerve cells or their bodies into neural ganglia. The neurons are, however, concentrated about the tentacles and mouth. The distribution of neurons in the coelenterate gives the appearance of a net, but in many instances the neurons do not actually touch (Figure 35.1a). Presumably, such a system would be more versatile than a true net since it would permit pathways to become established. This in turn would allow for certain tissues or organs to react independently of others.

The most primitive flatworms have a netlike nervous system similar to that of the coelenterates, but in others the system is somewhat modified. For example, moderately primitive flatworms may have up to five nerve cords. In many planarians the nervous system resembles a ladder, consisting of only two parallel nerve cords with runglike cross connections and a concentration of nerve cells in the head region (Figure 35.1b). The anterior end of planarians has both a higher rate of cellular activity and greater sensitivity than the rest of the body. In addition, it is in the planarians that we first see large, anterior **ganglionic masses.** Ganglia are clusters of neurons or their cell bodies, often accompanied by supporting cells and surrounded by a sheath. Flatworms also boast cellular specializations in their nerve cells, having motor neurons, sensory neurons, and interneurons.

35.1 NETS, LADDERS, AND CEPHALIZATION

Representative invertebrate nervous systems. **(a)** The coelenterates, like this hydra, have a diffuse, rather uncentralized system. The netlike arrangement of nerve fibers is scattered over its body, with the greatest number near its mouth and tentacles. **(b)** In planaria, many cell bodies of neurons are accumulated in the head region, forming ganglia. Two nerves emerge from the brain and run in parallel down the body. Branches from these criss-cross the body, forming a ladderlike arrangement. **(c)** The earthworm has a complex, segmented nervous system with large paired ganglia above and below the esophagus. These are the largest groupings of cell bodies. It has a ventral, solid nerve cord running the length of its body. Paired ganglia appear in each segment, sending branching sensory and motor nerves around the body.

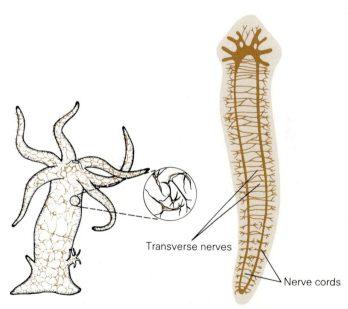

(a) **Hydra**—netlike arrangement   (b) **Planaria**—ladderlike arrangement

Transverse nerves

Nerve cords

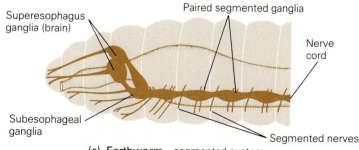

Superesophagus ganglia (brain)

Paired segmented ganglia

Nerve cord

Subesophageal ganglia

Segmented nerves

(c) **Earthworm**—segmented system

Massed Ganglia to Organized Brains

The annelids, we know, are segmented worms. In the earthworm, large ganglia surround the esophagus in the head of the worm (a common arrangement in annelids, arthropods, and other higher invertebrates), and each segment along the length of the worm has a pair of fused ganglia emanating from the nerve cord (see Figure 35.1c). Within the ventral nerve cord lie several **giant axons,** whose quick conduction of impulses permits the rapid burrowing and escape behavior of a startled worm.

If you have ever owned a pet clam, you may have noticed that it really wasn't much company. Clams and other bivalves are notoriously short on personality, perhaps because they are not very intelligent. Like many more sedentary mollusks, their nervous system is quite simple. The nervous system of the chiton takes on a ladder form roughly similar to that of a flatworm, except that it is much more complex. The greatest concentration of nerves form a nerve ring around the mouth and a dense ladderlike arrangement along the foot.

Cephalopods: Brainy Invertebrates. Surprisingly enough, another mollusk is rather bright, perhaps even smarter than some vertebrates. We're referring now to the octopus, an active and intelligent predator. It, like many invertebrates, has a ganglionic mass of neural tissue around its esophagus, but the mass is so differentiated that it may rightfully be called a brain. For example, a distinct and identifiable part of that complex serves as a respiratory center; another part controls the animal's rapid color changes; other parts deal with spatial associations and are hence important in learning; and other parts control eye movement (Figure 35.2a). The octopus, in fact, has been an important research animal in the study of learning and memory.

The squid, another cephalopod mollusk, is noted among physiologists for its giant axons, which form a complex over the mantle and are important to the fast "jet-propelled" escape movements for which the squid is noted. Much of what we know about neural propagation has come from studies of the giant axons of squid.

Another Look at the Arthropods. Arthropods are interesting from an evolutionary point of view because they have taken the segmentation we first saw in the annelids one step further—their segments are highly specialized. Not only are the body segments specialized, but so are the neural ganglia with which they are associated. In many groups the ganglia have massed to form a more centralized arrangement with a great reduction in the segmental ganglia. Compare the brain and segmental ganglia in the midge and blowfly in Figure 35.3. In the crayfish and other crustaceans capable of rapid darting movements, such movement is

35.2 NERVOUS SYSTEMS OF CEPHALOPOD, ECHINODERM, AND INSECT

(a) The octopus has a highly developed nervous system, complete with specialized regions related to keen eyesight, rapid movement, and intricate muscle control. (b) In the sea star, the nervous system consists of a central ring with major nerves radiating into each arm. There is little cephalization. (c) The insect nervous system includes supraesophageal and subesophageal ganglia; ventral, solid nerve cord; and paired ganglia serving most segments.

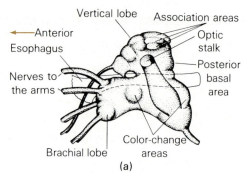

(a)
Octopus—complex centralized nervous system

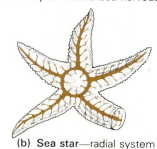

(b) Sea star—radial system

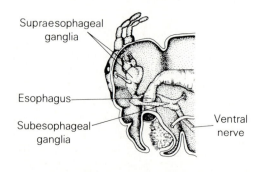

(c) Insect—segmented system

35.3 INSECT SPECIALIZATIONS

Evolutionary trends in many insects include consolidation of the segmental nervous system into a more centralized one. Although there are larger ganglia in the head and thorax of the midge (a), more primitive segmental ganglia remain in the abdominal region. The dancefly (b) shows some increase in consolidation in the ganglia, while this is even more apparent in the horsefly (c). Finally, segmental fusion of ganglia in the blowfly (d) is complete.

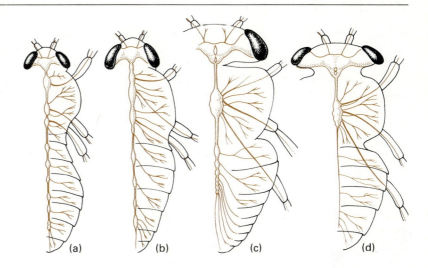

(a)   (b)   (c)   (d)

instigated by giant axons, as we saw in the annelids and mollusks.

The insect brain is composed of a large ganglionic mass above the esophagus and a smaller one below. Each of these brain parts has very specific roles with much of the larger mass devoted to integrating information from the eyes, antennas, and other sensory structures so abundant on the insect body (see Figure 35.2c). The ganglionic mass below the esophagus controls the mouthparts and salivary glands. Interestingly, in arthropods that have just the simple eye—one that crudely registers only light intensities—the visual centers take up about 3% of the brain. But in highly visual species, such as the housefly, the visual centers may occupy as much as 80% of the brain.

VERTEBRATE NERVOUS SYSTEMS

So far, the nerve cords we have seen have been ventral. Furthermore, these are all solid neural structures. Now, with the vertebrates, we find a dorsal hollow nerve cord and—at its bulbous anterior end—the large, complex, ganglionic mass that is the brain. We can begin with an overview of the nervous systems in various classes of vertebrates.

Among the vertebrates, there is a pronounced tendency toward increased cephalization. That is, the more recently evolved classes generally have larger brains in relation to body size. The living classes of vertebrates include the jawless lampreys and hagfish, the sharks and rays, bony fishes, amphibians, reptiles, birds, and mammals. Of these, the most recently evolved, and hence "higher" classes, are the birds and mammals. On the other hand, our lowly cephalochordate relative, the lancelet, evolved much earlier. It has a hollow spinal cord but no centralized brain at all, a trait it presumably shares with our own Precambrian ancestors.

All vertebrate brains have three parts: the **forebrain, midbrain,** and **hindbrain.** It is the forebrain that is associated with the cerebrum. In lower animals, the forebrain is reduced, and the posterior brain regions are disproportionately larger. The fish brain, for example, is dominated by the midbrain, but there is also a relatively large hindbrain. In many animals, the midbrain is important in analyzing visual and olfactory stimuli, but in humans it is greatly reduced and serves primarily as a communicating bridge between the forebrain and hindbrain. Compare the brains of the vertebrates in Figure 35.4. You can see that in higher vertebrates, the midbrain and hindbrain have yielded to the forebrain's cerebrum. It is in mammals that we find the most complex cerebral development. There are more neurons in the brain, and they interact in more ways than in any other group of animals. In addition, the wrinkled outer region of the brain—the **cortex**—is far more prominent in the mammals. For our in depth look at the mammalian nervous system, we will turn again to that brainiest of all creatures.

35.4 THE VERTEBRATE BRAIN

Comparing the anatomical structures of the brain in five classes of vertebrates (cartilaginous fish, amphibian, reptile, bird, and mammal) reveals general evolutionary trends and specific trends in specialization. **(a)** Note the relatively large olfactory lobes in the shark's brain. It clearly reflects the importance of chemical detection to this predator. **(b)** The frog feeds by visual means, as does the chicken. Note the relative sizes of their optic lobes. The trend toward increasing dominance of the cerebrum in vertebrate evolution is also apparent, beginning with the alligator **(c)** and becoming more pronounced in the bird **(d)**. **(e)** The trend is greatest in the mammal, with increased convolutions in the cortex. Convolutions or foldings are a way of increasing cerebral size without greatly increasing cranial size.

THE HUMAN CENTRAL NERVOUS SYSTEM

The human nervous system, like that of other vertebrates, is subdivided anatomically into two major parts, the **central nervous system** (CNS) and the **peripheral nervous system** (PNS) (Figure 35.5). The central nervous system is composed of the brain and spinal cord. (Embryologically, parts of the eye are out-pocketings of the brain, so these structures are also considered to be parts of the central nervous system). The peripheral nervous system includes all of the neural structures outside the brain and spinal cord, and these are organized into the **somatic system** and the **autonomic system.** The somatic system is largely voluntary and made up primarily of motor nerves that serve the muscles and the sensory nerves and sense organs, as well as the structures from which they arise (see Chapter 34). The autonomic system is largely involuntary and deeply involved in homeostasis, coordinating many of the regulatory activities of the body and operating below the conscious level. It contains many motor nerves and neurons that originate in the brain and spinal cord and that carry impulses to the internal organs.

The central nervous system develops from the familiar dorsal, hollow, nerve cord that is one of the defining characters of all chordates. This hollow nerve cord, the **neural tube,** is one of the first recognizable structures in the developing embryo (see Chapter 37). From it will develop the brain, the spinal cord, and much of the eye. We will begin our look at the central nervous system with the spinal cord.

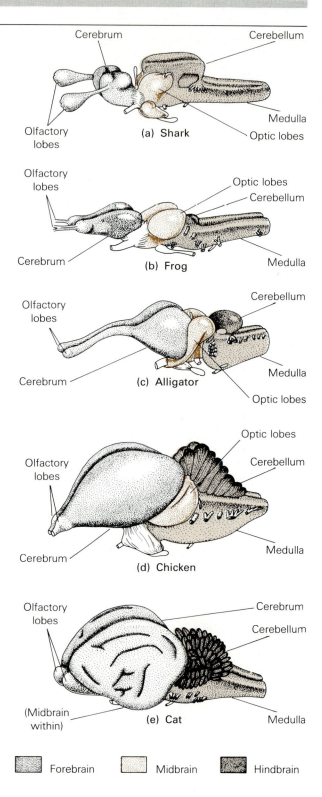

(a) Shark

(b) Frog

(c) Alligator

(d) Chicken

(e) Cat

Forebrain   Midbrain   Hindbrain

The Spinal Cord

The spinal cord (Figure 35.6) serves as a major link between the brain and other parts of the nervous system. Essentially, the spinal cord contains two regions which, in cross section, appear white and gray to the unaided eye. The white outer regions of the cord consists of astronomical numbers of myelinated axons running in neural pathways, or tracts, somewhat separated from their neural cell bodies. The cell bodies of the neurons form the gray butterfly-shaped areas of the spinal cord. The gray matter lacks myelinated neurons and consists mostly of the nerve cell bodies of motor and interneurons. Gray matter cannot repair itself if damaged, and white matter has only limited power of repair.

The spinal cord has a certain degree of autonomy in that it can carry out synaptic reflexes between sensory and motor neurons without consulting the brain (see Chapter 34). It is good to remove your hand from a hot stove as soon as possible, and letting the spinal cord control the situation reflexively helps preclude wasting time with cerebral debate.

The spinal cord begins as a narrow continuation of the brain, emerging from the **foramen magnum** ("big hole") at the base of the skull. It lies sheltered within the **vertebral canal,** a continuous channel formed by the neural arches of the vertebrae (see Figure 28.15). The spinal cord, like the brain, is surrounded by a tough, three-layered sheath, the **meninges,** which contains the cushioning **cerebrospinal fluid.**

Paired spinal nerves emerge from the great cord through the spaces that lie between adjacent vertebral arches. Each of these spinal nerves is formed from two roots in the cord, a dorsal (toward the back) **sensory root** and a ventral (toward the belly) **motor root** (see Figure 35.6). The cell bodies of the motor neurons, like the cell bodies of interneurons, lie in the gray matter of the cord, with their lengthy axons passing out through the spinal nerves and onward to the voluntary muscle. On the other hand, the cell bodies of sensory neurons are clumped in large ganglia just outside the cord.

The spinal nerves carry sensory messages that tell you that your foot is in cold water, and the vol-

35.5 HUMAN NERVOUS SYSTEM

(a) Organization of the human nervous system. **(b)** The central nervous system *(color)* includes the brain and spinal cord. The vast proliferation of branched sensory and motor nerves emerging from the brain and cord as cranial and spinal nerves make up the peripheral nervous system, which is again subdivided into its own divisions.

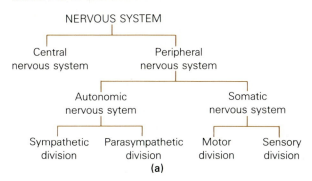

(a)

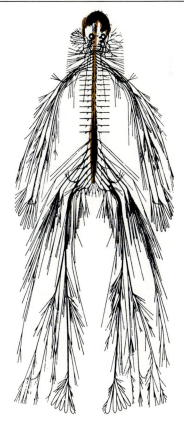

(b)

35.6 THE SPINAL CORD

The spinal cord (a) extends from the base of the brain into the lumbar region of the spine, where it begins branching into many descending nerves. The spinal nerves are major branches that emerge from between vertebrae. A cross section (b) reveals the spinal cord to be composed of two distinct regions. The gray, double-winged region consists primarily of cell bodies and nonmyelinated neural tracts, while the outer white region is largely composed of myelinated axons that form the major spinal tracts. The emerging spinal nerves contain motor and sensory neurons. Motor nerves emerge from a ventral (front) root; sensory nerves enter the cord through a dorsal (rear) root. The thickenings seen in the dorsal root are the cell bodies of numerous sensory neurons (the dorsal root ganglia).

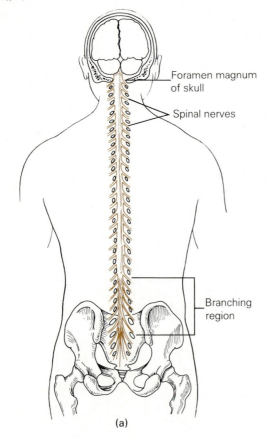

(a)

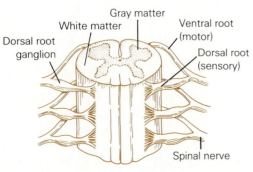

(b) Spinal cord
as seen from the back

untary motor messages that activate the appropriate muscles to get your foot out. Their neurons also form a large part of the autonomic nervous system—the constellation of nerves and nerve cells, which, as we mentioned, is involved in the involuntary activities of your internal organs.

The Human Brain

The human brain is an absolutely fascinating structure given to self-congratulation. (In these words we find one complimenting itself). An enlarged brain and an enlarged gluteus maximus are two of the most prominent traits of our species. One might wonder what the planet would look like had the human brain not motivated us to action and we had spent more time resting dolefully on our glutei maximi. The human brain has devoted a great deal of time to thinking about itself, but to this day many of its processes remain unknown. Furthermore, the things we do discover are often hard to believe. For example, there is some evidence that every word you have ever uttered or heard in your entire life may be filed way in your brain, even though you will go to your grave having recalled hardly any of that information. (How nice it would be to be able to retrieve the exact words of our last conversation with a loved one).

The human brain weighs about 1.4 kilograms (three pounds), has a volume of about 1300 to 1500 cubic centimeters, and contains some 100 billion or more neurons and at least 10 times that many supporting **glial cells.** Each neuron may form many synapses, and the number of alternative pathways of impulses, along with the coordination required for an appropriate response, produce a veritable harmonic neural symphony. The brain, like the spinal cord contains gray matter (cell bodies) and white matter (myelinated axons). But the gray matter of the brain lies on the outside, the reverse of what we find in the spinal cord. (Gray matter is what people like to accuse one another of lacking.) The delicate brain with its gelatinlike consistency requires protection. In addition to being covered by the skull, the brain, like the cord, is directly enclosed by the three tough protective meninges. The spaces within the meninges and the hollow spaces within the brain itself are filled with the pressurized, shock absorbing cerebrospinal fluid that also surrounds the spinal cord. Finally, the billions of fragile neurons themselves are embedded in even vaster numbers of glial cells, which make up much of the brain's mass.

You may be surprised to learn that our three-pound brain isn't the largest on earth; whales and porpoises have larger brains. But intelligence (since we invented the term) seems to be related to the *relative size* of the brain to the total body, and here we humans lead the pack. (Besides, whales and porpoises don't have opposable thumbs, so what would they do with great intelligence anyway?)

We have mentioned that in all vertebrates, the brain consists of three regions, elegantly named the forebrain, midbrain, and hindbrain. We will take a close look at these now, beginning with the hindbrain.

The Hindbrain.   The hindbrain consists of the **medulla,** the **cerebellum,** and the **pons,** and is continuous with the spinal cord. It contains the lower, more posterior, centers (Figure 35.7). As a rough generality, the unconscious, involuntary, and mechanical processes are directed by these lower regions. For example, the hindmost part, the **medulla oblongata,** or simply the medulla, contains centers that control breathing rate, heart rate, and blood pressure. In addition, all communication between the spinal cord and the brain must pass through the medulla.

The cerebellum is a small, bulbous, paired structure, about the size of and general appearance of the two halves of an enlarged walnut. It lies above the medulla and somewhat toward the back of the head. The cerebellum receives input from the eyes, balance organs, and proprioceptors,

which it uses in coordinating all complex voluntary muscle actions.

The pons (Latin for "bridge") is the portion of the brainstem just above the medulla. It contains ascending and descending neural tracts that run between the brain and spinal cord and between the cerebrum and cerebellum, linking the functions of the cerebellum with the more conscious centers of the forebrain. It also receives tracts to and from certain **cranial nerves,** the large nerves that emerge from the brain itself, to form connections in both the somatic and autonomic systems. The pons also aids the medulla in regulating breathing.

The Midbrain.   Essentially, the midbrain joins the hindbrain and forebrain, by numerous connecting tracts between the two. Certain parts of the midbrain receive sensory input from the eyes and ears. All auditory (sound) input of vertebrates is processed here before being sent to the forebrain. In most vertebrates, visual input is first processed in the midbrain, but in humans and other mammals, the visual information is sent directly to the forebrain. And while the midbrain is involved in complex behavior in fishes and amphibians, many of these same functions are assumed by the forebrain in reptiles, birds, and mammals.

The Forebrain.   The forebrain, the largest and most dominant part of the human brain, is respon-

---

35.7 ANATOMY OF THE BRAIN

(a) The outer surface of the brain contains the familiar convoluted (deeply wrinkled) cerebral cortex. **(b)** The brain here is divided along the midline from front to back. The dominant forebrain (thalamus, hypothalamus, and cerebrum) is obvious.

The areas of the hindbrain (medulla oblongata, pons, and cerebellum) are also quite distinctive, but the midbrain that connects them is not well defined. The corpus callosum connects the two halves of the brain.

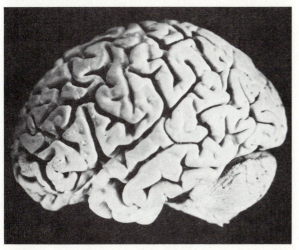

(a)

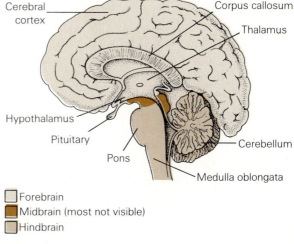

(b)

sible for conscious thought, reasoning, memory, language, sensory reception and decoding, and conscious movement of the skeletal muscles. The embryonic forebrain gives rise to such structures as the **thalamus,** the **reticular system,** the **hypothalamus,** and the **cerebrum.**

The Thalamus and the Reticular System. The thalamus is located at the base of the forebrain. It is a paired structure, with the two halves on either side of the central, fluid-filled **third ventricle** of the brain (see Figure 35.7). Since it specializes in making connections, the thalamus has been rather unpoetically called the "great relay station of the brain." It consists of densely packed clusters of neurons, which connect various parts of the brain; it relays sensory input to the cerbral cortex, motor impulses from the cortex to the spinal cord, and, curiously, relays impulses to the cortex that help maintain consciousness.

In the latter function, the thalamus relies on a portion of its tracts known as the reticular system, a region composed of interconnected neurons that are almost feltlike in appearance. These neurons run from the brainstem throughout the thalamus and as far as the lower part of the forebrain. The reticular system is still somewhat of a mystery, but several interesting facts are known about it. For

35.8 THE RETICULAR SYSTEM

The impulse originating at the lower right passes through the reticular system, a structure composed of untold millions of neurons. The smaller arrows indicate that, in this case, the entire cerebrum has been alerted, but a specific part (the shaded area) is the target of most of the impulses. It is likely, then, that this area of the cerebrum will be required to deal with whatever initiated the stimulus in the first place.

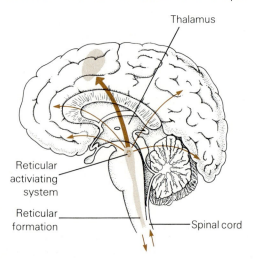

Thalamus

Reticular activiting system

Reticular formation

Spinal cord

example, it "bugs" your brain, tapping virtually all incoming and outgoing communications. The leading hypothesis is that the reticular network is something of an alarm system that serves to activate the appropriate parts of the brain upon receiving a stimulus. The more messages it intercepts, the more the brain is aroused (Figure 35.8).

The reticular system also seems to function importantly in sleep. You may have noticed that it is much easier to fall asleep when you are lying on a soft bed in a quiet room with the lights off. Under these conditions, there are fewer incoming stimuli; as a result, the reticular system receives fewer messages, and the brain is allowed to relax and initiate the processes associated with sleep.

The Hypothalamus. The hypothalamus is, as the name implies, below the thalamus. Actually it makes up the floor of the centrally located third ventricle. It is densely packed with cells that help regulate the body's internal environment and certain general aspects of behavior. In its homeostatic functions, the hypothalamus receives sensory input from many internal organs via the thalamus, and makes use of the input in coordinating such internal conditions as heart rate, appetite, water balance, blood pressure, and body temperature. It influences such basic drives as hunger, thirst, sex, and rage. Electrical stimulation of various centers in the hypothalamus can cause a cat to act hungry, angry, cold, hot, benign, or horny. In humans it is known that a tumor pressing against a "rage center" can cause the person to behave violently and even murderously.

You already know that a major function of the hypothalamus is the coordination of the nervous system with the endocrine system (see Chapter 33). In fact, the hypothalamus is pretty much in control of the pituitary (but subject to negative feedback, of course).

The Limbic System. The hypothalamus and the thalamus, along with certain pathways in the cortex, are functionally part of what is called the **limbic system** (Figure 35.9). The limbic system links the forebrain and midbrain and is composed of a number of **nuclei,** specific, sharply defined areas associated with a specific function. In the limbic system such nuclei control certain aspects of muscle tone, such as the positioning of an arm so that the fingers are in the best position to play the piano. Since it includes the hypothalamus, the limbic system encompasses its functions as well and influences emotions such as fear, rage, sexual

35.9 THE LIMBIC SYSTEM

The major components of the limbic system include the amygdala, hippocampus, thalamus, hypothalamus, and certain parts of the frontal and temporal lobes of the cortex.

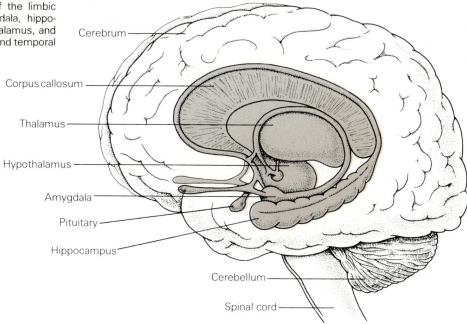

Cerebrum

Corpus callosum

Thalamus

Hypothalamus

Amygdala

Pituitary

Hippocampus

Cerebellum

Spinal cord

arousal, aggressiveness, and motivation. For example, the **amygdala,** lying within the limbic system, can produce rage if stimulated and docility if removed. The **hippocampus,** another limbic structure, figures importantly in memory, as will be discussed in Chapter 38.

The Cerebrum.   To most people, the word *brain* conjures up an image of two large, deeply convoluted gray lobes. Those lobes are the cerebrum, the largest and most prominent part of the human forebrain. Whereas the cerebrum is well known as the neural center of intelligence, its functions are actually quite diverse, as we will see.

The left and right halves of the cerebrum are the **cerebral hemispheres,** and the outer layer of gray cells is the **cerebral cortex.** The cerebral cortex consists of a thin but extremely dense layer of nerve cell bodies (about 15 billion in all) and their dendrites, overlying the more solid white matter below. The white region consists of myelinated nerve fibers, reaching in all directions throughout the brain (Figure 35.10). These fibers bring information to the cortex for processing and carry the integrated messages from the cortex to other parts of the brain.

Every vertebrate has a cerebrum, but the cerebrum differs markedly from class to class, especially in the degree of development of the cortex. In some

kinds of vertebrates the cerebrum may serve essentially to refine behavior that could be performed to some degree without it. These kinds of animals are not generally regarded as deep thinkers, but rely more on the genetically programmed behavior emanating from the "old brain"—that is, the noncerebral brain, the part that evolved first. In more advanced animals, the cerebrum takes on a greater importance and, as in the case of the visual and auditory centers, often takes over functions that were once the responsibility of other, older parts of the brain.

For example, if the cerebrum of a frog is removed, the frog will show relatively little change in behavior. If it is turned upside down, it will right itself; if it is touched with an irritant, it will scratch; it will even catch a fly. Also, sexual behavior in frogs can occur without the use of the brain—but it is probably best not to extrapolate from that. A rat is more dependent on its cerebrum. A rat that is surgically deprived of its cerebrum can visually distinguish only light and dark, although it seems to move normally. A cat with its cerebrum removed can meow and purr, swallow, and move to avoid pain, but its movements are sluggish and robotlike. Dogs treated this way are more helpless, and just stand around eventually starving unless food is thrust into their mouths. A monkey whose cerebrum has been removed is severely paralyzed and

35.10 THE CORTEX

(a) A cross section of the cerebrum reveals an outer layer of gray matter, the cortex, that contains several very dense layers of neurons. The underlying white matter consists of immense numbers of myelinated neural processes. The overlying cortex is highly convoluted. Such folds and fissures increase the cortical surface area within a confined space. (b) A cellular study of the cortex, made in 1888, but still generally valid.

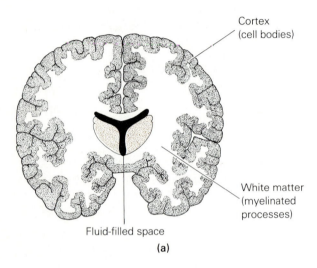

Cortex (cell bodies)

White matter (myelinated processes)

Fluid-filled space

(a)

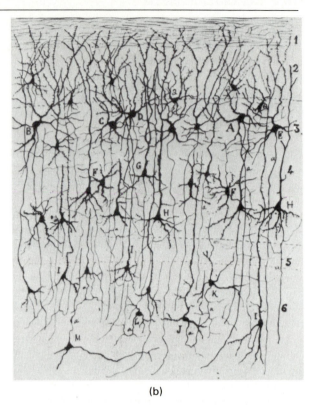

(b)

can barely distinguish light and dark. The result of massive cerebral damage in humans is total blindness and almost complete paralysis. Although such persons can breathe and swallow, they soon die.

It seems that, from an evolutionary standpoint, more and more of the functions of the lower brain are transferred to the cerebrum in the more "cerebrated," or intelligent species. Generally, the degree to which the cerebrum has taken over neural control is reflected in its size. Thus more "advanced" animals have relatively larger cerebrums (see Figure 35.4). However, size is not the only indicator of the cerebrum's complexity. Convolutions, for example, increase the surface area of the cortex without enlarging the braincase. The deep convolutions seen in the human brain are lacking in the brain of the rat (but note that rats, with their smooth and tiny brains, are notoriously clever). The highly touted and undoubtedly intelligent dolphin has a highly convoluted cerebrum, but with fewer layers than the human cerebrum (Figure 35.11).

Hemispheres and Lobes.   In humans each cerebral hemisphere is divided into four lobes (Figure 35.12). At the back is the **occipital lobe,** which receives and analyzes visual information. The

35.11 DOLPHIN BRAIN

The 3.5-pound dolphin brain is somewhat heavier than that of the human and its cortex is actually more convoluted. However, in proportion to body weight, humans—whose brain makes up about 2 percent—have the largest brain. The dolphin and chimpanzee are about even, with the brain making up about 1 percent of body weight. This proportion is only one of many factors taken into consideration where intelligence is concerned.

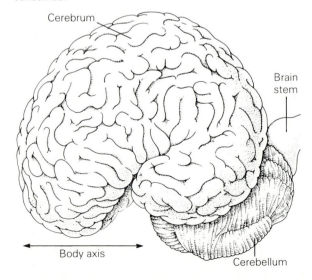

Cerebrum

Brain stem

Body axis

Cerebellum

visual area of the cortex is stimulated by the images formed on the retinas. If the occipital lobe is injured, black "holes" appear in the part of the visual field that is registered in that area.

The **temporal lobes** are at either side of the brain, under the temples. Each lobe roughly resembles the thumb on a boxing glove, and it is bound anteriorly by a deep fissure—the **fissure of Sylvius,** or **lateral sulcus.** The temporal lobe shares somewhat in the processing of visual information, but its main function is in hearing. The **frontal lobe** is right where you would expect to find it—at the front of the cerebrum. It underlies that part of the skull that people hit with the palm of their hand when they suddenly remember what they forgot. One part of the frontal lobe regulates precise voluntary movement. Another part controls the bodily movements that produce speech and is considered to be part of the speech center.

The area at the very front of the frontal lobe is called the **prefrontal area.** Whereas it was once believed that this area was the seat of the intellect, it is now apparent that its principal function is sorting out sensory information. In other words, it places information and stimuli into their proper context.

The gentle touch of a mate and the sight of a hand protruding from the bathtub drain might both serve as stimuli, but they would be sorted differently by the prefrontal area.

The **parietal lobe** lies directly behind the frontal lobe and the two are separated by a deep cleft called the **fissure of Rolando,** or **central sulcus.** The parietal lobe contains the sensory areas for the skin receptors and the cortical areas that detect body position. Even if you can't see your feet right now, you probably have some idea of where they are, thanks to receptors in your muscles and tendons that innervate centers in the parietal lobe (and the cerebellum as well). Damage to the parietal lobe can cause numbness and a sense that one's body is wildly distorted. In addition, the victim may be unable to perceive the spatial relationships of surrounding objects.

Sensory and Motor Regions of the Cortex. By probing the brain with electrodes, investigators have found it possible to determine exactly which areas of the cerebrum are involved in the body's various sensory and motor activities and to map these cortical functions (there is some individual

---

35.12 THE CEREBRUM

The human cerebrum is divided into four prominent lobes: frontal, temporal, parietal, and occipital. Each is somewhat specialized in its functions and all represent the so-called higher centers of the brain.

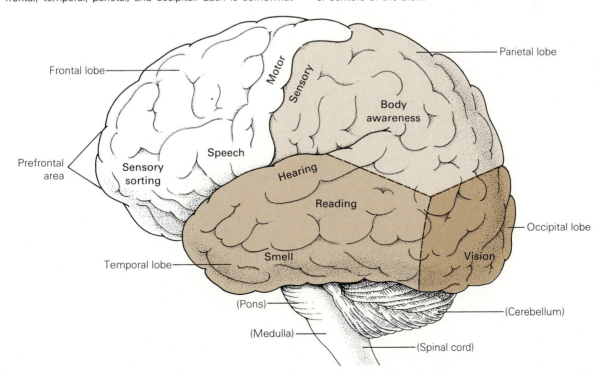

variation in where such functions are localized). Figure 35.13 shows the results of such mapping. The figures are distorted, not out of any appreciation of the macabre, but to demonstrate the relative area of the cerebrum devoted to each body part. The amount of gray matter devoted to a body part seems to depend on the importance that part has come to have in the evolution and natural history of humans. We see that the sensory areas of the cortex are largely devoted to integrating sensations from the face, tongue, hands, and genitals, while the larger parts of the motor control area are devoted to the muscle control in the tongue, face, and thumb. Thus, while our genitals are very sensitive, we have almost no conscious control over them, as you may know.

The sensory and motor areas are not randomly scattered through the cortex. There is a functional order in their spatial arrangement, depending on the location of the body part they control. Thus the index-finger control area lies near the thumb control area (even though the muscles that move the index finger are actually in the forearm), and the area that controls the movement of the elbow lies between the control areas of the wrist and the shoulder.

Right and Left Halves of the Brain.   Though the two cerebral hemispheres are roughly equal in size and potential, they are not identical in either form or function. Moreover, their differences apparently are accentuated by learning because the control of certain learned patterns takes place primarily in only one hemisphere. One example of the difference between the hemispheres is seen in handedness.

It is interesting that other species such as rats

35.13 CORTICAL MOTOR AND SENSORY REGIONS

Among the localized areas of the cortex are the sensory and motor regions. Each resides in a gyrus located on opposite sides of the fissure of Rolando, a prominent dividing line between the frontal and parietal lobes (see detail). The caricatures of the human body drawn over each region have their parts emphasized or deemphasized to indicate the degree of control present. Thus the face, and particularly the lips, is exaggerated in the figure overlying the sensory region. Note that the hand is similar in size in both, indicating that both sensory and motor functions are well developed in this organ.

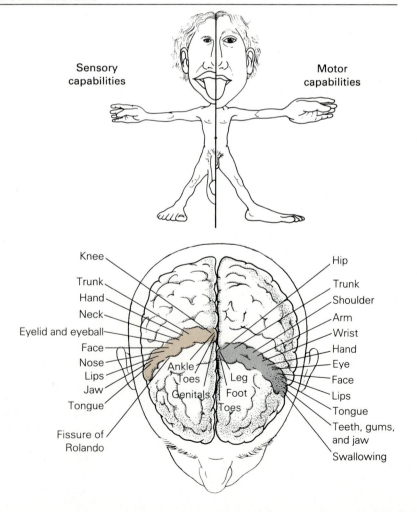

Sensory capabilities

Motor capabilities

Knee
Trunk
Hand
Neck
Eyelid and eyeball
Face
Nose
Lips
Jaw
Tongue
Fissure of Rolando

Ankle
Toes
Genitals

Leg
Foot
Toes

Hip
Trunk
Shoulder
Arm
Wrist
Hand
Eye
Face
Lips
Tongue
Teeth, gums, and jaw
Swallowing

and parrots also show right- and left-handedness. But whereas 89% of humans are right-handed, rats and parrots show equal frequencies of left and right dominance. Because neural tracts cross from one side to the other, the right side of the body is controlled by the left side of the brain and vice versa; thus in right-handed people, the left half of the brain is dominant. It is also slightly larger than the right half. There are many mysteries to handedness. For example, none of the numerous facile "explanations" adequately account for the unusual prevalence of southpaws among professional tennis players and artists.

No one is completely sure of what causes some people to be left-handed. The data seem to indicate that there are two kinds of left-handedness: in the more common type, the left half the brain is still dominant. As usual it contains the functional speech center; but in this case the left hemisphere also controls the left hand. These left-handers are the ones who write by crooking their left hand around so as to write "upside down." In the less common type of left-handedness, the *right* half of the brain is dominant. It contains the functional speech center and controls the left hand, which is held at a more conventional writing angle. A few people have both traits, a sort of double negative: their right hands are controlled by dominant right hemispheres, so that they write with their right hands—which are held upside down!

Other functions, such as speech, perception, and different aspects of IQ test performance are also more likely to be located in one hemisphere than the other. The primary speech center of all right-handed people, and most left-handed people, as we have seen, is located in the left hemisphere. The left hemisphere also seems to be the seat of analytical thinking, while spatial perception is a right-hemisphere function. Left-hemisphere brain damage can result in **aphasia**—the inability to speak or understand language—while right-hemisphere brain damage may result in the inability to draw the simplest picture or diagram (but does not affect the ability to write letters and numbers, which is a left-hemisphere function). We will return to speech later in the chapter.

In spite of these specializations, the right and left hemispheres operate as an integrated functional unit. The primary route of communication between the right and left cerebral hemispheres is the **corpus callosum** (Figure 35.14). It seems that if one side of the brain learns something—for instance, by feeling an object with just one hand—the information will be transferred to the other hemisphere. However, if the corpus callosum has been severed, the left brain

35.14 THE CORPUS CALLOSUM

The corpus callosum is a neural bridge between the cerebral hemispheres, relaying information from one half to the other. This connective can be severed without obvious ill effects, as it has been in the treatment of violent forms of epilepsy. While there are few major problems after such drastic treatment, certain tests reveal an absence of the transfer of learning from one hemisphere to the other.

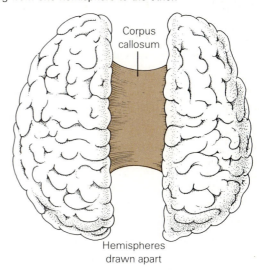

Corpus callosum

Hemispheres drawn apart

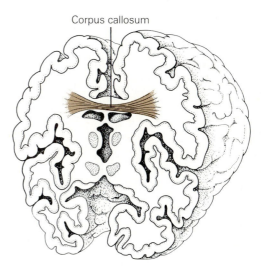

Corpus callosum

literally doesn't know what the left hand is doing (Essay 35.1).

It should be pointed out that a certain degree of compensation is possible between the two halves of the brain. For example, if the dominant hand is injured, it is possible to learn to use the other hand with almost equal facility. If one hemisphere of a young child's brain is severely injured, the other side will eventually take over some or all of its functions. This ability to shift, like so many other abilities, declines with age.

The Great Split-Brain Experiments

Some years ago, Roger W. Sperry and Robert E. Meyers experimentally separated the two hemispheres of the brains of cats by severing both the corpus callosum and the optic chiasma. This meant that the right eye was now connected only to the right half of the cerebral cortex and the left eye to the left cerebral cortex. Intensive testing then showed that the animals behaved as if they had two separate brains. A cat could be trained to perform a task by using one eye (the other covered) and when presented with the same task viewed by the other eye, the animal would respond as if no learning had occurred.

Later, Sperry and Michael Gazzaniga severed only the corpus callosum in human epilepsy patients as treatment for their condition. With their optic chiasmas intact they were tested in an apparatus that allowed a particular image to fall entirely on only one half of the retina (see figure). This meant an image could be effectively transmitted to only the right in the left hemisphere of the brain. The results were intriguing: the patients reported "seeing" only what was transmitted to the left hemisphere. If a word such as "heart" were shown but divided by a partition so that each eye could see only part of the word ("he" with the left eye and "art" with the right), then "he" fell on the right half of the brain, "art" on the left, and the subjects reported seeing "art."

It was quickly discovered that the results were not due to visual phenomena alone. The patients could only tell researchers what fell on the left halves of their brains because that is where the centers of *speech* are located. On the other hand, if they were allowed to express themselves nonverbally, they could immediately describe what the right half of the brain "saw." Because the right side of the cortex controls the left hand, and vice versa, a person can first feel an object, say with the left hand, and then pick out that object with the same hand from a collection hidden

Left hemisphere   Right hemisphere

If the corpus callosum is cut, the halves of the brain are capable of functioning as separate and roughly equivalent units, although each has its own special qualities. If the optic chiasma is also cut, images seen by one eye can't be transferred to the opposite cerebral hemisphere. When the right eye is covered, a person can learn tasks and make discriminations using only the left eye. But if the left eye is then covered, the person must relearn the same tasks.

behind a screen, simply by feel. However, because the information is traveling to the right half of the brain and that half cannot communicate with the left half, the person cannot name the object. Interestingly, the person will be able to pick out the object faster by using the left hand than the right because the right cortex is far superior at dealing with spatial relationships.

Further experimentation has revealed that the human brain is, in a sense, effectively divided into two halves—two brains, as it were. The left half indeed tends to deal more with rational, verbal, and logical information, while the right is more intuitive. However, the information has

fallen into the hands of pop psychologists, whom we now find breathlessly explaining that the problems of the world are due to too much dependence on the cold and logical left hemisphere and that we should be training ourselves to depend on the soulful and holistic right side. The secret, some say, is encompassed in the teachings of Eastern religions. Those who make such claims, however, seem to neglect the fact that the talents of each half of the brain are mutually dependent. They work together, and have historically, to produce that peculiar combination of behaviors that are so distinctively human.

Chemicals in the Brain

We know that the brain contains billions of neurons and that these neurons interact in a delicate and coordinated manner, integrating and shunting information from one place to another. This interaction includes both excitation and inhibition of adjacent neurons through very specific action in the trillions of synapses. While the peripheral nervous system uses only a few neurotransmitters—primarily acetylcholine and norepinephrine (see Chapter 34)—the brain has at least 50, with more being discovered all the time. This number probably shouldn't be too surprising considering the specificity that is necessary in orchestrating the brain's vast network of cells.

Some neurotransmitters are synthesized in the cell bodies of neurons, where the nucleus, ribosomes, endoplasmic reticulum, and the rest of the usual synthetic machinery is found (see Figure 34.2). Others are formed in the axon or axon terminal. The neurotransmitters are packaged into saclike vesicles, which then move along the axons to the terminal processes, where they are stored until their release into the synaptic cleft. Different regions of the brain have their own neurotransmitters and receptors, so that the brain is organized chemically as well as anatomically.

The brain's neurotransmitters may be simple **monoamines,** such as **norepinephrine, dopamine, histamine,** and **serotonin.** Each of the above monoamines is a modified amino acid, but in some instances unmodified amino acids, such as glycine and glutamate, act as neurotransmitters.

The **neuropeptides** are a more recently discovered class of neurotransmitters. As their name suggests, they are short chains of amino acids (from two to about 40). Some of the neuropeptides also occur as hormones. Evidence of this role is suggested when the neuropeptide **angiotensin II** is injected into cat brains, bringing on episodes of prolonged drinking. These substances, in most instances, function only if injected directly into the brain or its cavities. This is necessary because of the **blood-brain barrier** between the circulatory system and the brain. Unlike capillaries in the rest of the body, those in the brain lack the loosely organized, porous connections between the endothelial cells that make up the capillary walls. In fact, where the endothelial cells meet, they tend to overlap tightly by considerable margins. Transport out of the capillaries is assisted by certain glial cells called **astrocytes.** These cells form flattened, footlike processes against the capillary walls, and most materials must pass into the glial cells before entering intercellular spaces (Figure 35.15). Thus, the blood-brain barrier is highly selective in admitting substances into brain tissue, a protective measure that prevents toxic wastes from disrupting the delicate brain tissue.

Among the more interesting brain neuropeptides are the **enkephalins** and **endorphins.** These modify our perception of pain, and also have an elevating effect on mood. Anything acutely painful, such as running a marathon race or shooting oneself in the foot, will stimulate the release of enkephalins. The "runner's high," a slight euphoria that may appear after about a 10-mile run, is also attributed to the release of enkephalins (although some runners simply may be ecstatic at covering that distance without dying). These peptides are also called the **opioid** neurotransmitters because morphine and other opiates will also bind to the neural enkephalin receptors and mimic their action.

Our rapidly expanding knowledge of the neurotransmitters is also shedding some light on the action of certain other drugs that affect the central nervous system (and vice versa). Amphetamines, for example, are believed to accelerate the natural release of dopamine, a neurotransmitter known to be associated with arousal and pleasure systems. Cocaine has a different, more drastic effect. It amplifies the action of neurotransmitters such as serotonin, norepinephrine, and dopamine by block-

35.15 THE BLOOD-BRAIN BARRIER

Materials entering and leaving the brain tissue must pass across what is called the "blood-brain barrier." Note the tightly overlapping cells of the capillary wall and the footlike attachments of the astrocyte. Most materials are believed to pass through the astrocytes prior to entering the nervous tissue.

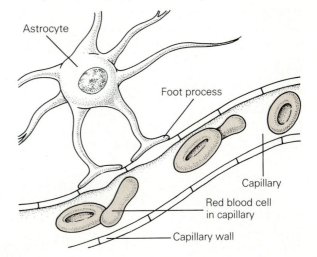

ing the enzymes and reuptake mechanisms that normally clear them from the synaptic cleft after they have acted. Drugs such as LSD, mescaline, and psilocybin mimic the role of other natural neurotransmitters. They produce their hallucinogenic effects by artificially stimulating action at the synapses.

Electrical Activity in the Brain

Each active neuron in the brain creates impulses that have electrical characteristics, although about a million neurons must fire simultaneously in order for this energy to be detected from outside the body. The instrument used in such detection is called an **electroencephalograph,** and the record obtained, an **electroencephalogram** (EEG). Electrodes leading to the device are fastened at various places on the head, and the feeble currents they pick up (primarily from the cortex) are amplified and recorded.

The electroencephalograph is not very useful in determining what, specifically, is going on in the brain, but it is useful in detecting abnormal electrical discharges and other gross changes in brain activity, such as the differences between normality and epilepsy, and even between wakefulness and sleep. Figure 35.16 compares the EEG of a normal person with that of a person with epilepsy. As the recording shows, epileptics are subject to sudden, random bursts of electrical activity. In a few instances, this abnormality can be traced to apparent physical defects, such as scars or lesions in the brain tissue, but most forms of epilepsy are simply not understood.

Sleep. Electrical activity in the brain does not cease with sleep. Quite the contrary. EEGs reveal that sleep is accompanied by a considerable amount of activity. Sleep has several distinct phases (Figure 35.17). By far the most interesting of these is called **paradoxical sleep,** or **REM sleep** (rapid eye movement sleep). During REM sleep the other skeletal muscles are in a highly relaxed state, but the eyes dart about beneath the eyelids. The EEG recording at this time is more similar to that produced by wakefulness. If a person is awak-

35.16 ELECTRICAL ACTIVITY IN THE BRAIN

The normal electrical activity of the brain is seen at the left, where several EEG tracings from different locations have been made. In the epileptic, normal electrical activity in the brain may occur between seizures, but when these episodes do occur, the electrical disturbances are quite obvious.

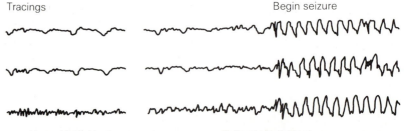

Tracings   Begin seizure

Normal individual   Epileptic individual

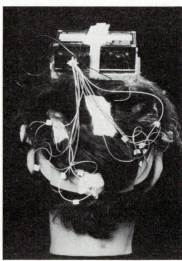

35.17   THE SLEEPING BRAIN

By studying the electrical activity of the brain and episodic eye movement, researchers have identified four sleep stages: drowsiness, light sleep, intermediate sleep, and deep sleep. Each has its own pattern of electrical activity as seen in the EEG. Typically, sleep is interrupted by bursts of electrical activity whose patterns resemble wakefulness. Since they are associated with rapid eye movement, they are called REM periods or REM sleep.

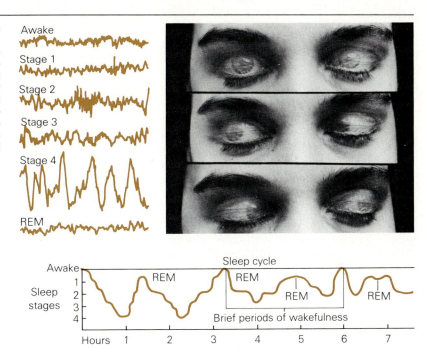

ened at this time, he or she will report vivid dreams (that will ordinarily be forgotten if the person is allowed to waken normally). REM sleep appears to be essential, although no one really knows why, just as no one really knows why dreaming might be useful. Experimental studies reveal that people deprived of REM sleep wake up tired, and in subsequent sleep periods they will extend the REM period (contrary to myth, you can make up for lost sleep). If the deprivation is continued, anxiety sets in and concentration becomes difficult. Eventually, personality disorders arise.

Psychologists and biologists generally agree that, for whatever reason, sleep appears to be highly essential and is somehow restoring. Theorists suspect that the sleep period, and REM sleep in particular, is a period of information sorting and storage, and may be essential to long-term memory storage.

Language Centers in the Human Brain

The human brain is unique in that it has areas devoted to the processing of language. (It would be surprising if other animals had language centers since as far as we know they don't have language.) Much of what we know about these centers comes from studies of stroke victims. When a stroke

occurs, part of the brain suffers an oxygen and nutrient loss, causing the death of brain tissue. Among the more typical results of stroke damage is aphasia, the loss or impairment of speech, especially when the damaged area is in the left side of the cortex. From stroke studies, neurophysiologists conclude that three regions of the left hemisphere are important to speech—**Broca's area, Wernicke's area,** and the **angular gyrus** (Figure 35.18). Of course, other regions of the cortex are also important to language, since linguistic articulation also involves hearing, seeing, even writing—not to mention thinking (although some people seem to have no trouble speaking without thinking much at all).

When we hear words, like all sounds, they are initially received by the auditory receptors, but unless the incoming neural information is processed in Wernicke's area its meaning will not be deciphered. The neural source of our words, however, is generated in Wernicke's area and transmitted to Broca's area along a special pathway. Broca's area then activates the motor cortex, providing information needed to properly activate the vocal cords, mouth, lips, tongue, and jaw, so that the proper sounds may be uttered.

Reading aloud involves the third region, the angular gyrus, as well as the visual and auditory areas of the cortex. Neural messages from the eyes reach the visual field of the cortex for processing, and this processed visual information is transferred to the angular gyrus, where the symbols are translated into words. In a manner of speaking, the angular gyrus "hears" the words we have just seen, by associating the visual message with auditory patterns in Wernicke's area. (Can you "hear" the words you see?) Speaking the words then requires the transfer of neural patterns to Broca's area and on to the motor cortex as usual. Poor readers mouth the words they read, but most of us learn to turn off our motor cortex unless we are deliberately trying to read aloud.

Damage to Broca's area results in badly impaired grammatical structure in both spoken and written sentences. Verbs and pronouns are commonly absent, although, with some effort, the sentences can be understood. Lesions in Wernicke's area result in an inability to decipher incoming messages properly. Speech, in this case, will include the verbs and pronouns and recognizable phrases strung together in a nonsensical manner. Neurologists conclude, then, that Wernicke's area produces the basic format of the spoken sentence, but Broca's area refines the structure and coordinates its actual vocalization.

THE PERIPHERAL NERVOUS SYSTEM

The peripheral nervous system is a vast network of nerves that spreads from the central nervous system to all parts of the body. It consists of motor neurons and sensory neurons, running side-by-side within the same nerves. The peripheral nervous system is further divided into the somatic nervous system and the autonomic nervous system. There are motor neurons in both systems, but only the somatic system has sensory neurons. The somatic nervous system carries the impulses that our conscious minds are most aware of: sensations that inform us of events in the world around us and the commands to our voluntary muscles that enable us to do something about those perceptions. The autonomic nervous system is more concerned with our internal workings and with our less conscious, less voluntary reactions.

The autonomic nervous system itself has two parts, the **sympathetic division** and the **parasympathetic division.** While the somatic, sympathetic, and parasympathetic entities have their own neurons, they often share the major peripheral nerves

35.18 LANGUAGE AND THE BRAIN

Using language requires the cooperation of neural centers in two cortical regions known as Broca's area and Wernicke's area, along with visual, hearing, and motor centers. The principal pathways used in repeating what one has just heard, and in reading aloud, are indicated by the arrows.

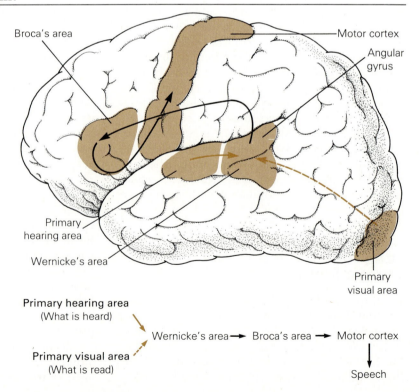

from the brain and spinal cord. Thus, a cranial nerve may carry elements of both the autonomic system and peripheral system. The vagus nerve, for instance, carries sensory and motor neurons of the pharynx, yet is also contains many motor neurons of the parasympathetic division of the autonomic nervous system.

The Autonomic Nervous System

The autonomic nervous system (ANS) is essentially a *motor* system, carrying impulses from the brain and spinal cord to the organs it serves (Figure 35.19). In doing this, it works in concert with the central nervous system in regulating the activity of the viscera, the internal organs. These include not only the prominent contents of the thoracic and abdominal cavities (the heart, lungs, digestive organs, kidneys, and bladder), but also the smooth muscle of the blood vessels, the iris, nasal lining, sweat glands, salivary glands, and even the tiny **arrector pili muscles** that cause our hairs to stand on end when we see who's been elected.

Generally, the ANS promotes homeostasis. Working under the direction of the central nervous system, especially the medulla and hypothalamus, it coordinates the adjustments needed to maintain an overall stability or constancy in the face of changing conditions. For such a system to function efficiently requires a considerable amount of sensory feedback from the organs served. This is carried out by sensory nerves of the somatic nervous system, which, by keeping the brain informed, also play an important role in the process of homeostasis.

The two divisions of the autonomic nervous system, the sympathetic and parasympathetic, have opposite effects and operate in a highly coordinated manner to produce an overall adaptive effect. One familiar example is seen in heart rate. The human heart, without outside influence, contracts about 70 to 80 times per minute but will speed up when stimulated by a sympathetic nerve called the **cardioaccelerator,** and slow down on a signal from a parasympathetic nerve (the vagus; see Figure 35.19). How can neural impulses from two different nerves have opposing effects on heart rate? The answer has to do with the specific neurotransmitters released by those neurons. In many species, most, but not all, of the sympathetic neurons secrete the neurotransmitter norepinephrine at their target organs, while neurons of the parasympathetic system secrete acetylcholine (as do motor neurons that move skeletal muscle). Norepinephrine accelerates heart rate, while acetylcholine slows it down.

An interesting example of how the two ANS divisions effect such changes in the viscera and muscles is seen in the "fight or flight" response discussed in Chapter 33. In an emergency, the response is instigated by the sympathetic nervous system. After the emergency has passed, the parasympathetic division takes over and conditions return to normal. (You may recall that the secretion of epinephrine and norepinephrine into the blood by the adrenal cortex produces many of the same effects.) The autonomic nervous system has a variety of effects on other organs as well (Table 35.1).

As you might expect, the divisions of the autonomic nervous system differ anatomically as well as physiologically. As seen in Figure 35.19, most of the nerves and nerve tracts that make up the parasympathetic division originate in the brain and toward the lower region of the spinal cord. Some go directly to their target organs; others end in external ganglia, where they relay impulses to nerves continuing toward the target organ. The sympathetic neurons all originate in the gray matter of the spinal cord. Furthermore, all of the sympathetic nerves leaving the cord pass through or synapse within rows of **sympathetic ganglia** just outside the cord. Some sympathetic nerves synapse once again in ganglia at various locations in the body. One of these, the **celiac plexus,** better known as the **solar plexus,** is often the target of a blow from a boxing or martial arts opponent, bringing about a temporary

TABLE 35.1

RESPONSES TO SYMPATHETIC AND PARASYMPATHETIC NERVES

Organ	Sympathetic effect	Parasympathetic effect
Pupil of eye	Dilates	Constricts
Heart	Accelerates	Slows
Intestine	Decreased movement	Increased movement
Salivary gland	Decreased secretion	Increased secretion
Stomach glands	Decreased secretion	Increased secretion
Lungs	Dilates air passages	Constricts air passages
Blood vessels		
Skin	Constricts	Dilates
Stomach	Constricts	Dilates
Metabolism	Increases	Decreases

35.19 THE AUTONOMIC NERVOUS SYSTEM

The organization of the autonomic nervous system, showing the various organs it serves. The autonomic system is motor in function, so its nerves transmit impulses from the brain and spinal cord to the organs served. Its principal role is regulato-ry—keeping the internal organs operating in a carefully coordinated manner in response to constantly shifting conditions and needs.

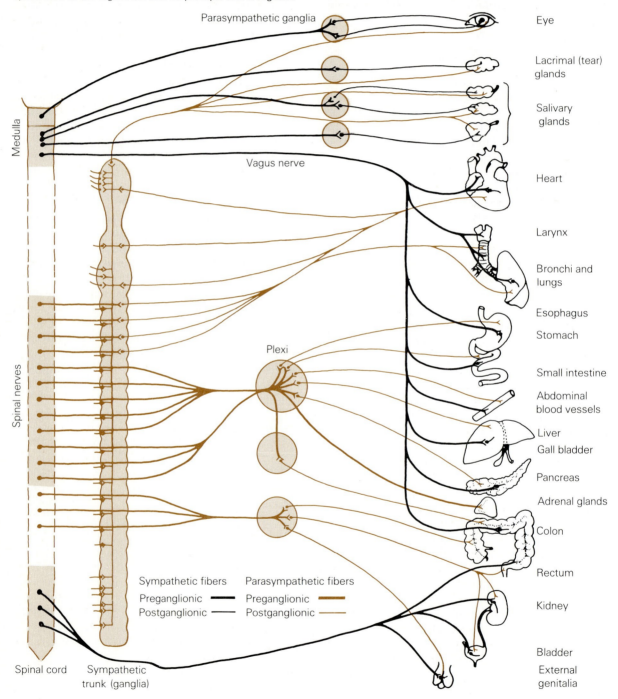

Parasympathetic ganglia

Eye

Lacrimal (tear) glands

Salivary glands

Medulla

Vagus nerve

Heart

Larynx

Bronchi and lungs

Esophagus

Stomach

Plexi

Small intestine

Abdominal blood vessels

Liver

Gall bladder

Spinal nerves

Pancreas

Adrenal glands

Colon

Rectum

Sympathetic fibers	Parasympathetic fibers
Preganglionic	Preganglionic
Postganglionic	Postganglionic

Kidney

Bladder

Spinal cord

Sympathetic trunk (ganglia)

External genitalia

paralysis of the diaphragm and a momentary adjustment of attitude.

The success of the animal kingdom on this planet has been, in large part, due to the very precise sensitivity and reactivity of its members. We have seen that this highly organized and adaptive responsiveness is based on the peculiar irritability of a constellation of cells called neurons. We know something about the kinds of neurons that exist, their supporting structures, and even certain details about how neurons work. Yet neural biology remains one of the most resistant challenges in all of science. Nonetheless, researchers continue to tell us more about just how animals are able to respond so precisely to both their internal and external environments in a very complex world.

APPLICATION OF IDEAS

1. Use the terms *primitive* and *advanced* to describe the major parts of the human brain. Explain the criteria on which you based your decisions.

2. The central nervous system, particularly the conscious centers, exercise what might be called "noisy control," as they receive, integrate, and respond to a barrage of sensory messages. The autonomic system, on the other hand, can be referred to as a "quiet" system, since its activity is generally below the conscious level. Elaborate on these ideas by considering how the two systems respond to problems in osmoregulation and in thermoregulation (in the latter include problems of both heating and cooling).

KEY WORDS AND IDEAS

INVERTEBRATE NERVOUS SYSTEMS

Simple Nerve Nets and Ladders

1. Coelenterates show no evidence of **cephalization.** They are radially symmetrical and contain a nerve net whose cells concentrate at the mouth and tentacles.

2. The flatworm nervous system resembles a ladder with two **ganglionic masses** (clusters of nerve cell bodies) at the anterior end and the ladder arrangement proceeding posteriorly.

Massed Ganglia to Organized Brains

1. The annelids have large ganglia above and below the esophagus and a ventral nerve cord with segmental ganglia. **Giant axons** in the cord support fast escape movements.

2. While the bivalve nervous system is fairly simple, that of the cephalopod is complex and features a functionally differentiated brain and complex behavior. Giant axons are found in the squid mantle.

3. Arthropod specializations include reduction in the number of segments and ganglia and much greater cephalization than is found in the annelids. The insect brain includes an enlarged ganglionic mass above the esophagus, which, in those with compound eyes, includes greatly enlarged visual integrating centers.

VERTEBRATE NERVOUS SYSTEMS

1. Vertebrate nervous systems include a pronounced trend toward cephalization and a dorsal, hollow nerve cord.

2. The vertebrate brain follows a three-part plan of **forebrain, midbrain,** and **hindbrain.** The **cerebrum** and surrounding **cortex** of the forebrain reach its greatest development in mammals.

THE HUMAN CENTRAL NERVOUS SYSTEM

1. The human nervous system is divided into the **central nervous system** (CNS)—the brain and spinal cord—and the **peripheral nervous system** (PNS)—the **somatic** and **autonomic systems.** The somatic system is largely composed of sensory receptors and sensory neurons and the voluntary motor nerves that move skeletal muscle, while the autonomic system is motor and involuntary, dedicated mainly to homeostasis. The central nervous system arises in the embryo from the **neural tube**.

The Spinal Cord

1. The spinal cord includes pathways between the brain and much of the peripheral nervous system. The white regions are myelinated neurons, while the gray are nerve cell bodies. Many reflexive acts occur at the cord level.

2. The cord emerges from the brain through the **foramen magnum** and lies within the **vertebral canal.** It is bathed in the **cerebrospinal fluid** and surrounded by the **meninges,** three layers of supporting connective tissue.

3. Motor neurons arise from cell bodies in the cord to pass out through the ventral **motor root** of the spinal nerves. Sensory neurons enter the cord via the dorsal **sensory root** from rows of ganglia outside, where their cell bodies cluster.

The Human Brain

1. The human brain weighs about 1.4 kg, has a volume of 1300 to 1500 cc, contains some 100 billion neurons and 10 times that number of **glial cells.** Cell bodies form its gray outer regions, while myelinated fibers make up its white inner mass. Cerebrospinal fluid and the meninges cushion and protect the brain, respectively.

2. The hindbrain consists of
   a. the **medulla oblongata** (or **medulla**), which controls breathing and heart rate and contains many pathways, including some from **cranial nerves;**
   b. the **cerebellum,** concerned with balance, equilibrium, and voluntary muscle coordination;
   c. the **pons,** which contains tracts traveling between the forebrain and cord and to and from the cerebellum.

3. The midbrain forms connections between the hind- and forebrain and receives sensory input from auditory and visual receptors.

4. The forebrain is involved in conscious thought, sensory reception, voluntary movement, and other voluntary acts. It consists of the following:
   a. The **thalamus,** which is a relay structure, connects various parts of the brain and includes the **reticular system,** which taps incoming and outgoing communications. It also acts as an alarm system and suppresses irrelevant stimuli, thus permitting sleep.
   b. The **hypothalamus,** which monitors many functions, acts as a homeostatic regulator (heart rate, blood pressure, body temperature, thirst, hunger, sex drive). It also stimulates hormonal activity in the pituitary and is subject to negative feedback.
   c. The **limbic system,** containing a number of specific **nuclei,** includes the hypothalamus, thalamus, and some cortical pathways. It links the fore- and midbrain, and is involved in emotion (for example, when areas of the **amygdala** are stimulated, we may experience rage).

The Cerebrum

1. The cerebrum, the largest region of the forebrain, is divided into left and right **cerebral hemispheres,** which are covered by the wrinkled **cerebral cortex.** The cortex contains some 15 billion cell bodies and dendrites. The white matter below includes the myelinated fibers.

2. The importance of the cerebrum to common voluntary acts is minimal in frogs but increasingly important in rats, cats, dogs, monkeys, and humans.

3. The cerebrum is made up of a number of lobes. The **occipital lobe,** receives and analyzes visual information, while the **temporal lobe** (bounded by the **fissure of Sylvius,** or **lateral sulcus**) processes auditory input and some visual information. The **frontal lobe** regulates voluntary movement, and speech. The **prefrontal area** of the frontal lobe sorts sensory input, putting it into proper context. The **parietal lobe** processes sensory information, including body position. It is separated from the frontal lobe by the **fissure of Rolando,** or **central sulcus.**

4. The processing of sensory input and the instigation of voluntary muscle action occurs in specific regions of the cortex.

5. The cerebral hemispheres are functionally distinct, and many learned patterns take place in just one hemisphere. The left hemisphere predominates in right-handed persons and in the more common type of left-handedness, but in the rare form of left-handedness, dominance is in the right. Speech is primarily a left hemisphere function, as is analytical thought. Damage to this hemisphere can provide a loss of language abilities.

6. Connections between hemispheres occur through the **corpus callosum.**

Chemicals in the Brain

1. The brain has at least 50 different neurotransmitters. The **monoamines,** modified amino acids, include **norepinephrine, dopamine, histamine,** and **seratonin. Neuropeptides,** short chains of amino acids, include **angiotensin II,** which plays a role in thirst-related functions.

2. Substances entering the brain must cross the **blood-brain barrier,** an area of tightly interwoven capillary endothelial cells. **Astrocytes** act as intermediaries in selecting and transporting materials into the brain tissue.

3. Other neuropeptides include the **enkephalins** and **endorphins,** mood elevators and pain modifiers. They are called **opioids** because their effects can be mimicked by opiates. Amphetamines accelerate neurotransmitter release, while cocaine blocks their normal enzymatic degradation, and LSD, mescaline, and psilocybin mimic neurotransmitters.

Electrical Activity in the Brain

1. Electrical activity in the brain can be detected

with an **electroencelphalograph** and recorded as an **electroencephalogram,** or EEG. Such recordings are useful for detecting gross abnormal patterns.

2. EEGs reveal that a considerable amount of electrical activity accompanies sleep. Electrical activity during **paradoxical sleep,** or **REM sleep** (rapid eye movement), a time of dreaming, is similar to that of wakefulness.

Language Centers in the Human Brain

1. Knowledge of language centers comes mainly from studies of stroke victims. Three left hemisphere regions, **Broca's area, Wernicke's area,** and the **angular gyrus,** are involved.

2. When heard, words are processed by Wernicke's area. Words to be spoken begin in this area but are transmitted to Broca's area, which activates the motor regions involved in voice.

3. Reading aloud involves all three areas. Visual images of the words go to the angular gyrus, which associates the words with auditory patterns in Wernicke's area. Wernicke's area produces the basic sentence format, while Broca's area refines sentence structure and coordinates vocalization.

THE PERIPHERAL NERVOUS SYSTEM

1. The peripheral system includes the somatic and autonomic systems. The autonomic system is made up of the **sympathetic** and **parasympathetic divisions.**

The Autonomic Nervous System

1. The ANS is essentially homeostatic in function. Its motor neurons carry impulses from the brain and spinal cord to the viscera, blood vessels, irises, secretory glands, and other involuntary structures, bringing about fine adjustments as necessary.

2. The sympathetic and parasympathetic divisions have opposing functions, with the former generally increasing an action and the latter decreasing it. For example, the sympathetic **cardioaccelerator** nerve releases neurotransmitters that increase heart rate, while the parasympathetic vagus nerve releases those that slow the heart rate.

3. In "fight or flight" events, the sympathetic division increases heart and breathing rate and blood pressure, and sends more blood to the muscles and brain. Afterward, these actions are reversed by the parasympathetic division.

4. Most parasympathetic nerves originate in the brain and a few in the lower spinal cord. Some go to external ganglia, where they synapse with neurons that innervate the target organ. Sympathetic nerves all emerge from the cord and all pass through, or synapse within, the **sympathetic ganglia** along the cord. Secondary ganglia also occur, including those of the **celiac plexus** or **solar plexus.**

REVIEW QUESTIONS

1. Describe the arrangement and concentration of neurons in a coelenterate. What is such an arrangement called? (899)

2. What evidence of bilateral symmetry can you find in the flatworm's nervous system? Is there any cephalization? Explain. (899)

3. Name two invertebrates with giant axons. What role do such axons play? (900)

4. What characteristics of the cephalopod central nervous system qualify the usual invertebrate ganglionic mass as a full-fledged brain? What capabilities does this provide to the cephalopod? (900)

5. Name the three parts of the vertebrate brain and describe the general evolutionary trends seen in each part. (901)

6. Describe the organization of the spinal cord, including its covering membranes, its gray and white regions, and the manner in which spinal nerves enter and leave. (903)

7. In addition to neurons, what other cells are common in the brain? What is the approximate ratio of these cells to neurons? (904)

8. What is the function of the cerebrospinal fluid? Where is it found in the brain? (904)

9. List the three regions of the hindbrain and state their general functions. (905)

10. In which vertebrates is the midbrain prominent? What is its significance to humans? (905)

11. What are the main functions of the thalamus and its reticular system? (906)

12. List five processes that are regulated by the hypothalamus. With what other system does it closely interact? (906)

13. What structures, in addition to the hypothalamus and thalamus, make up the limbic system? Name some activities of these. (906–907)

14. Using the terms *cerebral hemisphere, cerebral cortex, convolutions,* and *gray and white matter,* describe the organization of the cerebrum. (907)

15. Compare the importance of the cerebrum in the rat, frog, and primate. What general trend in the localizing of functions does your answer suggest? (907–908)

16. State a function that is localized in each of the following cerebral lobes: prefrontal, occipital, temporal, and frontal. (909)

17. Describe the organization of sensory and motor areas about the central sulcus. Compare the amount of sensory and motor control area devoted to the hand, genitals, leg, tongue, and toes. (909–910)

18. What is the general rule about the cerebral hemispheres and control of right and left sides of the body? How does this affect right- and left-handedness? (910–911)

19. What evidence is there that the left hemisphere contains the speech centers? (911)

20. Explain, theoretically, how it came to be that there are about 50 neurotransmitters in the central nervous system and only a few in the peripheral nervous system. (913)

21. What is the blood-brain barrier? How are materials actually transferred from capillaries to the brain cells? (913)

22. What is the role of enkephalins and endorphins? Why are they called opioid neurotransmitters? (913)

23. Compare the action of amphetamines, cocaine, and LSD in the brain. (913–914)

24. During which sleep episode is brain activity close to that of wakefulness? Why is this period of sleep so important? (914–915)

25. Explain the manner in which the three speech areas of the cortex interact when one reads aloud. In general, what are the roles of the Wernicke and Broca areas? (915–916)

26. What characteristic clearly distinguishes between the somatic and autonomic nervous systems? (916)

27. Describe the organization of the autonomic nervous system, naming its divisions and explaining their anatomical arrangement. (916–917)

28. Compare the role of the autonomic divisions in the "fight or flight" response. (917)

Sexual Reproduction

As we've mentioned before, the primary advantage of sex is that it permits new combinations of genes to form, thus, increasing variation among offspring. This means our offspring are different from us. We may be very proud of ourselves and think we fit into our world very well indeed. Some might even say that we humans are splendid. However, we are only splendid in our present world—it would be presumptuous to assume that we would fit into some future (or past) world with equal ease. And, even as we have our "moment," the world changes. Thus, if we were to make replicas of ourselves, each of our offspring bearing 100% of our genes, they would do well as long as the world remained as it is now. But they might be in a lot of trouble if it should change. In fact, they might not fit a different kind of world well at all and might die out.

But if we mix our genes randomly with someone else's, we will produce a motley crew of offspring—all of them different. Thus, should new conditions arise, at least some of our offspring would be likely to survive and to propel our genes into yet other generations. This is essentially the biological explanation for the evolution of sex. The idea was discussed in more detail in Chapter 11.

Even if such advantages are bestowed by sex, our best theoreticians still have trouble weighing them against the disadvantage of gene dilution. Obviously, our observations are not complete—or something is amiss with the theory. But holding any judgments in abeyance for now, let's see how this gene mixing works in the animal kingdom as we take a close look at animal reproduction.

SEX AND REPRODUCTION IN ANIMALS

The Aquatic Environment

Animals that practice sexual reproduction are faced with the basic problem of how to get eggs and sperm together. You may think you know a way, but it is surprising how chancy sex generally is. Under the best of circumstances, so many things can go wrong that one wonders how fertilization could *ever* occur. The chanciest conditions are found in those species that simply discharge eggs and sperm into the environment and then hope for the best. But that is precisely what many species do.

External Fertilization. The process is called **external fertilization** and is carried on by a number of aquatic invertebrates as well as some fishes and amphibians. In such species the eggs are usually discharged into the water and sperm immediately shed over them (Figure 36.1).

36.1 EXTERNAL FERTILIZATION

External fertilization is common in aquatic animals. Sessile bottom dwellers (a) must rely on huge numbers of gametes to assure fertilization, but in motile animals, more elaborate behavior is seen. (b) The female fish here is courted and induced into egg laying, whereupon the attentive male quickly releases his sperm on the egg mass. (c) After a lengthy period of vocal courting, male frogs and toads provoke egg laying by clasping and squeezing the female (known as amplexus). The eggs are then fertilized as they are released.

(a)

(b)

(c)

A number of animals play this game of chance, from coelenterates, which depend on water currents to bear their gametes, to salmon, which return from the open sea to shed their eggs and sperm in the very stream where they were born.

Most frogs and toads and some salamanders leave less to chance as males clasp females in a mating behavior known as **amplexus,** shedding their sperm as the egg mass is released. Thus these amphibians have taken a step toward one-on-one fertilization, although the union of gametes is still external. A specific member of the opposite sex is selected and gametes are joined with those of that individual (although males may mate indiscriminately). While external fertilization, wasteful of gametes as it seems, works well enough in the water, it wouldn't do for terrestrial life. As animals ventured hesitantly farther and farther onto the land, a different method was required. The risk of desiccation was simply too great. The sperm had to be placed where fertilization was assured and where the egg was protected. So on land, **internal fertilization** developed.

Internal Fertilization. Actually, internal fertilization is not restricted to terrestrial forms; it is also fairly common among aquatic invertebrates and vertebrates. For example, it occurs in barnacles and fishes alike. In fact, internal fertilization seems to have evolved in the earth's waters. It could have begun as a simple variation, one in which fertilization was simply more assured for at least some groups. With the safer introduction of sperm into the female body there would certainly be less risk of unsuccessful matings. Thus, internal fertilization may have arisen through constant selection for means to get sperm closer to the eggs ready to be fertilized. If so, then some animals were reproductively prepared for terrestrial life long before they ventured out of the comforting waters.

Copulation. Let's see how some species go about getting the male's sperm to the female's eggs inside her body. In almost every case, it involves the process known as **copulation.** This is the use of a specialized intromittent organ—a penis or its equivalent—to introduce sperm into a similarly specialized, but oppositely arranged receptacle—a vagina or its equivalent. This definition is somewhat anthropomorphic, since many animals simply bring ducted body openings together, transferring sperm reasonably well in the absence of an intromittent organ.

As we have seen, copulation is primarily a device of land dwellers, but is found in many aquatic organisms. Even stationary barnacles, like other arthropods, rely exclusively on internal fertilization. In their case, this requires a long, roving and probing penis that can reach the barnacle next door, or even one several doors down (Figure 36.2). It doesn't matter which one, because the barnacle is hermaphroditic. Another aquatic example, the planarian flatworm, is well equipped for copulation (see Figure 25.11b). It is also hermaphroditic, so it doesn't have to find a planarian of the opposite sex; it needs only another planarian. They then exchange sperm. The sperm moves up the lengthy oviducts to fertilize the eggs, which are waiting in the ovaries. As the fertilized eggs pass down the oviducts, they build up yolk stores as they go. An egg case forms around the eggs just before they are released.

36.2 SEX IN BARNACLES

Lepas, the barnacle, is hermaphroditic but not self-fertilizing. Since it is sessile, the problem of sperm transfer and fertilization is solved by the use of a long extensible penis. When it mates, *Lepas* sends its penis over to a neighboring barnacle (any one will do) and probing down into the scutum, releases its sperm. The sperm enter the nearby oviduct, where fertilization take place.

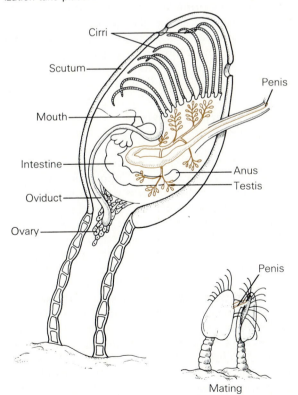

Cirri
Scutum
Penis
Mouth
Intestine
Anus
Testis
Oviduct
Ovary
Penis
Mating

36.3 OVOVIVIPARITY

Mollies are well known live-bearers—bearing fully developed young rather than laying eggs as many fishes do. Unlike live-bearing in mammals, where nourishment, gas exchange, and waste removal are provided by the placenta, the live-bearing fishes simply provide a safe place for the young to develop within the egg.

Modes of Fertilization and Development in Fishes. A number of fishes also copulate as a mode of reproduction, but their offspring follow three different developmental routes. First, we should note that in external fertilization, fish development occurs entirely outside the female's body—a developmental state long known as **oviparity.** Where fertilization in fishes is internal, however, development is commonly **ovoviviparous;** that is, the fertilized eggs are retained within the uterus while they mature. In fact, they hatch there and the female gives birth to live babies (Figure 36.3). You may have seen this in aquarium fishes such as guppies **Ovoviviparity** may seem similar to **viviparity,** the third mode of development, but there is a fairly clear distinction. Ovoviviparous species develop without any apparent exchanges with the mother. In mammals, which provide the best examples of viviparity, the young are in intimate association with the mother as they develop. They draw their sustenance from her through a placenta, and are not isolated in egg cases or shells. They, like guppies, are born alive, but their development is drastically different from that of the fish.

The distinction between ovoviviparity and viviparity becomes less distinct in some of the sharks and rays as we saw in Chapter 27 (and because of this, some biologists are dropping the terminology). Sharks often fertilize internally. The male is equipped with external **claspers**—long, tapering modifications of the pelvic fins that can be erected (Figure 36.4). During copulation, the erect claspers are held together and pushed into the female's cloacal opening (the common reproductive, excretory,

36.4 REPRODUCTION IN SHARKS

Male sharks can be readily distinguished by the presence of paired claspers near the cloaca. These are erectile structures used in copulation. The leathery egg cases of egg-laying rays and sharks are often found washed up along beaches.

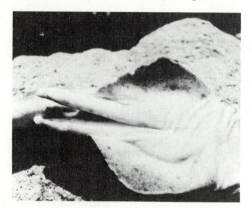

Claspers of male shark

Fifteen-minute-old shark among egg cases

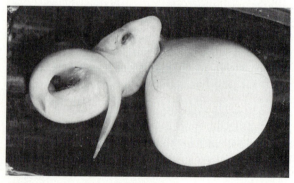

Shark embryo

and digestive waste passageway). The male then ejaculates between his grooved claspers, directly into her cloaca.

In oviparous sharks and rays, the fertilized eggs become enclosed in leathery **egg cases,** in which the young complete their development. In clearly ovoviviparous species, the young are maintained within the uterus, later emerging through the urogenital opening, in a manner reminiscent of the mammals. Some species are very likely viviparous. They are clearly like mammals, with the young provided the

same kind of support that is provided by the placenta. For example, at least three genera of sharks and rays may actually nourish their young in the uterus by exposing them to placentalike beds of capillaries—showing again that living things don't always fit into our neat categories.

Reproduction and the Survival Principle.
One reason we have included some diverse examples is that it brings up an important reproductive principle, readily seen in the fishes, but also true of other animals. The sharks don't produce many offspring, but, since most are ovoviviparous, they offer a lot of protection to their embryos. However, there is no evidence of maternal shark care after birth. On the other hand, the herring (a bony fish) produces about 50,000 eggs each season but offer no protection during development. The eggs sink to the ocean floor and most of them perish. But some sift into concealed places where predators miss them. The eggs of turbot and cod are in an even more precarious position—they float. Thus, they are highly exposed in the open surface waters. In order to leave any offspring at all, each female must produce up to nine million eggs per season. But whether a fish produces half a dozen eggs or 9 million—in the long run—the average number of offspring surviving to adulthood is two for each pair of parents. The point is, the more a parent of any species protects or cares for its offspring, the fewer offspring it produces (or can energetically *afford* to produce). We will explore this idea in greater detail in Chapter 42.

Reproduction in the Terrestrial Environment

As we have seen, the transition to land brought its own reproductive problems. Copulation was necessary and unless the egg was to be housed in the female during development, it would require substantial modification to survive the rigors of a dry atmosphere and varying temperatures. Solutions to these problems have provided interesting reproductive modes and variations thereon.

Reproduction in Terrestrial Arthropods.
The insects have been described as one of nature's greatest success stories. One reason for their success in the terrestrial environment is their tremendous reproductive capacity. Perhaps because their life spans are often short and the reproductive risks great, females generally produce large numbers of young. This ability can, in fact, set the stage for great fluctuations in numbers. Predation and

limited food supplies tend to suppress insect populations, but when food is abundant, populations increase phenomenally. This is notably true in agricultural areas where predators are few (often a result of human behavior) and where food is almost unlimited (always a result of human behavior).

All insects practice internal fertilization, and they have developed an array of devices to accomplish this end (Figure 36.5). Males are commonly equipped with clasping organs for holding the female in position while the penis is introduced and sperm released. Also, the penis is usually barbed, hooked, or braced in such a bizarre fashion that it can only fit into one kind of vagina—that of females of the correct species. Students of evolutionary theory note that this lock-and-key arrangement obviously has evolved to prevent inappropriate matings and unwanted hybridization between similar species.

Female insects are commonly equipped with an **ovipositor,** which, as the name implies, is used to deposit the eggs (Figure 36.6). In grasshoppers, the ovipositor is used to dig holes in the ground where the eggs will be buried. Many insects lay eggs in the soft ovary of flowers, assuring the young of an adequate food supply as they develop along with the fruit. Females of some species, such as the cicada and gall wasps, are even capable of boring into wood with the ovipositor and then filling the space with eggs. Variation in egg-laying behavior in arthropods is indeed seemingly endless.

Many female insects mate only once in their lifetime and then store the sperm in a saclike structure

36.5 "LOCK AND KEY" GENITALIA IN INSECTS

Many male insects are equipped with accessory reproductive structures that are used in grasping the female and holding her in position for copulation. The complex genitalia often contain barbs and hooks that bear little resemblance to what we think of as a penis and each species has its own unique structures. The hooked apparatus assures the successful transfer of sperm.

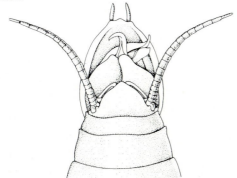

36.6 THE OVIPOSITOR IN GRASSHOPPERS

The ovipositor of the female grasshopper is a digging device used to make a hole in the earth for the eggs. She will bore about ten holes, depositing twenty or so eggs in each. Each group of eggs is enclosed in a sticky egg pod.

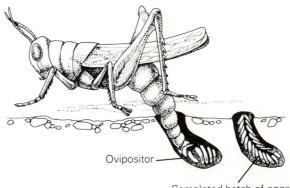

Ovipositor

Completed batch of eggs

called the **spermathecum.** They simply release the sperm when they are needed. Insect eggs are well protected by a tough covering, the **chorion,** which forms before the eggs are laid. Each egg is fertilized through the **micropyle,** a small pore in its chorion, as it passes the spermatheca on its way through the vagina (Figure 36.7).

Mating in Spiders. Spiders are all carnivores and they are all venomous. Many species are solitary animals. And a venomous, carnivorous hermit must not be taken lightly. In particular, the habits of spiders add up to problems in mating, especially for the small, weak males who must initiate the process. In some species, they timorously approach the females at mating time, waving special appendages as though they were signal flags (Figure 36.8). The waving seems to mesmerize the female, at least long enough for the male to mate with her. The orb spider males signal their presence in an even more cautious manner. The male climbs on the female's intricate web and "strums" out a signal that inhibits her feeding behavior. Males of other species even prepare a gift—usually a paralyzed insect, neatly wrapped in silk—anything to distract or occupy the cannibalistic female while he dashes in and accomplishes his task.

After all this trouble, most species of spiders don't actually copulate. Instead, the male may deposit his sperm on the ground and then scoop it up in a modified appendage. He then either draws near the female or climbs on her back and rakes his sperm-laden appendage over the plates covering her vagina. Then he clears out. He doesn't always

36.7 INSECT EGGS

As you might expect, insect eggs occur in a variety of shapes and sizes, representing specific adaptations. General features shown include a central yolky region harboring the nucleus, surrounded by a region of dense cytoplasm. A protective case (the chorion) has a pore (micropyle) at one end through which sperm enter. Fertilization generally occurs as eggs pass a sperm storage place (the spermatheca) on their way through the oviduct.

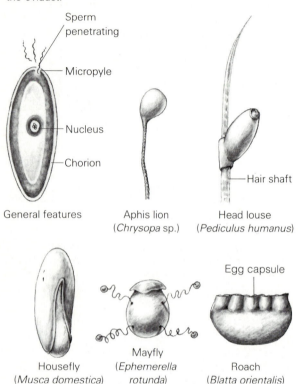

Sperm penetrating

Micropyle

Nucleus

Chorion

General features

Aphis lion
(*Chrysopa* sp.)

Hair shaft

Head louse
(*Pediculus humanus*)

Egg capsule

Housefly
(*Musca domestica*)

Mayfly
(*Ephemerella rotunda*)

Roach
(*Blatta orientalis*)

make it, but in some species the chances are increased if he pinches the female hard as he leaves. She will then immediately tuck into a defensive ball, and before she can recover, he's gone.

Reproduction in Reptiles and Birds. Reptiles and birds share many reproductive similarities. They practice internal fertilization and are nearly all oviparous. Exceptions include a few species of ovoviviparous snakes and the sea snakes, which have evolved a placentalike structure and can be said to be viviparous. Interestingly, the crocodilians, which in so many ways are quite advanced, are also egg-layers.

In male reptiles, there is a well-developed, but hidden, copulatory organ. (With the copulatory organ hidden it's often difficult to distinguish sexes in reptiles.) Actually, whereas turtles have only one penis, lizards and snakes have *two* penises or, more correctly, two **hemipenes,** which can be everted (turned inside out, like the fingers of a rubber glove). However, only one is used at a time, depending on which side of the male the female's body is on, since lizards and snakes mate more or less side-by-side. (Actually, since each hemipenis is connected to only one testis, the males prefer to alternate sides, using the hemipenis on the side with the larger store of sperm.) Each hemipenis is covered with backwardly directed barbs and hooks and, whereas it can be inserted into the female cloaca easily enough, it must be carefully withdrawn. The female is often larger than the male and if she

36.8 MATING IN SPIDERS

The smaller male spider may use specific recognition signals **(a)** so that the female recognizes him as a mate and not a meal. **(b)** But even when all is well, males do not actually copulate. Instead they deposit sperm in packets near the female, using bulblike pedipalps **(c)** on the forelegs.

(a)

(b)

(c)

36.9 THE "CLOACAL KISS"

Most male birds lack a penis, so copulation is accomplished through simple cloacal contact. It may sound inefficient, but it works well. A single barnyard rooster can inseminate a sizeable flock of hens.

should be startled during the act, she may take off, dragging the helpless male away by his organ.

A large part of the reproductive success of reptiles depends on the female's successful concealment of the eggs. Sea turtles, for example, bury their eggs on sandy beaches, where they are relatively safe from predators and are not exposed to the direct heat of the sun. But upon hatching, the young are completely on their own. As they scramble toward the sea, many fall prey to the voracious gulls wheeling above, and those that make it to the water face a battery of marine predators. Because of the high mortality rate of the young, then, the reproductive success of a female sea turtle often depends on her laying a lot of eggs each season and living through a lot of seasons. Sea turtles that make it through those first few years, can look forward to a long life span, for which, in part, they can thank their stable aquatic environment.

Male birds, with the exception of a few species, including ostriches and some ducks, lack a penis and their mating techniques can best described as somewhat crass and disorganized—at least from a human perspective. Insemination is accomplished by the male pushing his cloaca against the female's and quickly ejaculating during what has been poetically called "a cloacal kiss." He usually does this first by treading on her tail, causing her to crouch and raise her rear. His tail then cups under hers—cloacal contact is made—and in a flurry of wing-flapping, the deed is done (Figure 36.9). Incidentally, the swifts, which are known as great fliers, are even better fliers than we thought—they mate in mid-air.

Birds employ a reproductive strategy different from that of most reptiles. Birds are well known for the care they give their offspring. Typically, only a few eggs are produced each season, and then one or both parents lavish them with warm attention. In species of ground-foraging fowl, quail, and in most water birds, the young are born in what is called a **precocial state:** the eyes are functional, feathers are present and they can immediately walk about and feed themselves. In others, particularly the song birds such as robins and meadow larks, the young may hatch in a completely helpless, semideveloped **altricial** state, doddering weakly about and demanding feeding and protection for many days (Figure 36.10). The reproductive success of these birds depends, again, on the successful rearing of just a few well-protected offspring.

36.10 ALTRICIAL AND PRECOCIAL YOUNG

Birds fall into two groups according to the developmental state of the newly hatched young. (a) Altricial birds hatch in a completely helpless, blind state, nearly featherless and requiring a lengthy period of constant care if they are to survive. (b) Precocial young hatch with feathers, are well developed and able to feed independently. Nevertheless, they are usually sheltered and protected from predators for a time.

(a)

(b)

Reproduction in Mammals. All mammals produce milk, and with very few exceptions, they are viviparous, producing a placenta or pseudoplacenta for the metabolic support of the embryo. It is through the placenta that viviparous embryos receive their oxygen and nourishment, and discharge wastes and carbon dioxide. Let's turn to the exceptions first, and then look into the details of reproduction of a more typical mammal, a eutherian that produces a true placenta.

Monotremes and Marsupials. True placentas are not found in the monotremes and marsupials, two very different kinds of mammals discussed earlier (see Chapter 27). One monotreme, the duckbilled platypus, is an unusual creature that lays eggs, but to confuse the taxonomist it has fur and gives milk. So it's a mammal. (Isn't it?) Its curious features do not end there. The females lack the uterus and vagina, retaining the primitive cloaca which serves in copulatory, excretory and defecatory functions. The tiny eggs (1.3 × 2 cm), usually one to three, are laid in a nest within the duckbill's burrow, where they are periodically incubated. Males also retain the cloaca, and they have a penis which only conducts sperm. The testes do not descend into a saclike scrotum as in other mammals, but remain within the abdomen. Add to this the ducklike beak and toothless mouth (of adults), the poison conducting spurs on the hind limbs, and it is small wonder that when the first stuffed specimen was sent from Australia to the British Museum, it was considered to be a crude hoax.

The marsupials, such as opossums and kangaroos, are also taxonomically distinct. They are mammals and the young develop, for a time, in the uterus, but with the help of a **pseudoplacenta,** a dense capillary bed in the embryo's yolk sac. In marsupials, the pseudoplacenta provides for the supply of nourishment, the exchange of gases and removal of wastes. But a fully developed placenta is not necessary anyway, because the young don't remain in the uterus for long. While they are still hardly more than embryos—except for well-developed front limbs—they squirm out of the vagina almost unnoticed and, using a sort of Australian crawl, make their way to the mother's pouch and attach to a nipple to begin their second stage of growth.

The adaptive significance of this strange mode of development is, at the moment, anyone's guess. Importantly, the kangaroo can conceive again while young are developing in the pouch, so it can support young in different stages. If it already has an offspring in its pouch, the development of the second embryo appears to be restricted until the pouch is free. Sometimes both of her two nipples are in use, with a new and an older offspring in the pouch at the same time. Whatever the ins and outs of marsupial reproduction, one point is very clear: it works. Kangaroos can reproduce prolifically if populations are not decimated for dog food and trendy clothing.

The rest of the mammals are placental. Like bird hatchlings, the newborn mammals are either precocial or altricial. Precocial mammals include the hooved mammals and of course aquatic mammals such as whales and porpoises. As anyone raised on a farm knows, the hooved animals are born with the eyes functional, and the young ready to trot along after the mother within a few hours of birth. Altricial young, such as primates, carnivores, and rodents, are relatively helpless at birth for varying lengths of time afterwards. Mammals are noted for the care they take with their young, which may include a great deal of teaching and always a brave defense when the young are threatened. In most other ways sexual reproduction is similar in all placental mammals, so we'll focus on humans, making comparisons with some other animals along the way.

HUMAN REPRODUCTION

Humans are reproductively unique in some ways, and we will deal with some of our unusual behavior since it is biologically interesting and essential to our understanding of our own sexuality. We'll begin with anatomy of males and females, and then go on to some of the physiological and behavioral effects of hormones.

The Male Reproductive System

External Anatomy. The genitalia of the human male is rather typical of mammals in general (Figure 36.11). The external organs include the **penis** and the **testes.** Shortly before birth, the testes descend from the abdomen, where they developed, moving over the pubic bone and down into the saclike **scrotum.** If the testes fail to descend, sterility results since the scrotum is cooler than body temperature and sperm will not develop properly at the internal body temperature.

The testes are paired oval bodies that produce both sperm and sex hormones. Much of the testis consists of the highly coiled **seminiferous tubules,** which contain a dense **germinal epithelium,** where sperm develop through meiosis. Outside the seminiferous tubules are found clusters of **interstitial cells** in a connective tissue matrix—endocrine tissue that secretes testosterone, the male sex hormone. Following their development in the seminiferous tubules, the immature sperm are swept by ciliary action into a network of tubules known as the **rete testis,** which leads to a number of ducts that finally enter the **epididymis.** The epididymis, a sperm storage structure, is a highly coiled tubule that lies perched above and extending down one side of each testis. Each entire testis is surrounded by a tough sheath and contains a rich supply of sensory neurons (which as most males will attest include a seemingly inordinate number of pain receptors).

Internal Anatomy: The Sperm Route. Let's follow the route taken by sperm as they leave the testis during ejaculation. Leaving the epididymis, sperm enter the **vas deferens,** which will carry them to the urethra through which they will exit the body. Along the way, the sperm will receive secretions from three different sources.

The vas deferens passes through its surrounding **spermatic cord** along with the blood vessels and nerves of the testes. Each spermatic cord—one from each testis—extends upward, over the bones of the pubic arch, and then into the body cavity, over the path taken by the testes in their previous descent into the scrotum. The paired vas deferens continue to the underside of the bladder where they unite, joined by the duct of the paired **seminal vesicles,** the first accessory glands. Unlike this structure in some invertebrates, vertebrate seminal vesicles are not used to store sperm. Instead, they produce most of the volume of the sperm-bearing

36.11 HUMAN MALE GENITALIA AND SPERM ROUTE

(a) The human male genitalia includes the penis, paired testes within the saclike scrotum, and accessory ducts and glands which produce and carry the semen. The route of sperm-bearing semen travel is seen in color. **(b)** The testis consists of highly coiled seminiferous tubules—the site of sperm production—and, where the tubules join at the upper region, the sperm-storing epididymis. The latter is continuous with the spermatic cord.

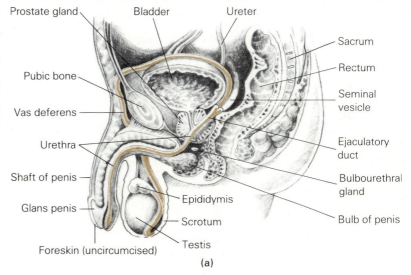

(a)

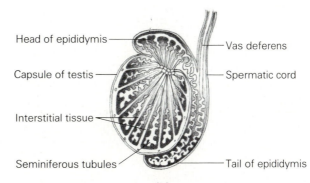

(b)

semen. The fluids added from the seminal vesicles contain nutritive and buffering materials. The misleading name "seminal vesicle" is a legacy from early anatomists who named the structure before its function was understood.

After being joined by fluids from the seminal vesicles, the newly formed, sperm-bearing semen passes from the vas deferens to the **urethra.** Surrounding this junction is the fleshy tissue of the second accessory, the **prostate gland.** From the prostate, the semen receives an alkaline secretion that raises its pH and produces its characteristic viscosity, color, and odor. Raising the pH of the semen activates the immobile sperm and, following ejaculation, protects them by neutralizing the normally acidic environment of the vagina. The late activation—just before ejaculation—is vital since each sperm has limited energy resources and, once in the female genital tract, has a lengthy journey to complete.

Following the addition of fluids from the prostate, the semen receives a final secretion from the **bulbourethral glands,** or **Cowper's glands.** While their function isn't clear, they do add slippery mucous secretions to the semen, that may help wash residual urine from the urethra, and since these secretions also precede ejaculation they probably have some lubricating value. The semen in its final state follows the route of the urethra through the penis, from which it is forcefully ejected. As you may have surmised, the urethra has the dual role of conducting both semen and urine—at different times.

Erectile Tissue.   In mammals, the penis consists of very specialized tissue with an erectile capability. The penis must be more or less erect and turgid before it can be introduced into the female vagina. In mammals such as horses and humans, the **erection** occurs when spaces in the penis fill with blood, causing it to increase in length and girth and to become turgid. In some mammals, such as cattle and sheep, erection may be brought about by extending the penis, already stiffened by cartilage or bone, from deep within the body where it usually rests.

In humans the penis is an erectile shaft tipped by an enlarged region, the **glans.** The penis contains three cylinders of very spongy tissue (Figure 36.12). Two of these are the paired **corpora cavernosa** and the third is the **corpus spongiosum,** which extends into the glans. During sexual excitement,

36.12 THE HUMAN PENIS

The penis contains three regions of erectile tissue which are capable of retaining blood during an erection. The largest are the two corpora cavernosa. Note the large arteries at their midregions. The third region, the corpus spongiosum, is below these and contains the urethra. During erection, the spongy passages of these bodies become engorged with blood because its exit has been restricted. This enlarges, curves, and firms the penis in preparation for copulation.

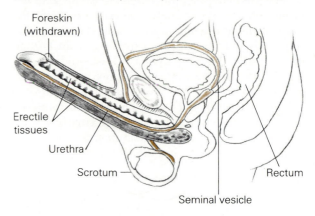

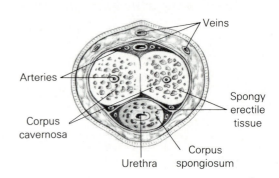

the blood flow into the spongy tissue increases, and venous flow back to the body is retarded. As the penis fills with blood, it becomes firm and erect. A complete or partial loss of erection almost invariably follows ejaculation, the normal result of a more restricted arterial blood flow and subsequent drainage of the spongy tissue. We will return to the subject in Essay 36.1, where the human sexual response is taken up.

Sperm and Their Production.   Spermatogenesis (sperm production) occurs in the walls of the seminiferous tubules (Figure 36.13). The cells in the outermost region of the tubule continually divide by mitosis. Some (roughly half) of the mitotic daughter cells begin the series of meiotic divisions that will produce the haploid sperm. The uncommitted cells—**stem cells** as they are known—continue to provide a source of new cells for the meiotic process.

When one of the spermatogonia enters the meiotic process it becomes known as a **primary spermatocyte.** Then meiosis I separates homologous chromosomes into daughter cells known as **secondary spermatocytes.** Meiosis II follows, with the production of four **spermatids,** each with the haploid chromosome number (see Chapter 11). Each spermatid will then undergo the complete reorganization that is necessary to form mature **spermatozoa,** its development guided by genetic instructions from the Y-chromosome. The spermatids at this time are apparently nourished by groups of specialized "nurse" cells called **Sertoli cells.**

In the mature spermatozoa, the chromatin will be condensed into a minute **sperm head.** The cytoplasm itself will differentiate into a midpiece and tail, and excess cytoplasm will be sloughed off. The midpiece contains a peculiar spiral-shaped mitochondrion and a centriole. The tail is a very long, tapered flagellum, which propels the sperm along. At the tip of the sperm head is the **acrosome,** a caplike structure that arises from an enzyme-laden lysosome and overlays the compact nucleus. We'll see in the next chapter how the enzymes of the acrosome function in egg penetration and fertilization. Figure 36.14 compares the sperm from several animals.

36.13 HUMAN SPERMATOGENESIS

Sperm are produced in the walls of the seminiferous tubules of the testis, seen here in both the illustration and SEM. Larger cells of the germinal layer, the spermatogonia, enter meiosis as primary spermatocytes. These are readily recognized by their large size and huge nuclei. After the first meiotic division occurs, the daughter cells are the secondary spermatocytes. These enter meiosis II, becoming spermatids. From this stage, nursed by Sertoli cells, each spermatid becomes greatly reduced in size, the flagellum develops, and the fully differentiated spermatozoan emerges.

Photo from *Tissues and Organs: A Text-Atlas of Scanning Electron Microscopy*, by Richard G. Kessel and Randy H. Kardon, W. H. Freeman and Company, copyright © 1979.

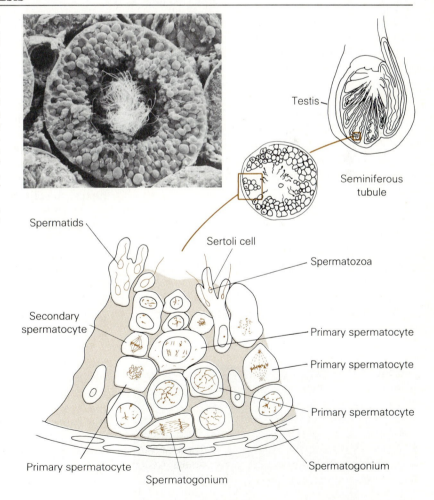

The Female Reproductive System

External Anatomy. The most prominent part of the external genitalia, or **vulva,** of women is the **mons pubis,** also called the **mons veneris** ("mountain of love,") a hair-covered mound of fatty tissue overlying the pubic arch (Figure 36.15). Below the mons pubis lie the larger outer folds, or **labia majo-**

36.14 ANIMAL SPERMATOZOA

(a) These various types of sperm cells are all drawn to the same scale, indicating that the human sperm is rather small. Perhaps the most unusual sperm is that of the rat, with its sickle-shaped head. However, each sperm consists of essentially the same parts: head, midpiece, and tail. (b) The head of the human sperm cell contains highly condensed haploid chromatin containing the 23 human chromosomes that have completed meiosis. At the tip is the enzyme-laden acrosome, which functions in egg penetration. The midpiece contains a centriole and a lengthy, spiraled mitochondrion. The tail is a flagellum, the propulsive structure of the sperm.

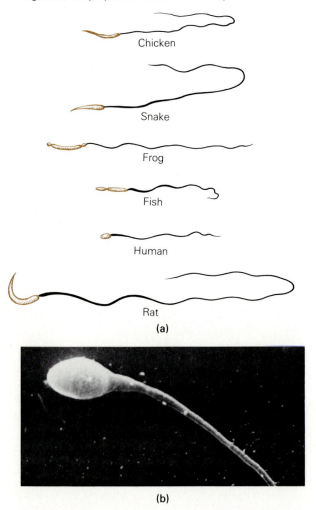

Chicken

Snake

Frog

Fish

Human

Rat

(a)

(b)

36.15 HUMAN FEMALE GENITALIA

The external features of the female reproductive system include the labia majora and minora, the clitoris, introitus of the vagina, and the meatus (opening) of the urethra. Prior to sexual experience, the introitus is sometimes partially occluded by membranes known as the hymen.

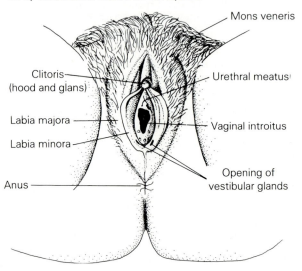

Mons veneris

Clitoris (hood and glans)

Urethral meatus

Labia majora

Vaginal introitus

Labia minora

Anus

Opening of vestibular glands

ra ("major lips"), of the vulva. In an unaroused state, these folds cover a number of rather sensitive structures. Just inside the labia majora are the less prominent, thinner, inner folds, or **labia minora** ("minor lips"). These join together at their upper portion to form a hood over a small prominence, the **clitoris.** This partially hooded, cylindrical organ is composed of highly sensitive, spongy erectile tissue. It is, in fact, homologous with the penis and develops from the same embryonic structures as the glans of the penis. Although it is in no way a miniature of the male structure (it does not enfold the urethra, for instance) the clitoral body contains a similar corpus cavernosum and a terminal enlargement called the **glans clitoridis,** and its sensitivity is similar to that of the penis.

Two openings are located between the labia minora. The upper, much smaller opening is the **urethral meatus** (the external opening of the urethra), through which the urine is voided. The second is the vaginal opening, or **introitus.** In virgins, the vaginal opening may be partially blocked by membranes known collectively as the **hymen.** While there is much variation, a certain amount of discomfort and bleeding is commonly experienced when the hymen is ruptured during the first intercourse. In a few instances, an overgrown hymen must be opened surgically (a hymenectomy).

Internal Anatomy. The **vagina** is a distendable, muscular tube about three inches long in its relaxed state (Figure 36.16). It is the female copulatory organ, receiving the erect penis and semen during copulation and it is also the birth passageway. The vagina is marvelously adapted for such diverse functions. Its highly folded walls consist of an inner mucous membrane (a stratified squamous epithelium; see Chapter 28), a middle muscular layer, and an outer layer of fibrous connective tissue. The inner layer presents a moistened, yet firm, stimulatory surface for copulation. During sexual arousal the vagina releases lubricating mucous secretions from glands in its inner recesses and through the **vestibular glands**, or **Bartholin's glands**, whose ducts open into the lower margin of the vaginal opening. They correspond somewhat to the bulbourethral glands in the male. The vaginal muscle and connective tissue layers are capable of great extension during passage of the infant at birth, accommodating a head about four inches in diameter. What is more remarkable is the rapid return to the original state from this strained expansion.

Other internal organs of the female reproductive system include the **uterus**, the **oviducts**, or **fallopian tubes**, and the **ovaries**, along with their supporting structures. As you can see in Figure 36.16, the uterus begins at the upper end of the vagina. Its opening, the **cervix**, is surrounded by a ring of dense tissue. The uterus is a hollow, pear-shaped organ that tilts its larger end forward in a slightly folded manner (if you can imagine a folded pear). Most mammals have paired uteri; the single uterus of humans and other primates (and bats as well) is an evolutionary specialization for species that customarily give birth to only one offspring at a time.

36.16 HUMAN FEMALE INTERNAL ANATOMY

Internally, the major organs of reproduction in the female are the vagina, uterus, oviducts (fallopian tubes), and ovaries. The vagina, a tubular structure about three inches long, ends at the cervix. The oviducts terminate in the fingerlike fimbriae, which do not connect to the ovaries.

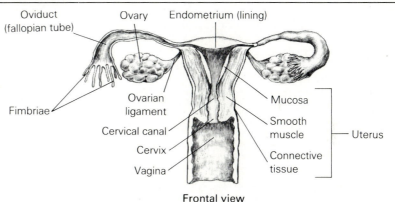

Frontal view

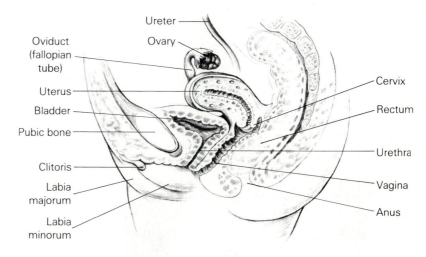

Midsaggital view

The Human Sexual Response

Much of what we know about human sexuality today has grown out of the pioneering efforts of William H. Masters and Virginia E. Johnson, of the Reproductive Biology Research Foundation in St. Louis. Using human volunteers, Masters and Johnson developed observing and measuring techniques for studying many of the physiological aspects of human copulation that were heretofore only known anecdotally, if at all. Not insignificantly, they also laid to rest many erroneous ideas on the subject.

In their analysis, Masters and Johnson divided the sex act, perhaps somewhat arbitrarily, into four phases: *arousal, plateau, orgasm,* and *resolution,* in that order. In their usual context, the four phases refer to actual copulation, but they can of course occur in other ways. We had better add also that not all the events listed occur every time, thus keeping in mind that when it comes to humans, there are variations on variations. Further, the phases are generally continuous, not clearly separated, and the time involved may take anywhere from a few minutes to many minutes.

In women, the preliminary acts of sexual arousal, such as kissing and caressing (variations here are endless) bring on increased heartrate, faster breathing, and a rise in blood pressure. This is accompanied by more outward manifestations such as firming of the nipples, spontaneous muscle contractions, and often what is known as the *sex flush.* This amounts to a reddening of the skin, particularly about the genitals, face, breasts, and abdomen, but not restricted to those areas. Changes occur in the glans clitoridis (analogous to erection in the male), general moistening of the vaginal walls, a marked elevation and parting of the labia majora and a size increase in the labia minora, such increases brought about by the accumulation of blood in these organs. These responses are physiological preparations for insertion of the penis; however, such changes do not necessarily signal psychological readiness of the woman.

In the male, arousal is characterized by somewhat similar events in circulation and respiration. Genital changes include erection, accompanied by a general contraction and elevation of the scrotum, and a moistening of the glans penis. Such events can occur notoriously fast in a male, often preceding correlating events in the female. But for the sake of this idealized discussion, let's assume arousal is simultaneous and copulation has begun.

As women enter the phase known as plateau, sensation and movement heightens and vigorous pelvic thrusting is common. The swollen clitoris withdraws into its hood rather suddenly at this time, and is now exquisitely sensitive to touch. The labial swelling increases as does swelling in the lower portion of the vagina, especially the opening. The uterus may elevate, tilting backward somewhat. As intensity grows, the genitals reach the *orgasmic platform*—preparatory physiological and anatomical changes leading to orgasm. Physical activity may peak now, with intense thrusting of the pelvis, although there is a great deal of variation here. Characteristically, the facial muscles relax, producing a slack appearance and almost an absence of expression.

In men, the intensity of the arousal phase has also increased steadily. The glans reaches its greatest size and the penis achieves its greatest curvature and rigidity. The testes increase in size and further elevate in the scrotum. Pelvic thrusting increases and secretions from the bulbourethral glands increase just prior to orgasm.

The term *climax,* often used to describe orgasm, is fitting. The events in the previous phase, as their intensity increases, now come to a climax, and orgasm begins. In both men and women, the orgasm is a matter of complex muscular contractions, usually accompanied by intensely pleasurable sensations. These contractions are reflexive and involuntary, occurring in a steady series.

In women they emanate from the muscles and thickened tissue about the vaginal opening—joined to a varying extent by activity in the vagina and the uterus. During these contractions, *tenting* may occur: the cervix is drawn upward, increasing the diameter of the surrounding inner vagina. Some authorities claim that tenting creates a suction that causes semen to pool around the cervix, and subsequent contractions in the uterus are thought somehow to draw the semen in, although certainly orgasm is not a prerequisite to fertilization. Uterine and cervical contractions are caused at least in part by the hormone oxytocin, which is released from the pituitary at this time.

The involuntary contractions of orgasm occur in 0.8 second intervals, spreading through the lower pelvis. In an intense orgasm, many of the other muscles of the body may begin to contract spasmodically. The actual origin of an orgasm is a subject of controversy at this time, but the preponderance of opinion is that clitoral stimulation triggers this phase. Stimulation preceding and during orgasm is probably indirect, involving the clitoral hood, which is apparently moved rhythmically against the clitoris by the thrusting action of the penis and by shoving of the two pelvises. Thrusting also tugs against the labia minora which, as you can see in Figure 36.16, is continuous with the hood.

The duration of the orgasm in women is highly variable. Several common reactions are shown in the accompanying graph. Orgasms may occur with one intense peak or several, or the peaks may be far more numerous but less intense. The ability of women to experience several orgasmic peaks in a single coitus is sharply contrasted to the usual single orgasmic peak to which men are limited.

In men, ejaculation is a part of orgasm, however orgasm can be divided into two distinct parts. The first is *emission*. Emission occurs with the rhythmic, peristaltic contractions in the vas deferens, prostate, and seminal vesicles. As a result, the elements of semen, spermatozoa, and prostatic fluids, collect at the base of the urethra prior to expulsion. The second part, *ejaculation,* a spinal reflex action triggered by the presence of semen in the urethra, will immediately follow. This forceful expulsion of semen occurs as powerful striated muscles at the base of the penis contract, also in intervals of about 0.8 sec. The semen is forced through the urethra and out of the penis in several spurts, the first usually bearing more semen than the rest. Ejaculatory contractions usually continue for a short time (several seconds). Involuntary contractions of many muscles of the lower pelvis are usually involved also. On the border between voluntary and involuntary movements are the intense, even violent pelvic thrustings and muscular contractions involving the whole body that commonly accompany this phase of the sex act.

Following orgasm, both males and females retreat from the state of sexual excitement into what is known as *resolution,* a phase that has also been referred to as *afterglow*. Men usually lose erection rapidly and enter a *refractory* period in which rearousal can be quite difficult—at least for some minutes and usually longer. The refractory period varies, typically increasing with age. Resolution in women is somewhat different, since immediate arousal is often possible. In any case, the quiet, relaxed, and often tender resolution phase may have an intense pleasure of its own.

(a) In the highly variable female orgasm, we see three types. In 1, it occurs in several peaks of intense response. The number varies considerably in such multiple orgasms. In 2, there are a greater number of multiple events which are far less intense—just above the plateau level. In 3, the female orgasm is similar to that in males, with one intense period that lasts somewhat longer than the male's. Note the difference in resolution time in the three types of female orgasm. (b) Compared to this, the male orgasm is much simpler and more stereotyped, although as the graph indicates, orgasm can occur again in males after a period of resolution. The recovery time between orgasms in males increases with age.

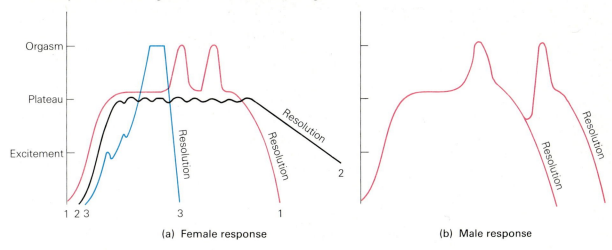

(a) Female response   (b) Male response

The walls of the uterus consist of three specialized layers. The inner layer is the versatile **mucosa**. It produces the soft, highly vascularized **endometrium** that will receive and support a zygote if fertilization occurs. (We will look at the cyclic production of the endometrium a little later in this chapter.) Below the mucosa lies the middle layer, a region of smooth muscle whose fibers run in several directions. The uterus has an enormous capability to expand, as it must when accommodating a fetus. The outermost layer is connective tissue, which is continuous with the **broad ligament**. This latter structure forms two wings on either side of the uterus, supporting it as well as the ovaries and the oviducts.

The oviducts emerge from each side of the upper end of the uterus and extend for a few inches outward and then downward. They terminate in a cluster of fingerlike processes known as the **fimbriae**. Curiously, there is no direct connection between the oviduct and the ovary; eggs are shed into the body cavity and later picked up by the highly moveable, ciliated fimbriae.

Ovary and Oocyte Development. The ovaries are located about in the center of the fanlike broad ligament and are additionally supported by the **ovarian ligament**. They are oval in shape and about an inch in length. The outer tissues contain the germinal epithelium where all the eggs are stored, arrested in prophase I of meiosis during fetal life (Figure 36.17). The number of oocytes in the ovaries at birth—some 2 million—is astounding, especially considering how few can ever mature. By puberty this number has shrunk considerably, to approximately 200,000 oocytes in each ovary. Of course, nowhere near that many will mature. In her reproductive lifetime, each human female has about 450 opportunities to become pregnant (about once every four weeks for about 35 years).

Gamete production represents an important difference between males and females. Females are born with all the eggs they are ever going to have. The male, however, continues to produce new sperm cells throughout his reproductive lifetime—literally trillions of them. Since each successful fertilization involves one egg and one sperm, such prodigious production seems unnecessary, but we will see.

Hormonal Control of Human Reproduction

The reproductive activities of humans, as with all higher animals, are under the influence of hormones. The primary sites of human reproductive hormonal activity are the pituitary gland, the

36.17 HUMAN OOGENESIS

The outer region of the ovary is the germinal epithelium that contains all of the oocytes, each arrested in prophase I of meiosis. **(a)** Each month, an oocyte becomes surrounded by a cluster of supporting cells—the follicle—and resumes meiosis. **(b, c)** The follicle enlarges, becoming fluid-filled, with a number of cells directly surrounding the oocyte itself. **(d)** At midcycle, the oocyte completes meiosis I, and is ejected from the ovary at ovulation. **(e)** The vacated follicle will form a hormone-secreting body—the corpus luteum.

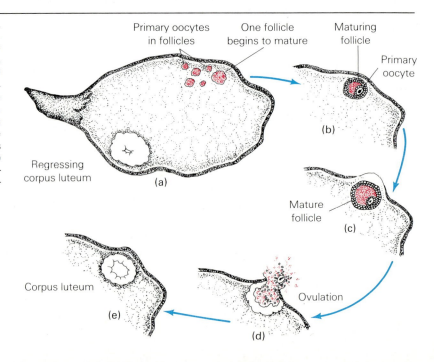

gonad, and the placenta. These centers of hormone production are, in turn, under the influence of the hypothalamus. The role of the hypothalamus was described in detail in Chapter 33, but a review here will be helpful. Essentially, the hypothalamus does two things: it initiates the release of certain hormones and it reads the blood level of other hormones. In this manner it can act as a precise regulator whose operation is subject to delicate feedback mechanisms.

When prompted by gonadotropic releasing hormones from the hypothalamus, the anterior pituitary secretes sex hormones known as **gonadotropins** (see Chapter 33). As their name implies, gonadotropins stimulate growth or activity in the gonads (both the ovaries and the testes). In addition, they are indirectly responsible for the surge of growth during puberty and help direct the development of those familiar secondary sex characteristics. More directly, the gonadotropins stimulate the production of the sex hormones and initiate the development of sperm and egg cells.

Two specific gonadotropins of the pituitary are the follicle-stimulating hormone (FSH) and the luteinizing hormone (LH). The targets of both are the ovaries in women and the testes in men. LH is known as **interstitial cell-stimulating hormone** (ICSH) in males. Its target is the testosterone-producing interstitial cells of the testes. Let's look further into the action of gonadotropins in the simpler system of males. Then we'll go on to hormones of females.

Male Hormonal Action. The specific target of FSH is the sperm-producing seminiferous tubules of the testes, and the hormone is continuously released throughout the life of the male. In addition to its role in sperm production, FSH acts synergistically with ICSH to induce the testes to produce **testosterone**, the primary male sex hormone.

Testosterone is produced throughout the testes in the interstitial cells that lie outside the seminiferous tubules. There are indications that the production of testosterone may rise and fall with increased and decreased sexual activity—or even with social interactions with the opposite sex. Testosterone levels commonly fall in all-male crews of ships at sea, for instance, as evidenced by lower rates of beard growth. (More recently, researchers claim that the mere anticipation of sex can produce an elevation in body testosterone.) But even if the male's social life follows the pattern of feast and famine, his testosterone levels will be far more constant than the corresponding ovarian hormones in a female, as we shall see.

36.18 TESTOSTERONE CONTROL

The male sex hormone, testosterone, is released by the testes through a delicate feedback mechanism involving the hypothalamus and anterior pituitary.

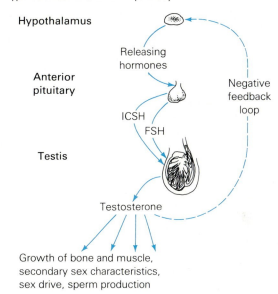

Testosterone is vital to the development of the characteristics of "maleness." Although the appearance of the embryonic gonads is genetic, their subsequent development requires testosterone. If it is absent during the eighth or ninth month of development, the testes fail to descend into the scrotum. Later on, testosterone is essential for those perturbing events associated with puberty, such as voice changes, the growth of body hair, bone and muscle development, and the enlargement of the testes and penis. Abnormalities resulting from low testosterone levels in the developing male are readily corrected by injections of the hormone. Finally, the sex drive itself appears to be greatly influenced by the presence of testosterone. As we will see, testosterone is responsible for both pubic hair growth and sex drive in women as well as men (also see Essay 33.1).

High levels of testosterone inhibit the release of ICSH. The critical level is apparently detected by the hypothalamus, which then retards its stimulation of the pituitary. The pituitary, in response, slows its release of ICSH, thus lowering the production of testosterone by the interstitial cells. So, as you can see, a feedback mechanism regulates testosterone levels (Figure 36.18). Actually, things aren't quite that simple. For example, we mentioned earlier that sexual and social activity can influence hormonal levels. Apparently, the frequency of ejaculation and/or sexual arousal has some influence on the hypothalamus, or perhaps directly on the pituitary, or the testes—or all three.

Now let's see if we can simplify some of this. Under the influence of the hypothalamus, the pituitary secretes the gonadotropins FSH and ICSH. These hormones stimulate sperm production and initiate the production and release of testosterone by the testes. Levels of testosterone are sensed by the hypothalamus, which responds by adjusting the release of gonadotropins. The result is a fairly steady level of hormone production. As you will see, all this is quite straightforward when compared to what happens in the human female.

Female Hormonal Action.   The onset of puberty in girls most often occurs between 9 and 12 years of age. The beginning of fertility follows in two to three years. Both events are initiated by a rise in the pituitary gonadotropin, FSH (follicle-stimulating hormone), which acts on the ovaries. The ovaries respond to FSH by producing estrogens. In turn, estrogens influence the growth of the breasts and nipples, broadening of the hips, and in general, the development of the usual adult female contours. Less conspicuously, estrogen influences the growth of the uterus, the vaginal lining, the labia, and the clitoris. The appearance of pubic and axillary (underarm) hair, another signal of puberty, is initiated by the combined effects of estrogen and ACTH (adrenocorticotrophic hormone from the pituitary) on the adrenal glands (see Chapter 33). In response to these hormones, the adrenals produce androgens (essentially, male hormones), which stimulate the growth of the coarser hair of the genitals and underarms.

Fertility in the female is marked by the start of the **ovarian**, or **menstrual cycle**, in which the gonadotropins and ovarian hormones rise and fall with some regularity. Other female vertebrates may experience somewhat analogous **estrous cycles** with a regular frequency or perhaps only once or twice each year. The remainder of the time they are sexually unresponsive and infertile. The human female is considered to be unusual among mammals in always being potentially responsive (or at least in not having predictable cycles of responsiveness).

The Ovarian or Menstrual Cycle.   On the average, a mature ovum is released from one of the two ovaries about every 28 days. Coinciding with this event is a peak in preparatory activities in the uterus. Its lining, the endometrium, is prepared for receiving the fertilized ovum within a short time after its release and fertilization. As we'll see, these events are closely correlated through the action of pituitary and ovarian hormones, which are themselves subject to timely negative feedback events in the hypothalamus.

In discussing the menstrual cycle, it is common to place the essential event of ovulation (the release of an ovum) at midcycle, or the 14th day, and begin counting on the first day of menstrual flow. The accompanying figures will help in understanding how the sequence is organized (Figures 36.19 and 36.20).

Days 1 to 14: Proliferation.   The first half of a cycle is known as the **proliferative,** or **preovulatory phase.** During the first three or four days (again, there is some variation), blood and tissues of the most recently produced uterine lining—the endometrium—are shed in what is referred to as the **menses,** or **menstruation.** Since menstruation is cyclic, we will return to it, but first let's see what the hormones are doing.

In the proliferative phase, the pituitary, under the influence of the hypothalamus, releases more FSH. Its targets are the immature ova in the germinal epithelium of the ovary. Under the influence of increasing levels of FSH, the cell clusters around several immature ova will grow, forming **follicles.** Usually, only one ovum and its surrounding follicle matures. Its follicular cells increase in number and secrete fluids, and the follicle grows steadily through the first 14 days. The cluster of cells will thus form a fluid-filled cavity—the **Graffian follicle**—which will eventually enlarge to protrude like a blister from the ovary's surface. Sometime during the fourteenth day, the ovum, clustered within a layer of surrounding cells, will burst from the ovary and begin its journey to the oviduct (see Figure 36.17).

As a follicle grows, it secretes estrogen which steadily increases in the bloodstream. Estrogen's principal target is the lining of the uterus, which responds by undergoing rapid cell division, building dense layers of cells over the entire surface, layers permeated by a profusion of blood vessels. The result is a highly vascularized lining we have called the endometrium. The cervix, also in response to rising estrogen levels, secretes an alkaline mucus, changing the usually acid vaginal environment to one more hospitable to sperm cells.

As you can see, hormones tightly coordinate

36.19 HORMONAL CONTROL OF THE OVARIAN CYCLE

(a) From day 1 to 14, FSH and estrogen dominate the cycle, bringing on follicle development and endometrial growth and repair. At midcycle **(b),** negative feedback brings on a pituitary shift to LH secretion. LH stimulates ovulation and the development of a corpus luteum. From day 14 to 28 **(c),** LH, estro-gen, and progesterone dominate the cycle, bringing the endometrium to a fully receptive, glandular state. Without the implantation of an embryo, a second negative feedback event diminishes the hormones and the cycle ends.

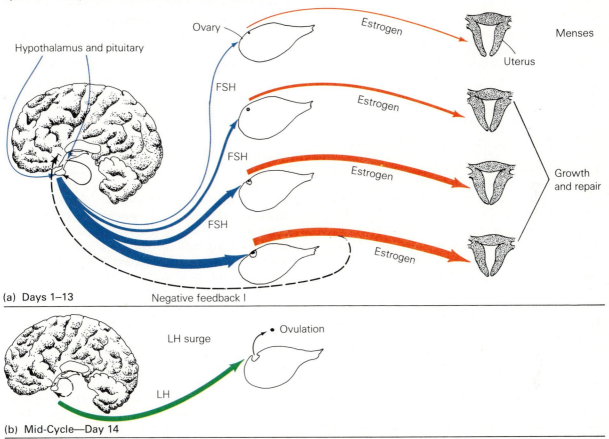

Hypothalamus and pituitary

Ovary

Estrogen

Uterus

Menses

FSH

Estrogen

FSH

Estrogen

Growth and repair

FSH

Estrogen

(a) Days 1–13 Negative feedback I

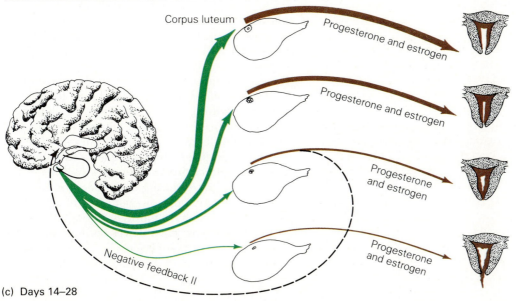

LH surge

Ovulation

LH

(b) Mid-Cycle—Day 14

Corpus luteum

Progesterone and estrogen

Progesterone and estrogen

Progesterone and estrogen

Negative feedback II

Progesterone and estrogen

(c) Days 14–28

the events in the ovary with those in the uterus. One result is that at the time of ovulation the uterus is nearing a fully receptive state. Should an egg be fertilized, it will reach the uterus in a few days and there it will find itself in a highly receptive environment.

36.20 PEAKS AND VALLEYS IN THE MENSTRUAL CYCLE

Note in each case the events of day 14 and ovulation. Both LH and FSH levels peak at this time, and the follicle reaches maturity. Although estrogen levels peak shortly before day 14, progesterone levels increase *after* ovulation. The endometrium undergoes growth and repair during the first 14 days and reaches its fullest development about a week later. At midcycle (day 14) body temperature is slightly elevated. This is a fairly reliable indicator of ovulation, an important part of the rhythm method of birth control.

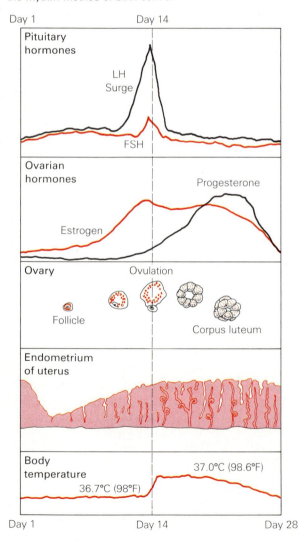

Day 14: Ovulation. As the fourteenth day of the cycle approaches, the hormone production changes. The rising blood estrogen level triggers a new action by the hypothalamus. Its prompting of the pituitary include two changes. First, FSH secretion is inhibited, or at least dramatically lowered. Second, the pituitary rapidly increases its release of luteinizing hormone, bringing on what is called the "LH surge."

The target of LH is the enlarged, fluid-filled follicle. Under LH influence, the fluid content rapidly increases and the follicle ruptures, releasing the ovum in what is called ovulation. The event is followed by a slight, temporary elevation in body temperature (about 1°F.). This temperature elevation is an important indicator for women trying to establish the "fertile period," either in the interest of conception or contraception.

Days 15 Through 28: Secretion. Following ovulation, the **secretory,** or **postovulatory phase** of the cycle begins. The vacated ovarian follicle, under the continued influence of LH and FSH, forms a new structure called the **corpus luteum (corpus,** body; **luteum,** yellow). The corpus luteum continues estrogen secretion and, in addition, secretes a second hormone, **progesterone.** The target of both is the endometrium, where progesterone brings on the secretory phase, prompting blood vessel growth and proliferation and glycogen accumulation. Thus progesterone instigates the final preparation of the uterus for the reception of a zygote. And should pregnancy occur, it is responsible for maintaining the condition of the uterine lining until this function is taken over by the placenta itself, as we will see in the next chapter. Progesterone also stimulates a temporary thickening and an increase in secretion in the vaginal lining and the growth of milk ducts in the breasts. Finally, the appearance of progesterone is believed to bring on the slight temperature rise following ovulation.

Hormonal levels rise in the blood for about 6 days following ovulation. During this time, the endometrium of the uterus will have reached its greatest development. Its loosely packed cell layers become fluid-filled sinuses, penetrated with glandular tissue that produces other secretions. If fertilization occurred during midcycle, as it normally does, a tiny embryo, now just a ball of cells about six days old, will have formed. Through the action of proteolytic enzymes, it will become implanted deep in this receptive tissue, literally digesting its way deep into the fluid-filled spaces where it continues its development. The slight bleeding caused by the embryo settling into the uterus may cause the woman to believe her period has started and that she isn't pregnant.

The uterus remains receptive for about 10 to 12 days following ovulation. Unless an embryo is snugly implanted by that time and putting out its own progesterone-prompting signals to the corpus luteum, that body will start to degenerate. Its production of estrogen and progesterone will decrease and the endometrium will lose its receptive state. In a few days it will slough away, signaling the beginning of menstruation and the start of the next 28-day cycle.

Why does the corpus luteum fail? While definitive answers aren't available there is some consensus on the following. Unless pregnancy has occurred, the signals from the pituitary start to fall off during the fourth week of the cycle. The decrease in LH and FSH may well occur because of negative feedback, primarily that of progesterone. Figure 36.19c shows that the levels of progesterone peak on about day 21. The high blood hormone levels may produce a negative effect on the hypothalamus. It slows its stimulation of the pituitary which, in response, slows its LH secretion. The resulting drop in LH causes its own chain reaction. Without LH, the corpus luteum recedes, which causes a drop in the secretion of progesterone and estrogen. And finally, without their support, the endometrium simply breaks down.

Now that we have explained how the cycle ends, we are left with one more question. How does the next cycle start? What we are really asking is why does FSH production begin again? If you recall, at midcycle, FSH production slowed, presumably as a result of high estrogen levels. Now, at the end of the cycle, estrogen (and progesterone) levels fall drastically. Clearly, this could remove any inhibitory effect that prevented FSH production. The signal "produce FSH" is again transmitted from the hypothalamus and FSH secretion follows. New follicles in the ovary respond and the cycle repeats.

However, if pregnancy has occurred, things are very different. The extraembryonic tissue of the fetus becomes an endocrine gland, secreting **human chorionic gonadotropin.** The latter constitutes the progesterone-secreting signal mentioned earlier— the signal directed to the corpus luteum. So the aging corpus luteum continues to secrete hormones that keep the uterine lining hospitable for the voracious little embryo. (The presence of HCG in the urine, incidentally, is the basis for many pregnancy tests.) Later—about the third month of pregnancy—the HCG will diminish as the placenta begins secreting its own progesterone and estrogen.

Let us now summarize the important events of the menstrual cycle. It involves the production of a mature ovum along with the simultaneous preparation of the uterus to receive it. The first half of the cycle is under the influence of FSH and estrogen which, respectively, cause follicle and uterine growth. The second half is principally under the influence of LH, progesterone, and estrogen. The effects are ovulation and the continued preparation of the uterus for implantation. Rising blood estrogen levels bring on the hormone-switching negative feedback effect at midcycle, and rising blood progesterone levels create the negative feedback effect that terminates the cycle.

Contraception

Conception, at least in the more developed nations, has increasingly involved decision-making processes. The general acceptance of birth control has produced a great deal of technological interest. As a result, there are many options in **contraception** today. We will consider some of these for two reasons. First, it is obviously relevant to people of reproductive age. Second, there is a biological principle at work behind each method. We will organize our discussion of birth control under four major headings, although you will notice that there are instances when the groups overlap or are used in combination.

Natural Methods. Some "natural methods" may seem somewhat unnatural in their application. We call them this simply because they don't require chemical, mechanical, or surgical intervention.

Coitus interruptus, withdrawing the penis just before ejaculation, is a surprisingly common method of contraception, particularly among novices, the poor, and the unprepared. But it is risky. There are the problems of premature release of small amounts of semen, and then, couples may engage in intercourse more than once, ignorant of the presence of residual sperm cells in the male urethra. But most significantly, there is the psychological frustration of a man's having to do exactly what he is least inclined to do, (and some men are less dependable than others). Even when he succeeds, withdrawal may deprive his partner of her

TABLE 36.1

METHODS OF BIRTH CONTROL

Method	Application	Effectiveness	Drawbacks
Surgical intervention (by qualified medical persons)			
Abortion (vacuum aspiration, up to 12 wks)	Cervix dilated, embryo and placenta removed by gentle suction	Virtually 100%, hospital stay not required	Psychological effect common; strongly controversial
Abortion (D&C) (up to 12 wks)	Cervix dilated, embryo and placenta removed by scraping	Virtually 100%, hospital stay generally not required	Psychological effect common; 1%-2% have medical complications
Abortion (saline injection, after 16 wks)	Saline (NaCl) solution injected into uterine cavity; labor and expulsion of fetus and placenta ensues	Usually 100%	Psychological effect; some risk from saline poisoning
Sterilization			
Vasectomy	Incisions through scrotum, section of vas deferens removed and remainder tied off; sperm cannot leave testes	Virtually 100%, now 45% reversible	Some psychological effects occur
Tubal ligation	Incision through abdominal wall, oviducts cut and tied or just tied; sperm cannot reach egg	Nearly 100%	Hospital stay generally required; costly; simply tying is somewhat reversible; newer microtechniques produce less surgical risk
Chemical and/or mechanical intervention (with advice of qualified physician when needed)			
Oral contraceptives ("the Pill")	Combination or sequential estrogen and progestins or progestin alone (minipill); taken daily, prevents ovulation	95%-99%	Costly; must be prescribed and monitored by physician; temporary side effects; greater risk to heavy smokers
Intrauterine devices (IUDs)	Plastic and/or metal devices inserted into uterus; may contain slow-release progestin; prevents implantation	95%-99%	Must be inserted and monitored by physician; temporary discomfort in some; some expulsion; complications if pregnancy occurs; recommended only after at least 1 child

own orgasm. Thus coitus interruptus is considered to rank only as "fair" in terms of dependability (Table 36.1).

Douching after intercourse was a fairly common method of contraception in past years. It involves flushing the vagina with plain water or various household or commercial solutions. Diluted vinegar has traditionally been the most common douche; its acid properties were intended to kill the sperm cells. We mention this crude method only because of its great failure rate (about 40%; failure rate in contraception studies refers to percentage of pregnancies per year of dependence on one method). Even fastidious douching fails because sperm cells are known to pass safely through the cervix into the uterus only moments after ejaculation.

The **rhythm method** is perhaps the most biolog-

ically interesting method of birth control. It is called the "most natural" method by its proponents, yet its tools are papers, pencils, thermometers, and calendars. Its success depends on a woman's precise knowledge of her menstrual cycle, including an accurate prediction of ovulation and the ability to analyze key postovulatory changes in the density of mucus secretions. The principle itself is simple: abstinence, that is, no intercourse during the period in which fertility is possible. Theoretically, this includes four days each month, two days on either side of ovulation. Usually, since no one is sure of how long sperm survive in the female genital tract, an additional day on each side is recommended. A review of Figure 36.21 adds another dimension to the problem, however. As you can see, a peak in sexual receptiveness and orgasm for many women

TABLE 36.1 (continued)

METHODS OF BIRTH CONTROL

Method	Application	Effectiveness	Drawbacks
Chemical and/or mechanical intervention continued			
Diaphragm with spermicide	Rubber dome; fits over cervix; blocks sperm; spermicide kills sperm	98% with spermicide and completely regimented use	Must be fitted by physician; strong motivation required; somewhat messy
Condom	Rubber sheath worn over penis; blocks sperm entry	70%-93%, depending on quality and strict usage; safer with spermicide; withdrawal immediately after ejaculation is essential	Requires strong motivation since it is interruptive and some sensation is lost for men; can break; cost fairly high
Spermicide alone (foams, jellies, creams, or suppositories)	Placed in vagina with applicator before each intercourse; kills sperm	About 90% when properly used	Generally messy and short-lived (newer foams somewhat better); interruptive
Rhythm (natural method)	Intercourse avoided during carefully determined fertile period	Under ideal conditions, about 80%	Requires great motivation; fails when cycles are irregular
Coitus interruptus (withdrawal)	Penis withdrawn just before ejaculation	About 70%	Mental stress for both partners; great self-control required; sperm can leak prior to ejaculation
Douche	Vagina flushed with water or chemical solution (vinegar common) after intercourse	Below 60%	Slightly better than no precautions—in other words, worthless; sperm can enter the cervix within 1-2 seconds after ejaculation
Sponge	Polyurethane sponge saturated with spermicide placed in vagina before intercourse	Believed to be same as diaphragm (98%), but data not yet available; no fitting necessary	Interruptive; reliability not proven

NOTE: Experimental innovations still being tested or developed include Silastic implants for slow release of progestin [1-6 years]; long-lasting injections of progestin [3 mos.], now used outside the U.S.; vaccine that brings on menstruation if implantation occurs; morning-after prostaglandin treatment; and male hormonal sperm suppressors.

occurs just prior to ovulation, a decidedly high-risk time, and a very difficult time to abstain. Under the best of conditions, this system fails 20% of the time. Some of the problems of the rhythm method are summarized in Figure 36.22.

Mechanical Devices. Mechanical devices include any physical object used to block sperm travel or embryo implantation. We include the **condom, diaphragm,** and **intrauterine device (IUD).** Condoms are very thin rubber or animal-membrane sheaths designed to fit over the penis. Condoms are probably the most common means of contraception and data indicate a failure rate of 7 to 30%. The lower figure represents *faithful* use of a quality product. The diaphragm is a rubber dome with a thick rim that is worn over the cervix during intercourse. Generally, the inner side is coated with a spermicidal jelly or cream before it is inserted. Data on the diaphragm's reliability are inconsistent, with failure rates varying from 2 to 20%. The lower figures represent faithful use with a spermicide.

The IUD is a widely used, very effective contraceptive device. It is also the least understood. Although first developed in the early 1930s, interest in the IUD didn't really grow in the United States until the late 1960s, when the very promising reports of a 30-year study were published. Since that time, a variety of forms of the IUD has found use all over the world. It appears that the IUD interferes with implantation of the embryo. Whatever its action, its failure rate today is reported 1 to 4%, a very respectable range. It is commonly used as an alternative to the pill for females who fall into a high risk pill-user group.

36.21 SEXUAL RESPONSE CYCLES IN WOMEN

The results of one study of female sexual response and orgasm through the 28-day menstrual cycle. From these data several observations can be made about sexual behavior: First, receptivity occurs throughout the cycle, but has peaks and valleys. Second, an activity peak occurs just before ovulation and it is accompanied by an orgasmic peak. (This is to be expected if fertilization is to be maximized.) Third, the second set of peaks is a little more difficult to account for since it is apparently too late in the cycle to accomplish fertilization.

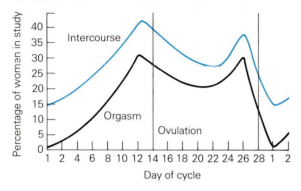

36.22 THE RHYTHM METHOD

The rhythm method of contraception assumes that there are definite periods of time when conception is highly unlikely. However, there are serious problems in setting exact limits. For instance, ovulation on day 14 is only an average, and can be offset by illness or individual variability. In addition, no one knows exactly how long sperm live in the female genital tract, although 48 hours seems to be the consensus. Use of this system requires that people read thermometers well, keep good records, and, most difficult, perhaps, abstain during mid-cycle, when studies indicate the maximum receptivity and orgasmic peak in females occurs. The system illustrated here is probably conservative, and allows for some slippage.

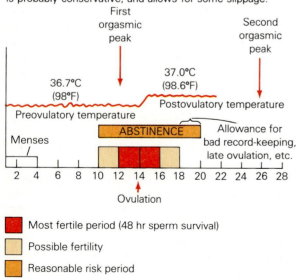

The primary advantage of the IUD is that it is convenient. Once it is inserted, it can more or less be forgotten, and fertility is restored when it is removed. Therefore, it is often the contraceptive of choice for people interested in family planning. However, IUDs cause cramps and pain in almost a third of women using them, and frequently they must be removed. Some have also been associated with uterine infections, and should pregnancy occur anyway, their presence can create complications.

Chemical Methods: The Pill.   One of the most widely used contraceptives today is the birth control pill. At present, an estimated 100 million women around the world are using this form of birth control. Its effectiveness rates highest among the various methods—short of surgical sterilization or abortion.

The active substances in the pill are synthetic estrogen and progesterone, used in combination or sequentially. The record of combination pills is slightly better because of its greater hormone content. Both function in the same manner. They inhibit ovulation by overriding the normal rise and fall of estrogen and progesterone produced by the ovary. In a sense, this simulates pregnancy, in which high levels of the two hormones also suppress ovulation (Figure 36.23). To understand this overriding effect, recall the discussion of the interactions of LH and estrogen.

Birth control pills receive some bad press occa-

36.23 ACTION OF THE PILL

The effect of the 20-day birth control regimen is to override the normal ebb and flow of ovarian hormones, as seen here. It is believed that the delicate feedback mechanism between ovary and pituitary, which initiates ovulation is blocked, thus preventing ovulation. In addition, the presence of a steady high level of progesterone in combination pills is believed to simulate pregnancy, thus overriding the normal cycle. Actually, there is much to learn about the precise action of the synthetic hormones of birth control.

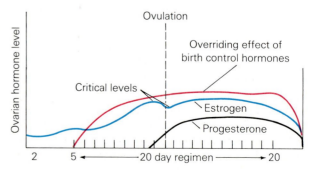

36.24 SURGICAL STERILIZATION

Both the vasectomy and the tubal ligation are common surgical methods of contraception. Both result in sterilization by preventing gametes from following their natural pathway. **(a)** Vasectomy involves a simple surgical procedure which is often done in a doctor's office. After an incision is made in the scrotum, the spermatic cord is located and cut open to reveal the vas deferens. The vas deferens is then cut and a section removed. The ends are then tied tightly with suturing thread and subsequently grow closed, preventing sperm from reaching the urethra. **(b)** Tubal ligation is more complex, surgically, than vasectomy, requiring abdominal surgery. More recently, it has been possible to reduce the size of the incision and use a viewing device and slender surgical instruments to complete the procedure. Tubal ligation consists of cutting and tying the oviducts, thus preventing the egg and sperm from meeting.

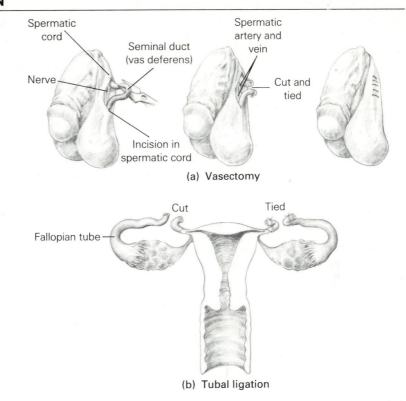

(a) Vasectomy

(b) Tubal ligation

sionally because of limited, but critical, side effects in some women. These include a higher risk to women who suffer from blood-clotting problems, high blood pressure, diabetes, and family histories of certain other conditions. Further, heavy smokers are under substantially greater risk. Thus women should be carefully screened by a physician before using the pill. The debate on side effects is by no means resolved, but one point about the risks stands out very clearly: in all age groups, the risks associated with pregnancy significantly exceed the risks in taking the pill. Deaths from complications in pregnancy are highly age-dependent. For every 100,000 pregnancies in the United States, there are about seven deaths in women between the ages of 15 and 30. This increases steadily to 21 deaths per 100,000 as women approach age 40. Deaths attributed to the pill in women below 35 are one to two per 100,000, increasing to about four as women approach age 40.

The possibility of chemically controlling fertility in males has been repeatedly considered. In spite of the fact that efforts to control male fertility hormonally have not met with great success, the search goes on. Apparently, the regularity of male hor-

mone production and the comparative simplicity of male reproductive physiology doesn't present an easy target.

In addition to fertility control chemicals, there are several other methods of chemical contraception. They are generally far less effective than hormonal control, however. For the most part, they consist of sperm-killing preparations such as jellies, foams, and suppositories, all used in the vagina. The most effective of these is undoubtedly the aerosol vaginal foam, since it can cover the entire cervix and vaginal lining with the spermicide. A more recent chemical contraceptive is the "sponge," a device containing spermicides that is inserted into the vagina prior to intercourse. It is reportedly about as effective as the diaphragm.

Surgical Intervention

We now find that surgical intervention in the reproductive process is gaining popularity throughout the Western world. Surgical intervention, of course, refers to sterilization and abortion. Male sterilization involves a minor operation known as a **vasectomy** (Figure 36.24a). The procedure can be carried out in a doctor's office with the use of local anesthetic. Essentially, the operation consists of two small incisions in the scrotum and spermatic cords, removal of a small segment of the vas deferens, tying off the cut ends, and sewing up the incisions. As a result of the surgery, sperm travel is blocked just beyond the epididymis. Sperm production continues, but they disintegrate and are cleared away by phagocytes. Other than loss of sperm, semen volume and ejaculation are normal, and many men, their anxieties reduced, report heightened sexual pleasure.

Sterilization in women, **tubal ligation,** is similar in that the surgery involves cutting and tying the oviducts, or just tying them off, thus preventing the sperm and eggs from joining. The surgery can be done through small abdominal incisions or by working through the vaginal wall just below the cervix (Figure 36.24b). While the cutting and tying procedure is essentially irreversible, tying alone can be readily reversed. While almost nothing is known about physiological side effects, sterilization brings many women a new attitude of freedom, greatly enhancing their enjoyment of sex. Because of the difficulties in restoring fertility no one should contemplate complete sterilization without thoughtful consideration and preferably counselling. Restoration can be a major operation, requiring full anesthesia and hospitalization, with only about a 50% success rate.

Abortion as a means of birth control is generally considered to be more of an "if all else fails" procedure, although in a few modern nations the abortion rate has come to exceed the birth rate! Abortion is the surgical removal of an embryo or fetus, or its expulsion by the inducement of labor. In early pregnancies (less than three months), surgical removal is commonly done with a suction or vacuum device, or through **dilation and curettage,** ("D&C"). The latter is a bit more severe since a curette (a spoon-shaped surgical knife) is used to scrape the uterine wall, dislodging the embryo.

If the fetus has passed its third month, a saline treatment may be used to induce labor. A 20% saline (NaCl) solution is injected into the uterus. Labor and delivery of the fetus and its supporting placenta follow. Saline treatments are riskier than the other means. There is an occasional problem of salt poisoning and also a greater risk of infection. No one takes saline abortion very lightly. The performance of an abortion after the second trimester is illegal in most places, and a bad idea in any case.

The field of reproductive biology, we see, has a number of aspects that have attracted researchers, particularly in recent years. Theorists are interested in the biological advantage of sex. Physiologists are fascinated by the remarkable hormonal control of reproductive processes. Anatomists wonder at animals with two penises. And many of the rest of us are drawn to such discussions because we want to better understand ourselves and to gain more control over our own reproductive activity. The advances and technical information are increasingly hard-won, but each step seems to set up yet another compelling question relating to this complex and critical aspect of life.

APPLICATION OF IDEAS

1. In evolutionary terms, hatching in the precocial state is considered by ornithologists to be primitive, while the altrical state is thought to be advanced. Offer reasons why such a hypothesis might be logical, taking into consideration the evolutionary history of birds and that precocious species are usually ground foragers.

2. Male and female pairing in humans is often described as a matter of "bonding"—that is, the pleasure bond, the pair bond, and so on. Both cultural and biological explanations of bonding have been proposed, and these are often in conflict. In the first, bonding is often considered to be a learned human behavior, approved and reinforced by society. In the second, bonding is considered to be genetic, a product of evolution, and fairly commonplace among higher animal species. Select one or the other of these points of view and offer supporting or opposing arguments. Can either assertion be tested? Can both be correct?

KEY WORDS AND IDEAS

SEX AND REPRODUCTION IN ANIMALS

The Aquatic Environment and External Fertilization

1. In **external fertilization,** eggs and sperm are shed into water, where fertilization and development usually occur. This method of reproduction is seen in aquatic species such as coelenterates and bony fishes. Complex mating and nesting behavior help assure successful fertilization, cutting down on the loss of gametes (for example, **amplexus,** clasping behavior with synchronous release of gametes in frogs, toads, and salamanders).

The Aquatic Environment and Internal Fertilization

1. **Internal fertilization** is almost universal on land and is also common in the aquatic realm.

2. Most internal fertilization occurs through **copulation,** usually, but not always, involving the introduction of some form of a penis into a vagina.

3. Both barnacles and planarians are hermaphroditic, with a mutual exchange of sperm occurring between partners.

4. Fertilization in fishes may be external or internal. All three developmental states are seen to some extent in the aquatic environment.
   a. **Oviparity** involves independent, external development of an embryo, usually in an egg case. Almost always follows external fertilization, but may also follow internal.
   b. **Ovoviviparity,** which almost always follows internal fertilization, refers to the retention of the embryo within the uterus, where little if any exchange goes on between mother and offspring.
   c. **Viviparity** follows internal fertilization and is the retention of the embryo in the uterus, where the mother's circulatory system provides nourishment, gas exchange, and waste removal.

5. Male sharks and rays have paired appendages called **claspers** that transfer sperm into the female cloaca. The embryos develop within **egg cases** following all three modes. A placentalike structure provides for viviparous development.

6. The number of offspring produced relates inversely to the mode of development and degree of maternal care and protection. The range is enormous with the available energy either invested in numbers or in care.

Reproduction in the Terrestrial Environment

1. Copulation and internal fertilization are common prerequisites to land-dwelling.

2. Insect species often produce great numbers of offspring with little if any care of the young, except in some social species. Insect genitalia are often complex and unique in some way to each species. Females often have a specialized **ovipositor** for boring into objects where the eggs are hidden. Eggs have a tough surrounding **chorion,** and a porelike **micropyle** that admits sperm stored in the **spermathecum.**

3. Because of the spider's solitary and predatory habits, females may attack and kill males during mating. Males often follow elaborate courtship behavior to escape this fate.

4. Reptiles and birds utilize internal fertilization and oviparity is common. Some reptiles are ovoviviparous.
   a. Male reptiles have copulatory organs. Lizards have two barbed **hemipenes,** either of which may be used. Except for ostriches and some ducks, the penis is absent in birds and sperm are exchanged through contact of cloacas.
   b. Egg production in usually greater in reptiles than in birds, and maternal care is much more common, although not exclusive, in species of the latter.
   c. Ground-nesting and foraging birds are usually **precocial,** hatching in a well-developed, self-feeding state. The young of other species are **altricial,** hatching in an incompletely developed, helpless, and often blind state.

5. Nearly all mammals are viviparous, and except for monotremes and marsupials, support the embryo via a well-developed placenta.
   a. Monotremes fertilize internally, but lay eggs in which the young develop oviparously.
   b. Marsupials fertilize internally, and the young begin development in the uterus, nourished by a **pseudoplacenta.** The young emerge semideveloped from the vagina and enter the pouch where they fasten to a nipple. Two or more young in different states of development can be supported.
   c. Among the placental mammals, precocial young include those of many hoofed species and all marine species (for example, whales and porpoises), while altricial species include rodents, carnivores, and primates.

HUMAN REPRODUCTION

The Male Reproductive System

1. Male genitalia include the **penis** and **testes,** which occupy the saclike **scrotum.**

2. Most of testes is made up of highly coiled **seminiferous tubules,** in which a **germinal epitheli-**

um produces sperm, and **interstitial cells** produce testosterone. Maturing sperm are swept by cilia to the **rete testes** and then to the **epididymis** for storage.

3. The path of sperm during ejaculation is from the epididymis to the **vas deferens,** which passes through the **spermatic cord.** Nutritive secretions are received from the **seminal vesicles** as the sperm enter the **urethra.** Next, the **prostate** adds alkaline secretions and fully formed semen is then forcefully ejected through the urethra.

4. In an **erection** the spongy spaces of the penis, the **corpora cavernosa** and **corpus spongiosum** (which includes the **glans**), become engorged with blood.

5. Sperm cells originate from undifferentiated **stem cells,** first becoming **primary spermatocytes,** which then enter meiosis I to form **secondary spermatocytes,** and after meiosis II, **spermatids.** The latter associate with supporting **Sertoli cells,** whereupon they differentiate into spermatozoa.

The Female Reproductive System

1. The external female genitalia, or **vulva,** include the **mons veneris,** a fatty mound overlying two folds or lips, the larger **labia majora** and smaller **labia minora.** The minor lips join to produce a hood overlying the **clitoris,** a sensitive, erectile organ tipped by the **glans clitoridis** (homologous to the penis). The labia minora also enclose the **urethral meatus** (opening of the urethra), and the **introitus** (opening of the vagina), the latter of which, in the virgin state, is often partially blocked by a membranous **hymen.**

2. The **vagina** is a distendable, muscular tube containing a squamous epithelial lining, a middle layer of smooth muscle and an outer layer of fibrous connective tissue. Its inner wall is a moist mucous membrane and additional moisture and lubrication is provided at the orifice by the **vestibular glands (Bartholin's glands).**

3. The internal female reproductive anatomy includes the **uterus, oviducts,** and **ovaries.**

4. The uterus opens in a muscular ring called the **cervix.** The uterine wall consists of an innermost **mucosa** which produces the vascular **endometrium,** a middle region of smooth muscle, and an outer fibrous connective tissue that merges with the **broad ligament,** a structure that supports the ovaries.

5. The oviducts emerge from the uterus to terminate in fingerlike, movable **fimbriae.** Its ciliated surface sweeps the oocytes into the oviduct.

6. The ovaries are supported by the dense **ovarian ligament.** Each ovary has an outer germinal epithelium that contains the oocytes, all of which formed before birth.

Hormonal Control of Reproduction

1. The hormonal control of reproduction involves an interaction between the hypothalamus, pituitary and gonads in both sexes, with the addition of placenta in females. Gonadotropin releasing hormones from the hypothalamus prompts the anterior pituitary to release the **gonadotropins,** follicle stimulating hormone (FSH) and luteinizing hormone (LH) (in males, **interstitial cell-stimulating hormone** or ICSH), which act on the ovaries or testes.

2. In males, FSH stimulates sperm production and works with ICSH to initiate **testosterone** production. Testosterone levels in the blood create a negative feedback loop back to the hypothalamus. Testosterone influences the development of maleness, the onset of puberty, and sex drive.

3. Puberty in females is associated with FSH secretion which prompts estrogen production and release by the ovaries. Estrogens influence development in the breasts, reproductive organs, and uterus, and general body growth. Androgens from the adrenal cortex influence the growth of pubic and axillary hair. The onset of fertility begins with the monthly **ovarian** or **menstrual cycle.**

4. In other female mammals, cycles of fertility, **estrous cycles,** occur less frequently, usually annually or semiannually.

5. The menstrual cycle includes the release of an ovum every 28 days, an event that corresponds to growth and thickening of the uterine endometrium, a preparation for the reception of an embryo. The cycle includes the following.
   a. *Days 1-14:* The **proliferative** or **preovulatory phase,** begins with the **menses or menstruation.** The pituitary secretes FSH which prompts **follicle** development and estrogen secretion. Follicle cells surround the oocyte, supporting its growth into a mature **Graffian follicle.** Under estrogen's influence, the endometrium enlarges and becomes vascularized.
   b. *Day 14:* Estrogen forms a negative feedback loop to the hypothalamus, pituitary FSH secretions slow, and LH increases ("LH surge"). Upon reaching the follicle, LH brings about ovulation—the release of the ovum.
   c. *Days 15-28:* In the **secretory** or **postovulatory phase,** LH and FSH maintain the **corpus luteum,** which secretes both estrogen and **progesterone,** further influencing endometrial growth. During the days following ovulation, the endometrium is ideally suited to receive the embryo. Terminating events in the cycle are still unclear, but it is suspected that ovarian hormones produce a second negative

feedback, suppressing LH and FSH secretion. The corpus luteum then fails, and without hormonal support, the endometrium breaks down.

d. The next cycle may be prompted by the lessening of FSH and LH suppression. But, if fertilization and implantation of an embryo has occurred, cells associated with the embryo secrete **human chorionic gonadotropin,** which supports the corpus luteum for the first two months. Later, progesterone and estrogen from the placenta maintain the uterus.

Contraception

1. Natural methods of **contraception** include **coitus interruptus** (withdrawal prior to ejaculation), **douching** (attempting to flush semen from the vagina), and the **rhythm method,** in which intercourse is avoided during fertile periods. None are very reliable. The variables in the rhythm method include difficulties in determining the fertile period, and the longevity of egg and sperm.

2. Mechanical methods include blocking sperm movement with rubber, sheathlike **condoms,** worn over the penis, rubber **diaphragms** inserted with or without **spermicide** over the cervix, and **intrauterine devices** or **IUDs.** The latter is a device placed in the uterus where it is thought to prevent implantation, but in ways not understood. Some IUDs contain contraceptive chemicals.

3. Chemical methods of birth control include spermicides and the pill. The pill contains synthetic estrogen and progesterone which are believed to suppress ovulation by overriding the natural rise and fall of fertility hormones. Side effects are noted in certain risk users, and heavy smoking increases the risks. However, in the general population, the risk of pregnancy is greater.

4. Surgical intervention includes the **vasectomy** (cutting and tying the vas deferens, and **tubal ligation** (cutting and tying—or just tying the oviducts).

5. The most controversial method of birth control is **abortion,** surgically removing an embryo or fetus or inducing its expulsion. Surgical methods in the early months of pregnancy include suction or vacuuming the uterus, and **dilation and curettage** (D&C), scraping the uterus. In later pregnancy, a saline (salt) solution is injected into the uterine cavity to induce labor and expulsion of the fetus.

REVIEW QUESTIONS

1. In what way does the number of eggs or offspring produced relate to the risk of survival? (926)

2. Define external fertilization and provide an example. (923–924)

3. How does the mating behavior of nesting fishes and of frogs and toads help assure successful fertilization? (924)

4. What advantages, if any, are seen in internal fertilization in the aquatic environment? (924)

5. Compare sexual reproduction in the barnacles and flatworms. What similarities exist? (925)

6. Define and provide examples of oviparity, ovoviviparity and viviparity. What provides the dividing line between the latter two? (925–926)

7. According to the survival principle described in the text, does any particular reproductive mode hold advantages of any other mode? Why are there so many diverse modes? Explain carefully. (926)

8. Why is it adaptive for insect species to have highly specific genital structure? (927)

9. Describe the typical mating behavior of spiders. Why is their courtship sometimes very elaborate? (927)

10. Cite two ways in which reptilian and avian reproduction are similar and two ways in which they are different. (928–929)

11. Explain the reproductive strategy of birds. In what aspects of reproduction is most of their reproductive energy invested? (929)

12. In which category of birds would you expect nesting behavior and nest structure to be most significant, altricial or precocial? Explain. (929)

13. Is there any parallel in mammals to the altrical and precocial modes of development in birds? Explain. (930)

14. Describe sexual reproduction in the kangaroo. In what way might their peculiar way of raising up young be adaptive? (930)

15. List the structures and events in the passage of sperm during ejaculation in humans, beginning with the epididymis. (931–932)

16. Describe the steps and name the stages in spermatogenesis in the human testis. (933)

17. In what two ways are the prostate gland secretions essential? (932)

18. In what way is the erectile tissue of the human penis analogous to the hydrostatic skeleton of some invertebrates? (932)

19. Describe the main anatomical structures of a spermatozoan. (933–934)

20. List the five structures that make up the human vulva. Which is unrelated to reproduction? (934)

21. In what way is the clitoris homologous to the penis? (934)

22. Describe the makeup of the vagina starting with the outermost layer of tissue. (933)

23. Discuss the manner in which the human ovum finds its way into the oviduct, listing the structures involved. (935)

24. List three structures that play a part in the hormonal control of reproduction, starting with one located in the brain. (939)

25. Explain how negative feedback works in controlling testosterone production in males. (939)

26. In what three aspects of life is testosterone important to males? (939)

27. What two events does the human menstrual cycle closely coordinate? Why is such coordination needed? (940)

28. Describe events in the ovary, hypothalamus, uterus and anterior pituitary, through the first thirteen days of the human menstrual cycle. Include any negative feedback loops that may form. (940)

29. Explain how shifting hormones bring on the process of ovulation. (942)

30. Describe the output of hormones that maintains the uterus through the first part of the postovulatory phase. What events end the cycle? (942)

31. Once fertilization and implantation have occurred, what is the source of hormonal support during the early and later stages of pregnancy? (943)

32. Describe the content of one type of birth control pill and suggest how it prevents conception. (946)

33. List several reasons why the rhythm system of birth control is subject to failure. (944–946)

34. List five methods of birth control that are unacceptable from the standpoint of failure. (944–945)

35. Discuss the risks of oral contraceptives, but put them in the perspective of normal risks. (947)

36. List an advantage and a disadvantage of vasectomy and tubal ligation. (947)

37. Describe the three methods of clinical abortion. Which involves the greatest risk? (948)

Development

One cell divides into two, and two into four. The four again divide and the divisions continue until eventually there is a great number of cells. All the cells have stemmed from that one primordial cell, but yet, many divisions later, far down the line, descendant cells may come to be very different from one another. One descendant may become an elongated nerve cell, while another becomes a pigmented cell within the eye's retina, and yet another an unprepossessing liver cell with an impressive battery of biochemical tricks. But since they arise by mitotic divisions from a single fertilized egg, they all (with certain notable exceptions, such as lymphocytes) carry the same genetic information in their chromosomes.

How does this seeming miracle of development and differentiation occur? This, it turns out, is not a rhetorical question; we don't quite know. It seems that there should be some ready answer, some satisfying explanation.

Part of the problem is that the study of development has two aspects: the descriptive and the experimental—the familiar "what" and "how." The descriptive work—the story of "what" actually can be seen in a developing embryo—is already rather complete. There are serious, unanswered questions in the "how it happened" department, however.

In this chapter we will discuss cell differentia-

tion in the context of development. We will also focus on the equally intriguing organized movements of cells and tissues; as well as the control of cleavage; the orientation and interaction of cells and tissues; programmed cell deaths; and, of course, the emergence of morphological shape and form, what is known as **morphogenesis.** In order to get at such questions, however, scientists have had to go beyond descriptive biology. They have experimented. We will review some of the most notable and fascinating of the embryological experiments.

But let's begin at the beginning—with the gametes and fertilization.

GAMETES AND FERTILIZATION

The Sperm

The sperm is one of the most highly specialized of cells, and its role is straightforward—it must encounter and penetrate an egg. It thereby delivers its half of the offspring's required chromosome complement. The sperm's entire structure is devoted to this one brief mission.

The three major parts of the sperm are the head, midpiece, and tail (Figure 37.1). The sperm head contains the haploid chromosome complement in

37.1 HUMAN SPERMATOZOAN

In this detailed reconstruction, the human spermatozoan is revealed as an efficient means of delivering a haploid nucleus to a receptive ovum.

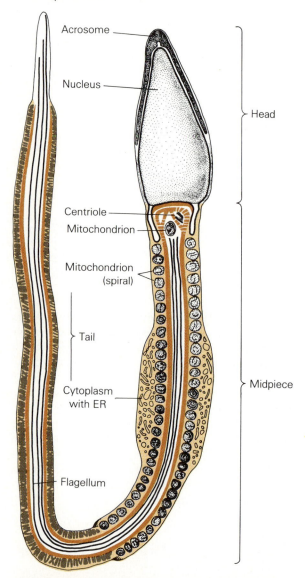

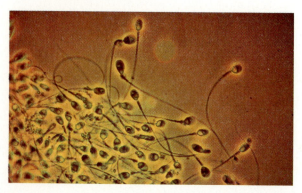

the form of highly condensed chromatin. At its tip is found the **acrosome,** a caplike structure that surrounds an enzyme-laden lysosome. If you recall the general nature of lysosomes, you can anticipate its role in penetrating the egg, but we will get to that shortly. The sperm midpiece includes an unusually long, spiralling mitochondrion, which provides the ATP energy needed by the lashing flagellum that propels the sperm along. As with other flagella, it has the familiar 9 + 2 microtubular construction. The midpiece also contains other typical organelles in its scant cytoplasm, including a centriole. Upon fertilization, the sperm centriole will join up with the egg centriole, producing the usual paired condition.

The Egg

Animal egg cells are somewhat more difficult to generalize about because they differ so greatly among major animal groups. Whereas a human egg is about the size of a period, you would be hard put to hide an ostrich egg behind this entire book. We can generalize a bit, however, by saying that each egg cell at the time of fertilization consists of a haploid nucleus surrounded by a region of cytoplasm containing yolky substances and all the usual cellular components.

But an egg is more than just a cell. In nearly all instances, animal eggs have developed some degree of **polarity.** That is, the egg constituents are not homogeneous, but are distributed in such a manner as to give the egg some form of organization. Typically there is a more metabolically active **animal pole** and a less active, often yolky, **vegetal pole.** Because of this kind of organization, the planes in which cell divisions occur can have a profound effect on the cytoplasmic content of daughter cells, and this in turn can affect the course of development.

Vertebrate eggs can also be characterized on the basis of their yolk quantity and distribution. The terms **macrolecithal, mesolecithal,** and **microlecithal** refer respectively to large, medium, and small amounts of yolk (*lecithal,* referring to egg yolk). The yolk quantity, we will find, also influences early development. Birds and reptiles boast the greatest relative amounts of yolk, but we shouldn't be fooled by the size of macrolecithal eggs. The **blastodisc,** the cytoplasmic region destined to become the embryo, is still relatively small.

The amount of yolk is related to the specific mode of nourishment of the embryo. The large quantity of yolk in the bird egg is necessary because development is completed within the confines of the shell, so the independent embryo lives within its own life-support system. The mesolecithal egg of amphibians has a fairly sizable yolk mass, but less than those of birds, since they need only enough nutrient to get the embryo to the larval stage, when it begins feeding itself. Placental mammals, such as humans, produce microlecithal eggs, whose comparatively minute amount of yolk contains only enough nutrients to support the embryo for the brief period required for it to implant in the nutrient-rich endometrium of the uterine wall. The eggs of many marine invertebrates, such as echinoderms and lancelets, are also small, since they proceed very rapidly into a tiny but self-sustaining larval stage.

The outer coverings of eggs—that is, the surrounding membranes and supplemental parts—may also vary in different species. Birds' eggs have the largest number of membranes, including a fibrous **vitelline membrane** surrounding the cytoplasm, followed by the dense "egg white," a watery protein suspension, two keratin-rich shell membranes, and finally the porous, calcium-hardened shell itself. Vitelline membranes are also found on the eggs of insects, mollusks, amphibians, and birds. Just below the vitelline membrane is the egg proper and a region of granules and microvilli. These features are unique to egg cells and each plays an important role in fertilization. Observa-

tions of the echinoderm egg reveal an interesting array of supplemental parts, such as a dense transparent substance, the **jelly coat,** surrounding the egg and vitelline membrane.

Mammalian egg cells, or, more properly, oocytes, are about the same size as echinoderm eggs, but rather than being enclosed in a jelly layer, they emerge from the ovary surrounded by a dense layer of follicle cells known collectively as the **corona radiata** ("radiating crown"). The corona cells contain interlocking borders, which, along with jellylike **hyaluronic acid** holds things together. Below the corona radiata lies a thick, glassy membrane, the **zona pellucida** (clear zone), which is secreted by the **follicle cells.** Below this is the **plasmalemma**—the mammalian oocyte's equivalent of a plasma membrane (Figure 37.2).

Fertilization

Our knowledge of human fertilization is far from complete, although there has been a new wave of interest associated with increasing successes with "in vitro" fertilization and implantation (producing the so-called test tube babies). In fact, we know much more about fertilization in echinoderms. Their gametes are readily available and very easy to observe. Let's follow the events of fertilization in

---

37.2 THE HUMAN OOCYTE

In humans the oocyte is surrounded by a large number of smaller follicle cells, the corona radiata, which emerges with the egg at ovulation. The cells are interconnected by cellular processes, and the mass is, in turn, connected to the egg by numerous microvilli. Just within the follicle cells is a region known as the zona pellucida, which is penetrated by the many microvilli. Following fertilization, the corona is shed and the entire surface changes significantly.

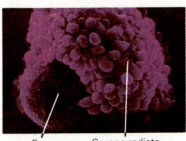

Egg   Corona radiata

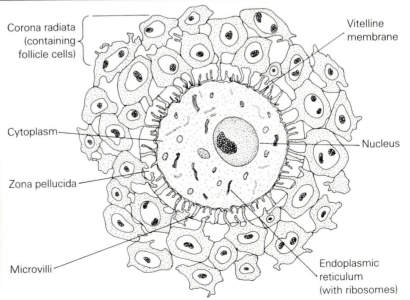

Corona radiata (containing follicle cells)

Cytoplasm

Zona pellucida

Microvilli

Vitelline membrane

Nucleus

Endoplasmic reticulum (with ribosomes)

37.3 FERTILIZATION IN A SEA URCHIN

The process of fertilization is easily observed in the sea urchin. **(a)** Typically, the jelly coat will contain many sperm but only one will penetrate the plasma membrane. When this hap- pens **(b)** a fertilization membrane will begin to form, and **(c)** upon its completion it will entirely surround the egg.

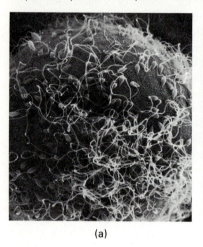
(a)

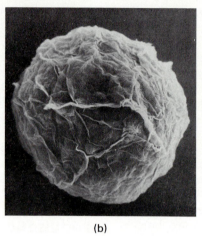
(b)

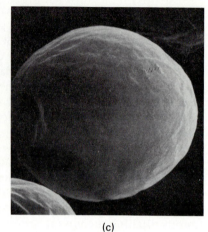
(c)

the sea urchin, keeping in mind that the details are not entirely representative of other species (Figure 37.3).

In sea urchins, as soon as a sperm cell contacts the jelly coat surrounding the egg, its acrosome releases enzymes that digest a path through the dense coat to the egg surface. During its passage, the acrosome extrudes a lengthy **acrosomal filament,** which penetrates the vitelline layer of the egg and attaches to the plasma membrane below. Investigators have discovered the presence of receptor proteins associated with the vitelline layer that identify with the acrosome. The match between acrosomal protein and egg receptors is believed to be quite specific and important in preventing interspecific fertilization (although it sometimes happens anyway). So far, it seems that the sperm is the only active participant, but this is far from true. Upon sperm attachment, the egg undergoes sweeping changes that both assist entry by one spermatozoan and act quickly to prevent penetration by others. Should such **polyspermy** occur, the embryo is almost always doomed to failure.

The first reaction is a sudden voltage shift across the plasma membrane, brought about by an influx of sodium ions. This is a first safeguard against polyspermy since it renders the plasma membrane impenetrable to sperm. As added insurance, a less instantaneous, chemical event, the **cortical reaction,** soon follows. Upon sperm attachment, **cortical granules** located just below the egg surface rupture, releasing their contents into the **perivitelline space** (around the vitelline membrane). The cortical reaction brings on the formation of a **fertilization membrane,** which starts as a small "blister" at the point of sperm attachment and enlarges in a rapidly expanding front around the entire egg—a minor tidal wave of chemical and physical activity. Any sperm that had managed to penetrate the vitelline membrane are detached at this time by the cortical reaction.

Meanwhile, a singularly amazing process occurs. The head of the successful sperm is actively engulfed and drawn into a mound of egg cytoplasm called the **fertilization cone.** The fertilization cone includes numerous fingerlike **microvilli,** which wrap themselves tightly about the sperm head, drawing it inward. Shortly afterward, the sperm head is detached from the midpiece and the sperm nuclear membrane breaks down, releasing its chromatin, the haploid male **pronucleus,** into the egg cytoplasm. There it will soon fuse with the haploid female pronucleus, completing fertilization (Figure 37.4).

THE EARLY DEVELOPMENTAL EVENTS

Embryogenesis, the "beginning of an embryo," refers to the early events following fertilization, where the embryo goes through a period of intense cell division, followed by a startling displacement and rearrangement of cells that creates the **triplo-blastic** (three-layered) condition. The events of embryogenesis, while varying considerably in detail, follow similar general patterns in many invertebrate and vertebrate animals.

The First Cleavages

Typically, the zygote will cleave into two equal halves in a manner that provides the newly emerging cells, or **blastomeres,** as they are known, with equal cytoplasmic (animal and vegetal) content. Where the cleavages completely divide the embryo the term **holoblastic cleavage** is used. The second cleavages usually occur more rapidly than the first. In sea urchins and frogs, the second cleavage plane follows the plane of the first, vertically dividing the zygote into four equal blastomeres, again, each with equal portions of the cytoplasm. The products thus take the form of four "orange-segment" blastomeres (Figure 37.5). The first cleavage in mammals also produces two equal blastomeres, but the second cleavage cuts across the animal-vegetal axis, dividing the cytoplasm unequally. Thus, in mammals, polarity occurs very early.

As cell division continues, the embryo remains

37.4 DETAILS OF SEA URCHIN FERTILIZATION

(a) After passing through the jelly coat, the sperm contacts a receptor site on the vitelline membrane, through which it passes, and binds to the plasma membrane. (b) Upon contact, the membranes rise up to encompass the sperm head in a fertilization cone. (c) The egg membranes are far from passive as numerous microvilli also rise up and (d) ensnare the sperm. (e) Simultaneously, cortical granules within the cyto-plasm rupture, creating a series of chemical events that (f) lead to the formation of a *fertilization membrane,* which will act as a barrier to other sperm. (g) The fertilization membrane rises up from the egg surface at the point of sperm penetration, and (h) moves in a wavelike front around the egg. Under ideal conditions, the egg will begin its first cleavage about an hour later. (i) TEM of sea urchin sperm entering egg.

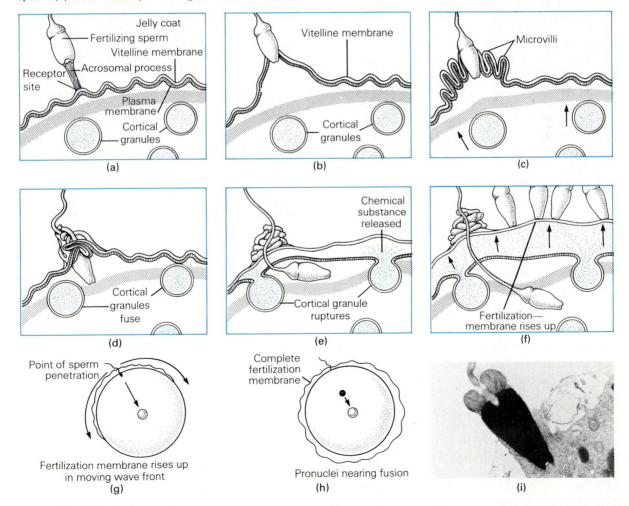

37.5 EARLY CLEAVAGES

In both **(a)** the sea urchin and **(b)** the frog, the first and second cleavages follow the plane of the animal-vegetal axis so that the first four cells are nearly identical in yolk content. The third cleavage, which produces the eight-celled stage, however, is perpendicular to that plane, so an unequal division of yolk and egg cytoplasmic elements occurs. In the sea urchin, the third and subsequent cleavages are consistently equal in size, or nearly so, through the blastula stage. This isn't true of the frog's embryo, where the third cleavage produces *micro-* *meres* (small cells) in the upper region and *macromeres* (large cells) in the lower region. This unequal cell division continues into the blastula stage, with a mass of small rapidly dividing cells clearly marking the amphibian animal pole and larger sluggishly dividing cells, the vegetal pole. The same pattern in the frog will continue as the embryo's form begins to emerge. The highly active animal pole represents the region of the egg that will produce most of the embryo's structure.

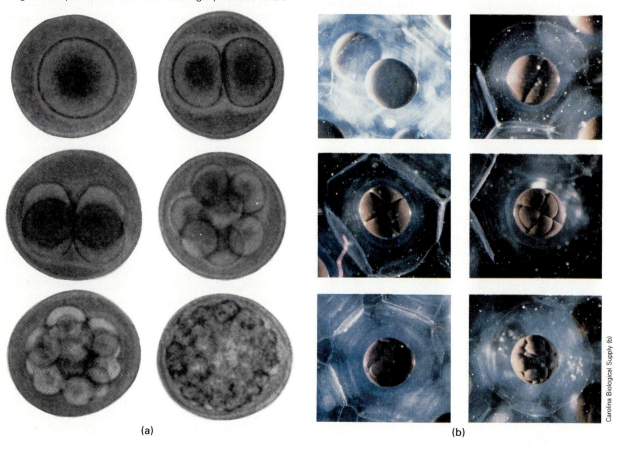

(a)   (b)

at about its original size, but the cells being produced get smaller and smaller with each division. In echinoderm embryos, each generation of blastomeres are of about equal size, and these then rearrange themselves until the embryo takes on the appearance of a hollow ball composed of small rounded cells. As we will see, this is the **blastula** stage, and the cavity within is the **blastocoel.** But in amphibian embryos with their greater yolk, the early symmetry of cleavage is soon lost. As seen in Figure 37.5, during the third cleavage, the furrow takes shape transversely (cutting across the other two) closer to one end, the more active animal pole. This produces four **micromeres,** smaller cells, at one end and signals the start of a trend that will soon produce a rapidly dividing animal pole and a slower dividing vegetal pole. As in the egg organization, vegetal pole cells, or **macromeres,** tend to be larger, more yolky, and metabolically sluggish.

Early Determination in the Frog.   After fertilization, cytoplasmic components of the frog egg become strategically rearranged, resulting in important implications for the future. At the opposite side of the egg from the point of sperm penetration, changes in the usual dark pigmentation result in the formation of the **gray crescent,** which looks exactly as it sounds. But more to the point, the area of the gray crescent is a **morphogenic determinant,** that is, its presence will influence future development. For instance, the crescent is bisected by the first cleavage, and the embryo immediately becomes bilateral—one blastomere is destined to form the embryo's left side and the other blastomere, the right side. The gray crescent also marks the site of a future ingrowth, the **blastopore** (introduced in Chapter 26). In a more general way, the gray crescent area also determines the future dorsal side of the embryo.

Cleavages in the Bird and Reptile, a Special Case.

Bird and reptile eggs, with their greater yolk reserves, undergo **meroblastic** (partial) **cleavage,** in which the cytoplasm is not completely divided. Actually, their yolky area doesn't divide at all; instead, cytoplasmic divisions are restricted to the blastodisc (Figure 37.6), which contains the nucleus and most of the cell's nonyolky cytoplasm. Unequal cleavages of the first four cells produce a central cluster, with the most rapid rate of cell division occurring toward the center of the disc. Eventually, the cells pinch off and become surrounded by membranes. All of this occurs in the blastodisc, which at this time has become a multicellular layer called the **blastoderm.** Before you go poking at your breakfast, remember that eggs sold for consumption are usually unfertilized.

The Blastula and Gastrulation

The Amphibian Blastula.

While the echinoderm blastula is a simple ball of cells surrounding a central blastocoel, that of the frog, as seen in Figure 37.7a, is asymmetrical with its blastocoel displaced toward the animal pole. Above is a thin rooflike region of small cells (animal pole), and a very dense floorlike region of larger, yolky cells (vegetal pole). Further, by the blastula stage, regions that will soon form the three germ cell layers are already identifiable. They are now referred to as **presumptive ectoderm, presumptive mesoderm,** and **presumptive endoderm,** with the term "presumptive" dropped once the layers take their final position in the upcoming events. Embryologists learned of this seemingly early commitment of cells through clever experiments in which harmless dyes were applied to the blastula surface, and the colored cells then tracked through their development. The three regions on the blastula's surface are graphically represented in the **fate map** seen in Figure 37.7b.

Gastrulation.

As the blastula stage in the amphibian progresses, the process of **gastrulation**

begins. Through a remarkable series of cell movements, and shape changes, the hollow blastula becomes a **gastrula,** a complex triploblastic embryo with its germ cell layers rearranged and poised to begin morphogenesis. Gastrulation involves several kinds of cell movement, including **invagination, involution,** and **epiboly.**

Gastrulation begins with invagination, an inward sinking of external endodermal cells. The first sign of impending events is a slight curved indentation at the site of the gray crescent, the beginning of a **blastopore.** The first invagination is the **dorsal lip** of the blastopore, which soon enlarges to form **lateral lips,** and, as the sinking process continues, the circular blastopore is completed as the **ventral lip** forms. At its center, a

37.6 CLEAVAGES IN THE CHICK EMBRYO

Because of the enormous quantity of yolk, cell divisions are restricted to a small patch of material at one side of the yolk—the blastodisc. Note that even here the first two cleavages are not complete. Cell division occurs more rapidly near the center of the blastodisc. In the bird embryo, there is no equivalent to the blastula stage of sea urchins and frogs, although the lower layer of cells will separate from the rest of the embryo during gastrulation.

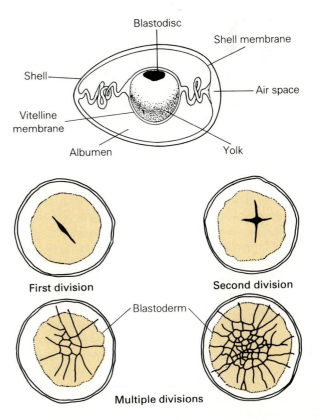

First division

Second division

Blastoderm

Multiple divisions

mass of endodermal cells forms the **yolk plug.** A close look at the site of invagination reveals that the presumptive endodermal cells responsible have taken on a teardrop or bottle shape, and, in fact, are referred to as **bottle cells.** As the blunt ends of these cells sink inward they draw tightly affixed cells in after them, forming a rolling, involuting front (Figure 37.8).

Involution, the continued inward movement of cells around the blastopore, produces a new cavity, the **archenteron** (primitive gut), that gradually replaces the old blastocoel. In addition, cells moving in to make a U-turn at the dorsal lip include underlying presumptive mesoderm, which come to occupy a "middle" position between outer and inner cell layers. As we will find later, the specific location and action of this presumptive mesoderm will have a profound effect on future development.

During all of this, ectodermal cells at the animal pole begin epiboly, a flattening and spreading process that results in an expansion in all directions around the embryo. Thus at the close of gastrulation the gastrula will contain, roughly, an outer ectoderm, middle mesoderm, and inner endoderm, along with a newly formed archenteron, the primitive gut (from which much of the gut will actually be derived).

Gastrulation in Birds.   The most obvious differences between the early embryos of birds and the early amphibian embryos lie in the cleavage patterns, and these, we've seen, are related to their yolk abundance. There is no blastula in the embryos of birds, but following the formation of the blastoderm, a flat disk of cells has formed which includes an upper layer of cells, the **epiblast,** and a lower layer, the **hypoblast.** The important layer is the epiblast, which, during gastrulation, will give rise to the three germ cell layers (Figure 37.9a).

Gastrulation in the bird embryo first becomes apparent as the cells of the epiblast begin to converge toward the midline of the blastoderm to produce a thickened region, extending toward the center of the disk. This line establishes the embryo's longitudinal axis, one that will later be represented by the dorsal nerve cord, the notochord, and eventually the vertebral column. As the cells converge from both sides of the line, they sink into the embryo, forming an elongated depression called the **primitive groove.** The groove and the two ridges bordering it are referred to as the **primitive streak** (Figure 37.9b). Upon turning under on each side of the groove, some of the migrating cells become organized into a middle mesodermal layer, while others displace or join the hypoblast to form the endoderm. The remaining cells of the epiblast,

37.7 FROG BLASTULA

(a) The frog blastula is a sphere of cells containing a cavity called the blastocoel. The blastocoel is offset towards the animal pole. This pole contains smaller, more active cells, while the vegetal pole is composed of large yolky cells. **(b)** An amphibian fate map, prepared by following dyed cells during development, reveals the presumptive ectoderm, mesoderm and endoderm.

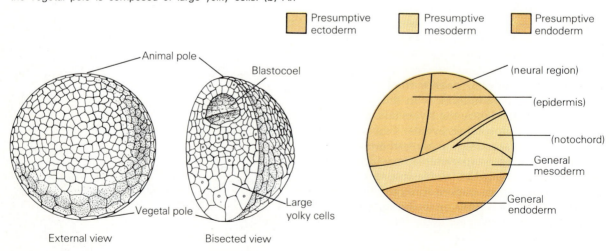

37.8 GASTRULATION IN THE FROG

(a) Gastrulation in the frog embryo begins with invagination, as cells from the overlying presumptive endoderm sink inward forming a crease that marks the dorsal lip of the blastopore. The inset shows the formation of bottle cells as invagination proceeds. (b) As gastrulation continues, the inward moving endoderm is joined by deeper layers of mesoderm which also involute, making a U-turn to relocate within the gastrula. The lateral lips of the blastopore form as the invagination spreads. Note the enlarging archenteron (primitive gut). (c) In a later stage, the involuting layers have extended back below the thin roof of the enlarging archenteron. On the surface, the ectodermal cells have also been active. They undergo epiboly, a flattening and spreading of the tissue over the gastrula's surface. (d) At the end of gastrulation, the massive migration has produced a triploblastic embryo, essentially consisting of an outer ectoderm, middle mesoderm and an inner endoderm. The completed blastopore contains a plug of yolk cells. The roof of the archenteron contains the layers that will form the embryo's axis.

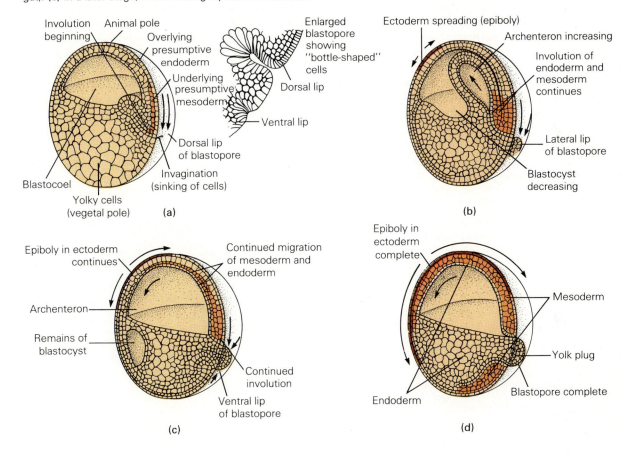

37.9 GASTRULATION IN THE BIRD

(a) Following a period of cleavage, the blastodisc of the bird embryo consists of an upper epiblast and lower hypoblast. (b) The embryo then forms the primitive streak through the merging of cells along a midline. Gastrulation begins as cells along the streak sink inward forming the primitive groove. Cells entering the cavity contribute to a newly forming endoderm and mesoderm. As in the frog gastrula, the inward movement creates a triploblastic embryo and forms the embryonic axis.

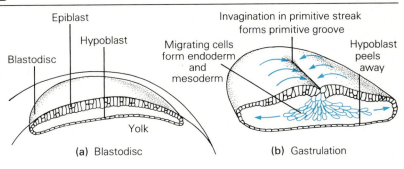

of course, constitute the outer ectoderm, the third layer. So, activity at the primitive streak is somewhat homologous to the invagination of cells around the blastopore in the amphibian embryo.

Gastrulation in Mammals. Mammals have evolved along a different line, but nevertheless, some of the early events are clearly reminiscent of events in the bird embryo. As we've seen, the mammalian egg has a meager yolk supply. Except for the first few days of development, the mammalian embryo will get nearly all its nourishment through maternal circulation.

In mammals, the union of sperm and egg occurs in the upper oviduct, following which the zygote begins its transit to the uterus, a journey that will require about six days (Figure 37.10). During this period the zygote undergoes active cell division, at first forming a **morula,** a solid sphere of cells. By the time the embryo reaches the uterus and begins implantation, it will be composed of about 32 cells and at this stage is called a **blastocyst.** The blastocyst is composed of two parts, a hollow sphere only one cell in thickness called a **trophoblast,** and a mass of cells extending inward, known as the **inner cell mass.** The fluid-filled space within the sphere is referred to simply as the **blastocyst cavity.** The inner cell mass will form the embryo as well as extraembryonic structures, such as the amnion and yolk sac, structures that will sustain the embryo. The trophoblast forms the early tissues that attach the blastocyst to the uterine wall during implantation and later will form most of the placenta. Upon implantation the trophoblast differentiates into an inner **syncytiotrophoblast,** which produces fingerlike growths that invade the

37.10 EARLY HUMAN EMBRYO

(a) After ovulation, the human oocyte is drawn into the oviduct, where fertilization can occur. From the time of fertilization, about six days are required for the transit to the receptive uterus. The first cleavage occurs about 36 hours after fertilization, but later divisions occur much faster. When the embryo enters the uterus, it will be a morula **(b).** This solid ball of cells will hollow out into a blastocyst, as seen in the photo.

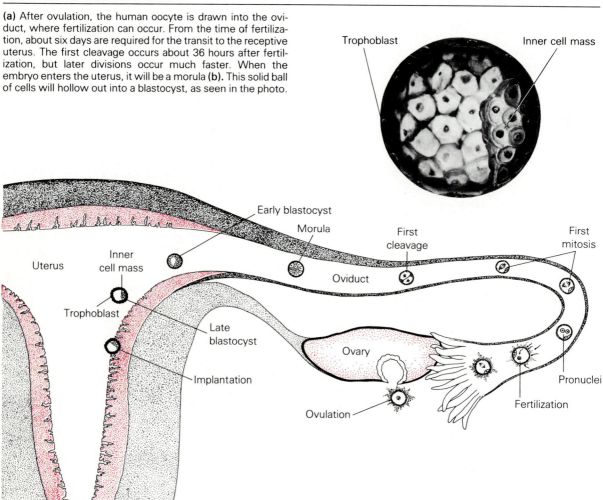

37.11 GASTRULATION IN THE MAMMAL

Gastrulation occurs after the blastocyst implants in the uterine lining, and supporting extraembryonic structures have begun their development. One part of the trophoblast forms the syncytiotrophoblast, whose growth penetrates the endometrium, anchoring the blastocyst and absorbing nutrients. The amniotic cavity forms as the cytotrophoblast separates from the inner cell mass. Next, cells from the cytotrophoblast form two membranes, the amnion about the amniotic cavity, and the yolk sac, below the embryo. It is then that the convergence of cells in the blastodisc forms the primitive streak and gastrulation finally begins as cells along the streak sink inward along its length. The separation of the lower endoderm, and the inward movement and formation of a mesoderm, create the triploblastic embryo.

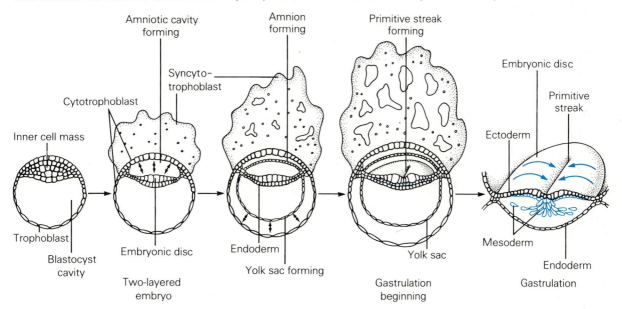

endometrium, and an outer **cytotrophoblast** surrounding the blastocyst.

As implantation proceeds, changes also occur in the inner cell mass, which forms the **bilaminar embryonic disc,** containing two distinct layers of cells, an epiblast and a hypoblast, just as we saw in the bird (Figure 37.11). In addition, two other structures emerge, the amnion above the epiblast (described later), and the yolk sac, below the hypoblast. In birds and reptiles, the yolk sac grows over the surface of the yolk and actively participates in the formation of blood cells and in the transport of yolk nutrients into the embryo. The mammalian yolk sac, on the other hand, is hollow and never contains yolk, but it remains as the site of the earliest blood cell formation.

Mammalian gastrulation is nearly identical to that of the bird and reptile. With the convergence of cells along its midline, the bilaminar disk enters the primitive streak stage. Then, with the sinking of epiblast cells into the primitive groove, cells contribute to the formation of mesoderm and endoderm layers, and a trilaminar embryo emerges (Figure 37.11). Later, we will return to the mammals with an in-depth look at the human embryo.

The Fate of Germ Layers. We have called the three layers of cells formed at gastrulation the "germ layers." Since *germ* refers to the beginnings, the term implies that these layers will give rise to new parts. Keep in mind, though, that the cells in the gastrula are not yet rigidly predisposed. In other words, they will retain a certain versatility for a while longer. The usual role of each layer, however, is summarized in Figure 37.12.

NEURULATION AND THE EMERGENCE OF FORM

With the conclusion of gastrulation, we begin to see the development of various organs and organ systems. The first system to develop in vertebrates is the central nervous system, which forms through a process called **neurulation.** In essence, the vertebrate embryo produces a dorsal, hollow tube extending the length of the developing body. That tube will become the central nervous system, the brain and spinal cord. Using Figure 37.13 as a guide, let's see how this vital system begins its formation in the amphibian.

Neurulation in the Amphibian

In the frog, the beginning of development of the central nervous system is marked by the formation of the **neural plate.** This is a thickened area of cells that extends from the dorsal lip of the blastopore, over the embryo along a line anticipating the spinal column, toward what will be the head. Soon, a depression called the **neural groove** forms along the length of the plate, producing a sort of valley with ridges on each side. The ectodermal ridges on either side of the valley are called **neural folds.** The "northern end" of the valley, the future head re-

gion, is somewhat wider than the rest. The "southern end" is marked by the blastopore. The neural ridges continue to thicken and to rise up, and gradually turn toward each other, finally converging at the midline. As the edges of the folds fuse together, they form the hollow **neural tube.** Since the head region closes last, it is much wider than the rest of the neural tube, and forms a bulbous structure containing the outlines of the brain and eyes.

Meanwhile, below the neural ectoderm, the mesoderm begins to form the long, central, rodlike supporting notochord. (Incidentally, the notochord is so special that some authorities consider it to comprise a unique, fourth germ layer.) All other chordates produce a notochord and rely on it for support in at least some stage of development.

As the nervous system and the notochord con-

37.12 THE THREE GERM LAYERS

As the diagram indicates, endoderm produces the linings of many internal structures. Mesoderm contributes to the formation of muscles, skeleton, and blood vessels. Ectoderm produces many of the outer structures and is very active in the earliest phases of development, as we saw in gastrulation and will see in the neurulation process.

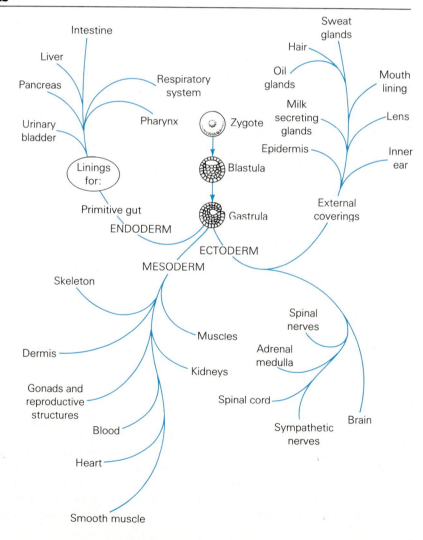

37.13 NEURULATION IN THE FROG EMBRYO

(a) The first indication of neurulation is the emergence of an elongated thickened plate in the region of the animal pole of the gastrula. The cells of this plate will form the primitive brain and spinal cord. (b) Next, ridges of cells rise up along either side of this plate to form the neural folds. (c) The neural folds continue to rise until the two sides lie prominently on either side of the neural groove. (d) As the process continues, the neural folds bend toward each other and then come together, first at the center, then at what will be the tail, and (e) finally at the future brain. Thus, a hollow tube is formed which will be the dorsal hollow nerve cord—a structure present in all vertebrates. The notochord will continue to lend some support, but will fade away as nearby islands of mesoderm begin to contribute to bones of the vertebral column.

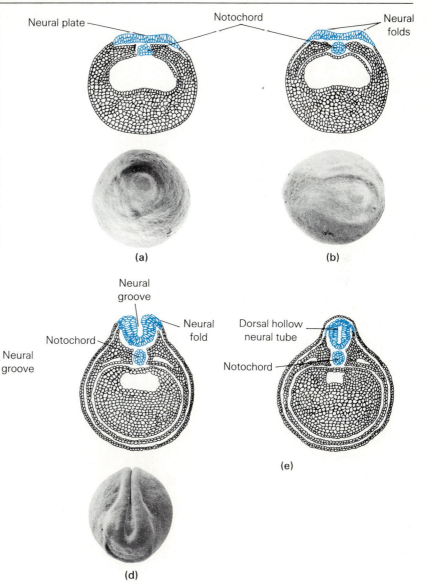

tinue their development, the recognizable features of the embryo gradually emerge. Through subtle changes, groups of cells in the anterior region of the neural tube begin to produce the first brain tissue. Shortly, the eyes will begin to form as cuplike extensions of the brain (Figure 37.14). Further back along the embryonic axis, a series of mesodermal clusters on either side of the notochord form the **somites.** These paired blocks of tissue are the forerunners of the body muscles and the axial skeleton. Their appearance reminds us again of the segmental plan common to so many animals.

These, then, are some of the basic events surrounding early vertebrate development. However, there are many supportive, peripheral or less obvious events that are also important to proper development. For instance, most vertebrate embryos produce an elaborate supporting system that aids continuing development. Some of the membranes involved in such support, as we learned earlier (see Chapter 27) are the **extraembryonic membranes.** We will briefly review their roles in the birds and reptiles and in the mammal, and then move on to some aspects of our own development.

37.14 LATER DEVELOPMENT IN THE FROG EMBRYO

Following neurulation **(a, b, c)**, the outlines of the frog embryo become more distinct. The bulbous brain and eye region, along with the gill pouches, are visible quite early **(d, e)**. As the body elongates **(f, g)** and the tail becomes prominent, the gills emerge and the head takes shape. Later, **(h, i)**, most of the yolky reserves have been absorbed, the eyes are well formed, and the larva (or tadpole) has reached its feeding stage.

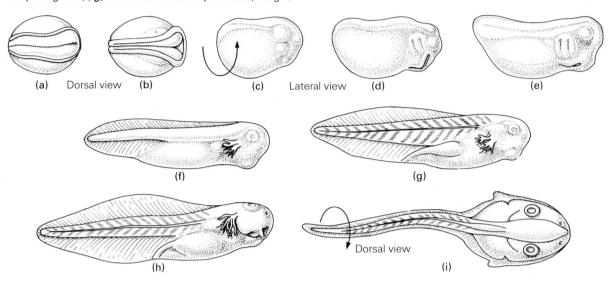

(a)   Dorsal view   (b)   (c)   Lateral view   (d)   (e)

(f)   (g)

(h)   (i)   Dorsal view

37.15 EXTRAEMBRYONIC MEMBRANES OF THE BIRD EMBRYO

(a) The amnion and part of the yolk sac are seen in an intact five-day-old chicken embryo (photo). **(b)** Following neurulation, a fold of mesoderm produces the amniotic folds which merge together over the embryo to complete the amnion. The same membrane continues outward forming the chorion, while below, a third membrane ventures out to become the yolk sac. **(c)** In a different view, the yolk sac is seen as a vascularized membrane spreading out over the yolk, from which it will bring in nutrients. The allantois begins as a simple sac that follows the path of the chorion as it enlarges. It, too, will carry blood vessels, since its function is gas exchange and nitrogen waste storage. The surrounding chorion is a simple sac that envelops the entire affair. The amnion provides a fluid medium surrounding the delicate embryo. **(d)** As development continues, the yolk sac covers the diminishing yolk; the allantois has grown along the route of the chorion, fusing in one area to form the chorioallantois.

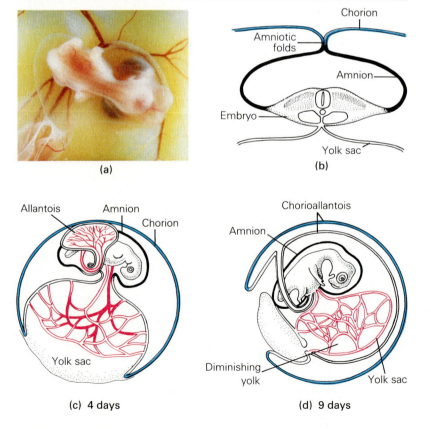

(a)

Chorion
Amniotic folds
Amnion
Embryo
Yolk sac

(b)

Allantois   Amnion   Chorion
Yolk sac

(c) 4 days

Chorioallantois
Amnion
Diminishing yolk   Yolk sac

(d) 9 days

Supporting Structures of Vertebrate Embryos

The Self-Contained Egg. Reptile and bird eggs consist essentially of four parts. The **shell** itself is composed of either a soft leathery covering (reptiles) or a hardened calcareous material (birds). A heavy **yolk mass** constitutes most of the stored food. The fluid **albumen** (egg white) acts primarily as a water supply, but also contains stored protein and some effective antibacterial agents. Finally, there is the **embryo,** which will produce four supporting, extraembryonic membranes as it develops—the **yolk sac,** the **chorion,** the **allantois,** and the **amnion** (Figure 37.15).

The yolk sac of birds and reptiles is the first extraembryonic membrane to develop. It is a product of endodermal growth and is actually an extension of the primitive gut. As it expands over the surface of the yolk, it becomes suffused with an extensive system of blood vessels. Food substances, absorbed into the blood as it circulates over the yolk's surface, are thus carried into the embryo.

The amnion and the chorion develop simultaneously, growing up over the embryo shortly after neurulation. The amnion forms a fluid-filled, protective sac that acts as a shock absorber and also keeps the body lubricated so that the growing parts don't fuse together. The chorion continues to grow around the entire egg contents, forming a continuous membrane just under the shell and later fusing with the allantois to form the **chorioallantois.**

The allantois, essentially endodermal in origin, begins as a peculiar little pouch but as it grows, it spreads out into a full-fledged membrane, coming to lie against the chorion, thus setting the stage for fusion with it. The network of blood vessels permeating the fused chorioallantois exchanges gases and functions as a waste receptacle.

With the evolution of the land egg, the vertebrate embryo was able to retain its watery ancestral environment. But this independent egg, with its supply of food, water, and extraembryonic membranes, was only one answer to the demands of the terrestrial environment.

Life Support of the Mammalian Embryo. The structures that support the mammalian embryo with food, water, and waste removal are collectively called the **placenta**—an entity clearly associated with the mammals.

In describing the development of the human placenta, we will return to about the sixth or seventh day after fertilization—the time when the blastocyst settles into the soft lining of the uterus (Figure 37.16). Fingerlike extensions of the innermost trophoblast, the syncytiotrophoblast, secrete enzymes that enable the embryo to digest its way into the endometrium, where it finally comes to rest in a pool of nutrient-rich blood. For a brief time, this blood will suffice to supply food and oxygen and carry off the meager wastes of the tiny embryonic mass.

Soon, the syncytiotrophoblast, joined by cells from the embryonic mesoderm, produces a series of fingerlike branches, known as the **primary chorionic villi,** which extend into the thick endometrium of the receptive uterus. Later, with the establishment of the chorion, the **secondary chorionic villi** will replace this early attachment. Then each villus will be penetrated by embryonic blood vessels, so a large maternal-embryonic surface area is created, across which materials can be readily exchanged.

As the many fingers of the primary chorionic villi probe deeply into the mother's tissues, additional membranes also develop. The amnion, which formed earlier, represents the second cavity in the blastocyst. A third cavity is produced with the formation of the hollow yolk sac. In addition to its role in producing the earliest red blood cells, endodermal cells from the roof of the yolk sac contribute to the formation of the gut as the digestive system begins to develop.

About this time, the allantois begins to take form (Figure 37.16d). Although it will not develop into the great extraembryonic structure it becomes in birds and reptiles, it will eventually form the urinary bladder. And later on the blood vessels that have grown into the allantois will become major vessels of the placenta. At this early stage of development, we see the human embryo, suspended by a **body stalk,** surrounded by a fluid-filled amniotic cavity, and dangling a yolk sac (Figure 37.16e). All of this is in turn surrounded by the **coelomic sphere** (earlier the blastocoel). The coelomic sphere, then, lies within the ring of secondary chorionic villi. At this time there is little about the embryo to indicate that it is a developing human (Figure 37.16e).

By the eighth week, the placenta has become well advanced, the heart is fully functional, and the umbilical vessels and circulation are well established. Within the chorionic villi, highly branched **chorionic microvilli** (not unlike the lung alveoli) have developed and are surrounded by small pools of maternal blood that have formed as sinuses in the placenta. It is across the fine walls of the microvilli that all materials will be exchanged between the mother and fetus for the next seven months (Figure 37.16f).

The Emergence of the Human Form

Finally, at about eight weeks, the embryo begins to take on a human appearance, and by custom, upon its eighth week anniversary, its status is elevated from humble embryo to full-fledged **fetus.** At this time the fetus is about 25 mm long—just about one inch. Even now, it takes some imagination to see a human resemblance, but as time passes, the fea-

tures will sharpen and assume their proper relationships. Let's look at this process step by step.

The First Trimester. The nine months of human development is divided into three equal parts, each called a **trimester.** For convenience, we will take the developing systems one at a time, but it's well to keep in mind that much of the system development is occurring simultaneously.

37.16 HUMAN EMBRYONIC SUPPORT

(a) The human blastocyst reaches the uterine lining about six days after fertilization. **(b)** Communication between embryo and mother occurs early as the trophoblast secretes a hormone, similar to LH, signalling the corpus luteum in the ovary to continue secreting the endometrium-supporting hormone, progesterone. As implantation begins **(c),** fingerlike growths of the syncytiotrophoblast penetrate the endometrium, secreting enzymes that will clear a small blood-filled cavity into which the blastocyst will sink. During this brief period, the embryo continues to change. The amnion, first of the extraembryonic structures, then begins to balloon out. Later, cells on the opposite side will form a second cavity, the yolk sac, which resides within the trophoblast cavity. At this time, the embryo itself is represented by an elongated plate, only two cell layers in thickness. **(d)** Following the mammalian version of gastrulation, the embryo takes on an elongated shape with a definite head and tail organization, the yolk sac and the amnion enlarge and the latter becomes fluid-filled. Next to appear is the allantois, which contributes to the body stalk where blood vessels will form. The primary chorionic villi also expand, increasing exchange surfaces. **(e)** At about five weeks, blood vessels, umbilical veins and arteries branch into the villi and a full-fledged placenta emerges. **(f)** At the eighth week, the embryo finally assumes a decidedly human form and is now a fetus. It floats in a pool of amniotic fluid in the greatly enlarged amnion. The amnion itself becomes surrounded by a newly formed chorionic membrane. The placenta has fully matured, and contains an extensive circulatory system. The fetal vessels form highly branched microvilli, each penetrating a sinus containing maternal blood, where all exchanges take place.

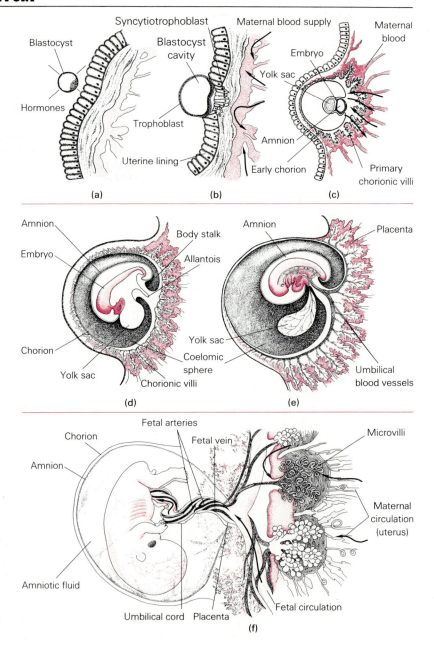

37.17 BRAIN DEVELOPMENT

Following the neural stage, the regions of the brain and other sensory structures become prominent. **(a)** At about 25 days, the neural structures already include the forebrain, optical vesicles, midbrain, hindbrain, and primitive ear. In addition, some of the cranial nerves are apparent. **(b)** At 38 days, regions of the brain have differentiated and the eye is forming. **(c)** The regions commonly associated with the adult brain begin to form by the end of the first trimester. The primitive features of the brain become overgrown by the massive cerebrum, which forms as hemispheres.

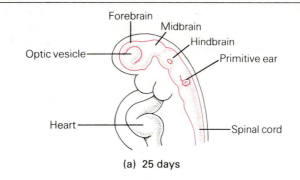

(a) 25 days

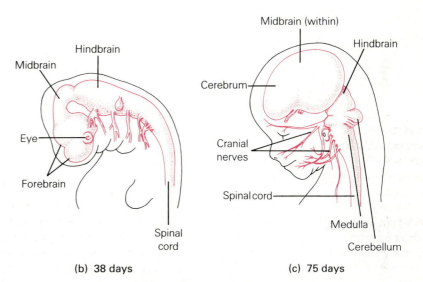

(b) 38 days   (c) 75 days

Nervous System. In typical vertebrate fashion, the nervous system develops early. Neurulation begins at about day 18 or 19, closely following gastrulation. As with the amphibian embryo discussed earlier, in humans the neural folds arise on the embryonic disk, reaching upward and then folding together to form the dorsal hollow nerve cord. Its anterior end enlarges to form the vesicles of the brain.

The nervous system continues its rapid progression in the first trimester. While it is the earliest system to begin development, it will not be completed until long after the birth process—perhaps not until the moment of death—since learning is in a sense a developmental process. By the fourth week, the major regions of the brain and spinal cord are recognizable. When the first trimester ends, these are already well-defined. The still-smooth cerebrum now extends over much of the embryonic brain, and the cerebellum and medulla have become distinct (Figure 37.17).

Circulatory System. As the neural ridges begin to break the contour of the human embryo, the heart and circulatory system make their earliest appearance. By day 22, you can make out the first timorous palpitations of the primitive heart. In vertebrates, this great organ is formed as cylinders of mesoderm converge, producing a single, tubelike structure. Within four to five days, the tube will have developed into a fully functional organ; it is primitive, but it moves blood. As the blood is pushed along it enters sinuses, forming channels that later become lined with endodermal cells, producing blood vessels.

Blood vessels begin to penetrate the chorionic villi at this time. Then, within another two weeks (or about 40 days from fertilization), the tubular heart will have looped back on itself, paving the way for its four-chambered pattern, which, by now, is nearly completed (Figure 37.18). An opening between the atrial chambers will remain until some time after birth.

37.18 DEVELOPMENT OF THE HEART

The human heart begins as a simple tubelike structure (a). By growing faster in the middle than either end, the tube loops (b) and thus the heart progresses from a single tube to a doubled one (c). In time, the atria and ventricles form (d).

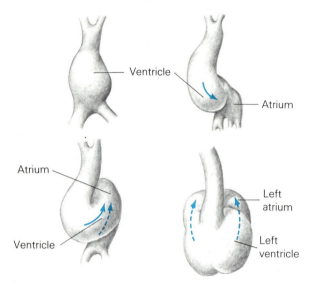

Respiratory and Digestive Systems. The respiratory and digestive systems develop fairly rapidly, so that by five to six weeks their basic patterns are clearly established. Interestingly, in human embryos, the primitive gut originates from an ingrowth of the yolk sac, rather than the other way around. Once the basic tube has been outlined, a few blind pouches form, then more and more outpocketings, until finally the indistinct outlines of the gut, liver, and pancreas can be seen. The trachea, bronchi, and lungs begin as a small outpocketing in the pharynx. The outpocketing then branches to form the two lung buds which, through morphogenesis, will give rise to the first individual lung lobes. At first, all of these respiratory structures are primarily endodermal, but later, both ectoderm and mesoderm will contribute to the final form (Figure 37.19).

Curiously, all mammals (and other chordates), including humans, produce embryos with a series of **pharyngeal arches** separated by **pharyngeal pouches** (Figure 37.20). In a fish, the genetic instructions for a pharyngeal arch might read, ''Produce gill arch.'' In mammals, however, these older instructions might have changed to, ''Modify pharyngeal arches into a jaw, a larynx, part of the middle ear, and tongue bone structures.''

Limbs. The limbs of humans appear as rounded buds during the fourth week (Figure 37.21). The arms and legs are distinguishable at six weeks, but fingers and toes require an additional week. The rudimentary hands and feet actually begin as simple webbed paddles that take form through a kind of developmental programmed cell death, as the tissue between the fingers and toes is broken down and absorbed.

Excretory System. The urinary and excretory system of the human develops in a series of distinct stages. Beginning on day 25, three successive sets of paired tubules develop, the **pronephros, mesonephros,** and **metanephros.** The pronephros, the most primitive, persists as a functional structure only in the hagfish; in all other vertebrates it is replaced during development. The mesonephros, a more advanced structure, persists in fishes and amphibian adults. As we will see, portions of the mesonephros are salvaged and put to use by the male reproductive system during its development. The final version, the metanephros, is the functional kidney of all amniotes—reptiles, birds, and mammals.

The metanephros becomes established as the paired kidneys in the human embryo by the 16th week. Actually the two versions that precede it are functional because the first helps develop permanent ducts and the second serves as a kidney for a few days during human development. (Figure 37.22 presents further details of this peculiar developmental story.)

Reproductive System. The reproductive system begins to develop during the first few weeks, but until the eighth week even a trained observer can't determine the sex of the embryo (without a chromosome test). It is true that before this time the genitals have begun to develop, but the genitals of the two sexes start off in much the same way. In the sexually ''indifferent'' state, three prominent parts of the reproductive anatomy are visible. They collectively make up the **genital tubercle** and include the **urogenital groove, urogenital folds,** and **genital swelling** (Figure 37.23). These structures already suggest the female genital system, but in a developing male they undergo great changes in response to male hormones.

The factors controlling sexual differentiation are complex. Certainly, the genetic sex of an individual is determined at fertilization by the chromosomal

37.19 DEVELOPMENT OF THE RESPIRATORY AND DIGESTIVE SYSTEMS

(a) The respiratory system makes its appearance in the third week as a simple outpocketing or lung bud from the embryonic pharynx. By the fourth week the lung bud has divided into right and left sides, which form the pattern for the two bronchi and lungs. By the fifth week of development, the lobes of the lung take on a crude form to be refined into a more definite shape one week later. **(b)** The digestive tube also takes on form during the fourth week. Its major regions, the esophagus, stomach, and intestine differentiate first. The tube ends in a cloaca, a temporary reminder of our chordate affiliation. By the fifth week, the tube has further differentiated, with the stomach enlarging and the small intestine, colon, and rectum clearly defined. In addition, the liver, gall bladder, and pancreas are clearly outlined. The small intestine will enter a coiling stage as development continues, and the stomach, shown at five and a half and eight weeks, will assume its final form.

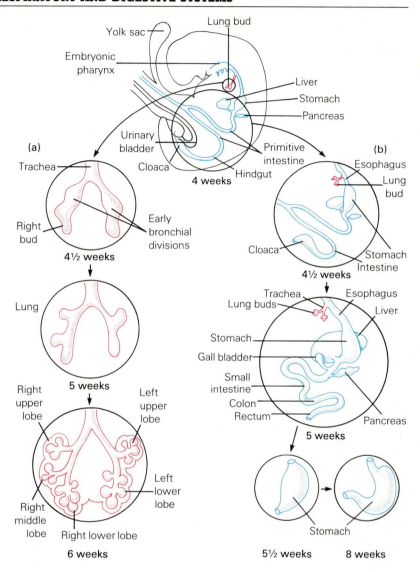

37.20 FATE OF THE PHARYNGEAL ARCHES

In the fifth week, the pharyngeal arches (note numbers) are prominent. In a fish, these structures form the jaw and gill structure, but in humans they contribute to numerous structures in the head and neck, including the lower jaw, tongue (and its hyoid bone), larynx, middle ear, pharynx, and many others. The incorporation of the pharyngeal arches into these structures is completed early in development, so this fishlike stage lasts only a few weeks.

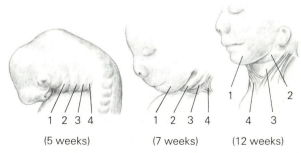

Pharyngeal arches

37.21 LIMB DEVELOPMENT

(a) By the fourth week the arms and legs appear as simple buds in the embryo. (b) By the fifth week they have grown significantly, forming paddlelike extensions. By six weeks, the finger and toe structure is visible, but a webbing remains. The tissue between fingers and toes will undergo a "programmed" death and by the eighth week separation will have become complete.

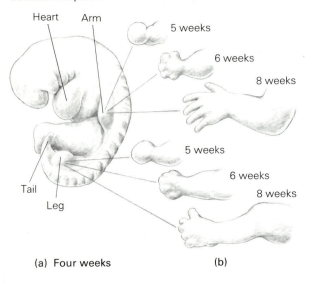

(a) Four weeks

(b)

complement of the sperm cell. Fertilizing sperm cells that carry X chromosomes produce females, while those bearing Y chromosomes produce males. However, this is only a beginning. A number of genes are involved in sexual development, and the control may stem from different levels (such as genetic, hormonal, or environmental). For example, it appears that the Y chromosome of mammals initiates the development of the testes and that the testes take over from there, producing hormones that determine the male's primary sexual characteristics.

In the male embryo, the mesonephric ducts are forerunners of the sperm-storing epididymis and sperm-conducting vas deferens (see Chapter 36). But this is only part of a peculiar story. During the indifferent state, all embryos develop a pair of **paramesonephric ducts** or **Mullerian ducts,** as they are also known, which lie, as their name implies, alongside the mesonephric ducts. As differentiation proceeds, the paramesonephric ducts in males degenerate, while the mesonephric ducts operate in the manner just described. In females, however, it is the mesonephric ducts that degenerate, while the paramesonephric ducts become the oviducts. The role taken on by the two ducts, like the rest of sexual

37.22 EXCRETORY SYSTEM DEVELOPMENT

Development of the excretory system in humans is quite interesting in that three versions of the kidney develop in a closely timed sequence. The earliest event is the completion of the pronephros, a primitive set of ducts that is very short lived (fourth to sixth weeks of development), and is replaced by the mesonephros (a) which utilizes the pronephros to form the mesonephric ducts. The mesonephros is also short lived and within a week of its completion (by the eighth week of development) begins to degenerate (b). Meanwhile, the most advanced version of the kidney, the metanephros, has formed paired buds that emerge from the mesonephric duct (a). Next, the buds enlarge and the metanephric ducts (future ureters) lengthen (b). By the eighth week, the kidney and its inner structures take on definite form (c).

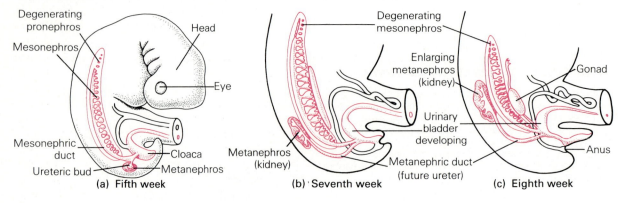

(a) Fifth week   (b) Seventh week   (c) Eighth week

37.23 SEXUAL DEVELOPMENT

At seven weeks, the embryo has produced a genital tubercle with a prominent glans and urogenital folds and groove. The tubercle is supported by a triangular genital swelling. Under the influence of male hormones from the embryonic gonad, these structures will produce typical male sex organs. In the absence of these hormones, female organs will emerge. In the genetic male (XY), the glans will enlarge to form the glans penis, with the shaft of erectile tissue emerging from the urethral folds. The urethral groove will close, while the genital swellings will differentiate into the scrotal sac to await the descent of the testes in the eighth month of life. In the genetic female (XX), the glans will become the glans clitoridis, and the urethral folds will remain open forming the labia minora. The genital swellings will form the labia majora.

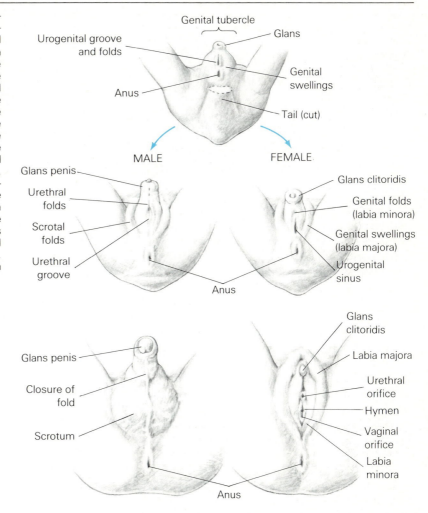

differentiation, is hormonally determined. (Recall the discussion of the role of sex hormones in sexual differentiation in Chapter 33.)

The Second and Third Trimesters.   The first trimester is marked by a great deal of activity and sweeping change—so much, in fact, that by its close, most major systems have approached their final form. In contrast, the second trimester is characterized by growth and refinement. Body length increases rapidly, catching up to the large head. Systems begin to approach a functional state. The fetus "breathes," but only amniotic fluid enters and leaves the tiny lungs. The digestive system becomes lined by secretory cells which will secrete enzymes. Bile from the liver joins these secretions. The fetus swallows, bringing in fluids and cellular debris that accumulate in the gut as dark, jellylike **meconium.** By 16 weeks, the dermis and epidermis of the skin reach a differentiated state and are penetrated by sweat glands, sebaceous glands, hair follicles, and sensory neurons. If stimulated, the 12-week-old fetus will respond with feeble, uncoordinated movements at first, but these movements soon grow more precise and brisk and by 16 weeks the mother feels the first fetal movements—usually as a pronounced "kick."

As the third trimester begins, the organ systems begin to be fully functional. A fetus may survive if born now, but its chances are very poor even under the most sophisticated medical supervision. At this time it will measure about 35 cm and weigh around 1000 g (14 in and 2.2 lbs), resembling a very thin and emaciated full-term baby. Growth

37.24 THE BIRTH PROCESS

In the first stage of labor, dilation of the cervix occurs and the baby's head eventually crowns (a). The second stage of labor involves the actual expulsion of the fetus (b). Expulsion of the placenta—the third stage—follows the birth of the baby.

(a)

(b)

and refinement continue but, in most cases, at a reduced rate. The fetus, no longer floating free, gradually fills the amniotic sac and uterine cavity. The brain and spinal cord continue to develop rapidly, and the cortex differentiates and begins to form the familiar fissures and convolutions. A layer of fat accumulates beneath the skin. In the skeletal framework, **ossification,** the hardening of bone, which began in the first trimester, increases rapidly. These activities increase the demand for protein and calcium, placing a substantial burden on the mother's reserves. As a final prelude to birth, antibodies cross the placenta during the last month. These will provide the newborn baby with immunity against viral and bacterial infections for a month or two. As the third trimester ends, the fetus, now crowding the mother's abdomen, will measure close to 50 cm and weigh about 3200 g (20 in and 7 lbs).

Birth.   In spite of our great interest in the process and the fact that many of us have experienced it, we still don't really understand what brings on the birth process. Currently, considerable research is focused on the changes in the placenta that occur at the end of gestation and the hormonal shifts that bring them about. Since the level of progesterone is important in maintaining the placental tissue, there may be a direct relationship between the aging of this structure and lowered progesterone levels. In addition, at term the uterus contracts more strongly in response to oxytocin.

Recent theories suggest that as pregnancy continues, the number of oxytocin receptors in the uterus increases, possibly brought about by increased estrogen levels and uterine stretching. Thus, while oxytocin levels do not rise appreciably, the uterus becomes increasingly sensitive to the hormone's presence. In addition, prostaglandin secretion by the placenta increases during labor, and prostaglandins are known to enhance the action of oxytocin.

The Stages of Birth.   In accordance with their penchant for pigeonholing, scientists have divided the birth process into three stages (Figure 37.24). The first is **dilation,** which involves a softening of the cervix. This is sometimes accompanied by a rupturing of the amnion, often referred to as "breaking the water," although this usually happens in the second stage. The period of dilation is unpredictable, and may vary from two to 16 hours. It is accompanied by periodic contractions called **labor pains.** These contractions increase in frequency through the first stage. When they begin to occur at least every three to four min, the fetus will **crown,** meaning that the head will begin its passage through the cervix and become visible. This is the start of stage two, the **fetal expulsion.** Expulsion may take just a few minutes or it may require hours. The mother may have to be given a spinal anesthetic at this time. The level of pain is highly variable, depending on the mother's preparation, her physical condition, her pain threshold, and her emotional state.

The final stage of birth is the **placental separation** and its expulsion from the uterus as afterbirth. Following this, the uterus rapidly contracts to its former state, which helps control the bleeding at the site of placental attachment.

Physiological Changes in the Newborn Baby.   The body systems of the fetus are in a state of readiness prior to birth, for it has been prepared for a new and more threatening kind of existence. For example, the vital exchange of gases has always been provided by the placenta. Now the infant's own respiratory system must function for the first time. There are the changes in the circulatory system (Figure 37.25). Oxygenated blood from the placental circulation had previously entered the fetus through the umbilical vein, proceeding directly into the vena cava and thence to the right side of the heart. A hole in the septum between the fetal atria, the **foramen ovale,** permitted the blood to flow from the right atrium to the left atrium, thus circumventing the route to the still uninflated and functionless embryonic lungs. In addition, a connecting vessel between the pulmonary artery and aorta, the **ductus arteriosus,** permitted blood from the right ventricle to bypass the lungs and go directly into the arterial distribution.

With the baby's first breaths, the ductus arteriosus constricts vigorously, closing that bypass. With the shortcut closed, the blood must move toward the rapidly expanding lungs. As a result of this increased pulmonary flow, the volume of blood returning directly to the left atrium suddenly increases. The foramen ovale is composed of two overlapping flaps of tissue which are forced closed when the atria contract. Actually, however, the foramen ovale will not seal itself completely until the baby is about a year old.

The first breath of a newborn infant is taken in response to a sudden increase in blood carbon dioxide when the placental exchange is lost. This increase stimulates the respiratory center of the medulla with the usual results, so the baby will breathe with or without the traditional swat on its rear.

Lactation.   Human mothers, like all mammals, are prepared to nurse their newborn young immediately. The breasts, under the influence of estrogen and progesterone, have enlarged during pregnancy, and with delivery, **colostrum** secretion begins. Colostrum, a clear, yellowish fluid, differs from milk in that it contains more protein, vitamins, and minerals and less sugar and fat. It also contains maternal antibodies that can be important to the infant in its first days.

In a few days, the colostrum is replaced by whiter, thicker milk. Human milk is sufficient in all required vitamins except vitamin D and contains

37.25 FETAL CIRCULATORY CHANGES

This is the circulatory system of the fetus shortly before birth. Note the foramen ovale, a septum between the right and left atria in the heart, the ductus arteriosis, connecting the pulmonary artery with the aorta and, of course, the umbilical circulation. The foramen ovale must close to prevent the mixing of oxygenated and deoxygenated atrial blood when the lungs become functional. The ductus arteriosis must constrict also, to prevent the mixing of deoxygenated and oxygenated blood. The umbilical circulation is disrupted when the cord is tied and cut.

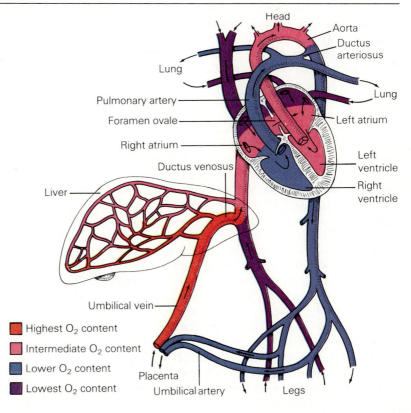

Highest O$_2$ content

Intermediate O$_2$ content

Lower O$_2$ content

Lowest O$_2$ content

about 700 calories per liter. Interestingly, it is bacteria-free (which in not true of cow's milk, either in grocery stores or at the source). There are additional benefits of breast feeding. Some studies reveal that breast-fed infants are less susceptible to disease, anemia, and vitamin deficiencies.

ANALYSIS OF DEVELOPMENT

Now that we have had a look at some of what happens during the development of some animals, we can focus on some of the mechanisms involved. Historically, we find a lively interest in development and, inevitably, the search for evidence in support of one point of view or another. That search has led to some of the milestone experiments in the science of embryology.

The Egg and Zygote in Differentiation

Late in the 19th century, Hans Driesch mechanically separated the two cells of a sea urchin embryo after it had undergone its first division. He cultured each half and found that each undamaged cell generally produced a normal, whole, but half-sized embryo. The data showed that at least in the two-celled stage, each cell had all that was required to produce a normal embryo.

This experiment, simple as it may seem in a more sophisticated time, probably marked the beginning of experimental embryology. It was incessantly repeated and elaborated upon, but experimenters finally turned to other experimental subjects, one of which was the amphibian embryo.

We have already noted that animal eggs show different degrees of polarity, and we pointed out that the fertilized amphibian (frog) egg forms a gray crescent after fertilization. The importance of this region to successful development was clearly demonstrated by Hans Spemann early in the 1900s. Using fine thread, Spemann ligated (tied off) and divided groups of newt zygotes through various planes in a way that emulated cleavage, producing pairs of blastomeres (Figure 37.26a). If the plane of the ligature bisected the gray crescent two normal embryos would develop. But if the gray crescent went entirely to one blastomere, that cell alone developed normally. The other would form an undifferentiated mass. Clearly much of the amphibian's future developmental organization is already determined in the zygote and the gray crescent plays an important role.

In the 1930s, similar experiments were carried out with sea urchin embryos by Sven Horstadius. He found that if sea urchin blastomeres were separated prior to the third cleavage, each would develop normal, though tiny, larvae. But after the third cleavage, the fate of separated blastomeres depended upon the plane of separation. Horstadius separated a number of eight-celled sea urchin embryos into quartets. When the embryo was divided along the animal-vegetal axis, normal larvae developed. But, separation of quartets in a plane running through (perpendicular to) the animal vegetal axis resulted in abnormal products (Figure 37.26b). Horstadius determined that organizational changes also begin early in the sea urchin's development.

But how early? Going one step further, Horstadius developed a technique for dividing unfertilized sea urchin eggs. Such eggs (like the human ovum) are about the size of the period at the end of this sentence, a fact that should generate some appreciation for his skill. Again, Horstadius found that after fertilization, successful development depended upon the plane in which the division was made (Figure 37.26c). It seems that in sea urchins, animal and vegetal poles are already established in the egg. Any separation that provided an equal distribution of both regions in artificially cleaved cells had a good chance of developing normally after fertilization. In all others, development failed. Since someone is bound to ask, let's bring it up: Only one of the artificially cleaved cells contained the egg nucleus, so after both cells were fertilized, the cell that got the egg nucleus was diploid, while the cell without the egg nucleus became haploid (contained the sperm nucleus only)—a condition that does not prevent development in the sea urchin. From these observations and many others, it becomes apparent that some animal eggs have already undergone a considerable amount of developmental organization prior to fertilization, and that this organization profoundly affects events to follow.

Experiments such as Horstadius conducted wouldn't work with other species that have more highly structured eggs. For example, in annelids and mollusks, and apparently insects as well, every definable part of the unfertilized egg has a strictly determined fate that cannot be affected by any kind of manipulation. Such eggs are called **mosaic eggs.** In these eggs, critical changes in the cytoplasm have obviously occurred very early—changes which unwaveringly send each part of the egg on its way to its developmental fate. Presumably, such changes occur during the long process of egg maturation. Think about this. If, in some species, differentiation is partly accomplished by the interac-

37.26 POLARITY

(a) The fertilized amphibian (newt) egg is far from homogeneous, its cytoplasmic constituents having undergone substantial organization prior to cleavage. Spemann ligated newt zygotes in two planes. Where the entire gray crescent went to one blastomere, it developed into a normal embryo, but the other blastomere produced only an amorphous mass. Ligating the zygote through the gray crescent resulted in two complete embryos. (b) Blastomeres in eight-celled sea urchin embryos can be separated into tetrads without producing developmental problems as long as such separations are not in a plane perpendicular to the animal-vegetal axis. (c) In sea urchins, polarity already exists in the unfertilized egg. When an unfertilized egg is divided in a plane perpendicular to what would be the animal-vegetal axis, and then fertilized, one of the zygotes fails. If, however, the division is made parallel to the animal-vegetal axis, both products successfully develop following fertilization.

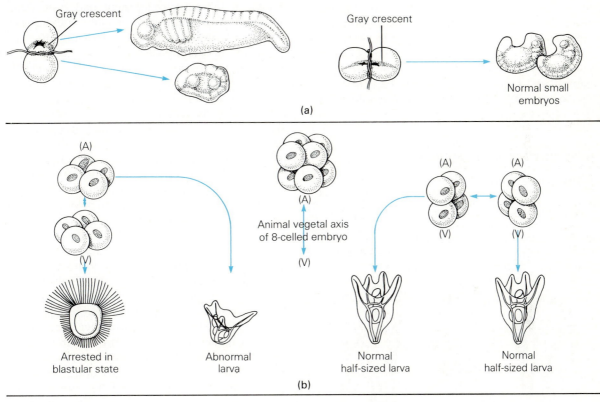

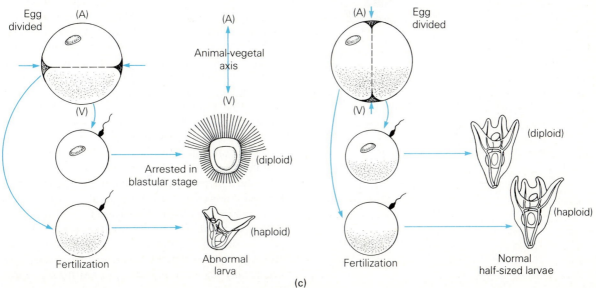

tion of cells, how do the cells of mosaic embryos develop? What triggers or controls their changes? No one knows.

Tissue Interaction: Organizers and the Induction Theory

The gastrula provides an excellent subject for the study of development—one that is fairly simple yet poised and readied for a series of incredibly complex events. We've seen that by the end of gastrulation, the cells of the ectoderm, mesoderm, and endoderm have become committed to producing specific tissue regions of the animals, yet their fates have not been irrevocably sealed. They can still be manipulated experimentally by altering their relationships to other cells. This kind of manipulation

37.27 THE PRIMARY INDUCER

Spemann and Mangold transplanted a part of the dorsal lip of the blastopore to a region of the embryo whose ectoderm would not normally produce a neural tube. The results, seen in whole embryos and in cross sections, reveal a supernumerary neural tube, notochord, and somites. Such transplants led Spemann to develop his induction theory of development. He postulated that the germ layers influence, or induce, each other to develop specific structures.

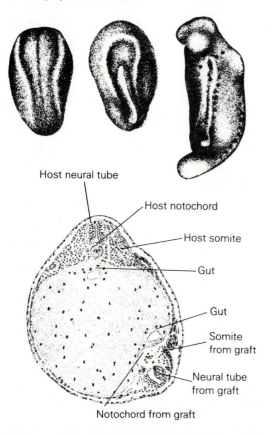

Host neural tube

Host notochord

Host somite

Gut

Gut

Somite from graft

Neural tube from graft

Notochord from graft

has given embryologists several important clues to how tissue interaction influences development.

A great deal of what we know about the interaction of the germ layers has emerged from the work of Hans Spemann and his protegee, Hilde Mangold. In the early 1920s they carried out a series of transplant experiments that other embryologists, seeking to repeat, could only describe as "formidable." Let's see what impressed everybody.

By Spemann and Mangold's time, many of the details of normal development had been clearly described for a number of animals, so there was a rather firm basis for designing experiments. They chose that part of the gastrula known as the dorsal lip of the blastopore (see Figure 37.8), which they knew was the site of the invagination and involution of the presumptive endoderm and underlying presumptive mesoderm during gastrulation. They also knew that as these actively migrating cells took up their new positions in the gastrula, they were to play a key role in organizing the embryonic axis. For example, cells originating as presumptive mesoderm would group themselves below the neural groove ectoderm and once in their new position, formed the notochord (Figure 37.27). The ectoderm above invariably proceeded through neurulation.

Early in their work, Spemann and Mangold found that the transplanting of a bit of dorsal lip containing endoderm and underlying mesoderm to some other region of the embryo brought about strange results. First, the transplant brought about invagination where it shouldn't have occurred, but just as it would have in its normal position. It then went on to produce a second embryonic axis, in which a gut, notochord, somites, and surprisingly, a neural tube developed. The latter was surprising in that it is not derived from dorsal lip endoderm or mesoderm, but, as we have seen in neurulation (see Figure 37.13), from ectoderm. Further, the ectoderm in the region of the transplant would normally have produced simple epidermis, unless it was somehow being influenced by the transplant.

Spemann and Mangold repeated the experiment, this time using donor dorsal lip that was highly pigmented. The color difference permitted them to follow the path of grafted tissue as development continued. Most of the graft invaginated and its pigmented cells could be found in the usual mesodermal derivatives. The neural tube, however, formed chiefly from unpigmented cells of the overlying ectoderm. The results were now clear. Somehow the dorsal lip mesoderm had *induced* the foreign ectoderm to form a neural tube. For this reason, Spemann referred to the dorsal lip as a **primary organizer**. The term *induction*, or **embryonic induc-**

tion, describes the ability of one group of cells to alter or influence the development of another.

In experiments to follow, Spemann and Mangold learned that by transplanting dorsal lip tissue mesoderm from younger gastrulas, they could produce a secondary head region, but when transplants were taken from older gastrulas, they would yield only a secondary trunk and tail. This concurs with what we know about gastrulation. The earliest presumptive mesoderm to involute migrates for a considerable distance along the inner wall of the gastrula, coming to rest at a region that will later form the head of the embryo. But in an older gastrula, the remaining dorsal lip presumptive mesoderm migrates only a short distance, ending up nearer the blastopore, thus making contributions to a tail region only (Figure 37.28).

Spemann's experiments reveal two major findings. First (as we mentioned earlier), he concluded that the ectodermal cells of the early gastrula are not yet totally determined. Determination in embryology, we have seen, means that the developmental fate of a tissue is set or fixed. We emphasize this discovery mainly because it led to a barrage of experiments that tested the determination of embryonic tissues. We will summarize a few of these later on. Second, Spemann was able to conclude that germ layers in the embryo influence each other. As we have seen, the mesoderm from the dorsal lip, which Spemann called a primary organizer, directs the course of events in the overlying ectoderm through the process of induction. This discovery was one of the first substantial clues to the basis of differentiation.

In 1935 Spemann received the Nobel Prize for his work. An era of experimental embryology followed in which the limits on transplanting were set only by the imagination and technical skill of the researcher and the toughness and recuperative powers of the experimental embryo. Many examples of embryonic induction have been discovered since that time. One of the most noteworthy of these was related to the development of the lens of the eye. This added a new twist since it involves a "triple-inductive" system, one involving induction and counterinduction—a veritable dialogue among **optic vesicles, lens placodes,** and **optic cups,** embryonic structures described in Figure 37.29.

The discovery that mesoderm can influence the fate of ectoderm raises the usual barrage of questions: How does the mesoderm influence the ectoderm? What is the relationship between the two layers? Is there some sort of chemical message transmitted across the cells? The last question, of course, leads us back to the molecular level. And if

37.28 TIMING OF THE TRANSPLANTS

Spemann and Mangold found that the results of dorsal lip transplants varied with the age of the gastrula. **(a)** When a source of the graft was an early gastrula, it produced the head region of the embryo, while grafts from the dorsal lip of substantially older gastrulas **(b)** produced only tail regions. These observations agree with our understanding of involution, the continued inward migration of mesoderm from the dorsal lip region to the future head region of the embryo.

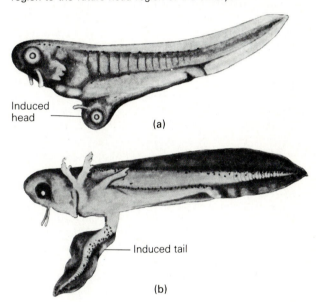

Induced head

(a)

Induced tail

(b)

chemical messengers are involved in induction, then what kinds of molecules are they?

Testing the assumption that a chemical messenger passes between mesoderm and ectoderm turned out to be relatively easy. (It's often more difficult to formulate the right questions than to perform the experiments that will answer them.) One procedure consisted of placing dorsal lip tissue on a laboratory growth medium for a brief period and, after removing it, placing a segment of ectoderm on the same medium. In those first experiments, as the isolated ectoderm began to change—under the wide-eyed observations of the experimenter—it produced crude structures strongly reminiscent of a neural apparatus. It was enough to encourage them to redefine the experiment by placing mesoderm and ectoderm on either side of an agar block. The results were the same. Obviously, something passed through the agar. But what was it?

Experimenters tested the media and the agar block to determine what substance was being released from the mesoderm. This is the point at which developmental biologists found themselves in the early 1960s.

Unfortunately, it turned out that many substances pass between mesoderm and ectoderm,

37.29 **TRIPLE INDUCTION**

Experiments on the development of the eye in frogs showed that there is a complex two-way interplay among the embryonic tissues. **(a)** As the brain begins its development, two bulges of tissue, the *optic vesicles*, begin their growth outward toward the surrounding ectoderm. **(b)** When contact is made, the ectoderm responds by forming the thickened *lens placode*. **(c)** The lens placode then becomes an inducer to the optic vesicle which responds by forming the curved *optic cup*. **(d)** In a final episode, the optic cup then induces the lens placode to invaginate, and it forms the sphere that will later become the lens **(e)**. Interestingly, transplanting the optic vesicle to other regions of the embryo will begin the multiple induction similar to that described above.

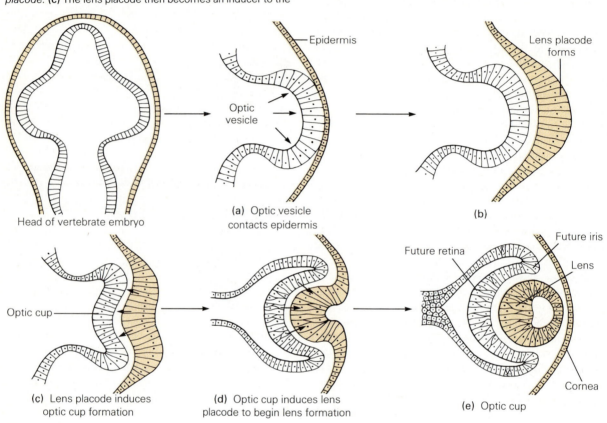

Epidermis

Optic vesicle

Lens placode forms

Head of vertebrate embryo

(a) Optic vesicle contacts epidermis

(b)

Optic cup

Future retina

Future iris

Lens

(c) Lens placode induces optic cup formation

(d) Optic cup induces lens placode to begin lens formation

(e) Optic cup

Cornea

and as far as is known, the inducer is simply nonspecific. Some hold that the responding ectodermal tissue has the capacity to form neural structure as well as epidermis, and all that the inducing substance or condition does is simply allow it to happen. It is suggested that there are many inducers acting alone or in concert, perhaps even changing from day to day as the embryo develops.

The Cell Nucleus in Differentiation

In their search for clues to explain differentiation, some embryologists in the 1950s turned to the nucleus, which isn't surprising since the nucleus controls heredity and great research emphasis was directed to determining the molecular basis of heredity. This was the era when the riddles of DNA and RNA structure, genetic coding, and protein synthesis were being solved in quick succession. For developmental biologists, the versatility of ectodermal cells in the early gastrula, established by Spemann, was of great concern. One line of research, reported in the 1950s by American embryologists R. W. Briggs and T. J. King, involved the transplanting of isolated nuclei from frog blastula cells to frog eggs whose own nucleus had been destroyed. Briggs and King predicted that if the nucleus had undergone irreversible changes during the formation of a blastula, then cells with such transplanted nuclei would fail to develop. The experimental data were clear enough. Most—about 80%—of the eggs, when activated, reached the blastula stage, and about half of these went on to form tadpoles. Allowing for error and irreparable injury, it was safe to conclude that blastula nuclei had lost none of their potency (Figure 37.30).

Other experiments, notably those done by John Gurdon at Oxford, revealed that even older cells maintained totipotency. Thus development did not alter the genetic capability of the cell. Developmental biologists would have to turn to other possibilities, perhaps to the environment of the cell or tissue as it undergoes its programmed changes.

Nerve Cells and Developmental Patterns

Developmental biologists, as modern embryologists call themselves, have taken a keen interest in the nervous system and its neurons as their "experimental organism." They have found some surprising things about the way neurons assume their role in the developing vertebrate embryo.

The longest cells of the vertebrate body are the neurons, or nerve cells. The cell body (nucleus and most of the cytoplasm) of the cell are often located in or near the brain or spinal cord, while long extensions or processes travel to a distant part of the body where they branch into treelike endings (see Chapter 34).

In embryonic development, the neural processes follow along defined nerve pathways from their location in the brain or spinal cord. How they are guided to their destination is unclear, however, and there may even be an element of chance determining which part of the body any given neuron will eventually reach. It is important that a developing neuron reach a target, since only those that make connections with appropriate target tissues will live; those that do not will perish during embryonic development. It appears that perhaps half of the neurons that grow out of the central nervous system do not survive.

The Sprouting Factor and Antisprouting Factor.
Once the growing neuron reaches its intended target area, it begins to sprout branches. Other neurons in the area will also branch, until the target area (say, the toe) has the proper density of neuron endings. How do the nerves know when the proper density has been reached? As occurs so often in development, regulation depends on negative feedback. In this case, two hormonelike substances are involved. The tissues receiving the neuron produce a **sprouting factor.** At the same time, an **antisprouting factor,** manufactured back in the body of the nerve cell, is transmitted down through the process and released at the ends of the neuron branches. This substance inhibits further sprouting of that neuron, and also of any other

neurons in the area. As more and more branches of the neuron sprout, a balance is reached between sprouting factor and antisprouting factor, and further branching ceases. But if some of the terminal processes in the area are killed and removed, the balance shifts and the remaining neurons will again put out new sprouts until the proper density of nerve endings is reestablished. If transmission of the antisprouting factor is experimentally blocked (for example, by colchicine, which inhibits cytoplasmic movement), all the neurons in the target area will sprout new branches.

How does the neuron know when its tip reached its own target area—or has failed to do so? The answer is complex and not fully developed. We do know that the target area itself has rather little effect on nerve growth patterns. This can be seen in experiments in which a patch of embryonic skin is removed prior to nerve growth and is replaced with a patch from another area. Regard-

37.30 NUCLEAR POTENCY

In their experiments, Briggs and King transplanted blastula nuclei into frog's eggs whose nuclei had been destroyed with ultraviolet light. The treated egg was then activated to instigate cleavage and development. Most of the treated eggs developed normally.

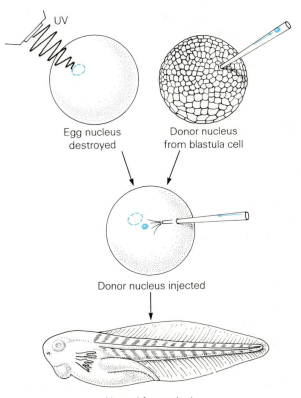

Egg nucleus destroyed

Donor nucleus from blastula cell

Donor nucleus injected

Normal frog tadpole

37.31 SKIN TRANSPLANT

In this simple transplant experiment, a patch of skin is excised from a developing frog embryo and reversed. Later when the patch has grown in and its neuronal connections made, a simple test is carried out. The lower region of the patch is stimulated with an irritant. The frog responds by trying to wipe the area where the skin would have been had it not been reversed.

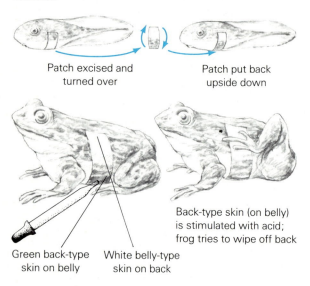

Patch excised and turned over

Patch put back upside down

Back-type skin (on belly) is stimulated with acid; frog tries to wipe off back

Green back-type skin on belly

White belly-type skin on back

less of the source of the transplanted skin, neurons reaching it will sprout and branch in a regular pattern, as if the original skin were still in place. But, as stated above, the eventual survival of the neuron depends on the nature of the tissue. The growing process, once it reaches its target area, apparently sends some unknown kind of message back to the body of the cell, informing it that contact has been made—*and with what kind of tissue.* The cell is then given the OK to continue living. If the appropriate contact is not made, the cell dies.

An Experimental Investigation of Nerve Specificity in Development. Consider the classic experiment illustrated in Figure 37.31. A piece of epidermis is removed from the side of a developing frog embryo, at a time when its own future as back skin or belly skin has been irreversibly determined, but before the axons of the nervous system have yet reached their target tissues. The piece of epidermis is rotated 180° and replaced. When the frog finally becomes an adult, a patch of skin on one side is upside down. The frog has white belly skin on part of its back and green back skin on part of its belly. Now, it is

known that vinegar irritates the skin of frogs. If an experimenter puts filter paper soaked in vinegar on the white patch of belly skin on the frog's back, however, the animal kicks itself in the stomach with its hind foot. If the experimenter then applies the stimulus to the green patch on the frog's belly, the confused animal reaches its hind foot up over its back, repeatedly hitting the spot where that back skin ought to be. The experiment indicates that the central nervous system has received messages from neurons that correctly indicate exactly what target tissue had been innervated—but it behaves as though the skin was in the right place.

If the frog is then killed and dissected, the visible nerve pathways are found to be just as they would have been if the ectoderm hadn't been reversed. The neurons going down the nerve pathway to the belly found themselves actually innervating back skin. Those innervating the patch on the back also followed normal nerve pathways. In both cases, the developing neurons recognized the tissue they terminated in, and passed this information on. The tissue-selection hypothesis would suggest that these particular neurons would have degenerated if the epidermis had not been rotated. But as luck would have it, some neurons targeted for belly skin actually made it to the misplaced belly skin, and other neurons—targeted for back skin—found their target tissue, even though it was in the wrong place. So when the back was touched with vinegar, belly neurons were activated, and the spinal column and brain deciphered the information accordingly.

How can the growing nerve cells "know" they have terminated in, say, back skin? The answer is not clear. Presumably each kind of target tissue has its identifying mix of chemical signals. We know even less about the programming and control of nerve cell interconnections in the brain, except that any such system must be phenomenally complex. Every brain neuron has connections—usually multiple connections—with the processes of many other neurons. The patterns are complex and surely not random. Does each neuron or class of neurons put out its own identifying set of chemical stimuli? We do know that mouse brains have many more kinds of mRNA than all other mouse tissues put together, which indicates that a substantial proportion of the genes and proteins of vertebrates are involved with brain function. And human brains are probably even more complex than mouse brains.

APPLICATION OF IDEAS

1. Fertilization has been intensely studied in echinoderms simply because the gametes lend themselves so well to manipulation and observation. Do such studies help us to understand fertilization in other animals? In humans? What cautions might be kept in mind? How might such studies help the scientist in formulating the right questions about humans and other species?

2. The embryos of vertebrates go through stages in which certain structures resemble those formed by other vertebrates but are soon modified into tissues and organs not seen in those animals.

Suggest a molecular basis for this observation, and consider how it might be of evolutionary significance.

3. Sometime during vertebrate evolution, the switch was made from the self-sustained, independent egg case to placental support and development. Considering the scenario of vertebrate evolution (see Chapter 27), suggest what some of the intermediate stages might have been. Suggest reasons why natural selection might have favored this shift in the evolving mammal. Why didn't the same selection occur in birds?

KEY WORDS AND IDEAS

1. A vast store of information on the visible events of development is available, but far less is known on how it occurs. Development in animals includes mitosis, cleavage, cell movement, tissue orientation and interaction, programmed cell death, and **morphogenesis,** the final emergence of form.

GAMETES AND FERTILIZATION

The Sperm

1. Structures in the sperm include the head, which contains the haploid chromatin and an enzyme laden **acrosome;** the midpiece, with a centriole and spiralling mitochondrion; and a tail which is a typical flagellum.

The Egg

1. Egg cells, or oocytes, may contain a degree of polarity, with the cytoplasmic elements organized in a gradient from the metabolically active **animal pole** to a less active **vegetal pole.** Even before fertilization, portions of some eggs are already committed to certain developmental fates.

2. In terms of yolk quantity, eggs are **macrolecithal** (large amount—bird), **mesolecithal** (intermediate—amphibian), and **microlecithal** (minute—echinoderm, mammal). The yolk quantity relates directly to the amount of development supported by materials in the egg. In the bird only, a minute **blastodisc** represents the embryo.

3. Many species produce a fibrous **vitelline membrane** just outside the plasma membrane. Echinoderm eggs are surrounded by a dense **jelly coat,** while those of mammals are surrounded by the **corona radiata,** which contains interlocking **follicle cells** and **hyaluronic acid,** overlying a clear **zona pellucida** and **plasmalemma** (plasma membrane).

Fertilization

1. Upon contact with the jelly coat, sea urchin sperm release acrosomal enzymes which digest a path to the vitelline membrane. The **acrosomal filament** identifies with a receptor protein in the vitelline membrane which it penetrates, attaching to the plasma membrane proper.

2. **Polyspermy** is avoided in echinoderms by electrical potential changes, which make penetration impossible, and the **cortical reaction,** which detaches extra sperm. **Cortical granules** release substances into the **perivitelline space,** initiating the formation of a **fertilization membrane.**

3. Upon sperm attachment in echinoderms, **microvilli** ensnare the sperm head, forming a **fertilization cone** and drawing the head into the cytoplasm, in which its chromatin forms the male **pronucleus.**

THE EARLY DEVELOPMENTAL EVENTS

1. **Embryogenesis** is a period of intensive cell division and cell rearrangement which produces a **triploblastic** (three-layered) embryo, preparing it for morphogenesis.

The First Cleavages

1. The time required for the first cleavage varies with species. Complete or **holoblastic cleavage** is typical in microlecithal and mesolecithal eggs, with equal-sized **blastomeres** forming in the first and second cleavages. In humans the sec-

ond cleavage in one blastomere is perpendicular to the first, initiating early changes in polarity.

2. In echinoderms, continued cleavages produce increasingly smaller but similar-sized cells that then form the **blastula,** a ball of cells surrounding a hollow called the **blastocoel.**

3. In amphibians **micromeres,** smaller cells, at the animal hemisphere also become progressively smaller, but in the vegetal hemisphere large and yolky **macromeres** occur. In the highly polarized frog egg, a pigmentation change following fertilization produces the **gray crescent,** which is a **morphogenic determinant.** Its presence and the manner in which it is divided provide symmetry to the embryo. It also marks the site at which a **blastopore** will form.

4. Cleavages in birds are restricted to the blastodisc and are at first **meroblastic** (partially complete). The blastodisc becomes a multicellular **blastoderm.**

The Blastula and Gastrulation

1. The frog blastula has its blastocoel displaced toward the animal pole. Three regions, the **presumptive ectoderm, mesoderm,** and **endoderm,** are already designated. **Fate mapping** suggests the predetermined role of blastula cells.

2. The three-germ layered embryo or **gastrula** is produced through **gastrulation,** a massive rearrangement of cells occurring through **invagination, involution,** and **epiboly.**
   a. Invagination, the sinking in of outer cells, first forms the **blastopore,** beginning with the **dorsal lip,** which widens to form the **lateral lips** and finally, a **ventral lip.** The blastopore contains a endodermal **yolk plug.** Invaginating cells take on a teardrop form and are called **bottle cells.**
   b. Continued inward movement or involution produces the **archenteron** (primitive gut). The presumptive mesoderm comes to lie just under the animal pole ectoderm.
   c. Through epiboly, a flattening and spreading action, ectodermal cells migrate around and enclose the gastrula. At the close of gastrulation, the embryo contains an outer ectoderm, middle mesoderm and inner endoderm.

3. Prior to gastrulation in birds, the blastoderm consists of an **epiblast** and **hypoblast.** Upon gastrulation, a thickened region, the **primitive streak** forms, followed by the sinking in of its cells, forming the **primitive groove.** Following this inward movement, these cells contribute to mesoderm and endoderm, while the remaining epiblast forms ectoderm.

4. In humans the zygote enters cell division, forming a spherical **morula** and then a **blastocyst.**
   a. The blastocyst stage is reached at about six days, just prior to implantation. It consists of a thin-walled sphere, the **trophoblast,** surrounding the **blastocyst cavity,** and a denser **inner cell mass.** Following implantation, part of the trophoblast forms the fingerlike **syncytiotrophoblast** and the rest forms the **cytotrophoblast.**
   b. The inner cell mass forms a **bilaminar embryo** (two-layered), with an epiblast and hypoblast. Gastrulation occurs in a manner similar to that in the bird.

5. As development proceeds, each germ layer is a primary contributor to specific tissues and organs of the body.

NEURULATION AND THE EMERGENCE OF FORM

1. **Neurulation** is the formation of the rudimentary central nervous system. A **neural plate** forms along the embryonic axis, its center forming a **neural groove** and its edges, as **neural folds,** rising up to close, completing the **neural tube.** A widening at one end marks the brain. The tube becomes overgrown by ectoderm.

2. Following neurulation, the basic form of the embryo emerges. Below the neural tube, mesodermal elements form the **notochord,** a temporary supporting rod. Alongside the notochord, mesodermal clusters called **somites** arise, later forming body wall muscles and the vertebral column.

Supporting Structures of Vertebrate Embryos

1. The bird and reptile egg contains a **shell,** a **yolk mass, albumen** and an **embryo. Extraembryonic membranes** include the **yolk sac, chorion, allantois,** and **amnion.**
   a. The yolk sac, which absorbs and transports food from yolk, is an endodermal extension of primitive gut.
   b. The amnion and chorion originate together as a mesodermal outgrowth, with the amnion closing first, and then the chorion. The amniotic cavity is fluid-filled.
   c. The chorion is joined later by the allantois to form the **chorioallantois,** which is important in gas exchange. The allantois is endodermal in origin.

2. In mammals, food, water, and waste removal is provided by the **placenta.**
   a. The first supporting structures are fingerlike **primary chorionic villi** of the syncytiotrophoblast. They will later be replaced by **secondary chorionic villi** as the placenta takes form.
   b. The amnion and yolk sac are the first membranes to form. The latter contributes to the digestive system and forms blood cells but never contains yolk.

c. The allantois remains small, housing placental blood vessels and later contributing to the urinary bladder.

d. Upon the formation of its extraembryonic membranes, the formless embryo is suspended by a **body stalk,** and the blastocoel has been replaced by the **coelomic sphere.** By eight weeks, the advanced placenta is well penetrated by blood vessels and contains great numbers of **chorionic microvilli.** They lie in placental blood sinuses where all exchanges take place.

The Emergence of Human Form

1. At eight weeks, the embryo is called a **fetus.**

2. During the first three **trimesters** the body systems will emerge, become refined, and begin functioning.

   a. *Nervous system.* Neurulation occurs by day 19 and by the fourth week, each region of the brain is recognizable. The cortex will be clearly visible by the trimester's end.

   b. *Circulatory system.* By day 22 the simple tubular heart functions and blood vessels form. By the 40th day the tube has looped back and formed a four-chambered heart.

   c. *Digestive and respiratory systems.* The gut, a simple tube at first, undergoes refinements and with the growth of outpocketings, forms the liver and pancreas. Similar outgrowths in the pharynx form the bronchi and lungs. While **pharyngeal arches** and **pharyngeal pouches** form gill structures in fishes, in humans they contribute to the jaw, larynx, middle ear, and hyoid apparatus.

   d. *Limbs.* Limbs first appear as simple buds, then take on a webbed, paddle appearance before cells in the webbing die off, separating the fingers and toes.

   e. *Excretory system.* The excretory system traces its own evolutionary history, first with the **pronephros,** then the more advanced **mesonephros.** The mesonephros contributes to the reproductive system, but finally the **metanephros** emerges and becomes consolidated into the paired kidneys.

   f. *Reproductive system.* In the early sexually indifferent state of the eight-week embryo, the external genitalia consist of a **genital tubercle,** which includes the **urogenital groove, urogenital folds,** and **genital swellings.** While sex is primarily chromosomally derived, unless testosterone is present, these structures will differentiate into female organs. Under the influence of testosterone, the mesonephric ducts give rise to the male epididymis and vas deferens, while the **para-**

mesonephric or **Mullerian duct** disintegrates. In the hormone's absence, the mesonephric ducts disintegrate and the paramesonephric ducts form the female oviducts.

3. The second trimester includes substantial growth and refinement, putting the final touches on the organs and systems. **Meconium** builds up in the gut, the specialized skin regions form, and movement becomes more coordinated. A fetus may survive if born at the end of the second trimester. In the third trimester, brain development continues, the cartilaginous skeleton intensifies **ossification,** and later, maternal antibodies cross the placenta.

4. Birth occurs when the placenta reaches a certain aging state and progesterone declines. The uterine contractile response to oxytocin increases, possibly because of an increase in oxytocin receptors. An increase in prostaglandin also enhances oxytocin's action. The fetal role is passive.

5. The birth process occurs in three stages, beginning with **dilation,** a softening and opening of the cervix sometimes accompanied by a rupture of the amnion. Contractions increase, bringing on **labor pains.** Stage two, **fetal expulsion,** begins when the fetus **crowns** (appears in the cervix). The third, **placental separation,** involves expulsion of the **afterbirth** or placenta.

6. With the loss of placental function during birth, the newborn's respiratory system must become functional and changes from fetal circulation must occur. Closure of the **foramen ovale** between atria and constriction of the **ductus arteriosus** greatly increase pulmonary circulation. Rising blood CO_2 levels stimulate respiratory activity and breathing begins.

7. The first milk, a clear, yellowish **colostrum,** is rich in maternal antibodies. With the drop in progesterone, prolactin secretion increases and milk is produced.

ANALYSIS OF DEVELOPMENT

The Egg and Zygote in Differentiation

1. Driesch determined that the two blastomeres from a sea urchin zygote's first cleavage produces two normal embryos.

2. Spemann ligated frog zygotes, finding that the plane of "cleavage" determined whether the success of development would occur, indicating that the frog egg is highly polarized.

3. By separating eight-celled sea urchin embryos into quartets in various ways, Horstadius found that the animal-vegetal axis is well established by the third cleavage. He further determined that polarity exists in the sea urchin egg, and successful development of artificially divided

eggs also depends on the "cleavage" plane. Such experiments will not work in the **mosaic eggs,** or zygotes, of some invertebrates because they are too highly organized to survive manipulation or disturbance of their polarity.

Tissue Interaction:
Organizers and the Induction Theory

1. Spemann and Mangold transplanted tissue from the dorsal lip of the blastopore (presumptive mesoderm and endoderm) to other regions in the late gastrula. In addition to forming expected endodermal and mesodermal derivatives, these regions also underwent neurulation.

2. The finding was later supported by the use of highly pigmented donor dorsal lip tissue which could be tracked through subsequent stages. It was unpigmented recipient ectoderm overlying the graft that differentiated into the neural structures.

3. Spemann called the dorsal lip a **primary organizer,** indicating that it induces the ectoderm to form the neural structures. This is known as **embryonic induction.**

4. In later experiments, Spemann and Mangold found that the age of the dorsal lip graft determines whether a supernumerary head or tail would form in the recipient region. They also concluded that cells in the gastrula have not reached an inrreversible state, and are still versatile.

5. Many examples of induction have been found. A classic is the triple induction system in the eye, where the **optic vesicle** induces formation of a **lens placode,** which in turn, induces formation of the **optic cup.**

6. In spite of intense investigation the inducing mechanism or substances remain unknown. There may be many inducers and these may constantly change as development proceeds.

The Cell Nucleus in Differentiation

1. Looking to the nucleus for clues, Briggs and King inactivated frog egg nuclei and replaced them with blastula nuclei. Many such transplants underwent normal development, indicating that in this stage, nuclei remain totipotent.

2. Gurdon refined the experiments, using nuclei from various advanced stages of frog development. His results indicated that the nuclei remains indefinitely totipotent, so theories of development cannot be based on irreversible changes in the nucleus.

Nerve Cells and Developmental Patterns

1. Neurons from the developing brain and spinal cord are extended out to all parts of the body but how they form the right connections isn't clear. Some simple chance may be involved.

2. Theoretically, neurons reaching their target will sprout until the proper density is reached, whereupon, they stop. The target cells release a **sprouting factor** while the neuron releases an **antisprouting factor.** When a balance is reached, the neuron stops further sprouting and inhibits other nearby neurons. Cells that do not reach their target tissues generally die.

3. In an experiment a patch of skin is reversed in a frog embryo, prior to the time when it's being penetrated by neurons. Later when one end or the other of the patch is stimulated, the frog responds as though the patches were never reversed, indicating that the neurons had made the right connections after all.

REVIEW QUESTIONS

1. Prepare a simple outline drawing of an animal sperm cell, labeling chromatin, flagellum, acrosome, mitochondrion, midpiece, head, and centriole. (954)

2. Explain what the term egg polarity means. In which of the animals discussed is such polarity most expressed? (954)

3. Provide examples of animals with microlecithal, mesolecithal, and macrolecithal eggs. What seems to determine the amount of food supply needed? (954–955)

4. Name the main supplemental parts of the human oocyte, beginning with the plasma membrane and working outward. (955)

5. Describe the action of the following during fertilization in sea urchins: acrosome, acrosomal filament, egg membrane, microvilli. (956)

6. Describe the manner in which a fertilization membrane is produced. What is its function? (956)

7. Compare the first two cleavages among the zygotes of the frog, sea urchin and human. Which of the three undergoes the most drastic shift in polarity? (957–958)

8. How does the early cleavage in the bird differ from that of the mammal and amphibian? What is the basis for the difference? (959)

9. What is the significance of the amphibian gray crescent to future developmental events? (958)

10. Describe a fate map of the frog blastula, listing the presumptive regions. How are such fate maps developed? (959)

11. Using the terms invagination, involution and epiboly, generally describe what happens to the embryo during gastrulation. (959)

12. Prepare a simple drawing of a late gastrula, labeling endoderm, mesoderm, ectoderm, archenteron and blastopore. What became of the blastocoel? (961)

13. Describe the events of gastrulation in the bird embryo, beginning with organization of the blastoderm. (960, 962)

14. Explain the organization of the mammalian blastocyst. What structures do each of its parts contribute? (962)

15. Compare the events of gastrulation in the bird and mammal. (963)

16. List several kinds of tissues derived from each embryonic germ layer. (964)

17. Describe the steps involved in neurulation. What germ layer is primarily involved? (964)

18. What germ layer forms the notochord? What is its apparent function in the embryo? (965)

19. Using a brief outline or table, compare the roles of the amnion, chorion, allantois and yolk sac in the mammal and bird. (967)

20. Explain the organization and functioning of the chorionic microvilli. (967)

21. Briefly describe the appearance of the earliest heart. How does it become four-chambered? (969)

22. What is the significance of the pharyngeal arches and pouches to the fish? To the human? (970)

23. Using the terms metanephros, pronephros and mesonephros, summarize the peculiar way in which the human excretory system develops. What parallel is seen in the pharyngeal structures? (970)

24. Name the structures that make up the human external genitalia at the sixth week of development. Why is this called an indifferent state? (970)

25. Explain the role of the sex chromosomes and of the male hormone in determining sex. (972)

26. Discuss the developmental fate of the mesonephric ducts and paramesonephric ducts in male and female embryos. (972–973)

27. List three specific events of the second trimester. In general what goes on in the last two trimesters? (973–974)

28. Name the three stages of birth and explain what happens in each. (974)

29. Name three structures of the fetal circulatory system that must undergo drastic change at birth. (975)

30. Briefly describe the cleavage experiments of Driesch, Spemann, and Horstadius. What common conclusions were the three able to make? (976)

31. Explain what Spemann meant when he called the dorsal lip of the blastopore a "primary organizer." Summarize the experiment that led to this conclusion. (979)

32. Briefly summarize the triple induction mechanism in eye development. (979)

33. What conclusion did Briggs and King reach about the potency of nuclei from the frog blastula? (980)

34. How far into development did Gurdon venture to determine when, if ever, the nucleus loses its potency? What general answer is there to the question of totipotency? (981)

35. Briefly summarize the manner in which axons emerge from the central nervous system. How might chance be involved? (981)

36. Suggest how sprouting and antisprouting factors might regulate the way axons innervate target organs. (981)

37. In what way does the frog skin transplant experiment support the idea that target tissues influence the proper organization of neurons? (982)

SUGGESTED READING

Alexander, T. 1975. "A Revolution Called Plate Techtonics Has Given Us a Whole New Earth." *Smithsonian* 5:30.

Balinsky, B. I. 1974. "Development, animal." In *The New Encyclopedia Brittanica*, 15th ed., 6:625. Although this article is almost entirely devoted to vertebrate embryology, it does that well.

Cornejo, D. 1982. "Night of the Spadefoot Toad." *Science 82*. September.

Griffin, D. R., ed. 1974. *Animal Engineering*. Readings from *Scientific American*. W. H. Freeman, San Francisco. Engineering majors with an interest in biology or biologists with an engineering background will find themselves at home here.

Johanson, D. C., and M. A. Edey. 1981. "Lucy: The Inside Story." *Science 81*, March.

Johanson, D. C., and T. D. White. 1979. "A Systematic Assessment of Early African Hominids." *Science* 203:321. Since our ideas about early hominid relationships are rapidly changing, it pays to keep up with the news. Here the authors take a fresh look at some old bones from East Africa.

Katchadourian, H. 1974. *Human Sexuality: Sense and Nonsense.* W. H. Freeman, San Francisco.

Kimble, D. P. 1973. *Psychology as a Biological Science.* Goodyear Publishing Co., Santa Monica, Calif. Kimble lives up to his impudent title: biology students who have been baffled by the orientation of standard psychology courses will find some relief here.

Kolata, G. 1982. "New Theory of Hormones Proposed." *Science*, 215:1383.

Landau, B. R. 1980. *Essential Human Anatomy and Physiology.* 2d ed. Scott, Foresman and Co., Glenview, 11.

MacArthur, R. H. 1972. *Geographical Ecology: Patterns in the Distribution of Species.* Harper and Row, New York.

Mayr, E. 1963. *Animal Species and Evolution.* Harvard University Press, Cambridge, Mass. A truly great work uniting biogeography, natural history, evolution, and genetics. Here Mayr introduces and defends "the biological species concept."

McMenamin, M. A. S. 1982. "A Case for Two Late Proterozoic—Earliest Cambrian Faunal Province Loci." *Geology*, June.

Miller, W. H. 1974. "Photoreception." In *The New Encyclopedia Brittanica*, 15th ed., 14:353. In far more detail than found in most *Brittanica* articles, Miller gives a full account of the morphology, physiology, and biochemistry of eyes and other vertebrate and invertebrate photoreceptors.

Money, John, and Anke A. Ehrhardt. 1972. *Man and Woman, Boy and Girl: The Differentiation and Dimorphism of Gender Identity from Conception to Maturity.* Johns Hopkins University Press, Baltimore. From their studies of patients who have been misdiagnosed as to gender because of congenital abnormalities of the genitalia, the authors conclude that personal sexual identity is confirmed by 18 months of age. On the other hand, prenatal sex hormone levels are shown to have a decided influence on personality.

Morton, J. E. 1967. *Guts.* St. Martin's Press, New York.

Oppenheimer, J. H. 1979. "Thyroid Hormone Action at the Cellular Level." *Science* 203:971.

Petit, C., and L. Ehrman. 1969. "Sexual Selection in *Drosophila.*" *Evolutionary Biology* 3:177. Perhaps you thought those little flies wouldn't *care.* But they do.

Rensberger, B. 1981. "Facing the Past." *Science 81,* October. (Neanderthal Man.)

Rugh, R., and L. B. Shettles. 1971. *From Conception to Birth: the Drama of Life's Beginnings.* Harper and Row, New York.

Shell, E. R. 1982. "The Guinea Pig Town." *Science 82,* December. (Framingham, Mass.)

Shodell, M. 1983. "The Prostaglandin Connection." *Science 83,* March.

Simmons, J. A., M. B. Fenton, and M. J. O'Farrell. 1979. "Echolocation and Pursuit of Prey by Bats." *Science* 203:16.

Sperry, R. 1982. "Some Effects of Disconnecting the Cerebral Hemispheres." *Science,* 217:1223.

Storer, T. I., R. L. Usinger, R. C. Stebbins, and J. W. Nybakken. 1979. *General Zoology,* 6th ed. McGraw-Hill, New York. An updated—complete, but still somewhat old-fashioned in style—version of the classic 1943 zoology textbook.

Tanner, J. M. 1974. "Development, human." In *The New Encyclopedia Brittanica*, 15th ed., 6:650. Tanner's emphasis is not on embryology but on the growth and development of human beings between birth and adulthood.

Waddington, C. H. 1974. "Development, biological." In *The New Encyclopedia Brittanica*, 15th ed., 6:643. Waddington takes an overview of the *principles* of animal development.

Wallace, R. A. 1980. *How They Do It.* Morrow, New York. Your author's paperback description of how a wide variety of species from bacteria to whales . . . do it.

Weisman, I. L., L. E. Hood, and W. B. Wood. 1978. *Essential Concepts in Immunology.* Benjamin, Menlo Park, Calif. There is no way to make the complexities of immunology simple. Advanced students will at least find this paperback a good place to get started.

West, S. 1983. "One Step Behind a Killer." *Science 83,* March (AIDS.)

Wilson, A. C., S. S. Carlson, and T. J. White. 1977. "Biochemical Evolution." *Annual Review of Biochemistry* 46:573. Each kind of protein appears to evolve (change its amino acid sequence) at a steady rate, making it possible to use protein sequence divergence as an "evolutionary clock."

Woolacott, R. M., and R. L. Zimmer. 1979. *Biology of Bryozoans.* McGraw-Hill, New York.

Scientific American articles. San Francisco; W. H. Freeman and Co.

Baker, M. A. 1979. "A Brain-Cooling System in Mammals." *Scientific American*, April.

Beaconsfield, P. et al. 1980. "The Placenta." *Scientific American*, August.

Bloom, F. E. 1981. "Neuropeptides." *Scientific American*, October.

Buisseret, P. D. 1982. "Allergy." *Scientific American*, August.

Cairns, J. 1975. "The Cancer Problem." *Scientific American*, November. Cairns argues that almost all cancers could be prevented by eliminating cancer-associated factors from the environment.

Caplan, A. 1984. "Cartilage." *Scientific American*, October.

Dautry-Yarsat, A., and H. Lodish. 1984. "How Receptors Bring Proteins and Particles into Cells." *Scientific American*, May.

Degabriele, R. 1980. "The Physiology of the Koala." *Scientific American*, July.

Edelman, G. M. 1970. "The Structure and Function of Antibodies." *Scientific American*, August.

Epel, D. 1978. "The Program of Fertilization." *Scientific American*. Includes the most dramatic scanning electron micrographs of eggs and sperm that we've seen yet.

Garcia-Bellido, A., P. A. Lawrence, and G. Morata. 1979. "Compartments in Animal Development." *Scientific American*, July.

Gurdon, J. B. 1968. "Transplanted Nuclei and Cell Differentiation." *Scientific American*, December.

Heimer, L. 1971. "Pathways in the Brain." *Scientific American*, January.

Hoyle, G. 1970. "How Is Muscle Turned On and Off?" *Scientific American*, April.

Hudspeth, A. J. 1983. "The Hair Cells of the Inner Ear." *Scientific American*, January.

Jarvik, R. K. 1981. "The Total Artificial Heart." *Scientific American*, January.

Lazarides, E., and J. P. Revel. 1979. "The Molecular Basis of Cell Movement." *Scientific American*, March. Some superb photographs of the cytoskeleton and the incredible "cellular geodome," as revealed by immunoflorescence microscopy.

Leder, P. 1982. "The Genetics of Antibody Diversity." *Scientific American*, May.

Lerner, R. A. 1983. "Synthetic Vaccines." *Scientific American*, February.

Llinas, R. R. 1982. "Calcium in Synaptic Transmission." *Scientific American*, October.

Moog, F. 1981. "The Lining of the Small Intestine." *Scientific American*, November.

Morris, S. C., and H. B. Whittington. 1979. "The Animals of the Burgess Shale." *Scientific American*, July. An exciting look at an accidentally preserved nearshore community in the early Cambrian era, including primitive members of most major phyla, some members of mysterious and bizarre long-extinct phyla, a marine onychophoran, and one amphioxuslike chordate.

Morrison, A. 1984. "A Window on the Sleeping Brain." *Scientific American*, April.

Morrell, P., and Norton, W. T. 1980. "Myelin." *Scientific American*, May.

Mossman, D., and W. Sarjeant. 1983. "The Footprints of Extinct Animals." *Scientific American*, January.

Newman, E., and P. Harline. 1982. "The Infrared 'Vision' of Snakes." *Scientific American*, March.

Pilbeam, D. 1984. "The Descent of Hominoids and Hominids." *Scientific American*, March.

Roper, C., and K. Boss. 1982. "The Giant Squid." *Scientific American*, April.

Rose, N. R. 1981. "Autoimmune Disease." *Scientific American*, February.

Rukang, W., and L. Shenglong, 1983. "Peking Man." *Scientific American*, June.

Russell, D. 1982. "The Mass Extinctions of the Late Mesozoic." *Scientific American*, January.

Schmidt-Nielsen, K. 1981. "Countercurrent Systems in Animals." *Scientific American*, May.

Short, R. V. 1984. "Breastfeeding." *Scientific American*, April.

Smith, C. A. 1963. "The First Breath." *Scientific American*, April. The dramatic changes that occur when fetal circulation and respiration suddenly shift to the adult patterns in mammals.

Van Dyke, C., and R. Byck. 1982. "Cocaine." *Scientific American*, March.

Winfree, A. 1983. "Sudden Cardiac Death: A Problem of Topology." *Scientific American*, May.

Wurtman, R. 1982. "Nutrients That Modify Brain Function." *Scientific American*, April.

Zucker, M. 1980. "The Functioning of Blood Platelets." *Scientific American*, June.

(a) By the fifth week the human embryo is just ½ inch long. The hands have taken on a paddle shape, and darkened lines mark future centers of bone formation. (b) In another few days the limbs have become better defined and blood vessels in the umbilical cord have progressed, as have surface vessels over the head and body. (c) At seven weeks, the embryo is 1 inch in length, weighing in at 2 grams. The fleecy outlines of a well-formed placenta are clearly seen. (d) The limbs have lengthened in the 9-week fetus and its fingers and toes are clearly defined. The eyes and ears are also taking form. (e) At the first trimester's end, all systems in the 2½-inch-long fetus have neared a functional state. (f) Although it looks as if it could survive on its own, the 16-week fetus must still undergo a lot of growth and refinement.

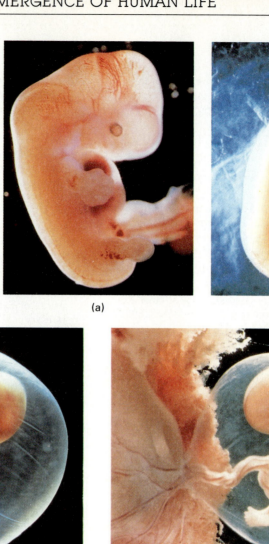

(a)

(b)

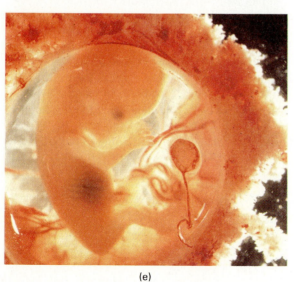

(c)

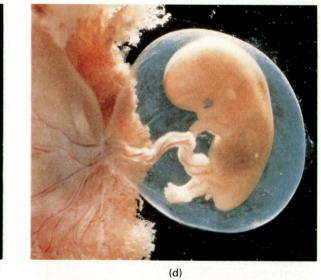

(d)

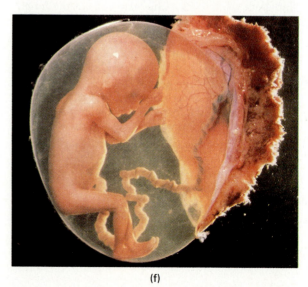

(e)

(f)

The Mechanisms and Development of Behavior

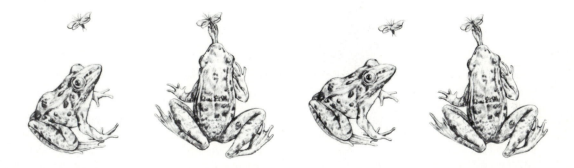

The field of study loosely called **behavior** can't exactly be called rowdy, but it certainly hasn't been as sedate or circumspect as some other areas of science either. One reason is that when it began to be studied seriously in this century, research branched off in several different directions. As each branch developed more or less independently, and as researchers in each group became vaguely aware of what the others were up to, the air became filled with criticism. At first, the lines were rather sharply drawn. Some people considered behavior in general to be genetically determined; others answered in outrage that all behavior was a result of learning. Some believed that behavior should be studied under natural conditions; others believed that it should be studied in the laboratory where the environment could be carefully controlled. No one had much use for anyone else's ideas. It would be nice if we could say that the dissension is past and that we are now beyond all that, but many basic disagreements remain to this day.

Much of the disagreement, of course, comes from the complexity of the problem. How can we study the behavior of other cultures and other species when our own behavior so often surprises us? We simply don't know what roams the cellars of our minds. We sometimes try to answer questions about ourselves by studying other species. How often are our self-conceptions based on the behavior of bright-eyed and bewhiskered beasties with an intense interest in sunflower seeds? But perhaps we are too quick in extrapolating from their behavior to ours. In fact, some say that we not only cannot use animals to understand ourselves, but we can't learn about *any* animal without studying that animal specifically. What motivates a crayfish may not even exist in the world of the rat, and one strain of rat may behave differently from another strain.

Nonetheless, the study of behavior has come of age. It has attracted armies of hard-nosed young biologists who not only tromp around in muddied boots but also speak algebra and converse with computer terminals. They are drawn partly by the challenges that remain. We may be able to synthesize a moth attractant, but we don't know what motivates moths, or what it feels like to be a moth, or what a moth senses and knows. Nonetheless, the only way to learn about behavior is to study it. The alternative is not to study it until we know enough about it to study it.

ETHOLOGY AND COMPARATIVE PSYCHOLOGY

People have been interested in animal behavior for centuries. The ancients had theories to account for such events as migration and territorial behavior, but they tended to rely on preconceived philosophy rather than on hard data gained through research. In fact, few advances were made in the study of animal behavior until the 1930s, when new dimensions were added to the discipline as a compelling theory of **instinct** was developed. The idea originated in Europe, although it is difficult to trace to its precise origins. Its development is attributed primarily to the Austrian Konrad Lorenz and his co-workers, who set forth their theory in 1937, and again in 1950 (Figure 38.1). The Dutch zoologist Niko Tinbergen, at Oxford University, amended the scheme in 1942 and 1951, and the resulting theory, in spite of its imperfections, demanded serious consideration. Then the arguments began.

Lorenz and Tinbergen, who shared the 1973 Nobel Prize in medicine and physiology with the German biologist Karl von Frisch, were the founders of an approach to behavior called **ethology,** which incorporates the instinct idea. Ethology traditionally concerns itself with **innate** (inborn) behavioral patterns. Ethological research originally was done in the field, under conditions as natural as possible, although ethologists also relied on observations of the many tame or half-tame animals that so often share their surroundings.

The strongest disagreements with the ethologists' instinct theory came from American **comparative psychologists,** led in recent years by B. F. Skinner. Their research had traditionally dealt with learning processes in laboratory animals, primarily the Norway rat. Although many of the arguments continue, the gap between ethologists and comparative psychologists has been steadily closing. The final integration of the two approaches probably depends on further data from a third area of study, **neurophysiology,** which is concerned with how the nervous system works. Such information, however, is hard won and we haven't yet filled in those nebulous gray areas. But let's take a look at the ethologist's concept of instinct and see what all the argument has been about.

INSTINCT

It was once believed that, whereas humans learn their behavior, other animals respond only to unalterable "instincts" that are indelibly stamped into their nervous systems at birth. Since mounting evidence over the years didn't bear out this sweeping generalization, the concept of instinct fell into some disrepute. In fact, many behavioral scientists declined even to use the word. The problem was

38.1 NOBEL PRIZE-WINNING ETHOLOGISTS

(a) Konrad Lorenz and gosling. Lorenz, a rather aloof Austrian, has time and again proven himself to be one of this century's most astute observers and synthesizers in the area of animal behavior. Drawing upon the work of his predecessors and coupling it with his intimate knowledge of animals, he was, with Niko Tinbergen, primary in developing the modern *instinct* theory. He has been strongly criticized for his assertion that humans are instinctively aggressive.

(b) Niko Tinbergen, a humanistic and imaginative researcher, shared the Nobel Prize with Lorenz and Karl von Frisch, an impeccable experimenter who worked primarily with insects. Tinbergen, at Oxford, worked in close association with Lorenz in the early years of ethology and summarized the concept, with his own modifications, in his landmark book *The Study of Instinct.*

(a)

(b)

38.2 A LEOPARD FROG ORIENTING

The tongue flick, a fixed action pattern, is not released until the central nervous system is stimulated by the sight of an insect in the proper position (close and in the midline). The orientation is obvious as the frog shifts its position, but once the tongue flick has started, no adjustment is possible; it will continue to its completion.

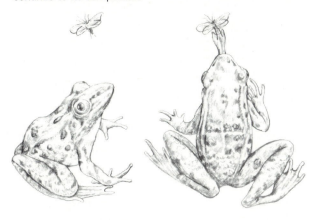

compounded by the fact that while ethologists were seeking to clarify the concept, the word *instinct* found its way into the public's vocabulary. Thus, while scientists were attempting to define instinct more precisely, others were using the term with wild abandon to explain everything from a baby's grip to swimming. We can avoid casual use of the term by viewing the idea through the eye of the ethologist.

Fixed Action Patterns and Orientation

It can't be denied that animals are born with certain behavioral repertoires. The first time a tern chick is given a small fish, it jerks its head and snaps in such a way that the fish lands head down and can be swallowed head first, with the spines safely flattened. Other birds build nests by using peculiar sideways swipes of their heads with which they jam twigs into the nest mass. And all dogs scratch their ears the same way, by moving their rear leg outside the foreleg. Such precise and identifiable patterns, which are innate and characteristic of a given species, are called **fixed action patterns.** In order to be directed properly, fixed action patterns must be coupled with **orienting movements,** and the result is an **instinctive pattern.**

To illustrate the relationship between a fixed action pattern and its orientation, consider the fly-catching movements of a frog (Figure 38.2). For the fixed action pattern (the tongue flick) to be effective, the frog must carefully orient itself with respect to

the fly's position. If the fixed action pattern is performed properly, the frog gains a fly. It is important to realize, however, that in most cases, once the fixed action pattern is initiated it cannot be altered. If the fly should move after the tongue begins its motion, the frog will miss, since the fixed sequence of movements will be completed whether the fly is there or not. Thus the frog first *orients* and then *performs* the fixed action pattern.

Whereas the frog first orients and then flicks its tongue, in other instinctive patterns the orientation may be indistinguishable from the fixed action pattern. For example, if a goose sees an egg lying outside her nest, she will roll it back under her chin in a straight line, using lateral movements of the head (Figure 38.3). Though the fixed and orienting components are difficult to separate, they were distinguished in a simple but classic experiment. The orienting movements, it turns out, depend on the direction taken by the rolling egg. If it rolls to the right, the goose moves her head to the right as she continues to draw it toward her. If an experimenter removes the egg during the retrieval, the goose will continue to draw her head back (the fixed component)—but in a straight line, without the orienting side-to-side movements. Thus we see that the fixed action pattern, once initiated, is independent of any environmental cue, so that once it starts, it continues. It might help to compare the relationship between this fixed action pattern and its orientation component to that between a car's engine and its steering wheel. Once the engine is started, the car continues to run, but each change of direction requires a new movement of the steering wheel

38.3 INSTINCTIVE RETRIEVAL BEHAVIOR

In a classic experiment, Lorenz and Tinbergen showed that the behavior of a greylag goose rolling an egg back to her nest had two components: fixed and orienting. When they removed the egg while she was retrieving it, she continued the chin-tucking movements (the fixed component) but stopped making the slight side-to-side movements that kept the egg moving in a straight line (the orienting component).

depending on the specific environmental circumstance.

An animal may not be ready to perform all its fixed action patterns at birth, however. Many of these patterns, like morphological characteristics, appear as the animal matures. For example, wing flapping or other patterns associated with flight normally don't appear until about the time a bird is actually ready to begin flying, although there may be some preliminary flapping just prior to this time. This delay once led some researchers to assume that these patterns were learned. In addition, instinctive actions may improve with time, as if they were being learned, but it is important to realize that in some cases, instinctive patterns simply mature with age, like many physical structures. It is also interesting that the fixed action pattern and its orientation component do not always mature simultaneously. A baby mouse may scratch vigorously

with its leg flailing thin air; only later will it apply the scratch to the itch.

Now let's consider another drama—one that might pull some of these ideas together. In this example, highly variable behavior (with strong learning components) becomes increasingly stereotyped until the final fixed action pattern is performed. The last orienting move is still variable, of course, but only within strict limits.

A peregrine falcon begins its hunt at daybreak. At first its behavior is highly variable; it may fly high or low, or bank to the left or the right. It is driven by hunger and would be equally pleased by the sight of a flitting sparrow or a scampering mouse. But suddenly it sees a flock of teal flying below. Its behavior now becomes less random and more predictable. It swoops toward them in what is now clearly "teal-hunting behavior" (Figure 38.4).

Almost all falcons perform this dive in just about

38.4 PEREGRINE HUNTING BEHAVIOR

The phenomenally swift and violent attack of a peregrine falcon. At this point, the bird's behavior is highly stereotyped, and if the teal makes a last minute change, the falcon will miss. This last stereotyped effort, however, was preceded by a series of modifiable acts, the first and most variable being the search for prey. Each step after that was performed in increasingly inflexible ways until the final strike. A successful strike may then initiate the next instinctive sequence—the one that results in swallowing. And that provides the greatest and longest-lasting relief.

the same way, although there are some variations. At the sight of the falcon, the teal close ranks for mutual protection. (It is hard to pick a single individual out of a tightly packed group. A predator is likely to shift its attention from one individual to another, and in failing to concentrate on a single prey, it may come up empty-handed.) The falcon first makes a sham pass to scatter the flock. As the teal scatter, the falcon immediately beats its way upward, and after picking out an isolated target below, it begins a second dive.

If we were to watch this same falcon perform such a dive on several occasions, we might notice that the procedure is much the same every time. And if we were to watch several different falcons, we might notice that they all perform this action in much the same way. The greatest amount of variation occurs during the earliest part of such dives since, at this stage, what the falcon does will be dictated in part by what the teal does. As the falcon descends, its options become fewer and its behavior becomes more and more stereotyped—that is, more rigid, distinctive, and unvarying. Finally, the falcon initiates the last part of the attack, the part that is no longer influenced by anything the teal does. At this instant the falcon's feet are clenched, and it may be traveling over 150 miles an hour. Unless the teal makes a last split-second change in direction, it will be knocked from the sky, and the falcon's hunt will have been successful.

The performance of this foot-clenched fixed action pattern then triggers the *next* instinctive sequence. The falcon may follow the teal to the ground over any of several paths and approach the senseless teal a number of ways, perhaps fluttering or hopping. With increasingly unvarying and stereotyped movements, it will grasp and pluck the teal, tear away its flesh with specific movements of its head, and swallow the meat. The sequence of muscle actions in swallowing, another fixed action pattern, is very precise and always the same. Theoretically, it is the desire to perform swallowing movements that brought the falcon out to hunt in the first place. In other words, the falcon was searching for a situation that would enable it to release the fixed action pattern of swallowing. Repeated swallowing actions seem to bring the falcon a measure of relief, and it doesn't resume hunting for a while.

We can derive several points from this example. First, the state of the animal is significant. The falcon's hunger provides the impetus, drive, or motivation. In general, if an instinctive pattern has not been performed for a time, there is an increasing

likelihood of its appearance. Second, the earlier stages of this instinctive sequence were highly variable, but the behavior became more and more stereotyped until the final fixed action (swallowing) was accomplished. Third, we assume that the performance of this final action provided some relief because the falcon does not immediately seek out conditions that will permit it to do so again. It has probably not escaped you that the performance of this fixed action pattern provides the animal with a biologically important element—in this case, food.

Let's take another case, one that tells us something about the desire to perform fixed action patterns. Courting behavior in birds is believed to be instinctive. In an experiment, Daniel Lehrman of Rutgers University found that when a male blond ring dove was isolated from females, it soon began to bow and coo to a stuffed model of a female—a model it had previously ignored. When the model was replaced by a rolled-up cloth, he began to court the cloth; when this was removed, the sex-crazed bird directed his attention to a corner of the cage, where he could at least focus his gaze. It seems that the threshold for release of the fixed courtship pattern became increasingly lower as time went by without the sight of a live female dove. It is almost as though some specific energy for performing courting behavior were building up within the male ring dove. As the energy level increased, the **response threshold,** the minimum stimulus necessary to elicit a response, lowered to a point at which almost anything would stimulate the dove. Maybe you can relate all this to something in your own experience.

Hypothesizing the buildup of some mysterious "energy" seems less necessary to explain the hunting behavior of the falcon, since the reason the falcon hunts seems apparent: it is hungry. But how do you explain hunting behavior in well-fed animals such as house cats? Perhaps the hunt is an instinctive action pattern in itself, whose performance provides a measure of relief.

Appetitive and Consummatory Behavior

There are labels for the actions we have been describing. Instinctive behavior is usually divided into two parts, the **appetitive stage** and **consummatory stage**. Perhaps the clearest examples of appetitive behavior are seen in feeding patterns, from which the name is probably derived. Appetitive behavior is usually variable, or nonstereotyped, and involves searching—for example, for food. Some stages of appetitive behavior may be more variable than others, as we saw in the increasingly stereotyped hunting behavior of the falcon. The specific objective of this appetitive behavior is the performance of consummatory behavior. Consummatory behavior is highly stereotyped and involves the performance of a fixed action pattern. In feeding behavior, swallowing is the consummatory pattern and involves a complex and highly coordinated sequence of contractions of throat muscles (Figure 38.5). These contractions are the components of a fixed action pattern. The performance of consummatory behavior is followed by a period of rest or relief, during which the appetitive patterns can't be so easily initiated again.

Numerous experiments have demonstrated that the performance of consummatory behavior is, in fact, the actual goal of appetitive behavior. This means, in effect, that a hungry animal isn't looking for something to fill its stomach but for something to swallow. In an experimental demonstration, the food swallowed by a hungry dog was shunted away from the stomach by tubes leading to a collecting pan on the floor. As soon as the dog had swallowed

the proper amount of food, it stopped eating, apparently satisfied, although its stomach was still empty. The performance of the consummatory swallowing patterns had apparently provided relief. The dog, however, began eating again sooner than it normally would have.

Releasers

Instinctive behavior is released by certain very specific signals from the environment. In the case of the falcon, the diving sequence was released by the sight of the teal flying below. The eating sequence was released by the sight of the downed bird. Courtship bowing is normally released in male ring doves by the sight of an adult female. Thus, environmental factors that evoke, or release, instinctive patterns are called **releasers.** The releaser itself may be only a small part of any appropriate situation. For example, fighting behavior may be released in territorial male European robins, not only by the sight of another male, but even by the sight of a tuft of red feathers at a certain height within their territories (Figure 38.6). Of course, such a response is usually adaptive because tufts of red feathers at that height are normally on the breast of a competitor. The point is that the instinctive act may be triggered by only *certain parts* of the environment.

The exact mechanism by which releasers work isn't known, but one theory is that there are certain neural centers called **innate releasing mechanisms** (IRMs), which, when stimulated by impulses set up by the perception of a releaser, trigger a chain of neuromuscular events. It is these events that comprise instinctive behavior. With some complex patterns, many such IRMs are involved, and specific releasers must be presented in sequence. And, as we have learned, the performance of one consummatory act (such as the falcon's knocking a teal to the ground) may act as the releaser for the next part of the pattern (such as plucking the teal before eating it).

Action-Specific Energy and Vacuum Behavior

The energy that is believed to build up for each instinctive pattern is called **action-specific energy.** The continued buildup of this energy in the absence of a releaser may cause a progressive lowering of the threshold necessary to release a pattern, as we saw in the case of the male ring dove. Eventually, if no releaser is provided at all, the instinctive pattern may "go off" on its own. The performance of an

38.5 SWALLOWING PATTERN IN THE DOG

Shown here is a diagrammatic summary of the sequence, timing, and intensity of the muscular contractions involved in swallowing in the dog. A fixed pattern such as swallowing, then, may have many components. In order for the pattern to be performed correctly, each component must function in a precise way. Such precision and complexity builds a certain conservatism into the overall pattern, since one part can't change drastically without affecting the other parts.

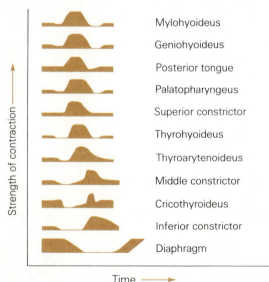

Mylohyoideus

Geniohyoideus

Posterior tongue

Palatopharyngeus

Superior constrictor

Thyrohyoideus

Thyroarytenoideus

Middle constrictor

Cricothyroideus

Inferior constrictor

Diaphragm

Strength of contraction

Time

38.6 INSTINCTIVE ATTACK

A male European robin in breeding condition will attack a tuft of red feathers placed in his territory. Since red feathers are usually on the breast of a competitor, it is to his reproductive advantage to behave aggressively at the very sight of them.

The phenomenon illustrates that releasers of instinctive behavior need to meet only certain criteria—in other words, need to represent only a specific part of the total situation.

instinctive pattern in the absence of a releaser is called **vacuum behavior.** Konrad Lorenz once noticed one of his tame starlings flying up from perches on the furniture and catching insects on the wing. He hadn't noticed any insects in his house, so he climbed up to see what the starling was catching. He found nothing. The well-fed starling had no chance to release its insect-catching behavior, and the behavior was apparently being released in vac-uo—in absence of any stimulus. Neurophysiologists have pointed out that there is no physical evidence of any such energy in the body, and have suggested that this weakens the entire ethological concept of instinct.

The Behavioral Hierarchy

Niko Tinbergen proposed a **behavioral hierarchy** to define the relationships of the different levels of an instinctive behavioral pattern. According to this model, increasingly specific stimuli are needed to release the next lowest level of an instinctive pattern on a hierarchical scale. As an example, he described the pattern of breeding behavior in the male stickleback fish—a common but fascinating denizen of European ditches (Figure 38.7). The breeding behavior is triggered by the lengthening days of spring, at which time the visual stimulus of a suitable territory may elicit either aggression or nest building. Nest building will commence unless another male appears; then aggression is released.

The sight of a competitor will elicit any of a number of aggressive responses, depending on what that male does. If the intruder bites, the defender will bite back; if the intruder runs, the defender will chase, and so on.

According to Tinbergen's model, the energy for an instinct passes from higher centers to lower ones (from left to right in Figure 38.7b). This happens only when a block, or inhibitor, for a particular action is removed. Hence, the energy of the general reproductive instinct passes to a "fighting" center (an IRM) at the sight of another male. The sight of that male fleeing then removes a block at the next level of behavior so that energy flows to the centers for chasing behavior.

It is thus apparent that the term *instinct* carries with it very specific implications. For a behavioral pattern to qualify as instinctive, it should meet the kinds of conditions we have described. For example, does the pattern show an appetitive phase? Does it ever appear as vacuum behavior? Is there a period immediately following its performance when the threshold for its release is raised? A rapid blinking of the eyes in response to a loud noise is an innate behavior but does not involve the specific components that would qualify it as instinctive.

You may have noted that there are certain problems with the instinct model. For example, a strictly hierarchical organization does not account for the fact that different appetitive actions can lead to the same consummatory act. Also, how is it that an ani-

38.7 STICKLEBACK COURTSHIP

The sequence in courtship and mating in the stickleback *Gasterosteus*, a fish found in European ditches and streams. The male is attracted by the sight of the female swollen with eggs. He is torn between chasing her and leading her to the nest he has built (a), and the result is a peculiar zigzag dance. Once she is maneuvered into his "tunnel of love" (b) he pokes her tail with his snout and thus induces her to lay eggs (c). He then promptly chases her off (d), returns to fertilize the eggs (e), and then spends days fanning oxygen-laden water over them until they hatch (f). *(right)* The hierarchical organization of the reproductive instinct of the male stickleback, according to the model devised by Tinbergen in 1942. The overall reproductive instinct is composed of a number of different patterns. The pattern that is expressed is determined by what happens in the environment. These patterns are, in turn, composed of yet more specific ones, and their expression is also environmentally determined.

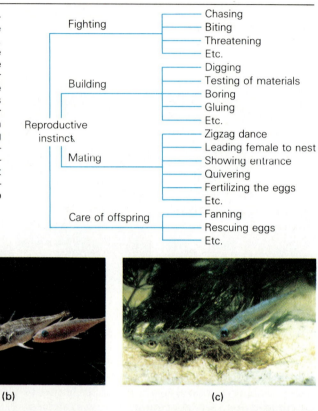

Reproductive instinct

Fighting — Chasing / Biting / Threatening / Etc.

Building — Digging / Testing of materials / Boring / Gluing / Etc.

Mating — Zigzag dance / Leading female to nest / Showing entrance / Quivering / Fertilizing the eggs / Etc.

Care of offspring — Fanning / Rescuing eggs / Etc.

(a)

(b)

(c)

(d)

(e)

(f)

mal can perform a complex chain of fixed action patterns without any intervening appetitive patterns? Finally, what we have presented is the classical ethological interpretation of instinct. It is important to understand that *innate* behavior (inborn, appearing without learning) is not synonymous with *instinctive* behavior. Put another way, not all innate patterns are instinctive. The instinct concept described here is probably most useful when describing the behavior of insects and the lower, less brainy vertebrates. But when not applied too rigidly, it can help explain certain kinds of behavior in other animals. Perhaps, then, the model's greatest usefulness lies in behavioral prediction and as a springboard for further research.

LEARNING

As we leave the realm of instinct to discuss **learning**, there are many diehard souls who would say that we are now moving to the other end of the behavioral spectrum, that instinct and learning have little in common, and that the two kinds of actions are clearly separate. However, as we will see, much adaptive behavior consists of both innate and learned components. In fact, we will also see that it is no longer considered acceptable to ask whether a behavior is learned or instinctive, since the two are often inextricably woven to produce an adaptive result.

Reward and Reinforcement

Before we consider the various types of learning, it would be well to mention a much debated concept that enters into many types of learning—the concept of **reward** or **reinforcement.**

Reinforcement is defined as a result of an action that increases the probability of that action being repeated. Reward, on the other hand, is considered to be something "good" that is gained through the performance of some action. But what does the term *good* mean in this context? It has been said that if an act fulfills or relieves a perceived need, then it is good. That act is thereby reinforced so that it will probably be repeated when the same body need appears again. "Good" has also been defined as that which reduces some physiological tension. That tension has been described as a **drive** (such as the hunger drive). It should be pointed out, however, that the reward doesn't actually have to be biologically advantageous (as we see in the case of addicting drugs).

The **drive-reduction hypothesis** is based on the notion that learning generally occurs as the animal seeks to satisfy its drives (or lower its physiological tension). However, there are a number of problems associated with the traditional concept of drive. In fact, like *instinct*, this concept has been so broadly and frequently used that it has fallen under strong criticism from researchers interested in tightening our thinking in such areas.

None of this is to downplay the importance of reward in learning situations. Most rewards probably do reduce anxiety or tension and rewards have been shown time and again to be critical in learning. However, the exceptions do not really permit drive reduction to be considered synonymous with reward.

The question arises, at what level does reinforcement operate? Lorenz proposed that the mechanism that directly brings about the good, or adaptive, effect is the fixed action pattern—the consummatory act. He said that the completion of an act (such as eating) has a reinforcing effect through **reafference,** or sensory feedback. As an example of such a learning-instinct interaction, consider nest building in crows. A crow, standing on a suitable nest site with nest material such as a twig in its beak, performs a downward and sideways sweep of the head that forces the material against the branches and later against the partially completed nest. When the twig meets resistance, the bird shoves harder, thrusting repeatedly, as would a man trying to push a pipe cleaner through a pipe-

Several complex patterns are innate and performed more or less correctly without learning. However, their adaptive sequence must be learned.

stem. When the twig is wedged tight, its resistance is increased and the efforts of the bird are heightened until, when it sticks fast, the bird consummates its activities in an orgiastic maximum of effort. Then it loses interest in that twig. Some species apparently possess no innate mechanism to guide their selection of nest material, so young birds will attempt to build with anything they can carry, which may include light bulbs or icicles. However, most of these unlikely materials will not wedge firmly into the nest, preventing the birds from reaching the consummatory stage that leads to both relief and biological success. Such failure quickly extinguishes the bird's acceptance of inadequate material until, finally, the bird becomes a true connoisseur of twigs. In such situations, then, the reinforcer of a learning situation is the performance of an innate consummatory act.

Learning may also function in placing innate patterns into their proper adaptive sequence through reinforcement mechanisms. Nest building in rats is accomplished by the performance of three motor patterns. First, the rat brings nest material to a selected site. Then, it stands in the center of the material and, turning to and fro, stacks the material into a roughly circular wall. Finally, it pats the wall with its paws, tamping it down and smoothing the inside (Figure 38.8). However, inexperienced rats, when offered paper strips, may go into a frenzy of all three activities—running around with nest material, patting, and tamping all at once. After carrying a few paper strips, it may stand above them performing heaping movements in the air and tamping down a wall that doesn't exist. Each act may be performed perfectly—but no nest results. Finally, the failure of the pattern to produce resistance to the rat's paws teaches the rat not to attempt heaping before the nest material is carried in, or patting before the wall is built.

Lorenz also pointed out that the structure and functional properties of the consummatory act were never fully understood until their teaching function was realized. He noted that there does not seem to be any case in which conditioning by reinforcement could be demonstrated in any behavioral system that does not include appetitive behavior. However, rats with electrodes implanted in their brains that are allowed to press a bar and send an impulse to their own pleasure centers will learn a task in order to be able to press the bar. Superficially, this seems to refute Lorenz's notion that appetitive behavior must be involved, since there is not likely to be an appetitive pattern for bar-pressing or pleasure-center stimulation. However, we don't know what pleasure-center stimulation feels like. Does it feel like sexual relief? Eating sumptuously? We only know that the rat will frantically press the bar up to 7000 times per hour, until it collapses in total exhaustion. Also, since learning plays such an important part in appetitive behavior, can the goal sometimes be an acquired one? Must it always be innate? At this point we simply don't know enough about what appetitive action really is to be able to support or deny Lorenz's assumption.

Three Types of Learning

The importance of learning varies widely from one species to the next. For example, there are probably very few clever tapeworms. But then, why should they be clever? They live in an environment that is soft, warm, moist, and filled with food. The matter of leaving offspring is also simplified; they merely lay thousands and thousands of eggs and leave the rest to chance—to sheer blind luck. In contrast, chimpanzees live in changeable and often dangerous environments, and they must learn to cope with a variety of complex conditions. Unlike tapeworms, they are long-lived and highly social, and the young mature slowly enough to give them time to accumulate information.

Apparently, different species may learn in different ways. However certain kinds of learning seem to be common to a number of species. Here, we will consider three of these: **habituation, classical conditioning,** and **operant conditioning.**

Habituation.   Habituation, in a sense, involves learning not to respond to a stimulus. In some cases, the first time a stimulus is presented, the response is immediate and vigorous. But if the stimulus is presented over and over again, the response to it gradually lessens and may disappear altogether. Habituation is not necessarily permanent, however. If the stimulus is withheld for a time after the animal has become habituated to it, the response may reappear when the stimulus is later presented again.

Habituation is important to animals in several ways. For example, a bird must learn not to waste energy by taking flight at the sight of every skittering leaf. A reef fish holding a territory may come to accept and pretty much ignore its neighbors, but will immediately drive away a strange fish wandering through the area. The wandering fish is likely

38.9 PAVLOV

The Russian biologist Ivan Pavlov formulated the notion of *classical conditioning*. Pavlov is best remembered for his development of the concept of classical conditioning, but his genius lay in his experiment design. He managed to overcome the formidable number of variables involved in behavioral experiments. Interestingly, his findings were eagerly embraced by Soviet communism, becoming an integral part of the party's methodology in molding the "new Soviet man," because they helped overcome the stereotyped view of class structure in society.

to be searching for a territory and therefore be more of a threat.

It may also help explain why animals continue to avoid predators (which they are likely to see only rarely), while ignoring more common, harmless species. Habituation is often ignored in discussions of learning, perhaps because it seems so simple, but it may well be one of the more important learning phenomena in nature.

Classical Conditioning.   Classical conditioning was first described through the well-known experiments of the Russian biologist Ivan Pavlov (Figure 38.9). In classical conditioning, a behavior that is normally released by a certain stimulus comes to be elicited by a substitute stimulus. For example, Pavlov found that a dog would normally salivate at the sight of food. He then experimentally presented a dog with a signal light five seconds before food was dropped into a feeding tray. After every few trials, the light was presented without the food. Pavlov found that the number of drops of saliva elicited by the light alone was in direct proportion to the number of previous trials in which the light had been followed by food.

This conditioning process also worked in reverse. When the conditioned animal was presented repeatedly with a light signal that was not followed by food, the salivary response to the light diminished until it finally was extinguished (Figure 38.10).

Pavlov's experiments demonstrated two other important properties of classical conditioning: **generalization** and **discrimination.** It was found that if a dog had been conditioned to salivate in response to a green light, it would also respond to a blue or red light. In other words, the dog was able to generalize regarding the qualities of lighted bulbs. However, it was found that with careful conditioning the dog could be taught to respond only to light with certain properties. (Dogs do not see color well and are more likely to be responding to some differences in intensity.) If food was consistently presented only with a green light and never with a red or blue light, the dog would come to respond only to green light, thus demonstrating clearly that it was able to discriminate differences among stimuli (discrimination) as well as the properties they had in common (generalization).

Operant Conditioning.   Operant conditioning differs from classical conditioning in several important ways. Whereas in classical conditioning the reinforcers (or rewards, such as food) follow the

This apparatus was devised to demonstrate classical conditioning. Upon presentation of a light, meat powder would be blown into the dog's mouth, causing it to salivate at the sight of a light alone. The salivation, then, was *conditional* on the light. Note in the first graph that the dog salivated at maximal levels after only eight trials. When the experiment was reversed and food no longer followed the light, the dog stopped salivating after only nine more trials.

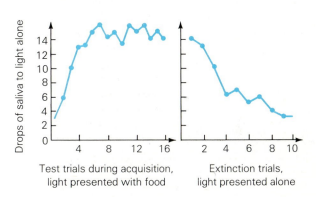

Test trials during acquisition, light presented with food

Extinction trials, light presented alone

stimulus, in operant conditioning reinforcers follow the behavioral response (Figure 38.11). Also, in classical conditioning the experimental animal has no control over the situation. In Pavlov's experiment all the dog could do was wait for lights to go on and food appear. There was nothing the dog could do one way or the other to make it happen. In operant conditioning the animal's own behavior determines whether or not the reward appears.

In the 1930s, B. F. Skinner developed an apparatus that made it possible to demonstrate operant conditioning. This device, now called a **Skinner box,** differed from earlier arrangements involving mazes and boxes from which the animal had to escape in order to reach the food.

Once inside the Skinner box, an animal has to press a small bar in order to receive a pellet of food

38.11 DIFFERENCES IN CLASSICAL AND OPERANT CONDITIONING

Note that in classical conditioning, an unconditioned stimulus (such as food) becomes paired with a conditioned stimulus (such as a flash of light). In time, the conditioned stimulus becomes a substitute for the unconditioned stimulus and is then able to produce the unconditioned response, such as salivation. In operant conditioning, after receiving a conditioned stimulus, there is an opportunity for the animal to respond in various ways. However, only the "correct" response is reinforced. Finally, the stimulus produces a specific response.

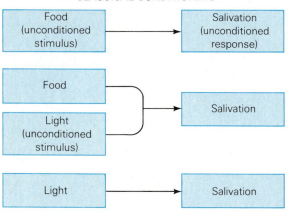

CLASSICAL CONDITIONING

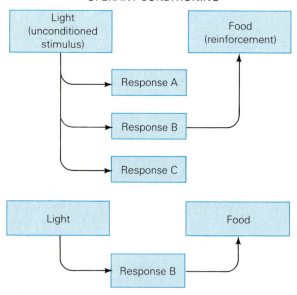

OPERANT CONDITIONING

from an automatic dispenser (Figure 38.12). When the experimental animal (usually a rat or a hamster) is first placed in the box, it ordinarily responds to hunger with random investigation of its surroundings. When it accidently presses the bar, lo! a food pellet is delivered. The animal doesn't immediately show any signs of associating the two events, bar-pressing and appearance of food, but in time its searching behavior becomes less random. It begins to press the bar more frequently. Eventually, it spends most of its time just sitting and pressing the bar. Skinner called this sort of learning *operant conditioning*. It is based on the principle that if a behavior is rewarded, then the probability of that pattern reappearing is increased. (In other words, reward is reinforcing.)

It is possible to build "stimulus control" into operant conditioning situations. For example, a pigeon in a Skinner box may be required to peck a bar in order to receive food. If food is delivered only when the bar is pecked within a certain period after a "Peck" sign lights up, the pigeon will come to peck the bar mostly at that time. If the pigeon pecks after a "Don't peck" sign lights up, and that peck is not followed by food, the pigeon will come to ignore that sign through discrimination learning. (It should be pointed out that pigeons can also be trained to peck in response to the "Don't peck" sign—pigeons can't really read.)

The relative importance of each type of learning to animals in the wild isn't known at this point. However, it is likely that most behavior patterns

38.12 RAT IN A SKINNER BOX

Skinner boxes are designed to promote operant conditioning. For example, a hungry rat may move randomly, searching for food, until it accidently presses a bar that delivers a food pellet. Each delivery means a greater probability that the rat will press the bar again, until finally the rat learns simply to press the bar each time it wants a pellet.

Sociobiology

Sociobiology burst upon the academic scene in the mid-1970s, prompted and defined by a monumental book written by Edward O. Wilson at Harvard University. His book dealt almost entirely with other organisms, especially insects, with only the last 27 pages dealing with humans. But it was enough. Wilson was attacked with a vigor usually reserved for the worst sorts of charlatans. But Wilson was no charlatan. Why, then, the attack?

Wilson, in including humans in his discussion, had suggested what most biologists had always taken for granted—that human social behavior was molded in part by evolutionary processes. This, of course, means that we are not *tabulae rasae*, blank slates waiting to be defined and molded solely by experience, by culture. Instead, suggested Wilson, we are born with certain tendencies and proclivities that have been shaped by our long processes of evolution. He suggested, further, that we are born with a certain genetically based foundation upon which we can build any number of social structures. However, as in building a house, those social structures can't exceed the limits of the foundation without grave risks.

The howls of denial began from down the hall. Some of Wilson's colleagues at Harvard held the view that human social systems were infinitely malleable and "perfectable." Some, of definite political persuasion, further noted that the sociobiological view flew in the face of Marxist doctrine. Many of Wilson's detractors were activists and formed "study groups" that denounced, hounded, and even physically attacked Wilson as he continued to bring his ideas to public attention. Because some of these detractors were among the best minds in biology, the academic community was at first perplexed. Why was Wilson being attacked? Was there something in his statements that had escaped the rest of us? When it dawned on the scientific establishment that the attacks were politically inspired, many people began to turn away from the noisemakers.

Wilson's detractors, of course, may have had the best of motives. After all, could one not argue that if we are genetically "compelled" to behave in certain ways, then isn't change impossible, even unnatural? The sociobiologist's position, they feared, could be used to promote racism and sexism. Perhaps they confused the attempt to describe "what is" with "what ought to be." A careful reading of sociobiology shows that it, in fact, does not defend the status quo; it simply seeks to understand it. If changes are in order, we must know what prompts us to behave as we do. It does no good to pretend that people are as we *wish* they were.

Obviously, there had been a grave misreading of sociobiology. The discipline is merely one more tool we can use in the examination of ourselves. If we are biologically constrained to behave in this way or that, we need to know it. Changes are indeed called for, but to bring them about we need information. After all, we didn't overcome gravity by simply denying its existence.

---

arise through the interaction of several types of learning.

The learning capabilities of an animal are largely related to its evolutionary heritage. For example, M. E. Bitterman of Bryn Mawr College demonstrated that there are strong species differences in **habit reversal** ability. In these experiments, once an animal had learned to discriminate between two choices for its reward, the reward was transferred to the opposite choice. Monkeys and rats showed the best performance, fish the worst, and turtles scored somewhere in the middle. The turtle scores were rather peculiar. They progressively improved in making spatial choices (for example, in differentiating high and low) but not in visual choices (such as in differentiating between a circle and a square). Thus, it appears that habit reversal requires a specific kind of learning ability that is present in mammals but not in fish, and in turtles only with respect to a certain class of problems.

HOW INSTINCT AND LEARNING CAN INTERACT

Much of the argument in the field of behavior has been over one simple question: Is a certain behavior innate or learned? After many years of bickering, the question itself was rapidly losing its validity, but in recent times the controversy has been renewed (see Essay 38.1). Nevertheless, in general, the supposition that a particular behavior stems from one source or the other neglects the myriad ways in which behavioral components interact. In a sense, it's like asking whether the area of a triangle is due to its height or its base. A better question is, how do innate and learned patterns interact in the development of a particular behavior? (Notice that we are not equating *innate* and *instinctive*. Instinctive patterns, you recall, are a special kind of innate behavior.)

As an example of such interaction, consider the development of flight in birds. Flight is a largely innate pattern. A bird must be able to fly pretty well on the first attempt or it will crash to the ground as surely as would a launched mouse. It was once believed that the little fluttering hops of nestling songbirds were beginning flight movements, and that the birds were, in effect, learning to fly before they left the nest. But in a series of experiments, one group of nestlings was allowed to flutter and hop up and down, while another group was reared in boxes that prevented any such movement. At the time when the young birds would have normally begun to fly, both groups were released. Surprise! The restricted birds flew just as well as the ones that had "practiced."

Flight behavior, then, must be an innate, or unlearned, pattern. It is important to realize, though, that generally young birds don't fly as well as adults. Many innate patterns can be improved upon by learning through practice (and strengthening the muscles involved).

In other instances, the innate and learned components of a behavior can be more clearly differentiated. The process by which a red squirrel opens a hazelnut is a rather complex behavior. The squirrels first cut a groove along the growth lines on one or both of the flat sides of the nut, where the shell is

thinnest. Then they insert their incisors into the groove and break the shell open. In one study, baby squirrels were reared without any solid food. When they were finally presented with a hazelnut, they correctly performed the gnawing actions and even inserted their incisors into the grooves correctly. Thus, these patterns are apparently innate. The problem was, they tried to gnaw all parts of the shell, even the thickest areas. Eventually they were usually able to break open the shell but only after great expenditure of energy. In time, however, their performance improved. After they had handled a few nuts, they began to make their grooves at the thinnest part of the shell and the shell opened more easily. Most of the squirrels soon became proficient at opening hazelnuts, as their innate gnawing pattern was modified through learning to produce the complex adaptive result (Figure 38.13).

The relative importance of innate and learned components of behavior may vary widely. For example, whereas reproductive behavior in stickleback fish may be largely innate, the nut opening in squirrels is perhaps less so. Furthermore, in some songbird species, the song is largely learned; in others it is almost completely innate. White-crowned sparrows demonstrate a very interesting interaction of innate and learned behavior. In order to learn their song properly, they must hear the song of their species during a brief "sensitive period" when they are too young to sing at all. If they are not exposed to the song until after this period, they will produce abnormal songs, lacking the finer details (Essay 38.2).

38.13 LEARNING IN YOUNG SQUIRRELS

Young squirrels instinctively gnaw at the surfaces of nuts. Later they will learn to bite through the thinnest part of the shell and deftly open the nut.

MEMORY AND LEARNING

Many years had elapsed during which nothing of Cambray had any existence for me, when one day in winter my mother offered me some tea. . . . I raised to my lips a spoonful of the tea in which I had soaked a morsel of cake. No sooner had the warm liquid, and the crumbs with it, touched my palate than a shudder ran through my whole body. . . . And suddenly the memory returns. The taste was that of the little crumb of madeleine which on Sunday mornings at Cambray my Aunt Leonie used to give me, dipping it first in her own cup of real or of lime-flower tea. . . . Once I had recognized the taste, all the flowers in our garden and in M. Swann's park, and all the water lilies on the Vivonne and the good folk of the village and their little dwellings and the parish church and the whole of Cambray sprang into being.

Marcel Proust
Remembrance of Things Past

Imprinting

Some years ago, Konrad Lorenz discovered that newly hatched goslings would follow whatever moving object they saw and that they would continue to identify with this object throughout the rest of their lives. Of course, under most circumstances any such object was likely to be a parent, and in this way they learned to identify their own species. Later, as they approached their first breeding season, they would seek out an individual with the traits of the individual they had followed soon after hatching. If they somehow were exposed to something else at that time, they would focus on whatever resembled that thing when it came time to breed. A group of goslings hatched in an incubator saw Lorenz first and, as they grew, would often dutifully fall into single file, following after him as he walked around the farm.

Many animals also learn sexual identification during this critical period. Lorenz once had a tame jackdaw that he had hand reared and it would try to "courtship feed" him during mating season. On occasion when Lorenz turned his mouth away, he would receive an earful of worm pulp! The story of Tex, the dancing whooping crane, provides another example of this type of learning (see photo).

The sort of learning that takes place in such a brief period of a young animal's life has been called *imprinting*. It was once believed to take place through special learning processes but most researchers now believe that it occurs through the same neural mechanisms that are involved in other types of learning.

Imprinting is especially important in many kinds of bird song. If male white-crowned sparrows are deafened while very young, they will sing disconnected notes, but no real song. The birds apparently must be able to hear themselves in order to learn the song of their species. Normally, these birds must be exposed to the song at about the age of three months. They will not begin to sing, however, until some months later. Then, even a bird that has been experimentally isolated after hearing the song of its group will sing the correct song—not only the basic song of the species, but the variation (dialect) of the particular local population whose song it heard.

To further illustrate the specificity of the imprinting process, whereas white-crowned sparrows are able to learn dialects of their own species during their critical period, they are completely unable to learn anything about songs of other species to which they have been exposed.

Imprinting, then, is a kind of learning that takes place at a specific period in an animal's life and that is very strongly influenced by genetic constraints on the type of learning that is possible.

Tex, the only female whooping crane at the International Crane Foundation breeding area in 1982, had been hand reared and therefore had imprinted on humans. She rejected the mate provided for her but could be enticed to lay eggs (artifically fertilized) by "dancing" with humans. She preferred Caucasian men of average size with dark hair.

There are indeed not only bits of information tucked away in our memories, but overall settings, moods, and impressions that can be recalled if properly summoned. In some cases, we can recall things from our memory at will, at times we simply cannot, and at other times, remembrances flock to our conscious thought that we would rather have suppressed. Obviously, we have not learned to use our memories as well as we might like. But our chances are better if we know more about how they work.

Memory is the storage and retrieval of information, and its characteristics are important in any consideration of learning. Learning is accompanied by changes in the central nervous system. Some change in the neural apparatus produces a more or less permanent record of the learned thing. The physical change that is presumed to occur in the brain when something is learned is called the **memory scar, memory trace,** or **engram.**

It is suspected that virtually everything we encounter is learned—stored away in the brain.

Wilder Penfield (Figure 38.14) and his group, working in Montreal, used electrodes to probe the brain (which has no pain receptors) of conscious patients, and asked them to describe their sensations as various parts of the brain were stimulated. The results were startling. Some patients "heard" conversations that had taken place years before, some heard music or seemed to find themselves with old friends, long deceased. They seemed almost to "relive" the experiences, rather than simply remember them. The powers of recall under such circumstances are phenomenal. A hypnotized bricklayer described markings on every brick he had laid in building a wall many years before. His recollections were carefully recorded, then the wall was found and examined. The markings were there! (Our memory is not infallible, however. We also have a tendency to embellish partially recalled events.)

Although we may have entered every jot and tittle of our lives into our mind's ledger, we are not consciously able to recall very much of it. Sometimes we can't even recall what we have tried to memorize—as you well know. But apparently the information is there, just the same. The fault lies with our recall techniques. The implications of this finding are enormous. If our research ever enables us to recall or relive earlier events at will, we might "reread" novels on long train trips and all our exams would be virtually open-book tests. Perhaps terminally ill and pain-racked patients could be stimulated to relive happier, youthful days with loved ones.

The Advantages of Forgetting

There are all sorts of strange abilities associated with memory in humans. **Idiot savants,** a special class of retardates, have very low IQs but are able to accomplish incredible feats, such as multiplying two five-figure numbers in their heads, or immediately giving the day of the week on which Christmas fell in 1492—or any day in any year (although this is probably not a memory feat). A Russian man of normal intelligence made his living giving stage performances as a **mnemonist.** He would sometimes memorize lines of 50 words. Once, in just a few moments he memorized this nonsense formula:

$$N \cdot \sqrt{\frac{d^2 \cdot 85^3}{vx}} \cdot \sqrt[3]{\frac{276^2 \cdot 86x}{n^2v \cdot 264}} \cdot n^2b = sv \cdot \frac{1624}{32^2} \cdot r^2s$$

Fifteen years later, upon request, he repeated the entire formula without a single mistake.

And what about those people with "photo-

38.14 WILDER PENFIELD

Penfield was a pioneer in the study of mapping the functional areas of the brain. In one group of experiments he elicited responses from conscious patients by electrically stimulating different parts of the brain.

graphic memories"? They exist, and they are called **eidetikers.** Proof of their abilities has been demonstrated with **stereograms.** These are apparently randomized dot patterns on two different cards; when they are superimposed by use of a stereoscope, a three-dimensional image appears. One person was asked to look at a 10,000-dot pattern with her right eye for one minute. After 10 seconds, she viewed another "random" dot pattern. She then recalled the positions of the dots on the first card and conjured up the image of a "T."

If such abilities are possible for our species, why haven't they been selected for so that by now we can all perform such feats? On the surface, the advantages seem enormous. But there are also serious drawbacks to remembering everything. The Russian mnemonist recalled everything so well that it seemed to him as if every event of his life had just occurred. He could look at a barren tree, "recall" its leaves, and, when he looked away, not be sure whether the tree was leafy or not. The mnemonist could watch the hands on a clock and not notice they had moved, remembering (or "seeing") only where they had been. Moreover, what about all those insignificant events it is of no advantage to remember? If we could not forget, the energetically expensive neural apparatus would be wasted retaining such information (assuming such retention takes more energy than the storage of material we normally can't recall).

Thus, the reason we can't remember as well as the people in these examples is that we, as a species, have not found it necessary or useful. We generally don't need to recall every stone on the path to the place where we found food yesterday; we only need to remember the location of the path and the food. In fact, remembering too much about our physical environment might cause us to relate to the environment on the basis of remembrances. Thus, we could be expected to adjust to environmental changes more rapidly if we were never quite sure what to expect because with our faulty memories, we must be prepared for change.

Theories on Information Storage

Long-Term and Short-Term Memory.   It now seems that there are two separate mechanisms by which information is stored, as evidenced in the differences between **long-term memory** and **short-term memory.** The existence of these two types of memory was first noticed in humans with brain concussions who were unable to recall what happened just before an accident, but could remember what happened much earlier. On the basis of such findings, researchers postulated two memory systems that could process information independently. Information stored in the long-term memory system is relatively permanent (subject to very slow decay). Once information is processed into the long-term system, it is not easily disrupted and therefore may be recalled with minimum confusion after long periods of time. Short-term memory, on the other hand, is relatively easily disrupted and subject to rather rapid decay. (This is why cramming for exams is not a good idea. If the exam is postponed, all is lost.) Also, there is evidence that the short-term memory can be overloaded, whereas the long-term memory can't be.

Apparently, short-term storage is necessary before long-term storage can occur. In a series of experiments, it was found that if an inedible object is presented to one arm of an octopus, over and over in rapid succession, the arm will come to reject it after learning that it is inedible. If the object is then immediately presented to other arms they will accept it. But if some time is allowed to elapse before the second presentation, all the arms will reject it. This finding suggests that the information acquired by one arm is only locally adaptive. It must filter from a hypothesized short-term center to a long-term system before a general adaptive pattern (total rejection) can appear. The experiment actually only suggests that something filters from neural centers associated with one arm to general centers. The short- and long-term phenomena are postulated because the learning arm acquires no more information than the other arms once the information has become integrated.

In another experiment, a mouse was put into an arena (enclosed area) containing only a box with a hole. If the mouse ran into the hole, its feet received a terrific electric shock, and it never went near that hole again. But if the mouse was given electroconvulsive or chemical shock treatment within about an hour after the unpleasant incident, it would forget the whole experience and helplessly run into the same hole the next day, and every day as long as it survived the experiment. On the other hand, if the electroconvulsive treatment was delayed for two hours, it was not effective and the mouse did not forget its experience. Mouse memory, then, is labile for up to about an hour, and then becomes fixed. We infer that transfer is made from short-term "holding" memory to permanent memory in that time.

The Consolidation Hypothesis.   Such observations have led to the development of a **consolidation hypothesis,** which states that memory is stored in two ways. First, any registered experience enters a short-term memory system where its neurological effects last about an hour; then the system returns to normal (forgets). But while events are still in the short-term system and subject to recall by memory, they are being shifted into the long-term memory system, where they form an engram. The transfer from short-term to long-term is called **consolidation,** and during this process, the memory is susceptible to all sorts of disturbances or even obliteration.

Though we have been speculating about how engrams are formed, we don't know for sure whether engrams *exist*, at least in the sense that we have been discussing them. Admittedly, there is a danger in constructing an imaginary working picture to account for what we've heard or observed, and then assuming this imaginary thing exists. However, if we hope to ever gain some real understanding of behavior, we must keep in mind the level of supportive evidence for each assumption we encounter.

APPLICATION OF IDEAS

1. Clearly distinguish between innate behavior and instinctive behavior.

2. What are the main factors (or information) that have drawn ethology and comparative psychology closer together?

3. Choose what you consider to be an instinctive act and break it down into its appetitive and consummatory behaviors. (Remember that there may be several appetitive or consummatory patterns involved in any complex behavior.)

KEY WORDS AND IDEAS

ETHOLOGY AND COMPARATIVE PSYCHOLOGY

1. The study of animal behavior was principally limited to philosophy until the theory of **instinct** was developed in the 1930s.

2. The founders of **ethology** were Konrad Lorenz, Niko Tinbergen, and Karl von Frisch. Ethology concerns itself with **innate** behavioral mechanisms, observed primarily in the field.

3. The field of **comparative psychology,** largely developed by B. F. Skinner, has traditionally focused on laboratory studies primarily involving learning experiments. The developing science of **neurosphysiology** shows promise in further uniting the two disciplines.

INSTINCT

1. Instinct refers to certain classes of innate responses that have various specific aspects.

Fixed Action Patterns and Orientation

1. **Fixed action patterns** are precise, identifiable behavior patterns that are innate and characteristic of a given species. When coupled with **orienting movements** they form **instinctive patterns.** Orienting movements and fixed action patterns are not always clearly separable.

2. A complex instinctive pattern can be demonstrated by a hunting falcon that is motivated by hunger to begin the appetitive searching phase. Each instinctive phase of the hunt grows increasingly stereotyped until it ends in a consummatory act that brings a degree of satisfaction, namely, the relief of hunger.

3. In the absence of appropriate releasers, the **response threshold** may become lower with less specific releasers required.

Appetitive and Consummatory Behavior

1. **Appetitive stages** of behavior are usually variable and nonstereotyped, while **consummatory stages** of behavior are fixed. Consummatory behavior may involve the execution of one or several fixed action patterns.

Releasers

1. **Releasers** are specific signals from the animal's environment that bring on instinctive behavior. There may be neural centers called **innate releasing mechanisms** that trigger neuromuscular events. The performance of one consummatory act may provide the releaser for the next.

Action-Specific Energy and Vacuum Behavior

1. The hypothetical buildup of energy for an instinctive pattern is known as **action-specific energy.** As it builds, the threshold for its release lowers. In the absence of a specific releaser, it may eventually go off on its own, producing **vacuum behavior.**

The Behavioral Hierarchy

1. Tinbergen noted different levels of an instinctive behavior pattern. He arranged these in a **behavioral hierarchy.** Stickleback mating behavior is an example. Tinbergen suggested that the energy for instincts passes from higher centers to lower. As inhibitors for each act are removed, behavior moves from one level down to another.

LEARNING

1. Adaptive behavior is generally composed of both innate and learned components.

Reward and Reinforcement

1. **Reward** involves the fulfillment of a need, whereas **reinforcement** refers to an increased probability of an action being repeated.

2. The **drive reduction hypothesis** states that reward involves the relief of the physiological or psychological tension that produces **drive.**

3. Because many exceptions to the theory have been found, drive-reduction is no longer considered synonymous with reward. Reward or reinforcement can operate through **reafference** (sensory feedback).

Three Types of Learning

1. **Learning** involves changes in behavior due to experience. Some animals exhibit little or no

need to learn, while others depend heavily on learning. Animals learn in three ways: by **habituation, classical conditioning,** and **operant conditioning.**

Habituation

1. Habituation involves learning not to respond to a stimulus. Animals learn to sort stimuli and to stop responding to those that are irrelevant to their survival.

Classical Conditioning

1. Pavlov's classic experiment demonstrated that stimuli can be substituted. Thus, the dog learned to salivate at a signal that it had associated with food, even when the food was not present.

2. The dog also learned **generalization,** so that a similar signal would provoke the same response. The dog also was able to learn **discrimination** between similar signals.

Operant Conditioning

1. In classical conditioning the reward follows the stimulus, while in operant conditioning the reward follows the behavioral response.

2. Operant conditioning was investigated by using the **Skinner box,** a device that rewarded the subject for the correct response.

3. Some animals undergo rapid **habit reversal** when rewards are withheld.

HOW INSTINCT AND LEARNING CAN INTERACT

1. The question of whether a particular behavior is instinctive or learned is no longer considered particularly useful. It is better to ask how the two interact.

Memory and Learning

1. The physical change that is presumed to occur in the brain when something is learned is known as a **memory scan, memory trace,** or **engram.** Memories of virtually everything we encounter may be stored in the brain. Studies by Wilder Penfield reveal that electrical stimulation can bring on recall of experiences presumably forgotten.

2. Persons capable of phenomenal memory tasks include **idiot savants, mnemonists,** and **eidetikers.**

The Advantage of Forgetting

1. There are serious disadvantages to total memory recall. There may be no advantage in terms of how we function in an ever-changing environment.

Theories on Information Storage

1. From experiments it was concluded that information is first filtered through **short-term memory** centers and later finds its way to **long-term memory** centers. This explanation is called the **consolidation hypothesis.**

REVIEW QUESTIONS

1. Carefully define the term *ethology* and explain how it differs from comparative psychology. (992)

2. Why did the term *instinct* fall into disrepute in the behavioral sciences? (992–993)

3. Describe the orientation and fixed action patterns in the feeding of a frog. (993)

4. Outline the steps involved in the falcon's hunt and indicate which are orientation movements and which are fixed action patterns. (994–995)

5. Explain what happens to the response threshold when a consummatory act is delayed. (995)

6. Carefully distinguish between appetitive and consummatory behavior. (996)

7. Describe an experiment that indicates that swallowing is a consummatory act. (996)

8. Using the term *IRM* (innate releasing mechanism), explain how releasers might work. (996)

9. Present an example of vacuum behavior and relate your example to the concept of action-specific energy. (997)

10. List the steps in the behavioral hierarchy in stickleback breeding and explain how one step might lead to another. (997)

11. Explain what the drive-reduction hypothesis is and what its drawbacks are. (999)

12. What is habituation and how is it of adaptive value to an animal? (1000–1001)

13. Using Pavlov's experiment as an example, define classical conditioning. (1001)

14. How does operant conditioning differ from classical conditioning? (1001)

15. Explain how the Skinner box is used to demonstrate operant conditioning. (1002)

16. In what groups of animals is habit reversal best demonstrated? (1003)

17. What is wrong with asking, ''Is a behavior innate or learned?'' (1003)

18. What is an engram? Review the evidence that suggests that engrams are rather permanent. (1005)

19. Describe the two types of memory and their characteristics. (1007)

Adaptiveness of Behavior

One of the most fundamental questions in the field of animal behavior is, why do animals do what they do? Sometimes the answer is obvious. That mouse in your kitchen scurries away when you walk in because it is in its best interest to avoid such large creatures. Tarantulas living in holes in the desert floor avoid drowning by leaving their burrows and climbing into bushes at the first sign of rain. Chickens spend the night on low-lying limbs where foxes can't reach them. We can indeed see why some animals behave in certain ways.

In other cases the explanations are far less apparent. Why do hammerhead sharks assemble in great numbers and then begin some long migration? Why do some predators hunt mainly at dusk? Why do some woodpecker pairs forage close together and others far apart? Why are some species colonial and others territorial?

Before going any further we should note that there are some problems with such "why" questions.

These problems are related to the assumption that the behavior is adaptive. Perhaps the behavior is not adaptive at all. For example, some patterns may be "evolutionary hangovers," a behavior no longer useful and perhaps on its way out of the population's repertoire. Perhaps the animal is a bit quirky and is wasting its time performing some useless act (as a misguided scientist records every

detail). Perhaps the animal is merely experimenting with some behavior that will soon be abandoned. But perhaps the greater problem is the underlying philosophical assumption that anything an animal does is bound to be somehow useful or adaptive. Some scientists argue that the assumption itself is ill founded.

With this reminder, then, we will consider some of the ways that behavior can be adaptive and we can begin with one of the most fascinating puzzles in animal behavior today: the problem of orientation and navigation. These studies not only provide a marvelous detective story but they illustrate some of the ways that scientists, through their experimentation, are attempting to answer some of the most perplexing questions in biology.

ORIENTATION AND NAVIGATION

We can begin here by noting that on a variable and changing planet such as this, proper positioning and direction of movement can be critical. It is best for a sowbug to remain active in dry areas and to slow down when it encounters a more desirable moist area. It is best for certain aquatic insects to move to the protective, darkened bottom as larvae, but as they change to adults, to seek the brightly lit surface that signals a greater abundance of oxygen. Not only do many shorebirds stand so as to face cold winds that would otherwise ruffle their insulating feathers, but some of them also leave those cold winds to travel to warmer climes each fall.

There are undoubtedly people in South America who wonder where their robins go every spring. Those robins, of course, are "our" birds; they merely vacation in the south each winter. If you've been keeping a careful eye on the robins in your neighborhood, you may have noticed that a specific tree is often inhabited by the same bird or pair of birds each year. You may have wondered, since these birds winter thousands of miles away, how they find their way back to that very tree each spring. Such questions have perplexed people for years. And the more we learn, it seems, the more strange the story becomes.

The migratory behavior of birds is truly astounding, but other animals may have similar, if less dramatic, adaptive patterns. The impassive toad in your garden sits in a hole and gorges on bugs, but when the urge hits, the toad heads through grass jungles for the pond where it was born. In fact, thousands of toads suddenly begin to move together, hopping slowly and generally unnoticed. In contrast with the robins' 5000-mile journey, the toads may migrate perhaps two or three miles, but it is an epic trek that may take weeks. We don't notice them much until they try to cross highways, where their flattened evidence becomes more apparent. But many make the journey, and after the males set up small territories around their ponds and begin chorusing, the females move toward specific males (usually the larger ones) and they mate. After mating and egg laying, the majority of the toads will make their way back to their own holes and drainpipes. How do they find their way and why is it important that they do?

Orientation

One aspect of navigation involves **orientation,** which simply refers to the ability to face in the right direction. **Navigation,** on the other hand, involves starting at point A and finding one's way to point B. More is known about orientation than about other aspects of navigation. Much of the recent work with birds began in the 1950s when a young German scientist, Gustav Kramer, provided some answers and, in the process, raised new questions.

Kramer found that caged migratory birds became very restless at about the time they would normally have begun migration in the wild. Furthermore, he noticed that as they fluttered around in the cage, they usually launched themselves in the direction of their normal migratory route. Kramer devised experiments with caged starlings and found that their orientation was, in fact, in the proper migratory direction—except when the sky was overcast. When they couldn't see the sun, there was no clear direction to their restless movements. Kramer surmised, therefore, that they were orienting by the sun. To test this idea, he blocked their view of the sun and used mirrors to change its apparent position. He found that under these circumstances the birds oriented with respect to the position of the new "suns." This seems preposterous. How could a stupid bird navigate by the sun when we often lose our way with the road maps? Obviously, more testing was in order.

In another set of experiments Kramer put identical food boxes around the cage, with food in only one of the boxes (Figure 39.1). The boxes were sta-

39.1 KRAMER'S ORIENTATION CAGE

The birds can see only the sky through the glass roof. The apparent direction of the sun can be shifted with mirrors.

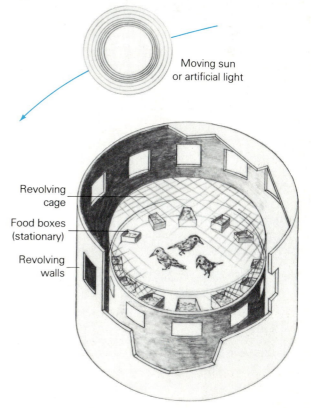

Moving sun
or artificial light

Revolving cage

Food boxes (stationary)

Revolving walls

tionary so that the one containing food was always at the same point of the compass. However, its position could be changed with respect to its surroundings by revolving either the inner cage containing the birds or the outer walls that served as the background. As long as the birds could see the sun through any window, no matter how their surroundings were altered, they went directly to the correct food box. It didn't matter where the box was placed; the birds were not confused. When the food was moved to a different box, they continued for a time to approach the old box, then finally gave up and searched around until they discovered the new food box. At that point they couldn't be easily fooled into approaching another box. On overcast days, however, the birds were completely disoriented and had great difficulty locating their food (Figure 39.2).

This sort of sun-compass orientation requires that starlings know both the time of day and the normal course of the sun. In other words, incredible as it seems, they apparently know where the sun is supposed to be in the sky at any given time. It seems that this ability is largely innate. In one experiment, a starling reared without ever having seen the sun was able to orient itself fairly well, although not as well as birds that had been exposed to the sun earlier in their lives (this would appear to be another example of the interaction of innate behavior, maturation, and learning). Evidence for

39.2 THE ANALYSIS OF SUN-COMPASS ORIENTATION IN STARLINGS

The dots in **(a)** and **(b)** represent directions the birds being tested in Kramer cages flew as they tried to migrate. The results in **(a)** show that the birds maintain the same heading even as the sun changes position. In **(b)** orientation is shifted as the apparent position of the sun is changed by mirrors. In **(c)** the birds were subjected to artificial light-dark schedules that were off by six hours (¼ circle). In **(d)** an artificial sun was held stationary. The birds then changed their direction at about the same rate the real sun would have moved.

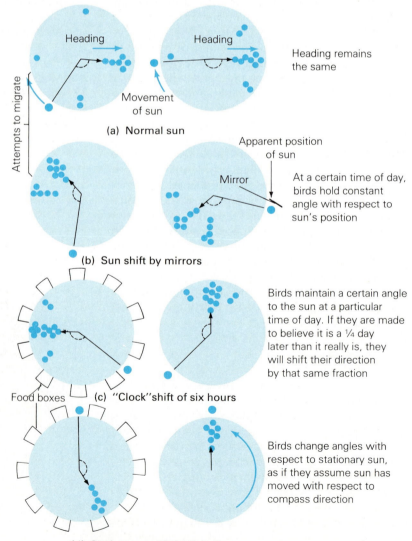

Attempts to migrate

Heading

Heading

Movement of sun

(a) Normal sun

Heading remains the same

Apparent position of sun

Mirror

At a certain time of day, birds hold constant angle with respect to sun's position

(b) Sun shift by mirrors

Food boxes

(c) "Clock" shift of six hours

Birds maintain a certain angle to the sun at a particular time of day. If they are made to believe it is a ¼ day later than it really is, they will shift their direction by that same fraction

(d) Stationary artificial "sun"

Birds change angles with respect to stationary sun, as if they assume sun has moved with respect to compass direction

39.3 A FAMOUS HOMING PIGEON

Two hundred fifty American soldiers in World War I owed their lives to this homing pigeon, which carried a distress message to its handler. Reinforcements were sent to rescue the beleaguered battalion.

sun-compass orientation has also been found in a variety of other animals, including insects, fish, and reptiles.

Navigation

The sun compass can't be the whole answer, of course. Possession of a compass may be necessary for navigation, but it isn't sufficient. Suppose you were blindfolded, driven out into the desert, and released with only a compass. Finding north would be easy; finding Bakersfield wouldn't. Even if you had a compass and a map, you would need to know your position on the map. And, as you sat on a rock, distraught and staring hopelessly at your devices, birds might be passing overhead on their way to a certain tree in Argentina.

As night drew on and you hadn't budged from your rock, other bird species might pass overhead. As you strained to see their dim shapes through teary eyes, you might wonder: How do *they* navigate? Since it is night, they would obviously not be using the sun. Apparently, some birds navigate the way sailors once did—by the stars. In experiments somewhat analogous to the sun-compass work, night-migrating birds were brought in cages into a planetarium. A planetarium is essentially a theater with a domelike ceiling onto which can be projected simulated night skies. Sure enough, the fly-by-night birds oriented in their cages according to the sky that was projected on the ceiling, even if it was

different from the sky outside. Again, they needed to know the time, just as the old sailors needed decent timepieces. In one experiment, birds that normally migrate north and south between western Europe and Africa were shown a simulated sky as it would appear that moment over Siberia. The birds, behaving as though they were thousands of miles off course, oriented toward the "west."

Homing Pigeons. The abilities of homing pigeons have fascinated people for years. (All pigeons are technically homing pigeons.) A homing pigeon can be put into a dark box and transported to a distant location by a circuitous route. When it is released, there's a good chance it will soon show up in its home loft. Some make it quickly, others never turn up—there is a great deal of variation in abilities (Figure 39.3).

Studies with homing pigeons have proven inconclusive but instructive. Researchers have asked a wide range of questions. Do pigeons use a sun compass? If so, how do they home successfully on cloudy days? In fact, pigeons fitted with frosted goggles so that the sun can't be pinpointed find their way back to the vicinity of their home loft, where they may flutter about directly overhead, or sit on the ground near the loft, unable to see it. Do pigeons use a magnetic compass? Small electromagnets have been attached to the heads of homing pigeons (Figure 39.4), and by producing a magnetic

39.4 HOMING EXPERIMENT

Pigeon carrying coils with reversed polarity on the head tended to home in opposite directions, indicating that pigeons can detect slight magnetic fields and that magnetism can influence their orientation.

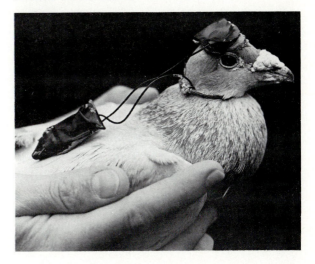

field across the bird's head that was deflected from the normal magnetic field, the birds could be made to fly off course by the same degree that the artificial field deviated from normal. If the sun was visible, some returned home anyway, though not as many as those birds who carried a similar but nonfunctional apparatus.

Evidently, then, pigeons use a variety of clues, including the sun compass, magnetic force fields, and visual details. Researchers have been foiled in trying to find *the* mechanism; the pigeon apparently has backup systems.

Navigation, as we've noted, requires a map as well as a compass. So do pigeons have built-in maps? The late William Keeton at Cornell University concluded that they do, although pigeon maps apparently are very different from human maps. He noted that released pigeons do not fly home in a straight line, but take indirect routes. For instance, pigeons released from one location due north of Cornell usually took off due east, and later make a series of sharply angled turns to take themselves home. Even those few pigeons that did fly straight south made the same turns, and they ended up lost somewhere in western New York.

It should be pointed out that the study of animal navigation and orientation is an extremely vigorous field at present, and new data are appearing almost daily. Any synthesis at this point would be extremely tenuous. However, these examples provide some idea of the value of innate characteristics in orientation, as well as the surprising sensitivity of some species to environmental cues.

Let's now turn to another vigorous area in the study of animal behavior, one that also has stimulated a great deal of first-class detective work, but one that is more generally adaptive across a wide range of animals. This area is loosely referred to as **communication.**

COMMUNICATION

If you should come upon a large dog eating a bone, walk right up and reach out as if to take the bone away. You may notice several changes in the dog's appearance. The hair on its back may rise, its lips may curl back to expose its teeth, and it may utter peculiar guttural noises. The dog is communicating with you. To see what the message is, grab the bone.

It should also be noted that in successful communication, both sender and receiver benefit. In the example above, if the message is properly understood, the dog gets to keep the bone and you get to keep your arm.

It would be a remarkable understatement to say that communication is an important phenomenon in the animal world. It is accomplished in many ways and with innumerable adaptive effects. Ultimately, however, it increases reproductive success in those animals able to send and receive the signals.

Communication may be directly involved with reproductive success as a component of mating behavior or precopulatory displays. It may also indirectly increase reproductive output by helping the offspring avoid danger when parents give warning cries, or by simply helping the reproducing animal live better or longer so that it may successfully mate again. Remember, the charge to all living things is "reproduce or your genes will be lost." Communication helps animals to carry out that reproductive imperative.

Let's consider a few general methods of communication, and in so doing perhaps we can learn something about the ways in which animals influence the behavior (or the probability of behavior) of other animals. That, after all, is what communication is all about.

39.5 GRADED DISPLAY

At left the bird *Fringilla coelebs* is showing a high-intensity threat. The posture at center is a medium-intensity threat and below is a low-intensity threat.

39.6 PERMANENT DISPLAY

A male mandrill has permanent markings that advertise his sex, but his signals of anger are temporary and dependent on his mood. It is probably often to his advantage to be identified as a male, but less frequently as an angry male.

Visual Communication

Visual communication is particularly important among certain fish, lizards, birds, and insects—and among some primates as well. Visual messages may be communicated by a variety of means, such as color, posture or shape, or movement and its timing. As was mentioned in Chapter 38, the color red can release territorial behavior in European robins. The female quickly solicits the attention of the male bird by assuming a head-up posture while fluttering her drooping wings. Some female butterflies attract male butterflies by the way they fly. As an example of communication by timing, fireflies are attracted to each other on the basis of their flash intervals, each species having its own frequency. (One predatory species "taps" communication lines by flashing at another species' frequency and then eating whomever comes to call. See Essay 39.1 for other examples of deception.)

Because visual signals carry high information content, subtle variations in the message can be conveyed by gradations in the display (Figure 39.5). Of course, **graded displays** are useful only to the species that are sensitive and intelligent enough to be able to recognize such subtleties. Another advantage of the high information load in visual sig-

nals is that the same message may be conveyed by more than one means. Such redundancy may be used either to modify the message or to emphasize it in order to reduce the chance of error in interpretation.

A visual signal may be a permanent part of the animal, as in the elaborate coloration of the male pheasant and the striking facial markings of the male mandrill (Figure 39.6). These animals advertise their maleness at all times and are continually responded to as males by members of their own species. On appropriate occasions they may emphasize their "machismo" through behavior such as the strutting of the pheasant and the glare of the mandrill. In other cases, a visual signal may be of a more temporary nature, such as the reddened rump areas in female chimpanzees and baboons during estrus. As a more short-term signal, a male baboon may expose his long canine teeth as a threat (Figure 39.7).

Short-term visual signals have the advantage that they can be started or stopped immediately. If a displaying bird suddenly spots a hawk, it can freeze, and its former position won't be given away by any lingering images. Also, the recipient of a visual message is notified of the exact location of the sender. The recipient can then respond in terms of the sender's precise location, as well as its general presence and behavioral state (aggressive, romantic, or whatever).

Visual signals also have certain disadvantages. For example, the sender must be seen, and all sorts

39.7 TEMPORARY DISPLAY

A male baboon displays his large canine teeth as a threat signal. These males are extremely powerful and dangerous.

Deception in the Animal World

It was once thought that animal communication is essentially a way for one animal to let another know what it is about to do. Now, however, we are aware that this notion is a bit naive. Communication is more realistically thought of as a kind of "enabling device"; it enables the communicator to increase its likelihood of success at survival and reproduction. In other words, communication can enable an animal to more effectively manipulate its environment. In some cases, this manipulation involves deception.

For example, a fish may develop an "eyespot" near its tail while concealing its own eye with a stripe; an animal seeking to head it off moves in one direction while the fish moves off in another. A variety of dangerous species take on common warning coloration such as yellow and black patterns, as we see in the wasps of the genera *Vespula* and *Bembex*. (This kind of convergent evolution is called *Müllerian mimicry*.) The harmless banded king snake mimics the red, yellow, and black stripes of the dangerous coral snake (an example of *Batesian mimicry*), and is also avoided by predators.

The orchid *Cryptostylis* emits a

Vespula

Bembex

Coral snake

Banded king snake

of things can block vision, from mountains and trees to fog and big hats. Visual signals are generally useless at night or in dark places (except for light-producing species). Since visibility weakens with distance, such signals are useless at long distances. Furthermore, as distance increases, the signal must become bolder and simpler; hence it can carry less information.

Sound Communication

Sound plays such an important part in communication in our own species that it may surprise you to learn that it is limited, for the most part, to arthropods and vertebrates. If you are familiar with the songs of the cricket and the cicada, you are probably aware that insect sounds are usually produced by some sort of friction, such as rubbing the wings together or rubbing the legs against the wings. The key aspect of these sounds is cadence, whereas with most birds and mammals, pitch or tone is more important.

Vertebrates use a number of different forms of sound communication. Fish may produce sound by means of frictional devices in the head area or by manipulation of the air bladder. Land vertebrates, on the other hand, usually produce sounds by forcing air through vibrating membranes in the respiratory tract. They use sound communication in other ways as well: rabbits thump the ground, gorillas

scent like that of the female Ichneumon wasp. The male wasps thus pollinate the orchids by *pseudocopulation* (shown here). In another example, the *Ophrys* orchids resemble the female black wasp, and they are pollinated as male wasps try to copulate with first one, then another.

The coloration and marking of some animals suggests "I am not here." The leopard, for example, easily blends into its grassy surroundings. Others essentially say "I am something else." A praying mantis captures visiting insects as it appears to be part of a flower.

In some cases, animals deceive others about their conditions. Some birds, such as the common killdeer, feign injury and flutter along the ground in the "broken wing display," drawing predators away from their nesting areas. A great deal of communication, we see, involves living things deceiving each other in a variety of fascinating, unexpected, and bizarre ways.

Ichneumon (male) and *Cryptostylis*

Leopard

Praying mantis

Killdeer

pound their chests, and woodpeckers hammer hollow trees and drainpipes early on Sunday mornings.

Most of the lower vertebrates rely on signals other than sound, but some species of salamanders can squeak and whistle, and there is even one that barks. Frogs and toads advertise territoriality and choose mates at least partly by sound. Few reptiles communicate by sound, but large territorial bull alligators can be heard roaring in more remote areas of the swamps in the southern United States. Darwin described the roaring and bellowing of mating tortoises when he visited the Galapagos Islands.

Sound may vary in pitch (low and high), in volume, and in tonal quality. The last is apparent as two people hum the same note, yet their voices remain distinguishable. It is possible to show the characteristics of sounds graphically by use of a **sound spectrogram.** In effect, this is a translation of sound into visual markings, which makes possible a more precise analysis of the sound.

The function of the message may dictate the characteristics of the sound. For example, Figure 39.8 shows a sound spectrogram of mobbing cries of several bird species. Such calls appear as brief, low-pitched *chuk* sounds, and their source is easy to pinpoint. When a bird hears the repetitious mobbing call of a member of its own species, it is able to locate the caller quickly and join it in driving away the object of concern—usually a hawk, crow, or

39.8 MOBBING CALLS

Shown here are sound spectograms of the mobbing calls of several species of British birds while attacking an owl. The sounds have qualities that make them easy to locate. Such calls are low-pitched *chuk* sounds. Different species of birds have developed similar calls through convergent evolution. That is, their calls serve much the same purpose and were developed under relatively similar conditions; thus the qualities of the sounds came to be somewhat alike.

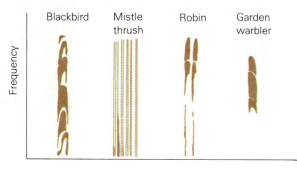

produce a sound. And of course, sounds, unlike visual images, can go around or through many kinds of environmental objects.

One disadvantage of sound communication is that it is rather useless in noisy environments. Thus, some sea birds that live on pounding, wave-beaten shorelines rely primarily on visual signaling. Sound also weakens with distance. And the source of a sound is sometimes difficult to locate, especially underwater.

Chemical Communication

You have probably seen ants rushing along single file as they sack your cupboard. You may also have taken perturbingly slow walks with dogs that stop to urinate on every bush. The behavior in both is based on chemical communication. Ants have laid down chemical trails that the others can follow and the dogs are gloriously advertising their presence.

Insects make extensive use of chemical signals, and certain chemicals elicit very rigid and stereo-typed behavior. An **alarm chemical** is produced when an ant encounters some sort of threat. As the chemical permeates the area, ants of that species react by rushing about in a very agitated manner, apparently ready to attack or protect the group in some other way. Female moths produce a *sex attrac-*

other marauder. If an aerial predator is spotted flying overhead, the warning cry of songbirds is usually a high-pitched, extended *tweeeeee,* a sound that is difficult to locate (Figure 39.9). The response of a bird hearing this call is quite different from its response to the mobbing call. The warning cry sends the listener heading for cover, often diving into deep, protective foliage, from where it may also take up the plaintive, hard-to-locate cry.

Sound signals have the advantage of potentially high information load through subtle variations in frequency, volume, timing, and tonal quality. They are distinguishable at low levels, but louder sounds, with their higher energy levels, can carry over greater distances. In addition, sounds are transitory; they don't linger in the environment after they have been emitted. Thus, an animal can cut off a sound signal should its situation suddenly change—for example, with the appearance of a predator. A further advantage is that an animal doesn't ordinarily have to stop what it is doing to

39.9 ALARM CALLS

The sound spectograms shown here are of five species of British birds when a hawk flies over. Such calls are high-pitched, drawn-out, and difficult to locate. They, too, have achieved their similarity through convergent evolution.

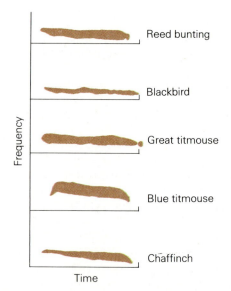

tant that attracts males from astonishing distances. You may recall our earlier discussion of one such substance called *bombykol;* a male moth can detect one molecule of this chemical (see Chapter 34).

Chemical signals are so immediate and invariable that they can be compared to hormonal messages. But whereas hormones work within an individual, these chemicals work between individuals and are called **pheromones** (from the Greek *pherein,* ''to carry,'' and *horman,* ''to excite'').

It was at first believed that pheromones are found only in insects, species whose behavior is largely genetically programmed. Later it was found that pheromones also exist in mammals. For example, if pregnant rats of certain species smell the urine of a strange male, some component of that urine will cause them to abort their fetuses and become sexually receptive again. And, as we have seen, the females of most mammal species signal their sexual receptivity by some chemical means.

Honeybees provide another interesting example of chemical communication. The queen produces a ''queen substance,'' which, when fed to developing worker females, inhibits sexual development. When the queen dies and the inhibiting substance is no longer present, new queens begin to develop (the first one to hatch may massacre the rest). At this time, the undeveloped ovaries of some adult workers also suddenly begin to mature. Their eggs will produce drones—royal consorts, one of which will mate with the new queen on her nuptial flight.

Careful molecular analysis of several chemical signals has revealed that the backbone of the molecule usually consists of 5 to 20 carbon atoms. The molecular weight of the chains ranges between 80 and 300. It is believed that molecules with at least 5 carbons are needed to provide the variation and specificity for different messages. However, the molecules cannot be too large either, since they are usually carried in the air and because more energy is expended in the manufacture of larger molecules.

Chemical signals have the advantage of being extremely potent in very small amounts. Also, because of their persistence in the environment, the sender and receiver do not have to be precisely situated in order to communicate. In addition, chemicals can move around many sorts of environmental obstacles.

The specificity of chemicals, however, limits their information load. Moreover, because chemicals do linger in the environment, they may advertise the signaler to arriving predators as well as to the intended recipient of the message.

WHY COMMUNICATE?

Now that we have some idea of how communication can be accomplished, we might ask, why do animals communicate at all? You can probably think of a host of reasons, but let's see if we can't organize the ideas a bit.

Species Recognition

First, communication may permit one animal to know that another animal is of the same species. This may not seem too important—unless one is interested in reproduction (Essay 39.2). After all, if two animals of different species mate, they cannot normally produce healthy offspring. Many species are very similar, and in the absence of some precise means of identification an animal might waste a lot of time and energy in trying to mate with a member of the wrong species. Therefore, species identification must somehow be quickly established.

For example, the golden-fronted woodpecker and the red-bellied woodpecker often share the same woods along a rather narrow band in Texas (Figure 39.10). The two species look very similar to

39.10 SIMILAR SPECIES

(a) The golden-fronted woodpecker *(Centurus aurifrons)* looks a bit scruffy while molting, but is still remarkably similar to **(b)** the red-bellied woodpecker *(Centurus carolinus).* The range of the golden-fronted extends through dry areas from Honduras northward to central Texas, where it rather abruptly stops and that of the red-bellied begins. The red-bellied's range extends to the eastern coast, where it lives primarily in wooded habitats. The birds overlap in a narrow range in Texas, but in spite of strong similarities in appearance, vocalization, and behavior, they don't interbreed. Whereas they don't recognize each other as sexual partners, each does see the other as a competitor and they mutually exclude each other from their territories as though they were of the same species. Obviously, they use different cues in the recognition of mates and competitors.

Animal Courtship

Albatrosses may bow and preen and clatter their bills together for days before breeding.

The reproductive success of many species of animals involves persuasion: one animal must persuade another to be its mate. Thus courtship behavior has evolved, and its array of manifestations never ceases to amaze even the most jaded biologists. The elaborate ceremonies that have resulted are often beautiful and distinctive, but they all serve the same function—to maximize the reproductive success of the participants.

In some cases, courtship involves only species recognition. These patterns are usually relatively brief, and mating occurs between individuals that need never see each other again. In species in which the male is not important in protection or in feeding the young, the sexes join only to mate, and it is in these species that we generally find the larger, more garish, conspicuous males. In such species, *sexual dimorphism* (different appearances of the sexes) is associated with a different strategy. The males maximize their reproductive success by attracting and inseminating as many females as possible. Of course, their appearance and behavior may attract predators and cut their reproductive life short.

The courtship ritual is more extended in species in which the male will remain with the female. In these cases, ceremonies may last several days; because the participants tend to remain together for long periods of time, they have more invested in the relationship and therefore must be highly selective. In many of these species (such as the albatross), the males and females may look very much alike (sexually monomorphic). It has been suggested that, in contrast to males in sexually dimorphic species, the drab male coloration may be an adaptation that reduces the possibility of predators. The drab and sexually monomorphic prairie chickens show another strategy. These birds gather at traditional leks (mating areas), where the males whirl and strut and females make their choices. When the dancing stops, however, the males inconspicuously go their way. This, then, is a form of *behavioral dimorphism*.

Western grebes not only display to each other in elaborate swimming rituals, but may also bring each other gifts of seaweed. In yet other species, the male may bring food to the female. The advantages of this behavior are obvious to a reproductive

the casual observer. The most conspicuous difference is that the golden-fronted woodpecker has a small yellow band along the base of its red cap, which is absent on the red-bellied woodpecker. Also, the red-bellied woodpecker has slightly more white in its tail. The habits of the species are much alike and their calls are very similar, at least to the human ear. Nevertheless, each of these birds is able to recognize its own species. In any case, cross-mating has never been observed, nor have any hybrids. No one knows exactly what cue the birds use, but to them, apparently, the signals are clear.

Interestingly enough, whereas the two species ignore each other sexually, they have come to treat each other as competitors and will eject a member of

the other species from a territory as quickly as they would a member of their own species. Apparently, the stimuli that release sexual and territorial behavior are quite different.

Individual Recognition

It may be important for an individual to be able to recognize specific individuals within its own population. If you have ever watched gulls at the beach, you may have noticed that the adults all look alike. This is because you aren't a gull (all people probably look alike to them). If you should visit a gullery, the place where they nest, you might see thousands of "identical" gulls. Color banding and recording playback experiments, however, have shown that

female. In essence, courtship behavior is a means of finding not only a mate, but a healthy mate that has demonstrated the ability to either recognize the appropriate signals or perform the elaborate and often demanding rituals of the species. In some species, advertising is less important. Male manatees simply follow estrous females until they are accepted, and powerful male hippos may force themselves on the smaller females.

Prairie chickens, a form of grouse, gather on leks to choose mates. The sexes have similar appearances until the males begin the elaborate strutting, whirling, and flapping that attracts receptive females.

A female manatee is physiologically ready to mate before she is behaviorally ready, eluding the cumbersome males for days before accepting one or more of them.

Western grebes go through elaborate rituals of swimming, diving, and "gift-swapping" before they mate.

Hippos mate beneath the water, where the male's great weight can easily be supported. Should the female resist his advances or be unresponsive to his preliminary displays, the male may attack and force her to submit.

gulls can recognize their own mates on sight and are able to filter out the calls of their mates from the raucous din of the clamoring gulls wheeling above.

Individual mate recognition is important only in species that establish pairs. In species in which the sexes come together only for copulation, the mate-recognition problem simply becomes one of sexual identification.

Individual recognition is important for two primary reasons. First, one should be able to recognize one's own mate, especially when both parents rear the offspring. Mate recognition ensures that pairs will attend to the nest that harbors their own offspring, thus increasing their chances of reproductive output. This is an essential factor in species that show a high level of coordination while rearing offspring. For example, in the African hunting dog, the male is more likely to regurgitate food to his own mate when she is nursing pups than to any other nursing females.

Second, individual recognition is important in the maintenance of **dominance hierarchies.** Once animals in a group know their rank with respect to the others, they will fight less over food or other commodities. In any group where rank is unknown, or not yet established, the incidence of fighting is likely to be high. Rank, however, can only be maintained when each animal can recognize the other individually.

AGGRESSION

The old image of "Nature, red in tooth and claw" has recently been superseded by the popular notion that other animals get along with their own kind and that humans are the only animals that kill each other. It might be a good idea to take a look around us to see what is actually going on.

First of all, we should note that whereas *aggressiveness* can be a mood, **aggression** is belligerent behavior, an action that usually arises as a result of *competition*. A lioness may spring toward a group of wild pigs, single out one in the chase, catch it, and quickly smother it or break its neck. However, she is not acting out of aggression. She is being about as aggressive as you are toward a hamburger. However, if one of the big boars should turn and charge her, the lioness may momentarily show a component of aggression—fear. In fact, fear is an important component of aggression; it is simply at one end of the behavorial spectrum of aggressive behavior. At the other is sheer, unbridled belligerence.

Fighting

Let's consider the most obvious form of aggression—fighting. You can discount the old films you have seen of leopards and pythons battling to the death. Such fights simply aren't likely to happen. What is a python likely to have that a leopard needs badly enough to risk its life for, and vice versa? Although such fighting might occur in the unlikely event that one should try to eat the other, fighting is much more likely to occur between two animals that are competing for the same commodity. The closest competitor is one that uses the same habitat in the same way, and the animal most likely to do this would be a member of the same species. Moreover, if there is strong competition for mates, fighting is even more likely between members of the same sex within that species. And this is in fact where most fighting occurs—between members of the same sex of a given species.

Of course, fighting between species sometimes occurs. The golden-fronted and red-bellied woodpeckers exclude each other from their respective territories. And lions may attack and kill African cape dogs at the site of a kill. The lions don't eat the dogs; they just exclude them.

Animals may fight in a number of ways, but combatants of the same species usually manage to avoid injuring each other (Figure 39.11). There are several apparent benefits to such a system. First, no

Each male *(Crotalus horridus)* tries to push the other to the ground, but the deadly poisonous snakes never bite each other.

one is likely to get hurt. The competitor is permitted to continue its existence, it is true, but the possibility of having to compete again entails less risk than does serious fighting. Also, since animals are most likely to breed with the individuals around them, the opponent may be a relative that is carrying some of the same genes; hence, sparing the opponent has a reproductive advantage. In fact, if the population is confined to a small area, the competitor might even be one's own mature offspring; it could also be a prospective mate. So even though the "motive" is selfish, not benevolent, it's best not to hurt each other.

When fighting occurs between potentially dangerous combatants, the fights are usually stylized and relatively harmless. For example, a horned antelope may gore an attacking lion, but when they fight each other, the horns are almost never directed toward the exposed flank of the opponent (Figure 39.12). Such stylized fighting does, however, enable the combatants to establish which is the stronger animal. Once dominance is established, the loser is usually permitted to retreat.

On the other hand, all-out fighting may occur between animals that are unequipped to injure each other seriously, such as hornless female antelope (Figure 39.13), or between animals that are so fast that the loser can usually escape before serious injury, as in house cats.

We might ask ourselves, since it is useless to ask an antelope, why they don't gore each other. An incurable romantic might assume that they simply don't want to hurt one another. In all likelihood, what the two antelopes "want" has very little to do with it. The fact is, they can't hurt each other. When the system works, an antelope could no more gore an opponent than fly! In terms of the actual mechanism, it may be that the sight of an opponent's exposed flank acts as an inhibitor of butting behavior. Conversely, a facing view, under certain conditions, might serve as a release of very stereotyped fighting behavior.

But regardless of what we may or may not surmise about animal motivation, such standardized behavior must have evolved because it was beneficial to the ancestors of the present combatants. Thus, we return to the original question: How does the animal benefit by refusing to do serious injury to the opponent?

One benefit is really not that subtle. The antelope might not gore its opponent because if it did so, it might get gored back. The situation is similar to that of two toughs in a barroom brawl. They slug and punch, and each is really trying to win the fight. On the other hand, each has a jackknife in his pocket—and both of them know it. Neither is willing to pull his knife, because then the other would be obliged to retaliate, and someone could get hurt that way. Either brawler would rather risk a loss in a fistfight than a draw in a knife fight.

Animal evolution has proceeded according to the same kind of logical accounting. John Maynard Smith, of the University of Sussex, points out that

39.12 MALE PRONGHORN ANTELOPES FIGHTING

Although the horns of these medium-sized antelopes are formidable weapons, neither animal will attack the vulnerable flank of the other. Instead, a harmless pushing contest ensues as the tips of the ridged horns are engaged. These animals effectively employ horns and hoofs against species other than their own, but when confronted with a member of their own species, they are genetically constrained to behave in very circumscribed ways.

39.13 HORNLESS FEMALE ANTELOPE FIGHTING

Hornless females of the Nilgai antelope have no inhibitions against attacking the flank of the competitor, but their butts are quite harmless, at least in the immediate sense. Though the butt itself may not be dangerous, it establishes dominance. A loss, however, usually just means a temporary setback, so it behooves the loser to accept it gracefully and attempt to breed another time. Interestingly, horned males of the same species almost never attack in this way, nor do horned females of other species.

the antlers of deer are used solely for intraspecific fights, and are constructed so as *not* to injure the opponent. The "points," or side branches, cause the antlers to lock in head-to-head combat so that the sharp tips do not actually reach the opponent's flesh. When they are seriously fighting a predator, they use their hooves with dazzling effect. Occasionally there are male deer with antlers that lack the side branches. When these males engage in what should be a ritual fight, their antlers slip through and gore their opponent. Does this give the mutant deer an advantage? Not at all. The gored opponent may be bigger and stronger and, once gored, may become very angry. Ritual combat then becomes real combat, and the sharp-antlered deer doesn't always win.

Maynard Smith has analyzed fighting strategies with mathematical models and computer simulations. What he strove to determine was the **evolutionarily stable strategy,** which he defined as an innate behavioral pattern that would outcompete all other behavioral patterns and would be stable against the invasion of a new mutant pattern of behavior. He asked: Is it best never to fight and always retreat? Then a born fighter will always win. Is it better always to fight? No, because there is too much risk of being beaten. Is it best to bluff consistently? No, the bluff will be called (it's better to bluff inconsistently and unpredictably). Maynard Smith found two behavioral "strategies" that proved to be the most stable in populations, depending on the circumstances. The first strategy was called the **retaliator strategy;** a retaliator engages only in ritual display and mock battle unless it is seriously attacked, whereupon it will retaliate with just as much seriousness. In a mathematical model, this behavior was always the most successful in the long run so it should not be surprising to see ritual display and mock battle so common in nature. There's always the threat of real injury to any animal that breaks the rules.

The second strategy, the **bourgeois strategy,** employs quite a different set of rules. The bourgeois approach, according to the modeling studies, was even more effective than retaliation, but it required following a peculiar rule. In each encounter between adversaries, one retreats immediately, so that neither wastes time and energy fighting. This strategy might work in territorial conflicts (discussed shortly) and in dominance hierarchies, or **pecking orders.** It could operate in nature, for example, with the quite reasonable "rule" that the smaller of two potential combatants concedes immediately. But it also works if the determination of who concedes is completely arbitrary. Some of our traffic rules work this way. The first car to the intersection has the right of way; if two cars arrive simultaneously, the car on the right has the right of way and the car on the left concedes (but don't count on it). As long as everyone knows the rules, it is advantageous to everyone to follow them.

Can we assume, then, that animals do not fight to the death with their own kind? As a general rule they don't, but there are many exceptions. Accidents may occur in normally harmless fighting, and this can lead to retaliation and escalation. There are also some species that normally engage in dangerous fighting. If a strange rat is placed in a cage with a group of established rats, the group may sniff at the newcomer carefully for a long time, but eventually they will begin to attack it and to do so repeatedly until they kill it. If escape is impossible, male guinea pigs and mice often fight to the death. The males of a pride of lions may kill any strange male they find within their hunting area, and a pack of hyenas may kill any of another pack that they can catch. Even gangs of male chimpanzees have been seen to ambush and kill isolated males from other troops.

Is Aggression Instinctive?

Our discussion of aggression so far suggests that aggressive behavior in many animals is largely innate. But there are many who contend that aggressive behavior, especially among mammals, is directly attributable to learning. Others, of course, stress that aggressive behavior has both learned and innate components. The argument has extreme significance for our species. If aggression is socially disruptive in an increasingly crowded world, then we must ask how it can be controlled. As a practical example, does television violence serve to release aggression, and hence dissipate it, or does it serve as a behavioral model for aggression, and hence encourage it? Since there is a tendency in most academic disciplines to focus on aggression as a cultural, or learned, phenomenon, let's consider the evidence that suggests an instinctive component.

In certain species of highly aggressive cichlid fish, the males must fight before they are able to mate. If a male's reproductive state is appropriate and a female is available, the male will frantically seek an opponent upon which to release his fighting behavior. Finding none, he will often attack and kill the female. He is then ready to mate, but of course by then it is too late. The behavior can be described in terms of Tinbergen's hierarchical model. In removing the inhibitory block for mating (by fighting), the male can move to the next level of reproductive behavior, but unfortunately he has destroyed the releaser of that behavioral level, the female. Among mammals, rats will learn mazes in order to be able to kill mice, and it has thus been inferred that the killing is a consummatory act that reduces tension.

39.14 TERRITORIAL LIZARDS

Anolis lizards are among the vast array of territorial species. The males may hold areas in which a number of females take residence, but all other males are excluded.

Does an animal become aggressive when it is shielded from all opportunities to learn such behavior? Rats and mice have been reared in isolation so that there was no opportunity for them to learn aggression. But when other members of the same species were introduced into their cages, the orphan rodents attacked, showing all the normal threat and fighting patterns. The evidence from all sides is essentially circumstantial, but we would do well to focus more research attention on the roots of aggression in these critical times.

TERRITORIALITY

Aggression is the tool with which animals are able to hold **territories.** (Territories are most broadly defined as any defended area.) People first observed that animals held territories many hundreds of years ago, but our attention to the topic was not focused until 1920 when Eliot Howard wrote *Territory in Bird Life.* The book was based on his observations of birds expelling each other from certain plots of land. Since one bird successfully defended one place and another bird was able to win in another place, it seemed as if the birds "owned" the land in that a particular bird seemed to have certain rights when it was on a certain plot of land. In fact, when on that land, the bird was virtually undefeatable, and with apparent confidence would immediately drive out any intruder of the same species and sex. Of the same species and sex? Obviously, the bird was driving out competitors. Members of the same species would compete for commodities, such as food and nest sites, and members of the same sex would compete for mates. So here was a working hypothesis: Territorial behavior is a means of reducing competition.

Since Howard's time, territoriality has been discovered in a vast range of species. Many very different species, it turns out, exhibit the same kind of territorial behavior. Some *Anolis* lizards (Figure 39.14) have territories very similar to those of many birds, and stickleback fish behave similarly to these. It is tempting, then, to look for encompassing underlying themes, broad generalizations to cover all the cases. Alas, such generalizations are hard to come by. For example, a hummingbird may defend different bushes at different times of the day, taking a proprietary interest in each as its flowers open. Several tomcats may hold the same territory at different times of the night, each taking ownership in turn. Even closely related animals may show differ-

39.15 INDIVIDUAL DISTANCE

Starlings sit peacefully next to each other on a power line. They are regularly spaced according to their tolerance of each other's proximity.

ent territorial patterns. The Norway rat *(Rattus norvegicus)* chases strange rats only on special paths it has marked, whereas the black rat *(Rattus rattus)* defends the entire area crossed by its paths. Some animals, in a sense, carry their territories with them as *individual distances.* This is the radius that must not be violated by other members of the species. Starlings sitting on a wire sit evenly distributed (Figure 39.15), each a circumscribed distance from the others, perhaps having to do with how far its neighbor can reach with its bill. (You can carry out a quick study of this aspect of human behavior by observing people on elevators. Notice how uneasy people become if you don't move to readjust "your space" when someone gets off—and you can make them even more insecure by riding facing the rear of the elevator.)

In spite of such variation, however, there are certain basic characteristics of territorial behavior. First, territories are rather specific and do not, as a rule, include all the places where an animal may wander. The area in which an animal may be found at any given time is called its **home range.** Animals meeting within home ranges, but outside territories, normally don't fight. Baboons defend their sleeping trees and feeding grounds, but they meet peacefully at water holes.

To emphasize the variation in territorial behavior, let's consider a few examples from birds. In species that live as pairs, the territory is usually defended by the male against other males, although the female may join the fray. Whereas the male will not usually attack an intruding female, the resident female may attack her on sight. At certain times, such as in the winter, the sexes of some species may hold separate territories, but in spring they lower their defenses and come together to mate.

In some species, pairs may join together to defend mutual territory in a community effort against members of other groups. In other species, pairs may live close together but may only defend small areas around their nests from other pairs. In yet other species, males may come together at the beginning of the mating season to form small, temporary territories on a communal display ground. Territoriality, then, is clearly a complex phenomenon. What follows is a list of some of the adaptive features that various researchers have attributed to territoriality:

1. *Assurance of a food supply.* Territories often, perhaps usually, hold food.
2. *Division of resources among dominant and subordinate individuals.* Once the resources have been distributed by the establishment of territories, less time will be spent deciding who gets what. Losers have a better opportunity of finding food if they look elsewhere than if they challenge a territory holder.
3. *Provision of a mating and nesting place.* The territory may hold a nest site, or in the case of communal breeding grounds, or **leks,** (Figure 39.16), a place to breed, since the male has signaled his vigor to other males and to females by possession of the territory.
4. *Selection of the most vigorous to breed.* A female should ideally choose the male who has successfully competed against other males. In this way she increases the probability of her offspring receiving "good" genes.
5. *Stimulation of breeding behavior.* In some species males without territories cannot attract mates, and, in fact, males without territories may not even develop sexually.
6. *Limitation of population and the assurance of space.* Territories obviously ensure space, but if space is limited and needed for breeding, they may also reduce the population size (see Chapter 42).
7. *Increased efficiency of habitat use.* Since an animal is likely to spend most of its time in or near its territory, it reserves a place for itself with which it can become very familiar, learning the best areas to hide and find food.
8. *Protection against predators, disease, and parasitism.* Disease and parasites travel with more difficulty in a sparsely distributed population, but spacing also may be advantageous if it insures against a predator finding all members of the population just because it finds one.

By the way, Maynard Smith's bourgeois theory of evolutionarily stable strategies (ESSs) has thrown some light on the adaptiveness of territorial behavior, and in particular on that unwritten rule that the "owner" of a territory almost always wins an encounter within the confines of its turf. In every case of territoriality, the rule is *stay off my turf.* The defender "knows" it (that is, it behaves as though it knows it); the intruder "knows" it; and both of them "know" that it is to their own advantage to follow the rule. According to the theory, it may be an arbitrary rule, but since it works to the advantage of all parties, it has persisted and been stabilized by evolution.

39.16 LEK DISPLAY

Male *Philomachus pugnax*, a species of ruff, are seen here displaying on their lek (breeding ground). A female is at right. These birds are unusual in that the males are highly variable in plumage.

39.17 COOPERATION IN VASTLY DIFFERENT SPECIES

(a) These social insects show extremely high levels of cooperation, coordination, and self-sacrifice. The result is a hive in which the watchword is cold, sometimes brutal, efficiency.

(b) Musk-oxen form a defensive circle to protect the females and young.

(a)

(b)

COOPERATION

Cooperation seems much "nicer" than aggression, and most of the animal stories of our youth (our pre-Jack London youth, that is) involved animals that helped each other in some way. Certainly, cooperation is highly developed in some species, and its complexity and coordination in certain instances surpasses imagination. But again we have the problem of sifting fact from fiction—a difficult, if not impossible, task for those with a casual interest in the subject.

Cooperative behavior occurs both within species and between species. As an example of *interspecific* (between species) cooperation we have the relationship of the rhinoceros and the tickbird. The little birds get free food while the rhinoceros rids itself of ticks and harbors a wary lookout. Such relationships are well known in nature because of their inherent interest, but the highest levels of cooperation are most likely to exist between members of the same species.

Let's consider a few examples of *intraspecific* (within species) cooperative behavior. Porpoises are air-breathing mammals, much vaunted in the popular press for their intelligence. In fact, certain of their actions support the claim. Groups of porpoises will swim around a female in the throes of birth and will drive away any predatory sharks that might be attracted by the blood. They will also carry a wounded comrade to the surface so that it can breathe. Their behavior in such cases is highly flex-

ible, rather than stereotyped. Such flexibility indicates that their behavior is not a blind response to innate genetic influences.

Group cooperation among mammals is probably most common in defensive and hunting behavior. For example, yaks of the Himalayas form a defensive circle around the young at the approach of danger, standing shoulder to shoulder with their massive horns directed outward. This defense is effective against all predators except humans, since it provides no defense against high-powered rifles. Wolves, African cape dogs, jackals, and hyenas often hunt in packs and sometimes cooperate in bringing down their prey. In addition, they may bring food to members of the group that were unable to participate in the hunt.

We might expect mammals, with their high intelligence, to cooperate closely. But social behavior and cooperation are most highly developed in the lowly insects (Figure 39.17). The complex and highly coordinated behavior patterns of insects are usually considered to be genetically programmed, highly stereotyped, and generally not influenced by learning. Some of the best examples of insect cooperation are found among the honeybees.

In honeybee colonies, the queen lays the eggs and all other duties are performed by the workers, which are sterile females. Each worker has a specific job, but that job may change with time. For example, newly emerged workers prepare cells in the hive to receive eggs and food. After a day or so, their **brood glands** develop, and they begin to feed larvae. Later, they begin to accept nectar from field workers and pack pollen loads into cells. At about this time their wax glands develop, and they begin to build combs. Some of these "house bees" may become guards that patrol the area around the hive. Eventually, each bee becomes a field worker, or forager. She flies afield to collect nectar, pollen, or water, according to the needs of the hive. Apparently, these needs are indicated by the "eagerness" with which the field bees' different loads are accepted by the house bees.

If a large number of bees with a particular duty are removed from the hive, the normal sequence of duties can be altered. Young bees may shorten or omit certain duties and begin to fill in where they are needed. Other bees may revert to a previous job where they are now needed again.

The watchword in a beehive is *efficiency*. In the more "feminist" species, the drone (males) exist only as sex objects. Once the queen has been inseminated, the rest of the drones are quickly killed off by the workers; they are of no further use. The females themselves live only to work. They tend the queen, rear the young, and maintain and defend the hive. When their wings are so torn and battered that they can no longer fly, they either die or are killed by their sisters. But the hive goes on.

ALTRUISM

You probably won't be shattered to learn that most of the "Lassie stories" aren't true. Consider what would happen to the genes of any dog that was given to rushing in front of speeding trains to save baby chickens. The reproductive advantages would be considerable to chickens, but dogs with those tendencies might be selected out of the population by the action of fast trains. In contrast, the genes of a "chicken" dog that spent its energy, not in chivalrous deeds, but in seeking out suitable mates, would be expected to increase in the population.

Altruism may be defined in a biological sense as an act by one individual that benefits another, but at the first individual's expense. There are many apparent cases of altruism in the animal world, but our job here will be to ask if these are really altruistic acts, after all, and if so, how the behavior might have evolved.

It is easy to see how certain forms of altruism are maintained in a population. For example, pregnancy, in a sense, is altruistic. The prospective mother is swollen and slowed. Much of her energy goes to the maintenance of the developing fetus. At birth she is not only almost completely incapacitated, but

is in marked danger. Pregnancy is clearly detrimental to her. So why do females so willingly take the risk? It may help to understand the enigma if we remember that the population at any time is composed entirely of the offspring of individuals who have made such a sacrifice. Thus, the females in the population are the descendants of generations who did reproduce, so they too are likely to be predisposed to make such a sacrifice. It makes little difference whether the tendency is genetic or learned. The tendency to have offspring or *to do the things that result in having offspring* can be innate (genetically based) or, especially in humans, transmitted culturally.

However, altruism on this basis doesn't explain why a bird may feed the young of another pair, or an African hunting dog will regurgitate food to almost any puppy in the group. Why, also, would a bird that may have no offspring of its own give a warning cry at the approach of a hawk, alerting other birds at the risk of attracting the hawk's attention to itself? To answer such questions we must look past the answers that first come to mind. It may seem cynical, but we must start with the premise that birds don't give a hoot about each other. A bird that issues a warning call isn't thinking, "I must save the others." At least, there is a simpler explanation of its behavior.

Keep in mind that the biologically "successful" individual is the one that maximizes its reproductive output. One way of accomplishing this is for the organism to leave its own offspring, but another way is to leave as many individuals as possible carrying that type of genes in the next generation. This latter point may not be so readily apparent—it implies that an individual can also leave its type of genes in the next generation by helping a relative's offspring survive. We need only keep in mind that an individual also shares genes with a cousin, albeit fewer than with a son or a daughter. So it is reproductively beneficial for an individual not only to bear offspring, but to assist relatives, as long as the cost is not too high. In fact, there is theoretically a point at which an individual could increase the success of its genes by saving its nieces and nephews (provided there were enough of them), rather than its own offspring. From the standpoint of effective reproductive output, the organism would be better off leaving 100 nieces than one daughter.

To illustrate, suppose a gene for altruism appears in a population (notice that this sets up a mechanism for the continuance of the behavior). As you can see from Figure 39.18, altruistic behavior would most likely be maintained only in groups in which the individuals are related (that is, have some kinds of genes in common). Altruism might be expected, then, when there is a high probability that proximity indicates kinship, as we see in relatively stationary populations.

Keep in mind that no conscious decision on the part of the altruist is necessary. It simply works out that when conditions are right, those individuals that behave altruistically increase their kinds of genes in the population, including the "altruism gene." Nonrelatives would benefit from the behavior of altruists, of course, but there is an increased likelihood of individuals nearer an altruist to be related to it.

It has been determined mathematically that the probability of altruism increasing in a population depends on how closely the altruist and benefitee are related (as well as on the risk to the individual, of course). In other words, the advantages to the benefitee must increase as the kinship becomes more remote. For instance, an altruistic act that results in a risk of death of the altruist will be selected for if the net genetic gain to brothers and sisters is more than twice the loss to the altruist; to half-brothers, four times the loss; and so on. To put it another way, an altruistic animal would gain reproductively if it sacrificed its life for more than two brothers, but not for fewer, or for more than four half-brothers, but not for fewer, and so on. Therefore we can deduce that in highly related groups, such as a small troop of baboons, a male might fight a leopard to the death in defense of the troop.

This model, developed by J. B. S. Haldane in 1932 and expanded by W. D. Hamilton in 1963, helps explain the extreme altruism shown by social insects such as honeybees. Since workers are sterile, their only hope of propagating their own genotype is to maximize the egg-laying output of the queen. In some species, the queen is inseminated only once (by a haploid male), so all the workers in a hive are sisters and have three fourths of their genes in common. In such a system, then, almost any sacrifice is worth any net gain to the hive and to the queen.

In a brilliant essay, Robert Trivers expanded our understanding of the evolution of altruism by suggesting that most apparent altruism (outside of parental behavior) depends on the expectation of reciprocation: "I'll scratch your back if you'll scratch mine." **Reciprocal altruism** is an evolutionarily stable strategy with some complex rules. Help (altruistic acts) is given to others—even offered to strangers—when the cost or risk is not too great to the giver. The expectation is that some kind of help will be reciprocated at another time. If the expected reciprocation is not forthcoming, an evolutionarily

39.18 A GENETIC MECHANISM OF ALTRUISM

In this population, A and A' are nonaltruists. They behave in such a way as to maximize their own reproductive success but do nothing to benefit the offspring of other individuals. In another segment of the population (B and B'), a gene for altruism has appeared that results in individuals benefiting the offspring of others in some way. It can be seen that, assuming the altruistic behavior is only minimally disadvantageous to the altruist, generations springing from B and B' are likely to increase in the population over those from A and A'. The altruistic behavior is likely to be greatest where B and B' are most strongly related, so that B shares the maximum number of genes in common with the offspring of B' and vice versa. The idea is that B, for example, can increase its own reproductive success by caring for the offspring of a relative with whom it has some genes in common. After all, reproduction is simply a way of continuing one's own kinds of genes.

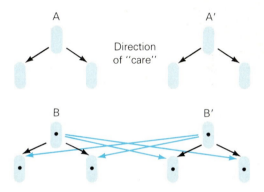

derived emotion is experienced: **moral indignation.** The individual that fails to reciprocate is scorned, turned out of the social group, no longer aided. To get ahead, everyone has to play by the rules—or appear to.

Some evidence of intraspecific reciprocal altruism has been reported in troops of social mammals, such as hunting dogs and higher primates, but the evidence is not very strong for such behavior in any animals but humans. Trivers implies, in fact, that reciprocal altruism is the key to human evolution. The complexity of such behavior, entailing as it does memory of past actions, the calculation of risk, the foreseeing of the probable consequences of present actions, the possibility of advantageous cheating, and the need to be able to detect such cheating, all require a level of intelligence that is beyond most species. In the opinion of some anthropologists, it is exactly for the management of these elaborate social interactions that the human brain—and the conscious human mind—evolved.

The discussion of the development of altruistic behavior is admittedly somewhat esoteric. However, it seems important to consider even our most cherished and most despised behaviors in the context of their evolutionary development.

APPLICATION OF IDEAS

1. Why is a biological clock (a precise sense of time) necessary for solar or stellar navigation?

2. Under what sort of competitive conditions would population recognition be particularly important?

3. What line of evidence suggests that aggression is instinctive? Learned? Both?

4. What is an evolutionarily stable strategy? Can you think of another one in addition to those described by J. Maynard Smith?

5. Would you expect competition between males to be stronger under monogamous or polygamous conditions? Why?

KEY WORDS AND IDEAS

ORIENTATION AND NAVIGATION

Orientation

1. Gustav Kramer's experiments with the migratory behavior of caged birds have provided strong evidence that **orientation** to the sun's position is important in **navigation.** He made the following observations:
   a. In their cage movements, daytime migratory birds oriented themselves according to their usual migratory direction if the sky was visible.
   b. When their surroundings were changed, they were still able to find their food box as long as its compass direction was not changed and they could see the sun.

Navigation

1. The sun-orientation studies indicate that some birds can determine the time of day and the normal course of the sun.

2. Studies of night-migrating birds indicate that they are capable of stellar (star) navigation.

3. Pigeons use a sun compass, but when the sky is overcast they have backup cues such as the earth's magnetic field and landmarks.

COMMUNICATION

1. The ultimate effect of **communication** is to increase reproductive success.

Visual Communication

1. Animals communicate visually through color, posture, and movement and its timing.

2. Visual communication carries a great information potential. Such signals can become more complex by graded displays.

3. Coloration may be a permanent part of visual signaling, as with male birds. Mating coloration, on the other hand, may be temporary, as in changes in seasonal coloration in fish.

4. Visual signals can be blocked by obstructions and except for light signals are generally ineffective at night.

Sound Communication

1. Sound communication is limited to arthropods and vertebrates. Arthropod sound is usually produced by friction (rubbing parts together). Vertebrates use a variety of devices, including vocal sounds. Fish use the air bladder (swim bladder). Sound signals vary in pitch, volume, and tonal quality. Sound characteristics can be shown graphically by a **sound spectrogram.**

2. Bird sounds can vary. Low-pitched sounds used in mobbing are easy to locate. High-pitched, extended sounds, used in certain predator alarms, are difficult to locate.

3. The advantages of sound communication are high information load, distance, amplification, and transience. The disadvantages include problems in noisy environments, weakening with distance, and difficulty of localization.

Chemical Communication

1. Chemical signaling is vital in a number of species, such as many vertebrates and the social insects.

2. The chemical substances used in communication are called **pheromones.** They are commonly used by insects, but are also important in mammals. Ants, for example, leave chemical trails and produce **alarm chemicals,** while rats rely on chemical signals in urine.

3. The advantages of chemical signals are their potency, their residual effect, and their ability to move around obstacles. A disadvantage is also found in the residual effect, which can alert a predator to the prey's presence.

WHY COMMUNICATE?

Species Recognition

1. Species recognition is vital for reproduction in that it decreases the probability of infertile hybrids produced by mating errors.

Individual Recognition

1. In situations where parental care is important, species must be able to recognize their mates and their offspring.

2. In some social species, individual recognition is important in the establishment and maintenance of **dominance hierarchies.** In hierarchical populations there is a minimum of discord and fighting.

AGGRESSION

1. **Aggression** is belligerent behavior that frequently arises from competition. Fear is an important component of aggression.

Fighting

1. Fighting between species is very unlikely simply because different species are not commonly in competition. The greatest competition occurs among members of the same species, especially those of the same sex.

2. In fighting within most species, there is usually very little injury. It is more adaptive to avoid injuring a competitor through stylized fighting.

3. J. Maynard Smith hypothesizes that engaging in ritual combat, but being willing to retaliate if injured, is an **evolutionarily stable strategy** (ESS). Included is the **retaliator strategy,** in which an adversary only fights if attacked. The **bourgeois strategy** permits retreat, whereupon no chase is made. This is the response in dominance hierarchies or **pecking orders.**

4. Fighting to the death or to serious injury does occur in some species. The serious attacks are directed most commonly to strangers.

Is Aggression Instinctive?

1. The origin of aggression is uncertain. Is it innate, learned, or both? Some animal behavior indicates a strong innate basis for aggression.

Territories

1. Animals may employ aggressive behavior to maintain **territories.** The strongest attacks are against others of their species and in particular against others of their own sex.

2. Territorial behavior varies widely in different species. A territory is defended and may be only a part of the animal's **home range.**

3. Some of the proposed adaptive features of territoriality are:
   a. insurance of food supply;
   b. equitable division of resources;
   c. provision of mating and nesting places;
   d. selection of vigorous breeders;
   e. stimulation of breeding behavior;
   f. assurance of required space;
   g. increasing efficiency of habitat utilization;
   h. protection against predators, disease, and parasitism.

4. Territorial defenders appear to know that others must stay out, and intruders also appear to know they are at a disadvantage. The behavior has become evolutionarily stabilized.

COOPERATION

1. Cooperative behavior is most highly developed within species. For example, porpoises protect females during the birth process and often aid injured members and Himalayan yaks form defensive circles to resist predators.

ALTRUISM

1. **Altruism** may be defined as an activity that benefits another organism at the altruist's expense.

2. Altruism probably has its greatest adaptiveness when it maximizes the reproductive output of one's own genes. The genetic value of altruistic behavior is directly related to the closeness of kinship.

3. This hypothesis explains why extreme altruism is common in social insects such as honeybees. Because of their extremely close relatedness, any sacrifice by members directly benefits the genes that determine the behavior.

4. **Reciprocal altruism** is an evolutionarily stable strategy. The rules are strict; there must be reciprocal behavior.

REVIEW QUESTIONS

1. Describe Kramer's experimental procedure and results in using mirrors to test the sun orientation hypothesis. (1011)

2. What did Kramer discover from his second experiment using the food box in changing surroundings? (1012)

3. Homing pigeons are excellent navigators. What kinds of environmental cues do they use and how are some of these redundant? (1014)

4. What are graded displays and how do they increase the information load of a visual signal? (1015–1016)

5. List three examples of permanent visual displays. (1015)

6. What are some advantages and disadvantages of visual communication? (1015–1016)

7. In general, how do insects make sounds? (1016)

8. Using the mobbing and warning cries of birds as examples, show how bird sounds can either be easy or difficult to locate. (1018)

9. List three aspects of sound that provide variability and great information load potentially. (1017)

10. List several advantages of sound communication. (1018)

11. Describe two examples of chemical communication in insects. (1018–1019)

12. Describe two examples of pheromones at work in mammals. (1019)

13. How are pheromones responsible for controlling the sex of developing bees? Of what advantage is this to the hive? (1019)

14. What are the molecular requirements of a pheromone? (1019)

15. List several advantages and disadvantages of chemical communication. (1019)

16. Why is it important for individuals of one species to be able to recognize one another? (1019)

17. How is individual recognition important in dominance hierarchies? (1021)

18. Define the term *aggression* in its biological sense. (1022)

19. Among what individuals is fighting most likely to occur, and why is this so? (1022–1023)

20. Why is it adaptive for fighting animals to avoid injuring or killing each other? (1023)

21. Define and give an example of stylized fighting. (1022)

22. Review J. Maynard Smith's concept of an evolutionarily stable strategy. What kind of fighting is most adaptive? (1023)

23. Are there any animals that fight to kill? Cite examples. (1024)

24. What are the three possible bases for aggressive behavior? (1024)

25. Describe an example of innate aggressive behavior. (1024)

26. List several adaptive features of territoriality. (1025)

27. What is the general rule about behavior in territories? (1026)

28. Briefly describe two examples of cooperative behavior in mammals. (1027)

29. In which group of animals has cooperative behavior reached its most intricate and coordinated level? Describe an example. (1028)

30. Define altruism. Under what circumstances would sacrifice for cousins be more adaptive than for one's own offspring? (1029)

SUGGESTED READING

Barash, D. P. 1977. *Sociobiology and Behavior*. Elsevier, New York and Amsterdam.

Brown, J. L. 1975. *The Evolution of Behavior*. W. W. Norton, New York.

Christian, J. J. 1970. "Social Subordination, Population Density, and Mammalian Evolution." *Science* 168:84. Mammals have marked endocrine responses to crowding and to aggressive encounters.

Darwin, C. 1871. *The Descent of Man, and Selection in Relation to Sex*. Appleton, New York. Darwin has a great deal to say about the evolutionary importance of sexual behavior.

———. 1872. *The Expression of the Emotions in Man and Animals*. Appleton, New York. The first great work of comparative ethology.

Dawkins, R. 1976. *The Selfish Gene*. Oxford University Press, New York and Oxford. The selfish gene concept is applied to animal behavior in the second half of this popular paperback.

Ehrman, L., and P. A. Parsons. 1976. *The Genetics of Behavior*. Sinauer Associates, Sunderland, Mass. A textbook approach.

Gardner, R. A., and B. T. Gardner. 1971 "Two-way Communication with an Infant Chimpanzee." In *Behavior of Non-Human Primates*, ed. A. Schrier and F. Stollnitz, p. 117. Academic Press, New York.

Gould, J. L. 1982. *Ethology of Behavior*. W. W. Norton, N.Y.

Jansen, D. H. 1966. "Coevolution of Mutualisms between Ants and Acacias in Central America."

Evolution 20:249. One of biology's best gee-whiz stories.

Johnsgard, P. A. 1967. "Dawn Rendezvous on the Lek." *Natural History* 76:16. A *lek* is a special place where males compete with one another for territory and prestige and wait for the arrival of sexually interested females.

Klopfer, P. H. and P. J. Hailman. 1973. *Behavioral Aspects of Ecology.* Prentice-Hall, Englewood Cliffs, N.J.

Marler, P., and W. Hamilton. 1966. *Mechanisms of Animal Behavior.* Wiley, New York.

Maynard Smith, J. 1964. "Group Selection and Kin Selection." *Nature* 201:1145. The problem of group selection and the evolution of altruistic, group-directed behavior.

————. 1974. "The Theory of Games and the Evolution of Animal Conflicts." *Journal of Theoretical Biology* 47:209.

Mech. L. D. 1970. *The Wolf: The Ecology and Behavior of an Endangered Species.* Natural History Press, Garden City, N.Y. Fascinating reading for all you wolf fans.

O'Keefe, J., and L. Nadel. 1978. *The Hippocampus as a Cognitive Map.* Clarendon (Oxford University Press), New York and Oxford. The mammalian brain stores "pictures" and "maps"—any two- or three-dimensional memory—in the hippocampus.

Oster, G. F., and E. O. Wilson. 1978. *Caste and Ecology in the Social Insects.* Princeton University Press, Princeton, N.J. Hedging bets against rare disasters demands very different allocations of scarce resources from what would be feasible if the world were predictable.

Skinner, B. F. 1938. *The Behavior of Organisms: An Experimental Analysis.* Appleton-Century-Crofts, New York.

Trivers, R. L. 1971. "The Evolution of Reciprocal Altruism." *Quarterly Review of Biology* 46:35. Frequently behavior that appears to be altruistic is really selfish in the long run: it is insurance for similar help when it might be needed.

Wallace, R. A. 1979. *The Ecology and Evolution of Animal Behavior,* 2d ed. Scott, Foresman and Company, Glenview, Ill. The shorter version of Wallace's behavior book. The previous edition was among the first texts to bring together ecological and evolutionary aspects of animal behavior with attention to a range of behavior studies.

————. 1979. *Animal Behavior: Its Development, Ecology, and Evolution.* Scott, Foresman and Company, Glenview, Ill. In this treatment, Wallace brings to the field of animal behavior a strong evolutionary viewpoint and a lively style.

————. 1979. *The Genesis Factor.* William Morrow, New York. Wallace considers the evolutionary influences on human behavior in this enjoyable and thought-provoking popular book on the implications of sociobiology.

Wilson, E. O. 1971. *The Insect Societies.* Harvard University Press, Cambridge, Mass.

————. 1975. *Sociobiology, The New Synthesis.* Harvard Press, Cambridge, Mass.

————. 1976. "Academic Vigilantism and the Political Significance of Sociobiology." *Bioscience* 26:183.

————. 1978. *On Human Nature.* Harvard University Press, Cambridge, Mass. The three books listed above form an "unplanned trilogy" in which Wilson successively considers insect societies, animal social behavior in general, and the application of evolutionary sociobiology to the study of human affairs.

Scientific American articles. San Francisco, W. H. Freeman Co.

Alkon, D. 1983. "Learning in a Marine Snail." *Scientific American,* July.

Heinrich, B. 1981. "The Regulation of Temperature in the Honeybee Swarm." *Scientific American,* June.

Lloyd, J. 1981. "Mimicry in the Sexual Signals of Fireflies." *Scientific American,* July.

Maynard Smith, J. 1978. "The Evolution of Behavior." *Scientific American,* September. More on the key problem of altruism and on the logic of formalized, nonlethal aggressive conflicts.

Partridge, B. 1982. "The Structure and Function of Fish Schools." *Scientific American,* July.

Wilson, E. O. 1975. "Slavery in Ants." *Scientific American,* June. Some slave-making species have become so specialized that they are no longer capable of feeding themselves.

Biosphere and Biomes

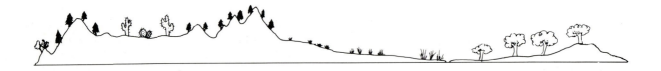

The word "ecology" has made its way into the public vocabulary and has proved to be remarkably resilient there. Often, once a word enters that "great consciousness" it becomes overused, misinterpreted, restructured, and finally battered into uselessness. ("Instinct" is an example.) But *ecology* has weathered the attacks in a remarkable fashion. The reasons, it seems, are twofold. One, it has been freely interchanged with the word *environment* ("What are we doing to the ecology?"), a word with which people seem to be somewhat comfortable, so it has escaped intellectual massage. Two, people have found that it just isn't possible to deal with the idea of ecology in a simplistic fashion. It is a manifold and cumbersome concept, covering too much and touching too much to be handled tidily. We feel comfortable with it only if we don't know much about it.

In this chapter, we certainly want to convey an appreciation of the immensity of the problems associated with ecology. But more than that, we hope to show the great challenge of it all. Ecologists may one day be recognized as the most important scientists on earth.

What, then, do ecologists study? In the original Greek the root word, *oikos*, means "house." Thus ecology is "the study of the house"—the place where we live, or the **environment.** The environment technically includes all of those factors, both nonliving and living, that affect an organism. Ecology, then, is the study of the interaction between organisms and their environment. The key word is *interaction.*

We saved the study of interaction until late in our discussions of life because a certain amount of basic information about organisms is necessary in order to understand their interactions. Certainly, ecology is where the sciences come together. It is a relatively new field, probably emerging as much from what was called **natural history** as from anything else. Natural historians are those crusty old souls who tramp around in fields and who know where rabbits live. Unfortunately, natural history became unfashionable too soon, trampled in the rush to the altar of the molecule and the equation. It is being reborn as a valid topic for ecologists who sometimes find themselves a bit embarrassed by what they don't know about rabbits.

Let's begin the discussion now by looking at ways in which ecologists have managed to organize and categorize the earth's environment. We will begin by considering that thin veil over the earth wherein the wondrous properties of light and water interact to permit life.

THE BIOSPHERE

The **biosphere** is that part of the earth that supports life. It includes much of the terrestrial surface, the subterrestrial realm, and the fresh and marine waters. When we consider the biosphere of the earth as a whole, we find that we're dealing with a lot of square footage but not much depth. Things can't live very far above or below the earth's surface. To be precise, the habitable regions of the earth lie within an amazingly thin layer of approximately 14 miles. This includes the highest mountains and the deepest ocean trenches. If the earth were the size of a basketball, the biosphere would be about the thickness of one coat of paint.

Physical Characteristics of the Biosphere

Conditions within the biosphere are quite special. Our space probes have found no other place in the solar system where these conditions exist. The conditions here result from our distance from the sun, the presence of water, the makeup of our atmosphere, and the earth's solid crust (Jupiter and Saturn, in fact, are gaseous balls). Let's look at some of these factors a little more closely and then consider how various life forms are distributed over the earth.

Water. You'll recall that much of the biochemistry of life is centered on the peculiar chemical traits of water (see Chapter 2). In addition, you may remember that water has a high **specific heat**— that is, it requires a relatively great input of energy to raise its temperature. Likewise, water loses energy very slowly (due to the hydrogen bonds between water molecules). Water, then, acts as a great stabilizer of temperature. It absorbs heat slowly and retains it well. Water vapor in the atmosphere is one of the reasons for the earth's comparatively moderate climates. This moisture helps hold heat and slows the radiation of heat from the earth's surface. We'll return to this point in a moment.

The Atmosphere. The earth's atmosphere may indeed be wispy and ethereal, but all life depends on this fragile veil. Chemically, it is a protective envelope of gases—78% nitrogen, 21% oxygen, slightly less than 1% argon, and about 0.04% carbon dioxide, with varying amounts of water vapor

and a number of other quite rare gases (neon, helium, sulfur dioxide, hydrogen). Most of the earth's atmosphere clings close to the planet, not extending more than five to seven miles above the surface. In addition to its reservoir of useful gases, the atmosphere is also vital to life in that it screens out much of the dangerous ultraviolet radiation that would otherwise make the earth's surface inhospitable to life.

As we also have noted, the atmosphere absorbs heat, and in so doing, acts as a great "heat sink," temporarily holding heat close to the earth's surface. In its role as a heat sink, the atmosphere can be compared to a florist's greenhouse. In a greenhouse, light energy is readily admitted through the glass or plastic windows. The absorbed light energy is radiated back as heat energy, but unlike light, the escape of heat is retarded. For this reason, greenhouses remain warm without added heating—even in the winter—but only as long as ample sunlight is available. The heat striking the earth is trapped in this manner by atmospheric carbon dioxide and water. (People sometimes learn about the "greenhouse effect" the hard way— when they return from shopping after having left their pets in a closed car.)

It is important to note that while the total energy the earth receives from the sun has historically been equaled by the escape of radiant energy, this equilibrium is being seriously disrupted as humans continue to pollute the atmosphere. It's interesting that some well-developed nations have taken considerable strides to curb the problem of atmospheric pollution. But one pollutant we cannot control is carbon dioxide. Every form of combustion in which we involve ourselves—from burning fossil fuels to clearing forests and fields—releases carbon dioxide. The added carbon dioxide in the atmosphere contributes to the earth's greenhouse effect as it traps radiant heat, producing what is believed to be a gradual but inexorable increase in atmospheric temperature. How serious is this? The CO_2 problem is examined in further detail in the next chapter (see Essay 41.1).

Solar Energy. Only about half of the incoming solar radiation ever reaches the earth's surface. About 30% is reflected back into space by the earth and its atmosphere—described as the **albedo effect.** Another 20% is absorbed by gases in the atmosphere. The remaining half reaches the earth's surface as light energy, where it is absorbed

40.1 THE EARTH'S ENERGY BUDGET

The relative constancy of conditions in the biosphere depends ultimately on an equilibrium between energy entering and energy leaving. Of the solar energy reaching the upper atmosphere, 30% is immediately reflected back into space. Another 20% is absorbed as heat by water vapor in the atmosphere. The remaining 50% reaches the earth's surface. Most of this energy reenters the atmosphere through evaporation from the earth's waters. The thin arrow represents energy released by organisms.

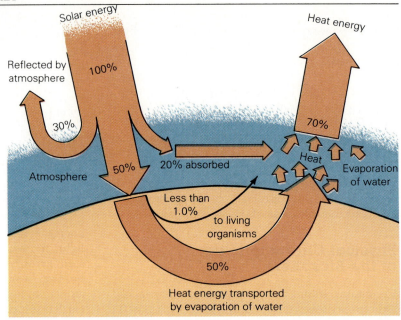

by the land and waters, and is then radiated back into space as heat (Figure 40.1). But a great deal of work is accomplished by the 50% reaching the biosphere.

Surprisingly enough, of this light, only about 1% will enter photosynthetic processes. Far more of that energy is used in shuffling water around. After all, the heat from solar energy is responsible for most of the evaporation from the oceans, lakes, rivers, and not insignificantly, from the leaves of plants through transpiration. As water absorbs this energy, its molecules gain heat—they move more rapidly—and are finally lifted into the atmosphere as water vapor. There, the heat is lost as the water condenses into its liquid phase, falling as precipitation (such as rain or snow). So the absorption of solar heat is vital in the distributing of water over the earth, which, as we have seen, is also a way of redistributing heat, a factor that moderates our climate. The constant shifting of water between its liquid and gaseous phases over the earth is called the **hydrologic cycle** (Figure 40.2).

Climate in the Biosphere. One of the most striking effects of solar energy reaching the earth is seen in the great annual seasons of the planet. The cyclic seasonal changes occur as the earth follows its orbit, exposing different parts of its surface to the direct rays of the sun as it moves. This changing exposure occurs because the earth somehow ended up with a rotational axis that is not perpen-

dicular to the sun's rays, but is 23.5° from vertical (Figure 40.3). Thus most of the earth's creatures are blessed with fluctuating but moderate surface conditions, and not the alternative: great heat at the equator, and huge expanses of perpetually frozen belts where the temperate zones are now located. This tilted axis also produces less dramatic phenomena, such as the tradewinds and patterns of rainfall. How do such changes arise?

The sun's rays fall more directly on the equator than any other part of the earth, so the equatorial regions are hot. This heat causes great warm air masses to rise, carrying with them large amounts of water vapor. The rising of such air masses creates a void below, and lower, colder, and drier air masses rush in from the north. These masses, in turn, heat up, become moisture-laden, and rise. Thus we have air cells both north and south of the equator circulating in opposite directions. Because of the earth's rotational force, the cells are thrown off at an angle as you see in Figure 40.4. This is called the Coriolis effect. These moving air cells produce what are called the **tradewinds,** the consistent winds so long used by sailing vessels. You can see in Figure 40.4 that there are other air cells in more northerly and southerly latitudes.

The movement of moisture-laden air from the equator creates the equatorial rainfall patterns. The rising air cells cool rapidly and lose most of their water as rainfall near the equator. As the air in the cells moves northward, it becomes increasingly

40.2 THE HYDROLOGIC CYCLE

Water is constantly being redistributed about the earth through the ongoing hydrologic cycle. The cycle, driven by solar energy, includes evaporation and precipitation. The distribution of water on the earth is quite uneven, with the greatest percentage—almost 98%—occurring in the oceans. Visible fresh water—that of rivers, lakes, streams, etc.—makes up far less than 1% of the earth's H₂O, an amount that is well exceeded by water locked into the polar ice caps, snow-capped mountains and glaciers—some 1.8%. The water vapor seems to be a paltry amount, but it is nevertheless quite significant to the earth's moderate climate.

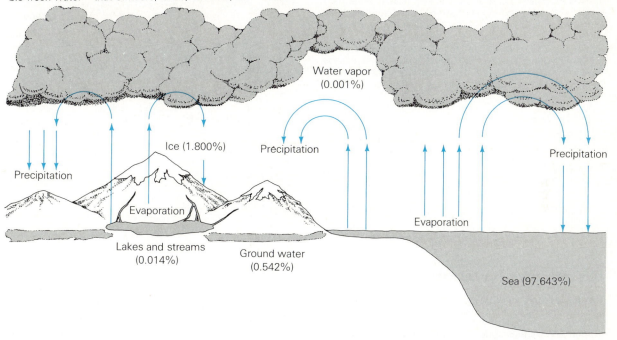

40.3 THE SEASONS IN NORTH AMERICA

(a) Because of its tilted rotational axis, the earth presents a constantly changing face to the sun as it continues its orbit. (b) The tilting produces these seasons in the northern hemisphere.

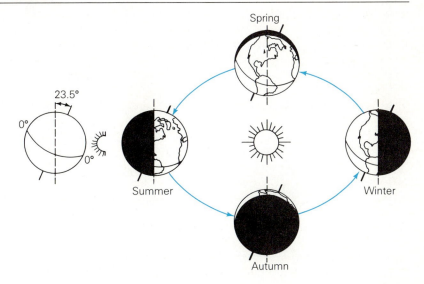

40.4 MOVING AIR MASSES AND RAINFALL DISTRIBUTION

The distribution of precipitation on the earth is determined to a large measure by the formation of several groups of air cells. Air nearest the equator rises, carrying abundant moisture with it. As it rises, cooler, drier air rushes in underneath and the cell rotates. The rising air mass cools and dumps most of its water in belts north and south of the equator. Rainfall is far more limited just above and below these belts and here are found some of the earth's great deserts. Note that the great rotating equatorial cells are thrown off center by the spinning of the earth—the Coriolis effect—so the winds do not simply blow north and south, but occur in easterly and westerly directions. These are the reliable trade winds upon which the old sailing vessels depended. Additional cells are seen further north and south.

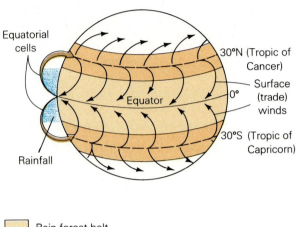

Rain forest belt

Desert belt

40.5 THE OCEAN CURRENTS

The major ocean currents are produced by the earth's winds and modified by its rotational forces. The two great Pacific currents, the Japan Current in the north and the Humboldt Current in the south, carry cold water south and north along the west coasts of North America and South America respec-tively. Note that the Atlantic Gulf Stream is quite different, originating in a more tropical region, and carrying warm water north along the east coast of North America before heading out across the Atlantic.

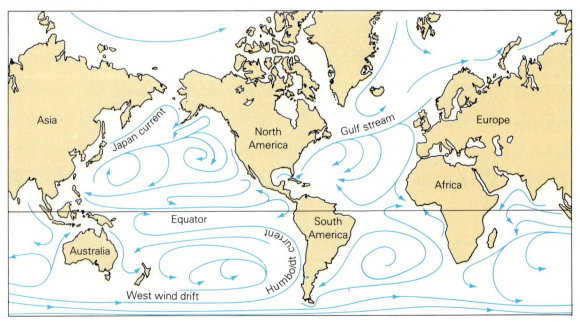

drained of moisture and thus yields less precipitation. Finally, at about 30° latitude, north and south, the lack of rain produces deserts.

Other factors may also contribute to the formation of deserts. One is called a **rain shadow.** Wherever moisture-laden prevailing winds encounter high mountain ranges, most of the precipitation falls on the windward slopes of the mountains. As the air masses move up the mountain slopes, they cool, and lose most of their moisture as rain or snow. The drier air mass moves down the leeward slopes, then scurries over the desert and gathers up whatever moisture there is, intensifying the situation. As an example, the great American desert is a product of the rain shadow produced by the Sierra Nevada mountains. Moisture-laden air, moving eastward from the Pacific, is confronted by the mountains. As the air masses rise, they cool and most of the water is dumped on the western slopes with very little to the leeward. In South America, where prevailing winds blow to the west, the rain shadow is produced by the Andes range, and the deserts that form west of the Andes along the coasts of Peru and Chile are some of the driest known.

The moving air masses also help create the ocean currents (Figure 40.5). Ocean currents are, in turn, important because of their effects on the climate of nearby land masses and because they help mix the waters and distribute the nutrients and gases required by aquatic organisms. Major oceanic currents called **gyres** are found in the northern and southern hemisphere. They are chiefly surface phenomena with the flow extending down just 100 to 200 meters. The deepest flow of a major current is found in the Gulf Stream, extending as far down as 1000 meters. The movement of the major gyres transfers equatorial heat northward and southward, a feat that profoundly affects the climate of coastal regions.

THE DISTRIBUTION OF LIFE: TERRESTRIAL ENVIRONMENT

The earth's surface varies markedly from place to place, as does the life it supports. Vast, distinct, and recognizable associations of life are called **biomes** (Figure 40.6). More precisely, a biome is a particular array of plants and animals within a geographic area brought about by distinctive climatic conditions. Biomes are usually identified more by their plant associations than those of animals, not only because the first is far more obvious but because it determines the second. The specific plant associations are, as you would expect, a product of adaptation to several climatic factors, including precipitation, temperature, and light.

Biomes in turn may be subdivided into **ecological communities,** referred to simply as **communities.** Ecologically speaking, a community is a number of interacting populations of organisms. As in biomes, communities are generally recognized according to their predominating plant species. Several communities may exist within any biome. If you were to begin a walk across a biome, you might notice that although the general climate may remain the same, specific groups of plants and animals change somewhat. Thus, we might find both a mesquite community and a sage community in a desert biome. If you were somehow able to return to that biome hundreds of years later, you would find that biomes also change over time. We will return to community structure and interaction in the next chapter, but for now let's make a few more general observations about biomes and then describe the major ones.

As the biomes change from one place to another, we find some transitions are gradual—forming gradients of species in what are called **transition zones.** For example, we find a gradual transition across the United States from the east coast and Appalachian Mountains to the western coastline (Figure 40.7). The moist forests of the Appalachians slowly give way to drier oak-hickory forests, and then to forests consisting almost entirely of oak. The forests become less luxuriant as they fade into the great American grasslands. That is, they did before the grasses yielded to the intensive agriculture of America's "breadbasket." The prairies were once seas of tall grasses, which gave way (where there was less precipitation) to shorter and shorter grasses. Finally these yield to the Great American Desert, which is followed by the great Sierra Nevada and coastal mountain ranges and finally the Pacific shore. Coniferous forests are common on the mountain slopes, and minor grasslands and deserts are found between some ranges. We see, then, that the borders of biomes are usually indistinct, with the mixture of plants seemingly engaged in an endless tug-of-war over boundaries. The principal determining factors in transitions are precipitation and temperature (see the biome map in Figure 40.6).

The distribution of biomes on the earth generally follows latitude, but this pattern is more obvious in the Northern Hemisphere than in the Southern

Hemisphere. Going from the equator northward we tend to move from tropical rain forest through desert, grassland, deciduous forest, the taiga (coniferous forest), and finally tundra and ice cap. This latitudinal arrangement of biomes is not entirely orderly for several reasons, one of which is the terrain, or topography. Mountain ranges interrupt the orderly distribution of biomes and, as we have seen, are often responsible for the presence of deserts because they can block the movement of moisture.

Mountains also influence biomes in another interesting way: biome distribution is a product of altitude as well as latitude. This is because temperature decreases as altitude increases. Thus, high mountains with permanent snow or ice are found in the tropics, and the usual biome transitions may be represented as well. Further, the *altitudinal* transition may resemble the latitudinal one (grassland into deciduous forest, and so on; see Figure 40.8).

One of the best examples of altitudinal transition is the Ruwenzori mountain range in east central Africa. One can ascend from tropical rain forest through broad-leaved evergreens, to deciduous and coniferous forests, then to alpine meadows, and finally to the barren, snowswept peaks. Once again we see that the factors determining the distribution of life are complex.

Keeping in mind that the dividing lines are often indistinct and arbitrary, and that the complexities are greater than any brief discussion can convey, we will now consider the nature of the earth's great biomes, beginning with the driest and perhaps most fabled.

40.6 THE BIOMES

Each of the biomes can be identified by its dominant form of plant life. These forms are adapted to the climatic conditions, including precipitation, the availability of light, and, of course, temperature. Both latitude and altitude affect all these conditions.

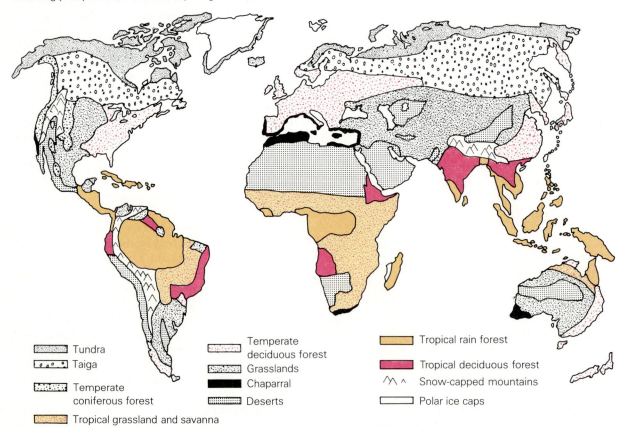

40.7 PROFILE ALONG THE 39TH PARALLEL CROSSING THE UNITED STATES

The line transects several of the major biomes and helps indicate the conditions by which they were produced. For example, the American deserts lie in the rain shadow of the western mountain ranges. Rain and snow fall on the western slopes only. Moisture-laden air from the Atlantic, at the other end of the profile, drops its burden as it moves west, the total rainfall diminishing as it moves inland. Forests occur west to the Mississippi, but from there, prairie extends westward, dwindling into short grasses toward the Rocky Mountains. The Great American Desert extends between the Rocky and Sierra Nevada Mountains.

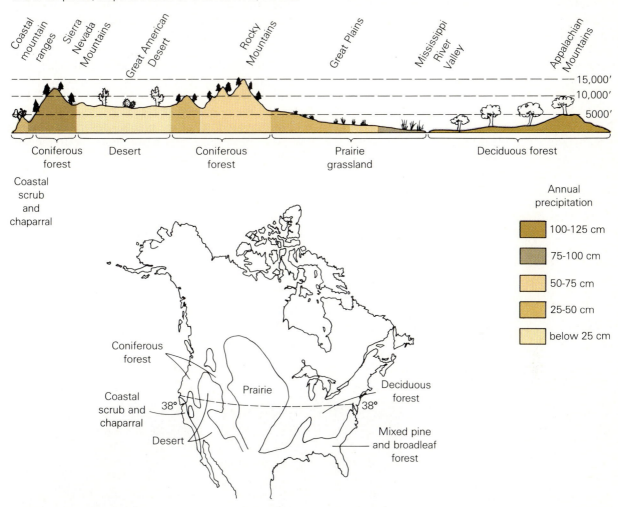

The Desert Biome

Deserts are dry. By definition, they receive less than 25 cm (10 in) of rain each year. You may be surprised to learn that deserts are not necessarily hot (even in the day) and not necessarily tropical. The largest desert is the Sahara, which, as you can see on our biome map, covers nearly half of the African continent (and for reasons we will get to shortly, is getting bigger). Other large expanses of desert are found in Australia, Asia, western North America, and South America (in fact, every continent but Europe). Temperatures in these regions undergo dramatic day-night fluctuations, and may vary as much as 30°C (54° F) in a 24-hour period. The reason for such extremes is the lack of buffering, heat-retaining moisture in the desert air. The surface heats up rapidly in the daytime, but cools just as rapidly by evening. In spite of the long periods of drought, much of the actual topography of the desert floor is determined by water. Very seasonal but torrential rains cause flash flooding that continually remakes the face of the desert.

If you have never seen a desert, you might have an image of lifeless regions of drifting sand studded with a few palm-covered oases. Actually, there are such places, but most deserts are alive with plants and animals. Since dry areas are called **xeric** (Greek

40.8 AN ALTITUDINAL DISTRIBUTION OF BIOMES

The altitudinal variation provided by mountains often mimics latitudinal distribution. In this hypothetical model, a rain forest at sea level gives way to a deciduous forest at a higher elevation. Conifers replace the deciduous trees farther up, which yield to alpine meadows, and finally a rather typical tundra situation near the glacier. These transitions are a product of decreasing precipitation and temperatures. On the other side of the mountain, air masses, with little moisture remaining, rush down the slopes and across the barren terrain, leaving typical desert conditions behind as they pick up much of whatever moisture exists.

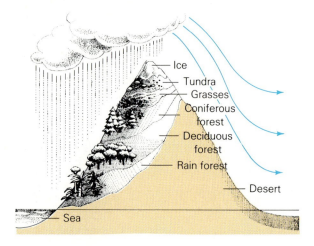

for "dry"), plants adapted to deserts are called **xerophytes.** Their adaptations can take a number of routes, but they have one common imperative: Save water. Native American perennials, such as cactus, ocotillo, joshua, creosote and palo verde (Figure 40.9), are adapted to living long periods on what little water they contain, while waiting for rain. Through evolution, the leaves of the cacti have been reduced to spines, and photosynthesis occurs mainly in the green stems. The stems are covered with a thick, waxy cuticle, and water is stored in the oversized cells of deeper tissues. The spines of cacti, and the thorns of many other desert plants discourage animal browsing and prevent subsequent water loss by the injured stems.

Compared to other plants, desert perennials tend to have fewer and more widely scattered stomata. This helps in conserving water, but the price is a considerably slower growth rate, since CO_2 uptake and photosynthesis are retarded. Other desert perennials have deep root systems able to tap whatever water seeps into the porous soil, very small leathery leaves, and the ability to lie metabolically dormant, growing very little, for long periods. Some desert cacti and other succulents have reversed their gas exchange cycles, taking advantage of the cooler and moister nights. This latter group is composed of the *CAM* plants discussed in Chapter 23.

40.9 DESERT PLANTS

Both of the North American desert plants shown here (the saguaro, left, and the Joshua tree, right) have adapted to limited water; these are called xerophytes. Their thorny epidermis discourages browsing and helps shield the green surfaces from the harsh, direct sunlight. Many xerophytes have pulpy water-storing tissues within, while others simply slow their metabolic processes, and become scraggy and leafless between rainy seasons.

40.10 DAY AND NIGHT DESERT FOOD WEBS

Animals in the desert forage in two shifts. The day feeders are fewer in number but are much more active. These include birds such as the African ground squirrel (a), roadrunner (b), and many insects. But even these hardy foragers often remain in shadows at midday. At sundown the second shift begins as carnivores like the scorpion (c) and the Fennec fox (d) go quietly about their deadly business.

(a)
(b)
(c)
(d)

The desert annuals have a different strategy for survival. After a spring rain (or *the* spring rain), the desert may become transformed as tiny flowers of all descriptions erupt into full bloom, a trembling and riotous offering of vibrant color and delicate forms. Their life cycle is short and they soon die. But their seeds will be highly resistant, lying in the sand until subsequent rains propel them into their surge of growth and reproduction.

Just as plants have had to adapt to the rigors of the desert, so have the animals. Animals have the advantage of being able not only to adapt anatomically and physiologically, but in having more behavioral options. The desert is rich in animal life, including a variety of arthropods, some resident birds, and often many seasonal migratory bird species. There are also reptiles, and even some mammals. The primary problems of all these species include escaping daytime heat and avoiding water loss. Many desert animals beat the heat by simply staying out of the sun and becoming nocturnal. In fact, if you should visit the desert you may see only a few lizards, a bird or two, and a few insects— unless you want to traipse around at night with the snakes.

Desert mammals are largely represented by rodents, animals for which nocturnality has very clear advantages. They tend to lose water quickly because they breathe rapidly and have a large surface area relative to their volume. Thus, rodents spend the desert days deep in insulating burrows, venturing out only at night to forage for plant food. Rodent-hunting predators such as owls and rattlesnakes, follow suit, confining most of their activity to the cool of evening or night.

The few daytime animals, such as long-legged and fast-moving lizards, are preyed upon by hawks and roadrunners, which are also adapted to daylight conditions. Even the daytime animals, however, restrict most of their activity to the morning and evening hours (Figure 40.10).

Perhaps we can best gain some insight into the adaptive strategies of desert animals by considering a mammal and, of all things, an amphibian, each of which has made unique adjustments to desert life. These are the versatile kangaroo rat and the spadefoot toad. The kangaroo rat (*Dipodomys deserti*) of the southern California desert is particularly interesting because it doesn't drink. It survives on the water content in its food and supplements this with metabolic water (produced as a waste product in cell respiration; see Chapter 8). The kangaroo rat is also a great water miser. Its remarkably efficient kidney produces only highly concentrated urine, and very little at that (23% urea and 7% salt, compared to 6% urea and 2.2% salt in humans). The feces are also dry and crumbly. Most water loss, in fact, occurs through simple breathing. Even these special physiological features, however, wouldn't permit desert survival, were they not coupled with nocturnality. The rat spends its day in a humid, hair-lined burrow, venturing forth only in the cooler evening.

While amphibians are notably scarce in the desert, the fascinating spadefoot toad is a year round resident of the American deserts. It escapes heat and drying by burrowing, and while entombed it is also capable of **estivation,** a sort of hibernation, but under reversed climatic conditions: it remains buried during long, hot, dry intervals. When sufficient rain occurs, it springs (hops?) into action. For this creature, time is short. Like other toads, the spadefoot must reproduce in water. Finding a temporary pond, the male begins at once to call prospective mates in the briefest of courtships. Eggs, fertilized in the shallow ponds, hatch in a day or two. The young tadpoles complete their metamorphosis and emerge as adults in a few short weeks.

Although deserts seem tough and unyielding, they actually represent one of the more fragile biomes. Simple systems are always the most vulnerable, and deserts are essentially simple places, harboring few species compared to many other places. Because the desert is so vulnerable, the impact of humans can be great.

The recent experience of the region to the south of the Sahara is sobering. People of the Sahel region long supported themselves and their small herds of cattle by hand-drawn water from a few scattered deep wells. In recent decades, gasoline pumps were introduced, making water suddenly abundant. The result was an explosive increase in the cattle population, followed by a substantial increase in the human population. The cattle placed great pressure on the vegetation of the area until a recent drought—not unlike the many droughts the region has previously survived—killed most of the cattle and many of the people. As a result of such destructive patterns, the Sahara itself has expanded by thousands of square miles, probably permanently, demonstrating that well-intentioned but ill-conceived aid programs can backfire with tragic results.

The Grassland Biome

Grasslands of the Northern Hemisphere exist as huge inland plains and include such areas as the Asian steppes and (in times past) the prairies of North America. There are similarities between grassland and desert and, in fact, in these two regions the grassland gradually fades off into desert. The chief climatic difference between the two is precipitation; grasslands, of course, get more rain—roughly 25 to 75 cm (10 to 30 in)—but the rain is of a seasonal nature.

In the Southern Hemisphere, grasslands are known under various names: the pampas of South America, and the veldt and savanna in Africa. (As we will see, the savanna is not considered a true grassland because it is dotted with groves of trees.) Grassland in Australia is very extensive, occupying over half the continent (Figure 40.11).

We are compelled to say that the dominant plants of grassland are grasses. Fortunately, we may be able to regain your interest by also noting that grasslands are very different from one place to the next. For example, in the former American prairie, grasses east of the Mississippi reportedly grew 10 feet tall, while those in the west rarely surpassed a foot or two. Again, the difference was principally due to variation in rainfall. But it is important to realize that the grassland biome is also maintained by fire and by the action of grazing animals. Enor-

40.11 GRASSLANDS

Natural grasslands such as the South American pampas support enormous numbers of herbivores, some of which are among the largest of all mammals. These regions are slowly changing with the encroachment of civilization.

40.12 GRASSLAND HERBIVORES

The highly efficient producers of the grassland support an extensive food web, often including incredible numbers of large herbivores, for example, these Australian sheep. These mammals, in turn, support a sizable predator population.

mous fires, often started by lightning, periodically sweep through all natural grasslands today. But the rhizomes and extensive root systems survive and regrowth begins with the next rains. Were it not for fire and grazing animals, the deciduous forest—whose plants are often not fire-adapted—would undoubtedly encroach on many grasslands.

Since rains are usually seasonal in grasslands, the plants have developed important strategies for the dry periods. In lowland regions, root systems may penetrate a permanent water table as far as 3 meters below the surface. More commonly, the grasses rely on vast, spreading **diffuse root systems** in which no root dominates; the great surface area provided by so many finely branching roots enables the grasses to quickly absorb water from light rains. In addition, grasses readily become dormant, reviving when water is once again available. Some grasses produce underground stems (rhizomes) that remain alive after all the foliage has died. The rhizomes, as well as the above-ground horizontal stems (stolons), form the dense sods which also actively prevent the growth of trees where rainfall would otherwise permit it. This matlike growth prevented agricultural intrusion in the American prairie until there were improvements in plowing implements. (The first successful grassland farmers were dubbed "sodbusters.")

Grasslands are highly efficient at rapidly converting solar energy into the chemical-bond energy of their living matter. They are able to support larger populations of animals than any biome on earth. So it is not surprising to find huge herds of grazing animals (Figure 40.12). In fact, grasslands are the habitats of many of the world's large hooved herbivores. The original herbivores of the American prairie—the bison and the pronghorns—have been displaced by cattle and sheep, but in the plains of Africa there are still vast herds of wildebeest, zebra, and other natural grazers. These native grazers quickly move on without destroying the grasses and the plants continue growing from their cropped bases.

Where wildlife is protected or nurtured in the United States, herds of bison and pronghorn antelope are the conspicuous grazers. But the fields are full of unobtrusive grass eaters, such as jackrabbits, rodents, and prairie dogs, as well as insects and seed-eating birds.

With such a food reserve, we can expect the grassland to support large numbers of predators. Unfortunately, they may prey on domestic herds as well as wild species and this has presented problems for both us and them. Obviously, we're not going to allow predators to get away with eating herbivores we have raised for profit, so large predators are generally restricted to protected grasslands. These animals include wolves, cougars, coyotes, and even foxes.

Snakes and carnivorous birds range a bit wider, but they don't eat many animals from our domesticated herds. (In spite of the lack of supporting evidence some ranchers still consider birds of prey to be a threat and thus justify shooting them—including the endangered bald eagle, our national emblem.) One particular and grisly exception is a parrot in New Zealand, the kea (*Nestor notabilis*), that swoops out of the forests, lands on a sheep's back, tears into its flesh and eats the fat from around the kidneys of the living animal.

Overgrazing, or grazing by herbivores that kill grass by cropping it too short (such as sheep may do), can irreparably destroy grassland. The mesquite- and cactus-covered wastelands area in Texas (55 million acres) was stable, productive grassland before the cattle barons subjected it to heavy grazing. Much of the Sahara and the deserts of the Middle East, in fact, have been created by domestic grazing in past centuries. The barren Middle East, remember, was once called "the land of milk and honey."

The Tropical Savanna Biome

The **tropical savanna biome** is a special kind of grassland that forms at the borders of the tropical rain forest. Unlike other grasslands, the savanna is frequently interrupted by scattered trees or groves

40.13 THE SAVANNA

Large hooved mammals are prominent among the savanna's herbivore population. On the African plains, predators include the large cats.

(Figure 40.13). And unlike the tropical forests, the savannas have a prolonged dry season with an annual rainfall of 100 to 150 cm (40 to 60 in). It is during the dry season that the savannas are subject to great fires. The great annual droughts are also responsible for the movements of great migratory herds of grazers in search of food.

The largest savannas occur in Africa, but they are also found in South America and Australia. While grasses are the dominant form of plant life in the African savanna, the drab landscape is brought to life by palms, colorful acacias, and the strange, misshapen baobab tree, which appears to be growing upside down. The number and variety of hooved animal species exceeds that of all other biomes, and includes the familiar zebras, giraffes, wildebeests, and numerous antelopes of the African plains. The African savanna is also the domain of familiar predators such as the lion and cheetah. Like the grassland, the natural fauna of the savanna is being replaced by domestic grazers, especially cattle.

Tropical Rain Forest

Forests are scattered over much of the world, but only in those places where the water supply is adequate. Rarely is water more available than in the tropical rain forest. Typically, this biome receives about 250 to 450 cm (100 to 180 in) of rainfall per year. Rain falls throughout the year but is somewhat heavier during the "rainy season." The largest tropical rain forest is in the Amazon River basin in South America (Figure 40.14). The second largest is in the wilds of the Indonesian archipelago. And

then there are those of the Congo basin in Africa, parts of India, Burma, Central America, and the Philippines.

Tropical rain forests are best described as lush, with a very large number of tree species growing to great heights. The floor is dark and wet, and the air is often cool and laden with rich smells. Unlike what we find in the other forest biomes, no single kind of plant dominates the forest. Any tree is likely to be a different species from its neighbors, and nearest trees of the same species are often miles apart. The exceedingly tall trees, ranging in height from 30 to 45 meters (about 100 to 150 ft) form a dense, continuous canopy overhead (Figure 40.15). The crowns of lesser trees form a subcanopy below their taller neighbors. The trees are invaded by large numbers of vines, and both trees and vines may be festooned with **epiphytes,** plant species that have evolved ways of joining the keen competition for sunlight. Epiphytes live on the stems and branches of tall trees in the canopy, with no contact with the soil. They absorb water directly from the surrounding humid air. (One species surrounds its roots with

40.14 THE TROPICAL RAIN FOREST

Tropical rain forests receive enormous amounts of rain throughout the year. There is little seasonal change. No single plant species dominates the terrain, but a number of kinds form a dense canopy over the sodden earth. The humidity on the forest floor can be stifling.

a bucketlike base which collects water and insects, the decay of the latter assuring it of a continuing supply of nitrogen compounds.) The forest floor may have little to moderate foliage, but it is teeming with fungal and bacterial decomposers and insect scavengers. The darkness, warmth, and blanketing humidity there are ideal for rapid decomposition.

If the rain forest floors often don't have much foliage, then what about those reports of the "impenetrable jungle"? The answer is, jungles are special kinds of tropical forests. Essentially, they contain a low-growing tangle of plants, which is, in fact, almost impenetrable. Jungles arise where light reaches the forest floor. Jungle areas may be scattered through rain forests, but they are particularly common along river banks and steep slopes, and in disturbed areas such as clearings and deserted farms. Along river banks the forest is called *wet jungle* and it abounds with insects and reptiles. The idea that jungles represent tropical rain forests grew from descriptions by river travelers who weren't about to get out of the boat and so they missed seeing the relatively clear forest floor, often just a few hundred yards from the river bank.

The tropical rain forest harbors a great number of different animal species, more than any other biome. Insects and birds are particularly abundant, and reptiles, small mammals, and amphibians are common. Many of the animals are arboreal (tree dwellers) and many species are stratified according to the layers established by the plants, becoming specialists at occupying certain levels of the canopy and subcanopy. In one study of the Costa Rican rain forest, ecologists found 14 species of ground foraging birds, 59 species occupying the subcanopy, and 69 in the upper canopy. They further found that about two thirds of the mammals there were arboreal, as were a number of frogs, lizards, and snakes (Figure 40.16).

Tropical seasonal forests occupy a considerable portion of the tropical biomes. These differ from true tropical rain forests in that rains are, as you might expect, seasonal. A familiar example is the monsoon forest of Southeast Asia, although such forests exist in other tropical regions. In many instances, the trees there are deciduous, losing their leaves during the dry season. When the monsoons arrive with their torrential rains, the forest takes on some of the characteristics of the tropical rain forest. Some of the more highly valued hardwoods, such as Burmese teak, are found in the tropical seasonal forests.

The tropical rain forests of the world are rapidly disappearing. The Amazon Basin rain forest of Bra-

40.15 STRATIFICATION IN THE TROPICAL RAIN FOREST

In the struggle for sunlight, plants of the tropical rain forest form a multistoried canopy. The uppermost consists of trees that often reach 30 to 45 m above the floor. Below these, the foliage of a subcanopy layer is found, and there may exist an even lower subcanopy. The layered arrangement of trees is further complicated by climbing vines (the lianas), adding their foliage. The stratified foliage provides several types of animal habitats. Brightly colored birds abound in the leafy canopies, as well as reptiles, amphibians, and mammals. In fact, most of the vertebrates in such places live in the trees as opposed to the forest floor.

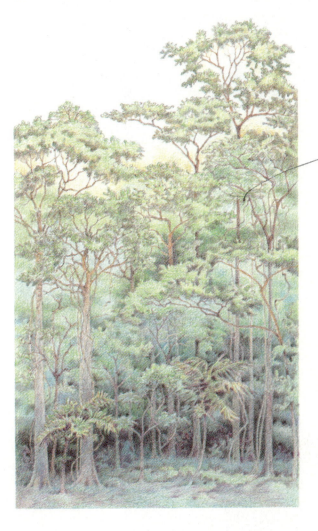

zil, the last really large area of undisturbed forest, is now in the process of being rapidly cleared for timber and farming and exploited for minerals. The results may prove to be catastrophic for several reasons.

The extremely heavy rainfall that gave rise to this rain forest in the first place is almost entirely composed of moisture sent aloft from the steaming forest itself. Little additional moisture blows in from the ocean. Cutting down the trees greatly reduces the amount of moisture returned to the atmosphere within the Amazon Basin, and this will ultimately reduce the rainfall there by a substantial amount. The rain that does fall onto the cleared land will tend to run off into the river system and then to the ocean.

In addition, the tropical rain forests, which are being cleared at an average rate of 1 percent per year, represent an enormous carbon dioxide "sink." This means that the lush growth absorbs a great deal of carbon dioxide from the atmosphere, locking it up in the carbon compounds of the plants. While there is a steady turnover because of death and decay, the sudden removal of plants means that the normal cycling is interrupted. Further, since the plants are burned, the carbon stored in their molecules is suddenly released as carbon dioxide. As we have mentioned, atmospheric scientists are particularly alarmed by this steadily increasing carbon dioxide burden.

Finally, you may be surprised to learn that tropical soils are very poor and infertile. The nutrients formed by decomposition are immediately recycled back into plant growth, so no reservoir of humus remains. In addition, the rains continually leach precious nutrients from the porous soil. This fact has become painfully clear to proponents of jungle agriculture. Once the native plants have been removed, the soil may rapidly change into a hard, water-resistant crust known as **laterite.** The term, which means "brick," rather aptly describes the reddened crust.

Chaparral: Mediterranean Scrub Forest

Our biome map indicates that the Mediterranean scrub forest, or **chaparral** as it is known in California, is rather insignificant among the forests of the world. It does, however, have unique characteristics and its own peculiar plant associations. Note, for example, that chaparral is exclusively coastal, found mainly along the Pacific coast of North America and the coastal hills of Chile, the Mediterranean, southernmost Africa, and southern Australia. This forest is unique in that it consists of broad-leaved evergreens, growing in subtropical regions marked by a marine air flow, low rainfall and a long summer drought. Depending on the altitude, California's coastal chaparral receives between 25 and 75 cm (about 10 to 30 in) of rainfall, almost all of it in the short winter rainy season. The seasonality of the rains means that the plants experience drought through most of the year.

Plants and animals adapted to chaparral have adopted strategies similar to those of desert dwell-

40.16 ANIMALS OF THE CANOPY

Tropical forest animals are generally arboreal, each species specialized for life in specific parts of the canopy and subcanopy.

Squirrel monkeys

Jaguar

Kinkajou

40.17 THE CHAPARRAL

The chaparral in southern California may lack the lushness of many other forests, but its plants are tenacious and hardy. These scrubby plants resist an annual drought that would discourage most other plants. Many parts of the chaparral actually qualify as deserts in terms of total precipitation, but they are not called deserts because of their cool, moist marine air.

ers. In the chaparral, plants are chiefly represented by shrubs, mostly with dwarfed, gnarled, and scrubby stems and scattered succulents (Figure 40.17). Leaves are generally small, with very waxy and tough cuticles. Many plants become dormant after the seasonal rains, remaining that way through the dry summers. Insects abound in the chaparral, particularly an army of beetles that lives in the leaf litter covering the forest floor. Other inhabitants include vertebrates such as mule deer, rabbits, bobcats, many rodents, lizards, snakes, and a considerable number of birds. Wrentits and towhees are typical residents, and large numbers of other birds pass through on their annual migrations.

Chaparral has a peculiar problem related to drought. It is often seared by fast-moving brush fires. Controlling the fires along the California coast is complicated by the rough, hilly terrain. Little of the chaparral escapes the periodic fires, but the plants appear to have adapted to this violence. Some sprout quickly from burned stumps, and others scatter fire-resistant seeds over the burned ground. Ecologists have described the chaparral as a **fire-disclimax community.** This means that the chaparral virtually never reaches maturity, and any given stand is always in some stage of recovery. It shows, too, that fires were common before humans came on the scene. The floor of chaparral is often heavily layered with leaf litter. Decomposition is

slow because of the dryness. The accumulation of those leaves (the plants may shed leaves year-round) magnifies the fire problem.

The Temperate Deciduous Forest

If you live east of the Mississippi, the temperate deciduous forests may dominate your surroundings. The trees here shed their leaves each fall and rake the gnarly fingers of their bare branches at the winter sky. Each spring they blush anew with fresh faint buds that will change to the deep greens of summer. Temperate deciduous forests, then, are characterized by both leaf fall and changing seasons (Figure 40.18). Such forests extend over much of the

40.18 THE CHANGING DECIDUOUS FOREST

The deciduous forest changes its appearance at different times of the year. The lovely green hillside of summer will explode in a riot of colors when autumn arrives. Both are in sharp contrast to the starkness of winter. Animals here must adapt to the long cold winters. Many will sleep or hibernate, while others simply migrate. The remainder are faced with problems of food and shelter for the winter months.

40.19 ANIMALS OF THE DECIDUOUS FOREST

Herbivores of the deciduous forest are numerous, including the gray squirrel, cardinal (left), and white tail deer (center), in places where they still remain. Carnivores once included large cats, wolves, and bears, but in most of this disturbed biome these animals are now rare. Typical carnivores today include the raccoon (right), eastern garter snake, and the insect-eating redheaded woodpecker, and an occasional fox is still seen.

eastern United States, northward into southeast Canada. They are also found in the central and northern parts of Europe, including Great Britain and reaching into southern Norway and Sweden. A long finger of the forest pushes into the center of the Soviet Union. In eastern Asia, deciduous forests are found in China, the eastern Soviet Union, Korea, and Japan; in the Southern Hemisphere, although much less conspicuous, they do exist in coastal Brazil, east Africa, the eastern coast of Australia, and across most of New Zealand.

Precipitation in the deciduous forest is rather evenly distributed throughout the year, rainfall often averaging over 100 cm (39 in)—enough to support a variety of trees. Typically, the larger trees such as oak, maple, beech, birch, and hickory form the canopy. In northerly deciduous forests, communities of beech and maple dominate, while in the south, the oak is common. In times past the oak was joined by the chestnut, but this majestic tree has been all but obliterated by a bark fungus (*Endothia parasitica*), accidentally introduced from China at the turn of the century. Interestingly, whereas the adult trees are gone, shoots still emerge from the persistent roots, and some seeds are produced, but further growth is stopped by the fungus.

Moderate levels of light reach the forest floor and encourage the growth of younger trees, shrubs, ground-hugging plants, and a variety of annuals. The annuals begin to grow in early spring, or even late winter. They rather quickly produce their seeds and die off with the autumn frosts, contributing to the rich leaf litter. The litter and humus harbor an abundance of scavengers and decomposers, especially in the warm summer months. The forest floor is like a giant soft sponge, soaking up rain and contributing to the luxurious forest growth.

There is a variety of animal life in deciduous forests. Although we have displaced or killed off most of the larger mammals, in some areas they have been protected by those interested in hunting them. The forests were once the home of deer, wolves, bears, foxes, and mountain lions. Mammals in these forests today are largely represented by rabbits, squirrels, raccoons, opossums, and rodents (Figure 40.19). The largest predators are likely to be humans, owls, hawks, a few black bears, and occasional bobcats and badgers. In the northern deciduous forests, a few wolves have escaped having their skins trim cheap coats. Finally, as any camper knows, arthropods inhabit the forest in great numbers, primarily rummaging through the litter and foliage, and crawling into tents.

Winter in the deciduous forest is heralded by one of the most beautiful and moving events in nature: autumn. As the abscission layers in the leaf petioles prepare to separate, and the green chlorophyll wanes, other colors come to dominate the woodland scene. The leaves almost seem to celebrate the cool days, as browns, reds, and yellows mix with the greens of conifers. It is too soon over. And the fading light of the shorter days, which changed the leaves, also warns the animals of the approach of winter. Some birds leave, others change color and give up territories. Some mammals, grown heavy with fat, find holes in which to sleep or to hibernate. Others, unable to escape either behaviorally or physiologically, simply face the coming cold; some will not survive. Soon, the stark, lifeless days of winter settle in.

The deciduous forest biome produces trees with strong and flexible cell walls, and much of it has been cleared in order to satisfy our needs not only for wood but our extraordinary demand for paper. The forests have also fallen simply because we had something else to put in that space. In the American Southeast, native hardwoods have been replaced by faster-growing, more marketable pines, planted in orderly rows. In any case, we have removed so much forest that today almost all American hardwood forests are, at best, second-growth areas. As such, they are in transition, with many regions dominated by scrub oak and other uninspiring invaders.

The Taiga and Other Coniferous Forests

The taiga, or **boreal** (northern) **forest,** is almost exclusively confined to the Northern Hemisphere. Great coniferous forests of pine, spruce, hemlock, and fir extend across the North American and Asian

40.20 THE TAIGA

The taiga is an extensive biome, found almost exclusively in the Northern Hemisphere. Its dominant plant is the conifer, although communities of birch, willow, poplar, and alder are not unusual. Since the conditions of the taiga are duplicated in high mountains, similar communities are found there. The harsh climate includes a short growing season, limited rainfall, and a severe, long winter.

40.21 MUSKEGS

Where bogs and marshes form, the taiga is interrupted. These areas are often referred to as muskegs. They represent a perpetual tug-of-war between aquatic and terrestrial environment, as plants continually invade the marshes, some failing, others becoming established.

continents, with narrower bands reaching into Norway and Sweden. They are also found at higher elevations in many mountain ranges (Figure 40.20). The taiga is unmistakable; there is nothing else like it. It is subject to long, cold, moist winters and short summer growing seasons. It is difficult to generalize about rainfall because the taiga is so extensive; some places get a great deal of rain, some don't. But in many parts, there is much less rainfall than in deciduous forests. (In the Canadian interior, for example, rainfall is between 50 and 100 cm—20 to 40 in.) The taiga is interrupted in places by extensive bogs, or **muskegs,** the remnants of large ponds. These wet regions are commonly dominated by spruce, but low-lying shrubs, mosses, and grasses form the spongy ground cover (Figure 40.21). The northernmost taiga is touched by sunlight for only six to eight hours each day through much of the winter, and bathed in sunlight for 19 hours in the summer.

The dominant plants of the taiga are conifers, but occasional communities of hardy, broad-leaved poplar, alder, willow, and birch may be found in disturbed places. Conifer leaves, which are usually needle- or scalelike with dense, waxy surfaces, represent adaptations to the dry conditions common in the taiga. Such adaptations enable conifers to survive the winter with many of their leaves intact; thus they are sometimes referred to as evergreens. The leaves are adapted for water conservation since much of the water reaching the taiga may be bound up (and made unavailable) in ice. (One is reminded of the spines and waxy cuticles of desert plants.)

Leisurely strolls can be quite pleasant in a coniferous forest because there is very little underbrush. One reason is because those soft needles that feel so good under your boots are hard on other plants. They make the soil notoriously acid (and thus the conifers reduce their level of competition with other species). The straight trunks and dark, unobstructed forest floor lend a cathedrallike atmosphere to taiga forests, an atmosphere that some find compelling and others forbidding.

The taiga harbors such large herbivores as moose, elk, and deer. It is the last refuge of the grizzly bear and black bear, and wolves still roam here, as do lynx and wolverines. While rabbits, porcupines, hares, and rodents abound, insect populations aren't as large as in deciduous forests, although there is an abundance of mosquitoes and flies in boggy regions during the summer (Figure 40.22).

There are other large coniferous forests that are not considered taiga. A number of temperate coniferous forests are to be found throughout the Northern Hemisphere. For example, the Olympic forest in western Washington is, on its western slopes, a bona fide temperate rain forest with 500 cm (200 in) of rain per year near the glacial peaks. Here, the Sitka spruce reaches its greatest size, approaching that of the famed California redwood. Fingers of the western forests follow the Cascade Mountains into Oregon and California.

Perhaps the most fascinating coniferous forests are the California coastal redwoods. Their gigantic demands for water are not met by rain alone, but also by coastal fogs. Of equal stature, on the western slopes of the Sierra Nevada Mountains, we find the magnificent forests of giant sequoias. These awesome and splendid trees are among the world's oldest living organisms. Finally, not to be ignored are the expanses of pines that dominate the coastal plain forest in the southeastern United States. These vast forests of longleaf, loblolly, and slash pine are probably a temporary invasion, replacing the usual oak-hickory communities that have been disturbed by human intervention and disease.

The coniferous forest is a continuing target of the lumber industry, but its very vastness has protected its more distant regions. The latest and most controversial of foresting methods is called "clearcutting." Unlike traditional selective foresting, clear-cutting involves clearing *every* tree from the land (Figure 40.23). As this practice continues, the timbering industry is careful to advertise that it is replanting the areas at great expense.

There is a problem with such reforestation,

40.22 ANIMALS OF THE TAIGA

Herbivores of the taiga include the porcupine (top) and large mammals such as the elk (bottom) and caribou. The lynx and grizzly bear are still found in the taiga, protected somewhat from human intervention by the vastness of the area.

however. Mixed forests of genetically diverse and resilient plants are replaced by artificially developed and genetically homogenous trees that are fast-growing and can quickly be reharvested. Because some kinds of conifers are now being routinely cloned through developments in tissue culture techniques, absolute genetic uniformity in some forests looms just over the horizon. Biologists are quite concerned with the dangers of genetic uniformity in any system, since without variation, destruction by an unforseen invasion of a mutated fungus, bacterium, or virus could virtually destroy an extensive forest. All of these eventualities aside, where will the jay, the bear, and the raccoon live while the clearcut forest recovers?

A number of studies have shown that other kinds of damage result from clear-cutting forests. Following clear-cutting, runoff from rainfall increases substantially, as you might expect, but this is accompanied by what ecologists call a "nutrient flush." There is a three- to twentyfold increase in the loss of mineral nutrients from the soil over what is normally recorded in forests. Most dramatic is a substantial loss of critical nitrogen.

The Tundra

The tundra is the northernmost land biome. Except for certain alpine meadows, it has no equivalent in the Southern Hemisphere. Tundra is actually located in a narrow band between northern taiga and the Arctic ice extending from the tip of the Alaskan peninsula around the earth and back to the Bering Sea (Figure 40.24). Anyone traveling in the tundra may be struck by the absence of tall trees and shrubs. The annual precipitation is often less than 15 cm (5 in), and much of this occurs as snow. So it's a dry place—at least during the long winter when everything is frozen. However, you wouldn't think it dry if you visited the tundra during its brief, damp summer. When spring and summer finally arrive, the upper few feet of soil thaw, leaving the frozen soil below in a state of **permafrost.** The surface thaw produces an unusual condition for a "desert." Ponds begin to form everywhere. Since there is no way for the water to percolate down, the plains of the tundra become a veritable bog.

The tundra receives less energy from the sun than does any other biome. Because it is near the

40.23 CLEAR-CUTTING

Clear-cutting—a controversial, potentially disastrous foresting practice—involves the widespread clearing of trees in large stretches of land, as can be seen in this view of an Oregon forest.

40.24 THE TUNDRA

In summer, the treeless tundra becomes a veritable marsh as the snow melts. With little runoff, the landscape becomes dotted with innumerable small ponds. Below the water the soils of the tundra are perpetually frozen. The plants dotting the landscape include a number of dwarfed trees, grasses, and abundant reindeer moss.

pole, any sunlight there strikes at an acute angle, losing much of its energy after having traversed diagonally through the atmosphere. The lack of sunlight drastically affects the growing season. During the brief six-to-eight-week summer, plants must photosynthesize and store enough food to last through the rest of the year. And they must compress their reproductive period into this brief season, all of which means they can't put much energy into growth. So the tundra is carpeted by low-lying plants. Another factor which dictates that plants hug the ground is the constant high winds, particularly in regions of higher elevation. A tall plant would simply be buffeted to death or even ripped away.

Because of the permafrost, the plants of the tundra can't form deep anchoring root systems; so they form shallow, diffuse roots that often become entangled with the roots of their neighbors to form a continuous mat. In the wet season, any disturbance on hilly slopes can break the entire root mat loose from the wet soil underneath and send a great mass of plant conglomerate sliding down, exposing the barren soil below. Because of the limited growth of its plants, tundra recovery is a very long, very slow process.

Lichens and mosses are common in the tundra, but so are willows and birches. These latter, however, are almost unrecognizable, being only dwarfed symbols of their more southerly relatives. They, with the lichens and mosses, are joined by

40.25 PLANTS OF THE TUNDRA

The plant seen here is actually a willow. The harsh tundra climate produces a stunted growth in woody plants. Willows that are 25 years old may only be 1 m high.

grasses, rushes, sedges, and other annuals to complete the summer ground cover (Figure 40.25).

Animal life isn't as rare as one might expect in this most northerly biome. Where autotrophs are at work capturing and storing energy there will always be opportunistic heterotrophs ready to harvest. In fact, year-round residents of the tundra include some rather large herbivores such as, in North America, the caribou and musk oxen, and in Europe and Asia, reindeer. Other browsing animals found here are the ptarmigan, the snowshoe hare, and the ever-present and legendary lemmings. Lemming populations are clear indicators of how good the season is for producers. In good years, their populations soar.

Lemmings, in turn, determine the success of a number of predators, including the arctic fox, lynx, snowy owl, weasel, and arctic wolf. Also, the jaeger, a migratory bird, travels great distances to feed on the tiny rodents. The duration of the predators' stay, as well as their reproductive success, will depend on the number of lemmings. In years when lemmings are few, many of the predatory birds will migrate early while the permanent residents will survive by switching to other prey (Figure 40.26).

A number of waterfowl and shore birds also migrate to the tundra. When they arrive in the spring they must mate and rear their broods as quickly as possible, before the brief summer is over. Why would some species travel long distances to breed in such a dismal and risky place? Perhaps because the long days permit extended feeding periods and less competition for food than might be present in the winter feeding grounds. If winters are harsh, resident competitors are few.

While invertebrates are certainly more limited in the frigid tundra, remarkable swarms of mosquitoes and tiny flies abound. They enter a frantic race to complete their reproductive activities before summer ends, since the adults only live for one season. Such insects survive by producing highly resistant immature stages that remain dormant through the long winter.

Winter comes early in the tundra and with the rapidly shortening days the migratory animals dis-

40.26 ANIMALS OF THE TUNDRA

Because of the paucity of species in the tundra, the food webs may be comparatively simple (and, therefore, easy to disrupt). Common herbivores in the tundra include large animals such as the caribou and small ones such as the lemming. The grizzly bear is one of the many carnivores of the tundra.

appear. The caribou, for example, leave for the forested taiga where winter food is more plentiful. Those that remain prepare for survival by whatever strategy they have developed. Lemmings retreat to food-laden burrows, while ptarmigans tunnel into snow banks to emerge periodically on foraging expeditions. Since the larger herbivores and predators don't hibernate, they must rove the barren, windswept landscape to feed on mosses, lichens, and each other.

Our own species, with the exception of a few Laplanders in northern Scandinavia and Finland, avoids the tundra. Even the hardy Eskimos prefer the Arctic coastline. For this reason, humans have had little effect on this most fragile ecosystem. This is fortunate because the links in the food chains here are few and important. The sudden loss of a single predator or producer through human intervention could be disastrous. We know that recovery from damage is agonizingly slow and costly to the tundra communities. Wheel ruts left by a single wagon that passed over the soil 100 years ago are still clearly visible.

North of the Tundra (or in other places at higher altitudes), the vegetation gives way to barren and rocky soil similar to that of the Antarctic. Plant life is sparse and patchy. These dry windswept plains extend to the coastal ice floes and glaciers. Here the marine environment, in sharp contrast, is amazingly rich in life, even in its colder waters. Of course, the perpetual search for more petroleum threatens these rich waters as well.

THE WATER ENVIRONMENT

About three quarters of the earth's surface is covered with water, and most of it is salty (Table 40.1). But not only is the marine environment vast, it's complex; it contains many kinds of communities. They fall into two major divisions: those in the relatively shallow areas along continents, or the **neritic province,** and those in the deep-water, open sea or **oceanic province** (Figure 40.27). The marine environment can be further subdivided into the **euphotic zone,** through which light penetrates, and the **aphotic zone,** which is in perpetual darkness.

The Oceanic Province

There's less life in the open sea than in most other places. Much of the open sea is devoid of life forms. Nonetheless, the oceanic province is the home of the **pelagic** organisms, those that drift in the open

TABLE 40.1

OCEAN SALINITY

The ocean averages about 3.5% salts, usually expressed as 35 parts per thousand. Six elements, occurring as ions, comprise 99% of the minerals in sea water. Salinity varies particularly in surface waters, but also in the depths, so the 35 parts per thousand applies to measurements taken at 300 m (about 985 ft).

Element	Ion	% of salts	Parts per 1000 of sea water
Chlorine	Cl^-	55.0	18.98
Sodium	Na^+	30.6	10.56
Magnesium	Mg^{2+}	3.7	1.27
Sulfur (sulfate)	SO_4^{2-}	7.7	0.88
Calcium	Ca^{2+}	1.2	0.40
Potassium	K^+	1.1	0.38

Less than 1% of salts:	
Bromine	0.065
Carbon	0.028
Strontium	0.013
Boron	0.005
Silicon	0.003
Fluorine	0.001
Nitrogen	0.0007
Aluminum	0.0005
Rubidium	0.0002
Lithium	0.0001
Phosphorus	0.0001
Barium	0.00005
Iodine	0.00005

40.27 ORGANIZATION OF THE MARINE ENVIRONMENT

The marine environment is subdivided into the oceanic and neritic provinces, which in turn are divided into lighted and unlighted zones. Most marine life is concentrated in the neritic province where more abundant mineral nutrients are found.

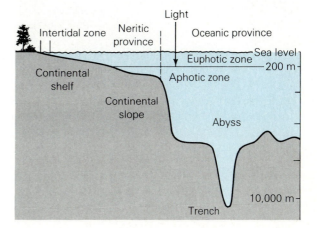

sea. Most of these are restricted to the euphotic zone—regions where light can penetrate. Almost all wavelengths of light, depending on the turbidity, are absorbed by the upper 100 m of water, although some shorter wavelengths (blue) may be detected at 300 m. Oceanic waters that are nearly devoid of particles (and thus, of nutrients) are quite blue, while ocean waters with suspended particles (nutrient rich) may be green, brownish and sometimes even reddish. In such turbid waters, light may only penetrate 10 or so meters.

Perhaps the greatest mysteries of the sea lie in the dark waters of the oceanic province, in the area called the **abyssal region.** Depths here can vary from 300 to nearly 11,000 m. The famed Marianas Trench is 10,680 m deep (over 6 miles). Little was known about the ocean abyss until new technology permitted its direct exploration. In 1960, the floor of the Marianas Trench was reached by Don Walsh and Jacques Piccard in the bathyscaph, *Trieste.* It was a momentous event, but scientists have visited these depths many times since then, and we are beginning to understand more about what is going on down there.

We know, for instance, that the ocean depths are places of darkness, tremendous pressure, and numbing cold. Nevertheless, these formidable deep abyssal regions support a surprisingly large number of peculiar **benthic** (bottom-dwelling) scavengers. Tethered cameras focused on bait, miles deep, have photographed primitive hagfish, many species of bony fish, crustaceans, mollusks, echinoderms, and even an occasional shark. For years oceanographers and marine biologists assumed that since producer populations could not exist at such depths, benthic creatures of the aphotic zone had to rely for their energy on the continual "rain" of the remains of pelagic (surface-swimming) organisms as they settled to the ocean floor. But more recently, biologists have discovered that there are unusual communities of organisms living along rifts and vents in the seabed, and that these have producer populations. Obviously, the producers aren't phototrophs, but are instead, chemotrophic organisms. We will learn more about the chemotrophs and the ocean rift community in the next chapter.

The waters below the photosynthesizers, but above the ocean floor, harbor some of the most peculiar creatures on earth: the deep-sea fish (Figure 40.28). They are usually small and dark and keen of sight, with big toothy mouths. These predators often produce their own light with which they

40.28 FISHES OF THE ABYSS

The deep-sea fishes are among the most bizarre of all creatures. They live where there is no sunlight and very few other animals. Thus, they must be equipped to capitalize on just about any living thing they might come across. *Chauliodus* uses the lengthy first ray of its dorsal fin as a lure, dangling it in front of its toothy mouth.

Chauliodus

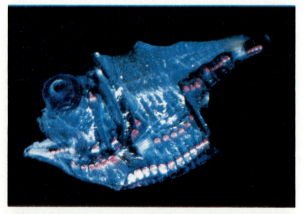

Argyropelecus

Eutaeniophorus

signal each other or attract prey. Some species cultivate luminescent patches of bacteria beneath their eyes which become visible when the fish rolls down a specialized eye covering.

The Neritic Province

In the neritic province, the land masses extend outward below the sea, forming the highly variable **continental shelf.** The neritic province ends where the shelf drops off, often abruptly, at the **continental slope.** In many shallower areas of the shelf, light penetrates to the ocean bottom. Such regions are continually stirred by waves, winds, and tides, keeping nutrients suspended and supporting many forms of swimming and bottom-dwelling marine life. Giant kelps and other seaweeds form extensive beds, offering hiding places for many fish species (see Figure 19.26). Let's look briefly at the organization of life and the flow of energy in this province.

The seas are home to many small, drifting species called **plankton.** The photosynthetic species are called the **phytoplankton.** Included are algal protists—mainly diatoms and dinoflagellates, all of which are microscopic. Also included are the more recently discovered, minute, flagellated **nanoplankton,** which are clearly important ecologically, but poorly understood. Phytoplankton, along with the marine plants—the kelps and seaweeds—are the **primary producers** of the sea. Thus, energy enters the marine ecosystem via these species. It has been estimated that 80 to 90% of the earth's photosynthetic activity is carried on by marine organisms. Thus, the chain of life in the sea begins with tiny floating plants capturing the energy of the sun to build energy-containing compounds within their fragile bodies. But where there is energy, there is likely to be something trying to utilize it, and the tiny plants are fed upon by animals only a little larger than themselves—the **zooplankton.** A variety of animals, from tiny fish to the great baleen whales, feed upon plankton of all sorts (Figure 40.29).

Among the more productive regions of the marine environment are the colder offshore waters, where **upwellings** occur (Table 40.2). Upwellings are the movement of deep, nutrient laden colder waters to the surface. While little is known about the vertical movement of water in the open sea, coastal upwellings are better understood. They are generally a seasonal phenomenon, where coastal winds blow either seaward or parallel to the coast, moving the surface layers which are then replaced by deeper layers. This stirring brings up nutrients

40.29 THE FOOD PYRAMID OF THE OCEAN

The tiny phytoplankton at the base capture the energy of the sun. These are eaten by animals larger than themselves, which are, in turn, eaten by still larger animals. At the top are the largest carnivores of the sea. (There are no large herbivores in the open oceans as there are on land.) Note that each level consists of far fewer organisms than the one below it.

TABLE 40.2

GROSS PRODUCTIVITY IN THE MARINE ENVIRONMENT, BY REGIONS OF INTEREST

Region	Percentage of total area	Gross productivity (kcal/m²/yr)
Oceanic province	90.0	1,000
Neritic province		
Coastal	9.4	2,000
Upwellings	0.1	6,000
Coral reefs and estuaries	0.5	20,000

that would otherwise be forever locked in the bottom sediments. The nutrients support photosynthetic organisms, which provide the base for marine food chains. The vast anchovy fisheries off the coast of Peru are dependent on such upwellings.

Thus, the ocean communities that are richest in life exist within the neritic province. Unfortunately, the world's neritic fishing grounds are being exploited at a rate that makes the likelihood of their recovery questionable. When increased fishing effort is not accompanied by increased catches, then it is fairly certain that the resource is in danger of depletion. Today's fishing technology is so sophisticated that even in areas where catches have diminished, the last remnants of once-abundant species are being captured.

Coastal Communities. Coastal communities nearer the shore—in what is called the **littoral zone**—are diverse, as is apparent by a simple drive along the coast. These areas include sandy beaches, rocky coasts, bays, estuaries, and tidal mud flats. Such places have one thing in common—they abound with a variety of life. The richness of life in the littoral zone is made possible by the availability of light, shelter (seaweeds, rocks,

40.30 ROCKY TIDE POOLS

The rocky coast is home for numerous marine animals. Each organism of the rocky tide pool is adapted in some way to withstand both the surging and pounding of waves and intermittent periods of exposure to the air at low tide. Some live in burrows; others cling tenaciously by bristles, sucker, or foot; others come in and out with the tide.

40.31 A MARINE MUD FLAT

Mud flats are saltwater areas in bays and estuaries that appear during intermittent low tides. At such times, their inhabitants are exposed to air, sunlight, and changing salinity, and must adapt to such harsh conditions. Burrowing mollusks and worms are common, and predators from the shore often stalk these areas in search of such prey. The estuarine mud flat may harbor a great variety of organisms if the river has not been heavily polluted.

kelp beds), and nutrient runoff from rivers entering the sea.

Each region supports its own particular type of plant and animal community. The rocky tidal communities, for example, consist of a large variety of organisms, all adapted in some way for holding on or burrowing in. Their biggest problem is to avoid being swept away or beached by the tremendous force of the waves. If you have dived in such places, you know about the nearly irresistible ebb and flow of the surging water. Like other littoral zone inhabitants, animals of the rocky coast have the added problem of being left "high-and-dry" by low tides, so each must adapt to such intermittent periods.

Visitors to rocky tide pools are usually amazed at the variety of life found there (Figure 40.30). That life may include plants such as seaweeds, eel grass (which is not really grass), and microscopic phytoplankton. Animal life from nearly all phyla are represented, as we mentioned in earlier chapters. Inhabitants include paper-thin flatworms, sponges, bryozoans, tiny but tenacious mollusks, echinoderms, crustaceans, tunicates, and small fishes (often visible only when they dart from one tiny crevice to another). The more sedentary of these species are adapted to regular exposure to the wind and sun as the tides fall.

The greatest productivity in the marine environment is in the estuaries and reefs (see Table 40.2).

Estuaries, places where rivers enter the sea, are fascinating for humans and demanding for marine life because of the special problems of osmoregulation. The problems arise because of the constantly changing salinity, which varies with the changes in the tides and river flow. Estuaries commonly have silty or muddy bottoms, and their backwaters may form mud flats at low tide (Figure 40.31). The animals inhabiting the mud flats usually survive such drastic changes by burrowing to escape the drying sun. The mud flats are inhabited by polychaete worms, and mollusks such as clams and scallops, by sea cucumbers, and by crustacean arthropods, such as crabs. With such rich food sources available, it's no wonder that many species of shore birds visit the mud flats at low tide.

Estuaries are also regions of serious pollution, and many have been drastically altered by humans. As long as rivers are used as dumping places for manufacturing wastes and sewage, the problem will persist. Our policies are often ill advised because bays and estuaries are rich sources of nutrients and in many cases are the very places where marine animals reproduce. Thus, the young of many marine animals—even those that will eventually move to deeper waters—are frequently endangered by polluted conditions.

Another kind of shallow area that is among the richest of marine habitats is the coral reef, but you haven't seen one unless you have been to tropical regions where the ocean temperatures remain above 20°C. Corals, of course, are coelenterates that live in huge colonies, building heavy exoskeletons of calcium carbonate. As years pass, the mass of exoskeletons grows, with the newer members of the colonies growing on top of the old. Coral atolls, common in the South Pacific, were shown by Darwin to represent coral growth at the top of submerged volcanoes. Barrier reefs, on the other hand, are coral deposits that form along coastlines, usually a short distance from the beach. The largest and most spectacular of these is the Great Barrier Reef, which extends some 1200 miles along the east coast of Queensland, Australia.

Reefs, by their irregular growth, form natural refuges for marine animals and thus set the stage for complex food chains. The complexity is due to the diversity of food types and a multitude of hiding places which forces predators to be able to exploit a number of different kinds of food, specializing in none. Coral reefs usually shelter sponges, encrusting algae, and bryozoans, as well as mollusks such as the octopus. Fishes abound in reefs and sharks patrol the deeper waters alongside.

APPLICATION OF IDEAS

1. Contrast the distribution of plant species in the tropical rain forest with that in the taiga and temperate deciduous forest. Suggest reasons for differences.

2. List the ways various animals have adapted to the desert biome. What general rules can be made for desert dwellers? Why are environmentalists more concerned with the human impact on this biome than they might be with most others?

3. Among the greatest environmental concerns today are those associated with the human encroachment on the tropical forests in South America and Asia. What is going on in these places, and how can the effects of this activity possibly be of significant global importance?

KEY WORDS AND IDEAS

Ecology, which has its roots in **natural history,** is the study of the "house," the interaction of organisms with each other and with their **environment.**

THE BIOSPHERE

1. The **biosphere** includes those portions of the earth that support life.

Physical Characteristics of the Biosphere

1. Unique characteristics of the earth's biosphere include the presence of water, a gaseous atmosphere, tilted axis, and moderate climate.

2. Water has very significant chemical and physical characteristics which include **specific heat,** the resistance to temperature change. Moisture in the atmosphere provides for a moderate climate.

3. The atmosphere contains variable amounts of water, is 78% N_2, 21% O_2, and about 1% rarer gases, including argon and CO_2. The atmosphere is vital in that it screens out harmful ultraviolet radiation. It also acts as a heat sink, producing conditions similar to those in a greenhouse.

4. About 30% of incoming **solar energy** is reflected from the atmosphere—the **albedo effect,** while 20% is absorbed by the atmosphere. The remaining 50% reaches the earth and radiates back into space as heat. Less than 1.0% is captured in photosynthesis, and most becomes involved in the evaporation of water. Evaporation and condensation constitute the **hydrologic cycle.**

5. The earth's changing seasons, and the resulting varying climate, is attributed to its tilted rotational axis (23.5 degrees), which presents a constantly shifting face to the sun.

6. While the equator receives much of the solar radiation, equatorial heat causes moisture laden air to rise, forming giant rotating air cells, that carry moisture and heat northward and southward from the equator. Such air cell movement is thrown westward by rotational forces, producing the **tradewinds.** Most of the earth's large deserts occur north of the precipitation zone.

7. Deserts also form as a result of **rain shadow,** in which high mountains in the path of moisture laden air receive the precipitation, while the land masses receive little.

8. The air masses also produce the great ocean currents, including the oceanic **gyres,** the circular surface movement of entire oceans.

THE DISTRIBUTION OF LIFE: THE TERRESTRIAL ENVIRONMENT

1. **Biomes** are major, recognizable associations of plants and animals maintained by specific climatic conditions.

2. Biomes subdivide into **ecological communities**—interacting populations of organisms—which are also usually identified by dominating plant life.

3. Biome distribution often follows latitude with transition zones from one biome to the next. In a northerly direction from the equator, tropical rain forests are followed by desert, grassland, deciduous forests, taiga, and tundra. Similar distributions are seen in altitudinal transitions, where increasing altitude emulates northerly changes.

The Desert Biome

1. **Deserts** receive less than 25 cm of highly seasonal precipitation per year, and occur in all continents except Europe. Without heat retaining water, night-day temperature shifts are drastic. Most deserts occur at about 30 degrees north or south latitude, or to the lee of high mountains.

2. Desert-adapted plants are called **xerophytes.** Specific adaptations include water storage, fewer stomata, small leathery leaves, spines, long dormant periods, and specialized cycles as seen in the CAM plants—**crassulacean acid metabolism**—all of which result in slower growth rates. Many annuals have rapid flowering cycles, and the seeds of some species resist germination unless the quantity of rainfall is enough to assure success.

3. Most animal species are **nocturnal,** becoming active at night, but some are active in the evening and early morning. Rodents, such as the desert kangaroo rat, survive by remaining in humid burrows. The kangaroo rat has a highly efficient kidney and does not require water. Burrowing desert spadefoot toads undergo **estivation,** remaining dormant through dry periods and rapidly undergoing reproduction when water is available.

The Grassland Biome

1. Major grasslands, which occur in inland plains (steppes, prairies, pampas, and veldts) have more rainfall than deserts, but it is also seasonal so drought is common. Fires are a significant factor in preventing trees from becoming established.

2. Grasses recover from fires and drought because of their protected stems (underground **rhizomes),** and extensive **diffuse root systems.** Rhizomes and **stolons** (surface runners) form sods that prevent the invasion of other plants.

3. Grasslands support huge grazing populations and survive by basal regrowth although overgrazing has destroyed many marginal grasslands. Most grasslands today have been converted to agriculture.

The Tropical Savanna Biome

1. The tropical savanna is characterized by intermittent groves of trees in otherwise typical grassland. It occurs in Africa, South America and Australia. Prolonged dry seasons and fire maintain the grassy state. Huge migratory populations of grazing mammals and large carnivores are common.

Tropical Rain Forest

1. Tropical rain forests receive from 250 to 450 cm of evenly distributed rainfall. The largest are in the Amazon, Indonesia, and the Congo. In addition there are tropical seasonal rain forests, typified by the monsoon forests of Southeast Asia which include many deciduous trees with leaf fall occurring in the dry season.

2. The foliage of trees in the tropical rain forest forms strata with an overlying canopy 30 to 45 m high, below which is a subcanopy of smaller trees and vines. Other plants, the **epiphytes,** live entirely within the trees, receiving water mainly from the humid air. Great numbers of species occur, but individuals in each species are widely dispersed.

3. Animal species also occur in great numbers and each species generally occupies a specific region in the canopy and subcanopy.

4. Clearing of the world's tropical rain forests may produce dramatic changes in climate. Transpired water is lost and rather than being absorbed, rainfall runs off into the rivers. Carbon dioxide is no longer absorbed by the plant "sink" and even more is released during burning as the land is cleared. The soil, which has a meager nutrient store, is rapidly depleted by agriculture, and such exposed soil forms rocklike laterite.

Chaparral: Mediterranean Scrub Forest

1. The broadleafed evergreens that typify the **chaparral** and **Mediterranean scrub** are restricted to coastal regions in western South and North America, Mediterranean, and southernmost Africa and Australia. While rainfall is meager and seasonal (25 to 75 cm), the droughtlike conditions are modified by a moist marine air flow.

2. Plant are typically xerophytic, with well protected small leaves that are often waxy, leathery, and spined. Decomposition is slow and ground litter is extensive.

3. Rodents abound, as do migratory birds, but there are few large herbivores or carnivores.

4. The chaparral is a **fire-disclimax community,** that is, it is always in a state of recovery from fires that periodically sweep through.

Temperate Deciduous Forest

1. Deciduous forests, those with trees that shed their leaves seasonally, occur mainly in the north temperate zones of North America, Europe and Asia. They are all characterized by warm summers and freezing winters, where the principal plant adaptation is leaf-shedding and dormancy. Precipitation averages about 100 cm and is evenly distributed.

2. Typical species are oaks, maples, beeches, birches, and hickory. The forest floor is densely populated by younger trees and many kinds of shrubs and ferns.

3. A rich humus and deep litter support large soil populations. Larger animals adapt to the seasons by hibernation, migration, or simply facing the cold.

4. Because of human encroachment in much of the deciduous forest biome, large mammalian herbivores and predators have been largely displaced.

The Taiga and Other Coniferous Forests

1. The taiga is the northernmost coniferous forest, extending in a northerly belt around the world. The winters are long and cold. Summers are short. Rainfall is often less than in the deciduous forest. Deciduous communities interrupt the conifers, particularly around boggy **muskegs.** In the northern taiga, light, as well as water, becomes a growth-limiting factor.

2. The needle or scaly conifer leaf with its waxy surface and recessed stomata is well adapted to dryness. The forest floor is relatively clear because of the heavy leaf litter that produces an acidic condition in the soil.

3. Taiga herbivores include moose, elk, deer, rabbits, hares, porcupines and numerous rodents. Carnivores include bears, wolves, lynx and wolverines.

4. Temperate coniferous forests in the U.S. are found in the Pacific Northwest, coastal and inland California, and the southeastern Atlantic coast. All are quite accessible and the target of powerful lumbering interests. Clear-cutting, a cost-effective way of lumbering, is apparently a serious ecological threat. Studies reveal that significant soil nutrients are lost by newly increased runoff and that water conditions are drastically changed downstream.

The Tundra

1. The northernmost biome is the tundra, an assemblage of dwarfed trees, grasses, and lichens found north of the taiga and at high elevations. Precipitation is less than 15 cm, much of which is snow. Winters are dry, but in the short summer, surface water collects in countless ponds. Below is the **permafrost,** permanently frozen soil.

2. A lack of sunlight results in a short six to eight-week growing season so plant growth is retarded. High winds in some regions, and the presence of permafrost prevent trees from becoming established. Lichens abound and plant life includes mosses, grasses, sedges, and dwarf willows and birches.

3. Larger mammals include caribou, musk oxen and reindeer, along with smaller hares and lemmings. Lemmings make up an important link in the food chain, supporting arctic fox, lynx, snowy owl, weasel, jaegers, and arctic wolf populations. Many waterfowl and shore birds also migrate in to nest in summer, a time when insects also abound.

THE WATER ENVIRONMENT

1. The marine environment is subdivided into the shallower, coastal **neritic province** and the deeper, open sea, **oceanic province.** Each includes a light-penetrable **euphotic zone** and a dark **aphotic zone.**

The Oceanic Province

1. Life is sparse in the nutrient poor oceanic province, and includes floating and drifting **pelagic** organisms, most of which are restricted to the upper 100 to 300 m, the euphotic zone.
2. The perpetually dark **abyssal region** (300 to 11,000 m) supports a few forms of **benthic** (bottom-dwelling) life that are adapted to the cold and immense pressure.

The Neritic Province

1. The neritic province includes the **continental shelf,** which falls off at the **continental slope.**
2. Nutrients from upwellings support the phytoplankton—algal protists—and poorly understood **nanoplankton,** the **primary producers** of the sea. They capture the sun's energy and form the base which makes all other marine life possible. Minute protists and animals (often larval forms) make up the zooplankton, upon which many other marine animals feed.

3. The greatest productivity is in colder offshore waters where upwellings bring nutrients up to surface waters. Coastal upwellings are brought about by seasonal winds blowing seaward or parallel to the coast.
4. The **littoral zone** of the neritic province contains the coastal communities: sandy beaches, rocky coasts, bays, estuaries, and reefs. Light, shelter, and nutrients abound.
   a. Along rocky coasts organisms must adapt to wave surge and to exposure at low tide. At low tide numerous tide pools form, each with a mixed population of marine plants, algae, and animals.
   b. Estuaries produce their own problems, including shifts in salinity that require osmoregulatory adaptations. Marine animals in tidal mud flats burrow in to escape drying and predatory shore birds. Estuarine life, which often includes immature forms of deeper water life, is constantly threatened by human encroachment and pollution. Reefs form natural refuges for a variety of marine populations.

REVIEW QUESTIONS

1. Define the term *ecology*. Where does it have its roots? What nonbiological disciplines might be important to ecologists? (1034)
2. Describe the physical extent of the biosphere. (1035)
3. List several life supporting characteristics of the earth. (1035)
4. Explain the relationship between water in the biosphere and temperature stability. (1035)
5. Explain what happens to the total amount of solar energy impinging upon the earth's atmosphere. What keeps the earth from continually heating up because of solar energy? (1036)
6. Fully explain what the earth's rotational axis has to do with the distribution of heat over the surface. What would the climates be like if the axis were not tilted? (1036)
7. Describe the formation and movement of the equatorial air cells. What do they have to do with the formation and location of the major deserts? (1036, 1039)
8. Explain how the presence of mountains can produce desert conditions. (1039)

9. Explain what causes oceanic gyres. How do they contribute to the distribution of heat over the globe? (1039)
10. List the biomes one might cross when moving from the equator to the North Pole. How might this distribution be mimicked by high mountains in the tropics? (1040)
11. Describe the physical conditions of the desert biome, including precipitation, its distribution, day-night temperature changes, and the effects of rainstorms. (1041)
12. List four significant ways in which plants are adapted to desert life. (1041–1043)
13. Briefly discuss three ways in which animals in hot deserts adapt to the heat. (1043)
14. What is the spadefoot toad's special survival strategy in the harsh desert climate? (1044)
15. List regions of the world where major grasslands are located and give their regional names. (1044)
16. What characteristics of grasses enable them to recover from both grazing and fire? (1045)

17. Characterize by kind and number the major animal life in grasslands. What does this tell you about the grasses' efficiency of energy capture? (1045)

18. How does the tropical savanna biome differ from the grassland? List some of the major mammals in the African savanna. (1046)

19. Characterize the conditions that support the tropical rain forest of the Amazon basin. In what significant way does the Southeast Asian rain forest differ from the Amazonian forest? (1046–1047)

20. Describe the organization of the tropical rain forest, including numbers of species, vertical stratification, forest floor, and nature of epiphytes. (1046)

21. What are the unique physical conditions of the chaparral? What keeps it from becoming a desert? (1048)

22. List several climatic adaptations made by chaparral plants. What is the general effect of fire on the chaparral? (1049)

23. List the major countries of the world where deciduous forests are found. What one climatic characteristic do they all share? (1050)

24. What is the major adaptation of broadleafed plants to seasonal changes in the deciduous forest? What three responses do the animals make to seasonal change? (1051)

25. What has been the fate of the deciduous forest throughout most of the United States and Europe? (1051)

26. Where is the taiga? Is this the only region of large populations of conifers? Explain. (1051)

27. What are three growth-limiting conditions in the taiga? (1051)

28. Describe the special adaptations of the conifer leaf. (1051)

29. List three large herbivores and three relatively large carnivores of the taiga. (1052)

30. What is clear-cutting? What specific studies and data present arguments against this practice? (1052)

31. Characterize the physical conditions of the tundra, including precipitation, temperatures, seasons, and special soil conditions. (1053)

32. List five prominent plants of the tundra. What two conditions prevent the growth of tall trees? (1053–1054)

33. Describe the short food chain of the tundra. (1054)

34. Name and describe the two major divisions of the marine environment. In what way is each divided vertically? (1055)

35. What is a pelagic organism? What determines the depths in which these organisms occur? (1055–1056)

36. What are three prominent physical conditions of the abyss? List five phyla of animals represented in the abyssal region. What is their primary food source? (1056)

37. Describe the primary producers and primary consumers of the marine environment. What do these organisms further support? (1058)

38. What is an upwelling? How does this phenomenon affect the distribution of marine life? (1057)

39. What two conditions in the rocky coast demand special adaptations by the inhabitants? What are three such adaptations? (1058)

40. What is the role of the coelenterate in forming atolls? Barrier reefs? Of what ecological significance are reefs? (1059)

Ecosystems and Communities

One of the great tasks of environmental biologists is to try to make some sense of the unimaginably diverse and complex world around us. To gain some appreciation of the problem, one need only look out the window. That view is but a single facet of a vast mosaic and, furthermore, such a view changes from season to season, and even from second to second—at least in the mind of the scientist. The task of making sense of it is overwhelming and the scientist has been forced to make sweeping and perhaps somewhat simplistic generalizations.

The concept of the biome is one such generalization, but it is a way to organize the living world into general categories that are conceptually somewhat manageable and, one would hope, that indeed reflect something of the nature of the real world. Biomes represent a grand and general view of life in the biosphere, and whereas such conceptual manipulations make our task of understanding the planet somewhat more manageable, they are rather unwieldy divisions and so we must devise yet new ways of categorizing, separating and dividing what we see around us. One way of doing this is to try to consider the living world in terms of its **interactions.** Obviously, here we must focus on smaller ecological units where interactions are more easily described.

In ecology, the complete unit of interaction takes place in what is known as an **ecosystem,** a handy operational term that can be applied at any level, from the entire biosphere to a tiny pool of microorganisms. Ideally, ecosystems are highly organized, self-contained entities, where the flow of energy and the cycling of essential substances occurs through clearly recognizable **trophic** ("feeding") **levels.** The ecosystem structure is assumed to occur in all biotic communities, no matter what the makeup of their organisms. We can begin, then, with this system of interaction. Our plan will be to first discuss the common characteristics of ecosystems, and then to learn how they apply to biotic communities. A logical place to start is by understanding something about energy flow.

ENERGETICS IN ECOSYSTEMS

Tracing energy flow through a living system sounds like a rather esoteric task, but it also involves such mundane behavior as eating. Since this is the case, we can begin with the synthesis of food and a reminder that the ultimate source of energy for most forms of life on the earth is the visible light from the sun. Of course, energy from visible light enters the living realm of the biosphere through photosynthesis. In the last chapter we saw that about 50% of the solar energy reaching the atmosphere finds its way to the earth's surface. Of this, less than one percent (actually one tenth of one percent as a worldwide average) is captured by photosynthesizers. But meager as it seems, this is enough to produce 150 to 200 billion tons of dry organic matter each year. (Such estimates are based on the dry weight of organic matter because water is not organic and makes up a substantial part of the weight of living organisms.) Such matter is often referred to as **biomass**—the total weight of organisms per unit of area (often stated as grams or kilograms per square meter [g/m^2, or kg/m^2] and recorded as either dry or wet). On a much smaller scale, this reduces down to several kilograms of dry organic matter per year for each square meter. (We should add that chemotrophs, organisms using the chemical energy of molecules spewing from vents in the seabed, also contribute to the earth's organic matter. However, this contribution is far less than that of the photosynthesizers.)

Trophic Levels

Energy flowing through the ecosystems is first captured by autotrophs such as photosynthesizers. These, in ecological terms, are the **producers,** or as we described them earlier, **primary producers.**

From the producers energy passes to various **consumers,** heterotrophic organisms that cannot utilize light energy and that rely on other organisms for their required organic compounds. In the last chapter we described the passage of food through the marine ecosystem in terms of food chains, where phytoplankton, the primary producers, were fed upon by zooplankton, and these by several levels of larger and larger marine creatures (see Figure 40.29). In the energetics of ecosystems, these steps in energy transfers are called trophic levels. Figure 41.1 traces the flow of energy through several trophic levels. The passage of energy, we should emphasize, involves the passage of food molecules. With these points fresh in mind, let's have a closer look at each level.

Producers. Producers include plants, algal protists, and phototrophic bacteria, all photosynthesizers. As you know, these organisms use light energy, usually captured by chlorophyll pigments in molecular complexes known as photosystems, to produce organic food materials from carbon dioxide, water, and a few minerals (see Chapter 7). The chemotrophs include bacteria that obtain energy from inorganic substances in the earth's

41.1 ENERGY FLOW IN AN ECOSYSTEM

Sunlight energy is first captured by producers during photosynthesis. It then passes, as chemical bond energy in foods, from one consumer to the next and from one trophic level to another. Eventually the energy passes to reducers. Each transfer is about 10% efficient. The remaining energy escapes as heat, a product of the respiratory activities that support the organisms.

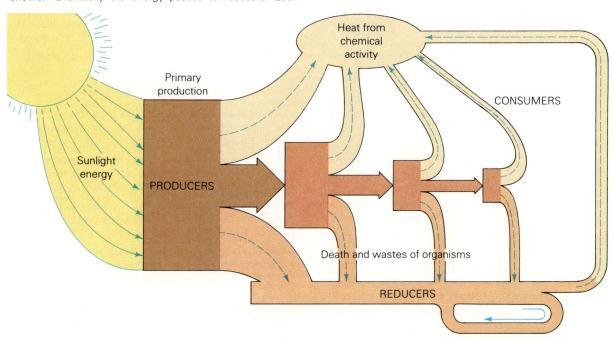

The Changing Carbon Cycle and the Greenhouse Effect: A Destabilized Equilibrium

In recent years, both physical and biological scientists have become increasingly concerned with the status of the carbon cycle. In the past 80 years alone, atmospheric carbon dioxide has increased by about 15%, and at the present rate of increase, human activities could easily double the present level over the next 40 years. Recently, an aroused group of scientists from the prestigious National Academy of Sciences informed Congress of the matter and initiated, of course, hearings. The hearings go on and the CO_2 continues to rise, but at least the problem is becoming recognized. The increasingly conservative Environmental Protection Agency concurs with the findings of the NAS, and they too are beginning to address the problem. However, there will be no quick fixes or easy answers to the global problems of increased atmospheric carbon dioxide.

Some of the more hopeful solutions once suggested have not materialized. For example, not long ago some scientists said that the oceans would act as a great buffer by absorbing any overburden of atmospheric CO_2. However, it turns out that the oceans are not very good at absorbing the gas—the mixing of water and gas occurs only in the top 80 meters of the sea.

So what does an increasing level of CO_2 mean to us? It means trouble. The trouble arises because carbon dioxide does not absorb or reflect short (ultraviolet) lightwaves, but it does absorb and reflect the longer (infrared) lightwaves. Thus, it freely admits solar energy into the biosphere but it slows the escape of heat that radiates from the surface of the sunlit earth. Carbon dioxide is thus a factor in the balance between energy entering and leaving the biosphere. So, an increase in atmospheric carbon dioxide results in an increase in heat in the biosphere. The principle is similar to that in a greenhouse. Greenhouses also work by letting in the short lightwaves and retaining the longer waves of heat energy. Thus, the carbon dioxide effect has become known as the *greenhouse effect*. It has been much publicized in the press but so far the effect seems to have been small. However, what otherwise might have been a detectable warming period may possibly have been dampened by a short-term, cyclic cooling period.

According to the experts, when this phase of the temperature cycle eventually passes, we may see some severe greenhouse effects. Warming of the earth would result in a warming of the ocean. If the ocean water gets warmer, the solubility of carbon dioxide will decrease. There are much greater reserves of dissolved carbon dioxide in the ocean water than there are in the atmosphere, and an ocean temperature rise of only 1 or 2°C would unload more carbon dioxide into the atmosphere. This in turn would increase the greenhouse effect and further raise the world's tem-

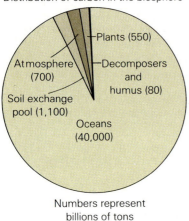

Distribution of carbon in the biosphere

Plants (550)
Atmosphere (700)
Decomposers and humus (80)
Soil exchange pool (1,100)
Oceans (40,000)

Numbers represent billions of tons

crust (see Essay 41.2). The biomass of the earth's producers is enormous, about 99% of the total present in the biosphere.

Consumers. Consumers include animals, fungi, many protists, and most bacteria—in a word—heterotrophs. Since some consumers eat producers, while others eat consumers, the flow of energy through the consumers may occur in several steps or trophic levels. Thus this group has been divided into **primary consumers,** the herbivores that feed directly on producers, **secondary consumers,** carnivores that feed on primary consumers, and so on through **tertiary** and **quaternary,** and even higher (but rare) consumer levels.

Organizing the trophic levels in this manner is obviously reminiscent of the food chains we described earlier but, in reality, nothing is ever that simple. Some consumers—primarily carnivores—are likely to cross trophic levels as they feed. Consider humans as an example. How many trophic levels do *you* occupy? Such complicated feeding patterns in a community are better represented by **food webs,** such as we see in Figure 41.2. However, such diagrams can only suggest the complexity of most food webs; to be accurate, they would have to include every species found in an ecosystem.

Reducers. The saprobic fungi and bacteria are called **reducers.** Reducers assure that essential

perature, which would cause the release of even more carbon dioxide. To make matters worse, the accelerated melting and subsequent decreases in the size of the polar ice caps would further decrease the amount of solar energy that is reflected back into space. Again we see what is known variously as a positive feedback loop, a destabilized equilibrium, or a vicious circle.

Of particular concern is the Antarctic ice pack, much of which is actually above sea level, either in floating masses or over the Antarctic continent itself. Obviously, the melting of such ice will raise the present sea level. By some estimates, this could be as great as 6 meters (almost 20 ft) but more conservative estimates suggest that the mean sea level will increase by about 3 meters (almost 10 ft) by the year 2100. The most immediate effects will of course be felt in coastal regions and low lying inland plains that have access to the sea. But quite possibly flooding along the continents is not the most serious consequence to consider. A far greater problem may be the effect of slight increases in the earth's mean temperature on climate patterns.

Studies of model systems help in our predicting patterns of weather change. For example, in the western United States, a change of just a few degrees will alter precipitation enough to reduce the flow of the Colorado River by half. This could be disastrous to the great population centers of the Southwest which are completely dependent on the Colorado River water for consumption and irrigation of crops. In other parts of the globe, there would be drastically increased river flow, for example, in the Niger and Nile in Africa, the Mekong, Volta, and Tigris-Euphrates in Asia and Sao Francisco in Brazil. As far as agriculture per se is concerned, Americans have long enjoyed the favorable climatic conditions that make us the world's leading producer of grains. Consider the political and economic ramifications of any significant climatic changes affecting this capability. For example, according to one of the predictions, the rain belt supporting American grain production could well move north, establishing itself in Canada and leaving the American grain belt in a semiarid condition.

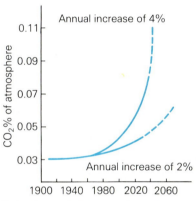

Projections of future carbon dioxide increases

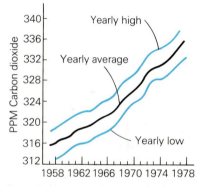

Recorded increases in atmospheric carbon dioxide

molecules cycle back to the producer and so they represent a vital link between consumers and producers. Essentially, they break down the organic matter, the remains of organisms, into simple products such as ammonia, sulfates, nitrites, nitrates, phosphates, and the usual carbon dioxide and water. We will look into their activities in more detail shortly, but for now keep in mind that reducers are the recyclers of organic nutrients, and without them the world would be a far different, and life-limited, place.

It may have occurred to you that in feeding on dead organisms, reducers do not differ essentially from carnivores or herbivores. One difference, however, is that reducers normally do not kill the organisms they feed upon. But then, how about scavengers or carrion eaters such as crows, crabs, jackals, vultures, and veritable armies of beetles and ants? Also, reducers may specialize in the eating of wastes. There are specific differences between the groups, however. For example, reducers commonly feed by absorption through the body wall. They secrete digestive enzymes into their food and absorb the products of digestion (Figure 41.3). Ecologists sometimes refer to organisms that feed on the already dead as **saprophages** (eaters of the dead). The other consumers are known as **biophages** (eaters of the living).

41.2 A SIMPLIFIED FOOD WEB

Food webs reflect the complex feeding patterns in an ecosystem. The arrows indicate the direction of the flow of energy and matter. In nature, some animals feed from more than one trophic level, particularly during shortages when usual food sources become reduced.

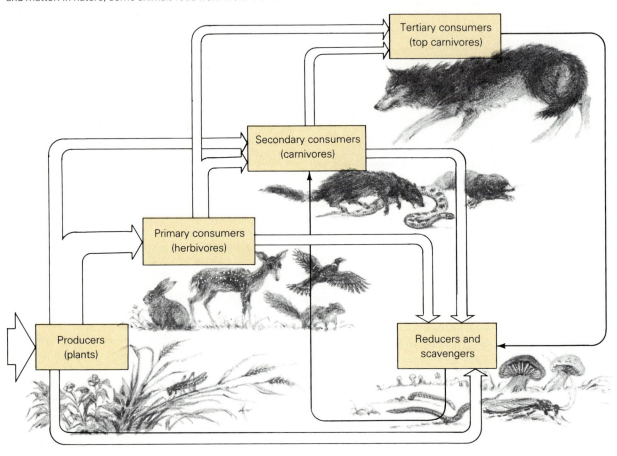

41.3 REDUCER ORGANISMS

Reducers decompose wastes and the remains of plants and animals, making nutrients available to the ecosystem. Most are microscopic in size, but the bracket fungi that live in fallen trees may be quite large.

Trophic Levels as Pyramids

The relationships among trophic levels are much easier to visualize if they are represented graphically, something ecologists are prone to do with their data. Pioneering ecologist Charles Elton first conceived of the idea of the **ecological pyramid,** or **Eltonian pyramids** as they are appreciatively known. In ecological pyramids, the producer populations form the base, followed by primary, secondary, tertiary, and higher levels of consumers. Three types of ecological pyramids are commonly used: *pyramids of numbers*, *pyramids of biomass* (total dry weight of all living matter) and *pyramids of energy* or *pyramids of energy flow.*

Pyramids of Numbers. Graphing numbers of individuals at each trophic level in, let's say, a grassland community, produces a typical "stepped" pyramidal shape as seen in Figure 41.4a. But number pyramids can take markedly different forms. For example, if we count trees in a forest and then count the parasites, herbivorous insects, and birds that feed from the tree, a single kind of producer will be supporting many consumers. Thus, the ecological pyramid then takes on an inverted form as seen in Figure 41.4b.

Pyramids of Biomass. Graphic depictions of biomass in an ecosystem can be quite revealing. Obviously, one normally cannot weigh a community (or, for that matter, count all of its individuals), so elaborate sampling techniques have been developed to estimate the actual weight or count. For example, if an ecologist were estimating biomass, he or she might choose to "harvest" several randomly selected square meters of a much larger area, or perhaps harvest in narrow swaths across the designated area. The ecologist is able to weigh all the individuals in these smaller areas and to extrapolate those measurements to the larger area. If the organisms are then sorted according to known trophic levels, dried and weighed, the data can be plotted. Usually, in such a case, the expect-

ed stepped pyramid emerges. Typically, the biomass of the producers is far greater than that of the consumers (Figure 41.4c). As we indicated earlier, 99% of the earth's biomass is to be found in the primary producer level. Not much of this is actually stored following its transfer to the primary consumer level, for several reasons. Of the plant matter eaten by herbivores, a considerable amount is consumed (converted to carbon dioxide, water, nitrogen waste, and so on) during metabolism. In addition, some is not actually absorbed, appearing in the feces. For the same reasons, the primary consumer level can support less biomass than does the producer. Thus, less secondary biomass can be sustained, and we see the stepwise narrowing of the pyramid.

Pyramids of Energy. Energy pyramids represent the flow of energy from one trophic level to the next and show the loss that occurs in such transfers. That energy is present in the chemical bonds of the molecules taken in as each group of organisms feeds. They have the usual stepped configuration (Figure 41.4e), but they are not as radical in appearance as are biomass pyramids. In this

41.4 ECOLOGICAL PYRAMIDS

(a) Pyramid of numbers. Plotting the numbers of organisms in each trophic level of a grassland community produces the stepped pyramid. (b) In a forest community, the pyramid becomes inverted, since most of the producers (P) are large trees and shrubs. Their numbers are far exceeded by the numbers of consumers (C) they support. The numbers represent counts for each 0.1 hectare (quarter acre). (c) Pyramid of biomass. Measurements of biomass (grams per square meter) in a tropical rain forest reveal a great difference between the producer and consumer levels, where 40,000 g/m² supports a total consumer mass of only 5 g/m². The larger reducer and scavenger biomass should be expected in a community where nutrient cycling is very rapid. (d) Pyramids of biomass in the marine environment seem to represent the impossible—the weight of producers is less than that of consumers. (e) An energy pyramid from a freshwater aquatic community at Silver Springs, Florida. Energy flow here is expressed as kcal/m²/year.

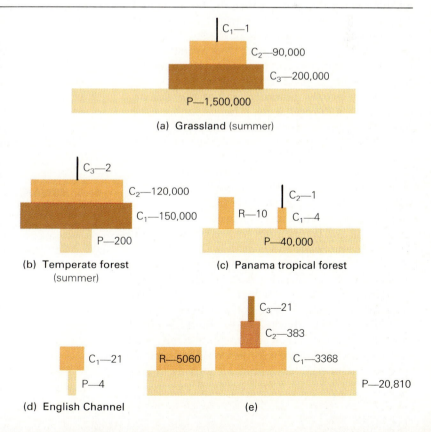

case, the stepped pyramid would be predicted by the second law of thermodynamics, which, as you may recall (Chapter 6), states that energy transfers are never perfect—there is a loss of energy with each transfer. This is true in living as well as non-living systems. In fact, that loss can be considerable; the energy transfers from one trophic level to the next average about 10% in efficiency. Thus, not much energy handled by any trophic level is available at all to the next level. It turns out that 90% of the energy in the food eaten by consumers is not stored but is used in maintaining the animal. Consumers do more than eat, and every act requires energy. (People who measure such things say that about 7 percent of the chemical bond energy of foods taken in by dairy cattle passes out of their bodies in the chemical bonds of gas!) Eventually, all of the energy entering the earth's ecosystems is released as heat. Heat, of course, is a low grade of energy, and there is no way for organisms to gather heat energy for useful work.

Humans and Trophic Levels

Energy pyramids may seem abstract and academic, but they apply dramatically to human populations and suggest some fundamental lessons in economics. (We might point out that economics and ecology share many things other than the prefix, including certain premises and theories.) Humans are basically **omnivores**—capable of feeding at several trophic levels. In large part, the availability of resources and one's economic state determines the trophic level most used. Under the worst economic conditions, people generally feed at the primary consumer (herbivore) level. The reason is clear: more energy is stored at the primary level in the form of more calories. Feeding at the secondary level or higher involves costly (or wasteful) energy transfers as energy from the primary level is converted, for example, to the bodies of cattle, sheep, hogs, and other animal stocks (Figure 41.5). But in most instances, while affluent people may choose to eat steak and lamb, impoverished people tend to eat cereals and beans and peas. It is also efficient to utilize fishes and other seafood (as do many Asians). Although there are costs in harvesting the high-protein sea food, the energy transfers up the chain are not costly in terms of human effort.

It seems, then, that the earth could support a much larger population if all humans ate only plant food. It's true that more calories would be available without the energy losses that accompany energy transfers from one trophic level the next. But humans feeding at the herbivore level face a problem with nutritional quality. For example, many common plant foods cannot provide certain essential amino acids (such as tryptophan, methionine, and lysine; see Chapter 29) that are critical to our own synthesis of proteins. The absence of such essential amino acids from the daily diet can lead to such infamous protein deficiency diseases as Kwashiorkor, a chronic problem in some parts of Africa (Figure 41.6). The critical amino acids can be provided if legumes such as soybeans and peas are included in the diet, but this must be done on a daily basis. Humans do not store amino acids very well, and excesses are rapidly converted to fats or carbohydrates in a daily turnover.

Energy and Productivity

The rate of energy storage in organic substances per unit of space by producers in an ecosystem is referred to as **primary productivity,** a concept which has two main aspects: **gross primary productivity** (GP) and **net primary productivity** (NP). GP is the total rate at which energy is assimilated by producers over a certain period of time. GP measurements are useful when comparing one ecosystem or community with another, or making comparisons throughout some specified time period. However, GP doesn't tell us how much energy the photosynthesizers are actually storing or how much growth is occurring. As we now know, a considerable amount of the captured energy must be expended for synthesis, transport, and numerous other activities. Table 41.1 compares the gross primary productivity—GP—for several major aquatic and terrestrial regions and biomes.

NP takes energy use into account. It is deter-

41.5 HUMANS AS SECOND LEVEL CONSUMERS

Where humans live as secondary consumers, energy must first flow from a producer to a primary consumer, and the change entails a considerable loss in the energy level. For each 1000 calories stored by a beef-eating human, the producers will have stored nearly 2 million calories.

Human—1050 calories

Steer—150,000 calories

Alfalfa—1,882,867 calories

41.6 PROTEIN DEFICIENT DIET

Kwashiorkor victims suffer from serious dietary amino acid deficiencies, although their calorie intake may be sufficient to provide energy. Symptoms include discolored, often thin, reddish or rust-colored and brittle hair, wasting of muscles, flaking skin, and puffiness from water retention. Often simple dietary supplements, such as a daily ration of peas or beans is all that is required to provide the missing amino acids.

mined by subtracting the rate of respiration by photosynthesizers (energy utilization) from GP. In other words, NP is the rate of energy stored minus the rate of energy released. NP would be reflected in new growth, seed production, and simple storage of energy rich compounds such as lipids and carbohydrates.

A second way of recording measurements of productivity is called **net community productivity** or **NCP.** This measurement takes the entire community into consideration, so respiration in heterotrophs as well as in the photosynthetic autotrophs is subtracted from gross productivity. Productivity is expressed in calories or grams (or kcal and kg) per unit of area (usually square meters or some multiple of this unit) over some period of time. Using GP, NP, and NCP to represent gross, net, and net community productivity, and R_a and R_h to represent respiration in autotrophs and heterotrophs respectively, the three aspects of productivity can be summarized as:

$$GP = \text{Energy assimilated by producers}$$
$$NP = GP - R_a$$
$$NCP = NP - R_h$$

Ecologists have devised many techniques for measuring the net productivity in communities and

TABLE 41.1

ESTIMATED GROSS PRIMARY PRODUCTION (ANNUAL BASIS) OF THE BIOSPHERE AND ITS DISTRIBUTION AMONG MAJOR ECOSYSTEMS

Ecosystem	Area, Millions of km²	Gross Primary Productivity, kcal/m²/yr	Total Gross Production, 10¹⁶ kcal/yr
Marine			
Open ocean	326.0	1,000	32.6
Coastal zones	34.0	2,000	6.8
Upwelling zones	0.4	6,000	0.2
Estuaries and reefs	2.0	20,000	4.0
Subtotal	362.4	—	43.6
Terrestrial			
Deserts and tundras	40.0	200	0.8
Grasslands and pastures	42.0	2,500	10.5
Dry forests	9.4	2,500	2.4
Northern coniferous forests	10.0	3,000	3.0
Cultivated lands with little or no energy subsidy	10.0	3,000	3.0
Moist temperate forests	4.9	8,000	3.9
Fuel-subsidized (mechanized) agriculture	4.0	12,000	4.8
Wet tropical and subtropical (broad-leaved evergreen) forests	14.7	20,000	29.0
Subtotal	135.0	—	57.0
Total for biosphere (round figures; not including ice caps)	500.0	2,000	100.0

SOURCE: From *Fundamentals of Ecology,* 3rd Edition by Eugene P. Odum. Copyright © 1971 by W. B. Saunders Company. Reprinted by permission of Holt, Rinehart and Winston, CBS College Publishing.

from this information they can determine the community's status. Is it growing? Has it reached a **climax state** (full maturity)? Is it declining? Net productivity accumulates only in communities that are in a growth stage. As they approach the climax state, an equilibrium between the rates of energy assimilation and energy use is established.

NUTRIENT CYCLING IN ECOSYSTEMS

While energy flows *through* an ecosystem, emerging eventually as heat, the nutrients essential to life tend to *cycle* in the usual "eat and be eaten" relationships. Nutrients are taken up by organisms as molecules and ions, and while some remain unchanged, others are incorporated into new molecules and structure, and some are metabolized for their chemical bond energy. But eventually the elements reappear in metabolic wastes, or, when death occurs, in the products of decay. These elements include those of the familiar SPONCH series (Chapter 2), sulfur, phosphorus, oxygen, nitrogen, carbon, and hydrogen, along with a host of others such as iron, cobalt, sodium, and chlorine. Most of the elements cycle back to the producers as mineral ions, or mineral nutrients as we have called them before (Chapter 23). Since the cycling of such nutrients involves both geological and biological activity in the ecosystem, the pathways are often called **biogeochemical cycles.** As we mentioned earlier, reducers play a key role in recycling essential substances.

The biochemical cycles include three aspects. In one, the elements are integrated into the bodies of living organisms, while in a second they are readily available as free water-soluble molecules and ions in **exchange pools.** In the third, mineral nutrients are locked away in less available **reservoirs,** usually in a less soluble form such as the elements in animal bones or shells.

Some of the biogeochemical cycles are rather short, involving only a few steps. For example, some of the water taken in from the environment by the plant simply passes through its vascular system to be released back into the environment through transpiration (see Chapter 23). However, the cycling of water can be more complex. For example, the water may be used in photosynthesis. Here, the molecules are disrupted, with the oxygen released as a gas into the atmosphere and the hydrogen used in the Calvin cycle to reduce carbon during the synthesis of carbohydrates (see Chapter 7). Aerobic organisms (including both autotrophs and heterotrophs) then take in the oxygen and use it in cell respiration (see Chapter 8). There, at the very end of the long mitochondrial process, the oxygen is reunited with protons and spent electrons, forming water. The water is then recycled to the environment. (As you can see, there's no telling where our drinking water has been.)

In describing biogeochemical cycles, we can again begin with the producers. Once they incorporate simple ions and molecules into their bodies, the substances are then available to be passed from one trophic level to the next as food, eventually reaching the ever-waiting reducers. (Of course, if the plant dies, its ions and molecules may go *directly* to the reducer.)

The Nitrogen Cycle

We have mentioned the nitrogen cycle several times before, so let's use it for our primary example of biogeochemical cycling (Figure 41.7). Nitrogen, as you know, is highly essential to life since it is a principal constituent of proteins, nucleic acids, chlorophyll, coenzymes, and many other molecules of life. Like other essential elements, nitrogen is found in both readily available exchange pools and in the less available reservoirs. The largest nitrogen reservoir is the atmosphere were molecular nitrogen (N_2) accounts for about 78% of the dry atmospheric gases. But, with the exception of certain nitrogen-fixing bacteria, atmospheric nitrogen as such is not available to the earth's organisms. Plants and nearly all other producers must have their nitrogen primarily in the form of nitrate ions in order to incorporate it into amino acids. Nitrate ions (NO_3^-) are produced by soil and water bacteria in two complex pathways: nitrogen fixation and nitrification—a part of decomposition. Nitrates in the soil—along with other nitrogen-containing ions—make up the readily available exchange pools.

The Reducers.   We can begin, here, to trace the flow of nitrogen from plants to animals to reducers, and back to plants. First, plants take in nitrate ions, process them, and then incorporate the nitrogen into amino acids which are finally assembled into plant protein. (For simplicity we will ignore other nitrogen-containing molecules.) When plants are eaten, the amino acids pass into the consumer levels where some are used to produce animal protein (and, of course, other nitrogen-containing molecules.) The rest may be metabolized and the nitrogen waste product excreted in various forms (urea, uric acid, or ammonia—see Chapter 32). Eventually, all producers and consumers (and the nitrogen wastes) enter the province of the reducers.

In the next stages, several populations of microorganisms, each with its role in a multi-stepped process, recycle the protein and nitrogen wastes. The reducers include the bacteria and fungi of decay or **decomposition.** The product of this decomposition is ammonia (NH_3), so this step is sometimes called **ammonification.** Ammonia readily ionizes in water, forming ammonium ions (NH_4^+). In the next step, **nitrification,** the ammonium ions are acted upon by bacteria such as the autotroph *Nitrosomonas,* which converts ammonium ions to nitrites (NO_2^-). A second group of nitrifiers, represented by *Nitrobacter,* then converts the

nitrites to nitrates (NO_3^-). The nitrates then join the exchange pools and cycle back to the plant. Nitrification is vital because ammonium ions are far less available to plants than are nitrates. Although the ammonium ion is chemically easier to incorporate into amino acids than nitrate, while in the soil, these positively charged ions tend to interact with and cling tenaciously to negatively charged clay particles. The negatively charged nitrate ion remains mobile in the soil and is more readily available for uptake by the plant root.

So far, the nitrogen cycle may seem smooth-running and efficient, but there are complications which lead to losses from the exchange pools. For example, the soluble nitrogen-containing ions can be carried out of the producer's reach by leaching—removal by the downward percolation of water. Further, should anaerobic conditions pre-

41.7 THE NITROGEN CYCLE

The nitrogen cycle can be viewed as two cycles. The inner cycle (dark arrows) includes the uptake of nitrate by producers and its passage as protein, nucleic acid, and other organic molecules through the usual trophic levels. At the reducer level, several major steps are involved in making nitrates available. In the outer cycle, (light arrows), nitrogen enters an ecosystem through the action of nitrogen fixers (although a minor amount of nitrogen fixation has always occurred during lightning storms and, more recently, through the action of sunlight on air pollutants). This input is roughly balanced by nitrogen loss through denitrification, which is carried out by anaerobic soil bacteria. Not seen in the cycle is the extensive human input of synthetic ammonia and nitrates through agriculture. In recent years, the human input has exceeded losses through denitrification.

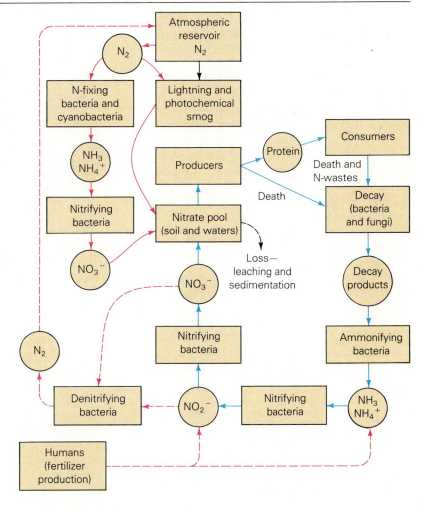

vail, organisms such as *Pseudomonas denitrificans*—anaerobic soil bacteria that act as *denitrifiers,* converting the nitrites and nitrates to nitrous oxide (N_2O, laughing gas) and nitrogen gas (N_2). The two gases escape to enter the atmospheric reservoir. Obviously denitrifiers can drastically deplete soil fertility. It is for this reason that anaerobic swamps and bogs are notoriously nitrate-poor.

The Nitrogen-Fixers.   In a balanced ecosystem, the losses through denitrification can be recovered by the gain from **nitrogen fixation** (see Chapters 18 and 23). Nitrogen-fixing bacteria (including many cyanobacteria), take in atmospheric nitrogen which they send through a complex biochemical pathway to combine with hydrogen, producing ammonia. Their excesses are released into the soil or water where nitrifying bacteria convert the ammonia to nitrite and nitrate. As you can see in Figure 41.7, denitrification and nitrogen fixation are not part of the main cycle of nitrification.

Knowledgeable farmers, aware of nitrogen fixation for years, have commonly rotated crops to include periodic alfalfa plantings. ("For years" is an understatement—Greek writings from the third century indicate that legumes were used even then to enrich the soil.) Nitrogen fixing bacteria of the genus *Rhizobium* invade the roots of alfalfa and other leguminous plants, which respond by forming cystlike nodules around the bacterial colony (see Figure 2.17). In a mutualistic relationship, the bacteria absorb organic nutrients produced by the plant, and the plant gains usable nitrogen fixed by its guest bacteria. In rice paddies, and other aquatic ecosystems, much of the nitrogen fixation is carried out by aquatic cyanobacteria. In natural terrestrial ecosystems, nitrogen-fixing bacteria live in mutualistic associations with the roots of wild legumes, alders, buckthorns and locust. But as far as agriculture is concerned, it is the leguminous plant that is most vital in harboring nitrogen-fixing bacteria.

Nitrogen may also be fixed the hard way, by lightning, and more gently, by the photochemical action of the sun on certain pollutants such as oxides of nitrogen. However, nitrogen fixed through such atmospheric action amounts to less than 10% of that fixed by organisms.

A problem has rather recently arisen regarding our manipulation of nitrogen. It turns out that we are not using up all our nitrate as one might have expected. In fact, we are overloading the environment with nitrogenous products. The problem has

41.8 NUTRIENT ENRICHMENT THROUGH POLLUTION

A dramatic increase in algal growth occurs when excess nitrate suddenly enriches a body of water. Conditions like this often result in the destruction of other precariously balanced life forms that inhabit the water. Masses of dead and decaying algae will support huge populations of bacteria, which, in turn, will deplete the oxygen supply. The result is the formation of an anaerobic water "desert," with only a few species of anaerobes and highly tolerant aerobes surviving.

only become apparent since we began synthesizing ammonia through industrial means.

Synthetic fertilizers are applied liberally to the soil and thus enter the natural nitrogen cycle. When we add the nitrogen from synthetic fertilizers to the nitrogen compounds produced by natural nitrogen fixation and by automobile exhaust (another major new source), the total amount of available nitrogen is astounding. C. C. Delwiche, at the University of California at Davis, has calculated that there is a net gain of 9 million metric tons of fixed nitrogen to the biosphere each year. Where is the surplus going? In California's central valley, which receives more synthetic fertilizer than any place on earth, the answer is: right into the water table and river systems and from there into San Francisco Bay and on to the Pacific.

How does excess nitrogen affect the water supplies in the soil? What is its effect on river systems and their estuarine life? One visible effect is the sporadic choking of waterways by uncontrolled, runaway algal growth (Figure 41.8). Such rapid growth always accompanies what is called **cultural eutrophication,** the sudden nutrient enrichment of lakes—to be discussed later. We have yet to measure the effect of the nitrogen load on the marine environment, but there, even more nitrogen is added from sewage dumped into the world's oceans.

Are we altering the nitrogen cycle in other ways? It's difficult to know just what's happening out there, but obviously we need to find out. It is quite conceivable that we can feed the world's burgeoning populations without calling the environment down around us. By coming to understand the nitrogen cycle better, perhaps we will find it vital and even economically feasible to keep our agricultural practices within the limits of natural systems.

The Phosphorus and Calcium Cycles

Now let's briefly consider two more cycles, those of phosphorus and calcium (Figure 41.9). Their cycles do not involve the atmosphere, but they do involve water. Phosphorus enters the roots of plants as the soluble phosphate ion HPO_4^{2-}. Phosphates are required for the production of many familiar molecules, such as ADP and ATP, phospholipids, nucleic acids, and the coenzymes of photosynthesis and respiration. Since ATP appears in the list, a shortage of phosphates will dramatically affect the energy-requiring activities of plants.

Calcium, another mineral nutrient, is critical to the proper functioning of cell membranes and many enzymatic reactions. It enters the root as the cation Ca^{2+}. A shortage of calcium can disrupt transport processes in the plant cells, causing the plant to die. Although it functions in transport systems, the ion itself is rather immobile once in place. When local

41.9 THE PHOSPHORUS CYCLE

Usable phosphorus in the form of soluble phosphates is found in soil water and in aquatic systems. Some phosphates pass to consumers through the trophic levels, while some are taken in through drinking water. During decomposition, reducers make some phosphates available again, but some locked in animal remains (bones, teeth, shells, etc.) is unavailable for long periods. The loss of phosphates through leaching (from soil water) and through runoff (to the sea) is considerable. Some end up as insoluble phosphorus in deep sediments. A gain in available phosphates occurs through erosion and from pollutants introduced by humans.

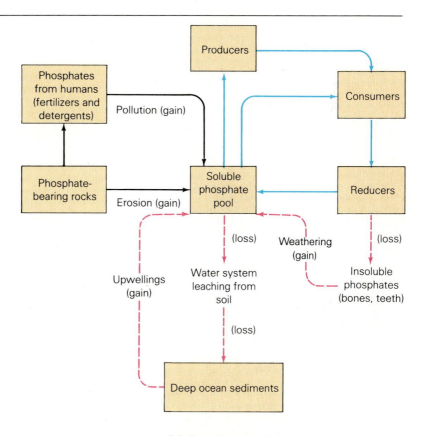

(a) Phosphorus cycle

shortages of other ions occur, they can simply be shifted from other parts of the plant. But not calcium; it must be taken in constantly by the roots.

The cycles of phosphorus and calcium are rather straightforward. Cellular phosphates, for example, are released into the phosphate pool by the action of decomposers. Since they are soluble, they may be recycled at once. Any phosphate incorporated into skeletal, tooth, or shell material, however, is released very slowly by weathering. Calcium may also cycle very slowly, since it is commonly bound up in skeletons and shells. Calcium from such dense structures is very slowly leached into the soil. In freshwater biomes, the reservoirs of phosphorus and calcium may lie bound in the bottom sediments for long periods of time until currents agitate those

murky depths. In marine biomes, occasional upwellings bring the reservoir sediments and dissolved ions to the surface, where they reenter the cycle via the phytoplankton.

The Carbon Cycle

We are aware that life is based on carbon, and as with any key element, its very availability may determine the size of populations. In the terrestrial realm, the main pool of carbon for photosynthesis is atmospheric carbon dioxide (CO_2). In the oceans and other waters, plants and algae use the carbonate ion (HCO_3^-) from dissolved carbon dioxide (and carbonate rock) as their principal carbon source.

41.10 THE CARBON CYCLE

The atmosphere and ocean are the earth's greatest reservoirs of carbon. However, a sizable amount is present in carbonate rock and fossil fuel. The cycling of carbon begins with CO_2 entering plants during photosynthesis and becoming incorporated in carbohydrates. Plants then carry on respiration, releasing some carbon back into the atmosphere and soil as

carbon dioxide. Some of the plant carbohydrates enter animals and other heterotrophs, where respiration returns more CO_2 to the atmosphere. The cycle includes one more significant source of CO_2—the burning of fossil fuels by humans.

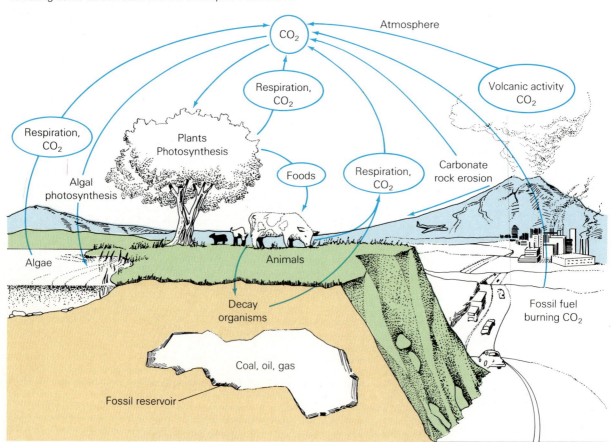

Paradoxically, as important as atmospheric carbon dioxide is to life, it is present in such small proportions (slightly less than 0.04%) that it can almost be called a "rare gas." The quantity is admittedly small when compared to other gases, but it still represents an enormous amount in absolute terms.

In addition to the carbon found in the atmosphere, the earth has a sizable reservoir in the form of carbonate rock (such as limestone) and fossil fuels (natural gas, oil, coal, peat). However, this reservoir is being steadily altered by human and geological processes. Since the industrial revolution began, CO_2 from fossil fuels has been released into the atmosphere in steadily increasing amounts (Essay 41.1). Today, atmospheric scientists estimate that 5 to 6 billion metric tons of CO_2 from fossil fuels are released into the air annually. Fortunately, as we shall see, about two thirds of this amount is quickly removed by the oceans and by photosynthesizers. For a look at how CO_2 is shuffled about in its cycle, see Figure 41.10. Since we have already covered critical aspects of the oxygen and water cycles in our discussions of photosynthesis, respiration, and the biosphere, let's move along now to community organization.

COMMUNITY ORGANIZATION AND DYNAMICS

Communities are populations of organisms that interact with each other and with their physical environment within a specific area. As usual the key term is *interaction.* Such definitions are admittedly vague, and it is often difficult to determine the boundaries of communities. As with biomes, some communities simply blend gradually into others and for this reason are called **open communities.** Forest communities are like that, as different vegetation types blend together in mixed associations at borders. There are more definite borders in the **closed communities.** An example might be a spruce forest community bordering Puget Sound where the presence of salt water ends the forest rather abruptly. Caves also tend to contain distinct communities since they are isolated from those surrounding, and the inhabitants have had to adapt to an entirely different range of conditions. Actually, in terms of energetics, there are no completely closed communities, since there is always some degree of interaction between neighboring communities. The cave community, for instance, must rely on organic materials carried in by streams or rain

41.11 AN OAK-PINE FOREST COMMUNITY

Oak-pine forests in Long Island, New York, are generally in a state of regrowth since they are periodically ravaged by fires.

water, although when an occasional surface animal blunders in, it provides a temporary energy source for a whole host of cave dwellers.

As we consider community life we will try to accomplish several things. First, we will compare several kinds of communities. We will then discuss the questions of energetics in a forest community, and the organization and conditions in an aquatic community. Finally we will look at communities over time; specifically, the succession of events as communities change.

Productivity in a Forest Community

In 1969, ecologist George Woodwell and his colleagues completed a decade long study of productivity in a scrub oak-pine forest community near the Brookhaven National Laboratory in Long Island, New York. Oak-pine communities are common in this area (Figure 41.11), the product of disturbed deciduous forests whose lush elegance once graced the Long Island landscape. Such an undertaking can be enormous in scope and serves as a clear illustration of what studies in community ecology can be like. While productivity is generally expressed in calories or kilocalories (per unit of area per unit of time), in this study the unit used was the gram.

In their preliminary efforts, the ecologists first described the makeup and structure of the community. In addition, they periodically sampled and estimated the total organic matter present. From measurements of the biomass, they determined the gross productivity of the forest. But, as we have seen, the net productivity can only be calculated if the rate of respiration is known. To find this, Woodwell and his associates used a direct indicator: the carbon dioxide output of the forest. As you can imagine, this is easier said than done.

One problem was that such measurements must be made in the dark in order to eliminate the problem of carbon dioxide uptake during photosynthesis. Then, of course, there is the problem of the usual air movement which would carry away any carbon dioxide diffusing from the plants. Fortunately for the study, the region experienced frequent

nighttime temperature inversions, such as the ones that produce smoggy days in the Los Angeles area. In an inversion, cool surface air becomes trapped close to the ground by warmer layers of air above. The nighttime inversions in the oak-pine forest prevented the usual vertical movement of air, and permitted the accumulation and measurement of respiratory carbon dioxide.

From the lengthy study, it was concluded that the *annual* gross productivity of the forest community was 2650 grams (about 5.8 lbs) per square

41.12 PRODUCTIVITY IN AN OAK-PINE FOREST COMMUNITY

Measurements in this oak-pine forest reveal that the total matter being produced through photosynthesis exceeds the total being metabolized in respiration. Thus forest growth is on the rise, although the actual matter being stored is only about 20%. The numbers themselves represent a rate, the grams of dry matter per square meter per year. The total amount of matter (dry) produced by the forest in this time period was 2650 g/m². The total metabolized in respiration to meet the energy requirements of life in the forest was 2100

g/m². This includes respiration in both autotrophs (RS_A = 1450) and in heterotrophs (RS_H = 650). The difference between matter fixed during the year and that consumed in respiration, 550 g/m,² represents the total stored energy (designated NEP: net energy production or NCP: net community productivity) is reflected in new growth of the forest dwellers along with some stored in humus. Other measurements (left) indicate the distribution of forest biomass among the foliage, soil litter, and decaying humus.

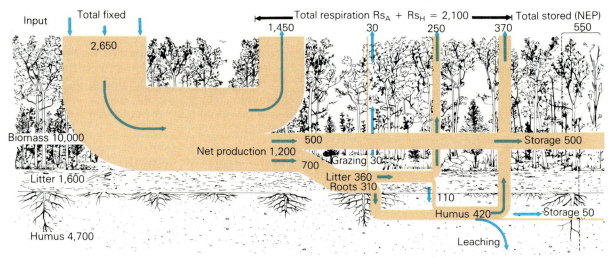

TABLE 41.2

ANNUAL PRODUCTION AND RESPIRATION AS KCAL/M²/YEAR IN GROWING AND CLIMAX ECOSYSTEMS

	Alfalfa Field (USA)	Young Pine Plantations (England)	Medium-aged Oak-Pine Forest (NY)	Large Flowing Spring (Silver Springs, FL)	Mature Rain Forest (Puerto Rico)	Coastal Sound (Long Island, NY)
Gross primary production	24,400	12,200	11,500	20,800	45,000	5,700
Autotrophic respiration	9,200	4,700	6,400	12,000	32,000	3,200
Net primary production	15,200	7,500	5,000	8,800	13,000	2,500
Heterotrophic respiration	800	4,600	3,000	6,800	13,000	2,500
Net community production	14,400	2,900	2,000	2,000	Very little or none	Very little or none

meter, while the annual net productivity, after the respiration correction was made, was 1200 g/m². Further determinations established that the rate at which organic material was accumulating—the net community productivity (net production minus heterotroph respiration)—was 550 g/m²/year. The forest study is summarized in Figure 41.12.

Compared to some communities, the net productivity of this forest is modest. Annual net productivity in each square meter of some tropical rain forest communities, for example, can reach several thousand grams. However, the greatest annual productivity is not in natural communities, but in agricultural communities. For instance, in tropically grown sugarcane (an efficient C4 plant; see Chapter 7), the annual net productivity can exceed 9000 g/m², while grain fields range between 6000 and 10,000 g/m². Of course, if we factor in the energy of fossil fuels used to operate farm machinery, and the energy used in manufacturing and applying pesticides and fertilizers, the true net yields fall drastically. In fact, on a calorie for calorie basis—the caloric yield of food versus the energy needed to produce it—there is often a loss. Table 41.2 compares productivity (expressed in kcal) in six quite different ecosystems.

One more observation—perhaps an obvious one—was made in the Woodwell study: the oak-pine community was growing—it had not reached its ecological climax state. As we mentioned earlier, communities that have reached their climax state have no net productivity. Their respiratory output equals their photosynthetic input. We will look more closely at the climax state and community growth shortly.

The Lake: A Freshwater Community

So far, we have concentrated on terrestrial and marine environments. However, the freshwater aquatic communities are equally interesting and in some ways, unique. Freshwater communities include ponds, streams, rivers, and marshes, and each has its unique conditions and organisms, but here we will focus on lakes. We will encounter familiar themes as we discover that common elements appear in the flow of energy and the cycling of nutrients through trophic levels of all ecosystems.

Lakes are usually of recent geological origin. Most occur in northerly regions as products of the last glacial retreats (10,000 to 12,000 years ago). But they are also produced through volcanic activity, as was the famed Crater lake in Oregon, and through uplifting of the land, as were many lakes in Florida.

41.13 LAKE ZONATION

Deeper lakes contain three zones, the littoral, limnetic and profundal, each with its own unique physical characteristics and populations of organisms.

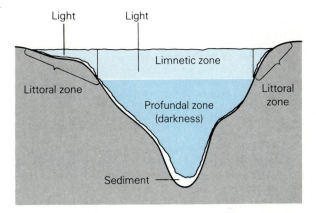

Lake Baikal in Russia, the world's deepest at 1750 m (5742 ft, well over a mile), is exceptionally ancient, having formed during the Mesozoic era.

Limnologists (*limne*, "pool" or "lake") are ecologists who study freshwater communities, and they have divided lakes into zones, each with its own physical features and each harboring a characteristic array of life. These zones are called **littoral, limnetic,** and **profundal** (Figure 41.13). Although some of the terminology is different, the organization is similar to that of the marine environment.

Conditions and Life in the Lake Zones. The littoral zone includes the shore and adjacent waters in which light penetrates to the lake bottom. Producers in the littoral zone include a variety of free-floating plants, plants that are rooted and submerged, and those rooted and emergent (starting above the water's surface). The plants form a progression of types as the water deepens. Like the marine waters, fresh water producers also include a phytoplankton component—numerous species of photosynthetic bacteria and algal protists (Figure 41.14). Consumers in the littoral zone include protozoan protists, snails, mussels, aquatic insects and insect larvae. Salamanders and frogs also prefer the littoral zone, as do both herbivorous and carnivorous fish and turtles, along with a number of birds.

The limnetic zone, like the euphotic zone of the ocean, is defined as the region of open water extending down to the depth of effective light penetration. The actual depth varies considerably with turbidity, but in general the limnetic zone ends where the rate of respiration catches up with that of

41.14 THE LITTORAL ZONE

Producers of the littoral zone include **(a)** bottom-rooted, aquatic plants such as cattails, bulrushes, and arrowheads living nearer the shore, and floating lilies and pond weeds farther out. **(b)** Microscopic producers (phytoplankton) include many filamentous and single-celled green algae, along with diatoms and cyanobacteria. **(c)** Primary consumers include the familiar freshwater snails, bottom-dwelling water mites, crustaceans, insect larva, and nymphs. **(d)** Smaller aquatic predators (secondary and tertiary consumers) of the littoral zone include the diving beetles, water scorpions, and nymphs of the damsel and dragon flies. There are also larger carnivorous species such as fish, frogs, and wading birds.

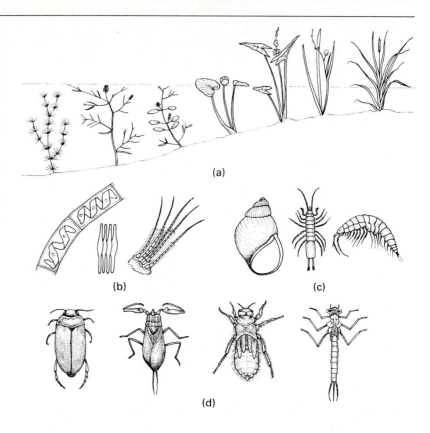

41.15 THE LIMNETIC ZONE

Limnetic zone producers **(a)** encompass several species, including phytoplankton common to the littoral zone. **(b)** The zooplankton include rotifers as well as crustaceans of principally two groups, the "cyclops" (one-eyed) copepods and the strange, flattened cladocerans. **(c)** The plankton support predatory insects and plankton-straining fishes, which in turn are fed upon by larger fishes **(d)** such as bass and pike.

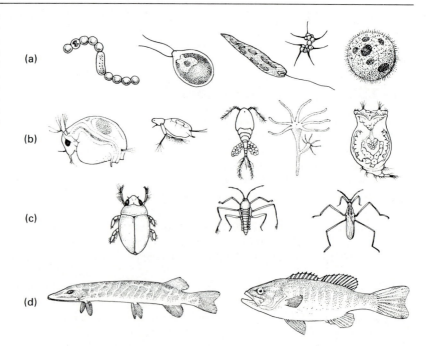

photosynthesis. This zone is notably absent in shallow lakes where effective light penetrates all the way to the lake bed. Producers in the limnetic zone are not obvious since they are nearly all microscopic. They include many of the phytoplankton found in the littoral zone along with flagellated algal forms, such as *Euglena* and *Volvox*. In northern lakes phytoplankton populations undergo seasonal blooms, during which their productivity exceeds the plants of the littoral zone. These blooms closely correspond to the availability of nutrients, light and favorable temperatures, and often follow a preceding bloom of nitrogen-fixing cyanobacteria that enrich the waters with their excess nitrogen compounds. Zooplankton form large populations of only a few species. Included are minute freshwater crustaceans such as the copepods and cladocerans along with dense blooms of rotifers. These populations rise and fall in response to the numbers of producers. Higher consumer levels are made up principally of the lake fishes: plankton feeding species such as shad, and the carnivorous species, such as bass and pike. The food web of the limnetic zone is often simple and direct (Figure 41.15).

The profundal zone begins where the effective penetration of light ceases—or where respiration begins to exceed photosynthesis. Again, turbidity is a determining factor in deep lakes. The profundal zone includes the muddy, sediment-rich lake floor. There is, of course, no producer trophic level here, and life is restricted primarily to bacteria and fungi, reducers of the lake ecosystem, and a few detritus-feeding clams and wormlike insect larvae. All of the profundal species are adapted to periods of very low oxygen concentrations. As sparse as life is in the profundal zone, the activities of these organisms are critical to the organisms above. As scavengers, they, along with the bacteria, convert a virtual rain of corpses from above into nitrates, phosphates, sulfates, and so on—the usual mineral nutrients. How well these nutrients become distributed in the lake depends, as it did in the marine environment, on the occasional vertical movement of the waters.

Thermal Overturn and Lake Productivity. The deeper waters of tropical Lake Tanganyika, which is 1450 meters deep (nearly a mile) are devoid of aerobic (oxygen using) life, while temperate Lake Baikal, about 300 meters deeper, supports animal life throughout much of its depths. The difference between the two is attributed to the availability of oxygen, which, in turn, is a product of *thermal overturn*, the deep circulation of lake water brought on by seasonal temperature

changes. Such changes, as you might expect, are far more significant in the earth's temperate regions where climate extremes are experienced. Thermal overturn carries dissolved oxygen to the lake depths, and brings nutrients to the surface. In the absence of seasonal overturns, the lake bottom becomes highly anaerobic, greatly restricting the life there.

The conditions preventing overturn in Lake Tanganyika and some other tropical lakes are similar to those occurring in temperate zone lakes during the summer. In both, a temperature gradient exists between top and bottom waters. The lighter, warm upper waters may be well mixed by wind action, but they cannot mix with the dense, colder waters below. Many temperate zone lakes have three distinct summer temperature regions, an upper warm water region called the **epilimnion** ("upper lake") and a lower, cold water region called the **hypolimnion** ("lower lake")—each of which experiences little temperature variation throughout its depths. Between the two is a middle region that reveals a steep **thermocline** (an area of rapidly changing temperature). Typically, oxygen depletion begins just below the thermocline.

Thermal overturn in temperate lakes occurs in the fall and spring when surface waters undergo drastic changes in temperature. When this happens, surface layers sink to the bottom, displacing lighter layers there, which rise to the surface. You see, it is a peculiar characteristic of water that as its temperature decreases, its density increases—it gets heavier—until it reaches 4° C (39.2° F), its maximum density. Below 4° C, the density of water decreases sharply—it gets lighter, and at 0° C (32° F, the freezing point), water is lightest. In addition to explaining thermal overturn, these peculiarities of water explain the reversed temperature gradients of lakes in summer and winter, both of which resist overturn. The effects of changing temperatures and density are further explained in Figure 41.16 where each season's condition is described.

So, while all communities are shaped by surrounding climate, we see that it affects the lake communities in unique ways. Rarely do we find that oxygen is a life-limiting factor in the terrestrial communities.

Before leaving the subject of community life, we should remind ourselves of one of the strangest on earth. This is the **Oceanic rift community** composed of the bizarre and recently discovered chemotrophs. The inhabitants of this type of community are subjected to the severest conditions, but they manage to thrive in great numbers, as we see in Essay 41.2 (p. 1086).

41.16 TEMPERATURE GRADIENTS IN THE LAKE

(a) In summer the highest water temperature is in the epilimnion and the lowest is in the hypolimnion. Between the two, a steep temperature gradient, or thermocline, occurs. Movement in the less dense epilimnion cannot disturb the denser cooler layers below, so overturn cannot occur. **(b)** In the fall, cooling surface waters approach 4°C (maximum density), sink to the bottom, and permit a wind-driven overturn to occur. **(c)** In winter, the summer temperature gradient is reversed, with the coldest temperatures at the surface (no thermocline forms), yet the density gradient is roughly similar to that of summer, preventing overturn. **(d)** In early spring, warming surface waters reach 4°C (maximum density) and sink, starting another wind-driven overturn.

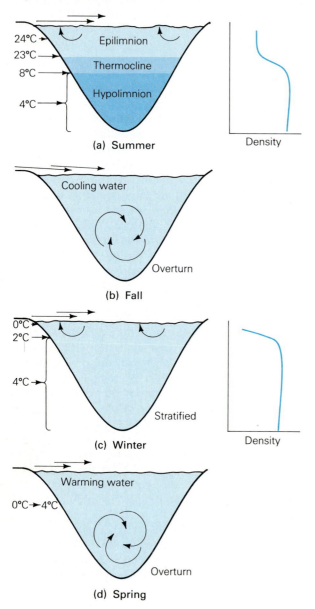

24°C
23°C
8°C
4°C

Epilimnion
Thermocline
Hypolimnion

Density

(a) Summer

Cooling water

Overturn

(b) Fall

0°C
2°C
4°C

Stratified

Density

(c) Winter

Warming water

0°C→4°C

Overturn

(d) Spring

COMMUNITY DEVELOPMENT OVER TIME: ECOLOGICAL SUCCESSION

Our view of biotic communities has for the most part been static—suspended in time—as a convenient way to describe their organizational components. However, they do tend to change as part of their normal development. They change not only in response to climatic and geological forces, but also in response to the activities of their inhabitants. In some cases the inhabitants will alter the environment, which then influences the community in new ways in a form of feedback loop.

Community change over time is known as **community development,** or, more traditionally, as **ecological succession.** Where ecological succession is the product of the organisms themselves, it is known as **autogenic succession. Allogenic succession** occurs where outside forces—particularly physical forces such as fire or flood—regularly affect change. In most instances, succession is a result of both autogenic and allogenic factors, although one or the other may have triggered the process.

As populations alter their environment, they set the stage for an invasion by yet other species. In this way, community structure continues to change. Such a series of sequential changes, in its entirety, is known as a **sere,** and each stage a **seral stage.** Allogenic succession is less predictable than autogenic succession. For example, an orderly progression of species during succession is often interrupted by the sudden bloom of unexpected opportunistic species such as weeds. In addition, we would not like to leave the impression that one population gracefully gives up its place for the next. On the contrary, species are often quite persistent, seemingly resisting their own displacement.

Ecological succession includes both **primary succession** and **secondary succession.** Primary succession occurs where no community previously existed, such as rocky outcroppings, newly-formed deltas, sand dunes, emerging volcanic islands, and lava flows. (Scientists are now studying the emerging succession on the slopes of Mount St. Helens.) Secondary succession occurs where a community has been disrupted, such as neglected farms reverting to the wild, or in a forest community that has been subjected to "clear-cutting," the controversial lumbering practice where all trees are removed.

41.17 PIONEER PLANTS

Lichens and mosses are able to exist under conditions that discourage most other life. Their success in colonizing bare rock surfaces has earned them the name "pioneer."

Primary Succession

In primary succession, such as occurs on a rocky outcropping (Figure 41.17), the sere begins with hardy, drought resistant species, often called **pioneer organisms.** Lichens are often the first to invade rocky outcroppings, held fast by their tenacious water-seeking, fungal component (see Chapter 20). Lichens are soil builders, producing weak acids that very gradually erode the rock surface. As organic products and sand particles accumulate in tiny fissures, opportunities arise for plants such as grasses and mosses to establish themselves and begin a new seral stage (Figure 41.18).

Plant roots penetrate the rocky crevices, exerting a remarkable turgor pressure, prying at the rocks and gradually widening the fissures. By then, certain insect and reducer populations will also have established themselves. In time, the lichens that made the penetration of plant roots possible are no longer able to compete for light, water and minerals and they give way to the plants. Similarly, these plants will contribute to the soil building for a

41.18 PRIMARY SUCCESSION

Succession begins here with a bare rock outcropping and ends with a fir-birch-spruce community. Pioneering lichens and mosses begin the soil-building process, followed by the invasion of increasingly larger plants until a more stable long-lived, climax forest community emerges.

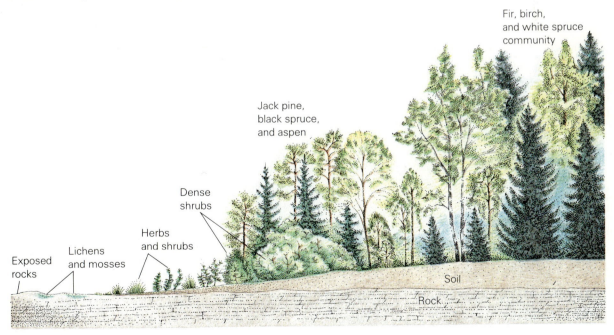

Fir, birch, and white spruce community

Jack pine, black spruce, and aspen

Dense shrubs

Herbs and shrubs

Lichens and mosses

Exposed rocks

Soil

Rock

time, and then they too will be replaced by fast-growing shrubs, and new populations of animals will invade.

The succession on bare rock outcroppings is, at first, an extremely slow process, with a sere often lasting hundreds of years or more. But once soil formation has begun, the process can accelerate. Succession in other sorts of places can also be slow. It is estimated that succession from sand dune to climax forest community on the shores of Lake Michigan took about a thousand years (Figure 41.19).

Secondary Succession

In secondary succession, the principles are similar to those of primary succession, but the seres occur at a more rapid pace. This is possible because the soil is already in place. In a deserted farm, weeds, grasses, shrubs, and saplings are often the first to

appear. Weeds are fast-growing, opportunistic plants that quickly invade disturbed communities, but they are often held in check where communities are undisturbed. Weeds are often imported from distant places where they may be less of a nuisance. For example, in Southern California, the large ball-shaped tumbleweed so prevalent there is a native of Russia, and the wild oat and mustard are natives of the Mediterranean region.

As secondary succession progresses, the initial invaders are eventually replaced by plants from the surrounding community. Larger, fast growing trees such as pines block the sunlight, and a new generation of shade-tolerant shrubs emerge below the canopy. Eventually there is a general blending with the surrounding community (Figure 41.20). Such a simple transition may require well over 100 years, depending upon the community. Secondary succession in grassland communities, as you might expect, is much faster, taking perhaps 20 to 40

41.19 SUCCESSION ALONG THE GREAT LAKES

Sand dunes bordering Lake Michigan have long offered ecologists an excellent laboratory for the study of succession.

About 1000 years are required for succession to be completed.

(a) Beach grass and cottonwoods

(b) Oak forest

(c) Oak-hickory-pine forest

(d) Climax beech-maple forest

41.20 SECONDARY SUCCESSION

The stages of secondary succession are revealed in a series of photos taken in the same section of the Bitterroot National Forest in Montana over a 70-year period—(a) 1909, (b) 1925, (c) 1937, and (d) 1979.

(a) (b) (c) (d)

41.21 SUCCESSION IN A POND

(a) Early in succession, aquatic plants begin to spread from the edges of the pond. (b) Eventually, these plants extend across the open water. (c) As the pond's waters disappear, invading marsh grasses, cattails, and sedges replace the floating plants, converting the pond into a marsh.

(a) (b) (c)

years. At the other extreme, fragile, disturbed tundra may require many hundreds of years to recover, if it ever does.

Succession in the Aquatic Community

Aquatic communities also undergo community development or succession although such changes may be held in check by shortages of mineral nutrients. Succession in lakes and ponds occurs as a product of **natural eutrophication,** changes brought about by the natural increase in nutrients carried in by streams and runoff from the land. Lakes and ponds that are rich in nutrients and high in productivity are called **eutrophic** ("true foods") **lakes,** while those that have limited nutrient supply and little productivity are called **oligotrophic** ("few foods") **lakes.** The general trend in freshwater bodies is toward increased eutrophication and thus, increased community growth, but the loss of any of the essential nutrients can reverse the latter trend.

As community growth in a lake progresses, the sediments increase and the depth decreases. Littoral zone plants crowd the shores, extending further and further into the lake, followed by increasing numbers of water tolerant shore plants (Figure 41.21). Unless the trend is interrupted, the lake will eventually convert to a marsh, and with the invasion of terrestrial plants from the surrounding community, the last traces of the lake will be lost. Interestingly, the ancient and oligotrophic Lake Baikal has shown alarming indications of eutrophication,

An Unusual Community: The Galapagos Rift

The Galapagos rift community, a bizarre assemblage of animals and bacteria, centers around the vents of sulfide hot springs in an area of active sea floor spreading some 612 km (380 mi) from Darwin's islands. It was discovered in 1977 by geologists aboard the submarine *Alvin,* which was cruising 2500 m deep (8202 ft) at the time. The geothermal hot springs, spewing boiling-hot solutions of hydrogen sulfide and carbon dioxide, support an entire ecosystem based on autotrophic chemosynthesis. It is one of the most dense and productive communities on earth—near the vents, the mass of living tissue approaches 50 to 100 kg per square meter.

Prominent among these denizens are enormous, blood-red tube worms of a previously unknown group, apparently belonging to the pogonophores (an obscure phylum now thought to be related to annelids). Like all pogonophores, they lack all trace of a mouth, anus, or gut. They are chemosynthetic autotrophs, deriving all their energy and carbon needs directly from the oxidation of the hydrogen sulfide and the reduction of the carbon dioxide, probably with the help of intracellular bacterial symbionts. The redness of their flesh is due to heavy concentrations of oxygen-binding hemoglobin; oxygen from the surrounding cold seawater is needed both for the oxidation of H_2S and for the worm's own oxidative metabolism. The tube worms range up to nearly 3 m in length (nearly 10 ft) and are about the girth of a man's wrist.

The entire ecosystem, in fact, is based on hydrogen sulfide chemosynthesis and is entirely cut off from other photosynthesis-based ecosystems. Apart from the tube worms, the primary producers are chemosynthetic bacteria, which swarm in enormous numbers where the often superheated, carbon dioxide- and hydrogen sulfide-laden vent water mixes with the near freezing, oxygenated abyssal seawater. At least 200 bacterial species proliferate near the vents and were visible to observers in *Alvin* as milky clouds. Filter-feeding crabs, clams, mussels, smaller worms (dubbed "spaghetti" by their geologist discoverers), and barnacles live off the bacteria. A little higher on

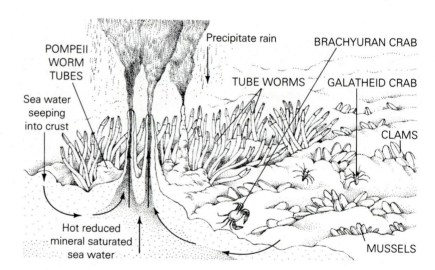

but not through natural means. The source of nutrient enrichment is the surrounding human community.

Eutrophication is an extremely gradual process in lakes, since nutrients are often whisked away by streams or buried deep in the bottom sediments. Typically, growth is checked by a scarcity of two nutrients, phosphates and nitrates. In recent times, however, such nutrients have become readily available through cultural eutrophication, in the form of increased mineral runoff produced by humans, as we dump sewage into water systems or permit runoff from heavily fertilized farms, cattle feed-lots, and denuded (clear-cut) forest regions. These practices vastly increase the availability of nitrates. Furthermore, the addition of phosphates to laundry detergents in recent years has added a heavy phosphate burden to natural water systems, rapidly speeding the succession and aging process of lakes and ponds.

the food chain, larger crabs and a variety of fish scavenge on clumps of bacteria as well as on animal remains. Whelks, leeches, limpets, and miscellaneous worms complete the community. Some species were previously known, but others, such as the tube worms, are new discoveries with uncertain affinities. Similar communities have now been found on the East Pacific Rise Rift, and other deep-sea geothermal communities probably occur wherever there are appropriate sulfide springs. In fact, the rift ecosystems may be so widespread as to constitute a major earth community.

The Galapagos rift community contrasts markedly with the rest of the deep-sea benthic communities which are characterized by constant cold and a severely limited energy input. The usual benthic communities survive on what little organic material drifts down from the surface waters and are thus based ultimately on energy from distant photosynthesis. The animals of the cold water deep-sea bottom are generally slow-moving and slow-growing, playing a variety of refrigerated waiting games: scavengers patrol listlessly for the chance of a dead fish or a fecal pellet, and predators lie motionless in ambush for wandering scavengers. In the hot springs communities, however, food is virtually unlimited, the temperature is not so uniformly cold, and both growth and metabolism are relatively rapid. The Galapagos rift clams, for instance, grow up to a third of a meter (13 inches) long at a rate of 4 cm per year, some 500 times faster than their smaller cold water relatives. The flesh of these clams, like that of the giant tube worms, is bright red with hemoglobin. This factor, along with their phenomenal growth rate, indicates a high metabolic rate.

Rich and active as they are, the deep-sea rift communities are ephemeral. The hot springs eventually die down, like volcanoes, leaving behind ghostly communities of empty clam shells. As the earth's crust shifts and new hot springs form, immigrants from established or dying geothermal communities arrive to begin the unusual chemosynthetic ecosystem anew. Rapid ecological succession is one more unusual characteristic of the rift community.

Tube worms

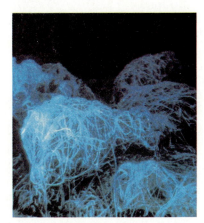
Spaghetti worms

Clam field

General Characteristics of Succession

Communities in succession tend to produce more organic material than they use, while in climax communities, an equilibrium is reached between net production and utilization. In the early stages of succession the exchange rate between organisms and the environment is slow because mineral nutrients are largely stored in environmental reservoirs. But as the climax state is reached, more of the nutrients cycle directly, through exchange pools, between the organisms and the decomposing material. The organisms themselves tend to become more diverse as the community enters the climax state, so a greater number of increasingly specialized niches are developed. Concurrently, some ecologists believe, feeding relationships go from a simple chainlike structure to the intricate food web. There is also some evidence that climax communities are much more stable than their transition stages, less susceptible to external influence, such as human intervention.

About Ecosystems and Human Intervention.

As you are aware, humans can and do change natural communities. We are forever accidentally or deliberately altering the intricate and myriad factors that maintain the delicate equilibrium of ecosystems. Following this we often make well-intentioned but uninformed efforts to make things right—but our efforts often falter for lack of basic information. All of this proves that we just don't know how to live in harmony with the ecosystems of which we are a part. Our technology has far outstripped our basic understanding of our environment. As we turn to the scientific community for answers, ecologists will play an increasingly important role in changing the ways in which we interact with the natural world.

APPLICATION OF IDEAS

1. While ecological communities may appear vastly different, each has many basic organizational features in common. Name three such different communities, and then discuss each in terms of these organizational features. As a start consider such aspects as energetics, trophic levels, and the cycling of mineral nutrients.

2. Considering how environmentally disruptive some mining and industrial practices can be (shale coal mining, open pit mining, copper smelting, and so on), there is a vitally important need for industry to reasonably restore the land following operations. In what ways would it be sensible for industry to bring in ecologists for consultation before beginning restorative programs? What specific knowledge might ecologists have to offer, and what kinds of preliminary studies might ecologists make before beginning?

KEY WORDS AND IDEAS

1. A complete unit of interaction, regardless of extent, can be called an **ecosystem**. Within ecosystems, the flow of energy and cycling of essential substances occurs through **trophic levels**.

ENERGETICS IN ECOSYSTEMS

1. About one tenth of one percent (0.1%) of incoming solar energy is captured in photosynthesis, with 150 to 200 billion tons of dry organic matter or **biomass** produced annually.

Trophic Levels

1. Photosynthesizers and other autotrophs are **producers** or **primary producers**. From the producer, energy and molecules go to the consumer trophic levels.

2. Photosynthetic producers use chlorophyll and light energy along with carbon dioxide, water, and a few minerals to produce essential organic compounds.

3. Consumer organisms occur at several trophic levels and are identified as **reducers** and **primary, secondary, tertiary,** or **quaternary consumers**. It is not uncommon for organisms to live at more than one level. Interaction among trophic levels can be represented by diagrams called **food webs**.

4. **Reducers** are chiefly fungi and bacteria, saprobic organisms, or **saprophages**. Other consumers are called **biophages**. Saprophages break organic matter down into inorganic ions, carbon dioxide, and water, thus recycling them in a form useful to producers.

Trophic Levels as Pyramids

1. **Ecological pyramids** (or **Eltonian pyramids**) can be used to represent **numbers, biomass,** or **energy flow** within communities.
   a. Producers generally represent the great bulk of biomass with other trophic levels greatly reduced from level to level.
   b. Energy pyramids reveal the total amount of energy stored in each trophic level. Typically, they are radically stepped with an average transfer of only 10% from one level to the next.

2. As **omnivores,** humans feed at all trophic levels, but the impoverished tend to live at the primary

consumer level. Feeding at higher levels is prohibitive because of the great losses involved in transfers.

Energy and Productivity

1. The rate of energy storage is called **primary productivity,** recorded as **gross primary productivity (GP), net primary productivity (NP),** and **net community productivity (NCP).**

2. GP is the total rate of energy assimilated by producers, while NP is determined by subtracting the rate of energy utilization by the producers.

3. NCP includes the same measurements, but also takes into account heterotrophs as well as autotrophs, so real community growth is known.

4. Such measurements determine whether a community is declining, growing or in a nonchanging or **climax state.**

NUTRIENT CYCLING IN ECOSYSTEMS

1. The movement of mineral ions and molecules in and out of ecosystems occurs through **biogeochemical cycles.** Most ions enter the living realm at the producer level.

The Nitrogen Cycle

1. Outside of life, nitrogen occurs in **exchange pools** and **reservoirs.** The largest reservoir is N_2 in the atmosphere, but it is only available to nitrogen fixers. Nitrate ions are made available in soil exchange pools.

2. Plants incorporate nitrate or ammonium ions into protein. When plants die or are consumed or eaten, the protein goes to the consumer or reducer level, but eventually all of the incorporated nitrogen goes to the reducers.
   a. **Ammonification:** Bacteria of decomposition reduce the nitrogen to ammonia which forms ammonium ions.
   b. **Nitrification:** In two steps, ammonium ions are converted to nitrite and then to nitrate which enters the exchange pools.

3. Where anaerobic conditions prevail, losses from the exchange pools occur through the action of **denitrifiers.**

4. During **nitrogen fixation,** bacteria (including cyanobacteria) convert atmospheric nitrogen to ammonia which then undergoes nitrification, forming nitrite and nitrate. Nitrogen fixers include symbionts that live in the roots of leguminous plants.

5. Synthetic nitrogen fertilizers are produced from methane gas and atmospheric nitrogen. Their liberal use in support of crops has produced worldwide soil and water excesses and the balance between nitrogen fixation and denitrification has been lost. Two products of the nitrogen load are **eutrophication**—nutrient enrichment of waters, and the pollution of soil water supplies.

The Phosphorus and Calcium Cycles

1. Cycles of phosphorus and calcium occur between living organisms and water. The two elements are taken up in soluble phosphate and calcium ions. Phosphates are used in producing ATP, nucleic acids, phospholipids and tooth and shell materials, while calcium is essential to bone and shell development and in membrane activity.

2. Cellular phosphates cycle directly from living organisms to water and return, while phosphorus in skeletal material and teeth is freed very slowly. Calcium is also freed very slowly from skeletons and shells. Both accumulate in deep ocean and lake bottom sediments until upwellings and overturns redistribute them.

The Carbon Cycle

1. Carbon is an essential part of nearly all the molecules of life. The principal exchange pool on land consists of carbon dioxide gas while the source in the waters is dissolved carbon dioxide gas and the carbonate ion. A large reservoir occurs in the form of limestone and fossil fuels.

2. Carbon enters the producer level during photosynthesis where it is used initially to form carbohydrate. Producers, consumers, and reducers all release carbon during cell respiration.

COMMUNITY ORGANIZATION AND DYNAMICS

1. While open communities blend together at indistinct borders, closed communities have definite borders.

Productivity in a Forest Community

1. In determining the productivity of an oak-pine forest community, Woodwell first determined its structure, including a determination of species, organic matter and biomass. Next, taking advantage of nighttime atmospheric inversions, he determined the output of respiratory carbon dioxide gas.

2. Net community productivity was determined to be 550 $g/m^2/yr$, a modest figure compared to communities that reach ten to twenty times that figure. The oak-pine community studied was determined to be in a state of growth and not in climax.

The Lake: A Freshwater Community

1. Most lakes are recent, having formed less than 12,000 years ago, although a few are of Mesozoic origin.

2. **Limnologists** recognize three lake zones, which are characterized as follows.
   a. In the **littoral zone,** light penetrates to the lake bottom. Producers include algae, bacteria, and submerged, emergent, and floating plants. Consumers include protists, snails, mussels, aquatic insects, fishes, frogs and turtles.

b. The **limnetic zone** is open water penetrated by light. Producers are mainly phytoplankton including green algae and cyanobacteria, whose numbers increase in seasonal "blooms" that correspond to the availability of bottom nutrients. Zooplankton are primary consumers, while lake fishes make up most of the higher level consumers.

c. The **profundal zone** lacks light, extending to the bottom. Profundal organisms usually tolerate low oxygen concentrations and are made up primarily of reducers, a few clams, and wormlike insect larvae.

3. Oxygen and nutrients are redistributed in lakes that undergo seasonal, wind-driven **thermal overturns**—the total circulation of lake waters. In summer, temperate zone lakes form a warm, upper **epilimnion,** a cold lowermost, **hypolimnion,** and a steep **thermocline** between. No mixing occurs below the thermocline because of the greater density of hypolimnion waters.

4. Temperate zone lakes undergo two seasonal thermal overturns when cold surface water sinks, displacing the bottom layers and bringing oxygen down from surface waters and nutrients up from the sediments.

Community Development Over Time: Ecological Succession

1. While communities exist for long periods in a state of climax, they change over time under the influence of geological and climatic changes and through biological factors and human intervention. Such change is called **ecosystem development** (or **ecological succession**). When brought about by living inhabitants, the process is called **autogenic succession,** while change brought about by outside forces is **allogenic succession.**

2. During community development, the organisms themselves produce changes that bring about their own replacement by other species. Each sequential stage is a **seral stage,** part of a **sere.**

3. Ecological succession includes **primary succession** and **secondary succession**

a. Primary succession occurs where no living organisms have yet become established. **Pioneer organisms,** such as lichens and mosses are often the first to become established, followed by grasses, and more sun-tolerant plants. Eventually, the community will reach a climax state. The entire sere may require hundreds of years to develop.

b. Secondary succession occurs wherever a climax community is disturbed (such as deserted farms, burns, or clearings). It is common for the first invaders to include opportunistics, fast-growing weeds. Initial invaders are soon replaced by fast-growing shrubs and trees, and a gradual blending with the surrounding community occurs.

Succession in the Aquatic Community

1. Succession in aquatic communities is accelerated by **natural eutrophication.** Lakes are rich in nutrients and are described as **eutrophic.** In **oligotrophic lakes,** nutrients are limited, with minute amounts in exchange pools, and succession is extremely slow.

2. Succession in lakes begins with increased nutrient sediments, encroachment by shore plants, and a general increase in both numbers and kinds of organisms. The continued trend will result in the lake filling in and later blending in with the surrounding terrestrial community.

3. Human activities often result in **cultural eutrophication,** nutrient enrichment through the pollution of lakes. Of particular significance are increases in phosphates and nitrates which are usually delicately balanced in freshwater communities.

4. Succession is characterized by the following: increased productivity, the shift of nutrients from reservoirs to exchange pools and an increase in the cycling rate, increased diversity in organisms with increased niche development, and a general increase in the complexity of food webs.

REVIEW QUESTIONS

1. Carefully distinguish between the terms biosphere, community, ecosystem, and biome. (1064)

2. Approximately what part of incoming solar energy is trapped by producers? In what form is this energy found in the producer? (1065)

3. Describe and provide examples of each of the following: primary producers, primary consumers, secondary consumers, and reducers. List two foods we eat that place us at the quaternary level of consumption. (1065–1066)

4. How does a saprophage differ from a biophage? (1067)

5. Construct a typical pyramid of biomass or energy. Describe the shape and explain the decrease from one level to the next. (1069)

6. Approximately what part of the energy moving through trophic levels is stored in each level? Can it ever be 100%? Why? What happens to the rest? (1070)

7. From an energetics viewpoint, which trophic level might support the greatest number of humans? Very specifically, what problems arise when we feed exclusively from this level? (1070)

8. How are the gross primary productivity and net primary productivity determined? Which represents real growth in the producer population? What becomes of the difference between the two? (1070–1071)

9. What does net community productivity tell us that GP and NP cannot? If NCP is a positive number, what does this tell us about the community? (1071)

10. How does a nutrient reservoir differ from a nutrient exchange pool? (1072)

11. Briefly summarize the events of the "inner nitrogen cycle," including mention of decomposition, ammonification, and nitrification. Why is nitrification desirable when plants can readily incorporate ammonium ions into amino acids? (1072–1073)

12. Briefly describe how nitrogen is constantly lost and gained by the nitrogen cycle. Under what specific conditions do losses occur? How have humans offset the balance between loss and gain? (1074)

13. What are two problems stemming from excessive use of synthetic fertilizers? (1074)

14. What must happen to phosphates and calcium before they can cycle directly from organisms to exchange pool and back to organisms again? (1076)

15. Using simple formulas for photosynthesis and respiration, explain how carbon dioxide, oxygen, and water cycle among living things. In addition to carbon dioxide gas, what other sources of carbon are available to life? (1077)

16. Distinguish between an open and a closed community. (1077)

17. Reviewing the Woodwell study of the oak-pine forest community, describe how gross productivity was determined. What was the problem with determining net productivity? How was this overcome? (1077–1078)

18. How did net productivity in the oak-pine community compare with that in the tropical rain forest? A field of grain? (1078–1079)

19. In which of the three lake zones would one expect to find the greatest net community productivity? Which has the greatest species diversity? The least? (1079, 1081)

20. Using a simple diagram, illustrate the organization of the lake into its three zones. What one factor distinguishes the three zones? (1079, 1081)

21. Briefly explain why the deep waters of Lake Tanganyika are permanently anaerobic, while the waters of Lake Baikal are aerobic at all depths. (1081)

22. Using the terms thermocline, epilimnion and hypolimnion, describe temperature conditions in a temperate zone lake in summer. What effect does such an organization have on the distribution of oxygen and nutrients? (1081)

23. Briefly describe how the density of water changes as its temperature decreases. How does such winter cooling in lake waters explain thermal overturn? (1081)

24. Distinguish between and provide examples of autogenic and allogenic succession. (1082)

25. Suggest several seral stages that might occur during primary succession. How do organisms actually produce the sere? (1083–1084)

26. List several indicators one can use to distinguish between an oligotrophic and a eutrophic lake? What kinds of human activities tend to increase eutrophication? (1085–1086)

27. List four characteristics of a community in succession. (1087)

SUMMARY FIGURE INTERACTION IN THE EARTH ECOSYSTEM

The interaction of organisms with each other and with their environment is seemingly endless. As we see, the source of energy for most life is the sun. Its boundless radiance drives photosynthesis, providing chemical bond energy that, in turn, supports the earth's heterotrophs. Sunlight energy also powers the great hydrologic cycle, an atmospheric heat machine that continually redistributes precious warmth and moisture over the globe. Within the terrestrial and aquatic realms, energy captured in photosynthesis flows from primary producer to primary consumer and on through higher consumer levels, thus making animal life possible. In a far more subtle interaction, the molecules of all organisms finally reach the reducer, where the last atoms and calories are extracted. Of greater significance, the vital ions and molecules are soon freed to recycle back to waiting producers for another turn of the cycle of life. In the ongoing energy flow, each participant releases its quota of low-energy waste heat, which escapes back into the physical realm, adding to spent energy radiating back into space. Thus the elements and the energy of life reach a state of equilibrium between the physical and biological world. Or do they? Off to the side of our scenario we are reminded of those activities that threaten to upset the precarious equilibrium of chemicals and energy. As human populations burgeon and technology expands, the carbon reservoirs are tapped at an incredible rate, and strange new chemicals never before known by the planet, along with extraordinary volumes of carbon dioxide, spew outward.

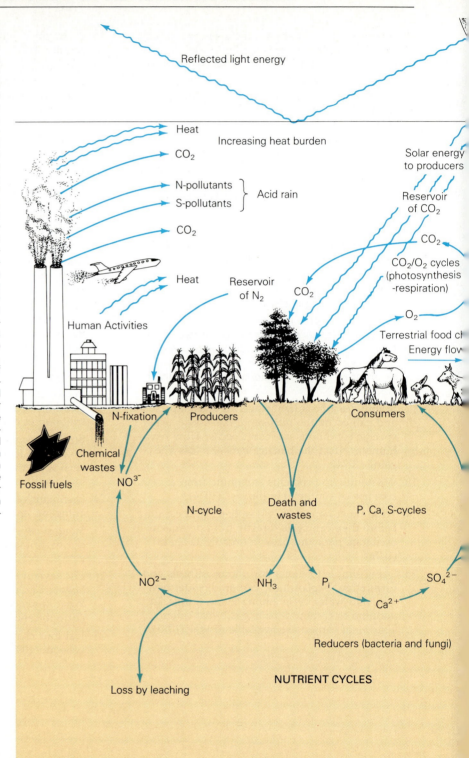

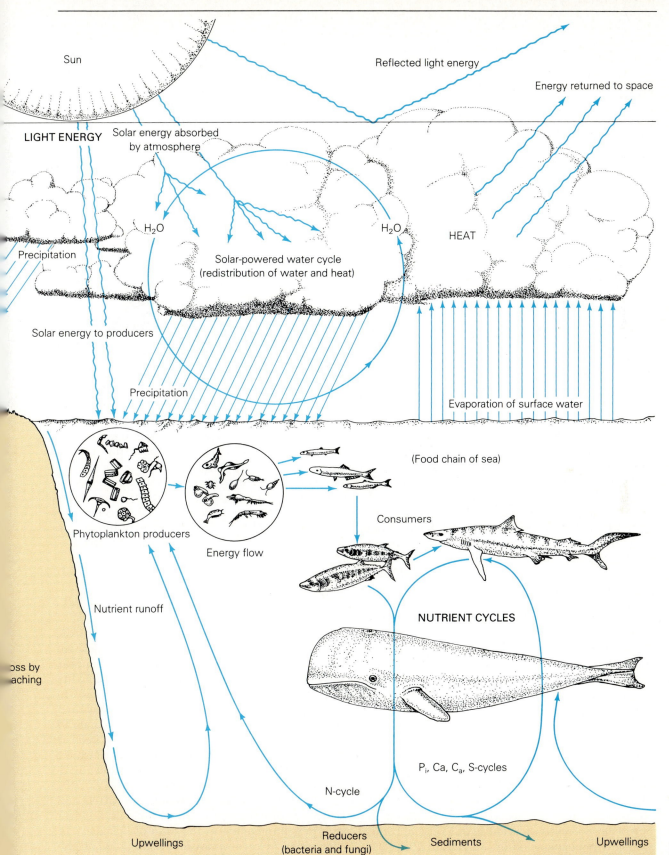

Sun

Reflected light energy

Energy returned to space

LIGHT ENERGY
Solar energy absorbed by atmosphere

H_2O

H_2O

HEAT

Precipitation

Solar-powered water cycle (redistribution of water and heat)

Solar energy to producers

Precipitation

Evaporation of surface water

(Food chain of sea)

Phytoplankton producers

Energy flow

Consumers

NUTRIENT CYCLES

Nutrient runoff

oss by aching

P_i, Ca, C_a, S-cycles

N-cycle

Upwellings

Reducers (bacteria and fungi)

Sediments

Upwellings

Interaction in Populations

The howling of wolves once accented the nights over most of the United States. Hunters and hikers regularly came across their tracks, and the remains of their kills were a part of the American wilderness experience. But now wolves are entirely absent over much of their former range and rare over most of their present range. Today, only a few well-publicized bands precariously persist in scattered areas. What has happened to their numbers? Why have wolf populations declined so precipitously? And what about other species? Why are bluebirds and vultures seen less frequently in Arkansas, just as the beaver is apparently making a comeback there? Sometimes we know the reasons for such population changes; sometimes we don't. Wolves, for example, have traditionally been fair targets for the rifles of farmers and game shooters. But what about places where they are not hunted, for example, protectorates in Michigan and Minnesota? Are the wolf populations in these areas also changing? If so, why? If not, what is keeping them stable? The point is, the numbers of living things change, and usually we don't know precisely why.

Because population numbers reflect a myriad of environmental, evolutionary, and physiological effects, most of which are not very well understood, any question about populations tends to be a tough one. A host of researchers are, at this very moment, watching populations of tiny animals in little glass tanks—just trying to fit more small pieces into the great puzzle. Others are traipsing around in fields or spending long hours watching a hole in the ground, waiting for some small face to appear and be counted. A great many people, in fact, are asking very basic questions about how populations change. So our goals here must remain modest. We can only describe, in general terms, a few of the factors that can cause populations to change.

POPULATION CHANGES

Some population changes are short-term or periodic and easily explainable. Grasshoppers, which are plentiful in summer, reproduce and then disappear by winter, leaving their fertilized eggs hidden in the soil. The disappearance of the adults is easy to understand—they die. Some birds disappear from American forests in the winter as they migrate to more hospitable southern areas. These are familiar and rather predictable phenomena. We will expect the next generation of grasshoppers and the migratory birds to show up in spring.

Other sorts of changes are less predictable and more puzzling. As some species dwindle, we hear of others burgeoning in the form of "plagues" or "invasions." However, the inordinate attention given to such changes may not accurately reflect what is going on. This is because *most species seem to maintain rather stable numbers*. They seem to be under some sort of control or regulating influence (Figure 42.1). In order to consider any such control, we must first understand something about how populations generally behave. What determines the numbers of a population? What are the characteristics of its growth? What causes populations to fluctuate, and what sort of stabilizing influences might be at work? Finally, what influences the distribution of populations? We should keep in mind that, technically, a population is not just a group of organisms, but a group of the same species that lives in close enough association to be able to interbreed.

Population Dynamics and Growth Patterns

A good place to begin looking at population change is through an idealized mathematical model. Consider first the simple population equation:

$$I = (b - d)N$$

which translates into, "*I*, the rate of change in the number of individuals in a population, is equal to *b*, the average birth rate, minus *d*, the average death rate, times *N*, the number of individuals in the population." As we will soon see, this straightforward equation describes only one aspect of change, one that is unrestricted and does not take into account the many population-limiting factors. But let's explore further.

One can see from the equation that for populations to increase, that is, for *I* to become larger, the value of *b* must exceed the value of *d*—births must exceed deaths. Should they be the same, *I* will remain unchanged, or should the average death rate, *d*, exceed the average birth rate, then *I* will become smaller. So, the value of *b* − *d* seems to determine what happens to population size. The simple difference between the two values alone is called the **intrinsic rate of increase** or *r* (also called the **instantaneous rate of increase**). The determination of this rate is simply, $r = b - d$. It may have occurred to you that we can now rewrite *I* to incorporate the new term, *r*. Thus,

$$I = rN$$

This new equation reminds us of something important in population dynamics. In a growing population, the growth rate is determined not only by *r*, but by *N*. For instance, when *r* is a positive number, *N* will grow with each new generation of

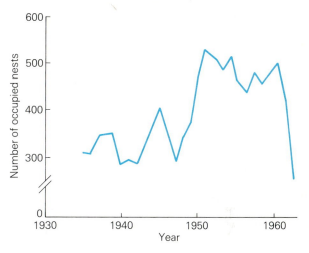

42.1 A STABLE POPULATION

This population of herons has fluctuated around a long-term mean. The data were collected from a study of occupied nests in the area over a period of 30 years.

offspring. Thus *I* increases in each generation. The story of population increase and decrease can be summarized in two simple curves, the **J-shaped curve** and the **S-shaped curve.** The letters *J* and *S* represent the direction taken by lines when population growth patterns are plotted. The J-shaped curve is the one environmentalists talk about when they try to interest people in the human population growth problem, so let's consider this curve first.

Exponential Increase and the J-shaped Curve. If a few reproductively active organisms are placed in an idealized environment—one with unlimited resources and space and without danger from disease, predation, or any other hazards—they may be expected to reproduce at their maximum physiological rate. This theoretical rate is referred to as a species' **biotic potential,** and it implies maximum gamete production, mating, fertilization, and survival of offspring. (If we allow a 30-year reproductive period for humans, then our species' biotic potential would be a little over 30 offspring for each female.) Biotic potential is also referred to as r_{max}—the maximum value of *r*, or the maximum intrinsic rate of increase.

Under r_{max} numbers may rise slowly at first, simply because we are dealing with low numbers—from 2 to 4, to 8, to 16, and so on. In plotting such numbers, at first we see a gently increasing slope. But soon, as the numbers continue to double, the curve arches sharply upward until it seems to approach (but never reaches) the vertical (Figure

42.2 THE J-SHAPED GROWTH CURVE

The J-shaped growth curve is characteristic of exponential growth that occurs when species reproduce at or near their biotic potential. In such growth the numbers increase in doubling increments, such as 2, 4, 8, 16, 32, and so on.

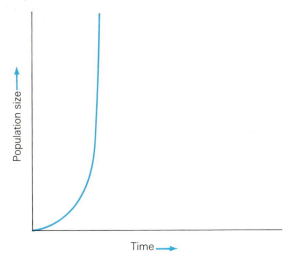

42.2). Such increases are called **exponential increases.** They are unlike linear or **arithmetic increases,** those that progress from 1 to 2 to 3.

We can visualize exponential increases in ways other than the J-shaped curve. You might consider, for instance, "inverted pyramids with ever-widening sides," but another example may be more familiar. Consider the way interest is compounded in a savings account. As you make weekly deposits (and refrain from withdrawing), your savings grow linearly, that is, by simple addition. In addition, money in the form of interest is added to your savings, and in subsequent interest-calculating periods, the interest paid is based on the new combined total. So, although the interest rate may remain constant, say at 5.5%, N (your wealth) accumulates in an ever-increasing fashion, at least until the government hears about it.

One of the clearest biological examples of exponential growth is seen when a small number of bacteria are introduced into a rich laboratory culture medium. Under such ideal conditions, the familiar *E. coli*, for example, will divide every 20 minutes. At this rate, after only 24 hours, just one bacterium would give rise to 40 septillion descendants (2^{72}: where the exponent equals the numbers of generations in 24 hours). But as simple as *E. coli's* growth requirements are, such rapid expansion could not be sustained. Very likely, sometime midway through the 24-hour period, the resources that were

supporting such phenomenal growth would be reduced to a point where the biotic potential could no longer be reached. *E. coli* would then have encountered **environmental resistance,** and the rapid rate of increase would begin to slow.

Environmental Resistance and Population Crashes. Growth-inhibiting factors, collectively called environmental resistance, can take many forms, but in a laboratory-grown bacterial population, such resistance is commonly in the form of food shortage and the accumulation of toxic waste products. As these things occurred, the rate of population increase would rapidly slow, and then, for a brief period, the cells would die as fast as they were produced ($b - d = 0$). Finally, the bacterial environment would be so fouled and depleted that the cells would begin to die far more rapidly than they were being produced. Such a sudden decrease in numbers is called a **population crash** (Figure 42.3). Logically, environmental resistance is often measured as the difference between the theoretical biotic potential (r_{max}) and the growth that is actually observed.

Carrying Capacity and the S-shaped Curve. Population crashes occur when a population exceeds the **carrying capacity** of its environment. The carrying capacity can be defined, in its sim-

42.3 A POPULATION CRASH

In populations that have undergone a period of unlimited, exponential growth, population crashes are not uncommon. The dieback usually places the numbers somewhere well below the ability of the environment to support the population, although in some instances, the dieback is complete.

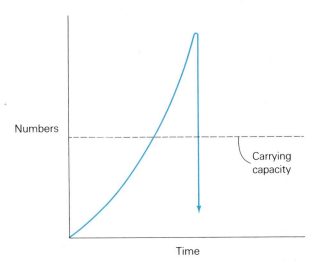

plest terms, as the number of individuals that can be supported by the resources of an environment without damage to those resources. The carrying capacity is not fixed; it may change drastically from one time to another. Further, it is constantly affected by many factors, both **biotic** (living) and **abiotic** (nonliving), which we will elaborate upon later. In the case of the bacterial colony, the number of individuals far exceeded the environmental carrying capacity.

The bacterial population crash is an extreme example, since there is no way for such an artificial environment to recover on its own. Typically, however, the resources that maintain a population are renewable: the plants sustaining a population of rabbits will recover; the fieldmouse population depressed by a population of predatory owls will recover as the owls seek new prey sources, and the rich supply of nitrates supporting a phytoplankton bloom will return at the next season's thermal overturn. It's important to realize that population booms and crashes are the natural way of life for many species.

While population crashes can and do occur under natural conditions, for many species another phenomenon occurs with less drastic results. In such cases, as the population approaches the carrying capacity, growth slows. The nearly vertically rising "J" tapers off as a gentler slope, forming the less radical S-shaped curve. Typically, the population numbers then fluctuate around the carrying capacity (Figure 42.4).

The S-shaped curve is also known as the **logistic growth curve.** Logistic growth curves start out similarly to exponential curves, with the birth rate greatly exceeding the death rate, but then at some point, acceleration ceases and deceleration begins. During deceleration the birth rate generally begins to decrease while the death rate increases. Finally, the two reach equality and $r = 0$, completing the S-shaped curve. The population may then hover about this level.

The carrying capacity is represented in population equations as K, with the following expression or term:

$$\left(\frac{K-N}{K}\right)$$

which, is incorporated into our equation as:

$$I = r\left(\frac{K-N}{K}\right)N$$

42.4 DYNAMICS OF POPULATION GROWTH

The dynamic factors in population change include the biotic potential, environmental resistance and carrying capacity. Typically, the interaction of these factors produce the S-shaped curve. Sustained growth at the biotic potential cannot persist because of environmental resistance which includes limited resources. Such resources help determine the carrying capacity—the numbers any environment can sustain over long periods without serious damage.

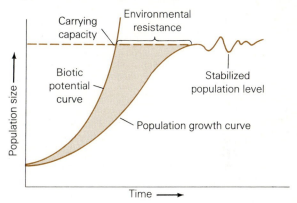

The effect of K increases as the population size, N, increases. Early in the population's growth, the value of N is small compared to K, and subtracting it from K in the numerator has little effect. In fact, at this time the value of $K - N/K$ approximates 1 and of course, multiplying the other factors in the equation by the 1 has no effect at all. As a result, early growth is exponential and rapid. But as the population grows and N becomes larger, the value of K decreases from 1, and rN, now multiplied by a fraction, becomes smaller. In fact, as you may have anticipated, as the value of N reaches K, the new quantity becomes zero and, theoretically, no further population increase is possible. To illustrate, simply substitute zero for K and multiply:

$$I = r \times 0 \times N = 0$$

In reality, populations often exceed their carrying capacity, and when they do K becomes a negative number, and we can predict a temporary population decline.

Population growth is certainly more complex than we have portrayed with these theoretical equations, and while populations of some species approximate the neat S-shaped, logistic growth curve, most only approximate such idealized situations (Figure 42.5). Many kinds of populations tend

42.5 THE S-SHAPED GROWTH CURVE

A long term study of sheep in Tasmania reveals a fairly typical S-shaped curve. Each dot represents an averaging of numbers over a five year period.

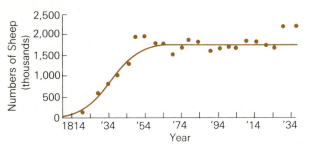

to fluctuate about some density, regulated in part by the apparent negative feedback effects of the carrying capacity. But it is important to keep in mind that the value of K can also change, perhaps due to permanent environmental damage, or cyclically, as seasons change and the availability of resources diminish or increase. K can also change through the influence of other species, as we will see.

Age, Life Span, and the Population. A number of factors can influence populations, such as the ages of its members and their longevity. In particular, one might ask: How many individuals within a population are below, within, or beyond the reproductive age? How long is their expected life span? What percentage of individuals complete the life span? As we will find, these are highly significant factors in predicting the future course of human population growth. For example, the 1983 census revealed that 34% of the human population was below 15 years of age—many having yet to participate in reproduction. How will this affect our future rate of increase?

We have seen that the death rate is a vital part of population growth dynamics, but knowing the death rate doesn't give us much information. What is important is knowing the *age groups* that are subjected to the greatest mortality. In many natural populations, we find that death occurs most frequently in the young, and once an individual has survived the rigors of early life, the life expectancy is significantly extended. Obviously, mortality increases again in the oldest segment of the population. This characterization applies to many species, but certainly not all. Modern humans, for example, have overcome what was formerly a large infant mortality and have succeeded in extending

the life span. In Figure 42.6, **survivorship curves** depict the relative average life span of five species, including humans.

Survivorship curves are prepared by plotting the number of survivors on the vertical axis (usually a logarithmic scale) and the percent of the life span on the horizontal axis. From such curves we can learn some interesting things about species. For example, in the comparison of humans with oysters, we find exaggerated convex and concave curves respectively. Oysters apparently have an extremely high "infant mortality," while there is little loss of young humans. (Actually, it's the oysters' swimming larvae that meet with misfortune, so we have to know something about life history to interpret the curve. With humans, for instance, as with most mammals and with birds and some social insects, parental care accounts for the small loss of offspring.) But, at the other end of the spectrum,

42.6 SURVIVORSHIP CURVES

Survivorship curves for six species, each beginning with a population of 1000 (vertical logarithmic scale). Points along each line represent the percent of the life span reached as the populations age (horizontal axis). In black-tail deer, a high death rate is experienced among the young, the numbers drop off rapidly with the average individual completing only 6 percent of its potential life span. Loss during the early period is even more dramatically seen with oysters where great numbers of offspring are devoured by other marine animals. But most surviving oysters can expect to complete their life span. The average human and rotifer survive the rigors of early life to complete about two thirds of their potential life span, while the average *Hydra* and gull complete about 37% and 24% respectively.

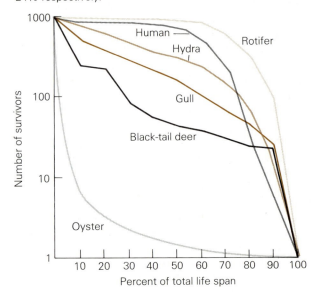

TABLE 42.1

PROJECTED POPULATIONS OF THE HOUSEFLY
(*MUSCA DOMESTICA*)[a]

Generation	Numbers if all survive
1	120
2	7,200
3	432,000
4	25,920,000
5	1,555,200,000
6	93,312,000,000
7	5,598,720,000,000

[a]In 1 year, about seven generations are produced. The numbers are based on each female laying 120 eggs per generation, each fly surviving just one generation, and half of these being females. Adapted from E. V. Kormondy, *Concepts of Ecology* (New Jersey: Prentice-Hall, Inc., 1969), p. 63.

when humans get old they die off rapidly, while death comes more slowly and steadily to the fewer surviving members of the oyster population. So that's how to tell people from oysters.

Reproductive Strategies

The reproductive potential of most species, even the slowest breeding ones, is incredibly high. Charles Darwin estimated that a single pair of elephants could leave over 19 million descendants in only 750 years. The fact that the entire world is not teeming with elephants indicated to him that some elephants were not reproducing, and that the ones that were successful were somehow "selected" by the environment.

Whereas elephants are very slow breeders, other species are more prolific. For example, the reproductive potential of the housefly for a *single year* is shown in Table 42.1. Again, we are pleased to report that not all survive—it just seems that way.

What determines the rate at which an organism reproduces? Why haven't flies or elephants taken over the entire surface of the earth? It should be apparent that simple laws of energetics would not allow an elephant to become pregnant several times a year. There is just no way that a female could find enough food to produce that many offspring. And what would happen to the helpless baby elephants already born as new brothers and sisters appeared on the scene?

It may seem puzzling, then, to learn that each animal theoretically reproduces to its biological limit. This limit does not depend on the number of births (or hatchings) then, but on the number of

young that can safely make it to reproductive age. So the elephant, with its single young, may not measure up to the fly's reproductive output, but because of its many restraints, it can manage no more. Let's take a closer look at these restraints in terms of evolutionary theory.

The Theory of "K" and "r" Selection. One attempt to answer the difficult questions about populations has led to the theory of **K-selection** and **r-selection.** (The letters *K* and *r* are borrowed from formulas in population statistics, described earlier, where they represent the carrying capacity and the intrinsic rate of increase, respectively.) K-selected species are those whose population growth curves closely resemble the idealized S-shape. Their numbers hover about the carrying capacity, to which they are highly responsive. Thus, they have slower rates of population growth. K-selected species tend to be large animals, such as mammals and birds, with long life spans, continuous reproduction—year after year—and lengthy growth periods. At the other extreme are the r-selected species, sometimes referred to as "boom and bust" types. In their adaptation to large population size, they tend to have short life spans, are small in size, mature early, and reproduce only once but with prodigious numbers of offspring—which they generally take no part in raising. Thus their populations suddenly boom, shooting up beyond the carrying capacity, following the familiar J-shaped growth curve before crashing. While population crashes can be quite dramatic in r-selected species, they are really a routine part of life and recovery is just as routine and predictable (Table 42.2). You might want to keep in mind, also, that K and r adaptations do not necessarily follow general taxonomic lines—thus while one might expect insects to be r-selected, and the locusts certainly are, bees are excellent examples of K-selected species. Perhaps we can make the distinction between r- and K-selected species more clear with a couple of examples.

The Tapeworm, an r-Selected Species. The tapeworm living so comfortably in the idyllic environs of your intestine is lucky to be there—although its luck is the inverse of yours. It is lucky because the life of a tapeworm is fraught with risk. There are incredible odds against its ever finding a host at all. In the case of the beef tapeworm, the egg must pass out with the feces of the previous human host. Then, not only must it survive the elements, but it must also be swallowed by a calf,

TABLE 42.2

SOME CHARACTERISTICS OF r- AND K-SELECTION

	r-selection	K-selection
Climate	Variable and/or unpredictable	Fairly constant and/or predictable
Mortality	Density independent	Density dependent
Survivorship	High mortality when young; high survivorship afterwards	Either little mortality until a certain age, or constant death rates over a period of time
Intraspecific and interspecific competition	Variable, lax	Usually keen
Selection favors	Rapid development; high rate of population increase; early reproduction; small body size; single reproduction	Slower development; greater competitive ability; delayed reproduction; larger body size; repeated reproductions
Length of life	Usually less than one year	Usually more than one year
Leads to	Productivity	Efficiency

SOURCE: Adapted from E. R. Pianka, "On r- and K-selection." *American Naturalist* 104:592-97 (1970).

so already the odds are against it. Once eaten, the larval tapeworm changes its appearance, moves through the intestinal wall of its host, and on through the bloodstream until it comes to lie as a "bladderworm" in the muscle. Here it forms a protective capsule around itself and waits. It waits for a human to eat the beef without cooking it too well. When this happens, the worm attaches to the intestine of its human host, grows into an adult, and begins to lay eggs. Actually, this is a rather simple cycle; some intestinal parasites have three hosts, and take different forms in each of these (see Figure 25.12).

There is obviously a strong probability that not all the necessary conditions will be met, and thus only a remarkably small number of tapeworm eggs ever develop into adult tapeworms. So what is the tapeworm's answer to such a demanding life cycle? It has become capable of self-fertilization (although it will also cross fertilize if another worm is present) and it lays thousands and thousands of eggs. Like other r-selected species, its whole life seems to be devoted to reproduction. It exists only

to lay eggs, eggs, EGGS! A few of these will wend their way, by chance, through the complex maze that is the tapeworm's life cycle.

K-Selected Species. We can consider the chimpanzee as our representative K-adapted species, and then draw on an example from birds. Like other such species, the chimpanzee tends to live in a stable environment, has a long life span and reproduces only after a prolonged period of development. When the female chimpanzee is sexually receptive, she may copulate with almost any male who signals his desire. Once she has become pregnant, she may not become sexually receptive again for years. Jane Goodall, who spent more than 20 years among the chimps at the Gombe Stream Preserve in Africa, tells us that Flo, the aging but sexy female did not become sexually receptive again for five years after giving birth. During this time, she attended carefully to her baby.

In the first months, Flo carried her baby everywhere and diligently guarded it against danger. Later, the baby was permitted brief forays on its own, but it was not allowed to stray far from its mother's vigilant eye. During this time, the youngster would scurry back to Flo at any real or imagined sign of danger. As the young chimpanzee gradually became able to care for itself during those first few years, the association between mother and offspring relaxed, until finally Flo was free to mate again and rear another baby. So, among chimpanzees, the female does not maxi-

42.7 THE K-ADAPTED CHIMPANZEE

Chimpanzee mothers carefully attend to their young. Because of their great investment in each of their offspring, they cannot afford to produce them in great numbers.

42.8 PREDATION AND BIRD REPRODUCTION

(a) The ground-nesting gulls, such as the herring gull, have been forced to assume a different reproductive strategy than **(b)** the cliff-nesting kittiwake. The nests of the ground-nesting species are vulnerable to marauding predators, so they compensate by laying more eggs than they will be likely to rear. The ground nesters, for example, lay three eggs, whereas they can probably rear only two young, but at least one of the eggs is likely to fall to a predator. The kittiwake, on the other hand, nests on cliffs safe from predators. Since all their young are likely to survive, they lay only two eggs.

mize her reproductive success by giving birth to large numbers of offspring. Instead, few offspring are produced and these receive very careful attention until they become independent (Figure 42.7).

A quick look at some data from birds provides a clearer picture of how the dictates of a particular environment can influence the number of offspring a given species attempts to produce. For example, food supply often influences the number of eggs laid by birds in a particular season. The number will be roughly equivalent to the number of hatchlings they can feed successfully.

The number of offspring attempted by some bird species may also reflect the vulnerability of the young to predators. For example, certain species of gulls nest on accessible beaches where the young are in real danger of being found by prowling foxes and other marauders. These gulls often lay three eggs, although usually only two young can be fed successfully. The odds are that at least one of the young will be eaten, so quite possibly the three egg nest is an adaptation to high predation. On the other hand, the kittiwake, a species of gull that nests on steep cliffs, normally lays only two eggs. Since any marauding fox would be likely to break its neck trying to climb the cliffs, all the young are safe from foxes (Figure 42.8). Thus the parents need only to produce the number of young that they can feed.

As a final example of an extremely K-selected

species, we need only consider the reproductive strategy of our own group.

A History of Human Reproductive Strategies.

Since the reproductive strategy of any species is a response to its own evolutionary history and immediate environment, let's take a closer look at the factors that might have determined our own reproductive propensity. The well-nourished human female has the ability to produce one child each year for over 30 years. Does this mean that humans are prepared, either physiologically or psychologically, to rear 30 children? Or is our reproductive ability an evolutionary adjustment to a historically rigorous life in which most, or even all, of our offspring were not likely to survive? In other words, has the ability of the human female to reproduce prolifically evolved in response to a traditionally high death rate among children?

What about today? As Table 42.3 shows, a child born in a modern, developed nation such as the United States is very likely to live through its first critical years. In humans today, the mortality rate decreases after the first year and is very low

---

TABLE 42.3

ESTIMATED AVERAGE LIFE SPAN IN HUMAN POPULATIONS

It seems that our life span, as a group, was fixed by evolutionary processes early in our history. Our attempts at increasing our longevity probably have resulted only in our coming closer to reaching the limits set by our physical constitutions. The examination of the remains of our primitive ancestors shows that many of the conditions we associate with old age affected them in much the same way that they affect us today. Also, in classic Rome, a person who lived to the age of 75 was more likely to reach 90 than someone in the United States today. (However, it might be argued that that is because it took a sturdier constitution to reach 75 in those days.)

Population	Years
Neanderthal	29.4
Upper Paleolithic	32.4
Mesolithic	31.5
Neolithic Anatolia	38.2
Austrian Bronze Age	38
Classic Greece	35
Classic Rome	32
United States, 1900–1902	48
United States, 1950	70
United States, 1980	73.7

SOURCE: Adapted from "The Probability of Death" by E. S. Deevey. Copyright © 1950 by Scientific American, Inc. All rights reserved. Reprinted by permission.

through the teens; then it increases gradually until about age 60, when it rises rather rapidly. (Actually, this sort of curve doesn't apply very well to the poorer countries.)

The means by which the human species has increased the likelihood of survival are almost universally viewed as good and desirable. For example, we have specific medicines to combat various maladies which, in earlier days, would have proven fatal. We also have therapeutic and corrective devices to aid the sick. If someone in our midst is unable to provide for him- or herself, that person will usually be cared for, however minimally. Our society often provides for those who, in harsher days, would have been selected against. A person doesn't have to be keen of wit and physically agile in order to cross a busy street. He or she simply waits for a light—a light that means it's safe to cross. The result of our social care has been a negation of many of the usual influences of natural selection. However, at the same time that we have reduced selection for swiftness and strength in our species, we have increased the level of variation. Thus we find among us all sorts of interests, talents, tendencies, and appearances. The point is that, with its highly developed social programs, our society is attempting to ensure that every individual will live and reproduce. This raises some important questions. With our reproductive potential so high and so many of the natural curbs removed, are we placing our species in a precarious position? Are we setting the stage for generations of ill-adapted people? Are we psychologically prepared for a great worldwide surge in our numbers?

In essence, then, the reproductive potential of humans has been established through the eons of our development. But although the direction and strength of natural selection have changed as we have altered the environment, we are left with a reproductive capacity better suited to earlier cultures.

Death as a Population Growth-Controlling Mechanism

Have you ever wondered why death occurs at all? Why hasn't natural selection resulted in organisms that simply live forever? If life is better than death, then why is there no marked selection for longevity? Why do humans have so much trouble surviving past their "three score years and ten"? Why death?

All sorts of people have struggled with the "meaning" of death—theologians, poets, novelists, philosophers, drunks, and others of bad habit. Even aboriginal societies deal with the meaning of death in one way or another—some very casually, not fearing it at all but treating it as though it were simply the natural extension of life. So let's join the fray and consider death from a biological point of view.

Programmed Death.   Annual plants germinate and grow in the early spring, flower in the late spring, undergo seed development in the summer, and disperse their seeds and die in the fall. There are "annuals" among the animals, also. One species of Brazilian fishes live in temporary ponds that exist only during the rainy season. Shortly before the ponds are due to dry up, the fish spawn and lay cystlike, drought-resistant eggs. The fish die when the ponds dwindle, but the eggs hatch the next year. If these fish are netted and kept in an aquarium, they continue to follow their natural life cycle. After the females have laid their eggs and the males have spawned, the fish literally begin to fall apart. Why do they die after reproducing?

The life spans of squids and octopuses are also intimately associated with their reproductive efforts. Put simply, after they reproduce, they die. They are programmed to die at this time because both reproduction and death are under hormonal control. If the cephalic gland of a mature octopus is removed, the animal won't engage in sexual activity or reproduce—and also won't die; that is—it won't die on schedule. The experimentally altered animal continues to feed and engage in normal activities. Sooner or later, however, death comes anyway. The altered animals produce tumors and infections, otherwise unknown in the species, and the long-lived but barren animals ultimately succumb.

Do humans have a similar kind of programmed death? There are some reasons for thinking that they do, to some extent at least, although the program is by no means as absolute as that of the squid. Usually between ages 60 and 70 the death rates suddenly begin a precipitous rise and continue to accelerate due to a host of apparently unrelated diseases. It seems that all kinds of systems wear out more or less at the same time. The time of this rapid aging may differ from individual to individual, which smooths out the human death rate curves, but when you're old, you're old. It has been observed that people who have been saved from cancer soon die of heart disease or some other infirmity of age. In fact, although cancers claim about 25% of all humans, the combined human death rates for all causes increases so rapidly that if we were to eliminate cancer, say at noon today, we would only extend our average life expectancy by less than three years. So why do humans seem to wear out all at once?

Humans are among the longest-lived of all animal species. Our normal life span exceeds that of all other mammals and most vertebrates and invertebrates—possibly only giant tortoises and certain sea anemones regularly outlive us. Correspondingly, we have very effective cancer-suppressing and disease-fighting mechanisms. Our bodies mature very slowly, but they last a long time. We have quite efficient defenses against the wear and tear of time, against infection and our own cancerous cells. We pay a price in physiological energy, in developmental time, and in our rather low rate of reproduction. However, after perhaps 15 to 20 years of maturation, another 25 to 30 years of peak reproduction, and another 20 years to see our youngest children reach their own maturity and independence, our bodies have had it. Once the body has served its allotted time, it wears out—all at once. There doesn't seem to be any strong selection in favor of metabolic systems that protect against the depredations of postreproductive degeneration. From the relentless viewpoint of natural selection, there's just not enough of a payoff involved.

Now, in any system in which the commodities are limited, such as the planet earth, as the numbers of individuals increase, so will the competition for the available commodities, such as food, shelter, land, and natural resources. Consider, then, an organism that finishes its period of reproduction and then hangs around to compete with its offspring. Such an individual could be expected ultimately to leave fewer offspring than a parent that, after having reproduced, died and thus did not interfere with the success of its offspring.

We've tossed around some concepts fairly loosely here, as you are aware. However, these are legitimate questions for evolutionary biologists. As scientists, we are not in a position to uncover whole truths, but we can begin to accumulate certain kinds of evidence. Anyhow, this sort of approach makes as much sense as dealing with the concept in subjective, artistic, abstract, or mystical terms.

POPULATION-REGULATING MECHANISMS

We can now begin to ask what, precisely, regulates populations. What are the effects of the physical environment? Of other species? In some instances the answers are obvious while in others they may be quite subtle or completely unknown. Let's begin by listing some such influences. We will see that they fall into two categories, nonliving and living.

Abiotic Control

Just as seasonal weather changes can alter population numbers, so can irregular or unusual weather. Drought may kill many kinds of plants and animals. Many birds perish in some years because they begin their northward migration in the spring only to be caught by a late cold spell. Because such population-depressing factors are of a physical nature rather than biological, they are called **abiotic population controls.**

It should be kept in mind that population-depressing influences such as severe weather are usually **density independent** (their effects not influenced by population density—the number of individuals within an area). In a severe drought, the parching sun doesn't care how many corn plants are struggling in the field below. It kills them all. In an area saturated with DDT, most of the insects die—whether there are few or many. Only their individual resistance counts—their numbers mean nothing. Thus, mortality brought about by such means is independent of the density of the population.

Nevertheless, there are instances where the severity of the effect of an abiotic factor clearly varies with density—that is, they are **density dependent.** For instance, if a killing frost strikes and life or death for a population of insects depends upon the availability of sheltering nooks in their surroundings, then the percentage of the population lost will depend on the number of individuals in the area and the number of available nooks.

Biotic Control

Now let's look at a much more complex phenomenon regarding how living organisms may influence each other's numbers. **Biotic population controls** refer to any influences on a population brought about by a living agent. For example, the organism that causes bubonic plague can reduce populations. So can a tiger. We know how these work, but biotic influences can operate in more subtle ways as well. If a territorial bird drives a competitor into an area where there is less food, when winter comes, the underfed competitor may be more likely to succumb to the rigors of the season. Thus, the territory holder has indirectly brought about the reduction of the population.

Unlike abiotic controls, which are usually density independent, biotic controls on populations are likely to be density dependent. This means that as population density rises, there will be increasing pressures on it that tend to reduce that density. As the density falls, the pressures will lessen, thus permitting the numbers to increase again.

We have generally been discussing factors that bring about death, but "population regulation" actually refers to the effects of factors that keep population size *within certain limits*. It is a specific term and intuitively we can see that this sort of regulation is likely to be achieved primarily through density-dependent mechanisms. Density-dependent effects could theoretically keep population sizes within more or less defined ranges, because they generally involve negative feedback control, the familiar stabilizing effect seen so frequently in physiological systems.

So, most density-dependent mortality is under the influence of biotic factors—either directly or indirectly. The shape of the idealized population growth curve indicates that, as the population density increases, its growth declines, and eventually reaches zero. This is no abstraction; it really happens. It occurs because density-dependent effects increase mortality, or reduce reproduction, or both. Let's consider some important biological effects on population size, paying particular attention to when these effects are density dependent and when they are not.

We'll consider six categories of biotic population-controlling mechanisms, each of which has prominent density-dependent aspects:

1. competition within species;
2. competition among species;
3. emigration;
4. self-poisoning;
5. disease and parasitism;
6. predation.

Competition Within Species. Competition within a species or population—**intraspecific competition**—can take many forms and involve any of

42.9   THE RED-WING, A TERRITORIAL BIRD

Among some birds, such as red-winged blackbirds, males may take territories of varying qualities. Females can assess territories and tend to choose the males with the best real estate. In some cases, a female will choose a male who already has a mate over a bachelor with an inferior status.

the resources of life. By definition, competition usually results in some harm coming to the loser of a struggle for any commodity that is in short supply. That harm can result in reduced population numbers. As an example, populations are often limited by the amount of food available. Biologists have gone to great lengths to try to understand how this might occur. For example, certain (truly dedicated) researchers have learned that only so many maggots can survive in a cow dropping and the density of a population of flies may be limited to the density and size of cow piles in the area. Observers report that when female flies find a cow pile, whether they lay eggs or fly off to find a fresher pile, depends on how many maggots or eggs are already present, an adaptation that helps assure that their own offspring will not be crowded out. Usually, though, food limitations exert their influence much more bluntly. As competition for food intensifies, those who can't find enough weaken and die. Here, then, competition for resources is an important factor in regulating the size of the fly populations. Let's consider how some species organize around this principal.

Territoriality and Dominance Hierarchies. In many species, the level of purely competitive activity is limited through organization. This doesn't imply any kind of benevolent sharing of resources, but as we'll see, assures that some will always have enough. In territorial species, for instance, some animals are excluded from certain areas held by other animals. If the habitat is variable, some animals will end up with "better" areas than others. If those areas hold more food, then in times of food shortage the most successful territory holders are more likely to survive. In some species of birds, the male must have a good territory in order to attract females. A splendid male holding inferior territory will not find a mate. If there aren't enough good territories to go around, than not all individuals will be able to reproduce, and thus population growth is restrained (Figure 42.9).

Some species form hierarchies, or pecking orders. The ranking is usually first established early in life through interactions such as fighting or play. In such hierarchies certain animals have freer access to commodities than others. In such a system, each animal must be able to recognize others individually and to respond to them according to their rank. Hierarchies reduce conflict within groups as an animal acquiesces when confronted by a dominant individual. In times of shortage of critical commodities, then, the higher-ranking animals are more likely to survive since subordinates give way before them (Figure 42.10).

Competition Between Species and the Ecological Niche. We will preface our look at **interspecific** (between species) **competition** with a look at the ecological niche. This is necessary since we can only judge the degree of competition between species by considering how much of an overlap exists between one niche and another. It is in the overlap that competition really exists.

The term *niche* is notoriously difficult to define; in fact, cynics say there is no record of any two

42.11 THE CROW, A BROAD-NICHED SPECIES

Crows are notoriously broad-niched, feeding on a variety of food sources ranging from bird eggs to armadillos to garbage.

biologists ever completely agreeing on a definition. Some ecologists have attempted to clarify matters by distinguishing between the **fundamental niche** and the **realized niche.** This distinction carries with it the notion that an animal's potential niche is seldom fully utilized at any point in time or space. In other words, the organism *can* usually interact with the environment in more ways than it does. What it *does* do is fill its realized niche. For purposes of simplicity, here, then, we will concentrate on the realized niche. We will say that the niche is the role a species plays in a community of interacting species—how it affects them and how they affect it. It also includes its effect on its physical surroundings and their effect on it. It has been said that if the habitat is the organism's address, the niche is its profession.

Some niches can be very narrow, while others are quite extensive. For example, some crows are **broad-niched.** In other words, they interact with their environment in a number of ways. They walk along plowed fields eating newly planted corn, they fly into trees to sample berries, and they don't turn up their beaks at a dead squirrel along the roadside. They may eat the eggs of other birds and so they are attacked on sight by smaller, quicker, nesting birds. They retreat from these tiny tormenters, but they may turn around and mob an owl. Because they interact with their environment in a number of complex ways, they are called broad-niched (Figure 42.11). On the other hand,

42.10 SOCIAL ORGANIZATION IN THE BABOON TROOP

Complex social interaction characterizes the baboon troop, providing a measure of both safety and order. Individuals and subgroups are organized in a hierarchy with dominant members receiving first priority in feeding and reproduction.

most microbes are **narrow-niched.** For example, such reducers as decomposing bacteria operate within very rigid and narrow limits. That rotting squirrel was not decomposed by the action of a single kind of bacteria. Instead, a variety of microbes were at work, each utilizing the waste products and by-products of others and each incapable of doing anything else. The decomposition process, then, is complex, each step carried out by a microbial specialist. All specialists are regarded as having narrow niches. Let's turn now to how different species interact when their niches overlap.

Competition leading to out and out physical aggression between members of different species is relatively rare. Instead, usually one species is simply and quietly repressed. This repression, in fact, may be so severe that the losing population completely dies out. The classic experiment demonstrating this principle was conducted by the Russian biologist, A. F. Gause. According to what is now called **Gause's law,** or the **principle of competitive exclusion,** two species cannot indefinitely live together and interact with the environment in the same way. The results of the key experiment that led him to this conclusion are shown in Figure 42.12. As you see, the two *Paramecium* species do well alone, but when grown together one—*P. aure-*

lia—outdoes the other. We are not certain about the specific nature of the competitive edge *P. aurelia* has over *P. caudatum*, but it certainly exists. There is apparently a critical overlap in the niche of the two protists.

As another example, when two species of duckweed, *Lemna gibba* and *Lemna polyrhiza* were grown alone in a tank, each did well, but when they were grown in the same tank, *L. polyrhiza* died out. The reason is, *L. gibba* is a better competitor for light. It has air-filled sacs that cause it to float higher in the water, blocking light from its competitor. Had the experiment been carried out under a variety of conditions, the experimenters would most likely, sooner or later, have come across a situation where *L. polyrhiza* was the winner.

Different species are most likely to rely on the same kind of food when that food is plentiful. For example, in the tundra, two predatory birds, snowy owls and jaegers, or skuas as the latter are also known, both feed on lemmings, but only when the lemmings are in abundance. The owls and jaegers switch to alternate prey if lemmings are sparse or simply cut short their annual visit. A study of 11 species of Panamanian stream fishes revealed that each species had a very specialized diet during the dry season when food was scarce. But in the wet

42.12 COMPETITIVE EXCLUSION AT WORK

Part of Gause's evidence for competitive exclusion was the study of population growth in two species of *Paramecium* that utilize the same kinds of foods. When grown separately the numbers of *P. aurelia* and *P. caudatum* produced a typical

S-shaped growth curve, each reaching a stable population size. When the two were grown together, *P. caudatum* was displaced by the smaller, but apparently more efficient *P. aurelia.*

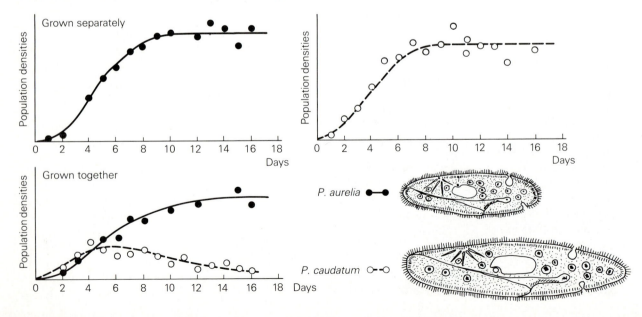

season, the same, abundant food sources were used by all the fishes.

From many such observations grows the hypothesis that competition in the wild rarely results in extermination, but rather it encourages a subdivision of the habitat, each species coming to live where it does best. This point is well illustrated in a classical observation by the noted ecologist R. H. MacArthur. As Figure 42.13 reveals, five species of warblers that utilize the resources and shelter of spruce trees do not actually occupy the same niche. Although some overlap is inevitable, their feeding and nesting habits differ enough so that they are not in serious competition.

Emigration. Naturally, populations can be reduced if some animals just ship out. Biologically, (and geographically) emigration refers to a one-way trip out of a population. In some species emigration is a common result of increasing population density and the accompanying intensification of competition. As such, it may represent an important safety valve in the regulation of density by some species.

The classic example of population reduction by emigration is seen in the Arctic lemming (see Figure 40.26). Populations in northern Europe periodically begin to behave erratically and to emigrate in great numbers. In their travels, they frequently swim across streams and ponds, and occasionally try the sea, apparently unaware of the difference. So the legendary (and somewhat mythical) "suicidal" drowning of lemmings may be no more than a fatal navigation error.

From Biblical times we have heard of plagues of locusts (Figure 42.14), which have swarmed across the land, devouring entire crops that lay in their paths. It seems that when the population density of locusts reaches a certain critical level, the locusts begin to produce offspring with a different developmental hormone condition. These hormones stimulate changes that result in longer wings, lighter bodies and darker colors, and it is this generation that emigrates in the dreaded swarms.

We don't know much about what causes emigrations, but we do know that the result is always the same: a reduction in the local population. Further, we can assume that the emigrators derive some benefit from it all. They may simply end up in new areas where competition is lessened because of a sparser population.

If animals that normally undergo cycles of population growth and dispersal are not allowed to disperse—for example, if they are kept in cages or confined to islands—the unnaturally high density of

42.13 SUBDIVIDING THE HABITAT

Five species of North American warblers use spruce trees for feeding and nesting, but each has its own zone. The darkened areas indicate where each species spends at least half of its feeding time. By exploiting different parts of the tree, the species avoid direct competition; thus they can occupy the same habitat.

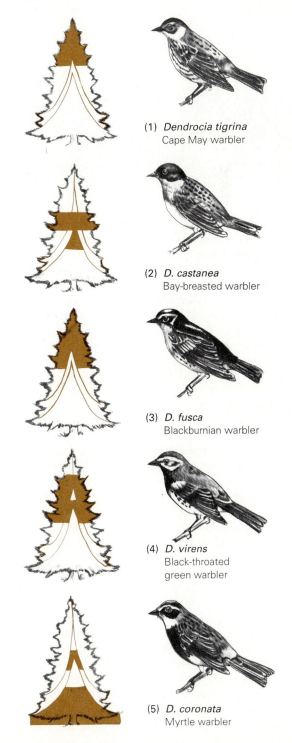

(1) *Dendrocia tigrina*
Cape May warbler

(2) *D. castanea*
Bay-breasted warbler

(3) *D. fusca*
Blackburnian warbler

(4) *D. virens*
Black-throated
green warbler

(5) *D. coronata*
Myrtle warbler

42.14 EMIGRATING LOCUSTS

Locusts have periodically ravaged African crops as the insects move in great swarms over the land. Virtually all the plant material in their path is devoured.

the population that results may cause social and behavioral breakdowns. Confined mice or rats in high population densities have shown abnormal interactions—such as fighting constantly or completely ignoring one another, gorging on food to the point of obesity or wasting away even in the presence of abundant food. In particular, normal sexual and maternal functions break down, and mothers frequently abandon their litters. Death rates become very high in the absence of any obvious factor other than emotional (and endocrine) upsets.

Self-Poisoning. Under high population densities, many species experience a literal "fouling of their nests." In the growth curve of bacteria in a pure culture or in a rotting squirrel corpse, it might seem that food is again the limiting factor for the microbes. However, even after growth has fallen to zero and the population has reached its plateau—chemical analyses often show that there were ample foodstuffs available. The increased mortality and decreased rates of cell division, it turns out, have more to do with the accumulation of toxic waste products than with lack of food. This is especially true in the pure culture, because in a rotting squirrel, where many species of decomposers are at work, one microbe's poison may be another microbe's food. Thus the numbers of one might swell as its predecessor died out. You may recall evidence of this in the succession of organisms in the nitrogen cycle (see Chapter 41).

Some poisons are more than by-products. They may be used as weapons against competitors. For instance, bacteria may produce toxins that kill fun-

gi that may otherwise compete for the same food, and fungi can produce substances, such as penicillin, that kill bacteria. However, these chemical weapons can backfire and eventually poison the very species that produced them. After all, microbes produce such poisonous metabolic breakdown products as ammonia, organic acids, and alcohol—none of which they can tolerate well themselves. (Daniel Janzen, one of our more imaginative ecologists, has suggested that the ability of microorganisms to render dead matter putrid and unpalatable is quite adaptive. It helps them avoid being incidentally eaten by scavenging vertebrates.)

Microbes are not the only organisms capable of controlling other species through toxic secretions. Many plant species secrete poisonous substances into the soil that inhibit other plants. It is not without cost, however, since it may also inhibit the growth of the plant that manufactured it (Figure 42.15).

Humans have, in recent years, managed to escape most other biological checks on our population growth. Now we, like the yeast and bacteria, are faced with the possibility of polluting ourselves into extinction. Never mind that the poisons are made by mines, factories, and automobiles, and not directly by our simple primate bodies; the principle

42.15 CHEMICAL INHIBITION AMONG PLANT SPECIES

The sandy, grass-free zone bordering the cluster of purple sage is a product of volatile, growth-inhibiting chemicals released by the sage.

42.16 THE THREAT OF DIOXIN

Residents of Times Beach, Missouri, were recently forced to vacate their community when it was learned that the area was contaminated by the highly toxic industrial waste, dioxin. Such events, the result of negligent toxic waste disposal, are becoming all too commonplace.

is the same (Figure 42.16). It's difficult to say whether the effects on humans will be (or are) density dependent or density independent. If, in the future, finding an appropriate shelter with filtered air or bottled oxygen spells the difference between life and death during "smog alerts," then the number of such spaces and the population density could indeed become significant factors. If we are caught in a wave of some manufactured poison, then its effects will likely be density independent. In 1985, thousands of people were killed and maimed by the accidental release of poisonous gases from a Union Carbide plant in Bhopal, India. The population density of the adjacent slum had no influence on the gas's effect.

The awesome impact of the human species on the life-support system may soon bring the reality of human self-poisoning into sharp focus. Recently, we have witnessed the strange phenomenon of **acid rain.** Because of high levels of oxides of sulfur and nitrogen in the air, the rain in many parts of North America and Europe contains startling concentrations of sulfuric and nitric acids. Sweden and Canada are deluged with such rains, although ironically, both are leaders in environmental pollution control. The Swedes can blame the industrial centers in England and the Ruhr Valley for their corrosive rains, while the Canadians know the source of their trouble to be their southern neighbor. Interestingly, the heavy industries often comply with local pollution restrictions by building enormous smokestacks, some a quarter of a mile high, from which the toxic fumes can be carried "safely" away from the local inhabitants. We cannot yet predict what

the final results of acid rain will be, but an indicator is the visible damage seen in buildings and monuments. Of far-reaching ecological significance, acid rains leach nutrients from the soil, carrying them into water systems where they wreak havoc on aquatic life. And even now, the crystal waters of those lovely northern glacial lakes—remote, pristine bodies that have so far escaped the heavy human touch—are beginning to quietly submit to irreversible chemical change.

Disease and Parasitism. Both disease and parasitism can certainly affect the sizes of populations, often through density-dependent mechanisms. Some of the mechanisms are obvious. For example, the more closely that susceptible individuals are packed together, the more opportunity there is for disease transmission. Also, the more individuals there are in a population, the greater will be the number of potential reservoirs in which more virulent mutant strains of the disease microorganisms can develop.

In some cases, disease may interact with predation to control populations. For example, a two-week-old caribou fawn can already outsprint a full-grown timber wolf, and healthy caribou seldom fall prey to wolves. However, caribou are subject to a hoof disease that lames them before it affects other parts of the body, and it is these lamed animals that a wolf is likely to cull out of a herd. In areas where wolves were poisoned in order to protect the migrating caribou, this hoof disease spread unchecked, and in a few seasons decimated entire caribou herds.

The relationship between a parasite and its host is often an interesting one. In most cases, the parasite is **host-specific;** it can infect only a single species, and thus its welfare is intimately tied to the welfare of its host. If the host species goes into a population decline, so does the parasite.

Thus, a parasite that can permanently infect a long-lived, relatively healthy host has a selective advantage over one that destroys its host. Still, there are parasites that regularly kill their hosts, and others that weaken their hosts so that they are nearly certain to die. Some parasites render their hosts permanently sterile, which of course will have an effect on population dynamics too. These debilitating parasites are believed to be rather newly associated with their hosts, not yet having adapted in such a way as to minimize their effects.

A special category of parasitism includes the parasitoids. A **parasitoid infection** is like a parasitic one except that it always kills its host. In many respects, a parasitoid-host relationship is more like a predator-prey relationship with the important distinction that the parasitoid is host-specific. The most common and best-known parasitoids are the tiny wasps that parasitize the larvae and pupae of other insects. The parasitoid wasp locates its victim—perhaps a caterpillar—and uses its long ovipositor to inject an egg or several eggs into the desperately squirming host. As the host develops, so does the parasitoid larvae within its tissues. The sickened host eventually dies at about the time that adult parasitoid wasps eat their way out of its body and fly off. Parasitoid wasps are very effective in population control, and have been used commercially in the biological control of insect pests of agricultural crops.

Predation.   One of the interesting aspects of predation as a population regulator is that a predator species normally doesn't eliminate its prey. For example, when the vigorous and intelligent dingo, a predatory wild dog, was introduced to Australia by aborigines some 10,000 years ago, it didn't kill off the new prey it found there. However, its hunting prowess proved so superior to that of its competitors, the Tasmanian devil and the Tasmanian wolf, that those animals disappeared from the Australian continent. But why didn't it wipe out its prey?

Let's examine the question by reviewing the classical theory that accounts for the density-dependent effect on predators. According to the **Lotka-Volterra theory** (simplified here), under undisturbed conditions, prey numbers rise steadily, thus providing more and more food for predators. Then the predator numbers begin to rise.

Their numbers do not rise immediately, however, since it takes time for the energy from food to be converted into successful reproductive efforts. Because of this time lag, the prey may be well on the road to recovery before the predator population begins to rise. When the predator numbers finally do rise, though, there is increasing pressure on the prey. Then, as the prey begin to be killed off, the predators find themselves with less food, and so their own numbers soon fall off due to starvation or simply a failure to reproduce. Then the prey begin to recover.

As an example of predator-prey population dynamics, consider the curves produced by the predation of one kind of mite on another in Figure 42.17. We should add, however, that predator-prey relationships are not often as straightforward as this. For example, if the predator is able to hunt more than one prey type, the story can be complicated by **prey-switching** in which the predators seek other, more available, food.

Our own species is providing us with clear examples of how density-dependent regulation of predation can fail. The great whales have been hunted to the brink of oblivion over the past few decades as modern whaling methods have increased efficiency and reduced personal risk. Although there is nothing that whales provide that can't be obtained elsewhere, the demand for whale products continues at a marginal but still profitable level.

As a result of such increased predatory efficiency, the whales become scarcer. Then, instead of relaxing their pressure and allowing the whale population to recover, whaling fleets intensify their depressing effect on populations of the great mam-

42.17 PREDATOR—PREY INTERACTIONS

The oscillation of population numbers of the predatory mite, (*Typhlodromus occidentalis*), and its prey, the six–spotted mite (*Eotetranychus sexmaculatus*), is almost classic in its characteristics. Prey number increase first, followed at once by predator numbers. Then as the predator increase continues, the prey species diminishes in what is reminiscent of a population crash. The predator's fate is quite similar as its numbers fall rapidly.

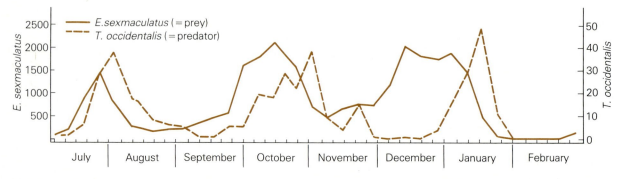

42.18 SOME EFFECTS OF HUMAN PREDATION

The history of whale hunting dramatically illustrates the effects of human predation. As modern hunting methods were developed, predatory efficiency increased and the whale population declined. But since there was no corresponding decrease in the human food supply, whaling fleets did not, until recently, restrict their hunting activities. It remains to be seen whether the populations of various whale species will fully recover.

mals (Figure 42.18), relenting only when the profits wane. As the populations of one whale species become critically depressed, the hunters switch to other, less desirable but more available species, and the hunt goes on. It seems that humans may be exempt from some of the usual predator-prey restraints. If humans actually starved when they couldn't catch whales (which might have been the case among the Eskimos) both populations might eventually stabilize (or cycle). But the decline in whale numbers has not appreciably lessened the efforts to hunt them.

THE HUMAN POPULATION

Now we turn our attention entirely to humans and to a fascinating set of numbers: "4.7, 29, 11, 1.8, 39, and 34." These, we see in Table 42.4, represent the (1983) total world population in billions, the **crude birth rate, crude death rate, percent annual growth,**

TABLE 42.4

WORLD POPULATION DATA: 1970 AND 1980

Region	Year	Total (millions)	Crude Birth Rate	Crude Death Rate	Annual % Inc.	Doubling Time (years)	% Below 15 yrs. of Age
World	1970	3632	34	14	2.0	35	37
	1983	4677	29	11	1.8	39	34
Africa	1970	344	47	20	2.6	27	44
	1983	513	46	16	3.0	23	45
Asia	1970	2045	38	15	2.3	31	40
	1983	2730	30	11	1.9	36	36
North America	1970	228	18	9	1.1	63	20
	1983	259	16	8	0.7	94	23
Latin America	1970	283	38	9	2.9	24	42
	1983	390	31	8	2.3	30	39
Europe	1970	462	18	10	0.8	88	25
	1983	489	14	10	0.4	199	22
Nations of Special Interest							
United States	1970	205	17.5	9.6	1.0	70	30
	1983	234	16.0	9.0	0.7	95	23
Soviet Union	1970	243	17.9	7.7	1.0	70	28
	1983	272	19.0	10.0	0.8	83	25
People's Rep. of China (estimate)	1970	760	34.0	15.0	1.8	39	?
	1983	1023	23.0	8.0	1.5	46	32
India	1970	554	42.0	17.0	2.6	27	41
	1983	730	36.0	15.0	2.1	33	39

SOURCES: Population Reference Bureau and the Environmental Defense Fund.

The Wolf as Predator: A Close-Up View

One day I watched a long line of wolves heading along the frozen shore line of Isle Royale in Lake Superior. Suddenly they stopped and faced upwind toward a large moose. After a few seconds the wolves assembled closely, wagged their tails, and touched noses. Then they started upwind single file toward the moose.

L. David Mech
The Wolf (1970)

In recent years, there has been increasing attention to large predators, their life histories, how they kill, and the effects of their predation. For a number of reasons—some scientific, some emotional—North Americans have focused on one of their own, the fabled and mysterious wolf. The hard information, however, was not easy to come by, and so a long-term study by L. David Mech was met with enthusiasm. The story he told was based on many years of watching wolves in the northern United States, especially on Isle Royale in Lake Superior and it answered a number of questions. Is the wolf an effective killer? Yes. Does it ever kill more than it can eat? Sometimes. Does it attack humans? No. Is it always successful in its hunts? No. (In fact, fewer than one moose hunt in ten yielded prey.)

One of the things we know from Mech's work and the studies that followed it is that wolves are continually on the hunt. They may attack a prey animal only hours or minutes after a successful kill, and any time they are on the move, they are hunting. When they are successful, they can gorge. Their stomachs can hold up to 20 pounds of food which is apparently rapidly (or incompletely) digested.

Wolves probably locate prey most often by scent (although they may stumble across prey). Mech found 42 cases in 51 hunts in which he could tell that the wolves were trailing their prey (moose) through the snow by direct scenting. In fact, they once detected a cow and her twins 1.5 miles away. He found that when the lead animals catch wind of the prey, they stop and all pack members stand alert with eyes, ears, and nose focused toward the prey. Then they carry out a peculiar ceremony, standing nose-to-nose and wagging their tails. If they are in deep snow, they just pile up behind the leader, who then sets out straight for the prey. As they near quarry, they quicken the pace and seem anxious to lunge ahead. But they hold themselves in check, alert, and tails wagging. The restrained approach permits the wolves to draw near their prey without sending it into flight.

As soon as wolves realize their quarry has spotted them but is not running, they stop stalking. If they proceed, they do so cautiously. There may be an advantage in this wariness. First of all, a moose is a strong and dangerous animal, easily capable of killing a wolf. If the animal doesn't run, then, it may be out of a spirit of confidence in its own prowess. Perhaps it has successfully dominated other moose—or other wolves—and hasn't developed the habit of retreat. Also, if it doesn't run, it isn't wasting its energy and can therefore probably put up a better fight.

Smaller or weaker animals such as deer don't have this option; they could certainly not fend off a wolf pack and so they tend to flee. Interestingly, a wolf may need the sight of a running animal in order to stimulate its closing rush. Mech believes that the sight of a standing animal inhibits the rush response. The wolves may still approach, but they do it more slowly.

If the prey bolts and runs, the wolves give chase. If the quarry is fast enough in its initial run, it usually gets away. Wolves are very perceptive about such matters. If the chase looks hopeless, they stop. If they do give chase, they may run for miles, but this is unusual; they normally chase only a very short distance. Interestingly enough, the prey (whether moose, deer, caribou, or Dall sheep) is also perceptive. It runs no farther than necessary. When the wolves give up, the prey will stop and turn around to watch the wolves. It doesn't waste energy running needlessly.

It is known that wolves attack different animals in different ways. For example, moose are usually bitten on the rump area, with one wolf sometimes clinging to the large, rubbery nose. If the kill is not at first successful, the wounded moose may stiffen and weaken so that it can be brought down some days later. On the other hand, caribou are usually attacked at the shoulder and neck area. When larger prey is not available, wolves will pounce on field mice, landing with all four feet on the hapless rodents.

It is important to realize that wolves are opportunists. They will eat not only large game and mice, but insects, fish, rabbits, birds, or just about anything they can catch. They will also eat certain kinds of berries.

Wolves have changed their diets in some cases when they have come into contact with humans. They raid garbage dumps, for example. More important, where humans have replaced wild game with domestic animals, the wolf's diet has changed in accordance. To the farmer's dismay, whereas wolves almost always dispose of all the remains of wild game, they sometimes eat only parts of domestic animals such as cattle and sheep but this may be because human activity drives them from such kills.

A lingering question has involved the "sanitizing" effects of wolves on their prey populations. Do they take the weak and infirm and elevate the quality of the prey populations in general? Since they make so many attempts and their success rate is so low, it seems that primarily disadvantaged prey are taken. Such prey could be newborn, inexperienced, malformed, sick, old, wounded, parasitized, starving, crippled, or just plain stupid. Several studies have shown that wolves kill primarily animals less than 1 year old, or those that have lived at least half the usual life span for that species in the wild. In a sample of 93 deer killed by wolves in Minnesota, 59% of the adults were relatively old—at least 4.5 years old. (In contrast, of most deer killed by hunters, only 20% were over 4.5 years old.)

Do wolves control the population of their prey? For a number of reasons it is difficult to know. For example, in many places where such studies have been made, there have been unusually high numbers of ungulates. Possibly, at one time, when ungulate numbers were lower and the number of wolves was higher, wolf activity may have been a major factor in the control of ungulates. But because of the artificial controls on wolves and their prey, it is difficult to know just what sort of population control was imposed by the predators under natural conditions.

There are a few documented cases in which wolves have strongly influenced prey populations. For example, on Isle Royale there were about 30 moose (averaging 800 pounds each) per wolf (from 1959 through 1966). So there were about 24,000 pounds of moose per wolf. The 210 square mile island supports about 600 moose in late winter, and about 225 calves are born each spring, but only about 85 survive to be recruited into the herd. There are about 23 wolves on the island and these kill about 140 calves and 83 adults each year for a total of 223 animals. Thus, the annual production of the moose herd is balanced by the average number of wolf kills. The implication is that the wolves are taking enough moose to control the herd.

As further evidence of the regulatory impact of the wolves, the moose have lived on the island since about the first of the century, but the wolves didn't arrive until 1949. In the early 1930s, A. Murie estimated the moose herd at 1000 to 3000 animals. They apparently overbrowsed their area and their numbers fell drastically through the effects of starvation and disease only a few years later. The herd recovered, but began to starve again in the late 1940s. However, for decades since the wolves have arrived, moose numbers were lower and more stable than ever before, and the vegetation recovered. In recent years, the Isle Royale moose population has again been rapidly increasing, this time despite the presence of wolves.

The caribou has been singled out, separated from the herd, and systematically attacked by wolves. Each type of prey is handled differently by the wolf pack. In this case, the prey is brought down by frontal attacks on the shoulder and neck.

population doubling time in years, and the percentage of humans below the age of 15. (Table 42.5 explains how such numbers are derived.) These are the haunting numbers that tell the status of our own population, and they are indeed unsettling.

The need to know about the history and status of human numbers becomes apparent when we realize that the population of the world will have doubled between the time that most of today's college students were born, and the year 2000 A.D. Such gloomy statistics are abundant, and everyone who is at all interested has probably heard enough of them by now. But, just in case, here is one more: a net gain of 27 new humans joined the world in the 10 seconds you have spent so far in reading this sentence. By tomorrow at this time, 231,000 more will have been added, and in the next year, some 84,000,000—about the population of Mexico—will appear, and most of the new hopefuls can realistically expect to achieve a lifestyle similar to the average citizen of that struggling nation (Figure 42.19).

What is behind such unprecedented growth? What problems are we bringing upon ourselves? To gain some insight into the situation let's briefly consider the history of our numbers.

A Brief Look at Human Population History

By all accounts, throughout most of human existence, our numbers have remained rather stable. However, in the past 50,000 to 100,000 years, there were three significant growth surges (Figure 42.20). But before that, say about 1 million years ago, there were only about 125,000 wandering hominids on the grassy East African plains, and they were not having a great deal of impact on their environment. These early hunters and gatherers were strongly subjected to the same biotic and abiotic factors that influence most populations of herbivores and carnivores today. Infants and children probably suffered high mortality rates, but were quickly replaced in a species where, we believe, fertility was not a seasonal event. The average life span is estimated to have been 30 years, but the high infant mortality hides the possibility that some individuals lived much longer (see Table 42.3). Then, however, something happened and the population underwent a sudden, rapid increase.

The first growth surge in the human population was probably due to the development of increasingly efficient tools, leading to more efficient ways of killing game and, in general, easier utilization of whatever was available in the surroundings. In addition, humans—inveterate wanderers—had by then penetrated and established themselves on all of the continents. Thus about 10,000 years ago, the earth probably supported only about 5 million hunting and gathering humans who still had relatively little impact on their surrounding environment.

Then the human population began its second growth surge, one that is attributed to the rise of agriculture. The increased quantity and dependability of the food supply assured a lower death rate, and better nutrition may have increased the birth rate. By the time of Christ, the human population had risen from 5 million to about 133 million, and by 1650 A.D., there were an estimated 500 million people. Of course there were setbacks due to crop failures, disease and war. In the 14th century, for example, one fourth of the European population was killed by the "black death"—the bubonic plague. Other diseases, such as typhus, influenza, and syphilis also took their toll in the crowded and incredibly filthy hamlets and towns of the medieval period. Thus, in spite of a more assured food supply, density dependent factors still maintained a hold over population growth.

The third population growth surge began in Europe around 1650, as swelling populations there found some relief in the opening of the New World. But more significantly, by the mid 1800s, the con-

---

TABLE 42.5

BASIC POPULATION ARITHMETIC

Crude birth rate = number of births per year per 1000 population

$$\left(\text{determined by: } \frac{\text{total births}}{\text{midyear population}} \times 1000\right)$$

Crude death rate = number of deaths per year per 1000 population

$$\left(\text{determined by: } \frac{\text{total deaths}}{\text{midyear population}} \times 1000\right)$$

Rate of natural increase (or decrease) = crude BR − crude DR

$$\text{Percent annual growth} = \frac{\text{rate of natural increase}}{10}$$

$$\text{Doubling time (approximately)} = \frac{70}{\text{percent annual growth}}$$

(example: % annual growth in the world in 1983 was 1.8: $\frac{70}{1.8} = 38.89$ years

Chapter 42 Interaction in Populations **1115**

42.19 EXPONENTIAL GROWTH, THE HUMAN AFTERMATH

A scene of squalor in Mexico City, the world's fastest growing metropolis. As the population approaches 20 million, drastic unemployment and the continued shortage of dwellings, adequate water, and sewage facilities all point to impending disaster unless heroic efforts to find a solution begin soon.

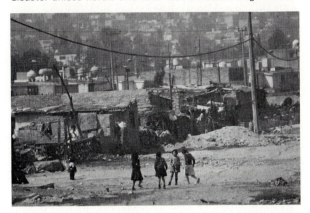

quest of disease had begun. Soon there were vast improvements in public sanitation, vaccines were developed, and rapid advances in technology led to a marked increase in food supplies. Between 1750 and 1850, the European population doubled, while that of the New World increased fivefold. In China, the most heavily populated nation at that time, agriculture had made great gains and a long period of comparative political stability followed the overthrow of the Ming dynasty in 1644. In spite of periodic famines in India (in 1770, the greatest of all famines reportedly killed half the population of Bengal), its population burgeoned. African populations remained stable until about 1850 after which the impact of European medical advances depressed the death rate there.

The third rise continues into modern times. It is marked by an increasing acceleration which is largely due to a decreasing death rate sustained by innovations in industry, agriculture and public health. Many of these innovations are benevolently exported from the developed regions of the world to the heavily populated developing regions where they contribute to unusually large population increases.*

---

*Developed nations are those that have a slow rate of population growth, stable industrialized economy, low percentage of agricultural labor, high per capita income, and a high degree of literacy. Developing nations are marked by the opposite characteristics.

In summary, the human population grew from about 5 million at the dawn of agriculture to 500 million by 1650. It doubled to its first billion in the next 200 years, and doubled again in 80 years and next in 50 years reaching two billion by 1930. In 1975 there were four billion of us. The next doubling—a growth to 8 billion—was predicted for 2005, only 35 years later. Let's see what has happened in the years since those dire predictions were made.

42.20 TWO WAYS OF VIEWING HUMAN POPULATION HISTORY

(a) The arithmetic plot of the past 10,000 years clearly reveals the soaring near-vertical rise of the J-shaped curve, a danger signal to most species. Hidden in the gradually rising line preceding this is the earlier population surge brought on by the switch from hunting and gathering to agriculture. The logarithmic plot (b), where time is seemingly compressed, covers human history over the past million years. Here, three growth surges are seen, the products, respectively, of advances in hunting and gathering technology, the rise of agriculture, and the recent era of modern health, industrial and agricultural practices.

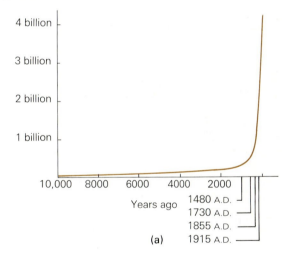

(a)

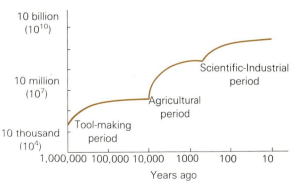

(b)

The Human Population Today

Since the late 1970s, the rate of natural increase in the world population has been slowing (see Table 42.4). The percent annual growth rate has fallen slightly and the doubling time has increased from 35 to 39 years. These data suggest some headway in the attempt to control population. Part of this headway is attributable to changing attitudes about the role of women in society, preferences in family size, advances in methods of birth control and the liberalization of abortion laws in the more developed nations.

One may feel relief and optimism because of such numbers, but a close look may dampen that view. Except for China, the depressed growth rates occurred primarily in the most developed nations, those that could best support increased populations: the United States, Japan, western Europe, and the Soviet Union. Most developing regions of the world show few signs of population growth abatement. For example, the annual rate of increase in those regions is an ominous 2.4% and the doubling time is 30 years. The doubling time is simply the time it takes for a growing population to double in size. In itself, the calculation is not significant, but it can trigger an intuitive understanding of the enormity of the problem of population growth when one considers that a doubling of food production, schools, roads, houses, manufacturing, and so on is required just in order to march in place. (To determine how doubling time is calculated see Table 42.5.)

While the crude birth and death rates in the United States in 1983 were 16 and 9, those numbers in Africa were 46 and 16. Further, the population doubling time both in Africa and politically and economically troubled Latin America is an alarmingly brief 23 and 30 years, respectively.

What of Asia? It is instructive to keep in mind that well over half of the people walking the earth today are Asian, and nearly half of these are Chinese. Because of successful, if harsh, legislation on birth control in the Peoples Republic of China, the doubling time for Asian populations increased from 31 years in the 1970s to 36 years in 1983. Similar efforts in India have been only partially successful, and India's annual increase is still over 2%; the doubling time there is only 33 years. (Doubling times in selected nations are shown in Table 42.6.) But, in spite of success in parts of Asia, keep in mind the problem of momentum. *Forty-one* percent of the Southeast Asians are under the age of 15 and have yet to make their reproductive impact felt. (In Africa this percentage is 45.)

TABLE 42.6

DOUBLING TIME IN SELECTED NATIONS

% Annual Growth	Doubling Time (yrs)	Nation		
		(Africa)	(Eurasia)	(Latin America)
4	23 or less	Algeria; Egypt; Ghana; Kenya; Libya; Nigeria	Iran; Iraq; Syria	Ecuador; Guatemala; Honduras; Nicaragua
3	23-35	Chad; Congo; Ethiopia; Niger; Somalia; S. Africa	Afghanistan; Burma; India; Lebanon; N. Korea; Pakistan; Philippines	Bolivia; Brazil; Mexico; Panama; Paraguay; Peru
2	35-69		China; Iceland; Indonesia; Israel; S. Korea; Taiwan; Thailand	Argentina; Chile
			(All nations)	
1	70 +	Australia; Canada; Japan; New Zealand; U.S.A.; U.S.S.R.		Most of Europe (Austria: 3465 years) (Luxembourg: 770 years) (Sweden: 3465 years)

42.21 A RECENT FERTILITY STUDY IN THE U.S.

The two plots include the general fertility rate (top) and total fertility rate (bottom). The Depression years are seen as a valley, followed by a reproductive peak—the "baby boom" years—that followed World War II. The recent valley, following the boom years, is good news to those alarmed by population growth, but keep in mind that attitudes toward family size change and women born in the baby boom years are still in the reproductive period of life.

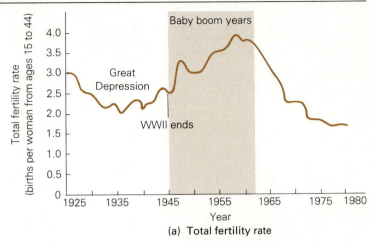

(a) Total fertility rate

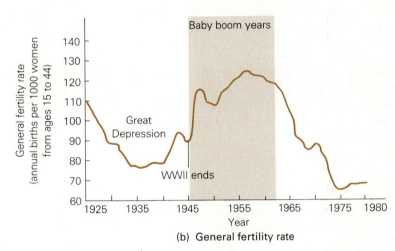

(b) General fertility rate

The Future of the Human Population

One of our best indicators for future growth is the **fertility rate.** This rate actually includes two statistics: the **general fertility rate** and the **total fertility rate** (Figure 42.21). The first is the number of live births per 1000 women of reproductive age (15 to 44 yrs in the United States). The general fertility rate gives us a clear idea of what is happening in terms of family size, but like the crude birth and death rates, it comes after the fact. The total fertility rate is a *prediction* of the average number of children women will have over their reproductive lifetime. It is based on information from many sources, including economists, governmental agencies, churches,

and family planning groups, but of particular importance is the *attitude* of couples in their reproductive period. For example, the attitude of Americans toward domestic life and family size changed radically in the 1960s and 1970s, as did that of people in many other developed nations. Women entered the labor force, relying heavily on newly developed birth control agents to assure a smaller family size.

Population Structure.   We can tell a great deal about future trends from graphs known as **age structure pyramids.** Here, the population is divided into groups by age and sex and these data are plotted as seen in Figure 42.22. Note the marked differences in the shapes of such pyramids in developed and developing nations. In developed

42.22 AGE STRUCTURE PYRAMIDS

Age structure pyramids for Mexico, the United States, and Sweden are quite different. Future growth rates can be predicted by a study of the lower levels of the three pyramids, since these include females who have yet to reproduce. The middle region of a pyramid contains the reproducers who are also the working labor force. These individuals have the burden of supporting both the lower and upper levels.

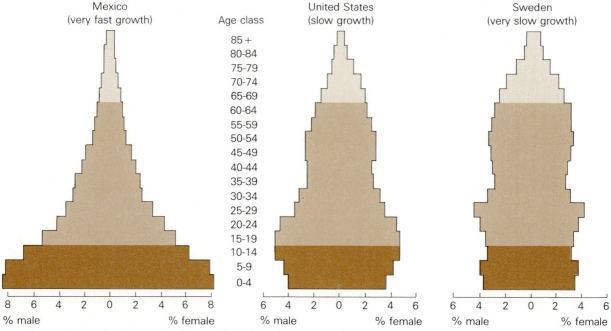

Mexico (very fast growth) | United States (slow growth) | Sweden (very slow growth)

Age class: 85+, 80-84, 75-79, 70-74, 65-69, 60-64, 55-59, 50-54, 45-49, 40-44, 35-39, 30-34, 25-29, 20-24, 15-19, 10-14, 5-9, 0-4

% male   % female

From *Living in the Environment*, Third Edition by G. Tyler Miller, Jr. © 1982 by Wadsworth, Inc. © 1979, 1975 by Wadsworth Publishing Company, Inc. Used by permission of the publisher.

countries, population increases have been comparatively slow, so the pyramid's base is not very wide. Also, people tend to live longer, and thus to occupy the upper levels in greater numbers, so the slope of the sides is quite steep. In developing nations, the rate of increase is great, so the pyramid is broad-based and, because of a higher death rate and a shorter average life span, fewer people occupy the apex. The base, of course, represents future reproducers. A broad base can be bad news since it signals an impending rapid population increase. But for the developed nations, a top-heavy pyramid can also spell economic difficulties since it means a larger group of dependent old people.

Demographic Transition. Studies of human population growth trends reveal that as nations undergo economic and technological development, their population growth decreases. Interpretations of this observation have led to the theory of *demographic transition*. Undeveloped nations experience a high birth and death rate and slow growth.

Then, as development begins, the birth rate remains high, but the death rate falls, thus the population begins to grow rapidly. Finally, with greater industrial development, the birth rate falls, nearing the death rate and the population reaches the equilibrium state we see in some developed nations today (Figure 42.23).

The major prediction to be made from the theory of demographic transition is clear. Population growth in developing parts of Asia, Africa, and Latin America will slow as these areas become further industrialized. Can the answer to the enigma of world population growth be that simple? Not quite. First, it is the developed nations that must pay the enormous costs of any such massive developmental program, and this would be at great risk to their own economic stability. Second, if massive development programs should stall in the second and most difficult phase, soaring populations would quickly wipe out any potential advances. Thus, any such developmental program must be approached with extreme caution.

Growth Predictions and the Earth's Carrying Capacity

The fundamental question of how great the human population *can* become is irrevocably tied to what the earth can support: its human carrying capacity. If we know anything from population studies of other species, it is that this capacity cannot be exceeded for long without environmental damage. Such damage acts to lower the carrying capacity for that species, escalating the inevitable population crash.

The range of estimates on the earth's carrying capacity is very wide. In other words, the experts cannot agree. Conservative estimates, based on many factors, suggest that the human population could be sustained very temporarily at 8 to 15 billion, but that programmed decreases must then occur until a lower stable number is reached. In other words 8 to 15 billion is actually an "overshoot," and should the programmed decrease fail to occur, a crash may well follow. Such a crash simply means a dramatic increase in the death rate. It has been calculated that such a crash might kill 50% to 80% of our numbers. Population biologist Paul Erlich, writing in 1977, emphasized that this is probably the way our population *will* stabilize. The die-back, says Ehrlich, will likely be due to a combination of famine, war, disease, and ecological disruption. With

the exception of war—which is rare among other species—we should note that these are common biotic, density-dependent regulators.

To leave the subject on a note of cautious optimism, we should reiterate that world population growth is on a slowing trend, and that some of the more heavily populated nations have joined this trend. Further, it is within the power of the world's family of nations to see that the decreasing trend is continued and further accentuated. The goal, if we are to avoid chaos, is **ZPG—zero population growth,** where the crude birth rate equals the crude death rate. (Contrary to popular opinion, the United States has not yet achieved this goal—while our birth rate has declined, our death rate has fallen even faster. The birth rate and death rate are inexorable quantities that must reach equality for population growth to cease. ZPG means b = d.) Highly optimistic souls anticipate ZPG at about 2020 A.D. If this happens, the world's population should have reached about 8 billion.

We must assume that, in the long run, our species is not exempt from the principles that govern the populations of other species. We are now aware of some of those principles, and we must carefully and wisely apply our knowledge so that we do not overburden our resources and ultimately call down some great disaster on perhaps the most remarkable species this good planet has yet produced.

42.23 DEMOGRAPHIC TRANSITION

According to the theory of demographic transition, nations or regions go through three population phases. Many developing third world nations are in the second or transitional stage, where falling death rates bring about rapid population increases.

From *Living in the Environment*, Third Edition by G. Tyler Miller, Jr. © 1982 by Wadsworth, Inc. © 1979, 1975 by Wadsworth Publishing Company, Inc. Used by permission of the publisher.

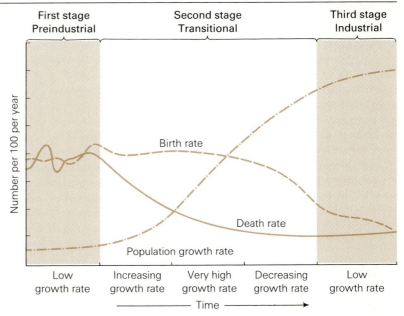

APPLICATION OF IDEAS

1. Consider what we know about population control in other species, and apply these factors to the future of the human species if voluntary population control measures fail. Considering such control, in what forms might we expect population reduction?

2. Why are the three historical human population growth surges usually understood in terms of decreased death rates?

KEY WORDS AND IDEAS

POPULATION CHANGES

1. In spite of noticeable short-term changes in population size, in the long run, the numbers of most species tend to remain about the same. Central questions concern size, growth fluctuation, stabilizing influences, and distribution.

Population Dynamics and Growth Patterns

1. Population changes are represented in the equation $I = (b - d)N$, where I, the rate of increase in individuals, equals average births minus average deaths times population size. The equation describes unlimited growth.

2. Births minus deaths $(b - d)$ is the **intrinsic rate of increase,** r, or the **instantaneous rate of increase;** thus $r = b - d$. Therefore $I = rN$. Because N grows with each generation, I, the rate also increases.

3. Population growth can be characterized with J-**shaped curves** and **S-shaped curves.** The first is a maximal rate when growth approaches the **biotic potential** or r_{max}. The curve rises slowly at first but increases until it approaches the vertical, due to an **exponential increase** (constantly increasing rate) as opposed to a linear, **arithmetic increase.**

4. Growing populations eventually encounter **environmental resistance,** the total factors that tend to slow such growth, sometimes then undergoing a **population crash.**

5. Crashes generally occur when the environmental **carrying capacity** is drastically exceeded. **Biotic** (living) and **abiotic** (nonliving) factors both determine the carrying capacity—the number that the environment can support. Population booms and crashes are part of the natural cycle of some species.

6. Populations undergoing an S-shaped growth may first follow what seems to be a J-shaped growth that then level off to fluctuate near the carrying capacity.

7. The carrying capacity is growth limiting and is represented by K in the expression $K - N/K$. Inserted into the exponential equation it appears as

$$I = r \left(\frac{K - N}{K} \right) N$$

The growth-limiting effects of K increase as the population nears the established carrying capacity, which acts through negative feedback in modulating growth.

8. Age structure and longevity also are important in determining the characteristics of a population. **Survivorship curves** reveal how aging affects populations. Some species experience a drastic loss of young, with most survivors living out their normal span, while in others the losses occur more equally along the normal span. Still others have light early losses, with few adults reaching the potential life span.

Reproductive Strategies

1. In general, organisms tend to leave as many offspring as possible, but adaptive strategies differ.
   a. **K-selected species** tend to describe the S-shaped curve, are highly responsive to carrying capacity, and tend to be large with long life spans and a long reproductive life.
   b. Growth of **r-selected species** ("boom or bust") follows the J-shaped curve. Typically, growth exceeds the carrying capacity, individuals are small with short life spans, and most reproduce but once.
   c. Birds and mammals are K-selected, producing few offspring but investing much of their energy into assuring survival.
   d. Examples of r-selected species include the tapeworm, a species that overcomes a high mortality rate in its offspring by laying great numbers of eggs.

2. The reproductive capacity of humans appears to be a response to a historically high rate of infant mortality. Today, with more assured survival such a capacity may no longer be adaptive.

Death as a Growth-Controlling Mechanism
1. For many species, death closely follows reproduction. Examples include annual plants, certain fishes, and squid. The adaptiveness of such a strategy involves the reduction of competition with one's variable offspring in a system of limited commodities.
2. With humans the emphasis is on a long period of growth, development, reproduction and rearing the young, followed by marked deterioration.

POPULATION-REGULATING MECHANISMS
Abiotic Control
1. Abiotic controls include nonliving environmental factors such as temperature, rainfall, and drought. Most abiotic control is **density independent**—its effect has nothing to do with population size and distribution. **Density-dependent** control is modified by the population size and distribution.

Biotic Control
1. **Biotic controls** involve other organisms. They are likely to be density dependent. They are much more likely to have a stabilizing effect than density-independent factors because they often operate through negative feedback.
2. Competition arises when individuals use common resources and one's gain is another's loss. **Intraspecific competition** is within species and is quite common.
3. Intraspecific competition may be limited through social organization such as territoriality. **Territorial species** defend certain areas within their range. **Hierarchical organization** (producing pecking orders) results in a system of rank in which some individuals have greater access to commodities than do others.
4. **Interspecific competition** (between or among species) is limited because of the different ways species utilize the environment or establish **ecological niches.** The **fundamental niche** is the potential use of the environment while the **realized niche** is what a species actually uses.
   a. Some species are **broad-niched,** which means they interact with their environment in a number of ways and are rather flexible in such interactions. Others are **narrow-niched** with very limited, specific, unvarying requirements.
   b. **Gause's law,** or the **principle of competitive exclusion** states that two species cannot share

the same niche indefinitely. One will replace the other, or their niches will change.
   c. Darwin's finches reveal **character displacement,** physical changes that tend to further increase divergence between similar species. In the finches, differences in bill size result in taking different foods, thus theoretically reducing competition.
5. **Emigrations** reduce population density, but neither their evolutionary history nor adaptiveness is well understood. In some cases, when emigration has been experimentally restricted, resulting in increased density, social pathologies have arisen, such as in rat populations where fighting, overeating, withdrawal, sexual dysfunction, and death rates have increased.
6. In laboratory cultures of microorganisms, the production of toxic wastes increases with density, eventually limiting population growth. Humans increase toxic wastes in the environment in a number of ways. A recent problem is the formation of **acid rain,** a product of air pollution by heavy industry.
7. Disease and parasitism are examples of density-dependent biotic controls. High density increases the chances of contagion, and the usual problems of crowding exacerbate the conditions. Predation may limit the spread of disease since the sick are more likely to fall victim to predators. Parasites evolve toward a coexistence with the host, since killing the host may not be adaptive. In **parasitoid infections,** the parasite is likely to kill the host, as when certain wasps lay their eggs in caterpillars, which then become a food source for the emerging young.
8. Predation can depress prey populations, but its effects are subtle. According to the **Lotka-Volterra theory,** where predator-prey relationships are well established, predator numbers closely track and respond to changes in prey numbers. Such idealized relationships are often complicated by **prey-switching.** Hunters do not always respond to the Lotka-Volterra model, partly because technology enables them to hunt low-density prey.

THE HUMAN POPULATION
1. Human population numbers have remained relatively stable throughout much of our evolutionary history but recently has begun to follow the J-shaped curve.

A Brief Look at Human Population History
1. Early hominids lived by hunting and gathering, their numbers regulated by the usual biotic and abiotic factors. Infant mortality was high, but replacement was rapid.

2. The first population surge took place in response to increasingly efficient tools and weapons, better assuring the food supply and depressing the death rate. By 10,000 years ago, humans probably numbered about 5 million.

3. Agriculture began roughly 10,000 years ago, resulting in a further decreased death rate. By the time of Christ, the human population was 133 million. By 1650, the population was about 500 million.

4. The last population surge was brought about by: colonization of the New World, conquest of many bacterial diseases, improved public sanitation, improved technology in industry and agriculture, modern medicine and public health practices, and most recently, by the development of antibiotics and pesticides. These were brought to developing nations as well, depressing the death rate there. Population doubled between 1650 and 1850, again by 1930 and once more by 1975.

The Human Population Today

1. The rate of human population growth began decreasing by the late 1970s, and doubling time increased from 35 to 39 years.

The Future of the Human Population

1. An important population growth indicator is the **fertility rate.** The **general fertility rate** is the number of live births per 1000 women in their reproductive years, while the **total fertility rate** is a prediction of the average number of children women will bear.

2. Information about population trends can be obtained through the study of **age structure pyramids,** in which the population is plotted by groups according to age and sex. A broad-based pyramid is typical of developing nations where much of the population is young, and the death rate rapidly increases with age.

3. According to the theory of **demographic transition,** in undeveloped nations, both birth and death rates are high and growth slow. As development begins, a population's death rate decreases and numbers rapidly increase. But as development peaks, the birth rate decreases markedly, more closely approximating the death rate, so population growth slows dramatically. Today, much of the world is in the middle category.

Growth Predictions
and the Earth's Carrying Capacity

1. Many biologists conclude that current human population growth is affecting the earth's carrying capacity, and that a crash looms ahead. The crash can come as a product of famine, war, disease, and ecological disruption.

2. The slowing trend is encouraging, but the ideal goal is **ZPG—zero population growth,** where the birth rate and death rate are equal.

REVIEW QUESTIONS

1. What generalization do biologists make about readily visible increases and decreases in natural populations? How do we intuitively suspect that this generalization is valid? (1094)

2. Explain the meaning of the population symbol r. How is r obtained? Is it a positive number? Explain. (1095)

3. What factors account for the J-shaped growth curve? Can a population actually reach its biotic potential? (1095)

4. Discuss the meaning of exponential or geometric growth. How does this differ from linear or arithmetic growth? (1096)

5. List five factors that might contribute to environmental resistance. (1096)

6. What is the carrying capacity of an environment? What generally happens to a population that exceeds this point and what happens to the carrying capacity itself? (1096)

7. Is the J-shaped growth a rare phenomenon? Where might it be a common occurrence? (1096)

8. Draw three radically different survivorship curves and explain what they mean. Which most closely approximates the human condition? (1098)

9. What seems to be the general rule concerning the rate of reproduction in species? Why is this rate never realized in the long run? (1099)

10. Select an example of a K-adapted species and an r-adapted species and discuss the following: growth curves, size, life span, and reproductive lifetime. (1099–1100)

11. What evidence suggests that the human reproductive capacity was once adaptive but perhaps is no longer so? (1101–1102)

12. List two examples of programmed death in organisms. Give an example of what happens to such organisms when life is extended experimentally. (1102)

13. Is human death programmed? In what way(s) might it be adaptive for humans to live far beyond their reproductive years? In the final analysis, how is human death adaptive? (1103)

14. Provide an example of a density-dependent, abiotic, population control factor. Why are most abiotic factors density independent? (1103)

15. Using an example, describe a way in which density-dependent, biotic controls are regulative in nature. (1104)

16. What is territoriality? How might it reduce intraspecific competition? (1105)

17. Define the ecological niche and compare a broad-niched species with a narrow-niched species. (1105)

18. Using Gause's law explain how interspecific competition may be reduced. What is likely to happen when the niches of two species strongly overlap? (1106)

19. Cite an example of emigration acting as a population regulating factor. Give an example of what happens to a species that is experimentally restricted from migrating. (1107)

20. Provide two examples of populations being limited by their own wastes. What, if anything, does this suggest to our own species? (1108)

21. Explain the problem of acid rain, its source and possible long-term effect. (1109)

22. What generalization can we make about disease, parasitism and population density? In what way do disease and parasitism tend to be self-controlling in caribou herds? (1109)

23. What appears to be the most adaptive kind of strategy for a parasite? Cite an example of such a strategy with human parasites. (1109)

24. Draw a simple graph depicting the Lotka-Volterra theory of prey regulation by predators. Explain what is happening. (1110)

25. Typify the modern human predator. Do his or her predatory habits follow the trends described in the Lotka-Volterra theory? Explain. (1111)

26. Briefly describe the causes of the first two human population growth surges. What population growth statistic did they most strongly change? (1114)

27. The last human population growth surge began in the 1600s. List at least five factors that might have been responsible. Is this growth surge continuing? (1115)

28. In what world regions has population growth continued unabated? State the doubling times, percentage of people under 15, and growth rates of these regions. Why is the percentage of persons under 15 so important to future predictions? (1116)

29. Distinguish between the general and total fertility rate. Which has the most predictive value? Why is this? (1117)

30. Describe population age structure pyramids for India, the U.S., and Sweden. Characterize each in terms of future population size, problems for the producing portion of the population, problems of the aged. (1118)

31. Describe population growth in the three stages of demographic transition: undeveloped, developing, and developed. What is the danger of helping nations emerge from an undeveloped condition? (1118)

32. Should all restraints to population growth fail, how might this affect the earth's carrying capacity for humans? What inevitable event awaits us should this happen? (1119)

33. What two factors must reach equality if human populations are to reach ZPG? Has the U.S. reached this point? (1119)

SUGGESTED READING

Borgstrom, G. 1976. "Never Before Has Humankind Had to Face the Problem of Feeding So Many People with So Little Food." *Smithsonian* 7:70.

Connell, J. H. 1978. "Diversity in Tropical Rain Forests and Coral Reefs." *Science* 199:1302. What makes some habitats so species-rich? Connell suggests that unpredictable natural disasters, such as typhoons, prevent them from ever approaching species equilibrium.

Corliss, J. B., et al. 1979. "Submarine Thermal Springs on the Galápagos Rift." *Science* 203:1073. *Alvin* takes a look at a novel ecosystem built on sulfur-oxidizing chemosynthetic bacteria.

Ehrlich, P. R., and A. H. Ehrlich. 1979. "What Happened to the Population Bomb?" *Human Nature*, January. Everyone expected an unprecedented population explosion in the United States in the 1970s, as the children of the 1945–1960 baby boom came of age. The bomb's failure to explode offers some hope for humanity.

Gressit, J. L. 1977. "Symbiosis Runs Wild in a Lilli-putian 'Forest' Growing on the Backs of High-living Weevils in New Guinea." *Smithsonian* 7:135. With a title like that, what more can we say?

Hutchinson, G. E. 1959. "Homage to Santa Rosalia, or Why Are There So Many Kinds of Animals?" *American Naturalist* 93:145. A masterpiece by one of the century's most prestigious ecologists.

National Academy of Sciences (U.S.). 1975. *Underexploited Tropical Plants with Promising Economic Value*. Instead of continuing to grow the standard "world crops" in new tropical nations, why not take a closer look at the native plants?

Odum, E. P. 1983. *Basic Ecology*. Saunders. Philadelphia, Pa.

Ricklefs, R. E. 1978. *Ecology*, 2d ed. Chiron Press, Newton, Mass. Our favorite all-around ecology textbook, which also has an excellent coverage of the basics of population genetics.

Schaller, G. B. 1972. *The Serengeti Lion: A Study of Predator—Prey Relations*. University of Chicago Press, Chicago.

A Scientific American book. 1970. *The Biosphere*. W. H. Freeman, San Francisco.

Terborgh, J. 1974. "Preservation of Natural Diversity, the Problem of Extinction-prone Species." *Bioscience* 24:715. How much responsibility should we accept for endangered species—and at what expense?

Wilson, E. O., ed. 1974. *Ecology, Evolution and Population Biology*. Readings from *Scientific American*. W. H. Freeman, San Francisco.

Wilson, E. O., and W. H. Bossert. 1971. *A Primer of Population Biology*. Sinauer, Sunderland, Mass. (paperback).

Scientific American articles. San Francisco, W. H. Freeman, Co.

Beddington, J., and R. May. 1982. "The Harvesting of Interacting Species in a Natural Ecosystem." *Scientific American*, November.

Bergerud, A. 1984. "Prey Switching in a Simple Ecosystem." *Scientific American*, December.

Calhoun, J. R. 1962. "Population Density and Social Pathology." *Scientific American*, February. Rats kept in crowded conditions mistreat their babies and lose interest in sex.

Clutton-Brock, T. 1985. "Reproductive Success in Red Deer." *Scientific American*, February.

Cloud, P. 1970. "The Biosphere." *Scientific American*, September.

Edmond, J., and K. Von Damm. 1984. "Hot Springs on the Ocean Floor." *Scientific American*, April.

Hauser, P. 1981. "The Census of 1980." *Scientific American*, November.

Horn, H. H. 1975. "Forest Succession." *Scientific American*, May. The succession of trees in a New Jersey forest depends primarily on soil moisture and the geometry of leaves.

Ingersoll, A. 1983. "The Atmosphere." *Scientific American*, September.

Perry, D. 1984. "The Canopy of the Tropical Rain Forest." *Scientific American*, November.

Revelle, R. 1982. "Carbon Dioxide and World Climate." *Scientific American*, July.

The classification scheme was drawn from various commonly used sources in microbiology, botany, and zoology, but has been modified to accommodate the most recent ideas. Most taxa mentioned are also described in Chapters 18–21 and 25–27. Notably absent from this scheme are the viruses. Their taxonomic relationships have never been clarified, although some authors include them with the Monera while others have invented "Kingdom Virus." In our opinion there is too little known about viral evolution and phylogeny to permit their consideration as a taxon.

THE PROKARYOTES

KINGDOM MONERA (also PROKARYOTA): Includes the prokaryotes or bacteria, single-celled and colonial organisms that generally lack membrane-bounded organelles, including an organized nucleus; and whose DNA is organized into a single, circular, main chromosome, sometimes supplemented by minute circular plasmids, all of which lack protein. Cell division is through fission and sexual recombination is limited to conjugation and plasmid transfer. Where present, the flagellum is solid and rotating. Spore formation is common and absorption is the usual mode of feeding by heterotrophs, including the parasites. Autotrophic bacteria include both chemotrophs and phototrophs. Classification in this kingdom is unsettled and undergoing intensive revision where new techniques in biochemistry and molecular biology are being applied. Older classifications divided Monera into two subkingdoms, the Schizophyta and Cyanophyta. The former included phyla such as the Eubacteriae (true bacteria), Myxobacteriae (slime or gliding bacteria), Chlamydobacteriae (mycelium bacteria), Spirochaetae, and Mycoplasmae. The latter included only the cyanobacteria. The new taxonomies favor replacing Kingdom Monera with two new kingdoms, Archaebacteria and Eubacteria, basing the new distinction in part on ribosomal RNA and cell wall chemistry.

THE EUKARYOTES

All other kingdoms are eukaryotic. All have cells containing membrane-bounded organelles, their DNA complexed with histone and other proteins, forming chromatin. Mitotic cell division is common and the eukaryotic flagellum and cilium, where occurring, follows the $9 + 2$ microtubular organization. Many undergo meiosis and exchange gametes in sexual reproduction.

KINGDOM PROTISTA: A polyphyletic kingdom with many unresolved taxonomic problems. Includes the protozoans, the fungallike protists, and the algae. Protists include unicellular, colonial, and multicellular levels of organization. Nutrition is both heterotrophic and phototrophic. Sexual reproduction commonly occurs through meiosis and the union of gametes, although it has not been observed in many species.

Protozoan Protists

PHYLUM MASTIGOPHORA: Flagellated, unicellular heterotrophs that feed by phagocytosis and absorption. Includes parasites such as *Trypanosoma gambiense*, the agent of African sleeping sickness.

PHYLUM SARCODINA: Unicellar heterotrophs with ameboid movement that feed by phagocytosis and absorption. Includes marine radiolaria and foraminiferans and *Endamoeba histolytica*, the parasite of ameboid dysentery.

PHYLUM SPOROZOA: Unicellar, nonmotile heterotrophs that feed by absorption. Mainly parasitic, including *Plasmodium vivax*, an agent of human malaria.

PHYLUM CILIOPHORA: Ciliates. Unicellular, ciliated heterotrophs that feed by phagocytosis and absorption. Often large with complex organelles and sexual reproduction by meiosis and conjugation. Includes *Paramecium.*

Funguslike Protists

PHYLUM ACRASIOMYCETES: Cellular slime molds. Individual myxameba that fuse to form a pseudoplasmodium and later a compound sporangia. Feed by phagocytosis. *Dictyostelium discoideum.*

PHYLUM MYXOMYCETES: Acellular slime molds. Unicellular individuals fuse to form a true plasmodium and sporangia. Feed by phagocytosis. Some reproduce sexually. Includes *Physarum polycephalum.*

Algal Protists

PHYLUM PYRROPHYTA: Dinoflagellates. Flagellated phototrophs with chitinous cell walls and a primitive fission and one-step meiosis. Chlorophylls *a* and *c* and carotenoids, starch storage. Red tide organisms include *Gonyaulax* and *Gymnodinium.*

PHYLUM EUGLENOPHYTA: Euglenoids. Unicellular, flagellated phototrophs and heterotrophs with red "eye spot." Chlorophylls *a* and *b* and carotenoids, paramylon starch storage. Includes *Euglena gracilis* and *Phacus.*

PHYLUM CHRYSOPHYTA: Yellow-green and golden-brown algae. Unicellular and colonial phototrophs with pectin and glassy cell walls. Chlorophylls *a* and *c* and carotenoids and leucosin storage. Diatoms.

PHYLUM RHODOPHYTA: Red algae, seaweeds. Multicellular phototrophs with phycocyanin and phycoerythrin, chlorophyll *a,* and carotenoid pigments; floridean or carageenan storage. Separate sporophyte and gametophyte with varying dominance. *Polysiphonia, Porphyra.*

PHYLUM PHAEOPHYTA: Brown algae, seaweeds and kelps. Multicellular phototrophs with fucoxanthin, chlorophylls *a* and *c,* and carotenoid pigments; laminarin and mannitol storage. Separate sporophytes and gametophytes in some. Considerable tissue specialization. *Macrocystis, Nereocystis,* and *Sargassum.*

PHYLUM CHLOROPHYTA: Green algae. Unicellular, colonial, and multicellular phototrophs, with chlorophylls *a* and *b* and carotenoids, starch storage. *Chlamydomonas, Ulva, Volvox.*

KINGDOM FUNGI: Multicellular heterotrophs with saprobic feeding. Many are parasites. Fungi usually have simple septate or nonseptate vegetative mycelia, but many form complex, specialized, multicellular reproductive structures. The mycelia have chitin or cellulose walls. Fungi are primarily haploid with brief diploid interludes. Asexual reproduction is by mitotic spore formation and sexual reproduction is often by conjugation, followed by a lengthy dikaryotic state.

PHYLUM ZYGOMYCOTA: Bread molds. Extensive, simple, nonseptate mycelium, sexual reproduction through conjugation, and zygospore formation. *Rhizopus, Neurospora.*

PHYLUM ASCOMYCOTA: Sac fungi. Extensive, septate mycelium, asexual reproduction by conidiospore, sexual reproduction through conjugation, and dikaryotic state. Fertilization followed by meiosis and development of ascospores in saclike asci. *Saccharomyces cerevisiae* (baker's yeast), *Claviceps purpurea, Piziza.*

PHYLUM BASIDIOMYCOTA: Club fungi. Extensive septate mycelium, often with a large, raised, fruiting body, the basidiocarp. In sexual reproduction conjugation leads to the dikaryotic state. Fertilization occurs in the clublike basidia, followed by meiosis and the production of basidiospores. Mushrooms, shelf and bracket fungi, *Puccinia graminis* (wheat rust).

PHYLUM DEUTEROMYCOTA: The Fungi Imperfecti. Those fungi whose sexual reproduction has not been observed and have yet to be classified among the fungi. *Penicillium rocquefortii.*

KINGDOM PLANTAE: A monophyletic kingdom containing nonmotile, multicellular, phototrophs, most with protected embryos and great tissue specialization and organ development. Pigments include chlorophylls *a* and *b* and carotenoids and the chief storage polysaccharide is starch. The sporophyte is clearly dominant in nearly all divisions. (Note: In plant taxonomy, division replaces phylum.)

Nonvascular plants
(Plants without vascular tissue)

DIVISION BRYOPHYTA: Mosses, liverworts, hornworts. Nonvascular plants with simple tissue organization, no true (vascular) roots, stems or leaves. Dominant gametophyte, motile sperm, fertilization requires water. Bryophytes include three major classes, Musci (mosses), Hepaticae (liverworts), and Anthocerotae (hornworts).

Vascular Plants (Seedless)
(Vascular tissue, highly dominating sporophyte, embryo develops within separate gametophyte)

DIVISION PSILOPHYTA: Whisk ferns. Primitive group of four species, vascular stem but nonvascular scalelike leaves and rhizoids. Motile sperm; fertilization requires water. *Psilotum nudum.*

DIVISION LYCOPHYTA: Club mosses. Widely distributed, vascular roots, stems, leaves. Motile sperm; fertilization requires water. Homosporous and heterosporous species. *Lycopodium, Selaginella.*

DIVISION SPHENOPHYTA: Horsetails. Limited distribution, leaves highly reduced, slender stems bearing sporangia at tips. Motile sperm; fertilization requires water. *Equisetum.*

DIVISION PTEROPHYTA: Ferns. Widely distributed, foliage, often treelike with complex leaves emerging from rhizomes. Motile sperm; fertilization requires water. Tree ferns, bracken ferns, sword ferns, and staghorn ferns.

Vascular Plants with Seeds: Gymnosperms

(Cones rather than flowers, seeds naked—not enclosed in fleshy fruits)

DIVISION GINKGOPHYTA: Maidenhair tree. One species, a large tree with bilobed, fan-shaped leaves. Motile sperm; plant provides fluids for sperm to swim to egg. *Ginkgo biloba.*

DIVISION CYCADOPHYTA: Cycads. Widespread, tropical, palmlike leaves, and very prominant cones. Swimming sperm; plant provides fluids for sperm to swim to egg. *Zamia, Cycas.*

DIVISION GNETOPHYTA: Widespread in deserts and warm temperate regions. Advanced, angiosperm characteristics in some: xylem with vessels, bladelike leaf, and flowerlike pollen cones. *Gnetum, Welwitschia, Ephedra.*

DIVISION CONIFEROPHYTA: Conifers. Widespread in temperate, subarctic regions and high altitudes; well adapted to dryness. Commonly with needlelike leaves. Nonmotile sperm; no fluids required for fertilization. Pines, spruce, hemlock, cedar, larch, fir, juniper, redwood.

Vascular Plants with Seeds: Angiosperms

(Seed plants with flowers, fleshy fruits, and nonmotile sperm)

DIVISION ANTHOPHYTA: Flowering plants. Seeds surrounded by fruit.

CLASS DICOTYLEDONAE: Dicots. Worldwide distribution, floral parts in fours and fives or their multiples, two cotyledons in embryo, net-veined leaves. Perennials with pronounced secondary growth. Magnolia, oak, willow, maple, apple, rose, cucumber, bean and pea.

CLASS MONOCOTYLEDONAE: Monocots. Worldwide distribution, floral parts in threes or its multiples, one cotyledon in embryo. Chiefly herbivorous with debatable secondary growth in some. Grasses, lily, palm, orchids.

KINGDOM ANIMALIA: A polyphyletic kingdom made up of motile, multicellular heterotrophs, generally with tissue specialization and organ and organ-system organization. Feeding commonly by engulfing and extracellular digestion, but phagocytosis and intracellular digestion is also common. Sexual reproduction by the fusion of haploid gametes that are produced directly through meiosis.

SUBKINGDOM PARAZOA
(Origin believed to be the flagellate protists, primarily cell-level organization)

PHYLUM PORIFERA: Sponges. Nonmotile, filter-feeding, solitary or colonial adults with cellular level of organization and few cell types. Asexual reproduction by budding and sexual reproduction by internal fertilization. Swimming planula larva.

CLASS CALCARIA: Calcium carbonate skeleton. *Leucosolenia.*

CLASS HEXACTINELLIDA: Silicon dioxide skeleton. *Euplectella* (Venus flower basket).

CLASS DEMOSPONGIAE: Skeleton of silicon dioxide or the protein spongin. *Cliona*, bath sponges.

SUBKINGDOM METAZOA
(Origin believed to be the ciliate protists: tissue, organ, and organ-system levels of organization)

Radiate Phyla
(Diploblastic animals with radial symmetry)

PHYLUM COELENTERATA: Cnidarians. Radial animals with thin-walled bodies, tentacles armed with stinging cells. Saclike gastrovascular cavity, extracellular and intracellular digestion. Individuals exist as polyps, medusae, or alternating cycles of the two.

CLASS HYDROZOA: Solitary hydroids or those with colonial feeding polyps that bud reproductive medusae in which fertilization occurs. *Obelia, Hydra.*

CLASS SCYPHOZOA: Includes the jellyfish, a medusa in which feeding and sexual reproduction occur. Swimming larva forms a polyp stage that buds off young jellyfish. *Aurelia.*

CLASS ANTHOZOA: Anemones and corals, polyps that feed and reproduce. Development is usually direct from zygote to polyp. *Metridium*, sea fan coral, brain coral.

PHYLUM CTENOPHORA: Radial, marine animals with thin-walled bodies, tentacles armed with glue cells. Saclike bodies like coelenterates, but with eight rows of combs consisting of fused cilia. Comb jelly, Venus girdle.

Bilateral, Protostome, Acoelomate Phyla

(Bilateral symmetry, triploblastic, spiral cleavage, blastopore area gives rise to mouth, solid body lacking a coelom)

PHYLUM PLATYHELMINTHES: Flatworms. Dorsoventrally flattened, solid-bodied, marine, freshwater, and terrestrial worms, most with a highly branched gastrovascular cavity and some organ-system development.

CLASS TURBELLARIA: Free-living planarians, with extensive organ-system development, hermaphroditic. *Dugesia.*

CLASS TREMATODA: Parasitic flukes with oral suckers, often with two or more hosts. *Clonorchis sinensis* (Asian liver fluke), *Schistosoma* (blood flukes).

CLASS CESTODA: Parasitic tapeworms with small scolex, numerous proglottids in the mature adults, often with two hosts. *Taenia solium* (pork tapeworm), *Taeniarhynchus* (beef tapeworm).

PHYLUM RHYNCHOCOELA: Proboscis worms. Thin, ribbonlike worms with a complete, one-way digestive tract. Protrusible pharynx with piercing stylet. Marine.

Bilateral, Protostome, Pseudocoelomate Phyla

(Body cavity a false coelom, incompletely lined by mesodermal tissue)

PHYLUM NEMATODA: Roundworms. Slender, threadlike worms with a fluid-filled body cavity and a complete digestive tract. Terrestrial, free-living predators and internal parasites. Widespread in soil and water—outnumbered only by arthropods. *Rhabditis, Ascaris lumbricoides* (giant roundworm), *Necator americanus* (hookworm), *Enterobias vermicularis* (pinworm).

PHYLUM ROTIFERA: Rotifers or "wheel animals." Complex, free-living, minute, ciliated animals with a complex, complete digestive tract and well-developed organ systems. Sexes separate with parthenogenesis common. *Philodina.*

Bilateral, Protostome, Coelomate Phyla

(Triploblastic animals with a true—mesodermally lined—body cavity)

PHYLA ECTOPROCTA, PHORONIDA, AND BRACHIOPODA: Lophophorate phyla. Ciliated tentacles attached to a ridgelike lophophore. Moss animals, lampshells, and wormlike *Phoronis.*

PHYLUM ANNELIDA: Segmented worms. Most with body divided into numerous metameres (segments), well-developed coelom, complete and specialized digestive tract, and closed circulatory system.

CLASS OLIGOCHAETA: Terrestrial, free-living and burrowing. Hermaphroditic. *Lumbricus terrestris* (earthworm).

CLASS POLYCHAETA: Marine, free-living burrowing, tube-dwelling and swimming forms. *Neries virens* (clam worm), fanworm, peacock worm.

CLASS HIRUDINEA: Freshwater and terrestrial, parasitic, sucker mouth. *Hirudo* (medicinal leech).

PHYLUM MOLLUSCA: Mollusks. Highly reduced coelom and segmentation, diversified through adaptive variations in foot, shell, and mantle. Open circulatory system. Chiefly marine and freshwater, some terrestrial.

CLASS AMPHINEURA: Marine, shell in eight parts, lengthy foot surrounded by simple gills and mantle. *Mopalia* (a chiton).

CLASS GASTROPODA: Aquatic and terrestrial, large muscular foot, often retractable foot and spiral shell, feed by radula. *Helix* (land snail), *Haliotis* (an abalone), whelks, limpets, slugs.

CLASS BIVALVIA: Hinged shells, digging retractable foot, filter feeders. *Mytilus* (a mussel), *Tagelus* (a clam), oyster, scallop.

CLASS CEPHALOPODA: Marine, enlarged head and brain, image-forming eyes, tentacles, internalized shell in most. *Octopus, Loligo* (a squid), and *Nautilus* (chambered nautilus).

PHYLUM ONYCOPHORA: Intermediate between annelids and arthropods, terrestrial with jointed appendages, claws, wormlike body. *Peripatus.*

PHYLUM ARTHROPODA: Largest animal phylum. Jointed-legged animals with chitinous exoskeleton, modified segmentation.

Subphylum Chelicerata (no jaws).

CLASS MEROSTOMATA, ARACHNIDA, AND PYCNOGONIDIA: Terrestrial and aquatic, four pairs of legs, fangs rather than mandibles, two body regions and book lungs common. Horseshoe crab, spiders, mites, ticks, scorpions, and sea spiders.

Subphylum Mandibulata (jaws).

CLASS CRUSTACEA: Mainly aquatic, two pairs of antennas, compound eye, hardened exoskeleton. Crabs, lobsters, copepods, barnacles, water fleas.

CLASS INSECTA: Terrestrial and aquatic, three body regions, one pair of antennas, compound eye, three pairs of walking legs, wings are common. Beetles, moths, bees, ants, flies, bugs, grasshoppers, lice, termites, fleas, dragonflies, and others. (For the major insect orders see Table 26.1, pp. 660-661.)

CLASS CHILOPODA AND DIPLOPODA: Centipedes and millipeds.

Bilateral, Deuterostome Phyla
(Radial cleavage, blastopore area gives rise to anus)

PHYLUM ECHINODERMATA: Spiny-skinned stars. Marine, pentaradial as adults, endoskeleton of calcareous plates, commonly with spines, and pedicellaria. All with a water vascular system.

CLASS CRINOIDIA: Cup-shaped body with branched arms, no suckers on tube feet. Sea lily.

CLASS HOLOTHUROIDIA: Cylindrical body, spines reduced, no pedicellaria. *Cucumaria* (a sea cucumber).

CLASS ECHINOIDEA: Spherical or round, flattened body without arms, prominent movable spines, tube feet with suckers. *Lytechinus* (a sea urchin), *Echinarachnius* (a sand dollar).

CLASS ASTEROIDEA: Prominent arms, short, blunt spines, numerous pedicellaria, tubefeet with suckers. *Asterias* (a sea star).

CLASS OPHIUROIDEA: Prominent arms, movable spines, tube feet without suckers, no pedicellaria. *Ophiura* (a brittle star), *Gorgonocephalus* (a basket star).

PHYLUM HEMICHORDATA: Hemichordates. Marine, burrowing, acorn-shaped proboscis, gill slits. Larvae similar to echinoderm's. *Saccoglossus* (an acorn worm).

PHYLUM CHORDATA: Chordates. Aquatic and terrestrial, gill slits, notochord, dorsal hollow nerve cord, postanal tail.

Subphylum Urochordata: Marine filter feeders, saclike body with gill slits, bilateral larva with all chordate characteristics. *Ciona* (a tunicate), *Thetys* (a salp).

Subphylum Cephalochordata: Marine filter feeder, fishlike body form, notochord in adults, prominent gill slits and segmented trunk muscles, sensory cirri. *Branchiostoma* (lancelet).

Subphylum Vertebrata: Vertebral column of bone or cartilage, increasing cephalization, ventral heart and dorsal aorta, no more than two pairs of limbs, separate sexes.

CLASS AGNATHA: Fishes with jawless mouth, cartilaginous skeleton. Lamprey and hagfish.

CLASS CHONDRICHTHYES: Marine fishes with cartilaginous skeleton and jaws, placoid scales and rows of replacement teeth, dorsoventrally flattened body. Sharks, rays, chimera.

CLASS OSTEICHTHYES: Marine and freshwater fishes with bony skeleton and jaws. Includes Subclass Crossopterygii (lobe-finned fishes), Subclass Dipneusti (lungfishes), and Subclass Actinopterygii (ray-finned fishes), which include Superorder Chondrostei (sturgeons), Superorder Holostei (gar pike and bowfin), and Superorder Teleostei (perch, bass, tuna, eel, sunfish).

CLASS AMPHIBIA: Freshwater and terrestrial amphibians, chiefly air breathers, moist vascularized skin, three-chambered heart, most requiring water for reproduction. Includes Order Anura (frogs and toads), Order Urodela (salamanders), and Order Apoda (caecilians).

CLASS REPTILIA: Aquatic and terrestrial air breathers with dry, scaly skin and three-chambered hearts in most. Fertilization is internal and development occurs in the independent, amniotic, land egg. Includes Order Chelonia (turtles), Order Squamata (snakes and lizards), Order Crocodilia (crocodiles and alligators), and Order Rhynchocephalia (tuatara).

CLASS AVES: Birds, principally terrestrial and flying; body modifications adapting for flight include lightweight skeleton with fusion in the vertebrae, no teeth, covering of feathers, highly modified forelimbs, four-chambered heart, endothermic, and crosscurrent air-blood flow in lung. Hawks, penguins, peafowl, jays, bluebirds, ducks, turkey.

CLASS MAMMALIA: Mammals, terrestrial and aquatic air-breathers, body covering of hair, glandular skin, milk produced in mammaries, highly specialized, socketed teeth, skull bones highly fused, enlarged cerebrum, and greatly increased learning capacity.

Subclass Prototheria: Egg-laying mammals. Spiny anteater and duckbilled platypus.

Subclass Metatheria: Marsupial mammals, pseudoplacenta, immature young housed and nourished in marsupium (pouch), nearly all restricted to Australia and New Zealand. Opossums, kangaroos, wallabies, koalas, wombats.

Subclass Eutheria: Placental mammals, true placenta supports embryo development.

Order Insectivora: Smallest mammals, primitive, pointed teeth, tapered snout. Shrew and mole.

Order Dermaptera: Fur-covered patagium, gliding flight. Flying lemur.

Order Chiroptera: Greatly extended fingers with flight membrane, true flight, sonar navigation. Bat.

Order Primates: Enlarged cerebrum, hands and feet with nailed digits, eyes forward, opposable thumb in some. Lemur, monkey, ape, human.

Order Edentata: Toothless or molars only, no enamel. Sloth, anteater.

Order Pholidota: Overlapping horny scales. Scaly anteater.

Order Lagomorpha: Rodentlike, but four upper incisors. Rabbit, hare.

Order Rodentia: Largest order, small, continuous growth in incisors. Squirrel, mouse, gopher, beaver.

Order Cetacea: Forelimbs modified to flippers, hindlimbs absent, lateral tail fluke, nostrils as blowhole, teeth without enamel. Porpoise, whale.

Order Carnivora: Enlarged canines, an adaptation for predation. Bear, cat family, dog family, raccoon, weasel.

Order Pinnipedia: Similar to carnivores, but feet modified for swimming. Seal, sea lion, walrus.

Order Tubulidentata: Piglike snout, no incisors or canines, teeth without enamel. Aardvark.

Order Proboscidea: Largest terrestrial mammal, upper lip a prehensile trunk, incisors enlarged to tusks. Elephant.

Order Hyracoidea: small, similar to guinea pig. Hyrax.

Order Sirenia: Forelimbs are paddles, hindlimbs absent, lateral tail flukes, heavy, blunt muzzle. Manatee.

Order Perissodactyla: Hooved feet with odd numbers of toes, (tapirs have even-numbered toes on forelimbs, odd on hindlimbs). Horse, ass, zebra, tapir, rhinoceros.

Order Artiodactyla: Hooved feet with even numbers of toes, many with antlers, some ruminants. Pig, hippopotamus, camel, deer, giraffe, cattle.

A list of Greek and Latin prefixes, suffixes, and word roots commonly used in biological terms, alphabetized by the most common combining form in English; with examples illustrating each usage.

a-, an- [Gk. *an-*, not, without, lacking]: anaerobic, abiotic, anorexia, anesthesia, aseptic, asexual

acro- [Gk. *akros*, highest]: acrophobia, acromegaly

ad- [L. *ad-*, toward, to]: adhesion, adrenal, adventitious, adsorption

allo- [Gk. *allos*, other]: allele, allopatric, allosteric, allotetraploid

amphi- [Gk. *amphi-*, two, both, both sides of]: amphibian, Amphineura, amphipod

ana- [Gk. *ana-*, up, up against]: anaphase, anatomy, anabolic

andro- [Gk. *andros*, an old man]: androecium, androgen, androgynous, polyandry

anti- [Gk. *anti-*, against, opposite, opposed to]: antibiotic, antibody, antigen, antidiuretic hormone, antihistamine, antiseptic, antipathy, antiparallel, antihistamine

apo- [Gk., *apo-*, different]: apomoxis, apodeme, aponeurosis

archeo- [Gk. *archaios*, beginning]: archegonium, archenteron

arthro- [Gk. *arthron*, a joint]: arthropod, arthritis

auto- [Gk. *auto-*, self, same]: autoimmune, autotroph, autosome

auxo- [Gk. *aux*, to grow or increase]: auxin, auxospore, auxotroph

bi-, bin- [L. *bis*, twice; *bini*, two-by-two]: binary fission, binocular vision, binomial

bio- [Gk. *bios*, life]: biology, biomass, biome, biosphere, biotic

blasto-, -blast [Gk. *blastos*, sprout; now "pertaining to the embryo"]: blastoderm, blastopore, blastula, trophoblast

brachi- [Gk. *brachion*, arm]: brachiation, brachiopod

brachy- [Gk. *brachys*, short]: brachydactyly, brachycardia

branchio-, -branch [Gk. *branchia*, a gill]: branchiopod, elasmobranch, branchial arch

broncho- [Gk. *bronchos*, windpipe]: bronchus, bronchiole, bronchitis

carcino- [Gk. *karkin*, a crab, cancer]: carcinogen, carcinoma

cardio- [Gk. *kardia*, heart]: cardiac, cardiology, myocardium, electrocardiogram

cephalo- [Gk. *kephale*, head]: cephalization, cephalochordate, encephalitis, encephalogram

chloro- [Gk. *chloros*, green]: chlorobacterium, chlorophyll, chloroplast, chlorine

chole- [Gk. *chole*, bile; Gk. cholecyst, gall bladder]: cholesterol, cholinesterase, cholecystokinin

chondro- [Gk. *chondros*, cartilage]: achondroplasia, chondroitin, chondroblast

chromo- [Gk. *chroma*, color]: chromosome, chromatophore, chromatin

coelo-, -coel [Gk. *koilos*, hollow, cavity]: coelenteron, coelenterate, coelom, pseudocoelom

com-, con-, col-, cor-, co- [L. *cum*, with, together]: commensal, conjugation, covalent

cranio- [Gk. *kranios*, L. *cranium*, skull]: cranial, craniotomy, cranium

cuti- [L. *cutis*, skin]: cutaneous, cuticle, cutin

cyclo-, -cycle [Gk. *kyklos*, circle, ring, cycle]: cyclosis, cyclostome, pericycle

cyto-, -cyte [Gk. *kytos*, vessel or container; now, "cell"]: cytoplasm, cytology, erythrocyte, cytosine

de- [L. *de-*, "away, off"; deprivation, removal, separation, negation]: deciduous, decomposer, deoxyribose

derm-, dermato- [Gk. *derma*, skin]: dermatitis, dermis, epidermis

di- [Gk. *dis*, twice]: dikaryon, dicotyledon, diencephalon, diatomic

dia- [Gk. through, passing through, thorough, thoroughly]: diabetes, dialysis, diapause, diaphragm

diplo- [Gk. *diploos*, two-fold]: diplotene, diploid, diploblastic

eco- [Gk. *oikos*, house, home]: ecology, androecium, ecosystem

ecto- [Gk. *ektos*, outside]: ectoderm, ectoplasm, ectoparasite

en- [Gk., L. in, into]: encephalon, encephalitis, entropy, environment

endo- [Gk. *endon*, within]: endocrine, endoderm, endodermis, endoparasite, endoskeleton

entero- [Gk. *enteron*, intestine]: archenteron, enteric, entameba

epi- [Gk. *epi*, on, upon, over]: epiboly, epicotyl, epidermis, epiphyte, epithelium

eu- [Gk. *eus*, good; *eu*, well; now "true"]: eubacterium, eukaryote

ex-, exo-, ec-, e- [Gk., L. out, out of, from, beyond]: emission, ejaculation, exhale, exocytosis, exoskeleton, tonsillectomy

extra- [L. outside of, beyond]: extracellular, extraembryonic

gam-, gameto- [Gk. gamos, marriage; now usually in reference to gametes (sex cells)]: gamete, gametogenesis, isogamete, heterogamete

gastro- [Gk. *gaster*, stomach]: gastric, gastrula, gastrin, gastrovascular cavity

gen- [Gk. *gen*, born, produced by; Gk. *genos*, race, kind; L. *genus, generare*, to beget]: gene, polygenic, genotype, glycogen, florigen, pyrogen

gluco-, glyco- [Gk. *glykys*, sweet; now pertaining to sugar]: glucose, glycogen, glycerol, glycolipid

gyn-, gyno-, gyneco- [Gk. *gyne*, woman]: gynecology, gynoecium, epigyny, hypogyny

hemo-, hemato-, -hemia, -emia [Gk. *haima*, blood]: hematology, hemoglobin, hemorrhoid, hemotoxin, leukemia, toxemia

hepato- [Gk. *hepar, hepat-*, liver]: hepatitis, hepatic

hetero- [Gk. *heteros*, other, different]: heterosexual, heteromorphic, heterozygote

histo- [Gk. *histos*, web of a loom, tissue; now pertaining to biological tissues]: histology, histocompatibility, histamine

homo-, homeo- [Gk. *homos*, same; Gk. *homios*, similar]: homeostasis, homeothermy, homosexual, homology

hydro- [Gk. *hydor*, water; now, confusingly, pertaining either to water or to hydrogen]: dehydration, hydraulic, carbohydrate, hydrophobia

hyper- [Gk. *hyper*, over, above, more than]: hyperacidity, hyperglycemia, hypertension, hypertonic

hypo- [Gk. *hypo*, under, below, beneath, less than]: hypocotyl, hypodermic, hypoglycemia, hypothalamus, hypotonic

inter- [L. *inter*, between, among, together, during]: interbreed, intercellular, interphase, interstitial

intra-, intro- [L. *intra*, within]: intracellular, intravenous, introitus

-itis [L., Gk. *-itis*, inflammation of]: arthritis, bronchitis, dermatitis

leuko-, leuco- [Gk. *leukos*, white]: leukocyte, leukemia, leukoplast

-logue, -logy [Gk. *-logos*, word, language, type of speech]: analogy, homology, dialogue

-logy [Gk. *-logia*, the study of, from *logos*, word]: biology, cytology, embryology

-lysis, lys-, lyso-, -lyze, -lyte [Gk. *lysis*, a loosening, dissolution]: lyse, lysis, lysogeny, hydrolysis, catalysis, catalytic, electrolyte

macro- [Gk. *makro-*, now "great," "large"]: macromere, macromolecule, macronucleus

mega-, megalo-, -megaly [Gk. *megas*, large, great, powerful]: megagamete, megalocephalic, megaspore, acromegaly

-mere, -mer, mero- [Gk. *meros*, part]: blastomere, centromere, dimer, polymer, meroblastic, merozoite

meso-, mes- [Gk. *mesos*, middle, in the middle]: mesencephalon, mesentery, mesoderm, mesophyll

meta-, met- [Gk. *meta*, after, beyond; now often denoting change]: metabolism, metacarpal, metamorphosis, metastasis

micro- [Gk. *mikros*, small]: microbe, microbiology, microgametophyte, micropyle

myo- [Gk. *mys*, mouse, muscle]: myasthenia gravis, myocardial infarction, myosin

neuro- [Gk. *neuron*, nerve, sinew, tendon]: neurofibril, neuron, aponeurosis

oligo- [Gk. *oligos*, few, little]: Oligocene, oligochaete, oligotrophic

-oma [Gk. *oma*, tumor, swelling]: carcinoma, glaucoma, hematoma

oo- [Gk. *oion*, egg]: oogenesis, oogonium, oospore

-osis [Gk. *-osis*, a state of being, condition]: arteriosclerosis, cirrhosis, halitosis

osteo-, oss- [Gk. *osteon*, bone; L. *os, ossa*, bone]: ossification, Osteichthyes, teleost

para- [Gk. *para-*, alongside of, beside, beyond]: paramedic, parapatric, paraphyletic, parathyroid

patho-, -pathy, -path [Gk. *pathos*, suffering; now often disease or the treatment of disease]: pathogen, osteopath, homeopathy

peri- [Gk. *peri*, around]: pericarp, pericycle, perinium, periosteum, photoperiod, peristome, peritoneum

phago-, -phage [Gk. *phagein*, to eat]: phagocyte, phage, bacteriophage

plasm-, -plasm, -plast, -plasty [Gk. *plasm*, something molded or formed; Gk. *plassein*, to form or mold]: plasma, plasmid, thromboplastin, cytoplasm, plastid, chloroplast, dermoplasty

-pod [Gk. *pod*, foot]: amphipod, Apoda, arthropod, cephalopod

poly- [Gk. *poly-, polys,* many]: polyandry, polychaete, polydactyly

-rrhea [Gk. *rhoia,* flow]: amenorrhea, diarrhea, gonorrhea

septi-, -sepsis, -septic [Gk. *septicos,* rotten, infected]: septic, septicemia, antiseptic

-some, somat- [Gk. *soma,* body; Gk. *somat-,* of the body]: somatic cell, psychosomatic, centrosome, acrosome

-stat, -stasis, stato- [Gk. *stasis,* stand]: statocyst, metastasis, hydrostatic

stoma-, stomato-, -stome [Gk. *stoma,* mouth, opening]: stoma, stomatoplasty, stomodeum, deuterostome

sym-, syn- [Gk. *syn,* with, together]: symbiont, symbiosis, synapsis, synaptinemal complex

taxo-, -taxis [Gk. *taxis,* to arrange, put in order; now often referring to ordered movement]: taxonomy, taxon, chemotaxis, phototaxis

tomo-, -tome, -tomy [Gk. *tome,* a cutting; Gk. *tomos,* slice]: tomography, microtome, anatomy

trich-, tricho-, -trich [Gk. *tricho-,* combining form of *thrix,* hair]: trichina, trichocyst, gastrotrich

trop-, tropo-, -tropy, -tropism [Gk. *tropos,* to turn, to turn toward]: tropism, entropy, phototropism

tropho-, -troph, -trophy [Gk. *trope,* nutrition]: trophic level, autotroph, heterotroph

ur-, -uria [Gk. *ouron,* urine]: urea, ureter, uric acid, urine, alkaptonuria, phenylketonuria

uro-, -uran [Gk. *oura,* tail]: urochordate, uropod, anuran

A band one of the bands of striated muscle, corresponding to the fixed length of the myosin filament.

abdomen 1. in mammals, the body cavity between the diaphragm and the pelvis; 2. in other vertebrates, the body cavity containing the stomach, intestines, liver, and reproductive organs. 3. in arthropods, the posterior section of the body.

abiotic 1. characterized by the absence of life. 2. nonbiological; factors independent of living organisms.

abiotic control control of population numbers by nonbiological factors, such as weather.

ABO blood group system a genetically controlled polymorphic cell surface polysaccharide antigen.

abomasum see *ruminant*.

abortion the spontaneous or *induced* expulsion of the human fetus before it is viable.

abscisic acid a plant hormone that suppresses dormancy.

abcission the normal separation, through an abscission zone, of flowers, fruit, and especially leaves from plants.

abscission zone also *abscission layer*, a layer of specialized, cutinized parenchyma cells at the base of a leaf petiole, fruit stalk, or branch, in which normal separation occurs.

absorption spectrum a graph indicating the relative light absorption of a molecule as a function of the wave length of light.

abyssal region the lowest depths of the ocean, especially the bottom waters; also, *abysmal region*.

accessory parts in flowers, any parts not directly involved in reproduction, such as the receptacle, calyx, and corolla.

accessory pigment a pigment other than chlorophyll that absorbs energy from light and is capable of transferring that energy to chlorophyll for photosynthesis.

acellular not composed of cells; not divisible into smaller cellular units; also, *noncellular*.

acetabulum the socket in the hip bone that receives the head of the thigh bone.

acetyl CoA a key intermediate in metabolism, consisting of an acetyl group covalently bonded to coenzyme A.

acetylcholine a neurotransmitter released and hydrolyzed in certain synaptic transmission and in the initiation of muscle contraction.

acetylcholinesterase a membrane-bound enzyme that hydrolyzes acetylcholine in the course of synaptic nerve impulse transmission.

achondroplasia a genetic pathology, usually due to a dominant allele, in which the conversion of cartilage into bone is defective, resulting in extremely short appendages, dwarfism, and facial anomalies.

acid a compound capable of neutralizing alkalis and of lowering the pH of aqueous solutions. Acids are proton donors that yield hydronium ions in water solutions.

acid-base buffering system in body fluids such as blood, substances that aid the body in resisting changes in pH.

acidosis an abnormally acidic condition of body fluids.

acoelomate a lower invertebrate animal, one that does not develop a pseudocoelom or coelom.

acrosomal filament a filament produced by rupture of the acrosome of the sperm of echinoderms, which penetrates the egg and initiates fertilization.

acrosome an organelle in the tip of a spermatozoan, which ruptures in fertilization with the release of enzymes and (in echinoderms) of an acrosomal filament.

actin a cytoplasmic protein known in both globular and fibrous forms. See *actin filament*.

actin filament also *myofilament*, a cytoplasmic protein fiber found in the cytoskeleton and in muscle contractile units, associated with cellular movement.

action potential during a neural impulse, a brief shift in the voltage potential across the membrane of a neuron from –60 mV to +40 mV.

action-specific energy (ASE), a hypothetical endogenous tension that increases in the nervous system until it is discharged, either on its own or due to the perception of an appropriate stimulus.

action spectrum a graph relating light-induced biological activity (e.g., carbon fixation or oxygen release in photosynthesis) per photon as a function of wave length.

activation gate see *sodium ion gate*.

active site that part of an enzyme directly involved in specific enzymatic activity.

active transport energy-requiring transport of a substance across the cell membrane usually against the concentration gradient.

adaptation 1. an adjusting to conditions. 2. a change that improves function. 3. any alteration in the structure or function of an organism or any of its parts that results from natural selection and by which the organism becomes better able to survive and multiply in its environment.

adaptive radiation the spread of a population into new and differing environments, accompanied by adaptive evolutionary changes.

additive law in mathematical probability theory, the relation that the probability of the occurrence of one of two or more mutually exclusive events is equal to the sum of the individual probabilities.

adenine a purine, one of the nitrogenous bases found in both DNA and RNA as well as in several coenzymes.

adenosine a nucleoside formed of adenine and ribose covalently linked.

adenylate cyclase an enzyme, usually incorporated into the cell membrane, that is capable of transforming ATP into cyclic AMP and pyrophosphate.

adhesion the molecular force of attraction in the area of contact between unlike substances.

ADP adenosine diphosphate, a compound of adenine, ribose, and two phosphate groups.

adrenal also *adrenal gland*, one of a pair of endocrine (ductless) glands located above the kidneys, each consisting of two distinct parts, a central *medulla* and an outer *cortex*.

adrenal cortex the outer portion of the adrenal gland producing steroid hormones.

adrenaline see *epinephrine*.

adrenal medulla the inner portion of the adrenal gland producing epinephrine and norepinephrine (adrenalin and noradrenalin) as its principle hormonal products.

adrenocorticotropic hormone (ACTH), a pituitary hormone that stimulates the production of hormones of the adrenal cortex.

advanced relatively later in evolutionary origin or state, as opposed to *primitive*. See also *derived*.

adventitious in plants, growth of a root or bud in an unusual place; *adventitious root*, a secondary root growing from stem tissue.

aerobe an organism that uses oxygen as a hydrogen acceptor in cellular respiration forming water.

aerobic utilizing oxygen in respiration.

afferent vessel any blood vessel carrying blood toward a specific structure such as a gill filament or kidney nephron.

African sleeping sickness a disease of cattle and humans caused by a parasitic protozoan, transmitted by the tsetse fly and eventually affecting the central nervous system.

agar also *agar-agar*, a polysaccharide produced by red algae; also, a gel made from this material, used as a moist semi-solid base for the experimental growth of microorganisms.

age profile see *population pyramid.*

age structure pyramid, also *age profile* and *population pyramid*, a pyramidal graph of a population, divided into age groups. Each age group is represented by a horizontal bar with that of the youngest forming the base.

agglutination the clumping of bacteria, erythrocytes, or other cells by antibody/antigen association.

aggregation a number of independent organisms grouped together either in a casual, temporary way or for more permanent mutual reasons.

aggression hostility, attack, or threat, especially unprovoked, usually against a competitor or potential competitor.

agricultural revolution also *green revolution*, the introduction of advanced agricultural technology and improved strains into the third world economy.

AIDS (acquired immune deficiency syndrome) a recently discovered disease in which the immune system fails.

air sac in birds and insects, any of numerous extensions of the respiratory system as air-filled, membranous sacs into various body parts.

alarm chemicals pheromones released as alarm signals.

albedo effect the reflection of a portion of solar radiation by the atmosphere.

albinism a marked and abnormal, genetically based deficiency in pigmentation.

albino a human being, plant, or animal lacking or abnormally deficient in pigmentation.

albumin 1. any of a class of clear, water-soluble plant or animal proteins. 2. *serum albumin*, a clear, water-soluble constituent of blood plasma, thought to serve detoxifying and osmotic functions. 3. *albumen* (egg white protein).

alcohol 1. ethyl alcohol, C_2H_5OH. 2. any of a class of organic compounds analogous to ethyl alcohol in having a $-CH_2OH$ group and lacking aldehyde or carboxyl groups.

aldehyde any organic compound with the reactive group $-CHO$.

aldosterone a steroid hormone produced by the adrenal cortex, involved in potassium reabsorption by the kidney.

aleurone also *aleurone layer*, a single, outermost layer of cells in the endosperm of cereals.

alga (pl. *algae*) any photosynthetic member of the kingdom *Protista.*

algin a structural polysaccharide of the brown algae.

alkaloid any organic, nitrogenous, alkaline, water-soluble, bitter-tasting compound produced by plants, usually with pharmacological effects; e.g., morphine, nicotine, quinine, caffeine.

alkalosis an alkaline condition in the entire body characterized by abnormally high levels of bicarbonate in the blood.

alkaptonuria a genetic pathology caused by the lack of an essential enzyme and characterized by the excretion of homogentisic acid.

allantois one of the extraembryonic membranes. In birds and reptiles it serves as a repository for nitrogenous wastes of the embryo.

allele a particular form of a gene at a locus.

allele frequency the proportion (relative number) of alleles of a given type at a specific gene locus.

allogenic succession succession brought about by physical forces generally beyond the influence of living organisms.

allopatric speciation the rise of new species from populations that have become geographically isolated.

allosteric control the control of enzyme activity through the binding of small metabolites at one or more secondary (allosteric) binding sites.

allosteric site on certain enzymes, a secondary binding site for small metobolites involved in the regulation of enzymatic activity.

allotetraploidy adj., derived from the hybridization of two distinct species and carrying the full diploid chromosome complements of both parental species; n., an allotetraploid organism. Also, *amphidiploid.*

alpha amylase a starch-digesting enzyme.

alpha cell see *Islets of Langerhans.*

alpha helix the right-handed helical configuration spontaneously formed by certain polymers. See *secondary level of organization.*

alternate factors see *factors.*

alternation of generations in algae and plants, the existence of two phases in the life of a single individual, including a diploid, spore-producing phase and a haploid, gamete-producing phase.

altricial of birds, helpless at hatching and requiring parental feeding and care. Compare *precocial.*

altruism behavior that is directly beneficial to others at some cost or risk to the altruistic individual.

alula a small feathered, clawed-spike (the thumb) at the leading edge of the wing that functions in the flight control of birds.

Alvarez hypothesis a hypothesis that proposes that much of the massive extinction of life that accompanied the end of the Mesozoic era was produced by the aftereffects of a gigantic asteroid's collision with the earth.

alveolus (pl. *alveoli*), the air cells of the vertebrate lung, formed by the terminal dilation of tiny air passages.

ameba also *amoeba*, 1. any protozoan of the large genus *Amoeba*, characterized by lobose pseudopods and the lack of permanent organelles or supporting structures; 2. any ameboid protist such as the ameboid stage of a flagellate or sporozoan.

amebic dysentery a contagious disease caused by an enteric ameba.

amebocyte in sponges, an amoeboid cell involved in reproduction, digestion, and spicule formation.

ameboid also *amoeboid*, like an ameba in moving or changing shape by protoplasmic flow.

ameboid movement also *amoeboid movement* or change in the shape of a cell by cytoskeleton deformation, the formation of pseudopods, and protoplasmic flow.

amino acid also *alpha-amino acid*. 1. any organic molecule of the general formula $R—CH(NH_2)COOH$, having both acidic and basic properties; 2. any of the 20 subunits found as normal constituents of polypeptides.

amino group the univalent group $-NH_2$, often ionized as $-NH_3^+$.

aminopeptidase a peptidase (proteolytic enzyme) that attacks peptide bonds adjacent to the amino end of a polypeptide chain.

ammonia 1. a colorless, pungent, poisonous, high soluble gas, NH_3. 2. this compound dissolved in water; ammonium hydroxide.

ammonification the production of ammonia by soil organisms, particularly by the reduction of nitrates or nitrites.

ammonium ion the univalent ion NH_4^+. Ammonia (NH_3) reacts with water to form an ammonium ion and a hydroxyl ion.

amnion the innermost of the extraembryonic membranes of reptiles, birds, and mammals, and thus the sac in which the embryo itself is suspended.

amniote egg as in reptiles and birds, an egg that forms an amnion within the egg case.

amoeba see *ameba.*

AMP *adenosine monophosphate*, a molecule consisting of adenine, ribose and one phosphate group.

amplexus the clasping of the female amphibian by the male during mating.

ampulla (pl. *ampullae*) a dilated portion of a canal or duct, such as the ampulae of the semicircular canals of the inner ear, and the ampullae of the echinoderm water vascular system, which are muscular sacs located above the tube feet.

amylase any enzyme that digests starch.

amylopectin a form of plant starch consisting of branched chains of alpha glucose subunits.

amylose a form of plant starch consisting of unbranched chains of alpha glucose subunits.

anabolic see *anabolism.*

anabolism in cells, synthetic (or building) chemical activity that produces a more highly ordered chemical organization and a higher free energy state. Compare with *catabolism.*

anaerobe an organism that does not require free oxygen to live *Facultative anerobe*, an organism that can utilize free oxygen when available but does not require it (e.g., yeast). *Obligate anerobe*, an organism that cannot tolerate free oxygen.

anaerobic 1. living or functioning in the absence of oxygen. 2. *anaerobic respiration*, respiration not requiring or utilizing oxygen; glycolysis.

analog also *analogue*, 1. an analogous relationship. 2. an artificial biochemical sharing some properties (e.g., enzyme binding) with a natural biochemical.

analogous in comparative morphology, similar in form or function but derived from different evolutionary or embryonic precursors (e.g., the wings of insects and birds).

analogy similarity in function, and perhaps appearance, but stemming from different evolutionary origins.

anaphase the stage of mitosis or meiosis in which the centromeres divide and separate and the two daughter chromosomes of each chromosome travel to opposite poles of the spindle; and in which the spindle elongates while the centromeric spindle fibers shorten.

anaphase I of meiosis the stage at which the chromosomal tetrads part and the homologous centromeres of each pair travel to opposite poles of the spindle.

anaphase II of meiosis the stage of which the centromeres divide and daughter centromeres separate, moving to opposite poles of the spindle.

anastomoses a netlike arrangement, as in neural fibers or blood vessels.

androecium a whorl of stamens; the stamens of a flower taken together.

androgen any substance that promotes masculine characteristics; male sex hormone.

anemone *sea anemone*, a sedentary coelenterate having a largish, cylindrical body and one or more whorls of tentacles.

angiosperm a plant in which the seeds are enclosed in an ovary; a flowering plant; a monocot or dicot.

angular gyrus a region of the cerebrum concerned with the translation of word symbols relayed from the visual center.

animal a member of the animal kingdom, consisting solely of multicellular forms, almost entirely diploid except for the gamete stage.

animal pole 1. the part of the ovum having the most cytoplasm and containing the nucleus. 2. in the early embryo, the comparable region and the region in which the greatest amount of cell division occurs. Compare *vegetal pole.*

anion a negatively charged ion.

annual yearly; a plant that completes its life cycle within a year; see *perennial.*

annual growth rings the concentric rings seen in the cross section of a woody stem, each ring corresponding to one year's growth.

annulus (pl. *annuli*), ("little ring"), 1. a pore in the nuclear membrane. 2. the dense wall of a sporangium, within a sorus (ferns).

antagonist one of a pair of skeletal muscles (or groups of muscles) the actions of which oppose one another.

antenna 1. one of the paired, jointed, movable sensory appendages of the heads of insects or other arthropods. 2. *light antenna*, a cluster of photosynthetic pigments (chlorophyll and accessory pigments) which receives energy from photons and transfers that energy to a single photocenter.

anterior (adj.) toward the front or head end of an organism.

anterior chamber see *eyeball.*

anterior (pituitary) lobe see *pituitary gland.*

anther the pollen-producing organ of the stamen.

antheridium (pl. *antheridia*) in lower plants and fungi, a male reproductive organ containing motile sperm.

antheridogen a hormone inducing the formation of antheridia.

anthocyanin a class of water-soluble plant pigments, including most of those that give blue and red flowers their colors.

anthropoid 1. resembling a human. 2. *anthropoid ape*, any tailless ape of the family *Pongidae*, comprising the gorillas, chimpanzees, orangutans, gibbons and simiang.

antibiotic any of a large number of substances, produced by various microorganisms and fungi, capable of inhibiting or killing bacteria and usually not harmful to higher organisms; e.g., *penicillin, streptomycin.*

antibiotic resistance the ability of an organism to survive in the presence of an antibiotic. Antibiotic resistance is sometimes transmitted from one bacterium to another by means of a plasmid vector.

antibody a protein molecule that can recognize and bind to a foreign molecule. See also *immunoglobulin.*

antibody-antigen complex a molecular complex consisting of an antibody bound noncovalently to one or more specific antigens.

anticodon a region of a transfer RNA molecule consisting of three sequential nucleotides capable of antiparallel Watson-Crick pairing with the three sequential nucleotides of a codon.

anticodon loop the region of a transfer RNA molecule containing the anticodon; one of the three or four loops of a transfer RNA molecule.

antidiuretic any substance that helps the body conserve water by increasing water reabsorption in the nephrons.

antidiuretic hormone (ADH) also *vasopressin*, a polypeptide hormone secreted by the posterior pituitary, the action of which is to increase the resorption of fluid from the kidney filtrate.

antigen 1. any large molecule, such as a cell-surface protein or carbohydrate, that stimulates the production of specific antibodies, or that binds specifically with such antibodies. 2. any antibody-specific site on such a molecule.

antigenic determinant see *variable region.*

antigen-presenting cell an activated macrophage carrying an antigen that will activate a matching helper T-cell.

antihemophilic factor a blood factor necessary for normal clotting, congenitally lacking in hemophilics. Antihemophilic factor can be isolated from normal blood and administered to hemophilics.

antiparallel (adj.) running parallel but in the opposite direction; specifically, the two strands of DNA, which form parallel interwound helices in which the 5'-to-3' direction in one strand is the 3'-to-5' direction in the other.

antiserum (pl. *antisera*), blood serum containing antibodies specific to some particular antigen.

antisprouting factor a local hormone produced by the axons of nerves that inhibits the growth of other nerve axons, in a developmental negative feedback control loop.

anus the opening of the lower end of the alimentary canal (gut), through which solid wastes are voided.

aorta in vertebrates, the principle or largest artery, conveying blood away from the heart.

aortic arch in embryology, one of a series of five paired, curved blood vessels that arise in the embryo from the ventral aorta, pass through the branchial arches, and unite to form the dorsal aorta.

aortic body sensory structures in the aorta that respond to changes in blood CO_2 and pH levels by sending neural impulses to the respiratory control center.

aortic semilunar valve in the heart, a one-way valve at the base of the aorta that prevents backflow into the left ventricle during systole.

aphotic zone in the aquatic environment, those depths below the penetration of effective light, where respiratory activity exceeds photosynthesis.

apical pertaining to the apex or tip, as of a shoot or root.

apical dominance the uppermost growing tip of a plant stem that hormonally inhibits the upward growth or formation of other branches.

apical meristem the meristem (actively dividing cells) of the growing tip of a root or shoot.

apodeme an ingrowth of the arthropod exoskeleton that serves as the point of attachment for a muscle.

apomixis the production of seeds without the union of gametes.

aponeurosis a flattened sheet of dense connective tissue covering certain muscles.

appendicular skeleton the bones of the limbs (two pairs), along with the pelvis and pectoral girdles.

appendix also *vermiform appendix*, a hollow, fingerlike, blind extension of the cecum, having no known function in humans.

appetitive behavior a variable, nonstereotyped part of instinctive behavior involving searching (for food, water, a mate) for the opportunity to perform.

appetitive stage see *appetitive behavior.*

aqueous pertaining to water; dissolved in water; watery.

aqueous humor see *eyeball.*

arboreal tree-dwelling.

archegonium the female reproductive organs of ferns and bryophytes.

archenteron in embryology, the primitive digestive cavity of the gastrula.

arithmetic increase also *linear increase*, in populations, increases that occur by simple addition, thus 1, 2, 3, 4, 5, 6, 7, etc. Compare *exponential increase.*

arteriole a small artery.

arteriosclerosis inelasticity and thickening of the arterial walls.

artery a vessel carrying blood away from the heart and toward a capillary bed.

articulate (v.), to form a joint (with).

artificial selection the deliberate selection for breeding by humans of domesticated animals or plants on the basis of desired characteristics.

ascending colon first region of the large intestine.

ascocarp a cuplike or saclike body in the Ascomytes in which asci are produced.

ascogonium female reproductive organ of ascomycetes which receives the antheridial nuclei in fertilization, and from which dikaryotic hyphae emerge.

ascomycete a sac fungus.

ascospore a haploid spore produced after sexual reproduction within the ascus of a sac fungus.

ascus (pl. *asci*), in ascomycetes, the sac in which meiosis occurs and in which four or eight ascospores are subsequently formed.

aseptic free of microorganisms.

asexual reproduction reproduction not involving the union of genetic material from two sources.

assay system in biology, a situation in which the amount or character of growth or other biological activity is an indicator of the presence or activity of a micronutrient, toxin, hormone, etc.

aster one of a pair of structures, formed in mitosis and meiosis of all animal cells and many other eukaryotic cells, each consisting of *astral rays* (of microtubules and microfilaments) radiating from a centriole.

astrocyte cells of the central nervous system specialized for transporting substances in and out of the capillaries. See also *blood brain barrier*.

atherosclerosis a form of arteriosclerosis characterized by fat deposits (plaques) in the inner lining of the arterial wall.

atlas the first cervical vertebra, which supports the head.

atom the smallest indivisible unit of an element still retaining the element's characteristics.

atomic mass (also, *atomic weight*), the average mass of the atoms of an element, given in daltons; the exact mass of a specific isotope.

atomic number the number of protons in an atomic nucleus.

ATP, adenosine triphosphate a ubiquitous small molecule involved in many biological energy exchange reactions, consisting of the nitrogenous base adenine, the sugar ribose, and three phosphate residues.

atrioventricular node a small mass of specialized muscle fibers at the base of the wall between the atria of the heart, conducting impulses to the bundle of His.

atrioventricular valve a valve between an atrium and the corresponding ventricle.

atrium also *auricle*, either of the two upper chambers of the heart, each of which receives blood from veins and in turn forces it into the corresponding ventricle.

attachment site one of three regions on the intact ribosome, two of which are pockets that interact with tRNA and mRNA during translation, and one that acts as an insertion point for the leading end of mRNA.

auditory canal the open, bony canal from the outer ear to the eardrum.

auditory receptor 1. a structure specialized for the reception of sound.

australopithecine pertaining to members of the extinct hominid genus *Australopithecus*.

autogenic succession community development or succession as a result of the activities of the organisms themselves.

autoimmune disease a disease in which the organism's immune system attacks and destroys one or more of the organism's own tissues.

autoimmune response see *autoimmune disease*.

autonomic learning any learned response mediated by the autonomic nervous system.

autonomic system the system of motor nerves and ganglion which innervates blood vessels, heart, smooth muscles, and glands and controls their involuntary functions. See *sympathetic* and the *parasympathetic divisions*.

autophagy "self-eating," destruction of aging or damaged cellular organelles by powerful hydrolyzing enzymes contained in lysosomes.

autoradiography the production of pictures (*audioradiograph*) revealing the presence of radioactive material in a thin object or section, in which a film or photographic emulsion is laid directly on the object to be tested, is exposed to radiation for a period of time, and is developed photographically.

autosome any chromosome other than the sex (X and Y) chromosomes.

autotetraploid (adj.) having four homologues of each chromosome type, derived from chromosomal duplication without cell division in a diploid organism; n., an autotetraploid organism.

autotroph 1. a microorganism capable of using carbon dioxide as its only source of carbon (and thus receiving its energy from sources other than organic compounds). 2. a microorganism capable of growth on minimal medium (as opposed to *auxotroph*).

auxin a class of natural or artificial substances that act as the principle growth hormone in plants. The naturally occurring auxin is primarily, or solely, *indoleacetic acid*.

auxospore the zygote of a diatom, consisting of a naked unflagellated cell capable of considerable growth before becoming encapsulated.

auxotroph a mutant organism requiring at least one growth factor not normally required of its species, and thus unable to grow on minimal medium.

axial pertaining to the axis. Axial fruit and flowers grow close to the stem as opposed to terminal fruit and flowers, which grow on the tips of branches.

axial skeleton in vertebrates, the skull, vertebral column, and bones of the chest, opposed to appendicular skeleton.

axil the angle between a petiole or branch and the supporting stem.

axillary bud see *lateral branch bud*.

axis the second cervical vertebra.

axon the (usually) long extension of a neuron that conducts nerve impulses away from the body of the cell.

axoneme in cilia and flagella, the active central unit, including the microtubules and their interconnections.

axon tree the branching, terminal region of an axon. See *neuron*.

axopod (pl. *axopodia*) in sarcodines, slender, microtubular spines upon which feeding pseudopods move. See also *pseudopod*.

bacillus (pl. *bacilli*) 1. an aerobic, rod-shaped, spore-producing bacterium of the genus *Bacillus*. 2. any rod-shaped bacterium.

backcross a cross between an individual of the first filial generation (i.e., an individual produced by the cross between two true-breeding strains) and an individual of either parental strain; see *testcross*.

bacteriochlorophyll a purple photosynthetic pigment used by various anaerobic photosynthetic bacteria.

bacteriodopsin a light-sensitive pigment used by certain archaebacteria for capturing light energy to be used in chemiosmosis.

bacteriophage a virus that infects and lyses bacteria.

bacterium (pl. *bacteria*), any of numerous prokaryotic organisms.

bailer in the crayfish, an appendage that beats rapidly, moving water below the carapace and anteriorly over the gills.

ball-and-socket joint a joint allowing maximal rotation and flexion, consisting of a ball-like termination on one part, held within a concave, spherical socket on the other; e.g., the hip joint.

ball-and-stick model 1. a solid model of a molecule, in which atomic centers are represented by colored balls and covalent bonds are represented by sticks. 2. a drawing of such a model.

bark the portion of a stem outside of the wood (xylem), consisting of cambium, phloem, cortex, epidermis, cork cambium, and cork; everything from the vascular cambium outward.

baroreceptor see *sensory receptor*.

Barr body a dark-staining feature of the nucleus of the cells of female mammals, representing the condensed X chromosome.

Bartholin's glands see *vestibular glands*.

basal body a structure found beneath each eukaryotic flagellum or cilium, consisting of a circle of nine short triplets of microtubules.

basal cell a bulbous cell at the base of the suspensor, believed to anchor and transport food into the embryo.

basal meristem a meristematic reserve in grasses that is below the leaves and stem, and thus is protected from grazing.

basal metabolism rate a measure of oxygen consumption per minute per kilogram of an individual at rest.

base 1. a compound that reacts with an acid to form a salt; a substance that releases hydroxyl ions when dissolved in water. 2. a nitrogenous base (purine or pyrimidine) of nucleic acids.

base deletion a mutation of DNA in which a single nucleotide or nucleotide pair is removed and the phosphate-sugar backbone is rejoined.

base insertion mutation in DNA in which a single nucleotide is inserted into the chain.

base pairing the specific pairing of purines and pyrimidines in DNA, occurring between adenine and thymine and between guanine and cytosine, and an essential part of replication, transcription, and translation.

base substitution a mutation in DNA in which one nucleotide is replaced by another, or modified into another.

basement membrane in animals, a sheet of collagen that underlies and supports the cells of a tissue; also *basal membrane, basilar membrane.*

basidiocarp the spore-producing organ in the *Basidiomycetes;* a mushroom.

basidiospore an aerial spore produced by meiosis in *Basidiomycetes.*

basidium *(pl. basidia),* the meiotic cell of *Basidiomycetes* that produces basidiospores by budding.

basilar membrane in the vertebrate ear, a membrane that conducts sound waves; also *basement membrane, basal membrane.*

basking to expose one's body to warmth, as the warmth of the sun; a frequent component of *behavioral thermoregulation.*

Batesian mimicry coloration or structural configuration in harmless species, rendering it similar to that of a dangerous species.

B cell a lymphocyte that matures in the bone marrow (mammals) or bursa of Fabricius (birds), and later circulates in the blood; involved in the immune response, especially in the production of free antibodies.

behavior observable activity of an organism; anything an organism does that involves action and/or response to stimulation.

behavioral hierarchy a sequence of fixed action patterns in animals that proceeds from the general to the specific.

behavioral thermoregulation regulating one's internal body temperature by behavioral means, such as *basking,* seeking shade, and taking cold showers.

bell-shaped curve see *normal distribution.*

belly 1. the abdomen. 2. of a muscle, the fleshy, bulging portion.

benthic referring to the benthos, or bottom-dwelling community of organisms.

benthos the aggregation of organisms living on or at the bottom of a body of water.

beta cell see *Islets of Langerhans.*

beta pleated sheet the zig-zag pattern assumed by many structural proteins as they assume their tertiary level of organization.

bicarbonate ion HCO_3^-.

bicuspid valve also *mitral valve,* the two-flapped valve between the left atrium and the left ventricle of the heart.

bilaminar embryonic disc also *embryonic shield,* in the early mammalian embryo, the disklike, two-cell-layered stage just prior to gastrulation.

bilateral symmetry having left and right sides that are approximate mirror images; having a single plane of symmetry.

bile a bitter, highly pigmented, alkaline, fat-emulsifying liquid secreted by the liver, containing bile salts and bile pigments.

bile duct the duct that carries bile from the cystic duct of the gall bladder to the small intestine.

bile salts cholesterol and other steroid compounds secreted by the liver and essential for the emulsifying, digestion and absorption of dietary fats.

bilirubin a reddish-yellow pigment occurring in bile, blood, and urine; a breakdown product of hemoglobin.

binary fission fission (splitting) into two organisms of approximately the same size; cell division (asexual reproduction) in prokaryotes.

binocular having or involving two eyes; *binocular vision,* stereoscopic vision, the ability to perceive depth through the integration of two overlapping fields of vision.

binomial system the tradition, introduced by Linnaeus, that each organism is given a taxonomic name in Latin consisting of a generic term and a specific term.

binucleate having two nuclei.

biochemical pathway a series of enzymatic steps by which an organic molecule is progressively modified.

biochemistry the science dealing with the chemistry of living organisms.

biodegradable capable of being rendered harmless upon exposure to the elements and organisms of the soil or water.

biogeochemical cycle the pathway of elements (e.g., carbon, nitrogen) or compounds (water) as they are taken up and released by organisms into the physical environment.

biogeography the study of the geographical distribution of living things.

biological clock an innate mechanism by which living organisms are able to perceive the lapse of time.

biological species concept see *species.*

biomass 1. the total weight of living organisms, usually expressed as dry weight, per unit area of volume in a particular habitat. 2. the weight of organic material produced in a unit time period under specified conditions.

biome a complex of ecological communities characterized by a distinctive type of vegetation, as determined by the climate.

biophage an organism that kills to feed.

biosphere the entire part of the earth's land, soil, waters and atmosphere in which the living organisms are found.

biotic 1. pertaining to life. 2. ecological factors due to the interactions of living organisms, as opposed to abiotic factors such as climate.

biotic control population control by living factors, including both intraspecific and interspecific influences. Compare *abiotic control.*

biotic potential the maximum growth rate of a population when it is unrestricted by environmental resistance.

biotin a water-soluble vitamin essential to most organisms.

bipedal literally two-footed; walking on two feet, as birds, humans, kangaroos and some dinosaurs.

bipinnaria larva a larval form in some echinoderms, similar to the tornaria larva of hemichordates.

birth control see *conception.*

birth control pill an oral steroid contraceptive that inhibits ovulation, fertilization or implantation, causing temporary infertility in women.

birth rate crude birth rate, the number of births per year per 1000 individuals of all ages in the population. Other birth rates may be expressed in terms of births per year per number of women of a specific age range, the expected lifetime reproduction per female and so on.

bivalve a mollusk with two shells (valves) hinged together; a pelecypod.

bladder 1. any membraneous sac serving as a receptacle for fluid or gas. 2. in brown algae, the hollow spheres that act as floats to keep the algae at the surface of the water. 3. *gall bladder,* a muscular sac-like organ that is the temporary receptacle of bile. 4. *urinary bladder,* the storage organ for urine in many vertebrates.

bladderworm (technically, *cysticerci*) the encystment stage following the migration of newly hatched tapeworm larvae, often found in skeletal muscle of infected steers and hogs.

blade 1. *leaf blade,* broad part of a leaf, as distinguished from petiole or midrib. 2. any broad, thin part of the thallus (body) of a red, green or brown alga.

blastocoel the cavity of a blastula.

blastocyst the early preimplantation stage in the mammalian embryo, in which the embryo consists of an *inner cell mass,* a *blastocyst cavity,* and an outer *trophoblast.*

blastocyst cavity see blastocyst.

blastoderm the single upper layer of cells of a blastula in embryos that develop in telolecithal eggs (fish, birds, and reptiles).

blastodisc the small disc of cytoplasm on the yolk of a bird or reptile egg, containing the egg nucleus, that becomes the early embryo.

blastomere any cell produced during cleavage, through the blastula stage.

blastopore in a gastrula, the opening of the archenteron produced by the invagination of cells during gastrulation.

blastula the early embryonic stage, consisting of a single layer of cells that form a hollow ball enclosing a central cavity, the blastocoel.

blending inheritance a theory of inheritance, which states that parental characters blend to produce an intermediate character in the offspring.

blight invasion of a plant by a parasitic water mold.

blood a circulating fluid in animals that helps distribute gases, nutrients, etc., and often collects cellular wastes.

blood brain barrier in the brain, a state of highly selective permeability to many substances that readily move into or out of other tissues, attributed in part to a lack of the usual looseness of capillary structure.

blood fluke parasitic flatworm of humans and other mammals.

blood type also *blood group,* one of a group of categories into which an individual can be categorized depending on his or her blood cell surface antigens.

body stalk in embryology, the connection between the early embryo and the extraembryonic tissues; later the body stalk becomes the umbilical cord.

Bohr effect the competitive effect of carbon dioxide in reducing the affinity of hemoglobin for oxygen.

bolus a clump of chewed food.

bombykol the sex attractant produced by the female silkworm moth.

bond see *chemical bond, covalent bond, hydrogen bond, salt linkage, van der Waals forces.*

bone the hard connective tissue forming the skeleton of most vertebrates, consisting principally of a collagen matrix impregnated with calcium phosphate.

bone compact, see *compact bone;* bone, spongy, see *spongy bone.*

bone marrow see *marrow.*

book lung the respiratory organ of a spider, scorpion or other terrestrial arachnid, consisting of thin, membranous structures arranged like the leaves of a book; *book gill,* the similar structure in horseshoe crabs.

botanist a scientist who studies plants.

botany the scientific study of plants.

bottle cell during involution in animal gastrulation, cells that take on a peculiar bottlelike appearance when drawn into the embryo.

bottleneck *population bottleneck,* a relatively short period of time during which the size of a population becomes unusually small, resulting in a random change in gene frequencies.

botulism a bacterial disease caused by the presence in foods (usually canned foods) of *Clostridium botulinum,* whose nerve toxins are among the most powerful poisons known.

bound ribosome a ribosome attached to the surface of a membrane, as of the endoplasmic reticulum. See also *endoplasmic reticulum, rough.*

bounded (adj.), limited by: e.g., *a membrane-bounded vesicle.*

bourgeois strategy in animal behavior, the immediate retreat from an adversary without physical encounter, as though through prearranged rules.

bowel see *colon.*

Bowman's capsule one of numerous double-walled membraneous capsules in the nephron, each surrounding a glomerulus (ball of capillaries).

brain 1. in vertebrates, the anterior enlargement of the central nervous system, encased in the cranium. 2. in invertebrates, any anterior concentration of neurons more or less corresponding in function to the vertebrate brain

braincase the primitive bony covering of the vertebrate brain as found in most fish and in the embryos of other vertebrates.

brain hormone (BH), also *prothoracicotrophic hormone (PTTH),* in insects, a hormone originating in the brain that stimulates the prothoracic gland to release ecdysone.

branch primordium see *primordium.*

branch root also *secondary root* in plants, a root that arises from an older primary root.

broad ligament one of a pair of supporting ligaments of the uterus, bearing the ovaries.

broad-niched an organism able to live under a wide variety of conditions.

Broca's area a region of the cerebrum concerned with motor aspects of speech.

bronchiole a small branch of a bronchus, part of the respiratory tree of the lungs.

bronchus (pl. *bronchi*), either of the two main branches of the trachea.

brown alga any alga of the *Phaeophyta,* usually brown due to fucoxanthin pigment; e.g., kelp.

Brownian motion also *Brownian movement,* the irregular motion of microscopic particles in a liquid or gas, caused by random thermal agitation of molecules in the medium.

bud a small protuberance on the stem of a plant, containing meristematic tissue and covered with overlapping rudimentary foliage.

bud scale an often hairy, waxy, or resinous scale enclosing an immature bud.

bud scale scar external scar marking the extent of each year's growth in a woody stem.

budding 1. asexual cell reproduction with unequal cytoplasmic division, as of yeasts. 2. similar asexual reproduction of a multicellular animal, as in hydras.

buffer a solution of chemical compounds capable of neutralizing both acids and bases, and thus able to maintain an equilibrium pH.

bulbourethral glands (Cowper's glands), a pair of small glands that secrete a mucous substance into the urethra in males, especially on sexual arousal.

bulk flow the net movement of water brought about by gravity or pressure.

bundle of His, also atrioventricular bundle, a bundle of specialized muscle fibers that conduct impulses along the ventricular septum.

bundle sheath one or more layers of compactly arranged parenchyma cells that enclose the small veins in a leaf, in C_4 plants, the site of the Calvin cycle.

bursa of Fabricius in birds, a discrete abdominal organ in which B-cell thymocytes mature.

calcaneous the protruding heel bone, receives the *achilles tendon* from the calf muscle.

calcareous containing or composed of calcium carbonate.

calcareous sponge, a sponge with a skeleton of calcium carbonate spicules.

calcitonin one of the two major hormones of the thyroid gland, whose major action is to inhibit the release of calcium from the bone.

calcium salt a salt in which calcium is the metallic element; in nature, most often calcium carbonate or calcium phosphate.

callus tissue an undifferentiated tissue that forms over wounds in plants.

calorie 1. also *small calorie* (calorie proper), the amount of heat (or equivalent chemical energy) needed to raise the temperature of one gram of water by 1°C. 2. also *large calorie, kilocalorie,* the heat needed to raise the temperature of a kilogram of water by 1°C; 1000 small calories.

Calvin cycle the cycle in C_3 photosynthesis in which NADPH and ATP reduce carbon dioxide to glyceraldehyde phosphate (PGAL).

calyx (pl. calyces) 1. the outermost whorl of floral parts (the sepals), usually green and leaflike. 2. in the kidney, cup-shaped regions that receive urine from the collecting ducts.

cambium undifferentiated meristematic tissue in a plant. See *cork cambium, primary growth, secondary growth, procambium, vascular cambium.*

canaliculus (pl. *canaliculi*), a tiny canal, as in bone, where canals communicate between osteocytes.

canine tooth in many mammals, any of four pointed, often elongated teeth used for tearing in feeding and in defense.

canopy the upper leafy area of a tree or especially of a forest

capillary the smallest blood vessels; the fine channel between the arteriole and venule.

capillary action the tendency of a liquid to rise in a small tube due to adhesion to its inner surfaces and cohesion among water molecules.

capillary bed tissue rich in capillaries; a body of capillaries taken together.

capsid in viruses, the protein coat surrounding the nucleic acid core.

capsomere in viral capsids, an individual protein subunit whose varied arrangements determine shape of different viruses.

capsule 1. a type of fruit which becomes dry and hard before rupturing to release seeds. 2. in animals, any membraneous sac or covering.

carapace a hard, shield-like covering of the dorsal surface of an animal, as a lobster, turtle, armadillo, etc.

carbaminohemoglobin deoxygenated hemoglobin complexed with carbon dioxide.

carbohydrate a class of organic compounds with multiple hydroxyl side groups and an aldehyde or ketone group; sugars, starches, cellulose, and chitin; empirical formula $(CH_2O)n$.

carbon dioxide a colorless gas of the formula CO_2, readily dissolved in water and capable of reacting with water to form a hydroxyl ion (OH^-) and a bicarbonate ion (HCO_3^+).

carbonate rock rock rich in calcium carbonate, formed by deposits of marine shells; included are limestone and marble.

carbonic anhydrase an enzyme that catalyses the reversible conversion of carbonic acid to carbon dioxide gas and water.

carboxyhemoglobin see *carbaminohemoglobin*.

carboxylation the enzymatic addition of carbon to substrate. Compare *decarboxylation*.

carboxypeptidase a protease that cleaves peptide bonds near the carboxy terminal (C-terminal) end of a polypeptide.

cardiac pertaining to the heart; near or toward the heart.

cardiac circuit the blood vessels of the heart, including the coronary arteries and veins and the capillaries.

cardiac muscle specialized muscle of the heart that is both striated and involuntary.

cardiac output the volume of blood pumped by the heart each minute.

cardiac sphincter the ring of muscle that closes the passageway between the lower esophagus and the stomach.

cardinal vein veins of fish, running anteriorly and posteriorly along the vertebral column and discharging into the right and left common cardinal veins.

carnivorous (adj.) flesh eating.

carotene a red or orange hydrocarbon, $C_{40}H_{56}$, found in most plants as an accessory photosynthetic pigment.

carotenoid any of a group of red, yellow and orange plant pigments chemically and functionally similar to carotene.

carotid artery either of a pair of large arteries on each side of the neck and head.

carotid body a mass of cells and nerve endings on either carotid artery that senses blood CO_2 and pH levels and responds by affecting the rate of breathing and the heart beat.

carpal any wrist bone.

carpel in flowers, a simple pistil, or a single member of a compound pistil; one sector or chamber of a compound fruit.

carposporangium see *carpospore*.

carpospore a diploid spore in members of the genus *Polysiphonia*, formed in the pouchlike carposporangium after fertilization.

carrageenan a polysaccharide produced by Irish moss, a red alga; used as a thickening agent.

carrier a heterozygote for a mutant or rare recessive allele.

carrier molecule a protein that binds a small molecule in facilitated diffusion or active transport.

carrying capacity a property of the environment defined as the size of a population that can be maintained indefinitely.

cartilage a firm, elastic, flexible, translucent type of connective tissue; in development, a precursor of bone formation.

cartilaginous fishes fish having skeletons of cartilage, comprising sharks, rays, and chimeras; one of the eight vertebrate classes.

casein the principal protein component of milk.

Casparian strip a waxy strip on cell walls of endoderm that serves as a barrier to the conduction of moisture in roots and stems.

Castle-Hardy-Weinberg law also *Hardy-Weinberg law, binomial law;* a statement of the mathematical expectations of genotype frequencies given allele frequencies in a population of random mating diploid individuals.

catabolism in cells, chemical activity that decreases chemical organization and free energy therein. Compare with *anabolism*.

catalase an enzyme that decomposes hydrogen peroxide to molecular oxygen and water, in the reaction $2 H_2O_2 \rightarrow 2 H_2O + O_2$.

catalyst an agent that causes or accelerates a chemical reaction, while not being permanently altered itself.

cation a positively charged ion.

CCA stem a sequence of three nucleotides added enzymatically to the 3' end of all tRNA molecules.

cecum in vertebrates, a blind pouch or diverticulum of the intestine, at the juncture of the small and large intestines.

cell 1. the structural unit of plant and animal life, consisting of cytoplasm and a nucleus, enclosed in a semipermeable membrane. 2. any similar organization, as that of a protist or prokaryotic organism.

cell body see *neuron*.

cell cycle the cycle of events in the life of a cell, including G_1 (Gap 1, synthesis and growth), S (DNA replication), G_2 (Gap 2, synthesis of spindle proteins), and M (mitosis and cell division).

cell division the division of a cell into two daughter cells.

cell-free system a preparation of cytoplasmic constituents extracted from cells by the disruption of the plasma membrane.

cell plate in plant cell division, the forming plasma membrane between nascent daughter cells.

cell respiration the energy-yielding metabolism of foods in which oxygen is used. Also see *respiration* and *glycolysis*.

cell sap the fluid that fills large vacuoles in plant cells; primarily water with various substances in solution or suspension.

cell streaming see *cyclosis*.

cell-surface antigen any constituent of the cell surface capable of eliciting a specific antibody response.

cell theory the universally accepted proposal that cells are the functional units of organization in living organisms and that all cells today come from preexisting cells.

cellulase an enzyme that digests cellulose.

cellulose an inert, insoluble carbohydrate, a principal constituent of plant cell walls, consisting of unbranched chains of beta glucose.

cell wall the semirigid extracellular encasement of a plant, fungal, algal, cyanophyte, or bacterial cell that gives it a definite shape.

central canal 1. also *Haversian canal*. See *Haversian system*. 2. See *water vascular system*.

central cell or *endosperm mother cell,* in flowering plants, the binucleate cell of the embryo sac, containing two polar nuclei; following fertilization, becomes the triploid primary endosperm nucleus.

central cylinder the supporting structure of the moss plant, which some botanists believe to have specialized conductive cells.

central dogma the proposition that all biological information is encoded in DNA, transmitted by DNA replication, transcribed into RNA, and translated into protein; together with several exceptions for certain misbehaving viruses. The term was coined by Francis Crick.

central lymphoid tissue lymphoid tissue of bone marrow and thymus.

central nervous system the brain and spinal cord.

centriole one of a pair of organelles in animal, protist, fungi, and lower plant cells, each centriole in turn consisting of a pair of short cylinders of nine triplet microtubules.

centrolecithal of an egg, having a central yolk entirely surrounded by cytoplasm, as in insects.

centromere also *kinetochore*, also *spindle fiber attachment*, the specialized region of a chromosome to which spindle fibers are attached, visible under the light microscope as a light-staining body, and under the electron microscope as a pair of shield-like plates. A centromere is considered to be a single object until the actual physical separation of daughter centromeres in anaphase.

centromeric pertaining to the centromere.

centromeric heterochromatin a heterochromatin region adjacent to the centromere.

centromeric spindle fiber any of a group of microtubules attached to each centromere and proceeding to a spindle pole in mitosis or meiosis.

cephalic pertaining to the head.

cephalization the evolutionary tendency to concentrate neural and sensory functions in an anterior end.

cephalochordate ("head cord animal"), a lancelet (*Branchiostoma*) of a chordate subphylum in which the permanent notochord extends through what would be the head if it had one, which it doesn't.

cercaria see *human liver fluke*.

cerebellum a portion of the brain serving to coordinate voluntary movement, posture, and balance; it is located behind the cerebrum.

cerebral cortex the outermost region of the cerebrum, the "gray matter," consisting of several dense layers of neural cell bodies and including numerous conscious centers, as well as regions specialized in voluntary movement and sensory reception.

cerebral hemisphere either the right or left half of the cerebrum.

cerebrospinal fluid a cushioning fluid found in the ventricles of the brain and in the central canal of the spinal cord.

cerebrum the anterior, dorsal portion of the vertebrate brain, the largest portion in humans, consisting of two *cerebral hemispheres* and controlling many localized functions, among them voluntary movement, perception, speech, memory and thought.

cervical 1. pertaining to the neck. 2. pertaining to the cervix of the uterus.

cervical cap a birth control device consisting of a small cap that encloses the cervix.

cervical vertebra any vertebra of the neck region.

cervix the opening to the uterus and the surrounding ring of firm, muscular tissue.

CF1 particle an ultramicroscopic structure on the outer surface of the thylakoid, the site of chemiosmotic phosphorylation in photosynthesis.

C4 plant a plant in which the initial product of CO_2 fixation is a four-carbon compound.

chain reaction 1. a reaction which produces a product necessary for the continuation of the reaction. 2. any series of events in which each event causes the next.

chain termination codon any of the three codons of the genetic code causing the termination of polypeptide synthesis; in RNA these are UGA, UAG, and UAA.

chain termination mutation a base substitution mutation in which the new codon created is a chain termination codon.

chaparral a vegetation type common in California, characterized by a dense growth of low, evergreen shrubs and trees.

character a classifiable feature, trait or characteristic of an individual organism; a specific component of a phenotype.

character displacement where similar species share a niche, the tendency for physical differences to become emphasized through natural selection.

character state a specific state of morphological character, defined in terms of presence or absence.

Chargaff's rule in DNA structure, the observation that the quantity of adenine in DNA is equal to the quantity of thymine while that of guanine is equal to that of cytosine.

charging enzyme any of a group of specific enzymes that covalently attach amino acids to their appropriate tRNA's.

chelicera (pl. *chelicerae*), one of the first pair of appendages in the *Chelicerata*; a spider's fang.

chemical bond any of several forms of attraction between atoms in a molecule.

chemical bond energy the potential energy invested in the formation of a chemical bond or that released upon its dissolution. See also *covalent bond, hydrogen bond, salt linkage, van der Waals forces*.

chemical communication the transmission of information between individuals by the use of phenomones.

chemical formula 1. representation of a molecule or compound, using chemical symbols and indicating the ratio of one element to another. 2. a mathematical statement including an equation, rule, principle, or answer.

chemically gated channel an ion channel whose gate is activated by chemical means rather than by voltage. See also *ion channel*.

chemical reaction the reciprocal action of chemical agents on one another; chemical change.

chemiosmosis the process in mitochondria, chloroplasts and aerobic bacteria in which an electron transport system utilizes the energy of photosynthesis or oxidation to pump hydrogen ions across a membrane, resulting in a proton concentration gradient that can be utilized to produce ATP.

chemiosmotic differential also known as *electrochemical proton gradient*. See *chemiosmosis*.

chemiosmotic phosphorylation the production of ATP using the energy of protons passing across a membrane and through F_1 and CF_1 particles.

chemoreceptor a neural receptor sensitive to a specific chemical or class of chemicals.

chemosynthesis the synthesis of organic compounds with energy derived from inorganic chemical reactions.

chemotroph an organism that lives on the energy of inorganic chemical reactions; *chemotrophic*, adj.

chew to grind with the teeth, especially utilizing sideways motion.

chiasma (pl. *chiasmata*) as first seen in diplotene of the first meiotic prophase, the cross- or X-shaped configurations taken by homologous chromatids as repulsion occurs. The regions still fused indicate where crossing over occurred during pachytene.

chief cell a stomach cell that secretes pepsinogen.

chitin a structural carbohydrate that is the principal organic component of arthropod exoskeletons.

chlamydobacterium a filamentous sheathed bacterium.

chlorenchyma chloroplast-containing parenchymal tissue in leaves.

chloride ion the anion of the element chlorine (Cl^-).

chloride ion gate see *ion gate*.

chlorobacterium a green sulfur bacterium.

chlorophyll a green photosynthetic pigment found in chloroplasts, cyanobacteria, and chloroxybacteria. It occurs in several forms, *chlorophyll a, b*, and *c*.

chloroplast a plastid containing chlorophyll.

chlorosis an abnormally pale or yellow condition of plants caused by lack of chlorophyll, due to disease or mineral deficiency.

chloroxybacterium a recently discovered photosynthetic prokaryote apparently more closely related to the ancestor of the plant chloroplast than are cyanobacteria; a symbiont of certain tunicates.

choanocyte also *collar cell*, a type of cell in all sponges and certain protists in which a single flagellum is surrounded at its base by a screen of fused cilia (collar), that filters food from the water current created by the flagellum.

cholecystekinin-pancreozymin a hormone that stimulates the contraction of the gall bladder.

cholesterol a common sterol occurring in all animal fats, a vital component of plasma membranes, an important constituent of bile for fat absorption, a precursor of vitamin D, and too much of which is not good for you.

chondroitin the principal structural protein of cartilage.

chordae tendineae in the heart, tough, cordlike tendons that prevent backflow by holding the tricuspid and bicuspid (auricular-ventricular) valves in place during systole.

chorioallantois a highly vascular extraembryonic membrane of birds, reptiles, and some mammals, formed by the fusion of the chorion and the allantois.

chorion 1. the tough covering on insect eggs; 2. the outermost of the extraembryonic membranes of birds, reptiles and mammals, contributing to the formation of the placenta in placental mammals.

chorionic villi small fingerlike processes of the chorion of the early mammalian embryo, especially before the formation of the placenta.

choroid see *eyeball*.

chromatid in a G_2 chromosome, one of the two identical strands of the chromosome following replication and prior to cell division.

chromatin the substance of chromosomes, a molecular complex consisting of DNA, histones, nonhistone chromosomal proteins, and usually some RNA of unknown function.

chromomere one of the beadlike clumps or granules arranged in a linear array of a chromonema, especially when visible in partially condensed prophase chromosomes. Chromomeres may be more or less representative of functional genetic units (genes).

chromoplast a plastid containing primarily pigments other than chlorophyll, as certain flower pigments.

chromosomal mutation a massive spontaneous change in DNA, generally breakage involving a whole chromosome that has not been repaired or has been repaired improperly.

chromosomal replication see *DNA replication*.

chromosome 1. in eukaryotes, an independent nuclear body carrying genetic information in a specific linear order, and consisting of one linear DNA molecule (in G_1) or two DNA molecules (in G_2), one centromere, and associated proteins. 2. in prokaryotes, an analogous circular DNA molecule. 3. the analogous DNA or genetic RNA molecule of a virus.

chromosome puff in polytene chromosomes, an enlargement of one band associated with transcriptional activity (mRNA production).

chrysophyte a yellow-green alga.

chylomicron in digestion, a minute, protein-coated fat droplet formed during lipid transport in the intestinal villi.

chyme churned, semiliquified food in the stomach.

chymotrypsin a proteolytic enzyme produced by the pancreas.

cilia (sing. *cilium*), fine, hairlike, motile organelles found in groups on the surface of some cells; shorter and more numerous than flagella, but similar in structure, they exhibit coordinated oarlike movement.

ciliary basal body see *basal body*.

circadian rhythm any recurrent sequence of physiological or behavioral activities repeated on a daily basis.

circulatory system the vascular system, consisting of blood-forming organs or tissues, vessels, the heart, and blood.

cisterna slender, membranous channels making up much of the Golgi apparatus.

cistron 1. also *structural gene*, a sequence of DNA specifying the sequence of a polypeptide chain. 2. any continuous genetic unit in which different recessive mutant lesions fail to complement one another in a double heterozygote.

citric acid cycle also *Krebs cycle, tricarboxylic acid cycle*, a cyclic series of chemical transformation in the mitochondrion by which pyruvate is degraded to carbon dioxide, NAD, and FAD are reduced to $NADH_2$ and $FADH_2$, and ATP is generated.

clasper one of a pair of specialized grooved caudal fins used by male sharks and rays as a penis.

class a major taxonomic grouping intermediate between *phylum* (or *division*) and *order*.

classical conditioning see *conditioning, classical.*

clavicle also *collar bone*, one of a pair of bones of the pectoral girdle, articulating with the sternum and scapula.

cleaning symbiont a usually conspicuous organism that interacts mutualistically with others by removing ectoparasites, necrotic tissue, etc., receiving nourishment and inhibiting aggressive behavior of the individual being cleaned.

cleavage the total or partial division of a zygote into smaller cells (blastomeres).

cleavage furrow the initial depression indicating cytoplasmic cleavage.

climax 1. that stage in ecological succession of a plant or animal community that is stable and self-perpetuating. 2. an orgasm.

cline a regular change in allele frequency over geographic space.

clitellum a thickened, ring-like glandular portion of the body wall of earthworms, which secretes mucus to form a cocoon for eggs.

clitoris in female mammals, an erectile, erotically sensitive organ of the vulva, homologous embryologically to most of the penis.

cloaca the common cavity into which the intestinal, urinary, and reproductive canals open in vertebrates other than placental and marsupial mammals.

clonal selection theory in immunology, the proposition that all potential antibody specificity is present in differentiated cells early in development and that the specific immune response consists of inducing appropriately differentiated cells to proliferate clonally.

clone 1. a group of genetically identical organisms derived from a single individual by asexual reproduction. 2. a group of identically differentiated cells derived mitotically from a single differentiated cell.

cloning, gene see *gene cloning.*

closed circulatory system a system in which blood is enclosed within arteries, veins, and capillaries throughout, and is not in direct contact with cells other than those lining these vessels.

closed community biotic communities that are abruptly circumscribed, often because of a confining physical barrier such as a river, bay, or gorge.

closed system 1. in thermodynamics, a system in which matter and energy neither enter nor leave. 2. see *closed circulatory system.*

cnidoblast in coelenterates, a specialized cell containing a stinging or snaring nematocyst.

cnidocyte in coelenterates, a stinging cell.

coacervate droplets with membranelike surface layers that form spontaneously when certain substances are attracted together in water, a hypothetical step in the earliest formation of cells.

coat 1. *seed coat*, the outer protective covering of a seed, composed of one or more layers of tissue derived from the parent sporphyte. 2. *viral coat*, the outer protective protein jacket of a virus.

coccus (pl. *cocci*), any spherical bacterium (principally eubacteria), a condition found in many distantly related groups.

coccyx, the tailbone or lowest portion of the vertebral column in humans and tailless apes, consisting usually of four fused vertebrae.

cochlea in mammals, a spiral cavity of the inner ear containing fluid, vibrating membranes, and sound-sensitive neural receptors.

cocoon in some invertebrates, a covering used to protect offspring during a stage in development, for example, the silky cocoon of moths and spiders.

codominance the individual expression of both alleles in a heterozygote; see *dominance relationships.*

codon also code group, 1. a series of three nucleotides in mRNA specifying a specific amino acid (or chain termination) in protein synthesis. 2. the colinear, complementary series of three nucleotides or necleotide pairs in the DNA from which mRNA codon is transcribed.

coelenteron the saclike cavity of coelenterates, which carries out ingestion, digestion, phagocytosis and absorption of food, and also has circulatory functions; also, gastrovascular cavity.

coelom *true coelom*, a principal body cavity, or one of several such cavities between the body wall and gut, entirely lined with mesodermal epithelium; compare *pseudocoelom.*

coelomate having a true coelom.

coenocyte an organism made up of a multinucleate mass of cytoplasm without subdivision into cells, as certain algae and fungi.

coenocytic consisting of a multinucleate mass of cytoplasm without subdivision into cells.

coenzyme a small organic molecule required for an enzymatic reaction.

coevolution evolution of two closely interacting species, such as predator and prey, where changes in the first determine those of the second.

cofactor 1. any organic or inorganic substance, especially an ion, that is required for the function of an enzyme. 2. in blood clotting, any of the many active participating elements.

cohesion the attraction between the molecules of a single substance.

coitus the act of sexual intercourse, especially between people.

coitus interruptus coitus that is intentionally interrupted by withdrawal before ejaculation.

colchicine a chemical agent that arrests mitosis and meiosis at metaphase.

coleorhiza a protective sheath over the radicle in the embryo of grasses and grains.

coleoptile in the embryos and early growth of grasses, a specialized tubular structure, completely enclosing and protecting the pulmule during emergence.

colinearity *principle of colinearity*, the finding that corresponding parts of a structural gene, mRNA, and polypeptide occur in the same linear order.

collagen in animals, a widely distributed fibrous protein of connective tissue that forms much of the structure of tendons and ligaments.

collar cell see *choanocyte.*

collecting duct also *collecting tubule*, the part of a nephron that collects fluids from distal convoluted tubules and discharges it into the renal pelvis.

collenchyma in plants, a strengthening tissue, a modified parenchyma consisting of elongated cells with greatly thickened cellulose walls.

colon in mammals, the large intestine from the cecum to the rectum; including the ascending, transverse, descending, and sigmoid regions, and the rectum.

colonial 1. generally existing in colonies. 2 *colonial organism*, an organism of semidependent parts, derived by asexual reproduction.

colony 1. a group of animals or plants of the same kind living in a close semidependent association. 2. an aggregation of bacteria growing together as the descendants of a single individual, usually on a culture plate.

color blindness the inability to distinguish colors. *Red-green color blindness*, the inability to distinguish certain shades of red from corresponding shades of green.

color vision the ability to discern colors.

coloration appearance as to color and patterns of color.

colostrum clear, yellowish milk secreted a few days before and after giving birth; colostrum is rich in maternal antibodies.

columnar epithelium epithelium consisting of one (*simple columnar*) or more (*stratified columnar*) layers of elongated, cylindrical cells.

comb a locomotor structure in ctenophorans, one of several slender plates on the body surface, bearing numerous cilia.

comb jelly see *ctenophore.*

combination pill a birth control pill that combines two or more female hormones.

common cardinal vein see *cardinal vein.*

common descent descent of two or more species (or individuals) from a common ancestor; e.g., the similarity in blood chemistry of apes and humans is due to common descent.

communication in its most general sense, any activity that functions to alter the behavior of another animal.

community as assemblage of interacting plants and animals forming an identifiable group within a biome, as in salt marsh or sage desert community.

community development also *ecological succession* and *succession* 1. the process of change in the populations of an area as competing organisms alter their environment. 2. the sequence of identifiable ecological stages or communities occurring over time in the progress of bare rock to a climax community.

compact bone dense, hard bone with spaces of microscopic size.

companion cell in plants, a nucleated cell adjacent to a sieve tube member, believed to assist it in its functions.

comparative anatomy the science of comparing the anatomy of animals, tracing homologies and drawing evolutionary and phylogenetic inferences; see also *allometry*.

comparative psychology the study of mental processes (behavior) in animals from a comparative point of view, usually in a laboratory environment.

competition 1. seeking to gain what another is seeking to gain at the same time; a common struggle for the same object. 2. in ecology, the utilization by two or more individuals or species of the same limiting resource.

competitive exclusion principle see *Gause's law.* .

complement (n.) 1. a group of blood proteins that interact with antibody-antigen complexes to destroy foreign cells. 2. (v.t.) the production of a normal phenotype by two recessive mutations in a double heterozygote; see *complementation test*.

complete flower see *flower*.

complete digestive tract a tubular digestive tract with an anal as well as an oral opening.

complete metamorphosis of insects, development includes distinct larval, pupal, and adult stages.

composite flower inflorescence in which the individual flowers are tightly clustered into a disklike head, often with a border of ray flowers that include petals only.

compound eye an arthropod eye consisting of many simple eyes closely crowded together, each with an indivdual lens and a restricted field of vision, so that a mosaic image is formed.

compound microscope an optical instrument for forming magnified images of small objects, consisting of an *objective lens* that can be brought close to the object being examined, aligned with an *ocular lens* mounted at the other end of a body tube of fixed length.

compound in chemistry, a pure substance consisting of two or more elements in a fixed ratio; consisting of a single molecular type.

concentration gradient 1. a slow, consistent decrease in the concentration of a substance along a line in space. 2. for any spatial difference in concentration, the direction away from the region of greater concentration.

conceptacles in *Fucus*, a strongly heteromorphic brown alga, regions within the blade where the minute gametophyte phase arises.

condensation of chromosomes, the coiling and supercoiling that transforms diffuse chromatin into a compact, discrete body in mitosis.

conditioned response an involuntary response that becomes associated with an arbitrary, previously unrelated stimulus through repeated presentation of the arbitrary stimulus simultaneously with a stimulus normally yielding the response.

conditioning, classical the process by which a conditional response is learned and elicited; compare *operant conditioning*.

condom a thin sheath of rubber or animal membrane worn over the penis during sexual intercourse to prevent conception or venereal infection (after Dr. Condom, an 18th-century English physician); also, *rubber, prophylactic*.

conduction 1. the transfer of movement of fluid, heat, ions, impulses, etc. through or along a channel or medium; e.g., the conduction of impulses through the central nervous system. 2. the transfer of heat through a solid, as opposed to *convection* or *radiation*.

conducting tissue see *vascular tissue*.

cone 1. one of a class of conical photoreceptors in retina that detect color, consisting of a highly modified cilium with specialized proteinaceous pigments for detecting red, green, or blue wavelengths. 2. a male or female reproductive structure of conifers, consisting of a cluster of scalelike modified leaves and either pollen or ovules, or the seed.

conidium (pl. *conidia*), an asexual spore borne on the tip of a fungal hypha; also *conidiospore*.

conidiophore a specialized branch of the mycelium bearing conidia.

conifer an evergreen gymnosperm of the order *Coniferales*, bearing ovules and pollen in cones; included are spruce, fir, pine, cedar, and juniper.

conjugated protein a compound of one or more polypeptides with one or more nonprotein substance; e.g., hemoglobin, which consists of four polypeptide chains and four heme groups.

conjugation in ciliates, a temporary cytoplasmic union in pairs, accompanied by meiosis and the exchange of haploid nuclei.

connective tissue a principal type of vertebrate supporting tissue, often with an extracellular matrix of collagen. Included are bone, cartilage, ligaments, and blood.

connective tissue matrix a noncellular, secreted matrix in which the cells of some connective tissues are embedded or immersed, e.g., bone, cartilage, and blood.

conservative replication replication of a molecule in which the original molecule remains intact. Compare *semiconservative replication*.

consolidation the transfer of engrams from short-term memory to long-term memory.

consolidation hypothesis the presumption that memory is first stored in short-term centers from where it is rapidly lost unless it is transferred to long-term centers where its loss is more gradual.

constant region the C-terminal portion of an immunoglobulin (antibody) light or heavy chain, not involved in antigen-specific binding and coded by a constant-region cistron.

consumer one that consumes; in ecology, an animal that feeds on plants (primary consumer) or other animals (secondary consumer).

consummatory behavior the satisfying, stereotyped behavior that completes an instinctive act. Usually preceded by *appetitive behavior*.

consummatory stage see *consummatory behavior*.

contact receptor see *sensory receptor*.

continental drift a theory proposing that today's continents or land masses were once parts of a supercontinent called Pangaea, which started to divide and drift apart some 200 million years ago. Also, *plate tectonics*.

continental shelf the sea bottom of shallow oceanic waters adjacent to the shores of a continent; considered to be a submerged part of the continent.

continental slope see *continental shelf*

continuous distribution see *distribution*.

continuously varying trait see *polygenic trait*.

continuous spindle fibers see *polar spindle fibers*.

continuous variation see *polygenic inheritance*.

contraception, also *birth control,* any process or method intended to prevent the sperm from reaching and fertilizing the egg, or preventing ovulation or implantation.

contractile capable of contracting (shortening, drawing in, or becoming smaller).

contractile unit (of muscle), see *sarcomere*.

contractile vacuole an osmoregulatory, water-containing vacuole in protists, capable of filling and emptying through the contraction of microfilaments.

control a standard of comparison in a scientific experiment; a replicate of the experiment in which a possibly crucial factor being studied is omitted.

controlled experiment an experiment involving several replicates, each differing by a single variable factor.

convection the transfer of heat by the movement of a circulating fluid or gas; compare *radiation, conduction*.

convergent evolution the independent evolution of similar structures in distantly related organisms; often found in organisms that have adopted similar ecological niches, as marsupial moles and placental moles.

copulation sexual union or sexual intercourse in animals involving internal fertilization.

CoQ Complex III a major membranal proton-pumping complex in the mitochondrial electron/hydrogen transport system.

coral 1. a colonial anthozoan that secretes a calcareous skeleton. 2. the skeleton itself.

coriolis effect the deflection of a moving body of water or air caused by the rotation of the earth, resulting in clockwise motion in the Northern hemisphere and counterclockwise motion in the Southern hemisphere, as seen in tornados, ocean currents, and prevailing winds.

cork 1. in plants, secondary tissue, produced by the *cork cambium*, consisting of cells that become heavily suberized and die at maturity, resistant to the passage of moisture and gasses; the outer layer of bark. 2. such a tissue from the cork oak, used to seal wine bottles.

cork cambium also *phellogen*, in plants, the outermost meristematic layer of the stem of woody plants, from which the outer layer of bark is produced.

cornea see *eyeball*.

corneal lens see *compound eye*.

corn smut a fungal disease of corn; see *smut*.

corolla in flowers, the whorl of petals surrounding the carpels or carpel.

corona radiata ("radiating crown"), an aggregation of follicle cells surrounding the mammalian egg at ovulation.

coronary thrombosis the formation of a blood clot that clings to a coronary vessel.

corpus allatum in insects, one of the paired endocrine bodies that secretes juvenile hormone, which prompts retention of juvenile stages.

corpus callosum a broad, white neural tract that connects the cerebral hemispheres and correlates their activities.

corpus cavernosum (pl. *corpora cavernosa*), in the penis or clitoris, a mass of erectile tissue with large interspaces capable of being distended with blood.

corpus luteum (pl. *corpus lutea*), ("yellow body"), a temporary yellow endocrine body on the surface of the ovary, consisting of secretory cells filling a follicle after ovulation; regressing quickly if the ovum is not fertilized or persisting throughout pregnancy.

cortex 1. (*zoology*) the outer layer or rind of an organ, as *adrenal cortex, kidney cortex*. 2. (*botany*) the portion of stem between the epidermis and the vascular tissue.

cortical (adj.), referring to the cortex.

cortical dominance increase of the cortex at the expense of the medulla. In the embryonic development of the vertebrate gonad, the early undifferentiated gonad consists of an inner *medulla* and an outer *cortex*; in females, cortical dominance results in the formation of an ovary, while medullary dominance results in the formation of a testis in males.

cortical granules membranous vesicles beneath the surface of unfertilized echinoderm eggs; at fertilization these rupture to form the *fertilization membrane*, during the *cortical reaction*.

cortical reaction during fertilization in echinoderms, the rupture of cortical granules, detachment of supernumerary sperm, and subsequent events leading to the formation of a fertilization membrane.

cortisol also *hydrocortisone*, a steroid hormone of the adrenal cortex that inhibits immune responses, lessening inflamation, and is also active in carbohydrate and protein metabolism, response to stress and other processes.

cotyledon also *seed leaf*, a food-storing structure in dicot seeds, sometimes emerging as first leaves; food-digesting organ in most monocot seeds; first leaves in a gymnosperm embryo.

cotylosaur a group of very early Carboniferous reptiles from which most later reptiles evolved.

countercurrent heat exchanger a heat-conserving mechanism in endothermic animals; e.g., the presence of closely paralleled arteries and veins that permits an efficient heat exchange as the outward flow of warmer blood opposes the inward flow of cooler blood from the extremities.

countercurrent multiplier an arrangement such that two fluids (or gases) flow in opposite directions on either side of a membrane, increasing the efficiency of heat exchange, solute transfer, etc.

covalent bond a relatively strong chemical bond in which an electron pair is shared by two atoms, simultaneously filling the outer electron shells of both.

Cowper's gland also *bulbourethral gland*, either of a pair of small exocrine glands discharging lubricating mucus into the male urethra at times of sexual arousal; after William Cowper, 1666–1709.

coxa (pl. *coxae*), 1. also *innominate bone, hip bone*; either half of the pelvis; created by the fusion of the ilium, ischium and pubis.

cranial nerve any of the 12 large, paired nerves that emerge directly from the brain to serve both somatic and autonomic nervous systems. Compare *spinal nerves*.

cranium 1. the skull. 2. the part of the skull enclosing the brain.

crash population crash, a sudden die-off or diminution in numbers, particularly after the carrying capacity of the environment has been exceeded and resources have been depleted.

crassulacean acid metabolism (CAM) in desert plants, a means of obtaining carbon dioxide without the risk of opening the stomata during the day. CAM plants utilize a biochemical system in which carbon dioxide is admitted into the plant at night and fixed into organic acids which can be broken back down during daylight, releasing carbon dioxide for the Calvin cycle.

creatine phosphate also *phosphocreatine*, in vertebrate muscles, a source of high-energy phosphate for restoring ATP expended during muscle contraction.

creationism the doctrine that all things, especially plant and animal species, were created fairly recently and substantially as they now exist, by an omnipotent creator and not by gradual evolution.

cretin a person affected by cretinism (human being; derived from *Christian*).

cretinism a recessive genetic abnormality that results in the inability to produce thyroxine; affected persons are extremely retarded physically and mentally unless thyroxine is administered from early infancy.

Crick strand either of the two strands of DNA, the other being designated the *Watson strand*.

crinoid a type of sessile, stalked echinoderm.

crista (pl. *cristae*), a shelflike fold of the inner mitochondrial membrane into the central cavity of the mitochondrion.

crop also *craw*, 1. an enlargement of the gullet of many birds, which serves as a temporary storage organ. 2. an analogous organ in certain insects and earthworms.

crosscurrent flow in the bird lung, the flow of blood at right angles to the passage of air.

crossing over 1. the exchange of chromatid segments by enzymatic breakage and reunion during meiotic prophase. 2. a specific instance of such an exchange; a crossover.

crossover 1. a crossing over. 2. the result of crossing over. 3. the place on a cromatid where crossing over has occurred.

crustecdysone the molting hormone of crustaceans.

cryptic coloration coloration that serves to conceal or to render inconspicuous.

ctenophore also *comb jelly*, any marine swimming invertebrate of the phylum *Ctenophora*, with a gelatinous body and eight rows of plates (combs) of fused cilia.

C-terminal end also *carboxy-terminal end*, the end of a polypeptide chain in which the final amino acid has its primary amino group fixed in a peptide linkage and its primary carboxyl group free; the last part of the chain to be synthesized.

C_3 plant a plant in which the initial product of carbon dioxide fixation is a 3-carbon compound of the Calvin cycle. C_3 *pathway*, the Calvin cycle.

cuticle a tough, often waterproof, nonliving covering, usually secreted by epidermal cells.

cutin a waxy, water-resistant substance covering the epidermis of leaves and stems.

cuttlebone the skeleton of a cuttlefish, filled with thin, hollow chambers and used as a flotation device.

cyanobacterium formerly *blue-green alga*, any of a large group of blue-green photosynthetic prokaryotes having as photopigments chlorophyll a, phycocyanin and phycoerythrin; and producing oxygen as a photosynthetic waste product.

cycad a plant of an order of palmlike gymnosperms, known from the Triassic to the present.

cyclic AMP (cAMP) adenosine monophosphate in which the phosphate is linked between the 3' and 5' carbons of the ribose group; serves as an intracellular gene regulator under a variety of circumstances.

cyclic phosphorylation light reactions employing only photosystem I, during which electrons from P700 reaction center pass through the associated electron transport system and cycle back to P700, thus serving chemiosmosis only.

cyclosis also *cell streaming* or *cytoplasmic streaming*, the movement of cytoplasm within the cell, often in more or less regular, circular pathways.

cyst 1. a bladder, sac or vesicle. 2. a closed sac, containing fluid, embedded in a tissue. 3. a heavy protective covering of a dormant animal or protist. 4. an encysted organism.

cytochrome 1. any of a group of iron heme enzymes or carrier proteins in the oxidative and photosynthetic electron transport chains. 2. any iron heme protein other than hemoglobin or myoglobin.

cytokinesis cytoplasmic cell division; actual division of the cell into two daughter cells.

cytokinin a plant cell hormone, mitogen, and plant tissue culture growth factor, which interacts with other plant hormones in the control of cell differentiation.

cytological map a map locating genes on the physical chromosome by means other than recombination or genetic mapping.

cytology the scientific study of cells.

cytoplasm the semisolid, protein-rich matrix of a cell exclusive of the plasma membrane, the nucleus, or other large inclusions.

cytoplasmic binding protein a cytoplasmic protein believed to bind with certain hormones prior to their entry into the nucleus, where the complex activates a gene.

cytoplasmic streaming see *cyclosis.*

cytopyge the fixed position on the surface of a protist from which wastes are discharged by exocytosis; the "anus" of a ciliate protozoan.

cytosine a pyrimadine, one of the four nucleotide bases of DNA, and also one of the four bases of RNA.

cytoskeleton the internal structure of animal cells, composed of microtubules and actin microfilaments; it controls the size, shape, and movement of the cell.

cytostome the mouth of a protist.

cytotoxic T-cell see *effector T-cells.*

cytotrophoblast an inner layer of trophoblast cells.

dalton a unit of atomic and molecular mass, equivalent to one twelfth the mass of an atom of carbon 12.

D and C see *dilation* and *curettage.*

Danielli model a historically important (and largely validated) conception of the cell membrane as a phospholipid bilayer stabilized by proteins with channels for the passage of water-soluble compounds.

dark clock an unknown mechanism by which plants are able to measure a period of darkness with fair accuracy, and thus produce flowers at the appropriate season.

dark reaction see *light-independent reaction.*

Darwinian fitness see under *fitness.*

daughter cell either of the two cells created when one cell divides.

deamination the removal of an amino group.

death control the reduction of a population growth rate by an increase in the death rate, as opposed to achieving the same end by a reduction in the birth rate.

death rate 1. in human populations, the number of deaths per 1000 population per year. 2. any analogous rate.

decarboxylation the enzymatic removal of carbon to a substrate. Compare *carboxylation.*

deciduous plant a perennial plant that seasonally drops its leaves.

decomposer also *reducer,* an organism that breaks down organic wastes and the remains of dead organisms into simpler compounds, such as carbon dioxide, ammonia, and water.

dedifferention the return of a differentiated cell to an undifferentiated, plastic state.

dehydration linkage a covalent bond formed between two compounds by the removal of one oxygen atom and two hydrogen atoms.

dehydration reaction also *dehydration synthesis,* an enzymatic reaction during which water is lost and a covalent bond forms between the reactants. See also *dehydration linkage.*

deletion 1. the removal of any segment of a chromosome or gene. 2. the site of such a removal after chromosome healing, considered as a mutation. See also *base deletion.* 3. the deleted chromosome.

deletion map a genetic map constructed from recombination tests between various deletion chromosomes.

delta cell see *Islets of Langerhans.*

demographic transition or *theory of demographic transition,* the proposal that in the development of nations, their population growth has distinct stages: (a) high birth and death rates and slow growth; (b) high birth rate, low death rate, and very rapid growth; and (c) low birth and death rates, and very slow growth.

demography the statistical study of human population (or other) with reference to growth, birth, and death rates, size, density, and migration.

denaturation the alteration of a protein so as to destroy its properties, through heating or chemical treatment (protein denaturation may be reversible or irreversible).

dendrite an extension of a neuron that conducts impulses toward the cell body and axon.

dendrograph a device for recording fluctuations in the diameter of a tree trunk.

denitrification the conversion of nitrates and nitrites to nitrogen gas; *dentrifying bacteria,* common soil and manure bacteria that are responsible for denitrification.

density-dependent effects factors affecting population parameters (growth, birth and death rates) in different ways depending on competition and population density, e.g., food availability, nesting site availability, disease, and predation.

density-independent effects those factors affecting population parameters independently of population size, as temperature, salinity, and meteorites.

denticle 1. a small tooth. 2. a small pointed projection, developmentally homologous with teeth, found in shark skin.

deoxyhemoglobin hemoglobin not carrying oxygen.

deoxynucleotide any nucleotide of DNA.

deoxyribonucleotide a nucleotide in which the sugar portion is deoxyribose rather than ribose.

deoxyribose a 5-carbon sugar identical to ribose except that the 2' hydroxyl group is replaced by a hydrogen.

depolarizing wave see *neural impulse.*

derived a character that has undergone an evolutionary change in the group being considered; the opposite of *primitive.* See also *advanced.*

dermal (adj.) pertaining to the skin or dermis; *dermal bone,* bone originating in evolution and embryology as flat plates beneath the skin, as certain bones of the cranium.

dermal branchiae "skin gills" of echinoderms, consisting of ciliated outpocketings of the coelom and body wall.

dermal system various plant tissues derived from protoderm, including epidermis, guard cells, leaf hairs, root hairs, and periderm.

dermis in animals, the inner mesodermally-derived layer of the skin, beneath the ectodermally-derived epidermis.

descending colon a part of the large intestine that descends to the rectum.

desert a region characterized by scanty rainfall (especially less than 25 cm annually).

desiccate (vt.) to dry up, cause to dry up, dehydrate; desiccated; (adj.) dried up, dehydrated.

desmotubule within the plasmodesmata of plant cells, a continuation of the endoplasmic reticulum of one cell with that of the adjacent cell.

deuterium an isotope of hydrogen, present in "heavy water" and designated ^{2}H.

deuterostome a bilateral animal with radial cleavage and a mouth that does not arise, developmentally or phylogenetically, from the blastopore; compare *protostome.*

developed nation a nation that is industrialized, has a mechanized, labor-free agriculture, high per capita income, and high literacy. Compare *developing nation.*

developing nation a nation that is characterized by low industrialization, labor-intensive agriculture, low per capita income, and low literacy. Compare *developed nation.*

development 1. the whole process of growth and differentiation by which a zygote, spore or embryo is transformed into a functioning organism. 2. any part of this process, as the development of the kidney. 3. evolution.

dextrose see *glucose.*

diabetes mellitus a genetic disease of carbohydrate metabolism characterized by abnormally high levels of glucose in the blood and urine, and the inadequate secretion or utilization of insulin.

diakinesis final stage of meiotic prophase, in which the chromosomes are maximally condensed and homologs are attached only at their ends.

diapause in insects, a period of dormancy, interrupting developmental activity of the pupa or other stage, frequently occurring during hibernation or estivation.

diaphragm 1. in mammals, a dome-shaped, muscularized body partition separating the chest and abdominal cavities, involved in breathing movement. 2. a dome-shaped, rubber, contraceptive device fitted over the cervix, often used with a spermicide.

diastole the period of expansion and dilation of the heart during which it fills with blood; the period between forceful contractions (systole) of the heart. See *systole.*

diatom an acellular or colonial yellow-green photosynthetic protist, having a silicon-impregnated cell wall in two parts.

diatomaceous earth geological deposits consisting largely of the cell walls of diatoms.

dicot a flowering plant of the angiosperm class *Dicotyledonae,* characterized by producing seeds with two cotyledons; compare *monocot.*

dicotyledon see *dicot* and *cotyledon.*

dictyosome in plants, the Golgi apparatus. Also see *Golgi apparatus.*

differential 1. *concentration differential,* the difference in concentration of a solute in two fluids, especially the fluids on either side of a semipermeable membrane. 2. *pH differential,* the difference in pH of two fluids, especially the fluids on either side of a semipermeable membrane.

differentiation in development, the process whereby a cell or cell line becomes morphologically, developmentally, or physiologically specialized. *Terminal differentiation,* differentiation to a form in which the cell will normally not undergo further cell division. See *dedifferentiation.*

diffuse root system also *fibrous root system,* a root system of a plant in which there are many roots all about the same size.

diffusion the random movement of molecules of a gas or solute under thermal agitation, resulting in a net movement from regions of higher initial concentration to regions of lower initial concentration.

digestion the process of making food absorbable by breaking it down into simpler chemical compounds, chiefly through the action of enzymes, acid, and emulsifiers in digestive secretions; see *extracellular* and *intracellular digestion.*

digestive vacuole in animals and protists, intracellular vacuoles that arise by the phagocytosis of solid food materials and in which digestive processes occur, nutrient molecules being absorbed and undigested materials eventually being expelled by exocytosis.

digit in terrestrial vertebrates, any of the terminal divisions of the limbs; a finger, thumb or toe.

dihybrid an organism that is heterozygous at two different loci. *Dihybrid cross,* a cross between two genotypically identical dihybrids.

dikaryotic in many fungi, a condition arising after conjugation, wherein parental plus and minus nuclei remain separated for a time before fusion occurs.

dilation an enlargement or widening; in the human birth process, the initial stage, in which the body of the cervix softens and its opening widens.

dilation and curettage (D and C) a surgical operation in which the cervix is forcefully dilated and the uterine mucosa scraped with a curette; a common method of induced abortion as well as a procedure for removing cysts and polyps.

dimer a molecule consisting of two similar subunits.

dinokaryote a dinoflagellate, considered as being other than a eukaryote; the term emphasizing the fact that cell and nuclear organization in the dinoflagellates is different from that of either prokaryotes or eukaryotes.

dinoflagellate a flagellated, photosynthetic, marine protist of the *Dinoflagellata,* a group considered by various authorities to be an order, class, division, kingdom, or even higher group *(dinokaryotes),* on a level with prokaryotes and eukaryotes.

dinosaur 1. any of a taxon of large, extinct reptiles, widely distributed from the Triassic to the Mesozoic, including the largest known land animals. 2. any large, extinct reptile.

dioecious 1. in flowering plants, having separate sexes in the sporophyte generation. 2. in algae plants and bryophytes, having separate sexes in the gametophyte generation; that is, producing only eggs or sperm in any one thallus.

dipeptidase an enzyme of the small intestine that hydrolyzes the peptide bond of a dipeptide. See also *dipeptide.*

dipeptide two amino acids united by a peptide linkage; a common intermediate product of digestion. *Dipeptidase,* an enzyme that hydrolyzes dipeptides but not longer polypeptides.

diplococcus (pl. *diplococci*), 1. a coccus (spherical) in which the cells are arranged in colonies of two. 2. a bacterium of the genus *Diplococcus,* which includes several serious human pathogens, causing pneumonia and gonorrhea.

diploid having a double set of genes and chromosomes—one set from each parent.

diplotene also *diplonema,* the stage in meiotic prophase during which chiasmata become visible and separation of homologs begins.

directional selection selection favoring one extreme of a continuous phenotypic distribution.

disaccharide a carbohydrate consisting of two simple sugar subunits.

discrimination in behavior, the ability to perceive the difference between similar stimuli.

disruptive selection natural selection during which extreme phenotypes receive favorable selection, whereas average phenotypes are selected against.

dissociate to come apart into discrete units; *dissociation constant,* of a chemical reaction, a constant depending on the equilibrium between the combined and dissociated forms of a chemical or chemicals.

distal away from the center of the body, heart, or other reference point.

distal convoluted tubule a portion of the nephron between the loop of Henle and the collecting duct.

distance receptor a sensory receptor of stimuli arising at a distance, such as sight and sound.

distribution in statistics, a set of values or measurements of a population, each measurement being associated with one individual. *Continuous distribution,* the distribution of a character that varies continuously or by small increments, such as heights and bristle number. *Discrete distribution,* the distribution of a character that exists in a finite number of distinct character states; see also *polymorphism.*

disulfide linkage, also *disulfide bond* a covalent link formed by the oxidation of two sulfhydryl groups: $R_1—SH + HS—R_2 \rightarrow R_1—S—S—R_2 + H_2$.

divergence see *divergent evolution.*

divergent evolution following speciation, the continued accumulation of differences between or among species, attributable to adaptive radiation.

diversity 1. variety; variability. 2. the range of types in a major taxon: *plant diversity.* 3. in ecology, a measure of the number of species coexisting in a community.

division a major primary category or taxon of the plant, fungal, and sometimes moneran and protist kingdoms, equivalent to *phylum.*

DNAase an enzyme capable of hydrolyzing DNA sugar-phosphate bonds.

DNA, deoxyribonucleic acid the genetic material of all organisms (except RNA viruses); in eukaryotes DNA is confined to the nucleus, mitochondria and plastids.

DNA polymerase any of several enzymes or enzyme complexes that catalyze the replication of DNA.

DNA repair system any of several systems of enzymes that detects and excises primary lesions in the Watson or Crick DNA strand, replacing them with a corrected segment.

DNA replication also *replication* or *DNA synthesis,* the semiconservative synthesis of DNA in which the double helix opens, the two strands separate and each is used as a template for producing a new opposing strand.

Dollo's law a generalization in biology; characters lost in the course of evolution are never regained in their original form.

dominance the phenotypic expression of only one of the two alleles in a heterozygote. See *dominance relationships.*

dominance, behavioral dominance a behavioral relationship between two animals where the *subordinate* individual withdraws or behaves submissively toward the *dominant* individual in any conflict, potential conflict, or interaction.

dominance, cortical see *cortical dominance.*

dominance hierarchy 1. also *pecking order,* behavioral interactions established in a troop, flock, or other species group, in which every individual is dominant to those lower on the order and submissive to those above. 2. in genetics, the relationship among a group of multiple alleles in which each ordered series behaves as a dominant to those below it and as a recessive to those above it.

dominance, medullary see *medullary dominance.*

dominance relationships the interaction between two alleles of a given locus in the development and expression of a specific phenotypic trait. One allele of the pair is *dominant* if its typical phenotype is expressed in both the homozygote and the heterozygote; in this case the other allele, which is expressed only when it is homozygous, is said to be recessive. If each of two alleles has a unique phenotypic expression and both are independently expressed in the heterozygote, both are said to be *codominant.* For other dominance relationships consult the text.

dominant 1. a phenotypic trait that is always expressed when a certain allele is present. 2. an allele that expresses a given phenotypic trait whether homozygous or heterozygous. See *dominance relationships.*

dominant brown spotting a human condition produced by a rare X-linked dominant gene, the result of which is brown spots on the teeth.

dormancy the temporary suspension of growth, development, or other biological activity; suspended animation, *dormant*, in a state of dormancy.

dorsal toward the back or (usually) upper surface of a bilateral animal.

dorsal aorta see under *aorta*.

dorsal hollow nerve cord a defining characteristic of chordates, arising in development as a flat dorsal plate of neurectoderm that rounds up and sinks beneath the surface to form, in vertebrates, the *brain* and *spinal cord (central nervous system)*.

dorsal lip in the animal gastrula, tissue associated with the early blastopore, the site of involution by presumptive mesoderm and endoderm.

dorsoventrally from top to bottom, or back to front, in humans.

double bond a bond between two atoms consisting of two covalent bonds, with a total of four electrons shared.

double fertilization in flowering plants, fertilization of egg cell and central cell by two sperm entering from a pollen tube.

double helix the configuration of the native DNA molecule, which consists of two antiparallel strands wound helically around each other.

doubling time the time it takes a growing population to double in numbers.

Down's syndrome see *trisomy 21*.

downy mildew 1. a protistan parasite of plant leaves. 2. the disease caused by the fungus.

drift *random drift*, also *genetic drift*, 1. the chance fluctuation of allele frequencies from generation to generation in a finite population. 2. the long-term consequences of such fluctuations, such as the loss or fixation of selectively neutral alleles.

drive an urgent, basic or instinctual need pressing for satisfaction; a physiological or psychological tension, lack or imbalance that the organism actively seeks to redress, quench or fulfull.

drive-reduction hypothesis the idea that most behavior aims at (or is rewarded by) the reduction of specific innate drives.

drumstick a small projection (not unlike a chicken leg) of the nucleus of the polymorphonuclear leucocytes of human females, attributed to the condensed second X chromosome.

ductless gland see *endocrine gland*.

ductus arteriosis in the mammalian fetus, a short broad vessel conducting blood from the pulmonary artery to the aorta, thus bypassing the fetal lungs.

duodenum the first region of small intestine, just posterior to the stomach, receiving the common bile duct.

dwarf 1. any abnormally short person. 2. also *achondroplastic dwarf*, a genetic abnormality, an abnormally short person with markedly atypical body proportions. 3. any plant or animal unusually small for its species or type. *Pituitary dwarf*, a dwarf whose small stature is due to a lack of growth hormone during development.

dynein arm in cilia and flagella, protein connections between the nine paired groups of microtubules.

ear ossicle one of the bones of the middle ear, in mammals, the malleus, incus, and stapes, that transfers sound from the eardrum to the internal ear.

ecdysial gland see *prothoracic gland*.

ecdysis also *molting*, 1. in arthropods, the shedding of the outer, noncellular cuticle. 2. this shedding together with associated changes in size, shape or function.

ecdysone the molting hormone of insects.

echinoderm any organism of the marine, coelomate, deuterostome phylum *Echinodermata*.

ecological community an assemblage of plants, animals, and other organisms forming an interacting unit within a biome, generally identifiable by dominant plant type or types, as seen in a spruce or sage desert community.

ecological niche see *niche*.

ecological succession see *community development*.

ecologist a scientist who studies ecology.

ecology the branch of biology dealing with the relationships between organisms and their environment.

Eco-R1 one of a family of restriction enzymes used in gene splicing.

ecosystem the biotic and abiotic factors of an ecological community considered together.

ectoderm in animal development, the outermost of the three primary germ layers of an embryo: the source of all neural tissue, sense organs, the outer cellular layer of the skin and associated organs of the skin; also, any of the tissues derived from embryonic ectoderm; also, the outer cellular layer of a coelenterate.

ectoparasite a parasite that feeds on or attaches to the surface tissue of the host; fleas, lice, ticks and athlete's foot fungus are human ectoparasites.

ectoproct a lophophorate, coelomate, sessile, colonial bryozoan ("moss animal") distinguished from the outwardly similar but unrelated pseudocoelomate entoproct by having the anus outside of the whorl of tentacles.

ectotherm also *poikilotherm* lacking the ability to thermoregulate; often called "cold-blooded," but body temperature is usually that of the surroundings. See also *endothermic*.

edema an abnormal accumulation of fluid in connective tissues, causing a general puffiness.

effector also *effector organ*, a bodily organ actively used in behavior, especially in communication—distinguished from *receptor*; e.g., the human eyebrows and associated muscles constitute an effector organ, and a rather effective one at that.

effector T-cells a group including cytotoxic T-cell and encapsulating T-cells: lymphocytes believed to destroy or encapsulate invading cells, virus infected cells, or cancer cells.

efferent neuron a neuron conducting impulses from the brain or spinal cord to an effector, such as a *motor neuron*.

efferent vessel any blood vessel carrying blood away from a structure under consideration, specifically in a gill filament or kidney nephron.

egg cell the larger gamete of an anisogametic organism; in plants, the one haploid cell of the embryo sac that will participate as the female gamete; in animals, the functional product of meiosis in females.

ejaculate (v.i.), 1. to eject semen by involuntary muscular contractions of the vas deferens and urethra in an orgasm. 2. (n.) the amount of semen ejaculated in one orgasm. *Ejaculation*, the act of ejaculating.

elastin an extracellular structural protein that forms long, elastic fibers in connective tissue.

elator in liverworts, a springlike structure in the sporangium that ejects the spore outward.

electroencephalograph a clinical device for detecting and recording *electronencephalograms* (EEGs) of electrical activity within the brain.

electromagnetic radiation waves of energy propagated by simultaneous electric and magnetic oscillations, including radio waves, microwaves, infrared, visible light, ultraviolet radiation, X rays, and gamma rays. *Electromagnetic spectrum*, the range of electromagnetic radiation from low-energy, low frequency radio waves to high-energy, high frequency gamma rays.

electron one of the three common constituents of an atom, a *lepton* with a mass of 1/1837 of that of a proton and an electrostatic charge of $^-1$.

electron acceptor a molecule that accepts one or more electrons in an oxidation-reduction reaction (and thus becomes reduced).

electron carrier a molecule that behaves cyclically as an electron acceptor and an electron donor.

electron donor a molecule that loses one or more electrons to an electron acceptor in an oxidation-reduction reaction (and thereby becomes oxidized).

electronegativity the relative ability of any element to hold electrons or to attract them from other elements.

electron microscope a device for creating magnified images of small specimens through bombarding it with an electron beam and by subsequent magnetic focusing.

electron orbit also *electron orbital*, 1. the state of an electron as determined by its energy as it moves within an atom. 2. the space within which an electron pair moves within an atom.

electron shell 1. the space occupied by the orbits of a group of electrons of approximately equal energy; 2. an energy level of a group of electrons in an atom.

electron transport system also *electron transport chain*, a series of cytochromes and other proteins, bound within a membrane of a thylakoid, mitochondrion, or prokaryote cell, that passes electrons and/or hydrogen atoms in a series of oxidation-reduction reactions that result in the net movement of hydrogen ions across the membrane.

electrostatic (adj.) pertaining to the attraction or repulsion of electric charges independent of their motion.

element a substance that cannot be separated into simpler substances by purely chemical means.

elongation lengthening; in plant development, the stretching in one direction of stem or root cells under turgor.

Eltonian pyramid also *pyramid of numbers*, a concept in ecology that, because of thermodynamic inefficiency, organisms forming the base of a food chain are numerically abundant and comprise a large total biomass, while organisms of each succeeding level of the chain are successively less abundant and of smaller total biomass, and the top predator is always numerically rare and of relatively small total biomass.

embolism 1. the sudden blockage of a blood vessel by a dislodged blood clot, air bubble or other obstruction. 2. also *embolus*, the clot or obstruction itself.

embryo an animal or seed plant sporophyte in an early state of development. In animals this stage begins with cleavage, includes the laying down of germ layers, organ systems, and basic tissue types, and grades imperceptibly into the *fetal* stage; in seeds the embryo stage lasts from fertilization to germination, the "mature embryo" comprising a rudimentary plant with plumule, radicle, and cotyledons.

embryo axis the plant embryo not including cotyledons.

embryogenesis in animals, the early period of development when cell division and gastrulation prepare the embryo for morphogenesis.

embryology the scientific study of early development in plants and animals.

embryonic disc 1. also *embryonic shield*, the part of the inner cell mass from which a mammalian embryo develops. 2. *blastodisc*.

embryonic induction an influencing action, employing chemical or physical agents, during which one embryonic tissue influences the developmental fate of another.

embryo sac in flowering plants, the mature megagametophyte after division into six haploid cells and one binucleate cell, enclosed in a common cell wall.

emigrate to move out of an area; the moving out of an area by former residents.

empirical based on observation rather than theory. *Empirical law,* a statement that describes a relationship but offers no explanation of its mechanism.

encapsulating T-cell see *effector T-cells*

endergonic (adj.) of a biochemical reaction or half-reaction, requiring work or the expenditure of energy; moving from a state of lower potential energy to one of higher potential energy; compare *exergonic*.

endocrine gland also *ductless gland*, a discrete gland that secretes hormones into the blood system.

endocrine system the endocrine glands taken together, and their hormonal actions and interactions.

endocytosis the process of taking food or solutes into the cell by engulfment (a form of active transport); see also *phagocytosis, pinocytosis*; compare *exocytosis*.

endoderm also *entoderm* 1. the innermost of the three primary germ layers of a metazoan embryo and the source of the gut epithelium and its embryonic outpocketings (in vertebrates, the liver, pancreas, lung). 2. any tissue derived from endoderm. 3. in coelenterates, the epithelium of the gastrovascular cavity.

endodermis a single layer of cells around the stele of vascular plant roots that forms a moisture barrier in that the lateral cell walls are closely oppressed and waterproofed in bands, forming the Casparian strip.

endometrium in mammals, the mucous membrane tissue lining the cavity of the uterus, responds cyclically to ovarian hormones by thickening as preparation for implantation of the blastocyst.

endomysium the delicate connective tissue sheath surrounding the individual muscle fibers.

endonuclease a family of enzymes capable of hydrolyzing sugar-phosphate bonds anywhere within DNA strands, an essential enzyme to crossing over. See also *recombination nodule*.

endopeptidase a proteolytic enzyme that cleaves peptide bonds in the middle portions of polypeptides.

endoplasmic reticulum (ER) internal membranes of the cell, usually a site of synthesis; *rough endoplasmic reticulum,* ER with bound ribosomes, the site of synthesis of noncytoplasmic proteins; *smooth endoplasmic reticulum,* ER without bound ribosomes, usually the site of synthesis of nonprotein materials.

endoskeleton a mesodermally derived supporting skeleton inside the organism, surrounded by living tissue, as in vertebrates and echinoderms.

endosperm a nutritive tissue of seeds, formed around the embryo in the embryo sac by the proliferation of the (usually) triploid endosperm nucleus to form a starch-rich mass; the endosperm may persist until germination or be resorbed by the cotyledon(s) during seed maturation.

endosperm mother cell a large, central, usually diploid or binucleate cell formed in the megagametophyte by fusion of haploid nuclei; when fertilized in double fertilization, forms the *endosperm nucleus.*

endospore an asexual resistant spore formed within a bacterial cell; see *spore*.

endosymbiont a mutualistic symbiont that resides within the cells of its symbiotic partner.

endosymbiosis hypothesis also *symbiosis hypothesis* or *serial endosymbiosis hypothesis* three interrelated hypotheses attributable to Lynn Margulis to the effect that the eukaryotic cell evolved from the mutualistic endosymbiosis of various prokaryotic organisms, one of which gave rise basically to the cytoplasm, nucleus, and motile membranes; a second to mitochondria; a third to chloroplasts and other plastids; and a fourth to cilia, eukaryotic flagella, basal bodies, centrioles, the spindle, and all other microtubule structures.

endothelium an epithelial tissue that forms the inner lining of blood and lymph vessels.

endotherm also *homeotherm* and *homoiotherm* an organism with the ability to metabolically thermoregulate, also called "warm-blooded." See also *ectotherm*.

energetic tendencies the probable behavior of interacting atoms as they proceed to new energetically stable configurations, e.g., the tendency for electrons to pair, for outer shells to fill, and to reach a balance of plus and minus charges.

energy a fundamental concept of physics, either being associated with physical bodies (potential energy, kinetic energy) or with electromagnetic radiation; under some conditions interconvertable with mass by the relation $E=mc^2$ or with entropy but not capable of being created or destroyed.

energy pyramid the Eltonian pyramid with regard to chemical energy rather than numbers or biomass; see *Eltonian pyramid*.

energy of activation see *heat of activation*.

energy-rich bond see *hich energy bond*.

energy state also *energy level,* a quantum state of an electron in an atom, a form of potential energy as the electron changes from a higher energy state to a lower one; a measure of the free energy in a system.

engram (also *memory scar, memory trace*) the physical change that is presumed to occur in the brain when something is learned.

entropy in thermodynamics, the amount of energy in a closed system that is not available for doing work; also defined as a measure of the randomness or disorder of such a system. *Negative entropy,* free energy in the form of organization. See *energy, free energy.*

enucleate without a nucleus.

environment the surrounding conditions, influences, or forces that influence or modify an organism, population, or community.

environmental resistance the sum of environmental factors (e.g., limited resources, drought, disease, predation) that restrict the growth of a population below its biotic potential (maximum possible population size).

enzyme a protein that catalyzes chemical reactions.

enzyme-substrate complex the unit formed by an enzyme bound by non-covalent bonds to its substrate.

ephyra a young, newly budded scyphozoan jellyfish.

epiblast the upper layer of cells in the early bird, reptilian, or mammalian embryo that, following gastrulation, contributes to the three-layered embryo.

epiboly in gastrulation, the spread of ectoderm down over the surface of the embryo.

epicotyl in the embryo of a seed plant, the part of the stem above the attachment of the cotyledon(s); forms the epicotyl hook in the growth of some seeds.

epidermis 1. in plants, the outer protective cell layer in leaves and in root and stem primary growth, one cell thick and made waterproof with an outer layer of cutin; replaced in secondary growth by *periderm*. 2. in animals, the outer epithelial layer, derived from ectoderm; lacking innervation or vascularization in vertebrates.

epididymis in mammals, an elongated soft mass lying alongside each testis, consisting of convoluted tubules; the site of sperm maturation.

epiglottis in the lower pharynx, a flexible cartilaginous flap that folds over the glottis during swallowing.

epinephrine also *adrenaline*, a neurohormone and systemic hormone with numerous effects, including raising of blood pressure, increasing heart beat, constricting blood flow to internal organs and skin and increasing blood flow to muscles, together with general arousal and excitement; produced by the adrenal medulla and by nerve synapses of the autonomic nervous system.

epiphysis in vertebrates, a part of a bone that ossifies separately and later becomes fused to the rest. *Epiphyseal cartilage*, also *epiphyseal plate*, a plate of cartilage separating the body of a long bone from its terminal epiphysis, being the site of bone growth and elongation.

epiphyte a plant that grows nonparasitically on another plant or sometimes an object.

episome a separable genetic unit in a prokaryote; see *plasmid*.

epistasis the masking of a trait ordinarily determined by one gene locus by the action of a gene or genes at another locus.

epithelial (adj.) of cells or tissue, covering a surface or lining a cavity.

epithelium (pl. *epithelia*) a tissue consisting of tightly adjoining cells that cover a surface or line a canal or cavity, and that serves to enclose and protect.

equilibrium frequency in population genetics, the idealized allele frequency at which selection, mutation and other forces balance, so that there is no expected net change; chance fluctuations may cause oscillations around this idealized equilibrium.

ER see *endoplasmic reticulum*.

era one of the major divisions of geologic time; see the geologic timetable inside the front cover.

erection 1. the becoming stiff and firm (turgid), by dilation under blood pressure, of a penis, clitoris, or nipple. 2. the stiffened, turgid state of such an organ.

ergot a toxic fungal infective state of rye.

ergotism an illness in humans brought about by eating bread made with ergoted flour, that is, flour milled from grain infected with the fungus *Claviceps purpurea*.

erythroblastosis also *erythroblastosis fetalis*, during Rh incompatibility, the destruction of red blood cells in an Rh^+ fetus or newborn through the action of maternal Rh^- antibodies that have crossed the placenta.

erythrocyte also *red blood cell*, a hemoglobin-filled, oxygen-carrying, circulating blood cell; enucleate in mammals, but having a physiologically inactive, condensed nucleus in other vertebrates.

erythropoietin a hormone that stimulates the production and differentiation of erythrocytes during erythropoiesis.

esophagus the anterior part of the digestive tract; in mammals it is muscularized and leads from the pharynx to the stomach.

ester one of a class of chemical compounds formed by a dehydration linkage between a hydroxyl group and a carboxyl group: $R_1 - CH_2OH + HOOCH - R_2 \rightarrow R_1 - CH_2 - O - CO - R_2 + H_2O$. *Ester linkage*, the dehydration linkage forming an ester.

estivation a state of dormancy or torpor induced by heat or dryness; summer dormancy.

estradiol see *estrogen*.

estrogen 1. any female hormone or feminizing hormone. 2. *Estradiol*, *estriol*, or *estrone*, steriod female hormones produced primarily in vertebrate ovaries but also in the adrenal cortex, able to promote estrus and the development of secondary sexual characteristics.

estrous cycle the sequence of endocrine and other reproductive system changes that occur from one estrus period to the next.

estrus also *heat*, in the females of most mammalian species, the regularly recurring state of sexual excitement around the time of ovulation; the only time during which the female will accept the male and is capable of conceiving; the term is not applicable to people.

ethanol ethyl alcohol.

ethology the scientific study of animal behavior as it occurs in the organism's natural environment, or among free-ranging animals in the ethologist's environment.

ethyl alcohol fermentation see *fermentation*.

ethylene a colorless, sweet-smelling, unsaturated hydrocarbon gas, $CH_2 = CH_2$, emitted by ripening fruit and inducing further ripening.

eukaryote a nucleated organism.

euchromatin chromatin other than heterochromatin.

eucoelomate also *coelomate* also *coelom*, any animal with a coelom, a body cavity entirely lined by mesodermally derived tissue. See also *coelom*.

euphotic zone in the aquatic environment, those depths penetrated by effective light, where photosynthetic activity exceeds respiration.

eutrophic (adj.) a rapidly changing lake or other body of fresh water, rich in nutrients but periodically depleted of oxygen.

eutrophication the process in which nutrients increase in a lake, through evapoaration and silting, or artificially by pollution.

evolution 1. any gradual process of formation, growth, or change. 2. descent with modification. 3. long-term change and speciation (division into discrete species) of biological entities. 4. the continuous genetic adaptation of organisms or populations through mutation, hybridization, random drift, and natural selection.

evolutionary stable strategy (ESS) an innate behavioral strategy (of conflict, etc.) that confers greater fitness on individuals than can any alternative behavior that might arise by mutation or recombination.

exchange pool in biogeochemical cycles, the readily available reserves of a mineral nutrient, such as a soluble phosphate or nitrate pool in the soil water; compare *reservoir*.

excision repair see *DNA repair system*.

excretion the removal of metabolic wastes from the body.

excretory system an organ system involved in the removal of metabolic wastes; e.g., the kidney, urinary bladder, and associated ducts and valves.

excurrent siphon also *exhalent siphon*, see *incurrent siphon*.

exergonic (adj.) of a biochemical reaction or half-reaction, producing work, heat, or other energy; moving from a higher energy state to a lower energy state; as opposed to *endergonic*.

exocrine gland also *ducted gland*, gland that releases its secretions externally, into the gut, or anywhere other than the bloodstream; usually exocrine glands are connected to ducts, but glandular secretory cells of the gut epithelium are considered exocrine.

exocytosis the process of expelling material from the cell through the plasma membrane by the fusion of vacuoles, secretion granules, etc. with the plasma membrane, and their subsequent eversion.

exonuclease a nucleic acid hydrolyzing enzyme that operates at the ends of the nucleic acid strand.

exopeptidase a proteolytic enzyme that hydrolyzes peptide linkages near the two ends of a polypeptide, see *carboxypeptidase* and *aminopeptidase*; compare *dipeptidase*, *endopeptidase*.

exoskeleton an external skeleton or supportive covering, as in arthropods and armadillos; arthropod exoskeletons are formed by secretions of epidermal cells, largely chitin, proteins and calcium carbonate.

exponential growth also *geometric growth*, population growth in which the population size increases by a fixed proportion in each time period, successive values forming an exponential series, $P_t = P_o r^t$ where P_o is the initial population size, t is the time, and r is the Malthusian parameter.

exponential growth curve also *exponential increase* and *J-shaped curve*, population growth that when plotted takes on a J-shaped curve, growth that approaches the *biotic potential* of a species and is not immediately responsive to *environmental resistance*. Growth increments may be 2, 4, 8, 16, 32, etc. Compare *arithmetic increase*.

expressed sequence or *exon* the portion of mRNA remaining after the *introns* or *unexpressed sequences* have been removed; see also *intron*.

external ear all parts of the ear external to the eardrum, comprising the external ear canal, the external auditory meatus, and the pinna (pl. *pinnae*—what you wriggle when you wriggle your ears).

external fertilization fertilization outside the body, as in echinoderms and most bony fish.

extinction a coming to an end or dying out of a species or other taxon (group of related organisms).

extracellular outside, between, or among cells.

extracellular digestion digestion outside of cells (usually in the gut).

extraembryonic outside of the embryo proper. *Extraembryonic membrane*, any of several membranes (*amion, chorion, allantois, yolk sac*) produced by the zygote but not part of the embryo proper.

extraembryonic membrane see *extraembryonic*.

extra-Y syndrome see *XYY syndrome*.

extreme halophile an organism, often an archaebacterium, that thrives in an excessively salty environment.

extreme thermophile an organism, often an archaebacterium, that thrives in an excessively hot environment, above 55° C.

eyeball in the vertebrate, the entire eye, consisting of an outer *sclera* containing the transparent *cornea*, a middle *choroid*, and an inner light-sensitive *retina*, along with the *iris, lens, ciliary muscles,* and fluid-filled *anterior chamber (aqueous humor), posterior chamber (vitreous humor),* and *optic nerve,* which emerges from the retina.

eyespot 1. a simple visual organ in many invertebrates, consisting of a pigmented cup with light-sensitive receptors. 2. a simple pigmented visual organelle in various flagellate protists, including euglenoids and *Chlamydomonas*.

facilitated diffusion diffusion of specific molecules across a plasma membrane that is facilitated by reversible association with carrier molecules able to traverse the membrane. Facilitated diffusion differs from active transport in that no energy is expended and net movement follows the concentration gradient.

factor in Mendelian language, the alternative expression of a gene and the equivalent of an allele.

facultative anaerobe an organism such as yeast that can use oxygen when it is available but is also able to live anaerobically.

FAD, FADH$_2$ flavin adenine dinucleotide, a hydrogen carrier in metabolism. FAD is the oxidized form and FADH$_2$ is the reduced form.

fallopian tube see *oviduct*.

family in classification, a grouping smaller than *order* but larger than *genus*.

fascia (pl. *fasciae* or *fascias*), a heavy sheet of connective tissue covering or binding together muscles or other internal structure of the body, often connecting with ligaments or tendons.

fascicle a distinct bundle of muscle fibers, surrounded by connective tissue and carrying nerves and blood vessels.

fat 1. *neural fat,* also *triglyceride,* any of the solid, semisolid, or liquid lipid-soluble compounds consisting of glycerol bound by ester linkages to three fatty acid molecules. 2. also *adipose tissue,* parts of an animal consisting largely of cells distended with triglycerides.

fate map a plan of a zygote, blastula or early embryo indicating the normal developmental fates of the cellular descendants of embryonic ectoderm, mesoderm, and endoderm.

fatty acid an organic acid consisting of a linear hydrocarbon "tail" and one terminal carboxyl group.

feces bodily waste discharged through the anus.

feedback of any system, the return to the input of a part of the output, which serves to regulate or affect the functioning of the system. *Negative feedback:* the output damps the activity of the system, reducing the flow of output. *Positive feedback:* the output stimulates the activity of the system, further increasing the flow of output.

feedback inhibition see *negative feedback*.

femur 1. also *thigh bone,* the upper bone of the hind leg of a vertebrate. 2. in insect legs, the third segment from the body, often enlarged.

fermentation 1. anaerobic breakdown of glucose by yeast to form alcohol and carbon dioxide. 2. any controlled enzymatic transformation of an organic substrate by microorganisms.

fertility factor also *F episome,* in *E. coli,* a plasmid that codes for the production of a sex pilus, thus permitting conjugation.

fertility rate the number of births per 1000 women from 15 to 44 years of age; a clearer indicator of reproductive activity in a population than the birth rate.

fertilization the process of the union of gametes or gamete nuclei.

fertilization cone a cone-shaped mound of egg cytoplasm that, upon activation of the egg by a fertilizing sperm, rises to engulf the sperm.

fertilization membrane in echinoderms, a membrane formed upon egg activation by the rupture of cortical granules, preventing further sperm penetration.

fetus in vertebrates, an unborn or unhatched individual past the embryo stage; in humans, a developing, unborn individual past the first 8 weeks of pregnancy.

Feulgen staining a specific procedure for staining and identifying DNA in the cell.

fiber 1. in plants a lengthy, slender, tapering sclerenchyma cell, often thick walled and nonliving at maturity. 2. see *muscle fiber*.

fibril also *myofibril,* cylindrical, striated, contractile units of muscle, containing myofilament's of myosin and actin.

fibrin an insoluble fibrous protein forming blood clots and also contributing to the viscosity of blood; see *fibrinogen*.

fibrinogen a globular blood protein (*globulin*) that is converted into fibrin by the action of thrombin as part of the normal blood clotting process.

fibroblast a cell in the connective tissue group that produces fibers and matrix substances such as collagen.

fibrocartilage slightly compressible cartilage containing thick bundles of collaginous fibers, as in intervertebral disks, the knee joint, and pubic symphysis.

fibroin the insoluble fibrous protein of silk.

fibrous protein a protein in which the molecules tend to align in linear arrays, in contrast to globular proteins.

fibrous root system also *diffuse root system,* the roots collectively of a plant in which there are many equivalent roots rather than a prominent central root.

fibula in land vertebrates, the smaller of the two bones of the lower hind leg; fused with the *tibia* in birds and many mammals.

fiddlehead a young unfurling frond of a fern, resembling the head of a violin.

filament 1. a long fiber. 2. see *muscle filament*. 3. in flowers, the slender stalk of the stamen, on which the anther is situated.

filamentous algae a group of green algae that tend to form lengthy filaments of cells.

filial generation in Mendelian genetics, 1. *first filial generation,* F_1, the offspring of a cross between two homozygotes for different alleles (or individuals of two true-breeding strains). 2. subsequent filial generations (F_2, F_3 etc.) The progeny of selfing, of inbreeding, or of crosses between individuals of a given filial generation.

filterable virus any virus that passes through filters designed to remove bacteria or their spores from a liquid.

filter-feeder an animal that obtains its food by filtering minute organisms from a current of water.

fimbria (pl. *fimbriae*), in vertebrates, a fringe of ciliated tissue surrounding the opening of the oviduct into the peritoneal cavity.

fire-disclimax community an ecological community in continual transition because of recurring fires (e.g., chaparral).

first messenger a hormone, as distinguished from a *second messenger*.

first law of thermodynamics the physical law that states that energy cannot be created or destroyed; later amended to allow for the interconversion of matter and energy.

fission the division of an organism into two (binary fission) or more organisms, as a process of asexual reproduction.

fissure of Rolando also *central sulcus,* the deep groove separating the frontal lobe from the parietal lobe of the cerebral cortex.

fissure of Sylvius also lateral fissure, a deep groove in the human cerebral cortex separating the temporal lobe from the frontal and parietal lobes.

fit in biology, adapted to the environment and thus well able to survive and reproduce.

fitness 1. the state of being adapted or suited (e.g., to the environment). 2. also *relative fitness, Darwinian fitness,* the relative expectation of surviving and reproducing of an individual or of a specific genotype, compared with that of the general population or of a standard genotype. 3. one half the expected number of offspring of a diploid genotype.

five-prime end (5′ *end*,) of a nucleic acid, the end at which the 5′ carbon of ribose or deoxyribose is free to form a phosphate linkage; the first end of the chain in the direction of synthesis.

five-prime leading region the end of mRNA to act first in translation, which in eukaryotic mRNA is capped through the addition of methylated triphosphate guanine nucleotide.

fixation 1. *nitrogen fixation,* the energy-consuming transformation of nitrogen gas to soluble nitrogen compounds. 2. the preparation of specimens for microscopic viewing by treatment that precipitates proteins and leaves spatial relationships intact; *to fix* (v.t.) to so treat a specimen.

fixed action pattern a precise and identifiable set of movements, innate and characteristic of a given species; see also *instinctive pattern*.

flagellin the crystalline protein comprising the eubacterial flagellum.

flagellum (pl. *flagella*) 1. a long, whiplike, motile eukaryote cell organelle, projecting from the cell surface but enclosed in a membrane continuous with the plasma membrane; longer and fewer in number than cilia, it propels the cell by undulations. 2. an analogous, nonhomologous organelle of bacteria, consisting of a solid, helical protein fiber that passes through the cell wall and propels the cell by rotating like a propeller.

flame bulb see *flame cell system*.

flame cell in flatworms, rotifers, and lancelets, a cup-shaped cell with a tuft of cilia in its depression, the flamelike beating of which creates suction that draws waste materials into an excretory duct.

flame cell system in several invertebrate phyla, an osmoregulatory/ excretory system consisting of ciliated flame bulbs and associated conducting tubules in which fluids and nitrogen wastes are swept to external pores. See also *protonephridium*.

flatus gas expelled through the anus.

flatworm a platyhelminth, especially a turbellarian such as the common planaria.

floridean starch a storage starch or carbohydrate of the red algae.

florigen a hypothetical flower-inducing hormone.

flower the part of a seed plant that bears the reproductive organs, especially around the time of reproduction or during the time in which some or all of its parts are conspicuously white or brightly colored. *Complete flower*, a flower having all four sets of floral organs; pistils, stamens, petals, sepals. *Incomplete flower*, a flower lacking one or more of the four sets of floral organs. *Regular flower*, a radially symmetrical flower. *Irregular flower*, a bilaterally symmetrical flower. *Peloric flower*, an abnormally regular flower in a species in which flowers are normally irregular.

fluid mosaic model also *Singer model*, a description of the plasma membrane as a phospholipid bilayer stabilized by specifically oriented proteins, with some proteins extending through to both surfaces and other proteins specific for the inner or outer surfaces.

fluke a parasitic trematode flatworm; see also *blood fluke* and *Schistosome*.

follicle see *hair follicle*, *Graafian follicle*, or *corpus luteum*.

follicle-stimulating hormone (FSH) a peptide hormone of the mammalian anterior pituitary, which stimulates the growth of ovarian follicles; the same peptide activates sperm production in males.

food chain a sequence of organisms in an ecological community, each of which is food for the next higher organism, from the primary producer to the top predator.

food pyramid a graphic arrangement of trophic levels with producers forming the base, followed by first, second, third, and subsequent levels of consumers.

food vacuole see *digestive vacuole*.

food web also *food cycle*, a group of interacting food chains; all of the feeding relations of a community taken together; the flow of chemical energy between organisms.

foot in the moss sporophyte, the anchoring base.

F-one particle usually **F1 particle**, an ultramicroscopic mitochondrial organelle attached to the inner surface of the crista; the site of chemiosmotic phosphorylation.

foramen (pl. *foramina*), an opening or perforation.

foramen magnum the large opening at the base of the skull through which the spinal cord passes into the brain.

foramen ovale in the fetal mammalian heart, an opening in the septum between the atria; it normally closes at the time of birth.

foraminiferan a shelled marine sarcodine protist with one or more nuclei in a single plasma membrane; the carbonate shell having numerous minute openings (foramina) through which slender branching pseudopods are extended.

force filtration in the nephron, the movement of smaller molecules out of the blood and into Bowman's capsule by hydrostatic pressure in the glomerulus.

forebrain 1. the anterior of the three primary divisions of the vertebrate brain. 2. the parts of the brain developed from the embryonic forebrain.

fossil any remains, impression, or trace of an animal or plant of a former geological age, such as mineralized skeleton, a footprint, or a frozen mammoth.

fossil fuel coal, petroleum, or natural gas.

founder effect the population genetic effect of the chance assortment of genes carried by the successful founder or by a few founders of a subsequently large population.

four-chambered heart the heart of mammals, birds and crocodilians (and probably of dinosaurs), in which the atrium and the ventricle are both separated into left and right chambers by the evolutionary development of septa.

fourth ventricle a fluid-filled space in the brain stem that is continuous with the central canal of the spinal cord.

fovea (pl. *foveae*), the most sensitive part of the retina; a small central depression rich in cones and devoid of rods.

F$^+$ (fertility-positive) bacterium containing the plasmid responsible for producing a sex pilus through which a replica of the plasmid can pass to an F$^-$ bacterium.

frame shift mutation a small insertion or deletion in a structural gene such that all mRNA codons downstream are misread in translation.

free energy energy theoretically available to be used for work.

free energy of evaporation the energy involved in the evaporation of a liquid.

free radical a highly reactive atom or molecular unit with an unpaired electron.

freeze-etching also *freeze-fracturing*, the slicing of freeze-dried materials for observation through the transmission electron microscope, particularly useful for dividing membranes along their lipid core.

freeze-fracturing see *freeze-etching*

frequency dependent selection natural selection in which a phenotype is more fit than average when it is numerically rare, and is less fit than average when its relative number surpasses some equilibrium frequency value.

frontal lobe an anterior division of the cerebral hemisphere, believed to be a site of higher cognition.

fructose a 6-carbon ketose sugar, occurring free in honey, and combined with glucose to form the disaccharide *sucrose*.

fruit the seed-bearing ovary of a flowering plant.

fruiting body in fungi, slime molds, and myxobacteria, a structure specialized for the production of spores.

fucoxanthin a brown carotenoid pigment characteristic of brown algae.

functional group in biochemistry, a side group with a characteristic chemical behavior or function.

fundamental niche the potential niche of an organism. Compare *realized niche*.

galactose a hexose of sugar, one of the two subunits of lactose, the disaccharide of milk sugar.

galactoside a compound of galactose; e.g., lactose. *Galactosidase*, an enzyme that dissociates galactose from a galactoside.

galactoside permease in *E. coli*, a membrane protein involved in the facilitated diffusion of galactosides into the cell.

gall bladder, see under *bladder*.

gametangium (pl. *gametangia*), in algae and fungi, the structure (cell or organ) in which gametes are formed.

gametocyte also *meiocyte*, the cell that undergoes meiosis to form gametes; a spermatocyte or oocyte.

gametogenesis the process through which haploid gametes (sex cells: eggs and sperm) are produced, commonly including meiosis.

gametophyte in plants with alteration of generations, the haploid form, in which gametes are produced.

gamma radiation electromagnetic radiation of very high energy per photon, beyond X-radiation.

ganglion a mass of nerve tissue containing the cell bodies of neurons.

gap junction dense structures that physically connect plasma membranes of adjacent cells along with channels for cell-to-cell transport.

gas gangrene a bacterial disease caused by *Clostridium perfringins*, the symptoms of which include rotting of the affected flesh.

gas gland in bony fishes, a gas-producing gland associated with the swim bladder and the maintenance of buoyancy. See also *reabsorptive area*.

gastric lipase a fat-digesting enzyme secreted by the stomach lining.

gastrin a hormone, produced by the stomach lining, that induces the secretion of gastric juice (and may be identical with histamine).

gastrointestinal tract 1. the stomach and intestines as a functional unit. 2. the entire passageway from the mouth to and including the anus.

gastrovascular cavity also *coelenteron*, the cavity of coelenterates, ctenophores and flatworms, which opens to the outside only via the mouth, and serves the functions of a digestive cavity, crude circulatory system, and coelom.

gastrula an early metazoan embryo consisting of a hollow, two-layered cup with an inner cavity (archenteron) opening out through a blastopore.

gastrulation the process usually involving invagination, involution, and epiboly, whereby a blastula becomes a gastrula.

Gause's law also, the *competitive exclusion principle*, the ecological principle that no two genetically isolated kinds of organisms can coexist (exist at the same time and place) while occupying the same ecological niche; the principle that each species of a group of coexisting species must be specialized in some way so as to occupy a unique niche.

gel electrophoresis an analytical procedure in biochemistry, whereby a mixture of biomolecules enters a gel and separates by the differential migration of its constituents in an imposed electrical field.

gemma (pl. *gemmae*), in liverworts, a multicellular bud that becomes detached from the parent thallus in a form of asexual reproduction.

gene (variously defined), 1. the unit of heredity. 2. the unit of heredity transmitted in the chromosome and that, through the interaction with other genes and gene products, controls the development of hereditary character. 3. a continuous length of DNA with a single genetic function. 4. the unit of transcription of DNA. 5. a *cistron* or *structural gene*, that is, the sequence of DNA coding for a single polypeptide sequence.

gene cloning recent techniques whereby pieces of DNA from any source are spliced into plasmid DNA, cultured in growing bacteria, purified and recovered in quantity. Also *gene splicing, DNA cloning, recombinant DNA.*

gene flow the exchange of genes between populations through migration, pollen dispersal, chance encounters and the like.

gene frequency see *allele frequency*.

gene pool all the genetic information of a population considered collectively.

gene splicing the use of recombinant DNA techniques to form covalent bonds between DNA of different sources. See *gene cloning*.

general fertility rate see *fertility rate*.

generalization the process whereby a response is made to stimuli similar to, but not identical with, a reference stimulus.

generative cell a cell in the microgametophyte of a seed plant, capable of dividing to form two sperm cells.

generative nucleus one of the haploid nuclei of a pollen grain; when successful pollination occurs, the generative nucleus divides to form two sperm nuclei.

gene sequencing determining the specific sequence of nucleotides in a gene.

gene splicing using recombinant DNA techniques to insert selected DNA fragments or genes into bacterial plasmids, which will then be reintroduced into bacterial subjects for cloning.

genetic code the relationship by which the 64 possible codons (sequences of three nucleotides in DNA or RNA) each specify an amino acid or chain termination in protein synthesis.

genetic drift see *drift*.

genetic dwarf a dwarf whose condition is due to genetic rather than environmental causes. See *achondroplasia*.

genetic engineering the manipulation of genes through recombinant DNA techniques.

genetic equilibrium the genetic state of a population wherein the frequency of certain alleles remains constant generation after generation.

genetic map a chart of the sequence of and relative distances between specific genes along a chromosome, based on recombination frequencies observed between gene loci. Also *recombination map*.

genetic marker a standard mutant allele used in genetic tests in which it serves only to trace the fate or distribution of a chromosome, chromosome region, or gene locus, especially in recombination studies.

genetic recombination any process, such as crossing over, that leads to new gene combinations on the chromosomes.

genetic variability a broad term indicating the presence of different genetic constitutions in a population or populations.

genetics 1. the science of heredity, dealing with the resemblances and differences of related organisms resulting from the interaction of their genes and the environment. 2. the study of the structure, function and transmission of genes.

genital swelling in mammalian embryos, a paired swelling beneath the *genital tubercle*; it will form the scrotum in males and the *labia majora* in females.

genital tubercle in the sexually undifferentiated mammalian embryo, three small swellings, the glans and two urogenital folds; in the male these will form the head and shaft of the penis, and in the female they will form the *glans clitoridis* and the *labia majora*.

genitals also *genitalia*, 1. the organs of the reproductive system. 2. also *external genitalia*, the external reproductive organs; the penis and scrotum in human males and the vulva in females.

genome the full haploid complement of genetic information of a diploid organism, usually viewed as a property of the species.

genotype 1. the genetic constitution of an organism with respect to the gene locus or loci under consideration. 2. also *total genotype*, the sum total of genetic information of an organism.

genotype frequency the proportion of individuals in a population having a specified genotype.

genus (pl. *genera*), a group of similar and related species; a taxon smaller than *family* and larger than *species*.

geological era one of five divisions of the earth's history (Cenozoic, Mesozoic, Paleozoic, Proterozoic, and Archeozoic); see the geological timetable inside the front cover.

geology the scientific study of the physical history of the earth, the rocks and soil of which it is composed, and the physical changes that the earth has undergone or is undergoing.

geotaxis also *positive geotaxis*, the tendency to move downward. *Negative geotaxis*, the tendency to move upward.

geotropism also *positive geotropism*, the tendency of a growing root tip to turn downward. *Negative geotropism*, the tendency of a growing shoot to turn upward.

germinal tissues epithelial tissues specialized for gametogenesis, located in the gonad (testis or ovary) of animals.

germ layer any of the three layers of cells formed at gastrulation, the partially differentiated precursors of specific body tissues and organs; see *ectoderm, endoderm,* and *mesoderm*.

germ line *germ line cell*, the gametes and any cell or tissue that will or might give rise to gametocytes, or that otherwise contains the physical precursor of DNA that will be transmitted to offspring. *Germ-line DNA*, the DNA of germ line cells.

germ line theory the proposal that the genes needed to produce any antibody are already encoded in the DNA of the organism. Compare to *somatic variation theory*.

germinal adj. 1. pertaining to the germ line. 2. pertaining to reproduction. 3. giving rise to embryonic structures.

germinal epithelium 1. the epithelium of the gonads, which through mitosis gives rise to gametocytes. 2. also *germinal layer*, the innermost layer of an epithelium, containing mitotic cells.

germination the process whereby a seed and embryo ends dormancy, the stored materials of the seed being digested, the seed coat ruptured, and the plumule and hypocotyl begin to grow; spore germination, the end of dormancy in a spore.

giant chromosome see *polytene chromosome*.

giantism 1. also *gigantism*, growth to abnormal or unusual size. 2. see *pituitary giant*.

gibberellin any of a family of plant growth hormones that control cell elongation, bud development, differentiation, and other growth effects.

gill 1. an organ for obtaining oxygen from water. 1a. any of the highly vascular, filamentous processes of the pharyns of fishes, between the gill clefts. 1b. any of the analogous but not homologous organs of invertebrates. 2. the thin, laminar structures on the underside of a mushroom (basidocarp), bearing the spore-forming basidia.

gill arches also *pharyngeal arches* in fish, a row of bony or cartilaginous curved bars extending vertically between the gill slits on either side of the pharynx, supporting the gills. 2. in land vertebrate embryos, a row of corresponding homologous rudimentary ridges that give rise to jaw, tongue and ear bones.

gill basket the cartilaginous structure supporting the gills in urochordates, cephalochordates, and larval lampreys, serving as a sieve for filter feeding.

gill chamber in fishes, one of the two chambers housing the gills; covered by a bony flap, the operculum.

gill clefts also *gill slits, pharyngeal clefts*, 1. in fish, the openings between the gills. 2. in vertebrate embryos, the corresponding and homologous grooves in the neck region, between the branchial arches.

gill filament one of the many feathery, leaf-like or filamentous structures making up the working surfaces of the gill.

gill pouch also *pharyngeal pouch,* one of the hollowed-out cavities corresponding to the gill clefts in cyclostomes and some sharks.

gill raker in fish, one of the bony processes on the inner side of a gill arch, serving to prevent solids from passing out through the gill clefts.

gill slit see *gill cleft.*

gizzard 1. in birds, a thick-walled muscular enlargement of the alimentary canal, filled with small stones and used to grind seeds or other food. 2. any analogous structure in invertebrates, as in earthworms, rotifers, and gastrotrichs.

glandular epithelium simple columnar epithelial tissue that lines glands and the intestine, specializing in synthesis and secretion.

glans (L. *glans,* acorn), 1. also *glans, penis,* the conical, vascular, highly innervated body forming the end of the penis. 2. also *glans clitoridis,* a homologous body forming the tip of the clitoris.

glassy sponge sponges with a skeletal element of silicon dioxide or glass.

glial cell see *neuroglia.*

gliding joint a skeletal joint in which the articulating surfaces glide over one another without twisting.

globulin any of a large class of globular proteins occurring in plant and animal tissues and in blood.

glomerulus (pl. *glomeruli*), the mass or tuft of capillaries within a Bowman's capsule.

glucagon a polypeptide hormone secreted by the pancreatic islets of Langerhans, whose action increases the blood glucose level by stimulating the breakdown of glycogen in the liver.

glottis the opening between the vocal chords in the larynx.

gluconeogenesis in the liver, a biochemical pathway in which, at the expense of ATP, lactate from active muscle is converted back to glucose or glycogen.

glucose also *dextrose, blood sugar, corn sugar, grape sugar,* a 6-carbon sugar, occuring in an open chain form or either of two ring forms; the subunit of which the polysaccharides starch, glycogen, and cellulose are composed, and a constituent of most other polysaccharides and disaccharides.

glucosamine an amino derivative of glucose, a constituent of chitin.

glue cell a glandular, thread-bearing cell, found only in ctenophores; used to capture prey by adhesion.

glycerol formerly *glycerin,* an organic compound composed of a 3-carbon backbone with an alcohol (hydroxyl) group on each carbon; a component of neutral fats and of phospholipids. Pure glycerol is a sweet, sticky, and disgusting liquid.

glycocalyx in plasma membranes, the complex of glycoprotein, glycosaccharides, and/or glycolipids that form a frilly-appearing surface.

glycogen also *animal starch,* a highly branched polysaccharide consisting of alpha glucose subunits; a carbohydrate storage material in the liver, muscle and other animal tissues.

glycolipid a compound with lipid and carbohydrate subunits; e.g., cerebrosides and gangliosides.

glycolysis the enzymatic, anaerobic breakdown of glucose in cells, yielding ATP (from ADP), and either lactic acid, or alcohol and CO_2, or pyruvate and NADH (from NAD).

glycoprotein a compound containing polypeptide and carbohydrate subunits.

G_1 also *G_1 phase, gap one* see **cell cycle**.

goblet cell a goblet-shaped, mucus-secreting epithelial cell; found in the lining of the nasal cavity, bronchi, and elsewhere.

goiter a visible enlargement of the thyroid gland on the front and sides of the neck resulting from iodine deficiency, associated with normal or subnormal levels of thyroid hormone.

Golgi apparatus also *Golgi complex, Golgi body,* a cytoplasmic, membranous, subcellular structure found especially in secretory cells, and which is involved in the packaging of cell products from the endoplasmic reticulum into secretion granules.

Golgi vesicle, also **coated vesicle** or **secretory vesicle** spherical, membrane-bounded bodies formed from the Golgi membranes and containing Golgi-modified substances for transport and for secretion.

gonad in animals, the primary sex gland, with endocrine functions, and the site of meiosis; an ovary or testis.

gonadotropin a hormone that stimulates growth or activity in the gonads; in vertebrates, a specific peptide hormone of the anterior pituitary.

gonorrhea a sexually transmitted bacterial disease caused by the diplococcus *Neisseria gonorrhoeae,* which thrives in the mucous membranes of the male urethra and genital tract; in women infection occurs in the cervix, often spreading to the fallopian tubes.

graded display a communication pattern in which there exist intermediate states such as a behavioral continuum from low to high levels of motivation.

gradient a continuous change in concentration (or of any other quantity such as temperature, partial pressure, or allele frequency) over space. *With the gradient* (in the movement of solutes), in the direction from higher to lower concentration. *Against the gradient,* in the direction from lower to higher concentration, i.e., in the direction opposite to that of expected net diffusion.

gradualism the Darwinian proposal that the pace of evolution is slow but steady with an ongoing accumulation of minor changes leading eventually to the formation of new species. Compare to *punctuated equilibrium.*

Graffian follicle or *follicle,* a fluid-filled body in the ovary during the preovulatory period, containing the maturing oocyte within a cluster of follicular cells.

graft 1. the act of uniting two plants in such a way that they grow together. 2. the resulting growth. 3. v.t., to make a graft.

grafting an artificial method of propagation in which the bud, stem, or shoot of a plant is inserted into a cut in a recipient stem such that the cambium of both plants comes into contact and will eventually grow together.

grain also *cereal grain,* the hard fruit of a cereal grass, such as wheat, rye, corn, rice, or barley, consisting of a single seed in a tough covering.

Gram stain a solution of iodine and iodide used to stain certain large classes of bacteria. *Gram negative bacterium,* a bacterium that cannot be stained by gram stain. *Gram positive bacterium,* a bacterium that can be stained with Gram stain.

granum (pl. *grana*), a stack of thylakoid disks in a chloroplast; seen as a series of minute, multiple green bodies containing most of the chlorophyll of the plant.

gray crescent in the amphibian embryo, a pigment-free, crescent-shaped region that appears in the egg surface following fertilization.

greenhouse effect the warming principle of greenhouses, in which high-energy solar rays enter easily, while less energetic heat waves are not radiated outward; now especially applied to the analogous effect of increasing atmospheric concentrations of carbon dioxide through the burning of fossil fuels and forest biomass.

gross productivity the amount of biochemical energy captured by photosynthesis in a particular area per unit time.

ground meristem primary plant tissue from which the ground tissues, collenchyma, parenchyma, and sclerenchyma, are derived.

growth curve a graph of population numbers per unit time, especially in a period of rapid initial growth leading into population size stabilization or to a population crash.

growth factor a hormone, usually a mitogen, necessary for the proliferation of cells in tissue culture.

growth hormone (GH) 1. a polypeptide hormone of the anterior pituitary that regulates growth in vertebrates. 2. any hormone that regulates growth, e.g., auxin or gibberellin in plants.

G_2 also *G_2 phase, gap two* see *cell cycle.*

guanine one of the nitrogenous bases of RNA and DNA.

guanosine a nucleoside of guanine; guanine linked to ribose.

gullet 1. the esophagus. 2. an invagination in the surface of certain protozoan protists, especially when used for the intake of food.

guard cell in leaf and stem epidermis, either of a pair of crescent-shaped cells that with the pore make up the stoma.

gustation the sense of taste; the act of tasting.

guttation the normal exuding of moisture from the tip of a leaf or stem, presumably due to root pressure.

gynoecium in a flower, the whorl or carpels taken as a group; the pistil or pistils.

habit reversal a test involving the ease with which an animal learns to respond in a manner opposite to that to which it has previously been trained.

habitat the place where a plant or animal species naturally lives and grows.

habituation learning not to respond to environmental stimuli that may have no relevance to the organism.

hair cell see *organ of Corti.*

half life. 1. of a radioisotope, the time it takes for half of the atoms in a sample to undergo spontaneous decay. 2. the time it takes for half of the amount of an introduced substance (such as a drug) to be eliminated by a natural system, either by excretion or by metabolic breakdown.

Haldane-Oparin hypothesis the proposal that life arose spontaneously in the sea following a period of organic synthesis and under conditions that are no longer present on the earth.

haploid having a single set of genes and chromosomes; compare *diploid, polyploid.*

Hardy-Weinberg law see *Castle-Hardy-Weinberg law.*

Haversian canal in bone, a small canal through which a blood vessel runs.

Haversian system a Haversian canal together with its surrounding, concentrically arranged layers of bone, canaliculi, lacuna, and osteocytes.

heat of activation the energy input needed before an exergonic chemical reaction can proceed.

heat receptor a sense organ that perceives warmth or radiant heat.

helicase or *unwinding enzyme* see *replication complex.*

heliozoan also *sun animacule,* a protist of a group in the phylum Sarcodina, consisting of free-living, freshwater forms looking rather like tiny suns, with multiple thin, stiff, radiating pseudopods.

helper T-cell also called *inducer T-cells* see *regulatory T-cells.*

heme group an iron-containing, oxygen-binding porphyrin ring present in all hemoglobin chains and in myoglobin; see *porphyrin group.*

hemicellulose any of a number of complex polysaccharides found in wood and other plant cell walls, having as subunits not only glucose but also uronic acids, xylose, galactose, and various pentose sugars.

hemichordate an animal of the deuterostome phylum *Hemichordata.*

hemipenis either of the paired hemipenes or copulatory organs of a male lizard or snake.

hemizygous a term used to describe the condition of X-linked genes in males, which are neither homozygous nor heterozygous.

hemocoel a body cavity, especially in arthropods and mollusks, formed by expansion of parts of the blood vascular system.

hemochromatosis a genetic disease in human males, characterized by widespread deposition of iron in the tissues, resulting in bronzing of the skin, cirrhosis of the liver and pancreas, diabetes, and loss of armpit hair; being a disease of males, it affects neither females nor children.

hemocyanin a copper-containing respiratory pigment occurring in solution in the blood plasma of various arthropods and mollusks.

hemoglobin an iron-containing respiratory pigment consisting of one or more polypeptide chains, each associated with a heme group.

hemophilia a genetic tendency to uncontrolled bleeding in humans, due to the lack of a necessary constituent of the blood clotting process; caused by recessive alleles at either of two sex-linked loci, it affects males almost exclusively.

hepatic portal vein in all vertebrates, a large blood vessel that collects blood from the capillaries and venules of the esophagus, stomach and intestine and carries it to the liver, where it once more divides into capillaries. *Hepatic portal system,* the hepatic portal vein and associated blood vessels.

herbivorous plant-eating, *Herbivore,* an animal that feeds only on plants.

heredity the transmission of genetic characters from parents to offspring, and the effects of this transmission. See also *genetics.*

hermaphrodite an animal or plant that is equipped with both male and female reproductive organs.

Herpes simplex see *Type I* and *Type II Herpes simplex.*

heterochromtin chromatin that is relatively heavily condensed at times other than mitosis or meiosis, or that condenses early in prophase, and is genetically largely inert.

heterocyst in filamentous cyanophytes, a relatively large, transparent, thick-walled cell specialized for nitrogen fixation.

heteromorphic alternation in alternation of generations, the condition wherein the sporophyte and gametophyte are quite different in appearance, usually with one dominating the life cycle.

heteromorphic chromosomes homologous chromosomes that can be distinguished from one another by some visible difference, as a variable knob or constriction; e.g., X and Y chromosomes.

heterophagy "other-eating," phagocytosis and digestion of aging or damaged cells and cellular debris by phagocytic cells.

heterosporous a state in plants in which morphologically distinct spores, microspores, and megaspores are produced.

heterotroph a microorganism that requires organic compounds as an energy or carbon source.

heterozygous having two different alleles at a specific gene locus on homologous chromosomes of a diploid organism.

Hfr (or *high frequency of recombination*) a strain of *E. coli* in which the F plasmid is incorporated into the main host chromosome, thus increasing the frequency of sexual recombination involving the host chromosome.

high-energy bond a molecular bond that releases a large amount of energy when it is broken (that is, replaced with a different type of bond).

hilum (pl. *hila*), a scar left on a seed, marking the point of separation from the ovary; a bean's belly button.

hindbrain 1. the posterior of the three primary divisions of the embryonic vertebrate brain. 2. the parts of the adult brain derived from the embryonic hindbrain, including the *cerebellum, pons,* and *medulla oblongata.*

hinge joint a joint that moves in one plane, like a door hinge.

hip bone see *coxa.*

hippocampus 1. a long, curved ridge of grey and white matter on the inner surface of each lateral ventricle of the brain, involved in memory consolidation, visual memory and other functions. 2. a sea horse.

hirudin an anticoagulant secreted by leeches.

histocompatibility antigen any of several extremely active cell-surface antigens involved in normal cell-to-cell interactions in the immune response, and highly variable in all mammalian populations; these antigens are the principal cause of rejection of tissue and organ transplants between individuals.

histocompatibility locus any of the four genetic loci responsible for producing the highly varied histocompatibility antigens present on the surfaces of cells.

histology the scientific study of tissues.

histone one of a class of small, highly basic proteins that complex with the nuclear DNA of higher eukaryotes, forming nucleosomes that presumably protect the DNA from degradation; one histone (histone I) appears to have a role in chromatin condensation and thus in one level of gene control.

holdfast a rhizoidal base of a seaweed, serving to anchor the algal thallus to the ocean floor or rock substratum.

holoblastic cleavage in the early animal embryo, cytokinesis that completely divides the embryo. Compare to *meroblastic cleavage.*

homeostasis the tendency toward maintaining a stable internal environment in the body of a higher animal through interacting physiological processes involving negative feedback control.

homeostatic mechanism any mechanism involved in maintaining a stable internal equilibrium, especially one involving self-correcting negative feedback.

homeotherm see *endotherm.*

home range the area to which an animal confines its activities; compare *territory.*

hominids a primate group composed of humans and related extinct forms.

hominoid a primate group composed of humans, apes, and related extinct forms.

homologous 1. similar because of a common evolutionary origin. 2. derived by independent evolutionary modification from a corresponding body part of a common ancestor. 3. derived from a common embryological precursor; e.g., the *glans clitoridis* is homologous with the *glans penis.* 4. *homologous chromosomes,* chromosomes which, because of common descent, have the same kinds of genes in the same order and which pair in meiosis.

homology similarity due to common descent.

homologue also *homologous chromosome,* either of the two members of each pair of chromosomes in a diploid cell; see *homologous.*

homosporous a state in plants where all spores are the same. See also *heterosporous.*

homozygous having the same allele at a given locus in both homologous chromosomes of a diploid organism.

hormone a chemical messenger transmitted in body fluids or sap from one part of the organism to another, that produces a specific effect on target cells often remote from its point of origin, and that functions to regulate physiology, growth, differentiation, or behavior.

hornwort a nonvascular terrestrial plant allied to mosses and liverworts.

host a living organism that harbors or sustains a parasite, pathogen, or other symbiont.

host-specific a parasite or pathogen that can infect only a single host species.

human chorionic gonadotropin (HCG) a hormone secreted by the human chorion, prompting the corpus luteum to continue manufacturing progesterone.

human leukocyte-associated antigen (HLA antigen) the major histocompatibility complex of humans. See also *major histocompatibility complex* and *histocompatibility antigen.*

human liver fluke also *Asian liver fluke* and *Chinese liver fluke (Chlonorchis sinensis),* a flatworm parasite of humans, the life cycle of which occurs in three hosts and includes five stages: sexually mature *metacercaria* that reproduce in the human host; *miracidia* that form a *sporocyst* and then *redia* in a snail; redia that develop into burrowing *cercaria* and encyst in a fish; metacercaria that reproduce sexually in humans who eat the fish without having cooked it sufficiently.

humerus the long bone that forms the upper portion of the vertebrate forelimb, e.g., the bone of the human upper arm.

humus partially decayed organic materials in the soil.

Huntington's disease also *Huntington's chorea,* a lethal hereditary disease of the nervous system developing in adult life and attributable to a dominant allele.

H-Y cell-surface antigen also *H-YA* an antigen found on the surface of all cells of all male mammals, but not in female mammals; and also on the cells of all female birds, but not male birds. It is believed to be controlled by a Y-linked gene in mammals, and to be responsible for the differentiation of the mammalian testis.

hyaline cartilage translucent, bluish-white connective tissue consisting of cells embedded in a homogeneous matrix.

hyaluronic acid a viscous mucopolysaccharide occurring chiefly in connective tissues, but also as a cell-binding agent in other tissues, e.g., it holds together the cells of the *corona radiata.*

hyaluronidase an enzyme that dissolves hyaluronic acid and thus disrupts connective tissue; produced by invading bacteria, and also by mammalian spermatozoa.

hybrid the offspring of two animals or plants of different races, breeds, varieties, species, or genera.

hybrid swarm genetically diversified populations produced by introgressive hybridization, that is, the breeding of hybrids with parent stock, thus creating numerous partial hybrids.

hydration shell the layer or layers of water surrounding and loosely bound to an ion in solution.

hydrocarbon 1. a chemical compound consisting of hydrogen and carbon only, often in chains or rings: e.g., gasoline, benzene, paraffin; 2. a portion of an organic molecule consisting of hydrogen and carbon only, e.g., the hydrocarbon portion of a fatty acid.

hydrocortisone a trade name for the hormone *cortisone.*

hydrogenation chemically saturating an unsaturated lipid with hydrogen through the elimination of double carbon-to-carbon bonds.

hydrogen bond a weak, noncovalent, usually intramolecular electrostatic attraction between the positively polar hydrogen of a side group and a negatively polar oxygen of another side group.

hydrogen carrier a certain membranal element within an electron transport system that accepts both electrons and protons, transporting hydrogen across the membrane, whereupon its proton is released and its electron continues in the system. See *chemiosmosis.*

hydrogen ion 1. a proton. 2. as a convenient fiction, a proton in solution and bound to a water molecule; actually a *hydronium ion* (H_3O^+).

hydroid in colelenterates, a member of class Hydrozoa.

hydrologic cycle also *water cycle,* the cycle of water between its liquid and gaseous phases brought about by heat in the atmosphere.

hydrolysis see *hydrolytic cleavage.*

hydrolytic cleavage also *hydrolysis,* the reaction of a compound with water such that the compound is split into two parts by the breaking of a covalent bond; and water is added in the place of the bond, an −OH group going to one subunit and an −H group going to the other.

hydronium ion a symmetrical charged ion, H_3O^+, that can dissociate into water and a hydrogen ion; most hydrogen ions in solution are actually in the form of hydronium ions.

hydrophilic "water loving," a characteristic of charged molecules in which they readily interact with water molecules.

hydrophobic with regard to a molecule or side group, tending to dissolve readily in organic solvents but not in water; resisting wetting; not containing polar groups or subgroups.

hydrostatic pressure the pressure exerted in all directions within a liquid at rest.

hydrostatic skeleton a supporting and locomotory mechanism involving a fluid confined in a space within layers of muscle and connective tissue; movement occurring when muscle contractions increase or decrease the hydrostatic pressure of the fluids.

hydroxyl group −OH, consisting of an oxygen and hydrogen covalently bonded to the remainder of the molecule; a constituent of alcohols, sugars, glycols, phenols, and other compounds.

hydroxyl ions also *hydroxide ion,* the univalent anion OH^-, one of the dissociation products of water in the reaction $2H_2O \rightarrow H_3O^+ + OH^-$.

hymen also *maidenhead,* fold of thin mucous membrane closing the orifice of the vagina, especially in virgins.

hyperglycemia an abnormal excess of glucose in the blood.

hyperosmotic see *hypertonic.*

hyperpolarization in a neuron, an inhibitory state caused by an influx of negative ions, whereby the threshold of stimulation is greater than during the usual resting state.

hyperthyroidism a condition caused by an excess of circulating thyroid hormone; symptoms in humans include nervousness, sleeplessness, hyperactivity, weight loss, and, if prolonged, a bulging of the eyeballs.

hypertonic (adj.), having a higher osmotic potential (e.g., higher solute concentration) than the cytoplasm of a living cell (or other reference solution).

hyperventilate to breathe deeply deliberately, purging the body of CO_2.

hypha (pl. *hyphae*), one of the individual filaments that make up a fungal mycelium.

hypoblast the lower layer of cells in the bird or reptile blastoderm.

hypocotyl the part of a plant embryo below the point of attachment of the cotyledon.

hypocotyl hook in some seedlings, the curved region of the hypocotyl that acts as a bumper as the seedling emerges from the soil.

hypoglycemia 1. an abnormally low level of blood glucose. 2. a medical condition involving recurrent episodes of low blood sugar, due to the overproduction of insulin.

hypophysis see *pituitary gland.*

hypothalamus a portion of the floor of the midbrain, containing vital autonomic regulatory centers, and closely associated functionally with the pituitary gland.

hypothesis a proposition set forth as an explanation for a specified group of phenomena, either asserted merely as a provisional conjecture to guide investigation (e.g., working hypothesis) or accepted as highly probable in the light of established facts; see also *theory.*

hypothyroidism a condition caused by abnormally low levels of circulating thyroid hormone; symptoms include physical and mental sluggishness and weight gain.

hypotonic (adj.), having a lower osmotic potential (e.g., a lower concentration of solutes) than the cytoplasm of a living cell (or other reference solution).

hysterectomy the surgical removal of the uterus.

H zone clearer, centermost zone of the relaxed contractile unit consisting of myosin alone.

IAA indoleacetic acid; see *auxin.*

I band one of the striations of striated muscle, of variable width, corresponding to the distance between the ends of the myosin filaments of adjacent units.

ileocecal valve a sphincter separating the large intestine from the small intestine.

ileum a region of small intestine.

ilium the hip bone, one of the three pairs of fused bones forming the pelvis.

imbibition the taking up (absorption) of a fluid (often water) by a hygroscopic gel, colloid, or fibrous matrix; such as the ultramicroscopic spaces in a plant cell wall.

immigration moving into a given area of population and taking up residence.

immune response the entire array of physiological and developmental responses involving specific protective actions against a foreign substance; including phagocytosis, the production of antibodies, complement fixation, lysis, agglutination, and inflammation.

immune system in vertebrates, widely-dispersed tissues that respond to the presence of the antigens of invading microorganisms or foreign chemical substances.

immunoglobin a protein antibody produced by thymocytes in response to specific foreign substances, consisting of constant regions and two specific antigen binding sites; or a polymer of such molecules.

imperfect flowers lacking either pistils or anthers.

implantation the act of attachment of the mammalian embryo (blastocyst) to the uterine endometrium.

impulse neural impulse, 1. a wave of excitement (transitory membrane depolarization) transmitted through a neuron; 2. the same wave as it travels a nerve pathway, including its transmission across synapses.

inactivation gate see *sodium ion gate*.

inborn error or metabolism a genetic defect in which an individual lacks one of the enzymes of a biochemical pathway.

incisors in humans and most mammals, the four upper and four lower front teeth, which are chisel-shaped for biting and cutting; in rodents the medial incisors are greatly exaggerated and the lateral incisors are absent.

incomplete flower see under *flower*.

incomplete penetrance see *penetrance*.

incurrent siphon also *inhalent siphon*, in bivalve mollusks, tubular structures used to bring in water for respiration and food filtration, which exits through a similar *excurrent siphon*.

incus see *middle ear*.

indeterminate life span in perennial plants, a life span hypothetically without end.

indeterminate or **day-neutral** plants whose flowering is independent of the length of day or night.

indeterminate plant one that shows no photoperiodicity in its flowering.

individual distance the space around an individual within which intrusion by another individual will elicit flight or attack.

indoleacetic acid also *IAA*, the naturally-occurring form of auxin, the plant growth hormone; see *auxin*.

induced fit hypothesis the proposition that enzymatic actions proceed because their active site forms an inexact fit with their substrate, producing physical stress.

inducer 1. in molecular genetics, a small molecule that triggers the activity of an inducible enzyme. 2. in embryology, a substance that stimulates the differentiation of cells or the development of a particular structure.

inducer T-cell also called *helper T-cells*, see *regulatory T-cells*.

inducible enzyme an enzyme produced by an organism only when induced by an appropriate stimulus. *Inducible operon*, the entire operon involved in the control of a battery of inducible enzymes.

induction the process by which the fate of embryonic cells of tissue is determined, especially as due to tissue interactions.

indusium in ferns, a covering over the sorus (spore-forming organ).

inferior situated below or beneath; e.g., *inferior vena cava*, inferior ovary in a flower (epigyny).

inflorescence a clustering of flowers with a specific arrangement.

information load a measure of the level of information that can be communicated in a single display.

ingestion swallowing.

inhibition the restraining, diminishing, or preventing of activity.

inhibitor a substance that interferes with a chemical or biological process.

inhibitory block according to the classical definition of instinct, the neurological inhibitors of behavior that are selectively removed by the perception of the appropriate releaser.

initiation the start of translation or polypeptide synthesis. See also *initiation complex*.

initiation complex the elements needed to begin initiation or polypeptide synthesis, a process requiring the presence of mRNA, methionine-charged tRNA, and the two ribosomal subunits.

initiator codon or *start codon*, a codon (sequence of three mRNA nucleotides) that initiates translation of a polypeptide; specifically, AUG, which also codes for methionine and N-formyl methionine.

initiator tRNA transfer RNA charged with methionine.

innate inborn, not due to learning.

innate behavioral pattern a genetically programmed behavior pattern.

innate releasing mechanism (IRM) a hypothetical center in the central nervous system that, when stimulated by the proper environmental cue, activates neural pathways that result in an *instinctive behavior pattern*.

inner cell mass in a mammalian blastocyst, the portion that is destined to become the embryo proper.

inner compartment see *matrix*.

inner ear the portion of the ear enclosed within the temporal bone, consisting of a complex fluid-filled labyrinth and the associated auditory nerve; included are the three *semicircular canals*, the *cochlea*, and the *round* and *oval windows*; compare *external ear*, *middle ear*.

inner sheath in cilia and flagella, a sheath surrounding the two central pairs of microtubules.

innominate bone (''nameless bone'') see *coxa*.

inoculum a small quantity of living cells or organisms introduced into a suitable medium for growth.

inoperative allele see *dominance relationships*.

inorganic molecule one not containing carbon, generally one found occurring outside of living organisms.

inorganic phosphate also P_i, the anion of phosphoric acid; the phosphate group once separated from a nucleotide triphosphate such as ATP or ADP.

insectivore an eater of insects.

insectivorous habitually eating insects; *insectivorous plant*, one with special adaptations for trapping and digesting insects for their nitrogen and mineral content.

insertion 1. in genetics, the addition of extra genetic material into the middle of a chromosome, gene or other DNA sequence; *base insertion*, the insertion of a single base pair into a DNA sequence. Also *muscle insertion*, 2. in anatomy, the distal attachment of a tendon or muscle; compare *origin*.

instinct innate behavior involving appetitive and consummatory phases, the latter usually in response to an environmental releaser but sometimes occurring as vacuum behavior.

instinctive pattern a *fixed-action pattern* that is coupled with orienting movements.

insulin a polypeptide hormone secreted by the islets of Langerhans of the pancreas, whose principal action is to facilitate the transport of glucose into cells and stimulate the conversion of glucose to glycogen, but having many other functions; it is an essential growth factor in mammalian cell culture.

integument a covering or envelope; in the flower, the covering that encloses the nucellus of an ovule, and later forms part of the seed coat; in animals, the skin, exoskeleton, tunic, cuticle, or other covering.

interbreeding 1. commonly breeding together. 2. breeding together; hybridizing.

intercalated disk in heart muscle, the highly convoluted double plasma membrane between two adjacent cells in a heart muscle fiber.

intercostal muscles the muscles of the rib cage, involved in breathing.

interecdysis also *intermolt*, *instar*, in arthropods, the period between successive molts.

interferon cellular substance that interferes with replication by viruses, generally produced by virus-infected cells.

intermediate host in parasitic relationships, a host in which a parasite may undergo asexual reproduction or at least some degree of development.

internal fertilization in animals, fertilization involving copulation, in which sperm release and fertilization occur within the female.

internal nares see *nares*.

interneuron a neuron typically found in the spinal cord that synapses with sensory and motor neurons and other interneurons.

internode the interval between two nodes, see *nodes*.

interphase all of the cell cycle between successive mitoses; included are G_1, S phase, and G_2.

interstitial cells cells of the testis that have an endocrine function.

interstitial cell-stimulating hormone (ICSH) (identical to LH), in males, an anterior pituitary hormone that stimulates testosterone production in the interstitial cells of the testis.

intertidal zone the zone between the mean high tide line and the mean low tide line, intermittently covered by the sea or exposed to air.

intervertebral disk one of the tough, elastic, fibrous disks situated between the centra of adjacent vertebrae.

intestine also *alimentary canal* or *gut*, essentially tubular organ in which the digestion and absorption of foods occur.

intracellular within cells.

intracellular digestion digestion within cells, in digestive vacuoles following phagocytosis.

intraspecific with a species; between members of the same species.

intrauterine device metallic contraceptive device that inhibits implantation when in place in the uterus.

intrinsic rate of increase *(i)* also *instantaneous rate of increase*, the rate of population increase determined by subtracting the average death rate from the average birth rate.

introgression backcrossing of a hybrid with one or the other parental types. Common means of diversification in flowering plants.

introitus the external opening of the vagina.

intron, also *intervening sequence*, a region of DNA separating two parts of a structural gene; transcribed into HnRNA but deleted from mRNA.

invagination the folding inward of a surface or tissue to make a cavity; specifically, the formation of a gastrula by the inward folding of part of the wall of the blastula.

in vitro ("in glass"), in a test tube or other artificial environment; compare *in vivo*.

in vivo in the living body of a plant or animal; compare *in vitro*.

involuntary muscle see *smooth muscle*.

involution an inward roll or curve, in gastrulation, the movement of cells toward the blastopore, over the blastopore lip, and into the archenteron.

ion any electrostatically charged atom or molecule.

ion channel in the neural membrane, sodium and potassium (and chloride) channels that, through the opening and closing of gates, selectively admit or reject ions.

ionic bond a chemical attraction between ions of opposite charge.

ionization 1. the dissociation of a molecule into oppositely charged ions in solution. 2. the creation of ions by the energy of ionizing radiation.

ionizing radiation energetic radiation that produces ions in air and free hydroxyl radicals in water, the latter being responsible for induced mutation and tissue damage; included are X rays, gamma rays, and streams of charged particles from the breakdown of radioactive isotopes.

iris see *eyeball*.

ischium one of the bones of the pelvis, specifically the one you sit on. See also *coxa*.

islets of Langerhans also *islets*, beta, alpha, and delta endocrine cells within the pancreas that secrete the hormones insulin, glucagon, and somatostatin, respectively.

isolecithal of an egg, having a small amount of yolk that is evenly dispersed in the cytoplasm.

isogametes gametes that are identical in size and appearance, such as the (+) and (−) isogametes of *Chlamydomonas*.

isomorphic alternation in alternation of generations, the condition wherein sporophyte and gametophyte are virtually identical in appearance and no dominance exists.

isoosmotic see *isotonic*.

isotonic having the same osmotic potential (e.g., the same concentration of solutes) as the cytoplasm of a living cell, or of some other reference fluid.

isotope a particular form of an element in terms of the number of neutrons in the nucleus; *radioactive isotope,* also *radioisotope,* an unstable isotope that spontaneously breaks down with the release of ionizing radiation.

Jacobson's organ a pouchlike olfactory organ in the roof of the mouth of most mammals and other vertebrates; in lizards and snakes it receives the ends of the forked tongue; absent in primates.

jaundice a yellowish pigmentation of the skin and other tissues caused by the deposition of bile pigments, following bile duct obstruction, liver disease, or the excessive breakdown of red blood cells.

jejunum a region of small intestine.

jelly coat a layer of transparent substance surrounding the egg of an echinoderm, fish or amphibian.

joint a point where two bones articulate, often movable.

juvenile hormone (JH), an insect hormone that prevents the differentiation of adult characters; secreted by the corpora allata of larval and subadult insects.

juxtaglomerular complex in the nephron, a physical connection between the afferent arteriole and the distal convoluted tubule, believed to coordinate the reabsorption of sodium.

karyotype 1. a mounted display of enlarged photomicrographs of all the stained chromosomes of an individual, arranged in order of decreasing size. 2. the total chromosome constitution of an individual. 3. the chromosomal makeup of a species.

keel the enlarged breastbone to which the flight muscles of birds are attached.

kelp any of various large brown algal seaweeds.

keratin a cysteine-rich, fibrous, insoluble, intracellular structural protein making up most of the substance of the dead cells of hair, horn, nails, claws, feathers, and the outer epidermis.

kidney 1. in vertebrates, one of a pair of ducted excretory organs situated in the body cavity beneath the dorsal peritoneum, serving to excrete nitrogenous wastes and to regulate the balance of body ions and fluids. 2. any analogous organ in invertebrate metazoans.

kilocalorie see *calorie*.

kinetic energy the energy of motion.

kinetochore see *centromere*.

kingdom one of the primary divisions of life forms; to the traditional plant and animal kingdoms Whittaker has added fungal, protist, and moneran (prokaryote) kingdoms.

Klinefelter's syndrome in humans, an abnormal condition of males, caused by the chormosomal condition XXY; characterized by increased stature, moderate retardation, small testes, low testosterone and secondary effects of low testosterone.

Krause corpuscle also *Krause end bulb,* a sensory receptor in the skin that responds to temperature changes.

Krebs cycle see *citric acid cycle*.

krill planktonic crustaceans that constitute the principal food of baleen (toothless) whales.

K-selection population growth characterized as logistic in form, responding to environmental resistance and carrying capacity. Compare *r-selection*.

kwashiorkor a syndrome of severe protein deficiency in human infants and children, including failure to grow, deficiency of melanin pigment, edema, degeneration of the liver, anemia, and retardation.

labia majora (sing. *labium majorum*), in human females, the outer, fatty, often hairy pair of folds bounding the vulva.

labia minora (sing. *labium minorum*), in the human female, the inner, thinner, highly vascular pair of folds surrounding the introitus, enfolding and extending downward from the clitoris.

labial palps a pair of leaflike, fleshy appendages on either side of the mouth of a bivalve mollusk which remove food from the gills in feeding.

labile readily undergoing chemical breakdown, dissociation, or denaturation.

labium 1. see *labia majora, labia minora*. 2. the lower lip of a snapdragon or similar flower. 3. the single lower lip of an insect or crustacean.

labrum the single upper lip of an insect or crustacean.

lac operon (an inducible operon) in the chromosome of *E. coli*, composed of genes that code for three lactose-metabolizing enzymes and the region that controls their transcription. See also *operon*.

lactase a digestive enzyme that hydrolyzes lactose to glucose and galactose; absent in most non-Caucasian adults.

lactate fermentation see *fermentation*.

lacteal a lymphatic vessel of the intestinal villi.

lactose also *milk sugar*, a disaccharide consisting of glucose and galactose subunits.

lactose intolerance the inability of certain persons to readily digest the milk sugar lactose, which leads to problems in the bowel.

lacuna (pl. *lacunas, lacunae*), a minute cavity in bone or cartilage that holds an osteocyte or chondrocyte.

lagging end in DNA replication, the 5' end of a DNA strand, where nucleotides are first assembled into Okazaki fragments and then added in using the enzyme ligase.

lamella (pl. *lamellae*), 1. one of the bony concentric layers (ring-like in cross section) that surround a Haversian canal in bone. 2. one of the thin plates composing the gills of a bivalve mollusk. 3. any thin, plate-like structure.

laminarin a storage carbohydrate of the brown algae.

lampbrush chromosome a chromosomal region commonly seen in oocytes that becomes temporarily active, spinning out loops of DNA which become active in transcription.

lanolin a greasy sterol found in untreated wool and expensive cosmetics.

large intestine also colon or bowel, a division of the alimentary canal, primarily functioning in the resorption of water.

larva in animals, an early, active, feeding stage of development during which the offspring may be quite unlike the adult.

laryngopharynx the lower part of the pharynx at the region of the larynx.

larynx in terrestrial vertebrates, the expanded part of the respiratory passage just below the glottis and at the top of the trachea; in mammals it contains the vocal cords and constitutes a resonating *voice box* or *Adam's apple*.

late blight the protistan disease caused by *Phytophthora infestans*, responsible for the Irish potato famine of the 1840s.

latency Darwin's term for the gross phenomenon of recessivity.

lateral branch bud also *axillary bud*, a dormant bud in the leaf axil or above a leaf scar, capable of giving rise to new stems.

lateral line organ in nearly all fish, a sensory organ considered to be responsive to water currents and vibrations; thought to be distantly homologous with the inner ear.

laterally from side to side.

laterite a hard, water-resistant, crusty soil, largely aluminum oxides leached of mineral nutrients, common in tropical rainforest regions.

law of alternate segregation also *Mendel's first law*, the observation, based on the regularity of meiosis, that heterozygous alleles separate in gamete formation, each allele going to approximately half the gametes produced.

law of independent assortment also *Mendel's second law*, the observation, based on the regularity of meiosis, that genetic elements on different chromosomes behave independently in the production of gametes.

Laws of Mass Action law stating that, in reversible chemical reactions, all other factors being equal, the rate and direction of the reaction depends on the concentration of the reactants.

laws of thermodynamics in physics, laws governing the interconversions of energy. See also *thermodynamics*.

leader sequence see *signal peptide*.

leading end in DNA replication, the 3' end of a DNA strand, where nucleotides are added one at a time to the growing strand.

leaf in vascular plants, a lateral outgrowth from stem functioning primarily in photosynthesis, arising in regular succession from the apical meristem, consisting typically of flattened blade jointed to the stem by a petiole.

leaf blade the typically flattened, photosynthetic region of a leaf only a few cell layers thick and containing a supporting petiole, midrib, and veins.

leaf hair see *lower epidermis*.

leaf mesophyll the photosynthetic parenchymal cells within a leaf, including an uppermost palisade layer and a lower spongy layer.

leaf mesophyll cell photosynthetic parenchymal cells within the leaf.

leaf primordium see *primordium*.

leaf scale a flattened, photosynthetic, leaflike structure in bryophytes.

leaf scar the mark left on a stem by a fallen leaf.

learning in animal behavior, the process of acquiring a persistent change in a behavioral response as a result of experience.

leech an annelid ectoparasite known to attach to and draw blood from warmblooded hosts.

lens any clear, biconvex organelle that functions to focus or collect light.

lenticel a pore in stem of plant for the passage of gases between the atmosphere and stem tissues.

leptotene (adj. and n), also *leptonema* (n.), the first stage of meiotic prophase, at which time the still unsynapsed chromosomes appear as elongated, thin threads.

leucoplast a colorless plastid; see also *plastid*.

leucosin a form of starch produced by protists of the phylum Chrysophya.

leukocyte also *leucocyte*, a vertebrate white blood cell; including, eosinophils, neutrophils, basophils, monocytes, and lymphocytes.

LH see *luteinizing hormone*.

lethal–recessive dominant an allele that has prominent visible effect as a heterozygote, and is lethal as a homozygote; in humans such an allele is also called a *rare dominant* since it may never occur as a homozygote.

lichen a combination of a fungus and an alga growing in a symbiotic relationship.

ligament a tough, flexible, but inelastic band of connective tissue that connects bones, or that supports an organ in place; compare *tendon*.

ligase an enzyme that heals nicks (single-strand breaks) in DNA.

light-harvesting antenna a cluster of photosynthetic pigments (chlorophyll and accessory pigments) which receives energy from photons and transfers that energy to a single *reaction center*.

light-independent reaction that part of photosynthesis not immediately involved in chemiosmosis, specifically the fixation of CO_2 into carbohydrate from the NADPH and ATP produced by the light reaction (Calvin cycle).

light reaction that part of photosynthesis directly dependent on the capture of photons; specifically the photolysis of water, the thylakoid electron transport system, and the chemiosmotic synthesis of ATP and NADPH.

lignin an amorphous substance that gives wood its rigidity.

limbic system a region of the brain concerned primarily with emotions.

limiting factor any factor in the environment which, by its presence or relative absence, limits the growth of a population; e.g., disease, predation, food supply, mineral nutrient.

limiting resource a resource, such as food, shelter, or mineral nutrient, that by its scarcity limits the growth of a population or determines the carrying capacity of the environment.

limnetic zone in freshwater bodies, the open water deep enough for a photic and an aphotic zone to occur and where food chains begin with phytoplankton.

limnologist an ecologist who specializes in the study of interaction in the freshwaters.

linkage group a group of gene loci shown to be linked by recombination tests; ultimately, a chromosome.

linked (adj.), of two gene loci, not segregating independently.

linoleic acid an essential polyunsaturated fatty acid, necessary for cell membrane function.

lipase any fat-digesting enzyme.

lip cell in the fern sporangium, thinner walled cells that split open through changes in the mature dry annulus, releasing the spores. See also *annulus*.

lipid an organic molecule that tends to be more soluble in non-polar solvents (such as gasoline) than in polar solvents (such as water).

lipoprotein a conjugated protein with a lipid subgroup.

litter also *leaf litter, duff*, the uppermost slightly decayed layer of organic material on a forest floor.

littoral in reference to the lake or seashore. *Littoral zone*, 1. a coastal region including both the land along the coast, the water along the shore and the intertidal. 2. a similar area in lakes.

liver 1. in vertebrates, a large, glandular, highly vascular organ that serves many metabolic functions including detoxification, the production of blood proteins, food storage, the biochemical alteration of food molecules, and the production of bile. 2. in mollusks, a digestive organ opening into the gut, in which intracellular digestion of food by phagocytosis occurs.

liverwort a kind of ground-hugging bryophyte.

loam a highly fertile soil defined ideally as a mixture of three-quarters silt and sand and one-quarter clay.

lobe a curved or rounded projection or division, as of the liver or lung.

lobe-finned fish lung fishes and crossopterygians.

locus (pl. *loci*), also *gene locus*, specific place on a chromosome where a gene is located.

logarithmic spiral a form of spiralling curve that keeps its overall shape constant as it increases in length.

logistic growth curve also *logistic increase* and *S-shaped curve*, population growth plot that takes the form of a sigmoid curve, growth that is eventually subject to *environmental resistance* and hovers about the environment's *carrying capacity*.

long-day plant also *short-night plant* a plant that begins flowering at some specific time before the summer solstice when day length exceeds night, flowering being triggered by a critical, established period of darkness.

long-term memory 1. learning that persists more than a few hours, the memory trace of which is physically located in a different part of the brain than short-term memory. 2. the part of the brain and the general neural function with which such persistent memory traces are associated.

loop of Henle the hairpin loop of the vertebrate nephron, between the proximal and distal convoluted tubules, which leaves the cortex and descends into the medulla, then loops back to the cortex; it is involved in water resorption.

lophophorate invertebrate animals bearing the lophophore, a ridge-like feeding structure bearing ciliated tentacles.

lophophore in brachiopods, ectoprocts and phoronids, a filter-feeding organ consisting of a spiral or horseshoe-shaped ridge surrounding the mouth, and bearing tentacles.

Lotka-Volterra theory the proposal that predator population size rises and falls according to the population size of the prey.

lower epidermis an outer layer of leaf cells, commonly containing numerous stomata and projecting leaf hairs.

lumbar pertaining to the lower trunk; *lumbar region* of the spine, the vertebrae between the thorax and the sacrum.

lumen the cavity or channel of a hollow tubular organ or organelle.

lung in land vertebrates, one of a pair of compound, saccular organs that function in the exchange of gases between the atmosphere and the bloodstream. 2. any of the several analogous organs in invertebrates, e.g., *book lung*.

luteinizing hormone (LH) a vertebrate polypeptide hormone of the anterior pituitary that in the female stimulates ovulation and the development and maintenance of the corpus luteum; the same polypeptide functions in the male as the interstitial cell stimulating hormone.

luteining hormone releasing hormone (LHR hormone) see *releasing hormone*.

lymphatic collecting duct larger lymphatic vessels that receive lymph from the lymphatic capillaries.

lymphatic system the system of lymphatic vessels, lymph nodes, lymphocytes, the thoracic duct and the thymus, which together serve to drain body tissues of excess fluids and to combat infections.

lymphatic vessel also *lymphatic*, a thin-walled vessel that conveys lymph, originating as an open intercellular cleft and draining eventually into the *thoracic duct*.

lymph capillary one of the many tiny, blind endings in the lymphatic system, including the lacteals of the small intestine.

lymph duct see *thoracic duct*.

lymph node a rounded, encapsulated mass of lymphoid tissue through which lymph ducts drain, consisting of a fibrous mesh containing numerous lymphocytes and phagocytes.

lymphoid tissue tissues in which lymphocytes are activated and aggregate.

lymphocyte any of several varieties of similar-appearing leukocytes involved in the production of antibodies and in other aspects of the immune response; see *B-cell, T-cell*.

Lyon effect in human females, the genetic mosaic created by the random manner in which either the paternal or the maternal X chromosome is selected for permanent inactivation.

lyse (v.t.), to destroy a cell by rupturing the plasma membrane.

lysis (n.), the destruction or lysing of a cell by rupture of the plasma membrane.

lysogeny the incorporation of a bacteriophage chromosome into the bacterial host chromosome, together with mechanisms preventing further infection and lysis; in later cell generations the incorporated DNA may excise, replicate and eventually cause cell lysis.

lysosome a small membrane-bounded cytoplasmic organelle, generally containing strong digestive enzymes or other cytotoxic materials.

lysozyme an enzyme that lyses bacteria by dissolving the bacteria cell wall; a natural constituent of eggwhite and tears.

lytic cycle the short cycle following bacteriophage invasion during which viral replication, capsid synthesis, viral assembly, and cell lysis occur, the latter releasing new infective phage particles. See also *lysogenic cycle*.

macrolecithal in animal eggs, the presence of a large amount of concentrated yolky material, as seen in bird and reptile eggs.

macromere a relatively large, yolk-filled blastomere of the lower half (vegetal pole) of a blastula.

macromolecule any large biological polymer, as a protein, carbohydrate, or nucleic acid.

macronucleus the larger of the two nuclei of *Paramecium* and certain other ciliate protozoans, carrying somatic line DNA; compare *micronucleus*.

macronutrient a plant nutrient required in relatively substantial quantities, as nitrogen, phosphorus, potassium sulfur, magnesium, and calcium; compare *micronutrient*.

macrophage a large phagocyte.

madreporite see *water vascular system*.

major histocompatibility complex (MHC) groups of glycoproteins coating the plasma membrane of cells, making individuals biochemically unique. See also *histocompatibility antigen*.

malaria disease caused by parasitic infection by sporozoan parasites (*Plasmodium*) of the red blood cells; transmitted by *Anopheles* mosquitos.

malleus see *middle ear*.

Malpighian tubule any of numerous blind, hollow, tubular structures that empty into the insect midgut and function as a nitrogenous excretory system.

maltase an enzyme that hydrolyzes maltose.

maltose a disaccharide consisting of two glucose subunits in an alpha linkage; a digestion product of starch.

mammal any vertebrate of the class *Mammalia*.

mammary gland also *milk gland, mamma, mammary*, the organ that, in female mammals, secretes milk for the nourishment of the young.

mandible 1. the vertebrate lower jaw or jaw bone. 2. either of two laterally paired anterior mouth appendages of a mandibulate arthropod, which form strong biting jaws.

mannitol a storage carbohydrate of the brown algae.

mantle a fleshy covering of mollusks that secretes material to form the external shell.

mantle cavity in mollusks, a cavity between the mantle and the body proper, in which the respiratory organs lie; a highly vascularized mantle cavity serves as a lung in pulmonate snails.

manubrium see *sternum*.

marker see *genetic marker*.

marrow *bone marrow*, the soft tissue in the interior cavity of a bone; *blood marrow*, vascularized bone marow in which white bood cells of all types are produced; *fatty marrow*, bone marrow consisting of adipose tissue.

marsupial a mammal of the subclass Metatheria, usually with a pouch (*marsupium*) in the female; included are the kangaroo, wombat, koala, Tasmanian devil, opossum, and wallaby.

mass action also *law of mass action*, in reversible chemical reactions, the rate and net direction of the reaction depends on the relative concentrations of the reactants and products.

masturbation manipulation of one's sexual organs for erotic gratification

mating type in fungi and various protists, a grouping of organisms incapable of sexual reproduction with one another but capable of such reproduction with members of one or more other groups or mating types.

matrix also *inner compartment*, in the mitochondrion, the enzyme-laden region within the highly convoluted inner membrane, the site of the citric acid or Krebs' cycle.

maturation the process of coming to full development.

maxilla (pl. *maxillae*), 1. the upper jaw of a vertebrate, *anterior* to the mandible. 2. in insects, one of the first or second pair of mouth parts *posterior* to the mandibles.

meatus 1. the opening of any passageway to the exterior. 2. *urinary meatus*, the opening of the urethra into the vulva in human females. 3. *penile meatus*, the opening of the urethra at the end of the penis in males. 4. *external auditory meatus*, the opening of the external ear canal.

mechanism 1. the agency or means by which an effect is produced or a purpose is accomplished; 2. the theory that everything in the universe is produced by matter in motion through definable physical laws; *materialism.*

mechanoreceptor see *sensory receptor.*

median eye 1. a simple photoreceptive organ, usually lacking a lens, found in lancelets and certain reptiles; apparently used in responses to photoperiodicity. 2. also *pineal body*, a homologous organ found in birds and various other vertebrates. 3. also *parietal eye*, a similar dorsal outgrowth of the brain of various vertebrates. 4. a midline ocellus in insects.

medulla 1. the inner portion of a gland or organ; compare *cortex.* 2. also *medulla oblongata*, a part of the brainstem developed from the posterior portion of the hindbrain and tapering into the spinal cord.

medulla oblongata see *medulla.*

medullary dominance see *cortical dominance.*

medusa the motile, free–swimming jelly fish body type of coelenterate.

megagametophyte the female gametophyte produced from a megaspore in flowering plants; the *embryo sac* consisting of eight haploid nuclei or cells.

megaphyll a large veined leaf, believed to have evolved from primitive branching stems.

megaspore a single large cell with one haploid nucleus formed after meiosis and a megaspore mother cell in plants, and from which the megagametophyte develops by mitosis.

megaspore mother cell in the ovule, a large diploid cell that will give rise to the megaspore by meiosis and the degeneration of three of the four haploid nuclei.

megasporogenesis the formation of megaspores.

megasporophyll a sporophyte structure specialized for producing megaspores, e.g., the ovules of a flowering plant.

meiosis also *reduction division*, in all sexually reproducing eukaryotes, the process in which a diploid cell or cell nucleus is transformed into four haploid cells or nuclei through one round of chromosome replication and two rounds of nuclear division, the first of which involves the unique pairing of chromosome homologues.

meiosis I all of the events leading up to and including the first of the two meiotic divisions. See also *meiosis.*

meiosis II all of the events of meiosis following meiosis I, the first meiotic division. See also *meiosis.*

meiospore a haploid spore produced by meiosis.

meiotic division. See also *meiosis.*

meiotic interphase the period, which may be prolonged or so brief as to be virtually nonexistent, between telophase I of meiosis and prophase II of meiosis, during which there is no DNA replication.

Meissner's corpuscle in mammals, a small ellipsoidal touch-responsive neural end organ.

melanin the characteristic animal surface pigmentation, a class of polymers of tyrosine or dopa, giving black, brown, orange, or red coloration depending on chemical composition and pigment granule size; also found in plants.

melanism an unusual development of a black or nearly black color in a species not generally so pigmented.

melatonin a hormone produced in the mammalian pineal body that has a role in sexual development and maturation through its effect on the hypothalamus; it also causes color changes in frogs, and appears to be related to photoperiodicity responses.

membranal pump any mechanism within the plasma membrane that actively transports molecules or ions in or out of the cell.

membrane 1. any thin, soft, pliable sheet, as of connective tissue. 2. *plasma membrane*, the semipermeable, lipid bilayer, proteinaceous sac enclosing the cytoplasm. 3. any internal cell structure of a similar construction, such as the nuclear membrane, thylakoid membrane, and others; see also *fluid-mosaic model.*

membrane-bound 1. bound to a membrane, e.g., *membrane-bound ribosomes*. 2. also *membrane-bounded*, having an outer boundary consisting of a membrane, e.g., a lysosome is a membrane-bounded organelle.

membrane-bounded having an outer boundary consisting of a membrane.

memory 1. the ability to retain a learned response or to recognize a stimulus previously encountered. 2. the part or function of the brain in which learned responses are maintained. See *engram, long-term memory, short-term memory.*

memory cell a mature, long-lived, B- or T-cell lymphocyte, specialized in retaining specific antigen information.

memory trace also *memory scar*, see *engram.*

menarche in human females, the onset of menstruation; the first menses, taken as the beginning of puberty.

meninges tough, protective, connective tissues covering the brain and spinal cord.

menopause in human females, 1. the cessation of menstruation, usually occurring between the ages of 45 and 50. 2. the whole group of physical, physiological and behavioral occurrences and changes associated with the cessation of menstruation.

menstrual cycle also *ovarian cycle*, the entire cycle of hormonal and physiological events and changes involving the growth of the uterine mucosa, ovulation, and the subsequent breakdown and discharge of the uterine mucosa in menses; normally occurring at three to five week periods in nonpregnant women.

menstruation also *menses*, in nonpregnant females of the human species only, the periodic discharge of blood, secretions, and tissue debris resulting from the normal, temporary breakdown of the uterine mucosa in the absence of implantation following ovulation.

menses see *menstruation, menstrual cycle.*

meristem also *meristematic tissue*, a plant tissue consisting of small undifferentiated cells that give rise by cell division both to additional meristematic cells and to cells that undergo terminal differentiation; all postembryonic plant growth depends on meristem.

meristematic pertaining to meristem.

meristematic tissue undifferentiated cells from which the plant produces new tissues.

meroblastic cleavage in the bird and reptile embryo, cell division that does not completely divide the embryo.

merozoite a stage of *Plasmodium* that reproduces asexually in a mammalian red blood cell, releasing the poisons that cause malarial symptoms.

mesoderm the middle of the three primary germ layers of the gastrula, giving rise in development to the skeletal, muscular, vascular, renal, and connective tissues, and to the inner layer of the skin and to the epithelium of the coelom (*peritoneum*).

mesoglea the loose, gelatinous middle layer of the bodies of sponges and coelenterates, between the outer ectoderm and the inner endoderm.

mesohyl in the sponge, a gelatinous matrix in which amebocytes, spicules, and various fibrils are arranged.

mesolecithal of an egg, having the moderate yolk displaced primarily toward the vegetal pole but with cleavage complete, as in amphibia.

mesonephros 1. the adult kidney of an amphibian or fish. 2. the second of three pairs of kidneys formed in reptile, bird and mammalian embryos, superceded in later development by the *metanephros.*

mesophyll the chloroplast-rich parenchyma of a leaf, between the upper and lower epidermal cells.

mesosome in many bacteria, an inward extension of the plasma membrane that forms a spherical membranous network; function not resolved.

messenger RNA *mRNA*, 1. in prokaryotes, RNA directly transcribed from an operon or structural gene, containing one or more contiguous regions (cistrons) specifying a polypeptide sequence. 2. in eukaryotes, RNA transcribed from a structural gene, tailored and usually capped and polyadenylated in the nucleus and transported to the cytoplasm; containing a single contiguous region specifying a polypeptide sequence as well as leader and follower sequences.

metabolic defect see *inborn error of metabolism.*

metabolic pathway an orderly series or progression of enzyme-mediated chemical reactions leading to a final product, each step catalyzed by its own specific enzyme.

metabolism the chemical changes and processes of living cells, including but not limited to respiration, the synthesis of biochemicals, and the breakdown of wastes; see also *anabolism, catabolism, basal metabolism, biochemical pathway.*

metabolite 1. a metabolic waste, especially one that is toxic. 2. an intermediate in a biochemical pathway.

metacarpals one of the usually five bones of the hand or forefoot, between the carpals and digits.

metamere segment of a segmented animal. See also *segmentation*.

metamerism see *segmentation*.

metanephros in the vertebrate embryo, the third and most posterior of the paired kidney structures to develop, and the adult kidney of reptiles, birds, and mammals; see *mesonephrose, pronephros*.

metaphase the stage of mitosis or meiosis in which the centromeres of the chromosomes are brought to a well-defined plane in the middle of the mitotic spindle prior to separation in anaphase.

metaphase plate the equatorial plane of the mitotic spindle on which the centromeres are oriented in mitosis.

metatarsal one of the usually five bones of the foot or hindfoot, between tarsals and toes, forming the instep in humans and part of the rear leg in ungulates and birds.

Metazoa in Whittaker's five-kingdom scheme, all animals other than sponges.

methanogen any archaebacterium from the anaerobic group known to produce methane gas (CH_4) as a metabolic byproduct.

methemoglobinemia, a pathological condition in which methemoglobin polypeptides from which the heme group has been detached is present in the blood (a symptom of nitrite poisoning).

mica a mineral that forms thin, clear, waterproof crystals.

microbe also *microorganisms* any very small organism, such as a bacterium, yeast, or protist.

microbial genetics the scientific study of inheritance and DNA function in bacteria, bacteriophage, yeast, fungi, or protists.

microevolution evolutionary change below the species level, including changes in gene frequencies brought about by natural selection and random drift.

microfibril one of the submicroscopic elongated bundles of cellulose in a plant cell wall.

microfilament 1. a submicroscopic filament of the cytoskeleton, involved in cell movement and shape; the principal component is fibers of the protein *actin*. 2. see *myofilament*.

microgametophyte a sperm-producing body developed from a microspore.

microlecithal in animal eggs, the presence of minute amounts of yolky material, as seen in mammalian eggs.

micromere in cleavage and in the blastula, a relatively small cell of the animal pole; compare *macromere*.

micrometer 1. symbol μm, one millionth of a meter. 2. an instrument for making very precise or very small measurements.

micronucleus (pl. *micronuclei*), in ciliate protists, the smaller of the two nuclei and the one carrying germ-line DNA; compare *macronucleus*.

micronutrient also *trace element*, an element necessary for plant growth but needed only in vanishingly small quantities.

microorganism also *microbe*, any organism too small to be seen readily without the aid of a microscope; such as bacterium, protist, or yeast.

microphyll a small leaf believed to have evolved from outgrowths of the plant epidermal cells. See also *megaphyll*.

micropyle 1. in insect eggs, a depression or differentiated area through which the sperm enters. 2. in seed plants, a minute opening in the integument of an ovule through which the pollen tube enters.

microspore in seed plants, one of the four haploid cells formed from meiosis of the microspore mother cell, which undergoes mitosis and differentiation to form a pollen grain.

microspore mother cell in the anther of a flowering plant, the diploid cells that will undergo meiosis to form the haploid microspores.

microsporogenesis in flowering plants the process of producing microspores, which develop into the male gametophyte.

microsporophyll a sporophyte structure specialized for producing microspores, e.g., the anthers of a flowering plant.

microtrabecular lattice the newly discovered cytoskeletal organization within the cytoplasm of cells, consisting of an interwoven array of microtubules.

microsporogenesis the formation of microspores.

microtubule a cytoplasmic hollow tubule composed of spherical molecules of tubulin, found in the cytoskeleton, the spindle, centrioles, basal bodies, cilia, and flagella.

microvilli (sing. *microvillus*), also *brush border*, tiny fingerlike outpocketings of the plasma membrane of various epithelial secretory or absorbing cells, such as those of kidney tubule epithelium and the intestinal epithelium.

midbrain the middle of the three divisions of the vertebrate embryonic brain; the adult structures derived from the embryonic midbrain.

middle ear in mammals, the part of the ear between the eardrum and the oval window, consisting of a chain of three ear ossicles; the *malleus, incus,* and *stapes,* in an air-filled chamber-communicating with the pharynx by way of the eustacian tube.

middle lamella a layer of cementing material between adjacent plant cell walls.

midpiece in a spermatozoan, the segment between the head and the tail, containing one or more mitochondria.

midrib the large central vein of a dicot leaf.

milk teeth the temporary, deciduous first teeth in a mammal, replaced by permanent teeth during growth and maturation.

millipede also *myriopod*, nonpoisonous, herbivorous terrestrial arthropod of the class *Diplopoda,* with a long, cylindrical, segmented body, hard integument, and two pairs of legs per segment.

mineral nutrient an inorganic compound, element or ion needed for normal growth of all organisms.

mineralocorticoid any mineral-regulating steroid hormone of the adrenal cortex.

minimal medium broth or agar gel consisting of glucose, mineral salts, water, and glycine or other nitrogen source; the least complex medium capable of sustaining the growth of a specific microorganism.

minimum mutation tree in systematics, a hypothetical phylogenetic tree selected because it represents the smallest number of evolutionary changes needed to account for known relationships.

missense mutation a base change in a structural gene that alters the coding property of the codon in which it appears, producing an abnormal polypeptide with one amino acid difference.

mitochondrion (pl. *mitochondria*), threadlike, self-replicating, membrane-bounded organelle found in every eukaryotic cell, functioning in oxidative chemiosmotic phosphorylation, apparently derived in evolution from a bacterial endosymbiont.

mitogen a hormone that stimulates cell division.

mitosis cell division or nuclear division in eukaryotes, involving chromosome consensation, spindle formation, precise alignment of centromeres, and the regular segregation of daughter chromosomes to daughter nuclei.

mitotic apparatus the centrioles, spindle, spindle fibers, and asters; in plants, the spindle and spindle fibers.

mitotic spindle or *spindle*, a spindle-shaped system of microtubules appearing in the cell during mitosis and meiosis, including both *centromeric spindle fibers* (pole to centromere) and *polar spindle fibers* (pole to midpoint overlapping fibers from opposite pole), and in animals, lower plants, and some protists, the *asters* or *astral rays*.

mitotic spindle apparatus see *mitotic apparatus*.

model a contrived biological mechanism, based on a minimal number of assumptions that, when applied, is expected to yield numerical data consistent with past observations; a biological hypothesis with mathematical predictions, e.g., *laws of thermodynamics, Mendel's first and second laws.*

modifier gene a gene or allele that modifies the expression of genotypes at another gene locus.

molar also *molar tooth,* a cheek tooth adapted for grinding; in human adults, the three most posterior pairs of teeth in each jaw.

mole 1. *gram molecule,* the quantity of a chemical substance that has a mass in grams numerically equal to its molecular mass in daltons; 6.023×10^{23} molecules of a substance.

molecular biology a branch of biology concerned with the ultimate physiochemical organization of living matter; the study of biological systems using biochemical methods.

molecular mass also, *molecular weight,* the mass of a molecule expressed in daltons.

molecule a unit of a chemical substance, consisting of atoms bound to one another by covalent bonds.

molt (v.i.), to cast off an outer covering in a periodic process of growth or renewal.

molt (n.) 1. the process of molting; see *ecdysis*. 2. the cast-off covering. 3. the period of a molt; an intermolt in interecdysis.

molting hormone a hormone that initiates or causes molting.

monera the prokaryotes, comprised of the bacteria, cyanobacteria, and chloroxybacteria.

mongolism an archaic term for *trisomy 21*.

monocotyledon see *monocot*.

monocot a flowering plant of the angiosperm class *Monocotyledonae*, characterized by producing seeds with one cotyledon; included are palms, grasses, orchids, lilies, irises, and others.

monoclonal antibody a specific immunoglobin produced by hybrid cells artificially cloned in the laboratory.

monoecious 1. of a gametophyte, producing both egg and sperm in the same thallus. 2. of a seed plant sporophyte, producing both ovules and pollen. Compare *dioecious*.

monomer one of the subunits or potential subunits of a polymer.

monophyletic (adj.), of a taxonomic group, deriving entirely from a single ancestral species that possessed the defining characters of the group, while not having given rise to organisms outside of the group.

monosaccharide a sugar not composed of smaller sugar subunits (e.g., glucose, fructose).

monotreme a platypus, echidna, or extinct egg-laying mammal of the order *Monotremata*, subclass *Protheria*.

monozygotic twins also *identical twins*, twins derived from a single fertilized egg.

mons veneris in women, a rounded, usually hairy bulge of fatty tissue over the pubic symphysis and above the vulva.

morph one of the particular forms of an organism that exists in two or more distinct forms in a single population.

morphogenesis the development of form and structure.

morphogenic determinate a developmental event early in the embryo's history that influences later events of development.

mortality death, considered as an aspect of population dynamics.

mortality rate also *death rate*, the number of deaths per unit time occurring among a specified number of individuals (usually per 1000) in a given area or population.

morula in the early embryo, a simple ball preceding the blastula state.

mosaic egg an egg in which the developmental fates of different parts of the cytoplasm or surface are fairly rigidly determined prior to fertilization.

motor end-plate the terminal branching of the axon of a motor neuron as it forms multiple synapses with a muscle fiber.

motor neuron a neuron that innervates muscle fibers, and impulses from which cause the muscle fibers to contract.

motor unit a motor neuron together with the muscle fibers it innervates, which contract as a unit.

M phase see *cell cycle*.

mRNA *messenger ribonucleic acid*, see *messenger RNA*.

mucin any of a class of mucoproteins that bind water and form thick, slimy, viscid fluids in various secretions; e.g., *mucin* secreted by the stomach lining protects it from the action of gastric juices.

mucopolysaccharide a polysaccharide that binds water to form a thick, gelatinous material that serves to cement animal cells together; a constituent of mucoproteins, e.g., *chondroitin, hyaluronic acid, heparin*.

mucoprotein a combination of mucopolysaccharide and a polypeptide to form *mucin*.

mucosa also *uterine mucosa, endometrium*, the highly glandular mucous membrane lining of the uterus; inner lining of the gut.

mucous (adj.), covered with or secreting mucus. *mucous membrane*, 1. any epithelium rich in mucus-secreting cells. 2. the epithelium that lines those passages and cavities of the body that communicate directly with the exterior such as the mouth, nasal cavity, and genital tract.

mucus a viscid, slippery secretion rich in mucins, that is secreted by mucous membranes and that serves to moisten and protect such membranes.

Mullerian ducts or *paramesonephric ducts*, the embryological origin of oviducts in some female vertebrates.

Mullerian mimicry in a number of dangerous species, a common coloration or configuration serving as a common warning.

multicellular (adj.), consisting of a number of specialized cells that cooperatively carry out the functions of life.

multinucleate (adj.), having many nuclei within a common cytoplasm.

multiple alleles the alleles of a gene locus when there are more than two alternatives in a population.

multiplicative law in mathematical probability theory, the statement that the probability that all of a group of independent events will occur is equal to the product of their individual probabilities.

muscle a contractile tissue, in vertebrates including skeletal, visceral, and cardiac muscle.

muscle fiber 1. one of the multinucleate cells of striated muscle, which takes the form of a long, unbranched cylinder. 2. cardiac muscle fiber, one of the long, branched, cylindrical subunits of heart muscle, consisting of individual cells joined end to end by intercalated disks.

muscular dystrophy an incurable hereditary disease characterized by the progressive wasting of muscle tissue and eventual death. *Duchenne-type muscular dystrophy*, a common sex-linked variety affecting primarily preteen boys.

muscularis the smooth muscle layer of the wall of a hollow contractile organ such as the uterus, gall bladder, or urinary bladder.

mushroom a basidiocarp, especially but not necessarily an edible one.

muskeg a North American sphagnum bog, particularly in the taiga.

mutagen a chemical or physical agent that causes mutations.

mutant 1. a new or abnormal type of organism produced by a mutation. 2. a mutated gene. 3. an individual that bears and is affected by a mutated gene, 4. (adj.), mutated.

mutate 1. (v.t.), to alter, cause a change, or cause a mutation or DNA change to occur in; to mutagenize. 2. (v.i.), to change in state or genetic condition, to become altered, to undergo a mutation.

mutation 1. any change in DNA. 2. a change in DNA that is not immediately and properly repaired. 3. any abnormal, heritable change in genetic material. 4. the act or process of mutating.

mutation rate the rate at which new mutations occur, generally in terms of mutations per locus per gamete per generation.

mutualism a mutually beneficial association between different kinds of organisms; symbiosis in which both partners gain fitness; compare *symbiosis, parasitism*.

mycelium (pl. *mycelia*), 1. the mass of interwoven hyphae that forms the vegetative body of a fungus. 2. the analogous filaments formed by certain filamentous bacteria.

mycoplasma a tiny, nonmotile, wall-less bacterium of irregular shape, occurring as an intracellular parasite of animals and plants.

mycorrhizae a mutualistic fungus-root association with the fungal mycelium either surrounding or penetrating the roots of a plant.

myelinated having a myelin sheath.

myelin sheath fatty sheath surrounding the axons of many vertebrate neurons; also see *Schwann cell* and *oligodendrocyte*.

myofibril a tubular suburit of muscle fiber structure, consisting of many sarcomeres end-to-end.

myofilament the highly organized microfilaments of striated muscle. *Actin myofilaments*, thin filaments of the protein actin bound to the Z-line of a sarcomere; *myosin myofilament*, thicker filaments of the protein myosin interspersed among the actin myofilaments in a regular hexagonal array.

myopia nearsightedness.

myosin a protein involved in cell movement, and structure; especially in muscle cells; see also *myofibril, sarcomere, cytoskeleton*.

myosin bridge the more or less globular head of the generally fibrous protein myosin, which forms a movable bridge between a myosin myofilament and an actin myofilament.

myotome 1. the portion of a vertebrate embryonic somite from which skeletal musculature is developed. 2. one of the muscular segments of the body wall of a fish or lancelet.

myxameba the free-living, unicellular, ameboid stage of a slime mold.

myxobacterium a slime or gliding bacterium, many species of which are capable of forming spores in stalked fruiting bodies.

NAD also *NAD⁺*, nicotine adenine dinucleotide, a hydrogen carrier in respiration. *NADH*, also *NADH + H⁺*, the reduced form of NAD. *NADP, NADPH*, nicotine adenine dinucleotide phosphate, a similar hydrogen carrier in photosynthesis.

nanoplankton a newly discovered group of microscopic, flagellated phytoplanton, believed to be important in marine food chains.

nares (sing. *naris*), 1. *external nares*, also *nostrils*, the paired external openings of the nasal cavity. 2. *internal nares*, the paired openings of the nasal cavity into the pharynx.

natural selection the differential survival and reproduction in nature of organisms having different heritable characteristics, resulting in the perpetuation of those characteristics and/or organisms that are best adapted to a specific environment. Also, *survival of the fittest*.

navigation the ability to locate or maintain reference to a particular place without the use of landmarks.

Neanderthal human an extinct, highly variable type or race of humans (*Homo sapiens neanderthalensis*) of the middle Paleolithic.

nectary a plant gland that secretes a sweet fluid (nectar), most often at the base of a flower, that functions as a reinforcing reward for the visits of pollinating animals.

negative control in a genetic control system, control such that transcription occurs at all times except when prevented by an inhibitor molecule or an inhibitor complex.

negative feedback see under *feedback*.

negative phototropism in plants, growth away from the light.

nematocyst one of the minute stinging cells of the coelenterates, consisting of a hollow thread coiled within a capsule, and an external hair trigger.

neoteny the condition in which animals become developmentally arrested in an immature or larval state, while becoming sexually mature and able to reproduce, as in the *Larvaceae*.

nephridiopore the exterior orifice of a nephridium.

nephridium an excretory organ primitive to many coelomate phyla and still found in annelids, brachipods, mollusks and some arthropods, occurring paired in each body segment in segmented animals, and typically consisting of a ciliated funnel (*nephrostome*) draining the coelom through a convoluted, glandular duct to the exterior.

nephron a single excretory unit of a kidney, consisting of a glomerulus, Bowman's capsule, proximal convoluted tubule, loop of Henle, distal convoluted tubule, and collecting duct discharging into the renal pelvis.

nerve (n.), 1. a filamentous band of nerve cell axons and dendrites, and protective and supporting tissue, that connect parts of the nervous system with other parts of the body. 2. (adj.), pertaining to the nerve or nervous system, e.g., *nerve cell, nerve net, nerve fiber*.

nerve cell see *neuron*.

nerve cord 1. *dorsal tubular nerve cord*, the spinal cord of chordates. 2. *ventral nerve cord*, the solid, paired, segmentally ganglionated, longitudinal central nerve of many invertebrates.

nerve fiber an axon or dendrite.

nerve impulse see under *impulse*.

nerve net in colenterates, the arrangement of nerve fibers into a net-like pattern with little concentration.

nerve synapse see under *synapse*.

neritic province the coastal sea from the low-tide line to a depth of 100 fathoms, generally waters of the continental shelf.

nervous system the brain, spinal cord, nerves, ganglia and the neural parts of receptor organs, considered as an integrated whole.

net community productivity (NCP) the rate of community productivity by autotrophs after their energy utilization through respiration has been subtracted.

net productivity in ecology, the rate of production of energy or biomass by plants, being the gross productivity less the energy used by the plants in their own activities.

neural arch the open ring formed by vertebral processes rising from the centrum and closing above the spinal cord.

neural canal 1. the canal formed by the series of vertebral neural arches, through which the spinal cord passes. 2. the lumen (cavity) of the spinal cord.

neural crest the ridge of a neural fold, migrating cells of which will give rise to spinal ganglia, the adrenal medulla, the autonomic nervous system, and melanocytes (pigment cells).

neural folds in early vertebrate embryology, a pair of longitudinal ridges that arise from the neural plate on either side of the neural groove, which fold over and fuse to give rise to the *neural tube*.

neural groove in early vertebrate embryology, the linear depression in the neural plate between the neural folds that will invaginate to form the neural tube.

neural impulse a transient membrane depolarization, followed by immediate repolarization, traveling in a wavelike manner along a neuron.

neural plate in the vertebrate gastrula, a thickened plate of ectoderm along the dorsal midline that gives rise to the neural tube, neural crests and ultimately to the nervous system.

neural tube 1. in early vertebrate embryology, the hollow dorsal tube formed by the fusion of the neural folds over the neural groove. 2. the spinal cord.

neuroglia also *glia* supporting cells of the central nervous system.

neuromuscular junction the synapse between a neural motor end plate and a muscle fiber.

neuron also *nerve cell*, a cell specialized for the transmission of nerve impulses, consisting of one or more branched *dendrites*, a *nerve cell body* in which the nucleus resides, and a terminally branched *axon*.

neurophysiology the physiology of nerve cells, neural receptors, and other aspects of the nervous system.

neurosecretion 1. a secretion of a nerve cell, especially hormonal secretions into the general circulation, usually produced in the nerve cell body, transmitted along the axon, and released at the terminus of the axon. 2. the process of secretion by neurons.

neurotransmitter a short-lived, hormone-like chemical (e.g., *acetylcholine, norepinephrine*) released from the terminus of an axon into a synaptic cleft, where it stimulates a dendrite in the transmission of a nerve impulse from one cell to another; see also *synapse, synaptic cleft, synaptic knob*.

neurula an early vertebrate embryo slightly older than a gastrula, in which the neural tube has formed.

neurulation the development of a neurula from a gastrula; equivalently, the development of the neural plate, neural folds, and neural tube.

neutral mutation also *selectively neutral mutation*, a mutational change in DNA that has no measurable effect on the fitness of an organism, and which may sometimes be incorporated into the genome of a species by genetic drift.

neutralist a population geneticist or evolutionary theorist who argues that a substantial proportion of protein and DNA changes in evolution have been due to selectively neutral mutations fixed by random drift; compare *selectionist*.

neutrophil the most common mammalian phagocytic leukocyte.

neutron one of the two common constituents of an atomic nucleus, a hadron having no charge and no effect on chemical reactions.

New-World monkeys the primate superfamily Ceboidea native to South America, with wide-set nostrils and prehensile tails.

niche the position or function of an organism in a community of plants and animals. It has been said that if a habitat is an organism's address, a *niche* is its occupation. *Niche* has also been defined as the totality of the adaptations, specializations, tolerance limits, functions, biological interactions, and behavior of a species.

nick 1. (n.), a single-strand break in DNA. 2. (v.t.), to cause a single-strand break in DNA by the hydrolysis of a phosphate linkage.

nicotinic acid a 5-carbon nitrogenous acid, a constituent of NAD and NADP in respiration and photosynthesis; a vitamin (dietary requirement) of animals.

nictitating membrane in many vertebrates, including many mammals and most birds, a thin translucent membrane that can be folded into the inner angle of the eye or drawn across the front of the eyeball; vestigial in humans.

nitrification the chemical oxidation of ammonium salts into nitrites and nitrates by the action of soil bacteria.

nitrogen fixation the chemical change of nitrogen from the stable, generally unavailable form of atmospheric nitrogen ($N \equiv N$) to soluble and more readily utilized forms such as ammonia, nitrate, or nitrite; usually by the action of cyanophytes or nitrogen-fixing bacteria, but sometimes by lightning, automobile engines, or synthetic fertilizer factories.

nitrogenase an enzyme that catalyzes nitrogen fixation utilizing cell energy.

nitrogenous base 1. a purine or pyrimadine used in the synthesis of nucleic acids; specifically, adenine, cytosine, guanine, thymine, or uracil. 2. any related heterocyclic compound, such as xanthine or uric acid.

noble element also *noble gas*, an element in which the uncharged atoms have all electron shells filled and all electrons paired and which consequently are chemically unreactive.

node 1. in a plant stem, any point at which one or more leaves emerge. 2. in a growing stem tip, a region of potential leaf growth, containing meristematic tissue. 3. see *node of Ranvier*. 4. see *lymph node*.

node of Ranvier a constriction in a myelin sheath, corresponding to a gap between successive Schwann cells. See *saltatory propagation*.

nodule 1. any small lump. 2. see *root nodule*.

noncellular see *acellular*.

nexin in cilia and flagella, protein connections between the nine-paired groups of microtubules.

noncoding strand during transcription, the DNA strand not involved. See also *transcription* and *transcribed strand*.

noncyclic phosphorylation light reactions employing both photosystem II and photosystem I, during which electrons from water pass through both photosystems, reducing NADP to NADPH + H+, thus serving both chemiosmosis and carbohydrate synthesis.

non-Darwinian evolution decent with modification attributable to genetic drift rather than natural selection; the fixation of selectively neutral mutations in evolutionary lineages.

nondisjunction 1. the failure of a pair of homologous chromosomes to segregate to opposite poles in anaphase I of meiosis. 2. the failure of a pair of daughter chromosomes or daughter centromeres to segregate to opposite poles in mitotic anaphase or in anaphase II of meiosis.

nonhistone chromosomal protein a large class of proteins other than histones associated with the chromosomes, constituting about one-third of the substance of chromatin in animals and plants, highly variable from tissue to tissue and known in at least some cases to be involved in gene control.

nonpolar of a molecule or part of a molecule, not forming hydrogen bonds with water, uncharged.

nonseptate of a fungal hypha, not having septa; not being separated into individual cells.

nontranscribed strand the DNA strand of a cistron opposite the strand that is transcribed into RNA.

norepinephrine also *noradrenalin*, a compound that serves as a synaptic neurotransmitter, and also as a systemic hormone where it acts as a vasoconstrictor.

normal distribution also *normal curve, Gaussinan distribution, bell-shaped curve*, the idealized, symmetrical distribution taken by a population of values centering around a mean, when departures from the mean are due to the chance occurrence of a large number of individually small independent effects; often approached in real populations.

notochord in all chordates at some point in development, a turgid, flexible rod running along the back beneath the nerve cord and serving as a skeletal support; replaced during development by the vertebral column in most vertebrates but persistent in adult coelocanths, cyclostomes, lancelets, and larvacean urochordates.

N-terminal end also *amino-terminal end*, the end of a polypeptide chain in which the end amino acid has its primary carboxyl group fixed in a peptide linkage and its primary amino group free (or formylated); the first part of the chain to be synthesized.

nucellus in flowers, a mass of thin-walled parenchymal cells that composes most of the ovule, encloses the embryo sac, and is enclosed by one or more integuments.

nuclear envelope also *nuclear membrane* the double cellular membrane surrounding the eukaryote nucleus, the outermost of which is continuous with the endoplasmic reticulum.

nuclear force also *strong nuclear force*, the mutually attractive forces within the atomic nucleus.

nuclear membrane see *nuclear envelope*.

nuclear pore a structure in the nuclear envelope that appears as a hole in electron micrographs although probably filled with a protein mass in life; through which RNA, nuclear proteins, and other large molecules are presumably able to pass. Also *annulus*.

nuclease any enzyme that hydrolyzes a nucleic acid; see also *endonuclease, exonuclease*.

nucleic acid either DNA or RNA, DNA being a double polymer of deoxynucleotides and RNA being a polymer of nucleotides.

nucleic acid base see *base*; also see *purine, pyrimidine,* and *nitrogenous base*.

nucleolar organizer region within the nucleolus, multiple copies of a DNA loop that transcribes ribosomal RNA. See also *lampbrush chromosomes*.

nucleolus (pl. *nucleoli*) a dark-staining body of RNA and protein found within the interphase nucleus of a cell, the site of synthesis and storage of ribosomes and ribosomal materials; disperses during mitosis and is reconstituted following nuclear envelope reorganization.

nucleoplasm the viscous fluid matrix of the nucleus, contrasted with *cytoplasm*.

nucleoside 1. a compound consisting of a nitrogenous base linked to the 1' carbon of ribose. 2. also *deoxynucleoside*, a compound consisting of a nitrogenous base linked to the 1' carbon of deoxyribose. Compare *nucleotide*.

nucleosome in the chromosome, globular bodies of histone about which eukaryotic DNA is wound.

nucleotide 1. a compound consisting of a nitrogenous base and a phosphate group linked to the 1' and 5' carbons of ribose respectively; the repeating subunit of RNA. 2. also *deoxynucleotide*, a compound consisting of a nitrogenous base and a phosphate group linked to the 1' and 5' carbons of deoxyribose respectively, the repeating subunit of DNA.

nucleoside triphosphate also *triphosphonucleoside*, a nucleoside with a chain of three phosphate groups linked to the 5' carbon of ribose or deoxyribose.

nucleus in all eukaryote cells, a prominent, usually spherical or ellipsoidal membrane-bounded sac containing the chromosomes and providing physical separation between transcription and translation.

nutrient (n.), 1. any substance required for growth and maintenance of an organism. 2. a chemical element or inorganic compound needed for normal growth of a plant, see also *macronutrient, micronutrient, mineral nutrient*. 3. (adj.) furnishing nourishment, e.g., *nutrient broth*, a broth in which microorganisms may readily be grown.

nutrition the process of being nourished, particularly the steps through which an organism obtains food and uses it for bodily processes.

obligate restricted to a particular condition of life; e.g., obligate aerobe, obligate anaerobe, obligate parasite, obligate symbiont.

obligate anaerobe an organism, most commonly a bacterium, that cannot utilize oxygen, and often is killed by its presence.

occipital condyle a rounded portion of the occipital bone at the base of the skull, which articulates with the first vertebra.

occipital lobe the posterior lobe of each cerebral hemisphere.

oceanic province (n.), the open sea as distinguished from the neritic province; *oceanic* (adj.), pertaining to the open sea.

Oceanic rift community a newly discovered kind of community along ocean floor rifts, where a variety of animal life is supported by chemosynthetic bacteria, producers that thrive near vents releasing heated water and hydrogen sulfide gas.

ocellus (pl. *ocelli*), a minute simple eye or eyespot of any organism. 2. one of the elements of an arthropod compound eye; an *ommatidium*. 3. one of the three simple eyes that form a triangle dorsally between the compound eyes.

oil a common term for any fat or viscous lipid that is liquid at ambient temperatures.

Okazaki fragment in DNA synthesis, the original form of a newly synthesized single strand, being a polynucleotide of some 200-300 bases produced at the lagging ends of each DNA strand and added in by ligase.

Old-World monkey a primate of the superfamily *Cercopithecoidea*, found only in Asia, Africa and Gibraltar, characterized by close-set nostrils on a somewhat dog-like snout; included are baboons, Barbary ape, rhesus, and macaques.

olfaction 1. the sense of smell. 2. the process of smelling.

olfactory bulb an etension of the brain that receives neurons from the olfactory epithelium within the nasal cavity.

olfactory organ a sensory organ that responds to chemicals or mixes of chemicals, especially to trace quantities dispersed in air or water.

olfactory receptor the odor-sensitive cells of the mucous membrane of an olfactory organ.

oligodendrocyte a neuroglial cell that forms the myelin sheath over axons of the central nervous system.

oligotrophic of a lake, rich in dissolved oxygen and poor in plant nutrients and algal growth, with clear water and no marked stratification.

omasum see *ruminant*.

ommatidium (pl. *ommatidia*), one of the elements of a compound eye, consisting of a corneal lens, crystalline cone, rhabdome, light-sensitive retinula, and sheathing pigment cells.

omnivorous (adj.), feeding on both animal and plant material; literally, eating everything. *omnivore* (n.), an omnivorous animal e.g., pigs, people.

oncogene a cancer-causing gene, often of viral origin.

oocyst a stage in the life cycle of the *Plasmodium* that increases asexually to form numerous sporozoites.

oocyte an egg cell before maturation; *primary oocyte*, a diploid cell precursor of an egg, before meiosis; *secondary oocyte*, an egg cell after the formation of the first polar body.

oogonium (pl. *oogonia*) a cell that gives rise to oocytes, the large, spherical, unicellular female sex organ of water molds and egg-producing algae in which egg cells are produced.

open circulatory system a circulatory system in which the arterioles end openly into intercellular space, allowing blood to percolate directly through nonvascular tissues.

open communities biotic communities that blend into each other in a gradual transition.

open growth, also *indeterminant growth* in perennial plants the capability of continuous growth.

operant conditioning see *conditioning, operant.*

operator see *operon.*

operator locus in an operon, the binding site of an inhibitor protein or inhibitor complex.

operculum a body process functioning as a lid or cover, e.g., (a) the horny plate on the foot of certain gastropods, which serves to close off the opening of the shell; (b) the covering flap of a moss spore capsule; (c) the skin-covered bony plates that cover the gills of a fish.

operon in prokaryotes, a region of DNA that includes structural genes and the genes controlling them; transcription may be *inducible*, remaining shut down until activated by an inducer substance, or *repressible*, remaining active until shut down by a repressor substance. Control regions generally consist of a *promotor region (p)*, an *operator region (o)*, and a *regulator gene (i)* which produces a *repressor protein*. Also see *lac operon* and *tryptophan operon.*

optic nerve see *eyeball.*

oral cavity also *mouth cavity*, the cavity between the mouth and the pharynx.

oral-facial-digital syndrome a genetic condition of human females, caused by an allele that is dominant with regard to abnormalities of the mouth, face and fingers, and is lethal in the hemizygous state.

oral groove in ciliates, a ciliated fold in the body wall, leading into the cytostome or mouth.

orbit 1. see *electron orbit.* 2. in vertebrates, the bony cavity of the eye socket.

order a taxonomic level between class and family.

organ an organized assembly of various tissues performing some major body function; e.g., the heart, brain, and liver.

organelle a functionally and morphologically specialized part of a cell.

organic compound a chemical compound containing carbon.

organic molecule a molecule containing carbon and generally produced by living organisms.

organism 1. a form of life composed of mutually dependent parts that maintain various vital processes. 2. any form of life. 3. an individual plant, animal or microorganism.

organ-system level of organization in organisms, a complex structural organization, including the presence of specific organs performing in a coordinated manner as systems.

organ of Corti also *spiral organ of Corti*, on the basilar membrane within the cochlea, an organ containing the neural receptors for hearing, including sensory hair cells and associated neural fibers.

orgasm in humans, the climax of sexual excitement typically occurring toward the end of coitus, usually accompanied in men by ejaculation and in women by rhythmic contractions of the cervix.

orientation the directing of bodily position according to the location of a particular stimulus; may be part of an instinctive action.

orienting movements see *orientation.*

origin 1. evolutionary ancestry. 2. the fixed skeletal attachment of a muscle or tendon; compare *insertion.*

ornithologist a scientist who studies birds.

osculum an opening in a sponge through which water exits.

osmoconformer an aquatic organism that does not regulate the osmotic potential of its body tissues, but allows it to fluctuate with that of the environment.

osmoregulation in aquatic organisms, the homeostatic regulation of the osmotic potential of body fluids.

osmoregulator an aquatic organism that maintains a constant internal osmotic potential in spite of fluctuations in the salinity of its environment.

osmosis the tendency of water to diffuse through a semipermeable membrane in the net direction from its higher to its lower concentration.

osmotic gradient any difference in the concentration of water molecules across a membrane, the difference between two fluids in terms of osmotic potential.

osmotic potential the tendency or capacity for water to move across a selectively permeable membrane into a second solution, such movement occurring because of the presence of a relatively high solute concentration (or less water) on the other side.

osmotic pressure the actual hydrostatic pressure that builds up in a confined fluid because of osmosis.

ossification the process of bone formation and especially of the replacement of cartilage by bone.

osteoblast a bone-forming cell.

osteoclast a bone-destroying ameboid cell that dissolves calcium phosphate and releases Ca^{2+} into the bloodstream.

osteocyte a bone cell isolated in a lacuna of bone tissue.

osteon see *Haversian system.*

osteoporosis a condition involving the decalcification of bone, producing bone porosity and fragility.

ostracoderm any of a group of extinct jawless fish.

outer compartment in the mitochondrion, the region between the outer and inner membranes into which protons are transported.

oval window in the cochlea, a membrane articulating with the stapes that moves in response to its vibration, subsequently creating movement in the fluid perilymph within.

ovary 1. in animals, the (usually paired) organ in which oogenesis occurs and in which eggs mature. 2. in flowering plants, the enlarged, rounded base of a pistil, consisting of a carpel or several united carpels, in which ovules mature and megasporogenesis occurs.

overdominance if the phenotype of the heterozygote is more fit than those of both homozygotes, both alleles are said to be *overdominant.*

oviduct a tube, usually paired, for the passage of eggs from the ovary toward the exterior or to a uterus, often modified for the secretion of a shell or protective membrane; in humans it is sometimes known as a *fallopian tube.*

oviparous (adj.), producing eggs that develop and hatch outside of the mother's body; compare *ovoviviparous, viviparous.*

oviparity (n.), the condition of being *oviparous.*

ovipositor a female insect organ specialized for the depositing of eggs and often for boring holes in which eggs may be deposited.

ovoviviparity the condition of being ovoviviparous.

ovoviviparous producing eggs that are fertilized internally, develop within the mother's body but without any direct connection with the maternal circulation; the young being released shortly before or after hatching. Compare *viviparous, oviparous.*

ovulation the release of one or more eggs from an ovary.

ovule 1. in animals, a small egg; an egg in the process of growth and maturation. 2. in seed plants, a rounded outgrowth of the ovary, consisting of the embryo sac surrounded by maternal tissue including a stalk, the nucellus, and one or more integuments.

oxidation 1. the loss of electrons from an element or compound. 2. also dehydrogenation, the loss of hydrogens from a compound.

oxidation-reduction reaction a chemical reaction in which one reactant is oxidized and another is reduced.

oxidative phosphorylation the production of ATP from ADP and phosphate in a process consuming oxygen, as by mitochondria or aerobic bacteria; see also *chemiosmosis, chemiosmotic, phosphorylation, oxidation.*

oxidative respiration the breakdown of biochemicals to produce cellular energy, utilizing oxygen as the final electron acceptor; see *oxidative phosphorylation, respiration.*

oxygen carrier a molecule specialized for the transport or storage of oxygen, e.g., *hemoglobin, myoglobin.*

oxygen debt a state of oxygen depletion after extreme physical exertion; measured by the amount of oxygen required to restore the system to its original state.

oxygen sink in the primitive earth, elements and compounds in the earth's crust that readily reacted with oxygen.

oxyhemoglobin hemoglobin carrying four oxygen molecules; the bright red arterial form of hemoglobin.

oxytocin a polypeptide hormone of the posterior pituitary that stimulates the contraction of the uterus and the release of milk.

pacemaker also *sinoatrial* or *SA node*, the portion of the heart in which the impulse for the heartbeat originates and is regulated.

pachytene a phase of meiosis in which crossing over occurs and in which the synapsed chromosomes appear as solid, thickened threads; between zygotene and diplotene.

Pacinian corpuscles oval pressure receptors containing the termini of sensory nerves, especially in the skin of hands and feet.

pairing see *synapsis*.

palate in mammals, the roof of the mouth cavity, separating the mouth cavity from the nasal cavity; consisting of an anterior, bony *hard palate* and a posterior fleshy *soft palate*.

paleobiology the branch of paleontology dealing with the origin, growth, biological functioning and ecological relationships of fossil organisms.

Paleolithic also *old stone age*, a period in human cultural evolution characterized by rough or chipped stone implements.

paleontology the scientific study of the forms of life existing in former geological periods, as represented by fossil animals, plants, and microorganisms.

palisade parenchyma see *leaf mesophyll*.

palp see *labial palps*.

pampas (sing, *pampa*), the extensive, grassy plains of southern South America.

pancreas a large digestive and endocrine gland of vertebrates, which secretes various digestive enzymes into the duodenum by way of the *pancreatic duct*, and which also contains endocrine tissues in the form of interspersed *islets of Langerhans*, responsible for the production of the hormones *insulin* and *glucagon*.

pancreatic amylase a starch-digesting enzyme secreted by the pancreas. *pancreatic lipase*, a fat-digesting enzyme of the pancreas. *pancreatic proteases*, the protein-digesting enzymes *trypsin, chymotrypsin* and *elastin. pancreatic hormones*, insulin and glucagon.

pancreatic duct a tube in the pancreas that directs pancreatic secretions to the duodenum.

paradoxical sleep see *REM (rapid eye movement) sleep*.

parallax the apparent displacement of objects seen from different points of view, especially from the slightly different placement of the eyes in binocular vision.

parallel evolution similar evolutionary change occurring simultaneously in separated lines of descent.

paramesonephric duct see *Mullerian ducts*.

parapatric living in separate but adjacent geographic regions not separated by geographical barriers; compare *allopatric, sympatric*.

paraphyletic of a taxonomic group, sharing defining characters because of common descent from a single ancestral species having those characters, but having also given rise to other organisms not now included in the group.

parapod also *parapodium*, one of the short, unsegmented, paired, leglike or finlike locomotive organs borne on either side of each body segment in *Nereis* and certain other polycheate worms.

parasite an organism living in or on another living organism from which it obtains its organic sustenance to the detriment of its host; see also *endoparasite, ectoparasite*; compare *symbiont, parasitoid, mutualism*.

parasitic castration a common form of parasitism in which the parasite invades the host's reproductive system, rendering it sterile; sometimes involving hormonal changes that lead to phenotypic sex reversal.

parasitism a relationship in which an organism of one kind (the parasite) lives in or on an organism of another kind (the host) at the expense of which it obtains food and shelter, causing some degree of damage but usually not killing the host directly; compare *symbiosis, mutualism*.

parasitoid an organism that invades the body of a host organism and eventually but regularly causes its death; specifically certain insects that lay one or more eggs in a host organism, the resulting larvae developing within and eventually killing the host at about the time that larval development is complete.

parasympathetic division one of the two divisions of the vertebrate autonomic nervous system, the one that utilizes acetylcholine as a neurotransmitter, that increases the activity of smooth muscle and digestive glands, slows the heart, and dilates blood vessels; compare *sympathetic division*.

parasympathetic nerve in the autonomic nervous system, any nerve of the parasympathetic division.

parathormone also *parathyroid hormone*, the internal secretion of the parathyroid glands, involved in maintaining normal calcium balance.

parathyroid glands four small endocrine glands embedded in or adjacent to the thyroid gland and involved in the regulation of calcium ion levels in the blood.

Parazoa a monophyletic group consisting of the sponges, usually included with the Metazoa in the animal kingdom but not directly related to any other animal groups.

parenchyma in higher plants, a tissue consisting of thin-walled living cells that remain capable of cell division even when mature, and function in photosynthesis or food storage.

parietal cells large cells of the stomach lining that secrete hydrochloric acid.

parietal eye a median light-sensitive organ of ancient vertebrates, or its vestige in certain contemporary vertebrates; analogous to the pineal eye but apparently not homologous with it.

parietal lobe the middle division of each cerebral hemisphere.

parotid glands a pair of ducted salivary glands situated below the ears.

parthenogenesis asexual reproduction in which gametes develop without fertilization, either with or without having undergone meiosis; e.g., male hymenopterans are produced parthenogenetically from haploid eggs, and *Daphnia* of either sex may be produced parthenogenetically from unreduced diploid eggs.

partial dominance a phenotype of a heterozygote in which the expressed alleles are half-way between the homozygous phenotypes. Also see *dominance relationships*.

partial pressure 1. the independent pressure exerted by each gas in a mixture of gases. *Total pressure* is equal to the sum of the partial pressures.

particulate inheritance inheritance consisting of discrete, separable units of information, that is, of genes; compare *blending inheritance*.

passive transport movement of fluids, solutes or other materials without the expenditure of energy, e.g., by diffusion, especially across a membrane.

pasteurization heating of wine or milk to a certain temperature for a given amount of time in order to kill pathogenic or otherwise undesirable bacteria without destroying the quality of the wine or milk.

patella the kneecap.

pathway see *biochemical pathway*.

paved teeth also *pavement teeth*, teeth that form a continuous, flat, roughened surface, and that function in crushing or grinding.

pecking order, also *peck order*, in many gregarious animals, a specific linear order of relative dominance and submissiveness; see *dominance hierarchy*.

pecten 1. any comblike organ. 2. a comblike, pigmented, highly vascularized projection from the retina into the vitreous humor of the eyes of most birds; function unknown.

pectin any of a group of complex methylated polysaccharides occurring in plant tissues, especially of fruits and succulent leaves, that bind water and sugar to make viscous solutions or gels.

pectoral girdle the bones or cartilage supporting and articulating with the vertebrate forelimb; in most vertebrates not connected with the axial skeleton, but in humans connecting to the sternum by way of the clavicle and consisting of the clavicle, coracoid process, and scapula.

pelagic (adj.), living in the open sea; *pelagic zone*, the region of the open sea beyond the littoral and neritic zones and extending from the surface to the depth of light penetration; compare *abyssal, benthic, neritic, littoral*.

pellicle 1. the semirigid, proteinaceous integument of many protists. 2. the plasma membrane and associated cytoskeleton of an animal cell.

peloric of a flower, having radial symmetry abnormally in a species in which flowers are normally irregular (bilaterally symmetrical); see also *flower*.

pelvic girdle the bones or cartilage supporting and articulating with the vertebrate hind limbs; in humans consisting of the fused bones of the *pelvis*.

pelvis 1. the bones of the pelvic girdle, consisting of the *sacrum, coccyx* and the paired and fused *ischium, ilium* and *pubis*; see also *coxa*. 2. *renal pelvis*, the main cavity of the kidney, into which the nephrons discharge urine.

penicillin a group of closely related powerful antibiotics produced by fungi of the genus *Penicillium*, that function by disrupting the synthesis of bacterial cell walls.

penis 1. the copulatory organ of the male in any species in which internal fertilization is achieved by the insertion of a male body part into the female genital tract. 2. the erectile intromittant organ of male mammals, which also serves as a channel for the discharge of urine.

pentaradial symmetry radial symmetry with a five-part organization. See also *radial symmetry*.

PEP cycle also known as *Hatch-Slack Pathway* in C4 plants, a biochemical cycle during which carbon dioxide is first incorporated into 4-carbon compounds in the leaf mesophyll and later released into the Calvin cycle in the bundle sheath cells.

pepsin a proteolytic enzyme secreted as *pepsinogen* by glands of the stomach lining, that is active only at low pH and which acts primarily to reduce complex proteins to simple polypeptides.

pepsinogen the initial, inactive form of pepsin as occurring in and secreted by gastric glands, that readily converts to pepsin in an acid medium.

peptidase a proteolytic enzyme that hydrolyzes peptide bonds; see *amino-peptidase, carboxypeptidase, dipeptidase, endopeptidase, exopeptidase.*

peptide a chain of two or more amino acids linked by peptide bonds, too short to be coagulated by heat or precipitated by saturated ammonium sulfate; most often seen as a partial digestion product of a protein or polypeptide.

peptide bond also *peptide linkage,* the dehydration linkage formed between the carboxyl group of one amino acid and the amino group of another: $R_1—COOH + NH_2—R_2 \rightarrow R_1—CO—NH—R_2 + H_2O$.

peptidoglycan the primary substance in the cell walls of eubacteria, consisting of N-acetylmuramic acid, N-acetylglucosamine, and certain amino acids.

peptidyl transferase during translation, the enzyme involved in the formation of peptide bonds between adjacent amino acids in the growing polypeptide.

perception response or responsiveness to sensory stimuli, e.g., *color perception, depth perception,* perception of movement.

perennial (adj.), 1. continuing or lasting for several years. 2. (n.), a plant that lives for an indefinite number of years, as compared with *annual* or *biennial.*

perfect flower see under *flower.*

pericardium 1. the membraneous sac surrounding the vertebrate heart. 2. the cavity or hemocoel containing the arthropod heart.

pericarp the ripened and variously modified wall of a plant ovary (fruit) such as a pea pod, seed capsule, berry, or nutshell.

pericycle a layer of parenchyma or sclerenchyma that sheathes the stele of the root, and is associated with the formation of vascular cambium and lateral roots.

periderm in plants, a protective layer of secondary tissue derived from epidermal cells and consisting of cork, cork cambium and underlying parenchyma.

perineum the area between the anus and the exteria genitalia, especially in the female.

perinuclear space the space between the two nuclear membranes, continuous with the endoplasmic reticulum.

periosteum tough connective tissue covering of bone.

peripheral lymphoid tissue widely dispersed lymphoid tissues where lymphocytes are activated and aggregate, as in the tonsils, lymph nodes, spleen and adenoids.

peripheral nervous system nerves and receptors outside the central nervous system, including sensory and motor nerves of the somatic system and autonomic nervous system.

peristalsis successive waves of involuntary contractions passing along the walls of the esophagus, intestine or other hollow muscularized tube, forcing the contents onward.

peristome 1. the fringe of teeth around the opening of a moss spore capsule. 2. the lip of a spiral shell. 3. the area surrounding the mouth of an invertebrate.

peritoneum the smooth, transparent membrane lining the abdominal cavity of a mammal. *Peritoneal cavity,* the principal body cavity (abdominal coelom) of a mammal.

peritubular capillaries in the kidney, a capillary bed surrounding the nephron and involved in tubular reabsorption and secretion, related to osmoregulation and excretion.

permafrost the permanently frozen layer of soil and/or subsoil in arctic and subarctic regions.

permanent tissues plant tissues that have become specialized; in *simple permanent tissue,* one type of cell predominates; in *complex permanent tissue,* two or more types of cell are in close association.

permeable allowing materials to pass through. *Permeability,* the degree to which materials are able to pass through a substance, membrane, or barrier.

permease an enzymelike carrier that functions in facilitated transport of a specific substrate across a plasma membrane.

peroxisome a cytoplasmic organelle involved in the detoxification of peroxides.

petal one of the usually white or brightly colored leaflike elements of the corolla of a flower.

petiole the small stalk that supports the blade of a leaf; a *leafstalk.*

Petri dish a small shallow dish of thin glass or plastic with a loosely fitting overlapping cover, used for plate cultures in bacteriology.

pH the negative logarithm of the hydronium ion concentration of a solution and a common measure of the acidity or alkalinity of a liquid; pH values of less than 7 indicating acidity, and values greater than 7 indicating alkalinity.

phage also *bacteriophage,* a virus that infects and lyses bacteria.

phagoctye 1. any leucocyte that engulfs particulate matter. 2. any cell that characteristically engulfs foreign matter, e.g., cells of the reticuloendothelial system.

phagocytosis taking solid materials into the cell by engulfment and the subsequent pinching off of the plasma membrane to form a digestive vacuole.

phalanges (sing. *phalanx*), in land vertebrates, the bones of the fingers and toes.

pharyngeal arch one of several archlike structures in the embryo, some of which form jaws and gill arches in fishes but contribute to a variety of structures in the head and neck in mammals.

pharyngeal gill cleft see *gill cleft.*

pharyngeal pouch in vertebrate embryos, any of the series of outpocketings of the ectoderm of the pharynx opposite the gill clefts, which may or may not break through; in mammals, one of the pairs of pharyngeal pouches becomes the eustachian tubes and the cavities of the middle ears.

pharynx 1. in fishes, the portion of the alimentary canal in which the gills and gill slits are located. 2. in land vertebrates, the corresponding region, posterior to the oral and nasal cavities and anterior to the esophagus. 3. an analogous region in the alimentary canals of various invertebrates, including some in which it is eversible and toothed.

pH differential the difference in acidity between two solutions.

phenotype the visible or otherwise detectable physical and chemical traits of an organism, as influenced by heredity and by the environment. Compare with *genotype.*

phenotype frequency the proportion of individuals in a population falling into a specific phenotypic category.

phenylalanine one of the 20 amino acids of protein structure.

phloem a complex vascular tissue of higher plants consisting of sieve tubes, companion cells, and phloem fibers, and functioning in transport of sugars and nutrients.

phloem fiber a fiber cell associated with phloem, with great strength and pliability, e.g., the linen fibers of flax.

phloem ray the part of a vascular ray located in the phloem; see *vascular ray.*

phosphate linkage also *phosphate diester,* a phosphate that is covalently linked to two organic residues.

phosphate group phosphate linked by a dehydration linkage to an organic molecule: HPO_4^-.

phosphate ion any of the four ionization steps of phosphate.

phosphodiesterase a phosphatase, common in snake venom, that hydrolyzes phosphate links of nucleic acids.

phospholipid also *phosphatide,* any of a class of phosphate-esterified lipids, including *lecithins, cephalins, sphingomyelins;* a major component of cell membranes.

phosphorylase an enzyme that catalyzes phosphorolysis.

phosphorolysis also *phosphorolytic cleavage,* a reversible reaction analogous to hydrolysis, in which a dehydration linkage is broken by the addition of phosphate: $R_1—O—R_2 + HPO_4^{2-} \rightarrow R_1PO_4^{2-} + HO—R_2.$

phosphorylation the addition of a phosphate group to a compound, e.g., the addition of phosphate to ADP to produce ATP and water.

photoevent the unitary event of the light-dependent reaction of photosynthesis, conceived as the pathway followed by two electrons released by the photolysis of water.

photolysis of water oxidation of water through the removal of hydrogen by highly oxidizing elements of photosystem 680, which in turn are regenerated with the energy of captured photons; the net reaction of a complex series of events being $2H_2O \rightarrow 4H^+ + 4e^- + O_2\uparrow$.

photon a quantum of electromagnetic radiant energy.

photoperiod the relative lengths of alternating periods of lightness and darkness, which change with the season especially in temperate climates.

photoperiodism the response of an organism to photoperiods, involving sensitivity to the onset of light or darkness and a capacity to measure time.

photoreceptor a receptor of light stimuli.

photorespiration a puzzling phenomenon in which abundant light energy is captured by photosynthesis but little or no net carbon dioxide fixation occurs; common in C_3 plants in bright sunlight on hot days.

photosynthesis the organized capture of light energy and its transformation into usable chemical energy in the synthesis of organic compounds.

photosynthetic pigment any pigment involved in the capture of light; see *chlorophyll, bacteriochlorophyll, accessory pigments.*

photosystem also *photosynthetic unit, pigment system,* an organized array of proteins and pigments bound within the thylakoid membrane, consisting of a light antenna, a photocenter, and an associated electron transport system.

photosystem I also *photosystem 700, P700,* the second in the two photosystems in the electron pathway of photosynthesis in cyanobacteria and chloroplasts, and the one involving the reduction of NADP to $NADPH_2$; it may be evolutionarily more ancient than photosystem II.

photosystem II also *photosystem 680, P680,* the first of the two photosystems in the electron pathway of photosynthesis in cyanobacteria and all photosynthetic eukaryotes, and the one involving the photolysis of water; it probably occurred later in evolution than photosystem I.

phototaxis a tendency to move toward light. *Negative phototaxis,* a tendency to move away from light.

phototroph a photosynthetic organism; compare *autotroph, heterotroph, chemotroph.*

phototropism 1. the turning toward light of a growing plant stem. 2. phototaxis. 3. *negative phototropism,* the turning away from light, as of a root tip.

phragmoplast in plants, the cytoplasmic organelle initiating the formation of a cell plate following mitosis.

phycobilin a photosynthetic accessory pigment of cyanobacteria and of red algal chloroplasts (which are apparently derived from symbiotic cyanobacteria).

phycocyanin a blue-green photosynthetic pigment of cyanobacteria and red algal chloroplasts.

phycoerythrin a red photosynthetic pigment of cyanobacteria and red algal chloroplasts.

phylogenetic tree a graphical representation of the interrelations and evolutionary history of a group of organisms, indicating the relative order of successive divisions of the line of descent, coincident with past speciation events.

phylogeny 1. the evolutionary history of an organism or group of organisms. 2. a phylogenetic tree.

phylum 1. a major taxonomic unit of related, similar classes of animals, e.g. Phylum Annelida. 2. a division of the plant kingdom.

physiological characteristic to an organism's healthy functioning, as contrasted with *pathological;* e.g., *physiological pH,* the range of pH found in the tissues of a healthy organism.

physiology 1. a branch of biology dealing with the processes, activities, and phenomena of individual living organisms, organs, tissues, and cells. 2. the normal functioning of an organism.

phytochrome a red-light-sensitive protein complex of certain plant cell membranes, involved in many light-induced phenomena, including phototropism, photoperiodicity, and others.

phytoplankton photosynthesizing planktonic organisms.

P_i inorganic phosphate.

pigment any chemical substance that absorbs light, whether or not its normal function involves light absorption: e.g., chlorophyll, cytochrome c, hemoglobin, melanin.

piloerection the lifting up of hair by tiny involuntary muscles; bristling.

pilus (pl. *pili*) see *sex pilus.*

pineal body also *pineal organ, pineal gland,* a small body arising from the roof of the third ventricle in all vertebrates, forming a small red cone in most but forming an eyelike photoreceptive organ in larval lampreys, tuatara and some lizards; directly sensitive to light in reptiles and birds but not in mammals, although it appears to be the center of photoperiod responses in all vertebrates; it secretes the hormone *melatonin.*

pinna in common terms, the "ear."

pinocytosis taking dissolved molecular food materials, such as proteins, into the cell by adhering them to the plasma membrane and invaginating portions of the plasma membrane to form digestive vacuoles; a form of active transport.

pioneer organism 1. an organism that successfully establishes residence and produces offspring in an area not previously inhabited by its kind. 2. a type of organism specialized for the initial invasion of a disturbed area, such as a landslide or burned-out region.

pistil in flowering plants, a unit comprised of one or more ovaries, a style and a stigma; see also *gynocecium.*

pistillate having pistils but no stamens.

pit also *pit pair,* thin, porelike regions in adjacent plant cell walls, where only the primary cell wall remains.

pith thin-walled parenchymous tissue in the central strand of the primary growth of a stem; the dead remains of such tissue at the center of a woody stem.

pituitary gland also *hypophysis,* a small double gland lying just below the brain and intimately associated with the hypothalamus in all vertebrates, consisting of an anterior lobe and a functionally distinct posterior lobe; *anterior pituitary,* also *adenohypophysis,* a glandular body communicating with the hypothalamus only by way of a small vascular portal system; secretes growth hormone, prolactin, melanocyte-stimulating hormone, thyrotrophic hormone, ACTH, FSH, and LH; *posterior pituitary,* also *neurohypophysis,* not actually a gland but the enlarged, glandlike termini of axons of cell bodies in the hypothalamus, the hormones being synthesized in the cell bodies and translocated to the posterior pituitary within the axons, to be stored pending release; among these are *oxytocin* and *antidiuretic hormone.*

pituitary dwarf a small individual resulting from the failure of the anterior pituitary to secrete growth hormone.

pituitary giant an abnormally large individual whose excessive growth is due to excessive secretions of growth hormone, usually because of tumorous overgrowth of the anterior pituitary.

pivotal joint also *pivot joint,* 1. a joint that permits rotation. 2. a joint that permits only rotation.

placebo a substance having no pharmacological effect but administered as a control in testing experimentally or clinically the efficacy of a biologically active preparation.

placenta 1. in mammals other than monotremes and marsupials, the organ formed by the union of the uterine mucosa with the extraembryonic membranes of the fetus, which provides for the nourishment of the fetus, the elimination of waste products, and the exchange of dissolved gases. 2. in flowering plants, the part of the ovary to which the ovule and seeds attach.

placental mammals also *eutheria,* mammals that form chorioallantoic placentas; apparently a monophyletic group originating about 65 million years ago, and including all living mammals other than marsupials and monotremes.

placoid scale see *denticle* (2).

plankton minute, drifting plant, algal, and animal life in marine and freshwaters.

planula the early, ciliated, free-swimming larva of a coelenterate.

plaque 1. a pathological deposit of lipid, fibrous material and often calcium salts in the inner wall of a blood vessel. 2. a film of bacterial polysaccharide, mucus, and detritus harboring dental bacteria.

plasma the fluid matrix of blood tissue, 90% water and 10% various other substances; distinguished from *blood serum* by the presence of fibrinogen and the absence of certain platelet-derived hormones.

plasma membrane the external semipermeable limiting layer of the cytoplasm; see also *membrane, fluid-mosaic model.*

plasmid in bacteria, a small ring of DNA that occurs in addition to the main bacterial chromosome and is transferred from host to host by direct contact through pili. *Plasmid plus,* a bacterial strain carrying a plasmid and immune to further transfer. *Plasmid minus,* a bacterial strain not harboring a plasmid and susceptible to plasmid infection.

plasma cell a mature, short-lived B-cell lymphocyte, specialized in secreting antibodies.

plasmodesma (pl. *plasmodesmata*) in plants, minute cytoplasmic junctions between cells, occurring at pores in the cell wall through which the plasma membrane of one cell becomes continuous with that of the next. Highly significant to transport.

plasmodium (pl. *plasmodia*), 1. a motile, multinucleate mass of protoplasm produced by the fusion of uninucleate slime mold ameboid cells. 2. also *Plasmodium,* the malarial parasite.

plastid any of several forms of a self-replicating, semiautonomous plant cell organelle, primarily as a *chloroplast* specialized for photosynthesis, a *chromoplast* specialized for pigmentation, or a *leucoplast* specialized for starch storage.

plasticity the capacity of an organism for adjustments to changes in the environment; *genetic plasticity*, the capacity of a population to respond rapidly to changes in the environment through short-term natural selection.

plastiquinone a lipid-soluble hydrogen carrier that moves within the membrane in an electron transport chain.

plate (v.t.), to spread a thin suspension of living microorganisms over the surface of an agar gel, so as to be able to isolate and count colonies that result from subsequent growth.

platelet mother cell or **megakaryocyte** see also *platelet.*

plate tectonics the movement of great land and ocean floor masses (plates) on the surface of the earth relative to one another occurring largely in the Cenozoic era; see also *continental drift.*

platelets minute, cell-like but enucleate, fragile, membrane-bounded cytoplasmic disks present in the vertebrate blood as the result of the programmed fragmentation of thrombocytes; rupture of platelets releases factors that initiate blood clotting, produce or enhance pain, and induce the proliferation of fibroblasts.

pleura (pl. *pleurae*), in mammals, the tough, clear, serous connective tissue membrane covering a lung and lining the cavity in which the lung lies.

pleural cavity in mammals, either of two divisions of the coelom constituting the thoracic cavities, harboring the lungs and lined with the pleurae.

plumule 1. the apex of the plant embryo, consisting of immature leaves and an epicotyl that forms the primary stem. 2. a down feather. 3. a feathery extension of certain air-borne fruit, such as that of the dandelion.

point mutation a small spontaneous change in DNA involving individual nucleotides, such as a single base change, a single base addition or deletion.

polar of a chemical or a part of a molecule, capable of forming hydrogen bonds with water and with other polar molecules, charged portion.

polar body either of two small cells produced by meiosis of an egg cell: *first polar body*, the recipient of the chromosomes going to one pole in anaphase I; *second polar body*, the recipient of the chromosomes going to one pole in anaphase II of the egg. Neither polar body divides again.

polarity in animal egg cells, the establishment of metabolic gradients that will later reveal themselves as the active animal and sluggish vegetal poles.

polarized state in a neuron, the period when the region outside a neuron is positive and the region inside is negative, producing a membrane voltage potential of −60mv. See also *resting state.*

polar microtubule or **polar spindle fiber** two partially overlapping spindle fibers (not actually continuous), each originating near a pole and extending across the chromosomes without attaching. Also see *mitotic spindle.*

polar nuclei see *central cell.*

pollen grain one of the microscopic particles making up pollen, each being the microgametophyte of a seed plant containing a generative nucleus and a tube nucleus.

pollen sac one of two or four chambers in an anther in which pollen develops and is held.

pollen tube a tube that extends from a germinating pollen grain and grows down through style to the embryo sac, into which it releases sperm nuclei.

pollination the transfer of pollen from a stamen to a stigma, preceding fertilization of a flowering plant.

poly-A tail *polyadenylic acid tail*, a string of about 200 adenosine nucleotides added enzymatically to the 3′ end of HnRNA transcripts in the nucleus during the maturation of mRNA; thought to protect mRNA from degradation by cytoplasmic exonucleases.

polyadenylization the enzymatic addition of a poly-A tail to an RNA transcript.

polymer a generally large molecule, consisting of chemically bonded subunits, as in a polypeptide or polysaccharide.

polymerase an enzyme that causes polymerization; e.g., *RNA polymerase*, *DNA polymerase.*

polydactyly the genetically derived state of having extra fingers or toes.

polygenic inheritance inheritance involving many interacting variable genes, each having a small effect on a specific trait; also, *quantitative inheritance.*

polygenic trait also *continuously varying trait, quantitative trait, metric trait*, a trait in which variation in a population is expressed in continuous increments about a mean, and is attributable to the action and interaction of many variable gene loci and usually also to multiple variable effects of the environment; see also *distribution, continuous.*

polymorphism 1. the existence within a population of two or more discrete, genetically determined forms other than variations in sex or maturity and apart from rare mutant forms; e.g., black and spotted leopards; see also *distribution, discrete*. 2. the existence within a population of two or more alleles at a locus, at allele frequencies greater than some arbitrary value such as 1% or 5%; e.g., **ABO** and **MN** blood group polymorphisms.

polymorphonuclear leukocyte a large white blood cell with permanently condensed, highly lobed nucleus.

polyp the typical sessile form of a coelenterate; compare *medusa.*

polypeptide a continuous string of amino acids in peptide linkage, longer than a peptide; compare *protein.*

polyphyletic of a taxonomic group, having two or more ancestral lines of origin; having defining characteristics in common because of convergent evolution rather than because of common descent; having a last common ancestor not belonging to the group or sharing its defining characters; e.g., *Pachyderma*, a discarded group originally containing elephants, rhinoceroses, hippopotamuses, pigs, and horses; also the *animal kingdom* inasmuch as Metazoa and Parazoa are derived from different protist ancestors.

polyploid having more than two complete sets of chromosomes; *autopolyploid*, having three or more complete sets of a specific complement of chromosomes; *allopolyploid*, having two or more full diploid sets of chromosomes derived from different ancestral diploid species.

polysaccharide a polymer of sugar subunits.

polysome also *polyribosome*, a small group of ribosomes held together by their common attachment to a messenger RNA molecule.

polyspermy an abnormal condition in which two or more sperm successfully penetrate the egg cytoplasm.

polytene chromosome in certain tissues of certain insects, enlarged chromosomes created by chromatin replication without chromosome division, see also *salivary chromosome.*

polyunsaturated of a fatty acid, having more than one carbon-carbon double bond.

pons a broad mass of nerve fibers running across the ventral surface of the mammalian brain.

population 1. (demography) the total number of persons inhabiting a given geographical or political area. 2. (genetics) an aggregate of individuals of one species, interbreeding or closely related through interbreeding and recent common descent, and evolving as a unit. 3. (ecology), the assemblage of plants and/or animals living in a given area; or all of the individuals of one species in a given area. 4. (statistics), a finite or infinite aggregation of individuals under study.

population bottleneck see *bottleneck.*

population genetics the scientific study of genetic variation within populations, of the genetic correlation between related individuals in a population, and of the genetic basis of evolutionary change.

population recognition the ability to recognize members of one's own particular population or subgroup of a species; see also *recognition.*

porocyte a pore-bearing cell in the body wall of a sponge, specialized for admitting water.

porphyria variegata a rare genetic disease characterized by excess porphyrins in the blood, accompanied by red urine, light sensitivity, and liver damage; its high incidence in Afrikaaners is considered an example of the *founder effect.*

porphyrin group also *porphrin ring*, a complex organic compound consisting of a flat circle of nitrogenous rings, usually with iron or another metal complexed in the center; a constituent of hemoglobin, myoglobin, chlorophyll, and the cytochromes.

positive control of a gene or operon, control such that a gene remains inactive until stimulated by an appropriate controlling protein, hormone or other signal.

positive feedback see *feedback.*

positive geotropism see *geotropism.*

positive phototropism see *phototropism.*

postanal tail tail that extends from the anus posteriorly.

postecdysis in arthropods, the period following *ecdysis* (shedding), during which the new exoskeleton becomes hardened.

posterior 1. in most bilateral organisms, away from the head end of an organism, the opposite of *anterior*. 2. in human anatomy, in view of the upright position of the organism at least in the daytime, *anterior* and *posterior* are often used in place of *ventral* and *dorsal* to mean toward the front and toward the back respectively, while *superior* and *inferior* are used in place of *anterior* and *posterior* meaning toward the head end and toward the tail end respectively.

posterior chamber see *eyeball*.

posterior pituitary the posterior lobe of the pituitary, a stalked extension of neural tissue extending from the hypothalamus.

posterior vena cava in humans *inferior vena cava*, either of two large veins draining the body posterior to the heart; see *vena cava*.

postsynaptic membrane the receptive surface of a dendrite or cell body adjacent to a synapse; see also *synapse*.

posttranscriptional modification the entire series of enzymatic changes made in RNA in the nucleus after transcription, including *capping, polyadenylation,* and *tailoring*.

potassium ion gate see *ion channel*.

potential energy energy stored in chemical bonds, in nonrandom organization, in elastic bodies, in elevated weight or any other static form in which it can theoretically be transformed into another form or into work.

powdery mildew 1. a fungal plant disease characterized by the presence of powdery conidia. 2. the fungus that causes the disease.

prairie the originally extensive tract of level or rolling treeless grassland of the Mississippi-Missouri valley, characterized by deep fertile soil and tall, coarse perennial grasses.

preadaptation the appearance of some trait that precedes its eventual usefulness.

precapillary sphincter in arterioles, rings of smooth muscle capable of regulating flood flow into capillaries; controlled by the autonomic nervous system.

precocial capable of walking and a high degree of independent activity from hatching or birth; compare *altricial*.

precursor a chemical substance from which another substance is formed, especially in a specific step of a biochemical pathway.

predation 1. the act of catching and eating. 2. being caught and eaten; e.g., *subject to predation*. 3. a mode of life in which food is primarily obtained by killing and eating other animals.

predator an animal that habitually preys on other animals, a carnivorous animal.

preecdysis see *premolt*.

prehensile adapted for seizing, grasping, or wrapping around, as the tail of a New World monkey or the upper lip of a rhinoceros.

premolar in humans, any of the eight bicuspid teeth, located in groups of two anterior to the molars.

premolt also *preecdysis*, the preparatory events of molting or ecdysis during which mineral salts are withdrawn from the old exoskeleton and rapid cell division occurs in the cells below.

prepotence Darwin's term signifying dominance. See also *dominance*.

pressure receptor also *baroreceptor*, see *sensory receptor*.

presumptive in embryology, designating the future fate of a tissue as in *presumptive ectoderm, mesoderm* and *endoderm*.

presynaptic membrane the membrane of an axon or of the synaptic knob of an axon in the region of a synaptic cleft, into which it secretes neurotransmitters in the course of the transmission of a nerve impulse; see *synapse, synaptic cleft*.

prey 1. (n.), an animal hunted or seized for food by a predator. 2. (v.t.), (followed by *upon*), to hunt or seize for food.

prey switching behavior of a predator in response to the relative densities of two prey species, in which the predator ceases hunting one prey and begins to hunt the other.

primary bronchus the left or right division of the trachea. See also *respiratory tree*.

primary consumer a herbivore; see *consumer*.

primary endosperm nucleus in flowering plants, a triploid nucleus resulting from fertilization in the binucleate central or endosperm mother cell.

primary growth of a plant, the initial growth or elongation of a stem or root, resulting primarily in an increase in length and the addition of leaves, buds and branches. *Primary growth pattern*, the distinctive pattern of xylem, phloem and other tissues in primary growth; compare *secondary growth, secondary growth pattern*.

primary host also *definitive host*, in parasitic relationships, a host in which a parasite may undergo sexual reproduction or at least reaches a mature state.

primary immune response the slower, initial response against invasion of the body by organisms or foreign molecules, during which immature, inactive lymphocytes are activated into specialized B- and T-cell lymphocytes. See *secondary immune response*.

primary lesion a damaged or mismatched segment of DNA, subject to repair. See *DNA repair system*.

primary meristem any of the three primary plant tissues—protoderm, ground meristem, and procambium—derived from apical meristem.

primary mRNA in transcription, the raw mRNA product prior to posttranscriptional modification and tailoring.

primary mutation an alteration in DNA subject to repair by repair enzymes, as distinguished from mutations that cannot be repaired.

primary oocyte a diploid cell of egg-producing potential prior to its first meiotic division.

primary organizer in development, any tissue that induces or influences the future development of other subservient tissues.

primary phloem phloem developed from apical meristem, that is, the phloem of primary growth.

primary producer in ecology, plants, algae or other photosynthetic (or chemosynthetic) organisms that produce the food of a food chain.

primary productivity see *gross productivity*.

primary spermatocyte a diploid meiotic cell of sperm-producing potential prior to its first meiotic division.

primary structure or *level of organization* the linear sequence of amino acids in a polypeptide, or of nucleotides in a nucleic acid.

primary succession the succession of vegetational states that occurs as an area changes from bare earth to a climax community.

primary xylem xylem in primary growth, which is produced by apical meristem rather than vascular cambium.

primitive a character or character state, characteristic of the original condition of the group under consideration; ancestral; not derived; of or like the earliest state within the group considered; old; compare *advanced, derived*.

primitive groove in bird, reptile, and mammalian embryos, a lengthy indentation formed as gastrulation begins.

primitive gut see *archenteron*.

primitive streak in bird, reptile, and mammalian embryos, a thickening in the blastoderm formed by convergence of cells in preparation for gastrulation.

primordium the rudiment or commencement of a part or organ, often appearing as a mass lump of undifferentiated cells; e.g., *leaf primordium, branch primordium, liver primordium*.

principle of colinearity see *colinearity*.

probability 1. the relative likelihood of the occurrence of a specific outcome or event based on the proportion of its occurrence among the number of trials or opportunities for its occurrence, viewed as indefinitely extended. 2. the field of mathematics dealing with probability relationships.

probability cloud a graphic representation of a three-dimensional electron orbital or group of orbitals, in which the relative probability that an electron will occupy a specific location at a given instant is represented by shading or stippling.

proboscis in vertebrates, a snout or trunk. In invertebrates, any tubular feeding organ that is often protrusible with a mouth at the end, sometimes used as a sensory or defensive structure.

procambium the part of a meristem that gives rise to cambium.

prokaryote also *procaryote*, any organism of the kingdom Monera, having no nucleus and a single circular chromosome of nearly naked DNA; a *eubacterium*, or *archaebacterium*.

producer see *primary producer*.

product something produced; *enzyme product*, one of the resulting compounds of an enzymatic reaction; *gene product*, either RNA directly transcribed from a gene or, more commonly, a polypeptide translated from the mRNA of an active gene.

productivity see *gross productivity, net productivity*.

profundal zone in freshwater bodies, the dark waters below the photic zone, often having a low oxygen content and limited aerobic life, that consists primarily of reducers and scavengers.

progeny testing determination of an organism's genotype by crossing it with a known recessive homozygous individual and observing the resulting offspring.

progesterone a steroid hormone produced by the placenta and corpus luteum, that functions in maintaining the uterine mucosa and the pregnant state; a frequent constituent of birth control pills.

proglottid any segment of a mature tapeworm formed by strobilation, containing male and female organs and being shed when full of mature fertilized eggs.

prolactin a hormone of the anterior pituitary that in mammals induces lactation and helps maintain the corpus luteum; in pigeons and related birds, stimulates the crop gland to produce a milklike substance fed to nestlings.

proliferation rapid and repeated production of new cells by a succession of cell divisions.

proliferative phase also *preovulatory phase*, in women, that portion of the menstrual cycle prior to ovulation when estrogen levels rise and endometrial growth and repair begins.

promoter a DNA sequence to which RNA polymerase must bind in order for transcription to begin. See *operon*.

pronephros the most primitive form of kidney in vertebrate development, later replaced by the *mesonephros,* which is itself replaced by the *metanephros* from which the final adult kidney forms.

pronucleus the nucleus of either gamete after the entry of a sperm into an egg and before nuclear fusion; *male pronucleus,* the sperm nucleus in the egg cytoplasm; *female pronucleus,* the egg nucleus after sperm entry.

prophase the first phase of mitosis and meiosis, the events of which include chromosomal condensation, dismantling of the nuclear envelope, organization of the spindle apparatus, and, in meiosis, synapsing and crossing over. See also *mitosis* and *meiosis*.

prophylactic 1. (adj.), preventing or contributing to the prevention of disease. 2. (n.) any drug or other substance used as a disease preventive measure. 3. a condom.

proprioception the sum of neural mechanisms involved in sensing, integrating and responding to stimuli from within the body, including the integration of the movement of body parts, balance, and stance.

proprioceptive sensor see *proprioceptor.*

proprioceptor a sensory receptor that is located in internal tissues (as in skeletal, smooth and heart muscle, tendons, carotid body) and responds to conditions of body positions, muscle tension, or internal chemistry.

prostacyclin a prostaglandin.

prostaglandin any of a group of hormonelike substances derived from long-chain fatty acids and produced in most animal tissues by a variety of stimuli including endocrine, neural, and mechanical stimuli as well as inflammation and oxygen deprivation; present in tissues, sometimes transported by the blood stream but more often found in other body fluids, e.g., prostaglandins in semen, which stimulate uterine contractions; appear to act by adenyl cyclase activation.

prostate gland a pale, firm, partly muscular and partly glandular organ that surrounds and connects with the base of the urethra in male mammals; its viscid, opalescent secretion is a major component of semen.

prosthetic group a part of a functional protein not part of a polypeptide, such as a heme group, coenzyme, carbohydrate or lipid.

protein a naturally occurring functional macromolecule consisting of one or more polypeptides held together by hydrogen bonds and van der Waals forces and often sulfhydryl linkages, as well frequently including one or more prosthetic groups.

proteinaceous sponge highly complex sponges with skeletal elements formed from the protein spongin.

protein kinase a class of enzymes that activates other enzymes.

proteinoid spherical aggregations of amino acid polymers with a membranelike surface formation, known to grow and divide in a manner reminiscent of cells.

protein-sequencing a determination of the primary structure of proteins or polypeptides—the number, kinds, and order of amino acids—useful in establishing evolutionary relationships.

prothallium also *prothallus*, the fern gametocyte, typically a small, flat, green thallus with rhizoids, bearing numerous antheridia and/or archegonia.

prothoracic gland an endocrine gland located in the prothorax of an insect, the source of the molting hormone *ecdysone*.

prothoracicotrophic hormone or **PTTH** see *brain hormone*.

prothrombin a blood plasma protein that is converted to the protein thrombin during the clotting process.

protocell the earliest form to exhibit characteristics associated with life. A coacervatelike body containing autocatalytic properties that assure faithful replication of specific chemical properties.

protoderm primary plant tissue from which the epidermis is derived.

protoeukaryote a hypothetical predatory microorganism, capable of the engulfment of prey, that by the successive engulfment of a protomitochondrion, a protochloroplast, and perhaps a protocilium evolved into the ancestral eukaryotic cell, perhaps a billion years ago.

Proto-Gondwana see *Proto-Laurasia*.

Proto-Laurasia according to the continental drift theory, one of two land masses, the other, *Proto-Gondwana,* preceding and giving rise to the supercontinent *Pangaea.* The two earlier masses, according to one hypothesis, are the origin of the deuterostomes and proterostomes.

proton 1. one of the two hadrons composing the atomic nucleus in ordinary matter, having an electrostatic charge of +1, and a mass 1837 times that of an electron, and by its numbers determining the chemical properties of the atom. 2. a hydrogen ion.

protonephridium literally, "first kidney," in both a phylogenetic and embryological sense; a primitive osmoregulatory and excretory tubule containing flame bulbs; also a unit in the flame cell system.

proton pump an active transport system using energy to move hydrogen ions from one side of a membrane to the other against a concentration gradient, as in chemiosmosis.

protonema (pl. *protonemata*), the threadlike and usually transitory stage of a moss or liverwort gametophyte as it emerges from a germinating spore; indistinguishable from a filamentous green alga in structure and appearance.

protoplasm the inner living substance of cells, including cytoplasm and nucleoplasm.

protoplast 1. a plant or bacterial cell artificially deprived of its cell wall. 2. a plant or bacterial cell, considered apart from its cell wall.

protostome an animal in which the mouth derives (developmentally or phylogenetically) from the blastopore; compare *deuterostome.*

protozoan 1. any of a large group of protists. 2. any nonphotosynthetic protist.

proventriculus in the bird, a glandular region of the stomach between the crop and the gizzard; in the insect foregut, a muscular region fitted with chitinous teeth.

proximal nearer the point of attachment or origin; nearer the center of the body; compare *distal.*

proximal convoluted tubule see *nephron.*

pruning cutting off of the growing tips of branches in order to produce a better shaped or more fruitful growth.

pseudocoelom also *pseudocoel*, "false coelom"; the body cavity in nematodes, rotifers and certain other invertebrates, between the body wall and the intestine, which is not entirely lined with mesodermal epithelium; in which the gut is entirely endodermal and not muscularized.

pseudoplacenta 1. also *yolk-sac placenta*, the placentalike structure of marsupials. 2. any placentalike structure, as in sea snakes, some cartilaginous fishes and a few insects.

pseudopod also *pseudopodium*, any temporary protrusion of the protoplasm of a cell serving as an organ of locomotion or engulfment; often having a fairly definite filamentous form, and sometimes fusing with others to form a network.

P700 see *photosystem I.*

psilophytes 1. the earliest known vascular plants; simple dichotomously branching plants of the Silurian and Devonian. 2. see *psilopsid.*

psilopsid a simple plant of the tracheophyte subdivision *Psilopsida,* vascular plants lacking roots, cambium, leaves and leaf traces, and have a reduced, subterranean, nonphotosynthetic gametophyte.

P680 see *photosystem II.*

pterophyte any member of the plant division sphenophyta, the ferns.

pterosaur also *pterodactyl*, any of numerous extinct flying reptiles of the lower Jurassic to the close of the Mesozoic.

ptyalin also *salivary amylase*, a starch-digesting enzyme found in the saliva of humans and of certain other mammals.

pubic bone also *pubis*, one of the constituent bones of the coxa or innominate bone.

pubic symphysis in mammals, the usually semirigid fibrous articulation of the two pubic bones in the midline above the genitalia; subject to considerable hormone-induced loosening toward the end of pregnancy.

pubis see *pubic bone.*

puff a temporary enlargement of a band of a polytene chromosome indicating gene transcriptional activity.

pulmonary artery a branching artery that conveys venous (i.e., deoxygenated) blood from the right ventricle to the lungs, consisting of a common pulmonary artery, right and left pulmonary arteries, and further subdivisions.

pulmonary circuit the passage of venous blood from the right side of the heart through the pulmonary arteries to the capillaries of the lung where it is oxygenated and from which it returns by way of the pulmonary veins to the left atrium of the heart.

pulmonary semilunar valve also *pulmonary valve* in the heart, a one-way valve at the base of the pulmonary artery that prevents backflow of blood into the right ventricle during systole.

pulmonary valve a three-cusped valve of the heart separating the cavity of the right ventricle from the common pulmonary artery.

pulmonary vein in birds and mammals the only vein that carries arterial (oxygenated) blood, returning it from the lungs to the left atrium.

pulp cavity the central cavity of a tooth, containing the nerves and blood vessels comprising *dental pulp.*

punctuated equilibrium a recently proposed theory that evolution does not occur gradually but that life exists over long periods of time with little evolutionary change, interrupted periodically by great changes. Compare to *gradualism.*

pupa a metamorphic insect in a specific quiescent intermolt between the larval and adult stages, during which time extensive body transformations occur, the pupa generally being enclosed in a hardened pupal case or cocoon.

pupil the contractile aperture in the iris of the eye.

purine a heterocyclic nitrogenous base consisting of conjoined 5-membered and 6-membered rings; e.g., adenine, guanine, uric acid.

Purkinje fiber in the heart, specialized conducting fibers that carry contractile impulses from the bundle of His to the outer ventricular walls.

pus a thick, yellowish-white accumulation of dead phagocytes, dead or living bacteria, and tissue debris in a fluid exudate.

pyloric sphincter also *pyloric valve,* a ring of muscle capable of closing off the opening between the stomach and the small intestine.

pyloric stenosis an inherited but surgically correctable defect of the human gut, characterized by a thickening and restriction in the pyloric orifice (lower valve) of the stomach.

pylorus the opening between the stomach and the small intestine in a vertebrate.

pyramid 1. see *population pyramid.* 2. also *pyramid of numbers:* see *Eltonian pyramid.* 3. any of various anatomical features shaped like the Egyptian pyramids, including certain structures of the brain and kidney.

pyrimidine a nitrogenous base consisting of a six-membered heterocyclic ring; notably cytosine, uracil, and thymine, constituents of nucleic acids.

pyrogen any fever-producing substance or hormone.

pyrophosphate an inorganic ion consisting of two phosphate groups joined by a phosphate-phosphate high energy bond.

pyruvate the anion of pyruvic acid, the final organic product of glycolysis in animal cells and the initial substrate of the citric acid cycle.

quadrate also *quadrate bone,* the bony element of the skull articulating with the lower jaw in most vertebrates other than mammals.

quantum (pl. *quanta*), one of the very small increments or parcels into which many forms of energy are subdivided; directly proportional to the wave frequency of a photon, electron, or other particle.

quantum mechanics also *quantum theory,* an extensive branch of physics based on Planck's concept of radiant and other energy being subdivided into discrete quanta.

quaternary structure or **level of organization** the interaction of two or more polypeptides through disulfide linkages.

radial have parts arranged like rays eminating from a center.

radial canal 1. in a sponge, one of the numerous choanocyte-lined canals radiating from the central cavity. 2. in a jellyfish, one of the gastro vascular canals extending from the central cavity to the marginal circular canal. 3. in an echinoderm water vascular system, a tube extending outward from the ring canal.

radial spoke in cilia and flagella, protein connections between the outer ring of microtubules and the inner sheath surrounding the central pair.

radial symmetry the condition of having similar parts regularly arranged as radii from a central axis, as in a starfish.

radiation 1. the transfer of heat or other energy as particles or waves; compare *convection, conduction.* 2. heat or other energy transmitted as particles or waves; see also *ionizing radiation.*

radiation, electromagnetic see *electromagnetic radiation.*

radiation, ionizing see *ionizing radiation.*

radical 1. also *moiety, residue,* a part or functional group of a molecule. 2. see *free radical.*

radicle the lower portion of the axis of a plant embryo, especially the part that will become the root.

radioactive in elements, a state of instability in which radiation is emitted by the atomic nucleus.

radioactive isotope see *isotope.*

radioactive tracer a radioactive element or a compound containing a radioactive element, that is put into a biological system and transformed or translocated, its eventual fate being determined by physiochemical means.

radiolarian ooze a siliceous mud of deep sea bottoms, composed largely of the skeletal remains of radiolarians.

radioisotope see *isotope.*

radius one of the two bones of the forearm of land vertebrates, rotating about the ulna and articulating with the wrist.

radula in all mollusks other than bivalves, a toothed, chitinous band that slides backward and forward over a protrusible prominance from the floor of the mouth, scraping and tearing food and bringing it into the mouth.

rain shadow a region of low rainfall on the prevailing lee side of a mountain or mountain range.

random drift see *drift.*

range the geographical area occupied by a species.

rare dominant 1. an allele of low frequency leading to a definable abnormal phenotype in the heterozygote, and unknown or lethal in the homozygote. 2. a lethal-recessive dominant.

ray see *vascular ray.*

reabsorptive area in bony fishes, a structure that absorbs gases from the swim bladder, this aiding in the maintenance of buoyancy. Also see *swim bladder* and *gas gland.*

reactant any element, ion, or molecule participating in a chemical reaction.

reaction center, also *photocenter* the part of a photosystem in which light-activated chlorophyll *a* transfers an electron to the electron transport system.

reaction see *chemical reaction.*

reading-frame shift a serious point mutation involving the loss or gain of a single base-pair, whereby all codons of mRNA beyond or "downstream" from the change will be misread.

reafference the presumed reward associated with the performance of a particular movement when an animal is performing an adaptive or instinctive pattern; behavior as its own reward.

reagent 1. a chemical used in preparative, experimental, or diagnostic work because of the chemical reactions it undergoes. 2. any laboratory chemical.

realized niche the niche of an organism as it actually is, generally only a portion of the fundamental niche. Compare *fundamental niche* and see *niche.*

receptacle the end of a floral stalk, which forms the base on which the flower parts are borne.

receptor 1. a specialized neuron involved in the detection of environmental stimuli. 2. any sense organ or group of sensitive cells that transmit impulses.

receptor-mediated endocytosis the transport of substances into the cell by vesicles formed from the plasma membrane, following the recognition of and binding of such substances to highly specific membranal receptor sites.

recessive (adj.), of an allele, not expressed in a heterozygote; see *dominance interaction.*

recessive-lethal dominant an allele that has a pathological or visible effect as a heterozygote and is lethal as a homozygote.

recessivity the failure of one allele of a heterozygote to have any discernable effect on a specific phenotypic trait; see *dominance interaction*.

reciprocal altruism behavior that appears to be altruistic in individual instances but in fact functions to increase the fitness of the individual insofar as it increases the likelihood that the individual will be the recipient of beneficial behavior at another time.

reciprocal testcross in seed plants, a cross in which pollen is exchanged between the flowers of two individual plants, e.g., pollen from a certain tall pea plant is used to pollinate a certain short plant, and then pollen from that short plant is used to pollinate the tall plant.

recognition 1. the ability to respond appropriately to the sight, sound or smell of another organism or a situation, whether on the basis of genetic propensities or of individual experience. 2. an act of appropriate response.

recombinant chromosome a chromosome emerging from meiosis with a combination of alleles not present on the chromosomes entering meiosis.

recombinant DNA a recent general term for the laboratory manipulation of DNA in which DNA molecules or fragments from various sources are severed and combined enzymatically and reinserted into living organisms; see also *gene cloning*.

recombinant DNA technology the techniques used in genetic engineering. See *genetic engineering*.

recombination frequency the proportion of gametes bearing recombinant chromosomes for two specific gene markers that were heterozygous in the parent.

recombination nodule during crossing over in meiosis I, large enzyme complexes within which strands of DNA are exchanged between homologues.

rectal valve any of several folds of the rectal wall.

rectum the terminal part of the intestine, used for the temporary storage of feces.

red alga any alga of the division *Rhodophyta*.

red blood cell see *erythrocyte*.

redia see *human liver fluke*.

red marrow regions within the ribs, sternum, vertebrae, and hip bones where red blood cells are produced.

red tide seawater discolored by a dinoflagellate bloom in a density fatal to many forms of life.

reducer in ecology, a fungus or microorganism that breaks down plant or animal matter into small molecules.

reducing power the relative amount of energy released by a substance for each electron or hydrogen transferred from it in an oxidation-reduction reaction; the relative ability of one substance to transfer electrons or hydrogen atoms to another.

reduction of a substance, the addition of electrons or hydrogen atoms.

reduction division the first division of meiosis.

reductionism see reductionist.

reductionist in science, one who attempts to divide phenomena and mechanisms to their most elemental parts, isolating as far as possible the effects of individual factors and testing these effects in separate controlled experiments.

redundancy having information present in more than one copy.

reflex arc in the vertebrate nervous system, a simple neural pathway involving as few as two or three neurons that sense and react to a stimulus.

refractory state a brief period when a neuron cannot generate a second impulse.

regular flower see *flower*.

regulator gene see *operon*.

regulatory T-Cell including helper T-cells (also called inducer T-cells) and suppressor T-cells; the first activates elements of the immune system and the second suppresses immune reactions.

releaser in animal behavior, any stimulus that evokes the release of instinctive behavioral patterns.

releasing hormone also *releasing factor*, a chemical messenger released by the hypothalamus that stimulates hormonal release by the pituitary.

REM sleep also *paradoxical sleep*, a normal period of sleep during which the muscles are very relaxed but the eyes move rapidly under closed lids; accompanied by high brain electrical activity.

renal circuit in mammals, the circuit of the branching *renal arteries* arising from the abdominal aorta and continuing through the glomerulus, the capillary beds of the nephron, and the *renal veins* to return to the vena cava.

renal pelvis the cavity of the kidney into which the collecting ducts empty.

renin a hormone of the juxtaglomerular complex, believed to stimulate the adrenal cortex to release aldosterone.

rennin an enzyme produced by the stomach linings of young mammals, the action of which is to coagulate milk.

repair enzyme any of several different complexes of enzymes that recognize improper base pairing in DNA, excise a region of one of the strands, and rebuild the DNA according to the rules of Watson-Crick pairing; or that otherwise repair mutational damage, including double-strand chromosome breaks.

replacement rate the birth rate (in terms of the lifetime production of offspring per woman) which, at age structure equilibrium, results in a stable population size; e.g., about 2.1 infants per adult woman.

replica plating a screening technique in which an array of colonies on a nutrient-supplemented agar plate are transferred as a group to one or more additional plates of nutrient-deficient agar to test for nutritional mutants.

replication see DNA replication.

replication complex during replication, a grouping of the essential enzymes of that process including the unwinding enzyme, helicase, and DNA polymerase.

replication fork the point at which unwinding proteins separate the two DNA strands in the course of DNA replication.

repolarization in the neuron, reestablishment of the resting potential or polarized state following an action potential. See also *neural impulse*.

repressible operon an operon governing a synthetic pathway, and which is generally active but which can be inactivated by the presence of its normal metabolic product in the medium.

repressor-inducer complex in an inducible operon, the repressor protein bound to and temporarily inactivated by the small molecule that induces operon activity; see also *operon*.

repressor molecule in bacterial operons, a small metabolite that combines with an inactive repressor protein to form an active complex that binds with an operator to prevent transcription of an operon.

repressor protein in bacterial operons, a protein that binds the operator and prevents transcription either when bound to an inducing molecule (inducible operon) or when not bound (repressible operon).

reproductive isolation 1. the state of a population in which there is no mating between members of the group and members of other groups, and no immigration or emigration of individuals. 2. the state of a population or species in which successful matings outside of the group are biologically impossible because of physical mating barriers, hybrid inviability, or hybrid sterility.

reproduction see *sexual reproduction, asexual reproduction*.

reservoir in biogeochemical cycles, the less readily available reserves of mineral nutrients, such as atmospheric nitrogen, which is only useful to nitrogen-fixing organisms. Compare *exchange pool*.

residual air or **residual volume** the volume of air remaining in the lungs following maximum, forceful exhalation.

resolution 1. the ability to distinguish detail in stimuli, specifically the ability to distinguish two closely spaced stimuli from a single stimulus. 2. in optics, the minimum distance between two points at which they still appear to be separate entities. 3. in sexual activity, a retreat from the excitement states, generally following orgasm.

resolving power the ability of the eye to distinguish objects near each other as distinct and separate; also the chief factor limiting useful magnification in light microscopes.

respiration 1. breathing. 2. the physical and chemical process by which an organism supplies oxygen to its tissues and removes carbon dioxide. 3. also *aerobic respiration*, any energy-yielding reaction in living matter involving oxygen. 4. the metabolic transformation of food or food storage molecules yielding energy, *aerobic respiration* (e.g, citric acid cycle and electron transport).

respiratory center a region in the medulla oblongata that regulates breathing movements.

respiratory system in animals, that organ system responsible for the exchange of gases with the environment.

respiratory tree in air-breathing vertebrates, much of the respiratory system, including the trachea, paired bronchi, highly branched tracheoles, and alveoli.

resting potential the charge difference across the membrane of a neuron or muscle fiber while not transmitting an impulse.

resting state a state of seeming inactivity in a neuron, but one in which the membrane activity maintains a polarized state in preparation for conduction.

restriction enzyme in bacteria, an enzyme that recognizes and serves a specific, short DNA sequence, thus protecting the cell from all but a few highly adapted, host-specific viruses; such enzymes have proven useful for experimental DNA manipulation.

retaliator an animal that displays and threatens but does not attack others of its kind unless attacked first, but always fights back if attacked.

retaliatory strategy see *retaliator*.

rete mirabile a small but very dense network of blood vessels in a counter-current heat exchange.

reticular system also *reticular formation*, a major neural tract in the brainstem containing neural pathways to other parts of the brain and to the *reticular activating system* (RAS), and arousal center.

reticulum see *ruminant*.

retina see *eyeball*.

retinal cell see *compound eye*.

retrovirus a single-stranded RNA virus that, after undergoing reverse transcription and producing double-stranded DNA from its RNA coding, inserts into the host chromosome where it is replicated through many host cell generations. It may later escape from the chromosome to enter a period of viral reproduction and cell lysis.

reverse transcriptase also *RNA-dependent DNA polymerase*, and enzyme of certain RNA viruses that copies RNA sequences into single-stranded and double-stranded DNA sequences in a minor reversal of the central dogma.

R-group a chemistry shorthand where R stands for the variable part of an amino acid.

Rh blood group system also *rhesus factor, Rh factor*, a polymorphic blood cell antigen system consisting of three variable antigenic sites controlled by eight alleles at a single locus.

Rh negative (adj.), homozygous for the absence of the most antigenic of the three Rh antigen sites; not responding to anti-Rh antibodies.

Rh positive homozygous or heterozygous for the most antigenic of the three Rh antigen sites; responding with blood cell agglutination to anti-Rh antibodies.

rhizoid 1. a rootlike structure that serves to anchor the gametophyte of a fern or bryophyte to the soil and to absorb water and mineral nutrients. 2. a portion of a fungal mycelium that penetrates its food medium.

rhizome an underground horizontal shoot specialized for food storage.

rhynophyte an extinct plant group thought to represent the first vascular plants.

rhythm method also *natural birth control*, a method of contraception whereby copulation is avoided during periods when conception is likely.

riboflavin a derivative of vitamin B_2 and an active group in the coenzyme FAD, flavin adenine dinucleotide, capable of being reduced during an oxidation-reduction reaction.

ribose a 5-carbon aldose sugar, a constituent of many nucleosides and nucleotides.

ribosomal RNA also *rRNA*, the RNA that forms the matrix of ribosome structure, consisting of a large, intermediate, and two relatively small sequences.

ribosome an ultramicroscopic cytoplasmic organelle or very large molecule consisting of a matrix of RNA and numerous specific proteins, binding a messenger RNA molecule, a charged tRNA and a growing polypeptide chain during protein synthesis.

rib one of the numerous curved bony or cartilaginous rods that stiffen the lateral walls of most vertebrates and articulate with the vertebral column.

ring canal 1. also *circumoral canal*, a circular tube of the water vascular system that surrounds the esophagus in echinoderms. 2. a circular, tubular extension of the gastrovascular cavity in the edge of the umbrella of a jellyfish; see also *radial canal*.

ritual combat intraspecific combat following a rigidly defined format that precludes serious or permanent damage to either participant.

RNA also *ribonucleic acid*, a single-stranded nucleic acid macromolecule consisting of adenine, guanine, cytosine and uracil on a backbone of repeating ribose and phosphate units; divided functionally into rRNA (ribosomal RNA), mRNA (messenger RNA), tRNA (transfer RNA), and viral RNA.

RNAase an enzyme capable of hydrolyzing RNA sugar-phosphate bonds.

RNA-dependent DNA polymerase also *reverse transcriptase*, a viral enzyme that catalyzes the copying of an RNA sequence into single-stranded or double-stranded DNA.

RNA-dependent DNA synthesis see *reverse transcription*.

RNA polymerase the enzyme or enzyme complex catalyzing transcription.

RNA synthetase see RNA polymerase.

rRNA see *ribosomal RNA*.

rod one of the numerous long, rod-shaped sensory bodies in the vertebrate retina, containing many membrane layers bearing visual pigments, responsive to faint light but not to detail or to variations in color; compare *cone*.

root the portion of a seed plant, originating from the radicle, that functions as an organ of absorption, anchorage and sometimes food storage, and differs from the stem in lacking nodes, buds, and leaves.

root apical meristem see *apical meristem*.

root cap a protective mass of parenchymal cells that covers the root apical meristem.

root hair one of the many tiny tubular outgrowths of root epidermal cells, especially just behind the root apex, that function in absorption.

root meristem see *apical meristem*.

root nodule one of the multiple swellings on the roots of a leguminous plant, developed in response to a symbiotic nitrogen fixing bacterium and harboring nourishing colonies of such bacteria.

root pressure the active force by which water rises into the stems from the roots.

root tip in plants, one of many delicate root endings, containing the apical meristem responsible for primary growth in the root.

rough endoplasmic reticulum see *endoplasmic reticulum*.

round window in the inner ear, a membranous window, similar to the oval window, but at the cochlear's opposite end. It moves in response to perilymph movement started by the oval window. See also *oval window*.

r-selection a form of natural selection characterized by exponential population growth. Usually characterized by high reproductivity with little parental care.

R6 plasmid a recently discovered plasmid containing genes that confer resistance to several antibiotics.

Ruffini corpuscle sensory receptors of the skin that respond to temperature changes.

rumen the large first compartment of the stomach of cattle and other ruminants, used as a temporary repository for food which is partly broken down by bacterial and protozoan symbionts and later regurgitated for rumination.

ruminant hooved grazing mammals that digest cellulose through the action of microorganisms in a four-part stomach, which includes the *rumen, reticulum, omasum,* and *abomasum*.

saccharide any compound consisting of carbon, hydrogen, and oxygen, in which the ratio of hydrogen to oxygen is two to one, or $(CH_2O)_n$.

saccule 1. a small sac. 2. the smaller of two chambers of the membranous labyrinth of the ear; compare *utricle*.

sacroiliac 1. the juncture of the sacrum of the vertebral column and the iliac bones of the pelvis. 2. the fibrous cartilage between the sacrum and the iliac.

sacrum the part of the vertebral column that connects with the pelvis; in people it consists of five fused vertebrae with transverse processes fused into a solid bony mass on either side.

saggital (adj.), 1. the median plane of the body of a bilateral animal. 2. any plane parallel to the median plane.

saline 1. also *saline solution, physiological saline*, a solution of sodium, potassium and sometimes magnesium salts isotonic with physiological fluids. 2. (adj.), salty.

salinity the concentration of salt in a solution; saltiness.

saliva 1. in land vertebrates, a viscous, colorless, mucoid fluid secreted into the mouth by ducted salivary glands. 2. in invertebrates, any analogous secretion into or from the mouth.

salivary chromosome also *polytene chromosome*, any of the gigantic interphase chromosomes of the salivary glands of larval dipterans (e.g., *Drosophila*), each consisting of 1024 chromonemata aligned side by side, with thousands of prominently stainable bands corresponding with condensed and folded genes and occasional puffs corresponding to unfolded, actively transcribing genes.

salivary gland 1. in land vertebrates, any of several glands secreting saliva into the mouth; in mammals these are the paired *parotid, sublingual* and *submaxillary glands;* in snakes the salivary glands are modified into venom glands. 2. in invertebrates, any analogous gland ducting secretions into or from the mouth.

salt linkage an electrostatic bond formed between a negatively charged ion or side group and a positively charged ion or side group.

saltatory proceeding by leaps and bounds rather than by gradual transitions. *Saltatory evolution*, also *macroevolution, saltation*, evolution by sudden variation or by periods of active change with intervening periods of morphological stability. *Saltatory propagation*, the skipping movement of an impulse from one node of a myelinated neuron to another.

salting out also *saline abortion*, a method of abortion used in the second trimester, in which the fetus is killed with an injection of a concentrated salt solution into the amniotic cavity.

sap the watery fluid transported by phloem that circulates dissolved gases, sugars, other organic compounds, and mineral nutrients from one part of the plant to another.

saprobe an organism that reduces dead plant and animal matter.

saprobic see *saprobe*.

saprophage an organism that feeds on the dead. Including reducers such as bacteria and fungi, and carrion-feeders or scavengers such as crows, beetles, and vultures.

sarcolemma the membraneous sheath enclosing a muscle fibril.

sarcomere the contractile unit of striated muscle bounded by Z-line partitions; consisting of regular hexagonal arrays of actin filaments bound to the Z-line partitions and parallel myosin filaments regularly interspersed between them.

sarcoplasmic reticulum membranous, hollow tubules in the cytoplasm of a muscle fiber, similar to the *endoplasmic reticulum* of other cells.

saturated having accepted as many hydrogens as possible. *Saturated fat*, a triglyceride lacking carbon-carbon double bonds.

savanna a tropical or subtropical grassland with scattered trees or shrubs, usually maintained by such human activities as burning and by foraging for firewood.

scale a multicellular, anchoring growth on the underside of liverworts.

scale insect any of various homopteran insects in which adult females become scale-like and permanently attached to a host plant.

scanning electron microscope (SEM), a device for visualizing microscopic objects by scanning them with a moving beam of electrons, recording impulses from scattered electrons and displaying the image by means of the synchronized scan of an electron beam in a cathode ray (television) tube.

scapula in land vertebrates, either of a pair of large, flat, triangular bones that slide dorsally above the rib cage and bear opposing muscles of the upper arm and the socket articulating with the humerus; the scapula forms the bone of a chuck roast.

schistosome also *blood fluke*, 1. any of several parasitic trematode flatworms, the primary host of which is a human, the secondary host being a fresh-water snail. 2. any related trematode parasite.

schistosomiasis any of several diseases caused by various schistosomes parasitic on humans.

Schwann cell also *neura lemma cell*, one of the cells that constitute the myelin sheath, each cell being greatly flattened so as to consist almost entirely of cell membrane, and being repeatedly wrapped around an axon so as to form a sheath many layers thick.

scientific method (variously defined). 1. any method used by scientists to investigate phenomena. 2. any of several more or less formal schemes of research methodology, in which a problem is identified, relevant data are gathered, a hypothesis and its null alternative are formulated from these data, and the null hypothesis is empirically tested for possible rejection.

sclera the heavy, white connective tissue enclosing most of the eyeball; the white of the eye.

sclerenchyma a protective or supporting plant tissue composed of cells with greatly thickened, lignified, and often mineralized cell walls.

scolex the hook-bearing head of a larval or adult tapeworm, from which the proglottids are produced by strobilation.

scrotum the external pouch of skin that contains the testes in most adult male mammals.

scutellum in monocots, the cotyledon in its specialized form as a digestive and absorptive organ.

scyphistoma a highly reduced polyp stage in the life cycle of a scyphozoan (jellyfish).

sea anemone see *anemone*.

seaweed a multicellular marine alga, of the red, green or brown algal groups.

secondary cambium or **vascular cambium** meristematic cells derived from procambium as secondary growth ensues; produces secondary xylem and phloem.

secondary consumer a carnivore; see *consumer*.

secondary growth growth in dicot plants that results from the activity of secondary meristem, producing primarily an increase in diameter of stem or root; compare *primary growth*.

secondary growth pattern the arrangement of xylem, cambium, phloem, cork cambium, cork and other tissues in dicot secondary growth, in which the various tissues form concentric sheaths that appear as rings in cross-section; compare *primary growth pattern*.

secondary immune response a rapid response to a second or subsequent invasion of the body by organisms or foreign molecules, during which memory cells quickly produce large numbers of active, specialized B- and T-cell lymphocytes. See also *primary immune response*.

secondary phloem in plants, phloem originating from secondary or vascular cambium.

secondary oocyte a developing egg cell after the first polar body has been produced and before the second polar body appears, after meiosis I.

secondary sex characteristic structural and psychological changes that occur at puberty, or seasonally in seasonal breeders, usually in one sex only, and is not directly part of reproduction; in humans, enlarged breasts and hips in women, beard growth, enlarged larynx, and increased muscle development in men, pubic and axillary hair in both sexes; breeding plumage in adult male birds.

secondary spermatocyte a sperm-forming cell after the first meiotic division and before the second division.

secondary structure or **level of organization** of a macromolecule, the pattern of folding of adjacent residues, usually in the form of helices or sheets.

secondary succession succession occurring in abandoned croplands. See also *community development*.

secondary tissue see *secondary growth*.

secondary xylem in plants, xylem originating from the secondary or vascular cambium.

second law of thermodynamics the statement in physics that all processes occurring spontaneously within a closed system result in an increase of total entropy.

second messenger an intracellular chemical compound transferring a hormonal message from the cell membrane to the nucleus or cytoplasm.

secretory phase or *postovulatory phase* in women, that portion of the menstrual cycle following ovulation, during which estrogen and progesterone levels rise and the thickening endometrium takes on a glandular appearance.

seed the fertilized and ripened ovule of a seed plant, comprised of an embryo (miniature plant) including one, two or more cotyledons, and usually a supply of food in a protective seed coat; capable of germinating under proper conditions and developing into a plant.

seed coat see *coat*.

seed plant a plant of the tracheophyte subdivision *Spermophyta*, consisting of gymnosperms and angiosperms.

segmented body plan see *segmentation*.

segmentation also *metamerism* 1. the condition of being divided into segments, originally repetitions of nearly identical parts (as still seen in some annelids, chilopods, diplopods, and *Peripatus*), but frequently followed in evolution by the specialization of different segments, as seen in most arthropods, some annelids, and vertebrates. 2. the serial repetition of body parts, such as ganglia or somites.

selectionists biologists that attribute most, if not all, evolutionary change to natural selection; compare to *neutralists*

selectively permeable in cellular membranes, the characteristic of permitting selected substances to pass through while rejecting others.

self-fertilization fertilization of ovules or eggs effected by pollen or sperm from the same individual.

self-pollination the transfer of pollen from the anther to the stigma of the same flower.

self-replicating a molecular entity that, with the aid of replicating enzymes, is capable of producing identical copies of itself. Also see *replication*.

SEM see *scanning electron microscope*.

semen 1. in mammals, a viscous white fluid produced in the male reproductive tract and released by ejaculation, consisting of spermatozoa suspended in the secretions of accessory glands, principally the prostate and seminal vesicle. 2. any fluid containing sperm and released by male animals.

semicircular canals in vertebrates, a group of three loop-shaped tubular portions of the membranous labyrinth of the inner ear. Serves to sense balance, orientation and movement.

semiconservative replication replication of a DNA molecule in which the original molecule divides into two complementary parts, both halves being preserved while each half promotes the synthesis of a new complement to itself.

seminal vesicle 1. in various invertebrates, a pouch in the male reproductive tract serving for the temporary storage of sperm. 2. in male mammals, paired outpocketings of the vas deferens producing much of the fluid substance of semen.

seminiferous tubule any of the coiled, thread-like tubules that make up the bulk of a testis, and are lined with germinal epithelium from which sperm are produced.

semipermeable membrane a membrane that allows some substances to pass freely through it but which restricts or prevents the passage of other substances.

senescence 1. aging. 2. changes, especially deterioration, associated with increasing age.

sensillium in the moth antenna, pored, hairlike structures bearing the olfactory neural endings.

sensory 1. pertaining to the senses. 2. receptive of stimuli. 3. conveying nerve impulses from a sense organ to the central nervous system.

sensory cell a peripheral nerve cell that is the primary receptor of a sensory stimulus.

sensory hair an innervated arthropod hair that serves as a sense organ.

sensory neuron also *afferent neuron*, a neuron that conducts impulses carrying sensory information from a receptor to the brain or spinal cord.

sensory palp also *palpus* a segmented process on either side of the mouth of an arthropod, occurring in insects in pairs on the maxillae and on the labium, having tactile and gustatory functions.

sensory receptor a cell or tissue, specialized in responding to specific kinds of stimuli.

sepals green, leaflike parts of flowers that surround the flower bud before it opens and later forms a whorl (the *calyx*) beneath and outside the whorl of petals.

septate divided by septa

septum (pl. *septa*), any dividing wall or membrane, such as those dividing the cavities of a compound fruit; the transverse partitions dividing the chambers of a chambered nautilus, the transverse partitions dividing the coelom of an annelid, the cross-walls of the hyphae of septate fungus, the partition dividing the left and right chambers of the heart, etc.

seral stage one stage in a sere. See also *sere* and *community development*.

sere a recognizable sequence of changes, sometimes predictable, occurring during community development or succession.

serial dilution a technique for measuring the density of suspended bodies (e.g., bacteria, viruses) in a fluid by progressively, repeatedly and accurately diluting the fluid until the density is small enough for accurate counts to be made.

serosa 1. see *chorion*. 2. also *serous membrane*, any thin, clear membrane consisting of a single layer of flattened, moist cells covering a connective tissue base, lining a body cavity or covering an internal organ.

serotonin a vasoconstricting hormone released from damaged tissues.

Sertoli cell also *nurse cell* one of the elongated cells of the tubules of the testis to which spermatids become attached while they mature.

seta 1. in the moss sporophyte, the stalklike growth that supports the sporangium. 2. the bristlelike, chitinous structure in the body wall of annelids, arthropods, and certain other invertebrates.

sex chromosome an X or Y chromosome, or any chromosome involved in the determination of sex.

sexduction the transfer of bacterial host genes from an F^+ bacterium to an F^- bacterium following its fortuitous incorporation into the genome of an F episome.

sex-influenced trait a congenital, gene-influenced or chronic abnormality that can occur in either sex but is more common in one than the other; e.g., breast cancer, ulcers, pyloric stenosis.

sex-limited trait a variable trait that affects members of one sex only; e.g., prostate cancer, pattern baldness, endometriosis.

sex pilus in some prokaryotes, an enlarged pilus (tube) through which a DNA replica can presumedly pass from one cell to another.

sexual intercourse also *coitus, copulation*, the entry of the erect penis into the vagina.

sexual recombination genetic recombination as the result of meiosis and biparental reproduction.

sexual reproduction reproduction involving the union of gametes; biparental reproduction.

shelf fungus also *bracket fungus*, a basidiomycete that forms a shelf-like basidiocarp on the sides of infected tree trunks.

shell 1. see *electron shell*. 2. a hard, rigid, usually calcaneous covering of an animal. 3. the calcaneous covering of a bird egg. 4. by extension, the leathery covering of a reptile or monotreme egg. 5. the hard, rigid outer covering of a fruit or seed, e.g., walnut shell, sunflower seed shell.

shoot apical meristem see under *apical meristem*.

short-day plant also *long night plant*, a plant that begins flowering after the summer solstice and in which flowering is triggered by periods of dark longer than some innately determined minimum.

short-term memory see under *memory*.

sickle-cell an abnormally crescent-shaped erythrocyte. *Sickle-cell anemia*, a severe recessive condition attributable to homozygosity for an allele producing sickel-cell hemoglobin. *Sickle-cell hemoglobin*, symbol Hb^S, adult hemoglobin in which the beta chain has valine rather than glutamic acid in the sixth position, resulting in the formation of sickle-shaped crystals. *Sickle-cell trait*, the heterozygous condition for the allele specifying Hb^S which entails an increased risk of mild anemia but also an increased resistance to malaria.

sieve plate 1. partition or cell wall between sieve tube members, bearing multiple perforations allowing for the continuity of sieve tube cytoplasm. 2. also *madreporite*, the perforated calcareous disk through which the echinoderm water vascular system communicates with the exterior water.

sieve tube a thin-walled tube consisting of an end-to-end series of enucleate, living cells joined by sieve plates, forming the channels through which sap flows in phloem.

sieve-tube element also *sieve-tube member*, a thin-walled phloem cell having no nucleus at maturity and forming a continuous cytoplasm with other such cells to form *sieve tubes*, functioning in the transport of organic solutes, hormones, and mineral nutrients.

sigmoid S-shaped; *sigmoid colon*, an S-shaped part of the large intestine; *sigmoid curve*, an S-shaped graphed function.

sigmoid colon see *colon*.

signal peptidase in translation by bound ribosomes, an enzyme that cleaves the leader sequence or signal polypeptide from a polypeptide entering the endoplasmic reticulum. See also *signal peptide*.

signal peptide in translation by bound ribosomes, the *leader sequence* of amino acids in the polypeptide being synthesized. Before entering the endoplasmic reticulum, the signal peptide must first interact with a signal recognition protein and then with a *receptor site* on the rough endoplasmic reticulum.

signal recognition protein see *signal peptide*.

silent mutation a change in a DNA codon in which a synonymous codon forms (third letter change), having no effect on the amino acid sequence of the polypeptide.

siliceous containing silica; *siliceous sponge*, a sponge having a skeleton of silica spicules.

silicon one of the most abundant elements in the earth's crust, a major component of glass, found in certain animal spicules, diatom shells, nettle spines, the outer cell walls of grasses and elsewhere.

simple eye also *ocellus*, in invertebrates, light receptors resembling single ommatidial units within the compound eye, specialized for detecting changes in light intensity.

Singer model see *fluid-mosaic model*.

sinoatrial node also *SA node*, *pacemaker*, a small mass of conducting tissue embedded in the musculature of the right atrium of a land vertebrate, representing the vestige of the sinus venosus of ancestral forms, serving as a source of regular contraction impulses to the heart, and transmitting the impulse to the auricles, the atrioventricular node, the bundle of His, and thus to the ventricles.

sinus a cavity that forms part of an animal body, e.g., any of the several air-filled, mucous-membrane lined cavities of the skull; a hemocoel.

sinus venosus in fish and in the embryos of land vertebrates, a distinct chamber of the heart formed by the union of the large systemic veins, opening into the single auricle.

sinusoid a minute, endothelium-lined space or passage for blood in the tissues of an organ, such as the liver or spleen.

siphonous alga coenocytic green algae, an evolutionary line of green algae characterized by a multinucleate cytoplasm, without the usual division by cell walls or membranes, e.g., *Codium*.

site a specific place; a part of a large molecule; e.g., the *active site* of an enzyme.

skeletal muscle also *voluntary muscle*, *striated muscle*, muscle attached to the skeleton or in some cases to the dermis, under direct control of the central nervous system, characterized by distinct striations at the subcellular level, with multinucleate unbranched muscle fibers; compare *smooth muscle*, *cardiac muscle*.

skin-breather any organism in which a significant proportion of the exchange of respiratory gasses occurs through a vascularized moist skin; e.g., most amphibia.

Skinner box a device for investigating operant conditioning, after B. F. Skinner.

skull the skeleton of the head of a vertebrate, being a cartilaginous braincase in cyclostomes, sharks and the embryos of all forms; in most vertebrates the cartilage is replaced by bone and the structure made more complete by its union with other bones developed in dermal membranes; consisting of the cranium, the bony capsules of the nose, ear, and eye, and the jaws and teeth.

slime mold a funguslike protist possibly related to fungi, consisting of an ameboid, saprophytic *plasmodium* that on maturity coalesces into structured fruiting bodies; two subgroups, the *acellular slime molds* and the *cellular slime molds*, though sharing many features may also be only distantly related to one another.

small intestine in vertebrates, the region of the alimentary canal and the cecum; the region in which most food absorption occurs.

smooth endoplasmic reticulum see *endoplasmic reticulum*.

smooth muscle also *involuntary muscle*, the muscle tissue of the glands, viscera, iris, piloerection and other involuntary functions, consisting of masses of uninucleate, unstriated, spindle-shaped cells occurring usually in thin sheets.

smut 1. a fungal disease of plants characterized by black, powdery masses of spores. 2. the fungus that causes the disease.

sodium ion gate either of the two gates controlling sodium ion passage through an ion channel, including an *activation gate* and *inactivation gate*.

sodium/potassium ion exchange pump also *sodium pump*, a poorly understood molecular entity in the plasma membrane, capable of actively transporting sodium out of the cell and potassium in, at a cost of ATP energy.

soft palate in the oval cavity, the membranous, non-bony posterior extension of the hard palate.

solute something dissolved in solution.

somatic cells cells of the body; any cells other than germinal cells, especially of terminally differentiated tissues.

somatic nervous system the voluntary nervous system, as distinguished from the visceral (autonomic) nervous system.

somatic variation theory the proposal that variable regions of immunoglobins are coded by a limited number of genes that can be rearranged in a vast number of ways during clonal selection. Compare to *germ line theory*.

somatostatin a hormonal secretion of the hypothalamus and delta cells in the *islets of Langerhans*.

somite 1. in the early vertebrate embryo, one of a longitudinal series of paired blocks of tissue that are the forerunners of body muscles and the axial skeleton. 2. one of the segments of any segmented animal.

sorus (pl. *sori*), one of the clusters of sporangia on the underside of a fern frond.

sound spectrograph a device for visually displaying the elements of a complex sound; *sound spectrogram*, one such display.

source in nutrient transport in plants, any region from which highly concentrated sugars are being transported.

space-filling model a solid model of a molecule, in which the outer orbitals of atoms are represented proportionally by large intersecting spheres.

speciation 1. the division of a species into two or more species. 2. the process or processes whereby new species are formed.

species (p., *species*) (variously defined), 1. a class of things or individuals having some common characteristics; a distinct type. 2. the major subdivision of a genus or subgenus, regarded as the basic category of biological classification and composed of related individuals that resemble one another through recent common ancestry and that share a single ecological niche. 3. in sexual organisms, a group whose members are at least potentially able to breed with one another but are unable to breed with members of any other group; a reproductively isolated group.

specific heat the amount of heat required to raise the temperature of a substance a specified number of units. With water, the standard reference at standard conditions, the amount of heat required to raise the temperature of one cubic centimeter by 1°C.

specific name the second part of a formal scientific name, indicating species within a genus, always italicized and never capitalized.

spectrum see *absorption spectrum*, *action spectrum*, *electromagnetic spectrum*, *sound spectrograph*.

sperm 1. a male gamete. 2. a spermatozoan, 3. a spermatozoid. 4. (adj.), pertaining to a male gamete or male gamete function, as in *sperm head*, *sperm nucleus*, *sperm cell*.

sperm nucleus 1. the nucleus of a sperm. 2. either of the two nuclei that arise from the generative nucleus of a pollen tube and function in the double fertilization characteristic of seed plants.

spermatheca a sac connected with the female reproductive tract of most insects, many other invertebrates, and some vertebrates, that receives and stores spermatozoa after insemination and retains them, often for a long time, until egg maturation and fertilization.

spermatic cord in mammals, the structure that suspends the testis in the scrotum, consisting of a tough fibrous coat containing an extension of the coelom, the vas deferens, the nerves and blood vessels supplying the testis, and a retractory muscle, the last being vestigial in humans.

spermatid one of the four haploid cells formed by meiosis of a spermatocyte, prior to maturation into a spermatozoan.

spermatocyte a cell giving rise to spermatozoa; see also *primary spermatocyte*, *secondary spermatocyte*.

spermatogonium (pl. *spermatogonia*), a diploid testicular cell from which spermatocytes and ultimately sperm are produced; a germ line cell of an adult male.

spermatophore a packet containing sperm, produced by spiders and certain other invertebrates, and by certain salamanders.

spermatozoan (pl. *spermatozoa*) also *spermatozoon*, *sperm*, *sperm cell* a motile male gamete of an animal, produced in great numbers in the male gonad, consisting of a sperm head containing the nucleus and an acrosome, a midpiece and a single flagellum (sperm tail).

sperm receptacle also *spermatheca* in invertebrates, a pouch or chamber for receiving and storing sperm.

S phase (synthesis phase), one of the phases of the cell cycle, occurring in interphase between G_1 and G_2; the period of chromosome *replication* and DNA synthesis.

sphenophyte any member of the plant division sphenophyta, the horse tails.

sphincter a ring of muscle surrounding a bodily opening or channel and able to close it off; e.g., *oral sphincter*, *anal sphincter*.

spicule 1. one of the many small to minute calcareous or siliceous pointed bodies embedded in and serving to stiffen and support the tissues of various invertebrates, including sponges, radiolarians, and sea cucumbers. 2. any small, stiff, spike-like or needle-like body part.

spike an action potential, in a traveling nerve impulse, the period of most extreme depolarization.

spinal column see *vertebral column*.

spinal cord see *dorsal hollow nerve cord*.

spinal nerve any of the many nerves that enter and leave the spinal cord, including both somatic and autonomic. Compare *cranial nerve*.

spindle a cytoplasmic body present in mitosis and meiosis, in shape resembling the spindle of a primitive loom, and composed of tubulin microtubules and actin microfilaments; serving to segregate daughter chromosomes in anaphase.

spindle apparatus a functional unit consisting of the continuous and centromeric fibers of the spindle, the centromeres, the centrioles (or the analogous *polar ground substance*) and, when present, the asters.

spindle fiber one of the microtubule filaments constituting the mitotic spindle. *Centromeric spindle fiber*, one of the several microtubules attaching firmly to a centromere and extending to a spindle pole. *Fiber*, a microtubule extending from one pole, interlacing with continuous spindle fibers of the other pole, and ending blindly somewhat short of the other spindle pole.

spinneret 1. any organ for producing threads of silk from the secretion of a silk gland. 2. one of the six nipplelike, jointed processes near the end of a spider's abdomen, each bearing numerous minute orifices of silk gland ducts.

spiracle 1. the blowhole of a cetacean. 2. one of the external openings of an insect tracheal system. 3. one of the pair of openings on the upper back part of the head of an elasmobranch or sturgeon, communicating with the pharynx, derived from gill clefts but serving as incurrent respiratory openings in rays.

spiral valve a helical fold of the intestinal wall in the short intestine of a shark, which serves to slow the passage of food and provides additional absorptive surface.

spleen a discrete organ of the reticuloendothelial system, consisting of lymphoid, reticular and endothelial tissues with blood supply and red blood cells circulating freely in intercellular spaces; all enclosed in a fibroelastic, muscularized capsule; functions include the scavenging of debris and the maintenance of blood volume.

spontaneous mutation a chemical change in DNA not induced by external agents; compare *induced mutation, mutation, mutagen*.

SPONCH an mnemonic acronym of the six most common elements of living matter, in the order of decreasing atomic mass: Sulfur, Phosphorus, Oxygen, Nitrogen, Carbon, and Hydrogen.

spongin a tough, insoluble protein that makes up the skeleton of a class of sponges including the once commercially important bath sponge.

spongocoel the body cavity of a simple sponge (not a true coelom).

spongy bone bone with a network of thin, hard walls surrounding open, non-bony pockets.

spongy parenchyma see *leaf mesophyll*.

sporangiophore in some protists and most fungi, an erect growth upon which pore-forming bodies develop.

sporangium a case, cell, or organ in which asexual spores are borne, in algae, fungi, bryophytes and ferns.

spore a minute unicellular reproductive or resistant body, specialized for dispersal, for surviving unfavorable environmental conditions, and for germinating to produce a new vegetative individual when conditions improve.

sporocyst see *human liver fluke*.

sporophyll a modified leaf capable of producing spores. See also *megasporophyll, microsporophyll*.

sporophyte in algae and plants having an alternation of generations, a diploid individual capable of producing haploid spores by meiosis; the prominent form of ferns and seed plants; compare *gametophyte*.

sporozoite a small, motile and elongate infective stage of a sporozoan, a product of a sexual fusion that initiates a new asexual cycle when transmitted to a new host.

sprouting factor a local hormone produced by tissues that induces nerve axon sprouting; compare *antisprouting factor*.

squamous epithelium stratified epithelium that consists, at least in its outer layers, of small, flattened, scale-like cells, generated from an underlying germinal layer.

stabilizing selection natural selection that tends to maintain the status quo, selection for the average and against extremes.

stamen the male reproductive structure of the flower, consisting of a pollen-bearing anther and the filament on which it is borne.

staminate having stamens but no pistils; an exclusively male flower.

standard conditions or *standard temperature and pressure*, in chemistry: 0.0° C or 273° K and 1.0 atmosphere of pressure. Conditions to which all gas volumes measured under the environmental conditions are corrected, permitting one measurement to be accurately compared to another.

stapes see *middle ear*.

starfish see *sea star*.

starch a polysaccharide of glucose subunits in alpha glycoside linkages, produced by cyanobacteria and by plastids in protists and plants. See *amylose, amylopectin*; compare *glycogen, cellulose*.

statistics 1. the science that deals with the classification, analysis and interpretation of numerical data, and that uses mathematical probability theory to seek order from such data. 2. the use of mathematical probability theory in hypothesis testing, specifically in determining whether or not observed correlations or departures from expectation can reasonably be attributed to chance.

statocyst 1. a cell containing one or more statoliths. 2. a sense organ of equilibrium, orientation and movement consisting of a fluid-filled chamber containing a statolith; widely distributed among invertebrate animals.

statolith 1. the calcareous body within a statocyst which, by its greater specific gravity, tends to move downward or against the direction of acceleration. 2. a similar body in the inner ear of a fish or amphibian. 3. a small stone taken into the statocyst of certain invertebrates, such as the lobster. 4. dense crystals or vacuolar inclusions that are believed to function in gravity-sensing in plant cells.

steady state a controlled condition of physiological stability in an open system, differing from an *equilibrium* in that it is a function of input and output as well as internal interactions.

stele the cylindrical central portion of the axis of a vascular plant, including pith, xylem, and phloem, surrounded by a pericycle.

stem cell also *hemocytoblast*, generalized cell of red bone marrow from which all blood cells form.

stem reptile also *cotylosaur*, a Permian reptile of the group believed to be ancestral to all reptiles.

stereoscopic vision the ability to perceive depth through the integration of two overlapping fields of vision.

sterile jacket cell nonreproductive cell surrounding moss antheridium and archegonium.

sternum a median ventral bone or cartilage in land vertebrates, connecting with the ribs, shoulder girdle or both.

steroid any of a class of lipid-soluble compounds on four interlocking saturated hydrocarbon rings; included are all *sterols* (alcoholic steroids), e.g., cholesterol, estradiol, testosterone, cortisol.

steroid hormone a class of hormones consisting of the steroid molecule with various side group substitutions, believed to freely pass across the cell membrane and, once bound by a specific carrier protein, interact directly with the chromatin in gene control; included are the vertebrate sex hormones.

stigma the top, slightly enlarged and often sticky end of the style, on which pollen adhere and germinate.

stimulus an aspect of the environment that influences the activity of a living organism or part of an organism, especially through a sense organ.

stimulus generalization the tendency to accept a variety of stimuli that have traits in common with a previously rewarded stimulus.

stipe the stemlike structure in red or brown algae that supports the blades or blade.

stolon 1. in plants, a horizontal branch that produces new plants from buds; also *runner*. 2. a long horizontal fungal hypha that spreads across the surface and periodically sends down a mycelium into the substrate and a group of conidia into the air.

stoma (pl. *stomata*), structure in the epidermis of leaves, stems, and other plant organs, made up of the guard cells and pore, allowing the diffusion of gases in and out of intercellular spaces.

stomach a muscular dilation of the alimentary canal in vertebrates, between the esophagus and the duodenum, that functions in temporary storage, preliminary digestion, sterilization, and physical breakdown of ingested food.

stone canal a calcified tube of the echinoderm water vascular system, leading from an external sieve plate (madreporite) to the ring canal.

stop codon also *termination codon*, in MRNA, one or more of the codons UAA, UAG, or UGA, signalling the end of polypeptide translation. See also *start codon*.

storage disease a pathological condition characterized by the accumulation within cells of partially metabolized substances that accumulate because of the absence of a critical enzyme.

stratified squamous epithelium multilayered, flattened epithelial cells as seen in the epidermis of the skin, esophagus, and vagina.

strep throat a bacterial disease caused by *Streptococcus pyogenes*, characterized by an inflamed throat, which may progress to scarlet fever, kidney inflammation, and rheumatic fever.

stretch receptor a proprioceptive sensory receptor that is stimulated by stretching, as in a tendon, muscle, or bladder wall.

striated muscle see *skeletal muscle*.

strobilation asexual reproduction by transverse division of the body into segments which break free as independent organisms; occurring in certain coelenterates and flatworms (tapeworms).

strobilus (pl. *strobili*), a conelike aggregation of sporophylls in a club moss or horsetail.

stroke volume the amount of blood passing through the heart per heartbeat.

stroma 1. the matrix or supporting connective tissue of an organ. 2. the cytoskeleton. 3. the proteinaceous matrix surrounding the thylakoids of a chloroplast.

stromatolite a macroscopic geological structure of layered domes of deposited material attributed to the presence of shallow water photosynthetic prokaryotes.

strong nuclear force one of the basic forces in physics within a nucleus.

structural formula a two-dimensional representation of the topological relationships of atoms and bonds in a molecule; sometimes including information on spatial relationships as well.

structural protein protein that is incorporated into cellular or extracellular structures.

style the stalk of the pistil in a flower, connecting the stigma with the ovary.

stylized fighting see *ritual combat*.

suberin a complex fatty substance of cork cell walls and other waterproofed cell walls.

sublingual under the tongue; see *salivary gland*.

submaxillary or **submaxillary gland** see *salivary gland*.

submucosa the layer of aveolar connective tissue beneath a mucous membrane.

subspecies a more or less clearly defined, morphologically distinct, named geographic variety of a species; when named, the name forms a third part of the scientific name (*genus, species, subspecies*).

substitution 1. see *base substitution*. 2. *amino acid substitution*, a base substitution that changes the identity of one of the amino acids of a protein.

substrate 1. the base on which an organism lives. 2. a substance acted upon by an enzyme. 3. a nutrient source or medium.

substrate-level phosphorylation the production in one or more steps of ATP from ADP, P_i, and an appropriate organic substrate; the capture of high-energy phosphate bonds directly from metabolic transformations; compare *chemiosmostic phosphorylation*.

succession see *community development*.

sucrose also *table sugar*, a sweet, 12-carbon disaccharide consisting of glucose and fructose subunits.

sugar diabetes see *diabetes mellitus*.

sulfhydryl group the side group R-SH.

sulfur bacterium 1. *purple sulfur bacterium*, a photosynthetic bacterium utilizing H_2S as a source of hydrogen for photosynthesis; 2. *nonpurple sulfur bacterium*, a nonphotosynthetic chemotroph utilizing sulfur compounds as a source of energy.

sun compass an innate mechanism utilizing the angle of the sun and the time of day to compute direction for navigation.

supercoiling bending of a helical linear structure into larger orders of helices, as occurs in the mitotic condensation of chromosomes.

superior vena cava the anterior vena cava in humans; see *vena cava*.

supplemented medium a growth medium providing metabolites not normally required by the organism, and thus able to sustain the growth of certain nutritional mutants that would not be recovered on minimal medium.

suppressor lymphocyte a lymphocyte specialized in preventing the immune system from reacting against self.

surface tension a condition that exists on the free surface of water or other liquid by reason of intermolecular forces unsymmetrically disposed around individual surface molecules, tending to make the surface layer behave in some respects as an elastic membrane.

surface-volume hypothesis the proposal that cells are restricted to a size that assures a surface-volume ratio that provides a sufficient membrane area to support the transport needed to maintain metabolic activity.

survivorship curve a graph with age on the X axis and zero to one on the Y axis, the monotonic curve presenting the probability of surviving to age X for each X.

suspensor in a developing seed, a group or chain of large cells, swollen with water, produced by the zygote and serving to anchor the embryo in place and to push it into contact with the endosperm.

suture in anatomy, immovable joints formed by the articulation of skull bones.

swarm cell also *swarm spore*, any of various minute, motile, sexual or asexual spores produced in large numbers by certain algae, fungi and protists.

swim bladder also *air bladder*, a gas-filled sac giving controlled bouyancy to most bony fish; homologous with the lungs of land vertebrates and lungfish and probably derived from the primitive lung of an ancestral air-breathing fish.

symbiont a symbiotic organism.

symbiosis 1. the living together in intimate association of two species. 2. *parasitic symbiosis*, also *parasitism*, symbiosis in which one organism gains fitness at some expense to the other. *Commensalism*, symbiosis in which one organism gains at no cost to the other. *Mutualistic symbiosis*, also *mutualism*, symbiosis in which both individuals gain fitness.

symbiosis hypothesis see *endosymbiosis hypothesis*.

sympathetic division see *autonomic nervous system*.

sympathetic nerve in the autonomic nervous system, any nerve of the sympathetic division.

sympatric speciation speciation occurring within a continuous population.

synapse 1. (n.), the place at which nerve impulses pass from the axon or the synaptic knobs of an axon of one neuron to the dendrite or cell body of another. 2. (v.i.), of homologous chromosomes, to pair together, chromomere by chromomere, in zygotene of meiosis.

synapsis the pairing up and fusing of homologous chromosomes during zygotene of the first meiotic prophase, whereby preparations are made for crossing over.

synaptic cleft the minute space between synaptic knob of one neuron and the dendrite or cell body of another, into which neurotransmitters are released in the transmission of nerve impulses between cells.

synaptic knob one of the multiple bulbous swellings at ends of the branching terminus of an axon, containing neurotransmitters in secretion granules and forming one side of the synaptic cleft.

synaptinemal complex a complex structure composed of protein and RNA, the material of which is first formed between sister strands in leptotene of meiotic prophase and combines with a like structure to form the complex and accomplish the specific zipperlike pairing of homologous chromosomes in zygotene.

syncytiotrophoblast in the newly implanted mammalian embryo, fingerlike growths of the trophoblast that form connections with the endometrium.

syncytium 1. a multinucleate mass of protoplasm resulting from the fusion of cells, as in the plasmodium of a slime mold. 2. also *coenocyte*, a multinucleate mass created by repeated nuclear mitosis without cytoplasmic cell division.

synonymous codon in the genetic code, two or more codons that specify the same amino acid.

synovial capsule a completely enclosed capsule containing *synovial fluid*, formed by the smooth cartilages covering the articular surface of the bones of a synovial joint and a surrounding *capsular ligament*.

synovial fluid also *synovia*, a transparent, viscid lubricating fluid secreted by the synovial membranes of joints, bursae, and tendon sheaths.

synovial joint a freely movable joint surrounded by a fibrous capsule lined by a synovial membrane that secretes lubricating synovial fluid.

synovial membrane see *synovial joint.*

synthesist in science, one who attempts to deduce broad general principles from widely disparate observations, or who draws upon such observations to substantiate or refute such principles; one who reinterprets established relationships in the support of a new general idea or *synthesis.*

syphilis a sexually transmitted bacterial disease caused by the spirochete *Treponema pallidum;* symptoms in the primary infection include local chancre on the genitals; in the secondary stages, body rash, mouth lesions, and runny nose; in the tertiary stages, which may occur years later, the widespread infection may damage the blood vessels and eyes and lead to insanity and paralysis.

systole the period of heart contraction particularly of the ventricles; compare *diastole.*

systolic pressure the highest arterial blood pressure of the cardiac cycle.

systematics 1. the science of classifying organisms or groups of organisms on the basis of their evolutionary relatedness. 2. the science of determining the evolutionary relatedness of groups of organisms, particularly at higher taxonomic levels.

systematist a biologist who tries to determine the evolutionary relationships of organisms.

TACT in plants, an acronym representing the water transporting forces, *t*ranspiration, *a*dhesion, *c*ohesion and *t*ension.

tactile receptor, a sensory receptor responsive to light touch; compare *pressure receptor.*

taiga moist, subarctic forest biome of Europe and North America dominated by spruces and firs.

tailpiece the portion of a bacteriophage between the head and the end plate.

talus also *astragalus, anklebone,* the principal bone of the human ankle.

tap root also *taproot,* 1. a root having a prominent central portion giving off smaller lateral roots in succession. 2. the central root of such a system, especially when it grows vertically and deep.

tap root system a root system consisting of a large primary root and its secondary and lateral branches.

tapeworm a parasitic flatworm of the class *Cestoda.*

target cell a cell acted upon by a specific hormone, generally containing or bearing specific hormone receptor proteins not found in other cells.

tarsal bone one of the smaller bones of the ankle, between the talus and the metatarsals.

tarsus 1. the tarsal bones taken together. 2. the vertebrate ankle. 3. the distal portion of an arthropod leg.

taste bud a sensory receptor sensitive to taste, found chiefly in the epithelium of the tongue, being a conical mass of cells, made up of supporting cells and *taste cells,* which terminate in modified sensory cilia that project into a porelike space in the overlying epithelium.

tautology a statement that excludes no logical possibilities.

taxon 1. any taxonomic category, such as *species, genus, subspecies, phylum* etc. 2. any named group of related organisms.

taxonomist an individual skilled in identifying, describing and classifying organisms, usually specializing in a particular group.

taxonomy the science dealing with the identification, naming, and classification of organisms.

Tay-Sachs disease a disease due to homozygosity for a rare recessive autosomal allele, characterized by inclusion bodies in neurons, juvenile idiocy and early death. The recessive allele and the disease unusually common in persons of Eastern European Jewish descent.

T-cell a lymphocyte of a variety that matures in the thymus, produces immunoglobulins bound tightly to its exterior surface, and interacts with invading cells, and with other cells of the immune system.

tectorial membrane a membrane of the cochlea, overlying and contacting the hair cells of the organ of Corti.

teichoic acid in the cell walls of Gram positive bacteria, an acidic polymer of alternating glycerol and phosphate groups with sidechains such as *N*-acetyl glucosamine.

telolecithal of an egg, having a large yolk, with active cytoplasm confined to a small disk on the upper surface, as in reptiles, birds, and monotremes.

telophase the stage of mitosis or meiosis in which new nuclear membranes form around each group of daughter chromosomes, the nucleoli appear and the chromosomes decondense; at which time the plasma membrane and cytoplasm of the cell usually divide to form two daughter cells.

TEM *transmission electron microscope;* see *electron microscope.*

temperate deciduous forest a forest biome of the temperate zone, usually in areas of significant winter snowfall, in which the dominant tree species and most other trees are deciduous and are bare in winter months.

temperate rain forest a cool woodland of the temperate zone in an area of heavy rainfall but little or no snow, usually including many different kinds of trees as well as a single dominant tree species.

temperate zone the region of the earth between the Tropic of Cancer and the Arctic Circle together with the region between the Tropic of Capricorn and the Antarctic Circle.

temporal bone a compound bone of the side of the mammalian skull.

temporal lobe a large lobe on the lateral portion of each cerebral hemisphere.

tendon a tough dense cord of fibrous connective tissue that is attached at one end to a muscle and at the other to the skeleton, and transmits the force exerted by the muscle in moving the skeleton.

tendril a slender, fast-growing tip of a climbing plant that grows in spirals around an external supporting tree, bush, or trellis.

tensile strength the ability to be pulled without breaking.

tentacle an elongate, flexible, fleshy, sometimes branched process borne by animals chiefly in groups or pairs on the head or surrounding the mouth, and serving as a tactile and/or prehensile organ.

tenting the expansion of the internal vagina (cul-de-sac) and elevation of the uterus on approaching orgasm.

terminal bud the dormant bud at the stem tip, representing the next season's potential growth.

terminal differentiation cell differentiation in which the cell remains in G_1 and does not normally undergo subsequent cell division.

termination in transcription, a point when the ribosome reaches a stop codon, whereupon the polypeptide is released and the ribosomal subunits separate.

termination signal a DNA sequence that signals the end of a transcription sequence, dislodging RNA polymerase.

terminator codon (also *termination codon, stop codon, nonsense codon*), one of the codons UAA, UAG, or UGA in mRNA, or the corresponding codons in DNA, signalling the end of polypeptide transcription.

territorial behavior also *territoriality,* a species-specific pattern of behavior associated with the defense of a territory, in most territorial species by the male, and often including territorial marking, singing or other displays, threats, attacks on trespassing conspecifics, and patrolling by the defending animal, as well as avoidance and submissive behavior by others when in or around a defended territory.

territory the area defended by a territorial animal.

tertiary consumer a consumer at the third level; see *consumer.*

tertiary structure or **level of organization** of a protein, the pattern of folding of a polypeptide upon itself, which is generally quite specific for each protein type.

testcross 1. a cross of an individual of unknown genotype with a homozygous recessive to determine zygosity by progeny testing. 2. the cross of a known double heterozygote with a double homozygote to determine linkage relationships.

testis (pl. *testes*), the male gonad, in which spermatozoa are produced by meiosis.

testosterone a steroid hormone, the principal androgen of vertebrates.

tetanus a bacterial disease characterized by sustained, painful, muscle contraction caused by the toxins of *Clostridium tetani,* which interfere with clearing of the neurotransmitter acetylcholinesterase from neuromuscular junctions.

tetrad 1. a group of four. 2. a group of four equal cells, such as microspores or spermatids, produced by meiosis. 3. also *bivalent,* the synapsed homologues in pachytene through diakinesis, considered as a group of four synapsed chromatids.

tetrahedron a perfect geometric solid with four equal sides, four equal edges, and four corners.

tetramer a molecule composed of four subunits, e.g., hemoglobin, a protein which is composed of two alpha chains and two beta chains.

tetrapod also *land vertebrate*, any vertebrate of the classes Amphibia, Reptilia, Mammalia, and Aves, including some that don't have four feet (e.g., birds, people, whales, dugongs).

tetraploid having four complete genomes in each nucleus; *autotetraploid*, having four genomes (i.e., twice the diploid number) of one normally diploid species; *allotetraploid*, also *amphidiploid*, having two genomes (i.e., the usual diploid number) of each of two different, normally diploid species.

T-even phage one general type of *E. coli* bacteriophage, with a double-stranded DNA genome and a complex body consisting of a head, tailpiece, end plate and attachment fibers; arbitrarily designated T_2, T_4, T_6, etc.

thalamus a large subdivision of the diencephalon, consisting of a mass of nuclei in each lateral wall of the centrally-located third ventricle of the brain.

thalassemia also *Cooley's anemia*, a severe congenital anemia due to homozygosity for a variant allele of the beta hemoglobin locus, producing little or no adult hemoglobin. *alpha thalassemia*, an analogous partial deficiency of the hemoglobin alpha chain.

thallose a leafy growth form seen in some liverworts.

thallus a plant body of a multicellular alga, that does not grow from an apical meristem, shows no differentiation into distinct tissues, lacks stems, leaves, or roots; may be simple or branched, consisting of filaments, thin plates or solid bodies of cells, and may include fairly complex blades, bladders and holdfasts.

theory a proposed explanation whose status is still conjectural, in contrast to well-established propositions that are often regarded as facts. *Theory* and *hypothesis* are terms often used colloquially to mean an untested idea or opinion. A *theory* properly is a more or less verified explanation accounting for a body of known facts or phenomena, whereas a *hypothesis* is a conjecture put forth as a possible explanation of a specific phenomenon or relationship that serves as a basis for argument or experimentation.

therapsid also *mammallike reptile*, any Permian or Triassic reptile of the extinct order *Therapsida*, with mammallike upright locomotion rather than crawling or bipedal locomotion.

thermal energy 1. energy in the form of heat. 2. the heat equivalent of some other form of energy.

thermal efficiency also *thermodynamic efficiency*, in a chemical reaction or series of reactions, the proportion of the original potential chemical bond energy of the substrate still retained in the potential bond energy of the product and *not* dissipated as thermal energy or entropy.

thermoacidophile an organism, often an archaebacterium, that thrives in a hot, acidic environment.

thermodynamics 1. the branch of physics that deals with the interconversions of energy as heat, potential energy, kinetic energy, radiant energy, entropy and work. 2. the processes and phenomena of energy interconversions.

thermoreceptor a sensory receptor that responds to temperature or to changes in temperature.

thermoregulation 1. an animal's control over its internal temperature. 2. the physiological mechanisms that maintain a body at a particular temperature in an environment with a fluctuating temperature. *Behavioral thermoregulation*, maintaining a body temperature within acceptable limits by behavioral means, as by basking and by seeking out appropriate microenvironments.

thigmotropism in plants, a sudden reaction to touch, associated with rapid ion transport and sudden changes in turgor.

thoracic (adj.), pertaining to the thorax.

thoracic region the portion of vertebral column that articulates with ribs, between the cervical and lumbar regions.

thorax 1. in mammals, the part of the body anterior to the diaphragm and posterior to the neck, containing the lungs and heart. 2. the middle of the three parts of an insect body, bearing the legs and wings.

threshold of stimulation in neurons or muscles, the lowest level of stimulation that will initiate a response.

thrombin in the blood clotting reactions, a proteolytic enzyme that catalyzes the conversion of fibrinogen to fibrin by the removal of two short peptide segments, and in turn is produced from *prothrombin* by the action of *thromboplastin*.

thromboplastin a complex phosphoprotein, released from damaged platelets, that catalyzes the conversion of the inactive serum component *protrombin* into the active enzyme *thrombin* in the initation of blood clotting.

thromboxane a prostaglandin.

thylakoid a membranous structure of chloroplasts and cyanophytes consisting of a *thylakoid membrane* containing chlorophyll, accessory pigments, photocenters, and the photosynthetic electron transport chain; an inner lumen that becomes acidic during active photosynthesis; and CF_1 particles attached to the outer sides of the thylakoid membrane, which are the sites of chemiosmotic phosphorylation.

thymidine a nucleoside consisting of the pyrimidine *thymine* and a deoxyribose residue.

thymine a pyrimidine, one of the four nitrogenous bases of DNA and the only nitrogenous base specific for DNA.

thymosin see *thymus*.

thymus a glandular body above the lungs, believed to stimulate T-cell lymphocyte development through the secretion of thymosin.

thyroid also *thyroid gland*, a large endocrine gland in the lower neck region of all vertebrates, arising as a ventral median outpocketing of the pharynx and believed to be homologous with the endostyle of cephalochordates, lamprey larvae, and urochordates; secretes *throxine* and *calcitonin*.

thyroid-stimulating hormone (TSH) also *thyrotropin*, a peptide hormone of the anterior pituitary, the action of which is to stimulate the release of thyroxin from the thyroid.

thyrotropic hormone releasing hormone (THRH) a small peptide produced in the hypothalamus and carried by the pituitary portal vein to the anterior lobe of the pituitary, where it stimulates the release of TSH.

thyroxin also *thyroxine*, also *thyroid hormone*, a specialized amino acid with four iodine residues, $C_{15}H_{11}I_4NO_4$, a hormone with an effect of increasing general body metabolism and the stimulation of growth and also serves as a stimulus to metamorphosis in amphibia.

tibia 1. the larger of the two bones of the vertebrate lower hind leg, between the femur and the talus. 2. an analogous segment in an insect leg.

tide pool a pool of sea water left behind by the receding tide and covered at high tide.

tissue a group of associated cells, identical in structure and function.

toadstool a mushroom especially if poisonous, distasteful, or suspicious.

tobacco mosaic virus a common infective agent of tobacco and the first virus to be seen by microscopy.

tolerance the ability of the body to resist immunological reactions against self. Also the body's ability to withstand (fail to respond to) an agent or stimulus.

total fertility rate the predicted general fertility rate, determined through attitudinal studies and other forecasting indicators.

torsion an event of gastropod development in which the upper part of the body is rotated 180 degrees, bringing the anus to a position above the head.

totipotent of a cell, able to undergo development and differentiation along any of the lines inherently possible for the species; undifferentiated.

touch receptor see *sensory receptor*.

toxin a poisonous substance produced by a biological organism.

trace element any element essential to life, but only in minute or "trace" quantities. See also *micronutrient*.

tracer see *radioactive tracer*.

trachea (pl. *tracheae*), 1. in land vertebrates, the main trunk of the air passage between the lungs and the larynx, usually stiffened with rings of cartilage. 2. one of the air-conveying chitinous tubules comprising the respiratory system of an insect.

tracheal gill in insect nymphs, a feathery apparatus used in exchanging gases with the surrounding water.

tracheal system respiratory system of insects, composed of thin-walled air conducting tubules opening to *spiracles* and extending to finer, branched *tracheoles*, terminating in *air sacs*.

tracheid a long tubular xylem cell or its lignified, empty cell wall, which functions in support and the conduction of water, distinguished from *xylem vessels* by having tapered, closed ends communicating with other tracheids through pits.

tracheole see *tracheal system*

tracheophyte a vascular plant.

tradewind a persistent, directional, global wind created by the rotational displacement of great rising air cells.

transduction the transfer of a host DNA fragment from one bacterium to another by a viral particle.

transcribed strand the DNA strand of a structural gene that is physically involved in transcription.

transcription the process of RNA synthesis as the RNA nucleotide sequence is directed by specific base pairing with the nucleotide sequence of the transcribed strand of a DNA cistron.

transducer an apparatus that converts nonelectrical forms of energy into electricity. In sensory receptors, the conversion of a physical stimulus such as pressure or heat into a neural impulse.

transfer RNA (tRNA), any of a class of relatively short ribonucleotides with a common secondary and tertiary structure consisting of three or four loops and a double-stranded stem, with numerous modified bases in addition to the four normally found in RNA; specific varieties become covalently linked to specific amino acids by specific enzymes, then function in transcription to recognize appropriate mRNA codons and to transfer the amino acid to the growing polypeptide chain.

transformation in a bacterium, the direct incorporation of a DNA fragment from its medium into its own chromosome.

transition zone a region of overlap between two ecological communities in which plants and animals of both communities are found.

translation polypeptide synthesis as it is directed by an mRNA on a ribosome; the transfer of linear information from a nucleotide sequence to an amino acid sequence according to the strictures of the genetic code.

translocation 1. the step in protein synthesis in which a transfer RNA molecule that is covalently attached to the growing polypeptide chain is moved (translocated) from one ribosomal tRNA attachment site to the other. 2. a chromosome rearrangement in which a terminal segment of one chromosome replaces a terminal segment of another, nonhomologous chromosome. *Reciprocal translocation,* a chromosome rearrangement in which terminal segments of two nonhomologous chromosomes are exchanged. 3. the directed movement of materials from one part of an organism to another part, especially the directed movement of solutes through phloem.

translocation Down's syndrome also *trisomy 14/21,* a rare inherited form of Down's syndrome where the chromosome number 14 in cells of the carrier are fused to a number 21 chromosome; see also *trisomy 21.*

transmembranal protein plasma membrane protein that extends entirely through the membrane, often forming a pore.

transmission electron microscope (TEM), see *electron microscope.*

transpiration 1. the evaporation of water vapor from leaves, especially through the stomata. 2. the physical effects of such evaporation taken together. *Transpiration pull,* the pulling of water up through the xylem of a plant utilizing the energy of evaporation and the tensile strength of water.

transverse colon see *colon.*

transverse tubule also *T tubule,* a specialized system of tubules in a muscle fiber, transmitting the contractile impulse from the sarcolemma to the sarcomeres.

trichocyst in some ciliates, minute harpoonlike bodies below the pellicle that can be extruded.

tricuspid valve the valve between the right atrium and the right ventricle consisting of three triangular membranous flaps.

triglyceride, a compound of glycerol with each of its three hydroxyl side groups bound in an ester linkage to a fatty acid or other residue.

triiodothyronine a hormonally active substance similar to thyroxin but lacking one of the four iodines of thyroxin, usually made synthetically.

trimester one of three-month periods of the nine months of human gestation, which thus is divided into first, second, and third trimesters.

triploblastic in animals, the existence in the embryo of three primary germ layers, the ectoderm, mesoderm and endoderm, from which all other tissues are formed.

trisomy the condition in which an otherwise diploid individual has a third homologue of one chromosome.

trisomy 21 also *Down's syndrome,* originally *mongolian idiocy,* a severe human congenital pathology attributable to the presence of three rather than two homologues of chromosome 21 attributable to either nondisjunction or translocation.

trisomy X syndrome also *XXX female,* a mild congenital human abnormality attributable to the presence of three rather than two X chromosomes, characterized by moderately increased stature, moderately decreased mental functioning, and frequently by amenorrhea and sterility.

tRNA see *transfer RNA.*

tRNA charging enzyme see *charging enzyme.*

trochophore also *trochophore larva,* a rather specific body form of a free-swimming ciliated larva of marine invertebrates of several phyla, including annelids, mollusks and rotifers.

trophic relating to nutrition. *Trophic level,* a level in a food pyramid.

trophic level see *trophic.*

trophoblast. see *blastocyst.*

tropical rain forest a tropical woodland biome that has an annual rainfall of at least 250 cm and often much more, typically restricted to lowland areas, characterized by a mix of many species of tall broad-leaved evergreen trees forming a continuous canopy, with vines and woody epiphytes, and a dark, nearly bare forest floor.

tropism a turning toward or away from a stimulus, usually accomplished by differential growth; see *geotropism, phototropism.*

tropomyosin a low-molecular-weight filamentous protein that accompanies the globular protein actin in making up actin microfilaments.

troponin a protein of low molecular weight that binds calcium in muscle contraction, a specific variant of the ubiquitous calcium-binding molecule *calmodulin.*

true bacteria see *eubacteria.*

true-breeding (adj.), of an organism or strain, when mated with individuals like itself producing offspring like itself; specifically, homozygous for all relevant loci.

trypanosome a parasitic, flagellated protozoan of the genus *Trypanosoma,* infecting mammals and transmitted by insect vectors, responsible for *chagas disease,* and *sleeping sickness.*

trypsin a powerful proteolytic enzyme secreted by the pancreas in the inactive form *trypsinogen* and activated in the intestine.

tryptophan one of the twenty amino acids of polypeptide synthesis.

tryptophan operon (a repressible operon) a region of DNA in *E. coli* that includes structural genes coding for enzymes that synthesize tryptophan, and a region that controls their transcription; see also *operon.*

tsetse fly also *tsetse,* a biting fly of Africa that is the carrier of the trypanosome responsible for sleeping sickness.

T tubule see transverse tubule.

tubal ligation sterilization of a female mammal by cutting and tying the oviducts.

tube cell a cell in the microgametophyte of a seed plant, responsible for pollen tube growth.

tube foot one of the numerous specialized, hollow, extensible, flexible organs of locomotion and grasping in echinoderms formed as extensions of the water vascular system and each usually bearing a terminal sucker.

tube nucleus in seed plants, one of the two haploid nuclei of a pollen grain, which controls the subsequent growth and enzyme production of the pollen tube upon germination.

tubular secretion the active transport of certain wastes from the peritubular capillaries into the nephron.

tubule any slender, elongated channel in an anatomical structure; *kidney tubule,* the tubular part of a nephron, being all of the nephron except Bowman's capsule; see also *proximal convoluted tubule, distal convoluted tubule.*

tubulin the protein that comprises the spherical subunit of microtubules.

tundra a biome, characterized by level or gently undulating treeless plains of the arctic and subarctic, supporting a dense growth of mosses and lichens and dwarf herbs and shrubs, seasonally covered with snow and underlain with permafrost.

tunic in urochordates the outer portion of the sac-like sea squirt body.

turgid stiffened and distended by hydrostatic pressure from within.

turgor the normal state of turgidity and tension in living plant cells, created by osmosis. *Turgor pressure,* the actual hydrostatic pressure developed by the fluid in a turgid plant cell.

Turner's syndrome also *XO syndrome*, a congenital human abnormality due to the presence of a single X chromosome and no other sex chromosome; resulting in immature female development, short stature, sterility, certain anatomical peculiarities and normal mental functioning except for a marked deficiency in certain narrowly delimited spatial tasks.

tympanal organ see *tympanum*.

tympanic membrane also *tympanum, eardrum,* a thin, clear, tense double membrane of connective tissue covered with living epithelium, dividing the middle ear from the external auditory canal, vibrating sympathetically with received sound and transmitting the impulses to the inner ear by way of the ossicles of the middle ear; deep within a bony recess in mammals, on the surface in frogs and toads, and intermediate in birds and reptiles.

tympanum 1. see *tympanic membrane.* 2. an analogous thin tense membrane of an organ of hearing in an insect. 3. the membrane of a sound-producing organ that acts as a resonator.

Type I herpes simplex a relatively benign herpes simplex virus that causes cold sores or fever blisters around the mouth.

Type II herpes simplex also *genital herpes,* a highly contagious, sexually transmitted virus, causing intermittent outbreaks of painful, infective blisters on the genitals, followed by a period of remission.

typhlosole 1. in the earthworm, an infolding in the roof of the intestine that greatly increases the digestive and absorptive surface. 2. a similar fold projecting into the gut of certain pelecypod mollusks and starfish.

ulna one of the two bones of the forearm or the corresponding part of the forelimb of land vertebrates.

ultracentrifuge a device capable of creating great rotational force, used to separate fluid-suspended cellular and subcellular materials according to density.

ultraviolet light also *ultraviolet radiation, u.v.,* electromagnetic radiation having a shorter wavelength than visible light and longer than X-rays, including wavelengths normally invisible to humans but visible to bees and hummingbirds; interacts destructively with DNA and thus is mutagenic, carcinogenic, and destructive of skin tissues.

umbilical cord in placental mammals, a vascular cord connecting the fetus with the placenta, containing two *umbilical arteries,* an *umbilical vein* and the blind remnant of the yolk sac, in a connective tissue matrix.

ungulate a mammal that has hoofs, especially of the orders *Perissodactyla* and *Artiodactyla.*

unicellular also *acellular,* 1. consisting of a single cell. 2. having a single cell nucleus.

unsaturated capable of accepting hydrogens; *unsaturated fat,* a triglyceride with one or more carbon-carbon double bonds.

unstable in the isotopes of elements, the tendency to undergo radioactive decay. Also see *radioactive.*

upper epidermis the upper (light exposed) layer of cutinized epidermal cells in a leaf.

upwelling the coming to the surface of water from the depths of the ocean, which is associated with the introduction of mineral nutrients and a consequent richness in productivity and biomass.

uracil a pyramidine, one of the four nitrogenous bases of RNA and the only one specific to RNA.

urea a highly soluble nitrogenous compound, the principal nitrogenous waste of the urine of mammals; retained in high concentrations in the blood of elasmobranchs and the coelocanth as a means of osmoregulation.

ureter either of the pair of ducts that carry urine from the kidney to the bladder in mammals, or from the kidney the cloaca in other vertebrates; compare *urethra.*

urethra the canal that carries urine from the mammalian bladder to the exterior, and in males serves also for the transmission of semen.

urethral meatus in humans, the external opening of the female urethra.

uric acid a relatively insoluble purine, a principal nitrogenous excretion product of reptiles, birds and insects; excreted in small quantities as a product of nucleic acid breakdown in mammals; sometimes crystalizing in the tissues of men causing a painful inflammation *(gout).*

urinary bladder see *bladder.*

urogenital folds a pair of early prominences in the undifferentiated stage of the development of the mammalian external genitalia; form the shaft of the penis in males and the *labia minora* in females.

urogenital groove in the undifferentiated stage of the embryonic development of the mammalian genitalia, a longitudinal depression between the urogenital folds; becomes the *penile urethra* in males and the *vestibule* between the labia minora in females.

uterine mucosa see *endometrium.*

uterus 1. in female mammals, a muscular, vascularized, mucous-membrane-lined organ for containing and nourishing the developing young prior to birth and for expelling them at the time of birth, consisting of enlargements of the paired oviducts or of a single median enlargement of the fused oviducts. 2. an enlarged section of the oviduct of various vertebrates and invertebrates modified to serve as a place of development of the young or of eggs.

utricle the chamber of the membranous labyrinth of the middle ear into which the semicircular canals open.

uv light *uv radiation,* see *ultraviolet light.*

vacuole a general term for any fluid-filled membrane-bounded body within the cytoplasm of a cell, in higher plants occupying most at the volume of most differentiated living cells, in protozoa performing numerous functions; see *contractile vacuole, digestive vacuole.*

vacuum behavior instinctive actions that are released spontaneously, without the stimulus of a releaser, presumably due to the build-up of action-specific energy that finally demands to be discharged.

vagina 1. an expandable canal that leads from the cervix of the uterus of a female mammal to the external orifice *(introitus),* serving to receive the penis in copulation and as a *birth canal* in parturition. 2. any canal of similar function in other animals.

vagus nerve either of the tenth pair of cranial nerves, with sensory and autonomic motor fibers innervating the heart and viscera.

van der Waals forces also *nonpolar attraction,* relatively weak attractive forces acting between nonpolar atoms and molecules, serving to bind together lipid-soluble portions of molecules and important in the stability of membranes, the specific folding of proteins, and the attachment of proteins to or within membranes.

variability *genetic variability,* the general qualitative term for the presence of genetic differences between individuals in a population.

variable or **experimental variable** the focus of an experiment, to be tested and compared with a control.

variable age of onset of a genetic pathology, not being observed at birth but becoming manifest at some variable time in later development.

variable expressivity of a genetic pathology, having different degrees of abnormality in different individuals with the same or similar genotype.

variable region the portion of an immunoglobulin polypeptide concerned with the binding of an antigen, which varies from antigen to antigen.

variety a particular breed or genetic type of a domesticated plant or animal. any group of organisms below the species level having some genetic characteristic or group of characteristics in common.

vascular (adj.), pertaining to vessels or tissues of conduction, as phloem and xylem in plants or the blood vessels in animals.

vascular bundle a unit of the vascular system of plants, a strand of xylem associated with a strand of phloem and usually a sheath of fibrous sclerenchyma, in the primary growth pattern and in leaves and petioles.

vascular cambium the cylinder of lateral meristem that produces xylem on its inner side and phloem on its outer side in secondary growth, contributing thus to growth in circumference.

vascularized having many blood vessels, especially capillaries.

vascular plant a plant with xylem and phloem; a tracheophyte.

vascular ray spokelike layers of parenchyma tissue that extend through the wood, vascular cambium, and phloem.

vascular system 1. the circulatory system of an animal. 2. the xylem and phloem of a vascular plant.

vascular tissue plant tissues specialized for conducting water and foods; see also *xylem* and *phloem*.

vas deferens (pl. *vasa deferentia)*, the male genital duct or spermatic duct.

vasectomy sterilization of the male by severance and ligature of the vasa deferentia (see vas deferens).

vasoconstriction contraction of sphincters in arterioles resulting in reduced blood flow.

vasodilation relaxation of arteriolar sphincters, allowing increased blood flow.

vector an organism that transmits a parasite, virus, bacterium or other pathogen from one host to another.

vegetal pole the lower, more yolk-filled end of a zygote or early cleavage blastula, determining the ventral side in development.

vegetative reproduction asexual reproduction in plants, in which offspring are produced from portions of the roots, stems, or leaves of the parent.

vein 1. a vessel returning blood toward the heart at relatively lower velocity and pressures; such blood is usually relatively deoxygenated, but in the case of the *pulmonary veins* it is freshly oxygenated. 2. a vascular bundle in a leaf or petiole.

veldt eastern and southern African grassland, usually level, often with scattered trees or shrubs; where the lions live (see savanna).

vena cava (pl. *venae cavae)*, any one of three large veins by which blood is returned to the right atrium of land vertebrates.

vent in vertebrates the external opening of the cloaca.

ventilation breathing.

ventral (adj.), in bilateral animals, toward the belly; downward, opposite the back; compare *dorsal.* In human anatomy, in view of the peculiar upright stance of the organism, *ventral* is frequently replaced by *anterior* (toward the front).

ventral aorta see *aorta.*

ventral nerve cord a common feature of many invertebrate phyla, the main longitudinal nerve cord of the body, being solid, paired, and ventral, with a series of ganglionic masses.

ventricle a cavity of a body part or organ: 1. one of the large muscular chambers of the four-chambered heart. 2. one of the systems of communicating cavities of the brain, consisting of two *lateral ventricles* and a median *third ventricle.*

ventriculus see *gizzard.*

venule a small vein

vertebra (pl. *vertebrae)*, one of the bony or cartilaginous elements that together make up the vertebrate spinal column, in land vertebrates being a complex bone consisting of a cylindrical centrum articulating with adjacent centra by a fibrous cartilaginous *intervertebral disk*, a dorsal *neural arch* providing a passage for the spinal cord, and various small spines, articular processes, and muscle attachments.

vertebral canal the pathway of the spinal cord through the vertebral column.

vertebral column also *spinal column*, the articulated series of vertebrae connected by ligaments and separated by intervertebral disks that in vertebrates forms the supporting axis of the body, and of the tail in most forms.

vesicle a small cavity, fluid-filled sac, blister, cyst, vacuole, etc.

vessel 1. a tube or canal in which a body fluid is contained or circulated within the interior, e.g., *blood vessel, lymph vessel.* 2. a conducting tube in an angiosperm formed in the xylem by the end-to-end fusion of a series of *vessel elements.* Compare *tracheid.*

vessel element see *vessel.*

vestibular apparatus that portion of the inner ear involved in sensing position and movement of the head, containing the semicircular canals, saccule, and utricle.

vestibular gland also *gland of Bartholin*, a pair of oval, ducted glands lying on either side of the lower part of the vagina, secreting a lubricating mucus especially on appropriate occasions.

villus (pl. *villi)*, a small, slender, vascular, fingerlike process: 1. one of the minute processes that cover and give a velvety appearance to the inner surface of the small intestine, that contain blood vessels and a lacteal, are covered with microvilli and serve in the absorption of nutrients. 2. one of the branching processes of the surface of the implanted blastodermic vesicle of most mammals.

viron an individual viral particle.

virulence 1. disease-producing capability. 2. the capacity of an infective organism to overcome host defenses.

virus a noncellular organism transmitted as DNA or RNA enclosed in a membrane or protein coat, often together with one or a few enzymes; replicating only within a host cell, utilizing host ribosomes and enzymes of synthesis.

visceral muscle also *smooth muscle* or *involuntary muscle*, muscle of the gut, uterus, and blood vessels.

visible light electromagnetic wavelengths longer than about 400 nm and shorter than about 750 nm, which can serve as visual stimuli to most photoreceptive organisms.

visual cliff 1. a dropoff placed beneath a plate of glass. 2. a flat pattern creating the optical illusion of a steep dropoff.

visual receptor a sense receptor stimulated by light or by patterns of light.

vital capacity the maximum volume of air that can be forcefully exhaled.

vitalism a doctrine in biology that ascribes the functions of a living organism to a *vital principle* or *vital force* distinct from chemical and physical forces; the doctrine is no longer taken seriously by practicing biologists.

vitamin any organic substance that is essential to the nutrition of an organism, usually by supplying part of a coenzyme.

vitreous humor see *eyeball.*

viviparity the condition of producing live young.

viviparous (adj.), producing live young from within the body of the female, following development of the young within a uterus in intimate association with the tissues of the mother; compare *ovoviviparous, oviparous.*

vocal cord in many vertebrates, one of the paired, mucous membranes stretched across the glottis, capable of producing varied sounds when vibrated by air.

voice box see *larynx.*

voluntary muscle see *skeletal muscle.*

vulva the external genitalia of a woman.

water mold an aquatic protist of the oomycetes.

water potential the potential energy of water to move, as a result of concentration, gravity, pressure, or solute content.

water vacuole see *contractile vacuole.*

water vascular system a system of vessels in echinoderms, containing sea water, used in the movement of tentacles and tube feet and possibly in excretion and respiration.

Watson strand either of the two strands of DNA, the other being designated the *Crick strand.*

wave length the physical distance from one point of maximum intensity to the next in a propagated wave, especially of electromagnetic radiation in a vacuum; inversely proportional to *frequency*, the number of waves passing a point per second, and also inversely proportional to the energy per photon.

wax 1. a dense, hard, lipid-soluble ester of a long-chain alcohol and a fatty acid. 2. beeswax, one such compound. 3. a natural mixture of such monohydroxy esters and other lipid substances.

Weberian ossicles a set of tiny bones that connects the swim bladder to the inner ear in some fish.

wheat rust 1. any of three distinct fungal diseases of wheat. 2. a rust fungus responsible for such a disease.

Wernicke's area a region of the cerebrum concerned with deciphering speech.

whorl 1. a group of parts repeated in a circle. 2. any of the four basic radially repeated groups of flower parts; see *gynoecium, androecium, corolla, calyx.* 3. a fingerprint in which the ridges form concentric circles.

wild type 1. of a phenotype, normal or typical in appearance; not mutant, rare or abnormal. 2. of an allele, normal and common in the population; not mutant; giving rise to a typical phenotype when homozygous.

wilt (v.i.), to lose turgor, especially because of an inadequate supply of water to a plant or plant part.

wood the hard, fibrous xylem of secondary growth, especially that of the central stem of a tree, consisting of lignified cellulose cell walls.

woody containing wood or wood fibers; consisting largely of secondary growth with hard lignified tissues.

work in physics, the transfer of energy resulting in the movement of an object formerly at rest.

X chromosome 1. one of the two heteromorphic sex chromosomes of mammals, flies and certain other insects, such that the possession of two X chromosomes without a Y chromosome results in the development of a normal female, while the possession of one X and one Y chromosome results in a normal male and other combinations are rare and abnormal. 2. the sex chromosome of grasshoppers and certain other insects such that the possession of two X chromosomes results in female development while the possession of only one results in male development. 3. see *Z chromosome*.

X-linked also *sex-linked*, the condition, common to many genes of functions unrelated to sex, of being present on the X chromosome; thus males have only one copy rather than the normal diploid two and recessive X-linked alleles are always expressed in males, thus radically affecting patterns of inheritance.

xanthophyll any of the bright yellow carotenoid pigments of flowers, leaves, or fruit.

xeric (adj.), of an environment, low in moisture available for plant life; dry.

xeroderma pigmentosum a rare autosomal recessive condition of humans in which the DNA repair system specific for u.v.-induced damage is defunct, resulting in a pathological sensitivity to sunlight that usually leads to multiple skin cancers.

xerophyte a plant adapted to dry areas.

xiphoid process see *sternum*.

X-organ an endocrine gland, found in the eyestalk of crayfish, that produces *molt-inhibiting hormone*; compare *Y-organ*.

X-ray a highly energetic form of electromagnetic radiation with wavelengths of .01 nm to 10 nm, between gamma radiation and u.v. radiation.

X-ray crystallography a procedure of directing X rays at a crystal of a protein or other large molecule, recording the pattern produced as the rays are bent by the regular, repeating molecular structures within the crystal, and using this pattern to reconstruct the 3-dimensional structure of the molecule.

XXY syndrome see *Klinefelter's Syndrome*.

xylem one of the two complex tissues in the vascular system of vascular plants, consisting of the dead cell walls of vessels, tracheids or both, often together with sclerenchyma and parenchyma cells; functioning chiefly in the conduction of water and in support, but sometimes also in food storage; the substance of wood in secondary growth as well as a part of the vascular bundles of primary growth; see *tracheid, vessel, lignin;* compare *phloem*.

xylem vessel water-conducting tissue in plants, consisting of vessel elements arranged end to end.

XYY syndrome also called extra-Y syndrome, an abnormal genetic condition of males due to nondisjunction, characterized by markedly excessive height, a tendency to severe acne and mild mental retardation that in itself tends to get affected individuals into trouble of various kinds.

Y chromosome 1. one of the two heteromorphic sex chromosomes of mammals, flies, and certain other insects such that XY individuals are male and XX individuals are female; in mammals carrying the primary male sex determinant for the H-Y antigen, a determinant for increased height, and few other genes; in general nearly devoid of genes not concerned with maleness; compare *X chromosome*.

yeast a unicellular sac fungus, especially *Saccharomyces cerevisiae*, which is used in the production of bread, beer, wine, and raised doughnuts.

yellow marrow yellow, fatty material within the central cavity of long bones.

yolk the material stored in an ovum that supplies food material to the developing embryo, consisting chiefly of *vitellin* (a lecithin-containing phosphoprotein), nucleoproteins, other proteins, lecithin (a phospholipid) and cholesterol.

yolk gland also *vitellarium*, 1. a modified part of the ovary of flatworms and rotifers that produces yolk-filled cells that nourish the true eggs. 2. the organ of an insect in which the egg cells grow to mature size.

yolk mass the mass of yolk in an egg cytoplasm; see *yolk*.

yolk plug a pluglike mass of yolk-filled endoderm cells left protruding from the blastopore of an amphibian embryo after the cresentic blastopore enlarges to form a complete circle.

yolk sac 1. one of the extraembryonic membranes of a bird, reptile or mammal, that in birds and reptiles grows over and encloses the yolk mass and in placental mammals encloses a fluid-filled space; the first site of blood cell and circulatory system formation. 2. an analogous structure in elasmobranchs and cephalochordates.

Y-organ one of the endocrine glands of a crayfish eyestalk, secreting *crustecdysone*, the crustacean molt-inducing hormone.

Z chromosome one of the two heteromorphic sex chromosomes of birds and butterflies, such that ZZ individuals are male and ZW individuals are female; sometimes considered to be an X chromosome; compare *W chromosome, X chromosome, Y chromosome*.

zeatin a plant cytokinin hormone and growth factor extracted from corn.

zero population growth (ZPG) a point in population dynamics where the birth rate equals the death rate.

Z line in striated muscle, the partition between adjacent contractile units to which actin filaments are anchored.

zona pellucida a thick, transparent, elastic membrane or envelope secreted around an ovum by follicle cells.

zooplankton animal life drifting at or near the surface of the open sea.

zoospore 1. any independently motile spore. 2. the flagellated gamete of a foraminiferan.

zooxanthela intracellular, photosynthetic, symbiotic dinoflagellates common in radiolarians, corals, marine flatworms and various other invertebrates.

Z protein an enzyme in the thylakoid that is involved in a complex series of reactions resulting in the oxidation of water, releasing molecular oxygen and hydrogen ions into the thylakoid lumen and passing electrons to oxidized chlorophyll *a*.

Z scheme a graphic presentation of the oxidation-reduction reactions occurring in the light reaction and electron transport chain in photosynthesis.

Zwitterion an ion with both positively charged and negatively charged molecular side groups; e.g., a free amino acid, $+NH_3$—HCR—COO−.

zygomatic arch the arch of bone that extends along the front and side of the skull below the orbit, passing outside of the jaw muscle; in humans also called the *cheekbone*.

zygospore a diploid fungal or algal spore formed by the union of two similar sexual cells, having a thickened and usually ornamented wall, that serves as a resistant resting spore.

zygote a cell formed by the union of two gametes; a fertilized egg.

zygotene the stage in meiotic prophase in which homologous chromosomes synapse; see *synapse, synaptinemal complex*.

ACKNOWLEDGMENTS

Unless otherwise acknowledged, all photos are the property of Scott, Foresman and Company.
Abbreviations: P. R. Photo Researchers BPS Biological Photo Supply (T) top (C) center (B) bottom (L) left (R) right (U) upper

Contents
p. xi (TL) Dr. Frank Carpenter
p. xi (TR) Macnab, R., and L. Ornston. 1977. *J. Mol. Biol.* 112:1–30. © 1977 by Academic Press, Inc.
p. xi (B) Dr. Thomas Eisner
p. xii (T) Robert Frerck/Odyssey Productions
p. xii (C) Courtesy Tsuyoshi Kakefuda, National Institutes of Health
p. xii (BL) Miller, O. L., Jr., and B. R. Beatty. 1969. *Science* 164:955–957
p. xii (BR) Dr. G. Schatten, SPL/Science Source, P.R.
p. xiv (TL) Courtesy Marshall A. Lichtman, M.D., U. of Rochester School of medicine
p. xiv (TR) Courtesy Carmen Raventos-Suarez and Ronald L. Nagel, M.D.
p. xiv Dan McCoy/Rainbow
p. xiv (B) Wolfgang Kaehler
p. xv (T) Wayne Lankinen
p. xv (B) Larry West
p. xvi (TL) M. Dworkin and H. Reichenback/Phototake
p. xvi (TR) Dr. Rudolf Rottger
p. xvi (C) Dwight R. Kuhn
p. xvi (B) N. Allin and G. L. Barron. Courtesy N.S.E.R.C., Canada. *Can. J. Bot.* 57:187–193
p. xix (T) James Moore/Anthro-Photo
p. xix (B) Omikron/Science Source, P.R.
p. xxi (T) Ed Reschke
p. xxi (U) William Roston
p. xxi (C) Dwight R. Kuhn
p. xxi (B) Tegner, M. J., and D. Epel. 1973. *Science* 179:685–688, figs. 2c, f, and g. © 1973 by AAAS
p. xxii (T) Kyoko Archibald/International Crane Foundation
p. xxii (U) Gordon Wiltsie/Bruce Coleman, Inc.
p. xxii (C) Mrs. Lorrimer Armstrong
p. xxi (B) David C. Fritts
xviii Dudley Foster/WHOI

Chapter 1
1.1 © Down House and The Royal College of Surgeons of England
1.2 National Maritime Museum, London
1.3 Des and Jan Bartlett/Bruce Coleman Inc.
Es. 1.1 p.6(T), 7(TR) George H. Harrison
Es. 1.1 p.6(CL,CR), 7(TL,B) © Elliott Varner Smith
Es. 1.1 p.6(BL,BR) David M. Stone
Es. 1.1 p.7(C) E. R. Degginger (4)
1.6(C) © 1985 Judith J. McClung/Click, Chicago Ltd.
1.6(B) © 1985 Karl Knize/Click, Chicago Ltd.
1.8 E. R. Degginger
1.10(L) © 1985 Brian Seed/Click, Chicago Ltd.
1.10(R) Lynn M. Stone/Bruce Coleman Inc.
Es. 1.3 p.18(T) E. R. Degginger
Es. 1.3 p.18(C,BL), 19 (BR) Dwight R. Kuhn
Es. 1.3 p.18(BR) Rajesh Bedi/Life Magazine © 1979 Time Inc.

Es. 1.3 p.19(TL) Baron Hugo van Lawick, © National Geographic Society
Es. 1.3 p.19(TR) Stephen J. Krasemann/DRK Photo
Es. 1.3 p.19(C) Dr. Frank Carpenter
Es. 1.3 p.19(BL) Manfred Kage/Peter Arnold, Inc.

Chapter 2
2.1 The University of Chicago
2.19(T) Ronald F. Thomas/Bruce Coleman Inc.
2.19(B) Martin W. Grosnick/Bruce Coleman Inc.
2.20 Dwight R. Kuhn (2)
2.25 Lynn M. Stone

Chapter 3
3.4 Eric V. Gravé
3.8 Brown, R. M., and J. H. M. Willison, 1977. In *International Cell Biology 1976–1977*, ed., B. R. Brinkley and K. R. Porter, pp. 267–283. © 1977 by The Rockefeller Univ. Press
3.9 Jane Burton/Bruce Coleman Inc.
3.10 U.S. Dept. of Agriculture (2)
3.12 National Institutes of Health (2)

Chapter 4
4.3 Ed Reschke (4)
4.6 Courtesy Dr. Eva Frei and Prof. R. D. Preston
4.8 Courtesy Manfred Schliwa
Es. 4.2 p.86 Virus Lab, U. of California
Es. 4.2 p.87 Dr. Daniel Branton
4.9a,c Dr. William E. Barstow
4.9b Dr. Daniel Branton
4.10 Dr. Keith R. Porter
4.11 B. F. King, U. of California School of Medicine/BPS
4.12 Dr. John Taylor
4.13 Dr. William E. Barstow
4.14 Richard Chao
4.15 Frederick, S. E., and E. H. Newcomb. 1969. *J. Cell Biol.* 43:343
4.16 G. F. Leedale/Science Source, P.R.
4.17a Hugh Spencer
4.17b Courtesy Dr. William E. Barstow
4.18 Dr. Keith R. Porter
4.19a McGill et al. 1976. *J. Ultrastruct. Res.* 57:43–53. Micrograph courtesy Dr. B. R. Brinkley
4.19b Dr. William E. Barstow
4.20 Macnab, R., and L. Ornston. 1977. *J. Mol. Biol.* 112:1–30. © 1977 by Academic Press, Inc.
4.21a Bouck, G. B. 1971. *J. Cell Biol.* 50:362–384. Reprinted by copyright permission of The Rockefeller Univ. Press
4.21b Dr. William E. Barstow
4.22(L) L. E. Roth, U. of Tennessee/BPS
4.22(R) P. R. Burton, U. of Kansas/BPS
4.23 Courtesy Dr. R. Wyckoff, National Institutes of Health

Chapter 5
5.1 Dr. J. David Robertson
5.3 Giuseppina Raviola, courtesy Don Fawcett/Science Source, P.R.
5.4 Dr. Henry C. Aldrich
5.5a,b Courtesy M. H. Moscona and A. A. Moscona
5.7 David R. Frazier
5.13 H. E. Buhse, Jr., and R. C. Holsen, U. of Illinois at Chicago
5.15a Dr. Thomas Eisner (2)
5.17c,d E. H. Newcomb, U. of Wisconsin—Madison/BPS

Chapter 6
6.1 Charles G. Summers, Jr.
6.2 Lynn M. Stone
6.3 UPI/The Bettmann Archive
Es. 6.1 Wide World

Chapter 7
7.1 Robert Frerck/Odyssey Productions
Es. 7.1 Dr. Kenneth R. Miller
7.14a,c Wayne A. Bladholm
7.14b Alan Pitcairn/Grant Heilman Photography
7.16b S. E. Frederick, courtesy of E. H. Newcomb, U. of Wisconsin—Madison

Chapter 8
8.6 Alain Nogues/Sygma
Es. 8.1 Candee Productions

Chapter 9
9.8 Courtesy Tsuyoshi Kakefuda, National Institutes of Health
9.11 Omikron/Science Source, P.R.
9.12 Lee D. Simon/Science Source, P.R.
9.14 X-ray diffraction of DNA by Rosalind Franklin. In Watson, J. D. 1968. *The Double Helix*. New York: Atheneum, p. 168. © 1968 by J. D. Watson
9.15 Watson and Crick in front of the DNA model. Photographer: A. C. Barrington Brown. In Watson, J. D. 1968. *The Double Helix*. New York: Atheneum, p. 215. © 1968 by J. D. Watson

Chapter 10
10.4a Dr. Joseph Gall
10.5 Miller, O. L., Jr., and B. R. Beatty. 1969. *Science* 164:955–957
10.15 Miller, O. L., Jr., B. A. Hamkalo, and C. A. Thomas, Jr. 1970. *Science* 169:392–395
10.17b Olins, D. E., and A. L. Olins. 1978. *Am. Scientist* 66:704–711. Copyright by Sigma Xi, The Scientific Research Society of North America, Inc.

Chapter 11
11.2 Whaley, W. G., H. H. Mollenhauer, and J. H. Leech, 1960. *Am. J. Botany* 47:425
11.4a Ed Reschke (7)
11.4b Ed Reschke (7)
11.6 Du Praw, E. J., and G. F. Bahr. 1969. *Acta Cytol.* 13:188
11.7 J. Pickett-Heaps/Science Source, P.R.
11.8 Dr. G. Schatten, SPL/Science Source, P.R.
11.9 Kessel, R. G., and C. Y. Shih. 1974. *Scanning Electron Microscopy in Biology: A Student's Atlas on Biological Organization*. New York, Heidelberg, Berlin: Springer-Verlag © 1974 (6)
11.10 Giménez-Martín, G, C. de la Torre, and J. F. López-Sàez. In *Mechanisms and Control of Cell Division*, ed. T. L. Rost and E. M. Gifford, Jr., pp. 267–283 (4)
11.12a–e J. Kezer. In Novitski, E. 1982. *Human Genetics*. 2d ed. New York: Macmillan Publishing Co.
11.14 J. Kezer. In Novitski, E. 1982. *Human Genetics*. 2d ed. New York: Macmillan Publishing Co.
11.15 J. Kezer. In Novitski, E. 1982. *Human Genetics*. 2d ed. New York: Macmillan Publishing Co. (2)
11.16 Dr. James L. Walters
11.17a,b,e J. Kezer. In Novitski, E. 1982. *Human Genetics*. 2d ed. New York: Macmillan Publishing Co.
11.17c,d Dr. James L. Walters
1.19 L. B. Shettles/Science Source, P.R.

Chapter 12
12.2 Wellcome Institute Library, London
12.11 Bernadette Cullen/Animals Animals
Es. 12.1(L) Courtesy Marshall A. Lichtman, M.D., U. of Rochester School of Medicine
Es. 12.1(R) Courtesy Carmen Raventos-Suarez and Ronald L. Nagel, M.D.
12.15 Eric V. Gravé

Chapter 13
13.5a,b Courtesy Dr. Murray L. Barr
13.6 Courtesy Macmillan Science Co., Inc.
Es. 13.2 Courtesy the New York Public Library, Astor, Lenox, and Tilden Foundations

13.8 Walter Chandoha
13.9(T) Keystone Press Agency
13.9(B) Wide World
13.17a Edström, J., and W. Beermann. 1962. *J. Cell Biol.* 14:374. Reprinted by copyright permission of The Rockefeller Univ. Press

Chapter 14
14.1 Dan McCoy/Rainbow
14.7 Drs. N. Yamamoto and T. F. Anderson
14.10 Charles C. Brinton, Jr., and Judith Carnahan
Es. 14.1(L) Douglas Kirkland/Sygma
Es. 14.1(R) Dan McCoy/Rainbow

Chapter 15
15.3 The Bettmann Archive
Es. 15.1 Walter Chandoha

Chapter 16
16.4 Dr. H. B. D. Kettlewell (2)
16.6 Wolfgang Kaehler
16.9 Blakeslee, A. F. 1914. *J. of Heredity* 5:512
16.13 Jeff Foott/Bruce Coleman Inc.
16.16 Michael Tennesen

Chapter 17
17.1(TL) J. Van Wormer/Bruce Coleman Inc.
17.1(TR) Bob and Clara Calhoun/Bruce Coleman Inc.
17.1(CL) Russ Kinne/Nat'l Audubon Society Collection, P.R.
17.1(C) Mark S. Carlson/Tom Stack & Associates
17.1(CR) Leonard Lee Rue III/Bruce Coleman Inc.
17.1(BL) Jen and Des Bartlett/Nat'l Audubon Society Collection, P.R.
17.1(BC) Clem Haagner/Bruce Coleman Inc.
17.1(BR) Warren Garst/Tom Stack & Associates
17.2(L) Tom McHugh/Nat'l Audubon Society Collection, P.R.
17.2(R) Stephen Maslowski/Nat'l Audubon Society Collection, P.R.
17.4(L) Richard Ellis
17.4(C) Merlin D. Tuttle, Bat Conservation International
17.4(R) Sonja Bullaty and Angelo Lomeo
17.11 U.S. Dept. of Agriculture
17.17(TL) Wayne Lankinen
17.17(TR, UC, CR) The Photographic Library of Australia
17.17(UL) Francisco Erize/Bruce Coleman Inc.
17.17(CL) Annie Griffiths
17.17(BL) Hans and Judy Beste/Animals Animals
17.17(BC) Wayne Lankinen/Bruce Coleman Inc.
17.17(BR) Tom McHugh/Nat'l Audubon Society Collection, P.R.
17.18(L) Dr. L. E. Gilbert, U. of Texas at Austin/BPS
17.18(R) Larry West
17.18(B) James L. Castner

Chapter 18
18.2 UCSD Photo
18.5 Courtesy Sidney W. Fox. In Fox, S. W., and K. Dose. 1977. *Molecular Evolution and the Origins of Life*. 2d ed. New York: Marcel Dekker Inc.
18.6 Dr. A. Oparin, Bakh Institute of Biochemistry, Moscow
18.9 Rick Smolan
18.12 NASA/Ames Research Center
18.13(L) Dr. G. Cohen-Bazire
18.13(R) Courtesy Dr. R. Wyckoff, National Institutes of Health
18.14a Centers for Disease Control, Atlanta
18.14b Clay Adams
18.14c,e Martin M. Rotker/Taurus Photos
18.14d,h Manfred Kage/Peter Arnold, Inc.
18.14f Courtesy Dr. C. F. Robinow
18.14g Ed Reschke
18.15 Dr. G. de Haller, courtesy Dr. C. F. Robinow
18.17(L) Manfred Kage/Peter Arnold, Inc.
18.17(R) Centers for Disease Control, Atlanta
18.18 Centers for Disease Control, Atlanta (2)
18.19 M. Dworkin and H. Reichenback/Phototake
18.20 T. J. Beveridge, U. of Guelph/BPS
18.22a Frieder Sauer/Bruce Coleman Ltd.
18.22b Kim Taylor/Bruce Coleman Ltd.
18.22c J. R. Waaland, U. of Washington/BPS

18.23 Courtesy D. L. Findley, P. L. Walne, and R. W. Holton, U. of Tennessee—Knoxville. From *J. Phycology* 6:182–188, 1970
18.25 Dr. Ralph Lewin
18.28a L. J. LeBeau, U. of Illinois Hospital/BPS
18.28b Esau, K. 1968. *Viruses in Plant Hosts.* Madison, WI: Univ. of Wisconsin Press
18.28c Gene M. Milbrath, Ph.D.
18.29a Dennis Kunkel/Phototake
18.29b Phototake
18.29c Valentine, R. C., and H. G. Pereira. 1965. *J. Mol. Biol.* 13:13–20. © 1965 by Academic Press, Inc.
18.29d Martin M. Rotker/Phototake
18.29e Dr. Michael Moody
18.30a Courtesy Sally Hensen, Pediatrics Dept., and John Hay, Microbiology Dept., U.S.U.H.S., Bethesda, MD (2)
18.30b Simons, K., H. Garoff, and A. Helenius. 1982 *Sci. American* 246(2):58–66
18.31a Simons, K., H. Garoff, and A. Helenius. 1982. *Sci. American* 246(2):58–66

Chapter 19
19.2 Steven T. Brentano, U. of Iowa
19.4 Dr. Kwang W. Jeon
19.5 Eric V. Gravé
19.6a Manfred Kage/Peter Arnold, Inc.
19.6b Dr. Rudolf Röttger
19.7(T) Oxford Scientific Films/Animals Animals
19.7(B) Ed Reschke
19.8(TL, CR) E. R. Degginger
19.8(TR) Ed Reschke
19.8(CL) P. W. Johnson and J. McN. Sieburth, U. of Rhode Island/BPS
19.8(BL) Centers for Disease Control, Atlanta
19.8(BR) Eric V. Gravé
19.11(L) Ed Reschke
19.12(L) E. R. Degginger
19.12(C, R) John Shaw
19.13 Eric V. Gravé
19.16 Russ Kinne/Nat'l Audubon Society Collection, P.R.
19.17(TL) Dennis Kunkel/Phototake
19.17(TR) Dr. Paul E. Hargraves
19.17(BL) J. Pickett-Heaps/Science Source, P.R.
19.17(BR) Gordon Leedale, Biophoto Associates/Science Source, P.R.
19.18 N. Murayama, Murayama Research Lab./BPS
19.19 Frieder Sauer/Bruce Coleman Ltd. (2)
19.21a(TL) E. R. Degginger
19.21a(TR) Eric V. Gravé/Phototake
19.21b Dennis Kunkel/Phototake
19.22 Doug Wechsler
19.24 Nancy Sefton (2)
19.26a Jeff Foott/Bruce Coleman Ltd.
19.26b Kathleen Olsen/Courtesy Monterey Bay Aquarium
19.28 Kim Taylor/Bruce Coleman Ltd.
19.29a Nancy Sefton
19.29b Anne Hubbard/Nat'l Audubon Society Collection, P.R.
19.30 Courtesy of Macmillan Science Co., Inc.

Chapter 20
20.1 Heather Angel/Biofotos
Es. 20.1 Gravé, E. V. 1984. *Discover the Invisible.* Englewood Cliffs, NJ: Prentice-Hall, Inc. Used by permission of Prentice-Hall, Inc.
20.7a,c John Shaw
20.7b Rod Planck
20.8 Eric V. Gravé/Phototake
20.9a E. R. Degginger
20.9b Doug Wechsler
20.9c Don and Pat Valenti
20.9d Peter Ward/Bruce Coleman Ltd.
20.11 N. Allin and G. L. Barron. Courtesy N.S.E.R.C., Canada. *Can. J. Bot.* 57:187–193.

Chapter 21
21.1 Dr. Linda E. Graham (2)
21.3(L) Doug Wechsler
21.3(R) William E. Ferguson
21.4 John Shaw
21.5(L) John Shaw

21.5(R) Dwight R. Kuhn
21.7 Doug Wechsler
21.8 Rod Planck
21.9 The Age of Reptiles mural (reversed). The Peabody Museum of Natural History, Yale University
21.10 William E. Ferguson
21.11a Anne Hubbard/Nat'l Audubon Society Collection, P.R.
21.11b G. R. Roberts
21.11c Rod Planck
21.11d Doug Wechsler
21.12b Ed Reschke
21.12d From *Living Images* by Gene Shih and Richard Kessel. Reprinted courtesy of Jones and Bartlett Publishers, Inc., Boston, MA
21.12e J. N. A. Lott, McMaster U./BPS
21.14 Center for Ultrastructure Studies, SUNY College of Environmental Science and Forestry
21.15(L) William H. Allen, Jr.
21.15(R) William E. Ferguson
21.16 William E. Ferguson
21.17 Field Museum of Natural History, Chicago
21.18 Lynn M. Stone
21.23 William E. Ferguson
21.24a,b Dr. Thomas Eisner
21.24c Eric Crichton/Bruce Coleman Ltd.
21.24f Frieder Sauer/Bruce Coleman Ltd.

Chapter 22
22.2 W. K. Fletcher/Nat'l Audubon Society Collection, P.R.
22.3(L) William E. Ferguson
22.3(R) Rod Planck
22.4 G. R. Roberts
22.5 From *Living Images* by Gene Shih and Richard Kessel. Reprinted courtesy of Jones and Bartlett Publishers, Inc., Boston, MA
Es. 22.2 p. 545(TC), 547(TC) G. R. Roberts
Es. 22.2 p. 546(TL), 547(C) Robert P. Carr
Es. 22.2 p. 547(CL) John Shaw
22.7 J. N. A. Lott, McMaster U./BPS
22.12 U.S. Forest Service
22.13a From *Living Images* by Gene Shih and Richard Kessel. Reprinted courtesy of Jones and Bartlett Publishers, Inc., Boston, MA
22.13b Dr. Alice Belling (2)
22.13c Center for Ultrastructure Studies, SUNY College of Environmental Science and Forestry (2)
22.17 Ed Reschke
22.18 Omikron/Science Source, P.R. (2)
22.20(L) J. R. Waaland, U. of Washington/BPS
22.20(R) Ed Reschke
22.21 Ed Reschke (2)
22.23a(L), b(L) John Shaw
22.23a(R) William H. Mullins
22.23b(R) Wayne A. Bladholm
22.24 Ed Reschke (2)
22.26(L) Mitchell Bleier/Peter Arnold, Inc.
22.26(R) Ed Reschke
22.27 Ed Reschke

Chapter 23
23.1 William H. Mullins
23.7b E. Zeiger and N. Burstein, Stanford U./BPS
23.13 Lynn M. Stone
23.15 S. A. Wilde
23.16 Gerald P. Sanders
23.17a,b Dwight R. Kuhn
23.17c Rod Planck

Chapter 24
24.1 Dwight R. Kuhn
24.6 Biophoto Associates
24.7 Rod Planck
24.8a Rod Planck
24.8b Eric Crichton/Bruce Coleman Ltd.
24.11a D. R. Paulson, U. of Washington/BPS
24.11b(C) Dwight R. Kuhn
24.11b(R) Peter Parks, OSF/Animals Animals
24.12 Sylvan H. Wittwer, Michigan State U. Agricultural Experiment Station
24.16a,c Eric Crichton/Bruce Coleman Ltd.
24.16b Wayne A. Bladholm

Chapter 25
25.5 Nancy Sefton
25.7 Ed Reschke
25.8b Nancy Sefton
25.9a Nancy Sefton (2)
25.9b Doug Wechsler
25.10 Jeffrey L. Rotman
25.11b Ed Reschke
25.13a(L) Ed Reschke
25.15(L) Frieder Sauer/Bruce Coleman Ltd.
25.15(R) U. S. Dept. of Agriculture
25.16 Eric V. Gravé
25.17 Russ Kinne/Science Source, P.R.
25.18 Roland Birke/Peter Arnold, Inc.
25.19 Tom McHugh/Nat'l Audubon Society Collection, P.R.

Chapter 26
26.2a Frieder Sauer/Bruce Coleman Ltd.
26.2b Fred Bavendam/Peter Arnold, Inc.
26.2c H. Chaumeton/Nature
26.4 Jan Taylor/Bruce Coleman Ltd.
26.5(L) Jeff Goodman/NHPA
26.5(R) Oxford Scientific Films/Animals Animals
26.6(TL) Doug Wechsler
26.6(TR) Alex Kerstitch/Sea of Cortez Enterprises
26.6(BL) Lynn M. Stone
26.6(BR) Doug Wechsler/Animals Animals
26.8a,c Doug Wechsler
26.8b Alex Kerstitch/Sea of Cortez Enterprises
26.10 G. I. Bernard, OSF/Animals Animals
26.12(L) Rod Planck
26.12(R) Hans Pfletschinger/Peter Arnold, Inc.
26.13a,c Alex Kerstitch/Sea of Cortez Enterprises
26.13b,d,e,f James L. Castner
26.17 Dwight R. Kuhn
26.18 Kim Taylor/Bruce Coleman Ltd.
26.19(L) Fred Bavendam/Peter Arnold, Inc.
26.19(C) William E. Ferguson
26.19(R) C. B. and D. W. Frith/Bruce Coleman Ltd.
26.20a Alex Kerstitch/Sea of Cortez Enterprises
26.20b,c Doug Wechsler
26.20d Philip Alan Rosenberg
26.20e Jeffrey L. Rotman
26.21 Jeffrey L. Rotman
26.22(T) Jeffrey L. Rotman
26.22(BL, BR) H. Chaumeton/Nature

Chapter 27
27.3 Nancy Sefton
27.4 Heather Angel/Biofotos
27.7a(L) Dr. Giuseppe Mazza
27.7a(C) Heather Angel/Biofotos
27.7b Tom McHugh, Courtesy Steinhart Aquarium/Nat'l Audubon Society Collection, P.R.
27.10b Wolf H. Fahrenbach
27.11(TL, TR, BL, BR) Alex Kerstitch/Sea of Cortez Enterprises
27.11(TC) Jeffrey L. Rotman
27.11(BC) Peter David/Planet Earth Pictures—Seaphot
27.13 Bruce Coleman Ltd.
27.14 Peter Scoones/Planet Earth Pictures—Seaphot
27.15(L) Alex Kerstitch/Sea of Cortez Enterprises
27.15(C, R) Dr. Giuseppe Mazza
27.17a,c, Dr. Giuseppe Mazza
27.17b Charles G. Summers, Jr.
27.21 Paläontologisches Museum, Museum für Naturkunde der Humboldt-Universität, Berlin, DDR
27.22 Dr. Carl Welty
27.24(TL, TR, BL) G. R. Roberts
27.24(BC) William E. Ferguson
27.24(BR) Dwight R. Kuhn
27.26a Graham Pizzey/Bruce Coleman Ltd.
27.26b Leonard Lee Rue III/Nat'l Audubon Society Collection, P.R.
27.26c Don and Pat Valenti
27.30(L) Bruce Coleman Ltd.
27.30(R) David Agee/Anthro-Photo
27.31a James Moore/Anthro-Photo
27.32(L, C) © Margo Crabtree. Courtesy AAAS
27.32(R) Institute for Human Origins

Chapter 28
28.1(L) James Carmichael/NHPA
28.1(R) Alan G. Nelson/Animals Animals
28.2 Dr. Giuseppe Mazza
28.4 Stephen Dalton/NHPA
28.5(T) Dr. Giuseppe Mazza
28.5(B) Russ Kinne/Nat'l Audubon Society Collection, P.R.
28.6(TL) Lynn M. Stone
28.6(TR) Dr. Giuseppe Mazza
28.7 Dr. Ralph Buchsbaum
28.8 Ed Reschke (5)
28.9 Ed Reschke (2)
28.10b Ed Reschke
28.11(L) John Watney Photo Library
28.11(C) Nilsson, L. 1974. *Behold Man*. Boston: Little, Brown & Co.
28.20 Ed Reschke (3)
28.24 Biophoto Associates (2)
28.25a Omikron/Science Source, P.R.

Chapter 29
29.7 Jeffrey L. Rotman
29.9 Michael Fogden/Bruce Coleman Inc.
29.10(TL) Doug Wechsler
29.10(TC) Lynn M. Stone
29.10(TR) Rod Planck
29.10(BL) George H. Harrison
29.10(BC) Larry West
29.10(BR) Don and Pat Valenti
29.15 Omikron/Science Source, P.R.
29.19 D. W. Fawcett, Omikron/Science Source, P.R.

Chapter 30
30.8(TL) H. R. Duncker
30.10b B. F. King, U. of California/BPS
30.12 Netland, P. A., and B. R. Zetter, 1984. *Science* 224:cover. © 1984 by the AAAS

Chapter 31
31.8(TL, TR) McAlpine, W. A. 1975. *Heart and Coronary Arteries*. New York, Heidelburg, Berlin: Springer-Verlag © 1975
31.8(B) Nilsson, L. 1974. *Behold Man*. Boston: Little, Brown & Co.
31.10 Ed Reschke
31.13 Biophoto Associates/Science Source, P.R.
31.15 D. W. Fawcett/Science Source, P.R.
31.16 Ed Reschke
31.17 Bessis, M. 1974. *Corpuscles*. New York, Heidelberg, Berlin: Springer-Verlag © 1974
31.19 Ed Reschke
31.20 Zucker-Franklin, D., M. F. Greaves, C. E. Grossi, and A. M. Marmont. 1981. *Atlas of Blood Cells: Function and Pathology*, vol. 2. Philadelphia: Lea & Febiger. © 1981 by Edi.Ermes s.r.l.-Milan, Italy (3)

Chapter 32
32.5 William E. Ferguson (2)
32.6a Lynn M. Stone
32.6b Rod Planck
32.10 Herbert Biebach/Max Planck Gesellschaft (3)

Chapter 33
33.1 Cosmo Blank/Nat'l Audubon Society Collection, P.R.
33.2 G. J. Cambridge/NHPA
33.9a UPI/The Bettmann Archive
33.9b Wide World
33.11a Science Source, P.R.
33.11b Paul Almasy, Omikron/Science Source, P.R.
33.14 Ed Reschke

Chapter 34
34.1 William Roston
34.3 Courtesy Dr. Cedric S. Raine
34.10b Dr. John Heuser
34.12 Rod Planck
34.16 Dr. Edward S. Ross

Es. 34.1 Jane Burton and Kim Taylor/Bruce Coleman Ltd.
34.20(L) Rod Planck
34.20(R) Biophoto Associates/Science Source, P.R.
34.22 Omikron/Science Source, P.R.

Chapter 35
35.5(L) Prepared by Dr. Rufus B. Weaver, M.D., 1888. Courtesy of
        Dr. Peter S. Amenta, Hahnemann University, Phila., PA
35.7a Manfred Kage/Peter Arnold, Inc.
35.10b Museo Cajal, Madrid
35.16 Dan McCoy/Rainbow (2)
35.17 Courtesy DREAMSTAGE Scientific Catalog. © J. Allan Hobson and Hoffmann-LaRoche Inc.

Chapter 36
36.1a Nancy Sefton/Nat'l Audubon Society Collection, P.R.
36.1b William Roston
36.1c R. Andrew Odum/Peter Arnold, Inc.
36.3 Ken Libby
36.4(T) Neville Coleman/Bruce Coleman Ltd.
36.4(C, B) Alex Kerstitch/Sea of Cortez Enterprises
36.8(L) Lynn M. Stone
36.8(R) Tom McHugh/Nat'l Audubon Society Collection, P.R.
36.9 Charles G. Summers, Jr.
36.10a Dwight R. Kuhn
36.10b Rod Planck
36.14 Dr. G. P. Schatten, SPL/Science Source, P.R.

Chapter 37
37.1 John Walsh, SPL/Science Source, P.R.
37.2 Per Sundström/Gamma-Liaison
37.3a,b,c Tegner, M. J., and D. Epel. 1973. *Science* 179:685–688,
          figs. 2c, f, and g. © 1973 by the AAAS
37.4 Anderson, E. *J. Cell Biol.* 37:514–539. Reproduced by copyright permission of The Rockefeller Univ. Press
37.5a Stephens, R. E. 1972. *Biol. Bull.* 142:132–144 (6)
37.10b Russ Kinne/Science Source, P.R.
37.13a,c Kessel, R. G., H. W. Breams, and C. Y. Shih. 1974. *Am. J. Anatomy* 141:341–359
37.13b,d Richard G. Kessel and Harold W. Breams
37.15 Oxford Scientific Films/Animals Animals
37.24 Maternity Center Association (3)
37.27 From Spemann, 1936
37.28 After Mangold and Tiedemann, from Balinsky, 1975
Sum. Fig. p. 980 Nilsson, L. 1977. *A Child Is Born.* New York: Dell Publishing Co., Inc. (6)

Chapter 38
38.1a,b Nina Leen/Life Magazine © Time Inc.
38.6 BBC Natural History Unit. From Sparks, J. 1982. *The Discovery of Animal Behavior.* London: a Collins Publishers/BBC co-production (4)
38.7 Dwight R. Kuhn (6)
38.8 Peter A. Hinchliffe/Bruce Coleman Ltd.
38.9 Culver Pictures
38.12 Courtesy Charles Pfizer and Co., Inc.
38.14 Dr. I. Eibl-Eibesfeldt
Es. 38.2 Kyoko Archibald/International Crane Foundation
38.15 Reproduced with permission of Curator, Penfield Papers, Montreal Neurological Institute

Chapter 39
39.3 UPI/The Bettman Archive
39.4 Charles Walcott
39.6 Toni Angermayer/Nat'l Audubon Society Collection, P.R.
39.7 Toni Angermayer/Nat'l Audubon Society Collection, P.R.
Es. 39.1 p. 1016(TL,TR) Dr. Edward S. Ross
Es. 39.1 p. 1016(BL,BR) E. R. Degginger
Es. 39.1 p. 1017(TL) Oxford Scientific Films/Animals Animals
Es. 39.1 p. 1017(TR) Charles G. Summers, Jr.
Es. 39.1 p. 1017(BL) William E. Ferguson
Es. 39.1 p. 1017(BR) Lynn M. Stone
39.10a Robert P. Carr
39.10b Lynn M. Stone
Es. 39.2 p. 1020(T,B) Candace Bayer/Wolfgang Bayer Productions
Es. 39.2 p. 1021(TL) Lynn M. Stone
Es. 39.2 p. 1021(TR) Charles G. Summers, Jr.
Es. 39.2 p. 1021(BL) Fred Bavendam/Peter Arnold, Inc.

Es. 39.2 p. 1021(BC) Robert C. Fields/Animals Animals
Es. 39.2 p. 1021(BR) Mrs. Lorrimer Armstrong
39.11 Gordon Wiltsie/Bruce Coleman Inc.
39.12 Charles G. Summers, Jr.
39.14 Gwen Fidler
39.15 Manfred Danegger/NHPA
39.16(L) Gunter Ziesler/Bruce Coleman Ltd.
39.16(R) Charles G. Summers, Jr.
39.17a Stephen Dalton/NHPA
39.17b Stephen J. Krasemann/DRK Photo

Chapter 40
40.9 Rod Planck (2)
40.10a K. G. Preston-Mafham/Animals Animals
40.10b Sullivan and Rogers/Bruce Coleman, Inc.
40.10c Alex Kerstitch/Sea of Cortez Enterprises
40.10d E.P.I. Nancy Adams
40.11 Clem Haagner/Bruce Coleman Inc.
40.12 G. R. Roberts
40.13 Peter Johnson/NHPA
40.14 David C. Fritts
40.16(L) Wolfgang Kaehler
40.16(C) Fulvio Eccardi/Bruce Coleman Inc.
40.16(R) James H. Carmichael, Jr./Bruce Coleman Inc.
40.17 Tom McHugh/P.R.
40.18(T) William E. Ferguson
40.18(C,B) Rod Planck
40.19(L) John Shaw
40.19(C) Rita Summers
40.19(R) Don and Pat Valenti
40.20 David C. Fritts
40.21 G. R. Roberts
40.22(T) Charles G. Summers, Jr.
40.22(B) Rita Summers
40.23 David R. Frazier
40.24 David C. Fritts
40.25 William E. Ferguson
40.26(T) David C. Fritts
40.26(C) Steve McCutcheon
40.26(B) Stephen J. Krasemann/DRK Photo
40.28(L) J. M. Bassot, H. Chaumeton/Nature
40.28(TR) H. Chaumeton/Nature
40.28(BR) Peter David/Planet Earth Pictures—Seaphot
40.30 Jeffrey L. Rotman
40.31 Steve McCutcheon

Chapter 41
41.3 G. R. Roberts
41.4 From *Fundamentals of Ecology*, 3rd Edition by Eugene P. Odum. Copyright © 1971 by W. B. Saunders Company. Reprinted by permission of CBS College Publishing.
41.6 Joseph P. Shapiro/U.S. News & World Report
41.8 William E. Ferguson
41.11 Woodwell, G. M. 1970, *Sci. American* 223(3):64–74
41.17 Don and Pat Valenti
41.20 U.S. Forest Service, Bitterroot National Forest (4)
41.21a,b,c John Ebeling
Es. 41.2(L) Dudley Foster/WHOI
Es. 41.2(C) James Childress, UC-SB/WHOI
Es. 41.2(R) Alvin External Camera/WHOI

Chapter 42
42.7 Helmut Albrecht/Bruce Coleman Ltd.
42.8a John Shaw
42.8b Steve McCutcheon
42.9 Rod Planck
42.10 Zig Leszczynski/Animals Animals
42.11(TL) Robert P. Carr
42.11(TR) Tom Brakefield/Bruce Coleman Inc.
42.11(B) James H. Carmichael, Jr./Bruce Coleman Inc.
42.14 Gianni Tortoli/Nat'l Audubon Society Collection, P.R.
42.15 Muller, C. H. 1966. *Bulletin of The Torrey Botanical Club* 93:332–351
42.16 Gilbert Dupuy/Black Star
42.18 B. and C. Alexander/Black Star
Es. 42.1 Jim Brandenburg
42.19 Alon Reininger/Contact Stock Images, Woodfin Camp & Associates